HANDBOOK OF OPTICS

Other McGraw-Hill Books of Interest

HANDBOOK OF OPTICS

Volume II
Devices, Measurements, and Properties

Second Edition

Sponsored by the
OPTICAL SOCIETY OF AMERICA

Michael Bass   Editor in Chief

The Center for Research and
Education in Optics and Lasers (CREOL)
University of Central Florida
Orlando, Florida

Eric W. Van Stryland   Associate Editor

The Center for Research and
Education in Optics and Lasers (CREOL)
University of Central Florida
Orlando, Florida

David R. Williams   Associate Editor

Center for Visual Science
University of Rochester
Rochester, New York

William L. Wolfe   Associate Editor

Optical Sciences Center
University of Arizona
Tucson, Arizona

McGRAW-HILL, INC.

New York   San Francisco   Washington, D.C.   Auckland   Bogotá
Caracas   Lisbon   London   Madrid   Mexico City   Milan
Montreal   New Delhi   San Juan   Singapore
Sydney   Tokyo   Toronto

Library of Congress Cataloging-in-Publication Data

Handbook of optics / sponsored by the Optical Society of America ;
  Michael Bass, editor in chief. — 2nd ed.
     p.   cm.
    Includes bibliographical references and index.
    Contents: — 2. Devices, measurement, and properties.
    ISBN 0-07-047974-7
    1. Optics—Handbooks, manuals, etc. 2. Optical instruments—
Handbooks, manuals, etc. I. Bass, Michael. II. Optical Society
of America.
QC369.H35   1995
535—dc20   94-19339
                                           CIP

3 4 5 6 7 8 9 DOC/DOC 9 0 9 8 7 6

ISBN 0-07-047974-7

The sponsoring editor for this book was Stephen S. Chapman, the editing
supervisor was Paul R. Sobel, and the production supervisor was Suzanne
W. Babeuf. It was set in Times Roman by The Universities Press (Belfast)
Ltd.

Printed and bound by R.R. Donnelly & Sons Company.

This book is printed on acid-free paper.

CONTENTS

Chapter 8. Binary Optics   *Michael W. Farn and Wilfrid B. Veldkamp*   8.1

Chapter 9. Gradient Index Optics   *Duncan T. Moore*   9.1

Chapter 10. Optical Fibers and Fiber-Optic Communications   *Tom G. Brown*   10.1

Chapter 11. X-Ray Optics   *James E. Harvey*   11.1

Chapter 16. Camera Lenses   *Ellis Betensky, M. Kreitzer, and J. Moskovich*   16.1

Chapter 17. Microscopes   *Shinya Inoué and Rudolf Oldenboug*   17.1

Chapter 18. Reflective and Catadioptric Objectives   *Lloyd Jones*   18.1

Chapter 19. Scanners   *Leo Beiser and R. Barry Johnson*   19.1

Chapter 20. Optical Spectrometers   *Brian Henderson*   **20.1**

Chapter 21. Interferometers   *P. Hariharan*   **21.1**

Chapter 22. Polarimetry   *Russell A. Chipman*   **22.1**

Chapter 23. Holography and Holographic Instruments *Lloyd Huff*   23.1

Part 3.   Optical Measurements   24.1

Chapter 24. Radiometry and Photometry *Edward F. Zalewski*   24.3

Chapter 25. The Measurement of Transmission, Absorption, Emission, and Reflection *James M. Palmer*   25.1

Chapter 26. Scatterometers *John C. Stover*   26.1

Chapter 27. Ellipsometry *Rasheed M. A. Azzam*   27.1

Chapter 28. Spectroscopic Measurements *Brian Henderson*   25.1

Chapter 29. Optical Metrology *Daniel Malacara and Zacarias Malacara*   29.1

Chapter 35. Properties of Metals *Roger A. Paquin* **35.1**

Chapter 36. Optical Properties of Semiconductors *Paul M. Amirtharaj and David G. Seiler* **36.1**

Chapter 37. Black Surfaces for Optical Systems *Stephen M. Pompea and Robert P. Breault* **37.1**

Part 5. Nonlinear and Photorefractive Optics **38.1**

Chapter 38. Nonlinear Optics *Chung L. Tang* **38.3**

Chapter 39. Photorefractive Materials and Devices *Mark Cronin-Golomb and Marvin Klein* **39.1**

CONTRIBUTORS

Paul M. Amirtharaj *Materials Technology Group, Semiconductor Electronics Division, National Institute of Standards and Technology, Gaithersburg, Maryland* (CHAP. 36).

Rasheed M. A. Azzam *Department of Electrical Engineering, College of Engineering, University of New Orleans, New Orleans, Louisiana* (CHAP. 27).

Leo Beiser *Leo Beiser Inc., Flushing, New York* (CHAP. 19).

Jean M. Bennett *Research Department, Michelson Laboratory, Naval Air Warfare Center, China Lake, California* (CHAP. 3).

Ellis Betensky *Opcon Associates, Inc., West Redding, Connecticut* (CHAP. 16).

Glenn D. Boreman *The Center for Research and Education in Optics and Lasers (CREOL), University of Central Florida, Ornando, Florida* (CHAP. 32).

Robert P. Breault *Breault Research Organization, Inc., Tucson, Arizona* (CHAP. 37).

Tom G. Brown *The Institute of Optics, University of Rochester, Rochester, New York* (CHAP. 10).

I. C. Chang *Aurora Associates, Santa Clara, California* (CHAP. 12).

Russell A. Chipman *Physics Department, University of Alabama in Huntsville, Huntsville, Alabama* (CHAP. 22).

Katherine Creath *Optical Sciences Center, University of Arizona, Tucson, Arizona* (CHAP. 31).

Mark Cronin-Golomb *Electro-Optics Technology Center, Tufts University, Medford, Massachusetts* (CHAP. 39).

Michael W. Farn *MIT/Lincoln Laboratory, Lexington, Massachusetts* (CHAP. 8).

Norman Goldberg *Madison, Wisconsin* (CHAP. 15).

P. Hariharin *Division of Applied Physics, CSIRO, Sydney, Australia* (CHAP. 21).

Terry J. Harris *Applied Physics Laboratory, Johns Hopkins University, Laurel, Maryland* (CHAP. 33).

James E. Harvey *The Center for Research and Education in Optics and Lasers (CREOL), University of Central Florida, Orlando, Florida* (CHAP. 11).

Brian Henderson *Department of Physics and Applied Physics, University of Strathclyde, Glasgow, United Kingdom* (CHAPS. 20, 28).

Lloyd Huff *Research Institute, University of Dayton, Dayton, Ohio* (CHAP. 23).

Shinya Inoué *Marine Biological Laboratory, Woods Hole, Massachusetts* (CHAP. 17).

R. Barry Johnson *Optical E.T.C., Inc., Huntsville, Alabama and Center for Applied Optics, University of Alabama in Huntsville, Huntsville, Alabama* (CHAPS. 1, 19).

Lloyd Jones *Optical Sciences Center, University of Arizona, Tucson, Arizona* (CHAP. 18).

Marvin Klein *Hughes Research, Malibu, California* (CHAP. 39).

Thomas L. Koch *AT&T Bell Laboratories, Holmdel, New Jersey* (CHAP. 6).

M. Kreitzer *Opcon Associates, Inc., Cincinnati, Ohio* (CHAP. 16).

F. J. Leonberger *United Technologies Photonics, Bloomfield, Connecticut* (CHAP. 6).

John D. Lytle *Advanced Optical Concepts, Santa Cruz, California* (CHAP. 34).

Daniel Malacara *Centro de Investigaciones en Optica, A.C., León, Gto, Mexico* (CHAPS. 29, 30).

Zacarias Malacara *Centro de Investigaciones en Optica, A.C., León, Gto, Mexico* (CHAP. 29).

Theresa A. Maldonado *Department of Electrical Engineering, The University of Texas at Arlington, Arlington, Texas* (CHAP. 13).

Tom D. Milster *Optical Sciences Center, University of Arizona, Tucson, Arizona* (CHAP. 7).

Duncan T. Moore *The Institue of Optics and Gradient Lens Corporation, Rochester, New York* (CHAP. 9).

J. Moskovich *Opcon Associates, Inc., Cincinnati, Ohio* (CHAP. 16).

Rudolf Oldenbourg *Marine Biological Laboratory, Woods Hole, Massachusetts* (CHAP. 17).

James M. Palmer *Optical Sciences Center, University of Arizona, Tucson, Arizona* (CHAP. 25).

Roger A. Paquin *Advanced Materials Consultants, Tucson, Arizona and Optical Sciences Center, University of Arizona, Tucson, Arizona* (CHAP. 35).

Stephen M. Pompea *S. M. Pompea and Associates, Tucson, Arizona and Steward Observatory, University of Arizona, Tucson, Arizona* (CHAP. 37).

David G. Seiler *Materials Technology Group, Semiconductor Electronics Division, National Institue of Standards and Technology, Gaithersburg, Maryland* (CHAP. 36).

John C. Stover *TMA Technologies, Bozeman, Montana* (CHAP. 22).

P. G. Suchoski *United Technologies Photonics, Bloomfield, Connecticut* (CHAP. 6).

Chung L. Tang *School of Electrical Engineering, Cornell University, Ithaca, New York* (CHAP. 38).

Michael E. Thomas *Applied Physics Laboratory, Johns Hopkins University, Laurel, Maryland* (CHAP. 33).

William J. Tropf *Applied Physics Laboratory, Johns Hopkins University, Laurel, Maryland* (CHAP. 33).

Wilfrid B. Veldkamp *MIT/Lincoln Laboratory, Lexington, Massachusetts* (CHAP. 8).

William B. Wetherell *Optical Research Associates, Framington, Massachusetts* (CHAP. 2).

William L. Wolfe *Optical Sciences Center, University of Arizona, Tucson, Arizona* (CHAP. 4).

Shin-Tson Wu *Exploratory Studies Laboratory, Hughes Research Laboratories, Malibu, California* (CHAP. 14).

James C. Wyant *Optical Sciences Center, University of Arizona, Tucson, Arizona and WYKO Corporation, Tucson, Arizona* (CHAP. 31).

Edward F. Zalewski *Hughes Danbury Optical Systems, Danbury, Connecticut* (CHAP. 25).

George J. Zissis *Environmental Research Institue of Michigan, Ann Arbor, Michigan* (CHAP. 5).

PREFACE

The *Handbook of Optics,* Second Edition, is designed to serve as a general purpose desktop reference for the field of Optics yet stay within the confines of two books of finite length. Our purpose is to cover as much of optics as possible in a manner enabling the reader to deal with both basic and applied problems. To this end, we present articles about basic concepts, techniques, devices, instruments, measurements, and optical properties. In selecting subjects to include, we also had to select which subjects to leave out. The criteria we applied when excluding a subject were: (1) was it a specific application of optics rather than a core science or technology and (2) was it a subject in which the role of optics was peripheral to the central issue addressed. Thus, such topics as medical optics, laser surgery, and laser materials processing were not included. The resulting *Handbook of Optics,* Second Edition, serves the long-term information needs of those working in optics rather than presenting highly specific papers of current interest.

The authors were asked to prepare archival, tutorial articles which contain not only useful data but also descriptive material and references. Such articles were designed to enable the reader to understand a topic sufficiently well to get started using that knowledge. They also supply guidance as to where to find more in-depth material. Most include cross references to related articles within the Handbook. While applications of optics are mentioned, there is not space in the Handbook to include articles devoted to all of the myriad uses of optics in today's world. If we had, the Handbook would have been many volumes long and would have been too soon outdated.

The *Handbook of Optics,* Second Edition, contains 83 chapters organized into 17 broad categories or parts. The categorization enables the reader to find articles on a specific subject, say Vision, more easily and to find related articles within the Handbook. Within the categories the articles are grouped to make it simpler to find related material.

Volume I presents tutorial articles in the categories of Geometric Optics, Physical Optics, Quantum Optics, Optical Sources, Optical Detectors, Imaging Detectors, Vision, Optical Information and Image Processing, Optical Design Techniques, Optical Fabrication, Optical Properties of Films and Coatings, and Terrestrial Optics. This material is, for the most part, in a form which could serve to teach the underlying concepts of optics and its implementation. In fact, by careful selection of what to present and how to present it, the contents of Volume I could be used as a text for a comprehensive course in Optics.

The subjects covered in Volume II are Optical Elements, Optical Instruments, Optical Measurements, Optical and Physical Properties of Materials, and Nonlinear and Photorefractive Optics. As can be seen from these titles, Volume II concerns the specific devices, instruments, and techniques which are needed to employ optics in a wide variety of problems. It also provides data and discussion to assist one in the choice of optical materials.

The *Handbook of Optics*, Second Edition, would not have been possible without the support of the staff of the Optical Society of America and in particular Mr. Alan N. Tourtlotte and Ms. Kelly Furr.

For his pivotal roles in the development of the Optical Society of America, in the development of the profession of Optics, and for his encouragement to us in the task of preparing this Handbook, the editors dedicate this edition to Dr. Jarus Quinn.

Michael Bass, Editor-in-Chief
Eric W. Van Stryland, Associate Editor
David R. Williams, Associate Editor
William L. Wolfe, Associate Editor

GLOSSARY AND FUNDAMENTAL CONSTANTS

Introduction

This glossary of the terms used in the Handbook represents to a large extent the language of optics. The symbols are representations of numbers, variables, and concepts. Although the basic list was compiled by the author of this section, all the editors have contributed and agreed to this set of symbols and definitions. Every attempt has been made to use the same symbols for the same concepts throughout the entire handbook, although there are exceptions. Some symbols seem to be used for many concepts. The symbol α is a prime example, as it is used for absorptivity, absorption coefficient, coefficient of linear thermal expansion, and more. Although we have tried to limit this kind of redundancy, we have also bowed deeply to custom.

Units

The abbreviations for the most common units are given first. They are consistent with most of the established lists of symbols, such as given by the International Standards Organization ISO[1] and the International Union of Pure and Applied Physics, IUPAP.[2]

Prefixes

Similarly, a list of the numerical prefixes[1] that are most frequently used is given, along with both the common names (where they exist) and the multiples of ten that they represent.

Fundamental Constants

The values of the fundamental constants[3] are listed following the sections on SI units.

Symbols

The most commonly used symbols are then given. Most chapters of the Handbook also have a glossary of the terms and symbols specific to them for the convenience of the reader. In the following list, the symbol is given, its meaning is next, and the most customary unit of measure for the quantity is presented in brackets. A bracket with a dash in it indicates that the quantity is unitless. Note that there is a difference between units and dimensions. An angle has units of degrees or radians and a solid angle square degrees or steradians, but both are pure ratios and are dimensionless. The unit symbols as recommended in the SI system are used, but decimal multiples of some of the dimensions are sometimes given. The symbols chosen, with some cited exceptions are also those of the first two references.

RATIONALE FOR SOME DISPUTED SYMBOLS

The choice of symbols is a personal decision, but commonality improves communication. This section explains why the editors have chosen the preferred symbols for the Handbook. We hope that this will encourage more agreement.

Fundamental Constants

It is encouraging that there is almost universal agreement for the symbols for the fundamental constants. We have taken one small exception by adding a subscript B to the k for Boltzmann's constant.

Mathematics

We have chosen i as the imaginary almost arbitrarily. IUPAP lists both i and j, while ISO does not report on these.

Spectral Variables

These include expressions for the wavelength, λ, frequency, ν, wave number, σ, ω for circular or radian frequency, k for circular or radian wave number and dimensionless frequency x. Although some use f for frequency, it can be easily confused with electronic or spatial frequency. Some use $\bar{\nu}$ for wave number, but, because of typography problems and agreement with ISO and IUPAP, we have chosen σ; it should not be confused with the Stephan Boltzmann constant. For spatial frequencies we have chosen ξ and η, although f_x and f_y are sometimes used. ISO and IUPAP do not report on these.

Radiometry

Radiometric terms are contentious. The most recent set of recommendations by ISO and IUPAP are L for radiance [$\mathrm{Wcm}^{-2}\,\mathrm{sr}^{-1}$], M for radiant emittance or exitance [Wcm^{-2}], E for irradiance or incidance [Wcm^{-2}], and I for intensity [Wsr^{-2}]. The previous terms, W, H, N and J respectively, are still in many texts, notably Smith and Lloyd[4] but we have used the revised set, although there are still shortcomings. We have tried to deal with the vexatious term *intensity* by using *specific intensity* when the units are $\mathrm{Wcm}^{-2}\,\mathrm{sr}^{-1}$, *field intensity* when they are Wcm^{-2}, and *radiometric intensity* when they are Wsr^{-1}.

There are two sets of terms for these radiometric quantities, that arise in part from the terms for different types of reflection, transmission, absorption, and emission. It has been proposed that the *ion* ending indicate a process, that the *ance* ending indicate a value associated with a particular sample, and that the *ivity* ending indicate a generic value for a "pure" substance. Then one also has reflectance, transmittance, absorptance, and emittance as well as reflectivity, transmissivity, absorptivity, and emissivity. There are now two different uses of the word emissivity. Thus the words *exitance, incidance,* and *sterance* were coined to be used in place of emittance, irradiance, and radiance. It is interesting that ISO uses radiance, exitance, and irradiance whereas IUPAP uses radiance, excitance [*sic*] and irradiance. We have chosen to use them both, i.e., emittance, irradiance, and radiance will be followed in square brackets by exitance, incidance, and sterance (or vice versa). Individual authors will use the different endings for transmission, reflection, absorption, and emission as they see fit.

We are still troubled by the use of the symbol E for irradiance, as it is so close in meaning to electric field, but we have maintained that accepted use. The spectral concentrations of these quantities, indicated by a wavelength, wave number, or frequency subscript (e.g., L_λ) represent partial differentiations; a subscript q represents a photon quantity; and a subscript v indicates a quantity normalized to the response of the eye. Thereby, L_v is luminance, E_v illuminance, and M_v and I_v luminous emittance and luminous intensity. The symbols we have chosen are consistent with ISO and IUPAP.

The refractive index may be considered a radiometric quantity. It is generally complex and is indicated by $\bar{n} = n - ik$. The real part is the relative refractive index and k is the extinction coefficient. These are consistent with ISO and IUPAP, but they do not address the complex index or extinction coefficient.

Optical Design

For the most part ISO and IUPAP do not address the symbols that are important in this area.

There were at least 20 different ways to indicate focal ratio; we have chosen FN as symmetrical with NA; we chose f and efl to indicate the effective focal length. Object and image distance, although given many different symbols, were finally called s_o and s_i since s is an almost universal symbol for distance. Field angles are θ and ϕ; angles that measure the slope of a ray to the optical axis are u; u can also be $\sin u$. Wave aberrations are indicated by W_{ijk}, while third order ray aberrations are indicated by σ_i and more mnemonic symbols.

Electromagnetic Fields

There is no argument about **E** and **H** for the electric and magnetic field strengths, Q for quantity of charge, ρ for volume charge density, σ for surface charge density, etc. There is no guidance from References 1 and 2 on polarization indication. We chose $\perp$ and $\parallel$ rather than p and s, partly because s is sometimes also used to indicate scattered light.

There are several sets of symbols used for reflection, transmission, and (sometimes) absorption, each with good logic. The versions of these quantities dealing with field amplitudes are usually specified with lower case symbols: r, t, and a. The versions dealing with power are alternately given by the uppercase symbols or the corresponding Greek symbols: R and T vs ρ and τ. We have chosen to use the Greek, mainly because these quantities are also closely associated with Kirchhoff's law that is usually stated symbolically as $\alpha = \epsilon$. The law of conservation of energy for light on a surface is also usually written as $\alpha + \rho + \tau = 1$.

Base SI Quantities

length	m	meter
time	s	second
mass	kg	kilogram
electric current	A	ampere
Temperature	K	kelvin
Amount of substance	mol	mole
Luminous intensity	cd	candela

Derived SI Quantities

energy	J	joule
electric charge	C	coulomb
electric potential	V	volt
electric capacitance	F	farad
electric resistance	Ω	ohm
electric conductance	S	siemens

magnetic flux	Wb	weber
inductance	H	henry
pressure	Pa	pascal
magnetic flux density	T	tesla
frequency	Hz	hertz
power	W	watt
force	N	newton
angle	rad	radian
angle	sr	steradian

Prefixes

Symbol	Name	Common name	Exponent of ten
E	exa		18
p	peta		15
T	tera	trillion	12
G	giga	billion	9
M	mega	million	6
k	kilo	thousand	3
h	hecto	hundred	2
da	deca	ten	1
d	deci	tenth	−1
c	centi	hundredth	−2
m	milli	thousandth	−3
μ	micro	millionth	−6
n	nano	billionth	−9
p	pico	trillionth	−12
f	femto		−15
a	atto		−18

Constants

c	speed of light in vacuo [299792458 ms^{-1}]
c_1	first radiation constant = $2\pi c^2 h = 3.7417749 \times 10^{-16}$ [Wm^2]
c_2	second radiation constant = hc/k = 0.01438769 [mK]
e	elementary charge [$1.60217733 \times 10^{-19}$C]
g_n	free fall constant [9.80665 ms^{-2}]
h	Planck's constant [$6.6260755 \times 10^{-34}$ Ws]
k_B	Boltzmann constant [1.380658×10^{-23} JK^{-1}]
m_e	mass of the electron [$9.1093897 \times 10^{-31}$ kg]
N_A	Avogadro constant [6.0221367×10^{23} mol^{-1}]
R_∞	Rydberg constant [10973731.534 m^{-1}]
ϵ_o	vacuum permittivity [$\mu_o^{-1} c^{-2}$]
σ	Stefan Boltzmann constant [5.67051×10^{-8} Wm^{-1} K^{-4}]
μ_o	vacuum permeability [$4\pi \times 10^{-7}$ NA^{-2}]
μ_B	Bohr magneton [$9.2740154 \times 10^{-24}$ JT^{-1}]

General

B	magnetic induction [Wbm^{-2}, kgs^{-1} C^{-1}]
C	capacitance [f, C^2 s^2 m^{-2} kg^{-1}]
C	curvature [m^{-1}]

c	speed of light in vacuo $[\mathrm{ms}^{-1}]$
c_1	first radiation constant $[\mathrm{Wm}^2]$
c_2	second radiation constant $[\mathrm{mK}]$
D	electric displacement $[\mathrm{Cm}^{-2}]$
E	incidance [irradiance] $[\mathrm{Wm}^{-2}]$
e	electronic charge [coulomb]
E_v	illuminance $[\mathrm{lux},\ \mathrm{lmm}^{-2}]$
E	electrical field strength $[\mathrm{Vm}^{-1}]$
E	transition energy $[\mathrm{J}]$
E_g	band-gap energy $[\mathrm{eV}]$
f	focal length $[\mathrm{m}]$
f_c	Fermi occupation function, conduction band
f_v	Fermi occupation function, valence band
FN	focal ratio (f/number) $[-]$
g	gain per unit length $[\mathrm{m}^{-1}]$
g_{th}	gain threshold per unit length $[\mathrm{m}^1]$
H	magnetic field strength $[\mathrm{Am}^{-1},\ \mathrm{Cs}^{-1}\,\mathrm{m}^{-1}]$
h	height $[\mathrm{m}]$
I	irradiance (see also E) $[\mathrm{Wm}^{-2}]$
I	radiant intensity $[\mathrm{Wsr}^{-1}]$
I	nuclear spin quantum number $[-]$
I	current $[\mathrm{A}]$
i	$\sqrt{-1}$
$\mathrm{Im}()$	Imaginary part of
J	current density $[\mathrm{Am}^{-2}]$
j	total angular momentum $[\mathrm{kg\,m}^2\,\mathrm{sec}^{-1}]$
$J_1()$	Bessel function of the first kind $[-]$
k	radian wave number $= 2\pi/\lambda$ $[\mathrm{rad\,cm}^{-1}]$
k	wave vector $[\mathrm{rad\,cm}^{-1}]$
k	extinction coefficient $[-]$
L	sterance [radiance] $[\mathrm{Wm}^{-2}\,\mathrm{sr}^{-1}]$
L_v	luminance $[\mathrm{cdm}^{-2}]$
L	inductance $[\mathrm{h},\ \mathrm{m}^2\mathrm{kgC}^{-2}]$
L	laser cavity length
L, M, N	direction cosines $[-]$
M	angular magnification $[-]$
M	radiant exitance [radiant emittance] $[\mathrm{Wm}^{-2}]$
m	linear magnification $[-]$
m	effective mass $[\mathrm{kg}]$
MTF	modulation transfer function $[-]$
N	photon flux $[\mathrm{s}^{-1}]$
N	carrier (number) density $[\mathrm{m}^{-3}]$
n	real part of the relative refractive index $[-]$
$\tilde{n}$	complex index of refraction $[-]$
NA	numerical aperture $[-]$
OPD	optical path difference $[\mathrm{m}]$
P	macroscopic polarization $[\mathrm{C\,m}^{-2}]$
$\mathrm{Re}()$	real part of $[-]$
R	resistance $[\Omega]$
r	position vector $[\mathrm{m}]$
r	(amplitude) reflectivity
S	Seebeck coefficient $[\mathrm{VK}^{-1}]$
s	spin quantum number $[-]$
s	path length $[\mathrm{m}]$

s_o	object distance [m]
s_i	image distance [m]
T	temperature [K, C]
t	time [s]
t	thickness [m]
u	slope of ray with the optical axis [rad]
V	Abbé reciprocal dispersion [—]
V	voltage [V, $m^2\,kgs^{-2}\,C^{-1}$]
x, y, z	rectangular coordinates [m]
Z	atomic number [—]

Greek Symbols

α	absorption coefficient [cm^{-1}]
α	(power) absorptance (absorptivity)
ϵ	dielectric coefficient (constant) [—]
ϵ	emittance (emissivity) [—]
ϵ	eccentricity [—]
ϵ_1	Re (ϵ)
ϵ_2	Im (ϵ)
τ	(power) transmittance (transmissivity) [—]
ν	radiation frequency [Hz]
ω	circular frequency $= 2\pi\nu$ [$rads^{-1}$]
ω_p	plasma frequency [H_2]
λ	wavelength [μm, nm]
σ	wave number $= 1/\lambda$ [cm^{-1}]
σ	Stefan Boltzmann constant [$Wm^{-2}\,K^{-1}$]
ρ	reflectance (reflectivity) [—]
θ, ϕ	angular coordinates [rad,°]
ξ, η	rectangular spatial frequencies [m^{-1}, r^{-1}]
ϕ	phase [rad, °]
ϕ	lens power [m^{-1}]
Φ	flux [W]
χ	electric susceptibility tensor [—]
Ω	solid angle [sr]

Other

$\Re$	responsivity
$\exp(x)$	e^x
$\log_a(x)$	log to the base a of x
$\ln(x)$	natural log of x
$\log(x)$	standard log of x: $\log_{10}(x)$
Σ	summation
Π	product
Δ	finite difference
δx	variation in x
dx	total differential
∂x	partial derivative of x
$\delta(x)$	Dirac delta function of x
δ_{ij}	Kronecker delta

REFERENCES

1. Anonymous, *ISO Standards Handbook 2: Units of Measurement,* 2dcd., International Organization for Standardization, 1982.
2. Anonymous, *Symbols, Units and Nomenclature in Physics,* Document U.I.P. 20, International Union of Pure and Applied Physics, 1978.
3. E. Cohen and B. Taylor, "*The Fundamental Physical Constants,*" *Physics Today,* 9, *August* 1990.
4. W. J. Smith, *Modern Optical Engineering,* 2d ed., McGraw-Hill, 1990; J. M. Lloyd, *Thermal Imaging Systems,* Plenum Press, 1972.

William L. Wolfe
Optical Sciences Center
University of Arizona
Tucson, Arizona

CONTENTS IN BRIEF: VOLUME I

CONTENTS IN BRIEF: VOLUME II

P · A · R · T · 1

OPTICAL ELEMENTS

CHAPTER 1
LENSES

R. Barry Johnson
Optical E.T.C., Inc.
Huntsville, Alabama
and
Center for Applied Optics
University of Alabama in Huntsville
Huntsville, Alabama

1.1 GLOSSARY

AChr	axial chromatic aberration
AST	astigmatism
b	factor
bfl	back focal length
C_o	scaling factor
c	curvature
C_1	scaling factor
C_2	scaling factor
CC	conic constant
CMA_s	sagittal coma
CMA_t	tangential coma
D_{ep}	diameter of entrance pupil
d_o	distance from object to loupe
d_e	distance from loupe to the eye
E	irradiance
efl	effective focal length
ep	eyepiece
FN	F-number
f	focal length
h	height above axis
H_i	height of ray intercept in image plane

$\mathcal{K}$	shape factor
i	image
$J_1()$	Bessel function of the first kind
k	$2\pi/\lambda$
L	length
MP	magnifying power [cf. linear lateral longitudinal magnification]
m	linear, lateral magnification
$\bar{m}$	linear, longitudinal, magnification
n	refractive index
$\mathcal{M}$	factor
MTF	modulation transfer function
NA	numerical aperture
a,b	first and second lenses
o	object
obj	objective
P	partial dispersion
P_i	principal points
p	$= s_d/f_a$
$\tilde{\mathcal{R}}$	peak normalized spectral weighting function
$\mathcal{S}$	object to image distance
SA3	third-order spherical aberration
SAC	secondary angular spectrum
s_i	image distance
s_{ot}	optical tube length
s_o	object distance
TPAC	transverse primary chromatic aberration
t	thickness
u	slope
V	Abbe number or reciprocal dispersion
υ	ϕ-normalized reciprocal object distance $1/s_o\phi$
x, y, z	cartesian coordinates
β	angular blur diameter
δ	depth of focus
ζ	sag
$\Delta\theta$	angular blur tolerance

θ	field of view
λ	wavelength
v	spatial frequency
ϕ	lens power
ρ	radius
σ	standard deviation of the irradiance distribution
τ	transmission
Ω	normalized spatial frequency

1.2 INTRODUCTION

This section provides a basic understanding of using lenses for image formation and manipulation. The principles of image formation are reviewed first. The effects of lens shape, index of refraction, magnification, and F-number on the image quality of a singlet lens are discussed in some detail. Achromatic doublets and more complex lens systems are covered next. A representative variety of lenses is analyzed and discussed. Performance that may be expected of each class of lens is presented. The section concludes with several techniques for rapid estimation of the performance of lenses. Refer to Chap. 1 "Geometric Optics" in Vol. I, for further discussion of geometrical optics and aberrations.

1.3 BASICS

Figure 1 illustrates an image being formed by a simple lens. The object height is h_o and the image height is h_i, with u_o and u_i being the corresponding slope angles. It follows from the Lagrange invariant that the *lateral magnification* is defined to be

$$m \equiv \frac{h_i}{h_o}$$

$$= \frac{(nu)_o}{(nu)_i} \tag{1}$$

where n_o and n_i are the refractive indices of the medium in which the object and image lie, respectively. By convention, a height is positive if above the optical axis and a ray angle is positive if its slope angle is positive. Distances are positive if the ray propagates left to right. Since the Lagrange invariant is applicable for paraxial rays, the angle nu

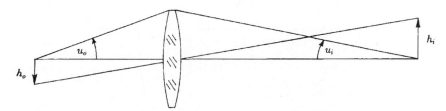

FIGURE 1   Imaging by a simple lens.

should be understood to mean $n \tan u$. This interpretation applies to all paraxial computations. For an *aplanatic* lens, which is free of spherical aberration and linear coma, the magnification can be shown by the *optical sine theorem* to be given by

$$m \equiv \frac{h_i}{h_o}$$

$$= \frac{n_o \sin u_o}{n_i \sin u_i} \tag{2}$$

If the object is moved a small distance ∂s_o longitudinally, the corresponding displacement of the image ∂s_i can be found by the differential form of the basic imaging equation and leads to an equation analogous to the Lagrange invariant. The *longitudinal magnification* is then defined as

$$\bar{m} \equiv \frac{\partial s_i}{\partial s_o}$$

$$= \frac{(nu^2)_o}{(nu^2)_i}$$

$$= m^2 \left[\frac{n_i}{n_o} \right] \tag{3}$$

The following example will illustrate one application of m and $\bar{m}$. Consider that a spherical object of radius r_o is to be imaged as shown in Fig. 2. The equation of the object is $r_o^2 = y_o^2 + z^2$, where z is measured along the optical axis and is zero at the object's center of curvature. Letting the surface sag as measured from the vertex plane of the object be denoted as ζ_o, the equation of the object becomes $r_o^2 = (r_o - \zeta_o)^2 + y_o^2$ since $z = r_o - \zeta_o$. In the region near the optical axis, $\zeta_o^2 \ll r_o^2$, which implies that $r_o \approx y_o^2/2\zeta_o$. The image of the object is expressed in the transverse or lateral direction by $y_i = my_o$ and in the longitudinal or axial direction by $\zeta_i = \bar{m}\zeta_o = \zeta_o m^2(n_i/n_o)$. In a like manner, the image of the spherical object is expressed as $r_i \approx (y_i)^2/2\zeta_i$. By substitution, the sag of the image is expressed by

$$r_i \equiv \frac{n_o y_o^2}{2n_i \zeta_o}$$

$$= r_o \left[\frac{n_o}{n_i} \right] \tag{4}$$

Hence, in the paraxial region about the optical axis, the radius of the image of a spherical

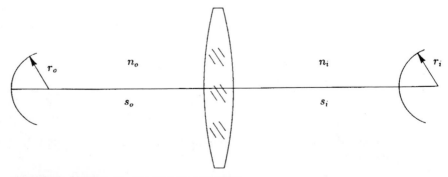

FIGURE 2 Imaging of a spherical object by a lens.

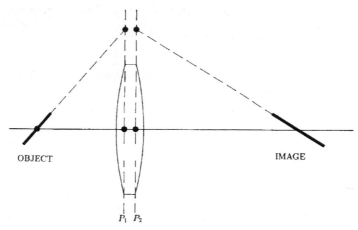

FIGURE 3 Imaging of a tilted object illustrating the Scheimpflug condition.

object is independent of the magnification and depends only on the ratio of the refractive indices of the object and image spaces.

When an optical system as shown in Fig. 3 images a tilted object, the image will also be tilted. By employing the concept of lateral and longitudinal magnification, it can be easily shown that the intersection height of the object plane with the first principal plane P_1 of the lens must be the same as the intersection height of the image plane with the second principal plane P_2 of the lens. This principle is known as the *Scheimpflug condition.*

The object-image relationship of a lens system is often described with respect to its *cardinal points,* which are as follows:

- *Principal points*: the axial intersection point of conjugate planes related by unit lateral magnification
- *Nodal points*: conjugate points related by unit angular magnification ($m = u_i/u_0$)
- *Focal points*: front (f_1) and rear (f_2)

The focal length of a lens is related to the power of the lens by

$$\phi = \frac{n_o}{f_o} = \frac{n_i}{f_i} \tag{5}$$

This relationship is important in such optical systems as underwater cameras, cameras in space, etc. For example, it is evident that the field of view is decreased for a camera in water.

The *lens law* can be expressed in several forms. If s_o and s_i are the distance from the object to the first principal point and the distance from the second principal point to the image, then the relationship between the object and the image is given by

$$\phi = \frac{n_i}{s_i} + \frac{n_o}{s_o} \tag{6}$$

Should the distances be measured with respect to the nodal points, the imaging equation becomes

$$\phi = \frac{n_o}{s_i} + \frac{n_i}{s_o} \tag{7}$$

When the distances are measured from the focal points, the image relationship, known as the *Newtonian imaging equation,* is given by

$$f_1 f_2 = s_o s_i \tag{8}$$

The power of a spherical refracting surface, with curvature c and n being the refractive index following the surface, is given by

$$\phi = c(n - n_o) \tag{9}$$

It can be shown that the power of a single thick lens in air is

$$\phi_{\text{thick}} = \phi_1 + \phi_2 - \phi_1 \phi_2 \frac{t}{n} \tag{10}$$

where t is the thickness of the lens. The distance from the first principal plane to the first surface is $-(t/n)\phi_2 f_1$ and the distance from the second principal point to the rear surface is $(-t/n)\phi_1 f_2$. The power of a thin lens $(t \to 0)$ in air is given by

$$\phi_{\text{thin}} = (n - 1)(c_1 - c_2) \tag{11}$$

1.4 STOPS AND PUPILS

The *aperture stop* or *stop* of a lens is the limiting aperture associated with the lens that determines how large an axial beam may pass through the lens. The stop is also called an *iris.* The *marginal ray* is the extreme ray from the axial point of the object through the edge of the stop. The *entrance pupil* is the image of the stop formed by all lenses preceding it when viewed from object space. The *exit pupil* is the image of the stop formed by all lenses following it when viewed from image space. These pupils and the stop are all images of one another. The *principal ray* is defined as the ray emanating from an off-axis object point that passes through the center of the stop. In the absence of pupil aberrations, the principal ray also passes through the center of the entrance and exit pupils.

As the obliquity angle of the principal ray increases, the defining apertures of the components comprising the lens may limit the passage of some of the rays in the entering beam thereby causing the stop not to be filled with rays. The failure of an off-axis beam to fill the aperture stop is called *vignetting.* The ray centered between the upper and lower rays defining the oblique beam is called the *chief ray.* When the object moves to large off-axis locations, the entrance pupil often has a highly distorted shape, may be tilted, and/or displaced longitudinally and transversely. Due to the vignetting and pupil aberrations, the chief and principal rays may become displaced from one another. In some cases, the principal ray is vignetted.

The *field stop* is an aperture that limits the passage of principal rays beyond a certain field angle. The image of the field stop when viewed from object space is called the *entrance window* and is called the *exit window* when viewed from image space. The field stop effectively controls the field of view of the lens system. Should the field stop be coincident with an image formed within or by the lens system, the entrance and exit windows will be located at the object and/or image(s).

A *telecentric stop* is an aperture located such that the entrance and/or exit pupils are located at infinity. This is accomplished by placing the aperture in the focal plane. Consider a stop placed at the front focal plane of a lens. The image is located at infinity and the principal ray exits the lens parallel to the optical axis. This feature is often used in metrology since the measurement error is reduced when compared to conventional lens systems because the centroid of the blur remains at the same height from the optical axis even as the focus is varied.

1.5 F-NUMBER AND NUMERICAL APERTURE

The focal ratio or F-number (FN) of a lens is defined as the effective focal length divided by the entrance pupil diameter D_{ep}. When the object is not located at infinity, the effective FN is given by

$$\text{FN}_{eff} = \text{FN}_\infty(1 - m) \tag{12}$$

where m is the magnification. For example, for a simple positive lens being used at unity magnification ($m = -1$), the $\text{FN}_{eff} = 2\text{FN}_\infty$. The *numerical aperture* of a lens is defined as

$$\text{NA} = n_i \sin U_i \tag{13}$$

where n_i is the refractive index in which the image lies and U_i is the slope angle of the marginal ray exiting the lens. If the lens is aplanatic, then

$$\text{FN}_{eff} = \frac{1}{2\text{NA}} \tag{14}$$

1.6 MAGNIFIER OR EYE LOUPE

The typical magnifying glass, or *loupe,* comprises a singlet lens and is used to produce an erect but virtual magnified image of an object. The magnifying power of the loupe is stated to be the ratio of the angular size of the image when viewed through the magnifier to the angular size without the magnifier. By using the thin-lens model of the human eye, the magnifying power (MP) can be shown to be given by

$$\text{MP} = \frac{25 \text{ cm}}{d_e + d_o - \phi d_e d_o} \tag{15}$$

where d_o is the distance from the object to the loupe, d_e is the separation of the loupe from the eye, and $\phi = 1/f$ is the power of the magnifier. When d_o is set to the focal length of the lens, the virtual image is placed at infinity and the magnifying power reduces to

$$\text{MP} = \frac{25 \text{ cm}}{f} \tag{16}$$

Should the virtual image be located at the near viewing distance of the eye (about 25 cm), then

$$\text{MP} = \frac{25 \text{ cm}}{f} + 1 \tag{17}$$

Typically simple magnifiers are difficult to make with magnifying powers greater than about 10×.

1.7 COMPOUND MICROSCOPES

For magnifying power greater than that of a simple magnifier, a compound microscope, which comprises an objective lens and an eyepiece, may be used. The objective forms an aerial image of the object at a distance s_{ot} from the rear focal point of the objective. The distance s_{ot} is called the *optical tube length* and is typically 160 mm. The objective magnification is

$$\text{MP}_{obj} = \frac{s_{ot}}{f_{obj}} \tag{18}$$

The image formed is further magnified by the eyepiece which has a $\text{MP}_{ep} = 250 \text{ mm}/f_{ep}$. The total magnifying power of the compound microscope is given by

$$\begin{aligned} \text{MP} &= \text{MP}_{obj}\text{MP}_{ep} \\ &= \frac{160}{f_{obj}} \cdot \frac{250}{f_{ep}} \end{aligned} \tag{19}$$

Typically, $f_{ep} = 25$ mm, so its $\text{MP} = 10$. Should the objective have a focal length of 10 mm, the total magnifying power of the microscope is $16\times$ times $10\times$, or $160\times$.

1.8 FIELD AND RELAY LENSES

Field lenses are placed at (or near) an image location for the purpose of optically relocating the pupil or to increase the field of view of the optical system. For example, a field lens may be used at the image plane of an astronomical telescope such that the field lens images the objective lens onto the eyepiece. In general, the field lens does not contribute to the aberrations of the system except for distortion and field curvature. Since the field lens must be positive, it adds inward curving Petzval. For systems having a small detector requiring an apparent increase in size, the field lens is a possible solution. The detector is located beyond the image plane such that it subtends the same angle as the objective lens when viewed from the image point. The field lens images the objective lens onto the detector.

Relay lenses are used to transfer an image from one location to another such as in a submarine periscope or borescope. It is also used as a means to erect an image in many types of telescopes and other such instruments. Often relay lenses are made using two lens groups spaced about a stop, or an image of the system stop, in order to take advantage of the principle of symmetry, thereby minimizing the comatic aberrations and lateral color. The relayed image is frequently magnified.

1.9 APLANATIC SURFACES AND IMMERSION LENSES

Abbe called a lens an aplanat that has an equivalent refractive surface which is a portion of a sphere with a radius r centered about the focal point. Such a lens satisfies the Abbe sine condition and implies that the lens is free of spherical and coma near the optical axis. Consequently, the maximum possible numerical aperture (NA) of an aplanat is unity, or an $\text{FN} = 0.5$. In practice, an FN less than 0.6 is difficult to achieve. For an aplanat,

$$\text{FN} = \frac{1}{2 \cdot \text{NA}} \tag{20}$$

It can be shown that three cases exist where the spherical aberration is zero for a spherical surface. These are: (1) the trivial case where the object and image are located at the surface, (2) the object and image are located at the center of curvature of the surface, and (3) the object is located at the aplanatic point. The third case is of primary interest. If

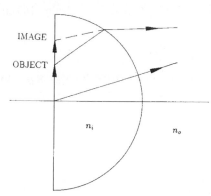

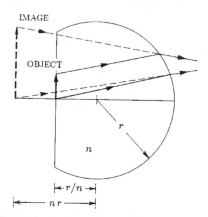

FIGURE 4 Aplanatic hemispherical magnifier with the object and image located at the center of curvature of the spherical surface. This type of magnifier has a magnification of n_i/n_o which can be used as a contact magnifier or as an immersion lens.

FIGURE 5 Aplanatic hyperhemispherical magnifier or Amici lens has the object located at the aplanatic point. The lateral magnification is $(n_i/n_0)^2$.

the refractive index preceding the surface is n_o and following the surface is n_i, then the object is located a distance s_o from the surface as expressed by

$$s_o = \frac{r(n_o + n_i)}{n_o} \tag{21}$$

and the image is located at

$$s_i = \frac{r(n_o + n_i)}{n_i} \tag{22}$$

An immersion lens or contact lens can be formed from an aplanatic surface and a plano surface. Figure 4 illustrates a hemispherical magnifier that employs the second aplanatic case. The resultant magnification is n_i if in air or n_i/n_o otherwise. A similar magnifier can be constructed by using a hyperhemispherical surface and a plano surface as depicted in Fig. 5. The lateral magnification is n_i^2. This lens, called an *Amici lens*, is based upon the third aplanatic case. The image is free of all orders of spherical aberration, third-order coma, and third-order astigmatism. Axial color is also absent from the hemispherical magnifier. These magnifiers are often used as a means to make a detector appear larger and as the first component in microscope objectives.

1.10 SINGLE ELEMENT LENS

It is well known that the spherical aberration of a lens is a function of its shape factor or bending. Although several definitions for the shape factor have been suggested, a useful formulation is

$$\mathcal{H} = \frac{c_1}{c_1 - c_2} \tag{23}$$

where c_1 and c_2 are the curvatures of the lens with the first surface facing the object. By adjusting the lens bending, the spherical aberration can be seen to have a minimum value.

The power of a thin lens or the reciprocal of its focal length is given by

$$\phi = \frac{(n-1)c_1}{\mathcal{H}} \tag{24}$$

When the object is located at infinity, the shape factor for minimum spherical aberration can be represented by

$$\mathcal{H} = \frac{n(2n+1)}{2(n+2)} \tag{25}$$

The resultant third-order spherical aberration of the marginal ray in angular units is

$$SA3 = \frac{n^2 - (2n+1)\mathcal{H} + (1 + 2/n)\mathcal{H}^2}{16(n-1)^2(FN)^3} \tag{26}$$

or after some algebraic manipulations,

$$SA3 = \frac{n(4n-1)}{64(n+2)(n-1)^2(FN)^3} \tag{27}$$

where, for a thin lens, the FN is the focal length f divided by the lens diameter, which in this case is the same as entrance pupil diameter D_{ep}. Inspection of this equation illustrates that smaller values of spherical aberration are obtained as the refractive index increases.

When the object is located at a finite distance s_o, the equations for the shape factor and residual spherical aberration are more complex. Recalling that the magnification m is the ratio of the object distance to the image distance and that the object distance is negative if the object lies to the left of the lens, the relationship between the object distance and the magnification is

$$\frac{1}{s_o\phi} = \frac{m}{1-m} \tag{28}$$

where m is negative if the object distance and the lens power have opposite signs. The term $1/s_o\phi$ represents the reduced or ϕ-normalized reciprocal object distance v, i.e., s_o is measured in units of focal length ϕ^{-1}. The shape factor for minimum spherical aberration is given by

$$\mathcal{H} = \frac{n(2n+1)}{2(n+2)} + \frac{2(n^2-1)}{n+2}\left(\frac{m}{1-m}\right) \tag{29}$$

and the resultant third-order spherical aberration of the marginal ray in angular units is

$$SA3 = \frac{1}{16(n-1)^2(FN)^3}\left[n^2 - (2n+1)\mathcal{H} + \frac{n+2}{n}\mathcal{H}^2 + (3n+1)(n-1)\left(\frac{m}{1-m}\right)\right.$$

$$\left. - \frac{4(n^2-1)}{n}\left(\frac{m}{1-m}\right)\mathcal{H} + \frac{(3n+2)(n-1)^2}{n}\left(\frac{m}{1-m}\right)^2\right] \tag{30}$$

where FN is the effective focal length of the lens f divided by its entrance pupil diameter. When the object is located at infinity, the magnification becomes zero and the above two equations reduce to those previously given.

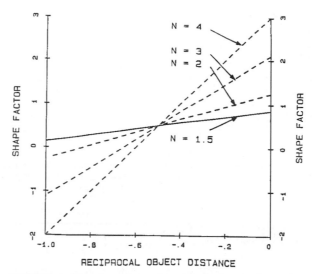

FIGURE 6 The shape factor for a single lens is shown for several refractive indexes as a function of reciprocal object distance v where the distance is measured in units of focal length.

Figure 6 illustrates the variation in shape factor as a function of v for refractive indices of 1.5–4 for an FN = 1. As can be seen from the figure, lenses have a shape factor of 0.5 regardless of the refractive index when the magnification is -1 or $v = -0.5$. For this shape factor, all lenses have biconvex surfaces with equal radii. When the object is at infinity and the refractive index is 4, lenses have a meniscus shape towards the image. For a lens with a refractive index of 1.5, the shape is somewhat biconvex, with the second surface having a radius about 6 times greater than the first surface radius.

Since the minimum-spherical lens shape is selected for a specific magnification, the spherical aberration will vary as the object-image conjugates are adjusted. For example, a lens having a refractive index of 1.5 and configured for $m = 0$ exhibits a substantial increase in spherical aberration when the lens is used at a magnification of -1. Figure 7 illustrates the variation in the angular spherical aberration as both a function of refractive index and reciprocal object distance v when the lens bending is for minimum spherical aberration with the object located at infinity. As can be observed from Fig. 7, the ratio of the spherical aberration, when $m = -0.5$ and $m = 0$, increases as n increases. Figure 8 shows the variation in angular spherical aberration when the lens bending is for minimum spherical aberration at a magnification of -1. In a like manner, Fig. 9 presents the variation in angular spherical aberration for a convex-plano lens with the plano side facing the image. The figure can also be used when the lens is reversed by simply replacing the object distance with the image distance.

Figures 7–9 may provide useful guidance in setting up experiments when the three forms of lenses are available. The so-called "off-the-shelf" lenses that are readily available from a number of vendors often have the convex-plano, equal-radii biconvex, and minimum spherical shapes.

Figure 10 shows the relationship between the third-order spherical aberration and coma, and the shape factor for a thin lens with a refractive index of 1.5, stop in contact, and the object at infinity. The coma is near zero at the minimum spherical aberration shape. The shape of the lens as a function of shape factor is shown at the top of the figure.

For certain cases, it is desirable to have a single lens with no spherical aberration. A

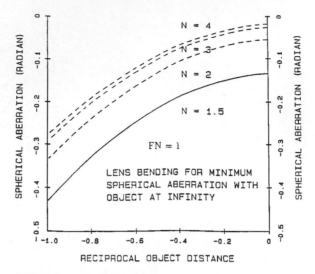

FIGURE 7 Variation of angular spherical aberration as a function of reciprocal object distance v for various refractive indices when the lens is shaped for minimum spherical aberration with the object at infinity. Spherical aberration for a specific FN is determined by dividing the aberration value shown by $(FN)^3$.

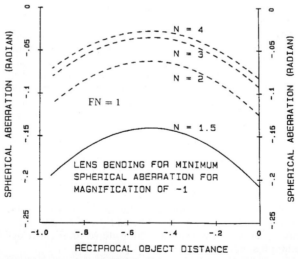

FIGURE 8 Variation of angular spherical aberration as a function of reciprocal object distance v for various refractive indices when the lens is shaped for minimum spherical aberration for a magnification of -1. Spherical aberration for a specific FN is determined by dividing the aberration value shown by $(FN)^3$.

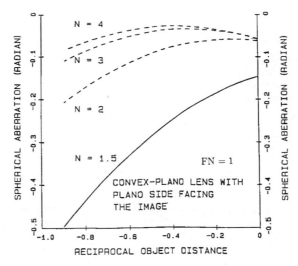

FIGURE 9 Variation of angular spherical aberration as a function of reciprocal object distance v for various refractive indices when the lens has a convex-plano shape with the plano side facing the object. Spherical aberration for a specific FN is determined by dividing the aberration value shown by $(FN)^3$.

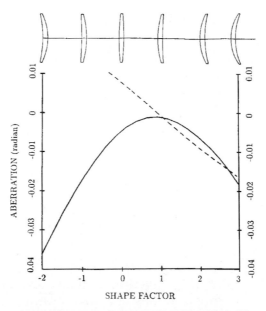

FIGURE 10 Variation of spherical aberration (solid curve) and coma (dashed line) as a function of shape factor for a thin lens with a refractive index of 1.5, stop in contact with the lens, and the object at infinity. The shape of the lens as the shape factor changes is shown at the top of the figure.

useful form is the plano-convex, with the plano side facing the object, if the convex side is figured as a conic surface with a conic constant of $-n^2$. Caution should be exercised when using this lens form at other than infinite object distances; however, imaging at finite conjugates can be accomplished by using two lenses with their plano surfaces facing one another and the magnification being determined by the ratio of the focal lengths. It should be noted that for this lens form, the actual thickness of the lenses is not important and that the inclusion of the conic surface does not alter the focal length.

The off-axis performance of a lens shaped for minimum spherical aberration with the object at infinity can be estimated by using the following equations. Assuming that the stop is in contact with the lens, the third-order angular sagittal coma is given by

$$\text{CMA}_s = \frac{\theta}{16(n+2)(\text{FN})^2} \tag{31}$$

where the field angle θ is expressed in radians. The tangential coma is three times the sagittal coma or $\text{CMA}_t = 3 \cdot \text{CMA}_s$. The diameter of the angular astigmatic blur formed at best focus is expressed by

$$\text{AST} = \frac{\theta^2}{\text{FN}} \tag{32}$$

The best focus location lies midway between the sagittal and tangential foci. An estimate of the axial angular chromatic aberration is given by

$$\text{AChr} = \frac{1}{2V(\text{FN})} \tag{33}$$

where V is the Abbe number of the glass and $V = (n_2 - 1)/(n_3 - n_1)$, with $n_1 < n_2 < n_3$.

If a singlet is made with a conic or fourth-order surface, the spherical aberration is corrected by the aspheric surface, and the bending can be used to remove the coma. With the stop in contact with the lens, the residual astigmatism and chromatic errors remain as expressed by the preceding equations. Figure 11 depicts the shapes of such singlets for

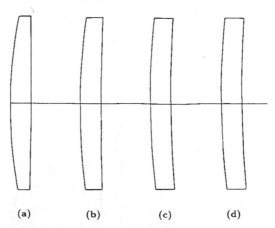

(a)   (b)   (c)   (d)

FIGURE 11 Variation of shape of singlets when the spherical aberration is corrected by the conic constant and the coma by the bending.

TABLE 1. Prescription of Singlets Corrected for Both Spherical Aberration and Coma

Lens	R_1	Thickness	R_2	Index	CC_2
a	0.55143	0.025	−5.27966	1.5	−673.543
b	0.74715	0.025	2.90553	2.0	23.2435
c	0.88729	0.025	1.56487	3.0	0.86904
d	0.93648	0.025	1.33421	4.0	0.24340

refractive indices of 1.5, 2, 3, and 4. Each lens has a unity focal length and an FN of 10. Table 1 presents the prescription of each lens where CC_2 is the conic constant of the second surface.

1.11 LANDSCAPE LENSES AND THE INFLUENCE OF STOP POSITION

The first lens used for photography was designed in 1812 by the English scientist W. H. Wollaston about a quarter of a century before the invention of photography. He discovered that a meniscus lens with its concave surface towards the object could produce a much flatter image field than the simple biconvex lens commonly used at that time in the camera obscuras. This lens became known as the landscape lens and is illustrated in Fig. 12. Wollaston realized that if the stop was placed an appropriate amount in front of the lens and the F-number was made to be modest, the image quality would be improved significantly over the biconvex lens.

The rationale for this can be readily seen by considering the influence on the residual aberrations of the lens by movement of the stop. Functionally, the stop allows certain rays in the oblique beam to pass through it while rejecting the rest. By simple inspection, it is clear that the movement of the stop (assuming a constant FN is maintained) will not affect the axial aberrations, while the oblique aberrations will be changed. In order to understand the influence of stop movement on the image quality, a graphical method was devised by

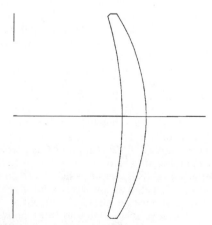

FIGURE 12 Landscape lens with the aperture stop located to the left of the lens.

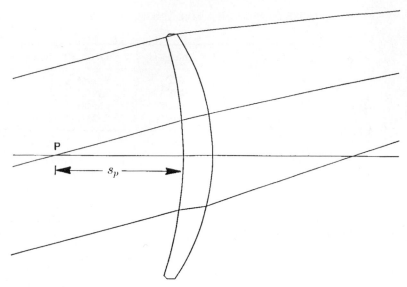

FIGURE 13   Rays traced at a given obliquity where the intersection of a given ray with the optical axis is P, located a distance s_p from the front surface of the lens.

R. Kingslake in which he traced a number of rays in the meridional plane at a given obliquity angle as illustrated in Fig. 13. A plot is generated that relates the intercept height of each real ray at the image plane H_i to the distance s_p from the intersection of the ray with optical axis P to the front surface of the lens. Each ray can be viewed as the principal ray when the stop is located at the intersection point P. This $H_i - s_p$ plot provides significant insight into the effect upon image quality incurred by placement of the stop. The shape of the curve provides information about the spherical aberration, coma, tangential field curvature, and distortion. Spherical aberration is indicated by an S-shaped curve, while the curvature at the principal ray point is a gauge of the coma. The coma is zero at inflection points. When the curve is a straight line, both coma and spherical aberration are essentially absent. The slope of the curve at the principal ray point is a measure of the tangential field curvature or the sag of the tangential field, i.e., astigmatism. The difference in height of the real and Gaussian principal rays in the image plane is distortion. For situations where the curve does not exhibit spherical aberration, it is impossible to correct the coma by shifting the stop.

Since a simple meniscus lens has stop position and lens bending as degrees of freedom, only two aberrations can be corrected. Typically, coma and tangential field curvature are chosen to be corrected, while axial aberrations are controlled by adjusting the FN of the lens. The $H_i - s_p$ plot for the lens shown in Fig. 13 is presented in Fig. 14, where the field angle is 10° and the image height is expressed as a percent of the Gaussian image height. The lens has a unity focal length, and the lens diameter is 0.275. Table 2 contains the prescription of the lens. Examination of this graph indicates that the best selection for stop location is when the stop is located at $s_p = -0.1505$ (left of the lens). For this selection, the coma and tangential astigmatism will be zero since the slope of the curve is zero and an inflection point is located at this stop position. Figure 15 shows the astigmatic field curves which clearly demonstrate the flat tangential image field for all field angles. Other aberrations cannot be controlled and must consequently be tolerated. When this lens is used at F/11, the angular blur diameter is less than 300 μradians. It should be noted that this condition is generally valid for only the evaluated field-angle obliquity and will likely

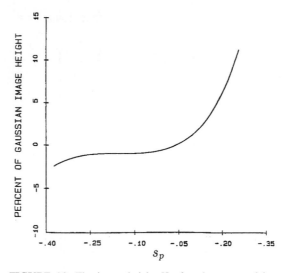

FIGURE 14 The image height H_i of each ray traced in Fig. 13 is plotted against the intersection length s_p to form the $H_i - s_p$ plot. H_i is expressed as a percent of the Gaussian image height as a direct measure of distortion.

TABLE 2. Prescription of Landscape Lens Shown in Fig. 13

Surface no.	Radius	Thickness	Index	Comment
1	Infinite	0.15050	1.0	Stop
2	−0.45759	0.03419	1.51680	BK7
3	−0.24887	0.99843	1.0	
4	Infinite			Image

be different at other field angles. Nevertheless, the performance of this lens is often acceptable for many applications.

An alternate configuration can be used where the lens is in front of the stop. Such configuration is used to conserve space since the stop would be located between the lens and the image. The optical performance is typically poorer due to greater residual spherical aberration.

The principle demonstrated by the $H_i - s_p$ plot can be applied to lenses of any complexity as a means to locate the proper stop position. It should be noted that movement of the stop will not affect the coma if spherical aberration is absent nor will astigmatism be affected if both spherical aberration and coma have been eliminated.

1.12 TWO-LENS SYSTEMS

Figure 16 illustrates the general imaging problem where an image is formed of an object by two lenses at a specified magnification and object-to-image distance. Most imaging

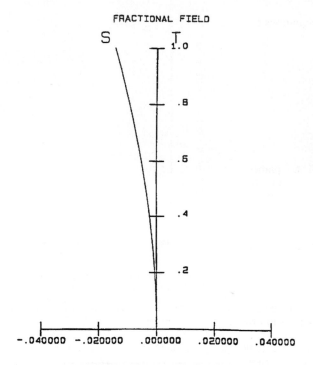

FRACTIONAL FIELD

LONGITUDINAL ABERRATION

FIGURE 15 Astigmatic field curves for the landscape lens having the stop located at the zero slope location on the $H_i - s_p$ plot in Fig. 14, which is the flat tangential field position. S represents the sagittal astigmatic focus while T indicates the tangential astigmatic focus.

problems can be solved by using two *equivalent* lens elements. An equivalent lens can comprise one lens or multiple lenses and may be represented by the principal planes and power of a single thick lens. All distances are measured from the principal points of each equivalent lens element. For simplicity, the lenses shown in Fig. 16 are thin lenses. If the magnification m, object-image distance $\mathscr{S}$, and lens powers ϕ_a and ϕ_b are known, then the

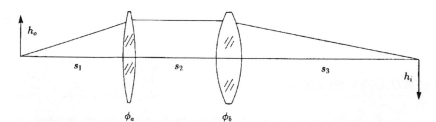

FIGURE 16 General imaging problem where the image is formed by two separated lenses.

equations for s_1, s_2, and s_3 are given by

$$s_1 = \frac{\phi_b(\mathscr{S} - s_2) - 1 + m}{m\phi_a + \phi_b}$$

$$s_2 = \frac{\mathscr{S}}{2}\left[1 \pm \sqrt{1 - \frac{4[\mathscr{S}m(\phi_a + \phi_b) + (m-1)^2]}{\mathscr{S}^2 m\phi_a\phi_b}}\right]$$

$$s_3 = \mathscr{S} - s_1 - s_2 \tag{34}$$

The equation for s_2 indicates that zero, one, or two solutions may exist.

If the magnification and the distances are known, then the lens powers can be determined by

$$\phi_a = \frac{\mathscr{S} + (s_1 + s_2)(m - 1)}{ms_1 s_2}$$

and $\tag{35}$

$$\phi_b = \frac{\mathscr{S} + s_1(m - 1)}{s_2(\mathscr{S} - s_1 - s_2)}$$

It can be shown that only certain pairs of lens powers can satisfy the magnification and separation requirements. Commonly, only the magnification and object-image distance are specified with the selection of the lens powers and locations to be determined. By utilizing the preceding equations, a plot of regions of all possible lens power pairs can be generated. Such a plot is shown as the shaded region in Fig. 17 where $\mathscr{S} = 1$ and $m = -0.2$.

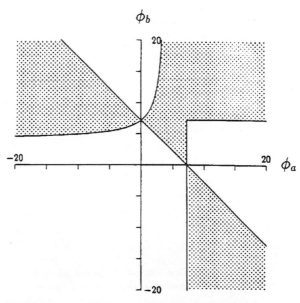

FIGURE 17 Shaded regions indicate all possible power pairs for the two lenses used for imaging. The solution space may be limited by physical considerations such as maximum aperture.

Examination of this plot can assist in the selection of lenses that may likely produce better performance by, for example, selecting the minimum power lenses. The potential solution space may be limited by placing various physical constraints on the lens system. For example, the allowable lens diameters can dictate the maximum powers that are reasonable. Lines of maximum power can then be plotted to show the solution space.

When s_1 becomes very large compared to the effective focal length *efl* of the lens combination, the optical power of the combination of these lenses is expressed by

$$\phi_{ab} = \phi_a + \phi_b - s_2 \phi_a \phi_b \qquad (36)$$

The effective focal length is ϕ_{ab}^{-1} or

$$f_{ab} = \frac{f_a f_b}{f_a + f_b - s_2} \qquad (37)$$

and the back focal length is given by

$$bfl = f_{ab}\left(\frac{f_a - s_2}{f_a}\right) \qquad (38)$$

The separation between lenses is expressed by

$$s_2 = f_a + f_b - \frac{f_a f_b}{f_{ab}} \qquad (39)$$

Figure 18 illustrates the two-lens configuration when thick lenses are used. The principal points for the lens combination are denoted by P_1 and P_2, P_{a1} and P_{a2} for lens *a*, and P_{b1} and P_{b2} for lens *b*. With the exception of the back focal length, all distances are measured from the principal points of each lens element or the combined lens system as shown in

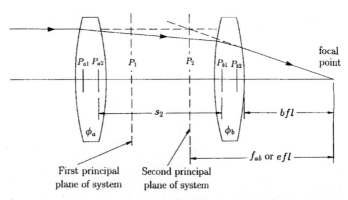

FIGURE 18 Combination of two thick lenses illustrating the principal points of each lens and the system, the f_{ab} or *efl*, and the *bfl*. Distances are measured from the principal points with the exception of the *bfl*.

the figure. For example, s_2 is the distance from P_{a2} to P_{b1}. The *bfl* is measured from the final surface vertex of the lens system to the focal point.

1.13 ACHROMATIC DOUBLETS

The singlet lens suffers from axial chromatic aberration, which is determined by the Abbe number V of the lens material and its FN. A widely used lens form that corrects this aberration is the achromatic doublet as illustrated in Fig. 19. An achromatic lens has equal focal lengths in c and f light. This lens comprises two lens elements where one element with a high V-number (crown glass) has the same power sign as the doublet and the other element has a low V-number (flint glass) with opposite power sign. Three basic configurations are used. These are the cemented doublet, broken contact doublet, and the widely airspaced doublet (dialyte). The degrees of freedom are two lens powers, glasses, and shape of each lens.

The resultant power of two thin lenses in close proximity, $s_2 \to 0$, is $\phi_{ab} = \phi_a + \phi_b$ and the transverse primary chromatic aberration TPAC is

$$\text{TPAC} = -yf_{ab}\left[\frac{\phi_a}{V_a} + \frac{\phi_b}{V_b}\right] \tag{40}$$

where y is the marginal ray height. Setting $\text{TPAC} = 0$ and solving for the powers of the lenses yields

$$\phi_a = \frac{V_a}{f_{ab}(V_a - V_b)} \tag{41}$$

and

$$\phi_b = \frac{-V_b\phi_a}{V_a} \tag{42}$$

The bending or shape of a lens is expressed by $c = c_1 - c_2$ and affects the aberrations of the lens. The bending of each lens is related to its power by $c_a = \phi_a/(n_a - 1)$ and $c_b = \phi_b(n_b - 1)$. Since the two bendings can be used to correct the third-order spherical and coma, the equations for these aberrations can be combined to form a quadratic equation in terms of the curvature of the first surface c_1. Solving for c_1 will yield zero, one, or two solutions for the first lens. A linear equation relates c_1 to c_2 of the second lens.

While maintaining the achromatic correction of a doublet, the spherical aberration as a function of its shape (c_1) is described by a parabolic curve. Depending upon the choices of

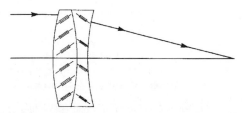

FIGURE 19 Typical achromatic doublet lens.

glasses, the peak of the curve may be above, below, or at the zero spherical aberration value. When the peak lies in the positive spherical aberration region, two solutions with zero spherical aberration exist in which the solution with the smaller value of c_1 is called the left-hand solution (Fraunhofer or Steinheil forms) and the other is called the right-hand solution (Gaussian form). Two additional solutions are possible by reversal of the glasses. These two classes of designs are denoted as crown-in-front and flint-in-front designs. Depending upon the particular design requirements, one should examine all four configurations to select the most appropriate. The spherical aberration curve can be raised or lowered by the selection of the V difference or the n difference. Specifically, the curve will be lowered as the V difference is increased or if the n difference is reduced. As for the thin singlet lens, the coma will be zero for the configuration corresponding to the peak of the spherical aberration curve.

Although the primary chromatic aberration may be corrected, a residual chromatic error often remains and is called the secondary spectrum, which is the difference between the ray intercepts in d and c. Figure 20a illustrates an F/5 airspaced doublet that exhibits well-corrected spherical light and primary chromatic aberrations and has notable secondary color. The angular secondary spectrum for an achromatic thin-lens doublet is given by

$$\text{SAC} = \frac{-(P_a - P_b)}{2(\text{FN})(V_a - V_b)} \tag{43}$$

where $P = (n_\lambda - n_c)/(n_f - n_c)$ is the partial dispersion of a lens material. In general, the ratio $(P_a - P_b)/(V_a - V_b)$ is nearly a constant which means little can be done to correct the SAC. A few glasses exist that allow $P_a - P_b \approx 0$, but the $V_a - V_b$ is often small, which results in lens element powers of rather excessive strength in order to achieve achromatism. Figure 20b shows an F/5 airspaced doublet using a relatively new pair of glasses that have a small $P_a - P_b$ and a more typical $V_a - V_b$. Both the primary and secondary chromatic aberration are well corrected. Due to the relatively low refractive index of the crown glass, the higher power of the elements results in spherical aberration through the seventh order. Almost no spherochromatism (variation of spherical aberration with wavelength) is

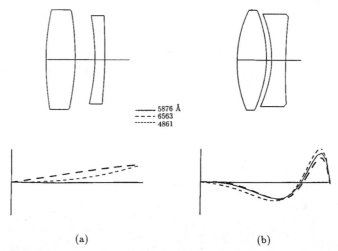

$$\begin{array}{l}\text{——— 5876 Å}\\ \text{– – – 6563}\\ \text{······· 4861}\end{array}$$

(a)   (b)

FIGURE 20 An F/5 airspaced doublet using conventional glasses is shown in a and exhibits residual secondary chromatic aberration. A similar lens is shown in b that uses a new glass to effectively eliminate the secondary color.

TABLE 3. Prescriptions for Achromatic Doublets Shown in Fig. 20

	Achromatic doublet—1		
Surface no.	Radius	Thickness	Glass
1	49.331	6.000	BK7 517:642
2	−52.351	4.044	Air
3	−43.888	2.000	SF1 717:295
4	−141.706		Air

	Achromatic doublet—2		
Surface no.	Radius	Thickness	Glass
1	23.457	6.000	FK03 439:950
2	−24.822	1.059	Air
3	−22.516	3.000	BK7 517:642
4	94.310		Air

observed. The 80 percent blur diameter is almost the same for both lenses and is 0.007. Table 3 contains the prescriptions for these lenses.

When the separation between the lens elements is made a finite value, the resultant lens is known as a *dialyte* and is illustrated in Fig. 21. As the lenses are separated by a distance s_d, the power of the flint or negative lens increases rapidly. The distance s_d may be expressed as a fraction of the crown-lens focal length by $p = s_d/f_a$. Requiring the chromatic aberration to be zero implies that

$$\frac{y_a^2}{f_a V_a} + \frac{y_b^2}{f_b V_b} = 0 \tag{44}$$

By inspection of the figure and the definition of p, it is evident that $y_b = y_a(1 - p)$ from which it follows that

$$f_b V_b = -f_a V_a (1 - p)^2 \tag{45}$$

The total power of the dialyte is

$$\phi = \phi_a + \phi_b(1 - p) \tag{46}$$

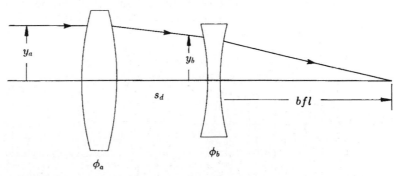

FIGURE 21 Widely separated achromatic doublet known as the dialyte lens.

Solving for the focal lengths of the lenses yields

$$f_a = f_{ab}\left[1 - \frac{V_b}{V_a(1-p)}\right] \tag{47}$$

and

$$f_b = f_{ab}(1-p)\left[1 - \frac{V_a(1-p)}{V_b}\right] \tag{48}$$

The power of both lenses increases as p increases.

The typical dialyte lens suffers from residual secondary spectrum; however, it is possible to design an airspaced achromatic doublet with only one glass type that has significantly reduced secondary spectrum. Letting $V_a = V_b$ results in the former equations becoming

$$f_a = \frac{pf_{ab}}{p-1} \qquad f_b = -pf_{ab}(p-1) \qquad s_d = pf_a \qquad bfl = -f_{ab}(p-1) \tag{49}$$

When $f_{ab} > 0$, then p must be greater than unity, which means that the lens is quite long. The focal point lies between the two lenses, which reduces its general usefulness. This type of lens is known as the Schupmann lens, based upon his research in the late 1890s. Several significant telescopes, as well as eyepieces, have employed this configuraton. For $f_{ab} < 0$, the lens can be made rather compact and is sometimes used as the rear component of some telephoto lenses.

1.14 TRIPLET LENSES

In 1893, a new type of triplet lens for photographic applications was invented by the English designer H. Dennis Taylor. He realized that the power of two lenses in contact of equal, but opposite, power is zero, as is its Petzval sum. As the lenses are separated, the system power becomes positive since the negative lens contributes less power. The Petzval sum remains zero, since it does not depend upon the marginal ray height. In order to overcome the large aberrations of such a configuration, Taylor split the positive lens into two positive lenses and placed one on each side of the negative lens. A stop is often located between the negative and rear-positive lenses. Figure 22 illustrates a typical triplet lens. The triplet can be used at reasonably large apertures ($>F/4$) and moderately large fields of view ($>\pm25°$).

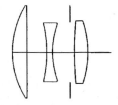

FIGURE 22 Typical triplet lens.

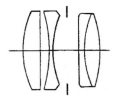

FIGURE 23 Typical Tessar lens.

The triplet has eight degrees of freedom which are the three powers, two airspaces, and three lens bendings. The lens powers and airspaces are used to control the axial and lateral chromatic aberrations, the Petzval sum, the focal length, and the ratio of the airspaces. Spherical aberration, coma, and astigmatism are corrected by the lens bendings. Distortion is usually controlled by the airspace ratio or the choice of glasses. Consequently, the triplet has exactly the number of degrees of freedom to allow correction of the basic aberrations and maintain the focal length.

The design of a triplet is somewhat difficult since a change of any surface affects every aberration. The choice of glass is important and impacts the relative aperture, field of view, and overall length. For example, a large ΔV produces a long system. It should be noted that a triplet corrected for third-order aberrations by using the degrees of freedom almost always leads to a lens with poor performance. A designer normally leaves a certain amount of residual third-order aberrations to balance the higher-order terms. The process for thin-lens predesign is beyond the scope of this handbook; however, it may be found in various references comprising the bibliography.

A few years later, Paul Rudolph of Zeiss developed the Tessar, which resembles the triplet, with the rear lens replaced by an achromatic doublet. The Tessar shown in Fig. 23 was an evolution of Rudolph's anastigmats which were achromatic lenses located about a central stop. The advantage of the achromatic rear component is that it allows reduction of the zonal spherical aberration and the oblique spherical aberration, and reduces the separation of the astigmatic foci at other than the design maximum field angle. Performance of the Tessar is quite good and has generally larger relative apertures at equivalent field angles than the triplet. A variety of lenses were derived from the triplet and the Tessar in which the component lenses were made into doublets or cemented triplets.

1.15 SYMMETRICAL LENSES

In the early 1840s, it was recognized that lenses that exhibit symmetry afford various benefits to the lens designer. The first aberration acknowledged to be corrected by the symmetry principle was distortion. It can also be shown that coma and lateral color are necessarily corrected by a symmetrical lens construction. Although the principle of symmetry implies that the lens be operated at a magnification of -1, the degree to which the aberrations are upset by utilizing the lens at other conjugates is remarkably small. This principle forms the basis of most wide-field-of-view lenses.

One of the earliest symmetrical lenses was the Periscopic (Periskop) lens invented by C. A. Steinheil in 1865. Figure 24 shows an F/11 Periscopic lens constructed from the landscape lens discussed previously. Symmetry corrects for coma and distortion, while the

FIGURE 24 The periscopic lens illustrates the earliest form of symmetrical lenses. It is formed by placing two landscape lenses about a central stop. Symmetry removes the aberrations of coma, distortion, and lateral color.

spacing of the lenses and their shapes are selected to produce a flat tangential astigmatic field. Since the stop position for the landscape lens was chosen to yield a flat tangential astigmatic field, essentially no change in the lens separation is necessary even though the Periscopic lens is being used at infinite conjugates. No correction for spherical aberration can be made. When used at other than unit magnification, some optical improvement can be achieved by making the stop slightly asymmetrical and/or having a different shape for the front or rear lens. This lens has continued to find application throughout this century.

By 1966, Dallmeyer in England and Steinheil and von Seidel in Germany both invented the Rapid Rectilinear lens that could be used at apertures of up to F/6. The lens has two cemented achromats about a central stop. Use of the doublet allows correction of the axial chromatic and spherical aberrations. Glass selection is of importance in the design. Typically, the Δn between the glasses should be large while the ΔV should be relatively small. The positive lens is located nearest the stop and has the lower refractive index. A notable characteristic of the lens is that the aberrations are reasonably stable over a broad range of object distances.

It should be noted that vignetting is often used in these and other lens types to control the higher-order aberrations that are often observed at large field angles. Although a loss in illumination occurs, the gain in resolution is often worthwhile.

The airspaced dialyte lens comprises four lenses symmetrically arranged about a central stop. The rear portion of the lens is an achromatic doublet that has five degrees of freedom (an air space, two powers, and two bendings) which may be used to control the focal length, spherical aberration, axial chromatic aberration, astigmatism, and the Petzval sum. With a like pair of lenses mounted in front of the stop, the symmetry corrects the coma, distortion, and lateral color. When used at infinite conjugates, the resultant residuals of the aberrations can be controlled by deviating somewhat from perfect symmetry of the air spaces about the stop. Lenses of this type can provide useful performance with apertures approaching F/4 and fields of view of about $\pm 20°$ or so.

1.16 DOUBLE-GAUSS LENSES

In the early 1800s, Gauss described a telescope objective comprising a pair of meniscus lenses with one having positive power and the other negative power. An interesting aspect of his lens is that the spherochromatism is essentially constant. Although this lens found little acceptance, in 1888, Alvan Clark of Massachusetts placed a pair of the Gauss lenses around a central stop to create a high-aperture, wide-field-of-view lens. This lens form is known as the Double-Gauss lens and is the basis of almost every high-aperture lens developed to date. An example of this lens was patented by Richter in 1933 and can cover a field of view of $\pm 45°$ at F/6.

In 1896, Paul Rudolph of Zeiss developed the Planar which reduces the often serious oblique spherical aberration and the separation of the astigmatic foci at intermediate field angles. Rudolph placed a buried surface into the thick negative elements to control the chromatic aberration. A buried surface is defined as the interface between two glasses that have the same refractive index n_d at the central wavelength, but have significantly different Abbe numbers. Such a surface has no effect upon the monochromatic aberrations or the lens system power, but does allow the inclusion of a wide range of chromatic aberration to compensate for that caused by the rest of the lens.

Many Double-Gauss lenses are symmetrical; however, it was discovered that if the lens was made unsymmetrical, then an improvement in performance could be realized. This lens form is often called the Biotar. A large portion of 35-mm camera lenses are based

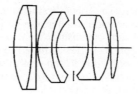

FIGURE 25 Unsymmetrical Double-Gauss or Biotar lens introduced as the Leica Summitar in 1939.

upon this design form or some modification thereof. Figure 25 shows the configuration of the Leica Summitar introduced in 1939.

It is the general nature of meniscus lens systems of this type to exhibit little coma, distortion, or lateral color; however, oblique spherical aberration is often observed to increase to significant levels as the field angle increases. Oblique spherical aberration can be recognized in transverse ray plots as the S shape of spherical aberration, but with the S becoming increasingly stronger as the field angle increases. As the aperture is increased beyond about F/8, the outer negative elements must be thickened dramatically and achromatic surfaces must necessarily be included.

1.17 PETZVAL LENSES

In 1839, Petzval designed a new type of lens that comprises a front objective with an achromatic, airspaced doublet as the rear elements. The Petzval lens has found great application in projectors and as a portrait lens. Both spherical aberration and coma can be well-corrected, but the lens configuration causes the Petzval sum to be undercorrected, which results in the field of view being limited by the astigmatism. The Petzval field curves inward and may be corrected by including a *field flattener lens* in close proximity to the image plane. A typical example of a Petzval lens is shown in Fig. 26.

1.18 TELEPHOTO LENSES

A telephoto lens provides an effective focal length *efl* that is longer than its overall length s_{ol} as measured from the front of the lens to the image plane. The telephoto ratio is defined as s_{ol}/efl, thus a lens with a ratio less than one is a telephoto lens. The basic concept of a telephoto lens is illustrated by the dialyte lens configuration in which a negative lens is inserted between the objective lens and the image plane. This concept goes back to Kepler, but Peter Barlow developed the idea in the early 1800s by including a negative achromat

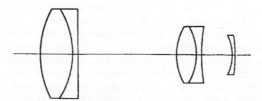

FIGURE 26 Typical Petzval lens.

in telescopes to increase their magnification. Barlow type lenses are widely used today. As the telephoto ratio is made smaller, the design of the lens becomes more difficult, primarily due to the Petzval sum increasing.

When most telephoto lenses are used to view objects that are relatively close, the image quality degrades rapidly due to the typical unsymmetrical lens configuration. Some modern telephoto lenses include one or more elements that move as the lens is focused for the purpose of aberration correction.

1.19 INVERTED OR REVERSE TELEPHOTO LENSES

A reverse telephoto lens has a telephoto ratio greater than unity and exhibits a shorter focal length than its overall length, a larger *bfl* than is provided by normal lenses of the same *efl*, lenses with generally large apertures and wide fields of view, and lens elements of physically larger size that allow easier manufacture and handling. The basic configuration has a large negative lens located in front of a positive objective lens. Since the negative lens makes the object appear closer to the objective lens, the resultant image moves beyond the focal point, thereby making the *bfl* greater than the *efl*.

An extreme form of the reverse telephoto lens is the fish-eye or sky lens. Such lenses have a total field of view of 180° or more. The image formed by these lenses has very large barrel distortion. Recalling that the image height for a distortionless lens on a flat image surface is $f \tan \theta$, the reverse telephoto lens has mapping relationships such as $f\theta$ and $f \sin \theta$. When the barrel distortion is given by $f \sin \theta$, the illumination across the image will be constant if such effects as vignetting and stop distortion are absent. Barrel distortion has the effect of compressing the outer portions of the image towards the central portion, thereby increasing the flux density appropriately.

After World War II, the Russian designer M. M. Roosinov patented a double-ended reverse-telephoto lens that was nearly symmetrical with large negative lenses surrounding a pair of positive lenses with a central stop. Although the back focal length is quite short, it provides relatively large aperture with a wide field of view and essentially no distortion. Lenses of this type have found significant use in aerial photography and photogrammetry.

1.20 PERFORMANCE OF REPRESENTATIVE LENSES

Figures 27–38 present the performance of lenses, selected generally from the patent literature, representing a variety of lens types. The measures of performance provided in each figure have been selected for utilization purposes. Diffraction effects have not been included.

Each figure is divided into four sections *a–d*. Section *a* is a drawing of the lens showing the aperture stop. Section *b* contains two set of plots. The *solid* line is for the distortion versus field of view (θ) in degrees while the *dashed* lines show the transmission of the lens versus field of view for three F-numbers. Transmission in this case is *one minus the fractional vignetting.* No loss for coatings, surface reflection, absorption, etc., is included. The rms diameter of the geometric point source image versus field of view for three F-numbers is presented in section *c*. The spot sizes are in angular units and were calculated for the central wavelength only, i.e., monochromatic values. Note that the ordinate is logarithmic. The final section, *d*, contains angular transverse ray plots in all three colors for both the on-axis and near-extreme field angles with y_{ep} being measured in the entrance pupil. The lower right plot shows the axial aberrations while the upper left plot represents the tangential/meridional aberrations and the upper right plot presents the sagittal

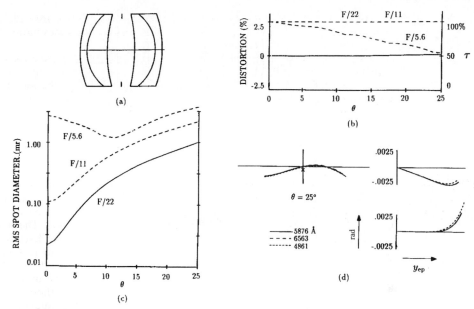

(a)

(b)

(c)

(d)

FIGURE 27   Rapid Rectilinear: This lens is an aplanat which is symmetrical with the rear half corrected for spherical aberration and flat tangential field. A compact configuration is realized by having a large amount of coma in each half. Symmetry removes the lens system coma, distortion, and lateral color. This type of lens is one of the most popular camera lenses ever made.

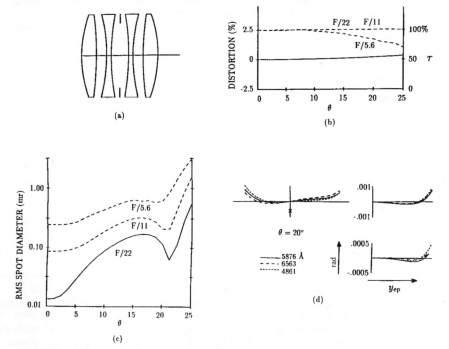

(a)

(b)

(c)

(d)

FIGURE 28   Celor: F/5.6 with 50° total field of view. Also known as an airspaced dialyte lens. *After R. Kingslake,* Lens Design Fundamentals, *Academic Press, New York, 1978, p. 243.*

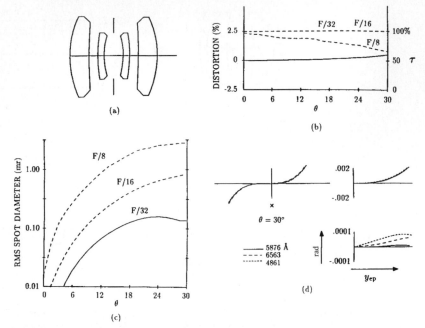

FIGURE 29 Symmetrical double anastigmat or Gauss homocentric objective: basic form of Double-Gauss lens using a pair of Gauss telescope objectives. First patented by Alvan Clark in 1888, USP 399,499. After *R. Kingslake*, Lens Design Fundamentals, *Academic Press, New York, 1978, pp. 244–250.*

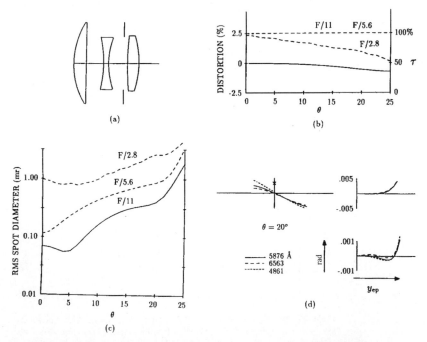

FIGURE 30 Triplet: F/2.8 with 50° total field of view. (*Tronnier, USP 3,176,582.*)

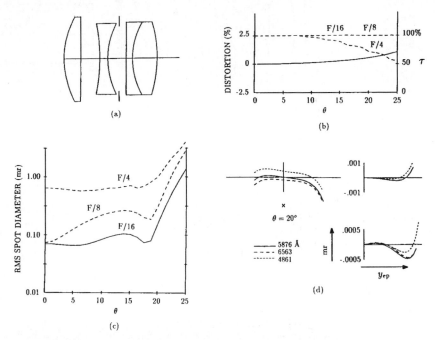

(a)

(b)

(c)

(d)

FIGURE 31   Tessar: F/4 with 50° total field of view. (*Tronnier, USP 2,084,714, 1937.*)

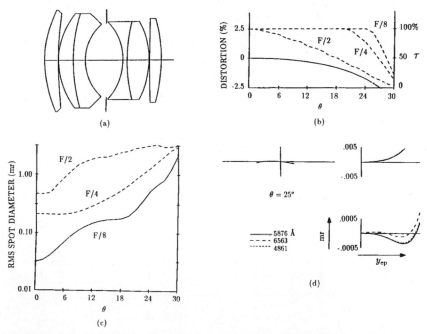

(a)

(b)

(c)

(d)

FIGURE 32   Unsymmetrical Double-Gauss: This lens was designed in 1933 for Leitz and was called the Summar. F/2 with 60° total field of view. This lens was replaced by the Leitz Summitar in 1939, due to rapidly degrading off-axis resolution and vignetting. Compare this lens with the lens shown in Fig. 33. (*Tronnier, USP 2,673,491.*)

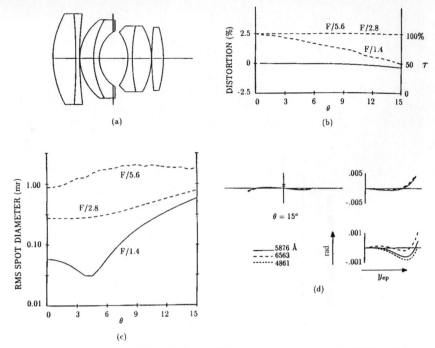

FIGURE 33 Unsymmetrical Double-Gauss: This lens type was designed in 1939 for Leitz and was called the F/2 Summitar. Kodak had a similar lens called the F/1.9 Ektar. A later example of this design form is shown and operates at F/1.4 with 30° total field of view. (*Klemp, USP 3,005,379.*)

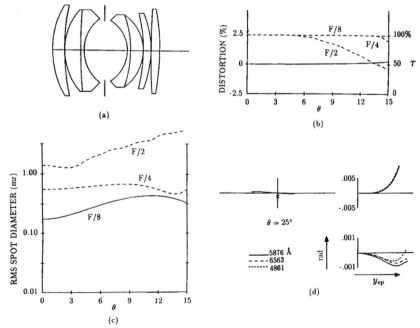

FIGURE 34 Unsymmetrical Double-Gauss: F/1.75 with 50° total field of view. Similar to the 1949 Leitz F/1.5 Summarit. This lens has a split rear element which produces improved resolution of the field of view and less vignetting than the earlier Summar type lens. (*Cook, USP 2,959,102.*)

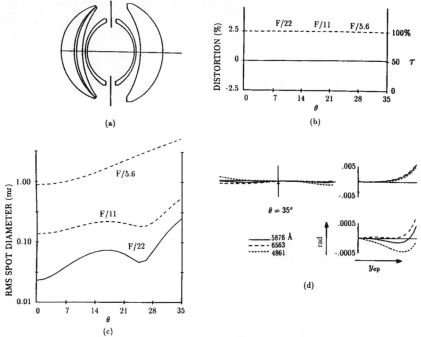

FIGURE 35 Unsymmetrical Double-Gauss: F/5.6 with 70° field of view. This lens is a variant of the 1933 Zeiss F/6.3 Topogon (*USP 2,031,792*) and is the Bausch & Lomb Metrogon. The principal difference is the splitting of the front element. (*Rayton, USP 2,325,275.*)

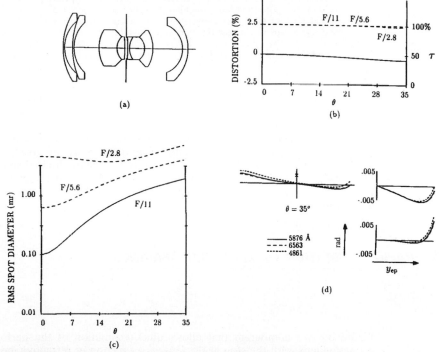

FIGURE 36 Reverse Telephoto: This lens was developed by Zeiss in 1951 and is known as the Biogon. It operates at F/2.8 with 70° field of view. This lens comprises two reverse-telephoto objectives about a central stop. (*Bertele, USP 2,721,499.*)

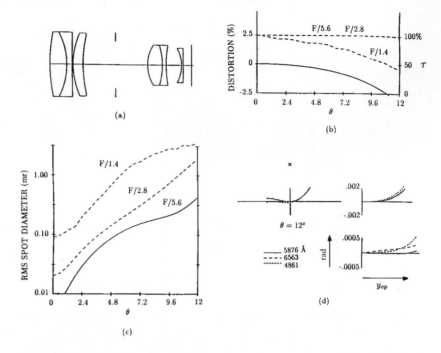

FIGURE 37   Petzval: Example of Kodak projector lens operating at F/1.4 with 24° total field of view. The front lens group has its power shared between a cemented doublet and a singlet for aberration correction. Note that the aperture stop is located between the front and rear groups rather than the more common location at the front group. Resolution in the region near the optical axis is very good although it falls off roughly exponentially. The limiting aberrations are oblique spherical and cubic coma. (*Schade, USP 2,541,484.*)

aberrations. The X included on some of the tangential plots represents the location of the paraxial principal ray. The legend indicating the relationship between line type and wavelength is included.

The linear spot size is computed by multiplying the *efl* by the angular spot size. This value can be compared against the diffraction-limited spot size given by $2.44(\lambda/D_{ep})$. If the geometric spot is several times *smaller* than the diffraction-limited spot, then the lens may be considered to be diffraction-limited for most purposes. If the geometric spot is several times *larger*, then the lens performance is controlled by the geometric spot size for most applications.

1.21 *RAPID ESTIMATION OF LENS PERFORMANCE*

Singlet

Figure 39 is a nomogram that allows quick estimation of the performance of a single refracting lens, with the stop at the lens, as a function of refractive index N, dispersion V, F-number, and field of view θ. Chart A estimates the angular blur diameter β resulting

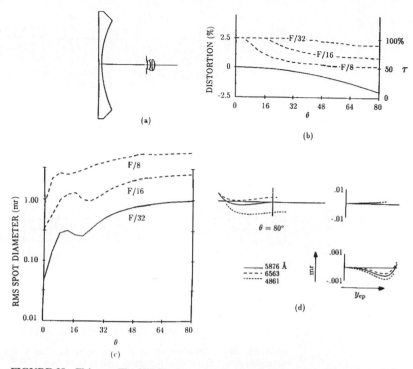

FIGURE 38 Fish-eye: The Hill Sky lens was manufactured by Beck of London in 1924. The lens has moderate resolution and enormous distortion characteristic of this type of lens. (*Merte, USP 2,126,126.*)

from a singlet with bending for minimum spherical aberration. The angular chromatic blur diameter is given by Chart B. The three rows of FN values below the chart represent the angular blur diameter that contains the indicated percentage of the total energy. Chart C shows the blur diameter due to astigmatism. Coma for a singlet bent for minimum spherical aberration with the stop at the lens is approximately

$$\frac{\theta}{16 \cdot (N+2) \cdot (\text{FN})^2} \tag{50}$$

Depth of Focus

The *depth of focus* of an optical system is expressed as the axial displacement that the image may experience before the resultant image blur becomes excessive. Figure 40 shows the geometric relationship of the angular blur tolerance $\Delta\theta$ to the depth of focus $\delta_\pm$. If the entrance pupil diameter is D_{ep} and the image distance is s_i, then the depth of focus is

$$\delta_\pm = \frac{s_i^2 \, \Delta\theta}{D_{ep} \pm s_i \, \Delta\theta} \tag{51}$$

or when $\delta \ll s_i$, the depth of focus becomes

$$\delta = \frac{s_i^2 \, \Delta\theta}{D_{ep}} \tag{52}$$

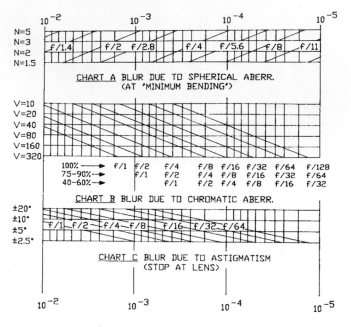

FIGURE 39 Estimation of single lens spot size as a function of refractive index, dispersion, F-number, and field of view. (*Smith,* Modern Optical Engineering, *McGraw-Hill, New York, 1990, p. 458.*)

When $s_i = f$, then

$$\delta = f \, \Delta\theta \, \text{FN} \tag{53}$$

The *depth of field* is distance that the object may be moved without causing excessive image blur with a fixed image location. The distance at which a lens may be focused such that the depth of field extends to infinity is $s_o = D_{ep}/\Delta\theta$ and is called the hyperfocal distance.

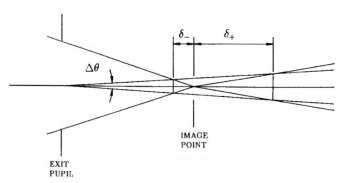

FIGURE 40 Geometric relationships for determining the geometric depth of focus of a lens.

If the lens system is diffraction-limited, then the depth of focus according to the Rayleigh criterion is given by

$$\delta = \pm \frac{\lambda}{2 n_i \sin^2 u_i} \tag{54}$$

Diffraction-Limited Lenses

It is well known that the shape of the image irradiance of an incoherent, monochromatic point-source formed by an aberration-free, circularly-symmetric lens system is described by the Airy function

$$E(r) = C_0 \left[\frac{2 J_1 (k D_{ep} r / 2)}{k D_{ep} r} \right]^2 \tag{55}$$

where J_1 is the first order Bessel function of the first kind, D_{ep} is the diameter of the entrance pupil, k is $2\pi/\lambda$, r is the radial distance from the center of the image to the

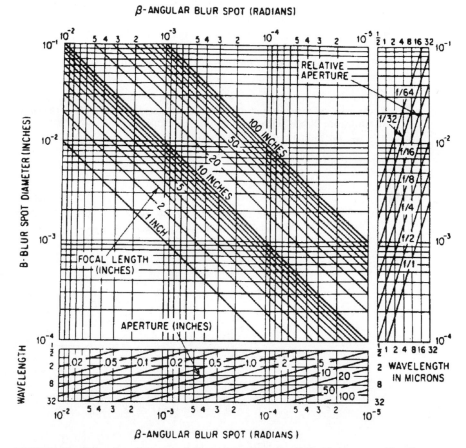

FIGURE 41 Estimation of the spot diameter for a diffraction-limited lens system. The diameter is that of the first dark ring of the Airy disk. (*Smith*, Modern Optical Engineering, *McGraw-Hill, New York, 1990, p. 458.*)

observation point, and C_0 is a scaling factor. The angular radius β_{DL} of the first dark ring of the image is $1.22(\lambda/D_{ep})$. A common measure for the resolution is Lord Rayleigh's criterion that asserts that two point sources are just resolvable when the maximum of one Airy pattern coincides with the first dark ring of the second Airy pattern, i.e., an angular separation of β_{DL}. Figure 41 presents a nomogram that can be used to make a rapid estimate of the diameter of angular or linear blur for a diffraction-limited system.

The modulation transfer function (MTF) *at a specific wavelength* λ for a circular entrance pupil can be computed by

$$\mathrm{MTF}_\lambda(\Omega) = \frac{2}{\pi}[\arccos\Omega - \Omega\sqrt{1-\Omega^2}] \quad \text{for} \quad 0 \le \Omega \le 1 \tag{56}$$

where Ω is the normalized spatial frequency (ν/ν_{co}) with the maximum or cut-off frequency ν_{co} being given by $1/\lambda_o$ FN.

Should the source be polychromatic and the lens system be aberration-free, then the perfect-image irradiance distribution of a point source can be written as

$$E(r) = C_1 \int_0^\infty \tilde{\mathscr{R}}(\lambda)\left[\frac{2J_1(kD_{ep}r/2)}{kD_{ep}r}\right]^2 d\lambda \tag{57}$$

where $\tilde{\mathscr{R}}(\lambda)$ is the peak normalized spectral weighting factor and C_1 is a scaling factor.

A quick estimation of this ideal irradiance distribution can be made by invoking the central limit theorem to approximate this distribution by a Gaussian function, i.e.,

$$E(r) \approx C_2 e^{-(r^2/2\sigma^2)} \tag{58}$$

where C_2 is a scaling constant and σ^2 is the estimated variance of the irradiance distribution. When $\tilde{\mathscr{R}}(\lambda) = 1$ in the spectral interval λ_S to λ_L and zero otherwise with $\lambda_s < \lambda_L$, an estimate of σ can be written as

$$\sigma = \frac{\mathscr{M}\lambda_L}{\pi D_{ep}} \tag{59}$$

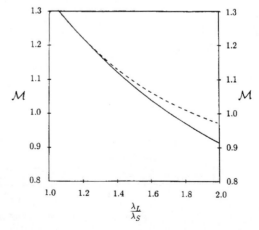

FIGURE 42 Variation of $\mathscr{M}$ with λ_L/λ_S or $b+1$ for $\tilde{\mathscr{R}}(\lambda) = 1$ as the solid curve and $\tilde{\mathscr{R}}(\lambda) = \lambda/\lambda_S$ as the dashed curve.

where $\mathcal{M} = 1.335 - 0.625b + 0.25b^2 - 0.0465b^3$ with $b = (\lambda_L/\lambda_S) - 1$. Should $\tilde{\mathcal{R}}(\lambda) = \lambda/\lambda_L$ in the spectral interval λ_S to λ_L and zero otherwise, which approximates the behavior of a quantum detector, $\mathcal{M} = 1.335 - 0.65b + 0.385b^2 - 0.099b^3$. The Gaussian estimate residual error is less than a few percent for $b = 0.5$ and remains useful even as $b \to 0$. Figure 42 contains plots of $\mathcal{M}$ for both cases of $\tilde{\mathcal{R}}(\lambda)$, where the abscissa is λ_L/λ_S.

A useful estimation of the modulation transfer function for this *polychromatic* lens system is given by

$$\text{MTF}(\nu) \approx e^{-2(\pi\sigma\nu)^2} \tag{60}$$

where ν is the spatial frequency. This approximation overestimates the MTF somewhat at lower spatial frequencies, while being rather a close fit at medium and higher spatial frequencies. The reason for this is that the central portion of the irradiance distribution is closely matched by the Gaussian approximation, while the irradiance estimation beyond several Airy radii begins to degrade, therefore impacting the lower spatial frequencies. Nevertheless, this approximation can provide useful insight into expected performance limits.

1.22 BIBLIOGRAPHY

Douglas S. Goodman, "Basic Optical Instruments," Chap. 4 in *Geometrical and Instrumental Optics,* Daniel Malacara ed., *Methods of Experimental Physics,* **25,** Academic Press, San Diego (1988).

R. E. Hopkins, "Geometrical Optics," Chap. 2 in *Geometrical and Instrumental Optics,* Daniel Malacara ed., *Methods of Experimental Physics,* **25,** Academic Press, San Diego (1988).

R. E. Hopkins, "The Components in the Basic Optical Systems," Chap. 3 in *Geometrical and Instrumental Optics,* Daniel Malacara ed., *Methods of Experimental Physics,* **25,** Academic Press, San Diego (1988).

R. Barry Johnson and C. Feng, "A History of IR Lens Designs," SPIE Critical Reviews **CR37,** 3–18 (1991).

Rudolf Kingslake, *A History of the Photographic Lens,* Academic Press, San Diego (1989).

Rudolf Kingslake, *Lens Design Fundamentals,* Academic Press, New York (1978).

Rudolf Kingslake, *Optical System Design,* Academic Press, New York (1983).

Rudolf Kingslake, *Optics in Photography,* SPIE Press, Bellingham, (1992).

Rudolf Kingslake, "Basic Geometrical Optics," Chap. 6 in *Applied Optics and Optical Engineering,* **1,** Academic Press, New York (1965).

Milton Laikin, *Lens Design,* Marcel Dekker, New York (1991).

MIL-HDBK-141, *Optical Design,* Defense Supply Agency, Washington (1962).

Warren J. Smith, *Modern Lens Design, A Resource Manual,* McGraw-Hill, New York (1992).

Warren J. Smith, *Modern Optical Engineering,* second edition, McGraw-Hill, New York (1990).

CHAPTER 2
AFOCAL SYSTEMS

William B. Wetherell
Optical Research Associates
Framingham, Massachusetts

2.1 GLOSSARY

BFL	back focal length
D	pupil diameter
ER_{cp}	eye relief common pupil position
ER_k	eye relief keplerian
e	exit pupil; eye space
F, F'	focal points
FFL	front focal length
h, h'	object and image heights
l, l'	object and image distances
M	angular magnification
m	linear, lateral magnification
n	refractive index
OR	object relief
o	entrance pupil; object space
P, P'	principal points
R	radius
TTL	total length
$\tan \alpha$	slope
x, y, z	cartesian coordinates
Δz	axial separation

2.2 INTRODUCTION

If collimated (parallel) light rays from an infinitely distant point source fall incident on the input end of a lens system, rays exiting from the output end will show one of three characteristics: (1) they will converge to a real point focus outside the lens system, (2) they will appear to diverge from a virtual point focus within the lens system, or (3) they will emerge as collimated rays that may differ in some characteristics from the incident collimated rays. In cases 1 and 2, the paraxial imaging properties of the lens system can be modeled accurately by a characteristic focal length and a set of fixed principal surfaces. Such lens systems might be called *focusing* or *focal* lenses, but are usually referred to simply as *lenses*. In case 3, a single finite focal length cannot model the paraxial characteristics of the lens system; in effect, the focal length is infinite, with the output focal point an infinite distance behind the lens, and the associated principal surface an infinite distance in front of the lens. Such lens systems are referred to as *afocal*, or without focal length. They will be called *afocal lenses* here, following the common practice of using "lens" to refer to both single element and multielement lens systems. They are the topic of this chapter.

The first afocal lens was the galilean telescope (to be described later), a visual telescope made famous by Galileo's astronomical observations. It is now believed to have been invented by Hans Lipperhey in 1608.[1] Afocal lenses are usually thought of in the context of viewing instruments or attachments to change the effective focal length of focusing lenses, whose outputs are always collimated. In fact, afocal lenses can form real images of real objects. A more useful distinction between focusing and afocal lenses concerns which optical parameters are fixed, and which can vary in use. Focusing lenses have a fixed, finite focal length, can produce real images for a wide range of object distances, and have a linear magnification which varies with object distance. Afocal lenses have a fixed magnification which is independent of object distance, and the range of object distances yielding real images is severely restricted.

This chapter is divided into six sections, including this introduction. The second section reviews the Gaussian (paraxial) image forming characteristics of afocal lenses and compares them to the properties of focusing lenses. The importance of the optical invariant in designing afocal lenses is discussed. The third section reviews the keplerian telescope and its descendants, including both infinite conjugate and finite conjugate variants. The fourth section discusses the galilean telescope and its descendants. Thin-lens models are used in the third and fourth sections to define imaging characteristics and design principles for afocal lenses. The fifth section reviews relay trains and periscopes. The final section reviews reflecting and catadioptric afocal lenses.

This chapter is based on an earlier article by Wetherell.[2] That article contains derivations of many of the equations appearing here, as well as a more extensive list of patents illustrating different types of afocal lens systems.

2.3 GAUSSIAN ANALYSIS OF AFOCAL LENSES

Afocal lenses differ from focusing lenses in ways that are not always obvious. It is useful to review the basic image-forming characteristics of focusing lenses before defining the characteristics unique to afocal lenses.

Focusing Lenses

In this chapter, all lens elements are assumed to be immersed in air, so that object space and image space have the same index of refraction. Points in object space and image space are represented by two rectangular coordinate systems (x, y, z) and (x', y', z'), with the prime indicating image space. The z- and z'-axes form a common line in space, the *optical axis* of the system. It is assumed, unless noted otherwise, that all lens elements are rotationally symmetric with respect to the optical axis. Under these conditions, the imaging geometry of a focusing lens can be defined in terms of two principal points P and P', two focal points F and F', and a single characteristic focal length f, as shown in Fig. 1. P, P', F, and F' all lie on the optical axis.

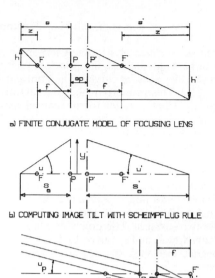

a) FINITE CONJUGATE MODEL OF FOCUSING LENS

b) COMPUTING IMAGE TILT WITH SCHEIMPFLUG RULE

c) INFINITE CONJUGATE MODEL OF FOCUSING LENS

FIGURE 1 Imaging geometry of focusing lenses.

The focal points F and F', will be the origins for the coordinate systems (x, y, z) and (x', y', z'). If the origins are at P and P', the coordinates will be given as (x, y, s) and (x', y', s'), where $s = z - f$ and $s' = z' + f$. Normal right-hand sign conventions are used for each set of coordinates, and light travels along the z-axis from negative z toward positive z', unless the optical system has internal mirrors. Figure 1a illustrates the terminology for finite conjugate objects.

Object points and image points are assumed to lie in planes normal to the optical axis, for paraxial computations. *Object distance* is specified by the axial distance to the object surface, z or s, and *image distance* by z' or s'. The two most commonly used equations relating image distance to object distance are

$$\frac{1}{s'} - \frac{1}{s} = \frac{1}{f} \tag{1}$$

and

$$zz' = -f^2 \tag{2}$$

For infinitely distant object points, $z' = 0$ and $s' = f$, and the corresponding image points will lie in the focal plane at F'.

To determine the actual distance from object plane to image plane, it is necessary to know the distance sp between P and P'. The value of sp is a constant specific to each real lens system, and may be either positive [moving object and image further apart than predicted by Eqs. (1) or (2)] or negative (moving them closer together).

For rotationally symmetric systems, off-axis object and image coordinates can be expressed by the *object height h* and *image height h'*, where $h^2 = x^2 + y^2$ and $h'^2 = x'^2 + y'^2$. Object height and image height are related by the *linear magnification m*, where

$$m = \frac{h'}{h} = \frac{s'}{s} = \frac{z' + f}{z - f} \tag{3}$$

Since the product zz' is a constant, Eq. (3) implies that magnification varies with object distance.

The *principal surfaces* of a focusing lens intersect the optical axis at the principal points P and P'. In paraxial analysis, the principal surfaces are planes normal to the optical axis; for real lenses, they may be curved. The principal surfaces are conjugate image surfaces for which $m = +1.0$. This property makes the raytrace construction shown in Fig. 1a possible, since a ray traveling parallel to the optical axis in either object or image space must intersect the focal point in the conjugate space, and must also intersect both principal surfaces at the same height.

In real lenses, the object and image surfaces may be tilted or curved. Planes normal to the optical axis are still used to define object and image positions for off-axis object points, and to compute magnification. For tilted object surfaces, the properties of the principal surfaces can be used to relate object surface and image surface tilt angles, as shown in Fig. 1b. Tilted object and image planes intersect the optical axis and the two principal planes. The tilt angles with respect to the optical axis, u and u', are defined by meridional rays lying in the two surfaces. The points at which conjugate tilted planes intercept the optical axis are defined by s_a and s'_a, given by Eq. (1). Both object and image planes must intersect their respective principal surfaces at the same height y, where $y = s_a \tan u = s'_a \tan u'$. It follows that

$$\frac{\tan u'}{\tan u} = \frac{s_a}{s'_a} = \frac{1}{m_a} \tag{4}$$

The geometry of Fig. 1b is known as the *Scheimpflug condition*, and Eq. (4) is the *Scheimpflug rule*, relating image to object tilt. The magnification m_a applies only to the axial image.

The height off axis of an infinitely distant object is defined by the principal ray angle u_p measured from F or P, as shown in Fig. 1c. In this case, the image height is

$$h' = f \tan u_p \tag{5}$$

A focusing lens which obeys Eq. (5) for all values of u_p within a specified range is said to be *distortion-free*: if the object is a set of equally spaced parallel lines lying in an object plane perpendicular to the optical axis, it will be imaged as a set of equally spaced parallel lines in an image plane perpendicular to the optical axis, with line spacing proportional to m.

Equations (1) through (5) are the basic Gaussian imaging equations defining a perfect focusing lens. Equation (2) is sometimes called the *newtonian* form of Eq. (1), and is the more useful form for application to afocal lens systems.

Afocal Lenses

With afocal lenses, somewhat different coordinate system origins and nomenclature are used, as shown in Fig. 2. The object and image space reference points RO and RE are at conjugate image points. Since the earliest and most common use for afocal lenses is as an aid to the eye for viewing distant objects, image space is referred to as *eye space*. Object position is defined by a right-hand coordinate system (x_o, y_o, z_o) centered on reference point RO. Image position in eye space is defined by coordinates (x_e, y_e, z_e) centered on RE.

Because afocal lenses are most commonly used for viewing distant objects, their imaging characteristics are usually specified in terms of *angular magnification M, entrance pupil* diameter D_o, and total field of view. Figure 2a models an afocal lens used at infinite conjugates. Object height off axis is defined by the principal ray angle u_{po}, and the corresponding image height is defined by u_{pe}. Objects viewed through the afocal lens will appear to be magnified by a factor M, where

$$\tan u_{pe} = M \tan u_{po} \tag{6}$$

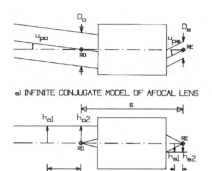

a) INFINITE CONJUGATE MODEL OF AFOCAL LENS

b) FINITE CONJUGATE MODEL OF AFOCAL LENS

FIGURE 2   Imaging geometry of afocal lenses.

If M is negative, as in Fig. 2a, the image will appear to be inverted. [Strictly speaking, since RO and RE are separated by a distance S, the apparent magnification seen by an eye at RE with and without the afocal lens will differ slightly from that indicated by Eq. (6) for nearby objects.][2]
   The imaging geometry of an afocal lens for finite conjugates is illustrated in Fig. 2b. Since a ray entering the afocal lens parallel to the optical axis will exit the afocal lens parallel to the optical axis, it follows that the linear magnification m relating object height h_o and image height h_e must be invariant with object distance. The linear magnification m is the inverse of the angular magnification M:

$$m = \frac{h_e}{h_o} = \frac{1}{M} \tag{7}$$

The axial separation Δz_e of any two images h_{e1} and h_{e2} is related to the separation Δz_o of the corresponding objects h_{o1} and h_{o2} by

$$\Delta z_e = m^2 \, \Delta z_o = \frac{\Delta z_o}{M^2} \tag{8}$$

It follows that any convenient pair of conjugate image points can be chosen as reference points RO and RE. Given the location of RO, the reference point separation S, and the magnifications $m = 1/M$, the imaging geometry of a rotationally symmetric distortion-free afocal lens can be given as

$$x_e = mx_o = \frac{x_o}{M}; \qquad y_e = my_o = \frac{y_o}{M}; \qquad z_e = m^2 z_o = \frac{z_o}{M^2} \tag{9}$$

Equation (9) is a statement that coordinate transformation between object space and eye space is rectilinear for afocal lenses, and is solely dependent on the *afocal magnification* M and the location of two conjugate reference points RO and RE. The equations apply (paraxially) to all object and image points independent of their distances from the afocal lens. Any straight line of equally spaced object points will be imaged as a straight line of equally spaced image points, even if the line does not lie in a plane normal to the optical axis. Either RO or RE may be chosen arbitrarily, and need not lie on the axis of symmetry of the lens system, so long as the z_o- and z_e-axes are set parallel to the axis of symmetry.
   A corollary of invariance in lateral and axial linear magnification is invariance in

angular magnification. Equation (6) thus applies to any ray traced through the afocal system, and to tilted object and image surfaces. In the latter context, Eq. (6) can be seen as an extension of Eq. (4) to afocal lenses.

The *eye space pupil* diameter D_e is of special importance to the design of visual instruments and afocal attachments: D_e must usually be large enough to fill the pupil of the associated instrument or eye. The *object space pupil* diameter D_o is related to D_e by

$$D_e = \frac{D_o}{M} = mD_o \tag{10}$$

(The more common terminology *exit pupil* and *entrance pupil* will be used later in this chapter.)

Subjective Aspects of Afocal Imagery

The angular magnification M is usually thought of in terms of Eq. (6), which is often taken to indicate that an afocal lens projects an image which is M-times as large as the object. (See, for example, Fig. 5.88 in Hecht and Zajac.)[3] Equation (9) shows that the image height is actually $1/M$-times the object height (i.e., smaller than the object when $|M| > 1$). Equation 9 also shows, however, that the image distance is reduced to $1/M^2$-times the object distance, and it is this combination of linear height reduction and quadratic distance reduction which produces the subjective appearance of magnification. Equation (6) can be derived directly from Eq. (9).

$$\tan u_{pe} = \frac{y_e}{z_e} = \frac{y_o/M}{z_o/M^2} = M \tan u_{po}.$$

Equation (9) is therefore a more complete model than Eq. (6) for rotationally symmetric, distortion-free afocal lenses.

Figure 3 illustrates two subjective effects which arise when viewing objects through afocal lenses. In Fig. 3a, for which $M = +3\times$, Eq. (9) predicts that image dimensions normal to the optical axis will be reduced by 1/3, while image dimensions along the optical axis will be reduced by 1/9. The image of the cube in Fig. 3a looks three times as tall and wide because it is nine times closer, but it appears compressed by a factor of 3 in the axial direction, making it look like a cardboard cutout. This subjective compression, most apparent when using binoculars, is intrinsic to the principle producing angular magnification, and is independent of pupil spacing in the binoculars.

Figure 3a assumes the optical axis is horizontal within the observer's reference framework. If the axis of the afocal lens is not horizontal, the afocal lens may create the

a) IMAGE OF 18 mm CUBE FORMED BY +3X AFOCAL LENS

b) HORIZONTAL SURFACE IMAGED BY +7X AFOCAL LENS
WHOSE AXIS IS TILTED 10 DEGREES TO HORIZON

FIGURE 3 Subjective aspects of afocal imagery.

illusion that horizontal surfaces are tilted. Figure 3b represents an $M = +7\times$ afocal lens whose axis is tilted 10° to a horizontal surface. Equation (6) can be used to show that the image of this surface is tilted approximately 51° to the axis of the afocal lens, creating the illusion that the surface is tilted 41° to the observer's horizon. This illusion is most noticeable when looking downward at a surface known to be horizontal, such as a body of water, through a pair of binoculars.

Afocal Lenses and the Optical Invariant

Equations (6) and (7) can be combined to yield

$$h_e \tan u_{pe} = h_o \tan u_{po} \tag{11}$$

which is a statement of the optical invariant as applied to distortion-free afocal lenses. Neither u_{po} nor u_{pe} is typically larger than 35°–40° in distortion-free afocal lenses, although there are examples with distortion where $u_{po} \rightarrow 90°$. Given a limit on one angle, Eq. (11) implies a limit on the other angle related to the ratio $h_o/h_e = D_o/D_e$. Put in words, *the ratio D_o/D_e cannot be made arbitrarily large without a corresponding reduction in the maximum allowable field of view.* All designers of afocal lens systems *must* take this fundamental principle into consideration.

2.4 KEPLERIAN AFOCAL LENSES

A simple afocal lens can be made up of two focusing lenses, an *objective* and an *eyepiece,* set up so that the rear focal point of the objective coincides with the front focal point of the eyepiece. There are two general classes of simple afocal lenses, one in which both focusing lenses are positive, and the other in which one of the two is negative. Afocal lenses containing two positive lenses were first described by Johannes Kepler in *Dioptrice,* in 1611,[4] and are called *keplerian.* Lenses containing a negative eyepiece are called *galilean,* and will be discussed separately. Generally, afocal lenses contain at least two powered surfaces. The simplest model for an afocal lens consists of two thin lenses.

Thin-Lens Model of a Keplerian Afocal Lens

Figure 4 shows a thin-lens model of a keplerian telescope. The focal length of its objective is f_o and the focal length of its eyepiece is f_e. Its properties can be understood by tracing two rays, ray 1 entering the objective parallel to the optical axis, and ray 2 passing through

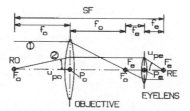

FIGURE 4 Thin-lens model of keplerian afocal lens.

F_o, the front focal point of the objective. Ray 1 leads directly to the linear magnification m, and ray 2 to the angular magnification M:

$$m = -\frac{f_e}{f_o}; \quad M = -\frac{f_o}{f_e} = \frac{\tan u_{pe}}{\tan u_{po}} \tag{12}$$

Equation (12) makes the relationship of afocal magnification to the Scheimpflug rule of Eq. (4) more explicit, with focal lengths f_o and f_e substituting for s_a and s'_a.

The second ray shows that placing the reference point RO at F_o will result in the reference point RE falling on F'_e, the rear focal point of the eyepiece. The reference point separation for RO in this location is

$$SF = 2f_e + 2f_o = 2(1 - M)f_e = 2(1 - m)f_o \tag{13}$$

Equation (13) can be used as a starting point for calculating any other locations for RO and RE, in combination with Eq. (9).

One additional generalization can be drawn from Fig. 4: the ray passing through F_o will emerge from the objective parallel to the optical axis. It will therefore also pass through F'_e even if the spacing between objective and eyepiece is increased to focus on nearby objects. Thus the angular magnification remains invariant, if u_{po} is measured from F_o and u_{pe} is measured from F'_e, even when adjusting the eyepiece to focus on nearby objects makes the lens system depart from being strictly afocal.

The simple thin-lens model of the keplerian telescope can be extended to systems composed of two real focusing lenses if we know their focal lengths and the location of each lens' front and rear focal points. Equation (12) can be used to derive M, and SF can be measured. Equation (9) can then be used to compute both finite and infinite conjugate image geometry.

Eye Relief Manipulation

The earliest application of keplerian afocal lenses was to obtain magnified views of distant objects. To view distant objects, the eye is placed at RE. An important design consideration in such instruments is to move RE far enough away from the last surface of the eyepiece for comfortable viewing. The distance from the last optical surface to the exit pupil at RE is called the *eye relief ER*. One way to increase eye relief ER is to move the entrance pupil at RO toward the objective. Most telescopes and binoculars have the system stop at the first surface of the objective, coincident with the entrance pupil, as shown in Fig. 5a.

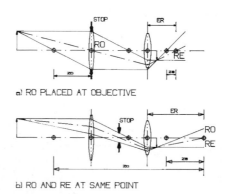

a) RO PLACED AT OBJECTIVE

b) RO AND RE AT SAME POINT

FIGURE 5 Increasing eye relief ER by moving stop.

In the thin-lens model of Fig. 5a, RO is moved a distance $zo = f_o$ to place it at the objective. Thus RE must move a distance $ze = f_o/M^2 = -f_e/M$, keeping in mind that M is negative in this example. Thus for a thin-lens keplerian telescope with its stop at the objective, the eye relief ER_k is

$$ER_k = \frac{(M-1)}{M} f_e \qquad (14)$$

It is possible to increase the eye relief further by placing the stop inside the telescope, moving the location of RO into virtual object space. Figure 5b shows an extreme example of this, where the virtual location of RO has been matched to the real location of RE. For this common-pupil-position case, the eye relief ER_{cp} is

$$ER_{cp} = \frac{(M-1)}{(M+1)} f_e \qquad (15)$$

A price must be paid for locating the stop inside the afocal lens, in that the elements ahead of the stop must be increased in diameter if the same field of view is to be covered without vignetting.

The larger the magnitude of M, the smaller the gain in ER yielded by using an internal stop. To increase the eye relief further, it is necessary to make the objective and/or the eyepiece more complex, increasing the distance between F_o and the first surface of the objective, and between the last surface of the eyepiece and F'_e. If this is done, placing RO at the first surface of the objective will further increase ER.

Figure 6 shows a thin-lens model of a telephoto focusing lens of focal length f_t. For convenience, a zero Petzval sum design is used, for which $f_1 = f$ and $f_2 = -f$. Given the telephoto's focal length f_t and the lens separation d, the rest of the parameters shown in Fig. 6 can be defined in terms of the constant $C = d/f_t$. The component focal length f, back focal length bfl, and front focal length ffl, are given by

$$f = f_t C^{1/2}; \quad bfl = f_t(1 - C^{1/2}); \quad ffl = f_t(1 + C^{1/2}) \qquad (16)$$

and the total physical length ttl and focal point separation sf are given by

$$ttl = f_t(1 + C - C^{1/2}); \quad sf = f_t(2 + C) \qquad (17)$$

The maximum gain in eye relief will be obtained by using telephoto designs for both objective and eyepiece, with the negative elements of each facing each other. Two cases are of special interest. First, ttl can be minimized by setting $C = 0.25$ for both objective and eyepiece. In this case, the eye relief ER_{ttl} is

$$ER_{ttl} = 1.5 \frac{(M-1)}{M} f_e = 1.5 ER_k \qquad (18)$$

Second, sf can be maximized by setting $C = 1.0$ for both objective and eyepiece. This

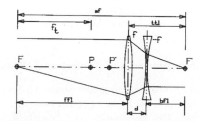

FIGURE 6 Zero Petzval sum telephoto lens.

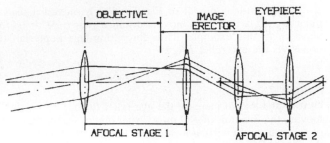

FIGURE 7 Terrestrial telescope.

places the negative element at the focal plane, merging the objective and eyepiece negative elements into a single negative field lens. The eye relief in this case, ER_{sf}, is

$$ER_{sf} = 2.0\frac{(M-1)}{M} = 2.0ER_k \qquad (19)$$

Placing a field lens at the focus between objective and eyepiece can be problematical, when viewing distant objects, since dust or scratches on the field lens will be visible. If a reticle is required, however, it can be incorporated into the field lens. Equations (14), (18), and (19) show that significant gains in eye relief can be made by power redistribution. In the example of Eq. (18), the gain in *ER* is accompanied by a reduction in the physical length of the optics, which is frequently beneficial.

Terrestrial Telescopes

Keplerian telescopes form an inverted image, which is considered undesirable when viewing earthbound objects. One way to make the image erect, commonly used in binoculars, is to incorporate erecting prisms. A second is to insert a relay stage between objective and eyepiece, as shown in Fig. 7. The added relay is called an *image erector,* and telescopes of this form are called *terrestrial telescopes.* (The keplerian telescope is often referred to as an *astronomical telescope,* to distinguish it from terrestrial telescopes, since astronomers do not usually object to inverted images. *Astronomical* has become ambiguous in this context, since it now more commonly refers to the very large aperture reflecting objectives found in astronomical observatories. *Keplerian* is the preferred terminology.) The terrestrial telescope can be thought of as containing an objective, eyepiece, and image erector, or as containing two afocal relay stages.

There are many variants of terrestrial telescopes made today, in the form of binoculars, theodolites, range finders, spotting scopes, rifle scopes, and other military optical instrumentation. All are offshoots of the keplerian telescope, containing a positive objective and a positive eyepiece, with intermediate relay stages to perform special functions. Patrick[5] and Jacobs[6] are good starting points for obtaining more information.

Field of View Limitations in Keplerian and Terrestrial Telescopes

The maximum allowable eye space angle u_{pe} and magnification M set an upper limit on achievable fields of view, in accordance with Eq. (11). MIL-HDBK-141[7] lists one eyepiece design for which the maximum $u_{pe} = 36°$. If $M = 7×$, using that eyepiece allows a 5.9° maximum value for u_{po}. It is a common commercial practice to specify the total field of view FOV as the width in feet which subtends an angle $2u_{po}$ from 1000 yards away, even when the pupil diameter is given in millimeters. FOV is thus given by

$$\text{FOV} = 6000 \tan u_{po} = \frac{6000}{M} \tan u_{pe} \tag{20}$$

For our 7× example, with $u_{pe} = 36°$, FOV = 620 ft at 1000 yd. For commercial 7×50 binoculars ($M = 7\times$ and $D_o = 50$ mm), FOV = 376 ft at 1000 yd is more typical.

Finite Conjugate Afocal Relays

If an object is placed in contact with the front surface of the keplerian telescope of Fig. 5, its image will appear a distance ER_k behind the last surface of the eyepiece, in accordance with Eq. (14). There is a corresponding *object relief* distance $OR_k = M^2 ER_k$ defining the position of an object that will be imaged at the output surface of the eyepiece, as shown in Fig. 8. OR_k and ER_k define the portions of object space and eye space within which real images can be formed of real objects with a simple keplerian afocal lens.

$$OR_k = M(M - 1)f_e \tag{21}$$

Object relief is enlarged by the power redistribution technique used to extend eye relief. Thus there is a minimum total length design corresponding to Eq. (18), for which the object relief OR_{ttl} is

$$OR_{ttl} = 1.5M(M - 1)f_e \tag{22}$$

and a maximum eye relief design corresponding to Eq. (19), for which OR_{sf}

$$OR_{sf} = 2.0M(M - 1)f_e \tag{23}$$

is also maximized.

Figure 9 shows an example of a zero Petzval sum finite conjugate afocal relay designed to maximize OR and ER by placing a negative field lens at the central infinite conjugate image. Placing the stop at the field lens means that the lens is *telecentric* (principal rays parallel to the optical axis) in both object and eye space. As a result, magnification, principal ray angle of incidence on object and image surface, and cone angle are all invariant over the entire range of OR and ER for which there is no vignetting. Magnification and cone angle invariance means that object and image surfaces can be tilted with respect to the optical axis without introducing keystoning or variation in image irradiance over the field of view. Having the principal rays telecentric means that object and image position can be adjusted for focus without altering magnification. It also means that the lens can be defocused without altering magnification, a property very useful for unsharp masking techniques used in the movie industry.

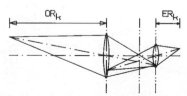

FIGURE 8 Finite conjugate keplerian afocal lens showing limits on usable object space and image space.

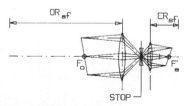

FIGURE 9 Finite conjugate afocal relay configured to maximize eye relief ER and object relief OR. Stop at common focus collimates principal rays in both object space and eye space.

One potential disadvantage of telecentric finite conjugate afocal relays is evident from Fig. 9: to avoid vignetting, the apertures of both objective and eyepiece must be larger than the size of the associated object and image. While it is possible to reduce the diameter of either the objective or the eyepiece by shifting the stop to make the design nontelecentric, the diameter of the other lens group becomes larger. Afocal relays are thus likely to be more expensive to manufacture than focusing lens relays, unless object and image are small.

Finite conjugate afocal lenses have been used for alignment telescopes,[8] for laser velocimeters,[9] and for automatic inspection systems for printed circuit boards.[10] In the last case, invariance of magnification, cone angle, and angle of incidence on a tilted object surface make it possible to measure the volume of solder beads automatically with a computerized video system. Finite conjugate afocal lenses are also used as Fourier transform lenses.[11] Brief descriptions of these applications are given in Wetherell.[2]

Afocal Lenses for Scanners

Many optical systems require scanners, and if the apertures of the systems are large enough, it is preferable to place the scanner inside the system. Although scanners have been designed for use in convergent light, they are more commonly placed in collimated light (see the chapter on scanners in this volume, Marshall,[12] and chapter 7 in Lloyd,[13] for descriptions of scanning techniques). A large aperture objective can be converted into a high magnification keplerian afocal lens with the aid of a short focal length eyepiece collimator, as shown in Fig. 10, providing a pupil in a collimated beam in which to insert a scanner. For the polygonal scanner shown, given the desired scan angle and telescope aperture diameter, Eq. (11) will define the combination of scanner facet size and number of facets needed to achieve the desired scanning efficiency. Scanning efficiency is the time it takes to complete one scan divided by the time between the start of two sequential scans. It is tied to the ratio of facet length to beam diameter, the amount of vignetting allowed within a scan, the number of facets, and the angle to be scanned.

Two limitations need to be kept in mind. First, the optical invariant will place an upper limit on M for the given combination of D_o and u_{po}, since there will be a practical upper limit on the achievable value of u_{pe}. Second, it may be desirable in some cases for the keplerian afocal relay to have enough barrel distortion so that Eq. (6) becomes

$$u_{pe} = Mu_{po} \tag{24}$$

An afocal lens obeying Eq. (24) will convert a constant rotation rate of the scan mirror into a constant angular scan rate for the external beam. The same property in "f-theta"

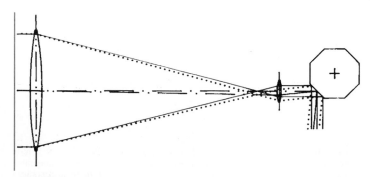

FIGURE 10   Afocal lens scanner geometry.

focusing lenses is used to convert a constant angular velocity scanner rotation rate into a constant linear velocity rate for the recording spot of light.

The above discussion applies to scanning with a point detector. When the detector is a linear diode array, or when a rectangular image is being projected onto moving film, the required distortion characteristics for the optical system may be more complex.

Imaging in Binoculars

Most commercial binoculars consist of two keplerian afocal lenses with internal prismatic image erectors. Object and image space coordinates for binoculars of this type are shown schematically in Fig. 11. Equation (9) can be applied to Fig. 11 to analyze their imaging properties. In most binoculars, the spacing S_o between objectives differs from the spacing S_e between eyepieces, and S_o may be either larger or smaller than S_e. Each telescope has its own set of reference points, ROL and REL for the left telescope, and ROR and RER for the right. Object space is a single domain with a single origin O. The object point at z_o, midway between the objective axes, will be x_{oL} units to the right of the left objective axis, and x_{oR} units to the left of the right objective axis. In an ideal binocular system, the images of the object formed by the two telescopes would merge at one point, z_e units in front of eye space origin E. This will happen if $S_o = MS_e$, so that $x_{eL} = x_{oL}/M$ and $x_{eR} = x_{oR}/M$. In most modern binoculars, however, $S_o \ll MS_e$, and separate eye space reference points EL and ER will be formed for the left and right eye. As a result, each eye sees its own eye space, and while they overlap, they are not coincident. This property of binoculars can affect stereo acuity[2] and eye accommodation for the user.

It is normal for the angle at which a person's left-eye and right-eye lines of sight converge to be linked to the distance at which the eyes focus. (In my case, this linkage was quite strong before I began wearing glasses.) Eyes focused for a distance z_e normally would converge with an angle β, as shown in Fig. 11. When $S_o \ll MS_e$, as is commonly the case, the actual convergence angle β' is much smaller. A viewer for whom focus distance is strongly linked to convergence angle may find such binoculars uncomfortable to use for extended periods, and may be in need of frequent focus adjustment for different object distances.

A related but more critical problem arises if the axes of the left and right telescopes are not accurately parallel to each other. Misalignment of the axes requires the eyes to twist in unaccustomed directions to fuse the two images, and refocusing the eyepiece is seldom

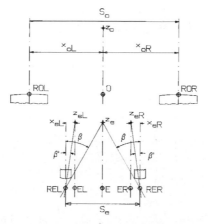

FIGURE 11 Imaging geometry of binoculars.

able to ease the burden. Jacobs[6] is one of the few authors to discuss this problem. Jacobs divides the axes misalignment into three categories: (1) misalignments requiring a divergence D of the eye axes to fuse the images, (2) misalignments requiring a convergence C, and (3) misalignments requiring a vertical displacement V. The tolerance on allowable misalignment in minutes of arc is given by Jacobs as

$$D = 7.5/(M-1); \qquad C = 22.5/(M-1); \qquad V = 8.0/(M-1) \qquad (25)$$

Note that the tolerance on C, which corresponds to convergence to focus on nearby objects, is substantially larger than the tolerances on D and V.

2.5 GALILEAN AND INVERSE GALILEAN AFOCAL LENSES

The combination of a positive objective and a negative eyepiece forms a *galilean* telescope. If the objective is negative and the eyepiece positive, it is referred to as an *inverse galilean* telescope. The galilean telescope has the advantage that it forms an erect image. It is the oldest form of visual telescope, but it has been largely replaced by terrestrial telescopes for magnified viewing of distant objects, because of field of view limitations. In terms of number of viewing devices manufactured, there are far more inverse galilean than galilean telescopes. Both are used frequently as power-changing attachments to change the effective focal length of focusing lenses.

Thin-Lens Model of a Galilean Afocal Lens

Figure 12 shows a thin-lens model of a galilean afocal lens. The properties of galilean lenses can be derived from Eqs. (9), (12), and (13). Given that f_e is negative and f_o is positive, M is positive, indicating an erect image. If RO is placed at the front focal point of the objective, RE is a virtual pupil buried inside the lens system. In fact, galilean lenses cannot form real images of real objects under any conditions, and at least one pupil will always be virtual.

Field of View in Galilean Telescopes

The fact that only one pupil can be real places strong limitations on the use of galilean telescopes as visual instruments when $M \gg 1x$. Given the relationship $\Delta z_o = M^2 \Delta z_e$, moving RE far enough outside the negative eyepiece to provide adequate eye relief moves RO far enough into virtual object space to cause drastic vignetting at even small field

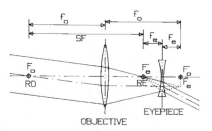

FIGURE 12 Thin-lens model of galilean afocal lens.

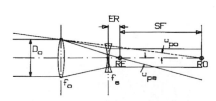

FIGURE 13 Galilean field-of-view limitations.

angles. Placing RE a distance ER behind the negative lens moves RO to the position shown in Fig. 13, SF' units behind RE, where

$$SF' = (M^2 - 1)ER - (M - 1)^2 f_e \qquad (26)$$

In effect, the objective is both field stop and limiting aperture, and vignetting defines the maximum usable field of view. The maximum acceptable object space angle u_{po} is taken to be that for the principal ray which passes just inside D_o, the entrance pupil at the objective. If the F-number of the objective is $\text{FN}_{ob} = f_o / D_o$, then

$$\tan u_{po} = \frac{-f_e}{2\text{FN}_{ob}(MER + f_e - Mf_e)} \qquad (27)$$

For convenience, assume $ER = -f_e$. In this case, Eq. (27) reduces to

$$\tan u_{po} = \frac{1}{2\text{FN}_{ob}(2M - 1)} \qquad (28)$$

For normal achromatic doublets, $\text{FN}_{ob} \geq 4.0$. For $M = 3x$, in this case, Eq. (28) indicates that $u_{po} \leq 1.43°$ (FOV ≤ 150 ft at 1000 yd). For $M = 7x$, $u_{po} \leq 0.55°$ (FOV ≤ 57.7 ft at 1000 yd). The effective field of view can be increased by making the objective faster and more complex, as can be seen in early patents by von Rohr[14] and Erfle.[15] In current practice, galilean telescopes for direct viewing are seldom made with M larger than $1.5x$–$3.0x$. They are more typically used as power changers in viewing instruments, or to increase the effective focal length of camera lenses.[16]

Field of View in Inverse Galilean Telescopes

For inverse galilean telescopes, where $M \ll 1x$, adequate eye relief can be achieved without moving RO far inside the first surface of the objective. Inverse galilean telescopes for which $u_{po} \rightarrow 90°$ are very common in the form of security viewers[17] of the sort shown in Fig. 14, which are built into doors in hotel rooms, apartments, and many houses. These may be the most common of all optical systems more complex than eyeglasses. The negative objective lens is designed with enough distortion to allow viewing of all or most of the forward hemisphere, as shown by the principal ray in Fig. 14.

Inverse galilean telescopes are often used in camera view finders.[18] These present reduced scale images of the scene to be photographed, and often have built in arrangements to project a frame of lines representing the field of view into the image.

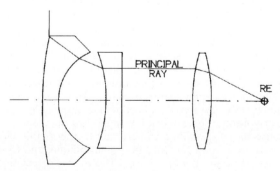

FIGURE 14 Inverse galilean security viewer with hemispheric field of view.

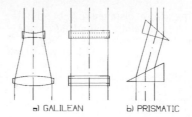

FIGURE 15   Anamorphic afocal attachments.

Inverse galilean power changers are also used to increase the field of view of submarine periscopes and other complex viewing instruments, and to reduce the effective focal length of camera lenses.[19]

Anamorphic Afocal Attachments

Afocal attachments can compress or expand the scale of an image in one axis. Such devices are called *anamorphosers,* or *anamorphic afocal attachments.* One class of anamorphoser is the cylindrical galilean telescope, shown schematically in Fig. 15a. Cox[20] and Harris[21] have patented representative examples. The keplerian form is seldom if ever used, since a cylindrical keplerian telescope would introduce image inversion in one direction. Anamorphic compression can also be obtained using two prisms, as shown in Fig. 15b. The adjustable magnification anamorphoser patented by Luboshez[22] is a good example of prismatic anamorphosers. Many anamorphic attachments were developed in the 1950s for the movie industry for use in wide-field cameras and projectors. An extensive list of both types will be found in Wetherell.[2]

Equation (9) can be modified to describe anamorphic afocal lenses by specifying separate afocal magnifications M_x and M_y for the two axes. One important qualification is that separate equations are needed for object and image distances for the x and y planes. In general, anamorphic galilean attachments work best when used for distant objects, where any difference in x-axis and y-axis focus falls within the depth of focus of the associated camera lens. If it is necessary to use a galilean anamorphoser over a wide range of object distances, it may be necessary to add focus adjustment capabilities within the anamorphoser.

2.6   *RELAY TRAINS AND PERISCOPES*

There are many applications where it is necessary to perform remote viewing because the object to be viewed is in an environment hostile to the viewer, or because the object is inaccessible to the viewer without unacceptable damage to its environment. Military applications[5] fall into the former category, and medical applications[23] fall into the latter. For these applications, instrumentation is needed to *collect* light from the object, *transport* the light to a location more favorable for viewing, and *dispense* the light to the viewing instruments or personnel. Collecting and dispensing optical images is done with focusing lenses, typically. There are three image transportation techniques in common use today: (1) sense the image with a camera and transport the data electronically, (2) transport the light pattern with a coherent fiber optics bundle, and (3) transport the light pattern with a relay lens or train of relay lenses. The first two techniques are outside the scope of this chapter. *Relay trains,* however, are commonly made up of a series of unit power afocal lenses, and are one of the most important applications of finite conjugate afocal lenses.

Unit Power Afocal Relay Trains

Several factors are important in designing relay trains. First, it is desirable to minimize the number of relay stages in the relay train, both to maximize transmittance and to minimize the field curvature caused by the large number of positive lenses. Second, the outside diameter of the relay train is typically restricted (or a single relay lens could be used), so the choice of image and pupil diameter within the relay is important. Third, economic considerations make it desirable to use as many common elements as possible, while minimizing the total number of elements. Fourth, it is desirable to keep internal images well clear of optical surfaces where dust and scratches can obscure portions of the image. Fifth, the number of relay stages must be either odd or even to insure the desired output image orientation.

Figure 16 shows thin-lens models of the two basic afocal lens designs which can be applied to relay train designs. Central to both designs is the use of symmetry fore and aft of the central stop to control coma, distortion, and lateral color, and matching the image diameter D_i and stop diameter D_s to maximize the stage length to diameter ratio. In paraxial terms, if $D_i = D_s$, then the marginal ray angle u matches the principal ray angle u_p, in accordance with the optical invariant. If the relay lens is both aplanatic and distortion free, a better model of the optical invariant is

$$D_i \sin u = D_s \tan u_p \tag{29}$$

and either the field of view $2u_p$ or the numerical aperture $\mathrm{NA} = n \sin u$ must be adjusted to match pupil and image diameters. For some applications, maximizing the optical invariant which can pass through a given tube diameter D_t in a minimum number of stages is also critical.

If maximizing the ratio $D_i \sin u / D_t$ is not critical, Fig. 16a shows how the number of elements can be minimized by using a keplerian afocal lens with the stop at the common focus, eliminating the need for field lenses between stages. The required tube diameter in this example is at least twice the image diameter. If maximizing $D_i \sin u / D_t$ is critical, field lenses FL must be added to the objectives OB as shown in Fig. 16b, and the field lenses should be located as close to the image as possible within limits set by obstructions due to dirt and scratches on the field lens surfaces. Symmetry fore and aft of the central stop at 1 is still necessary for aberration balancing. If possible within performance constraints,

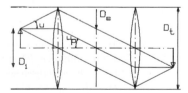

a) MINIMUM NUMBER OF LENSES

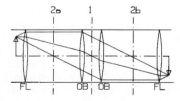

b) MINIMUM TUBE DIAMETER

FIGURE 16 Basic unit power afocal relay designs.

a) McKINLEY TYPE 1 ROD RELAY

b) BAKER FLAT FIELD RELAY

FIGURE 17 Improved unit power afocal relays.

symmetry of OB and FL with respect to the planes 2a and 2b is economically desirable, making OB and FL identical.

For medical instruments, where minimizing tube diameter is critical, variants of the second approach are common. The rod lens design[24] developed by H. H. Hopkins[25] can be considered an extreme example of either approach, making a single lens so thick that it combines the functions of OB and FL. Figure 17a shows an example from the first of two patents by McKinley.[26,27] The central element in each symmetrical cemented triplet is a sphere. Using rod lenses does maximize the optical invariant which can be passed through a given tube diameter, but it does not eliminate field curvature. It also maximizes weight, since the relay train is almost solid glass, so it is most applicable to small medical instruments.

If larger diameter relays are permissible, it is possible to correct field curvature in each relay stage, making it possible to increase the number of stages without adding field curvature. Baker[28] has patented the lens design shown in Fig. 17b for such an application. In this case, field lens and objective are identical, so that an entire relay train can be built using only three different element forms. Pupil and image diameters are the same, and pupil and image are interchangeable.

For purposes of comparison, the two designs shown in Fig. 17 have been scaled to have the same image diameter (2.8 mm) and numerical aperture (0.10), with component focal lengths chosen so that $D_s = D_i$. Minimum tube diameter is 4.0 mm for the rod lens and 5.6 mm for the Baker relay. The image radius of curvature is about 20 mm for the rod relay and about -368 mm for the Baker relay (i.e., field curvature is overcorrected). Image quality for the rod relay is 0.011 waves rms on axis and 0.116 waves rms at full field, both values for best focus, referenced to 587 nm wavelength. For the Baker relay, the corresponding values are 0.025 and 0.056 waves rms, respectively. The Baker design used for this comparison was adapted from the cited patent, with modern glasses substituted for types no longer available. No changes were made to the design other than refocusing it and scaling it to match the first order parameters of the McKinley design. Neither design necessarily represents the best performance which can be obtained from its design type, and both should be evaluated in the context of a complete system design where, for example, the field curvature of the McKinley design may be compensated for by that of the collecting and dispensing objectives. Comparing the individual relay designs does, however, show the price which must be paid for either maximizing the optical invariant within a given tube diameter or minimizing field curvature.

Periscopes

Periscopes are relay trains designed to displace the object space reference point *RO* a substantial distance away from the eye space reference point *RE*. This allows the observer to look over an intervening obstacle, or to view objects in a dangerous environment while the observer is in a safer environment. The submarine periscope is the archetypical example.[5] Many other examples can be found in the military and patent[2] literature.

The simplest form of periscope is the pair of fold mirrors shown in Fig. 18a, used to

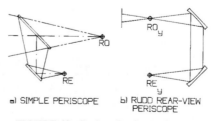

a) SIMPLE PERISCOPE   b) RUDD REAR-VIEW PERISCOPE

FIGURE 18   Basic reflecting periscopes.

allow the viewer to see over nearby obstacles. Figure 18b shows the next higher level of complexity, in the form of a rear-view vehicle periscope patented[29] by Rudd.[30] This consists of a pair of cylindrical mirrors in a roof arrangement. The cylinders image one axis of object space with the principal purpose of compensating for the image inversion caused by the roof mirror arrangement. This could be considered to be a keplerian anamorphoser, except that it is usually a unit power magnifier, producing no anamorphic compression. Beyond these examples, the complexity of periscopes varies widely.

The optics of complex periscopes such as the submarine periscope can be broken down into a series of component relays. The core of a submarine periscope is a pair of fold prisms arranged like the mirrors in Fig. 18a. The upper prism can be rotated to scan in elevation, while the entire periscope is rotated to scan in azimuth, typically. The main optics is composed of keplerian afocal relays of different magnification, designed to transfer an erect image to the observer inside the submarine, usually at unit net magnification. Galilean and inverse galilean power changers can be inserted between the upper prism and main relay optics to change the field of view. Variants of this arrangement will be found in other military periscopes, along with accessories such as reticles or image intensifiers located at internal foci. Optical design procedures follow those for other keplerian afocal lenses.

2.7 *REFLECTING AND CATADIOPTRIC AFOCAL LENSES*

Afocal lenses can be designed with powered mirrors or combinations of mirrors and refractors. Several such designs have been developed in recent years for use in the photolithography of microcircuits. All-reflecting afocal lenses are classified here according to the number of powered mirrors they contain. They will be reviewed in order of increasing complexity, followed by a discussion of catadioptric afocal systems.

Two-powered-mirror Afocal Lenses

The simplest reflecting afocal lenses are the variants of the galilean and keplerian telescopes shown in Fig. 19a and 19b. They may also be thought of as afocal cassegrainian and gregorian telescopes. The galilean/cassegrainian version is often called a *Mersenne* telescope. In fact, both galilean and keplerian versions were proposed by Mersenne in 1636,[31] so his name should not be associated solely with the galilean variant.

Making both mirrors parabolic corrects all third order aberrations except field curvature. This property of *confocal parabolas* has led to their periodic rediscovery,[32,33] and to subsequent discussions of their merits and shortcomings.[34,35,36] The problem with both designs, in the forms shown in Fig. 19a and 19b, is that their eyepieces are buried so

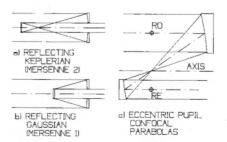

FIGURE 19 Two-powered-mirror afocal lenses.

deeply inside the design that their usable field of view is negligible. The galilean form is used as a laser beam expander,[37] where field of view and pupil location is not a factor, and where elimination of internal foci may be vital.

Eccentric pupil versions of the keplerian form of confocal parabolas, as shown in Fig. 19c, have proven useful as lens attachments.[38] *RO, RE,* and the internal image are all accessible when *RO* is set one focal length ahead of the primary, as shown. It is then possible to place a field stop at the image and pupil stops at *RO* and *RE,* which very effectively blocks stray light from entering the following optics. Being all-reflecting, confocal parabolas can be used at any wavelength, and such attachments have seen use in infrared designs.

Three-powered-mirror Afocal Lenses

The principle which results in third-order aberration correction for confocal parabolas also applies when one of the parabolas is replaced by a classical cassegrainian telescope (parabolic primary and hyperbolic secondary), as shown in Fig. 20, with two important added advantages. First, with one negative and two positive mirrors, it is possible to reduce the Petzval sum to zero, or to leave a small residual of field curvature to balance higher-order astigmatism. Second, because the cassegrainian is a telephoto lens with a remote front focal point, placing the stop at the cassegrainian primary puts the exit pupil in a more accessible location. This design configuration has been patented by Offner,[39] and is more usefully set up as an eccentric pupil design, eliminating the central obstruction and increasing exit pupil accessibility.

Four-powered-mirror Afocal Lenses

The confocal parabola principle can be extrapolated one step further by replacing both parabolas with classical cassegrainian telescopes, as shown in Fig. 21a. Each cassegrainian is corrected for field curvature independently, and the image quality of such *confocal cassegrainians* can be quite good. The most useful versions are eccentric pupil. Figure 21b shows an example from Wetherell.[40] Since both objective and eyepiece are telephoto designs, the separation between entrance pupil *RO* and exit pupil *RE* can be quite large. An afocal relay train made up of eccentric pupil confocal cassegrainians will have very

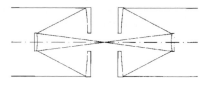

a) CENTERED PUPIL CONFOCAL CASSEGRAINS

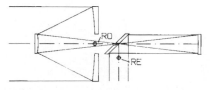

FIGURE 20   Three-powered-mirror afocal lenses.

b) ECCENTRIC PUPIL CONFOCAL CASSEGRAINS

FIGURE 21   Four-powered-mirror afocal lenses.

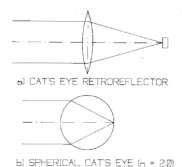

a) CAT'S EYE RETROREFLECTOR

b) SPHERICAL CAT'S EYE (n = 2.0)

FIGURE 22   Afocal retroreflector designs.

long collimated paths. If the vertex curvatures of the primary and secondary mirrors within each cassegrainian are matched, the relay will have zero field curvature, as well. In general, such designs work best at or near unit magnification.

Unit Power Finite Conjugate Afocal Lenses

The simplest catadioptric afocal lens is the cat's-eye retroreflector shown in Fig. 22*a*, made up of a lens with a mirror in its focal plane. Any ray entering the lens will exit parallel to the incoming ray but traveling in the opposite direction. If made with glass of index of refraction $n = 2.00$, a sphere with one hemisphere reflectorized (Fig. 22*b*) will act as a perfect retroreflector for collimated light entering the transparent hemisphere. Both designs are, in effect, unit power ($M = -1.00$) afocal lenses. Variations on this technique are used for many retroreflective devices.

Unit power relays are of much interest in photolithography, particularly for microcircuit manufacturing, which requires very high resolution, low focal ratio unit power lenses. In the *Dyson lens*,[41] shown in Fig. 23*a*, the powered surfaces of the refractor and the reflector are concentric, with radii R and r given by

$$\frac{R}{r} = \frac{n}{(n-1)} \tag{30}$$

where n is the index of refraction of the glass. At the center point, spherical aberration and coma are completely corrected. In the nominal design, object and image are on the surface intersecting the center of curvature, displaced laterally to separate object from image

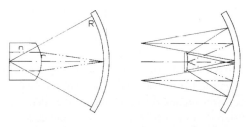

a) DYSON LENS   b) OFFNER CONCENTRIC

FIGURE 23   Concentric spheres unit power afocal lenses.

sensor (this arrangement is termed *eccentric field,* and is common to many multimirror lens systems). In practice, performance of the system is limited by off-axis aberrations, and it is desirable to depart from the nominal design to balance aberrations across the field of view.[42]

The unit power all-reflecting concentric design shown in Fig. 23*b* is patented[43] by Offner.[44] It was developed for use in manufacturing microcircuits, and is one of the most successful finite conjugate afocal lens designs in recent years. The spheres are concentric and the plane containing object and image surfaces passes through the common center of curvature. It is an all-reflecting, unit power equivalent of the refracting design shown in Fig. 9. Object and image points are eccentric field, and this is an example of the *ring field* design concept, where axial symmetry ensures good correction throughout a narrow annular region centered on the optical axis. As with the Dyson lens, having an eccentric field means performance is limited by off-axis aberrations. Correction of the design can be improved at the off-axis point by departing from the ideal design to balance on-axis and off-axis aberrations.[45]

2.8 REFERENCES

1. A. van Helden, "The Invention of the Telescope," *Trans. Am. Philos. Soc.* **67,** part 4, 1977.

2. W. B. Wetherell, In "Afocal Lenses," R. R. Shannon and J. C. Wyant (eds.), *Applied Optics and Optical Engineering,* vol. X, Academic Press, New York, 1987, pp. 109–192.

3. E. Hecht and A. Zajac, *Optics,* Addison-Wesley, Reading, Mass., 1974, p. 152.

4. H. C. King, *The History of the Telescope,* Dover, New York, 1979, pp. 44–45.

5. F. B. Patrick, "Military Optical Instruments," In R. Kingslake (ed.), *Applied Optics and Optical Engineering,* vol. V, Academic Press, New York, 1969, pp. 183–230.

6. D. H. Jacobs, *Fundamentals of Optical Engineering,* McGraw-Hill, New York, 1943.

7. Defense Supply Agency, *Military Standardization Handbook*: *Optical Design,* MIL-HDBK-141, Defense Supply Agency, Washington, D.C., 1962, section 14, p. 18.

8. A. König, Telescope, U.S. Patent 1,952,795, March 27, 1934.

9. D. B. Rhodes, Scanning Afocal Laser Velocimeter Projection Lens System, U.S. Patent 4,346,990, August 31, 1982.

10. J. C. A. Chastang and R. F. Koerner, Optical System for Oblique Viewing, U.S. Patent 4,428,676, January 31, 1984.

11. A. R. Shulman, *Optical Data Processing,* Wiley, New York, 1970, p. 325.

12. G. F. Marshall, (ed.), *Laser Beam Scanners,* Marcel Dekker, Inc., New York, 1970.

13. J. M. Lloyd, *Thermal Imaging Systems,* Plenum, New York, 1975.

14. M. von Ruhr, Galilean Telescope System, U.S. Patent 962,920, June 28, 1910.

15. H. Erfle, Lens System for Galilean Telescope, U.S. Patent 1,507,111, September 2, 1924.

16. H. Köhler, R. Richter, and H. Kaselitz, Afocal Lens System Attachment for Photographic Objectives, U.S. Patent 2,803,167, August 20, 1957.

17. J. C. J. Blosse, Optical Viewer, U.S. Patent 2,538,077, January 16, 1951.

18. D. L. Wood, View Finder for Cameras, U.S. Patent 2,234,716, March 11, 1941.

19. H. F. Bennett, Lens Attachment, U.S. Patent 2,324,057, July 13, 1943.

20. A. Cox, Anamorphosing Optical System, U.S. Patent 2,720,813, October 18, 1955.

21. T. J. Harris, W. J. Johnson, and I. C. Sandbeck, Wide Angle Lens Attachment, U.S. Patent 2,956,475, October 18, 1960.

22. B. J. Luboshez, Prism Magnification System Having Correction Means for Unilateral Color, U.S. Patent 2,780,141, February 5, 1957.

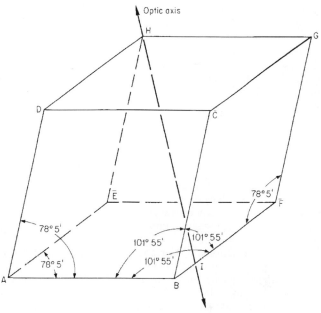

FIGURE 1   Schematic representation of a rhombohedral calcite crystal showing the angles between faces. The optic axis passes through corner *H* and point *I* on side *BF*.

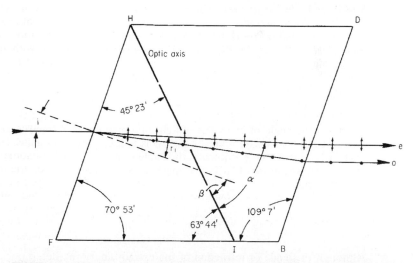

FIGURE 2   Side view of a principal section for the calcite rhomb in Fig. 1. The direction of the optic axis and the angles of the principal section are indicated. The angle of incidence is *i*, angle of refraction is *r*, angle between the *e* ray and the optic axis is *a*, and angle between the normal to the surface and the optic axis is *β*. The directions of vibration of the *e* and *o* rays are in the plane of the paper and perpendicular to it, respectively.

will precess about the o ray. However, unlike the o ray, it will not remain in the plane of incidence unless this plane coincides with the principal section.

The plane containing the o ray and the optic axis is defined as the *principal plane of the o ray,* and that containing the e ray and the optic axis as the *principal plane of the e ray.* In the case discussed above, the two principal planes and the principal section coincide. In the general case, they may all be different. However, in all cases, the o ray is polarized with its plane of vibration perpendicular to its principal plane and the e ray with its plane of vibration in its principal plane (see Fig. 2). In all cases, the vibration direction of the e ray remains perpendicular to that of the o ray.

The value of the index of refraction of the e ray which differs most from that of the o ray, i.e., the index when the e ray vibrations are parallel to the optic axis, is called the *principal index for the extraordinary ray n_e.* Snell's law can be used to calculate the path of the e ray through a prism for this case. Snell's law can always be used to calculate the direction of propagation of the ordinary ray.

Table 1 lists values of n_o and n_e for calcite, along with the two absorption coefficients a_o and a_e, all as a function of wavelength. Since $n_e < n_o$ in the ultraviolet, visible, and infrared regions, calcite is a negative uniaxial crystal. However, at wavelengths shorter than 1520 Å in the vacuum ultraviolet, the birefringence $n_e - n_o$ becomes positive, in agreement with theoretical predictions.[5,10] For additional data in the 0.17 to 0.19-μm region, see Uzan *et al.*[11] The range of transparency of calcite is approximately from 0.214 to 3.3 μm for the extraordinary ray but only from about 0.23 to 2.2 μm for the ordinary ray.

If the principal plane of the e ray and the principal section coincide (Fig. 2), the wave normal (*but not the e ray*) obeys Snell's law, except that the index of refraction n_ϕ of this wave is given by[12,13]

$$\frac{1}{n_\phi^2} = \frac{\sin^2\phi}{n_e^2} + \frac{\cos^2\phi}{n_o^2} \tag{1}$$

where ϕ is the angle between the direction of the *wave normal* and the optic axis ($\phi \le 90°$). When $\phi = 0°$, $n_\phi = n_o$, and when $\phi = 90°$, $n_\phi = n_e$. The angle of refraction for the wave normal is $\phi - \beta$, where β is the angle the normal to the surface makes with the optic axis. Snell's law for the extraordinary-ray *wave normal* then becomes

$$n \sin i = \frac{n_e n_o \sin(\phi - \beta)}{(n_o^2 \sin^2\phi + n_e^2 \cos^2\phi)^{1/2}} \tag{2}$$

where i is the angle of incidence of light in a medium of refractive index n. Since all other quantities in this equation are known, ϕ is uniquely determined but often must be solved for by iteration. Once ϕ is known, the angle of refraction r for the extraordinary ray can be determined as follows. If α is the angle the ray makes with the optic axis ($\alpha \le 90°$), then $r = \alpha - \beta$ and[13]

$$\tan \alpha = \frac{n_o^2}{n_e^2} \tan \phi \tag{3}$$

Although the angle of refraction of the extraordinary ray determines the path of the light beam through the prism, one must use the angle of refraction of the *wave normal*, $\phi - \beta$, in Fresnel's equation [Eq. (21) in Chap. 5, "Polarization," in Vol. I of this handbook] when calculating the reflection loss of the e ray at the surface of the prism.

For the special case in which the optic axis is parallel to the surface as well as in the plane of incidence, α and ϕ are the complements of the angles of refraction of the ray and wave normal, respectively. If the light is normally incident on the surface, ϕ and α are

TABLE 1   Refractive Indices[a] and Absorption Coefficients[a] for Calcite

λ, μm	n_o	α_o	n_e	α_e	λ, μm	n_o	α_o	n_e	α_e
0.1318	1.56[b]	534,000[b]	1.80[b]	477,000[b]	0.3195	—	0.059		
0.1355	1.48	473,000	1.84	380,000	0.327	—	0.028		
0.1411	1.40	561,000	1.82	196,000	0.330	1.70515	—	1.50746	
0.1447	1.48	669,000	1.80	87,000	0.3355	—	0.028		
0.1467	1.51	711,000	1.75	20,500	0.340	1.70078	—	1.50562	
0.1478$_5$	1.54	722,000	1.75	17,000	0.3450	—	0.0170		
0.1487	1.58	735,000	1.75	14,400	0.346	1.69833	—	1.50450	
0.1495$_5$	1.62	714,000	1.75	12,600	0.3565	—	0.0112		
0.1513	1.68	756,000	1.75	8,300	0.361	1.69316	—	1.50224	
0.1518$_5$	1.72	753,000	1.74	10,700	0.3685	—	0.0056		
0.1536	1.80	761,000	1.74	9,000	0.3820	—	0.0056		
0.1544$_5$	1.87	748,000	1.74	6,500	0.394	1.68374	—	1.49810	
0.1558$_5$	1.92	766,000	1.74	8,100	0.397	—	0.000	1.49640[c]	
0.1581$_5$	2.02	715,000	1.73	11,100	0.410	1.68014[c]	—	1.49430	
0.1596	2.14	669,000	1.72	12,600	0.434	1.67552	—	1.49373	
0.1608	2.20	594,000	1.70	13,300	0.441	1.67423	—	1.48956	
0.1620	2.10	566,000	1.65	14,000	0.508	1.66527	—	1.48841	
0.1633	2.00	608,000	1.65	10,800	0.533	1.66277	—	1.48736	
0.1662	2.00	559,000	1.64	7,500	0.560	1.66046	—	1.48640	
0.1700	1.94	414,000	1.63	≤4,400	0.589	1.65835	—	1.48490	
0.1800	1.70	391,000	1.61	≤1,400	0.643	1.65504	—	1.48459	
0.1900	1.72	278,000	1.59	≤321[d]	0.656	1.65437	—	1.48426	
0.198	—	—	1.57796[c]		0.670	1.65367	—	1.48353	
0.200	1.90284[c]	257,000	1.57649	133	0.706	1.65207	—	1.48259	
0.204	1.88242	—	1.57081		0.768	1.64974	—	1.48215	
0.208	1.86733	149,000	1.56640		0.795	1.64886	—	1.48216	
0.211	1.85692	—	1.56327		0.801	1.64869	—	1.48176	
0.214	1.84558	—	1.55976	~0.1	0.833	1.64772	—	1.48137	
0.219	1.83075	—	1.55496		0.867	1.64676	—	1.48098	
0.226	1.81309	—	1.54921		0.905	1.64578	—	1.48060	
0.231	1.80233	—	1.54541		0.946	1.64480	—	1.48022	
0.242	1.78111	—	1.53782		0.991	1.64380	—	1.47985	
0.2475	—	0.159[e]			1.042	1.64276	—	1.47948	
0.2520	—	0.125			1.097	1.64167	—	1.47910	
0.256	—	0.109			1.159	1.64051	—	1.47870	
0.257	1.76038	—	1.53005		1.229	1.63926	—		
0.2605	—	0.102			1.273	1.63849		1.47831	
0.263	1.75343	—	1.52736		1.307	1.63789	—		
0.265	—	0.096			1.320	1.63767			
0.267	1.74864	—	1.52547		1.369	1.63681			
0.270	—	0.096			1.396	1.63637	—	1.47789	
0.274	1.74139	—	1.52261		1.422	1.63590			
0.275	—	0.102			1.479	1.63490			
0.2805	—	0.096			1.497	1.63457	—	1.47744	
0.286	—	0.102			1.541	1.63381			
0.291	1.72774	—	1.51705		1.6	—	0.05[f]		
0.2918	—	0.109			1.609	1.63261			
0.2980	—	0.118			1.615	—	—	1.47695	
0.303	1.71959	—	1.51365		1.682	1.63127			
0.305	—	0.118			1.7	—	0.09		
0.312	1.71425	0.096	1.51140		1.749	—	—	1.47638	

TABLE 1 Refractive Indices[a] and Absorption Coefficients[a] for Calcite (*Continued*)

λ, μm	n_o	α_o	n_e	α_e	λ, μm	n_o	α_o	n_e	α_e
1.761	1.62974				2.4	—	2.3	—	0.09
1.8	—	0.16			2.5	—	2.7	—	0.14
1.849	1.62800				2.6	—	2.5	—	0.07
1.9	—	0.23			2.7	—	2.3	—	0.07
1.909	—	—	1.47573		2.8	—	2.3	—	0.09
1.946	1.62602				2.9	—	2.8	—	0.18
2.0	—	0.37			3.0	—	4.0	—	0.28
2.053	1.62372				3.1	—	6.7	—	0.46
2.100	—	0.62	1.47492	0.02[f]	3.2	—	10.6	—	0.69
2.172	1.62099				3.3	—	15.0	—	0.92
2.2	—	1.1	—	0.05	3.324	—	—	1.47392	
2.3	—	1.7	—	0.07	3.4	—	19.0	—	1.2

[a] Refractive indexes n_o and n_e are the ordinary and extraordinary rays, respectively, and the corresponding absorption coefficients are $\alpha_o = 4\pi k_o/\lambda$ cm^{-1} and $\alpha_e = 4\pi k_e/\lambda$ cm^{-1}, where the wavelength λ is in centimeters. In the table, the wavelength is in micrometers.
[b] Uzan *et al.*, Ref. 5; α_o and α_e were calculated from the reported values of k_o and k_e.
[c] Ballard *et al.*, Ref. 6.
[d] Schellman *et al.*, Ref. 7; α_e was calculated from the optical density for the extraordinary ray.
[e] Bouriau and Lenoble, Ref. 8; reported absorption coefficient in this paper was for both *o* and *e* rays. α_o was calculated by assuming $\alpha_e = 0$.
[f] Ballard *et al.*, Ref. 9.

both 90° and the extraordinary ray is undeviated and has its minimum refractive index n_e. In other cases for which the optic axis is not parallel to the surface, the extraordinary ray is refracted even for normal incidence.

If the plane of incidence is neither in a principal section nor perpendicular to the optic axis, it is more difficult to determine the angle of refraction of the extraordinary ray. In such cases, Huygens' construction is helpful.[13–15]

2. Types of Polarizing Prisms and Definitions In order to make a polarizing prism out of calcite, some way must be found to separate the two polarized beams. In wavelength regions where calcite is absorbing (and hence only a minimum thickness of calcite can be used), this separation has been made simply by using a very thin calcite wedge cut so that the optic axis is parallel to the faces of the wedge to enable the *e* and *o* rays to be separated by a maximum amount. The incident light beam is restricted to a narrow pencil. Calcite polarizers of this type can be used at wavelengths as short as 1900 Å.[16] In more favorable wavelength regions, where the amount of calcite through which the light passes is not so critical, more sophisticated designs are usually employed. Such prisms can be divided into two main categories, *conventional polarizing prisms* (Pars. 3 to 22) and *polarizing beam-splitter prisms* (Pars. 23 to 31), and a third category, *Feussner prisms* (Par. 32).

In conventional polarizing prisms, only light polarized in one direction is transmitted. This is accomplished by cutting and cementing the two halves of the prism together in such a way that the other beam suffers total internal reflection at the cut. It is usually deflected to the side, where it is absorbed by a coating containing a material such as lampblack. Since the ordinary ray, which has the higher index, is the one usually deflected, the lampblack is often mixed in a matching high-index binder such as resin of aloes ($n_D = 1.634$) or balsam of Tolu ($n_D = 1.628$) to minimize reflections.[17] When high-powered lasers are used, the coating is omitted to avoid overheating the prism, and the light is absorbed externally.

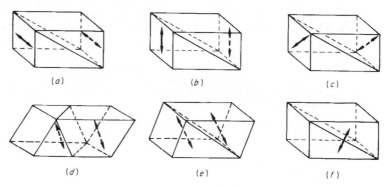

FIGURE 3 Types of conventional polarizing prisms. Glan types: (*a*) Glan-Thompson, (*b*) Lippich, and (*c*) Frank-Ritter; Nicol types: (*d*) conventional Nicol, (*e*) Nicol, Halle form, and (*f*) Hartnack-Prazmowsky. The optic axes are indicated by the double-pointed arrows.

Conventional polarizing prisms fall into two general categories: *Glan types* (Pars. 3 to 13) and *Nicol types* (Pars. 14 to 21), which are illustrated in Fig. 3. Glan types have the optic axis in the plane of the entrance face. If the principal section is parallel to the plane of the cut, the prism is a Glan-Thompson design (sometimes called a Glazebrook design); if perpendicular, a Lippich design; and if 45°, a Frank-Ritter design. In Nicol-type prisms, which include the various Nicol designs and the Hartnack-Prazmowsky, the principal section is perpendicular to the entrance face, but the optic axis is neither parallel nor perpendicular to the face.

Air-spaced prisms can be used at shorter wavelengths than cemented prisms, and special names have been given to some of them. An air-spaced Glan-Thompson prism is called a Glan-Foucault, and an air-spaced Lippich prism, a Glan-Taylor. In common practice, either of these may be called a Glan prism. An air-spaced Nicol prism is called a Foucault prism. Double prisms can also be made, thus increasing the prism aperture without a corresponding increase in length. Most double prisms are referred to as double Frank-Ritter, etc., but a double Glan-Thompson is called an Ahrens prism.

In polarizing beam-splitter prisms, two beams, which are polarized at right angles to each other, emerge but are separated spatially. The prisms have usually been used in applications for which both beams are needed, e.g., in interference experiments, but they can also be used when only one beam is desired. These prisms are also of two general types, illustrated in Fig. 10 in Par. 23; those having the optic axis in the two sections of the prism perpendicular and those having them parallel. Prisms of the first type include the Rochon, Sénarmont, Wollaston, double Rochon, and double Sénarmont. Prisms of the second type are similar to the conventional polarizing prisms but usually have their shape modified so that the two beams emerge in special directions. Examples are the Foster, the beam-splitting Glan-Thompson, and the beam-splitting Ahrens.

The Feussner-type prisms, shown in Fig. 12 in Par. 32, are made of isotropic material, and the film separating them is birefringent. For negative uniaxial materials the ordinary ray rather than the extraordinary ray is transmitted. These prisms have the advantage that much less birefringent material is required than for the other types of polarizing prisms, but they have a more limited wavelength range when calcite or sodium nitrate is used because, for these materials, the extraordinary ray is transmitted over a wider wavelength range than the ordinary ray.

The amount of flux which can be transmitted through a prism or other optical element depends on both its angular aperture and its cross-sectional area. The greater the amount of flux which can be transmitted, the better the *throughput* or *light-gathering power* (sometimes called *étendue* or *luminosity*) of the system.[18,19] If a pupil or object is magnified, the convergence angle of the light beam is reduced in direct ratio to the increase in size of the image. The maximum throughput of a prism is thus proportional to the product of the prism's solid angle of acceptance and its cross-sectional area perpendicular to the prism axis. Hence, a large Glan-Taylor prism having an 8° field angle may, if suitable magnification is used, have a throughput comparable to a small Glan-Thompson prism with a 26° field angle. In general, to maximize prism throughput in an optical system, both the angular aperture and clear aperture (diameter of the largest circle perpendicular to the prism axis which can be included by the prism) should be as large as possible.

The quantities normally specified for a prism are its clear aperture, field angle, and length-to-aperture (L/A) ratio. The *semi-field angle* is defined is the maximum angle to the prism axis* at which a ray can strike the prism and still be completely polarized *when the prism is rotated about its axis*. The field angle is properly twice the semi-field angle.† (Some manufacturers quote a "field angle" for their polarizing prisms which is not symmetric about the prism axis and is thus in most cases unusable.) The *length-to-aperture* (L/A) *ratio* is the ratio of the length of the prism base (parallel to the prism axis) to the minimum dimension of the prism measured perpendicular to the prism base. For a square-ended prism, the L/A ratio is thus the ratio of prism length to width.

In determining the maximum angular spread a light beam can have and still be passed by the prism, both the field angle and the L/A ratio must be considered, as illustrated in Fig. 4. If the image of a point source were focused at the center of the prism, as in Fig. 4a, the limiting angular divergence of the beam would be determined by the field angle $2i$ of the prism.‡ However, if an extended source were focused there (Fig. 4b), the limiting angular divergence would be determined by the L/A ratio, not the field angle.

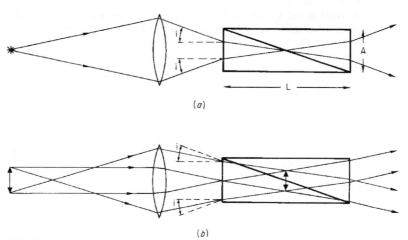

FIGURE 4 The effect of field angle and length-to-aperture ratio of a prism polarizer on the maximum angular beam spread for (a) a point source and (b) an extended source. The field angle is $2i$, and $L/A = 3$. The field angle is exaggerated for clarity.

---

* The prism axis, which is parallel to its base, is not to be confused with the optic axis of the calcite.

† In many prism designs, there is asymmetry about the prism axis, so that although light transmitted at a given angle to the prism axis may be completely polarized for one prism orientation, it will not be completely polarized when the prism is rotated about its axis. Thus, the semi-field angle is not necessarily the largest angle at which completely polarized light can be transmitted by the prism in any orientation.

‡ We are assuming that the prism is wide enough to ensure that the sides of the prism do not limit the angular width of the beam.

The field angle of a polarizing prism is strongly wavelength-dependent. For example, a Glan prism having an 8° field angle at 0.4 µm has only a 2° field angle at 2 µm. In designing optical systems in which polarizing prisms are to be used, the designer must allow for this variation in field angle. If he does not, serious systematic errors may occur in measurements made with the system.

GLAN-TYPE PRISMS

3. Most prisms used at the present time are of the Glan type. Although they require considerably more calcite than Nicol types of comparable size, they are optically superior in several ways. (1) Since the optic axis is perpendicular to the prism axis, the index of the extraordinary ray differs by a maximum amount from that of the ordinary ray. Thus, a wider field angle or a smaller L/A ratio is possible than with Nicol types. (2) The light is nearly uniformly polarized over the field; it is not for Nicol types. (3) There is effectively no lateral displacement in the apparent position of an axial object viewed through a (perfectly constructed) Glan-type prism. Nicol types give a lateral displacement. (4) Since off-axis wander results in images which have astigmatism when the prism is placed in a converging beam, Glan types have slightly better imaging qualities than Nicol types.

Two other often-stated advantages of Glan-type prisms over Nicol types appear to be fallacious. One is that the slanting end faces of Nicol-type prisms have higher reflection losses than the square-ended faces of Glan types. Since the extraordinary ray vibrates in the plane of incidence and hence is in the p direction, increasing the angle of incidence toward the polarizing angle should decrease the reflection loss. However, the index of refraction for the extraordinary ray is higher in Nicol-type prisms (Glan types have the minimum value of the extraordinary index), so the reflection losses are actually almost identical in the two types of prisms. The second "advantage" of Glan-type prisms is that the slanting end faces of the Nicol type supposedly induce elliptical polarization. This widely stated belief probably arises because in converging light the field in Nicol-type polarizers is not uniformly polarized, an effect which could be misinterpreted as ellipticity (see Par. 22). It is possible that strain birefringence could be introduced in the surface layer of a calcite prism by some optical polishing techniques resulting in ellipticity in the transmitted light, but there is no reason why Nicol-type prisms should be more affected than Glan types.

Glan-Thompson-Type Prisms

4. Glan-Thompson-type prisms may be either cemented or air-spaced. Since, as was mentioned previously, an air-spaced Glan-Thompson-type prism is called a Glan-Foucault or simply a Glan prism,* the name Glan-Thompson prism implies that the prism is cemented. Both cemented and air-spaced prisms, however, have the same basic design. The cemented prisms are optically the better design for most applications and are the most common type of prisms in use today. The Glan-Thompson prism is named for P. Glan,[20] who described an air-spaced Glan-Thompson-type prism in 1880, and for S. P. Thompson,[21] who constructed a cemented version in 1881 and modified it to its present

---

* An air-spaced Lippich prism, the Glan-Taylor (Par. 9), has similar optical properties to the Glan-Foucault prism but better transmission. It is also called a Glan prism.

square-ended design in 1882.[22] These prisms are also sometimes called Glazebrook prisms because R. T. Glazebrook[23] demonstrated analytically in 1883 than when rotated about its axis, this prism gives the most uniform rotation of the plane of polarization for a conical beam of incident light. The cut in a Glan-Thompson-type prism is made parallel to the optic axis, which may either be parallel to two sides, as in Fig. 3a, or along a diagonal. The end faces are always perpendicular to the axis of the prism and contain the optic axis.

The extinction ratio* obtainable with a good Glan-Thompson-type prism equals or exceeds that of any other polarizer. Ratios of 5 parts in 100,000 to 1 part in 1 million can be expected, although values as high as 1 part in 3×10^7 have been reported for small selected apertures of the prism.[24] The small residuals result mainly from imperfections in the calcite or from depolarizaton by scattering from the prism faces,[24] although if the optic axis is not strictly in the plane of the end face, or if the optic axes in the two halves of the prism are not accurately parallel, the extinction ratio will be reduced. Also, the extinction ratio may depend strongly upon which end of the prism the light is incident. When prisms are turned end for end, changes in the extinction ratio of as much as a factor of 6 have been reported.[24]

When measuring the extinction ratio, it is essential that none of the unwanted ordinary ray, which is internally reflected at the interface and absorbed or scattered at the blackened side of the prism, reach the detector. King and Talim[25] found that they had to use two 4-mm-diameter apertures and a distance of 80 mm between the photomultiplier detector and prism to eliminate the o-ray scattered light. With no limiting apertures and a 20-mm distance, their measured extinction ratio was in error by a factor of 80.

The field angle of the prism depends both on the cement used between the two halves and on the angle of the cut, which is determined by the L/A ratio. Calculation of the field angle is discussed in Par. 6 and by Bennett and Bennett.[1] Very large field angles can be obtained with Glan-Thompson prisms. For example, if the L/A ratio is 4, the field angle can be nearly 42°. Normally, however, smaller L/A ratios are used. The most common types of cemented prisms are the long form, having an L/A ratio of 3 and a field angle of 26°, and the short form, having an L/A ratio of 2.5 and a field angle of 15°.

5. Transmission In Fig. 5 the transmission of a typical Glan-Thompson prism is compared with curves for a Glan-Taylor prism and a Nicol prism. The Glan-Thompson is superior over most of the range, but its transmission decreases in the near ultraviolet, primarily because the cement begins to absorb. Its usable transmission range can be extended to about 2500 Å by using an ultraviolet-transmitting cement. Highly purified glycerin, mineral oil, castor oil, and Dow Corning DC-200 silicone oil, which because of its high viscosity is not as subject to seepage as lighter oils, have been used as cements in the ultraviolet, as have dextrose, glucose, and *gédamine* (a urea formaldehyde resin in butyl alcohol). Transmission curves for 1-mm thicknesses of several of these materials are shown in Fig. 6, along with the curve for Canada balsam, a cement formerly widely used for polarizing prisms in the visible region.[8] *Gédamine,* one of the best of the ultraviolet-transmitting cements, has an index of refraction $n_D = 1.465_7$ and can be fitted to the dispersion relation[8]

$$n = 1.464 + \frac{0.0048}{\lambda^2} \tag{4}$$

where the wavelength λ is in micrometers.

Figure 7 shows ultraviolet transmission curves for Glan-Thompson prisms with L/A

---

* The extinction ratio is the ratio of the maximum to the minimum transmittance when a polarizer is placed in a plane polarized beam and is rotated about an axis parallel to the beam direction.

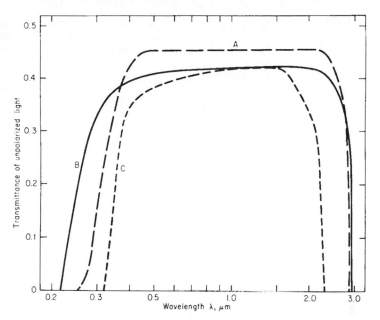

FIGURE 5 Transmittance curves for typical polarizing prisms: *A*, Glan-Thompson, *B*, Glan-Taylor, and *C*, Nicol prism. (*Measured by D. L. Decker, Michelson Laboratory.*) In the visible and near infrared regions the Glan-Thompson has the best energy throughput. In the near ultraviolet the Glan-Thompson may still be superior because the Glan-Taylor has such an extremely small field angle that it may cut out most of the incident beam.

ratios of 2.5 and 3 which are probably cemented with *n*-butyl methacrylate, a low-index polymer that has largely replaced Canada balsam. Better ultraviolet transmission is obtained with a Glan-Thompson prism cemented with DC-200 silicone oil. Air-spaced prisms can be used to nearly 2140 Å in the ultraviolet, where calcite begins to absorb strongly. Transmission curves for two such prisms are shown in Fig. 7. The Glan-Taylor, which is an air-spaced prism of the Lippich design, has a higher ultraviolet transmission than the Glan-Foucault, an air-spaced Glan-Thompson prism. The reason for this difference is that multiple reflections occur between the two halves of the Glan-Foucault prism, resulting in a lowered transmission, but are largely absent in the Glan-Taylor design (see Par. 9).

The infrared transmission limit of typical Glan-Thompson prisms is about 2.7 μm although they have been used to 3 μm.[26] The same authors report using a 2.5-cm-long Glan-Thompson prism in the 4.4- to 4.9-μm region.

6. Field Angle Since many prism polarizers are used with lasers that have parallel beams of small diameter, field-angle effects are not as important as previously when extended area sources were used. Extensive calculations of the field angles for a Glan-Thompson prism are included in the earlier Polarization chapter.[1]

7. Other Glan-Thompson-Type Prisms Other types of Glan-Thompson-type prisms include the Ahrens prism (two Glan-Thompson prisms placed side-by-side), Glan-Foucault prism (an air-spaced Glan-Thompson prism), Grosse prism (an air-spaced Ahrens prism),

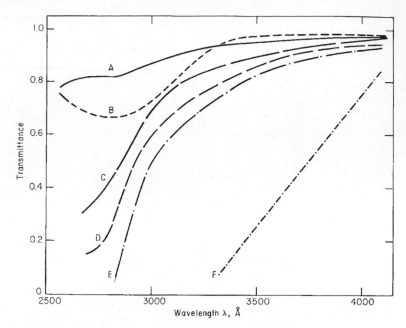

FIGURE 6 Transmittance curves for 1-mm thicknesses of various cements: *A,* crystalline glucose, *B,* glycerine, *C,* gédamine (urea formaldehyde resin in butyl alcohol), *D,* Rhodopas N60A (polymerized vinyl acetate in alcohol), *E,* urea formaldehyde, and *F,* Canada balsam. (*Modified from Bouriau and Lenoble, Ref. 8*). The transmittance of these materials is adequate at longer wavelengths.

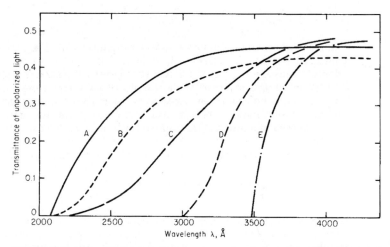

FIGURE 7 Ultraviolet transmittance curves for various Glan-Thompson and air-spaced prisms: *A,* Glan-Taylor (air-spaced Lippich-type prism), *B,* Glan-Foucault (air-spaced Glan-Thompson prism), *C,* Glan-Thompson prism with L/A ratio of 2 cemented with DC-200 silicone oil. *D,* Glan-Thompson prism with L/A ratio of 2.5 probably cemented with *n*-butyl methacrylate, and *E,* Glan-Thompson prism similar to *D* except with $L/A = 3$. (*Modified from curves supplied by Karl Lambrecht Corporation, Chicago.*)

and those constructed of glass and calcite. Information about these prisms can be found in the earlier Polarization chapter.[1]

Lippich-Type Prisms

8. Lippich[27] (1885) suggested a polarizing-prism design similar to the Glan-Thompson but with the optical axis in the entrance face and at right angles to the intersection of the cut with the entrance face (Fig. 3b).* For this case, the index of refraction of the extraordinary ray is a function of angle of incidence and can be calculated from Eq. (1) after ϕ, the complement of the angle of refraction of the wave normal is determined from Eq. (2). In the latter equation, β, the angle the normal to the surface makes with the optic axis, is 90° since the optic axis is parallel to the entrance face. Since the directions of the ray and the wave normal no longer coincide, the ray direction must be calculated from Eq. (3). Lippich prisms are now little-used because they have small field angles, except for two; the air-spaced Lippich, often called a Glan-Taylor prism, and the Marple-Hess prism (two Glan-Taylor prisms back-to-back) that is described in Par. 10. Further information about all Lippich-type prisms is given in the earlier Polarization chapter.[1]

9. Glan-Taylor Prism The Glan-Taylor prism, first described in 1948 by Archard and Taylor,[29] has substantial advantages over its Glan-Thompson design counterpart, the Glan-Foucault prism (Par. 7). Since air-spaced prisms have a very small field angle, the light must be nearly normally incident on the prism face, so that the difference in field angles between the Glan-Taylor and Glan-Foucault prisms (caused by the difference in the refractive index of the extraordinary ray) is negligible.

The major advantages of the Glan-Taylor prism are that its calculated transmission is between 60 and 100 percent higher than that of the Glan-Foucault prism and the intensity of multiple reflections between the two sides of the cut, always a principal drawback with air-spaced prisms, is reduced to less than 10 percent of the value for the Glan-Foucault prism.

The calculated and measured transmittances of a Glan-Taylor prism are in reasonable agreement, but the measured transmittance of a Glan-Foucault prism (Fig. 7) may be considerably higher than its theoretical value.[29] Even so, the transmission of the Glan-Taylor prism is definitely superior to that of the Glan-Foucault prism, as can be seen in Fig. 7. Extinction ratios of better than 1 part in 10^3 are obtainable for the Glan-Taylor prism.[30]

A final advantage of the Glan-Taylor prism is that it can be cut in such a way as to conserve calcite. Archard and Taylor[29] used the Ahrens method of spar cutting described by Thompson[22] and found that 35 percent of the original calcite rhomb could be used in the finished prism.

In a modified version of the Glan-Taylor prism becoming popular for laser applications, the cut angle† is increased, the front and back faces are coated with antireflection coatings, and portions of the sides are either covered with absorbing black glass plates or highly polished to let the unwanted beams escape.[30] The effect of increasing the cut angle is twofold: a beam normally incident on the prism face will have a smaller angle of incidence

---

* The Lippich prism should not be confused with the Lippich half-shade prism, which is a device to determine a photometric match point. The half-shade prism consists of a Glan-Thompson or Nicol prism placed between the polarizer and analyzer such that it intercepts half the beam and is tipped slightly in the beam. The prism edge at the center of the field is highly polished to give a sharp dividing line. The eye is focused on this edge; the disappearance of the edge gives the photometric match point.[28]

† The cut angle is the acute angle the cut makes with the prism base.

on the cut and hence a smaller reflection loss at the cut than a standard Glan-Taylor prism, but, at the same time, the semi-field angle will be reduced throughout most of the visible and near-infrared regions.

A new type of air-spaced prism[31] has a very high transmittance for the extraordinary ray. It resembles the Glan-Taylor prism in that the optic axis is parallel to the entrance face and at right angles to the intersection of the cut with the entrance face. However, instead of striking the prism face at normal incidence, the light is incident at the Brewster angle for the extraordinary ray (54.02° for the 6328-Å helium-neon laser wavelength), so that there is no reflection loss for the *e* ray at this surface. Since the ordinary ray is deviated about 3° more than the extraordinary ray and its critical angle is over 4° less, it can be totally reflected at the cut with tolerance to spare while the extraordinary ray can be incident on the cut at only a few degrees beyond its Brewster angle. Thus this prism design has the possibility of an extremely low light loss caused by reflections at the various surfaces. A prototype had a measured transmission of 0.985 for the extraordinary ray at 6328 Å.[31] If the prism is to be used with light sources other than lasers, its semi-field angle can be calculated.[1]

A major drawback to the Brewster angle prism is that since the light beam passes through a plane-parallel slab of calcite at nonnormal incidence, it is displaced by an amount that is proportional to the total thickness of the calcite. Some of the prisms are made with glass in place of calcite for the second element. In this case, the beam will usually be deviated in addition to being displaced. Measurements on a calcite-glass prototype at 6328 Å showed that the output beam was laterally displaced by several millimeters with an angular deviation estimated to be less than 0.5°.[31]

10. Marple-Hess Prism If a larger field angle is required than can be obtained with a Glan-Taylor prism, a Marple-Hess prism may be used. This prism, which was first proposed in 1960 as a double Glan-Foucault by D. T. F. Marple of the General Electric Research Laboratories and modified to the Taylor design by Howard Hess of the Karl Lambrecht Corporation,[32] is effectively two Glan-Taylor prisms back-to-back. The analysis for this prism is made in the same way as for the Glan-Taylor prism (Par. 9) and Lippich-type prisms in general, keeping in mind that the refractive index of the "cement" is 1 since the components are air-spaced.

Since the ordinary ray is totally reflected for all angles of incidence by one or the other of the two cuts, the field angle is symmetric about the longitudinal axis of the prism and is determined entirely by the angle at which the extraordinary ray is totally reflected at one of the two cuts. This angle can be readily calculated.[1] The field angle is considerably larger than for the Glan-Foucault or Glan-Taylor prism and does not decrease as the wavelength increases.

Unlike the Glan-Foucault or Glan-Taylor prisms, which stop being efficient polarizers when the angle of incidence on the prism face becomes too large, the Marple-Hess prism continues to be an efficient polarizer as long as the axial ordinary ray is not transmitted. If the prism is used at a longer wavelength than the longest one for which it was designed (smallest value of n_o used to determine the cut angle), the value of n_o will be still smaller and the critical angle for the axial ordinary ray will not be exceeded. Thus the axial *o* ray will start to be transmitted before off-axis rays get through. When this situation occurs, it only makes matters worse to decrease the convergence angle. Thus, there is a limiting long wavelength, depending on the cut angle, beyond which the Marple-Hess prism is not a good polarizer. At wavelengths shorter than the limiting wavelength, the Marple-Hess prism has significant advantages over other air-spaced prism designs.

It is not easy to make a Marple-Hess prism, and the extinction ratio in the commercial model is given as between 1×10^{-4} and 5×10^{-5}, somewhat lower than for a Glan-Taylor prism.[30] On the other hand, even though the Marple-Hess prism has an increased L/A

ratio, 1.8 as compared to 0.85 for a Glan-Taylor prism, its ultraviolet transmission is still superior to commercially available ultraviolet transmitting Glan-Thompson prisms of comparable aperture.

Frank-Ritter-Type Prisms

11. The third general category of Glan-type polarizing prisms is the Frank-Ritter design. Prisms of this type are characterized by having the optic axis in the plane of the entrance face, as in other Glan-type prisms, but having the cut made at 45° to the optic axis (Fig. 3c) rather than at 0°, as in Glan-Thompson prisms, or at 90°, as in Lippich prisms. Frank-Ritter prisms are particularly popular in the Soviet Union, and over 80 percent of the polarizing prisms made there have been of this design.[33] Usually double prisms comparable to the Ahrens modification of the Glan-Thompson are used,[1] primarily because from a rhombohedron of Iceland spar two Frank-Ritter double prisms can be obtained but only one Ahrens of the same cross section or one Glan-Thompson of smaller cross section.[33] However, this apparent advantage can be illusory since Iceland spar crystals often are not obtained as rhombs. For example, if the natural crystal is in the form of a plate, it may be less wasteful of material to make a Glan-Thompson or Ahrens prism than a Frank-Ritter prism.[33]

Optically, Frank-Ritter prisms should be similar to Glan-Thompson and Ahrens types, although the acceptance angle for a given L/A ratio is somewhat smaller since the refractive index of the extraordinary ray is larger than n_e in the prism section containing the longitudinal axis and perpendicular to the cut. In practice, the degree of polarization for a Frank-Ritter prism seems to be quite inferior to that of a good Glan-Thompson or even an Ahrens prism.[33]

Use of Glan-Type Prisms in Optical Systems

12. Several precautions should be taken when using Glan-type prisms in optical systems: (1) the field angle of the prism should not be exceeded, (2) there should be an adequate entrance aperture so that the prism does not become the limiting aperture of the optical system, and (3) baffles should be placed preceding and following the prism to avoid incorrect collection of polarized light or extraneous stray light. The reason why these precautions are important are discussed in the earlier Polarization chapter.[1]

Common Defects and Testing of Glan-Type Prisms

13. Several common defects are found in the construction of Glan-type prisms and limit their performance:

1. The axial beam is displaced as the prism is rotated. This defect, called *squirm,* results when the optic axes in the two halves of the prism are not strictly parallel. A line object viewed through the completed prism will oscillate as the prism is turned around the line of sight.[34]

2. The axial ray is deviated as the prism is rotated. This defect is caused by the two

prism faces not being parallel. A residual deviation of 3 minutes of arc is a normal tolerance for a good Glan-Thompson prism; deviations of 1 minute or less can be obtained on special order.

3. The optic axis does not lie in the end face. This error is often the most serious, since if the optic axis is not in the end face and the prism is illuminated with convergent light, the planes of vibration of the transmitted light are no longer parallel across the face of the prism. This effect, which in Nicol-type prisms gives rise to the Landolt fringe, is illustrated in the following practical case.[35] For a convergent beam of light of semi-cone angle i, the maximum variation of the plane of vibration of the emergent beam is $\pm\gamma$, where, approximately,

$$\tan \gamma = n_e \sin i \tan \phi \qquad (5)$$

and ϕ is the angle of inclination of the optic axis to the end face, caused by a polishing error. For $i = 3°$ and $p = 5°$, the plane of vibration of the emergent beam varies across the prism face by ±23 minutes of arc. Thus, good extinction cannot be achieved over the entire aperture of this prism even if nearly parallel light is incident on it. The field angle is also affected if the optic axis is not in the end face or is not properly oriented in the end face, but these effects are small.

4. The cut angle is incorrect or is different in the two halves of the prism. If the cut angle is slightly incorrect, the field angle may be decreased. This error is particularly important in Glan-Foucault or Glan-Taylor prisms, for which the angular tolerances are quite severe, and a small change in cut angle for these prisms may greatly alter the field angle, as discussed in Par. 9 and Ref. 1. If the cut angles are different in the two halves of the prism, the field angle will change when the prism is turned end-for-end. The field angle is determined by the cut angle in the half of the prism toward the incident beam. Differences in the two cut angles may also cause a beam deviation. If the angles in the two halves differ by a small angle α that makes the end faces nonparallel, the beam will be deviated by an angle $\delta = \alpha(n_e - 1)$.[35] If instead, the end faces are parallel and the difference in cut angle is taken up by the cement layer which has a refractive index of approximately n_e, there will be no deviation. However, if the prism is air-spaced, the deviation δ' caused by a nonparallel air film is approximately $\delta' = \alpha n_e$, illustrating one reason why air-spaced prisms are harder to make then conventional Glan-Thompson prisms.[35]

5. The transmittance is different when the prism is rotated through 180°. A potentially more serious problem when one is making photometric measurements is that the transmission of the prism may not be the same in two orientations exactly 180° apart.[36] This effect may be caused by the presence of additional light outside the entrance or exit field angle, possibly because of strain birefringence in the calcite.

Two factors which limit other aspects of polarizer performance in addition to the extinction ratio are *axis wander*, i.e. variation of the azimuth of the transmitted beam over the polarizer aperture, and the ellipticity of the emergent polarized beams[25] caused by material defects in the second half of the prism. Further details are discussed in the earlier Polarization chapter.[1]

In order to determine the cut angle, field angle, parallelism of the prism surfaces, thickness and parallelism of the air film or cement layer, and other prism parameters, one can use the testing procedures outlined by Decker et al.,[37] which require a spectrometer with a Gauss eyepiece, laser source, and moderately good polarizer. (Other testing procedures have been suggested by Archard.[35]) Rowell et al.[38] have given a procedure for determining the absolute alignment of a prism polarizer. However, they failed to consider some polarizer defects, as pointed out by Aspnes[39] who gives a more general alignment procedure that compensates for the prism defects. (There is also a response from

Rowell.[40]) Further information about testing Glan-type prisms and reasons why prism errors are important can be found in the earlier Polarization chapter.[1]

NICOL-TYPE PRISMS

14. Nicol-type prisms are not generally used at the present time, as Glan types are optically preferable. However, they were the first kind made and were once so common that Nicol became a synonym for polarizer. There is much more calcite wastage in making Glan-type prisms than in making the simpler Nicol types so that, even though Glan polarizers were developed in the nineteenth century, it was only following the recent discoveries of new calcite deposits that they became popular. Many of the older instruments are still equipped with Nicol prisms so they will be briefly described here.

15. Conventional Nicol Prism The first polarizing prism was made in 1828 by William Nicol[41] a teacher of physics in Edinburgh. By cutting a calcite rhomb diagonally and symmetrically through its blunt corners and then cementing the pieces together with Canada balsam, he could produce a better polarizer than any known up to that time. A three-dimensional view of Nicol's prism is shown in Fig. 3d. The cut is made perpendicular to the principal section (defined in Par. 1), and the angle is such that the ordinary ray is totally reflected and only the extraordinary ray emerges. When the rhomb is intact, the direction of polarization can be determined by inspection. However, the corners are sometimes cut off, making the rhomb difficult to recognize.

The principal section of Nicol's original prism is similar to that shown in Fig. 2 except that the ordinary ray is internally reflected at the cut along diagonal BH. The cut makes an angle of 19°8' with edge BF in Fig. 2 and an angle of about 90° with the end face of the rhomb. Since the obtuse angle is 109°7' (Fig. 3d), the angle between the cut and the optic axis is 44°36'. The field of the prism is limited on one side by the angle at which the ordinary ray is no longer totally reflected from the balsam film, about 18.8° from the axis of rotation of the prism, and on the other by the angle at which the extraordinary ray is totally reflected by the film, about 9.7° from the axis. Thus the total angle is about 28.5° but is not by any means symmetric about the axis of rotation; the field angle (Par. 2) is only $2 \times 9.7° = 19.4°$.

In order to produce a somewhat more symmetric field and increase the field angle, the end faces of Nicol prisms are usually trimmed to an angle of 68°. This practice was apparently started by Nicol himself.[22] If the cut is made at 90° to the new face, as shown in Fig. 8, the new field angle is twice the smaller of θ_1 and θ_1'. The field angles are computed as described in the earlier Polarization chapter.[1]

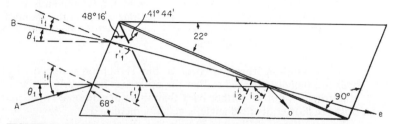

FIGURE 8 Principal section of a conventional Nicol prism with slightly trimmed end faces. Ray A gives the limiting angle θ_1 beyond which the ordinary ray is no longer totally internally reflected at the cut; ray B gives the limiting angle θ_1' for which the extraordinary ray starts to be totally internally reflected at the cut.

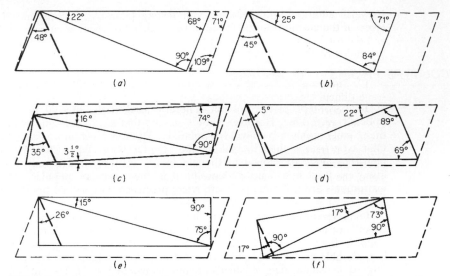

FIGURE 9 Principal sections of various types of trimmed cemented Nicol prisms shown superimposed on the principal section of a cleaved calcite rhomb (see Fig. 2): (*a*) conventional trimmed Nicol; (*b*) Steeg and Reuter shortened Nicol (*Thompson, Ref. 22*); (*c*) Ahrens Nicol (*Thompson, Ref. 22*); (*d*) Thompson reversed Nicol (*Thompson, Ref. 42*); (*e*) square-ended Nicol; and (*f*) Hartnack-Prazmowski reversed Nicol. In all cases, the angle between the prism face and the optic axis (*heavy dashed line*), the angle of the cut, and the acute angle of the rhomb are indicated.

Trimmed Nicol-Type Prisms

16. The angle at which the cut is made in a Nicol-type prism is not critical. The field angle is affected, but a useful prism will probably result even if the cut is made at an angle considerably different from 90°. The conventional trimmed Nicol, discussed in Par. 15, is shown again in Fig. 9a. In this and the other five parts of the figure, principal sections of various prisms are shown superimposed on the principal section of the basic calcite rhomb (Fig. 2). Thus, it is clear how much of the original rhomb is lost in making the different types of trimmed Nicols.

In the Steeg and Reuter Nicol shown in Fig. 9b, the rhomb faces are not trimmed, and the cut is made at 84° to the faces instead of 90°, giving a smaller L/A ratio. The asymmetry of the field which results is reduced by using a cement having a slightly higher index than Canada balsam.

Alternately, in the Ahrens Nicol shown in Fig. 9c, the ends are trimmed in the opposite direction, increasing their angles with the long edges of the rhomb from 70°53′ to 74°30′ or more. By also trimming the long edges by 3°30′, the limiting angles are made more symmetric about the prism axis.

17. Thompson Reversed Nicol In the Thompson reversed Nicol shown in Fig. 9d, the ends are heavily trimmed so that the optic axis lies nearly in the end face. As a result, the blue fringe is thrown farther back than in a conventional Nicol, and although the resulting prism is shorter, its field angle is actually increased.

18. Nicol Curtate, or Halle, Prism The sides of the calcite rhomb may also be trimmed so that they are parallel or perpendicular to the principal section. Thus, the prism is square (or sometimes octagonal). This prism is of the Halle type[43,44] and was shown in Fig. 3e.

Halle, in addition, used thickened linseed oil instead of Canada balsam and altered the angle of the cut. In this way he reduced the length-to-aperture ratio from about 2.7 to 1.8 and the total acceptance angle from 25° to about 17°. Such shortened prisms cemented with low-index cements are often called Nicol curtate prisms (curtate means shortened).

19. Square-ended Nicol The slanting end faces on conventional Nicol prisms introduce some difficulties, primarily because the image is slightly displaced as the prism is rotated. To help correct this defect, the slanting ends of the calcite rhomb can be squared off, as in Fig. 9e, producing the so-called square-ended Nicol prism. The angle at which the cut is made must then be altered since the limiting angle θ_1 for an ordinary ray depends on the angle of refraction at the end face in a conventional prism, in which the limiting ray travels nearly parallel to the prism axis inside the prism (ray A in Fig. 8). If the cut remained the same, the limiting value of θ_1 would thus be zero. However, if the cut is modified to be 15° to the sides of the prism, the total acceptance angle is in the 24 to 27° range, depending on the type of cement used.[22]

Some image displacement will occur even in square-ended Nicol prisms since the optic axis is not in the plane of the entrance face. Therefore, the extraordinary ray will be bent even if light strikes the entrance face of the prism at normal incidence. There is considerable confusion on this point in the literature.[22,45]

20. Hartnack-Prazmowski Prism A reversed Nicol which has the cut at 90° to the optic axis[46] is shown in Figs. 3f and 9f. If it is cemented with linseed oil, the optimum cut angle calculated by Hartnack is 17° to the long axis of the prism, giving a total acceptance angle of 35° and an L/A ratio of 3.4.[22] If Canada balsam is used, the cut should be 11°, in which case the total acceptance angle is 33° and the L/A ratio is 5.2.

21. Foucault Prism A modified Nicol prism in which an air space is used between the two prism halves instead of a cement layer[47] consists of a natural-cleavage rhombohedron of calcite which has been cut at an angle of 51° to the face. The cut nearly parallels the optic axis. Square-ended Foucault-type prisms, such as the Hofmann prism, have also been reported.[22] The angle at which the cut is made can be varied slightly in both the normal Foucault prism and the Hofmann variation of it. In all designs the L/A ratio is 1.5 or less, and the total acceptance angle about 8° or less. The prisms suffer somewhat from multiple reflections, but the principal trouble, as with all Nicol prisms, is that the optic axis is not in the plane of the entrance face. This defect causes various difficulties, including nonuniform polarization across the field and the occurrence of a Landolt fringe (Par. 22 and Ref. 1) when two Nicol-type prisms are crossed.

22. Landolt Fringe If an intense extended light source is viewed through crossed polarizing prisms, careful observation will reveal that the field is not uniformly dark. In Nicol-type prisms the darkened field is crossed by a darker line whose position is an extremely sensitive function of the angle between the polarizer and analyzer. Other types of polarizing prisms also exhibit this anomaly but to a lesser extent. The origin of the Landolt fringe is given in the earlier Polarization chapter[1] and the references cited therein.

POLARIZING BEAM-SPLITTER PRISMS

23. The three classic polarizing beam-splitter prisms are the Rochon, Sénarmont, and Wollaston, shown in perspective in Fig. 10a to c and in side view in Fig. 11a to c. In

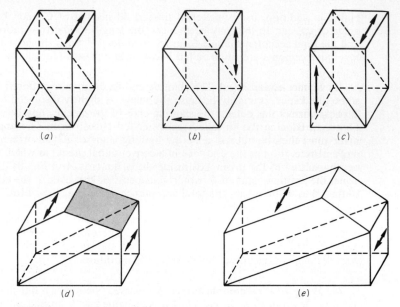

FIGURE 10   Three-dimensional views of various types of polarizing beam-splitter prisms: (*a*) Rochon; (*b*) Sénarmont; (*c*) Wollaston; (*d*) Foster (shaded face is silvered); and (*e*) beam-splitting Glan-Thompson.

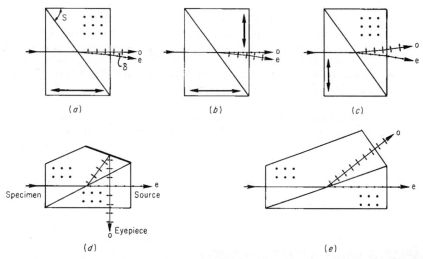

FIGURE 11   Side views of the polarizing beam-splitter prisms in Fig. 10. The directions of the optic axes are indicated by the dots and the heavy double-pointed arrows. The angle of the cut for the Rochon prism is *S*. When the Foster prism is used as a microscope illuminator, the source, specimen, and eyepiece are in the positions indicated.

addition, any polarizing prism can be used as a polarizing beam splitter by changing the shape of one side and removing the absorbing coating from its surface. Two examples of such prisms are the Foster prism, in which the ordinary and extraordinary rays emerge at right angles to each other, and the beam-splitting Glan-Thompson prism, in which the ordinary ray emerges normal to one side (Figs. 10*d* and *e* and 11*d* and *e*). Another prism of this type, the beam-splitting Ahrens prism, is a double beam-splitting Glan-Thompson prism (see Par. 7).

In polarizing prisms, the optic axes are always parallel to each other in the two halves of the prism. By contrast, the optic axes in the two halves of the Rochon, Sénarmont, and Wollaston polarizing beam-splitter prisms are at right angles to each other. Crystal quartz is often used to make these beam splitters, and such prisms can be used down to the vacuum ultraviolet. In applications not requiring such short wavelengths, calcite is preferable because it gives a greater angular separation of the beams (typically 10° as compared to 0.5° for quartz) and does not produce optical rotation.

24. Rochon Prism   The Rochon prism, invented in 1783,[48] is the most common type of polarizing beam splitter. It is often used in photometric applications in which both beams are utilized. It is also used as a polarizing prism in the ultraviolet, in which case one of the beams must be eliminated, e.g., by imaging the source beyond the prism and blocking off the deviated image.

The paths of the two beams through the prism are shown in Fig. 11*a*. A ray normally incident on the entrance face travels along the optic axis in the first half of the prism, so that both ordinary and extraordinary rays are undeviated and have the same refractive index n_0. The second half of the prism has its optic axis at right angles to that in the first half, but the ordinary ray is undeviated since its refractive index is the same in both halves. The extraordinary ray, however, has its minimum index in the second half, so that it is refracted at the cut according to Snell's law (see Par. 1). Since the deviation angle depends on the ratio n_e/n_0, it is a function of wavelength. If the angle of the cut is S, to a good approximation the beam deviation δ of the extraordinary ray depends on the cut angle in the following manner, according to Steinmetz *et al.*,[49]

$$\tan S = \frac{n_o - n_e}{\sin \delta} + \frac{\sin \delta}{2n_e}. \tag{6}$$

This relation holds for light normally incident on the prism face. The semifield angle i_{max} is given by[49]

$$\tan i_{max} = \tfrac{1}{2}(n_e - n_o) \cot S. \tag{7}$$

If the prism is to be used as a polarizer, the light should be incident as shown. Rochon prisms also act as polarizing beam splitters when used backward, but the deviation of the two beams is then slightly less.

When a Rochon prism is used backward, both the dispersion and the optical activity (for quartz) will adversely affect the polarization. Thus, one generally uses a Rochon in the normal manner. However, an exception occurs when a quartz Rochon is to be used as an analyzer. In this case it is best to reverse the prism and use a detector that is insensitive to polarization to monitor the relative intensities of the two transmitted beams.

A Rochon prism is achromatic for the ordinary ray but chromatic for the extraordinary ray. Since total internal reflection does not occur for either beam, the type of cement used between the two halves of the prism is less critical than that used for conventional polarizing prisms. Canada balsam is generally used, although the two halves are sometimes optically contacted for high-power laser applications or for use in the ultraviolet at wavelengths shorter than 3500 Å. Optically contacted crystalline-quartz Rochon prisms can be used to wavelengths as short as 1700 Å, and a double Rochon of MgF$_2$ has been used to 1300 Å in the vacuum ultraviolet.[49] Optically contacted single Rochon prisms of MgF$_2$ have also been constructed, and the transmission of one has been measured from 1400 Å to 7 μm.[50] Ultraviolet-transmitting cements such as *gédamine* can be used to extend the short-wavelength limit of calcite prisms to about 2500 Å (see Par. 5).

25. Defects Quartz and calcite Rochon prisms suffer from several defects. Quartz exhibits optical activity when light is transmitted through it parallel to the optic axis, and although two mutually perpendicular, polarized beams will emerge from a quartz Rochon prism used in the conventional direction, their spectral composition will not faithfully reproduce the spectral compositions of the horizontal and vertical components of the input. If such a prism is used backward, different wavelengths emerge from the prism vibrating in different planes. Hence the output consists of many different polarizations instead of the desired two.[51]

Calcite Rochon prisms do not exhibit optical activity but are difficult to make, since when calcite surfaces are cut normal to the optic axis, small tetrahedra tend to cleave out from the surface during pitch polishing. These tetrahedra may also cleave out during attempts to clean the prisms, and occasionally glass plates are cemented to such surfaces to prevent damage. Some image distortion will occur in calcite prisms; if nonnormally incident rays pass through the prism, both beams will be distorted along their directions of vibration; i.e., the undeviated beam (*o* ray), which vibrates in a vertical plane, will be distorted vertically, and the deviated beam (*e* ray), which vibrates in a horizontal plane, will be distorted horizontally.[51]

26. Glass-Calcite Rochons Some of the difficulties mentioned in Par. 25 can be minimized or eliminated by making the entrance half of the Rochon prism out of glass of matching index instead of quartz or calcite. Both *o* and *e* rays travel along the same path and have the same reflective index in this half of the prism, so that the birefringent qualities of the quartz or calcite are not being used and an isotropic medium would serve just as well. By properly choosing the index of the glass, either the ordinary or the extraordinary ray can be deviated, and glasses are available for matching either index of calcite reasonably well over much of the visible region.[51] The extraordinary ray always suffers some distortion in its direction of vibration, but the distortion of the ordinary ray can be eliminated in the glass-calcite construction. By properly choosing the refractive index of the glass, we can determine whether the *e* ray will be the deviated or the undeviated beam. (Some distortion also arises for deviated beams in the direction of the deviation because of Snell's law and cannot be corrected in this way.) Another method of obtaining an undeviated beam was used by Hardy;[52] unable to find a glass with refractive index and dispersion matching those of calcite, he selected a glass with the correct dispersive power and then compensated for the difference in refractive index by putting a slight wedge angle on the calcite surface. Now a wider selection of glasses is available, but glass-calcite prisms cannot be made strictly achromatic over an extended wavelength range, and thermally induced strains caused by the difference in expansion coefficients in the two parts of the prism may be expected unless the cement yields readily.

27. Total Internal Reflection in Rochons When normal Rochon prisms are used as polarizers, one of the beams must be screened off and eliminated. This restriction might be removed by making the cut between halves of the prism at a sufficiently small angle for the extraordinary ray to be totally reflected. Calculations indicate that this approach should be feasible,[53] but it has apparently not been followed.

28. Sénarmont Prism The Sénarmont polarizing beam splitter, shown in Figs. 10*b* and 11*b*, is similar to the Rochon prism except that the optic axis in the exit half of the prism is coplanar with the optic axis in the entrance half, i.e., at right angles to the Rochon configuration. As a result, light whose plane of vibration is initially vertical is deviated in the Sénarmont prism, while in the Rochon prism the deviated beam has its plane of vibration horizontal (assuming no optical activity in either case) (compare Fig. 11a and b). The amount of the deviation in the Sénarmont prism is slightly less than in the Rochon because the extraordinary ray does not have its minimum refractive index [Eq. (1)].

An alternate form of Sénarmont prism, the right-angle Sénarmont or Cotton polarizer,[54] consists of only the first half of the Sénarmont prism. Unpolarized light normally incident on the prism face is totally internally reflected at the hypotenuse and is then resolved into two planes of vibration, one parallel to the optic axis and the other perpendicular to it. Double refraction will then occur just as in a normal Sénarmont prism. Such a prism has a transmission equivalent to that of an optically contacted Sénarmont or Rochon but is much less expensive.

29. Wollaston Prism The Wollaston prism (Figs. 10c and 11c) is a polarizing beam splitter, also used as a polarizing prism in the vacuum ultraviolet,[55] that deviates both transmitted beams. The deviations, indicated in Fig. 11c, are nearly symmetrical about the incident direction, so that the Wollaston has about twice the angular separation of a Rochon or Sénarmont prism. A normally incident beam is undeviated upon entering the prism, but the o ray, vibrating perpendicular to the optic axis, has a refractive index n_0 while the e ray, vibration parallel to the optic axis has its minimum (or principal) index n_e. At the interface the e ray becomes the o ray and vice versa because the direction of the optic axis in the second half is perpendicular to its direction in the first half. Thus the original o ray enters a medium of lower refractive index and is refracted away from the normal at the cut, while the original e ray passes into a medium of higher refractive index and is refracted toward the normal. On leaving the second half of the prism, both rays are refracted away from the normal, so that their divergence increases.

The deviation of each beam is chromatic in Wollaston prisms, which are most commonly used to determine the relative intensities of two plane-polarized components. Since the light never travels along the optic axis, optical activity does not occur and the relative intensities of the two beams are always proportional to the intensities of the horizontal and vertical polarization components in the incident beam. For an L/A ratio of 1.0, the angular separation between beams is about 1° for a crystalline-quartz Wollaston prism; it can be as high as 3°30' for an L/A ratio of 4.0. With a calcite prism, the beams would have an angular separation of about 19° for an L/A ratio of 1.0, but severe image distortion and lateral chromatism results when such large angular separations are used. These effects can be minimized or the angular separation can be increased for a given L/A ratio by using a three-element Wollaston prism, a modification, apparently suggested by Karl Lambrecht.[30] Divergences as large as 30° can be obtained.[1]

The ellipticity in the emergent polarized beams has been measured by King and Talim.[25] For calcite Wollaston prisms, the ellipticities were in the 0.004 to 0.025° range, comparable to those of Glan-Thompson prisms (Par. 13). Larger values, between 0.12 and 0.16°, were measured for crystalline-quartz Wollaston prisms. The major contribution, which was from the combined optical activity and birefringence in the quartz rather than from defects within the crystal, cannot be avoided in quartz polarizers.

30. Foster Prism This prism, shown in a three-dimensional view in Fig. 10d and in cross section in Fig. 11d, can be used to form two plane-polarized beams separated by 90° from each other.[56] Its construction is similar to that of a Glan-Thompson prism except that one side is cut at an angle and silvered to reflect the ordinary ray out the other side.

The Foster prism is often used backward as a polarizing microscope illuminator for observing reflecting specimens. For this application, the light source is at e in Fig. 11d, and unpolarized light enters the right-hand face of the prism. The ordinary ray (not shown) is reflected at the cut and absorbed in the blackened side of the prism, while the extraordinary ray is transmitted undeviated out the left face of the prism. It then passes through the microscope objective and is reflected by the specimen, returning on its same path to the prism. Light that is unchanged in polarization will be transmitted undeviated by the prism along the path to the light source. If, however, the plane of vibration has been rotated so that it is at right angles to the optic axis (in the plane of the figure), the light will be reflected into the eyepiece. The prism thus acts like a crossed polarizer-analyzer combination.

If a correctly oriented quarter-wave plate is inserted in the beam between the prism and the microscope objective, the light striking the sample will be circularly polarized, and, after being reflected back through the quarter-wave plate, it will be linearly polarized again but with the plane of vibration rotated by 90°. This light is vibrating perpendicular to the optic axis and will be reflected into the eyepiece, giving bright-field illumination. Foster prisms used in this manner introduce no astigmatism since the light forming the image enters and leaves the prism normal to the prism faces and is reflected only by plane surfaces.

31. Beam-splitting Glan-Thompson Prism If a prism design similar to the Foster is used but the side of the prism is cut at an angle so that the ordinary ray, which is deflected, passes out normal to the surface of the prism rather than being reflected, the prism is called a beam-splitting Glan-Thompson prism (Figs. 10*e* and 11*e*). Since no refraction occurs for either beam, the prism is achromatic and nearly free from distortion. The angle between the two emerging beams is determined by the angle of the cut between the two halves of the prism and hence depends on the L/A ratio of the prism. For an L/A ratio of 2.414, the angle is 45°. The field angle around each beam is calculated for different L/A ratios just as for a conventional Glan-Thompson prism. By making the prism double, i.e., a beam-splitting Ahrens prism, the incident beam can be divided into three parts, one deflected to the left, one to the right, and one undeviated.

32. Feussner Prisms The polarizing prisms discussed so far require large pieces of birefringent material, and the extraordinary ray is the one usually transmitted. Feussner[57] suggested an alternate prism design in which only thin plates of birefringent material are required and the ordinary ray rather than the extraordinary ray is transmitted for negative uniaxial materials. A similar suggestion was apparently made by Sang in 1837, although he did not publish it until 1891.[58] In essence, Feussner's idea was to make the prisms isotropic and the film separating them birefringent, as shown in Fig. 12. The isotropic prisms should have the same refractive index as the higher index of the birefringent material so that for negative uniaxial materials, e.g., calcite or sodium nitrate, the ordinary ray is transmitted and the extraordinary ray totally internally reflected. Advantages of this design are (1) since the ordinary ray is transmitted, the refractive index does not vary with angle of incidence and hence the image is anastigmatic, (2) large field angles or prisms of compact size can be obtained, and (3) the birefringent material is used economically.

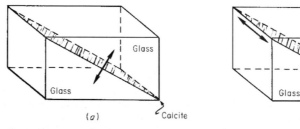

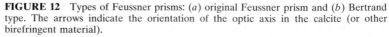

FIGURE 12   Types of Feussner prisms: (*a*) original Feussner prism and (*b*) Bertrand type. The arrows indicate the orientation of the optic axis in the calcite (or other birefringent material).

Furthermore, because the path length of the ray through the birefringent material is short, a lower-quality material can be used.

Disadvantages are (1) for both calcite and sodium nitrate, the extraordinary ray is transmitted over a larger wavelength range than the ordinary ray so that Feussner prisms do not transmit over as as large a wavelength range as conventional prisms, and (2) the thermal-expansion coefficients of the isotropic and birefringent materials are different, making thermally induced strains likely. Solutions to the second problem are to use a thixotropic cement, which flows more readily with increasing stress, or to enclose the system in a metal sleeve and use oil instead of cement. If the ordinary index is matched by the oil, the birefringent material does not even need to be polished very well. Even a cleavage section of calcite can be used, with only a tolerable loss in angular field.[59]

Feussner suggested orienting the optic axis of the birefringent slab perpendicular to the cut, as indicated in Fig. 12a. Since the thermal expansion of the slab is the same in all directions perpendicular to the optic axis, thermally induced strains are minimized in this way. Field angles for Feussner prisms employing calcite and sodium nitrate slabs are given in the earlier Polarization chapter.[1]

Shortly after Feussner's article was published, Bertrand[60] pointed out that the optic axis of the birefringent slab should be parallel to the entrance face of the prism to give the maximum difference between the refractive indices of the ordinary and extraordinary rays. A prism made in this way, sometimes called a Bertrand-type Feussner prism, is shown in Fig. 12b.

Since sodium nitrate is easily obtainable and has a birefringence even larger than that of calcite, attempts have been made to produce polarizing prisms of this material by Wulff,[61] Stöber,[62–64] Tzekhovitzer,[65] West,[66] Huot de Longchamp,[67] and Yamaguti.[68,69] However, it is not only deliquescent but also very soft, so that although large single crystals can be obtained, they are difficult to work. They can be crystallized in the desired orientation from a melt using a technique discovered by West.[66] When sodium nitrate crystallizes from a melt on a mica cleavage surface, one of its basal planes is oriented parallel to the mica cleavage and hence its optic axis is perpendicular to the mica surface. West reports growing single crystals as large as $38 \times 19 \times 2$ cm using this technique. Yamaguti[68,69] has produced polarizing prisms of sodium nitrate by placing thin, closely spaced glass plates on edge on a mica sheet and then immersing the assembly in a melt of sodium nitrate. The thin single crystal thus formed was annealed and cemented between glass prisms to form a Bertrand-type Feussner prism. Conceivably, the sodium nitrate could have been grown directly between the glass prisms themselves, but when such thick pieces of glass are used, it is difficult to avoid setting up strains in the crystal and consequently reducing the polarization ratio. Yamaguti used SK5 glass prisms ($n_D = 1.5889$) cut at an angle of $23°$ to form his polarizing prism and reports a field of view of $31°$, symmetric about the normal to the entrance face.

Another possible birefringent material suitable for a Feussner prism is muscovite mica, and such prisms have actually been constructed and tested.[70,71] A $6°$ field angle can be obtained,[59] which is adequate for many optical systems illuminated by lasers.

Noncalcite Polarizing Prisms

33. Polarizing prisms made of materials other than calcite have been used primarily in the ultraviolet region at wavelengths for which calcite is opaque. Prism materials used successfully in this region include crystalline quartz, magnesium fluoride, sodium nitrate, and ammonium dihydrogen phosphate. Rutile polarizing prisms have been used beyond the calcite cutoff in the infrared. A new prism material, yttrium orthovanadate, has been used to make high-transmission polarizers for the visible and near-infrared spectral regions.[72] Properties of this material were described in the earlier Polarization chapter.[1]

Rochon or Wollaston prisms (Pars. 24 and 29) are sometimes made of crystalline quartz for use in the far ultraviolet. The short-wavelength cutoff of the quartz is variable, depending on the impurities present, but can be as low as 1600 Å.

By utilizing magnesium fluoride instead of quartz for the polarizing prisms, the short-wavelength limit can be extended to 1300 Å. Magnesium fluoride transmits to about 1125 Å, but below 1300 Å its birefringence decreases rapidly and changes sign at 1194 Å.[55,73] Although it is the most birefringent material available in this region, MgF_2 has a much smaller birefringence than that of calcite; hence, a small cut angle and large L/A ratio for the prism are unavoidable. Since absorption does occur, it is desirable to minimize the length of the prism. Johnson[55] solved this problem by constructing a MgF_2 Wollaston prism which requires only half the path length necessary for a Rochon prism. However, both beams are deviated, creating instrumental difficulties.

Steinmetz et al.[49] constructed a double Rochon prism of MgF_2 which has the same L/A ratio as the Wollaston prism but does not deviate the desired beam. Problems with the prism included fluorescence, scattered light, and nonparallelism of the optic axes.[1] In principle, however, a MgF_2 double Rochon polarizing prism should be an efficient, high-extinction-ratio, on-axis polarizer for the 1300- to 3000-Å wavelength range and should also be useful at longer wavelengths. Morris and Abramson[50] reported on the characteristics of optically contacted MgF_2 single Rochon prisms.

A different type of polarizer suggested by Chandrasekharan and Damany[74] to take the place of a Rochon or Wollaston prism in the vacuum ultraviolet consisted of a combination of two MgF_2 lenses, one planoconcave and the other planoconvex of the same radius of curvature, combined so that their optic axes were crossed. The combination acted as a convergent lens for one polarization and as a divergent lens for the other. It had the advantage that the polarized beam remained on axis and was focused. A measured degree of polarization of 98.5 percent was obtained at 1608 Å, in good agreement with the calculated value.

Prism polarizers can also be constructed for use in the infrared at wavelengths longer than those transmitted by calcite. Rutile, TiO_2, a positive uniaxial mineral with a large birefringence and good transmittance to 5 μm in the infrared, has been used by Landais[75] to make a Glan-Foucault-type crystal polarizer. Since rutile has a positive birefringence (in contrast to the negative birefringence of calcite), the ordinary ray is transmitted undeviated and the extraordinary ray is reflected out one side. Other characteristics are given in the earlier Polarization chapter.[1]

DICHROIC AND DIFFRACTION-TYPE POLARIZERS

34. Some of the most useful polarizers available employ either dichroism or diffraction effects. These polarizers come in sheet form, sometimes in large sizes, are easily rotated, and produce negligible beam deviation. Also, they are thin, lightweight, and rugged, and most can be made in any desired shape. The cost is generally much less than that of a prism-type polarizer. Furthermore, both types are insensitive to the degree of collimation of the beam, so that dichroic or diffraction-type polarizers can be used in strongly convergent or divergent light.

A dichroic* material is one which absorbs light polarized in one direction more strongly than light polarized at right angles to that direction. Dichroic materials are to be distinguished from birefringent materials, which may have different refractive indexes for the two electric vectors vibrating at right angles to each other but similar (usually

---

* The term *dichroic* is also used in three other ways: (1) to denote the change in color of a dye solution with change in concentration, (2) to denote a color filter that has two transmission bands in very different portions of the visible region and hence changes color when the spectral distribution of the illuminating source is changed, and (3) to denote an interference filter that appears to be of a different color when viewed in reflected or transmitted light.

negligible) absorption coefficients. Various materials are dichroic, either in their natural state or in a stretched condition. The most common materials used as dichroic polarizers are stretched polyvinyl alcohol sheets treated with absorbing dyes or polymeric iodine, commonly marketed under the trade name Polaroid. These and similar materials are discussed in Par. 35. Another type of dichroic polarizer is prepared by rubbing a glass or plastic surface in a single direction and then treating it with an appropriate dye. Polarizers of this type are sold under the trade name Polacoat and will be described in Par. 36. In certain portions of the infrared spectral region, calcite is strongly dichroic and makes an excellent high-extinction polarizer.[76] Pyrolytic graphite is electrically and optically aniso-tropic and has been successfully used as an infrared polarizer; it is described in Par. 37. Other materials which exhibit dichroism in the infrared include single-crystal tellurium,[77] ammonium nitrate,[78] mica, rubber under tension, polyvinyl alcohol, and polyethylene.[79] In the visible region, gold, silver, and mercury in the form of microcrystals,[80] needles of tellurium,[81] graphite particles,[82] and glasses containing small elongated silver particles[83] are all dichroic.

A sodium nitrate polarizer described by Yamaguti[84] is not dichroic in the strict sense of the word but acts like a dichroic polarizer. Roughened plates of SK5 glass are bonded together by a single crystal of sodium nitrate, which has a refractive index for the ordinary ray nearly equal to that of the glass. The extraordinary ray has a much lower index, so that it is scattered out of the beam by the rough surfaces, leaving the ordinary ray to be transmitted nearly undiminished. (Yamaguti has also made Feussner prisms out of single-crystal sodium nitrate described in Par. 32.)

Diffraction-type polarizers include diffraction gratings, echelettes, and wire grids. These are all planar structures that have properties similar to those of dichroic polarizers except that they transmit one component of polarization and reflect the other when the wavelength of the radiation is much longer than the grating or grid spacing. Wire grid and grating polarizers are covered in Par. 38.

None of these polarizers has as high a degree of polarization as the prism polarizers of Pars. 1 to 33. Thus it is frequently necessary to measure the polarizing properties of the particular polarizer used. A source of plane-polarized light is desirable for such a measurement. Lacking that, one of the procedures described in Par. 39 can be followed if there are two identical imperfect polarizers. Alternate methods are also described which are applicable to two nonidentical imperfect polarizers.

35. Sheet Polarizers Various types of sheet polarizers have been developed by Edwin H. Land and coworkers at the Polaroid Corporation, Cambridge, Mass. Sheet polarizers are also available from several European companies. The J sheet polarizer, the first type available in America (around 1930), consisted of submicroscopic needles of herapathite oriented parallel to one another in a sheet of cellulose acetate. Since this type of polarizer, being microcrystalline, had some tendency to scatter light, it was superseded by H and K sheet molecular polarizers, which exhibit virtually no scattering. The most widely used sheet polarizer is the H type, which consists of a sheet of polyvinyl alcohol that has been unidirectionally stretched and stained with iodine in a polymeric form. The K type is made by heating a sheet of polyvinyl alcohol in the presence of a catalyst to remove some of the water molecules and produce the dichromophore polyvinylene. It was developed primarily for applications where resistance to high temperature and high humidity are necessary. Another type of polarizing sheet, made from a combination of the H and K types, has an absorption maximum at about 1.5 μm in the infrared and is designated as HR Polaroid.

The history of the development of the various kinds of sheet polarizers has been given by Land,[81] their chemical composition by Land and West,[80] and their optical performance by Shurcliff,[82] Baumeister and Evans,[85] Land and West,[80] and Land.[81] In addition, Blake et al.[86] mention the HR infrared polarizer, and Makas[87] describes the modified H-film polarizer for use in the near ultraviolet. Baxter et al.[88] describe a technique for measuring

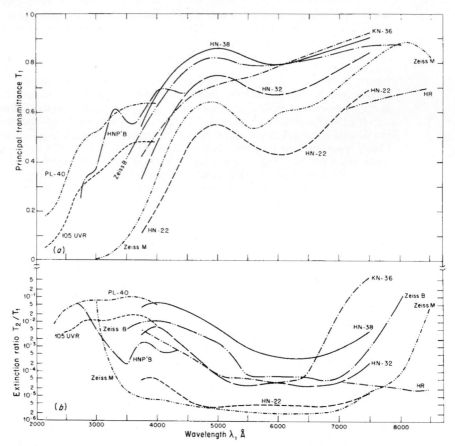

FIGURE 13 (*a*) Principal transmittance and (*b*) extinction ratio for various types of dichroic polarizers: Polaroid sheet polarizers HN-22, HN-32, HN-38, and KN-36; Zeiss (Oberkochen) Bernotar and Micro Polarization filters; and Polacoat PL-40 and 105 UVR polarizing filters. The last is stated to have a transmittance (for unpolarized light) of 32 percent at 5460 Å. (*Modified from curves of Shurcliff, Ref. 82, Baumeister and Evans, Ref. 85, Jones, Ref. 89, Haase, Ref. 90, and McDermott and Novick, Ref. 91.*)

the optical density of high-extinction polarizers in the presence of instrumental polarization.

Figure 13 shows the principal transmittance T_1 and extinction ratio T_2/T_1 of various types of H and K sheet polarizers used in the visible and near ultraviolet.[82,85,89] In addition, curves for two sheet polarizers manufactured by Zeiss and two types of polarizing filters from Polacoat (Par. 36) are shown. The letter N in the designation of the Polaroid sheets stands for neutral (to distinguish them from sheet polarizers prepared from colored dyes), and the number 22, 32, etc., indicates the approximate transmittance of unpolarized visible light. Figure 14 gives the principal transmittance and extinction ratio of a typical plastic laminated HR infrared polarizer.[82,89] Sometimes the optical density D of a polarizer is plotted instead of its transmittance. The relation between these two quantities is

$$D = \log \frac{1}{T} \qquad (8)$$

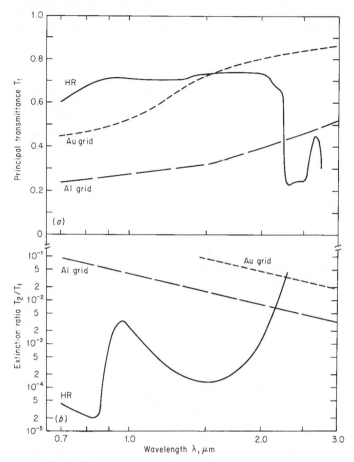

FIGURE 14 (*a*) Principal transmittance and (*b*) extinction ratio for plastic laminated HR infrared polarizer (*modified from curves of Shurcliff, Ref. 82, and Jones, Ref. 89*) and two wire grid polarizers with 0.463-μm grating spacings (*Bird and Parrish, Ref. 92*).

The extinction ratio of the HN-22 Polaroid compares favorably with that of Glan-Thompson prisms throughout the visible region, but the transmission of the Glan-Thompson is superior. In the ultraviolet, the new HNP′B material has a reasonably good extinction ratio (about 10^{-3} or better) for wavelengths longer than 3200 Å. It is a specially purified form of HN-32, and its properties match those of the standard HNT-32 Polaroid at wavelengths longer than 4500 Å. Optical properties of various types of Polaroid dichroic polarizers have been described by Trapani.[93] According to West and Jones,[48] the extinction ratio for a dichroic polarizer of the Polaroid type has a practical limit of about 10^{-5} because, as the concentration of dichromophore is increased beyond a certain value, the optical density no longer increases proportionately. Gunning and Foschaar[94] have described a method for the controlled bleaching of the iodine dichromophore in iodine-polyvinyl alcohol polarizers to achieve an increased internal transmission of up to 95 percent for the principal transmittance of linearly polarized light in the 5000- to 6000-Å wavelength region. This is achieved at the expense of degrading the extinction ratio and drastically affecting the short wavelength performance of the polarizer. Baum[95] describes the application of sheet polarizers to liquid crystal displays and problems encountered in this application.

If Polaroids are used in applications where beam deviation is important, they should be checked for possible deviation. Most Polaroids, which are laminated in plastic sheets, do produce a slight beam deviation that can be observed through a telescope as a shift in the image position when the Polaroid is rotated. The amount of the deviation varies from point to point on the Polaroid and can be much worse if the material is mounted between glass plates. It is possible to order specially selected sheet Polaroid laminated between polished glass plates that deviates the beam by only about 5 seconds of arc.

Sheet polarizers made of stretched polyvinyl alcohol that has been stained with iodine or various dyes are also made in countries outside the United States, as described in the earlier Polarization chapter.[1]

King and Talim[25] have measured the axis wander and ellipticity of beams transmitted by various types of sheet polarizers in the same way as for Glan-Thompson prisms, (Par. 13). They found considerable variations from one type of sheet polarizer to another and also over a single sheet. Details are given in the earlier chapter on Polarization.[1]

36. Dichroic Polarizing Coatings Beilby-layer polarizers[82] are dichroic coatings that can be applied to the surface of glass or plastic. The process was developed by Dreyer,[96] who founded the company which manufactures Polacoat polarizing filters. There are three main steps in the production of these polarizers. First, the substrate (quartz, glass, plastic, etc.) is rubbed along parallel lines with filter paper, cotton, or rouge to produce a preferred surface orientation. (The affected region of minute scratches extends to a depth of less than 1 μm.) Then the sheet is rinsed and treated with a solution of dichroic molecules, e.g., a 0.5 percent solution of methylene blue in ethanol or one or more azo dyes, and then dried in a controlled fashion. Presumably the molecules line up preferentially along the rubbing direction, resulting in a greater absorption for light, polarized in that direction. As a final step, the surface is treated with an acidic solution, often that of a metallic salt such as stannous chloride, which can increase the dichroism and produce a more neutral color. A protective coating over the polarized surface provides mechanical protection for the fragile layer with no loss in transmission. McDermott and Novick[91] give a somewhat more complete description of the Polacoat process, and Anderson[97] has investigated the absorption of methylene blue molecules on a unidirectionally polished surface. References to patents and related work are given by Shurcliff.[82]

The principal transmittance and extinction ratio of two standard Polacoat coatings, PL-40 and 105 UVR (32 percent transmission of unpolarized light at 5460 Å), are shown in Fig. 13. These curves are taken from the data of McDermott and Novick.[91] Polacoat 105 UVR coating comes in various densities; the data shown are for the highest-density material with the best extinction ratio.* A major advantage of Polacoat over sheet Polaroid is that it does not bleach upon exposure to intense ultraviolet radiation.

Kyser[99] tested a stock PL40 polarizing filter on fused quartz and found that it produced a large quantity of scattered light of the unwanted component. This light was dispersed spectrally and was scattered at angles up to about 20° as though the scratches on the rubbed surface were acting like rulings on a diffraction grating. There was relatively little of the unwanted component on axis; most of it was scattered at larger angles. Despite these difficulties, Polacoat PL40 polarizers appear to be the best large-aperture transmission-type polarizers available for work in the 2000- to 3000-Å wavelength range in the ultraviolet.

---

* The company literature[98] is somewhat misleading in that the transmittance of this material is stated to be 35 percent, but the transmission curve (for unpolarized light) given in the bulletin does not rise above 30 percent until the wavelength becomes longer than 6500 Å.

37. Pyrolytic-Graphite Polarizers Pyrolytic graphite has a large anisotropy in both the electric conductivity and in the optical properties. If the E vector of an electromagnetic wave is pointing in the direction of the c-axis of the graphite, the absorption coefficient is a minimum, the reflectance is also a minimum, and hence the transmittance is a maximum. If the E vector lies in the plane perpendicular to the c direction, the absorption is a maximum, reflectance is a maximum, and transmittance is a minimum. Thus, pyrolytic graphite should be a good material from which to make a dichroic polarizer if a thin foil is cut and polished to contain the c-axis. Several such polarizers have been made by Rupprecht et al.[100]; two had thicknesses of 9.2 μm, and a third was 4.2 μm thick. The transmittances T_1 of the thinner one and T_1 and T_2 of the two thicker ones were determined using one of the methods described in Par. 49 of the earlier Polarization chapter.[1] The principal transmittance and extinction ratio for one of the 9.2-μm-thick ones are shown in Fig. 15 for infrared wavelengths from 2 to 16 μm, along with curves for various wire-grid polarizers (Par. 38). In the far infrared out to 600 μm, T_1 gradually increases to 0.50, and T_2/T_1 drops down to the 10^{-3} range.[100] The transmittance of the thinner pyrographite polarizer was larger than the curve shown, but its extinction ratio, although not given, was probably poorer. Pyrolytic-graphite polarizers have the advantages of being planar and thus easily rotatable, having large acceptance angles, and having reasonably high

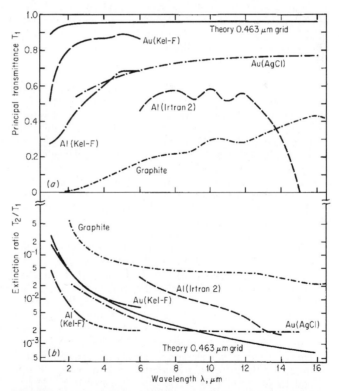

FIGURE 15 (*a*) Principal transmittance and (*b*) extinction ratio for a pyrolytic-graphite polarizer (*Rupprecht et al., Ref. 100*) and various wire-grid polarizers (*Bird and Parrish, Ref. 92, Perkin-Elmer, Ref. 101 and Young et al., Ref. 102*). The substrate materials and metals used for the grids are indicated. Theoretical curves (*solid lines*) calculated from relations given in Ref. 1 with $n = 1.5$ and $d = 0.463$ are also shown for comparison.

transmittances and good extinction ratios in the far infrared. However, in the shorter-wavelength region shown in Fig. 15, they are inferior to all the wire-grid polarizers. In addition, they are fragile, and the largest clear aperture obtained by Rupprecht et al.[100] was about 12 mm diameter.

38. Wire-Grid and Grating Polarizers Wire grids* have a long history of use as optical elements to disperse radiation and detect polarization in far-infrared radiation and radio waves.[92] They transmit radiation whose E vector is vibrating perpendicular to the grid wires and reflect radiation with the E vector vibrating parallel to the wires when the wavelength λ is much longer than the grid spacing d. When λ is comparable to d, both components are transmitted. For grids made of good conductors, absorption is negligible. Various aspects of the theory of reflection and transmission of radiation by wire grids are summarized in the earlier Polarization chapter.[1] In addition to that theoretical treatment, Casey and Lewis[104,105] considered the effect of the finite conductivity of the wires on the transmission and reflection of wire-grid polarizers when the light was polarized parallel to the wires. Mohebi, Liang, and Soileau[106] extended the treatment to the case for which light was polarized both parallel and perpendicular to the wires; they also calculated the absorption of the wire grids as a function of d/λ. In addition, they measured the absorption and surface damage of wire-grid polarizers consisting of aluminum strips (0.84 μm period) deposited on ZnSe substrates at 10.6 μm, 1.06 μm, and 0.533 μm. Stobie and Dignam[107] calculated the amplitude transmission coefficients for parallel and perpendicular components and relative phase retardation between them, both as a function of λ/d. Burton[108] proposed using wire-grid polarizers in the form of cylinders and paraboloids instead of planar structures in infrared interferometers, but did not show any experimental measurements.

Figure 16 shows values of the calculated principal transmittance and extinction ratio for various values of the refractive index n as a function of λ/d. These curves were calculated from relations given in the earlier Polarization chapter.[1] It is clear that the shortest wavelength for which a given grid will act as a useful polarizer is $\lambda \approx 2d$. Also, the best performance is obtained with the lowest refractive index substrate. Since absorption in the substrate material has been neglected, principal transmittances measured for real materials will be lower than the calculated values, but the extinction ratios should be unaffected. If one must use a high refractive index substrate such as silicon or germanium, the performance of the grid can be considerably improved by applying an antireflection coating to the substrate *before* depositing the conducting strips, since a perfectly antireflected substrate acts like an unsupported grid.[109] However, if the antireflecting layer is laid down *over* the grid strips, the performance of the wire grid polarizer is degraded.[1]

Many people have built and tested wire-grid polarizers including Bird and Parrish,[92] Young et al.,[102] Hass and O'Hara,[110] Hilton and Jones,[111] Auton,[109] Vickers et al.,[112] Cheo and Bass,[113] Auton and Hutley,[114] Costley et al.,[115] Beunen et al.,[116] Leonard,[117] Sonek et al.,[118] Eichhorn and Magner,[119] and Novak et al.[120] In addition, two types of wire grids are manufactured commercially by Buckbee Mears (see Ref. 110) and Perkin-Elmer,[101] and a third type composed of 152-μm-diameter tungsten wires spaced 800 to the inch has been mentioned, but no performance characteristics have been given.[121] Hwang and Park[122] measured the polarization characteristics of two-dimensional wire mesh (64 μm and 51 μm spacings) at a laser wavelength of 118.8 μm. The different wire-grid polarizers are listed in Table 2, and the principal transmittances and extinction ratios of several are shown in Figs 14 and 15.

The polarizers with grid spacings of 1.69 μm and less were all made by evaporating the grid material at a very oblique angle onto a grating surface which had been prepared

---

* *Wire grid* is being used here, as is customary, to denote a planar structure composed of a series of parallel wires or strips. Renk and Genzel[103] and a few others use the term to designate a two-dimensional array with two series of elements arranged at right angles to each other. They call a one-dimensional array a wire or strip grating.

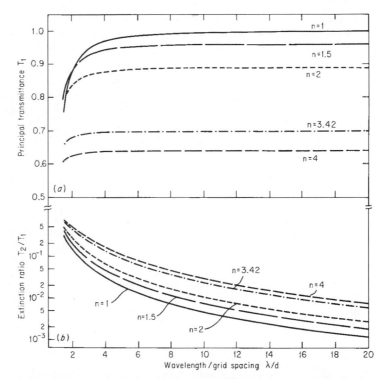

FIGURE 16 (*a*) Principal transmittance and (*b*) extinction ratio as a function of λ/d calculated from relations given in Ref. 1 for various values of n for the substrate. Substrate indexes correspond approximately to an antireflected substrate of air, organic plastic, silver chloride, silicon, and germanium.

either by replicating a diffraction grating with the appropriate substrate material (silver bromide, Kel-F, polymethyl methacrylate, etc.) or by ruling a series of lines directly onto the substrate (Irtran 2 and Irtran 4). The oblique evaporation (8 to 12° from the surface) produced metallic lines on the groove tips which acted like the conducting strips of the theory, while the rest of the surface was uncoated and became the transparent region between strips. Larger grid spaces (4 to 25.4 μm) were produced by a photoetching process, and one 25.4-μm grid was made by an electroforming process. Still larger grid spacings were achieved by wrapping wires around suitable mandrels.

If a wire-grid polarizer is to be used in the near infrared, it is desirable to have the grid spacing as small as possible. Bird and Parrish[92] succeeded in obtaining a very good extinction ratio in the 2- to 6-μm wavelength region with an aluminum-coated Kel-F substrate (Figs. 14 and 15). Unfortunately, Kel-F $(CF_2CFCl)_n$, has absorption bands at 7.7 to 9.2 and 10.0 to 11.0 μm, making the polarizer useless in these regions, but it can be used at longer wavelengths out to 25 μm.[92] Polyethylene would be an excellent substrate material since it has fewer absorption bands than Kel-F, but its insolubility in common solvents makes it much more difficult to use for replicating gratings.[110] It does, however, make an excellent substrate material for photoetched grids.[109]

For infrared wavelengths longer than about 24 μm, a photoetched grid with 1-μm-wide lines (close to the present limit for the photoetching process) and a 2-μm spacing should have an extinction ratio of 5×10^{-3} or better if the refractive index of the substrate is about 1.5—for example, polyethylene. The extinction ratio would continue to decrease;

TABLE 2 Types of Wire-Grid Polarizers

Grid spacing, μm	Grid material	Substrate	Wavelength range, μm	Reference
0.115	Evaporated Al	Quartz	0.2–0.8	Sonek et al., Ref. 118
0.347	Evaporated Au	Silver chloride	2.5–30	Perkin-Elmer, Ref. 101
0.22–0.71	Evaporated Al	KRS-5	3–15, 3.39, 10.6	Auton and Hutley, Ref. 114
0.22–0.45	Evaporated Al	CaF$_2$	3–10, 3.39	Auton and Hutley, Ref. 114
0.42	Evaporated Al	Glass	3–5	Auton and Hutley, Ref. 114
0.463	Evaporated Au	Kel-F	1.5–10*	Bird and Parrish, Ref. 92
0.463	Evaporated Al	Kel-F	0.7–15*	Bird and Parrish, Ref. 92
0.463	Evaporated Al	Polymethyl methacrylate	1–4000†	Hass and O'Hara, Ref. 110
1.67	Evaporated Al	Irtran 2	6–14	Young et al., Ref. 102
1.67	Evaporated Al	Irtran 4	8–19	Young et al.,, Ref. 102
1.69	Evaporated Al	Polyethylene	2.9–200‡	Hass and O'Hara, Ref. 110
2	Evaporated Cr	Silicon	10.6	Cheo and Bass, Ref. 113
?	?	BaF$_2$	2–12	Leonard, Ref. 117
?	?	ZnSe	3–17	Leonard, Ref. 117
4	Photoetched Al	Polyethylene	>16	Auton, Ref. 109
5.1	Photoetched Al	Silicon	54.6	Hilton and Jones, Ref. 111
10	Photoetched Al	Polyethylene	>16	Auton, Ref. 109
25.4	Photoetched Al	Silicon	54.6	Hilton and Jones, Ref. 111
25.4	Evaporated Au	Mylar	>60	Hass and O'Hara, Ref. 110
25	Stainless steel wire 8 μm diam	Air	80–135	Novak et al., Ref. 120
32.4	Gold-coated W wire 21 μm diam	Air	100–10,000	Eichhorn and Magner, Ref. 119
64, 51	Wire mesh (2D)	Air	118.8	Hwang and Park, Ref. 122
?	Stainless steel wire 50 μm diam	Air	200–1000	Vickers et al., Ref. 112
317	W wire 152 μm diam	Air	40–300	Roberts and Coon, Ref. 121
25–1800	W wire 10 μm diam	Air	>50	Costley et al., Ref. 115
30–65	W wire 10 μm diam	Air	22–500,337	Beunen et al., Ref. 116

* Strong absorption bands near 8.3 and 10.5 μm.

† Strong absorption bands between 5.7 and 12.5 μm.

‡ Absorption bands between 6 and 15.5 μm.

i.e., the polarization properties would improve as the wavelength increased. At very long wavelengths, grids with a larger spacing would have a high degree of polarization. The important factor is the ratio of wavelength to grid spacing, which should be kept as large as possible (Fig. 16b).

One definite advantage of the wire-grid polarizer is that it can be used in sharply converging beams, i.e. systems with high numerical apertures. Young et al.[102] found no decrease in percent of polarization for an Irtran 2 polarizer at 12 μm used at angles of incidence from 0 to 45°. They did find, however, that the transmittance decreased from 0.55 at normal incidence to less than 0.40 at 45° incidence.

If a grid were to be used at a single wavelength, one might possibly make use of interference effects in the substrate to increase the transmission.[109] If the substrate has perfectly plane-parallel surfaces, it will act like a Fabry-Perot interferometer and transmit a maximum amount of light when twice the product of the thickness and refractive index is equal to an integral number of wavelengths. The 0.25-mm-thick pressed polyethylene substrates used by Auton[109] were not uniform enough to show interference effects, but the Mylar film backing on the Buckbee Mears electroformed grid did show interference effects.[110]

Lamellar eutectics of two phases consist of thin needles of a conducting material embedded in a transparent matrix. The material is made by a controlled cooling process in which there is a unidirectional temperature gradient. This method of cooling orients conducting needles parallel to the temperature gradient, and hence the material can act like a wire-grid polarizer. Weiss and coworkers[123–125] have grown eutectic alloys of InSb and NiSb in which the conducting needles of NiSb are approximately 1 µm in diameter and approximately 50 µm long. A degree of polarization of more than 99 percent has been reported. Other eutectic alloys of InAs, GaSb, and InSb containing conducting needlelike crystals of Ni, Fe, Mn, Cr, and Co (or their compounds) have also been investigated. An advantage of this type of polarizer is that its performance can be optimized at a specific wavelength, e.g., that of a CO_2 laser line, by choosing the thickness of the crystalline film so that there will be an interference maximum at the desired wavelength.[126] Recently, Saito and Miyagi[127] have proposed using a thin film of anodized aluminum with implanted metallic columns to make a high-performance polarizer. Their theoretical calculations suggest that this type of polarizer should have a large extinction ratio and low loss in the infrared.

In summary, wire grids are very useful infrared polarizers, particularly for wavelengths much greater than the grid spacing. They are compact and easily rotatable and can be used with sharply converging beams. A major advantage is the extreme breadth of the wavelength band over which they have good polarizing properties. The long-wavelength limit is set by the transmission of the substrate material rather than by the loss of polarization of the grid. The short-wavelength limit is determined by the grid spacing; if gratings with smaller spacings could be successfully replicated and coated, the short-wavelength limit could be pushed closer to the visible region.

Another possible method of producing plane-polarized light is by using diffraction gratings or echelette gratings. Light reflected from diffraction gratings has long been known to be polarized, but the effect is generally small and extremely wavelength-dependent.[128,129] However, Roumiguieres[130] predicted that under certain conditions (rectangular groove grating with equal groove and land spacings and small groove depth), a high polarizing efficiency could be obtained. For wavelengths in the range $1.1 < \lambda/d < 1.7$, over 80 percent of the light polarized parallel to the grooves should be reflected in the zero order at a 50° angle of incidence and less than 5 percent of the other polarization. His predictions were verified by Knop[131] who fabricated gold-coated photoresist gratings as well as an electroplated nickel master grating. Knop's measured reflectances of the two polarized components were within ±3 percent of the predicted values. In general, one tries to avoid polarization in the diffracted light to obtain high efficiencies in a blazed grating since polarization effects are frequently associated with grating anomalies.[132,133]

In contrast to diffraction gratings, echelette gratings have been found to produce an appreciable amount of plane-polarized light. Experimental studies have been made by Peters et al.,[134] Hadni et al.,[135,136] and Mitsuishi et al.,[137] as discussed in the earlier Polarization chapter.[1] The theory of the polarization of light reflected by echelette gratings in the far-infrared and microwave regions has been given by Janot and Hadni[138] and Rohrbaugh et al.[139] A general numerical technique published by Kalhor and Neureuther[140] should be useful for calculating the polarization effects of echelette gratings of arbitrary groove shape used in the visible region.

39. Measuring Polarization of Imperfect Polarizers In determining the principal transmittance, extinction ratio, and other properties of an imperfect polarizer, the effects of source polarization, instrumental polarization, and sensitivity of the detector to the plane of polarization must either be measured or eliminated from the calculations. This is easy if an auxiliary polarizer is available that has a much higher degree of polarization than the one to be measured. In such a case, the "perfect" polarizer can be placed in the beam, and the transmittances T_1 and T_2 for the unknown polarizer can be measured directly.* Source polarization, instrumental polarization, and variation of detector response with plane of polarization can all be lumped together as a product. If this product is different in the horizontal and vertical planes, the ratio of the signals obtained when the "perfect" polarizer is oriented horizontally and vertically will not equal unity. One should always take more than the minimum number of measurements, i.e., introduce redundancy, to make sure that no systematic errors are present.

If a high-quality polarizer is not available, two polarizers having unknown properties may be used instead. Several procedures have been described in detail in the earlier Polarization chapter.[1] The method of Hamm et al.[141] which yields the extinction ratio of each polarizer and the instrumental polarization was described in detail and a brief summary of the method of Kudo et al.[142] was given. The methods of Hamm et al.,[141] Horton et al.,[143] and Schledermann and Skibowski[144] were specifically developed for non-normal incidence reflection polarizers (see Par. 41).

NON-NORMAL-INCIDENCE REFLECTION AND TRANSMISSION POLARIZERS

40. By far the largest class of polarizers used in the infrared and ultraviolet spectral regions (where dichroic sheet polarizers and calcite polarizing prisms cannot be used) is the so-called *pile-of-plates polarizers* from which light is reflected (or transmitted) at non-normal incidence. Since most of these polarizers operate at angles near the Brewster or polarizing angle [see Eq. (48) in Chap. 5, "Polarization," in Vol. I of this handbook], they are frequently called Brewster angle polarizers. The plane-parallel plates which are used for Brewster angle transmission polarizers (Par. 42) are generally thick enough to ensure that although multiple reflections occur within each plate, the coherence of the light beam is lost and there are no interference effects. However, another class of non-normal-incidence transmission polarizers makes use of interference effects to enhance their polarizing properties, (Pars. 43 and 44). These include interference polarizers (Par. 43) and polarizing beam splitters (Par. 44). These thin-film devices are discussed in much more detail in the chapter on Optical and Physical Properties of Films and Coatings by J. A. Dóbrowolski (Chap. 42, Vol. I). A relation which is frequently used in connection with non-normal-incidence reflectance measurements is the Abelès condition, discussed in Par. 41.

41. Brewster Angle Reflection Polarizers Most reflection-type polarizers are made of plates which are either nonabsorbing or only slightly absorbing. The angle of incidence most often used is the Brewster angle at which the reflection of the p component, light polarized parallel to the plane of incidence, goes to 0. Thus the reflected light is completely plane polarized with the electric vector vibrating perpendicular to the plane of incidence (s component). Curves showing the reflectance and extinction ratio for various materials and angles near the Brewster angle are given in Fig. 5 of Chap. 5, "Polarization," in Vol. I of this handbook. The polarizing efficiency of reflection-type polarizers can be experimentally determined using any of the methods given in Par. 49 of the earlier Polarization chapter;[1] the methods of Hamm et al.,[141] Horton et al.,[143] and Schledermann and Skibowski[144] were specifically developed for polarizers of this type.

Brewster angle reflection polarizers for the infrared are made from the semiconductors silicon, germanium, and selenium which are transparent beyond their absorption edges

---

* When using an air-spaced polarizing prism, extreme care should be taken not to exceed the acceptance angle of the prism.

TABLE 3 Infrared Brewster Angle Reflection Polarizers

Material	Description	Reference
Ge-Hg	Multiple internal reflections in Ge immersed in Hg	Harrick, Ref. 145
Ge	Single external reflection from 1-cm-thick polished Ge single crystal	Edwards and Bruemmer, Ref. 146
Ge	Proposed parallel and antiparallel arrangements of two Ge plates	Krízek, Ref. 147
Ge	Double-beam system: beam 1, single reflection; beam 2, one transmission, one reflection	Craig *et al.*, Ref. 148
Ge	Axial arrangement with reflections from two Ge wedges and two Al mirrors	Bor and Brooks, Ref. 149
Se	Reflections from two cast-Se films on roughened glass plates	Pfund, Ref. 150
Se	Axial arrangement with reflections from two Se films evaporated on NaCl and one Ag mirror	Barchewitz and Henry, Ref. 151
Se	Large-aperture, axial, venetian-blind arrangement with one or two reflections from evaporated Se films on roughened glass plates (additional reflections from Al mirrors)	Takahashi, Ref. 152
Si	Single reflection from polished single crystal Si	Walton and Moss, Ref. 153
Si	Axial arrangement with reflection from two Al mirrors and polished Si plate with roughened back	Baumel and Schnatterly, Ref. 154
PbS	Axial arrangement with reflections from two chemically deposited PbS films and one Al film	Grechushnikov and Petrov, Ref. 155
CdTe	Single plate	Leonard, Ref. 117
$Al + Al_2O_3$	Multiple reflections from Al_2O_3 coated with metal at 10.6 μm (calculations only)	Cox and Hass, Ref. 156
$Ti + SiO_2$	Multiple reflections from dielectric coated Ti at 2.8 μm (calculations only)	Thonn and Azzam, Ref. 157

and have high refractive indexes. Table 3 lists various infrared polarizers which have been described in the literature. All involve external reflections except the Ge-Hg polarizer described by Harrick,[145] in which light undergoes two or four reflections within a bar of germanium. While Harrick's polarizer has attractive features, it depends on maintaining polarization in the germanium, so that great care must be taken to obtain material with a minimum of strain birefringence.

In the ultraviolet, materials such as LiF, MgF_2, CaF_2, and Al_2O_3, can be used as polarizers. Biotite, a form of mica, has also been found to perform very well in the 1000- to 6000-Å region. In the extreme ultraviolet, metallic films, particularly Au, Ag, and Al, have been used as polarizers. Table 4 lists various non-normal-incidence ultraviolet reflection polarizers as well as authors who have made calculations and measurements on various materials for ultraviolet polarizers.

The most versatile non-normal-incidence reflection polarizer would be one which does not deviate or displace the beam from its axial position. One convenient arrangement would be a symmetric three-reflection system in which the light is incident on one side of a triangle, reflected to a plane mirror opposite the apex, and back to the other side of the triangle, as was done by Horton *et al.*,[143] and Barchewitz and Henry.[151] If the polarizer must have a good extinction ratio and the light beam is highly convergent, two of the reflections could be from the polarizing material and the third from a silvered or aluminized mirror. If the beam is highly collimated or more throughput is required, only one reflection may be from the polarizing material. The throughput can also be increased by using a plane-parallel plate for the polarizing reflection. The major drawback to a reflection polarizer is the extreme length of the device required to accommodate a beam of large cross-sectional area. For example, if a germanium polarizer were used at the

TABLE 4 Ultraviolet Reflection Polarizers and Polarization Measurements

Material	Description	Wavelength range, Å	Reference
Al$_2$O$_3$, Al, Au, ZnS, glass, and others	Calculated values of R_s and $(R_s/R_p)_{max}$ vs. wavelength for a single reflection	500–2000	Hunter, Ref. 158
Al$_2$O$_3$, Al, glass, and others	Calculated values of R_s and $(R_s - R_p)/(R_s + R_p)$ vs. angle of incidence for a single reflection; also principal angle and related angles	584	Damany, Ref. 159
Al$_2$O$_3$, CaF$_2$, LiF, and Pyrex	Measured optical constants; calculated R_s and $(R_s - R_p)/(R_s + R_p)$ vs. angle of incidence and wavelength for a single reflection	200–2000	Stephan et al., Ref. 160
Al$_2$O$_3$ and CaF$_2$	Measured $(R_s - R_p)/(R_s + R_p)$ vs. angle of incidence at selected wavelengths for a single reflection; used both materials as single-reflection polarizers	1026–1600	de Chelle and Merdy, Ref. 161
LiF, Al$_2$O$_3$, MgF$_2$, SiO, ZnS	Used single-reflection LiF polarizer at Brewster angle to measure R_s and R_s/R_p for various materials; best polarizers were Al$_2$O$_3$ and MgF$_2$	1216	McIlrath, Ref. 162
Al, Ag, Au, MgF$_2$, SiO, ZnS	Measured polarization of uncoated aluminum grating and optical constants of all materials listed	304–1216	Cole and Oppenheimer, Ref. 163
Al, Au	Determined polarization of Au- and Al-coated gratings by measuring reflectance of Au and fused-silica mirrors at 45°	600–2000	Uzan et al., Ref. 164
Al, Au, glass	Measured average reflectance and degree of polarization of Al, Au, and glass as a function of angle of incidence; measured polarization of a glass grating and an Al-coated grating	584	Rabinovitch et al., Ref. 165
MgF$_2$	Measured R_p/R_s at 60° for a single reflection and compared it with calculated values	916, 1085, 1216	Sasaki and Fukutani, Ref. 166
MgF$_2$ + Al	Calculated performance, constructed axial triple-reflection polarizer and analyzer of MgF$_2$-coated Al, and measured transmission	1216	Winter et al., Ref. 167
MgF$_2$, MgF$_2$ + Al	Calculated performance, constructed triple-reflection polarizer of a MgF$_2$ plate and two MgF$_2$-coated Al mirrors, and measured transmission	300–2000	Hass and Hunter, Ref. 168

Material	Wavelength (Å)	Description	Reference
MgF$_2$, MgF$_2$ + Al	1150–visible	Constructed four-reflection polarizers of a MgF$_2$ plate and three MgF$_2$-coated Al mirrors, no performance properties measured; polarizer part of the UV spectrometer and polarimeter for the NASA Solar Maximum Mission	Spencer et al., Ref. 169
MgF$_2$, Au and other metals	300–2000	Calculated values of R_s and $(R_s/R_p)_{max}$ vs. angle of incidence and wavelength for one and more reflections for a variety of materials	Hunter, Ref. 170
Au, Ag	500–1300	Measured R_p/R_s at 45° for a single reflection and compared it with calculated values; measured R_p/R_s for a platinized grating	Hamm et al., Ref. 141
Au	600–1200	Used single-reflection Au mirror at 60° as polarizer (Brewster angle about 55°)	Ejiri, Ref. 171
Au	500–1000	Used axial arrangement of eight Au mirrors at 60° as polarizer and analyzer to measure polarization of synchrotron radiation; determined polarizing properties of each polarizer	Rosenbaum et al., Ref. 172
Au	500–5000	Constructed axial triple-reflection Au polarizer, measured extinction ratio and transmission for different angles of incidence on Au plates	Horton et al., Ref. 143
Au	1200–3000	Reflection from two cylindrical gold mirrors in a Seya-Namioka monochromator; measured polarization ratio	Rehfeld et al., Ref. 173
Au	584	Calculated performance of a polarizer made of 2 concave Au-coated spherical mirrors used off axis, constructed polarizer, no measurements made of polarization or transmission	Van Hoof, Ref. 174
Au	400–1300	Constructed a 4-reflection Au-coated polarizer of Van Hoof's design (2 plane, 2 spherical mirrors), measured transmission and degree of polarization	Hibst and Bukow, Ref. 175
Au	584	Constructed a single-reflection Au-coated polarizer and measured the polarizing efficiency	Khakoo et al., Ref. 176
Biotite	1100–6000	Constructed axial polarizer and analyzer each with 61° Brewster angle reflection from biotite and two reflections from MgF$_2$-coated Al mirrors; measured transmission and extinction ratio	Robin et al., Ref. 177
Biotite	1000–2000	Constructed two polarizers: (1) axial polarizer with two 60° reflections from biotite and 30° reflection from MgF$_2$-coated Al mirror; (2) displaced-beam polarizer with 60° reflections from two biotite plates; measured degree of polarization of various gratings	Matsui and Walker, Ref. 178

Brewster angle (76°) and the beam width were about 25 mm, each Ge plate would have to be about 25 by 100 mm and the overall length of the polarizer would be greater than 200 mm if a three-reflection axial arrangement such as that described above were used.

The Abelès condition,[179] which applies to the amplitude reflectance at 45° angle of incidence (see Par. 4 in Chap. 5, "Polarization," in Vol. I) is useful for testing the quality of reflection polarizers. Schulz and Tangherlini[180] apparently rediscovered the Abelès condition and used the ratio $R_s^2/R_p = 1$ as a test to evaluate their reflecting surfaces. They found that surface roughness made the ratio too small but annealing the metal films at temperatures higher than 150°C made the ratio larger than unity. Rabinovitch *et al.*[165] made use of the Abelès condition to determine the polarization of their Seya-Namioka vacuum-ultraviolet monochromator. They measured the reflectance at 45° of a sample whose plane of incidence was perpendicular or parallel to the exit slit. From these measurements they deduced the instrumental polarization by assuming the Abelès condition. Values of instrumental polarization obtained using carefully prepared gold and fused-silica samples were in excellent agreement, showing that neither of these materials had surface films which invalidated the Abelès condition. Surface films usually have relatively little effect on the Abelès condition in the visible region[181] but become important in the vacuum ultraviolet. Hamm *et al.*[141] eliminated the effect of instrumental polarization from their measurements of the reflectance of a sample in unpolarized light at 45° angle of incidence by making use of the Abelès condition. Although McIlrath[162] did not refer to the Abelès condition as such, he used it to determine the instrumental polarization of his vacuum-ultraviolet apparatus so he could measure the absolute reflectance of a sample at 45° angle of incidence. Thonn and Azzam[157] have calculated the polarizing properties of dielectric-coated metal mirrors at 2.8 μm in the infrared. Reflections from 2, 3, or 4 such mirrors at the Brewster angle should give excellent performance, although the polarizer would be quite long.

42. Brewster Angle Transmission Polarizers   To help overcome the beam-deviation problem and the extreme length of reflection-type polarizers, Brewster angle polarizers are often used in transmission, particularly in the infrared, where transparent materials are available. At the Brewster angle, all of the *p* component and an appreciable fraction of the *s* component are transmitted. Thus, several plates must be used to achieve a reasonable degree of polarization. The higher the refractive index of the plates, the fewer are required.

Tables 1 and 2 in Chap. 5, "Polarization," in Vol. I of this handbook give equations for the transmittances and degree of polarization for a single plate and multiple plates at any angle of incidence in terms of R_s and R_p for a single surface, as well as these same quantities at the Brewster angle. Conn and Eaton[182] have shown that the formulas which assume incoherent multiple reflections within each plate and none between plates give the correct degree of polarization for a series of Zapon lacquer films ($n = 1.54$) and also for a series of eight selenium films, whereas the formula of Provostaye and Desains[183] predicted values which were much too low. These authors also point out that the number of multiply reflected beams between plates that enter the optical system depends on the spacing between plates and the diaphragm used to limit the number of beams. One can use a fanned arrangement, as suggested by Bird and Shurcliff,[184] to eliminate these multiply reflected beams. Internal reflections within each plate can be removed by wedging the plates.[184]

Most of the infrared Brewster angle transmission polarizers described in the literature have been made of selenium, silver chloride, or polyethylene sheet; they are listed in Table 5. For wavelengths longer than 3 μm, where calcite polarizing prisms become highly absorbing, to about 10 μm, beyond which wire-grid polarizers have good extinction ratios, Brewster angle transmission polarizers are the most useful, since the better-extinction, on-axis reflection-type polarizers (Par. 41) are impossibly long. Some of the interference polarizers described in Pars. 43 and 44 are superior if the beam-convergence angle is

TABLE 5 Infrared Brewster Angle Transmission Polarizers

Material	Description	Wavelength range, μm	Reference
Se	5 or 6 unbacked films (4 μm thick) at 65° angle of incidence (Brewster angle 68.5°)	2–14	Elliott and Ambrose, Ref. 185; Elliott et al., Ref. 186
Se	5 unbacked films at 65° incidence (different method of preparation from above)		Ames and Sampson, Ref. 187
Se	8 unbacked films at the Brewster angle	1–15	Conn and Eaton, Ref. 182
Se	Se films (3–8 μm thick) evaporated on one side of collodion films; 68.5° angle of incidence		Barchewitz and Henry, Ref. 151
Se	1 to 6 unbacked films (1.44–8 μm thick) at 68° angle of incidence	Visible–20	Duverney, Ref. 188
Se	3 unbacked films (0.95 μm thick) at 71° angle of incidence	6–17	Hertz, Ref. 189
Se	5 Formvar films coated on both sides with Se (various thicknesses) at 65° angle of incidence	1.8–3.2	Buijs, Ref. 190
Se	Unbacked films (different method of preparation from Elliott et al. Ref. 186)		Bradbury and Elliott, Ref. 191
Se	4 to 6 pleated unsupported films (4–8 μm thick) at the Brewster angle	2.5–25	Greenler et al., Ref. 192
AgCl	3 plates (1 mm thick) at 63.5° angle of incidence	Visible–15	Wright, Ref. 193
AgCl	6 to 12 plates (0.05 mm thick) at 60–75° angle of incidence	2–20	Newman and Halford, Ref. 78
AgCl	6 plates (0.5 mm thick) at 63.5° stacked in alternate directions		Makas and Shurcliff, Ref. 194
AgCl	Suggest 6 wedge-shaped plates at 68° stacked in alternate directions in a fanned arrangement		Bird and Shurcliff, Ref. 184
AgCl	2 V-shaped plates (3.2 mm thick) at 77° angle of incidence; large aperture		Bennett et al., Ref. 195
KRS-5	1 to 3 thallium bromide-iodide plates (1 and 4 mm thick) at polarizing angle	1–15	Lagemann and Miller, Ref. 196
ZnS	4 glass plates (0.1 mm thick) coated on both sides with uniform ZnS films of same thickness (several sets of plates to cover extended wavelength range)	Visible–6	Huldt and Staflin, Ref. 197
ZnSe	6 plates, extinction ratio of 800 at 4 μm	4	Leonard et al., Ref. 198
Ge	1 single-crystal Ge plate (0.8 mm thick) at 76° angle of incidence		Meier and Günthard, Ref. 199
Ge	2 plates (1 mm thick) in an X-shaped arrangement at 76° angle of incidence		Harrick, Ref. 200
Ge	3 plates (2 wedged) at the Brewster angle	2–6	Murarka and Wilner, Ref. 201
Polyethylene	12 sheets (8 μm thick) at the Brewster angle	6–20 (absorption bands 6–14)	Smith et al., Ref. 202
Polyethylene	9 to 15 sheets (20–50 μm thick) at the Brewster angle (55°)	30–200	Mitsuishi et al., Ref. 137
Polyethylene	4 sheets at the Brewster angle	200–350	Hadni et al., Ref. 136
Polyethylene	12 sheets (5 μm thick) at the Brewster angle	1.5–13 (selected wavelengths)	Walton and Moss, Ref. 153
Polyethylene	1–15 stretched sheets (12.7 μm thick) at the Brewster angle	10.6	Rampton and Grow, Ref. 203
Polyethylene	20 sheets (30 μm thick) at the Brewster angle	45–200	Munier et al., Ref. 204
Polyethylene	25 to 30 sheets at the Brewster angle	54.6	Hilton and Jones, Ref. 111
Melinex	11 to 13 polyethylene terephthalate (Melinex) sheets (4.25–9 μm thick) at the Brewster angle	1–5	Walton et al., Ref. 205

TABLE 6 Ultraviolet Brewster Angle Transmission Polarizers

Material	Description	Wavelength range, Å	Reference
LiF	4 to 8 plates (0.3–0.8 mm thick) at 60° angle of incidence (Brewster angle 55.7–58.7°) stacked in alternate directions	1200–2000	Walker, Ref. 206
LiF	8 plates	1100–3000	Hinson, Ref. 207
LiF	8 plates (0.25–0.38 mm thick) at 60° angle of incidence stacked in groups of 4 in alternate directions	1200–2000	Heath, Ref. 208
CaF$_2$	4 to 8 wedged plates stacked in alternate directions in fanned arrangement at 65° angle of incidence (Brewster angle 56.7)	1500–2500	Schellman *et al.*, Ref. 209
Al	Calculations of polarizing efficiency for 1000-Å-thick unbacked Al film, 1000- and 500-Å Al films each covered with 30-Å Al$_2$O$_3$ and 100-Å Au films	300–800	Hunter, Ref. 158

small. Ultraviolet Brewster angle transmission polarizers are not nearly as common; LiF and CaF$_2$ have mainly been used from about 1500 to 2500 Å (see Table 6). In the wavelength region where calcite polarizing prisms are usable (>2140 Å), Brewster angle polarizers have the advantage of a larger linear aperture and less absorption.

Low-absorption glass pile-of-plates polarizers have been used in the visible spectral region by Weiser,[210] in preference to more absorbing Glan-Thompson prism polarizers, to increase the power output of giant-pulse ruby lasers. Weinberg[211] calculated the degree of polarization of glass and silver chloride plates, but he did not calculate the transmittance of his polarizers.

Interference Polarizers

43. When the sheets or films constituting a non-normal-incidence transmission polarizer are thin and have very smooth surfaces, the internally reflected beams can interfere constructively or destructively. In this case, the transmittance of the *p* component remains unity at the Brewster angle (where $R_p = 0$) and only oscillates slightly (with respect to wavelength) for angles close to the Brewster angle. However, the *s* transmittance varies from a maximum of unity to a minimum of $(1 - R_s)^2/(1 + R_s)^2$ whenever λ changes by an amount that will make the quantity $(nd \cos \theta_1)/\lambda$ in Eq. (26) in Chap 5., "Polarization," in Vol. I change by $\frac{1}{2}$.* These transmittance oscillations are only ± 0.225 for a single film of refractive index 1.5 but can become as large as ± 0.492 when $n = 4.0$. Since the *p* transmittance remains essentially constant, the extinction ratio will vary cyclically with the *s* transmittance, as can be seen in the upper curve of Fig. 17 for a 2.016-μm-thick selenium film.

If a transmission polarizer with a good extinction ratio is needed for use over a limited wavelength range, it can be made of several uniform films of a thickness that yields a minimum extinction ratio in the given wavelength region. The extinction ratio for a series of *m* films is $(T_s/T_p)^m$ when there are no multiple reflections between them. In this way only *half* as many films would be needed to achieve a given extinction ratio as would be necessary if interference effects were not present. This rather surprising result can be seen from the expressions for $(T_s)_{\text{sample}}$ for *m* plates with and without interference effects in Table 2 in Chap. 5, "Polarization," in Vol. I of this handbook. Assuming no multiple reflections between plates, the expressions are $[2n^2/(n^4 + 1)]^{2m}$ and $[2n^2/(n^4 + 1)]^m$,

---

* The approximate expression for this wavelength interval $\Delta\lambda$ (assuming that the oscillations are sufficiently close together for $\lambda_1\lambda_2 \approx \lambda_2$) is given in Eq. (27) in Chap. 5, "Polarization," in Vol. I of this handbook.

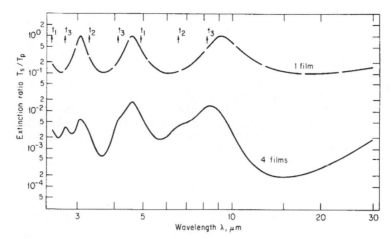

FIGURE 17 Calculated extinction ratios for a series of selenium films ($n = 2.46$) as a function of wavelength from 2.5 to 30 μm. Light is incident at the Brewster angle, 67.9°, and multiply reflected beams interfere within the film. The upper curve is for a single film 2.016 μm thick; arrows indicate positions of maxima for three thinner films: $t_1 = 1.080$ μm, $t_2 = 1.440$ μm, and $t_3 = 1.800$ μm. The lower curve is the extinction ratio for the four films in series assuming no reflections between films. The calculated p transmittance for each film (and for four films in series) is unity at the Brewster angle.

respectively. Hertz[189] achieved a degree of polarization of 99.5 percent in the 6- to 17-μm region using three unbacked selenium films 0.95 μm thick. Conn and Eaton[182] obtained only a slightly better performance with eight thicker nonuniform selenium films.

As can be seen in Fig. 17, the calculated extinction ratio for the 2.016-μm-thick film goes to unity at 3.0, 4.6, and 9.2 μm, indicating that the s as well as the p transmittance at these wavelengths is unity. This ratio will remain unity at the above wavelengths if there are several nonabsorbing films of the same thickness. Even if the films have slightly different thicknesses, or if their surfaces are somewhat rough, interference effects may still persist, adversely affecting polarizer performance. Such effects have been observed by Elliott et al.,[186] Barchewitz and Henry,[151] Duverney,[188] Mitsuishi et al.,[137] and Walton et al.[205]

By choosing films of appropriate thicknesses, interference effects can be used to advantage. The lower curve in Fig. 17 shows the extinction ratio obtained if four selenium films of thicknesses 1.08, 1.44, 1.80, and 2.02 μm are used at the Brewster angle as a transmission polarizer. (The wavelengths at which maxima occur for the three thinner films are indicated by arrows in the upper portion of the figure.) In this example the extinction ratio for the four films in series is better than 2×10^{-2} from 2.5 to 30 μm and at most wavelengths is better than 10^{-2} (corresponding to a degree of polarization in excess of 98 percent). In the 11- to 27-μm wavelength region the extinction ratio is better than 10^{-3}. Four thick or nonuniform selenium films without interference effects have a calculated extinction ratio of about 10^{-2}, and six films are required to change this ratio to 10^{-3}. Thus, in the 11- to 27-μm wavelength region, *four* selenium films of appropriate thicknesses *with interference* have a superior extinction ratio to *six* selenium films *without interference*. If one wishes to optimize the extinction ratio over a more limited wavelength range, the film thicknesses can be adjusted accordingly and the extinction ratio improved. Unfortunately, the gain in extinction ratio is offset by a more sensitive angular function than that shown in Fig. 7 in Chap. 5., "Polarization," in Vol. I, so that the incident beam must be very well collimated.

Interference effects can also be used to advantage in other types of non-normal-incidence polarizers. Bennett et al.[212] made a transmission polarizer from a series of four germanium films (ranging in thickness from 0.164 to 0.593 μm), evaporated onto strain-free plates of sodium chloride. The plates were inclined at the Brewster angle for germanium and arranged in the form of an X so that the polarizer would have a large square aperture and would not deviate the beam. An extinction ratio better than 3×10^{-3} was measured at 2.5 μm, and the plates transmitted from 2 to 13 μm. (Calculated extinction ratios in this wavelength range vary from 1×10^{-3} to 2×10^{-4} for radiation incident at the Brewster angle.)

Polarizers consisting of a high refractive index transparent film on a lower refractive index transparent substrate have been suggested for use in the visible wavelength region by Schröder[213] and Abelès.[214] These still have a Brewster angle where $R_p = 0$, and further-more R_s at this angle is greatly increased over its value for an uncoated low refractive index substrate. Thus, a large-aperture, high-efficiency polarizer with no absorption losses is possible, which should find numerous applications in laser systems. One polarizer of this type, suggested independently by Schröder and by Abelès, would consist of high refractive index titanium dioxide films ($n \approx 2.5$) evaporated onto both sides of a glass substrate ($n = 1.51$). At the Brewster angle, 74.4°, $R_s \approx 0.8$, making this polarizer equivalent to one made from a material of refractive index 4 ($\theta_B = 76.0°$ as shown in Fig. 4 in Chap. 5, "Polarizers," in Vol. I of this handbook).* Two glass plates coated on both sides with TiO_2 films should have an extinction ratio of about 1.6×10^{-3} at 5500 Å and about twice that value at the extreme ends of the visible region, according to Abelès.[214] Schröder[213] measured the degree of polarization as a function of angle of incidence for one such TiO_2-coated glass plate and found values comparable to the calculated ones. Kubo[215] calculated the degree of polarization, reflectance, and transmittance (as a function of angle of incidence and wavelength) of a glass plate ($n = 1.50$) covered with a thin transparent film of index, 2.20. His results are similar to those of Abelès and Schröder.

Schopper,[216] Ruiz-Urbieta and Sparrow,[207–219] and Abelès[220] have also investigated making non-normal-incidence reflection polarizers from a thin transparent or absorbing film deposited onto an absorbing substrate. Zaghloul and Azzam[221] proposed using silicon films on fused silica substrates as reflection polarizers for different mercury spectral lines in the visible and ultraviolet regions. Abelès designed some specialized reflection polarizers for use in the vacuum ultraviolet. Unfortunately the wavelength range covered by such a polarizer is very narrow; for one polarizer it was 25 Å at a wavelength of 1500 Å. However, the spectral range could possibly be increased by using several thin films instead of one.

Multilayer film stacks have also been used to produce non-normal-incidence reflection or transmission polarizers by Buchman et al.[222] Buchman[223] later improved the design performance of his polarizers by adding antireflection layers between the repeating groups of layers. Although this type of polarizer has a relatively narrow operating bandwidth, a small angular acceptance, tight wavelength centering, and layer thickness uniformity requirements, it can be used successfully in high power laser systems as shown by Refermat and Eastman.[224] Songer[225] described how to design and fabricate a Brewster angle multilayer interference polarizer out of a titanium dioxide, silicon dioxide multilayer on BK 7 glass for use in a 1.06-μm laser beam. Blanc, Lissberger, and Roy[226] designed, built, and tested multilayer zinc sulfide–cryolite-coated glass and quartz polarizers for use with a pulsed 1.06-μm laser. Recently, Maehara et al.[227] have reported excellent performance for a pair of polarizers coated with 21 ruthenium and silicon films on a silicon wafer over a wide wavelength range in the soft x-ray region. In several designs of multilayer film stacks, both the reflected and transmitted beams are used; they are discussed in Par. 44.

44. Polarizing Beam Splitters Polarizing beam splitters are a special form of non-normal-incidence interference polarizer in which the beam is incident on a multilayer

---

* We are assuming no multiply reflected beams within the substrate in either case.

dielectric stack at 45°. The transmitted beam is almost entirely plane-polarized in the p direction, while the reflected beam is nearly all plane-polarized in the s direction. Generally the alternating high and low refractive index dielectric layers are deposited onto the hypotenuses of two right-angle prisms, which are then cemented together to form a cube. The beam enters a cube face normally and strikes the multilayers on the hypotenuse (the high refractive index layer is next to the glass), and the reflected and transmitted beams emerge normal to cube faces, being separated by 90°. Clapham et al.[228] have a good discussion of polarizing beam splitters, which were invented by S. M. MacNeille[229] and developed by Banning.[230] Banning's beam splitter was made with three zinc sulfide and two cryolite layers on each prism; the polarization for white light was greater than 98 percent over a 5°-angle on each side of the normal to the cube face for both the reflected and transmitted beams. Variations on this design have since been proposed by Dobrowolski and Waldorf,[231] Monga et al.,[232] and Mouchart et al.,[233] primarily to improve the laser damage resistance of the device and increase the angular field of view. Dobrowolski and Waldorf[231] designed and built a polarizing beam splitter consisting of a multilayer coating of HfO_2 and SiO_2 deposited onto fused silica and immersed in a water cell that acted like the MacNeille cube. Tests with a 0.308 μm excimer laser showed a high laser damage threshold. The multi-wavelength polarizing beam splitters designed by Monga et al.[232] could be made in large sizes and could withstand high laser power levels. The modified MacNeille cube polarizers designed by Mouchart et al.[233] had angular fields of view that could be increased to about ±10° when the polarizers were used with monochromatic light sources.

Lees and Baumeister[234] designed a frustrated total internal reflection beam splitter that had a multilayer dielectric stack deposited onto the hypotenuse of a prism. Their designs, for use in the infrared spectral region, consisted of multilayer stacks of PbF_2 and Ge deposited onto a germanium prism and covered by a second germanium prism. Azzam[235] designed polarization independent beam splitters for 0.6328 μm and 10.6 μm using single-layer coated zinc sulfide and germanium prisms. The devices were found to be reasonably achromatic and their beam-splitting ratio could be varied over a wide range with little degradation in polarization properties. Azzam[236] also proposed coating a low-refractive-index dielectric slab on both sides with high-refractive-index dielectric films to make an infrared polarizing beam splitter.

Various high- and low-refractive-index materials have been successfully used in the multilayer stacks. In addition to zinc sufilde and cryolite on glass by Banning[230] and Schröder and Schläfer,[237] layers of a controlled mixture of silicon dioxide and titanium dioxide have been alternated with pure titanium dioxide on fused-silica prisms by Pridatko and Krylova,[238] thorium dioxide and silicon dioxide have been used on fused-silica prisms by Sokolova and Krylova,[239] chiolite (a mixture of sodium and aluminum fluorides) and lead fluoride have been used on fused-silica prisms by Turner and Baumeister,[240] bismuth oxide and magnesium fluoride have been used on EDF glass prisms by Clapham et al.,[228] and zirconium oxide and magnesium fluoride have been used on dense flint-glass prisms by Clapham et al.[228] The calculations involved in optimizing these beam splitters for good polarizing characteristics, achromaticity, and relative insensitivity to angle of incidence are quite involved. Clapham et al.[228] and Turner and Baumeister[240] discuss various calculational techniques frequently used. Clapham[241] also gives the measured characteristics of a high-performance achromatic polarizing beam splitter made with zirconium oxide and magnesium fluoride multilayers.

Although polarizing beam splitters are generally designed so that the s and p polarized beams emerge at right angles to each other, Schröder and Schläfer[237] have an ingenious arrangement in which a half-wave plate and mirror are introduced into the path of the reflected beam to make it parallel to the transmitted beam and of the same polarization. Other optical schemes to accomplish the same purpose have been described in a later paper.[242]

For some purposes it is desirable to have a beam splitter that is insensitive to the polarization of the incident beam. Baumeister[243] has discussed the design of such beam splitters made from multilayer dielectric stacks of alternating low- and high-refractive-index materials. One of his designs is composed of six dielectric layers for which the extinction ratio T_s/T_p varies from 0.93 to 0.99 in a bandwidth of about 800 Å, with a $\pm 1°$ variation in the angle of incidence. In principle, any multilayer filter which is nonreflecting at normal incidence will be nonpolarizing at all angles of incidence, according to Baumeister.[244] Costich[245] has described filter designs for use in the near infrared which are relatively independent of polarization at 45° angle of incidence.

RETARDATION PLATES

45. Introduction The theory of retardation plates and especially quarter-wave retarders is given in Chap. 5, "Polarization," in Vol. I of this handbook. The basic relation for retardation plates, Eq. (73) in that section, is

$$N\lambda = d(n_e - n_o) \qquad (9)$$

where n_o = refractive index of the ordinary ray, n_e = refractive index of the extraordinary ray, d = physical thickness of the plate, and λ = wavelength.

Retardation plates are generally made of mica, stretched polyvinyl alcohol, and quartz, although other stretched plastics such as cellophane, Mylar, cellulose acetate, cellulose nitrate, sapphire, magnesium fluoride, and other materials can also be used (see West and Makas[246]). Polyvinyl alcohol in sheet form transmits well into the ultraviolet beyond the cutoff for natural mica and is thus particularly useful for ultraviolet retardation plates, according to McDermott and Novick.[91] As suggested by Jacobs et al.,[247] permanent birefringence can be thermomechanically induced in the borosilicate optical glass ARG-2, making it an attractive alternate to natural crystalline quartz and mica for large aperture wave plates for laser systems. Refractive indexes and birefringences of some materials are listed in Tables 7 and 8. The birefringences reported for mica and apophyllite should be

TABLE 7 Refractive Indices of Selected Materials at 5893 Å (Billings, Ref. 248)

Material	n_o	n_e
Positive uniaxial crystals		
Ice, H_2O	1.309	1.313
Sellaite, MgF_2	1.378	1.390
Apophyllite, $2[KCa_4Si_8O_{20}(F, OH)\cdot 8H_2O]$	1.535±	1.537±
Crystalline quartz, SiO_2	1.544	1.553
Dioptase, $CuSiO_3\cdot H_2O$	1.654	1.707
Zircon, $ZrSiO_4$	1.923±	1.968±
Rutile, TiO_2	2.616	2.903
Negative uniaxial crystals		
Beryl (emerald), $Be_3Al_2(SiO_3)_6$	1.581±	1.575±
Sodium nitrate, $NaNO_3$	1.584	1.336
Muscovite mica (complex silicate)	1.5977±	1.5936±
Apatite, $Ca_{10}(F, Cl)_2(PO_4)_6$	1.634	1.631
Calcite, $CaCO_3$	1.658	1.486
Tourmaline (complex silicate)	1.669±	1.638±
Sapphire, Al_2O_3	1.768	1.760

TABLE 8 Birefringence $n_e - n_o$ of Various Optical Materials[a]

Wavelength, μm	Rutile[b] TiO2	CdSe[b]	Crystalline quartz[c-g]	MgF2[c,d,h,i]	CdS[i-l]	Apophyllite[m]	ZnS[j] (Wurtzite)	Calcite[c]	LiNbO[n]	BaTiO[o]	AdP[p]	KDP[p]	Sapphire[d,h,q,r] (Al2O3)	Mica[s]
0.15	—	—	0.0214	0.0143	—	—	—	—	—	—	—	—	—	—
0.20	—	—	0.0130	0.0134	—	—	—	-0.326	—	—	-0.0613	-0.0587	-0.0111	—
0.25	—	—	0.0111	0.0128	—	—	—	-0.234	—	—	-0.0543	-0.0508	-0.0097	—
0.30	—	—	0.0103	0.0125	—	—	—	-0.206	—	—	-0.0511	-0.0474	-0.0091	—
0.35	—	—	0.0098	0.0122	—	—	—	-0.193	—	—	-0.0492	-0.0456	-0.0087	—
0.40	—	—	0.0096	0.0121	—	—	0.004	-0.184	—	—	-0.0482	-0.0442	-0.0085	—
0.45	0.338	—	0.00937	0.0120	—	0.0019	0.004	-0.179	-0.1049	-0.097	-0.0473	-0.0432	-0.0083	-0.00457
0.50	0.313	—	0.00925	0.0119	—	0.0022	0.004	-0.176	-0.0998	-0.079	-0.0465	-0.0424	-0.0082	-0.00468
0.55	0.297	—	0.00917	0.0118	0.014	0.0024	0.004	-0.173	-0.0947	-0.070	-0.0458	-0.0417	-0.0081	-0.00476
0.60	0.287	—	0.00909	0.0118	0.018	0.0026	0.004	-0.172	-0.0919	—	-0.0451	-0.0410	-0.0081	-0.00480
0.65	0.279	—	0.00903	0.0117	0.018	0.0028	0.004	-0.170	-0.0898	-0.064	-0.0444	-0.0403	-0.0080	-0.00482
0.70	0.274	—	0.00898	0.0117	0.018	—	0.004	-0.169	-0.0882	—	-0.0438	-0.0396	-0.0080	-0.00483
0.80	0.265	—	0.0089	0.0116	0.018	—	0.004	-0.167	-0.0857	—	-0.0425	-0.0382	-0.0079	—
0.90	0.262	0.0195	0.0088	0.0115	0.018	—	0.004	-0.165	-0.0840	—	-0.0411	-0.0367	-0.0079	—
1.00	0.259	0.0195	0.0088	0.0114	0.018	—	0.004	-0.164	-0.0827	—	-0.0396	-0.0350	—	—
1.10	0.256	0.0195	0.0087	0.0114	0.018	—	0.004	-0.162	-0.0818	—	-0.0379	-0.0332	—	—
1.20	0.254	0.0195	0.0087	0.0114	0.017	—	0.004	-0.161	-0.0810	—	-0.0361	-0.0313	—	—
1.30	0.252	0.0195	0.0086	0.0113	0.017	—	0.004	-0.161	-0.0804	—	-0.342	-0.0292	—	—
1.40	0.251	0.0195	0.0085	0.0113	0.017	—	0.004	-0.160	-0.0798	—	-0.0321	-0.0269	—	—
1.50	0.250	0.0195	0.0085	0.0113	0.017	—	0.004	-0.158	-0.0793	—	-0.0298	-0.0245	—	—
1.60	0.249	0.0195	0.0084	0.0112	—	—	—	-0.157	-0.0788	—	-0.0274	-0.0219	—	—
1.70	0.248	0.0195	0.0084	0.0112	—	—	—	-0.156	-0.0782	—	-0.0248	-0.0191	—	—
1.80	0.247	0.0195	0.0083	0.0112	—	—	—	-0.154	-0.0777	—	-0.0221	-0.0162	—	—
1.90	0.246	0.0195	0.0082	0.0112	—	—	—	-0.153	-0.0774	—	-0.0192	-0.0130	—	—
2.00	0.246	0.0195	0.0081	0.0111	—	—	—	-0.151	-0.0771	—	-0.0161	-0.0097	—	—
2.10	0.245	0.0195	0.0081	0.0111	—	—	—	-0.150	-0.0766	—	—	—	—	—
2.20	0.245	0.0195	0.0080	0.0111	—	—	—	-0.148	-0.0761	—	—	—	—	—
2.30	0.244	0.0195	0.0079	0.0110	—	—	—	—	-0.0752	—	—	—	—	—
2.40	0.243	0.0195	0.0078	0.0110	—	—	—	—	-0.0744	—	—	—	—	—
2.50	0.241	0.0195	0.0077	0.0110	—	—	—	—	-0.0739	—	—	—	—	—
2.60	—	0.0195	0.0076	0.0110	—	—	—	—	-0.0734	—	—	—	—	—

[a] Calculated values at 24.8°C obtained from analytical expressions are given for crystalline quartz, MgF2, calcite, ADP, and KDP by Beckers, Ref. 249.
[b] Bond, Ref. 250.
[c] Shields and Ellis, Ref. 251.
[d] Chandrasekharan and Damany, Ref. 260.
[e] Ballard et al., Ref. 255.
[f] Ennos and Opperman, Ref. 256.
[g] Maillard, Ref. 261.
[h] Chandrasekharan and Damany, Ref. 73.
[i] Palik, Ref. 257.
[j] Bieniewski and Czyzak, Ref. 262.
[k] Palik and Henvis. Ref. 252.
[l] Gobrecht and Bartschat, Ref. 258.
[m] Françon et al., Ref. 263.
[n] Boyd et al., Ref. 253.
[o] Shumate, Ref. 258a.
[p] Zernike, Ref. 264.
[q] Jeppeson, Ref. 254.
[r] Loewenstein, Ref. 259.
[s] Einsporn, Ref. 265.

considered as approximate, since they are measurements made on single samples. There is good reason to believe that the birefringence of apophyllite may be different for other samples, (see Par. 53). Although calcite would seem at first to be a good material for retardation plates, its birefringence is so high that an extremely thin piece, less than 1 μm, would be required for a single λ/4 retardation plate. If a "first-order" or multiple-order plate were constructed (Pars. 48 and 50), or if calcite were used as one component of an achromatic retardation plate (Par. 53), the tolerance on the thickness would be very stringent.

Retardation plates are generally made of a single piece of material, although when the thickness required for a plate is too small, two thicker pieces may be used with the fast axis of one aligned parallel to the slow axis of the other to cancel out all but the desired retardation. Plates which are a little too thin or a little too thick may be rotated about an axis parallel or perpendicular to the optic axis to change the retardation to the desired amount, as suggested by Gieszelmann et al.,[266] and Daniels.[267] There are also some novel circular polarizers and polarization rotators for use in the far ultraviolet (see the papers by McIlrath,[162] Saito et al.,[268] and Westerveld et al.[269]), far infrared (Richards and Smith,[270] Johnston,[271] and Gonates et al.[272]), and visible region (Lostis,[273] and Greninger[274]).

Achromatic retardation plates which have the same retardation over a range of wavelengths can be made from two or more different materials or from two or more plates of the same material whose axes are oriented at appropriate angles with respect to each other. These latter devices are known as composite plates (Par. 55 and the earlier Polarization chapter[1]), and although they can change plane-polarized light into circularly polarized light, they do not have all the other properties of true retardation plates. By far the most achromatic λ/4 retarders are devices, such as the Fresnel rhomb, which obtain their retardation from internal reflections at angles greater than the critical angle.

Mica retardation plates are mentioned in Par. 46 and are discussed in detail in the earlier Polarization chapter,[1] which includes the theory of multiple reflections; Pars. 47 to 52 are devoted to various types of crystalline-quartz retardation plates, and Par. 53 covers all achromatic retardation plates, except those of the rhomb-type; the latter are mentioned in Par. 54 and in detail by Bennett[275] and also in the earlier Polarization chapter.[1] Various types of composite plates and unusual retardation plates are also described in detail in Ref. 1.

Methods for making and testing quarter-wave plates including ways of splitting mica, how to distinguish between fast and slow axes, methods for measuring retardations close to λ/4, and the tolerance on plate thickness have all been described in detail in the earlier Polarization chapter.[1] An additional paper by Nakadate[276] shows how Young's fringes can be used for a highly precise measurement of phase retardation.

Waveplates are all sensitive to some degree to temperature changes, variations in the angle of incidence, coherence effects in the light beam, and wavelength variations. Multiple-order plates are much more sensitive than "first-order" or single-order plates. Hale and Day[277] discuss these effects for various types of waveplates and suggest designs that are less sensitive to various parameters.

Most retardation plates are designed to be used in transmission, generally at normal incidence. However, there are also reflection devices that act as quarter-wave and half-wave retarders and polarization rotators. In the vacuum ultraviolet, Westerveld et al.[269] produced circularly polarized light by using Au-coated reflection optics. Saito et al.[268] used an evaporated Al mirror as a retardation plate at 1216 Å, Lyman α radiation, following earlier work by McIlrath.[162] Greninger[274] showed that a three-mirror device could be used in place of a half-wave plate to rotate the plane of polarization of a plane-polarized beam and preserve the collinearity of input and output beams. Johnston[271] used a different three-mirror arrangement for the same application in the far-infrared. Thonn and Azzam[157] designed three-reflection half-wave and quarter-wave retarders from single-layer dielectric coatings on metallic film substrates. They showed calculations for ZnS-Ag film-substrate retarders used at 10.6 μm. Previously Zaghloul, Azzam, and

Bashara[278,279] had proposed using a SiO_2 film on Si as an angle-of-incidence tunable reflection retarder for the 2537-Å mercury line in the ultraviolet spectral region. Kawabata and Suzuki[280] showed that a film of MgF_2 on Ag was superior to Zaghloul *et al.*'s design at 6328 Å. They also performed calculations using Al, Cu, and Au as the metals and concluded that Ag worked best.

46. Mica Retardation Plates Mica quarter-wave plates can be made by splitting thick sheets of mica down to the appropriate thickness, as described by Chu *et al.*,[281] and in the earlier Polarization chapter.[1] Since the difference between the velocities of the ordinary and extraordinary rays is very small, the mica sheets need not be split too thin; typical thicknesses lie in the range 0.032 to 0.036 mm for yellow light. The fast and slow axes of a mica quarter-wave plate can be distinguished using Tutton's test, as mentioned in Strong's book,[282] and the retardation can be measured using one of several rather simple tests.[1]

If the mica sheets are used without glass cover plates, multiply reflected beams in the mica can cause the retardation to oscillate around the value calculated from the simple theory, as described in the earlier Polarization chapter.[1] Fortunately this effect can be eliminated in one of several ways.[1] Mica does have one serious drawback. There are zones in the cleaved mica sheets which lie at angles to each other and which do not extinguish at the same angle, as noted by Smith.[283] Thus, extinction cannot be obtained over the whole sheet simultaneously. In very critical applications such as ellipsometry, much better extinction can be obtained using quarter-wave plates made of crystalline quartz (Pars. 47 to 50), which do not exhibit this effect. Properties of mica quarter-wave plates and methods for making and testing all $\lambda/4$ plates are discussed in detail in the earlier Polarization chapter.[1]

Crystalline-Quartz Retardation Plates

47. Crystalline quartz is also frequently used for retardation plates, particularly those of the highest quality. It escapes the problem of zones with different orientations like those found in mica. The thickness of quartz required for a single quarter-wave retardation at the 6328-Å helium-neon laser line is about 0.017 mm, much too thin for convenient polishing. If the plate is to be used in the infrared, single-order quarter-wave plates are feasible (Par. 49). Two types of quartz retardation plates are generally employed in the visible and ultraviolet regions: so-called "first-order" plates made of two pieces of material (Par. 48), which are the best for critical applications, and multiple-order plates made of one thick piece of crystalline quartz (Pars. 50 to 52). The multiple-order plates are generally not used for work of the highest accuracy since they are extremely sensitive to small temperature changes (Par. 51) and to angle of incidence. Also, they have $\lambda/4$ retardation only at certain wavelengths; at other wavelengths the retardation may not even be close to $\lambda/4$.

When using any of the different types of retardation plates at a single wavelength, the methods for measuring the retardation and for distinguishing between fast and slow axes given in the earlier Polarization chapter[1] can be used.

48. "First-Order" Plates A so-called "first-order" plate is made by cementing together two nearly equal thicknesses of quartz such that the fast axis of one is aligned parallel to the slow axis of the other (both axes lie in planes parallel to the polished faces). The plate is then polished until the difference in thickness between the two pieces equals the thickness of a single $\lambda/4$ plate. The retardation of this plate can be calculated from Eq. (9) by setting *d* equal to the *difference in thickness* between the two pieces. The "first-order" plate acts strictly like a single-order quarter-wave plate with respect to the variation of retardation with wavelength, temperature coefficient of retardation, and angle of incidence.

The change in phase retardation with temperature at 6328 Å, as calculated from equations given in the earlier Polarization chapter,[1] is 0.0091°/°C, less than one-hundredth that of the 1.973-mm multiple-order plate discussed in Par 51. The change in retardation with angle of incidence* at this wavelength is also small: $(\Delta N)_{10^\circ} = 0.0016$, as compared with 0.18 for the thick plate (see Par. 52).

A "first-order" quartz $\lambda/4$ plate has several advantages over a mica $\lambda/4$ plate. (1) Crystalline quartz has a uniform structure, so that extinction can be obtained over the entire area of the plate at a given angular setting. (2) Since the total plate thickness is generally large, of the order of 1 mm or so, the coherence of the multiple, internally reflected beams is lost and there are no oscillations in the transmitted light or in the phase retardation. (3) Crystalline quartz is not pleochroic, except in the infrared, so that the intensity transmitted along the two axes is the same. (4) Crystalline quartz transmits farther into the ultraviolet than mica, so that "first-order" plates can be used from about 0.185 to 2.0 μm (see Table 8).

49. Single-Order Plates in the Infrared Although a crystalline-quartz retardation plate which is $\lambda/4$ in the visible is too thin to make from a single piece of material, the thickness required for such a plate is larger in the infrared. Jacobs and coworkers[266] describe such a $\lambda/4$ plate for use at the 3.39-μm helium-neon laser line. They measured the birefringence of quartz at this wavelength and found it to be 0.0065 ± 0.0001, so that the thickness required for the plate was 0.1304 mm. The actual plate was slightly thinner (0.1278 mm), so that it was tipped at an angle of 10° (rotating it about an axis parallel to the optic axis) to give it exactly $\lambda/4$ retardation (see Par. 52). Maillard[261] has also measured the birefringence of quartz at 3.39 and 3.51 μm and obtained values of 0.00659 and 0.00642, respectively (both ± 0.00002), in agreement with Jacobs' value. These data lie on a smooth curve extrapolated from the values of Shields and Ellis.[251]

A problem encountered when using crystalline quartz in the infrared is that, in general, the ordinary and extraordinary rays have different absorption coefficients; thus it may be impossible to construct a perfect wave plate regardless of the relative retardation between the rays. For an absorbing wave plate to have a retardation of exactly $\lambda/4$, the requirement

$$\left(\frac{n_o + 1}{n_e + 1}\right)^2 \exp\left[-\frac{(\alpha_e - \alpha_o)\lambda}{8(n_e - n_o)}\right] = 1 \tag{10}$$

must be met;[266] α_e and α_o are the absorption coefficients for the extraordinary and ordinary rays, respectively. At wavelengths shorter than 3.39 μm, the birefringence is small enough for it to be possible to approximate the condition in Eq. (10) closely whenever $\alpha_e \approx \alpha_o$. Values of these quantities are given by Drummond.[285] Gonatas et al.[272] concluded that, in the far infrared and submillimeter wavelength region, the effect of different absorption coefficients in the crystalline quartz was small and could be corrected for.

Another problem which occurs for crystalline quartz and also for sapphire[272] in the infrared is that the Fresnel reflection coefficients are slightly different for the ordinary and extraordinary rays since the refractive indexes and absorption coefficients are in general different. One possible solution is to deposit isotropic thin films on the crystal surfaces.[272] The refractive index of these films is chosen to balance the anisotropic absorption effect by making the Fresnel reflection coefficients appropriately anisotropic. On the other hand, if anisotropic Fresnel reflection proves to be undesirable, it can be greatly diminished by using an antireflection coating, as suggested by Gieszelmann et al.[266]

---

* Grechushnikov[284] has an incorrect relation for the change in phase retardation with angle of incidence [his eq. (2)]. He assumed that the retardations in the two halves of the plate add rather than subtract, yielding a retardation comparable to that of a thick quartz plate.

If a single-order, crystalline-quartz plate is to be used for a continuous range of wavelengths, both the phase retardation and the transmittance of the ordinary and extraordinary rays will oscillate as a function of wavelength because of multiple coherent reflections in the quartz. The separation between adjacent maxima in the phase retardation can be calculated from Eq. (144) in the earlier Polarization chapter.[1] Using $\lambda = 3.3913$ μm, $n \approx 1.4881$, and $d = 127.8$ μm, $\Delta\lambda = 0.03024$ μm, an amount which should be well-resolved with most infrared instruments. Thus, if a wave plate is to be used over a range of wavelengths, it would be well to antireflect the surfaces to eliminate the phase oscillations.

50. Multiple-Order Plates Thick plates made from crystalline quartz are sometimes used to produce circularly polarized light at a single wavelength or a discrete series of wavelengths. The plate thickness is generally of the order of one or more millimeters so that the retardation is an integral number of wavelengths plus $\lambda/4$, hence the name multiple-order wave plate. This plate acts like a single $\lambda/4$ plate providing it is used only at certain specific wavelengths; at other wavelengths it may not even approximate the desired retardation. For example, a 1.973-mm-thick quartz plate was purchased which had an order of interference $N = 28.25$ at 6328 Å. From Eq. (9) and Table 8, this plate would have $N = 30.52$ at 5890 Å, and would thus be an almost perfect half-wave plate at this latter wavelength.

If a multiple-order plate is used to produce circularly polarized light at unspecified discrete wavelengths, e.g., to measure circular or linear dichroism, it can be placed following a polarizer and oriented at 45° to the plane of vibration of the polarized beam. When the wavelengths are such that N calculated from Eq. (9) equals 1/4, 3/4, or in general $(2M - 1)/4$ (where M is a positive integer), the emerging beam will be alternately right and left circularly polarized. The frequency interval Δv between wavelengths at which circular polarization occurs is

$$\Delta v = \frac{1}{2d(n_e - n_o)} \tag{11}$$

where $v = 1/\lambda$. If the birefringence is independent of wavelength, the retardation plate will thus produce circularly polarized light at equal intervals on a frequency scale and can conveniently be used to measure circular dichroism, as described by Holzwarth.[286]

In order to approximately calibrate a multiple-order retardation plate at a series of wavelengths, it can be inserted between crossed polarizers and oriented at 45° to the polarizer axis. Transmission maxima will occur when the plate retardation is $\lambda/2$ or an odd multiple thereof; minima will occur when the retardation is a full wave or multiple thereof. If the axes of the two polarizers are parallel, maxima in the transmitted beam will occur when the plate retardation is a multiple of a full wavelength. The birefringence of the retardation plate can be determined by measuring the wavelengths at which maxima or minima occur if the plate thickness is known. Otherwise d can be measured with a micrometer, and an approximate value of $n_e - n_o$ can be obtained.

Palik[287] made and tested a 2.070-mm-thick CdS plate for the 2- to 15-μm infrared region and also made thick retardation plates of SnSe, sapphire, and crystalline quartz to be used in various parts of the infrared. Holzwarth[286] used a cultured-quartz retardation plate 0.8 mm thick to measure circular dichroism in the 1850- to 2500-Å region of the ultraviolet; Jaffe et al.[288] measured linear dichroism in the ultraviolet using a thick quartz plate and linear polarizer.

51. Sensitivity to Temperature Changes Small temperature changes can have a large effect on the retardation of a multiple-order plate. The method for calculating this effect was given in the earlier Polarization chapter.[1] For the 1.973-mm-thick quartz plate mentioned in Par. 50 ($N = 28.25$ at 6328 Å), the phase retardation will decrease 1.03° for each Celsius degree increase in temperature. If the temperature of the wave plate is not controlled extremely accurately, the large temperature coefficient of retardation can introduce sizable errors in precise ellipsometric measurements in which polarizer and analyzer settings can be made to ±0.01°.

52. Sensitivity to Angle of Incidence The effect of angle of incidence (and hence field angle) on the retardation was calculated in the earlier Polarization chapter.[1] It was shown there that the change in phase retardation with angle of incidence, $2\pi(\Delta N)_\theta$, is proportional to the total thickness of the plate (which is incorporated into N) and the square of the angle of incidence when the rotation is about an axis parallel to the optic axis. If the 1.973-mm-thick plate mentioned previously is rotated parallel to the optic axis through an angle of 10° at a wavelength of 6328 Å, the total retardation changes from 28.25 to 28.43, so that the $\lambda/4$ plate is now nearly a $\lambda/2$ plate.

If the plate had been rotated about an axis *perpendicular* to the direction of the optic axis, in the limit when the angle of incidence is 90°, the beam would have been traveling along the optic axis; in this case the ordinary and extraordinary rays would be traveling with the same velocities, and there would have been *no retardation* of one relative to the other. For any intermediate angle of incidence the retardation would have been *less than* the value at normal incidence. The relation for the retardation as a function of angle of incidence is not simple, but the retardation will be approximately as angle-sensitive as it was in the other case. An advantage of rotation about either axis is that, with care, one can adjust the retardation of an inexact wave plate to a desired value. Rotation about an axis *parallel* to the optic axis will *increase* the retardation, while rotation about an axis *perpendicular* to the optic axis will *decrease* the retardation.

Achromatic Retardation Plates

53. Achromatic retardation plates are those for which the phase retardation is independent of wavelength. The name arose because when a plate of this type is placed between polarizers, it does not appear colored and hence is achromatic, as shown by Gaudefroy.[289] In many applications, a truly achromatic retardation plate is not required. Since the wavelength of light changes by less than a factor of 2 across the visible region, a quarter- or half-wave mica plate often introduces only tolerable errors even in white light. The errors that do occur cancel out in many kinds of experiments.

Achromatic retardation plates can be made in various ways. The most achromatic are based on the principle of the Fresnel rhomb, in which the phase retardation occurs when light undergoes two or more total internal reflections (Par. 54 and Ref. 1). A material with the appropriate variation of birefringence with wavelength can also be used. Such materials are uncommon, but plates of two or more different birefringent materials can be combined to produce a reasonably achromatic combination. Composite plates, consisting of two or more plates of the same material whose axes are oriented at the appropriate angles, can be used as achromatic circular polarizers or achromatic polarization rotators,[1] although they do not have all the properties of true $\lambda/4$ or $\lambda/2$ plates. One unusual achromatic half-wave plate is described in the earlier Polarization chapter.[1]

The simplest type of achromatic retardation plate could be made from a single material if its birefringence satisfied the requirement that $(n_e - n_o)/\lambda$ be independent of wavelength, i.e., that $n_e - n_o$ be directly proportional to λ. This result follows from Eq. (9) since $d(n_e - n_o)/\lambda$ must be independent of λ to make N independent of wavelength. (The plate thickness d is constant.) The birefringences of various materials are listed in Table 8 and

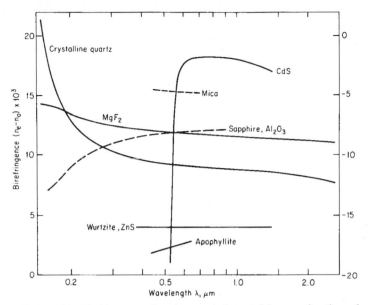

FIGURE 18 Birefringence of various optical materials as a function of wavelength. The scale at the left is for materials having a positive birefringence (*solid curves*), and the scale at the right is for materials with a negative birefringence (*dashed curves*).

plotted in Figs. 18 and 19. Only one material, the mineral apophyllite, has a birefringence which increases in the correct manner with increasing wavelength.[263] * A curve of the phase retardation vs. wavelength for a quarter-wave apophyllite plate is shown as curve *D* in Fig. 20. Also included are curves for other so-called achromatic $\lambda/4$ plates as well as for simple $\lambda/4$ plates of quartz and mica. The phase retardation of apophyllite is not as constant with λ as that of the rhomb-type retarders, but it is considerably more constant than that of the other "achromatic" $\lambda/4$ plates. Since the birefringence of apophyllite is small, a $\lambda/4$ plate needs a thickness of about 56.8 μm, which is enough for it to be made as a single piece rather than as a "first-order" plate. Unfortunately optical-grade apophyllite is rare, the sample for which data are reported here having come from Sweden. There is some indication that the optical properties of other apophyllite samples may be different. Isotropic, positive, and negative-birefringent specimens have been reported by Deer *et al.*.[290] According to them, the optical properties of apophyllite are often anomalous, some specimens being isotropic, uniaxial negative, or even biaxial with crossed dispersion of optic axial planes. Whether many samples have the favorable birefringence of the Swedish sample is uncertain.

Certain types of plastic film stretched during the manufacturing process have birefringences which are nearly proportional to wavelength and can serve as achromatic retardation plates if they have the proper thickness, as pointed out by West and Makas.[246] Curve *C* in Fig. 20 is the retardation of a stretched cellulose nitrate film as measured by West and Makas.[246] A combination of stretched cellulose acetate and cellulose nitrate sheets with their axes parallel will also make a reasonably achromatic $\lambda/4$ plate over the visible region. The advantages of using stretched plastic films for retardation plates are that they are cheap, readily available, have a retardation which is uniform over large areas,

---

* For materials having a negative birefringence the requirement is that $-(n_e - n_o)$ be proportional to λ.

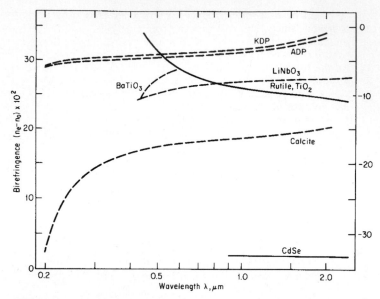

FIGURE 19 Birefringence of various optical materials which have larger birefringences than those shown in Fig. 18. The scale at the left is for materials having a positive birefringence (*solid curves*), and the scale at the right is for materials with a negative birefringence (*dashed curves*).

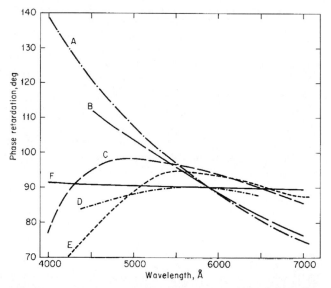

FIGURE 20 Curves of the phase retardation vs. wavelength for $\lambda/4$ plates; *A*, quartz; *B*, mica; *C*, stretched plastic film; *D*, apophyllite; and *E*, quartz-calcite achromatic combination. Curve *F* is for a Fresnel rhomb but is representative of all the rhomb-type devices. (*Bennett, Ref. 275.*)

and can be used in strongly convergent light. However, each sheet must be individually selected since the birefringence is a strong function of the treatment during the manufacturing process and the sheets come in various thicknesses, with the result that their retardations are not necessarily $\lambda/4$ or $\lambda/2$. Also, Ennos[291] found that while the magnitude of the retardation was uniform over large areas of the sheets, the direction of the effective crystal axis varied from point to point by as much as 1.5° on the samples he was testing. Thus, film retarders appear to be excellent for many applications but are probably not suitable for measurements of the highest precision.

A reasonably achromatic retardation plate can be constructed from pairs of readily available birefringent materials such as crystalline quartz, sapphire, magnesium fluoride, calcite, or others whose birefringences are listed in Table 8. Assume that the plate is to be made of materials a and b having thicknesses d_a and d_b, respectively (to be calculated), and that it is to be achromatized at wavelengths λ_1 and λ_2. From Eq. (9) we can obtain the relations

$$N\lambda_1 = d_\alpha \, \Delta n_{1a} + d_b \, \Delta n_{1b}$$

$$N\lambda_2 = d_a \, \Delta n_{2a} + d_b \, \Delta n_{2b}$$

$$(12)$$

where $N = \frac{1}{4}$ for a $\lambda/4$ plate, $\frac{1}{2}$ for a $\lambda/2$ plate, etc., and the Δn's are values of $n_e - n_o$ for the particular materials at the wavelengths specified; Δn will be positive for a positive uniaxial crystal and negative for a negative uniaxial crystal. (A positive uniaxial material can be used with its fast axis crossed with that of another positive uniaxial material; in this case the first material will have a negative Δn.) Equations (12) can be solved for d_a and d_b:

$$d_\alpha = \frac{N(\lambda_1 \, \Delta n_{2b} - \lambda_2 \, \Delta n_{1b})}{\Delta n_{1a} \, \Delta n_{2b} - \Delta n_{1b} \, \Delta n_{2a}} \qquad d_b = \frac{N(\lambda_2 \, \Delta n_{1a} - \lambda_1 \, \Delta n_{2a})}{\Delta n_{1a} \, \Delta n_{2b} - \Delta n_{1b} \, \Delta n_{2a}} \qquad (13)$$

As an example of a compound plate, let us design a $\lambda/4$ plate of crystalline quartz and calcite and achromatize it at wavelengths $\lambda_1 = 0.508 \, \mu m$ and $\lambda_2 = 0.656 \, \mu m$. Quartz has a positive birefringence and calcite a negative birefringence (Table 8) so that Δn_{1a} and Δn_{2a} (for quartz) are positive and Δn_{1b} and Δn_{2b} (for calcite) are negative. Equations (13) are satisfied for $d_{qtz} = 426.2 \, \mu m$ and $d_{calc} = 21.69 \, \mu m$; thus the phase retardation is exactly 90° at these two wavelengths. An equation of the form of those in Eqs. (12) is now used to calculate N for all wavelengths in the visible region using birefringence values listed in Table 8, and the results are plotted as curve E in Fig. 20. Although the achromatization for this quartz-calcite combination is not as good as can be obtained with a rhomb-type device or apophyllite, the phase retardation is within ±5° of 90° in the wavelength region 4900 to 7000 Å and is thus much more constant than the retardation of a single mica or quartz $\lambda/4$ plate. Better two-plate combinations have been calculated by Beckers,[249] the best being MgF_2-ADP and MgF_2-KDP, which have maximum deviations of ±0.5 and ±0.4 percent, respectively, compared with ±7.2 percent for a quartz-calcite combination over the same 4000- to 7000-Å wavelength region. The thicknesses of the materials which are required to produce $\lambda/4$ retardation* are $d_{MgF_2} = 113.79 \, \mu m$, $d_{ADP} = 26.38 \, \mu m$, and $d_{MgF_2} = 94.47 \, \mu m$, $d_{KDP} = 23.49 \, \mu m$. Since the ADP and KDP must be so thin, these components could be made in two pieces as "first-order" plates.

Other two-component compound plates have been proposed by Chandrasekharan and Damany,[260] Gaudefroy,[289] Ioffe and Smirnova,[292] and Mitchell.[293] The paper by Ioffe and Smirnova describes a quartz-calcite combination similar to the one illustrated above, but it contains various numerical errors which partially invalidate the results.

---

* Beckers' tables II to V give the thickness of materials required to produce one full-wave retardation. To obtain values of thicknesses for $\lambda/4$ retardation, for example, multiply all d values in the table by 0.25. The percent deviations should remain unchanged.

If better achromatization is desired and one does not wish to use a rhomb-type $\lambda/4$ device, three materials can be used which satisfy the relations

$$N\lambda_1 = d_a \, \Delta n_{1a} + d_b \, \Delta n_{1b} + d_c \, \Delta n_{1c}$$

$$N\lambda_2 = d_a \, \Delta n_{2a} + d_b \, \Delta n_{2b} + d_c \, \Delta n_{2c} \qquad (14)$$

$$N\lambda_3 = d_a \, \Delta n_{3a} + d_b \, \Delta n_{3b} + d_c \, \Delta n_{3c}$$

where the Δn's are birefringences of the various materials at wavelengths λ_1, λ_2, and λ_3.

Instead of using only three wavelengths, Beckers[249] suggested that the thicknesses can be optimized such that the maximum deviations from achromatization are minimized over the entire wavelength interval desired. In this way, he obtained a three-component combination of quartz, calcite, and MgF_2 which has a retardation of a full wavelength and a maximum deviation of only ±0.2 percent over the 4000- to 7000-Å wavelength region. The maximum deviation of slightly different thicknesses of these same three materials rises to ±2.6 percent if the wavelength interval is extended to 3000 to 11,000 Å. Chandrasekharan and Damany[260] have designed a three-component $\lambda/4$ plate from quartz, MgF_2, and sapphire for use in the vacuum ultraviolet. Title[294] has designed achromatic combinations of three-element, four-element, nine-element, and ten-element waveplates using Jones matrix techniques. The nine-element combination is achromatic to within 1° from 3500 to 10,000 Å. He constructed and tested several waveplate combinations, and they performed as designed.

Rhombs as Achromatic $\lambda/4$ Retarders

54. The simplest stable, highly achromatic $\lambda/4$ retarder with a reasonable acceptance angle and convenient size appears to be a rhomb-type retarder. Several types are available; the choice of which one to use for a specific application depends on (1) the geometry of the optical system (can a deviated or displaced beam be tolerated?), (2) wavelength range, (3) degree of collimation of the beam, (4) beam diameter (determining the aperture of the retarder), (5) space available, and (6) accuracy required. Table 9 summarizes the properties of the various achromatic rhombs. This subject has been covered in detail by

TABLE 9 Properties of Achromatic Rhombs (Bennett, Ref. 275)

				Refractive index		With wavelength		With angle of incidence	
								Variation of phase retardation	
Name	Light path	Internal angle of incidence, deg	Material	n	Wavelength, Å	Var., deg	Wavelength, Å	Var., deg.	Angle, deg
Fresnel rhomb	Translated	54.7	Crown glass	1.511	5893	2.5	3650–7682	9.1	−7 to +7
Coated Fr. rhomb	Translated	51.5	Crown glass	1.5217	5461	0.4	3341–5461	2.5	−4 to +6
	Translated	54.0	Fused quartz	1.4880	3000	0.7	2148–3341	<0.5	−1.5 to +1.5
Mooney rhomb	Deviated	60.0	Flint glass	1.650	5893	1.9	4047–6708	0.7	−7 to +7
AD-1	Undeviated	74.3	Fused quartz	1.4702	4000	2.0	3000–8000	0.7	−7 to +7
AD-2	Undeviated	73.2, 56.4	Fused quartz	1.4702	4000	2.9	3000–8000	13.2	−3 to +3
Coated AD-2	Undeviated	72.2	Fused quartz	1.4601	5461	0.3	2140–5461	6.0	−1.5 to +1.5
AD	Undeviated	53.5	Crown glass	1.511	5893	1.6	3650–7682	9.4	−7 to +7

Bennett.[275] and is condensed from that reference in the earlier Polarization chapter.[1] Anderson[295] has compared the retardation of a CdS $\lambda/4$ plate and a Fresnel rhomb in the 10-μm CO_2 laser emission region. Wizinowich[296] used a Fresnel rhomb along with some additional optics to change an unpolarized light beam from a faint star object into linearly polarized light to improve the throughput of a grating spectrograph and make it independent of the input polarization.

Composite Retardation Plates

55. A composite retardation plate is made up of two or more elements of the same material combined so that their optic axes are at appropriate angles to each other. Some of the composite plates have nearly all the properties of a true retardation plate, whereas others do not. In the earlier Polarization chapter,[1] composite plates were described which produced circularly polarized light at a given wavelength, those which acted as achromatic circular polarizers, and those which acted as achromatic polarization rotators or pseudo $\lambda/2$ plates. The effect of combining several birefringent plates with their axes at arbitrary angles to each other can be easily understood using the Poincaré sphere. A general treatment of this subject has been given by Ramachandran and Ramaseshan.[297]

VARIABLE RETARDATION PLATES AND COMPENSATORS

56. Variable retardation plates can be used to modulate or vary the phase of a beam of plane-polarized light, to measure birefringence in mineral specimens, flow birefringence, or stress in transparent materials, or to analyze a beam of elliptically polarized light such as might be produced by transmission through a birefringent material or by reflection from a metal or film-covered surface. The term compensator is frequently applied to a variable retardation plate since it can be used to compensate for the phase retardation produced by a specimen. Common types of variable compensators include the Babinet and Soleil compensators, in which the total thickness of birefringent material in the light path is changed, the Sénarmont compensator,[1] which consists of a fixed quarter-wave plate and rotatable analyzer to compensate for varying amounts of ellipticity in a light beam, and tilting-plate compensators,[1] with which the total thickness of birefringent material in the light beam is changed by changing the angle of incidence. Electro-optic and piezo-optic modulators can also be used as variable retardation plates since their birefringence can be changed by varying the electric field or pressure. However, they are generally used for modulating the amplitude, phase, frequency, or direction of a light beam, in particular a laser beam, at frequencies too high for mechanical shutters or moving mirrors to follow. Information on electro-optic materials and devices is contained in the chapter on Electro-Optic Modulators by T. A. Maldonado (Chap. 13, Vol II) and in the earlier Polarization chapter.[1]

57. Babinet Compensator There are many devices which compensate for differences in phase retardation by having a variable thickness of a birefringent material (such as crystalline quartz) in the light beam, as discussed by Johansen,[298] and Jerrard.[299] One such device, described by Hunt,[300] can compensate for a residual wedge angle between the entrance and exit faces of birefringent optical components such as optical modulators and waveplates.

The most common variable retardation plates are the Babinet compensator and the Soleil compensator. The Babinet compensator was proposed by Babinet in 1837 and later

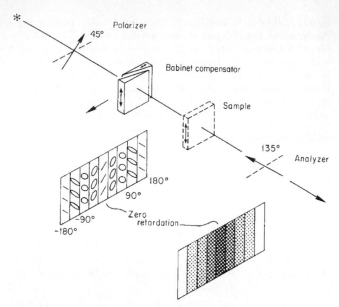

FIGURE 21 Arrangement of a Babinet compensator, polarizer, and analyzer for measuring the retardation of a sample. The appearance of the field after the light has passed through the compensator is shown to the left of the sample position. Retardations are indicated for alternate regions. After the beam passes through the analyzer, the field is crossed by a series of dark bands, one of which is shown to the left of the analyzer.

modified by Jamin; references to the voluminous early literature are given by Partington.[301] Ellerbroek and Groosmuller[302] have a good description of the theory of operation (in German), and Jerrard[303-305] and Archard[306] describe various optical and mechanical defects of Babinet compensators.

The Babinet compensator, shown schematically in Fig. 21, consists of two crystalline-quartz wedges, each with its optic axis in the plane of the face but with the two optic axes exactly 90° apart. One wedge is stationary, and the other is movable by means of a micrometer screw in the direction indicated by the arrow, so that the total amount of quartz through which the light passes can be varied uniformly. In the first wedge, the extraordinary ray vibrates in a horizontal plane and is retarded relative to the ordinary ray (crystalline quartz has a positive birefringence; see Table 8). When the rays enter the second wedge, the ray vibrating in the horizontal plane becomes the ordinary ray and is advanced relative to the ray vibrating in the vertical plane. Thus, the total retardation is proportional to the difference in thickness between the two wedges:

$$N\lambda = (d_1 - d_2)(n_e - n_o) \tag{15}$$

where N = retardation in integral and fractional parts of a wavelength

d_1, d_2 = thickness of the first and second wedges where light passes through

n_o, n_e = ordinary and extraordinary refractive indexes for crystalline quartz

If light polarized at an angle of 45° to one of the axes of the compensator passes through it, the field will appear as shown in Fig. 21; the wedges have been set so there is zero retardation at the center of the field. (If the angle α of the incident plane-polarized beam were different from 45°, the beam retarded or advanced by 180° in phase angle would make an angle of 2α instead of 90° with the original beam.) When an analyzer whose axis is crossed with that of the polarizer is used to observe the beam passing through the compensator, a series of light and dark bands is observed in monochromatic light. In white light only one band, that for which the retardation is zero, remains black. All the other bands are colored. These are the bands for which the retardation is multiples of 2π (or, expressed in terms of path differences, integral numbers of wavelengths). On one side of the central black band one ray is advanced in phase relative to the other ray; on the other side it is retarded. If one wedge is moved, the whole fringe system translates across the field of view. The reference line is scribed on the stationary wedge so that it remains in the center of the field. Information on calibrating and using a Babinet compensator is given in the earlier Polarization chapter.[1]

58. Soleil Compensator The Soleil compensator (see Wood[307] and Ditchburn[308]), sometimes called a Babinet-Soleil compensator, is shown in Fig. 22. It is similar to the Babinet compensator in the way it is used, but instead of having a field crossed with alternating light and dark bands in monochromatic light, the field has a uniform tint if the compensator is constructed correctly. This is because the ratio of the thicknesses of the two quartz blocks (one composed of a fixed and a movable wedge) is the same over the entire field. The Soleil compensator will produce light of varying ellipticity depending on the position of the movable wedge. Calibration of the Soleil compensator is similar to that of

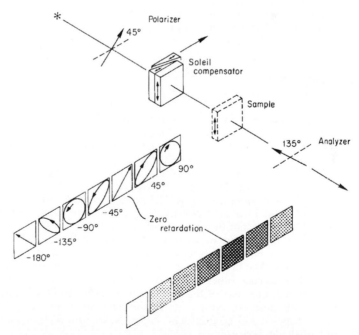

FIGURE 22 Arrangement of a Soleil compensator, polarizer, and analyzer for measuring the retardation of a sample. The appearance of the field after the light has passed through the compensator is shown to the left of the sample position. After the beam passes through the analyzer, the field appears as one of the shades of gray shown to the left of the analyzer.

the Babinet compensator.[1] The zero-retardation position is found in the same manner except that now the entire field is dark. The compensator is used in the same way as a Babinet compensator with the uniformly dark field (in white light) of the Soleil corresponding to the black zero-retardation band in the Babinet.

The major advantage of the Soleil compensator is that a photoelectric detector can be used to make the settings. The compensator is offset a small amount on each side of the null position so that equal-intensity readings are obtained. The average of the two drum positions gives the null position. Photoelectric setting can be much more precise than visual setting, but this will not necessarily imply increased accuracy unless the compensator is properly constructed. Since Soleil compensators are composed of three pieces of crystalline quartz, all of which must be very accurately made, they are subject to more optical and mechanical defects than Babinet compensators. Jerrard[309-311] has described many of these defects in detail. Ives and Briggs[312] found random departures of about $\pm 1.5°$ from their straight-line calibration curve of micrometer reading for extinction vs. wedge position. This variation was considerably larger than the setting error with a half-shade plate and was attributed to variations in thickness of the order of $\pm \lambda/4$ along the quartz wedges.

Soleil compensators have been used for measurements of retardation in the infrared. They have been made of crystalline quartz, cadmium sulfide, and magnesium fluoride (see the work of Palik[257,313] and Palik and Henvis[252]). A by-product of this work was the measurement of the birefringence of these materials in the infrared.

Two other uniform-field compensators have been proposed. Jerrard,[310] following a suggestion by Soleil, has taken the Babinet wedges and reversed one of them so the light passes through the thicker portions of each wedge. This reversed Babinet compensator is less subject to mechanical imperfections than the Soleil compensator but does produce a small deviation of the main beam. Hariharan and Sen[314] suggest double-passing a Babinet compensator (with a reflection between the two passes) to obtain a uniform field.

HALF-SHADE DEVICES

59. It is sometimes necessary to measure accurately the azimuth of a beam of plane-polarized light, i.e., the angle the plane of vibration makes with a reference coordinate system. This can be done most easily by using a polarizer as an analyzer and rotating it to the position where the field appears the darkest. The analyzer azimuth is then exactly 90° from the azimuth of the plane-polarized beam. A more sensitive method is to use a photoelectric detector and offset on either side of the extinction position at angles where the intensities are equal. The average of these two angles is generally more accurate than the value measured directly, but care must be taken to keep the angles small so that asymmetries will not become important.

Before the advent of sensitive photoelectric detectors, the most accurate method of setting on a minimum was to use a half-shade device as the analyzer or in conjunction with the analyzer. The device generally consisted of two polarizers having their axes inclined at an angle α to each other (angle fixed in some types and variable in others). As the device was rotated, one part of the field became darker while the other part became lighter. At the match position, both parts of the field appeared equally bright. The Jellett-Cornu prism, Lippich and Laurent half shades, Nakamura biplate, and Savart plate are examples of half-shade devices.[1]

Ellipticity half-shade devices are useful for detecting very small amounts of ellipticity in a nominally plane-polarized beam and hence can indicate when a compensator has completely converted elliptically polarized light into plane-polarized light. Two of these devices are the Bravais biplate and the Brace half-shade plate. Half-shade devices for both plane and elliptically polarized light are described in detail in the earlier Polarization chapter.[1]

MINIATURE POLARIZATION DEVICES

60. Polarization Devices for Optical Fibers Single-mode optical fiber-type polarizers are important devices for optical fiber communication and fiber sensor systems. These polarizers have been made by a variety of techniques. Polarizers have been made by bending[315] or by tapering[316] a birefringent fiber to induce differential attenuation in the orthogonal modes. In most cases a fiber was polished laterally and some device was placed in contact with the exposed guiding region of the fiber to couple out the unwanted polarization. Bergh et al.[317] used a birefringent crystal as the outcoupling device and obtained a high extinction ratio polarizer. Optical fiber polarizers made with a metal film coated onto the polished area to eliminate the unwanted polarization state seem to be preferred because they are stable and rugged. The original version by Eickhoff[318] used the thin cladding remaining after polishing as the buffer layer, but it had an insufficient extinction ratio. Other designs using metal coatings were suggested by Gruchmann et al.,[319] and Hosaka et al.[320] Feth and Chang[321] used a fiber polished into its core to which a superstrate coated with a very thin metal layer was attached by an index-matching oil. Yu and Wu[322] gave a theoretical analysis of metal-clad single-mode fiber-type polarizers. Dyott et al.[323] made a metal-fiber polarizer from an etched D-shaped fiber coated with indium.

In the above approaches, either expensive components are used or the structure of the polarizer is complicated and fragile. Lee and Chen[324] suggested a new way of fabricating high-quality metal-clad polarizers by polishing a fiber ~0.4 µm into its core and then overcoating it with a 265-nm MgF_2 film as the buffer layer followed by a 100-nm Al film. Polarizers fabricated in this way had an average extinction ratio of 28 dB with a 2-dB insertion loss at a 0.63-µm wavelength or a 34-dB extinction ratio with a 3-dB insertion loss at 0.82 µm.[324]

Other devices for optical fibers have also been designed. Ulrich and Johnson[325] made a single-mode fiber-optical polarization rotator by mechanically twisting successive half-wave fiber sections in alternating directions; Hosaka et al.'s fiber circular polarizer[326] was composed of a metal-coated fiber polarizer and a λ/4 platelet fabricated on a birefringent fiber; polished-type couplers acting as polarizing beam splitters were made by Snyder and Stevenson.[327] The patent literature contains references to other polarization devices for optical fibers.

61. Polarization Devices for Integrated Circuits Small and highly efficient polarization devices are also needed for integrated circuits. Some such devices have been proposed and fabricated. Uehara et al.[328] made an optical waveguiding polarizer for optical fiber transmission out of a plate of calcite attached to borosilicate glass into which a three-dimensional high-index region had been formed by ion migration to act as the waveguide. Mahlein[329] deposited a multilayer dielectric film onto a glass superstrate which was then contacted to a planar waveguide to couple out the TM polarization. This paper contains a good description of the polarizer design as well as extensive references. Suchoski et al.[330] fabricated low-loss, high-extinction polarizers in $LiNbO_3$ by proton exchange. Noé et al.[331] achieved automatic endless polarization control with integrated optical $Ti:LiNbO_3$ polarization transformers. This was a better method of matching polarization states between two superposed waves than techniques that had been used previously. Finally, Baba et al.[332] proposed making a polarizer for integrated circuits out of periodic metal-dielectric laminated layers (Lamipol structures). Their experiments with $Al-SiO_2$ structures were encouraging. Patents have been filed for other polarization devices for integrated circuits.

REFERENCES*

1. H. E. Bennett and J. M. Bennett, "Polarization," in W. G. Driscoll and W. Vaughan, (eds.). *Handbook of Optics,* 1st ed., McGraw-Hill, New York, 1978, pp. 10-1–10-164.

2. H. J. Nickl and H. K. Henisch, *J. Electrochem. Soc.* **116,** pp. 1258–1260, 1969.

3. R. N. Smartt, *J. Sci. Instrum.* **38,** p. 165, 1961.

4. E. E. Wahlstrom, *Optical Crystallography,* 4th ed. Wiley, New York, 1969, pp. 236–267.

5. E. Uzan, H. Damany, and V. Chandrasekharan, *Opt. Commun.* **1,** pp. 221–222, 1969.

6. S. S. Ballard, J. S. Browder, and J. F. Ebersole, in D. E. Gray (ed.), *American Institute of Physics Handbook,* 3d ed., McGraw-Hill, New York, 1972, p. 6-20.

7. J. Schnellman, V. Chandrasekharan, H. Damany, and J. Romand, *C. R. Acad. Sci.* **260,** pp. 117–120, 1965.

8. Y. Bouriau and J. Lenoble, *Rev. Opt.* **36,** pp. 531–543, 1957.

9. S. S. Ballard, J. S. Browder, and J. F. Ebersole, in D. E. Gray, (ed.), *American Institute of Physics Handbook,* 3d ed., McGraw-Hill, New York, 1972, p. 6-65.

10. H. Damany, Laboratoire des Hautes Pressions, Bellevue, France, private communication, 1970.

11. E. Uzan, H. Damany, and V. Chandrasekharan, *Opt. Commun.* **2,** pp. 273–275, 1970.

12. M. Born and E. Wolf, *Principles of Optics,* 6th ed. Pergamon Press, New York, 1980, p. 680.

13. L. C. Martin, *Technical Optics,* vol. 1, Pitman, London, 1948, pp. 196–198.

14. R. W. Ditchburn, *Light,* 2d ed. Interscience, New York, 1963, pp. 595–616.

15. ·A. Schuster, *Theory of Optics,* 2d ed, Arnold, London, 1920, pp. 168–187.

16. R. Müller, *Optik* **20,** pp. 510–511, 1963.

17. C. Dévé, *Optical Workshop Principles,* T. L. Tippell (trans.), Hilger & Watts, London, 1954, p. 295.

18. P. Jacquinot, *J. Opt. Soc. Am.* **44,** pp. 761–765, 1954.

19. L. Mertz, *Transformations in Optics,* Wiley, New York, 1965, pp. 15–16.

20. P. Glan, *Carl's Repert.* **16,** p. 570, 1880.

21. S. P. Thompson, *Phil. Mag.,* ser. 5, **12,** p. 349, 1881.

22. S. P. Thompson, *Proc. Opt. Conv.,* 1905, pp. 216–235.

23. R. T. Glazebrook, *Phil. Mag.,* ser. 5, **15,** p. 352, 1883.

24. C. E. Moeller and D. R. Grieser, *Appl. Opt.* **8,** pp. 206–207, 1969.

25. R. J. King and S. P. Talim, *J. Phys.* (GB), ser. E, **4,** pp. 93–96, 1971.

26. J. W. Ellis and J. Bath, *J. Chem. Phys.* **6,** pp. 221–222, 1938.

27. F. Lippich, *Wien Akad. Sitzungsber.,* ser. III, **91,** p. 1059, 1885.

28. A. C. Hardy and F. H. Perrin, *The Principles of Optics,* McGraw-Hill, New York, 1932, p. 611.

29. J. F. Archard and A. M. Taylor, *J. Sci. Instrum.* **25,** pp. 407–409, 1948.

30. Karl Lambrecht Corp., Bull. P-73, Chicago, 1973.

31. J. Swartz, D. K. Wilson, and R. J. Kapash, *High Efficiency Laser Polarizers,* Electro-Opt. 1971 West Conf., Anaheim, Calif., May, 1971.

32. A. Lambrecht, Karl Lambrecht Corp., Chicago, Ill., private communication, 1969.

33. A. V. Shustov, *Sov. J. Opt. Technol.* **34,** pp. 177–181, 1967.

34. F. Twyman, *Prism and Lens Making,* 2d ed., Hilger & Watts, London, 1952, pp. 244, 599.

35. J. F. Archard, *J. Sci. Instrum.* **26,** pp. 188–192, 1949.

* In all references to the Russian literature, volume and pages cited are for the English translation.

36. H. E. Bennett and J. M. Bennett, "Precision Measurements in Thin Film Optics," in G. Hass and R. E. Thun (eds.), *Physics of Thin Films,* vol. 4, Academic Press, New York, 1967, pp. 69–78.

37. D. L. Decker, J. L. Stanford, and H. E. Bennett, *J. Opt. Soc. Am.* **60,** p. 1557A, 1970.

38. R. L. Rowell, A. B. Levit, and G. M. Aval, *Appl. Opt.* **8,** p. 1734, 1969.

39. D. E. Aspnes, *Appl. Opt.* **9,** pp. 1708–1709, 1970.

40. R. L. Rowell, *Appl. Opt.* **9,** p. 1709, 1970.

41. W. Nicol, *Edinb. New Phil. J.* **6,** p. 83, 1828–1829, as quoted in A. Johannsen, *Manual of Petrographic Methods,* 2d ed. Hafner, New York, 1968, p. 158; (originally published in 1918).

42. S. P. Thompson, *Phil. Mag.,* ser. 5, **21,** p. 476, 1886.

43. B. Halle, *Dtsch. Mech. Z.,* no. 1, pp. 6–7, Jan. 1, 1908.

44. B. Halle, *Dtsch. Mech. Z.,* no. 2, pp. 16–19, Jan. 15, 1908.

45. A. B. Dale, in R. Glazebrook (ed.), *A Dictionary of Applied Physics,* vol. 4, Macmillan, London, 1923, pp. 496–497.

46. Hartnack and Prazmowski, *Ann. Chim. Phys.,* ser. 4, **7,** p. 181, 1866.

47. L. Foucault, *C. R. Acad. Sci.* **45,** p. 238, 1857.

48. C. D. West, and R. C. Jones, *J. Opt. Soc. Am.* **41,** pp. 976–982, 1951.

49. D. L. Steinmetz, W. G. Phillips, M. Wirick, and F. F. Forbes, *Appl. Opt.* **6,** pp. 1001–1004, 1967.

50. G. C. Morris and A. S. Abramson, *Appl. Opt.* **8,** pp. 1249–1250, 1969.

51. E. O. Ammann and G. A. Massey, *J. Opt. Soc. Am.* **58,** pp. 1427–1433, 1968.

52. A. C. Hardy, *J. Opt. Soc. Am.* **25,** pp. 305–311, 1935.

53. C. Bouhet and R. LaFont, *Rev. Opt.* **28,** pp. 490–493, 1949.

54. A. Cotton, *C. R. Acad. Sci.* **193,** pp. 268–271, 1931.

55. W. C. Johnson, Jr., *Rev. Sci. Instrum.* **35,** pp. 1375–1376, 1964.

56. L. V. Foster, *J. Opt. Soc. Am.* **28,** pp. 124–126, 127–129, 1938.

57. K. Feussner, *Z. Instrumentenkd.* **4,** p. 41, 1884.

58. A. Johannsen, *Manual of Petrographic Methods,* 2d ed., Hafner, New York, 1968, pp. 169, 283–285; (originally published in 1918).

59. W. L. Hyde, New York Univ., Bronx N.Y., 1970, private communication.

60. E. Bertrand, *C. R. Acad. Sci.* **49,** p. 538, 1884.

61. L. Wulff, *Sitz. Preuss. Akad. Wiss.* **135,** p. 879, 1896.

62. P. Stöber, *Z. Krist.* **61,** p. 299, 1924.

63. P. Stöber, *Neues Jahrb. Mineral.* **A57,** p. 139, 1928.

64. P. Stöber, *Chem. Erde* **6,** p. 357, 453, 1930.

65. E. Tzekhnovitzer, *J. Phys. Chem.* (USSR) **5,** p. 1452, 1934.

66. C. D. West, *J. Opt. Soc. Am.* **35,** pp. 26–31, 1945.

67. M. Huot de Longchamp, *Rev. Opt.* **26,** pp. 94–98, 1947.

68. T. Yamaguti, *J. Phys. Soc. Jap.* **10,** pp. 219–221, 1955.

69. T. Yamaguti, I. Makino, S. Shinoda, and I. Kuroha, *J. Phys. Soc. Jap.* **14,** p. 199–201, 1959.

70. F. J. Dumont and R. N. Smartt, *J. Opt. Soc. Am.* **59,** p. 1541A, 1969

71. F. J. Dumont, *J. Opt. Soc. Am.* **60,** p. 719A, 1970

72. L. G. DeShazer, Dept. of Electrical Engineering, Univ. Southern California, Los Angeles, private communication, 1971.

73. V. Chandrasekharan and H. Damany, *Appl. Opt.* **8,** pp. 675–675, 1969.

74. V. Chandrasekharan and H. Damany, *Appl. Opt.* **10,** pp. 681–682, 1971.

75. E. Landais, *Bull. Soc. Fr. Mineral. Cristallogr.* **91,** pp. 350–354, 1968.

76. T. J. Bridges and J. W. Kluver, *Appl. Opt.* **4,** pp. 1121–1125, 1965.

77. J. J. Loferski, *Phys. Rev.* **87,** pp. 905–906, 1952.

78. R. Newman and R. S. Halford, *Rev. Sci. Instrum.* **19,** pp. 270–271, 1948.

79. W. L. Hyde, *J. Opt. Soc. Am.* **38,** p. 663A, 1948.

80. E. H. Land and C. D. West, in J. Alexander (ed.), *Colloid Chemistry,* vol. 6, Reinhold, New York, 1946, pp. 160–190.

81. E. H. Land, *J. Opt. Soc. Am.* **41,** pp. 957–963, 1951.

82. W. A. Shurcliff, *Polarized Light,* Harvard University Press, Cambridge, Mass., 1962, pp. 43–64.

83. S. D. Stookey and R. J. Araujo, *Appl. Opt.* **7,** pp. 777–779, 1968.

84. T. Yamaguti, *J. Opt. Soc. Am.* **45,** pp. 891–892, 1955.

85. P. Baumeister and J. Evans, in D. E. Gray (ed.), *American Institute of Physics Handbook,* 3rd ed., McGraw-Hill, New York, 1972, pp. 6-171–6-172.

86. R. P. Blake, A. S. Makas, and C. D. West, *J. Opt. Soc. Am.* **39,** p. 1054A, 1949.

87. A. S. Makas, *J. Opt. Soc. Am.* **52,** pp. 43–44, 1962.

88. L. Baxter, A. S. Makas, and W. A. Shurcliff, *J. Opt. Soc. Am.* **46,** p. 229, 1956.

89. R. C. Jones, Polaroid Corporation, Cambridge, Mass., private communication, 1970.

90. M. Haase, *Zeiss-Mitt.* **2,** p. 173, 1961.

91. M. N. McDermott and R. Novick, *J. Opt. Soc. Am.* **51,** pp. 1008–1010, 1961.

92. G. R. Bird and M. Parrish, Jr., *J. Opt. Soc. Am.* **50,** pp. 886–891, 1960.

93. G. B. Trapani, *Proc. Soc. Photo-Opt. Instrum. Eng.* **88,** pp. 105–113, 1976.

94. W. J. Gunning and J. Foschaar, *Appl. Opt.* **22,** pp. 3229–3231, 1983.

95. S. J. Baum, *Proc. Soc. Photo-Opt. Instrum. Eng.* **88,** pp. 50–56, 1976.

96. J. F. Dreyer, *J. Opt. Soc. Am.* **37,** p. 983A, 1947.

97. S. Anderson, *J. Opt. Soc. Am.* **39,** pp. 49–56, 1949.

98. Polacoat Bull. P-108 and P-112, Polacoat, Inc., Cincinnati, Ohio, 1967.

99. D. S. Kyser, Michelson Laboratory, Naval Weapons Center, China Lake, Calif., private communication, 1970.

100. G. Rupprecht, D. M. Ginsberg, and J. D. Leslie, *J. Opt. Soc. Am.* **52,** pp. 665–669, 1962.

101. Perkin-Elmer Corp., Instrument Division, Norwalk, Conn., "Wire Grid Polarizer Accessory," Sheet no. D-454, 1966.

102. J. B. Young, H. A. Graham, and E. W. Peterson, *Appl. Opt.* **4,** pp. 1023–1026, 1965.

103. K. F. Renk and L. Genzel, *Appl. Opt.* **1,** pp. 643–648, 1962.

104. J. P. Casey and E. A. Lewis, *J. Opt. Soc. Am.* **42,** pp. 971–977, 1952.

105. E. A. Lewis and J. P. Casey, *J. Appl. Phys.* **23,** pp. 605–608, 1952.

106. M. Mohebi, J. Q. Liang, and M. J. Soileau, *Appl. Opt.* **28,** pp. 3681–3683, 1989.

107. R. W. Stobie and J. J. Dignam, *Appl. Opt.* **12,** pp. 1390–1391, 1973.

108. C. H. Burton, *Appl. Opt.* **18,** pp. 420–422, 1979.

109. J. P. Auton, *Appl. Opt.* **6,** pp. 1023–1027, 1967.

110. M. Hass and M. O'Hara, *Appl. Opt.* **4,** pp. 1027–1031, 1965.

111. A. R. Hilton and C. E. Jones, *J. Electrochem. Soc.* **113,** pp. 472–478, 1966.

112. D. G. Vickers, E. I. Robson, and J. E. Beckman, *Appl. Opt.* **10,** pp. 682–684, 1971.

113. P. K. Cheo and C. D. Bass, *Appl. Phys. Lett.* **18,** pp. 565–567, 1971.

114. J. P. Auton and M. C. Hutley, *Infrared Phys.* **12,** pp. 95–100, 1972.

115. A. E. Costley, K. H. Hursey, G. F. Neill, and J. M. Ward, *J. Opt. Soc. Am.* **67,** pp. 979–981, 1977.

116. J. A. Beunen, A. E. Costley, G. F. Neill, C. L. Mok, T. J. Parker, and G. Tait, *J. Opt. Soc. Am.* **71,** pp. 184–188, 1981.

117. T. A. Leonard, *Soc. Photo-Opt. Instrum. Eng.* **288,** pp. 129–135, 1981.

118. G. J. Sonek, D. K. Wanger, and J. M. Ballantyne, *Appl. Opt.* **22,** pp. 1270–1272, 1983.

119. W. L. Eichhorn and T. J. Magner, *Opt. Eng.* **25,** pp. 541–544, 1986.

120. G. Novak, R. J. Pernic, and J. L. Sundwall, *Appl. Opt.* **28,** pp. 3425–3427, 1989.

121. S. Roberts and D. D. Coon, *J. Opt. Soc. Am.* **52,** pp. 1023–1029, 1962.

122. Y. S. Hwang and H. K. Park, *Appl. Opt.* **28,** pp. 4999–5001, 1989.

123. H. Weiss and M. Wilhelm, *Z. Phys.* **176,** pp. 399–408, 1963.

124. B. Paul, H. Weiss, and M. Wilhelm, *Solid State Electron.* (GB) **7,** pp. 835–842, 1964.

125. A. Mueller and M. Wilhelm, *J. Phys. Chem. Solids* **26,** p. 2029, 1965.

126. N. M. Davis, A. R. Clawson, and H. H. Wieder, *Appl. Phys. Lett.* **15,** pp. 213–215, 1969.

127. M. Saito and M. Miyagi, *Appl. Opt.* **28,** pp. 3529–3533, 1989.

128. A. Hidalgo, J. Pastor, and J. M. Serratosa, *J. Opt. Soc. Am.* **52,** pp. 1081–1082, 1962.

129. T. G. R. Rawlins, *J. Opt. Soc. Am.* **54,** pp. 423–424, 1964.

130. J.-L. Roumiguieres, *Opt. Commun.* **19,** pp. 76–78, 1976.

131. K. Knop, *Opt. Commun.* **26,** pp. 281–283, 1978.

132. G. W. Stroke, *Phys. Lett.* (Neth.) **5,** pp. 45–48, 1963.

133. G. W. Stroke, *J. Opt. Soc. Am.* **54,** p. 846, 1964.

134. C. W. Peters, T. F. Zipf, and P. V. Deibel, *J. Opt. Soc. Am.* **43,** p. 816A, 1953.

135. A. Hadni, E. Décamps, and P. Delorme, *J. Phys. Radium,* ser. 8, **19,** 793–794, 1958.

136. A. Hadni, E. Décamps, D. Grandjean, and C. Janot, *C. R. Acad. Sci.* **250,** pp. 2007–2009, 1960.

137. A. Mitsuishi, Y. Yamada, S. Fujita, and H. Yoshinaga, *J. Opt. Soc. Am.* **50,** pp. 433–436, 1960.

138. C. Janot and A. Hadni, *J. Phys. Radium,* ser. 8, **24,** pp. 1073–1077, 1963.

139. J. H. Rohrbaugh, C. Pine, W. G. Zoellner, and R. D. Hatcher, *J. Opt. Soc. Am.* **48,** pp. 710–711, 1958; [see also R. D. Hatcher and J. H. Rohrbaugh, *J. Opt. Soc. Am.* **46,** pp. 104–110, 1956 and J. H. Rohrbaugh and R. D. Hatcher, *J. Opt. Soc. Am.* **48,** pp. 704–709, 1958].

140. H. A. Kalhor and A. R. Neureuther, *J. Opt. Soc. Am.* **61,** pp. 43–48, 1971.

141. R. N. Hamm, R. A. MacRae, and E. T. Arakawa, *J. Opt. Soc. Am.* **55,** pp. 1460–1463, 1965.

142. K. Kudo, T. Arai, and T. Ogawa, *J. Opt. Soc. Am.* **60,** pp. 1046–1050, 1970.

143. V. G. Horton, E. T. Arakawa, R. N. Hamm, and M. W. Williams, *Appl. Opt.* **8,** pp. 667–670, 1969.

144. M. Schledermann and M. Skibowski, *Appl. Opt.* **10,** pp. 321–326, 1971.

145. N. J. Harrick, *J. Opt. Soc. Am.* **49,** pp. 376–379, 379–380, 1959.

146. D. F. Edwards and M. J. Bruemmer, *J. Opt. Soc. Am.* **49,** pp. 860–861, 1959.

147. M. Krízek, *Czech. J. Phys.* **13B,** pp. 599–610, 683–691, 1963.

148. J. P. Craig, R. F. Gribble, and A. A. Dougal, *Rev. Sci. Instrum.* **35,** pp. 1501–1503, 1964.

149. J. Bor, and L. A. Brooks, *J. Sci. Instrum.* **43,** p. 944, 1966.

150. A. H. Pfund, *J. Opt. Soc. Am.* **37,** pp. 558–559, 1947.

151. P. Barchewitz and L. Henry, *J. Phys. Radium,* ser. 8, **15,** pp. 639–640, 1954.

152. S. Takahashi, *J. Opt. Soc. Am.* **51,** pp. 441–444, 1961.

153. A. K. Walton and T. S. Moss, *Proc. Phys. Soc.* (GB) **78,** pp. 1393–1407, 1961

154. R. T. Baumel and S. E. Schnatterly, *J. Opt. Soc. Am.* **61,** pp. 832–833, 1971.

155. B. N. Grechushnikov and I. P. Petrov, *Opt. Spectrosc.* (USSR) **14,** pp. 160–161, 1963.

156. J. T. Cox and G. Hass, *Appl. Opt.* **17,** pp. 1657–1658, 1978.

157. T. F. Thonn and R. M. A. Azzam, *Opt. Eng.* **24,** pp. 202–206, 1985.

158. W. R. Hunter, *Jap. J. Appl. Phys.* **4,** suppl. 1, p. 520, 1965; (*Proc. Conf. Photogr. Spectrosc. Opt.,* 1964).

159. H. Damany, *Opt. Acta* **12**, pp. 95–107, 1965.

160. G. Stephan, J.-C. Lemonnier, Y. LeCalvez, and S. Robin, *C. R. Acad. Sci.* **262B**, pp. 1272–1275, 1966.

161. F. de Chelle and H. Merdy, *C. R. Acad. Sci.* **265B**, pp. 968–971, 1967.

162. T. J. McIlrath, *J. Opt. Soc. Am.* **58**, pp. 506–510, 1968.

163. T. T. Cole and F. Oppenheimer, *Appl. Opt.* **1**, pp. 709–710, 1962.

164. E. Uzan, H. Damany, and J. Romand, *C. R. Acad. Sci.* **260**, pp. 5735–5737, 1965.

165. K. Rabinovitch, L. R. Canfield, and R. P. Madden, *Appl. Opt.* **4**, pp. 1005–1010, 1965.

166. T. Sasaki and H. Fukutani, *Jap. J. Appl. Phys.* **3**, pp. 125–126, 1964.

167. H. Winter, H. H. Bukow, and P. H. Heckmann, *Opt. Commun.* **11**, pp. 299–300, 1974.

168. G. Hass and W. R. Hunter, *Appl. Opt.* **17**, pp. 76–82, 1978.

169. R. S. Spencer, G. J. Bergen, C. M. Fleetwood, H. Herzig, L. Miner, S. H. Rice, E. Smigocki, B. E. Woodgate, and J. J. Zaniewski, *Opt. Eng.* **24**, pp. 548–554, 1985.

170. W. R. Hunter, *Appl. Opt.* **17**, pp. 1259–1270, 1978.

171. A. Ejiri, *J. Phys. Soc. Jap.* **23**, p. 901, 1967.

172. G. Rosenbaum, B. Feuerbacher, R. P. Godwin, and M. Skibowski, *Appl. Opt.* **7**, pp. 1917–1920, 1968.

173. N. Rehfeld, U. Gerhardt, and E. Dietz, *Appl. Phys.* **1**, pp. 229–232, 1973.

174. H. A. Van Hoof, *Appl. Opt.* **19**, pp. 189–190, 1980.

175. R. Hibst and H. H. Bukow, *Appl. Opt.* **28**, pp. 1806–1812, 1989.

176. M. A. Khakoo, P. Hammond, and J. W. McConkey, *Appl. Opt.* **26**, pp. 3492–3494, 1987.

177. M. B. Robin, N. A. Kuebler, and Y.-H. Pao, *Rev. Sci. Instrum.* **37**, pp. 922–924, 1966.

178. A. Matsui and W. C. Walker, *J. Opt. Soc. Am.* **60**, pp. 64–65, 1970.

179. F. Abelès, *C. R. Acad. Sci.* **230**, pp. 1942–1943, 1950.

180. L. G. Schulz and F. R. Tangherlini, *J. Opt. Soc. Am.* **44**, pp. 362–368, 1954.

181. D. K. Burge and H. E. Bennett, *J. Opt. Soc. Am.* **54**, pp. 1428–1433, 1964.

182. G. K. T. Conn and G. K. Eaton, *J. Opt. Soc. Am.* **44**, pp. 553–557, 1954.

183. M. F. de la Provostaye and P. Desains, *Ann. Chim. Phys.,* ser. 3, **30**, p. 158, 1850.

184. G. R. Bird and W. A. Shurcliff, *J. Opt. Soc. Am.* **49**, pp. 235–237, 1959.

185. A. Elliott and E. J. Ambrose, *Nature* **159**, pp. 641–642, 1947.

186. A. Elliott, E. J. Ambrose, and R. Temple, *J. Opt. Soc. Am.* **38**, pp. 212–216, 1948.

187. J. Ames and A. M. D. Sampson, *J. Sci. Instrum.* **26**, p. 132, 1949.

188. R. Duverney, *J. Phys. Radium,* ser. 8, **20**, suppl. 7, p. 66A, 1959.

189. J. H. Hertz, *Exper. Tech. der Phys.* **7**, pp. 277–280, 1959.

190. K. Buijs, *Appl. Spectrosc.* **14**, pp. 81–82, 1960.

191. E. M. Bradbury and A. Elliott, *J. Sci. Instrum.* **39**, p. 390, 1962.

192. R. G. Greenler, K. W. Adolph, and G. M. Emmons, *Appl. Opt.* **5**, pp. 1468–1469, 1966.

193. N. Wright, *J. Opt. Soc. Am.* **38**, pp. 69–70, 1948.

194. A. S. Makas and W. A. Shurcliff, *J. Opt. Soc. Am.* **45**, pp. 998–999, 1955.

195. H. E. Bennett, J. M. Bennett, and M. R. Nagel, *J. Opt. Soc. Am.* **51**, p. 237, 1961.

196. R. T. Lagemann and T. G. Miller, *J. Opt. Soc. Am.* **41**, pp. 1063–1064, 1951.

197. L. Huldt and T. Staflin, *Opt. Acta* **6**, pp. 27–36, 1959.

198. T. A. Leonard, J. Loomis, K. G. Harding, and M. Scott, *Opt. Eng.* **21**, pp. 971–975, 1982.

199. R. Meier and H. H. Günthard, *J. Opt. Soc. Am.* **49**, pp. 1122–1123, 1959.

200. N. J. Harrick, *J. Opt. Soc. Am.* **54**, pp. 1281–1282, 1964.

201. N. P. Murarka and K. Wilner, *Appl. Opt.* **20,** pp. 3275–3276, 1981.

202. S. D. Smith, T. S. Moss, and K. W. Taylor, *J. Phys. Chem. Solids* **11,** pp. 131–139, 1959.

203. D. T. Rampton and R. W. Grow, *Appl. Opt.* **15,** pp. 1034–1036, 1976.

204. J.-M. Munier, J. Claudel, E. Décamps, and A. Hadni, *Rev. Opt.* **41,** pp. 245–253, 1962.

205. A. K. Walton, T. S. Moss, and B. Ellis, *J. Sci. Instrum.* **41,** pp. 687–688, 1964.

206. W. C. Walker, *Appl. Opt.* **3,** pp. 1457–1459, 1964.

207. D. C. Hinson, *J. Opt. Soc. Am.* **56,** p. 408, 1966.

208. D. F. Heath, *Appl. Opt.* **7,** pp. 455–459, 1968.

209. J. Schellman, V. Chandrasekharan, and H. Damany, *C. R. Acad. Sci.* **259,** pp. 4560–4563, 1964.

210. G. Weiser, *Proc. IEEE* **52,** p. 966, 1964.

211. J. L. Weinberg, *Appl. Opt.* **3,** pp. 1057–1061, 1964.

212. J. M. Bennett, D. L. Decker, and E. J. Ashley, *J. Opt. Soc. Am.* **60,** p. 1577A, 1970.

213. H. Schröder, *Optik* **3,** pp. 499–503, 1948.

214. F. Abelès, *J. Phys. Radium.* ser. 8, **11,** pp. 403–406, 1950.

215. K. Kubo, *J. Sci. Res. Instrum.* (Tokyo Inst. Phys. Chem. Res.) **47,** pp. 1–6, 1953.

216. H. Schopper, *Optik* **10,** pp. 426–438, 1953.

217. M. Ruiz-Urbieta and E. M. Sparrow, *J. Opt. Soc. Am.* **62,** pp. 1188–1194, 1972.

218. M. Ruiz-Urbieta and E. M. Sparrow, *J. Opt. Soc. Am.* **63,** pp. 194–200, 1973.

219. M. Ruiz-Urbieta, E. M. Sparrow, and G. W. Goldman, *Appl. Opt.* **12,** pp. 590–596, 1973.

220. F. Abelès, *Jap. J. Appl. Phys.* **4,** suppl. 1, p. 517, 1965; (Proc. Conf. Photogr. Spectrosc. Opt., 1964).

221. A.-R. M. Zaghloul and R. M. A. Azzam, *Appl. Opt.* **16,** pp. 1488–1489, 1977.

222. W. W. Buchman, S. J. Holmes, and F. J. Woodberry, *J. Opt. Soc. Am.* **61,** pp. 1604–1606, 1971.

223. W. W. Buchman, *Appl. Opt.* **14,** pp. 1220–1224, 1975.

224. S. Refermat and J. Eastman, *Proc. Soc. Photo-Opt. Instrum. Eng.* **88,** pp. 28–33, 1976.

225. L. Songer, *Optical Spectra* **12** (10), pp. 49–50, October 1978.

226. D. Blanc, P. H. Lissberger, and A. Roy, *Thin Solid Films* **57,** pp. 191–198, 1979.

227. T. Maehara, H. Kimura, H. Nomura, M. Yanagihara, and T. Namioka, *Appl. Opt.* **30,** pp. 5018–5020, 1991.

228. P. B. Clapham, M. J. Downs, and R. J. King, *Appl. Opt.* **8,** pp. 1965–1974, 1969. [See also P. B. Clapham, *Thin Solid Films* **4,** pp. 291–305, 1969].

229. S. M. MacNeille, U.S. Patent 2,403,731, July 9, 1946.

230. M. Banning, *J. Opt. Soc. Am.* **37,** pp. 792–797, 1947.

231. J. A. Dobrowolski and A. Waldorf, *Appl. Opt.* **20,** pp. 111–116, 1981.

232. J. C. Monga, P. D. Gupta, and D. D. Bhawalkar, *Appl. Opt.* **23,** pp. 3538–3540, 1984.

233. J. Mouchart, J. Begel, and E. Duda, *Appl. Opt.* **28,** pp. 2847–2853, 1989.

234. D. Lees and P. Baumeister, *Opt. Lett.* **4,** pp. 66–67, 1979.

235. R. M. A. Azzam, *Opt. Lett.* **10,** pp. 110–112, 1985.

236. R. M. A. Azzam, *Appl. Opt.* **25,** pp. 4225–4227, 1986.

237. H. Schröder and R. Schläfer, *Z. Naturforsch.* **4a,** pp. 576–577, 1949.

238. G. Pridatko and T. Krylova, *Opt.-Mekh. Prom.* **3,** p. 23, 1958.

239. R. S. Sokolova and T. N. Krylova, *Opt. Spectrosc.* (USSR) **14,** pp. 213–215, 1963.

240. A. F. Turner and P. W. Baumeister, *Appl. Opt.* **5,** pp. 69–76, 1966.

241. P. B. Clapham, *Opt. Acta* **18,** pp. 563–575, 1971.

242. H. Schröder, *Optik* **13,** pp. 158–168, 169–174, 1956.

243. P. Baumeister, *Opt. Acta* **8,** pp. 105–119, 1961.

244. P. Baumeister, Institute of Optics, Univ. Rochester, Rochester, N.Y., private communication, 1971.

245. V. R. Costich, *Appl. Opt.* **9,** pp. 866–870, 1970.

246. C. D. West and A. S. Makas, *J. Opt. Soc. Am.* **39,** pp. 791–794, 1949.

247. S. D. Jacobs, Y. Asahara, and T. Izumitani, *Appl. Opt.* **21,** pp. 4526–4532, 1982.

248. B. H. Billings, in D. E. Gray, (ed.), *American Institute of Physics Handbook,* 3d ed., McGraw-Hill, New York, 1972, pp. 6-37,6-40, 6-46, 6-112, and 6-113.

249. J. M. Beckers, *Appl. Opt.* **10,** pp. 973–975, 1971.

250. W. L. Bond, *J. Appl. Phys.* **36,** pp. 1674–1677, 1965.

251. J. H. Shields and J. W. Ellis, *J. Opt. Soc. Am.* **46,** pp. 263–265, 1956.

252. E. D. Palik and B. W. Henvis, *Appl. Opt.* **6,** pp. 2198–2199, 1967.

253. G. D. Boyd, W. L. Bond, and H. L. Carter, *J. Appl. Phys.* **38,** pp. 1941–1943, 1967.

254. M. A. Jeppesen, *J. Opt. Soc. Am.* **48,** pp. 629–632, 1958.

255. S. S. Ballard, J. S. Browder, and J. F. Ebersole, in D. E. Gray (ed.), *American Institute of Physics Handbook,* 3d ed., McGraw-Hill, New York, 1972, pp. 6-20, 6-27, and 6-35.

256. A. E. Ennos and K. W. Opperman, *Appl. Opt.* **5,** p. 170, 1966.

257. E. D. Palik, *Appl. Opt.* **7,** pp. 978–979, 1968.

258. H. Gobrecht and A. Bartschat, *Z. Phys.* **156,** pp. 131–143, 1959.

258a. M. S. Shumate, *Appl. Opt.* **5,** pp. 327–332, 1966.

259. E. V. Loewenstein, *J. Opt. Soc. Am.* **51,** pp. 108–112, 1961.

260. V. Chandrasekharan and H. Damany, *Appl. Opt.* **7,** pp. 939–941, 1968.

261. J.-P. Maillard, *Opt. Commun.* **4,** pp. 175–177, 1971.

262. T. M. Bieniewski and S. J. Czyzak, *J. Opt. Soc. Am.* **53,** pp. 496–497, 1963.

263. M. Françon, S. Mallick, and J. Vulmière, *J. Opt. Soc. Am.* **55,** p. 1553, 1965.

264. F. Zernicke, Jr., *J. Opt. Soc. Am.* **54,** pp. 1215–1220, 1964, [erratum in *J. Opt. Soc. Am.* **55,** pp. 210–211, 1965].

265. E. Einsporn, *Phys. Z.* **37,** pp. 83–88, 1936.

266. E. L. Gieszelmann, S. F. Jacobs, and H. E. Morrow, *J. Opt. Soc. Am.* **59,** pp. 1381–1383, 1969 [erratum in *J. Opt. Soc. Am.* **60,** p. 705, 1970].

267. J. M. Daniels, *Rev. Sci. Instrum.* **38,** pp. 284–285, 1967.

268. T. Saito, A. Ejiri, and H. Onuki, *Appl. Opt.* **29,** pp. 4538–4540, 1990.

269. W. B. Westerveld, K. Becker, P. W. Zetner, J. J. Corr, and J. W. McConkey, *Appl. Opt.* **14,** pp. 2256–2262, 1985.

270. P. L. Richards and G. E. Smith, *Rev. Sci. Instrum.* **35,** pp. 1535–1537, 1964.

271. L. H. Johnston, *Appl. Opt.* **16,** pp. 1082–1084, 1977.

272. D. P. Gonatas, X. D. Wu, G. Novak, and R. H. Hildebrand, *Appl. Opt.* **28,** pp. 1000–1006, 1989.

273. P. Lostis, *J. Phys. Radium,* ser. 8, **18,** p. 51S, 1957.

274. C. E. Greninger, *Appl. Opt.* **27,** pp. 774–776, 1988.

275. J. M. Bennett, *Appl. Opt.* **9,** pp. 2123–2129, 1970.

276. S. Nakadate, *Appl. Opt.* **29,** pp. 242–246, 1990.

277. P. D. Hale and G. W. Day, *Appl. Opt.* **27,** pp. 5146–5153, 1988.

278. A.-R. M. Zaghloul, R. M. A. Azzam, and N. M. Bashara, *Opt. Commun.* **14,** pp. 260–262, 1975.

279. A.-R. M. Zaghloul, R. M. A. Azzam, and N. M. Bashara, *J. Opt. Soc. Am.* **65,** pp. 1043–1049, 1975.

280. S. Kawabata and M. Suzuki, *Appl. Opt.* **19,** pp. 484–485, 1980.

281. S. Chu, R. Conti, P. Bucksbaum, and E. Commins, *Appl. Opt.* **18,** pp. 1138–1139, 1979.

282. J. Strong, *Procedures in Experimental Physics,* Prentice-Hall, Englewood Cliffs, N.J., 1938, pp. 388–389.

283. P. H. Smith, *Proc. Symp. Recent Dev. Ellipsometry, Surf. Sci.* **16,** pp. 34–66, 1969.

284. B. N. Grechushnikov, *Opt. Spectrosc.* (USSR) **12,** p. 69, 1962.

285. D. G. Drummond, *Proc. Roy. Soc.* (Lond.) **153A,** pp. 318–339, 1936.

286. G. Holzwarth, *Rev. Sci. Instrum.* **36,** pp. 59–63, 1965.

287. E. D. Palik, *Appl. Opt.* **2,** pp. 527–539, 1963.

288. J. H. Jaffe, H. Jaffe, and K. Rosenbeck, *Rev. Sci. Instrum.* **38,** pp. 935–938, 1967.

289. C. Gaudefroy, *C. R. Acad. Sci.* **189,** pp. 1289–1291, 1929.

290. W. A. Deer, R. A. Howie, and J. Zussman, *Rock-forming Minerals,* vol. 3; *Sheet Silicates,* Wiley, New York, 1962, pp. 258–262.

291. A. E. Ennos, *J. Sci. Instrum.* **40,** pp. 316–317, 1963.

292. S. B. Ioffe and T. A. Smirnova, *Opt. Spectrosc.* (USSR) **16,** pp. 484–485, 1964.

293. S. Mitchell, *Nature* **212,** pp. 65–66, 1966.

294. A. M. Title, *Appl. Opt.* **14,** pp. 229–237, 1975.

295. R. Anderson, *Appl. Opt.* **27,** pp. 2746–2747, 1988.

296. P. L. Wizinowich, *Opt. Eng.* **28,** p. 157–159, 1989.

297. G. N. Ramachandran and S. Ramaseshan, "Crystal Optics," in S. Flügge (ed.), *Handbuch der Physik,* vol. 25/1, Springer, Berlin, 1961, pp. 156–158.

298. A. Johannsen, *Manual of Petrographic Methods,* 2d ed. Hafner, New York, 1968, pp. 369–385 (originally published in 1918).

299. H. G. Jerrard, *J. Opt. Soc. Am.* **38,** pp. 35–59, 1948.

300. R. P. Hunt, *Appl. Opt.* **9,** pp. 1220–1221, 1970.

301. J. R. Partington, *An Advanced Treatise on Physical Chemistry,* vol. 4, Wiley, New York, 1953, pp. 173–177.

302. J. Ellerbroek and J. T. Groosmuller, *Phys. Z.* **27,** pp. 468–471, 1926.

303. H. G. Jerrard, *J. Opt. Soc. Am.* **39,** pp. 1031–1035, 1949.

304. H. G. Jerrard, *J. Sci. Instrum.* **26,** pp. 353–357, 1949.

305. H. G. Jerrard, *J. Sci. Instrum.* **27,** pp. 62–66, 1950.

306. J. F. Archard, *J. Sci. Instrum.* **27,** pp. 238–241, 1950.

307. R. W. Wood, *Physical Optics,* 3d ed., Macmillan, New York, 1934, pp. 356–361.

308. R. W. Ditchburn, *Light,* 2d ed., Interscience, New York, 1963b, pp. 483–485.

309. H. G. Jerrard, *J. Sci. Instrum.* **27,** pp. 164–167, 1950.

310. H. G. Jerrard, *J. Sci. Instrum.* **28,** pp. 10–14, 1951.

311. H. G. Jerrard, *J. Sci. Instrum.* **30,** pp. 65–70, 1953.

312. H. E. Ives and H. B. Briggs, *J. Opt. Soc. Am.* **26,** pp. 238–246, 1936.

313. E. D. Palik, *Appl. Opt.* **4,** pp. 1017–1021, 1965.

314. P. Hariharan and D. Sen, *J. Sci. Instrum.* **37,** pp. 278–281, 1960.

315. M. P. Varnham, D. N. Payne, A. J. Barlow, and E. J. Tarbox, *Opt. Lett.* **9,** pp. 306–308, 1984.

316. C. A. Villarruel, M. Abebe, W. K. Burns, and R. P. Moeller, in *Digest of the Seventh Topical Conference on Optical Fiber Communication,* vol. 84.1, Optical Society of America, Washington, D.C., 1984.

317. R. A. Bergh, H. C. Lefevre, and H. J. Shaw, *Opt. Lett.* **5,** pp. 479–481, 1980.

318. W. Eickhoff, *Electron. Lett.* **16,** pp. 762–763, 1980.

319. D. Gruchmann, K. Petermann, L. Satandigel, and E. Weidel, in *Proceedings of the European Conference on Optical Communication,* North Holland, Amsterdam, 1983, pp. 305–308.

320. T. Hosaka, K. Okamoto, and T. Edahiro, *Opt. Lett.* **8,** pp. 124–126, 1983.

321. J. R. Feth and C. L. Chang, *Opt. Lett.* **11,** pp. 386–388, 1986.

322. T. Yu and Y. Wu, *Opt. Lett.* **13,** pp. 832–834, 1988.

323. R. B. Dyott, J. Bello, and V. A. Handerek, *Opt. Lett.* **12,** pp. 287–289, 1987.

324. S. C. Lee and J.-I. Chen, *Appl. Opt.* **29,** pp. 2667–2668, 1990.

325. R. Ulrich and M. Johnson, *Appl. Opt.* **18,** pp. 1857–1861, 1979.

326. T. Hosaka, K. Okamoto, and T. Edahiro, *Appl. Opt.* **22,** pp. 3850–3858, 1983.

327. A. W. Snyder and A. J. Stevenson, *Opt. Lett.* **11,** pp. 254–256, 1986.

328. S. Uehara, T. Izawa, and H. Nakagome, *Appl. Opt.* **13,** pp. 1753–1754, 1974.

329. H. F. Mahlein, *Opt. Commun.* **16,** pp. 420–424, 1976.

330. P. G. Suchoski, T. K. Findakly, and F. J. Leonberger, *Opt. Lett.* **13,** pp. 172–174, 1988.

331. R. Noé, H. Heidrich, and D. Hoffmann, *Opt. Lett.* **13,** pp. 527–529, 1988.

332. K. Baba, K. Shiraishi, K. Obi, T. Kataoka, and S. Kawakami, *Appl. Opt.* **27,** pp. 2554–2560, 1988.

CHAPTER 4
NONDISPERSIVE PRISMS[1,2]

William L. Wolfe
Optical Sciences Center
University of Arizona
Tucson, Arizona

4.1 GLOSSARY

δ	angular deviation
ϕ	phase
ω	radian frequency of rotation
subscripts	
A, B, C, D, d	prism dimensions
t	time
x, y	rectangular components
α	angle
$1, 2$	prism number

4.2 INTRODUCTION

Prisms of various shapes and sizes are used for folding, inverting, reverting, displacing, and deviating a beam of light, whether it be collimated, converging, or diverging.

Prisms, rather than mirrors, are often used for the applications discussed here, since they make use of reflecting coatings at what amounts to an interior surface. The coatings can be protected on their backs by other means, and do not tarnish with age and exposure. Even better, some prisms do not need such coatings if the (internal) angle of incidence exceeds the critical angle.

In these applications, chromatism is to be avoided. Thus, the arrangements either make use of perpendicular incidence or compensating angles of incidence.

Almost all of these prisms are meant to be used with collimated beams. Most of the operations are somewhat equivalent to the use of plane parallel plates, which displace but do not deviate a collimated beam. However, such plates have both chromatism and spherical aberration in a convergent beam.

Dispersing prisms are discussed by Zissis and polarizing prisms by Bennett, both in this handbook.

4.3 INVERSION, REVERSION

A reverted image shifts the image left for right. An inverted image is upside down. A reinverted image or inverted-reverted image does both.

The best two references on this subject are the Frankford Arsenal book called *Design of Fire Control Optics*,[1] and Jacobs' book called *Optical Engineering*.[2] Many of the diagrams shown here have been taken from the former since they provide direct design information as well as descriptions of the performance of the prism.

4.4 DEVIATION, DISPLACEMENT

The beam, in addition to being inverted and/or reverted, can also be displaced and/or deviated. Displacement means that the beam has been translated in x or y, but it has not changed the direction in which it was traveling. Deviation indicates that the beam has been caused to change its direction. Deviation is measured in angular measure; displacement in linear measure. If a beam has been deviated, displacement is not important. If a beam has been displaced, it usually has not been deviated. Although the two can occur simultaneously, it seldom happens in optical prisms.

4.5 SUMMARY OF PRISM PROPERTIES

Table 1 is a listing of the prisms that are described in this section. The first column has the name of the prism. The second column indicates whether or not the image has been reverted; the second column, whether it has been inverted. The third column indicates the extent to which the prism displaces the beam. This value is given in terms of the dimension A that is indicated for each of the prisms. It is a characteristic, normalizing dimension.

The next column indicates the angular deviation of the beam in degrees. Some entries have two angles. This indicates that the beam is deviated in two directions. The first is the horizontal deviation; the second is the vertical deviation, h and v. These are the conventional planes. Of course, if the prism is rotated, the directions are reversed. The general deviation prisms have a range of deviation angles that can be obtained by reasonable changes in the prism angles. Thus, each of them is a representative design for a range of angles. The final column is for comments.

A prism that has neither deviation nor displacement is a *direct-vision* prism.

4.6 PRISM DESCRIPTIONS

Each diagram shows at least one view of the prism and a set of dimensions. The A dimension is a reference dimension. It is always 1.00, and the rest of the dimensions are related to it. The refractive index is almost always taken as 1.5170, a representative value for glass in the visible. Prism dimensions can change somewhat if the refractive index is modestly different from the chosen value. If, for instance, germanium is used, however, the prism might be drastically different.

TABLE 1 Summary of Prism Properties

Prism	Reverts	Inverts	Displaces	Deviates	Comments
Right angle	Yes	No	—	90	Simplest
	No	Yes			
Porro A	Yes	Yes	1.1A	0	Binocs
Porro B	Yes	Yes	1.1A	0	Binocs
Abbe A	Yes	Yes	0	0	
Abbe B	Yes	Yes	0	0	
Dove, double dove	Yes	No	0	0	Parallel light
Pcchan	Yes	No	0	0	Nonparallel
Amici	Yes	No	—	90	Roof
Schmidt	Yes	Yes	—	45	
Leman	Yes	Yes	3A	0	
Penta	No	No	—	90	exactly
Reversion	Yes	Yes	0	0	Nonparallel
Wollaston	No	No	—	90	tracing
Zeiss	Yes	Yes	Design		
Goerz	Yes	Yes	Design		
Frank 1	Yes	Yes	—	115	
Frank 2	Yes	Yes	—	60	
Frank 3	Yes	Yes	—	45v 90h	
Frank 4	Yes	No	—	45v 90h	
Frank 5	No	Yes	—	60v 90h	
Frank 6	Yes	Yes	—	60v 90h	
Frank 7	No	No	—	45v 90h	
Hastings	Yes	Yes	0	0	
Rhomboid	No	No	A	0	
Risleys	No	No	No	0–180	
Retro	Yes	Yes	No	180	
D40	No	No	—	40–50	
D60	No	No	—	50–60	
D90	No	No	—	80–100	
D120	No	No	—	110–130	

Right-angle Prism

Perhaps the simplest of the deviating prisms is the right-angle prism. Light enters one of the two perpendicular faces, as shown in Fig. 1*a*, reflects off the diagonal face, and emerges at 90° from the other perpendicular face. The beam has been rotated by 90°, and the image has been inverted. If the prism is used in the other orientation, shown in Fig. 1*b*, then the image is reverted. The internal angle of incidence is 45°, which is sufficient for total internal reflection (as long as the refractive index is greater than 1.42).

Porro Prism

A Porro prism, shown in Fig. 2, has a double reflection and may be considered to be two right-angle prisms together. They are often identical. Often, two Porro prisms are used together to invert and revert the image. The incidence angles are the same as with the right-angle prism, so that total internal reflection takes place with refractive indices larger than 1.42. It is a direct-vision prism.

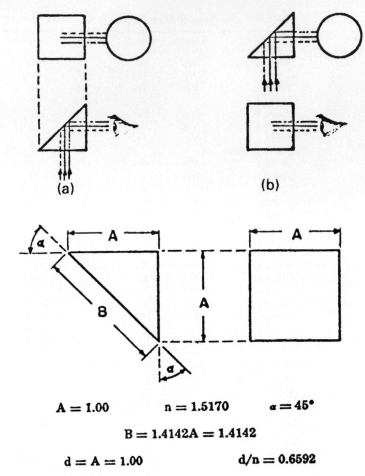

$$A = 1.00 \qquad n = 1.5170 \qquad \alpha = 45°$$

$$B = 1.4142A = 1.4142$$

$$d = A = 1.00 \qquad d/n = 0.6592$$

FIGURE 1 Right-angle prism.

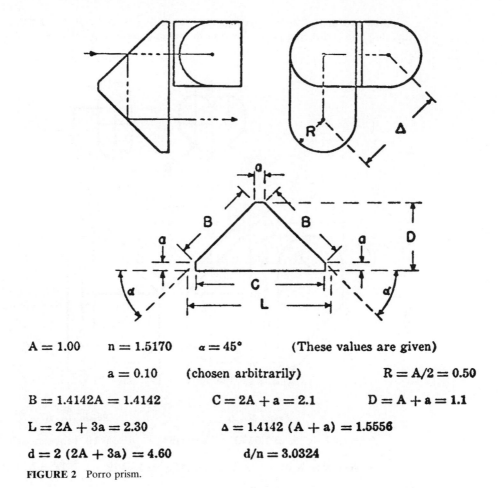

$$A = 1.00 \qquad n = 1.5170 \qquad \alpha = 45° \qquad \text{(These values are given)}$$

$$a = 0.10 \qquad \text{(chosen arbitrarily)} \qquad R = A/2 = 0.50$$

$$B = 1.4142A = 1.4142 \qquad C = 2A + a = 2.1 \qquad D = A + a = 1.1$$

$$L = 2A + 3a = 2.30 \qquad \Delta = 1.4142\,(A + a) = 1.5556$$

$$d = 2\,(2A + 3a) = 4.60 \qquad d/n = 3.0324$$

FIGURE 2 Porro prism.

Abbé's version of the Porro prism is shown in Fig. 3. The resultant beam is inverted and reverted and is directed parallel and in the same direction as the incident beam.

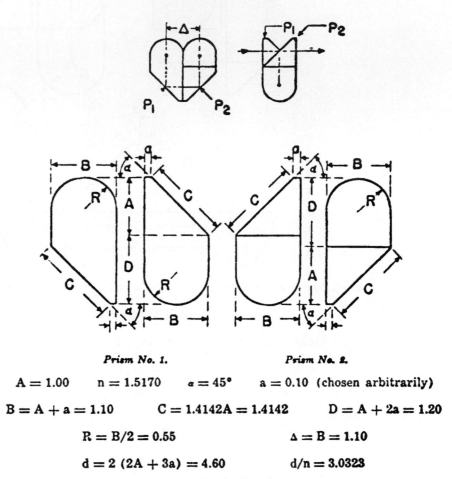

Prism No. 1.   Prism No. 2.

$A = 1.00$ $n = 1.5170$ $a = 45°$ $a = 0.10$ (chosen arbitrarily)

$B = A + a = 1.10$ $C = 1.4142A = 1.4142$ $D = A + 2a = 1.20$

$R = B/2 = 0.55$ $\Delta = B = 1.10$

$d = 2(2A + 3a) = 4.60$ $d/n = 3.0323$

FIGURE 3 Abbe modification of Porro prisms for binoculars.

Abbé's Prisms

Two versions of prisms invented by Abbé are shown. they both are direct-vision prisms that revert and invert the image. One version is symmetrical; the other uses three different prism segments. They are shown in Figs. 4 and 5.

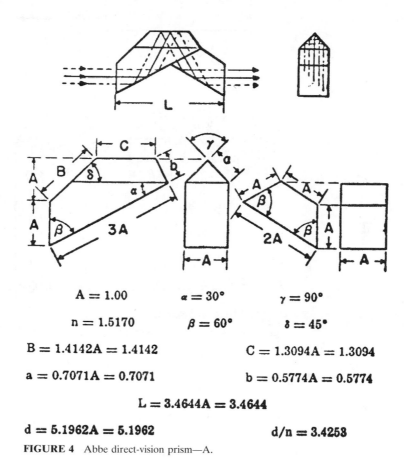

$$A = 1.00 \qquad \alpha = 30° \qquad \gamma = 90°$$

$$n = 1.5170 \qquad \beta = 60° \qquad \delta = 45°$$

$$B = 1.4142A = 1.4142 \qquad C = 1.3094A = 1.3094$$

$$a = 0.7071A = 0.7071 \qquad b = 0.5774A = 0.5774$$

$$L = 3.4644A = 3.4644$$

$$d = 5.1962A = 5.1962 \qquad d/n = 3.4253$$

FIGURE 4 Abbe direct-vision prism—A.

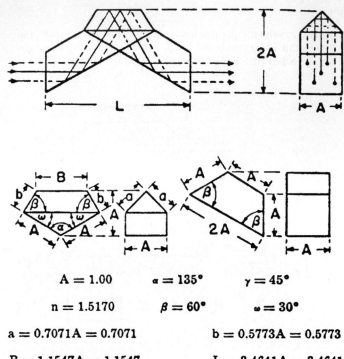

$$A = 1.00 \qquad \alpha = 135° \qquad \gamma = 45°$$

$$n = 1.5170 \qquad \beta = 60° \qquad \omega = 30°$$

$$a = 0.7071A = 0.7071 \qquad\qquad b = 0.5773A = 0.5773$$

$$B = 1.1547A = 1.1547 \qquad\qquad L = 3.4641A = 3.4641$$

$$d = 5.1962A = 5.1962 \qquad\qquad d/n = 3.4253$$

FIGURE 5 Abbe direct-vision prism—B.

Dove Prism

A Dove prism (also known as a Harting-Dove prism) does not deviate or displace an image but it can be used to either invert or revert an image. It must be placed in parallel light. Such a prism is shown in Fig. 6.

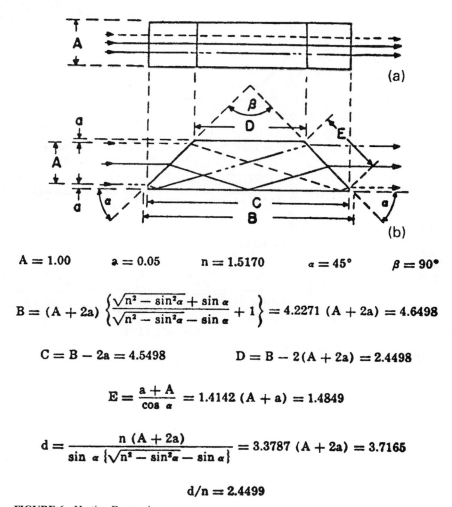

$$A = 1.00 \qquad a = 0.05 \qquad n = 1.5170 \qquad \alpha = 45° \qquad \beta = 90°$$

$$B = (A + 2a) \left\{ \frac{\sqrt{n^2 - \sin^2\alpha} + \sin\alpha}{\sqrt{n^2 - \sin^2\alpha} - \sin\alpha} + 1 \right\} = 4.2271 \, (A + 2a) = 4.6498$$

$$C = B - 2a = 4.5498 \qquad\qquad D = B - 2(A + 2a) = 2.4498$$

$$E = \frac{a + A}{\cos\alpha} = 1.4142 \, (A + a) = 1.4849$$

$$d = \frac{n\,(A + 2a)}{\sin\alpha \left\{ \sqrt{n^2 - \sin^2\alpha} - \sin\alpha \right\}} = 3.3787 \, (A + 2a) = 3.7165$$

$$d/n = 2.4499$$

FIGURE 6   Harting-Dove prism.

Double Dove

Two Dove prisms are glued together. The length is halved, but the height is doubled. It performs the same functions as a single Dove in almost the same way. It is shown in Fig. 7.

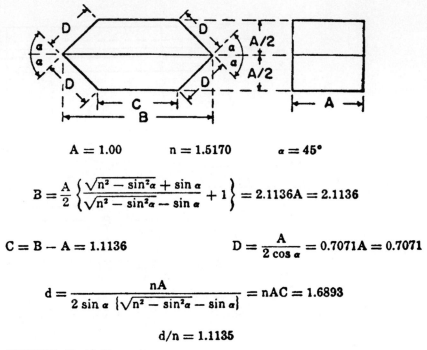

$$A = 1.00 \qquad n = 1.5170 \qquad \alpha = 45°$$

$$B = \frac{A}{2} \left\{ \frac{\sqrt{n^2 - \sin^2\alpha} + \sin\alpha}{\sqrt{n^2 - \sin^2\alpha} - \sin\alpha} + 1 \right\} = 2.1136A = 2.1136$$

$$C = B - A = 1.1136 \qquad\qquad D = \frac{A}{2\cos\alpha} = 0.7071A = 0.7071$$

$$d = \frac{nA}{2\sin\alpha \left\{ \sqrt{n^2 - \sin^2\alpha} - \sin\alpha \right\}} = nAC = 1.6893$$

$$d/n = 1.1135$$

FIGURE 7 Double Dove prism.

Pechan Prism

The Pechan prism (shown in Fig. 8) performs the same function as the Dove, but it can do it in converging or diverging beams. The surfaces marked B are silvered and protected. The surfaces bordering the air space are unsilvered.

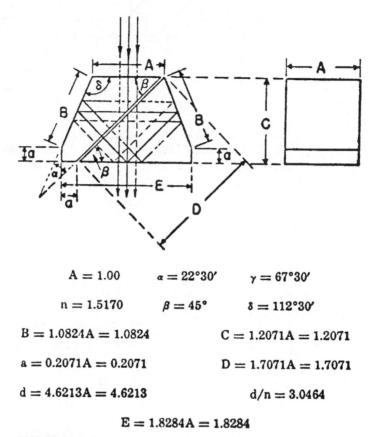

$$A = 1.00 \qquad \alpha = 22°30' \qquad \gamma = 67°30'$$

$$n = 1.5170 \qquad \beta = 45° \qquad \delta = 112°30'$$

$$B = 1.0824A = 1.0824 \qquad\qquad C = 1.2071A = 1.2071$$

$$a = 0.2071A = 0.2071 \qquad\qquad D = 1.7071A = 1.7071$$

$$d = 4.6213A = 4.6213 \qquad\qquad d/n = 3.0464$$

$$E = 1.8284A = 1.8284$$

FIGURE 8   Pechan prism.

Amici (Roof) Prism

This more complex arrangement of surfaces inverts the image, reverts it, and deviates it 90°. It is shown in Fig. 9. Since this prism makes use of the roof effect, it has the same angles as both the right-angle and Porro prisms, and exhibits total internal reflection for refractive indices larger than 1.42.

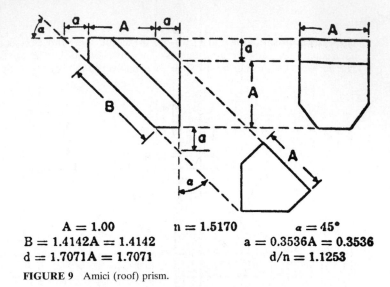

$$A = 1.00 \qquad n = 1.5170 \qquad \alpha = 45°$$
$$B = 1.4142A = 1.4142 \qquad a = 0.3536A = 0.3536$$
$$d = 1.7071A = 1.7071 \qquad d/n = 1.1253$$

FIGURE 9 Amici (roof) prism.

Schmidt Prism

The prism will invert and revert the image, and it will deviate it through 45°. It is shown in Fig. 10.

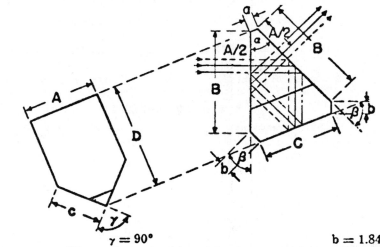

$A = 1.00$	$\gamma = 90°$	$b = 1.8478a = 0.1848$
$a = 0.10$ (chosen at will)	$B = 1.4142A + 0.5412a = 1.4683$	$c = 0.7071A = 0.7071$
$n = 1.5170$	$C = 1.0824A = 1.082$	$d = 3.4142A = 3.4142$
$\alpha = 45°$	$D = 1.4142A + 2.3890a = 1.6531$	$d/n = 2.2506$
$\beta = 67°30'$		

FIGURE 10 Schmidt prism.

Leman Prism

This rather strange looking device, shown in Fig. 11, reverts, inverts, and displaces by 3A, an image.

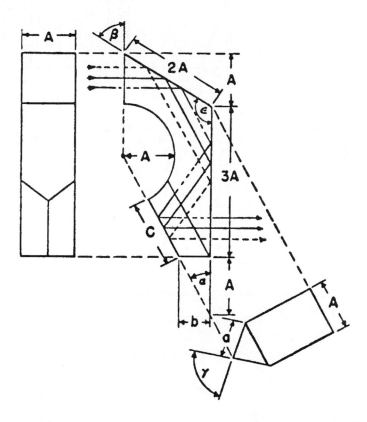

$$A = 1.00$$
$$n = 1.5170$$
$$\alpha = 30°$$
$$\beta = 60°$$
$$\gamma = 90°$$
$$\epsilon = 120°$$
$$d = 5.1962A = 5.1962$$

$$B = 1.7321A = 1.7321$$
$$a = 0.7071A = 0.7071$$
$$C = 1.3099A = 1.3099$$
$$b = 0.5774A = 0.5774$$
$$d/n = 3.4253$$

FIGURE 11 Leman prism.

Penta Prism

A penta prism has the remarkable property that it always deviates a beam by exactly 90° in the principal plane. This is akin to the operation of a cube corner. The two reflecting surfaces of the penta prism, shown in Fig. 12, must be reflectorized, as the angles are 22.5° and therefore require a refractive index of 2.62 or greater for total internal reflection. Some penta-prisms are equipped with a roof to revert the image.

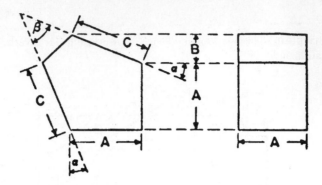

A = 1.00   n = 1.5170   α = 22°30′   β = 45°
    B = 0.4142A = 0.4142   C = 1.0824A = 1.0824
    a = 3.4142A = 3.4142   d/n = 2.2506

FIGURE 12   Penta prism.

Reversion Prism

This prism operates like an Abbé prism, type A, but does not require parallel light. It is shown in Fig. 13.

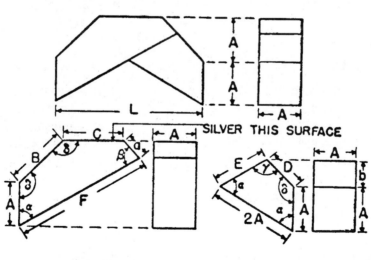

A = 1.00   C = 1.4641A   β = 75°   a = 0.5176A
B = 1.4142A   E = 1.2679A   γ = 105°   d/n = 3.4253A
D = 0.8966A   d = 5.1962A   δ = 135°   L = 3.4641A
n = 1.5170   α = 60°   F = 3.2679A   b = 0.6340A

FIGURE 13   Reversion prism.

Goerz Prism System

This is an alternate to the Zeiss system. It does the same things. It is shown in Fig. 16.

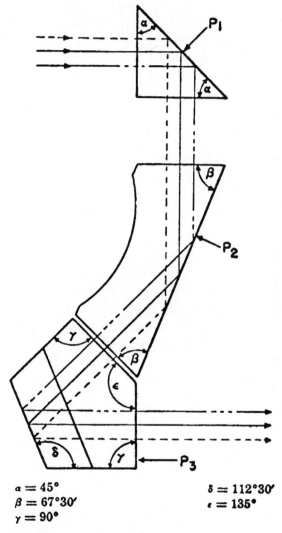

$\alpha = 45°$
$\beta = 67°30'$
$\gamma = 90°$

$\delta = 112°30'$
$\epsilon = 135°$

FIGURE 16   C. P. Goerz prism system.

Frankford Arsenal 1

This prism, shown in Fig. 17, reverts, inverts, and deviates through 115°.

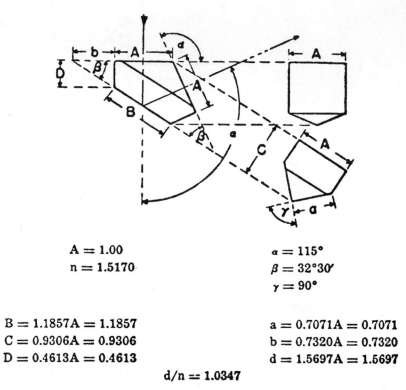

$$A = 1.00$$
$$n = 1.5170$$

$$\alpha = 115°$$
$$\beta = 32°30'$$
$$\gamma = 90°$$

$$B = 1.1857A = 1.1857$$
$$C = 0.9306A = 0.9306$$
$$D = 0.4613A = 0.4613$$

$$a = 0.7071A = 0.7071$$
$$b = 0.7320A = 0.7320$$
$$d = 1.5697A = 1.5697$$

$$d/n = 1.0347$$

FIGURE 17   Frankford Arsenal prism 1.

Frankford Arsenal 2

This prism reverts, inverts, and deviates through 60°. It is shown in Fig. 18.

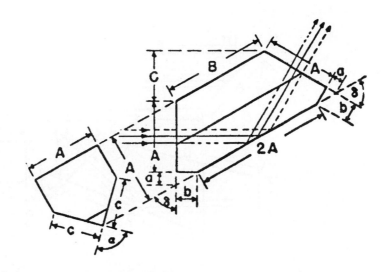

$A = 1.00$ $n = 1.5170$ $\delta = 60°$ $\alpha = 90°$

$a = 0.1547A = 0.1547$ $b = 0.2680A = 0.2680$

$B = 1.4641A = 1.4641$ $C = 0.7321 = 0.7321$

$d = 2.2690A = 2.2680$ $d/n = 1.4951$

FIGURE 18 Frankford Arsenal prism 2.

Frankford Arsenal 3

This prism reverts, inverts, and deviates through an angle of 45° upward and 90° horizontally. It is shown in Fig. 19.

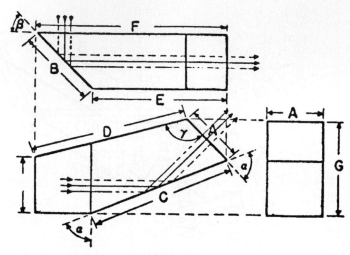

A = 1.00	n = 1.5170	α = 67°30′	β = 45°	γ = 120°21′40″

$$B = 1.4142A = 1.4142 \qquad\qquad C = 2.6131A = 2.6131$$

$$D = 2.7979A = 2.7979 \qquad\qquad E = 2.4142A = 2.4142$$

$$F = 3.4142A = 3.4142 \qquad\qquad G = 1.7071A = 1.7071$$

$$d = 3.4142A = 3.4142 \qquad\qquad d/n = 2.2506$$

FIGURE 19 Frankford Arsenal prism 3.

Frankford Arsenal 4

This prism reverts the image and deviates it 45° upward and 90° sidewards, like Frankford Arsenal 3. It is shown in Fig. 20.

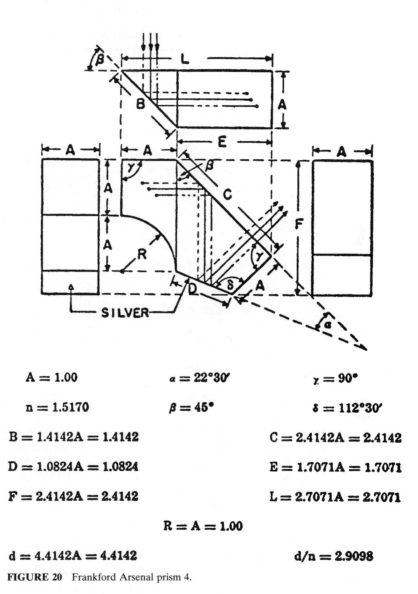

A = 1.00	$a = 22°30'$	$\gamma = 90°$
n = 1.5170	$\beta = 45°$	$\delta = 112°30'$
B = 1.4142A = 1.4142		C = 2.4142A = 2.4142
D = 1.0824A = 1.0824		E = 1.7071A = 1.7071
F = 2.4142A = 2.4142		L = 2.7071A = 2.7071

$$R = A = 1.00$$

d = 4.4142A = 4.4142   d/n = 2.9098

FIGURE 20 Frankford Arsenal prism 4.

Frankford Arsenal 5

This prism inverts the image while deviating it 90° sideways and 60° upwards. It is shown in Fig. 21.

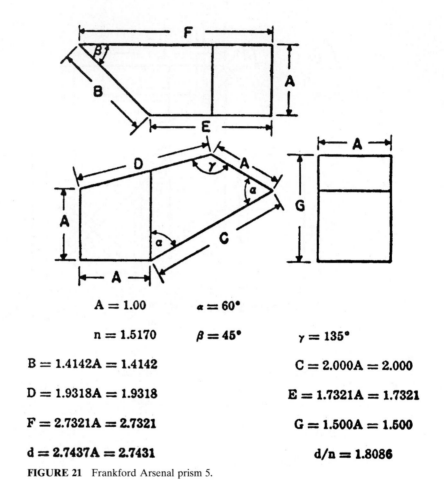

A = 1.00	α = 60°	
n = 1.5170	β = 45°	γ = 135°
B = 1.4142A = 1.4142		C = 2.000A = 2.000
D = 1.9318A = 1.9318		E = 1.7321A = 1.7321
F = 2.7321A = 2.7321		G = 1.500A = 1.500
d = 2.7437A = 2.7431		d/n = 1.8086

FIGURE 21   Frankford Arsenal prism 5.

Frankford Arsenal 6

This prism inverts, reverts, and deviates 90° horizontally and 60° vertically. It is shown in Fig. 22.

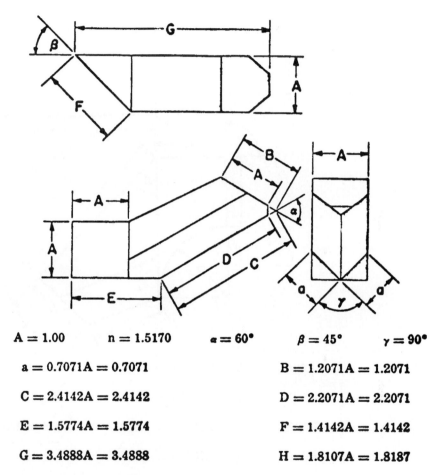

A = 1.00	n = 1.5170	α = 60°	β = 45°	γ = 90°

$$a = 0.7071A = 0.7071 \qquad\qquad B = 1.2071A = 1.2071$$

$$C = 2.4142A = 2.4142 \qquad\qquad D = 2.2071A = 2.2071$$

$$E = 1.5774A = 1.5774 \qquad\qquad F = 1.4142A = 1.4142$$

$$G = 3.4888A = 3.4888 \qquad\qquad H = 1.8107A = 1.8187$$

$$d = 3.6681A = 3.6681 \qquad\qquad d/n = 2.4180$$

FIGURE 22 Frankford Arsenal prism 6.

Frankford Arsenal 7

This prism neither reverts nor inverts, but deviates 90° horizontally and 45° vertically. It is shown in Fig. 23.

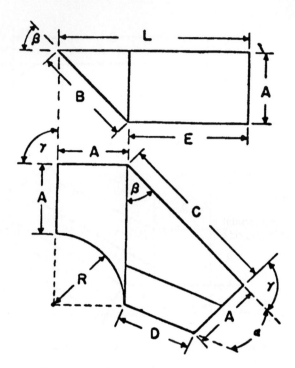

A = 1.00	**B = 1.4142A = 1.4142**	**R = A = 1.00**
n = 1.5170	**C = 2.4142A = 2.4142**	**d = 4.4142A = 4.4142**
α = 22°30′	**D = 1.0824A = 1.0824**	**d/n = 2.9098**
β = 45°	**E = 1.7071A = 1.7071**	
γ = 90°	**L = 2.7071A = 2.7071**	

FIGURE 23 Frankford Arsenal prism 7.

Brashear-Hastings Prism

This device, shown in Fig. 24, inverts an image without changing the direction of the beam. Since this is a relatively complicated optical element, it does not see much use.

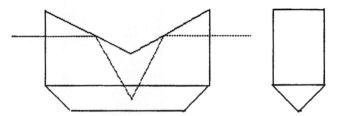

FIGURE 24   Brashear-Hastings prism.

Rhomboidal Prism

A rhomboidal prism, as shown in Fig. 25, displaces the beam without inverting, reverting, deviating, or otherwise changing things. The reflecting analog is a pair of mirrors at 45°.

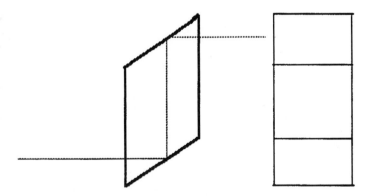

FIGURE 25   Rhomboidal prism.

Risley Prisms

Risley prisms are used in two ways. If they are slightly absorbing, they can be used as variable attenuators by translating one with respect to the other perpendicular to their apexes.[3] They can also be rotated to generate a variety of angular deviations.[4] A single prism deviates the beam according to its wedge angle and refractive index. If rotated in a circle about an axis perpendicular to its face, it will rotate the beam in a similar circle. A second, identical prism in series with it, as shown in Fig. 26, can double the angle of the beam rotation and generate a circle of twice the radius. If they rotate in opposite directions, one motion is canceled and a line is generated. In fact, all sorts of Lissajous-type figures can be obtained; some are shown in Fig. 27. The equations that

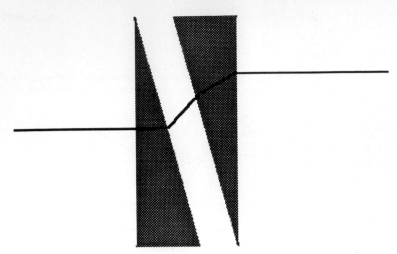

FIGURE 26 Risley prisms.

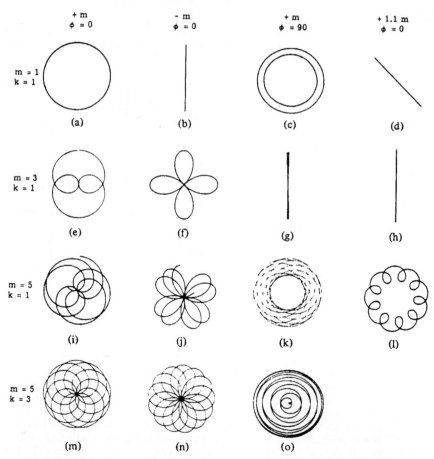

FIGURE 27 Risley prism patterns.

govern the patterns are

$$\delta_x = \delta_1 \cos \omega_1 t + \delta_2 \cos (\omega_2 t + \phi) \tag{1}$$

$$\delta_y = \delta_1 \sin \omega_1 t + \delta_2 \sin (\omega_2 t + \phi) \tag{2}$$

where δ_x and δ_y are the beam deviations, δ_1 and δ_2 are the individual prism deviations, ω is the rotation rate, t is time, and ϕ is the phase of the prism position. For relatively monochromatic applications, the prisms can be "Fresnelled," as shown in Fig. 28, and the mirror analogs, shown in Fig. 29, can also be used.

FIGURE 28   Fresnel Risleys.

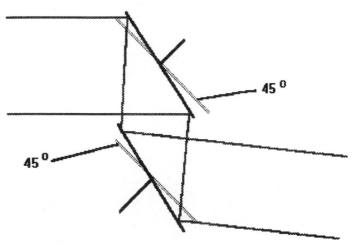

FIGURE 29   Risely mirrors.

Retroreflectors

The familiar reflective cube corner (not corner cube), that sends a ray back in the direction from which it came, has its refractive analog, as shown in Fig. 30. The angles are so that total internal reflection occurs. The angular acceptance range can be large.

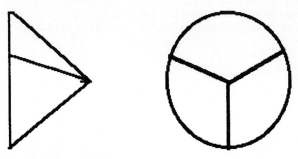

FIGURE 30 Retroreflectors.

General Deviation Prisms

Figure 31 shows a 60°-deviation prism. Other angles are obtainable with appropriate changes in the prism manufacture, as shown for example in Figs. 32 and 33.

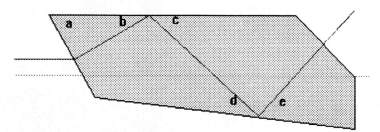

FIGURE 31 40°-deviation prism—D40.

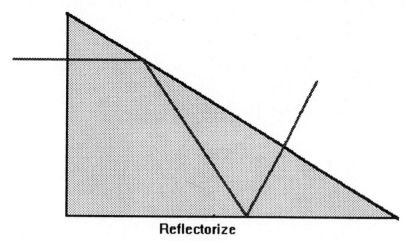

FIGURE 32   60°-deviation prism—D60.

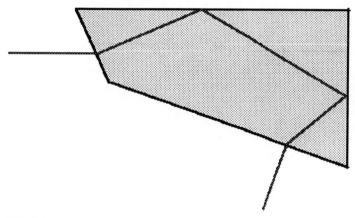

FIGURE 33   120°-deviation prism—D120.

4.7 REFERENCES

1. Frankford Arsenal, *Design of Fire Control Optics,* 1952, U.S. Government Printing Office.
2. D. H. Jacobs, *Fundamentals of Optical Engineering,* McGraw-Hill, 1943.
3. F. A. Jenkins and H. E. White, *Fundamental Optics,* 3d ed., McGraw-Hill, 1957.
4. W. L. Wolfe and G. J. Zissis, *The Infrared Handbook,* U.S. Government Printing Office, 1978.

CHAPTER 5
DISPERSIVE PRISMS AND GRATINGS

George J. Zissis
Environmental Research Institute of Michigan
Ann Arbor, Michigan

5.1 GLOSSARY

A_p prism angle

B prism base

D_p angle of minimum derivation

d grating constant

E irradiance

N number of slits

n refractive index

p order number

RP resolving power

r angles

W prism width

β angle

γ angle

5.2 INTRODUCTION

Spectroradiometers (Fig. 1) are radiometers designed specifically to allow determination of the wavelength distribution of radiation. This category of measurement systems usually consists of those in which separation of the radiation into its spectral components, or *dispersion,* is accomplished by the use of an optical element possessing a known functional dependence on wavelength—specifically, prisms and diffraction gratings. (Interferometers can also provide spectral dispersion as is discussed in the chapter on Interferometer Instruments by P. Hariharan.)

5.3 PRISMS[1,2,3]

The wavelength dependence of the index of refraction is used in prism spectrometers. Such an optical element disperses parallel rays or collimated radiation into different angles from

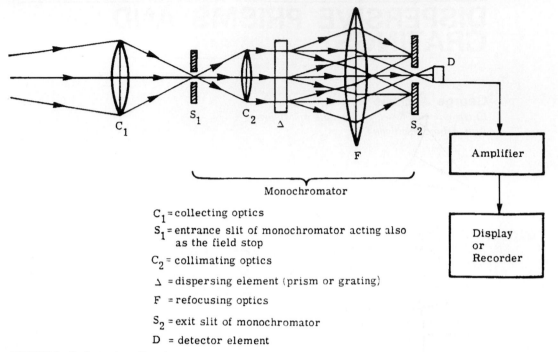

C_1 = collecting optics

S_1 = entrance slit of monochromator acting also
as the field stop

C_2 = collimating optics

$\triangle$ = dispersing element (prism or grating)

F = refocusing optics

S_2 = exit slit of monochromator

D = detector element

FIGURE 1   Basic spectroradiometer.

the prism according to wavelength. Distortion of the image of the entrance slit is minimized by the use of plane wave illumination. Even with plate wave illumination, the image of the slit is curved because not all of the rays from the entrance slit can traverse the prism in its principal plane. A prism is shown in the position of minimum angular deviation of the incoming rays in Fig. 2. At minimum angular deviation, maximum power can pass through the prism. For a prism adjusted to the position of minimum deviation,

$$r_1 = r_2 = A_p/2 \tag{1}$$

and

$$i_1 = i_2 = [D_p + A_p]/2 \tag{2}$$

where D_p = angle of minimum deviation for the prism
A_p = angle of the prism
r_1 and r_2 = internal angles of refraction
i_1 and i_2 = angles of entry and exit

The angle of minimum deviation D_p varies with wavelength. The angular dispersion is defined as $dD_p/d\lambda$, while the linear dispersion is

$$dx/d\lambda = F\, dD_p/d\lambda \tag{3}$$

where F is the focal length of the camera or imaging lens and x is the distance across the image plane. It can be shown[1] that

$$dD_p/d\lambda = (B/W)(dn/d\lambda) \tag{4}$$

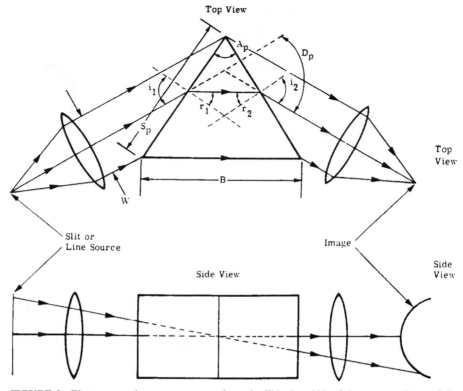

FIGURE 2 Elementary prism spectrometer schematic. W is the width of the entrance beam; S_p is the length of the prism face; and B is the prism base length.

where B = base length of the prism
            W = width of the illumination beam
            n = index of refraction

$$dx/d\lambda = F[B/W][dn/d\lambda]$$

The resolving power RP of an instrument may be defined as the smallest resolvable wavelength difference, according to the Rayleigh criterion, divided into the average wavelength in that spectral region. The limiting resolution is set by diffraction due to the finite beam width, or effective aperture of the prism, which is rectangular. Thus,

$$\mathrm{RP}_p = B[dn/d\lambda] \tag{5}$$

If the entire prism face is not illuminated, then only the illuminated base length must be used for B.

5.4 GRATINGS

A grating is an n-slit system used in Fraunhofer diffraction with interference arising from division of the incident, plane wave front. Thus it is a multiple beam interferometer.

$$p\lambda = d(\sin \theta + \sin \phi) \tag{6}$$

where p = order number ($=0, 1, 2, \ldots$) of the principal maxima
$\quad\quad d$ = the grating constant or spacing (the distance between adjacent slits)
$\quad\quad \phi$ = angle of incidence
$\quad\quad \theta$ = angle of diffraction
$\quad\quad w$ = width of any one slit

The most common case is $\phi = 0$, so that

$$p\lambda = d \sin \theta \tag{7}$$

and the irradiance distribution is

$$E = E_o\{\sin ((\pi w \sin \theta)/\lambda)/((\pi w \sin \theta)/\lambda)\}^2$$
$$\times \{\sin ((N\pi d \sin \theta)/\lambda)/\sin ((\pi d \sin \theta)/\lambda)\}^2 \tag{8}$$

where N is the number of slits or grooves. This equation is more often written as:

$$E = E_0[(\sin \beta)/\beta]^2[(\sin N\gamma)/\sin \gamma]^2 \tag{9}$$

which can be considered to be

$$E = (\text{constant}) \times (\text{single-slit diffraction function})$$
$$\times (N\text{-slit interference function}) \tag{10}$$

These considerations are for unblazed gratings. For a diffraction grating, the angular dispersion is given (for angle ϕ constant) by

$$dD_g/d\lambda \quad \text{or} \quad d\theta/d\lambda = p/(d \cos \theta) \tag{11}$$

The resolving power is given by

$$\text{RP}_g = pN \tag{12}$$

5.5 PRISM AND GRATING CONFIGURATIONS AND INSTRUMENTS

Classical

There are several basic prism and grating configurations and spectrometer designs which continue to be useful. One of the oldest spectrometer configurations is shown in Fig. 3.[1] Reflective interactions and prism combinations are used in Figs. 4, 5, and 6. Dispersion without deviation is realized in Figs. 7 and 8, while half-prisms are used in Fig. 9 in an arrangement which uses smaller prisms but still attains the same beam width. A few classical prism instrumental configurations are shown in Figs. 10, 11, and 12. Multiple-pass prism configurations are illustrated in Figs. 13 and 14.[4,5]

A well-known example of a single beam double-pass prism infrared spectrometer was the Perkin-Elmer Model 112 instrument shown in Fig. 15. Infrared radiation from a source is focused by mirrors M_1 and M_2 on the entrance slit S_1 of the monochromator. The radiation beam from S_1, path 1, is collimated by the off-axis paraboloid M_3 and a parallel beam traverses the prism for a first refraction. The beam is reflected by the Littrow mirror M_4, through the prism for a second refraction, and focused by the paraboloid, path 2, at the corner mirror M_6. The radiation returns along path 3, traverses the prism again, and is returned back along path 4 for reflection by mirror M_7 to the exit slit S_2. By this double dispersion, the radiation is spread out along the plane of S_2. The radiation of the frequency interval which passes through S_2 is focused by mirrors M_8 and M_9 on the thermocouple

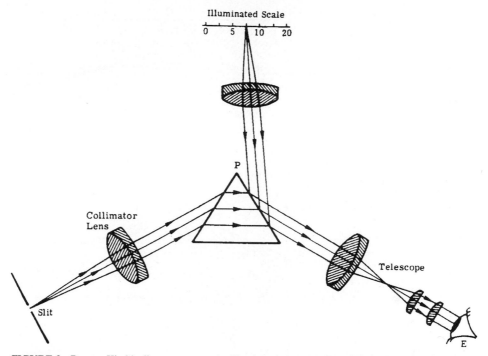

FIGURE 3   Bunsen-Kirchhoff spectrometer. An illuminated scale is reflected from the prism face into the telescope.

TC. The beam is chopped by CH, near M_6, to produce a voltage (at the thermocouple) which is proportional to the radiant power or intensity of the beam. This voltage is amplified and recorded by an electronic potentiometer. Motor-driven rotation of Littrow mirror M_4 causes the infrared spectrum to pass across exit slit S_2 permitting measurement of the radiant intensity of successive frequencies.

Gratings can be used either in transmission or reflection.[6] Another interesting variation comes from their use in plane or concave reflection form. The last was treated most completely by Rowland, who achieved a useful combination of focusing and grating action. He showed that the radius of curvature of the grating surface is the diameter of a circle (called the Rowland circle). Any source placed on the circle will be imaged on the circle,

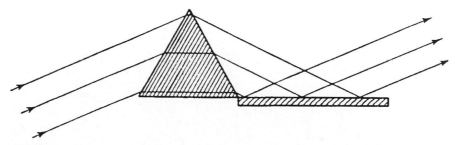

FIGURE 4   Wadsworth constant-deviation, prism-mirror arrangement. The beam enters the prism at minimum deviation and emerges displaced but not deviated from its original direction.

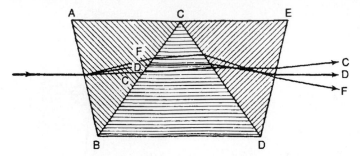

FIGURE 5 Amici prism. The central ray *D* enters and leaves parallel to the base. The *C* and *F* rays are deviated and dispersed.

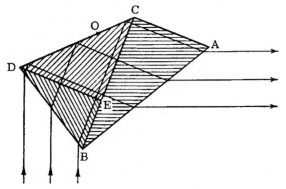

FIGURE 6 Pellin–Broca prism. The prism is equivalent to two 30° prisms, *ABC* and *BED,* and one 45° prism, *DEC,* but is made in one place. The beam shown, entering at minimum deviation, emerges at 90° deviation to its entrance direction.

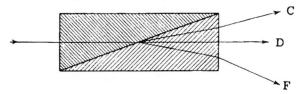

FIGURE 7 Zenger prism. The central ray *D* is undeviated. The *C* and *F* rays are deviated and dispersed.

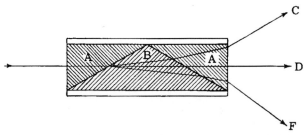

FIGURE 8 Wernicke prism. This arrangement is essentially two Zenger prisms, back-to-back.

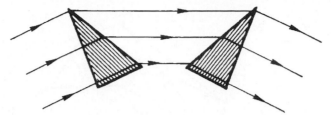

FIGURE 9 Young–Thollon half prisms. The passage of a beam at minimum deviation is shown.

with dispersion, if the rulings are made so that d is constant on the secant to the grating-blank (spherical) surface. The astigmatism acts so that a point source on a Rowland circle is imaged as a vertical line perpendicular to the plane of the circle. Rowland invented and constructed the first concave grating mounting, illustrated in Fig. 16.[1]

If dispersion is sufficiently large, one may find overlapping of the lines from one order with members of the spectra belonging to a neighboring order. Errors and imperfections in the ruling of gratings can produce spurious images which are called "ghosts." Also, the grooves in a grating can be shaped so as to send more radiation along a preferred direction corresponding to an order other than the zero order. Such gratings are said to be blazed in that order. These issues and many more involved in the production of gratings by ruling engines were thoroughly discussed by Harrison in his 1973 paper "The Diffraction Grating—An Opinionated Appraisal."[7]

Six more grating configurations[1] which are considered to be "classics" are:

1. *Paschen-Runge*, illustrated in Fig. 17. In this argument, one or more fixed slits are placed to give an angle of incidence suitable for the uses of the instrument. The spectra are focused along the Rowland circle $P\ P'$, and photographic plates, or other detectors, are placed along a large portion of this circle.

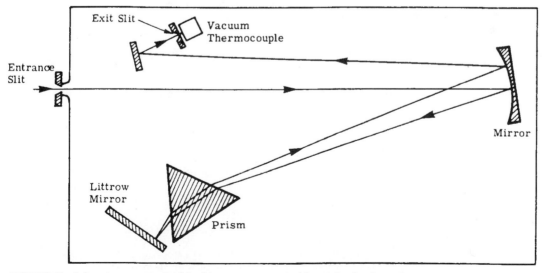

FIGURE 10 Infrared spectrograph of the Littrow-type mount with a rock salt prism.

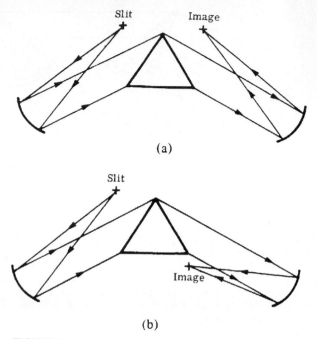

(a)

(b)

FIGURE 11   Mirror spectrometer with two choices of the location of the image. Arrangement (*b*) leads to smaller aberrations than arrangement (*a*) and is used in the Czerny-Turner mount.

2. *Eagle,* shown in Fig. 18. This is similar to the Littrow prism spectrograph. The slit and plate holder are mounted close together on one end of a rigid bar with the concave grating mounted on the other end.

3. *Wadsworth,* shown in Fig. 19. The Rowland circle is not used in this mounting in which the grating receives parallel light.

4. *Ebert-Fastie,* shown in Fig. 20. The Ebert-Fastie features a single, spherical,

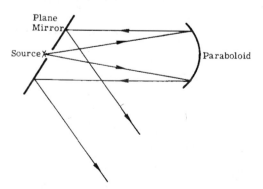

FIGURE 12   Pfund mirror. The use of a plane mirror to avoid astigmatism in the use of a paraboloidal mirror.

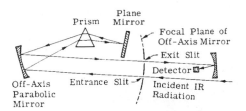

FIGURE 13 Double-pass monochromator.

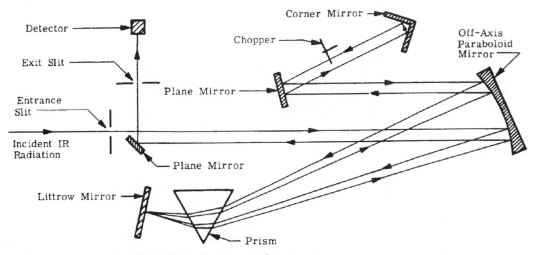

FIGURE 14 Perkin-Elmer Model 99 double-pass monochromator.

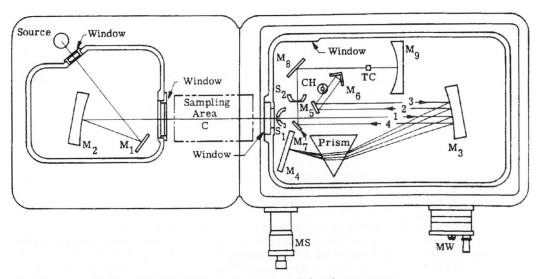

FIGURE 15 Perkin-Elmer Model 112 single-beam double-pass infrared spectrometer.

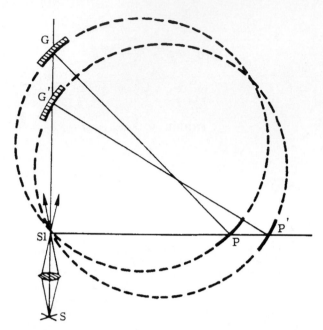

FIGURE 16 Rowland mounting of the concave grating. The grating plate-holder bar, which slides on the two perpendicular ways, is shown in two positions, *GP* and *G'P'*. The slit *SI* and Source *S* remain fixed.

collimating mirror and a grating placed symmetrically between the two slits. The major advantage of the Ebert system is the fact that it is self-correcting for spherical aberration. With the use of curved slits, astigmatism is almost completely overcome.

5. *Littrow,* shown in Fig. 10. The Littrow system has slits on the same side of the grating to minimize astigmatism. An advantage of the Littrow mount, therefore, is that straight slits can be used. In fact, such slits may be used even for a spherical collimating mirror if the aperture is not too large. Its greatest disadvantage is that it does not correct for spherical aberration—not too serious a defect for long focal-length/small-aperture instruments. If an off-axis parabola is used to collimate the light, aberrations are greatly reduced.

6. *Pfund,* shown in Figs. 12 and 21. This is an on-axis, Pfund-type grating instrument.[5] Incident infrared radiation, focused by a collimating lens on the entrance slit and modulated by a chopper, passes through the central aperture of plane mirror M_1. Reflected by the paraboloidal mirror P_1, it emerges as a parallel beam of radiation, which is reflected by mirror M_1 to the grating. The grating is accurately located on a turntable, which may be rotated to scan the spectrum. From the grating, the diffracted beam, reflected by mirror M_2, is focused by a second paraboloid P_2 through the central aperture of mirror M_2 to the exit slit. The emerging beam is then focused by the ellipsoidal mirror M_3 on the detector.

An off-axis, double-pass grating instrument is illustrated in Fig. 22.[6]

Combinations of prisms and gratings are not uncommon. An illustrative and complex prism-grating, double-monochromator spectrometer designed by Unicam Instruments, Ltd. is shown in Fig. 23.[5] The prism monochromator had four interchangeable prisms, and the grating monochromator had two interchangeable gratings. The two monochromators,

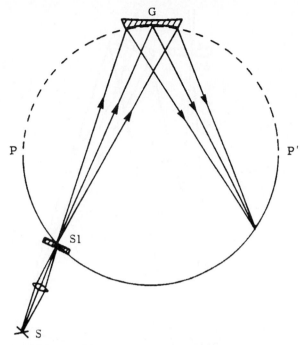

FIGURE 17   Paschen-Runge mounting of the concave grating. *Sl* is the slit, *G* is the grating, and *S* is the light source.

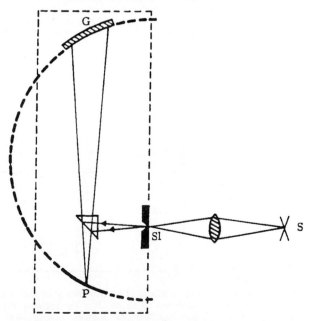

FIGURE 18   Eagle mounting on the concave grating. *Sl* is the slit, *G* is the grating, *S* is the light source, and *P* is the plate holder.

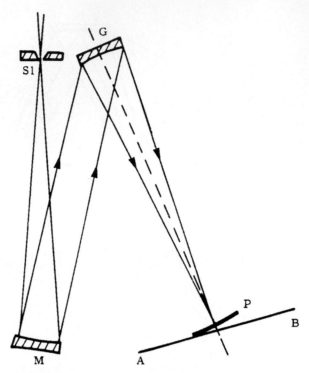

FIGURE 19 Wadsworth mounting of the concave grating. *Sl* is the entrance slit, *G* is the concave grating, *M* is the concave mirror, *P* is the plate holder, and *AB* is the rail for the plate holder. To minimize aberrations, one must locate the slit close beside the grating.

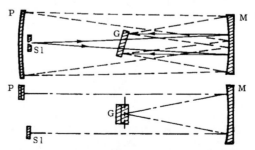

FIGURE 20 Ebert mounting of the plane grating designed by Fastie. *Sl* is the entrance slit, *G* is the grating, *M* is the concave mirror, and *P* is the photographic plate. The horizontal section is at the top and the vertical section is at the bottom.

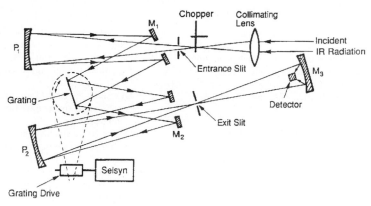

FIGURE 21   On-axis Pfund grating spectrograph.

ganged by cams which are linear in wave number, were driven by a common shaft. The instrument could be used either as a prism-grating double monochromator, or as a prism spectrometer by blanking the grating monochromator. Gratings, prisms, and cams could be automatically interchanged by means of push buttons. Magnetically operated slits, programmed by a taped potentiometer, provided a constant energy background. A star-wheel, time-sharing, beam attenuator was used in the double-beam photometer.

Contemporary

In recent years there has been more attention paid to total system design and integration for specific purposes and applications, as for analytical atomic and molecular spectroscopy in analytical chemistry. Thus the conventional dispersive elements are often used in the classical configurations with variations. Innovations have come especially in designs tailored for complete computer control; introduction of one- and two-dimensional detector arrays as well as new detector types (especially for signal matching); the use of holographic optical elements either alone or combined with holographic gratings; and special data-processing software packages, displays, and data storage systems. This is the case also for interferometric systems as discussed in the chapter on Interferometer Instruments by P. Hariharan.

Some examples found by a brief look through manufacturers' literature and journals

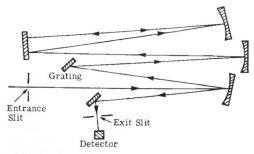

FIGURE 22   Off-axis, double-pass grating spectrograph.

TABLE 1 Examples of Prism/Grating Spectroradiometers

Manufacturer	Comments
ARC (Acton Research Corp.), Acton, Mass.	Czerny-Turner or Rowland systems with triple indexable Vac UV/IR gratings
ARIES (Acton Research Instrument & Equipment Services Inc.), QEI (Quantum Electronics Instruments Inc.), Concord, Mass.	Czerny-Turner variation with double or triple selectable gratings for 165-nm to 40-μm regions
Beckman Instruments Inc., Fullerton, Calif.	DU Series 60 and 70 modular construction, computer-controlled spectrophotometers for analytical applications
C VI Laser Corp., Albuquerque, N. Mex.	Digikrom Monochrometers, 1/8-, 1/4-, and 1/2-m Czerny-Turner grating systems, 186 nm–20 μm
Cary/Varian Instrument Group, San Fernando, Calif.	Cary 1, 3, 4, and 5 spectrophotometers for UV-Vis-IR; double beam, dual chopper/grating Littrow systems; attachments (e.g., reflectance) and applications software
CI Systems Ltd., New York City, N.Y. and Israel	CVF spectroradiometers for 0.4- to 20-μm scan
Infrared Systems, Inc., Orlando, Fla.	CVF spectroradiometer
Instruments SA, Inc., J-Y Optical Systems, Edison, N.J.	Monochrometers, spectrometers for UV-Vis-IR, holographic gratings in Czerny-Turner or concave aberration-corrected holographic gratings and Rowland mounts; single and double pass; imaging spectrographs
LECO Corp., St. Joseph, Mich.	ICP (Inductively Coupled Plasma) spectrometer system with Pachen-Runge mount concave grating followed by an Echelle and a linear detector array
Leeman Labs, Inc., Lowell, Mass.	ICP system with a fixed echelle grating followed by a prism with crossed order dispersion and scanned photomultipliers or detector arrays
McPherson, Division of SI Corp., Acton, Mass.	Double/triple monochrometers, spectroradiometers using gratings and/or prisms in Seya-Namioka, Czerny-Turner (C-T), crossed C-T, or Rowland configurations
Minirad Systems, Inc., Fairfield, Conn.	CVF and discrete filters in spectroradiometers for field measurements, 0.2 to 30 μm
Optometrics Corp., Ayer, Mass.	Monochrometers, prism or grating, Ebert-Fastie systems for UV-Vis-NIR
Optronic Laboratories, Inc., A Subsidiary of Kollmorgen Corp., Orlando, Fla.	Spectroradiometers, UV-Vis-IR for precision measurements; filter wheels, gratings, and prisms in single/double monochrometer configurations
Oriel Corporation, Stratford, Conn.	Scanning monochrometers, rotation filter wheels, and detector array instruments
Perkin-Elmer Corporation, Norwalk, Conn.	Complete sets of UV-Vis-IR spectroscopic systems using gratings and prisms, or FT-IR, with software and hardware for computer control, and accessories for microscopy, reflectance measurement, etc.
Shimadzu Scientific Instruments, Inc., Columbia, Md.	UV-Vis-NIR spectroscopic systems using holographic gratings in Czerny-Turner mounts in single- and double-beam configurations, computer-controlled, with accessories for analyses
SPEX Industries, Inc., Edison, N.J.	UV through IR grating spectrometers, 1/2- and 1/4-m, with CCD or PDA multichannel detectors
Thermo Jarrell Ash Corp., A Subsidiary of Thermo Instrument Systems, Inc., Franklin, Mass.	Monochromators and spectroscopic systems for analyses, UV-Vis-IR with gratings (in 1942 in Wadsworth, then in 1953, Ebert, and now Paschen-Runge and crossed Czerny-Turner mounts); complete systems

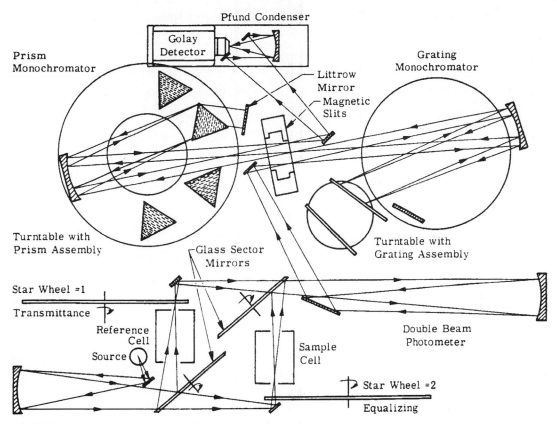

FIGURE 23 Unicam prism-grating double monochromator spectrometer.

such as *Spectroscopy, Physics Today, Laser Focus, Photonics Spectra,* and *Lasers & Optronics,*[8] are presented in Table 1. Most of these systems are designed for analytical spectroscopy with techniques described in many texts such as Robinson's *Atomic Spectroscopy.*[9]

5.6 REFERENCES

1. R. A. Sawyer, *Experimental Spectroscopy,* 3d ed. esp. Chapters 4, 6, 7, and 11, Dover Press, New York, 1963.

2. F. A. Jenkins and H. E. White, *Fundamentals of Optics,* 4th ed., McGraw-Hill, New York, 1976.

3. E. Hecht, *Optics: Second Edition,* Addison-Wesley, Reading, MA, reprinted April 1988.

4. A. Walsh, "Multiple Monochromators II. Application of a Double Monochromator to Infrared Spectroscopy," *Journal of the Optical Society of America,* Optical Society of America, Washington, DC, vol. 42, 1952, p. 95.

5. H. L. Hackforth, *Infrared Radiation,* McGraw-Hill, New York, 1960, pp. 209, 211, 214.

6. A. H. Nielsen, "Recent Advances in IR Spectroscopy," Tech. Memo 53-2, Office of Ordnance Research, Durham, NC, December 1953.

7. G. R. Harrison, "The Diffraction Grating—An Opinionated Appraisal," *Applied Optics,* vol. 12, no. 9, 1973, p. 2039.

8. See:

   Spectroscopy, especially issues of May and June 1990, Aster Publishing Corp., Eugene, OR.

   Physics Today, Annual Buyers' Guide, 7 August 1990, American Institute of Physics, 335 East 45th St., New York.

   Laser Focus World and LFW's *The Buyers' Guide,* 25th ed., 1990, PennWell Publishing Co., Westford, MA.

   Photonics Spectra and *The Photonics Directory,* 4 vols., 36th ed., 1990, Laurin Publishing Company Inc., Pittsfield, MA.

   Lasers & Optronics and *L & O's 1990 Buying Guide,* Gorden Publications, Inc., Morris Plains, NJ.

9. J. W. Robinson, *Atomic Spectroscopy,* Marcel Dekker, Inc., New York, 1990.

CHAPTER 6
INTEGRATED OPTICS

Thomas L. Koch
AT&T Bell Laboratories
Holmdel, New Jersey

F. J. Leonberger and P. G. Suchoski
United Technologies Photonics
Bloomfield, Connecticut

6.1 GLOSSARY

APE	annealed proton exchange
CATV	cable television
CVD	chemical vapor deposition
DBR	distributed Bragg reflector
DFB	distributed feedback
$\vec{E}$	electric field of propagating light
FOG	fiber optic gyroscope
Gb/s	gigabits per second
$\vec{H}$	magnetic field of propagating light
IOC	integrated optic circuit
L_c	coupling length of directional coupler
Mb/s	megabits per second
MMIC	monolithic millimeter-wave integrated circuit
MZ	Mach-Zehnder
n	index of refraction
OEIC	optoelectronic integrated circuit
r_{ij}	electro-optic tensor element
PIC	photonic integrated circuit
t_{cutoff}	waveguide thickness for cutoff of first odd mode
TE	transverse electric mode
TM	transverse magnetic mode

V_π	voltage for π radian phase shift in electrooptic modulator
VLSI	very large scale integration
WDM	wavelength division multiplexing
β	propagation constant of waveguide mode
ε_m	field amplitude of mode m
η	coupling efficiency between modes
θ_{crit}	critical angle for total internal reflection
Λ	spatial period of periodic feature along waveguide
λ	vacuum wavelength of propagating light
λ_{PL}	photoluminescence peak wavelength of semiconductor

6.2   INTRODUCTION

The field of integrated optics is concerned with the theory, fabrication, and applications of guided wave optical devices. In these structures, light is guided along the surface region of a wafer by being confined in dielectric waveguides at or near the wafer surface; the light is confined to a cross-sectional region having a typical dimension of several wavelengths. Guided wave devices that perform passive operations analogous to classical optics (e.g., beam splitting) can be formed using microelectronic-based fabrication techniques. By fabricating devices in active materials such as ferroelectrics, modulators and switches based on the classical electro-optic effect can be formed. By fabricating the devices in compound semiconductors, passive and active guided wave devices can be monolithically combined with lasers, detectors, and optical amplifiers. The combination of both passive and active devices in a multicomponent circuit is referred to as an integrated optic circuit (IOC) or a photonic integrated circuit (PIC). In semiconductor materials, purely electronic devices can be integrated as well to form what is often referred to as an optoelectronic integrated circuit (OEIC).

Progress in the field of integrated optics has been rapid since its inception in 1970. Much of this progress is due to the availability of high-quality materials, microelectronic processing equipment and techniques, and the overall rapid advancement and deployment of fiber optic systems. The interest in integrated optics is due to its numerous advantages over other optical technologies. These include large electrical bandwidth, low power consumption, small size and weight, and improved reliability. Integrated optics devices also interface efficiently with optical fibers, and can reduce cost in complex circuits by eliminating the need for separate, individual packaging of each circuit element.

The applications for integrated optics are widespread. Generally these applications involve interfacing with single-mode fiber optic systems. Primary uses are in digital and analog communications, sensors (especially fiber optic gyroscopes), signal processing, and instrumentation. To a lesser extent, IO devices are being explored in nonfiber systems for laser beam control and optical signal processing and computing. IOCs are viewed in the marketplace as a key enabling technology for high-speed digital telecommunications, CATV signal distribution, and gyros. As such they will have a significant impact on commercially deployed fiber systems and devices in the 1990s.

This chapter reviews the IO field, beginning with a brief review of IO device physics and fabrication techniques. A phenomenological description of IO circuit elements, both passive and active, is given, followed by a discussion of IO applications and system demonstrations. The chapter concludes with a short look at future trends. Due to the brevity of this chapter relative to the work in the field, much of the coverage is necessarily limited. The reader is referred to Refs. 1–13 for more detailed information at a variety of levels.

6.3 DEVICE PHYSICS

Optical Waveguides

Central to integrated optics is the concept of guiding light in dielectric waveguide structures with dimensions comparable to the wavelength of the guided light. In this section we present only a brief survey of the relevant physics and analysis techniques used to study their properties. The reader is referred to a number of excellent texts dealing with this topic for a more comprehensive treatment.[1–3]

A dielectric waveguide confines light to the core of the waveguide by somehow reflecting power back towards the waveguide core that would otherwise diffract or propagate away. While any means of reflection can accomplish this end (for example, glancing-incidence partial reflections from interfaces between different media can serve as the basis for *leaky* waveguides), the most common technique employs a 100 percent *total internal reflection* from the boundary of a high-index core and a lower-index cladding material. As light propagates down the axis of such a structure, the waveguide cross section can also be viewed as a lens-like phase plate that provides a larger retardation in the core region. Propagation down the guide then resembles a continuous refocusing of light that would otherwise diffract away.

The pedagogical structure used to illustrate this phenomenon is the symmetric slab waveguide, composed of three layers of homogeneous dielectrics as shown in Fig. 1. It is well known that propagation in slab structures can be analyzed using either a ray-optics approach or through the use of interface boundary conditions applied to the simple solutions of Maxwell's equations in each homogeneous layer of the structure.[1–3] In the ray-optics description, the rays represent the phase fronts of two intersecting plane-waves propagating in the waveguide core region. Since the steady-state field has a well-defined phase at each point, a finite set of discrete modes arises from the self-consistency condition that, after propagation and two reflections from the core-cladding boundaries, the phase front must rejoin itself with an integral multiple of a 2π phase shift. For a given core thickness, there will be a limited discrete number of propagation angles in the core that

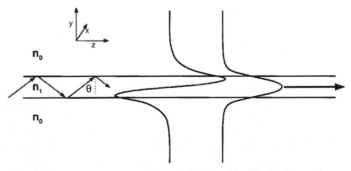

FIGURE 1   A symmetric three-layer slab waveguide. The fundamental even and first odd mode are shown.

satisfy this criterion, with the lower bound on the angle given by the critical angle for total internal reflection, $\theta_{\text{crit}} = \sin^{-1}(n_0/n_1)$. In general, a thicker and higher-index waveguide core will admit a larger number of confined solutions or *bound* modes. Figure 1 shows both the fundamental even mode and the first higher-order odd mode. If the dimensions are small enough, only one bound mode for each polarization state will exist and the guide is termed a single-mode waveguide. Care must be exercised to include the angle-dependent phase shift upon total internal reflection, referred to as the Goos-Hanchen shift, that can be viewed as a displaced effective reflection plane.[1] The quantity $\beta = (2\pi n_1/\lambda) \cdot \sin\theta$, referred to as the propagation constant, is the z projection of the wave-vector and thus governs the phase evolution of the field along the guide. In addition to the discrete set of bound modes, plane waves can also enter from one side and pass vertically through such a structure, and form a continuous set of *radiation* modes. In an asymmetrical structure, some of the radiation modes may be propagating on one side of the guide but evanescent on the other. From a mathematical point of view, the set of all bound and radiation modes forms a complete set for expansion of any electromagnetic field in the structure. Analysis techniques will subsequently be discussed in more detail.

The slab waveguide in Fig. 1 employed total internal reflection from an abrupt index discontinuity for confinement. Some fabrication techniques for waveguides, particularly in glasses or electro-optic materials such as $LiNbO_3$, achieve the high-index core by impurity diffusion or implantation, leading to a *graded* index profile. Here a field solution will usually be required to properly describe the modes and the ray paths become curved, but total internal reflection is still responsible for confinement.

Most useful integrated optics devices require waveguide confinement in a stripe or channel geometry. While recognizing that the vector nature of the electromagnetic field makes the rigorous analysis of a particular structure quite cumbersome, the reader can appreciate that the same phenomenon of confinement by reflection will be operative in two dimensions as well. Figure 2 shows the cross sections of the most common stripe or channel waveguide types used in integrated optics. Common to the cross section for all

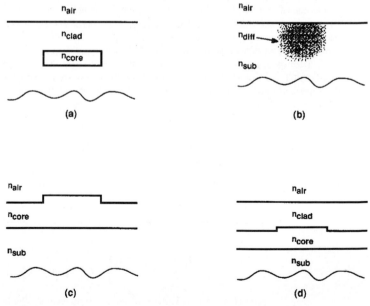

FIGURE 2   Various types of channel or stripe waveguides.

these structures is a region on the waveguide axis containing more high-index material than the surrounding cladding areas. The diffused waveguide may require a full two-dimensional analysis, but a common technique for the approximate analysis of high-aspect-ratio channel guides such as (a), (c), and (d), is the *effective index method*[1,2,14] In this technique, a slab waveguide analysis is applied sequentially to the two dimensions. First, three separate vertical problems are solved to obtain the modal phase index $n_{\mathrm{mode}} \equiv \beta \cdot \lambda / 2\pi$ for each lateral region as if it were an infinite slab. These indices are then used as input to a final "effective" slab waveguide problem in the lateral dimension. Since the properties of multilayer slab waveguides play an important role in waveguide analysis, a more comprehensive general formulation is outlined below. This task is more tractable using the field solutions of Maxwell's equations than the ray-optics approach.

A general multilayer slab is shown in Fig. 3. Since Maxwell's equations are separable, we need only consider a two-dimensional problem in the y direction perpendicular to the layers, and a propagation direction z. The concept of a mode in such a structure is quantified in physical terms as a solution to Maxwell's equations whose sole dependence on the coordinate in the propagation direction z is given by $e^{i\beta z}$. This translates to a requirement that the *shape* of the field distribution in the y direction, perpendicular to layers, remain unchanged with propagation. If we generalize to leaky structures or materials exhibiting loss or gain, β may be complex, allowing for a scaling of the mode amplitude with propagation, but the relative mode profile in the perpendicular y direction still remains constant. These latter solutions are not normalizable or "proper" in the sense of mathematical completeness, but are very useful in understanding propagation behavior in such structures.

Since the field in each homogeneous layer m is well known to be $e^{\pm i\vec{k}_m \cdot \vec{r}}$, with

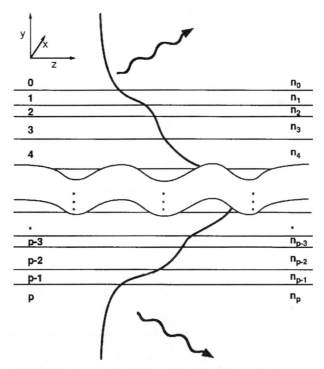

FIGURE 3 A general multilayer slab waveguide structure.

$|\vec{k}_m| = 2\pi n_m/\lambda$ for the (generally complex) index of refraction n_m, the general solution to the field amplitude in each layer m is

$$\varepsilon_m = [a_m e^{iq_m y} + b_m e^{-iq_m y}]e^{i\beta z} \tag{1}$$

where $q_m \equiv [(2\pi n_m/\lambda)^2 - \beta^2]^{1/2}$. Inspection of the vector Maxwell's equations reveals that the general vector solution in the multilayer slab can be broken down into the superposition of a TE (transverse electric) and a TM (transverse magnetic) solution.[1,2] The TE (TM) solution is characterized by having only one component of the electric (magnetic) field that points in the x direction, parallel to the layers and perpendicular to the propagation direction z. The mode field amplitude ε_m in Eq. (1) refers to the E_x or the H_x field for the TE and TM case, respectively.

In a very simple exercise, for each of these cases one can successively match boundary conditions for continuous tangential $\vec{E}$ and $\vec{H}$ across the interfaces to provide the coefficients a_{m+1} and b_{m+1} in each layer $m + 1$ based upon the value of the coefficients in the preceding layer m,

$$\begin{bmatrix} a_{m+1} \\ b_{m+1} \end{bmatrix} = \begin{bmatrix} \frac{1}{2}\left(1 + \frac{q_m \gamma_m}{q_{m+1}\gamma_{m+1}}\right)e^{-i(q_{m+1}-q_m)y_m} & \frac{1}{2}\left(1 - \frac{q_m \gamma_m}{q_{m+1}\gamma_{m+1}}\right)e^{-i(q_{m+1}+q_m)y_m} \\ \frac{1}{2}\left(1 - \frac{q_m \gamma_m}{q_{m+1}\gamma_{m+1}}\right)e^{i(q_{m+1}+q_m)y_m} & \frac{1}{2}\left(1 + \frac{q_m \gamma_m}{q_{m+1}\gamma_{m+1}}\right)e^{i(q_{m+1}-q_m)y_m} \end{bmatrix} \cdot \begin{bmatrix} a_m \\ b_m \end{bmatrix} \tag{2}$$

where y_m are the coordinates of the interfaces between layers m and $m + 1$, and $\gamma_m \equiv 1$ for TE modes and $\gamma_m \equiv n_m^{-2}$ for TM modes. The wave is assumed evanescently decaying or outward leaking on one initial side of the arbitrary stack of complex-index layers, i.e., $b_0 = 0$ on the uppermost layer. When the lowermost "cladding" layer $m = p$ is reached, one again demands that only the coefficient b_p of the evanescently decaying, or possibly the outward leaking, component be nonzero, which recursively provides the eigenvalue equation $a_p(\beta) = 0$ for the eigenvalues β_j. Arbitrarily letting $a_0 = 1$, this can be written explicitly as

$$a_p(\beta) = [1 \quad 0] \cdot \left[\prod_{m=p-1}^{m=0} \mathbf{M}_m(\beta)\right] \cdot \begin{bmatrix} 1 \\ 0 \end{bmatrix} = 0 \tag{3}$$

where $\mathbf{M}_m(\beta)$ is the matrix appearing in Eq. (2). In practice this is solved numerically in the form of two equations (for the real and imaginary parts of a_p) in two unknowns (the real and imaginary parts of β). Once the complex solutions β_j are obtained using standard root-finding routines, the spatial profiles are easily calculated for each mode j by actually evaluating the coefficients for the solutions using the relations above with $a_0 = 1$, for example.

Application of Eq. (3) to the simple symmetric slab of Fig. 1 with thickness t, and real core index n_1 and cladding index n_0 can be reduced with some trigonometric half-angle identities to a simple set of equations with intuitive solutions by graphical construction.[15] Defining new independent variables $r \equiv (t/2)[(2\pi n_1/\lambda)^2 - \beta^2]^{1/2}$ and $s \equiv (t/2)[\beta^2 - (2\pi n_0/\lambda)^2]^{1/2}$, one must simultaneously solve for positive r and s the equation

$$r^2 + s^2 = (\pi t/\lambda)^2(n_1^2 - n_0^2) \tag{4}$$

and either one of the following equations:

$$s = \frac{\gamma_1}{\gamma_0} \cdot r \cdot \begin{cases} \tan(r) & \text{(even modes)} \\ -\cot(r) & \text{(odd modes)} \end{cases} \tag{5}$$

where again $\gamma_m \equiv 1$ for TE modes and $\gamma_m \equiv n_m^{-2}$ for TM modes.

By plotting the circles described by Eq. (4) and the functions in Eq. (5) in the (r, s)

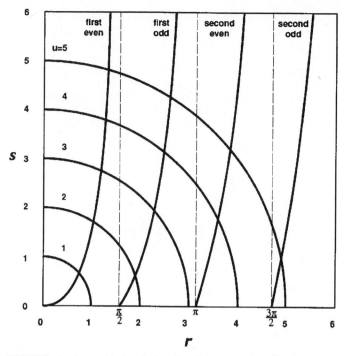

FIGURE 4 A graphical solution for the symmetric three-layer slab waveguide. For an arbitrary value of the parameter u, solutions are found at the intersections of the circular arcs and the transcendental functions as shown.

plane, intersections provide the solutions (r_j, s_j) for mode j, yielding β_j from the definition of either r or s. This construction is shown in Fig. 4 for TE modes, where Eq. (4) has been parametrized with $u \equiv (\pi t / \lambda)(n_1^2 - n_0^2)^{1/2}$. Due to the presence of γ_m in Eq. (5), the TE and TM modes will have different propagation constants, leading to waveguide birefringence.

It is easy to see from the zero-crossing of Eq. (5) at $r = \pi/2$ that the cutoff of the first odd mode occurs when the thickness reaches a value

$$t_{\text{cutoff}} = \left(\frac{\lambda}{2}\right)(n_1^2 - n_0^2)^{-1/2} \tag{6}$$

Another important feature of the symmetric slab is the fact that neither the TE nor TM fundamental (even) mode is ever cutoff. This is not true for asymmetric guides. More complicated structures are easily handled using Eq. (2), and one can also approximate graded-index profiles using the multilayer slab. However, analytical solutions also exist[1] for a number of interesting graded-index profiles, including parabolic, exponential, and $\cos h^{-2}$.

Once the modes of a waveguide are known, there are many physical phenomena of interest that can be easily calculated. Quite often the propagation constant is required to evaluate the phase evolution of guided-wave components. In other cases, the designer may need to know the fraction of the propagating energy that lies within a certain layer of the waveguide system. Perhaps the most important application is in evaluating waveguide coupling phenomena that describe how the light distribution evolves spatially as light propagates down a waveguide system. For example, it is easy to show from the mathematical completeness of the waveguide modes[1–3] that the energy efficiency η of the coupling into a mode ε_m of a waveguide from a field ε_{inj} injected at the input facet of the waveguide is given by $\eta = |\int \varepsilon_m^*(y)\varepsilon_{inj}(y)\,dy|^2$. Here the fields have been assumed normalized such that $\int_{-\infty}^{\infty} |\varepsilon_m|^2\,dy = 1$.

Coupled-mode theory[1,2,7] is one of the most important design tools for the guided-wave device designer. This formalism allows for the calculation of the coupling between parallel waveguides as in a directional coupler. It also allows for the evaluation of coupling between different modes of a waveguide when a longitudinal perturbation along the propagation direction ruins the exact mode orthogonality. An important example of the latter is when the perturbation is periodic as in a corrugated-waveguide grating. Waveguide gratings are used to form wavelength selective coupling as in Bragg reflectors, distributed feedback lasers, and other grating-coupled devices. The comprehensive treatment of these phenomena is beyond the scope of this chapter, and the reader is referred to the references cited above. In some instances, evaluating the performance of devices where radiation plays a significant role may be tedious using a modal analysis, and techniques such as the *beam propagation method (BPM)* are used to actually launch waves through the structure to evaluate radiation losses in waveguide bends, branches, or complicated splitters.[16]

Index of Refraction and Active Index-changing Mechanisms

Index of Refraction. Waveguide analysis and design requires precise knowledge of the material index of refraction. One of the most common electro-optic materials is $LiNbO_3$, a uniaxial birefringent crystal whose index can be characterized by providing the wavelength-dependent ordinary and extraordinary indices n_o and n_e. They are given by[17]

$$n_{o,e}^2 = A_{o,e} + \frac{B_{o,e}}{D_{o,e} - \lambda^2} + C_{o,e}\lambda^2 \tag{7}$$

where $A_o = 4.9048$ $\quad B_o = -0.11768$ $\quad C_o = -0.027169$ $\quad D_o = 0.04750$
$A_e = 4.5820$ $\quad B_e = -0.099169$ $\quad C_e = -0.021950$ $\quad D_e = 0.044432$

Glasses are also common substrate materials, but compositions and indices are too varied to list here; indices usually lie in the range of 1.44 to 1.65, and are mildly dispersive. Commonly deposited dielectrics are SiO_2 and Si_3N_4 with indices of ~1.44 and ~2.0 at 1.55 μm. The reader is referred to various tables in the literature for more details.[17]

InP and GaAs are by far the most common substrates for IO devices in semiconductors. The usual epitaxial material on GaAs substrates is $Al_xGa_{1-x}As$, which is nearly lattice-matched for all values of x, with GaAs at $x = 0$ providing the narrowest bandgap. For photon energies below the absorption edge, the index of refraction of this material system is given by[18]

$$n_{AlGaAs}(E, x) = \left[1 + \gamma(E_f^4 - E_\Gamma^4) + 2\gamma(E_f^2 - E_\Gamma^2)E^2 + 2\gamma E^4 \cdot ln\left(\frac{E_f^2 - E^2}{E_\Gamma^2 - E^2}\right)\right]^{1/2} \tag{8}$$

where $E = 1.2398/\lambda$ is the incident photon energy,

$$\gamma = \frac{E_d}{4E_0^3(E_0^2 - E_\Gamma^2)} \quad \text{and} \quad E_f = (2E_0^2 - E_\Gamma^2)^{1/2} \tag{9}$$

where $E_0(x) = 3.65 + 0.871x + 0.179x^2$
$E_d(x) = 36.1 - 2.45x$
$E_\Gamma(x) = 1.424 + 1.266x + 0.26x^2$

For devices fabricated on InP substrates, common in telecommunications applications for devices in the 1.3 μm and 1.55 μm bands, the most common epitaxial material is a quaternary alloy composition $In_xGa_{1-x}As_yP_{1-y}$. In this case, the material is only lattice matched for the specific combination $y = 2.917x$, and this lattice-matched alloy can be characterized by its photoluminescence wavelength λ_{PL} under low-intensity optical excitation. The index of this quaternary allow is given by[19]

$$n_Q(E, E_{PL}) = \left(1 + \frac{A_1}{1 - \left(\dfrac{E}{E_{PL} + E_1}\right)^2} + \frac{A_2}{1 - \left(\dfrac{E}{E_{PL} + E_2}\right)^2} \right)^{1/2} \tag{10}$$

where $E = 1.2398/\lambda$ and $E_{PL} = 1.2398/\lambda_{PL}$ are, respectively, the incident photon energy and photoluminescence peak photon energy for λ in μm and $A_1(E_{PL})$, $A_2(E_{PL})$, E_1, E_2 are fitted parameters given by

$$A_1 = 13.3510 - 5.4554 \cdot E_{PL} + 1.2332 \cdot E_{PL}^2$$

$$A_2 = 0.7140 - 0.3606 \cdot E_{PL} \tag{11}$$

$$E_1 = 2.5048 \, eV \qquad \text{and} \qquad E_2 = 0.1638 \, eV$$

For application to the binary InP, the value of the photoluminescence peak should be taken as $\lambda_{PL} = 0.939$ μm.

Many integrated optics devices rely on active phenomena such as the electro-optic effect to alter the real or imaginary index of refraction. This index change is used to achieve a different device state, such as the tuning of a filter, the switching action of a waveguide switch, or the induced absorption of an electroabsorption modulator. A brief survey is provided here of the most commonly exploited index-changing phenomena.

Linear Electro-optic Effect. The linear electro-optic or Pockels effect refers to the change in the optical dielectric permittivity experienced in noncentrosymmetric ordered materials that is *linear* with applied quasi-static electric field. This relation is commonly expressed using the dielectric *impermeability* $(1/n^2)_{ij} \equiv \varepsilon_0 \, \partial E_i/\partial D_j$ appearing in the *index ellipsoid* equation for propagation in anisotropic crystals.[20] Symmetry arguments allow $(1/n^2)_{ij}$ to be contracted to a single subscript $(1/n^2)_i$ for $i = 1, \ldots, 6$. In the *principal axes* coordinate system, the impermeability is diagonalized and $(1/n^2)_i = 0$ for $i = 4, 5, 6$ in the absence of an applied electric field, with the value of $(1/n^2)_i$ providing the inverse square of the index for optical fields polarized along each axis $i = 1, 2, 3$. For an electric field expressed in the principal axes coordinate system, the changes in the coefficients are evaluated using the 6×3 *electro-optic tensor* $\mathbf{r}$

$$\Delta \left(\frac{1}{n^2} \right)_i = \sum_{j=1}^3 r_{ij} E_j \qquad i = 1, \ldots, 6 \tag{12}$$

With an applied field, the equation for the index ellipsoid in general must be *rediagonalized* to again yield $(1/n^2)_i = 0$ for $i = 4, 5, 6$. This provides a new set of principal axes and the coefficients in the new index ellipsoid equation provide the altered value of the refractive index along each new principal axis.

For a particular crystal, symmetry also limits the number of nonzero r_{ij} that are possible. In the cubic zinc-blend III-V compounds there are only three equal nonzero components

$r_{63} = r_{52} = r_{41}$ and the analysis is relatively easy. As an example, consider a static field **E** applied along the (001) direction, surface normal to the wafers commonly used for epitaxial growth. The rediagonalized principal axes in the presence of the field become the (001) direction (z-axis), the (011) direction (x-axis), and the (01$\bar{1}$) direction (y-axis); the latter two directions are the cleavage planes and are thus common directions for propagation. The respective index values become

$$n_x = n_0 - \tfrac{1}{2}n_0^3 r_{41} \mathbf{E}$$

$$n_y = n_0 + \tfrac{1}{2}n_0^3 r_{41} \mathbf{E} \tag{13}$$

$$n_z = n_0$$

For a slab guide on a (001) substrate and propagation in the (011) direction, the applied field would produce a phase retardation for TE-polarized light of $\Delta\phi = (\pi/\lambda)n_0^3 r_{41}\mathbf{E} \cdot L$ after a propagation length L. With values of $r_{41} \sim 1.4 \times 10^{-10}$ cm/V, micron-scale waveguide structures in GaAs or InP lead to retardations in the range of $10°/\text{V} \cdot \text{mm}$. This TE retardation could be used as a phase modulator, or in a Mach-Zehnder interferometer to provide intensity modulation. For fields applied in other directions such as the (011), the principal axes are rotated away from the (011) and (01$\bar{1}$) directions. Propagation along a cleavage direction can then alter the polarization state, a phenomenon that also has device implications as will be discussed in more detail later.

In the case of LiNbO$_3$ and LiTaO$_3$, two of the most common integrated optic materials for electro-optic devices, the dielectric tensor is more complex and the materials are also birefringent in the absence of an applied field. There are eight nonzero components to the electro-optic tensor, $r_{22} = -r_{12} = -r_{61}$, $r_{51} = r_{42}$, $r_{13} = r_{23}$, and r_{33}. For LiNbO$_3$, the largest coefficient is $r_{33} \sim 30.8 \times 10^{-10}$ cm/V. Both retardation and polarization changes are readily achieved, and the reader is referred to Vol. II, Chap. 13, "Electro-Optic Modulators" of this handbook or the literature for a comprehensive treatment of the properties of these and other materials.[20,21]

The electro-optic effect in these materials is associated with field-induced changes in the positions of the constituent atoms in the crystal, and the resulting change in the crystal polarizability. The absorption induced by the conventional electro-optic effect is thus negligible.

Carrier Effects. In semiconductors, other powerful index-changing mechanisms are available related to the interaction of the optical field with the free electrons or holes. The simplest of these is the plasma contribution resulting from the polarizability of the mobile carriers. According to the simple Drude model,[22] this is given for each carrier species by $\Delta n \approx -N \cdot e^2 \lambda^2/(8\pi^2 \varepsilon_0 n c^2 m^*)$ in MKS units, where N and m^* are the carrier concentration and effective mass, e is the electronic charge, and ε_0 is the free-space permittivity. This can produce index changes approaching $\Delta n \sim -0.01$ at $10^{18}/\text{cm}^3$ electron/hole doping of the semiconductor, and can be exploited for waveguide design. Near the bandgap of the semiconductor, there are additional strong index changes with variations in the carrier density that arise from the associated dramatic changes in the optical loss or gain. Since these correspond to changes in the imaginary index, the Kramers-Kronig relation dictates that changes also occur in the real index. These effects are comparable in magnitude and are of the same sign as the free-carrier index contribution, and dramatically impact the performance of semiconductor devices and PICs that emply gain media.

In addition to the effects described here arising from changing carrier populations, the electronic transitions that generate the free carriers can be modified by an applied electric field. For optical frequencies close to these transitions, this can give rise both to electroabsorption and to an enhanced electro-optic effect, which shall be termed electrorefraction, due to the Stark effect on the carrier-generating transitions. In bulk

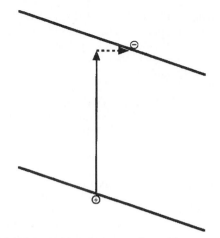

FIGURE 5 Franz-Keldysh effect. Electron can complete transition to the tilted conduction band by tunneling.

material, the induced absorption is termed the Franz-Keldysh effect,[23] and can be viewed as a tunneling effect. For an electron in the valence band with insufficient energy to complete a transition to the conduction band, the field can be viewed as adding a potential to the bands that effectively tilts them in space as shown in Fig. 5. If the excited carrier also traversed a short distance down-field from its initial location, it would have sufficient energy to complete the transitions. This distance depends on the tilt, and thus the strength of the applied field. Since carriers cannot be precisely localized according to the Heisenberg uncertainty principle, there is a finite amplitude for completing the transition that is an increasing function of electric field. For fields on the order of 10^5 V/cm, absorption values of $\sim 100 \, \mathrm{cm}^{-1}$ can be readily achieved in material that is quite transparent at zero field. According to the Kramers-Kronig relations, in addition to the absorption described above, this will also induce a change in the real index that will be positive below the band edge.

In the case of quantum-wells, carrier-induced effects can be enhanced due to the confinement in the wells. *Excitonic* effects, resulting from the Coulombic attraction between electrons and holes, produce sharp features in the absorption spectrum near the band gap that can be dramatically altered by an applied electric field. This quantum-confined Stark effect (QCSE) can enhance both electroabsorptive and electrorefractive effects.[24,25] This suggests that more compact, lower-drive voltage devices are possible when quantum wells are employed, a fact that has been confirmed experimentally. However, in both the bulk and especially the quantum-well case, care must be taken to operate at an optical frequency where strong electroabsorptive or electrorefractive effects are operative but the zero-field background absorption is not prohibitively high. Another issue that impacts the design of devices based on electroabsorption is the requirement for removal of the photogenerated carriers to prevent saturation.

Thermal Effects. In virtually all materials, the optical path length of a given section of waveguide will increase with temperature. This is the combination of both the physical expansion of the material and the change of the index of refraction with temperature. While both are significant, in most integrated-optic materials the latter effect is dominant. In SiO_2 on Si, for example, this mechanism provides a useful means of index change, and numbers on the order of $\Delta n \sim 10^{-5}/°C$ are observed. This effect has been used to form a variety of thermo-optic switches and filters, but a significant disadvantage for many applications is the inherent slow speed and high power dissipation. In semiconductors, this index change is more than an order of magnitude larger, and leads to frequency shifts in filter devices and single-longitudinal-mode lasers of $\Delta f \sim 10\ GHz/°C$.

Nonlinear Effects. Another class of index changes results from the nonlinar optical effects caused by the incident optical fields themselves. This is treated in depth in other portions of this handbook,[25] but two phenomena will be mentioned here. The first is closely related to the electrooptic effect discussed earlier, where the field giving rise to the index change is no longer "quasi-static" but is in fact the optical field itself. The response of the medium is in general not the same at optical frequencies, but the same symmetry arguments and contracted tensor approach are employed. The polarization resulting from the incident field at ω multiplied by the index oscillating at ω generates second harmonic components at 2ω.

This frequency doubling can occur in waveguides, but great care must be taken to achieve *phase matching* where each locally generated frequency-doubled field propagates in such a way as to add coherently to frequency-doubled fields generated farther along the guide. This requires either that the dispersive properties of the materials and guide geometry allow $n(\omega) = n(2\omega)$, or that the frequency-doubled light radiates away from the guide at an angle to allow phase-matching of the z-component of the wave-vector, or that a periodic domain reversal be introduced into the crystal to allow phase matching. This latter approach, often referred to as quasi phase matching, has generated considerable interest recently. In this approach, the optic axis of the crystal is periodically reversed with a period equal to the difference in the wave vectors of the fundamental and second harmonic. To date, the most promising results are in $LiTaO_3$, $LiNbO_3$, and KTP. In $LiNbO_3$, periodic domain reversal has been obtained by the application of 100 μsec pulsed electric fields of 24 kV/mm using a 2.8-μm-period segmented electrode that is subsequently removed.[27] The domain reversal in $LiTaO_3$ can be obtained on a few-micron scale by proton exchange or electron bombardment. KTP has a higher nonlinear coefficient, but the material is not as well developed. Lower efficiencies have been obtained.

A second application of nonlinear optics to integrated structures involves a higher order of nonlinearity referred to as four-wave mixing, or in some situations as self-phase modulation. The change in index in these cases arises from the *product* of two optical fields. If all fields are the same frequency, this is termed *degenerate,* and if only one field is present, it becomes self-phase modulation with the index change driven by the *intensity* of the incident wave. This nonlinearity has been of interest in research aimed at creating all-optical logic devices. Here the intensity of either the input signal or a separate gating signal can determine the output port of a Mach-Zehnder or directional coupler switch, for example.[28]

6.4 INTEGRATED OPTICS MATERIALS AND FABRICATION TECHNOLOGY

Ion-exchanged Glass Waveguides

Passive integrated optic devices can be fabricated in certain glass substrates using the ion-exchange technique.[29,30] In this fabrication process, a sodium-rich glass substrate is placed in a mixture of molten nitrate salts containing alkali cations, such as Cs^+, Rb^+, Li^+, K^+, Ag^+, and Tl^1. During this process, sodium ions at the surface of the host glass are replaced with the alkali cations, resulting in a local increase in the refractive index. Channel waveguides are realized by selectively masking the glass surface. The index change and the diffusion depth are a function of host material composition, the specific alkali cation being used, the temperature, and the diffusion time. The exchange process can be substantially enhanced by applying an electric field across the wafer while the substrate is immersed in the salt bath.

Multimode devices are typically fabricated using thallium ion exchange in borosilicate glasses.[31] The high polarizability of the thallium ions results in a large index change ($>$0.1) while providing low propagation losses (0.1 dB/cm). However, thallium-sodium ion exchange has two significant drawbacks. Thallium is extremely toxic, and it also has a large ionic radius compared to sodium (1.49 Å compared to 0.95 Å), resulting in low diffusion coefficients and low mobility. It is therefore necessary to process the waveguides at high bath temperatures approaching the glass transition temperature of the host material (500°C) for long diffusion times (10 h) with high applied electric fields ($>$100 V/cm) to achieve deep multimode waveguides that efficiently couple to commercially available multimode fiber (50 to 100 micron core size). Finding suitable masking materials is a challenge.

Single-mode devices are typically realized using either Ag^+-Na^+, K^+-Na^+, or Cs^+-K^+ exchange.[29,30] The first two processes have been extensively studied and are well understood; however, they each appear to have drawbacks. Ag^+-Na^+ exchanged waveguides are somewhat lossy (0.5 dB/cm) due to a tendency for silver reduction in the host material. K^+-Na^+ exchanged waveguides are typically highly stressed and prone to surface scattering that increases the propagation loss. Although not as extensively studied, Cs^+-K^+ exchanged waveguides show great promise. These waveguides are nearly stress-free, low-loss ($<$0.1 dB/cm), reproducible, and can be buried using a two-step exchange process. The two-step process further reduces the propagation loss and results in efficient fiber-waveguide coupling ($<$0.1 dB loss per interface).

Thin Film Oxides

In recent years there has been substantial interest in IO devices fabricated in thin film dielectrics on silicon substrates. This is due in part to the excellent surface quality, large-area wafers, and mechanical integrity of silicon itself. However, this interest also stems in some cases from the availability of mature silicon processing technology developed by the electronic integrated circuit industry. IO technology on silicon substrates is usually carried out in SiO_2, and there are two generic approaches to the Si/SiO_2 fabrication that have proven capable of producing very high performance IO devices. IO devices using both approaches are characterized by waveguides that are extremely low-loss and are easily matched in mode characteristics to optical fibers used for transmission, thereby providing very efficient coupling.

The first approach borrows more from the technology of optical fiber manufacture than it does from the Si electronics industry.[32] Using a technique known as flame hydrolysis (FHD), a "soot" of SiO_2 is deposited on a Si wafer to a depth of 50–60 μm, followed by a thinner layer of a SiO_2/GeO_2 mix to form what will become the high-index waveguide core. This material is consolidated at ~1300°C for several hours down to roughly half its original thickness, and then the waveguide core layer is patterned using reactive ion etching to form square cross section waveguide cores. Then FHD is again used, followed by more consolidation, to form the upper cladding layers. Typical index differences for the core material are in the range of 0.25 percent to 0.75 percent, with core dimensions of 6–8 μm^2.

A measure of the material quality obtained using this approach is given by some of the extraordinary devices results obtained. IO power splitters have been fabricated to sizes of 1×128 using seven stages and a total of 127 Y-branch 1×2 splitters and a total device length of 5 cm. Total *fiber-to-fiber* excess loss for this device was 3.2 dB with a standard deviation of 0.63 dB.[33] A large variety of devices has been made using this technique.

Another technique for Si/SiO_2 fabrication employs film deposition technology borrowed from silicon electronics processing.[33] First a base SiO_2 layer is deposited using high-pressure steam to a thickness of ~15 μm to prevent leakage to the high-index Si substrate. The waveguide and cladding layers are deposited using low-pressure chemical vapor deposition, either from silane and oxygen, or from tetraethylorthosilane and ammonia. Phosphine is added to increase the index, with guide cores typically containing 6.5–8 percent P. The wafer is usually annealed at 1000°C to relieve strain and to densify the films. Waveguide losses below 0.05 dB/cm have been reported[34] using this technique, and a large variety of devices have been demonstrated using this approach, including splitters, couplers, and WDM devices. One of the interesting features of this approach to fabrication is that it readily lends itself to the inclusion of other thin films common in Si processing. One such film is Si_3N_4 and this has been used as a high-index core for waveguides with much larger core-cladding index step. Such waveguides can generate tightly confined modes that are a much closer match to the modes commonly found in active semiconductor components such as lasers. This feature has been used in a novel mode converter device[35] that adiabatically transforms from the smaller mode into the larger, fiber-matched mode commonly employed in Si/SiO_2 IOCs.

In some instances, slow-response *active* devices have been fabricated in Si/SiO_2 technology using thermal effects to achieve local index changes in one arm of a Mach-Zehnder interferometer. This can either be used as a thermo-optic switch or as a tuning element in WDM components. The heating element in these devices comprises a simple metal film resistive heater deposited directly on the upper surface of the wafer.

Another characteristic feature of IOCs in Si/SiO_2 is a degree of birefringence that results from the compressive stress induced in the film by the Si substrate after cooling down from the high-temperature film deposition or consolidation. Typical amounts of birefringence are $n_{TE} - n_{TM} = 3 \times 10^{-4}$. This birefringence can cause wavelength shifts with input polarization in WDM components, and techniques to counteract it include stress removal by adding strain-relief grooves in the film, or stress compensation by adding a counteracting stress-inducing film on the surface of the guide.

The Si/SiO_2 technology has not fully matured, but it certainly promises to play an important role in high-performance passive components.

LiNbO₃ and LiTaO₃

The majority of the integrated optics R&D from 1975 to 1985 and the majority of the currently commercial integrated optics product offerings utilize $LiNbO_3$ as the substrate material. A number of excellent review papers detailing R&D efforts in $LiNbO_3$ are available.[36–40] $LiNbO_3$ is an excellent electro-optic material with high optical transmission in the visible and near infrared, a relatively large refractive index ($n = 2.15–2.2$), and a large electro-optic coefficient ($r_{33} = 30.8 \times 10^{-10}$ cm/V). Probably most important, but frequently overlooked, is the widespread availability of high-quality $LiNbO_3$ wafers. Hundreds of tons of $LiNbO_3$ are produced annually for the fabrication of surface acoustic wave (SAW) devices. This large volume has resulted in well-developed crystal growth and wafer processing techniques. In addition, $LiNbO_3$ wafers are at least an order of magnitude less expensive than they would be if integrated optics was the only application of this material. High-quality three- and four-inch optical-grade $LiNbO_3$ wafers are now available from multiple vendors.

$LiNbO_3$ is a uniaxial crystal which is capable of supporting an extraordinary polarization mode for light polarized along the optic axis (z-axis) and an ordinary polarization mode for light polarized in the $x - y$ plane. LiNbO$_3$ is slightly birefringent with $n_e = 2.15$ and $n_o = 2.20$. LiNbO$_3$ devices can be fabricated on x-, y-, and z-cut wafers. Phase modulators, fiber gyro circuits, and Mach-Zehnder interferometers are typically fabricated on x-cut, y-propagating wafers, and operate with the TE (extraordinary) mode. Push-pull devices, such as delta-beta directional coupler switches, are typically fabricated on z-cut, y-propagating wafers and operate with the TM (extraordinary) mode. Both configurations utilize the strong r_{33} electro-optic coefficient. Devices that require two phase-matched modes for operation are typically fabricated on x-cut, z-propagating wafers.

The majority of LiNbO$_3$ integrated optic devices demonstrated to date have been fabricated using the titanium in-diffusion process.[41] Titanium strips of width 3–10 microns and thickness 500–1200 angstroms are diffused into the LiNbO$_3$ at 950–1050°C for diffusion times of 5–10 h.[42,43] The titanium diffusion results in a local increase in both the ordinary and extraordinary refractive indices so that both TE and TM modes can be supported for any crystal orientation. Titanium thickness and strip width typically need to be controlled to ±1 percent and ±0.1 microns, respectively, for reproducible device performance. Due to the high processing temperatures that approach the Curie temperature of LiNbO$_3$, extreme care must be taken to prevent Li$_2$O out-diffusion[44,45] and ferroelectric domain inversion, both of which significantly degrade device performance. Photorefractive optical damage[46] also needs to be considered when utilizing Ti-diffused devices for optical wavelengths shorter than 1 micron. Optical damage typically prevents the use of Ti-diffused devices for optical power greater than a few hundred μW at 800-nm wavelength, although the problem can be reduced by utilizing MgO-doped LiNbO$_3$ wafers. Optical damage is typically not a problem at 1300 and 1550 nm for optical powers up to 100 mW.

An alternative process for fabricating high-quality waveguides in LiNbO$_3$ is the annealed proton exchange (APE) process.[47,48] In the APE process, a masked LiNbO$_3$ wafer is immersed in a proton-rich source (benzoic acid is typically used) at temperatures between 150 and 245°C and times ranging from 10 to 120 min. The wafer is then annealed at temperatures between 350 and 400°C for 1 to 5 h. During the initial acid immersion, lithium ions from the wafer are exchanged with hydrogen ions from the bath in the unmasked region, resulting in a stress-induced waveguide that supports only the extraordinary polarization mode. Proton-exchanged waveguides that are not subjected to further processing are practically useless due to temporal instabilities in the modal propagation constants, high propagation loss, DC drift, and a much-reduced electro-optic coefficient. However, it has been demonstrated[48] that proper post annealing results in extremely high-quality waveguides that are stable, low-loss, and electro-optically efficient.

The APE process has recently become the fabrication process of choice for the majority of applications currently in production. Since the APE waveguides only support the extraordinary polarization mode, they function as high-quality polarizers with polarization extinction in excess of 60 dB.[49] As described later in this chapter, high-quality polarizers are essential for reducing the drift in fiber optic gyroscopes and minimizing nonlinear distortion products in analog links. APE waveguides exhibit low propagation losses of 0.15 dB/cm for wavelengths ranging from 800 to 1550 nm. APE LiNbO$_3$ devices exhibit stable performance for optical powers of 10 mW at 800 nm and 200 mW at 1300 and 1550 nm. The APE process can also be used to fabricate devices in LiTaO$_3$ for applications requiring higher optical powers (up to 200 mW) at 800 nm.[50] In addition to offering performance advantages, the APE process also appears to be the more manufacturable process. It is relatively easy to scale the APE process so that it can handle 25-wafer lots with no degradation in device uniformity. The fiber pigtailing requirements are also substantially reduced when packaging APE devices since these devices only support a single polarization mode.

After the waveguides have been fabricated in the LiNbO$_3$ wafer, electrodes need to be deposited on the surface. One-micron-thick gold is typically used for lumped-electrode

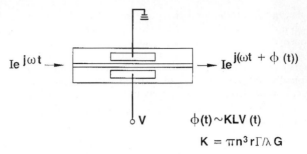

FIGURE 6 Top-down view of a typical LiNbO$_3$ phase modulator. Field is applied laterally across the guide by surface electrodes on each side.

devices while five-micron-thick gold is typically used for traveling-wave devices to reduce RF resistive losses. The lift-off process and electron-beam deposition is typically used for lumped-electrode devices while up-plating is typically used for realizing the thicker gold electrodes. Better than 0.5-micron layer-to-layer registration is required for optimum device performance. As shown in Fig. 6, electrodes on x-cut LiNbO$_3$ are usually placed alongside the waveguide so that the horizontal component of the electric field interacts with the propagating TE mode. Electrodes on z-cut LiNbO$_3$ are placed on top of the waveguide so that the vertical component of the electric field interacts with the propagating TM mode. An SiO$_2$ buffer layer (0.1 to 1 micron thick) is required between the optical waveguide and the electrode on all z-cut devices to reduce metal-loading loss. A thick (1 micron) SiO$_2$ buffer layer is also utilized on some x- and z-cut devices to reduce the velocity mismatch between the microwave and optical waves in high-speed traveling-wave modulators. A thin layer of amorphous silicon is also utilized on some z-cut devices to improve device stability over temperature.

III-V Materials and Fabrication Technology

In this section we will briefly review some of the epitaxial growth and fabrication techniques that are used to make PICs in III-V materials, with a primary focus on InP-based devices.

III-V Epitaxial Crystal Growth. The epitaxial growth of III-V optoelectronic materials has evolved rapidly during the last decade from nearly exclusive use of manually controlled liquid-phase epitaxial (LPE) growth to a variety of highly versatile computer-automated vapor and beam growth techniques. These include atmospheric-pressure and low-pressure metal-organic vapor-phase epitaxy (MOVPE), hydride and chloride vapor-phase epitaxy (VPE), molecular beam epitaxy (MBE), chemical beam epitaxy (CBE), and metal-organic molecular beam epitaxy (MOMBE). Detailed descriptions of reactor design and growth chemistry are beyond the scope of this section, and the interested reader is referred to recent texts and conference proceedings for the most current information.[51]

One of the critical criteria for evaluating crystal growth is the uniformity, both in thickness and in epitaxial composition. Layer thickness changes of several percent can lead to nm-scale wavelength changes in grating-based lasers and filter devices. Similarly, compositional changes leading to a 10-nm shift in the photoluminescence peak wavelength of the guide layers, which is not at all uncommon, can also result in nm-scale wavelength shifts in distributed feedback (DFB) laser emission wavelengths, in addition to potential undesirable gain-peak mismatches that may result from the λ_{PL} shift itself.

Proper reactor geometry, sometimes with substrate rotation, have been shown capable of percent-level uniformity both in MOVPE and in the beam techniques. One difficulty associated with the latter lies in the ballistic "line-of-sight" growth which prevents regrowth over re-entrant mesa geometries or overhanging mask surfaces often encountered in PIC and laser fabrication, while MOVPE and especially VPE offer outstanding coverage over a wide range of morphologies. Other criteria to be considered are the doping capabilities. The lower growth temperatures associated with MBE, CBE, and MOMBE enable very abrupt changes in doping level, and highly concentrated doping sheets that are desirable for high-speed transistors in OEICs, for example. Both the vapor and beam techniques have successfully grown semi-insulating Fe-doped InP, a material that is playing an increasingly pivotal role in photonic devices.

The typical PIC processing involves the growth of a base structure that is followed by processing and regrowths. During both the base wafer and regrowths, selective area growth is often employed where a patterned dielectric film is used to prevent crystal growth over protected areas. This film is typically SiO_2 or Si_3N_4 deposited by CVD or plasma-assisted CVD. This technique is readily used with MOVPE, but care must be taken to keep a substantial portion of the field open for growth to avoid the formation of polycrystalline deposits on the dielectric mask. Caution must be exercised during regrowths over mesas or other nonplanar geometries, as well as in the vicinity of masked surfaces. Gross deviations from planarity can occur due to overshoots of crystal growth resulting from crystal-orientation-dependent growth rates on the various exposed surfaces.

III-V Etching Technology.

A fundamenal step in III-V PIC processing is mesa etching for definition of the optical waveguides. This is usually accomplished by patterning a stripe etch mask on a base wafer that has a number of epitaxial layers already grown, and removing some of the layers in the exposed regions to leave a mesa comprised of several of the epitaxial layers. The etching process can either be a "wet" chemical etchant, or a "dry" plasma-type etch.

Wet etching refers to the use of an acid bath to attack the unprotected regions of a surface. The acids that are commonly used to etch III-V materials[52] also tend to significantly undercut a photoresist pattern, and hence photoresist is usually used only in broad-area features or in shallow etches where undercutting is not a concern. For precise geometries such as waveguide stripes, another masking material such as SiO_2 or Si_3N_4 is first deposited and patterned with photoresist and plasma etching, or HF etching (for SiO_2).

In some instances, it is required that the etchants be nonselective, uniformly removing layers regardless of composition. This is usually the case when etching a mesa through a multilayer active region to form a buried heterostructure laser. Br-based etchants such as bromine in percent-level concentration in methanol, tend to be very good in this regard. This etchant, along with many of the nonselective etchants, will form a re-entrant 54.7° (111A) face mesa for stripes along the (011) direction (with a nonundercutting mask) and will form an outward-sloping 54.7° walled mesa for stripes along the (01$\bar{1}$) direction. Other etchants, with varying degrees of nonselectivity and crystallographic behavior, include mixtures of HBr, CH_3COOH, or HCl, CH_3COOH, and H_2O_2.[53]

In fabricating precise geometries in III-V integrated optic or PIC devices, it is often desirable to remove specific layers while leaving others, or control mesa heights to a very high degree of precision. The combination of material-selective etchants and the inclusion of special etch-stop layers offers a convenient and precise means of achieving this. 100-Å-thick layers of InGaAsP can easily halt InP etches even after many microns of etching. Extensive compilations have been made of etches for the InP-based compounds,[52] and the most common selective InP etches are HCl-based. Typical mixtures are HCl and H_3PO_4 in ratios ranging from $3:1$ to $1:3$, with the lower HCl content leading to less undercutting and slower etch rates. The HCl-based etchants are highly crystallographic in nature,[54] and can produce mesas with nearly vertical walls or outward-sloping walls, depending on the mesa stripe orientation.

A common selective etch for removing InGaAsP or InGaAs while only weakly attacking InP are mixtures of H_2SO_4, H_2O_2 and H_2O, in a ratio of $X:1:1$ with X typically ranging from 3 to 30. Selectivities in the range of $10:1$ and typically much higher are readily achieved. Other selective etchants for InGaAsP are based on HNO_3 or mixtures of KOH, $K_3Fe(CN)_6$, and H_2O.

Dry etching techniques, such as reactive ion etching (RIE), or other variants, such as chemically assisted reactive ion beam etching (CAIBE), also play a key role in III-V PIC processing. These have often been carried out using Cl_2-based mixtures with O_2 and Ar,[55] while in other cases the reactive chlorine is derived from compounds such as CCl_2F_2. Recent work has demonstrated excellent results with methane/hydrogen mixtures or ethane/hydrogen.[56] In these latter cases, Ar is also often used as a sputtering gas to remove interfering redeposited compounds. Reactive ion etching has been used both to form mesa and facet structures as well as in transferring grating patterns into semiconductors through an etch mask.

The appeal of reactive ion etching is the lack of mask undercutting that can usually be achieved, allowing very high lateral precision with the promise of reproducible submicron mesa features. In addition, the ability to create vertical-wall etched facets through a variety of different composition epitaxial layers suggests the possibility of integrated resonator or reflecting and coupling structures without the use of gratings. This approach has been used to form corner reflectors,[57] square-geometry ring-resonators,[58] and a variety of complex waveguide patterns using beam splitters.[59] Another recent application has been the use of etched-facet technology to create gratings, not as an interfacial corrugation *along* the waveguide, but as a grating in the other dimension *at the end surface* of a waveguide for two-dimensional "free-space" grating spectrometers.[60,61]

Grating Fabrication. Many of the PICs employ corrugated-waveguide grating-based resonators or filters, and the most common technique for fabricating these gratings involves a "holographic" or interferometric exposure using a short-wavelength laser source. Here a thin (typically 500–1000-Å-thick) layer of photoresist is spun on a wafer surface and exposed with two collimated, expanded beams from a blue or UV laser at an appropriate angle to form high contrast fringes at the desired pitch. Since the illuminating wavelength is precisely known, and angles are easily measured in the mrad range, the typical corrugation in the 2000-Å-period range can be fabricated to Å-level precision in period. The resist is developed and then functions as an etch mask for the underlying layers. This etching can be either a wet etch (commonly using HBr-based etchants), or a dry reactive ion etch. Commonly used lasers are HeCd at 325 nm or one of the UV lines of an argon ion laser at 364 nm. Electron-beam lithography has also been successfully applied to the generation of gratings for III-V integrated optic devices.

Active-Passive Transitions. Compound semiconductors are appealing for PICs in large part due to their ability to emit, amplify, and detect light. However, waveguide elements that perform these functions are not low-loss without excitation, and are generally not suitable for providing passive interconnections between circuit elements. One of the most fundamental problems to overcome is the proper engineering and fabrication of the coupling between active waveguides, containing lower bandgap material, and passive waveguides composed of higher bandgap material.

Most PICs demonstrated to date have employed some form of butt-coupling, where an active waveguide of one vertical and/or lateral structure mates end-on with a passive waveguide of a different vertical and/or lateral structure. Butt-coupling offers design simplicity, flexibility, and favorable fabrication tolerances. The most straightforward approach for butt-coupling involves the selective removal of the entire active waveguide core stack using selective wet chemical etching, followed by a regrowth of a mated, aligned passive waveguide structure. The principle advantage of such an approach is the independent selection of compositional and dimensional design parameters for the two guides.

Another approach to butt-coupling employs a largely continuous passive waveguide structure with a thin active layer residing on top, which is selectively removed on the portions of the structure which are to be passive. Using material-selective wet chemical etches, the thin active layer (often a thin MQW stack) can be removed with very high reproducibility and precision, and the dimensional control is thus placed in the original computer-automated MOVPE growth of the base wafer. The removal of the thin active layer constitutes only a small perturbation of the continuous guide core constituted by the lower, thicker layer, and efficient coupling can be achieved.

Yet another approach to coupling between two different waveguides employs directional coupling in the vertical plane between epitaxial layers serving as the cores of the two distinct waveguides. This type of vertical coupling can either be accomplished using the principle of intersecting dispersion curves, or through the use of a corrugated-waveguide grating to achieve phase matching. Vertical coupler structures may be useful for wide-tuning applications, since a small change of effective index for one mode can lead to a large change in coupling wavelength.[62]

Organic Polymers

Polymer films are a relatively new class of materials for integrated optics.[63] Polymers offer much versatility, in that molecular engineering permits many different materials to be fabricated; they can be applied by coating techniques to many types of substrates, and their optical and electro-optical properties can be modified in a variety of ways. Applications range from optical interconnects, in which passive guides are used in an optical PC board arrangement, to equivalents of IOCs and OEICs. Polymer devices are also being explored for third-order nonlinear applications.

While numerous methods for fabricating polymer waveguide electro-optic devices have been reported, the most attractive technique consists of spin-coating a three-layer polymer sandwich over a metal film, often on a semiconductor (Si) substrate. The three polymer layers form a symmetric planar waveguide; the middle layer is electro-optic, due to the presence of a guest molecule that imparts the electro-optic property, or the use of a side-chain polymer. The sample is overcoated with metal and the entire structure is heated near the glass transition temperature and poled at an electric field of typically $150\,V/\mu m$. The poling aligns the nonlinear molecules in the middle polymer layer, thereby inducing the Pockels effect and a birefringence. Typical values of index and birefringence are 1.6 and 0.05 respectively. Electro-optic coefficients are in the 16–38-pm/V range. Channel waveguides are subsequently defined by a variety of methods. An attractive technique is photobleaching, in which the waveguide region is masked with a metal film and the surrounding area exposed to UV light. This exposure alters the molecules/linking in the active layer, thereby reducing the refractive index and providing lateral confinement. Losses in such guides are typically in the 1-dB/cm range.

The basic IO modulators have been demonstrated in a variety of polymers. Of particular note is a traveling wave modulator with a 3-dB bandwidth of 40 GHz and a low-frequency V pi of 6 V.[64] Relative to LiNbO$_3$, polymer modulators can have higher overlap factors because the lower metal layer provides vertical, well-confined signal fields. However, the relatively low index of polymers and their comparable electro-optic coefficient to LiNbO$_3$ implies a lower electro-optic efficiency. Polymers do provide a better velocity match of optical and electromagnetic velocities, which can result in very high frequency performance as described above.

For polymers to fulfill their potential, a number of material and packaging issues must be addressed. First, it is highly desirable to develop polymers that overcome the long-term relaxation of the electro-optic effect typical of many of the materials reported to date. Development of polymers with transition temperatures in the 300°C range (so they can withstand the temperatures typical of device processing and packaging) is also highly desirable. Work on polyimide is particularly promising in this area. Finally, techniques to polish device end faces, pigtail, and package these devices need to be developed.

6.5 CIRCUIT ELEMENTS

Passive Devices

This section provides a phenomenological description of the most common passive and active IO devices. Detailed descriptions of the device theoretical properties as well as typical characteristics can be found in the literature.[1-3]

Passive guided wave devices are the fundamental building blocks and interconnection structures of IOCs and OEICs. Passive devices are here defined as those dielectric waveguide structures which involve neither application of electrical signals nor nonlinear optical effects. This section will focus on the most important structures: waveguide bends, polarizers, and power splitters, and on the closely related issue of fiber-to-chip coupling.

Waveguide bends, such as those illustrated in Figs. 10, 14, and 18, are needed to laterally offset modulators and increase device-packing density. The most widely used bend is based on an S-bend geometry described by a raised cosine function.[1] This structure minimizes the tendency of light in a dielectric waveguide to "leak" as the guide's direction is altered by starting with a small bend (large effective bend radius) and then increasing the bend rate until the midpoint of the offset, then following the pattern in reverse through the bend completion.

Since the index difference between the guide and surrounding dielectric material is usually small (10^{-3} to 10^{-4}), bends must be gradual (effectively a few degrees) to keep losses acceptably (<0.5 dB) small. In LiNbO$_3$, offsets of 100 microns require linear distances of typically 3 mm. In semiconductor research device work, designs with high index steps are sometimes used to form small-radius bends, and selective etching has been utilized to form reflective micro mirrors[57] at 45° to the guide to create a right-angle bend. To date this latter approach is relatively lossy.

Polarizers are necessary for polarization-sensitive devices, such as many electro-optic modulators, and in polarization-sensitive applications such as fiber gyroscopes. Polarizers can be formed on dielectric waveguides that support both TE and TM propagation by forming overlays that selectively couple one polarization out of the guide. For example, a plasmon polarizer formed on LiNbO$_3$ by overcoating the guide with a Si$_3$N$_4$/Au/Ag thin-film sandwich selectively attenuates the TM mode.[65] In some materials it is possible to form waveguides that only support one polarization (the other polarization is not guided and any light so polarized radiates into the substrate). By inserting short (mm) lengths of such guides in circuits or alternatively forming entire circuits from these polarizing guides, high extinction can be obtained. For example, annealed proton exchange waveguides (APE) in LiNbO$_3$ exhibit polarization extinction ratios of at least 60 dB.[66]

Guided wave devices for splitting light beams are essential for most IOCs. Figure 7 illustrates the two common splitters: a directional coupler and a Y junction. The figure illustrates 3-dB coupling (1X2); by cascading such devices and using variations on the basic designs it is possible to fabricate N × N structures. IO splitters of complexity 8 × 8 are commercially available in glass.

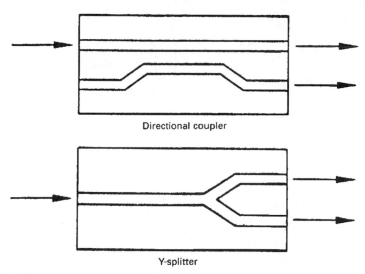

Directional coupler

Y-splitter

FIGURE 7   Passive directional coupler and Y-branch IO splitter devices.

The operation of the directional coupler is analogous to the microwave coupler and is described by the coupled mode equations. The coupling strength is exponentially dependent of the ratio of the guide spacing and the effective tail length of the guided mode. Thus, when guides are far apart (typically greater than 10 microns), as in the left-most portion of the structure in Fig. 7, there is negligible coupling. When the guides are close together (typically a few microns), power will couple to the adjacent guide. The fraction of power coupled is sinusoidally dependent on the ratio of the interaction length to the coupling length L_c. L_c is typically 0.5–10 mm and is defined as that length for full power transfer from an incident guide to a coupled guide. The 3-dB coupling illustrated requires an interaction length of half L_c.[1-3] Operation of this device is symmetric; light incident in any one of the four inputs will result in 3-dB splitting of the output light. However, if coherent light is incident on both input guides simultaneously, the relative power out of the two output guides will depend on the phase and power relationship of the incident signals.

The Y splitter illustrated in Fig. 7 operates on a modal evolution principle. Light incident on the junction from the left will divide symmetrically so that the fundamental mode of each output branch is excited. Branching circuit design follows closely from the design of waveguide bends. The Y-junction angle is effectively a few degrees and the interaction length is a few mm. Operation of this device is not symmetric with respect to loss. If coherent light is incident on both guides from the right, the amount of light exiting the single guide will depend on the power and phase relationship of the optical signals as they enter the junction area. If coherent light is only incident in one arm of the junction from the right, it will experience a fundamental 3-dB loss in propagation to the left to the single guide. This is due to the asymmetric modal excitation of the junction.

An extremely important issue in integrated optics is the matching of the waveguide mode to the mode of the fiber coupled to the guide. Significant mode mismatch causes high insertion loss, whereas a properly designed waveguide mode can have coupling loss well under 1 dB. To design the guide, one must estimate the overlap integral of the optical fields in the two media. It is reasonable to assume that some sort of index matching between the two materials is also employed. Figure 8 illustrates the issue with mode profiles in the two transverse directions for a Ti indiffused guide. In general the IO mode is

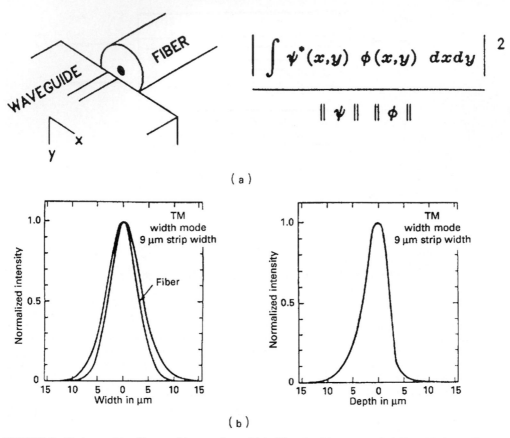

$$\frac{\left| \int \psi^*(x,y) \; \phi(x,y) \; dx\,dy \right|^2}{\| \psi \| \; \| \phi \|}$$

(a)

(b)

FIGURE 8 Mode-matching illustrated by coupling a Ti indiffused guide to an optical fiber. Mode profiles are shown both in width and depth for the waveguide.

elliptical and often asymmetrical relative to the fiber mode. It should be noted that the loss obtained on a pigtailed fiber-chip joint is also highly determined by the mechanical method of attaching the fiber to the chip. Most techniques use some sort of carrier block for the fiber (e.g., a Si V-groove) and attach the block to the IO chip. Performance on commercially available devices is typically <1 dB coupling loss per interface with robust performance over at least the −30 to 60°C range.

Active Devices

Active IO devices are those capable of having their state altered by an external applied voltage, current, or other stimulus. This may include electro-optic devices, or devices that generate, amplify, or detect light. Active IO devices in non-semiconducting dielectrics generally depend on the linear electro-optic effect, or Pockels effect.[20] The electro-optic effect results in a change of the index of refraction of a material upon the application of an electric field. Typical values for a variety of materials is about 10^{-4} for a field of 10^{-4} V/cm. This results in a phase change for light propagating in the field region and is the basis for a large family of modulators.

The fundamental guided wave modulator is a phase modulator, as illustrated in Fig. 6. In this device, electrodes are placed alongside the waveguide, and the lateral electric field determines the modulation. In other modulator designs, the vertical field component is used. For the geometry shown, the phase shift is KLV, where K is a constant, L is the electrode length, and V is the applied voltage. For LiNbO$_3$, $K = \pi n^3 r_{63} \Gamma / g\lambda$ for the preferred orientation of field along the z (optic) axis and propagation along the y-axis. Here, n is the index, r_{63} is the electro-optic coefficient, λ is the wavelength, g is the electrode gap, and Γ is the overlap of the electrical and optical fields. In general, the value of K is anisotropic and is determined by the electro-optic tensor.[21] It should be noted that modulators are often characterized by their V_π value. This is the voltage required for a pi-radian phase shift; in this nomenclature, phase shift is written as $\Phi = \pi \cdot V/V_\pi$ where $V_\pi \equiv \pi/KL$. Due to the requirement that the optical field be aligned with a particular crystal axis (e.g., in LiNbO$_3$ and III-V semiconductors), the input fiber on modulators is generally polarization maintaining.

Modulators in LiNbO$_3$ typically have efficiencies at 1.3 microns of 50°/volt-cm, a V_π of 5 V for a 1-GHz 3-dB bandwidth, and a fiber-to-fiber insertion loss of 2-3 dB. In semiconductors, modulation efficiencies can be significantly higher if one designs a tightly guided mode (i.e., one well-suited for on-chip laser coupling, but having a relatively high fiber-to-chip mismatch coupling loss).

The modulation bandwidth of phase and intensity modulators is determined by the dielectric properties of the electro-optic material and the electrode geometry. For structures in which the electrode length is comparable to or shorter than a quarter RF wavelength, it is reasonable to consider the electrodes as lumped and to model the modulator as a capacitor with a parasitic resistance and inductance. In this case, the bandwidth is proportional to $1/L$. For most IO materials, lumped-element modulators have bandwidths less than 5 GHz to maintain reasonable drive voltages. For larger bandwidths, the electrodes are designed as transmission lines and the RF signal copropagates with the optical wave. This is referred to as a traveling wave modulator. The microwave performance of this type of structure is determined by the degree of velocity match of the optical and RF waves, the electrode microwave loss, the characteristic impedance of the electrodes, and a variety of microwave packaging considerations. In general, semiconductor and polymer modulators fundamentally have better velocity match than LiNbO$_3$ and thus are attractive for highest frequency operation; however, in recent years, techniques and structures have been developed to substantially improve the velocity match in LiNbO$_3$ and intensity modulators with 50 GHz bandwidth have been reported.[67]

To achieve intensity modulation, it is generally necessary to incorporate a phase modulator into a somewhat more complex guided wave structure. The two most common devices are the Mach-Zehnder (MZ) and the directional coupler modulator. Figure 9 illustrates the MZ modulator. This device is the guided wave analog of the classical MZ interferometer. The input and output Y-junctions serve as 3-dB splitters, and modulation is achieved in a push-pull manner by phase-modulating both arms of the interferometer. The interferometer arms are spaced sufficiently that there is no coupling between them. When the applied voltage results in a pi-radian phase shift in light propagating in the two arms when they recombine at the output junction, the resultant second-order mode cannot be guided and light radiates into the substrate. The output intensity I of this device is given by $I = I_0/2[1 + \cos(KLV)]$. The sinusoidal transfer characteristic is unique in IO modulators and provides the unique capability to "count fringes" by applying drive signals that are multiples of V_π. This feature has been exploited in a novel analog-to-digital converter.[68] The device can be operated about its linear bias point $\pi/2$ for analog applications and can also be used as a digital switch. A variation on this device is a balanced bridge modulator. In this structure the two Y-junctions are replaced by 3-dB directional couplers. This structure retains a sinusoidal transfer characteristic, but can function as a 2×2 switch.

A directional coupler switch is shown in Fig. 10. In the embodiment illustrated, a set of electrodes is positioned over the entire coupler region. The coupler is chosen to be L_c, a

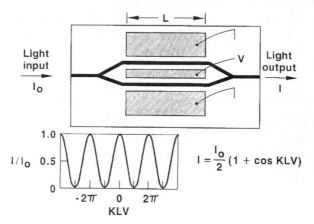

$$I = \frac{I_0}{2}(1 + \cos KLV)$$

FIGURE 9 Geometry of lumped-element Mach-Zehnder modulator and transfer characteristic.

coupling length long, so that in the absence of an applied voltage, all light incident in one guide will cross over and exit the coupled guide. The performance of the directional coupler switch can be modeled by coupled mode theory. The application of an electric field spoils the synchronism of the guides, resulting in reduced coupling, and a shorter effective coupling length. For application of a voltage such that $KLV = \pi\sqrt{3}$, all light will exit the input guide. In general, the transfer characteristic is given by

$$I = \frac{I_0}{2(1 + (KLV/\pi)^2)} \cdot (1 - \cos(\pi\sqrt{1 + (KLV/\pi)^2})) \tag{14}$$

Directional coupler switches can also be used for analog or digital modulation. They have also been fabricated in matrix arrays for applications in $N \times N$ switch arrays (see **Optical Storage and Display** in this chapter). To increase the fabrication tolerance of directional

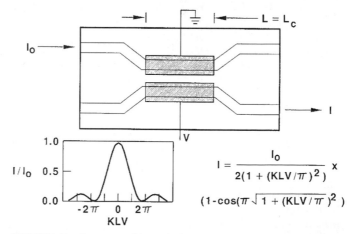

$$I = \frac{I_0}{2(1 + (KLV/\pi)^2)} \times (1 - \cos(\pi\sqrt{1 + (KLV/\pi)^2}))$$

FIGURE 10 Geometry of lumped-element directional coupler switch and transfer characteristic.

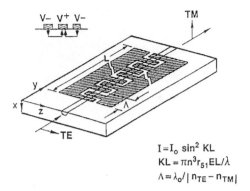

$$I = I_0 \sin^2 KL$$
$$KL = \pi n^3 r_{51} EL / \lambda$$
$$\Lambda = \lambda_0 / |n_{TE} - n_{TM}|$$

FIGURE 11 TE-TM mode converter using periodic electrodes to achieve phase matching.

coupler switches, designs based on reversing the sign of index change (delta beta) periodically along the coupler have been developed. The most common device consists of a device 1–2 coupling lengths long and a single reversal of the voltage formed by a two-section electrode.

Both Mach-Zehnder and directional coupler devices have been developed in semiconductors, LiNbO$_3$, and polymers. Devices are commercially available in LiNbO$_3$. Drive voltages and bandwidths achieved are similar to the values quoted above for phase modulators. Additional effort has been focused in LiNbO$_3$ to make devices that are polarization insensitive so that they are compatible with conventional single-mode fiber.[69] Mach-Zehnder modulators have also been formed in glass waveguides. Here a resistive pad is heated to vary the index of the waveguide via the thermo-optic effect.

Another important IO component is the TE-to-TM mode converter. This device, illustrated in Fig. 11, depends on an off-diagonal component r_{51} of the electro-optic tensor in LiNbO$_3$ to convert incident TE (TM) light to TM (TE) polarization. In the converter, a periodic electrode structure is used to create a periodic index change along the waveguide to provide phase matching, and thus coupling, between the TE and TM wave. The period Λ of this index change is given by $\Lambda = \lambda/(n_{TE} - n_{TM})$. The coupling efficiency at the phase-matched wavelength is given by $\sin^2(\kappa L)$ where $\kappa = \pi n^3 r_{51} E / \lambda$ and E is the applied field. This type of device can be used for polarization control. In a polarization controller, a phase modulator is placed before and after the converter so that signals of arbitrary input polarization can be converted into any desired output polarization. The converter also serves as the basis for a tunable wavelength filter.[70]

There are numerous other types of IO intensity modulators that have been reported. These include a crossing channel switch, a digital optical switch, an acousto-optic tunable wavelength switch, and a cut-off modulator. The first two devices depend on modal interference and evolution effects. The acousto-optic switch utilizes a combination of acoustically induced TE-to-TM mode conversion and TE-TM splitting couplers to switch narrow-optical-band signals. The cut-off modulator is simply a phase modulator designed near the cutoff of the fundamental mode such that an applied field effectively eliminates the guiding index change between the guide and the substrate. This results in light radiating into the substrate.

In addition to the electro-optic devices described above, another common modulation technique employed in III-V materials employs the electroabsorption of electrorefraction effects discussed previously. Here the bandgap energy of a bulk medium or an appropriately engineered quantum-well medium is chosen to be somewhat higher than the

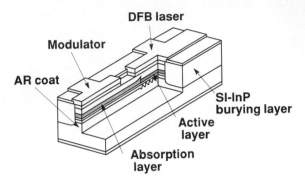

FIGURE 12 Integrated semiconductor laser/electroabsorption modulator PIC.

energy of the propagating photons. An applied field directly induces absorption, or a large index shift associated with the change in absorption at higher energy. The latter effect is used interferometrically in directional couplers, Mach-Zehnder modulators, or other designs as an enhanced substitute for the conventional electro-optic effect. The former is used as a single-pass absorptive waveguide modulator.

To achieve low operating voltages, such modulators are usually designed with small waveguide dimensions for tight confinement of the optical mode. This usually leads to a significant insertion loss of ~2–3 dB/cm when coupling to optical fibers. However, the tight waveguide mode is very similar to the waveguides employed in semiconductor lasers, and hence the primary appeal of the waveguide electroabsorption modulators lies in their potential for integration with semiconductor lasers on a single PIC chip.[71]

A particular implementation used by Soda et al.[72] is shown schematically in Fig. 12. A 1.55-μm DFB laser structure is mated to an electroabsorption modulator with an InGaAsP core layer having a photoluminescence wavelength of $\lambda_{PL} \sim 1.40 \, \mu$m. The entire structure uses a buried heterostructure waveguide with semi-insulating InP lateral cladding to provide good current blocking with low capacitance for high modulator bandwidth. Optimization of the modulator core λ_{PL} is very important in this device. With proper design, devices have yielded a good compromise between high output power and high modulator extinction ratios with low voltage drive. Typical device designs exhibit mW-level fiber-coupled output power with a −10-dB extinction ratio at drive levels of 2–4 V.

The integrated DFB/electroabsorption modulator mentioned previously provided one illustration of laser integration. Common to most laser/detector/waveguide PICs is the inclusion of a guide containing an amplifying or gain medium, or an absorptive medium for detection, and the requirements of current drive or extraction. The design and processing associated with these guided-wave components relies heavily on a relatively mature technology associated with semiconductor lasers.[73]

The gain section of a semiconductor laser is usually fabricated in a buried heterostructure guide as shown in Fig. 2a, and is driven through a forward biased p-n junction where the layers are usually doped during the crystal growth. With zero or reverse bias, this same structure can function as a waveguide photodetector. In a DFB laser or a distributed Bragg reflector (DBR) laser, this feature can be used to provide an integrated detector, on the back end of the device external to the cavity, for monitoring laser power. Alternatively, a separate gain medium external to the laser cavity can be located external to the cavity on the *output* side to function as an integrated power amplifier for the laser. Such a structure is shown in Fig. 13, where a DBR laser is followed by a fan-shaped amplifier to keep the amplifier medium at a relatively constant state of saturation. These PICs are termed

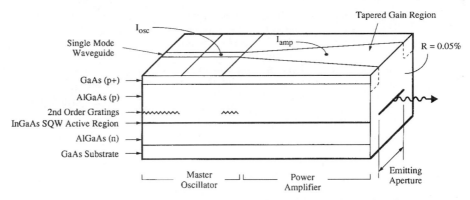

FIGURE 13 Integrated semiconductor master-oscillator/power-amplifier (MOPA) PIC.

master-oscillator/power-amplifiers (MOPAs), and can provide Watt-level single-frequency, diffraction-limited output beams from a single chip.[74]

The challenge of laser integration is to fabricate other guided-wave components without compromising the intricate processing sequence required to make high-performance lasers. Figure 14 shows a balanced heterodyne receiver PIC suitable for coherent optical communications links.[75] Here a tunable local oscillator is tuned to an optical frequency

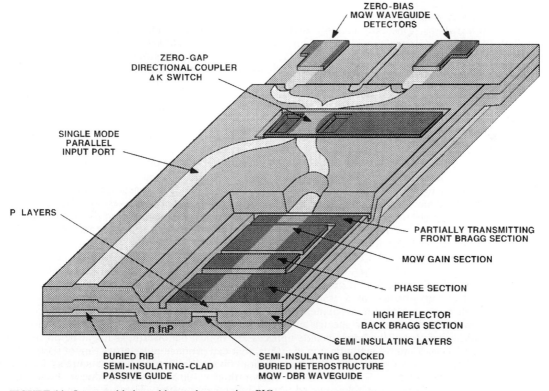

FIGURE 14 Integrated balanced heterodyne receiver PIC.

offset by a predetermined amount from one of potentially many incoming signals. The beat signals are generated in the integrated photodetectors, whose signals can be subtracted for noise reduction, and then electrically amplified, filtered, and fed to a decision circuit. This PIC combined five different types, and a total of seven guided-wave optical devices: two tunable Bragg reflection filters, an MQW optical gain section, an electrically adjustable phase shifter, a zero-gap directional coupler switch, and two MQW waveguide photodetectors. It also demonstrates self-aligned connections between the buried heterostructure guides, which offer current access and semi-insulating InP lateral current blocking, and the low-loss semi-insulating InP-clad rib guides used in the S-bends and input port. The processing sequence for PICs of this complexity has been described in some detail in the literature, and can be carried out following most of the same steps used in commercial semiconductor laser fabrication.[76]

Tuning of the DBR lasers, as used in the PIC above, is accomplished by injecting current into the (transparent) Bragg reflectors, shifting their index via the plasma and anomalous dispersion effects discussed under *Carrier Effects* in this chapter. This shifts the wavelength of peak Bragg reflectivity, thereby selecting different longitudinal modes for the laser. The laser can also be *continuously* tuned by shifting the frequency of any particular longitudinal mode by injecting current to provide an index shift in the (transparent) phase section of the laser. Detectors in PICs of this type often employ for absorption the same layers used for gain in the laser, and typically have capacitance of several pF dominated by the contact pads rather than the depletion capacitance of the p-n junction.

Early experimental prototypes of PICs of this type have demonstrated total on-chip losses including propagation losses, bending losses, radiation losses at the coupler and at the active/passive detector transitions, and any departures from 100 percent quantum efficiency in the detectors, of ~4 dB, providing encouragement for even more complex integrations. This PIC demonstrated error-free heterodyne reception of frequency-shift-keyed digital signals with sensitivities of −40 dBm at 200 Mb/s measured in free-space outside the chip. PICs such as this may in the future offer high selectivity in multichannel optical broadcasts, much in the same way radio receivers operate today.

6.6 *APPLICATIONS OF INTEGRATED OPTICS*

Digital Transmission

The performance metric in digital optical fiber transmission is the ability of a transmitter to deliver a signal to the receiver on the end of the link in a manner such that the receiver can clearly distinguish between the "0" and "1" state in each time period or bit slot. Binary amplitude-shift-keyed transmission (ASK) is by far the most common format in commercial systems, but research systems also employ frequency-shift-keyed (FSK) and phase-shift-keyed (PSK) formats as well. A decision circuit at the receiver must distinguish between "0" and "1," and this circuit will be more susceptible to noise when the "0" and "1" level difference is reduced, or when the time over which this difference is maintained is reduced below the full bit period.

The performance of a transmitter is thus governed by its rise and fall times, its modulation bandwidth or flatness of response to avoid pattern-effects, and its output power. Furthermore, the spectral characteristics of its optical output can impair transmission. Examples of the latter include unnecessary optical bandwidth, as might be present in an LED or a multi-longitudinal-mode laser, that can produce pulse spreading of the digital pulses due to dispersion in the transmission fiber. While transmission sources include current-modulated LEDs, for speeds higher than ~100 Mb/s semiconductor lasers are used, and IO technology has played a fundamental role in the evolution of semiconductor laser technology. In very high speed systems (typically >1 Gb/s), dispersive pulse distortion can cause severe degradation with directly modulated lasers unless devices which emit only one longitudinal mode are employed. The incorporation of gratings in DFB and DBR lasers has produced sources with exceptional spectral purity and allow multi-Gb/s transmission over intermediate distances (<100 km) in conventional fiber.

The advent of fiber amplifiers has raised interest in longer transmission spans, and here the unavoidable frequency excursions that result from directly modulating even a single-longitudinal-mode laser again lead to pulse spreading and distortion. In these instances, a CW laser followed by an external modulator is a preferred source. The integrated DFB/electroabsorption modulator, as discussed previously, provides such a source. These PICs have demonstrated error-free transmission in excess of 500 km in dispersive fiber at 2.5 Gb/s.[77] However, even these devices impose a small residual dynamic phase shift on the signal due to electrorefractive effects accompanying the induced absorption in the modulator. This can be especially problematic with typical percent-level antireflection coatings on the output facet, since this will provide time-varying optical feedback into the laser and further disrupt its frequency stability.

The highest performance digital transmission has been achieved using external LiNbO$_3$ Mach-Zehnder modulators to encode a CW semiconductor laser. Modulators similar to that in Fig. 9 have been constructed to allow separate traveling-wave voltages to be applied to each arm of the modulator in a push-pull configuration. This device design can produce a digitally encoded signal with *zero* residual phase shift or chirp.[78] Such a source has only its information-limited bandwidth and generally provides nearly optimized performance in a dispersive environment. The Mach-Zehnder can also be driven to intentionally provide positive or negative chirping, and transmission experiments over distances up to 256 km in dispersive fiber at bit rates of 5 Gb/s have revealed a slight increase in performance when a slightly negative chirp parameter is used.[79] Recent work has focused on the integration of semiconductor lasers with Mach-Zehnder modulators using electrorefraction from the QCSE.[80] Such a source is expected to provide a convenient and high-performance digital transmitter.

Analog Transmission

A second application area that is expected to use a large number of IOCs is analog fiber optic links. Analog fiber optic links are currently being used to transmit cable television (CATV) signals at the trunk and supertrunk level. They are also being used for both commercial and military antenna remoting. Analog fiber optic links are being fielded for these applications because of their low distortion, low loss, and low life-cycle cost when compared to more conventional coaxial-cable-based transmission systems.

An analog fiber optic link using IOCs is typically configured as shown in Fig. 15. A high-power CW solid-state laser, such as a 150-mW diode-pumped YAG laser operating at 1319 nm, is typically used as the source in order to maximize dynamic range, carrier-to-noise ratio, and link gain. An interferometric modulator, such as a Mach-Zehnder interferometer or a Y-fed balanced bridge modulator, is typically used to modulate the light with the applied RF or microwave signal via the linear electro-optic effect. Current analog links for CATV signal distribution utilize a 1-GHz Y-fed balanced bridge modulator biased at the quadrature point (linear 3-dB point).[81,82] A predistortion circuit is required to minimize third-order distortion associated with the interferometric modulator response. The CATV modulated signal can be transmitted on both output fibers of the device. Analog links for antenna remoting typically fit into one of two categories. Certain applications require relatively narrow passbands in the UHF region while other microwave-carrier applications require very broadband (several GHz) performance. Y-fed balanced bridge modulators biased at the quadrature point are again used to perform the electrical-to-optical conversion, with the electrode structure tailored to the application. In

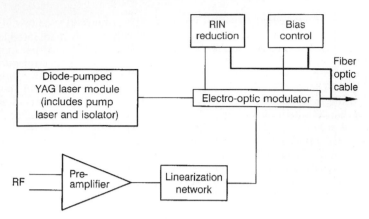

FIGURE 15   Standard configuration for externally modulated analog fiber optic link.

both narrowband and broadband applications, 20–30 dB preamplifiers are typically utilized to minimize the noise figure and maximize the RF gain of the link.

Two important modulator parameters are the insertion loss and the half-wave drive voltage, both of which impact the link gain and dynamic range. Fully-packaged Y-fed balanced bridge modulators with 2.5–4.0-dB insertion loss are now readily achieved in production for both UHF and microwave bandwidths. A trade-off typically needs to be made between half-wave voltage and bandwidth for both lumped-element and traveling-wave electrode structures. Commercially available lumped-element LiNbO$_3$ interferometric modulators typically have half-wave voltages of ~5 V for 600-MHz, 1-dB bandwidths. Commercially available traveling-wave LiNbO$_3$ interferometric modulators typically have half-wave voltages of ~8 V for 12-GHz, 3-dB bandwidths. The half-wave voltages of LiNbO$_3$ traveling-wave modulators can be reduced by more than a factor of two using a velocity-matched electrode structure as described in Ref. 67.

In order to be used in practical systems, it is critical that the integrated optical modulators have well-behaved flat frequency responses in the band of interest. Modulators for CATV signal transmission and UHF antenna remoting typically required that the amplitude response and the phase response be flat to ±0.25 dB and ±2 degrees, respectively. The frequency response of an integrated optical modulator is a function of both the device design and packaging parasitics. Care must be exercised in designing modulators since LiNbO$_3$ is both a piezoelectric and an acousto-optic material. Early LiNbO$_3$ modulators typically had 1–3 dB of ripple in the passband due to acoustic mode excitation. When packaging lumped-electrode devices, it is also critical to keep terminating resistors and wire bonds short to minimize stray inductance and capacitance. When packaging traveling-wave modulators, it is critical to control the impedance of the launch, the transitions, the device, and the termination. Through proper device design and packaging, it is possible to achieve well-behaved frequency responses in both lumped-electrode and traveling-wave devices as shown in Figs. 16 and 17.

An additional issue that impacts IOC modulator design for analog links is harmonic and intermodulation distortion. Most modulators used in analog links are interferometric in nature with a sinusoidal transfer function. By operating the device precisely at the quadrature point, all even harmonics can be suppressed. Second-harmonic distortion less than −75 dBc is easily achieved using an electronic feedback loop for modulator bias. Alternative techniques are being investigated to laser trim one leg of a Mach-Zehnder to bring the modulator to quadrature. Third-order distortion due to the sinusoidal transfer

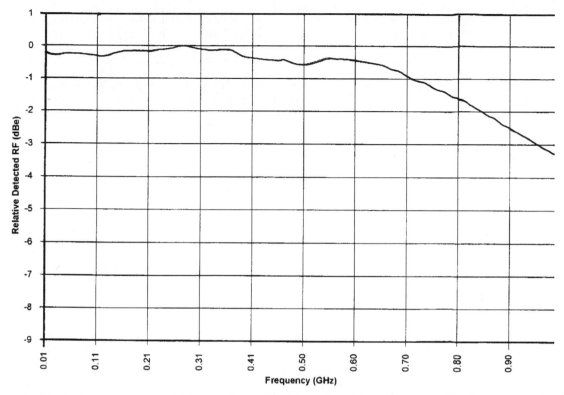

FIGURE 16 Frequency response of APE LiNbO$_3$ UHF interferometric modulator for CATV and UHF antenna remoting.

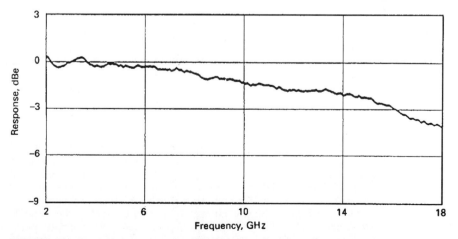

FIGURE 17 Frequency response of APE LiNbO$_3$ microwave interferometric modulator for broadband antenna remoting.

function of the interferometric modulator also poses a problem, but the transfer functions are very well-behaved and predictable, and this distortion can be suppressed to acceptable levels using electronic predistortion or feed forward techniques.

Forty- and eighty-channel CATV transmitters operating at 1300-nm wavelengths with APE LiNbO$_3$ modulators are currently being fielded. Compared to coaxial transmission, which requires transmitters every 500 meters, the fiber optic systems can transmit over distances up to 50 km without repeaters. Similarly, externally modulated analog fiber optic links are currently being fielded for military and commercial applications. UHF links with 115 dB/Hz$^{2/3}$ dynamic range, 4-dB noise figure, and unity gain have been demonstrated using commercially available hardware. These systems will maintain this quality of transmission over temperature ranges of −25 to +50°C. Microwave links with 2–18-GHz frequency response, 114-dB/Hz$^{2/3}$ spurious-free dynamic range, and input noise figure of 22 dB can also be achieved using commercially available hardware.

Switching

Arrays of IO switches have been proposed and demonstrated for a variety of space switching and time-multiplexed switching (TMS) applications. In space switching, it is generally necessary to formulate the switches in a nonblocking geometry, and the reconfiguration time can be relatively slow. This requirement led to the development of crossbar switches in which an N × N switch contains N^2 IO switches and $2N − 1$ stages and from 1 to $2N − 1$ cross points. Typically $N = 4$ in LiNbO$_3$ and in InP. More recently, much attention in IO switch arrays has shifted to the dilated Benes architecture which is only rearrangeably nonblocking but reconfigurable in short (nsec) times suitable for TMS, and has the advantage of requiring substantially fewer switches and a constant number $2 \log_2 N$ of crosspoints.

A schematic of an 8 × 8 dilated Benes switch composed of two separate IO chips is shown in Fig. 18.[83] This device uses waveguide crossovers and is representative of the state of the art in IO integration in 1992. The performance of switch arrays of the dilated Benes architecture are much more forgiving than crossbar switches to the degradation of individual switches. The device shown contains 48 delta beta switches driven by a single-voltage arrangement. The switching voltage at 1.3 microns was 9.4 + −0.2 V, the insertion loss varied from −8 to −11 dB (93 percent of the 256 paths through the switch were within ±1 dB), the crosstalk levels in individual switches ranged from −22 dB to −45 dB, and the reconfiguration time was 2.5 nsec.

An advantage of these types of IO switch arrays is that they are data-rate transparent. That is, once the switch is reconfigured, the data stream through the device is simply the passage of light and can easily be multi-Gbit/sec. As of 1992, crossbar switches were commercially available and other types of arrays were under advanced development.

Fiber Optic Gyroscopes

Another application that may require large quantities of integrated optical circuits is the fiber optic gyroscope (FOG).[84–89] A FOG is one form of a Sagnac interferometer, in which a rotation rate results in a phase shift between clockwise- and counterclockwise-propagating optical fields. The most frequently utilized FOG configuration, which was first proposed by workers at Thompson CSF in the mid-1980s,[90] is presented in Fig. 19.

FOG IOCs are typically fabricated in LiNbO$_3$ using the annealed proton exchange (APE) process,[66] although titanium-diffused IOCs with surface plasmon polarizers have also been utilized. The IOC performs four primary functions in the fiber gyroscope. First, the Y-junction serves as the loop coupler splitting and recombining the clockwise- and

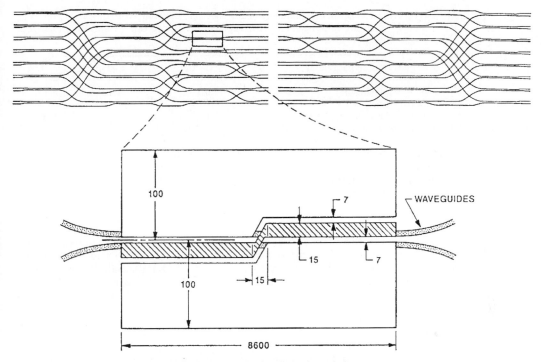

FIGURE 18   Architecture of 8×8 dilated Benes directional coupler switch array.

counterclockwise-propagating optical fields. Second, the IOC functions as a high-quality polarizer. Third, a ninety-degree phase dither (at the eigen frequency of the fiber coil) is typically applied to one of the integrated optical phase modulators. This approach keeps the Sagnac interferometer biased at the 3-dB point where it is linear and most sensitive to rotation. Finally, in a closed-loop FOG configuration, one of the phase modulators functions as a frequency shifter. A serrodyne signal (sawtooth wave) is applied to the phase modulator to effectively cancel the shift due to the rotation.

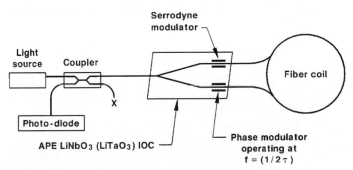

FIGURE 19   Standard configuration for fiber optic gyroscope incorporating a three-port integrated optical circuit.

TABLE 1 Critical Performance Parameters for 800- and 1300-nm APE LiNbO$_3$ FOG IOCs. Listed Values are Maintained Over a Temperature Range of -55 to $+95°C$ and During Vibration up to 15 Grms

Performance parameter	1300-nm IOCs	800-nm IOCs
Insertion loss (pigtailed)	3 dB	4 dB
Polarization extinction	70 dB	60 dB
Y-junction split ratio (pigtailed)	48/52 to 52/48	45/55 to 55/45
Polarization crosstalk at fiber-waveguide interfaces	< -30 dB	< -25 dB
Optical back reflection	< -65 dB	< -65 dB
Half-wave voltage	4.5 V	3.5 V
Residual intensity modulation	0.02%	0.05%

The output signal from a fiber gyro at rest is the sum of white receiver noise, primarily dependent on the amount of optical power arriving at the detector, and an additional long-term drift of the mean value. The long-term drift in a FOG associated with a residual lack of reciprocity typically limits the sensitivity of the FOG to measure low rotation rates. Another important characteristic of a gyro is the scale factor, which is a measure of the linearity between the actual rotation rate and the gyro response. The critical performance parameters for a FOG IOC are presented in Table 1. The performance of 800- and 1300-nm APE LiNbO$_3$ FOG IOCs that are currently in production is also presented in this table.

One application of the FOG is inertial guidance, requiring a FOG with a bias drift <0.01 deg/h and a scale factor accuracy <5 ppm. A 1300-nm LED or an erbium-doped fiber is typically used as the light source. A large coil of polarization-maintaining fiber (typically 1 km of fiber wound in a 15–20 cm diameter coil) and precise source spectral stability are required to achieve the desired sensitivity. The fiber is typically wound in a quadrupole configuration to minimize sensitivity to temperature gradients. With recent improvements in optical sources, integrated optics, and fiber coil winding technology, it is now possible to achieve inertial grade FOG performance over a temperature range of -55 to $+95°C$.

A second tactical-grade FOG design is more typical to aerospace applications, with bias drift and scale factor accuracy requirements ranging from 0.1 to 10 deg/h and 10 to 1000 ppm, respectively. These systems are typically designed for operation at 810–830 nm to make use of low-cost multimode 830-nm AlGaAs laser diodes as used in consumer electronic products. These systems typically utilize 2- to 5-cm-diameter fiber coils with 100–200 meters of either polarization-maintaining or single-mode fiber. A third very low-cost, low-grade FOG design for automotive navigation is also nearing production. The required bias drift is only 1000 deg/h, and a closed-loop configuration is unnecessary since the scale factor accuracy is only 0.1 percent. Current designs to achieve low cost include low performance IOCs, laser, and fiber couplers, fully automated FOG assembly and test procedures, and only ~50 m of single-mode fiber. More advanced IOCs, including four-port designs that integrate the source/detector coupler into the IOC, are also being considered to reduce component count.

WDM Systems

Wavelength division multiplexing (WDM), by encoding parallel data streams at different wavelengths on the same fiber, offers a technique to increase transmission capacity, or

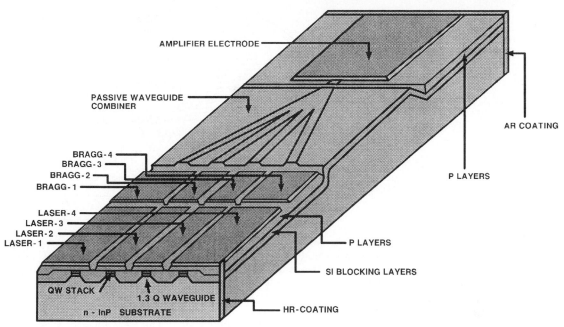

AMPLIFIER ELECTRODE

PASSIVE WAVEGUIDE
COMBINER

AR COATING

BRAGG-4
BRAGG-3
BRAGG-2
BRAGG-1

P LAYERS

LASER-4
LASER-3
LASER-2
LASER-1

P LAYERS

SI BLOCKING LAYERS

QW STACK   1.3 Q WAVEGUIDE

n - InP SUBSTRATE

HR-COATING

FIGURE 20   Four-channel WDM transmitter PIC.

increase networking or switching flexibility, without requiring higher speed electronics to process each channel. IO will be an enabling technology in both WDM transmitter design and in demultiplexer design.

Figure 20 shows a WDM transmission PIC, following the concept introduced by Aiki et al,[91] where the outputs of a number of single-frequency lasers are combined into a single waveguide output port. The PIC shown here combines the outputs of four independently modulatable and independently tunable MQW-DBR lasers through passive power combining optics, and also includes an on-chip MQW optical output amplifier to partially recover the inherent losses of the power combining operation.[92] This WDM source PIC has successfully demonstrated a high-speed WDM transmission capability. The PIC was mounted in a fixture with SMA connectors to 50Ω microstrip leading to ceramic stand-offs with bond wires to the PIC contacts, and a thermoelectric cooler to provide frequency stability. One concern in PICs employing optical amplifiers, just as in the integrated laser/modulator, is the need for good anti-reflection coatings since the source sees the facet reflection after a double-pass through the amplifier. Another area of concern is the cross-talk between channels, either through electrical leakage on chip or through amplifier saturation effects since all sources share the same amplifier.

Cross-talk in this PIC has been evaluated a large-signal digital transmission environment. The lasers were each simultaneously modulated with a pseudo-random 2 Gb/s signal, and the combined signal at an aggregate bit-rate of 8 Gb/s was sent over a 36 km transmission path of conventional 1.3 µm dispersion-zero fiber. A fiber Fabry-Perot interferometer was used for channel selection. The PIC used in this experiment had two MQW-DBR lasers at each of two distinct zero-current Bragg frequencies shifted by ~50 Å, and one of each pair was then tuned by ~25 Å to yield four channels with a spacing of ~25 Å. There was virtually no cross-talk penalty without fiber, indicating that the level of cross-talk is not significant in a direct sense, i.e., in its impact on the intensity modulation waveforms at 2 Gb/s. When the 36 km link of dispersive fiber is inserted, a

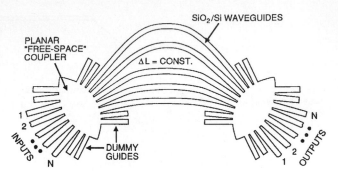

FIGURE 21 Si/SiO$_2$ demultiplexer using star couplers and rastered-length waveguide interconnects.

small penalty of ~1 dB was observed, suggesting a small dispersion penalty probably resulting from electrical cross-modulation from the other current drives on the PIC. This result indicates that cross-talk is not a severe problem, and simple design improvements could reduce it to inconsequential levels for digital applications. PICs of similar design have been extended to include 16 lasers with 8-Å spacing and an individual electroabsorption modulator for each laser.[93]

IO demultiplexers are also expected to be important for WDM. Figure 21 shows a demultiplexer design that employs star couplers and arrayed waveguides, and it has been successfully executed both in the Si/SiO$_2$ and InP-based technologies. In this design[94] each input to a primary star coupler expands in the free-space (in-the-plane) portion of the star to uniformly illuminate each output waveguide of the primary star. The path lengths of each guide in the array between the primary and secondary star are incremented in length by an integral multiple of some base wavelength. At this wavelength, upon arrival at the input to the second star, each wave has the same phase relation to its neighbors that it had at the output to the first star coupler, and reciprocity demands that this focus the beam, as a phased array, back to the corresponding single waveguide at the secondary star output. However, a slight change in wavelength will produce a phase tilt across this phased array, and the focus will go to a different output waveguide. In this manner, the device operates as a demultiplexer. The incremental length difference between adjacent guides in the region between the two stars functions just as a grating would in a bulk-optic equivalent of this device.

The performance of these devices has been extraordinary.[94–96] Out-of-band rejection in excess of 25 dB has been obtained, with individual channel bandwidths in the 1-nm-scale and fiber-to-fiber insertion losses of only several dB. Device sizes up to 15 channels have been demonstrated, and these devices are also amenable to fiber ribbon array connections. IO devices that perform this function will be instrumental for cost-effective WDM deployment.

Optical Storage and Display

IO devices are of special interest for propagation and manipulation of visible light for storage and display applications. For optical data storage, there is interest in operating at wavelengths shorter than obtainable from GaAs lasers to increase storage capacity (which is proportional to diffraction limited laser spot size). For display applications, multiple colors are needed. The method receiving most interest is to frequency double diode lasers in materials with high nonlinear optical coefficients. Waveguide doubling can be quite efficient because of the long interaction length and high optical power densities. Both channel and slab guide doublers have been demonstrated. As discussed under *Nonlinear Effects* in this chapter, periodic domain reversal has been obtained in LiNbO$_3$ by the application of pulsed electric fields with segmented contacts. This approach has recently yielded waveguides capable of producing 20.7 mW of blue light from 195.9 mW of fundamental at 851.7 nm.[27] For these devices to find practical application, considerable effort in robust packaging of laser-doubling waveguide systems will be required.

6.7 FUTURE TRENDS

Shift from R&D to Manufacturing

Integrated optical circuit technology has advanced to the point where it is now moving from R&D to manufacturing. There are now several companies producing LiNbO$_3$ and silicon integrated optical circuits in moderate volumes (several thousand devices each per year). In the semiconductor area, integrated DFB/electroabsorption modulators and laser/amplifier MOPA PICs are appearing commercially. In the more mature integrated optic technologies, the majority of the development effort is now focusing on reducing manufacturing costs and understanding reliability and device lifetime issues.

Early results on LiNbO$_3$ wafer level processing are very encouraging. The majority of the LiNbO$_3$ devices described in this chapter can be fabricated using either existing or slightly modified semiconductor processing equipment. Clean room requirements are not nearly as tight as what is required for VLSI and MMIC wafer fabrication. Production yields of LiNbO$_3$ integrated optical circuits have been obtained well in excess of 95 percent. The majority of the defects are mechanical in nature (probes scratching electrodes or fibers damaging polished end faces). These should be minimized as the processes become more automated. At this time, it appears that all wafer processing operations should be easily scalable to wafer batch processing, including the end-face optical polishing. The majority of the cost of an integrated optical circuit is currently associated with fiber attachment, packaging, and final testing. Analysis indicates that these operations are not fundamentally expensive, but instead are expensive because of limited production volumes. Efforts are currently underway to automate these processes by incorporating robotics and machine vision.

The second area that needs to be addressed is device reliability. First-generation commercial integrated optic products were plagued by premature failures and poor performance over temperature. The majority of these problems can be traced to poor fiber attachment and packaging techniques. These problems have recently been remedied by carefully selecting compatible material systems and incorporating proven hybrid electronic packaging techniques. Commercially available LiNbO$_3$ integrated optical curcuits can be thermally cycled hundreds of times from −65 to +125°C with less than 1-dB variation in insertion loss. The devices can also withstand 30 Grms random vibration testing. Efforts are now underway to better understand and model long-term aging effects on the device performance. This is accomplished by first identifying potential failure mechanisms and then by developing physical models of these failures. Early results on LiNbO$_3$ integrated optical circuits are encouraging. No failure mechanism has been identified that will limit device lifetime to less than 25 years, assuming the device is properly fabricated and assembled. Obviously, this can only be guaranteed by fabricating the devices in well-controlled, well-documented production environments.

In the semiconductor area, reliability certification technology has reached a more mature level as far as semiconductor lasers are concerned, and this methodology is being carried over to PIC devices as well. This includes sequences of purges at high currents or voltages combined with high temperatures, together with extended accelerated aging to

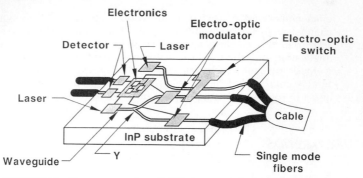

FIGURE 22 Future transceiver OEIC that combines optical signal routing switch, sources lasers, detectors, and receiver and transmitter drive electronics.

identify activation energies and aging rates for various degradation mechanisms. The early III-V PIC commercial offerings are just entering this stage of development.

Advanced Integration

Active research continues in $LiNbO_3$. The high performance and low fiber-to-fiber insertion loss are unparalleled and make this the technology of choice for extremely demanding applications. One area of active research is in ever-larger switching fabrics, with dimensions of 16×16 already demonstrated.[97] As this work migrates to semiconductors with the potential for several stages of amplification to counter losses and smaller overall node size, even larger dimensions should become feasible.

As stated in the introduction, integrated optics can be carried out in semiconductors in combination with electronics to form OEICs. The majority of OEIC work to date has been on receivers where the first stages of electrical amplification are integrated with a photodetector. These OEICs usually have little or no IO technology and employ surface-illuminated photodetectors, and are clearly dominated by electronics. The III-V IO technology combining active devices has mostly been categorized as PICs since traditional electronics have been virtually nonexistent in the guided-wave integrations as described in this chapter.

However, the appeal of a material that supports high-performance transistor circuitry, electro-optic waveguide devices, and active laser sources, amplifiers, and detectors is too strong to dismiss, and substantial worldwide research continues in this area. It is envisioned that entire transceiver units can be made lithographically on a single chip, such a futuristic OEIC is shown in Fig. 22. This circuit might perform detection of incoming signals, subsequent electronic processing, and retransmission with several external modulators. As of 1992, a variety of circuits that embody portions of this circuit have been demonstrated in numerous research labs.

6.8 REFERENCES

1. T. Tamir (ed.), *Guided-Wave Optoelectronics,* 2d ed., Springer-Verlag, New York, 1990.

2. R. G. Hunsperger, *Integrated Optics: Theory and Technology,* Springer-Verlag, Berlin, 1985.

3. T. Tamir (ed.), *Integrated Optics, 2nd Edition,* Springer-Verlag, New York, 1979.

4. L. D. Hutcheson (ed.), *Integrated Optical Circuits and Components,* Marcel Dekker, New York, 1987.

5. H. Nishihara, M. Haruna, and T. Suhara, *Optical Integrated Circuits,* McGraw-Hill, New York, 1985.

6. D. Marcuse, *Theory of Dielectric Optical Waveguide,* 2d ed., Academic Press, San Diego, 1991, pp. 307–318.

7. A. Yariv and P. Yeh, *Optical Waves in Crystals,* John Wiley & Sons, Inc., New York, 1984, pp. 177–201, pp. 425–459.

8. S. E. Miller and I. P. Kaminow (ed.), *Optical Fiber Telecommunications II,* Academic Press, New York, 1985.

9. Special issue on photonic devices and circuits, *IEEE J. Quantum Electron.,* vol. QE-27 (3), 1991.

10. Special issue on integrated optics, *J. Lightwave Technology,* vol. 6 (6), 1988.

11. Special issue on integrated optics, *IEEE J. Quantum Electron.,* vol. QE-22 (6), 1986.

12. *Digests of IEEE/OSA Conferences on Integrated Photonics Research,* 1990–present, OSA, Washington, D.C.

13. *Digests of IEEE Conference on Integrated and Guided Wave Optics,* 1972–1989, OSA, Washington, D.C.

14. G. B. Hocker and W. K. Burns, "Mode Dispersion in Diffused Channel Waveguides by the Effective Index Method," *Appl. Optics,* vol. 16, 1977, pp. 113–118.

15. J. F. Lotspeich, "Explicit General Eigenvalue Solutions for Dielectric Slab Waveguides," *Appl. Opt.,* vol. 14, 1975, pp. 327–335.

16. M. D. Feit and J. A. Fleck, Jr., "Light Propagation in Graded-Index Optical Fibers," *Appl. Optics,* vol. 17, 1978, pp. 3990–3998.

17. M. J. Weber (ed.), *CRC Handbook of Laser Science and Technology, Vol. IV, Optical Materials, Part 2,* CRC Press, Inc., Boca Raton, 1986.

18. M. A. Afromowitz, "Refractive Index of $Ga_{1-x}Al_xAs$," *Solid State Commun.,* vol. 15, 1974, pp. 59–63.

19. C. H. Henry, L. F. Johnson, R. A. Logan, and D. P. Clarke, "Determination of the Refractive Index of InGaAsP Epitaxial Layers by Mode Line Luminescence Spectroscopy," *IEEE J. Quantum Electron.,* vol. QE-21, 1985, pp. 1887–1892.

20. T. Maldonado, "Electro-Optics Modulators," in vol. II, chap. 13 of this handbook.

21. I. P. Kaminow, *An Introduction to Electrooptic Devices,* Academic Press, Orlando, 1974.

22. T. S. Moss, G. S. Burrell, and B. Ellis, *Semiconductor Opto-Electronics,* John Wiley & Sons, Inc., New York, 1973.

23. K. Tharmalingam, "Optical Absorption in the Presence of a Uniform Field," *Phys. Rev.,* 130, 1963, pp. 2204–2206.

24. D. A. B. Miller, J. S. Weiner, and D. S. Chemla, "Electric-Field Dependence of Linear Optical Properties Quantum Well Structures: Waveguide Electroabsorption and Sum Rules," *IEEE J. Quantum Electron.,* vol. QE-22, 1986, pp. 1816–1830.

25. J. E. Zucker, I. Bar-Joseph, G. Sucha, U. Koren, B. I. Miller, and D. S. Chemla, "Electrorefraction in GaInAs/InP Multiple Quantum Well Heterostructures," *Electron. Lett.,* vol. 24, 1988, pp. 458–460.

26. C. L. Tang, "Nonlinear Optics," in vol. II, chap. 38 of this handbook.

27. M. Yamada, N. Nada, M. Saitoh, and K. Watanabe, "First-order Quasi-Phase Matched $LiNbO_3$ Waveguide Periodically Poled by Applying an External Field for Efficient Blue Second-Harmonic Generation," *Appl. Phys. Lett.,* 62, 1993, pp. 435–436.

28. H. A. Haus, E. P. Ippen, and F. J. Leonberger, in J. L. Horner (ed.), *Optical Signal Processing,* Academic Press, Orlando, 1987.

29. T. K. Findakly, "Glass Waveguides by Ion-Exchange: A Review," *Opt. Eng.,* 24, 1985, p. 244.

30. R. V. Ramaswamy and R. Srivastava, "Ion-exchanged Glass Waveguides: A Review," *J. Lighwave Technology,* LT-6, 1988, pp. 984–1002.

31. T. Izawa and H. Nakogome, "Optical Waveguide Formed by Electrically Induced Migration of Ions in Glass Plates," *Appl. Phys. Lett.,* 21, 1972, p. 584.

32. M. Kawachi, "Silica Waveguides on Silicon and Their Applications to Integrated-optic Components," *Opt. and Quant. Electron.,* 22, 1990, pp. 391–416.

33. H. Takahashi, Y. Ohnori, and M. Kawachi, "Design and Fabrication of Silica-based Integrated-optic 1 × 128 Power Splitter," *Electron. Lett.,* vol. 27, 1991, pp. 2131–2133.

34. C. H. Henry, G. E. Blonder, and R. F. Kazarinov, "Glass Waveguides on Silicon for Hybrid Optical Packaging," *J. Lightwave Tech.,* 7, 1989, pp. 1530–1539.

35. Y. Shani, C. H. Henry, R. C. Kistler, K. J. Orlowski, and D. A. Ackerman, "Efficient Coupling of a Semiconductor Laser to an Optical Fiber by Means of a Tapered Waveguide on Silicon," *Appl. Phys. Lett.,* 56, 1989, pp. 2389–2391.

36. R. C. Alferness, "Guided-wave Devices for Optical Communications," *IEEE J. Quantum Electron.,* vol. QE-17 (6), 1981, pp. 946–959.

37. H. F. Taylor, "Applications of Guided-wave Optics in Signal Processing and Sensing," *Proc. IEEE,* 75 (11), 1987, pp. 1524–1535.

38. E. Voges and A. Neyer, "Integrated-optic Devices on LiNbO$_3$ for Optical Communication," *J. Lightwave Technology,* LT-5 (9), 1987, pp. 1229–1238.

39. L. Thylen, "Integrated Optics in LiNbO$_3$: Recent Developments in Devices for Telecommunications," *J. Lightwave Technology,* LT-6 (6), 1988, pp. 847–861.

40. R. C. Alferness, "Waveguide Electro-optic Switch Arrays," *IEEE J. Selected Areas in Communications,* 6 (7), 1988, pp. 1117–1130.

41. R. V. Schmidt and I. P. Kaminow, "Metal Diffused Optical Waveguides in LiNbO$_3$," *Appl. Phys. Lett.,* 15, 1974, pp. 458–460.

42. M. Fukuma and J. Noda, "Optical Properties of Ti-diffused LiNbO$_3$ Strip Waveguides and Their Coupling-to-a-Fiber Characteristics," *Appl. Opt.,* vol. 19, 1980, pp. 591–597.

43. S. K. Korotky and R. C. Alferness, *Integrated Optical Circuits and Components,* L. D. Hutcheson (ed.), Marcel Dekker, New York, 1987.

44. R. J. Esdaile, "Closed-tube Control of Out-diffusion During Fabrication of Optical Waveguides in LiNbO$_3$," *Appl. Phys. Lett.,* 33, 1978, pp. 733–734.

45. J. L. Jackel, V. Ramaswamy, and S. P. Lyman, "Elimination of Out-diffused Surface Guiding in Titanium-diffused LiNbO$_3$," *Appl. Phys. Lett,* 38, 1981, pp. 509–511.

46. A. M. Glass, "The Photorefractive Effect," *Opt. Eng.,* 17, 1978, pp. 470–479.

47. J. L. Jackel, C. E. Rice, and J. J. Veselka, "Proton Exchange for High-index Waveguides in LiNbO$_3$," *Appl. Phys. Lett.,* 47, 1982, pp. 607–608.

48. P. G. Suchoski, T. K. Findakly, and F. J. Leonberger, "Stable Low-loss Proton-exchanged LiNbO$_3$ Waveguide Devices with no Electro-optic Degradation," *Opt. Lett.,* 13, 1988, pp. 1050–1052.

49. P. G. Suchoski, T. K. Findakly, and F. J. Leonberger, "Low-loss High Extinction Polarizers Fabricated in LiNbO$_3$ by Proton Exchange," *Opt. Lett.,* 13, 1988, pp. 172–174.

50. T. K. Findakly, P. G. Suchoski, and F. J. Leonberger, "High-quality LiTaO$_3$ Integrated Optical Waveguides and Devices Fabricated by the Annealed-Proton-Exchange Technique," *Opt. Lett.,* 13, 1988, pp. 797–799.

51. See, for example, *Conference Proceedings of 3rd Int. Conf. on Indium Phosphide and Related Materials,* Cardiff, 1991, IEEE Cat. #91 CH2950-4.

52. A. R. Clawson, "Reference Guide to Chemical Etching of InGaAsP and In$_{0.53}$Ga$_{0.47}$As Semiconductors," NOSC Tech. Note 1206, San Diego, CA, 1982.

53. S. Adachi and H. Kawaguchi, "Chemical Etching Characteristics of (001)InP," *J. Electrochem. Soc.: Sol. St. Sci. and Tech.,* vol. 128, 1981, pp. 1342–1349.

54. L. A. Coldren, K. Furuya, B. I. Miller, and J. A. Rentschler, "Etched Mirror and Groove-Coupled GaInAsP/InP Laser Devices for Integrated Optics," *IEEE J. Quantum Electron.,* vol. QE-18, 1982, pp. 1679–1688.

55. L. A. Coldren and J. A. Rentschler, "Directional Reactive-Ion-Etching of InP with Cl$_2$ Containing Gases," *J. Vac. Sci. Technol.,* 19, 1981, pp. 225–230.

56. J. W. McNabb, H. G. Craighead, and H. Temkin, "Anisotropic Reactive Ion Etching of InP in Methane/Hydrogen Based Plasma," paper TuD15 in *Tech. Digest of Integrated Photonics Research,* Monterey, 1991, pp. 26–27.

57. P. Buchmann and H. Kaufmann, "GaAs Single-Mode Rib Waveguides with Reactive Ion-Etched Totally Reflecting Corner Mirrors," *IEEE J. Lightwave Tech.,* vol. LT-3, 1985, pp. 785–788.

58. S. Oku, M. Okayasu, and M. Ikeda, "Low-Threshold CW Operation of Square-Shaped Semiconductor Ringe Lasers (Orbiter Lasers)," *IEEE Phot. Tech. Lett.,* vol. 3, 1991, pp. 588–590.

59. W. J. Grande, J. E. Johnson, and C. L. Tang, "AlGaAs Photonic Integrated Circuits Fabricated Using Chemically Assisted Ion Beam Etching," paper OE10.4/ThUU4 in *Conf. Digest of IEEE LEOS Annual Meeting,* Boston, 1990. p. 169.

60. J. B. D. Soole, A. Scherer, H. P. LeBlanc, N. C. Andreadakis, R. Bhat, and M. A. Koza, "Monolithic InP-Based Grating Spectrometer for Wavelength-Division Multiplexed Systems at 1.5 μm," *Electron. Lett.,* 27, 1991, pp. 132–134.

61. C. Cremer, G. Ebbinghaus, G. Heise, R. Muller-Nawrath, M. Shienle, and L. Stoll, "Grating Spectrograph in InGaAsP/InP for Dense Wavelength Division Multiplexing," *Appl. Phys. Lett.,* 59, 1991, pp. 627–629.

62. R. C. Alferness, U. Koren, L. L. Buhl, B. I. Miller, M. G. Young, T. L. Koch, G. Raybon, and C. A. Burrus, "Broadly Tunable InGaAsP/InP Laser Based on a Vertical Coupler Filter with 57-nm Tuning Range," *Appl. Phys. Lett.,* 60, 1992, pp. 3209–3211.

63. E. Van Tomme, P. P. Van Daele, R. G. Baets, and P. E. Lagasse, "Integrated Optic Devices Based on Nonlinear Optical Polymers," *IEEE J. Quan Electron.,* vol. QE-27, 1991, p. 778.

64. C. C. Teng, M. A. Scature, and T. K. Findakly, *OFC '92 Technical Digest,* OSA, Washington, D.C. 1992, p. 333.

65. J. N. Polkay and G. L. Mitchell, "Metal-clad Planar Dielectric Waveguide for Integrated Optics," *J. Opt. Soc. Am.,* 64, 1974, pp. 274–279.

66. P. G. Suchoski, T. K. Findakly, and F. J. Leonberger, "Stable Low-Loss Proton-Exchanged $LiNbO_3$ Waveguide Devices with no Electro-Optic Degradation," *Optics Lett.,* 13, 1988, p. 1050.

67. D. W. Dolfi and T. R. Ranganath, "50 GHz Velocity-matched Broad Wavelength $Ti:LiNbO_3$ Mach-Zehnder Modulator with Multimode Active Section," *Electron. Lett.,* vol. 28, 1992, pp. 1197–1198.

68. A. Becker, C. E. Woodward, F. J. Leonberger, and R. C. Williamson, "Wide-Band Electro-Optic Guided-Wave Analog-to-Digital Converters," *IEEE Proc.* 72, 802, 1984.

69. R. C. Alferness, "Polarization Independent Optical Directional Coupler Switch Using Weighted Coupling," *Appl. Phys, Lett.,* 35, 1979, pp. 748–750.

70. F. Heismann, L. L. Buhl, and R. C. Alferness, "Electro-Optically Tunable Narrowband $Ti:LiNbO_3$ Wavelength Filters," *Electron. Lett.,* 23, 1987, pp. 572–573.

71. M. Suzuki, Y. Noda, H. Tanaka, S. Akiba, Y. Kushiro, and H. Isshiki, "Monolithic Integration of InGaAsP/InP Distributed Feedback Laser and Electroabsorption Modulator by Vapor Phase Epitaxy," *IEEE J. Lightwave Tech.,* LT-5, pp. 1277–1285.

72. H. Soda, M. Furutsu, K. Sato, N. Okazaki, Y. Yamazaki, H. Nishimoto, and H. Ishikawa, "High-Power and High-Speed Semi-Insulating BH Structure Monolithic Electroabsorption Modulator/DFB Laser Light Source," *Electron Lett.,* 26, 1990, pp. 9–10.

73. G. P. Agrawal and N. K. Dutta, "Long-Wavelength Semiconductor Lasers," Van Nostrand Reinhold Company, Inc., New York, 1986.

74. R. Parke, D. F. Welch, A. Hardy, R. Lang, D. Muhuys, S. O'Brien, K. Dzurko, and D. Scifres, "2.0 W CW, Diffraction-Limited Operation of a Monolithically Integrated Master Oscillator Power Amplifier," *IEEE Phot. Tech. Lett.,* vol. 5, 1993, pp. 297–300.

75. T. L. Koch, F. S. Choa, U. Koren, R. P. Gnall, F. Hernandez-Gil, C. A. Burrus, M. G. Young, M. Oron, and B. I. Miller, "Balanced Operation of an InGaAs/InGaAsP Multiple-Quantum-Well Integrated Heterodyne Receiver," *IEEE Phot. Tech. Lett.,* vol. 2, 1990, pp. 577–580.

76. T. L. Koch and U. Koren, "Semiconductor Photonic Integrated Circuits," *IEEE J. Quantum Electron.,* vol. QE-27, 1991, pp. 641–653.

77. P. D. Magill, K. C. Reichman, R. Jopson, R. Derosier, U. Koren, B. I. Miller, M. Young, and B. Tell, "1.3 Tbit · km/s Transmission Through Non-dispersion Shifted Fiber by Direct Modulation of a Monolithic Modulator/Laser," paper PD9 in *Technical Digest of OFC '92*, Optical Society of America, San Jose, 1992, pp. 347–350.

78. S. K. Korotky, J. J. Vaselka, C. T. Kemmerer, W. J. Minford, D. T. Moser, J. E. Watson, C. A. Mattoe, and P. L. Stoddard, "High-speed, low power optical modulator with adjustable chirp parameter," paper TuG2 in *Tech. Digest of 1991 Integrated Photonics Research Topical Meeting*, Optical Society of America, Monterey, 1991, p. 53.

79. A. H. Gnauck, S. K. Korotky, J. J. Vaselka, J. Nagel, C. T. Kemmerer, W. J. Minford, and D. T. Moser, "Dispersion Penalty Reduction Using an Optical Modulator with Adjustable Chirp," *IEEE Phot. Techn. Lett.*, vol. 3, 1991, pp. 916–918.

80. J. E. Zucker, K. L. Jones, M. A. Newkirk, R. P. Gnall, B. I. Miller, M. G. Young, U. Koren, C. A. Burrus, and B. Tell, "Quantum Well Interfermontric Modulator Monolithically Integrated with 1.55 μm Tunable Distributed Bragg Reflector Laser," *Electron. Lett.*, 28 1992, pp. 1888–1889.

81. G. S. Maurer, G. T. Tremblay, and S. R. McKenzie, "Transmitter Design Alternatives for CATV Distribution in FITL Deployment," in *Proc. SPIE OE/Fibers*, 1992.

82. R. B. Childs and V. A. O'Byrne, "Multichannel AM Video Transmission Using a High Power Nd: YAG Laser and a Linearized External Modulator," *IEEE J. On Selected Areas of Comm.*, vol. 8, 1990, p. 1369.

83. J. E. Watson, M. A. Milbrodt, K. Bahadori, M. F. Dautartas, C. T. Kemmerer, D. T. Moser, A. W. Schelling, T. O. Murphy, J. J. Veselka, and D. A. Herr," A Low-Voltage 8 × 8 TiLiNbO₃ Switch with a Dilated-Benes Architecture," *IEEE/OSA J. Lightwave Tech*, 8, 1990, pp. 794–801.

84. H. Lefevre, *The Fiber-Optic Gyroscope*, Artech House, Inc., Norwood, Mass., 1993.

85. R. B. Smith, "Fiber-optic Gyroscopes 1991: A Bibliography of Published Literature," *SPIE Proceedings*, vol. 1585, 1991, pp. 464–503.

86. S. Ezekiel and H. J. Arditty (eds.), "Fiber-optic Rotation Sensors and Related Technologies," *Proceedings of the First International Conference*, Springer Series in Optical Sciences, vol. 32, 1981.

87. E. Udd (ed.), "Fiber Optic Gyros: 10th Anniversary Conference," *SPIE Proceedings*, vol. 719, 1986.

88. S. Ezekiel and E. Udd (eds.), "Fiber Optic Gyros: 15th Anniversary Conference," *SPIE Proceedings*, vol. 1585, 1991.

89. R. A. Bergh, H. C. Lefevre, and H. J. Shaw, "An Overview of Fiber-optic Gyroscopes," *J. Lightwave Technology*, LT-2, 1984, pp. 91–107.

90. H. C. Lefevre, S. Vatoux, M. Papuchon, and C. Puech, "Integrated Optics: A Practical Solution for the Fiber-optic Gyroscope," *SPIE Proceedings*, vol. 719, 1986, pp. 101–112.

91. K. Aiki, M. Nakamura, and J. Umeda, "A Frequency-Multiplexing Light Source with Monolithically Integrated Distributed-Feedback Diode Lasers," *IEEE J. Quantum Electron.*, vol. QE-13, 1977, pp. 220–223.

92. A. H. Gnauck, U. Koren, T. L. Koch, F. S. Choa, C. A. Burrus, G. Eisenstein, and G. Raybon, paper PD26 in *Tech. Digest of Optical Fiber Communications Conference*, San Francisco, 1990.

93. M. G. Young, U. Koren, B. I. Miller, M. A. Newkirk, M. Chien, M. Zirngibl, C. Dragone, B. Tell, H. M. Presby, and G. Raybon," A 16X1 WDM Transmitter with Integrated DBR Lasers and Electroabsorption Modulators," paper IWA3 in *Tech. Digest of 1993 Integrated Photonics Research Topical Meeting*, Optical Society of America, Palm Springs, 1993, pp. 414–417.

94. C. Dragone, "An N × N Optical Multiplexer Using a Planar Arrangement of Two Star Couplers," *IEEE Photon. Tech. Lett.*, 3, 1991, pp. 812–815.

95. H. Takahashi, Y. Hibino, and I. Nishi, "Polarization-insensitive Arrayed-waveguide Grating Wavelength Multiplexer on Silicon," *Opt. Lett.*, 17, 1992, pp. 499–501.

96. M. Zirngible, C. Dragone, and C. H. Joyner, "Demonstration of a 15X15 Arrayed Waveguide Multiplexer on InP," *IEEE Phot. Tech. Lett.*, 1992, pp. 1250–1252.

97. S. S. Bergstein, A. F. Ambrose, B. H. Lee, M. T. Fatehi, E. J. Murphy, T. O. Murphy, G. W. Richards, F. Heismann, F. R. Feldman, A. Jozan, P. Peng, K. S. Liu, and A. Yorinks, "A Fully Implemented Strictly Non-blocking 16X16 Photonic Switching System," paper PD30 in *Tech. Digest of OFC/IOOC '93*, San Jose, 1993, pp. 123–126.

CHAPTER 7
MINIATURE AND MICRO-OPTICS

Tom D. Milster
Optical Sciences Center
University of Arizona
Tucson, Arizona

7.1 GLOSSARY

A, B, C, D	constants
$A(r, z)$	converging spherical wavefront
c	curvature
D	diffusion constant
d	diffusion depth
EFL	effective focal length
f	focal length
g	gradient constant
h	radial distance from vertex
i	imaginary
k	conic constants
k	wave number
LA	longitudinal aberration
l_0	paraxial focal length
M	total number of zones
NA	numerical aperture
n	refractive index
r	radial distance from optical axis
r_{mask}	mask radius
r_m	radius of the mth zone
t	fabrication time
$\bar{u}$	slope
W_{ijk}	wavefront function
X	shape factor
x, y	Cartesian coordinates
y	height

Z	sag
z	optical axis
Δ	relative refractive difference
ρ	propagation distance
λ	wavelength
$\bar{\sigma}$	$\sigma_{rms}/2y$
σ_{rms}	rms wavefront error
Φ	phase
ψ	special function

7.2   INTRODUCTION

Optical components come in many sizes and shapes. A class of optical components that has become very useful in many applications is called micro-optics. We define micro-optics very broadly as optical components ranging in size from several millimeters to several hundred microns. In many cases, micro-optic components are designed to be manufactured in volume, thereby reducing cost to the customer. The following paragraphs describe micro-optic components that are potentially useful for large-volume applications. The discussion includes several uses of micro-optics, design considerations for micro-optic components, molded glass and plastic lenses, distributed-index planar lenses, Corning's SMILE™ lenses, microFresnel lenses, and, finally, a few other technologies that could become useful in the near future.

7.3   USES OF MICRO-OPTICS

Micro-optics are becoming an important part of many optical systems. This is especially true in systems that demand compact design and form factor. Some optical fiber-based applications include fiber-to-fiber coupling, laser-diode-to-fiber connections, LED-to-fiber coupling, and fiber-to-detector coupling. Microlens arrays are useful for improving radiometric efficiency in focal-plane arrays, where relatively high numerical aperture (NA) microlenslets focus light onto individual detector elements. Microlens arrays can also be used for wavefront sensors, where relatively low-NA lenslets are required. Each lenslet is designed to sample the input wavefront and provide a deviation on the detector plane that is proportional to the slope of the wavefront over the lenslet area. Micro-optics are also used for coupling laser diodes to waveguides and collimating arrays of laser diodes. An example of a large-volume application of micro-optics is data storage, where the objective and collimating lenses are only a few millimeters in diameter.[1]

7.4   MICRO-OPTICS DESIGN CONSIDERATIONS

Conventional lenses made with bulk elements can exploit numerous design parameters, such as the number of surfaces, element spacings, and index/dispersion combinations, to achieve performance requirements for NA, operating wavelength, and field of view. However, fabricators of micro-optic lenses seek to explore molded or planar technologies, and thus the design parameters tend to be more constrained. For example, refractive microlenses made by molding, ion exchange, mass transport, or the SMILE™ process resemble single-element optics. Performance of these lenses is optimized by manipulating one or possibly two radii, the thickness, and the index or index distribution. Index choices are limited by the available materials. Distributed-index and graded-index lenses have a limited range of index profiles that can be achieved. Additional performance correction is possible by aspherizing one or both surfaces of the element. This is most efficiently done with the molding process, but molded optics are difficult to produce when the diameter of the lens is less than 1.0 mm. In general, one or two aberrations may be corrected with one or two aspheres, respectively.

Due to the single-element nature of microlenses, insight into their performance may be gained by studying the well-known third-order aberrations of a thin lens in various configurations. Lens bending and stop shift are the two parameters used to control aberrations for a lens of a given power and index. Bending refers to distribution of power between the two surfaces, i.e., the shape of the lens, as described in R. Barry Johnson's Chap. 1, (Vol. II) on "Lenses." The shape is described by the shape factor X which is

$$X = \frac{C_1 + C_2}{C_1 - C_2} \tag{1}$$

where C_1 and C_2 are the curvatures of the surfaces. The third-order aberrations as a function of X are shown in Fig. 1. These curves are for a lens with a focal length of 10.0 mm, an entrance pupil diameter of 1.0 mm, field angle $\bar{u} = 20°$, an optical index of refraction of 1.5, $\lambda = 0.6328 \ \mu m$, and the object at infinity. For any given bending of the lens, there is a corresponding stop position that eliminates coma,[2] and this is the stop position plotted in the figure. The stop position for which coma is zero is referred to as the *natural* stop shift, and it also produces the least curved tangential field for the given bending. Because the coma is zero, these configurations of the thin lens necessarily satisfy the Abbe sine condition. When the stop is at the lens (zero stop shift), the optimum shape to eliminate coma is approximately convex-plano ($X = +1$) with the convex side toward the object. The optimum shape is a function of the index, and the higher the index, the more the lens must be bent into a meniscus. Spherical aberration is minimized with the stop at the lens, but astigmatism is near its maximum. It is interesting to note that biaspheric objectives for data storage tend toward the convex-plano shape.

Astigmatism can be eliminated for two different lens-shape/stop-shift combinations, as

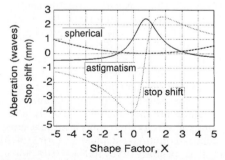

FIGURE 1 Third-order aberrations as a function of the shape factor, or bending, of a simple thin lens with focal length 10.0 mm, entrance pupil diameter of 1.0 mm, field angle 20°, $n = 1.5$, and object at infinity. The stop position shown is the *natural* stop shift, that is, the position that produces zero coma.

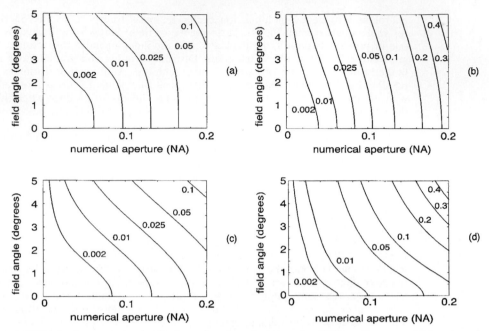

FIGURE 2   Contours of normalized rms wavefront deviation, $\bar{\sigma} = 1000\lambda\sigma_{rms}/2y$, versus field angle and NA, where $2y$ is the diameter of the stop. The stop is located at the lens. The focus is adjusted to give minimum rms deviation of the wavefront, so effects of Petzval curvature are not included. A: $X = 1$, $n = 1.5$; B: $X = -1$, $n = 1.5$; C: $X = 1$, $n = 3.0$; D: $X = -1$, $n = 3.0$.

shown in Fig. 1. The penalty is an increase in spherical aberration. Note that there is no lens shape for which spherical, coma, and astigmatism are simultaneously zero in Fig. 1, that is, there is no aplanatic solution when the object is at infinity. The aplanatic condition for a thin lens is only satisfied at finite conjugates.

The plano-convex shape ($X = -1$) that eliminates astigmatism is particularly interesting because the stop location is in front of the lens at the optical center of curvature of the second surface. All chief rays are normally incident at the second surface. Thus, the design is monocentric.[3] (Obviously, the first surface is not monocentric with respect to the center of the stop, but it has zero power and only contributes distortion.)

Two very common configurations of micro-optic lenses are $X = +1$ and $X = -1$ with the stop at the lens. Typically, the object is at infinity. In Fig. 2, we display contours of normalized rms wavefront deviation, $\bar{\sigma} = \sigma\lambda_{rms}/2y$, versus $\bar{u}$ and NA, where $2y =$ diameter of the stop. Aberration components in σ_{rms} include third-order spherical, astigmatism, and coma. The focus is adjusted to give minimum rms deviation of the wavefront, so effects of Petzval curvature are not included. Tilt is also subtracted. As NA or field angle is increased, rms wavefront aberration increases substantially. The usable field of view of the optical system is commonly defined in terms of Maréchal's criterion[4] as field angles less than those that produce $2y\bar{\sigma}/1000\lambda \leq 0.07$ wave. For example, if the optical system operates at $2y = 1.0$ mm, $\lambda = 0.6328\,\mu$m, NA $= 0.1$, $X = +1$, $n = 1.5$, and $\bar{u} = 2°$, the wavefront aberration due to third-order contributions is

$$\sigma_{rms} = \frac{2y\bar{\sigma}}{1000\lambda} \approx \frac{(1.0 \times 10^{-3}\,\text{m})(0.015)}{(10^{3})(0.6328 \times 10^{-6}\,\text{m/wave})} = 0.024 \text{ wave} \qquad (2)$$

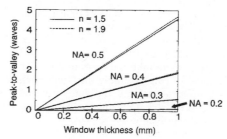

FIGURE 3 Effect of a window on wavefront distortion at $\lambda = 830$ nm.

which is acceptable for most situations. Note that the configuration for $X = -1$ yields $\sigma_{rms} \approx 0.079$ wave, which is beyond the acceptable limit. When large values of σ_{rms} are derived from Fig. 2, care must be taken in interpretation of the result because higher-order aberrations are not included in the calculation. Also, if field curvature is included in the calculation, the usable field of view is significantly reduced.

Coma and astigmatism are only significant if the image field contains off-axis locations. In many laser applications, like laser diode collimators, the micro-optic lens is designed to operate on axis with only a very small field of view. In this case, spherical aberration is very significant. A common technique that is used to minimize spherical aberration is to aspherize a surface of the lens. Third-, fifth-, and higher orders of spherical aberration may be corrected by choosing the proper surface shape. In some lens design codes, the shape is specified by

$$ Z = \frac{ch^2}{1 + \sqrt{1 - (1 + k)c^2 h^2}} + Ah^4 + Bh^6 + Ch^8 + Dh^{10} \tag{3} $$

where Z is the sag of the surface, c is the base curvature of the surface, k is the conic constant ($k = 0$ is a sphere, $k = -1$ is a paraboloid, etc.), and $h = \sqrt{x^2 + y^2}$ is the radial distance from the vertex. The A, B, C, and D coefficients specify the amount of aspheric departure in terms of a polynomial expansion in h.

When a plane-parallel plate is inserted in a diverging or converging beam, such as the window glass of a laser diode or an optical disk, spherical aberration is introduced. The amount of aberration depends on the thickness of the plate, the NA of the beam, and to a lesser extent the refractive index of the plate,[5] as shown in Fig. 3. The magnitude of all orders of spherical aberration is linearly proportional to the thickness of the plate. The sign is opposite that of the spherical aberration introduced by an $X = +1$ singlet that could be used to focus the beam through the plate. Therefore, the aspheric correction on the singlet compensates for the difference of the spherical aberration of the singlet and the plate. This observation follows the fact that minimum spherical aberration without aspheric correction is achieved with the smallest possible air gap between the lens and the plate. For high-NA singlet objectives, one or two aspheric surfaces are added to correct the residual spherical aberration.

7.5 MOLDED MICROLENSES

Molded micro-optic components have found applications in several commercial products, which include compact disk players, bar-code scanners, and diode-to-fiber couplers. Molded lenses become especially attractive when one is designing an application that requires aspheric surfaces. Conventional techniques for polishing and grinding lenses tend to be time-expensive and do not yield good piece-to-piece uniformity. Direct molding, on the other hand, eliminates the need for any grinding or polishing. Another advantage of direct molding is that useful reference surfaces can be designed directly into the mold. The reference surfaces can take the form of flats.[6] The reference flats are used to aid in aligning the lens element during assembly into the optical device. Therefore, in volume applications that require aspheric surfaces, molding becomes a cost-effective and practical solution. The molding process utilizes a master mold, which is commonly made by single-point diamond turning and post polishing to remove tooling marks and thus minimize scatter from the surface. The master can be tested with conventional null techniques, computer-generated null holograms,[7] or null Ronchi screens.[8] Two types of molding technology are described in the following paragraphs. The first is molded glass technology. The second is molded plastic technology.

Molded Glass

One of the reasons glass is specified as the material of choice is thermal stability. Other factors include low birefringence, high transmission over a broad wavelength band, and resistance to harsh environments.

Several considerations must be made when molding glass optics. Special attention must be made to the glass softening point and refractive index.[9] The softening point of the glass used in molded optics is lower than that of conventional components. This enables the lenses to be formed at lower temperatures, thereby increasing options for cost-effective tooling and molding. The refractive index of the glass material can influence the design of the surface. For example, a higher refractive index will reduce the surface curvature. Smaller curvatures are generally easier to fabricate and are thus desirable.

An illustration is Corning's glass molding process.[9] The molds that are used for aspheric glass surfaces are constructed with a single-point diamond turning machine under strict temperature and humidity control. The finished molds are assembled into a precision-bored alignment sleeve to control centration and tilt of the molds. A ring member forms the outside diameter of the lens, as shown in Fig. 4. The glass material, which is called a preform, is inserted between the molds. Two keys to accurate replication of the aspheric surfaces are forming the material at high glass viscosity and maintaining an isothermal environment. After the mold and preform are heated to the molding temperature, a load is applied to one of the molds to press the preform into shape. After molding, the assembly

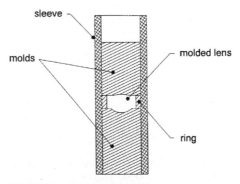

FIGURE 4   Mold for glass optics.

is cooled to below the glass transformation point before the lens is removed. Optical performance characteristics of the finished lens are determined by the quality of the mold surfaces, the glass material, and the preform volume, which also determines the thickness of the lens when pressed.

An alternative process is used at Kodak, Inc., where molded optics are injection molded and mounted into precision lens cells.[10] In this process, a tuned production mold can reproduce intricate mounting datum features and extremely well-aligned optics. It can also form a stop, baffle, or a film-plane reference in the system. Table 1 lists preferred and possible tolerances for molded glass components. The Kodak process has been tested with over 50 optical glasses, which include both crowns and flints. This provides a wide index-of-refraction range, $1.51 < n < 1.85$, to choose from.

Most of the molded glass microlenses manufactured to date have been designed to operate with infrared laser diodes at $\lambda = 780$–830 nm. The glass used to make the lenses is transparent over a much broader range, so the operating wavelength is not a significant factor if designing in the visible or near infrared. Figure 5 displays a chart of the external transmission of several optical materials versus wavelength. LaK09 (curve B) is representative of the type of glass used in molded optics. The external transmission from 300 nm to over 2200 nm is limited primarily by Fresnel losses due to the relatively high index of refraction ($n = 1.73$). The transmission can be improved dramatically with antireflection coatings. Figure 6 displays the on-axis operating characteristics of a Corning 350110 lens, which is used for collimating laser diodes. The rms wavefront variation and effective focal length (EFL) are shown versus wavelength. The highest aberration is observed at shorter wavelengths. As the wavelength increases, the EFL increases, which decreases the NA slightly. Table 2 lists several optical properties of molded optical materials. The trend in molded glass lenses is to make smaller, lighter, and higher NA components.[11] Reduction in mass and size allows for shorter access times in optical data storage devices, and higher NA improves storage density in such devices.

Molded Plastic

Molded plastic lenses are an inexpensive alternative to molded glass. In addition, plastic components are lighter than glass components. However, plastic lenses are more sensitive to temperatures and environmental factors. The most common use of molded plastic lenses is in compact disk (CD) players.

Precision plastic microlenses are commonly manufactured with injection molding equipment in high-volume applications. However, the classical injection molding process typically leaves some inhomogeneities in the material due to shear and cooling stresses.[12] Improved molding techniques can significantly reduce variations, as can compression molding and casting. The current state of the art in optical molding permits master surfaces to be replicated to an accuracy of roughly one fringe per 25 mm diameter, or perhaps a bit better.[13] Detail as small as 5 nm may be transferred if the material properties and processing are optimum and the shapes are modest. Table 3 lists tolerances of injection-molded lenses.[14] The tooling costs associated with molded plastics are typically less than those associated with molded glass because of the lower transition temperature of the plastics. Also, the material cost is lower for polymers than for glass. Consequently, the costs associated with manufacture of molded plastic microlenses are much less than those for molded glass microlenses. The index of refraction for the plastics is less than that for the glass lenses, so the curvature of the surfaces must be greater, and therefore harder to manufacture, for comparable NA.

The glass map for molded plastic materials is shown in Fig. 7. The few polymers that have been characterized lie mainly outside the region containing the optical glasses and particularly far from the flint materials.[15] Data on index of refraction and Abbe number are particularly difficult to obtain for molded plastic. The material is supplied in pelletized

TABLE 1 Preferred and Possible Tolerances for Molded Glass Components[10]

	Preferred	Possible
Center thickness (mm)	10.00 max	25.00
	0.40 min	0.35
	±0.030 tol	±0.015
Diameter (mm)	25.00 max	50.00
	4.00 min	2.00
	±0.10 tol	±0.01
Diameter of lens beyond clear aperture (mm)	2.00	0.50
Surface quality	80–50	40–20
Axis alignment	3×10^{-3} radians	2×10^{-3} radians
Radius (mm)— best fit sphere	5 to ∞	2 to ∞
Slope (λ/mm)	50 max	100 max
Wavelengths (λ) departure from BFS	≤250	≤500

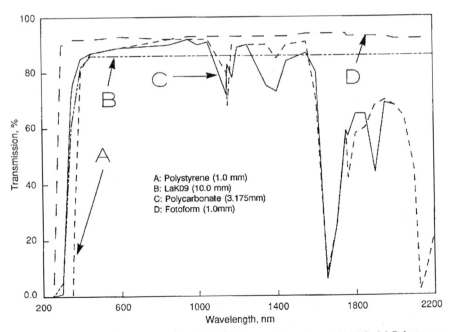

FIGURE 5 External transmission of several optical materials versus wavelength. (*a*) Polystyrene 1.0 mm thick, which is used for molded plastic lenses,[12] (*b*) LaK09 10.0 mm thick, which is used for molded glass lenses;[65] (*c*) Polycarbonate 3.175 mm thick, which is used for molded plastic lenses,[12] (*d*) Fotoform glass 1.0 mm thick, which is used in the production of SMILE[TM] lenses.[38]

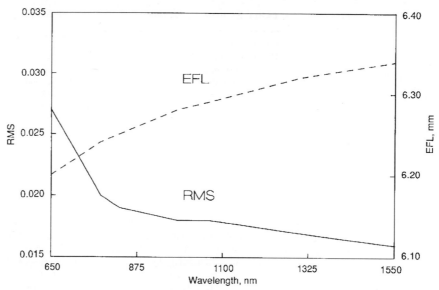

FIGURE 6 On-axis operating characteristics versus wavelength of a Corning 350110 lens, which is a molded glass aspheric used for collimating laser diodes.[66]

TABLE 2 Properties of Materials Used for Molding Micro-optics

Property	PMMA (acrylic)	PMMA (imide)	SSMA	Poly-carbonate	Poly-styrene	LaK09	BK7
Index (n_d)	1.491	1.528	1.564	1.586	1.589	1.734	1.517
Abbe # (V_d)	57.4	48	35	30	31	51.5	64.2
Density (g/mm³)	1.19	1.21	1.09	1.20	1.06	4.04	2.51
Max service temp (°C)	72	142	87	121	75	500	500
Thermal expansion coefficient (1E-6 mm/mm °C)	67.9	—	56.0	65.5	50.0	5.5	7.1
Thermal index coefficient (1E-6/°C)	−105	—	—	−107	—	6.5	3
Young's modulus (10E4 kg/cm²)	3.02	—	3.30	2.43	3.16	11.37	83.1
Impact strength	2	—	3	5	4	—	1
Abrasion resistance	4	—	3	1	2	—	5
Cost/lb	3	—	2	4	2	—	5
Birefringence	2	—	4	3	5	—	1

(1 = lowest/5 = highest).

TABLE 3 Injection Molding Tolerances for Plastic Lenses

Focal length	±0.5%
Radius of curvature	±0.5%
Spherical power	2 to 5 f*
Surface quality	60/40 (40/20 possible)
Vertex thickness (in)	±0.0005
Diameter (in. per in. DIA.)	±0.002 to 0.0005
Repeatability lens-to-lens	0.1% to 0.3%

* Tolerances given in optical fringes abbreviated by "f".
Vertex-to-edge thickness ratio
4:1 Difficult to mold
3:1 Moderately easy to mold
2:1 Easy to mold

form, so it must first be molded into a form suitable for measurement. The molding process subjects the material to a heating and annealing cycle that potentially affects the optical properties. Typically, the effect of the additional thermal history is to shift the dispersion curve upward or downward, leaving the shape unchanged. A more complete listing of optical plastics and their properties is given in Ref. 12. Additional information can be obtained from the *Modern Plastics Encyclopedia,*[16] the *Plastics Technology, Manufacturing Handbook and Buyer's Guide,*[17] and in John D. Lytle's Chap. 34, (Vol. II) on "Polymeric Optics."

Changes in dimension or refractive index due to thermal variations occur in both molded glass and molded plastic lenses. However, the effect is more pronounced in polymer optical systems because the thermal coefficients of refractive index and expansion are ten times greater than for optical glasses, as shown in Table 2. When these changes are modeled in a computer, a majority of the optical systems exhibit a simple defocus and a change of effective focal length and corresponding first-order parameters. An experimental study[18] was made on an acrylic lens designed for a focal length of 6.171 mm at $\lambda = 780$ nm and 20°C. At 16°C, the focal length changed to 6.133 mm. At 60°C, the focal length changed to 6.221 mm. Thermal gradients, which can introduce complex aberrations, are a more serious problem. Therefore, more care must be exercised in the design of athermalized mounts for polymer optical systems.

The transmission of two common optical plastics, polystyrene and polycarbonate, are

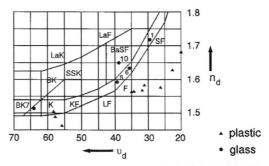

FIGURE 7 Glass map for molded plastic materials, which are shown as triangles in the figure. The few polymers that have been characterized lie mainly outside the region containing the optical glasses and particularly far from the flint materials.[13]

shown in Fig. 5. The useful transmittance range is from 380 to 1000 nm. The transmission curve is severely degraded above 1000 nm due to C-H vibrational overtone and recombination bands, except for windows around 1300 nm and 1500 nm. Sometimes, a blue dye is added to the resins to make the manufactured part appear "water clear," instead of slightly yellowish in color. It is recommended that resins be specified with *no* blue toner for the best and most predictable optical results.[12]

The shape of the lens element influences how easily it can be manufactured. Reasonable edge thickness is preferred in order to allow easier filling. Weak surfaces are to be avoided because surface-tension forces on weak surfaces will tend to be very indeterminate. Consequently, more strongly curved surfaces tend to have better shape retention due to surface-tension forces. However, strongly curved surfaces are a problem because it is difficult to produce the mold. Avoid clear apertures that are too large of a percentage of the physical surface diameter. Avoid sharp angles on flange surfaces. Use a center/edge thickness ratio less than 3 for positive lenses (or 1/3 for negative lenses). Avoid cemented interfaces. Figure 8 displays a few lens forms. The examples that mold well are C, E, F, and H. Form A should be avoided due to a small edge thickness. Forms A and B should be avoided due to weak rear surfaces. Form D will mold poorly due to bad edge/center thickness ratio. Form G uses a cemented interface, which could develop considerable stress due to the fact that thermal differences may deform the pair, or possibly even destroy the bond.

Since polymers are generally softer than glass, there is concern about damage from ordinary cleaning procedures. Surface treatments, such as diamond films,[19] can be applied that greatly reduce the damage susceptibility of polymer optical surfaces.

A final consideration is the centration tolerance associated with aspheric surfaces. With spherical optics, the lens manufacturer is usually free to trade off tilt and decentration tolerances. With aspheric surfaces, this tradeoff is no longer possible. The centration

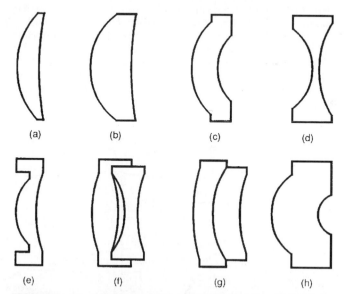

FIGURE 8 Example lens forms for molded plastic lenses. Forms C, E, F, and H mold well. Form A should be avoided due to small edge thickness. Forms A and B should be avoided due to weak rear surfaces. Form D will mold poorly due to bad edge/center ratio. Form G uses a cemented interface, which could develop stress.[15]

tolerance for molded aspherics is determined by the alignment of the mold halves. A common specification is 4 to 6 μm, although 3 to 4 μm is possible.

7.6 MONOLITHIC LENSLET MODULES

Monolithic lenslet modules (MLMs) are micro-optic lenslets configured into close-packed arrays. Lenslets can be circular, square, rectangular, or hexagonal. Aperture sizes range from as small as 25 μm to 1.0 mm. Overall array sizes can be fabricated up to 68 × 68 mm. These elements, like those described in the previous section, are fabricated from molds. Unlike molded glass and plastic lenses, MLMs are typically fabricated on only one surface of a substrate, as shown in the wavefront sensing arrangement of Fig. 9. An advantage of MLMs over other microlens array techniques is that the fill factor, which is the fraction of usable area in the array, can be as high as 95 to 99 percent. Applications for MLMs include Hartman testing,[20] spatial light modulators, optical computing, video projection systems, detector fill-factor improvement,[22] and image processing.

There are three processes that have been made used to construct MLMs.[22] All three techniques depend on using a master made of high-purity annealed and polished material. After the master is formed, a small amount of release agent is applied to the surface. In the most common fabrication process, a small amount of epoxy is placed on the surface of the master. A thin glass substrate is placed on top. The lenslet material is a single-part polymer epoxy. A slow-curing epoxy can be used if alignment is necessary during the curing process.[23] The second process is injection molding of plastics for high-volume applications. The third process for fabrication of MLMs is to grow infrared materials, like zinc selenide, on the master by chemical vapor deposition. Also, transparent elastomers can be used to produce flexible arrays.

MLMs are advertised[24] to be diffraction-limited for lenslets with NA < 0.10. Since the lens material is only a very thin layer on top of the glass substrate, MLMs do not have the same concerns that molded plastic lenses have with respect to birefringence and transmission of the substrate. For most low-NA applications, individual lenslets can be analyzed as plano-convex lenses. Aspheres can be fabricated to improve imaging performance for higher NAs. Aspheres as fast as NA = 0.5 have been fabricated with spot sizes about twice what would be expected from a diffraction-limited system. The residual error is probably due to fabrication imperfections observed near the edges and corners of the lenslets.[25]

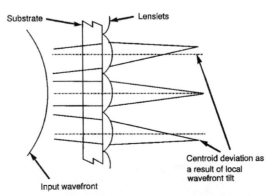

FIGURE 9 Monolithic lenslet modules (MLMs) configured for wavefront sensing.[28]

7.7 DISTRIBUTED-INDEX PLANAR MICROLENSES

A distributed-index planar microlens, which is also called a Luneberg lens,[26] is formed with a radially symmetric index distribution. The index begins at a high value at the center and decreases to the index value of the substrate at the edge of the lens. The function that describes axial and radial variation of the index is given by[27]

$$n(r, z) \approx n(0, 0) \sqrt{1 - g^2 r^2 - \frac{2g \, \Delta n^2(0, 0)}{d} z^2} \qquad (4)$$

where r is the radial distance from the optical axis, z is the axial distance, $n(0, 0)$ is the maximum index at the surface of the lens, g is a constant that expresses the index gradient, d is the diffusion depth, and $\Delta = (n(0, 0) - n_2)/n(0, 0)$, where n_2 is the substrate index. Typical values are $\Delta = 0.05$, $d = 0.4$ mm, $r_{max} = 0.5$ mm, $n_2 = 1.5$, and $g = \sqrt{2\Delta}/r_{max} = 0.63$ mm^{-1}. These lenses are typically fabricated on flat substrates and yield hemispherical index profiles, as shown in Fig. 10. Two substrates placed together will produce a spherical lens. Several applications of light coupling with distributed-index microlenses have recently been demonstrated.[28] These include coupling laser diodes to fibers, LEDs to fibers, fibers to fibers, and fibers to detectors. In the future, arrays of lenslets might aid in parallel communication systems.

One way to introduce the index gradient is through ion exchange.[29] As shown in Fig. 10, a glass substrate is first coated with a metallic film. The film is then patterned with a mask that allows ions to diffuse from a molten salt bath through open areas of the mask. Ions in the glass substrate are exchanged for other ions in the molten salt at high temperatures. The diffused ions change the refractive index of the substrate by an amount that is proportional to their electric polarizability and concentration. To increase the index, diffusing ions from the salt bath must have a larger electronic polarizability than that of the ions involved in the glass substrate. Since ions that have larger electron polarizability also have larger ionic radius, the selective ion exchange changes the index distribution and creates local swelling where the diffusing ion concentration is high. The swelling can be removed with polishing for a smooth surface. Alternatively, the swelling can be left to aid in the lensing action of the device. To obtain the proper index distribution, the mask radius and diffusion time must be chosen carefully.[30] If the mask radius, r_{mask}, is small compared to the diffusion depth, the derivative of the index distribution with respect to radial

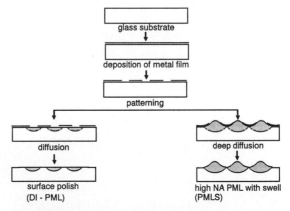

FIGURE 10 Planar distributed-index microlens array and fabrication process.[28]

TABLE 4 Summary of Diffusion Times for Planar DI Lenses[32]

Materials	W_n/n	D (m^2/sec)	t (sec)*
Plastics (DIA-MMA)	0.05	3×10^{-10}	3×10^2
Glass (TI) ion-exchange	0.05	4×10^{-13}	9×10^4
Glass (TI) electromigration	0.05	—	3×10^4†

* t = $(r_m^2/D) \times 0.4$.
† Experimental data with radius of 0.6 mm.

distance r monotonically decreases. Since the curvature of a light ray passing through the medium is proportional to the gradient of the logarithm of the refractive index, the rays tend not to focus. A suitable combination of diffusion time t and mask radius is given by $Dt/r_{mask}^2 \approx 0.4$, where D is the diffusion constant of the dopant in the substrate. Table 4 displays the diffusion time necessary for making a planar microlens with a radius of 0.5 mm. Typically, the paraxial focal length in the substrate is $l_0 \approx 20\ r_{mask}$, and the numerical aperture is NA $\approx n_2/20$.

Other fabrication techniques can also be used. Planar lens arrays in plastics are fabricated with monomer-exchange diffusion.[31,32] Plastics are suitable for making larger-diameter lenses because they have large diffusion constants at relatively low temperatures (100°C). The electromigration technique[33] is more effective for creating devices with short focal length. For example, by applying an electric field of 7 V/mm for 8 h, it is possible to obtain a planar microlens with radius of 0.6 mm and focal length of 6.8 mm.[30] A distributed-index microlens array using a plasma chemical vapor deposition (CVD) method has also been reported.[34] In this process, hemispherical holes are etched into a planar glass substrate. The holes are filled with thin layers of a combination of SiO_2 and Si_2N_4. These materials have different indices of refraction, and the composition is varied from the hemispherical outside shell to the center to provide a Luneburg index distribution.

Shearing interferometry can be used to measure the index distribution from thinly sliced samples of lenslets. Samplets are acquired laterally or longitudinally, as shown in Fig. 11. Results of the measurement on a lateral section are shown in Fig. 12 for the

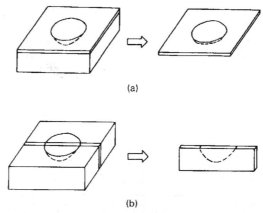

(a)

(b)

FIGURE 11 Slicing a lens to obtain a thin sample for interferometric characterization.[30] (*a*) lateral slice; (*b*) longitudinal slice.

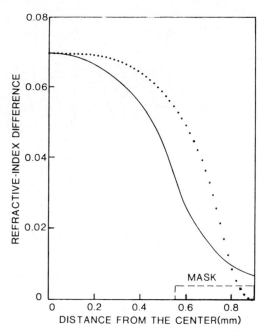

FIGURE 12 Surface index distribution of a planar microlens. Theoretical (——) and experimental (• • •).[67]

ion-exchange technique. The solid line is the theoretical prediction, and the dotted line corresponds to measured data. The large discrepancy between measured and theoretical results is probably due to concentration-dependent diffusion or the interaction of the dopants. Figure 13 shows the two-dimensional index profile resulting from a deep electromigration technique.[35] These data correspond much more closely to the theoretical values in Fig. 12.

The ray aberration of a distributed-index lens is commonly determined by observing the longitudinal aberration at infinite conjugates, as shown in Fig. 14. The paraxial focusing length, l_0, is given by

$$l_0 = d + \frac{\sqrt{1 - 2\Delta}}{g} \cot \left[\frac{gd \sin^{-1}\sqrt{2\Delta}}{\sqrt{2\Delta}} \right] \tag{5}$$

The amount of longitudinal aberration is defined by $LA = (l - l_0)/l_0$, where l is the distance at which a ray crosses the optical axis. LA increases with the radius r of the ray. In order to display the effects of different Δ and n_2 parameters, we define a normalized numerical aperture that is given by

$$\overline{NA} = \frac{NA}{n_2\sqrt{2\Delta}} \tag{6}$$

and is plotted in Fig. 15 versus diffusion depth for several values of LA. Notice that, for small values of LA, the maximum $\overline{NA}$ occurs at a diffusion depth of $d \approx 0.9/g$.

Wave aberration of a planar distributed index microlens is shown in Fig. 16. The large departure at the maximum radius indicates severe aberration if used at full aperture. Swelled-structure lenses can exhibit much improved performance.[36] It has been determined that the index distribution contributes very little to the power of the swelled-surface

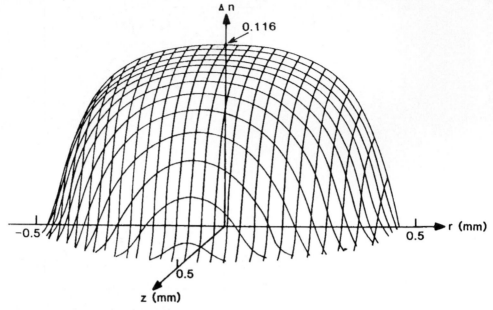

FIGURE 13 Two-dimensional index distribution of a distributed-index planar microlens prepared with the deep electromigration technique.[35]

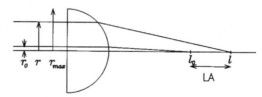

FIGURE 14 Longitudinal ray aberration, LA, of a distributed-index planar microlens. The object is at infinity. l_0 is the paraxial focal distance. LA increases with r.

element. Most of the focusing power comes from the swelled surface-air interface. A few characteristics of ion-exchanged distributed-index microlenses are shown in Table 5.

7.8 SMILE™ MICROLENSES

Spherical Micro Integrated Lenses (SMILE™) are also micro-optic lenslets configured into arrays. Unlike MLMs, SMILE™ lenses are formed from a photolytic technique in photosensitive glass. Applications for SMILE™ lenses include facsimile machines and photocopiers, LED/LCD enhancers, autofocus devices in video and SLR cameras, optical waveguide connectors, and others.

The process used for construction of SMILE™ lenses involves an exposure of the glass to ultraviolet light through a chrome mask.[37] The mask is patterned so that circular areas,

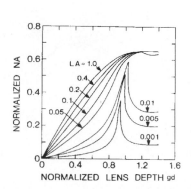

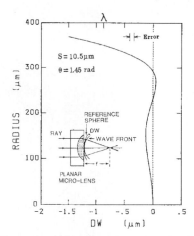

FIGURE 15 Normalized NA versus normalized depth of the distributed index region at several values of LA.[27]

FIGURE 16 Wave aberration of a distributed-index planar microlens.[68]

which correspond to the lens diameter, are opaque on a clear background. The arrangement of the opaque circles on the mask determines the layout of the lenses in the final device. The exposure is followed by a thermal development schedule that initiates the formation of noble metal particles, which in turn serve as nuclei for the growth of a lithium metasilicate microcrystalline phase from the homogeneous glass. The thermo-optically developed crystallized region is slightly more dense than the unexposed homogeneous glass. The exposed region contracts as the crystalline phase develops. This squeezes the soft undeveloped glass and forces it beyond the plane of the original surface. Minimization of surface energy determines the spherical nature of the surface. The spherical eruption constitutes the lenslet. In addition to the lens-forming process, the exposed and developed region surrounding the lens is rendered optically opaque, thus providing optical isolation. Figure 17 displays an electron photomicrograph showing the spherical protrusions in perspective. Figure 18 shows an optical micrograph of the lenses in a close-packed geometry. The lens sag, as well as the optical density and color of the exposed region, is a function of the excitation light exposure and the thermal schedule. The lens sag is also determined to some extent from the distance between the lenslets. The glass can be exposed from both sides simultaneously through a pair of precisely aligned masks. This permits a lens pattern that has symmetric lens curvatures on opposing sides. It is advertised[38] that SMILE™ lens substrates can be as thin as 0.25 mm with diameters from 75 to 1000 μm. The minimum separation of the lenses must be greater than 15 μm. Effective focal lengths are available between 50 and 200 μm, with NA ≤ 0.35.

SMILE™ lenses suffer from aberration near the lenslet edges, as shown in the interferogram of Fig. 19. Over most of the lenslet area, straight and equally spaced fringes are observed. However, near the edges, a strong curve indicates the presence of spherical

TABLE 5 Fundamental Characteristics of the Planar DI Microlens[28]

	Diameter	NA	Focal length
Planar	10–1000 μm	0.02–0.25	20–4000 μm
Swelled	50–400 μm	0.4–0.6	55–500 μm

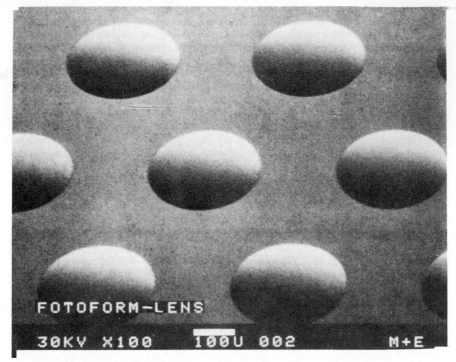

FIGURE 17 Electron photomicrograph showing spherical protrusion in perspective of a SMILE™ microlens. Each lenslet is 400 μm in diameter and has a maximum height of ~20 μm.

aberration. An irradiance spot profile is displayed in Fig. 20 that results from focusing a 0.2-NA lenslet at $\lambda = 0.6238$ μm. The full-width-at $1/e^2$ is 3.7 μm. It has been shown that, for facsimile applications, arrays of SMILE™ lenses exhibit better irradiance uniformity and have an enhanced depth of focus when compared to a rod-lens array of GRIN lenses.[39]

7.9 MICRO-FRESNEL LENSES

The curvature of an optical beam's wavefront determines whether the beam is converging, diverging, or collimated. A bulk lens changes the wavefront curvature in order to perform a desired function, like focusing on a detector plane. The micro-Fresnel lens (MFL) performs the same function as a bulk lens, that is, it changes the curvature of the wavefront. In a simple example, the MFL converts a plane wavefront into a converging spherical wavefront, $A(x, y, z)$, as shown in Fig. 21. The difference between an MFL and a bulk lens is that the MFL must change the wavefront over a very thin surface.

A Fresnel lens is constructed of many divided annular zones, as shown in Fig. 22. Fresnel lenses are closely related to Fresnel zone plates.[40,41] Both zone patterns are the same. However, unlike a Fresnel zone plate, the Fresnel lens has smooth contours in each zone, which delay the phase of the optical beam by 2π radians at the thickest point. In the central zone, the contour is usually smooth enough that it acts as a refractive element. Toward the edges, zone spacing can become close to the wavelength of light, so the Fresnel lens exhibits diffractive properties. Also, due to the quasi-periodic nature of the zones and the diffractive properties, Fresnel lenses have strong wavelength dependencies.

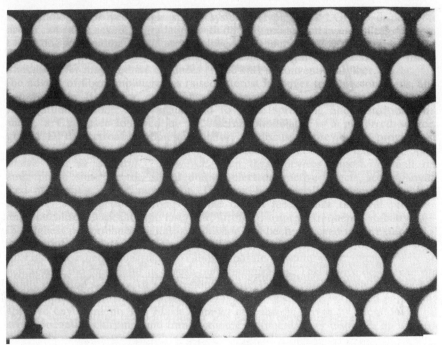

FIGURE 18   Optical micrograph showing SMILE™ lenses surrounded by an optically opaque region.[39]

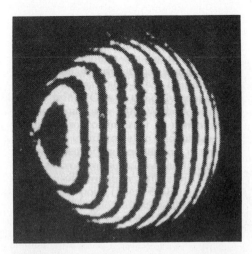

FIGURE 19   Interferogram of a Corning 160-μm-diameter F/1.4 lenslet. The central two-thirds of the lenslet is nearly perfect.[69]

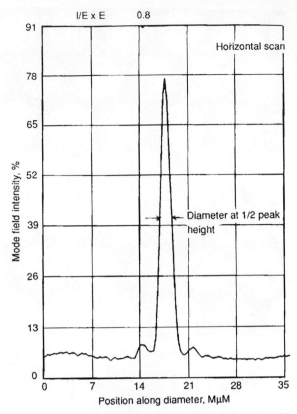

FIGURE 20   Image-plane irradiance profile as measured with Leitz-TAS microscope system and a He-Ne source.[70]

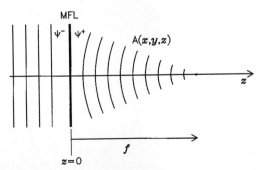

FIGURE 21   A micro-Fresnel lens (MFL) is often used to convert a planar wavefront into a converging spherical wave, $A(x, y, z)$, which focuses a distance f away from the MFL. The phase of the light in a plane on either side of the MFL is described by ψ^- and ψ^+.

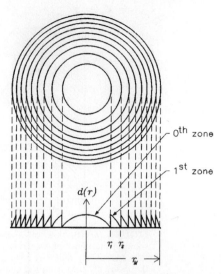

FIGURE 22 Fresnel lens construction. M divided
annular zones occur at radii r_i in the same manner as
a Fresnel zone plate. The profiles of each zone are
given by $d(r)$, and they are optimized to yield the
maximum efficiency in the focused beam.

Advantages of the Fresnel lens are that they can be made small and light compared to bulk
optical components. Note that binary optics, which are described in Farn and Veldkamp's
Chap. 8, (Vol. II) on "Binary Optics" are stepped approximations to the MFL
smooth-zone contour.

To understand the zonal profiles of the MFL, we return to our example problem
illustrated in Fig. 21. Our development is similar to that described by Nishihara and
Suhara.[42] The converging spherical wavefront is given by

$$A(r, z) = \frac{A_0}{\rho} \exp\left[-i(k\rho + \omega t)\right] = \frac{A_0}{\rho} \exp\left[i\phi(r, z)\right] \tag{7}$$

where A_0 is the amplitude of the wave, $\rho^2 = (z - f)^2 + r^2$, $r = \sqrt{x^2 + y^2}$, f is the focal length,
and $k = 2\pi/\lambda$. The phase of $A(x, y, z)$ at $t = 0$ and in a plane just behind the MFL is given
by

$$\phi(x, y, 0^+) = -k\sqrt{f^2 + r^2} \tag{8}$$

We could add a constant to Eq. (8) and not change any optical properties other than a dc
phase shift. Let

$$\psi^+(r) = \phi(x, y, 0^+) + kf + 2\pi = 2\pi + k(f - \sqrt{f^2 + r^2}) \tag{9}$$

Zone radii are found by solving

$$k(f - \sqrt{f^2 + r_m^2}) = -2\pi m \tag{10}$$

where $m = 1, 2, 3, \ldots$ is the zone number. The result is

$$r_m = \sqrt{2\lambda f m + (\lambda m)^2} \tag{11}$$

Equation (9) becomes

$$\psi_m^+(r) = 2\pi(m + 1) + k(f - \sqrt{f^2 + r^2}) \tag{12}$$

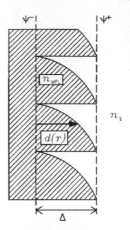

FIGURE 23 Portion of a Fresnel lens profile showing the thickness variation of the pattern. The thickness at any radius is given by $d(r)$, where r is the radial distance from the center of the lens. The phase shift that is added to wavefront ψ^- is determined by $d(r)$, the index of refraction of the substrate, n_{MFL}, and the index of refraction of the image space, n_i. The maximum thickness of the pattern is given by Δ. The resulting phase in a plane just after the MFL is given by ψ^+.

The job of the MFL is to provide a phase change so that the incident wavefront phase, $\psi^-(r)$, is changed into $\psi^+(r)$. The phase introduced by the MFL, $\psi_{\text{MFL}}(r)$, must be

$$\psi_{\text{MFL}}(r) = \psi^+(r) - \psi^-(r) \tag{13}$$

A phase change occurs when a wave is passed through a plate of varying thickness, as shown in Fig. 23. $\psi^+(r)$ is given by

$$\psi^+(r) = \psi^-(r) + kn_{\text{MFL}}d(r) + kn_i[\Delta - d(r)]$$
$$= \psi^-(r) + k(n_{\text{MFL}} - n_i)d(r) + kn_i\Delta \tag{14}$$

where $d(r)$ is the thickness profile, n_i is the refractive index of the image space, n_{MFL} is the refractive index of the substrate, and Δ is the maximum thickness of the MFL pattern. $d(r)$ is found by substituting Eq. (14) into Eq. (12). Note that the factor Δ is a constant and only adds a constant phase shift to Eq. (14). Therefore, we will ignore Δ in the remainder of our development. If $n_i = 1$, the result is

$$d_m(r) = \frac{\lambda(m+1)}{n_{\text{MFL}} - 1} - \frac{\sqrt{f^2 + r^2} - f}{n_{\text{MFL}} - 1} \tag{15}$$

where we have arbitrarily set $\psi^-(r) = 0$. The total number of zones M for a lens of radius r_M is

$$M = \frac{r_M(1 - \sqrt{1 - \text{NA}^2})}{\lambda \text{NA}} \tag{16}$$

The minimum zone period, $\Lambda_{\min}$, occurs at the outermost part of the lens and is given by

$$\Lambda_{\min} = r_M - r_{M-1} = r_M\left(1 - \sqrt{1 - \frac{2\lambda f + (2M-1)\lambda^2}{2M\lambda f + (M\lambda)^2}}\right) \tag{17}$$

The following approximations may be used without significant error if $NA < 0.2$ and $M \gg 1$:

$$d_m(r) \approx \frac{m\lambda f - 0.5r^2}{f(n_{MFL} - 1)} \tag{18}$$

$$r_m \approx \sqrt{2m\lambda f} \tag{19}$$

$$M \approx \frac{r_M}{2\lambda} NA \tag{20}$$

and

$$\Lambda_{min} \approx \frac{\lambda}{NA} \tag{21}$$

The consequence of using Eqs. (18) and (19) for $NA > 0.2$ is that a small amount of spherical aberration is introduced into the system.

The aberration characteristics of the MFL and the Fresnel zone plate are very similar. Aberrations of Fresnel zone plates have been discussed by Young.[43] For convenience, we describe a zone plate with the stop at the lens that is illuminated with a plane wave at angle α. For an MFL made according to Eq. (15) and used at the proper conjugates, there will be no spherical aberration or distortion. Coma, astigmatism, and field curvature are given by $W_{131} = \alpha r_M^3/2\lambda f^2$, $W_{222} = \alpha^2 r_M^2/2\lambda f$, and $W_{220} = \alpha^2 r_M^2/4\lambda f$, respectively. When $M \gg 1$ and α is small, the dominant aberration is coma, W_{131}. If the substrate of the zone plate is curved with a radius of curvature equal to the focal length, coma can be eliminated.[44] Chromatic variations in the focal length of the MFL are also similar to a Fresnel zone plate. For $NA < 0.2$,

$$\lambda f \approx \frac{r_M^2}{2M} \tag{22}$$

A focal-length-shift versus wavelength comparison of a Fresnel (hologram) lens and some single-element bulk-optic lenses are shown in Fig. 24. Note that the dispersion of the MFL is much greater than the bulk lens, and the dispersion of the MFL is opposite in sign to that of the bulk lenses. These facts have been used to design hybrid achromats by combining bulk lenses and diffractive lenses into the same system.[45] The thermal variations in MFLs primarily result in a change of focal length given by[46]

$$\Delta f = 2f\alpha_g \Delta T \tag{23}$$

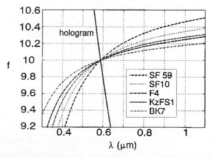

FIGURE 24 Single-element dispersions for a Fresnel (hologram) lens and refractive singlets. The focal lengths (arbitrary units) of thin lenses are plotted versus wavelength for refractive lenses of various optical glasses. Each lens was constructed to have a focal length of 10 at $\lambda_g = 0.5876\ \mu m$.[45]

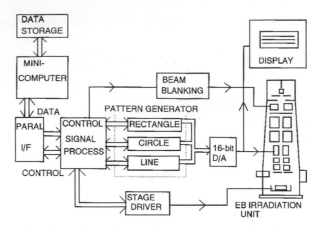

FIGURE 25 Block diagram of the computer-controlled electron-beam writing system.[48]

where f is the nominal focal length, α_g is the coefficient of thermal expansion for the substrate material, and ΔT is a uniform temperature change of the element. For most optical glasses, α_g ranges from $5 \times 10^{-4}\,°C^{-1}$ to $10 \times 10^{-4}\,°C^{-1}$.

There are several technologies that have been used to fabricate MFLs. These include electron-beam writing in resist,[47,48] laser writing in resist,[49,50,51] diamond turning,[52] and molding.[53]

Electron-beam writing in resist usually involves complicated translation stages under computer control in a high-vacuum environment, as shown in Fig. 25. The focused electron beam is scanned over the sample, and the amount of exposure is controlled by varying the electron-beam current, dwell time, or number of repetitive scans. After exposure, the resist is developed. Smooth-zone profiles are obtained by properly varying the exposure. An example of a depth-versus-dose curve for PMMA resist is given in Fig. 26. Notice that this exposure differs from that used in an electronic semiconductor device process, where only binary (fully exposed or unexposed) patterns are of interest. For a 0.2-μm-thick resist, this technique has produced 0.1 μm patterning of grating profiles.[48] However, according to Eq. (15), the medium must be at least $\lambda/(n_{\mathrm{MFL}} - 1)$ deep, which is greater than 1.0 μm for most applications. If the pattern is to be transferred into a glass substrate with ion milling, the resist thickness must be adjusted according to the differential etching rate between the resist and the glass.[54] It has been argued that, due to the relatively thick resist requirements, the resolution of electron-beam writing is similar to that of optical writing. The minimum blazed zone width that can be fabricated reliably with either technique is 2 to 3 μm. The problem of using thick resists can be avoided to some degree if one uses a reflection MFL rather than a transmission MFL, because less surface-height relief is required.[55]

Optical writing in photoresist requires similar positioning requirements to those used in electron-beam writing. However, it is not necessary to perform the exposure in a high-vacuum environment. Instead of using an electron beam, optical writing uses a focused laser beam, as shown in Fig. 27. Usually, some form of autofocus control is implemented to keep the spot size as small as possible. Standard photoresists, like Shipley 1400-27, may be used, so the wavelength of the exposure laser is typically 442 nm (HeCd) or 457 nm (Ar$^+$). Exposure is controlled by varying the power of the laser beam or by varying the number of repetitive scans.

The diffraction efficiency, η, of an MFL is defined as the ratio of the power in the focused spot to the power in the unfocused beam transmitted through the lens. At best,

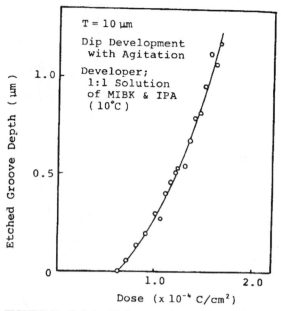

FIGURE 26   Relationship between electron dose and depth obtained in developed PMMA for 10 μm period grating profiles.[47]

Fresnel zone plates exhibit $\eta = 40.5\%$.[56] Blazing the grating profile can significantly increase the efficiency of the lens. Theoretically, η of an MFL can be 100 percent with the proper profile. However, there are several process parameters that limit η, as shown in Fig. 28, where a perfect zone profile has width T and height $d_{MAX} = \lambda/(n_{MFL} - 1)$. Variation of film thickness, over-etching, and swell all exhibit sinc-squared dependency on the errors. Shoulder imperfection is the most critical parameter, with η proportional to $(s/T)^2$. For $\eta > 90\%$, $s/T \geq 0.95$, which implies that the falling edge of the zone profile must take no more than 5 percent of the grating period. This is possible with low NA systems, where the zone spacing is large compared to the resolution of the exposure system, but it becomes difficult in high NA systems, where the zone spacing is on the order of several microns. Analysis of the three remaining parameters indicates fairly loose tolerances are acceptable. For $\eta > 98\%$, tolerance on individual parameters are: $|d(n-1)/\lambda - 1| < 0.25$, $a/T > 0.50$, and $\Delta d(n-1)/\lambda < 0.50$. Due to the increasing difficulty in fabricating correct zone profiles with decreasing zone width, most MFLs exhibit a variation in diffraction efficiency versus radius. In the center of the zone pattern, where the zone spacing is large, the measured diffraction efficiency can be in excess of 90 percent. At the edge of the zone pattern, where the zone spacing can be on the order of a few wavelengths, the measured diffraction efficiency is much lower. One possible solution to this problem is to use "superzone" construction,[57] in which the zone radii are found from a modified form of Eq. (10), that is

$$k(f - \sqrt{f^2 + r_M^2}) = 2\pi Nm \qquad (24)$$

where N is the superzone number. This results in a maximum thickness of $d_{MAX} = N\lambda/(n_{MFL} - 1)$. Note that $N = 1$ corresponds to the standard MFL. $N = 2$ implies that zones are spaced at every 4π phase transition boundary instead of at every 2π phase

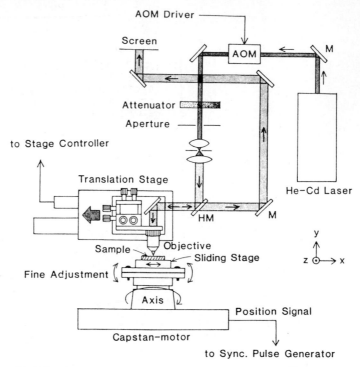

FIGURE 27   System configuration for laser beam lithography.[50]

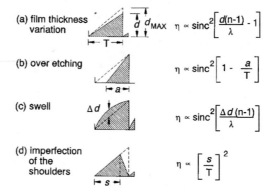

(a) film thickness
    variation

$$\eta \propto \text{sinc}^2\left[\frac{d(n-1)}{\lambda} - 1\right]$$

(b) over etching

$$\eta \propto \text{sinc}^2\left[1 - \frac{a}{T}\right]$$

(c) swell

$$\eta \propto \text{sinc}^2\left[\frac{\Delta d\,(n-1)}{\lambda}\right]$$

(d) imperfection
    of the
    shoulders

$$\eta \propto \left[\frac{s}{T}\right]^2$$

FIGURE 28 Four parameters that influence the diffraction efficiency of MFLs are: (*a*) film thickness variation; (*b*) over etching; (*c*) swell of the resist, and (*d*) imperfection of the shoulders. A profile of one zone is illustrated for each parameter. The ideal profile is shown as a dotted line, where d_{MAX} is the ideal height and T is the ideal period. The diffraction efficiency η of each profile is determined from extrapolating the result obtained from an infinite blazed grating.[47]

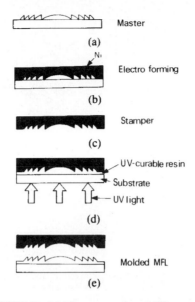

FIGURE 29 Molding process for a MFL on a glass substrate. First, a master is made by electron-beam lithography, then a stamper is electroformed from the master. MFLs are molded by potting a UV-curable resin between the stamper and the substrate and then exposing through the substrate.[53]

transition boundary. Although this makes the zones wider apart, the surface relief pattern must be twice as thick.

Molding provides a potentially valuable process for fabricating large quantities of MFLs economically. MFLs can be produced with conventional injection molding, but due to the large thermal expansion coefficient of polymers, the lenses are sensitive to thermal variations. An alternative MFL molding process is shown in Fig. 29, where a glass substrate is used to avoid large thermal effects. First, a master lens is formed with electron-beam or laser writing. A stamper is prepared using conventional nickel electro-forming methods.[58] After potting a UV-curable resin between the stamper and the glass substrate, the replica lenses are molded by the photopolymerization (2P) process.[59] The wavefront aberration versus temperature for a $\lambda = 780$ nm, NA $= 0.25$, diameter $= 0.5$ mm lens formed with this technique is shown in Fig. 30. A variation on this technique is to use the stamper as a substrate in an electron-beam evaporation device.[60] Inorganic materials of various refractive indices can be deposited on the stamper, resulting in a thin lens of high refractive index. The high refractive index of a material like ZnS ($n = 2.35$) can be used to lower the maximum thickness requirement of the lens, which makes fabrication of the master with electron-beam writing easier.

7.10 OTHER TECHNOLOGIES

There are several other technologies that are potentially valuable for micro-optic components. Four particularly interesting technologies are melted-resin arrays,[61] laser-assisted chemical etching,[62] mass transport,[63] and drawn preform cylindrical lenses.[61]

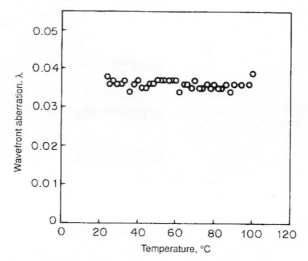

FIGURE 30 Wavefront aberration versus substrate temperature for a 0.25 NA molded MFL on a glass substrate designed to operate at $\lambda = 780$ nm.[53]

Melted-resin arrays are formed with the process shown in Fig. 31. First, an Al film is deposited on a quartz substrate and patterned with holes that serve as aperture stops for the array. Next, circular pedestals are formed on top of the aperture holes. The pedestals are hardened so that they are insoluble and stable for temperatures in excess of 180°C. Cylinders of resin are then developed on top of the pedestals. The device is heated to 140°C to melt the resin. The pedestals serve to confine the melting resin. The lenses form

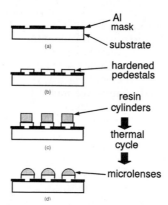

FIGURE 31 Process used to form melted-resin microlenses: (*a*) an AL film is deposited on the substrate and patterned with holes to serve as aperture stops for the array; (*b*) circular pedestals are formed on top of the aperture holes and hardened; (*c*) cylinders of resin are developed of the pedestals; (*d*) pedestals are melted to form spherical surfaces.[61]

into hemispherical shapes due to surface tension forces. Lens diameters have been demonstrated at 30 μm with good wavefront performance and uniformity.[61]

Laser-assisted chemical etching (LACE) can be used to make arrays of F/0.7 to F/10 lenslets with spacings of 50 to 300 μm. Microlenses have been fabricated in glass, silicon, CdTe, and sapphire with 95 percent fill factors and figure quality better than 1/10th wave.[62] In the LACE process, a focused laser beam is scanned over a thick layer of photoresist. The irradiance of the laser beam is modulated in order to vary the exposure and thus the thickness of the developed resist. With the proper irradiance mapping, accurate lens profiles can be produced in the developed resist. If lenslet material other than photoresist is required, a pattern can be exposed in the photoresist and transferred into the new material by ion milling.

Our discussion of the mass-transport process follows discussion presented in Ref. 63. In the mass-transport process, a multilevel mesa structure is first etched into a semiconductor, as shown in Fig. 32a. The semiconductor must be a binary compound in which the evaporation rate of one element is negligible compared to that of the other. For example, InP has been used successfully. The mesa structure is placed in a furnace at an elevated temperature. Since some surface decomposition occurs with InP, a minimum phosphorus vapor pressure must be maintained in the gas ambient to prevent the sample from being transformed into metallic In. The decomposition produces free In atoms located at the crystal surface, which are in equilibrium with phosphorus in the vapor and InP in the crystal. The concentration of In in the vapor is negligible. The equilibrium concentration of free In atoms increases with increasing positive surface curvature, since the higher surface

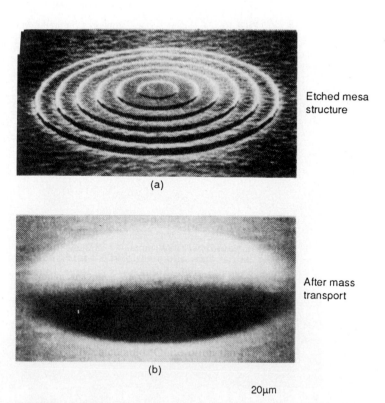

Etched mesa structure

(a)

After mass transport

(b)

20μm

FIGURE 32 SEM photographs showing perspective views of (a) etched multilevel mesa structure and (b) the microlens formed after mass transport.[63]

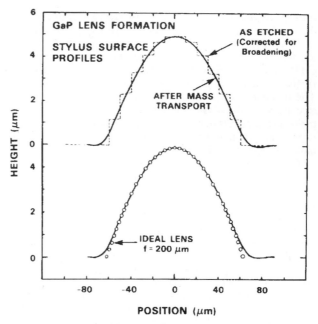

FIGURE 33 Stylus surface profiles of the multilevel mesa structure and the microlens formed after mass transport (upper half) and the comparison of the measured lens profile with an ideal one (lower half).[63]

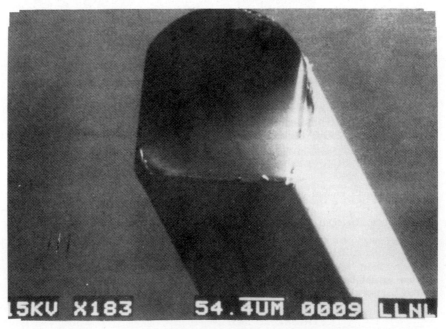

FIGURE 34 Scanning electron microscope photo of an elliptical cylindrical microlens. The lens width is 200 μm.[64]

energy of the high-curvature regions translates into a lower bonding energy in the decomposition process. Consequently, a variation in curvature across the surface will result in diffusion of free In atoms from regions of high positive curvature, where the concentrations are high, to low-curvature regions, where the In-diffused atoms exceed the equilibrium concentration and have to be reincorporated into the crystal by reaction with P to form InP. (The diffusion of P in the vapor phase is presumably much faster than the diffusion of free In atoms on the surface. The latter is therefore assumed to be the rate-limiting process.) The mass transport of InP, resulting from decomposition in high-curvature regions, will continue until the difference in curvature is completely eliminated. After mass transport, a smooth profile is obtained, as shown in Fig. 32b. The design of the mesa structure can result in very accurate control of the lens profile, as shown in Fig. 33. Mass-transport lens arrays have been used to collimate arrays of laser diodes, with diffraction-limited performance at NA ~ 0.5.

Very accurate cylindrical lenses can be drawn from preforms.[64] An SEM photo of an elliptical cylindrical lens is shown in Fig. 34. The first step in the process is to make a preform of a suitable glass material. Since the cross-sectional dimensions of the preform are uniformly reduced (typically ~ 50–100X) in the fiber drawing process, small manufacturing errors become optically insignificant. Therefore, standard numerically controlled grinding techniques can be utilized to generate a preform of any desired shape. Besides maintaining the preform shape, the drawing process also polishes the fiber. Results are presented[61] that demonstrate a 200-μm-wide elliptical cylindrical lens. The SFL6 preform was about 0.75 cm wide. The lens has a nominal focal length of 220 μm at $\lambda = 800$ μm. The lens is diffraction-limited over about a 150-μm clear aperture, or NA ~ 0.6. The application is to collimate the fast axis of laser diodes.

7.11 REFERENCES

1. Glenn T. Sincerbox, "Miniature Optics for Optical Recording," *Proc. SPIE,* **935,** 1988, pp. 63–76.

2. W. J. Smith, *Modern Optical Engineering,* McGraw-Hill, New York, 1966, p. 61.

3. D. K. Towner, "Scanning Techniques for Optical Data Storage," *Proc. SPIE,* **695,** 1986, pp. 172–180.

4. M. Born and E. Wolf, *Principles of Optics,* Pergamon Press Ltd., Oxford, 1980, pp. 468–469.

5. J. F. Forkner and D. W. Kurtz, "Characteristics of Efficient Laser Diode Collimators," *Proc. SPIE,* **740,** 1987, 27–35.

6. R. O. Maschmeyer, U.S. Patent 4,537,473, 1985.

7. D. Malacara, *Optical Shop Testing,* John Wiley and Sons, New York, 1978, Chap. 12, pp. 381–397.

8. G. W. Hopkins and R. N. Shagam, "Null Ronchi Gratings from Spot Diagrams," *Appl. Opt.,* **16,** 1977, p. 2602.

9. M. A. Fitch, "Molded Optics: Mating Precision and Mass Production," *Photonics Spectra,* October 1991, pp. 84–87.

10. T. Aquilina, D. Richards, and H. Pollicove, "Finished Lens Molding Saves Time and Money," *Photonics Spectra,* September 1986, pp. 73–80.

11. M. A. Fitch and Y. K. Konishi, "Technical Directions in Molded Optics," *Proc. SPIE,* **1139,** 1989, pp. 187–190.

12. Donald L. Keyes, "Fundamental Properties of Optical Plastics," *CRC Handbook of Laser Science and Technology,* CRC Press, Inc., Boca Raton, Fl, in press.

13. J. D. Lytle, "Aspheric Surfaces in Polymer Optics," *Proc. SPIE,* **381,** 1983, pp. 64–68.

14. D. Keyes, private communication.

15. R. M. Altman and J. D. Lytle, "Optical-design Techniques for Polymer Optics," *Proc. SPIE,* **237,** 1980, pp. 380–385.

16. *Modern Plastics Encyclopedia,* **68,** (11), McGraw-Hill, New York, 1991/1992.

17. *Plastics Technology, Manufacturing Handbook and Buyer's Guide,* vol. 37(8), Bill Communications Inc., New York, 1991/1992.

18. Diverse Optics, Inc., commercial literature, 1992.

19. A. H. Deutchman, R. J. Partyka, and J. C. Lewis, "Dual Ion Beam Deposition of Diamond Films on Optical Elements," *Proc. SPIE,* **1146,** 1989, pp. 124–134.

20. L. E. Schmutz, "Hartman Sensing at Adaptive Optics Associates," *Proc. SPIE,* **779,** 1987, pp. 13–17.

21. D. D'Amato and R. Centamore, "Two Applications for Microlens Arrays: Detector Fill Factor Improvement and Laser Diode Collimation," *Proc. SPIE,* **1544,** 1991, pp. 166–177.

22. Adaptive Optics Associates, private communication.

23. D. D'Amato, S. Barletta, P. Cone, J. Hizny, R. Martinsen, and L. Schmutz, "Fabricating and Testing of Monolithic Lenslet Module (MLM)," in *Optical Fabrication and Testing,* pp. 74–77, 1990 Technical Digest Series, vol. 11, Conference Edition, June 12–14, 1990, Monterey, California.

24. Adaptive Optics Associates, commercial literature.

25. T. D. Milster and J. N. Wong, "Modeling and Measurement of a Micro-Optic Beam Deflector," to be published in *Proc. SPIE,* 1625.

26. R. K. Luneburg, *A Mathematical Theory of Optics,* University of California Press, Berkeley, 1964, pp. 164–195.

27. K. Iga, M. Oikawa, and J. Banno, "Ray Traces in a Distributed-index Planar Microlens," *Appl. Opt.,* **21,** 1982, pp. 3451–3455.

28. M. Oikawa, H. Nemoto, K. Hamanaka, H. Imanishi, and T. Kishimoto, "Light Coupling Characteristics of Planar Microlens," *Proc. SPIE,* **1544,** 1991, pp. 226–237.

29. K. Iga and S. Misawa, "Distributed-index Planar Microlens and Stacked Planar Optics: A Review of Progress," *Appl. Opt.,* **25,** 1986, p. 3388.

30. K. Iga, Y. Kokobun, and M. Oikawa, *Fundamentals of Microoptics,* Academic Press, Inc., Tokyo, 1984.

31. Y. Ohtsuka, "Light-focusing Plastic Rod Prepared from Diallyl Isophthalate-methyl Methacrylate Copolymerization," *Appl. Phys. Lett.,* **23,** 1973, p. 247.

32. K. Iga and N. Yamanamoto, "Plastic Focusing Fiber for Imaging Applications," *Appl. Opt.,* **16,** 1977, p. 1305.

33. T. Izawa and H. Nakagome, "Optical Waveguide Formed By Electrically Induced Migration of Ions in Glass Plates," *Appl. Phys. Lett.,* **21,** 1972, p. 584.

34. G. D. Khoe, H. G. Kock, J. A. Luijendijk, C. H. J. van den Brekel, and D. Küppers, "Plasma CVD Prepared SiO_2/Si_3N_4 Graded Index Lenses Integrated in Windows of Laser Diode Packages," in *Technical Digest,* Seventh European Conference on Optical Communication, Copenhagen, 1981, pp. 7.6-1–7.6-4.

35. K. Iga, M. Oikawa, and T. Sanada, *Electron. Lett.,* **17,** 1981, p. 452.

36. M. Oikawa, H. Nemoto, K. Hamanaka, and E. Okuda, "High Numerical Aperture Planar Microlenses with Swelled Structure," *Appl. Opt.,* **29,** 1990, p. 4077.

37. N. F. Borrelli, D. L. Morse, R. H. Bellman, and W. L. Morgan, "Photolytic Technique for Producing Microlenses in Photosensitive Glass," *Appl. Opt.,* **24,** 1985, pp. 2520–2525.

38. Corning, Inc., commercial literature.

39. R. H. Bellman, N. F. Borrelli, L. G. Mann, and J. M. Quintal, "Fabrication and Performance of a One-to-One Erect Imaging Microlens Array for Fax," *Proc. SPIE,* **1544,** 1991, pp. 209–217.

40. O. E. Meyres, "Studies of Transmission Zone Plates," *Am. J. Phys.,* **19,** 1951, p. 359.

41. M. Sussman, "Elementary Diffraction Theory of Zone Plates," *Am. J. Phys.,* **28,** 1960, p. 394.

42. H. Nishihara and T. Suhara, in *Progress in Optics,* North-Holland, New York, 1987, Vol. 24, pp. 1–37.

43. M. Young, "Zone Plates and Their Aberrations," *J. Opt. Soc. Am.,* **62,** 1972, pp. 972–976.

44. T. D. Milster, R. M. Trusty, M. S. Wang, F. F. Froehlich, and J. K. Erwin, "Micro-Optic Lens for Data Storage," *Proc. SPIE,* **1499,** 1991, pp. 286–292.

45. T. Stone and N. George, "Hybrid Diffractive-Refractive Lenses and Achromats," *Appl. Opt.,* **27,** 1988, pp. 2960–2971.

46. G. P. Behrmann and J. P. Bowen, "Thermal Effects in Diffractive Lenses," in *Diffractive Optics: Design, Fabrication, and Applications Technical Digest, 1992,* Optical Society of America, Washington, D.C., 1992, Vol. 9, pp. 8–10.

47. T. Fujita, H. Nishihara, and J. Koyama, "Blazed Gratings and Fresnel Lenses Fabricated by Electron-beam Lithography," *Opt. Lett.,* **7,** 1982, pp. 578–580.

48. T. Shiono, K. Setsune, O. Yamazaki, and K. Wasa, "Computer-controlled Electron-beam Writing System for Thin-film Micro-optics," *J. Vac. Sci. Technol.,* B5 (1), 1987, pp. 33–36.

49. W. Goltsos and S. Liu, "Polar Coordinate Laser Writer for Binary Optics Fabrication," *Proc. SPIE,* **1211,** 1990, pp. 137–147.

50. M. Harana, M. Takahashi, K. Wakahayashi, and H. Nishihara, "Laser Beam Lithographed Micro-Fresnel Lenses," *Appl. Opt.,* **29,** 1990, pp. 5120–5126.

51. H. Nishihara and T. Suhara, "Micro Fresnel Lenses," in *Progress in Optics,* E. Wolf (ed.), North-Holland, Physics Publishing, The Netherlands, 1987, vol. 24, chap. 1, pp. 3–37.

52. P. P. Clark and C. Londono, "Production of Kinoforms by Single Point Diamond Machining," *Opt. News,* **15,** 1989, pp. 39–40.

53. M. Tanigami, S. Aoyama, T. Yamashita, and K. Imanaka, "Low Wavefront Aberration and High Temperature Stability Molded Micro-Fresnel Lens," *IEEE Photonics Technology Letters,* **1** (11), 1989, pp. 384–385.

54. P. J. Revell and G. F. Goldspink, "A Review of Reactive Ion Beam Etching for Production," *Vacuum,* **34,** 1984, pp. 455–462.

55. T. Shiono, M. Kitagawa, K. Setsune, and T. Mitsuyu, "Reflection Microlenses and Their Use in an Integrated Focus Sensor," *Appl. Opt.,* **28,** 1989, pp. 3434–3442.

56. R. Magnusson and T. K. Gaylord, "Diffraction Efficiencies of Thin Phase Gratings with Arbitrary Grating Shape," *J. Opt. Soc. Am.,* **68,** 1978, pp. 806–809.

57. J. Futhey and M. Fleming, "Superzone Diffractive Lenses," in *Diffractive Optics: Design, Fabrication, and Applications Technical Digest, 1992,* Optical Society of America, Washington, D.C., 1992, vol. 9, pp. 4–6.

58. R. W. Schneek, "Process Factors for Electroforming Video Disks," *Plating and Surface Finishing,* **71** (1), 1984, p. 38.

59. K. Goto, K. Mori, G. Hatakoshi, and S. Takahashi, "Spherical Grating Objective Lenses for Optical Disk Pickups," *Proc. International Symposium on Optical Memory, 1987, Jpn. J. Appl. Phys.,* **26,** suppl. 26-4, 1987, p. 135.

60. H. Hosokawa and T. Yamashita, "ZnS Micro-Fresnel Lens and Its Uses," *Appl. Opt.,* **29,** 1990, pp. 5706–5710.

61. Z. D. Popovic, R. A. Sprague, and G. A. Neville Connell, "Technique for Monolithic Fabrication of Microlens Arrays," *Appl. Opt.,* **27,** 1988, p. 1281–1284.

62. E. J. Gratrix, and C. B. Zarowin, "Fabrication of Microlenses by Laser Assisted Chemical Etching (LACE)," *Proc. SPIE,* **1544,** 1991, pp. 238–243.

63. V. Diadiuk, Z. L. Liau, and J. N. Walpole, "Fabrication and Characterization of Semiconductor Microlens Arrays," *Proc. SPIE,* **1354,** 1990, pp. 496–500.

64. J. J. Snyder, P. Reichert, and T. M. Baer, "Fast Diffraction-limited Cylindrical Microlenses," *Appl. Opt.,* **30,** 1991, pp. 2743–2747.

65. Ohara Optical Glass Catalog, Ohara Corporation, Somerville, New Jersey, 1990.

66. Corning, Incorporated, Precision Molded Optics Department, Corning, New York, commercial literature, 1990.

67. Y. Kokuban and K. Iga, *Appl. Opt.,* **21,** 1982, p. 1030.

68. Y. Kokuban, T. Usui, M. Oikawa, and K. Iga, "Wave Aberration Testing System for Microlenses by Shearing Interference Method," *Jap. J. of Appl. Phys.,* vol. 23, no. 1, 1984, pp. 101–104.

69. P. de Groot, F. D'Amato, and E. Gratrix, "Interferometric Evaluation of Lenslet Arrays for 2D Phase-locked Laser Diode Sources," *SPIE,* **1333,** 1990, pp. 347–355.

70. N. F. Borrelli and D. L. Morse, "Microlens Arrays Produced by a Photolytic Technique," *Appl. Opt.,* **27** (3)**,** 1988, pp. 476–479.

CHAPTER 8
BINARY OPTICS

Michael W. Farn and Wilfrid B. Veldkamp
MIT/Lincoln Laboratory
Lexington, Massachusetts

8.1 GLOSSARY

A	aspheric
C	describes spherical aberration
C_m	Fourier coefficients
c	curvature
$c(x, y)$	complex transmittance
D	local period
f	focal length
k, l	running indices
l_i	paraxial image position
L, M	direction cosines
m	diffraction order
P	partial dispersion
s	spheric
t	thickness
V_d	Abbe number
x, y, z	Cartesian coordinates
λ	wavelength
η	diffraction efficiency
ξ_i	paraxial image height
$\phi(x, y)$	phase
$0, i$	iterative points
$'$	diffracted

8.2 INTRODUCTION

Binary optics is a surface-relief optics technology based on VLSI fabrication techniques (primarily photolithography and etching), with the "binary" in the name referring to the binary coding scheme used in creating the photolithographic masks. The technology allows the creation of new, unconventional optical elements and provides greater design freedom and new materials choices for conventional elements. This capability allows designers to create innovative components that can solve problems in optical sensors, optical communications, and optical processors. Over the past decade, the technology has advanced sufficiently to allow the production of diffractive elements, hybrid refractive-diffractive elements, and refractive micro-optics which are satisfactory for use in cameras, military systems, medical applications, and other demanding areas.

The boundaries of the binary optics field are not clearly defined, so in this section, the concentration will be on the core of the technology: passive optical elements which are fabricated using VLSI technology. As so defined, binary optics technology can be broadly divided into the areas of optical design and VLSI-based fabrication. Optical design can be further categorized according to the optical theory used to model the element: geometrical optics, scalar diffraction theory, or vector diffraction theory; while fabrication is composed of two parts: translation of the optical design into the mask layout and the actual micromachining of the element. The following sections discuss each of these topics in some detail, with the emphasis on optical design. For a more general overview, the reader is referred to Refs. 1 for many of the original papers, 2 and 3 for a sampling of applications and research, and 4–6 for a lay overview.

Directly related areas which are discussed in other sections but not in this section include micro-optics and diffractive optics fabricated by other means (e.g., diamond turning, conventional manufacturing, or optical production), display holography (especially computer-generated holography), mass replication technologies (e.g., embossing, injection molding, or epoxy casting), integrated optics, and other micromachining technologies.

8.3 DESIGN—GEOMETRICAL OPTICS

In many applications, binary optics elements are designed by ray tracing and "classical" lens design principles. These designs can be divided into two classes: broadband and monochromatic. In broadband applications, the binary optics structure has little optical power in order to reduce the chromatic aberrations and its primary purpose is aberration correction. The device can be viewed as an aspheric aberration, corrector, similar to a Schmidt corrector, when used to correct the monochromatic aberrations and it can be viewed as a material with dispersion an order of magnitude greater than and opposite in sign to conventional materials when used to correct chromatic aberrations. In monochromatic applications, binary optics components can have significant optical power and can be viewed as replacements for refractive optics.

In both classes of designs, binary optics typically offers the following key advantages:

- Reduction in system size, weight, and/or number of elements

- Elimination of exotic materials

- Increased design freedom in correcting aberrations, resulting in better system performance

- The generation of arbitrary lens shapes (including micro-optics) and phase profiles

Analytical Models

Representation of a Binary Optics Element. As with any diffractive element, a binary optics structure is defined by its phase profile $\phi(x, y)$ (z is taken as the optical axis), design wavelength λ_0, and the surface on which the element lies. For simplicity, this surface is assumed to be planar for the remainder of this section, although this is commonly not the case. For example, in many refractive/diffractive systems, the binary optics structure is placed on a refractive lens which may be curved. The phase function is commonly represented by either explicit analytical expression or decomposition into polynomials in x and y (e.g., the HOE option in CODE V).

Explicit analytic expressions are used in simple designs, the two most common being lenses and gratings. A lens used to image point (x_o, y_o, z_o) to point (x_i, y_i, z_i) at wavelength λ_0 has a phase profile

$$\phi(x, y) = \frac{2\pi}{\lambda_0}[z_o(\sqrt{(x - x_o)^2/z_o^2 + (y - y_o)^2/z_o^2 + 1} - 1)$$

$$- z_i(\sqrt{(x - x_i)^2/z_i^2 + (y - y_i)^2/z_i^2 + 1} - 1)] \tag{1}$$

where z_o and z_i are both taken as positive to the right of the lens. The focal length is given by the Gaussian lens formula:

$$1/f_0 = 1/z_i - 1/z_o \tag{2}$$

with the subscript indicating that f_0 is the focal length at λ_0. A grating which deflects a normally incident ray of wavelength λ_0 to the direction with direction cosines (L, M) is described by

$$\phi(x, y) = \frac{2\pi}{\lambda_0}(xL + yM) \tag{3}$$

Axicons are circular gratings and are described by

$$\phi(x, y) = \frac{2\pi}{\lambda_0}(\sqrt{x^2 + y^2}L) \tag{4}$$

where L now describes the radial deflection.

For historical reasons, the polynomial decomposition of the phase profile of the element commonly consists of a spheric term and an aspheric term:

$$\phi(x, y) = \phi_S(x, y) + \phi_A(x, y) \tag{5}$$

where

$$\phi_A(x, y) = \frac{2\pi}{\lambda_0}\sum_k \sum_l a_{kl}x^k y^l$$

and the spheric term $\phi_S(x, y)$ takes the form of Eq. (1). Since the phase profiles produced by binary optics technology are not constrained to be spheric, $\phi_S(x, y)$ is often set to zero by using the same object and image locations and the aspheric term alone is used to describe the profile. The binary optics element is then optimized by optimizing the polynomial coefficients a_{kl}. If necessary, the aspheric term can be forced to be radially symmetric by constraining the appropriate polynomial coefficients.

It is possible to describe the phase profile of a binary optics element in other ways. For example, $\phi(x, y)$ could be described by Zernicke polynomials or could be interpolated from a two-dimensional look-up table. However, these methods are not widely used since lens design software currently does not support these alternatives.

Ray Tracing by the Grating Equation. A binary optics element with phase $\phi(x, y)$ can be ray traced using the grating equation by modeling the element as a grating, the period of which varies with position. This yields

$$L' = L + \frac{m\lambda}{2\pi} \frac{\partial \phi}{\partial x} \tag{6}$$

$$M' = M + \frac{m\lambda}{2\pi} \frac{\partial \phi}{\partial y} \tag{7}$$

where m is the diffracted order, L, M are the direction cosines of the incident ray, and L', M' are the direction cosines of the diffracted ray.[7] In geometrical designs, the element is usually blazed for the first order ($m = 1$). Note that it is the phase gradient $\nabla\phi(x, y)$ (a vector quantity proportional to the local spatial frequency) and not the phase $\phi(x, y)$ which appears in the grating equation. The magnitude of the local period is inversely proportional to the local spatial frequency and given by

$$D(x, y) = 2\pi / |\nabla\phi| \tag{8}$$

where $|\,|$ denotes the vector magnitude. The minimum local period determines the minimum feature size of the binary optics structure, a concern in device fabrication (see "Fabrication" later in this chapter).

Ray Tracing by the Sweatt Model. The Sweatt model,[8] which is an approximation to the grating equation, is another method for ray tracing. The Sweatt approach models a binary optics element as an equivalent refractive element and is important since it allows results derived for refractive optics to be applied to binary optics. In the Sweatt model, a binary optics element with phase $\phi(x, y)$ at wavelength λ_0 is replaced by a refractive equivalent with thickness and refractive index given by

$$t(x, y) = \frac{\lambda_0}{n_0 - 1} \frac{\phi(x, y)}{2\pi} + t_0 \tag{9}$$

$$n(\lambda) - 1 = \frac{\lambda}{\lambda_0}(n_0 - 1) \tag{10}$$

Here, t_0 is a constant chosen to make $t(x, y)$ always positive and n_0 is the index of the material at wavelength λ_0. The index n_0 is chosen by the designer and as $n_0 \to \infty$, the Sweatt model approaches the grating equation. In practice, values of $n_0 = 10,000$ are sufficiently high for accurate results.[9]

In the special case of a binary optics lens described by Eq. (1), the more accurate Sweatt lens[10] can be used. In this case, the element is modeled by two surfaces of curvature

$$c_o = 1/[(1 - n_0)z_o] \tag{11}$$

$$c_i = 1/[(1 - n_0)z_i] \tag{12}$$

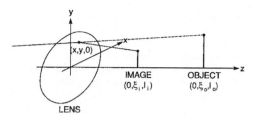

FIGURE 1 Primary aberrations of a binary optics lens.[7]

and conic constant $-n_0^2$, with the axis of each surface passing through the respective point source. The refractive index is still modeled by Eq. (10).

Aberration Correction

Aberrations of a Binary Optics Singlet. As a simple example of a monochromatic imaging system, consider a binary optics singlet which is designed to image the point $(0, 0, z_o)$ to the point $(0, 0, z_i)$ at wavelength λ_0. The phase profile of this lens can be derived from Eq. (1) and the focal length f_0 from Eq. (2). Now consider an object point of wavelength λ located at $(0, \xi_o, l_o)$. The lens will form an image at $(0, \xi_i, l_i)$ (see Fig. 1), with the paraxial image position l_i and height ξ_i given by[7]

$$\frac{1}{l_i} = \frac{\lambda}{f_0 \lambda_0} + \frac{1}{l_o} \tag{13}$$

$$\xi_i / l_i = \xi_o / l_o \tag{14}$$

Note that the first equation is just the Gaussian lens law but using a wavelength-dependent focal length of

$$f(\lambda) = f_0 \frac{\lambda_0}{\lambda} \tag{15}$$

The focal length being inversely proportional to the wavelength is a fundamental property of diffractive lenses. In addition, due to the wavelength shift and position change of the object point, the lens will form a wavefront with a primary aberration of[7]

$$W(x, y) = \frac{1}{8} \left[\left(\frac{1}{l_i^3} - \frac{1}{l_o^3} \right) - \frac{\lambda}{\lambda_0} \left(\frac{1}{z_i^3} - \frac{1}{z_o^3} \right) \right] (x^2 + y^2)^2$$

$$- \frac{1}{2l_i} \left(\frac{1}{l_i^2} - \frac{1}{l_o^2} \right) \xi_i y (x^2 + y^2)$$

$$+ \frac{3}{4l_i^2} \left(\frac{1}{l_i} - \frac{1}{l_o} \right) \xi_i^2 y^2 + \frac{1}{4l_i^2} \left(\frac{1}{l_i} - \frac{1}{l_o} \right) \xi_i^2 x^2 \tag{16}$$

where the ray strikes the lens at (x, y). The first term is spherical aberration, the second is coma, and the last two are tangential and sagittal field curvature. As noted by Welford, all the off-axis aberrations can be eliminated if and only if $l_i = l_o$, a useless configuration. In most systems of interest, the limiting aberration is coma.

The performance of the binary optics singlet can be improved by introducing more degrees of freedom: varying the stop position, allowing the binary optics lens to be placed on a curved surface, using additional elements, etc. For a more detailed discussion, see Refs. 1, 7, and 11.

Chromatic Aberration Correction. Binary optics lenses inherently suffer from large chromatic aberrations, the wavelength-dependent focal length [Eq. (15)] being a prime example. By themselves, they are unsuitable for broadband imaging and it has been shown that an achromatic system consisting only of diffractive lenses cannot produce a real image.[12]

However, these lenses can be combined successfully with refractive lenses to achieve chromatic correction (for a more detailed discussion than what follows, see Refs. 4, 13, and 14). The chromatic behavior can be understood by using the Sweatt model, which states that a binary optics lens behaves like an ultrahigh index refractive lens with an index which varies linearly with wavelength [let $n_0 \to \infty$ in Eq. (10)]. Accordingly, they can be used to correct the primary chromatic aberration of conventional refractive lenses but cannot correct the secondary spectrum. For the design of achromats and apochromats, an effective Abbe number and partial dispersion can also be calculated. For example, using the $C, d,$ and F lines, the Abbe number is defined as $V_d = [n(\lambda_d) - 1]/[n(\lambda_F) - n(\lambda_C)]$. Substituting Eq. (10) and letting $n_0 \to \infty$ yields

$$V_d = \lambda_d/(\lambda_F - \lambda_C) = -3.45 \qquad (17)$$

In a similar fashion, the effective partial dispersion using the g and F lines is

$$P_{gF} = (\lambda_g - \lambda_F)/(\lambda_F - \lambda_C) = 0.296 \qquad (18)$$

By using these effective values, the conventional procedure for designing achromats and apochromats[15] can be extended to designs in which one element is a binary optics lens.

Figure 2 plots the partial dispersion P_{gF} versus Abbe number V_d for various glasses. Unlike all other materials, a binary optics lens has a negative Abbe number. Thus, an achromatic doublet can be formed by combining a refractive lens and a binary optics lens, both with positive power. This significantly reduces the lens curvatures required, allowing for larger apertures. In addition, the binary optics lens has a position in Fig. 2 which is not collinear with the other glasses, thus also allowing the design of apochromats with reduced lens curvatures and larger apertures.

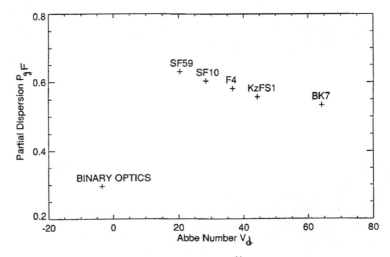

FIGURE 2 Partial dispersion vs. Abbe number.[14]

Monochromatic Aberration Correction. For a detailed discussion, the reader is referred to Refs. 1 and 11. As a simple example,[4] consider a refractive system which suffers from third-order spherical aberration and has a residual phase given by

$$\phi_r(x, y) = \frac{2\pi}{\lambda} C(x^2 + y^2)^2 \tag{19}$$

where C describes the spherical aberration. Then, a binary optics corrector with phase

$$\phi_b(x, y) = -\frac{2\pi}{\lambda_0} C(x^2 + y^2)^2 \tag{20}$$

will completely correct the aberration at wavelength λ_0 and will reduce the aberration at other wavelengths to

$$\phi_r + \phi_b = \frac{2\pi}{\lambda} C(1 - \lambda/\lambda_0)(x^2 + y^2)^2 \tag{21}$$

The residual aberration is spherochromatism.

Micro-optics

Binary optics technology is especially suited for the fabrication of micro-optics and micro-optics arrays, as shown in Fig. 3. The advantages of binary optics technology include the following:

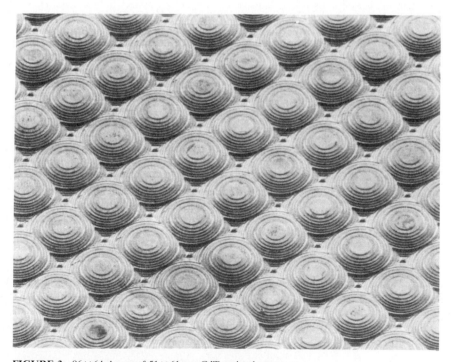

FIGURE 3 96×64 Array of 51×61 μm CdTe microlenses.

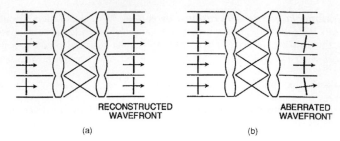

RECONSTRUCTED
WAVEFRONT

ABERRATED
WAVEFRONT

(a)

(b)

FIGURE 4 Micro-optic telescope using (*a*) coherent arrays; (*b*) incoherent arrays.

- *Uniformity and coherence.* If desired, all micro-optics in an array can be made identical to optical tolerances. This results in coherence over the entire array (see Fig. 4).

- *Refractive optics.* Binary optics is usually associated with diffractive optics. This is not a fundamental limit but results primarily from fabrication constraints on the maximum achievable depth (typically, 3 μm with ease and up to 20 μm with effort). However, for many micro-optics, this is sufficient to allow the etching of refractive elements. For example, a lens or radius R_0 which is corrected for spherical aberration[15] and focuses collimated light at a distance z_0 (see Fig. 5) has a thickness of

$$t_{max} = n[\sqrt{R_0^2 + z_0^2} - z_0]/(n - 1) \qquad (22)$$

where n is the index of the material.

- *Arbitrary phase profiles.* Binary optics can produce arbitrary phase profiles in micro-optics just as easily as in macro-optics. Fabricating arrays of anamorphic lenses to correct the astigmatism of semiconductor lasers, for example, is no more difficult than fabricating arrays of conventional spherical lenses.

- *100 percent fill factor.* While many technologies are limited in fill factor (e.g., round lenses on a square grid yield a 79 percent fill factor), binary optics can achieve 100 per cent fill factor on any shape grid.

- *Spatial multiplexing.* Each micro-optic in an array can be different from its neighbors and the array itself can compose an arbitrary mosaic rather than a regular grid. For example, a binary optics array of individually designed micro-optics can be used to optimally mode match one-dimensional laser arrays to laser cavities or optical fibers.[16]

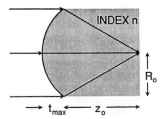

INDEX n

R_0

t_{max} z_0

FIGURE 5 Thickness of a refractive lens.[15]

Optical Performance

Wavefront Quality.   The wavefront quality of binary optics components is determined by the accuracy with which the lateral features of the element are reproduced. Since the local period (typically several μm) is usually much larger than the resolution with which it can be reproduced (of order 0.1 μm), wavefront quality is excellent. In fact, wavefront errors are typically limited by the optical quality of the substrate rather than the quality of the fabrication.

Diffraction Efficiency.   The diffraction efficiency of a device is determined by how closely the binary optics stepped-phase profile approximates a true blaze. The theoretical efficiency at wavelength λ of an element with I steps designed for use at λ_0 is:[4]

$$\eta(\lambda, I) = \left| \operatorname{sinc} (1/I) \frac{\sin (I\pi\alpha)}{I \sin \pi\alpha} \right|^2 \tag{23}$$

where $\operatorname{sinc} (x) = \sin (\pi x)/(\pi x)$

$$\alpha = (\lambda_0/\lambda - 1)/I$$

This result is based on scalar theory, assumes perfect fabrication, and neglects any material dispersion. Figure 6 plots the efficiency $\eta(\lambda, I)$ for different numbers of steps I;

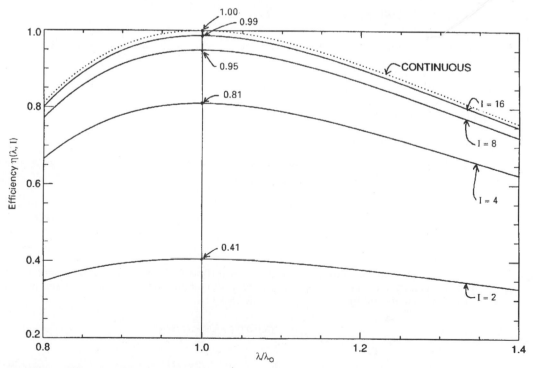

FIGURE 6   Diffraction efficiency of binary optics.[4]

TABLE 1   Average Diffraction Efficiency for Various Bandwidths[4]

$\Delta\lambda/\lambda_0$	$\bar{\eta}$
0.00	1.00
0.10	1.00
0.20	0.99
0.30	0.98
0.40	0.96
0.50	0.93
0.60	0.90

while Table 1 gives the average efficiency over the bandwidth $\Delta\lambda$ for a perfectly blazed element $(I \to \infty)$.[4] The efficiency equation is asymmetric in λ but symmetric in $1/\lambda$.

The use of scalar theory in the previous equation assumes that the local period $D(x, y)$ [see Eq. (8)] is large compared to the wavelength. As a rule of thumb, this assumption begins to lose validity when the period dips below 10 wavelengths (e.g., a grating with period less than $10\lambda_0$ or a lens faster than F/5) and lower efficiencies can be expected in these cases. For a more detailed discussion, see Ref. 17.

The efficiency discussed here is the diffraction efficiency of an element. Light lost in this context is primarily diffracted into other diffraction orders, which can also be traced through a system to determine their effect. As with conventional elements, binary optics elements will also suffer reflection losses which can be minimized in the usual manner.

8.4   DESIGN—SCALAR DIFFRACTION THEORY

Designs based on scalar diffraction theory are based on the direct manipulation of the phase of a wavefront. The incident wavefront is generally from a coherent source and the binary optics element manipulates the phase of each point of the wavefront such that the points interfere constructively or destructively, as desired, at points downstream of the element. In this regime, binary optics can perform some unique functions, two major applications being wavefront multiplexing and beam shaping.

Analytical Models

In the scalar regime, the binary optics component with phase profile $\phi(x, y)$ is modeled as a thin-phase screen with a complex transmittance of

$$c(x, y) = \exp\left[j\phi(x, y)\right] \tag{24}$$

The phase screen retards the incident wavefront and propagation of the new wavefront is modeled by the appropriate scalar formulation (e.g., angular spectrum, Fresnel diffraction, Fraunhofer diffraction) for nonperiodic cases, or by Fourier series decomposition for periodic cases.

The design of linear gratings is an important problem in the scalar regime since other problems can be solved by analogy. A grating with complex transmittance $c(x)$ and period D can be decomposed into its Fourier coefficients C_m, where

$$C_m = \frac{1}{D}\int_0^D c(x)\exp\left(-j2\pi mx/D\right)dx \tag{25}$$

$$c(x) = \sum_{m=-\infty}^{\infty} C_m \exp\left(j2\pi mx/D\right) \tag{26}$$

The relative intensity or efficiency of the mth diffracted order of the grating is

$$\eta_m = |C_m|^2 \tag{27}$$

Due to the fabrication process, binary optics gratings are piecewise flat. The grating transmission in this special case can be expressed as $c(x) = c_i$ for $x_i < x < x_{i+1}$, where c_i is the complex transmission of step i of I total steps, $x_0 = 0$, and $x_I = D$. The Fourier coefficients then take the form

$$C_m = \sum_{i=0}^{I-1} c_i \delta_i \exp\left(-j2\pi m \Delta_i\right) \operatorname{sinc}\left(m\delta_i\right) \tag{28}$$

where $\delta_i = (x_{i+1} - x_i)/D$

$\Delta_i = (x_{i+1} + x_i)/(2D)$

The sinc term is due to the piecewise flat nature of the grating. If, in addition to the above, the grating transition points are equally spaced, then $x_i = iD/I$ and Eq. (28) reduces to

$$C_m = \exp\left(-j\pi m/I\right) \operatorname{sinc}\left(m/I\right) \left[\frac{1}{I} \sum_{i=0}^{I-1} c_i \exp\left(-j2\pi mi/I\right)\right] \tag{29}$$

The bracketed term is the FFT of c_i, which makes this case attractive for numerical optimizations. If the complex transmittance is also stepped in phase by increments of ϕ_0, then $c_i = \exp\left(ji\phi_0\right)$ and Eq. (29) further reduces to[18]

$$C_m = \exp\left[j\pi((I-1)\alpha - m/I)\right] \operatorname{sinc}\left(m/I\right) \frac{\sin\left(I\pi\alpha\right)}{I \sin \pi\alpha} \tag{30}$$

where $\alpha = \phi_0/(2\pi) - m/I$

This important case occurs whenever a true blaze is approximated by a stepped-phase profile. The efficiency equation [Eq. (23)] is a further specialization of this case.

Wavefront Multiplexers

Grating Designs. Grating multiplexers (also known as beam-splitter gratings) split one beam into many diffracted beams which may be of equal intensity or weighted in intensity.[19] Table 2 shows some common designs. In general, the designs can be divided into two categories: continuous phase and binary. Continuous phase multiplexers generally have better performance, as measured by the total efficiency and intensity uniformity of the diffracted beams, while binary multiplexers are easier to fabricate (with the exception

TABLE 2 Grating Multiplexers of Period D, $0 < x < D$

Phase profile	η_{-1}	η_0	η_1	Remarks
$\phi(x, y) = \begin{cases} 0 & x < D/2 \\ \pi & D/2 < x \end{cases}$	0.41	0	0.41	Binary 1:2 splitter
$\phi(x, y) = \begin{cases} 0 & x < D/2 \\ 2.01 & D/2 < x \end{cases}$	0.29	0.29	0.29	Binary 1:3 splitter
$\phi(x, y) = \pi x/D$	×	0.41	0.41	Continuous 1:2 splitter
$\phi(x, y) = \arctan\left[2.657 \cos\left(2\pi x/D\right)\right]$	0.31	0.31	0.31	Continuous 1:3 splitter

of several naturally occurring continuous phase profiles). Upper bounds for the efficiency of both continuous and binary types are derived in Ref. 20.

If the phase is allowed to be continuous or nearly continuous (8 or 16 phase levels), then the grating design problem is analogous to the phase retrieval problem and iterative techniques are commonly used.[21] A generic problem is the design of a multiplexer to split one beam into K equal intensity beams. Fanouts up to 1:50 with perfect uniformity and efficiencies of 90–100 percent are typical.

The complex transmittance of a binary grating has only two possible values [typically +1 and −1, or $\exp(j\phi_0)$ and $\exp(-j\phi_0)$], with the value changing at the transition points of the grating. By nature, the response of these gratings have the following properties:

- The intensity response is symmetric; that is, $\eta_m = \eta_{-m}$.
- The relative intensities of the nonzero orders are determined strictly by the transition points. That is, if the transition points are held constant, then the ratios η_m/η_n for all $m, n \neq 0$ will be constant, regardless of the actual complex transmittance values.
- The complex transmittance values only affect the balance of energy between the zero and nonzero orders.

Binary gratings are usually designed via the Dammann approach or search methods and tables of binary designs have been compiled.[22,23] Efficiencies of 60 to 90 percent are typical for the 1:K beam-splitter problem.

Multifocal Lenses. The concepts used to design gratings with multiple orders can be directly extended to lenses and axicons to design elements with multiple focal lengths by taking advantage of the fact that while gratings are periodic in x, paraxial lenses are periodic in $(x^2 + y^2)$, nonparaxial lenses in $\sqrt{r^2 + y^2 + f_0^2}$, and axicons in $\sqrt{x^2 + y^2}$. For gratings, different diffraction orders correspond to plane waves traveling in different directions, but for a lens of focal length f_0, the mth diffraction order corresponds to a lens of focal length f_0/m. By splitting the light into different diffraction orders, a lens with multiple focal lengths (even of opposite sign if desired) can be designed.

As an example, consider the paraxial design of a bifocal lens, as is used in intraocular implants. Half the light should see a lens of focal length f_0, while the other half should see no lens. This is a lens of focal length f_0, but with the light split evenly between the 0 and +1 orders. The phase profile of a single focus lens is given by $\phi(r) = -2\pi r^2/(2\lambda_0 f_0)$, where $r^2 = x^2 + y^2$. This phase, with the 2π ambiguity removed, is plotted in Fig. 7a as a function of r and in Fig. 7b as a function of r^2, where the periodicity in r^2 is evident. To split the light between the 0 and +1 orders, the blaze of Fig. 7b is replaced by the 1:2 continuous splitter of Table 2, resulting in Fig. 7c. This is the final design and the phase profile is displayed in Fig. 7d as a function of r.

Beam Shapers and Diffusers

In many cases, the reshaping of a laser beam can be achieved by introducing the appropriate phase shifts via a binary optics element and then letting diffraction reshape the beam as it propagates. If the incoming beam is well characterized, then it is possible to deterministically design the binary optics element.[24] For example, Fig. 8a shows the focal spot of a Gaussian beam without any beam-forming optics. In Fig. 8b, a binary optics element flattens and widens the focal spot. In this case, the element could be designed using phase-retrieval techniques, the simplest design being a clear aperture with a π phase shift over a central region. If the beam is not well-behaved, then a statistical design may be more appropriate.[25] For example, in Fig. 8c, the aperture is subdivided into randomly phased subapertures. The envelope of the resulting intensity profile is determined by the subaperture but is modulated by the speckle pattern from the random phasing. If there is

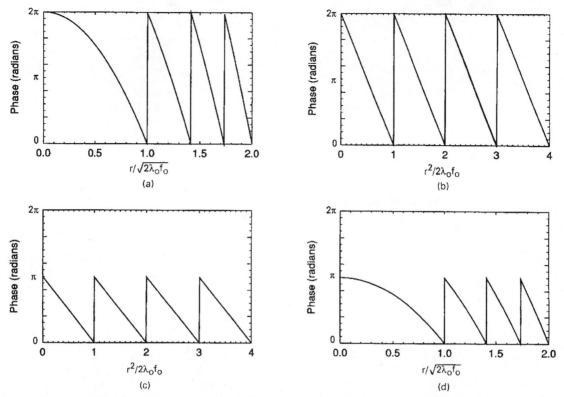

FIGURE 7 Designing a bifocal lens: (a) lens with a single focus; (b) same as (a), but showing periodicity in r^2; (c) substitution of a beam-splitting design; (d) same as (c), but as a function of r.

some randomness in the system (e.g., changing laser wavefront), then the speckle pattern will average out and the result will be a design which reshapes the beam and is robust to variations in beam shape.

Other Devices

Other Fourier optics-based applications which benefit from binary optics include the coupling of laser arrays via filtering in the Fourier plane or other means,[26] the fabrication of phase-only components for optical correlators,[27] and the implementation of coordinate transformations.[16,28] In all these applications, binary optics is used to directly manipulate the phase of a wavefront.

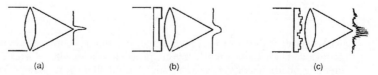

FIGURE 8 Reshaping a focused beam: (a) Gaussian focus; (b) deterministic beam-shaper; (c) statistical diffuser.

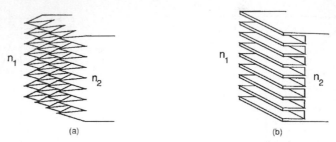

FIGURE 9 Artificial index designs: (*a*) antireflection layer; (*b*) form birefringence.

8.5 DESIGN—VECTOR DIFFRACTION THEORY

Binary optics designs based on vector diffraction theory fall into two categories: grating-based designs and artificial index designs.

Grating-based designs rely on solving Maxwell's equations for diffraction from the element. At present, this is practical only for periodic structures. Two major methods for this analysis are the expansion in terms of space harmonics (coupled wave theory) and the expansion in terms of modes (modal theory).[29] In this category, optical design is difficult since it can be both nonintuitive and computationally intensive.

Artificial index designs are based on the following premise. When features on the component are small compared to the wavelength, then the binary optics element will behave as a material of some average index. Two common applications are shown in Fig. 9. In Fig. 9*a*, the device behaves as an antireflection coating (analogous to anechoic chambers) since, at different depths, the structure has a different average index, continuously increasing from n_1 to n_2. In Fig. 9*b*, the regular, subwavelength structure exhibits form birefringence.[30] For light polarized with the electric vector perpendicular to the grooves, the effective index is

$$\frac{1}{n_{\text{eff}}^2} = p\,\frac{1}{n_1^2} + (1-p)\,\frac{1}{n_2^2}$$ (31)

where p is the fraction of total volume filled by material 1. However, for light polarized with the electric vector parallel to the grooves,

$$n_{\text{eff}}^2 = pn_1^2 + (1-p)n_2^2$$ (32)

In both these cases, the period of the structure must be much less than the wavelength in either medium so that only the zero order is propagating.

8.6 FABRICATION

Mask Layout

At the end of the optical design stage, the binary optics element is described by a phase profile $\phi(x, y)$. In the mask layout process, this profile is transformed into a geometrical layout and then converted to a set of data files in a format suitable for electron-beam pattern generation. From these files, a mask maker generates the set of photomasks which are used to fabricate the element.

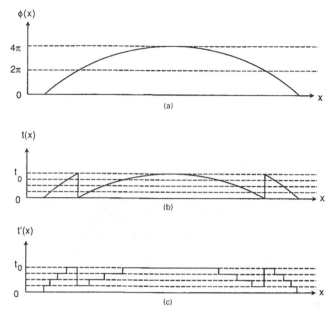

FIGURE 10 Translation from $\phi(x, y)$ to micromachined surface: (a) phase $\phi(x, y)$; (b) thickness $t(x, y)$; (c) binary optics profile $t'(x, y)$.

The first step is to convert the phase profile $\phi(x, y)$ into a thickness profile (see Fig. 10a,b) by the relation

$$t(x, y) = \frac{\lambda_0}{2\pi(n_0 - 1)} (\phi \bmod 2\pi) \tag{33}$$

where λ_0 is the design wavelength and n_0 is the index of the substrate at λ_0. The thickness profile is the surface relief required to introduce a phase shift of $\phi(x, y)$. The thickness varies continuously from 0 to t_0, where

$$t_0 = \lambda_0 / (n_0 - 1) \tag{34}$$

is the thickness required to introduce one wave of optical path difference.

To facilitate fabrication, $t(x, y)$ is approximated by a multilevel profile $t'(x, y)$ (Fig. 10c), which normally would require one processing cycle (photolithography plus etching) to produce each thickness level. However, in binary optics, a binary coding scheme is used so that only N processing cycles are required to produce

$$I = 2^N \tag{35}$$

thickness levels (hence the name binary optics).

The photomasks and etch depths required for each processing cycle are determined from contours of the thickness $t(x, y)$ or equivalently the phase $\phi(x, y)$, as shown in Table 3. The contours can be generated in several ways. For simple phase profiles, the contours are determined analytically. Otherwise, the contours are determined either by calculating the thickness at every point on a grid and then interpolating between points[31] or by using a numerical contouring method,[32] analogous to tracing fringes on an interferogram.

To generate the photomasks, the geometrical areas bounded by the contours must be described in a graphics format compatible with the mask vendor (see Fig. 11a,b). Common formats are GDSII and CIF,[33] both of which are high-level graphics descriptions which

TABLE 3   Processing Steps for Binary Optics

Layer	Etch region, defined by $t(x, y)$	Etch region, defined by $\phi(x, y)$	Etch depth
1	$0 < t \bmod (t_0) < t_0/2$	$0 < \phi \bmod 2\pi < \pi$	$t_0/2$
2	$0 < t \bmod (t_0/2) < t_0/4$	$0 < \phi \bmod \pi < \pi/2$	$t_0/4$
3	$0 < t \bmod (t_0/4) < t_0/8$	$0 < \phi \bmod \pi/2 < \pi/4$	$t_0/8$
4	$0 < t \bmod (t_0/8) < t_0/16$	$0 < \phi \bmod \pi/4 < \pi/8$	$t_0/16$

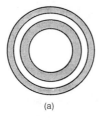  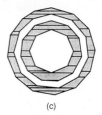

(a)   (b)   (c)

FIGURE 11 Mask layout descriptions: (a) mathematical description based on thickness contours; (b) high-level graphics description; (c) MEBES.

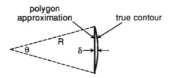

FIGURE 12 Quantization angle.

use the multisided polygon (often limited to 200 sides) as the basic building block. Hierarchical constructions (defining structures in terms of previously defined structures) and arraying of structures are also allowed.

The photomasks are usually written by electron-beam generators using the MEBES (Moving Electron Beam Exposure System) format as input. Most common high-level graphics descriptions can be translated or "fractured" to MEBES with negligible loss in fidelity via existing translation routines. Currently, commercial mask makers can achieve a minimum feature size or "critical dimension" (CD) of 0.8 μm with ease, 0.5 μm with effort, and 0.3 μm in special cases. The CD of a binary optics element is determined by the minimum local period [see Eq. (8)] divided by the number of steps, $D_{\min}/I$. For lenses,

$$D_{\min} \doteq 2\lambda_0 F \tag{36}$$

where F is the F-number of the lens; while, for gratings, $D_{\min}$ is the period of the grating.

In MEBES, all geometrical shapes are subdivided into trapezoids whose vertices lie on a fixed rectangular grid determined by the resolution of the electron-beam machine (see Fig. 11c). The resolution (typically 0.05 μm) should not be confused with the CD achievable by the mask maker.

In summary, the description of the photomask begins as a mathematical description based on contours of the thickness profile and ends as a set of trapezoids whose vertices fall on a regular grid (see Fig. 11). This series of translations results in the following artifacts. First, curves are approximated by straight lines. The error introduced by this approximation (see Fig. 12) is

$$\delta = R(1 - \cos \theta/2) \doteq R\theta^2/8 \tag{37}$$

Normally, the maximum allowable error is matched to the electron-beam resolution. Second, all coordinates are digitized to a regular grid. This results in pixelization artifacts (which are usually negligible), analogous to the ziggurat pattern produced on video monitors when plotting gently sloped lines. Finally, the MEBES writing process itself has a preferred direction since it uses electrostatic beam deflection in one direction and mechanical translation in the other.

In addition to the digitized thickness profile, photomasks normally include the following features which aid in the fabrication process. Alignment marks[34] are used to align successive photomasks, control features such as witness boxes allow the measurement of etch depths and feature sizes without probing the actual device, and labels allow the fabricator to easily determine the mask name, orientation, layer, etc.

Micromachining Techniques

Binary optics uses the same fabrication technologies as integrated circuit manufacturing.[34–35] Specifically, the micromachining of binary optics consists of two steps: replication of the photomasks pattern into photoresist (photolithography) and the subsequent transfer of the pattern into the substrate material to a precise depth (etching or deposition).

The replication of the photomasks onto a photoresist-covered substrate is achieved primarily via contact, proximity, or projection optical lithography. Contact and proximity printing offer lower equipment costs and more flexibility in handling different substrate sizes and substrate materials. In contact printing, the photomask is in direct contact with the photoresist during exposure. Vacuum-contact photolithography, which pulls a vacuum between the mask and photoresist, results in the highest resolution (submicron features) and linewidth fidelity. Proximity printing, which separates the mask and photoresist by 5 to 50 μm, results in lower resolution due to diffraction. Both contact and proximity printing require 1:1 masks. In projection printing, the mask is imaged onto the photoresist with a demagnification from $1\times$ to $20\times$. Projection printers are suitable for volume manufacturing and can take advantage of magnified masks. However, they also require expensive optics, strict environmental controls, and can only expose limited areas (typically $2\,\text{cm} \times 2\,\text{cm}$).

Following exposure, either the exposed photoresist is removed (positive resist) or the unexposed photoresist is removed (negative resist) in a developer solution. The remaining resist serves as a protective mask during the subsequent etching step.

The most pertinent etching methods are reactive ion etching (RIE) and ion milling. In RIE, a plasma containing reactive neutral species, ions, and electrons is formed at the substrate surface. Etching of the surface is achieved through both chemical reaction and mechanical bombardment by particles. The resulting etch is primarily in the vertical direction with little lateral etching (an anisotropic etch) and the chemistry makes the etch attack some materials much more vigorously than others (a selective etch). Because of the chemistry, RIE is material-dependent. For example, RIE can be used to smoothly etch quartz and silicon, but RIE of borosilicate glasses results in micropatterned surfaces due to the impurities in the glass. In ion milling, a stream of inert gas ions (usually Ar) is directed at the substrate surface and removes material by physical sputtering. While ion milling is applicable to any material, it is usually slower than RIE.

For binary optics designed to be blazed for a single order (i.e., designs based on geometrical optics), the major effect of fabrication errors is to decrease the efficiency of the blaze. There is little or no degradation in the wavefront quality. Fabrication errors can be classified as lithographic errors, which include alignment errors and over/underexposure of photoresist, and etching errors, which include depth errors and nonuniform etching of the substrate. As a rule of thumb, lithographic errors should be held to less than 5 per cent of the minimum feature size ($<0.05\,D_{\min}/I$), which can be quite challenging; while etching errors should be held to less than 5 percent of t_0, which is usually not too difficult. For binary optics designed via scalar or vector diffraction theory, manufacturing tolerances are estimated on a case-by-case basis through computer simulations.

8.7   REFERENCES

1. "Holographic and Diffractive Lenses and Mirrors," *Proc. Soc. Photo-Opt. Instrum. Eng.,* Milestone Series 34, 1991.

2. "Computer and Optically Generated Holographic Optics" series, *Proc. Soc. Photo-Opt. Instrum. Eng.,* 1052, 1989; 1211; 1990; 1555, 1991.

3. "Miniature and Microoptics" series, *Proc. Soc. Photo-Opt. Instrum. Eng.,* 1544, 1991 and 1751, 1992.

4. G. J. Swanson, "Binary Optics Technology: The Theory and Design of Multi-level Diffractive Optical Elements," M.I.T. Lincoln Laboratory Technical Report 854, NTIS Publ. AD-A213-404, 1989.

5. M. W. Farn and W. B. Veldkamp, "Binary Optics: Trends and Limitations," *Conference on Binary Optics,* NASA Conference Publication 3227, 1993, pp. 19–30.

6. S. H. Lee, "Recent Advances in Computer Generated Hologram Applications," *Opt. and Phot. News,* **16**:7, 1990, pp. 18–23.

7. W. T. Welford, *Aberrations of Optical Systems,* Adam Hilber, Ltd., Boston, 1986, pp. 75–78, 217–225.

8. W. C. Sweatt, "Mathematical Equivalence between a Holographic Optical Element and an Ultra-high Index Lens," *J. Opt. Soc. Am.,* **69,** 1979, pp. 486–487.

9. M. W. Farn, "Quantitative Comparison of the General Sweatt Model and the Grating Equation," *Appl. Opt.,* 1992, pp. 5312–5316.

10. W. C. Sweatt, "Describing Holographic Optical Elements as Lenses," *J. Opt. Soc. Am.,* **67,** 1977, pp. 803–808.

11. D. A. Buralli and G. M. Morris, "Design of Diffractive Singlets for Monochromatic Imaging," *Appl. Opt.,* **30,** 1991, pp. 2151–2158.

12. D. A. Buralli and J. R. Rogers, "Some Fundamental Limitations of Achromatic Holographic Systems," *J. Opt. Soc. Am.,* **A6,** 1989, pp. 1863–1868.

13. C. W. Chen, "Application of Diffractive Optical Elements in Visible and Infrared Optical Systems," *Proc. Soc. Photo-Opt. Instrum. Eng.* CR41, 1992, pp. 157–172.

14. T. Stone and N. George, "Hybrid Diffractive-refractive Lenses and Achromats," *Appl. Opt.,* **27,** 1988, 2960–2971.

15. R. Kingslake, *Lens Design Fundamentals,* Academic Press, Inc., New York, 1978, pp. 77–78, 112–114.

16. J. R. Leger and W. C. Goltsos, "Geometrical Transformation of Linear Diode-laser Arrays for Longitudinal Pumping of Solid-state Lasers," *IEEE J. of Quant. Elec.,* **28,** 1992, pp. 1088–1100.

17. G. J. Swanson, "Binary Optics Technology: Theoretical Limits on the Diffraction Efficiency of Multilevel Diffractive Optical Elements," M.I.T. Lincoln Laboratory Technical Report 914, 1991.

18. H. Dammann, "Spectral Characteristics of Stepped-phase Gratings," *Optik,* **53,** 1979, pp. 409–417.

19. A. Vasara, et al., "Binary Surface-Relief Gratings for Array Illumination in Digital Optics," *Appl. Opt.* **31,** 1992, pp. 3320–3336.

20. U. Krackhardt, et al., "Upper Bound on the Diffraction Efficiency of Phase-only Farnout Elements," *Appl. Opt.,* **31,** 1992, pp. 27–37.

21. D. Prongue, et al., "Optimized Kinoform Structures for Highly Efficient Fan-Out Elements," *Appl. Opt.* **31,** 1992, pp. 5706–5711.

22. U. Killat, G. Rabe, and W. Rave, "Binary Phase Gratings for Star Couplers with High Splitting Ratios" *Fiber and Integrated Optics,* **4,** 1982, pp. 159–167.

23. U. Krackhardt, "Binaere Phasengitter als Vielfach-Strahlteiler," *Diplomarbeit, Universitaet Erlangen-Nuernberg,* Erlangen, Germany, 1989.

24. J. Hossfeld, et al., "Rectangular Focus Spots with Uniform Intensity Profile Formed by Computer Generated Holograms," *Proc. Soc. Photo-Opt. Instrum. Eng.,* 1574, 1991, pp. 159–166.

25. C. N. Kurtz, "Transmittance Characteristics of Surface Diffusers and the Design of Nearly Band-Limited Binary Diffusers," *J. Opt. Soc. Am.,* **62,** 1972, pp. 982–989.

26. J. R. Leger, et al., "Coherent Laser Beam Addition: An Application of Binary-optics Technology," *The Lincoln Lab Journal,* **1,** 1988, pp. 225–246.

27. M. A. Flavin and J. L. Horner, "Amplitude Encoded Phase-only Filters," *Appl. Opt.* **28,** 1989, pp. 1692–1696.

28. O. Bryngdahl, "Geometrical Transforms in Optics," *J. Opt. Soc. Am.,* **64,** 1974, pp. 1092–1099.

29. T. K. Gaylord, et al., "Analysis and Applications of Optical Diffraction by Gratings," *Proc. IEEE* 73, 1985, pp. 894–937.

30. D. H. Raguin and G. M. Morris, "Antireflection Structured Surfaces for the Infrared Spectral Region," *Appl. Opt.* **32,** 1993, pp. 1154–1167.

31. J. Logue and M. L. Chisholm, "General Approaches to Mask Design for Binary Optics," *Proc. Soc. Photo-Opt. Instrum. Eng.,* 1052, 1989, pp. 19–24.

32. A. D. Kathman, "Efficient Algorithm for Encoding and Data Fracture of Electron Beam Written Holograms," *Proc. Soc. Photo-Opt. Instrum. Eng.,* 1052, 1989, pp. 47–51.

33. S. M. Rubin, *Computer Aids for VLSI Design,* Addison-Wesley Publishing Co., Reading, MA, 1987.

34. N. G. Einspruch and R. K. Watts (eds.), *Lithography for VLSI,* VLSI Electronics Series 16, Academic Press, Inc., Boston, MA, 1987.

35. N. G. Einspruch and D. M. Brown (ed.), *Plasma Processing for VLSI,* VLSI Electronics Series 8, Academic Press, Inc., Boston, MA, 1984.

CHAPTER 9
GRADIENT INDEX OPTICS

Duncan T. Moore
The Institute of Optics
and
Gradient Lens Corporation
Rochester, New York

9.1 GLOSSARY

A	constant
a, b	constants
g	constant
h_i	constants
n	refractive index
r	radius
V_{ij}	Abbe numbers
z	cartesian coordinate (optical axis direction)
Φ	power

9.2 INTRODUCTION

Gradient index (GRIN) optics[1] refers to the field of optics in which light propagates along a curved path. This contrasts with normal homogeneous materials in which light propagates in a rectilinear fashion. Other terms that have been used to describe this field are inhomogeneous optics, index of refraction gradients, and distributed index of refraction. The most familiar example of a gradient index phenomenon is the mirage when a road appears to be wet on a hot summer day. This can be understood by the fact that the road is absorbing heat, thus slightly raising the temperature of the air relative to the temperature a few meters above the surface. By the gas law, the density decreases, and therefore the index of refraction decreases. Light entering this gradient medium follows a curved path. The ray path, as shown in Fig. 1, is such that the ray propagates downward towards the road and then gradually upwards to the observer's eye. The observer sees two images. One is the normal image propagating through the homogeneous material and the second is an image that is inverted and appears below the road surface. Thus, the index of refraction gradient acts as a mirror by gradual light refraction rather than reflection.

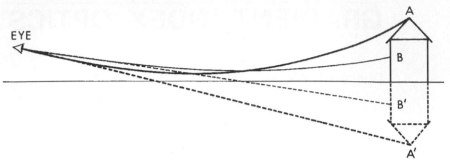

FIGURE 1 Light from point A emits or reflects in all directions. Light propagating several meters above the heated road travels in a straight line. Light passing through the lower index of refraction region near the road undergoes a bending. This light appears to have come from below the road.

9.3 ANALYTIC SOLUTIONS

Over the approximately 150 years that gradient index optics has been studied, a wealth of very interesting analytic solutions has been published.[2] A classic example was published by James Clerk Maxwell in 1850. Maxwell[3] showed through geometrical optics that the ray paths in a spherically symmetric material whose index of refraction is given by

$$n(r) = a/(b^2 + r^2) \tag{1}$$

are circles. The object and the image lie on the surface of the sphere but, otherwise, the imaging is perfect between the conjugate points on the sphere. The medium between the object and the image is continuous with no discreet surfaces. A century later, Luneburg[4] modified the system to allow for discontinuities of the index of refraction. While these have not been implemented in ordinary optical systems, they have, however, been shown to be useful in integrated optics.[5]

A final example of a numerical solution is that of a radial (cylindrical) gradient in which the index of refraction varies perpendicular to a line. In 1954, Fletcher[6] showed that if the index of refraction is given by

$$n(r) = n_o\, sech\,(ar^2) \tag{2}$$

then the ray paths inside the material in the meridional plane are sinusoidal. Nearly fifty years earlier, Wood[7] had shown experimentally that the paths appeared to be sinusoidal. This solution has several important commercial applications. It is the basis of the Selfoc® lens used in arrays for facsimile and photocopying machines and in endoscopes used for medical applications.

9.4 MATHEMATICAL REPRESENTATION

Most of the gradient index profiles are represented by a polynomial expansion. While these expansions are not necessarily the most desirable from the gradient materials manufacturing standpoint, they are convenient for determining the aberrations of systems embodying GRIN materials. There are basically two major representations for gradient index materials. The first, used by the Nippon Sheet Glass, is used exclusively by representing radial gradient components. In this case, the index of refraction is written as a function of the radial coordinate r

$$N(r) = N_0(1 - Ar^2/2 + h_4r^4 + h_6r^6 + \cdots) \tag{3}$$

The second method of representing index of refraction profiles is a polynomial expansion in both the radial coordinate r and the optical axis coordinate z. In this case, the representation is

$$N(r, z) = \sum_{j=0} \sum_{i=0} N_{ij} \, r^{2i} z^{j} \qquad (4)$$

where the coefficients N_{ij} are the coefficients of the index of refraction polynomial. A pure axial gradient (in the z direction) has coefficients only in the form of N_{0j} and those in the radial would be of the form of N_{i0}. These representations for the index of refraction polynomial have been the basis of the aberration theory which was first developed for gradient index materials with discreet surfaces by Sands.[8] These coefficients are wavelength dependent and are typically defined at three wavelengths. A gradient dispersion is defined by using a general Abbe number

$$V_{ij} = N_{ij,d} / (N_{ij,F} - N_{ij,C}) \qquad (5)$$

except for i and j both equal to zero. In the case for $i = j = 0$, then the Abbe number becomes the standard form, namely

$$V_{00} = (N_{00,d} - 1) / (N_{00,F} - N_{00,C}) \qquad (6)$$

The subscripts $d, F,$ and C refer to the wavelengths 0.5876, 0.4861, and 0.6563 μm, respectively. Unlike the normal dispersion of glasses where V_{00} is between 20 and 90, the V_{ij} can have negative and positive values or can be infinite (implying that the gradient is the same as both the red and the blue portions of the spectrum).

9.5 *AXIAL GRADIENT LENSES*

When the index of refraction varies in the direction of the optical axis (the z direction), the bending of the light within the material is very small. Thus, the main feature of an axial gradient is its ability to correct aberrations rather than to add power to the lens. Sands showed that the effect of an axial gradient on monochromatic aberrations is exactly equivalent to that of an aspheric surface. In fact, one could convert any aspherical surface to an axial gradient with a spherical surface and have the same image performance to the third-order approximation. There is, however, one very important difference between aspheric surfaces and axial gradients, i.e., the variation of the index of refraction profile with wavelength. Since an aspheric is the same for all wavelengths, its effect on spherochromatism is established once the aspheric has been determined. Further, an asphere has no effect on paraxial axial or lateral chromatic aberrations. This is not the case for axial gradients. Since the index of refraction profile varies with wavelength, it is possible to significantly modify the spherochromatism of the lens and, in the case where the gradient extends from the front to the back surface, to affect the paraxial chromatic aberrations. Depending upon the dispersion of the gradient index material, the sphero-chromatism can be increased or decreased independent of the monochromatic correction. The effect of an axial gradient on paraxial axial chromatic aberration is best understood by placing a surface perpendicular to the optical axis in the middle of a single lens dividing it into two parts. The gradient dispersion implies that the medium will have one dispersion at the front surface and a different dispersion at the second surface. Thus, if one were to design a material in which the dispersion of the front surface is 60 and at the rear surface is 40, then the combination of a positive (convex surface on the front) and negative lens (concave surface on the back), reduces the chromatic aberration. This can only be done if the lens is meniscus. In that case, the theoretical front lens is plano-convex while the back

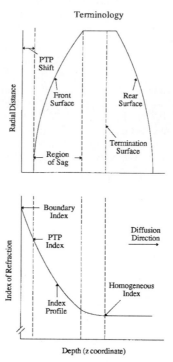

Terminology

FIGURE 2 Diagram of axial gradient terminology. The effective region of the gradient is in the "region of sag."

one is plano-concave. If the negative element has the higher dispersion (lower V numbers), then it is possible to chromatize the lens by a proper bending of the lens surfaces. This was first shown in the infrared part of the spectrum using a zinc sulfide-zinc selenide gradient material.[9]

The simplest example of an axial gradient is the linear profile in which the index of refraction is written as

$$N(z) = N_{00} + N_{01}z \tag{7}$$

The coefficient N_{01} is an additional degree of freedom which can be used to correct any of the third-order monochromatic aberrations except Petzval curvature of field. There are two ways to approach the design of these lenses. In the case where the index of refraction profile does not continue to the rear surface (see Fig. 2), a simple formula can be used to relate the amount of index change to the F-number of the lens surface if the third-order spherical aberration and coma are to be correct to zero,[10] namely,

$$\Delta n = (0.0375/(N_{00} - 1)^2)/f\#^2 \tag{8}$$

in this formula, the important parameters are the index of refraction of the base material, N_{00}, the change in index of index of refraction, Δn, from the polar tangent to the maximum sag point, and the F-number of the lens. One sees that if the F-number of the lens is doubled, then the amount of index change necessary to correct the spherical aberration and coma to zero increases by a factor of 4. Thus, while it is possible to correct the spherical aberration of the singlet operating at F/4 with an index change of only 0.0094, that same lens operating at F/1 will require an index change of 0.15. In most lenses, one never corrects the spherical aberration of individual elements to zero, but corrects the total amount of spherical aberration of all lens elements to zero.

Axial gradients have been used in a number of lens designs. Most of the work in this field has occurred in photographic objectives.[11,12] In these cases, they offer a slight advantage over aspherics because of the chromatic variation of the gradient.

9.6 RADIAL GRADIENTS

In the most generalized case for radial gradients (one in which all coefficients are nonzero), it is possible not only to use the gradient for aberration correction, but also to modify the focal length of the lens. Independent of which representation is used, the coefficient of the parabolic term [Eq. (2) or Eq. (3)] dictates the amount of power that is introduced by the radial gradient component. Assuming only a radial gradient component, Eq. (4) can be expanded as

$$N(r) = N_{00} + N_{01}r^2 + N_{02}r^4 + \cdots \tag{9}$$

Equating the terms in Eq. (3) and Eq. (9) gives

$$N_{00} = N_0 \qquad \text{and} \qquad N_{10} = -N_0 A/2 \tag{10}$$

In the most general form, the power ϕ, due to the radial gradient component, is written as

$$\phi = -N_0 A^{0.5} \sin (A^{0.5}t) \tag{11}$$

From Eq. (11), the length of the material t determines the focal length of the system. In fact, depending on the choice of length, the power can be positive, negative, or zero. See Fig. 3b. A convenient variation on this formula is to determine the length at which light

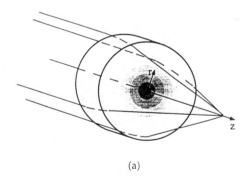

(a)

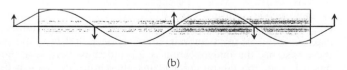

(b)

FIGURE 3 Diagram of radial gradient: (a) wood lens; (b) a long radial gradient lens illustrating period ray path.

entering the material collimated will be focused on the rear surface. This length is called the quarter pitch length of the rod and is given by

$$P_{1/4} = (\pi/2)(-N_{00}/(2N_{10}))^{0.5} \tag{12}$$

The full period length of the rod is simply four times Eq. (12).

For the case where the focal length (the reciprocal of the power) is long compared to its thickness, this can be approximated by the formula (see Fig. 3a)

$$\phi = -2N_{10}t \tag{13}$$

This simplifying formula was derived by the entomologist Exner[13] in 1889 while he was analyzing insect eyes and found them to have radial gradient components. Since the dispersion of a gradient material can be positive, negative, or infinity, the implication is that the paraxial axial chromatic aberration can be negative, positive, or zero. This leads to the possibility of an achromatized singlet with flat surfaces; or, by combining the dispersion of the gradient with that of the homogeneous materials, to single element lenses with curved surfaces that are color-corrected.

The radial gradient lens with flat surface is a very important example, both from a theoretical and a commercial standpoint. Consider such a lens with an object of infinity where the lens is thin relative to its focal length. As has already been shown, the value of N_{10} and the thickness determine the focal length of such a lens. According to third-order aberration theory,[8] the only other term that can influence the third-order monochromatic aberrations is the coefficient N_{20}. This term can be used to correct any one of the third-order aberrations except Petzval curvature of field. The coefficient N_{20} is normally used to correct the spherical aberration; however, once this choice is made, there are no other degrees of freedom to reduce other aberrations such as coma. It can be shown that the coma in such a single element lens is very large if the lens is used at infinite conjugates. Of course, if such a lens is used at unit magnification in a system which is symmetric about the aperture stop, the coma (as well as the distortion and paraxial lateral color) is zero. As the length of the rod increases, the approximation for the focal length becomes inaccurate and the more rigorous formula given by Eq. (11) is appropriate. However, the rules governing the aberration correction remain the same. That is, the choice of the value of N_{20}, or in the Nippon Sheet Glass representation, in h_4 coefficient, corrects the third-order spherical aberration to zero. In Fletcher's original paper, he showed that rays propagating in a material whose index of refraction is given by Eq. (2) would focus light in the meridional plane periodically with no aberration along the length of such a rod. If one expands a hyperbolic secant in a polynomial expansion, one obtains

$$N_{20} = 5N_{10}^2/6N_{00} \tag{14}$$

The implication is that if N_{20} is chosen according to Eq. (14), then not only is the spherical aberration corrected, but so is the tangential field (that is, the sum of three times the astigmatism plus the field curvature). Rawson[14] showed that a more appropriate value for N_{20} was $3N_{10}^2/2N_{00}$. This is a compromise for the correction of sagittal and tangential fields.

The second limiting case is to use these rods with arbitrary length but at unit magnification. This has important commercial applications in photocopying and fax machines, for couplers for single-mode fibers, and in relays used in endoscopes. In all of these systems, the magnification is ± 1 and thus there is no need to correct the coma, the distortion, or the lateral color. Thus, the choice of N_{20} can be used to either correct the spherical aberration or to achieve a compromise between the tangential and sagittal fields.

In one of the most common applications, a series of lenses is assembled to form an

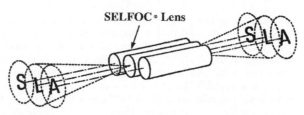

FIGURE 4 An array of radial gradient lenses (only three shown) can be used to form an image of an extended object. This principle is used in photocopying and fax machines.

array (see Fig. 4). In this case, the magnification between the object and the image must be a +1 with an inverted image halfway through the gradient index rods. Light from an object point is imaged through multiple GRIN rods depending on the numerical aperture of each of the rods. The effective numerical aperture of the array is significantly higher than that of a single rod. Theoretically, a full two-dimensional array can be constructed to image an entire two-dimensional object. In practice, to reduce costs the object is scanned by moving the object across the fixed lens array with either a charged couple device or a transfer drum used to record the image.

9.7 RADIAL GRADIENTS WITH CURVED SURFACES

While the radial gradient with flat surfaces offers tremendous commercial applications today, it has limited applications because of the large amount of coma that is introduced unless the lens system is used at unit magnification. Thus, it is often desirable to introduce other degrees of freedom that may improve the imagery. The simplest way to do this is to make one or both of the end cases curved. The ability to chromatize such a lens is not lost so long as the power resulting from the curved surfaces and that of the radial gradient maintain the same ratio (but with opposite sign) as that of the Abbe number of homogeneous material and the Abbe number of N_{10}. Thus, the lens shape can be determined to reduce the coma to zero and the value of the N_{20} coefficient is chosen to eliminate the spherical aberration. An example of a curved lens with a radial gradient was developed by Nippon Sheet Glass for a compact disc player.[15] In that case, it is not necessary to achromatize the lens since the source is a monochromatic laser diode, but it was necessary to extend the field and reduce the amount of spherical aberration simultaneously. It is also often desirable to place part of the power on the curvature rather than using the gradient to refract all of the light. This reduces the magnitude of the index change and makes the lens easier to manufacture.

In a radial gradient material with curved surfaces, it is possible to eliminate four out of five monochromatic aberrations,[16] and any four can be chosen. However, these lenses tend to be very sensitive to slight manufacturing errors, as they require a very delicate balance between the coefficients of the gradient profile and typically have very large amounts of higher-order aberrations.

9.8 SHALLOW RADIAL GRADIENTS

An interesting compromise between an axial gradient and a radial gradient with power is the shallow radial gradient (SRGRIN). In this type of gradient, there is no power generated by the gradient (i.e., $N_{10} = 0$). Like the axial gradient, it has no effect on Petzval curvature of field, but its aberration correction is significantly different than that of axial gradients. Sands[8] showed that in the case of an axial gradient, the important parameter is the differential refraction of the ray at the surface which causes an additional surface contribution. In the shallow radial gradient there is no surface contribution, since the N_{10} coefficient is zero. All of the aberration correction is from the transfer contribution through the material.[8] The implication of this fact is that the thickness of the shallow radial gradient is very important and, in fact, the most important parameter is the product of the thickness and the N_{20} coefficient. Thus, if only a small index change can be manufactured, the same amount of aberration correction can be achieved by increasing the thickness of the element. The other significant difference between this gradient and a normal radial gradient is the sign of the index change. In most lenses designed to date, the index of refraction of a conventional radial gradient should be lower at the periphery than it is at the center, thus creating a positive lens. However, in the shallow radial gradient, the index of refraction should be higher at the periphery than at the center. This has also normally been the case in the axial gradient in which the index of refraction should be higher at the polar tangent plane than at the maximum sag point. This has important implications for the manufacturing process. Furthermore, the amount of index change necessary for shallow gradient correction is usually very small compared to the amount of index change needed in a regular radial gradient.

9.9 MATERIALS

While several materials systems have been proposed for forming gradient index materials, gradients have only been made for commercial applications in glasses and polymers. However, research has been conducted in zinc selenide-zinc sufilde,[17] and germanium-silicon[18] for the infrared portion of the spectrum, and in fluoride materials for the ultraviolet.[19] However, none of these have reached the stage, at this writing, which can be commercialized. For glasses, several processes have been proposed. The most common method of making gradient index materials is by the ion exchange process. In this case, a glass containing a single valence ion (such as sodium, lithium, or potassium) is placed in a molten salt bath at temperatures between 400 and 600°C. The molten salt bath contains a different ion than that in the glass. The ions from the salt diffuse into the glass and exchange for an ion of equal valence in the glass. The variation in composition leads to a variation in index of refraction. The variation in index of refraction occurs due to the change of polarizability between the two ions and the slight change in the density of the material. In some cases, these two phenomena can cancel one another, producing a composition variation, but no corresponding change in index of refraction. A model for predicting the index refraction change as well as the chromatic variation of the gradient has been developed.[20] In this system, it is clear that the maximum index change is limited by the changes in the properties of single valence ions. While very large index changes have been made (approaching 0.27), these gradients suffer from large amounts of chromatic aberration. In axial gradients, a large amount of chromatic aberration is desirable, as it normally improves the spherochromatism. In the case of radial gradients, however, it creates large paraxial axial chromatic aberration which is normally not desirable.

The manufacturing method is quite simple. If one wishes to make axial gradients, a sheet of glass is placed in a molten salt bath. Typical times for diffusion are a few days for diffusion depths of 3 to 7 mm at temperatures around 500°C. The higher the temperature, the faster the diffusion; however, at high temperature the glass will begin to deform. Lower temperatures increase the diffusion times. For radial gradients, one simply starts with glass with cylindrical symmetry and places the rods inside an ion exchange bath. In order to form good parabolic profiles, it is necessary for the ions to diffuse through the center.

Two other methods have been proposed for making gradients in glass. In the first, the gradient is formed by leaching or by stuffing in a sol-gel formed glass. This system has only shown to be applicable to radial gradients. After the glass is formed by the sol-gel (solution gelatin process), the glass is in a porous state where one of the components can be dissolved out in an acid bath[21] or molecules can be stuffed into the glass to form the index of refraction gradient.[22] By the leaching method, gradients have been formed in either titanium or zirconium. Index changes of up to 0.03 have been formed by this method. Alternatively, the glass can be stuffed with ions such as lead. The lead precipitates on the walls of the porous material whereupon it is included in the glass during the sintering step. While it is possible to get much larger changes using the method based on lead, both of these techniques suffer from large amounts of chromatic aberration.

A new method shown to be very useful for axial gradients is based on the fusion of glass slabs.[23] The index of refraction of each slab is slightly different than its adjacent slab. Very large index of refraction changes can be formed by this technique ($\Delta n = 0.4$). Further, these materials can be made in apertures up to 100 mm.

Two basic methods for manufacturing of polymers for gradient index have been demonstrated. In the first, an exchange of one monomer for a monomer in a partially polymerized material forms a profile in the same way as the ion exchange method.[24] In the second, ultraviolet light is used to induce photocopolymerization to form an index of refraction in the material.[25]

9.10 REFERENCES

1. For a source of over 100 articles on Gradient-Index(GRIN) Optics, the reader is referred to a series of special issues in *Applied Optics,* GRIN I (April 1, 1980), GRIN II (March 15, 1982), GRIN III (Feb. 1, 1983), GRIN IV (June 1, 1984), GRIN V (December 15, 1985), GRIN VI (October 1, 1986), GRIN VII (February 1, 1988), GRIN VIII (October 1, 1990 and December 1, 1990), and GRIN IX (September 1, 1992).

2. S. Cornbleet, *Microwave Optics,* Academic Press, New York (1976), pp. 108–187.

3. J. C. Maxwell, *The Scientific Papers of James Clerk Maxwell,* W. D. Niven (ed.), New York, 1965, pp. 76–78.

4. R. K. Luneburg, *Mathematical Theory of Optics,* University of California Press, 1966, pp. 182–195.

5. W. H. Southwell, "Planar Optical Waveguide Lens Design," *Appl. Opt.,* **21,** 1982, p. 1985.

6. A. Fletcher, T. Murphy, and A. Young, "Solutions of Two Optical Problems," *Proc R. Soc. Lond. A,* **223,** 1954, pp. 216–225.

7. R. W. Wood, *Physical Optics,* Macmillan New York, 1905, pp. 86–91.

8. P. J. Sands, "Third-Order Aberrations of Inhomogeneous Lenses," *J. Opt. Soc. Am.,* **60,** 1970, pp. 1436–1443.

9. J. W. Howard and D. P. Ryan-Howard, "Optical Design of Thermal Imaging Systems Utilizing Gradient-Index Optical Materials," *Opt Eng.,* **24,** 1985, p. 263.

10. D. S. Kindred, "Development of New Gradient Index Glasses for Optical Imaging Systems," Ph.d. thesis, Univ. of Rochester, 1990, pp. 207–210.

11. D. S. Kindred and D. T. Moore, "Design, Fabrication, and Testing of a Gradient-Index Binocular Objective," *Appl. Opt.,* **27,** 1988, pp. 492–495.

12. L. G. Atkinson III, et al., "Design of a Gradient-Index Photographic Objective," *Appl. Opt.,* **21,** 1984 p. 1735.

13. S. Exner, "The Retinal Image of Insect Eyes" (in Ger.), *Sb. Akad. Wiis Wien,* **98,** 1889, p. 13.

14. E. G. Rawson, D. R. Herriott, and J. McKenna, "Analysis of Refractive Index Distributions in Cylindrical Graded-Index Glass Rods Used as Image Relays," *App. Optl.,* **9,** pp. 753–759.

15. H. Nishi, H. Ichikawa, M. Toyama, and I. Kitano, "Gradient-Index Objective for the Compact Disk System," *Appl. Opt.,* **25,** 1986, p. 3340.

16. D. T. Moore and R. T. Salvage, "Radial Gradient-Index Lenses with Zero Petzval Aberration," *Appl Opt.,* **19,** 1980, pp. 1081–1086.

17. M. A. Pickering, R. L. Taylor, and D. T. Moore, "Gradient Infrared Optical Material Chemical Vapor Deposition Process," *Appl. Opt.,* **25,** 1986, pp. 3364–3372.

18. J. J. Miceli, "Infrared Gradient-Index Optics: Materials, Fabrication and Testing," Ph.D. thesis, University of Rochester, New York, 1982.

19. M. T. Houk, "Fabrication and Testing of Index Gradients in Flouride Materials," Ph.D. thesis, University of Rochester, New York, 1990.

20. S. D. Fantone, "Refractive Index and Spectral Models for Gradient-Index Materials", *Appl. Opt.,* **22,** 1983, pp. 432–440.

21. T. M. Che, J. B. Caldwell, and R. M. Mininni, "Sol-gel Derived Gradient-Index Optical Materials," *SPIE,* **1328,** 1990, p. 145.

22. M. Yamane, H. Kawazoe, A. Yasumori, and T. Takahashi, "Gradient-Index Glass Rods of PbO-K_2O-B_2O_3-SiO_2 System Prepared by the Sol-Gel Process," *J. Non-Cryst. Solids,* **100,** [1–3] 1988, pp. 506–510.

23. J. J. Hagerty, "Glass Plate Fusion for Macro-Gradient-Index Materials," U.S. Patent 4,929,065, 1990.

24. Y. Ohtsuka and T. Sugano, "GRIN Rod of CR 39-Trifluoroethyl Methacrylate Copolymer by Vapor-Phase Transfer Process," *Appl. Opt.,* **22,** 1983, pp. 413–417.

25. Y. Koike and Y. Ohtsuka, "Studies on the Light Focusing Plastic Rods: Control Of Refractive-Index Distribution of Plastic Radial Gradient-Index Rod by Photocopolymerization," *Appl Opt.,* **24,** 1985, pp. 4316–4320.

CHAPTER 10
OPTICAL FIBERS AND FIBER-OPTIC COMMUNICATIONS

Tom G. Brown
The Institute of Optics
University of Rochester
Rochester, New York

10.1 GLOSSARY

A	open loop gain of receiver amplifier
A	pulse amplitude
a	core radius
a_P	effective pump area
A_{eff}	effective (modal) area of fiber
A_i	cross-sectional area of ith layer
B	data rate
B_n	noise bandwidth of amplifier
c	vacuum velocity of light
D	fiber dispersion (total)
E_i	Young's modulus
e_{LO}, e_S	polarization unit vectors for signal and local oscillator fields
F	tensile loading
F_e	excess noise factor (for APD)
g_B	Brillouin gain
g_R	Raman gain
i_d	leakage current (dark)
I_m	current modulation
$I(r)$	power per unit area guided in single mode fiber
k	Boltzmann's constant
J_m	Bessel function of order m
K_m	modified Bessel function of order m
k_0	vacuum wave vector
l	fiber length

l_0	length normalization factor
L_D	dispersion length
m	Weibull exponent
M	modulation depth
N	order of soliton
n	actual number of detected photons
N_{eff}	effective refractive index
N_p	average number of detected photons per pulse
n_0	core index
n_1	cladding index
$n(r)$	radial dependence of the core refractive index for a gradient-index fiber
P	optical power guided by fiber
P_E	error probability
P_f	probability of fiber failure
P_s	received signal power
P_s	signal power
P_R	power in Raman-shifted mode
P_0	peak power
P_0	peak power of soliton
R	detector responsivity (A/W)
RIN	relative intensity noise
R_L	load resistor
$R(r)$	radial dependence of the electric field
S	failure stress
SNR	signal-to-noise ratio measured in a bandwidth B_n
S_0	location parameter
T	temperature (Kelvin)
t	time
T_0	pulse width
U	normalized pulse amplitude
z	longitudinal coordinate
$Z(z)$	longitudinal dependence of the electric field
α	profile exponent
α_f	frequency chirp
α_R	attenuation of Raman-shifted mode
$\tilde{\beta}$	complex propagation constant
β_1	propagation constant
β_2	dispersion (2d order)
Δ	peak index difference between core and cladding
Δf	frequency deviation

ΔL change in length of fiber under load

$\Delta \phi$ phase difference between signal and local oscillator

$\Delta \nu$ source spectral width

$\Delta \tau$ time delay induced by strain

ε strain

η_{HET} heterodyne efficiency

θ_c critical angle

$\Theta(\theta)$ azimuthal dependence of the electric field

λ vacuum wavelength

λ_c cut-off wavelength

ξ normalized distance

$\Psi(r, \theta, z)$ scalar component of the electric field

Ψ_S, Ψ_{LO} normalized amplitude distributions for signal and LO

σ_A^2 amplifier noise

$\sigma_d^2 = 2ei_d B_n$ shot noise due to leakage current

$\sigma_J^2 = \dfrac{4kT}{R_L} B_n$ Johnson noise power

$\sigma_R^2 = R^2 P_s^2 B_n \times 10^{-(\text{RIN}/10)}$ receiver noise due to source RIN

$\sigma_s^2 = 2eRP_s B_n Fe$ signal shot noise

τ time normalized to moving frame

r, θ, z cylindrical coordinates in the fiber

10.2 INTRODUCTION

Optical fibers were first envisioned as optical elements in the early 1960s. It was perhaps those scientists well-acquainted with the microscopic structure of the insect eye who realized that an appropriate bundle of optical waveguides could be made to transfer an image, and the first application of optical fibers to imaging was conceived. It was Charles Kao[1] who first suggested the possibility that low-loss optical fibers could be competitive with coaxial cable and metal waveguides for telecommunications applications. It was not, however, until 1970 when Corning Glass Works announced an optical fiber loss less than the benchmark level of 10 dB/km[2,3] that commercial applications began to be realized. The revolutionary concept which Corning incorporated and which eventually drove the rapid development of optical fiber communications was primarily a materials one—it was the realization that low doping levels and very small index changes could successfully guide light for tens of kilometers before reaching the detection limit. The ensuing demand for optical fibers in engineering and research applications spurred further applications. Today we see a tremendous variety of commercial and laboratory applications of optical fiber technology. This chapter will discuss important fiber properties, describe fiber fabrication and chemistry, and discuss materials trends and a few commercial applications of optical fiber.

While it is important, for completeness, to include a treatment of optical fibers in any handbook of modern optics, an exhaustive treatment would fill up many volumes all by itself. Indeed, the topics covered in this chapter have been the subject of monographs, reference books, and textbooks; there is hardly a scientific publisher that has not published several books on fiber optics. The interested reader is referred to the "Further Reading" section at the end of this chapter for additional reference material.

Optical fiber science and technology relies heavily on both geometrical and physical optics, materials science, integrated and guided-wave optics, quantum optics and optical physics, communications engineering, and other disciplines. Interested readers are referred to other chapters within this collection for additional information on many of these topics.

The applications which are discussed in detail in this chapter are limited to information technology and telecommunications. Readers should, however, be aware of the tremendous activity and range of applications for optical fibers in metrology and medicine. The latter, which includes surgery, endoscopy, and sensing, is an area of tremendous technological importance and great recent interest. While the fiber design may be quite different when optimized for these applications, the general principles of operation remain much the same. A list of references which are entirely devoted to optical fibers in medicine is listed in "Further Reading".

10.3 PRINCIPLES OF OPERATION

The optical fiber falls into a subset (albeit the most commercially significant subset) of structures known as dielectric optical waveguides. The general principles of optical waveguides are discussed elsewhere in Chap. 6 of Vol. II, "Integrated Optics"; the optical fiber works on principles similar to other waveguides, with the important inclusion of a cylindrical axis of symmetry. For some specific applications, the fiber may deviate slightly from this symmetry; it is nevertheless fundamental to fiber design and fabrication. Figure 1 shows the generic optical fiber design, with a core of high refractive index surrounded by a low-index cladding. This index difference requires that light from inside the fiber which is incident at an angle greater than the critical angle

$$\theta_c = \sin^{-1}\left(\frac{n_1}{n_0}\right) \tag{1}$$

be totally internally reflected at the interface. A simple geometrical picture appears to allow a continuous range of internally reflected rays inside the structure; in fact, the light (being a wave) must satisfy a self-interference condition in order to be trapped in the waveguide. There are only a finite number of paths which satisfy this condition; these are analogous to the propagating electromagnetic modes of the structure. Fibers which support a large number of modes (these are fibers of large core and large numerical aperture) can be adequately analyzed by the tools of geometrical optics; fibers which support a small

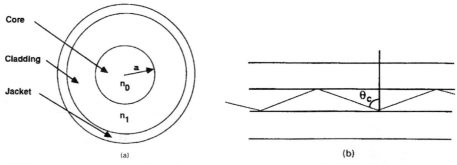

FIGURE 1   (a) Generic optical fiber design, (b) path of a ray propagating at the geometric angle for total internal reflection.

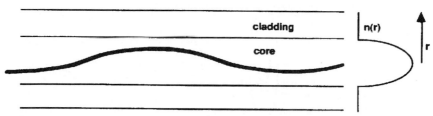

FIGURE 2   Ray path in a gradient-index fiber.

number of modes must be characterized by solving Maxwell's equations with the appropriate boundary conditions for the structure.

Fibers which exhibit a discontinuity in the index of refraction at the boundary between the core and cladding are termed *step-index fibers.* Those designs which incorporate a continuously changing index of refraction from the core to the cladding are termed *gradient-index fibers.* The geometrical ray path in such fibers does not follow a straight line—rather it curves with the index gradient as would a particle in a curved potential (Fig. 2). Such fibers will also exhibit a characteristic angle beyond which light will not internally propagate. A ray at this angle, when traced through the fiber endface, emerges at an angle in air which represents the maximum geometrical acceptance angle for rays *entering* the fiber; this angle is the numerical aperture of the fiber (Fig. 3). Both the core size and numerical aperture are very important when considering problems of fiber-fiber or

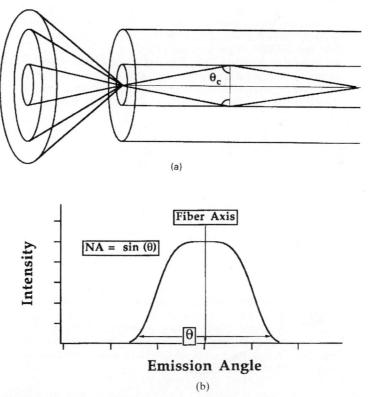

(a)

(b)

FIGURE 3   The numerical aperture of the fiber defines the range of external acceptance angles.

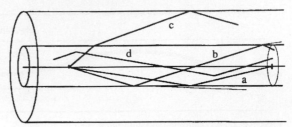

FIGURE 4 Classification of geometrical ray paths in an optical fiber. (a) Meridional ray; (b) leaky ray; (c) ray corresponding to a cladding mode; (d) skew ray.

laser-fiber coupling. A larger core and larger numerical aperture will, in general, yield a higher coupling efficiency. Coupling between fibers which are mismatched either in core or numerical aperture is difficult and generally results in excess loss.

The final concept for which a geometrical construction is helpful is ray classification. Those geometrical paths which pass through the axis of symmetry and obey the self-interference condition are known as *meridional rays*. There are classes of rays which are nearly totally internally reflected and may still propagate some distance down the fiber. These are known as *leaky rays* (or modes). Other geometrical paths are not at all confined in the core, but internally reflect off of the cladding-air (or jacket) interface. These are known as *cladding modes*. Finally, there exists a class of geometrical paths which are bound, can be introduced outside of the normal numerical aperture of the fiber, and do not pass through the axis of symmetry. These are often called *skew rays*. Figure 4 illustrates the classification of geometrical paths.

Geometrical optics has a limited function in the description of optical fibers, and the actual propagation characteristics must be understood in the context of guided-wave optics. For waveguides such as optical fibers which exhibit a small change in refractive index at the boundaries, the electric field can be well described by a scalar wave equation,

$$\nabla^2 \Psi(r, \theta, z) + k_0^2 r^2(r) \Psi(r, \theta, z) = 0 \tag{2}$$

the solutions of which are the modes of the fiber. $\Psi(r, \theta, z)$ is generally assumed to be separable in the variables of the cylindrical coordinate system of the fiber:

$$\Psi(r, \theta, z) = R(r)\Theta(\theta)Z(z) \tag{3}$$

This separation results in the following eigenvalue equation for the radial part of the scalar field:

$$\frac{d^2 R}{dr^2} + \frac{1}{r}\frac{dR}{dr} + \left(k_0^2 n^2(r) - \beta^2 - \frac{m^2}{r^2}\right)R = 0 \tag{4}$$

in which m denotes the azimuthal mode number, and β is the propagation constant. The solutions must obey the necessary continuity conditions at the core-cladding boundary. In addition, guided modes must decay to zero outside the core region. These solutions are readily found for fibers having uniform, cylindrically symmetric regions but require numerical methods for fibers lacking cylindrical symmetry or having an arbitrary index gradient. A common form of the latter is the so-called α-*profile* in which the refractive index exhibits the radial gradient[4]

$$m(r) = \begin{cases} n_1\left[1 - \Delta\left(\dfrac{r}{a}\right)^\alpha\right] & r < a \\ n_1[1 - \Delta] = n_2 & r \geq a \end{cases} \tag{5}$$

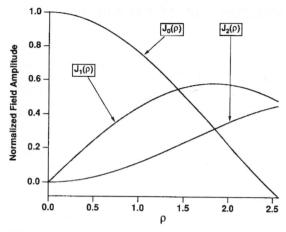

FIGURE 5 Bessel functions $J_m(\rho)$ for $m = 0$, 1, and 2.

The step-index fiber of circular symmetry is a particularly important case, because analytic field solutions are possible and the concept of the "order" of a mode can be illustrated. For this case, the radial dependence of the refractive index is the step function

$$n(r) = \begin{cases} n_1 & r < a \\ n_2 & r \geq a \end{cases} \tag{6}$$

The solutions to this are Bessel functions[5] and are illustrated in Fig. 5. It can be seen that only the lowest-order mode ($m = 0$) has an amplitude maximum at the center. Its solution in the (core) propagating region ($r < a$) is

$$R(r) = \frac{J_0\left((n_1^2 k_0^2 - \beta^2)^{1/2}\left(\dfrac{r}{a}\right)\right)}{J_0((n_1^2 k_0^2 - \beta^2)^{1/2})} \tag{7}$$

while the solution in the cladding ($r > a$) is the modified Bessel function

$$R(r) = \frac{K_0\left((\beta^2 - n_2^2 k_0^2)^{1/2}\left(\dfrac{r}{a}\right)\right)}{K_0((\beta^2 - n_2^2 k_0^2)^{1/2})} \tag{8}$$

Higher-order modes will have an increasing number of zero crossings in the cross section of the field distribution.

Fibers which allow more than one bound solution for each polarization are termed *multimode* fibers. Each mode will propagate with its own velocity and have a unique field distribution. Fibers with large cores and high numerical apertures will typically allow many modes to propagate. This often allows a larger amount of light to be transmitted from incoherent sources such as light-emitting diodes (LEDs). It typically results in higher attenuation and dispersion, as discussed in the following section.

By far the most popular fibers for long distance telecommunications applications allow only a single mode of each polarization to propagate. Records for low dispersion and attenuation have been set using single-mode fibers, resulting in length-bandwidth products exceeding 10 Gb-km/s. In order to restrict the guide to single-mode operation, the core diameter must typically be 10 μm or less. This introduces stringent requirements for connectors and splices and increases the peak power density inside the guide. As will be discussed, this property of the single-mode fiber enhances optical nonlinearities which can act to either limit or increase the performance of an optical fiber system.

10.4   *FIBER DISPERSION AND ATTENUATION*

Attenuation

In most cases, the modes of interest exhibit a complex exponential behavior along the direction of propagation z.

$$Z(z) = \exp(i\tilde{\beta}z) \tag{9}$$

β is generally termed the propagation constant and may be a complex quantity. The real part of β is proportional to the phase velocity of the mode in question, and produces a phase shift on propagation which changes rather rapidly with optical wavelength. It is often expressed as an effective refractive index for the mode by normalizing to the vacuum wave vector:

$$N_{\text{eff}} = \frac{Re\{\tilde{\beta}\}}{k_0} \tag{10}$$

The imaginary part of β represents the loss (or gain) in the fiber and is a weak (but certainly not negligible) function of optical wavelength. Fiber attenuation occurs due to fundamental scattering processes (the most important contribution is Rayleigh scattering), absorption (both the OH-absorption and the long-wavelength vibrational absorption), and scattering due to inhomogeneities arising in the fabrication process. Attenuation limits both the short- and long-wavelength applications of optical fibers. Figure 6 illustrates the attenuation characteristics of a typical fiber.

The variation of the longitudinal propagation velocity with either optical frequency or path length introduces a fundamental limit to fiber communications. Since signaling necessarily requires a nonzero bandwidth, the dispersion in propagation velocity between different frequency components of the signal or between different modes of a multimode fiber produces a signal distortion and intersymbol interference (in digital systems) which is unacceptable. Fiber dispersion is commonly classified as follows.

Intermodal Dispersion

The earliest telecommunications links as well as many modern data communications systems have made use of multimode fiber. These modes (which we have noted have some connection to geometrical ray angles) will typically have a broad range of propagation velocities. An optical pulse which couples to this range of guided modes will tend to broaden by an amount equal to the mean-squared difference in propagation time among

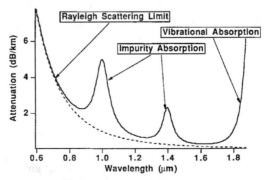

FIGURE 6   Attenuation characteristics of a typical fiber: (*a*) schematic, showing the important mechanisms of fiber attenuation.

the modes. This was the original purpose behind the gradient-index fiber; the geometrical illustrations of Figs 1 and 2 show that, in the case of a step-index fiber, a higher-order mode (one with a steeper geometrical angle or a higher mode index m) will propagate by a longer path than an axial mode. A fiber with a suitable index gradient will support a wide range of modes with nearly the same phase velocity. Vassell was among the first to show this,[6] and demonstrated that a hyperbolic secant profile could very nearly equalize the velocity of all modes. The α-profile description eventually became the most popular due to the analytic expansions it allows (for certain values of α) and the fact that it requires the optimization of only a single parameter.

Multimode fibers are no longer used in long distance ($>10\,km$) telecommunications due to the significant performance advantages offered by single-mode systems. Many short-link applications, for which intermodal dispersion is not a problem, still make use of multimode fibers.

Material Dispersion

The same physical processes which introduce fiber attenuation also produce a refractive index which varies with wavelength. This intrinsic, or material, dispersion is primarily a property of the glass used in the core, although the dispersion of the cladding will influence the fiber in proportion to the fraction of guided energy which actually resides outside the core. Material dispersion is particularly important if sources of broad spectral width are used, but narrow linewidth lasers which are spectrally broadened under modulation also incur penalties from material dispersion. For single-mode fibers, material dispersion must always be considered along with waveguide and profile dispersion.

Waveguide and Profile Dispersion

The energy distribution in a single-mode fiber is a consequence of the boundary conditions at the core-cladding interface, and is therefore a function of optical frequency. A change in frequency will therefore change the propagation constant independent of the dispersion of the core and cladding materials; this results in what is commonly termed *waveguide dispersion*. Since dispersion of the core and cladding materials differs, a change in frequency can result in a small but measurable change in index profile, resulting in *profile dispersion* (this contribution, being small, is often neglected). Material, waveguide, and profile dispersion act together, the waveguide dispersion being of opposite sign to that of the material dispersion. There exists, therefore, a wavelength at which the total dispersion will vanish. Beyond this, the fiber exhibits a region of anomalous dispersion in which the real part of the propagation constant increases with increasing wavelength. Anomalous dispersion has been used in the compression of pulses in optical fibers and to support long distance soliton propagation.

Dispersion, which results in a degradation of the signal with length, combines with attenuation to yield a length limit for a communications link operating at a fixed bandwidth. The bandwidth-length product is often cited as a practical figure of merit which can include the effects of either a dispersion or attenuation limit.

Normalized Variables in Fiber Description

The propagation constant and dispersion of guided modes in optical fibers can be conveniently expressed in the form of normalized variables. Two common engineering problems are the determination of mode content and the computation of total dispersion. For example, commonly available single-mode fibers are designed for a wavelength range

TABLE 1 Normalized Variables in the Mathematical Description of Optical Fibers

Symbol	Description
$k_0 = \dfrac{2\pi}{\lambda}$	Vacuum wave vector
a	Core radius
n_0	Core index
n_1	Cladding index
$\tilde{\beta} = \beta' + i\beta''$	Mode propagation constant
$\alpha = 2\beta''$	Fiber attenuation
$N_{\text{eff}} = \beta'/k_0$	Effective index of mode
$\Delta = \dfrac{n_0^2 - n_1^2}{2n_1^2}$	Normalized core-cladding index differences
$V = \sqrt{2k_0}\, an_1\Delta$	Normalized frequency
$b = \left(\dfrac{N_{\text{eff}}}{n_1} - 1\right)\Big/\Delta$	Normalized effective index
$f(r)$	Gradient-index shape factor
$\Gamma = \dfrac{\displaystyle\int_0^a f(r)\Psi^2(r)r\,dr}{\displaystyle\int_0^a \Psi^2(r)r\,dr}$	Profile parameter ($\Gamma = 1$ for step-index)

of 1.3 to 1.55 μm. Shorter wavelengths will typically support two or more modes, resulting in significant intermodal interference at the output. In order to guarantee single-mode performance, it is important to determine the single-mode cut-off wavelength for a given fiber. Normalized variables allow one to readily determine the cut-off wavelength and dispersion limits of a fiber using universal curves.

The normalized variables are listed in Table 1 along with the usual designations for fiber parameters. The definitions here apply to the limit of the "weakly guiding" fiber of Gloge,[7] for which $\Delta \ll 1$. The cutoff for single-mode performance appears at a normalized frequency of $V = 2.405$. For values of V greater than this, the fiber is multimode. The practical range of frequencies for good single-mode fiber operation lie in the range

$$1.8 < V < 2.4 \tag{11}$$

An analytic approximation for the normalized propagation constant b which is valid for this range is given by

$$b(V) \approx \left(1 - 1.1428 - \frac{0.996}{V}\right)^2 \tag{12}$$

Operation close to the cutoff $V = 2.405$ risks introducing higher-order modes if the fiber parameters are not precisely targeted. A useful expression which applies to step-index fibers relates the core diameter and wavelength at the single-mode cutoff[5]:

$$\lambda_{\text{cutoff}} = \left(\frac{\pi}{2.405}\right)(2a)n_0\sqrt{2\Delta} \tag{13}$$

Evaluation of Fiber Dispersion

Evaluation of the fiber dispersion requires:

1. Detailed material dispersion curves such as may be obtained from a Sellmeier formula.[4] The Sellmeier constants for a range of silica-based materials used in fiber fabrication are contained in Chap. 33 of Vol. II, "Crystals and glasses."

2. Complete information about the fiber profile, including compositional as well as refractive index information.

3. Numerical evaluation of the effective indices of the modes in question *and their first and second derivatives.* Several authors have noted the considerable numerical challenge involved in this,[8,9] particularly since measurements of the refractive index/composition possess intrinsic uncertainties.

Figure 7 shows an example of the dispersion exhibited by a step-index single-mode fiber. Different components of the dispersion are shown in order to illustrate the point of zero dispersion near 1.3 μm. The section devoted to fiber properties will describe how profile control can shift the minimum dispersion point to the very low-loss window near 1.55 μm.

10.5 *POLARIZATION CHARACTERISTICS OF FIBERS*

The cylindrical symmetry of an optical fiber leads to a natural decoupling of the radial and tangential components of the electric field vector. These polarizations are, however, so nearly degenerate that a fiber of circular symmetry is generally described in terms of orthogonal linear polarizations. This near-degeneracy is easily broken by any stresses or imperfections which break the cylindrical symmetry of the fiber. Any such symmetry breaking (which may arise accidentally or be introduced intentionally in the fabrication

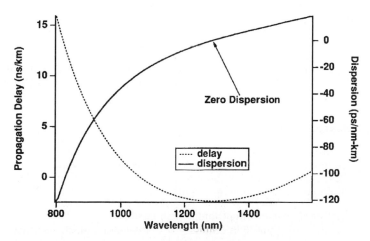

FIGURE 7 Dispersion of a typical single-mode fiber. The opposite contributions of the waveguide and material dispersion cancel near $\lambda = 1.3$ μm. (*Courtesy of Corning, Inc.*)

process) will result in two orthogonally polarized modes with slightly different propagation constants. These two modes need not be linearly polarized; in general, they are two elliptical polarizations. Such polarization splitting is referred to as *birefringence.*

The difference in effective index between the two polarizations results in a state of polarization (SOP) which evolves through various states of ellipticity and orientation. After some propagation distance, the two modes will differ in phase by a multiple of 2π, resulting in a state of polarization identical to that at the input. This characteristic length is called the *beat length* between the two polarizations and is a measure of the intrinsic birefringence in the fiber. The time delay between polarizations is sometimes termed *polarization dispersion,* because it can have an effect on optical communication links which is similar to intermodal dispersion.

If this delay is much less than the coherence time of the source, coherence is maintained and the light in the fiber remains fully polarized. For sources of wide spectral width, however, the delay between the two polarizations may exceed the source coherence time and yield light which emerges from the fiber in a partially polarized or unpolarized state. The orthogonal polarizations then have little or no statistical correlation. The state of polarization of the output can have an important impact on systems with polarizing elements. For links producing an unpolarized output, a 3-dB power loss is experienced when passing through a polarizing element at the output.

The intentional introduction of birefringence can be used to provide polarization stability. An elliptical or double-core geometry will introduce a large birefringence, decoupling a pair of (approximately) linearly polarized modes.[10,11] It also will tend to introduce loss discrimination between modes. This combination of birefringence and loss discrimination is the primary principle behind polarization-maintaining fiber. As will be discussed in the description of optical fiber systems, there is a class of transmission techniques which requires control over the polarization of the transmitted light, and therefore requires polarization-maintaining fiber.

10.6 OPTICAL AND MECHANICAL PROPERTIES OF FIBERS

This section contains brief descriptions of fiber measurement methods and general information on fiber attenuation, dispersion, strength, and reliability. It should be emphasized that nearly all optical and mechanical properties of fibers are functions of chemistry, fabrication process, and transverse structure. Fibers are now well into the commercial arena and specific links between fiber structure, chemistry, and optical and mechanical properties are considered highly proprietary by fiber manufacturers. On the other hand, most fiber measurements now have established standards. We therefore give attention to the generic properties of fibers and the relevant evaluation techniques.

Attenuation Measurement

There are two general methods for the measurement of fiber attenuation. Source-to-fiber coupling must be taken into account in any scheme to measure attenuation, and destructive evaluation accomplishes this rather simply. The *cut-back method*[12,13] for attenuation measurement requires

1. Coupling light into a long length of fiber
2. Measuring the light output into a large area detector (so fiber-detector coupling remains constant)
3. Cutting the fiber back by a known distance and measuring the change in transmitted intensity

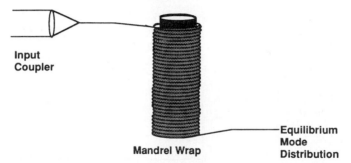

FIGURE 8 Mandrel wrap method of achieving an equilibrium mode distribution.

For single-mode fiber, the fiber can be cut back to a relatively short length provided that the cladding modes are effectively stripped. The concept of "mode stripping" is an important one for both attenuation and bandwidth measurements[14] (since modes near or just beyond cutoff can propagate some distance but with very high attenuation). If these modes are included in the measurement, the result yields an anomalously high attenuation. Lossy modes can be effectively stripped by a mandrel wrap or a sufficiently long length of fiber well-matched to the test fiber (see Fig. 8).

For multimode fiber (whether step-index or gradient-index) the excitation conditions are particularly important. This is because the propagating modes of a multimode fiber exhibit widely varying losses. If the laser used for performing the measurement is focused to a tight spot at the center of the core, a group of low-order modes may be initially excited. This group of lower-order modes will have lower loss and the first 10 to 1000 meters will show an anomalously low attenuation. As the propagation distance increases, lower-order modes gradually scatter into higher-order modes and the mode volume "fills up." The high-order modes are substantially lossier, so the actual power flow at equilibrium is that from the lower-order modes to the higher-order and out of the fiber. This process is illustrated in Fig. 9. It is easy to see that if the excitation conditions are set so that all modes guide approximately the same power at the input, the loss in the first hundred meters would be much higher than the equilibrium loss.

With modern single-mode splices, connectors, and couplers, it is sometimes possible to make nondestructive attenuation measurements simply by assuring that the connector loss is much less than the total loss of the fiber length being measured. With this method, care must be taken that the connector design exhibits no interference between fiber endfaces.

Connector loss measurements must have similar control over launch conditions. In addition, it is important to place a sufficiently long length of fiber (or short mandrel wrap) *after* the connector to strip the lossy modes. A slightly misaligned connector will often

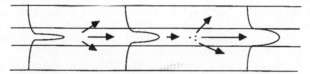

FIGURE 9 In a multimode fiber, low-order modes lose power to the high-order modes, and the high-order modes scatter into cladding and other lossy modes.

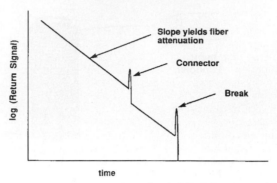

FIGURE 10 Typical OTDR signal. OTDR can be used for attenuation measurement, splice and connector evaluation, and fault location.

exhibit an extremely low loss prior to mode stripping. This is because power is coupled into modes which, while still guided, have high attenuation. It is important, in evaluation of fibers, to properly attribute this loss to the connector and not to the length of fiber which follows.

Another method of nondestructive evaluation of attenuation is optical time domain reflectometry (OTDR). The excitation of a fiber with a narrow laser pulse produces a continuous backscatter signal from the fiber. Assuming a linear and homogeneous scattering process, the reduction in backscattered light with time becomes a map of the round-trip attenuation versus distance. Sudden reductions in intensity typically indicate a splice loss, while a narrow peak will usually indicate a reflection. A typical OTDR signal is shown in Fig. 10. OTDR is extremely sensitive to excitation conditions—a fiber which is not properly excited will often exhibit anomalous behavior. Control of the launch conditions is therefore important for all methods of attenuation measurement.

A major theme of research and development in optical telecommunications has been the elimination of troublesome reflections from optical networks. In particular, high-return loss connectors have been developed which exhibit 30 to 40 dB of reflection suppression.[15–18] OTDR can be used to assess the reflection at network connections as well as perform on-line fault monitoring.

Dispersion and Bandwidth Measurement

The fiber has often been presented as the "multi-TeraHertz bandwidth transmission channel." While it is true that the total attenuation window of the fiber is extremely large by communications standards, the actual information bandwidth at any given wavelength is limited by the various sources of dispersion. The bandwidth of the fiber can be measured either in the time or frequency domain. Both measurements assume the fiber to be linear in its baseband (intensity) transfer characteristics. This assumption breaks down at very high powers and short pulses, but is nevertheless useful in most system applications.

The *time domain measurement*[19] measures the temporal broadening of a narrow input pulse. The ratio of the Fourier transform of the output intensity to that of the input yields a baseband transfer function for the fiber. If the laser and detector are linear, this transfer function relates the drive current of the laser to the photocurrent of the receiver and treats the fiber simply as a linear transmission channel of limited bandwidth. The use of the Fourier transform readily allows the phase to be extracted from the baseband transfer function. For intermodal pulse broadening in multimode fibers, this phase can be a nonlinear function of frequency, indicating a distortion as well as a broadening of the optical pulse.

Swept-frequency methods[20] have also been used for fiber evaluation. A pure sinusoidal modulation of the input laser is detected and compared in amplitude (and phase, if a network analyzer is available). In principle, this yields a transfer function similar to the pulse method. Both rely on the linearity of the laser for an accurate estimation, but since the swept-frequency method generally uses a single tone, the harmonics produced by laser nonlinearities can be rejected. Agreement between the two methods requires repeatable excitation conditions, a nontrivial requirement for multimode fibers.

The usual bandwidth specification of a multimode fiber is in the form of a 3-dB bandwidth (for a fixed length) or a length-bandwidth product. A single-mode fiber is typically specified simply in terms of the measured total dispersion. This dispersion can be measured either interferometrically, temporally, or using frequency domain techniques.

The *interferometric measurement*[21,22] is appropriate for short fiber lengths, and allows a detailed, direct comparison of the optical phase shifts between a test fiber and a reference arm with a suitable delay. This approach is illustrated in Fig. 11, which makes use of a Mach-Zehnder interferometer. This requires a source which is tunable, and one with sufficient coherence to tolerate small path differences between the two arms. The advantage of the approach is the fact that it allows measurements of extremely small absolute delays (a shift of one optical wavelength represents less than 10 fs time delay). It

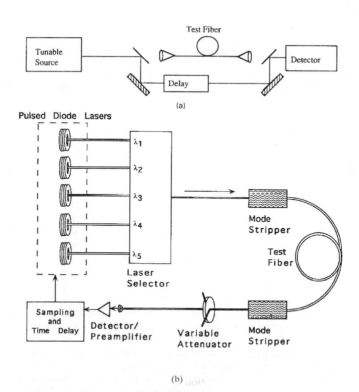

FIGURE 11 (*a*) Interferometric measurement of fiber dispersion; (*b*) time delay measurement of fiber dispersion.

tends to be limited to rather short lengths of fiber; if a fiber is used in the reference arm to balance the interferometer, the properties of that fiber must be known with some accuracy.

Time-domain measurements[23] over a broad spectral range can be made provided a multiwavelength source is available with a sufficiently short optical pulse. One can make use of a series of pulsed diode lasers spaced at different wavelengths, use Raman scattering to generate many wavelengths from a single source, or make use of a tunable, mode-locked solid state laser. The relative delay between neighboring wavelengths yields the dispersion directly. This technique requires fibers long enough to adequately measure the delay, and the optical pulses must be weak enough not to incur additional phase shifts associated with fiber nonlinearities.

Frequency-domain or phase-shift measurements attempt to measure the effects of the dispersion on the baseband signal. A sinusoidally modulated signal will experience a phase shift with propagation delay; that phase shift can be readily measured electronically. This technique uses a filtered broadband source (such as an LED) or a CW, tunable, solid state source to measure the propagation delay as a function of wavelength.

Shifting and Flattening of Fiber Dispersion

A major dilemma facing system designers in the early 1980s was the choice between zero dispersion at 1.3 μm and the loss minimum at 1.55 μm. The loss minimum is an indelible consequence of the chemistry of silica fiber, as is the material dispersion. The waveguide dispersion can, however, be influenced by suitable profile designs.[24] Figure 12 illustrates a generic design which has been successfully used to shift the dispersion minimum to 1.55 μm.

The addition of several core and cladding layers to the fiber design allows for more complicated dispersion compensation to be accomplished. *Dispersion-flattened* fiber is designed for very low dispersion in an entire wavelength range; the spectral region from 1.3 to 1.6 μm is the usual range of interest. This is important for broadband WDM applications, for which the fiber dispersion must be as uniform as possible over a wide spectral region.

Reliability Assessment

The reliability of an optical fiber is of paramount importance in communications applications—long links represent large investments and require high reliability. There will, of course, always be unforeseen reliability problems. Perhaps the most famous such example was the fiber cable design on the first transatlantic link—the designers had not quite appreciated the Atlantic shark's need for a high-fiber diet. The sharks, apparently attracted by the scent of the cable materials, made short work of the initial cable installations. However, most of the stresses which an optical fiber will experience in the field can be replicated in the laboratory. A variety of accelerated aging models (usually relying on temperature as the accelerating factor) can be used to test for active and passive component reliability. In this section, we will review the reliability assessment of the fiber itself, referring interested readers to other sources for information on cable design.

Among the most important mechanical properties of the fiber in a wide range of applications is the tensile strength.[25] The strength is primarily measured destructively, by finding the maximum load just prior to fracture.[26] Full reliability information requires a knowledge of the maximum load, the relation between load and strain, a knowledge of the strain experienced by the fully packaged fiber, and some idea of how the maximum tolerable strain will change over long periods of time. One must finally determine the strain and associated failure probability for fibers with finite bends.

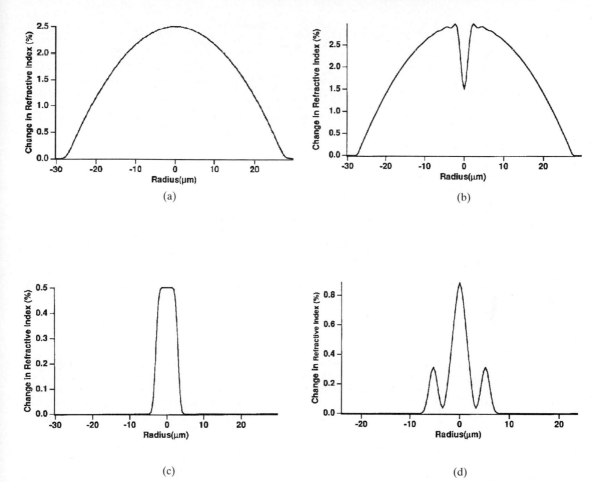

FIGURE 12 Typical index profiles for (*a*), (*b*) gradient-index multimode fiber; (*c*) step-index single-mode fiber; (*d*) dispersion-shifted fiber.

The tensile strength typically decreases slowly over time as the material exhibits fatigue, but in some cases can degrade rather rapidly after a long period of comparative strength. The former behavior is usually linked to fatigue associated with purely mechanical influences, while the latter often indicates chemical damage to the glass matrix. The strain ε and tensile loading F are related through the fiber cross section and Young's modulus:[27]

$$\varepsilon = \frac{F}{\sum_i E_i A_i} \tag{14}$$

E_i and A_i represent the Young's modulus and cross-sectional area of the i_{th} layer of the fiber-jacketing combination. Thus, if the Young's moduli are known, a measurement of the load yields the strain.

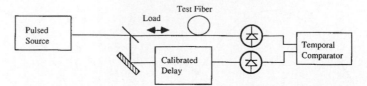

FIGURE 13 Single-pass technique for time-domain measurement of fiber strain.

It is sometimes helpful to measure the fiber strain directly in cases where either the load or Young's moduli are not known. For example, a fiber does not necessarily have a uniform load after jacketing, cabling, and pulling; the load would (in any case) be a difficult quantity to measure. Using the relation between the strain and the optical properties of the fiber it is possible to infer the fiber strain from optical measurements. These techniques have been successful enough to lead to the development of fiber strain gauges for use in mechanical systems.

Optical measurements of strain make use of the transit time of light through a medium of refractive index N_{eff}. (We will, for simplicity, assume single-mode propagation.) A change in length ΔL produced by a strain $\Delta L/L$ will yield a change in transit time

$$\frac{\Delta \tau}{\Delta L} = \frac{N_{\text{eff}}}{c}\left(1 + \frac{L}{N_{\text{eff}}}\frac{dN_{\text{eff}}}{dL}\right) \tag{15}$$

For most cases of interest, the effective index is simply taken to be the value for that of the core. The ratio $\Delta \tau/\Delta L$ can be calculated (it is about 3.83 ns/m for a germania-silica fiber with $\Delta = 1\%$) or calibrated by using a control fiber and a measured load. It is important to note that this measurement yields only information on the *average* strain of a given fiber length.

There are three categories of optoelectronic techniques for measuring $\Delta \tau$; these are very similar to the approaches for dispersion measurement. A single-pass optical approach generally employs a short-pulse laser source passing through the fiber, with the delay of the transmitted pulse deduced by a comparison with a reference (which presumably is jitter-free). This is shown in Fig. 13. Figure 14a shows a multipass optoelectronic scheme, in which an optoelectronic oscillator circuit is set up with the fiber as a delay loop. The Q of the optoelectronic oscillator determines the effective number of passes in this measurement of optical delay. Finally, one can use an all-optical circuit in which the test fiber is placed in a fiber loop with weak optical taps to a laser and detector/signal processor (Fig. 14b). This "ring resonator" arrangement can also be set up with a fiber amplifier in the resonator to form the all-optical analog of the multipass optoelectronic scheme of Fig. 14a.

If the strain is being used to gain information about fiber reliability, it is necessary to understand how strain, load, and fiber failure are related. Fatigue, the delayed failure of the fiber, appears to be the primary model for fiber failure. One experimental evaluation of this process is to measure the mean time to failure as a function of the load on the fiber with the temperature, the chemical environment, and a host of other factors serving as control parameters.

Since the actual time to failure represents only the average of a performance distribution, the reliability of manufactured fibers is sometimes specified in terms of the two-parameter Weibull distribution[25,27–30]

$$P_f = 1 - \exp\left\{\left(\frac{l}{l_0}\right)\left(\frac{S}{S_0}\right)^m\right\} \tag{16}$$

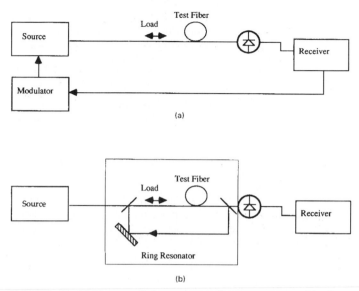

FIGURE 14 Multipass techniques for strain measurement. (*a*) Opto-electronic oscillator; (*b*) optical ring resonator.

where P_f denotes the cumulative failure probability and the parameters are as defined in Table 2. The Weibull exponent m is one of the primary descriptors of long-term fiber reliability. Figure 15 shows a series of Weibull plots associated with both bending and tensile strength measurements for low, intermediate, and high values of m.

One factor which has been shown to have a strong impact on reliability is the absolute humidity of the fiber environment and the ability of the protective coating to isolate the SiO_2 from the effects of H_2O. A recent review by Inniss, Brownlow, and Kurkjian[31] pointed out the correlation between a sudden change in slope, or "knee," in the time-to-failure curve and the H_2O content—a stark difference appeared between liquid and vapor environments. Before this knee, a combination of moisture and stress are required for fiber failure. In the case of fiber environments with a knee, a rather early mean time to failure will exist even for very low fiber stresses, indicating that chemistry rather than mechanical strain is responsible for the failure. The same authors investigated the effects of sodium solutions on the strength and aging of bare silica fibers.

10.7 OPTICAL FIBER COMMUNICATIONS

The optical fiber found its first large-scale application in telecommunications systems. Beginning with the first LED-based systems,[32,34,35] the technology progressed rapidly to longer wavelengths and laser-based systems of repeater lengths over 30 km.[36] The first applications were primarily digital, since source nonlinearities precluded multichannel analog applications. Early links were designed for the 800- to 900-nm window of the optical fiber transmission spectrum, consistent with the emission wavelengths of the GaAs-AlGaAs materials system for semiconductor lasers and LEDs. The development of sources and detectors in the 1.3- to 1.55-μm wavelength range and the further improvement in optical fiber loss over those ranges has directed most applications to either the 1.3-μm window (for low dispersion) or the 1.55-μm window (for minimum loss). The

TABLE 2 Variables Used in the Weibull Distribution

l	Fiber length
l_0	Length normalization factor
S	Failure stress
S_0	Location parameter
m	Weibull exponent

design of dispersion-shifted single-mode fiber along the availability of erbium-doped fiber amplifiers has solidified 1.55 µm as the wavelength of choice for high-speed communications.

The largest currently emerging application for optical fibers is in the local area network (LAN) environment for computer data communications, and the local subscriber loop for telephone, video, and data services for homes and small businesses. Both of these applications place a premium on reliability, connectivity, and economy. While existing systems still use point-to-point optical links as building blocks, there is a considerable range of networking components on the market which allow splitting, tapping, and multiplexing of optical components without the need for optical detection and retransmission.

Point-to-Point Links

The simplest optical communications system is the single-channel (no optical multiplexing) point-to-point digital link. As illustrated in Fig. 16, it consists of a diode laser (with associated driver circuitry and temperature control), optical fiber (with associated splices, connectors, and supporting material), and a detector (with appropriate electronics for signal processing and regeneration). The physics and principles of operation of the laser and detector are covered elsewhere in this collection (see Chap. 11 of Vol. I, "Lasers" Chap. 15 of Vol. I, "Photodetectors"), but the impact of certain device characteristics on the optical fiber communications link is of some importance.

Modulation and Source Characteristics. For information to be accurately transmitted, an appropriate modulation scheme is required. The most common modulation schemes employ direct modulation of the laser drive current, thereby achieving a modulation depth of 80 percent or better. The modulation depth is defined as

$$m = \frac{P_{\max} - P_{\min}}{P_{\max} + P_{\min}} \tag{17}$$

where $P_{\min}$ and $P_{\max}$ are the minimum and maximum laser power, respectively. The modulation depth is limited by the requirement that the laser always remain above threshold, since modulation near the lasing threshold results in a longer turn-on time, a broader spectrum, and higher source noise.

The transmitting laser contributes noise to the system in a fashion that is, generally speaking, proportional to the peak transmitted laser power. This noise is always evaluated as a fraction of the laser power and is therefore termed *relative intensity noise* (RIN). The RIN contribution from a laser is specified in dB/Hz, to reflect a spectral density which is

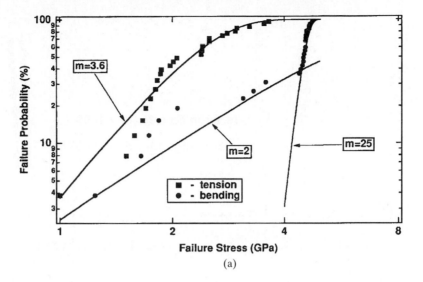

(a)

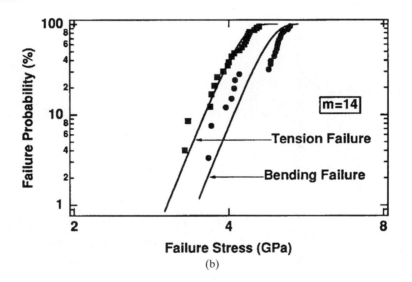

(b)

FIGURE 15 A series of Weibull plots comparing bending and tensile strength for (*a*) low, (*b*) intermediate, and (*c*) high values of the Weibull exponent *m*; (*d*) shows a typical mean time to failure plot. Actual fibers will often exhibit slope discontinuities, indicating a change in the dominant failure mechanism. (Data Courtesy of Corning, Inc.)

approximately flat and at a fixed ratio (expressed in dB) to the laser power. Figure 17 shows a typical plot of the relative intensity noise of a source. The specification of RIN as a flat noise source is valid only at frequencies much less than the relaxation oscillation frequency and in situations where reflections are small.

The relative intensity noise is affected rather dramatically by the environment of the

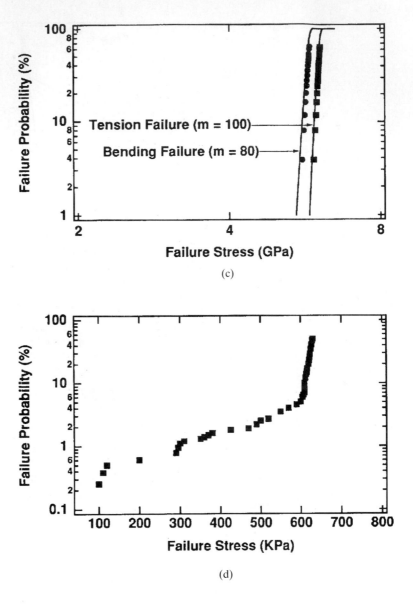

(c)

(d)

FIGURE 15 (*Continued*)

diode laser. A rather weak reflection back into the laser will both increase the magnitude of the relative intensity noise and modify its spectrum. As the reflection increases, it can produce self-pulsations and chaos in the output of the laser, rendering it useless for communications applications.[37] Thus, the laser cannot be thought of as an isolated component in the communications system. Just as RF and microwave systems require impedance matching for good performance, an optical communications system must minimize reflections. This is relatively easily accomplished for a long distance telecommunications link which makes use of low-reflection fusion splices. However, in a short

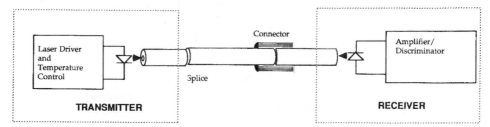

FIGURE 16   Typical point-to-point optical fiber communications link.

link-network environment which must be modular, a small number of connectors can cause severe problems unless those connectors are designed to minimize reflections. It is now widely accepted that optical fiber connectors must be specified both in terms of insertion loss and reflection. A 1 percent reflection from a fiber connector can have far more serious implications for an optical fiber link than a 1 percent loss which is not reflected back to the laser. Optical isolators are available but only at considerable expense and are not generally considered economically realistic for network environments.

Impact of Fiber Properties on a Communications Link.   For moderate power levels, the fiber is a passive, dispersive transmission channel. Dispersion can limit system performance in two ways. It results in a spreading of data pulses by an amount proportional to the spectral width of the source. This pulse spreading produces what is commonly termed "intersymbol interference." This should not be confused with an optical interference effect, but is simply the blurring of pulse energy into the neighboring time slot. In simple terms, it can be thought of as a reduction in the modulation depth of the signal as a function of link length. The effects of dispersion are often quantified in the form of a power penalty. This is simply a measure of the additional power required to overcome the effects of the dispersion, or bring the modulated power to what it would be in an identical link without dispersion. It is commonly expressed as a decibel ratio of the power required at the receiver compared to that of the ideal link.

Modulation-induced frequency chirp of the laser source will also result in pulse

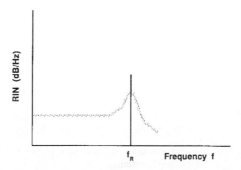

FIGURE 17   Typical RIN spectrum for a diode laser. The peak corresponds to the relaxation resonance frequency, f_R, of the laser.

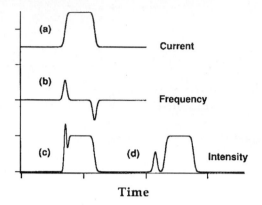

FIGURE 18 Modulation of the drive current in a semiconductor laser (*a*) results in both an intensity (*b*) and a frequency modulation (*c*). The pulse is distorted after transmission through the fiber (*d*).

distortion. This is illustrated in Fig. 18, in which the drive current of the laser is modulated. The accompanying population relaxation produces a frequency modulation of the pulse. Upon transmission through a dispersive link, these portions of the pulse which are "chirped" will be advanced or retarded, resulting in both pulse distortion and intersymbol interference.

System Design. The optical receiver must, within the signal bandwidth, establish an adequate signal-to-noise ratio (SNR) for accurate regeneration/retransmission of the signal. It must accomplish this within the constraints of the fiber dispersion and attenuation, the required system bandwidth, and the available source power. First-order system design normally requires the following steps:

1. Determine the maximum system bandwidth (or data rate for digital systems) and the appropriate transmission wavelength required for the system.

2. Find the maximum source RIN allowable for the system. For analog systems, in which a signal-to-noise ratio (SNR) must be specified in a bandwidth B_n, the RIN (which is usually specified in dB/Hz, indicating a measurement in a 1-Hz bandwidth) must obey the following inequality:

$$|\text{RIN(dB/Hz)}| \ll 10 \log (\text{SNR} \cdot B_n) \qquad (18)$$

The SNR is specified here as an absolute ratio of carrier power to noise power. For an SNR specified as a decibel ratio,

$$|\text{RIN(dB/Hz)}| \ll \text{SNR(dB)} + 10 \log (B_n) \qquad (19)$$

For digital systems, a Gaussian assumption allows a simple relationship between error probability (also termed bit error rate) and the signal-to-noise ratio:

$$P_E = 0.5 erfc[0.5(0.5\text{SNR})^{1/2}] \qquad (20)$$

Where *erfc* denotes the complementary error function and the decision threshold is assumed to be midway between the on and off states. The maximum error probability due to source noise should be considerably less than the eventual target error probability. For system targets of 10^{-9} to 10^{-12}, the error probability due to source RIN should be considerably less than 10^{-20}. This will allow at least a 3-dB margin to allow for increases in RIN due to device aging.

3. Establish a length limit associated with the source frequency chirp and spectral width. The frequency chirp α_f is specified in GHz per milliampere change in the drive current of the laser. A total current modulation I_m therefore yields a frequency deviation Δf of

$$\Delta f = I_m \alpha_f \tag{21}$$

This frequency deviation translates into a propagation delay via the fiber intramodal dispersion D. This delay must be kept less than the minimum pulse width (data rate). With D specified in ps/nm-km, the length in kilometers must obey the following inequality to avoid penalties due to frequency chirp:

$$L \ll \frac{c}{B \Delta f D \lambda_0^2} = \frac{c}{\alpha_f I_m B D \lambda_0^2} \tag{22}$$

where B denotes the data rate and is the reciprocal of the pulse width for data pulses that fill up an entire time slot. (These signals are designated non-return-to-zero, or NRZ.)

The length limit due to source spectral width $\Delta \nu$ obeys a similar inequality—in this case, the delay associated with the spectral spread of the source must remain much less than one pulse width:

$$L \ll \frac{c}{\Delta \nu B D \lambda_0^2} \tag{23}$$

If the chirp is low and the unmodulated source bandwidth is less than the system bandwidth being considered, one must require that the delay distortion of the signal spectrum itself be small compared to a pulse width, requiring

$$L \ll \frac{c}{B^2 D \lambda_0^2} \tag{24}$$

For multimode fiber systems, the limiting length will generally be associated with the intermodal dispersion rather than the material and waveguide dispersion. A length-bandwidth product is generally quoted for such fibers. With the length and bandwidth limits established, it is now possible to design, within those limits, a receiver which meets the necessary specifications.

4. Determine the minimum power required at the receiver to achieve the target SNR or error probability. This minimum acceptable power (MAP) is first computed assuming an ideal source (no RIN contribution). A correction for the RIN can be carried out later. A computation of the MAP requires a knowledge of the noise sources and detector bandwidth. It is conventional to express the noise sources in terms of equivalent input noise current sources. The noise sources of importance for such systems are: the shot noise of the photocurrent, dark current, and drain current (in the case of a field effect transistor (FET) preamplifier); the Johnson noise associated with the load resistor or equivalent amplifier input impedance; 1/f noise from certain classes of FETs. The noise contributions from amplifiers other than the first stage are generally second-order corrections. Figure 19 shows a schematic of the receiver and relevant noise sources. Table 3 gives expressions for, and definitions of the important physical quantities which determine the receiver sensitivity.

Figure 20 illustrates two possible configurations for the detector/amplifier combination.

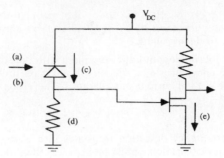

FIGURE 19 Schematic of the receiver, showing the introduction of noise into the system. Noise sources which may be relevant include (*a*) signal shot noise; (*b*) background noise (due to thermal background or channel crosstalk); (*c*) shot noise from the leakage current; (*d*) Johnson noise in the load resistor; (*e*) shot noise and 1/f noise in the drain current of the field effect transistor.

Of these, the integrating front end is the simplest (particularly for high-frequency operation) but tends to be slower than a transimpedance amplifier with an equivalent load resistance. This is because the transimpedance amplifier reduces the effective input impedance of the circuit by $(A + 1)$, where A denotes the open loop gain of the amplifier.

For equivalent bandwidth, the transimpedance amplifier exhibits a lower Johnson noise contribution since a higher feedback resistance is possible. It is worth mentioning that the transimpedance design tends to be much more sensitive to the parasitic capacitance which appears across the feedback resistor—small parasitics across the load resistor tend to be less important for the integrating front end.

The excess noise factor F_e is determined by the choice of detector. There are several choices over the wavelength range generally of interest for optical fiber transmission. (A detailed discussion of the principles of operation can be found in Chaps. 15–17 of Vol. I.)

TABLE 3 Symbols and Expressions for Receiver Noise

Symbol	Description
R_L	Load resistor
k	Boltzmann's constant
T	Temperature (Kelvin)
$\sigma_J^2 = \dfrac{4kT}{R_L} B_n$	Johnson noise power
R	Detector responsivity (A/W)
$P_g P$	Signal power
B_n	Noise bandwidth of amplifier
$\sigma_s^2 = 2eRP_s B_n F e$	Signal shot noise
i_d	Leakage current (dark)
$\sigma_d^2 = 2ei_d B_n$	Shot noise due to leakage current
$\sigma_R^2 = R^2 P_s^2 B_n \times 10^{-(\text{RIN}/10)}$	Receiver noise due to source RIN
F_e	Excess noise factor (for APD)
σ_λ^2	Amplifier noise

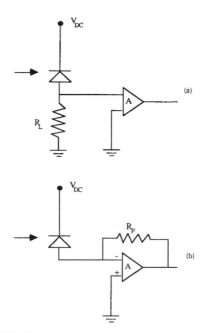

FIGURE 20 Two possible configurations for the detector/amplifier: (*a*) the integrating front end yields the simplest design for high speed operation; (*b*) the transimpedance amplifier provides an expansion of the receiver bandwidth by a factor of $A + 1$, where A is the open loop gain of the amplifier.

1. The p-i-n photodiode is described in some detail in Chap. 15 of Vol. I ("Photodetectors"). It can provide high quantum efficiencies and speeds in excess of 1 GHz. Dark currents range from less than 1 nA for silicon devices to 1 μA or more for Ge diodes. The dark current increases and the device slows down as the active area is increased.

2. The avalanche photodiode is a solid state device which exhibits internal multiplication of the photocurrent in a fashion that is sometimes compared with the gain in photomultiplier tubes. The multiplication does not come without a penalty, however, and that penalty is typically quantified in the form of an excess noise factor which multiplies the shot noise. The excess noise factor is a function both of the gain and the ratio of impact ionization rates between electrons and holes.

Figure 21 shows the excess noise factor for values of k ranging from 50 (large hole multiplication) to 0.03 (large electron multiplication). The former is claimed to be typical of certain III–V compounds while the latter is typical of silicon devices. Germanium, which would otherwise be the clear detector of choice for long wavelengths, has the unfortunate property of having k near unity. This results in maximum excess noise, and Ge avalanche photodiodes must typically be operated at low voltages and relatively small gains. The choice of a p-i-n detector, which exhibits no internal gain, yields $F_e = 1$.

3. The need for very high speed detectors combined with the fabrication challenges present in III-V detector technology has led to a renewed interest in Schottky barrier detectors for optical communications. A detector of considerable importance today is the metal-semiconductor-metal detector, which can operate at extremely high speed in an

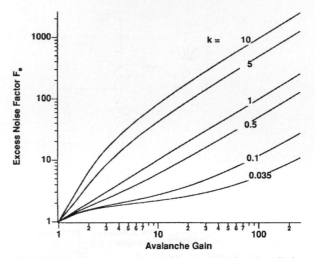

FIGURE 21 Excess noise factor for an avalanche photodiode with the electron/hole multiplication ratio k as a parameter. Small values of k indicate purely electron multiplication while large values of k indicate purely hole multiplication.

interdigitated electrode geometry. Chapter 17 of Vol. I provides further discussion of MSM detectors.

With all noise sources taken into account (see Table 3 for the relevant expressions), the signal-to-noise ratio of an optical receiver can be expressed as follows:

$$\text{SNR} = \frac{R^2 P_S^2}{\sigma_S^2 + \sigma_T^2} \tag{25}$$

where σ_T^2 denotes the total signal-independent receiver noise:

$$\sigma_T^2 = \sigma_D^2 + \sigma_J^2 + \sigma_A^2 \tag{26}$$

and σ_S^2 is the signal shot noise as in Table 3. If the effects of RIN are to be included, the following correction to the SNR may be made:

$$\text{SNR}^{-1} = \text{SNR}^{-1} + \sigma_R^2 \tag{27}$$

With the signal-to-noise ratio determined, the error probability may be expressed in terms of the signal-to-noise ratio

$$P_E = 0.5 erfc[0.5(0.5 \times \text{SNR})^{1/2}] \tag{28}$$

The above expressions assume Gaussian distributed noise sources. This is a good assumption for nearly all cases of interest. The one situation in which the Gaussian assumption underestimates the effects of noise is for avalanche photodiodes with large excess noise. It was shown by McIntyre[38,39] and Personick[40] that the avalanche multiplication statistics are skewed and that the Gaussian assumption yields overly optimistic results.

5. Given the MAP of the receiver, the fiber attenuation and splice loss budget, and the available pigtailed laser power (the maximum power coupled into the first length of fiber by the laser), it is possible to calculate a link loss budget. The budget must include a substantial power margin to allow for device aging, imperfect splices, and a small measure of stupidity. The result will be a link length which, if shorter than the dispersion limit, will provide an adequate signal-to-noise ratio.

For further link modeling, a variety of approaches can be used to numerically simulate the link performance and fully include the effects of fiber dispersion, a realistic detector-preamplifier combination, and a variety of other factors which the first-order design does not include. Nevertheless, a first-order design is necessary to reduce the range of free parameters used in the simulation.

The ultimate goal of the point-to-point link is to transparently transmit the data (or the analog signal) in such a way that standard communications techniques may be used in the optical link. Examples include the use of block or error-correcting codes in digital systems, standard protocols for point-to-point links between nodes of a network, or frequency allocation in the case of a multichannel analog link.

Advanced Transmission Techniques

The optical bandwidth available in either of the low-loss transmission windows of the fiber exceeds 10^{13} Hz. Two ways of taking full advantage of this bandwidth are through the use of ultrashort pulse transmission combined with time-division multiplexing or the use of wavelength/frequency-division multiplexing. Either technique can overcome the limits imposed by the channel dispersion, but both techniques have their limitations. The first technique seeks to turn fiber dispersion to advantage; the second attempts to simply reduce the negative effects of dispersion on a broadband optical signal.

Ultrashort Pulse Transmission. The most common form of multiplexing in digital communication systems is the combination of a number of low data rate signals into a single, high data rate signal by the use of time-division multiplexing. This requires much shorter optical pulses than are used in conventional transmission. As mentioned earlier, the normal (linear) limitation to the data rate is imposed by the fiber attenuation and dispersion. Both of these limits can be exceeded by the use of soliton transmission and optical amplification.

The physics of soliton formation[41–45] is discussed in "Nonlinear Optical Properties of Fibers," later in this chapter. Solitons, in conjunction with fiber amplifiers, have been shown to promise ultralong distance transmission without the need for optoelectronic repeaters/regenerators. Time-division multiplexing of optical solitons offers the possibility of extremely long distance repeaterless communications.

No communication technique is noise-free, and even solitons amplified by ideal amplifiers will exhibit phase fluctuations which broaden the spectrum and eventually cause the soliton to break up. This spontaneous-emission noise limit is known as the Gordon-Haus limit,[46] and had been thought to place a rather severe upper limit on the bit rate distance product for optical fiber systems. It has recently been noted,[47] that a unique series of linear filters can prevent the buildup of unwanted phase fluctuations in the soliton, thereby justifying amplified soliton transmission as a viable technology for undersea communications.

Such a communications system puts great demands on the signal processing both at the input and the output. For very high bit rates, one needs either all-optical demultiplexing or extremely fast electronic logic. Current limits on silicon logic are in the range of several

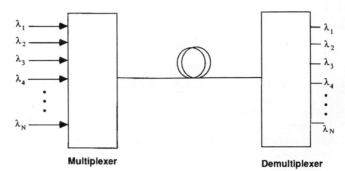

FIGURE 22 Schematic of a WDM transmission system. The main figures of merit are insertion loss (for both the multiplexer and demultiplexer) and channel crosstalk (for the demultiplexer).

Gb/s, which may be adequate for the first implementations of soliton transmission. It is anticipated that all-optical multiplexing and demultiplexing will be required in order to fully exploit the optical fiber bandwidth.

Solitons supported by an optical fiber bear a very specific relationship between pulse width T_0, peak power P_0, fiber dispersion D, effective area A_{eff}, and the intensity-dependent refractive index n_2. For a lowest-order ($N = 1$) soliton,

$$T_0^2 = \frac{\lambda^3 D}{(2\pi)^2 n_2 (P_0/A_{\text{eff}})} \tag{29}$$

Under normal operation, a fiber will propagate lowest-order solitons of about 10 ps in duration. Even for a pulse train of comparatively high duty cycle, this represents less than 100 GHz of a much larger fiber bandwidth. To fully span the bandwidth requires wavelength-division multiplexing.

Wavelength-division Multiplexing (WDM).   The troublesome delay between frequencies which is introduced by the fiber dispersion can also be overcome by division of the fiber transmission region into mutually incoherent (uncorrelated) wavelength channels. It is important for these channels to be uncorrelated in order to eliminate any worry about dispersion-induced delay between channels. Figure 22 shows a schematic picture of a WDM transmission system. The concept is quite simple, but reliable implementation can be a considerable challenge.

An attractive feature of WDM is the fact that the only active components of the system remain the optical sources and detectors. The multiplexers/demultiplexers are passive and are therefore intrinsically more reliable than active multiplexers. These schemes range from simple refractive/reflective beam combiners to diffractive approaches and are summarized in Fig. 23. For a multiplexing scheme, the key figure of merit is the insertion loss per channel. A simple 50-50 beam splitter for a two-channel combiner offers simple multiplexing with high insertion loss. If the beam splitter is coated to provide high reflectivity at one wavelength and high transmissivity at the other, the insertion loss is reduced, the coupler becomes wavelength-specific, and the element can act either as a multiplexer or demultiplexer.

Grating combiners offer an effective way to maximize the number of channels while still controlling the insertion loss. The grating shape must be appropriately designed—a problem which is easily solved for a single-wavelength, single-angle geometry. However, the diffraction efficiency is a function both of wavelength and angle of incidence. The optimum combination of a range of wavelengths over a wide angular range will typically

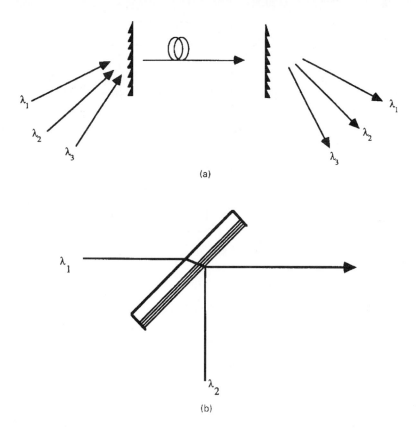

FIGURE 23 Multiplexing/demultiplexing schemes for WDM; (*a*) grating combiner (bulk optics); (*b*) wavelength selective beamsplitter (bulk optics); (*c*) directional coupler (integrated optics); (*d*) all-fiber multiplexer/demultiplexer.

require a tradeoff between insertion loss, wavelength range, and angular discrimination. Wavelength-division multiplexing technology has been greatly aided by the rapid advances in diffractive optics, synthetic holography, and binary optics in recent years. More on these subjects is included in Chap. 8 of Vol. II.

There have been considerable accomplishments in the past ten years in the fabrication of integrated optical components for WDM applications. Much of these involve the waveguide equivalent of bulk diffractive optical elements. Since the optical elements are passive and efficient fiber coupling is required, glass waveguides have often been the medium of choice. A great variety of couplers, beam splitters, and multiplexer/demultiplexers have been successfully fabricated in ion-exchanged glass waveguides. Further details on the properties of these waveguides is contained in Chap. 36 of Vol. I. There has also been a major effort to fabricate low-cost polymer-based WDM components. These can be in the form of either waveguides or fibers.

From the point of view of connectivity and modular design, all-fiber WDM components are the most popular. Evanescent single-mode fiber couplers are inherently wavelength-sensitive and can be designed for minimum insertion loss. As with the bulk approaches, all-fiber components become more difficult to design and optimize as the number of

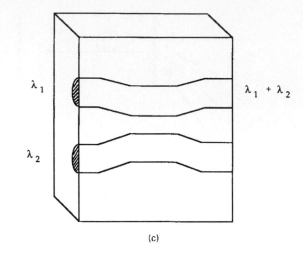

(c)

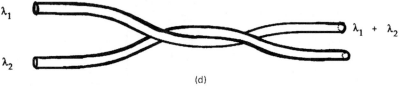

(d)

FIGURE 23 *(Continued)*

channels increases. Most commercially available all-fiber components are designed for widely separated wavelength channels. For example, Corning, Inc. currently offers multiplexers designed for combining signals from 1.5-μm, 1.3-μm, and 0.8-μm sources.

Advances in source fabrication technology in recent years have offered the possibility of fabricating diode laser arrays equipped with a controlled gradient in emission wavelength across the array. Such an array, equipped with appropriate beam-combining optics, could greatly reduce the packaging and alignment requirements in a large-scale WDM system. Minimizing crosstalk for closely spaced wavelength channels presents a significant challenge for demultiplexer design.

Coherent Optical Communications. Intensity modulation with direct detection remains the most popular scheme for optical communications systems. Under absolutely ideal transmission and detection conditions (no source RIN, perfect photon-counting detection, no background radiation), the probability of detecting n photons in a pulse train having an average of N_P photons per pulse would obey the Poisson distribution

$$p(n) = \frac{N_p^n e^{-N_P}}{n!} \tag{30}$$

The probability of an "error" P_E would be the detection of no photons during the pulse,

$$P_E = \exp{(-N_P)} \tag{31}$$

If we choose the benchmark error probability of 10^{-9}, we require an average of about 21 photons per pulse. This represents the quantum limit for the direct detection of optical

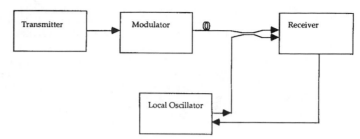

FIGURE 24   Generic coherent optical fiber communication link.

signals. This limit can scarcely be reached, since it assumes no dark count and perfectly efficient photon counting.

Current optical communication[48–54] offers a way to achieve quantum-limited receiver sensitivities even in the presence of receiver noise. By using either amplitude, phase, or frequency modulation combined with heterodyne or homodyne detection, it is possible to approach, and even exceed, the quantum limit for direct detection.

A generic coherent optical communication link is shown in Fig. 24. The crucial differences with direct detection lie in the role of the modulator in transmission and the presence of the local oscillator laser in reception. To understand the role of the modulator, we first consider the method of heterodyne detection. We will then discuss the component requirements for a coherent optical fiber communication link.

Heterodyne and Homodyne Detection.   We consider the receiver shown in Fig. 25, in which an optical signal arriving from some distant point is combined with an intense local oscillator laser by use of a 2×2 coupler. The power $I(r)$ guided in the single-mode fiber due to the interfering amplitudes can be expressed as

$$I(r) = P_S(t) |\Psi_S(r)|^2 + P_{LO} |\Psi_{LO}(r)|^2$$
$$+ 2e_S(t) \cdot e_{LO} \Psi_S(r) \Psi_{LO}(r) \sqrt{P_S(t) P_{LO}} \cos (\omega_{IF} t + \Delta\phi) \quad (32)$$

in which $e_{LO}(t)$ and $e_S(t)$ denote the polarizations of the local oscillator and signal, P_{LO} and $P_S(t)$ denote the powers of the local oscillator and signal, $\Psi_S(r)$ and $\Psi_{LO}(r)$ are the spatial amplitude distributions, and $\Delta\phi(t)$ denotes the phase difference between the two sources. The two sources may oscillate at two nominally different frequencies, the difference being labeled the *intermediate frequency* ω_{IF} (from heterodyne radio nomenclature). If the

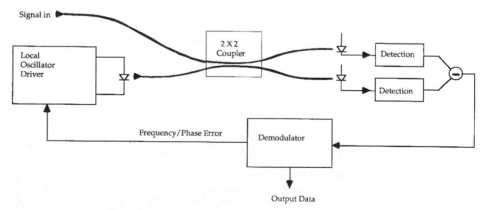

FIGURE 25   Heterodyne/homodyne receiver.

intermediate frequency is zero, the detection process is termed *homodyne* detection; if a microwave or radio carrier frequency is chosen for postdetection processing, the detection process is referred to as *heterodyne* detection.

If the local oscillator power is much larger than the signal power, the first term is negligible. The second represents a large, continuous signal which carries no information but does provide a shot noise contribution. The third term represents the signal information. If the signal is coupled to a detector of responsivity R and ac-coupled to eliminate the local oscillator signal, the photocurrent $i(t)$ can be expressed as follows:

$$i(t) = 2R\eta_{\mathrm{HET}}\sqrt{P_s(t)P_{LO}} \cos{(\omega_{IF}t + \Delta\phi)} \tag{33}$$

The heterodyne efficiency η_{HET} is determined by the spatial overlap of the fields and the inner product of the polarization components:

$$\eta_{\mathrm{HET}} = (e_S(t) \cdot e_{LO}) \int_{\substack{\text{Fiber} \\ \text{Area}}} \Psi_S(r)\Psi_{LO}(r)\, d^2r \tag{34}$$

These results illustrate four principles of coherent optical fiber communications:

1. The optical frequency and phase of the signal relative to those of the local oscillator are preserved, including the phase and frequency fluctuations.

2. The local oscillator "preamplifies" the signal, yielding a larger information-carrying component of the photocurrent than would be detected directly.

3. The local oscillator and signal fields must occupy the same spatial modes. Modes orthogonal to that of the local oscillator are rejected.

4. Only matching polarization components contribute to the detection process.

The first principle allows the detection of frequency or phase information, provided the local oscillator has sufficient stability. The second provides an improvement of the signal-to-noise ratio in the limit of large local oscillator power. Both the first and fourth lead to component requirements which are rather more stringent than those encountered with direct detection. The following sections will discuss the source, modulator, fiber, and receiver requirements in a coherent transmission system.

Receiver Sensitivity. Let σ_T^2 represent the receiver noise described in Eq. 26. The signal-to-noise ratio for heterodyne detection may be expressed as

$$\mathrm{SNR} = \frac{2\eta_{\mathrm{HET}}R^2P_sP_{LO}}{2eRP_{LO}B_n + \sigma_T^2} \tag{35}$$

where B_n denotes the noise bandwidth of the receiver. (B_n is generally about half of the data rate for digital systems.) For homodyne detection, the signal envelope carries twice the energy, and

$$\mathrm{SNR} = \frac{4\eta_{\mathrm{HET}}R^2P_sP_{LO}}{2eRP_{LO}B_n + \sigma_T^2} \tag{36}$$

For a given modulation scheme, homodyne detection will therefore be twice as sensitive as heterodyne.

Modulation Formats. The modulation formats appropriate for coherent optical communications can be summarized as follows:

1. *Amplitude-Shift Keying (ASK).* This technique is simply on-off keying (similar to simple intensity modulation) but with the important constraint that the frequency and phase of the laser be kept constant. Direct modulation of ordinary semiconductor lasers produces a frequency chirp which is unacceptable for ASK modulation. An external modulator such as an electro-optic modulator, a Mach-Zehnder modulator, or an electroabsorption modulator would therefore be appropriate for ASK.

2. *Phase-Shift Keying (PSK).*   This technique requires switching the phase between two or more values. Any phase modulator can be suitable for phase-shift keying. Direct modulation of semiconductor lasers is not suitable for PSK for the same reasons mentioned for ASK.

3. *Frequency-Shift Keying (FSK).*   FSK has received a good deal of attention[55] because it can be achieved by direct modulation of the source. It is possible to make use of the natural frequency chirp of the semiconductor laser to frequency modulate the laser simply by a small modulation of the drive current.

All of the modulation techniques can operate between two states (binary) or extend to four or more levels. The only technique which benefits from an increase in the number of channels is FSK. The sensitivity of PSK to source phase noise generally precludes higher-level signaling. Multilevel FSK, being a bandwidth expansion technique, offers a receiver sensitivity improvement over binary FSK without placing severe constraints on the source.

Table 4 gives expressions for the receiver error probability as a function of received power for each modulation technique. The right-hand column gives, for comparison purposes, the number of photons required per pulse to assure an error rate of better than 10^{-9}. PSK modulation with homodyne detection is the most sensitive, requiring only nine photons per pulse, which is below the quantum limit for direct detection.

Source Requirements.   One of the ways coherent optical communications systems differ from their microwave counterparts is in the comparatively large phase noise of the source. Since the detection system is sensitive to the frequency and phase of the laser, the source linewidth is a major consideration. This is very different from intensity modulation/direct detection, in which the source spectral width limits the system only through the channel dispersion. When two sources are heterodyned to generate an intermediate frequency in the microwave region, the spectral spread of the heterodyned signal is the combined spectral spread of the signal and local oscillator. Thus, the rule of thumb for high-quality coherent detection is that the sum of the linewidths of the signal and local oscillator be much less than the receiver bandwidth.

TABLE 4   Receiver Sensitivities for a Variety of Modulation/Detection Schemes

Modulation/Detection Scheme	P_E	Photons per pulse @ $P_E = 10^{-9}$
ASK heterodyne	$0.5 erfc\left(\sqrt{\dfrac{\eta P_S}{4h\nu B}}\right)$	72
ASK homodyne	$0.5 erfc\left(\sqrt{\dfrac{\eta P_S}{2h\nu B}}\right)$	36
FSK heterodyne	$0.5 erfc\left(\sqrt{\dfrac{\eta P_S}{2h\nu B}}\right)$	36
PSK heterodyne	$0.5 erfc\left(\sqrt{\dfrac{\eta P_S}{h\nu B}}\right)$	18
PSK homodyne	$0.5 erfc\left(\sqrt{\dfrac{2\eta P_S}{h\nu B}}\right)$	9
Direction detection quantum limit	$0.5 \exp\left(\dfrac{-\eta P_S}{h\nu B}\right)$	21

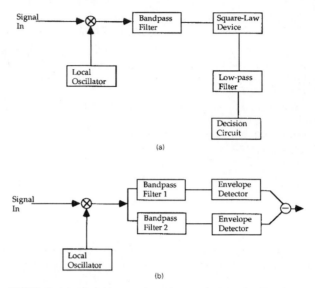

FIGURE 26 Noncoherent (asynchronous) demodulation schemes: (*a*) ASK envelope detection; (*b*) FSK dual filter detection, in which the signal is separated into complementary channels for ASK envelope detection.

Precisely how narrow the linewidth must be has been a topic of many papers.[49–52] The result varies somewhat with modulation scheme and varies strongly with the demodulation process. The general trends can be summarized as follows:

Incoherent demodulation (envelope detection). Either ASK or FSK can be demodulated simply by using an appropriate combination of filters and nonlinear elements. The basic principle of incoherent ASK or dual-filter FSK detection is illustrated in Fig. 26. This type of detection is, in general, least sensitive to the spectral width of the source. The primary effect of a broad source is to broaden the *IF* signal spectrum, resulting in attenuation but not a catastrophic increase in bit error rate. Further, the receiver bandwidth can always be broadened to accommodate the signal. This yields a penalty in excess receiver noise, but the source spectral width can be a substantial fraction of the bit rate and still keep the receiver sensitivity within tolerable limits.

There are two approaches to PSK detection which avoid the need for a phase-locked loop. The first is differential phase-shift keying (DPSK), in which the information is transmitted in the form of phase *differences* between neighboring time slots. The second is phase diversity reception, in which a multiport interferometer is designed to yield signals proportional to the power in different phase quadrants.

Coherent demodulation with electronic phase-locked loop. Some PSK signals cannot be demodulated incoherently and require careful receiver design for proper carrier recovery. Suppressed carrier communications schemes such as PSK require a nonlinear recovery circuit. The phase estimation required in proper carrier recovery is far more sensitive to phase noise than is the case with envelope detection. In contrast to incoherent demodulation, source spectral widths must generally be kept to less than 1 percent of the bit rate (10 percent of the phase-locked loop bandwidth) to maintain reliable detection.

Coherent demodulation with optoelectronic phase-locked loop. Homodyne detection requires that an error signal be fed back to the local oscillator; phase and frequency errors must be corrected optically in order to maintain precise frequency and phase matching between the two signals. This generally results in a narrower phase-locked loop bandwidth and a much narrower spectral width requirement for the transmitter and local oscillator. Homodyne systems therefore require considerably narrower linewidths than their heterodyne counterparts.

Fiber Requirements.   Heterodyne or homodyne reception is inherently single-mode, and it is therefore necessary for coherent links to use single-mode fibers. Single-mode couplers can then be used to combine the signal and local oscillator lasers for efficient heterodyne detection.

As with other forms of fiber communications, fiber dispersion presents a degradation in the signal-to-noise ratio due to differential delay between different components of the signal spectrum. The power penalty associated with fiber dispersion is determined entirely by the dispersion, the fiber length, and the bit rate. Because of the stringent source linewidth requirements for coherent detection, the spectral broadening is entirely due to the signal itself. The coherent detection system is therefore inherently less sensitive to fiber dispersion.

One important requirement of any fiber that is to be used for coherent transmission is polarization control. As was discussed briefly under "Polarization Characteristics of Fibers" earlier in this chapter, the transmitted polarization of light from a single-mode fiber varies randomly with temperature, stress on the fiber, and other environmental influences. If heterodyning is attempted under these circumstances, the heat signal will fade in and out as the polarization of the signal changes.

Polarization fading can be controlled either by external compensation,[56] internal control,[11] or polarization diversity reception.[57] External compensation seeks to actively control the polarization of the output by sensing the error through a polarizer-analyzer combination and feeding back to correct the polarization. The latter can be accomplished through mechanical, electro-optical, or magneto-optical means.

There are classes of optical fiber sensors which have source and fiber requirements very similar to those of a coherent communication link. One of the most widely studied has been the optical fiber gyro, in which counterpropagating waves in a rotating fiber coil interfere with one another; the resulting beat frequency between the waves is proportional to the angular velocity of the coil. There are other interferometric sensors which make use of optical fibers. Most of them require polarization control and a high degree of frequency stability for the source. The relatively low frequencies and small bandwidths which are required for sensing represent the major difference between these applications and coherent data transmission.

10.8 NONLINEAR OPTICAL PROPERTIES OF FIBERS

Chapters and entire books have been devoted to the subject of optical nonlinearities in fibers. A selection of these are included in "Further Reading" at the end of this chapter. We will content ourselves with an overview of the subject, and consider nonlinear effects which are most important in either limiting or enhancing the performance of fibers. To date, most of the *applications* of nonlinear optics in fibers are in the area of ultralong distance telecommunications.[41,58–60] However, nonlinearities can limit the power-handling ability of fibers and can be an important limitation for certain medical/surgical applications.

Stimulated Scattering Processes

The low loss and long interaction length of an optical fiber makes it an ideal medium for stimulating even relatively weak scattering processes. Two important processes in fibers are: (1) stimulated Raman scattering, the interaction of the guided wave with high-frequency optical phonons in the material, and (2) stimulated Brillouin scattering, the emission, amplification, and scattering of low-frequency acoustic waves. These are distinguished by the size of the frequency shift and the dynamics of the process, but both can act to limit the power available for transmission.

Stimulated Raman scattering (SRS) produces a frequency shift of about $400 \, \text{cm}^{-1}$ from the incident laser line. The equation governing the power growth of the Raman-shifted mode is as follows

$$\frac{dP_R}{dz} = -\alpha_R P_R + \frac{g_R}{a_P} P_P P_R \tag{37}$$

where P_R denotes the power of the Stokes-shifted light, P_P is the pump power (this is the power in the initially excited mode), and a_P is the effective area of the pump. The Raman gain g_R ultimately determines the SRS-limited light intensity. For typical single-mode silica fibers, g_R is about $10^{-11} \, \text{cm/W}$, and yields a power limit of

$$P_{CR} = \frac{16\alpha a_P}{g_R} \tag{38}$$

beyond which the guided wave power will be efficiently Raman-shifted and excess loss will begin to appear at the pump wavelength.

Stimulated Brillouin scattering (SBS) can yield an even lower stimulated scattering threshold. Acoustic waves in the fiber tend to form a Bragg index grating, and scattering occurs primarily in the backward direction. The Brillouin gain g_B is much higher than Raman gain in fibers ($g_B = 5 \times 10^{-9} \, \text{cm/W}$) and leads to a stimulated scattering threshold of

$$P_{CR} = \frac{25\alpha a_P}{g_B} \tag{39}$$

for a narrowband, CW input.

Either type of stimulated scattering process can be used as a source of gain in the fiber. Injecting a signal within the frequency band of the stimulated scattering process will provide amplification of the input signal. Raman amplification tends to be the more useful of the two because of the relatively large frequency shift and the broader-gain bandwidth. SBS has been used in applications such as coherent optical communications[48] where amplification of a pilot carrier is desired.

The gain bandwidth for SBS is quite narrow—100 MHz for a typical fiber. SBS is therefore only important for sources whose spectra lie within this band. An unmodulated narrow-band laser source such as would be used as a local oscillator in a coherent system would be highly susceptible to SBS, but a directly modulated laser with a 1-GHz linewidth under modulation (modulated laser linewidths can extend well into the GHz range due to frequency chirp) would have an SBS threshold about ten times that of the narrow linewidth source.

Pulse Compression and Soliton Propagation

A major accomplishment in the push toward short pulse propagation in optical fibers was the prediction and observation of solitary wave propagation. In a nonlinear dispersive medium, solitary waves may exist provided the nonlinearity and dispersion act to balance one another. In the case of soliton propagation, the nonlinearity is a refractive index which follows the pulse intensity in a nearly instantaneous fashion:

$$n(t) = n_0 + n_2 I(t) \tag{40}$$

For silica fibers, $n_2 = 3 \times 10^{-16} \, \text{cm}^2/\text{W}$.

TABLE 5 Normalized Variables of the Nonlinear Schrödinger Equation

A	Pulse amplitude
z	Longitudinal coordinate
t	Time
P_0	Peak power
T_0	Pulse width
U	$A/\sqrt{P_0}$ normalized pulse amplitude
β_1	Propagation constant
β_2	Dispersion (2d order)
L_D	$T_0^2/|\beta_2|$ dispersion length
n_2	Nonlinear refractive index
τ	$\dfrac{t-\beta_1 z}{T_0}$ time normalized to moving frame
ξ	$\dfrac{z}{L_D}$ normalized distance
N	$n_2\beta_1 P_0 T_0^2/|\beta_2|$ Order of soliton

The scalar equation governing pulse propagation in such a nonlinear dispersive medium is sometimes termed the *nonlinear Schrödinger equation*

$$i\frac{dU}{d\xi}+\frac{1}{2}\frac{d^2 U}{d\tau^2}+N^2\,|U|^2 U = 0 \tag{41}$$

where the symbols are defined in Table 5. Certain solutions of this equation exist in which the pulse propagates without change in shape; these are the soliton solutions. Solitons can be excited in fibers and propagate great distances before breaking up. This is the basis for fiber-based soliton communication.

Figure 27 illustrates what happens to a pulse which propagates in such a medium. The local refractive index change produces what is commonly known as *self phase modulation.* Since n_2 is positive, the leading edge of the pulse produces a local increase in refractive index. This results in a red shift in the instantaneous frequency. On the trailing edge, the pulse experiences a blue shift. If the channel is one which exhibits normal dispersion, the red-shifted edge will advance while the blue-shifted edge will retard, resulting in pulse spreading If, however, the fiber exhibits anomalous dispersion (beyond 1.3 μm for most single-mode fibers), the red-shifted edge will retard and the pulse will be compressed. Fibers have been used in this way as pulse compressors for some time. In the normal dispersion regime, the fiber nonlinearity is used to chirp the pulse, and a grating pair supplies the dispersion necessary for compression. In the anomalous dispersion regime, the fiber can act both to chirp and compress the pulse. Near the dispersion minimum,

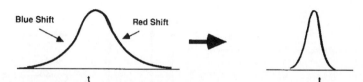

FIGURE 27 A pulse propagating through a medium with an intensity-dependent refractive index will experience frequency shifts of the leading and trailing edges of the pulse (*left*). Upon transmission through a fiber having anomalous dispersion, the pulse compresses (*right*).

higher-order dependence of the group delay on wavelength becomes important, and simple pulse compression does not take place.

Pulse compression cannot continue indefinitely, since the linear dispersion will always act to spread the pulse. At a critical shape, the pulse stabilizes and will propagate without change in shape. This is the point at which a soliton forms. The lowest-order soliton will propagate entirely without change in shape, higher order solitons (which also carry higher energy) experience a periodic evolution of pulse shape.

A soliton requires a certain power level in order to maintain the necessary index change. Distortion-free pulses will therefore propagate only until the fiber loss depletes the energy. Since solitons cannot persist in a lossy channel, they were long treated merely as laboratory curiosities. This was changed by several successful demonstrations of extremely long distance soliton transmission by the inclusion of gain to balance the loss. The gain sections, which initially made use of stimulated Raman scattering, now consist of rare-earth doped fiber amplifiers. The record for repeaterless soliton transmission is constantly being challenged. At the time of this writing, distance of well over 10,000 km have been demonstrated in recirculating loop experiments.

In the laboratory, solitons have most often been generated by mode-locked laser sources. Mode-locked solid state laser sources are generally limited to low duty-cycle pulses, with repetition rates in the 1-GHz range or less. The mode-locked pulse train must then be modulated to carry data, a process which must be carried out externally. There is a high level of current interest in Erbium-doped fiber lasers as mode-locked sources for ultralong distance data communications. Despite the capability of high duty cycle, directly modulated semiconductor lasers are generally rendered unsuitable for soliton communications by the spectral broadening that occurs under modulation.

Four-Wave Mixing

The nonlinear refractive index is simply a degenerate case of a third-order optical nonlinearity, in which the polarization of the medium responds to the cube of the applied electric field. It is possible for widely separated frequencies to phase modulate one another via the fiber nonlinearity, generating sidebands which interfere with neighboring channels in a multiplexed system. This represents an important limit to channel capacity in either WDM or FDM systems. The simplest picture of the four-wave mixing process in fibers can be illustrated by the transmission and cross-phase modulation of four equally spaced channels shown in Fig. 28. Channels 1 and 2 interfere, producing an index of refraction which oscillates at the difference frequency. This modulation in refractive index modulates channel 4, producing sidebands at channels 3 and 5. This is only the simplest combination of frequencies. Four-wave mixing allows any combination of three frequencies beating together to produce a fourth. If the fourth frequency lies within a communication band, that channel can be rendered unusable.

This channel interference can effect either closely spaced channels, as one encounters

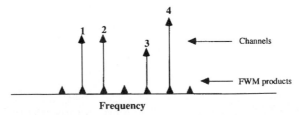

FIGURE 28 The effects of four-wave mixing on multichannel transmission through an optical fiber.

with coherent communications, or the rather widely separated channels of a WDM system. Efficient four-wave mixing requires phase matching of the interacting waves throughout the interaction length—widely separated channels will therefore be phase matched only in a region of low-fiber dispersion.

The communications engineer will recognize this as little more than the intermodulation products which must always be dealt with in a multichannel communications system with small nonlinearities. Four-wave mixing merely produces intermodulation products over an extremely wide bandwidth. Just as with baseband nonlinearities in analog communications systems, judicious allocation of channels can minimize the problem, but at the expense of bandwidth. The cumulative effect of the nonlinearities increases with interaction length and therefore imposes an important limit on frequency or wavelength-division multiplexed systems.

Photorefractive Nonlinearities in Fibers

There also exists a class of integrating, photorefractive nonlinearities in optical fibers which have been of some interest in recent years. We use the word photorefractive loosely here, simply to indicate a long-term change in either the first- or second-order susceptibility with light exposure. The effects appear strongest in fibers with a germania content, but the precise role of the glass constituents in these processes is still an area of active research.

Bragg Index Gratings. Photons of energy near a UV absorption edge can often write permanent phase gratings by photoionizing certain constituents or impurities in the material. This is the case for $LiNbO_4$ and certain other ferroelectric materials, and such effects have also been observed in germania-silica fibers. The effects were first observed in the process of guiding relatively high power densities of green light—it was found that a high backscatter developed over a period of prolonged exposure. The fiber then exhibited the transmission characteristics of a Bragg grating, with extremely high resonant reflectivities.

The writing of permanent gratings in fibers using UV exposure is now relatively commonplace. Bragg gratings can be used as filters in WDM systems, reflectors on fiber lasers, and possibly optical switches. For short lengths, the gratings are most easily formed holographically, by using two interfering beams from a pulsed UV source such as an excimer laser. The fiber is exposed from the side; by controlling the angle of the two interfering beams, any grating period may be chosen.

Frequency Doubling in Germania-Silica Fibers. While it is not surprising that UV exposure could produce refractive index changes, a rather unexpected discovery was the fact that strong optical fields inside the fiber could produce a second-order susceptibility, resulting in efficient frequency doubling. Electro-optic effects such as frequency doubling require that a crystalline material lack a center of symmetry while an amorphous material must lack a statistical center of symmetry. It has long been known that certain materials will develop an electro-optic effect under a suitable applied field. This process, known as *poling,* provides the necessary microscopic alignment of dipoles for development of the nonlinear susceptibility. In optical fibers, a type of self-poling occurs from the strong fundamental beam, resulting in a second-order susceptibility and efficient frequency doubling.

Efficient frequency doubling requires both a noncentrosymmetric material and adequate phase matching between the fundamental and second harmonic waves. The mechanism by which the fiber is both poled and phase matched is still not fully understood at the time of this writing, and it remains to be seen whether this represents an exciting, new application of germania-silica fibers or simply an internal damage mechanism which limits the ultimate power delivery of the fiber.

10.9 *OPTICAL FIBER MATERIALS*: *CHEMISTRY AND FABRICATION*

What is arguably the most important breakthrough in the history of optical fiber technology occurred in the materials development. Until 1970, many scientists felt that glasses of moderate softening points and smooth phase transitions would allow for easier drawing and better control. The choice of Corning Glass Works (now Corning, Inc.) to go to (what was then) the somewhat more difficult chemistry of the nearly pure silica fiber allowed both a dramatic reduction in fiber attenuation and a better understanding of the role of the chemical constituents in fiber loss. Researchers soon found that the best dopants for altering the refractive index were those which provided a weak index change without providing a large shift in the UV absorption edge. Conventional fiber chemistry consists of dopants such as GeO_2, P_2O_5 (for raising the refractive index) and B_2O_3 or SiF_4 (for lowering the refractive index).

Silica has both UV and mid-IR absorption bands; these two bands result in a fundamental limit to the attenuation which one can achieve in the silica system. This occurs despite the fact that the Rayleigh scattering contribution decreases as λ^{-4}, and the ultraviolet Urbach absorption edge decreases even faster with increasing λ. The infrared absorption increases with long wavelengths, and becomes dominant beyond wavelengths of about 1.6 μm, resulting in a fundamental loss minimum near 1.55 μm.

The promise of achieving a lower Rayleigh scattering limit in the mid-infrared (as well as the possible applications of fiber to the CO_2 laser wavelength range) have spurred a great deal of research in fiber materials which exhibit better infrared transparency. Two important representative materials are the heavy-metal fluoride glasses and the chalcogenide glasses. While both classes exhibit better infrared transparency, neither has yet improved to the point of serious competition with silica materials.

For a number of years, attenuation in optical fibers was limited by a strong absorption band near $\lambda = 1.4$ μm. (An examination of attenuation curves of early telecommunications-grade fiber shows it nearly unusable at what are now the wavelengths of prime interest—1.3 μm and 1.55 μm.) This absorption, which was linked to the presence of residual OH ions, grew steadily lower with the improvement of fiber fabrication techniques until the loss minimum at $\lambda = 1.55$ μm was eventually brought close to the Rayleigh scattering limit.

The low-cost, low-temperature processes by which polymers can be fabricated has led to continued research into the applications of plastic fiber to technologies which require low cost, easy connectivity, and that are not loss-limited. The additional flexibility of these materials makes them attractive for large-core, short-length applications in which one wishes to maximize the light insertion. Hybrid polymer cladding-silica core fibers have also received some attention in applications requiring additional flexibility.

The final triumph of fiber chemistry in recent years has been the introduction and successful demonstration of extremely long distance repeaterless fiber links using rare-earth doped fiber amplifiers. This represented the climax of a long period of research in rare-earth doped glasses which went largely unnoticed by the optics community. As a result, there has been an explosion of work in the materials science, materials engineering, and applications of rare-earth doped optical fibers.

Fabrication of Conventional Optical Fibers

Conventional fabrication of low-loss optical fibers requires two stages. The desired refractive index profile is first fabricated in macroscopic dimensions in a preform. A typical preform is several centimeters in width and a meter in length, maintaining the dimensions and dopant distribution in the core and cladding that will eventually form in the fiber.

Chemical vapor deposition (CVD) is the primary technology used in fiber manufacturing. The fabrication process must satisfy two requirements: (1) high purity, and (2) precise control over composition (hence, refractive index) profiles. Manufacturing considerations favor approaches which provide a fast deposition rate and comparatively large preforms. In CVD processes, submicron silica particles are produced through one (or both) of the following chemical reactions

$$SiCl_4 + O_2 \rightarrow SiO_2 + 2Cl_2$$

$$SiCl_4 + 2H_2O \rightarrow SiO_2 + HCl$$

The reactions are carried out at a temperature of about 1800°C. The deposition leads to a high-purity silica soot which must then be sintered in order to form optical quality glass.

Modern manufacturing techniques, generally speaking, use one of two processes.[61] In the so-called "inside process," a rotating silica substrate tube is subjected to an internal flow of reactive gases. The two inside processes which have received the most attention are modified chemical vapor deposition (MCVD) and plasma-assisted chemical vapor deposition (PCVD). Both techniques require a layer-by-layer deposition, controlling the composition at each step in order to reach the correct target refractive index. Oxygen, as a carrier gas, is bubbled through $SiCl_4$, which has a relatively high vapor pressure at room temperature.

The PCVD process provides the necessary energy for the chemical reaction by direct RF plasma excitation. The submicron-sized particles form on the inner layer of the substrate, and the composition of the layer is controlled strictly by the composition of the gas. PCVD does not require the careful thermal control of other methods, but requires a separate sintering step to provide a pore-free preform. A final heating to 2150°C collapses the preform into a state in which it is ready to be drawn.

The MCVD process (Fig. 29) accomplishes the deposition by an external, local application of a torch. The torch has the dual role of providing the necessary energy for oxidation and the heat necessary for sintering the deposited SiO_2. The submicron particles are deposited on the "leading edge" of the torch; as the torch moves over these particles, they are sintered into a vitreous, pore-free layer. Multiple passes result in a layered, glassy deposit which should approximate the target radial profile of the fiber. As with PCVD, a final pass is necessary for collapse of the preform before the fiber is ready to be drawn. MCVD requires rather precise control over the temperature gradients in the tube but has the advantage of accomplishing the deposition and sintering in a single step.

In the "outside process," a rotating, thin cylindrical target (or mandrel) is used as the substrate for a subsequent chemical vapor deposition, and requires removal before the

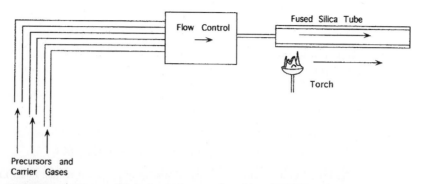

FIGURE 29 The modified chemical vapor deposition (MCVD) process for preform fabrication.

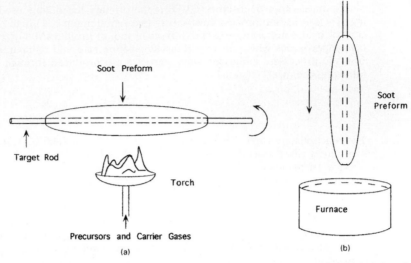

FIGURE 30 Outside method of preform fabrication. The soot deposition (*a*) is followed by sintering (*b*) to cast the preform.

boule is sintered. Much of the control in these deposition techniques lies in the construction of the torch. For an outside process, the torch supplies both the chemical constituents and the heat for the reaction.

Two outside processes which have been used a great deal are the outside vapor deposition (OVD) and the vapor axial deposition (VAD) techniques. Figure 30 illustrates a generic outside process. In the OVD process the torch consists of discrete holes formed in a pattern of concentric rings. The primary chemical stream is at the center, followed by O_2 (acting as a shield gas), premixed methane/oxygen, and another shield ring. The torch itself translates along the rotating boule and the dopants are dynamically controlled to achieve the necessary profiling.

The VAD torch is comprised of a set of concentric annular apertures, with a chemical sequence similar to the OVD. In contrast to the OVD method, the VAD torch is held stationary during the deposition; the rotating target is held vertically, and is lifted as the deposition continues.

Dopant Chemistry

Standard dopants for silica fiber include GeO_2, P_2O_5, B_2O_3, and SiF_4. The former two are used to increase the refractive index (and are therefore used in the core), while the latter decrease the index of refraction (and are therefore used in the cladding). The CVD processes will often use oxygen as a carrier gas with the high vapor pressure liquids $GeCl_4$, $POCl_3$, or SiF_4. The reaction which produces the dopant "soot" is then

$$GECl_4 + O_2 \rightarrow GeO_2 + 2Cl_2$$

$$4POCl_3 + 3O_2 \rightarrow 2P_2O_5 + 6Cl_2$$

As noted in a recent article by Morse et al.,[62] "Nature has been kind in the creation of the high vapor pressure liquid precursors used in the fabrication of optical fibers for the transmission of telecommunication signals." This has been an extremely important factor in the success of CVD fiber fabrication techniques. The problem of introducing more exotic dopants, such as the rare-earth elements, is not quite so straightforward and there does not appear to exist, at this time, a single, widely used technique. The problem of control over the rare-earth dopant profile is compounded by the fact that research in laser and amplifier design is ongoing, and the optimum dopant profile for rare-earth doped fibers and amplifiers is, in many cases, still unknown. Despite these uncertainties, rare-earth doped fibers have already been introduced into commercial products and promise to spearhead the next generation of long distance telecommunications systems.

Other Fabrication Techniques

There are other preform fabrication and fiber drawing techniques. These are not generally used in telecommunications-grade silica fiber, but can be of advantage for glass chemistries which do not easily lend themselves to chemical vapor deposition. Several examples of this will be described in the following section on infrared fiber fabrication.

CVD materials, while the most popular, are not the only methods for preform fabrication. Alternative methods of preform fabrication include both bulk casting and a class of non-CVD tubular casting techniques. One such technique is the "rod-in-tube" method, in which the core and cladding materials are cast separately and combined in a final melting/collapsing step. This method assures a homogeneous, low-impurity content core but risks introducing defects and bubbles into the core/cladding interface.

The most well-known method of *preform-free drawing* is the double crucible method, in which the core and cladding melts are formed separately and combined in the drawing process itself. This eliminates the need for a very large preform in the case of long lengths of fiber. The index profile is established in the drawing process itself, and index gradients are therefore difficult to establish unless additional crucibles are added. Another difficulty of the crucible method is the sometimes inadequate control of the concentricity of the core and cladding.

Infrared Fiber Fabrication

The major applications of interest for infrared optical fibers are as follows:

1. Ultra-low-loss communication links

2. CO_2 laser transmission for medical applications

3. Thermal imaging and remote temperature monitoring

4. Gas sensing

These may differ rather dramatically in their attenuation requirements and spectral region of interest. For example, an ultra-low-loss communications link requires attenuation somewhat less than 0.1 dB/km in order to be competitive with silica fiber. Typical medical applications simply require high-power handling capabilities of a CO_2 laser over meter lengths. All of these applications require a departure from the silica-based chemistry which has been so successful for applications in the near infrared and visible. Much of the generic chemistry of these glasses is covered in Chap. 33 of Vol. II, "Crystals and glasses". Our intent here is to give an overview of the fiber types and the general state of the materials technology in each case.

Chalcogenide Fibers. Sulfide, selenide, and telluride glasses have all been used for bulk infrared optics—particularly for applications involving CO_2 ($\lambda = 10.6\,\mu\text{m}$) or CO laser transmission ($\lambda = 5.4\,\mu\text{m}$). Infrared fibers have been drawn from these materials and yielded transmission losses of the order of 1 dB/meter in the 5- to 7-μm region.[63] The preform fabrication and drawing of chalcogenide fibers is much more difficult than that of silica due primarily to its sensitivity both to oxygen and moisture. Both oxidation and crystallization can occur at the temperatures necessary to draw the fiber. Either will result in catastrophically high losses and fiber weakness.

Fluoride Fibers. Fluoride fibers have received the most attention for low-loss telecom-munications applications, because the theoretical limit for Rayleigh scattering is con-siderably lower. This is due both to a higher-energy UV absorption edge and better infrared transparency. The difficulty is that excess absorption has proven rather difficult to reduce, and the lowest published losses to date have been near 1 dB/km for long fiber lengths.[64,65] The state-of-the-art in fluoride fiber fabrication is still well above the Rayleigh scattering limit but does show the expected improvement over silica fiber in wavelengths beyond 1.6 μm. Fabrication of very short fiber lengths has been somewhat more successful, with reported losses as low as 0.025 dB/km at 2.55 μm.[66]

The residual loss for longer fibers has been largely linked to extrinsic impurity/defect content. Recent articles by Takahashi and Sanghera[64,65] have noted the role of transition metal and rare-earth ions, submicron platinum particles, oxyfluoride particles, fluoride microcrystals, and bubbles as both extrinsic absorbers and scatterers. The defects of interest originate from a variety of sources, and there has been much discussion on which defects dominate the scattering process. To date, the consensus appears to be that impurity absorption does *not* adequately account for the current loss limits, but does account for residual losses in the neighborhood of 0.2 dB/km.

The classes of defects which have been blamed for the current loss limits are as follows:

Platinum particles. These arise from the use of platinum crucibles. The use of vitreous carbon crucibles eases this contamination.

Core bubbles. This is clearly a problem in the preform fabrication and appears in some of the bulk casting techniques.

Interfacial bubbles. Bubbles appearing at the core-cladding interface have been named as being a major cause of excess scattering. These appear to be a particular problem for those techniques which employ separate core and cladding melts. This unfortunately negates some of the advantages offered by the crucible techniques in fluoride fiber fabrication.

Fluoride microcrystals. Crystals can nucleate at a variety of defect sites. Many of these sites appear at the core-cladding interface, producing interface roughness and scatter-ing. Since, for step-index fibers, the integrity of the core-cladding interface is essential to the confinement of the optical wave, a small amount of interface roughness can produce rather high excess losses.

Chapter 33 of Vol. II, "Crystals and glasses", gives information on the composition and properties of a single-mode fiber grade fluoride glass. This class of compositions has received the designation ZBLAN, after its heavy-metal constituents. The large number of components makes it immediately obvious that both phase separation and crystallization are important issues in fabrication. Either can produce catastrophic increases in loss as well as mechanical weakening of the fiber, and it is clear that many materials science challenges remain in the area of fluoride fiber fabrication.

10.10 REFERENCES

1. K. C. Kao and G. A. Hockham, "Dielectric Fibre Surface Waveguides for Optical Frequencies", *Proc. IEE,* **113,** 1966, pp. 1151–1158.
2. D. B. Keck, P. C. Schultz, and F. W. Zimar, U.S. Patent 3,737,393.

3. F. P. Kapron, D. B. Keck, and R. D. Maurer, "Radiation Losses in Glass Optical Waveguides" *Appl. Phys. Lett.,* **17,** 1970, p. 423.

4. M. J. Adams, *An Introduction to Optical Waveguides,* John Wiley and Sons, Chichester, 1981.

5. D. Davidson, "Single-Mode Wave Propagation in Cylindrical Optical Fibers", in E. E. Basch (ed.), *Optical Fiber Transmission,* Howard W. Sams, Indianapolis, 1987, pp. 27–64.

6. M. O. Vassell, "Calculation of Propagating Modes in a Graded-Index Optical Fiber", *Optoelectronics,* **6,** 1974, pp. 271–286.

7. D. Gloge, "Weakly Guiding Fibers", *Appl. Opt.,* **10,** 1971, pp. 2252–2258.

8. R. W. Davies, D. Davidson, and M. P. Singh, "Single Mode Optical Fiber with Arbitrary Refractive Index Profile: Propagation Solution by the Numerov Method", *J. Lightwave Tech.,* LT-3, 1985, pp. 619–627.

9. P. C. Chow, "Computer Solutions to the Schroedinger Problem", *Am. J. of Physics,* **40,** 1972, pp. 730–734.

10. A. Kumar, R. K. Varshney, and K. Thyagarajan, "Birefringence Calculations in Elliptical-core Optical Fibers", *Electron. Lett.,* **20,** 1984, pp. 112–113.

11. K. Sano and Y. Fuji, "Polarization Transmission Characteristics of Optical Fibers With Elliptical Cross Section", *Electron. Commun. Japan,* **63,** 1980, p. 87.

12. EIA-FOTP-46, *Spectral Attenuation Measurement for Long-Length, Graded-Index Optical Fibers, Procedure B,* Electronic Industries Association (Washington, D.C.).

13. EIA-FOTP-78, *Spectral Attenuation Cutback Measurement for Single Mode Optical Fibers,* Electronic Industries Association (Washington, D.C.).

14. EIA-FOTP-50, *Light Launch Conditions for Long-Length, Graded-Index Optical Fiber Spectral Attenuation Measurements, Procedure B,* Electronic Industries Association (Washington, D.C.).

15. A. W. Carlisle, "Small Size High-performance Lightguide Connector for LAN's, *Proc. Opt. Fiber Comm.,* 1985, paper TUQ 18, p. 74–75.

16. E. Sugita, et al., "SC-Type Single-Mode Optical Fiber Connectors", *J. Lightwave Tech.,* **LT-7,** 1989, pp. 1689–1696.

17. N. Suzuki, M. Saruwatari, and M. Okuyama, "Low Insertion- and High Return-loss Optical Connectors with Spherically Convex-polished Ends." *Electron. Lett.,* **22**(2)**,** 1986, pp. 110–112.

18. W. C. Young, et al., "Design and Performance of the Biconic Connector Used in the FT3 Lightwave System," in *30th IWCS,* 1981.

19. EIA-FOTP-168, *Chromatic Dispersion Measurement of Multimode Graded-Index and Single-Mode Optical Fibers by Spectral Group Delay Measurement in the Time Domain,* Electronic Industries Association (Washington, D.C.).

20. R. Rao, "Field Dispersion Measurement—A Swept-Frequency Technique," in *NBS Special Publication 683,* Boulder, 1984, p. 135.

21. L. G. Cohen and J. Stone, "Interferometric Measurements of Minimum Dispersion Spectra in Short Lengths of Single-Mode Fiber," *Electron. Lett.,* **18,** 1982, p. 564.

22. L. G. Cohen, et al., "Experimental Technique for Evaluation of Fiber Transmission Loss and Dispersion," *Proc. IEEE,* **68,** 1980, p. 1203.

23. R. A. Modavis and W. F. Love, "Multiple-Wavelength System for Characterization of Dispersion in Single-Mode Optical Fibers", in *NBS Special Publication 683,* Boulder, 1984, p. 115.

24. T. Miya, et al., "Fabrication of Low-dispersion Single-Mode Fiber Over a Wide Spectral Range", *IEEE J. Quantum Electronics,* **QE-17,** 1981, p. 858.

25. R. Olshansky and R. D. Maurer, "Tensile Strength and Fatigue of Optical Fibers", *J. Applied Physics,* **47,** 1976, pp. 4497–4499.

26. EIA-FOTP-28, *Method for Measuring Dynamic Tensile Strength of Optical Fibers,* Electronic Industries Association (Washington, D.C.).

27. M. R. Brininstool, "Measuring Longitudinal Strain in Optical Fibers," *Optical Engineering,* **26,** 1987, p. 1113.

28. M. J. Matthewson, C. R. Kurkjian, and S. T. Gulati, "Strength Measurement of Optical Fibers by Bending," *J. Am. Ceramic. Soc.,* **69,** 1986, p. 815.

29. W. Weibull, "A Statistical Distribution Function of Wide Applicability," *J. Appl. Mech.*, **24,** 1951, pp. 293–297.

30. J. D. Helfinstine, "Adding Static and Dynamic Fatigue Effects Directly to the Weibull Distribution," *J. Am. Ceramic Soc.*, **63,** 1980, p. 113.

31. D. Inniss, D. L. Brownlow, and C. R. Kurkjian, "Effect of Sodium Chloride Solutions on the Strength and Fatigue of Bare Silica Fibers," *J. Am. Ceramic Soc.*, **75,** 1992, p. 364.

32. J. S. Cook and O. I. Scentesi, "North American Field Trials and Early Applications in Telephony," *IEEE J. Selected Areas in Communications*, **SAC-1,** 1983, pp. 393–397.

33. H. Ishio "Japanese Field Trials and Early Applications in Telephony." *IEEE J. Selected Areas in Communications*, **SAC-1,** 1983, pp. 398–403.

34. A. Moncolvo and F. Tosco, "European Field Trials and Early Applications in Telephony", *IEEE J. Selected Areas in Communications*, **SAC-1,** 1983, pp. 398–403.

35. E. E. Basch, R. A. Beaudette, and H. A. Carnes, "Optical Transmission for Interoffice Trunks," *IEEE Trans. on Communications*, **COM-26,** 1978, pp. 1007–1014.

36. G. P. Agrawal, *Fiber-Optic Communication Systems,* Wiley Series in Microwave and Optical Engineering, K. Chang (ed.), John Wiley and Sons, New York, 1992.

37. K. Petermann, *Laser Diode Modulation and Noise,* Kluwer Academic, Dordrecht, The Netherlands, 1991.

38. R. J. McIntyre, "Multiplication Noise in Uniform Avalanche Diodes," *IEEE Trans. Electron Devices,* **ED-13,** 1966, p. 164.

39. R. J. McIntyre, "The Distribution of Gains in Uniformly Multiplying Avalanche Photodiodes: Theory", *IEEE Trans. Electron Devices,* **ED-19,** 1972, pp. 703–713.

40. S. D. Personick, "Statistics of a General Class of Avalanche Detectors with Applications to Optical Communications," *Bell System Technical Journal,* **50,** 1971, pp. 167–189.

41. R. K. Dodd, et al., *Solitons and Nonlinear Wave Equations,* Academic Press, Orlando, Fl., 1984.

42. A. Hasegawa, *Solitons in Optical Fibers,* Springer-Verlag, Berlin, 1989.

43. L. F. Mollenauer, R. H. Stolen, and J. P. Gordon, "Experimental Observation of Picosecond Pulse Narrowing and Solitons in Optical Fibers," *Phys. Rev. Letters,* **45,** 1980, p. 1095.

44. L. F. Mollenauer, et al., "Extreme Picosecond Pulse Narrowing by Means of Soliton Effect in Single-Mode Optical Fibers", *Opt. Lett.,* **8,** 1983, p. 289.

45. L. F. Mollenauer, R. H. Stolen, and M. N. Islam, "Experimental Demonstration of Soliton Propagation in Long Fibers: Loss Compensated by Raman Gain", *Opt. Lett.,* **10,** 1985, p. 229.

46. J. P. Gordon and H. A. Haus, *Opt. Lett.,* **11,** 1986, p. 665

47. L. F. Mollenauer, J. P. Gordon, and S. G. Evangelides, "The Sliding Frequency Guiding Filter—an Improved Form of Soliton Jitter Control," *Opt. Lett.,* **17,** 1992, p. 1575.

48. A. R. Chraplyvy and R. W. Tkach, *Electron. Lett.,* **22,** 1986, p. 1084.

49. I. Garrett and G. Jacobsen, "Theoretical Anaylsis of Heterodyne Optical Receivers using Semiconductor Lasers of Non-negligible Linewidth," *J. Lightwave Technology,* **4,** 1986, p. 323.

50. B. Glance, "Performance of Homodyne Detection of Binary PSK Optical Signals," *J. Lightwave Technology,* **4,** 1986, p. 228.

51. L. G. Kazovsky, "Performance Analysis and Laser Linewidth Requirements for Optical PSK Heterodyne Communications," *J. Lightwave Technology,* **4,** 1986, p. 415.

52. K. Kikuchi, et al., *J. Lightwave Technology,* **2,** 1984, p. 1024.

53. T. Okoshi and K. Kikuchi, *Coherent Optical Fiber Communications,* Kluwer, Boston, 1988.

54. N. A. Olsson et al., "400 Mbit/s 372 = Km Coherent Transmission Experiment," *Electron. Lett.,* **24,** 1988, p. 36.

55. E. G. Bryant et al., "A 1.2 Gbit/s Optical FSK Field Trial Demonstration," *British Telecom Technology Journal,* **8,** 1990, p. 18.

56. T. Okoshi, "Polarization-State Control Schemes for Heterodyne of Homodyne Optical Fiber Communications," *J. Lightwave Technology,* **3,** 1985, pp. 1232–1237.

57. B. Glance, "Polarization Independent Coherent Optical Receiver," *J. Lightwave Technology,* **5,** 1987, p. 274.

58. R. H. Stolen, "Nonlinear Properties of Optical Fibers," in S. E. Miller and A. G. Chynowth, eds., *Optical Fiber Telecommunications,* Academic Press, New York.

59. G. P. Agrawal, "Nonlinear Interactions in Optical Fibers," in G. P. Agrawal and R. W. Boyd, eds., *Contemporary Nonlinear Optics,* Academic Press, San Diego, CA, 1992.

60. G. P. Agrawal, *Nonlinear Fiber Optics,* Academic Press, San Diego, CA, 1989.

61. J. R. Bautista and R. M. Atkins, "The Formation and Deposition of SiO_2 Aerosols in Optical Fiber Manufacturing Torches", *J. Aerosol Science,* **22,** 1991, pp. 667–675.

62. T. F. Morse, et al., "Aerosol Transport for Optical Fiber Core Doping: A New Technique for Glass Formation," *J. Aerosol Science,* **22,** 1991, pp. 657–666.

63. J. Nishii, et al., "Recent Advances and Trends in Chalcogenide Glass Fiber Technology: A Review," *J. Noncrystalline Solids,* **140,** 1992, pp. 199–208.

64. J. S. Sanghera, B. B. Harbison, and I. D. Aggarwal, "Challenges in Obtaining Low Loss Fluoride Glass Fibers," *J. Non-Crystalline Solids,* **140,** 1992, pp. 146–149.

65. S. Takahashi, "Prospects for Ultra-low Loss Using Fluoride Glass Optical Fiber," *J. Non-Crystalline Solids,* **140,** 1992, pp. 172–178.

66. I. Aggarwal, G. Lu, and L. Busse, *Materials Science Forum,* **32 & 33,** Plenum, New York, 1988, p. 495.

10.11 FURTHER READING

Agrawal, G. P., *Fiber-Optic Communication Systems,* John Wiley and Sons, New York, 1992.

Baack, C. (ed.), *Optical Wideband Transmission Systems,* CRC Press, Boca Raton, Fla., 1986.

Baker, D. G., *Fiber Optic Design and Applications,* Reston Publishing, Reston, Va., 1985.

Barnoski, M. K. (ed.), *Fundamentals of Optical Fiber Communications,* Academic Press, New York, 1981.

Basch, E. E. (ed.), *Optical Fiber Transmission,* Howard W. Sams, Indianapolis, 1987.

Chaffee, C. D., *The Rewiring of America*: *The Fiber Optics Revolution,* Academic Press, Boston, 1988.

Chaimowitz, J. C. A., *Lightwave Technology,* Butterworths, Boston, 1989.

Cheo, P. K., *Fiber Optics*: *Devices and Systems,* Prentice-Hall, Englewood Cliffs, NJ, 1985.

Cheo, P. K., *Fiber Optics and Optoelectronics,* Prentice-Hall, Englewood Cliffs, NJ, 1990.

Cherin, A. H., *An Introduction to Optical Fibers,* McGraw-Hill, 1983.

Culshaw, B., *Optical Fibre Sensing and Signal Processing,* Peter Peregrinus, London, 1984.

Daly, J. C., (ed.), *Fiber Optics,* CRC Press, Boca Raton, Fla., 1984.

Day, G. W., *Measurement of Optical-Fiber Bandwidth in the Frequency Domain,* NBS Special Publication, No. 637, National Bureau of Standards, Boulder, 1983.

Edwards, T. C., *Fiber-Optic Systems*: *Network Applications,* John Wiley and Sons, New York, 1989.

Geckeler, S., *Optical Fiber Transmission Systems,* Artech House, Norwood, Mass., 1987.

Gowar, J., *Optical Communications Systems,* Prentice-Hall, London, 1984.

Howes, M. J. and D. V. Morgan (ed.), *Optical Fiber Communications,* John Wiley and Sons, New York, 1980.

Jones, W. B., Jr., *Introduction to Optical Fiber Communications,* Holt, Rinehart and Winston, New York, 1988.

Kaiser, G. E., *Optical Fiber Communications,* McGraw-Hill, New York, 1991.

Kao, C. K., *Optical Fibre,* Peter Pereginus, London, 1988.

Karp, S., R. Gagliardi, et al., *Optical Channels*: *Fibers, Clouds, Water, and the Atmosphere,* Plenum Press, New York, 1988.

Killen, H. B., *Fiber Optic Communications,* Prentice-Hall, Englewood Cliffs, NJ, 1991.

Li, T., (ed.), *Optical Fiber Data Transmission,* Academic Press, Boston, 1991.

Lin, C. (ed.), *Optoelectronic Technology and Lightwave Communications Systems,* Van Nostrand Reinhold, New York, 1989.

Mahlke, G. and P. Gossing, *Fiber Optic Cables,* John Wiley and Sons, New York, 1987.

Miller, S. E. and J. P. Kaminow (eds.), *Optical Fiber Telecommunications II,* Academic Press, San Diego, Calif., 1988.

Okoshi, T. and K. Kikuchi, *Coherent Optical Fiber Communications,* Kluwer, Boston, 1988.

Palais, J. C., *Fiber-Optic Communications,* Prentice-Hall, Englewood Cliffs, NJ, 1988.

Personick, S. D., *Optical Fiber Transmission Systems,* Plenum, New York, 1981.

Personick, S. D., *Fiber Optics*: *Technology and Applications,* Plenum Press, New York, 1985.

Runge, P. K. and P. R. Trischitta (eds.), *Undersea Lightwave Communications,* IEEE Press, New York, 1986.

Senior, J. M., *Optical Fiber Communications,* Prentice-Hall, London, 1985.

Sharma, A. B., S. J. Halme, et al., *Optical Fiber Systems and Their Components,* Springer-Verlag, Berlin, 1981.

Sibley, M. J. N., *Optical Communications,* Macmillan, London, 1990.

Taylor, H. F. (ed.), *Fiber Optics Communications,* Artech House, Norwood, Mass., 1983.

Taylor, H. F. (ed.), *Advances in Fiber Optics Communications,* Artech House, Norwood, Mass., 1988.

Tsang, W. T. (ed.), *Lightwave Communications Technology,* Semiconductors and Semimetals. Academic Press, Orlando, Fla., 1985.

Fibers in Medicine

Harrington, J. A. (ed.), *Infrared Fiber Optics III,* SPIE, Bellingham, Wash., 1991.

Joffe, S. N., *Lasers in General Surgery,* Williams & Wilkins, Baltimore, 1989.

Katzir, A. (ed.), *Selected papers on optical fibers in medicine,* SPIE, Bellingham, Wash., 1990.

Katzir, A. (ed.), *Proc. Optical Fibers in Medicine VII,* Bellingham, Wash., SPIE, 1992.

Nonlinear Properties of Fibers

Agrawal, G. P., *Nonlinear Fiber Optics,* Academic Press, San Diego, Calif., 1989.

Agrawal, G. P., "*Nonlinear Interactions in Optical Fibers,*" in *Contemporary Nonlinear Optics,* Academic Press, San Diego, Calif., 1992.

Hasegawa, A., *Solitons in Optical Fibers,* Springer-Verlag, Berlin, 1989.

CHAPTER 11
X-RAY OPTICS

James E. Harvey
The Center for Research and Education in Optics and Lasers
(CREOL)
University of Central Florida
Orlando, Florida

11.1 GLOSSARY

Acronyms

ACV	autocovariance function
AXAF	Advanced X-ray astrophysical facility
BBXRT	broadband x-ray telescope
EUV	extreme ultraviolet
FWHM	full-width at half-maximum
HPR	half-power-radius
HPR_G	geometrical half-power-radius
MTF	modulation transfer function
OTF	optical transfer function
PSD	power spectral density function
PSF	point spread function

Symbols

I	irradiance
E	electric field
ε	linear obscuration ratio
f	focal length
λ	wavelength

n	index of refraction
r_o	diffraction-limited (linear) image radius
θ_o	diffraction-limited (angular) image radius
D	aperture diameter
P	radiant power
A_n, A_t	effective collecting area
α, β, γ	direction cosines
σ_s, σ	root-mean-square surface roughness
σ_w	root-mean-square wavefront error
t	substrate thickness
l	autocovariance length
$P(x, y)$	complex pupil function
$H(x, y)$	optical transfer function
$H_S(x, y)$	surface transfer function
$C_S(x, y)$	surface autocovariance function
$\mathscr{S}(\alpha, \beta)$	angle spread function

11.2 *INTRODUCTION*

The following fundamental difficulties have served as a deterrent to the development of x-ray imaging systems: adequate sources and detectors have not always been available; no suitable refractive material exists from which conventional x-ray lenses can be fabricated; mirrors have traditionally exhibited useful reflectances only for grazing incident angles; grazing incidence optical designs are cumbersome and difficult to fabricate and align; scattering effects from imperfectly polished surfaces severely degrade image quality for these very short wavelengths; and finally, the absorption of x-rays by the atmosphere limits practical applications to evacuated propagation paths in the laboratory or to space applications.

New technologies such as synchrotron sources, free electron lasers, and laser-generated plasmas are being developed.[1-5] Improved optical fabrication techniques and the rapidly emerging technology of enhanced reflectance x-ray multilayers are therefore resulting in a resurgence of activity in the field of x-ray and extreme ultraviolet (EUV) imaging systems.[6-9] These significant advances are stimulating renewed efforts in the areas of x-ray/EUV astronomy, soft x-ray microscopy, and x-ray microlithography.[10,11]

In this chapter we will first review the historical background of x-ray optics. The reader will then be warned about several potential pitfalls in the design of x-ray/EUV imaging systems; in particular, it will be emphasized that scattering effects from optical fabrication errors almost always dominate residual optical design errors at these very short wavelengths. The optical performance of x-ray/EUV optical systems will be discussed from a systems engineering viewpoint and a technique of error-budgeting will be discussed. Diffraction effects will be shown to be a significant source of image degradation for certain grazing incidence x-ray optics applications. Scattering effects will be discussed in detail and characterized by an effective surface transfer function. And x-ray multilayers will be shown to act as a surface power spectral density (PSD) filter function. Our

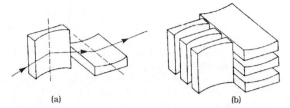

FIGURE 1 (*a*) The Kirkpatrick-Baez telescope consists of two orthogonal grazing incidence parabolic sheet mirrors; (*b*) a multiplate stack of several mirrors can be used to substantially increase the collecting area.

understanding of the combined effect of diffraction, geometrical aberrations, and scattering will then be demonstrated by making accurate image quality predictions for a variety of applications of interest. Finally, a brief summary and conclusion will be presented.

11.3 HISTORICAL BACKGROUND

Grazing Incidence X-ray Imaging Systems

The first two-dimensional image produced by deflecting rays in a controlled manner was obtained by Kirkpatrick and Baez in 1948 with two grazing incidence mirrors as illustrated in Fig. 1(*a*).[12] The extremely small collecting area of such an imaging system can be alleviated by constructing the multiplate Kirkpatrick-Baez mirror of Fig. 1(*b*).

In 1952 Wolter published a paper in which he discussed several rotationally symmetric grazing incidence x-ray telescope systems.[13] The Wolter Type I telescope consists of a coaxial paraboloid (primary mirror) and hyperboloid (secondary mirror) as illustrated in Fig. 2(*a*). The focus of the paraboloid is coincident with the rear focus of the hyperboloid

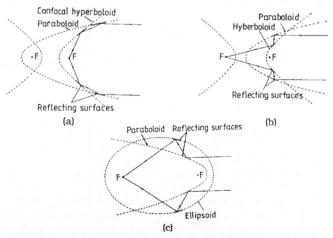

FIGURE 2 (*a*) Wolter Type I telescope; (*b*) Wolter Type II telescope; (*c*) Wolter Type III telescope.

and the reflection occurs on the inside of both mirrors. The Wolter Type II telescope also consists of a coaxial paraboloid and hyperboloid. However, the focus of the paraboloid is coincident with the front focus of the hyperboloid and the reflection occurs on the inside of the paraboloid and the outside of the hyperboloid. This system is the grazing incidence analog to the classical Cassegrain telescope. The Wolter Type III telescope consists of a paraboloid and an ellipse. The focus of the paraboloid is coincident with the front focus of the ellipse, and the reflection occurs on the outside of the paraboloid and on the inside of the ellipse. The Wolter Type I telescope typically has a grazing angle of less than a degree and is used for hard x-rays (greater than 1 keV). The Wolter Type II telescope typically has a grazing angle of approximately 10 degrees and is used for soft x-rays and the extreme ultraviolet (EUV).

These grazing incidence optical configurations are free of spherical aberration; however, they exhibit severe field curvature, coma, and astigmatism. They are also quite cumbersome to fabricate and require a huge surface area to achieve a very small collecting area. Furthermore, at these extremely short wavelengths, very smooth surfaces are required to prevent scattering effects from severely grading the resulting image quality. Aschenbach has presented a nice review of these scattering effects in x-ray telescopes.[14]

Primarily due to much improved optical surface metrology capabilities, the conventional optical fabrication techniques of grinding and polishing glass substrates are resulting in major advances in the resolution of grazing incidence x-ray telescopes.[15,16] The European ROSAT (Röntgensatellit) telescope[17] consisting of four nested Wolter Type I grazing incidence telescopes provides substantial improvement in both effective collecting area and resolution over the Einstein Observatory which was launched in 1978,[18,19] and the technology mirror assembly (TMA)[20] for NASA's advanced X-ray Astrophysical Facility (AXAF)[21] has demonstrated that the technology exists to produce larger Wolter Type I grazing incidence x-ray telescopes with subarcsecond resolution. This progress in grazing incidence x-ray optics performance is illustrated graphically in Fig. 3.[22] A Wolter Type II telescope capable of achieving a half-power-radius of 0.5 arcsec at a wavelength of 1000 Å is currently being studied for use in NASA's Far Ultraviolet Spectroscopic Explorer (FUSE) program.[23]

The very smooth surfaces required of high-resolution x-ray optics have been achieved by tedious and time-consuming optical polishing efforts of skilled opticians. AXAF and

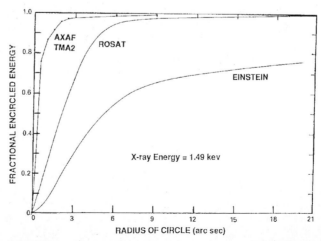

FIGURE 3 Fractional encircled energy plots at 1.49 keV for the Einstein Observatory, the European ROSAT telescope, and the AXAF technology mirror assembly (TMA2) are compared.

ROSAT are thus very expensive to produce. Less labor-intensive (and therefore less expensive) optical fabrication techniques such as plasma-assisted chemical etching (PACE) are therefore being developed.[24]

When such high resolution is not required, other optical materials and fabrication techniques may be applicable. The extreme ultraviolet explorer (EUVE) is a NASA-funded astronomy mission intended to perform an all-sky survey in the 70- to 760-Å spectral region. The deep survey and spectroscopic portion of the mission utilizes a Wolter Type II grazing incidence telescope built at the Space Sciences Laboratory at the University of California, Berkeley. Its mirrors were fabricated from aluminum substrates by forging, rough machining, diamond turning to the desired figure, nickel plating, polishing, and coating with gold. An image half-power-width of approximately 15 arcsec was achieved.[25]

Smooth x-ray mirror surfaces have been achieved without any labor-intensive polishing by merely dipping diamond-turned metal substrates in lacquer, then depositing tungsten or gold coatings to yield the desired high reflectance.[26,27]

The apparent smoothness of the lacquer-coated surfaces and a desire for light weight and a high throughput or filling factor led Petre and Serlemitsos to develop the concept of tightly nested conical foil x-ray telescopes.[28] Conical x-ray imaging mirrors represent the long focal length limit of Wolter Type I grazing incidence mirrors; i.e., the departure of the paraboloid and hyperboloid from simple cones diminishes with increasing focal length. The technology to fabricate and assemble many concentric, nested conical foil x-ray telescopes for high-throughput, moderate resolution, spectroscopic applications has been described in detail. This technique consists of dipping metal foil substrates in lacquer to obtain the optical surface, coating them with gold, then precisely positioning them in a fixture to comprise a tightly packed concentric array of individual telescopes to achieve a relatively high aperture filling factor. The NASA/GSFC broadband x-ray telescope (BBXRT), consisting of 101 concentric conical foil mirror pairs, has recently provided the first flight data from an instrument of this type.[29] Scientists at the Danish Space Research Institute (DSRI) are currently fabricating and testing similar mirrors for the XSPECT telescope being developed for the Soviet Spectrum-X-Gamma mission.[30] This telescope will have 154 concentric shells with grazing angles varying from 9 to 33 arcmin. The Japanese are planning an x-ray telescope mission called SXO in which 89 concentric foil shells will be utilized.[31]

Still other novel optical fabrication concepts for grazing incidence x-ray optical surfaces include a variety of replication techniques.[32] The Italian x-ray astronomy satellite (SAX) will consist of 30 nested coaxial mirrors electroformed over conical mandrels to a thickness ranging from 0.2 to 0.4 min.[33] Finally, the European Space Agency (ESA) will provide a dramatic increase in collecting area with its high throughput x-ray spectroscopy XMM mission featuring several modules of 58 tightly nested confocal Wolter Type I telescopes fabricated with a metal/epoxy replication technique.[34]

Normal Incidence X-ray Multilayers

The possibility of obtaining enhanced reflectance of soft x-rays by multilayer structures was recognized soon after the discovery of x-ray diffraction by crystals. The first successful x-ray multilayers were obtained by Du Mond and Youtz in 1935 by vapor deposition of gold and copper with periods of approximately 100 Å.[35–36] These multilayers turned out to be metallurgically unstable, interdiffusing over a period of a few weeks. In the 1960s, Dinklage and Frerichs fabricated lead/magnesium, gold/magnesium, and iron/magnesium multilayers with periods of 30 to 50 Å which were stable for approximately a year.[37–38]

These multilayer structures were intended for use as dispersion elements in soft x-ray spectroscopy. In 1972, Spiller proposed that quarter-wave stacks of scattering/absorbing materials deposited by thermal-source vapor deposition techniques could be used to develop normal-incidence imaging sysems in the EUV.[39–41] In 1976, Barbee and Keith reported upon sputter deposition techniques for producing multilayer structures on the atomic scale at a Workshop on X-ray instrumentation for Synchrotron Radiation.[42]

The field of normal-incidence x-ray multilayer optics has advanced at a very rapid pace during the last 15 years. There are now over 40 groups worldwide actively working on x-ray multilayers for applications in soft x-ray astronomy, microscopy, microlithography, and synchrotron source experiments.[43–47] Perhaps the most dramatic demonstration of the capabilities of normal-incidence x-ray multilayers to date was provided by the solar corona images obtained with the Stanford/MSFC Multispectral Solar Telescope Array. This 63.5-mm-diameter spherical Cassegrain telescope produced spectroheliograms with approximately 1 arcsec resolution at a wavelength of 171 Å.[48]

These developments permit the return from the Wolter Type grazing incidence x-ray telescope designs to "classical" design forms such as the newtonian (prime focus paraboloid) and cassegrainian telescopes.[49]

11.4   *OPTICAL PERFORMANCE OF X-RAY / EUV IMAGING SYSTEMS*

There has long been a desire to come up with a simple single-number merit function for characterizing the optical performance of imaging systems. However, the proper criterion for evaluating optical performance depends upon a number of different factors: the nature of the source or object to be imaged, the nature of the detector or sensor to be used, and the goal of the particular application.

Image Quality Criteria

The Strehl ratio is a common image quality criterion for near diffraction-limited imaging systems, but is completely inappropriate for some applications. The full width at half maximum (FWHM) of the point spread function (PSF) has been the astronomers' classical definition of *resolution* and is quite appropriate when observing bright point sources; however, the optical transfer function (OTF) is a much more appropriate image quality criterion if the application involves studying fine detail in extended images (such as x-ray solar physics experiments and many soft x-ray microscopy applications). Soft x-ray projection lithography deals with imaging very specific object features which are best evaluated in terms of the square wave response (modulation in the image of a three-bar target) at a particular wavelength and spatial frequency. Fractional encircled energy of the PSF has become a very common image quality requirement imposed upon optical system manufacturers in recent years. Fractional encircled energy is a particularly relevant image quality criterion for some x-ray astronomy programs and for many x-ray synchrotron beamline applications where the imaging system is used to collect radiation and concentrate it on the slit of a spectrographic instrument.

Since the PSF is the squared modulus of the Fourier transform of the complex pupil function, the autocorrelation theorem from Fourier transform theory allows us to define

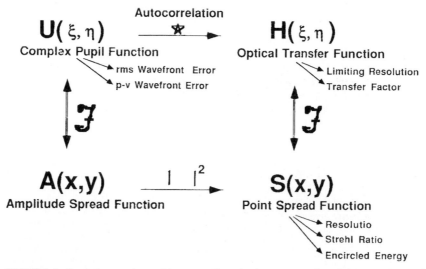

FIGURE 4 Some frequently used image quality criteria are properties of the complex pupil function. Others are obtained from the PSF or the OTF. This figure shows the relationship between these different image quality criteria.

the OTF as the normalized autocorrelation of the complex pupil function and to draw the relationship illustrated in Fig. 4.[50]

The complex pupil function describes the wavefront aberrations that degrade image quality and, furthermore, these wavefront aberrations are rendered observable and measurable by interferometric techniques. Single-number merit functions derivable from interferometric data include the rms wavefront error and the peak-to-valley wavefront error. The amplitude spread function is not an observable quantity with ordinary sensors. The PSF is the irradiance distribution making up the image of an ideal point source. Frequently used single-number merit functions obtained from the PSF are the resolution, Strehl ratio, and the fractional encircled energy. The OTF contains all of the information about the spatial frequency content of the image. Limiting resolution and the transfer factor at a specific spatial frequency are single-number merit functions derivable from the OTF.

Potential Pitfalls in the Design and Analysis of X-ray/EUV Optics

There are many error sources in addition to residual design errors (geometric aberrations) that can limit the performance of high-resolution, short-wavelength imaging systems. Potential pitfalls in the design and analysis of x-ray/EUV imaging systems include: (1) assuming negligible diffraction effects at x-ray wavelengths; (2) overlooking the effects of ghost images in grazing incidence x-ray/EUV systems; (3) assuming that residual design errors dominate optical fabrication errors at these very short wavelengths; and (4) assuming that high reflectance implies negligible scattering in x-ray multilayers. Each of these potential pitfalls in the design and analysis of high-resolution x-ray/EUV imaging systems will be discussed and illustrated by examples in current programs of major interest.

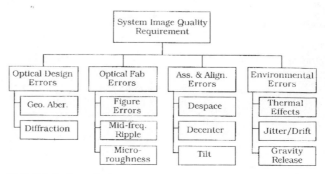

FIGURE 5 Error budget tree form.

Systems Engineering and the Error Budget Tree

The best way to avoid the above pitfalls is to take a systems engineering approach in the design and analysis of x-ray/EUV imaging systems and develop an error budget tree which includes all possible error sources. Figure 5 illustrates one form of an error budget tree that shows these error sources grouped into four main categories: optical design errors (including diffraction effects), optical fabrication errors (including scattering effects from surface irregularities over the entire range of relevant spatial frequencies), assembly and alignment errors, and environmental errors.

The top-level requirement should be expressed in terms of an appropriate image quality criterion for the particular application. Allocations for the individual error sources should then be determined by detailed analysis so that an equitable distribution of difficulty is achieved.

In general, as the wavelength becomes shorter and shorter, diffraction effects will diminish and scattering effects will be enhanced, while the geometrical aberrations are independent of wavelength. Since even the best optical surfaces are not always "smooth" relative to these very short wavelengths, optical fabrication errors will frequently dominate geometrical design errors in the degradation of image quality. Traditional optical design and analysis techniques (geometrical ray tracing) are therefore frequently inadequate for predicting the performance of high-resolution imaging systems at these very short wavelengths.

An error budget tree of this form is useful not only for the initial allocation of allowable errors from which specified tolerances can be derived, but also as a living tool for tracking achievements during the fabrication and alignment process. Occasional reallocation of the remaining error sources can then be made to assure that the system performance goals are met while minimizing the cost and schedule to complete the project.

11.5 DIFFRACTION EFFECTS OF GRAZING INCIDENCE X-RAY OPTICS

Diffraction effects of x-ray optical systems are often (justifiably) ignored due to the small wavelength of the x-ray radiation. However, the extremely large obscuration ratio inherent to grazing incidence optical systems produces profound degradation of the diffraction image over that produced by a moderately obscured aperture of the same diameter.[51]

Diffraction Behavior of Highly Obscured Annular Apertures

The diffraction-limited imaging performance of annual apertures has been discussed in detail by Tschunko.[52] The irradiance distribution of an aberration-free image formed by an annular aperture with a linear obscuration ratio of ε is given by the expression

$$I(x) = \frac{1}{(1-\varepsilon^2)^2} \left[\frac{2J_1(x)}{x} - \varepsilon^2 \frac{2J_1(\varepsilon x)}{\varepsilon x} \right]^2 \qquad (1)$$

For large obscuration ratios, the cross terms in the above squared modulus represent a dominant interference effect that produces an irradiance distribution made up of ring groups as illustrated in Fig. 6.

Tschunko shows that, for $\varepsilon > 0.8$, the number of rings in each ring group is given by

$$n = \frac{2}{(1-\varepsilon)} \qquad (2)$$

Furthermore, 90 percent of the energy is contained within the first ring group and 95 percent within the second ring group, independent of the obscuration ratio. This is compared to 84 percent of the energy in the central lobe of the Airy pattern produced by

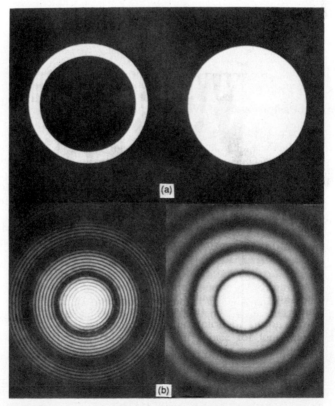

FIGURE 6 (*a*) A highly obscured annular aperture and an un-obstructed circular aperture; (*b*) their respective Fraunhofer diffraction patterns.

an unobscured circular aperture. The central ring group clearly replaces the Airy disc as the meaningful image size. The central lobe itself contains a very small fraction of the energy and in no way represents a meaningful image size or resolution. This is particularly true for the incoherent superposition of images from a nested array of annular subapertures where the spacing of the ring structure varies for each element of the array. Also, small amounts of jitter or optical fabrication error will tend to smear the ring structure without significantly affecting the size of the central ring group or the fraction of energy contained within it.

Tschunko also points out that the minima between the ring groups are regularly spaced and the radii are equal to the number of rings multiplied by π, or

$$\text{Diffraction-limited image radius} = x_o = \pi n = \frac{2\pi}{(1 - \varepsilon)} \tag{3}$$

However, the actual radius of the diffraction image is obtained by setting

$$x = \frac{\pi r}{\lambda f / D} \tag{4}$$

hence

$$r_o = \frac{2\lambda f / D}{1 - \varepsilon} \tag{5}$$

or the angular radius containing 90 percent of the energy is

$$\theta_o = \frac{2\lambda / D}{1 - \varepsilon} \tag{6}$$

Diffraction-limited Performance of Wolter Type I X-ray Telescopes

NASA's Advanced X-ray Astrophysical Facility (AXAF) consists of six concentric Wolter Type I grazing incidence x-ray telescopes. Due to the severe field curvature and off-axis geometrical aberrations characteristic of grazing incidence telescopes,[53,54] several alternative AXAF optical designs that were shorter and more compact and would therefore exhibit more desirable geometric image characteristics were proposed.[55,56] However, to maintain a viable effective collecting area, many more nested shells had to be incorporated, as shown in Fig. 7.

Since the obscuration ratio of each Wolter Type I shell in the AXAF baseline design is 0.98, the diffraction-limited angular radius of the resulting image is given by

$$\theta_n = 100 \, \lambda / D_n \tag{7}$$

The effective collecting area of grazing incidence telescopes varies substantially with both grazing angle and x-ray energy; hence, the composite image made up of the incoherent superposition of the images produced by the individual shells is given by

$$\theta_o = \frac{1}{A_t} \sum_{n=1}^{N} A_n \theta_n \tag{8}$$

where A_t is the total effective collecting area for a given x-ray energy.[57]

At least three different alternative optical designs with 18, 26, and 36 nested Wolter Type I shells were proposed for AXAF. The obscuration ratio, and therefore the diffraction-limited optical performance of these alternative optical designs, are substantially different. Figure 8 illustrates the predicted rms image radius as a function of x-ray

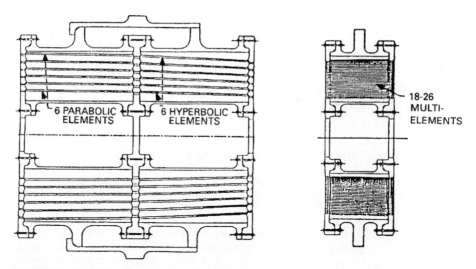

FIGURE 7   NASA baseline AXAF optical design and an alternative design with 18 to 26 concentric shells.

energy for each of these alternative designs plus the baseline design. The rms image radius was calculated from the radius θ_o of the first ring group (which contains 90 percent of the energy) by assuming a Gaussian image distribution with an encircled energy of 90 percent at $\theta = \theta_o$. This is a reasonable assumption since the final image distribution can be portrayed as the convolution of the image distribution due to a variety of error sources, and the central limit theorem of Fourier transform theory states that the convolution of N arbitrary functions approaches a Gaussian function as N approaches infinity.

The entire AXAF telescope error budget allocation (which must include optical

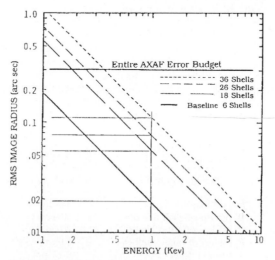

FIGURE 8   Diffraction-limited image quality of various AXAF optical designs.

fabrication, assembly, alignment, environmental, and long-term material stability errors as well as optical design errors) is also indicated in Fig. 8. The diffraction-limited image size for the NASA baseline design is negligible for x-ray energies greater than 1 keV: below 1 keV, diffraction rapidly becomes a dominant error source; and for the x-ray energy of 0.1 keV, it consumes a very significant fraction of the entire error budget. Note also that diffraction alone will prevent all three of the alternative designs from satisfying the top-level AXAF image quality requirement at the low-energy end of the AXAF spectral range.

Diffraction Effects of Nested Conical Foil Imaging Mirrors

Conical x-ray imaging mirrors represent the long focal length limit of Wolter Type I grazing incidence mirrors; i.e., the departure of the paraboloid and hyperboloid from simple cones diminishes with increasing focal length. A conical design and fabrication technique for the many nested thin-walled cones necessary to achieve a relatively high aperture filling factor has been described in Ref. 28.

This conical approximation to the ideal surface figure of the conventional Wolter Type I design results in a field-independent image degradation due to spherical aberration. By going to a short, compact design, the field curvature, coma, and astigmatism can be reduced to the point that they are dominated by the spherical aberration, thus resulting in a constant spatial resolution across the field of view. In Ref. 28, Petre and Serlemitsos have determined the geometrical optical performance of several conical x-ray telescope designs by utilizing a Monte Carlo ray-tracing procedure. In their conclusion they state that it is possible to design a conical foil telescope with arbitrarily high intrinsic spatial resolution by merely reducing the length of the mirror elements and increasing their number to maintain an acceptable effective collecting area. This, of course, increases the obscuration ratio and the corresponding image degradation due to diffraction.

Table 1 lists the design parameters taken from Ref. 28 for the NASA/GSFC BBXRT prototype mirror and a conical mirror assembly with the AXAF diameter and focal length.

The five parameters f, W, R_o, R_i, and t (defined in Table 1) completely define the conical mirror design, and the number of concentric shells N_t is then determined by the geometry; alternatively, given N_t, either R_o or R_i becomes constrained. Also tabulated is the predicted spatial resolution expressed as the angle within which half of the geometrically reflected rays fall. Petre and Serlemitsos also state that the axial spatial resolution may be expressed as

$$\text{HRP}_G = KWR_o^2/f^2 \tag{9}$$

where K depends weakly upon t and R_i.

From the above design parameters, one can calculate the minimum obscuration ratio of the individual nested shells making up these two x-ray telescope designs. The obscuration ratio for the BBXRT prototype is 0.9934 and that for the AXAF application is 0.9988.

TABLE 1 Nested Conical Foil Design Parameters[28]

Design parameter	BBXRT prototype	AXAF application
f (focal length)	380 cm	1000 cm
W (length of each mirror)	10.0 cm	5.0 cm
R_o (radius of outer shell)	20 cm	60 cm
R_i (radius of inner shell)	8.9 cm	30 cm
t (substrate thickness)	0.17 mm	0.17 mm
N_t (number of shells)	101	900
HRP_G (half-power-radius)	12 arcsec	3 arcsec

These are extremely high obscuration ratios and may result in significant diffraction effects for some x-ray energies of interest.

Harvey has shown that diffraction effects are negligible compared to the geometrical HPR for the BBXRT design.[51] However, suppose that we want to achieve a higher resolution. Petre and Serlemitsos state that virtually any intrinsic spatial resolution is possible by merely reducing W and increasing N_t to maintain an acceptable effective collecting area. This, of course, increases the obscuration ratio and the corresponding image degradation due to diffraction. Since the intrinsic image size is obtained by convolving these two independent contributions for each shell and calculating the effective area weighted average for all of the shells, a crossover point, beyond which diffraction is the dominant mechanism limiting the intrinsic resolution of the nested conical imaging telescope, is reached. We refer to *intrinsic resolution* as the image size produced by design parameters alone, assuming no image degradation due to fabrication errors, assembly and alignment errors, environmental errors, etc. Figure 9 illustrates the angular resolution (expressed as a half-power-radius) obtained by root-sum-squaring the geometrical half-power radius and the diffraction-limited half-power radius for several different conical mirror element lengths (W) as a function of x-ray energy. Clearly it is impossible to produce a telescope with these design parameters resulting in a resolution better than approximately 2.5 arcsec for 1 keV x-rays.

There is widespread international activity in the design and fabrication of grazing incidence x-ray telescopes, with particular interest in tightly nested conical foil designs. Furthermore, diffraction effects do not appear to be widely recognized as a potential limitation to spatial resolution in x-ray imaging systems. We have shown here that diffraction alone can prevent grazing incidence x-ray telescopes from meeting the desired optical performance requirements. However, just as diffraction can dominate geometrical aberrations in these grazing incidence imaging systems, we will soon see that scattering effects and alignment errors will almost always dominate both diffraction and geometrical aberrations.

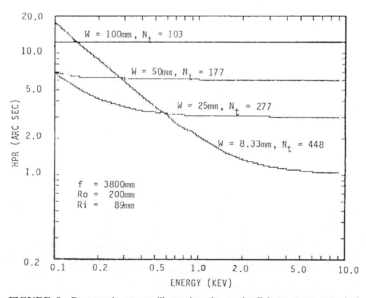

FIGURE 9 Parametric curves illustrating the tradeoff between geometrical effects and diffraction effects in the image size as the BBXRT mirror elements are shortened in an attempt to impove angular resolution.

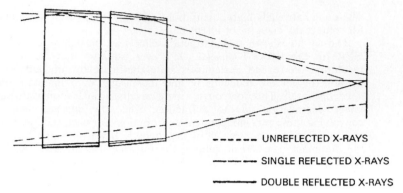

- - - - - - UNREFLECTED X-RAYS

— - — - — SINGLE REFLECTED X-RAYS

————→ DOUBLE REFLECTED X-RAYS

FIGURE 10 Ghost images in Wolter Type I telescopes are caused by x-rays which originate from out-of-field sources and reach the focal plane without being reflected from both mirrors.

11.6 *GHOST IMAGES IN GRAZING INCIDENCE X-RAY TELESCOPES*

Images are formed in Wolter Type I grazing incidence telescopes by x-rays which are reflected from both the primary and the secondary mirrors before reaching the focal plane. Ghost images result from rays which originate from out-of-field sources and reach the focal plane without being reflected from both mirrors as illustrated in Fig. 10.

When ghost images strike the focal plane near the optical axis, they can seriously degrade the quality of the desired image. Because of the extreme grazing angles of the Wolter Type I telescopes, ghost images are difficult to control by conventional aperture plates and baffles.

Aperture plates are usually required in a Wolter Type I system for the structural purpose of mounting and supporting the cylindrical mirrors. There are frequently conflicting demands placed upon the design of these aperture plates by requirements on field of view, vignetting, and ghost image control. Moran and Harvey have discussed the five different positions of aperture plates illustrated in Fig. 11 for use in the control of ghost images in Wolter Type I telescopes.[58]

The spot diagrams in Fig. 12 were generated as aperture plates, sequentially added to a Wolter Type I system. The system focal length was 1000 cm and the mirrors were 80 cm

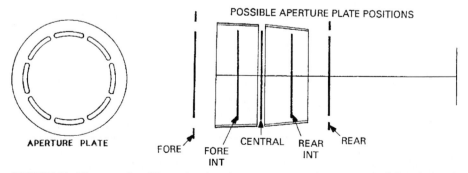

APERTURE PLATE

POSSIBLE APERTURE PLATE POSITIONS

FORE FORE INT CENTRAL REAR INT REAR

FIGURE 11 There are five different locations for aperture plates in the control of ghost images in Wolter Type I telescopes.

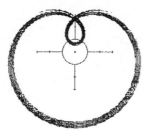

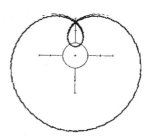

(a) NO APERTURE PLATES   (b) CENTRAL APERTURE PLATE ONLY

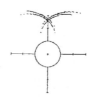

(c) CENTRAL AND REAR   (d) FRONT, CENTRAL, AND REAR

FIGURE 12 Ghost image characteristics as aperture plates are sequentially added to a Wolter Type I telescope.

long with a nominal diameter of 60 cm. This results in a grazing angle of approximately 0.43 degrees. These spot diagrams qualitatively demonstrate the relationship between ghost image characteristics and aperture plate position. The distribution of singly reflected ghost images from a point source at a field angle of 40 min of arc when no aperture plates are present is shown in Fig. 12(a). The circle centered on the optical axis in the focal plane is 15 min of arc in radius and represents the field within which we would like to exclude all ghost images.

The addition of a central aperture plate which does not introduce any vignetting thins out the ghost image distribution by blocking many of the singly-reflected rays, as shown in Fig. 12b. If we now add an aperture plate at the rear of the hyperboloid that satisfies a given vignetting requirement, we obtain the ghost image distribution shown in Fig. 12c. Notice that the ghost images furthest from the optical axis have been blocked. Finally, if we add a front aperture plate in front of the paraboloid which satisfies the same vignetting requirement, the ghost image distribution shown in Fig. 12d is produced. This fore aperture plate has eliminated the ghost images closest to the optical axis.

The position of the fore aperture plate (which blocks the ghost images closest to the optical axis in the focal plane) is a critical parameter in ghost image control. To obtain the ghost image pattern shown in Fig. 12d, the fore aperture plate was placed 100 cm in front of the primary mirror. This long distance would result in rather cumbersome and unwieldy aperture plate support structure. Figure 13 illustrates the ghost image behavior as a function of this fore aperture plate position. Note that 50 cm is the shortest distance in front of the primary mirror that the front aperture plate can be placed in order to eliminate all ghost images from a 20-arcmin semifield angle.

It is impractical to place aperture plates in front of each concentric shell of the tightly nested, high-throughput conical foil x-ray telescopes. A full analysis of the geometrical ghost imaging characteristics has thus been performed for the Soviet XSPECT and it was determined that it would be impossible to completely eliminate ghost images within a 10-arcmin radius of the optical axis in the focal plane.[59]

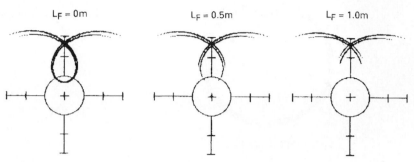

FIGURE 13 Illustration of ghost image distribution for various positions of the fore aperture plate.

Ghost image behavior is also a major concern for Wolter Type II telescopes. They typically have grazing angles of approximately 10 degrees and also require unconventional techniques for stray radiation control. Mangus has developed a strategy and calculations for the design of baffles for Wolter Type II telescopes.[60]

11.7 SCATTERING EFFECTS FROM OPTICAL FABRICATION ERRORS

When light is reflected from an imperfect optical surface, the reflected radiation consists of a specularly reflected component and a diffusely reflected component as illustrated in Fig. 14. The light scattered from optical surface irregularities degrades optical performance in several different ways: (1) it reduces optical throughput since some of the scattered radiation will not even reach the focal plane, (2) the wide-angle scatter will produce a veiling glare which reduces image contrast or signal-to-noise ratio, and (3) the small-angle scatter will decrease resolution by producing an image blur.

It is customary to present angular scattering data as scattered intensity (flux per unit solid angle) versus scattering angle. Figure 15a illustrates several scattered light profiles from a polished and aluminized fused quartz sample. The results confirm the well-known fact that the scattered light distribution changes shape drastically with angle of incidence, becoming quite skewed and asymmetrical at large angles of incidence. However, if we take these scattered intensity curves and divide by the cosine of the scattering angle, then replot as a function of $\beta - \beta_o$, where $\beta = \sin \theta$ and $\beta_o = \sin \theta_o$, the five curves with the incident angle varying from 0 to 60 degrees coincide almost perfectly, as shown in Fig. 15b.

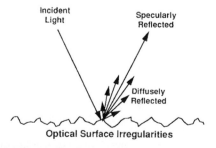

FIGURE 14 Optical surface irregularities produce a specularly reflected beam with a diffusely reflected component that can degrade optical performance in several different ways.

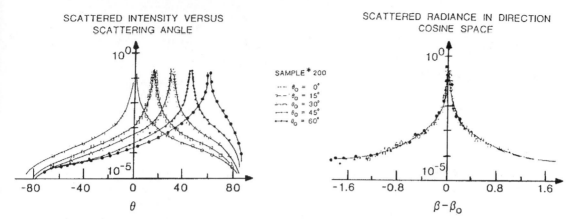

SCATTERED INTENSITY VERSUS
SCATTERING ANGLE

SAMPLE * 200

- $\theta_0 = 0°$
- $\theta_0 = 15°$
- $\theta_0 = 30°$
- $\theta_0 = 45°$
- $\theta_0 = 60°$

SCATTERED RADIANCE IN DIRECTION
COSINE SPACE

FIGURE 15 (*a*) Scattered intensity versus scattering angle; (*b*) scattered radiance in direction cosine space.

Transfer Function Characterization of Scattering Surfaces

The *shift-invariant* scattering behavior shown above implies the existence of a *surface transfer function* which can be shown to relate the scattering properties to the surface characteristics.[61] By describing surface scatter phenomena as a diffraction process in which the rough surface introduces random phase variations into the effective pupil function of the system, an analytical expression can be obtained for the transfer function of the scattering surface. Harvey has shown that if we assume a stationary process (i.e., a random, homogeneous, isotropic mirror surface), and a Gaussian surface height distribution function as shown in Fig. 16, this transfer function is described by the following expression:[62]

$$H_s(\hat{x}, \hat{y}) = \exp\left\{-(4\pi\hat{\sigma}_s)^2\left[1 - \hat{C}_s\left(\frac{\hat{x}}{\hat{l}}, \frac{\hat{y}}{\hat{l}}\right)\Big/\hat{\sigma}_s^2\right]\right\} \tag{10}$$

This transfer function is described only in terms of the rms surface roughness and the surface autocovariance function and therefore provides a simple solution to the inverse scattering problem. Note that a scaled coordinate system is utilized in which the spatial coordinates are normalized by the wavelength of light; i.e., $\hat{x} = x/\lambda$, $\hat{y} = y/\lambda$, and $\hat{\sigma} = \sigma/\lambda$, etc.

Considerable insight into the scattering process can now be obtained by considering the

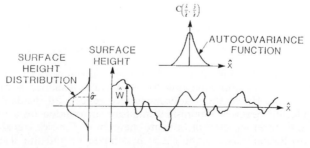

FIGURE 16 A surface profile and the relevant statistical parameters.

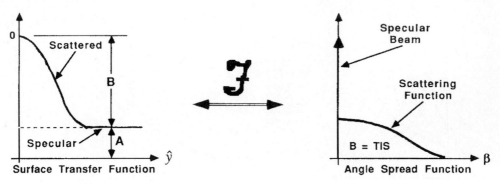

FIGURE 17 (*a*) The surface transfer function, and (*b*) the associated angle spread function are related by the Fourier transform operation just as are the optical transfer function (OTF) and the point spread function (PSF) of modern image formation theory.

nature of this transfer function. The autocovariance function approaches the value σ^2 as the displacement approaches zero. The equivalent transfer function thus approaches unity as expected. As the displacement approaches infinity, the autocovariance function approaches zero and the equivalent transfer function approaches a plateau of height $\exp\left[-(4\pi\hat{\sigma}_s)^2\right]$. The equivalent transfer function of the scattering surface can thus be regarded as the sum of a constant component and a bell-shaped component as shown in Fig. 17*a*. Equation (10) can therefore be rewritten as

$$H(\hat{x}, \hat{y}) = A + BQ(\hat{x}, \hat{y}) \tag{11}$$

where

$$A = \exp\left[-(4\pi\hat{\sigma})^2\right] \tag{12}$$

$$B = 1 - A = 1 - \exp\left[-(4\pi\hat{\sigma})^2\right] \tag{13}$$

$$Q(\hat{x}, \hat{y}) = \frac{\exp\left[(4\pi\hat{\sigma}_s)^2\hat{C}_s\left(\frac{\hat{x}}{\hat{l}}, \frac{\hat{y}}{\hat{l}}\right)\Big/\hat{\sigma}_s^2\right] - 1}{\exp\left[-(4\pi\hat{\sigma}_s)^2\right] - 1} \tag{14}$$

The significance of this interpretation of the equivalent transfer function of the scattering surface is shown by the inferred properties of the corresponding *angle spread function*. Since the transfer function is the sum of two separate components, the angle spread function of the scattering surface is the sum of the Fourier transforms of the two component functions.

$$\mathscr{S}(\alpha, \beta) = \mathscr{F}\{H_s(\hat{x}, \hat{y})\} = A\,\delta(\alpha, \beta) + S(\alpha, \beta) \tag{15}$$

where $\mathscr{F}$ denotes the Fourier transform operator and the scattering function is given by

$$S(\alpha, \beta) = B\mathscr{F}\{Q(\hat{x}, \hat{y})\} \tag{16}$$

The constant component of the transfer function transforms into a delta function, and the bell-shaped component transforms into a bell-shaped scattering function as illustrated in Fig. 17*b*. Hence the scattering surface reflects an incident beam of light as a specularly reflected beam of diminished intensity surrounded by a halo of scattered light. Furthermore, from the central ordinate theorem of Fourier transform theory, the relative power distribution between the specular component and the scattered component of the angle spread function are given by the quantities A and B, respectively.

By describing the transfer function in terms of the normalized spatial variables, the

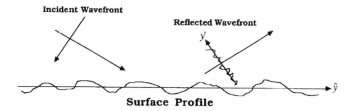

FIGURE 18 Grazing incidence reduces the rms wavefront error induced by the surface irregularities and foreshortens the wavefront features in the plane of incidence.

reciprocal variables in Fourier transform space are the direction cosines of the propagation vectors of the scattered radiation.

Effects of Grazing Incidence. The profile of an optical surface determines the detailed phase front of the reflect wave. At normal incidence, the surface irregularities are merely replicated onto the reflected wavefront with the wavefront error W being twice the surface error. For an arbitrary angle of incidence, this wavefront error is given by $W = 2h \cos \theta$, where h is the surface height deviation and θ is the incident angle measured from the surface normal. The grazing angle ϕ is the complement of the incident angle θ; hence, the rms wavefront error is

$$\sigma_w = 2\sigma_s \sin \phi \tag{17}$$

where σ_s is the rms surface roughness. There is also a foreshortening of the wavefront features in the plane of incidence by an amount equal to the sine of the grazing angle as illustrated in Fig. 18.

An isotropic surface with a rotationally symmetric surface autocovariance function of width l will thus produce a wavefront autocovariance function which is attenuated by the factor $\sin^2 \phi$ and foreshortened in the plane of incidence by the factor $\sin \phi$[63]

$$C_W(\hat{x}, \hat{y}) = (4 \sin^2 \phi)C_s\left(\frac{\hat{x}}{\hat{l}}, \frac{\hat{y}}{\hat{l} \sin \phi}\right) \tag{18}$$

The transfer function of a scattering surface at grazing incidence can thus be written as

$$H_S(\hat{x}, \hat{y}) = \exp\left\{-(4\pi(\sin \phi)\sigma_s)^2\left[1 - \hat{C}_s\left(\frac{\hat{x}}{\hat{l}}, \frac{\hat{y}}{\hat{l} \sin \phi}\right)\bigg/\hat{\sigma}_s^2\right]\right\} \tag{19}$$

and the previous expression, Eq. (11), for the surface transfer function in terms of the quantities A and B is still valid:

$$H(\hat{x}, \hat{y}) = A + BQ(\hat{x}, \hat{y}) \tag{20}$$

However, now

$$A = \exp\left[-(4\pi(\sin \phi)\hat{\sigma}_s)^2\right] \tag{21}$$

$$B = 1 - A = 1 - \exp\left[-(4\pi(\sin \phi)\hat{\sigma}_s)^2\right] \tag{22}$$

and

$$Q(\hat{x}, \hat{y}) = \frac{\exp\left[(4\pi \sin \phi \hat{\sigma}_s)^2 \hat{C}_s\left(\frac{\hat{x}}{\hat{l}}, \frac{\hat{y}}{\hat{l} \sin \phi}\right)\bigg/\hat{\sigma}_s^2\right] - 1}{\exp\left[-(4\pi(\sin \phi)\hat{\sigma}_s)^2\right] - 1} \tag{23}$$

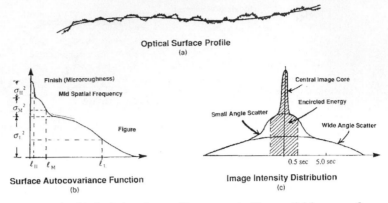

Optical Surface Profile
(a)

Surface Autocovariance Function
(b)

Image Intensity Distribution
(c)

FIGURE 19 (*a*) Optical surface profile composed of low spatial frequency figure errors, mid spatial frequency surface irregularities, and high spatial frequency microroughness; (*b*) composite surface autocovariance function; (*c*) point spread function consisting of a narrow image core, a small-angle scatter function, and a wide-angle scattered halo.

The quantities A and B still correspond to the fraction of the total reflected energy which is contained in the specular beam and the total integrated scatter (TIS), respectively.

Bridging the Gap Between Figure and Finish. It is finally being recognized that residual surface roughness over the *entire range of relevant spatial frequencies* must be specified and controlled in many precision optical systems. This includes the "mid" spatial frequency surface errors that span the gap between the traditional "figure" and "finish" errors.[64] For many high-resolution grazing incidence x-ray imaging systems, the small-angle scatter due to mid spatial frequency surface ripple is often a dominant source of image degradation.

If we schematically represent a hypothetical optical surface as a low spatial frequency surface figure error overlaid with an uncorrelated mid spatial frequency ripple, which in turn is modulated by a high spatial frequency microroughness as shown in Fig. 19*a*; we can express the optical fabrication transfer function of this surface as the product of three separate transfer functions representing each of the three spatial frequency regimes

$$H_{\text{fab}} = H_L + H_M + H_H \tag{24}$$

or

$$
\begin{aligned}
H_{\text{fab}} = {}& \exp\left\{-\left(4\pi(\sin\phi)\hat{\sigma}_L\right)^2[1 - \hat{C}_L/\hat{\sigma}_L^2]\right\} \\
& \times \exp\left\{-\left(4\pi(\sin\phi)\hat{\sigma}_M\right)^2[1 - \hat{C}_M/\hat{\sigma}_M^2]\right\} \\
& \times \exp\left\{-\left(4\pi(\sin\phi)\hat{\sigma}_H\right)^2[1 - \hat{C}_H/\hat{\sigma}_H^2]\right\}
\end{aligned}
\tag{25}
$$

The expression for the total fabrication transfer function can be simplified by noting that the total surface autocovariance function can be expressed as a sum of the individual autocovariance functions representing the different spatial frequency regimes as illustrated in Fig. 19*b*

$$C_S = C_L + C_M + C_H \tag{26}$$

The component variances sum to a total surface variance of

$$\sigma_s^2 = \sigma_L^2 + \sigma_M^2 + \sigma_H^2 \tag{27}$$

hence

$$H_{\text{fab}} = \exp\left\{-\left(4\pi(\sin\phi)\hat{\sigma}_s\right)^2[1 - \hat{C}_s/\hat{\sigma}_s^2]\right\} \tag{28}$$

The microroughness will produce a very wide-angle scatter function, the mid spatial frequency ripple will produce a small-angle scatter function, and the low spatial frequency figure error will contribute to the central image core as depicted in Fig. 19c. The shaded portion of the image intensity distribution represents the fractional encircled energy.

A requirement on fractional encircled energy is widely used as a specification on image quality. Since most x-ray/EUV imaging systems are far from diffraction-limited, the image core is usually completely contained within the specified circled. Virtually all the radiation scattered at wide angles due to microroughness falls outside the specified circle. However, the small-angle scatter distribution due to mid spatial frequency surface errors may have a width comparable to the specified circle diameter. Hence, the fractional encircled energy will be quite sensitive to both the amplitude and spatial frequency of the optical surface errors in this intermediate domain between the traditional surface figure and surface finish.

Short-wavelength Considerations. At these very short x-ray wavelengths, the smooth-surface approximation is frequently not valid, and conventional perturbation techniques cannot always be used for calculating the effects of scatter due to residual optical fabrication errors. It is important that we thoroughly understand the relationship between these surface characteristics and the associated image degradation.

The transfer function characterization of scattering surfaces expressed by Eq. (10) was based upon a scalar diffraction formulation; however, no explicit smooth surface approximation has been made. As illustrated in Fig. 20, the Fourier transform of the surface autocovariance function is the surface power spectrum or power spectral density (PSD) function. The image characteristics (generalized point spread function) are related to the complex pupil function in exactly the same way as the surface PSD is related to the surface profile; however, the surface profile appears in the phase term of the complex pupil function

$$P(\hat{x}, \hat{y}) = A(\hat{x}, \hat{y}) \exp [2\pi \hat{W}(\hat{x}, \hat{y})], \qquad \hat{W}(\hat{x}, \hat{y}) = 2(\sin \phi)\hat{h}(\hat{x}, \hat{y}) \tag{29}$$

The surface transfer function, expressed previously in Eq. (28), becomes one term of the system optical transfer function (OTF). The other term is the conventional OTF representing the effects that determine the width of the image core (diffraction, aberrations, misalignments, environmental errors, etc.)

$$H(\hat{x}, \hat{y}) = H_c(\hat{x}, \hat{y})H_{\text{fab}}(\hat{x}, \hat{y}) \tag{30}$$

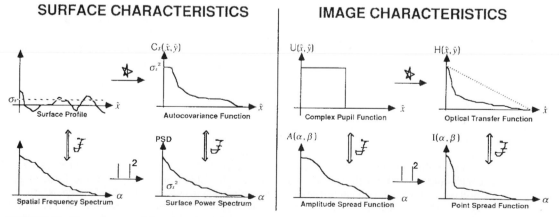

FIGURE 20 The point spread function of an imaging system is related to the complex pupil function in exactly the same way that the surface PSD is related to the surface profile.

The generalized point spread function will thus be given by convolving this image core with the angle spread function discussed earlier in Eq. (15)

$$I(\alpha, \beta) = I_c(\alpha, \beta) * \mathscr{S}(\alpha, \beta) \tag{31}$$

It should again be emphasized that the smooth surface approximation is generally not valid at these very short wavelengths; however, if we do assume that $\hat{\sigma} \ll 1$, the specular reflectance is given by

$$A \approx 1 - (4\pi\hat{\sigma}_s)^2 \tag{32}$$

the total integrated scatter (TIS) is given by

$$B \approx (4\pi\hat{\sigma}_s)^2 \tag{33}$$

and the function $Q(\hat{x}, \hat{y})$ reduces to the normalized surface autocovariance function

$$Q(\hat{x}, \hat{y}) \approx \hat{C}_s\left(\frac{\hat{x}}{\hat{l}}, \frac{\hat{y}}{\hat{l}}\right) \Big/ \hat{\sigma}_s^2 \tag{34}$$

This results in the often quoted and widely used statement that the scattering function is directly proportional to the surface power spectral density (PSD) function.[65,66]

$$S(\alpha, \beta) = \text{scattering function} = B\mathscr{F}\{Q(\hat{x}, \hat{y})\} = (4\pi/\lambda)^2 \text{PSD} \tag{35}$$

However, one must remember that this is only true when the smooth surface approximation is valid, and even for the best optical surfaces available, this is frequently not the case when dealing with x-ray/EUV radiation.

Multilayer Scattering Characteristics

It is widely recognized that scatter effects from interface microroughness can decimate the reflectance of multilayer coatings in normal-incidence soft x-ray imaging systems.[67–71] The (spectral) reflectance has thus become the most common measure of performance in the evaluation of x-ray multilayers. However, high x-ray reflectance is a necessary but not sufficient condition for producing high-quality images. A second and equally important condition is the ability to concentrate the reflected radiation in a very small region in the focal plane. We will proceed to describe the concept of an enhanced reflectance multilayer as a low-pass spatial frequency filter acting upon the substrate PSD. This concept allows us to apply conventional linear systems techniques to the evaluation of image quality, and to the derivation of optical fabrication tolerances, for applications utilizing multilayer coatings.

The reflectance, at normal incidence, from a single interface can be calculated using the following Fresnel reflectance formula, if the refractive indices of the respective media separated by the interface are known.

$$R = |r|^2 = (n_2 - n_1)^2/(n_2 + n_1)^2 \tag{36}$$

This value for reflectance is valid whether the interface is rough or not. In the case of a rough interface, the reflected wavefront takes on a disturbance of the same form as the interface but of twice the amplitude. As a result, some fraction of the light will be scattered away from the specular direction. We represent this fact by writing the specularly reflected power as

$$P_s = RAP_o \tag{37}$$

and the diffusely scattered power as

$$P_d - RBP_o \tag{38}$$

where R is the Fresnel reflectance, P_o is the incident power, and A and B are again given by

$$A = \exp\left[-(4\pi\sigma/\lambda)^2\right] \tag{39}$$

and

$$B = \text{total integrated scatter} = 1 - A \tag{40}$$

and are the fraction of reflected power in the specular and scattered beams, respectively. Here σ is the rms surface roughness over the entire range of relevant spatial frequencies, from high spatial frequency microroughness, through mid spatial frequency surface irregularities, and including low spatial frequency figure errors

$$\sigma^2 = \sigma_L^2 + \sigma_M^2 + \sigma_H^2 \tag{41}$$

Similarly, the radiation reflected by an enhanced reflectance multilayer will consist of a specular and diffuse part as shown in Fig. 21. The specularly reflected and diffusely reflected components are still given by equations of the form of Eqs. (37) and (38); however, the relative strength of the specular beam will now depend upon the *effective* roughness, σ', which in turn depends upon the degree of *correlation* between the various interfaces. Hence, for the multilayer

$$A' = \exp\left[-(4\pi\sigma'/\lambda)^2\right] \tag{42}$$

and

$$B' = \text{total integrated scatter} = 1 - A' \tag{43}$$

where

$$\sigma'^2 = \sigma_L'^2 + \sigma_M'^2 + \sigma_H'^2 \tag{44}$$

Eastman (1974), Carniglia (1979), Elson (1980), and Amra (1991) have also contributed to our understanding of scattering from optical surfaces with multilayer coatings.[72–75] However, in most cases they have dealt only with microroughness and have assumed that the interfaces are either perfectly correlated or perfectly uncorrelated.

We are considering the surface irregularities over the entire range of relevant spatial frequencies from the very low spatial frequency *figure* errors to the very high spatial

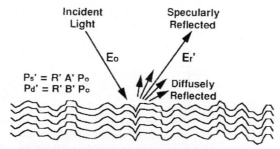

FIGURE 21 An enhanced reflectance multilayer also produces a specularly reflected beam and a diffusely reflected component whose relative strengths depend upon the degree of correlation between the various interfaces.

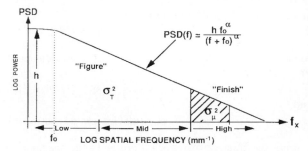

FIGURE 22 The substrate surface PSD is assumed to obey an inverse power law which spans the entire range of spatial frequencies from low spatial frequency *figure* errors to high spatial frequency *finish*.

frequency *finish* errors. It therefore seems rather intuitive that for any reasonable thin-film deposition process, the low spatial frequency figure errors will *print through* and be correlated from layer to layer, while the high spatial frequency microroughness inherent to the deposition process itself will be uncorrelated from layer to layer. Let us assume that each interface making up the multilayer has the same surface statistics.

Let us also assume that the interfaces can be characterized by an inverse power law surface PSD as illustrated in Fig. 22.

Note that this is a radial profile of a two-dimensional PSD plotted as log power expressed in waves squared per spatial frequency squared versus log spatial frequency. The units are thus consistent with a volume under the PSD of σ^2 as required. The low, mid, and high spatial frequency domains are indicated in Fig. 22. Furthermore, knowing the functional form of the surface PSD now enables one to calculate the total rms surface error σ from a band-limited measurement of the microroughness σ_μ with an instrument such as a micro phase-measuring interferometer.

It has been pointed out by Spiller that spatially uncorrelated microroughness in x-ray/EUV multilayers will yield an effective rms surface roughness reduced by a factor of $1/\sqrt{N}$, where N is the number of layer pairs.[69] From this and the knowledge that the interfaces will be highly correlated at mid and low spatial frequencies, it is clear that the multilayer will act as a low-pass spatial frequency filter which has a value of unity for correlated low spatial frequencies and drops to a value of $1/N$ for uncorrelated high spatial frequencies. The exact location and shape of this cutoff depends upon the material and deposition process. However, the PSD filter function is illustrated qualitatively in Fig. 23.

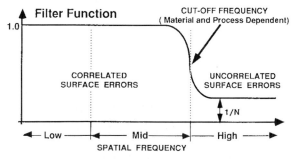

FIGURE 23 Enhanced reflectance multilayer coatings behave as a low-pass spatial frequency filter acting upon the interface PSD.

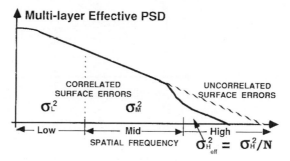

FIGURE 24 The filtered interface PSD is the effective PSD of the multilayer.

Figure 24 illustrates the effective multilayer PSD obtained by multiplying the interface PSD by the multilayer filter function. Note that the effective microroughness (and hence the wide-angle scatter) has been substantially reduced.

This effective PSD of the multilayer completely characterizes the scattering effects which degrade the image quality of normal-incidence x-ray imaging systems.

Other useful references dealing with surface scatter for general optical applications include *Introduction to Surface Roughness and Scattering* by J. M. Bennett and L. Mattsson,[76] *Optical Scattering, Measurement and Analysis* by John C. Stover,[77] and a myriad of excellent papers published over the years by Eugene Church.[78–85]

11.8 IMAGE QUALITY PREDICTIONS FOR VARIOUS APPLICATIONS

A scalar diffraction treatment of surface scatter phenomena can now be used to make image quality predictions as degraded by residual optical fabrication errors. The effective PSD (including the effects of multilayer scattering) can be Fourier transformed to obtain the effective surface autocovariance function. This effective surface autocovariance function can be substituted into Eq. (28) to obtain the transfer function of the scattering surface. The product of this surface transfer function and the conventional optical transfer function describing the effects of diffraction, geometrical aberrations, misalignments, etc. can then be Fourier transformed to obtain a generalized point spread function including the effects of scattering.[86]

In the absence of good metrology data, parametric calculations can be made to determine the performance sensitivity to assumed optical fabrication tolerances and to wavelength. In fact, these parametric curves can be used to derive the optical fabrication tolerances necessary for a given application. A few examples follow.

Extreme Ultraviolet (EUV) Astronomy

NASA's Far Ultraviolet Spectroscopic Explorer (FUSE) telescope design is the Wolter Type II grazing incidence configuration shown in Fig. 25. The FUSE program requires a fractional encircled energy of 0.5 in a 1.0 arcsec diameter circle from an on-axis point source with a wavelength of 1000 Å. Figure 26 illustrates the severe image degradation that would occur in the 100- to 350-Å EUV region of the electromagnetic spectrum for optical fabrication tolerances chosen to meet the above requirement.[87] The scattering theory discussed in the previous section of this chapter has been used to make exhaustive image

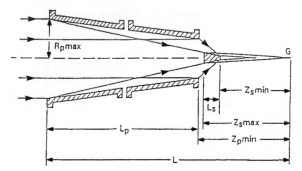

FIGURE 25 Schematic illustration of the FUSE telescope design.

quality predictions as a function of various optical surface parameters. The resulting sensitivity curves were used to determine the optical fabrication tolerances necessary to achieve considerably enhanced EUV performance.[88] These parametric performance predictions can then be used as the basis of a detailed cost versus performance trade study.

High Energy X-ray Astrophysics with Conical Foil X-ray Telescopes

Scientists at the Danish Space Research Institute (DSRI) are currently fabricating and testing conical foil grazing incidence mirrors for the XSPECT telescope being developed for the Soviet Spectrum-X-Gamma mission.[30] This telescope will have 154 concentric shells with grazing angles varying from 9 to 33 arcmin. Image degradation due to diffraction effects and geometrical aberrations caused by the conical shape of these mirrors were discussed previously in this chapter; however, detailed image quality predictions indicate that scattering effects and alignment errors will limit the achievable resolution to

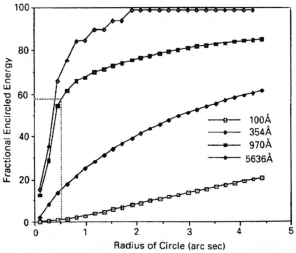

FIGURE 26 Fractional encircled energy at different wavelengths including optical fabrication and alignment errors.

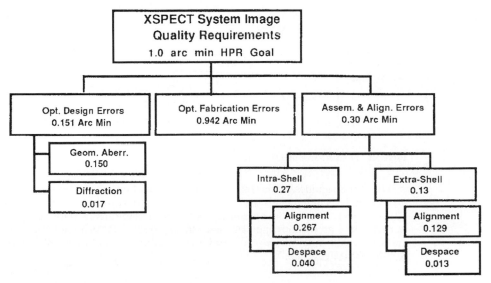

FIGURE 27 XSPECT system error budget.

approximately two minutes of arc.[89] Figure 27 shows an error budget tree that illustrates the relative contributions of the various error sources for x-ray telescopes of this type. Note that scattering effects from optical fabrication errors are by far the dominant error source.

X-ray Solar Physics

Figure 28 illustrates fractional encircled energy predictions for the Stanford/MSFC normal-incidence Cassegrain solar telescope reported to have a resolution of approximately 1.0 arcsec.[48] In order to obtain super-smooth surfaces, spherical mirrors were used rather than the paraboloid and hyperboloid of a classical Cassegrain configuration. This

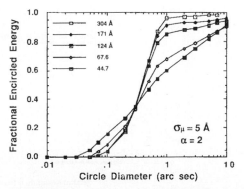

FIGURE 28 Encircled energy predictions for the Stanford/MSFC Cassegrain X-ray Solar Telescope. A surface PSD obeying an inverse-square law and a band-limited microroughness of 5 Å were assumed.

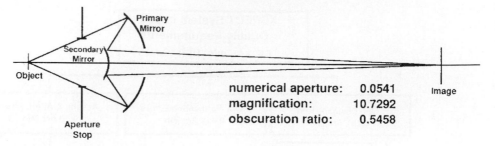

FIGURE 29 Schwarzschild soft x-ray microscope configuration.

resulted in a significant amount of spherical aberration. Note that our predictions indicate a fractional encircled energy of 0.9 in a 1.0-arcsec circle at the operational wavelength of 171 Å, which is in good agreement with the experimental data. However, parametric performance predictions indicate that current optical fabrication technology for aspheric surfaces will not allow the order-of-magnitude improvement projected by the Stanford/MSFC group.[90]

Soft X-ray Microscopy

The Schwarzschild microscope configuration consists of two concentric spherical mirrors with an aperture stop at their common center of curvature as shown in Fig. 29.

The curves in Fig. 30 illustrate the predicted optical performance of this Schwarzschild microscope with state-of-the-art surfaces at the soft x-ray wavelengths of 300 Å, 130 Å, and 50 Å. For general microscopy applications where fine details in extended objects are being studied, some property of the modulation transfer function (MTF) is probably the most appropriate image quality criterion. Let us assume that a modulation of 0.20 is just acceptable for our application. Image plane *resolution* will thus be defined as the reciprocal of the spatial frequency at which the modulation drops to a value of 0.20 and

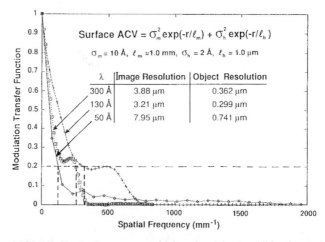

FIGURE 30 Performance sensitivity of a Schwarzschild soft x-ray microscope upon surface scatter effects due to mid spatial frequency optical fabrication errors.

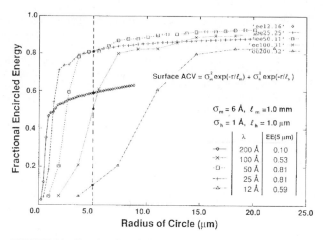

FIGURE 31 Fractional encircled energy of a soft x-ray synchrotron beamline consisting of five grazing incidence mirrors with state-of-the-art fabrication tolerances.

the object plane resolution is this number divided by the microscope magnification. Note that the intermediate wavelength has the best resolution. The long wavelength (300 Å) resolution is limited by diffraction with negligible degradation from scattering. At 130 Å there is a modest 16 percent degradation in the object resolution due to scattering. And at 50 Å, mid spatial frequency scattering effects are by far the dominant image degradation mechanism, with the smallest resolvable object 12 times as large as the diffraction-limited value.[91] Clearly, high resolution soft x-ray microscopy of biological specimens in the water window (25 Å < λ < 50 Å) will require ultrasmooth (rms microroughness < 1 Å) surfaces.

Detailed analysis has shown that a 0.1-μm object resolution with a 1.0-μm depth-of-field over an 850-μm object field diameter can be obtained at a wavelength of 43 Å with a Schwarzschild microscope configuration. This includes the effects of geometric aberrations, assembly and alignment errors, diffraction effects, and scattering effects from optical fabrication errors.[92]

X-ray Synchrotron Source Applications

Figure 31 illustrates the predicted performance of a soft x-ray synchrotron beamline consisting of five grazing incidence mirrors ($\alpha = 3°$) with state-of-the-art optical fabrication tolerances. We have chosen fractional encircled energy in a 5-μm radius circle (corresponds to a 10-μm monochrometer slit width) as our image quality criterion. Again we see that diffraction dominates image degradation at the long wavelengths ($\lambda > 100$ Å), and scattering dominates image degradation at the short wavelength ($\lambda = 12$ Å). The best performance is obtained for intermediate wavelengths (25 Å > λ > 50 Å).[93]

11.9 SUMMARY AND CONCLUSIONS

We first reviewed recent progress in both grazing incidence x-ray imaging systems and the rapidly emerging technology of enhanced reflectance x-ray multilayers. Four specific potential pitfalls in the design and analysis of x-ray/EUV imaging systems were then identified. These included: (1) assuming negligible diffraction effects at x-ray wavelengths; (2) overlooking the effects of ghost images in grazing incidence x-ray/EUV imaging systems; (3) assuming that residual design errors dominate optical fabrication errors at these very short wavelengths; and (4) assuming that high reflectance implies negligible scattering in x-ray multilayers. The importance of utilizing a systems engineering approach and an exhaustive error budget analysis was also emphasized.

The diffraction behavior of highly obscured annular apertures was shown to offset the popular notion that diffraction effects are negligible at short wavelengths.

Grazing incidence imaging systems are particularly difficult to control stray radiation from out-of-field sources. Ghost image behavior in Wolter Type I telescopes was discussed in detail.

It was then emphasized that, even our best optical surfaces are not always "smooth" relative to these very short wavelengths. Image degradation due to scattering effects from surface irregularities in several different spatial frequency regimes was discussed. Small angle scatter from mid-spatial frequency optical fabrication errors that bridge the gap between traditional figure and finish errors were shown to dominate residual design errors in several x-ray/EUV imaging applications of interest. X-ray multilayers were shown to behave as a surface PSD filter function which can conveniently be used to predict image degradation due to interface roughness. Uncorrelated microroughness in x-ray multilayers decimates reflectance but does not significantly degrade image quality.

Finally, parametric image quality predictions for a variety of applications including x-ray/EUV astronomy, soft x-ray microscopy, soft x-ray microlithography, and x-ray synchrotron beamline mirrors were presented to demonstrate the wavelength dependence of optical fabrication tolerances for x-ray imaging systems.

11.10 REFERENCES

1. N. M. Ceglio, "Revolution in X-ray Optics," *J. X-ray Sci. and Tech.* **1:**7–78 (1989).

2. V. Rehn, "Grazing Incidence Optics for Synchrotron Radiation Insertion-device Beams," *Proc SPIE* **640:**106–115 (1986).

3. H. Rarback et al., "Coherent Radiation for X-ray Imaging—The Soft X-ray Undulator and the X1A Beamline at the NSLS," *J. X-ray Sci. and Tech.* **2:**274–296 (1990).

4. D. R. Gabardi and D. L. Shealy, "Optical Analysis of Grazing Incidence Ring Resonators for Free-electron Lasers," *Opt. Eng.* **29:**641–648 (1990).

5. A. G. Michette, "Laser-generated Plasmas: Source Requirements for X-ray Microscopy," *J. X-ray Sci. and Tech.* **2:**1–16 (1990).

6. T. W. Barbee, Jr., "Multilayers for X-ray Optics," *Opt. Eng.* **25:**898–915 (1986).

7. J. F. Osantowski and L. Van Speybroeck (eds.), Twenty-three Papers Presented at the 1986 SPIE Conference "Grazing Incidence Optics," *Proc. SPIE* **640** (1986).

8. Fourteen Papers on X-ray/EUV Optics in a Special Issue of Optical Engineering, *Opt. Eng.*, **29:**576–671 (June 1990).

9. Eleven Papers on X-ray/EUV Optics in a Special Issue of Optical Engineering, *Opt. Eng.*, **29:**698–780 (July 1990).

10. R. B. Hoover (ed.), Seventy-one Papers Presented at the 1989 SPIE Conference "X-ray/EUV Optics for Astronomy and Microscopy," *Proc. SPIE* **1160** (1989).

11. J. Boker (ed.), Thirty-four Papers Presented at a Symposium on "Soft X-ray Projection Lithography," *Proc. OSA* **12** (1991).

12 P. Kirkpatrick and A. V. Baez, "Formation of Optical Images by X-rays," *J. Opt. Soc. Am.* **38:**776 (1948).

13. H. Wolter, "Spiegelsysteme streifenden Einfalls als abbildende Optiken für Röntgenstrahlen," *Ann. Phys.*, NY **10:**94 (1952).

14. B. Aschenbach, "X-ray Telescopes," *Rep. Prog. Phys.,* **48:**579–629 (1985).

15. L. P. Van Speybroeck, "Grazing Incidence Optics for the U.S. High resolution X-ray Astronomy Program," *Opt. Eng.* **27:**1398–1403 (1988).

16. A. Slomba, R. Babish, and P. Glenn, "Mirror Surface Metrology and Polishing for AXAF/TMA," *Proc. SPIE* **597:**40 (1985).

17. B. Aschenbach, "Design, Construction, and Performance of the ROSAT High-resolution Mirror Assembly," *Appl. Opt.* **27:**1404–1413 (1988).

18. R. Giacconi et al., "The Einstein (HEAO 2) X-ray Observatory," *Astrophys. J.* **230:**540 (1979).

19. L. P. Van Speybroeck, "Einstein Observatory (HEAO B) Mirror Design and Performance," *Proc. SPIE* **184:**2 (1979).

20 L. P. Van Speybroeck, P. Reid, D. Schwartz, and J. Bilbro, "Predicted and Preliminary Evaluation of the X-ray Performance of the ACAF Technology Mirror Assembly," *Proc. SPIE* **1160:**94 (1989).

21. M. V. Zombeck, "Advanced X-ray Astrophysics Facility (AXAF)-Performance Requirements and Design Considerations", *Proc. SPIE,* **184** (1979).

22. J. E. Harvey, "Recent Progress in X-ray Imaging", presented at the *AIAA Space Programs and Technologies Conference* in Huntsville, AL (Sept. 1990).

23. T. T. Saha, D. A. Thomas, and J. F. Osantowski, "OSAC Analysis of the Far Ultraviolet Explorer (FUSE) Telescope," *Proc. SPIE* **640:**79–84 (1986).

24. C. B. Zarowin, "A Theory of Plasma-assisted Chemical Vapor Transport Processes," *J. Appl. Phys.* **57**(3):929–942 (1985).

25. S. Bowyer and J. Green, "Fabrication, Evaluation, and Performance of Machined Metal Grazing Incidence Telescopes," *Appl. Opt.* **27:**1414–1422 (1988).

26. R. C. Catura, E. G. Joki, D. T. Roethig, and W. J. Brookover, "Lacquer Coated X-ray Optics," *Proc. SPIE* **640:**140–144 (1986).

27. J. A. Nousek et al., "Diamond-turned Lacquer-coated Soft X-ray Telescope Mirrors," *Appl. Opt.* **27:**1430–1432 (1988).

28. R. Petre and P. J. Serlemitsos, "Conical Imaging Mirrors for High-speed X-ray Telescopes," *Appl. Opt.* **24:**1833 (1985).

29. R. Petre, P. J. Serlemitsos, F. E. Marshall, K. Jahoda, and H. Kunieda, "In Flight Performance of the Broad-Band X-ray Telescope," *Proc. SPIE* **1546:**72–81 (1991).

30. N. J. Westergaad, B. P. Byrnak, F. E. Christensen, P. Grundsoe, A. Hornstrup, S. Henrichsen, U. Henrichsen, E. Jespersen, H. U. Norgaard-Nielsen, and J. Polny, "Status of the Development of a Thin Foil High Throughput X-ray Telescope for the Soviet Spectrum X-gamma Mission," *Proc. SPIE* **1160:**488–499 (1989).

31. Y. Tanaka and F. Makino, "Grazing Incidence Optics for the X-ray Astronomy Mission SXO," *Proc. SPIE* **830:**242 (1987).

32. Y. Matsui, M. P. Ulmer, and P. Z. Takacs, "X-ray and Optical Profiler Analysis of Electroformed X-ray Optics," *Appl. Opt.* **27:**1558–1563 (1988).

33. O. Citterio, et al., "Optics for the X-ray Imaging Concentrators Aboard the X-ray Astronomy Satellite SAX," *Proc. SPIE* **830:**139 (1987).

34. W. Egle, H. Bulla, P. Kaufmann, B. Aschenbach, and H. Brauninger, "Production of the First Mirror Shell for ESA's XMM Telescope by Application of a Dedicated Large Area Replication Technique," *Opt. Eng.* **29:**1267 (1990).

35. J. Du Mond and J. P. Youtz, "Selective X-ray Diffraction from Artificially Stratified Metal Films Deposited by Evaporation," *Phys. Rev.* **48:**703 (1935).

36. J. Du Mond and J. P. Youtz, "An X-ray Method for Determining Rates of Diffusion in the Solid State," *J. Appl. Phys.* **11:**357 (1940).

37. J. Dinklage and R. Frerichs, "X-ray Diffraction and Diffusion in Metal Film Layered Structures," *J. Appl. Phys.* **34:**2633 (1963).

38. J. Dinklage, "X-ray Diffraction by Multilayered Thin Film Structures and Their Diffusion," *J. Appl. Phys.* **38:**3781 (1967).

39. E. Spiller, "Low-loss Reflection Coatings Using Absorbing Materials," *Appl. Phys. Lett.* **20:**365 (1972).

40. E. Spiller, "Multilayer Interference Coatings for the Vacuum Ultraviolet," in Proc. ICO-IX, *Space Optics,* p. 525, Natl. Acad. Science, Washington, D.C. (1974).

41. E. Spiller, "Reflective Multilayer Coatings in the Far UV Region," *Appl. Opt.* **15:**2333 (1976).

42. T. W. Barbee, Jr. and D. L. Kieth, "Synthetic Structures Layered on the Atomic Scale," H. Winick and G. Brown (eds.), in Workshop on *X-ray Instrumentation for Synchrotron Radiation Research,* p. III-26, Stanford SSRL Report 7804 (1978).

43. Ten Papers on X-ray Multilayered Optics in a Special Issue of Optical Engineering, *Opt. Eng.* **25:**897–978 (August 1986).

44. G. F. Marshall (ed.), Forty-eight Papers Presented at the 1985 SPIE Conference "Applications of Thin-film Multilayered Structures to Figured X-ray Optics," *Proc. SPIE* **563** (1985).

45. F. E. Christensen (ed.), Thirty-three Papers Presented at the 1988 SPIE Conference on "*X-ray Multilayers for Diffractometers, Monochromators , and Spectrometers,*" *Proc. SPIE* **984** (1988).

46. R. B. Hoover (ed.), Twenty-three Papers Presented at the 1989 SPIE Conference on "X-ray/EUV Optics for Astronomy and Microscopy," *Proc. SPIE* **1160** (1989).

47. N. M. Ceglio (ed.), Thirty-one papers presented at the 1991 SPIE Conference on "Multilayer Optics for Advanced X-ray Applications," *Proc. SPIE* **1547** (1991).

48. A. B. C. Walker, Jr., T. W. Barbee, Jr., R. B. Hoover, and J. F. Lindblom, "Soft X-ray Images of the Solar Corona with a Normal-Incidence Cassegrain Multilayer Telescope," *Science,* vol. 241:1781 (Sept. 1988).

49. W. J. Smith, *Modern Optical Engineering,* 2nd ed., McGraw-Hill, New York, 1990.

50. J. W. Goodman, *Introduction to Fourier Optics,* McGraw-Hill, New York, 1968.

51. J. E. Harvey, "Diffraction Effects in Grazing Incidence X-ray Telescopes," *J. X-ray Sci. and Tech.* **3:**68–76 (1991).

52. H. F. A. Tschunko, "Imaging Performance of Annular Apertures," *Appl. Opt.* **13:**1820 (1974).

53. W. Werner, "Imaging Properties of Wolter I Type X-ray Telescopes," *Appl. Opt.* **16:**764 (1977).

54. C. E. Winkler and D. Korsch, "Primary Aberrations for Grazing Incidence," *Appl. Opt.* **16:**2464 (1977).

55. United Technologies Research Center, *AXAF Technology Briefiing,* NASA/MSFC, August 7, 1984.

56. D. Korsch, "Near Anastigmatic Grazing Incidence Telescope", *Proc. SPIE* **493** (1984).

57. M. V. Zombeck, "AXAF Effective Area Studies," *SAO-AXAF-83-015* (1983).

58. E. C. Moran and J. E. Harvey, "Ghost Image Behavior in Wolter Type I X-ray Telescopes," *Appl. Opt.* **27:**1486 (15 April 1988).

59. E. C. Moran, J. E. Harvey, F. E. Christensen, N. J. Westergaard, H. W. Schnopper, B. P. Byrnak, and H. U. Noergaard-Nielsen, "Ghost Image Analysis for XSPECT High-throughput X-ray Telescope Mission," presented at the *173rd Meeting of the American Astronomical Society* in Boston, MA (Jan. 1989).

60. J. D. Mangus, "Strategy and Calculations for the Design of Baffles for Wolter Type II telescopes," *Proc. SPIE* **830:**245–253 (1987).

61. J. E. Harvey, "Surface Scatter Phenomena: a Linear, Shift-invariant Process," *Proc. SPIE* **1165:**87–99 (1989).

62. J. E. Harvey, "Light-Scattering Characteristics of Optical Surfaces," Ph.D. Dissertation, Univ. Arizona (1976).

63. J. E. Harvey, E. C. Moran, and W. P. Zmek, "Transfer Function Characterization of Grazing Incidence Optical Systems," *Appl. Opt.* **27:**1527–1533 (1988).

64. R. J. Noll, "Effect of Mid and High Spatial Frequencies on Optical Performance," *Opt. Eng.* **18:**137 (1979).

65. J. C. Stover, "Roughness Characterization of Smooth Machined Surfaces by Light Scattering," *Appl. Opt.* **14:**1796 (1975).

66. E. L. Church, H. A. Henkinson, and J. M. Zavada, "Relationship Between Surface Scattering and Microtopographic Features," *Opt. Eng.* **18:**125 (1979).

67. J. H. Underwood, T. W. Barbee, and D. L. Shealy, "X-ray and Extreme Ultraviolet Imaging Using Layered Synthetic Microstructures," *SPIE* vol. 316, High Res. Soft X-ray Optics (1981).

68. A. E. Rosenbluth and J. M. Forsyth, "The Reflecting Properties of Soft X-Ray Multilayers," *SPIE,* vol. 563, Applications of Thin-Film Multilayered Structures to Figured X-Ray Optics, 284 (1985).

69. E. Spiller and A. E. Rosenbluth, "Determination of Thickness Errors and Boundary Roughness from the Measured Performance of a Multilayer Coating," *Opt. Eng.* **25:**898 (1986).

70. D. L. Windt and R. C. Catura, "Multilayer Characterization at LPARL," *SPIE,* vol. 984, X-ray Multilayers for Diffractometers, Monochromaters, and Spectrometers, 82 (Aug. 1988).

71. D. G. Stearns, "The Scattering of X-rays from Nonideal Multilayer Structures," *J. Appl. Phys.* **65**(2)**:**498 (15 Jan. 1989).

72. J. M. Eastman, "Surface Scattering in Optical Interface Coatings," Ph.D. dissertation, Univ. of Rochester, Rochester, NY (1974). (Available from University Microfilms, Ann Arbor, MI 48106.)

73. C. K. Carniglia, "Scalar Scattering Theory for Multilayer Optical Coatings," *Opt. Eng.* **18:**104 (1979).

74. J. M. Elson, J. P. Rahn, and J. M. Bennett, "Light Scattering from Multilayer Optics; Comparison of Theory and Experiment," *Appl. Opt.* **19:**669 (1980).

75. C. Amra, J. H. Apfel, and E. Pelletier, "The Role of Interface Correlation in Light Scattering by a Multilayer," *Appl. Opt.* **31:**3134–3151 (1992).

76. J. M. Bennett and L. Mattsson, *Introduction to Surface Roughness and Scattering,* Opt. Soc. of Am., Washington, D.C., 1989.

77. J. C. Stover, *Optical Scattering, Measurement and Analysis,* McGraw-Hill, New York, 1990.

78. E. L. Church, "The Role of Surface Topography in X-ray Scattering," *Proc. SPIE* **184:**196 (1979).

79. E. L. Church, "Small-Angle Scattering from Smooth Surfaces," *J. Opt. Soc. Am.* **70:**1592 (1980).

80. E. L. Church, "Interpretation of High-Resolution X-ray Scattering Measurements," *Proc. SPIE* **257:**254 (1980).

81. E. L. Church, "Fractal Surface Finish," *Appl. Opt.* **27:**1518–1526 (1988).

82. E. L. Church and P. Z. Takacs, "Instrumental Effects in Surface Finish Measurements," *Proc. SPIE* **1009:**46–55 (1988).

83. E. L. Church and P. Z. Takacs, "Prediction of Mirror Performance from Laboratory Measurements," *Proc. SPIE* **1160:**323–336 (1989).

84. E. L. Church and P. Z. Takacs, "The Optical Estimation of Surface Finish Parameters," *Proc. SPIE* **1530:**71–86 (1991).

85. E. L. Church and P. Z. Takacs, "Specification of the Surface Figure and Finish of Optical Elements in Terms of Systems Performance," *Proc. SPIE* **1791:**118–130 (1992).

86. J. E. Harvey and K. L. Lewotsky, "Scattering from Multilayer Coatings: a Linear Systems Model," *Proc. SPIE* **1530:**35–46 (1991).

87. Anita Kotha and James E. Harvey, "Enhanced EUV Performance of Wolter Type II Telescopes," *Proc. SPIE* **2011:**34–46 (1993).

88. Anita Kotha, "EUV Performance of Wolter Type II Telescopes for Space Astronomy Applications," M.S. thesis, Dept. of Physics, Univ. of Central Florida (1992).

89. William J. Gresslor, "Conical Foil X-ray Telescope Performance Predictions for Space Astronomy Applications," M.S. thesis, Dept. of Electrical and Computer Engineering, Univ. of Central Florida (1993).

90. J. E. Harvey, William P. Zmek, and C. Ftaclas, "Imaging Capabilities of Normal-incidence X-ray Telescopes," *Opt. Eng.* **29:**603–608 (1990).

91. K. L. Lewotsky, A. Kotha, and J. E. Harvey, "Performance Limitations of Imaging Microscopes for Soft X-ray Applications," *Proc. SPIE* **1741** (1992).

92. Kristin L. Lewotsky, "Performance Limitations of Imaging Microscopes for Soft X-ray Applications," M.S. thesis, Dept. of Electrical and Computer Engineering, Univ. of Central Florida (1992).

93. K. L. Lewotsky, A. Kotha, and J. E. Harvey, "Optical Fabrication Tolerances for Synchrotron Beamline Optics," presented at the (1992) *Annual Meeting of the Optical Society of America*, Albuquerque, NM (Sept 1992).

CHAPTER 12
ACOUSTO-OPTIC DEVICES AND APPLICATIONS

I. C. Chang
Aurora Associates
Santa Clara, California

12.1 GLOSSARY

α	acoustic attenuation
$\delta\theta_o, \delta\theta_a$	divergence: optical, acoustic
ΔBm	impermeability tensor
$\Delta f, \Delta F$	bandwidth, normalized bandwidth
Δn	birefringence
$\Delta\theta$	deflection angle
λ_o	optical wavelength (in vacuum)
f, F	acoustic frequency, normalized acoustic frequency
Λ	acoustic wavelength
ρ	density
τ	acoustic transit time
ψ	phase mismatch function
a	divergence ratio
D	optical aperture
$\mathbf{E}_i, \mathbf{E}_d$	electric field: incident, diffracted light acoustic wave
$\mathbf{k}_i, \mathbf{k}_d, \mathbf{k}_a$	wavevector: incident, diffracted light, acoustic wave
L, ℓ	interaction length, normalized interaction length
L_o	characteristic length
M	figure of merit
n_o, n_e	refractive index: ordinary, extraordinary

P_a, P_d	acoustic power, acoustic power density
p, p_{mn}, p_{ijkl}	elasto-optic coefficient
$\mathbf{S}, S_{ij}$	strain, strain tensor components
t_r	rise time
T	scan time
V	acoustic velocity

12.2   INTRODUCTION

When an acoustic wave propagates in an optically transparent medium, it produces a periodic modulation of the index of refraction via the elasto-optical effect. This provides a moving phase grating which may diffract portions of an incident light into one or more directions. This phenomenon, known as the acousto-optic (AO) diffraction, has led to a variety of optical devices that perform spatial, temporal, and spectral modulations of light. These devices have been used in optical systems for light-beam control and signal-processing applications.

Historically, the diffraction of light by acoustic waves was first predicted by Brillouin[1] in 1922. Ten years later, Debye and Sears[2] and Lucas and Biquard[3] experimentally observed the effect. In contrast to Brillouin's prediction of a single diffraction order, a large number of diffraction orders were observed. This discrepancy was later explained by the theoretical work of Raman and Nath.[4] They derived a set of coupled wave equations that fully described the AO diffraction in unbounded isotropic media. The theory predicts two diffraction regimes; the Raman-Nath regime, characterized by the multiple of diffraction orders, and the Bragg regime, characterized by a single diffraction order. Discussion of the early work on AO diffraction can be found in Refs. 5 and 6.

Although the basic theory of AO diffraction in isotropic media was well understood, there had been relatively few practical applications prior to the invention of the laser. It was the need of optical devices for laser beam control that stimulated extensive research on the theory and practice of acousto-optics. Significant progress of the AO devices has been made during the past two decades, due primarily to the development of superior AO materials and efficient broadband transducers. By now, acousto-optics has developed into a mature technology and is deployed in a wide range of optical system applications.

It is the purpose of this chapter to review the theory and practice of bulkwave acousto-optic devices and their applications. The review emphasizes design and implementation of AO devices. It also reports the status of most recent developments. Previous review of acousto-optics may be found in references 7–11.

In addition to bulkwave acousto-optics, there have also been studies on the interaction of optical guided waves and surface acoustic waves (SAW). However, the effort has remained primarily at the research stage and has not yet resulted in practical applications. As such, the subject of guided-wave acousto-optics will not be discussed here. The interested reader may refer to a recent review article.[12]

This chapter is organized as follows: The next section discusses the theory of acousto-optic interaction. It provides the necessary background for the design of acousto-optic devices. The important subject of acousto-optic materials is discussed in the section following. Then a detailed discussion on three basic types of acousto-optic devices is presented. Included in the discussion are the topics of deflectors, modulators, and tunable filters. The last section discusses the use of AO devices for optical beam control and signal processing applications.

12.3 *THEORY OF ACOUSTO-OPTIC INTERACTION*

Elasto-optic Effect

The elasto-optic effect is the basic mechanism responsible for the AO interaction. It describes the change of refractive index of an optical medium due to the presence of an acoustic wave. To describe the effect in crystals, we need to introduce the elasto-optic tensor based on Pockels' phenomenological theory.[1]

An elastic wave propagating in a crystalline medium is generally described by the strain tensor **S**, which is defined as the symmetric part of the deformation gradient

$$S_{ij} = \left(\frac{\partial u_i}{\partial x_j} + \frac{\partial u_j}{\partial x_i} \right) \Big/ 2 \qquad i,j = 1 \text{ to } 3 \tag{1}$$

where u_i is the displacement. Since the strain tensor is symmetric, there are only six independent components. It is customary to express the strain tensor in the contracted notation

$$S_1 = S_{11}, \qquad S_2 = S_{22}, \qquad S_3 = S_{33}, \qquad S_4 = S_{23}, \qquad S_5 = S_{13}, \qquad S_6 = S_{12} \tag{2}$$

The conventional elasto-optic effect introduced by Pockels states that the change of the impermeability tensor, ΔB_{ij}, is linearly proportional to the symmetric strain tensor.

$$\Delta B_{ij} = p_{ijkl} S_{kl} \tag{3}$$

where p_{ijkl} is the elasto-optic tensor. In the contracted notation,

$$\Delta B_m = p_{mn} S_n \qquad m,n, = 1 \text{ to } 6 \tag{4}$$

Most generally, there are 36 components. For the more common crystals of higher symmetry, only a few of the elasto-optic tensor components are non-zero.

In the above classical Pockels' theory, the elasto-optic effect is defined in terms of the change of the impermeability tensor ΔB_{ij}. In the more recent theoretical work on AO interactions, analysis of the elasto-optic effect has been more convenient in terms of the nonlinear polarization resulting from the change of dielectric tensor $\Delta \varepsilon_{ij}$. We need to derive the proper relationship that connects the two formulations.

Given the inverse relationship of ε_{ij} and B_{ij} in a principal axis system $\Delta \varepsilon_{ij}$ is:

$$\Delta \varepsilon_{ij} = -\varepsilon_{ii} \Delta B_{ij} \varepsilon_{jj} = -n_i^2 n_j^2 \Delta B_{ij} \tag{5}$$

where n_i is the refractive index. Substituting Eq. (3) into Eq. (5), we can write:

$$\Delta \varepsilon_{ij} = \chi_{ijk\ell} S_{k\ell} \tag{6}$$

where we have introduced the elasto-optic susceptibility tensor,

$$\chi_{ijk\ell} = -n_i^2 n_j^2 p_{ijk\ell} \tag{7}$$

For completeness, two additional modifications of the basic elasto-optic effect are discussed as follows.

Roto-optic Effect. Nelson and Lax[14] discovered that the classical formulation of elasto-optic was inadequate for birefringent crystals. They pointed out that there exists an additional roto-optic susceptibility due to the antisymmetric rotation part of the deformation gradient.

$$\Delta B'_{ij} = p'_{ijkl} R_{kl} \qquad (8)$$

where $R_{ij} = (S_{ij} - S_{ji})/2$.

It turns out that the roto-optic tensor components can be predicted analytically. The coefficient of $\mathbf{p}_{ijkl}$ is antisymmetric in kl and vanishes except for shear waves in birefringent crystals. In a uniaxial crystal the only nonvanishing components are $p_{2323} = p_{2313} = (n_o^{-2} - n_e^{-2})/2$, where n_o and n_e are the principal refractive indices for the ordinary and extraordinary wave, respectively. Thus, the roto-optic effect can be ignored except when the birefringence is large.

Indirect Elasto-optic Effect. In the piezoelectric crystal, an indirect elasto-optic effect occurs as the result of the piezoelectric effect and electro-optic effect in succession. The effective elasto-optic tensor for the indirect elasto-optic effect is given by[15]

$$p_{ij}^* = p_{ij} - \frac{r_{im} S_m e_{jn} S_n}{\varepsilon_{mn} S_m S_n} \qquad (9)$$

where p_{ij} is the direct elasto-optic tensor, r_{im} is the electro-optic tensor, e_{jn} is the piezoelectric tensor, ε_{mn} is the dielectric tensor, and S_m is the unit acoustic wave vector. The effective elasto-optic tensor thus depends on the direction of the acoustic mode. In most crystals the indirect effect is negligible. A notable exception is $LiNbO_3$. For instance, along the z axis, $r_{33} = 31 \times 10^{-12}$ m/v, $e_{33} = 1.3$ c/m^2, $E_{33}^s = 29$, thus $p^* = 0.088$, which differs notably from the contribution $p_{33} = 0.248$.

Plane Wave Analysis of Acousto-optic Interaction

We now consider the diffraction of light by acoustic waves in an optically transparent medium in which the acoustic wave is excited. An optical beam is incident onto the cell and travels through the acoustic beam. Via the elasto-optical effect, the traveling acoustic wave sets up a spatial modulation of the refractive index which, under proper conditions, will diffract the incident beam into one or more directions.

In order to determine the detailed characteristics of AO devices, a theoretical analysis for the AO interaction is required. As pointed out before, in the early development, the AO diffraction in isotropic media was described by a set of coupled wave equations known as the Raman-Nath equations.[4] In this model the incident light is assumed to be a plane wave of infinite extent. It is diffracted by a rectangular sound column into a number of plane waves propagating along different directions. Solution of the Raman-Nath equations gives the amplitudes of these various orders of diffracted optical waves.

In general, the Raman-Nath equations can be solved only numerically and judicious approximations are required to obtain analytic solutions. Solutions of these equations can be classified into different regimes that are determined by the ratio of the interaction length L to a characteristic length $L_o = n\Lambda^2/\lambda_o$[10] where n is the refractive index and Λ and λ_o are wavelengths of the acoustic and optical waves, respectively.

The Raman-Nath equations admit analytic solutions in the two limiting cases. In the Raman-Nath regime, where $L \ll L_o$, the AO diffractions appear as a large number of different orders. The diffraction is similar to that from a thin phase grating. The direction of the various diffraction orders are given by the familiar grating equation, $\sin \theta_m = m\lambda_o/n\Lambda$, where m is the diffraction order. Solution of the Raman-Nath equations shows that the amplitude of the mth-order diffracted light is proportional to the mth-order Bessel functions. The maximum intensity of the first-order diffracted light (relative to the incident light) is about 34 percent. Due to this relatively low efficiency, AO diffractions in the Raman-Nath regime are of little interest to device applications.

In the opposite limit, $L > L_o$, the AO diffraction appears as a predominant first order and is said to be in the Bragg regime. The effect is called Bragg diffraction since it is similar to that of the x-ray diffraction in crystals. An important feature of the Bragg diffraction is that the maximum first-order diffraction efficiency obtainable is 100 percent. Therefore, practically all of today's AO devices are designed to operate in the Bragg regime.

In the intermediate case, $L \le L_o$, the AO diffractions now appear as a few dominant orders. The relative intensities of these diffraction orders can be obtained by numerically solving the Raman-Nath equations. This problem has been studied in detail by Klein and Cook,[16] who calculated numerically the diffracted light intensities for the different regimes. Their analysis shows that when $L = L_o$, the maximum first-order diffraction efficiency is about 90 percent. Thus $L = L_o$ may be used as a criterion for insuring the AO device operated in the Bragg regime.

Many modern high-performance devices are based on the light diffraction in anisotropic media.[17] In this case, the indices of the incident and diffracted light may be different. This is referred to as birefringent diffraction. The classical Raman-Nath equations are no longer adequate and a new formulation is required. We have previously presented a plane wave analysis of AO interaction in anisotropic media.[10] Results of the plane wave analysis for the Bragg diffraction are summarized as follows.

The AO interaction can be viewed as a parametric process where the incident optical wave mixes with the acoustic wave to generate a number of polarization waves. The polarization waves in turn generate new optical waves at various diffraction orders. Let the angular frequency and wave vector of the incident optical wave be denoted by ω_i and $\mathbf{k}_i$, respectively, and those of the acoustic waves by ω_a and $\mathbf{k}_a$. In the Bragg limit, only the first-order diffracted light grows to a finite amplitude. The polarization wave is then characterized by the angular frequency $\omega_d = \omega_i + \omega_a$ and wavevector $\mathbf{K}_d = \mathbf{k}_i + \mathbf{k}_a$. The total electric fields of the optical wave can be expressed as

$$\bar{E}(r, t) = \tfrac{1}{2}(\hat{\mathbf{e}}_i E_i(z) e^{j(\omega_i t - \bar{k}_o \cdot \bar{r})} + \hat{\mathbf{e}}_d E_d(z)^{j(\omega_d t - \mathbf{K}_d \cdot \bar{r})}) + \text{c.c.} \tag{10}$$

where $\hat{\mathbf{e}}_i$ and $\hat{\mathbf{e}}_d$ are unit vectors in the directions of electric fields for the incident and diffracted optical waves, respectively, $\mathbf{E}_i(z)$ and $\mathbf{E}_d(z)$ are the corresponding slowly varying electric field amplitudes, and c.c. stands for complex conjugate. The electric field of the optical wave satisfies the wave equation,

$$\nabla \times \nabla \times \mathbf{E} + \frac{1}{c^2}\left(\bar{\bar{\varepsilon}} \cdot \frac{\partial^2 \bar{E}}{\partial t^2}\right) = -\mu_o \frac{\partial^2 \mathbf{P}}{\partial t^2} \tag{11}$$

where $\bar{\bar{\varepsilon}}$ is the relative dielectric tensor and $\mathbf{P}$ is the acoustically induced polarization. Based on the Pockel's theory of the elasto-optic effect,

$$\mathbf{P}(\mathbf{r}, t) = \varepsilon_o \bar{\bar{\chi}} \cdot \mathbf{S}(\mathbf{r}, t) \cdot \mathbf{E}(\mathbf{r}, t) \tag{12}$$

where χ is the elasto-optical susceptibility tensor given in Eq. (7). $\mathbf{S}(r, t)$ is the strain of the acoustic wave

$$S(\bar{r}, t) = \tfrac{1}{2}\hat{s}Se^{j(\omega_a t - \bar{\mathbf{k}}_a \cdot \bar{r})} + \text{c.c.} \tag{13}$$

where $\hat{s}$ is a unit strain tensor of the acoustic wave and S is the acoustic wave amplitude. Substituting Eqs. (10), (12), and (13) into Eq. (11) and neglecting the second-order derivatives of electric field amplitudes, we obtain the coupled wave equations for AO Bragg diffraction.

$$\frac{dE_i}{dz} = j(Cn_d/n_i)S^*E_d \tag{14}$$

$$\frac{dE_d}{dz} - j\,\Delta k E_d = jCSE_i \tag{15}$$

where $C = \pi n_i^2 n_d p/2\lambda_o$, n_i and n_d are the refractive indices for the incident and diffracted light, p is the effective elasto-optic constant for the particular mode of AO interaction, and Δk is the magnitude of momentum mismatch between the polarization wave and free wave of the diffracted light.

$$\Delta k = |\mathbf{K}_d - \mathbf{k}_d| = |\mathbf{k}_i + \mathbf{k}_a - \mathbf{k}_d| \tag{16}$$

Eqs. (14) and (15) admit simple analytic solutions. At $z = L$, the intensity of the first-order diffracted light (normalized to the incident light) is

$$I_1 = \frac{I_d(L)}{I_i(0)} = \eta \operatorname{sinc}^2 \frac{1}{\pi}\left(\eta + \left(\frac{\Delta kL}{2}\right)^2\right)^{1/2} \tag{17}$$

where $\operatorname{sinc}(x) = (\sin \pi x)/\pi x$, and

$$\eta = \frac{\pi^2}{2\lambda_0^2}\left(\frac{n^6 p^2}{2}\right)S^2 L^2 = \frac{\pi^2}{2\lambda_0^2} M_2 P_a \left(\frac{L}{H}\right) \tag{18}$$

In the preceding equation, we have used the relation $P_a = \tfrac{1}{2}\rho V^3 S^2 LH$ where P_a is the acoustic power, H is the acoustic beam height, ρ is the mass density, V is the acoustic wave velocity, and $M_2 = n^6 p^2/\rho V^3$ is a material figure of merit.

Equation (17) shows that for sufficiently long interaction length the diffracted light builds up only when the momentum is nearly matched. For exact phase matching, ($\Delta \mathbf{k} = 0$), the peak intensity of the diffracted light is given by

$$I_p = \sin^2\sqrt{\eta} \tag{19}$$

When $\eta \ll 1$, $I_p \approx \eta$. The diffraction efficiency is thus linearly proportional to acoustic power. This is referred to as the weak interaction (or small signal) approximation which is valid when the peak efficiency is below 70 percent. As acoustic power increases, the diffraction efficiency saturates and approaches 100 percent. Thus in the Bragg regime, complete depletion of the incident light is obtainable. However, the acoustic power required is about 2.5 times that predicted by the small signal theory.

When the momentum is not matched, the fractional diffracted light I_1, in Eq. (17) can be approximated by

$$I_1 = I_p \operatorname{sinc}^2 \psi \tag{20}$$

where $\psi = \Delta k L / 2\pi$ is the phase mismatch (normalized to 2π). We shall show later that Ψ determines the frequency and angular characteristics of AO interaction. In the next section we shall first consider the case $\psi = 0$; i.e., when the AO wavevectors are exactly momentum matched.

Phase Matching

It was shown in the preceding section that significant diffraction of light occurs only when the exact momentum matching is met.

$$\mathbf{k}_d = \mathbf{k}_i + \mathbf{k}_a \tag{21}$$

In the general case of AO interaction in an anisotropic medium, the magnitudes of the wave vectors are given by:

$$k_i = \frac{2\pi n_i}{\lambda_o} \qquad k_d = \frac{2\pi n_d}{\lambda_o} \qquad k_a = \frac{2\pi}{\Lambda} \tag{22}$$

where $\Lambda = V/f$ is the acoustic wavelength, and V and f are the velocity and frequency of the acoustic wave. In Eq. (22), the small optical frequency difference (due to acoustic frequency shift) of the incident and diffracted light beams are neglected.

Isotropic Diffractions. Consider first the case of isotropic diffraction: $n_i = n_d = n$. At exact phase matching, the acoustic and optical wavevectors form an isosceles triangle, as shown in Fig. 1a. The loci of the incident and diffracted optical wavevectors fall on a circle of radius n. From the figure, it is seen that the incident and diffracted wave vectors make the same angle with respect to the acoustic wavefront

$$\sin \theta = \frac{\lambda_o}{2n\Lambda} = \frac{\lambda_o f}{2nV} \tag{23}$$

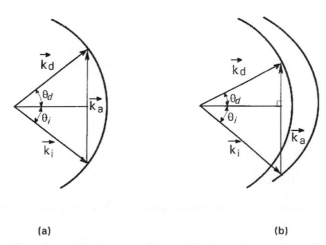

(a)   (b)

FIGURE 1 Wavevector diagram of acousto-optic interaction. (a) Isotropic diffraction, (b) birefringent diffraction.

For typical AO diffraction, small-angle approximation holds, i.e., $\sin \theta_b \approx \theta_b$. The deflection angle outside the medium is

$$\Delta\theta = \frac{\lambda_o f}{V} \tag{24}$$

The deflection angle is thus linearly proportional to the acoustic frequency. The linear dispersion relation forms the basis of the AO spectrum analyzer, an important signal processing application to be discussed later.

Referring to Fig. 1a, a change of incidence angle $\delta\theta_i$ will introduce a change of acoustic frequency (or optical wavelength) for exact phase matching. Thus, this phase matching condition for isotropic diffraction is critical to the angle of the incident light.

Anisotropic Diffraction.   Next, consider the Bragg diffraction in an optically anisotropic medium such as a birefringent crystal. The refractive index now depends on the direction as well as the polarization of the light beam. In general, the refractive indices of the incident and diffracted light beams are different. As an example, consider the Bragg diffraction in a positive uniaxial crystal. Figure 1b shows the wavevector diagram for the acoustic coupling from an incident extraordinary wave (polarized parallel to the c-axis) to a diffracted ordinary wave (polarized perpendicular to the c-axis). From the wave vector diagram shown in the figure and using the law of cosines one obtains:

$$\sin \theta_i = \frac{\lambda_o}{2n_i\Lambda}\left[1 + \frac{\Lambda^2}{\lambda_o^2}(n_i^2 - n_d^2)\right] \tag{25}$$

$$\sin \theta_d = \frac{\lambda_o}{2n_d\Lambda}\left[1 - \frac{\Lambda^2}{\lambda_o^2}(n_i^2 - n_d^2)\right] \tag{26}$$

Notice that the first term on the right-hand side of the preceding equations is the same as Eq. (23) and thus represents the usual Bragg condition for isotropic diffraction, while the remaining terms denote the modification due to the effect of anisotropy. Adding Eqs. (25) and (26) yields (in the small angle approximation) the same angle as the case of isotropic diffraction. However, addition of the second term in the above equations has significantly changed the angle-frequency characteristics of AO diffraction.

In order to show the distinct characteristics of birefringent diffraction, we shall consider AO diffraction in the constant azimuth plane of a uniaxial crystal. In general the refractive indices are functions and are dependent on the direction of the propagation. For a uniaxial crystal the refraction indices for the ordinary and extraordinary waves are n_o and $n_e(\theta_e)$, respectively.

$$n_e(\theta_e) = n_o\left(\cos^2 \theta_e + \frac{\sin^2 \theta_e}{e^2}\right)^{-1/2} \tag{27}$$

where θ_e is the polar angle of the e-wave, $e = n_E/n_o$, n_E is the refractive index for the e-wave polarized along the c-axis. Since the indices of refraction appear in both Eq. (25) and (26), the incident and diffraction angles are not separable. In the following analysis we shall decouple these equations and derive an explicit solution for the frequency angular relations.

Apply the momentum matching condition along the z-axis and use Eq. (22) to get

$$n_e(\theta_e) \cos (\theta_e - \theta_c) = n_o \cos (\theta_o - \theta_c) = n_z \tag{28}$$

where θ_c is the polar angle for the z-axis, n_z is the z-component of the refractive indices.

For an e-wave input, Eq. (28) can be readily solved to give the polar angle θ_o of the diffracted light.

$$\theta_o = \theta_c + \cos^{-1}\left(\frac{n_z}{n_o}\right) \tag{29}$$

The case of o-wave input is more complicated. After some tedious but straightforward algebra, the following formula for the diffracted e-wave are obtained. Introducing

$$\tan \phi_c = e \tan \theta_c \tag{30}$$

$$\phi_e = \phi_c + \cos^{-1}[n_z/n_e(\phi_c)] \tag{31}$$

yields the polar angle of the diffracted light,

$$\tan \theta_e = e \tan \phi_e \tag{32}$$

Once the directions of the incident and diffracted light are determined, the frequency-angular characteristics of AO diffraction can be determined by Eq. (25) or (26).

As an example, consider the AO diffraction in the polar plane of a shear wave TeO_2 crystal. Figure 2 shows the dependence of the incident angle θ_i and diffraction angle θ_d as a function of $n_a = \lambda_o/\Lambda$, the ratio of optical and acoustic wavelengths for a specific example where the polar angle θ_a of the acoustic waves is equal to 100°. The plots exhibit two operating conditions of particular interest. The point $\theta_i = \theta_1$, referred to as the tangential phase matching (TPM), shows at this angle there exists a wide range of acoustic frequencies that satisfy the phase matching condition. This operating condition provides the optimized design for wideband AO deflectors (or Bragg cells). The figure also shows that there exists two operating points, θ_2 and θ_3, where the phase matching is relatively insensitive to the changes of θ_i. The operating condition is referred to as noncritical phase matching (NPM). AO diffraction at these points exhibit a large angular aperture characteristic that is essential to tunable filter applications. Figure 4 shows incident light

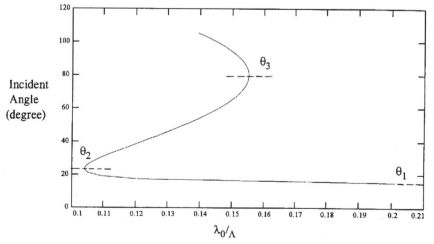

FIGURE 2 Dependence of incident and diffraction angles vs. ratio of optical and acoustic wavelengths.

for ordinary (θ_o) polarizations. The plot for incident light with extraordinary polarization is similar.

Frequency Characteristics of AO Interaction

The plane-wave analysis can be used to determine the frequency and angular characteristics of AO interaction. In this approach the acoustic wave is approximated as a single plane wave propagating normal to the transducer. The frequency or angular dependence is obtained from the phase mismatch caused by the change of acoustic frequency or incident optical wave direction. This is referred to as the phase mismatch method.[10]

Referring to the wavevector diagram shown in Fig. 3, the momentum match $\Delta\mathbf{k}$ is constrained to be normal to the boundary of the medium (i.e., along the z-axis).

An approximate expression of Δk is

$$\Delta k = ((k_i^2 + k_a^2 - k_d^2) - 2k_i k_a \sin\theta_i)/2k_d \tag{33}$$

where θ_i is the angle of incidence shown in Fig. 3.

According to Eq. (20) the bandshape of the AO interaction is a function of the phase mismatch.

$$W(\psi) = \text{sinc}^2(\psi) \tag{34}$$

where $\psi = \Delta\mathbf{k}L/2\pi$ is the phase mismatch (normalized to 2π). Substituting Eq. (22) into Eq. (33), we obtain the following expression for the phase mismatch.

$$\psi = \left(\frac{L}{2\lambda_o n_d}\right)\left(\left(\frac{\lambda_o}{\Lambda}\right)^2 - 2n_i\left(\frac{\lambda_o}{\Lambda}\right)\sin\theta_i + (n_i^2 - n_d^2)\right) \tag{35}$$

It is convenient to normalize the acoustic frequency to a center frequency f_o (wavelength Λ_o). In terms of the normalized acoustic frequency $F = f/f_o$, the phase mismatch function ψ can be written as,

$$\psi = \left(\frac{\ell}{2}\right)(F^2 - F_b F + F_c) \tag{36}$$

where

$$\ell = \frac{L}{L_o}, \qquad L_o = \frac{n_d \Lambda_o^2}{\lambda_o}, \qquad F_b = \frac{n_i \Lambda_o}{\lambda_o}\sin\theta_i, \qquad F_c = \left(\frac{\Lambda_o}{\lambda_o}\right)^2(n_i^2 - n_d^2) \tag{37}$$

Isotropic Diffraction. In this case $n_i = n_d = n_o$, $F_c = 0$. By choosing $F_b = 1 + (\Delta F/2)^2$, the phase mismatch function can be written as

$$\psi = \frac{\ell}{2}F\left(F - 1 - \frac{\Delta F^2}{2}\right) \tag{38}$$

where ΔF is the fractional bandwidth of the AO interaction, the diffraction efficiency reduces to 0.5 when $\psi = 0.45$. This corresponds to a fractional bandwidth

$$\Delta F = \frac{\Delta f}{f_o} = \frac{1.8}{\ell} \tag{39}$$

To realize octave bandwidth, ($\Delta F = 2/3$) for instance, the normalized interaction length ℓ is equal to 2.7. Figure 4a shows the bandshape of isotropic AO diffraction.

Birefringent Diffraction. Consider next the case of anisotropic AO diffraction in a

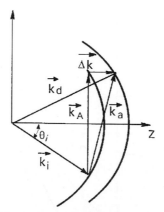

FIGURE 3   Acousto-optic wavevector diagram showing the construction of momentum mismatch.

birefringent crystal. We consider the case when $n_i > n_d$. We put $F_t = \sqrt{F_c} = (\Lambda_o/\lambda_o)\sqrt{n_i^2 - n_d^2}$. By choosing $F_b = 2$, the phase-mismatch function now takes the form

$$\psi = \frac{\ell}{2}(F^2 - 2F + F_t^2) \tag{40}$$

Equation (40) shows that there are two frequencies where the mismatch is zero. If we choose $F_t = 1$ the two frequencies coincide. This is referred to as the tangential phase matching, (TPM) since the acoustic wavevector is tangential to the locus of the diffracted light wavevector. At the tangential frequency the AO interaction exhibits wide bandwidth characteristics.

If the center frequency is chosen to be slightly off from the TFM frequency, the AO bandpass becomes a double peak response with a dip at the center frequency. It has an even larger bandwidth. If we choose $F_t^2 = 1 - \Delta F^2/8$, the phase mismatch at the center frequencies and the band edges will be equal. The 3 dB bandwidth is obtained by letting $\psi = 0.45$ and the fractional bandwidth of birefringent diffraction is then,

$$\Delta F = \frac{\Delta f}{f} = \left(\frac{7.2}{\ell}\right)^{1/2} \tag{41}$$

Figure 4b shows the AO bandshape of birefringent diffraction with octave bandwidth ($\Delta F = 2/3$). Notice that for octave bandwidth, the normalized interaction length is equal to 16.2. This represents an efficiency advantage factor of 6 compared to isotropic diffraction.

Acousto-optic Interaction of Finite Geometry

The plane wave analysis based on phase mismatch just presented provides an approximate theoretical description of AO interaction in the Bragg regime. However, applicability of this method may appear to be inadequate for real devices since the interaction geometry usually involves optical and acoustic beams of finite sizes with nonuniform amplitude distributions, e.g., a Gaussian optical beam and a divergent acoustic beam. One approach for taking into account the finite AO interaction geometry is to decompose the optical and

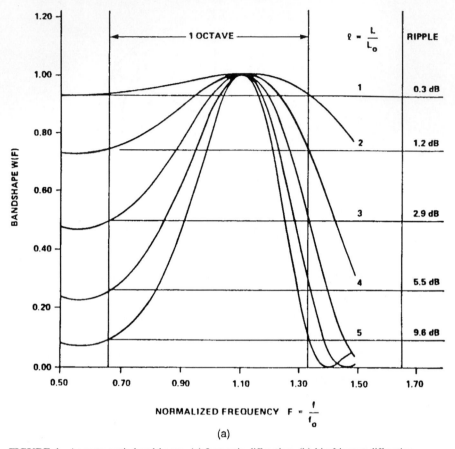

FIGURE 4 Acousto-optic bandshapes. (a) Isotropic diffraction, (b) birefringent diffraction.

acoustic beams into angular spectrum of plane waves and apply the plane wave solution as in a standard Fourier analysis. To make the approach analytically tractable, a practical solution is to impose the simplifying assumption of weak interaction.[18,19] Since the basic AO interaction is modeled as a filtering process in the spatial frequency domain, the approach is referred to as the frequency domain analysis.[20]

In the case of weak interaction, there is negligible depletion of the incident light. Using the approximation of constant amplitude of the incident optical wave, Eq. (15) can be integrated to yield the diffracted optical wave. At the far field, $(z \to \infty)$, the plane wave amplitude of diffracted light becomes

$$E_d(\mathbf{k}_a) = jCS(\mathbf{k}_a)E_i(\mathbf{k}_i)\delta(\mathbf{k}_d - \mathbf{k}_i - \mathbf{k}_a) \tag{42}$$

The preceding equation shows that in the far field the diffracted optical wave will be nonzero only when the exact momentum-matching condition is satisfied. Thus, assuming the incident light is a wide, collimated beam, the diffracted light intensity is proportional to the power spectra of the acoustic wave components which satisfies the exact phase matching condition.

As an example we use the frequency domain approach to determine the bandpass characteristics of the birefringent AO deflector. Suppose the AO device has a single

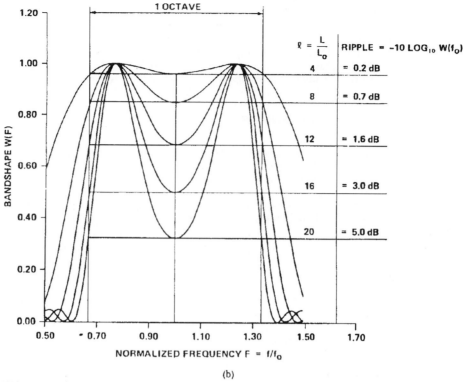

FIGURE 4 (*Continued*)

uniform transducer of length L in the z direction and height H in the y direction. The incident optical beam is assumed to be a plane wave propagating near the z axis in the $x\,z$ plane (referred to as the interaction plane). To take into account the finite size of the transducer, the acoustic beam is now modeled as an angular spectrum of plane waves propagating near the x axis.[20]

$$S(\bar{\sigma}) = S_o \cdot \text{sinc}\,(\sigma_x L)\,\text{sinc}\,(\bar{\sigma}_y H)\delta\left(\sigma_z - \frac{1}{\Lambda} - \frac{\Lambda}{2}(B\sigma_y^2 + C\sigma_z^2)\right) \qquad (43)$$

where $\sigma_x \sigma_y$ and σ_z are the spatial frequency components of the acoustic wavevector, B and C are the curvatures of acoustic slowness surface for the y and z directions.

We consider the AO diffraction only in the interaction plane where the phase-matching condition is satisfied. The diffracted light intensity distribution is proportional to the acoustic power spectra; i.e.,

$$I_d = \eta \,\text{sinc}^2\left(\frac{L}{\Lambda}\theta_a\right) = \eta \,\text{sinc}^2 \psi \qquad (44)$$

Referring to Fig. 3, we have

$$\psi = \frac{L}{\Lambda}(\sin\theta_i - \sin\bar{\theta}_i) \qquad (45)$$

where the angles θ_i and $\bar{\theta}_i$ are the optical incidence angle and Bragg angle (at exact phase matching), respectively. Substituting the Bragg angle θ_i from Eq. (25) into Eq. (45), we obtain the same result as Eq. (35). Thus, the frequency-domain analysis and the phase-mismatch method are equivalent.

In the above analysis the weak field approximation is used. The case of strong interaction has also been studied.[21] However, due to the mathematical complexities involved, only numerical solutions are obtainable.

12.4 ACOUSTO-OPTIC MATERIALS

The significant progress of AO devices in recent years has been largely due to the development of superior materials such as TeO_2 and GaP. In this section we shall review the material issues related to AO device applications. A comprehensive list of tables summarizing the properties of AO materials is presented at the end of this section. Discussion of AO material issues and tabulations of earlier material data have been given in several previous publications.[9,22–26]

The selection of AO materials depends on the specific device application. An AO material suited for one type of device may not even be applicable for another. For example, GaP is perhaps the best choice for making wideband AO deflectors or modulators. However, since GaP is optically isotropic, it cannot be used for tunable filter purposes.

Some of the requirements for materials' properties apply to the more general cases of optical device applications, e.g. high optical transparency over wavelength range of interest, availability in large single crystals, etc. We shall restrict the discussion to material properties that are particularly required for AO device applications.[26]

Acousto-optic Figures of Merit

A large AO figure of merit is desired for device applications. There are several AO figures of merit that have been used for judging the usefulness of an AO material. The relevant one to be used depends on the specific applications. Several AO figures of merit are defined in the literature. These include:

$$M_1 = \frac{n^7 p^2}{\rho V} \qquad M_2 = \frac{n^6 p^2}{\rho V^3} \qquad M_3 = \frac{n^7 p^2}{\rho V^2} \qquad M_4 = \frac{n^8 p^2 V}{\rho} \qquad M_5 = \frac{n^8 p^2}{\rho V^3} \qquad (46)$$

where n is the index of refraction, p is the relevant elasto-optic coefficient, ρ is the density, and V is the acoustic wave velocity. These figures of merit are generally listed as the normalized quantities M (normalized to values for fused silica).

Based on Eq. (18), the figure of merit M_2 relates the diffraction efficiency η, to the acoustic power P_a for a given device aspect ratio L/H.

M_2 is the AO figure of merit most often referred to in the literature and is widely used for the comparison of AO materials. This is a misconception, since from the viewpoint of device applications, M_2 is usually not appropriate. Comparison of AO materials (or modes) based on M_2 can lead to erroneous conclusions. M_2 is used only when efficiency is the only parameter of concern. In most practical cases, other parameters such as bandwidth and resolution must also be considered. To optimize the bandwidth product, the relevant figure of merit is M_1.

$$\eta = \frac{\pi^2}{2\lambda_o^3 f^2} M_1 \ell \frac{P_a}{H} \qquad (47)$$

where ℓ is the normalized interaction and is determined by the specified (fractional) bandwidth.

In the design of AO deflectors, or Bragg cells, besides efficiency and bandwidth, a third parameter of interest is the aperture time τ. A minimum acoustic beam height H must be chosen to insure that the aperture is within the near field of the acoustic radiation. Let $H = hH_o$ with $H_o = V(\tau/f)^{1/2}$.

$$\eta = \frac{\pi^2 M_3}{2\lambda_o^3 f^{3/2}\tau^{1/2}} P_a \frac{\ell}{h} \tag{48}$$

For wideband AO modulators, the acoustic power density P_d is often the limiting factor. The appropriate AO figure of merit is then M_4, i.e.,

$$\eta = \frac{\pi^2}{2\lambda_o^4 f^4} M_4 P_d \ell^2 \tag{49}$$

In the design of AO tunable filters, the parameters to be optimized are the product of efficiency η, the resolving power $\lambda_o/\Delta\lambda$, and the solid angular aperture $\Delta\Omega$. In this case the appropriate AO figure of merit is $M_5 = n^2 M_2$.

Acoustic Attenuation

The performance of AO devices also depends on the acoustic properties of the interaction medium. Low acoustic attenuation is desired for increased resolution of deflectors or aperture of tunable filters.

The theory of acoustic attenuation in crystals has been a subject of extensive study.[27] Generally, at room temperature $\omega\tau_{th} \ll 1$, where ω is the angular frequency of the acoustic wave and τ_{th} is the thermal phonon relaxation, the dominant contribution to acoustic attenuation is due to Akhieser loss caused by relaxation of the thermal phonon distribution toward equilibrium. A widely used result of this theory is the relation derived by Woodruff and Erhenrich. It states that acoustic attenuation measured in nepers per unit time is given by[27]

$$\alpha = \frac{\gamma^2 f^2 \kappa T}{\rho V^4} \tag{50}$$

where γ is the Grüneisen constant, T is the temperature, and κ is the thermal conductivity. Equation (50) shows that the acoustic attenuation has a quadratic frequency-dependence near to that observed in most crystals. In practice, in some crystals such as GaP, it has been found that the frequency dependence of attenuation $a \sim f^n$, where n varies in different frequency ranges and has an average value between 1 and 2. The deviation from a quadratic dependence may be attributed to the additional extrinsic attenuation caused by scattering from lattice imperfections.

Optical Birefringence

Optical birefringence is a requirement for materials used in AO tunable filters. The requirement is met by optically birefringent crystals in order that the phase matching for the AO filter interaction can be satisfied over a large angular distribution of incident light. For AO deflectors and modulators, optical birefringence is not necessary. AO devices with high efficiency, wide bandwidth, and large resolution are realizable with superior isotropic materials such as GaP. However, in an optically birefringent crystal it is possible to achieve tangential phase matching, which provides an enhancement of (normalized) interaction length ℓ for a given fractional bandwidth. This represents an increased interaction length advantage, by a factor of five or more, as compared to isotropic diffraction. The advantage is particularly significant when power density is the limiting factor, since the reduction of power density is proportional to ℓ^2. The optical birefringence in $LiNbO_3$ and TeO_2 have been largely responsible for the superior performance of these two AO materials.

The driving acoustic frequency corresponding to the passband wavelength in an AO tunable filter is proportional to the crystal birefringence. It is desirable to lower the acoustic frequency for simpler construction and improved performance. Therefore, for AO tunable filter applications, the birefringence is preferably small. For wideband AO Bragg cells using tangential phase matching in birefringent crystals, a larger value of birefringence is desirable since it limits the maximum frequency for wideband operations.

Tabulation of Acousto-optic Material Properties

To aid the selection of AO materials, the relevant properties of some promising materials are listed. Table 1 lists the values of elasto-optical tensor components. Table 2 lists the relevant properties of selected AO materials. The listed figures of merit M_1, M_2 and M_3 are normalized relative to that of fused silica, which has the following absolute values.

$$M_1 = 7.83 \times 10^{-7} \quad [cm^2sg^{-1}]$$
$$M_2 = 1.51 \times 10^{-18} \quad [s^3g^{-1}]$$
$$M_3 = 1.3 \times 10^{-12} \quad [cm^2s^2g^{-1}]$$

TABLE 1 Elasto-optic Coefficients of Materials

		(a) Isotropic		
Material	λ (μm)	p_{11}	p_{12}	Ref.
Fused silica (SiO_2)	0.63	+0.121	+0.270	24
As_2S_3 glass	1.15	+0.308	+0.299	24
Water	0.63	0.31	0.31	24
$Ge_{33}Se_{55}As_{12}$(glass)	1.06	0.21	0.21	24
Lucite	0.63	0.30	0.28	24
Polystyrene	0.63	0.30	0.31	24
SF-59	0.63	0.27	0.24	24
SF-8	0.63	0.198	0.262	28
Tellurite glass	0.63	0.257	0.241	29

		(b) Cubic: classes $\bar{4}3m$, 432, and m3m			
Material	λ (μm)	p_{11}	p_{12}	p_{44}	Ref.
CdTe	10.60	−0.152	−0.017	−0.057	25
GaAs	1.15	−0.165	−0.140	−0.072	24
GaP	0.633	−0.151	−0.082	−0.074	24
Ge	2.0–2.2	−0.063	−0.0535	−0.074	25
	10.60	0.27	0.235	0.125	24
NaCl	0.55–0.65	0.115	0.159	−0.011	25
NaF	0.633	0.08	0.20	−0.03	25
	0.589		−0.021	−0.10	25
Si	1.15	−0.101	0.0094		25
	3.39	−0.094	0.017	−0.051	25
$Y_3Fe_5O_{12}$ (YIG)	1.15	0.025	0.073	0.041	24
$Y_3Al_5O_{12}$ (YAG)	0.633	−0.029	+0.0091	−0.0615	24
KRS5	0.633	0.18	0.27	±0.15	25

TABLE 1   Elasto-optic Coefficients of Materials (*Continued*)

		(b) Cubic: classes $\bar{4}$3m, 432, and m3m			
Material	λ (μm)	p_{11}	p_{12}	p_{44}	Ref.
KRS6	0.633	0.28	0.25	±0.14	25
β-ZnS	0.633	0.091	−0.01	0.075	25
$Y_3Ga_5O_{12}$	0.63	0.091	0.019	0.079	25
Diamond	0.59	−0.31	−0.03		24
	0.59	−0.43	+0.19	−0.16	24
LiF	0.59	+0.02	+0.128	−0.064	24
MgO	0.59	−0.32	−0.08		24
KBr	0.59	+0.22	+0.71	−0.026	24
KCl	0.59	+0.17	+0.124		24
KI	0.59	+0.210	0.169		24

		(c) Cubic classes 23 and m3				
Material	λ (μm)	p_{11}	p_{12}	p_{13}	p_{44}	Ref.
$Ba(NO_3)_2$	0.63	0.15	0.35	0.29	0.02	24
$NaBrO_3$	0.59	0.185	0.218	0.213	−0.0139	24
$NaClO_3$	0.59	0.162	0.24	0.2	−0.198	24
$BA(NO_3)_2$	0.63	0.15	0.35	0.29	0.02	24
$Bi_{12}GeO_{20}$	0.63	0.12			0.04	30
$Bi_{12}SiO_{20}$	0.63	0.13			0.04	30

		(d) Hexagonal system: classes 6m2, 6mm, 622, and 6/mm						
Material	λ (μm)	p_{11}	p_{12}	p_{13}	p_{31}	p_{33}	p_{44}	Ref.
CdS	0.63	−0.142	−0.066	−0.057	−0.041	−0.20	±0.054	25
	10.60	0.104		0.011				25
SnO	0.63	0.222	0.099	−0.111	0.088	−0.235	−0.0585	25
α-ZnS	0.63	−0.115	0.017	0.025	0.0271	−0.13	−0.0627	25
ZnO	0.63	0.222	0.199	−0.111	0.088	−0.235	−0.061	31

		(e) Hexagonal system: classes 6, $\bar{6}$, and 6/m								
Substance	λ (μm)	p_{11}	p_{12}	p_{13}	p_{31}	p_{33}	p_{44}	p_{45}	p_{16}	Ref.
$LiIO_3$	0.63	0.32		0.31					0.03	24

		(f) Trigonal System: classes 3m, 32, and $\bar{3}$m								
Substance	λ (μm)	p_{11}	p_{12}	p_{13}	p_{14}	p_{31}	p_{33}	p_{41}	p_{44}	Ref.
Al_2O_3	0.644	−0.23	−0.03	0.02	0.00	−0.04	−0.20	0.01	−0.10	25
$LiNbO_3$	0.633	−0.026	0.090	0.133	−0.075	0.179	0.071	−0.151	0.146	25
$LiTaO_3$	0.633	−0.081	0.081	0.093	−0.026	0.089	−0.044	−0.085	0.028	25
SiO_2(quartz)	0.589	0.16	0.27	0.27	−0.030	0.29	0.10	−0.047	−0.079	24
Ag_3AsS_3	0.633	±0.10	±0.19	±0.22		±0.24	±0.20			25
(proustite)	1.15	±0.056	±0.082	±0.068		±0.103	±0.100	±0.01		25
Te	10.6	0.164	0.138	0.146		0.086	0.038			32
$CaCO_3$		±0.095	±0.189	±0.215	−0.006	+0.309	+0.178	+0.01	−0.090	24
HgS	0.63			0.445			0.115			33
Tl_3AsSe_3	3.39	0.4	0.22	0.24	0.04	0.2	0.22	0.018	0.15	34
Tl_3AsS_3	3.39	0.36	0.13	0.2		0.15	0.36	0.02		34

TABLE 1   Elasto-optic Coefficients of Materials (*Continued*)

(g) Tetragonal system: classes 4mm, $\bar{4}$2m, 422, and 4/mmm

Substance	λ (μm)	p_{11}	p_{12}	p_{13}	p_{31}	p_{33}	p_{44}	p_{66}	Ref.
$(NH_4)H_2PO_4$(ADP)	0.589	0.319	0.277	0.169	0.197	0.167	−0.058	−0.091	25
	0.63	0.296	0.243	0.208	0.188	0.228			25
KH_2PO_4(KDP)	0.589	0.287	0.282	0.174	0.241	0.122	−0.019	−0.064	25
	0.63	0.254	0.230	0.233	0.221	0.212		−0.0552	25
$Sr_{0.75}Ba_{0.25}Nb_2O_6$	0.63	0.16	0.10	0.08	0.11	0.47			24
$Sr_{0.5}Ba_{0.5}Nb_2O_6$	0.63	0.06	0.08	0.17	0.09	0.23			24
TeO_2	0.63	0.0074	0.187	0.340	0.0905	0.240	0.04*	−0.0463	24
TiO_2 (rutile)	0.514	−0.001	0.113	−0.167	−0.106	−0.064	0.0095	−0.066	25
	0.63	−0.011	0.172	−0.168	−0.0965	−0.058		±0.072	25
$ZrSiO_4$	0.63	0.06		0.13	0.07	0.09		0.10	24
$CdGeP_2$	0.63	0.21	−0.09	0.09		0.4	0.1	0.12	35
Hg_2Cl_2	0.63	0.551	0.44	0.256	0.137	0.01		0.047	36
Hg_2Br_2	0.63	0.262	0.175	0.148	0.177	0.116			37

(h) Tetragonal system: classes 4, $\bar{4}$, 422, and 4/m

Material	λ (μm)	p_{11}	p_{12}	p_{13}	p_{16}	p_{31}	p_{33}	p_{44}	p_{45}	p_{61}	p_{66}		Ref.
$PbMoO_4$	0.63	0.24	0.24	0.255	0.017	0.175	0.3	0.067	−0.01	−0.01	0.013	0.05	24
$CdMoO_4$	0.63	0.12	0.10	0.13	0.11	0.18							25
$NaBiMoO_4$	0.63	0.243	0.265	0.25		0.21	0.29						38
$LiBiMoO_4$	0.63	0.265	0.201	0.244		0.227	0.309						39
$CaMoO_4$	0.63	0.17	−0.15	−0.08	0.03	0.10	0.08	0.06	0.06	0.06	0.1	0.026	40

(i) Orthorhombic: all classes

Material	λ (μm)	p_{11}	p_{12}	p_{13}	p_{21}	p_{22}	p_{23}	p_{31}	p_{32}	p_{33}	p_{44}	p_{55}	p_{66}	Ref.
α-HIO_3	0.63	0.406	0.277	0.304	0.279	0.343	0.305	0.503	0.310	0.334			0.092	24
$PbCO_3$	0.63	0.15	0.12	0.16	0.05	0.06	0.21	0.14	0.16	0.12				24
$BaSO_4$	0.59	+0.21	+0.25	+0.16	+0.34	+0.24	+0.19	+0.27	+0.22	+0.31	+0.22	−0.012	+0.037	24
$Gd_2(MoO_4)_3$	0.63	0.19	0.31	0.175	0.215	0.235	0.175	0.185	0.23	0.115	−0.033	−0.028	0.035	41

TABLE 2   Selected AO materials

Material	Optical transmission (μm)	Density (g/cm³)	Acoustic mode	Acoustic velocity	Acoustic attenuation (dB/μs-GHz²)	Optical polarization	Refraction index	Figure of merit M_1	M_2	M_3
Fused silica	0.2–4.5	2.2	L	$5.96\,\dfrac{mm}{\mu s}$	$7.2\,\dfrac{dB}{\mu s\text{-}GHz}$	⊥	1.46	1.0	1.0	1.0
$LiNbO_3$	0.4–4.5	4.64	L[100]	6.57	1.0	35°Y Rot.	2.2	8.5	4.6	7.7
	0.4–4.5	0.4–4.5	S(100)35°	~3.6	~1.0	[100]	2.2	2.3	4.2	3.8
TiO_2	0.45–6.0	4.23	L[110]	7.93	~1.0	⊥	2.58	18.6	6.0	14
$PbMoO_4$	0.42–5.5	6.95	L[001]	3.63	5.5	⊥	2.39	14.6	23.9	24
BGO	0.45–7.5	9.22	L[110]	3.42	1.6	Arb.	2.55	3.8	6.7	6.7
TeO_2	0.35–5.0	6.0	L[001]	4.2	6.3	⊥	2.26	17.6	22.9	25
	0.4–4.5	6.0	S[110]	0.62	17.9	Cir.	2.26	13.1	795	127
GaP	0.6–10.0	4.13	L[110]	6.32	8.0	∥	3.31	75.3	29.5	71
	0.6–10.0	4.13	S[110]	4.13	2.0	⊥	3.31		16.6	26

TABLE 2   Selected AO materials (*Continued*)

Material	Optical transmission (μm)	Density (g/cm³)	Acoustic mode	Acoustic velocity	Acoustic attenuation (dB/μs-GHz²)	Optical polarization	Refraction index	Figure of merit M_1	Figure of merit M_2	Figure of merit M_3
Tl_3AsS_4	0.6–12.0	6.2	L[001]	2.15	5.0	∥	2.83	152	523	416
Tl_3AsSe_3	1.26–13.0	7.83	L[100]	2.05	14.0	∥	3.34	607	2259	1772
Hg_2Cl_2	0.38–28.0	7.18	L[100]	1.62	—	∥	2.62	34	337	125
			S[110]	0.347	8.0		2.27	4.3	703	73
$Ge_{33}As_{12}Se_{33}$	1.0–14.0	4.4	L	2.52	1.7	⊥	2.7	54.4	164	129
GaAs	1.0–11.0	5.34	L[110]	5.15	15.5	∥	3.37	118	69	137
As_2Se_3	0.9–11.0	4.64	L	2.25	27.5	∥	2.89	204	722	539
Ge	2.0–20.0	5.33	L[111]	5.5	16.5	∥	4.0	1117	482	1214

12.5 BASIC ACOUSTO-OPTIC DEVICES

In this section, we present in detail the theory and practices of AO devices. Following our previous classification,[10] three basic types of devices will be defined by the relative divergence of the optical and acoustic beams. Let

$$a = \frac{\delta\theta_o}{\delta\theta_a} \qquad (51)$$

be the divergence ratio characterizing the AO interaction geometry. In the limit $a \ll 1$, the device acts as a deflector or spatial modulator. For the intermediate value $a \approx 1$, the device serves as a (temporal) modulator. In the other limit $a \gg 1$, the device provides a spectral modulation, i.e. a tunable optical filter.

Table 3 summarizes the interaction geometry and appropriate figures of merit for the three basic AO devices.

Acousto-optic Deflector

Acousto-optic interaction provides a simple means to deflect an optical beam in a sequential or random-access manner. As the driving frequency of the acoustic wave is changed, the direction of the diffracted beam can also be varied. The angle between the first-order diffracted beam and the undiffracted beam for a frequency range Δf is approximately given by (outside the medium)

$$\Delta\theta_d = \frac{\lambda_o \Delta f}{V} \qquad (52)$$

TABLE 3   Figures of Merit for Acousto-Optic Devices

	Interaction plane	Transverse plane	Figure of merit
Deflectors (Bragg cells)	$a \ll 1$	$a \approx 1$	$M_3 = \dfrac{n^7 p^2}{\rho V^2}$
Modulators	$a \approx 1$	$a \approx 1$	$M_4 = \dfrac{n^8 p^2 V}{\rho}$
Tunable filters	$a \gg 1$	$a \gg 1$	$M_5 = \dfrac{n^8 p^2}{\rho V^3}$

In a deflector, the most important performance parameters are resolution and speed. Resolution, or the maximum number of resolvable spots, is defined as the ratio of the range of deflection angle divided by the angular spread of the diffracted beam, i.e.,

$$N = \frac{\Delta\theta}{\delta\theta_o} \tag{53}$$

where

$$\delta\theta_o = \xi\lambda_o/D \tag{54}$$

where D is the width of the incident beam and ξ is the a factor (near unity) that depends on the incident beam's amplitude distribution. For a nontruncated gaussian beam $\xi = 4/\pi$. From Eqs. (52), (53), and (54) it follows that

$$N \approx \tau\,\Delta f \tag{55}$$

Where $\tau = D/V \cos\theta_o$ is the acoustic-transit time across the optical aperture. Notice that the acoustic-transit time also represents the (random) access time and is a measure of the speed of the deflector. Equation (55) shows that the resolution is equal to time (aperture) bandwidth product. This is the basic tradeoff relation between resolution and speed (or bandwidth) of AO deflectors. In the design of AO deflectors, the primary goal is to obtain the highest diffraction efficiency for the specified bandwidth and resolution (or time aperture).

In the following we consider the design of AO deflectors [42]. Figure 5 shows the geometry of an AO deflector. A piezoelectric transducer is bonded to the appropriate crystal face, oriented for efficient AO interaction. A top electrode deposited on the transducer defines the active area with interaction length L and acoustic beam height, H. An acoustic wave is launched from the transducer into the interaction medium and produces a traveling phase grating. An optical beam is incident at a proper Bragg angle with respect to the acoustic wavefront. The incident beam is generally modeled as a gaussian profile in the interaction plane with a beam waist $2\omega_1$ at $/1e^2$ of the intensity. The optical beam in the interaction plane is truncated to an aperture width D. For sufficiently long interaction length L, the acoustic wave diffracts a portion of the incident light into the first order. The angular spectrum of the diffracted light is proportional to the acoustic power spectrum, weighted by the truncated optical beam profile.

The acoustic beam profile in the transverse plane is determined by the transducer height H, and the acoustic diffraction in the medium. The optical intensity distribution can be taken as Gaussian with waist $2\omega_2$ at $1/e^2$ intensity. The diffracted light intensity is then proportional to the overlapping integral of the optical beam and the acoustic diffraction profile.

Under momentum matching conditions, the peak diffraction efficiency of a Bragg cell is given by Eq. (18a)

$$\eta_o = \frac{\pi^2}{2\lambda_o^2} M_2\left(\frac{L}{H}\right) P_a$$

The diffraction efficiency can be increased by the choice of large L and small H. However, the acoustic beam width (i.e., interaction length) L defines the angular spread of the acoustic power spectrum and is thus limited by the required frequency bandwidth. The acoustic beam height H determines the transverse acoustic diffraction. A smaller value of H, however, will increase the divergence of the acoustic beam in the transverse plane. An optimum acoustic beam height in the Bragg cell design is chosen so that the AO diffraction occurs within the acoustic near field.[20,43]

$$H_o = \sqrt{BD} = V \sqrt{\frac{\tau}{f}} B \tag{56}$$

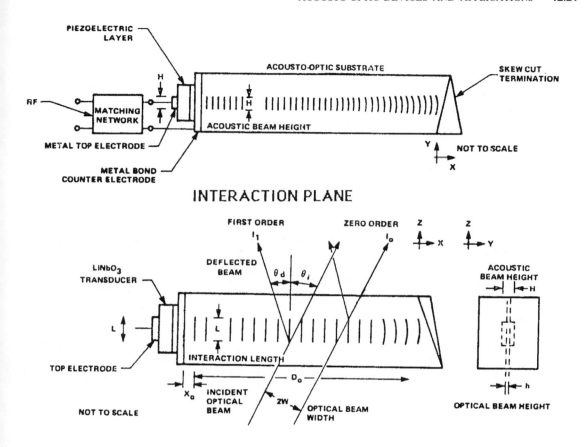

FIGURE 5   Acousto-optic deflector configuration.

where D is the total optical aperture and B is the curvature of the acoustic slowness surface. Substituting Eq. (48) into Eq. (18), we obtain:

$$\eta_o = \frac{\pi^2 P_a}{2\lambda_o^3 f^{3/2}\tau^{1/2}}\left(\frac{M_3}{\sqrt{B}}\right)\cdot \ell \tag{57}$$

where $\ell = L/L_o$ is the normalized acoustic beamwidth, and is determined by the specified fractional bandwidth. In Sect. 12.2 we have derived the frequency response of AO interaction. For the isotropic diffraction bandshape it was shown that ℓ is related to the fractional bandwidth $\Delta F = f/f_o$ by Eq. (39)

$$l \approx 1.8/\Delta F$$

Equation (57) shows increased diffraction efficiency can be obtained by the selection of AO materials and modes with large effective figure of merit $M_3^* = M_3/\sqrt{B}$ and applying techniques for increasing ℓ. The use of acoustic modes with minimum curvature is referred to as anisotropic acoustic beam collimation. A well-known example is a shear mode in GaP propagating along [110] direction.[43] In this case $B = 0.026$. Compared to an acoustically isotropic direction, the transducer height can be reduced by a factor of 6.2. Theoretically it is also possible to reduce the acoustic beam height by using acoustic focusing with cylindrical transducers. Since a deposited ZnO transducer is required, implementation of cylindrical acoustics is more complicated.

Another performance enhancement technique is to increase ℓ by using AO diffraction in birefringent crystals. In this case it is possible to choose the acoustic wave vector to be approximately tangential to the locus of the diffracted light vector.[17,44,45] As a result, a large band of acoustic frequencies will simultaneously satisfy the momentum-matching condition. Equivalently, for a given bandwidth, a larger interaction length can be used, thus yielding an enhancement of diffraction efficiency. The normalized interaction length is related to the fractional bandwidth by Eq. (33).

$$\ell = \frac{7.2}{\Delta F^2}$$

Compared to isotropic diffraction, the birefringent phase matching achieves an efficiency advantage factor of $4/\Delta F$, which becomes particularly significant for smaller fractional bandwidths.

Two types of birefringent phase matching are possible; these include the acoustically-rotated (AR)[46,47] and optically-rotated (OR)[42] phase matching. Figure 6a shows the wave vector diagram for AR tangential phase matching where the constant azimuth plane is chosen as the interaction plane and the acoustic wave vector is rotated in the plane to be tangential to the locus of the diffracted light wave vector. The wave vector diagram shown in Fig. 6b describes the OR type of tangential phase matching. In this case the acoustic wave vector is chosen to be perpendicular to the optic axis. The incident light wave vector is allowed to rotate in different polar angles to achieve tangential phase matching at desired frequencies. For both types of phase-matching schemes the tangential matching frequency, $f_t = F_t f_o$, is given by

$$f_t = \frac{V(\theta_a)}{\lambda_o} \sqrt{n_i^2(\theta_i) - n_d^2(\theta_d)} \tag{58}$$

where θ_i, θ_d and θ_a denote the directions of the incident optical wave, diffracted optical wave, and acoustic wave, respectively.

Another technique for increasing ℓ while maintaining the bandwidth is to use acoustic beam steering.[48,49] In this approach a phase array of transducers is used so that the composite acoustic wavefront will effectively track the Bragg angle. The simplest phase array employs fixed inter relevant phase difference that corresponds to an acoustic delay of $P\Lambda/2$, where P is an integer. This is referred to as first-order beam steering and can be realized in either a stepped array [48–50] or a planar configuration.[19,51] The two types of phased array configurations are shown in Fig. 7. The stepped array configuration is more efficient; but less practical due to the fabrication difficulty. The following analysis addresses only the planar configuration of first-order beam steering.

Consider the simplest geometry of a planar first-order beam-steered transducer array where each element is driven with an interelement phase difference of 180°. The acousto-optic bandpass response of this transmitter configuration is equal to the single-element bandshape multiplied by the interference (array) function;[42] i.e.,

$$W(F) = \left(\frac{\sin \pi X}{\pi X}\right)^2 \left(\frac{\sin N\pi Y}{N \sin \pi Y}\right)^2 \tag{59}$$

$$X = \frac{L_e F}{2L_o}(F_b - F) \tag{60}$$

$$Y = \frac{1}{2}\left\{\frac{D}{L_o} F(F_b - F) + 1\right\} \tag{61}$$

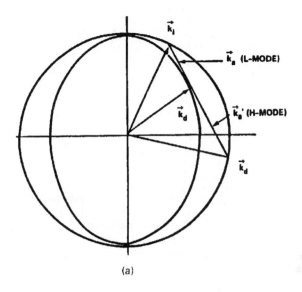

(a)

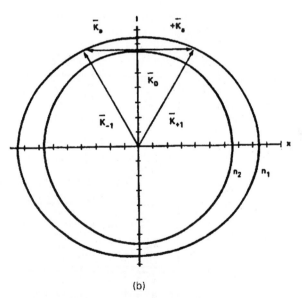

(b)

FIGURE 6 Wave vector diagram for tangential phase matching. (a) Acoustically rotated geometry, (b) optically rotated geometry.

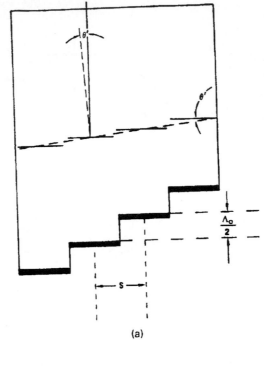

(a)

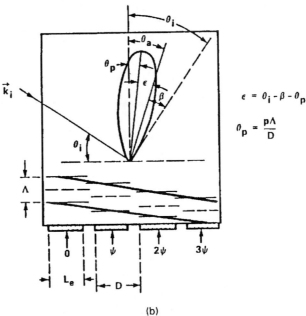

$$\epsilon = \theta_i - \beta - \theta_p$$

$$\theta_p \approx \frac{p\Lambda}{D}$$

(b)

FIGURE 7   Acoustic beam steering using phased array. (a) Stepped phased array, (b) planar phased array.

and L_e is the length of one element, N is the number of elements, where D is the center-to-center distance between adjacent elements.

For large N, the radiation pattern for a single element is broad, the bandpass function is primarily determined by the grating (array) functions. The array function can be approximately given by Eq. (34) except that the phase mismatch is given by:

$$\psi = \frac{Nd}{2}\left\{F(F_b - F) + \frac{1}{d}\right\}$$ (62)

where $d = D/L_o$. The bandpass characteristics of the grating loss are the same as the birefringent diffraction case with an equivalent interaction length $\ell = Nd$ and tangential matching frequency $F_1 = 1/\sqrt{d}$. The discussion on the interaction length-bandwidth relation of birefringent diffraction is thus directly applicable.

Referring to Eq. (59), notice that at the peak of the grating lobe for the phase-array radiation, the value of the single element radiation $\text{sinc}^2 x$ is approximately equal to 0.5. There is thus an additional 3-dB loss due to the planar phase array.

An interesting design is to combine the preceding techniques of tangential phase matching and acoustic beam steering in a birefringent phased-array Bragg cell.[52,53] The approach allows a higher degree of freedom in the choice of acoustic and optic modes for optimized $\mathbf{M}_3^*$, shifting of center frequency, and suppression of multiple AO diffractions.

Both techniques discussed here, decreasing acoustic beam height and increasing interaction length, allow the reduction of drive power required for obtaining a given diffraction efficiency. The increase of ℓ is particularly significant since the power density is proportional to ℓ^2. For instance, an increase of ℓ by a factor of 7 will reduce the power density by 50 times! In most wideband AO cells high power density has been the dominant factor limiting the device performance. The deterious effect due to high power density includes thermal, gradient nonlinear acoustics and possible transducer failure.

Besides bandwidth, there are other factors that limit the usable time aperture of AO deflectors: maximum available crystal size, requirement of large optics, and, most basically, the acoustic attenuation across the aperture. For most crystalline solids, the acoustic attenuation is proportional to f^2. If we allow a change of average attenuation of £ (dB) across the band, the maximum deflector resolution is given by

$$N_{\text{max}} = \frac{£}{\alpha_o f_o}$$ (63)

where α_o (dB/µsec GHz^2) is the acoustic attenuation coefficient.

It can be shown that the acoustic loss has negligible effects on the resolvable spot of the deflector. For most practical cases, a more severe problem associated with the acoustic loss is the thermal distortion resulting from heating of the deflector. The allowable acoustic loss thus depends upon other factors, such as the acoustic power level, thermal conductivity of the deflector material, etc.

It is instructive to estimate the maximum resolution achievable of AO deflectors. In the calculation, the following assumptions are made: octave bandwidth: $f_o = 1.5\Delta f$, maximum acoustic loss: £ = 4 dB, maximum aperture size: $D = 5$ cm. Results of the calculation are summarized in Fig. 8 where the deflector resolution is plotted for a number of selected AO materials. The figure clearly shows the basic tradeoff relation between the speed and the resolution of AO deflectors. A large number of resolvable spots is obtainable for low-bandwidth detectors using materials with slow acoustic velocities such as TeO$_2$ and Hg$_2$Cl$_2$. The maximum resolution of AO deflectors is limited to a few thousand.

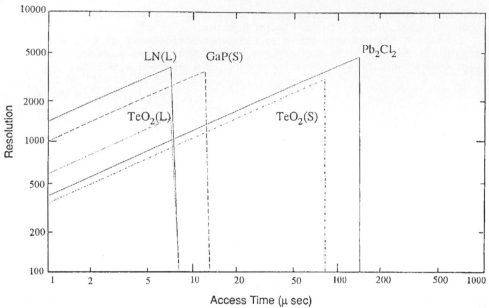

FIGURE 8 Resolution vs. access time for acousto-optic deflectors.

For certain applications such as laser scanning, a few thousand resolvable spots are insufficient, and further increase of deflector resolution is desirable. One early design for increased resolution involves the use of cascade deflectors.[54] Another technique is to use higher-order AO diffractions. One interesting design was to utilize the second-order diffraction in an OR-type birefringent cell.[55] Since both the first- and second-order diffraction are degenerately phase matched, efficient rediffraction of the first order into the second order was obtained. However, the use of the second-order diffraction allows the deflector resolution to be doubled for a given bandwidth and time aperture.

Acousto-optic Modulator

The acousto-optic interaction has also been used to modulate light. In order to match the Bragg condition over the modulator bandwidth, the acoustic beam should be made narrow, as in the case of deflectors. Unlike the case of deflectors, however, the optical beam should also have a divergence approximately equal to that of the acoustic beam, so that the carrier and the sidebands in the diffracted light will mix collinearly at the detector to give the intensity modulation. Roughly speaking, the optical beam should be about equal to that of the acoustic beam, i.e., $a = \delta\theta_o / \delta\theta_a \approx 1$. The actual value of the divergence ratio depends on the tradeoff between desired efficiency and modulation bandwidth.

The divergent optical beam can be obtained by using focusing optics. Figure 9 shows the diffraction geometry of a focused-beam AO modulator. In most practical cases the

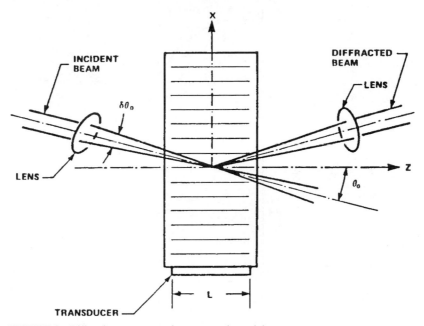

FIGURE 9 Diffraction geometry of acousto-optic modulator.

incident laser beam is a focused gaussian beam with a beam waist of diameter d. The corresponding optical beam divergence is

$$\delta\theta_o = \frac{4\lambda_o}{\pi n d} \tag{64}$$

The acoustic wave generated from a flat transducer is assumed to have a uniform amplitude distribution of width L and height H. The corresponding acoustic beam divergence in the interaction plane is

$$\delta\theta_a = \frac{\Lambda}{L} \tag{65}$$

The ratio of optical divergence to acoustic divergence becomes

$$a = \frac{\delta\theta_o}{\delta\theta_a} = \frac{4\ell}{\pi f_o \tau} \tag{66}$$

where $\tau = d/v$ is the acoustic transit time across the optical aperture. The characteristics of the AO modulator depend on a and ℓ, and thus the primary task in the design of the modulator is to choose these two parameters so that they will best meet the device specifications.

In the following we consider the design of a focused beam AO modulator for analog modulation.[10,19] We assume the incident laser beam profile is Gaussian, and calculate its diffraction by an amplitude-modulated (AM) acoustic wave. Three acoustic waves, the carrier, the upper, and the lower sidebands will generate three correspondingly diffracted

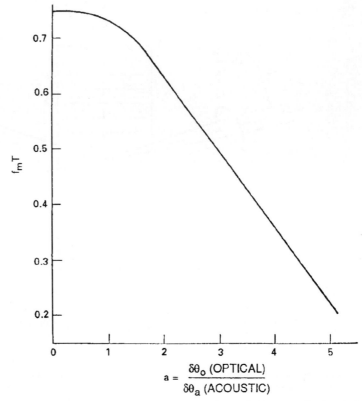

FIGURE 10 Modulation bandwidth × acoustic transit time vs. ratio of optical and acoustic beam divergence.

light waves traveling in separate directions. The modulated light intensity is determined by the overlapping collinear heterodyning of the diffracted optical carrier beam and the two sidebands. Using the frequency domain analysis, the diffracted light amplitudes can be calculated. Figure 10 shows the calculated AO modulation bandwidth as a function of optical-to-acoustic divergence ratio.[10] In the limit $a \ll 1$, the modulation bandwidth approaches the value

$$f_m \approx 0.75/\tau \qquad (67)$$

where $\tau = d/V$ is the acoustic transit time. Equation (60) shows that the modulation bandwidth may be increased by reducing the size of the optical beam. Further reduction of the beam size, however, will increase the angular spread of the optical beam $\delta\theta_o$ to become greater than $\delta\theta_a$; i.e., $a > 1$. The divergence angle of the diffracted beam is now determined by the acoustic divergence $\delta\theta_a$ through the momentum-matching condition. Thus, the modulation bandwidth starts to decrease as a increases. On the other hand, a larger value of a increases the interaction length l and thus the efficiency. A plot of the product of bandwidth and efficiency shows [10] a broad maximum near $a = 15$. In some laser systems it is desirable that the AO modulation does not introduce noticeable change to the guassian distribution of the incident beam. To satisfy this more restricted requirement an even smaller value for a should be chosen. The effect of the parameter a on the eccentricity of the diffracted beam was analyzed based on numerical calculation.[56]

The result shows that to limit the eccentricity to less than 10 percent the divergence ratio value for a must be about 0.67. In this region, $0.67 < a < 1.5$, the modulation bandwidth is approximately given by

$$f_m = 0.7/\tau \tag{68}$$

Another important case is the digital, or pulse modulation. Maydan[57] calculated the rise time and efficiency of pulsed AO modulators. His results show that an optimized choice of a is equal to 1.5, and that the corresponding rise time (10 to 90%) is

$$t_r = 0.85\tau \tag{69}$$

An additional constraint in the design of AO modulators is that the diffracted beam must be separated from the incident beam. To obtain an adequate extinction ratio, the angle of separation is chosen to be equal to twice that of the optical beam divergence. It follows that the acoustic frequency must be greater than

$$f_o = \frac{8}{\pi\tau} \tag{70}$$

Comparing to Eq. (68), the center frequency of the AO modulator should be about 4 times that of the modulation frequency.

The focused-beam-type AO modulator has certain disadvantages. The diffraction spread associated with the narrow optical beam tends to lower the diffraction efficiency. More importantly, the focusing of the incident beam results in a high peak intensity that can cause optical damage for even relatively low laser power levels. For these reasons, it is desirable to open up the optical aperture. Due to the basic issue of acoustic transit time, the temporal bandwidth of the wide-beam AO modulator will be severely degraded.

In certain applications, such as the laser display system, it is possible to use a much broader optical beam in the modulator than that which would be allowed by the transit time limitation. The operation of the wide-beam modulator is based on the ingenious technique of scophony light modulation.[58,59] A brief description of the scophony light modulator is presented as follows. A thorough treatment was given by Johnson.[60]

The basic idea, applicable to any system which scans a line at a uniform scan velocity, is to illuminate a number of picture elements, or pixels, in the modulator (window) onto the output line, such that the moving video signal in the modulator produces a corresponding image of the pixels which travels across the beam at sound velocity. The image can be made stationary by directing the image through a deflector that scans with equal and opposite velocity. Now, if the window contains N spots, then N picture elements can be simultaneously exposed in the image at any instant, and each picture element will be built up over the access time of the window which is equal to N times the spot time. Since the spots are immobilized, there is no loss of resolution in the image, provided that the modulator bandwidth is sufficient to produce the required resolution. The design of the wide-beam AO modulator is thus the same as that of the Bragg cell.

Acousto-optic Tunable Filter

The acousto-optic tunable filter (AOTF) is an all-solid-state optical filter that operates on the principle of acousto-optic diffraction in an anisotropic medium. The center wavelength of the filter passband can be rapidly tuned across a wide spectral range by changing the frequency of the applied radio frequency (RF) signal. In addition to the electronic tunability, other outstanding features of the AOTF include: large angular aperture while maintaining high spectral resolution, inherent intensity, and wavelength modulation capability.

The first AOTF, proposed by Harris and Wallace,[61] used a configuration in which the interacting optical and acoustic waves were collinear. Later, the AOTF concept was enlarged by Chang in a noncollinear configuration.[62] The theory and practice of the AOTF have been discussed in several review papers.[63,64]

Figure 11a shows the schematic of a transmissive-type collinear AOTF. It consists of a birefringent crystal onto which a piezoelectric transducer is bonded. When an RF signal is applied, the transducer launches an acoustic wave which is reflected by the acoustic prism and travels along a principle axis of the crystal. The incident optical beam, passing through the input polarizer, propagates along the crystal axis and interacts collinearly with the acoustic waves. At a fixed RF frequency, only a narrow band of optical waves is diffracted into the orthogonal polarization and is selected by the output analyzer. The center wavelength of the passband is determined by the momentum matching condition. Figure 11b shows the wave-vector diagram for the collinear AO interaction in a uniaxial crystal. In this case, Eq. (21) yields a relation between the center of the passband and the acoustic frequency

$$\lambda_o = \frac{V \, \Delta n}{f} \tag{71}$$

where Δn is the birefringence. Equation (71) shows that the passband wavelength can be tuned simply by changing the frequency of the RF signal.

Experimental demonstrations of the collinear AOTF were reported by Harris and coworkers in a series of papers.[65–67] For instance, a collinear AOTF was operated in the visible using $CaMoO_4$ as the interaction medium.[66] The full width at half-maximum (FWHM) of the filter passband was measured to be 8 Å with an input light cone angle of $\pm 4.8°$ (F/6). This angular aperture is more than one order of magnitude larger than that of a grating for the same spectral resolution.

The collinearity requirement limits the AOTF materials to rather restricted classes of crystals. Some of the most efficient AO materials (e.g., TeO_2) are not applicable for the collinear configuration. To utilize such materials, a new AOTF configuration was proposed in which the acoustic and optical waves are noncollinear.

Figure 12 shows the schematic of a noncollinear AOTF. The use of the noncollinear geometry has the significant advantage of simpler fabrication procedures. In addition, when the filtered narrowband beam is spatially separated from the incident broadband light, the noncollinear AOTF can be operated without the use of polarizers.

The basic concept of the noncollinear AOTF is shown by the wave vector diagram in Fig. 13 for an acoustically rotated AO interaction in a uniaxial crystal.[62] The acoustic wave vector is so chosen that the tangents to the incident and diffracted light wave vector loci are parallel. When the parallel tangents condition is met, the phase mismatch due to the change of angle incidence is compensated by the angular change of birefringence. The AO diffraction thus becomes relatively insensitive to the angle of light incidence, i.e., a process referred to as the noncritical phase matching (NPM) condition. Figure 13 also shows the NPM scheme for the special case of the collinear AOTF.

The first experimental demonstration of the noncollinear AOTF was reported for the visible spectral region using TeO_2 as the filter medium. The filter had a FWHM of 4 nm at an F/6 aperture. The center wavelength is tunable from 700 to 450 nm as the RF frequency is changed from 100 to 180 MHz. Nearly 100 percent of the incident light is diffracted with a drive power of 120 mW. The filtered beam is separated from the incident beam with an angle of about 6°.

The plane wave analysis of birefringent AO diffraction presented earlier can be used to determine the wavelength and angular bandpass characteristics of the AOTF. For proper operation, the requirement of NPM must be satisfied, i.e., the tangents to the optical wave vector surfaces are parallel. The parallel tangents condition is the equivalent of collinearity of ordinary and extraordinary rays,[64] i.e.,

$$\tan \theta_e = e^2 \tan \theta_o \tag{72}$$

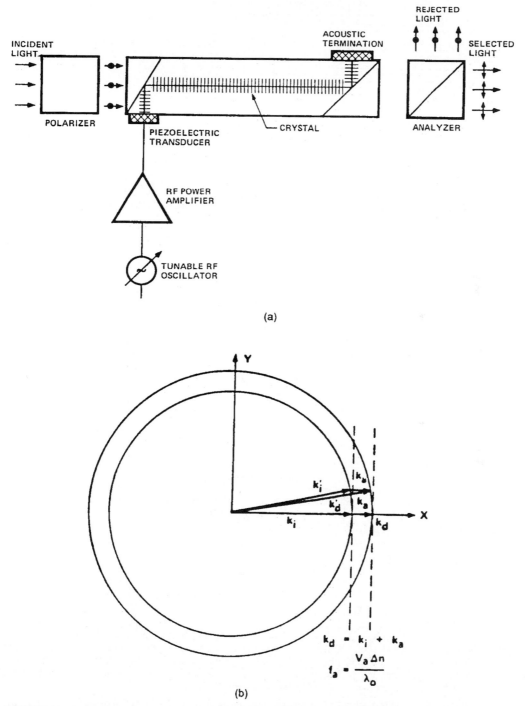

FIGURE 11   Collinear acousto-optic tunable filter. (a) Transmissive configuration, (b) wave vector diagram.

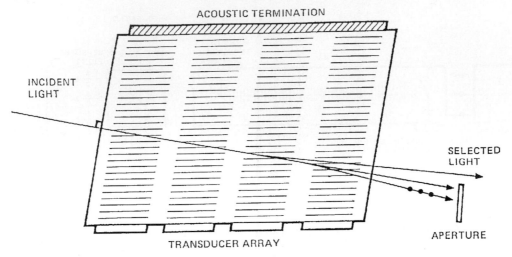

ACOUSTIC TERMINATION

INCIDENT
LIGHT

SELECTED
LIGHT

APERTURE

TRANSDUCER ARRAY

FIGURE 12 Schematic of noncollinear acousto-optic tunable filter.

where $e = n_E/n_o$, n_E is the refractive index of the extraordinary wave polarized along the optic c-axis. The momentum matching condition is

$$\tan \theta_a = (n_e \sin \theta_e - n_0 \sin \theta_o)/(n_e \cos \theta_e - n_o \cos \theta_o) \qquad (73)$$

$$f_a = (V/\lambda_o)(n_e^2 + n_o^2 - 2n_e n_o \cos(\theta_e - \theta_o))^{1/2} \qquad (74)$$

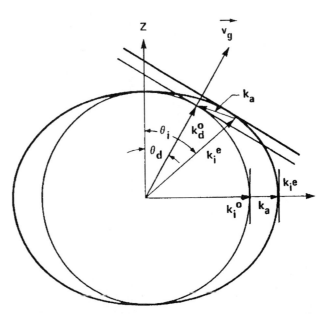

FIGURE 13 Wave vector diagram for noncollinear AOTF showing noncritical phase matching (NPM).

where $n_e(\theta_e)$ is the refractive index for the extraordinary wave with polar angle θ_e, i.e.,

$$n_e(\theta_e) = n_o(\cos^2 \theta_e + (\sin^2 \theta_e)/e^2)^{-1/2} \tag{75}$$

Substituting Eqs. (72) and (75) into Eq. (73) the acoustic wave angle, θ_a, can be written explicitly as a function of the optical wave angle, θ_o.[34]

$$\tan \theta_a = \frac{-(\cos \theta_o + \sqrt{1 + e^2 \tan^2 \theta_o})}{\sin \theta_o} \tag{76}$$

For small birefringence $\Delta n = |n_c - n_o| \ll n_o$, an approximate solution for the acoustic frequency is

$$f_a = (V \Delta n / \lambda_o)(\sin^4 \theta_e + \sin^2 2\theta_e)^{1/2} \tag{77}$$

where $\theta_e = 90°$. Equation (77) reduces to the simple expression for the collinear AOTF (Eq. (71)).

The bandpass characteristics of the AOTF are determined by the growth of optical waves that nearly satisfy the phase-matching condition over the finite interaction length. For an acoustic column of uniform amplitude, the bandpass response is given by

$$T(\lambda_o) = T_o \, \text{Sinc}^2 \, \sigma L \tag{78}$$

where T_o is the peak transmission at exact momentum matching, L is the interaction length, and σ is the momentum mismatch. It can be shown that[68]

$$\sigma = -b \sin^2 \theta_i (\Delta \lambda / \lambda_o^2) + (\Delta n / 2 \lambda_o)[F_i(\Delta \theta)^2 + F_2(\sin \theta_i \Delta \phi)^2] \tag{79}$$

where $\Delta \lambda$, $\Delta \theta$, and $\Delta \theta$ are deviations in wavelength, polar, and azimuth angles of the incident light beam, b is the dispersion constant defined by

$$b = \Delta n - \lambda_o \delta / \delta \lambda_o (\Delta n) \tag{80}$$

and $F_1 = 2 \cos^2 \theta_i - \sin^2 \theta_i$, $F_2 = 2 \cos^2 \theta_i + \sin^2 \theta_i$.

Resolution. Equation (78) shows that half peak transmission occurs when $\sigma L \approx 0.44$. The full width at half-maximum (FWHM) of the AOTF is

$$\Delta \lambda = 0.9 \lambda_o^2 / bL \sin^2 \theta_i \tag{81}$$

Angular Aperture. The acceptance (half) angles in the polar and azimuth planes are given by:

$$\Delta \theta = \pm n (\lambda_o / \Delta n L F_1)^{1/2} \tag{82}$$

$$\Delta \phi = \pm n (\lambda_o / \Delta n L F_2)^{1/2} \tag{83}$$

The frequency domain method described previously Sect. 2.2 can now be used to determine the AOTF bandpass characteristics for an incident cone of light. The filter transmission for normalized input light distribution $I(\theta_i, \phi_i)$ is obtained by integrating the plane-wave transmission [Eq. (78)] over the solid angle aperture

$$I_d(\Delta \lambda) = \int_{\alpha_i, \beta_i} T(\Delta \lambda, \theta_i, \phi_i) I(\theta_i, \phi_i) \sin \theta_i \, d\theta_i, \, d\phi_i \tag{84}$$

Angle of Deflection. For noncollinear AOTF the diffracted light is spatially separate from the incident light. The angle of deflection can be determined from Eq. (72). For small birefringence, an approximate expression for the deflection angle is given by

$$\Delta \theta_d = \Delta n \sin 2\theta_o \tag{85}$$

It reduces maximum when $\theta_o = 45°$. For instance, in a TeO_2 AOTF operated at 633 nm,

$\Delta n \approx 0.15$, the maximum deflection angle for TeO_2 AOTF is about 8.6 degrees. As long as the angular aperture $\Delta\theta$ is smaller than the deflection angle $\Delta\theta_d$, the noncollinear AOTF can realize the angular aperture without the use of polarizers. Notice that if the small birefringent dispersion is ignored, $\Delta\theta_d$ is independent of wavelength. This feature is important to spectral imaging applications.

Transmission and Drive Power.   An important parameter in the design of an AOTF is the required drive power. The peak transmission of an AOTF is given by

$$T_o = \sin^2\left(\frac{\pi^2}{2\lambda_o^2} M_2 P_d L^2\right)^{1/2} \qquad (86)$$

where P_d is the acoustic power density. Maximum transmission occurs when the drive power reaches the value

$$P_a = \frac{\lambda_o^2 A}{2M_2 L^2} \qquad (87)$$

where A is the optical aperture. The high drive power required for infrared AOTFs (particularly large-aperture imaging types) is perhaps the most severe limitation of this important device.

12.6   APPLICATIONS

Acousto-optic devices provide spatial, temporal, and spectral modulation of light. As such, they can be used in a variety of optical systems for optical beam control and signal processing applications. In this section we shall discuss the application of AO devices. The discussion will be limited to these cases where the AO devices have been successfully deployed and appear most promising for future development.

Acousto-optic Deflector

The early development of the AO deflector was aimed at laser beam scanning applications. The primary goal was to realize a large number of resolvable spots so that it may be used to replace mechanical scanners such as the rotating polygons. In recent years, however, most of the effort was directed toward Bragg cells for optical signal processing applications. In addition to the high-resolution AO deflectors, the most intensive activity has been the development of wideband Bragg cells. In the following we discuss in detail the various system applications of AO deflectors. First, we shall briefly discuss the present status of both types of AO deflectors (or Bragg cells). A summary of the representative devices is shown in Table 4.

High-resolution Cell.   One of the earlier designs of high-resolution cells was first demonstrated by Warner et al. using the low-shear [110] mode in TeO_2.[45] Sizable reduction of drive power was achieved by using tangential phase matching of the birefringent diffraction. Due to the slow velocity along [110] axis, $V = 0.62$ mm/μsec, the optical aperture size of a 70 μsec device is only 4.3 cm.

The on-axis has two drawbacks. First, the tangential frequency f_t, determined by Eq. (58), is relatively low. At 633 nm, f_t is equal to 37 MHz. This limits the usable bandwidth to below 25 MHz. Second, at the tangential phase-matching frequency, the second-order diffraction is also matched. The first-order diffracted light is rediffracted into the second order and results in a dip at the midband of the deflector. By utilizing the acoustically rotated (AR)-type birefringent diffraction, Yano et al.[46] demonstrated high efficient diffraction in a TeO_2 cell without the midband dip. The primary drawbacks of the AR-type

TABLE 4 Performance of Acousto-Optic Bragg Cells

Material (mode)	Center frequency (MHz)	Bandwidth (MHz)	Aperture (μs)	TB product	Efficiency (%/W)
TeO$_2$(S)	90	50	40	2000	110
TeO$_2$(S)	160	100	10	1000	95
GaP(S)	1000	500	2.0	1000	30
GaP(S)	2000	1000	1.0	1000	12
LiNbO$_3$(S)	2500	1000	1.0	1000	10
GaP(L)	2500	1000	0.25	250	44
GaP(L)	3000	2000	0.15	300	10
GaP(S)	3000	2000	0.25	500	8
LiNbO$_3$(S)	3000	2000	0.30	600	6

Note: $\lambda_o = 830$ nm

cell are the reduced aperture (and thus resolution) and significant optical aberration due to the large acoustic beam walkoff in TeO$_2$.

Another technique of raising the tangential phase-matching frequency is to use the optically rotated (OR) configuration.[42] Since the acoustic wave is along the principal[110] axis, in this case, there is no acoustic beam walkoff, and high resolution is obtainable. The upshift in f_t allows wider bandwidth to be realized. The optical aperture along the [110] axis is limited by the acoustic attenuation. The device shown in Table 4 is designed to operate at the tangential phase-matching frequency of 90 MHz with a 50-MHz bandwidth. For an aperture size of 2.5 cm (or time aperture of 40 μsec), the OR-type TeO$_2$ device shown in Table 4 has an overall resolution of about 2000 spots. A larger resolution can be realized with a longer optical aperture. Figure 8 shows a maximum resolution of 3500 spots for 5-cm aperture.

In recent years several new AO materials with exceptionally low acoustic velocities were developed. The acoustic velocities of slow shear waves in Hg$_2$Cl$_2$ and Hg$_2$Br$_2$ along the [110] direction are 0.347 mm/μsec and 0.282 mm/μsec, respectively. Very large time apertures are obtainable with crystals of moderate size. As shown in Fig. 9, an AO deflector resolution of 5300 spots is projected for a 5-cm-long Hg$_2$Cl$_2$ crystal along the [110] direction. Recently, an experimental Hg$_2$Cl$_2$ device was constructed that had a measured bandwidth of 30 MHz.[69] Due to difficulties in the device fabrication and the large curvature of the slow acoustic mode, relatively small time aperture was realized. The potential of the exceedingly large deflector resolution as predicted in Fig. 9 is yet to be experimentally demonstrated.

Wide Bandwidth Cell. Two types of wideband Bragg cells have been developed. These include the phased array GaP cell and the birefringent LiNbO$_3$ cell. The material GaP has very large figure of merit M_3 and relatively low acoustic attenuation. Two modes of particular interest are the L [111] and S [100] modes.[43,70,71] Both have low acoustic slowness curvature and thus allow efficiency enhancement due to anisotropic beam confinement. The effective figure of merit $M_3/\sqrt{B}$ for L [111] and S [100] are equal to 98 and 48, respectively.[72]

The wideband GaP device also achieves an efficiency enhancement by using acoustic beam steering.[70] To simplify fabrication of the transducer, planar phased-array configuration was used. The achievable efficiency and bandwidth of these phased array GaP devices are believed to be best state-of-the-art performance.

AO Bragg cells with even larger bandwidths have been demonstrated using the low attenuation materials. The best known design was a LiNbO$_3$ device using a y-z propagating off-axis x-polarized shear wave.[73,74] The device demonstrated an overall bandwidth of 2 GHz, a peak diffraction efficiency of 12 percent/watt and about 600 resolvable spots.

Multichannel Cell.   As an extension to usual single-channel configuration there has been considerable activity in the development of multichannel Bragg cells (MCBC).[75,76] The MCBC uses a pattern of multiple individually addressed transducer electrodes in the transverse plane. The use of the MCBC allows the implementation of compact two-dimensional optical signal-processing systems such as optical page composers, direction-of-arrival processors, etc. In addition to the design rules used for single-channel cells, other performance parameters such as crosstalk, amplitude, and phase tracking must be considered. Another critical issue is power handling. Since the drive power is proportional to the number of channels, the excessive amount of heat dissipation at the transducer will introduce a thermal gradient that severely degrades the Bragg cell performance.

Early work on MCBC uses the y-z cut shear-wave wideband $LiNbO_3$ cell design.[75] Good amplitude and phase tracking were obtained over the operating bandwidth of 1 GHz. The channel-to-channel isolation, typically about 25 dB, was limited by RF crosstalk. In a later development of GaP MCBC,[76] reduction of electrical crosstalk (to −40 dB) was obtained by using stripline interconnection structures. The use of the anisotropic collimating modes in GaP also brings the advantage of lowering acoustic crosstalk.

Laser Scanning.   There are a variety of laser applications that require laser beam scanning. Acousto-optic deflectors provide a simple solid state scanner which eliminates the inherent drawbacks of mechanical scanners due to moving parts such as facet errors and the requirement of realignment because of bearing wear. For certain applications such as beam-addressed optical memory, the rapid random access capability of AO deflectors offers a distinct advantage. One of the first practical AO deflector applications was a laser display system described by Korpel et al.[48] The AO device used water as the interaction medium. Due to its high acoustic attenuation, the water cell had limited bandwidth. By incorporating superior AO materials such as $PbMoO_4$ and TeO_2, a number of laser scanning systems were later developed that demonstrated significant performance improvements. A notable example was the TV rate laser scanner described by Goreg et al.[77,78] With a drive power as low as 50 mW, the TeO_2 deflector operated as a horizontal scanner and achieved the specified TV rate and 500 resolvable spots.

We have shown (Fig. 9) that using optimized designs, the AO deflector can achieve 2000 to 4000 resolvable spots with good efficiency and acceptable access time. In the late seventies, such AO deflectors became commercially available. Using a 2000-spot AO deflector, Grossman and Redder demonstrated a high-speed laser facsimile scanner with 240 pixel/in resolution.[79]

An intriguing approach to significantly increase the deflector resolution is the use of a traveling lens. The basic concept, proposed by Foster et al.,[80] utilizes the refractive effect of an acoustic pulse to form a traveling lens. The laser beam from a primary deflector passes through a second cell that has an acoustic pulse traveling across the optical aperture in synchronicity with the deflected beam from the first cell. The additional focusing by the traveling acoustic lens reduces the final spot size and thus increases the overall number of resolvable spots. The concept of the traveling wave acoustic lens was experimentally demonstrated with an enhancement factor from about 10 to 40.[81] Combined with a primary deflector with moderate resolution, this technique should be capable of achieving more than 10,000 pixels/scan line resolution. One major drawback of this device is the relatively high drive power required.

In the design of the AO deflector for linear scanning, one must consider the effect due to finite RF sweep rate. An important characteristic of linear frequency modulation (LFM) operation is the cylindrical focusing effect.[82] When an LFM acoustic pulse is passing through the cell, the incoming beam is diffracted by an angle $\theta = \lambda_o f / V$, where f is the frequency of the local acoustic wave. At a distance Δx away, the angle of diffraction will change by $\Delta\theta = \lambda_o / V \Delta f$ where Δf is the corresponding change of acoustic frequency. The diffracted rays will focus to a point located at a distance F from the center of the cell

$$F = \frac{\Delta x}{\Delta \theta} = V^2 \left/ \left(\lambda_o \frac{\Delta f}{T} \right) \right. \tag{88}$$

range Δf where $\Delta f/T$ is the linear scan rate and T is the time for sweeping over frequency. An additional lens may be used to compensate for the lens effect.

The finite scan rate also degrades the resolution of the AO deflector. When the acoustic transit time τ (which is equal to flyback time) becomes an appreciable portion of the scan time T, there is a reduction of the effective aperture by the factor $1 - \tau/T$. This results in a resolvable number of spots for the scanning mode,

$$N = \tau \Delta f (1 - \tau/T) \tag{89}$$

To avoid loss of resolution the general practice is to choose $\tau \ll T$. This limits the speed or resolution of the scanner. Another method is the AO chirp scanner.[83] The basic concept is to exploit the cylindrical lens effect described above. An LFM or chirp RF with a duration less than the acoustic transit time is applied to the transducer. As the acoustic chirp travels through the cell, due to the cylindrical lens effect, it produces a focused spot which moves down in the image plane with the acoustic wave velocity. To make efficient use of the laser power a prescanner is used to track the traveling chirp. For a chirp pulse of duration τ_c, the size of the focused spot is equal to the angular size of the incident beam multiplied by the focal length. Setting the sweep time equal to τ_c in Eq. (88) yields

$$d = \frac{F\lambda_o}{V\tau_c} = \frac{V}{\Delta f} \tag{90}$$

The total number of resolvable spots is approximately equal to $\Delta f \tau$ if $\tau_c \ll \tau$. The chirp scanner thus avoids the degradation resolution by reducing the duration of the chirp over the bandwidth Δf. The technical difficulty of the chirp scanner is the requirement of a stable fast chirp with extremely linear frequency scanning characteristics.

Despite these promising results, the AO deflector has not been able to compete with mechanical scanners such as rotating polygons for general laser scanning applications. Limited resolution, improved reliability, and, in particular, high cost have been the primary factors. As a result, in more recent developments, the AO deflectors are designed to be used in laser scanning systems for acquiring extremely wide bandwidth, random access capability, or to provide auxiliary functions to prime scanners such as wobble correction[84] and facet tracking.[85]

Acousto-optic Modulator

Among the three basic types of AO devices, the AO modulator has been most well developed in the marketplace. A variety of AO modulators is commercially available that is suited to external or intracavity applications. Compared to the competing electro-optic (EO) modulators, the AO modulator has many advantages that include low drive power, high extinction ratio, insensitivity to temperature change, simple drive electronics, and high safety factors. Since the AO effect occurs in all crystal types as well as amorphous solids, and high optical quality, low-loss AO materials are readily available for intracavity applications. Except in certain applications where very large bandwidths are required, AO modulators are generally preferred to their EO counterparts.

Most of the AO modulator applications had been well developed by the mid-70s. The early work has been discussed in a previous review paper.[10] In the following we shall primarily address some of the newer developments.

The AO modulators can be broadly classified into three categories: simple intensity modulators, intracavity modulators, and AO frequency shifters.

TABLE 5

Material (mode)	Wave-length (μm)	Center frequency (MHz)	RF bandwidth (MHz)	Rise time (nsec)	Efficiency (%)
PbMoO$_4$	0.633	80	40	25	80
TeO$_2$	0.633	110	50	20	75
TeO$_2$	0.633	200	100	7	65
GaP	0.83	500	250	4	50
GaP	0.83	1000	500	2	30

Intensity Modulators. Simple intensity AO modulators are used for general-purpose applications. The lower-cost types, usually made of glass (dense flint or Tellurite glass), are useful for bandwidths of up to about 10 MHz. The use of superior materials such as PbMoO$_4$ and TeO$_2$ has extended modulation bandwidth to about 50 MHz. Chang and Hecht[86] demonstrated the first GaP modulator with a risetime of 5 nsec. Later development has reduced the risetime to about 2 nsec, which corresponds to a modulation bandwidth of about 250 MHz. Table 5 lists a few selected AO modulators and some of the typical performances.

Historically, the most important market for AO modulators is for use in laser printers. Before the development of organic photoreceptors with spectral response extended to the near infrared, gas lasers such as HeCd or HeNe were used in laser printers. The requirement of an external modulator can be best met by the simple, low-cost AO modulator. The laser printer generally uses a flying-spot scanner configuration with the AO modulator operated in the focused beam mode in order to achieve the desired rise time.

Since the development of laser printers that use laser diodes as the optical source, the internal modulation capability of laser diodes has eliminated the need for external modulators. Therefore, for future laser printer applications, the use of AO modulators will probably be limited to cases where gas lasers are employed, or where the AO device offers a unique advantage. For example, the use of an AO device to perform combined deflector and modulation functions in the multibeam fast, dither scanners.[87]

The AO modulators have also been used for laser modulation in the infrared. The development of efficient infrared AO materials has resulted in various laser modulation uses, ranging from 1.06 μm to 10.6 μm. One important application of AO modulators is to serve as an external modulator in laser communication systems. For lower modulation bandwidth requirements, AO modulators are preferred to their EO counterparts due to lower optical insertion loss or greater contrast ratio. Efficient and fast miniature AO modulators suited to 1.06 μm can be constructed using GaP, GaAs, or GeAsSe.[88] Among these, GeAsSe glass is particularly suited to wideband uses due to its exceptionally low acoustic attenuations ($\alpha_o = 1.7$ dB/nsec GHz^2). Although it is slightly less efficient than both GaP and GaAs, it offers the important advantage of lower cost.

AO modulators have also been used to modulate CO$_2$ lasers for various applications such as precision machining, range finding, and communications. At 10.6 microns, the most popular material has been single-crystal Ge. Early work[89] of Ge modulators using longitudinal mode along [111] axes gave a high, large figure of merit ($M_2 = 540$) with respect to fused silica. Later, more accurate measurements[90,91] have shown a smaller figure of merit ($\mathbf{M}_2 = 120$). Typical performance of the Ge modulator shows a risetime of 30 nsec and a diffraction efficiency of 5 percent/watt.

Intracavity AO Modulators. Due to the high optical quality of AO materials such as fused silica, AO modulators have been used exclusively inside a laser cavity. These intracavity applications include Q-switching,[92] mode locking[93] and cavity dumping.[94]

Q-switching of YAG lasers has been an important requirement for industrial laser applications such as cutting, scribing, resistor trimming, and other material processing processes.

For Q-switching applications, an intracavity AO modulator is used to introduce an optical loss to keep the laser below threshold. When the loss is suddenly removed by switching off the acoustic pulse, the laser bursts into a short pulse with extremely high intensity. During the Q-switch period, the AO modulator should ideally add no additional loss to the laser cavity. Another important requirement for the AO Q-switch is due to special characteristics of the YAG laser; it achieves maximum gain when operated in the unpolarized state. Thus, it is essential that the AO Q-switch does not introduce any polarizing effect.

Presently all of the AO Q-switches use UV-grade fused silica as the interaction medium.[95] The low optical absorption, good homogeneity, and near strain-free material make it the only viable choice. In a longitudinal mode device, the diffraction efficiency for light polarized perpendicular to the acoustic wave is five times greater than that of parallel polarization. Thus, the YAG laser will tend to operate in the polarization state with the low diffraction efficiency. Because of this, today's AO Q-switches are designed to use acoustic shear waves since they are insensitive to the state of polarization. Due to the extremely low AO figure of merit of this mode, substantial RF power (~ 50 W) is required to hold off a laser with a moderate gain.

In the mode-locking applications, a standing wave AO modulator is used inside the cavity to introduce a loss modulation with a frequency that is equal to longitudinal mode spacing. $f_o = c/2L$ where L is the optical length of the laser cavity. The loss modulation provides a phase locking of the longitudinal modes and produces a train of short optical pulses.

In the Q-switching operation the maximum repetition rate is limited by the finite buildup time of population inversion, to about 50 KHz. To obtain higher repetition rates, the technique of cavity dumping is used. In the cavity dumping mode of operation, short acoustic pulses are fed into an intracavity AO modulator to couple the laser energy out of the cavity. The laser is always kept above the threshold, thus the switching rate is limited by the switching speed of the AO modulator. Most cavity dumpers are fused silica as the interaction medium. To reduce drive power, the TeO_2 modulator has been used for cavity dumping of YAG lasers.

There has been considerable activity in the development of intracavity AO devices that perform simultaneous Q-switching mode locking or cavity dumping.[96,97]

Acousto-optic Tunable Filter

Since the early work on AOTFs, there has been considerable effort to improve its performance for meeting system requirements. The key performance parameters include tuning range, spectral resolution, angular aperture, out-of-band rejection, drive power, and image degradation. Before a discussion of the applications, we shall first review the progress of these performance parameters. The performance characteristics of several typical AOTFs is given in Table 6.

Spectral Tuning Range. Early operation of the collinear AOTF was reported in the

TABLE 6 Performance of Acousto-optic Tunable Filters

Material	Configuration	Tuning range (μm)	Measured wavelength (μm)	Bandwidth (nm)	Angular aperture	Optical aperture	Efficiency
Quartz	Collinear	0.23–0.7	0.325	0.15	5°	7 mm × 7 mm	10%/W
Quartz	Noncollinear	0.23–0.7	0.325	1.0	10°	4 mm × 30 mm	10%/W
TeO$_2$	Noncollinear	0.45–0.75	0.633	2.0	10°	5 mm × 5 mm	75%/W
TeO$_2$	Noncollinear	1.2–2.5	1.3	1.5	10°	3 mm × 3 mm	35%/W
TeO$_2$	Noncollinear	2.0–5.0	3.39	15	2°	8 mm × 8 mm	8%/W
Tl$_3$AsSe$_3$	Noncollinear	7.0–11	10.6	600	35°	4 mm × 15 mm	6%/W

visible spectrum region. Later development extended the operating wavelength into the ultraviolet using crystal quartz[98,99] and into the infrared using Tl$_3$AsSe$_3$.[100,101]

The extension of spectral range is limited primarily by the availability of efficient AOTF materials that are transparent in the desired wavelength regions. Since the crystal classes applicable to collinear AOTF are more restricted, most of the effort for extending the spectral range has been focused on noncollinear AOTFs. Operation of noncollinear AOTF has been demonstrated in the UV region using crystal quartz,[102] sapphire, and MgF$_2$. Noncollinear AOTFs operated in the infrared up to about 11 microns have also been reported. The filter materials include TeO$_2$[103,104] Tl$_3$AsSe$_3$[101,105] and, most recently, HgCl$_2$.[69]

Drive Power. One of the major limitations of the AOTF is the relatively large drive power required. The problem is particularly severe in the infrared since the drive power is proportional to the square of the optical wave length. As an example, the acoustic power density of an infrared TeO$_2$ AOTF is estimated using a $\theta_i \approx 20°$ design ($M_2 \approx 10^{-12}$ m^2/W) and an interaction length of 1 cm. At $\lambda_2 = 4$ μm, the required acoustic power density is estimated to be about 8 W/cm^2. This relatively high power requirement is probably the most serious disadvantage of the AOTF and limits its use for certain important system applications (such as focal plane sensors). One technique to reduce the drive power is to utilize acoustic resonance enhancement.[106]

In an acoustic resonator structure, the peak acoustic field at resonance can be orders of magnitude greater than that of a traveling acoustic wave. It is thus possible to decrease the drive power by operating the AOTF in the acoustic resonance mode. Naturally, the resonant AOTF must be operated at the discrete resonant acoustic frequencies. For collinear AOTFs, the passband response (in acoustic frequency) consists of two to three modes. For noncollinear AOTFS, the passband response consists of many resonant modes. The passband response between successive resonant peaks will be sufficiently overlapped. Thus, the continuous tuning of the resonant AOTF is obtainable. The bandpass is essentially the same as the conventional traveling wave case.

The drive power of the resonant AOTF is reduced by the enhancement of diffraction efficiency due to acoustic resonance. At resonance the enhancement factor can be shown to be equal to the reciprocal of one-way acoustic loss. The total acoustic loss includes losses due to acoustic attenuation in the filter medium and reflection loss at the boundaries.

The reduction of drive power due to acoustic resonance was demonstrated in a collinear CaMoO$_4$ AOTF. Operating the filter at the peak of the acoustic resonance, a reduction in drive power of 22 dB was achieved. Acoustic resonance enhancement was also demonstrated using a noncollinear TeO$_2$ AOTF operated in the infrared region of 2 to 5 µm. About nine resonant modes were observed within one filter passband. Due to the relatively high acoustic attenuation in TeO$_2$, a power reduction factor of 16 was obtained.

Out-of-Band Rejection and Sidelobe Suppression.   The ability of spectral discrimination is determined by the ratio of peak transmission and out-of-band rejection of the AOTF. The out-of-band rejection for an AOTF is determined by two factors. Overall out-of-band rejection is determined by the contrast ratio; i.e., the fraction of undiffracted light leakage through crossed polarizers. In practice the contrast can also be limited by residual strain of the filter medium. A large contrast ratio is obtainable in the noncollinear AOTF, where the incident and diffracted light are spatially separated.

Near the filter band, out-of-band rejection is determined by the sidelobe structure of the filter passband response. The AOTF bandpass response is proportional to the power spectra of the acoustic field. For uniform acoustic excitation, the AOTF exhibits a sinc^2-type bandpass characteristic with the nearest sidelobe about 13 dB below the main lobe. Suppression of the high sidelobes can be obtained by techniques of amplitude apodization.[107] In the collinear AOTF, an acoustic pulse apodized in time is launched into the filter medium. By utilizing a triangular window, the first sidelobe was reduced by −22 dB with a collinear quartz AOTF. For the noncollinear AOTF, the apodization can be realized by weighted acoustic excitation at a transducer array. This was demonstrated in a noncollinear TeO$_2$ AOTF. A maximum sidelobe of −26 dB was achieved using a Hamming-type window.

Spectral Resolution.   For certain applications such as lidar receiver systems it is desirable to significantly narrow the passband of the AOTF. In principle, increased spectral resolution can be obtained by using long interaction length. However, there exists a basic tradeoff relation between the angular and spectral bandwidth. For instance, the angular aperture of a TeO$_2$ AOTF with 0.1 percent bandwidth (e.g., 5 Å at 0.5 µm) the angular is only ±3°.

One approach for realizing narrow passband while maintaining a large angular aperture is to use the wavelength dispersion of birefringence.[108] In the spectral range near the absorption band edge, certain uniaxial semiconductors such as CdS exhibit anomalous birefringence dispersion. The dispersion constant b in Eq. (74) can become orders of magnitude larger than the birefringence. Near the dispersive region, the AOTF would exhibit a large enhancement of resolution without degradation of angular aperture (which is not affected by the dispersion). Experimentally the concept of the dispersive AOTF was demonstrated using CdS as the filter medium. Over the wavelength 500 to 545 nm, a FWHM of 2 nm was obtained over an angular aperture of 38°.

The AOTF possesses many outstanding features that make it attractive for a variety of optical system applications. To proceed from experimental research to practical system deployment, the merits of the AOTF technology must be compared against other competing and usually more matured, technologies. Similar to the case of AO deflectors, the AOTF is most attractive to certain special applications that utilize its unique characteristics. The significant features of the AOTF include: rapid and random-access tuning, large optical throughput while maintaining high spectral resolution, and good image resolution, inherent amplitude, frequency modulation and wavelength multiplexing capability, and small rugged construction with no moving parts. In the next section we shall review the various applications of AOTFs.

Rapid-scan Spectrometers. The rapid, random-access tuning of the AOTF makes it well-suited to rapid-scan spectrometer applications, such as time-resolved spectra analysis. The random access speed of the AOTF is limited by the acoustic transit time and is typically on the order of a few seconds. Compared to the conventional grating spectrometer, the AOTF approach offers the advantages of higher resolution, larger throughput, and amplitude/wavelength modulation capability. Currently, inexpensive photodetector arrays with high performance are not yet available at wavelengths longer than 1.1 μm. Thus, the AOTF-based rapid-scan spectrometer will have clear advantages in the infrared. Other practical advantages, including easy computer interface, small size, and rugged construction, which make it suitable to field applications. These advantages were demonstrated in several experimental models of AOTF-based spectrometers. As an example, we briefly describe an infrared spectrometer operating over the tuning range from 2 to 5 micrometers.[109] The system used a noncollinear TeO_2 AOTF with a full width at half maximum (FWHM) of 7 cm^{-1} over an f/3 angular aperture. The AOTF device is capable of random accessing any wavelength within the tuning range in less than 10 microseconds. The actual scan speed is limited by the sweep of the RF synthesizer. The system has been used as a radiometer to measure the radiation spectrum of a rocket engine placed in a high-altitude simulation chamber. The measurement of the entire scan was completed in 20 milliseconds. Compared to conventional grating spectrometers, this represents a significant increase in the data acquisition speed. Another experimental type of AOTF spectrometer, also operated in the infrared, was developed for use as a commercial stack analyzer.[110] Current approaches based on in-situ spectrometers generally require absorption measurement at two selected wavelengths (differential spectroscopy) or the change of absorption with respect to wavelength change (derivative spectroscopy). The primary advantage of such a wavelength-modulation technique is that the degradation of system accuracy due to variations of the light source, intensity, optical misalignment, etc. can be self-compensated. Since the AOTF is inherently capable of performing wavelength modulation, it is ideally suited to the gas stack analyzer applications.

In spite of the encouraging results demonstrated by these early experimental systems, to date the AOTF technology has not been employed in any commercial spectrometers. This may be due to many factors that include reliability, cost, and performance deficiencies such as high sidelobes, fixed resolution, and high drive power. In view of some of the more recent technological progress, certain special-purpose AOTF spectrometers are expected to appear in the commercial market.

Multispectral Imaging. An interesting AOTF imaging application was demonstrated by Watson et al.[111] in the investigation of planetary atmospheres. A collinear $CaMoO_4$ AOTF operated in the near-infrared was used to obtain spectral images of Jupiter and Saturn. The filtered image was detected with a cooled silicon CID sensor array. The CID signals were read out into a microcomputer memory, processed pixel-by-pixel for subtraction of CID dark currents, and displayed on a color TV monitor. Spectral images of Saturn and Jupiter were obtained with this system in the near-infrared at wavelengths characteristic of CH_4 and NH_3 absorption bands.

When the first noncollinear TeO_2 AOTF was demonstrated, it was recognized that because of its larger aperture and simpler optical geometry, this new type of AOTF would be well-suited to spectral imaging applications. A multispectral spectral experiment using a TeO_2 AOTF was performed in the visible.[63] The white light beam was spatially separated from the filtered light and blocked by an aperture stop in the immediate frequency plane. A resolution target was imaged through the AOTF and relayed onto the camera.

Over the tuning range in the visible, good image resolution in excess of 100 lines/mm was obtained.

As an aid to the design of the AOTF-based imaging spectrometer, a ray-tracing program for optical imaging through AOTF has been described.[112]

Recently, the use of noncollinear AOTF imaging in astronomy has been demonstrated by spectropolarimetry of stars and planets.[113,114] To perform precise polarimetric imaging adds additional requirement on the design of the AOTFs. Since the angles for noncritical phase matching (NPM) are different for ordinary and extraordinary rays, AOTFs based on standard designs cannot be, in general, optimized for both polarizations. In another operation, AOTF spectral imaging has been shown to be a promising approach in the studies of biological materials. Treado et al.[115] demonstrated the use of a visible AOTF to obtain the absorption spectral images of human epithelial cells. The same authors have also described the same techniques for obtaining high fidelity Raman spectral images.[116]

Fiber Optic Communication.   One important application where the AOTF has shown great potential is in the area of fiber-optic communication. Due to its capability of selecting a narrow optical band over a wide spectral range within a time duration of microseconds, the AOTF appears to be well-suited to perform wavelength division multiplexing (WDM) networks. Both the bulkwave and guided wave type AOTF have been demonstrated for this application.[117] Compared to the guided wave version, the bulkwave device has the advantage of lower optical insertion loss; however, the drive power is orders of magnitude higher. For single-mode fiber applications, the laser beams are well-collimated, thus the stringent condition of NPM can be relaxed. A new type of noncollinear TeO_2 AOTF using a collinear beam geometry was described.[118,121] Because of the long interaction length of the new device geometry, narrower passband and lower drive power are obtainable.

Other applications of AOTFs include laser detection[119] and tuning of dye lasers.[120]

Acousto-optic Signal Processor

The inherent wide temporal bandwidth and parallel processing capability of optics makes it one of the most promising analog signal-processing techniques. It is particularly suited for performing special integral transforms such as correlation, convolution, and Fourier transforms. Implementation of the optical signal processing in real time requires a spatial light modulator (SLM) that impresses the electronic signal onto the optical beams. Up to now, the AO Bragg cell was by far the best developed and possibly the only practical one-dimensional SLM to use as the input electrical to optical transducer. In the past decade there was significant progress on the key relevant components that include laser diodes, photodetectors, and AO Bragg cells. A number of AO signal processors have been developed that demonstrated great potential for various system applications. In this section we shall review progress of AO signal processors. We shall focus our discussion on the specific topic of signal spectrum analysis. For more information on AO signal processing and the related subject of optical computing, the interested reader may refer to the references.[121,122]

The frequency-dispersion characteristics of the AO Bragg cell leads to the obvious application of signal spectrum analysis.[123] Referring to Fig. 6b, the momentum-matching condition yields the deflection angle for the diffracted light beam.

$$\Delta\theta = 2\sin^{-1}\left(\frac{\lambda_o f}{2V}\right) = \frac{\lambda_o f}{V} \tag{91}$$

for $\Delta\theta \leq 0.1$ radian, where λ_o is the optical wavelength, V is the acoustic velocity, and f is the acoustic frequency.

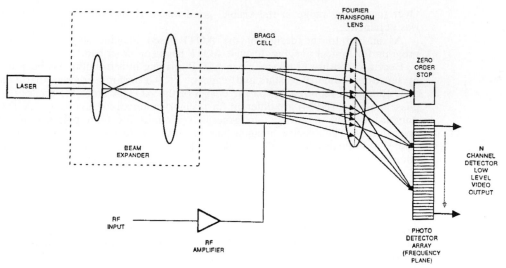

FIGURE 14 Acousto-optic power spectra analyzer.

If a lens is placed behind the Bragg cell, it will transform the angular deviation into a displacement in the back focal plane (referred to as the frequency plane). The displacement of the first-order deflected spot from the undeflected spot in the frequency plane is thus proportional to the frequency of the input signal, i.e.,

$$X = F\lambda_o f/V \tag{92}$$

where F is the transform lens focal length.

The frequency dispersion of AO diffraction described above can be used to perform RF channelization.[124,126] The simplest system architecture (shown in Fig. 14) is the power spectra analyzer (PSA) that measures the magnitude square of the Fourier transform. It consists of a laser, collimating/beamforming optics, AO Bragg cell, Fourier transform lens, and linear photodetector array for real-time electronic readout. The laser beam amplitude profile is tailored by the beamforming optics so as to match the interaction aperture of the Bragg cell. The Bragg cell diffracts a portion of the laser light into an angular distribution of intensities that are proportional to the input RF power spectra. The Fourier transform lens converts the angular diffracted beam into a linear density distribution coincident with the photodetector array. The output of the photodetector thus yields the real-time power spectra of the input RF signal.

Mathematically, the preceding operation yields a Fourier transform of the light distribution at the optical aperture of the Bragg cell

$$F(s) = \int w(x) \exp\left(-j2\pi sx\right) dx \tag{93}$$

where $w(x)$ is the weighting function, $s = f\tau$ is the normalized frequency (τ is the time aperture of the Bragg cell), and $F(s)$ is the desired Fourier transform. The photodetector array in the transform plane measured the magnitude square of $F(s)$, i.e., the instantaneous power spectra of the signal read into the Bragg cell.

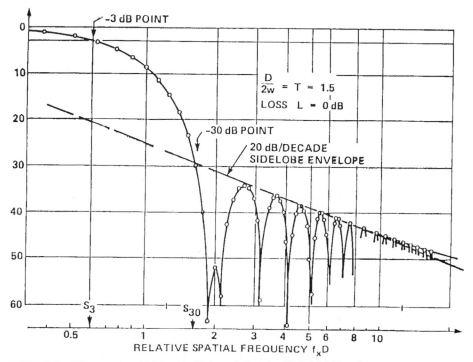

FIGURE 15   RF spectrum of PSA using Truncated Gaussian Weighting Function.

The frequency resolution and sidelobes inherent to the instantaneous power spectra are determined by the weighting function. In practice, the incident optical beam has a gaussian distribution modified by the finite aperture. The corresponding power spectra distribution can be obtained from a numerical calculation of Eq. (93). A typical spectrum is shown in Fig. 15 with a truncation ratio of 1.5[124] It shows a 3-dB width of $\delta f = 1.2/\tau$. This is the frequency resolution or minimum separation for resolving two equal signals.

In spite of its extreme simplicity, the AO channelized receiver described above is equivalent to a large number of contiguous narrowband receivers realized in small size. Unity POI is ensured for analysis of a large number of simultaneous signals over a wide instantaneous bandwidth.

An alternative approach for RF spectrum analysis is to implement a compressive receiver using acousto-optics. Functionally, the AO compressive receiver performs the Fourier transform via the chirp Z transform. Using the identity $2ft = (f - t)^2 - f^2 - t^2$, the Fourier transform can be written as:

$$F(f) = e^{-j\pi f^2} \int f(t)[e^{-j\pi t^2} \cdot e^{j\pi(f-t)^2}]\, dt = e^{-j\pi f^2}\{[e^{-j\pi t^2}f(t)] * e^{j\pi t^2}\} \tag{94}$$

where * denotes convolution. The Fourier transform is accomplished by premultiplying the signal with a forward chirp, convolving with a reverse chirp, and then post multiplying with another chirp. Since the algorithm involves premultiplication, correlation, and post multiplication, it is referred to as the MCM scheme. For power spectrum analysis, the second multiplication can be neglected.

The desired convolution can be implemented using either a space-integrating, or a time-integrating architecture. The space-integrating compressive receiver[127] is illustrated in

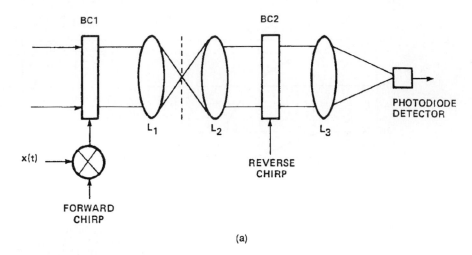

(a)

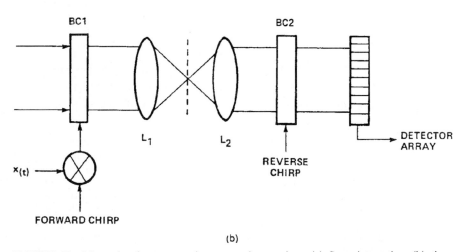

(b)

FIGURE 16 Schematic of acousto-optic compressive receiver. (a) Space-integrating, (b) time-integrating.

Fig. 16a. The signal to be analyzed, $x(t)$, is multiplied by the forward chirp and input into the first Bragg cell. The optical amplitude distribution is imaged onto the second Bragg cell, which is driven by the reverse chirp. The product is spatially integrated by lens $L3$ to produce the final output, $\tilde{x}(f)$, which is (except for a phase factor) the Fourier transform of $x(t)$. The single photodetector measures the desired power spectrum and displays it as a function of time. The frequency resolution of the space-integrating compressive receiver is set by the finite aperture of the Bragg cell.

An alternative method of implementing the compressive receiver is the time-integrating architecture[128] shown in Fig. 16b. Like the space-integrating system, the product of the input and the forward chirp from the first Bragg cell is imaged onto the second Bragg cell (driven by the reverse chirp). However, the product light distribution leaving the second cell is now time-integrated by the photodetector array (instead of the spatial integration due to a lens). The time-integrating architecture has significant advantages. Since the integration time is now limited by the photodetector and is orders of magnitude larger than that obtainable from spatial-integration, very long integration time is realizable, thereby achieving fine frequency resolution and large time bandwidth products. In addition, the frequency resolution can be varied by changing the integration time and the chirp rate.

Theoretically, the AO compressive receiver using the space- or time-integrating architecture appears to have great potential for advanced receiver applications. In practice, since the achievable dynamic range and bandwidth are limited by the performance of photodetectors, the AO implementation of compressive receiver offers little performance advantage compared to that based on the more mature SAW technology. As a result, most of the previous efforts have been directed toward RF signal channelization. Therefore, in the remainder of the chapter, the discussion will be limited to AO channelized receivers.

It has been pointed out that the major advantage of analog AO signal processor is its high throughput realized with small volume and low power consumption. A figure of merit for comparing different signal processors is $B \log_2 N$ where B is the bandwidth and N is the time-bandwidth product. Thus, the AO signal processor is most promising for wideband applications with the size and power constraint.

At present the major deficiency of AO signal processors such as the channelized receiver is the relatively low dynamic range. For channelized receivers, the dynamic range is the capability of analyzing a weak signal in the presence of strong signals. For signals that are close in frequency, the instantaneous dynamic range is limited by the sidelobes of the AO diffraction, which can be effectively suppressed by the choice of the weighting window for the illuminated optical beam (apodization). For signals far apart in frequency, the dynamic range is limited either by optical scattering or by the inmodulation products (IMPs) generated in the Bragg cell. The most dominant inband intermodulation products for two simultaneous signals at f_1 and f_2 are the third-order terms at $2f_1 - f_2$ and $2f_2 - f_1$.

The main contributions of IMPs are due to multiple AO diffractions occurring at high diffraction efficiencies.[129] For two equal intensity signals the two-tone third-order IMP is equal to:

$$I_3 = \frac{I_1^3}{36} \qquad (95)$$

Where I_1 is the first-order diffracted light efficiency.

The above theory assumes that exact momentum matching is met at each step of the multiple AO diffraction process. In an AO signal processor such as the PSA, the IMP in the Bragg cell becomes the dominant factor limiting the dynamic range only when the two signals are relatively far apart in frequency. A refined theory of IMPs in Bragg cells was developed for the case of when the two signals are separated.[130] It was shown that significant suppression of the IMPs can be obtained by using birefringent or phased-array Bragg cells.

In our recent experimental work on wideband Bragg cells, we have found[131] in many cases that the dominant contribution of IMPs are not due to multiple AO diffractions, but are instead due to acoustic nonlinearities. Further theoretical and experimental investigation on these acoustically generated IMPs were reported.[132–134] It was shown that the process could involve second- or third-order acoustic nonlinearities. The nonlinear acoustically-generated IMP in turn diffracts the incident optical beam to an angular position that appears as the optical spurious mode with the corresponding frequency shift.

This type of IMP becomes increasingly severe in wideband Bragg cells where the acoustic power densities are high. The dynamic range degradation due to acoustic nonlinearity thus sets a fundamental limit on the largest bandwidth obtainable. Most of the recent Bragg cells have employed birefringent or phased-array designs to increase the interaction length. These techniques have further extended the high frequency limit of wideband Bragg cells.

Besides the generation of IMPs in Bragg cells, the PSA also has limited dynamic range due to the use of direct detection. Since the photodetector current is proportional to the RF power, the input dynamic range is only one half that of the output dynamic range. One approach to overcome this deficiency is the use of an Interferometric Spectra Analyzer (ISA) configuration. The photodetector current is now proportional to the square root of the product of the signal and a constant reference. Therefore, theoretically the dynamic range of an ISA is twice that of a PSA.

VanderLugt[135] described a special ISA configuration which used two Bragg cells, a signal cell, and a reference cell. The reference cell is used to provide a spatially modulated reference beam. With proper arrangement, the interference term of the signal and reference beams will produce a heterodyning detector output with almost constant temporal IF (within one resolution) over the entire spatial frequency plane. The constant IF implementation scheme has the advantage of significantly lowering the required bandwidth of the detector arrays.

Based on the fixed offset IF scheme, a number of ISAs have been built and demonstrated increased dynamic range.[136,137] Experimental results obtained from these ISAs have shown that the dynamic range is limited primarily by the spurious modulation associated with the wideband RF reference.

Figure 17 shows the schematic of an ISA using a Mach-Zehnder type of interferometer. The optical bench is referred to as the AO interferometer since it uses two Bragg cells to modulate the optical path difference of the two equal path arms. The input signal is applied to one Bragg cell, the signal cell, through an RF amplifier. A wideband RF reference is fed to the second Bragg cell, the reference cell. The incident laser light is divided by the beam splitter and results in two diffracted laser beams out of the signal and reference cells. By appropriately adjusting the tilt between the two Bragg cells, the two diffracted beams are made to overlap at the Fourier plane and to mix collinearly at the desired intermediate frequencies (IF).

The signal cell is operated in the linear range at relatively low efficiency (about 5 percent or less) to avoid spurious signals due to intermodulation products (IMPs) and to

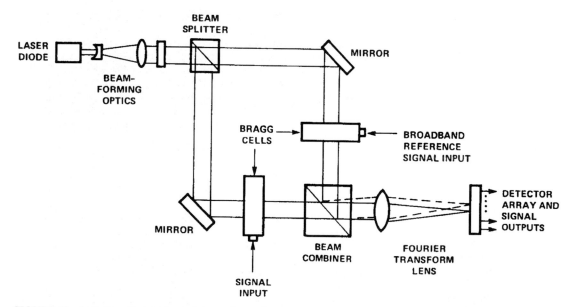

FIGURE 17   Acousto-optic interferometric spectra analyzer.

achieve a high two-tone dynamic range. The reference is usually driven at high efficiency in order to improve receiver sensitivity. The IMPs generated from the reference cell will not severely degrade the receiver performance.

Theoretically, the ISA can significantly improve the dynamic range of an AO channelized receiver. Measured results from an experimental model[138] reveals that the improvement is realizable only for purely CW signals. For pulsed signals, the sidelobe rejection of the IF bandpass is severely degraded due to the inherent frequency gradient in the leading and trailing pulse edges. This introduces false signals in the sidelobe region and greatly increases the difficulty of processing the optical processor outputs. Other critical problems include optical stability, complexity of wideband reference and IF subsystem, and most of all the implementation of wideband detector arrays.

12.7 REFERENCES

1. L. Brillouin, "Diffusion de la lumiére et des ray x par un corps transparent homogéne," *Ann. Phys.,* vol. 17, pp. 80–122, 1992.

2. P. Debye and F. W. Sears, "On the Scattering of Light by Supersonic Waves," *Proc. Nat. Acad. Sci. (U.S.),* vol. 18, pp. 409–414, 1932.

3. R. Lucas and P. Biquard, "Propriètès optiques des milieux solides et liquides soumis aux vibration èlastiques ultra sonores," *J. Phys. Rad.,* vol. 3, pp. 464–477, 1932.

4. C. V. Raman and N. S. Nagendra Nath, "The Diffraction of Light by High Frequency Sound Waves," *Proc. Ind. Acad. Sci.,* vol. 2, pp. 406–420, 1935; vol. 3, pp. 75–84, 1936; vol. 3, pp. 459–465, 1936.

5. M. Born and E. Wolf, "Principles of Optics," Third Edition, Pergamon Press, Ch. 12, New York, 1965.

6. C. F. Quate, C. D. W. Wilkinson, and D. K. Winslow, "Interaction of Light and Microwave Sound," *Proc. IEEE,* vol. 53, pp. 1604–1623, 1965.

7. R. W. Damon, W. T. Maloney, and D. H. McMahon, "Interaction of Light with Ultrasound: Phenomena and Applications in Physical Acoustics," W. P. Mason and R. N. Thurston (eds.), vol. 7, ch. 7, Academic Press, New York, 1970.

8. E. K. Sittig, "Elasto-Optic Light Modulation and Deflection," *Progress in Optics*, vol. X, E. Wolf (ed.), North-Holland, Amsterdam, 1972, Ch. VI.

9. N. Uchida and N. Niizcki, "Acoustooptic Deflection Materials and Techniques," *Proc. IEEE,* vol. 61, pp. 1073–1092, 1973.

10. I. C. Chang, "Acoustooptic Devices and Applications," *IEEE Trans. Sonics and Ultrasonics,* vol. SU-23, pp. 2–22, 1976.

11. A. Korpel, "Acousto-Optics," *Applied Optics and Optical Engineering,* R. Kingslake and B. J. Thompson (eds.) Academic Press, New York, vol. VI, 1980, Ch. IV.

12. C. S. Tsai, "Guided Wave Acousto-Optics," *Electronics and Photonics,* Springer Series, Springer, Verlag, Berlin, Heidelberg, Vol. 23, 1990.

13. J. F. Nye, "Physical Properties of Crystals," Clarendon Press, Oxford, England, 1967.

14. D. F. Nelson and M. Lax, "New Symmetry for Acousto-Optic Scattering," *Phys. Rev. Lett.* vol. 24, pp. 378–380, Feb. 1970—Theory of Photoelastic Interaction," *Phys. Rev.,* vol. B3, pp. 2778–2794, Apr. 1971.

15. J. Chapelle and L. Tauel, "Theorie de la diffusion de la lumiere par les cristeaux fortement piezoelectriques," *C. R. Acad. Sci.,* vol. 240, p. 743, 1955.

16. W. R. Klein and B. D. Cook, "Unified Approach to Ultrasonic Light Diffraction," *IEEE Transactions and Sonics on Ultrasonics,* vol. SU-14, pp. 123–134, 1967.

17. R. W. Dixon, "Acoustic Diffraction of Light in Anisotropic Media", *IEEE J. Quantum Electron.,* vol. QE-3, pp. 85–93, Feb. 1967.

18. M. G. Cohen and E. I. Gordon, "Acoustic Beam Probing Using Optical Techniques," *The Bell System Technical Journal,* pp. 693–721, 1965.

19. E. I. Gordon, "A Review of Acoustooptical Deflection and Modulation Devices," *Proceedings of the IEEE,* vol. 54, pp. 1391–1401, 1966.

20. I. C. Chang and D. L. Hecht, "Device Characteristics of Acousto-Optic Signal Processors," *Proc. SPIE* vol. 241, pp. 129–138, 1980. Also Opt. Eng. vol. 21, pp. 76–81, 1982.

21. L. N. Magdich and V. Y. Molchanov, "Diffraction of a Divergent Beam by Intense Acoustic Waves," *Optical Spectroscopy,* vol. 42, pp. 299–302, 1977.

22. R. W. Dixon, "Photoelastic Properties of Selected Materials and Their Relevance for Applications to Acoustic Light Modulators and Scanners," *J. App. Phys.,* vol. 38, pp. 5149–5153, Dec. 1967.

23. D. A. Pinnow, "Guided Lines for the Selection of Acousto-Optic Materials," *IEEE J. Quantum Electron.* vol. QE-6, pp. 223–238, Apr. 1970.

24. D. A. Pinnow, "Elasto-Optical Materials," in *CRC Handbook of Lasers,* R. J. Pressley (ed.) Cleveland, Ohio: The Chemical Rubber Co., 1971.

25. K. H. Hellwege, Landolt-Bornestein, vol. 11, Springer-Verlag, 1979.

26. I. C. Chang, "Selection of Materials for Acoustooptic Devices," *Optical Engineering,* vol. 24, pp. 132–137, 1985.

27. N. Pomerantz, "Ultrasonic Attenuation by Phonons is Insulators," *IEEE Ultrason. Symp.,* pp. 479–485, 1972.

28. E. Eschier and F. Weidinger, "Acousto-Optic Properties of Dense Flint Glasses," *J. Appl. Phys.,* vol. 42, pp. 65–70, 1975.

29. T. Yano, A. Fukomoto, and A. Watanabe, "Tellurite Glass: A New Acousto-Optic Material," *J. Appl. Phys.,* vol. 42, pp. 3674–3676, 1971.

30. Z. Kleszcewski, "Properties of Some Acousto-Optic Materials," *Archives of Acoustics,* vol. 3, pp. 175–184, 1978.

31. H. Sasaki, K. Tsubouchi, N. Chubachi, and N. Mikoshiba, "Photoelastic Effect in Piezoelectric Semiconductor: ZnO," *J. Appl. Phys.,* vol. 47, pp. 2046–2049, 1976.

32. S. Fukuda, T. Shiosuki, and A. Kawabatam, "Acousto-Optic Properties of Tellurium at 10.6 μm," *J. Appl. Phys.,* vol. 50, pp. 3899–3905, 1979.

33. J. Sapriel, "Cinnbar (α HgS), a Promising Acousto-Optic Material," *Appl. Phys. Lett.,* vol. 19, pp. 533–535, 1971.

34. I. C. Chang and P. Katzka, "Acousto-Optic Properties of Chalcogenide Compounds," *IEEE Ultrason. Symp. Proc.,* pp. 511–514, 1987.

35. P. Katzka and I. C. Chang, "Acousto-Optic Properties of Chalcopyrite Compounds," *IEEE Ultrason. Symp. Proc.,* p. 436, 1977.

36. I. M. Silvestrova, C. Barta, G. F. Dobrzhanskii, L. M. Belyaev, and Y. V. Pisarevsii, "Acousto-Optical Properties of Calomel Crystals Hg_2Cl_2," *Sov. Phys. Crystallogr.* vol. 20, pp. 649–651, 1975.

37. A. Zamkov, I. Kokov, and A. Anistratov, "Acousto-Optical Properties and Photoelasticity of $PbBr_2$ Crystals, "*Phys. Status Solidi A,* vol. 79, pp. K177–K180, 1983.

38. S. V. Akimov, T. M. Stolpakova, E. F. Didknik, and E. V. Sinyakov, "Photoelastic Properties of $NaBi(MoO_4)_2$," *Sov. Phys. Solid State,* vol. 19, p. 1600, Sept. 1977.

39. S. V. Akimov, T. M. Stolpakova, E. F. Didknik, and G. V. Dovchenko, "Acousto-Optic Characteristics of $LiBi(MoO_4)_2$," *Sov. Phys. Solid State,* vol. 20, p. 847, 1978.

40. L. P. Avakyants, V. V. Antipov, D. F. Kiselev, N. G. Sorokin, and K. U. Zakutailov, "Acoustooptic Parameters of Calcium Molybdate Crystals," *Sov. Phys. Solid State,* vol. 24, pp. 1799–1800, 1982.

41. J. Sapriel, "Photoelectric Tensor Components of $Gd_2(MoO_4)_3$," *J. Appl. Phys.,* vol. 48, pp. 1191–1194, 1977.

42. I. C. Chang, "Design of Wideband Acoustooptic Bragg Cells," *Proc. of SPIE,* vol. 352, pp. 34–41, 1983.

43. D. L. Hecht and G. W. Petrie, "Acousto-Optic Diffraction from Acoustic Anisotropic Shear Modes in GaP," *IEEE Ultrason. Symp. Proc.,* p. 474, Nov. 1980.

44. E. G. H. Lean, C. F. Quate, and H. J. Shaw, "Continuous Deflection of Laser Beams," *Appl. Phys. Letters,* vol. 10, pp. 48–50, 1967.

45. A. W. Warner, D. L. White, and W. A. Bonner, "Acousto-Optic Light Deflectors Using Optical Activity in Pratellurite," *Journal of Appl. Phys.,* vol. 43, pp. 4489–4495, 1972.

46. T. Yano, M. Kawabuchi, A. Fukumoto, and A. Watanabe," TeO_2 Anisotropic Bragg Light Deflector Without Midband Degeneracy," *Applied Physics Letters,* vol. 26, pp. 689–691, 1975.

47. I. C. Chang, R. Cadieux, and G. W. Petrie, "Wideband Acousto-Optic Bragg Cells," *IEEE Ultrason. Symp. Proc.,* p. 735, Oct. 1981.

48. A. Korpel, A. Adler, P. Desmares, and W. Watson, "A Television Display Using Acoustic Deflection and Modulation of Coherent Light," *Proc. IEEE,* vol. 54, pp. 1429–1437, 1966.

49. G. A. Couquin, J. P. Griffin, and L. K. Anderson, "Wide-band Acousto-Optic Deflectors Using Acoustic Beam Steering," *IEEE Trans. Sonics Ultrason.,* vol. SU-18, pp. 34–40, Jan. 1970.

50. D. A. Pinnow, "Acousto-Optic Light Deflection: Design Considerations for First Order Beamsteering Transducers," pp. 209–214, 1971.

51. G. A. Alphonse, "Broad-Band Acousto-Optic Deflectors: New Results," *Appl. Optics,* vol. 14, pp. 201–207, 1975.

52. I. C. Chang, "Birefringent Phased Array Bragg Cells," *IEEE Ultrason., Symp. Proc.,* pp. 381–384, 1985.

53. E. H. Young, H. C. Ho, S. K. Yao, and J. Xu, "Generalized Phased Array Bragg Interaction in Anisotropic Crystals, *Proc. SPIE,* vol. 1705, pp. 178–189.

54. W. H. Watson and R. Adler, "Cascading Wideband Acousto-Optic Deflectors," *IEEE Conf. Laser Engineering and Applications,* Wash., D.C., June 1969.

55. I. C. Chang and D. L. Hecht, "Doubling Acousto-Optic Deflector Resolution Utilizing Second Order Birefringent Diffraction," *Appl. Phys. Lett.,* vol. 27, pp. 517–518, 1975.

56. E. H. Young and S. K. Yao, "Design Considerations for Acousto-Optic Devices," *Proc. IEEE,* vol. 69, pp. 54–64, 1981.

57. D. Maydan, "Acousto-Optic Pulse Modulators," *J. Quantum Electronics,* vol. QE-6, pp. 15–24, 1967.

58. F. Okoliocsanyi, "The Waveslot, an Optical Television System," *Wireless Eng.,* vol. 14, pp. 527–536, 1937.

59. D. M. Robinson, "The Supersonic Light Control and its Application to Television with Spatial Reference to the Scophony Television Receiver," *Proc. IRE,* vol. 27, pp. 483–486, Aug. 1939.

60. R. V. Johnson, "Scophony Light Valve," *Appl. Opt.,* vol. 18, pp. 4030–4038, 1979.

61. S. E. Harris and R. W. Wallace, "Acousto-Optic Tunable Filter," *J. Opt. Soc. Am.,* vol. 59, pp. 744–747, June 1969.

62. I. C. Chang, "Noncollinear Acousto-Optic Filter with Large Angular Aperture," *Appl. Phys. Lett.,* vol. 25, pp. 370–372, Oct. 1974.

63. I. C. Chang, "Tunable Acousto-Optic Filters: An Overview," *Opt. Eng.,* vol. 16, pp. 455–460, 1977.

64. I. C. Chang, "Acousto-Optic Tunable Filters," *Opt. Eng.,* vol. 20, pp. 824–828, 1981.

65. S. E. Harris, S. T. K. Nieh, and D. K. Winslow, "Electronically Tunable Acousto-Optic Filter," *App. Phys. Lett.,* vol. 15, pp. 325–326, Nov. 1969.

66. S. E. Harris, S. T. K. Nieh, and R. S. Feigelson, "$CaMoO_4$ Electronically Tunable Acousto-Optical Filter," *Appl. Phys. Lett.,* vol. 17, pp. 223–225, Sept. 1970.

67. S. T. K. Nieh and E. Harris, "Aperture-Bandwidth Characteristics of Acousto-Optic Filter," *J. Opt. Soc. Am.,* vol. 62, pp. 672–676, May 1972.

68. I. C. Chang, "Analysis of the Noncollinear Acousto-Optic Filters," *Electron. Lett.,* vol. 11, pp. 617–618, 1975.

69. M. Gottlieb, A. P. Goutzoulis, and N. B. Singh, "Mercurous Chloride ($HgCl_2$) Acousto-Optic Devices," *IEEE Ultrason. Symp. Proc.,* pp. 423–427, 1986.

70. I. C. Chang et al., "Progress of Acousto-Optic Bragg Cells," *IEEE Ultrason. Symp. Proc.,* p. 328, 1984.

71. J. M. Bagshaw, S. E. Lowe, and T. F. Willats, "The Performance of Gallium Phosphide Bragg Cells," *GEC Journal of Research,* vol. 5, pp. 171–175, 1987.

72. I. C. Chang, "High Performance Wideband Bragg Cells," *IEEE Ultrason. Symp. Proc.,* p. 435, 1988.

73. I. C. Chang and S. Lee, "Efficient Wideband Acousto-Optic Bragg Cells," *IEEE Ultrason. Symp. Proc.,* p. 427, Oct. 1983.

74. J. M. Bagshaw and T. F. Willats, "Anisotropic Bragg Cells," *GEC Journal of Research,* vol. 2, p. 328, 1984.

75. I. C. Chang and R. Cadieux, "Multichannel Acousto-Optic Bragg Cells," *IEEE Ultrason. Symp.,* Proc., p. 413, 1982.

76. W. R. Beaudot, M. Popek, and D. R. Pape, Advances in Multichannel Bragg Cell Technology, *Proc. SPIE,* vol. 639, pp. 28–33, 1986.

77. I. Gorog, J. D. Knox, and P. V. Goerdertier, "A Television-Rate Laser Scanner. I. General Considerations," *RCA Rev.,* vol. 33, pp. 623–666, Dec. 1972.

78. I. Gorog, J. D. Knox, P. V. Goerdertier, and I. Shidlovskky, "A Television-Rate Laser Scanner. II. Recent Developments, *RCA Rev.* vol. 33, pp. 667–673, Dec. 1972.

79. Grossman and Redderson, High Speed Laser Facsimile Scanner, *Proc. SPIE* Symp., vol. 200, pp. 8–15, 1979.

80. L. C. Foster, C. B. Crumly, and R. L. Cohoon, "A High Resolution Linear Optical Scanner Using a Traveling Wave Acoustic Lens," *Appl. Optics,* vol. 9, pp. 2154–2160, Sept. 1970.

81. R. H. Johnson and R. M. Montgomery, "Optical Beam Deflection Using Acoustic Traveling Wave Technology," *Proc. SPIE,* vol. 90, 1976.

82. J. S. Gerig and H. Montague, "A Simple Optical Filter for Chirp Radar," *Proc. IEEE,* vol. 52, p. 1753, 1964.

83. J. Eveleth and L. Bademian, "Solid State Laser Beam Recorder for 875 Line TV," *Proc. Electro-Optical System Design,* pp. 76–81, 1975.

84. F. Bestenreiner, U. Greis, J. Helmburger, and K. Stadler, "Visibility and Corrections of Periodic Interference Structures in Line-by-Line Record Images," *J. Appl. Phot. Eng.,* vol. 2, pp. 86–102, 1976.

85. R. V. Johnson, "Facet Tracking Laser Scanning," *Tech. Dig. Conference on Lasers and Electro-Optics,* p. 78, 1980.

86. I. C. Chang and D. L. Hecht, "Efficient GaP Acousto-Optic Modulators," *IEEE Conf. on Laser and Electro-Optical Systems,* San Diego, CA, Feb. 1978.

87. G. Hrbeck and W. Watson, "A High Speed Laser Alphanumeric Generator," *Electro-Optical System Design Conf.,* New York, Sept. 1970.

88. A. W. Warner and D. A. Pinnow, "Miniature Acousto-Optic Modulators for Optical Communications," *J. Quantum Elect.,* vol. QE-9, pp. 1155–1157, 1973.

89. R. L. Abrams and D. A. Pinnow, "Efficient Acousto-Optic Modulator at 3.39 and 10.6 Microseconds in Crystalline Germanium," *IEEE J. Quantum Electron.* (Corresp.) vol. QE-7, pp. 135–136, Mar. 1971.

90. A. Feldman, R. M. Waxler, and D. Horowitz, "Photoelastic Constants of Germanium," *J. Appl. Phys.,* vol. 49, p. 2589, 1978.

91. A. J. Fox, "Acousto-Optic Figure of Merit for Single Crystal Germanium at 10.6 μm Wavelength," *Appl. Optics,* vol. 24, p. 2040, 1985.

92. A. J. DeMaria, R. Gagosz, and G. Barnard, "Ultrasonic-Refraction Shutter for Optical Master Oscillators," *J. Appl. Phys.* vol. 34, pp. 453–456, Mar. 1963.

93. L. E. Hargrove, R. L. Fork, and M. A. Pollack, "Locking of He-Ne Laser Modes Induced by Synchronous Intracavity Modulation," *Appl. Phys. Lett.,* vol. 5, pp. 4–5, Jul. 1964.

94. D. Maydan, "Fast Modulator for Extraction of Internal Laser Power," *J. Appl. Phys.,* vol. 41, pp. 1552–1559, Mar. 1970.

95. R. B. Chesler, M. A. Karr, and J. E. Geusic, "An Experimental and Theoretical Study of High Repetition Rate Q-Switched Nd: YAG Lasers," *Proc. IEEE* vol. 58, pp. 1899–1914, Dec. 1970.

96. R. H. Johnson, "Characteristics of Acousto-Optic Cavity Dumping in a Mode Locked Laser," *J. Quantum Electron.* (Corresp.), vol. QE-9, pp. 255–257, Feb. 1973.

97. D. J. Kuizenga, D. W. Phillion, T. Lund, and A. E. Siegman, "Simultaneous Q-Switching and Mode Locking in the CW Nd:YAG Laser," *Optics. Comm.,* vol. 9, pp. 221–226, Nov. 1973.

98. J. A. Kusters, D. A. Wilson, and D. L. Hammond, "Optimum Crystal Orientation for Acoustically Tuned Optic Filters", *J. Opt. Soc. Am.,* vol. 64, pp. 434–440, April 1974.

99. I. C. Chang, "Tunable Acousto-Optic Filter Utilizing Acoustic Beam Walk-off in Crystal Quartz," *Appl. Phys. Lett.,* vol. 25, pp. 323–324, Sept. 1974.

100. J. D. Feichtner, M. Gotlieb, and J. J. Conroy, "A Tunable Collinear Acousto-Optic Filter for the Intermediate Infrared Using Crystal Tl_3AsSe_3," *IEEE Conf. Laser Engineering and Applications,* Washington D.C., May 1975.

101. I. C. Chang and P. Katzka, "Tunable Acousto-Optic Filters at 10.6 Microns," *IEEE Ultrason. Symp. Proc.,* Sept., 1978.

102. P. Katzka and I. C. Chang, "Noncollinear Acousto-Optic Filter for the Ultraviolet," *SPIE Symp. Proc.,* Sept. 1978.

103. I. C. Chang, "Development of an Infrared Tunable Acousto-Optic Filter", *SPIE Symp. Proc.,* vol. 131, pp. 2–10, Jan. 1978.

104. M. Khoshnevisan, and E. Sovero, "Development of a Cryogenic Infrared Acousto-Optic Tunable Spectral Filter," *SPIE Symp. Proc.,* vol. 245, July, 1980.

105. J. D. Feichtner, M. Gottlieb, and J. J. Conroy, "Tl_3AsSe_3 Noncollinear Acousto-Optic Filter Operation at 10 Micrometers," *Appl. Phys. Lett.,* vol. 34, pp. 1–3, Jan. 1979.

106. I. C. Chang, P. Katzka, J. Jacob, and S. Estrin, "Programmable Acousto-Optic Filter," *IEEE Ultrason. Symp. Proc.,* Sept. 1979.

107. I. C. Chang and P. Katzka, "Tunable Acousto-Optic Filters and Apodized Acoustic Excitation," *J. Opt. Soc. Am.,* vol. 68, p. 1449, Oct. 1978.

108. I. C. Chang and P. Katzka, Enhancement of Acousto-Optic Filter Resolution Using Birefringent Dispersion in CdS, *Opt. Lett.,* vol. 7, 1982.

109. P. Katzka, S. Estrin, I. C. Chang, and G. W. Petrie, "Computer Controlled Infrared TAOF Spectrometer," *Proc. SPIE,* vol. 246, p. 121, 1980.

110. R. L. Nelson, "Role of a TAS AOTF in a Commercial Stack Analyzer," *Proc. SPIE,* vol. 753, pp. 103–113, 1987.

111. R. B. Watson, S. A. Rappaport, and E. E. Frederick, *Icarus,* vol. 27, p. 417, 1976.

112. I. C. Chang, "Electronically Tuned Imaging Spectrometer Using Acousto-Optic Tunable Filter," *Proc. SPIE,* vol. 1703, pp. 24–29, 1991.

113. W. H. Smith and K. M. Smith, "A Polarimetric Spectral Imager Using Acousto-Optic Tunable Filters," *Experimental Astronomy,* vol. 1, pp. 329–343, 1991.

114. D. A. Glenar, J. J. Hillman, B. Seif, and J. Bergstrahl, "POLARIS-II: An Acousto-Optic Imaging Spectropolarimeter for Ground Based Astronomy," *Proc. SPIE,* vol. 1747, pp. 92–101, 1992.

115. P. J. Treado, I. W. Levin and E. N. Lewis, "Near Infrared Acousto-Optic Filtered Spectroscopic Microscopy: A Solid-State Approach to Chemical Imaging," *Appl. Spectroscopy,* vol. 46, pp. 553–559, 1992.

116. P. J. Treado, I. W. Levin, and E. N. Lewis, "High Fidelity Raman Spectrometry: A Rapid Method Using Acousto-Optic Tunable Filter," *Appl. Spectroscopy,* vol. 46, pp. 1211–1216, 1992.

117. K. W. Cheung, M. M. Choy, and H. Kobrinski, Electronic Wavelength Tuning Using Acousto-Optic Tunable Filter with Broad Continuous Tuning Range and Narrow Channel Spacing, *IEEE Photonics. Tech. Lett.,* vol. 1, pp. 38–40, 1989.

118. I. C. Chang, "Collinear Beam Acousto-Optic Tunable Filters," *Electron. Lett.,* vol. 28, p. 1255, 1992.

119. I. C. Chang, Laser Detection Utilizing Tunable Acousto-Optic Filter, *J. Quantum Electron.*, vol. 14, p. 108, 1978.

120. D. J. Taylor, S. T. K. Neih, and T. W. Hansch, "Electronic Tuning of a Dye Laser Using the Acousto-Optic Filter," *Appl. Phys. Lett.*, vol. 19, pp. 269–271, Oct. 1971.

121. N. J. Berg and J. M. Pelligrino, *Acousto-Optical Signal Processing*, 2d ed., Marcel Dekker, New York, 1995.

122. J. N. Lee and A. VanderLugt, "Acousto-Optical Signal Processing and Computing," *Proc. IEEE*, vol. 77, No. 10, p. 1528, 1989.

123. L. B. Lambert, "Wide-Band, Instantaneous Spectrum Analyzers Employing Delay-Line Light Modulators," *IRE National Conv. Rec.*, vol. 10, pt. 6, pp. 69–78, Mar. 1962.

124. D. L. Hecht, "Spectrum Analysis Using Acousto-Optic Devices," *Optical Engineering*, vol. 16, No. 5, p. 461, 1977.

125. I. C. Chang, "Acousto-Optic Channelized Receiver," *Microwave Journal*, vol. 29, pp. 141–157, March, 1986.

126. G. W. Anderson et al., "Advanced Channelization Technology for RF, Microwave and Millimeter Wave Applications," *Proc. of IEEE*, vol. 79, No. 3, p. 355, 1991.

127. N. J. Berg et al., "Real Time Fourier Transformation via Acousto-Optics, "*Appl. Phys. Lett.*, vol. 34, p. 15, 1979.

128. T. M. Turpin, "Time Integrating Optical Processors," *Proc. of SPIE*, vol. 154, p. 196, 1978.

129. D. L. Hecht, "Multifrequency Acousto-Optic Diffraction," *IEEE Trans. Sonics and Ultrason.*, Vol. SU-24, p. 7, 1977.

130. I. C. Chang, "Intermodulation Products in Phased Array Bragg Cells," *IEEE Ultrason. Symp. Proc.*, pp. 505–508, 1989.

131. I. C. Chang, "IEEE/OSA Conference on Lasers and Electro-Optics," Baltimore, May, 1983.

132. I. C. Chang and R. T. Wererka, "Multifrequency Acousto-Optic Diffraction," *IEEE Ultrason. Symp. Proc*, p. 445, Oct. 1983.

133. G. Elston and P. Kellman, *IEEE Ultrason. Symp. Proc.*, pp. 449–453, Oct. 1983.

134. I. C. Chang, "Multifrequency Acousto-Optic Interaction in Bragg Cells," *Proc. SPIE*, vol. 753, p. 97, 1987.

135. A. VanderLugt, "Interferometric Spectra Analyzer," *Appl. Optics*, vol. 20, p. 2770, 1981.

136. M. L. Shah et al. "Interferometric Bragg Cell Spectrum Analyzer," *IEEE Ultrason. Sypm. Proc.*, p. 743, 1981.

137. I. C. Chang et al. "High Dynamic Range Acousto-Optic Receiver," *Proc. SPIE*, vol. 545, p. 95, 1985.

138. J. B. Y. Tsui, I. C. Chang, and E. Gill, "Interferometric Acousto-Optic Receiver," SPIE Proc., vol. 1102, *Optical Technology for Microwave Applications IV*, pp. 176–182, Oct. 1989.

CHAPTER 13
ELECTRO-OPTIC MODULATORS

Theresa A. Maldonado
Department of Electrical Engineering
The University of Texas at Arlington
Arlington, Texas

13.1 GLOSSARY

$\bar{A}$	general symmetric matrix
$\bar{a}$	orthogonal transformation matrix
b	electrode separation of the electro-optic modulator
$\mathbf{D}$	displacement vector
d	width of the electro-optic crystal
$\mathbf{E}$	electric field
$\mathbf{H}$	magnetic field
$IL:$	insertion loss
$\mathbf{k}$	wave vector, direction of phase propagation
L	length of the electro-optic crystal
L/b	aspect ratio
N	number of resolvable spots
n_m	refractive index of modulation field
n_x, n_y, n_z	principal indices of refraction
$\mathbf{R}$	global rotation axis of the index ellipsoid
r_{ijk}	third-rank linear electro-optic coefficient tensor
$\mathbf{S}$	Poynting (ray) vector, direction of energy flow
s_{ijkl}	fourth-rank quadratic electro-optic coefficient tensor
T	transmission or transmissivity
V	applied voltage
V_π	half-wave voltage
v	phase velocity
v_m	modulation phase velocity

v_s	ray velocity
w	beamwidth
w_o	resonant frequency of an electro-optic modulator circuit
$\mathbf{X}$	position vector in cartesian coordinates
$\mathbf{X}'$	electrically perturbed position vector in cartesian coordinates
(x, y, z)	unperturbed principal dielectric coordinate system
(x', y', z')	new electro-optically perturbed principal dielectric coordinate system
(x'', y'', z'')	wave vector coordinate system
(x''', y''', z''')	eigenpolarization coordinate system
β_1	polarization angle between x''' and x''
β_2	polarization angle between y''' and x''
Γ	phase retardation
Γ_m	amplitude modulation index
Δn	electro-optically induced change in the index of refraction or birefringence
$\Delta(1/n^2)$	electro-optically induced change in an impermeability tensor element
$\Delta\phi$	angular displacement of beam
Δv	bandwidth of a lumped electro-optic modulator
δ	phase modulation index
$\bar{\epsilon}$	permittivity tensor
ϵ_o	permittivity of free space
$\bar{\epsilon}^{-1}$	inverse permittivity tensor
$\epsilon_x, \epsilon_y, \epsilon_z$	principal permittivities
$\bar{\varepsilon}$	dielectric constant tensor
$\bar{\varepsilon}^{-1}$	inverse dielectric constant tensor
$\varepsilon_x, \varepsilon_y, \varepsilon_z$	principal dielectric constants
η_m	extinction ratio
θ	half-angle divergence
θ_k, ϕ_k	orientation angles of the wave vector in the (x, y, z) coordinate system
ϑ	optic axis angle in biaxial crystals
λ	wavelength of the light
$\bar{\lambda}$	diagonal matrix
v_{tw}	bandwidth of a traveling wave modulator
ξ	modulation efficiency
ρ	modulation index reduction factor

τ	transit time of modulation signal
Φ	global rotation angle of the index ellipsoid
ϕ	phase of the optical field
Ω	plane rotation angle
ϖ	beam parameter for bulk scanners
ω_d	frequency deviation
ω_e	stored electric energy density
ω_m	modulation radian frequency
$\overline{1/n^{2\prime}}$	electro-optically perturbed impermeability tensor
$\overline{1/n^2}$	inverse dielectric constant (impermeability) tensor

13.2 INTRODUCTION

The electro-optic effect is one of several means to impose information on, or modulate, a light wave carrier. Electro-optic devices have been developed for application in communications,[1-4] analog and digital signal processing,[5] information processing,[6] optical computing,[6,7] and sensing.[5,7] Example devices include phase and amplitude modulators, multiplexers, switch arrays, couplers, polarization controllers, deflectors,[1,2] Givens rotation devices,[8] correlators,[9] A/D converters,[10] multichannel processors,[11] matrix-matrix and matrix-vector multipliers,[11] and sensors for detecting temperature, humidity, and radio-frequency electrical signals.[5,7] The electro-optic effect allows for much higher modulation frequencies than other methods, such as mechanical shutters, moving mirrors, or acousto-optic devices, due to a faster response time.

The basic idea behind electro-optic devices is to alter the optical properties of a material with an applied voltage in a controlled way. The changes in the optical properties, particularly the permittivity tensor, translate into a modification of some parameter of a light wave carrier, such as phase, amplitude, frequency, polarization, or position, as it propagates through the device. Therefore, understanding how light propagates in these materials is necessary for the design and analysis of electro-optic devices. The following section gives an overview of light propagation in anisotropic materials that are homogeneous, nonmagnetic, lossless, optically inactive, and nonconducting. The third section, "The Electro-optic Effect," gives a geometrical and mathematical description of the linear and quadratic electro-optic effects. A geometrical approach using the index ellipsoid is presented to illustrate the details of how the optical properties change with applied voltage. A mathematical approach is offered to determine the perturbed principal dielectric axes and indices of refraction of any electro-optic material for any direction of the applied electric field as well as the phase velocity indices and eigenpolarization orientations for a given wave vector direction. Finally, basic bulk electro-optic modulators are described in the fourth section, "Modulator Devices," including some design considerations and performance criteria.

The discussion presented in this chapter applies to any electro-optic material, any direction of the applied voltage, and any direction of the wave vector. Therefore, no specific materials are described explicitly, although materials such as lithium niobate ($LiNbO_3$), potassium dihydrogen phosphate (KDP), and gallium arsenide (GaAs) are just a few of several materials commonly used. Emphasis is placed on the general fundamentals of the electro-optic effect and bulk modulator devices.

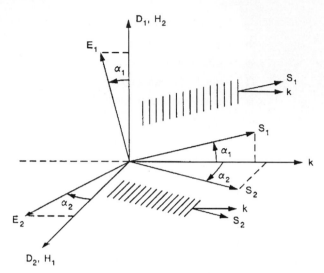

FIGURE 1 The geometric relationships of the electric quantities **D** and **E** and the magnetic quantities **B** and **H** to the wave-vector **k** and the ray vector **S** are shown for the two allowed extraordinary-like waves propagating in an anisotropic medium.[12]

13.3 CRYSTAL OPTICS AND THE INDEX ELLIPSOID

For any anisotropic (optically inactive) crystal class there are two allowed orthogonal lineraly polarized waves propagating with differing phase velocities for a given wave vector **k**. Biaxial crystals represent the general case of anisotropy. Generally, the allowed waves exhibit *extraordinary-like* behavior; the wave vector and ray (Poynting) vector directions differ. In addition, the phase velocity, polarization orientation, and ray vector of each wave change distinctly with wave vector direction. For each allowed wave, the electric field **E** is not parallel to the displacement vector **D** (which defines polarization orientation) and, therefore, the ray vector **S** is not parallel to the wave vector **k** as shown in Fig. 1. The angle α between **D** and **E** is the same as the angle between **k** and **S**, but for a given **k**, $\alpha_1 \neq \alpha_2$. Furthermore, for each wave $\mathbf{D} \perp \mathbf{k} \perp \mathbf{H}$ and $\mathbf{E} \perp \mathbf{S} \perp \mathbf{H}$, forming orthogonal sets of vectors. The vectors **D**, **E**, **k**, and **S** are coplanar for each wave.[12]

The propagation characteristics of the two allowed orthogonal waves are directly related to the fact that the optical properties of an anisotropic material depend on direction. These properties are represented by the constitutive relation $\mathbf{D} = \bar{\epsilon}\mathbf{E}$, where $\bar{\epsilon}$ is the permittivity tensor of the medium and **E** is the corresponding optical electric field vector. For a homogeneous, nonmagnetic, lossless, optically inactive, and nonconducting medium, the permittivity tensor has only real components. Moreover, the permittivity tensor and its inverse $\bar{\epsilon}^{-1} = 1/\epsilon_0(1/n^2)$, where n is the refractive index, are symmetric for all crystal classes and for any orientation of the dielectric axes.[13–15] Therefore, the matrix representation of the permittivity tensor can be diagonalized, and in principal coordinates the constitutive equation has the form

$$\begin{pmatrix} D_x \\ D_y \\ D_z \end{pmatrix} = \begin{pmatrix} \epsilon_x & 0 & 0 \\ 0 & \epsilon_y & 0 \\ 0 & 0 & \epsilon_z \end{pmatrix} \begin{pmatrix} E_x \\ E_y \\ E_z \end{pmatrix} \tag{1}$$

where reduced subscript notation is used. The principal permittivities lie on the diagonal of $\bar{\epsilon}$.

The index ellipsoid is a construct with geometric characteristics representing the phase velocities and the vibration directions of $\mathbf{D}$ of the two allowed plane waves corresponding to a given optical wave-normal direction $\mathbf{k}$ in a crystal. The index ellipsoid is a quadric surface of the stored electric energy density ω_e of a dielectric:[13,16]

$$\omega_e = \tfrac{1}{2}\mathbf{E}\cdot\mathbf{D} = \tfrac{1}{2}\sum_i\sum_j E_i\epsilon_{ij}E_j = \tfrac{1}{2}\epsilon_0\mathbf{E}^T\bar{\epsilon}\mathbf{E} \qquad i,j = x, y, z \tag{2a}$$

or

$$\omega_e = \tfrac{1}{2}\epsilon_0(E_x^2\varepsilon_x + E_y^2\varepsilon_y + E_z^2\varepsilon_z) \tag{2b}$$

in principal coordinates, where T indicates transpose. The stored energy density is positive for any value of electric field; therefore, the quadric surface is always given by an ellipsoid.[13,16-18]

With the constitutive equation, Eq. (2b) assumes the form $(D_x^2/\varepsilon_x) + (D_y^2/\varepsilon_y) + (D_z^2/\varepsilon_z) = 2\omega_e\epsilon_o$. By substituting $x = D_x/(2\omega_e\epsilon_o)^{1/2}$ and $n_x^2 = \varepsilon_x$, and similarly for y and z, the ellipsoid is expressed in cartesian principal coordinates as

$$\frac{x^2}{n_x^2} + \frac{y^2}{n_y^2} + \frac{z^2}{n_z^2} = 1 \tag{3}$$

Equation (3) is the general index ellipsoid for an optically biaxial crystal. If $n_x = n_y$, the surface becomes an ellipsoid of revolution, representing a uniaxial crystal. In this crystal, one of the two allowed eigenpolarizations will always be an *ordinary* wave with its Poynting vector parallel to the wave vector and $\mathbf{E}$ parallel to $\mathbf{D}$ for any direction of propagation. An isotropic crystal $(n_x = n_y = n_z)$ is represented by a sphere with the principal axes having equal length. Any wave propagating in this crystal will exhibit ordinary characteristics. The index ellipsoid for each of these three optical symmetries is shown in Fig. 2.

For a general direction of propagation, the section of the ellipsoid through the origin

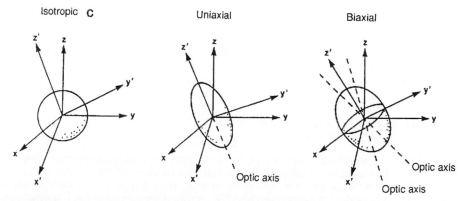

FIGURE 2 The index ellipsoids for the three crystal symmetries are shown in nonprincipal coordinates (x', y', z') relative to the principal coordinates (x, y, z). For isotropic crystals, the surface is a sphere. For uniaxial crystals, it is an ellipsoid of revolution. For biaxial crystals it is a general ellipsoid.[30]

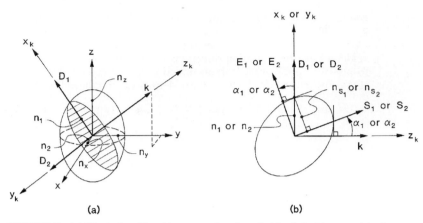

FIGURE 3   (*a*) The index ellipsoid cross section (crosshatched) that is normal to the wave vector **k** has the shape of an ellipse. The major and minor axes of this ellipse represent the directions of the allowed polarizations $\mathbf{D}_1$ and $\mathbf{D}_2$; (*b*) for each eigenpolarization (1 or 2) the vectors **D**, **E**, **S**, and **k** are coplanar.[30]

perpendicular to this direction is an ellipse, as shown in Fig. 3*a*. The major and minor axes of the ellipse represent the orthogonal vibration directions of **D** for that particular direction of propagation. The lengths of these axes correspond to the phase velocity refractive indices. They are, therefore, referred to as the "fast" and "slow" axes. Figure 3*b* illustrates the field relationships with respect to the index ellipsoid. The line in the $(\mathbf{k}, \mathbf{D}_i)$ plane ($i = 1$ or 2) that is tangent to the ellipsoid at $\mathbf{D}_i$ is parallel to the ray vector $\mathbf{S}_i$; the electric field $\mathbf{E}_i$ also lies in the $(\mathbf{k}, \mathbf{D}_i)$ plane and is normal to $\mathbf{S}_i$. The line length denoted by n_{s_i} gives the ray velocity as $v_{s_i} = c/n_{s_i}$ for $\mathbf{S}_i$. The same relationships hold for either vibration, $\mathbf{D}_1$ or $\mathbf{D}_2$.

In the general ellipsoid for a biaxial crystal there are two cross sections passing through the center that are circles. The normals to these cross sections are called the *optic axes* (denoted in Fig. 2 in a nonprincipal coordinate system), and they are coplanar and symmetric about the z principal axis in the x, z plane. The angle ϑ of an optic axis with respect to the z axis in the x, z plane is

$$\tan \vartheta = \frac{n_z}{n_x} \sqrt{\frac{n_y^2 - n_x^2}{n_z^2 - n_y^2}} \tag{4}$$

The phase velocities for $\mathbf{D}_1$ and $\mathbf{D}_2$ are equal for these two directions: $v_1 = v_2 = c/n_y$. In an ellipsoid of revolution for a uniaxial crystal there is one circular cross section perpendicular to the z principal axis. Therefore, the z axis is the optic axis, and $\vartheta = 0°$ in this case.

13.4   THE ELECTRO-OPTIC EFFECT

At an atomic level an electric field applied to certain crystals causes a redistribution of bond charges and possibly a slight deformation of the crystal lattice.[14] In general, these alterations are not isotropic; that is, the changes vary with direction in the crystal. Therefore, the inverse dielectric constant (impermeability) tensor changes accordingly. Crystals lacking a center of symmetry are noncentrosymmetric and exhibit a linear (Pockels) electro-optic effect. The changes in the impermeability tensor elements are linear in the applied electric field. On the other hand, all crystals exhibit a quadratic (Kerr) electro-optic effect. The changes in the impermeability tensor elements are quadratic in the applied field. When the linear effect is present, it generally dominates over the quadratic effect.

The linear electro-optic effect is represented by a third rank tensor r_{ijk}. The permutation symmetry of this tensor is $r_{ijk} = r_{jik}$, $i, j, k = 1, 2, 3$. Therefore, the tensor can be represented by a 6×3 matrix; i.e., $r_{ijk} \Rightarrow r_{ij}$, $i = 1, \ldots, 6$ and $j = 1, 2, 3$. Generally, the r_{ij} coefficients have very little dispersion in the optical transparent region of a crystal.[19] The electro-optic coefficient matrices for all crystal classes are given in Table 1. References 14, 19, 20, and 21, among others, contain extensive tables of numerical values for indices and electro-optic coefficients for different materials. The quadratic electro-optic effect is represented by a fourth rank tensor s_{ijkl}. The permutation symmetry of this tensor is $s_{ijkl} = s_{jikl} = s_{ijlk}$, $i, j, k, l = 1, 2, 3$. The tensor can be represented by a 6×6 matrix; i.e., $s_{ijkl} \Rightarrow s_{kl}$, $k, l = 1, \ldots, 6$. The quadratic electro-optic coefficient matrices for all crystal classes are given in Table 2. Reference 14 contains a table of quadratic electro-optic coefficients for several materials.

The Linear Electro-optic Effect

An electric field applied in a general direction to a noncentrosymmetric crystal produces a linear change in the constants $(1/n^2)_i$ due to the linear electro-optic effect according to

$$\Delta(1/n^2)_i = \sum_j r_{ij} E_j \qquad \begin{aligned} i &= 1, \ldots, 6 \\ j &= x, y, z = 1, 2, 3 \end{aligned} \qquad (5)$$

where r_{ij} is the ijth element of the linear electro-optic tensor in contracted notation. In matrix form Eq. (5) is

$$\begin{pmatrix} \Delta(1/n^2)_1 \\ \Delta(1/n^2)_2 \\ \Delta(1/n^2)_3 \\ \Delta(1/n^2)_4 \\ \Delta(1/n^2)_5 \\ \Delta(1/n^2)_6 \end{pmatrix} = \begin{pmatrix} r_{11} & r_{12} & r_{13} \\ r_{21} & r_{22} & r_{23} \\ r_{31} & r_{32} & r_{33} \\ r_{41} & r_{42} & r_{43} \\ r_{51} & r_{52} & r_{53} \\ r_{61} & r_{62} & r_{63} \end{pmatrix} \begin{pmatrix} E_x \\ E_y \\ E_z \end{pmatrix} \qquad (6)$$

E_x, E_y, and E_z are the components of the applied electric field in principal coordinates. The magnitude of $\Delta(1/n^2)$ is typically on the order of less than 10^{-5}. Therefore, these changes are mathematically referred to as perturbations. The new impermeability tensor $\overline{1/n^{2'}}$ in the presence of an applied electric field is no longer diagonal in the reference principal dielectric axes system. It is given by

$$\overline{1/n^{2'}} = \begin{pmatrix} 1/n_x^2 + \Delta(1/n^2)_1 & \Delta(1/n^2)_6 & \Delta(1/n^2)_5 \\ \Delta(1/n^2)_6 & 1/n_y^2 + \Delta(1/n^2)_2 & \Delta(1/n^2)_4 \\ \Delta(1/n^2)_5 & \Delta(1/n^2)_4 & 1/n_z^2 + \Delta(1/n^2)_3 \end{pmatrix} \qquad (7)$$

However, the field-induced perturbations are symmetric, so the symmetry of the tensor is not disturbed. The new index ellipsoid is now represented by

$$(1/n^2)_1' x^2 + (1/n^2)_2' y^2 + (1/n^2)_3' z^2 + 2(1/n^2)_4' yz + 2(1/n^2)_5' xz + 2(1/n^2)_6' xy = 1 \qquad (8)$$

TABLE 1 The Linear Electro-optic Coefficient Matrices in Contracted Form for All Crystal Symmetry Classes[14]

Centrosymmetric ($\bar{1}$, 2/m, mmm, 4/m, 4/mmm, $\bar{3}$, $\bar{3}$m6/m, 6/mmm, m3, m3m):

$$\begin{pmatrix} 0 & 0 & 0 \\ 0 & 0 & 0 \\ 0 & 0 & 0 \\ 0 & 0 & 0 \\ 0 & 0 & 0 \\ 0 & 0 & 0 \end{pmatrix}$$

Triclinic:

1*

$$\begin{pmatrix} r_{11} & r_{12} & r_{13} \\ r_{21} & r_{22} & r_{23} \\ r_{31} & r_{32} & r_{33} \\ r_{41} & r_{42} & r_{43} \\ r_{51} & r_{52} & r_{53} \\ r_{61} & r_{62} & r_{63} \end{pmatrix}$$

Cubic:

$\bar{4}$3m, 23 $\qquad$ 432

$$\begin{pmatrix} 0 & 0 & 0 \\ 0 & 0 & 0 \\ 0 & 0 & 0 \\ r_{41} & 0 & 0 \\ 0 & r_{41} & 0 \\ 0 & 0 & r_{41} \end{pmatrix} \begin{pmatrix} 0 & 0 & 0 \\ 0 & 0 & 0 \\ 0 & 0 & 0 \\ 0 & 0 & 0 \\ 0 & 0 & 0 \\ 0 & 0 & 0 \end{pmatrix}$$

Monoclinic:

$2\ (2 \parallel x_2)$ $\qquad$ $2\ (2 \parallel x_3)$

$$\begin{pmatrix} 0 & r_{12} & 0 \\ 0 & r_{22} & 0 \\ 0 & r_{32} & 0 \\ r_{41} & 0 & r_{43} \\ 0 & r_{52} & 0 \\ 6_{61} & 0 & r_{63} \end{pmatrix} \begin{pmatrix} 0 & 0 & r_{13} \\ 0 & 0 & r_{23} \\ 0 & 0 & r_{33} \\ r_{41} & r_{42} & 0 \\ r_{51} & r_{52} & 0 \\ 0 & 0 & r_{63} \end{pmatrix}$$

$m\ (m \perp x_2)$ $\qquad$ $m\ (m \perp x_3)$

$$\begin{pmatrix} r_{11} & 0 & r_{13} \\ r_{21} & 0 & r_{23} \\ r_{31} & 0 & r_{33} \\ 0 & r_{42} & 0 \\ r_{51} & 0 & r_{53} \\ 0 & r_{62} & 0 \end{pmatrix} \begin{pmatrix} r_{11} & r_{12} & 0 \\ r_{21} & r_{22} & 0 \\ r_{31} & r_{32} & 0 \\ 0 & 0 & r_{43} \\ 0 & 0 & r_{53} \\ r_{61} & r_{62} & 0 \end{pmatrix}$$

Tetragonal:

4 $\qquad$ $\bar{4}$ $\qquad$ 422

$$\begin{pmatrix} 0 & 0 & r_{13} \\ 0 & 0 & r_{13} \\ 0 & 0 & r_{33} \\ r_{41} & r_{51} & 0 \\ r_{51} & -r_{41} & 0 \\ 0 & 0 & 0 \end{pmatrix} \begin{pmatrix} 0 & 0 & r_{13} \\ 0 & 0 & -r_{13} \\ 0 & 0 & 0 \\ r_{41} & -r_{51} & 0 \\ r_{51} & r_{41} & 0 \\ 0 & 0 & r_{63} \end{pmatrix} \begin{pmatrix} 0 & 0 & 0 \\ 0 & 0 & 0 \\ 0 & 0 & 0 \\ r_{41} & 0 & 0 \\ 0 & -r_{41} & 0 \\ 0 & 0 & 0 \end{pmatrix}$$

4mm $\qquad$ $\bar{4}$2m $(2 \parallel x_1)$

$$\begin{pmatrix} 0 & 0 & r_{13} \\ 0 & 0 & r_{13} \\ 0 & 0 & r_{33} \\ 0 & r_{51} & 0 \\ r_{51} & 0 & 0 \\ 0 & 0 & 0 \end{pmatrix} \begin{pmatrix} 0 & 0 & 0 \\ 0 & 0 & 0 \\ 0 & 0 & 0 \\ r_{41} & 0 & 0 \\ 0 & r_{41} & 0 \\ 0 & 0 & r_{63} \end{pmatrix}$$

Orthorhombic:

222 $\qquad$ 2mm

$$\begin{pmatrix} 0 & 0 & 0 \\ 0 & 0 & 0 \\ 0 & 0 & 0 \\ r_{41} & 0 & 0 \\ 0 & r_{52} & 0 \\ 0 & 0 & r_{63} \end{pmatrix} \begin{pmatrix} 0 & 0 & r_{13} \\ 0 & 0 & r_{23} \\ 0 & 0 & r_{33} \\ 0 & r_{42} & 0 \\ r_{51} & 0 & 0 \\ 0 & 0 & 0 \end{pmatrix}$$

Trigonal:

3 $\qquad$ 32

$$\begin{pmatrix} r_{11} & -r_{22} & r_{13} \\ -r_{11} & -r_{22} & r_{13} \\ 0 & 0 & r_{33} \\ r_{41} & r_{51} & 0 \\ r_{51} & -r_{41} & 0 \\ -r_{22} & -r_{11} & 0 \end{pmatrix} \begin{pmatrix} r_{11} & 0 & 0 \\ -r_{11} & 0 & 0 \\ 0 & 0 & 0 \\ r_{41} & 0 & 0 \\ 0 & -r_{41} & 0 \\ 0 & -r_{11} & 0 \end{pmatrix}$$

Hexagonal:

6 $\qquad$ 6mm $\qquad$ 622

$$\begin{pmatrix} 0 & 0 & r_{13} \\ 0 & 0 & r_{13} \\ 0 & 0 & r_{33} \\ r_{41} & r_{51} & 0 \\ r_{51} & -r_{41} & 0 \\ 0 & 0 & 0 \end{pmatrix} \begin{pmatrix} 0 & 0 & r_{13} \\ 0 & 0 & r_{13} \\ 0 & 0 & r_{33} \\ 0 & r_{51} & 0 \\ r_{51} & 0 & 0 \\ 0 & 0 & 0 \end{pmatrix} \begin{pmatrix} 0 & 0 & 0 \\ 0 & 0 & 0 \\ 0 & 0 & 0 \\ r_{41} & 0 & 0 \\ 0 & -r_{41} & 0 \\ 0 & 0 & 0 \end{pmatrix}$$

$3m\ (m \perp x_1)$ $\qquad$ $3m\ (m \perp x_2)$

$$\begin{pmatrix} 0 & -r_{22} & r_{13} \\ 0 & r_{22} & r_{13} \\ 0 & 0 & r_{33} \\ 0 & r_{51} & 0 \\ r_{51} & 0 & 0 \\ -r_{22} & 0 & 0 \end{pmatrix} \begin{pmatrix} r_{11} & 0 & r_{13} \\ -r_{11} & 0 & r_{13} \\ 0 & 0 & r_{33} \\ 0 & r_{51} & 0 \\ r_{51} & 0 & 0 \\ 0 & -r_{11} & 0 \end{pmatrix}$$

$\bar{6}$ $\qquad$ $\bar{6}$m2 $(m \perp x_1)$ $\qquad$ $\bar{6}$m2 $(m \perp x_2)$

$$\begin{pmatrix} r_{11} & -r_{22} & 0 \\ -r_{11} & r_{22} & 0 \\ 0 & 0 & 0 \\ 0 & 0 & 0 \\ 0 & 0 & 0 \\ -r_{22} & r_{11} & 0 \end{pmatrix} \begin{pmatrix} 0 & -r_{22} & 0 \\ 0 & r_{22} & 0 \\ 0 & 0 & 0 \\ 0 & 0 & 0 \\ 0 & 0 & 0 \\ -r_{22} & 0 & 0 \end{pmatrix} \begin{pmatrix} r_{11} & 0 & 0 \\ -r_{11} & 0 & 0 \\ 0 & 0 & 0 \\ 0 & 0 & 0 \\ 0 & 0 & 0 \\ 0 & -r_{11} & 0 \end{pmatrix}$$

* The symbol over each matrix is the conventional symmetry-group designation.

TABLE 2 The Quadratic Electro-optic Coefficient Matrices in Contracted Form for all Crystal Symmetry Classes[14]

Triclinic:

$$1,\bar{1}$$

$$\begin{pmatrix} s_{11} & s_{12} & s_{13} & s_{14} & s_{15} & s_{16} \\ s_{21} & s_{22} & s_{23} & s_{24} & s_{25} & s_{26} \\ s_{31} & s_{32} & s_{33} & s_{34} & s_{35} & s_{36} \\ s_{41} & s_{42} & s_{43} & s_{44} & s_{45} & s_{46} \\ s_{51} & s_{52} & s_{53} & s_{54} & s_{55} & s_{56} \\ s_{61} & s_{62} & s_{63} & s_{64} & s_{65} & s_{66} \end{pmatrix}$$

Monoclinic:

$$2,m,2/m$$

$$\begin{pmatrix} s_{11} & s_{12} & s_{13} & 0 & s_{15} & 0 \\ s_{21} & s_{22} & s_{23} & 0 & s_{25} & 0 \\ s_{31} & s_{32} & s_{33} & 0 & s_{35} & 0 \\ 0 & 0 & 0 & s_{44} & 0 & s_{46} \\ s_{51} & s_{52} & s_{53} & 0 & s_{55} & 0 \\ 0 & 0 & 0 & s_{64} & 0 & s_{66} \end{pmatrix}$$

Orthorhombic:

$$2mm,222,mmm$$

$$\begin{pmatrix} s_{11} & s_{12} & s_{13} & 0 & 0 & 0 \\ s_{21} & s_{22} & s_{23} & 0 & 0 & 0 \\ s_{31} & s_{32} & s_{33} & 0 & 0 & 0 \\ 0 & 0 & 0 & s_{44} & 0 & 0 \\ 0 & 0 & 0 & 0 & s_{55} & 0 \\ 0 & 0 & 0 & 0 & 0 & s_{66} \end{pmatrix}$$

Tetragonal:

$$4,\bar{4},4/m$$

$$\begin{pmatrix} s_{11} & s_{12} & s_{13} & 0 & 0 & s_{16} \\ s_{12} & s_{11} & s_{13} & 0 & 0 & -s_{16} \\ s_{31} & s_{31} & s_{33} & 0 & 0 & 0 \\ 0 & 0 & 0 & s_{44} & s_{45} & 0 \\ 0 & 0 & 0 & -s_{45} & s_{44} & 0 \\ s_{61} & -s_{61} & 0 & 0 & 0 & s_{66} \end{pmatrix}$$

$$422,4mm,\bar{4}2m,4/mm$$

$$\begin{pmatrix} s_{11} & s_{12} & s_{13} & 0 & 0 & 0 \\ s_{12} & s_{11} & s_{13} & 0 & 0 & 0 \\ s_{31} & s_{31} & s_{33} & 0 & 0 & 0 \\ 0 & 0 & 0 & s_{44} & 0 & 0 \\ 0 & 0 & 0 & 0 & s_{44} & 0 \\ 0 & 0 & 0 & 0 & 0 & s_{66} \end{pmatrix}$$

Trigonal:

$$3,\bar{3}$$

$$\begin{pmatrix} s_{11} & s_{12} & s_{13} & s_{14} & s_{15} & -s_{61} \\ s_{12} & s_{11} & s_{13} & -s_{14} & -s_{15} & s_{61} \\ s_{31} & s_{31} & s_{33} & 0 & 0 & 0 \\ s_{41} & -s_{41} & 0 & s_{44} & s_{45} & -s_{51} \\ s_{51} & -s_{51} & 0 & -s_{45} & s_{44} & s_{41} \\ s_{61} & -s_{61} & 0 & -s_{15} & s_{14} & \frac{1}{2}(s_{11}-s_{12}) \end{pmatrix}$$

$$32,3m,\bar{3}m$$

$$\begin{pmatrix} s_{11} & s_{12} & s_{13} & s_{14} & 0 & 0 \\ s_{12} & s_{11} & s_{13} & -s_{14} & 0 & 0 \\ s_{13} & s_{13} & s_{33} & 0 & 0 & 0 \\ s_{41} & -s_{41} & 0 & s_{44} & 0 & 0 \\ 0 & 0 & 0 & 0 & s_{44} & s_{41} \\ 0 & 0 & 0 & 0 & s_{14} & \frac{1}{2}(s_{11}-s_{12}) \end{pmatrix}$$

Hexagonal:

$$6,\bar{6},6/m$$

$$\begin{pmatrix} s_{11} & s_{12} & s_{13} & 0 & 0 & -s_{61} \\ s_{12} & s_{11} & s_{13} & 0 & 0 & s_{61} \\ s_{31} & s_{31} & s_{33} & 0 & 0 & 0 \\ 0 & 0 & 0 & s_{44} & s_{45} & 0 \\ 0 & 0 & 0 & -s_{45} & s_{44} & 0 \\ s_{61} & -s_{61} & 0 & 0 & 0 & \frac{1}{2}(s_{11}-s_{12}) \end{pmatrix}$$

$$622,6mm,\bar{6}m2,6/mmm$$

$$\begin{pmatrix} s_{11} & s_{12} & s_{13} & 0 & 0 & 0 \\ s_{12} & s_{11} & s_{13} & 0 & 0 & 0 \\ s_{31} & s_{31} & s_{33} & 0 & 0 & 0 \\ 0 & 0 & 0 & s_{44} & 0 & 0 \\ 0 & 0 & 0 & 0 & s_{44} & 0 \\ 0 & 0 & 0 & 0 & 0 & \frac{1}{2}(s_{11}-s_{12}) \end{pmatrix}$$

Cubic:

$$23,m3$$

$$\begin{pmatrix} s_{11} & s_{12} & s_{13} & 0 & 0 & 0 \\ s_{13} & s_{11} & s_{12} & 0 & 0 & 0 \\ s_{12} & s_{13} & s_{11} & 0 & 0 & 0 \\ 0 & 0 & 0 & s_{44} & 0 & 0 \\ 0 & 0 & 0 & 0 & s_{44} & 0 \\ 0 & 0 & 0 & 0 & 0 & s_{44} \end{pmatrix}$$

$$432,m3m,\bar{4}3m$$

$$\begin{pmatrix} s_{11} & s_{12} & s_{12} & 0 & 0 & 0 \\ s_{12} & s_{11} & s_{12} & 0 & 0 & 0 \\ s_{12} & s_{12} & s_{11} & 0 & 0 & 0 \\ 0 & 0 & 0 & s_{44} & 0 & 0 \\ 0 & 0 & 0 & 0 & s_{44} & 0 \\ 0 & 0 & 0 & 0 & 0 & s_{44} \end{pmatrix}$$

Isotropic:

$$\begin{pmatrix} s_{11} & s_{12} & s_{12} & 0 & 0 & 0 \\ s_{12} & s_{11} & s_{12} & 0 & 0 & 0 \\ s_{12} & s_{12} & s_{11} & 0 & 0 & 0 \\ 0 & 0 & 0 & \frac{1}{2}(s_{11}-s_{12}) & 0 & 0 \\ 0 & 0 & 0 & 0 & \frac{1}{2}(s_{11}-s_{12}) & 0 \\ 0 & 0 & 0 & 0 & 0 & \frac{1}{2}(s_{11}-s_{12}) \end{pmatrix}$$

or equivalently, $\mathbf{X}^T \overline{1/n^{2\prime}} \mathbf{X} = 1$, where $\mathbf{X} = [x \; y \; z]$.[7,17,22] The presence of cross terms indicates that the ellipsoid is rotated and the lengths of the principal dielectric axes are changed. Determining the new orientation and shape of the ellipsoid requires that $\overline{1/\mathbf{n}^{2\prime}}$ be diagonalized, thus determining its eigenvalues and eigenvectors. The perturbed ellipsoid will then be represented by a square sum:

$$\frac{x'^2}{n_{x'}^2} + \frac{y'^2}{n_{y'}^2} + \frac{z'^2}{n_{z'}^2} = 1 \tag{9}$$

The eigenvalues of $\overline{1/n^{2\prime}}$ are $1/n_{x'}^2$, $1/n_{y'}^2$, $1/n_{z'}^2$. The corresponding eigenvectors are $\mathbf{x}' = [x_{x'} \; y_{x'} \; z_{x'}]^T$, $\mathbf{y}' = [x_{y'} \; y_{y'} \; z_{y'}]^T$, and $\mathbf{z}' = [x_{z'} \; y_{z'} \; z_{z'}]^T$, respectively.

The Quadratic Electro-optic Effect

An electric field applied in a general direction to any crystal, centrosymmetric or noncentrosymmetric, produces a quadratic change in the constants $(1/n^2)_i$ due to the quadratic electro-optic effect according to

$$
\begin{pmatrix}
\Delta(1/n^2)_1 \\
\Delta(1/n^2)_2 \\
\Delta(1/n^2)_3 \\
\Delta(1/n^2)_4 \\
\Delta(1/n^2)_5 \\
\Delta(1/n^2)_6
\end{pmatrix}
=
\begin{pmatrix}
s_{11} & s_{12} & s_{13} & s_{14} & s_{15} & s_{16} \\
s_{21} & s_{22} & s_{23} & s_{24} & s_{25} & s_{26} \\
s_{31} & s_{32} & s_{33} & s_{34} & s_{35} & s_{36} \\
s_{41} & s_{42} & s_{43} & s_{44} & s_{45} & s_{46} \\
s_{51} & s_{52} & s_{53} & s_{54} & s_{55} & s_{56} \\
s_{61} & s_{62} & s_{63} & s_{64} & s_{65} & s_{66}
\end{pmatrix}
\begin{pmatrix}
E_x^2 \\
E_y^2 \\
E_z^2 \\
E_y E_z \\
E_x E_z \\
E_x E_y
\end{pmatrix}
\tag{10}
$$

E_x, E_y, and E_z are the components of the applied electric field in principal coordinates. The perturbed impermeability tensor and the new index ellipsoid have the same form as Eqs. (7) and (8).

A Mathematical Approach

Although the eigenvalue problem is a familiar one, obtaining accurate solutions has been the subject of extensive study.[23–26] A number of formalisms are suggested in the literature to address the specific problem of finding the new set of principal dielectric axes relative to the zero-field principal dielectric axes. Most approaches, however, do not provide a consistent means of labeling the new axes. Also, some methods are highly susceptible to numerical instabilities when dealing with very small off-diagonal elements as in the case of the electro-optic effect. In contrast to other methods,[13,23,24,27–29] a similarity transformation is an attractive approach for diagonalizing a symmetric matrix for the purpose of determining its eigenvalues and eigenvectors.[23,24,26,30,31]

A symmetric matrix $\bar{A}$ can be reduced to diagonal form by the transformation $\bar{a}\bar{A}\bar{a}^T = \bar{\lambda}$, where $\bar{\lambda}$ is a 3×3 diagonal matrix and $\bar{a}$ is the orthogonal transformation matrix. Since the eigenvalues of $\bar{A}$ are preserved under similarity transformation, they lie on the diagonal of $\bar{I}\bar{\lambda}$, as in Eq. (1).

The problem of determining the new principal axes and indices of refraction of the index ellipsoid in the presence of an external electric field is analogous to the problem of finding the transformation matrix $\bar{a}$ that will diagonalize the perturbed impermeability tensor. The lengths of the semiaxes are the reciprocals of the square roots of the eigenvalues of $\overline{1/\mathbf{n}^{2\prime}}$. Generally, this matrix reduction requires a *sequence* of similarity transformations. Since similarity is a transitive property, several transformation matrices can be multiplied to generate the desired cumulative matrix.[23,30]

The Jacobi method is a form of similarity transformation which has been shown to provide both accurate eigenvalues and orthogonal eigenvectors. It produces reliable results for matrices with very small off-diagonal elements.[25,30] It is a systematic procedure for ordering the solutions to provide consistent labeling of the principal axes.

Principal Axes and Principal Refractive Indices

The Jacobi method utilizes the concepts of rigid-body rotation and the properties of ellipsoids to determine the principal axes and indices of a crystal.[30] From these concepts, a geometric interpretation of the electro-optic effect is developed. The Jacobi method is an iterative procedure that consists of a simple elementary plane rotation at each step to zero an off-diagonal element. The goal is to produce a diagonal matrix by minimizing the norm of the off-diagonal elements to within a desired level of accuracy. The transformation matrix is simply the product of the plane rotation matrices multiplied in the order in which they are applied. For step m the transformation is represented by

$$\overline{1/n_m^2} = \bar{a}_m \, \overline{1/n_{m-1}^2} \, \bar{a}_m^T \tag{11}$$

The perturbed impermeability matrix of Eq. (7) for a specific crystal is determined given the unperturbed principal refractive indices, the electro-optic coefficients, and the direction of the applied field defined in the principal coordinate system. The first step is to select the largest off-diagonal element $(1/n^2)_{ij}$ and execute a rotation in the (i, j) plane, $i < j$, so as to zero that element. The required rotation angle Ω is given by

$$\tan (2\Omega) = \frac{2(1/n^2)_{ij}}{(1/n^2)_{ii} - (1/n^2)_{jj}} \qquad i, j = 1, 2, 3 \tag{12}$$

For example, if the largest off-diagonal element is $(1/n^2)_{12}$, then the plane rotation is represented by

$$\bar{a} = \begin{pmatrix} \cos \Omega & \sin \Omega & 0 \\ -\sin \Omega & \cos \Omega & 0 \\ 0 & 0 & 1 \end{pmatrix} \tag{13}$$

If $(1/n^2)_{ii} = (1/n^2)_{jj}$, which can occur in isotropic and uniaxial crystals, then $|\Omega|$ is taken to be $45°$, and its sign is taken to be the same as the sign of $(1/n^2)_{ij}$. The impermeability matrix elements are updated with the following equations:

$$(1/n^2)'_{\Omega_{ii}} = (1/n^2)_{ii} \cos^2 \Omega + (1/n^2)_{jj} \sin^2 \Omega + 2(1/n^2)_{ij} \cos \Omega \sin \Omega$$

$$(1/n^2)'_{\Omega_{jj}} = (1/n^2)_{ii} \sin^2 \Omega + (1/n^2)_{jj} \cos^2 \Omega - 2(1/n^2)_{ij} \cos \Omega \sin \Omega$$

$$(1/n^2)'_{\Omega_{kk}} = (1/n^2)_{kk}$$

$$(1/n^2)'_{\Omega_{ij}} = [(1/n^2)_{jj} - (1/n^2)_{ii}] \cos \Omega \sin \Omega + (1/n^2)_{ij}(\cos^2 \Omega - \sin^2 \Omega) \tag{14}$$

$$= (1/n^2)'_{\Omega_{ji}} = 0$$

$$(1/n^2)'_{\Omega_{ik}} = (1/n^2)_{ik} \cos \Omega + (1/n^2)_{jk} \sin \Omega = (1/n^2)'_{\Omega_{ki}}$$

$$(1/n^2)'_{\Omega_{jk}} = -(1/n^2)_{ik} \sin \Omega + (1/n^2)_{jk} \cos \Omega = (1/n^2)'_{\Omega_{kj}}$$

Once the new elements are determined, the next iteration step is performed, selecting the new largest off-diagonal element and repeating the procedure. The process is terminated when all of the off-diagonal elements are reduced below the desired level (typically 10^{-10}).

The next step is to determine the cumulative transformation matrix $\bar{a}$ by either of two ways. The first way is to multiply the plane rotation matrices in order, either as $\bar{a} = \bar{a}_n \cdots \bar{a}_2 \bar{a}_1$ or equivalently for the transpose of $\bar{a}$ as

$$\bar{a}^T = \bar{a}_1^T \bar{a}_2^T \cdots \bar{a}_n^T \tag{15}$$

A much simpler way is to find each eigenvector individually be a series of matrix-vector multiplications and then to construct the cumulative matrix by placing these vectors in the rows of $\bar{a}$.[23] For example, to find the first eigenvector, the first matrix-vector multiplication is $\bar{a}_n^T [1 \ 0 \ 0]^T = \mathbf{a_n}$ (a vector), followed by $\bar{a}_{n-1}^T \mathbf{a_n} = \mathbf{a_{n-1}}$, etc., until all matrix-vector multiplications are performed to obtain

$$\bar{a}_1^T \bar{a}_2^T \cdots \bar{a}_n^T \begin{pmatrix} 1 \\ 0 \\ 0 \end{pmatrix} = \begin{pmatrix} a_{11} \\ a_{12} \\ a_{13} \end{pmatrix} \tag{16}$$

which is the first column of $\bar{a}^T$. Alternatively, a_{11}, a_{12}, and a_{13} are the elements of the first row of $\bar{a}$, and they represent the components of the first eigenvector $\mathbf{x'}$ of $\overline{1/n^2}'$. This eigenvector corresponds to the new x' axis. Likewise, the second ($\mathbf{y'}$) and third ($\mathbf{z'}$) eigenvectors may be found by multiplying the successive rotation matrixes by $[0 \ 1 \ 0]^T$ and $[0 \ 0 \ 1]^T$, respectively. All three eigenvectors are automatically in normalized form, since the rotation matrices are orthonormal.

The reorientation of the index ellipsoid can be described by a single rotation of Φ about a global rotation axis $\mathbf{R}$. Both Φ and $\mathbf{R}$ can be determined from $\bar{a}$ as shown in Fig. 4.[30,32] The cumulative transformation matrix always has an eigenvalue of $+1$, since it represents a rigid-body rotation. The global rotation axis is the eigenvector of $\bar{a}$ corresponding to this eigenvalue, and its direction cosines are unchanged by the transformation. Therefore, the rotation axis may be found by

$$(\bar{a} - \bar{I})\mathbf{R} = \mathbf{0} \tag{17}$$

where $\mathbf{R}$ is the vector of direction cosines for the rotation axis. The sense of $\mathbf{R}$ is not unique; $-\mathbf{R}$ is also a solution. To provide consistency in the orientation of $\mathbf{R}$, the z

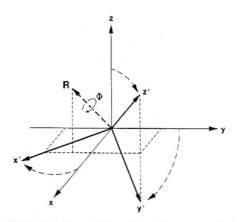

FIGURE 4   The transformation of the unperturbed (x, y, z) coordinate system to the (x', y', z') coordinate system can be described by one global rotation of angle Φ about a global rotation axis $\mathbf{R}$.[30]

component of **R** is set to +1 before normalizing. To determine the global rotation angle, a similarity transformation with orthogonal matrix $\overline{a'}$ is possible where the z axis is made to align with **R**, and the rotation is taken in the (x, y) plane perpendicular to **R**:

$$\overline{a'} = \begin{pmatrix} \cos \Phi & \sin \Phi & 0 \\ -\sin \Phi & \cos \Phi & 0 \\ 0 & 0 & 1 \end{pmatrix} \tag{18}$$

Since the trace of a matrix is invariant under similarity transformation, the magnitude of Φ may be found by

$$\sum_{i=1}^{3} a_{ii} = 1 + 2 \cos \Phi \tag{19}$$

where a_{ii} are the diagonal elements of $\overline{a}$. Using the a_{ii} elements as obtained by the general Jacobi method, Φ is automatically the *minimum* rotation angle required to reorient the zero-field principal axes to those of the perturbed index ellipsoid (x', y', z'). Other information that can be obtained from $\overline{a}$ is the set of Euler angles, which also defines the orientation of a rigid body.[32,33] These angles are given in the Appendix. Several examples for using the Jabobi method are given in Ref. 30.

Eigenpolarizations and Phase Velocity Indices of Refraction

For a general direction of phase propagation, there are two allowed linear orthogonal polarization directions. These waves are the only two that can propagate with unchanging orientation for the given wave vector direction. Figure 5a depicts these axes for a crystal in the absence of an applied field. Figure 5b depicts the x''' and y''' axes, which define the fast and slow axes, when a field is applied in a direction so as to reorient the index ellipsoid. A field, in general, rotates the allowed polarization directions in the plane perpendicular to the direction of phase propagation as shown in Fig. 5b.

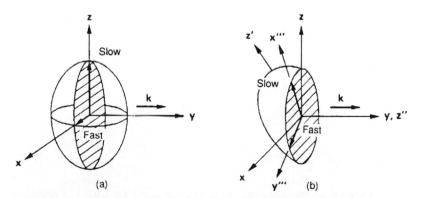

FIGURE 5 (*a*) The cross-section ellipse for a wave propagating along the y principal axis is shown with no field applied to the crystal; (*b*) with an applied electric field the index ellipsoid is reoriented, and the eigenpolarizations in the plane transverse to **k** are rotated, indicated by x''' and y'''.[30]

The perturbed index ellipsoid resulting from an external field is given by Eq. (8). For simplicity, the coefficients may be relabeled as

$$Ax^2 + By^2 + Cz^2 + 2Fyz + 2Gxz + 2Hxy = 1 \tag{20}$$

where x, y, and z represent the original dielectric axes with no applied field. The wave vector direction $\mathbf{k}$ may be specified by the spherical coordinates angles θ_k and ϕ_k in the (x, y, z) coordinate system as shown in Fig. 6. Given $\mathbf{k}$, the cross section ellipse through the center of the perturbed ellipsoid of Eq. (20) may be drawn. The directions of the semiaxes of this ellipse represent the fast and slow polarization directions of the two waves $\mathbf{D}_1$ and $\mathbf{D}_2$ that propagate independently. The lengths of the semiaxes are the phase velocity indices of refraction. The problem is to determine the new polarization directions x''' of $\mathbf{D}_1$ and y''' of $\mathbf{D}_2$ relative to the (x, y, z) axes and the corresponding new indices of refraction $n_{x'''}$ and $n_{y'''}$.

A new coordinate system (x'', y'', z'') may be defined with z'' parallel to $\mathbf{k}$ and x'' lying in the (z, z'') plane. The (x'', y'', z'') system is, of course, different from the (x', y', z') perturbed principal axes system. The (x'', y'', z'') system may be produced first by a counterclockwise rotation ϕ_k about the z axis followed by a counterclockwise rotation θ_k about y'' as shown in Fig. 6. This transformation is described by

$$x = x'' \cos \theta_k \cos \phi_k - y'' \sin \phi_k + z'' \sin \theta_k \cos \phi_k$$

$$y = x'' \cos \theta_k \sin \phi_k + y'' \cos \phi_k + z'' \sin \theta_k \sin \phi_k \tag{21}$$

$$z = -x'' \sin \theta_k + z'' \cos \theta_k$$

The equation for the cross section ellipse normal to $\mathbf{k}$ is determined by substituting Eqs. (21) into Eq. (20) and setting $z'' = 0$, giving

$$A''x''^2 + B''y''^2 + 2H''x''y'' = 1 \tag{22}$$

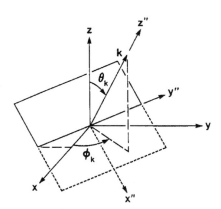

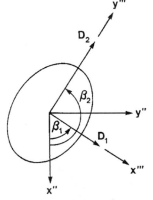

FIGURE 6 The coordinate system (x'', y'', z'') of the wave vector $\mathbf{k}$ is defined with its angular relationship (ϕ_k, θ_k) with respect to the unperturbed principal dielectric axes coordinate system (x, y, z).[30]

FIGURE 7 The polarization axes (x''', y''') are the *fast* and *slow* axes and are shown relative to the (x'', y'') axes of the wave vector coordinate system. The wave vector $\mathbf{k}$ and the axes z'' and z''' are normal to the plane of the figure.[30]

where $A'' = \cos^2 \theta_k [A \cos^2 \phi_k + B \sin^2 \phi_k + H \sin 2\phi_k] + C \sin^2 \theta_k$

$$- \sin 2\theta_k [F \sin \phi_k + G \cos \phi_k]$$

$$B'' = A \sin^2 \phi_k + B \cos^2 \phi_k - H \sin 2\phi_k \tag{23}$$

$$2H'' = \sin 2\phi_k \cos \theta_k [B - A] + 2 \sin \theta_k [G \sin \phi_k - F \cos \phi_k]$$

$$+ 2H \cos \theta_k \cos 2\phi_k$$

The polarization angle β_1 of x''' ($\mathbf{D}_1$) with respect to x'', as shown in Fig. 7, is given by

$$\beta_1 = \tfrac{1}{2} \tan^{-1} \left[\frac{2H''}{(A'' - B'')} \right] \tag{24}$$

The polarization angle β_2 of y''' ($\mathbf{D}_2$) with respect to x'' is $\beta_1 + \pi/2$. The axes are related by

$$x'' = x''' \cos \beta_1 - y''' \sin \beta_1$$
$$y'' = x''' \sin \beta_1 + y''' \cos \beta_1 \tag{25}$$

The refractive index $n_{x''}$ may be found by setting $y''' = 0$ and substituting Eqs. (25) into Eq. (22) and solving for $n_{x''} = x'''$ giving

$$n_{x''} = [A'' \cos^2 \beta_1 + B'' \sin^2 \beta_1 + 2H'' \cos \beta_1 \sin \beta_1]^{-1/2} \tag{26}$$

Similarly, the refractive index $n_{y''}$ may be found by setting $x''' = 0$ and solving for $n_{y''} = y'''$, giving

$$n_{y''} = [A'' \sin^2 \beta_1 + B'' \cos^2 \beta_1 - 2H'' \cos \beta_1 \sin \beta_1]^{-1/2} \tag{27}$$

The larger index corresponds to the slow axis and the smaller index to the fast axis.

13.5 MODULATOR DEVICES

An electro-optic modulator is a device with operation based on an electrically-induced change in index of refraction or change in natural birefringence. Depending on the device configuration, the following properties of the light wave can be varied in a controlled way: phase, polarization, amplitude, frequency, or direction of propagation. The device is typically designed for optimum performance at a single wavelength, with some degradation in performance with wideband or multimode lasers.[14,34,35]

Electro-optic devices can be used in analog or digital modulation formats. The choice is dictated by the system requirements and the characteristics of available components (optical fibers, sources/detectors, etc.). Analog modulation requires large signal-to-noise ratios (SNR), thereby limiting its use to narrow-bandwidth, short-distance applications. Digital modulation, on the other hand, is more applicable to large-bandwidth, medium to long distance systems.[34,35]

Device Configurations

An electro-optic modulator can be classified as one of two types, *longitudinal* or *transverse,* depending on how the voltage is applied relative to the direction of light propagation in the device. Basically, a bulk modulator consists of an electro-optic crystal sandwiched between a pair of electrodes and, therefore, can be modeled as a capacitor. In general, the

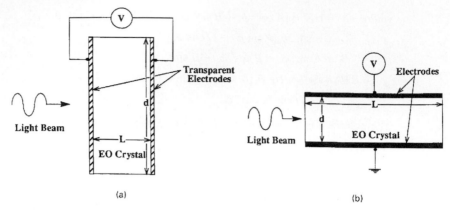

FIGURE 8 (*a*) A longitudinal electro-optic modulator has the voltage applied parallel to the direction of light propagation; (*b*) a transverse modulator has the voltage applied perpendicular to the direction of light propagation.[14]

input and output faces are parallel for the beam to undergo a uniform phase shift over the beam cross section.[14]

In the longitudinal configuration, the voltage is applied parallel to the wave vector direction as shown in Fig. 8*a*.[14,21,36–39] The electrodes must be transparent to the light either by the choice of material used for them (metal-oxide coatings of SnO, InO, or CdO) or by leaving a small aperture at their center at each end of the electro-optic crystal.[21,37–39] The ratio of the crystal length L to the electrode separation b is defined as the *aspect ratio*. For this configuration $b = L$, and, therefore, the aspect ratio is always unity. The magnitude of the applied electric field inside the crystal is $E = V/L$. The induced phase shift is proportional to V and the wavelength λ of the light but not the physical dimensions of the device. Therefore, for longitudinal modulators, the required magnitude of the applied electric field for a desired degree of modulation cannot be reduced by changing the aspect ratio, and it increases with wavelength. However, these modulators can have a large acceptance area and are useful if the light beam has a large cross-sectional area.

In the transverse configuration, the voltage is applied perpendicular to the direction of light propagation as shown in Fig. 8*b*.[14,36–39] The electrodes do not obstruct the light as it passes through the crystal. For this case, the aspect ratio can be very large. The magnitude of the applied electric field is $E = V/d$, ($b = d$), and d can be reduced to increase E for a given applied voltage, thereby increasing the aspect ratio L/b. The induced phase shift is inversely proportional to the aspect ratio; therefore, the voltage necessary to achieve a desired degree of modulation can be greatly reduced. Furthermore, the interaction length can be long for a given field strength. However, the transverse dimension d is limited by the increase in capacitance, which affects the modulation bandwidth or speed of the device, and by diffraction for a given length L, since a beam with finite cross section diverges as it propagates.[14,37,40]

Modulation of Light Parameters

The modulation of phase, polarization, amplitude, frequency, and position of light can be implemented using an electro-optic bulk modulator with polarizers and passive birefringent elements. Three assumptions are made in this section. First, the modulating field is *uniform* throughout the length of the crystal; the change in index or birefringence is uniform unless otherwise stated. Second, the modulation voltage is dc or very low frequency ($\omega_m \ll 2\pi/\tau$); the light experiences the same induced Δn during its transit time τ through the crystal of length L, and the capacitance is negligible. Finally, light propagation is taken to be along a principal axis, before and after the voltage is applied; therefore, the equations are presented in terms of the *optical electric field* **E**, rather than the *displacement vector* **D**, which is common practice in various optical references. For other general configurations the equations should be expressed in terms of the eigenpolarizations $\mathbf{D}_1$ and $\mathbf{D}_2$. However, the electric field will determine the direction of energy flow. References 14 and 37, among others, provide examples of modulator devices using potassium dihydrogen phosphate (KDP), lithium niobate ($LiNbO_3$), lithium tantalate ($LiTaO_3$), gallium arsenide (GaAs), and barium titanate ($BaTiO_3$).

Phase Modulation.   A light wave can be phase modulated, without change in polarization or intensity, using an electro-optic crystal and an input polarizer in the proper configuration. An example of a longitudinal device is shown in Fig. 9. In general, an applied voltage V will rotate the principal axes in the crystal cross section. For phase modulation, the input polarizer must be aligned parallel to one of the principal axes when the voltage is on or off. Figure 9 indicates a polarizer along x' with an input optical electric field $E_{i_{x'}}(t) = E_i \cos \omega t$.

The optical wave at the output of the crystal at $z = L$ is

$$E_o(t) = E_i \cos(\omega t - \phi) \tag{28}$$

where

$$\phi = \frac{2\pi}{\lambda}(n_{x'} + \Delta n_{x'})L = \phi_o + \Delta\phi_{x'} \tag{29}$$

is the total phase shift consisting of a natural phase term $\phi_o = (2\pi/\lambda)Ln_{x'}$, with $n_{x'}$ being the unperturbed index in the x' direction, and an electrically-induced phase term $\Delta\phi_{x'} = (2\pi/\lambda)L\Delta n_{x'}$ for a polarization along x'. The change in index is $\Delta n_{x'} \approx \frac{1}{2}n_{x'}^3 rE$, where r is the corresponding electro-optic coefficient.

For a longitudinal modulator the applied electric field is $E = V/L$, and the induced phase shift is $\Delta\phi_{x'} = \frac{\pi}{\lambda}n_{x'}^3 rV$, which is independent of L and is linearly related to V. For a

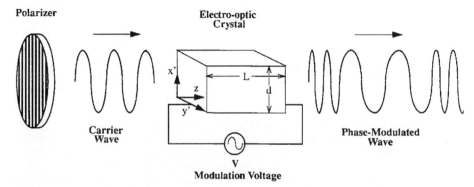

Polarizer

Electro-optic Crystal

x'

z

L

d

y'

Carrier Wave

Phase-Modulated Wave

V

Modulation Voltage

FIGURE 9   A longitudinal phase modulator is shown with the light polarized along the new x' principal axis when the modulation voltage V is applied.[14]

transverse modulator $E = V/d$, and the induced phase shift is $\Delta\phi_{x'} = \frac{\pi}{\lambda}n_{x'}^3 rV(L/d)$, which is a function of the aspect ratio L/d and V. The voltage that would produce an induced phase shift of $\Delta\phi_{x'} = \pi$ is the *half-wave voltage*. It is $V_\pi = \lambda/n_{x'}^3 r$ for a longitudinal modulator and $V_\pi = (\lambda/n_{x'}^3 r)(d/L)$ for a transverse modulator.

If a dc voltage is used, one of two possibilities is required for a crystal and its orientation. The first possibility is a crystal having principal axes which will not rotate with applied voltage V; an example is $LiNbO_3$ with V applied in the z direction and an input polarization along the $x' = x$ axis propagating along $z' = z$. The second possibility is a crystal having a characteristic plane perpendicular to the direction of propagation. If a field is applied such that the axes rotate in this plane, the input wave must be polarized along one of the new principal axes. Therefore, it will always be polarized along a principal axis, whether the voltage is on or off. An example is KDP with V along the z axis and an input wave polarized along the new principal axis x' and propagating along $z' = z$. Phase modulation is then achieved by turning the voltage on and off.

If the applied modulation voltage is sinusoidal in time ($V = V_m \sin \omega_m t$), the corresponding electric field can be represented by

$$E = E_m \sin \omega_m t \tag{30}$$

The magnitude of the field varies only with time, not space; it is a stationary wave applied in the same direction for all time. In other words, this time-varying voltage signal is to be distinguished from a traveling wave voltage which will be discussed in the next section. In this case

$$\phi = \left(\frac{2\pi}{\lambda}\right)(n_{x'} - \tfrac{1}{2}n_{x'}^3 rE_m \sin \omega_m t)L$$

$$= \left(\frac{2\pi}{\lambda}\right)n_{x'}L - \delta \sin \omega_m t \tag{31}$$

The parameter $\delta = (\pi/\lambda)n_{x'}^3 rE_m L = \pi V_m/V_\pi$, where V_π is the half-wave voltage for a given configuration, is the *phase modulation index* or *depth-of-phase modulation*. By neglecting the constant phase term ϕ_o, applying the identity $\cos(\delta \sin \omega_m t) + j \sin(\delta \sin \omega_m t) = \exp[j\delta \sin \omega_m t] = \sum_{l=-\infty}^{\infty} J_l(\delta) \exp[jl\omega t]$, and equating the real and imaginary parts, the output light wave becomes

$$E_o(t) = E_i[J_0(\delta) \cos \omega t + J_1(\delta) \cos(\omega + \omega_m)t - J_1(\delta) \cos(\omega - \omega_m)t$$

$$+ J_2(\delta) \cos(\omega + 2\omega_m)t + J_2(\delta) \cos(\omega - 2\omega_m)t + \cdots] \tag{32}$$

The output consists of components at frequencies ω and $(\omega + n\omega_m)$, $n = \pm 1, \pm 2, \ldots$. For no modulation, $\delta = 0$ and $J_0(0) = 1$, $J_n(0) = 0$ for $n \neq 0$ and $E_o(t) = E_i \cos \omega t = E_{i_{x'}}(t)$.[14] For $\delta \approx 2.4048$, $J_0(\delta) = 0$ and all the power is transferred to harmonic frequencies.[37]

Polarization Modulation (Dynamic Retardation).   Polarization modulation involves the coherent addition of two orthogonal waves, resulting in a change of the input polarization state at the output. As with a phase modulator, the basic components for an electro-optic polarization modulator (or dynamic retardation plate or polarization state converter) is an electro-optic crystal and an input polarizer. The crystal and applied voltage V (dc assumed) are configured to produce dynamically the fast and slow axes in the crystal cross section. In this case, however, the polarizer is positioned such that the input light wave is

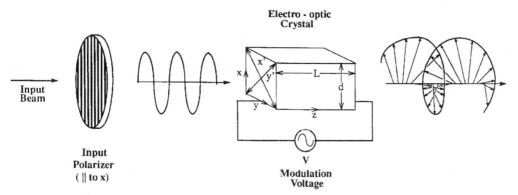

FIGURE 10 A longitudinal polarization modulator is shown with the input polarizer oriented along the x principal axis at 45° with respect to the perturbed x' and y' axes.

decomposed equally into the two orthogonal linear eigenpolarizations along these axes as shown in Fig. 10. If the light is polarized along the x axis and propagates along the z principal axis, for example, the propagating fields are

$$
\begin{aligned}
E_{x'} &= E_0 \cos\left[\omega t - (2\pi/\lambda)n_{x'}z\right] \\
E_{y'} &= E_0 \cos\left[\omega t - (2\pi/\lambda)n_{y'}z\right]
\end{aligned}
\tag{33}
$$

where the fast and slow axes are x' and y'. The corresponding refractive indices are

$$
\begin{aligned}
n_{x'} &\approx n_x - \tfrac{1}{2}r_x n_x^3 E = n_x - \Delta n_x \\
n_{y'} &\approx n_y - \tfrac{1}{2}r_y n_y^3 E = n_y - \Delta n_y
\end{aligned}
\tag{34}
$$

where n_x, n_y are the indices in the absence of an applied field and r_x, r_y are the appropriate electro-optic coefficients for the material being used and the orientation of the applied voltage. As the two polarizations propagate at different speeds through the crystal, a phase difference (relative phase) or *retardation* Γ evolves between them as a function of length:

$$
\begin{aligned}
\Gamma &= \frac{2\pi}{\lambda}(n_{x'} - n_{y'})L \\
&= \frac{2\pi}{\lambda}(n_x - n_y)L - \frac{\pi}{\lambda}(r_x n_x^3 - r_y n_y^3)EL = \Gamma_o + \Gamma_i
\end{aligned}
\tag{35}
$$

where Γ_o is the natural phase retardation in the absence of an applied voltage and Γ_i is the induced retardation linearly related to V.

For a longitudinal modulator the applied electric field is $E = V/L$, and the induced retardation is $\Gamma_i = \left(\dfrac{\pi}{\lambda}\right)(r_y n_y^3 - r_x n_x^3)V$, which is independent of L and linearly related to V.

For a transverse modulator $E = V/d$, and the induced retardation is $\Gamma_i = \left(\dfrac{\pi}{\lambda}\right)(r_y n_y^3 - r_x n_x^3)V\left(\dfrac{L}{d}\right)$, which is dependent on the aspect ratio L/d and V.

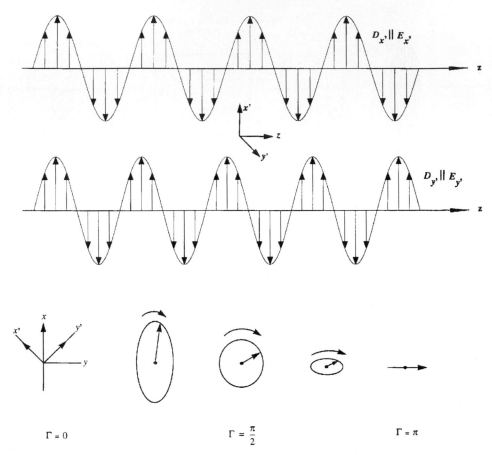

FIGURE 11 The polarization state of an input vertical linear polarization is shown as a function of crystal length L or applied voltage V. The retardation $\Gamma = \pi$ for a given length L_π in a passive $\lambda/2$ wave plate or applied voltage V_π in an electro-optic polarization modulator.[14]

The optical fields at the output can be expressed in terms of Γ:

$$E_{x'} = \cos \omega t$$
$$E_{y'} = \cos (\omega t - \Gamma) \tag{36}$$

Therefore, the desired output polarization is obtained by applying the appropriate voltage magnitude. Figure 11 illustrates the evolution of the polarization state as a function of propagation distance z. In terms of an active device, Fig. 11 also can be interpreted as a change in polarization state as a function of applied voltage for fixed length. The eigenpolarizations $E_{x'}$ and $E_{y'}$ are in phase at $z = 0$. They have the same frequency but different wavelengths. Light from one polarization gradually couples into the other. In the absence of natural birefringence, $n_x - n_y = 0$, the voltage that would produce a retardation of $\Gamma = \Gamma_i = \pi$, such that a vertical polarization input becomes a horizontal polarization output, is the *half-wave voltage* V_π. For a longitudinal modulator $V_\pi = \dfrac{\lambda}{r_x n_x^3 - r_y n_y^3}$, which is independent of L. For a transverse modulator $V_\pi = \dfrac{\lambda}{r_x n_x^3 - r_y n_y^3}\left(\dfrac{d}{L}\right)$, which is dependent on the aspect ratio L/d. The total retardation in terms of V_π (calculated assuming no birefringence) is

$$\Gamma = \Gamma_o + \pi\left(\frac{V}{V_\pi}\right) \tag{37}$$

To cancel the effect of natural birefringence, the phase retardation Γ_o can be made a multiple of 2π by slightly polishing the crystal to adjust the length or by applying a bias voltage. If birefringence is present, an effective V_π can be calculated that would give a total retardation of $\Gamma = \pi$.

To achieve polarization modulation, a birefringence must exist in the crystal cross section. If the cross section is a characteristic plane, then the input polarization propagates through the crystal unchanged when $V = 0$. If an applied voltage causes the axes to rotate 45° in this cross section with respect to the input polarization, as in Fig. 10, then the input will decompose into two equal components and change polarization state at the output. If the cross section has natural birefringence, then the input polarization state will change with $V = 0$ as well as with an applied voltage.

Amplitude Modulation.   The intensity (optical energy) of a light wave can be modulated in several ways. Some possibilities include using (1) a dynamic retarder configuration with a crossed polarizer at the output, (2) a dynamic retarder configuration with a parallel polarizer at the output, (3) a phase modulator configuration in a branch of a Mach-Zehnder interferometer, or (4) a dynamic retarder with push-pull electrodes. The intensity modulator parameter of interest is the *transmission* $T = I_o/I_i$, the ratio of output to input intensity.

An intensity modulator constructed using a dynamic retarder with crossed polarizers is shown in Fig. 12. The transmission for this modulator is

$$T(V) = \sin^2\left(\frac{\Gamma}{2}\right) = \sin^2\left(\frac{\Gamma_o}{2} + \frac{\pi V}{2V_\pi}\right) \tag{38}$$

using Eq. (37). For linear modulation, where the output is a replica of the modulating voltage signal, a fixed bias of $\Gamma_o = \pi/2$ must be introduced either by placing an additional phase retarder, a $\lambda/4$ wave plate (Fig. 12), at the output of the electro-optic crystal or by applying an additional dc voltage of $V_\pi/2$. This bias produces a transmission of $T = 0.5$ in the absence of a modulating voltage. If the crystal cross section has natural birefringence, then a variable compensator (Babinet-Soleil) or a voltage less than $V_\pi/2$ must be used to tune the birefringence to give a fixed retardation of $\pi/2$.

For a sinusoidal modulation voltage $V = V_m \sin \omega_m t$, the retardation at the output of the crystal, including the bias, is

$$\Gamma = \Gamma_o + \Gamma_i = \frac{\pi}{2} + \Gamma_m \sin \omega_m t \tag{39}$$

where $\Gamma_m = \pi V_m/V_\pi$ is the *amplitude modulation index* or *depth-of-amplitude modulation* and V_π is the half-wave voltage as presented in the previous section for polarization modulators. The transmission becomes

$$T(V) = \sin^2\left(\frac{\pi}{4} + \frac{\Gamma_m}{2}\sin \omega_m t\right)$$

$$= \frac{1}{2}\left[1 - \cos\left(\frac{\pi}{2} + \Gamma_m \sin \omega_m t\right)\right] \tag{40}$$

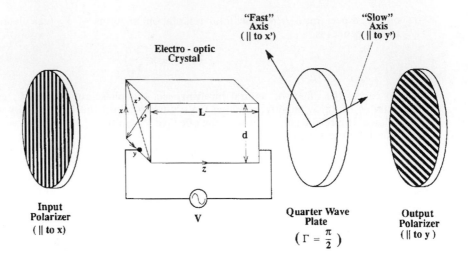

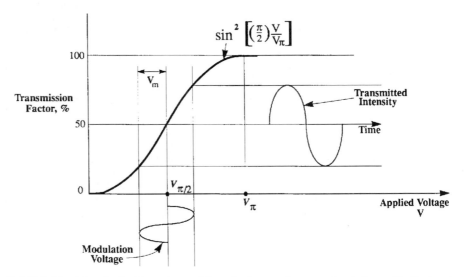

FIGURE 12 A longitudinal intensity modulator is shown using crossed polarizers with the input polarization along the x principal axis. A $\lambda/4$ wave plate is used as a bias to produce linear modulation.[14]

If the modulation voltage is small ($V_m \ll 1$), then the modulation depth is small ($\Gamma_m \ll 1$) and

$$T(V) = \tfrac{1}{2}[1 + \Gamma_m \sin \omega_m t] \tag{41}$$

Therefore, the transmission or output intensity is linearly related to the modulating voltage. If the signal is large, then the output intensity becomes distorted, and higher-order odd harmonics appear.[14]

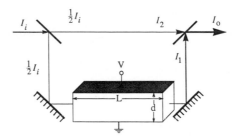

FIGURE 13 An intensity modulator is shown implementing a Mach-Zehnder interferometer configuration with a phase modulator in one branch.[38]

The dynamic retarder with parallel polarizers has a transmission of[14]

$$T = \cos^2\left(\frac{\Gamma}{2}\right) = \cos^2\left(\frac{\pi}{4} + \frac{\Gamma_m}{2}\sin\omega_m t\right)$$

$$= \frac{1}{2}\left[1 + \cos\left(\frac{\pi}{2} + \Gamma_m\sin\omega_m t\right)\right] \tag{42}$$

For small modulation, $T(V) = (1/2)[1 - \Gamma_m \sin \omega_m t]$, and again, the output is a replica of the modulating voltage.

Similarly, the output of a Mach-Zehnder interferometer is given by

$$I_o = I_1 + I_2 = \tfrac{1}{2}[I_i\cos\Gamma_o + I_i] = I_i\cos^2\left(\frac{\Gamma_o}{2}\right) \tag{43}$$

where Γ_o is the relative phase shift between the two branches. An intensity modulator is produced by placing a phase modulator in one branch[38] as shown in Fig. 13. The total retardation is $\Gamma = \Gamma_o + \Gamma_i$, as before. The transmission is

$$T = \frac{I_o}{I_i} = \cos^2\left(\frac{\Gamma}{2}\right) \tag{44}$$

The push-pull modulator is based on the Mach-Zehnder interferometer. In this case, a phase modulator is placed in each branch with opposite polarity voltages applied to the arms; the phase modulators are driven 180° out-of-phase. This configuration requires lower drive voltages and provides a shorter transit time for the light for a defined degree of modulation.[41]

Frequency Modulation. In frequency modulation a shift or deviation in the frequency by ω_d from the optical carrier instantaneous frequency ω is desired. One approach to achieve a shift in frequency is to use an intensity modulator configuration of an electro-optic crystal between left- and right-hand circular polarizers. The electrodes on the modulator must be designed to produce an applied circular electric field.[42]

A crystal and its orientation are selected such that there is no birefringence in the crystal cross section at zero voltage. When a circular electric field with frequency ω_m is applied, however, a birefringence is induced in the crystal cross section, and the induced principal axes rotate with angular velocity $\omega_m/2$ in the opposite sense with respect to the modulating field. The relative rotation between the axes and the modulating field creates a frequency shift in the optical electric field as it propagates through the crystal.

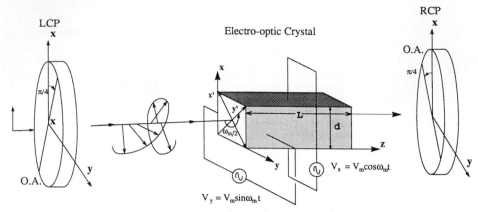

FIGURE 14 A frequency modulator using a phase modulator with two pairs of transverse electrodes set 90° out of phase to produce a circular applied electric field. The phase modulator is placed between left and right circular polarizers to create a frequency deviation in the output optical field.

An example of such a device[42] is shown in Fig. 14. There are two sets of electrodes in the transverse configuration. The applied voltages are 90° out of phase to produce a left circular modulating field:

$$E_x = E_m \cos \omega_m t$$
$$E_y = E_m \sin \omega_m t \tag{45}$$

The principal axes in the crystal cross section rotate through an angle

$$\beta_1(t) = -\tfrac{1}{2}(\omega_m t + \Phi) \tag{46}$$

where Φ is a fixed angle that depends on the electro-optic coefficients r_{ij} of the crystal and not on the electric field magnitude.

The input optical wave (after the left circular polarizer) has field components

$$E_{i_x} = E_i \cos \omega t$$
$$E_{i_y} = E_i \sin \omega t \tag{47}$$

The induced retardation $\Gamma_i = \Gamma = (2\pi/\lambda)\Delta nL$ is independent of time, since Δn is constant although the principal axes are rotating at constant angular velocity. The optical field components along the induced principal axes at the output of the crystal are

$$E_{o_{x'}} = E_i \cos\left(\omega t - \beta_1 + \frac{\Gamma}{2}\right)$$
$$E_{o_{y'}} = E_i \sin\left(\omega t - \beta_1 - \frac{\Gamma}{2}\right) \tag{48}$$

In terms of the original stationary x, y axes, the optical field components at the output are

$$E_{o_x} = E_i \cos(\Gamma/2) \cos \omega t - E_i \sin(\Gamma/2) \sin[(\omega + \omega_m)t + \Phi]$$
$$E_{o_y} = E_i \cos(\Gamma/2) \sin \omega t - E_i \sin(\Gamma/2) \cos[(\omega + \omega_m)t + \Phi] \tag{49}$$

The first terms in E_{o_x} and E_{o_y} represent left circular polarization at the original optical frequency ω with constant amplitude $E_i \cos(\Gamma/2)$ and phase independent of the rotating principal axes (i.e., of β_1). The second terms represent right circular polarization at frequency $(\omega + \omega_m)$ with constant amplitude $E_i \sin(\Gamma/2)$ and phase proportional to $2\beta_1 = \omega_m t$. Therefore, the frequency shift ω_d is the modulation frequency ω_m. If the retardation $\Gamma = \pi$, the frequency of the light has complete deviation to $(\omega + \omega_m)$. If Γ is very small, the component optical fields at frequency $(\omega + \omega_m)$ are linearly related to Γ and therefore, to the applied voltage.

A shift in frequency to $(\omega - \omega_m)$ is obtained if the optical and the applied modulating electric fields rotate in the opposite sense.

Scanners. The position of an optical beam can be changed dynamically by using an electro-optic deflecting device or scanner. Analog scanners are based on refraction phenomena: (1) refraction at a dielectric interface (prism) and (2) refraction by an index gradient that exists perpendicular to the direction of light propagation. Digital scanners or switches are based on birefringence.

One of the most important parameters characterizing the performance of a scanner is its resolution, the number of independent resolvable spots it can scan, which is limited by the diffraction occurring at the aperture of the device. The Rayleigh criterion states that two spots are just resolved when the angular displacement of the beam $\Delta\varphi$ is equal to the half-angle divergence θ due to diffraction.[13] Therefore, the total number of resolvable spots is given by the ratio of the total deflection angle φ to the half-angle divergence θ

$$N = \frac{\varphi}{\theta} \tag{50}$$

The half-angle divergence is $\theta = \varpi/w$, where w is the beamwidth, $\varpi = 1$ for a rectangular beam of uniform intensity, $\varpi = 1.22$ for a circular beam of uniform intensity, and $\varpi = 1.27$ for a beam of Gaussian intensity distribution.[43]

An analog scanner can be constructed by a prism of electro-optic material with electrodes on the crystal faces as shown in Fig. 15. The resolution is maximum in this isosceles-shaped prism when the beam is transmitted through at the minimum deviation angle and is[43,44]

$$N = \Delta n \left(\frac{l}{\varpi\lambda}\right)\left(\frac{w}{W}\right) \tag{51}$$

where l is the base length of the prism and w/W is the ratio of the beamwidth to input

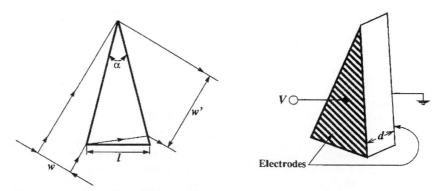

FIGURE 15 An analog scanner can be constructed using an isosceles-shaped prism with the beam transmitted at the minimum deviation angle.[44]

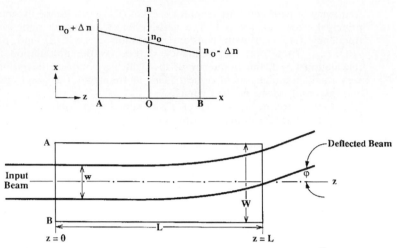

FIGURE 16   An analog scanner based on an index gradient is shown.[45]

aperture of the prism. This type of device typically requires a voltage much higher than the half-wave voltage V_π to resolve N spots.[38]

An analog scanner based on a gradient index of refraction[45] is shown in Fig. 16. The voltage is applied to a crystal such that the change in index is linear with distance perpendicular to the direction of light propagation; i.e., $n(x) = n_o + (2\Delta n/W)x$, where W is the width of the crystal. For linear gradient and small refraction angles, the wavefront remains planar. The (small) deflection angle of the ray after propagating a distance L in the crystal is approximated to be[43]

$$\varphi = L\frac{dn}{dx} \tag{52}$$

and the resolution is

$$N = \frac{\varphi}{\theta} = 2\Delta n\left(\frac{L}{\varpi\lambda}\right)\left(\frac{w}{W}\right) \tag{53}$$

A large crystal length L is needed to obtain appreciable deflection, since Δn is very small ($\sim 10^{-4}$). Laser beams, however, are very narrow to make such a device practical.

Digital light deflection can be implemented with a number of *binary units*,[38,46] each of which consists of a polarization modulator followed by a birefringent crystal (or discriminator) as shown in Fig. 17. The polarizer is oriented to give a horizontal polarization input to the modulator. With the voltage off, the light is polarized along a principal axis of the crystal, and it propagates through without change in polarization. It then enters the birefringent crystal, such as calcite, and passes through undeflected. With an applied voltage, the principal axes rotate 45° and the input is then decomposed into orthogonal components of equal amplitude. The voltage magnitude is set to produce a retardation of π, thereby rotating the polarization by 90°. The vertical polarization then enters the birefringent crystal which deflects it.[43,47]

The number of binary units n would produce a deflector of 2^n possible deflected positions at the output. An example of a three-stage deflector is shown in Fig. 18.[46-48] The on–off states of the three voltages determine at what position the output beam would be deflected. For example, if all three voltages are off, the input polarization remains horizontal through the system and is undeflected at the output. However, if all three voltages are on, the horizontal input is rotated 90° and becomes vertical after the first

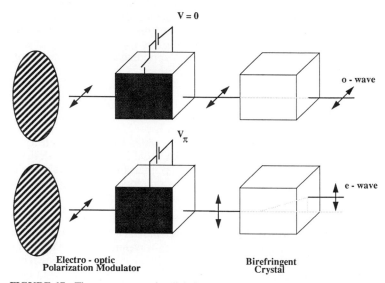

FIGURE 17   The component of a digital scanner is a binary unit. It consists of a polarization modulator followed by a birefringent crystal which serves as a discriminator.[38]

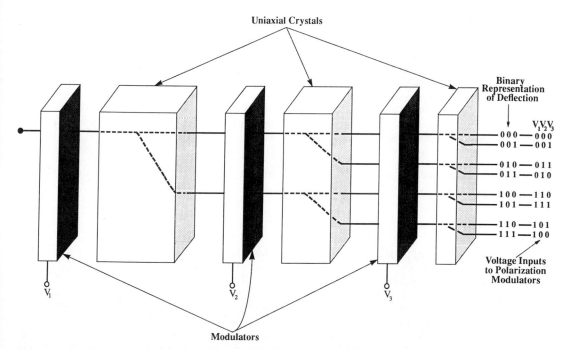

FIGURE 18   A three-stage digital scanner produces 2^3 possible deflected positions at the output.[46]

polarization modulator and is deflected. The polarization is rotated 90° after the second modulator, becoming horizontal, and therefore, propagates straight through the birefringent crystal. Finally, the polarization is rotated 90° after the third modulator, becoming vertical, and is deflected. The corresponding output represents a binary five; that is, 101.

Design Considerations

Two fundamental considerations by which to design an electro-optic modulator are signal voltage and frequency response. The goal is to achieve a desired depth of modulation with the smallest amount of power and at the same time obtain the largest possible bandwidth. In general, when designing bulk modulators, the field of the modulation signal and optical wave are assumed to have 100 percent overlap.

Choice of Material and Device Geometry.   An electro-optic material must be selected that would perform ideally at a given wavelength and in a given environment. The material should have high resistivity, a high electro-optic coefficient, good homogeneity, and temperature and mechanical stability. Some materials with natural birefringence, such as KDP and ferroelectric $LiTaO_3$ and $LiNbO_3$, are sensitive to temperature variations and would induce a temperature-dependent phase shift in the light. Some of these materials have acoustic resonances that cause undesirable peaks in the frequency response.[20,41,48,49] The half-wave voltage V_π required for a desired degree of modulation is an important parameter to consider when selecting a crystal, and V_π should be as small as possible. The material should have small dielectric dissipation, good optical quality in the proper size, and good ohmic contacts.[20,50]

When a modulator is operating in the transverse configuration, the half-wave voltage is inversely proportional to the aspect ratio L/d. Half-wave voltages on the order of tens of volts are typical for this configuration as compared to ~10,000 V for a longitudinal device.[37] However, the transverse dimension d is limited by diffraction of the finite cross-section light beam and by capacitance of the device. For optimum performance, the aperture should be just large enough to pass the light through. The dimensions L and d cannot be selected independently.[20,37,50]

For phase modulation, a crystal orientation is required that would give the maximum change in index of refraction. For amplitude modulation a crystal orientation must produce the maximum birefringence.

Transit Time Limitations.   The transit time of the light is the time for it to pass through the crystal:

$$\tau = \frac{nL}{c} \tag{54}$$

where n is the index seen by the light. This parameter has no relevance for modulation frequencies $\omega_m \ll 2\pi/\tau$; the modulation field appears uniform in the crystal at very low frequencies. A rule of thumb for defining the limiting frequency such that τ can be neglected is that the length of the crystal be less than $\frac{1}{10}$ the wavelength of the modulating field[51] or $L \ll 2\pi c/\omega_m\sqrt{\varepsilon}$.[37] The electro-optic crystal is modeled as a lumped capacitor at low frequencies.

As the modulating frequency becomes larger, the transit time must be taken into account in evaluating modulator performance. The modulation electric field has the form $E = E_m \sin \omega_m t$, and the optical phase can no longer follow the time-varying index of refraction adiabatically. The result is a reduction in the modulation index parameters, δ for phase modulation and Γ_m for amplitude modulation, by a factor[14,37,40]

$$\rho = \frac{\sin\left(\frac{1}{2}\omega_m\tau\right)}{\frac{1}{2}\omega_m\tau} \tag{55}$$

Therefore, the phase modulation index at high frequencies becomes

$$\delta_{RF} = \delta \cdot \rho = \delta \cdot \left[\frac{\sin\left(\frac{1}{2}\omega_m\tau\right)}{\frac{1}{2}\omega_m\tau}\right] \tag{56}$$

and the amplitude modulation index at high frequencies becomes

$$\Gamma_{m_{RF}} = \Gamma_m \cdot \rho = \Gamma_m \cdot \left[\frac{\sin\left(\frac{1}{2}\omega_m\tau\right)}{\frac{1}{2}\omega_m\tau}\right] \tag{57}$$

If $\tau = 2\pi/\omega_m$ such that the transit time of the light is equal to the time period of the modulation signal, then there is no retardation; the retardation produced in the first half of the crystal is exactly canceled by the retardation produced in the second half.[37] The maximum modulation frequency for a given crystal length L is determined by the allowable ρ parameter set by the designer.

Traveling Wave Modulators. The limitation of the transit time on the bandwidth of the modulator can be overcome by applying the voltage as a traveling wave, propagating collinearly with the optical wave. Figure 19 illustrates a transverse traveling wave configuration. The electrode is designed to be an extension of the driving transmission line to eliminate electrode charging time effects on the bandwidth. Therefore, the transit time problem is addressed by adjusting the phase velocity of the modulation signal to be equal to the phase velocity of the optical signal.[14,37,40]

The applied modulation electric field has the form

$$E_{RF}(t, z) = E_m \sin\left(\omega_m t - k_m z\right) \tag{58}$$

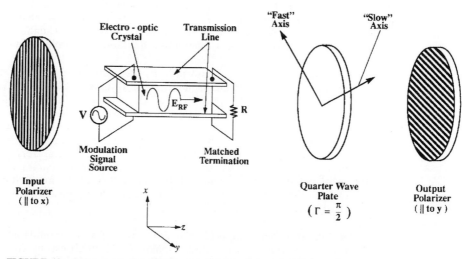

FIGURE 19 A transverse traveling wave modulator has a modulating voltage polarized in the same orientation as the input light and propagating collinearly.[14]

for propagation along the z axis of a crystal. The direction of field vibration is the direction of the applied field, not the direction it is traveling, and is along x in Fig. 19. The parameter k_m is the wave-vector magnitude of the modulation field and is

$$k_m = \frac{\omega_m}{v_m} = \frac{\omega_m n_m}{c} \tag{59}$$

where v_m and $n_m = \sqrt{\varepsilon}$ are the phase velocity and index of refraction of the modulating signal. A mismatch in the phase velocities of the modulating signal and optical wave will produce a reduction in the modulation index δ or Γ_m by a factor

$$\rho_{tw} = \frac{\sin\left[\dfrac{\omega_m}{2c}(n - n_m)L\right]}{\dfrac{\omega_m}{2c}(n - n_m)L} = \frac{\sin(\Delta L)}{\Delta L} \tag{60}$$

In the case of amplitude modulation Eq. (60) holds only if there is no natural birefringence in the cross section of the crystal. The eigenpolarization magnitudes are functions of time and space. Therefore, they satisfy the coupled wave equations, which are more complicated when birefringence is present.[14]

With no phase velocity mismatch, the phase modulation index is $\delta_{RF} = \delta$ and is linearly proportional to the crystal length L. Likewise, $\Gamma_{RF} = \Gamma_m$ for amplitude modulation. With a mismatch, the maximum possible phase modulation index is $\delta_{RF_{max}} = \delta/(\Delta L)$, likewise for the amplitude modulation index $\Gamma_{RF_{max}}$. The modulation index becomes a sinusoidal function of L. This maximum index can be achieved for crystal lengths $L = \pi/2\Delta$, $3\pi/2\Delta$, etc. The ratio n/n_m is approximately 1/2 for LiNbO$_3$, producing a walk-off between the optical and modulation waves.[40] Therefore, for a given length L the modulation frequency is greatly affected by the velocity mismatch.

Performance Criteria

The following parameters are indicators of modulator performance. Basically, the tradeoffs occur between the aspect ratio L/b, the drive voltage, and the electrode configuration (lumped or traveling wave) to achieve a desired depth of modulation with the smallest amount of power and at the same time obtain the largest possible bandwidth.

Modulation Bandwidth. The bandwidth is the difference between the two closest frequencies at which the modulation index, δ or Γ, falls to 50 percent of the maximum value.[52,53] Therefore, the 3 dB optical bandwidth is determined by setting the modulation reduction factor ρ or ρ_{tw} to 0.5.[40,52,53] Modulation speed depends on the electrode type, lumped or traveling wave, and the modulator capacitance per length, which depends on the RF dielectric constant ε and the geometry of the electrodes. For a lumped modulator the bandwidth is limited by the optical (c/Ln) or electrical ($c/L\sqrt{\varepsilon}$) transit time, whichever is smaller, or the time constant of the lumped-circuit parameters ($1/RC$), where R is the resistance in the circuit and C is the capacitance of the modulator. For a traveling wave modulator the bandwidth v_{tw} is limited by the velocity mismatch between the optical and modulation waves, $v_{tw} = c/Ln(1 - \sqrt{\varepsilon}/n)$.[14]

Power per Unit Bandwidth (Specific Energy). To apply the modulating signal efficiently to a *lumped* electro-optic modulator, a resonant RLC circuit can be created by adding an inductance and resistance in parallel to the modulator, which is modeled by a capacitor C.[14,37] The terminating resistance R ensures that more of the voltage drop occurs over the modulator rather than the internal resistance of the source R_s. The impedance of the resonant circuit is high over a bandwidth (determined by the RC time constant) of $\Delta v = \Delta \omega_m / 2\pi \simeq 1/2\pi RC$ centered at the resonant frequency of $\omega_o = 1/\sqrt{LC}$, where the impedance of the parallel RLC circuit is exactly equal to R. A peak voltage V_m must be applied to achieve a desired peak retardation $\Gamma_m = \pi V_m / V_\pi$. Therefore, $V_m = \Gamma_m (V_\pi / \pi)$. The power required to achieve Γ_m is $P = V_m^2 / 2R$, where $1/2R = \pi C \Delta v$, giving

$$P/\Delta v = \frac{1}{\pi} \Gamma_m^2 V_\pi^2 C \qquad (61)$$

The power per unit bandwidth (joules or mW/MHz) depends on the modulator capacitance C and the peak modulation voltage V_m to give the desired depth of modulation Γ_m. The required drive power increases with modulation frequency.[37,53] Since the capacitance is a function of the modulator active area dimensions, the required power is also a function of the modulator dimensions.[53] Furthermore, since V_π is directly proportional to the wavelength λ, a higher input power (voltage) is required at longer wavelengths for a given peak retardation.

Extinction Ratio.   The extinction ratio is the maximum depth of intensity modulation for a time-varying voltage at the output when the optical bias is adjusted properly.[49,52] If for no voltage the output intensity is I_0 and for maximum applied voltage the output intensity is I_m, then the extinction ratio is defined as[52]

$$
\begin{aligned}
\eta_m &= \frac{|I_m - I_0|}{I_0} \qquad I_m \leq I_0 \\
\eta_m &= \frac{|I_m - I_0|}{I_m} \qquad I_m \geq I_0
\end{aligned}
\qquad (62)
$$

Material effects, such as crystal imperfections, temperature sensitivities, and birefringence, can degrade the extinction ratio,[14, 34, 41] thereby affecting the signal-to-noise ratio at the detector.[52] Another definition is in terms of the transmission T, $\eta_m = T_{\max}/T_{\min}$.[38,54] In general, $T_{\max} < 1$ due to absorption, reflection, and scattering losses, and $T_{\min} > 0$ due to beam-divergence angle, residual crystal birefringence, crystal inhomogeneity, electric field uniformity, background scattered light, and polarizer-analyzer alignment.[36] Extinction ratio also can be applied to phase modulation, since phase changes can be related to equivalent changes in intensity.[52]

Maximum Frequency Deviation.   A similar figure of merit as exists for intensity modulators likewise exists for frequency modulators. The maximum deviation of a frequency modulator is defined as[53]

$$D_{\max} = \frac{|\omega_d - \omega|}{\omega} \qquad (63)$$

where ω_d is the frequency shift when the maximum voltage is applied.

Percent Modulation.   Percent modulation is an intensity modulation parameter. Basically, it is the transmission $T = I_o / I_i$ times 100 percent at a specific wavelength. For a device with total retardation of $\Gamma = \dfrac{\pi}{2}$ radians at no voltage, the transmission $T = 0.5$. Then 70 percent modulation is achieved with an analog signal if a voltage is applied such that $\Gamma = 2$ radians; i.e., $\sin^2(1) = 0.70$.[52] 100 percent intensity modulation is the output intensity varying between the input intensity and zero intensity.[37] A peak voltage of $V = V_\pi / 2$ is required to achieve this level of performance for a linear modulator. Another definition of percent modulation is the ratio of the peak output intensity at the modulation frequency to the maximum output intensity at dc voltage.[49] Reference 41 defines percent modulation as the peak modulation voltage output per dc voltage output, assuming no temperature variations.

Degree of Modulation.   For a sinusoidal reference of an optical electric field, $E(t) = E_i \sin \omega t$, an amplitude-modulated signal is typically expressed as

$$E(t) = E_i(1 + m \sin \omega_m t) \sin \omega t \tag{64}$$

where m is the degree of modulation at a specific wavelength.[17] For $m \ll 1$ the intensity is

$$I = \tfrac{1}{2}E_i^2(1 + 2m \sin \omega_m t) \tag{65}$$

The amplitude modulation index is then $\Gamma_m = 2m$, a function of the degree of modulation. The degree of modulation is often referred to as percent modulation for intensity modulation $(70 - 100$ percent nominal) and as modulation index for phase modulation (1 radian nominal).[34]

Modulation Efficiency.   Modulation efficiency is defined as the percentage of total power which conveys information.[55,56] The total power of an amplitude-modulated optical carrier is a function of the modulation frequency and is proportional to $1 + m^2 \langle \sin^2 \omega_m t \rangle = 1 + \tfrac{1}{2}m^2$. Therefore, the modulation efficiency is[55]

$$\xi = \frac{\tfrac{1}{2}m^2}{1 + \tfrac{1}{2}m^2} \tag{66}$$

The maximum efficiency is achieved when m is maximum; i.e., $m = \tfrac{1}{2}\Gamma_m = \pi V/2V_\pi$. In terms of the input power P the modulation efficiency is[56]

$$\xi = \frac{\Gamma^2}{P} \tag{67}$$

where $\Gamma = \pi V/V_\pi$ for maximum efficiency.

Optical Insertion Loss.   For an external modulator, in particular, the optical insertion loss must be minimized when coupling light into and out of the device. For an input light intensity I_{in} the insertion loss IL is[52]

$$IL = 1 - \frac{I_m}{I_{in}} \qquad I_m \geq I_0$$

$$IL = 1 - \frac{I_0}{I_{in}} \qquad I_m \leq I_0 \tag{68}$$

where I_0 and I_m are the output intensities at no voltage and maximum voltage, respectively. If a beam has a large cross-sectional area, the modulator must have a large acceptance area to couple all or most of the light into the device. A longitudinal modulator typically has a large acceptance area. Minimizing insertion loss in a transverse modulator is more of a challenge due to its narrow width, particularly for wide beams. However, a transverse modulator requires less voltage for a desired degree of modulation.

13.6 APPENDIX: EULER ANGLES

Euler angles represent a set of three independent parameters which specify the orientation of a rigid body, in this case the index ellipsoid. An orthogonal transformation matrix consisting of direction cosines and having a determinant of $+1$ corresponds to defining the orientation of a rigid body.

The transformation matrix is developed by a specific sequence of three plane rotations (not in principal planes) in a defined order. There are 12 possible conventions for defining a set of Euler angles in a right-handed coordinate system.[32] One convention that has been proposed as a standard is the y convention.[57] The transformation evolves by an initial counterclockwise rotation ζ about the z axis, followed by a counterclockwise rotation η about the intermediate y' axis, and finally a counterclockwise rotation ω about z''. The resulting transformation matrix is

$$\bar{a} = \begin{pmatrix} \cos\omega & \sin\omega & 0 \\ -\sin\omega & \cos\omega & 0 \\ 0 & 0 & 1 \end{pmatrix} \begin{pmatrix} \cos\eta & 0 & -\sin\eta \\ 0 & 1 & 0 \\ \sin\eta & 0 & \cos\eta \end{pmatrix} \begin{pmatrix} \cos\zeta & \sin\zeta & 0 \\ -\sin\zeta & \cos\xi & 0 \\ 0 & 0 & 1 \end{pmatrix} \quad (A.1)$$

To find the Euler angles (ζ, η, ω), Eq. (A.1) is rearranged as[33]

$$\begin{pmatrix} \cos\omega & -\sin\omega & 0 \\ \sin\omega & \cos\omega & 0 \\ 0 & 0 & 1 \end{pmatrix} \begin{pmatrix} a_{11} & a_{12} & a_{13} \\ a_{21} & a_{22} & a_{23} \\ a_{31} & a_{32} & a_{33} \end{pmatrix} = \begin{pmatrix} \cos\eta & 0 & -\sin\eta \\ 0 & 1 & 0 \\ \sin\eta & 0 & \cos\eta \end{pmatrix} \begin{pmatrix} \cos\zeta & \sin\zeta & 0 \\ -\sin\xi & \cos\zeta & 0 \\ 0 & 0 & 1 \end{pmatrix} \quad (A.2)$$

where $\bar{a}$ is the cumulative transformation matrix of the normalized eigenvectors. Multiplying the matrices, the Euler angles are related to the elements of $\bar{a}$:

$$a_{11}\cos\omega - a_{21}\sin\omega = \cos\eta\cos\zeta \quad (A.3a)$$

$$a_{12}\cos\omega - a_{22}\sin\omega = \cos\eta\sin\zeta \quad (A.3b)$$

$$a_{13}\cos\omega - a_{23}\sin\omega = -\sin\eta \quad (A.3c)$$

$$a_{11}\sin\omega + a_{21}\cos\omega = -\sin\zeta \quad (A.3d)$$

$$a_{12}\sin\omega + a_{22}\cos\omega = \cos\zeta \quad (A.3e)$$

$$a_{13}\sin\omega + a_{23}\cos\omega = 0 \quad (A.3f)$$

$$a_{31} = \sin\eta\cos\zeta \quad (A.3g)$$

$$a_{32} = \sin\eta\sin\zeta \quad (A.3h)$$

$$a_{33} = \cos\eta \quad (A.3i)$$

From Eq. (A.3f), the angle ω is $\omega = \tan^{-1}(-a_{23}/a_{13})$. From Eq. (A.3c), the angle η is $\eta = \sin^{-1}(a_{23}\sin\omega - a_{13}\cos\omega)$. From Eq. (A.3d), the angle ζ is $\zeta = \sin^{-1}(-a_{11}\sin\omega - a_{21}\cos\omega)$.[30]

13.7 REFERENCES

1. W. J. Tomlinson and C. A. Brackett, "Telecommunications Applications of Integrated Optics and Optoelectronics," *Proc. IEEE* **75**(11):1512–1523 (1987).

2. E. Vogues and A. Neyer, "Integrated-optic Devices on LiNbO₃ for Optical Communication," *IEEE/OSA J. Lightwave Technol.* **LT-5**:1229–1238 (1987).

3. L. Thylén, "Integrated Optics in LiNbO$_3$: Recent Developments in Devices in Telecommunications," *IEEE/OSA J. Lightwave Technol.* **6**(6):847–861 (1988).

4. R. C. Alferness, "Guided-wave Devices for Optical Communication," *IEEE J. Quantum Electron.* **QE-17**(6):946–959 (1981).

5. H. F. Taylor, "Application of Guided-wave Optics in Signal Processing and Sensing," *Proc. IEEE* **75**(11):1524–1535 (1987).

6. See, for example, Special Issue on Optical Computing, *Proc. IEEE* **72** (1984).

7. See, for example, "Special Feature on Integrated Optics: Evolution and Prospects," *Opt. News* **14** (1988).

8. T. K. Gaylord and E. I. Verriest, "Matrix Triangularization Using Arrays of Integrated Optical Givens Rotation Devices," *Computer* **20**:59–66 (1987).

9. C. M. Verber, R. P. Kenan, and J. R. Busch, "Design and Performance of an Integrated Optical Digital Correlator," *IEEE/OSA J. Lightwave Technol.* **LT-1**:256–261 (1983).

10. C. L. Chang and C. S. Tsai, "Electro-optic Analog-to-Digital Conversion Using Channel Waveguide Fabry-Perot Modulator Array," *Appl. Phys. Lett.* **43**:22 (1983).

11. C. M. Verber "Integrated-optical Approaches to Numerical Optical Processing," *Proc. IEEE* **72**:942–953 (1984).

12. T. A. Maldonado and T. K. Gaylord, "Light Propagation Characteristics for Arbitrary Wavevector Directions in Biaxial Crystals by a Coordinate-free Approach," *Appl. Opt.* **30**:2465–2480 (1991).

13. M. Born and E. Wolf, *Principles of Optics,* 6th ed., Pergamon Press, Oxford, UK, 1980.

14. A. Yariv and P. Yeh, *Optical Waves in Crystals,* Wiley, New York, 1984.

15. L. D. Landau and E. M. Lifshitz, *Electrodynamics of Continuous Media,* Pergamon, London, 1960.

16. A. I. Borisenko and I. E. Tarapov, *Vector and Tensor Analysis with Applications,* R. A. Silverman (ed.), Prentice-Hall, Englewood Cliffs, N.J., 1968.

17. I. P. Kaminow, *An Introduction to Electrooptic Devices,* Academic Press, New York, 1974.

18. T. C. Phemister, "Fletcher's Indicatrix and the Electromagnetic Theory of Light," *Am. Mineralogist* **39**:173–192 (1954).

19. I. P. Kaminow, in M. J. Weber (ed.), *Handbook of Laser Science and Technology,* vol. IV, part 2, CRC Press, Boca Raton, FL, 1986, pp. 253–278.

20. I. P. Kaminow and E. H. Turner, "Electro-optic Light Modulators," *Proc. IEEE* **54**(10):1374–1390 (1966).

21. J. M. Bennett and H. E. Bennett, "Polarization," in W. G. Driscoll and W. Vaughan (eds.), *Handbook of Optics,* McGraw-Hill, New York, 1978, chap. 10.

22. A. Yariv, *Optical Electronics,* Holt, Rinehart, and Winston, New York, 1976.

23. J. H. Wilkinson, *The Algebraic Eigenvalue Problem,* Oxford Univ. Press, London, 1965.

24. W. H. Press, B. P. Flannery, S. A. Teukolsky, and W. T. Vetterling, *Numerical Recipes,* Cambridge Univ. Press, New York, 1986.

25. J. H. Wilkinson and C. Reinsch, *Handbook for Automatic Computation,* Springer-Verlag, New York, 1971.

26. G. H. Golub and C. F. Van Loan, *Matrix Computations,* John Hopkins Univ. Press, Baltimore, 1983.

27. J. F. Nye, *Physical Properties of Crystals,* Oxford Univ. Press, London, 1957.

28. D. E. Sands, *Vectors and Tensors in Crystallography,* Addison-Wesley, Reading, MA, 1982.

29. D. R. Hartree, *Numerical Analysis,* Clarendon Press, Oxford, 1952.

30. T. A. Maldonado and T. K. Gaylord, "Electro-optic Effect Calculations: Simplified Procedure for Arbitrary Cases," *Appl. Opt.* **27**:5051–5066 (1988).

31. B. N. Parlett, *The Symmetric Eigenvalue Problem,* Prentice-Hall, Englewood Cliffs, N.J., 1980.

32. H. Goldstein, *Classical Mechanics,* Addison-Wesley, Reading, MA, 1981.

33. R. P. Paul, *Robot Manipulators,* The MIT Press, Cambridge, MA, 1981.

34. I. P. Kaminow and T. Li, "Modulation Techniques," in S. E. Miller and A. G. Chynoweth (eds.), *Optical Fiber Telecommunications,* Academic Press, New York, 1979, chap. 17.

35. D. F. Nelson, "The Modulation of Laser Light," *Scientific Am.* **218**(6):17–23 (1968).

36. E. Hartfield and B. J. Thompson, "Optical Modulators," in W. G. Driscoll and W. Vaughan (eds.), *Handbook of Optics,* McGraw-Hill, New York, 1978, chap. 17.

37. A. Ghatak and K. Thyagarajan, *Optical Electronics,* Cambridge Univ. Press, New York, 1989.

38. B. E. A. Saleh and M. C. Teich, *Fundamentals of Photonics,* Wiley, New York, 1991, chap. 18.

39. E. Hecht, *Optics,* 2d ed., Addison-Wesley, Reading, MA, 1990.

40. R. C. Alferness, in T. Tamir (ed.), *Guided-Wave Optoelectronics,* Springer-Verlag, New York, 1990, chap. 4.

41. W. H. Steier, "A Push-Pull Optical Amplitude Modulator," *IEEE J. Quantum Electron.* **QE-3**(12):664–667 (1967).

42. C. F. Buhrer, D. Baird, and E. M. Conwell, "Optical Frequency Shifting by Electro-optic Effect," *Appl. Phys. Lett.* **1**(2):46–49 (1962).

43. V. J. Fowler and J. Schlafer, "A Survey of Laser Beam Deflection Techniques," *Appl. Opt.* **5**:1675–1682 (1966).

44. F. S. Chen, et al., "Light Modulation and Beam Deflection with Potassium Tantalate Niobate Crystals," *J. Appl. Phys.* **37**:388–398 (1966).

45. V. J. Fowler, C. F. Buhrer, and L. R. Bloom, "Electro-optic Light Beam Deflector," *Proc. IEEE* (correspondence) **52**:193–194 (1964).

46. T. J. Nelson, "Digital Light Deflection," *B.S.T.J.,* 821–845 (1964).

47. W. Kulcke, et al., "A Fast, Digital-indexed Light Deflector," *IBM J.* **8**:64–67 (1964).

48. M. Gottlieb, C. L. M. Ireland, and J. M. Ley, *Electro-optic and Acousto-optic Scanning and Deflection,* Marcel Dekker, New York, 1983.

49. R. T. Denton, et al., "Lithium Tantalate Light Modulators," *Appl. Phys.* **38**(4):1611–1617 (1967).

50. F. S. Chen, "Modulators for Optical Communications," *Proc. IEEE* **58**:1440–1457 (1970).

51. D. M. Pozar, *Microwave Engineering,* Addison-Wesley, Reading, MA, 1990.

52. J. M. Hammer, in T. Tamir (ed.), "Integrated Optics," Springer-Verlag, Berlin, 1979, chap. 4, pp. 140–200.

53. R. G. Hunsperger, *Integrated Optics*: *Theory and Technology,* 2d ed., Springer-Verlag, Berlin, 1983.

54. R. Simon, *Optical Control of Microwave Devices,* Artech House, Boston, 1990.

55. R. E. Ziemer and W. H. Tranter, *Principles of Communications,* Houghton Mifflin Co., Boston, MA, 1976.

56. I. P. Kaminow and J. Liu, "Propagation Characteristics of Partially Loaded Two-conductor Transmission Line for Broadband Light Modulators," *Proc. IEEE* **51**(1): 132–136 (1963).

57. W. L. Bond "The Mathematics of the Physical Properties of Crystals," *Bell Sys. Tech. J.* **23**:1–72 (1943).

CHAPTER 14
LIQUID CRYSTALS

Shin-Tson Wu
Exploratory Studies Laboratory
Hughes Research Laboratories
Malibu, California

14.1 GLOSSARY

C_{ii}	reduced elastic constants
d	liquid crystal thickness
E	electric field
E_a	activation energy
e_c	electroclinic coefficient
F	Onsager reaction field
f_i	oscillator strength
f_c	crossover frequency
h	cavity field factor
K_{ii}	elastic constants
k	Boltzmann constant
l	aspect ratio
$n_{e,o}$	refractive indices
N	molecular packing density
P_4	order parameter of the fourth rank
P_s	spontaneous polarization
S	order parameter of the second rank
T	temperature
T_c	clearing temperature
V_n	mole volume
V	voltage
V_b	bias voltage
V_{th}	threshold voltage
Z	number of electrons
$\alpha_{l,t}$	molecular polarizability

α_l	Leslie viscosity coefficients
β	angle between dipole and molecular axis
δ	phase retardation
ε	dielectric constant
ε_o	vacuum permittivity
$\Delta\varepsilon$	dielectric anisotropy
Δn	birefringence
μ	dipole moment
γ_1	rotational viscosity
λ	wavelength
θ	twist angle
ϕ	tilt angle
ϕ_{ent}	entrance angle
ϕ_{exit}	exit angle
ϕ_s	pretilt angle
ϕ_m	maximum tilt angle
τ	response time

14.2 INTRODUCTION

Liquid crystals possess physical properties that are intermediate between conventional fluids and solids. They are fluidlike, yet the arrangement of molecules within them exhibits structural order.[1] Three types of liquid crystals have been studied extensively: (1) thermotropic, (2) lyotropic, and (3) polymeric. Among these three, the thermotropic liquid crystals have been studied extensively and their applications have reached mature stage. The potential applications are in areas such as flat panel displays,[2] light switches,[3] temperature sensors,[4] etc. Lyotropic liquid crystals are receiving increasing scientific and technological attention because of the way they reflect the unique properties of their constituent molecules.[5] Polymeric liquid crystals are potential candidates for electronic devices and ultra-high-strength materials.[6] Thermotropic liquid crystals can exist in three phases: (1) nematic, (2) cholesteric, and (3) smectic. Their structures and representative molecules are illustrated[7] in Fig. 1.

14.3 PHYSICAL PROPERTIES OF THERMOTROPIC LIQUID CRYSTALS

Optical Properties

Refractive indices and absorption are fundamentally and practically important parameters of an LC compound or mixture.[8] Almost all the light modulation mechanisms are involved with refractive index change. The absorption has a crucial impact on the photostability or lifetime of the liquid crystal devices. Both refractive indices and absorption in the visible spectral region are determined by the electronic structures of the liquid crystal studied.

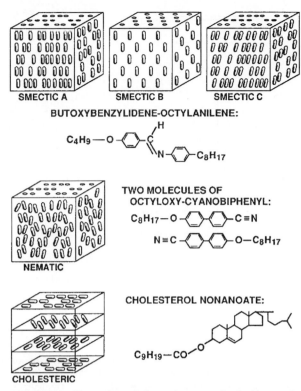

SMECTIC A   **SMECTIC B**   **SMECTIC C**

BUTOXYBENZYLIDENE-OCTYLANILENE:

C_4H_9 — O —◯— $\underset{\parallel}{\overset{H}{C}}$

— N —◯— C_8H_{17}

NEMATIC

TWO MOLECULES OF OCTYLOXY-CYANOBIPHENYL:

C_8H_{17} — O —◯—◯— $C \equiv N$

$N \equiv C$ —◯—◯— O — C_8H_{17}

CHOLESTERIC

CHOLESTEROL NONANOATE:

C_9H_{19} — CO — O

FIGURE 1 Illustration of three thermotropic liquid crystals: nematic, smectic, and cholesteric.[7]

Electronic Structures.   Three types of electronic transitions are often encountered in a liquid crystal compound. They are: (1) $\sigma \to \sigma^*$ (excited states of σ-electrons), (2) $n \to \pi^*$ (excited states of π-electrons), and (3) $\pi \to \pi^*$ transitions.

$\sigma \to \sigma^*$ *Transitions.*   The majority of liquid crystals discovered so far consist of either saturated cyclohexane rings, unsaturated phenyl rings, or a combination of both. In addition to affecting the mesogenic range, these rings also make the primary contribution to absorption and refractive indices of the liquid crystal molecule. For a cyclohexane ring, only σ-electrons are present. The $\sigma \to \sigma^*$ transitions take place at vacuum ultraviolet region ($\lambda < 180$ nm). Thus, the absorption of a totally saturated LC compound or mixture becomes negligible in the visible spectral region.

$n \to \pi^*$ *Transitions.*   In molecules containing a heteroatom (such as oxygen or nitrogen), the highest filled orbitals in the ground state are generally the nonbonding, essentially atomic n orbitals, and the lowest unfilled orbitals are π^*. Thus, the $n \to \pi^*$ transitions are the excitation mechanism for these special molecules. The oscillator strength of the $n \to \pi^*$ transition is not very strong. Thus, its contribution to refractive indices is not significant.

$\pi \to \pi^*$ *Transitions.*   The $\pi \to \pi^*$ electronic transitions of a benzene molecule have been studied extensively using the molecular-orbital theories.[9] These results can be used to explain the UV absorption of compounds containing a phenyl ring. From the Group Theory, a benzene molecule belongs to the D_{6h} point group, but a phenyl ring belongs to the C_{2v} group. Obviously, the degree of symmetry of a phenyl ring in a liquid crystal compound is lower than an unbounded benzene molecule due to the constraints from the

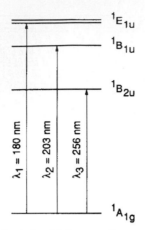

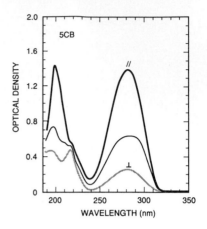

FIGURE 2 A simplified energy diagram of $\pi \rightarrow \pi^*$ electronic transitions of a benzene molecule. The transition intensity for the λ_1-band is very strong, λ_2-band is strong, and λ_3-band is weak. The λ_1-band has a twofold degeneracy.

FIGURE 3 Polarized absorption spectrum of 5CB. $\parallel$ and $\perp$ represent extraordinary and ordinary rays and the middle curve is for the unpolarized light. Sample: 1 percent 5CB dissolved in ZLI-2359. Cell thickness = 6 μm. $T_r = 0.865$.

connecting segments. If a liquid crystal molecule contains a phenyl ring, its absorption spectrum would be similar to that of a pure benzene molecule. The degree of similarity depends on how many conjugated bonds are attached to the phenyl ring.

The simplified $\pi \rightarrow \pi^*$ electronic transition diagram of a benzene molecule is shown in Fig. 2. From selection rules, the $^1A_{1g} \rightarrow {}^1E_{1u}$ is a spin- and symmetry-allowed transition. In a pure benzene molecule, this band has a twofold degeneracy and large extinction coefficient. But in a phenyl ring, this degeneracy gradually splits as the conjugation length increases. The $^1A_{1g} \rightarrow {}^1B_{1u}$ is a spin-allowed, symmetry-forbidden, but vibronically allowed transition. Its extinction coefficient is modest. Finally, the $^1A_{1g} \rightarrow {}^1B_{2u}$ is a spin-allowed but symmetry-forbidden transition. Its extinction coefficient is very weak. In the three-band model[10] for understanding the refractive index dispersion, only the two major $\pi \rightarrow \pi^*$ transitions, the $^1A_{1g} \rightarrow {}^1E_{1u}$ (designated as λ_1-band) and the $^1A_{1g} \rightarrow {}^1B_{1u}$ (λ_2-band), and the $\sigma \rightarrow \sigma^*$ transitions (λ_0-band) are included. The polarized absorption spectra of 5CB is shown in Fig. 3. Indeed, the λ_1-band consists of two closely separated bands; the averaged wavelength is $\lambda_1 \sim 210$ nm. The λ_2 is centered at 282 nm, and its dichroic ratio is large. As temperature increases, dichroic ratio decreases gradually.

Refractive Index Dispersions. Taking all three bands into consideration, the following expressions for refractive indices have been derived:[10]

$$n_e \cong 1 + g_{0e}\frac{\lambda^2\lambda_0^2}{\lambda^2 - \lambda_0^2} + g_{1e}\frac{\lambda^2\lambda_1^2}{\lambda^2 - \lambda_1^2} + g_{2e}\frac{\lambda^2\lambda_2^2}{\lambda^2 - \lambda_2^2} \tag{1a}$$

$$n_o \cong 1 + g_{0o}\frac{\lambda^2\lambda_0^2}{\lambda^2 - \lambda_0^2} + g_{1o}\frac{\lambda^2\lambda_1^2}{\lambda^2 - \lambda_1^2} + g_{2o}\frac{\lambda^2\lambda_2^2}{\lambda^2 - \lambda_2^2} \tag{1b}$$

where $g_i \sim NZ_if_i$ are proportionality constants; Z_0 is the number of responsible σ electrons and $Z_1 = Z_2$ is the number of π electrons in an LC compound. These g_i's determine the temperature effect of the refractive indices. In the off-resonant region, Eq. (1) can be expanded into the Cauchy-type formula:

$$n_{e,o} = A_{e,o} + \frac{B_{e,o}}{\lambda^2} + \frac{C_{e,o}}{\lambda^4} + \cdots \tag{2}$$

where $A_{e,o}$, $B_{e,o}$, and $C_{e,o}$ are expansion coefficients. For simplicity, we may just keep the first two terms for fitting the experimental data.

Molecular Vibrational Absorptions. In the visible spectral region, directors' fluctuation-induced light scattering surpasses the absorption loss.[11] Thus, absorption has to be measured at the isotropic state. Figure 4 shows the absorption of 5CB and MBBA at $T \sim 60°C$. At $\lambda \gg \lambda_2$, absorption decreases as λ increases. In the near IR region, some harmonics of molecular vibrational bands appear, resulting in enhanced absorption. As

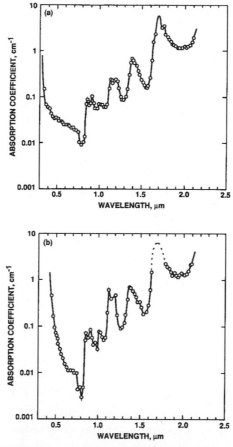

FIGURE 4 Absorption coefficient of (*a*) 5CB and (*b*) MBBA at $T \sim 60°C$ in the visible and near IR region. The clearing point of 5CB and MBBA is 35.2 and 46°C, respectively. Light scattering is neglected under this isotropic phase.

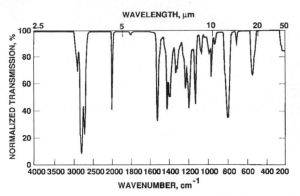

FIGURE 5 IR transmission of a 15-μm, parallel-aligned E-7 LC mixture. The truncation at the ~40-μm region is due to the KRS-5 substrates used.

wavelength extends to IR, molecular vibrational absorptions start at around 3.3 μm.[12] Figure 5 shows the normalized transmission of a ~15-μm-thick, parallel-aligned E-7 LC mixture using an unpolarized light. The truncated transmission at $\lambda \sim 40$ μm is due to the KRS-5 window, and not due to the liquid crystal. Although there exist some localized vibrational bands, there are broad regions where the transmission is reasonably high and useful electro-optic effect can be realized.

Dielectric Constants

Dielectric constants and their anisotropy affect the sharpness of the voltage-dependent optical transmission curve of an LV device and its threshold voltage. Maier and Meier[13] have developed a theory to correlate the microscopic molecular parameters with the macroscopic dielectric constants of anisotropic LCs:

$$\varepsilon_\parallel = NhF\{\langle\alpha_\parallel\rangle + (F\mu^2/3kT)[1 - (1 - 3\cos^2\beta)S]\} \tag{3a}$$

$$\varepsilon_\perp = NhF\{\langle\alpha_\perp\rangle + (F\mu^2/3kT)[1 + (1/2)(1 - 3\cos^2\beta)S]\} \tag{3b}$$

$$\Delta\varepsilon = NhF\{(\alpha_l - \alpha_t) - (F\mu^2/2kT)(1 - 3\cos^2\beta)\}S \tag{3c}$$

where N is the molecular packing density, $h = 3\varepsilon/(2\varepsilon + 1)$ is the cavity field factor, $\varepsilon = (\varepsilon_\parallel + 2\varepsilon_\perp)/3$ is the averaged dielectric constant, F is the Onsager reaction field, α_l and α_t are the principal elements of the molecular polarizability tensor, β is the angle between the dipole moment μ and the principal molecular axis, and S is the order parameter of the second rank. From Eq. (4), the dielectric constants of anisotropic liquid crystals are influenced by the molecular structure, temperature, and frequency. These individual effects are discussed separately.

Structural Effect. For a nonpolar liquid crystal compound, its dipole moment $\mu \sim 0$. Thus, its $\Delta\varepsilon$ is expected to be small and its magnitude is proportional to the differential molecular polarizability, similar to birefringence. On the other hand, for an LC molecule containing a polar group, such as the cyano,[14] isocyanate,[15] fluoro,[16] or chloro group,[17] its $\Delta\varepsilon$ can be positive or negative depending on the position(s) of the polar group(s). If $\beta \sim 0$, i.e., the dipole moment of the polar group is along the principal molecular axis, $\Delta\varepsilon$ is large and positive. Cyano-biphenyls are such examples. The $\Delta\varepsilon$ of 5CB is about 10 at $T = 20°C$ and $f = 1$ kHz. These positive $\Delta\varepsilon$ materials are useful for parallel or twist alignment. On the contrary, if $\beta > 55°$, $1 - 3\cos^2\beta > 0$ and $\Delta\varepsilon$ may become negative depending on the dipole moment as indicated in Eq. (3c). The negative $\Delta\varepsilon$ materials are useful for perpendicular alignment.

Temperature Effect. In general, as temperature rises, $\varepsilon_\parallel$ decreases, but $\varepsilon_\perp$ increases gradually resulting in a decreasing $\Delta\varepsilon$. From Eq. (3c), the temperature dependence of $\Delta\varepsilon$ is proportional to S for the nonpolar LCs and S/T for the polar LCs. At $T > T_c$, the isotropic phase is reached and dielectric anisotropy vanishes.

Frequency Effect. In an aligned LC, the molecular rotation around the short axis is strongly hindered. Thus, the frequency dispersion occurs mainly at $\varepsilon_\parallel$; $\varepsilon_\perp$ remains almost constant up to microwave region. Figure 6 shows the frequency-dependent dielectric constants of the M1 LC mixture (from Roche) at various temperatures.[18] As the frequency increases, $\varepsilon_\parallel$ decreases and beyond the crossover frequency f_c, $\Delta\varepsilon$ changes sign. The dielectric anisotropies of M1 are symmetric at low and high frequencies. The crossover frequency is sensitive to temperature. As temperature rises, the $\varepsilon_\parallel$ and $\varepsilon_\perp$ of M1 both decrease slightly. However, the frequency-dependent $\varepsilon_\parallel$ is strongly dependent on temperature, but $\varepsilon_\perp$ is inert. Thus, the crossover frequency increases exponentially with temperature; for example, $f_c \sim \exp(-E_a/kT)$; E_a is the activation energy. For M1 mixture, $E_a = 0.96$ eV.[18]

Dual frequency effect is a useful technique for improving the response times of an LC device.[19] In the dual frequency effect, a low frequency ($<f_c$ where $\Delta\varepsilon > 0$) electric field is used to drive the device to its on state, and during the decay period a high frequency ($>f_c$ where $\Delta\varepsilon < 0$) electric field is applied to speed up the relaxation time. From a material standpoint, an LC mixture with low f_c and large $\Delta\varepsilon$ at both low and high frequencies is beneficial. But for a single LC substance (such as cyano-biphenyls), its f_c is usually too high ($>10^6$ Hz) to be practically employed. In the microwave region, the LC absorption is small and birefringence relatively large.[20,21] Thus, liquid crystals are useful electro-optic media in the spectral range covering from UV, visible, and IR to microwave. Of course, in each spectral region, an appropriate LC material has to be selected.

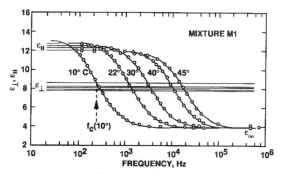

FIGURE 6 Experimental results on frequency and temperature dependences of $\varepsilon_\parallel$ and $\varepsilon_\perp$ of a Roche mixture M1. Note that the crossover frequency increases exponentially with temperature.[18]

Elastic Constants

Both threshold voltage and response time are related to the elastic constant of the LC used. Several molecular theories have been developed for correlating the Frank elastic constants with molecular constituents. Here we only introduce two theories: (1) the mean-field theory,[22,23] and (2) the generalized van der Waals theory.[24]

Mean-field Theory.   In the mean-field theory, the three elastic constants are expressed as:

$$K_{ii} = C_{ii} \cdot V_n^{-7/3} \cdot S^2 \qquad (4a)$$

$$C_{ii} = (3A/2)(L \cdot m^{-1} \cdot \gamma_{ii}^{-2})^{1/3} \qquad (4b)$$

where C_{ii} is called the reduced elastic constant; V_n is the mole volume; L is the length of the molecule; m is the number of molecules in a steric unit in order to reduce the steric hindrance; $\gamma_{11} = \gamma_{22} = z/x$ and $\gamma_{33} = (x/z)^2$ where x, $(y = x)$ and z are the averaged molecular distances in the x, y, and z directions; and $A = 1.3 \times 10^{-8}\,\text{erg} \cdot \text{cm}^6$. The temperature-dependent elastic and reduced constants for a liquid crystal PAA (4,4'-di(methoxy)-azoxybenzene) are shown in Fig. 7.

From Eq. (4), the ratio $K_{11} : K_{22} : K_{33}$ is equal to $1 : 1 : (z/x)^2$ and the temperature dependence of elastic constants is basically proportional to S^2. This S^2 dependence has been experimentally observed for many LCs. However, the prediction for the relative magnitude of K_{ii} is correct only to the first order. Experimental results indicate that K_{22} often has the lowest value, and the ratio of K_{33}/K_{11} can be either greater or less than unity. For PAA at 400°K,[25] $K_{11} : K_{22} : K_{33} = 2 : 1 : 3.2$.

Generalized van der Waals Theory.   Gelbart and Ben-Shaul[24] extended the generalized

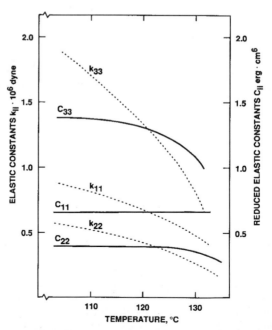

FIGURE 7   Temperature-dependent elastic and reduced elastic constants of PAA.[23]

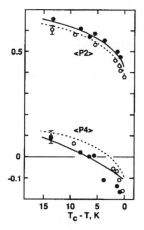

FIGURE 8 Temperature-dependent order parameters of 5CB. $P_2(\equiv S)$ and P_4 are the second- and fourth-rank order parameter, respectively. Data are obtained from the depolarization ratios of the Raman vibrational bands from the terminal $C \equiv N$ (*open circles*) and the central C-C (*filled circles*) biphenyl link of a 12-μm LC film. Solid and dashed lines are the corresponding results extrapolated to zero thickness.[26]

van der Waals theory for explaining the detailed relationship between elastic constants and molecular dimensions, polarizability, and temperature. They derived the following formula for nematic liquid crystals:

$$K_{ii} = a_i \langle P_2 \rangle \langle P_2 \rangle + b_i \langle P_2 \rangle \langle P_4 \rangle \tag{5}$$

where a_i and b_i represent sums of contributions of the energy and the entropy terms. They depend linearly on temperature; $\langle P_2 \rangle$ ($=S$) and $\langle P_4 \rangle$ are the order parameter of the second and the fourth rank, respectively. In general, the second term may not be negligible in comparison with the S^2 term depending on the value of $\langle P_4 \rangle$. Temperature-dependent $\langle P_2 \rangle$ and $\langle P_4 \rangle$ of 5CB[26] are shown in Fig. 8. As temperature increases, both S and $\langle P_4 \rangle$ decrease. The $\langle P_4 \rangle$ of 5CB changes sign at $T_c - T \approx 5°C$. The ratio of $\langle P_4 \rangle / S$ is about 15 percent at $T = 20°C$. If the $\langle P_4 \rangle$ of an LC is much smaller than S in its nematic range, Eq. (5) is reduced to the mean-field theory, or $K_{ii} \sim S^2$. The second term in Eq. (5) is responsible for the difference between K_{11} and K_{33}.

Viscosities

Rotational viscosity (γ_1) of an aligned liquid crystal represents an internal friction among LC directors during the rotation process. The magnitude of γ_1 depends on the detailed molecular constituents, structure, intermolecular association, and temperature. As temperature increases, γ_1 decreases rapidly. Rotational viscosity is an important parameter for many electro-optical applications employing liquid crystals, because the response time of the LC device is linearly proportional to γ_1. Several theories, including rigorous and semiempirical, have been developed in an attempt to account for the origin of the LC viscosity. However, owing to the complicated anisotropic attractive and steric repulsive interactions among LC molecules, these theoretical results are not yet completely satisfactory. Some models fit certain LCs, but fail to fit others.[27]

In the molecular theory developed by Osipov and Terentjev,[28] all six Leslie viscosity coefficients are expressed in terms of microscopic parameters. From the Osipov-Terentjev theory, the parameters affecting the rotational viscosity of an LC are:[28,29]

1. *Activation energy*:   An LC molecule with low activation energy leads to a low viscosity.

2. *Moment of inertia*:   An LC molecule with linear shape and low molecular weight would possess a small moment of inertia and exhibit a low viscosity.

3. *Intermolecular association*:   An LC with weak intermolecular association (e.g., not form dimer) would reduce the viscosity significantly.

4. *Temperature*:   Elevated temperature operation of an LC device may be the easiest way to lower viscosity. However, birefringence, elastic, and dielectric constants are all reduced as well.

14.4 *PHYSICAL MECHANISMS FOR MODULATING LIGHT*

Several mechanisms for modulating light using liquid crystals have been developed. These include: (1) dynamic scattering, (2) the guest-host effect, (3) field-induced nematic-cholesteric phase change, (4) field-induced director axis reorientation, (5) the laser-addressed thermal effect, and (6) light scattering by micron-sized droplets.

Dynamic Scattering

When a dc or low-frequency ac field is applied to a nematic liquid crystal cell,[30] the ionic motion due to the conductivity anisotropy induces electro-hydrodynamic flow which is coupled to the molecular alignment via viscous friction. The liquid crystal becomes turbulent and scatters light strongly. Usually a dc field of the order of 10^4 V/cm is required to generate such effects. The contrast ratio of the maximum and minimum transmitted light intensities is about 20 : 1, and the response time is about 200 ms.

For pure liquid crystal compounds, the conductivity should be very low. The observed nonnegligible conductivity of liquid crystals may originate from the impurities or dopants. Because of the finite current flow involved, the dynamic scattering mode encounters problems such as large power consumption, instability, and short lifetime. The dc field also tends to trigger undesirable electrochemical reactions among the liquid crystal molecules, thus generating more impurities and degrading their chemical stability. The dynamic scattering mode does have the advantage that it does not require a polarizer in order to modulate the light. It has been used in watches and photoaddressed light valves[31] in the early stage of liquid crystal device development, but has since been replaced by the field effect owing to the problems mentioned previously.

Guest-Host Effect

Guest-host systems[32] are formed by dissolving a few percent (1 to 5 percent, limited by the solubility) of dichroic dye in the liquid crystal. The host material should be highly transparent in the spectral region of interest. The dichroic dye molecules should have a strong absorption for one polarization and weak in another in order to enhance the contrast ratio. In the field-off state, the dye molecules are nearly parallel to the incident light polarization so that high absorption is obtained. When the liquid crystal directors are reoriented by the field, the dye molecules will also be reoriented, and the absorption is reduced.

Contrast ratio of the device employing the guest-host system is affected by the dichroic ratio and concentration of the dye molecules, and by the cell thickness. For a given cell thickness, a higher dye concentration leads to a higher contrast ratio, but the corresponding transmission is reduced and the response times are lengthened. Many dichroic dyes are available in the visible region. Due to the long conjugation of dye molecules, their viscosity is usually very large. The mixture containing merely 5 percent dyes is enough to result in a significantly larger viscosity. A typical contrast ratio of the guest-host system is 50 : 1. The guest-host effects in ferroelectric liquid crystals have also been investigated where a response time of less than 100 μs has been observed.[33]

Field-induced Nematic-Cholesteric Phase Change

Electric field-induced nematic-cholesteric phase change has been observed experimentally[34,35] and used for displays.[36] The liquid crystal is initially at the cholesteric phase where it has a helical structure whose axis is parallel to the glass substrates. The incident light is scattered and the cell appears milky white. When the applied electric field exceeds 10^5 V/cm, it will unwind the helix to give an aligned nematic phase. The cell then becomes transparent.

When the voltage is decreased, it has been observed that an intermediate metastable nematic phase could appear.[37] In this intermediate phase, the directors near the surfaces remain homeotropic but those in the bulk are slightly tilted. The optical output will therefore exhibit a hysteresis loop. This hysteresis effect can be used for optical storage. Once the voltage is removed completely, the liquid crystal directors return to their initial scattering states within a few ms.

Field-induced Director Axis Reorientation

Field-induced director axis reorientation on aligned nematic and ferroelectric liquid crystals is one of the most common electro-optic effects employed for modulating light. Many alignment methods have been developed for various applications employing nematic liquid crystals. Some examples include 90° twist,[38] homeotropic (also called perpendicular) alignment,[39] 45° twist,[40] π-cell,[41] and a variety of supertwist cells.[42–46] Each alignment exhibits its own unique features and also drawbacks.

Laser-induced Phase Change

Laser-induced phase change has been observed in cholesteric[47] and smectic-A[48] liquid crystals. In the case of a smectic-A liquid crystal, an IR laser (e.g., laser diode) is focused onto a small spot of the cell. Due to absorption, the irradiated area is heated over the smectic-nematic phase transition temperature. During the cooling process, this area may turn to the scattering or well-aligned nonscattering state, depending on the cooling rate and the applied electric field. Slow cooling permits the disordered molecules to reorganize themselves into the initial uniform alignment state which does not scatter light. Application of a sufficiently strong field during cooling also helps such a realignment process. Since the spot size of the writing laser can be as small as about 5 μm, this laser-induced phase change effect can be utilized to construct a light valve with a very high resolution. However, the laser addressing time is relatively slow; it is on the order of seconds, depending on the number of pixels involved.

Light Scattering by Micron-sized Droplets

Micron-sized liquid crystal droplets dispersed in a polymer matrix produce significant light scatterings in the visible spectral region.[49,50] The refractive index mismatch between the liquid crystal droplets and the host polymer is the physical mechanism responsible for such light scattering in the voltage-off state. At the voltage-on state, the droplets are reoriented along the field direction. The ordinary refractive index of the LC matches the index of the polymer. As a result, little scattering takes place and most of the light is transmitted. One advantage of this type of electro-optical modulation is that it does not require polarized light (i.e., no polarizer is needed). Therefore, optical efficienty is greatly enhanced.

14.5 ELECTRO-OPTICS OF NEMATIC LIQUID CRYSTALS

Field-induced Reorientational Effects

The appearance of liquid crystal molecules without alignment, such as in a bottle, is often milky. They scatter light strongly due to the fluctuations of liquid crystal clusters. A 5-mm-thick liquid crystal cluster in the mesogenic phase is enough to block out the transmission of a visible light. However, at the isotropic state, light scattering is suppressed dramatically and the liquid crystal clusters become clear. For realizing useful electro-optic effect, LC has to be aligned. Three basic alignments in nematic LCs have been developed and widely used for application. They are *parallel, perpendicular,* and *twist* alignments.

Parallel Alignments. In the parallel alignment, the directors in the front and back substrates are parallel. These substrates are first coated with an electrically conductive but optically transparent metallic film, such as indium-tin-oxide (ITO), and then deposited with a thin SiO_2 layer to create microgrooves for aligning LC molecules. The oblique-angle evaporation method[51,52] produces a high-quality alignment, well-defined pretilt angle.

Gently rubbing the ITO surface can produce microgrooves as well. The anchoring energy of the rubbed surfaces is found to be larger than the SiO_2 evaporation method.[53] Simplicity is the major advantage of the rubbing method. This technique has been widely employed for fabricating large panel LC devices. However, some problems of the rubbing technique exist, such as possible substrate contamination, creation of static charges, and variations of pretilt angle from substrate to substrate.

When a linearly polarized light impinges in a parallel-aligned cell, if the polarization axis is parallel ($\theta = 0°$) to the LC director, a pure phase modulation is achieved because the light behaves as an extraordinary ray. On the other hand, if $\theta = 45°$, then phase retardation occurs due to the different propagating speed of the extraordinary and ordinary rays in the LC medium. The phase retardation (δ) is determined by: (1) LC layer thickness d; (2) effective LC birefringence, $\Delta n(V, T, \lambda)$, which is dependent on the applied voltage, temperature, and wavelength λ of the incident light; and (3) wavelength as:

$$\delta(V, T, \lambda) = 2\pi d \, \Delta n(V, T, \lambda)/\lambda \tag{6}$$

At $V = 0$, Δn $(=n_e - n_o)$ has its maximum value; so does δ. In the $V > V_{th}$ regime, the effective birefringence decreases sharply as voltage increases, and then gradually saturates. The slope depends on the elastic constants, dielectric constants, and refractive indices of the LC material. At $V \gg V_{th}$ regime, basically all the bulk directors are aligned by the field to be perpendicular to the substrates, except the boundary layers. Thus, further increase in voltage only causes a small change in the orientation of boundary layers. In this regime, the effective birefringence is inversely proportional to the applied voltage.

The advantage of operating an LC device at low voltage regime is that a large phase change can be obtained merely with a small voltage swing. However, the directors' relaxation time is slow. On the other hand, at high voltage regime the phase change is inert to the applied voltage, but the response time is much faster.[54]

The transmission of a parallel cell is very sensitive to wavelength because the phase retardation depends strongly on wavelength and birefringence, as shown in Eq. (6). Thus, the single parallel-aligned cell is not suitable for color display using a broadband light source. A good contrast can be obtained only when a narrow laser line is used or the device is operated at high voltage regime. On the other hand, the parallel alignment is ideal for pure phase modulation and tunable phase retardation plate.

Perpendicular (or Homeotropic) Alignment. Perpendicular alignment is also known as homeotropic alignment. In such an alignment, LC directors are perpendicular to the substrate surfaces except for a small pretilt angle. Similar to parallel alignment, perpendicular alignment can be used for both pure phase modulation and phase retardation. But there exist three major differences between these two alignments. (1) For realizing useful electro-optical effects of a parallel-aligned cell, the LCs must have a positive dielectric anisotropy. But for a perpendicular alignment, LCs with negative dielectric anisotropy ($\varepsilon_\perp > \varepsilon_\parallel$) are required. (2) For achieving a stable perpendicular alignment, a certain alignment surfactant, such as DMOAP[53] or alcohol,[55] is needed. But for a parallel alignment, rubbed surfaces work well. (3) The voltage-off state of a parallel-aligned cell between crossed polarizers is dependent on the phase retardation of the cell and is normally bright unless δ happens to be $(2m + 1)\pi$. But for a perpendicular-aligned cell, this voltage-off state is dark. Moreover, this dark state is independent of LC thickness, LC birefringence, wavelength, and temperature.

Contrast ratio as high as $10^4 : 1$ for a laser beam at normal incidence has been observed in uniformly aligned perpendicular cells. For a broadband light, a contrast exceeding $200 : 1$ is easily achievable. However, the transmission of the voltage-off state slightly increases and contrast ratio decreases gradually as the incident angle tilts away from normal. A typical viewing angle of a homeotropic LC cell is about $\pm 30°$ in the horizontal direction.

Twist Alignment. In the twist alignment, both substrates are treated similarly to those of parallel alignment except that the back substrate is twisted with an angle. In order to realize the useful electro-optical effect, the positive nematic LC material has to be employed. This type of LC cell is abbreviated as a TN cell, standing for the twist nematic. For the twist angle greater than 90°, the cell bears a special name as supertwist nematic or STN. STN and TN are widely employed in the passive-matrix and active-matrix liquid crystal displays, respectively.

When a linearly polarized light traverses through a 90° TN cell, the plane of polarization follows the twist of the LC directors if the Mauguin's condition[56] is satisfied; i.e., $d\Delta n \gg \lambda$. Under this circumstance, the output beam remains linearly polarized except that its polarization axis is rotated by 90°. Thus, for a normally dark (no transmission at

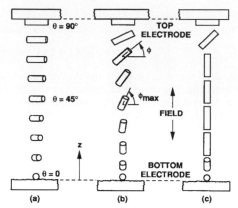

FIGURE 9 Schematic illustration of director's configuration of a 90° TN LC cell: (*a*) $V = 0$; (*b*) $V > V_{th}$; and (*c*) $V \gg V_{th}$. θ = twist angle; ϕ = tilt angle.[37]

$V = 0$) display using a 90° TN cell, the analyzer is set to be parallel to the polarizer. But when the applied voltage exceeds a threshold called optical threshold, the polarization guiding effect is interrupted and the light leaks through the analyzer.

Several twist alignments have been developed according to the twist angle, such as 45°,[40] 90°[38] twist, super birefringence effect (SBE, 270° twist),[42] optical mode interference (OMI, 180° twist),[45] and supertwist nematic with twist angle ranging from 200 to 270°.[43-46] For high-density information display applications using the multiplexing addressing technique, the sharp voltage-dependent optical transmission curve is highly desirable. The regular TN cells with twist angle smaller than 90° do not meet this challenge. Thus, several STN configurations have been developed for steepening the electro-optical transmission.

In the 90° TN cell, the director of the back surface is twisted 90° with respect to the front surface. The 90° TN cell is normally used in the transmissive mode. Figure 9 shows the directors' configurations of the 90° TN cell under three conditions: (1) $V < V_{th}$, (2) $V > V_{th}$, and (3) $V \gg V_{th}$. In Fig. 9, ϕ and θ represent the tilt and twist angles of the LC directors, respectively.[57] At a V not too far above V_{th}, the LC directors tilt collectively (but reserve the twist configuration) along the field direction. Thus, the output beam is modulated in phase only, but not in amplitude. However, as voltage further increases, the twist continuity of the LC directors is interrupted and its polarization guiding effect is broken. As a result, light leaks through the analyzer.

Optical Transmission of LC Cells

Optical transmission of an LC cell sandwiched between two polarizers is a fundamentally important subject for electro-optic application. Extensive efforts have been devoted to obtaining the analytical expressions and then optimizing the performance of various display modes. Three approaches are commonly employed: (1) the geometrical-optics approach,[58,59] (2) the 4×4 matrix method,[60] and (3) the extended Jones matrix method.[61] Both 4×4 matrix and extended Jones matrix methods are powerful in treating the oblique incidence which is particularly important for improving the viewing angle of the LC devices.

The generalized geometrical optics approach is developed for solving the optical transmission of a general TN cell with an arbitrary twist angle and for the double-layered

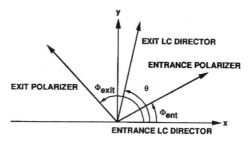

FIGURE 10 The coordinate system used in the Generalized Geometric Optics Approximation analysis for optical transmission of an LC cell.[59]

LC cell.[59,62] The geometry of an LC cell and polarizer orientation is depicted in Fig. 10. Here, a TN cell with thickness d is confined between the planes $Z_1 = 0$ and $Z_2 = d$ in a cartesian coordinate system. Both substrate surfaces are treated with alignment layers which give a uniform pretilt angle ϕ_S. ϕ_S is typically 1 to 5° for the twist angle $0 \le \theta \le 240°$, and ~20° for the 270° STN. This pretilt angle is necessary in order to avoid domain formation during director rotation. As shown in Fig. 10, the LC director at the entrance surface is directed along the x-axis and twisted by an angle θ at the exit surface. The entrance and exit polarizers are arranged at angles ϕ_{ent} and ϕ_{exit} with respect to the LC director, respectively.

After having considered the propagation of both ordinary and extraordinary rays in the LC medium, Ong[59] has derived the light transmission for an arbitrary twist LC cell under th normal incidence condition:

$$T = \cos^2(\theta - \phi_{exit} + \phi_{ent}) + \sin^2(\theta\sqrt{1 + u^2})\sin 2(\theta - \phi_{exit})\sin 2\phi_{ent}$$

$$+ \frac{1}{2\sqrt{1 + u^2}}\sin(2\theta\sqrt{1 + u^2})\sin 2(\theta - \phi_{exit} + \phi_{ent})$$

$$- \frac{1}{1 + u^2}\sin^2(\theta\sqrt{1 + u^2})\cos 2(\theta - \phi_{exit})\cos 2\phi_{ent} \quad (7)$$

where

$$u = (\pi d/\theta\lambda)(n_e/\sqrt{1 + v\sin^2\phi_s} \quad n_o) \quad (8)$$

and $v = (n_e/n_o)^2 - 1$. From this general expression, the transmission (under different polarizer-analyzer orientations) of parallel-aligned ($\theta = 0°$) TN and STN cells can be calculated.

LC Directors' Response to External Field

To solve the steady state directors' distribution, the total free energy needs to be minimized. The total free energy for nematics consists of two parts: (1) dielectric free energy which originates from the interaction between the applied field and anisotropic LC molecules, and (2) elastic free energy which originates from the elastic deformation. On the other hand, to derive the dynamic response of the LC directors, the elastic torque and electric field-induced torque need to balance with the viscous torque. Let us illustrate the detailed procedures using a parallel-aligned cell as an example. The treatment of a perpendicular cell is similar to that of a parallel one except for some parameter changes. Once the directors' distribution is obtained, the voltage-dependent phase retardation and capacitance changes can be calculated.

Directors' Distribution. Consider a parallel aligned nematic LC cell in which the directors are along the x axis and the two bounding surfaces are at $Z_1 = 0$ and $Z_2 = d$. When the applied voltage V (along the z-axis) exceeds the Freedericksz transition threshold V_{th},[63] the LC undergoes an elastic deformation. The directors then tilt in the xz plane, the amount of tilt $\phi(z)$ being a function of the distance from the aligning surface; $\phi(z)$ has a maximum value ϕ_m at $Z = d/2$ and $\phi(z) = 0$ at boundaries.

The static state directors' distribution is described by the Oseen-Frank equation:[64]

$$(K_{11} \cos^2 \phi + K_{33} \sin^2 \phi)\left(\frac{d^2\phi}{dz^2}\right) + (K_{33} - K_{11}) \sin \phi \cos \phi \left(\frac{d\phi}{dz}\right)^2 = \varepsilon_o \, \Delta\varepsilon E^2 \sin \phi \cos \phi \quad (9)$$

where K_{11}, K_{22}, and K_{33} stand for splay, twist, and bend elastic constant, respectively. After some algebra and introducing two new parameters, $\sin \phi = \sin \phi_m \sin \psi$ and $\eta = \sin^2 \phi_m$, the following important equations are obtained:[65,66]

$$\frac{V}{V_{th}} = \frac{2}{\pi} \sqrt{1 + \gamma\eta} \int_0^{\pi/2} \sqrt{\frac{1 + \kappa\eta \sin^2 \psi}{(1 + \gamma\eta \sin^2 \psi)(1 - \eta \sin^2 \psi)}} \, d\psi \quad (10)$$

and

$$\frac{2z}{d} \int_0^{\pi/2} \left[\frac{(1 + \kappa\eta \sin^2 \psi)(1 + \gamma\eta \sin^2 \psi)}{1 - \eta \sin^2 \psi}\right]^{1/2} d\psi$$

$$= \int^{\sin^{-1}(\sin\phi/\sqrt{\eta})} \left[\frac{(1 + \kappa\eta \sin^2 \psi)(1 + \gamma\eta \sin^2 \psi)}{1 - \eta \sin^2 \psi}\right] d\psi \quad (11)$$

where V_{th} is the threshold voltage given by $V_{th} = \pi[K_{11}/\varepsilon_o \, \Delta\varepsilon]^{1/2}$, $\gamma = \varepsilon_\parallel/\varepsilon_\perp - 1$, and $\kappa = K_{33}/K_{11} - 1$.

In principle, knowing the liquid crystal material constants, one can use Eq. (10) in an iterative fashion to evaluate ϕ_m for a given applied voltage. Once ϕ_m is obtained, the complete profile of $\phi(z)$ can be calculated from Eq. (11). Figure 11 shows the voltage-dependent directors' distribution of a parallel-aligned LC cell. At $V = 2 \, V_{th}$, $\phi_m \sim 60°$. But at $V = 4V_{th}$, $\phi_m > 85°$. Thus, voltage-dependent phase change is sensitive at

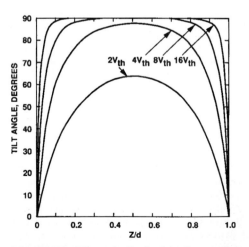

FIGURE 11 Director's distribution of a parallel-aligned LC cell for various normalized voltage (V/V_{th}).[66]

low voltage regime and gradually saturates as voltage increases. At high voltage regime, the central layers all reach their maximum tilt angle; only the surface layers tilt further. Thus, a small phase change is obtained with a large voltage increase.

Dynamic Response. In a static deformation of LC directors, the Oseen-Frank equation describes the balance between an elastic torque and an electric torque exerted by the applied field. But in the dynamic response, the viscous torque which opposes the directors' rotation has to be included. The most general treatment of the dynamic of LC directors is described by the Erickson-Leslie equation:[67]

$$\frac{\partial}{\partial z}\left[(K_{11}\cos^2\phi + K_{33}\sin^2\phi)\frac{\partial\phi}{\partial z}\right] + (K_{33} - K_{11})\sin\phi\cos\phi\left(\frac{\partial\phi}{\partial z}\right)^2$$

$$+ (\alpha_2\sin^2\phi - \alpha_3\cos^2\phi)\frac{\partial v}{\partial z} + \varepsilon_0\Delta\varepsilon E^2\sin\phi\cos\phi = \gamma_1\frac{\partial\phi}{\partial t} + I\frac{\partial^2\phi}{\partial t^2} \quad (12)$$

where ϕ is the deformation angle, $\alpha_{2,3}$ are Leslie viscosity coefficients, v is the flow velocity, $\gamma_1 = \alpha_3 - \alpha_2$ is the rotational viscosity, and I is the inertia of the LC directors. The Erickson-Leslie equation can be applied to both parallel, perpendicular, and twist alignments. We will deal with the simplest case, which is parallel alignment.

Neglecting the backflow and inertial effects in Eq. (12), the dynamic response of parallel-aligned LC directors is described as:

$$(K_{11}\cos^2\phi + K_{33}\sin^2\phi)\frac{\partial^2\phi}{\partial z^2} + (K_{33} - K_{11})\sin\phi\cos\phi\left(\frac{\partial\phi}{\partial z}\right)^2$$

$$+ \varepsilon_o\Delta\varepsilon E^2\sin\phi\cos\phi = \gamma_1\frac{\partial\phi}{\partial t} \quad (13)$$

The rise and decay times of the LC directors are described as:

$$\tau_{\text{rise}} = \tau_o/[(V/V_{th})^2 - 1] \quad (14)$$

$$\tau_{\text{decay}} = \tau_o/|(V_b/V_{th})^2 - 1| \quad (15)$$

where $\tau_o = \gamma_1 d^2/K_{11}\pi^2$ is the free relaxation time, V_b is the bias voltage of the LC cell, or final state of the relaxation. The absolute value in the denominator indicates that this formula is valid no matter if V_b is above or below V_{th}. It should be mentioned that the response times appearing in Eqs. (26) and (27) are directors' responses to the external field, but not the optical responses. Several methods[8] have been developed for improving the response times of nematic LC devices. Among them, dual frequency effect, bias voltage effect, transient nematic effect, temperature effect, Fabry-Perot effect, and molecular engineering method have been proven effective.

14.6 *ELECTRO-OPTICS OF POLYMER-DISPERSED LIQUID CRYSTALS*

A simple diagram illustrating the electro-optical effect of a polymer-dispersed liquid crystal (PDLC)[49,50] shutter is shown in Figs. 12a and 12b for the off and on states, respectively. Suppose the polymer matrix material is optically isotropic and has refractive index n_p. The LC directors within the droplets are determined by the polymer-LC interaction at each droplet boundary. They have no preferred orientation but vary nearly randomly from droplet to droplet in the absence of external field. The index mismatch $(n_e > n_p)$ between the LC droplets and the host polymer results in light scattering. Because the cell thickness

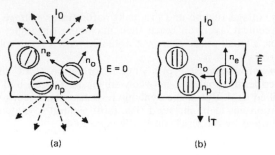

FIGURE 12 Light transmission characteristics of a normal-mode PDLC film at (*a*) voltage-off and (*b*) voltage-on state. n_e and n_o are the refractive indices of the LC droplets, and $n_p \sim n_o$ is the index of the isotropic polymer matrix.[50]

($\sim 10 \ \mu$m) is much larger than the droplet size ($\sim 0.5 \ \mu$m), the incident light will be scattered many times before emerging from the film. Thus, little light is transmitted through the cell. The degree of the off-state scattering depends on the size, birefringence, and concentration of the liquid crystalline droplets, and the film thickness. When the applied field is sufficiently strong, the directors in the droplets are reoriented along the field direction. Its effective refractive index is equal to n_o which is similar to the index of the polymer, n_p. Thus, the normally incident light acts as the ordinary ray and transmits through the cell. The PDLC film becomes clear. This mode of operation is called normal mode. Reversed mode operation has been demonstrated recently using nematic liquid crystals[68] and gels containing cholesteric liquid crystals.[69]

Physical Mechanisms

Optical transmission of a nonabsorbing PDLC film depends on the scattering properties of the micron-sized droplets and their distribution in space. The light scattering induced by the micron-sized nematic droplets is much more significant than that from the directors' fluctuation. To find an exact solution for the optical transmission of a PDLC system is a difficult task because the droplet size is comparable to the wavelength in the visible region. Thus, a certain approximation[70] (depending on the relative dimension of the droplet size as compared to the wavelength of the incident light) has to be taken. If the object size is much smaller than the wavelength and its optical anisotropy is not very large, the Rayleigh-Gans approximation[71] can be employed and the total scattering cross section is found to be proportional to λ^{-4}. On the other hand, if the object size is larger than the wavelength, the anomalous diffraction approach[72] can be used and the total cross section is proportional to λ^{-2}. If the object size is very large, the geometrical optics approach[73] can be taken.

Dynamic Response of PDLC

The dynamic response of a PDLC film has been solved by Wu et al.,[74] and the rise and decay times are shown as follows:

$$\tau_{\text{on}} = \frac{\gamma_1}{\varepsilon_o \, \Delta\varepsilon E^2 + K(l^2 - 1)/a^2} \tag{16}$$

$$\tau_{\text{off}} = \frac{\gamma_1 a^2}{K(l^2 - 1)} \tag{17}$$

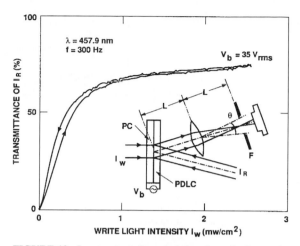

FIGURE 13 Input-output characteristics of a reflective-mode, photoactivated PDLC light valve. The photoconductor used is 500-μm-thick BSO. The PDLC film thickness is 20 μm and droplet size of about 1 μm.[76] (*Courtesy of Dr. H. Kikuchi of NHK, Japan.*)

The rise time can be shortened by increasing the applied voltage. However, the relaxation time is proportional to the viscosity, droplet size a, and effective elastic constant K in the same way as a nematic LC cell, except that the aspect ratio also plays an important role. For a PDLC that consists of E-7 LC droplets, a slightly shaped droplet of $l = 1.1$ leads to $\tau_{off} \sim 20$ ms. Equation (17) predicts that the more elongated LC droplets yield faster rise and decay times. A simple way to obtain elongated droplets is to shear the PDLC film during droplet formation.[75]

Photoactivated PDLC Light Valve

A reflective-mode, photoactivated PDLC light valve has been demonstrated[76] and results are shown in Fig. 13. This device is particularly useful for projection TV display. The photoconductor used is BSO ($Bi_{12}SiO_{20}$), whose photosensitivity is in the visible region.[77] In the beginning, an external voltage $V_b = 35\ V_{rms}$ is applied to the PDLC light valve. Due to the high impedance of the photoconductor, little voltage drops across the PDLC layer. The PDLC is in its scattering state; very little light is transmitted to the detector. As the photoconductor is illuminated by a wavelength ($\lambda = 457.9$ nm from an Ar^+ laser) which is within the bandgap of BSO, the photogenerated carriers cause the impedance of the photoconductor to decrease. As a result, more voltage drops across the PDLC film, which increases the transmittance of the readout light. The response time of the PDLC light valve is about 30 ms. Note that this light-scattering mechanism does not require a polarizer and, therefore, high brightness can be obtained, which is important for large screen display.

14.7 ELECTRO-OPTICS OF FERROELECTRIC LIQUID CRYSTALS

Surface-stabilized FLCs

The bookshelf geometry of the surface-stabilized ferroelectric liquid crystal cell[78,79] is sketched in Fig. 14. The molecules of Sm-C* LC form a layered structure located in the

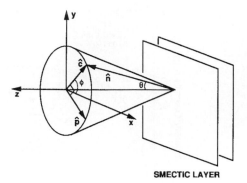

FIGURE 14   The coordinate system used for describing the dynamic response of a surface-stabilized FLC cell.[79]

x-y plane. The LC director $\mathbf{n}$ tilts away from the layer normal (z axis) by a constant angle θ. The projection of $\mathbf{n}$ onto the layer plane is defined as the C-director and ϕ is the azimuthal angle between the C-director and the x axis. The spontaneous polarization P_s is located in the x-y plane but perpendicular to both $\mathbf{n}$- and C-directors. The electric field is applied along the y axis and the sample is assumed to be uniform in the x and z axes without any deformation.

Using the similar treatment of free energy, Nakagawa et al.[80] have derived the dynamic response for the surface-stabilized ferroelectric LC:

$$\gamma_1 \frac{\partial \phi}{\partial t} = A(1 + v \sin^2 \phi)\left(\frac{\partial^2 \phi}{\partial y^2}\right) + \frac{v}{2} A \sin 2\phi \left(\frac{\partial \phi}{\partial y}\right)^2 + \frac{e}{2} \varepsilon_\perp E^2 \sin 2\phi + P_s E \sin \phi \qquad (18)$$

where $A = K_{11} \sin^2 \theta$, $v = (K_{22}/K_{11}) \cos^2 \theta + (K_{33}/K_{11}) \sin^2 \theta - 1$, and $e = (\varepsilon_\parallel/\varepsilon_\perp - 1) \sin^2 \theta$.

In the one elastic constant approximation ($K_{11} = K_{22} = K_{33} = K$, so that $v = 0$), Eq. (18) is greatly simplified:

$$K \sin^2 \theta \frac{\partial^2 \phi}{\partial y^2} + \frac{e}{2} \varepsilon_\perp E^2 \sin 2\phi + P_s E \sin \phi = \gamma_1 \frac{\partial \phi}{\partial t} \qquad (19)$$

For comparison, the magnitude of each term is estimated as follows. For a typical FLC material and device, $K \sim 10^{-11} N$, $\theta \sim 20°$, $\varepsilon_\perp \sim 10^{-11} F/m$, $E \sim 10^6 V/m$, and $P_s \sim 10^{-5} C/m^2$. The value e can be positive or negative depending on the dielectric anisotropy of the LC. If an FLC has its $\varepsilon_\parallel \sim \varepsilon_\perp$, then $e \sim 0$ and the dielectric (second) term in Eq. (19) can be neglected. Under this condition, Eq. (19) is reduced to that derived by Yamada et al.[81] On the other hand, if e is not too small, the magnitude of the dielectric term may be comparable to that of the spontaneous polarization (third) term. The elastic term becomes the smallest among the three and may be neglected. Under this assumption, Eq. (19) is then reduced to that derived by Xue et al.:[82,83]

$$\tfrac{1}{2}\Delta \varepsilon E^2 \sin 2\phi + P_s E \sin \phi = \gamma_1 \frac{\partial \phi}{\partial t} \qquad (20)$$

Xue et al.[83] derived the following analytical solution for Eq. (20):

$$\frac{t}{\tau} = \frac{1}{1 - a^2}\left\{\ln \frac{\tan (\phi/2)}{\tan (\phi_o/2)} + a \ln \frac{(1 + a \cos \phi) \sin \phi_o}{(1 + a \cos \phi_o) \sin \phi}\right\} \qquad (21)$$

where $\tau = \gamma_1/P_s E$, $\phi_o = \phi(t = 0)$, and $a = \Delta \varepsilon E \sin^2 \theta/(2P_s)$.

   In practical application, an SSFLC film is arranged such that the input light polarization is parallel to the director which is one of the bistable states—say, "up" state. The analyzer is crossed with respect to the polarizer. Thus, no light is transmitted and a good dark state is obtained. This dark state is independent of the incident wavelength. When the sign of the electric field is reversed, the director is changed from the up state to the down state (the total rotation angle is 2θ), and some light leaks through the analyzer. The normalized transmission is proportional to the tilt angle θ and phase retardation δ as

$$T_\perp = \sin^2(4\theta)\sin^2(\delta/2) \tag{22}$$

As seen from Eq. (22), the optimal transmission occurs at $\theta = 22.5°$ and $\delta = \pi$; i.e., the FLC film acts like a half-wave phase retardation plate. In order to achieve a uniform rotation, the SSFLC layer thickness is often limited to about $2\,\mu m$. This bright state transmission is dependent on the cell thickness, wavelength, and the birefringence of the FLC mixture employed. The response time of such an FLC cell is proportional to γ_1/P_sE and ranges from 1 to $100\,\mu s$. The lower rotational viscosity, larger spontaneous polarization, and higher switching field are helpful in improving the response times of the FLC devices.

Deformed Helix Ferroelectric (DHF) Effect

   The director configuration of the DHF LC cell is shown in Fig. 15. The applied dc electric voltage deforms the helix of the FLC.[84–89] As a result, the effective birefringence of the cell is changed and the transmitted light modulated. The DHF effect is complementary to the SSF effect in many ways. The basic features of the DHF effect are summarized as follows: (1) it uses short ($\sim 0.35\,\mu m$) pitch mixtures; (2) its surface anchoring energy is kept weak, just enough to generate the so-called "bookshelf structure"; (3) the field-induced tilt angle is linear to the applied voltage; (4) it has no inherent bistability; (5) it exhibits no optical threshold voltage; (6) it possesses greyscale; and (7) it operates at much lower voltage, and with response time shorter than the SSFLC devices. However, two problems remain to be overcome before the DHF effect can be used for displays: (1) the hysteresis effect, and (2) the relatively low contrast ratio. The hysteresis effect is more pronounced at higher voltages. This hysteresis makes the precise control of greyscale difficult. The contrast depends strongly on how the cell is prepared. The sheared cell seems to provide a reasonably good LC alignment. A contrast ratio of 40 : 1 has been reported.[85]

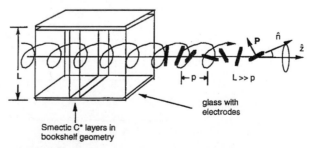

FIGURE 15 Cell configuration of the deformed helix effect. The incident light is normal to the cell. L = cell gap; p = helical pitch; Z = helical direction; n = long axis of the FLC; and P = spontaneous polarization vector.[89]

Soft-mode FLCs

There are similarities and differences between the soft-mode[90] and the surface-stabilized-mode ferroelectric effects. The similarities include the following. (1) Both effects are linear to the applied electric field. Thus, they are sensitive to the polarity of the applied field. This is different from the nematic, which is sensitive to the square of the field. (2) Both effects have fast response times. The major differences can be found in the following areas: (1) SSFLCs use changes in the azimuthal angle ϕ around the tilt cone; SMFLCs use changes in tilt θ. For SSFLC, θ remains constant and ϕ varies (called Goldstone mode). But for SMFLC, ϕ remains constant and θ varies (called soft mode). In general, the electric field-induced deflection angle in SMFLCs is small so that the associated optical change is small. (2) SSFLCs exhibit bistability, but SMFLCs show continuously controllable intensity change. (3) The SMFLCs employ smectic A* phase, but SSFLCs use smectic C* phase. Thus, the uniform alignment is much easier to obtain for SMFLCs than those for SSFLCs. The LC layer thickness in SSFLCs is usually limited to 2 μm in order to achieve uniform surface-stabilized states. But in SMFLCs, this requirement is greatly relieved. Good alignment can still be obtained even if the LC layer is as thick as 10 μm.

The experimental apparatus for realizing the soft-mode FLC effect is sketched as shown in Fig. 16. In the small angle approximation, the tilt angle θ is obtained[91] as

$$\theta = e_c E \tag{23a}$$

$$e_c = \frac{\mu}{\alpha(T - T_c)} \tag{23b}$$

where e_c is called the electroclinic coefficient, μ is the structure coefficient which is

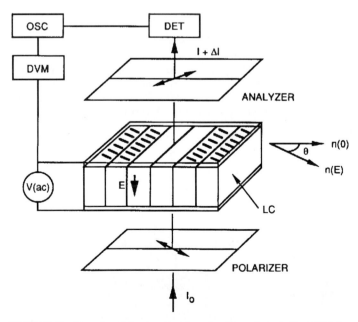

FIGURE 16   Experimental setup for observing the soft-mode ferroelectric LC effect.[90]

equivalent to the dipole moment per unit volume for unit tilt angle, and α is a proportionality constant. From Eq. (23*b*), e_c diverges when T approaches the Sm-A* to Sm-C* transition. At a given temperature, the field-induced tilt angle is linearly proportional to the applied electric field.

The decay time of the SMFLC cell has been derived[91] and expressed as

$$\tau = \frac{\gamma_\theta}{\alpha(T - T_c)} \tag{24}$$

To the first-order approximation, the response time is independent of the applied electric field. This prediction is also validated experimentally.[91] However, at near the smectic A-C* transition, the response time is dependent on temperature and the applied field.[92] As temperature approaches T_c, $\tau \sim E^{-2X}$ with $1/3 \leq X \leq 1$. Near transition, τ decreases rapidly as E increases. Far from the Sm A Sm-c* transition, τ is weakly dependent on E.

Response time of about 1 μs has been demonstrated although the modulation efficiency is low due to the small tilt angle. In the smectic-A* phase, it is difficult to obtain an induced tilt angle much greater than 10° unless one works near the phase transition. In this regime, the response time is comparatively slow and strongly dependent on the temperature. From Eq. (22), the maximum optical transmission for $\theta = 10°$ is calculated to be about 40 percent. One way to improve the optical transmission efficiency is to use two or more SMFLC cells in series. For example, if one arranges two electroclinic cells together and operates with reversed voltages, one can achieve a 100 percent modulation efficiency with tilt angle $\theta = 11.25°$ and phase retardation $\delta = \pi$. Under these circumstances, the response times remain the same as those of the single cell, but the optical modulation is greatly enhanced.

14.8 CONCLUSION

Liquid crystals are a unique electro-optic medium with spectral transmission spanning from UV, through visible and infrared to the microwave region. In the UV region, some electronic transitions take place. These absorptions determine the refractive index dispersions of the liquid crystals. Various physical mechanisms and devices, such as the dynamic scattering mode, the guest-host effect, the field-induced cholesteric-nematic phase transition, the field-induced director reorientation effect, light scattering induced by the micron-sized liquid crystal droplets, and the thermo-optic effect, have been developed for modulating light. The response times of the nematics are typically in the 10- to 100-ms range. For high-information-content displays using nematics, the device response time is marginal and further improvement is needed.

To achieve a faster response time, ferroelectric liquid crystals can be considered. Several electro-optic effects based on the ferroelectric liquid crystals have been developed. Among them, the surface-stabilization effect, deformed helix effect, and electroclinic effect hold great promise. The surface-stabilized FLCs exhibit bistability and fast response times. Thus, they are suitable for flat panel displays which don't require greyscale. On the other hand, both electronic effects and deformed helix exhibit greyscale and fast response times. The low transmission of the electroclinic FLC cell can be overcome by using two cells in series. However, the electroclinic coefficient is quite sensitive to temperature. Thus, the operation temperature of the device needs to be controlled precisely. Although the deformed helix effect is promising for display application, its hysteresis effect needs to be minimized and its contrast ratio improved.

14.9 REFERENCES

1. P. G. de Gennes, *The Physics of Liquid Crystals,* Clarendon, Oxford, 1974.

2. E. Kaneko, *Liquid Crystal TV Displays*: *Principles and Applications of Liquid Crystal Displays,* KTK Scientific Publishers, Tokyo, 1987.

3. L. M. Blinov, *Electro-optical and Magneto-optical Properties of Liquid Crystals.* Wiley & Sons, New York, 1983.

4. C. W. Smith, D. G. Gisser, M. Young, and S. R. Powers, Jr., *Appl. Phys. Lett.* **24:**453 (1974).

5. For a review, see P. S. Pershan, *Physics Today,* May 1982, p. 34.

6. For a review, see E. T. Samulski, *Physics Today,* May 1982, p. 40.

7. J. D. Lister and R. J. Birgeneau, *Physics Today,* May 1982, p. 26.

8. I. C. Khoo and S. T. Wu, *Optics and Nonlinear Optics of Liquid Crystals,* World Scientific, Singapore, 1993.

9. H. H. Jaffe and M. Orchin, *Theory and Applications of Ultraviolet Spectroscopy,* Wiley and Sons, New York, 1962.

10. S. T. Wu, *Phys. Rev.* **A33:**1270 (1986); Also, *J. Appl. Phys.* **69:**2080 (1991).

11. S. T. Wu and K. C. Lim, *Appl. Opt.* **26:**1722 (1987).

12. S. T. Wu, U. Efron, and L. D. Hess, *Appl. Phys. Lett.* **44:**1033 (1984).

13. W. Maier and G. Meier, *Z. Naturforsh,* Teil A **16:**262 (1961).

14. G. W. Gray, K. J. Harrison, and J. A. Nash, *Electron. Lett.* **9:**130 (1973).

15. M. Schadt, R. Buchecker, A. Villiger, F. Leenhouts, and J. Fromm, *IEEE Trans. Electron Devices* **33:**1187 (1986).

16. V. Reiffenrath, U. Finkenzeller, E. Poetsch, B. Rieger, and D. Coates, *SPIE* **1257:**84 (1990).

17. S. T. Wu, D. Coates, and E. Bartmann, *Liq. Cryst.* **10:**635 (1991).

18. M. Schadt, *Mol. Cryst. Liq. Cryst.* **89:**77 (1982).

19. H. K. Bucher, R. T. Klingbiel, and J. P. VanMeter, *Appl. Phys. Lett.* **25:**186 (1974).

20. D. Lippens, J. P. Parneix, and A. J. Chapoton, *J. Phys.* (Paris) **38:**1465 (1977).

21. T. K. Bose, B. Campbell, and S. Yagihara, *Phys. Rev. A* **36:**5767 (1987).

22. W. Maier and A. Saupe, *Z. Naturforsch.,* Teil A **15:**287 (1960).

23. H. Gruler, *Z. Naturforsch,* Teil A **30:**230 (1975).

24. W. M. Gelbart and A. Ben-Shaul, *J. Chem. Phys.* **77:**916 (1982).

25. W. H. de Jeu, "The Dielectric Permittivity of Liquid Crystals," *Solid State Phys.,* suppl. **14:** "Liquid Crystals", L. Liebert (ed.), Academic Press, New York, 1978; also, *Mol. Cryst. Liq. Cryst.* **63:**83 (1981).

26. K. Miyano, *J. Chem. Phys.* **69:**4807 (1978).

27. S. T. Wu and C. S. Wu, *Liq. Cryst.* **8:**171 (1990). Seven commonly used models (see the references therein) have been compared in this paper.

28. M. A. Osipov and E. M. Terentjev, *Z. Naturforsch.,* Teil A **44:**785 (1989).

29. S. T. Wu and C. S. Wu, *Phys. Rev. A* **42:**2219 (1990).

30. G. H. Heilmeier, L. A. Zanoni, and L. A. Barton, *Proc. IEEE* **56:**1162 (1968); also, *Appl. Phys. Lett.* **13:**46 (1968).

31. J. D. Margerum, J. Nimoy, and S.-Y. Wong, *Appl. Phys. Lett.* **17:**51 (1970).

32. G. H. Heilmeier and L. A. Zanoni, *Appl. Phys. Lett.* **13:**91 (1968).

33. H. J. Coles, H. F. Gleeson, and J. S. Kang, *Liq. Cryst.* **5:**1243 (1989).

34. J. J. Wysocki, J. Adams, and W. Haas, *Phys. Rev. Lett.* **20:**1024 (1968).

35. G. H. Heilmeier and J. E. Goldmacher, *Appl. Phys. Lett.* **13:**132 (1968).

36. A. Mochizuki, G. Gondo, T. Watanula, K. Saito, K. Ikegami, and H. Okuyama, *SID Technical Digest,* **16:**135 (1985).

37. C. G. Lin-Hendel, *Appl. Phys. Lett.* **38:**615 (1981); also, *J. Appl. Phys.* **53:**916 (1982).

38. M. Schadt and W. Helfrich, *Appl. Phys. Lett.* **18:**127 (1971).

39. M. F. Schiekel and K. Fahrenschon, *Appl. Phys. Lett.* **19:**391 (1971).

40. J. Grinberg, A. Jacobson, W. P. Bleha, L. Miller, L. Fraas, D. Bosewell, and G. Meyer, *Opt. Eng.* **14:**217 (1975).

41. P. J. Bos and K. R. Koehler/Beran, *Mol. Cryst. Liq. Cryst.* **113:**329 (1984).

42. T. J. Scheffer and J. Nehring, *Appl. Phys. Lett.* **45:**1021 (1984); also *J. Appl. Phys.* **58:**3022 (1985).

43. C. M. Waters, E. P. Raynes, and V. Brimmell, *Mol. Cryst. Liq. Cryst.* **123:**303 (1985).

44. K. Kinugawa, Y. Kando, M. Kanasaki, H. Kawakami, and E. Kaneko, *SID Digest* **17:**122 (1986).

45. M. Schadt and F. Leenhouts, *Appl. Phys. Lett.* **50:**236 (1987).

46. K. Kawasaki, K. Yamada, R. Watanabe, and K. Mizunoya, *SID Digest* **18:**391 (1987).

47. R. A. Soref, *J. Appl. Phys.* **41:**3022 (1970); also, A. Sasaki, K. Kurahashi, and T. Takagi, *J. Appl. Phys.* **45:**4356 (1974).

48. F. J. Kahn, *Appl. Phys. Lett.* **22:**111 (1973).

49. J. L. Fergason, *SID Digest* **16:**68 (1985).

50. J. W. Doane, N. A. Vaz, B. G. Wu, and S. Žumer, *Appl. Phys. Lett.* **48:**269 (1986).

51. J. L. Janning, *Appl. Phys. Lett.* **21:**173 (1972).

52. D. Meyerhofer, *Appl. Phys. Lett.* **29:**691 (1976).

53. J. Cognard, *Mol. Cryst. Liq. Cryst.,* Suppl. **1:**1 (1982).

54. P. D. Berezin, L. M. Blinov, I. N. Kompanets, and V. V. Nikitin, *Sov. J. Quantum Electron.* **3:**78 (1973).

55. A. M. Lackner, J. D. Margerum, L. J. Miller, and W. H. Smith, Jr., *Proc. SID.* **31:**321 (1990).

56. M. C. Mauguin, *Bull. Soc. Franc. Miner. Crist.* **34:**71 (1911).

57. D. W. Berreman, *Appl. Phys. Lett.* **25:**12 (1974).

58. E. P. Raynes, *Mol. Cryst. Liq. Cryst. Lett.* **4:**1 (1986).

59. H. L. Ong, *J. Appl. Phys.* **64:**614 (1988).

60. For example, H. Wöhler, G. Haas, M. Fritsch, and D. A. Mlynski, *J. Opt. Soc. Am.* A **5:**1554 (1988), and references therein.

61. A. Lien, *Appl. Phys. Lett.* **57:**2767 (1990); also, *SID Digest* **22:**586 (1991).

62. H. L. Ong, *J. Appl. Phys.* **64:**4867 (1988).

63. V. Freedericksz and V. Zolina, *Trans. Faraday Soc.* **29:**919 (1933).

64. C. W. Oseen, *Trans. Faraday Soc.* **29:**883 (1933); F. C. Frank, *Discuss. Faraday Soc.* **25:**19 (1958).

65. H. J. Deuling, "Elasticity of Nematic Liquid Crystals," *Solid State Phys.,* suppl. **14:** "Liquid Crystals", L. Liebert (ed.), Academic Press, New York, 1978.

66. K. R. Welford and J. R. Sambles, *Mol. Cryst. Liq. Cryst.* **147:**25 (1987).

67. J. L. Erickson, *Trans. Soc. Rheol.* **5:**23 (1961); F. M. Leslie, *Arch. Ration. Mechan. Anal.* **28:**265 (1968).

68. Y.-D. Ma, B.-G. Wu, and G. Xu, *Proc. SPIE,* **1257:**46 (1990).

69. D. K. Yang, L. C. Chien, and J. W. Doane, *Appl. Phys. Lett.* **60:**3102 (1992).

70. H. C. Van de Hulst, *Light Scattering by Small Particles,* Wiley, New York, 1957.

71. S. Žumer, A. Golemme, and J. W. Doane, *J. Opt. Soc. Am.* A **6:**403 (1989).

72. S. Žumer, *Phys. Rev.* A **37:**4006 (1988).

73. R. D. Sherman, *Phys. Rev.* A **40:**1591 (1989).

74. B. G. Wu, J. H. Erdmann, and J. W. Doane, *Liq. Cryst.* **5:**1453 (1989).

75. J. D. Margerum, A. M. Lackner, J. H. Erdmann, and E. Sherman, *Proc. SPIE* **1455:**27 (1981).

76. K. Takizawa, H. Kikuchi, H. Fujikake, Y. Namikawa and K. Tada, *Opt. Eng.* **32:**1781 (1993).

77. K. Tada, Y. Kuhara, M. Tatsumi, and T. Yamaguchi, *Appl. Opt.* **21:**2953 (1982).

78. N. A. Clark and S. T. Lagerwall, *Appl. Phys. Lett.* **36:**899 (1980).

79. Y. Ouchi, H. Takezoe, and A. Fukuda, *Jpn. J. Appl. Phys.* **26:**1 (1987).

80. M. Nakagawa, M. Ishikawa, and T. Akahane, *Jpn. J. Appl. Phys.* **27:**456 (1988).

81. Y. Yamada, T. Tsuge, N. Yamamoto, M. Yamawaki, H. Orihara, and Y. Ishibashi, *Jpn. J. Appl. Phys.* **26:**1811 (1987).

82. J. Z. Xue, M. A. Handschy, and N. A. Clark, *Liq. Cryst.* **2:**707 (1987).

83. J. Z. Xue, M. A. Handschy, and N. A. Clark, *Ferroelectrics* **73:**305 (1987).

84. B. I. Ostrovskii and V. G. Chigrinov, *Sov. Phys. Crystallogr.* **25:**322 (1980).

85. J. Funfschilling and M. Schadt, *J. Appl. Phys.* **66:**3877 (1989).

86. V. G. Chigrinov, V. A. Balkalov, E. P. Pozhidaev, L. M. Blinov, L. A. Beresnev, and A. I. Allagulov, *Sov. Phys. JETP* **61:**1193 (1985).

87. L. A. Beresnev, L. M. Blinov, and D. I. Dergachev, *Ferroelectrics* **85:**173 (1988).

88. L. A. Beresnev, V. G. Chigrinov, D. I. Dergachev, E. P. Poshidaev, J. Funfschilling, and M. Schadt, *Liq. Cryst.* **5:**1171 (1989).

89. M. D. Wand, R. Vohra, M. O'Callaghan, B. Roberts, and C. Escher, *Proc. SPIE* **1665:**176 (1992).

90. G. Andersson, I. Dahl, P. Keller, W. Kucynski, S. T. Lagerwall, K. Skarp, and B. Stebler, *Appl. Phys. Lett.* **51:**640 (1987); also, *J. Appl. Phys.* **66:**4983 (1989).

91. G. Andersson, I. Dahl, W. Kuczynski, S. T. Lagerwall, and K. Skarp, *Ferroelectrics* **84:**285 (1988).

92. S.-D. Lee and J. S. Patel, *Appl. Phys. Lett.* **55:**122 (1989).

OPTICAL INSTRUMENTS

CHAPTER 15
CAMERAS

Norman Goldberg
Madison, Wisconsin

15.1   INTRODUCTION

Thanks to technical progress and vigorous competition, the camera buyer faces a difficult challenge in making a choice. This chapter will attempt to reduce the difficulty by asking the buyer to consider the final image; its purpose, its audience, and its appearance.

Next, some of the more recent technical features are discussed. These include the intriguing ability to select objects in a scene for focus and/or exposure measurement by tracking the position of the user's eye. Finally, various types of cameras and their accessories are described.

In terms of technical sophistication, a moderately priced 35 mm snapshot camera made today would astonish a photographer who was suddenly time shifted from the 1950's. Consider the automation of exposure, focus, film loading, winding, rewinding, plus flash exposures from a tiny integral electronic flash unit no bigger than a spare roll of film.

The net result, for the snapshooter, is a higher percentage of "good" pictures per roll of film than ever before. The specialist also profits, particularly when the basis and limits of the feature are understood.

A good share of these technical features have been incorporated in the more advanced cameras; sometimes just because it can be done. Looking beyond this, the most basic technical camera ever made, the view camera, remains virtually unchanged for the past century. It is to photography what the wooden match is to fire making.

Portions of this chapter are adapted from the author's recent book, "Camera Technology: The Dark Side of The Lens" (Academic Press, 1992). The author acknowledges, with thanks, the permission granted by Academic Press to use certain material from that book in this chapter.

15.2   BACKGROUND

Imagine the first camera as nothing more than a tent with a small hole in the side casting an image upon the opposite wall. From this accidental version of a "pinhole" camera to today's "smart" cameras, we find a cornucopia of ingenuity embracing optics, mechanics, electronics, and chemistry.

The variety of cameras ranges from one tiny enough to be concealed in a man's ring to one large enough for several people to walk around in without obscuring the image. The price range of cameras stretches from under five dollars for a disposable model (complete with film) to several thousand dollars (without film).

Cameras have recorded images of the deepest ocean trenches and the surface features

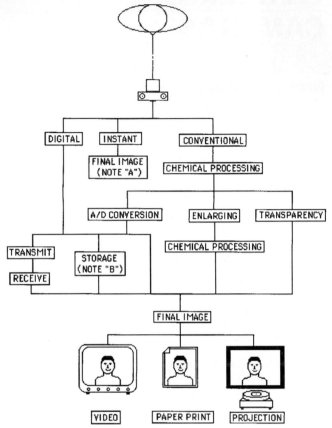

FIGURE 1 Final image flow chart. (A) Many instant photos can be manipulated just as the digital and conventional types, but are treated here in their primary use. (B) Storage means include magnetic tape and disks, optical disks, etc.

of Jupiter's moons. There are cameras that can freeze a bullet in midair or compress the germination of an acorn into a few minutes. From intimate portraits of bacteria to a 360° panoramic view of the Grand Canyon, there's a camera for any task.

Nonetheless, there is a common denominator: all cameras produce an image. This image may be the end product, or it may be converted in some way to the final image intended for viewing, as shown in Fig. 1. To choose the best camera for a given task, the properties of this final image should be determined first.

15.3 PROPERTIES OF THE FINAL IMAGE

1. Appearance
   a. Black-and-white
   b. Color
   c. High contrast
   d. Continuous tone

2. Smallest detail to be resolved
3. Type of display
    a. Audience population
    b. Viewing conditions
        (1) Viewing distance
            (*a*) Minimum
            (*b*) Maximum
        (2) Ambient illumination
    c. Display choices
        (1) Print
        (2) Projection
        (3) Self-luminous
4. Distribution

By considering the properties listed, we're obliged to visualize the final image through the viewer's eyes. Esthetics aside, we'll assume that the prime purpose of the final image is to convey information to the viewer.

15.4   FILM CHOICE

The appearance of the final image affects the choice of a camera by the kind of film required to produce that appearance. There are some films that are not available in all sizes. Other films are available in certain sizes only by special order. The availability of some films in some sizes changes over time, so check with your supplier before you select a camera for which film may be scarce.

Most film makers will be glad to send you their latest data on their current films, but be prepared for changes, because this is a very competitive field. New 35-mm color films in particular seem to come out with every change in the seasons.

15.5   RESOLVING FINE DETAIL

If the information in the final image is to be of any use, it must be legible to its detector, which we'll assume to be the human eye. Figure 2 shows that for high-contrast detail

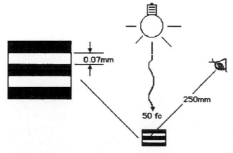

FIGURE 2   Visual resolution. Under ideal viewing conditions, we can resolve seven line-pairs per millimeter.

viewed under at least 50 foot-candles (office lighting), the eye has an angular resolution of about one minute of arc. This means that we can resolve about seven line-pairs per millimeter (LP/mm) at a distance of 250 mm. Since most photographic images exhibit moderate contrast and are viewed in moderate light, a more conservative limit of resolution would be 3.4 minutes of arc, which is good enough to resolve a pattern of two LP/mm at 250 mm.

In most cases, the final image is a magnification of the primary image formed in the camera. All else being equal, there is a practical limit to the extent of this magnification, after which the structure of the film, residual lens aberrations, focus inaccuracy, and/or diffraction effects begin to obscure fine image details.

Suppose then, that for some film we set a practical limit of magnification at 10×. Based on the visual resolution limit given previously, the smallest detail in the primary image could be 20 LP/mm, each line 0.025 mm wide.

Looking at it another way, if you want to photograph fine details and display the image legibly at a distance of 250 mm from the viewer, choose a film that will clearly resolve at least 20 LP/mm and is capable of being enlarged 10 diameters without its grain or other structure obscuring the image. Most films in common use today easily satisfy this criterion.

15.6 FILM SIZES

In terms of the widest variety of films available, 35-mm ranks number one. The most common format for this film is 24 × 36 mm. Although seldom used today, other 35-mm formats include 18 × 24 mm and 24 × 24 mm.

Next in line for a broad choice of film types is known as *medium-format* and is sold in 61.5-mm-wide rolls. The shortest rolls are paper-backed and are called 120. Many cameras that accept 120 film will also accept 220 film, which has an opaque paper leader and trailer, but no paper backing over the film. This permits a longer strip of film (more exposures per roll) and better film flatness than 120.

Common formats include (nominal dimensions) 45 × 60 mm, 60 × 60 mm, 60 × 70 mm, and 60 × 90 mm. Some medium-format cameras also accept 70-mm film that has a row of sprocket holes along each edge and may be loaded in special cassettes for use in the camera's large capacity, motorized, interchangeable film magazine.

The large formats, commonly referred to by their sheet film sizes in inches include 4 × 5, 5 × 7, and 8 × 10, to name the most well known. They may not offer as broad a choice of film as the smaller formats, but the most essential films are available for them.

15.7 DISPLAY

Choosing the best type of display for the final image should start with the number of people in the viewing audience. For large groups, a projected transparency has the advantage of being visible to the entire audience simultaneously. This is especially important if you want to use a pointer to single out detail in the image. Image detail should be clearly resolved by everyone in the audience, from the front row (image not too grainy) to the last row (image detail within the visual limits). In some cases, the best display is both a projected transparency that the lecturer can refer to with a pointer and a print for each viewer to examine closely, regardless of his position in the audience.

For the best viewing of projected images, the only light striking the screen should be that coming through the transparency. In other words, the room should be pitch black. Unless this condition is met, there is no possibility of reproducing the full tonal range, from deepest black to sparkling white, that the image could contain.

When this condition is difficult to satisfy, a self-luminous display may be best. One or more video monitors located at strategic points can provide good image contrast even under office illumination. The type of monitor may be the conventional cathode ray tube (CRT) or liquid crystal display (LCD). Of the two, the CRT produces a brighter image and is the least expensive. But it is bulky and fragile. The LCD has the virtue of minimal thickness; it's a flat screen display that can be hung on a wall like a framed photo.

At present, neither type can equal the fine detail and subtle color reproduction of a high-grade projected transparency viewed under the proper conditions. However, the gap in image quality is closing, especially now that high-definition television (HDTV) shows promise to become widely available in the near future.

15.8 DISTRIBUTING THE IMAGE

For many applications, the ease and speed with which an image can be distributed is crucial. Thanks to scanners, fax machines, modems, color photocopy machines, rapid photofinishing plants, self-processing "instant" films, etc., we can send practically any image to practically anyone who wants it in a matter of minutes. At present, our ability to do this depends on transforming the analog information in the subject into digital information for transmission, reception, manipulation, analysis, storage, and/or display, as indicated in Fig. 1.

15.9 VIDEO CAMERAS

If speed of acquisition and distribution is most important, we can capture the image on the charge coupled device (CCD) of the widely available camcorder, whose video and audio output signals are available in real time. These video cameras are versatile and moderately priced.

The still-picture counterpart to the camcorder seems to have come to a fork in the road. One path goes to a complete camera system, designed from scratch around the CCD chip and incorporating a miniature magnetic disk drive. The second path leads to a special video back, designed to replace the standard back of a conventional (film) camera. The video back contains a CCD chip and associated circuitry. In some cases the video back and the recorder, in which hundreds of images can be stored, require an "umbilical" cord between them. Some of the newer designs have integrated the back and recorder into a single (cordless) unit. Some of these backs can store up to 50 images internally.

As the capacity for image storage and/or manipulation grows, we see the emergence of systems within systems, where black box A converts black box B to communicate with computer C as long as you have the right adapter cables D, E, and F. This is typical of many rapidly expanding technologies.

Users of a video back on a conventional film camera will notice an unusually narrow angle of view for the lens in use if the light-sensitive area of the CCD chip is smaller than that of the film normally used in the camera. The reason is that only the central region of the camera's format is used. The result is that the camera's lenses perform as though their focal lengths have been "stretched" compared to their performance with conventional film that covers the whole format.

For example, Kodak's DCS 200 replaces the back of an unmodified 35-mm SLR camera, the Nikon N8008s. The "normal" lens for this camera's 24×36-mm film format has a 50 mm focal length, producing a (diagonal) angle of view of about 47°. The same lens used with the video back produces an angle of view of 37° because the CCD measures only 9.3×14 mm. To duplicate the 47° angle of view for this size CCD, a 19.3 mm focal length lens should be used.

Concerning the resolution from CCD images, Kodak's data for the DCS 200 gives a count of 1.54 million (square) pixels, arranged in a 1012×1524-pixel array that measures 9.3×14 mm. This gives a pixel spacing of 0.018 mm, which theoretically can resolve 54.4 monochromatic LP/mm. The color version uses a checkerboard pattern of red, green, and blue filters over the array, so divide the monochrome figure by three to come up with a color resolution of 18.1 LP/mm.

This is quite close to the criterion, discussed earlier, of two LP/mm for a 10× enlargement viewed at 250 mm. A 10× enlargement of the CCD image just described would measure 93×140 mm, about the size of a typical snapshot.

15.10 INSTANT PICTURES

For many applications, instant, self-processing film is the best choice. A familiar example is the oscilloscope camera loaded with high-speed film. With minimum, moderately priced equipment, a transient waveform on the scope screen can be captured on the film. Seconds later the print can be examined.

Polaroid dominates this field, which they spawned in 1948. Their range of camera models goes from snapshot to trucksize. They also have special backs which can be used on various cameras to adapt them for use with Polaroid films.

These films range from 35-mm color transparency to 8×10-inch (and larger) color print. Included in this variety are black-and-white sheet films that yield both a positive print and a negative. The negative must be stabilized, then washed and dried before being placed in an enlarger or contact printer.

15.11 CRITICAL FEATURES

In many cases, the availability of an accessory such as a Polaroid and/or digital image back is important enough to dictate the choice of a camera. Other factors that may tip the scales in favor of one camera over another might not be discovered until the chosen camera is used for some time.

For example, it may be very useful to have the kind of exposure automation that measures the light reflected from the film plane, before and during the exposure, thus being capable of responding instantly to any change in the scene luminance. There are some cameras that have this capability, yet they lack another feature that may be more valuable for some kinds of photography: the ability to observe the image through the viewfinder of an SLR not just before, but during the exposure.

Most SLRs employ a mirror that swings out of the way just before the exposure begins. This allows the image-forming light to reach the film, but it also blacks out the viewfinder, so that during the crucial instant of the exposure the photographer is momentarily blind.

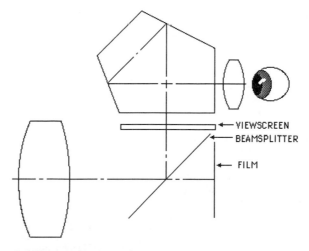

FIGURE 3 Beam-splitter SLR. The beam splitter eliminates the moving mirror, resulting in shorter time lag, reduced noise and vibration, plus the ability to monitor exposure and other image properties in real time.

Figure 3 illustrates that by using a beam splitter instead of a conventional mirror in an SLR, the problem is eliminated.

When the advantages of a beam-splitting system are considered, it seems strange that the feature isn't used more widely. Eliminating the swinging mirror reduces the noise and vibration generated each time an exposure is made. This can be crucial when the camera is attached to a microscope or telescope. Some SLRs provide for the mirror to be locked in its raised (shooting) position when desired.

15.12 TIME LAG

Even more important for some types of photography, substituting a beam splitter for a moving mirror in an SLR should reduce the camera's time lag. This is the interval between pressing the camera's trip button and the beginning of the exposure. It's a characteristic shared by all cameras and is rarely mentioned in a manufacturer's specifications for his camera. With few exceptions, time lag has increased in step with camera automation.

Testing 40 different 35-mm SLRs for their time lag resulted in a broad range, with the minimum of 46 ms and the maximum of 230 ms. The average was 120 ms. Figure 4 shows that during this interval, a walker moves about 0.8 ft, a runner about twice as far, a galloping horse about 7.0 ft, and a car going 60 mph moves 10.6 ft.

Various other cameras were also tested for their time lag, with these results:

Minox 35 EL (35-mm ultracompact): 8 ms

Leica M3 (35-mm coupled range finder classic): 17 ms

Hasselblad 500C (6×6-cm SLR classic): 82 ms

Kodak Disk 4000 (subminiature snapshot): 270 ms

Polaroid SX-70 Sonar (autofocus instant SLR): 600 ms

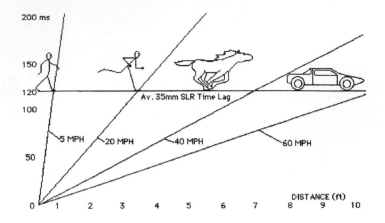

FIGURE 4   Time and motion.

Time mS		20	40	60	80	100	120	140	160	180	200
MOTION	MPH										
Walker	5	0.1	0.3	0.4	0.6	0.7	0.8	1.0	1.1	1.3	1.4
Runner	20	0.6	1.2	1.7	2.3	2.9	3.5	4.1	4.6	5.2	5.8
Horse	40	1.2	2.4	3.5	4.6	5.8	7.0	8.1	9.3	10.4	11.6
Car	60	1.8	3.5	5.3	7.0	8.8	10.6	12.3	14.1	15.4	17.6
Car	80	2.3	4.7	7.0	9.4	11.7	14.0	16.4	18.7	21.1	23.4

Distance Travelled (ft.)

15.13   AUTOMATION

Camera automation has taken full advantage of the miniaturization and economy of electronic devices, making two features, autoexposure and autofocus, available in all but the least expensive cameras. This increases the percentage of (technically) good photos per roll of film exposed by the typical amateur.

It's the amateur photographer that is first served when it comes to most of the significant camera automation features. Curious as this may seem, camera makers prefer to introduce a new concept by offering it first in a model intended for the casual snapshooter. This generally means large numbers will be produced. If problems with the feature show up, improvements are made and a "new, improved" model follows. Typically, the feature will be scoffed at by the more seasoned photographer who has learned to overcome the difficulties of making a technically good photograph with the most basic equipment. In time, the new feature is mature enough to be included in the camera maker's premier model. Eventually, even those that scoffed at the feature in its infancy learn to love it, but only after they discover how to recognize and compensate for its weaknesses, if any.

Autoexposure

Early autoexposure systems measured the average luminance of a scene with a selenium photocell, then regulated the shutter speed and/or f-stop based on the deflection of a galvanometer connected to the photocell. These were known as trapped needle systems and were successful in their prime mission: to produce acceptable exposures in snapshot cameras with the just-available color films, whose exposure error tolerance is much smaller than that of black-and-white film.

Most of the first generation autoexposure cameras using the trapped needle system relied on brute force, requiring a long, hard push to trip the camera. This caused camera motion, resulting in a (correctly exposed) smeared image. Nonetheless, many resourceful photographers used these early autoexposure cameras, bolted together with an intervalo-meter and electromagnetic tripping system, to create an unmanned camera for surveillance, traffic studies, etc.

Amateur movie cameras eagerly adopted autoexposure systems, which proved to be at least as much, if not more, of an improvement for them as they were in still cameras. The movie camera autoexposure systems work by regulating the lens opening (the f-stop), either with a galvonometer or a servomotor. With autoexposure, the movie maker can follow the subject as it moves from bright sunshine to deep shade without the distraction of manually adjusting the f-stop.

This same freedom to follow action without the distraction of manually resetting camera and/or lens controls explains the need for autofocus, a feature whose introduction enjoyed greater enthusiasm from amateur movie makers than from still photographers. Once again, the amateur models were the first to incorporate the feature, but in far less time than it took for autoexposure's acceptance, autofocus became a standard feature in both the amateur and front-line models from most of the makers of 35-mm cameras.

There are similarities between the automation of exposure and focusing. Both have become increasingly sophisticated as user expectations increase. Paradoxically, in the effort to perfect the making of a routine snapshot, some of the more sophisticated automation intrudes on the process by offering the user certain choices. Instead of simplifying photography, these technological marvels require the user to select a mode of operation from several available modes. For example, many cameras with autoexposure offer factory-programmed combinations of shutter speed and f-stop that favor:

- *Action*:   fast shutter speed, wide f-stop
- *Maximum depth of field*:   small f-stop, slow shutter speed
- *Average scenes*:   midway between the first two
- *Fill-flash*:   to illuminate portraits made against the light (backlit)

It comes down to this: if you know enough about photographic principles to choose the best autoexposure program, you will rarely need any of them. But when an unexpected change in the subject occurs, such as a cloud moving across the sun, some form of autoexposure can be valuable.

One of the more helpful refinements of autoexposure is the automatic shift of shutter speed with the focal length setting of a zoom lens. This is based on the time-honored guide that gives the slowest shutter speed that may be used without objectionable image motion from normal body tremor. The rule of thumb is to use the shutter speed given by the reciprocal of the lens's focal length. For example, if you're using a 35- to 105-mm zoom lens, the slowest shutter speed for arresting body tremor will shift as you zoom, from 1/35 s to 1/105 s (nominal). If the focal length's reciprocal doesn't coincide with a marked shutter speed, use the next faster speed. This guide applies to a hand-held camera, not for a camera mounted on a tripod.

Another autoexposure refinement combines a segmented silicon or gallium photocell with a microprocessor to automatically select the best exposure based on the distribution of light reflected from the subject. It amounts to making a series of narrow-angle "spot" readings of the subject, then assigning weighting factors to the different readings according to their relative importance. The weighting factors are determined by the camera maker based on the analysis of thousands of photographs.

Reduced to its most spartan form, a segmented photocell could have a very small central region, surrounded by a broad field. The user can flip a switch to select the desired reading—the center segment for spot readings, the broad segment for full field readings, or both segments for center-weighted full field readings.

To ensure optimum exposure for a subject, seasoned photographers "bracket" exposure settings by making at least three exposures of the subject. The first exposure obeys the meter's reading. The next two are one exposure step less and one greater than the first. This exposure bracketing, with some variations, has been incorporated as an on-demand automatic feature in some cameras.

Autofocus

Autofocus, in one form or another, has become a standard feature in camcorders and in most 35-mm cameras. The latter can be divided into two main types: (1) the snapshot "point–and–shoot," also known as "PHD" (press here, dummy) and (2) the SLR, spanning a wide range in price and sophistication. In between, there are several models which can be thought of as "PHDs on steroids." They have zoom lenses and elaborate viewfinders, making them too bulky to fit easily into a shirt pocket.

There are two main types of autofocus systems, the active and the passive. The active type emits a signal toward the subject and determines the subject's distance by measuring some property of the reflected signal. The passive type measures subject distance by analyzing the subject's image.

Active Autofocus Systems. Nearly every active system uses two windows, spaced some distance apart. The user centers the subject in the viewfinder's aiming circle and presses the shutter trip button. Figure 5 shows how a narrow infrared beam is projected from one of the windows, strikes the subject, and is reflected back to the second window. A photocell behind this window detects the reflected beam. The photocell is sensitive to the position of the beam on its surface and relays this information to its associated circuitry to regulate the camera's focus setting.

Initially, this was a straight-forward triangulation system, using a single infrared beam. But too many users were getting out-of-focus pictures of the main subject when it wasn't in the center of the picture. The camera's instruction book gives the solution: center the

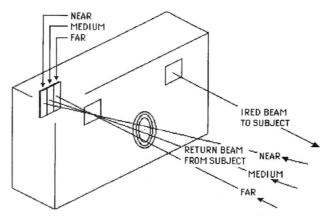

FIGURE 5 Active autofocus. Subject distance determines the angle of the reflected IR beam. The segmented photocell detects this angle, the AF system translates the angle to distance, moves the lens accordingly.

main subject in the finder's aiming circle, press the trip button halfway down, and hold it there, then recompose the scene and press the trip button all the way to make the exposure. This requires a fair amount of concentration and discipline, so it contradicted the purpose of having an automatic camera—to be free of cumbersome details, relying on the camera to make properly exposed, sharp photos.

A big improvement was made by projecting three beams from the camera, instead of one. The beams are divergent and the center beam coincides with the finder's aiming circle. Focus is set on the object closest to the camera.

A very different type of active autofocus is the ultrasonic system used by Polaroid in several models. Basically, it's a time-of-flight device that's been compared to sonar and bats. It uses an electrostatic transducer to emit an ultrasonic "chirp" towards the subject. Based on a round-trip travel time of about 5.9 milliseconds per meter, the time it takes for the chirp to reach the subject and be reflected back to the camera is translated into subject distance and a servomotor sets the focus accordingly.

A significant advantage of the active autofocus systems just described is their ability to work in total darkness. On the minus side is their inability to focus through a pane of glass or on a subject with an oblique glossy surface that reflects the signal away from the camera.

Passive Autofocus Systems.   Passive autofocus systems can be broadly characterized as acquiring two views of the subject, each view coming from a slightly different position, then focusing the lens to make the two views match. In this sense, the system operates just like a coincidence-type of optical range finder, but there are important differences.

With an optical range finder we rely on our ability to see when the two images are perfectly superimposed, so our focusing accuracy depends on our visual acuity. In a passive autofocus system, we relieve our eye of this burden and let the tireless electro-optical technology take over.

For the point-and-shoot camera, a passive autofocus system uses two windows, one whose line of sight coincides with that of the viewfinder's, and a second window, spaced some distance from the first. A simple, symmetrical optical system behind the windows includes a CCD for each window. The signal from the first CCD is taken as the reference against which the second CCD's signal is compared. Differences in the light distribution and/or differences in the relative location of the waveforms causes the control circuit to change the focus setting.

Autofocus SLRs.   Instead of the two windows just described, autofocus SLRs use two bean-shaped segments on opposite sides of the camera lens's exit pupil. Figure 6 shows how this is done. Two small lenslets are located a short distance behind the geometric equivalent of the camera's film plane. Each lenslet receives light only from its side of the exit pupil and projects it onto a CCD line array, one for each lenslet. The relative position of each image on its CCD strip is analyzed by the system's microcomputer which is programmed to recognize the focus condition as a function of the CCD's signals. If the signals deviate from the programmed values, the microcomputer issues the appropriate command to the focus motor.

For off-center subjects, it's necessary to prefocus on them by pressing the trip button halfway, holding it there as you recompose the scene, then pressing all the way on the trip button to make the exposure. This is asking too much of a photographer shooting any sort of action, and many of them mistrusted their autofocus SLRs. In response, camera makers offered new models with broader CCD arrays to provide a larger central region of autofocus sensitivity. Some of these can be switched between narrow and broad sensitivity regions.

Other refinements to SLR autofocusing include:

• Optimization of camera settings to maximize depth of field

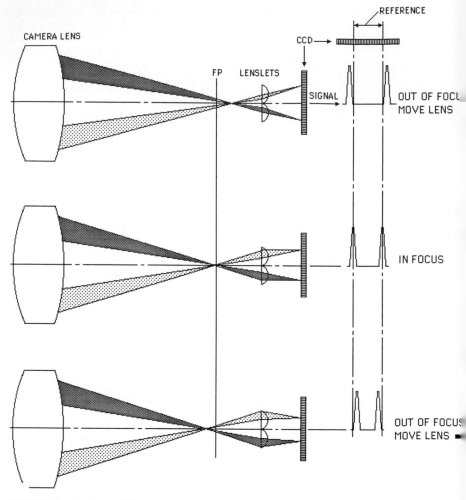

FIGURE 6   Autofocus SLR.

- Prediction of moving subject's distance at instant of exposure
- Accommodation for horizontal and vertical subject detail
- Focus priority according to position of user's eye

To optimize depth of field, the user aims the camera at the near point and presses the trip button halfway. This is repeated for the far point. Then the scene is recomposed in the viewfinder and the exposure is made with the actual focus set automatically to some midpoint calculated by the camera's microcomputer.

For predicting the distance of a moving subject, the subject's motion should be constant, both in direction and velocity. Under these conditions, the autofocus sensor's signals can be used to calculate where the subject will be when the exposure is made. The calculation must consider the camera's inherent time lag.

Early AFSLRs used focus sensors that were shaped to respond to vertical image detail,

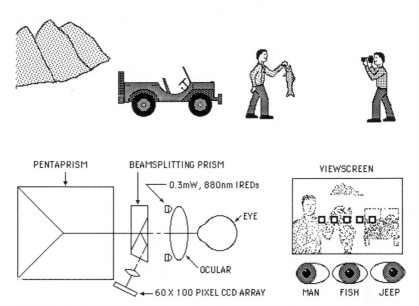

FIGURE 7 Canon's Eye Tracking SLR. The user's eye, illuminated by IREDs, is imaged on a CCD array. The resulting signal shifts in step with eye movements, causing a corresponding focus patch on the viewscreen to glow, indicating where the camera should focus.

with diminishing response as the detail approached the horizontal, where they were unable to respond. One solution incorporates three sets of lenslets and their CCD detector arrays. One set is laid out horizontally to respond to vertical detail, while the other two sets are vertical and straddle the first to form the letter "H." The two vertical sets respond to horizontal detail. Another solution has the individual, rectangular, detector segments (pixels) slanted to respond to both horizontal and vertical details.

By combining information from the autofocus detector and the focal length tracer in a zoom lens, some AFSLRs can maintain the image size (within the limits of the zoom range) chosen by the user, even as the subject distance changes.

Eye Tracking. Figure 7 shows how Canon's model EOS A2E overcomes the need for the subject to be centered in the viewfinder in order to be in focus. Canon devised an eye tracking system that detects what portion of the viewscreen the user is looking at. Using low-power infrared emitting diodes (IREDs) to illuminate the eye, the system is matched to the user by having him/her look at the extremities of the five autofocus aiming patches in the viewfinder's center. The reflections from the eye are detected by a 60×100-pixel CCD array and the resulting signals are stored in the camera's memory. At present, the five aiming patches occupy a 15-mm horizontal strip at the center of the finder, but it is possible that this could expand in future models. As it is, the camera's 16 user-selectable operational modes include one in which both the autofocus and autoexposure systems are commanded by the eye tracking feature.

If the user wants to preview the depth of field, all that's necessary is to look at a small patch near the finder's upper left corner (not shown here). This brief glance causes the lens to close down to the f-stop chosen by the autoexposure system.

Because this eye tracking feature is in an SLR, the user can see if it's working as expected just by looking at the viewscreen image. This indicates if, but not how, it works. To see how it works, I set up a simple experiment to measure the distribution of the light

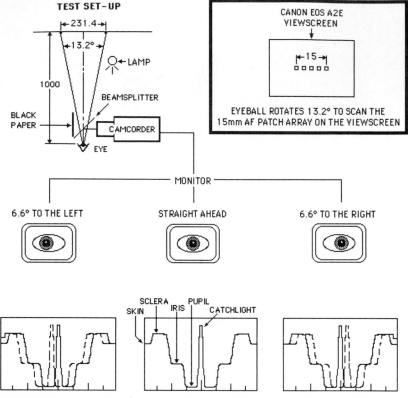

FIGURE 8 Eye tracking experiment. Line scan of the monitor's image at half screen height. (Dashed lines indicate image shift.)

reflected from my eye as I shifted my gaze between two marks on a wall. The separation between the marks and their distance from my eye were chosen to duplicate the angle swept by the eye when looking from one side to the other of the 15-mm focus patch array on the Canon EOS A2E viewscreen. As indicated by Fig. 8, the format was nearly filled with the image of my eye. Consistent eye placement was assured with a chin and head rest. Once the image of my eye was recorded on tape, I could play back and pause at any point, then select a line at half screen height and store its waveform in a storage oscilloscope. By superimposing line scan waveforms from the frames showing my gaze from one side to the other, I could easily see the difference and dismissed my skepticism. This novel feature has intriguing possibilities.

15.14 FLASH

Many 35-mm cameras feature a built-in electronic flash unit. Some are designed to flash every time the shutter is tripped, unless the user switches off the flash. Others fire only when the combination of scene luminance and film speed calls for flash. In some of the more advanced models with zoom lenses, the beam angle emitted by the flash changes in step with the focal length setting of the lens.

Red Eye

In the interest of compactness, the majority of cameras with built-in flash units have the flash close to the lens. The resulting flash photos of people frequently exhibit what is commonly known as "red eye," which describes the eerie red glow in the image of the pupils of a subject's eyes. The red glow is the light reflected from the retina, which is laced with fine blood vessels. Young, blue-eyed subjects photographed in dim light seem to produce the most intense red-eye images.

The effect is reduced by (1) increasing the angle subtended to the subject's eye by the separation between the centers of the lens and the flash; (2) reducing the subject's pupil diameter by increasing the ambient brightness or having the subject look at a bright light for a few seconds before making the exposure.

Examples of how some camera makers fight red eye include Kodak's Cobra Flash, used on several of their point-and-shoot models, and the "preflash," used on many different camera makes and models. The Cobra Flash describes a flash unit whose flashlamp/reflector unit is hinged at the camera's top. When the camera is not in use, the flash is folded down, covering the lens. To use the camera, the flash is swung up, positioning it further from the lens than would be possible if it had been contained in the camera's main body. One of their most compact cameras featuring the Cobra Flash is the Cameo motordrive model, which slips easily into a dress shirt pocket when the flash is folded down. When opened for use, the flash is 72 mm above the lens. Test shots were free of red eye when the subject was no more than seven feet away.

Another Kodak approach to the elimination of red eye is their single-use Fun Saver Portrait 35, whose integral electronic flash unit points upward, instead of forward. To use the camera, a simple white plastic panel, hinged at the camera's top rear edge above the flash is pulled open. It latches at a 45° angle to switch the flash circuit on and direct the light from the flash forward. The result is a diffused beam that appears to originate from a point 100 mm above the lens.

Other makes and models have integral flash units that pop up a short distance when put into play. This may only gain several millimeters of lens-to-flash separation, but my experiments indicate that, as sketched in Fig. 9, for every extra millimeter of separation between the lens and the flash, the (red-eye-free) subject distance can be increased about 30 mm.

Several 35-mm cameras use the *preflash* method to reduce red eye by emitting a brief, rapid burst of low intensity flashes just before the main flash goes off for the exposure. A variation uses a steady beam from an incandescent lamp in the flash unit. The beam switches on shortly before the flashlamp fires for the exposure. The purpose in both methods is to make the subject's pupils close down, reducing the light reflected from the eye during the exposure.

The preflash approach has two drawbacks: (1) it drains energy from the camera's battery, reducing the number of pictures per battery; (2) many times the subject reacts to the preflash and blinks, just in time for the exposure.

15.15 FLEXIBILITY THROUGH FEATURES AND ACCESSORIES

The seemingly endless combinations of operating modes with a camera like the Canon EOS A2E might be taken as an attempt to be all things to all photographers. Another way to look at it is to see it as a three-pound Swiss army knife: you'll never use all of the tools all of the time, but if there's the need for some tool, even just once, it might be nice to know you have it.

Many cameras have long lists of accessories. A typical camera system can be thought of as a box with an open front, top, and rear. For the front, the user may choose from as

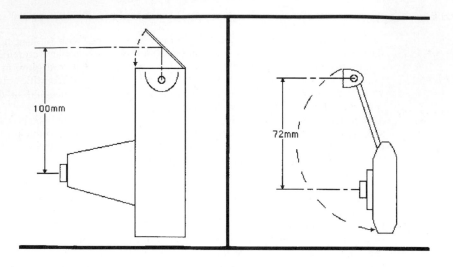

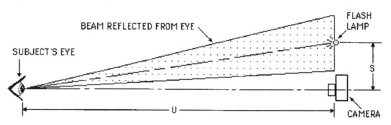

FIGURE 9 *Top left*: Kodak's "Fun Saver Portrait 35" bounces its flash from a folding reflector. *Top right*: Kodak's "Cameo Motordrive" uses the folding "Cobra Flash." *Bottom*: A beam reflected from the subject's eye misses the lens when the subject's distance "U" is not more than 20 S.

many as 40 different lenses. For the top, there may be three or more viewfinder hoods. For the back, choose one of perhaps five image receptacles.

Then there are the other groups, shown in Fig. 10: flash units, motor drives, close-up hardware, carrying cases, neck straps, lens hoods, filters, remote control cables, transmitters and receivers, mounting brackets, eyepiece magnifiers, corrective eyepiece lenses, cold weather heavy-duty battery packs, and more.

No matter how varied your photographic needs may be, the camera maker wants you to find everything you need in his or her catalog, Possibly, the availability of just one accessory, such as a wide-angle lens with tilt-shift controls for perspective correction, can decide which camera you choose.

15.16 ADVANTAGES OF VARIOUS FORMATS

In terms of versatility through a broad range of accessories plus the camera's intrinsic capabilities, it's hard to beat one of the major brands of 35-mm SLRs. No other type of camera has had as much ingenuity and as many refinements lavished on it for so many years. It's one of today's most highly evolved consumer-oriented products.

Accompanying the evolution in optics, mechanics, and electronics, film emulsions have

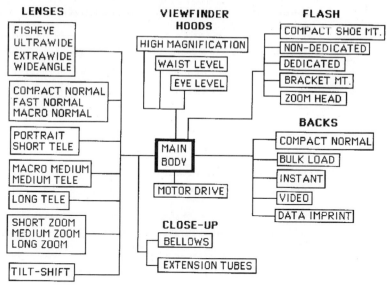

FIGURE 10   Camera system.

improved over the years, making the 35-mm format just as able as the larger formats for most applications. Even so, all else being equal, there is no substitute for "real estate"—the precious additional square millimeters of emulsion offered by the many 120-size medium formats. As the data in Fig. 11 shows, some of these are SLRs with systems as extensive as their 35-mm counterparts.

15.17 LARGE FORMAT: A DIFFERENT WORLD

When you make the jump from medium-format to large-format, you're in a different world. You use individual sheets of film, not rolls. Your camera will be used on a tripod or copy stand most of the time. Your photography will be contemplative, careful, and unhurried—perhaps better.

Scene composition and focusing are done with the lens at full aperture. Then the lens is stopped down, the shutter closed, the film holder inserted, its dark slide pulled, the shutter tripped, the dark slide replaced, and the film holder removed.

In a short time you'll realize that the large-format (view) camera can be thought of as a compact optical bench. As such, it lends itself to special applications that could be difficult for the smaller formats.

View Camera Versatility

To illustrate, suppose you need a picture of a picket fence at some obliquity, with every picket board, from near to far, in sharp focus and with the lens wide open. This calls for the use of the "Scheimpflug condition," shown in Fig. 12. It requires that the planes containing the lensboard, film, and subject all intersect on a common line. When this condition is satisfied, the entire surface of the subject plane will be in focus, even with the lens wide open.

CAMERA TYPE	35MM	MEDIUM FORMAT	LARGE FORMAT
FORMAT SIZES	18 X 24mm,	45 X 60mm,	2-1/4 X 3-1/4",
	24 X 24mm,	60 X 60mm,	3-1/4 X 4-1/4",
	24 X 36mm	60 X 70mm,	4 X 5", 5 X 7",
		60 X 90mm	8 X 10"
VIEWFINDER	Galilean, SLR	Galilean, TLR, SLR	Galilean, FP
FOCUSING	CRF, SLR, AF	CRF, TLR, SLR	CRF, FP
EXPOSURE	Manual, AE, TTL	Manual, AE, TTL	Manual
OPTIONS	Mot, Bulk, Pol,	Mot, Bulk, Pol,	Pol, Lens
	Dig, Data, Hood,	Dig, Data, Hood	
	Lens, Pgm, DEF	Lens, DEF	

Abreviations:

 AE = Autoexposure, **AF** = Autofocus, **Bulk** = Bulk Film Back,   **CRF** = Coupled Rangefinder,

Data = Data Recording, **DEF**= Dedicated Electronic Flash, **Dig** = Digital Image Detector,

FP = Film Plane, **Hood** = Interchangeable Viewfinder Hoods, **Lens** = Interchangeable Lenses,

Mot = Motor Drive, **Pgm** = Programable AE, AF, other functions, **Pol** = Polaroid Film Back,

SLR = Single Lens Reflex, **TLR** = Twin Lens Reflex, **TTL** = Through The Lens Metering.

FIGURE 11   Major camera features.

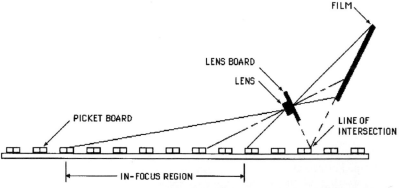

FIGURE 12   Scheimpflug condition. All of the picket boards within the field of view will be in focus when the planes of the lens board, film, and picket boards intersect on a common line.

The necessary camera movements, involving lensboard and film plane, are standard features of even the most spartan view cameras. These movements are known as swings and tilts. They take just a few seconds to adjust on a view camera and the job doesn't require a special lens. You can do it with a smaller format camera too, but you'll need one of their special (expensive) tilt-shift lenses or a bellows unit with articulated front and rear panels, plus a lens with a large enough image circle. The resulting combination may not retain all of the small-format camera's features, such as exposure metering, autofocus, etc.

The view camera's fully articulated front and rear provide for swing, tilt, rise, fall, and left-right shift. Thanks to this flexibility, objects such as boxes and buildings can be photographed without distortion, and distracting detail near the image borders can be omitted.

It takes first-time users a while to get used to the inverted and reversed image seen on the view camera's groundglass screen. This can be annoying when shooting a portrait, since an upside-down smile looks like a frown until you accept the fact that even though you understand the basic camera optics, it doesn't mean you have to enjoy coping with it. Worse, you'll ned to drape a dark cloth over the back of the camera and over your head in order to see the image if you're working in bright light. If you're claustrophobic, this may bother you.

On the plus side, large-format negatives are frequently contact-printed or only slightly enlarged for the final image. Because the image is large, depth of field and other image properties can be examined easily on the groundglass viewscreen with a small magnifier of modest power—a 4× loupe works well. The large negative has another attribute: it lends itself to retouching, masking, and other image manipulations, but these may be lost arts now that clever computer programs are available for doing the same things, provided your image is in digital form.

15.18 SPECIAL CAMERAS

Some photographic tasks call for cameras with special features, such as the ability to form images in near-total darkness or inside of a crowded mechanism. Among the long list of special cameras, we find:

- Aerial
- Clandestine
- Endoscopic
- High-speed
- Periphery
- Sewer
- Stereo (3-D)
- Streak
- Thermal imaging
- Underwater
- Wide-angle

Aerial Cameras

Aerial cameras come in a variety of sizes and features. Among the more common features are image motion compensation, where focal length, speed, and altitude are factored into the movement of the film during the exposure; a vacuum back to hold the film flat during the exposure; and a calibrated lens so that any rectilinear distortion can be factored into the measurements made of the image.

Clandestine Cameras

Clandestine, or "spy", cameras have been with us since photography was invented. In the broadest sense, any camera that is not recognized as such by the subject being photographed might be considered a successful spy camera. Many early box cameras were dubbed "detective" cameras because they were much smaller and more drab than a "real" camera with its prominent bellows and sturdy stand.

Cameras have been disguised as books, rings, binoculars, cigarette packs and lighters, matchboxes, portable radios, briefcases, canes, cravats, hats, even revolvers. Of all of them, the classic Minox is probably the best known. It can be concealed in an adult's fist, focuses down to eight inches, and is nearly silent. Its smooth exterior and gently rounded corners have inspired the belief among many that it was designed to be concealed in a body cavity with minimal discomfort.

Endoscopic Cameras

Endoscopic cameras use a tiny, short-focal-length lens to form an image that's transferred by a coherent, flexible fiber-optic bundle to a relay system that forms the image on the detector (film or CCD) in the camera. To illuminate the subject, the coherent bundle may be surrounded by an incoherent ring of fibers optically coupled to a light source at its free end, close to the camera.

Often fitted with a 90° prism on its tip, these cameras are used to photograph inside humans and machines. Another application is shown in Fig. 13: getting close-up views of architectural models from "ground" level. Variations include those without illumination optics but having a very small diameter image bundle to fit inconspicuously in some object for surveillance photography.

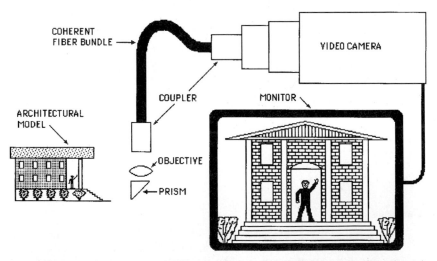

FIGURE 13 Endoscopic camera. While most often used for medical purposes, the endoscopic camera's properties make it valuable for photographing miniature scenes from the perspective of a miniature photographer.

High-speed Cameras

High-speed cameras were once defined as being able to make exposures of less than 1/1000 s. Today this would include many 35-mm SLRs which have a top speed of 1/10,000 s, a speed equaled by several consumer-grade camcorders. When shorter exposures are called for, a common, low-cost electronic flash unit can give flash durations as short as 1/32,000 s.

The next step includes the Kerr cell and Faraday shutters, both of which work by discharging a high-voltage capacitor across a medium located between crossed polarizers. This produces a momentary rotation of the plane of polarization within the medium, permitting light to pass through to the detector. Exposure times are in the nanosecond range for these electro-optical/magneto-optical devices.

For exposures in the picosecond range accompanied by image intensification, there's the electronic image tube. When a lens forms an image on the photocathode at the front of this tube, electrons are emitted. Their speed and direction are controlled by electrodes within the tube. A secondary image is formed by the electrons as they strike the phosphor screen at the rear of the tube. This image may be photographed, or, if the tube has a fiber-optic faceplate behind the screen, the image can be directly transferred to a film held against the faceplate.

By placing a microchannel plate in front of the phosphor screen, the image can be intensified by a factor of 10,000 or more. A microchannel plate is a thin glass disk riddled with microscopic holes that pierce the disk at an angle. In Fig. 14 the wall surface of each hole is coated with a substance that reacts to the impact of an electron by emitting more electrons. A high voltage across the disk accelerates the stream of electrons. For every electron that enters one of the angled holes, about 100,000 electrons emerge to strike the phosphor screen.

Periphery Cameras

A periphery camera is used to make photos of objects like gas engine pistons, bullets, and other cylindrical objects whose surface detail must be imaged as though the surface was "unrolled" and laid out flat before the camera. Depending on the size of the subject, either it or the camera is rotated about its longitudinal axis at a constant angular velocity. The image strikes the film moving behind a slit that's parallel to the axis of rotation. The film's velocity matches that of the image unless deliberate image compression or elongation is desired.

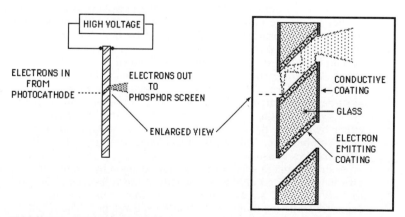

FIGURE 14 Microchannel plate.

Sewer Cameras

A sewer camera is designed to photograph the inside of pipes, tunnels, etc. It may be thought of as a small underwater camera on a sled. The camera's lens is encircled by an electronic flashtube and reflector to illuminate the scene. Pictures are made at regular intervals, as judged by distance marks on the cable attached to the sled. Other cables attached to the camera convey signals to and from the camera. With the miniaturization of video cameras, they have taken over this task, except where maximum resolution is required. This is where film cameras excel.

Stereo Cameras

Stereo cameras seem to come in and out of vogue with some mysterious rhythmic cycle. The root idea has been around since the dawn of photography and is based on the parallax difference between the views of our left and right eyes. The classic stereo camera mimics nature by using two lenses spaced about 65 mm apart to form two images of the subject.

The two images can be made in other ways. A simple reflection system using four small mirrors or an equivalent prism system placed in front of a normal camera's lens will form two images of the subject, as shown in Fig. 15. Another method requires that the subject is stationary because two separate exposures are made, with the camera being shifted 65 mm between exposures. In aerial stereo photography, two views are made of the ground, the views made so many seconds apart.

When the images are viewed in a manner that restricts the left and right images to their respective eye, the stereo effect is achieved. Various methods for viewing stereo pairs include projection, where the left and right views are polarized at 90° to one another. The viewer wears glasses with polarizing filters oriented to let each eye see the view intended for it.

Another viewing system is called a *parallax stereogram*. It (optically) slices the left and right images into narrow, interlaced strips. When viewed through a series of vertical lenticular prisms with a matching pitch, the 3-D effect is seen.

Streak Cameras

Streak cameras are useful for studying relative motion between the subject and camera. They share certain characteristics with the periphery camera described previously, insofar as they match the movement of the film to that of the image coming through a slit at the film plane. Exposure time is determined by how long it takes for a point on the film's surface to travel across the slit's width.

The basics of the streak camera are shown in Fig. 16. It would be pointless to use a streak camera without some relative motion between the image and film. Some photographers use a streak camera for creative effects, such as depicting motion tack-sharp at its beginning, then gradually elongating or compressing it, and ending in a smear. This is done by varying the relative velocity between the image and the film during the exposure, either by moving the camera, the subject, or the film, These motions may be made singly or in combination. Varying the focal-length setting of a zoom lens with the film moving also produces unusual images.

A streak camera's format has a width defined by the film it uses, but each picture has its own length, limited only by the length of the roll of film. One of the more critical factors to look for in a streak camera is freedom from *cogging,* a local density variation in exposure while the film is moving at a fixed velocity. The result of periodic or intermittent speed variations, the cause may be improperly meshed gears, a bad bearing, poor fit

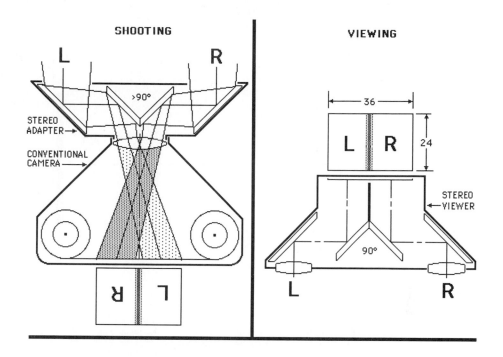

SHOOTING   VIEWING

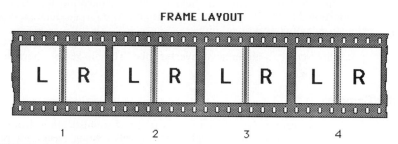

FRAME LAYOUT

FIGURE 15 Stereo adapter set.

between the film drive sprocket teeth and the film's sprocket perforations, or the magnetic pole effects of the drive motor.

Thermal Image Cameras

Thermal imaging cameras convert the intrinsic heat of a subject into a visible image. Among their many applications are detection of heat losses from buildings, blood circulation disorders, and surveillance. Some of these cameras produce false color images, in which each color represents a different temperature.

Among the various methods to form visible images of temperature variations, the most direct way is to use a normal camera loaded with film that's sensitive to the infrared (IR) portion of the electromagnetic spectrum.

In a more elaborate system, a moving mirror scans the subject and, line-by-line,

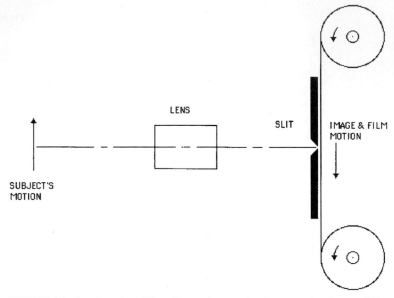

LENS

SLIT

IMAGE & FILM MOTION

SUBJECT'S MOTION

FIGURE 16 Streak camera. When film motion matches image motion, image will be free of distortion. If the film moves too fast, the image will be stretched. If the film moves too slow, the image will be compressed.

projects its image onto a heat-sensitive semiconductor device whose output is proportional to the IR intensity. The output is used to modulate a beam of light focused on the surface of conventional film.

Another version uses the semiconductor to modulate a stream of electrons striking a phosphor screen in an image converter tube. The image can then be photographed.

Underwater Cameras

Underwater cameras come in a wide variety of sophistication, from the disposable costing less than ten dollars, to the high-tech versions costing thousands of dollars. In between, there are dozens of underwater housings designed for specific cameras. Typically, these housings permit the user to change the camera's settings through watertight couplings. Most of the cameras used in such housings have motorized film advance, autoexposure, and autofocus, so the only external control needed is a pushbutton at one end of a simple electrical switch.

External attachments include flashguns, viewfinders, and ballast weights. The flashgun connections should be carefully examined because they are one of the leading sources of problems. In general, the simpler the connector, the better.

Wide-angle Photography

There are several 35-mm and medium-format cameras designed specifically for wide-angle photography. These include straightforward types which use lenses designed for wide-angle views on larger-format cameras. Essentially these cameras use only a rather long horizontal strip of the broad image circle the lens produces. This type of camera is uncomplicated and rugged.

Panoramic Cameras. A special kind of wide-angle camera is known as a panoramic camera, and there are two main types: one where the entire camera rotates; the other, where just the lens rotates.

The rotating camera type is capable of a full 360° vista. As the camera turns on its vertical axis, the film is moved past a narrow, stationary slit at the center of the film plane. The motion of the film is matched to that of the image. Because these cameras rotate slowly, a common prank in photos of large groups is for the prankster to stand at the edge of the group that's exposed first, then dash behind the group to the opposite edge in time for its exposure, with the result that the same person appears twice in the same photo, once at either edge of the group.

The rotating lens type shown in Fig. 17 produces images of about 140°. It works by rotating its lens on a vertical axis coinciding with its real nodal point. The image is swept across the film through a tubular image tunnel at the rear of the lens. The tunnel extends almost to the film surface and has a narrow slit at its end. The slit is parallel to the axis of rotation and extends over the width of the film. During the exposure the film is held stationary against a cylindrical film gate whose radius equals the focal length of the lens. The slit width, the rotating speed, and the lens opening may be adjusted for exposure control.

The panoramic cameras described here regulate their speed of rotation with precision governing systems to ensure edge-to-edge uniformity of exposure, so they should be kept as clean as possible. Also, to avoid unpleasant distortion, use care in leveling them and always use the best single camera accessory money can buy: a good, solid tripod.

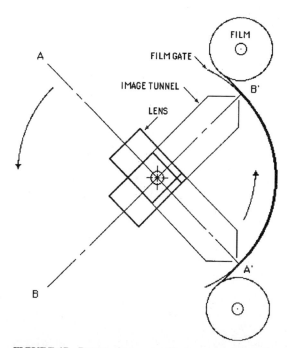

FIGURE 17 Panoramic camera. The lens rotates about its rear nodal point from A to B. Image-forming light reaches the film from A′ to B′ through a slit at the end of the image tunnel.

15.19 FURTHER READING

Clerc, *Photography,* vol. 1 and 2, Focal Press, 1970.

Edgerton, *Electronic Flash Strobe,* MIT Press, 1979.

Feynman, *The Feynman Lectures on Physics,* vol. 1 Addison, 1975.

Goldberg, *Camera Technology*: *The Dark Side of The Lens,* Academic Press, 1992.

Habell and Cox, *Engineering Optics,* Pitman, 1966.

Hyzer, *Engineering and Sci. Hi-Speed Photography,* McMillan, 1962.

Jenkins and White, *Fundamentals of Optics,* McGraw-Hill, 1957.

Kingslake, *Applied Optics and Opt. Eng.,* vol. IV, Academic Press, 1967.

Kingslake, *Optical System Design,* Academic Press, 1983.

Kingslake, *Optics in Photography,* SPIE Optical Engineering Press, 1992.

Morton, *Photography for the Scientist,* Academic Press, 1984.

Ray, *Applied Photographic Optics,* Focal Press, 1988.

Ray, *The Photographic Lens,* Focal Press, 1979.

Smith, *Modern Optical Engineering,* McGraw-Hill, 1966, 1990.

Spencer, *The Focal Dictionary of Photo Technologies,* Focal Press, 1973.

Stimson, *Photometry and Radiometry for Engineers,* Wiley, 1974.

Williamson and Cummins, *Light and Color in Nature and Art,* Wiley, 1983.

CHAPTER 16
CAMERA LENSES

Ellis Betensky
Opcon Associates, Inc.
West Redding, Connecticut

M. Kreitzer and J. Moskovich
Opcon Associates, Inc.
Cincinnati, Ohio

16.1 INTRODUCTION

Camera lenses have been discussed in a large number of books and articles. The approach in this chapter is to concentrate on modern types and to describe imaging performance in detail both in terms of digital applications and in terms of the optical transfer function. By modern types, we mean lens forms that were found on cameras in 1992. The chapter deals almost entirely with lenses for the 35-mm (24 × 36-mm) format. This limitation is unfortunate but not really inappropriate, given the widespread use of this format. Moreover, the different lens types that are described are used for applications ranging from 8-mm video to 6 × 9-cm roll film.

We have not included any specific design examples of lenses for large-format cameras, but the imaging capabilities of these lenses are described in terms of digital applications. By digital applications we mean the comparison of different lens types in terms of total pixels and pixels per unit solid angle. It is hoped that this feature will make comparisons between radically different imaging systems possible and also help to classify lenses in terms of information capability. See "Further Reading" at the end of this chapter for related information about photographic lenses, particularly with respect to older design types.

16.2 IMPOSED DESIGN LIMITATIONS

There are some limitations that are imposed on the design of camera lenses. The most significant ones are listed as follows.

Microprism focusing in single lens reflex cameras (SLRs) is difficult at apertures smaller than about F/4.5. Recent advances permit the use of microprisms at apertures down to F/5.6 and this is usually the smallest maximum aperture permitted in the specification of a lens for the SLR camera.

Depending on the camera type, there is a maximum rear lens opening allowable at the flange on SLR lenses. The limitation is approximately 33 to 36-mm diameter at flange to film plane distances of 40.5 to 46 mm. This affects the maximum possible aperture on normal lenses (typically to F/1.2) and also requires appropriate design of the exit pupil location on long-focal-length and high-speed retrofocus lenses in order to avoid excessive vignetting.

The minimum back focal length (BFL) allowable on SLR lenses (because of the swinging mirror) is about 38.5 mm. The BFL cannot be too short on non-SLRs because of in-focus dust or cosmetic problems on optical surfaces close to the film plane. The actual limitation depends on the minimum relative aperture that would be used but is rarely less than 4 mm and usually more than 8 mm.

Since most lens accessories such as filters and lens-shades are mounted on the front of a lens, there is a practical limitation to the allowable front diameter of most lenses. Filter sizes larger than 72 mm are not desirable, and smaller is always preferred. The actual clear aperture at the front of a lens is considerably smaller than the filter size, depending on the angular field and the mounting details of the filter. Obviously there are lenses such as 600-mm F/4 telephotos for which the 72-mm limitation is not possible. In these cases, the lens can be designed to use internal filters that are incorporated into the design.

Mechanical cams are still in widespread use for the practical realization of the required motions in zoom lenses. This technology requires that the motions themselves be controlled at the design stage to be reasonably monotonic and often to have certain mutual relationships. These requirements are particularly severe for the so-called "one-touch" zoom and focus manual control found on many SLR zoom lenses.

In general, size and weight restrictions pose the biggest problems for the designer of most camera lenses. Almost any lens can be designed if there are no physical limitations. These limitations are sometimes a consequence of ergonomic considerations but can equally be an effort to achieve a marketing advantage. Size restrictions almost always adversely affect the design, and exceptionally small lenses (for a given specification) should be regarded with suspicion.

16.3 MODERN LENS TYPES

Normal (with Aspherics) and Variations

35-mm SLR normal lenses are invariably Double-Gauss types. Refer to Fig. 1. This lens form is characterized by symmetry about a central stop to facilitate the correction of coma, distortion, and lateral color. These lenses are relatively easy to manufacture and a user can expect good quality in a production lens. Total angular coverage of about 45° is typical, and speeds as fast as F/1 are achievable. Extremely good optical performance is possible, particularly if the angular field and speed are reduced somewhat. Image quality generally deteriorates monotonically from axis to corner and improves dramatically as the lens aperture is reduced by about two F-numbers. With the addition of a fixed rear group, conjugate stability can be achieved over a wide range. Refer to Fig. 2.

Wide-angle

An interesting new wide-angle lens type is a four-component form found commonly on the so-called compact 35-mm cameras. This lens is characterized by a triplet construction followed by a rear element that is strongly meniscus-shaped, convex to the image plane. This lens has much less astigmatism than either conventional triplets or Tessars and can cover total fields of up to 75° at speeds of around F/4. Faster speeds are possible if the angular field is reduced. Most importantly, the rear meniscus component takes the burden

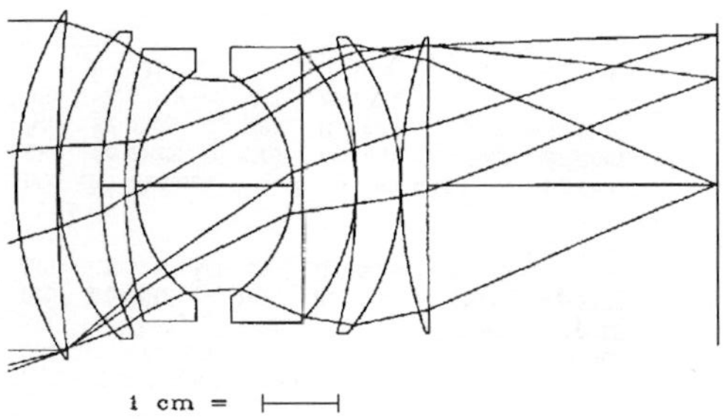

Optical Transfer Function

THROUGH FOCUS   BEST FOCAL PLANE

SN.	RADIUS	THICKNESS	GLASS	CLR. AP.
1	45.5018	5.57796	NBFD13	45.30
2	200.7383	0.12778	NBFD13	44.59
3	30.2250	5.87242	NBFD13	39.77
4	49.8199	3.16411		37.43
5	69.7190	1.46671	FDS3	34.94
6	18.9571	13.22338		29.04
7		7.70000		28.24
8	-17.8910	1.22226	FD11	28.13
9	-2598.6050	7.10023	NBFD12	35.42
10	-27.0475	0.10000		35.78
11	-78.2303	5.77241	NBFD11	37.30
12	-32.4479	0.12778		38.27
13	65.0998	3.42233	NBFD11	38.00
14	901.7260	38.39188		38.02

SYSTEM FIRST ORDER PROPERTIES

FIELD: 20.5 °   f/   1.24
STOP: 4.89 after surface 6. DIA: 29.273

EFL: 55.5383   FVD: 93.2693   ENP: 24.6604
BFL: 38.3919   BRL: 54.8774   EXP: -117.603

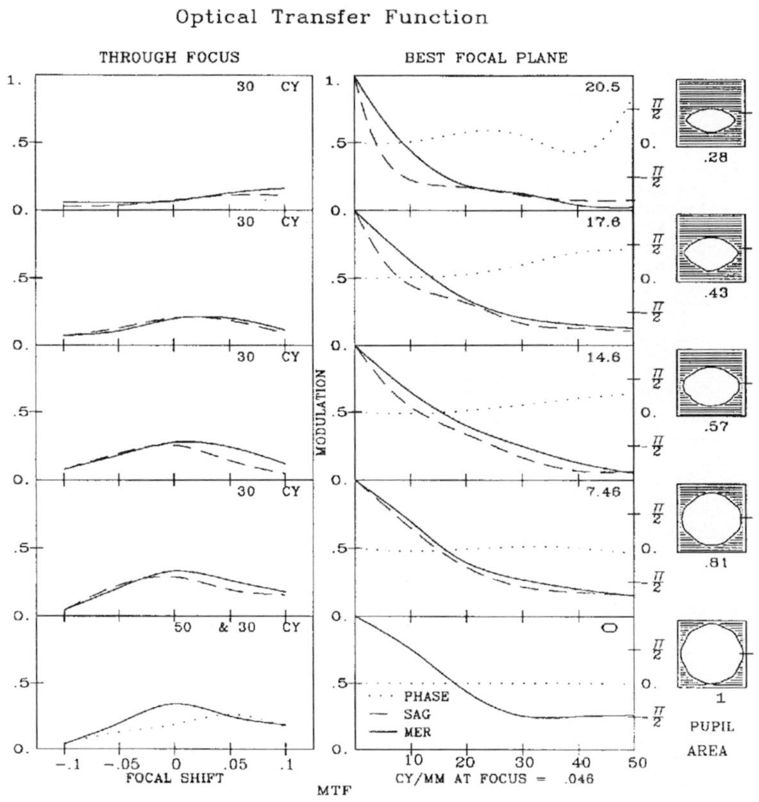

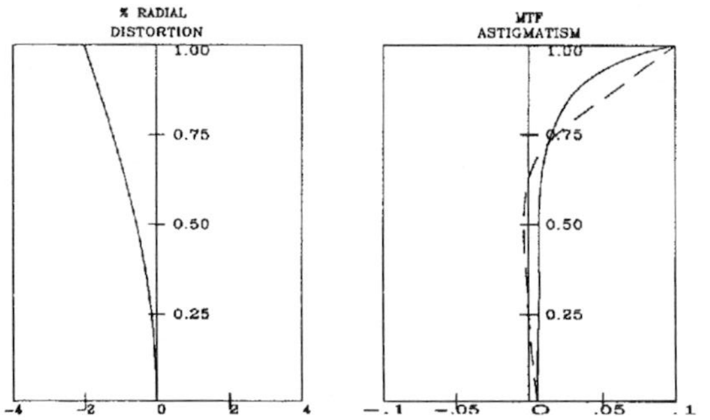

1 cm =

λ	WT
.558	.284
.483	.273
.634	.215
.433	.131
.685	.097

FIGURE 1 55-mm F/1.2 for 35-mm SLR.

Optical Transfer Function

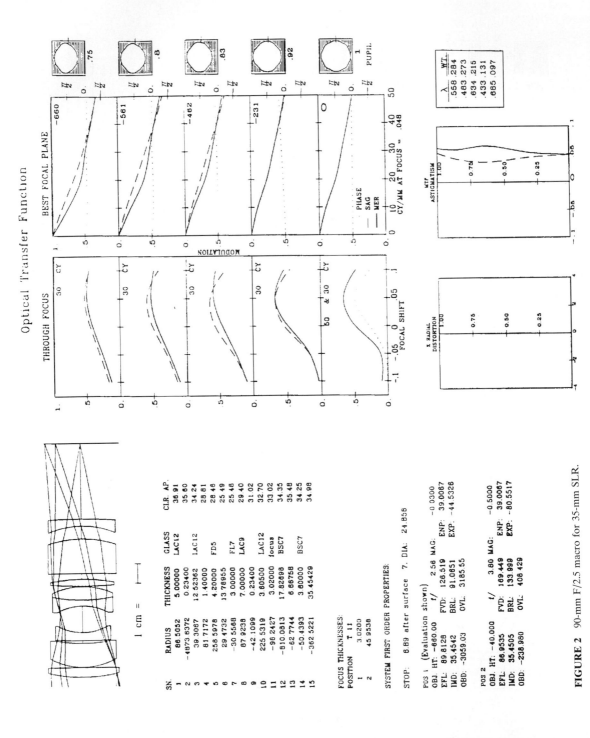

FIGURE 2 90-mm F/2.5 macro for 35-mm SLR.

of field flattening away from the triplet front part. This results in considerably lower individual element powers and correspondingly lower sensitivities to tilts and decentrations of the elements. It is this problem that makes conventional triplets extremely difficult to manufacture. Refer to Fig. 3.

Inverted Telephoto (Retrofocus)

These lens types, characterized by a long back focal length, are typically used for wide-angle applications for single lens reflex cameras having a swinging viewing mirror behind the lens. Inverted telephoto implies a front negative group followed by a rear positive group, just the reverse of a telephoto construction. This type of construction tends to result in relatively large front aperture sizes, and it is not easy to design small lenses without compromising on image quality. Retrofocus designs sometimes have a zone of poorer image quality in a field area between the axis and the corner. This zone is a by-product of the struggle to balance lower- and higher-order aberrations so that the outer parts of the field have acceptable image quality. These lenses have particularly good relative illumination both because the basic construction results in an exit pupil quite far from the image plane and also because it is possible for the size of the pupil to increase with field angle. In order to achieve conjugate stability, it is necessary to employ the use of so-called "floating elements" or variable airspaces that change with focusing. However, this feature does result in additional optomechanical complexity.

The newer forms of this lens type fall into four broad subcategories.

Very Compact Moderate Speed. These include six-element 35-mm F/2.8 with a front negative element and seven-element 28-mm F/2.8 with a leading positive element. Refer to Figs. 4 and 5, respectively. These relatively simple constructions are suitable for speeds of F/2.8 or slower and total angular coverages of up to 75°.

Highly Complex Extreme Speed. As the complexity of both the front and rear groups is increased, the inverted telephoto form can be designed to achieve speeds of F/1.4 and angular fields of 90°. The use of aspherical surfaces is essential in order to achieve these specifications. Refer to Figs. 6, 7, 8, and 9.

Highly Complex Extreme Wide-angle with Rectilinear Distortion Correction.

These are inverted telephoto designs covering total fields of up to 120°, often with speeds as fast as F/2.8. Distortion correction is rectilinear. The chromatic variations of distortion, astigmatism, and coma are usually the limiting aberrations and are virtually impossible to correct beyond a certain point. Refer to Figs. 10 and 11.

Extreme Wide-angle with Nonrectilinear Distortion ("Fish-eye Lenses"). Without the requirement of rectilinear correction of distortion, inverted telephoto designs can be achieved quite readily with total angular fields exceeding 180°. For these lenses, the image height h and focal length f are often related by $h = f \cdot \theta$, where θ is the semifield angle. See, for example, USP 4,412,726.

Telephoto Lenses

The term *telephoto* strictly applies to lenses having a front vertex length less than the focal length (telephoto ratio less than one). The classic telephoto construction has a front positive group followed by a rear negative group. This can lead to telephoto ratios that are as short as 0.7 or less. The term telephoto is often loosely used to refer to any

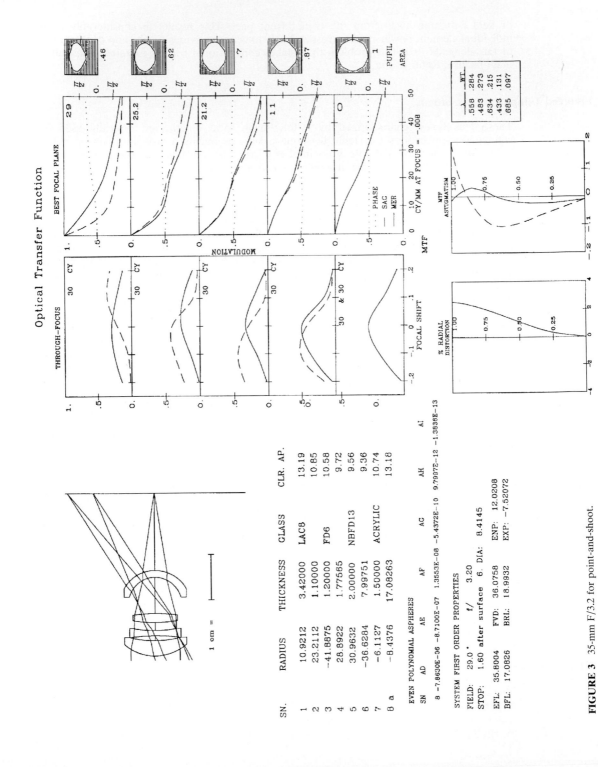

FIGURE 3 35-mm F/3.2 for point-and-shoot.

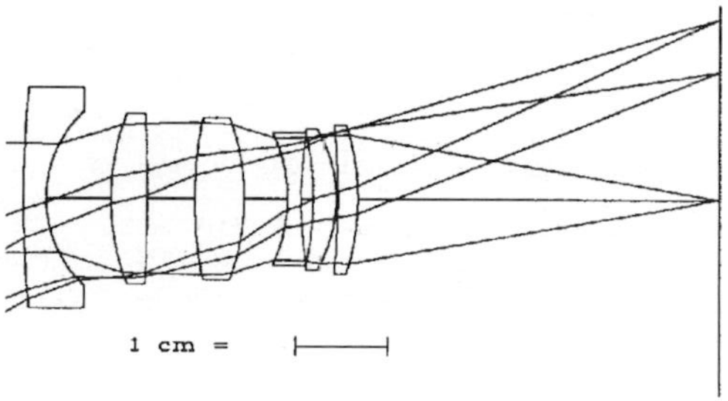

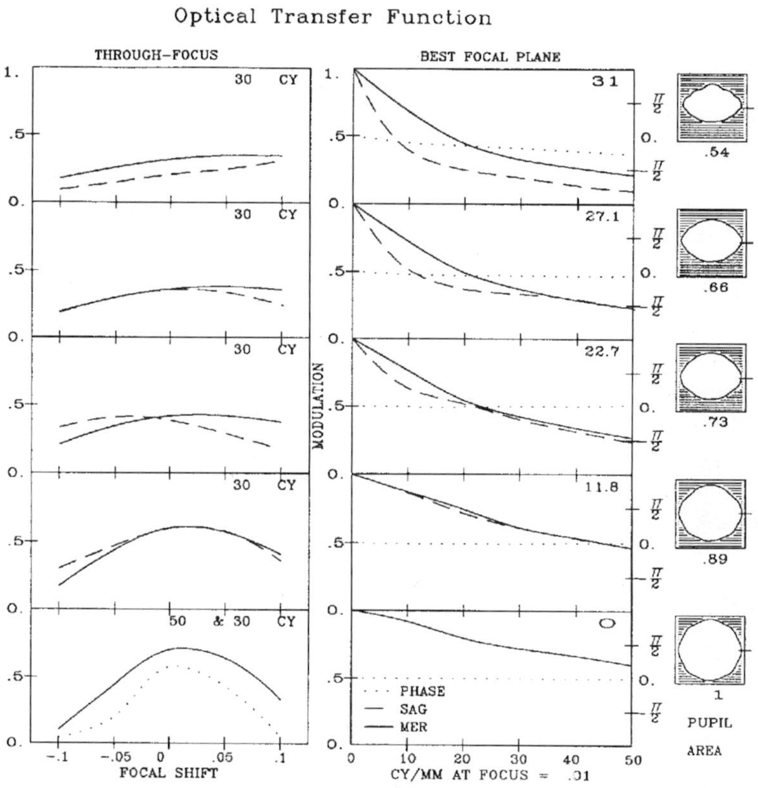

Optical Transfer Function

SYSTEM FIRST ORDER PROPERTIES

SN.	RADIUS	THICKNESS	GLASS	CLR. AP.
1	142.6758	2.50000	BACD2	22.86
2	13.6359	7.22000		16.58
3	26.8698	4.00000	LAC10	15.19
4	-305.5060	5.32000		17.83
5	45.5824	5.50000	BACED1	17.16
6	-26.9377	2.15000		16.80
7		2.63000		14.55
8	-17.7587	1.50000	FDS3	13.74
9	51.0640	1.31000		13.96
10	-34.5463	2.46000	LAC10	14.04
11	-16.8112	0.20000		14.86
12	-90.3640	2.20000	NBFD5	15.31
13	-28.3558	39.60170		15.91

FIELD: 31.0 °   f/ 2.85
STOP: 0.00 after surface 7. DIA: 14.546

EFL: 34.9990   FVD: 76.5917   ENP: 15.2855
BFL: 39.6017   BRL: 36.9900   EXP: -10.3446

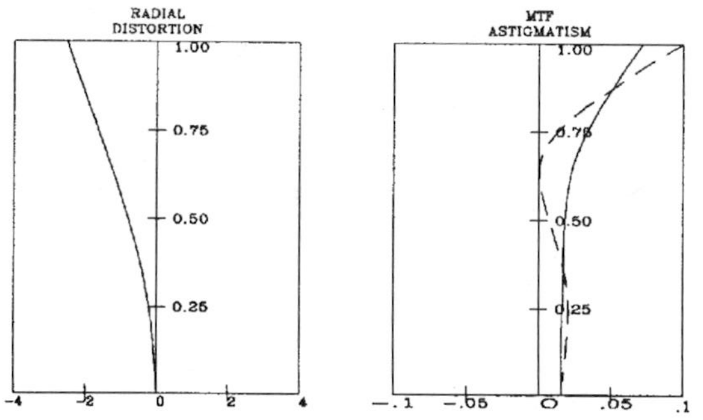

1 cm =

FIGURE 4   35-mm F/2.8 for 35-mm SLR.

16.7

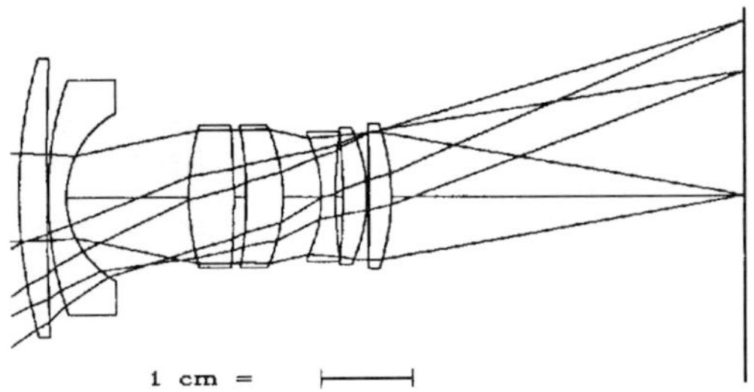

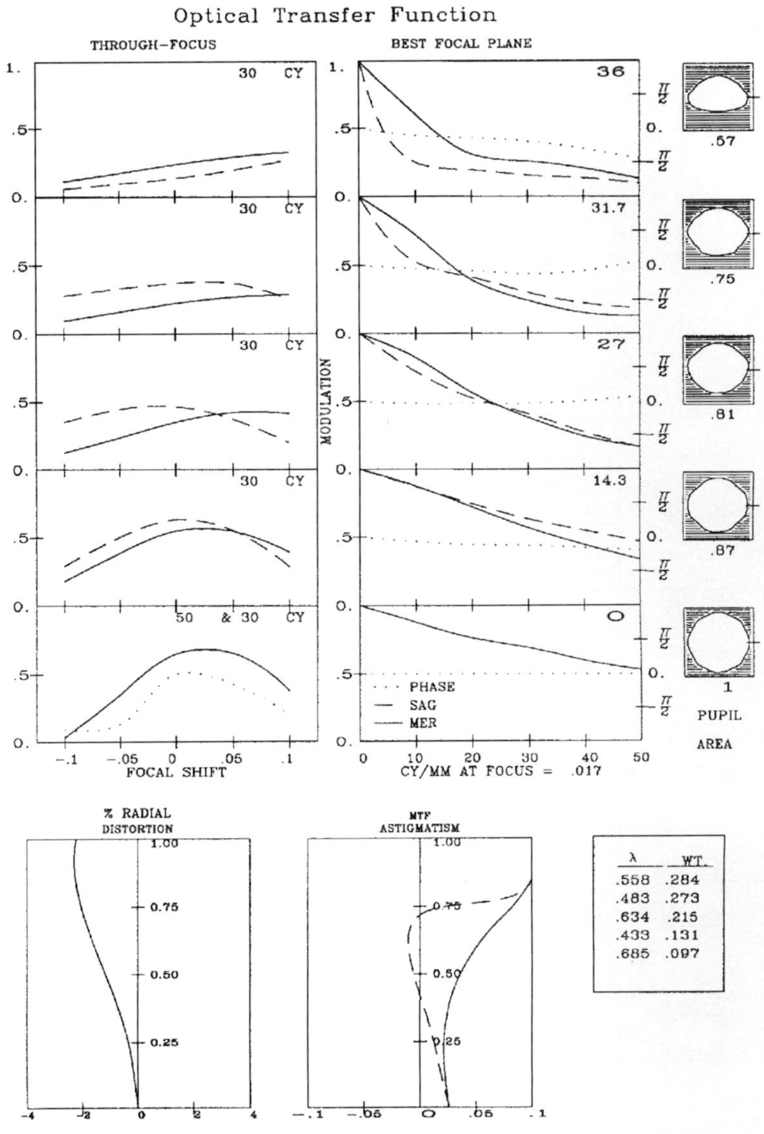

Optical Transfer Function

SN.	RADIUS	THICKNESS	GLASS	CLR. AP.
1	58.3632	3.08000	BACD11	29.56
2	358.8033	0.20000		28.01
3	40.8438	2.00000	LAC11	24.39
4	10.8112	13.63511		17.88
5	28.3296	5.00000	BAFD8	14.92
6	-101.4423	1.38000		14.92
7	-47.6164	4.00000	BACD4	14.80
8	-19.6821	1.00000		14.96
9		3.22000		13.84
10	-15.9894	1.50000	FDS3	13.01
11	54.4624	0.85000		13.53
12	-76.2485	2.60000	LAC10	13.59
13	-18.3952	0.20000		14.35
14	-426.2859	2.46000	LAC14	14.96
15	-25.6954	38.67632		15.28

SYSTEM FIRST ORDER PROPERTIES

FIELD: 36.0°   f/ 2.83
STOP: 0.00 after surface 9. DIA: 13.839

EFL: 27.9998   FVD: 79.8014   ENP: 17.5829
BFL: 38.6763   BRL: 41.1251   EXP: -11.5408

FIGURE 5 28-mm F/2.8 for 35-mm SLR.

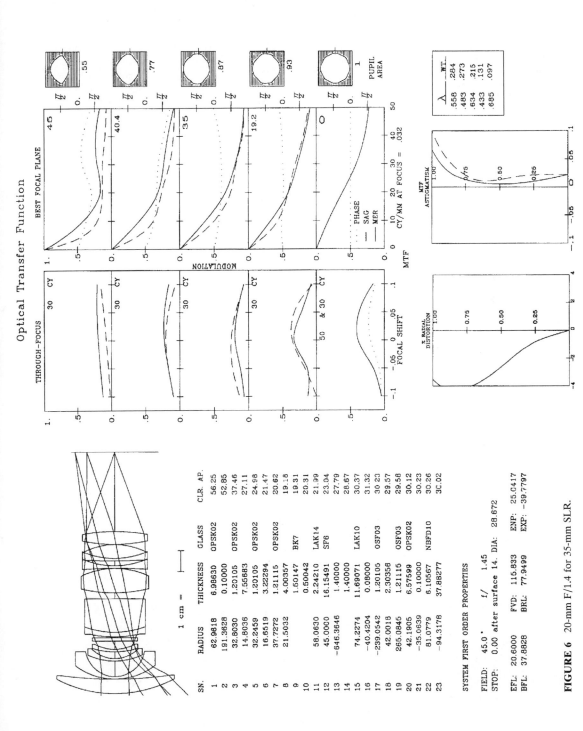

FIGURE 6 20-mm F/1.4 for 35-mm SLR.

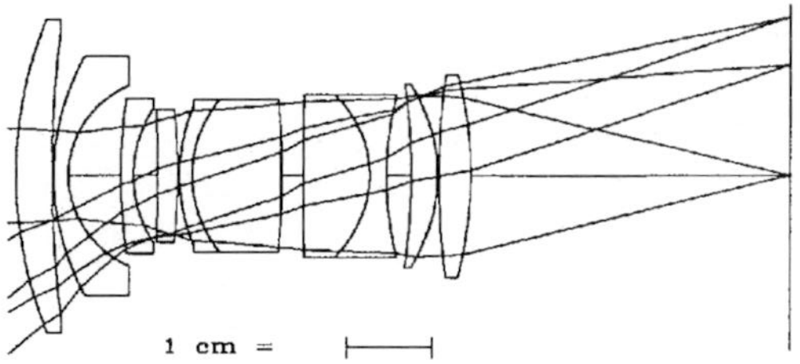

Optical Transfer Function

SN.	RADIUS	THICKNESS	GLASS	CLR. AP.
1	52.7393	4.45788	BAFD8	37.20
2	145.4670	0.20008		34.60
3	31.3749	1.79340	BACD14	27.97
4	12.3953	6.37329		20.57
5	72.5430	1.49328	BACD14	18.52
6	15.9588	2.98656		15.99
7	-182.9944	2.39120	CF6	15.65
8	-80.7948	0.20008		15.28
9	23.6565	1.49328	TAF1	16.45
10	15.5506	10.75553	F8	16.53
11	-114.3056	0.97600		18.00
12		1.61284		18.26
13	1383.4500	7.86657	TAF1	18.59
14	-14.5710	1.99104	FD15	19.04
15	43.4978	2.98900		19.43
16	-79.6178	2.98900	LACL4	20.38
17	-23.1138	0.20008		20.99
18	99.2167	3.78444	LACL7	22.62
19	-53.3986	38.25872		23.41

SYSTEM FIRST ORDER PROPERTIES

FIELD: 40.0 °   f/   2.06
STOP: 0.00 after surface 12. DIA: 18.261

EFL: 24.4997   FVD: 92.8123   ENP: 19.6895
BFL: 38.2587   BRL: 54.5536   EXP: -26.7172

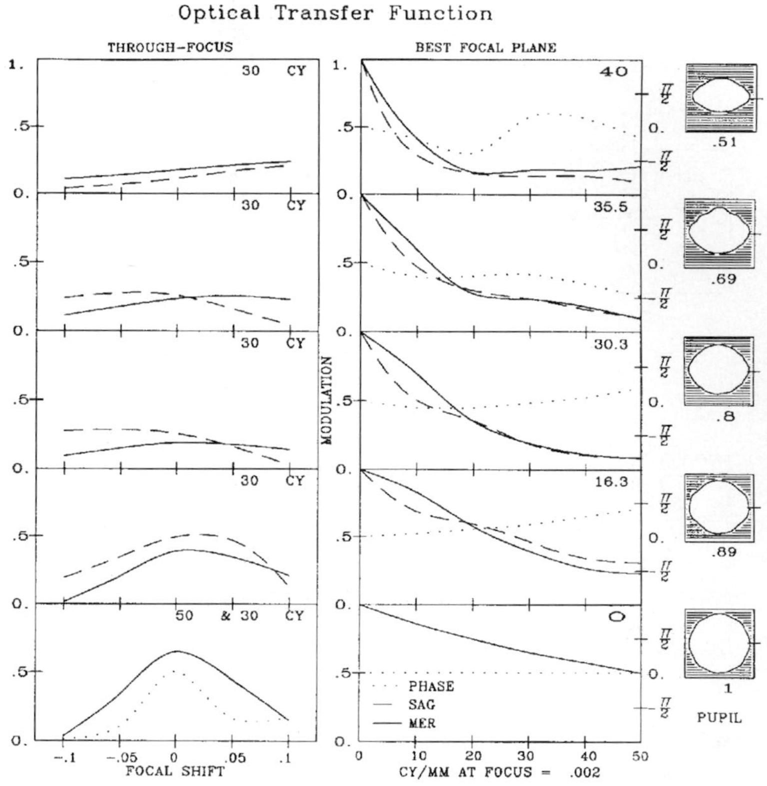

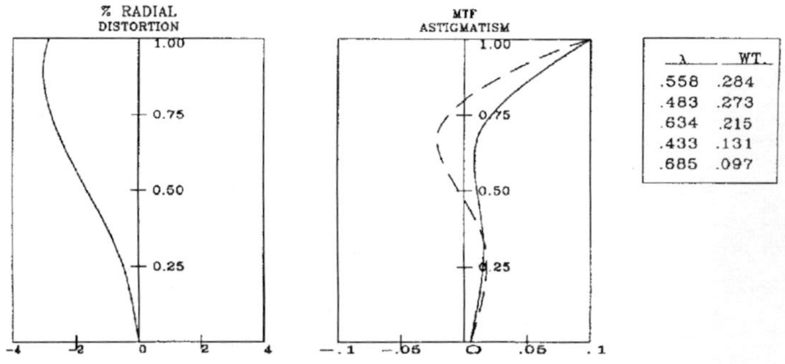

1 cm =

FIGURE 7 24-mm F/2 for 35-mm SLR.

Optical Transfer Function

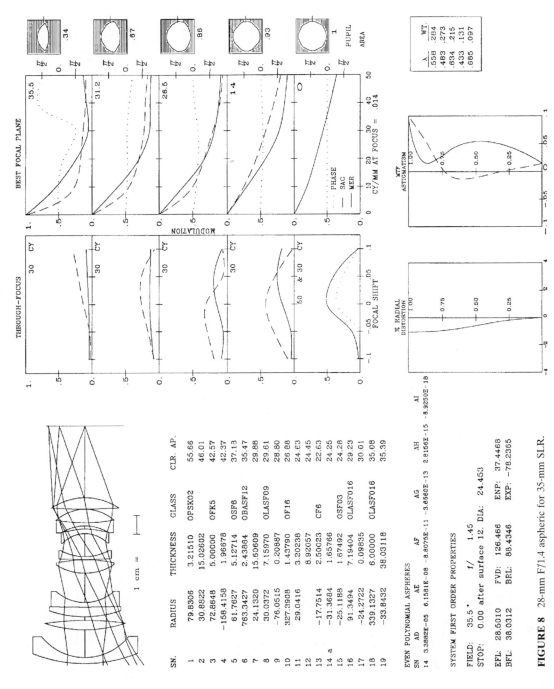

FIGURE 8 28-mm F/1.4 aspheric for 35-mm SLR.

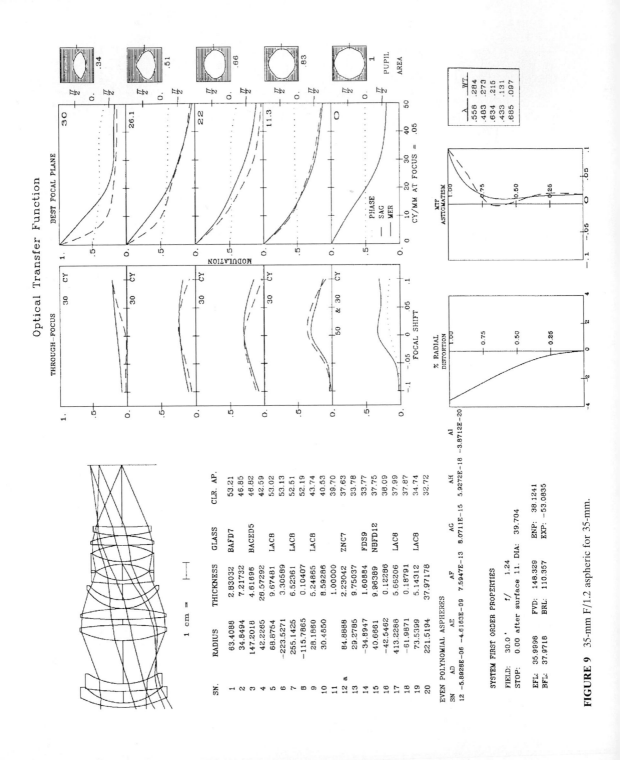

FIGURE 9 35-mm F/1.2 aspheric for 35-mm.

Optical Transfer Function

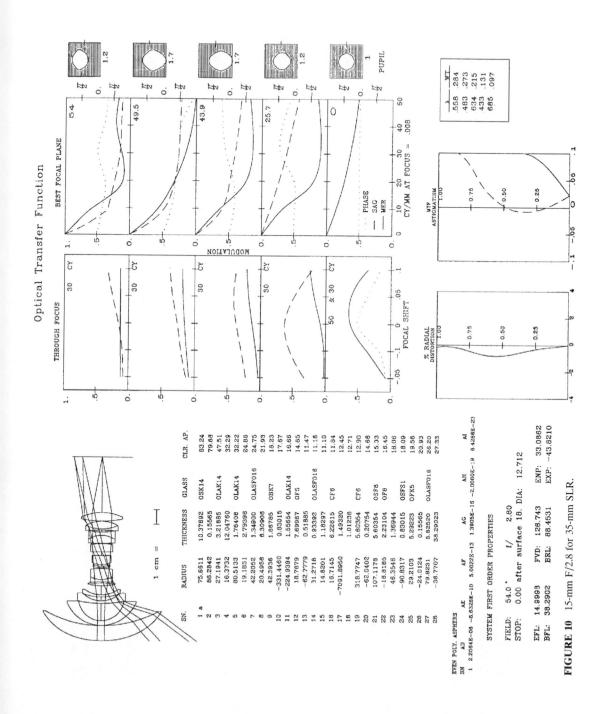

FIGURE 10 15-mm F/2.8 for 35-mm SLR.

16.13

Optical Transfer Function

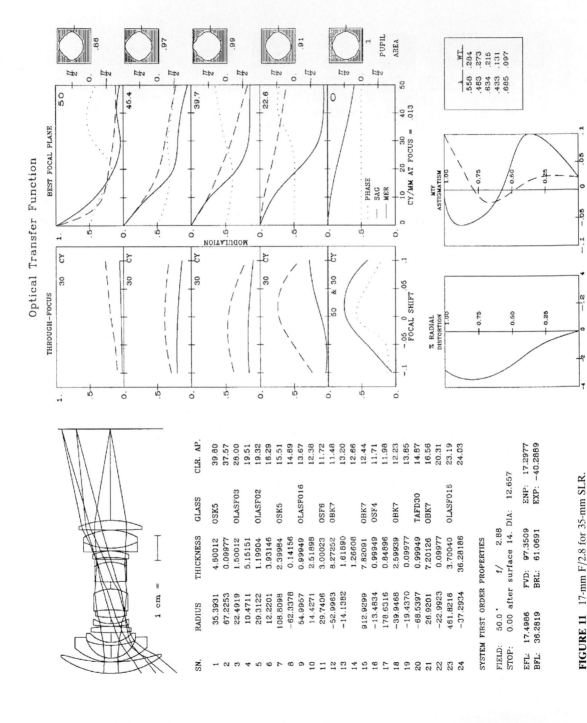

SN.	RADIUS	THICKNESS	GLASS	CLR. AP.
1	35.3931	4.80012	OSK5	39.80
2	67.2253	0.09977		37.57
3	22.4919	1.50012	OLASF03	28.00
4	10.4711	5.15151		19.51
5	29.3122	1.19904	OLASF02	19.32
6	12.2201	3.93146		16.29
7	108.8098	2.39984	OSK5	15.51
8	-62.3378	0.14156		14.89
9	54.9957	0.99949	OLASF016	13.67
10	14.4271	2.51898		12.38
11	29.7406	3.00023	OSF6	11.72
12	-52.9963	8.27252	OBK7	11.48
13	-14.1382	1.61890		13.20
14		1.26608		12.66
15	912.9299	7.82091	OBK7	12.44
16	-13.4834	0.99949	OSF4	11.71
17	178.6316	0.84896		11.98
18	-39.9468	2.59939	OBK7	12.23
19	-19.4370	0.09977		13.85
20	-68.5397	0.99949	TAFD30	14.87
21	26.9201	7.20126	OBK7	16.56
22	-22.9923	0.09977		20.31
23	461.8216	3.70040	OLASF016	23.19
24	-37.2934	36.28186		24.03

SYSTEM FIRST ORDER PROPERTIES

FIELD: 50.0 °   f/   2.88
STOP: 0.00 after surface 14. DIA: 12.657

EFL: 17.4986   FVD: 97.3509   ENP: 17.2977
BFL: 36.2819   BRL: 61.0691   EXP: -40.2889

λ	WT
.558	.284
.483	.273
.634	.215
.433	.131
.685	.097

FIGURE 11 17-mm F/2.8 for 35-mm SLR.

long-focal-length lens and one sometimes sees references made to the telephoto ratio of a wide-angle lens.

Two significant advances characterize the newer types of telephoto lenses, particularly those used for 35-mm SLR cameras. The first is the use of small internal groups for focusing, sometimes in conjunction with the front group. This feature has also led to significant improvement in the performance of these lenses with change of conjugate. This has been a problem with telephoto lenses, particularly with respect to attaining close focus with good optical quality. Internal focusing of a long-focal-length lens also has considerable advantages in terms of mechanical simplicity because a smaller mass is being moved over a significantly shorter distance.

The second advance is the employment of optical glasses having anomalous dispersion for the correction of chromatic aberrations. These newer glasses have anomalous dispersion characteristics similar to those of calcium fluorite, but with physical and chemical properties that make their use practical. These glasses are still expensive and more difficult to use than ordinary ones, but they do offer significant advantages in terms of reducing the chromatic aberrations that otherwise severely limit the imaging potential of all long-focal-length refracting optics. Typical available versions of these glass types are the FK Schott, FCD Hoya, FPL Ohara, and PFC Corning series of glasses.

These design types offer outstanding optical correction together with remarkable specifications, resulting in considerable size and cost. Commercial embodiments include 300-mm F/2 and 400-mm F/2.8 for 35-mm. They are widely used for sports and wildlife photography. Since secondary color increases as the front vertex distance is reduced, it is advisable to regard excessively short all-refractive telephotos with some caution. Refer to Figs. 12 and 13.

Zoom Lenses

Zoom lenses have evolved significantly in the past twenty years. In the early 1970s, there was basically only the classic four-group type of zoom lens. This four-group zoom has two moving groups between a front group used only for focusing and a stationary rear ("master") group. This type is still found on consumer video cameras. Figures 14 and 15 show a variation of this form with the rear group also moving for zooming. The master group could often be changed to yield a different zoom with the same ratio over a different range.

The second basic form, originating in the mid 1970s, was the two-group wide-angle zoom, typically 24 to 48 mm and 35 to 70 mm for the 35-mm format. Both the front negative group and the rear positive group move for zooming, and the front group is also used for focusing. This lens type has an inherently long back focal length, making it eminently suitable for the SLR camera. See, for example, USP 4,844,599. The maximum zoom range is about 3 : 1.

In order to achieve lens types such as a 28- to 200-mm zoom for 35-mm, new ideas had to be employed. The resulting lenses have up to five independent motions, including that of the diaphragm. These degrees of freedom allow for the location of the entrance pupil to be near the front of the lens at the short-focal-length position and also for the exit pupil to be located near the rear, particularly at the long-focal-length setting. These conditions result in acceptably small size. The extra zooming motions permit a large focal-length range to be achieved without any one motion being excessively long. There is a constant struggle in the design of these zooms to minimize the diameter of the front of the lens. This is not only to reduce size and weight, but also to permit the use of acceptably small filters. Some designs do have problems with relative illumination at the wide-angle end.

In the past, these lenses have been focused either by moving the front group or by moving the entire lens, the latter option leading to the so-called varifocal zoom. However, more recent developments in miniature electromechanical and autofocusing systems have led to the evolution of extended range zooms in which the distinction between a focusing

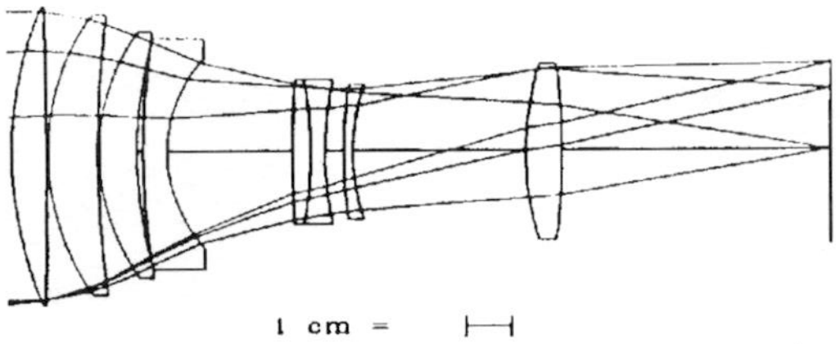

SN	RADIUS	THICKNESS	GLASS	CLR. AP.
1	94.7391	7.77937	FK51	68.50
2	−2172.6060	0.29673		68.17
3	60.2686	11.06877	FK51	64.27
4	252.3730	0.32992		61.84
5	52.0363	8.40407	LACL5	56.02
6	106.3570	1.32716		52.39
7	174.8780	5.38407	ADF455	52.36
8	34.3907	27.53793		42.51
9		0.39774z		32.21
10	544.8948	4.02731	FD4	31.89
11	−82.8590	3.02000	LAC10	31.33
12	84.3923	4.10159		29.96
13	121.2688	2.02830	FL5	29.65
14	53.3638	38.18467z		29.43
15	77.5462	7.99606	FC5	38.92
16	−149.3100	59.11712		39.34

SYSTEM FIRST ORDER PROPERTIES

FIELD: 6.0° f/ 2.88
STOP: 0.00 after surface 9. DIA: 32.215

EFL: 197.001 FVD: 181.001 ENP: 120.950
BFL: 59.1171 BRL: 121.884 EXP: −95.2419

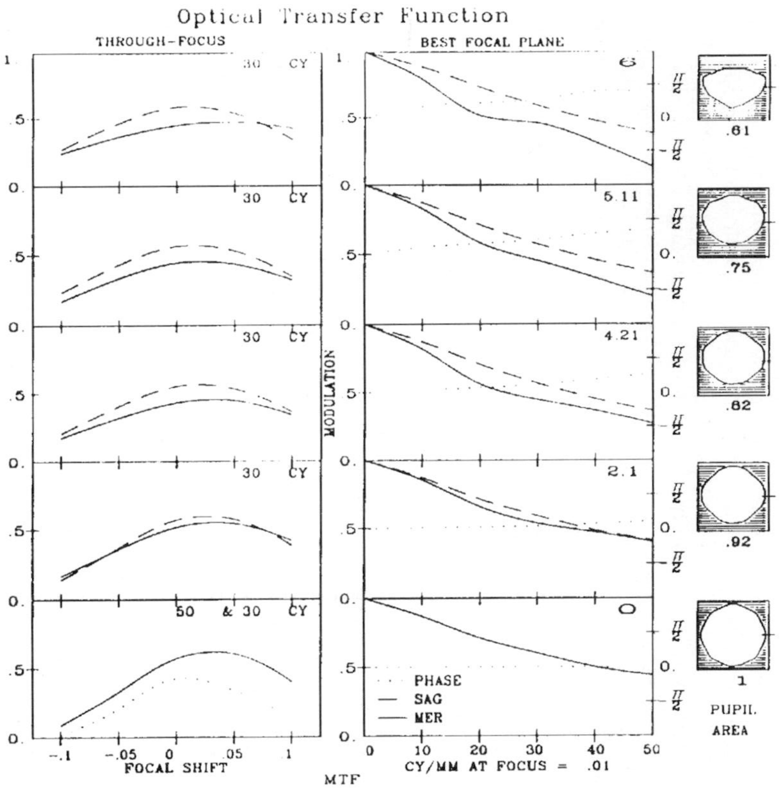

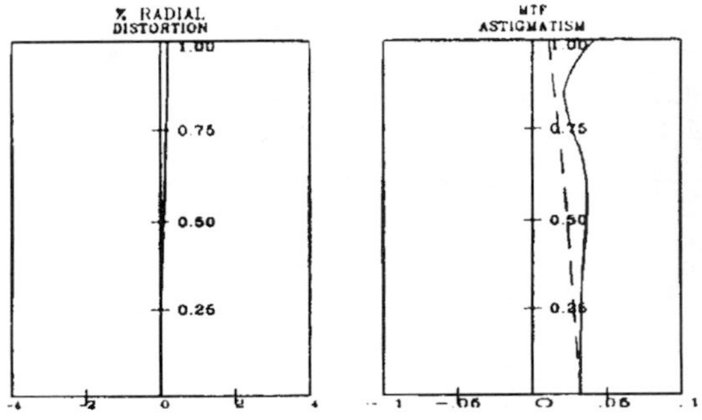

λ	WT.
.558	.284
.483	.273
.634	.215
.433	.131
.685	.097

FIGURE 12 200-mm F/2.8 for 35-mm SLR.

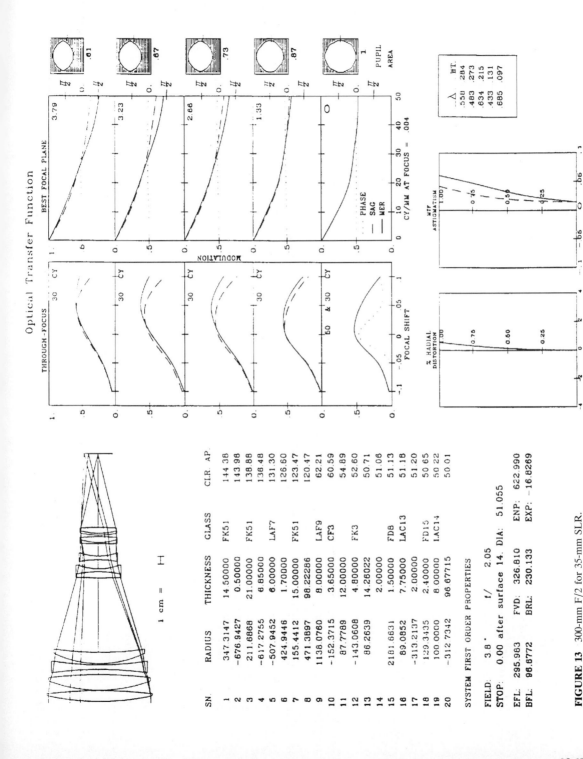

Optical Transfer Function

SN	RADIUS	THICKNESS	GLASS	CLR. AP.
1	347.3147	14.50000	FK51	144.38
2	-676.9427	0.50000		143.98
3	211.6868	21.00000	FK51	138.88
4	-617.2755	6.85000		136.48
5	-507.9452	6.00000	LAF7	131.30
6	424.9446	1.70000		126.60
7	155.4412	15.00000	FK51	123.47
8	471.3897	98.22286		120.47
9	1138.0760	8.00000	LAF9	62.21
10	-152.3715	3.65000	CF3	60.59
11	87.7789	12.00000		54.89
12	-143.0608	4.80000	FK3	52.60
13	86.2639	14.28022		50.71
14		2.00000		51.06
15	2181.6631	1.50000	FD8	51.13
16	89.0852	7.75000	LAC13	51.18
17	-313.2137	2.00000		51.20
18	129.3135	2.40000	FD15	50.65
19	100.0000	8.00000	LAC14	50.22
20	-312.7342	96.67715		50.01

1 cm = ⊢—⊣

SYSTEM FIRST ORDER PROPERTIES

FIELD: 3.8°   f/ 2.05
STOP: 0.00 after surface 14.   DIA: 51.055

EFL: 295.983   FVD: 326.810   ENP: 622.990
BFL: 96.8772   BRL: 230.133   EXP: -16.8269

FIGURE 13   300-mm F/2 for 35-mm SLR.

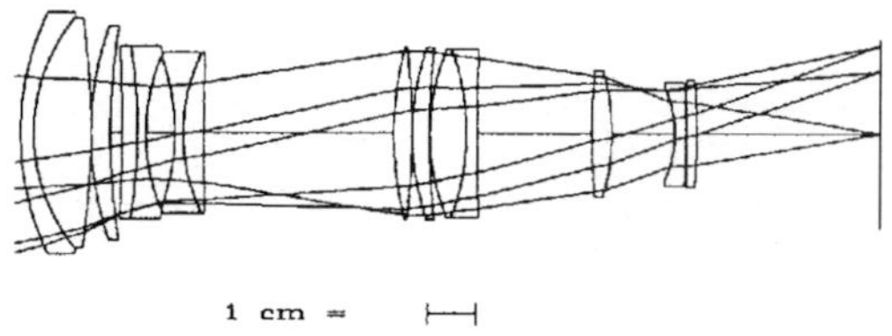

Optical Transfer Function

SN.	RADIUS	THICKNESS	GLASS	CLR. AP.
1	63.7262	2.90000	FD1	51.59
2	41.4392	12.20000	FCD1	49.08
3	-194.5870	0.10000		48.28
4	73.5061	4.20000	FC5	45.44
5	159.0812	18.25269z		44.23
6	-481.4192	3.40000	FD6	36.10
7	-119.3247	1.80000	TAC2	34.61
8	45.7271	6.21000		31.49
9	-50.2560	1.80000	LAC14	31.22
10	42.1790	4.40000	FD6	33.15
11	638.3038	0.50000z		33.36
12	83.0243	3.80000	FC5	36.06
13	-177.6418	0.10000		36.12
14	62.2167	3.60000	BACD5	36.14
15	192.0383	0.10000		35.75
16	50.9787	7.70000	FC5	35.17
17	-55.6299	2.20000	FD6	34.44
18	730.9876	12.17702		33.41
19		11.91055		29.50
20	179.5821	4.10000	F5	25.77
21	-55.0342	5.96406		25.30
	....	continued		

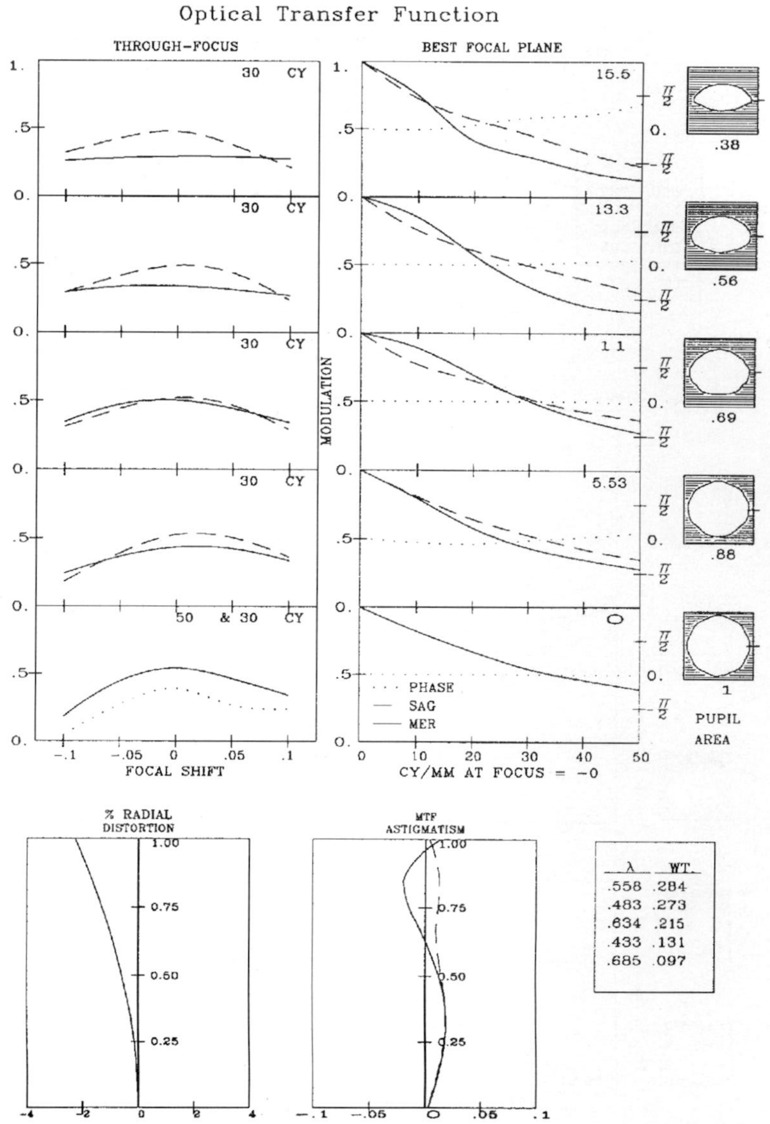

FIGURE 14 70–210-mm F/2.8–4 at $f = 70$.

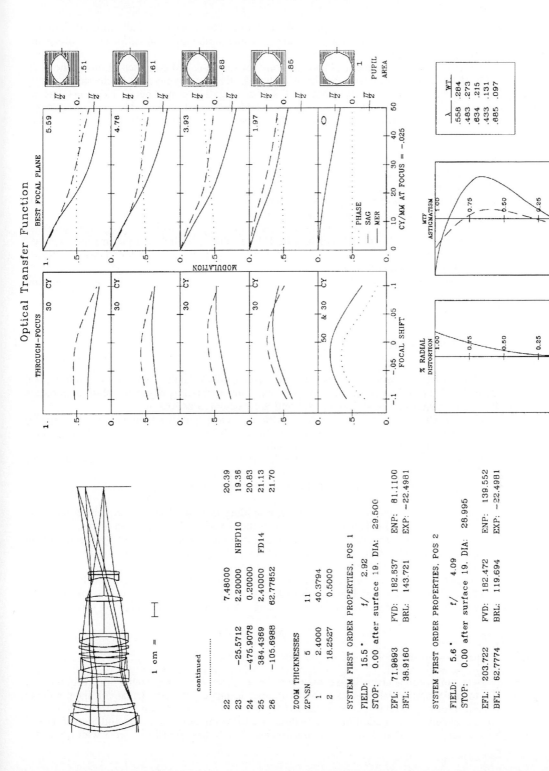

Optical Transfer Function

FIGURE 15 70–210-mm F/2.8–4 at f = 210.

and a compensating group has become academic. As a result, small internal zooming groups can serve a dual function as focusing groups under the control of an autofocusing system. Refer to Figs. 16, 17, and 18.

A recent new development in zooms is one for the so-called compact 35-mm camera. In its most basic form, this type can have as few as three elements and is characterized by having a front positive component and a rear negative component. This lens has an inherently short back focal length at the wide end, making it not suitable for SLR cameras with swinging mirrors. In more complicated versions, this idea can be extended to 28 to 160 mm or further, the main limitation being a small relative aperture at the long-focal-length end. A recent practical embodiment is a four-element 38- to 90-mm F/3.5 to 7.7 having three aspherical surfaces. See, for example, USP 4,936,661.

Zoom lenses are also found on most consumer video cameras. The classic fixed front- and rear-group type (with the aperture stop in the rear group) is still commonly used because the very small format sizes can permit acceptably small lenses. This lens form is also used for motion picture and television zooms. In many of these applications, it is desirable to have an exit pupil position that does not change with zooming. Telecentricity of the exit pupil is also sometimes required. In addition, the motion picture industry still prefers zoom lenses that have conventional front-group focusing in order to easily calibrate tape-measure focus measurements.

Very long range television zooms (often 30 : 1 or more) are also of the fixed front and rear type, with a succession of cascading zooming groups in between.

16.4 CLASSIFICATION SYSTEM

A wide variety of camera lenses has been classified in Table 1 in terms of total pixel capability P and pixels per steradian AD. Pixels are defined as digital resolution elements relative to a specified modulation level and are calculated as follows:

The polychromatic optical transfer function of each lens is calculated and the spatial frequencies at which the modulation falls to 0.5 and 0.2 is noted at each of five field points. The lower of the meridional and sagittal values is used.

The image field of the lens, assumed to be circular with diameter D, is divided into four annular regions. The outer boundaries of each region correspond respectively to $0.35H$, $0.7H$, $0.85H$, and $1.0H$, where H is the maximum field height. The area of each region is computed.

The average of the inner and outer boundary-limiting spatial frequency values is assigned to each region. This is done for both the 0.5 and 0.2 modulation levels.

The area of each annular region, in square millimeters, is multiplied by the square of the spatial frequency values from the previous step to yield regional pixel counts for both 0.5 and 0.2 modulation levels.

The pixel counts are summed over all regions to yield the D data in Table 1.

The AD data in Table 1 are obtained by dividing the total pixel values by the solid angle of the lens in object space. The solid angle S is given by the following formula:

$$S = 2\pi(1 - \cos W)$$

where W is the semifield angle of the lens in degrees.

In general, for a given image diameter D, a larger P implies higher image quality or greater information-gathering capability. A lens designed for a smaller D will have a lower P than a lens of similar quality designed for a larger D. These same generalizations hold for AD except that, in addition, a lens designed for a smaller field angle and a given D will have a larger AD than a lens of similar image quality designed to cover a wider field for

Optical Transfer Function

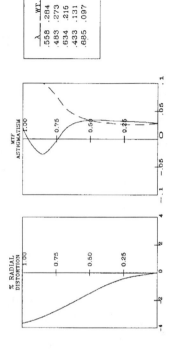

SN.	RADIUS	THICKNESS	GLASS	CLR. AP.
1	157.3035	1.90000	FDS9	53.12
2	59.0231	7.90000	BACED4	50.68
3	4807.9692	0.10000		50.21
4	49.8481	5.40000	LAC10	48.13
5	136.1973	35.88211z		47.31
6	65.6172	1.20000	TAF4	32.62
7	17.8072	7.80000		25.96
8	-68.2702	1.10000	TAF4	24.99
9	33.2828	2.30000	FDS9	23.67
10	54.4894	1.70433		23.24
11	33.7748	4.00000	FD11	22.73
12	-68.0717	1.00000		22.25
13	-43.1432	1.00000	TAF1	21.58
14	80.1552	1.69700z		20.72
15	133.2336	3.60000	FC5	19.77
16	-63.0384	0.10000		19.74
17	59.0801	4.00000	FC5	19.55
18	-326.5275	1.00000		19.66
19		1.20000		19.70
20	30.5400	1.30000	FD6	19.82
21	16.6772	8.00000	BACD15	19.08
22	-302.4610	13.16251z		18.54

.... continued

FIGURE 16 28–150-mm F/4.1–5.7 at $f = 28$.

1 cm =

λ	WT
.558	.284
.483	.273
.634	.216
.433	.131
.685	.097

BEST FOCAL PLANE

THROUGH-FOCUS

CY/MM AT FOCUS = .028

PUPIL AREA

.74   1.1   1.2   1.1   1

35   30.7   28.1   13.8   0

PHASE
SAG
MER

MODULATION

FOCAL SHIFT

MTF ASTIGMATISM

% RADIAL DISTORTION

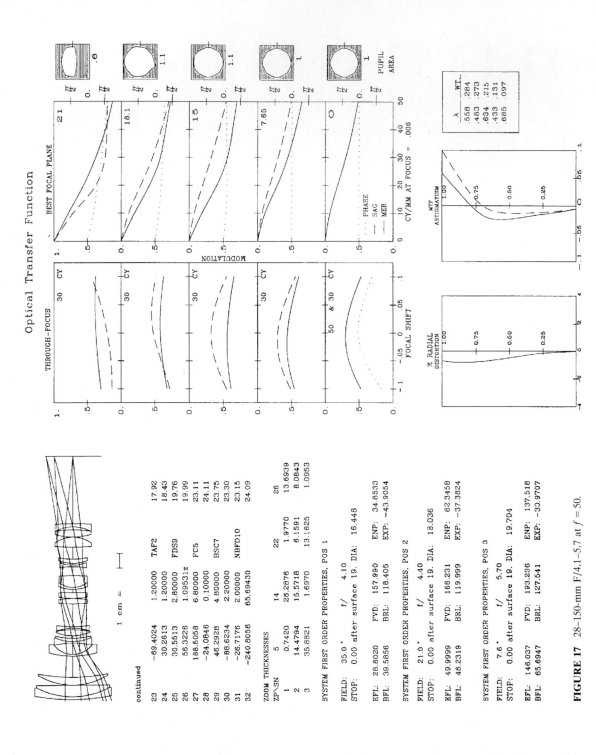

FIGURE 17 28–150-mm F/4.1–5.7 at $f = 50$.

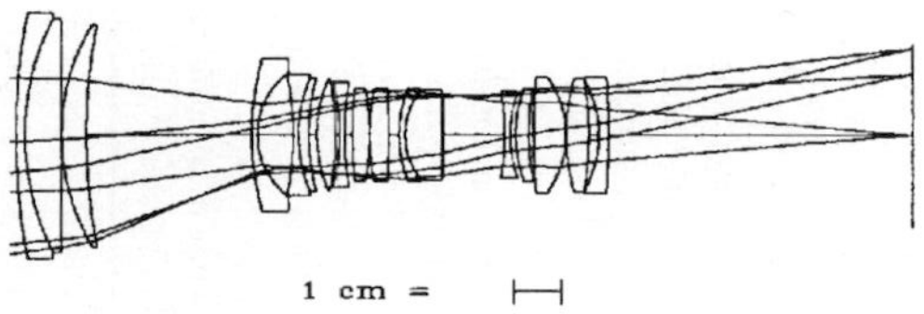

Optical Transfer Function

continued

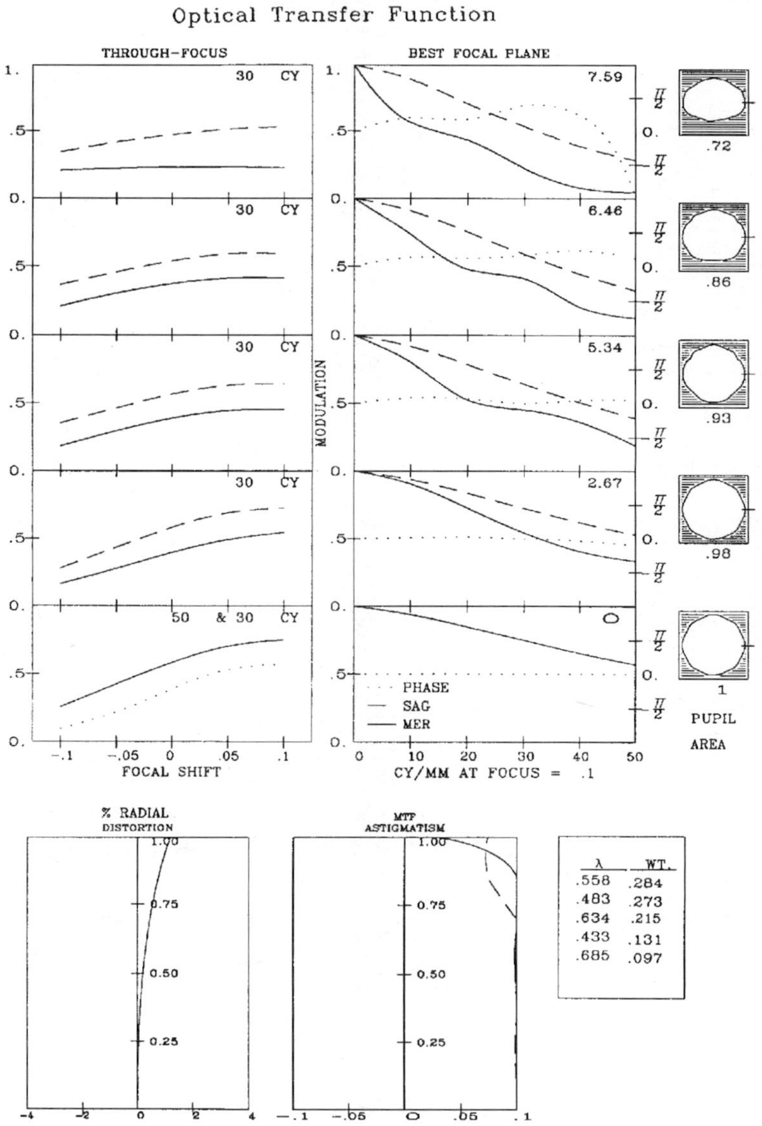

FIGURE 18 28–150-mm F/4.1–5.7 at $f = 150$.   —

TABLE 1

EFL	D	2W	F-no	$P_{0.5}$	$P_{0.2}$	$AD_{0.5}$	$AD_{0.2}$	Comments
5.1	11	94	1.8	0.27	0.93	0.14	0.47	Cinegon 2/3″ video
6.7	21	114	1.8	0.36	1.06	0.13	0.37	Xenoplan
10.2	16	76	1.8	0.66	1.59	0.50	1.19	Cinegon 1″ video
15.0	41	108	2.8	0.95	3.39	0.37	1.31	Ultrawide 35-mm SLR
17.1	11	36	1.0	0.25	0.66	0.81	2.15	Xenar 2/3″ video
17.5	42	100	2.9	1.48	3.14	0.66	1.40	Ultrawide 35-mm SLR
17.6	11	35	1.4	0.69	1.93	2.37	6.64	Xenon 2/3″ video
20.6	41	90	1.5	0.52	4.05	0.28	2.20	Ultrawide 35-mm SLR
24.5	41	80	2.1	1.24	4.86	0.84	3.30	Very wide 35-mm SLR
25.5	27	56	2.8	1.61	2.56	2.19	3.48	Panavision Primo zoom
28.0	30	56	2.8	1.32	8.74	1.79	11.88	Xenar 35-mm cine
28.0	41	72	2.8	2.32	7.28	1.93	6.06	Wide 35-mm SLR
28.5	41	71	1.5	0.82	3.14	0.70	2.69	Wide 35-mm SLR
28.8	40	70	4.1	3.35	10.31	2.95	9.08	Wide-tele 35-mm SLR
35.0	42	62	2.8	3.43	17.54	3.82	19.54	Wide 35-mm SLR
35.8	40	58	3.2	1.45	10.43	1.84	13.24	Snapshot 35-mm
36.0	42	60	1.2	2.63	5.27	3.12	6.26	Wide 35-mm SLR
50.0	38	42	4.4	3.23	14.28	7.74	34.21	Wide-tele 35-mm SLR
51.0	40	43	1.4	0.76	7.02	1.74	16.06	Normal 35-mm SLR
55.5	41	41	1.2	0.87	3.41	2.17	8.58	High-speed 35-mm SLR
55.5	41	41	1.2	0.54	4.22	1.35	10.60	Asph normal 35-mm SLR
57,5	40	37	1.4	0.98	3.67	3.05	11.41	Close focus 35-mm SLR
58.0	133	98	5.6	13.77	60.12	6.37	27.82	Super-Angulon large-format
72.0	40	31	2.9	3.22	10.44	14.11	45.70	Telezoom 35-mm SLR
73.8	16	12	2.8	0.31	1.26	9.01	36.61	Tele-xenar 1″ video
75.0	30	23	2.0	1.87	14.12	14.83	111.95	Xenar 35-mm cine
80.0	80	53	2.8	13.32	53.45	20.18	80.97	Xenotar medium-format
90.0	173	88	8.0	19.42	61.93	11.01	35.12	Super-Angulon large-format
90.0	207	98	5.6	30.07	75.59	13.92	34.98	Super-Angulon large-format
100.0	27	15	2.8	3.11	6.65	57.86	123.72	Panavision Primo zoom
100.0	116	60	5.6	33.11	85.72	39.34	101.84	APO-Symmar large-format
120.0	143	62	5.6	30.31	100.43	33.78	111.91	APO-Symmar large-format
120.0	168	70	5.6	23.78	113.14	20.93	99.57	Super-Symmar large-format
146.0	39	15	5.7	2.67	12.85	48.44	232.76	Wide-tele 35-mm SLR
150.0	80	30	4.0	12.35	87.89	57.69	410.55	Tele-Xenar medium-format
150.0	138	49	5.6	12.12	28.95	21.42	51.18	Xenar large-format
150.0	202	68	5.6	17.87	117.45	16.64	109.35	Super-Symmar large-format
180.0	300	45	5.6	100.80	368.45	210.77	770.41	Makro-Symmar 1 : 1
197.0	41	12	2.9	4.17	16.37	121.10	475.55	Telephoto 35-mm SLR
204.0	40	11	4.1	4.18	12.15	139.28	405.17	Telezoom 35-mm SLR
210.0	274	66	5.6	18.37	126.58	18.12	124.88	Super-Symmar large-format
210.0	400	87	8.0	22.62	130.62	13.11	75.70	Super-Angulon large-format
240.0	164	19	9.0	44.94	111.55	521.56	1294.60	Artar 1 : 1
250.0	126	28	5.6	38.05	89.71	203.88	480.70	Tele-Artar large-format
268.0	27	6	2.8	3.11	7.64	361.19	887.30	Panavision Primo zoom
296.0	39	8	2.1	5.36	20.36	388.37	1473.60	Telephoto 35-mm SLR
360.0	392	57	6.8	45.09	378.07	59.22	496.57	APO-Symmar large-format
400.0	250	35	5.6	33.77	160.01	116.13	550.27	APO-TXR large-format
480.0	327	19	11.0	76.29	256.34	885.40	2975.00	APO-Artar 1 : 1
480.0	400	45	8.4	86.48	509.51	180.83	1065.30	APO-Symmar large-format
800.0	500	35	12.0	108.62	460.37	373.54	1583.10	APO-TXR large-format

D—image diameter in mm
W—semifield angle in degrees
P_m—pixels $\times 10^6$ at modulation level m
AD_m—pixels $\times 10^6$ per steradian at modulation level m
F-no—F-number of the lens
EFL—effective focal length of the lens in mm

the same *D*. In other words, for the same image-quality level and format size, wide-angle lenses have lower *AD* values than do narrow angle lenses.

16.5 LENS PERFORMANCE DATA

A wide variety of camera lenses has been selected to show typical performance characteristics. In most cases, the data have been derived from the referenced published United States patents. The authors have taken the liberty of reoptimizing most of the data to arrive at what would, in our judgment, correspond to production-level designs. All performance data have been shown at maximum aperture. It is important to realize that photographic lenses are invariably designed so that optimum performance is achieved at F-numbers at least 2 stops slower than maximum. A general explanation of the data page follows.

The lens drawing shows the marginal axial rays together with the upper and lower meridional rays for seven-tenths and full field.

The lens prescription and all other data are in millimeters. Glass catalogs are Hoya, Ohara, and Schott. Distances to the right of a surface are positive. A positive radius means that the center of curvature is to the right of the surface. The thickness and glass data indicate the distance and medium immediately following the particular surface.

The optical transfer function (OTF) plots show the through-focus modulation transfer function (MTF) on the left and the OTF at best axial focus on the right. The data are shown for five field points, viz., the axis, 0.35*H*, 0.70*H*, 0.85*H*, and 1.0*H*, where *H* is the maximum field angle in object space. The actual field angles are indicated in the upper-right-hand corner of each best-focus OTF block and are in degrees. The through-focus data are at the indicated spatial frequency in cycles per millimeter with an additional frequency on-axis (dotted curve). Both the through-focus and best-focus data indicate meridional (solid curves) and sagittal (dashed curves) MTF. The modulus scale is on the left of each block and runs from zero to one. The phase of the OTF is shown as a dotted curve in the best-focus plots. The scale for the phase is indicated on the right of each best-focus block and is in radian measure. All the OTF data are polychromatic. The relative weights and wavelengths used appear in the lower-right-hand corner of each page. The wavelengths are in micrometers and the weights sum to one. The axial focus shift indicated beneath the best-focus plots is relative to the zero position of the through-focus plots. The best-focus plane is at the peak of the additional axial through-focus plot (dotted curve).

Vignetting for each field angle is illustrated by the relative pupil area plots on the right-hand side of each page. The distortion plots shows the percentage of radial distortion as a function of fractional field height. The MTF Astigmatism plot shows the loci of the through-focus MTF peaks as a function of fractional field height. The data can be readily determined directly from the through-focus MTF plots.

Certain acronyms are used in the System First-Order Properties:

Effective focal length (EFL)

Back focal length (BFL)

Front vertex distance (FVD)

Barrel length (BRL)

Entrance pupil distance (ENP)

Exit pupil distance (EXP)

The ENP and EXP data are measured from the front and rear vertices of the lens, respectively. A positive distance indicates that the pupil is to the right of the appropriate vertex.

16.6 ACKNOWLEDGMENTS

The authors would like to acknowledge data provided by R. Mühlschlag of Jos. Schneider Optische Werke and C. Marcin of Schneider Corporation of America. We also appreciate permission granted by Panavision Corporation to use data pertaining to the Primo zoom lens.

16.7 REFERENCES

Betensky, E. I., "Photographic Lenses," in R. Shannon and J. Wyant (eds.), *Applied Optics and Optical Engineering,* vol. 8, Academic Press, New York, 1980.

Cook, G. H., "Photographic Objectives," in R. Kingslake (ed.), *Applied Optics and Optical Engineering,* vol. 3, Academic Press, New York, 1965.

Figures 1–18 contain data that was originally derived from the following United States Patents. The patents are listed in order corresponding to the figure order. USP 4,431,273; 4,381,888; 4,095,873; 3,830,554; 4,770,512; 4,333,714; 4,136,931; 4,303,315; 4,792,216; 4,110,007; 3,942,875; 4,786,152; 4,732,459; 5,018,843 (Figs. 14–16); 4,758,073 (Figs. 17–18).

Kingslake, R., *A History of the Photographic Lens,* Academic Press, Inc., San Diego, 1989.

Kingslake, R., *Optics in Photography,* SPIE Press, Bellingham, WA, 1992.

CHAPTER 17
MICROSCOPES

Shinya Inoué and Rudolf Oldenbourg
Marine Biological Laboratory
Woods Hole, Massachusetts

17.1 GLOSSARY

f focal length

M magnification

n refractive index

NA   numerical aperture

z distance along optical axis

λ wavelength of light

17.2 INTRODUCTION

Historic Overview

The optical principles and basic lens design needed to generate a diffraction-limited, highly magnified image with the light microscope were already essentially perfected a century ago. Ernst Abbe demonstrated how a minimum of two successive orders of diffracted light had to be captured in order for a particular spacing to be resolved (see historical sketch about Abbe principle[1]). Thus, he explained and demonstrated with beautiful experiments the role of the wavelength of the imaging light and the numerical aperture (NA, $NA = n \sin \Theta$, Fig. 1) of the objective and condenser lenses on the resolving power of the microscope. In general, the minimum spacing δ for line gratings that can just be resolved cannot be smaller than

$$\delta = \frac{\lambda}{2\,NA} \tag{1}$$

when the NA of the condenser is equal to the NA of the objective.

With the NA of the Lister-type objective (Fig. 2a) extended by the addition of an aplanatic hyperhemisphere in the Amici-type objective (Fig. 2b), and further by the incorporation of homogeneous immersion (Fig. 2c), the numerical aperture of the objective lens was raised from around 0.25 to 0.65 and to 1.25. The apochromatic objective (Fig. 2d) was introduced in 1986. By incorporating low-dispersion fluorite lens

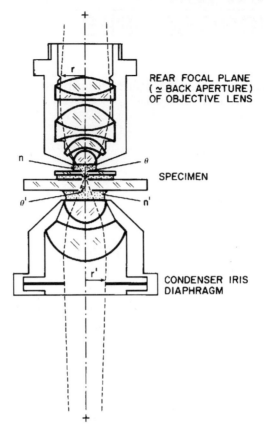

FIGURE 1   Definition of numerical aperture of objective ($NA_{obj} = n \sin \Theta$) and condenser ($NA_{cond} = n' \sin \Theta'$).[6]

elements, the apochromats provided good correction for lateral as well as longitudinal chromatic aberration up to an NA of 1.35 when used in conjunction with a compensating eyepiece, albeit with residual error of the secondary spectrum and appreciable curvature of field. Early high-resolution compound microscopes were equipped with an (oil-immersible) Abbe condenser whose iris diaphragm, placed at the front focal plane of the condenser, could be off-centered to achieve oblique illumination.

Much of the early use of the light microscope depended on the relatively high image contrast that could be generated by differential absorption, reflection, birefringence, etc., due to specimen composition or structure. Specimens, such as unstained living cells and other transparent objects introducing small optical path differences, were generally not amenable to direct microscopic observation for they would not produce detectable image contrast when brought to exact focus.

These impediments were removed by Zernicke who showed how contrast in the microscope image is generated by interference between the light waves that make up the direct rays (that are undeviated by the specimen) and those that were scattered and suffered a phase difference by the presence of the specimen. Using this principle, Zernicke invented the phase contrast microscope.[2] For the first time it became possible to see, in focus, the image of small, nonabsorbing objects. Zernicke's revelations, together with

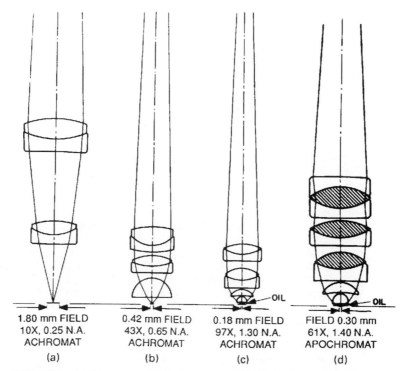

1.80 mm FIELD
10X, 0.25 N.A.
ACHROMAT
(a)

0.42 mm FIELD
43X, 0.65 N.A.
ACHROMAT
(b)

0.18 mm FIELD
97X, 1.30 N.A.
ACHROMAT
(c)

FIELD 0.30 mm
61X, 1.40 N.A.
APOCHROMAT
(d)

FIGURE 2 (*a*) Lister-type, (*b*) Amici-type, and (*c*) oil-contacted achromatic objectives. In (*d*) an apochromat is shown, with fluorite elements shaded.[57]

Gabor's further contributions,[3] not only opened up opportunities for the design of various types of interference-dependent image-forming devices but, even more importantly, improved our understanding of the basic wave optics involved in microscope image formation.

About the same time as Zernicke's contributions, perfection of the electron microscope made it possible to image objects down to the nanometer range, albeit necessitating use of a high-vacuum environment and other conditions compatible with electron imaging. Thus, for four decades following World War II, the light microscope in many fields took a back seat to the electron microscope.

During the last decade, however, the light microscope has reemerged as an indispensable, powerful tool for investigating the submicron world in many fields of application. In biology and medicine, appropriate tags, such as fluorescent and especially immunofluorescent tags, are used to signal the presence and location of selected molecular species with exceptionally high sensitivity. Dynamic behavior of objects far below the limit of resolution are visualized by digitally-enhanced video microscopy directly in their natural (e.g., aqueous) environment. Very thin optical sections are imaged by video microscopy, and even more effectively with confocal optics. Quantitative measurements are made rapidly with the aid of digital image analysis.

At the same time, computer chips and related information-processing and storage devices, whose availability in part has spurred the new developments in light microscopy, are themselves miniaturized to microscopic dimensions and packaged with increasingly higher density. These electronic and photonic devices in turn call for improved means for mass manufacturing and inspection, both of which require advanced microscope optics.

Driven by the new needs and aided in part by computerized ray tracing and the introduction of new optical materials, we see today another epochal advance in the quality of lens design. The precision and remote control capabilities of mechanical components are also steadily improving. Furthermore, we may expect another surge of progress, hand-in-hand with development of improved electro-optical and electromechanical devices, in regulated image filtration, contrast-generating schemes, as well as in optical manipulation of the specimen employing microscope optics.

Outline

In the following, we first review the general optical considerations that enter into the use and design of the light microscope, examine several commonly used modes of contrast generation, discuss additional modes of microscopy, list some applications in which the microscope is used to optically manipulate the specimen, and describe some accepted mechanical standards.

Many of the above subjects which are introduced here in the context of microscopy are discussed in more detail in other chapters of this Handbook. On general optical considerations consult Handbook chapters on Geometric Optics (Vol. I, Chap. 1) and on Optical Elements, such as Lenses (Vol. II, Chap. 1), Mirrors (Vol. II, Chap. 2), Polarizers (Vol. II, Chap. 3), etc., as well as chapters on Physical Optics for wave phenomena such as Interference (Vol. I, Chap. 2), Diffraction (Vol. I, Chap. 3), Coherence (Vol. 1, Chap. 4), and Polarization (Vol. I, Chap. 5) which, as phenomena, are essential to the workings of the various contrast modes of the microscope. Material on image detection and processing can be found in Handbook articles on Vision, Imaging Detectors, and Optical Information and Image Processing.

There are a number of excellent review articles and books discussing the optical principles of light microscopes[1,4,5] and microscopic techniques[6-8] and their applications.[9-11] The present article is intended in part to bridge the territories of the manufacturer and the user of the microscope, including those who incorporate microscope optics into other equipment or apply them in unconventional ways. Thus, the article emphasizes basic principles and does not attempt to discuss specialized instruments such as operating microscopes, specific inspection microscopes, etc.

17.3   GENERAL OPTICAL CONSIDERATIONS

Geometrical and Wave Optical Train, Magnification

In an optical train of a conventional microscope (Fig. 3), the objective lens L_{ob} projects an inverted, real, magnified image O' of the specimen O (or object) into the intermediate

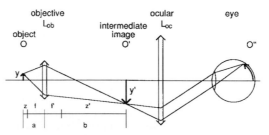

FIGURE 3   Ray path in the microscope from object to observer's eye (see text).

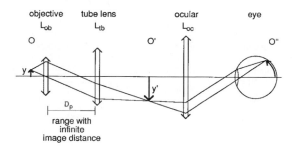

FIGURE 4 Ray path in microscope with infinity-corrected objective and tube lens.

image plane (or primary image plane). The intermediate image plane is located at a fixed distance $f' + z'$ behind L_{ob}, where f' is the back focal length of L_{ob} and z' is the optical tube length of the microscope. In general, O' is an aerial image for which an ocular L_{oc} (or the eyepiece) acts as a magnifier in front of the eye. Since L_{oc}, coupled with the corneal surface and lens of the eye, produces an erect image O'' of O on the retina, the object appears inverted to the observer. The ocular may conversely be used as a projector.

Continuing with the schematic diagram in Fig. 3, using thin-lens approximations, O is placed at a short distance z just outside of the front focus of L_{ob}, such that $z + f = a$, where f is the front focal length of L_{ob} and a is the distance between O and L_{ob}. O' is formed at a distance $b = (z' + f')$ behind L_{ob}. For a height y of O, the image height $y' = y \times b/a$. Thus, L_{ob} magnifies O by $M_{ob} = b/a$. Also, $M_{ob} = f/z = z'/f'$. M_{ob} is the transverse magnification (or lateral magnification) of L_{ob}. In turn, y' is magnified by L_{oc} by a factor $M_{oc} = 25$ cm$/f_{oc}$, where f_{oc} is the focal length of the ocular and 25 cm is the so-called near distance from the observer's eye (see section on Vision in this Handbook). Thus, the total transverse magnification of the microscope $M_{tot} = M_{ob} \times M_{oc}$.

Note that most microscope objectives are corrected for use only within a narrow range of image distances, and, in general, only in conjunction with specific groups of oculars. M_{ob}, which is the magnification inscribed on the barrel of the objective lens, is defined for its specified tube length (for high-power objectives, $M_{ob} = z'/f$). In the case of an infinity-corrected objective, M_{ob} is the ratio y'/y as used in conjunction with a specific tube lens L_{tb} (Fig. 4). These factors, as well as those mentioned under "Microscope Lenses, Aberrations," must be kept in mind when a microscope objective is used as a magnifying lens, or in reverse as a high-numerical-aperture reducing lens, to form a truly diffraction-limited image.

Conjugate Planes

Continuing the optical train back to the light source in a transilluminating microscope, Fig. 5a shows the ray paths and foci of the waves that focus on an on-axis point in the specimen. In Köhler illumination, the distance between the specimen and the condenser are adjusted so that the image of the field diaphragm in the illuminator is superimposed with the focused region of the specimen, and the lamp collector lens is adjusted so that the source image is focused in the plane of the condenser aperture diaphragm. Thus, $\bar{O}$, O, O', and O'' all lie in image planes that are conjugate with each other.

Tracing the rays emitted from a point in the light source (Fig. 5b), the rays are parallel between the condenser and the objective lenses, thus the specimen is illuminated by a plane wavefront. This situation arises because the iris diaphragm of the condenser is built into the front focal plane of the condenser. Also, since the pupil of an (experienced)

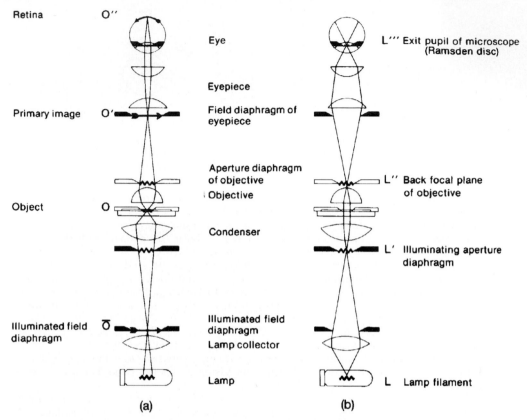

Retina — O''

Eye

Eyepiece

Primary image — O'

Field diaphragm of eyepiece

Aperture diaphragm of objective

Objective

Object — O

Condenser

Illuminated field diaphragm — Ō

Illuminated field diaphragm

Lamp collector

Lamp

(a)

L''' Exit pupil of microscope (Ramsden disc)

L'' Back focal plane of objective

L' Illuminating aperture diaphragm

L Lamp filament

(b)

FIGURE 5   Ray paths in a transmitted light microscope adjusted for Köhler illumination. Two sets of conjugate planes are shown: set O in (a) is conjugate with the object O and with the field diaphragm planes; set L in (b) is conjugate with the lamp filament L and with aperture diaphragm planes.[58]

observer's eye is placed at the eyepoint or back focal plane of the ocular, the four aperture planes L, L', L'', and L''' are again conjugate to each other.

As inspection of these two figures shows, the field planes and aperture planes are in reciprocal space relative to each other throughout the whole optical system. This reciprocal relationship explains how the various diaphragms and stops affect the cone angles, paths, and obliquity of the illuminating and image-forming rays, and the brightness, uniformity, and size of the microscope field. More fundamentally, a thorough grasp of these reciprocal relationships is needed to understand the wave optics of microscope image formation and for designing various contrast-generating devices and other microscope optical systems.

Airy Disk and Lateral Resolution

Given a perfect objective lens and an infinitely small point of light residing in the specimen plane, the image formed in the intermediate image plane by the objective lens is not another infinitely small point, but a diffraction image with a finite spread (Fig. 6a). This Airy diffraction image is the Fraunhofer diffraction pattern formed by the exit pupil of the

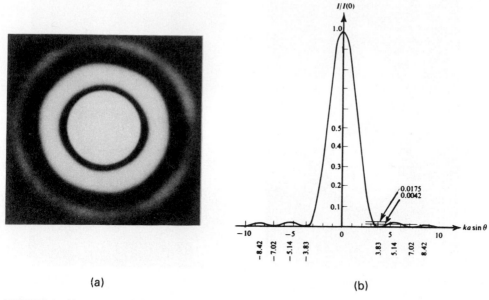

(a)

(b)

FIGURE 6 Airy pattern of circular aperture: image (a) of central Airy disk, first dark ring and subsidiary maximum; graph (b) of radial intensity distribution.[59]

objective lens from which spherical waves converge to the focal point. The distribution of irradiance of the diffraction image (Fig. 6b) is given by an expression containing the first order Bessel function $J_1(v)$:

$$I(v) = I_0 \left(\frac{2J_1(v)}{v} \right)^2 \qquad (2)$$

with v proportional to the diffraction angle. If the irradiance is calculated as a function of radius measured from the center of the Airy diffraction pattern located in the intermediate image plane, v takes on the form:

$$v = 2\pi \frac{\mathrm{NA}}{M\lambda} r_i \qquad (3)$$

where NA is the numerical aperture and M the magnification of the objective lens, λ the wavelength of light, and r_i the radial distances measured in the intermediate image plane. If we express r_i as a distance r_o in object space, with $r_i = Mr_o$, we obtain the more familiar relationship:

$$v = 2\pi \frac{\mathrm{NA}}{\lambda} r_o \qquad (4)$$

The central bright disk of the diffraction image is known as the Airy disk, and its radius (the radius from the central peak to the first minimum of the diffraction image) in object space units is given by:

$$r_{\mathrm{Airy}} = 0.61 \frac{\lambda}{\mathrm{NA}} \qquad (5)$$

When there exist two equally bright, self-luminous points of light separated by a small distance d in object space, i.e., the specimen plane, their diffraction images lie side-by-side

 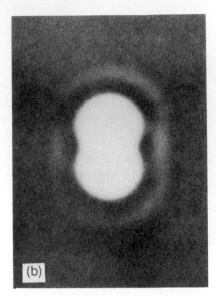

FIGURE 7 Overlapping Airy patterns: (*a*) clearly resolved; (*b*) center of Airy patterns separated by $d = r_{Airy}$, Rayleigh criterion.[59]

in the image plane. The sum of the two diffraction images, assuming the two points of light were mutually incoherent, appears as in Fig. 7a. As d becomes smaller so that the first minimum of one diffraction image overlaps with the central maximum of the neighboring diffraction image ($d = r_{Airy}$, Fig. 7b), their sum (measured along the axis joining the two maxima) still contains a dip of 26.5 percent of the peak intensities that signals the twoness of the source points (the Rayleigh criterion). Once d becomes less than this distance, the two diffraction images rapidly pass a stage where instead of a small dip, their sum shows a flat peak (the Sparrow criterion) at $d = 0.78 \, r_{Airy}$, and thereafter the sum of the diffraction images appears essentially indistinguishable from one arising from a single point source instead of two. In other words, we can no longer resolve the image of the two points once they are closer than the Rayleigh criterion, and we lose all cues of the periodicity at spacings below the Sparrow criterion. Since the diameter of the Airy diffraction image is governed by NA_{obj} and the wavelength of the image-forming light λ, this resolution limit normally cannot be exceeded (for exceptions, see the sections on "Confocal Microscopy" and "Proximity Scanning Microscopy" later in this chapter).

The consideration given here for two point sources of light applies equally well to two absorbing dots, assuming that they were illuminated incoherently. (Note, however, that it may, in fact, be difficult or impossible to illuminate the two dots totally incoherently since their spacing may approach the diameter of the diffraction image of the illuminating wave. For the influence of the condenser NA on resolution in transillumination, refer to the section on "Transillumination" later in this chapter. Also, the contrast of the diffraction images of the individual absorbing dots diminishes rapidly as their diameters are decreased, since the geometrical size of such small dots would occupy a decreasing fraction of the diameter of their diffraction images. For further detail see Ref. 12.)

The image of an infinitely small point or line thus acquires a diameter equal to that of the Airy disk when the total magnification of the image becomes sufficiently large so that we can actually perceive the diameter of the Airy disk. In classical microscopy, such a large magnification was deemed useless and defined as empty magnification. The situation is, however, quite different when one is visualizing objects smaller than the limit of resolution with video microscopy. The location of the Airy disk can, in fact, be established with very high precision. Distances between lines that are clearly isolated from each other can, therefore be measured to a precision much greater than the resolution limit of the microscope. Also, minute movements of nanometer or even Ångstrom steps have been measured with video-enhanced light microscopy using the center of gravity of the highly magnified diffraction image of marker particles.[13]

Three-dimensional Diffraction Pattern, Axial Resolution, Depth of Focus, Depth of Field

The two-dimensional Airy pattern that is formed in the image plane of a point object is, in fact, a cross section of a three-dimensional pattern that extends along the optical axis of the microscope. As one focuses an objective lens for short distances above and below exact focus, the brightness of the central spot periodically oscillates between bright and dark as its absolute intensity also diminishes. Simultaneously, the diameters of the outer rings expand, both events taking place symmetrically above and below the plane of focus in an aberration-free system (Fig. 8).

Figure 9c shows an isophot of the longitudinal section of this three-dimensional diffraction image. In the graph we recognize at $v = 1.22\pi$ (and $u = 0$, focal plane) the first minimum of the Airy pattern which we discussed in the previous section. The intensity distribution along u perpendicular to the focal plane has its first minima at $u = \pm 4\pi$ and $v = 0$ ($\pm z_1$ in Fig. 9a).

To find the actual extent of the three-dimensional diffraction pattern near the intermediate plane of the microscope, we express the dimensionless variables v and u of Fig. 9c as actual distances in image space. The relationship between v and the lateral distance r_i is given by Eq. (3). The axial distance z_i, oriented perpendicular to the image plane, is related to u by:

$$u = 2\pi \frac{NA^2}{M^2 \lambda} z_i \tag{6}$$

The first minimum ($u = 4\pi$) is at a distance $z_1 = (2M^2\lambda)/NA^2$. To transfer distance z_i in image space to distance z_o in object space, we use the relationship $z_i = z_o M^2/n$. (Note that for small axial distances, to a close approximation, the axial magnification is the square of the lateral magnification M divided by the refractive index n of the object medium.) The distance from the center of the three-dimensional diffraction pattern to the first axial minimum in object space is then given by:

$$z_{min} = 2 \frac{\lambda n}{NA^2} \tag{7}$$

z_{min} corresponds to the distance by which we have to raise the microscope objective in order to change from exact focus of a small pinhole to the first intensity minimum in the center of the observed diffraction pattern (see Fig. 8).

In correspondence to the lateral resolution limit which is taken as the Airy disk radius r_{Airy} [Eq. (5)], we can use z_{min} as a measure of the limit of axial resolution of microscope optics. Note that the ratio of axial to lateral resolution ($z_{min}/r_{Airy} = 3.28 \, n/NA$) is inversely proportional to the numerical aperture of the objective.

The axial resolution of the microscope is closely related to the depth of focus, which is the axial depth on both sides of the *image* plane within which the image remains acceptably sharp (e.g., when a focusing screen at the image plane is displaced axially without moving the object or objective). The depth of focus D is usually defined as $\frac{1}{4}$ of the axial distance between the first minima above and below focus of the diffraction image of a small pinhole. In the intermediate image plane, this distance is equal to $z_1/2$, with z_1 defined earlier. The depth of focus defined by z_1 is the diffraction-limited, or physical, depth of focus.

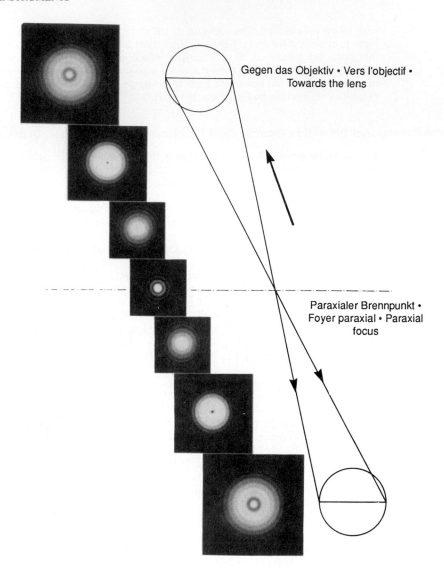

Gegen das Objektiv • Vers l'objectif • Towards the lens

Paraxialer Brennpunkt • Foyer paraxial • Paraxial focus

FIGURE 8 The evolution of the diffraction image of a circular aperture with differing planes of focus in an aberration-free system.[60]

A second and sometimes dominating contribution to the total depth of focus derives from the lateral resolution of the detector used to capture the image. This geometric depth of focus depends on the detector resolution and the geometric shape of the light cone converging to the image point. If the detector is placed in the intermediate image plane of an objective with magnification M and numerical aperture NA, the geometrical depth of focus D is given by

$$D = \frac{M}{\mathrm{NA}} e \qquad (8)$$

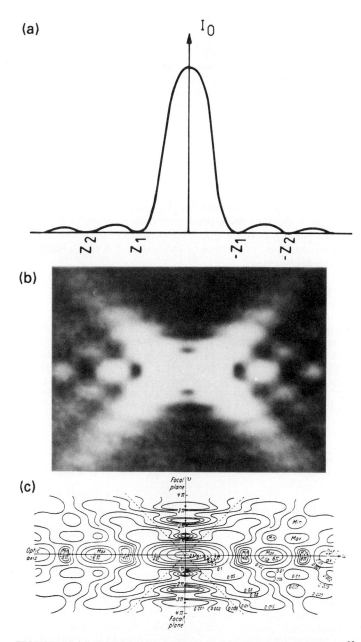

FIGURE 9 (*a*) Axial intensity distribution of irradiance near focal point;[15] (*b*) meridional section through diffraction pattern near focal point of a point source of light focused by lens with a uniform circular aperture;[15] (*c*) contour plot (isophot) of the same cross section as in (*b*).[35] The three-dimensional diffraction pattern is obtained by rotating the meridional section around the optic axis. The three-dimensional diffraction pattern is also called the intensity point spread function.

with e the smallest distance resolved by the detector (e is measured on the detector's face plate).

The depth in *specimen* space that appears to be in focus within the image, without readjustment of the microscope focus, is the depth of field (unfortunately often also called the depth of focus). To derive expressions for the depth of field, we can apply the same arguments as outlined above for the depth of focus. Instead of moving the image plane in and out of focus, we keep the image plane in the ideal focus position and move the small pinhole in object space. Axial distances in object space, however, are a factor n/M^2 smaller than corresponding distances in image space. Therefore, we apply this factor to the expression for the geometrical depth of focus [Eq. (8)] and add the physical depth of field [derived from Eq. (7)] for the total depth of field d_{tot}:

$$d_{tot} = \frac{\lambda n}{NA^2} + \frac{n}{MNA} e \qquad (9)$$

Notice that the diffraction-limited depth of field shrinks inversely proportionally with the square of the NA, while the lateral limit of resolution is reduced inversely proportionally to the first power of the NA. Thus, the axial resolution and thinness of optical sections that can be attained are affected by the system NA much more so than is the lateral resolution of the microscope.

These values for the depth of field, and the distribution of intensities in the three-dimensional diffraction pattern, are calculated for incoherently illuminated (or emitting) point sources (i.e., $NA_{cond} \geq NA_{obj}$). In general, the depth of field increases, up to a factor of two, as the coherence of illumination increases (i.e., as $NA_{cond} \rightarrow 0$). However, the three-dimensional point spread function with partially coherent illumination can depart in complex ways from that so far discussed when the aperture function is not uniform (see sections on "Phase Contrast and Other Aperture-modifying Contrast Modes," "Polarizing," and "Interference" later in this chapter). In a number of phase-based, contrast-generating modes of microscopy, the depth of field may turn out to be unexpectedly shallower than that predicted from Eq. (9) and may yield extremely thin optical sections.[14]

17.4   *MICROSCOPE LENSES, ABERRATIONS*

Designations of Objective Lenses

With few exceptions, microscope objective lenses are designed to form a diffraction-limited image in a specific image plane that is located at a fixed distance from the objective lens (or from the tube lens in the case of an infinity-focus system). The field of view is often quite limited, and the front element of the objective is placed close to the specimen with which it must lie in optical contact through a medium of defined refractive index (see Sects. "Coverslip Correction," "Tube Lengths and Tube Lenses for which Microscope Objectives Are Corrected," and "Mechanical Standards" later in this chapter) for standardized distances commonly used for microscope objectives).

Depending on the degree of correction, objectives are generally classified into achromats, fluorites, and apochromats with a plan designation added to lenses with low curvature of field and distortion (Table 1). Some of these characteristics are inscribed on the objective lens barrel, such as Plan Apo 60/1.40 oil 160/0.17, meaning 60 power/1.40 NA Plan Apochromatic objective lens designed to be used with oil immersion between the objective front element and the specimen, covered by an 0.17-mm-thick coverslip, and used at a 160-mm mechanical tube length. Another example might be Epiplan-Neofluar

TABLE 1   Objective Lens Types and Corrections

Type	Spherical	Chromatic	Flatness
Achromat	*	2λ	No
F-Achromat	*	2λ	Improved
Neofluar	3λ	$<3\lambda$	No
Plan-Neofluar	3λ	$<3\lambda$	Yes
Plan Apochromat	4λ	$>4\lambda$	Yes

* = corrected for two wavelengths at two specific aperture angles.
2λ = corrected for blue and red (broad range of visible spectrum).
3λ = corrected for blue, green, and red (full range of visible spectrum).
4λ = corrected for dark blue, blue, green, and red.
Source: Zeiss publication #41-9048/83.

$50\times/0.85 \infty/0$, which translates to Plan "Fluorite" objective designed for epi-illumination (i.e., surface illumination of specimen through the objective lens rather than through a separate condenser) with a $50\times$ magnification and 0.85 NA to be used in air (i.e., without added immersion medium between the objective front element and coverslip or specimen), with no coverslip, and an (optical) tube length of infinity. "Infinity corrected" objectives may require the use of a designated tube lens to eliminate residual aberration and to bring the rays to focus into the image plane. Achromats are usually not designated as such, and several other codes are inscribed or color-coded on microscope objectives (Tables 2 and 3).

Most objective lenses are designed to be used with a specified group of oculars or tube lenses that are placed at specific distances in order to remove residual errors. For example, compensation oculars are used in conjunction with apochromatic and other high-NA objectives to eliminate lateral chromatic aberration and improve flatness of field. However, some lenses such as Nikon CF (Chrome Free) and current Zeiss Jena objectives are fully corrected so as not to require additional chromatic correction by a tube lens or the ocular.

Coverslip Correction

For objective lenses with large NAs, the optical properties and thicknesses of the media lying between its front element and the specimen critically affect the calculations needed to satisfy the aplanatic and sine conditions and otherwise to correct for image aberrations. For homogeneous immersion objectives (that are designed to be used with the refractive indices and dispersion of the immersion oil, coverslip, and medium imbibing the specimen, all matched to that of the objective lens front element), the calculation is straightforward since all the media can be considered an extension of the front lens element.

However, with nonimmersion objectives, the cover glass can become a source of chromatic aberration which is worse the larger the dispersion and the greater the thickness of the cover glass. The spherical aberration is also proportional to the thickness of the cover glass. In designing objectives not to be used with homogeneous immersion, one assumes the presence of a standard cover glass and other specific optical media between the front lens element and the specimen. As one departs from these designated conditions, spherical aberration (and also coma) increases with the NA of the lens, since the difference between the tangent and sine of the angle of incidence is responsible for departure from the needed sine condition.

It should also be noted that oil immersion objectives fail to provide full correction, or full NA, when the specimen is mounted in an imbibing medium with a different refractive index, e.g., aqueous media, even with the objective and cover glass properly oil-contacted to each other. With such an arrangement, the diffraction image can degrade noticeably as one focuses into the specimen by as little as a few micrometers.[15] Special water immersion

TABLE 2 Common Abbreviations Designating Objective Lens Types

DIC, NIC	Differential (Nomarski) interference contrast
L, LL, LD, LWD, ELWD, ULWD	Long working distance (extra-) (ultra-)
FL, FLUOR, NEOFLUOR, FLUOTAR	With corrections as with "fluorite" objectives but no longer implies the inclusion of fluorite elements
PHASE, PHACO, PC, PH 1, 2, 3, etc.	Phase contrast, using phase condenser annulus 1, 2, 3, etc.
DL, DM, PLL, PL, PM, PH, NL, NM, NH	Phase contrast: dark low, dark medium, positive low low, low, medium, high contrast (regions with higher refractive index appear darker); negative low, medium, high contrast (regions with higher refractive index appear lighter)
PL, PLAN; EF	Flat field; extended field (larger field of view but not as high as with PLAN, achromats unless otherwise desginated)
PLAN APO	Flat field apochromat
NPL	Normal field of view plan
P, PO, POL	Low birefringence, for polarized light
UV	UV transmitting (down to approx. 340 nm), for UV-excited epifluorescence
ULTRAFLUAR	Fluorite objective for imaging down to approx. 250 nm in UV as well as in the visible range
CORR, W/CORR	With correction collar
I, IRIS, W/IRIS	Adjustable NA, with iris diaphragm built into back focal plane
M	Metallographic
NC, NCG	No coverslip
EPI	Surface illumination (specimen illuminated through objective lens), as contrasted to dia- or transillumination
BD, HD	For use in bright or darkfield (hell, dunkel)
CF	Chrome-free (Nikon: objective independently corrected longitudinal chromatic aberrations at specified tube length)
ICS	Infinity color-corrected system (Carl Zeiss: objective lens designed for infinity focus with lateral and longitudinal chromatic aberrations corrected in conjunction with a specified tube lens)
OIL, HI, H; WATER, W; GLY	Oil-immersion, Homogeneous-immersion, water-immersion, glycerol-immersion
U, UT	Designed to be used with universal stage (magnification/NA applies for use with glass hemisphere; divide both values by 1.51 when hemisphere is not used)
DI; MI; TI Michelson	Interferometry: noncontact; multiple-beam (Tollanski);
ICT; ICR	Interference contrast: in transillumination; in reflected light

objectives (e.g., Nikon Plan Apo 60×/1.2 NA and short-wavelength transmitting Fluor 40×/1.0 NA, both with collar to correct coverslip thickness deviation from 0.17 mm) overcome such aberrations, even when the specimen is imaged through an aqueous medium of 200-μm thickness.

For lenses that are designed to be used with a standard coverslip of 0.17-mm thickness (and $n_D = 1.515$), departure from standard thickness is not overly critical for objectives

TABLE 3   Color-coded Rings on Microscope Objectives

Color code (narrow colored ring located near the specimen
             end of objective)

Black	Oil immersion
Orange	Glycerol immersion
White	Water immersion
Red	Special

Magnification color code (narrow band located further away
    from specimen than immersion code)

Color	Magnification
Black	1, 1.25
Brown	2, 2.5
Red	4, 5
Yellow	10
Green	16, 20
Turquoise blue	25, 32
Light blue	40, 50
Cobalt (dark) blue	60, 63
White (cream)	100

with NA of 0.4 or less. However, for high-NA, nonhomogeneous immersion lenses, the problem becomes especially critical so that even a few micrometers' departure of the cover glass thickness degrades the image with *high-dry objectives* (i.e., nonimmersion objectives with high NA) of NA above 0.8 (Fig. 10). To compensate for such error, well-corrected, high-dry objectives are equipped with correction collars that adjust the spacing of their intermediate lens elements according to the thickness of the cover glass. Likewise, objective lenses that are made to be viewed through layers of silicon or plastic, or of different immersion media (e.g., water/glycerol/oil immersion lenses), are equipped with correction collars.

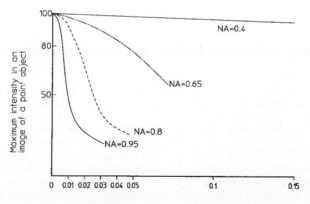

FIGURE 10   Calculated maximum intensity in the image of a point object versus the deviation of the coverglass thickness from the ideal thickness of 0.17 mm.[61]

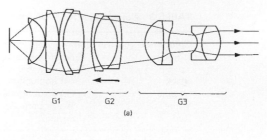

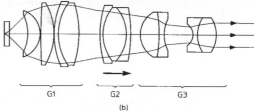

FIGURE 11 High-dry objective lens (60×/0.7 NA) equipped with a correction collar for (*a*) focusing at the surface or (*b*) through plane glass of up to 1.5-mm thickness. The lens group G_2 is moved forward to enhance the spherical and chromatic correction by G_1 and G_2 when focused on the surface, while it is moving backward to compensate for the presence of the glass layer when focusing deeper through the glass.[61] (U.S. Patent 4666256.)

The use of objective lenses with correction collars does, however, demand that the observer is experienced and alert enough to reset the collar using appropriate image criteria. Also, the focus tends to shift, and the image may wander, during adjustment of the correction collar. Figure 11 shows an example of a 60/0.7 objective lens equipped with a correction collar for focusing at the surface or through a cover glass of up to 1.5-mm thickness without altering the focal setting of the lens.

Tube Lengths and Tube Lenses for Which Microscope Objectives Are Corrected

For finite-focused "biological" objective lenses, most manufacturers have standardized the mechanical tube length to 160 mm (see Fig. 37). Most manufacturers use infinity focus for their metallurgical series (except Nikon uses a finite, 210-mm mechanical tube length). Carl Zeiss, Carl Zeiss Jena, Reichert and more recently Leica and Olympus have switched to infinity focus for both biological and metallurgical objectives.

For infinity-focused objective lenses, the rays emanating from a given object point are parallel between the objective and tube lens. Since the physical distance (D_p, Fig. 4) and optical path length between the objective and tube lens are not critical, optical components, such as compensators, analyzers, and beam splitters, can be inserted in this space without altering the objective's corrections. The tube lens focuses the parallel rays onto the intermediate image plane.

The magnification of an infinity-focused objective lens is calculated by dividing a *reference focal length* by the focal length of the objective lens. The reference focal lengths

TABLE 4 Reference Focal Lengths for Infinity-focused Objective Lenses

Leica	200 mm	B, M
Olympus	180 mm	B, M
Reichert	183.1 mm	B, M
Carl Zeiss	160 mm	B, M
Carl Zeiss Jena	250 mm*	B, M

B = biological, M = metallurgical.
* Subject to change in near future.

adopted by several manufacturers are listed in Table 4. For the ICS (Infinity Color-Corrected System optics) of Carl Zeiss, the reference focal length is fixed at 160 mm (both biological and metallographic) with the tube lens correcting the residual chromatic aberrations of the objective lenses. For current Carl Zeiss Jena infinity-corrected objectives (biological and metallographic), the objectives themselves are fully corrected for lateral and longitudinal chromatic aberrations, and tube lenses of various focal lengths are used to change magnification.

Working Distance

Microscope objectives are generally designed with a short free working distance, i.e., the distance from the front element of the objective lens to the surface of the cover glass or, in the case of lenses that are designed to be used without cover glass, to the specimen surface. For some applications, however, a long free working distance is indispensable, and special objectives are designed for such use despite the difficulty involved in achieving large numerical apertures and the needed degree of correction. Table 5 lists some objectives that provide extra-long working distances.

Field Size, Distortion

The diameter of the field in a microscope is expressed by the field-of-view number, or simply field number, which is the diameter of the field in millimeters measured in the intermediate image plane. The field size in the object plane is obviously the field number divided by the magnification of the objective. While the field number is often limited by

TABLE 5 Some Microscope Objectives with Long Working Distances Relative to NA

Manufacturer	Type	Mag/NA	FWD (mm)	CvrGl (mm)	Remarks*
Nikon	CF Plan Apo	4/0.20	15.0		TL = 160 mm
Leica	EF Achromat	4/0.12	24.0		Pol, TL = 160 mm
Zeiss	Plan Apo	5/0.16	12.2		ICO
Leica	Plan Fluotar	10/0.25	19.8		ICO, f.ref 200 mm, Pol
Leica	Plan	20/0.40	11.0	0	ICO, f.ref 200 mm
Olympus	Plan Achro	20/0.40	10.5	0–2	Corr, TL = 160 mm
Leica	Plan	40/0.60	6.8	0	ICO, f.ref 200 mm
Nikon	CF Plan	40/0.50	10.1	0.17	TL = 210 mm
Nikon	CF Plan	60/0.70	3.05	0.17	TL = 160 mm
Olympus	Neo S Plan	80/0.75	4.10	0	ICO, f.ref 180 mm
Olympus	Neo S Plan	100/0.80	3.20	0	ICO, f.ref 180 mm

FWD: free working distance; CvrGl: thickness of coverglass; TL: mechanical tube length; ICO: infinity-corrected objective, f.ref.: reference focal length.
* See Table 2 for other abbreviations.

the magnification and field stop of the ocular, there is clearly a limit that is also imposed by the design of the objective lens. In early microscope objectives, the maximum usable field diameter tended to be about 18 mm or considerably less, but with modern plan apochromats and other special flat field objectives, the maximum usable field can be as large as 28 mm or more. The maximum useful field number of objective lenses, while available from the manufacturers, is unfortunately not commonly listed in microscope catalogs. Acknowledging that these figures depend on proper combination with specific tube lenses and oculars, we should encourage listing of such data together with, for example, UV transmission characteristics (e.g., as the wavelength at which the transmission drops to 50 percent, or some other agreed upon fraction).

Design of Modern Microscope Objectives

Unlike earlier objective lenses in which the reduction of secondary chromatic aberration or curvature of field were not stressed, modern microscope objectives that do correct for these errors over a wide field tend to be very complex. Here we shall examine two examples, the first a 60/1.40 Plan Apochromat oil-immersion lens from Nikon (Fig. 12).

Starting with the hyperhemisphere at the front end (left side of figure) of the objective, this aplanatic element is designed to fulfill Abbe's sine condition in order to minimize off-axis spherical aberration and coma, while providing approximately half the total magnifying power of the objective (Fig. 13). In earlier designs, the hyperhemisphere has been made with as small a radius as possible in order to maximize its magnifying power and to minimize its spherical and chromatic aberrations, since these aberrations increase proportionally with the focal length of the lens. Modern demands for larger field size and reduced curvature of field, however, introduce a conflicting requirement, namely, the need to maintain as large a radius as practical in order to minimize the hyperhemisphere's contribution to the Petzval sum (the algebraic sum of the positive and negative curvatures multiplied by the refractive indices of the lens elements). The hyperhemisphere in these Plan Apochromats is made with a high-index, low-dispersion material to compensate for the greater radius. Additionally, a negative meniscus is generated in the front surface of the hyperhemisphere to which is cemented a minute, plano-convex lens. The negative curvature in the hyperhemisphere contributes to the reduction of the Petzval sum. At the same time the minute plano-convex lens protects the material of the hyperhemisphere which is less resistant to weathering. Index matching between the minute plano-convex lens and immersion oil eliminates or minimizes the refraction and reflection at the lens-oil interface and provides maximum transmission of the all-important high-NA rays into the objective lens. The index matching also reduces the influence of manufacturing errors of this minute lens element on the performance of the objective.

The low-dispersion-glass singlet behind the aplanatic hyperhemisphere further reduces the cone angle of the rays entering the doublets that follow, allowing these and the subsequent lenses to concentrate on correcting axial and lateral chromatic aberration as well as curvature of field. These errors, as well as residual spherical aberrations, are corrected by inclusion of low-dispersion positive and high-dispersion negative lens elements, use of thick-lens elements, appropriate placement of positive and negative lens

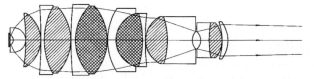

FIGURE 12 Design of Nikon Plan Apochromat oil-immersion objective with 60× magnification and 1.40 NA.[61]

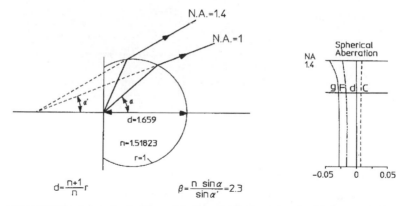

$$d = \frac{n+1}{n} r \qquad\qquad \beta = \frac{n \ \sin \alpha}{\sin \alpha'} = 2.3$$

FIGURE 13 Aplanatic condition of the hyperhemisphere placed at the front end of an oil-immersion objective. The front lens has the same refractive index as the coverglass and immersion oil. The aplanatic condition describes the necessary relationship between refractive index n, distance d between object and spherical surface, and radius r of the spherical surface, in order to make all rays emanating from an object point on the axis leave the hemispherical surface after refraction without introducing spherical aberration. On the right, the small amount of spherical aberration and deviation from the ideal focus point of the hyperhemisphere is shown for different wavelengths ($C =$ 656 nm, $d = 588$ nm, $F = 486$ nm, $g = 436$ nm).[61]

curvatures, and through extensive ray tracing. Near the exit pupil, the height of the ray paths through the concave surfaces is reduced in order to generate additional negative values that minimize the Petzval sum (to complement the inadequate negative contribution made by the concave surface in the hyperhemisphere), so that field flatness can be improved without overly reducing the objective lens' magnifying power or adding to its spherical aberration.

In reality, the Petzval sum of the objective as a whole is made somewhat negative in order to compensate for the inevitable positive Petzval sum contributed by the ocular. Thus, the image at the intermediate image surface, especially the sagittal surface of modern objectives, bows away from the object. Unless the image area is relatively small, one needs to use specified oculars in order to attain maximum field flatness combined with optimum correction otherwise.

Unlike earlier objective lenses (Figs. 2a to 2d) whose design did not appreciably vary from one manufacturer to another, the design of lenses in modern microscope objectives can vary considerably. For example, compare the Nikon Chrome Free 60/1.4 Plan Apo objective discussed above and the Zeiss Infinity Color-Corrected Systems 63/1.4 Plan Apo objective in Fig. 14. Both are excellent, state-of-the-art lenses. But in addition to general design philosophy, including the decision to avoid or to use tube lenses to achieve full chromatic corrections, other factors such as choice of optical elements with special dispersion characteristics; degrees of UV transmission; freedom from fluorescence, birefringence, aging loss of transmittance, etc., all affect the arrangement of choice.

While a modern research-grade microscope is corrected to keep the aberrations from spreading the image of a point source beyond the Airy disk, geometrical distortion of the image formed by microscope objectives tends not to be as well-corrected (e.g., compared to photographic objectives at the same picture angle). Thus, in objectives for biological use, pincushion distortions of up to 1 percent may be present. However, in objectives that are designed for imaging semiconductors, the distortion may be as low as 0.1 percent and they can be considered nearly distortion-free. To reduce stray light and flare, modern microscope objectives contain lens elements with carefully tuned, antireflection coatings,

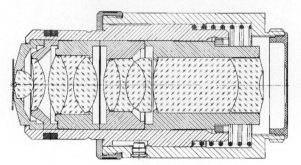

FIGURE 14 Carl Zeiss Infinity Color-Corrected 63/1.4 Plan Apo objective. (*Courtesy of E. Keller, Carl Zeiss, N.Y.*)

and lens curvatures are selected to minimize ghost images arising from multiple reflections.

Given the sophisticated design to provide a wide flat field, with spherical aberrations corrected over a broad wavelength range, and with low longitudinal as well as chromatic aberrations corrected at high NA, the aberration curves of these modern microscope objectives no longer remain simple cubic curves, but turn into complex combinations of higher-order curves (Fig. 15).

Oculars

As conventionally illustrated, the ocular in a light microscope further magnifies the primary (intermediate) image formed by the objective lens (Figs. 3 and 4). The ocular can also be viewed as the front elements of a macro (relay) lens system made up of the ocular plus the refractive elements of the viewer's eye (Fig. 5a) or a video or photographic camera lens. Special video and photo oculars combine these functions of the ocular plus the video or photo lenses into single units.

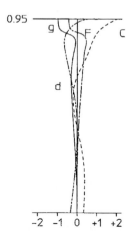

FIGURE 15 Spherical aberration of a highly corrected modern microscope objective with a high numerical aperture.[61]

The intermediate image plane (that lies between the lenses in many ocular types or precedes the lens elements in the Ramsden-type oculars), or its conjugate plane, is used to place field limiting stops, iris diaphragms, reticles, micrometer scales, comparator beam splitters, etc., that need to appear in the same focal plane as the specimen.

The Ramsden disk, the exit pupil of the objective lens imaged by the ocular, generally appears a short distance above the ocular (Fig. 5b). Since the Ramsden disk should lie in the observer's pupil, special high-eye-point oculars are provided for the benefit of observers wearing corrective eye glasses (especially those for astigmatism). High-eye-point oculars are also used for inserting beam-deviating devices (such as the scanning mirrors in laser scanning confocal microscopes) or aperture-modifying devices (such as aperture occluders for stereo viewing through single objective binocular microscopes[6]).

The magnification of an ocular is defined as 25 cm divided by the ocular's focal length. On the ocular, the magnification and field number are inscribed (e.g., as 10×/20, meaning 10-power or 25-mm focal length with a field of view of 20-mm diameter), together with manufacturer's name and special attributes of the ocular such as chromatic-aberration-free (CF), wide-field (W, WF, EWF), plan (P, Pl), compensation (Comp, C, K), high-eye-point (H, picture of glasses), with cross hair and orientation stub for crystallography (pol), projection (pro), photographic (photo), video (TV), etc. Also, special oculars provide larger and flatter fields of view (designated Wide Field, Extra Wide Field, Plan, Periplan, Hyperplan, etc., some with field numbers ranging up to 28 mm).

Compared to microscope objective lenses, fewer design standards have been adopted and fewer standard abbreviations are used to designate the performance or function of the oculars. Two physical parameters of the oculars have, however, become more or less standardized. The outside diameter of the ocular is either 23.2 mm or 30.0 mm, and the reference distance, or the parafocalizing distance of the ocular (i.e., the location of the intermediate image plane below the flange of the ocular) is now generally set to 10 mm.

In the past, oculars with wide ranges of incremental magnifications were provided to adjust the total image magnification of the microscope, but this practice is now replaced by the use of much fewer, better-corrected oculars coupled with a telan magnification changer in the microscope's body tube, or a zoom projection ocular.

Factors affecting choice of ocular focal length and magnification include optimizing the microscope total magnification and image resolution to match the MTF characteristics of the detector and to adjust the available field coverage. In video-enhanced fluorescence, DIC, polarizing, dark field, etc., microscopy, the total magnification often needs to be raised beyond the classical "empty magnification" limit, in order to be able to visualize minute objects whose diameters lie well below the limit of microscope resolution.[6] However, depending on the MTF characteristics, sensitivity, and total pixels available in the sensor, conflicts may arise between the need for greater magnification, image brightness, and field coverage. To optimize the total image magnification, fine trimming of the ocular magnification may be needed, in addition to choosing an objective with the appropriate magnification and NA-to-magnification ratio. Zoom oculars are especially suited for fine-tuning the magnification to optimize S/N ratio and image integration time in video microscopy. For very low light level images, e.g., in photon-counting imaging, ocular magnifications of less than one may be needed in order to sufficiently elevate the S/N ratio, albeit at a sacrifice to spatial resolution.

In addition to adjusting image magnification and placing the microscope's exit pupil at a convenient location, the ocular compensates for the aberrations that have not been adequately corrected in the objective and tube lens. Huygens oculars combined with lower-power achromatic objectives, and compensating oculars combined with higher-NA achromatic and apochromatic objectives, correct for lateral chromatic aberration. Some higher-NA achromatic objectives are purposely designed to provide residual aberrations (including field curvature) that are similar to those in the apochromats, so that the same compensation oculars can be used to compensate for both types of objectives.

Certain classes of modern objectives are sufficiently well corrected to require minimum compensatory correction by the oculars. For example, the Nikon CF objectives and the current Zeiss Jena objectives are designed to produce sufficiently well-corrected intermediate images so that the oculars themselves also are made independently free of lateral and longitudinal chromatic and some spherical aberrations. Regardless of the degree of correction relegated to the ocular, modern microscopes provide images with color corrections, fields of view, and flatness of field much superior to earlier models.

17.5 CONTRAST GENERATION

In microscopy, the generation of adequate and meaningful contrast is as important as providing the needed resolution. Many specimens are transparent and differ from their surroundings only in slight differences of refractive index, reflectance, or birefringence. Most objects that are black or show clear color when reasonably thick become transparent or colorless when their thickness is reduced to a few tenths of a micrometer (since absorption varies exponentially with thickness). Additionally, the specimen is illuminated at large cone angles to maximize resolution under the microscope, thus reducing the shadows and other contrast cues that aid detection of objects in macroscopic imaging. Furthermore, contrast is reduced at high spatial frequency because of an inherent fall-off of the contrast transfer function.

Many modes of contrast generation are used in microscopy partly to overcome these limitations and partly to measure, or detect, selected optical characteristics of the specimen. Thus, in addition to simply raising contrast to make an object visible, the introduction of contrast that reflects a specific physical or chemical characteristic of the specimen may provide particularly important information.

As a quantitative measure of expected contrast generation as functions of spatial frequencies, the modulation transfer functions (MTFs, of sinusoidal gratings) can be calculated theoretically for various contrast-generating modes assuming ideal lenses (Figs. 16 and 17[16]), or on the basis of measured point or line spread functions.[17] Alternatively, the contrast transfer function (CTF, of square wave gratings) can be measured directly using test targets made by electron lithography (Fig. 18).

In the remainder of this section, we survey microscope optical systems according to their modes of contrast generation.

Bright Field

Whether on an upright or inverted microscope, bright field is the prototypic illumination mode in microscopy (Fig. 5). In transmission bright-field illumination, image contrast commonly arises from absorption by stained objects, pigments, metal particles, etc., that possess exceptionally high extinction coefficients, or in the case of transparent objects, from the Becke lines introduced by refraction at object boundaries that are slightly out of focus. (The dark Becke line, which is used for immersion determination of refractive index of particles,[18] surrounds, or lies just inside, a boundary with a sharp gradient of refractive index when the boundary is slightly above or below focus. The Becke line disappears altogether when a thin boundary is exactly in focus.)

To gain additional contrast, especially in bright-field microscopy, the condenser NA is commonly reduced by closing down its iris diaphragm. This practice results in loss of resolution and superimposition of diffraction rings, Becke lines, and other undesirable optical effects originating from regions of the specimen that are not exactly in focus. The various modes of optical contrast enhancement discussed in following sections obviate this

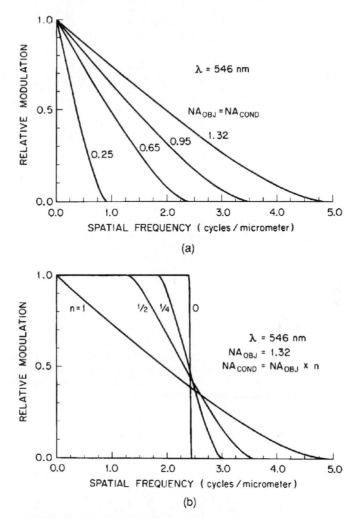

FIGURE 16   Modulation transfer function (MTF) curves for microscope lenses, calculated for periodic specimens in focus: (*a*) each curve represents a different numerical aperture (NA), which is the same for the objective and condenser lens in these curves. (*b*) These MTF curves all represent an objective lens of 1.32 NA, but with different condenser NAs; the conditions are otherwise the same as in (*a*). (*Courtesy of Dr. G. W. Ellis.*)[6]

limitation and provide images with improved lateral and axial resolution as well as improved contrast.

Before the advent of phase contrast and DIC microscopy, oblique illumination (that can be attained by off-centering a partially closed condenser iris diaphragm) was used to generate contrast of transparent objects. While this particular approach suffered from the problems listed in the previous paragraph, combination of oblique illumination at large condenser NA with video contrast enhancement proves to be an effective method for generating DIC-like thin optical sections.[19]

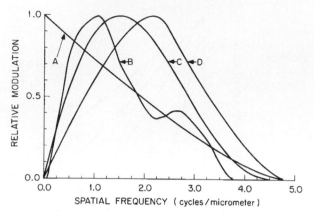

FIGURE 17 Modulation transfer function curves calculated for different modes of microscope contrast generation. A = bright field, B = phase contrast, C = differential interference contrast, and D = single-sideband edge enhancement. The curves are plotted with their peak modulation normalized to 1.0. (*Courtesy of Dr. G. W. Ellis.*)[6]

In reflection bright-field microscopy, the image is formed by the reflected or backscattered light of the specimen which is illuminated through the objective (see the sect. on "Epi-illumination"). Reflection contrast is used primarily for opaque and thick samples, especially for metals and semiconductors. Reflection contrast is also finding increasing applications in autoradiography and in correlative light and electron microscopy for detecting the distribution of colloidal gold particles that are conjugated to antibodies and other selective indicators.

Contrast Transfer Characteristics

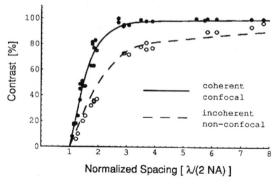

FIGURE 18 Measured contrast transfer values plotted as a function of spatial period in Airy disk diameter units, to normalize the values measured with different lenses and wavelengths. Data points were obtained with a laser spot scan microscope operating in the confocal reflection mode (*solid points*) and the nonconfocal transmission mode (*circles*). Curves are calculated contrast transfer values for the coherent confocal and the incoherent nonconfocal imaging mode.[62]

Total frustrated reflection microscopy[20] generates contrast due to objects that are present in a low-refractive-index medium located within the evanescent wave that extends only a few wavelengths' distance from the microscope coverslip surface. Regions of the specimen whose refractive index differs from its milieu produce interference fringes whose contrast sensitively reflects the refractive index difference and distance from the coverslip surface.

Dark Field

In dark field microscopy the illuminating beam is prevented from entering the image-forming ray paths. The background of the field is dark, and only light scattered by optical discontinuities in the specimen is designed to appear in the image as bright lines or dots. Thus, contrast can become extremely high, and diffraction images can be detected as bright points or lines even when the diameter of the scattering object becomes vanishingly small compared to the microscope's limit of resolution.[4,12,14]

For small objects that are not obscured by other light-scattering particles (a condition rather difficult to achieve) and are free in a fluid substrate, Brownian motion of the object and the time constant and sensitivity of the detector, rather than the object's absolute size, are more likely to set a lower limit to the size of the object that can be clearly visualized with dark field microscopy.

Phase Contrast and Other Aperture-modifying Contrast Modes

Microscopic objects, distinguished from their surround only by a difference of refractive index, lose their Becke line and disappear altogether when brought exactly into focus. Nevertheless, light diffracted by the small object still suffers a $\lambda/4$ phase shift relative to the undeviated background wave by the very act of being scattered (by a nonabsorbing object; the phase shift upon scattering by an absorbing object is $\lambda/2$[21]). As shown in Fig. 19, light s scattered by the small object and the undeviated light u, both originating from a common small point A of the condenser aperture, traverse different regions of the objective lens aperture. At the objective aperture, the undeviated light traverses only point B that is conjugate to A, while the scattered light passes those regions of the aperture defined by the spatial periods of the object.

Since light waves s and u arise from the same points in object space but traverse regions that are spatially separated in the objective aperture plane, a *phase plate* introduced in that plane can be used to modify the relative phase and amplitudes of those two waves. The phase plate is configured to subtract (or add) a $\lambda/4$ phase to u relative to s so as to introduce a $\lambda/2$ (or zero) phase difference between the two and, in addition, to reduce the amplitude of the u wave so that it approximates that of the s wave. Thus, when the two waves come to focus together in the image plane, they interfere destructively or constructively to produce a darker or brighter, in-focus image of the small, transparent object against a dark gray background.

As generally implemented, an annulus replaces the pinhole in the condenser aperture, and a complementary phase ring in the objective aperture plane or its conjugates (covering a somewhat larger area than the undisturbed image of the annulus in order to handle the u waves displaced by out-of-focus irregularities in the specimen) replaces the simple phase disk.

In the Polanret system, the phase retardation and effective absorbance of the phase ring can be modified by use of polarization optical components so that the optical path difference of a moderately small object can be measured by seeking the darkest setting of

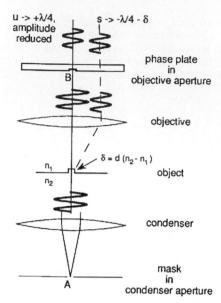

FIGURE 19 Optical principle of phase contrast microscopy illustrating the phase relationships between waves of the light s scattered by the specimen and the undeviated light u (see text).

the object.[22] Similarly, the Polanret system can be used to accentuate color or low contrast due to slight absorption by the object.

Several modes of microscopy, including phase contrast, specifically take advantage of the facts that (1) the condenser and objective lens apertures are conjugate planes, (2) the illuminating beam arising out of each point of the condenser aperture is variously deviated by the specimen structure according to its spatial frequency, and (3) the objective aperture is the Fourier plane.

In modulation contrast microscopy, the condenser aperture contains a slit mask with the slit placed towards the edge of the aperture. The objective aperture holds a second, complementary mask, called a modulator, which consists of two parts (Fig. 20). The dark part covers the smaller sector to one side of the projected slit and the gray part covers the slit area. The objective mask thus attenuates the zero-order light undeviated by the specimen and removes the light diffracted by the specimen to one side of the zero-order beam. The light deviated by specimen structure away from the dark sector of the mask passes unchanged, while the light deviated towards the dark sector is blocked. Thus, the image becomes shadow-cast, similar in appearance to differential interference contrast (DIC) that reflects gradients of refractive indices or of optical path differences in the specimen.

In single sideband edge enhancement (SSBEE) microscopy, a halfstop is placed at the front focal plane of the condenser to occlude one half of the condenser aperture. A complementary spatial filter in the objective aperture attenuates the undeviated beam and controls its phase displacement by adjustable amounts relative to the light scattered by the object into the aperture region obscured by the condenser halfstop (single sideband[23]). Thus, we again acquire an image similar in appearance to DIC, but which, unlike DIC, does not require polarization optical elements to sandwich the specimen, and thus can be used to detect very minute anisotropies of refractive index at high resolution. The SSBEE

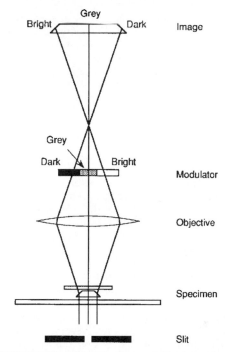

FIGURE 20 Schematic diagram indicating regions of the modulator that modify light from phase gradients in the object to enhance contrast.[63]

system takes advantage of the fact that one needs to capture only one of the two beams (sidebands) deviated by diffraction in order to gain the information regarding the spatial frequency in the specimen that gives rise to that particular diffraction angle.

Polarizing

The polarizing microscope generally differs from a standard transilluminating microscope by the addition of a polarizer below the condenser; a compensator slot and analyzer above the objective lens; strain-free optics; a graduated, revolving stage; centrable lens mounts; cross hairs in the ocular aligned parallel or at 45° to the polarizer axes; and a focusable Bertrand lens that can be inserted for conoscopic observation of interference patterns at the back aperture of the objective lens. In addition, the front element of the condenser can be swung into place for higher-NA conoscopic observations or swung out for low-NA orthoscopic observations of larger field areas.

Calcite polarizing prisms (which introduce astigmatism to all but collimated rays) have mostly been replaced in recent years by dichroic polarizing filters. The polarizing filters are thin and cost very much less, although inferior in transmittance and extinction compared to high-quality polarizing prisms. The lower transmittance through a pair of polarizing filters is especially detracting in high-extinction polarized light or DIC microscopy, since only a small fraction of light originating from a low birefringence or path difference in the sample reaches the observer's eyes or the video detector.

With standard polarizing microscopes, one can image and measure polarization optical parameters[18,24] on objects which are larger than a few micrometers and which introduce retardances greater than several tens of nanometers. However, as the dimension of the object or magnitude of retardance decrease below these ranges, one needs to use special techniques or devices for detecting and measuring birefringence or even for generating a reliable image with high-NA lenses.

The basic ingredients that are needed to detect low levels of birefringence (retardance ≤ 10 nm) are high-extinction optics, use of low-retardance compensator, light source with high irradiance, and high-sensitivity detector (e.g., dark adaptation for visual observation and measurements). These needs can be understood from the following formula:

$$I = I_p(\sin^2(R/2) + 1/\text{EF}) \tag{10}$$

where I is the irradiance of the specimen with a retardance of R (radians), L_p is the irradiance of the field when the polarizer and analyzer axes are parallel (i.e., maximum transmission), and the extinction factor $\text{EF} = I_p/I_c$, where I_c is the irradiance when $R = 0$.[25] For example, when the specimen retardance is 2 nm at a wavelength of 555 nm, then $R = 2\pi \times 2/555$, $\sin^2(R/2) = 1 \times 10^{-4}$. In other words, the specimen has only twice the irradiance as the background even when EF is as large as 10^4. The field is also only 10^{-4} of its irradiance compared to when the polarizer and analyzer axes are set parallel, or when observing a specimen with a full wavelength retardance between crossed polarizers. Inclusion of the compensator adds a bias compensation that increases contrast (the irradiance of the specimen relative to that of the background field) as well as field irradiance. An intense light source and dark adaptation of the observer also help shift the field and specimen irradiance into ranges that reduce the contrast threshold of the eye (or improve the S/N ratio of a photoelectric detector or for photography).

The EF of a polarizing microscope rapidly drops (to as low as 2×10^2 when $\text{NA}_{\text{obj}} = \text{NA}_{\text{cond}} = 1.25$) as the NA of the system is raised, even with polarizer $\text{EF} > 10^5$, and using carefully selected objective and condenser lenses that are free from strain birefringence or birefringent inclusions. The loss of EF originates from the rotation of polarization plane and birefringence that results from differential transmission of the P and S components at the optical interfaces. The depolarization results in four bright quadrants separated by a dark cross that is seen conoscopically for crossed polarizers in the absence of a specimen. The depolarization also gives rise to anomalous diffraction, and a four-leaved clover pattern replaces the Airy disk or each weakly birefringent image point[26] (see Ref. 6, Fig. III-23).

Anomalous diffraction and the loss of EF at high NAs are both eliminated or drastically reduced by introducing polarization rectifiers (Fig. 21). The rectifier corrects for the differential P vs. S transmission loss regardless of their absolute values, thus resulting in a uniformly high extinction for the full aperture (Fig. 22). The resulting uniform aperture function accounts for the absence of anomalous diffraction in a rectified system.[27] The image improvement achieved with rectifiers in a polarizing microscope is demonstrated in Fig. 23.

Interference

While all modes of contrast generation in light microscopy in fact depend on interference phenomena, a group of instruments is nevertheless known separately as interference microscopes. These microscopes form part of an interferometer, or contain an interferometer, that allows direct measurements of optical path difference (or generaton of contrast) based on inteference between the waves passing the specimen and a reference wave.

Among the many designs that have been manufactured or proposed, interference microscopes can be classified into three major groups: (1) the Mach-Zehnder type where two complete sets of microscope optics are placed, one in each arm of a Mach-Zehnder interferometer; (2) the beam-shearing type in which the reference wave is generated by

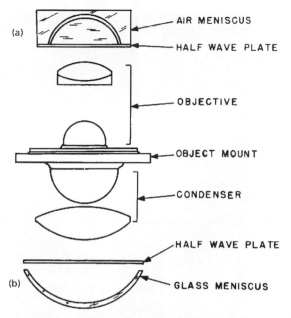

(a)

AIR MENISCUS

HALF WAVE PLATE

OBJECTIVE

OBJECT MOUNT

CONDENSER

HALF WAVE PLATE

(b)

GLASS MENISCUS

FIGURE 21 Arrangement of polarization rectifiers. Rectifiers are placed above the objective *A,* below the condenser *B,* or at both *A* and *B*. The glass or air meniscus introduces additional rotation to each beam of light which equals the amount introduced by the lenses and the specimen slide. The sense of rotation is, however, reversed by the half-wave plate so rectification is achieved over the whole aperture.[64]

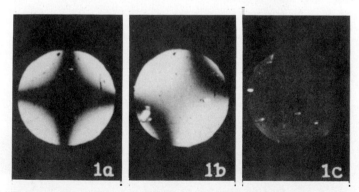

FIGURE 22 Appearance of the back aperture of a 1.25-NA strain-free objective with and without rectifiers. The condenser, which is identical with the objective, is used at full aperture. (*a*) Crossed polarizers, no rectifier; (*b*) polarizer turned 2°, no rectifier; (*c*) crossed polarizers, with rectifiers in both condenser and objective. Photographs were given identical exposures.[64]

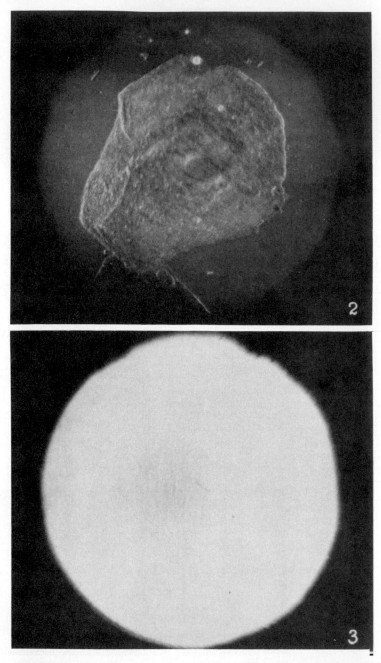

FIGURE 23 (*Top*) Photograph of an epithelial cell from a human mouth taken with a 43×/0.85-NA strain-free objective and a 43×/0.63 strain-free condenser equipped with a rectifier. Polarizers crossed, 1-nm background retardation (horizontal image width approx. 150 μm). (*Bottom*) Identical to *top* but without a rectifier. Notice the brighter background and the total lack of detail and contrast in the image.[64]

displacing a beam laterally within the field of a single microscope; and (3) the Mirau type in which the reference wave is focused to a different level than the specimen plane, again in a single microscope.

The Mach-Zehnder type (Fig. 24), while straightforward in principle, requires close

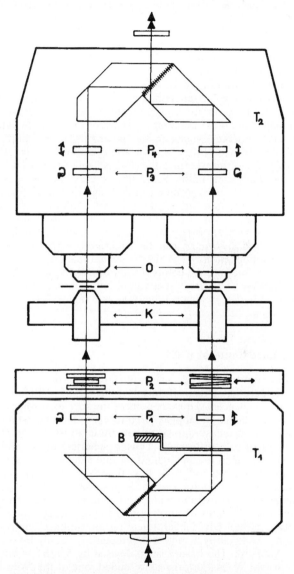

FIGURE 24 Mach-Zehnder type interference microscope with two complete sets of microscope optics, one in each arm of a Mach-Zehnder interferometer. The compensators $P_1, P_2, P_3,$ and P_4 introduced into the beam path permit optimum contrast setting and adjust for any variation of the interference picture. (*Reproduced from E. Leitz Inc., Catalog No. 500-101 Interference Systems, 1972.*)

matching of the optics in the two interferometer arms and a mechanical design that provides exceptional precision and stability. Thus, in addition to using one of the matched pairs of objectives and the corresponding matched pairs of condensers that are mounted on special sliders, and inserting a blank slide (that is similar to the specimen-containing slide) into the reference path, for each specimen change, one needs to carefully adjust the built-in beam deviators, path equalizers, and wedge components to correct for deviation of the beam paths and to precisely equalize the optical paths through the two microscopes. While unfortunately no longer manufactured, this type of microscope permits precise interferometric measurements of microscopic objects both in the uniform field mode and the fringe displacement mode, and can even be used to generate holograms (Dr. G. W. Ellis, personal communication).

Several variations exist of the two latter types of interference microscopes.[4,28] Many designs use a polarizer, polarized beam splitters, and a half-wave plate sandwiching the specimen to generate and recombine the two beams, and a quarter-wave plate and graduated rotatable analyzer to measure optical path difference (Fig. 25). Depending on the crystal arrangements in the polarizing beam splitter, the reference beam is either laterally displaced in the field or at a different focal level from the primary imaging beam. In this type of interference microscope, one needs to be wary of ghost images introduced by the dual beam paths. Some interference microscopes for noncontact surface profiling employ objectives according to the design of Mirau (Fig. 26).

With all the interference microscopes, both monochromatic and white light illumination can be used. Monochromatic illumination allows precise measurement of optical path difference and determination of the dry mass (i.e., reduced weight) of the specimen if its thickness is known or can be calibrated, e.g., by altering the refractive index of the immersion medium.[29,30] White light illumination allows determination of the order of the fringes or interference colors. In uniform field mode, the interference color has been used as a method of contrast generation, but such use of the interference microscope was soon replaced by DIC which can be used with higher-NA lenses and thus provide greater image resolution.

Differential Interference Contrast (DIC)

Differential interference contrast (DIC) microscopy provides a monochromatic shadow-cast image that effectively displays the gradient of optical paths for both low and high spatial frequencies. Those regions of the specimen where the optical paths increase along a reference direction appear brighter (or darker), while those regions where the path differences decrease appear in reverse contrast. Image contrast is greater the steeper the gradient of path difference. Thus, a spherical object with higher index than its surround would appear to be highlighted on one side with shadows cast on the other side. Objects whose refractive index is less than the surround appear with reverse shadow-cast appearance as though it were a depression. Very thin filaments or sharp interfaces likewise appear shadow-cast and with good contrast, even when their diameter or separation falls way below the limit of resolution of the optical system.

DIC is basically a beam-shearing interference contrast system in which the reference beam is sheared by a minuscule amount, generally by somewhat less than the diameter of the Airy disk. The basic system devised by Smith[31] is a modified polarizing microscope to which two Wollaston prisms are added, one at the front focal plane of the condenser and the other at the back focal plane of the objective lens (Fig. 27). The condenser Wollaston converts each ray illuminating the specimen into two, slightly displaced, parallel beams that are orthogonally polarized relative to each other. In the second Wollaston prism above the objective, the two beams become recombined again.

The phase difference between the two beams, introduced by a gradient of thickness or refractive index in the specimen, results in elliptical polarization for the recombined beam

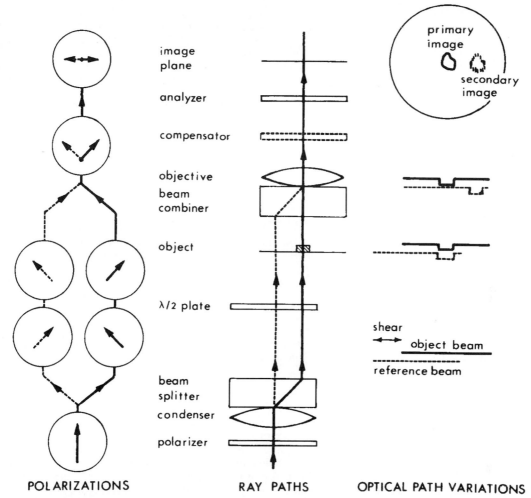

FIGURE 25 Jamin-Lebedeff interferometer microscope. The reference and object beam are polarized orthogonally by the birefringent beam splitter and are separated by a sizeable fraction of the field diameter. The half-wave plate switches the polarization of the two beams so that the second birefringent plate in front of the objective serves as a beam combiner. The compensator (quarter-wave plate and graduated rotatable analyzer) serves to measure optical path differences between reference and object beam.[65]

that leaves the second Wollaston. A deSenarmont compensator (a quarter-wave plate in extinction position and rotatable analyzer) is placed above the second Wollaston to extinguish light that has suffered a particular phase shift relative to the other beam. Alternatively, the quarter-wave plate can be inserted between a rotatable polarizer and the condenser Wollaston prism.

In medium- to high-power objective lenses, the back focal plane is usually inside the lens system and therefore not available for insertion of a Wollaston prism. To avoid this problem, Nomarski[32] introduced a modification of the Wollaston prism that can be placed outside of the objective lens (Fig. 28). By using crystal wedges with appropriately oriented axes, the Nomarski prism recombines the two beams that were separated by the

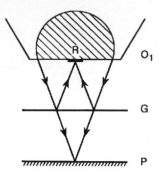

FIGURE 26 Mirau's interference microscope. The incident light beam, emerging from the objective O_1, is split in two parts in the semireflective plate G. One part is transmitted to the object P and the other is reflected to the reference area R extending over a small portion of the objective front surface. The wavefronts reflected by R and P are recombined at G to produce the interference pattern.[4]

condenser Wollaston as though a regular Wollaston prism were located in the proper plane in the objective lens. Regions of the specimen with selected phase shifts are compensated by translating the Nomarski prism. A second modification introduced by Nomarski (and for a while adopted by Nikon) uses a single crystal in place of the condenser Wollaston and a three-part prism in place of the Nomarski prism described above.[33]

Fluorescence

Fluorescence microscopy is one of the few modes of microscopy in which the illuminating wavelength differs from that of the emitted. In early designs, the exciting waves were prevented from contaminating the fluorescence image by a combination of (1) special illumination (such as the use of a dark-field condenser) that prevented the direct rays from entering the objective lens, and (2) the use of a barrier filter. The barrier filter absorbs the exciting light while transmitting much of the longer fluorescence wavelengths. (*Note*: For every fluorochrome, the longer wavelength portion of the absorption curve, i.e., the excitation wavelengths, overlaps with the shorter wavelength tail of its fluorescence emission curve.)

Today most fluorescence microscopes (or attachments) use epi-illumination incorporating interchangeable filter cubes (after Ploem, see Fig. 30) that are matched to the fluorochrome. The filter cube is placed in the collimated beam between the objective and a tube lens, at the intersection of the microscope axis and that of the excitation illuminator located on a side arm. The objective lens serves both as the condenser and the objective. A field diaphragm, and sometimes an aperture iris, are placed in the illuminating side arm together with the source collector at appropriate conjugate planes. The illuminating beam, commonly emitted by a xenon or mercury arc lamp, is filtered through a narrow band path interference filter and reflected down into the objective by a dichromatic beam splitter. The fluorescent imaging beam originating from the specimen passes straight through the dichromatic beam splitter and associated barrier filter and reaches the ocular. Each fluorescent cube contains the appropriate excitation interference filter, dichromatic beam splitter, and barrier filter so that they can be switched as a group, for example, to rapidly inspect specimens containing (or stained with) multiple fluorochrome.

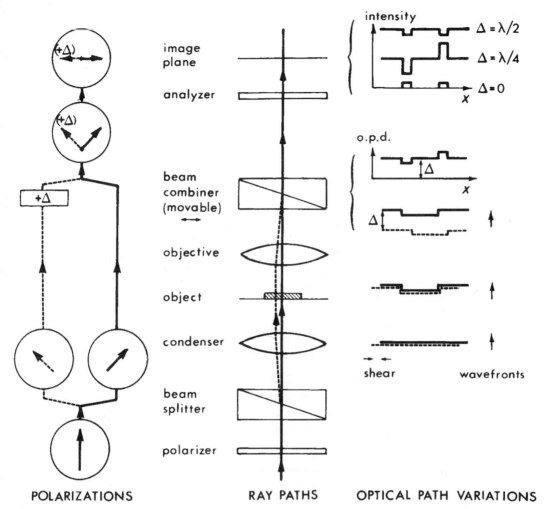

FIGURE 27 The optical system for differential interference contrast (DIC, see text).[65]

For fluorochrome requiring shorter-wave UV excitation, objective lenses must be designed for greater short-wavelength transmission and low autofluorescence. While aberrations for the shorter-UV exciting wavelengths are generally not as well-corrected as for the imaging wavelengths, it should be noted that such aberrations, or lack of parafocality, directly affect the resolution in the case of confocal fluorescence microscopes.

Also, it should be noted that, while little effort is commonly made to fill the objective aperture with the illuminating beam (presumably with the rationale that this should not affect image resolution because each fluorescent object is emitting incoherently relative to its close neighbor), one finds that in practice the fluorescent image is much improved by filling the aperture, for example, by use of an optical fiber light scrambler (see the section on "Transillumination" later in this chapter). (The reason for this improvement is still not fully understood—whether it is solely due to the increased level of illumination which, in turn, provides improved signal of the intrinsically low-light fluorescent image (but why

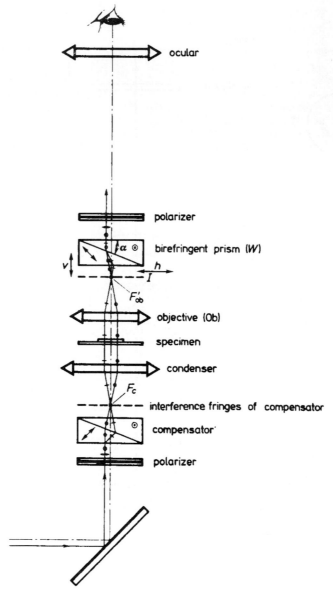

FIGURE 28 Standard Nomarski interference contrast with modified Wollaston prisms which are placed outside the focal planes of condenser and objective.[8]

should the scrambler improve the level of illumination to start with?) or whether NA_{cond} does affect resolution in fluorescence microscopy after all.)

While most fluorescence microscopes today use epi-illumination (since epi-illumination provides advantages such as avoiding loss of excitation by self-absorption by underlying fluorochrome layers, generating an image that more closely approximates an intuitive one when reconstructed in three dimensions, etc.), improvements in interference filters open up new opportunities for fluorescence microscopy using transillumination. New interference filters are available with exceptionally high extinction ($>10^5$) and sharp cutoff of the excitation wavelengths, coupled with high transmission of the pass band. With transillumination, one can more reliably combine fluorescence with polarization-based microscopy or carry out polarized fluorescence measurements with greater accuracy, since one can avoid the use of dichromatic beam splitters which tend to be optically anisotropic.

Fluorescence microscopy particularly benefits from video imaging, especially with the use of low-noise, chilled CCDs as imaging detectors, digital computers to enhance and rapidly process the signal (such as in ratio imaging), and the new fluorescence-conjugated chemical probes that provide incredible sensitivity and selectivity (see also pp. 17.46–17.48).[9–11,34]

17.6 ILLUMINATION AND IMAGING MODES

Transillumination

The full impact of the illumination system on the final quality of the microscope image is often not appreciated by the microscope user or designer. Undoubtedly, part of this neglect arises from a lack of understanding of the roles played by these components, in particular the condenser, and the common practice of closing down the condenser iris diaphragm to adjust image contrast for comfortable viewing. Also, it may in part stem from Zernicke's consideration that the microscope resolution is independent of the condenser's correction (cited, e.g., in Ref. 35). Regardless of the conventional view, critical examination of the microscope image or point spread function reveals the importance of the alignment, focus, tilt, NA, and effective aperture function of the condenser. The effects are especially noticeable when contrast is enhanced, e.g., by video microscopy.

The theoretical influence of condenser NA on image resolution has been calculated by Hopkins and Barham,[36] who showed that maximum resolution could be obtained at $NA_{cond} = 1.5 \times NA_{obj}$ (Fig. 29). Such high NA_{cond} is usually not achievable for high-NA objective lenses, and, in addition, with most objectives, flare due to internal reflection would reduce image contrast to an extent possibly unsalvageable even with video contrast enhancement. For $NA_{cond} \leq NA_{obj}$, the minimum resolvable distance between two point objects is commonly given by the formula:

$$d = \frac{1.22\lambda}{NA_{obj} + NA_{cond}} \tag{11}$$

which expresses the importance of NA_{cond} on resolution. Again, reduction of NA_{cond}, generally achieved by closing down the condenser iris diaphragm, tends to raise image contrast so that even experienced microscopists tend to use an $NA_{cond} \approx (0.3 \cdots 0.5) \times NA_{obj}$ to obtain a compromise between resolution and visibility. With video and other modes of electronic enhancement, the loss of contrast can be reversed so that improved lateral, and especially axial, resolution is achieved by using an NA_{cond} that equals, or nearly equals, the NA_{obj}. With a larger NA_{cond}, the S/N ratio is simultaneously improved by the increased photon flux that illuminates the specimen.

Under optimum circumstances, the light source and condenser should be focused for Köhler illumination (Fig. 5) to minimize flare and to improve the homogeneity of field illumination. Alternately, image brightness, especially in the middle of the field, can be maximized by *critical illumination* where the condenser is somewhat defocused from Köhler illumination to produce an image of the source rather than the field diaphragm

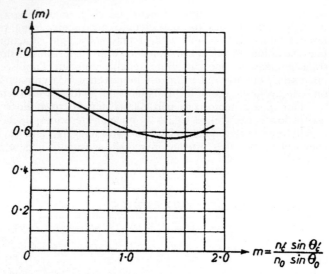

FIGURE 29 Effect of the condenser aperture on the resolution of two pinholes of equal brightness. m is the ratio of the numerical apertures of condenser to objective. L is the minimum resolved distance between the pinholes (Rayleigh criterion) in units of the wavelength divided by the objective aperture.[35]

superimposed on the specimen. Either mode of illumination can yield resolution approximately as given by Eq. (11).

The aperture function of the microscope can become nonuniform, or limited, for a number of reasons, These include misalignment between the objective and condenser lenses; misalignment of the condenser iris (relative to the condenser lens elements); misalignment of the illuminator and condenser axes; tilted objective or condenser lenses or lens elements; nonuniform illumination of the condenser aperture; limited source size; nonuniform intensity distribution in the source; and improper choice, or focusing, of the condenser or source collector. Whether intentional or accidental, these conditions can reduce the effective NA_{cond} and/or induce oblique illumination, thus sacrificing resolution and image quality. An improvement, using a single optical fiber light scrambler, that allows the filling of the full condenser aperture with uniform illumination and little loss of field brightness (especially when using concentrated arc lamps) was introduced by Ellis[37] (also see Figs. III-21, III-22 in Ref. 6).

Epi-illumination

In the epi-illumination mode, a beam splitter, part-aperture-filling mirror, or wavelength-discriminating dichromatic (unfortunately often called dichroic) mirror, placed behind the objective lens diverts the illuminating beam (originating from a light source placed in the side arm of the microscope) into the objective lens which also acts as the condenser (Fig. 30). Alternatively, a second set of lenses and a beam-diverting mirror (both of whose centers are bored out and are arranged coaxially around the objective lens) can provide a larger-NA illuminating beam, much as in dark field illumination in the transillumination mode.

This latter approach limits the maximum NA of the objective lens to around 1.25, but has the advantage that the illuminating beam traverses a path completely isolated from

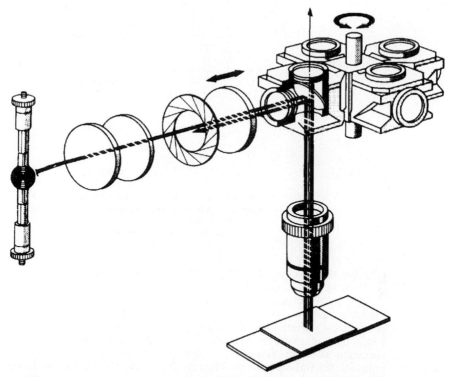

FIGURE 30 Schematic of epi-illuminating light path. The rotatable set of filter cubes with excitation filters, dichromatic mirrors, and barrier filters matched to specific fluorochromes are used in epifluorescence microscopy.[66]

the image-forming beam. When the two beams do pass through the same objective lens, as is the case with most epi-illuminating systems, the lens elements must be carefully designed (by appropriate choice of curvature and use of highly efficient antireflection coating) to reduce hot spots and flare introduced by (multiple) reflection at the lens surfaces. Modern microscope objectives for metallurgical and industrial epi-illuminating systems in particular are designed to meet these qualities. In addition, circular polarizers (linear polarizers plus $\lambda/4$ wave plates) and appropriate stops are used to further exclude light reflected from the surfaces of lens elements, cover glass, etc. For epi-illumination fluorescence microscopy, dichromatic beam splitters and barrier filters can reduce background contamination that arises from the exciting beam to less than one part in 10^4.

Orthoscopic vs. Conoscopic Imaging

The common mode of observation through a microscope is by orthoscopic observation of the focused image. For certain specific applications, particularly with polarizing microscopes, examination of the aperture plane, or conoscopic observation, sheds valuable complementary information.

Conoscopic observation can be made either by replacing the regular ocular with a telescope that brings the aperture plane into focus or by inserting a Bertrand lens (that serves as a telescope objective) in front of a regular ocular. Conversely, one can observe the aperture plane simply by removing the ocular and looking down the microscope body

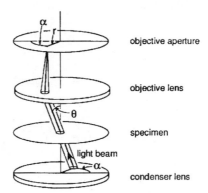

FIGURE 31 Parallel rays with inclination θ and azimuth orientation α, traversing the specimen plane, and focused by the objective lens at a point with radius r and same azimuth angle α in the aperture plane.

tube (in the absence of a Bertrand lens) or by examining the Ramsden disk above the ocular with a magnifier.

The polar coordinates of each point in the aperture plane, i.e., the radius r and azimuth angle α, are related to the rays traversing the specimen by: $r = \sin\theta$ and $\alpha = $ (azimuth orientation of the ray projected onto the aperture plane) (Fig. 31). Thus, conoscopic observation provides a plane projection of all of the rays traversing the specimen in three-dimensional space. For specimens, such as single crystal flakes or polished mineral sections in which a single crystal is illuminated (optically isolated) by closing down the field diaphragm, the conoscopic image reveals whether the crystal is uniaxial or biaxial, its optic axis angle and directions, as well as sign and strength of birefringence and other anisotropic or optically active properties of the crystal.[24]

Conoscopic observation also reveals several attributes of the condenser aperture plane and its conjugate planes (e.g., in Köhler illumination, the plane of the condenser iris diaphragm and the illuminating source). Thus, conoscopy can be used for checking the size, homogeneity, and alignment of the illuminating light source as well as the size and alignment of the condenser iris diaphragm and phase contrast annulus (located at the front focal plane of the condenser) relative to the objective exit pupil or the phase ring (located at the back focal plane of the objective). It also reveals the state of extinction in polarized light and interference contrast microscopy and provides a visual estimate of the aperture transfer function for the particular optical components and settings that are used.

The aperture plane of the microscope is also the Fourier plane of the image, so that diffraction introduced by periodic texture in the specimen can be visualized in the aperture plane by conoscopic observation. Depending on the NA of the objective and the spatial period in the specimen, the pattern of diffraction up to many higher orders can be visualized when the condenser iris is closed down and laterally displaced to illuminate the specimen with a narrow, oblique, coherent (monochromatic) beam, and the zero-order intensity is suppressed (e.g., by the use of appropriate polarizers and compensator).

Confocal Microscopy

In confocal microscopy, the specimen is scanned point by point either by displacing the specimen (stage scanning) or by scanning a minute illuminating spot (beam scanning), generally in a TV-raster fashion. In either case, the scanning spot is an Airy disk formed by a high-NA objective lens. An exit pinhole is placed conjugate to the spot being scanned so

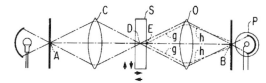

FIGURE 32 Optical path in simple confocal microscope. The condenser lens C forms an image of the first
pinhole A onto a confocal spot D in the specimen S.
The objective lens O forms an image of D into the
second (exit) pinhole B which is confocal with D and A.
Another point, such as E in the specimen, would not
be focused at A or B, so that the illumination would be
less and, in addition, most of the light g-h scattered
from E would not pass the exit pinhole. The light
reaching the phototube P from E is thus greatly
attenuated compared to that from the confocal point D.
In addition, the exit pinhole could be made small
enough to exclude the diffraction rings in the image of
D, so that the resolving power of the microscope is
improved. The phototube provides a signal of the light
passing through points D_1, D_2, D_3, etc. (*not shown*), as
the specimen is scanned. D_1, D_2, D_3, etc. can lie in a
plane normal to the optical axis of the microscope (as in
conventional microscopy), or parallel to it, or at any
angle defined by the scanning pattern, so that optical
sections can be made at angles tilted from the conventional image plane. Since, in the stage-scanning system,
D is a small spot that lies on the axis of the microscope,
lenses C and O can be considerably simpler than
conventional microscope lenses.[38,67]

that only the light originating from the scanned spot is transmitted through the exit
pinhole. Thus, light originating from other regions of the specimen or optical system are
prevented from reaching the photo detector (Fig. 32).

This optical arrangement reduces blurring of the image from out-of-focus light
scattering, fluorescence, etc., and yields exceptionally clear, thin optical sections. The
optical sections can then be processed and assembled electronically to yield three-
dimensional displays or tilted plane projections. Alternatively, the specimen itself can be
scanned through a tilted plane (e.g., by implementing a series of x scans with y, z
incremented) to yield a section viewed from any desired orientation, including that normal
to the microscope axis.

In addition to yielding optical sections with exceptional image (S/N) quality, confocal
microscopes of the stage-scanning type can be used to vastly expand the field of view. Here
the image area is not limited by the field of view of the optics but only by the range of
movement of the specimen and ability of the photo detector and processor to handle the
vast information generated at high speed. Furthermore, the objective lens needs only to
be corrected for a narrow field of view on axis.[5,38] Laser disk recorders are a form of
application that takes advantage of these attributes.

Beam-scanning confocal microscopes can be classified into two types. In the *disk-
scanning* type, the source and exit pinholes, arranged helically on a modified Nipkow disk,
concurrently scan (1) the field plane that is located in front of the collector lens of the light
source and (2) the image plane of the objective lens. By using epi-illumination, these two
planes lie at the same distance, both above the objective lens. Thus, a single spinning disk
(with symmetrically placed pinholes that alternatively serve as entrance and exit pinholes
in the Petráň type, or used with a beam splitter so that each pinhole serves both as

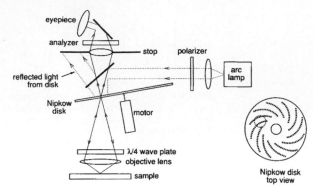

FIGURE 33   Kino-type real-time confocal microscope (see text).[40]

entrance and exit pinholes in the Kino type) can be made to provide synchronously scanning entrance and exit pinholes (Fig. 33).[39,40] The *disk-scanning* type confocal microscope suffers from light loss due to the small areas occupied by the pinholes (even though several pinholes are made to scan the field simultaneously) but has the advantage of permitting direct visual viewing of the confocal image.

In a confocal microscope, the exit pinhole can be made smaller than the diameter of the Airy diffraction image formed by the objective lens so that the Airy disk is trimmed down to regions near its central peak. With this optical arrangement, the unit diffraction pattern that makes up the image turns out to be the square of the Airy pattern given in Eq. (2). Thus, the radius at half maximum of the central bright area (Airy disk) is reduced by a factor of 1.36. (The radial position of the first minimum in both patterns is still equal to r_{Airy}.) The shape of the unit diffraction pattern is thus sharpened so that, compared to nonconfocal imaging, two points which radiate incoherently (as in fluorescence microscopy) can be expected to approach each other by up to a factor of $\sqrt{2}$ closer to each other before their diffraction patterns encounter the Rayleigh limit. Figure 18 compares the contrast transfer characteristics of a confocal microscope in the coherent imaging mode with the same lenses used in the nonconfocal, incoherent imaging mode.

In the laser-scanning epi-illuminating confocal microscope (which was developed into a practical instrument in the late 1980s and immediately adopted with great enthusiasm), a source pinhole is commonly omitted since the focused beam of the single-mode (TEM_{00}) laser itself produces a pointlike source. This source is expanded by relay lenses (or mirrors) and the ocular to fully cover the back aperture of a very well-corrected, flat-field objective lens and then focused onto the specimen where it forms a minute Airy disk defined by the NA of the objective and wavelength of the laser source. The Airy disk formed by the objective is scanned in the x, y image plane by two galvanometer mirrors placed at the eyepoint of the ocular (or its conjugate plane in order to insure that the direction of the rays alone is deviated at the aperture plane and that the objective aperture remains fully illuminated throughout the scanning). Viewed from the specimen towards the source, the two mirrors swing the source image (in sawtooth waves rapidly along the x direction and slowly along the y direction) so that the specimen is scanned along these orthogonal axes in a raster pattern by the Airy spot formed by the objective lens.

The scanned spot is imaged by the same set of lenses, and the image-forming beam is deviated by the same two sets of swinging mirrors, in reverse order, so that the image of the scanned spot becomes stationary at its image plane. The paths of the illuminating beam and the return imaging beam are shared until the two paths are separated by a (dichromatic) beam splitter into: (1) the path from the laser, and (2) the path to the exit pinhole. The location of the exit pinhole is adjusted to lie in the conjugate image plane coaxially with the source point, therefore, confocal with the source point and the scanning spot. The confocally adjusted exit pinhole intercepts the final Airy disk formed by the objective lens and trims away the surrounding rings as well as the outer foot of the disk. Depending on the amount of light available (or tolerable by the specimen) and the permissible scan duration for forming the image, a compromise is often needed between reduction of the exit pinhole diameter (that improves image resolution and optical sectioning capability) and expansion of the pinhole diameter to allow more light to reach the photodetector. When the need to acquire confocal images at high rate is frustrated by the limited amount of light reaching the photodetector, the exit pinhole can be replaced by a slit, with surprisingly little deterioration of optical sectioning capability.

Rather than using confocal optics to eliminate image blurring from out-of-focus planes, one can achieve the same end by computational deconvolution of a stack of serial optical sections obtained by video microscopy.[17,41] While computationally intensive and time consuming, this image restoration method allows one to isolate clean optical sections from a stack of images that can be acquired at a much higher speed than with laser-scanning confocal microscopy and in modes of contrast generation not accessible to confocal imaging.

Alternatively, thin optical sections can be obtained directly with digital enhanced video microscopy using high-NA condenser and objective lenses. Requiring little processing, this approach is especially convenient when many stacks of optical sections have to be acquired at high rates in succession, e.g., in order to record rapid, three-dimensional changes in microscopic domains over time.

Proximity-scanning Microscopy

A microscope's limit of resolution [Eq. (1)] can be exceeded by narrowing the field of illumination. In confocal microscopy, the resolution is expected to be improved by up to a factor of $\sqrt{2}$ by illuminating the field point by point with an Airy diffraction spot and by using a small confocal exit pinhole. Even in the absence of confocal optics, Harris[42] has argued that the diffraction pattern in the Fourier plane can be extrapolated beyond the spatial frequency that is cut off by the NA of the objective lens—in other words, that the limit of resolution can be exceeded by computational extrapolation of the diffraction orders as long as the specimen is illuminated in a narrowly limited field.

The field of illumination can be reduced beyond that defined by diffraction by placing the minute exit aperture of a tapered light guide or a minute pinhole closely adjacent to the specimen. By scanning such an aperture relative to the specimen, one obtains a proximity-scanned image whose resolution is no longer limited by the diffraction orders captured by the objective lens. Instead, only the size of the scanning pinhole and its proximity to the specimen limit the resolution.[43]

For non-optical microscopes, e.g., in scanning tunneling, force, and other proximity-scanning microscopes, resolution down to atomic dimensions can be obtained on images that reflect topological, electronic, ionic, and mechanical properties of the specimen surface.[44] In these types of proximity-scanning microscopes, a fine-tipped probe, mounted on a piezoelectric transducer that provides finely controlled $x, y,$ and z displacements of the probe, interacts with specific properties of the specimen surface (alternatively, the probe may be fixed and the sample mounted to the transducer). The resulting interaction signal is detected and fed back to the z-axis transducer, which generally induces the probe tip to rise and fall with the surface contour (that reflects the particular electrical or mechanical property of the surface) as the probe is scanned in a raster fashion along the x and y directions over an area several tens of Ångstroms to several tens of micrometers wide. A highly magnified contour image of the atomic or molecular lattices is generated on a monitor that displays the z signal as a function of the x, y position.

Aperture Scanning

In the aperture-scanning microscope devised by Ellis[45] for phase contrast microscopy, the tip of a flexible signal optical fiber, illuminated by an Hg arc, makes rapid circular sweeps at the periphery of the condenser aperture. This circular, scanning illumination spot replaces the conventional phase annulus in the condenser aperture plane. A quarter-wave plate and absorber, both covering only a small area conjugate to the illuminating spot, spins in synchrony with the fiber at the objective back aperture (or its projected conjugate). Thus, the specimen is illuminated by a narrow, coherent beam that enters the specimen obliquely at high NA, with the azimuth orientation of the beam swinging around and around to generate a full cone of illumination within the integration time of the detector. With this aperture-scanning approach, the specimen is illuminated by a large-NA cone of light which is temporally incoherent, with the phase disk covering only a small fraction of the area normally occupied by the phase ring in conventional phase contrast systems. The small size of the phase disk, while appropriately reducing the amplitude and introducing the requisite 1/4 wave phase retardation to the rays not scattered by the specimen, allows the transmission of a much larger fraction of the scattered rays that carry the high spatial frequency information. The aperture-scanning phase contrast microscope thus provides a very thin optical section. The image is also virtually free of the phase halo that obscures image detail adjacent to refractile boundaries in conventional phase contrast microscopy.

Extending this concept, modulation of the transfer functions of the condenser and objective apertures with electro-optical devices should open up intriguing new opportunities. Such modulation eliminates the need for mechanical scanning devices, the spatial distribution of the modulation function can be altered at will, and the amplitude and phase of light passing each point in the aperture can be adjusted rapidly, even coupled dynamically to the image signal through a feedback loop to generate dynamic spatial filters that enhance or select desired features in the image.

Video Enhancement and Digital Image Processing

Video and related forms of electronic imaging offer a number of advantages that vastly extend the utility of the light microscope. The electronic signal generated by the video camera can be readily amplified and biased to boost contrast in desired gray-value ranges of the image while suppressing unwanted background signal noise and light due to flare, inadequate extinction, etc. (Fig. 34). Through such analog processing, one gains the opportunity to use objective lenses at higher condenser NA, and in contrast modes that previously could not generate an image with high enough contrast for direct observation. Thus, in DIC, polarized light, and other modes of microscopy, one can now use

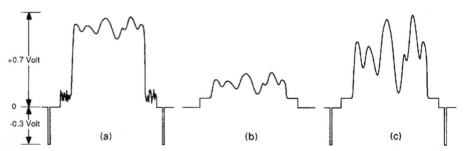

FIGURE 34 Analog contrast enhancement of video signal: (*a*) original signal; (*b*) sync pulses removed and signal level adjusted to suppress background noise; (*c*) signal amplified and sync pulses added back.

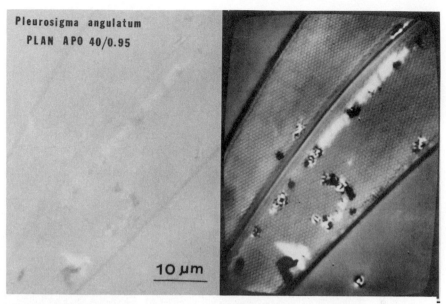

FIGURE 35 Diatom frustule viewed between crossed polarizers and a $\lambda/10$ compensator without (*left*) and with (*right*) analog video enhancement.[68]

better-corrected lenses, such as the 1.4 NA Plan Apo objective illuminated with matched-NA condenser, to generate images with greater resolution coupled with vastly improved contrast for objects that previously could barely be detected (Fig. 35).

Compared to visual observation or photography, which are based on logarithmic responses of photochemical detectors, an ideal video signal responds linearly to image intensity. Thus, in addition to analog enhancement, the video signal can be digitized in a straightforward fashion and processed with a digital computer. Today, the video signal can be digitized (e.g., to 640×480 pixels at 256 gray levels) and processed at video frame rate (30 frames/s) with a fairly simple personal computer to which is added an imaging board that contains digitizers, look-up tables, frame stores, video bus, logic units, and flexible operational and routing alternatives. These functions are controlled by software residing in the host computer. With such a digital image processor, one can carry out a large variety of operations, much of them on line, at video rate, including image capture, storage, grayscale stretching, binarizing; interimage manipulations such as averaging, subtraction, superposition, ratio imaging, selection of minimum or maximum value pixels; interpixel manipulations such as spatial filtering, gradient differentiation, unsharp masking; and many forms of (automatic) quantitative analyses of image geometrics and intensities. With such digital image processors, image defects due to uneven microscope illumination or fixed pattern noise can simply be subtracted away, noise in low-light-level images can be suppressed by frame averaging or spatial filtering, and contrast and desired spatial frequencies can be boosted or suppressed, etc. (Fig. 36).

Not only does analog and digital image processing permit the full use of the light microscope to deliver the theoretical maximum resolution in contrast modes and on specimens that hitherto escaped detection, but the clean, high contrast image provides opportunities for direct observation of the dynamic behavior of individual, unresolvable molecular filaments and membranes. The image of these nanometer-thick filaments (expanded to their Airy disk diameter) clearly depicts their mobility, polymerization-depolymerization, interaction with organelles in living cells, etc. Thus, new insights are

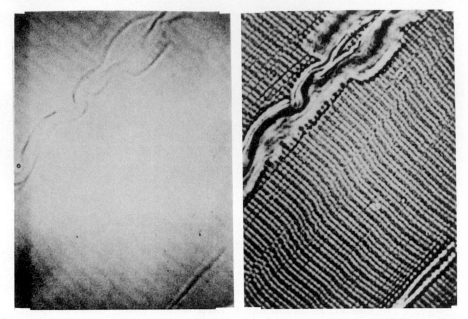

FIGURE 36 Muscle thin section in matched index medium (totally invisible without video enhancement). With analog video enhancement (*left*); with analog and digital enhancement (*right*). (*Reproduced from American Laboratory, April 1989.*)

gained by observing the behavior of individual macromolecular assemblies that were not anticipated from conventional chemical analyses that relied on statistical behavior of large numbers of sample molecules.

In addition to extending the applicable range of image detection and maximizing the attainable resolution in the x–y plane, the digitally processed, high-NA video images provide clearer and thinner optical sections. The effects of video enhancement on z-axis resolution and optical sectioning is all the more striking since they rise as the square of operative NA as contrasted to the linear rise of lateral resolution with NA.

Ratio Imaging

The concentration of minute quantities of specific ions and chemicals can be measured accurately in living cells and other irregularly shaped objects by fluorescence ratio imaging. For example the pH, and micromolar to nanomolar concentrations of calcium ions, can be measured in pico- to femtoliter volumes in active nerve and kidney cells preloaded with fluorescent reporter dyes such as SNARF and Fura-2.[10,34] The concentrations, as well as the spreading waves of the pH or calcium ion within the cell, can be followed second-by-second with a digital processor coupled to a fluorescence microscope equipped with low-light-level video camera and illuminated alternatively with, for example 340- and 380-nm excitation wavelengths. The reporter dye is formulated so as to emit fluorescence as a sensitive function of the ion concentration at one of the two exciting wavelengths, while at the other exciting wavelength the emission does not vary (isosbestic point) or is an inverse function of the ion concentration. Thus, for each pixel, the ratio of fluorescence intensities emitted by the reporter dye excited at the two wavelengths provides the concentration of the ionic species independent of sample thickness and local dye concentration.

Three-dimensional Imaging

For low-power observations, the three-dimensional features of a specimen can be viewed directly under a dissecting microscope equipped with pairs of tilted objectives, erecting prisms, and oculars (Greenough type), or through a single larger objective lens whose aperture is divided to provide the left- and right-eye images with appropriate stereoscopic parallax. With high-NA objectives, these approaches are not very effective owing to the shallow depth of field.

With the new capability to obtain serial optical sections through the highest-NA objectives using confocal or video-enhanced microscopy, and to rapidly store, retrieve, and manipulate the stored images in a digital computer, one can now generate stereoscopic and other three-dimensional views representing a substantial depth of the specimen at the highest resolution of the light microscope.[46] The stack of ultrathin optical sections can be used directly as acquired through the high-NA objectives or after digital deconvolution to further improve the quality of each optical section by reducing undesired contributions from out-of-focus fluorescence, light scattering, etc.[17,41]

17.7 OPTICAL MANIPULATION OF SPECIMEN WITH THE LIGHT MICROSCOPE

In confocal microscopy, light initially travels "in a reverse path" through an objective lens to form a diffraction-limited image of the source pinhole into the specimen plane. In a similar vein, the light microscope and microscope objectives are increasingly used to project reduced high-intensity images of source patterns into the object plane in order to manipulate minute regions of the specimen optically. Photolithography and laser disk recorders are examples of important industrial applications which have prompted the design of specially modified objective lenses for such purposes.

Microbeam Irradiation, Caged Compounds

Many applications are also found in the biomedical field, some using UV-transmitting, moderately high NA objectives that are parafocalized for visible light and UV down to approximately 250 nm (Zeiss Ultrafluar and catadioptric objectives from Cooke-A.E.I., also quartz monochromats from Leitz). In its extreme form, a concentrated image of a circular- or slit-shaped UV or laser source of selected wavelengths is imaged onto a biological specimen to locally ablate a small targeted organelle; for example, a part of a chromosome, the microtubules attached thereto, or tiny segments of cross-striated muscle, are irradiated with the microbeam in order to sever their mechanical connections and, for example, to analyze force transduction mechanisms.[47,48] In other cases, oriented chromophores can be selectively altered at the submolecular level, for example, by polarized UV microbeam irradiation. The stacking arrangement of the DNA nucleotide bases (which exhibit a strong UV dichroism, as well as birefringence in visible light) can be selectively altered and disclose the coiling arrangement of DNA molecules within each diffraction-limited spot in the nucleus of living sperm.[49] Brief microirradiation of slit- or grid-shaped patterns of UV are used to bleach fluorescent dyes incorporated into membranes of living cells. The time course of recovery of fluorescence into the bleached zone measures the rate of diffusion of the fluorescently tagged molecules in the membrane and reveals unexpected mobility patterns of cell membrane components.[50,51]

Also, selected target molecules within minute regions in living cells can be modified, tagged, or activated by focused beams of light. The target molecules can be naturally photosensitive species such as chlorophyll (which produces oxygen where illuminated with the appropriate visible wavelengths), rhodopsin (which isomerizes and triggers the release of calcium ions and action potentials in retinal cells), or artificially introduced photosensitive reagents such as the drug colchicine (whose antimitotic activity is abolished locally with 366-nm irradiation).

Of the photosensitive compounds, the *caged compounds* promise a far-reaching potential. These are compounds that are synthesized so as to "cage" and hide the active chemical group until a photosensitive part of the compound is altered (e.g., by long-wavelength UV irradiation) and unmasks the hidden active group. Thus, by preloading with the appropriate caged compound and irradiating the cell selectively in the region of interest, one can test the role of the uncaged compound. For example, the role of ATP can be tested using caged ATP and ATP analogs; response to subtle increase in calcium ions can be seen using caged calcium or caged calcium chelators.[52,53] Likewise, caged fluorescent dyes are irradiated to locally label and follow the transport of subunits within macromolecular filaments in a dividing cell.[54]

Optical Tweezers

Intense laser beams concentrated into a diffraction spot can generate a photon-driven force great enough to capture and suspend small particles whose refractive index differs from its surrounding.[55] Applied to microscopy, a single swimming bacterium or micrometer-sized organelles in live cells can be trapped and moved about at will at the focus of a near-infrared laser beam focused by an objective lens of high NA. While the energy density concentrated at the laser focus is very high, the temperature of the trapped object remains within a degree or so of its environment; biological targets typically exhibit low absorbance at near-infrared wavelengths and thermal diffusion through water from such minute bodies turns out to be highly effective. Thus, the bacterium continues to multiply while still trapped in the focused spot, and it swims away freely when the laser beam is interrupted.

The ability to use "optical tweezers," not only to capture and move about minute objects but to be able to instantly release the object, provides the microscopist with a unique form of noninvasive, quick-release micromanipulator.[56]

17.8 MECHANICAL STANDARDS

Some mechanical dimensions for the light microscope have gradually become standardized internationally (Fig. 37, Table 6). While the standards permit more ready interchange of

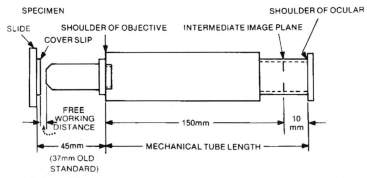

FIGURE 37 Dimensions for standard microscopes with finite focus objectives. See text section on "Tube Lengths and Tube Lenses for which Microscope Objectives Are Corrected" for further detail.[6]

TABLE 6 Royal Microscopical Society Standardized Objective Screw Thread Dimensions (in mm)

Thread		Ext. diam.	Pitch diam.	Core diam.	Calc. play male/female	Allowance	Tolerance
Female	Max	20.396	19.944	19.492		+0.076	
	Min	20.320	19.868	19.416	Min. 0.046	0	0.076
Male	Max	20.274	19.822	19.370	Max. 0.198	−0.046	
	Min	20.198	19.746	19.294		−0.122	0.076

Based on data given in the specification by the RMS [*J. R. Microsc. Soc.* **56:**377–380 (1936)] and now standardized in ISO Standard 8038 (1985) and in BS 7012, pt. 4.
Standard for ocular tube diameter: 23.2 mm or 30.0 mm.

objective lenses, oculars, etc., of different makes, in order to achieve optimum correction one still needs to be alert to the design constraint assumed by each manufacturer (finite or infinity focus, tube length, type and thickness of coverslip and immersion medium, aberrations corrected in objective alone or in combination with tube lens or with ocular, field size, and wavelength for correction, etc.: see section on"Microscope Lenses, Aberations" in this chapter). The safest approach is to use the objectives and oculars on the microscope body supplied, but for those well-versed in the art, standardized dimensions do allow for greater flexibility of choice and innovative application.

17.9 ACKNOWLEDGMENTS

The authors thank Yoshiyuki Shimizu and Hiroohi Takenaka from Nikon K. K. Japan, for providing a preprint and original figures of their most useful article on microscope lens design. The authors are also grateful to Ernst Keller of Carl Zeiss, Inc., Jan Hinsch of Leica, Inc., Mortimer Abramovitz of Olympus, Inc., and Lee Shuett and Mel Brenner of Nikon, Inc., for providing detailed data on microscope lenses. We are especially grateful to Gordon W. Ellis of the University of Pennsylvania and Katsuji Rikukawa of Nikon K. K., Yokohama, for extensive discussions regarding several contents of this article. The preparation of this article was supported in part by the National Institutes of Health grant R-37 GM31617 and National Science Foundation grant DCB-8908169 awarded to S. I. and National Institutes of Health grant R01 GM49210 awarded to R. O.

17.10 REFERENCES

1. L. C. Martin, *The Theory of the Microscope.* Blackie, London, 1966.

2. F. Zernicke, "Phase Contrast, A New Method for the Microscopic Observation of Transparent Objects, *Physica* **9:**686–693 (1942).

3. D. Gabor, "Microscopy by Reconstructed Wavefronts, *Proc. Roy. Soc. London* A **197:**454–487 (1949).

4. M. Françon, *Progress in Microscopy,* Row, Peterson, Evanston, Ill., 1961.

5. T. Wilson and C. Sheppard, *Theory and Practice of Scanning Optical Microscopy,* Academic Press, London, 1984.

6. S. Inoué, *Video Microscopy,* Plenum Press, New York, 1986.

7. M. Pluta, *Advanced Light Microscopy Vol I: Principles and Basic Properties,* Elsevier Science Publishing Co., Amsterdam, 1988.

8. M. Pluta, *Advanced Light Microscopy Vol. II: Specialized Methods,* Elsevier Science Publishing Co., Amsterdam, 1989.

9. Y.-L. Wang and D. L. Taylor, *Fluorescence Microscopy of Living Cells in Culture,* Part A, Academic Press, San Diego, 1989.

10. D. L. Taylor and Y.-L. Wang, *Flourescence Microscopy of Living Cells in Culture,* Part B, Academic Press, San Diego, 1989.

11. D. L. Taylor, M. Nederlof, F. Lanni, and A. S. Waggoner, "The New Vision of Light Microscopy," *American Scientist* **80:**322–335 (1992).

12. L. W. Smith and H. Osterberg, "Diffraction Images of Circular Self-radiant Disks," *J. Opt. Soc. Am.* **51:**412–414 (1961).

13. K. Svoboda, C. F. Schmidt, B. J. Schnapp, and S. M. Block "Direct Observation of Kinesin Stepping by Optical Trapping Interferometry," *Nature* **365:**72 (1993).

14. S. Inoué, "Imaging of Unresolved Objects, Superresolution and Precision of Distance Measurement with Video Microscopy, in D. L. Taylor and Y.-L. Wang (eds.), *Methods in Cell Biology,* Academic Press, New York, 1989, pp. 85–112.

15. H. E. Keller, "Objective Lenses for Confocal Microscopy" in J. B. Pawley (ed.), *Handbook of Biological Confocal Microscopy,* Plenum Publ. Corp., New York, 1990, pp. 77–86.

16. I. T. Young, "Image Fidelity: Characterizing the Imaging Transfer Function," in D. L. Taylor and Y.-L. Wang (eds.), *Methods in Cell Biology,* Academic Press, New York, 1989, pp. 1–45.

17. D. A. Agard, Y. Hiraoka, P. Shaw, and J. W. Sedat, "Fluorescence Microscopy in Three Dimensions," in D. L. Taylor and Y.-L. Wang (eds.), *Methods in Cell Biology,* Academic Press, San Diego, 1989, pp. 353–377.

18. E. M. Chamot and C. W. Mason, *Handbook of Chemical Microscopy,* 2d ed., John Wiley & Sons, New York, 1958.

19. B. Kachar, "Asymmetric Illumination Contrast: A Method of Image Formation for Video Microscopy," *Science* **227:**766–768 (1985).

20. C. S. Izzard and L. R. Lochner, "Formation of Cell-to-Substrate Contacts During Fibroblast Motility: An Interference-Reflection Study" *J. Cell Sci.* **42:**81–116 (1980).

21. F. Zernicke, "The Wave Theory of Microscopic Image Formation," in J. Strong (ed.), *Concepts of Classical Optics,* W. H. Freeman and Co., San Francisco, 1958, pp. 525–536.

22. A. H. Bennett, H. Osterberg, H. Jupnik, and O. W. Richards, *Phase Microscopy, Principles and Applications,* John Wiley & Sons, Inc., New York, 1951.

23. G. W. Ellis, "Advances in Visualization of Mitosis in vivo," in E. Dirksen, D. Prescott, and C. F. Fox (eds.), *Cell Reproduction*: *In Honor of Daniel Mazia,* Academic Press, New York, 1978, pp. 465–476.

24. N. H. Hartshorne and A. Stuart, *Crystals and the Polarizing Microscope*: *A Handbook for Chemists and Others,* 3d ed., Arnold, London, 1960.

25. M. M. Swann and J. M. Mitchison, "Refinements in Polarized Light Microscopy," *J. Exp. Biol.* **27:**226–237 (1950).

26. H. Kubota and S. Inoué, "Diffraction Images in the Polarizing Microscope," *J. Opt. Soc. Am.* **49:**191–198 (1959).

27. S. Inoué and H. Kubota, "Diffraction Anomaly in Polarizing Microscopes," *Nature* **182:**1725–1726 (1958).

28. C. J. Koester, "Interference Microscopy: Theory and Techniques," in G. L. Clark (ed.), *The Encyclopedia of Microscopy,* Reinhold, New York, 1961, pp. 420–434.

29. R. Barer, "Phase, Interference and Polarizing Microscopy," in R. C. Mellors (ed.), *Analytical Cytology,* 2d ed., McGraw-Hill, New York, 1959.

30. H. G. Davies, "The Determination of Mass and Concentration by Microscope Interferometry," in J. F. Danielli (ed.), *General Cytochemical Methods,* Academic Press, New York, 1958, pp. 55–161.

31. F. H. Smith, "Microscopic Interferometry," *Research* (London) **8:**385–395 (1955).

32. R. D. Allen, G. B. David, and G. Nomarski, "The Zeiss-Nomarski Differential Interference Equipment for Transmitted Light Microscopy," *Z. wiss. Mikroskopie* **69:**193–221 (1969).

33. T. Tsuruta, "Special Nomarski Prism," *Appl. Optics* **2:**98–99 (1990).

34. R. Y. Tsien, "Fluorescent Probes of Cell Signaling," in W. M. Cowas, E. M. Shooter, C. F. Stevens, and R. F. Thompson (eds.), *Annual Review of Neuroscience,* Annual Reviews Inc., Palo Alto, 1989, pp. 227–253.

35. M. Born and E. Wolf, *Principles of Optics,* 6th ed., Pergamon Press, Elmsford, N.Y., 1980.

36. H. H. Hopkins and P. M. Barham, "The Influence of the Condenser on Microscopic Resolution," *Proc. Phys. Soc. London* **63B:**737–744 (1950).

37. G. W. Ellis, "Microscope Illuminator with Fiber Optic Source Integrator," *J. Cell Biol.* **101:**83a (1985).

38. M. Minsky, Microscopy Apparatus, U.S. Patent #3013467, 1957.

39. M. Petráň, M. Hadravsky, D. Egger, and R. Galambos, "Tandem-scanning Reflected-light Microscope," *J. Opt. Soc. Am.* **58:**661–664 (1968).

40. G. S. Kino, "Intermediate Optics in Nipkow Disk Microscopes," in J. B. Pawley (ed.), *Handbook of Biological Confocal Microscopy,* Plenum Press, New York, 1990, pp. 105–111.

41. W. A. Carrington, K. E. Fogarty, L. Lifschitz, and F. S. Fay, "Three Dimensional Imaging on Confocal and Wide-field Microscopes," in J. B. Pawley (ed.), *Handbook of Biological Confocal Microscopy,* Plenum Press, New York, 1990, pp. 151–161.

42. J. L. Harris, "Diffraction and Resolving Power," *J. Opt. Soc. Am.* **54:**931–936 (1964).

43. E. Betzig, A. Lewis, A. Harootunian, M. Isaacson, and E. Kratschmer, "Near-field Scanning Optical Microscopy (NSOM); Development and Biophysical Applications," *Biophys. J.* **49:**269–279 (1986).

44. H. K. Wickramasinghe, "Scanned-probe microscopes," *Scientific American* **261:**98–105 (1989).

45. G. W. Ellis, "Scanned Aperture Light Microscopy," in G. W. Bailey (ed.), *Proceedings of the 46th Annual Meeting of the Electron Microscopy Society of America,* San Francisco Press, San Francisco, pp. 48–49 (1988).

46. J. K. Stevens, L. R. Mills, and J. E. Trogadis (eds.), *Three Dimensional Confocal Microscopy,* Academic Press, San Diego, 1993.

47. R. E. Stephens, "Analysis of Muscle Contraction by Ultraviolet Microbeam Disruption of Sarcomere Structure," *J. Cell Biol.* **25:**129–139 (1965).

48. M. W. Berns, *Biological Microirradiation, Classical and Laser Sources,* Prentice-Hall, Englewood Cliffs, N.J., 1974.

49. S. Inoué and H. Sato, "Deoxyribonucleic Acid Arrangement in Living Sperm," in T. Hayashi and A. G. Szent-Gyorgyi (eds.), *Molecular Architecture in Cell Physiology,* Prentice Hall, Englewood Cliffs, N.J., 1966, pp. 209–248.

50. D. E. Koppel, D. Axelrod, J. Schlessinger, E. L. Elson, and W. W. Webb, "Dynamics of Fluorescence Marker Concentration as a Probe of Mobility," *Biophys. J.* **16:**1315–1329 (1976).

51. H. G. Kapitza, G. McGregor, and K. A. Jacobson, "Direct Measurement of Lateral Transport in Membranes by Using Time-Resolved Spatial Photometry," *Proc. Natl. Acad. Sci. USA* **82:**4122–4126 (1985).

52. J. H. Kaplan and G. C. Ellis-Davies, "Photolabile Chelators for the Rapid Photorelease of Divalent Cations," *Proc. Natl. Acad. Sci. USA* **85:**6571–6575 (1988).

53. J. A. Dantzig, M. G. Hibberd, D. R. Trentham, and Y. E. Goldman, "Cross-bridge Kinetics in the Presence of MgADP Investigated by Photolysis of Caged ATP in Rabbit Psoas Muscle Fibres," *J. Physiol.* **432:**639–680 (1991).

54. T. J. Mitchison and E. D. Salmon, "Poleward Kinetochore Fiber Movement Occurs during both Metaphase and Anaphase-A in Newt Lung Cell Mitosis," *J. Cell Biol.* **119:**569–582 (1992).

55. J. M. Dziedzic, J. E. Bjorkholm, and S. Chu, "Observation of a Single Beam Gradient Force Optical Trap for Dielectric Particles," *Optics Letters* **11:**288–290 (1986).

56. S. M. Block, "Optical Tweezers: a New Tool for Biophysics," *Mod. Cell Biol.* **9:**375–402 (1990).

57. J. R. Benford, "Microscope Objectives," in R. Kingslake (ed.), *Applied Optics and Optical Engineering,* Academic Press, New York, 1965, pp. 145–182.

58. S. Bradbury, P. J. Evennett, H. Haselmann, and H. Piller, *Dictionary of Light Microscopy,* Oxford University Press, Oxford, 1989.

59. E. Hecht, *Optics,* Addison-Wesley Publ. Co., Reading, Mass., 1987.

60. M. Cagnet, M. Françon, and J. C. Thrierr, *Atlas of Optical Phenomena,* Springer Verlag, Berlin, 1962.

61. Y. Shimizu and H. Takenaka, "Microscope Objective Design", in C. Sheppard and T. Mulvey (eds.), *Advances in Optical and Electron Microscopy,* vol. 14, Academic Press, San Diego, 1994, pp. 249–334.

62. R. Oldenbourg, H. Terada, R. Tiberio, and S. Inoué, "Image Sharpness and Contrast Transfer in Coherent Confocal Microscopy," *J. Microsc.,* **172:**31–39 (1993).

63. R. Hoffman and L. Gross, "Modulation Contrast Microscopy," *Appl. Optics* **14:**1169–1176 (1975).

64. S. Inoué and W. L. Hyde, "Studies on Depolarization of Light at Microscope Lens Surfaces II. The Simultaneous Realization of High Resolution and High Sensitivity with the Polarizing Microscope," *J. Biophys. Biochem. Cytol.* **3:**831–838 (1957).

65. M. Spencer, *Fundamentals of Light Microscopy,* Cambridge University Press, London, 1982.

66. E. Becker, *Fluorescence Microscopy,* Ernst Leitz Wetzlar GmbH, Wetzlar, 1985.

67. S. Inoué, "Foundations of Confocal Scanned Imaging in Light Microscopy," in J. B. Pawley (ed.), *Handbook of Biological Confocal Microscopy,* Plenum Publ. Corp., New York, 1990, pp. 1–14.

68. S. Inoué, "Video Imaging Processing Greatly Enhances Contrast, Quality and Speed in Polarization-based Microscopy," *J. Cell Biol.* **89:**346–356 (1981).

CHAPTER 18
REFLECTIVE AND CATADIOPTRIC OBJECTIVES

Lloyd Jones
Optical Sciences Center
University of Arizona
Tucson, Arizona

18.1 GLOSSARY

A	4th-order aspheric deformation coefficient
AN	4th-order nonsymmetric deformation coefficient
B	6th-order aspheric deformation coefficient
C	8th-order aspheric deformation coefficient
c	surface base curvature
CON	conic constant
D	10th-order aspheric deformation coefficient
FN	focal ratio
GLA	glass type
h	radial surface height
INF	infinite radius of curvature
k	conic constant
n	index of refraction
R	radius of curvature
RDX	radius of curvature in the x dimension
RDY	radius of curvature in the y dimension
STO	stop surface
SUR	surface number
t	element thickness
THI	thickness of element or distance to next surface or element
Z	surface sag

18.2 INTRODUCTION

During the initial stages of an optical design, many optical engineers take advantage of existing configurations that exhibit useful properties. This chapter is a compilation of reflective and *catadioptric* objective designs that should help inform the reader of available choices and provide reasonable starting solutions.

The chapter also includes a cursory introduction to some of the more important topics in system analysis, such as angular and linear blur size, *image irradiance,* scaling, and stray light control.

An extensive list of referenced reading material and brief definitions of terms italicized throughout the text are included.

18.3 GLASS VARIETIES

Glasses used in the designs are represented in terms of index of refraction and Abbe number or *V* number, below. The *V* number indicates glass dispersion. Most glasses can be obtained from a number of vendors.

Glass	Index of refraction	*V*-number
BK7	1.516	64.2
F2	1.620	36.3
F9	1.620	38.1
FK51	1.487	84.5
FN11	1.621	36.2
Germanium	4.037	117.4
LLF1	1.548	45.8
LAK21	1.640	60.1
PSK2	1.569	63.2
Silica	1.445	27.7
Silicon	3.434	147.4
Sapphire	1.735	15.5
SK1	1.610	56.5
SK2	1.607	56.8
SK3	1.609	58.9
SK16	1.620	60.3
SF5	1.673	32.1
SF10	1.728	28.5
UBK7	1.517	64.3

18.4 INTRODUCTION TO CATADIOPTRIC AND REFLECTIVE OBJECTIVES

The variety of objectives presented in this chapter is large. Most of the intricate detail relating to each design is therefore presented with the design itself. In the following paragraphs, analysis of the general features of the catadioptric and reflective objectives is undertaken.

Conic Mirrors

It is apparent after a brief perusal of the designs that there are many surface types. Among these are the sphere, paraboloid, hyperboloid, prolate ellipsoid, and oblate ellipsoid. The

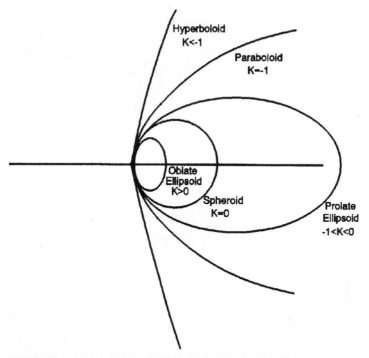

FIGURE 1 Relative shapes of conic surfaces in two dimensions.

oblate ellipsoid is a prolate ellipsoid turned on its side. The equation of a conic is given by the expression

$$Z = \frac{ch^2}{1 + \sqrt{1 - (1 + k)c^2h^2}} \tag{1}$$

where Z is the surface sag, k is the *conic constant*, C is the surface base curvature, and h is the radial height on the surface. The relative shapes of these surfaces are illustrated in Fig. 1.

Conic mirrors give perfect geometric imagery when an axial point object is located at one conic focus and the axial point image is located at the other conic focus. Figure 2 illustrates these ray paths.

General Aspheres

General aspheres are surfaces with fourth- and higher-order surface deformation on top of a flat or curved surface (see Schulz, 1988). The surface deformation of a rotationally symmetric general asphere is given by the relation

$$Z = \frac{ch^2}{1 + \sqrt{1 - (1 + k)c^2h^2}} + Ah^4 + Bh^6 + Ch^8 + Dh^{10} \tag{2}$$

where A, B, C, and D are 4th-, 6th-, 8th-, and 10th-order coefficients that determine the sign and magnitude of the deformation produced by that order. Although general aspheres allow correction of *third-* and *higher-order aberrations* and may reduce the number of elements in an optical system, general aspheres are more expensive than spheres or conics. If aspheric deformation is required, conic surfaces should be tried first, especially since a conic offers higher-order correction (Smith, 1992).

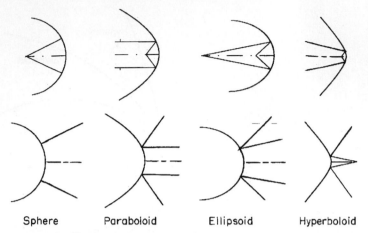

Sphere   Paraboloid   Ellipsoid   Hyperboloid

FIGURE 2   Ray paths for perfect axial imagery.

Obscurations

Obscurations that block portions of the entering beam reduce image irradiance and image contrast (Smith, 1990, Everhart, 1959) in reflective and catadioptric systems. Several methods are used to reduce or eliminate completely the effects of an obscuration (see Fig. 3).

Figure 3a illustrates a commonly employed technique for reducing large-mirror

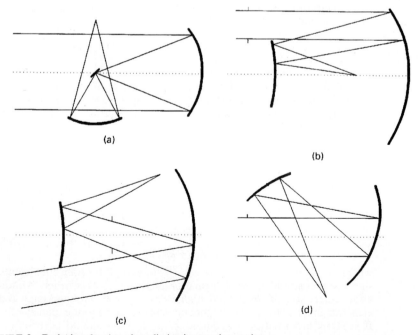

FIGURE 3   Reducing the size of or eliminating an obscuration.

obscuration: a small secondary mirror close to the intermediate image moves the larger tertiary mirror out of the beam path.

Figure 3*b* is an illustration of an eccentric pupil system. All elements are symmetric about the same axis and the aperture stop is decentered for a clear light path.

Figure 3*c* is an example of an off-axis objective with the field of view biased to direct the center of view away from any intervening elements. All elements and apertures are symmetric about the *optical axis.*

Figure 3*d* is an illustration of a tilted and decentered-component objective. Each element is rotationally symmetric about its own unique optical axis which may be tilted and/or decentered. The imaging behavior of this system is more complicated to deal with than the imaging behavior of eccentric pupil and off-axis systems. Vector aberration theory (Thompson, 1980, Sasian, 1990a) has been developed to properly model the imaging behavior of these systems.

Stray Light Suppression

Suppression of light diffracted from apertures and obscurations is facilitated with intermediate images and a real and accessible *Lyot stop.* Figure 4*a* illustrates a generic refractive configuration with an intermediate image and Lyot stop. Figure 4*b* illustrates where the diffracted light (shaded region) originates and terminates (at one edge of each aperture, for clarity).

A *field stop* is placed at the focus of the first lens to block diffracted light produced by the front light baffle. To block unwanted objects within the field of view, an occulting disc may be inserted at the focus of the first lens, as is done with a Lyot coronagraph in order to block the sun. By oversizing the field stop slightly, the light diffracted at the field stop falls just outside of the detector area.

Following the field stop is a second lens that reimages the intermediate image to the final image and the *entrance pupil* to the Lyot stop (the shaded region in Fig. 4*a* illustrates how the entrance pupil is imaged). Undersizing the Lyot stop blocks the light diffracted at the entrance pupil. In this way the Lyot stop becomes the *aperture stop* of the system.

Another application of the Lyot stop in the infrared (assuming the Lyot stop is located

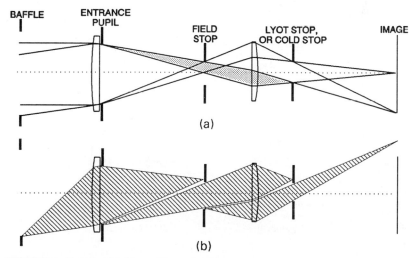

FIGURE 4 Generic objectives with apertures.

exterior to the objective optics) is as a cold stop (Fischer, 1992). The cold stop (Fig. 4*a*) is a baffle that prevents stray infrared light, radiated from the housing, from impinging upon the detector from outside its intended field.

Reflective and Catadioptric Objective Designs

The objectives to follow are listed according to *focal ratio* and design type. Objectives have a 20-cm diameter and catadioptric systems are optimized for a wavelength range from 480 to 680 nm unless otherwise stated. The angles of surface and element tilts are with respect to the horizontal optical axis. Decenters are also with respect to the optical axis. Since many of the designs are *aplanatic, anastigmatic,* and free of chromatic aberrations, the position of the stop does not affect third-order aberration correction and may be chosen to minimize *vignetting, distortion,* element size, or stray light contamination. All aberrations mentioned in this section are third-order unless otherwise stated.

Definitions of the abbreviated terminology used in the lens data are as follows:

SUR Surface number.

RDY Surface radius of curvature. A positive value means the center of curvature lies to the right of the surface; negative to the left.

THI Thickness of element or distance to next element. The value is positive if the next surface lies to the right of the surface.

GLA Glass-type or mirror surface, the latter referred to by the abbreviation REFL.

CON Conic constant *k*.

STO Stop surface.

INF A surface with an infinite radius of curvature; that is, a flat surface.

A, B, C, D The 4th-, 6th-, 8th-, and 10th-order aspheric deformation coefficients in Eq. (2).

A potential source of confusion is the terminology used to describe Mangin elements; that is, refractive elements with reflective back surfaces. This is illustrated in design 2 (F/4 Mangin): a ray enters the second refractive surface of the element (surface 2) and travels to the right where it intersects the mirror surface (surface 3). The thickness of surface 2 is therefore positive. The ray is reflected by the mirror surface (surface 3) and travels back through the glass element to surface 2; hence, the notation F9/REFL and the negative surface 3 thickness. Since surface 2 and 4 represent the same surface, the radii are the same.

1. *F/4 Paraboloid Objective*

(*a*) Newtonian; (*b*) Herschelian; (*c*) Pfund.

Comments: A single parabolic mirror objective can be arranged in a variety of forms, the best known being the Newtonian. Here a mirror flat diverts the image away from the light path. A tipped-mirror configuration is the Herschelian; a modern version is untipped and employs an eccentric-pupil to give an accessible image. A "backwards" Newtonian, the Pfund has a large flat-mirror primary. The Pfund has a smaller obscuration than the Newtonian and requires no diffraction-inducing support structure for the folding flat.

As has been mentioned, the on-axis performance of a paraboloid objective is perfect. Off-axis, *coma* quickly degrades image quality. For objectives slower than F/11, the easy-to-fabricate spherical mirror gives the same performance as a paraboloid when diffraction is also considered.

The paraboloid objective has image quality as good as a Cassegrain (design 3) of equivalent FN and aperture diameter, and is easier to align. The Cassegrain has the advantage of being compact.

2. F/4 Mangin

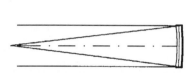

SUR	RDY	THI	GLA
1:	−75.15	1.0	BK7
2:	−307.1	1.4	F9
3:	−123.63	−1.4	F9/REFL
4:	−307.1	−1.0	BK7
5:	−75.15	−80.48	

Comments: The Mangin (1876) was invented by a French engineer of the same name to replace difficult-to-fabricate paraboloids in light houses. The objective relies upon the overcorrected *spherical aberration* produced by the negative first surface to cancel the undercorrected spherical aberration produced by the focusing, reflective surface. The chromatic aberration of the Mangin is reduced by achromatizing with two glasses of different dispersions. *Secondary spectrum* limits on-axis performance, and coma, one-half that of a spherical mirror, is the primary field-limiting aberration. Kingslake (1978) takes the reader through the design of a Mangin mirror.

Mangin mirrors are susceptible to ghost reflections from the refractive surfaces. Antireflection coatings are usually needed.

In some cases the overcorrected chromatic aberration of a Mangin is used to cancel undercorrected chromatic aberration produced by a refractive element. The Schupmann or medial objective (Schupmann, 1899, Olsen, 1986) has a positive refractive element with undercorrected chromatic aberration which is annulled by a Mangin element.

3. F/4 Cassegrain

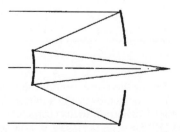

SUR	RDY	THI	GLA	CON
STO	−45.72	−16	REFL	−1
2:	−19.2	24.035	REFL	−3.236

Comments: The ubiquitous Cassegrain is predominant in situations where a small field of view, high resolution, compact size, long effective focal length, and accessible image are required. The classical Cassegrain is composed of a paraboloid primary and hyperboloid secondary, giving perfect imagery on-axis whenever the primary image coincides with the hyperboloidal focus. Coma and *field curvature* limit off-axis performance.

Many books discuss the first- and third-order properties of Cassegrain objectives. The Rutten (1988), Schroeder (1987), Korsch (1991), and Smith (1990) texts are among these.

4. F/4 Ritchey-Chretien

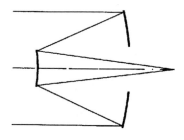

SUR	RDY	THI	GLA	CON
STO	−45.72	−16	REFL	−1.072
2:	−19.2	24.034	REFL	−3.898

Comments: The aplanatic Cassegrain or Ritchey-Chretien (Chretien, 1922) is also corrected for coma, leaving astigmatism and field curvature uncorrected. Both mirrors of the Ritchey-Chretien are hyperboloids.

Numerous modern telescope objectives are of Ritchey-Chretien form; among these are the Hubble space telescope and the infrared astronomical satellite IRAS.

5. F/9 Ritchey-Chretien Telescope with Two-Lens Corrector

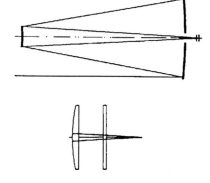

SUR	RDY	THI	GLA	CON
STO	−2139.7	−794.0	REFL	−1.0778
2:	−802.83	853.96	REFL	−4.579
3:	67.73	2.54	BK7	
4:	90.39	9.9		
5:	−1925.6	1.27	BK7	
6:	129.1	14.39		

Comments: This is a design by Wynne (1965) for the correction of the Cassegrain focus of a large (350-cm) Ritchey-Chretien. The corrector removes the inherent astigmatism and field curvature of the Ritchey-Chretien. Other Cassegrain focus correctors are discussed by Schulte (1966a), Rosin (1966), and Wilson (1968).

6. F/4 Dall-Kirkham

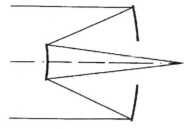

SUR	RDY	THI	GLA	CON
STO	−45.72	−16	REFL	−0.6456
6th order term:		0.593E-10		
2:	−19.2	24.035	REFL	

Comments: The Dall-Kirkham is another Cassegrain corrected for spherical aberration. The primary is an ellipsoid with 6th-order aspheric deformation and the secondary is spherical. An inverse Dall-Kirkham, or Carlisle, is just the reverse, with a spherical primary. There is *zonal spherical aberration* without the 6th-order deformation. Five times more coma is produced by the Dall-Kirkham than the classical Cassegrain, seriously limiting the field of view.

7. F/4 Cassegrain with Field Corrector and Spherical Secondary

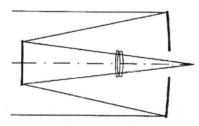

SUR	RDY	THI	GLA	CON
1:	−94.21	−27.937	REFL	−1
STO	−94.29	17.72	REFL	
3:	17.59	0.35	Silica	
4:	8.76	0.491		
5:	−64.15	0.6	Silica	
6:	−13.41	13.67		

Comments: By adding zero-power refractive correctors, the performance of a reflective objective is substantially enhanced. Zero power is maintained to prevent *axial color*. Such is the case with this objective similar to one designed by Rosin (1964). All third-order aberrations, with the exception of distortion, are corrected. The surfaces, with the exception of the primary, are spherical. One of the most attractive features of this design, in comparison to the Schmidt which will be discussed shortly, is the small size of the refractive elements. Add to this the capability of eliminating any remaining spherical aberration in an assembled objective by adjusting the axial positions of the lenses.

Zero *Petzval sum* and, hence, a flat image (in the absence of astigmatism) is ensured by giving the mirrors the same curvature and the lens elements equal and opposite power.

8. F/15 Spherical-primary Cassegrain with Reflective Field Corrector

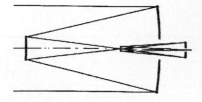

SUR	RDY	THI	GLA	CON
STO	−84.03	−30.69	REFL	
2:	−46.56	36.83	REFL	20.97
3:	−17.39	−14.77	REFL	−0.8745
4:	−20.87	16.26	REFL	−96.62

Comments: This well-corrected design from Korsch (1991) has an easily manufactured spherical primary and is intended for use as a large-aperture telescope objective. Another all-reflective corrector of Gregorian form has been developed for a fast (F/0.6) spherical primary (Meinel, 1984).

9. Afocal Cassegrain-Mersenne Telescope

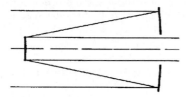

SUR	RDY	THI	GLA	CON
STO	−100	−35	REFL	−1
2:	−30	40	REFL	−1

Comments: The Mersenne contains two confocal paraboloids working at infinite conjugates. It is aplanatic, anastigmatic, and can be made distortion-free by choosing an appropriate stop location. The utility of confocal mirrors has been emphasized by Baker (1978) and Brueggeman (1968), and is illustrated in the following design.

10. Dual-magnification Cassegrain

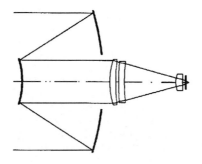

SUR	RDY	THI	GLA	CON
STO	−33.99	−11.69	REFL	−1
2:	−10.61	12.76	REFL	−1
3:	10.486	0.877	Silicon	
4:	25.673	0.462		
5:	48.33	0.798	Germanium	
6:	22.68	7.57		
7:	3.52	1.0	Silicon	
8:	4.22	0.377		
9:	INF	0.16	Sapphire	
10:	INF	0.396		

Comments: This IR design is related to one introduced by T. Fjeidsted (1984). The system offers two magnifications and fields of view. The high-magnification configuration is with the afocal Mersenne in the optical path. Removing the secondary lets light pass directly to the refractive assembly and a larger field of view is observed. The spectral range is from 3.3 to 4.2 μm.

11. *F/3.2 Three-lens Prime Focus Corrector*

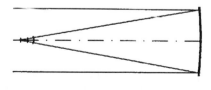

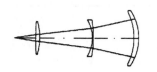

SUR	RDY	THI	GLA	CON
STO	−1494.57	−684.08	REFL	−1
2:	−26.98	−2.6	UBK7	
3:	−31.3	−22.43		
4:	−53.96	−0.586	UBK7	
5:	−19.0	−28.87		
6:	−33.36	−2.042	UBK7	
7:	236.7	−11.65		

Comments: This is a three-lens corrector for a 250-cm parabolic mirror. The corrector was developed by Wynne (1965, 1974) for the region of the spectrum extending from 365 to 1014 nm. It is used to extend the field of a parabolic mirror. Versions for a Ritchey-Chretien primary also exist. The corrector is able to correct spherical aberration, coma, astigmatism, and field curvature while keeping chromatic aberrations under control. The field of view can be extended considerably for smaller apertures.

The three-spherical lens corrector is one of the best large-optics prime-focus correctors to come along, both in terms of image quality and ease of fabrication. Other designs have either not performed as well or were heavily dependent on aspheric figuring.

This and other prime-focus correctors are surveyed in articles by Gascoigne (1973), Ross (1935), Meinel (1953), Schulte (1966b), Baker (1953), and Wynne (1972a).

12. *F/4 Gregorian*

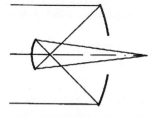

SUR	RDY	THI	GLA	CON
STO	−24.62	−16	REFL	−1
2:	6.4	24.1	REFL	−0.5394

Comments: The classical Gregorian is aberration-free on-axis when the paraboloidal mirror image coincides with one of the ellipsoidal-mirror foci; the other focus coincides with the final image. Like the Cassegrain, off-axis image quality is limited by coma and field curvature. The ellipsoidal secondary reimages the entrance pupil to a location between the secondary and final image. Thus, there exists the possibility of unwanted-light suppression at the primary-mirror image and *exit pupil.*

The Gregorian is longer than the Cassegrain and thus more expensive to support and house, but it produces an erect image and the concave secondary is easier to produce. In eccentric-pupil versions it has an accessible prime focus.

13. *F/4 Aplanatic Gregorian*

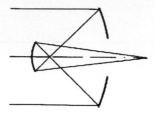

SUR	RDY	THI	GLA	CON
STO	−24.62	−16	REFL	−0.989
2:	6.4	24.1	REFL	−0.5633

Comments: The aplanatic Gregorian is corrected for spherical aberration and coma. Both mirrors are ellipsoids. Astigmatism and field curvature limit off-axis imagery.

14. *F/1.25 Flat-medial-field Aplanatic Gregorian*

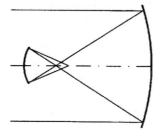

SUR	RDY	THI	GLA	CON
STO	−34.68	−22.806	REFL	−0.767
2:	6.47	7.924	REFL	−0.1837

Comments: The Gregorian's field performance is enhanced if image accessibility is sacrificed. This version of the Gregorian (Korsch, 1991) is aplanatic. A flat *medial image* is achieved by balancing Petzval curvature with astigmatism, which remains uncorrected.

15. *F/1.25 Flat-medial-field Aplanatic Gregorian with Spherical Primary*

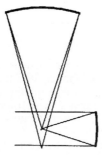

SUR	RDY	THI	GLA	CON
STO	−42.59	−21	REFL	
2:	INF	46.51	REFL	
		Tilt: 45°		
3:	−49.84	−54.08	REFL	−0.078

Comments: The field of this objective (Korsch, 1991) is larger than its cousins, the classical and aplanatic Gregorians, even with the spherical primary. Spherical aberration and coma are corrected, and the medial image is flat. The design has a real intermediate image and exit pupil. The obvious drawback is the size of the secondary in relation to the size of the entrance pupil, which is 15 cm in diameter.

Korsch (1991) analyzes two other designs that are loosely referred to as Gregorians.

16. *Afocal Gregorian-Mersenne Telescope*

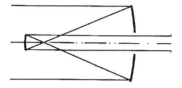

SUR	RDY	THI	GLA	CON
STO	−50	−30	REFL	−1
2:	10	40	REFL	−1

Comments: The Gregorian Mersenne, also composed of confocal paraboloids, is aplanatic, anastigmatic, and can be corrected for distortion. The Gregorian-Mersenne has an intermediate image and an accessible exit pupil.

17. *F/1.25 Couder*

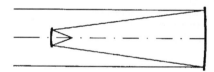

SUR	RDY	THI	GLA	CON
STO	−142.86	−52.9	REFL	−6.285
2:	23.08	7.1142	REFL	−0.707

Comments: The Couder (1926), composed of two conic mirrors, is corrected for third-order spherical aberration, coma, and astigmatism. Unfortunately, the Couder is long for its focal length and the image is not readily accessible.

18. *F/1.25 Aplanatic, Flat-medial-image Schwarzschild*

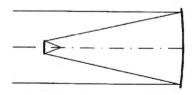

SUR	RDY	THI	GLA	CON
STO	−91.57	−38.17	REFL	−2.156
2:	23.67	4.637	REFL	5.256

Comments: The aplanatic, flat-medial-image Schwarzschild (1905) is similar in appearance to the Couder but the secondary mirror and image locations are different for identical *secondary magnifications.*

19. *F/1.25 Aplanatic, Anastigmatic Schwarzschild*

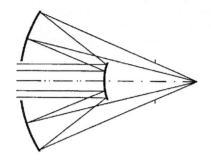

SUR	RDY	THI	GLA
1:	30.62	−49.44	REFL
2:	80.14	80.26	REFL
STO	INF	24.864	

Comments: The spherical-mirror Schwarzschild (1905) is aplanatic, *anastigmatic,* and distortion-free (Abel, 1980). The Schwarzschild relies on the principle of symmetry for its high level of aberration correction and a large field of view. All surfaces have the same center of curvature at the aperture stop. Hence, there are no off-axis aberrations. Spherical aberration is produced but each mirror produces an equal and opposite amount, thus canceling the effect of the aberration. Some higher-order aberrations are also corrected (Abel, 1980). Eccentric portions of this design—above and below the optical axis in the picture—form well-corrected, unobscured designs. Zonal spherical aberration from the mix of third- and higher-order terms limits on- and off-axis performance.

An aspheric plate positioned at the center-of-curvature of the mirrors removes this aberration as illustrated in the next design.

Wetherell and Rimmer (1972), Korsch (1991), Schroeder (1987), Linfoot (1955), and Gascoigne (1973) offer a general third-order analysis of two-mirror systems. The closed-form solutions described provide insight into third-order aberration theory of reflective systems.

20. *F/1 Aplanatic, Anastigmatic Schwarzschild with Aspheric Corrector Plate*

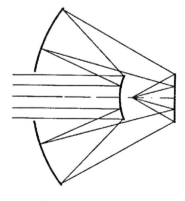

SUR	RDY	THI	GLA
1:	24.547	−39.456	REFL
2:	63.92	64.528	REFL
STO	INF	−19.098	REFL
	A:	−0.9998E-7	
	B:	−0.1269E-9	

Comments: With an aspheric plate at the aperture stop, spherical aberration is eliminated. The only aberrations still remaining are of higher order. To correct these, the mirrors must also be aspherized. Linfoot (1955) and Abel (1980) describe this design.

21. *F/1.25 Anastigmatic, Flat-image Schwarzschild*

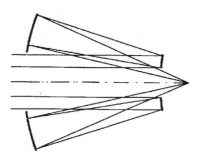

SUR	RDY	THI	GLA	CON
1:	69.7	−50.56	REFL	5.47
STO	71.35	61.26	REFL	0.171

Comments: With just two conics, this design type (Schwarzschild, 1905) achieves aplanatic and anastigmatic performance on a flat image surface. The flat field is attained by making the curvatures of the mirrors equal. Eccentric portions above or below the optical axis form unobscured versions; the design may alternatively be used off-axis. Sasian (1990a, 1990b) and Shafer (1988) have explored many of this design's features.

22. *F/1.25 Schmidt*

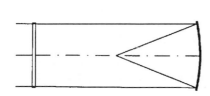

SUR	RDY	THI	GLA	CON
STO	1554	1	PSK2	
	A:	−0.2825E-5		
	B:	−0.1716E-8		
2:	INF	52.33		
3:	−52.95	−26.215	REFL	

Comments: The Schmidt (1932) also relies on the principle of symmetry; that is, the aperture stop is located at the center of curvature of the spherical mirror and hence the mirror produces no off-axis aberrations.

The Schmidt corrector is flat with aspheric deformation figured in to correct the spherical aberration produced by the mirror. It is positioned at the aperture stop because off-axis aberrations are independent of aspheric deformation when an aspheric surface coincides with the stop. Hence the Schmidt plate has no effect on off-axis aberrations, and the benefits of concentricity are preserved.

The corrector introduces chromatic variation of spherical aberration (spherochromatism). A small amount of positive power in the corrector introduces undercorrected axial color to reduce the effects of this aberration. Further improvement is obtained by achromatizing the corrector with two glasses of different dispersions.

Higher-order aberrations degrade image quality at low focal ratios and large field angles. Kingslake (1978), Schroeder (1987), Maxwell (1972), and Linfoot (1955) provide additional details of this and other catadioptric objectives.

23. *F/1.25 Field-flattened Schmidt*

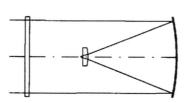

SUR	RDY	THI	GLA
STO	598.7	1.155	PSK2
	A: −0.273E-5		
	B: −0.129E-8		
2:	INF	40.38	
3:	−52.95	−24.06	REFL
4:	−10.35	−1.49	PSK2
5:	INF	−0.637	

Comments: As is known from third-order aberration theory, a thin element will be nearly aberration-free, except for Petzval curvature, and distortion when it is placed in close proximity to an image. Therefore, by properly choosing the lens power and index to give a Petzval curvature of equal and opposite sign to the Petzval curvature introduced by the other optics, the image is flattened.

The image in the Schmidt above has been flattened with the lens near the image plane. The only aberrations introduced by the lens are spherochromatism and *lateral color,* lateral color being the most noticeable aberration; this can be removed by achromatizing the field-flattening lens. The close proximity of the lens to the image can cause problems with light scattered from areas on the lens surfaces contaminated by dirt and dust particles. Clean optics are a must under these circumstances.

The field-flattening lens provides two more positive results. First, the lens introduces a small amount of coma which is compensated by moving the Schmidt corrector towards the mirror somewhat, thus reducing the overall length of the objective. Second, *oblique spherical aberration,* one of the primary field-limiting, higher-order aberrations of the Schmidt, is substantially reduced.

Besides its usual function as a telescope or photographic objective, the field-flattened Schmidt has also been used as a spectrograph camera (Wynne, 1977).

24. *F/1.25 Wright*

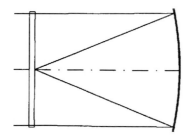

SUR	RDY	THI	GLA	CON
1:	INF	1.0	BK7	
2:	−699.8	26.1		
	A: 0.6168E-5			
	B: 0.5287E-8			
3:	−53.24	−26.094	REFL	1.026

Comments: The Wright (1935) is one-half the length of the Schmidt. It also relies on aspheric deformation of the corrector plate for the elimination of spherical aberration. Coma, introduced as the corrector is removed from the center of curvature of the mirror, is cancelled with conic deformation of the mirror; the surface figure is that of an oblate ellipsoid. The remaining astigmatism and Petzval curvature are balanced for a flat medial image. The only on-axis aberration, spherochromatism, is corrected by achromatizing the corrector.

25. F/4 Reflective Schmidt

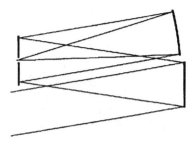

SUR	RDY	THI	GLA
STO	−66752	−67.37	REFL
	A: 0.5083E-7		
2:	INF	66.6	REFL
3:	−133.97	−66.85	REFL

Comments: Another way of defeating chromatic aberration is to eliminate it altogether with a reflective corrector (Abel, 1985). The elements are removed from the light path with a field bias (9°), and hence the objective is off-axis. Spherical aberration, coma, and astigmatism are all corrected. At large field angles, beyond about 12° half-field angle, oblique spherical aberration becomes evident, but otherwise this design provides excellent performance on a curved image over a 24° field of view. In order to avoid severe obstruction of the light path, the full 24° can be taken advantage of only in the plane that extends perpendicular to the picture of the design above. A considerably reduced field of view is allowed in the plane of the picture.

26. F/0.6 Solid Schmidt

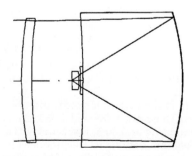

SUR	RDY	THI	GLA	CON
1:	62.69	1.69	BK7	
2:	103.38	8.39		
		A: 0.1492E-4		
		B: 0.1988E-7		
		C: 0.1013E-10		
		D: 0.35E-12		
3:	−169.36	16.47	BK7	
STO	−37.05	−16.47	REFL\BK7	
5:	INF	−0.3952	BK7	
6:	−45.86	−0.245		
7:	−10.295	−1.136	BK7	
8:	INF	−0.026		

Comments: All monochromatic aberrations, with the exception of distortion, are corrected by the appropriately-named solid Schmidt, a system used mostly as a spectrograph camera (Schulte, 1963). All chromatic aberrations, with the exception of lateral color, are corrected. The imaging theory behind the solid Schmidt is expounded by J. Baker (1940a). With a refractive index n, the solid Schmidt is n^2 times faster than the conventional Schmidt. Focal ratios of F/0.3 have been obtained. Schulte (1963) offers a design similar to the one given here.

27. F/1.25 Schmidt-Cassegrain

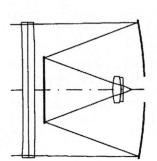

SUR	RDY	THI	GLA	CON
STO	INF	0.8	489.574	
		A: −0.1928E-4		
		B: 0.298E-7		
2:	2269.1	1.0	583.303	
3:	INF	16.49		
		A: −0.1085E-4		
		B: 0.2806E-7		
4:	−55.9	−15.0	REFL	1.077
5:	INF	10.267	REFL	
6:	9.1	1.2	489.574	
7:	−8.577	0.018		
8:	−8.59	0.3	583.303	
9:	−87.44	1.317		

Comments: The Schmidt-Cassegrain (Baker, 1940b) represents a successful attempt to resolve the difficulties related to the curved image, considerable length, and awkwardly located image of the Schmidt objective, without destroying the positive attributes of the design.

The Schmidt-Cassegrain comes in a wide variety of forms—too many to go into here. Linfoot (1955) performs an extensive exploration of the design, with one and two aspheric plate correctors. Warmisham (1941) has gone as far as three. Wayman (1944) has analyzed a *monocentric* Schmidt-Cassegrain.

In this fast version of the Schmidt-Cassegrain, the corrector is close to the flat secondary. Usually one or both mirrors are aspherics. An achromatized image-flattening lens has been introduced. An image-flattening lens is not usually required with a Schmidt-Cassegrain since enough degrees of freedom exist for aberration correction and a flat image. In this case, the secondary mirror is flat and one degree of freedom is lost. Additionally, the primary mirror power introduces a strong Petzval contribution which necessitates a field-flattening lens.

The first three digits of the six-digit code in the glass column identify the indices of the materials in the design, in this case plastics. These are 1.489 and 1.583; the *Abbe-numbers* are given by the last three digits and they are 57.4 and 30.3, respectively. The two plastics have nearly identical thermal coefficients and are very light. Buchroeder (1971a) analyzes designs of this variety with two aspheric correctors. Shafer (1981) offers a Schmidt-Cassegrain with an aspheric meniscus corrector. Only two elements are used since the secondary mirror surface is on the corrector. Rutten (1988) has examples of Schmidt-Cassegrains in a number of configurations.

28. *F/3.4 Reflective Schmidt-Cassegrain*

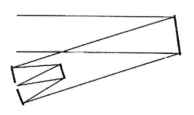

SUR	RDY	THI	GLA	CON
STO	INF	−92.16	REFL	
	A: 0.13344E-6	AN: −0.1255E-1		
2:	84	26	REFL	
3:	84	−25.848	REFL	−0.3318

Comments: The reflective Schmidt-Cassegrain exhibits all the nice properties of the Schmidt-Cassegrain and, in addition, is achromatic. Schroeder (1978) points out that, because the corrector is tilted (9° here), adequate aberration correction requires a nonrotationally symmetric corrector plate. The nonaxially symmetric surface deformation in this design is given by

$$Z = A[(1 - AN)X^2 + (1 + AN)Y^2]^2 \qquad (3)$$

where A is the fourth-order symmetric coefficient and AN is the fourth-order nonsymmetric coefficient. The y dimension is in the plane of the picture; the x dimension is perpendicular to it.

Since the corrector is tilted by 9°, the reflected rays are deviated by twice that amount. The element spacings (THI) are no longer measured along the horizontal optical axis after reflection off the corrector, but along the line of deviation. The secondary and tertiary are tilted by 18°.

29. *F/2 Shafer Relayed Virtual Schmidt*

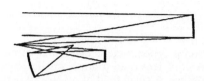

SUR	RDY	THI	GLA
STO	−320	−159.82	REFL
2:	106.7	80.0	REFL
3:	INF	−68.51	REFL
	A: 0.1882E-5		
	B: 0.1273E-8		
	C: −0.1757E-12		
	D: 0.1766E-14		
4:	63.967	40.774	REFL

Comments: Shafer (1978a) has introduced an eccentric-pupil (18-cm stop decenter), virtual Schmidt objective similar to this but with a decentered quaternary mirror. The center of curvature of the spherical primary is imaged by the concave secondary onto the flat Schmidt tertiary mirror. Now the Schmidt plate, which appears to be at the primary center of curvature, is aspherized to produce a well-corrected virtual image, hence the

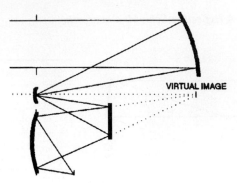

FIGURE 5 Picture of virtual Schmidt with decentered quaternary.

name (see Fig. 5). In this configuration, the Schmidt plate is one-half the size of the primary.

The Schmidt plate and the spherical quaternary mirror form a finite conjugate Schmidt system. Thus, the spherical aberration of this mirror is also corrected.

Figure 5 shows a pictorial representation of the Shafer design with the last mirror decentered to provide a more accessible image. Since the primary and quaternary mirrors no longer share the same axis of symmetry, a two-axis Schmidt corrector is required to remove the aberrations of both mirrors. The shape of this surface is described by Shafer, for an F/1, unobscured, wide-field design with an intermediate image and Lyot stop.

30. *F/2.2 Spherical-primary Objective That Employs the Schmidt Principle of Correction*

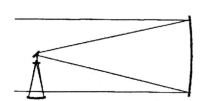

SUR	RDY	THI	GLA	CON
STO	−88.07	−42.51	REFL	
2:	INF	2.2	REFL	
		Tilt: 45°		
3:	−4.433	0.33	FK51	
4:	−2.527	9.217		
5:	−10.21	−9.217	REFL	−0.8631
6:	−2.527	−0.33	FK51	
7:	−4.433	−0.64		

Comments: Baker (1978) reports on a system where the center of curvature of a large, spherical primary is imaged by a positive lens onto a much smaller mirror where aspheric correction of spherical aberration occurs. A small field offset (0.25°) is required so that the one-to-one relay doesn't reimage the primary image back onto itself. To avoid overlap, this design is best used over a small or strip field of view.

Because of the geometry of the design, coma, astigmatism, image curvature, chromatic aberrations, and distortion are eliminated in addition to the spherical aberration correction from aspheric figuring of the tertiary mirror. Baker (1978) offers several other interesting designs in his article, including an F/0.8, 10.6-μm, 180° field-of-view Roesch (1950), a design that incorporates a Schmidt with a strong negative lens or lens group before the aspheric corrector. The strong divergence produced by this lens reduces the amount of light blocked by the image plane but increases the size of the spherical mirror.

31. F/2 Maksutov

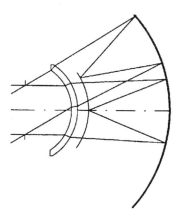

SUR	RDY	THI	GLA	CON
STO	INF	31.788		
2:	−23.06	3.5	BK7	
3:	−25.53	51.09		
4:	−83.9	−43.8	REFL	

Comments: The all-spherical Maksutov (1944) was intended as an inexpensive alternative to the Schmidt at slower speeds. In small sizes it is indeed less expensive. The meniscus corrector is "self-achromatic" when the following relationship is satisfied:

$$t = \frac{n^2}{n^2 - 1}(R_2 - R_1) \tag{4}$$

where R_1 and R_2 are the radii, t is the thickness, and n is the refractive index of the corrector.

Bouwers (1950) also developed a meniscus corrector. All elements of the Bouwers are concentric about the aperture stop. This ensures correction of third-order, off-axis aberrations over a nearly unlimited field of view. In exchange for the wide field, axial color is not well-corrected.

32. F/1.25 Solid Maksutov-Cassegrain

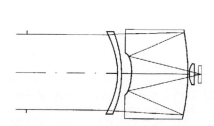

SUR	RDY	THI	GLA	CON
STO	INF	20.5		
2:	−20.5	0.955	Silica	
3:	−25.92	0.0313		
4:	138.58	15.3	Silica	
5:	−45.61	−12.973	REFL\Silica	
6:	−51.89	13.41	REFL\Silica	
7:	INF	0.0475		
8:	12.026	1.68	Silica	
9:	16.07	0.545		
10:	INF	0.394	Silica	
11:	INF	0.155		

Comments: The solid Maksutov-Cassegrain shown here and the solid Schmidt-Cassegrains have been studied extensively by Wynne (1971, 1972b). Lateral color is the most consequential aberration left uncorrected.

33. *F/1.2 Wide-field Objective with Maksutov Correction*

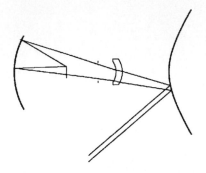

SUR	RDY	THI	GLA	CON
1:	12.467	−9.684	REFL	−3.243
2:	−4.81	−1.267	FK51	
3:	−3.762	−3.679		
STO	INF	−17.189		
5:	15.98	10.64	REFL	

Comments: This very wide field imaging system similar to one in Courtes et al. (1971) is essentially a Maksutov focused on the virtual image of the object produced by the hyperboloidal mirror. Both speed (F/1) and a very wide field of view (80 × 120 degrees) can be achieved with this design form on a flat image but only for small apertures—1.25 cm in this case. Courtes (1983) describes similar systems with refractive and reflective Schmidt plates instead of a Maksutov corrector.

34. *F/1 Gabor*

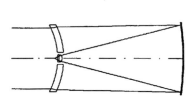

SUR	RDY	THI	GLA
STO	−23.3	2	SK1
2:	−25.468	39.5	
3:	−83.33	−40	REFL
4:	−1.67	−1	BK7
5:	9.85	−0.5	SF5
6:	−7.71	−0.942	

Comments: Another meniscus design was invented by Gabor (1941). The Gabor is more compact than the Maksutov or Bouwers, and has a smaller focal ratio and field of view.

The design shown here began without the field lens. The lens was introduced into the design with the surface closest to the image being concentric about the *chief ray* and the other surface being aplanatic (Welford, 1986). A surface concentric about the chief ray is free of coma, astigmatism, distortion, and lateral color. The aplanatic surface is free of spherical aberration, coma, and astigmatism with the result that the lens is coma- and astigmatism-free. The spherical aberration produced by the lens is balanced against the spherical aberration produced by the two other elements. The chromatic aberrations were corrected by achromatizing the lens.

Shafer (1980a, 1980b) offers interesting suggestions for design with aplanatic and concentric surfaces. Several varieties of field-flattening lens are described. Kingslake (1978) runs through the design procedure for a Gabor.

35. *F/4 Schmidt-meniscus Cassegrain*

SUR	RDY	THI	GLA
STO	787.7	1.4	BK7
2:	INF	32.69	
3:	−32.69	2.62	BK7
4:	−35.6	63.446	
5:	−81.97	−21.78	REFL
6:	−79.5	38.65	REFL

Comments: This system, originally by Bouwers, uses a slightly positive plate to compensate the overcorrected chromatic aberration produced by the meniscus. The Bouwers produces very good quality on a flat image, over a large field of view.

Fourth- and sixth-order deformation added to the plate eliminates any residual spherical aberration. Lateral color and oblique spherical aberration affect field performance, although both are small.

36. *F/1.2 Baker Super-Schmidt*

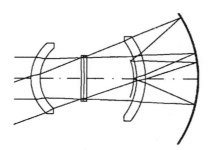

SUR	RDY	THI	GLA
1:	22.5	4.28	BK7
2:	19.13	19.75	
STO	−11,783	1.4	F2
	A:	−0.558E-6	
	B:	0.244E-8	
4:	−135.5	1.3	SK2
5:	INF	23.92	
6:	−29.31	4.28	BK7
7:	−32.42	25.1	
8:	−55	−25.1	REFL
9:	−32.42	−4.28	BK7
10:	−29.31	−1.45	

Comments: The Baker (1945) super-Schmidt, a design that incorporates both meniscus and Schmidt correction, achieves excellent performance over a wide field of view. The field-limiting aberration of a fast Schmidt, *oblique spherical aberration,* is controlled by adding a concentric meniscus lens which also introduces overcorrected spherical aberration, thus reducing the amount of overcorrection needed from the Schmidt plate. Since oblique spherical is proportional to the amount of overcorrection in the Schmidt plate, the effect of this aberration is reduced.

The most apparent aberration produced by the meniscus is axial color. This is minimized by achromatizing the Schmidt corrector. Spherochromatism is reduced since the magnitudes produced by the Schmidt corrector and meniscus are nearly equal and have opposite signs. Another meniscus element is added to further reduce aberrations.

37. *F/1 Baker-Nunn*

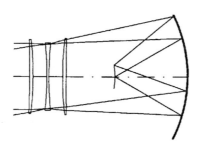

SUR	RDY	THI	GLA
1:	−491.9	1.06	LLF1
2:	−115.6	4.23	
	A:	−0.8243E-5	
	B:	0.1348E-8	
STO	−125.78	0.64	SK3
	A:	−0.1158E-4	
	B:	−0.427E-8	
	C:	−0.7304E-11	
4:	125.78	4.23	
	A:	0.1158E-4	
	B:	0.427E-8	
	C:	0.7304E-11	
5:	115.6	1.06	LLF1
	A:	0.8243E-5	
	B:	−0.1348E-8	
6:	491.87	36.77	
7:	−42.03	−21.961	REFL

Comments: The Baker-Nunn (Baker, 1962) was born of work by Houghton (1944) during World War II. Houghton wished to find a less expensive alternative to the Schmidt. The result was a zero-power, three-lens combination with easy-to-make spherical surfaces. Spherical aberration and coma can be eliminated for any position of the corrector. The surfaces have equal radii so they can be tested interferometrically against one another using the Newton ring method. Residual spherical aberration that remains after assembly is removed by altering the spacing between the lenses.

38. *F/10 Houghton-Cassegrain*

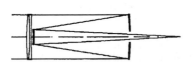

SUR	RDY	THI	GLA
STO:	145	1.2	BK7
2:	−172.1	0.164	
3:	−111.9	0.639	BK7
4:	264.7	44.61	
5:	−129.7	−43.16	REFL
6:	−63.94	66.84	REFL

Comments: A two-lens, afocal corrector developed by Houghton and Sonnefeld (1938) is used here as a corrector for a Cassegrain system. Sigler (1978) has written on the subject of Cassegrains with Houghton, Schmidt, and Maksutov refractive correctors. This Houghton-Cassegrain gives well-corrected imagery on a curved image surface. An afocal achromatized doublet corrector has also been tried (Hawkins and Linfoot, 1945).

39. F/3.6 Houghton-Cassegrain

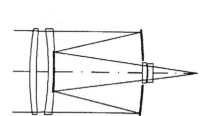

SUR	RDY	THI	GLA
STO	69.64	1.607	UBK7
2:	148.71	3.045	
3:	−61.43	1.607	LAK21
4:	−97.53	21.733	
5:	−85.11	−21.733	REFL
6:	−97.53	21.733	REFL
7:	70.44	1.2	UBK7
8:	−15.47	0.18	
9:	−15.23	1.3136	SK16
10:	−517.29	11.03	

Comments: Another Houghton corrector, with meniscus elements, is utilized in this design by D. Rao (1987). The spectral range is 550 to 850 nm. The design is similar to one introduced by Mandler (1951). Examples of other Houghton-Cassegrains of this form are studied by Gelles (1963).

40. F/1.25 Shenker

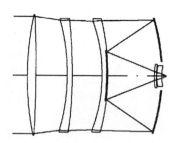

SUR	RDY	THI	GLA
STO	49.42	1.5	BK7
2:	−203.6	5.4	
3:	−34.7	0.863	BK7
4:	−79.25	5.08	
5:	−27	0.98	BK7
6:	−38.87	9.32	
7:	−31.96	−9.32	REFL
8:	−38.87	8.1	REFL
9:	13.73	0.39	BK7
10:	21.8	0.05	
11:	7.925	0.895	BK7
12:	8.56	0.856	

Comments: M. Shenker has studied a large number of variations on the theme of three-element correctors for a Cassegrain. This is related to one of the configurations developed by Shenker (1966). Note that the third corrector is also the secondary mirror. Zonal spherical aberration limits performance on-axis. This may be removed by aspherizing one or more surfaces. All elements are of the same glass. Laiken (1991) has a similar version of this design as well as other catadioptric objectives. Maxwell (1972) has design examples and catadioptric imaging theory.

41. *F/1.25 Mangin-Cassegrain with Correctors*

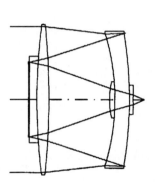

SUR	RDY	THI	GLA
STO	80.62	1.64	BK7
2:	−102.3	9.07	
3:	−30.43	2.02	BK7
4:	−54.52	−2.02	BK7/REFL
5:	−30.43	−9.07	
6:	−102.3	−1.64	BK7
7:	80.62	−1.01	BK7
8:	−526.4	1.01	BK7/REFL
9:	80.62	1.64	BK7
10:	−102.3	8.32	
11:	11.06	0.75	BK7
12:	−30.43	2.02	BK7
13:	−54.52	0.5	SF10
14:	52.92	1.445	

Comments: Mangin mirrors are evident in this design by L. Canzek (1985) and two elements are used twice. The design has exceptionally good on-axis performance. Lateral color and higher-order aberrations limit the field.

42. *F/1.25 Mangin-Cassegrain with Correctors*

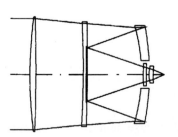

SUR	RDY	THI	GLA
STO	80.83	1.09	FN11
2:	−325.9	8.5	
3:	−191.4	0.728	FN11
4:	−440.3	9.69	
5:	−31.44	1.456	FN11
6:	−46.13	−1.456	FN11/REFL
7:	−31.44	−9.69	
8:	−440.3	10	REFL
9:	26.97	0.582	FN11
10:	38.33	0.544	
11:	8.44	0.728	FN11
12:	40.87	2.025	

Comments: Another short and fast catadioptric by Max Amon (1973) is shown here. The second corrector is also the secondary mirror.

43. *F/4 Eisenburg and Pearson Two-mirror, Three-reflection Objective*

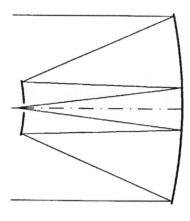

SUR	RDY	THI	GLA	CON
STO	−48.0	−17.289	REFL	−1.05
2:	−14.472	17.289	REFL	−1.539
3:	−48	−18.195	REFL	−1.05

Comments: This aplanatic, two-mirror, three-reflection configuration was first introduced by Rumsey (1969). The design presented here comes from Eisenburg and Pearson (1987). The first and third surface represent the same surface.

44. *F/4 Shafer Two-mirror, Three-reflection Objective*

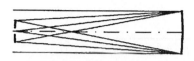

SUR	RDY	THI	GLA	CON
STO	−106.7	−80.01	REFL	−0.4066
2:	80.01	80.01	REFL	−5.959
3:	−106.7	−80.05	REFL	−0.4066

Comments: Shafer (1977) has documented numerous versions of multiple-reflection objectives. This is an aplanatic, anastigmatic design with field curvature. For optimum aberration correction, the primary is at the center of curvature of the secondary mirror. Shafer (1908b) suggests a ring field for a flat, accessible image on an annular surface, and a Lyot stop.

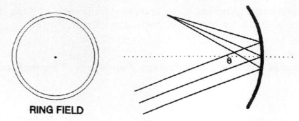

FIGURE 6 The ring field system.

A simple ring field design is depicted in Fig. 6. Only one field angle θ is required, easing the difficulties associated with off-axis aberration correction. The single viewing direction is rotated about the optical axis, forming, in this case, a ring image. In reality, less than half the ring image is used to avoid interference of the image with the entering beam.

45. *F/15 Two-mirror, Three-reflection Objective*

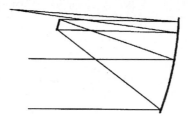

SUR	RDY	THI	GLA	CON
STO	−116.33	−46.53	REFL	−1.024
2:	−22.6	46.53	REFL	−1.0037
3:	−116.33	−67.05	REFL	−1.024

Comments: This is another aplanatic, anastigmatic, eccentric-pupil design which gives well-corrected imagery on a curved image. It has a 30-cm stop decenter.

46. *F/15 Yolo*

SUR	RDY	THI	GLA	CON
STO	−1015	−160.36	REFL	−4.278
		Tilt: −3.5°		
2:	1045.72	208.19	REFL	
		RDX: 1035.0		
		Tilt: −9.82°		
		Image tilt: −11.7°		

Comments: Arthur S. Leonard (1986a, 1986b) invented the Yolo (named after a scenic county in California) so that he could have an achromatic system without obscurations. The result is a tilted and decentered component objective that gives the high contrast of an unobscured refractive objective without the chromatic effects.

Spherical aberration is corrected by the hyperboloidal figuring of the first surface. The anamorphism introduced into the secondary (by a warping harness) corrects astigmatism; RDX is the surface radius of curvature perpendicular to the picture. Coma is eliminated by adjusting the curvatures and tilting the secondary.

Relatives of the two-mirror Yolo are the Solano, an in-line three-mirror Yolo, or the three-dimensional, three-mirror Yolo (Mackintosh, 1986). As in design 28, thickness (THI) is measured along the deviated ray paths. With the angle of reflection known, element decenter may be easily determined.

47. F/15 Schiefspiegler

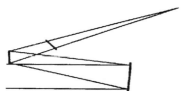

SUR	RDY	THI	GLA	CON
STO	−397.2	−101.4	REFL	−0.607
		Tilt angle: −4.5°		
2:	−552.5	35.84	REFL	
		Tilt angle: 3.64°		
3:	3411	0.52	BK7	
		Tilt angle: 50°		
4:	INF	111.11		
		Tilt angle: 50.0529°		
		Image tilt angle: 22.80°		

Comments: The Schiefspiegler ("oblique reflector" in German) was introduced about a century ago, and at the time was called a brachyt (or bent). The motivation behind the Schiefspiegler's design is essentially the same as the Yolo's. Like the Yolo, elements are tilted and decentered. Coma and astigmatism are corrected by tilting the secondary and corrector lens. The lens is thin and slightly wedged to minimize chromatic effects. Spherical aberration is corrected with the aspheric deformation of the primary.

A three-mirror Schiefspiegler, or Trischiefspiegler, has been developed by Anton Kutter (1975). This design is all-reflective and completely achromatic. Like the Schiefspiegler, aspheric deformation of the primary corrects spherical aberration; coma and astigmatism are corrected with element tilts.

A four-mirror Schiefspiegler was recently introduced by Michael Brunn (1992). For more on unusual telescope objectives, see Manly (1991).

48. *F/8 Catadioptric Herschelian Objective*

SUR	RDY	THI	GLA
STO	269.61	1.487	BK7
2:	INF	2.147	

Element tilt: 0.35°

3:	−269.61	1.321	BK7
4:	INF	151.97	

Element tilt: 5.38°

5:	−317.26	−158.69	REFL

Tilt angle: 3.0°
Image tilt angle: 0.174°

Comments: Several centuries ago, Herschel tilted his large parabolic mirror to give him access to the image. A spherical mirror in this design by D. Shafer (Telescope Making 41) has been tilted for the same reason. Element tilts in the Houghton corrector control the astigmatism introduced by tilting the mirror. The Houghton corrector also eliminates the spherical aberration of the mirror with lens bending. Note the smaller focal ratio of this design compared to either the Yolo or the Schiefspiegler.

Other catadioptric Herschelians, as well as Schiefspieglers and Yolos, have been studied by Buchroeder (1971b) and Leonard (1986b). Tilted, decentered, and unobscured Cassegrains are discussed by Gelles (1975).

49. *F/4 SEAL*

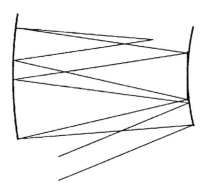

SUR	RDY	THI	GLA	CON
1:	181.2	−147.8	REFL	
2:	350.9	147.8	REFL	−0.404
STO	INF	−147.8	REFL	
4:	350.9	119	REFL	−0.404

Comments: For an all-reflective objective, this flat-image design provides an exceptionally wide, unobscured field of view—greater than 90° with a ring field. Referred to as the SEAL (Owen, 1990), it is derived from its cousin the WALRUS (Hallam, 1986); a related design has been described by Shafer (1978d). The SEAL is another Mersenne-Schmidt hybrid: primary and secondary form an inverse-Mersenne; tertiary and quaternary (also the secondary) form a reflective Schmidt. Residual spherical aberration limits the performance, but by aspherizing the flat, this residual aberration is corrected as well. Clearing all obscurations requires at least a 22° field offset. The SEAL shown here is optimized for a 20° strip field although a square, rectangular, annular, or almost any shape field is possible.

50. *F/4 Paul Three-mirror Objective*

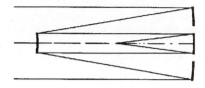

SUR	RDY	THI	GLA	CON
STO	−117.1	−42.87	REFL	−1
2:	−31.38	42.87	REFL	−.6076
3:	−42.87	−21.42	REFL	

Comments: This design is based on work by Paul (1935) and later by Baker (1969), who was looking for an achromatic field corrector for a parabolic primary. Their efforts culminated in a design similar to this one, which combines the properties of an afocal Cassegrain-Mersenne in the first two elements with the properties of an all-reflective Schmidt in the secondary and tertiary elements. Since both modules are corrected for spherical aberration, coma, and astigmatism to third order, the complete system is aplanatic and anastigmatic. Petzval curvatures are equal and opposite so a flat image is achieved. The conic deformation of the secondary is modified to give it an ellipsoidal shape. This gives the required Schmidt aspherization needed to correct the spherical aberration of the tertiary mirror.

Other all-reflective designs have been proposed by Meinel (1982, 1984) and Baker (1978). The Meinel-Shack objective (1966) exhibits similar performance and offers a more accessible image.

51. *F/4 Alternative Paul Three-mirror Objective*

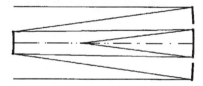

SUR	RDY	THI	GLA	CON
STO	−142.4	−49.28	REFL	−1
2:	−39.51	49.28	REFL	
3:	−54.69	−30.7	REFL	0.101

Comments: This Paul objective has an aspheric tertiary mirror, instead of an aspheric secondary.

52. *F/4 Off-axis, Eccentric-pupil Paul-Gregorian*

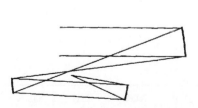

SUR	RDY	THI	GLA	CON
STO	INF	79.2		
2:	−158.4	−118.8	REFL	−1
3:	79.2	79.2	REFL	−1
	A: −0.2707E-6			
	B: −0.117E-9			
4:	−77.53	−38.768	REFL	

Comments: This eccentric-pupil (22-cm), off-axis (1°) design utilizes a Gregorian-Mersenne module in the primary and secondary mirrors. Spherical aberration produced by the spherical tertiary mirror is corrected by superimposing aspheric deformation on the paraboloid secondary, located at the tertiary mirror center of curvature. With all concave surfaces, field curvature is uncorrected. A real and accessible exit pupil and intermediate image offer possibilities for excellent stray-light suppression.

As is the case with the virtual Schmidt system, the tertiary mirror may be decentered to provide a more convenient image location. This requires two-axis aspheric deformation of the secondary mirror (Shafer, 1978c).

53. F/4 Three-mirror Cassegrain

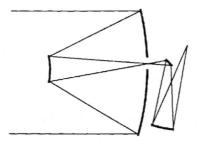

SUR	RDY	THI	GLA	CON
STO	−39.67	−15.814	REFL	−0.9315
2:	−10.66	21	REFL	−2.04
3:	INF	−9.05	REFL	
		Tilt: 45°		
4:	13.66	13.651	REFL	−0.4479

Comments: A design similar to the aplanatic, anastigmatic, flat-image design shown here was conceived by Korsch (1978) and is described by Williams (1979), Korsch (1977), and Abel (1985). The exit pupil is accessible and an intermediate image exists. A 1° field offset is needed to displace the image from the folding flat. Residual coma limits field performance. Small element tilts and decenters will improve the performance of this design.

54. Three-mirror Afocal Telescope

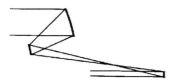

SUR	RDY	THI	GLA	CON
STO	−100.725	−33.514	REFL	−1
2:	−46.109	100	REFL	−3.016
3:	−74.819	−55.56	REFL	−1

Comments: This 5× afocal design from Smith (1992) is an eccentric-pupil Cassegrain and a parabolic tertiary combined. The design is aplanatic and anastigmatic. The entrance pupil is decentered by 32 cm.

55. Three-mirror Afocal Telescope

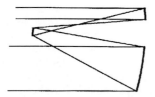

SUR	RDY	THI	GLA	CON
STO	−240	−200	REFL	−1
2:	−160	200	REFL	−9
3:	−480	−250	REFL	−1

Comments: A similar design by Korsch (1991) is also aplanatic and anastigmatic. The entrance pupil is decentered by 20 cm. Other afocal designs are described by Gelles (1974) and King (1974).

56. *F/4 Three-mirror Cassegrain*

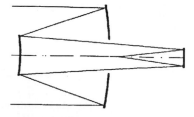

SUR	RDY	THI	GLA	CON
STO	−59.64	−18.29	REFL	−1.134
2:	−28.63	33.74	REFL	−2.841
3:	−55.05	−13.244	REFL	−5.938

Comments: Robb (1978) has introduced another aplanatic, anastigmatic, flat-image, three-mirror Cassegrain without an intermediate image.

57. *F/6.7 Spherical Primary Three-mirror Objective*

SUR	RDY	THI	GLA	CON
STO	−429.67	−149.87	REFL	
2:	−104.16	211.14	REFL	3.617
3:	−126.49	−73.0	REFL	−0.179

Comments: Making the largest element in an objective a spherical mirror reduces cost and may enhance performance. This aplanatic, anastigmatic, flat-image, eccentric-pupil design (−35 cm stop decenter) with an unobscured light path is similar to one described by Korsch (1991) and another developed for use as an astrometric camera by Richardson et al. (1986).

58. *F/4 Spherical Primary Three-mirror Objective*

SUR	RDY	THI	GLA	CON
STO	−194.58	−79.13	REFL	
2:	−64.42	113.68	REFL	12.952
3:	−38.47	−26.24	REFL	−0.4826

Comments: Here is another aplanatic, anastigmatic, flat-field, eccentric-pupil design with a 17-cm stop decenter and large spherical primary. There is an intermediate image and an accessible exit pupil.

59. *F/4 Three-mirror Korsch Objective*

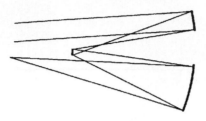

SUR	RDY	THI	GLA	CON
1:	−201.67	−133.36	REFL	−0.689
STO	−96.5	131.8	REFL	−1.729
3:	−172.54	−200.83	REFL	

Comments: This off-axis (5°) design by Korsch (1988) is aplanatic, anastigmatic, and has a flat image. The same configuration has been employed by Pollock (1987) as a collimator. Characteristics include a large field of view, low pupil magnification, accessible pupils, and an intermediate image.

The tertiary in this design is spherical. With reoptimization, the secondary may also be spherical.

60. *F/4 Three-mirror Cook Objective*

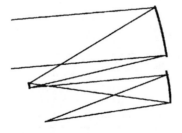

SUR	RDY	THI	GLA	CON
1:	−123.2	−57.38	REFL	−0.7114
2:	−37.46	57.45	REFL	−3.824
3:	−51.89	−35.87	REFL	−0.1185
STO	INF	−15.92		

Comments: This objective was introduced by L. Cook (1981, 1979, 1992). The aplanatic, anastigmatic, flat-image design shown here has a larger pupil magnification and a smaller field than the previous design. The eccentric-pupil, off-axis design has a −3.2-cm stop decenter and a 5° field bias. A space-based surveillance objective in this configuration has been developed and built by Wang et al. (1991).

61. *F/4 Three-mirror Wetherell and Womble Objective*

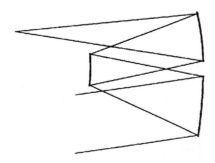

SUR	RDY	THI	GLA	CON
1:	−166.19	−38.78	REFL	−2.542
STO	−55.19	38.78	REFL	−0.428
3:	−82.46	−65.24	REFL	0.133

Comments: Another aplanatic, anastigmatic, flat-image, off-axis (9°) design has been introduced by Wetherell and Womble (1980). Figosky (1989) describes a variant of this form to be sent into orbit. The aperture stop is located at the secondary mirror; hence, this mirror is symmetric with respect to the optical axis.

62. *F/10 Korsch Three-mirror, Four-reflection Objective*

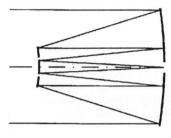

SUR	RDY	THI	GLA	CON
STO	−66.44	−22.15	REFL	−1.092
2:	−22.15	22.15	REFL	−1.295
3:	−66.44	−22.15	REFL	−1.092
4:	−44.29	21.96	REFL	0.8684

Comments: The three-mirror, four-reflection design shown here from Korsch (1974) is extremely compact for its 200-cm focal length, and the image is accessible.

63. *F/1.25 McCarthy*

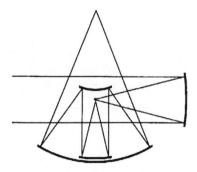

SUR	RDY	THI	GLA	CON
STO	−81.57	−40.21	REFL	−1
2:	INF	25.09	REFL	
		Tilt: 45°		
3:	−48.68	−29	REFL	−1
4:	−19.15	30.64	REFL	
5:	−49.85	−65.483	REFL	

Comments: McCarthy (1959) intended this design, which combines a Cassegrain-Mersenne primary and tertiary mirror with a quaternary and quintenary Schwarzschild arrangement, as a wide strip-field imager. Both the Mersenne and Schwarzschild groups are separately corrected for spherical aberration, coma, and astigmatism. The Petzval curvature of the Mersenne is equal and opposite in sign to the Petzval curvature of the Schwarzschild and hence there is no net Petzval curvature. The quaternary mirror may be moved out from the entering beam with only a slight reduction in performance.

64. F/2.2 Cassegrain Objective with Schwarzschild Relay

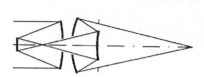

SUR	RDY	THI	GLA	CON
1:	−51.49	−19.01	REFL	−1.048
2:	−37.37	34.19	REFL	−20.35
3:	38.18	−10.493	REFL	−1.358
4:	29.94	11.27	REFL	
STO	INF	40.484		

Comments: S. Williams (1978) describes a technique for optimizing a high-resolution system similar to this one while maintaining proper clearances, obscuration sizes, and packaging requirements. An all-reflective zoom system of the above configuration, developed by Johnson et al. (1990), gives a 4× zoom range and a field-of-view range of 1.5 to 6.0°. The Schwarzschild module and image position change with zoom position, while the front Cassegrain module remains fixed.

65. F/4 Altenhof Objective

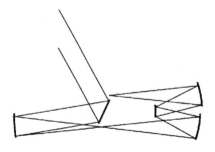

SUR	RDY	THI	GLA	CON
STO	INF	−80	REFL	
		Tilt: 25°		
2:	155.64	165.53	REFL	
3:	−77.26	−38.57	REFL	−0.0897
4:	−34.146	40.367	REFL	
5:	−80.65	−82.539	REFL	

Comments: This objective is similar to one designed by R. Altenhof (1975). Intended as a ring field system, the flat primary couples the light incident from a large azimuthal angle (60°) into the system where the spherical primary mirror focuses the light to a poorly corrected image.

The three-mirror Offner relay (Offner, 1975, Kingslake, 1978), a unit magnification relay which is corrected for spherical aberration, coma, and astigmatism, improves the degraded image in the process of reimaging it to a flat focal surface in the form of an annulus. A two-dimensional scene is imaged by rotating the flat mirror about an axis perpendicular to the picture so as to scan the other dimension. A two-dimensional mosaic image can also be produced by building up a series of one-dimensional annular strip images as the imaging system is moved along its trajectory.

66. *F/4.5 Shafer Four-mirror, Unobscured Objective*

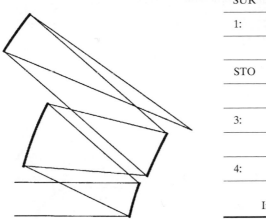

SUR	RDY	THI	GLA
1:	158.1	−71.21	REFL
	Tilt angle: −16.44°		
STO	186.8	74.25	REFL
	Tilt angle: −20.88°		
3:	337.4	−111.4	REFL
	Tilt angle: −24.82°		
4:	239.1	121.4	REFL
	Tilt angle: −34.76°		
	Image tilt angle: −24.29°		

Comments: This is a tilted and decentered-component infrared imaging system by David Shafer. Mirror tilts provide an unobscured path and help correct the aberrations. Thickness is measured along the deviated ray paths. With the reflection angle known, element decenter may be easily determined.

67. *F/4.5 Shafer Five-mirror, Unobscured Objective*

SUR	RDY	THI	GLA
1:	−239.5	−160.2	REFL
	Tilt angle: 6.4°		
2:	−228.9	48.69	REFL
	Tilt angle: −9.2°		
3:	−75.94	−37.24	REFL
	Tilt angle: −19.01°		
STO	−39.81	39.24	REFL
	Tilt angle: −28.82°		
5:	−78.72	−74.5	REFL
	Tilt angle: −40.55°		
	Image tilt angle: −11.28°		

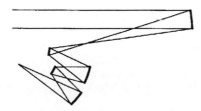

Comments: Another all-spherical, tilted, and decentered-component infrared imager by

D. Shafer is presented here. The entrance pupil is accessible and there is an intermediate image. A number of variations on this arrangement are described by Shafer (1978b).

68. *Korsch Two- and Three-mirror Objectives*

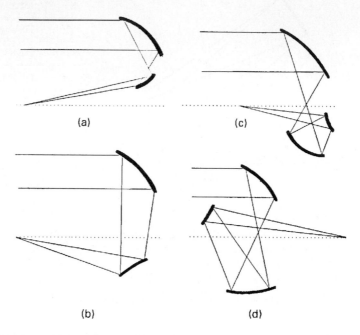

(a)

(c)

(b)

(d)

Comments: A new class of eccentric-pupil objectives has been introduced by Korsch (1991). Unlike most systems, which are conceived using third-order aberration theory, these systems are based upon the fulfillment of *axial stigmatism,* the *Abbé sine condition,* and the *Herschel condition*; meeting these three conditions guarantees a perfect axial point image, axially perpendicular image area, and axial line element, respectively.

Design examples are not given for two reasons. First, rays strike some mirror surfaces at angles greater than 90°, which can cause ray-trace errors. Second, some of the surface shapes are particularly complex and must be entered in design software as a user-defined surface.

Design (*c*) gives perfect imagery on-axis and less than one milliradian resolution at all other points over a 6° field of view, for an aperture diameter equal to F/6.0 (F is focal length).

18.5 FIELD-OF-VIEW PLOTS

The plots that follow give rms spot size and angular resolution as a function of half-field of view. The curves have been generated by calculating the resolution for a number of field angles and connecting them with smooth curves. The dashed horizontal line is the Airy disc diameter for 0.55-μm radiation.

The numbers in the plots correspond to the designs presented in the previous section. The aperture of each design is 20 cm and the spectral range 480 to 680 nm, unless stated otherwise in the previous section.

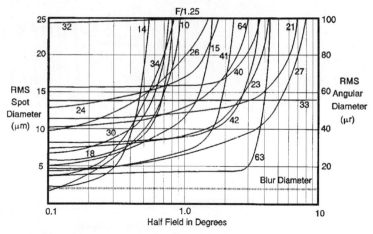

FIGURE 7 Field-of-view plots: F/1.25 on a flat image.

It should be kept in mind that these are representative designs: they have usually been optimized for a specific purpose; they are meant to be a starting place for the design of a new system that may have entirely different requirements.

Flat-field designs show consistent performance out to the field angle for which the objective is optimized. Beyond this point, the graph leaps upward. Reoptimization is needed if the field of view is to be extended further; a considerable increase in the average rms spot size may occur if this is attempted. The curved-image designs show a quadratic dependence with field angle.

Off-axis and eccentric-pupil designs have rectangular fields with most of the field of view in one dimension only. Data plotted for these designs are representative of the larger field.

In Figs. 8 and 10 plots for the curved image designs are provided. The curvature of the image is adjusted to give optimum performance. Figures 7 and 9 are for the flat image designs.

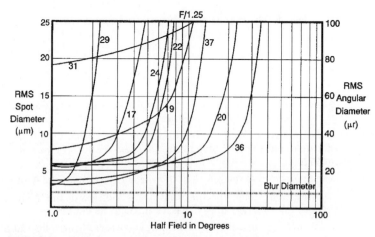

FIGURE 8 Field-of-view plots: F/1.25 on a curved image.

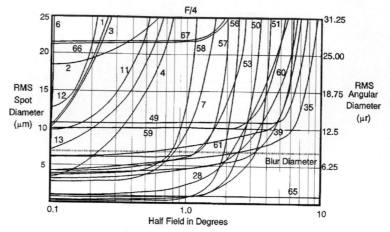

FIGURE 9   Field-of-view plots: F/4 on a flat image.

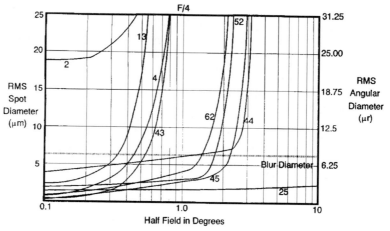

FIGURE 10   Field-of-view plots: F/4 on a curved image.

18.6   DEFINITIONS

Abbé number: A number that indicates the dispersion of a glass. Low dispersion glasses have a high Abbe number.

Abbé sine condition: A condition for zero coma, based on the requirement of equal marginal and paraxial magnifications. See Welford (1986), Kingslake (1978), or Korsch (1991).

anastigmatic: A surface or system free of astigmatism. Also stigmatic.

aperture stop: The aperture that limits the size of the axial beam passing through the system; the *chief ray* always passes through its center.

aplanatic: A surface or system that is corrected for spherical aberration and coma.

astigmatism: An aberration that generates two different focal positions for rays in two perpendicular planes centered on the optical axis. These are called the sagittal and tangential planes.

axial color: The variation in focal position with wavelength.

axial stigmatism: A characteristic of a surface which is able to produce a perfect image of a single point object on-axis.

catadioptric: An optical system composed of refractive and reflective elements: catoptric, reflective and dioptric, refractive.

chief ray: A ray that passes through the center of the aperture stop and the edge of the image or object.

coma: An aberration resulting from the change in magnification with ray height at the aperture, so that rays near the edge of the aperture are focused further from rays near the axis, for the same field point.

conic constant: A constant defined by

$$k = -\epsilon^2$$

where ϵ is the eccentricity of the conic.

distortion: The variation in magnification with field angle.

entrance pupil: The image of the aperture stop in object space. The chief ray passes or appears to pass through the center of the entrance pupil.

exit pupil: The image of the aperture stop in image space. The chief ray passes or appears to pass through the center of the exit pupil.

focal ratio: The effective focal length of an objective divided by its entrance-pupil diameter. Focal ratio is also referred to as the FN, F-number, and speed.

field curvature: Image curvature produced by the combined effects of astigmatism and Petzval curvature. When astigmatism is absent, the image surface coincides with the Petzval surface.

field stop: An aperture that limits the size of an intermediate or final image.

Herschel condition: A condition for invariance of aberrations with change in axial conjugates. See Welford (1986) and Korsch (1991).

higher-order aberrations: Aberrations defined by the higher-order terms in the aberration power series expansion. See Welford (1986) and Schulz (1988).

lateral color: An aberration that produces a dependence of image size on wavelength; also called chromatic difference of magnification.

Lyot stop: A real and accessible image of the aperture stop; used to block stray light.

marginal ray: A ray that passes through the center of the object or image and past the edge of the aperture stop.

medial image: The image halfway between the sagittal and tangential images. See Welford (1986).

monocentric system: An optical system in which all surfaces are concentric about the chief ray.

oblique spherical aberration: A *higher-order aberration* that is the variation of spherical aberration with field angle.

optical axis: The axis about which all optical elements are symmetric. In tilted and decentered systems, each element has a unique optical axis.

Petzval sum: The sum defined by

$$P = \sum \frac{\phi}{n}$$

where ϕ is element power and n is the index of refraction. The reciprocal of the Petzval sum is the image radius of curvature.

secondary magnification: System focal length divided by primary-mirror focal length.

secondary spectrum: The difference in focal position between two wavelengths corrected for axial color and one other wavelength which is not. For example, the blue and red focus coincide and the yellow focus is axially displaced.

spherical aberration: The only on-axis monochromatic aberration, spherical aberration results from rays at different heights coming to focus at different points along the optical axis. Smith (1990), Rutten (1988), Kingslake (1978), Mackintosh (1986), and Welford (1986) discuss aberrations. Welford specifically addresses aberrations.

third-order aberrations: Any of the Seidel aberrations: spherical aberration, coma, astigmatism, Petzval curvature, and distortion. See Welford (1986).

vignetting: The off-axis clipping of light by apertures in an optical system.

virtual image: A real image is visible when a screen is placed at its location. The image is visible because rays from the object converge at the image. A virtual image is not visible when a screen is placed at its location since real rays do not converge.

zonal spherical aberration: The incomplete correction of spherical aberration at radial zones in the aperture. For example, spherical aberration could be corrected for rays close to the center and edge of the aperture, but not corrected at other ray heights in the aperture.

18.7 REFERENCES

Abel, I. and M. Hatch, "The Pursuit of Symmetry in Wide-Angle Reflective Optical Designs," *SPIE* **237:**271 (1980).

Abel, I., "Mirror Systems: Engineering Features, Benefits, Limitations and Applications," *SPIE* **531:**121 (1985).

Altenhof, R., "The Design of a Large Aperture Infrared Optical System," *SPIE* **62:**129 (1975).

Amon, M., "Large Catadioptric Objective," U.S. Patent 3,711,184, 1973.

Baker, J., "The Solid Glass Schmidt Camera and a New Type of Nebular Spectrograph," *Proc. Am. Phil. Soc.* **82:** 323 (1940a).

Baker, J., "A Family of Flat-Field Cameras, Equivalent in Performance to the Schmidt Camera," *Proc. Amer. Phil. Soc.* **82**(3):339 (1940b).

Baker, J., "Schmidt Image Former with Spherical Abberation Corrector," U.S. Patent 2,458,132, 1945.

Baker, J., *Amateur Telescope Making, Book Three,* Scientific American, Inc., 1953, p. 1.

Baker, J., "Correcting Optical System," U.S. Patent 3,022,708, 1962.

Baker, J., "On Improving the Effectiveness of Large Telescopes," *IEEE Transact. AES-5,* (2):261 (1969).

Baker, J., "Explorations in Design," *SPIE* **147:**102 (1978).

Bouwers, A., *Achievements in Optics,* Elsevier, Amsterdam, 1950.

Brueggeman, H. *Conic Mirrors,* Focal Press, 1968.

Brunn, M., "Unobstructed All-reflecting Telescopes of the Schiefspiegler Type," U.S. Patent 5,142,417, 1992.

Buchroeder, R., "Catadioptric Designs," OSC, University of Arizona, Technical Report 68, May, 1971a.

Buchroeder, R., "Fundamentals of the TST," OSC, University of Arizona, Technical Report 68, May, 1971b.

Canzek, L., "High Speed Catadioptric Objective Lens System," U.S. Patent 4,547,045, 1985.

Chretien, M., "Le Telescope de Newton et le Telescope Aplanatique," *Rev. d'Optique* (2):49 (1922).

Cook, L., "Three-Mirror Anastigmat Used Off-Axis in Aperture and Field," *SPIE* **183:**207 (1979).

Cook, L., "Three-Mirror Anastigmatic Optical Systems," U.S. Patent 4,265,510, 1981.

Cook, L., "The Last Three-Mirror Anastigmat (TMA)?" *SPIE Proceedings* vol. CR**41:**310 (1992).

Couder, A., *Compt. Rend. Acad. Sci. Paris,* 183, II, 1276, p. 45 (1926).

Courtes, G., "Optical Systems for UV Space Researches," From *New Techniques in Space Astronomy,* D. Reidel Publishing Company, Dordrecht, Holland, 1971.

Courtes, G., P. Cruvellier, M. Detaille, and M. Saisse, *Progress in Optics* vol. XX, 1983, p. 3.

Eisenburg, S. and E. Pearson, "Two-Mirror Three-Surface Telescope," *SPIE* **751:**24 (1987).

Everhart, E. and J. Kantorski, "Diffraction Patterns Produced by Obstructions in Reflective Telescopes of Modest Size," *Astronomical Journal* **64**(Dec.):455 (1959).

Figosky, J., "Design and Tolerance Specification of a Wide-Field, Three-Mirror, Unobscured, High-Resolution Sensor," *SPIE* **1049:**157 (1989).

Fischer, R., "What's So Different About IR Lens Design?" *SPIE Proceedings* Vol. CR**41:**117 (1992).

Fjeidsted, T., "Selectable Field-of-View Infrared Lens," U.S. Patent 4,453,800, 1984.

Gabor, D., British Patent 544,694, 1941.

Gascoigne, S., "Recent Advances in Astronomical Optics," *Applied Optics* **12:**1419 (1973).

Gelles, R., "A New Family of Flat-Field Cameras," *Applied Optics* **2**(10):1081 (1963).

Gelles, R., "Unobscured Aperture Stigmatic Telescopes," *Optical Engineering* **13**(6):534 (1974).

Gelles, R., "Unobscured-Aperture Two-Mirror Systems," *Journal of the Optical Society of America* **65**(10):1141 (1975).

Hallam, K., B. Howell, and M. Wilson, "Wide-Angle Flat Field Telescope," U.S. Patent 4,598,981, 1986.

Hawkins, D. and E. Linfoot, "An Improved Type of Schmidt Camera," *Monthly Notices Royal Astronomical Soc.* **105:**334 (1945).

Houghton, J., "Lens System," U.S. Patent 2,350,112 (1944).

Johnson, R., J. Hadaway, and T. Burleson, "All-Reflective Four-Element Zoom Telescope: Design and Analysis," *SPIE* **1354:**669 (1990).

King, W., "Unobscured Laser-Beam-Expander Pointing System with Tilted Spherical Mirrors," *Applied Optics* **13**(1):21 (1974).

Kingslake, R., *Lens Design Fundamentals,* Academic Press, Inc., New York, 1978.

Korsch, D., "Two Well-Corrected Four-Mirror Telescopes," *Applied Optics* **13**(8):1767 (1974).

Korsch, D., "Anastigmatic Three-Mirror Telescope," *Applied Optics* **16**(8):2074 (1977).

Korsch, D., "Anastigmatic Three-Mirror Telescope," U.S. Patent 4,101,195 (1978).

Korsch, D., "Wide-Field Three-Mirror Collimator," U.S. Patent 4,737,021, 1988.

Korsch, D., *Reflective Optics,* Academic Press, San Diego, Calif. 1991.

Kutter, A., "A New Three-mirror Unobstructed Reflector," *Sky and Telescope,* Jan. 1975, p. 46; Feb. 1975, p. 115.

Laiken, M., *Lens Design,* Marcel Dekker, Inc., New York, 1991.

Leonard, A., "New Horizons for Tilted-Component Telescopes," in *Advanced Telescope Making Techniques,* Willmann-Bell Inc., Richmond, Va., 1986a, p. 110.

Leonard, A., "T.C.T.'s (Tilted Component Telescopes)," in *Advanced Telescope Making Techniques,* Willmann-Bell Inc., Richmond, Va., 1986b, p. 131.

Linfoot, E., *Recent Advances in Optics,* Oxford Univ. Press, London, 1955.

Mackintosh, A., *Advanced Telescope Making Techniques,* Willmann-Bell Inc., Richmond, Va., 1986.

Maksutov, D., *J. Opt. Soc. Am.,* **34:**270 (1944).

McCarthy, E., "Anastigmatic Catoptric Systems," U.S. Patent 3,062,101, 1959.

Mandler, W., "Reflecting Mirror and Lens System of the Cassegrain Type," U.S. Patent 2,726,574, 1951.

Mangin, A., "Memorial de L'officier du Genie (Paris)," **25**(2):10, 211 (1876).

Manly, P., *Unusual Telescopes,* Cambridge Univ. Press, Cambridge, England, 1991.

Maxwell, J., *Catadioptric Imaging Systems,* Elsevier, New York, 1972.

Meinel, A., "Aspheric Field Correctors for Large Telescopes," *Astrophysical Journal* **118:**335 (1953).

Meinel, A. and R. Shack, "A Wide-Angle All- Mirror UV Camera," Optical Sciences Center Tech. Rep. No. 6, 1966.

Meinel, A. and M. Meinel, *SPIE* **332:**178 (1982).

Meinel, A., M. Meinel, D. Su, and Ya-Nan Wang, "Four-Mirror Spherical Primary Submillimeter Telescope Design," *Applied Optics* **23**(17):3020 (1984).

Offner, A., "New Concepts in Projection Mask Aligners," *Optical Engineering* **14**(2):130 (1975).

Olsen, E., "A Schupmann for Amateurs," in *Advanced Telescope Making Techniques,* Willmann-Bell Inc., Richmond, Va., 1986, p. 223.

Owen, R., "Easily Fabricated Wide-Angle Telescope," *SPIE* **1354:**430 (1990).

Paul, M., *Rev. Opt.* **14:**169 (1935).

Pollock, D., "An All-Reflective, Wide Field of View Collimator," *SPIE* **751:**135 (1987).

Rao, D., "Design of a Near Diffraction-Limited Catadioptric Lens," *SPIE* **766:**220 (1987).

Richardson, E. and C. Morbey, "Optical Design of an Astrometric Space Telescope," *SPIE* **628:**197 (1986).

Robb, P., "Three-mirror Telescopes: Design and Optimization," *Applied Optics* **17**(17):2677 (1978).

Roesch, M., *Tr. Int. Astr. Union.,* Vol. VII:103 (1950).

Rosin, S., "Corrected Cassegrain System," *Applied Optics* **3**(1):151 (1964).

Rosin, S., "Ritchey-Chretien Corrector System," *Applied Optics* **5**(4):675 (1966).

Ross, F., "Lens Systems for Correcting Coma or Mirrors," *Astrophysics Journal,* **81:**156 (1935).

Rumsey, N., "Telescopic System Utilizing Three Axially Aligned Substantially Hyperbolic Mirrors," U.S. Patent 3,460,886, 1969.

Rutten, H. and M. Venrooij, *Telescope Optics,* Willmann-Bell, Inc., Richmond, Va., 1988.

Sasian, J., "Review of Methods for the Design of Unsymmetrical Optical Systems," *SPIE* **1396:**453 (1990a).

Sasian, J., "Design of a Schwarzschild Flat-Field, Anastigmatic, Unobstructed, Wide-Field Telescope," *Optical Engineering* **29**(1) January, p. 1 (1990b).

Schmidt, B., *Mitt. Hamburg. Sternwart* **7**(36) (1932).

Schroeder, D., "All-Reflecting Baker-Schmidt Flat-Field Telescopes," *Applied Optics* **17**(1):141 (1978).

Schroeder, D., *Astronomical Optics,* Academic Press, Inc., San Diego, Calif., 1987.

Schulte, D., "Auxiliary Optical Systems for the Kitt Peak Telescopes," *Applied Optics* **2**(2):141 (1963).

Schulte, D., "Anastigmatic Cassegrain Type Telescope," *Applied Optics* **5**(2):309 (1966a).

Schulte, D., "Prime Focus Correctors Involving Aspherics," *Applied Optics* **5**(2):313 (1966b).

Schulz, G., "Aspheric Surfaces," *Progress in Optics* **XXV:**351 (1988).

Schupmann, L., "Die Medial-Fernrehre," *Leipzig* (1899).

Schwarzschild, K., "Untersuchungen zur Geometrischen Optik, II; Theorie der Spiegelteleskope," *Abh. der Konigl. Ges. der Wiss. zu Gottingen, Math.-phys. Klasse,* 9, Folge, Bd. IV, No. 2 (1905).

Shafer, D., "Well Baffled Two-Mirror Corrector for a Parabola," *Applied Optics* **16**(5):1175 (1977).

Shafer, D., "Design With Two-Axis Aspheric Surfaces," *SPIE* **147:**171 (1978a).

Shafer, D., "Four-Mirror Unobscured Anastigmatic Telescopes with All-Spherical Surfaces," *Applied Optics* **17**(7):1072 (1978b).

Shafer, D., "20 Degree Field-of-View Strip Field Unobscured Telescope with Stray Light Rejection," presented Oct. 31 in San Francisco at *Optical Society of America Annual Meeting,* 1978c.

Shafer, D., "30 Degree F/1.0 Flat Field Three Mirror Telescope," presented Oct. 31 in San Francisco at *Optical Society of America Annual Meeting,* 1978d.

Shafer, D., "Simple Method for Designing Lenses," *SPIE* **237:**234 (1980a).

Shafer, D., "A Simple Unobstructed Telescope with High Performance," *Telescope Making* **41:**4 (TM 41).

Shafer, D., "Optical Design with Only Two Surfaces," *SPIE* **237:**256 (1980b).

Shafer, D., "Well-Corrected Two-Element Telescope with a Flat Image," *Optica Acta* **28**(11):1477 (1981).

Shafer, D., "An All-Reflective Infrared Target Simulator Design," *October OSA Workshop on Optical Fabrication and Testing,* 1988.

Shenker, M., "High Speed Catadioptric Objective in Which Three Corrector Elements Define Two Power Balanced Air Lenses," U.S. Patent 3,252,373, 1966.

Sigler, R., "Compound Catadioptric Telescopes with All-Spherical Surfaces," *Applied Optics* **17**(10):1519 (1978).

Smith, W., *Modern Optical Engineering,* McGraw-Hill, Inc., New York, 1990.

Smith, W., *Modern Lens Design,* McGraw-Hill, Inc., New York, 1992.

Sonnefeld, A., "Photographic Objective," U.S. Patent 2,141,884, 1938.

Thompson, K., "Aberration Fields in Tilted and Decentered Optical Systems," Ph.D. dissertation, Optical Sciences Center, University of Arizona, 1980.

Wang, D., L. Gardner, W. Wong, and P. Hadfield, "Space Based Visible All-Reflective Stray Light Telescope," *SPIE* **1479:**57 (1991).

Warmisham, A., British Patent 551,112, 1941.

Wayman, R., "The Monocentric Schmidt-Cassegrain Cameras," *Proc. Phys. Soc.* (London), **63:**553 (1944).

Welford, W., *Aberrations of Optical Systems,* Adam Hilger Ltd., Bristol, England, 1986.

Wetherell, W. and P. Rimmer, "General Analysis of Aplanatic Cassegrain, Gregorian and Schwarzschild Telescopes," *Applied Optics* **11:** no. 12, 2817 (1972).

Wetherell, W. and D. Womble, "All-Reflective Three Element Objective," U.S. Patent 4,240,707, 1980.

Williams, S., "Design Techniques for WFOV High-Resolution Reflecting Optics," *SPIE* **147:**94 (1978).

Williams, S., "On-Axis Three-Mirror Anastigmat with an Offset Field of View," *SPIE* **183:**212 (1979).

Wilson, R., "Corrector Systems for Cassegrain Telescopes," *Applied Optics* **7**(2):253 (1968).

Wright, F., "An Aplanatic Reflector With a Flat-Field Related to the Schmidt Telescope," *Astronomical Society of the Pacific* **47:**300 (1935).

Wynne, C., "Field Correctors for Large Telescopes," *Applied Optics* **4**(9):1185 (1965).

Wynne, C., "Maksutov Spectrograph Cameras," *Monthly Notices Royal Astronomical Society* **153:**261 (1971).

Wynne, C., "Field Correctors for Astronomical Telescopes," *Progress in Optics,* **X:**139 (1972a).

Wynne, C., "Five Spectrograph Camera Designs," *Monthly Notices Royal Astronomical Society* **157:**403 (1972b).

Wynne, C., "A New Wide-Field Triple Lens Paraboloid Field Corrector," *Monthly Notices Royal Astronomical Society* **167:**189 (1974).

Wynne, C., "Shorter than a Schmidt," *Monthly Notices Royal Astronomical Society* **180:**485 (1977).

CHAPTER 19
SCANNERS

Leo Beiser
Leo Beiser Inc.
Flushing, New York

R. Barry Johnson
Optical E.T.C., Inc.
Huntsville, Alabama
and
Center for Applied Optics
University of Alabama in Huntsville
Huntsville, Alabama

19.1 GLOSSARY

a	aperture shape factor; calibrates diffraction angle as function of aperture intensity distribution
A	area
C	capacitance, electrical
d	grating spacing
D	useful aperture width in scan (x) direction; D_m = enlarged aperture due to α
f_e	data bandwidth
f	focal length
F	F-number (F/D)
FOV	field of view
FWHM	full width (of δ) measured at half maximum intensity
H	vehicle height
I	resolution invariant; adaptation of Lagrange invariant, $I = \Theta D = \Theta' D'$
k	scanning constant (see m)
m	scan magnification ($d\Theta/d\Phi$) (for m = constant, $m = k = \Theta/\Phi$)
M	optical magnification (image/object)

n	number of facets, refractive index, diffractive order
N	number of resolution elements subtended by Θ or S
P	radiant power (watts)
PSF	point spread function (intensity distribution of focused spot)
Q	q-factor (electromechanical)
r	radius
R	reciprocity failure factor (≥ 1), scanner rpm
s	sensitivity (recording medium)
S	format width, no. of scans per revolution of scanner
T	time period, optical transfer factor (≤ 1)
V	vehicle velocity, electrical voltage
W	full aperture width, of which D is utilized (see ρ)
x	along-scan direction
y	cross-scan direction
α	angular departure from normal beam landing, spectral power transfer factor
γ	phosphor utilization factor (≤ 1)
δ	spot size; if gaussian, measured across $1/e^2$ intensity width or at FWHM
η	duty cycle, conversion efficiency (≤ 1)
Θ, θ	optical scan angle
Φ	mechanical scan angle, along-track optical scan angle, optical field angle
λ	wavelength
Λ	wavelength (acoustic grating)
ρ	truncation ratio (W/D)
τ	transit time (acoustic), retrace time, dwell time

19.2 *INTRODUCTION*

This chapter provides an overview of optical scanning techniques in context with their operational requirements. System objectives determine the characteristics of the scanner which, in turn, influence adjacent system elements. For example, the desired resolution, format, and data rate determine the scanner aperture size, scan angle, and speed, which then influence the associated optics. The purpose of this chapter is to review the diverse options for optical scanning and to provide insight to associated topics, such as scanned resolution and the reduction of spatial errors. This broad perspective is, however, limited to those factors which bear directly on the scanner. Referencing is provided for related system relationships, such as image processing and data display. Topics are introduced with brief expressions of the fundamentals. And, where appropriate, historical and technical origins are referenced.

The subject of scanning is often viewed quite differently in two communities. One is classified as *remote sensing* and the other, *input/output scanning.* Associated component nomenclature and jargon are, in many cases, different. While their characteristics are expanded in subsequent sections, it is useful to introduce some of their distinctions here. Remote sensing detects objects from a distance, as by a space-borne observation platform. An example is infrared imaging of terrain. Sensing is usually *passive* and the radiation *incoherent* and often multispectral. Input/output scanning, on the other hand, is *local.* A familiar example is document reading (input) or writing (output). Intensive use of the laser makes the scanning *active* and the radiation *coherent.* The scanned point is focused via finite-conjugate optics from a local fixed source.

While the scanning components may appear interchangeable, special characteristics and operational modes often preclude this option. This is most apparent for diffractive devices such as acousto-optic and holographic deflectors. It is not so apparent regarding the differently filled scanning apertures, imparting important distinctions in resolution and duty cycle. The unification of some of the historically separated parameters and nomenclature is considered an opportunity for this writing.

System Classifications

The following sections introduce the two principal disciplines of optical scanning, remote sensing and input/output scanning, in preparation for discussion of their characteristics and techniques.

Remote Sensing.   The applications for passive (noninvasive) remote sensing scanners are varied and cover many important aspects of our lives. A signature representative of the target is obtained to form a signal for subsequent recording or display. This process is operationally distinct from active scanning, as expressed further in this chapter. Table 1

TABLE 1   Representative Applications of Passive Scanning Sensors

Medical	Government
Cancer	Forest fires
Arthritis	Police
Whiplash	Smuggling
Industrial	Search and rescue
Energy management	Military
Thermal fault detection	Gun sights
Electronic circuit	Night vision
detection	Tactical
Nondestructive testing	Navigation
Scientific	Missiles
Earth resources	Strategic
Weather	Aircraft
Astronomy	ICBM
	Surveillance

lists typical applications of these techniques. Clearly, remote scanning sensors can be hand-held to satellite-borne.

A variety of scanning methods has been developed to accomplish the formation of image (or imagelike) data for remote sensing. These methods may be roughly divided into framing, pushbroom, and mechanical. Generally stated, frame scanning requires no physical scan motion and implies that the sensor has a two-dimensional array of detectors which are read out by use of electronic means (e.g., CCD), electron beam, or light beam. Such an array requires an optical system that has 2-D wide-angle capability. Pushbroom methods typically employ some external means to move the image of a linear array of detectors along the area to be imaged. Mechanical methods generally include one- and two-dimensional scanning techniques incorporating as few as one detector to multiple-detector arrays. As is the case for pushbroom methods, image formation by one-dimensional mechanical scanning requires that the platform containing the sensor (or in some cases the object) be moved to create the second dimension of the image. The latter two methods are discussed further in later sections of this chapter.

Mechanical scanners can be configured to perform either one- or two-dimensional scan patterns. In the case of a one-dimensional scanner, the second dimension needed to form an image is most often generated by movement of the sensor platform.

A number of optical scanning systems have been invented to satisfy the wide variety of applications. In the early years of passive scanning systems, the entire optical system was often moved in order to produce either a one- or two-dimensional scan pattern. The advent of airborne mapping or reconnaissance electro-optical systems began during the 1950s. Typically, the scanner performed a one-dimensional scan in object-space, as defined under the section "Object-space and Image-space Scanners," orthogonal to the flight direction of the aircraft while the motion of the aircraft generated the second dimension of the image. The resultant video information is stored on a recording medium such as photographic film, digital tape, etc. The design and resultant performance of a scanning system are governed by a variety of parameters that are related by tradeoff equations and considerations. The selection of the scanner type typically has a strong influence upon the ultimate system performance. In subsequent discussion, the more important parameters related to the scanner selection will be covered. The complexities of the total system design and optimization are not within the scope of this chapter.

Input/Output Scanning.   In contrast to remote sensing, which captures passive radiation, active input/output scanning illuminates an object or medium with a "flying spot," derived typically from a laser source. Some examples appear in Table 2, divided into two principal

TABLE 2   Examples of Input/Output Scanning

Input	Output
Image scanning/digitizing	Image recording/printing
Bar-code reading	Color image reproduction
Optical inspection	Medical image outputs
Optical character recognition	Data marking and engraving
Optical data readout	Microimage recording
Graphic arts camera	Reconnaissance recording
Scanning confocal microscopy	Optical data storage
Color separation	Phototypesetting
Robot vision	Graphic arts platemaking
Laser radar	Earth resources imaging
Mensuration	Data/Image display

functions: *input* (detecting radiation scattered from the scanning spot) and *output* (recording or display). Input is modulated by the target to form a signal; output is modulated by a signal.

Some merit clarification. Under input is laser radar—a special case of active remote sensing, using the same coherent and flying-spot scanning disciplines as the balance of those exemplified. Earth resources imaging is the *recording* of remotely sensed image signals. Finally, data/image display denotes the general presentation of information, which could include "hard copy" and/or actively projected and displayed images.

Active Scanning is synonymous with flying-spot scanning, the discipline most identified with the ubiquitous cathode-ray tube (CRT). While the utilized devices and their performance differ significantly, the distinctions between CRT and laser radiation are primarily their degrees of monochromaticity and coherence, as addressed later in this chapter.

Thus, most high-resolution and high-speed flying-spot scanning are now conducted using the laser as a light source. This work in input/output scanning concentrates on the control of laser radiation and the unique challenges encountered in deflecting photons, devoid as they are of the electric and magnetic fields accompanying the electron beam. Reference is provided[1,2,3] for pursuit of the CRT scanning discipline.

Scanner Classification. Following the nomenclature introduced in the early '70s,[4,5] laser scanners are designated as *preobjective, objective,* and *postobjective.* Figure 1 indicates the scan regions within a general conjugate optical transfer of a fixed reference (object) point P_o to a moving focal (image) point P_i. *The component which provides principal focusing of the wavefront identifies the objective lens.*

The scanner can perform two functions (see "Objective, Preobjective, and Postobjective Scanning" later in this chapter): one is *translation* of the aperture with respect to the information medium. This includes translation of the lens element(s) or translation of the object, or both, and is identified as an *objective* scan. The other is *angular change* of the optical beam with respect to the information medium. Angular scanners are exemplified by plane mirrors on rotating substrates. Although lenses can be added to an angular scanner, it is seldom so configured. The scanner is either preobjective or postobjective. In holographic scanning, however, the hologram can serve as an objective lens and scanner simultaneously.

Radial Symmetry and Scan Magnification. A basic characteristic of some angular scanners is identified as *radial symmetry.* When an illuminating beam converges to or diverges from the nodal or rotating axis of an angular scanner, it is said to exhibit radial

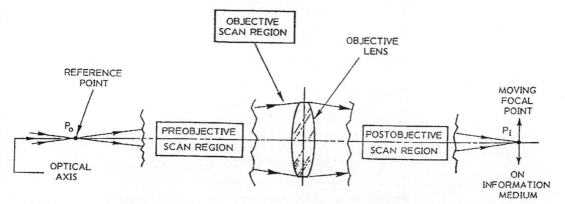

FIGURE 1 Conjugate imaging system, showing scan regions, as determined by position of scanning member relative to the objective lens. Translational (objective) scan and angular (pre/postobjective) scan can occur individually or simultaneously.[4]

symmetry.[6] The collimated beam which is parallel to the rotating axis is a special case of radial symmetry, in which the illuminating beam propagates to or from a very distant point on the axis. Scanners exhibiting radial symmetry provide unity angular optical change for unity mechanical change. That is, $m = d\Theta/d\Phi = 1$, where Θ is the optical scan angle and Φ is the mechanical change. The parameter m is called the *scan magnification*, discussed later under "Augmented Resolution" for Eq. (19). It ranges typically between 1 and approximately 2, depending on the scanner-illumination configuration, per Table 3. In remote sensing, $m = \Theta/\Phi = k$. (See "Compound Mirror Optics Configurations.")

The prismatic polygon (see "Monogon and Polygon Scanners") exhibits a variable m, depending on the degree of collimation or focusing of the output beam. When collimated, $m = 2$. When focusing, the value of m shifts from 2 according to

$$m' = 2 + r/f \tag{1}$$

where f and r are per Fig. 4 and Eq. (19). This is similar to the ratio of angular velocities of the scanned focal point along the arc of a limaçon,[5]

$$\dot{\Theta}/\dot{\Phi} = 2\left(1 + \frac{\cos \Phi}{1 + f/r}\right) \tag{2}$$

Note that when $r \to 0$ or when $f \to \infty$, $\dot{\Theta}/\dot{\Phi} \to 2$. In holographic scanners which are not radially symmetric, m depends on the angles of incidence and diffraction of the input and first-order output beam,

$$m = \sin \theta_i + \sin \theta_o = \lambda/d \tag{3}$$

where θ_i and θ_o are the input and diffracted angles (with respect to the grating normal) and d is the grating spacing. For example, when $\theta_i = \theta_o = 30°$, $m = 1$; when $\theta_i = \theta_o = 45°$, $m = \sqrt{2}$.

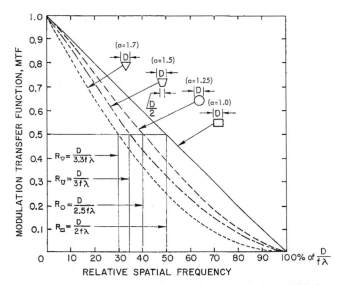

FIGURE 2 Modulation transfer function vs. relative spatial frequencies for uniformly illuminated rectangular, round, keystone, and triangular apertures. Spatial frequency at 50 percent modulation (relative to that of rectangular aperture) determines the a value.[5]

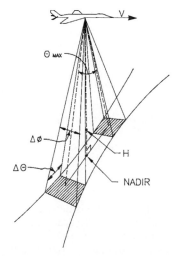

FIGURE 3 Scanning geometry for an airborne line-scanning system with a total scanned FOV of θ_{max}. The aircraft is flying at height H and velocity V. The cross-track and across-track instantaneous FOVs are $\Delta\theta$ and $\Delta\phi$, respectively. (*After Wolfe*, Proc. IRE, 1958).

19.3 SCANNED RESOLUTION

Remote Sensing Resolution and Data Rates

Figure 3 illustrates the scanning geometry for an airborne line-scanning system where the aircraft is flying along a "track" at a height H and velocity V. The total scanned field of view is θ_{max} and the cross-track and along-track instantaneous fields of view are $\Delta\theta$ and $\Delta\phi$, respectively.* The direction directly below the aircraft and normal to the scanned surface is called the *nadir*. The instantaneous field of view is defined as the geometrical projection of the detector with spatial dimensions d_{ct} and d_{at} by the optics having a focal length of F. Therefore, $\Delta\theta = d_{ct}/F$ and $\Delta\theta = d_{at}/F$. Figure 3 shows the "bow-tie" distortion of the scanning geometry which will be discussed further under "Image Consequences."

The basic equation relating the aircraft velocity to the angular velocity of the scanning system to produce contiguous scan lines at the nadir is $V/H = \dot{s} \cdot \Delta\phi$ where $\dot{s}$ is the scanning system's scan rate in scans/s. For a system with n detector elements aligned in the flight direction, $V/H = n\dot{s} \cdot \Delta\phi$.

The number of resolution elements or pixels in a single scan line is

$$N = \frac{\theta_{\mathrm{max}}}{\Delta\theta} \tag{4a}$$

$$= \frac{2\pi\theta_{\mathrm{max}}}{360° \cdot \Delta\theta} \tag{4b}$$

---

*Cross-track and along-track in remote sensing correspond to along-scan and cross-scan, respectively, in input/output scanning.

where $\Delta\theta$ is in radians, $\theta_{\max}$ is the total field of view measured in radians in Eq. (4a) and in degrees in Eq. (4b), for the scanning means employed, taking due regard for the duty cycle given in Eq. (23). The scan rate, in scans per second, may be expressed as a function of the scan mirror speed by

$$\dot{s} = \frac{R \cdot S}{60} \tag{5}$$

where R is the scan mirror rpm and S is the number of scans produced by the scanning mechanism per revolution of the optics. It follows that the number of resolution elements or pixels per second per scan line is

$$\dot{N} = \frac{2\pi\theta_{\max}RS}{60 \cdot 360 \cdot \Delta\theta} \tag{6}$$

The angle $\theta_{\max}$ (in degrees) is determined by the configuration of the scan mirror and is $\theta_{\max} = 360 \cdot k/S$ where k is the scanning constant or scan magnification* and can have values ranging from 1 to 2. The specific value is dependent upon the optical arrangement of the scanner as exemplified in Table 3. The pixel rate may now be written as

$$\dot{N} = \frac{2\pi kR}{60 \cdot \Delta\theta} \tag{7}$$

TABLE 3   Typical Features of Pyramidal and Prismatic Polygon Scanners

Item	Description	Pyramidal	Prismatic
1	Input beam direction[a]	Radially symmetric[b] (typically parallel to axis)	Perpendicular to axis[c]
2	Output beam direction[a]	Arbitrary angle to axis (typically perpendicular)	Perpendicular to axis[c]
3	Scan magnification[b] (scanning constant)	1	2[b]
3a	Along-scan error magnification	1	2[b]
3b	Max. scan angle, $\Theta_{\max}$ (n = no. of facets)	$2\pi/n$	$4\pi/n^b$
4	Output beam rotation about its axis[d]	Yes	No
5	Aperture shape[e] (overilluminated)	Triangular/keystone	Rectangular
6	Enlargement of along-scan beam width	No	Yes[f] $D_m = D/\cos\alpha$
7	Error due to axial polygon shift[g]	Yes	No
8	Error due to radial polygon shift[g]	Yes	Yes
9	Fabrication cost	Greater	Lower

[a] With respect to rotating axis.
[b] See sections on "Radial Symmetry and Scan Magnification" and "Augmenting and Scan Magnification." Output beam assumed collimated.
[c] All beams typically in same plane perpendicular to axis. See Figs. 26 and 28.
[d] Observable when beam is nonisotropic; e.g., elliptic, polarized, etc. Rotation of isotropic beam normally not perceived. See "Image Rotation and Derotation."
[e] See Table 4.
[f] α = angular departure from normal landing. See "Scanner-Lens Relationships."
[g] Shift of image focal point in noncollimated light. No error in collimated light.

---

* See "Radial Symmetry and Scan Magnification" regarding scan magnification m which represents a more general form of the scanning constant k.

The information retrieval rate of a system is often expressed in terms of the dwell time τ or the data bandwidth f_e as

$$f_e = \frac{1}{2\tau} = \frac{\dot{N}}{2} \tag{8}$$

By combining the preceding equations, the data bandwidth for a multiple-detector system can be expressed as

$$f_e = \frac{\pi k(V/H)}{nS\,\Delta\theta\,\Delta\phi} \tag{9}$$

which illustrates clearly the relationship between important system parameters such as f_e being inversely proportional to instantaneous field-of-view solid angle ($\Delta\theta/\Delta\phi$).

Input/Output Scanning

Resolution Criteria, Aperture Shape Factor. The resolution of an optical scanner is expressed[5,7] by the number N of spots or elements that can be conveyed along a contiguous spatial path. The path is usually (nearly) linear and traversed with uniform velocity. Although the elements δ are analogous to the familiar descriptors *pixels* or *pels* (picture elements), such identification is avoided, for pixels often denote spatially digitized scan, where each pixel is uniform in intensity and/or color. Active optical scan, on the other hand, is typically contiguous, except as it may be subjected to modulation. Normally, therefore, the scanned spots align and convolve to form a continuous spatial function that can be divided into elements by modulation of their intensity. To avoid perturbation of the elemental point spread function (PSF) by the modulating (or sampling) process, we assume that the scan function is modulated in intensity with a series of (Dirac) pulses of infinitesimal width, separated by a time t such that the spatial separation between spot centers is $w = vt$, where v is the velocity of the scanned beam. It is often assumed that the size δ of the thus-established elemental spot corresponds to w; that is, the width of the imaged spot is equal to the spacing from its neighbor.

To quantify the number N of such spots, those which exhibit a gaussian intensity distribution are usually considered overlapping at one of two widths; at their $1/e^2$ intensity points, or at their 50 percent intensity points (the latter denoted as FWHM; full width at half maximum). Their relationship is

$$\delta_{\text{FWHM}} = 0.589\delta_{1/e^2} \tag{10}$$

The resolution N is identified with its measurement criterion, for the same system will convey different apparent N, per Eq. (10). That is, it will count approx. $1.7\times$ as many spots at FWHM than at $1/e^2$ intensity.

These distinctions are accommodated[8] by their aperture shape factors a. For example, the above gaussian aperture distribution is represented by shape factors

$$a_{1/e^2} = \frac{4}{\pi} = 1.27 \tag{11a}$$

$$a_{\text{FWHM}} = 0.589a_{1/e^2} = 0.75 \tag{11b}$$

When adapted to the applicable equation for spot size

$$\delta = aF\lambda \tag{12}$$

in which $F = f/D$ is the F-number of the cone converging over the distance f from aperture width D, and λ is the radiation wavelength, the resulting gaussian spot size becomes

$$\delta_{1/e^2} = \frac{4}{\pi} \frac{f}{D} \lambda = 1.27F\lambda \qquad (13a)$$

when measured across the $1/e^2$ intensity points, and

$$\delta_{\text{FWHM}} = 0.75F\lambda \qquad (13b)$$

when measured across FWHM.

The factor a further accommodates the resolution changes due to changes in aperture shape, as for apodized and truncated gaussians. Ultimate truncation is that manifest when the illuminating spatial distribution is much larger than the limiting aperture (over-illumination or overfilling), forming the uniformly illuminated aperture.* Familiar analytic examples are the rectangular and round (or elliptic) apertures, which generate (for the variable x) the normalized intensity distributions $[\sin x/x]^2$ and $[2J_1(x)/x]^2$ respectively, in which $J_1(x)$ is the first-order Bessel function of the first kind.

Figure 2 illustrates† the MTFs[9] of several uniformly illuminated apertures. Their intersections with the 0.5 MTF value identifies the spatial frequency at which their modulation is 50 percent. With the rectangular aperture as a reference (its straight line intersects 0.5 MTF at 50 percent of the limit frequency, forming $a = 1$), the intersections of the others with MTF $= 0.5$ yield corresponding spatial frequencies and relative a-values. Since the spatial frequency bandpass is proportional to $D/f = 1/F$, the apertures of the others must be widened by their a-values (effectively lowering their F-numbers) to render equivalent response midrange.

Table 4 summarizes the aperture shape factors (a) for several useful distributions.[8] Truncated, when applied, is two-dimensional. Noteworthy characteristics are:

1. Scanning is in the direction of the aperture width D.

TABLE 4   Aperture Shape Factor a

Uniformly illuminated			Gaussian illuminated		
Shape	⊢D⊣	a	δ (Spot overlap)	a (Untruncated) $W \geqq 1.7\,D$	a (Truncated) $W = D$
Rectangular		1.0	@ $1/e^2$ Intensity	1.27	1.83
Round/elliptic		1.25	@ $\frac{1}{2}$-Intensity	0.75	1.13
Keystone		1.5	for 50% MTF	0.85	1.38
Triangular		1.7			
Width D for 50% MTF.			Beam width D @ $1/e^2$ intensity centered within aperture width W.		

---

* Although the illumination and resulting PSFs are of a coherent wave, scanning forms a sequence of incoherently related intensity measurements of the space-shifting function, yielding an incoherent MTF.

† See Figs. 23 and 24 and related discussion for reduced power throughput due to approaching uniform illumination within the aperture.

TABLE 5   Aperture Shape Factor a for One-dimensional Truncation of a Gaussian Intensity Distribution

Truncation ratio $\rho = W/D$	Shape Factor a for 50% MTF
0	1.0
0.5	1.05
1.0	1.15
1.5	1.35
2.0	1.75

W = Width of aperture
D = Width of gaussian beam at $1/e^2$ intensity points
W and D measured in scan direction

2. The a-value of 1.25 for the uniformly illuminated round/elliptic aperture corresponds closely to the Rayleigh radius value of 1.22.

3. The gaussian-illuminated data requires that the width D, measured at the $1/e^2$ intensity points be centered within the available aperture W. Two conditions are tabulated: untruncated[10] ($W \geq 1.7D$) and truncation at $W = D$.

4. The gaussian-illuminated data also provides the a-values for 50 percent MTF, allowing direct comparison with performance of the uniformly illuminated apertures.

This data relates to apertures which, if apodized, are truncated two-dimensionally. However, one-dimensional truncation of a gaussian beam by parallel boundaries is not uncommon, typical of that for acousto-optic scanning. There, the limiting aperture width W is constant, as determined by the device, while the gaussian width D is variable.[4,5,11,12] Table 5 tabulates the shape factor a for such conditions.

To relate to data in Table 4, the case of $\rho = W/D = 0$ represents illumination through a narrow slit. This corresponds to the uniformly illuminated rectangular aperture, whence $a = 1$. When $\rho = 1$, then $W = D$ and the parallel barriers truncate the gaussian beam at its $1/e^2$ intensity points. Compared to symmetric truncation, this allows more of the gaussian skirts to contribute to the aperture width, providing $a = 1.15$ vs. 1.38. When $\rho = 2$, the gaussian beam of half the width of the boundaries is effectively untruncated, halving the resolution ($a = 1.75$ vs. 0.85), but maximizing radiometric throughput. (See Fig. 24, observing nomenclature, in which $D = 2w_x$ and $W = 2r_o$.)

Fundamental Scanned Resolution.   The section on "Input/Output Scanning" introduced the two forms of optical scan: translation and angular deflection. Beam translation is conducted by objective scan, while angular deflection is either preobjective or postobjective. Examples of each are provided later in this chapter.

The resolution N_s of translational scan, by a beam focused to spot size δ executing a scanned path distance S, is simply,

$$N_s = \frac{S}{\delta} \tag{14}$$

Extremely high resolutions are practical, but are often limited to moderate speeds and bandwidths. Common implementations provide $N_s = 3000$ to $100{,}000$.*

---

* A high-resolution laser printer provides $N = 3000$ to 10,000, and a high-resolution graphic arts imager, $N = 10{,}000$ to 100,000.

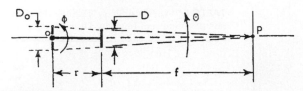

FIGURE 4 Deflecting element of width D (e.g., mirror of polygon) displaced by radius r from axis o, propagating a converging beam to p over focal distance f. Effective larger aperture D_o appears at rotating axis.[4]

The resolution N of angular scan,* represented schematically in Fig. 4, capable of much higher speeds, is given by[4,5,7]

$$N_\theta = \frac{\Theta D_o}{a\lambda} \tag{15}$$

in which Θ is the useful deflected optical angle and D_o is the effective aperture width at its nodal center, discussed in the next section. Common implementations provide $N_\theta = 2{,}000$ to $30{,}000$. Equation (15) is independent of spot size δ and dependent only on the aperture characteristics of D_o and a, and the wavelength λ. The beam could be converging, collimated, or diverging. When collimated, $D_o = D$, the actual width of the illuminated portion of the aperture. When converging or diverging, resolution augmentation occurs (see next section).

The numerator of Eq. (9) is a form of the Lagrange invariant,[13] expressed in this nomenclature as

$$n\Theta D = n'\Theta' D' \tag{16}$$

where the primed terms are the refractive index, (small) angular deviation, and aperture width, respectively, in the final image space. For the common condition of $n = n'$ in air, the ΘD product and resolution N are conserved, invariant with centered optics following the deflector.

Augmented Resolution, the Displaced Deflector. In general, a scanning system can accumulate resolution N by adding the two processes described previously, augmentation of angular scan with linear translation, forming

$$N = N_\theta + N_s \tag{17}$$

Augmentation occurs, for example, with conventional multielement scanners (such as polygons) having deflecting elements (facets) which are displaced from the rotating axis by a distance r, and whose output beam is noncollimated. One active element (of width D) and its focused output beam is illustrated in Fig. 4. For convenient analysis,[6] the deflecting element appears as overilluminated with an incident beam. The resulting resolution equations and focal spot positions are independent of over- or underillumination (see "Duty Cycle").

Augmentation for increased resolution is apparent in Fig. 4, in which the output beam is derived effectively from a larger aperture D_o which is located at o. By similar triangles, $D_o = D(1 + r/f)$, which yields from Eq. (15),

$$N = \frac{\Theta D}{a\lambda}\left(1 + \frac{r}{f}\right) \tag{18}$$

This corresponds to Eq. (17), for in the N_s term the aperture D executes a displacement component $S \simeq r\Theta$, which, with Eq. (12) forms Eq. (14).

---

* Derived from Eq. (4a) with $\sin \Delta\Theta \simeq \Delta\Theta = a\lambda/D_o$.

Following are some noteworthy observations regarding the parenthetic augmentation term:

1. Augmentation goes to zero when $r = 0$ (deflector on nodal axis) or when $f = \infty$ (output beam collimated).

2. Augmentation adds when output beam is convergent (f positive) and subtracts when output beam is divergent (f negative).

3. Augmentation adds when r is positive and subtracts when r is negative (output derived from opposite side of axis o).

The fundamental or nonaugmented portion of Eq. (18), $N = \Theta D/a\lambda$, has been transformed to a nomograph, Fig. 5, in which the angle Θ is represented directly in degrees. $D/a\lambda$ is plotted as a radius, particularly useful when $a\lambda = 1$ μm, whereupon $D/a\lambda$ becomes the aperture size D, directly in mm. The set of almost straight bold lines is the resolution N. Multiples of either scale yield corresponding multiples of resolution.

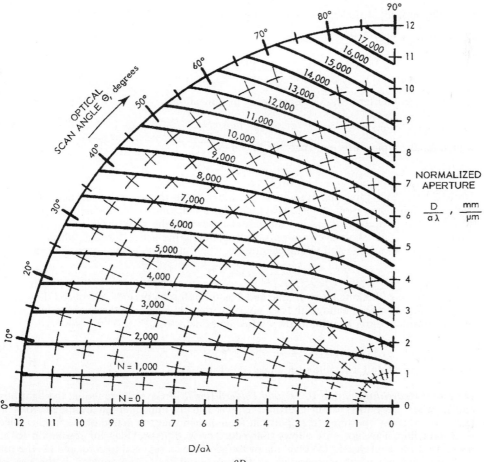

FIGURE 5 Nomograph of resolution equation $n = \dfrac{\theta D}{a\lambda}$.

Augmenting and Scan Magnification. Equation (18) develops from Fig. 4, assuming that the optimal scan angle Θ is equal to the mechanical angle Φ. This occurs only when the scanner exhibits radial symmetry (see "Radial Symmetry and Scan Magnification"). When, however, $m = d\Theta/d\Phi \neq 1$, as for configurations represented in the section on "Objective, Preobjective, and Postobjective Scanning," account must be taken of scan magnification m. Thus, the more complete resolution equation is represented by[6,14]

$$N = \frac{\Theta D}{a\lambda}\left(1 + \frac{r}{mf}\right)$$ (19)

in which, per Fig. 4, Θ = optical scan angle (active)

$\quad\quad D$ = scan aperture width

$\quad\quad \lambda$ = wavelength (same units as D)

$\quad\quad a$ = aperture shape factor

$\quad\quad m$ = scan magnification ($= d\Theta/d\Phi$)

$\quad\quad \Phi$ = mechanical scan angle about o

$\quad\quad r$ = distance from o to D

$\quad\quad f$ = distance from D to p

(∞ for collimated; $+$ for convergent; $-$ for divergent).

Considering $m = \Theta/\Phi$ as a constant, another useful form is

$$N = \frac{\Phi D}{a\lambda}\left(m + \frac{r}{f}\right)$$ (20)

whose augmenting term shows a composite magnification

$$m' = m + r/f$$ (21)

which, for the typical prismatic polygon becomes

$$m' = 2 + r/f$$ (21a)

Duty Cycle.   The foregoing resolution equations refer to the active portion of a scan cycle. The full scan period almost always includes a blanking or retrace interval. The ratio of the active portion to the full scan period is termed the duty cycle η. The blanking interval can include short overscan portions (straddling the active format), which are used typically for radiometric and timing calibration. The duty cycle is then expressed as

$$\eta = 1 - \tau/T$$ (22)

in which τ is the blanking interval time and T is the full scan period. A reduced duty cycle increases instantaneous bandwidth for a given average data rate. In terms of the scan angle of polygons, for example, it limits the useful component to

$$\theta = \eta\theta_{\text{max}}$$ (23)

where θ_{max} is the full available scan angle (see Table 3).

Over and Underillumination (Over and Underfilling).   In overillumination, the light flux encompasses the entire useful aperture. This is usually implemented by illuminating at least two adjacent apertures (e.g., polygon facets) such that the active one is always filled with light flux. This not only allows unity duty cycle, but provides for resolution to be maximized for two reasons: (1) blanking or retrace may be reduced to zero; and (2) the full available aperture width is operative as D throughout scan. The tradeoff is the loss of illuminating flux beyond the aperture edges (truncation) and attendant reduction in optical power throughput (see "Coherent Source" under "Scanning for Input/Output Imaging"). An alternative is to prescan[15] the light flux synchronously with the path of the scanning aperture such that it is filled with illumination during its entire transit.

In underillumination, the light flux is incident on a portion of the available aperture, such that this subtense delimits the useful portion D. A finite and often substantive blanking or retrace interval results, thereby depleting the duty cycle, but maximizing the transfer of incident flux to the scanned output.

19.4 SCANNERS FOR REMOTE SENSING

Early Single-Mirror Scanners

Early scanning systems comprised an object-space mirror followed by focusing optics and a detector element (or array). The first scanners were simple rotating mirrors oriented typically at 45° to the axis as illustrated in Fig. 6. The rotational axis of the scan mirror lies parallel to the flight direction. In Fig. 6a, the scan efficiency and duty cycle of the oblique or single ax-blade scanner (see monogon under "Monogon and Polygon Scanners") is quite low since only one scan per revolution ($S = 1$) is generated. The scan efficiency of the wedge or double ax-blade scanner shown in Fig. 6b is twice as great ($S = 2$), although the effective optical aperture is less than half that of the oblique scanner for the same mirror diameter. The scanning constant is $k = 1$ for both types (see "Remote Sensing Resolution and Data Rates").

Compound-Mirror-Optics Configurations

The aforementioned scanners suffered from a varying optical aperture as a function of view angle. To overcome this difficulty that causes severe variation in the video resolution during a scan line, several new line scanner configurations were developed. Most notable among these was the rotating prism scanner invented by Howard Kennedy[16] in the early 1960s and which forms the basis for most of the produced wide-field-of-view line scanners.

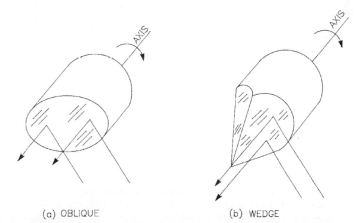

(a) OBLIQUE   (b) WEDGE

FIGURE 6 Early forms of scanners for remote sensing. The oblique or single ax-blade scanner is shown in (a) and the wedge or double ax-blade is shown in (b).

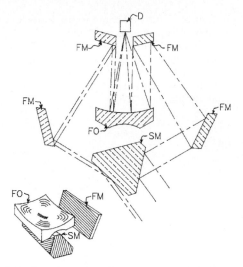 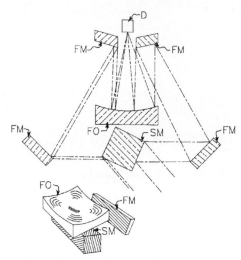

FIGURE 7 Basic split-aperture scanner with a three-sided scan mirror developed in the early 1960s. This scanner features wide FOV, constant optical aperture vs. scan angle, and compact mechanical configuration.

FIGURE 8 Basic split-aperture scanner with a four-sided scan mirror developed in the early 1960s. This scanner features wide FOV, constant optical aperture vs. scan angle, and compact mechanical configuration.

Figures 7 and 8 illustrate two configurations of this scanner. The three-sided scan mirror SM shown in Fig. 7 rotates about its longitudinal axis at a high rate and the concomitant folding mirrors FM are arranged such that the scanned flux is directed onto suitable focusing optics FO which focuses the flux at the detector D. As may be seen in the drawing of a four-sided scanner shown in Fig. 8, the effective optical aperture is split into two portions such that their sum is a constant value as a function of view angle. The width of each portion varies as the view angle is changed, with the portions being of equal value at the nadir position. The isometric view in Fig. 8 shows a portion of the scanner comprising the scan mirror, one folding mirror, and the focusing mirror. For this design, the number of scans per rotation of the scan mirror is equal to the number of faces on the scan mirror, and the scanning constant is $k = 2$, which is also known as optical doubling (see item 3 of the prismatic polygon in Table 3). Also, two faces of the scan mirror are always used to form the total optical aperture. Another advantage of this scanner configuration is that it produces a compact design for the total scanner system, a major reason for its popularity for over a quarter of a century.

Image Consequences

In airborne sensing, it is reasonable to assume that the earth is flat beneath the aircraft. When viewing along the nadir, the detector spatial footprint on the ground is $H \Delta\theta$ and $H \Delta\phi$ in the across- and along-track directions, respectively. As the view angle (θ) moves away from the nadir, the geometric resolution on the ground changes as illustrated in Fig. 3, which creates the bow-tie pattern. In the cross-track direction, it is easily shown that the footprint dimension is $H \Delta\theta \cdot sec^2 \theta$, while in the along-track direction, the footprint dimension is $H \Delta\phi \cdot sec \theta$. The change in footprint as a function of view angle can be significant. For example, if $\theta_{max} = 120°$, then the footprint area at the extremes of the scan line are about eight times greater than at the nadir.

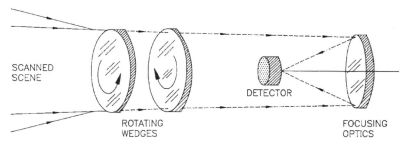

FIGURE 9 Basic geometry of a simple rotating wedge scanner. A wide variety of scan patterns can be produced by controlling the rotational rates and phasing of the wedges. In principal, the detector can view any point within the circular scanned FOV of the scanner. Figure 10 presents typical scan patterns for constant rotational rates. Two-dimensional raster scans can be generated if general positional control of each prism is allowed.

Image Relation and Overlap

When a linear array of n detectors is used, it is easily seen that the image of the detector array rotates by exactly the same amount as the view angle if the scanner is pyramidal as shown in Fig. 6. No such rotation occurs for the prismatic polygon, as in the Kennedy scanner, for which each scan comprises n adjacent detector footprints on the ground that form a segmented bow tie. The next scan footprint has significant overlap with the preceding scan(s) for $\theta \neq 0$. A means to compensate for the radiometric difficulties caused by the overlap of scans has been developed.[17] In a single detector system, this artifact is easily compensated by electronic means.

Rotating Wedge Scanner

Figure 9 shows a simple rotating wedge scanner that allows the generation of a wide variety of scan patterns, including a line scan. By controlling the rotational rates and phasing of the wedges, such patterns as included in Fig. 10 can be realized.[18]

Circular Scan

In some cases, a circular scan pattern has found utility. Typically, the entire optical system is rotated about the nadir with the optical axis inclined at an angle ψ to the nadir. Figure 11 depicts an object-plane scanner showing how the aircraft or satellite motion creates contiguous scans. Although the duty cycle is limited, an advantage of such a scanner is that, at a given altitude, the footprint has the same spatial size over the scanned arc.

Pushbroom Scan

A pushbroom scanner comprises typically an optical system that images onto the ground a linear array of detectors aligned in the cross-track direction or orthogonal to the flight direction. The entire array of detectors is read out every along-track dwell time which is $\tau_{at} = \Delta\phi/V/H$. Often, when a serial read-out array is employed, the array is rotated slightly such that the read-out time delay between detectors creates an image that is

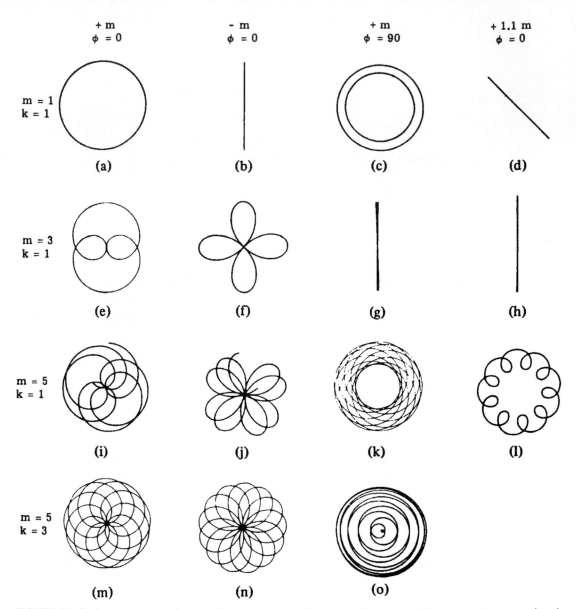

FIGURE 10 Typical scan patterns for a rotating wedge scanner. The ratio of the rotational frequencies of the two prisms is m, the ratio of the prism angles is k, and the phase relation at time zero is ϕ ($\phi = 0$ implies the prism apexes are oriented in the same direction). A negative value of m indicates that the prisms are counter-rotating. (*After Wolfe and Zissis, ONR, Washington, 1978, fig. 10-12, pp. 10–12.*)

properly aligned to the direction of motion. Some state-of-the-art arrays can transfer the image data in parallel to a storage register for further processing. The principal advantage of the pushbroom scanner is that no moving parts are required other than the moving platform upon which it is located.

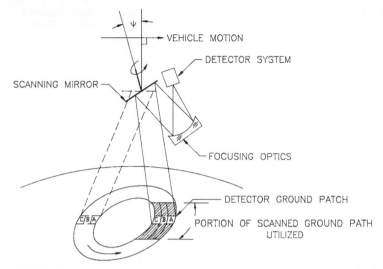

FIGURE 11 Basic configuration of a circular or conical scanner. The normal of the scanning mirror makes an angle Ψ with the nadir. The scan pattern of the ground forms a series of arcs illustrated by scans A, B, and C. (*After* Manual of Remote Sensing, *2d ed, Amer. Soc. of Photogrammetry, Falls Church, 1983, fig. 8-3, p. 340.*)

Two-dimensional Scanners

Two-dimensional scanners have become the workhorses of the infrared community during the past two decades even though line scanners still find many applications, particularly in the area of earth resources. Scanners of this category can be classified into three basic groups, namely, object-space scanner, convergent-beam or image-space scanner, and parallel-beam or intermediate space scanner. Figure 12 depicts the generic form of each group.

Object-space and Image-space Scanners

The earliest two-dimensional scanners utilized an object-space scan mechanism. The simplest optical configuration is a single flat-mirror (see Fig. 12a) that is articulated in such a manner as to form a raster scan. The difficulty with this scan mechanism is that movement of a large mirror with the necessary accuracy is challenging. The size of the mirror aperture when in object space must be greater than that of the focusing optics. By using two mirrors rotating about orthogonal axes, the scan can be generated by using smaller mirrors, although the objective optics must have the capability to cover the entire field of view rather than the FOV of the detector. Figure 13 illustrates such a scanner[19] where mirror SM1 moves the beam in the vertical direction at a slow rate while mirror SM2 generates the high-speed horizontal scan. Although the focusing optics FO is shown preceding the image-space scan mirrors, the optics could be placed following the mirrors which would then be in object-space. Although the high-F-number or low-numerical-aperture focusing lens before the mirrors must accommodate the FOV, it allows the use of smaller mirror facets. The left-hand side of Fig. 13 shows an integral recording mechanism that is automatically synchronized to the infrared receptor side. This feature is one of the more notable aspects of the configuration and sets the stage for other scanner designs

(a)   OBJECT SPACE SCANNER

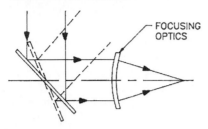

FOCUSING OPTICS

(c)   PARALLEL BEAM SCANNER (INTERMEDIATE SPACE)

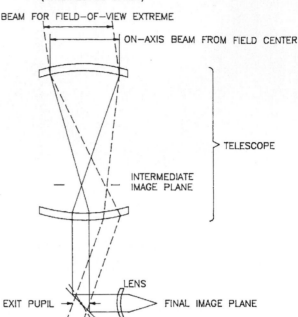

BEAM FOR FIELD—OF—VIEW EXTREME

ON—AXIS BEAM FROM FIELD CENTER

TELESCOPE

INTERMEDIATE IMAGE PLANE

(b)   CONVERGENT BEAM SCANNER (IMAGE SPACE)

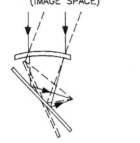

EXIT PUPIL

LENS

FINAL IMAGE PLANE

FIGURE 12   The three basic forms of two-dimensional scanners are shown in (a), (b), and (c). The object-space scanner in (a) comprises a scan mirror located in object-space where the mirror may be moved in orthogonal angular directions. Image-space or convergent-beam scanner in (b) forms a spherical image surface due to motion of the scan mirror (unless special compensation motion of the scan mirror is provided). The parallel-beam or intermediate-space scanner is illustrated in (c). It is similar to the scanner in (a) except that the scan mirror is preceded by an afocal telescope. By proper selection of the afocal ratio and FOV, the scan mirror and focusing lens become of reasonable size. The scan mirror forms the effective exit pupil. (*After J. M. Lloyd*, Thermal Imaging Systems, *Plenum Press, New York, 1983, figs. 7.1 and 7.10.*)

incorporating the integrated scene and display scanner. A disadvantage of this scanner is the large size and weight of the vertical scan mirror, in part, to accommodate both scene and display scan.

A variation of the two-mirror object-space scanner is known as the discoid scanner, which produces a raster scan at TV rates.[20] Figure 14 depicts the scanner configuration which uses a high-speed, multiple-facet scan mirror SM1 to generate the horizontal scan and a small, oscillating flat mirror SM2 to produce the vertical scan. An advantage of this scanner is that only a single detector is needed to cover the FOV, although a linear array oriented in the scan direction is sometimes used, with time-delay integration, to improve sensitivity. A feature of the "paddle" mirror scanner is the maintenance of a relatively stable aperture on the second deflector without the use of relay optics (see Figs. 21 and 32 and the section on the "Parallel Beam Scanner").

Figure 15 depicts a reflective polygon scanner that generates the high-speed horizontal scan (per facet) by rotation of mirror SM about its rotational axis and the vertical movement of the scan pattern by tilting the spinning mirror about pivots P1 and P2 using cam C and its follower F.[21] The path of the flux from the object reflects from the active facet A of the scan mirror to the folding mirror FM to the focusing mirror FO back through a hole in mirror FM to the detector located in dewar D. Almost all scanners of this type exhibit scanned-field distortion; i.e., the mapping of object to image space is non-rectilinear (e.g., see the target distortion discussion in the section "Image Consequences").

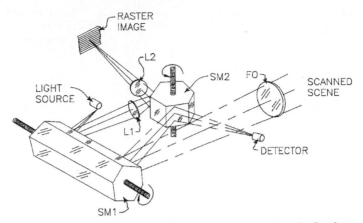

FIGURE 13 Early slow-scan-rate, image-space scanner where the flux from the scanned scene is imaged by the focusing objective lens FO onto the detector. The scene is scanned by mirrors SM1 (vertical) and SM2 (horizontal). A raster image of the scene is formed by imaging the light source using lenses L1 and L2. The display image and the scanned scene are synchronized by using the same pair of mirrors. The light source is modulated by the output of the detector.

In general, convergent-beam, image-space scanners suffer from severe defocus over the scanned field of view due to the field curvature produced by the rotation of the scan mirror in a convergent beam. The use of this type scanner is therefore rather limited unless some form of focus correction or curved detector array is employed. A clever invention by Lindberg[22,23] uses a high-speed refractive prism and a low-speed refractive prism to create the scanned frame. Figure 16 shows the basic configuration for a one-dimensional scanner

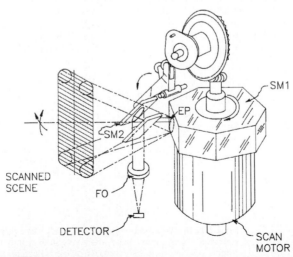

FIGURE 14 Real-time, object-space scanner that has a compact geometry. The exit pupil EP is located on the active facet of the horizontal scan mirror SM1.

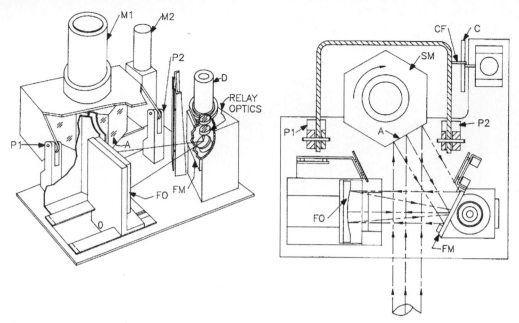

FIGURE 15 Object-space scanner that generates a raster scan using a single mirror SM which is driven by motor M1. The vertical motion of the mirror is accomplished by tilting the housing containing the scan mirror SM about pivots P1 and P2 using the drive mechanism comprising motor M2, cam C, and CF. The FOV of scanners of this type can exceed 30°.

where the cube P is rotated about the axis orthogonal to the page. By proper selection of the refractive index and the geometry of the prism, focus is maintained over a significant and useful field of view. As can be seen from the figure, flux from the object at a given view angle is focused by lens L onto surface I which is then directed to the detector D by the refraction caused by the rotated prism. Numerous commercial and military thermographic systems have utilized this principle for passive scanning. Since the field of view, maximum numerical aperture, optical throughput, and scan and frame rates are tightly coupled together, such scanners have a reasonably constrained design region.

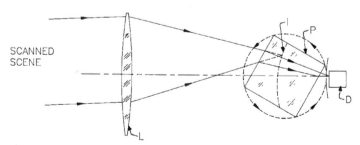

FIGURE 16 Basic configuration of a refractive prism scanner. The scan is generated by rotation of the prism. As shown, four scans are produced per complete rotation of the prism. By proper selection of the refractive index of the prism, reasonably wide FOV can be realized. Although only one prism is shown, a second prism can be included to produce the orthogonal scan direction.

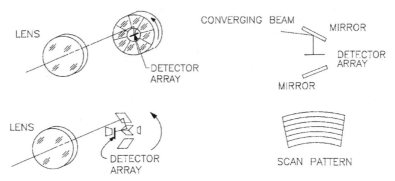

FIGURE 17   Rotating reflective or "soupbowl" scanner. (*After J. M. Lloyd,* Thermal Imaging Systems, *Plenum Press, New York, 1975, fig. 7.21, p. 309.*)

Other image-space scanners used in a convergent beam are the "soupbowl" and carousel scanners.[24] The soupbowl scanner shown in Fig. 17 uses a rotating array of mirrors to produce a circularly segmented raster scan. The mirror facets may be at the same angle to generate more frames per rotation, given a detector array that has adequate extent to cover the field of view. The facets could also be tilted appropriately with respect to one another to produce contiguous segments of the field of view if a small detector array is employed. Figure 18 illustrates the configuration of the carousel scanner which uses an array of mirrors arranged such that they create essentially a rectangular scan of the field of view. Another scanning means that has been used for certain forward-looking infrared systems (FLIRs) was to mechanically rotate a detector array of angular extent Φ about the optical axis of the focusing optics such that one end of the array was located an angular distance Φ_{os} from the optical axis. The rotating action generated a circular arc scan pattern similar to that of a windshield wiper. The inner radius of the scan pattern is Φ_{os} and the outer radius is $\Phi_{os} + \Phi$. Clearly, the scan efficiency is rather poor and the necessity to use slip rings or the equivalent to maintain electrical connections to the detector array complicated acceptance of this scanner. The windshield wiper scan can also be generated by rotating only the optics if the optics incorporates anamorphic elements. A pair of

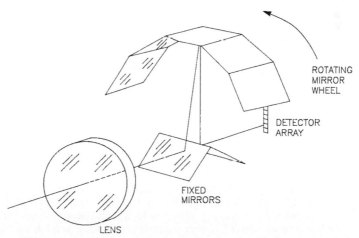

FIGURE 18   Rotating reflective carousel scanner. (*After J. M. Lloyd,* Thermal Imaging Systems, *Plenum Press, New York, 1975, fig. 7.22, p. 309.*)

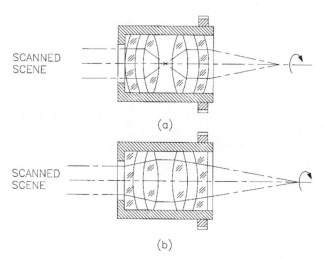

(a)

(b)

FIGURE 19 "Windshield wiper" scanner. The circular scan is generated by rotating the anamorphic optics about its optical axis. The detector array is located radially and offset from the optical axis. Two scans are produced for each complete rotation of the optics. The lens is shown in (*a*) at rotation angle θ and in (*b*) at rotation angle $\theta + 90°$. Although the lens shown is focal, the lens could be afocal and followed by a focusing lens.

cylindrical lenses placed in an afocal arrangement, as illustrated in Fig. 19 at rotational angles θ and $\theta + 90°$, will rotate the beam passing through it at twice the rotational rate of the optics.[25] See "Image Rotation in Derotation" in "Scanner Devices and Techniques."

Multiplexed Image Scanning

With the advent of detector arrays comprising significant numbers of elements, the use of a single scan mirror became attractive. Figure 20 presents the basic parallel-beam scanner configuration used for the common module FLIR and thermal night sights. The flat scan mirror SM is oscillated with either a sawtooth or a triangular waveform such that the detector array D (comprising 60, 120, or 180 elements) is scanned over the field of view in the azimuthal direction while the extent of the detector covers the elevation FOV. Since the detectors are spaced on centers two detector widths apart, the scan mirror is tilted slightly in elevation every other scan to produce a 2:1 interlaced scan of the field of view. As shown in Fig. 25, the back side of the scan mirror is used to produce a display of the scanned scene by coupling the outputs of the detectors to a corresponding array of LEDs which are projected to the user's eye by lenses L1, L2, L3, and L4.

Parallel-Beam Scanner

A more complex two-dimensional, parallel-beam scanner configuration of the type shown in Fig. 12*c* has been developed by Barr & Stroud and is illustrated in Fig. 21 which incorporates an oscillating mirror SM1, a high-speed polygon mirror SM2 driven by motor M2, and relay optics L1. (See discussion at end of section on "Scanning for Input/Optical Imaging".)[26] An afocal telescope is located before the scanner to change the FOV, as is typical of parallel-beam scanners. Another innovative and compact two-dimensional

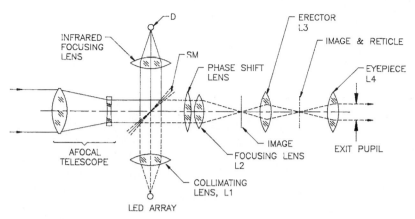

FIGURE 20 Basic configuration of the common module scanner. The front side of the flat scan mirror SM is used to direct the flux from the scanned scene to the detector D through the focusing lens. The outputs from the infrared detector array are used to modulate a corresponding LED array. The LED array is collimated by L1 and scanned over image space by the back side of SM. Lenses L2–L4 are used to project the image of the LED array to the observer's eye.

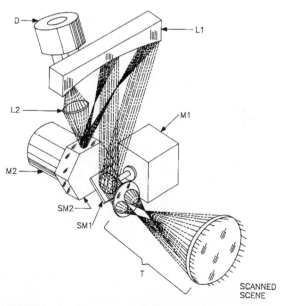

FIGURE 21 Compact real-time scanner. The horizontal scan mirror SM2 is shown in two positions to illustrate how the field of view is related to location on mirror L1. Mirror L1 serves as a relay mirror of the pupil on mirror SM1 to SM2.

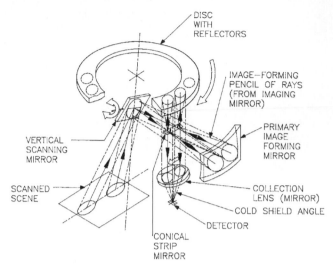

DISC
WITH
REFLECTORS

IMAGE-FORMING
PENCIL OF RAYS
(FROM IMAGING
MIRROR)

VERTICAL
SCANNING
MIRROR

PRIMARY
IMAGE
FORMING
MIRROR

SCANNED
SCENE

COLLECTION
LENS (MIRROR)

COLD SHIELD ANGLE

DETECTOR

CONICAL
STRIP
MIRROR

FIGURE 22 Extremely compact, real-time scanner. Diamond-turned optics are used throughout this configuration. [*After Thomas H. Jamieson, "Optical Design of Compact Thermal Scanner," Proc. SPIE* **518:**15, fig. 1 (1984).]

scanner design by Kollmorgen is depicted in Fig. 22 and features diamond-turned fabrication technology for ease of manufacture and alignment of the mirrors and mounts.[27] Another parallel-beam scanner that uses a simple scan mirror has been developed.[28] The scan mirror is multifaceted with each facet tilted at an angle that positions the detector array in a contiguous manner in elevation. By having the nominal tilt angle of the facets be 45° to the rotation axis, minimal scanned-field distortion is realized.

19.5 *SCANNING FOR INPUT/OUTPUT IMAGING*

Power Density and Power Transfer

Incoherent Source. This topic merits introduction as the predecessor to laser scanning— cathode-ray tube (CRT), flying-spot scanning and recording.[1,2,3] Adaptation to other forms of incoherent sources, such as light-emitting diodes (LEDs) will be apparent. Similarities and contrasts with the handling of coherent sources are expressed.

In a CRT, the electron beam power P (accelerating voltage · beam current) excites a phosphor of conversion efficiency η and utilization factor γ. The resulting radiant power is transferred through an imaging system of optical transmission efficiency T and spectral power transfer α to a photosensitive medium of area a during a time t. The resulting actinic energy density is given by[3]

$$E = \frac{\eta \alpha T \gamma P t}{A} \quad \text{joules/cm}^2 \tag{24}$$

(1 joule = 1 watt-sec = 10^7 ergs).

The first four terms are transfer factors (≤ 1) relating to the CRT, but adaptable to other radiant sources. They are determined[3] for a principal group of CRT recording phosphors having varying processes of deposition and aluminizing, and for two typical (silver halide) photosensitive spectral responses: noncolor sensitized and orthochromatic. The spectral transfer term α is determined from the relatively broad and nonanalytic spectral characteristics of the CRT phosphors and the photosensors,

$$\alpha \cong \frac{\sum\limits_{i=1}^{n} \frac{P_i}{P_{max}} \cdot \frac{S_i}{S_{max}} \Delta\lambda_i}{\sum\limits_{j=1}^{m} \frac{P_j}{P_{max}} \Delta\lambda_j} \quad (j \geq i) \tag{25}$$

where the Ps and the Ss are the radiant power and medium sensitivity respectively, taken at significant equal wavelength increments $\Delta\lambda$.

The optical transfer term T is composed of three principal factors, $T = T_r T_f T_v$ in which T_r is the fixed transmission which survives losses due to, for example, reflection and scatter, T_f is the fixed geometric transfer, and T_v is the spectrally variable transmission of, for example, different glass types. The fixed geometric transfer is given by[3]

$$T_f = \frac{cos^4\Phi \, V_\Phi}{1 + 4F^2(M+1)^2} \tag{26}$$

The numerator (≤ 1) is a transfer factor due to field angle Φ and vignetting[29] losses, F is the lens F-number, and M is the magnification, image/object. The variable component T_v requires evaluation in a manner similar to that conducted for the α. The resulting available energy density E is determined from Eq. (24) and compared to that required for satisfactory exposure of the selected storage material.

Coherent Source. Determination of power transfer is much simplified by utilization of a monochromatic (single-line laser) source. Even if it radiates several useful lines (as a multispectral source), power transfer is established with a finite number of relatively simple determinations. Laser lines are sufficiently narrow, compared to the spectral characteristics of most transmission and detection media, so that single point evaluations at the wavelengths of interest are usually adequate. The complexity due to spectral and spatial distributions per Eqs. (25) and (26) are effectively eliminated.

In contrast to the incoherent imaging system described above, which suffers a significant geometric power loss represented by T_f of Eq. (26), essentially all the radiant power from the laser (under controlled conditions discussed subsequently) can be transferred to the focal spot. Further, in contrast to the typical increase in radiating spot size with increased electron beam power of a CRT, the radiating source size of the laser remains essentially constant with power variation. The focused spot size is determined (per the previous section on "Resolution Criteria, Aperture Shape Factor") by the converging beam angle or corresponding numerical aperture or F-number, allowing for extremely high power densities. Thus, a more useful form of Eq. (24) expresses directly the laser power required to irradiate a photosensitive material as

$$P = \frac{sR}{T}\left(\frac{A}{t}\right) \text{ watts} \tag{27}$$

in which s = material sensitivity, J/cm^2
$\quad\quad R$ = reciprocity failure factor, ≥ 1
$\quad\quad T$ = optical throughput efficiency, ≤ 1
$\quad\quad A$ = exposed area, cm^2
$\quad\quad t$ = time over area A, sec.

The reciprocity failure factor R appears here,[30] since the exposure interval t (by laser) can be sufficiently short to elicit a loss in sensitivity of the photosensitive medium (usually registered by silver halide media). If the A/t value is taken as, for example, an entire

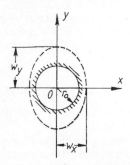

FIGURE 23 Irradiance of a single-mode laser beam, generalized to elliptical, centered within a circular aperture of radius r_o.[31] Glossary as published; D (as used here) $= 2w_x$ and w (as used here) $= 2r_o$.

frame of assumed uniform exposure interval (including blanking), then the two-dimensional values of η must appear in the denominator, for they could represent a significant combined loss of exposure time.

The optical throughput efficiency T is a result of loss factors including those due to absorption, reflection, scatter, diffraction, polarization, diffraction inefficiency in acousto-optic and holographic elements, and beam truncation or vignetting. Each requires disciplined attention. While the radiation from (fundamental mode) laser sources is essentially conserved in traversing sufficiently large apertures, practical implementation can be burdensome in scanners. To evaluate the aperture size consistent with throughput power transfer, Figs. 23 and 24 are useful. The data is generalized to elliptic, accommodating the irradiance of typical laser diodes.[31] Figure 23 shows an irradiance distribution having ellipticity $\epsilon = w_x/w_y$ (w @ $1/e^2$ intensity) apertured by a circle of radius r_o. Figure 24 plots the encircled power (percent) vs. the ellipticity, with the ratio r_o/w_x as a

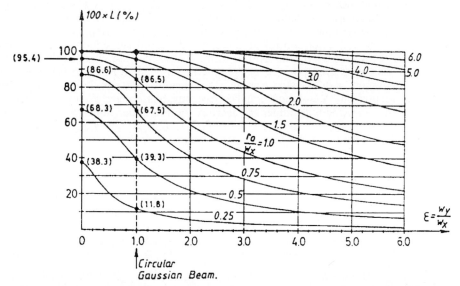

FIGURE 24 Variations of the encircled energy $100 \times L(\%)$ versus the ellipticity ε and the ratio r_o/w_x as a parameter.[31] Glossary as published; D (as used here) $= 2w_o$ and w (as used here) $= 2r_o$.

parameter. When $\epsilon = 1$, it represents the circular gaussian beam. Another parameter closely related to this efficiency is the aperture shape factor (discussed previously) affecting scanned resolution. (Note Glossary: $D = 2w_x$ and $w = 2r_o$.)

Objective, Preobjective, and Postobjective Scanning

Classification Characteristics. The scanner classifications designated as preobjective, objective, and postobjective were introduced previously and represented in Fig. 1 as a general conjugate optical transfer. This section expresses their characteristics.

Objective Scan (Transverse Translational). Translation of an objective lens transverse to its axis translates the imaged focal point on the information surface. (Axial lens translation which optimizes focus is not normally considered scanning.) Translation of the information medium (or object) with respect to the objective lens forms the same effect, both termed objective scan.* The two forms of objective scan appear in Fig. 25, the configuration of a drum scanner.

Preobjective Scan (Angular). Preobjective scan can provide a flat image field.† This is exemplified by angularly scanning a laser beam into a flat-field or f-θ lens,[32] as illustrated in Fig. 26, an important technique discussed further under "Pyramidal and Prismatic Facets" and "Flat Field Objective Optics."

Postobjective Scan (Angular). Postobjective scan which is radially symmetric per Fig. 27 generates a perfectly circular scan locus.‡ Departure from radial symmetry (e.g., focal point not on the axis of Fig. 27) generates noncircular (e.g., limaçon[5]) scan, except for the special case, as follows.
„special case, as follows.

A postobjective mirror with its surface on its axis generates a perfectly circular scan locus, illustrated in Fig. 28. The input beam is focused beyond the axis at point o. Scan magnification $m = 2$.

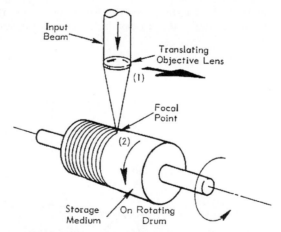

FIGURE 25 Drum configuration executing two forms of objective scan: (*a*) lens and its focal point translate with respect to storage medium; (*b*) storage medium translates with respect to lens during drum rotation.[4]

---

* Objective scan is limited in speed because the translating lens elements or storage medium must execute the desired final scan velocity, limited by the articulation of relatively massive components.

† Applies beyond the small scan angle $\theta \simeq \arctan \theta$, which may be considered linear. Also, no separate dynamic focus to aid forming a flat field.

‡ Placing the objective lens in the output beam and coupling it rigidly to the scanner (e.g., of Fig. 27) maintains the same characteristic. This is identified as Objective Scan (Angular). The scanner and lens may be combined, as in a hologram.

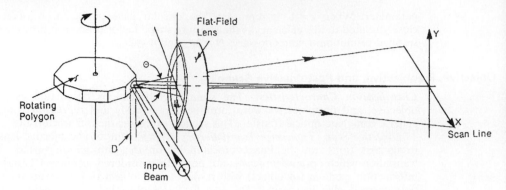

FIGURE 26 Polygon preobjective scan. The rotating polygon reflects and scans the input beam through the angle Θ. The flat-field lens transforms this Θ-change to (nominally) linear x displacement along a straight scan line. The input beam and the scanned beam reside nominally in the same plane.[4]

Objective Optics

The objective lens converges a scanned laser beam to a moving focal point. The deflector can appear before, at, or after the lens, forming preobjective, objective, and postobjective scanning, respectively (see previous discussion).

On-axis Objective Optics. The simplest objective lens arrangement is one which appears before the deflector, as in Fig. 27, where it is required only to focus a monochromatic beam on-axis. The (postobjective) deflector intercepts the converging beam to scan its focal point. Ideally, the process is conducted almost aberrationlessly, approaching

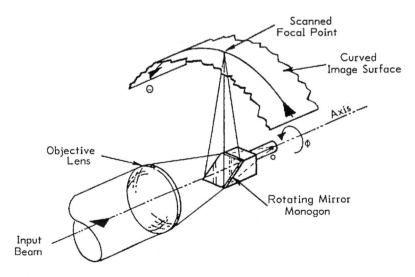

FIGURE 27 Monogon postobjective scan. Generates a curved image. When input beam is focused on the axis (at point o), then system becomes radially symmetric, and image locus forms section of a perfect circle.[4]

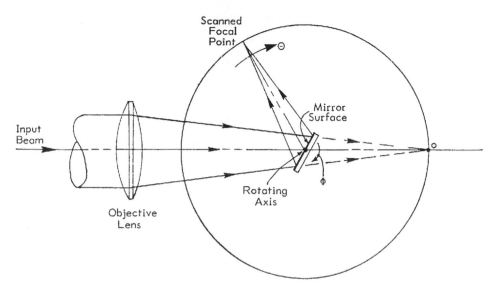

FIGURE 28 Postobjective scan with mirror surface on rotating axis. Input beam is focused by objective lens to point *o*, intercepted by mirror and reflected to scanning circular locus. Scan magnification $m = d\Theta/d\Phi = 2$.[4]

diffraction-limited performance. Since the lens operates on-axis only (accommodates no field angle) if the F-number of the converging cone is sufficiently high (see "Resolution Criteria, Aperture Shape Factor"), it can be composed of a single lens element. This simple arrangement can scan a perfectly circular arc [see "Postobjective Scan (Angular)]", the basis for the elegance of the internal drum scanner and the requirement for adapting the information medium to a curved surface.

Flat-field Objective Optics. Almost all other lens configurations are required to form a flat field by transforming the angular scan to a straight line.[32] The deflector appears before the lens—preobjective. The most common configuration is similar to that of Fig. 26, as detailed further in "Design Considerations" under "Monogon and Polygon Scanners," in which the scanned beam is shown collimated. Application is not limited to polygon scanners. Similar lenses adapt to a variety of techniques, including galvanometer, acousto-optic, electro-optic, and holographic scanners. The lens must accept the scanned angle θ from the aperture D and converge the beam throughout the scanned field to a best-focus along a straight-line locus. Depending on the magnitudes of θ and D, the F-number of the converging cone and the desired perfection of straight-line focus and linearity, the lens assembly can be composed of from 2 to 7 (or more) elements, with an equal number of choices of index of refraction, 4 to 14 (or more) surfaces and 3 to 8 (or more) lens spacings, all representing the degrees of freedom for the lens designer to accommodate performance. A typical arrangement of three elements is illustrated in Fig. 34.

  Telecentricity. A more demanding arrangement is illustrated in Fig. 29, showing six elements forming a high-performance scan lens in a telecentric configuration.[29] Telecentricity is represented schematically in Fig. 30, in which an ideal thin-lens element depicts the actual arrangement of Fig. 29. Interposed one focal length f between the scanning aperture D (entrance pupil) and the flat image surface, the ideal lens transforms the angular change at the input to a translation of the output cone. The chief ray of the ideal output beam

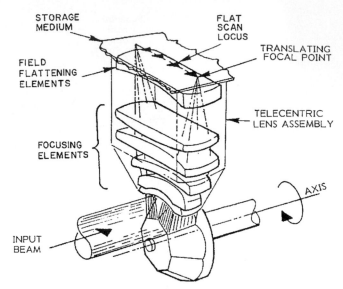

FIGURE 29   High-performance telecentric lens at output of pyramidal polygon scanner. see Fig. 34. (Lens elements shown cut for illustration only.)[4]

lands normal to the image surface. The degree of telecentricity is expressed by the angular departure from normal landing. Telecentricity is applied typically to restrict the spread of the landing beam and/or to retroreflect the probing beam efficiently for internal system calibration. This facility comes dearly, however, for the final lens elements must be at least as wide as the desired scan format. A further requirement is the need to correct the nonlinearity of the simple system of Fig. 30, in which the spot displacement is proportional to the tangent of the scan angle, rather than to the angle directly. As in all scan lenses, compensation to make displacement proportional to scan angle is termed the f-θ correction.

Double-pass and Beam Expansion.   Another variation of the objective lens is its

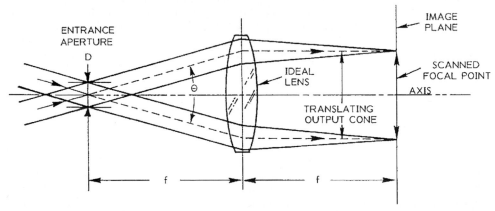

FIGURE 30   Telecentric optical system. Schematic illustration of ideal lens transforming angular scan Θ from aperture D to translational scan landing normal to the image plane.[4]

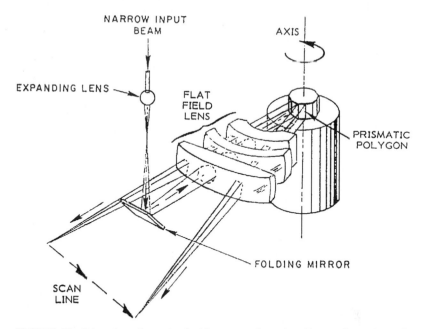

FIGURE 31 Prismatic polygon in double-pass configuration. Narrow input beam is focused by positive lens, expanded, and picked-off by folding mirror to be launched through flat-field lens (in reverse direction). Beam emerges collimated, illuminates facets and is reflected and scanned by rotating polygon to be reconverged to focus on scan line. Input and output beams are slightly skewed above and below lens axis to allow clear separation by folding mirror. (Flat-field lens elements shown cut for illustration.)[4]

adaptation to double-pass,[4,5,31] as depicted in Fig. 31. The lens assembly serves two functions: first, as the collimating portion of a lenticular beam expander*[32] and second, upon reflection by the scanner, as a conventional flat-field lens. This not only provides compaction, but since the illuminating beam is normal to the undeflected facet, the beam and facet undergo minimum enlargement, conserving the size of the deflector. A slight skew of the input and output planes, per Fig. 31, avoids obstruction of the input and scanned beams at the folding mirror. An alternate input method is to make the lens array wide enough to include injection of the input beam (via a small mirror) from the side; at an angle sufficiently off-axis to avoid obstruction of the reflected scanned beam.[31] This method imposes an off-axis angle and consequential facet enlargement and beam aberration, but allows all beams to remain in the same plane normal to the axis, avoiding the (typically) minor scanned bow which develops in the aforementioned center-skewed method. Other factors relating to increased surface scatter and reflection need be considered.

The requirement for beam expansion noted here is fundamental to the formation of the aperture width D which provides a desired scanned resolution. Since most gas lasers radiate a collimated beam which is narrower than that required, the beam is broadened by propagating it through an inverted telescope beam expander, that is, an afocal lens group having the shorter focal length followed by the longer focal length.* Operation may be reversed, forming *beam compression,* as required. In the previously described double-pass

---

* Beam expansion/compression can also be achieved nonlenticularly with prisms.[33] Introduction in 1964 of the phrase "beam expander" by Leo Beiser, and its dissemination to generic form, is summarized in App. 1 of Ref. 6.

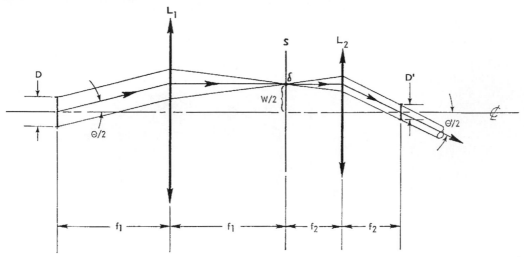

FIGURE 32   Illustration of invariance $I = \Theta D = \Theta'D'$ with telescopic transfer of scanned angle Θ from aperture D.[4]

system, the objective lens provides the collimating portion (long-focal-length group) of a beam expander.

Conservation of Resolution.   A most significant role of objective optics following the scanner is its determination of the integrity of scanned format, *not of scanned resolution,* as discussed under "Input/Output Scanning". Denoting N as the total number of scanned elements of resolution to be conveyed over a full format width, N is *invariant* with intervening ideal optics. In reasonably stigmatic systems, the lens determines the *size* of the spots, *not their total number.* The number of spots is determined *at the deflector,* whether it be galvanometer, acousto-optic, electro-optic, polygonal, holographic, phased array, or any other angular scanner. This invariance is expressed as

$$I = \theta D = \theta'D' \qquad (28)$$

an adaptation of the Lagrange invariant [see "Fundamental Scanned Resolution," Eq. (16)], which is illustrated effectively with telescopic operation. If the scanned beam is directed through a telescope (beam compression), as in Fig. 32, the demagnification of f_2/f_1 reduces D to D', but also expands θ to θ' by the same paraxial factor, sustaining resolution invariance. If D were the deflecting aperture width and L_1 were its objective lens (telecentric in this case), then the image along surface S would exhibit the same number of N spots as would appear if the output beam from D' were focused by another objective lens to another image plane. This schematic represents, effectively, a pupil-transferring optical relay.[4,5] (See R. B. Johnson, "Lenses" and W. B. Wetherell, "Afocal Lenses", this Handbook, vol. II, chaps. 1 and 2.)

19.6 *SCANNER DEVICES AND TECHNIQUES*

Many of the techniques addressed here for input/output imaging apply equally to remote sensing. Their reciprocal characteristic can be applied effectively by reversing the positions (and ray directions) of the light source(s) and detector(s). Preobjective and postobjective scanning have their counterparts in object-space and image-space scanning. A notable distinction, however, is in the option of underillumination or overillumination of the deflecting aperture in input/output imaging, while the aperture is most often fully subtended in collecting flux from a remote source. This leads to the required attention to aperture shape factor in input/output imaging, which is less of an issue in remote sensing. Another is the need to accommodate a relatively broad spectral range in remote sensing, while input/output operation can be monochromatic, or at least polychromatic. The frequent use of reflective optics in both disciplines tends to normalize this distinction.

Monogon and Polygon Scanners

The rotating mirrored polygon is noted for its capacity to render high data rate at high resolution. It is characterized by a multiplicity of facets which are usually plane and disposed in a regular array on a shaft which is rotatable about an axis. When the number of facets reduces to one, it is identified as a monogon scanner.

Pyramidal and Prismatic Facets.   Principal arrangements of facets are termed *prismatic* (Fig. 33) or *pyramidal* (Fig. 34). Figure 27 is a single-facet pyramidal equivalent, while Fig. 28 is a single-facet prismatic equivalent (common galvanometer mount).

The prismatic polygon of Fig. 33 is oriented typically with respect to its objective optics in a manner shown in Fig. 26, while Fig. 34 shows the relationship of the pyramidal polygon to its flat-field lens. The pyramidal arrangement allows the lens to be oriented close to the polygon, while, as in Fig. 26, the prismatic configuration requires space for clear passage of the input beam.* Design consideration for this most popular arrangement is provided later in this chapter.

Table 3 lists significant features and distinctions of typical polygon scanners. Consequences of Item 3, for example, are that the scan angle of the prismatic polygon is twice that of the pyramidal one for a given rotation. To obtain equal scan angles θ of equal

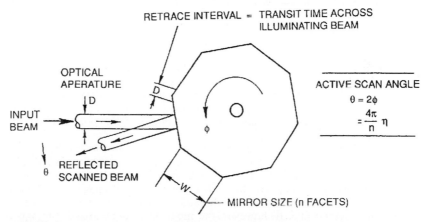

FIGURE 33   Prismatic polygon (underilluminated). Input beam perpendicular to axis. Its width D illuminates a portion of facet width W. Rotation through angle Φ yields scanned angle $\Theta = 2\Phi$, till facet corner encounters beam. Scan remains inactive for fraction D/W, yielding duty cycle $\eta = 1 - D_m/W$. See "Scanner-Lens Relationship."[4]

---

* In remote sensing, applying reciprocity, this is the *detected beam*.

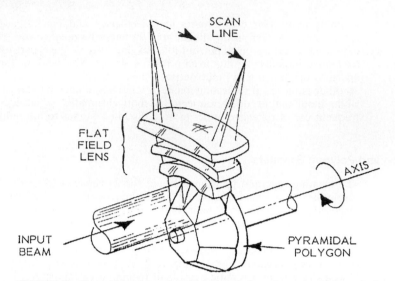

FIGURE 34 Pyramidal polygon (overilluminated). Input beam, parallel to axis, is scanned (by 45° pyramidal angle) in plane perpendicular to axis. When input beam illuminates two facets (as shown), one facet is active at all times, yielding (up to) 100 percent duty cycle. Facet width is full optical aperture *D*, minimizing polygon size for a given resolution, but wasting illumination around unused facet regions. Can operate underilluminated (per Fig. 33) to conserve throughput efficiency, but requires increased facet width (and polygon size) to attain high duty cycle. (Flat-field Lens elements shown cut for illustration.)[4]

beam width *D* (equal resolutions *N*) and to provide equal duty cycle (see "Augmenting and Scan Magnification") at equal scan rates, the prismatic polygon requires twice the number of facets, is almost twice the diameter, and rotates at half the speed of the pyramidal polygon. The actual diameter is determined with regard for the aperture shape factor (previously discussed) and the propagation of the beam cross section (pupil) across the facet during its rotation (see "Design Considerations").

Image Rotation and Derotation. When a beam having a round cross section is focused to an isotropic point spread function (psf), the rotation of this distribution about its axis is typically undetectable. If, however, the psf is nonisotropic (asymmetric or polarized), or if an array of 2 or more points is scanned to provide beam multiplexing,[6] certain scanning techniques can cause an undesired rotation of the point and the array of points on the image surface.

Consider a monogon scanner, per Fig. 27. As shown, the input beam *over*illuminates the rotating mirror. Thus, the mirror delimits the beam, establishing a rectangular cross section which maintains its relative orientation with respect to the image surface. Thus, if uniformly illuminated, the focal point on the image surface (having in this case a $\text{sinc}^2 x \cdot \text{sinc}^2 y$ psf, $x = $ along-scan and $y = $ cross-scan) maintains the same orientation along its scanned line. If, however, the input beam is polarized, the axis of polarization of the imaged point will rotate directly with mirror rotation within the rectangular psf. Similarly will be rotation for any radial asymmetry (e.g., intensity or ellipticity) within the aperture, resulting in rotation of the psf.

Consider, therefore, the same scanner *under*illuminated with, for example, an elliptical gaussian beam (with major axis horizontal). The axis of the imaged elliptic spot (intended major axis vertical) will rotate directly with the mirror. Similarly, if the scanner is illuminated with multiple beams displaced slightly angularly in a plane (to generate an in-line array of spots), the axis of the imaged array will rotate directly with mirror rotation.

This effect is transferrable directly to the pyramidal polygon which is illuminated per Fig. 34. It may be considered as an array of mirrors, each exhibiting the same rotation characteristics as the monogon of Fig. 27. The mirrors of Fig. 34 are also overilluminated, maintaining a stationary geometric psf during scan (if uniformly illuminated), but subject to rotation of, for example, polarization within the psf. Similarly, it is subject to rotation of an elliptical beam within the aperture, or of a multiple-beam array.

Not so, however, for the mirror mounted per Fig. 28 (galvanometer mount), or for the prismatic polygon of Figs. 26 & 33, which may be considered a multifacet extension of Fig. 28. When the illuminating beam and the scanned beam form a plane which is normal to the axis of mirror rotation, execution of scan does not alter the characteristics of the psf, except for the possible vignetting of the optical aperture and possible alteration of reflection characteristics (e.g., polarization) with variation in incident angle. It is noteworthy that in the prior examples, the angles of incidence remained constant, while the image is subject to rotation; and here, the angles of incidence change, while the image develops no rotation.

The distinction is in the symmetry of the scanning system with respect to the illumination. The prior examples (maintaining constant incidence while exhibiting image rotation) are radially symmetric. The latter examples (which vary incidence but execute no image rotation) represent the limit of radial asymmetry. While mirrored optical scanners seldom operate in regions between these two extremes, holographic scanners can, creating possible complications with, for example, polarization rotation. This is manifest in the variation in diffraction efficiency of gratings for variation in p and s polarizations during rotation. (See "Operation in the Bragg Regime.")

Image Derotation. Image derotation can be implemented by interposing another image rotating component in the optical path to cancel that caused by the scanner. The characteristic of an image rotator is that it inverts an image.[13] Thus, with continuous rotation, it rotates the image, developing two complete rotations per rotation of the component. It must, therefore, be rotated at half the angular velocity of the scanner.

While the Dove prism[13] is one of the most familiar components used for image rotation, other coaxial image inverters include[13]

- Three-mirror arrangement, which simulates the optical path of the Dove prism
- Cylindrical/spherical lens optical relay
- Pechan prism, which allows operation in converging or diverging beams

Design Considerations. A commonly encountered scanner configuration is the prismatic polygon feeding a flat-field lens in preobjective scan, illustrated schematically in Fig. 26. The size and cost of the flat field lens (given resolution and accuracy constraints) is determined primarily by its proximity to the scanner and the demand on its field angle. A larger distance from the scanner (pupil relief distance) imposes a wider acceptance aperture for a given scan angle, and a wider scan angle imposes more complex correction for off-axis aberration and field flattening. The pupil relief distance is determined primarily by the need for the input beam (Fig. 26) to clear the edge of the flat-field lens. Also, a wider scan angle reduces the accuracy requirement for pixel placement. Since the scan angle θ subtends the desired number N of resolution elements, a wider angle provides a larger angular subtense per element and correspondingly larger allowed error in angle $\Delta\theta$ for a desired elemental placement accuracy ΔN. This applies in both along-scan and cross-scan directions, $\Delta\theta_x$ and $\Delta\theta_y$, respectively (see "Scan Error Reduction").

Subsequent consideration of the scanner-lens relationships requires a preliminary estimate of the polygon facet count, in light of its diameter and speed. Its speed is determined by the desired data rates and entails considerations which transcend the optogeometric ones developed here. Diffraction-limited relationships are used throughout, requiring adjustment for anticipated aberration in real systems. The wavelength λ is a convenient parameter for buffering the design to accommodate aberration. For example, an anticipated fractional spot growth of 15 percent due to systematic aberration is accommodated by using $\lambda_+ = 1.15\lambda$.

Performance characteristics which are usually predisposed are the resolution N (elements per scan), the optical scan angle θ, and the duty cycle η. Their interrelationships are presented under "Input/Output Scanning," notably by Eqs. (15) and (23). The values of N and θ for a desired image format width must be deemed practical for the flat-field lens.

Following these preliminary judgments, the collimated input beam width D is determined from [see "Fundamental Scanned Resolution" Eq. (15)]

$$D = Na\lambda/\theta \tag{29}$$

in which a is the aperture shape factor and λ is the wavelength. For noncollimated beams, see "Augmented Resolution, the Displaced Deflector," notably Eq. (19). The number of facets is determined from Table 3 and Eq. (23),

$$n = 4\pi\eta/\theta \tag{30}$$

whereupon it is adjusted to an integer.

Scanner-Lens Relationships. The polygon size and related scan geometry into the flat-field lens may now be determined.[34] Figure 35 illustrates a typical prismatic polygon

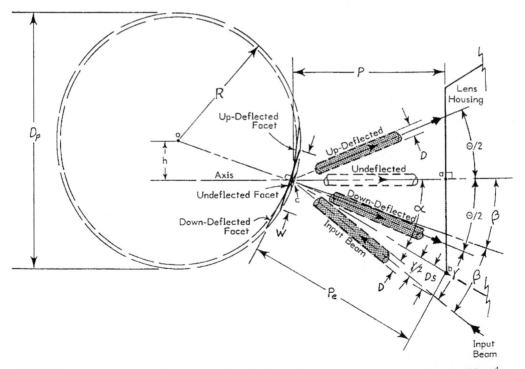

FIGURE 35 Polygon, beam, and lens relationships. Showing undeflected and limit facet and beam positions.[4]

and its input and output beams, all in the same plane. One of n facets of width W is shown in three positions: undeflected and in its limit-rotated positions. The optical beams are shown in corresponding undeflected and limit positions, deflected by $\pm\theta/2$. A lens housing edge denotes the input surface of a flat-field lens. Angle γ provides clear separation between the input beam and the down-deflected beam or lens housing. The pupil relief distance P (distance ac) and its slant distance P_e (distance bc) are system parameters which establish angle α such that $\cos\alpha = P/P_e$. Angle α represents the off-axis illumination on the polygon which broadens the input beam on the facet. The beam width D_m on the facet is widened due to α and due to an additional safety factor t ($1 \le t \le 1.4$) which limits one-sided truncation of the beam by the edge of the facet at the end of scan. Applying these factors, the beam width becomes

$$D_m = \frac{Dt}{\cos\alpha} \tag{31}$$

Following Eq. (15), the duty cycle is represented by $\eta = 1 - D_m/W$, yielding the facet width

$$W = D_m/(1 - \eta) \tag{32}$$

from which the outer (circumscribed) polygon diameter is developed;[34] expressed by

$$D_p = \frac{Dt}{(1 - \eta)\,\sin\pi/n\,\cos\alpha} \tag{33}$$

Solution of Eq. (33) or expressions of similar form[35] entails determination of α, the angle of off-axis illumination on the facet. This usually requires a detailed layout, similar to that of Fig. 35. Series approximation of $\cos\alpha$ allows transformation of Eq. (33) to replace α with more direct dependence on the important lens parameter P (pupil relief distance), yielding,

$$D_p = \frac{Dt}{(1 - \eta)\,\sin\pi/n} \cdot \frac{1 + \theta Ds/2P}{1 - \theta^2/8} \tag{34}$$

in which, per Fig. 35, $s \approx 2$ is a safety multiplier on D for secure input/output beam separation and clearance.

Orientation of the scanner and lens also requires the height h, the normal distance from the lens axis to the polygon center. This is developed[34] as

$$h = R_c \sin(\gamma/2 + \theta/4) \tag{35}$$

in which R_c is the radial distance oc, slightly shorter than the outer radius R, approximated to be

$$R_c = R[1 - \tfrac{1}{4}(\pi/n)^2] \tag{36}$$

Holographic Scanners

General Characteristics. Almost all holographic scanners comprise a substrate which is rotated about an axis, and utilize many of the concepts representative of polygons. An array of holographic elements disposed about the substrate serves as facets, to transfer a fixed incident beam to one which scans. As with polygons, the number of facets is determined by the optical scan angle and duty cycle (see "Duty Cycle"), and the elemental resolution is determined by the incident beam width and the scan angle (see Eq. 15). In radially symmetric systems, scan functions can be identical to those of the pyramidal polygon. While there are many similarities to polygons, there are significant advantages and limitations.[6] The most attractive features of holographic scanners are:

1. Reduced aerodynamic loading and windage with elimination of radial discontinuities of the substrate.

2. Reduced inertial deformation with elimination of radial variations.

3. Reduced optical-beam wobble when operated near the Bragg transmission angle.

Additional favorable factors are:

1. Operation in transmission, allowing efficient beam transfer and lens-size accommodation.

2. Provision for disk-scanner configuration, with facets disposed on the periphery of a flat surface, designable for replication.

3. No physical contact during exposure. Precision shaft indexing between exposures allows for high accuracy in facet orientation.

4. Filtering in retrocollection, allowing spatial and spectral selection by rediffraction.

5. Adjustability of focus, size, and orientation of individual facets.

Some limiting factors are:

1. Need for stringent design and fabrication procedures, with special expertise and facilities in diffractive optics, instrumentation, metrology, and processing chemistry.

2. Accommodation of wavelength shift: exposure at one wavelength (of high photosensitivity) and reconstruction at another (for system operation). Per the grating equation[6] for first-order diffraction,

$$sin\ \theta_i + sin\ \theta_o = \lambda/d \tag{37}$$

where θ_i and θ_o are the input and diffracted output angles with respect to the grating normal and d is the grating spacing, a plane linear grating reconstructs a collimated beam of a shifted wavelength at a shifted angle. Since wavefront purity is maintained, it is commonly employed,[6] although it requires separate focusing optics (as does a polygon). When optical power is added to the hologram (to provide self-focusing), its wavelength shift requires compensation for aberration.[6] Further complications arise when intended for multicolor operation, even if plane linear gratings. Further, even small wavelength shifts, as from laser diodes, can cause unacceptable beam misplacements, requiring corrective action.[6,36]

3. Departure from radial symmetry develops complex interactions which require critical balancing to achieve good scan linearity, scan-angle range, wobble correction, radio-

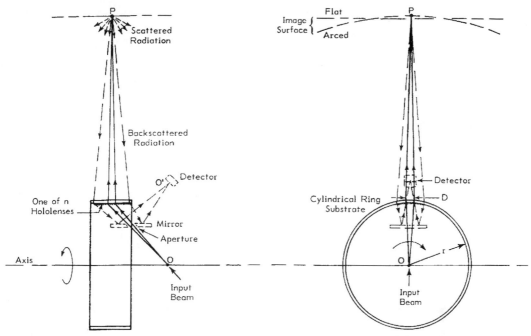

FIGURE 36 Transmissive cylindrical holographic scanner. Input beam underilluminates hololens which focuses diffracted beam to image surface. Dashed lines designate optional collection of backscattered radiation for document scanning.[6]

metric uniformity, and insensitivity to input beam polarization.[6,36] This is especially demanding in systems having optical power in the holograms.

4. Systems which retain radial symmetry to maintain scan uniformity may be limited in Bragg angle wobble reduction, and can require auxiliary compensation, such as anamorphic error correction.

Holographic Scanner Configurations. A scanner which embodies some of the characteristics expressed above is represented in Fig. 36.[37] A cylindrical glass substrate supports an array of equally spaced *hololenses* which image the input beam incident at *o* to the output point at *P*. Since point *o* intersects the axis, the scanner is radially symmetric, whereupon *P* executes a circular (arced) scan concentric with the axis, maintaining magnification $m = \theta/\Phi = 1$. A portion of the radiation incident on the image surface is backscattered and intercepted by the hololens, and reflected to a detector which is located at the mirror image *o'* of point *o*. The resolution of this configuration is shown to be analogous to that of the pyramidal polygon.[6]

An even closer analogy is provided by an earlier reflective form illustrated in Fig. 37, emulating the pyramidal polygon, Fig. 34. It scans a collimated beam which is transformed by a conventional flat-field lens to a scanned focused line. This is one of a family of holofacet scanners,[6] the most prominent of which tested to the highest performance yet achieved in combined resolution and speed—20,000 elements per scan at 200 Mpixels/s. This apparatus is now in the permanent collection of the Smithsonian Institution.

Operation in the Bragg Regime. The aforementioned systems are radially symmetric and utilize substrates which allow derivation of the output beam normal to the rotating axis.

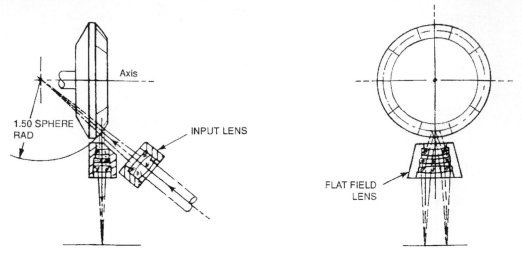

FIGURE 37 Reflective holofacet scanner, underilluminated. Flat-field microimage scanner (100 1p/mm over 11-mm format).

While operation with radial asymmetry was anticipated in 1967,[38] it was not until operation in the Bragg regime was introduced[6,39] that major progress developed in disk configurations. Referring to Fig. 38, the input and output beams I and O appear as principal rays directed to and diffracted from the holographic sector HS, forming angles θ_i and θ_o with respect to the grating surface normal.

For the tilt-error reduction in the vicinity of Bragg operation, the differential in output angle $d\theta_o$ for a differential in hologram tilt angle $d\alpha$ during tilt error $\Delta\alpha$ is given by

$$d\theta_o = \left[1 - \frac{\cos\,(\theta_i + \Delta\alpha)}{\cos\,(\theta_o - \Delta\alpha)}\right] d\alpha \qquad (38)$$

whence, when $\theta_i = \theta_o$, a small $\Delta\alpha$ is effectively nulled. While the θ_i and θ_o depart from perfect Bragg symmetry during hologram rotation and scan, the reduction in error

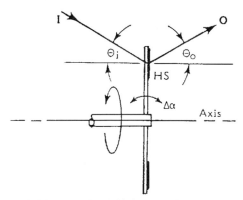

FIGURE 38 Holographic scanner in Bragg regime; $\Theta_i = \Theta_o$, in which output angle θ_o is stabilized against tilt error $\Delta\alpha$ of holographic segment *HS*.

remains significant. An analogy of this important property is developed for the tilting of a refractive wedge operating at minimum deviation.[6] When $\theta_i = \theta_o \simeq 45°$, another property develops in the unbowing of the output scanned beam: the locus of the output beam resides (almost) in a plane normal to that of the paper over a limited but useful range.[6,36] Further, the incremental angular scan for incremental disk rotation becomes almost uniform: their ratio m at small scan angles is shown to be equal to the ratio λ/d of the grating equation (see the section on "Radial Symmetry and Scan Magnification" and Eq. 3).[6] At $\theta_i = \theta_o = 45°$, $m = \lambda/d = \sqrt{2}$. This results in the output-scan angle to be $\sqrt{2}$ larger (in its plane) than the disk-rotation angle. While such operation provides the above attributes, it imposes two practical restrictions.

1. For high diffraction efficiency from relief gratings (e.g., photoresist), the depth-to-spacing ratio of the gratings must be extremely high, while the spacing $d = \lambda/\sqrt{2}$ must be extremely narrow. This is difficult to achieve and maintain, and difficult to replicate gratings which provide efficient diffraction.

2. Such gratings exhibit a high polarization selectivity, imposing a significant variation in diffraction efficiency with grating rotation (see "Image Rotation and Derotation").

Accommodation of these limitations is provided by reducing the Bragg angle and introducing a bow correction element to straighten the scan line. This is represented in Fig. 39; a high-performance scanner intended for application to the graphic arts. The Bragg angle is reduced to 30°. This reduces the magnification to $m = 1 = \lambda/d$ (as in radially symmetric systems), increases d to equal λ for more realizable deep-groove gratings, and reduces significantly the angular polarization sensitivity of the grating.

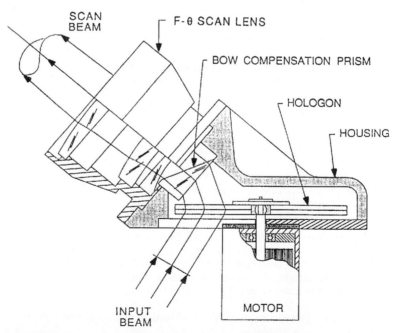

FIGURE 39 Plane linear grating (hologon) holographic disk scanner. 30° Bragg angle provides fabricatable and polarization-insensitive grating structure, but requires bow compensation prism. Useful scan beam is in and out of plane of paper. (*After Holotek Ltd, Rochester, NY product data.*)

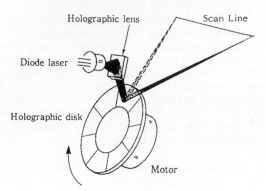

FIGURE 40 Holographic disk scanner with corrective holographic lens, both operating in approximately Bragg regime, providing complex error balancing.[41]

The elegance of the 45° Bragg configuration has been adapted[6] to achieve self-focusing in less demanding tasks (e.g., laser printing). This is exemplified in Fig. 40, which includes a holographic lens to balance the wavelength shift of the laser diode,[40,41,42] to shape the laser output for proper illumination of the scanner and to accommodate wavelength shift reconstruction. However, such multifunction systems are compounded by more critical centration requirements[6] and balancing of characteristics for achievement of a discrete set of objectives.

Galvanometer and Resonant Scanners

To avoid the scan nonuniformities which can arise from facet variations (see "Scan Error Reduction") of polygons or holographic deflectors, one might avoid multifacets. Reducing the number to one, the polygon becomes a monogon. This adapts well to the internal drum scanner (Fig. 27), which achieves a high duty cycle, executing a very large angular scan within a cylindrical image surface. Flat-field scanning, however, as projected through a flat-field lens, allows limited optical scan angle, resulting in a limited duty cycle from a rotating monogon. If the mirror is vibrated rather than rotated completely, the wasted scan interval may be reduced. Such components must, however, satisfy system speed, resolution, and linearity. Vibrational scanners include the familiar galvanometer and resonant devices[4,5,43,44] and the less commonly encountered piezoelectrically driven mirror transducer.[5,44]

The Galvanometer. Referring to Fig. 41*a*, a typical galvanometer driver is similar to a torque motor. Permanent magnets provide a fixed field which is augmented (±) by the variable field developed from an adjustable current through the stator coils. Seeking a new balanced field, the rotor* executes a limited angular excursion ($\pm\Phi/2$). With the mirror and principal ray per Fig. 28, the reflected beam scans through $\pm\theta/2$, twice that of the rotor.

The galvanometer is a broadband device, damped sufficiently to scan within a wide

---

* Rotor types include moving iron, moving magnet, or moving coil. Figure 41*a* illustrates the first two and Fig. 41*b* exemplifies the moving coil type. Moving magnet types (having NdFeB high-energy magnetic material) can exhibit some advantage in lower inertia and higher torque.

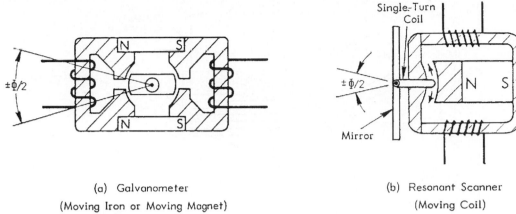

(a) Galvanometer

(Moving Iron or Moving Magnet)

(b) Resonant Scanner

(Moving Coil)

FIGURE 41 Examples of galvanometer and resonant scanner transducers. Fixed field of permanent magnet(s) is augmented by variable field from current through stator coils. (*a*) Galvanometer: torque rotates iron or magnetic core. Mirror surface (not shown) on extended shaft axis. (*b*) Resonant scanner: torque from field induced into single-turn armature coil (in plane perpendicular to paper) rotates mirror suspended between torsion bars. One stator coil may be nondriven and used for velocity pick-off.

range of frequencies, from zero to an upper value close to its mechanical resonance. Thus, it can provide the sawtooth waveform with a longer active linearized portion and shorter retrace time $= \tau$. This is represented in Fig. 42 (solid lines) showing rotation angle Φ vs. time. As a broadband device, it can also serve for random access, positioning to an arbitrary location within its access-time limitations. For this feature of waveform shaping, the galvanometer was categorized as a *low inertia scanner.*[5]

The Resonant Scanner. When damping is removed almost completely, large vibrations can be sustained only very near the resonant frequency of the oscillating system. The

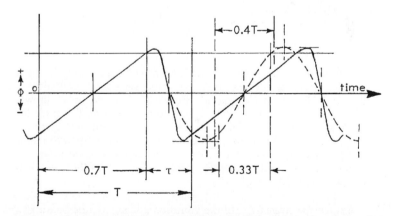

FIGURE 42 Waveforms (Φ vs. time) of vibrational scanners having same period and zero crossings. *Solid line:* galvanometer with linearized scan, providing 70 percent duty cycle. *Dashed line:* Resonant scanner providing 33.3 percent duty cycle (unidirectional) with 2 : 1 slope change, or 40 percent duty cycle with 3.24 : 1 slope change. Ratio of max. slopes: resonant/galvanometer $= 2.6/1$.[4]

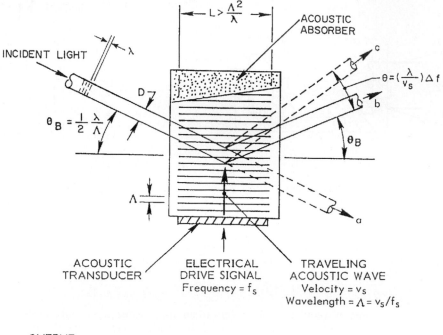

OUTPUT BEAM POSITION	BEAM CONDITION	ACOUSTIC FREQUENCY
a	Undiffracted Zero Order	f_s = Zero (or Amplitude = 0)
b	Nominal Bragg Angle Θ_B	$f_s = f_o$ (Center Frequency)
c	Added Diffraction, Scanned Through Angle Θ	$f_s = f_o + \Delta f$

FIGURE 43 Bragg acousto-optic deflector (angles exaggerated for illustration). Electrical drive signal generates traveling acoustic wave in medium which simulates a thick optical grating. Relationship between the output beam position and the electrical drive frequency is tabulated.[4]

resonant scanner is thus characterized by larger angular excursions at a fixed and usually higher frequency, executing near-perfect sinusoidal oscillations. A typical driver configuration is illustrated in Fig. 41b. Figure 42 (dashed lines) shows a sinusoid with the same zero-crossings as those of the sawtooth waveform. Contrary to its popular designation as "low-inertia," the resonant scanner provides rigid time increments, as though it exhibits a high inertia. While the rotary inertia of the suspension system is low to allow high repetition rates, it permits no random access and no scan waveform shaping, as do the galvanometer, acousto-optic, electro-optic, and other wideband scanners designated as low-inertia devices.[5]

Suspension Systems. In the vibrational scanners, the bearings and suspension systems are the principal determinants of scan uniformity. The galvanometer shaft must be sufficiently stiff and long to inhibit cross-scan wobble. However, to maximize the oscillating frequency, the armature is restricted in size and mass. Fortunately, its reciprocating motion tends to retrace its path (and its perturbations) faithfully over many cycles, making adjacent scans more uniform than if the same shaft rotated completely within the same bearings, as in a motor.

Some bearings are flexure, torsion, or taut-band devices which insert almost no along-scan perturbations.[43] Because of their low damping, these suspensions are most often applied to the resonant scanner. When damped, they can serve for the galvanometer, suffering a small sacrifice in bandwidth and maximum excursion, but gaining more uniform scan with very low noise and almost unlimited life. Some considerations are their low radial stiffness and possible coupling perturbation from torsion, shift of the axis of rotation with scan angle, and possible appearance of spurious modes when lightly damped. Most of these factors can be well-controlled in commercial instrument designs.

Adaptations and Comparisons.   Because the resonant scanner oscillates sinusoidally, and we seek typically a linearized scan, some significant adaptations are often required. As illustrated in Fig. 42 (dashed lines), we must select from the sine function a central portion which is sufficiently linear to be linearized further by timing the pixels or extracting them out of memory at a corresponding rate.[45] To limit the variation in pixel rate to 2:1 (i.e., velocity at zero crossover will be twice that at the same limit), then the useful excursion must be restricted to 60°/90° or 66.7 percent of its peak angle. When scanning with only one slope of the sinusoid (as for generation of a uniformly spaced raster), this represents a duty cycle of only 33.3 percent. To raise the duty cycle, one must accommodate a greater variation in data rate. If, for example, the useful scan is 80 percent of its full excursion (40 percent when using one slope), then the velocity variation rises to 3.24×. That is, the data rate or bandwidth at crossover is 3.24 times that at the scan limit. Also, its bandwidth at crossover is approximately $2\frac{1}{2}$ times that of the galvanometer, as represented by their relative slopes in Fig. 42

There is a corresponding variation in the dwell-time of the pixels, resulting in predictable but significant variation in image exposure or detectivity: 2:1 for 33.3 percent duty cycle and $3\frac{1}{4}:1$ for 40 percent duty cycle. This may require compensation over the full scan interval, using position sensing and control.[44,45,46,47] In contrast, the broadband galvanometer with feedback can provide a highly linearized scan[43] at a duty cycle of approximately 70 percent.[5]

Acousto-optic Scanners

Acousto-optic diffraction serves effectively for high-speed low-inertia optical deflection. It can provide random beam positioning within extremely short access times, or generate repetitive linear scans at very high rates, or divide a single beam into multiple beams for multiplexing applications. The tradeoff is, however, relatively low resolution, seldom providing more than $N = 1000$ elements per scan.

The principles of acousto-optics were formulated in 1932[48] and its attributes were applied only five years later to the Scophony TV projection system.[49] Its potential for laser scanning was explored in the mid-60s.[50,51] While various acousto-optic interactions exist, laser scanning is dominated by operation in the Bragg regime.[5,52]

Fundamental Characteristics.   Diffraction from a structure having a periodic spacing Λ is expressed as $\sin \theta_i + \sin \theta_o = n\lambda/\Lambda$, in which θ_i and θ_o are the input and output beam angles respectively, n is the diffractive order, and λ is the wavelength. Bragg operation requires that $\theta_i = \theta_o = \theta_B$. In a "thick" diffractor, length $L \gtrsim \Lambda^2/\lambda$, wherein all the orders are transferred efficiently to the first, and the Bragg angle reduces to

$$\theta_B = \frac{1}{2}\frac{\lambda}{\Lambda} \tag{39}$$

Per Fig. 43, the grating spacing Λ is synthesized by the wavefront spacing formed by an acoustic wave traveling through an elastic medium. An acoustic transducer at one end converts an electrical drive signal to a corresponding pressure wave which traverses the medium at the velocity v_s, whereupon it is absorbed at the far end to suppress standing waves. The varying pressure wave in the medium forms a corresponding variation in its refractive index. An incident light beam of width D is introduced at the Bragg angle (angle shown exaggerated). An electrical drive signal at the center frequency f_o develops a variable index grating of spacing Λ which diffracts the output beam at θ_B into position b. The drive signal magnitude is adjusted to maximize beam intensity at position b, minimizing intensity of the zero order beam at position a. When f_o is increased to $f_s = f_o + \Delta f$, the grating spacing is decreased, diffracting the output beam through a larger angle, to position c. The small scan angle θ is effectively proportional to the change in frequency Δf.

The scan angle is $\theta = \lambda/\Delta\Lambda = (\lambda/v_s)\Delta f$. The beam width, traversed by the acoustic wave over the transit time τ is $D = v_s\tau$. Substituting into Eq. (15) and accounting for duty cycle per Eq. (22), the resolution of the acousto-optic scanner (total N elements for total Δf) is

$$N = \frac{\tau\Delta f}{a}(1 - \tau/T) \qquad (40)$$

The $\tau\,\Delta f$ component represents the familiar time-bandwidth product, a measure of information-handling capacity.

Deflection Techniques. Because the clear aperture width W of the device is fixed, anamorphic optics is often used to illuminate W with an adjusted beam width D—encountering selective truncation by the parallel boundaries of W. The beam height (in quadrature to D) can be arbitrarily narrow to avoid apodization by the aperture. This one-dimensional truncation of the gaussian beam requires assignment of an appropriate aperture shape factor a, summarized in Table 5.

Additional topics in acousto-optic deflection are cylindrical lensing due to linearly swept f_s,[52] correction for decollimation in random access operation,[5] Scophony operation,[53] traveling lens or chirp operation,[54] correction for color dispersion,[55] polarization effects,[56] and multibeam operation.[57]

Electro-optic (Gradient) Scanners

The gradient deflector is a generalized form of beam scanner[4,5,58] in which the propagating wavefronts undergo increasing retardation transverse to the beam, thereby changing the wavefront spacing (wavelength) transverse to the beam. To maintain wavefront continuity, the rays (orthogonal trajectories of the wavefronts) bend in the direction of the shorter wavelength. Referring to Fig. 44a, this bend angle θ through such a deflection cell may be expressed as

$$\theta = k_o(dn/dy)l \qquad (41)$$

where n is taken as the number of wavelengths per unit axial length l, y is the transverse distance, and k_o is a cell system constant. For the refractive material form in which the wavefront traverses a change Δn in index of refraction and the light rays traverse the change in index over the full beam aperture D in a cell of length L, then the relatively small deflection angle becomes[4,5]

$$\theta = (\Delta n/n_f)L/D \qquad (42)$$

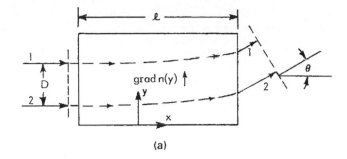

(a)

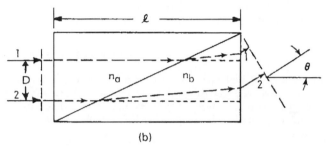

(b)

FIGURE 44 Equivalent gradient deflectors: (*a*) basic deflector cell composed of material having grad $n(y)$. Ray 1, propagating through a higher refractive index, is retarded more than Ray 2, tipping the wavefront through angle Θ (including boundary effect). (*b* Analogous prismatic cell, in which $n_a > n_b$, such that Ray 1 is retarded more than Ray 2, tipping the wavefront through angle Θ.[4]

where n_f is the refractive index of the final medium (applying Snell's law and assuming $\sin \theta = \theta$). Following Eq. (15), the corresponding resolution in elements per scan angle is expressed as

$$N = (\Delta n / n_f) L / a\lambda \tag{43}$$

The Δn is given by

$$\text{(for class I materials)} \quad \Delta n = n_o^3 r_{ij} E_z \tag{44a}$$

$$\text{(for class II materials)} \quad \Delta n = n_e^3 r_{ij} E_z \tag{44b}$$

where $n_{o,e}$ is the (ordinary, extraordinary) index of refraction, r_{ij} is the electro-optic coefficient, and $E_z = V/Z$ is the electric field in the z direction (see Fig. 45).

Methods of Implementation. An electroacoustic method of developing a time-dependent index gradient was proposed in 1963[59] utilizing the (harmonic) pressure variations in an acoustically driven cell (of a transparent elastic material). Although this appears similar to acousto-optic deflection (see "Acousto-optic Scanners"), it differs fundamentally in that the cell is terminated reflectively rather than absorptively (to support a standing wave). Also, the acoustic wavelength is much longer than the beam width, rather than much shorter for Bragg operation. A method of approaching a linearly varying index gradient utilizes a quadrupolar array of electrodes bounding an electro-optic material;[60,61] and is available commercially.[62]

A continuous index gradient can be simulated by the use of alternating electro-optic prisms.[58,63] A single stage biprism is illustrated in Fig. 44*b* and an iterated array for practical implementation appears in Fig. 45. Each interface imparts a cumulative differential in retardation across the beam. The direction and speed of retardation is controlled by the index changes in the electro-optic material. While resolution is limited primarily by available materials, significant experiment and test is reported for this form of deflector.[5]

Drive Power Considerations. The electrical power dissipated within the electro-optic material is given by[5]

$$P = \tfrac{1}{4}\pi V^2 C f / Q \tag{45}$$

where V is the applied (p-p sinusoidal) voltage in volts, C is the deflector capacitance in farads, f is the drive frequency in Hz, and Q is the material Q-factor [$Q = 1/$loss tangent $(\tan \delta) \simeq 1/$power factor, $(Q > 5)$].

The capacitance C for transverse electroded deflectors is approximately that for a parallel-plate capacitor of (rectangular) length L, width Y, and dielectric thickness Z (per Fig. 45)

$$C = 0.09\kappa L Y / Z \quad \text{pF} \tag{46}$$

where κ is the dielectric constant of the material (L,Y,Z in cm).

The loss characteristics of materials which determine their operating Q are often a strong function of frequency beyond 10^5 Hz. The dissipation characteristics of some electro-optic materials are provided,[5,64,65] and a resolution-speed-power figure of merit has been proposed.[66]

Unique Characteristics. Most electro-optic coefficients are extremely low, requiring high drive voltages to achieve even moderate resolutions (to $N \simeq 100$). However, these devices can scan to very high speeds (to 10^5/s) and suffer effectively no time delay (as do acousto-optic devices), allowing use of broadband feedback for position control.

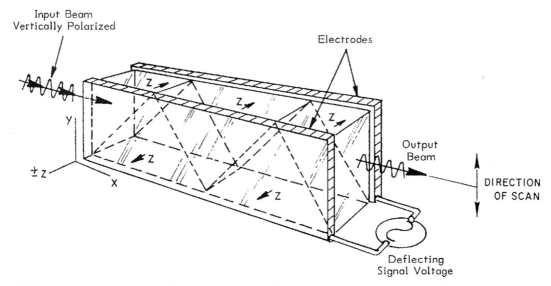

FIGURE 45 Iterated electro-optic prism deflector. Indicating alternating crystallographic (z) axes. Input beam polarization for class 1 (r_{63}) materials.[4]

19.7 SCAN-ERROR REDUCTION

High-resolution scanners often depend on precise rotation of a shaft about its axis, said shaft supporting a multiplicity of deflecting elements (facets, mirrors, holograms). The control of angular uniformity of these multielements with respect to the axis, and of the axis with respect to its frame, can create an imposing demand on fabrication procedures and consequential cost. Since uniformity of beam position in the cross-scan direction may not be approached by phasing and timing of the data (as can be the along-scan errors), several noteworthy techniques have been developed to alleviate this burden.

Available Methods

The general field of cross-scan error reduction is represented in Table 6. *Fabrication accuracy* may be selected as the only discipline, or it may be augmented by any of the auxiliary methods. The *active* ones utilize high-speed low-inertia (A-O or E-O) or piezoelectric deflectors[5,67] or lower-speed (galvanometer) deflectors which are programmed to rectify the beam-position errors. While open-loop programming is straightforward (while accounting for angular magnification/demagnification as a function of the accessed beam size), elegant closed-loop methods may be required to rectify pseudorandom perturbations. This must, however, be cost-effective when compared to the alternatives of increased fabrication accuracy and of the *passive* techniques.

Passive Methods

Passive techniques require no programming. They incorporate optical principles in novel configurations to reduce beam misplacement due to angular error in reflection or diffraction. Bragg-angle error reduction of tilted holographic deflectors is discussed in the section, "Operation in the Bragg Regime."

Anamorphic Error Control.   Anamorphic control, the most prominent treatment, may be applied to any deflector. The basics and operational characteristics[6] are summarized here.

Separating the nonaugmented portion of the resolution equation [Eq. (19)] into quadrature components and denoting the cross-scan direction as y, then the error, expressed in the number of resolvable elements, is

$$N_y = \frac{\theta y D y}{a\lambda} \tag{47}$$

TABLE 6   Techniques for Cross-Scan Error Reduction

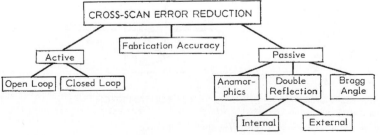

in which $a\lambda$ is assumed constant, θ_y is the angular error of the output beam direction, and D_y is the height of the beam illuminating the deflector. The objective is to make $N_y \rightarrow 0$. Mechanical accuracies determine θ_y, while anamorphics are introduced to reduce D_y; usually accomplished with a cylindrical lens focusing the illuminating beam in the y direction upon the deflector. [The quadrature (along-scan) resolution is retained by the unmodified D_x and scan angle θ_x.] As D_y is reduced, the y displacement error is reduced. Following deflection, the y direction scanned spot distribution is restored by additional anamorphics—restoring the nominal converging beam angle (via F_y, the F-number forming the scanning spot in the y direction).

The error reduction ratio is

$$R = D'_y/D_y \tag{48}$$

where D'_y is the compressed beam height and D_y is the original beam height on the deflector.

A variety of anamorphic configurations has been instituted, with principal variations in the output region, in consort with the objective lens, to reestablish the nominal F_y while maintaining focused spot quality and uniformity.

Double-Reflection Error Control. In double-reflection (Table 6), the deflector which creates a cross-scan error is reilluminated by the scanned beam in such phase as to tend to null the error. This can be conducted in two forms: internal and external.

An internal double-reflection scanner is exemplified by the pentaprism monogon[68] in Fig. 46a; a (glass) substrate having two of its five surfaces mirrored. This is an optically stabilized alternate to the 45° monogon of Fig. 27, operating preobjective in collimated light. Tipping the pentaprism cross-scan (in the plane of the paper) leaves the 90° output beam unaffected. A minor translation of the beam is nulled when focused by the objective lens. The pentamirror[68] per Fig. 46b, requires, however, significant balancing and support of the mirrors, since any shift in the nominal 45° included angle causes twice the error in the output beam. A stable double-reflector is the open mirror monogon[69] of Fig. 46c. Its nominal 135° angle serves identically to maintain the output beam angle at 90° from the axis, independently of cross-scan wobble. With a rigid included angle and simple balancing, it can provide high speed operation.

Two variations which double the duty cycle, as would a two-faceted pyramidal polygon or ax-blade scanner (see "Early Single-Mirror Scanners") appear in Fig. 47. Figure 47a is effectively two pentamirrors forming a *butterfly scanner*[70] and Fig. 47b is effectively a pair of open mirrors.[71] The absolute angles of each half-section must maintain equality to

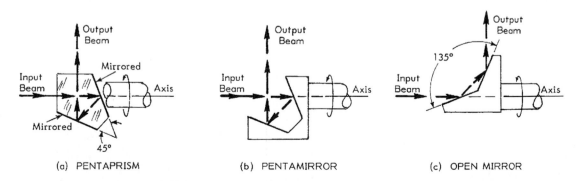

FIGURE 46 Monogon scanners employing double-reflection.[4]

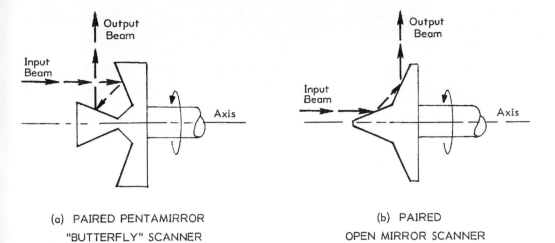

(a) PAIRED PENTAMIRROR
"BUTTERFLY" SCANNER

(b) PAIRED
OPEN MIRROR SCANNER

FIGURE 47   Paired scanners employing double-reflection.[4]

within half of the allowed error in the output beam. Also, the center section of (a) must be angularly stable to within one-quarter of the allowed error, because an increased included angle on one side forms a corresponding decrease on the other. Other dynamic considerations involve inertial deformation, and the beam displacements and mirror widths (not shown) to accommodate the distance of the input beam from the axis during rotation.

The need for near-perfect symmetry of the multiple double-reflectors can be avoided by transferring the accuracy requirement to an *external* element that redirects recurrent beam scans. One such form[72] is illustrated in Fig. 48. A prismatic polygon illuminated with a collimated beam of required width (only principal rays shown) deflects the beam first to a roof mirror, which returns the beam to the same facet for a second deflection toward the

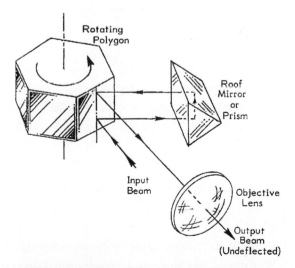

FIGURE 48   Method of external double-reflection—shown in undeflected position. Components and distances not to scale. Only principal rays shown.[4]

flat-field lens. The roof mirror phases the returned beam such as to null the cross-scan error upon the second reflection. Several characteristics are noteworthy:

1. The along-scan angle is doubled. That is, scan magnification $m = 4$ rather than 2.

2. This normally requires increasing the number of facets to provide the same angle with the same duty cycle.

3. However, during polygon rotation, the point of second reflection shifts significantly along the facet and sacrifices duty cycle.

4. The pupil distance from the flat-field lens is effectively extended by the extra reflections, requiring a larger lens to avoid vignetting.

5. The roof mirror and flat-field lens must be sized and positioned to minimize obstruction of the input and scanned beams. Allow for finite beam widths (see "Scanner-Lens Relationships").

19.8   REFERENCES

1. S. Sherr, *Fundamentals of Display System Design,* John Wiley and Sons, New York, 1970.

2. S. Sherr, *Electronic Displays,* John Wiley and Sons, New York, 1979.

3. L. Beiser, "Unified Approach to Photographic Recording from the Cathode-Ray Tube," *Photo. Sci. & Engr.* **7**(3):196–204 (1963).

4. L. Beiser, *Laser Scanning Notebook,* SPIE Press, Washington, vol. PM13, 1992.

5. L. Beiser, "Laser Scanning Systems," in *Laser Applications,* vol. 2. Academic Press, New York, 1974, pp. 53–159.

6. L. Beiser, *Holographic Scanning,* John Wiley and Sons, New York, 1988.

7. J. D. Zook, "Light Beam Deflector Performance: a Comparative Analysis," *Appl. Opt.* **13**(4):875–887 (1974).

8. L. Beiser, "A Guide and Nomograph for Laser Scanned Resolution," *Proc. SPIE* **1079:** 2–5 (1989), L. Beiser (ed.).

9. G. Boreman, "Modulation Transfer Techniques," *This Handbook,* vol. II, Chap. 32.

10. L. R. Dickson, "Characteristics of Propagating Gaussian Beams," *Appl. Opt.* **9**(8):1854–61 (1970).

11. J. Randolph and J. Morrison, "Rayleigh-Equivalent Resolution of Acoustooptic Deflection Cells." *Appl. Opt.* **10**(6):1383–1385 (1971).

12. L. R. Dickson, "Optical Considerations for Acoustooptic Deflectors," *Appl. Opt.* **11**(10):2196–2202 (1972).

13. L. Levy, *Applied Optics,* vol. 1, John Wiley and Sons, New York, 1968.

14. L. Beiser, "Generalized Equations for the Resolution of Laser Scanners," *Appl. Opt.* **22**(20):3149–51 (1983).

15. J. Urbach, et al., "Laser Scanning for Electronic Printing," *Proc. IEEE* **70**(6): 597–618 (1982).

16. H. V. Kennedy, Split Image, High Scanning Rate Optical System with Constant Aperture, U.S. Patent 3,211,046, 1965.

17. H. B. Henderson, Scanner System, U.S. Patent 3,632,870, 1972.

18. W. L. Wolfe, "Optical-Mechancial Scanning Techniques and Devices," in W. L. Wolfe and G. Zissis (eds.), *The Infrared Handbook,* chapt. 10, ERIM, Ann Arbor, MI, 1989.

19. R. B. Barnes, Infrared Thermogram Camera and Scanning Means Theoref, U.S. Patent 3,287,559, 1966.

20. B. A. Wheeler, Optical Raster Scan Generator, U.S. Patent 3,764,192, 1973.

21. R. B. Johnson, Target-Scanning Camera Comprising a Constant Temperature Source for Providing a Calibrated Signal, U.S. Patent 3,631,248, 1971.

22. P. J. Lindberg, "A Prism Line-Scanner for High Speed Thermography," *Optica Acta* **15:**305–316 (1966).

23. P. J. Lindberg, Scanning Mechanism for Electromagnetic Radiation, U.S. Patent 3,253,498, 1966.

24. J. M. Lloyd, *Thermal Imaging Systems,* Plenum Press, New York, 1975, pp. 308–309.

25. R. B. Johnson, Image Rotation Device for an Infrared Scanning System or the Like, U.S. Patent 3,813,552, 1974.

26. P. J. Berry and H. M. Runciman, Radiation Scanning System, U.S. Patent 4,210,810, 1980.

27. T. H. Jamieson, "Optical Design of Compact Thermal Scanner," *Proc. SPIE* **518:**15–21 (1984).

28. J. E. Modisette and R. B. Johnson, High Speed Infrared Imaging System, U.S. Patent 4,419,692, 1983.

29. D. C. O'Shea, *Elements of Modern Optical Design,* John Wiley and Sons, New York, 1985.

30. K. S. Petche, *Handbook of Optics,* OSA, W. G. Driscoll and W. Vaughn (eds.), McGraw Hill, 1978, pp. 5–10.

31. Y. Li and J. Katz, "Encircled Energy of Laser Diode Beams," *Appl. Opt.* **30**(30):4283 (1991).

32. R. E. Hopkins and D. Stephenson, "Optical Systems for Laser Scanners," in G. F. Marshall (ed.), *Optical Scanning,* Marcel Dekker, New York, 1991.

33. A. B. Marchant, U.S. Patent 4,759,616, 1988; and A. W. Lohman and W. Stork, "Modified Brewster Telescope," *Appl. Opt.* **28**(7):1318–19 (1989).

34. L. Beiser, "Design Equations for a Polygon Laser Scanner," *Proc. SPIE* **1454:**60–66 (1991), G. F. Marshall and L. Beiser (eds.)

35. D. Kessler, et al., "High Resolution Laser Writer," *Proc. SPIE* **1079:**27–35 (1989), L. Beiser (ed.).

36. C. J. Kramer, "Holographic Deflectors for Graphic Arts Systems," in G. F. Marshall (ed.), *Optical Scanning,* Marcel Dekker, New York, 1991.

37. R. V. Pole and H. P. Wollenmann, "Holographic Laser Beam Deflector," *Appl. Opt.* **14**(4):976–80 (1975).

38. I. Cindrich, "Image Scanning by Rotation of a Hologram," *Appl. Opt.* **6**(9):1531–34 (1967).

39. C. J. Kramer, Holo-Scanner for Reconstruction a Scanning Light Spot Insensitive to Mechanical Wobble, U.S. Patent 4,239,326, 1980.

40. D. B. Kay, Optical Scanning System with Wavelength Shift Correction, U.S. Patent 4,428,643, 1984.

41. H. Ikeda, et al., "Hologram Scanner," *Fujitsu Sci. Tech J.* **23:**3 (1987).

42. F. Yamagishi, et al., "Lensless Holographic Line Scanner," *Proc. SPIE* **615:**128–132 (1986).

43. J. Montagu, "Galvanometric and Resonant Low Inertia Scanners," in G. F. Marshall (ed.), *Optical Scanning,* Marcel Dekker, New York, 1991.

44. S. Reich, "Use of Electro-Mechanical Mirror Scanning Devices," *SPIE Milestone Series* **378:**229–238 (1985), L. Beiser (ed.).

45. D. G. Tweed, "Resonant Scanner Linearization Techniques," *Opt. Engr.* **24**(6):1018–22 (1985).

46. F. Blais, "Control of Galvanometers for High Precision Laser Scanning Systems," *Opt. Engr.* **27**(2):104–110 (1988).

47. G. F. Marshall and J. S. Gadhok, "Resonant and Galvanometer Scanners: Integral Position Sensing," *Photonics Spectra,* 155–160 (1991).

48. P. Debye and F. W. Sears, "On the Scattering of Light by Supersonic Waves," *Proc. Nat'l Acad. Sci.* (USA) **18:**409 (1932).

49. F. Okolicsanyi, "The Waveslot, an Optical TV System," *Wireless Engr.* **14:**536–572 (1937).

50. R. Adler, "Interaction between Light and Sound," *IEEE Spectrum* **4**(5):42–54 (1967).

51. E. I. Gordon, "Review of Acoustooptical Deflection and Modulation Devices," *Proc. IEEE* **54**(10):1391–1401 (1966).

52. A. Korpel, "Acousto-Optics," in R. Kingslake and B. J. Thompson (eds.), *Appl. Opt. and Opt. Engr.,* VI, Academic Press, New York, 1980, pp. 89–141.

53. R. V. Johnson, "Scophony Light Valve," *Appl. Opt.* **18**(23):4030–4038 (1979).

54. L. Bedamian, "Acousto-Optic Laser Recording," *Opt. Engr.* **20**(1):143–149 (1981).

55. W. H. Watson and A. Korpel, "Equalization of Acoustooptic Deflection Cells in a Laser Color TV System," *Appl. Opt.* **7**(5):1176–1179 (1970).

56. M. Gottlieb, "Acousto-Optical Scanners and Modulators," in G. Marshall (ed.), *Optical Scanning,* Marcel Dekker, New York, 1991.

57. D. L. Hecht, "Multibeam Acoustooptic and Electrooptic Modulators," *Proc. SPIE* **396:**2–9 (1983), L. Beiser (ed.).

58. L. Beiser, "Generalized Gradient Deflector and Consequences of Scan of Convergent Light," *J. Opt. Soc. Amer.* **57:**923–931 (1967).

59. A. J. Giarola and T. R. Billeter, "Electroacoustic Deflection of a Coherent Light Beam, *Proc. IEEE* **51:**1150 (1963).

60. V. J. Fowler, et al., "Electro-Optical Light Beam Deflection", *Proc. IEEE* **52:**193 (1964).

61. J. F. Lospeich, "Electro-Optic Light Beam Deflections," *IEEE Spectrum,* **5**(2):45–52 (1968).

62. R. J. Pizzo and R. J. Kocka, Conoptics Inc., Danbury, CT 06810.

63. T. C. Lee and J. D. Zook, "Light Beam Deflection with Electrooptic Prisms," *IEEE J. Quant. Electr.* **QE-4**(7):442–454 (1968).

64. C. S. Tsai and J. M. Kraushaar, "Electro-Optical Multiplexers and Demultiplexers for Time-Multiplexed PCM Laser Communication Systems," *Proc. Electro-Opt. Syst. Des. Conf.,* 176–182 (1972).

65. A. S. Vasilevskaya, *Soviet Physics Crystallography,* **12**(2):308 (1967).

66. J. D. Zook and T. C. Lee, "Design of Analog Electrooptic Deflectors," *Proc. SPIE* (of 1969), 281 (1970).

67. J. P. Donohue, "Laser Pattern Generator for Printed Circuit Board Artwork Generation," *Proc. SPIE* **200:**179–186 (1979). Also in *SPIE Milestone Series* vol. 378, 421–28 (1985), L. Beiser (ed.).

68. G. K. Starkweather, Single Facet Wobble Free Scanner, U.S. Patent 4,475,787, 1984.

69. L. Beiser, Light Scanner, U.S. Patent 4,936,643, 1990.

70. G. F. Marshall, et al., "Butterfly Line Scanner," *Proc. SPIE* **1454** (1991), G. F. Marshall and L. Beiser (eds.).

71. L. Beiser, Double Reflection Light Scanner, U.S. Patent 5,114,217, 1992.

72. D. F. Hanson and R. J. Sherman, U.S. Patent 4,433,894, Feb. 28, 1984: and R. J. Garwin, U.S. Patent 4,429,948, Feb. 7, 1984.

19.9 FURTHER READING

The following listing augments the text and its references with a select group of publications, many arcane and of historic value, representing substantive work in the field.

Barnes, C. W., "Direct Optical Data Readout and Transmission at High Data Rates," AFAL-TR-65-45, Wright Pat. A.F.B., AD-460486, 1965.

Becker, C. H., "Coherent Light Recording Techniques," RADC-TR-65-130, AFSC, Griffiss AFB, New York, 1965.

Beiser, L., "Laser Beam Scanning for High Density Data Extraction and Storage," *Proc. SPSE Symp. on Photography in Information Storage and Retrieval,* Washington, D.C., 1965.

Beiser, L., "Laser Beam and Electron Beam Extremely Wideband Information Storage and Retrieval," *Photo. Sci. & Engr.,* **10**(4).222 (1966).

Beiser, L., "Laser Scanning Systems," *Laser Applications,* vol. 2, M. Ross (ed.), Academic Press, New York, 1974.

Beiser, L. (ed.), *Selected Papers on Laser Scanning and Recording,* SPIE Milestone Series, Vol. 378, 1985.

Beiser, L., *Holographic Scanning,* John Wiley and Sons, New York, 1988.

Bousky, S., "Scanning Techniques for Light Modulation Recording," Wright Pat. AFB, Ohio, Contr. No. AF33(615)–2632, 1965.

Cindrich, I., et al., "Wide Swath Wide Bandwidth Recording," AFAL-TR-11-361, Wright Pat. AFB, Ohio, DDC No. 13217, AD 513171, 1971.

Dubovic, A. S., *Bases of Mirror Scanning Theory* (trans from Russian), AD-703569, Wright Pat. AFB, F.T.D. HT-23-654-69, 1969.

Dubovic, A. S., *The Photographic Recording of High Speed Processes,* John Wiley and Sons, New York, 1981.

Fowler, V. J., "Survey of Laser Beam Deflection Techniques," *Appl. Opt.* **5**(10):1675 (1966).

Johnson, R. B. and W. L. Wolfe (eds), *Selected Papers on Infrared Design,* SPIE Milestone Series, vol. 513, 1985.

Katys, G. P., *Automatic Scanning* (trans. from Russian, 1969), Nat'l Tech Info. Service, JPRS 51512, 1970.

Korpel, A., et al., "A TV Display Having Acoustic Deflection and Modulation of Coherent Light," *Appl. Opt.* **5**(10):667 (1966).

Klein, D. A., "Laser Imaging Techniques," Naval Air Syst. Comm. TR-1615 (AIR-52022), Naval Avionics Facility, Indiana 46218, 1970.

Laemmel, A. E., "The Scanning Process in Picture Transmission," Polytech. Inst. of B'klyn, RADC-TN-57-215, DDC AD-131112, 1957.

Marshall, G. F. (ed.), *Optical Scanning,* Marcel Dekker, New York, 1991.

Newgard, P. M. and Brain, E. E., "Optical Transmission at High Data Rates," AFAL-TR-67-217, Wright. Pat. AFB, Ohio, 1967.

Townsend, C. V. R., "Study of Advanced Scanning Methods," ASD-TDR-63-595, DDC 412794, Wright Pat. AFB, Ohio, 1963.

Wolfe, W. L. and G. J. Zissis (eds.), *The Infrared Handbook,* ERIM, Ann Arbor, MI, 1989.

Zook, J. D., "Light Beam Deflector Performance," *Appl. Opt.* **13:**875 (1974).

CHAPTER 20
OPTICAL SPECTROMETERS

Brian Henderson
Department of Physics and Applied Physics
University of Strathclyde
Glasgow, United Kingdom

20.1 GLOSSARY

A_{ba}	Einstein coefficient for spontaneous emission
a_o	Bohr radius
B_{if}	the Einstein coefficient for transition between initial state $|i\rangle$ and final state $|f\rangle$
e	charge on the electron
ED	electric dipole term
E_{DC}	Dirac Coulomb term
E_{hf}	hyperfine energy
E_n	eigenvalues of quantum state n
EQ	electric quadrupole term
$\mathbf{E}(t)$	electric field at time t
$\mathbf{E}(\omega)$	electric field at frequency ω
g_a	degeneracy of ground level
g_b	degeneracy of excited level
g_N	gyromagnetic ratio of nucleus
h	Planck's constant
H_{so}	spin-orbit interaction Hamiltonian
I	nuclear spin
$I(t)$	the emission intensity at time t
$\mathbf{j}$	total angular momentum vector given by $\mathbf{j} = 1 \pm \frac{1}{2}$
l_i	orbital state
m	mass of the electron
MD	magnetic dipole term
M_N	atom of nuclear mass

$n_\omega(T)$	equilibrium number of photons in a blackbody cavity radiator at angular frequency ω and temperature T
QED	quantum electrodynamics
$R_{nl}^{(r)}$	radial wave function
R_∞	Rydberg constant for an infinitely heavy nucleus
s	spin quantum number with the value $\frac{1}{2}$
s_i	electronic spin
T	absolute temperature
W_{ab}	transition rate in absorption transition between states a and b
W_{ba}	transition rate in emission transition from state b to state a
Z	charge on the nucleus
$\alpha = e^2/4\pi\varepsilon_0\hbar c$	fine structure constant
$\Delta\omega$	natural linewidth of the transition
$\Delta\omega_D$	Doppler width of transition
ε_o	permittivity of free space
$\zeta(r)$	spin-orbit parameter
μ_B	Bohr magneton
$\rho(\omega)$	energy-density at frequency ω
τ_R	radiative lifetime
ω	angular velocity
ω_k	mode k with angular frequency ω
$\langle f|\, V'\, |i \rangle$	matrix element of perturbation V

20.2   INTRODUCTION

This article outlines the physical basis of optical measurements in the wavelength/frequency and time domains. From the multiplicity of different apparatus, only simple examples are given of spectrometers designed for optical absorption, photoluminescence, and radiative decay measurements. Rather more detailed expositions are given of modern developments in laser spectrometers especially where high resolution is possible in both frequency and time domains. Included are specific developments for linewidth measurements down to tens of kilohertz using saturated absorption techniques as well as temporal decay characteristics in the sub-picosecond domain. A description is also given of a multiple resonance spectrometer including optically detected electron spin resonance and optically detected electron nuclear double resonance.

20.3   OPTICAL ABSORPTION SPECTROMETERS

General Principles

In optical absorption spectroscopy, electromagnetic radiation in the near-ultraviolet, visible, or near-infrared regions is used to excite transitions between the electronic states. Whereas atoms in low-pressure gas discharges exhibit very sharp lines, electronic centers in molecules and condensed matter display a variety of different bandshapes. In consequence, the absorbed intensity is a function of the photon wavelength (or energy). The most desirable experimental format plots the absorption coefficient α as a function of the radiation frequency v, because v is directly proportional to the energy separation between the states involved in the transition. Nevertheless, optical spectroscopists quote peak positions and linewidths in energy units E (eV, meV), in wave numbers $\bar{v}$ (in cm^{-1}), in frequency units (v or ω), or in wavelength λ [nanometers (nm) or micrometers (μm)]. The following approximate relationships exist: $1 cm^{-1} = 1.24 \times 10^{-4} eV$; $1 eV = 8066 cm^{-1}$; and $E(eV) = 1.24/\lambda$ (μm).

Very often the spectrometer output is given in terms of the specimen transmission, $T = I(v)/I_o(v)$ expressed as a percentage, or the optical density (or absorbance), $OD = \log_{10}(1/T)$, which are related to the absorption coefficient α by

$$OD = \log_{10}(1/T) = \alpha(v)l/2.303 \tag{1}$$

where l is the thickness of the sample. Typically one measures the absorption coefficient α over the wavelength range 185 to 3000 nm. Since α may be a function of both frequency v and polarization $\hat{\varepsilon}$, we may use the designation $\alpha(v, \hat{\varepsilon})$. For a center containing N noninteracting absorbing centers per unit volume, each absorbing radiation at frequency v and polarization $\hat{\varepsilon}$, the attenuation of a beam of intensity $I_o(v, \hat{\varepsilon})$ by a solid of thickness l is given by:

$$I(v, \hat{\varepsilon}) = I_o(v, \hat{\varepsilon}) \exp(-\alpha(v, \hat{\varepsilon})l). \tag{2}$$

Experimentally $I_o(v, \hat{\varepsilon})$ represents the transmission of the system in the absence of an absorbing specimen. In practice $I_o(v, \hat{\varepsilon})$ and $I(v, \hat{\varepsilon})$ are measured and the value of the absorption coefficient $\alpha(v, \hat{\varepsilon})$ at a particular frequency is obtained using the formula

$$\alpha(v, \hat{\varepsilon}) = \frac{1}{l} \ln \frac{I_o(v, \hat{\varepsilon})}{I(v, \hat{\varepsilon})} \tag{3}$$

$\alpha(v, \hat{\varepsilon})$ has units of cm^{-1} or m^{-1}. The variation of the absorption coefficient with frequency is difficult to predict. In general, the absorption transition has a finite width, and the absorption strength, $\int \alpha(v, \hat{\varepsilon}) \, dv$, is related to the density of absorbing centers and to the transition probability.

The value of the absorption coefficient in an isotropic material is related to the Einstein A coefficient for spontaneous emission by[1]

$$\alpha(v) = \left(N_a \frac{g_b}{g_a} - N_b\right) A_{ba} \frac{c^2}{8\pi v^2} \frac{1}{n^2} G(v) \tag{4}$$

where g_a and g_b are the statistical weights of the states, $G(v)$ is the lineshape function (defined such that $\int G(v) \, dv = 1$), v is the velocity of light in the medium, and n is the refractive index. In Eq. (4), the population densities in the ground and excited states, N_a and N_b respectively, have been assumed to be invariant with time and unaffected by the absorption process. Under conditions of weak excitation we can ignore the small value of N_b, and replace N_a by N so that

$$\alpha(v) = N A_{ba} \frac{c^2}{8\pi v^2} \frac{1}{n^2} \frac{g_b}{g_a} G(v) = N\sigma G(v) \tag{5}$$

where σ is the *absorption cross section per center*. The absorption strength, i.e. the area under the absorption band, is related to σ by

$$\int \alpha(v) \, dv = N\sigma \tag{6}$$

If we ignore the refractive index and local field correction factors and assume a gaussian-shaped absorption band, then

$$Nf_{ab} = 0.87 \times 10^{17} \, \alpha(v_o) \, \Delta v \tag{7}$$

where $\alpha(v_o)$ is measured in cm^{-1}, Δv, the full-width at half maximum absorption, is measured in eV, and N is the number of centers cm^{-3}. Equation (7) is often referred to as *Smakula's formula*. To obtain the oscillator strength from the area under the absorption band, one needs an independent determination of the density of centers. For impurity ions in solids, N may be determined by chemical assay or by electron spin resonance.

The Double-Beam Spectrophotometer

The first essential of an absorption spectrophotometer is a broadband source: deuterium, hydrogen, xenon, and tungsten lamps are commonly used. Their outputs cover different wavelength ranges: a hydrogen lamp is suitable for wavelengths in the range 150 to 350 nm whereas high-pressure xenon lamps have usable light outputs in the range 270 to 1100 nm. For Xe arc lamps the output is relatively continuous in the wavelength range 270 to 800 nm apart from some sharp lines near 450 nm. In the infrared region 800 to 1100 nm, much the most intense part of the output is in the form of sharp lines. In the arc lamp, radiation is due to the collision of Xe atoms with electrons which flow across the arc. Complete separation of the excited electrons from the atoms leads to ionization and the continuum output. The formation of Xe atoms in excited states leads to the sharp lines in the output from Xe arc lamps. Tungsten filament lamps may also be used in absorption spectrophotometers. The spectral output from such a heated filament is approximately that of a blackbody radiator at a temperature of 2000 K. In consequence, the emission intensity is a smooth function of wavelength with peak at ca. 1500 nm, with the detailed curve following Planck's thermal radiancy law. Accordingly, the peak in the distribution of light varies with filament temperature (and therefore current through the filament), being determined by $\lambda_{max} T = 2.898 \times 10^{-3} \, m \, °K$. This relationship expresses Wien's wavelength displacement law. Although containing all wavelengths from the ultraviolet into the infrared region, the total output is fairly modest compared with a high-pressure mercury lamp.

Accurate measurements of the absorption coefficient at different wavelengths are best made using a double-beam spectrophotometer: a schematic is shown in Fig. 1. The exciting beam from the broadband source passes through a grating monochromator: the resulting narrow band radiation is divided by a beam-splitting chopper into separate monochromatic beams which traverse the sample and a reference channel. Thus the light incident on the sample and that which passes through the reference channel have the same wavelengths and is square-wave modulated (on/off) at some frequency in the range 1 kHz to 5 kHz. The sample and reference beams are recombined at the phototube, and the magnitude and phase of sample and reference signals are amplified and compared by the lock-in detector. Chopping at a preselected frequency permits narrowband amplification of the detected signal. Thus any noise components in the signal are limited to a narrowband centered at the chopping frequency. The dc output from the lock-in detector is plotted as a function of wavelength using a pen recorder. Alternatively, the signal may be processed using a microcomputer, so that the absorbed intensity may be signaled as the transmission, the optical density (Eq. 1), or the absorption coefficient (Eq. 3) of the sample as a function of wavelength λ, wave number $\bar{v}$, or photon energy ($E = hv$).

Ensuring a high light throughput in both sample and reference channels usually limits the resolution of the monochromator used in the spectrophotometer (Fig. 1). In consequence, very narrow absorption lines, $\Delta\lambda < 0.1$ nm, are normally broadened instrumentally. Note that because in an absorption spectrophotometer one measures the light transmitted by the sample relative to that transmitted by the reference chamber

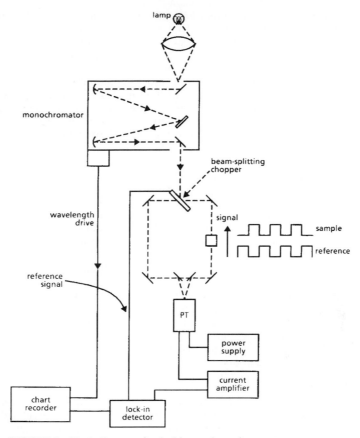

FIGURE 1   Block diagram of a dual-beam absorption spectrometer.

(Eqs. 2 and 3), the absorption coefficient is independent of the spectral dependences of the lamp, the monochromator, and the detection system. The measurement is also independent of the polarization properties of the monochromator system. By taking ratios, many nonideal behaviors of the components cancel.

20.4 *LUMINESCENCE SPECTROMETERS*

General Principles

To study luminescence it is necessary to optically pump into the absorption spectrum using high-intensity sources. Typical sources used in luminescence spectroscopy, which have broadbands in near ultraviolet and blue regions, include hydrogen and xenon arc lamps. The xenon arc lamp is particularly useful for exciting luminescence in the yellow-red region of the spectrum since xenon does not show interfering sharp line emission in this region. In general, high-pressure mercury (Hg) arc lamps have higher intensities than Xe arc lamps. However, the intensity is concentrated in sharp lines. Consequently, such lamps are utilized mainly with broadband absorbers or in situations that permit the individual lines to suit the absorption lines of the particular sample. In addition, a variety of lasers may be used, including Ar^+, Kr^+, He-Ne, and He-Cd lasers which have emissions at fixed wavelengths. Tunable dye lasers can be selected to closely match the absorption bands of particular materials. Because of their low intensity, tungsten filament lamps are not normally used in luminescence spectrometers.

The light emitted is resolved into its component lines/bands using a monochromator. For medium resolution a 1-m Czerny-Turner monochromator will give a spectral resolution of about 0.02 nm. Resolution an order of magnitude lower can be achieved using a grating spectrometer with focal length 0.25 m. The light emerging from the monochromator is detected using an electron multiplier phototube with associated high-voltage power supplies. Gallium arsenide phototubes operate with good quantum efficiency in the range 280 to 860 nm. For measurements in the near-infrared, a lead sulphide cell, cooled germanium photodetector, or special III-V compound photodiode may be used. Under steady-state optical pumping, a steady-state luminescence output is obtained and detected as a photocurrent which is amplified and converted to a voltage signal to be displayed on a pen recorder. Luminescence detection is inherently more sensitive than absorption measurements and sensitivities of 10^{11} centers cm^{-3} are routine.

Ideally, the excitation source should yield a constant light output at all wavelengths, the monochromator must pass all wavelengths with equal efficiency, and be independent of polarization. In addition, the detector should detect all wavelengths with equal efficiency. Unfortunately, such ideal light sources, monochromators, and phototubes are not available and it is necessary to compromise on the selection of components and to correct for the nonideal response of the luminescence spectrometer. Generally, luminescence spectra are recorded by selecting the excitation wavelength which results in the most intense emission and then scanning the wavelength of the emission monochromator. In consequence, techniques must be developed to allow for the wavelength-dependent efficiency of the emission monochromator and photomultiplier tube. This is not required in absorption spectrophotometers where the ratio of $I(v, \hat{e}) \,|\, I_o(v, \hat{e})$ are used to compute the values of $\alpha(v, \hat{e})$ from Eq. (3).

Modern spectrometers use diffraction gratings in monochromators rather than prisms. This results in less interference from stray light and in greater dispersion. Stray light may also be reduced using narrow entrance and exit slits as well as double monochromators (i.e., monochromators incorporating two gratings). Nevertheless, the transmission efficiency of the grating monochromator is a strong function of wavelength, which can be maximized at any given wavelength by choice of the blaze angle: the efficiency is less at other wavelengths as Fig. 2 shows. The stray light levels are to some extent controlled by

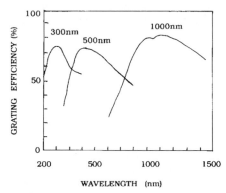

FIGURE 2 Showing how the grating efficiency varies with wavelength for gratings blazed at 300, 500, and 1000 nm.

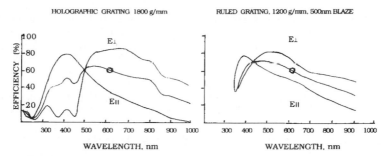

HOLOGRAPHIC GRATING 1800 g/mm   RULED GRATING, 1200 g/mm, 500nm BLAZE

FIGURE 3 Showing the effect of polarization on the efficiency of a ruled grating with 1200 grooves/mm and blazed at 500 nm.

exit and entrance slits. Smaller slit widths also yield higher resolution as do gratings with greater numbers of grooves per unit area. The efficiency of a grating monochromator also depends upon the polarization of the light. For this reason, the observed fluorescence intensities can be dependent upon the polarization of the emitted radiation. A typical plot of the wavelength dependence of the efficiency of a ruled grating as a function of polarization is shown in Fig. 3. As a consequence, the emission spectrum of a sample can be shifted in wavelength and altered in shape by the polarization properties of the monochromator. In modern spectrometers the monochromators can be calibrated using standard lamps and polarizers, the information stored in the memory of the control computer, and the detected intensities corrected at the data processing stage of the experiment. Most manufacturers also provide data sheets describing monochromator performance, and use can be made of such data for approximate corrections to the measured spectra.

Care must be taken with polarization anisotropy measurements. Thin-film polarizers have absorption properties which are strongly wavelength-dependent. Precise corrections can be made using computer-controlled facilities with provision for computerized data processing. However, it is preferable to use a Glan-Thompson prism made from quartz or calcite which has good transparency from the ultraviolet into the infrared. Furthermore, the polarization properties are not wavelength-dependent.

In general terms, the light signal is detected using a photomultiplier tube in which the photon flux produces an electrical current that is proportional to the light intensity. The basis of the device is the photoelectric effect. Incident photons cause photoelectrons to be emitted from a photocathode with an efficiency dependent upon the incident wavelength. The photocathode is held at a high negative potential of 1000 to 2000 V. The photoelectrons are incident upon a series of dynodes which are also held at negative potentials in order to accelerate electrons towards the next dynode. Each photoelectron arriving at the first dynode chain causes the ejection of a further 10 to 20 electrons, depending on the voltage difference between photocathode and first dynode. This process of electron multiplication and consequent current amplification continues down the dynode chain until a current pulse arrives at the anode. Although the photomultiplier tube responds to individual photons, the individual current pulses are generally detected as an average signal.

The anode current must be directly proportional to the light intensity. However, at wavelengths longer than the work function of the photocathode, the photomultiplier tube is no longer sensitive to the incident photons. Thus, different photocathodes are used in different wavelength ranges. For phototubes used in the ultraviolet region, quartz windows are used. For the ultraviolet-visible region (200 to 550 nm) a K-Cs bialkali photocathode may be used; such devices have high quantum efficiency, up to 25 percent between 350 to 500 nm, high gain, and low dark current. Typically, the operating anode currents are of the order of a few microamps, whereas the dark current is in the nanoamp range. A somewhat more useful device, in that the quantum efficiency is almost constant from 300 to 860 nm, uses a GaAs photocathode. For longer wavelength operation, 800 to 1250 nm, a germanium photodiode may be used. In other words, spectroscopic studies over a wide wavelength range may require several different photodetectors to be used. Techniques for correcting for the nonideal wavelength-dependent properties of the monochromator, polarizers, and photomultiplier tubes have been described at length by Lackowicz.[2]

Luminescence Spectrometers Using Phase-sensitive Detection

Where phase-sensitive detection techniques are used, the excitation intensity is switched on and off at a certain reference frequency so that the luminescence intensity is modulated at this same frequency. The detection system is then set to record signals at the reference frequency only. This effectively eliminates all noise signals except those closely centered on the modulation frequency. A typical luminescence spectrometer is shown in Fig. 4. The pumping light is modulated by a mechanical light chopper operating at frequencies up to 5 kHz. A reference signal is taken from the chopper to one channel of a lock-in detector. The magnitude and phase of the luminescence signal is then compared with the reference signal. Because of the finite radiative lifetime of the emission and phase changes within the electronics, the luminescence signal is not in phase with the reference signal. Hence, to maximize the output from the lock-in detector, the phase control of the reference signal is adjusted until input (luminescence) and reference signals to the lock-in detector are in phase. Of course, the phase of the reference signal may also be adjusted so that reference

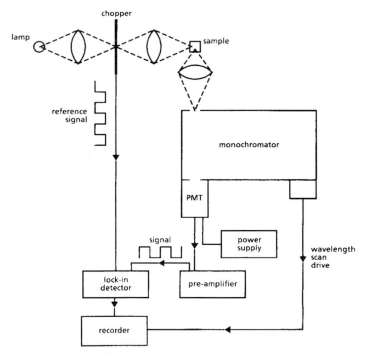

FIGURE 4   Schematic of a spectrometer for measuring luminescence spectra by phase-sensitive detection techniques.

and luminescence signals are in quadrature giving zero output from the lock-in. This method of phase adjusting may enable one to separate the overlapping luminescence bands from different centers. In such experiments, the chopping frequency is adjusted so that there is an appreciable reduction in the luminescence intensity during the "off" half-cycle. This effectively puts an upper limit on the rate at which the lock-in system can operate.

The use of a mechanical chopper restricts the maximum modulation frequency to ca 5 kHz. Essentially, the mechanical chopper consists of a rotating blade of metal into which slots are cut at regular angular intervals. When the excitation beam is incident on the metal section, the excitation intensity at the sample is zero, and when on the slot, the sample receives the full intensity in the excitation beam. If the blade is rotated at a frequency of 1 Hz and there are n slots cut in the blade, then the excitation beam is essentially switched on/off at a rate of n Hz. Obviously, the modulation rate can be increased either by increasing the number of slots cut in the blade and/or by increasing the revolution rate. If the excitation is well-focused onto the chopper, then the modulation is in the form of a square wave with maximum modulation index $M = 1$.

Other modulators impose sinusoidal variations in the excitation intensity, at frequencies up to 50 MHz.[2] There are various means for providing sinusoidal intensity variations, including Kerr cells and Pockels cells. Both types require high electric fields to obtain the desired modulation and such high driver voltages may interfere with the detection of weak signals. Kerr cells do not transmit ultraviolet light and so may only be used in the visible/near-infrared region. Currently available Pockels cells may be used in the ultraviolet region as well as at visible and infrared wavelengths. They may also be operated at variable frequencies. However, since they require highly collimated light sources for efficient operation, they require a laser for excitation. The ultrasonic Debye-Sears modulator overcomes the experimental difficulties associated with both Pockel cells and Kerr cells. A vibrating quartz crystal is used to set up standing waves in a tank containing an ethanol-water mixture. (The crystal restricts the device to operate at the fundamental and one or two harmonic frequencies only.) The standing waves act as a closely spaced refractive index diffraction grating normal to the incident exciting radiation. A slit permits only the undiffracted light to pass to the sample. The result is a sinusoidally varying light intensity with about 50 percent modulation index.

The emission signal is forced to respond to the modulated excitation at the same circular frequency ω as the excitation. However, the detected emission signal is delayed in phase by an angle ϕ relative to the excitation, and with reduced modulation depth. The radiative lifetime may be calculated from the measured phase angle ϕ and demodulation factor m. For a single exponential decay the appropriate relations are[3]

$$\tan \phi = \omega \tau_R \tag{8}$$

and

$$m = [1 + \omega^2 \tau_R^2]^{-1/2} \tag{9}$$

Even with more complex processes, where several decaying species are present, phase angles and demodulation factors can be measured and used to calculate actual lifetimes.[3,4]

Phase-sensitive detection techniques may also be used to "time-resolve" overlapping absorption/luminescence spectra with different decay characteristics. The phase-sensitive detector (PSD) yields a direct-current signal proportional to the modulated amplitude and to the cosine of the phase difference between the detector phase ϕ_D and the signal phase ϕ, i.e.

$$I(\lambda, \phi_D) = m_s I_o(\lambda) \cos (\phi_D - \phi) \tag{10}$$

where λ is the wavelength, $I_o(\lambda)$ is the steady-state excitation intensity, and m_s is the source modulation index. Now suppose that there are two components A and B with lifetimes $\tau_A < \tau_B$. The modulated emission measured with the PSD results in an unmodulated signal given by

$$I(\lambda, \phi_D) = m_s^A I_o^A(\lambda) \cos(\phi_D - \phi_A) + m_s^B I_o^B(\lambda) \cos(\phi_D - \phi_B) \qquad (11)$$

If the phase-control of the PSD is adjusted so that $\phi_D = \phi_B + 90°$, then the second term in Eq. (11) is zero, and the output intensity is given by

$$I(\lambda, \phi_D) = m_s^A I_o^A(\lambda) \sin(\phi_B - \phi_A) \qquad (12)$$

In other words, the emission output from species B has been suppressed. Species A can be suppressed at the detector phase angle $\phi_D = \phi_A + 90°$. If we now scan the wavelength, then the consequence of Eq. (12) is that the steady-state spectrum of species A is recorded, i.e., $I_o^A(\lambda)$, and conversely for species B.

In the example given in Fig. 5a, the steady-state fluorescence of a mixture of indole and dimethylindole dissolved in dodecane is shown.[5] With the detector phase angle set to $90° + 9.7$ and using a modulation frequency of 10 MHz in Fig. 5b, we resolve the indole emission with wavelength maximum at 306 nm. The phase angle of 9.7° corresponds to a radiative lifetime close to the isolated methylindole molecules in dodecane ($\tau_R = 5$ ns). The suppression of the indole signal gives the dimethylindole spectrum with peak at 323 nm at a phase angle of $28.6 - 90°$, giving the τ_R value of indole as 9.0 ns.

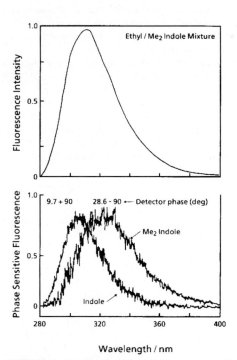

FIGURE 5 (a) Steady-state emission spectra of a mixture of indole and dimethylindole in dodecane; (b) shows the phase-resolved spectra of the indole and dimethylindole.[5]

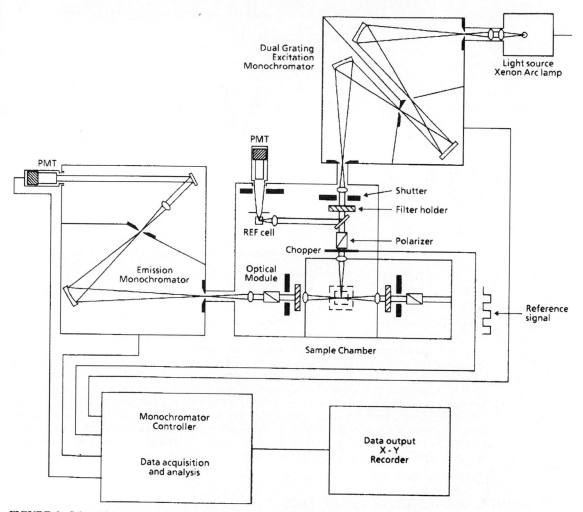

FIGURE 6 Schematic representation of a luminescence excitation spectrometer.

Luminescence Excitation Spectrometers

Some inorganic solids have strong overlapping absorption bands due to nonluminescent centers, which completely overwhelm the absorption spectrum related to a particular luminescence center. These difficulties are overcome by excitation spectroscopy, Fig. 6, in which the intensity of the luminescence output is recorded as a function of the wavelength of the excitation beam. Strong emission at a particular excitation wavelength signals that the emitting center absorbs strongly at that wavelength. In this way it is possible to determine the shape and position of the absorption bands which excite the emission process. A low-resolution scanning monochromator is placed immediately after the chopper, and light from its exit slit is then focused onto the sample. This monochromator may be of focal length only 250 mm and have a grating of area 5 cm × 5 cm, ruled with only 600 lines per mm. Alternatively, it may be a double monochromator chosen to reduce

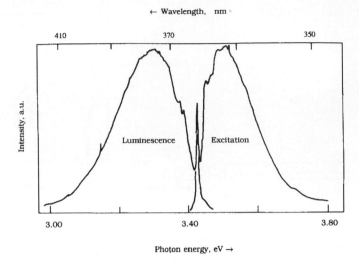

← Wavelength, nm

Intensity, a.u.

Luminescence   Excitation

Photon energy, eV →

FIGURE 7 Luminescence and excitation luminescence spectrum of F_2 centers in MgO.

stray light levels. In either case, an optical band pass filter may be used in front of the monochromator. Generally, the grating blaze is chosen to give high efficiency in the ultraviolet/blue/green regions for the excitation monochromator (e.g., gratings blazed at 300 nm or 500 nm), whereas the emission monochromator is chosen to give high efficiency at visible and near-infrared wavelength (i.e., gratings blazed at 500 nm or 750 nm). With such an apparatus, it is possible to distinguish absorption transitions from several centers whose absorption bands partially or completely overlap. The example given in Fig. 7 shows the luminescence pattern emitted by F_2-centers in magnesium oxide and the excitation spectrum associated with this emission. Other strong absorption bands due to Fe^{3+} ions and F centers which overlap the F_2-absorption bands are strongly discriminated against by selective detection of the F_2-center luminescence.

20.5 PHOTOLUMINESCENCE DECAY TIME

Radiative Lifetime

In order to measure the radiative lifetime of a transition it is necessary to use a sharp intense pulse of excitation in the absorption band together with some means of recording the temporal evolution of the luminescence signal. Suitable excitation sources include pulsed lasers, flash lamps, or stroboscopes. Laser systems may produce pulses of duration 0.1 to 100 ps; flash lamps and stroboscopes will produce pulses of order 10^{-8} s and 10^{-5} s, respectively. A possible spectrometer system is shown in Fig. 8. Usually the luminescence yield following a single excitation pulse is too small for good signal-to-noise throughout the decay period. In consequence, repetitive pulsing techniques are used together with signal averaging to obtain good decay statistics. The pulse reproducibility of the stroboscope is advantageous in the signal averaging process in which the output from the detector is sampled at equally spaced time intervals after each excitation pulse. If the pulse is repeated N times then there is an $N^{1/2}$ improvement in the signal-to-noise ratio. If a multichannel analyzer is used, the excitation pulse is used to trigger the analyser, and

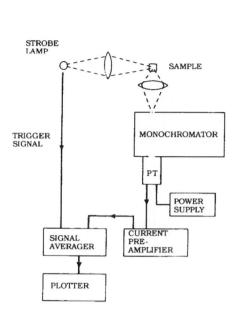

FIGURE 8 Spectrometer for measuring lumines-cence decay times.

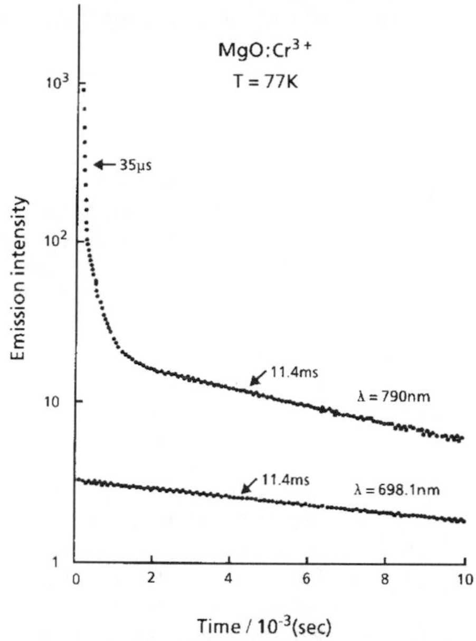

FIGURE 9 Decay in the intensity of Cr^{3+} lumin-escence in MgO. Detection of the broadband lumin-escence at 790 nm shows two components, one fast ($\tau_R = 35\ \mu s$) and one slow ($\tau_R = 11.4\ ms$). Detection of the R-line at 698.1 nm shows a single component with $\tau_R = 11.4\ ms$.

hence the time between pulses need not be constant. Of course the phase sensitive detection spectrometer may also be used to measure lifetimes, but only down to about 100 µs.

An illustration of the data obtainable using the stroboscope technique is shown in Fig. 9. The luminescence signal detected is the broadband emission with peak at 780 nm from Cr^{3+} ions in orthorhombic symmetry sites in magnesium oxide measured at 77 K. At low Cr^{3+} ion concentration, the radiative lifetime of this luminescence center is 35 µs. These data show that the evolution of the intensity during the pulse-decay cycle is not necessarily in the form of a single exponential decay. On sampling the emission at times long relative to τ_R there is a component with characteristic decay time of 11.4 ms, which is the lifetime of Cr^{3+} ions occupying octahedral symmetry sites in magnesium oxide and which emit a characteristic R-line emission at 698.1 nm. This result implies that excitation is being transferred from excited Cr^{3+} ions in octahedral sites to Cr^{3+} in orthorhombic sites.

For rather faster decay processes ($10^{-10} - 10^{-8}$ s), fast flashlamps are used to excite the luminescence. Modern gated flashlamps have extremely reproducible pulses, down to 0.8-ns width with repetition rates of up to 50 kHz.[2] The usual gases for such lamps are hydrogen, deuterium, and air. Hydrogen has several advantages, not the least being the continuum output in the ultraviolet and visible ranges, with pulse profiles which are independent of wavelength. The combination of pulse-sampling techniques and computer deconvolution of the decaying luminescence enables decay times to be measured down to 20 ps. However, judicious choice of photomultiplier tube and careful design of the photomultiplier dynode chain is necessary to eliminate signal noise. It is usual to use coincidence single-photon counting techniques to obtain good decay data.[2]

Picosecond and Sub-picosecond Relaxation

During the past two decades there have been quite remarkable developments in techniques for generating and measuring ultrashort pulses into the femtosecond domain. In semiconductors, a very wide range of ultrafast phenomena are being studied—electron-hole plasma formation, exciton and biexciton formation dynamics, hot electron effects, phase-conjugate self-defocusing and degenerate four-wave mixing. However, one very general optical phenomenon that may be addressed using ultrashort pulses involves nonradiative decay times in nonresonant fluorescence spectra. Such processes include ionic relaxations around a center in decaying from an excited state, sometimes including reorientations of anisotropic centers. Many picosecond phenomena, especially nonradiative decay processes, are studied by excite-and-probe techniques in which light pulses at wavelength λ_1 are used to excite a phenomenon of interest, and then a delayed optical pulse at wavelength λ_2 interrogates a change of some optical property of this phenomenon. Ideally, two sources of picosecond pulses at different, independently tunable wavelengths are required, which must be synchronized on the picosecond timescale.

A convenient experimental system for studying vibrational relaxation at color centers and transition metal ions in ionic crystals is shown in Fig. 10.[6] A mode-locked dye laser producing sub-picosecond pulses at wavelength λ_1 is used both to pump in the absorption band and to provide the timing beam. Such pumping leads to optical gain in the luminescence band and prepares the centers in their relaxed state. The CW probe beam, collinear with the pump beam, operates at a longer wavelength, λ_2. The probe beam and gated pulses from the pump laser are mixed in a nonlinear optical crystal and a filter allows only the sum frequency of the pump and probe beams, which is detected by a phototube. The photomultiplier tube actually measures the rise in intensity of the probe beam which signals the appearance of gain when the $F_A(Li)$-centers have reached the relaxed excited state. The pump beam is chopped at low frequency to permit phase-sensitive detection. The temporal evolution gain signal is measured by varying the time delay between pump

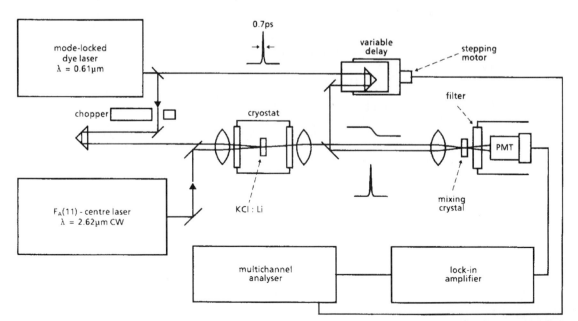

FIGURE 10 A sub-picosecond pump and probe spectrometer for measuring vibrational relaxation times in excited defects and transition metal ions.

and gating pulses. Although specifically used by Mollenauer et al.[6] to probe the relaxation dynamics of color centers, the spectrometer system shown in Fig. 10 is readily adapted to other four-level systems, including transition metal ions.

20.6 POLARIZATION SPECTROMETERS

General Principles

The absorbed intensity is sometimes dependent on the optical anisotropy of the sample. Whether or not a transition is allowed in a particular polarization is determined by examining the value of the square of the matrix element $\langle b| \boldsymbol{\mu} \cdot \boldsymbol{\varepsilon}_j |a\rangle$, see Optical Spectroscopy and Spectroscopic Lineshapes (Vol. 1, Chap. 8), Eqs. (12) and (13), where a and b are the states involved in the transition, $\boldsymbol{\mu} \cdot \boldsymbol{\varepsilon}_j$ is the appropriate component of the dipole operator summed over all $\mathbf{j}$ electrons involved in the transition. Optical transitions may be linearly or circularly polarized. For an electronic dipole transition, the dipole operator is $\boldsymbol{\mu}_e \cdot \boldsymbol{\varepsilon}_E$ where $\boldsymbol{\mu}_e = \Sigma_j e\mathbf{r}_j$ is summed over the $\mathbf{j}$ electrons and $\boldsymbol{\varepsilon}_E$ is the unit electric polarization vector parallel to the E-field of the radiation. The matrix element is evaluated using group theory, which shows how the symmetry properties of the states affect the transition rate.[1] From this matrix element the selection rules of the transition are determined. The polarization of the radiation is defined in Fig. 11 by reference to the $\hat{z}$ direction of the system, which itself is assumed to be parallel to an external perturbation (static electric or magnetic fields) or to unique symmetry direction in a crystal. For the π- and σ-senses of linear polarization, the radiation travels in a direction perpendicular to $\hat{z}$ with its electric field $\hat{\boldsymbol{\varepsilon}}_E$ either parallel to $\hat{z}$ (π-polarization) or perpendicular to $\hat{z}$ (σ-polarization). The electric dipole operators are then given by $\Sigma_j e\mathbf{r}_j \cdot \hat{z} = \Sigma_j ez_j$ for π-polarization and $\Sigma_j ex_j$ or $\Sigma_j ey_j$ for σ-polarization. The $\hat{x}$ and $\hat{y}$ directions have been assumed equivalent. In α-polarization the radiation propagates along the unique symmetry axis, $\hat{z}$, with $\hat{\boldsymbol{\varepsilon}}_E$ anywhere in the x-y plane: in this case the electric dipole operator is also $\Sigma_j ex_j$ or $\Sigma_j ey_j$. We define right circularly polarized (RCP) radiation as having electric (and magnetic) polarization vectors which rotate clockwise when viewed from behind the direction of propagation. For electric dipole absorption transitions, the electric dipole

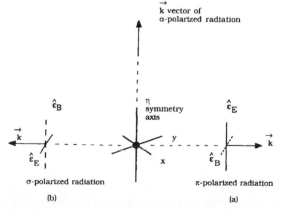

FIGURE 11 Definitions of the senses of α-, π- and σ-polarized light beams relative to a local symmetry axis.[1]

operator for RCP light propagating in the z direction is $\Sigma_j e(x + jy)_j/\sqrt{2}$. Accordingly, in the case of LCP light, where the sense of rotation is anticlockwise, the electric dipole operator is $\Sigma_j e(x - iy)_j/\sqrt{2}$.

Polarized Absorption

Although the selection rules of dipole transitions provide for polarized spectra, the optical spectra of atoms in the absence of external fields and of electronic centers with octahedral symmetry in solids are isotropic. Since the unit polarization vector, $\hat{\varepsilon}_E$, has direction cosines $\cos \alpha$, $\cos \beta$, and $\cos \gamma$, where the angles α, β, and γ are defined in Fig. 12a, the square of the electric dipole matrix element is

$$|\langle b| \mu_x \cos \alpha + \mu_y \cos \beta + \mu_z \cos \gamma |a\rangle|^2 \tag{13}$$

When the center has octahedral symmetry the cross terms in Eq. (13) are zero so that the squared matrix element becomes

$$\langle \mu_x \rangle^2 \cos^2 \alpha + \langle \mu_y \rangle^2 \cos^2 \beta + \langle \mu_z \rangle^2 \cos^2 \gamma \tag{14}$$

using $\langle \mu_x \rangle = \langle b| \mu_x |a\rangle$ with similar expressions for $\langle \mu_y \rangle$ and $\langle \mu_z \rangle$. Since in octahedral symmetry

$$\langle \mu_x \rangle^2 = \langle \mu_y \rangle^2 = \langle \mu_z \rangle^2 \tag{15}$$

$|\langle \mu \cdot \hat{\varepsilon} \rangle|^2$ becomes $\langle \mu_x \rangle^2$ and the strength of the transition is independent of the direction of the polarization of the incident radiation and the direction of propagation.

In octahedral solids, the local symmetry of an electronic center may be reduced by the application of an external perturbation or internally through the presence of a nearby

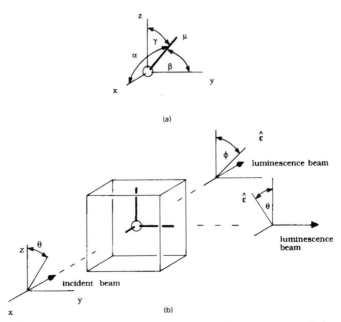

(a)

(b)

FIGURE 12   Showing (a) the orientation of the dipole moment μ relative to the x, y, and z axes; (b) two different geometrical arrangements used for polarized excitation luminescence spectroscopy.

defect. In tetragonal symmetry with the z axis parallel to the symmetry axis, the transition probability is again given by Eq. (14) but with $\langle\mu_x\rangle^2 = \langle\mu_y\rangle^2 \neq \langle\mu_z\rangle^2$. Since the transition probability for radiation at polarization $\hat{\varepsilon}_E$ is then proportional to

$$\langle\mu_x\rangle^2(\cos^2\alpha + \cos^2\beta) + \langle\mu_z\rangle^2\cos^2\gamma = A + B\cos^2\gamma \tag{16}$$

where A and B are constants, the spectroscopic properties of the center are anisotropic. In terms of the experimental situation referred to in Fig. 12b, in α-polarization the angle γ is always $\pi/4$ radians, and the intensity is proportional to A. For π-polarization, the angles are $\alpha = \pi/4$, $\beta = \pi/4 - \alpha$, and $\gamma = 0$ and the intensity is proportional to B. Similarly, for σ-polarization, the intensity is proportional to A. This shows that in tetragonal symmetry a rotation of the polarizer from the $\gamma = 0$ to $\gamma = \pi/4$ in, for example, the y-z plane determines the magnitudes of A and B. The linear dichroism D is then given by $D = (B - A)/(A + B)$.

To illustrate these ideas, Fig. 13 shows the polarization of the $^2S_{1/2} \rightarrow {}^2P_{1/2}$, $^2P_{3/2}$ lines of atomic sodium, i.e., the D_1 and D_2 absorption lines, in the presence of an applied magnetic field. The Zeeman splittings of energy levels are much smaller than the spin-orbit splitting between the $^2P_{1/2}$ and $^2P_{3/2}$ levels. The wave functions are labeled in Fig. 11 by the M_J-values: the relevant Clebsch-Gordan coefficients and theoretical intensities of the transitions for linear and circular polarizations are shown in Fig. 13, as are the theoretical intensities of the $^2S_{1/2} \rightarrow {}^2P_{1/2}$, $^2P_{3/2}$ right circularly polarized (RCP) and left circularly polarized (LCP) absorption transitions. The experimental pattern of lines for π- and σ-polarizations are in excellent agreement with the predicted Zeeman pattern.

An analysis of the polarization properties of the sample absorption requires that a polarizer be inserted in the light path immediately prior to the sample chamber. For accurate measurements of the absorption anisotropy, the polarizers must be accurately positioned relative to the beam and rotatable. The angle of rotation about the beam must be accurately indexed so that the orientation-dependence of the anisotropy may be determined. The polarizer should also be removable since it is unnecessary for measurements with optically isotropic solids. A sample which has different absorption coefficients in different crystallographic directions is said to be *dichroic*. The dichroism is defined as

$$D = \frac{\alpha(\pi) - \alpha(\sigma)}{\alpha(\pi) + \alpha(\sigma)} = \frac{1}{l}\left(\frac{I(\pi) - I(\sigma)}{I(\pi) + I(\sigma)}\right) \tag{17a}$$

in the limit of small absorption coefficients.

Although discussion has focused on radiative absorption transitions via electric dipole transitions, a similar analysis can be made for magnetic dipole transitions. In this case, the phase relationships for the magnetic fields $\mathbf{B}_x$ and $\mathbf{B}_y$ are exactly the same as those between $\mathbf{E}_x$ and $\mathbf{E}_y$, and the magnetic dipole operator is $\boldsymbol{\mu}_m \cdot \boldsymbol{\varepsilon}_B$; where $\boldsymbol{\mu}_m = \sum_j (e/2m) \times (l + 2s)_j$ and $\boldsymbol{\varepsilon}_B$ is the unit vector along the direction of the magnetic field of the radiation. If the absorption transitions used to excite the luminescence are unpolarized, so too will be the resulting luminescence spectrum. However, as discussed above, the absorption spectrum of an atomic system may be made anisotropic by the application of an external field or by using polarized exciting radiation. The resulting emission spectrum will be correspondingly polarized. Absorption and luminescence spectra from optically isotropic solids can also be made anisotropic using similar techniques.

Polarized Absorption/Luminescence

Just as the absorption spectra of free atoms and isotropic solids are unpolarized, so too are the luminescence spectra, at least when exciting with unpolarized radiation. This is shown by simple extensions to the arguments leading to Eq. (15) in which the electric dipole operators for luminescence are the complex conjugates of those for the appropriate

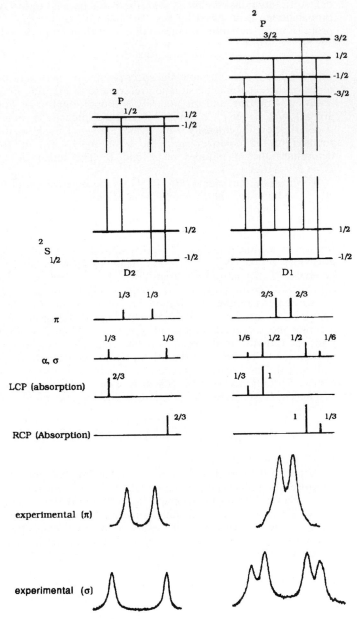

FIGURE 13 Zeeman splittings of $^2S_{1/2} \rightarrow {}^2P_{1/2}$, $^2P_{3/2}$ levels of sodium. The electric dipole matrix elements and relative intensities of linearly and circularly polarized absorption transitions are compared with some experimental spectra.[1]

absorption transitions. In practice, both absorption and emission properties are anisotropic. Although the host crystal may possess a cubic unit cell in which the electronic centers are anisotropic, a regular distribution of equivalent sites will still result in isotropic spectra. The use of polarized absorption/luminescence techniques can reveal details of the local site anisotropy. Such methods have been discussed by Henderson and Imbusch[1] and in more detail by Feofilev.[7]

To measure the effects of polarization on the absorption coefficient (i.e., $\alpha(v, \hat{\varepsilon})$) it is necessary to place a polarizer immediately before the beam splitter in the double-beam spectrophotometer, Fig. 1. In polarized luminescence measurements, linear polarizers are placed immediately before the sample in the absorption channel and just after the sample in the emission channel of a luminescence excitation spectrometer such as that shown in Fig. 6. The spectrometer may then operate in the "straight through" configuration or the emitted light may be collected in a direction at 90° to the direction of the excitation light, as illustrated in Fig. 13. Note that provision is made for rotatable polarizers in both excitation (θ) and detection channels (ϕ), and the measured emission signal will be a function of both θ and ϕ.

The circular dichroism may be defined in an analogous manner to the linear dichroism, i.e., Eq. (17a). Since circular dichroism has a specific relevance to the Zeeman effect, we use Fig. 14a and consider circularly polarized absorption transitions which are excited between two Kramers doublets. With light propagating along the direction of the magnetic field, the selection rule is that σ-polarized light induces $\Delta M_s = +1$ absorption transitions

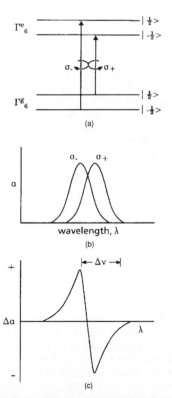

FIGURE 14 Circularly polarized dipole transitions excited between Kramers states.

and σ_+-polarized light induces absorption transitions in which $\Delta M_s = -1$. As a result of the Zeeman effect, the absorption peak splits into two overlapping bands (Fig. 14b) centered at different wavelengths with absorption coefficients $\alpha_\pm$ in σ_+- and σ_--polarizations that are different at particular frequencies. The peaks in the two oppositely polarized bands are separated in energy by $(g_e + g_g)\mu_B B$, where the g-values refer to the excited e and ground g states. The energy difference corresponds to a frequency splitting $\Delta\nu = (g_e + g_g)eB/4\pi m$, which for $g_e = g_g = 2.0$ and $B = 1\ T$ gives a separation between band peaks of $\simeq 0.04$ nm for a band centered at 500 nm. In a magnetic field, the difference $(\Delta\alpha(\nu))$ in the absorption coefficients for σ_+ and σ_- circularly polarized light is referred to as *magnetic circular dichroism* (MCD). In the limit of small absorption coefficient the circular dichroism is

$$\Delta\alpha(\nu) \simeq -\frac{2(I_+(\nu) - I_-(\nu))}{l(I_+(\nu) + I_-(\nu))} \tag{17b}$$

where l is the sample thickness and $I_+(\nu)$, $I_-(\nu)$ refer to the transmitted intensities of the σ_+ and σ_- circularly polarized light at frequency ν.

In most cases, a splitting of only 0.04 nm would be hard to resolve directly by Zeeman effect measurements on a broadband. However, this Zeeman structure may be resolved by measuring $\Delta\alpha(\nu)$ as a function of magnetic field as can be seen from a simple estimate. We approximate the MCD signal, $\Delta\alpha(\nu)$ for a sample of thickness l, as the product of the magnetic splitting, $\Delta\nu$, with the rate of change of the absorption coefficient with frequency which is given by $d\alpha(\nu)/d\nu \approx \alpha(\nu_o)/\Gamma$, for a symmetrical, structureless band. Hence

$$\Delta\nu = \frac{\Delta\alpha(\nu)}{l} \times \frac{\Gamma}{\alpha(\nu_o)} \tag{18}$$

In a typical experiment, $\Delta\alpha(\nu)l \simeq 10^{-5}$ and $\alpha(\nu_o)l \sim 1$, hence $\Delta\nu \simeq 10^{-5}\,\Gamma$. For a typical broadband, $\Gamma \simeq 0.25$ eV $\simeq 2000$ cm^{-1} and Eq. (18) yields $\Delta\nu \simeq 0.02$ cm^{-1} (i.e., $\Delta\lambda \sim 0.05$ nm) which is of the same order of magnitude as the Zeeman splitting calculated above. Although the intensity changes, which determine the magnitude of $\Delta\alpha(\nu)$, may be quite small, they may be assumed very precisely using lock-in techniques. This is done very efficiently by replacing the circular polarizer in the excitation system by a stress-modulated quarter-wave plate, a device which transmits circularly polarized light, the polarization of which is switched between σ_+ and σ_- at the vibration frequency of the plate, usually ≈ 50 kHz. Using this piezo-optic modulation, MCD signals as low as 10^{-6} can be measured.[8]

The MCD signal is strongly dependent on both frequency and temperature. Since at low temperatures the populations $N_\pm$ of the $M_s = \pm\frac{1}{2}$ levels of the spin $\frac{1}{2}$ ground state are different for a system in thermal equilibrium, the MCD signal (Eq. 17b) is given by

$$\Delta\alpha(\nu)l = \alpha_o(\nu)G(\Delta\nu)\tan h\left[\frac{g\mu_B B}{2kT}\right] \tag{19}$$

In this expression $\alpha_o(\nu)$ and the sample thickness, l, are experimental constants and, in consequence, the MCD signal only varies through the Brillouin function for the $s = \frac{1}{2}$ ground state (i.e., $\tan h(g\mu_B B/2kT)$). This MCD signal is paramagnetic, being strongest at high field and low temperature, and measurement of its magnitude probes the ground-state magnetization. In order to test Eq. (19) experimentally, it is best to work at either the positive or negative peak in Fig. 14 and so maximize the MCD signal. Having thus obtained a suitable MCD signal, its variation with temperature and magnetic field can then be measured. Excitation of the Kramers' system in Fig. 14 with circularly polarized radiation of appropriate frequency results in the emission of circularly polarized emission. The electric dipole operators for RCP and LCP emission are the complex conjugates of those for absorption. The consequent change in emission intensity, Eq. (17), is referred to as the magnetic circular polarization (MCP).

Optically Detected Magnetic Resonance (ODMR)

In optical absorption spectroscopy, electronic transitions (usually) out of the ground state may result in one of a rich tapestry of possible bandshapes, depending upon the strength of the electron-phonon coupling. Photoluminescence measurements involve transitions which originate on an excited electronic state and frequently the terminal state is the electronic ground state. Overlapping absorption and luminescence bands can cause difficulty in assigning individual optical absorption and luminescence bands to particular optical centers. Since the lifetimes of excited states are in the range 10^{-3} to 10^{-8} s, it is no trivial matter to measure excited-state electron spin resonance using the microwave detection techniques pioneered in ground-state studies. Geschwind et al.[9] developed techniques in which the excited-state ESR was detected optically. In favorable cases this method enables one to correlate in a single experiment ESR spectra in the ground state and in the excited state with particular optical absorption and luminescence spectra. The technique of measuring the effect of resonant microwave absorption on the MCD and/or MCP signal may be termed optically detected magnetic resonance (ODMR). In ODMR measurements involving the MCD signal, microwave-induced transitions between the Zeeman levels of the ground state are detected by a change in intensity of the absorption (i.e., MCD) spectrum. Electron spin resonance transitions in the excited state are signaled by microwave-induced changes in the MCP signal.

Figure 15 is a schematic drawing of an ODMR spectrometer. There are three necessary channels: a microwave channel and channels for optical excitation and detection. The microwave system is relatively simple, comprising a klystron or Gunn diode operating at some frequency in the range 8.5 to 50 GHz, followed by an isolator to protect the microwave source from unwanted reflected signals in the waveguide path. The microwave power is then square-wave modulated at frequencies up to 10 kHz, using a PIN diode. A variable attenuator determines the power incident upon the resonant cavity, although for high-power operation in a traveling-wave amplifier might be added to the waveguide system. The sample is contained in the microwave cavity, which is designed to allow optical access of the sample by linearly or circularly polarized light traveling either parallel or perpendicular to the magnetic field direction. The cavity is submerged in liquid helium to achieve as large a population difference as possible between the Zeeman levels. The magnetic field is provided either by an electromagnet ($B \simeq 0 - 2.0\ T$) or a superconducting solenoid ($B \simeq 0 - 6.5\ T$). Radiation from the sample is focused onto the detection system, which in its simplest form consists of suitable filters, a polarizer, and photomultiplier tube. A high-resolution monochromator may be used instead of the filters to resolve sharp features in the optical spectrum. The signal from the phototube is processed using a phase-sensitive detector, or alternatively using computer data collection with a multichannel analyzer or transient recorder. The recorded spectrum is plotted out using a pen recorder as a function of either magnetic field or of photon energy (or wavelength). With such an experimental arrangement one may examine the spectral dependence of the ODMR signal on the wavelength of the optical excitation or on the wavelength of the detected luminescence by use of one of the two scanning monochromators.

In order to carry out ODMR, microwave radiation of fixed frequency v is introduced while the optical wavelength is kept at the positive or negative peak in Fig. 14c. The magnetic field is then adjusted until the ESR condition, $hv = g\mu_B B$, is satisfied. Since ESR transitions tend to equalize the populations N_+ and N_-, resonance is observed as a decrease in $\Delta\alpha(v)$, and as the microwave power is increased, the MCD gradually tends to

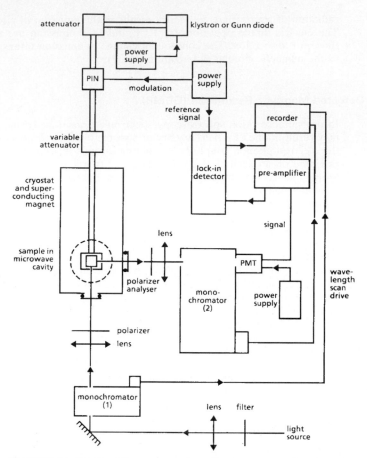

FIGURE 15 A schematic representation of a spectrometer for measuring optically detected magnetic resonance spectra via circularly polarized absorption or emission transitions.

zero. In certain circumstances the ground-state spin polarization may be used to monitor *excited-state* ESR transitions because of the selectivity of the transitions induced by circularly polarized radiation. This measurement technique is an example of *trigger detection* where one microwave photon in absorption triggers the detection of one optical photon emitted. The resulting enhancement in sensitivity relative to the normal ESR technique is approximately in the ratio of optical to microwave frequency (i.e., $10^{15}/10^{10} = 10^5$). At x-band ($\approx 10\,\text{GHz}$), the ESR sensitivity is about 10^{10} spins per gauss linewidth so that ODMR sensitivity is of order 10^5 atoms in the excited state. With the ODMR technique, one may gather information on a wide range of important solid-state processes including spin-lattice and cross relaxation, spin memory, energy transfer, electron-hole recombination, phonon bottlenecks, and spin coherence effects.

A major attribute of the ODMR technique is illustrated in Fig. 16, showing the optical characteristics of the ODMR spectrum of F centers in calcium oxide.[10] These spectra were measured at 18.7 GHz and 1.6 K with the magnetic field along a crystal $\langle 100 \rangle$ direction. A high-pressure xenon discharge lamp and monochromator (M_1 in Fig, 15) set at 400 nm was used to excite the fluorescence, which was detected through monochromator M_2. The

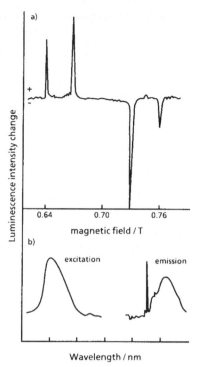

FIGURE 16 The ODMR spectrum of triplet state of *F*-centers in CaO.

spectrum consists of four equally spaced lines due to an $S = 1$ state of a center with tetragonal symmetry. Then with the magnetic field set at the strongest ODMR line, the excitation wavelength is scanned using monochromator M_1 (Fig. 15) over the visible and near-ultraviolet region. A single broad structureless excitation peak is observed at 400 nm corresponding to the $^1A_{1g} \rightarrow {}^1T_{1u}$ absorption band of the *F*-center (Fig. 16). Subsequently, the excitation monochromator is set at the peak of this excitation band and the same magnetic field while the detecting monochromator (M_2 in Fig. 15) is scanned over the fluorescence spectrum. This spectral dependence (Fig. 16) shows a sharp zero-phonon line at a wavelength of 574 nm with an accompanying broad vibronic sideband with peak at 602 nm. In a single experiment, a unique and unambiguous relationship has been established between the ESR spectrum, absorption, and fluorescence bands of an intrinsic lattice defect.

20.7 HIGH-RESOLUTION TECHNIQUES

Inhomogeneous broadening arises when individual atoms are distinguished by the frequency at which they absorb light. The absorption profile is then the sum of separate absorption lines. In atomic spectroscopy, the major source of the spectral linewidth is Doppler broadening; the frequency shift is $(\Delta v/v) = \pm(v_z/c)$ due to an atom moving with velocity component v_z towards $(+)$ or away from $(-)$ the observer. At thermal equilibrium, a gaussian lineshape is observed because of the Maxwell-Boltzmann velocity distribution. In solids, the distribution of internal strains is a source of inhomogeneous broadening. Because crystals contain imperfections, electronic centers experience crystal fields which vary slightly from site to site in the crystal; in consequence, zero-phonon lines may have linewidths of order 0.1 to 50 cm^{-1}. The use of narrow-band laser excitation makes it possible to eliminate inhomogeneous broadening and to realize a resolution limited only by the homogeneous width of the transition, which in crystals can vary from kilohertz to gigahertz. This factor of 10^3 to 10^4 improvement in resolution enables the spectroscopist to carry out high-resolution studies of the physical properties and electronic structures of centers and of the mechanisms responsible for homogeneous broadening. Contributions to homogeneous width come from population dynamics and random modulation of the optical frequency by phonons and nuclear spins.

Saturated Absorption and Optical Hole-burning

The experimental basis of recovering the homogeneous width of an inhomogeneously broadened optical spectrum, so-called saturated absorption or optical holeburning (OHB) spectroscopy, is illustrated in Fig. 17. An inhomogeneously broadened line of width Γ_{inh} is

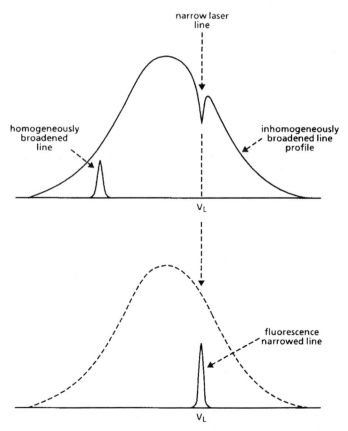

FIGURE 17 Optical holeburning (OHB) and fluorescence line narrowing (FLN) of an inhomogeneously broadened spectroscopic line.[1]

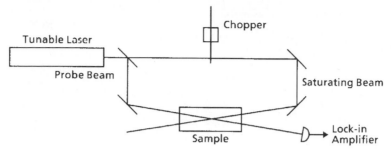

FIGURE 18 A spectrometer system for Doppler-free saturation spectroscopy.[12]

produced by many narrow components of homogeneous width $\Gamma_{hom} \ll \Gamma_{inh}$. Each component is centered at a different frequency within the inhomogeneous line profile. If a narrow laser line of frequency ν_L and bandwidth $\Gamma_L < \Gamma_{hom}$ is incident upon an atomic assembly having an inhomogeneously broadened linewidth Γ_{inh}, the resulting absorption of laser radiation depletes only that subassembly of excited centers whose energies are within Γ_{hom} of the laser line frequency ν_L. Consequently, a "hole" is burned in the lineshape in the neighborhood of ν_L. Resolution of the homogeneous width requires that $\Gamma_L < \Gamma_{hom} \ll \Gamma_{inh}$. In holeburning spectroscopy, the narrow laser linewidth and high power make it possible to maintain a significant fraction of those atoms with transition frequency ν_L in the excited state, where they no longer contribute to the absorption at this frequency. To observe holeburning experimentally requires that -5 percent of those centers within the pump laser bandwidth be transferred to the excited state.

The first measurements of optical holeburning or saturated absorption spectroscopy in atoms were made by the Stanford group on the H_α-line ($n = 2 \rightarrow n = 3$) in hydrogen using a pulsed dye laser.[11] A schematic diagram of an appropriate absorption spectrometer is shown in Fig. 18. A strong laser beam interacts with those atoms that are moving with the right velocity to Doppler shift them into resonance. If the laser beam is intense enough, it tends to equalize the population in the two levels, thereby reducing the intensity. The hole burned in the absorption profile, which extends over the natural width of the transition, is probed by a second beam at the same frequency but lower intensity and traveling in the opposite direction. This beam interacts with atoms having the same velocity but in the opposite direction to the saturating beam. When the laser is tuned to line center, both pump and probe beams interact with atoms moving with zero longitudinal velocity. The probe beam then measures the reduced absorption caused by the saturating beam. In experiments using pulsed lasers, very high intensity is required to achieve saturation and hence there must be very tight focusing and overlap of pump and probe beam. In consequence, CW lasers are preferred in both gas-phase and solid-state spectroscopy. Saturated absorption measurements on atomic hydrogen have been reviewed recently by Ferguson and Tolchard[12] and OHB in solids by Selzer[13] and by Yen.[14]

To burn a hole in a narrow absorption line in crystals requires that the laser be focused onto the sample for periods of order 10^2 to 10^3 s, depending upon the specific system. When the laser excitation is switched off, the holes recover on some timescale characteristic of the physical process responsible for holeburning. For short-lived holes the exciting beam is divided using a beam splitter into pump and probe beams. The weaker probe beam passes through an optoacoustic modulator which scans it backward and forward over the hole.[13] To observe long-lived holes, the sample is irradiated for a short time in the zero-phonon line with a few hundred milliwatts of single-mode dye laser light with a width of a few megahertz. The shape of the hole is then displayed by reducing the laser intensity to a few milliwatts and scanning the laser over the inhomogeneous line

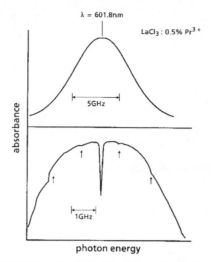

$\lambda = 601.8\,\text{nm}$

$LaCl_3 : 0.5\%\ Pr^{3+}$

5 GHz

1 GHz

absorbance

photon energy

FIGURE 19 Optical holeburning in the 601.28-nm line of Pr^{3+} ions in $LaCl_3$. The hole was burned using some $200\,\text{W cm}^{-2}$ of single-frequency laser light with a bandwidth of 2 MHz. The zero-phonon line has an inhomogeneous width of 7.5 GHz. (*After Harley and Macfarlane, 1986, unpublished.*)

profile. Figure 19 shows an example of holeburning in the 601.28-nm line of $Pr^{3+} : LaCl_3$. The homogeneous width measured in this holeburning experiment is $\Gamma_{hom} = 10$ MHz, which corresponds to a lifetime of 10 ns. There have been many reports of holeburning spectroscopy on transition metal ions, rare-earth ions, and color centers in inorganic materials. For rare-earth ions, holeburning with lifetimes determined by the nuclear spin relaxation processes have been reported to vary from 10 to 10^3 s. Many measurements are aimed at the mechanisms leading to the homogeneous width of optical transitions. In these cases, techniques have been developed for the detection of coherent transients (e.g., photon echo or free induction decay) because the measurements are made on the timescale of the dephasing and are not affected by spectral diffusion and other such processes.

Polarized Absorption Spectrometers

Polarized absorption spectroscopy is a technique related to sub-Doppler absorption spectroscopy. However, in this case use is made of the circularly polarized absorption properties of atomic transitions. In the Wieman-Hänsch experiment,[11] the probe beam passes through crossed polarizers immediately before and after the sample. If the pump beam is unpolarized, the sample is optically isotropic and no light falls on the detector. However, if the $s \rightarrow p$ transitions are excited using RCP light, the pump beam induces optical anisotropy in the medium with which it interacts. In consequence, as pump and probe beams are tuned to line center so that both interact with the same class of atoms, the weak probe beam becomes slightly elliptically polarized and light is transmitted through the crossed polarizers. The advantage of the method is a factor of about 10^3 enhancement in sensitivity relative to saturation absorption spectroscopy. Sub-Doppler two-photon absorption spectroscopy is also much used in atomic physics.[15] The selection rule for two-photon absorption is that $\Delta l = 0$ or 2. In consequence, for $1 = $ electron atoms $S \to S$ and $S \to D$ transitions are allowed.

Laser Stark Spectroscopy of Molecules

Sub-Doppler resolution enhancement is also used in studying the heterogeneously broadened rotational/vibrational spectra of molecules. Such spectra are generally observed in the mid-IR region and are studied using a wide variety of gas lasers (e.g., N_2O, CO, and CO_2). Such laser lines are not usually in exact resonance with the particular molecular transition: laser and molecular transition are brought into register using a variable electric field to tune the molecular system into resonance. In general, parallel-plate Stark cells are used in which free-space propagation of the short-wavelength infrared radiation occurs. This makes it easy to use both perpendicular and parallel polarization configurations in the electric resonance experiments so that both $\Delta M_J = 0$ and $\Delta M_J = \pm 1$ transitions are observed. The subject of laser Stark spectroscopy has been discussed at length by Duxbury.[16]

A schematic intracavity laser Stark spectrometer is shown in Fig. 20; the same basic principles obtain as with optical holeburning spectroscopy. The effects of the saturating laser field are confined to a narrow frequency region centered on the velocity component of those molecules whose absorption is Doppler-shifted into resonance. In a standing wave field, two holes are burned, one on either side of the line center, corresponding to molecules moving towards or away from the detector. The applied electric field is used to tune the two holes to line center where they coalesce to give a sharp dip in the absorption coefficient at line center. Since the resonance method relies on the use of an electric field for tuning, it is necessary both to generate high uniform fields and to study molecules with appreciable Stark tuning coefficients. In order to generate high electric fields, which may approach 90 kV cm^{-1}, narrow electrode spacings of from 1 to 4 mm are commonly used. With such narrow gaps, the plates must be flat to one or two fringes of visible light, and must be held accurately parallel. The gas pressure used must also be restricted to the low-pressure region below 100 mTorr. A potential difference of roughly 3000 V may be sustained without electrical breakdown across any gas at a pressure of 100 mTorr and below.

The electric field is then modulated at some convenient frequency to permit the use of phase-sensitive detection techniques. In order to get above the principal noise region of the electric discharge lasers used in the 5- and 10-μm regions and as pumps for the FIR

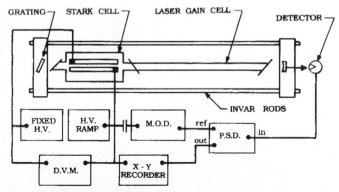

FIGURE 20 Schematic diagram of an intracavity laser Stark spectrometer. PSD stands for phase sensitive detector, DVM for digital voltmeter, HV for high voltage, and MOD for modulation source.[16]

lasers, it is necessary to use electric field modulation frequencies in the range from 5 to 100 kH. The amplitude of the electric field modulation used to detect the signals is usually small compared to the equivalent electric field linewidth of the transitions. The most common modulation waveform is sinusoidal. If the modulation amplitude is much smaller than the linewidth, detection at the fundamental modulation frequency results in a first derivative lineshape as in analogous electron spin resonance spectra. In order to remove the effects of sloping baselines produced by transitions with a slow Stark effect, it is common to use detection at either the second or third harmonic of the modulation frequency. Second-harmonic detection produces a second-derivative signal resembling a sharpened absorption line but with negative side lobes. Third-harmonic detection produces a third-derivative signal which resembles a sharpened first derivative, but which again possesses side lobes. Theoretical lineshapes are illustrated in Fig. 21. Second- and

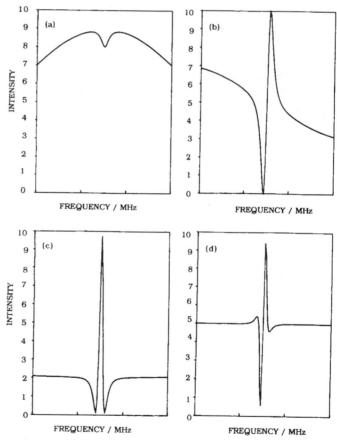

FIGURE 21 Lineshapes which occur when detecting at harmonics of the modulation frequency when small-amplitude field modulation is used. Doppler-broadened line showing the Lamb dip. (*a*) 30-MHz scan of a partially saturated Doppler-broadened line showing the Lamb dip; (*b*) 30-MHz scan with first-derivative detection; (*c*) 30-MHz scan with second-harmonic, second-derivative detection. The gain is increased by a factor of four from (*a*); (*d*) 30-MHz scan with third-harmonic, third-derivative detection.

third-harmonic detection are particularly useful for the observation of narrow saturation features free from background effects. The detectors used are quantum-limited liquid nitrogen cooled devices, PbSnTe or CdHgTe in the 10-μm region and InSb or Au doped Ge in the 5-μm region. In the far infrared, Golay cells have been used but in order to achieve a better signal-to-noise ratio it is necessary to use detectors cooled by liquid helium.

Just as in atomic spectroscopy one may use atomic beam spectroscopy as an alternative to absorption saturations, so too may one use molecular beam systems in high-resolution studies of the rotational-vibrational spectra of molecules.

Fluorescence Line Narrowing

Fluorescence line narrowing (FLN) is a technique complementary to that of OHB. It may also be understood by referring to Fig. 17. A narrow laser line is used to pump within the inhomogeneous linewidth, Γ_{inh}. The laser interacts only with the subset of levels spanning the bandwidth of the laser, Γ_L. These centers reradiate to some lower lying level, with a fluorescence linewidth much narrower than the inhomogeneous width. The fluorescence linewidth approaches the homogeneous width. In fact, for centers involved in a resonance fluorescence transition, the total FLN lineshape is a convolution of the laser lineshape and twice the homogeneous lineshape (once for the pump bandwidth and once for the fluorescence). The FLN linewidth Γ is then usually written as $\Gamma = \Gamma_L + 2\Gamma_h$. Experimentally, FLN requires a little more sophistication than does holeburning spectroscopy. Of course, one still requires a stable, high-resolution laser. Care must be used in extracting the true homogeneous linewidth, especially for nonresonant fluorescence. Many of the experimental problems relative to solid samples are discussed in the review by Selzer,[13] and numerous examples are given by Yen and Selzer.[17] The CW FLN spectrum shown in Fig. 22 is for the Cr^{3+} ion in aluminum oxide.[18] The fluorescence lifetime is 3.4 ms at 4.2 K. Hence the homogeneous width is of the order 0.3 kHz. A direct-phonon relaxation process between the two ^2E levels, $2\bar{A}$ and $\bar{E}$, separated in energy by 29 cm^{-1}, broadens the homogeneous width to $\approx$130 kHz. In early CW measurements, a homogeneous width in excess of 100 MHz was reported.[18] The problem is relaxations due to super-hyperfine interactions with neighboring aluminum nuclei. The application of a dc

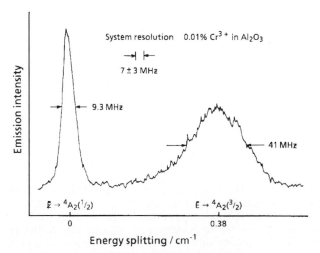

FIGURE 22   FLN in the R_1 transition of ruby.[18]

magnetic field of only $40\,mT$ has the effect of inhibiting relaxation due to local fields at the Cr^{3+} ions due to the ^{27}Al nuclear moments. A very considerable narrowing of the Cr^{3+} FLN spectrum is then achieved.

20.8 LIGHT SCATTERING

Light-scattering experiments are now a routine feature in many optical laboratories. The first observations of light scattering by small particles were reported by Tyndall.[19] Subsequently, theoretical work by Lord Rayleigh[20] showed both that the scattered intensity varied as the fourth power of the frequency and that the scattering was due to molecules rather than dust particles. Many of the early studies were concerned with the depolarization of the light after being scattered by the dust-free atmosphere. Of course, in the prelaser era, sufficient light intensity could only be achieved by use of strongly condensing lenses to focus light onto the gas cell. Very great care was then necessary to obtain reliable depolarization measurements. Even in the laser era it is still essential to avoid any effects due to parasitic light which often plague light-scattering experiments.

A significant early result from scattering of light by gases was that the scattered light intensity varied with the density of the gas being used as the sample. However, Lord Rayleigh discovered that the intensity scattered per molecule decreased by a factor of order 10 on condensation to the liquid phase. There is a somewhat smaller decrease in going from the liquid phase to the solid. Obviously, some scattering experiments become rather difficult in the solid state. The classical experimental geometry for studying Rayleigh scattering is in the 90° orientation for the scattered radiation. This is also the most useful orientation for Raman scattering in solids.[21]

One important feature of the structure of solids is the periodic disturbance of the crystal structure by the propagation of quantized elastic waves (i.e., phonons). Those elastic waves which travel at the velocity of sound (i.e., sonic waves) are essentially thermal density fluctuations in the elastic medium. Brillouin predicted that such fluctuations should give rise to fine structure in the Rayleigh scattered light when the Bragg coherence condition $\lambda_1 = 2\lambda_p \sin(\phi/2)$ is obeyed. Here λ_1 is the wavelength of light, λ_p is the wavelength of those phonons responsible for scattering the light, and ϕ is the scattering angle. Because the scattering centers are in motion, the scattered light is frequency shifted by the Doppler effect. It is an easy matter to show that the Doppler shift, Δv, is given by

$$\Delta v = \pm v_p = \pm 2v_1(v/c)\sin(\phi/2) \tag{20}$$

where v_p is the frequency of the density fluctuations in the medium and v is the velocity of sound in the medium. For light in the visible region then, that part of the phonon spectrum probed by the Brillouin scattering is in the gigahertz frequency region. In addition, the Brillouin components are completely polarized for 90° scattering. Before the advent of lasers, the study of Brillouin scattering effects in solids was exceedingly difficult. It remains a technique more used in gases than in condensed media.

C. V. Raman was one of numerous scientists engaged in research into light scattering during the decade 1920 to 1930. Much of his work was carried out using sunlight as a source. However, in experiments using monochromatic light, he observed in the spectrum of light scattered at 90° by liquid samples new lines at wavelengths not present in the original light.[21] The frequency displacement of these new lines from source frequency was found to be independent of the wavelength of the incident light. This was contrary both to fluorescence excitation and Brillouin scattering [Eq. (20)]; hence was born a new scattering phenomenon for which Raman was awarded the Nobel prize and which now bears his name. The frequency shifts in the Raman spectrum of a particular substance are related to but not identical to infrared absorption frequencies. In general, infrared transitions occur

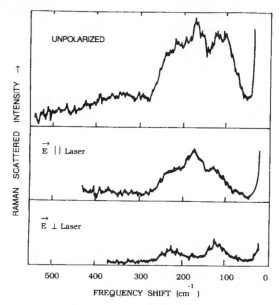

FIGURE 23 Raman spectra of *F*-centers in NaCl for different polarizations of the detected radiation. The shifts are measured relative to the 514.5 nm of the Ar[+] laser.[20]

when there is a change in the electric dipole moment of a center as a consequence of the local atomic vibrations. The Raman lines occur when a change in polarizability is involved during atomic vibrations. This usually means that infrared transitions occur only between states of opposite parity whereas Raman transitions occur between states of the same parity. Thus the infrared and Raman spectra give complementary information about the vibrational spectra of spectroscopic centers.

Raman scattering measurements have found wide application in condensed matter physics. The spectrometer systems have much in common with fluorescence spectrometers, although lasers provide the excitation source almost without exception. Single-frequency lasers (He-Ne, Ar[+], Ke[+]) and tunable dye lasers and solid-state lasers have all been used. Most lasers provide a polarized output and it is necessary to modify this to allow the excitation polarization to be varied. The scattered light is observed via a monochromator in a direction normal to the laser beam. Again, provision is made for the polarization of the scattered radiation to be analyzed. To permit observation closer to the laser line, double or triple monochromators are used to eliminate all traces of stray light. Furthermore, one must take trouble to separate out the Raman-scattered light from any fluorescence signal. Since the Raman signal is instantaneous, it is comparatively straight-forward to recover the desired signal from the decaying fluorescence signal using time-resolution techniques.

An example of the application of Raman spectroscopy in color center physics is shown in Fig. 23. The intensity of scattering versus wavelength shift from the Ar[+] laser excitation is shown for *F*-centers in NaCl,[22] in which crystals the longitudinal optic frequency is $270 \, \text{cm}^{-1}$. The major Raman-shifted spectrum occurs below $200 \, \text{cm}^{-1}$, showing that the vibrational interaction is due to ionic displacements close to the defect. These local modes have broad peak centers near $\hbar\omega = 175 \, \text{cm}^{-1}$. A comparison of the polarized and unpolarized excitation spectra shows that the local mode scattering is supplemented by a lattice vibrational contribution covering much of the 0 to $500 \, \text{cm}^{-1}$ frequency shift.

20.9 REFERENCES

1. B. Henderson and G. F. Imbusch, *Optical Spectroscopy of Inorganic Solids,* Oxford University Press, Oxford, England, 1989.

2. J. R. Lakowicz, *Principles of Fluorescence Spectroscopy,* Plenum Press, New York, 1983.

3. H. Engstrom and L. F. Mollenauer, *Phys. Rev.* B **7:**1616 (1973).

4. M. O. Henry, J. P. Larkin, and G. F. Imbusch, *Phys. Rev.* B **13:**1893 (1976).

5. J. R. Lakowicz and H. Charek, *J. Biol. Chem.* **256:**6348 (1981).

6. L. F. Mollenauer, J. M. Wiesenfeld, and E. P. Ippen, *Radiation Effects* **72:**73 (1983).

7. P. P. Feofilov, *The Physical Basis of Polarized Emission,* Consultants Bureau, New York, 1961.

8. J. C. Kemp, *J. Opt. Soc. Amer.* **59:**915 (1966).

9. S. Geschwind, R. J. Collins, and A. L. Schawlow, *Phys. Rev. Lett.* **3:**545 (1959).

10. P. Edel, C. Hennies, Y. Merle d'Aubigné, R. Romestain, and Y. Twarowski, *Phys. Rev. Lett.,* **28:**1268 (1972).

11. The Stanford group made a number of experimental improvements; see for example T. W. Hänsch, I. S. Shakin, and A. L. Shawlow, *Nature* (Lond.) **235:**63 (1972); T. W. Hänsch, M. H. Nayfeh, S. A. Lee, S. M. Curry, and I. S. Shahin, *Phys. Rev. Lett.* **32:**1336 (1974); and C. E. Wieman and T. W. Hänsch, *Phys. Rev. Lett.* **36:**1170 (1976).

12. A. I. Ferguson and J. M. Tolchard, *Contemp. Phys.* **28:**383 (1987).

13. P. M. Selzer, in W. M. Yen and P. M. Selzer (eds.), *Laser Spectroscopy of Solids I,* Springer-Verlag, Berlin, 1981, p. 113.

14. W. Yen, in W. M. Yen (ed.), *Laser Spectroscopy of Solids II,* Springer-Verlag, Berlin (1988).

15. B. Cagnac, G. Grynberg, and F. Biraben, *J. Phys. Paris* **34:**845 (1973).

16. G. Duxbury, *International Reviews in Physical Chemistry* **4:**237 (1985).

17. W. M. Yen and P. M. Selzer, in W. M. Yen and P. M. Selzer (eds.), *Laser Spectroscopy of Solids I,* Springer-Verlag, Berlin, 1981.

18. P. E. Jessop, T. Muramoto, and A. Szabo, *Phys. Rev.* B **21:**926 (1980).

19. J. W. Tyndall, *Notes of a Course of Nine Lectures on Light,* Royal Institution of Great Britain, Longmans, London, 1869.

20. Lord Rayleigh (Hon. J. W. Strutt), *Phil. Mag.* **x/i:**107 (1871).

21. C. V. Raman, *Indian J. Phys.* **2:**1 (1928).

22. J. M. Worlock and S. P. S. Porto, *Phys. Rev. Lett.* **15:**697 (1965).

CHAPTER 21
INTERFEROMETERS

P. Hariharan
Division of Applied Physics
CSIRO
Sydney, Australia

21.1 GLOSSARY

A	area
C	ratio of peaks to valleys
d	thickness
F	finesse
FSR	free spectral range
I	intensity
$J_i(\ )$	Bessel function
L	fiber length
m	integer
N	number of fringes
p	optical path difference
R	reflectance
r	radius
T	transmittance
λ_s	synthetic wavelength
θ	angle
ν	frequency
ϕ	phase

21.2 INTRODUCTION

Optical interferometers have made feasible a variety of precision measurements using the interference phenomena produced by light waves.[1,2] After a brief survey of the basic types of interferometers, this article will describe some of the interferometers that can be used for such applications as measurements of lengths and small changes in length; optical testing; studies of surface structure; measurements of the pressure and temperature distribution in gas flows and plasmas; measurements of particle velocities and vibration amplitudes; rotating sensing; measurements of temperature, pressure, and electric and magnetic fields; wavelength measurements, and measurements of the angular diameter of stars, as well as, possibly, the detection of gravitational waves.

21.3 BASIC TYPES OF INTERFEROMETERS

Interferometric measurements require an optical arrangement in which two or more beams, derived from the same source but traveling along separate paths, are made to interfere. Interferometers can be classified as *two-beam* interferometers or *multiple-beam* interferometers according to the number of interfering beams; they can also be grouped according to the methods used to obtain these beams. The most commonly used form of beam splitter is a partially reflecting metal or dielectric film on a transparent substrate; other devices that can be used are polarizing prisms and diffraction gratings. The best known types of two-beam interferometers are the Fizeau, the Michelson, the Mach-Zehnder, and the Sagnac interferometers; the best known multiple-beam interferometer is the Fabry-Perot interferometer.

The Fizeau Interferometer

In the Fizeau interferometer, as shown in Fig. 1, interference fringes of equal thickness are formed between two flat surfaces separated by an air gap and illuminated with a collimated beam. If one of the surfaces is a standard reference flat surface, the fringe pattern is a contour map of the errors of the test surface. Modified forms of the Fizeau interferometer are also used to test convex and concave surfaces by using a converging or diverging beam.[3]

The Michelson Interferometer

The Michelson interferometer, shown schematically in Fig. 2, uses a beam splitter to divide and recombine the beams. As can be seen, one of the beams traverses the beam splitter

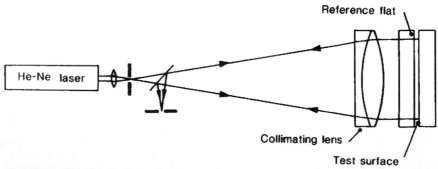

FIGURE 1 The Fizeau interferometer.

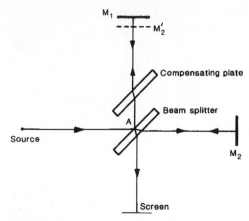

FIGURE 2 The Michelson interferometer.

three times, while the other traverses it only once. Accordingly, a compensating plate of the same thickness as the beam splitter is introduced in the second beam to equalize the optical paths in glass. With an extended source, the interference pattern is similar to that produced in a layer of air bounded by the mirror M_1 and M_2', the image of the other mirror in the beam splitter. With collimated light, fringes of equal thickness are obtained. The Michelson interferometer modified to use collimated light is known as the Twyman-Green interferometer and is used extensively in optical testing.[4]

The Mach-Zender Interferometer

The Mach-Zehnder interferometer uses two beam splitters and two mirrors to divide and recombine the beams. As shown in Fig. 3, the fringe spacing and the plane of localization

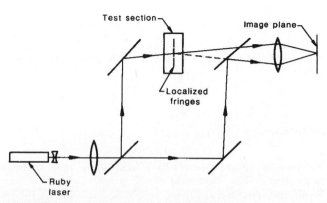

FIGURE 3 The Mach-Zehnder interferometer.

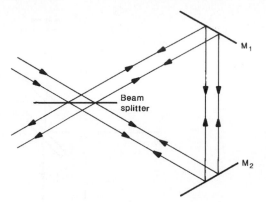

FIGURE 4 The Sagnac interferometer.

of the fringes obtained with an extended source can be controlled by varying the angle between the beams and their lateral separation when they emerge from the interferometer. This makes it possible to use a pulsed ruby laser, which may be operating in more than one transverse mode, as the source. Because the measurement path is traversed only once, and the separation of the beams can be made as large as desired, this interferometer is well suited to studies of gas flows, heat transfer, and the temperature distribution in flames and plasmas.

The Sagnac Interferometer

In the Sagnac interferometer, as shown in Fig. 4, the two beams traverse the same closed path in opposite directions. Because of this, the interferometer is extremely stable and easy to align, even with an extended broadband light source.

Modified versions of the Sagnac interferometer have been used for rotation sensing. When the interferometer is rotated with an angular velocity ω about an axis making an angle θ with the normal to the plane of the interferometer, a phase shift φ is introduced between the beams given by the relation

$$\varphi = (8\pi\omega A \cos \theta)/\lambda c \tag{1}$$

where A is the area enclosed by the light path, λ is the wavelength, and c is the speed of light.

Polarization Interferometers

Polarization interferometers have found their most extensive application in interference microscopy.[5] The Nomarski interferometer, shown schematically in Fig. 5, uses two Wollaston (polarizing) prisms to split and recombine the beams. If the separation of the beams in the object plane (the lateral shear) is small compared to the dimensions of the

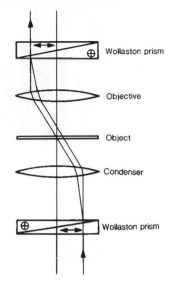

FIGURE 5 The Nomarski interferometer.

object, the optical path difference corresponds to the phase gradients in the test object. A similar setup is also used in reflection to study the surface structure of opaque objects.

Grating Interferometers

Gratings can be used as beam splitters in the Michelson and Mach-Zender interferometers. Such an arrangement is very stable, since the angle between the beams is affected only to a small extent by the orientation of the gratings. Figure 6 is a schematic of an interferometer that has been used to test fine-ground surfaces at grazing incidence utilizing two diffraction gratings to split and recombine the beams.[6]

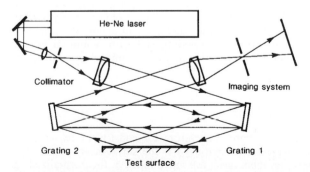

FIGURE 6 Grating interferometer used to test fine-ground surfaces at grazing incidence. [*P. Hariharan*, Opt. Eng. **14:**257 *(1975)*.]

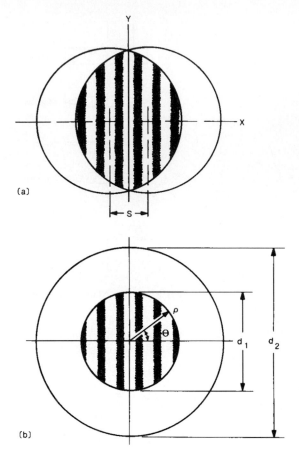

FIGURE 7 Fields of view in (a) lateral and (b) radial shearing interferometers.

Shearing Interferometers

Shearing interferometers are now widely used for optical testing, since they eliminate the need for a reference surface. The most commonly used types of shear are *lateral* shear and *radial* shear. As shown in Fig. 7, in a lateral shearing interferometer two images of the test wavefront are superimposed with a mutual lateral displacement, while in a radial shearing interferometer one of the images is contracted or expanded with respect to the other.[7,8]

The Fabry-Perot Interferometer

The Fabry-Perot interferometer consists of two parallel surfaces with highly reflecting, semitransparent coatings.[9] If the surfaces are separated by a distance d and the medium between them has a refractive index n, the normalized value of the transmitted intensity at a wavelength λ for rays traversing the interferometer at an angle θ is

$$I_T(\lambda) = T^2/(1 + R^2 - 2R \cos \varphi) \tag{2}$$

FIGURE 8 Ray paths in a confocal Fabry-Perot interferometer.

where T and R are, respectively, the transmittance and reflectance of the surfaces and $\varphi = (4\pi/\lambda)nd \cos \theta$. With an extended source of monochromatic light, the fringes seen by transmission are narrow, concentric rings. For a given angle of incidence, the difference in the wavelengths corresponding to successive peaks in the transmitted intensity (the *free spectral range*) is given by the relation

$$\text{FSR}_\lambda = \lambda^2/2nd \qquad (3)$$

The free spectral range corresponds to the range of wavelengths that can be handled without successive orders overlapping. The width of the peaks at half the maximum intensity corresponds to a change in φ given by the relation

$$\Delta\varphi = 2(1 - R)/R^{1/2} \qquad (4)$$

The ratio of the free spectral range to the width of the fringes at half maximum intensity is known as the *finesse F,* and is given by the relation

$$F = \pi R^{1/2}/(1 - R) \qquad (5)$$

Two useful variants of the Fabry-Perot interferometer are the multiple-passed Fabry-Perot interferometer and the confocal Fabry-Perot interferometer. With the conventional Fabry-Perot interferometer, the ratio of the intensity at the maxima to that at the minima between them is given by the relation

$$C = [(1 + R)/(1 - R)]^2 \qquad (6)$$

For typical values of reflectance ($R \approx 0.95$), the background due to a strong spectral line may mask a neighboring weak satellite. A much higher contrast factor may be obtained by double- or multiple-passing the interferometer.[10,11]

The confocal Fabry-Perot interferometer uses two spherical mirrors whose spacing is chosen, as shown in Fig. 8, so that their foci coincide. Any ray, after traversing the interferometer four times, then emerges along its original path.[12] The confocal Fabry-Perot interferometer has a higher throughput than the plane Fabry-Perot interferometer and produces a uniform output field. It is, therefore, the preferred form for operation in a scanning mode by using piezoelectric spacers to vary the separation of the mirrors.

21.4 THREE-BEAM AND DOUBLE-PASSED TWO-BEAM INTERFEROMETERS

Because of the sinusoidal intensity distribution in two-beam interference fringes, it is difficult to estimate their position visually to better than 1/20 of their spacing. However, it is possible to detect much smaller optical path variations using the intensity changes in a

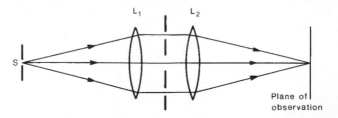

FIGURE 9   Zernike's three-beam interferometer.

uniform interference field. Two types of interferometers use photometric settings on a system of interference fringes to make very accurate measurements.

Three-beam Interferometers

Zernike's three-beam interferometer, shown schematically in Fig. 9, uses three beams produced by division of a wavefront at a screen containing three parallel, equidistant slits.[13] In this arrangement, the optical paths of all three beams are equal at a point in the back focal plane of the lens L_2. The two outer slits provide the reference beams, while the beam from the middle slit, which is twice as broad, is used for measurements. The intensity at any point in the interference pattern is then given by the relation

$$I = I_0[3 + 2 \cos 2\psi + 4 \cos \psi \cos \varphi] \qquad (7)$$

where ψ is the phase difference between the two outer beams, and φ is the phase difference between the middle beam and the two outer beams at the center of the field. As can be seen from the curves in Fig. 10, the intensities at adjacent maxima are equal only when φ is an odd multiple of $\pi/2$. Two positions of the plane of observation can be found that satisfy this condition, one inside and the other outside the focus, and any small change in the optical path of the middle beam can be measured from the shift in these positions.

Three-beam fringes can also be produced by amplitude division, using an optical system similar to that in the Jamin interferometer.[14] Settings are made by means of a compensator in the middle beam and can be repeated to $\lambda/200$ by visual observation, and to better than $\lambda/1000$ with a photoelectric detector.[15]

Double-passed Two-beam Interferometers

Fringes whose intensity is modulated in the same manner as three-beam fringes can also be produced by reflecting the beams emerging from a two-beam interferometer back through the interferometer.[16] In the arrangement shown in Fig. 11, the beams reflected back after a single pass are eliminated by the polarizer P_2, whose axis is at right angles to the axis of P_1, while the double-passed beams, which have traversed the $\lambda/4$ plate twice, are transmitted by P_2. In this case also, the intensity of the adjacent fringes is equal when the phase difference between the single-passed beams is

$$\varphi = (2m + 1)\pi/2 \qquad (8)$$

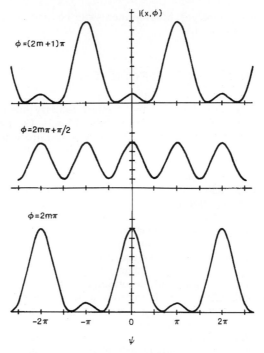

FIGURE 10 Intensity distribution in the fringes formed in a three-beam interferometer.

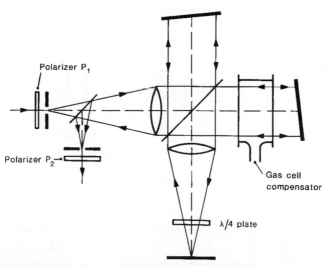

FIGURE 11 Double-passed two-beam interferometer. [*P. Hariharan and D. Sen,* J. Opt. Soc. Am. **50:***357 (1960)*.]

where m is an integer. Measurements can be made with a precision of $\lambda/1000$ using a gas cell as a compensator.

21.5 FRINGE-COUNTING INTERFEROMETERS

One of the main applications of interferometry has been in accurate measurements of length using the wavelengths of stabilized lasers. Because of the high degree of coherence of laser light, interference fringes can be obtained even with quite large optical path differences, and electronic fringe counting has become a very practical technique for length measurements.

The earliest fringe counting interferometers used an optical system giving two uniform interference fields, in one of which an additional phase difference of $\pi/2$ was introduced between the interfering beams. Two photodetectors viewing these fields provided signals in quadrature that were used to drive a bidirectional counter. Another arrangement used two orthogonally polarized beams which were converted by a $\lambda/4$ plate oriented at 45° into right- and left-handed circularly polarized beams, respectively. When these two beams were superposed, they produced a linearly polarized beam whose plane of polarization rotated through 360° for a change in the optical path diffrence of two wavelengths (an optical screw). Changes in the orientation of the plane of polarization and, hence, in the optical path difference were monitored by a polarizer controlled by a servo system.[17]

The very narrow spectral line widths of lasers make it possible to use a heterodyne system, which has the advantage that its operation is not significantly affected by variations in the intensity of the source. In one implementation of this technique, a He-Ne laser is forced to oscillate simultaneously at two frequencies, v_1 and v_2, separated by a constant frequency difference of about 2 MHz, by applying an axial magnetic field.[18] These two waves, which are circularly polarized in opposite senses, are converted to orthogonal linear polarizations by a $\lambda/4$ plate. Alternatively, two acousto-optic modulators driven at different frequencies can be used to introduce a suitable frequency difference between two orthogonally polarized beams derived from a single-frequency laser.

As shown in Fig. 12, a polarizing beam splitter reflects one beam to a fixed reflector, while the other is transmitted to a movable reflector. A differential counter receives the

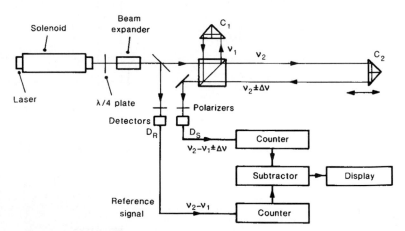

FIGURE 12 Heterodyne fringe-counting interferometer. [*After J. N. Dukes and G. B. Gordon,* Hewlett-Packard Journal, **21**(12) (1970). © Copyright Hewlett-Packard Company. Reproduced with permission.]

beat frequencies from the photodetector D_S and a reference photodetector D_R. If the two reflectors are stationary, the two beat frequencies are the same, and the net count is zero. However, if one of the optical paths is varied by translating one of the reflectors, the change in wavelength is given by the net count.

21.6 TWO-WAVELENGTH INTERFEROMETRY

If a length is known within certain limits, the use of a wavelength longer than the separation of these limits permits its exact value to be determined unambiguously by a single interferometric measurement. One way to synthesize such a long wavelength is by illuminating the interferometer simultaneously with two wavelengths λ_1 and λ_2. The envelope of the fringes then corresponds to the interference pattern that would be obtained with a synthetic wavelength

$$\lambda_s = \lambda_1 \lambda_2 / |\lambda_1 - \lambda_2| \tag{9}$$

This technique can be implemented very effectively with a carbon dioxide laser, since it can operate at a number of wavelengths that are known very accurately, yielding a wide range of synthetic wavelengths.[19]

Two-wavelength interferometry and fringe-counting can be combined to measure lengths up to 100 m by switching the laser rapidly between two wavelengths as one of the mirrors of a Twyman-Green interferometer is moved over the distance to be measured. The output signal from a photodetector is squared, low-pass filtered, and processed in a computer to yield the fringe count and the phase difference for the corresponding synthetic wavelength.[20]

21.7 FREQUENCY-MODULATION INTERFEROMETERS

New interferometric techniques are possible with laser diodes which can be tuned electrically over a range of wavelengths.[21] One of these is frequency-modulation interferometry.

Figure 13 shows a frequency-modulation interferometer that can be used to measure

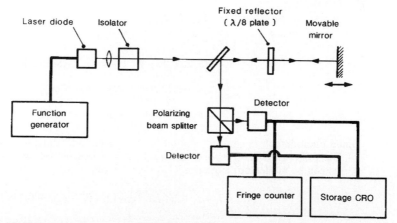

FIGURE 13 Frequency-modulation interferometer for measurements of distance. [*T. Kubota, M. Nara, and T. Yoshino,* Opt. Lett. **12:***310 (1987).*]

absolute distances, as well as relative displacements, with high accuracy.[22] In this arrangement, the signal beam reflected from the movable mirror returns as a circularly polarized beam, since it traverses the $\lambda/8$ plate twice. The reference beam reflected from the front surface of the $\lambda/8$ plate interferes with the two orthogonally polarized components of the signal beam at the two detectors to produce outputs that vary in quadrature and can be fed to a counter to determine the magnitude and sign of any displacement of the movable mirror.

To make direct measurements of the optical path between the reference surface and the movable mirror, the frequency of the laser is ramped linearly with time by using a function generator to vary the injection current of the laser. An optical path difference p introduces a time delay p/c between the two beams, so that they interfere at the detector to produce a beat signal with a frequency

$$f = (p/c)(dv/dt) \tag{10}$$

where dv/dt is the rate at which the laser frequency is varying with time. With this arrangement, distances of a few meters can be measured directly with an accuracy of $100 \ \mu m$.

21.8 HETERODYNE INTERFEROMETERS

In heterodyne interferometers, a frequency difference is introduced between the two beams by means of two acousto-optic modulators operated at slightly different frequencies. The electric fields due to the two beams can then be represented by the relations

$$E_1(t) = a_1 \cos{(2\pi v_1 t + \varphi_1)} \tag{11}$$

$$E_2(t) = a_2 \cos{(2\pi v_2 t + \varphi_2)} \tag{12}$$

where a_1 and a_2 are the amplitudes, v_1 and v_2 are the frequencies, and φ_1 and φ_2 are the phases of the two waves. The output from an ideal square-law detector would be, therefore,

$$
\begin{aligned}
I(t) &= [E_1(t) + E_2(t)]^2 \\
&= \tfrac{1}{2}a_1^2 + \tfrac{1}{2}a_2^2 \\
&\quad + \tfrac{1}{2}[a_1^2 \cos{(4\pi v_1 t + 2\varphi_1)} + a_2^2 \cos{(4\pi v_2 t + 2\varphi_2)}] \\
&\quad + a_1 a_2 \cos{[2\pi(v_1 + v_2)t + (\varphi_1 + \varphi_2)]} \\
&\quad + a_1 a_2 \cos{[2\pi(v_1 - v_2)t + (\varphi_1 - \varphi_2)]}
\end{aligned}
\tag{13}
$$

Since photodetectors cannot respond to the second and third terms on the right-hand side of Eq. (13), which correspond to signals at frequencies of $2v_1, 2v_2$, and $(v_1 + v_2)$, these terms can be neglected. Equation (13) then becomes

$$I(t) = I_1 + I_2 + 2(I_1 I_2)^{1/2} \cos{[2\pi(v_1 - v_2)t + (\varphi_1 - \varphi_2)]} \tag{14}$$

where $I_1 = \tfrac{1}{2}a_1^2$ and $I_2 = \tfrac{1}{2}a_2^2$. The output from the detector therefore contains an ac component at the difference frequency $(v_1 - v_2)$, whose phase is $(\varphi_1 - \varphi_2)$. The phase

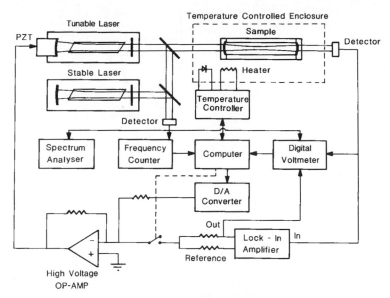

FIGURE 14 Heterodyne interferometer for measurements of thermal expansion.
[*S. F. Jacobs and D. Shough*, Appl. Opt. **20**:*3461 (1981).*]

difference between the interfering light waves can then be obtained from measurements of
the phase of the ac component with respect to a reference signal.[23]

Heterodyne techniques can also be used for measurements of very small changes in
length.[24,25] In the setup shown in Fig. 14, the frequency of a laser is locked to a
transmission peak of a Fabry-Perot interferometer formed by attaching two mirrors to the
ends of the sample. The beam from this slave laser is mixed at a photodetector with the
beam from a stable reference laser. Changes in the separation of the mirrors can be
evaluated from the changes in the beat frequency.

21.9 *PHASE-SHIFTING INTERFEROMETERS*

In phase-shifting interferometers, the phase difference between the two beams in the
interferometer is varied linearly with time and the values of intensity at any point in the
interference pattern are integrated over a number of equal segments covering one period
of the sinusoidal signal. Alternatively the phase difference between the two beams can be
changed in a number of equal steps, and the corresponding values of intensity at each data
point are measured and stored. In both cases, the values obtained can be represented by a
Fourier series, whose coefficients can be evaluated to obtain the original phase difference
between the interfering beams at each point.[26,27] Typically, four measurements are made at
each point, corresponding to phase intervals of 90°. If I_1, I_2, I_3, and I_4 are the values of
intensity obtained, the phase difference between the interfering beams is given by the
relation

$$\tan \varphi(x, y) = (I_1 - I_3)/(I_2 - I_4) \tag{15}$$

Phase-shifting interferometers are used widely in optical testing, since a detector array can
be used in conjunction with a microcomputer to make measurements simultaneously at a
large number of points covering the interference pattern.

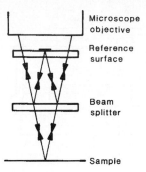

FIGURE 15 Schematic of a compact optical system (the Mirau interferometer) used for phase-stepping interference microscopy.

Figure 15 is a schematic of a compact optical system (the Mirau interferometer) used for phase-shifting interference microscopy. In this setup, the phase-steps are introduced by mounting the beam splitter on a piezoelectric transducer (PZT) to which an appropriately varying voltage is applied. In a Fizeau or Twyman-Green interferometer, it is possible to use a laser diode as the light source and vary its output frequency.[28] If the initial optical path difference between the beams in the interferometer is p, a frequency shift Δv in the output of the laser diode introduces an additional phase difference between the beams

$$\Delta\varphi = (2\pi p/v)\,\Delta v \tag{16}$$

21.10 PHASE-LOCKED INTERFEROMETERS

The output intensity from an interferometer depends on the phase difference between the beams. In phase-locked interferometers, any variation in the output intensity is detected and fed back to a phase modulator in the measurement path so as to hold the output intensity constant. The changes in the optical path can then be estimated from the changes in the drive signal to the phase modulator.[29]

Drifts can be eliminated by using an ac amplifier. In this case, the phase of one beam in the interferometer is subjected to a sinusoidal modulation

$$\Delta\varphi(t) = \Delta\varphi \sin \omega t \tag{17}$$

with an amplitude $\Delta\varphi \ll \pi$. The intensity in the interference pattern can then be written as

$$I(t) = I_1 + I_2 + 2(I_1 I_2)^{1/2} \cos\left[\varphi + \Delta\varphi(t)\right] \tag{18}$$

where I_1 and I_2 are the intensities due to the two beams taken separately, and φ is the mean phase difference between them. If the last term on the right-hand side of Eq. (18) is expanded as an infinite series of Bessel functions, we have

$$\begin{aligned} I(t) = I_1 + I_2 + 2(I_1 I_2)^{1/2}[J_0(\Delta\varphi) + 2J_2(\Delta\varphi) \cos 2\omega t + \cdots] \cos \varphi \\ - 2(I_1 I_2)^{1/2}[2J_1(\Delta\varphi) \sin \omega t + 2J_3(\Delta\varphi) \sin 3\omega t + \cdots] \sin \varphi \end{aligned} \tag{19}$$

The signal at the modulation frequency has an amplitude

$$I_\omega(t) = 4(I_1 I_2)^{1/2} J_1(\Delta\varphi) \sin \varphi \tag{20}$$

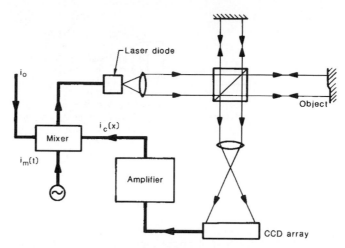

FIGURE 16 Schematic of a phase-locked interferometer using a laser diode source. [*T. Suzuki, O. Sasaki, and T. Maruyama,* Appl. Opt. **28**:*4407 (1989).*]

and drops to zero when $\varphi = m\pi$, where m is an integer. Since, at this point, both the magnitude and the sign of this signal change, it can be used as the input to a servo system that locks the phase difference between the beams at this point.

With a laser diode, it is possible to compensate for changes in the optical path difference by a change in the illuminating wavelength. A typical setup is shown in Fig. 16. The injection current of the laser then consists of a dc bias current i_0, a control current i_c, and a sinusoidal modulation current $i_m(t) = i_m \cos \omega t$ whose amplitude is chosen to produce the required phase modulation.[30]

Direct measurements of changes in the optical path are possible by sinusoidal phase-modulating interferometry, which uses a similar setup, except that in this case the amplitude of the phase modulation is much larger (typically around π radians). The modulation amplitude is determined from the amplitudes of the components in the detector output corresponding to the modulation frequency and its third harmonic. The average phase difference between the beams can then be determined from the amplitudes of the components at the modulation frequency and its second harmonic.[31]

21.11 LASER-DOPPLER INTERFEROMETERS

Light scattered from a moving particle undergoes a frequency shift, due to the Doppler effect, that is proportional to the component of its velocity in a direction determined by the directions of illumination and viewing. With laser light, this frequency shift can be evaluated by measuring the frequency of the beats produced by the scattered light and a reference beam.[32] Alternatively, it is possible to observe the beats produced by scattered light from two illuminating beams incident at different angles.[33]

Laser-Doppler interferometry can be used for measurements of the velocity of moving material,[34] as well as for measurements, at a given point, of the instantaneous flow velocity of a moving fluid to which suitable tracer particles have been added.[35] A typical optical

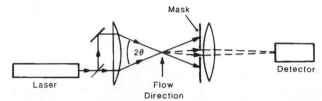

FIGURE 17 Optical arrangement used for laser-Doppler velocimetry.

system for measurements on fluids is shown in Fig. 17. If the two illuminating beams in this arrangement make equal but opposite angles $\pm\theta$ with the viewing direction, the frequency of the beat signal is given by the relation

$$f = (2\,|v|\sin\theta)/\lambda \tag{21}$$

where v is the component of the velocity of the particle in the plane of the beams at right angles to the direction of observation. To distinguish between positive and negative flow directions, the frequency of one of the beams is offset by a known amount by means of an acousto-optical modulator. Simultaneous measurements of the velocity components along two orthogonal directions can be made by using two pairs of beams in orthogonal planes. Interactions between the two pairs of beams are avoided by using different laser wavelengths.

Laser diodes and optical fibers can be used to build very compact laser-Doppler interferometers.[36,37] A frequency offset can be introduced between the beams either by using a piezoelectric fiber-stretcher driven by a sawtooth waveform in one path, or by ramping the injection current of the laser diode linearly.

Laser-Doppler interferometry can also be used to measure vibration amplitudes. Typically, one of the beams in an interferometer is reflected from a point on the vibrating specimen, while the other, whose frequency is offset, is reflected from a fixed reference mirror. The output from a photodetector then consists of a component at the offset frequency (the carrier) and two sidebands. The amplitude of the vibration can be determined from a comparison of the amplitudes of the carrier and the sidebands.[38] Modifications of this technique can be used to measure vibration amplitudes down to a few thousandths of a nanometer.[39]

21.12 LASER-FEEDBACK INTERFEROMETERS

Laser-feedback interferometers make use of the fact that the output of a laser is strongly affected if, as shown in Fig. 18, a fraction of the output beam is reflected back into the laser cavity by an external mirror M_3. The output of the laser then varies cyclically with the position of M_3, one cycle of modulation corresponding to a displacement of M_3 by half a wavelength.[40]

The operation of such an interferometer can be analyzed by considering the two

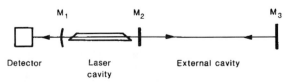

FIGURE 18 Schematic of a laser-feedback interferometer.

mirrors M_3 and M_2 as a Fabry-Perot interferometer that replaces the output mirror of the laser. A variation in the spacing of M_3 and M_2 results in a variation in the reflectivity of this interferometer for the laser wavelength and, hence, in the gain of the laser.

A very compact laser-feedback interferometer can be set up with a single-mode laser diode.[41] Small displacements can be detected by measuring the changes in the laser output when the laser current is held constant. Measurements can be made over a larger range by mounting the laser on a piezoelectric transducer and using an active feedback loop to stabilize the length of the optical path from the laser to the mirror.[42]

Laser-feedback interferometers can also be used for velocimetry. If the light reflected from the moving object is mixed with the original oscillating wave inside the laser cavity, the beat signal can be observed in the beam leaving the rear end of the laser.[43,44] Very high sensitivity can be obtained with a laser diode operated near threshold.[45] If a separate external cavity is used, as shown in Fig. 19, to ensure single-mode operation, measurements can be made at distances up to 50 m.

21.13 FIBER INTERFEROMETERS

Analogs of conventional two-beam interferometers can be built with single-mode optical fibers. High sensitivity can be obtained with fiber interferometers because it is possible to have very long optical paths in a small space. In addition, because of the extremely low noise level, sophisticated detection techniques can be used.

Fiber-interferometer Rotation Sensors

Fiber interferometers were first used for rotation sensing, by replacing the ring cavity in a conventional Sagnac interferometer with a closed, multiturn loop made of a single-mode fiber.[46] For a loop rotating with an angular velocity ω about an axis making an angle θ with the plane of the loop, the phase difference introduced between the two counterpropagating beams is

$$\Delta\varphi = (4\pi\omega Lr \cos\theta)/\lambda c \qquad (22)$$

where L is the length of the fiber, r is the radius of the loop, λ is the wavelength, and c is the speed of light. High sensitivity can be obtained by increasing the length of the fiber in the loop. In addition, very small phase shifts can be measured, and the sense of rotation determined, by introducing a nonreciprocal phase modulation in the beams and using a phase-sensitive detector.[47]

Figure 20 is a schematic of a typical all-fiber interferometric rotation sensor.[48] In this arrangement, the beam splitters are replaced by optical couplers, and a phase modulator consisting of a few turns of the fiber wound around a piezoelectric cylinder is located near one end of the optical fiber coil.

Fiber-interferometer rotation sensors are an attractive alternative to other types of rotation sensors and have the advantages of small size and low cost. If care is taken to minimize noise due to back scattering and nonreciprocal effects due to fiber birefringence, performance close to the limit set by photon noise can be obtained.[49]

Generalized Fiber-interferometer Sensors

The optical path length in a fiber is affected by its temperature and also changes when the fiber is stretched, or when the pressure changes. Accordingly, an optical fiber can be used in an interferometer to sense changes in these parameters.[50]

Figure 21 is a schematic of an all-fiber interferometer that can be used for such

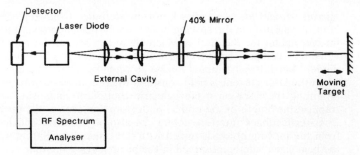

FIGURE 19   Feedback interferometer using a diode laser for velocimetry. [*P. J. de Groot and G. M. Gallatin, Opt. Lett.* **14:***165 (1989).*]

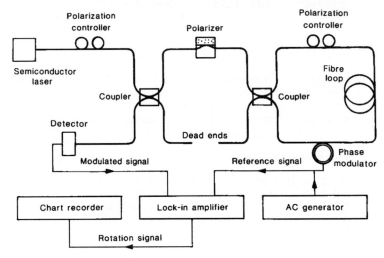

FIGURE 20   Fiber-interferometer for rotation sensing. [*R. A. Bergh, H. C. Lefevre, and H. J. Shaw, Opt. Lett.* **6:***502 (1981).*]

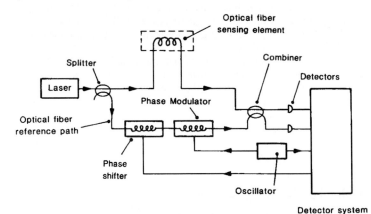

FIGURE 21   Schematic of a typical fiber-interferometer sensor. [*T. G. Giallorenzi, J. A. Bucaro, A. Dandridge, G. H. Sigel, Jr., J. H. Cole, S. Rashleigh, and R. G. Priest, IEEE J. Quantum Electron.* **QE-18:***626 (1982).*]

measurements.[51] A layout analogous to a Mach-Zehnder interferometer has the advantage that it avoids optical feedback to the laser. Optical fiber couplers are used to divide and recombine the beams, and measurements can be made with either a heterodyne system or a phase-tracking system. Detection schemes involving either laser-frequency switching or a modulated laser source can also be used. Optical phase shifts as small as 10^{-6} radian can be detected.

Fiber interferometers have been used as sensors for mechanical strains and changes in temperature and pressure. They can be used for measurements of magnetic fields by bonding the fiber sensor to a magnetostrictive element.[52] Electric fields can also be measured by using a single-mode fiber sensor bounded to a piezoelectric film, or jacketed with a piezoelectric polymer.[53]

Multiplexed Fiber-interferometer Sensors

Fiber-interferometer sensors can be multiplexed to measure different quantities at different locations with a single light source and detector and the same set of transmission lines. Techniques developed for this purpose include frequency-division multiplexing, time-division multiplexing, and coherence multiplexing.[54-57] These techniques can be combined to handle a larger number of sensors.[58]

21.14 INTERFEROMETRIC WAVE METERS

The availability of tunable lasers has created a need for instruments that can measure their output wavelength with an accuracy commensurate with their narrow line width. Two types of interferometric wave meters have been developed for this purpose: dynamic wave meters and static wave meters. In dynamic wave meters, the measurement involves the movement of some element; they have greater accuracy but can be used only with continuous sources. Static wave meters can also be used with pulsed lasers.

Dynamic Wave Meters

A dynamic wave meter typically consists of a two-beam interferometer in which the number of fringes crossing the field is counted as the optical path is changed by a known amount. In one form, shown in Fig. 22, two beams, one from the laser whose wavelength

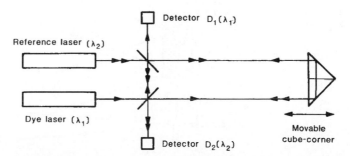

FIGURE 22 Optical system of a dynamic interferometric wave meter. [*F. V. Kowalski, R. T. Hawkins, and A. L. Schawlow*, J. Opt. Soc. Am. **66**:*965 (1976).*]

is to be determined and another from a frequency stabilized He-Ne laser, traverse the same two paths in opposite directions.[59] The fringe system formed by these two lasers are imaged on the two detectors D_1 and D_2, respectively. If, then, the end reflector is moved through a distance d, we have

$$\lambda_1/\lambda_2 = N_2/N_1 \tag{23}$$

where N_1 and N_2 are the numbers of fringes seen by D_1 and D_2, respectively, and λ_1 and λ_2 are the wavelengths in air. To obtain the highest precision, it is also necessary to measure the fractional order numbers. This can be done by phase-locking an oscillator to an exact multiple of the frequency of the ac signal from the reference channel, or by digitally averaging the two signal frequencies.[60] It is also possible to use a vernier method in which the counting cycle starts and stops when the phases of the two signals coincide.[61] With these techniques, a precision of 1 part in 10^9 can be obtained.

Another type of dynamic wave meter uses a scanning Fabry-Perot interferometer in which the separation of the mirrors is changed slowly. If this interferometer is illuminated with the two wavelengths to be compared, peak transmission will be obtained for both wavelengths at intervals given by the condition

$$m_1\lambda_1 = m_2\lambda_2 = p \tag{24}$$

where m_1 and m_2 are the changes in the integer order and p is the change in the optical path difference.[62] A precision of 1 part in 10^7 can be obtained with a range of movement of only 25 mm, because the Fabry-Perot fringes are much sharper than two-beam fringes.

Static Wave Meters

The simplest type of static wave meter is based on the Fizeau interferometer.[63] As shown in Fig. 23, a collimated beam from the laser is incident on two uncoated fused-silica flats separated by about 1 mm and making an angle of about 3 minutes of arc with each other. The intensity distribution in the fringe pattern formed in the region in which the shear between the two reflected beams is zero is recorded by a linear detector array.[64] In the first step, the integral interference order is calculated from the spatial period of the interference pattern: the exact value of the wavelength is then calculated from the positions of the

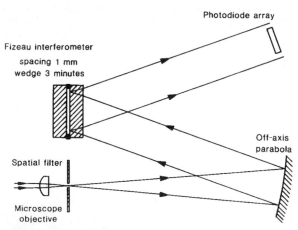

FIGURE 23 Schematic of a static interferometric wave meter. [*J. J. Snyder,* Proc. SPIE **288**:*258 (1981)*.]

maxima and minima. This wave meter can be used with pulsed lasers, and up to 15 measurements can be made every second, with an accuracy of 1 part in 10^7.

21.15 SECOND-HARMONIC AND PHASE-CONJUGATE INTERFEROMETERS

Nonlinear optical elements are used in second-harmonic and phase-conjugate interferometers.[65]

Second-harmonic Interferometers

One type of second-harmonic interferometer, shown in Fig. 24, is an analog of the Mach-Zehnder interferometer.[66] In this interferometer, the infrared beam from a Q-switched Nd:YAG laser ($\lambda_1 = 1.06\ \mu$m) is incident on a frequency-doubling crystal. The green ($\lambda_2 = 0.53\ \mu$m) and infrared beams emerging from this crystal traverse the test piece and are then incident on another frequency-doubling crystal. Fringes with good visibility are obtained, since very nearly the same fraction of the power in the infrared beam is converted into green light by each crystal.

The fringe number at any point in this interferometer is

$$N = (n_1 - n_2)d/\lambda_2 \qquad (25)$$

where n_1 and n_2 are the refractive indices of the test specimen at 1.06 and 0.53 μm, respectively, and d is its thickness.

Second-harmonic interferometers have potential applications in measurements of the electron density in plasmas because of the large dispersion due to the electrons.

Phase-conjugate Interferometers

The conjugate of a light wave can be generated by a range of nonlinear interactions. In one group of phase-conjugate interferometers, the wavefront that is being studied is made to interfere with its conjugate.[67] Such an interferometer has the advantage that a reference wavefront is not required; in addition, the sensitivity of the interferometer is doubled.

Figure 25 is a schematic of a phase-conjugate interferometer that is an analog of the Fizeau interferometer.[68] In this interferometer, the signal beam is incident on a conventional, partially reflecting mirror placed in front of a single crystal of barium titanate which functions as an internally self-pumped phase-conjugate mirror.

An interferometer in which both mirrors have been replaced by phase-conjugating mirrors has some unique properties.[69] Such an interferometer is unaffected by misalignment of the mirrors and the field of view is normally completely dark. However, because of the delay in the response of the phase conjugator, dynamic changes in the optical path difference are displayed.[70]

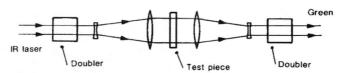

FIGURE 24 Second-harmonic interferometer: analog of the Mach-Zehnder interferometer. [*F. A. Hopf, A. Tomita, and G. Al-Jumaily, Opt. Lett.* **5:***386 (1980).*]

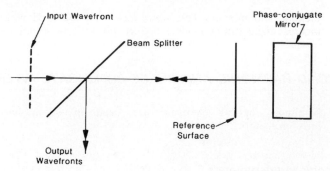

FIGURE 25 Schematic of a phase-conjugate Fizeau interferometer. [*D. J. Gauthier, R. W. Boyd, R. K. Jungquist, J. B. Lisson, and L. L. Voci,* Opt. Lett. **14:***323 (1989).*]

21.16 STELLAR INTERFEROMETERS

A star can be considered as an incoherent light source whose dimensions are small compared to its distance from the earth. Accordingly, the complex degree of coherence between the fields at two points on the earth's surface is given by the normalized Fourier transform of the intensity distribution over the stellar disc.

21.17 MICHELSON'S STELLAR INTERFEROMETER

Michelson used the interferometer shown schematically in Fig. 26 to make observations of the visibility of the fringes formed by light from a star, for different separations of the

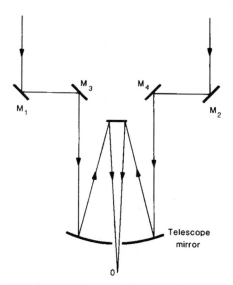

FIGURE 26 Michelson's stellar interferometer.

mirrors. The separation at which the fringes disappeared was used to determine the angular diameter of the star. The problems encountered by Michelson in making measurements at mirror separations greater than 6 m have been overcome in a new version of this interferometer using modern detection, control, and data-processing techniques, that is expected to make measurements over baselines extending up to 1000 m.[71]

The Intensity Interferometer

The intensity interferometer uses measurements of the degree of correlation between the fluctuations in the intensity at two photodetectors separated by a suitable distance.[72] With a thermal source, the correlation is proportional to the square of the modulus of the degree of coherence of the fields. The intensity interferometer has the advantage that atmospheric turbulence only affects the phase of the incident wave and has no effect on the measured correlation. In addition, since the spectral bandwidth is limited by the electronics, it is only necessary to equalize the optical paths to within a few centimeters. It was therefore possible to use light collectors with a diameter of 6.5 m separated by distances up to 188 m, corresponding to a resolution of 0.42×10^{-3} second of arc.

Heterodyne Stellar Interferometers

In heterodyne stellar interferometers, as shown in Fig. 27, light from the star is mixed with light from two lasers, whose frequencies are offset by 5 MHz with respect to each other, at two photodetectors, and the resulting heterodyne signals are multiplied in a correlator. The output signal from the correlator is a measure of the degree of coherence of the wave fields at the two photodetectors.[73]

As in the intensity interferometer, it is only necessary to equalize the two paths to within a few centimeters in the heterodyne stellar interferometer. However, higher sensitivity is obtained, because the output is proportional to the product of the intensities of the laser and the star.

The sensitivity of a heterodyne detector relative to a conventional detector increases with the wavelength.[73] Heterodyne interferometers therefore use a carbon dioxide laser operating at a wavelength of 10.6 μm. Another advantage of operating in the mid-infrared is that the useful diameter of a diffraction-limited telescope, which is limited by seeing conditions, increases as $\lambda^{6/5}$, permitting the study of faint objects. The largest heterodyne stellar interferometer constructed so far, with two 1.65-m telescopes, is expected to achieve a resolution of 0.001 second of arc.[74]

21.18 GRAVITATIONAL-WAVE INTERFEROMETERS

The general theory of relatively predicts the existence of gravitational waves, which can be thought of as an alternating strain that propagates through space, affecting the dimensions and spacing of all material objects. Potential sources of gravitational radiation are binary systems of neutron stars, collapsing supernovas, and black holes.

Figure 28 is a schematic of an interferometric setup that could be used to detect gravitational waves. It consists of a Michelson interferometer in which the beam splitter and the end mirrors are attached to separate, freely suspended masses. Because gravitational waves are transverse quadrupole waves, the arm lying along the direction of propagation of the wave is unaffected, while the other arm experiences a change in length at the frequency of the gravitational wave. For a gravitational wave propagating at right

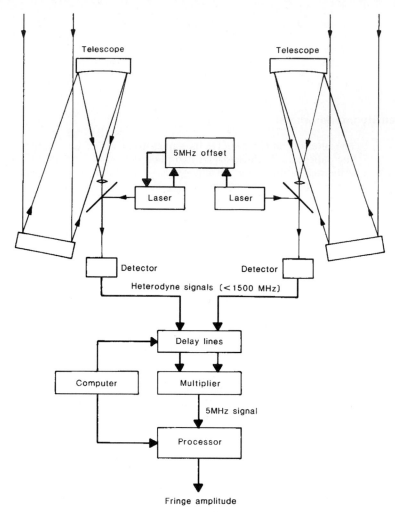

FIGURE 27 Schematic of an infrared heterodyne stellar interferometer. [*M. A. Johnson, A. L. Betz, and C. H. Townes,* Phys. Rev. Lett. **33:***1617 (1974).*]

angles to the plane of the interferometer, the changes in the lengths of the arms have opposite signs, so that the output is doubled.[75]

The ultimate limit to the sensitivity of an interferometer is set by photon noise. Since estimates of the intensity of bursts of gravitational radiation reaching the earth indicate that it would be necessary to detect strains of the order of 1 part in 10^{21}, such a simple setup would need to have unrealistically long arms (>100 km) with available laser powers.

Several methods are currently being explored for obtaining higher sensitivity.[76] One way is by using multiple reflections between two mirrors in each arm to increase the effective paths. Another involves using two identical Fabry-Perot interferometers at right angles to each other, with their mirrors mounted on freely suspended masses. The separations of the mirrors are compared by locking the frequency of a laser to a transmission peak of one interferometer, and using a servo system to adjust the length of the other interferometer continuously, so that its peak transmittance is at this frequency.

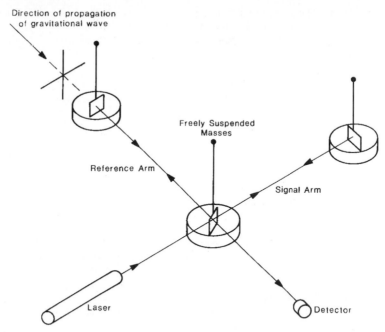

FIGURE 28   Schematic of an interferometer for detecting gravitational waves.

Even higher sensitivity can be obtained with limited laser power by making use of the fact that the interferometer is normally adjusted so that observations are made on a dark fringe, to avoid overloading the detector. Most of the light is then reflected back to the source. This light can be recycled if an additional mirror is used to reflect it back into the interferometer with the right phase. Another technique is to interchange the light paths in the arms during each half-cycle of the gravitational wave, so that the phase shifts accumulate.

Strain sensitivities of a few parts in 10^{18} have been achieved with prototype interferometers having arm lengths of 10 to 40 m, and gravitational-wave interferometers with arm lengths of up to 3 km are now under construction.

21.19   REFERENCES

1. W. H. Steel, *Interferometry,* Cambridge Univ. Press, Cambridge, 1983.

2. P. Hariharan, *Optical Interferometry,* Academic Press, Sydney, 1985.

3. M. V. Mantravadi, "Newton, Fizeau and Haidinger Interferometers," in D. Malacara (ed.), *Optical Shop Testing,* John Wiley, New York, 1991, pp. 1–49.

4. D. Malacara, "Twyman-Green Interferometer," in D. Malacara (ed.), *Optical Shop Testing,* John Wiley, New York, 1991, pp. 51–94.

5. M. Francon and S. Mallick, *Polarization Interferometers: Applications in Microscopy and Macroscopy,* Wiley-Interscience, London, 1971.

6. P. Hariharan, "Improved Oblique-Incidence Interferometer," *Opt. Eng.* **14**:257–258 (1975).

7. M. V. Mantravadi, "Lateral Shearing Interferometers," in D. Malacara (ed.), *Optical Shop Testing,* John Wiley, New York, 1991, pp. 123–172.

8. D. Malacara, "Radial, Rotational and Reversal Shear Interferometers," in D. Malacara (ed.), *Optical Shop Testing,* John Wiley, New York, 1991, pp. 173–206.

9. J. M. Vaughan, *The Fabry-Perot Interferometer,* Adam Hilger, Bristol, 1989.

10. P. Hariharan and D. Sen, "Double-Passed Fabry–Perot Interferometer," *J. Opt. Soc. Am.* **51:**398–399 (1961).

11. J. R. Sandercock, "Brillouin Scattering Study of SbSI Using a Double-Passed Stabilised Scanning Interferometer," *Opt. Commun.* **2:**73–76 (1970).

12. M. Hercher, "The Spherical Mirror Fabry-Perot Interferometer," *Appl. Opt.* **7:**951–966 (1968).

13. F. Zernike, "A Precision Method for Measuring Small Phase Differences," *J. Opt. Soc. Am.* **40:**326–328 (1950).

14. P. Hariharan and D. Sen, "Three-Beam Interferometer," *J. Sci. Instrum.* **36:**70–72 (1959).

15. P. Hariharan, D. Sen, and M. S. Bhalla, "Photoelectric Setting Methods for a Three-Beam Interferometer," *J. Sci. Instrum.* **36:**72–75 (1959).

16. P. Hariharan and D. Sen, "Double-Passed Two-Beam Interferometers," *J. Opt. Soc. Am.* **50:**357–361 (1960).

17. G. R. Hopkinson, "The Optical Screw as a Path Difference Measurement and Control Device: Analysis of Periodic Errors," *J. Optics* (Paris) **9:**151–155 (1978).

18. J. N. Dukes and G. B. Gordon, "A Two-Hundred Foot Yardstick With Graduations Every Microinch," *Hewlett-Packard Journal* **21**(12):2–8 (1970).

19. C. W. Gillard and N. E. Buholz, "Progress in Absolute Distance Interferometry," *Opt. Eng.* **22:**348–353 (1983).

20. H. Matsumoto, "Synthetic Interferometric Distance-Measuring System Using a CO_2 Laser," *Appl. Opt.* **25:**493–498 (1986).

21. P. Hariharan, "Interferometry with Laser Diodes," *Proc. SPIE* **1400:**2–10 (1991).

22. T. Kubota, M. Nara, and T. Yoshino, "Interferometer for Measuring Displacement and Distance," *Opt. Lett.* **12:**310–312 (1987).

23. R. Crane, "Interference Phase Measurement," *Appl. Opt.* **8:**538–542 (1969).

24. S. F. Jacobs and D. Shough, "Thermal Expansion Uniformity of Heraeus-Amersil TO8E Fused Silica," *Appl. Opt.* **20:**3461–3463 (1981).

25. S. F. Jacobs, D. Shough, and C. Connors, "Thermal Expansion Uniformity of Materials for Large Telescope Mirrors," *Appl. Opt.* **23:**4237–4244 (1984).

26. J. H. Bruning, D. R. Herriott, J. E. Gallagher, D. P. Rosenfeld, A. D. White, and D. J. Brangaccio, "Digital Wavefront Measuring Interferometer for Testing Optical Surfaces and Lenses," *Appl. Opt.* **13:**2693–2703 (1974).

27. K. Creath, "Phase-Measurement Interferometry Techniques," in E. Wolf (ed.), *Progress in Optics,* vol. XXVI, Elsevier, Amsterdam, 1988, pp. 349–393.

28. Y. Ishii, J. Chen, and K. Murata, "Digital Phase-Measuring Interferometry with a Tunable Laser Diode," *Opt. Lett.* **12:**233–235 (1987).

29. G. W. Johnson, D. C. Leiner, and D. T. Moore, "Phase-Locked Interferometry," *Proc. SPIE* **126:**152–160 (1977).

30. T. Suzuki, O. Sasaki, and T. Maruyama, "Phase Locked Laser Diode Interferometry for Surface Profile Measurement," *Appl. Opt.* **28:**4407–4410 (1989).

31. O. Sasaki and H. Okazaki, "Sinusoidal Phase Modulating Interferometry for Surface Profile Measurements," *Appl. Opt.* **25:**3137–3140 (1986).

32. Y. Yeh and H. Z. Cummins, "Localized Fluid Flow Measurements with an He-Ne Laser Spectrometer," *Appl. Phys. Lett.* **4:**176–178 (1964).

33. F. Durst and J. H. Whitelaw, "Integrated Optical Units for Laser Anemometry," *J. Phys. E: Sci. Instrum.* **4:**804–808 (1971).

34. B. E. Truax, F. C. Demarest, and G. E. Sommargren, "Laser Doppler Velocimetry for Velocity and Length Measurement of Moving Surfaces," *Appl. Opt.* **23:**67–73 (1984).

35. F. Durst, A. Melling, and J. H. Whitelaw, *Principles and Practice of Laser-Doppler Anemometry,* Academic Press, London, 1976.

36. O. Sasaki, T. Sato, T. Abe, T. Mizuguchi, and M. Niwayama, "Follow-Up Type Laser Doppler Velocimeter Using Single-Mode Optical Fibers," *Appl. Opt.* **19:**1306–1308 (1980).

37. J. D. C. Jones, M. Corke, A. D. Kersey, and D. A. Jackson, "Miniature Solid-State Directional Laser-Doppler Velocimeter," *Electron. Lett.* **18:**967–969 (1982).

38. W. Puschert, "Optical Detection of Amplitude and Phase of Mechanical Displacements in the Angstrom Range," *Opt. Commun.* **10:**357–361 (1974).

39. P. Hariharan, "Interferometry with Lasers," in E. Wolf (ed.), *Progress in Optics,* vol. XXIV, Elsevier, Amsterdam, 1987, pp. 123–125.

40. D. E. T. F. Ashby and D. F. Jephcott, "Measurement of Plasma Density Using a Gas Laser as an Infrared Interferometer," *Appl. Phys. Lett.* **3:**13–16 (1963).

41. A. Dandridge, R. O. Miles, and T. G. Giallorenzi, "Diode Laser Sensor," *Electron. Lett.* **16:**943–949 (1980).

42. T. Yoshino, M. Nara, S. Mnatzakanian, B. S. Lee, and T. C. Strand, "Laser Diode Feedback Interferometer for Stabilization and Displacement Measurements," *Appl. Opt.* **26:**892–897 (1987).

43. J. H. Churnside, "Laser Doppler Velocimetry by Modulating a CO_2 Laser with Backscattered Light," *Appl. Opt.* **23:**61–66 (1984).

44. S. Shinohara, A. Mochizuki, H. Yoshida, and M. Sumio, "Laser-Doppler Velocimeter Using the Self-Mixing Effect of a Semiconductor Laser Diode," *Appl. Opt.* **25:**1417–1419 (1986).

45. P. J. de Groot and G. M. Gallatin, "Backscatter-Modulation Velocimetry with an External-Cavity Laser Diode," *Opt. Lett.* **14:**165–167 (1989).

46. V. Vali and R. W. Shorthill, "Fiber Ring Interferometer," *Appl. Opt.* **15:**1099–1100 (1976).

47. S. Ezekiel, "An Overview of Passive Optical 'Gyros'," *Proc. SPIE* **487:**13–20 (1984).

48. R. A. Bergh, H. C. Lefevre, and H. J. Shaw, "All-Single-Mode Fiber-Optic Gyroscope with Long-Term Stability," *Opt. Lett.* **6:**502–504 (1981).

49. R. A. Bergh, H. C. Lefevre, and H. J. Shaw, "An Overview of Fiber-Optic Gyroscopes," *IEEE J. Lightwave Technol.* **TL-2:**91–107 (1984).

50. B. Culshaw, *Optical Fibre Sensing and Signal Processing,* Peregrinus, London, 1984.

51. T. G. Giallorenzi, J. A. Bucaro, A. Dandridge, G. H. Sigel Jr, J. H. Cole, S. C. Rashleigh, and R. G. Priest, "Optical Fiber Sensor Technology," *IEEE J. Quantum Electron.* **QE-18:**626–665 (1982).

52. J. P. Willson and R. E. Jones, "Magnetostrictive Fiber-Optic Sensor System for Detecting DC Magnetic Fields," *Opt. Lett.* **8:**333–335 (1983).

53. P. D. De Souza and M. D. Mermelstein, "Electrical Field Detection with a Piezoelectric Polymer-Jacketed Single-Mode Optical Fiber," *Appl. Opt.* **21:**4214–4218 (1982).

54. I. P. Gilles, D. Uttam, B. Culshaw, and D. E. N. Davies, "Coherent Optical-Fibre Sensors with Modulated Laser Sources," *Electron. Lett.* **19:**14–15 (1983).

55. J. L. Brooks, R. H. Wentworth, R. C. Youngquist, M. Tur, B. Y. Kim, and H. J. Shaw, "Coherence Multiplexing of Fiber-Optic Interferometric Sensors," *J. Lightwave Technol.* **LT-3:**1062–1071 (1985).

56. I. Sakai, G. Parry, and R. C. Youngquist, "Multiplexing Fiber-Optic Sensors by Frequency Modulation: Cross-Term Considerations," *Opt. Lett.* **11:**183–185 (1986).

57. J. L. Brooks, B. Moslehi, B. Y. Kim, and H. J. Shaw, "Time-Domain Addressing of Remote Fiber-Optic Interferometric Sensor Arrays," *J. Lightwave Technol.* **LT-5:**1014–1023 (1987).

58. F. Farahi, J. D. C. Jones, and D. A. Jackson, "Multiplexed Fibre-Optic Interferometric Sensing System: Combined Frequency and Time Division," *Electron. Lett.* **24:**409–410 (1988).

59. F. V. Kowalski, R. T. Hawkins, and A. L. Schawlow, "Digital Wavemeter for CW Lasers," *J. Opt. Soc. Am.* **66:**965–966 (1976).

60. J. L. Hall and S. A. Lee, "Interferometric Real-Time Display of CW Dye Laser Wavelength with Sub-Doppler Accuracy," *Appl. Phys. Lett.* **29:**367–369 (1976).

61. A. Kahane, M. S. O'Sullivan, N. M. Sanford, and B. P. Stoicheff, "Vernier Fringe Counting Device for Laser Wavelength Measurements," *Rev. Sci. Instrum.* **54:**1138–1142 (1983).

62. R. Salimbeni and R. V. Pole, "Compact High-Accuracy Wavemeter," *Opt. Lett.* **5:**39–41 (1980).

63. J. J. Snyder, "Fizeau Wavemeter," *Proc. SPIE* **288:**258–262 (1981).

64. J. L. Gardner, "Wave-Front Curvature in a Fizeau Wavemeter," *Opt. Lett.* **8:**91–93 (1983).

65. P. Hariharan, "Interferometry with Lasers," in E. Wolf (ed.), *Progress in Optics,* vol. XXIV, Elsevier, Amsterdam, 1987, pp. 144–151.

66. F. A. Hopf, A. Tomita, and G. Al-Jumaily, "Second-Harmonic Interferometry," *Opt. Lett.* **5:**386–388 (1980).

67. F. A. Hopf, "Interferometry Using Conjugate Wave Generation," *J. Opt. Soc. Am.* **70:**1320–1322 (1980).

68. D. J. Gauthier, R. W. Boyd, R. K. Jungquist, J. B. Lisson, and L. L. Voci, "Phase-Conjugate Fizeau Interferometer," *Opt. Lett.* **14:**323–325 (1989).

69. J. Feinberg, "Interferometer with a Self-Pumped Phase-Conjugating Mirror," *Opt. Lett.* **8:**569–571 (1983).

70. D. Z. Anderson, D. M. Lininger, and J. Feinberg, "Optical Tracking Novelty Filter," *Opt. Lett.* **12:**123–125 (1987).

71. J. Davis, "Long Baseline Optical Interferometry," in J. Roberts (ed.), *Proc. Int. Symposium on Measurement and Processing for Indirect Imaging,* Sydney, 1983, Cambridge University Press, Cambridge, 1984, pp. 125–141.

72. R. Hanbury Brown, *The Intensity Interferometer,* Taylor and Francis, London, 1974.

73. M. A. Johnson, A. L. Betz, and C. H. Townes, "10 μm Heterodyne Stellar Interferometer," *Phys. Rev. Lett.* **33:**1617–1620 (1974).

74. C. H. Townes, "Spatial Interferometry in the Mid-Infrared Region," *J. Astrophys. Astron.* **5:**111–130 (1984).

75. G. E. Moss, L. R. Miller, and R. L. Forward, "Photon-Noise-Limited Laser Transducer for Gravitational Antenna," *Appl. Opt.* **10:**2495–2498 (1971).

76. R. W. P. Drever, S. Hoggan, J. Hough, B. J. Meers, A. J. Munley, G. P. Newton, H. Ward, D. Z. Anderson, Y. Gursel, M. Hereld, R. E. Spero, and S. E. Whitcomb, "Developments in Laser Interferometer Gravitational Wave Detectors," in H. Ning (ed.), *Proc. Third Marcel Grossman Meeting on General Relativity,* vol. 1, Science Press, Beijing and North-Holland, Amsterdam, 1983, pp. 739–753.

CHAPTER 22
POLARIMETRY

Russell A. Chipman
Physics Department
University of Alabama in Huntsville
Huntsville, Alabama

22.1 GLOSSARY

A	analyzer vector
a	analyzer vector element
a	semimajor axis of ellipse
BRDF	bidirectional reflectance distribution function
b	semiminor axis of ellipse
D	diattenuation
DOCP	degree of circular polarization
DOLP	degree of linear polarization
DOP	degree of polarization
Dep	depolarization
E	extinction ratio
e	ellipticity
I	inhomogeneity of a Mueller matrix
ID	ideal depolarizer
J	Jones matrix
$j_{xx}, j_{xy}, j_{yx}, j_{yy}$	Jones matrix elements
LD	linear diattenuation
LP	linear polarizer
M	Mueller matrix
$\vec{\mathbf{M}}$	Mueller vector
MBRDF	Mueller bidirectional reflectance distribution function
$\mathbf{M}_\mathrm{D}$	diattenuator Mueller matrix
$\mathbf{M}_\mathrm{R}$	retarder Mueller matrix
$m_{00}, m_{01}, \ldots, m_{33}$	Mueller matrix elements
Δn	birefringence

n_1, n_2	refractive indices of a birefringent medium
$\mathbf{P}$	flux measurement vector
$\boldsymbol{P}$	polarizance
P	flux measurement
$\boldsymbol{PD}$	partial depolarizer
PDL	polarization-dependent loss
$\mathbf{QWLR}$	quarter-wave linear retarder
Q	index limit
q	index for a sequence of polarization elements
$\mathbf{R}_M$	rotational change of basis matrix for Stokes vectors
$\mathbf{S}$	Stokes vector
$\hat{\mathbf{S}}$	normalized polarized Stokes vector
$\mathbf{S}'$	exiting Stokes vector
$\mathbf{S}_{\mathrm{m}}$	measured Stokes vector
$\mathbf{S}_{\mathrm{max}}, \mathbf{S}_{\mathrm{min}}$	incident Stokes vectors of maximum and minimum intensity transmittance
s_0, s_1, s_2, s_3	Stokes vector elements
T	transpose, superscript
T	intensity transmittance
T_{avg}	intensity transmission averaged over all incident polarization states
T_{max}	maximum intensity transmittance
T_{min}	minimum intensity transmittance
Tr	trace of a matrix
t	thickness
$\mathbf{U}$	Jones/Mueller transformation matrix
$\boldsymbol{VD}$	variable partial depolarizer
$\mathbf{W}$	polarimetric measurement matrix
$\mathbf{W}^{-1}$	polarimetric data reduction matrix
$\mathbf{W}_P^{-1}$	pseudoinverse of $\mathbf{W}$
α, β	angles of incidence
γ, δ	angles of scatter
δ	retardance
η	azimuth of ellipse
ϵ	eccentricity

θ	orientation angle
$\Delta\theta$	angular increment
ρ	amplitudes of a complex number
ϕ	phases of a complex number
χ	angle between the eigenpolarizations on the Poincare sphere
$\otimes$	tensor product
$\dagger$	Hermitian adjoint

22.2 OBJECTIVES

This chapter surveys the principles of polarization measurements. Throughout this chapter all measured and derived quantities are formulated in terms of the Stokes vector and Mueller matrix, as these comprise the most appropriate representation of polarization for radiometric measurements. The Mueller matrix has a structure which is difficult to understand, so the interpretation has been discussed within this chapter. One of the primary difficulties in performing accurate polarization measurements is the systematic errors due to nonideal polarization elements. Therefore, a formulation of the polarimetric measurement and data reduction process is included which readily handles arbitrary polarization elements used in the polarimeter whose transmitted or analyzed Stokes vectors are determined through calibration. Finally, a survey of the literature on applications studies utilizing polarimeters has been included.

This chapter is closely related to the following *Handbook of Optics* chapters: "Polarization" (Vol. I, Chap. 5) by J. M. Bennett, "Polarizers" by J. M. Bennett (Vol. II, Chap. 3), and "Ellipsometry" by R. M. A. Azzam (Vol. II, Chap. 27).

22.3 POLARIMETERS

Polarimeters are optical instruments used for determining the polarization properties of light beams and samples. Polarimetry, the science of measuring polarization, is most simply characterized as radiometry with polarization elements. To perform accurate polarimetry, all the issues necessary for careful and accurate radiometry must be considered, together with many additional polarization issues. In this chapter, our emphasis is strictly on those additional polarization issues which must be mastered to accurately determine polarization properties from polarimetric measurements. Typical applications of polarimeters include the following: remote sensing of the earth and astronomical bodies, calibration of polarization elements, measuring the thickness and refractive indices of thin films (ellipsometry), spectroscopic studies of materials, and alignment of polarization-critical optical systems. We can broadly subdivide polarimeters into the several categories as discussed in succeeding sections.

22.4 LIGHT-MEASURING AND SAMPLE-MEASURING POLARIMETERS

Light-measuring polarimeters determine the polarization state of a beam of light or determine some of its polarization characteristics. These determinations may include the following: the direction of oscillation of the electric field vector for a linearly polarized beam, the helicity of a circularly polarized beam, or the elliptical parameters of an elliptically polarized beam, as well as the degree of polarization and other characteristics. A light-measuring polarimeter utilizes a set of polarization elements placed in a beam of light in front of a radiometer; or, to paraphrase, the light from the sample is analyzed by a series of polarization state analyzers, and a set of measurements is acquired. The polarization characteristics of the sample are determined from these measurements by a data reduction procedure. Measurement, calibration, and data reduction algorithms are treated under "Light-measuring Polarimeters."

22.5   SAMPLE-MEASURING POLARIMETERS

Sample-measuring polarimeters determine the relationship between the polarization states of incident and exiting beams for a sample. The term *exiting beam* is general and includes beams which are transmitted, reflected, diffracted, or scattered. The term *sample* is also an inclusive term used in a broad sense to describe a general light-matter interaction or sequence of such interactions and applies to practically anything. Measurements are acquired using a series of polarization elements located between a source and sample and the exiting beams are analyzed with a separate set of polarization elements between the sample and radiometer. Samples of great interest include surfaces, thin films on surfaces, polarization elements, optical elements, optical systems, natural scenes, biological samples, and industrial samples.

Accurate polarimetric measurements can be made only if the polarization generator and/or polarization analyzer are fully calibrated. To perform accurate polarimetry, the polarization elements do not need to be ideal. If the Mueller matrices of the polarization components are known, the systematic errors due to nonideal polarization elements can be removed during the data reduction (see "Polarimetric Measurement Equation and Polarimetric Data Reduction Equation").

22.6   COMPLETE AND INCOMPLETE POLARIMETERS

A light-measuring polarimeter is "complete" if it measures a Stokes vector or if a Stokes vector can be determined from its measurements. An "incomplete" light-measuring polarimeter cannot be used to determine a Stokes vector. For example, a polarimeter which employs a rotating polarizer in front of a detector does not determine the circular polarization content of a beam, and is incomplete. Similarly, a sample-measuring polarimeter is complete if it is capable of measuring the full Mueller matrix, and incomplete otherwise. Complete polarimeters are often referred to as Stokes polarimeters or Mueller polarimeters.

22.7   POLARIZATION GENERATORS AND ANALYZERS

A *polarization generator* consists of a source, optical elements, and polarization elements to produce a beam of known polarization state. A polarization generator is specified by the Stokes vector **S** of the exiting beam. A *polarization analyzer* is a configuration of polarization elements, optical elements, and a detector which performs a flux measurement of a particular polarization component in an incident beam. A polarization analyzer is characterized by a Stokes-like *analyzer vector* **A** which specifies the incident polarization state which is analyzed, the state which produces the maximal response at the detector. Sample-measuring polarimeters require polarization generators and polarization analyzers, while light-measuring polarimeters only require polarization analyzers. Frequently the terms "polarization generator" and "polarization analyzer" refer just to the polarization elements in the generator and analyzer. In this usage, it is important to distinguish between elliptical (and circular) generators and elliptical analyzers for a given state because they generally have different polarization characteristics and Mueller matrices (see "Elliptical and Circular Polarizers and Analyzers").

22.8 CLASSES OF LIGHT-MEASURING POLARIMETERS

Polarimeters operate by acquiring measurements with a set of polarization analyzers (and a set of polarization generators for sample-measuring instruments). The following sections classify polarimeters by the four broad methods by which these multiple measurements are acquired.

22.9 TIME-SEQUENTIAL MEASUREMENTS

In a time-sequential polarimeter, the measurements are taken sequentially in time. Between measurements, the polarization analyzer and/or polarization generator is changed. Time-sequential polarimeters frequently employ rotating polarization elements or filter wheels containing a set of analyzers. A time-sequential polarimeter generally employs a single source and detector.

22.10 POLARIZATION MODULATION

Polarimeters employing polarization modulation comprise a subset of time-sequential polarimeters. Here, the polarization analyzer contains a polarization modulator, a rapidly changing polarization element. The output of the analyzer is a rapidly fluctuating irradiance on which polarization information is encoded. Polarization parameters can then be determined by ac and vector voltmeters, by lock-in amplifiers, or by frequency-domain digital signal processing techniques. For example, a rapidly spinning polarizer produces a modulated output which allows the flux and the degree of linear polarization to be read with a dc voltmeter and an ac voltmeter. The most common high-speed polarization modulators in general use are the electro-optical modulator, the magneto-optical modulator, and the photoelastic modulator.

22.11 DIVISION OF APERTURE

Polarimeters based on division of aperture employ multiple polarization analyzers operating side-by-side. The aperture of the polarimeter is subdivided, with each beam going into a separate polarization analyzer and detector. The detectors are usually synchronized to acquire measurements simultaneously. This is similar in principle to the polarizing glasses used in 3-D movie systems, where a 45° polarizer is used for one eye and a 135° for the other, permitting two polarization measurements simultaneously in two eyes.

22.12 DIVISION OF AMPLITUDE

Division-of-amplitude polarimeters utilize beam splitters to divide beams and direct them toward multiple analyzers and detectors. A division-of-amplitude polarimeter can acquire its measurements simultaneously. Many division-of-amplitude polarimeters use polarizing beam splitters to simultaneously divide and analyze the polarization state of the beam. The four-detector photopolarimeter uses a sequence of detectors at nonnormal incidence to measure Stokes vectors (Azzam, 1985; Azzam, Elminyawi, and El-Saba, 1988).

22.13 DEFINITIONS

Analyzer—an element whose intensity transmission is proportional to the content of a specific polarization state in the incident beam. Analyzers are placed before the detector in polarimeters. The transmitted polarization state emerging from an analyzer is not necessarily the same as the state which is being analyzed.

Birefringence—a material property, the retardance associated with propagation through an anisotropic medium. For each propagation direction within a birefringent medium, there are two modes of propagation with different refractive indices n_1 and n_2. The birefringence Δn is $\Delta n = |n_1 - n_2|$.

Depolarization—a process which couples polarized light into unpolarized light. Depolarization is intrinsically associated with scattering and with diattenuation and retardance which vary in space, time, and/or wavelength.

Diattenuation—the property of an optical element or system whereby the intensity transmittance of the exiting beam depends on the polarization state of the incident beam. The intensity transmittance is a maximum P_{max} for one incident state, and a minimum P_{min} for the orthogonal state. The diattenuation is defined as $(P_{max} - P_{min})/(P_{max} + P_{min})$.

Diattenuator—any homogeneous polarization element which displays significant diattenuation and minimal retardance. Polarizers have a diattenuation close to one, but nearly all optical interfaces are weak diattenuators. Examples of diattenuators include the following: polarizers and dichroic materials, as well as metal and dielectric interfaces with reflection and transmission differences described by Fresnel equations; thin films (homogeneous and isotropic); and diffraction gratings.

Dichroism—the material property of displaying diattenuation during propagation. For each direction of propagation, dichroic media have two modes of propagation with different absorption coefficients. Examples of dichroic materials include sheet polarizers and dichroic crystals such as tourmaline.

Eigenpolarization—a polarization state transmitted unaltered by a polarization element except for a change of amplitude and phase. Every polarization element has two eigenpolarizations. Any incident light not in an eigenpolarization state is transmitted in a polarization state different from the incident state. Eigenpolarizations are the eigenvectors of the corresponding Mueller or Jones matrix.

Ellipsometry—a polarimetric technique which uses the change in the state of polarization of light upon reflection for the characterization of surfaces, interfaces, and thin films (after Azzam, 1993).

Homogeneous polarization element—an element whose eigenpolarizations are orthogonal. Then, the eigenpolarizations are the states of maximum and minimum transmittance and also of maximum and minimum optical path length. A homogeneous element is classified as linear, circular, or elliptical depending on the form of the eigenpolarizations.

Inhomogeneous polarization element—an element whose eigenpolarizations are not orthogonal. Such an element will display different polarization characteristics for forward and backward propagating beams. The eigenpolarizations are generally not the states of maximum and minimum transmittance. Often inhomogeneous elements cannot be simply classified as linear, circular, or elliptical.

Ideal polarizer—a polarizer with an intensity transmittance of one for its principal state and an intensity transmittance of zero for the orthogonal state.

Linear polarizer—a device which, when placed in an incident unpolarized beam, produces a beam of light whose electric field vector is oscillating primarily in one plane, with only a small component in the perpendicular plane (after Bennett, 1993).

Nonpolarizing element—an element which does not change the polarization state for arbitrary states. The polarization state of the output light is equal to the polarization state of the incident light for all possible input polarization states.

Partially polarized light—light containing an unpolarized component; cannot be extinguished by an ideal polarizer.

Polarimeter—an optical instrument for the determination of the polarization state of a light beam, or the polarization-altering properties of a sample.

Polarimetry—the science of measuring the polarization state of a light beam and the diattenuating, retarding, and depolarizing properties of materials.

Polarization—any process which alters the polarization state of a beam of light, including diattenuation, retardance, depolarization, and scattering.

Polarization coupling—any conversion of light from one polarization state into another state.

Polarized light—light in a fixed, elliptically (including linearly or circularly) polarized state. It can be extinguished by an ideal polarizer. For polychromatic light, the polarization ellipses associated with each spectral component have identical ellipticity, orientation, and helicity.

Polarizer—a strongly diattenuating optical element designed to transmit light in a specified polarization state independent of the incident polarization state. The transmission of one of the eigenpolarizations is very nearly zero.

Polarization element—any optical element which alters the polarization state of light. This includes polarizers, retarders, mirrors, thin films, and nearly all optical elements.

Pure diattenuator—a diattenuator with zero retardance and no depolarization.

Pure retarder—a retarder with zero diattenuation and no depolarization.

Retardance—a polarization-dependent phase change associated with a polarization element or system. The phase (optical path length) of the output beam depends upon the polarization state of the input beam. The transmitted phase is a maximum for one eigenpolarization, and a minimum for the other eigenpolarization. Other states show polarization coupling and an intermediate phase.

Retardation plate—a retarder constructed from a plane parallel plate or plates of linearly birefringent material.

Retarder—a polarization element designed to produce a specified phase difference between the exiting beams for two orthogonal incident polarization states (the eigenpolarizations of the element). For example, a quarter-wave linear retarder has as its eigenpolarizations two orthogonal linearly polarized states which are transmitted in their incident polarization states but with a 90° (quarter-wavelength) relative phase difference introduced.

Spectropolarimetry—the spectroscopic study of the polarization properties of materials. Spectropolarimetry is a generalization of conventional optical spectroscopy. Where conventional spectroscopy endeavors to measure the reflectance or transmission of a sample as a function of wavelength, spectropolarimetry also determines the diattenuating, retarding, and depolarizing properties of the sample. Complete characterization of these properties is accomplished by measuring the Mueller matrix of the sample as a function of wavelength.

Waveplate—a retarder.

22.14 STOKES VECTORS AND MUELLER MATRICES

Several calculi have been developed for analyzing polarization, including those based on the Jones matrix, coherency matrix, Mueller matrix, and other matrices (Shurcliff, 1962; Gerrard and Burch, 1975; Theocaris and Gdoutos, 1979; Azzam and Bashara, 1987; Coulson, 1988; Egan, 1992). Of these methods, the Mueller calculus is most generally suited for describing irradiance-measuring instruments, including most polarimeters, radiometers, and spectrometers, and is used exclusively in this paper.

In the Mueller calculus, the Stokes vector $\mathbf{S}$ is used to describe the polarization state of a light beam, and the Mueller matrix $\mathbf{M}$ to describe the polarization-altering characteristics of a sample. This sample may be a surface, a polarization element, an optical system, or some other light/matter interaction which produces a reflected, refracted, diffracted, or scattered light beam. All vectors and matrices are represented by bold characters. Normalized vectors have "hats" (i.e., $\hat{\mathbf{A}}$).

22.15 PHENOMENOLOGICAL DEFINITION OF THE STOKES VECTOR

The Stokes vector is defined relative to the following six flux measurements P performed with ideal polarizers in front of a radiometer (Shurcliff, 1962):

P_H horizontal linear polarizer (0°)

P_V vertical linear polarizer (90°)

P_{45} 45° linear polarizer

P_{135} 135° linear polarizer

P_R right circular polarizer

P_L left circular polarizer

Normally, these measurements are irradiance measurements (W/m^2) although other flux measurements might be used. The Stokes vector is defined as

$$\mathbf{S} = \begin{bmatrix} s_0 \\ s_1 \\ s_2 \\ s_3 \end{bmatrix} = \begin{bmatrix} P_H + P_V \\ P_H - P_V \\ P_{45} - P_{135} \\ P_R - P_L \end{bmatrix} \tag{1}$$

where s_0, s_1, s_2, and s_3 are the Stokes vector elements. The Stokes vector does not need to be measured by these six ideal measurements; what is required is that other methods reproduce the Stokes vector defined in this manner. Ideal polarizers are not required. Further, the Stokes vector is a function of wavelength, position on the object, and the light's direction of emission or scatter. Thus, a Stokes vector measurement is an average over area, solid angle, and wavelength, as is any radiometric measurement. Each Stokes vector element has units of watts per meter squared. The Stokes vector is defined relative to a local $x - y$ coordinate system defined in the plane perpendicular to the propagation vector. The coordinate system is right-handed; the cross product $\hat{x} \times \hat{y}$ of the basis vectors points in the direction of propagation of the beam.

22.16 POLARIZATION PROPERTIES OF LIGHT BEAMS

From the Stokes vector, the following polarization parameters are determined (Azzam and Bashara, 1977 and 1987; Kliger, Lewis, and Randall, 1990; Collett, 1992):

Flux
$$P = s_0 \tag{2}$$

Degree of polarization
$$\boldsymbol{DOP} = \frac{\sqrt{s_1^2 + s_2^2 + s_3^2}}{s_0} \tag{3}$$

Degree of linear polarization
$$\boldsymbol{DOLP} = \frac{\sqrt{s_1^2 + s_2^2}}{s_0} \tag{4}$$

Degree of circular polarization
$$\boldsymbol{DOCP} = \frac{s_3}{s_0} \tag{5}$$

The Stokes vector for a partially polarized beam ($\boldsymbol{DOP} < 1$) can be considered as a superposition of a completely polarized Stokes vector $\mathbf{S}_P$ and an unpolarized Stokes vector $\mathbf{S}_U$ which are uniquely related to $\mathbf{S}$ as follows (Collett, 1992):

$$\mathbf{S} = \mathbf{S}_P + \mathbf{S}_U = \begin{bmatrix} s_0 \\ s_1 \\ s_2 \\ s_3 \end{bmatrix} = s_0 \boldsymbol{DOP} \begin{bmatrix} 1 \\ s_1/(s_0\boldsymbol{DOP}) \\ s_2/(s_0\boldsymbol{DOP}) \\ s_3/(s_0\boldsymbol{DOP}) \end{bmatrix} + (1 - \boldsymbol{DOP})s_0 \begin{bmatrix} 1 \\ 0 \\ 0 \\ 0 \end{bmatrix} \tag{6}$$

The polarized portion of the beam represents a net polarization ellipse traced by the electric field vector as a function of time. The ellipse has a magnitude of the semimajor axis a, semiminor axis b, orientation of the major axis η (azimuth of the ellipse) measured counterclockwise from the x axis, and eccentricity (or ellipticity).

Ellipticity
$$e = \frac{b}{a} = \frac{s_3}{s_0 + \sqrt{s_1^2 + s_2^2}} \tag{7}$$

Orientation of major axis, azimuth
$$\eta = \tfrac{1}{2} \arctan \left[\frac{s_2}{s_1} \right] \tag{8}$$

Eccentricity
$$\epsilon = \sqrt{1 - e^2} \tag{9}$$

The ellipticity is the ratio of the minor to the major axis of the corresponding electric field polarization ellipse, and varies from 0 for linearly polarized light to 1 for circularly polarized light. The polarization ellipse is alternatively described by its eccentricity, which is zero for circularly polarized light, increases as the ellipse becomes thinner (more cigar-shaped), and becomes one for linearly polarized light.

22.17 MUELLER MATRICES

The Mueller matrix **M** for a polarization-altering device is defined as the matrix which transforms an incident Stokes vector **S** into the exiting (reflected, transmitted, or scattered) Stokes vector **S'**,

$$\mathbf{S}' = \begin{bmatrix} s_0' \\ s_1' \\ s_2' \\ s_3' \end{bmatrix} = \mathbf{MS} = \begin{bmatrix} m_{00} & m_{01} & m_{02} & m_{03} \\ m_{10} & m_{11} & m_{12} & m_{13} \\ m_{20} & m_{21} & m_{22} & m_{23} \\ m_{30} & m_{31} & m_{32} & m_{33} \end{bmatrix} \begin{bmatrix} s_0 \\ s_1 \\ s_2 \\ s_3 \end{bmatrix} \tag{10}$$

The Mueller matrix is a four-by-four matrix with real valued elements. The Mueller matrix $\mathbf{M}(k, \lambda)$ for a device is always a function of the direction of propagation k and wavelength λ. The Mueller matrix is an appropriate formalism for characterizing polarization measurements because it contains within its elements all of the polarization properties: diattenuation, retardance, depolarization, and their form, either linear, circular, or elliptical. When the Mueller matrix is known, then the exiting polarization state is known for an arbitrary incident polarization state. Table 1 is a compilation of Mueller matrices for common polarization elements, together with the corresponding transmitted Stokes vector. Other tables of Mueller matrices may be found in the following references: Shurcliff (1962), Gerrard and Burch (1975), Azzam and Bashara (1977), Theocaris and Gdoutos (1979), and Collett (1992). See detailed discussion of the polarization properties as related to the Mueller matrix elements later in this chapter.

The Mueller matrix **M** associated with a beam path through a sequence (cascade) of polarization elements $q = 1, 2, \ldots, Q$ is the right-to-left product of the individual matrices $\mathbf{M}_q$,

$$\mathbf{M} = \mathbf{M}_Q \mathbf{M}_{Q-1} \cdots \mathbf{M}_q \cdots \mathbf{M}_2 \mathbf{M}_1 = \sum_{q=Q,-1}^{1} \mathbf{M}_q \tag{11}$$

When a polarization element with Mueller matrix **M** is rotated about the beam of light by an angle θ such that the angle of incidence is unchanged (for example, for a normal-incidence beam, rotating the element about the normal), the resulting Mueller matrix $\mathbf{M}(\theta)$ is

$$\mathbf{M}(\theta) = \mathbf{R}_M(\theta)\mathbf{M}\mathbf{R}_M(-\theta) = \begin{bmatrix} 1 & 0 & 0 & 0 \\ 0 & \cos(2\theta) & -\sin(2\theta) & 0 \\ 0 & \sin(2\theta) & \cos(2\theta) & 0 \\ 0 & 0 & 0 & 1 \end{bmatrix} \begin{bmatrix} m_{00} & m_{01} & m_{02} & m_{03} \\ m_{10} & m_{11} & m_{12} & m_{13} \\ m_{20} & m_{21} & m_{22} & m_{23} \\ m_{30} & m_{31} & m_{32} & m_{33} \end{bmatrix}$$

$$\times \begin{bmatrix} 1 & 0 & 0 & 0 \\ 0 & \cos(2\theta) & \sin(2\theta) & 0 \\ 0 & -\sin(2\theta) & \cos(2\theta) & 0 \\ 0 & 0 & 0 & 1 \end{bmatrix} \tag{12}$$

where $\mathbf{R}_M$ is the rotational change of basis matrix for Stokes vectors and Mueller matrices.

TABLE 1   Example Mueller Matrices and Transmitted Stokes Vectors

Nonpolarizing element

$$
\begin{bmatrix} 1 & 0 & 0 & 0 \\ 0 & 1 & 0 & 0 \\ 0 & 0 & 1 & 0 \\ 0 & 0 & 0 & 1 \end{bmatrix}
\begin{bmatrix} s_0 \\ s_1 \\ s_2 \\ s_3 \end{bmatrix}
=
\begin{bmatrix} s_0 \\ s_1 \\ s_2 \\ s_3 \end{bmatrix}
$$

Absorber

$$
\begin{bmatrix} a & 0 & 0 & 0 \\ 0 & a & 0 & 0 \\ 0 & 0 & a & 0 \\ 0 & 0 & 0 & a \end{bmatrix}
\begin{bmatrix} s_0 \\ s_1 \\ s_2 \\ s_3 \end{bmatrix}
=
\begin{bmatrix} as_0 \\ as_1 \\ as_2 \\ as_3 \end{bmatrix}
$$

Linear polarizer, transmission axis 0°

$$
\frac{1}{2}\begin{bmatrix} 1 & 1 & 0 & 0 \\ 1 & 1 & 0 & 0 \\ 0 & 0 & 0 & 0 \\ 0 & 0 & 0 & 0 \end{bmatrix}
\begin{bmatrix} s_0 \\ s_1 \\ s_2 \\ s_3 \end{bmatrix}
=
\frac{1}{2}\begin{bmatrix} s_0 + s_1 \\ s_0 + s_1 \\ 0 \\ 0 \end{bmatrix}
$$

Linear diattenuator, axis 0°, intensity transmittances q, r

$$
\frac{1}{2}\begin{bmatrix} q+r & q-r & 0 & 0 \\ q-r & q+r & 0 & 0 \\ 0 & 0 & 2\sqrt{qr} & 0 \\ 0 & 0 & 0 & 2\sqrt{qr} \end{bmatrix}
\begin{bmatrix} s_0 \\ s_1 \\ s_2 \\ s_3 \end{bmatrix}
=
\frac{1}{2}\begin{bmatrix} q(s_0+s_1)+r(s_0-s_1) \\ q(s_0+s_1)-r(s_0-s_1) \\ 2s_2\sqrt{qr} \\ 2s_3\sqrt{qr} \end{bmatrix}
$$

Linear diattenuator, axis 45°, intensity transmittances q, r

$$
\frac{1}{2}\begin{bmatrix} q+r & 0 & q-r & 0 \\ 0 & 2\sqrt{qr} & 0 & 0 \\ q-r & 0 & q+r & 0 \\ 0 & 0 & 0 & 2\sqrt{qr} \end{bmatrix}
\begin{bmatrix} s_0 \\ s_1 \\ s_2 \\ s_3 \end{bmatrix}
=
\frac{1}{2}\begin{bmatrix} q(s_0+s_2)+r(s_0-s_2) \\ 2s_1\sqrt{qr} \\ q(s_0+s_2)-r(s_0-s_2) \\ 2s_3\sqrt{qr} \end{bmatrix}
$$

Linear diattenuator, axis θ, intensity transmittances q, r

$$
\frac{1}{2}\begin{bmatrix} q+r & (q-r)\cos 2\theta & (q-r)\sin 2\theta & 0 \\ (q-r)\cos 2\theta & (q+r)\cos^2 2\theta + 2\sqrt{qr}\sin^2 2\theta & (q+r-2\sqrt{qr})\sin 2\theta\cos 2\theta & 0 \\ (q-r)\sin 2\theta & (q+r-2\sqrt{qr})\sin 2\theta\cos 2\theta & (q+r)\sin^2 2\theta + 2\sqrt{qr}\cos^2 2\theta & 0 \\ 0 & 0 & 0 & 2\sqrt{qr} \end{bmatrix}
$$

Circular diattenuator, intensity transmittances q, r

$$
\frac{1}{2}\begin{bmatrix} q+r & 0 & 0 & q-r \\ 0 & 2\sqrt{qr} & 0 & 0 \\ 0 & 0 & 2\sqrt{qr} & 0 \\ q-r & 0 & 0 & q+r \end{bmatrix}
\begin{bmatrix} s_0 \\ s_1 \\ s_2 \\ s_3 \end{bmatrix}
=
\frac{1}{2}\begin{bmatrix} q(s_0+s_3)+r(s_0-s_3) \\ 2s_1\sqrt{qr} \\ 2s_2\sqrt{qr} \\ q(s_0+s_3)-r(s_0-s_3) \end{bmatrix}
$$

Linear retarder, fast axis 0°, retardance δ

$$
\begin{bmatrix} 1 & 0 & 0 & 0 \\ 0 & 1 & 0 & 0 \\ 0 & 0 & \cos\delta & \sin\delta \\ 0 & 0 & -\sin\delta & \cos\delta \end{bmatrix}
\begin{bmatrix} s_0 \\ s_1 \\ s_2 \\ s_3 \end{bmatrix}
=
\begin{bmatrix} s_0 \\ s_1 \\ s_2\cos\delta + s_3\sin\delta \\ -s_2\sin\delta + s_3\cos\delta \end{bmatrix}
$$

TABLE 1 *(Continued)*

Linear retarder, fast axis 45°, retardance δ

$$\begin{bmatrix} 1 & 0 & 0 & 0 \\ 0 & \cos\delta & 0 & -\sin\delta \\ 0 & 0 & 1 & 0 \\ 0 & \sin\delta & 0 & \cos\delta \end{bmatrix} \begin{bmatrix} s_0 \\ s_1 \\ s_2 \\ s_3 \end{bmatrix} = \begin{bmatrix} s_0 \\ s_1\cos\delta - s_3\sin\delta \\ s_2 \\ s_1\sin\delta + s_3\cos\delta \end{bmatrix}$$

Linear retarder, fast axis θ, retardance δ

$$\begin{bmatrix} 1 & 0 & 0 & 0 \\ 0 & \cos^2 2\theta + \sin^2 2\theta\cos\delta & \sin 2\theta\cos 2\theta(1-\cos\delta) & -\sin 2\theta\sin\delta \\ 0 & \sin 2\theta\cos 2\theta(1-\cos\delta) & \sin^2 2\theta + \cos^2 2\theta\cos\delta & \cos 2\theta\sin\delta \\ 0 & \sin 2\theta\sin\delta & -\cos 2\theta\sin\delta & \cos\delta \end{bmatrix}$$

Circular retarder, retardance δ

$$\begin{bmatrix} 1 & 0 & 0 & 0 \\ 0 & \cos\delta & \sin\delta & 0 \\ 0 & -\sin\delta & \cos\delta & 0 \\ 0 & 0 & 0 & 1 \end{bmatrix} \begin{bmatrix} s_0 \\ s_1 \\ s_2 \\ s_3 \end{bmatrix} = \begin{bmatrix} s_0 \\ s_1\cos\delta + s_2\sin\delta \\ -s_1\sin\delta + s_2\cos\delta \\ s_3 \end{bmatrix}$$

Linear diattenuator and retarder, fast axis 0°, intensity transmittance (q, r), retardance δ

$$\frac{1}{2}\begin{bmatrix} q+r & q-r & 0 & 0 \\ q-r & q+r & 0 & 0 \\ 0 & 0 & 2\sqrt{qr}\cos\delta & 2\sqrt{qr}\sin\delta \\ 0 & 0 & -2\sqrt{qr}\sin\delta & 2\sqrt{qr}\cos\delta \end{bmatrix}\begin{bmatrix} s_0 \\ s_1 \\ s_2 \\ s_3 \end{bmatrix} = \frac{1}{2}\begin{bmatrix} q(s_0+s_1) + r(s_0-s_1) \\ q(s_0+s_1) - r(s_0-s_1) \\ 2\sqrt{qr}\,(s_2\cos\delta + s_3\sin\delta) \\ 2\sqrt{qr}\,(-s_2\sin\delta + s_3\cos\delta) \end{bmatrix}$$

Ideal depolarizer

$$\begin{bmatrix} 1 & 0 & 0 & 0 \\ 0 & 0 & 0 & 0 \\ 0 & 0 & 0 & 0 \\ 0 & 0 & 0 & 0 \end{bmatrix}\begin{bmatrix} s_0 \\ s_1 \\ s_2 \\ s_3 \end{bmatrix} = \begin{bmatrix} s_0 \\ 0 \\ 0 \\ 0 \end{bmatrix}$$

Partial depolarizer

$$\begin{bmatrix} 1 & 0 & 0 & 0 \\ 0 & d & 0 & 0 \\ 0 & 0 & d & 0 \\ 0 & 0 & 0 & d \end{bmatrix}\begin{bmatrix} s_0 \\ s_1 \\ s_2 \\ s_3 \end{bmatrix} = \begin{bmatrix} s_0 \\ ds_1 \\ ds_2 \\ ds_3 \end{bmatrix}$$

Here $\theta > 0$ if the x axis of the device is rotated toward 45°. If the polarization element remains fixed but the coordinate system rotates by ϕ, the resulting Mueller matrix is $\mathbf{M}(\phi) = \mathbf{R}_M(-\phi)\mathbf{M}\mathbf{R}_m(\phi)$.

22.18 COORDINATE SYSTEM FOR THE MUELLER MATRIX

Consider a Mueller polarimeter consisting of a polarization generator which illuminates a sample, and a polarization analyzer which collects the light exiting the sample in a particular direction. We wish to characterize the polarization modification properties of the sample for a particular incident and exiting beam through the Mueller matrix. The incident polarization states are specified by Stokes vectors defined relative to an $\{\hat{x}, \hat{y}\}$ coordinate system orthogonal to the propagation direction of the incident light. Similarly, the exiting lights' Stokes vector is defined relative to an $\{\hat{x}', \hat{y}'\}$ coordinate system orthogonal to its propagation direction. For transmission measurements where the beam exits undeviated, the orientations of $\{\hat{x}, \hat{y}\}$ and $\{\hat{x}', \hat{y}'\}$ will naturally be chosen to be aligned, $(\hat{x} = \hat{x}', \hat{y} = \hat{y}')$. The global orientation of $\{\hat{x}, \hat{y}\}$ is arbitrary, and the measured Mueller matrix varies systematically if $\{\hat{x}, \hat{y}\}$ and $\{\hat{x}', \hat{y}'\}$ are rotated together.

When the exiting beam emerges in a different direction from the incident beam, orientations must be specified for both sets of coordinates. For measurements of reflection from a surface, a logical choice sets $\{\hat{x}, \hat{y}\}$ and $\{\hat{x}', \hat{y}'\}$ to the $\{\hat{s}, \hat{p}\}$ orientations for the two beams. Other Mueller matrix measurement configurations may have other obvious arrangements for the coordinates. All choices, however, are arbitrary, and lead to different Mueller matrices. Let a Mueller matrix $\mathbf{M}$ be defined relative to a particular $\{\hat{x}, \hat{y}\}$ and $\{\hat{x}', \hat{y}'\}$. Let another Mueller matrix $\mathbf{M}(\theta_1, \theta_2)$ for the same measurement conditions have its $\hat{x}$ axis rotated by θ_1 and x' axis rotated by θ_2, where $\theta > 0$ indicates a counterclockwise rotation looking into the beam ($\hat{x}$ into $\hat{y}$). These Mueller matrices are related by the equation

$$\mathbf{M}(\theta_1, \theta_2) = \begin{bmatrix} 1 & 0 & 0 & 0 \\ 0 & \cos 2\theta_2 & -\sin 2\theta_2 & 0 \\ 0 & \sin 2\theta_2 & \cos 2\theta_2 & 0 \\ 0 & 0 & 0 & 1 \end{bmatrix} \begin{bmatrix} m_{00} & m_{01} & m_{02} & m_{03} \\ m_{10} & m_{11} & m_{12} & m_{13} \\ m_{20} & m_{21} & m_{22} & m_{23} \\ m_{30} & m_{31} & m_{32} & m_{33} \end{bmatrix}$$

$$\times \begin{bmatrix} 1 & 0 & 0 & 0 \\ 0 & \cos 2\theta_1 & \sin 2\theta_1 & 0 \\ 0 & -\sin 2\theta_1 & \cos 2\theta_1 & 0 \\ 0 & 0 & 0 & 1 \end{bmatrix} \qquad (13)$$

When $\theta_1 = \theta_2$, the coordinates rotate together, the eigenvalues are preserved, the circular polarization properties are preserved, and the linear properties are shifted in orientation. When $\theta_1 \neq \theta_2$, the matrix properties are qualitatively different; the eigenvalues of the matrix change. If the eigenpolarizatons of $\mathbf{M}$ were orthogonal, they may not remain orthogonal. After we perform data reduction on the matrix, the basic polarization properties couple in a complex fashion. For example, linear diattenuation in $\mathbf{M}$ yields a circular retardance component in $\mathbf{M}(\theta_1, \theta_2)$, and a linear retardance component yields a circular diattenuation component. The conclusion is that the selection of the coordinate systems for the incident and exiting beams is not important for determining exiting polarization states, but is crucial for identifying polarization characteristics of the sample.

22.19 ELLIPTICAL AND CIRCULAR POLARIZERS AND ANALYZERS

There are few good and convenient circularly or elliptically polarizing mechanisms, whereas linear polarizers are simple, inexpensive, and of high quality. Therefore, most circular and elliptical polarizers incorporate linear polarizers to perform the polarizing, and retarders to convert polarization states. For such compound devices, the distinction between a polarizer and an analyzer becomes significant. This is perhaps best illustrated by three examples: (1) a left circular polarizer (which is also a horizontal linear analyzer) constructed from a horizontal linear polarizer $\mathbf{LP}(0°)$ followed by a quarter-wave linear retarder with the fast axis oriented at 135°, $\mathbf{QWLR}(135°)$ Eq. 14, (2) a left circular analyzer (which is also a horizontal linear polarizer) constructed from a $\mathbf{QWLR}(45°)$ followed by an $\mathbf{LP}(0°)$ Eq. 15, and, (3) a left circular analyzer and polarizer constructed from a

QWLR(135°), then an **LP**(0°), followed by a **QWLR**(45°) Eq. 16. The Mueller matrix equations and exiting polarization states for arbitrary incident states are as follows:

$$\mathbf{QWLR}(135°)\mathbf{LP}(0°)\mathbf{S} = \frac{1}{2}\begin{bmatrix} 1 & 1 & 0 & 0 \\ 0 & 0 & 0 & 0 \\ 0 & 0 & 0 & 0 \\ -1 & -1 & 0 & 0 \end{bmatrix}\begin{bmatrix} s_0 \\ s_1 \\ s_2 \\ s_3 \end{bmatrix} = \frac{1}{2}\begin{bmatrix} s_0 + s_1 \\ 0 \\ 0 \\ -s_0 - s_1 \end{bmatrix} \tag{14}$$

$$\mathbf{LP}(0°)\mathbf{QWLR}(45°)\mathbf{S} = \frac{1}{2}\begin{bmatrix} 1 & 0 & 0 & -1 \\ 1 & 0 & 0 & -1 \\ 0 & 0 & 0 & 0 \\ 0 & 0 & 0 & 0 \end{bmatrix}\begin{bmatrix} s_0 \\ s_1 \\ s_2 \\ s_3 \end{bmatrix} = \frac{1}{2}\begin{bmatrix} s_0 - s_3 \\ s_0 - s_3 \\ 0 \\ 0 \end{bmatrix} \tag{15}$$

$$\mathbf{QWLR}(135°)\mathbf{LP}(0°)\mathbf{QWLR}(45°)\mathbf{S} = \frac{1}{2}\begin{bmatrix} 1 & 0 & 0 & -1 \\ 0 & 0 & 0 & 0 \\ 0 & 0 & 0 & 0 \\ -1 & 0 & 0 & 1 \end{bmatrix}\begin{bmatrix} s_0 \\ s_1 \\ s_2 \\ s_3 \end{bmatrix} = \frac{1}{2}\begin{bmatrix} s_0 - s_3 \\ 0 \\ 0 \\ -s_0 + s_3 \end{bmatrix} \tag{16}$$

The device in Eq. (14) transmits only left circularly polarized light, because the zeroth and third elements have equal magnitude and opposite sign, making it a left circular polarizer. However, the transmitted flux $(s_0 + s_1)/2$ is the flux of horizontal linearly polarized light in the incident beam, making it a horizontal linear analyzer. Similarly, the transmitted flux from the example in Eq. (15), $(s_0 - s_3)/2$, is the flux of left circularly polarized light in the incident beam, making this combination a left circular analyzer. The final polarizer makes the device in Eq. (15) a horizontal linear polarizer, although this is not the standard Mueller matrix for horizontal linear polarizers found in tables. Thus an analyzer for a state does not necessarily transmit the state; its transmitted flux is proportional to the amount of the analyzed state in the incident beam. Examples in Eqs. (14) and (15) are referred to as inhomogeneous polarization elements because the eigenpolarizations are not orthogonal, and the characteristics of the device are different for propagation in opposite directions. The device in Eq. (16) is both a left circular polarizer and a left circular analyzer; it has the same characteristics for propagation in opposite directions, and is referred to as a homogeneous left circular polarizer.

22.20   LIGHT-MEASURING POLARIMETERS

This section presents a general formulation of the measurement and data reduction procedure for a polarimeter intended to measure the state of polarization of a light beam. Similar developments are found in Theil (1976), Azzam (1990), and Stenflo (1991). A survey of light-measuring polarimeter configurations is found in the Handbook, Chap. 27, "Ellipsometry" (Azzam, 1994).

Stokes vectors and related polarization parameters for a beam are determined by measuring the flux transmitted through a set of polarization analyzers. Each analyzer determines the flux of one polarization component in the incident beam. Since a polarization analyzer does not contain ideal polarization elements, the analyzer must be calibrated, and the calibration data used in the data reduction. This section describes data reduction algorithms for determining Stokes vectors which assume arbitrary analyzers; the algorithms allow for general calibration data to be used. Each analyzer is used to measure one polarization component of the incident light. The measured values are related to the incident Stokes vector and the analyzers by the polarimetric measurement equation. A set of linear equations, the data reduction equations, is then solved to determine the Stokes parameters for the beam.

Henceforth, the "polarization analyzer" is considered as the polarization elements used for analyzing the polarization state together with any and all optical elements (lenses, mirrors, etc.), and the detector contained in the polarimeter. The polarization effects from all elements are included in the measurement and data reduction procedures for the polarimeter. A polarization analyzer is characterized by an *analyzer vector* containing four elements and is defined in a manner analogous to a Stokes vector. Let P_H be flux measurement taken by the detector (the current or voltage generated) when one unit of horizontally polarized light is incident. Similarly P_V, P_{45}, P_{135}, P_R, and P_L are the detector's flux measurements for the corresponding incident polarized beams with unit flux. Then the analyzer vector **A** is

$$\mathbf{A} = \begin{bmatrix} a_0 \\ a_1 \\ a_2 \\ a_3 \end{bmatrix} = \begin{bmatrix} P_H + P_V \\ P_H - P_V \\ P_{45} - P_{135} \\ P_R - P_L \end{bmatrix} \tag{17}$$

Note that $P_H + P_V = P_{45} + P_{135} = P_R + P_L$. The response P of the polarization analyzer to an arbitrary polarization state **S** is the dot product

$$P = \mathbf{A} \cdot \mathbf{S} = a_0 s_0 + a_1 s_1 + a_2 s_2 + a_3 s_3 \tag{18}$$

A Stokes vector measurement consists of series of measurements taken with a set of polarization analyzers. Let the total number of analyzers be Q, with each analyzer $\mathbf{A}_q$ specified by index $q = 0, 1, \ldots, Q - 1$. We assume the incident Stokes vector is the same for all polarization analyzers and strive to ensure this in our experimental setup. The qth measurement generates an output $P_q = \mathbf{A}_q \cdot \mathbf{S}$. A polarimetric measurement matrix **W** is defined as a four-by-Q matrix with the qth row containing the analyzer vector $\mathbf{A}_q$,

$$\mathbf{W} = \begin{bmatrix} a_{0,0} & a_{0,1} & a_{0,2} & a_{0,3} \\ a_{1,0} & a_{1,1} & a_{1,2} & a_{1,3} \\ \vdots & & & \\ a_{Q-1,0} & a_{Q-1,1} & a_{Q-1,2} & a_{Q-1,3} \end{bmatrix} \tag{19}$$

The Q measured flux values are arranged in a measurement vector $\mathbf{P} = [P_0, P_1, \ldots, P_{Q-1}]^T \cdot \mathbf{P}$ is related to **S** by the polarimetric measurement equation

$$\mathbf{P} = \begin{bmatrix} P_0 \\ P_1 \\ \vdots \\ P_{Q-1} \end{bmatrix} = \mathbf{WS} = \begin{bmatrix} a_{0,0} & a_{0,1} & a_{0,2} & a_{0,3} \\ a_{1,0} & a_{1,1} & a_{1,2} & a_{1,3} \\ \vdots & & & \\ a_{Q-1,0} & a_{Q-1,1} & a_{Q-1,2} & a_{Q-1,3} \end{bmatrix} \begin{bmatrix} s_0 \\ s_1 \\ s_2 \\ s_3 \end{bmatrix} \tag{20}$$

If **W** is accurately known, then this equation can be inverted to solve for the incident Stokes vector. During calibration of the polarimeter, the principal objective is the determination of the matrix **W** or equivalent information regarding the states which the polarimeter analyzes at each of its analyzer settings. However, systematic errors, differences between the calibrated and actual **W**, will always be present.

To calculate the incident Stokes vector from the data, the inverse of **W** is determined and applied to the measured data. The measured value for the incident Stokes vector is designated $\mathbf{S}_m$ to distinguish it from the actual **S**. In principle, $\mathbf{S}_m$ is related to the data by the polarimetric data reduction matrix $\mathbf{W}^{-1}$,

$$\mathbf{S}_m = \mathbf{W}^{-1}\mathbf{P} \tag{21}$$

Three considerations in the solution of this equation are the existence, rank, and uniqueness of the matrix inverse $\mathbf{W}^{-1}$.

The simplest case occurs when four measurements are performed. If $Q = 4$ linearly independent measurements are made, $\mathbf{W}$ is of rank four, and the polarimetric measurement matrix $\mathbf{W}$ is nonsingular. Then $\mathbf{W}^{-1}$ exists and is unique. Data reduction is performed by Eq. 20 and the polarimeter measures all four elements of the incident Stokes vector.

The second case occurs when $Q > 4$. With more than four measurements, $\mathbf{W}$ is not square, $\mathbf{W}^{-1}$ is not unique, and $\mathbf{S}_m$ is overdetermined by the measurements. In the absence of noise in the measurements, the different $\mathbf{W}^{-1}$ would all yield the same value for $\mathbf{S}_m$. Because noise is always present, the optimum $\mathbf{W}^{-1}$ is desired. The least squares estimate for $\mathbf{S}_m$ utilizes the psuedoinverse $\mathbf{W}_P^{-1}$ of $\mathbf{W}$, $\mathbf{W}_P^{-1} = (\mathbf{W}^T\mathbf{W})^{-1}\mathbf{W}^T$. The best estimate of $\mathbf{S}$ in the presence of random noise is

$$\mathbf{S}_m = (\mathbf{W}^T\mathbf{W})^{-1}\mathbf{W}^T\mathbf{P} \tag{22}$$

The third case occurs when $\mathbf{W}$ is of rank three or less. The optimal matrix inverse is the pseudoinverse. However, only three or less of the Stokes vector elements can be determined from the data. The polarimeter is referred to as "incomplete." Figure 11 in Chap. 27, "Ellipsometry," in this Handbook, summarizes polarization element configurations for Stokes vector measurements listing the vector elements not determined by the incomplete configurations.

22.21 SAMPLE-MEASURING POLARIMETERS FOR MEASURING MUELLER MATRIX ELEMENTS

The polarization characteristics of a sample are characterized by its Mueller matrix. This section describes the particulars of measuring Mueller matrix elements. The section following contains a general formulation of Mueller matrix determination. Since the Mueller matrix is a function of wavelength, angle of incidence, and location on the sample, these are assumed fixed. Figure 1 is a block diagram of a sample-measuring polarimeter. The polarization state generator (PSG) prepares the polarization states which are incident on a sample. A beam of light exiting the sample is analyzed by the polarization state analyzer (PSA) and detected by a detector.

The objective is to determine several elements of a sample Mueller matrix $\mathbf{M}$ through a sequence $q = 0, 1, \ldots, Q - 1$ of polarimetric measurements. The polarization generator prepares a set of polarization states with a sequence of Stokes vectors $\mathbf{S}_q$. The Stokes vectors exiting the sample are $\mathbf{MS}_q$. These exiting states are analyzed by the qth polarization state analyzer $\mathbf{A}_q$, yielding the measured flux $P_q = \mathbf{A}_q^T\mathbf{MS}_q$. Each measured flux is assumed to be a linear function of the sample's Mueller matrix elements. From a set of polarimetric measurements, we develop a set of linear equations which can be solved for certain of the Mueller matrix elements.

Source   PSG   Sample   PSA   Detector

FIGURE 1 A sample-measuring polarimeter consists of a source, polarization state generator (PSG), the sample, a polarization state analyzer (PSA), and the detector. (*After Chenault, 1992.*)

For example, consider a measurement performed with horizontal linear polarizers for both the generator and analyzer. The measured flux depends on the Mueller matrix elements m_{00}, m_{01}, m_{10}, and m_{11} as follows:

$$P = \mathbf{A}^T \mathbf{M} \mathbf{S} = \frac{1}{2}[1 \quad 1 \quad 0 \quad 0] \begin{bmatrix} m_{00} & m_{01} & m_{02} & m_{03} \\ m_{10} & m_{11} & m_{12} & m_{13} \\ m_{20} & m_{21} & m_{22} & m_{23} \\ m_{30} & m_{31} & m_{32} & m_{33} \end{bmatrix} \frac{1}{2} \begin{bmatrix} 1 \\ 1 \\ 0 \\ 0 \end{bmatrix} = \frac{m_{00} + m_{01} + m_{10} + m_{11}}{4} \quad (23)$$

As another example, consider measuring the Mueller matrix elements m_{00}, m_{01}, m_{10}, and m_{11} using four measurements with ideal horizontal (H) and vertical (V) linear polarizers for the polarization state generators and analyzers. The four measurements P_0, P_1, P_2, and P_3 are taken with (generator/analyzer) settings of (H/H), (V/H), (H/V), and (V/V). The combination of Mueller matrix elements measured for each of the four permutations of these polarizers are as follows:

$$P_0 = (m_{00} + m_{01} + m_{10} + m_{11})/4, \qquad P_1 = (m_{00} + m_{01} - m_{10} - m_{11})/4$$
$$P_2 = (m_{00} - m_{01} + m_{10} - m_{11})/4, \qquad P_3 = (m_{00} - m_{01} - m_{10} + m_{11})/4 \qquad (24)$$

These four equations are solved for the Mueller matrix elements as a function of the measured intensities, yielding

$$\begin{bmatrix} m_{00} \\ m_{01} \\ m_{10} \\ m_{11} \end{bmatrix} = \begin{bmatrix} P_0 + P_1 + P_2 + P_3 \\ P_0 + P_1 - P_2 - P_3 \\ P_0 - P_1 + P_2 - P_3 \\ P_0 - P_1 - P_2 + P_3 \end{bmatrix} \qquad (25)$$

Other Mueller matrix elements are determined using other combinations of generator and analyzer states. For example, the four matrix elements at the corners of a rectangle in the Mueller matrix $\{m_{00}, m_{0i}, m_{j0}, m_{ji}\}$ can be determined from four measurements using a $\pm i$-generator and $\pm j$-analyzer. For example, a right and left circularly polarizing generator and 45° and 135° polarizing analyzer will determine $\{m_{00}, m_{02}, m_{30}, m_{32}\}$.

In practice, the data reduction equations are far more complex than the above examples because many more measurements are involved, and especially because the polarization elements are not ideal. The next section contains a method to sytematize the calculation of data reduction equations based on calibration data for the generator and analyzer.

22.22 POLARIMETRIC MEASUREMENT EQUATION AND POLARIMETRIC DATA REDUCTION EQUATION

This section develops equations which relate the measurements in a Mueller matrix polarimeter to the generator and analyzer states. The algorithm can use either ideal or calibrated values for the Stokes vectors of the polarization generator and analyzer. The data reduction equations then have the form of a straightforward matrix-vector multiplication on a data vector. This method is an extension of the matrix data reduction methods presented under "Light-Measuring Polarimeters" on Stokes vector measurement. This method corrects for systematic errors in the generator and analyzer, provided these are characterized in the calibration. A well-calibrated generator and analyzer are essential for accurate Mueller matrix measurements.

A Mueller matrix polarimeter takes Q measurements identified by the index $q = 0$, $1, \ldots, Q - 1$. For the qth measurement, the generator produces a beam with Stokes vector $\mathbf{S}_q$. The beam exiting the sample is analyzed by the polarization analyzer with an analyzer vector $\mathbf{A}_q$. The measured flux P_q is related to the sample Mueller matrix by

$$P_q = \mathbf{A}_q^T \mathbf{M} \mathbf{S}_q = \begin{bmatrix} a_{q,0} & a_{q,1} & a_{q,2} & a_{q,3} \end{bmatrix} \begin{bmatrix} m_{00} & m_{01} & m_{02} & m_{03} \\ m_{10} & m_{11} & m_{12} & m_{13} \\ m_{20} & m_{21} & m_{22} & m_{23} \\ m_{30} & m_{31} & m_{32} & m_{33} \end{bmatrix} \begin{bmatrix} s_{q,0} \\ s_{q,1} \\ s_{q,2} \\ s_{q,3} \end{bmatrix}$$

$$= \sum_{j=0}^{3} \sum_{k=0}^{3} a_{q,j} m_{j,k} s_{q,k} \tag{26}$$

This equation is now rewritten as a vector-vector dot product (Azzam, 1978; Goldstein, 1992). First, the Mueller matrix is flattened into a 16×1 *Mueller vector* $\vec{\mathbf{M}} = \begin{bmatrix} m_{00} & m_{01} & m_{02} & m_{03} & m_{10} & \cdots & m_{33} \end{bmatrix}^T$. A 16×1 polarimetric measurement vector $\mathbf{W}_q$ for the qth measurement is defined as follows

$$\mathbf{W}_q = \begin{bmatrix} w_{q,00} & w_{q,01} & w_{q,02} & w_{q,03} & w_{q,10} & \cdots & w_{q,33} \end{bmatrix}^T$$

$$= \begin{bmatrix} a_{q,0} s_{q,0} & a_{q,0} s_{q,1} & a_{q,0} s_{q,2} & a_{q,0} s_{q,3} & a_{q,1} s_{q,0} & \cdots & a_{q,3} s_{q,3} \end{bmatrix}^T \tag{27}$$

where $w_{q,jk} = a_{q,j} s_{q,k}$. The qth measured flux from Eq. (25) is rewritten as the dot product

$$P_q = \mathbf{W}_q \cdot \vec{\mathbf{M}} = \begin{bmatrix} a_{q,0} s_{q,0} \\ a_{q,0} s_{q,1} \\ a_{q,0} s_{q,2} \\ a_{q,0} s_{q,3} \\ a_{q,1} s_{q,0} \\ a_{q,1} s_{q,1} \\ \vdots \\ a_{q,3} s_{q,3} \end{bmatrix} \begin{bmatrix} m_{0,0} \\ m_{0,1} \\ m_{0,2} \\ m_{0,3} \\ m_{1,0} \\ m_{1,1} \\ \vdots \\ m_{3,3} \end{bmatrix} \tag{28}$$

The full sequence of measurements is described by the polarimetric measurement matrix $\mathbf{W}$, defined as the $Q \times 16$ matrix where the qth row is $\mathbf{W}_q$. The polarimetric measurement equation relates the measurement vector $\mathbf{P}$ to the sample Mueller vector by a matrix-vector multiplication,

$$\mathbf{P} = \mathbf{W} \vec{\mathbf{M}} = \begin{bmatrix} P_0 \\ P_1 \\ \vdots \\ P_{Q-1} \end{bmatrix} = \begin{bmatrix} w_{0,00} & w_{0,01} & \cdots & w_{0,33} \\ w_{1,00} & w_{1,01} & \cdots & w_{1,33} \\ \vdots & & & \\ w_{Q-1,00} & w_{Q-1,01} & \cdots & w_{Q-1,33} \end{bmatrix} \begin{bmatrix} m_{00} \\ m_{01} \\ \vdots \\ m_{33} \end{bmatrix} \tag{29}$$

If $\mathbf{W}$ contains sixteen linearly independent columns, all sixteen elements of the Mueller matrix can be determined. Then, if $Q = 16$, the matrix inverse is unique and the Mueller matrix elements are determined from the polarimetric data reduction equation $\vec{\mathbf{M}} = \mathbf{W}^{-1}\mathbf{P}$. More often, $Q > 16$, and $\vec{\mathbf{M}}$ is overdetermined. The optimal (least-squares) polarimetric data reduction equation for $\vec{\mathbf{M}}$ uses the pseudoinverse $\mathbf{W}_P^{-1}$ of $\mathbf{W}$, Eq. (21) where $\mathbf{W}_P^{-1}$ is a polarimetric data reduction matrix for the polarimeter. The polarimetric data reduction equation is then

$$\vec{\mathbf{M}} = (\mathbf{W}^T \mathbf{W})^{-1} \mathbf{W}^T \mathbf{P} = \mathbf{W}_P^{-1} \mathbf{P} \tag{30}$$

where $\mathbf{W}_P^{-1}$ operates on a set of measurements to estimate the Mueller matrix of the sample.

The advantages of this polarimetric measurement equation and polarimetric data reduction equation procedure are as follows. First, this procedure does not assume that the set of states of polarization state generator and analyzer have any particular form. For example, the polarization elements in the generator and analyzer do not need to be rotated in uniform angular increments, but can comprise an arbitrary sequence. Second, the polarization elements are not assumed to be ideal polarization elements or have any particular imperfections. If the Stokes vectors associated with the polarization generator and analyzer are determined through a calibration procedure, the effects of nonideal polarization elements are corrected in the data reduction. Third, the procedure readily treats overdetermined measurement sequences (more than sixteen measurements for the full Mueller matrix), providing a least-squares solution. Finally, a matrix-vector form of data reduction is readily implemented and understood.

The next two sections describe configurations of sample-measuring polarimeter with example data reduction matrices.

22.23 DUAL ROTATING RETARDER POLARIMETER

The dual rotating retarder Mueller matrix polarimeter is one of the most common Mueller polarimeters. Figure 2 shows the configuration: light from the source passes first through a fixed linear polarizer, then through a rotating linear retarder, the sample, a rotating linear retarder, and finally through a fixed linear polarizer. In the most common configuration, first described by Azzam (1978), the polarizers are parallel, and the retarders are rotated in angular increments of five-to-one. This five-to-one ratio encodes all 16 Mueller matrix elements onto the amplitudes and phases of 12 frequencies in the detected signal. The detected signal is Fourier analyzed, and the Mueller matrix elements are calculated from the Fourier coefficients.

This polarimeter design has an important advantage: the polarizers do not move. The polarizer in the generator accepts only one polarization state from the source optics, making the measurement immune to instrumental polarization from the source optics. If the polarizer did rotate, and if the beam incident on it were elliptically polarized, a systematic modulation of intensity would be introduced which would require compensation. Similarly, the polarizer in the analyzer does not rotate; only one polarization state is transmitted through the analyzing optics and onto the detector. Any diattenuation in the analyzing optics and any polarization sensitivity in the detector will not affect the measurements.

The data reduction matrix is presented here for a polarimeter with ideal linear retarders with arbitrary retardances δ_1 in the generator and δ_2 in the analyzer. Optimal values for the retardances are near $\lambda/4$ or $\lambda/3$, depending which characteristics of the

Source   Detector

Polarizer Retarder Sample Retarder Polarizer

FIGURE 2 The dual rotating retarder polarimeter consists of a source, a fixed linear polarizer, a retarder which rotates in steps, the sample, a second retarder which rotates in steps, a fixed linear polarizer, and the detector. This polarimeter measures the full Mueller matrix. It accepts only one polarization state from the source, and transmits only one polarization state to the detector. (*After Chenault, 1992.*)

Mueller matrix are chosen for a figure of merit. If $\delta_1 = \delta_2 = \pi$ rad, the last row and column of the sample Mueller matrix are not measured. Q measurements are taken, described by index $q = 0, 1, \ldots, Q - 1$. The angular orientations of the two retarders for measurement q are $\theta_{q,1} = q180°/Q$ and $\theta_{q,2} = 5q180°/Q$. The angular increment between settings of the generator retarder is $\Delta\theta = 180°/Q$. The data reduction matrix is a $16 \times Q$ matrix which multiplies a $Q \times 1$ data vector **P**, yielding the sample Mueller vector **M**. Table 2 lists the equations for the elements in each row q of the data reduction matrix assuming ideal polarization elements (Chenault, Pezzaniti, and Chipman, 1992).

Several data reduction methods have been published to account for additional imperfections in the polarization elements, leading to considerably more elaborate expressions than those presented here. Hauge (1978) developed an algorithm to compensate for the linear diattenuation and linear retardance of the retarders. Goldstein and Chipman (1990) treat five errors, the retardances of the two retarders, and orientation errors of the two retarders and one of the polarizers, in a small angle approximation good for small errors. Chenault, Pezzaniti, and Chipman (1992) extended this method to larger errors.

22.24 INCOMPLETE SAMPLE-MEASURING POLARIMETERS

Incomplete sample-measuring polarimeters do not measure the full Mueller matrix of a sample and thus provide incomplete information regarding the polarization properties of a sample. Often the full Mueller matrix is not needed. For example, many birefringent samples have considerable linear birefringence and minuscule amounts of the other forms of polarization. The magnitude of the birefringence can be measured, assuming all the other polarization effects are small, using much simpler configurations than a Mueller matrix polarimeter, such as the circular polariscope (Theocaris and Gdoutos, 1979). Similarly, homogeneous and isotropic interfaces, such as dielectrics, metals, and thin films, should only display linear diattenuation and linear retardance aligned with the $s - p$ planes. These interfaces do not need characterization of their circular diattenuation and circular retardance. Many categories of ellipsometer will characterize such samples without providing the full Mueller matrix (Azzam and Bashara, 1977, 1987; Azzam, 1993).

22.25 DUAL ROTATING POLARIZER POLARIMETER

This section describes the dual rotating polarizer polarimeter, a common polarimetric configuration capable of measuring nine Mueller matrix elements (Collins and Kim, 1990). Figure 3 shows the arrangement of polarization elements in the polarimeter. Light from the source passes through a linear polarizer whose orientation θ_1 is adjustable. This

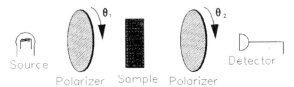

FIGURE 3 The dual rotating polarizer polarimeter consists of a source, a linear polarizer which rotated in steps, the sample, a second linear polarizer with a stepped angular orientation, and the detector. (*After Chenault, 1992.*)

TABLE 2 Elements of the Polarimetric Data Reduction Matrix for the Dual Rotating Retarder Polarimeter

---

$$w_{q,0} = 1$$

$$w_{q,1} = \cos^2\left(\frac{\delta_1}{2}\right) + \sin^2\left(\frac{\delta_1}{2}\right)\cos\left(4q\,\Delta\theta\right)$$

$$w_{q,2} = \sin^2\left(\frac{\delta_1}{2}\right)\sin\left(4q\,\Delta\theta\right)$$

$$w_{q,3} = \sin\left(\delta_1\right)\sin\left(2q\,\Delta\theta\right)$$

$$w_{q,4} = \cos^2\left(\frac{\delta_2}{2}\right) + \sin^2\left(\frac{\delta_2}{2}\right)\cos\left(20q\,\Delta\theta\right)$$

$$w_{q,5} = \cos^2\left(\frac{\delta_1}{2}\right)\cos^2\left(\frac{\delta_2}{2}\right) + \sin^2\left(\frac{\delta_1}{2}\right)\cos^2\left(\frac{\delta_2}{2}\right)\cos\left(4q\,\Delta\theta\right)$$

$$+ \cos^2\left(\frac{\delta_1}{2}\right)\sin^2\left(\frac{\delta_2}{2}\right)\cos\left(20q\,\Delta\theta\right)$$

$$+ \frac{1}{2}\sin^2\left(\frac{\delta_1}{2}\right)\sin^2\left(\frac{\delta_2}{2}\right)\left(\cos\left(16q\,\Delta\theta\right) + \cos\left(24q\,\Delta\theta\right)\right)$$

$$w_{q,6} = \sin^2\left(\frac{\delta_1}{2}\right)\cos^2\left(\frac{\delta_2}{2}\right)\sin\left(4q\,\Delta\theta\right)$$

$$+ \frac{1}{2}\sin^2\left(\frac{\delta_1}{2}\right)\sin^2\left(\frac{\delta_2}{2}\right)\left(-\sin\left(16q\,\Delta\theta\right) + \sin\left(24q\,\Delta\theta\right)\right)$$

$$w_{q,7} = \sin\left(\delta_1\right)\cos^2\left(\frac{\delta_2}{2}\right)\sin\left(2q\,\Delta\theta\right)$$

$$+ \frac{1}{2}\sin\left(\delta_1\right)\sin^2\left(\frac{\delta_2}{2}\right)\left(-\sin\left(18q\,\Delta\theta\right) + \sin\left(22q\,\Delta\theta\right)\right)$$

$$w_{q,8} = \sin^2\left(\frac{\delta_2}{2}\right)\sin\left(20q\,\Delta\theta\right)$$

$$w_{q,9} = \cos^2\left(\frac{\delta_1}{2}\right)\sin^2\left(\frac{\delta_2}{2}\right)\sin\left(20q\,\Delta\theta\right)$$

$$+ \frac{1}{2}\sin^2\left(\frac{\delta_1}{2}\right)\sin^2\left(\frac{\delta_2}{2}\right)\left(\sin\left(16q\,\Delta\theta\right) + \sin\left(24q\,\Delta\theta\right)\right)$$

$$w_{q,10} = \frac{1}{2}\sin^2\left(\frac{\delta_1}{2}\right)\sin^2\left(\frac{\delta_2}{2}\right)\left(\cos\left(16q\,\Delta\theta\right) - \cos\left(24q\,\Delta\theta\right)\right)$$

$$w_{q,11} = \frac{1}{2}\sin\left(\frac{\delta_1}{2}\right)\sin^2\left(\frac{\delta_2}{2}\right)\left(\cos\left(18q\,\Delta\theta\right) - \cos\left(22q\,\Delta\theta\right)\right)$$

$$w_{q,12} = -\sin\left(\delta_2\right)\sin\left(10q\,\Delta\theta\right)$$

$$w_{q,13} = -\cos^2\left(\frac{\delta_1}{2}\right)\sin\left(\delta_2\right)\sin\left(10q\,\Delta\theta\right)$$

$$- \frac{1}{2}\sin^2\left(\frac{\delta_1}{2}\right)\sin\left(\delta_2\right)\left(\sin\left(6q\,\Delta\theta\right) + \sin\left(14q\,\Delta\theta\right)\right)$$

$$w_{q,14} = -\frac{1}{2}\sin^2\left(\frac{\delta_1}{2}\right)\sin\left(\delta_2\right)\left(\cos\left(6q\,\Delta\theta - \cos\left(14q\,\Delta\theta\right)\right)\right)$$

$$w_{q,15} = -\frac{1}{2}\sin\left(\delta_1\right)\sin\left(\delta_2\right)\left(\cos\left(8q\,\Delta\theta\right) - \cos\left(12q\,\Delta\theta\right)\right)$$

---

linearly polarized light interacts with the sample and is analyzed by a second linear polarizer whose orientation θ_2 is also adjustable. This polarimeter is incomplete because measurement of the last column of the Mueller matrix requires elliptical states from the polarization generator. Similarly, elliptical analyzers are required in the polarization analyzer to measure the bottom row of the Mueller matrix.

The polarimetric data reduction matrix which follows is for a particular 16-measurement sequence. The most common defects of polarizers have been taken into consideration: less than ideal diattenuation, and transmission of less than unity. The polarizers are characterized by T_{max}, the maximum intensity transmittance for a single polarizer, and T_{min}, the minimum intensity transmittance, which are associated with orthogonal linear states. Let $a = (16(T_{max} + T_{min})^2)^{-1}$, $b = (8(T_{max}^2 - T_{min}^2))^{-1}$ and $c = (4(T_{max} - T_{min})^2)^{-1}$. Sixteen measurements are acquired with the generator polarizer angle $\theta_{q,1}$ and the analyzer polarizer angle $\theta_{q,2}$ oriented as follows: $\theta_{q,1} = (0°, 0°, 0°, 0°, 45°, 45°, 45°, 45°, 90°, 90°, 90°, 90°, 135°, 135°, 135°, 135°)$, $\theta_{q,2} = (0°, 45°, 90°, 135°, 0°, 45°, 90°, 135°, 0°, 45°, 90°, 135°, 0°, 45°, 90°, 135°)$. Since only nine Mueller matrix elements are measured, a nine-element Mueller vector is used:

$$\vec{\mathbf{M}} = [m_{00} \quad m_{01} \quad m_{02} \quad m_{10} \quad m_{11} \quad m_{12} \quad m_{20} \quad m_{21} \quad m_{22}]^T \tag{31}$$

The data reduction matrix $\mathbf{W}_P^{-1}$ which operates on the 16-element measurement vector $\vec{\mathbf{P}}$ yielding $\vec{\mathbf{M}}$ is

$$\mathbf{W}_P^{-1} = \begin{bmatrix} a & a & a & a & a & a & a & a & a & a & a & a & a & a & a & a \\ b & b & b & b & 0 & 0 & 0 & 0 & -b & -b & -b & -b & 0 & 0 & 0 & 0 \\ 0 & 0 & 0 & 0 & b & b & b & b & 0 & 0 & 0 & 0 & -b & -b & -b & -b \\ b & 0 & -b & 0 & b & 0 & -b & 0 & b & 0 & -b & 0 & b & 0 & -b & 0 \\ c & 0 & -c & 0 & 0 & 0 & 0 & 0 & -c & 0 & c & 0 & 0 & 0 & 0 & 0 \\ 0 & 0 & 0 & 0 & c & 0 & -c & 0 & 0 & 0 & 0 & 0 & -c & 0 & c & 0 \\ 0 & b & 0 & -b & 0 & b & 0 & -b & 0 & b & 0 & -b & 0 & b & 0 & -b \\ 0 & c & 0 & -c & 0 & 0 & 0 & 0 & 0 & -c & 0 & c & 0 & 0 & 0 & 0 \\ 0 & 0 & 0 & 0 & 0 & c & 0 & -c & 0 & 0 & 0 & 0 & 0 & -c & 0 & c \end{bmatrix} \tag{32}$$

The source is assumed to be unpolarized in this equation. Similarly, the detector is assumed to be polarization-insensitive. When this is not the case, the data reduction matrix is readily generalized to incorporate these and other systematic effects following the method shown under "Polarimetric Measurement Equation and Polarimetric Data Reduction Equation."

22.26 NONIDEAL POLARIZATION ELEMENTS

For use in polarimetry, polarization elements require a level of characterization beyond what is normally provided by vendors. For retarders, usually only the linear retardance is specified. For polarizers, usually only the two principal transmittances or the extinction ratio is given. For polarization elements used in critical applications such as polarimetry, this level of characterization is inadequate. In this section, defects of polarization elements are described, and the Mueller calculus is recommended as the most appropriate measure of performance.

22.27 POLARIZATION PROPERTIES OF POLARIZATION ELEMENTS

For ideal polarization elements, the polarization properties are readily defined. For real polarization elements, the precise description of the polarization properties is more complex. The handbook chapter "Polarizers" (Chap. 3) contains an extensive description of the various forms of polarizers and retarders and their characteristics (Bennett 1993). Polarization elements such as polarizers, retarders, and depolarizers have three general polarization properties: diattenuation, retardance, and depolarization, and a typical element displays some amount of all three. Diattenuation arises when the intensity transmittance of an element is a function of the incident polarization state (Chipman, 1989a). The diattenuation D of a device is defined in terms of the maximum T_{max} and minimum T_{min} intensity transmittances,

$$D = \frac{T_{max} - T_{min}}{T_{max} + T_{min}} \tag{33}$$

for an ideal polarizer, $D = 1$. When $D = 0$, all incident polarization states are transmitted with equal loss, although the polarization states in general change upon transmission. The quality of a polarizer is often expressed in terms of the related quantity, the extinction ratio E,

$$E = \frac{T_{max}}{T_{min}} = \frac{1 + D}{1 - D} \tag{34}$$

Retardance is the phase change a device introduces between its eigenpolarizations (eigenstates). For a birefringent retarder with refractive indices n_1 and n_2, and thickness t, the retardance δ expressed in radians is

$$\delta = \frac{2\pi(n_1 - n_2)t}{\lambda} \tag{35}$$

Depolarization describes the coupling by a device of incident polarized light into depolarized light in the exiting beam. For example, depolarization occurs when light transmits through milk or scatters from clouds. Multimode optical fibers generally depolarize the light. Depolarization is intrinsically associated with scattering and a loss of coherence in the polarization state. A small amount of depolarization is probably associated with the scattered light from all optical components. A depolarization coefficient e can be defined as the fraction of unpolarized power in the exiting beam when polarized light is incident. e is generally a function of the incident polarization state.

22.28 COMMON DEFECTS OF POLARIZATION ELEMENT

Here we list some common defects found in real polarization elements.

1. Polarizers have nonideal diattenuation since $T_{max} < 1$ and $T_{min} > 0$ (Bennett, 1993; King and Talim, 1971).

2. Retarders have the incorrect retardance. Thus, there will be some deviation from a quarter-wave or a half-wave of retardance, for example, because of fabrication errors or a change in wavelength.

3. Retarders usually have some diattenuation because of differences in absorption coefficients (dichroism) and due to different transmission and reflection coefficients at the interfaces. For example, birefringent retarders have diattenuation due to the difference of the Fresnel coefficients at normal incidence for the two eigenpolarizations since $n_1 \neq n_2$. This can be reduced by antireflection coatings.

4. Polarizers usually have some retardance; there is a difference in optical path length between the transmitted (principal) eigenpolarization and the small amount of the extinguished (secondary) eigenpolarization. For example, sheet polarizers and wire-grid polarizers show substantial retardance when the secondary state is not completely extinguished.

5. The polarization properties vary with angle of incidence; for example, Glan-Thompson polarizers polarize over only a 4° field of view (Bennett, 1994). Birefringent retarders commonly show a quadratic variation of retardance with angle of incidence which increases along one axis and decreases along the orthogonal axis (Title, 1979; Hale and Day, 1988). For polarizing beam-splitter cubes, the axis of linear polarization rotates for incident light out of its normal plane (the plane defined by the face normals and the beam-splitting interface normal).

7. The polarization properties vary with wavelength; for example, for simple retarders made from a single birefringent plate, the retardance varies approximately linearly with wavelength.

8. For polarizers, the *accepted state* and the *transmitted state* can be different. Consider a polarizing device formed from a linear polarizer oriented at 0° followed by a linear polarizer oriented at 2°. Incident light linearly polarized at 0° has the highest transmittance for all possible polarization states and is the accepted state. The corresponding exiting beam is linearly polarized at 2°, which is the only state exiting the device. In this example, the transmitted state is also an eigenpolarization. This "rotation" between the accepted and transmitted states of a polarizer frequently occurs, for example, when the crystal axes are misaligned in a birefringent polarizing prism assembly such as a Glan-Thompson polarizer.

9. A nominally "linear" element may be slightly elliptical (have elliptical eigenpolarizations). For example, a quartz linear retarder with the crystal axis misaligned becomes an elliptical retarder. Similarly a circular element may be slightly elliptical. For example, an (inhomogeneous) circular polarizer formed from a linear polarizer followed by a quarter-wave linear retarder at 45° [see Eq. (14)] becomes an elliptical polarizer as the retarder's fast axis is rotated.

10. The eigenpolarizations of the polarization element may not be orthogonal; i.e., a polarizer may transmit linearly polarized light at 0° without change of polarization while extinguishing linearly polarized light oriented at 88°. Such a polarization element is referred to as *inhomogeneous* (Shurcliff, 1962; Lu and Chipman, 1992). Sequences of polarization elements, such as optical isolator assemblies, often are inhomogeneous. The circular polarizer in Eq. 14 is inhomogeneous.

11. A polarization element may depolarize, coupling polarized light into unpolarized light. A polarizer or retarder with a small amount of depolarization, when illuminated by a completely polarized beam, will have a small amount of unpolarized light in the transmitted beam. Such a transmitted beam can no longer be extinguished by an ideal polarizer. Depolarization results from fabrication errors such as surface roughness, bulk scattering, random strains and dislocations, and thin-film microstructure.

12. Multiply reflected beams and other "secondary" beams may be present with undesired polarization properties. For example, the multiply reflected beams from a birefringent plate have various values for their retardance. Antireflection coatings will reduce this effect in one waveband, but may increase these problems with multiple reflections in other wavebands.

The preceding list of polarization element defects is by no means comprehensive. It should serve as a warning to those with demanding applications for polarization elements. In particular, the performance of polarizing beam-splitting cubes have been found to be quite different from the ideal (Pezzaniti and Chipman, 1991).

22.29 THE MUELLER MATRIX FOR POLARIZATION COMPONENT CHARACTERIZATION

The Mueller matrix provides the full characterization of a polarization element (Shurcliff, 1962; Azzam and Bashara, 1977). From the Mueller matrix, all of the performance defects listed previously and more are specified. Thus, when one is using polarization elements in critical applications such as polarimetry, it is highly desirable that the Mueller matrix of the elements be known. This is analogous to having the interferogram of a lens to ensure that it is of suitable quality for incorporation into a critical imaging system.

The optics community has been very slow to adopt Mueller matrices for the testing of optical components and optical systems, delaying a broad understanding of how real polarization elements actually perform. An impediment to the widespread acceptance of Mueller matrices for polarization element qualification has been that the polarization properties associated with a Mueller matrix (the diattenuation, retardance, and depolarization) are not easily "extracted" from the Mueller matrix. Thus, while the operational definition of the Mueller matrix, Eq. (10), is straightforward, determining the diattenuation, retardance, and depolarization from an experimentally determined Mueller matrix is a complex process (Gil and Bernabeau, 1987). This is described later in this chapter.

The following matrix element pairs indicate the presence of the various forms of diattenuation and retardance:

$$
\begin{bmatrix}
0 & a & b & c \\
a & 0 & -d & -e \\
b & d & 0 & -f \\
c & e & f & 0
\end{bmatrix}
\tag{36}
$$

Each pair of elements is related to the following properties:

a linear diattenuation oriented at 0° or 90°

b linear diattenuation oriented at 45° or 135°

c circular diattenuation

d linear retardance oriented at 0° or 90°

e linear retardance oriented at 45° or 135°

f circular retardance

For small amounts of these properties, the Mueller matrix elements indicated are linear in the diattenuation or retardance. Other degrees of freedom in the Mueller matrix, antisymmetry in a, b, or c or symmetry in d, e, or f, indicate the presence of depolarization and inhomogeneity.

22.30 *APPLICATIONS OF POLARIMETRY*

Polarimetry has found application in nearly all areas of science and technology with several tens of thousands of papers detailing various applications. The following summarizes a few of the principal applications and introduces some of the books, reference works, and review papers which provide gateways to the various applications.

Ellipsometry

Ellipsometry is the application of polarimetry for determining the optical properties of surfaces and interfaces. Example applications are refractive indices and thin-film thickness determination, and investigations of processes at surfaces such as contamination and corrosion. In this Handbook, Chap. 27, "Ellipsometry," by Azzam treats the fundamentals. A more extensive treatment is found in the textbook by Azzam and Bashara (1979 and 1986) which presents the mathematical fundamentals of polarization, determination of the properties of thin films, polarimetric instrumentation, and a myriad of applications. Azzam (1991) is a recent collection of historical papers. Calculation of the polarization properties of thin films is given a detailed presentation by Dobrowolski (1994) in Chap. 43 of Vol. I of this Handbook, and also in the text by Macleod (1986).

Spectropolarimetry for Chemical Applications

Spectropolarimeters are spectrometers which incorporate polarimeters for the purpose of measuring polarization properties as a function of wavelength. Whereas spectrometers measure transmission or reflectance as a function of wavelength, a spectropolarimeter also may measure dichroism (diattenuation), linear birefringence (linear retardance), optical activity (circular retardance), or depolarization, all as spectra. In physical chemistry, spectra of the linear dichroism and the linear retardance of a molecule permit the determination of the orientation of the electric dipole moment in three dimensions. Similarly, circular dichroism and optical activity provide information on the orbital magnetic moment. Schellman and Jensen (1987) and Johnson (1987) provide comprehensive surveys of the spectropolarimetry of oriented molecules and interpretation of the data in terms of molecular structure. The volumes by Michl and Thulstrup (1986), Samori and Thulstrup (eds.) (1988), and by Kliger, Lewis, and Randall (1990) cover the basics of polarimetry with an emphasis on spectroscopy with polarized light and interpretation of the resulting data. Texts and reviews on optical activity include the following: Jirgensons (1973), Mason (ed.) (1978), Mason (1982), Thulstrup (1982), Barron (1986), and Laktakia (1990). Chenault (1992) contains a survey of spectropolarimetric instrumentation.

Remote Sensing

Polarimetry has become an important technique in remote sensing, since it augments the limited information available from spectrometric techniques. Polarization in the scattered light from the earth has many subtle characteristics. The sunlight which illuminates the earth is essentially unpolarized, but the scattered light has a surprisingly large degree of polarization, which is mostly linear polarization (Egan, 1985; Konnen, 1985; Coulson, 1988; Coulson, 1989; Egan, 1992). Visible light scattered from forest canopy, cropland, meadows, and similar features frequently has a degree of polarization of 20 percent or greater in the visible (Curran, 1982; Duggin, 1989). Light reflecting from mudflats and water often has a degree of polarization of 50 percent or higher, particularly for light incident near

Brewster's angle. Light scattered from clouds is nearly unpolarized (Konnen, 1985; Coulson, 1988). The magnitude of the degree of linear polarization depends on many variables, including the angle of incidence, the angle of scatter, the wavelength, and the weather. The polarization from a site varies from day to day even if the angles of incidence and scatter remain the same; these variations are caused just by changes in the earth's vegetation, cloud cover, humidity, rain, and standing water. Polarization is complex to interpret but it conveys a great deal of useful information.

Astronomical Polarimetry

The polarization of light from astronomical bodies conveys considerable information regarding their physical state—information that generally cannot be acquired by any other means. Gehrels (1974) compiles information regarding the polarization of plants, stars, and other astronomical objects. Polarimetry is the principle technique for determining solar magnetic fields. Solar vector magnetographs are imaging polarimeters combined with narrowband tunable spectral filters which measure Zeeman splitting in magnetically active ions in the solar atmosphere, from which the magnetic fields can be determined. November (1991) is a recent survey of instrumentation and ongoing measurement programs for solar magnetic field study.

Polarization Light Scattering

Polarization light scattering is the application of polarimetry to scattered light (Van de Hulst, 1957; Stover, 1990). The scattering characteristics of a sample are generally described by its bidirectional reflectance distribution function, $BRDF(\alpha, \beta, \gamma, \delta, \lambda)$, which is the ratio of the scattered flux in a particular direction (γ, δ) to the flux of an incident beam from direction (α, β). The $BRDF$ function contains no polarization information, but is the m_{00} element of the Mueller matrix relating the incident and scattered beams. The $BRDF$ can be generalized to a Mueller bidirectional reflectance distribution function, or **MBRDF**$(\alpha, \beta, \gamma, \delta, \lambda)$, which is the Mueller matrix relating arbitrary incident and scattered beams. Scattered light is often a sensitive indicator to surface conditions; a small amount of surface roughness may reduce the specular power by less than a percent while increasing the scattered power by orders of magnitude. The retardance, diattenuation, and depolarization of the scattered light similarly provide sensitive indicators of light-scattering conditions, such as uniformity of refractive index, orientation of surface defects, texture, strain and birefringence at an interface, subsurface damage, coating microstructure, and the degree of multiple scattering.

Optical and Polarization Metrology

Polarimetry is useful in optical metrology for measuring the instrumental polarization of optical systems and for characterizing optical and polarization components. Optical systems, both common and exotic, modify the polarization state of light due to the reflections, refractions, and other interactions with optical materials. Each ray path through the optical system can be characterized by its polarization matrix (Chipman, 1989a; Chipman, 1989b). Polarization ray tracing is the technique of calculating the polarization matrices for ray paths from the optical and coating prescriptions (Waluschka, 1989; Bruegge, 1989; Wolff and Kurlander, 1990). Diffraction image formation of polarization-aberrated beams is then handled by vector extensions to diffraction theory (Kuboda and Inouè, 1959; Urbanczyk, 1984; Urbanczyk, 1986; McGuire and Chipman, 1990; McGuire and Chipman, 1991; Mansuripur, 1991). Polarimeters, particularly imaging polarimeters, are used to measure the Mueller matrices of ray paths through optical systems determining the polarization aberrations. These polarization aberrations frequently have the same functional forms as the geometrical aberrations, since they arise from similar geometrical considerations (Chipman, 1987; McGuire and Chipman, 1987, 1989, 1990a, 1990b, 1991; Hansen, 1988; Chipman and Chipman, 1989). Several conferences have surveyed these areas (Chipman, 1988; Chipman, 1989c; Goldstein and Chipman, 1992).

Radar Polarimetry

Polarimetric measurements are a standard and highly evolved technique in radar with broad application (Poelman and Guy, 1985; Holm, 1987; van Zyl and Zebker, 1990). Although radar is outside the scope of this handbook, several references to the radar literature are included since the optical community can greatly benefit from advances in radar polarimetry. The text by Mott (1992) develops the polarization properties of antennas and the techniques of radar polarimetry. Fundamental analyses of the Mueller matrix have been performed by Huynen (1965) and Kennaugh (1951), both of which have found broad application in the interpretation of radar polarization signatures. Morris and Chipman (eds.) (1990) and Boerner and Mott (eds.) (1992) are proceedings from meetings specifically intended to provide an exchange between the optical polarimetry and radar polarimetry communities.

22.31 INTERPRETATION OF MUELLER MATRICES

The Mueller matrix is defined as a matrix which transforms incident Stokes vectors into exiting Stokes vectors with each element seen as a coupling between corresponding Stokes vector elements. Despite this simple and elegant definition, the polarization properties associated with the Mueller matrix—the diattenuation, retardance, and depolarization—are not readily apparent from the matrix for two reasons. First, the Stokes vector has an unusual coordinate system in which the different elements do not represent orthogonal polarization components. Instead, positive and negative values on each component separately represent orthogonal polarization components. Second, the phenomenon of depolarization greatly complicates the matrix properties. It is not possible to analyze real arbitrary Mueller matrices measured by polarimeters without considering three tricky topics: physical realizability, depolarization, and inhomogeneity.

22.32 DIATTENUATION AND POLARIZATION SENSITIVITY

The intensity transmittance T for a given matrix $\mathbf{M}$ and incident polarization state $\mathbf{S}$ is defined as the ratio of exiting flux s_0' to incident flux s_0,

$$T(\mathbf{MS}) = \frac{s_0'}{s_0} = \frac{m_{00}s_0 + m_{01}s_1 + m_{02}s_2 + m_{03}s_3}{s_0} \tag{37}$$

The intensity transmittance averaged over all incident polarization states is $T_{\text{avg}} = m_{00}$. The maximum T_{max} and minimum T_{min} intensity transmittances are

$$T_{\text{max}} = m_{00} + \sqrt{m_{01}^2 + m_{02}^2 + m_{03}^2}, \qquad T_{\text{min}} = m_{00} - \sqrt{m_{01}^2 + m_{02}^2 + m_{03}^2} \tag{38}$$

and are associated with the unnormalized incident states

$$\mathbf{S}_{\max} = \begin{bmatrix} \sqrt{m_{01}^2 + m_{03}^2 + m_{03}^2} \\ m_{01} \\ m_{02} \\ m_{03} \end{bmatrix} \qquad \mathbf{S}_{\min} = \begin{bmatrix} \sqrt{m_{01}^2 + m_{03}^2 + m_{03}^2} \\ -m_{01} \\ -m_{02} \\ -m_{03} \end{bmatrix} \qquad (39)$$

The incident Stokes vectors of maximum $\mathbf{S}_{\max}$ and minimum $\mathbf{S}_{\min}$ intensity transmittance are always orthogonal. The term *diattenuation* refers to the two attenuations associated with these two orthogonal states. The diattenuation $D(\mathbf{M})$ of a Mueller matrix is a measure of the variation of intensity transmittance with incident polarization state,

$$D(\mathbf{M}) = \frac{T_{\max} - T_{\min}}{T_{\max} + T_{\min}} = \frac{\sqrt{m_{01}^2 + m_{02}^2 + m_{03}^2}}{m_{00}} \qquad (40)$$

When $D = 1$, the device is an ideal analyzer; it completely blocks one polarization component of the incident light, and only one Stokes vector exits the device. If this device is also nondepolarizing, then $\mathbf{M}$ represents a polarizer. When $D = 0$, all incident states have the same intensity transmittance: the device may be nonpolarizing, depolarizing, or a pure retarder. Diattenuation is also referred to as *polarization sensitivity*. *Linear polarization sensitivity* or *linear diattenuation* $LD(\mathbf{M})$ characterizes the variation of intensity transmittance with incident linear polarization states:

$$LD(M) = \frac{\sqrt{m_{01}^2 + m_{02}^2}}{m_{00}} \qquad (41)$$

Linear polarization sensitivity is frequently specified as a performance parameter in remote sensing systems designed to measure incident power independently of any linearly polarized component present in scattered earth-light (Maymon and Chipman, 1991). Note that $LD(\mathbf{M}) = 1$ specifies that $\mathbf{M}$ is a linear analyzer; $\mathbf{M}$ is not necessarily a linear polarizer, but may represent a linear polarizer followed by some other polarization element. Diattenuation in (fiber optical) components and systems is often characterized by the *polarization dependent loss*, given in decibels:

$$PDL(\mathbf{M}) = 10 \, \mathrm{Log}_{10} \frac{T_{\max}}{T_{\min}} \qquad (42)$$

22.33 POLARIZANCE

The polarizance $P(\mathbf{M})$ is the degree of polarization of the transmitted light when unpolarized light is incident (Bird and Shurcliff, 1959; Shurcliff, 1962).

$$P(\mathbf{M}) = \frac{\sqrt{m_{10}^2 + m_{20}^2 + m_{30}^2}}{m_{00}} \qquad (43)$$

The Stokes vector of the exiting light $\mathbf{S}_P$ is specified by the first column of $\mathbf{M}$,

$$\mathbf{S}_P(\mathbf{M}) = [m_{00} \quad m_{10} \quad m_{20} \quad m_{30}]^T \qquad (44)$$

and is not generally equal to $\mathbf{S}_{\max}$ when inhomogeneity or depolarization is present.

22.34   *PHYSICALLY REALIZABLE MUELLER MATRICES*

Mueller matrices form a subset of the four-by-four real matrices. A four-by-four real matrix is not a physically realizable Mueller matrix if it can operate on an incident Stokes vector to produce a vector with degree of polarization greater than one ($s_0^2 < s_1^2 + s_2^2 + s_3^2$), which represents a physically unrealizable polarization state. Similarly, a Mueller matrix cannot output a state with negative flux. Conditions for physical realizability have been studied extensively in the literature, and many necessary conditions have been published (Hovenier, van de Hulst, and van der Mee, 1986; Barakat, 1987; Cloude, 1989; Girgel, 1991; Xing, 1992; Kumar and Simon, 1992; van der Mee and Hovenier, 1992; Kostinski, Givens, and Kwiatkowski). A set of sufficient conditions for physical realizability is not known to this author. The following four necessary conditions for physical realizability are among the more general of those published:

1. $Tr\,(\mathbf{MM}^T) \leq 4m_{00}^2$

2. $m_{00} \geq |m_{ij}|$

3. $m_{00}^2 \geq b^2$

4. $(m_{00} - b)^2 \geq \sum\limits_{j=1}^{3} \left(m_{0,j} - \sum\limits_{k=1}^{3} m_{j,k} a_k \right)$

where $b = \sqrt{m_{01}^2 + m_{02}^2 + m_{03}^2}$, $a_j = m_{0,j}/b$, and Tr indicates the trace of a matrix.

Another condition for physical realizability is that the matrix can be expressed as a sum of nondepolarizing Mueller matrices. The Mueller matrix for a passive device $T_{\max} \leq 1$, a device without gain, must satisfy the relation $T_{\max} = m_{00} + \sqrt{m_{01}^2 + m_{02}^2 + m_{03}^2} \leq 1$.

In the 16-dimensional space of Mueller matrices, the matrices for ideal polarizers, ideal retarders, and other nondepolarizing elements lie on the boundary between the physically realizable Mueller matrices and the unrealizable matrices. Thus, a small amount of noise in the measurement of a Mueller matrix for a polarizer or retarder may yield a marginally unrealizable matrix.

22.35   *DEPOLARIZATION*

Depolarization is the coupling of polarized into unpolarized light. If an incident state is polarized and the exiting state has a degree of polarization less than one, then the sample has depolarization. Consider three Mueller matrices of the following forms:

$$\mathbf{ID} = \begin{bmatrix} 1 & 0 & 0 & 0 \\ 0 & 0 & 0 & 0 \\ 0 & 0 & 0 & 0 \\ 0 & 0 & 0 & 0 \end{bmatrix} \qquad \mathbf{PD} = \begin{bmatrix} 1 & 0 & 0 & 0 \\ 0 & a & 0 & 0 \\ 0 & 0 & a & 0 \\ 0 & 0 & 0 & a \end{bmatrix} \qquad \mathbf{VD} = \begin{bmatrix} 1 & 0 & 0 & 0 \\ 0 & a & 0 & 0 \\ 0 & 0 & b & 0 \\ 0 & 0 & 0 & c \end{bmatrix} \quad (45)$$

Matrix **ID** is the ideal depolarizer; only unpolarized light exits the depolarizer. Matrix **PD** is the partial depolarizer; all fully polarized incident states exit with their incident polarization ellipse, but with a degree of polarization **DOP(PD)** = a. Matrix **PD** represents a variable partial depolarizer; the degree of polarization of the exiting light is a function of the incident state. Physically, depolarization is closely related to scattering and usually has its origin in retardance or diattenuation which is rapidly varying in time, space, or wavelength.

The amount of depolarization is a function of the incident state, and is defined for polarized incident states as $1 - \mathbf{DOP}\{\mathbf{MS}\}$. To describe the depolarization characteristics of a Mueller matrix, two figures of merit are useful. The first is the Euclidian distance of the normalized Mueller matrix $\mathbf{M}/m_{00}$ from the ideal depolarizer:

$$\left\| \frac{\mathbf{M}}{m_{00}} - \mathbf{ID} \right\| = \frac{\sqrt{\left(\sum_{i,j} m_{i,j}^2 \right) - m_{00}^2}}{m_{00}} \tag{46}$$

This quantity varies from zero for the ideal depolarizer to $\sqrt{3}$ for nondepolarizing Mueller matrices, including all pure diattenuators, pure retarders, and any sequences composed from them. Another useful measure is the depolarization of the matrix $Dep(\mathbf{M})$:

$$Dep(\mathbf{M}) = 1 - \frac{\sqrt{\left(\sum_{i,j} m_{i,j}^2 \right) - m_{00}^2}}{\sqrt{3}\, m_{00}} \tag{47}$$

This index measures how close a Mueller matrix is to the set of nondepolarizing Mueller matrices, and is related to the average depolarization of the exiting light. If $Dep(\mathbf{M}) = 0$ and the matrix is physically realizable, then for incident polarized states, the exiting light is polarized. $Dep(\mathbf{M})$ is closely related to the depolarization index of Gil and Bernabeu, (1985, 1986).

If a polarized state becomes partially polarized and then is polarized again while interacting with a sequence of elements, depolarization is still present in the matrix, despite the fact the output beam is polarized. Consider a depolarizer followed by a horizontal linear polarizer:

$$\frac{1}{2} \begin{bmatrix} 1 & 1 & 0 & 0 \\ 1 & 1 & 0 & 0 \\ 0 & 0 & 0 & 0 \\ 0 & 0 & 0 & 0 \end{bmatrix} \begin{bmatrix} 1 & 0 & 0 & 0 \\ 0 & 0 & 0 & 0 \\ 0 & 0 & 0 & 0 \\ 0 & 0 & 0 & 0 \end{bmatrix} = \begin{bmatrix} 1 & 0 & 0 & 0 \\ 1 & 0 & 0 & 0 \\ 0 & 0 & 0 & 0 \\ 0 & 0 & 0 & 0 \end{bmatrix} \tag{48}$$

The exiting beam is horizontally polarized. All incident states, however, have equal intensity transmission due to the depolarizer. For this example, $Dep = 1 - 1/\sqrt{3}$.

22.36 *NONDEPOLARIZING MUELLER MATRICES AND JONES MATRICES*

A sample which does not display depolarization is nondepolarizing. A nondepolarizing Mueller matrix satisfies the condition

$$Tr(\mathbf{M}\mathbf{M}^T) = 4m_{00} \tag{49}$$

An incident beam with degree of polarization of one will exit with a degree of polarization of one. Many other necessary conditions for nondepolarization may be found in the literature (Abhyankar and Fymat, 1969; Barakat, 1981; Fry and Kattawar, 1981; Simon, 1982; Gil and Bernabeu, 1985; Cloude, 1986).

Jones matrices form an alternative and very useful representation of sample polarization, particularly because Jones matrices have simpler properties and are more easily manipulated and interpreted. It is desirable to be able to transform between these two matrix representations. The complication in mapping Mueller matrices onto Jones matrices and vice versa is that Mueller matrices cannot represent absolute phase and Jones matrices cannot represent depolarization. Thus, only nondepolarizing Mueller matrices have corresponding Jones matrices. All Jones matrices have a corresponding

Mueller matrix, but because the absolute phase is not represented, the mapping is many Jones matrices to one Mueller matrix. A Jones matrix $\mathbf{J}$ is transformed into a Mueller matrix by the relation

$$\mathbf{M} = \mathbf{U}(\mathbf{J} \otimes \mathbf{J}^*)\mathbf{U}^{-1} \tag{50}$$

in which $\otimes$ represents the tensor product and $\mathbf{U}$ is the Jones/Mueller transformation matrix (Simon, 1982; Kim, Mandel, and Wolf, 1987)

$$\mathbf{U} = \frac{1}{\sqrt{2}} \begin{bmatrix} 1 & 0 & 0 & 1 \\ 1 & 0 & 0 & -1 \\ 0 & 1 & 1 & 0 \\ 0 & i & -i & 0 \end{bmatrix} = (\mathbf{U}^{-1})^{\dagger} \tag{51}$$

where the Hermitian adjoint is represented by $\dagger$. All Jones matrices of the form $\mathbf{J}' = e^{j\phi}\mathbf{J}$ transform to the same Mueller matrix. Nondepolarizing Mueller matrices are transformed into Jones matrices using the following relations:

$$\mathbf{J} = \begin{bmatrix} j_{xx} & j_{xy} \\ j_{yx} & j_{yy} \end{bmatrix} = \begin{bmatrix} \rho_{xx}e^{i\phi_{xx}} & \rho_{xy}e^{i\phi_{xy}} \\ \rho_{yx}e^{i\phi_{yx}} & \rho_{yy}e^{i\phi_{yy}} \end{bmatrix} \tag{52}$$

where the amplitudes are

$$\rho_{xx} = \frac{1}{\sqrt{2}}\sqrt{m_{00} + m_{01} + m_{10} + m_{11}} \qquad \rho_{xy} = \frac{1}{\sqrt{2}}\sqrt{m_{00} - m_{01} + m_{10} - m_{11}}$$

$$\tag{53}$$

$$\rho_{yx} = \frac{1}{\sqrt{2}}\sqrt{m_{00} + m_{01} - m_{10} - m_{11}} \qquad \rho_{xy} = \frac{1}{\sqrt{2}}\sqrt{m_{00} - m_{01} - m_{10} + m_{11}}$$

and the relative phases are

$$\phi_{xy} - \phi_{xx} = \arctan\left(\frac{-m_{03} - m_{13}}{m_{02} + m_{12}}\right) \qquad \phi_{yx} - \phi_{xx} = \arctan\left(\frac{m_{30} + m_{31}}{m_{20} + m_{21}}\right)$$

$$\phi_{yy} - \phi_{xx} = \arctan\left(\frac{m_{32} - m_{23}}{m_{22} + m_{33}}\right) \tag{54}$$

The phase ϕ_{xx} is not determined; it represents the absolute phase relative to which the other phases are determined. If $j_{xx} = 0$, then both the numerator and denominator of the arctan are zero and the phase equations fail. The equations can then be recast in closely related forms to use the phase of another Jones matrix element as the reference for the "absolute phase."

22.37 *HOMOGENEOUS AND INHOMOGENEOUS POLARIZATION ELEMENTS*

This refers specifically to nondepolarizing Mueller matrices.

A nondepolarizing Mueller matrix is defined as *homogeneous* if the two eigenpolarizations are orthogonal, and *inhomogeneous* otherwise. A nondepolarizing Mueller matrix can be factored into a cascade of a diattenuator Mueller matrix $\mathbf{M}_D$ followed by a retarder Mueller matrix $\mathbf{M}_R$ or into a cascade of the same retarder followed by a diattenuator $\mathbf{M}'_D$,

$$\mathbf{M} = \mathbf{M}_R\mathbf{M}_D = \mathbf{M}'_D\mathbf{M}_R \tag{55}$$

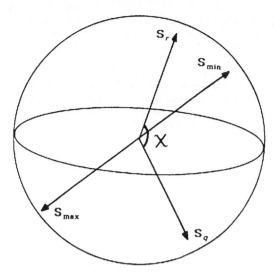

FIGURE 4 The principal Stokes vectors associated with an inhomogeneous polarization element mapped on the Poincarè sphere. The incident Stokes vectors of maximum $\mathbf{S}_{max}$ and minimum $\mathbf{S}_{min}$ intensity transmittance are diametrically opposite on the Poincare spherè (indicating orthogonal polarization states) while the eigenpolarizations $\mathbf{S}_q$ and $\mathbf{S}_r$ are separated by the angle χ.

where the diattenuation of $\mathbf{M}_D$ and $\mathbf{M}'_D$ are equal. We define the diattenuation of $\mathbf{M}$ as the diattenuation of $\mathbf{M}_D$, and the retardance of $\mathbf{M}$ as the retardance of $\mathbf{M}_R$. For a homogeneous device, $\mathbf{M}_D = \mathbf{M}'_D$ and the eiegenvectors of $\mathbf{M}_R$ and $\mathbf{M}_D$ are equal. Thus the retardance and diattenuation of a homogeneous Mueller matrix are "aligned," giving it substantially simpler properties than the inhomogeneous Mueller matrices. A necessary condition for a homogeneous Mueller matrix is $m_{01} = m_{10}$, $m_{02} = m_{20}$, $m_{03} = m_{30}$. Then, $P\{\mathbf{M}\} = D\{\mathbf{M}\}$.

The inhomogeneity of a Mueller matrix is characterized by an inhomogeneity index $I(\mathbf{M})$ which characterizes the orthogonality of the eigenpolarizations; $I(\mathbf{M})$ varies from zero for orthogonal eigenpolarizations to one for degenerate (equal) eigenpolarizations. Let $\hat{\mathbf{S}}_1$ and $\hat{\mathbf{S}}_2$ be normalized polarized Stokes vector eigenpolarizations of a Mueller matrix; then

$$I(\mathbf{M}) = \frac{\sqrt{\hat{\mathbf{S}}_1 \cdot \hat{\mathbf{S}}_2}}{2} = \cos{(\chi/2)} \tag{56}$$

where χ is the angle between the eigenpolarizations on the Poincare sphere measured from the center of the sphere as illustrated in Fig. 4.

22.38 REFERENCES

Abhyankar, K. D. and A. L. Fymat, "Relations Between the Elements of the Phase Matrix for Scattering," *J. Math. Phys.* **10:**1935–1938 (1969).

Azzam, R. M. A. and N. M. Bashara, *Ellipsometry and Polarized Light,* 1st ed., North-Holland, Amsterdam, 1977, 2d ed., North-Holland, Amsterdam, 1987.

Azzam, R. M. A., "Photopolarimetric Measurement of the Mueller Matrix by Fourier Analysis of a Single Detected Signal," *Opt. Lett.* **2:**148–150 (1978).

Azzam, R. M. A., "Arrangement of Four Photodetectors for Measuring the State of Polarization of Light," *Opt. Lett.* **10:**309–311 (1985).

Azzam, R. M. A., I. J. Elminyawi, and A. M. El-Saba, "General Analysis and Optimization of the Four-detector Photopolarimeter," *J. Opt. Soc. Am.* A, **5:**681–689 (1988).

Azzam, R. M. A., "Instrument Matrix of the Four-detector Photopolarimeter: Physical Meaning of its Rows and Columns and Constraints on its Elements," *J. Opt. Soc. Am.* A **7:**87–91 (1990).

Azzam, R. M. A. (ed.), "Ellipsometry," *Soc. Photo-Opt. Instrum. Eng. Milestone Series,* **MS27** (1991).

Azzam, R. M. A., "Ellipsometry," in M. Bass (ed.), *Handbook of Optics,* vol. 2, 2d ed., McGraw-Hill, New York, 1994.

Barakat, R., "Bilinear Constraints Between Elements of the 4×4 Mueller-Jones Transfer Matrix of Polarization Theory," *Opt. Commun.* **38:**159–161 (1981).

Barakat, R., "Conditions for the Physical Realizability of Polarization Matrices Characterizing Passive Systems," *J. Mod. Opt.* **34:**1535–1544 (1987).

Barron, L. D., *Molecular Light Scattering and Optical Activity,* Cambridge University Press, New York, 1982.

Bennett, J. M., "Polarization," in M. Bass (ed.), *Handbook of Optics,* vol. 2, 2d ed., McGraw-Hill, New York, 1994.

Bennett, J. M., "Polarizers," in M. Bass (ed.), *Handbook of Optics,* vol. 2, 2d ed., McGraw-Hill, New York, 1994.

Bird, G. R. and Shurcliff, W. A., "Pile-of-plates Polarizers for the Infrared: Improvement in Analysis and Design," *J. Opt. Soc. Am.* **49:** 235–240 (1959).

Boerner, W. and H. Mott (eds.), "Radar Polarimetry," *Proc. Soc. Photo-Opt. Instrum. Eng.* **1747** (1992).

Bruegge, T. J., "Analysis of Polarization in Optical Systems," in R. A. Chipman (ed.), "Polarization Considerations for Optical Systems II," *Proc. Soc. Photo-Opt. Instrum. Eng.* **1166:**165–176 (1989).

Chenault, D. B., "Infrared Spectropolarimetry," Ph.D. dissertation, University of Alabama in Huntsville, 1992.

Chenault, D. B., J. L. Pezzaniti, and R. A. Chipman, "Mueller Matrix Algorithms," in D. Goldstein and R. Chipman (eds.), "Polarization Analysis and Measurement," *Proc. Soc. Photo-Opt. Instrum. Eng.* **1746:**231–246 (1992).

Chipman, R. A., "Polarization Aberrations," Ph.D. Dissertation, Optical Sciences, University of Arizona, 1987.

Chipman, R. A. (ed.), "Polarization Considerations for Optical Systems," *Proc. Soc. Photo-Opt. Instrum. Eng.* **891** (1988).

Chipman, R. A., "Polarization Analysis of Optical System," *Opt. Eng.* **28:**90–99 (1989a).

Chipman, R. A., "Polarization Analysis of Optical Systems II," in Russell A. Chipman (ed.), "Polarization Considerations for Optical Systems II," *Proc. Soc. Photo-Opt. Instrum. Eng.* **1166:**79–99 (1989b).

Chipman, R. A. (ed.), "Polarization Considerations for Optical Systems II," *Proc. Soc. Photo-Opt. Instrum. Eng.* **1166** (1989c).

Chipman, R. A. and L. J. Chipman, "Polarization Aberration Diagrams," *Opt. Eng.* **28:**100–106 (1989).

Cloude, S. R., "Group Theory and Polarisation Algebra," *Optik* **75:**26–36 (1986).

Cloude, S. R., "Conditions for the Physical Realisability of Matrix Operators in Polarimetry," *Proc. Soc. Photo-Opt. Instrum. Eng.* **1166:**177–185 (1989).

Collett, E., *Polarized Light,* Marcel Dekker, Inc., New York, 1992.

Collins, R. W. and Y.-T. Kim, "Ellipsometry for Thin-film and Surface Analysis," *Ann. Chem.* **62:**887a–900a (1990).

Coulson, K. L., *Polarization and Intensity of Light in the Atmosphere,* A. Deepak, Hampton, Va., 1988.

Coulson, K. L., "Polarization of Light in the Natural Environment," in R. Chipman (ed.), *Polarization Considerations for Optical Systems II* **1166:**2–10 (1989).

Curran, P. J., "Polarized Visible Light as an Aid to Vegetation Classification," *Remote Sensing of the Environment* **12:**491–499 (1982).

Dobrowolski, G., "Optical Properties of Films and Coatings," in M. Bass (ed.), *Handbook of Optics,* Vol. 1, 2d ed., McGraw-Hill, New York, 1994.

Duggin, M. J., S. A. Israel, V. S. Whitehead, J. S. Myers, and D. R. Robertson, "Use of Polarization Methods in Earth Resource Investigations," in R. Chipman (ed.), *Polarization Considerations for Optical Systems II* **1166:**42–51 (1989).

Egan, W. G., *Photometry and Polarization in Remote Sensing,* Elsevier, New York, 1985.

Egan, W. (ed.), "Polarization in Remote Sensing," *Proc. Soc. Photo-Opt. Instrum. Eng.* **1748** (1992).

Fry, E. S. and G. W. Kattawar, "Relationships Between Elements of the Stokes Matrix," *Appl. Opt.* **20:**2811–2814 (1981).

Gehrels, T. (ed.), *Planets, Stars, and Nebulae Studied with Polarimetry,* Univ. Arizona Press, Tucson, 1974.

Gerrard, A. and J. M. Burch, *Introduction to Matrix Methods in Optics,* Wiley, London, 1975.

Gil, J. J. and E. Bernabeu, "A Depolarization Criterion in Mueller Matrices," *Opt. Acta* **32:**259–261 (1985).

Gil, J. J. and E. Bernabeu, "Depolarization and Polarization Indices of an Optical System," *Opt. Acta* **33:**185–189 (1986).

Gil, J. J. and E. Bernabeu, "Obtainment of the Polarizing and Retardation Parameters of a Non-depolarizing Optical System from the Polar Decomposition of Its Mueller Matrix," *Optik* **76:**67–71 (1987).

Girgel, S. S., "Structure of the Mueller Matrices of Depolarized Optical Systems," *Sov. Phys. Crystallogr.* **36:**890–891 (1991).

Goldstein, D. H., "Mueller Matrix Dual-rotating Retarder Polarimeter," *Appl. Opt.* **31,** 6676–6683 (1992).

Goldstein, D. H. and R. A. Chipman, "Error Analysis of a Mueller Matrix Polarimeter," *J. Opt. Soc. Am.* **A7:**693–700 (1990).

Goldstein, D. H. and R. A. Chipman (eds.), "Polarization Analysis and Measurement," *Proc. Soc. Photo-Opt. Instrum. Eng.* **1745** (1992).

Hale, P. D. and G. W. Day, "Stability of Birefringent Linear Retarders (Waveplates)," *Appl. Opt.* **27:**5146–5153 (1988).

Hansen, E. W., "Overcoming Polarization Aberrations in Microscopy," in R. Chipman (ed.), "Polarization Considerations for Optical Systems," *Proc. Soc. Photo-Opt. Instrum. Eng.* **891:**190–197 (1988).

Hauge, P. S., "Mueller Matrix Ellipsometry with Imperfect Compensators," *J. Opt. Soc. Am.* **68:**1519–1528 (1978).

Holm, W. A., "Polarimetric Fundamentals and Techniques," in J. L. Eaves and E. K. Reedy (eds.), *Principles of Modern Radar,* Van Nostrand Reinhold, New York, 1987.

Hovenier, J. W., H. C. van de Hulst, and C. V. M. van der Mee, "Conditions for the Elements of the Scattering Matrix," *Astron. Astrophys.* **157:**301–310 (1986).

Huynen, J. R., "Measurement of the Target Scattering Matrix," *Proc. IEEE* **53**(8):936–946 (1965).

Jirgensons, B., *Optical Activity of Proteins and Other Macromolecules,* Springer-Verlag, New York, 1973.

Johnson, W. C. Jr., "The Circular Dichroism of Carbohydrates," *Advances in Carbohydrate Chemistry and Biochemistry* **45:**73–124 (1987).

Kennaugh, E. M., "Effects of Type of Polarization on Echo Characteristics," Report No. 389-9, Antenna Laboratory, Ohio State University, 1951.

Kim, K., L. Mandel, and E. Wolf, "Relationship Between Jones and Mueller Matrices for Random Media," *J. Opt. Soc. Amer.* **A4:**433–437 (1987).

King, R. J. and S. P. Talim, "Some Aspects of Polarizer Performance," *J. Physics,* ser. E, **4:**93–96, (1971).

Kliger, D. S., J. W. Lewis, and C. E. Randall, *Polarized Light in Optics and Spectroscopy,* Academic Press, Boston, 1990.

Konnen, G. P., *Polarized Light in Nature,* Cambridge University, Cambridge, 1985.

Kostinski, A. B., B. D. James, and W.-M. Boerner, "Polarimetric Matched Filter for Coherent Imaging," *Canad. J. Phys.* **66:**871–877 (1988).

Kostinski, A. B., C. R. Clark, and J. M. Kwiatkowski, "Some Necessary Conditions on Mueller Matrices," in D. Goldstein and R. Chipman (eds.), "Polarization Analysis and Measurement, *Proc. Soc. Photo-Opt. Instrum. Eng.* **1746:**213–220 (1992).

Kuboda, H. and S. Inouè, "Diffraction Images in the Polarizing Microscope," *J. Opt. Soc. Am.* **49:**191–198 (1959).

Kumar, M. S. and R. Simon, "Characterizing of Mueller Matrices in Polarization Optics," *Opt. Commun.* **88:**464–470 (1992).

Lakhtakia, A. (ed.), "Natural Optical Activity," *Soc. Photo-Opt. Instrum. Eng. Milestone Series,* **MS15**, SPIE, Bellingham, WA, 1990.

-Lu, S. Y. and R. A. Chipman, "Generalized Diattenuation and Retardance for Inhomogeneous Polarization Elements," in D. Goldstein and R. Chipman (eds.), "Polarization Analysis and Measurement," *Proc. Soc. Photo-Opt. Instrum. Eng.* **1746:**197–200 (1992).

H. A. Macleod, *Thin Film Optical Filters,* 2d ed., Macmillan, New York, 1986.

Mansuripur, M., "Effects of High-numerical-aperture Focusing on the State of Polarization in Optical and Magneto-optical Data Storage Systems," *Appl. Opt.* **30:**3154–3162 (1991).

Mason, S. F., *Molecular Optical Activity and the Chiral Discrimination,* Cambridge University Press, Cambridge, 1982.

Mason, S. F. (ed.), *Optical Activity and Chiral Discrimination,* Reidel, Boston, 1978.

Maymon, P. W. and R. A. Chipman, "Linear Polarization Sensitivity Specification for Space-borne Instruments," in D. Goldstein and R. Chipman (eds.), "Polarization Analysis and Measurement," *Proc. Soc. Photo-Opt. Instrum. Eng.* **1746:**148–156 (1992).

McGuire, J. P. Jr. and R. A. Chipman, "Polarization Aberrations in Optical Systems," in R. Fischer and W. Smith, (eds.), "Current Developments in Optical Engineering II," *Proc. Soc. Photo-Opt. Instrum. Eng.* **818:**240–245 (1987).

McGuire, J. P. Jr. and R. A. Chipman, "Polarization Aberrations in the Solar Activity Measurements Experiments (SAMEX) Solar Vector Magnetograph," *Opt. Eng.* **28:**141–147 (1989).

McGuire, J. P. Jr. and R. A. Chipman, "Diffraction Image Formation in Optical Systems with Polarization Aberration I: Formulation and Example," *J. Opt. Soc. Am.* **A7:**1614–1626 (1990a).

McGuire, J. P. Jr. and R. A. Chipman, "Analysis of Spatial Pseudodepolarizers in Imaging Systems," *Optical Engineering* **29:**1478–1484 (1990b).

McGuire, J. P. Jr. and R. A. Chipman, "Diffraction Image Formation in Optical Systems with Polarization Aberration II: Amplitude Response Matrices for Rotationally Symmetric Systems," *J. Opt. Soc. Am.* **A8:**833–840 (1991).

Michl, J. and E. W. Thulstrup, *Spectroscopy with Polarized Light,* VCH, New York, 1986.

Morris, J. and R. A. Chipman (eds.), "Polarimetry: Radar, Infrared, Visible, Ultraviolet, and X-ray," *Proc. Soc. Photo-Opt. Instrum. Eng.* **1317** (1990).

Mott, H., *Antennas for Radar and Communications, a Polarimetric Approach,* J. Wiley and Sons, New York, 1992.

November, L. J. (ed.), "Solar Polarimetry," *Proceedings, National Solar Observatory,* Sunspot, N. M., 1991.

Pezzaniti, J. L. and R. A. Chipman, "Off-axis Polarizing Properties of Polarizing Beam Splitter Cubes, in D. Goldstein and R. Chipman (eds.), "Polarization Analysis and Measurement," *Proc. Soc. Photo-Opt. Instrum. Eng.* **1746:**343–357 (1992).

Poelman, A. J., and J. R. F. Guy, "Polarization Information Utilization in Primary Radar," in W.-M. Boerner (ed.), *Inverse Methods in Electromagnetic Imaging,* D. Reidel, Dordrecht, Holland, 1985.

Samori, B. and E. W. Thulstrup (eds.), *Polarized Spectroscopy of Ordered Systems,* Kluwer Academic, Boston, 1988.

Schellman, J. and H. P. Jensen, "Optical Spectroscopy of Oriented Molecules," *Chem. Rev.* **87:**1359–1399 (1987).

Schiff, T. F., D. J. Wilson, B. D. Swimley, M. E. Southwood, D. R. Bjork, and J. C. Stover, "Mueller Matrix Measurements with an Out-of-plane Polarimetric Scatterometer," in D. Goldstein and R.

Chipman (eds.), "Polarization Analysis and Measurement," *Proc. Soc. Photo-Opt. Instrum. Eng.* **1746:**295–306 (1992).

Shurcliff, W. A., *Polarized Light,* Harvard University Press, Cambridge, Mass., 1962.

Simon, R., "The Connection Between Mueller and Jones Matrices of Polarization Optics," *Opt. Commun.* **42:**293–297 (1982).

Simon, R., "Mueller Matrices and Depolarization Criteria," *J. Mod. Opt.* **34:**569–575 (1987).

Stenflo, J. O., "Optimization of the LEST Polarization Modulation System," in O. Engvold and O. Hauge (eds.), LEST Foundation Tech. Rep. 44, Institute of Theoretical Astrophysics, Univ. Of Oslo, 1991.

Stover, J. C., *Optical Scattering, Measurement and Analysis*, McGraw-Hill Inc., New York, 1990.

Theocaris, P. S. and E. E. Gdoutos, *Matrix Theory of Photoelasticity,* Springer-Verlag, Berlin, 1979.

Thiel, M. A. F., "Error Calculation of Polarization Measurements," *J. Opt. Soc. Am.* **66:**65–67 (1976).

Thulstrup, E. W., *Linear and Magnetic Circular Dichroism of Planar Organic Molecules,* Springer-Verlag, Berlin, 1980.

Title, A. M. and W. J. Rosenberg, "Improvements in Birefringent Filters. 5: Field of View Effects," *Appl. Opt.* **18:**3443–3456 (1979).

Urbanczyk, W., "Optical Imaging Systems Changing the State of Light Polarization," *Optik* **66:**301–309 (1984).

Urbanczyk, W., "Optical Transfer Functions for Imaging Systems Which Change the State of Light Polarization," *Opt. Acta* **33:**53–62 (1986).

van der Mee, C. V. M. and J. W. Hovenier, "Structure of Matrices Transforming Stokes Parameters," *J. Math. Phys.* **33:**3574–3584 (1992).

van de Hulst, H. C., *Light Scattering by Small Particles,* John Wiley and Sons, New York, 1957.

van Zyl, J. J. and H. A. Zebker, "Imaging Radar Polarimetry," in J. A. Kong (ed.), *Polarimetric Remote Sensing,* PIER 3, Elsevier, New York, 1990.

Waluschka, E., "A Polarization Ray Trace," *Opt. Eng.* **28:**86–89 (1989).

Wolff, L. B. and D. J. Kurlander, "Ray Tracing with Polarization Parameters," *IEEE Computer Graphics and Appl.,* pp. 44–55 (Nov. 1990).

Xing, Z.-F., "On the Deterministic and Non-deterministic Mueller Matrix," *J. Mod. Opt.* **39:**461–484 (1992).

CHAPTER 23
HOLOGRAPHY AND HOLOGRAPHIC INSTRUMENTS

Lloyd Huff
Research Institute
University of Dayton
Dayton, Ohio

23.1 GLOSSARY

A	wave amplitude
a	diameter of viewing lens in speckle imaging system
d	fringe spacing
d_{sp}	characteristic speckle diameter
$\mathbf{E}$	electric field vector
$\hat{\mathbf{e}}$	polarization unit vector
f	wave frequency
I	field irradiance
I_H	irradiance of the field in plane of hologram
K	proportionality constant
k	propagation constant
r	radial position coordinate
T	transmittance of the hologram
v	distance from lens to image plane in speckle imaging system
θ_1, θ_2	object and reference beam angles
λ	wavelength
ϕ	wave phase
Ψ	complex field amplitude
Ψ_H	complex field amplitude in plane of hologram
Ψ_0	complex amplitude of object wave field
Ψ_R	complex amplitude of reference wave field
Ψ_T	complex amplitude of field transmitted by hologram

23.2 INTRODUCTION

The three-dimensional imagery produced by holography accounts for most of the current popular interest in this technique. Conceptual applications, such as the holodeck seen on the television series *Star Trek: The Next Generation,* and actual applications, like the widespread use of embossed holograms on book and magazine covers and credit cards, have fascinated and captured the imagination of millions. Holography was discovered in 1947 by Gabor and revived in the early 1960s through the work of Leith and Upatnieks. Since that time, most practitioners in the field believe that technical applications, rather than imaging, have represented the utility of holography in a more significant way. This chapter is a brief overview of some of the more important technical applications, particularly as they relate to a variety of instrumentation problems. The discussion addresses several ways holography has been used to observe, detect, inspect, measure, or record numerous physical phenomena. The second section presents a brief review of the basic principles of wavefront reconstruction. The third section addresses what is undoubtedly the most important application of holography to date—holographic interferometry. This is followed by a review of electronic or television holography (the fourth section) which takes this powerful interferometric technique into the real-time domain. The fifth section addresses several instrumental applications of holographic optical elements (HOEs). The sixth and seventh discuss ways in which holography is being applied in the semiconductor industry. The final section briefly addresses the holographic storage of information.

23.3 BACKGROUND AND BASIC PRINCIPLES

Holography is a method of recording and reconstructing wavefronts residing anywhere in the electromagnetic spectrum or acoustic spectrum. This chapter addresses optical holography as practiced in or near the visible region of the electromagnetic spectrum. The principals of wavefront reconstruction were discovered by Gabor[1-3] in an attempt to improve the resolving power of the electron microscope. The original purpose was never accomplished, but this basic discovery evolved into one of the most significant new fields of study in this century. Gabor's early work received little attention because the lack of a light source with sufficient coherence severely limited the quality of the images produced. However, the invention of the laser in the early 1960s heralded a holographic renaissance. During this period, Leith and Upatnieks[4,5] recognized the parallels between their work in coherent radar and Gabor's wavefront reconstruction concepts. Their experiments in the optical region of the spectrum with the newly available HeNe laser produced the first high-quality, three-dimensional images. The publication of this work created an explosive interest in the field as well as many unrealistic predictions about what might be accomplished with holographic three-dimensional imagery. The work of numerous researchers established the medium's true capabilities and limitations; consequently, many successful applications ensued. Progress continues to be made in the development of new materials and techniques sustaining a high level of interest in holography and its technical, commercial, and artistic applications.

Holography is most often associated with its ability to produce striking three-dimensional images. Therefore, a logical place to start in understanding holography is to compare this imaging science with its two-dimensional predecessor—photography. A light wavefront is characterized by several parameters; the two most important of these are its intensity (or irradiance) and its local direction of propagation. Photography records only one of these parameters—intensity—in the plane of the recording medium or photographic film. The intensity distribution of a light wave emanating from an object may be recorded

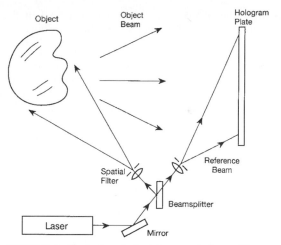

FIGURE 1 Typical optical arrangement for making a simple laser transmission hologram.

by simply exposing a film plate placed in proximity to the object; however, this will not produce a discernible image. Recording a photograph is accomplished by imaging the object onto the film with a lens, thereby establishing a correspondence between points on the object and points in the film plane.

Holography also records the intensity distribution of a wavefront; in addition, the local propagation direction (or phase) is recorded through the process of optical interference. The process in its simplest form is illustrated in Fig. 1. The light from a laser is split into two parts, expanded with a short-focal-length lens (usually a microscope objective), and spatially filtered with a pinhole to remove intensity variations caused primarily by nonuniformities in the lens. One of the split beams (object beam) is directed to the object; the other (reference beam) is incident directly on the recording medium (such as a high-resolution silver halide film). The light reflected from and scattered by the object combines with the reference beam at the plate to form an interference fringe field. These fringes are recorded by the film. The spacing of these fringes d is given by the grating equation

$$d - \frac{\lambda}{(\sin \theta_1 + \sin \theta_2)} \tag{1}$$

where λ is the wavelength of the light, and θ_1 and θ_2 are the angles made by the object and reference beams relative to the normal. For visible light and common recording geometries, the fringe frequency ($1/d$) can exceed 2000 fringes (or line-pairs) per millimeter. Therefore, the recording material must be of very high resolution relative to conventional photographic film which is usually in the range of 50 line-pairs per millimeter. The stability of the fringe is extremely sensitive to the mechanical motion of the object and optical components. To record holograms with good fringe stability, the optical system must be stable enough to prevent motions greater than a fraction of a wavelength. For this reason, the common practice is to use rigid optical components placed on a stable, vibration-isolated table.

Illumination of the developed hologram by the reference beam alone reveals a three-dimensional image which is essentially identical to the original object as viewed in laser light. Observing the holographic image of the object is exactly like looking at the object through the window formed by the plate with full parallax and look-around capability. The object wave is reconstructed when the illumination (reference) wave is diffracted by the grating formed in the recording medium. This grating is formed by variation of the optical transmittance or optical thickness of the material along the fringe lines. The amplitude hologram formed with silver halide film may be converted to a phase hologram by bleaching; this results in a significant increase in diffraction efficiency. Other materials (such as photopolymer film) produce phase holograms directly.

The holographic recording and reconstruction process may be described in general mathematical terms as follows. The object and reference fields satisfy the Helmholtz equation

$$\nabla^2 \mathbf{E} + k^2 \mathbf{E} = 0 \tag{2}$$

where $\mathbf{E}$ is the electric field vector and $k = 2\pi/\lambda$ is the propagation constant.

A spherical wave solution of this equation may be expressed in the form

$$\mathbf{E} = A e^{ikr} e^{i2\pi ft} \hat{\mathbf{e}} \tag{3}$$

where A is the amplitude of the wave, f is the frequency, and $\hat{\mathbf{e}}$ is the polarization unit vector.

The complex field amplitude Ψ is defined as

$$\Psi = A e^{i\phi} \tag{4}$$

where $\phi = kr$ is the phase. The irradiance of the field is given by

$$I = \mathbf{E} \cdot \mathbf{E}^* = \Psi\Psi^* \hat{\mathbf{e}} \cdot \hat{\mathbf{e}} = \Psi\Psi^*$$
$$= A^2 = |\Psi|^2 \tag{5}$$

where * denotes the complex conjugate.

The field in the plane of the hologram Ψ_H is the sum of the object and reference fields:

$$\Psi_H = \Psi_0 + \Psi_R \tag{6}$$

and the irradiance of the field at the hologram is given by (assuming parallel polarization of the waves):

$$I_H = \Psi_H \Psi_H^* = (\Psi_0 + \Psi_R)(\Psi_0 + \Psi_R)^*$$
$$= |\Psi_0|^2 + |\Psi_R|^2 + \Psi_0 \Psi_R^* + \Psi_R \Psi_0^* \tag{7}$$

After processing, and apart from a constant term, the transmittance T of the hologram is proportional to the irradiance of the field at the hologram;

$$T = K[|\Psi_0|^2 + |\Psi_R|^2 + \Psi_0 \Psi_R^* + \Psi_R \Psi_0^*] \tag{8}$$

When illuminated by the reference wave, the field transmitted by the hologram Ψ_T is given by the hologram transmittance multiplied by the reference wave field:

$$\Psi_T = K[\Psi_R(|\Psi_0|^2 + |\Psi_R|^2) + |\Psi_R|^2\Psi_0 + \Psi_R^2\Psi_0^*] \tag{9}$$

The first term in this equation for Ψ_T is simply the transmitted wave altered by an attenuation factor. The second term is the original object wave multiplied by an amplitude factor; this term represents the virtual holographic image of the object. The third term is the conjugate object wave. In off-axis holography, the real image formed by this wave is weak, lies out of the field of view, and does not make a significant contribution to the imaging process. However, for Gabor's original in-line holography, this term represented an objectionable twin image which overlapped and obstructed viewing of the desired image. An important contribution of the Leith and Upatnieks off-axis reference scheme was the elimination of this twin image.

Many different types of holograms can be made by varying the location of the object relative to the recording medium, the directions and relative angles of the object and reference beams, and the wavefront curvature of these beams. The properties of these hologram types vary greatly; much research has been performed to characterize and successfully apply the different formats. Vigorous work is still being pursued in both areas of imaging and technical applications. For a thorough explanation of holography and its many applications, the reader may consult any of several standard texts on the subject (e.g., Refs. 6–11).

23.4 HOLOGRAPHIC INTERFEROMETRY

Interferometry provides a means of measuring optical path differences through the analysis of fringe patterns formed by the interference of coherent light waves. Optical path differences of interest may be produced by mechanical displacements, variations in the contour of one surface relative to another, and variations in the refractive index of a material volume. Classical interferometry involves the interference of two relatively simple optical wavefronts which are formed and directed by optical components. These components must be of sufficient quality that they do not introduce random phase variations across the field that compete with or totally mask the optical path length differences of interest. Typical examples of classical interferometry include the use of configurations such as the Michelson or Twyman Green and Mach-Zehnder interferometers to determine the surface figure of optical components, study the refractive index variation in optical materials, and visualize the properties of flowing gases. The need for high-quality optical surfaces in classical interferometry is a consequence of the difficulty, using classical optical methods, of generating two separate but identical optical wavefronts of arbitrary shape. Although it was not immediately recognized by early holography researchers, the ability to holographically record then replay an arbitrary wavefront in a predictable fashion obviated this basic limitation of classical interferometry. With holographic interferometry, polished optical surfaces are not required and diffusely reflecting objects of any shape may be studied.

In holographic interferometry, a wavefront is stored in the hologram and later compared interferometrically with another wavefront. Phase differences between these two wavefronts produce fringes that can be analyzed to yield a wide range of both qualitative and quantitative information about the system originating these two wavefronts. Several researchers working independently made experimental observations related to this fact.[12–17] Once the full implication of this discovery was realized, a period of intense research activity began to develop a solid theoretical understanding of this powerful new technique. Holographic interferometry quickly became the most important application of the relatively young science of holography. Although other branches of holography have successfully matured, most notably HOEs, holographic interferometry remains today the area in which holography has probably made the greatest impact.

As stated earlier, holographic interferometry involves the interferometric comparison of two wavefronts separated in time. This comparison can be made in a variety of ways which constitute the basic methods of holographic interferometry: real-time, double-exposure, and time-average. Real-time interferometry is realized by the interference of a holographically reconstructed object wave with the wave emanating from the actual object. This is accomplished as follows. The holographic plate is exposed, developed, then replaced in its holder in its original position. Reference- and object-beam intensities are adjusted so that the illuminated object and its holographic image are of approximately equal brightness. Since the reconstructed object wavefront is 180° out of phase with the object wavefront, the object should be dark when viewed through the holographic plate. In practice, one or two broad fringes usually appear across the object due to emulsion shrinkage effects and lack of complete mechanical precision in returning the holographic plate to its original position. Any disturbance of the object which results in a mechanical displacement of its surface will now produce a fringe system which can be viewed in real time. The structure and periodicity of the fringes are related to the surface displacement. The mechanical surface deformation can result from an applied force, change in pressure, change in temperature, or any combination of the three. The quantitative details of this surface deformation can be derived from an analysis of this fringe system.

In double-exposure holographic interferometry, the two wavefronts to be compared are stored in the same hologram. This is done by holographically recording the image of the object under study in two exposures separated in time in the same holographic plate. If nothing is done to alter the object wavefront between these two exposures, the resulting image will appear as for a single-exposure hologram. However, if the object is perturbed in some way between these two exposures, an interference fringe system will appear in the final image. Again, this fringe system is related to the mechanical deformations of the object surface caused by the disturbance. Real-time holographic interferometry allows one to study the effects of object perturbation of varying types and degrees over any desired length of time and in real time. In contrast, double-exposure holographic interferometry examines a particular change of the state of the object between two particular points in time. A double-exposure holographic interferogram then might be thought of as a single data point record. The interferogram might be a record of the change of the object from one stable state to another, which might be recorded with a continuous wave laser. Or the interest might be to compare two states of a rapidly varying system most effectively recorded using a pulsed laser.

In some respects, the double-exposure holographic interferogram involves less experimental complexity, because both exposures are made in a single hologram plate held in a fixed position. Mechanical registration of the plate to a baseline position is not required. Emulsion shrinkage due to wet process development of silver halide film affects the holographic fringe systems for both exposures in the same way; therefore, it is not a problem for double-exposure interferometry. Another feature which may or may not be of benefit, depending on the parameters of the experiment, is that the superimposed images for the two exposures are in phase, thereby producing a bright baseline image. Double-exposure interferometry has been used very effectively to record fast events such as the flight of a bullet passing through a chamber.[18] The first exposure is made of the chamber alone before passage of the bullet; the second exposure is made with the bullet in midflight. The interferogram is an interference recording of the refractive index variations in the chamber created by the disturbance of the bullet.

Vibrating surfaces may be studied using either real-time or time-average holographic interferometry.[16,19] A vibrating object presents a continuum of surface configurations or surface deformations to an interferometric system. A unique interferometric fringe system is associated with each state of the surface for any particular point in time during the vibrating cycle. In real-time interferometry, the interference fringes are formed by the addition of wavefronts from the surface of the object at rest and from the vibrating surface at some point during its vibrating cycle. The fringe pattern observed is a visual time average of this continuum set of interferogram fringe systems. A time-average holographic interferogram is made by exposing the holographic plate while the object is vibrating; the exposure time is usually multiple vibration periods. This hologram may be thought of as a continuum set of exposures, each recording the interference of the object wave with its temporal counterparts in the rest of the continuum over one cycle. The end result is a fringe pattern, directly related to the surface vibration pattern, in which the fringe lines represent contours of constant vibration amplitude. A more physical view of the process derives from observing that holographic fringe movement in the recording medium due to object movement nullifies the holographic recording. Thus, regions of a vibrating surface in motion (antinodes) will appear dark while regions at rest (nodes) will appear bright.

Stroboscopic illumination has long been used to study objects in motion. Coupling this technique with holographic interferometry produces interferograms of vibrating objects with enhanced fringe visibility and greater information content. The technique may be used to make real-time observations or to record double-exposure holograms of vibrating surfaces. In the real-time configuration, the hologram of the test object at rest is made in the usual fashion. The object is then vibrated and strobed with a laser set to flash at a particular point in the vibrating cycle. The fringe system formed is produced by interference of the wavefront from the object at rest and the wavefront from the object at this particular point in the cycle. The timing of the laser flash may be varied to observe the evolution of the surface vibration throughout its entire cycle. A double-exposure interferogram is formed by exposing the hologram plate to two flashes from the strobed laser. In this manner, any two states of the vibrating surface may be compared during its cycle by varying the timing of the laser flashes and their separation. The actual exposure may extend over several vibration cycles. This method results in fringes of much higher contrast than those yielded by time-average holography.

An interesting variant of hologram interferometry is contour generation.[20-22] This is accomplished by making two exposures of an object with a refractive index change in the medium surrounding the object or a change in the laser wavelength between exposures. Either method yields fringes on the surface of the object. The fringe positions are related to the height of points on the object relative to a fixed plane. The two-wavelength method may be implemented in real-time by first making a hologram of an object in the usual manner at one wavelength, then illuminating both the object and the processed hologram (carefully placed in its original position) at a different wavelength. Both methods have been successfully applied in a variety of situations.

In many cases, the interpretation of the fringe pattern produced by a holographic interferogram is a simple matter of qualitative assessment. For example, defects or flaws may be identified by anomalous local variations in a background fringe pattern. The presence or absence of these anomalies may provide all the information required in a nondestructive evaluation experiment. However, a detailed quantitative assessment of the mechanical deformation of the surface may be desired—this can involve a complex mathematical analysis of the fringe system. Quantitative analysis of the fringe pattern is often complicated by the fact that the fringes are not necessarily localized on the surface of the object. The interpretation of holographic interferograms and analysis of the fringe data have been the subject of considerable study.[23-27] Numerous methods such as sandwich holographic interferometry,[28,29] fringe linearization interferometry,[30] difference holographic interferometry,[31] and fringe carrier techniques[32] have been developed to facilitate this interpretation. In addition, the development of automatic fringe reading systems and data reduction software have greatly aided this process. However, fringe data reduction remains a challenge for many situations despite the progress that has been made.

The basic methods of holographic interferometry (real-time, double-exposure, and time-average) are in widespread use and continue to be the mainstay of this technique. However, important refinements have been made in recent years which have greatly added to the power of holographic interferometry. These advances include the use of real-time recording media[33,34] and heterodyne holographic interferometry[35] Real-time holographic recording materials (such as photorefractive crystals) provide an adaptive feature that makes the interferometer less sensitive to vibration, air currents, and other instabilities. The reliability of the interferometric process in a hostile environment is thus improved. Heterodyne techniques using two separate reference waves and a frequency shift between these two waves upon reconstruction has greatly improved the accuracy of holographic interferometry. Measurements with accuracies as high as $\lambda/1000$ can be made using heterodyne methods. Holographic moiré,[36] infrared holographic interferometry,[37] and the use of optical fibers[38] have also significantly extended the capabilities of holographic interferometry.

In the laboratory, where conditions are well-controlled, silver halide film has been the recording material of choice for holographic interferometry due to its relatively high

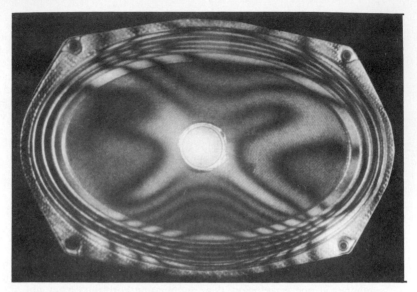

FIGURE 2 Time-average holographic interferogram of a loudspeaker vibrating at resonance. (*Photo courtesy of Newport Corporation.*)

sensitivity, low cost, and reliability. However, in field applications such as the factory floor, the wet processing requirements of silver halide film make this material much less attractive and, for some time, inhibited the use of holographic interferometry in many situations. Other materials that do not require wet processing (such as photopolymer films) are available, but these materials have very low sensitivity. The development of the thermoplastic recording material,[39] which does not require wet processing but retains the high sensitivity of silver halide film, made possible the much more convenient application of holographic interferometry in industrial situations. Several companies (Newport, Rottenkolber, and Micraudel) have commercially marketed holocamera systems using this material. The Rottenkolber and Micraudel systems are film-based with supply and take-up rollers. Only one exposure may be made on each area of film. The Newport holocamera has a smaller recording area than the film-based systems, but the plate may be erased and reused for several hundred exposures. A holographic interferogram made using the Newport holocamera is shown in Fig. 2.

Holographic interferometry has been applied to an enormous range of problems; this is a simple testimony to its utility and versatility. The classical interferometric testing of the figure of optical components during fabrication can be augmented with holographic interferometry to test for figure during the grinding process since the surface of the test object does not need to be polished.[21] The ability of holographic interferometry to make precise measurements of very small mechanical displacements has enabled it to be used in stress-strain measurements in materials, components, and systems. Mechanical displacements observed with holographic interferometry are often the result of thermal disturbances. Measurement of these thermally induced mechanical displacements with holographic interferometry can provide an accurate determination of the heat transfer properties of the material or system under study.[40] Similarly, diffusion coefficients in liquids can be determined using holographic interferometry.[41,42] Flow visualization and the accurate determination of fluid-flow properties using holographic interferometry has been an intense area of study.[43–45] As noted earlier, the technique can also be used to study high-speed events using short-pulse lasers in the double-pulse mode. Vibration analysis is

FIGURE 3 Time-average holographic interferogram displaying one of the vibration modes of a turbine blade. (*Photo courtesy of Karl Stetson, United Technologies Research Center.*)

one of the more powerful applications of holographic interferometry. In this area, the technique has been applied to a diverse array of problems including studies of the vibration properties of musical instruments,[46] vibration patterns in the human eardrum,[47] and vibration properties of mechanical parts such as turbine blades.[48] The application of holographic interferometry to turbine blade mechanics is illustrated in Fig. 3. One of the great virtues of holographic interferometry is that a tremendous wealth of information can be garnered from its application without destroying the test object or system. As a result, nondestructive evaluation or nondestructive testing has been one of the most important areas of application for this technique. As an example, holographic interferometry has been successfully used to observe subsurface defects in solid opaque objects. Even though the interference pattern is produced strictly by mechanical surface deformations, these surface variations are often indicative of subsurface changes (e.g., ply separations in automobile tires and interlayer delamination in composite materials).[49] Subsurface defects are usually manifest in local anomalies of the fringe pattern and a qualitative examination of the interferogram will often discern the effect. The literature is replete with articles describing these and many other applications of holographic interferometry. For the reader interested in an in-depth discussion of the theory and application of holographic interferometry, numerous textbooks and review articles are available (see, for example Refs. 6–8 and 50–57).

Electronic Holography

Even with the use of thermoplastic recording media, holographic interferometry remains a challenging and, in many cases, unacceptable technique for industrial applications—particularly those that involve on-line quality testing in a production environment. Speckle-pattern interferometry,[58] another technique closely associated with holographic interferometry, alleviates many shortcomings of the traditional holographic approach when combined with electronic image recording and processing equipment.

Speckle is the coarse granular or mottled intensity pattern observed when a diffuse surface is illuminated with coherent light. Wavelets reflected from the randomly oriented facets of an optically rough surface interfere to produce this effect when the size of the facets is on the order of a wavelength or larger. Although this interference occurs throughout the space occupied by the wave scattered by the surface, the interference that produces the observable pattern takes place in the plane of the detector or recording medium (i.e., the retina of the eye or the film plane of a camera). The speckle-pattern recorded by an imaging system (eye or camera) is known as subjective speckle, while the intensity variation detected by a scanning detector above a coherently illuminated diffuse surface is referred to as objective speckle. Objective speckle is the resultant sum of the waves scattered from all parts of the surface to a point in space; in subjective speckle, wave summation in the observation plane is limited to the resolution cell of the system. The objective-speckle scale depends only on the plane in space where it is viewed, not on the image system used to view it. The size of the image plane or subjective speckle depends on the aperture of the viewing or imaging system. For subjective speckle, the characteristic speckle diameter, d_{sp} is given by[59]

$$d_{sp} \approx \frac{2.4 \lambda v}{a} \tag{10}$$

where λ is the wavelength, v is the distance from the lens to the image plane, and a is the diameter of the viewing lens aperture.

A fringe pattern is formed when the speckle patterns of the diffuse surface in its original and displaced positions are properly combined. The formation of this fringe pattern is known as speckle-pattern interferometry, of which there are two basic types: speckle-pattern photography and speckle-pattern correlation interferometry. Both techniques, which form the basis for electronic speckle-pattern interferometry (ESPI), and other forms of electronic holography will be discussed in this section. The remarks made here are derived from Ref. 58, which contains a thorough discussion of the topic.

By varying the recording and viewing configurations, speckle-pattern interference fringes can be made sensitive to in-plane and out-of-plane displacements, displacement gradients, and the first derivative of displacement gradients. Speckle-pattern interferometry has two distinct advantages over holographic interferometry: (1) the direction of the magnitude sensitivity of the fringes can be varied over a larger range, and (2) the resolution of the recording medium for speckle-pattern interferometry does not need to be nearly as high. Therefore, speckle-pattern interferometry is a much more flexible technique, although the fringe definition is not nearly as good as with holographic interferometry due to the degradation of the images by the speckle pattern.

In speckle-pattern photography, the object is illuminated by a single light beam; no reference beam is involved. Some of the light scattered by the object is collected by a lens and recorded on photographic film. The film plane may be an image plane (in-focus) or some other plane (out-of-focus). The location of the film plane determines whether the resulting interferometric fringes are sensitive to in-plane or out-of-plane motion. Two exposures of the film are made: one with the object in its original position, the second with the object deformed or displaced. Proper illumination of the film negative with coherent light produces a fringe pattern in the observation plane which is related to the object motion. With the use of appropriate recording and viewing geometries, the fringes may be made to superimpose an image of the object. If the object is illuminated by a plane wavefront and the film is in the focal plane of a lens, the fringes are related to out-of-plane displacement gradients. Illumination of the object by a diverging wavefront in the proper geometry yields fringes related to the tilt of the object. Speckle-pattern photography can be used to make time-average stroboscopic and double-pulse measurements just as in holographic interferometry. In speckle-pattern correlation interferometry, a reference beam (either specular or diffuse) is incident upon the observation or recording plane in addition to the light scattered by the object. Interferometric fringes are produced by the correlation of the speckle patterns in the observation plane for the displaced and undisplaced object. Real-time or live-correlation fringes may be produced as follows. The object and reference beams are recorded with the object in its original position using photographic film. The film is developed and the film negative is replaced in its original position. The negative is illuminated with object and reference light, and the object is displaced. Correlation fringes are produced by the process of intensity multiplication. Because of the contrast reversal of the film negative, minimum transmission is found in areas of maximum correlation between the pattern recorded and the pattern produced by the displaced object. Unfortunately, the correlation fringes produced using this technique are of low contrast.

The variation in the correlation of the two speckle patterns which produces the fringes may be made sensitive to different components of surface displacement by using different object and reference beam geometries. One of the most important configurations uses a specular in-line reference beam introduced with a beam splitter or mirror with a pin hole. This configuration may be used to make dynamic displacement measurements or to observe the behavior of vibrating objects in real time. This particular optical geometry is also the most popular arrangement for ESPI.

In ESPI, the photographic film processing methods used for speckle-pattern photography and speckle-pattern correlation interferometry are replaced by video recording and display technology. The concept of using video equipment for this purpose was originated by several researchers working independently during the same period.[60-63] For speckle-pattern interferometry, the minimum speckle size is usually in the range of 5 to 100 μm so that standard television (TV) cameras can be used to record the pattern. The main advantage of using TV equipment is the high data rate. Real-time correlation fringes may be produced and displayed on the TV monitor at 30 frames per second. In addition, the full array of modern video image processing technology is available to further manipulate the image once it is recorded and stored in the system. Another advantage is the relatively high light sensitivity of TV cameras which operate at low light levels, thus enabling satisfactory ESPI measurements with relatively low power lasers.

The ready availability of advanced video recording and processing equipment, its ease of use, and its flexible adaption to various applications make ESPI a near-ideal measurement system in many instances—particularly, industrial situations (such as an assembly line) where rapid data generation and retrieval, and high throughput are required. ESPI overcomes many of the objections of holographic interferometry and has been used extensively for industrial measurements in recent years.

Intensity correlation fringes in ESPI are produced by a process of video subtraction or addition. In the subtraction process, the image of a displaced object is subtracted from an electronically stored image of the object in its original position to produce the correlation fringes. To observe these fringes, the subtracted video signal is rectified and high-pass filtered, then displayed on a video monitor. This video processing is analogous to the reconstruction step in holography.

For the addition method, both original and displaced images are added optically at the photo cathode of the TV camera. The TV camera detects the light intensity and, again, the signal is full-wave rectified, high-pass filtered, and displayed on the TV monitor. Because of the persistence of the TV tube, the two images need not be recorded simultaneously; however, they must be presented to the camera within its persistence time, usually on the order of 100 ms. The various optical configurations used in speckle-pattern correlation interferometry, employing both specular and diffuse reference beams, may be used in ESPI as well. The most popular of these configurations uses a specular in-line reference beam and may be used to make real-time vibration studies. ESPI has been used for this purpose

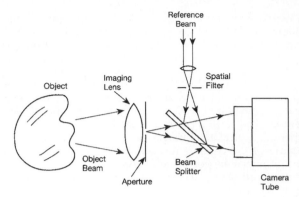

FIGURE 4 Typical ESPI optical arrangement with in-line reference beam.

more than any other application. This optical configuration is shown in Fig. 4. The object and reference beams in Fig. 4 are derived from the same laser with the use of a beam splitter, not shown for simplicity. In holographic interferometry, a high-resolution film is used and the reference beam angle may be any practical value desired. In ESPI, the recording medium (TV camera) has a resolution two orders of magnitude lower than holographic film (on the order of 30 line-pairs per millimeter). Therefore, in ESPI, an in-line reference beam must be used. Furthermore, the aperture of the system must be small enough to keep the interference angle below one degree. All the usual modes of holographic interferometry (real-time, time-average, stroboscopic, and double-exposure) may be performed with ESPI. A time-average ESPI interferogram of an object made with the system operating in the subtraction mode is shown in Fig. 5a. For comparison, a holographic interferogram of the same object is shown in Fig. 5b. ESPI has been applied to a wide range of measurement problems. These applications are discussed in numerous books, technical papers, and review articles.[64-71]

Despite the flexibility of ESPI and its ease of use, fringe definition is poor compared to holographic interferometry—this has somewhat limited its use. Speckle-averaging and video-processing techniques have provided some improvement in interferogram quality, but very fine interference fringes are still difficult to discern with ESPI. A significant improvement in interferogram quality has been achieved with a newer technique: electro-optic holography (EOH).[72-74] This technique uses the same speckle interferometer optical configuration as ESPI, but processes the video images in a different manner. In EOH, a phase-stepping mirror is added to the reference leg to advance the phase of the reference beam by 90° between successive video frames. Subsequent processing of these phase-stepped images combined with frame and speckle averaging provides interferograms in real-time with nearly the same resolution and clarity of the traditional film-based holographic interferogram. An interferogram made with an EOH system is shown in Fig. 6.

23.5 HOLOGRAPHIC OPTICAL ELEMENTS

An optical element with the power to direct and/or focus a light wave can be made by recording the fringe pattern formed by two interfering light beams. HOEs or diffractive optical elements, thus produced, have several features that distinguish them from

(a)

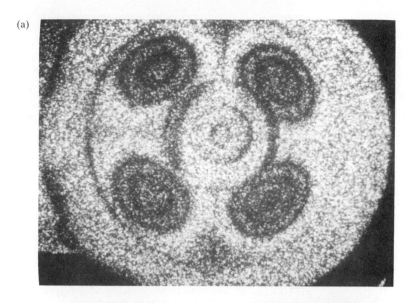

(b)

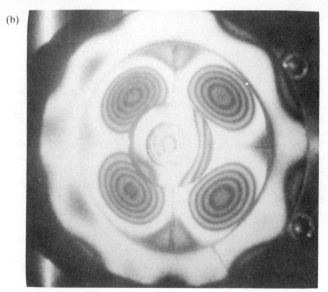

FIGURE 5 Time-average interferograms of a pulley made with (*a*) ESPI; (*b*) holographic interferometry.

conventional optics. A compilation of articles on diffractive optics may be found in Ref. 75. References 76 and 77 are recent reviews of the field.

Different types of elements may be produced by varying the curvature of the interfering wavefronts, interbeam angle, and configuration of the recording surface. Lenses with mild or very strong focusing power and plane or focusing mirrors are easily produced. The recording substrate may be plane or any arbitrarily curved shape. An element combining

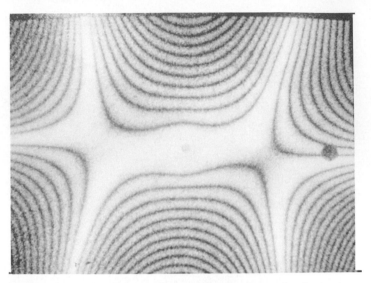

FIGURE 6 An interferogram illustrating a vibrating mode of a center-mounted rectangular plate made with an electronic holography system. (*Photo courtesy of Karl Stetson, United Technologies Research Center.*)

diffractive power with refractive power can be made by placing the recording medium on a curved surface. This method may be used to reduce the aberration of the combined optic since the achromatic dispersion of diffraction and refraction are of opposite sign. Although any of the many types of holographic recording materials may be used to make an HOE, dichromated gelatin and photoresist are preferred because of their high resolution and extremely low optical scatter. Dichromated gelatin has the additional advantage of forming HOEs of very high diffraction efficiency. Photoresist forms a surface relief hologram which makes possible the economical mass production of HOEs with straightforward mechanical replication means. Since the power of an HOE is derived from diffraction at the element's surface, the HOE may be very thin and lightweight. HOEs may be produced by the direct interference of physical light waves, or by calculating the desired interference pattern and printing this pattern onto a substrate by either photographic or electron beam lithographic means. Computer-generated HOEs are advantageous when the required optical wavefronts are difficult to create physically.

The use of the term "holographic optics" is not technically correct because the definition of holography implies that at least one of the wavefronts being combined to produce an interference record is an information carrier. Consequently, the term "diffractive optics" is gaining popularity and, when applied to gratings, the term "interference gratings" is certainly more appropriate. In this brief discussion, however, we will continue to use the currently popular "holographic" terminology.

Certainly one of the most common and successful applications of holographic optics is as a grating in spectrographic instruments.[78-80] The main advantage of holographic gratings* are that they can be made free of the random and periodic groove variation found in even the finest-ruled gratings, and they have low light scatter. This latter property is especially important when even a small amount of stray light is objectionable (such as in the study of Raman spectra of solid samples). To produce high-quality holographic

---

*The discussion here on holographic gratings is taken largely from Ref. 80. The author is indebted to Christopher Palmer of the Milton Roy Company for making an advance copy of this material available.

gratings, extreme care must be used in the fabrication process. Photoresist is the preferred recording material for reasons mentioned above; however, it is very insensitive and requires long exposure times, often many hours. Therefore, a highly stable optical system is essential. The recording room must be free of air currents, the air must be filtered and dust-free, and the photoresist coating must be defect-free. Stray light scatter from optical mounts and other objects in the recording setup must be eliminated by proper baffling and masking. The interfering beams must be appropriately conditioned by spatial filtering to ensure diffraction-limited performance. If beam-forming optics are used, for example, to produce collimated beams for making plane holographic gratings, these optics must be aberration-free and of diffraction-limited quality.

After exposure and chemical development, the surface relief pattern is metalized and the holographic grating is replicated as conventional master-ruled gratings are replicated. Both positive and negative photoresists are available for making holographic gratings, however, the negative resist is seldom used.

Grating-diffraction efficiency in the various orders is determined by the groove profile. In a ruled grating, the groove is profiled by appropriately shaping the diamond tool. Holographic gratings have a sinusoidal profile; blazing, in this case, is accomplished by ion etching. A wide range of groove spacing is possible with ruled gratings, however, holographic gratings offer more flexibility with respect to the groove pattern. For example, groove curvature may be used in holographic gratings to reduce aberrations in the spectrum, thereby improving the throughput and resolution of imaging spectrometers. The grooves in ruled gratings are produced by the traveling diamond tool, one after another. In holographic gratings, all grooves are produced in parallel; thus, the fabrication time for holographic gratings can be considerably shorter.

Another important instrumental HOE application is in optical beam scanning; the most common example of this is the supermarket scanner.[81] Laser beam scanning is usually accomplished by either mechanical means (e.g., rotating mirror) or with the use of some transparent medium whose optical properties are changed by some sort of stimulation (e.g., acousto-optic cell). Holographic scanners offer advantages over both of these more conventional methods.

The working principle of the holographic scanner may be illustrated by considering the translation of a focusing lens through an unexpanded laser beam. As the beam intercepts the lens from one side to the other, it is simultaneously deflected and focused at the lens focal point on the lens axis. Thus, moving the lens back and forth causes the laser beam to sweep back and forth in the focal plane of the lens. In its basic form, a holographic scanner is simply an HOE lens or mirror translating through the scan beam. The principal advantage of the holographic scanner over conventional scanning means is the ability to combine beam deflection and focusing into a single element. The form of both the deflection and focusing function can be tailored in a very flexible way by proper design of the HOE formation optical system. For example, the scan element may be easily made a line segment rather than a focal spot, and the locus of the scanned focal spot or line may be placed on either plane or curved surfaces. Multiple scan beams with multiple focal points may be generated by a scanner in the form of a segmented rotating disk. Each segment or facet of this disk has different deflection and focal properties. As the disk rotates through the beam, a multiplicity of scan beams is produced which can densely fill the desired scanned volume. This feature is especially important in a supermarket application because it allows products of varying sizes and shapes to be rapidly scanned. A further advantage of the HOE scanner is that the scanner disk can be small, thin, and lightweight, thereby greatly reducing the demands on the drive system. The disk format also produces little air movement due to windage and is very quiet in operation.

In addition to serving as the beam deflector, the holographic scanner also collects the light scattered from the laser spot on the product and directs it to the optical detector. This scattered light illuminates the holographic scanner along the conjugate object beam path and is diffracted into the fixed conjugate reference or primary scan beam path where it is

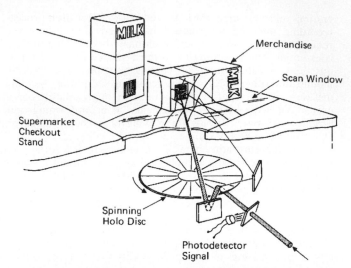

FIGURE 7 Schematic of a holographic supermarket scanner. (*Reprinted from Ref. 81, p. 9; courtesy of Marcel Dekker, Inc.*)

accessed by a beam splitter. The light scattered from all points in the scan volume is thus directed to a single, fixed detector position. An optical schematic for a point-of-sales scanner is shown in Fig. 7.

The ability to combine several optical functions into a single HOE makes this device attractive in many situations. Significant savings in space, weight, and cost can often be realized by replacing several conventional elements with a single HOE device. This feature has been incorporated into an optical head for compact disk applications with the use of a multifunctional HOE.[82] An optical diagram of the device is shown in Fig. 8. The objective lens images the light-emitting point of the laser diode (LD) to the compact disk. The light scattered from this focal point on the disk is reimaged by the objective lens through the HOE to the photodetector (PD). In this application, the HOE serves as a spherical lens, beam splitter, aberration-correcting lens, and cylindrical lens. In addition to simplifying the optical system, the HOE provides a better means of aligning the optical system.

Holography has even been applied to one of the oldest instrumental functions known—

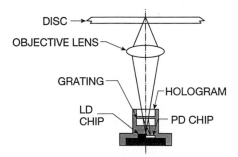

FIGURE 8 Diagram of compact disk optical head employing a holographic optical element. (*Diagram courtesy of Wai-Hon Lee, Hoetron, Inc.*)

the keeping of time. This has been accomplished by using an HOE as a holographic sundial.[83]

23.6 HOLOGRAPHIC INSPECTION

Quality assurance inspection and testing, important functions in any industrial manufacturing process, have also benefited from advances in holography. Holographic methods have been applied to the quality control problem in several areas, such as identifying and locating subsurface mechanical defects and determining the presence or absence of certain surface features. Holographic interferometry has been successfully applied to some of these problems,[84–86] and optical processing methods have also yielded good results in many cases.[87] Matched and spatial filtering in the Fourier transform plane of an optical processing system have proved to be especially powerful means of identifying features and determining surface detail. In this section, we describe a unique combination of holography and classical optical processing methods that made possible a very successful means of rapidly detecting defects in devices with highly regular and repetitive patterns.

Defects in integrated circuit photomasks and wafers at various stages of processing can greatly diminish the final device yield. Since the economics of wafer production is strongly influenced by yield, there has been an ever-present incentive to increase this yield by minimizing the number of defects introduced at various points in the production process. One way to increase yield is to identify these defects early, and eliminate them before they cascade in a multiplicative manner through the various stages of the process.

For years, the inspection of integrated circuit photomasks and wafers was performed by either manual microscopic examination or automated serial scanning using an optical detector. The latter method involves comparison of the detector signal from a magnified portion of the test pattern to a similar portion of a reference pattern, adjacent pattern, or digital design database. In either case, the inspection process was long and tedious, often requiring many hours or days to inspect a single photomask or wafer. These methods are very slow because of the large number of pixels involved and the serial pixel-by-pixel nature of the inspection. Clearly, a great advantage would be afforded by the ability to examine all the pixels in parallel rather than serial format.

This observation prompted a number of researchers to consider optical processing methods for integrated circuit inspection and for addressing other types of problems involving highly repetitive patterns such as cathode-ray tube masks and TV camera tube array targets.[88–90] The concept in all cases was to eliminate perfect pattern information and highlight defect areas of the image by spatial filtering in the Fourier transform plane. These methods met with only limited success because of the difficulty in fashioning effective blocking filters and the need for extremely high quality, large-aperture, low F-number Fourier transform lenses. Despite considerable work in this area, none of these efforts resulted in the development, production, and in-process use of integrated circuit inspection systems using Fourier optical processing concepts.

This situation was reversed by the adaption of a holographic documentation system used to document the surface microstructure of high-energy laser optical component test samples.[91,92] A schematic diagram of the holographic documentation optical system is shown in Fig. 9. An F/3.42 lens was used to image the test target onto the hologram film and an argon laser operating at a wavelength of 514.5 nm was the illumination source. A polarizing beam-splitting cube and a half-wave plate were used to split the beam into object and reference beams, and to adjust the beam ratios. A second polarizing beam-splitting cube and quarter-wave plate were used to efficiently illuminate the test target and direct the reflected light to the holographic plate. The holographic image was reconstructed with the conjugate reference beam by removing mirror M-4. The test target was removed and the holographic image of the sample was examined with the aid of a

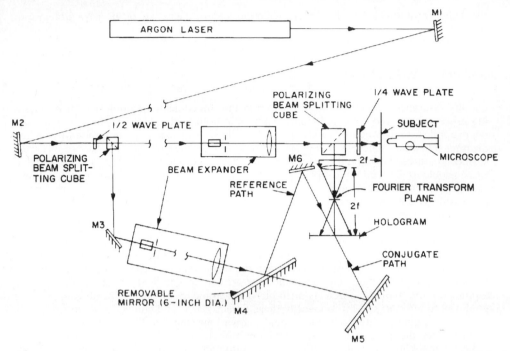

FIGURE 9 Holographic documentation optical system. (*Reprinted from Ref. 92, p. 87; courtesy of Oxford University Press.*)

microscope. The calculated resolution of this system (classic Rayleigh resolution limit) was 4.0 μm. A photograph of the holographic reconstruction of the standard Air Force resolution target made with this system is shown in Fig. 10. The smallest bars in this target are on 4.4-μm centers. Thus, the resolution observed with this system was comparable to the calculated value.

In considering how the documentation system might be adapted to other applications (including the integrated circuit inspection problem), Fusek et al. observed that because the real image of the test target was being examined by conjugate reconstruction, both functions of the classic Fourier optical system (i.e., transform and inverse transform) were performed in the documentation system.[93,94] Because of reverse ray tracing, a high-quality matched pair of specially designed Fourier transform lenses is no longer required to produce a diffraction-limited output image. As with previous work, the objective was to attenuate the image area where the pattern is defect-free and highlight the defects. This is done simply by placing the appropriate blocking filter in the Fourier transform plane. The method works effectively only if the filter efficiently blocks the light associated with defect-free areas of the image and efficiently transmits the defect light. Fortunately, this is the case for integrated circuit masks and wafers which consist of regular patterns of circuit elements repeated many times over the area of the wafer. For such patterns, the intensity distribution in the Fourier transform plane is a series of sharp spikes or bright points of light. Low spatial frequencies associated with slowly varying features (such as line-spacing) are represented by light points near the optical axis of the imaging lens. High spatial frequencies, representing such features as edge and corner detail, lie farther out in the Fourier transform plane. The Fourier transform plane intensity pattern for a production-integrated circuit photomask is shown in Fig. 11. Since defects are usually of a nonregular

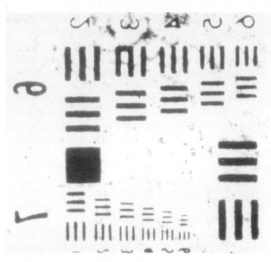

FIGURE 10 Magnified image of the holographic reconstruction of the Air Force resolution target. (*Reprinted from Ref. 92, p. 88; courtesy of Oxford University Press.*)

FIGURE 11 Optical Fourier transform of a production-integrated circuit photomask. (*Reprinted from Ref. 92, p. 98; courtesy of Oxford University Press.*)

nature, the light associated with defects is generally spread out fairly uniformly over the Fourier transform plane. Thus, a filter designed to block the regular-pattern light passes most of the light associated with defects.

The most straightforward way to produce the blocking filter is by photographic means, with the film placed in the Fourier transform plane. The filter is made first, then the hologram is exposed with the object beam passing through this filter. Thus, the reconstructed wave that produces the output image is effectively filtered twice. A problem with this method of filter fabrication is the extremely large range of intensities in the Fourier transform plane which can cover as many as ten orders of magnitude. If the filter is properly exposed near the axis where the transform light is brightest, high-spatial-frequency components away from the axis will be underexposed and inadequately filtered. However, if high frequencies are properly exposed, low frequencies will be overexposed, and too much defect light will be filtered. An effective means of alleviating this intensity-range problem is dynamic range compression by multistep filter generation. The process is as follows. With a mask in place, a holographic plate is exposed to the Fourier transform pattern with exposure parameters set to record the intensity distribution in the low-frequency region near the optical axis. After processing, this plate (Stage 1 filter) is replaced in the Fourier transform plane and a hologram of the mask is made through this filter. The hologram is illuminated by the conjugate reference beam and a second filter plate (Stage 2 filter) is exposed to the resulting Fourier transform intensity pattern. This pattern now has its low-frequency components attenuated because of the action of the Stage 1 filter. The Stage 2 filter can now be used to record a Stage 3 filter. The process may be repeated as many times as necessary to produce a filter with the desired attenuation properties. Because the defect light is of much lower intensity in the Fourier transform plane than the light corresponding to nondefect areas, defect light does not contribute significantly to the exposure of the filter, and the object under test (containing defects) may be used to generate the filter. Stage 1 and 2 filters for a defect calibration test mask (Master Images VeriMask™) are shown in Fig. 12.

Performance of the breadboard documentation system using the VeriMask™ is shown in Fig. 13. Figure 13a is a photo of a magnified region of the VeriMask™ containing a pinhole defect. Figure 13b shows the Stage 2 filter image of this same defect which is clearly enhanced. In addition, dimensional variations in the mask pattern from die to die are also highlighted.

The breadboard holographic inspection technology developed by Fusek and coworkers was further advanced and placed into production by Insystems of San Jose, California.[95,98] This company produced a series of mask and wafer inspection machines based on the holographic optical processing technique. The optical configuration of the Insystems Model 8800 Wafer Inspection System is shown in Fig. 14. The commercial instrument uses an argon ion laser as the light source, and the system functions in an optical manner identical to that of the original breadboard device. However, many refinements were incorporated into the commercial system which yielded substantially improved performance over the breadboard system. These refinements, which included a sophisticated Fourier transform lens design, made possible adequate performance without using the multistep filter generation technique. The wafer test piece and the hologram are placed on a rotating stage under a microscope and a video camera so that the filtered image of the defect and the microscopic image of the defect on the actual wafer can be viewed simultaneously. Figure 15 illustrates the advanced filtering capability of the commercial instrument. This instrument is sensitive to defects as small as 0.35 μm.

Disadvantages of the holographic defect detection system are the inconvenience and time delays associated with the wet processing of the silver halide holographic recording material. The use of photorefractive crystals, which operate in real time and do not require wet processing, has been studied as a means of eliminating these disadvantages.[97] The dual functions of image recording and spatial filtering are combined by placing the crystal in

(a)

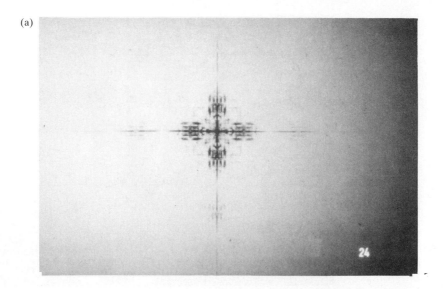

(b)

FIGURE 12 Fourier transform plane blocking filters for a defect calibration test mask: (*a*) first-stage initial filter; (*b*) second-stage dynamic range compressed filter. (*Reprinted from Ref. 92, p. 101; courtesy of Oxford University Press.*)

the Fourier transform plane and adjusting the reference beam intensity to the level of the defect light intensity. Since the light in the Fourier transform plane associated with defects is much lower in intensity than nondefect light, only the defects will be recorded with high diffraction efficiency. Practical use of photorefractive crystals in this application has not been realized, however, because these crystals have relatively low sensitivity and are not available in large sizes with good optical quality.

(a)

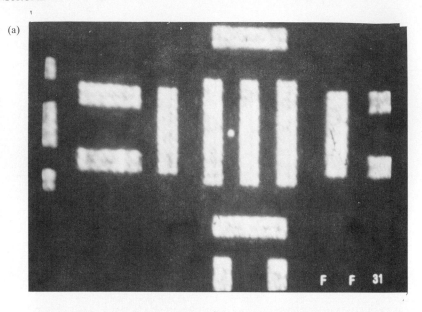

(b)

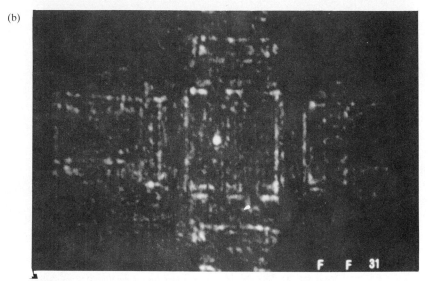

FIGURE 13 Images of the holographic reconstruction of a 2.01-μm pinhole on a calibration test mask; (*a*) unfiltered and (*b*) filtered. (*Reprinted from Ref. 92, pp. 102–103; courtesy of Oxford University Press.*)

23.7 *HOLOGRAPHIC LITHOGRAPHY*

The lithographic transfer of an integrated circuit photomask pattern to a resist-coated integrated circuit wafer has been accomplished by several methods, including contact printing, proximity printing, and step-and-repeat imaging. Each method has advantages and disadvantages. Contact printing is a simple, straightforward method suitable for

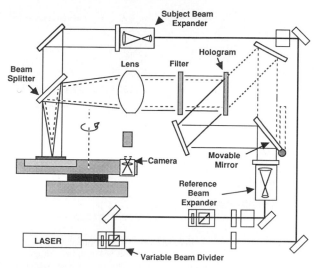

FIGURE 14 Optical schematic for the Insystems Model 8800 holographic wafer inspection system. (*Diagram courtesy of Insystems.*)

printing large wafer areas, but damage to the mask and contamination of the wafer are common problems. Proximity printing is resolution-limited because of near-field diffraction. The diffraction problem can be eliminated by imaging the mask onto the wafer, but full-field imaging systems do not provide the resolution required over the full area of the wafer. Stepper systems, which image only a small area of the mask at a time, provide the required resolution; however, they are complex and expensive because of the resolution demands placed on the optical imaging system and the mechanical difficulty in accurately stitching together the multiple image patterns over the full area of the wafer.

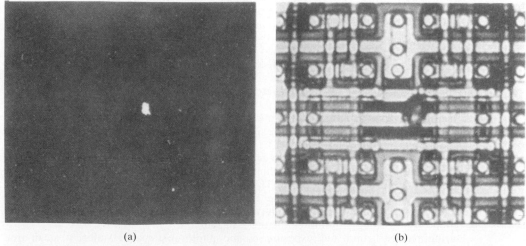

(a)   (b)

FIGURE 15 Metal layer defect in an integrated circuit wafer highlighted by the Insystems Model 8800 wafer inspection system: (*a*) filtered image and (*b*) unfiltered image. (*Photo courtesy of Insystems.*)

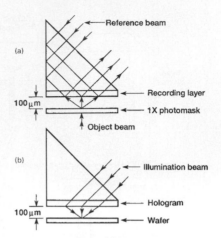

FIGURE 16 Basic optical arrangement for total internal reflection holographic lithography: (*a*) hologram exposure and (*b*) reconstruction onto a resist-coated substrate. (*Reprinted from Ref. 99; courtesy of PennWell Publications.*)

Clearly, a full-field method of printing the image onto the full area of the wafer with the required resolution is desirable. A holographic system capable of accomplishing this task has been developed by Holtronic Technologies Limited.[98–100] Rather than using a lens system to image the mask on to the wafer, the Holtronic holographic system uses real-image projection by conjugate illumination to overlay the mask image onto the wafer. Near-field holography is used to record the mask image by placing the mask in close proximity to the recording medium (100-μm separation) and illuminating the mask from the back with a collimated laser beam (364-nm line from an argon ion laser). To allow the introduction of an off-axis reference beam, an image of the mask could be relayed to the hologram plane with a lens. The approach taken was to eliminate the need for this lens with the use of the total internal reflection holography scheme of Stetson.[101] A diagram illustrating the optical principle is shown in Fig. 16.

Light transmitted and diffracted by the mask is incident directly on the holographic recording material which is deposited onto the opposing surface of a prism. Reference light is introduced through the diagonal surface of the prism and reflected at the holographic coating/air interface by total internal reflection. The recording medium is Dupont photopolymer. The holographic exposure is made with an expanded reference beam illuminating the entire hologram area.

Since three beams pass through the photosensitive material, three gratings are formed: (1) a reflection grating formed by the incident and reflected reference beams, (2) a reflection grating formed by the object beam and the incident reference beam, and (3) a transmission grating formed by the object beam and the reflected reference beam. To reconstruct the mask wavefront, an illumination beam conjugate to the reflected reference beam is introduced through the prism. This incident illumination beam interacts with the transmission object grating to form the conjugate real image of the mask. The totally internally reflected and the Lippmann-reflected illumination beams interact with the reflection object grating to reinforce the mask image.

During reconstruction and exposure of the photoresist-coated wafer, a small-area illumination beam is scanned over the hologram surface. Dynamic focusing of this scanned exposure beam eliminates the requirement that the mask and wafer substrates be ultraflat.

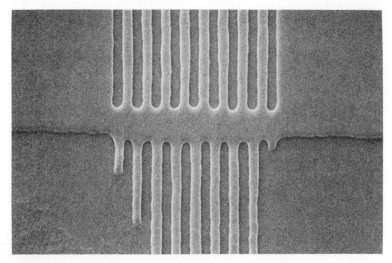

FIGURE 17   Scanning electron micrograph of 0.3-μm lines and spaces printed in photoresist by holographic lithography. (*Photo courtesy of Holtronic Technologies Ltd.*)

Figure 17 illustrates the 0.3-μm resolving capability of the holographic lithography process in 0.3-mm thick resist. The lines shown were printed on a silicon wafer coated with Olin Hunt HPR 204 i-line photoresist and developed using the Hunt HPRD 429 developer. The effective numerical aperture of the system is greater than 0.7.

23.8 HOLOGRAPHIC MEMORY

The information storage capability of holograms has been the subject of considerable study over the years with several applications in mind. With the parallel information storage and processing capability of holograms, and the promise of shorter access times, computer memory has received particular attention. Two review articles provide good summaries of this field of research through 1990.[102,103] However, little mention is made of the work of Russian scientists who have also been very active in this field (see, for example, Refs. 104–116).

Of all the holographic material recording possibilities available, volume storage in photorefractive crystals has received the most attention. There are two main reasons for this emphasis: (1) the large information storage capacity of these crystals, and (2) their capability to meet the write-read-erase requirement in real time with no wet-chemical or other material-processing delays. However, no commercial memory systems using holographic storage have been developed. One reason for this is the relatively large volume of space occupied by the laser beam-steering equipment and other optical components required in such a system, even though the actual holographic storage element may occupy a volume of less than a few cubic centimeters. However, the primary reason is the limitations of the recording material. Despite their distinct advantages over other recording material candidates, photorefractive crystals have some significant limitations. It is difficult to grow large crystals with good optical quality and to achieve stable, long-term storage without destructive readout.

Recent work by Redfield and Hesselink[117–119] shows promise for overcoming these previous limitations and leading the way to a practical holographic memory system. Rather than concentrating on developing large, high-quality crystals, their approach is to form a large-volume memory element by using an array of small crystallites of strontium barium niobate in the form of small cubes or crystalline fibers. Techniques have also been developed for accessing the holographic information stored in these crystallites without destructive readout. Information is stored in the memory structure by recording Fourier holograms of checkerboard patterns (pages) of digital information. Access times are projected to be 100 to 1000 times faster than with conventional magnetic disk drives.

23.9   CONCLUSION

This chapter has briefly reviewed some of the more important instrumental applications of holography and demonstrated how holographic methods have been used to creatively solve a variety of measurement and recording problems. These successful applications should pave the way for additional advances in this field. Consequently, we anticipate that the list of technical applications of holography will expand significantly in the future.

23.10   REFERENCES

1. D. Gabor, "A New Microscope Principle," *Nature* **161** (1948).
2. D. Gabor, "Microscopy by Reconstructed Wavefronts," *Proc. Roy. Soc.* **A197** (1949).
3. D. Gabor, "Microscopy by Reconstructed Wavefronts: II," *Proc. Phys. Soc.* **B64** (1951).
4. E. N. Leith and J. Upatnieks, "Reconstructed Wavefronts and Communication Theory," *J. Opt. Soc. Am.* **52**(10) (1962).
5. E. N. Leith and J. Upatnieks, "Wavefront Reconstruction with Diffused Illumination and Three-Dimensional Objects," *J. Opt. Soc. Am.* **54**(11) (1964).
6. R. J. Collier, C. B. Burchhardt, and L. H. Lin, *Optical Holography,* Academic Press, New York, 1971.
7. H. M. Smith, *Principles of Holography,* 2d ed., John Wiley & Sons, New York, 1975.
8. H. J. Caulfield (ed.), *Handbook of Optical Holography,* Academic Press, New York, 1979.
9. N. Abramson, *The Making and Evaluation of Holograms,* Academic Press, New York, 1981.
10. P. Hariharan, *Optical Holography,* Cambridge University Press, Cambridge, 1984.
11. G. Saxby, *Practical Holography,* Prentice-Hall, New York, 1988.
12. J. M. Burch, "The Application of Lasers in Production Engineering," *Prod. Eng.* (*London*) **44:**431 (1965).
13. K. A. Haines and B. P. Hildebrand, "Contour Generation by Wavefront Reconstruction," *Phys. Lett.* **19:**10 (1965).
14. R. J. Collier, E. T. Doherty, and K. S. Pennington, "Application of Moire Techniques to Holography," *Appl. Phys. Lett.* **7:**223 (1965).
15. R. E. Brooks, L. O. Heflinger, and R. F. Wuerker, "Interferometry with a Holographically Reconstructed Comparison Beam," *Appl. Phys. Lett.* **7:**248 (1965).
16. R. L. Powell and K. A. Stetson, "Interferometric Vibration Analysis by Wavefront Reconstruction," *J. Opt. Soc. Amer.* **55:**1593 (1965).

17. K. A. Stetson and R. L. Powell, "Interferometric Hologram Evaluation and Real-Time Vibration Analysis of Diffuse Objects," *J. Opt. Soc. Amer.* **55**:1694 (1965).

18. L. O. Heflinger, R. F. Wuerker, and R. E. Brooks, "Holographic Interferometry," *J. Appl. Phys.* **37**:642 (1966).

19. M. A. Monahan and K. Bromley, "Vibration Analysis by Holographic Interferometry," *J. Acoust. Soc. Amer.* **44**:1225 (1968).

20. B. P. Hildebrand and K. A. Haines, "Multiple-Wavelength and Multiple-Source Holography Applied to Contour Generation," *J. Opt. Soc. Amer.* **57**:155 (1967).

21. T. Tsuruta, N. Shiotake, J. Tsujiuchi, and K. Matsuda, "Holographic Generation of Contour Map of Diffusely Reflecting Surface by Using Immersion Method," *Jap. J. Appl. Phys.* **6**:661 (1967).

22. N. Shiotake, T. Tsuruta, Y. Itoh, J. Tsujiuchi, N. Takeya, and K. Matsuda, "Holographic Generation of Contour Map of Diffusely Reflecting Surface by Using Immersion Method," *Jap. J. Appl. Phys.* **7**:904 (1968).

23. J. D. Trolinger, "Automated Data Reduction in Holographic Interferometry," *Opt. Eng.* **24**(5) (1985).

24. R. J. Pryputniewicz, "Time Average Holography in Vibration Analysis," *Opt. Eng.* **24**(5) (1985).

25. R. J. Pryputniewicz, "Quantification of Holographic Interferograms: State of the Art Methods," *Topical Meeting on Holography Technical Digest* **86**(5), Opt. Soc. Am, Wash. D.C. (1986).

26. R. J. Pryputniewicz, "Review of Methods for Automatic Analysis of Fringes in Hologram Interferometry," *SPIE Proc.* **816** (1987).

27. R. J. Pryputniewicz, "Automated Systems for Quantitative Analysis of Holograms," *SPIE Institute Series*, vol. **IS 8** (1990).

28. N. Abramson, "Sandwich Hologram Interferometry: A New Dimension in Holographic Comparison," *Appl. Opt.* **13**(9) (1974).

29. H. Bjelkhagen, "Sandwich Holography for Compensation of Rigid Body Motion and Reposition of Large Objects," *SPIE Proc.* **215** (1980).

30. G. O. Reynolds, D. A. Servaes, L. Ramos-Izquierdo, J. B. DeVelis, D. C. Peirce, P. D. Hilton, and R. A. Mayville, "Holographic Fringe Linearization Interferometry for Defect Detection," *Opt. Eng.* **24**(5) (1985).

31. Z. Fuzessy and F. Gyimesi, "Difference Holographic Interferometry: An Overview," *SPIE Institute Series*, vol. **IS 8** (1990).

32. P. D. Plotkowski, Y. Y. Hung, J. D. Hovanesian, and G. Gerhart, "Improved Fringe Carrier Technique for Unambiguous Determination of Holographically Recorded Displacements," *Opt. Eng.* **24**(5) (1985).

33. A. A. Kamshilin, E. V. Mokrushina, and M. P. Petrov, "Adaptive Holographic Interferometers Operating Through Self-Diffraction of Recording Beams in Photorefractive Crystals," *Opt. Eng.* **28**(6) (1989).

34. V. I. Vlad, D. Popa, M. P. Petrov, and A. A. Kamshilin, "Optical Testing by Dynamic Holographic Interferometry with Photorefractive Crystals and Computer Image Processing," *SPIE Proc.* **1332** (1990).

35. R. Dandliker and R. Thalmann, "Heterodyne and Quasi-Heterodyne Holographic Interferometry," *Opt. Eng.* **24**(5) (1985).

36. X. Youren, C. M. Vest, and E. J. Delp, "Optical and Digital Moiré Detection of Flaws Applied to Holographic Nondestructive Testing," *Appl. Opt.* **8**:452–454 (1983).

37. M. Cormier, J. Lewandowski, B. Mongeau, F. Ledoyen, and J. Lapierre, "Infrared Holographic Interferometry," *Topical Meeting on Holography Technical Digest* **86**(5), Opt. Soc. Am., Wash., D.C. (1986).

38. J. A. Gilbert and T. D. Dudderar, "The Use of Fiber Optics to Enhance and Extend the Capabilities of Holographic Interferometry," *SPIE Institute Series,* vol. **IS 8** (1990).

39. T. C. Lee, "Holographic Recording on Thermoplastic Films," *Appl. Opt.* **13**(4) (1974).

40. N. G. Patil, C. R. Prasad, and V. H. Arakeri, "Holographic Interferometric Study of Heat

Transfer in Rectangular Cavities," *Topical Meeting on Holography Technical Digest* **86**(5), Opt. Soc. Am., Wash., D.C. (1986).

41. H. Fenichel and M. Lin, "Application of Holographic Interferometry to Investigations of Diffusion Processes in Liquid Solutions," *SPIE Proc.* **523** (1985).

42. H. Fenichel, G. E. Lohman, and D. Will, "Measurements of Diffusion Coefficients in Liquids Using Holographic Interferometry," *Topical Meeting on Holography Technical Digest* **86**(5), Opt. Soc. Am., Wash., D.C. (1986).

43. R. L. Perry and G. Lee, "Holographic Interferometry Applied to Symmetric Aerodynamic Models in a Wind Tunnel," *SPIE Proc.* **523** (1985).

44. V. A. Deason, L. D. Reynolds, and M. E. McIlwain, "Velocities of Gases and Plasmas from Real-Time Holographic Interferograms," *Opt. Eng.* **24**(5) (1985).

45. P. J. Bryanston-Cross, "Holographic Flow Visualization," *J. Phot. Sci.* **37**(1) (1989).

46. C. Agren and K. A. Stetson, "Measuring the Wood Resonance of Treble-Viol Plates by Hologram Interferometry," *J. Acoust. Soc. Amer.* **46**(1) (1969).

47. G. von Bally, "Otological Investigations in Living Man Using Holographic Interferometry," in G. von Bally (ed.), *Holography in Medicine and Biology,* Springer Series in Optical Sciences, vol. 18, Springer-Verlag, Berlin, 1979.

48. K. A. Stetson, "Holography as a Tool in the Gas Turbine Industry," *Topical Meeting on Holography Technical Digest* **86**(5), Opt. Soc. Am., Wash., D.C. (1986).

49. Y. Y. Hung, "Shearography Versus Holography in Nondestructive Evaluation of Tyres and Composites," *SPIE Proc.* **814** (1987).

50. G. M. Brown, R. M. Grant, and G. W. Stroke, "Theory of Holographic Interferometry," *J. Acoust. Soc. Amer.* **45**(5) (1969).

51. C. M. Vest, *Holographic Interferometry,* John Wiley & Sons, New York, 1979.

52. K. A. Stetson, "A Critical Review of Hologram Interferometry," *SPIE Proc.* **532** (1985).

53. C. M. Vest, "Holographic Metrology and Nondestructive Testing—Past and Future," *Proc. of the NATO Advanced Study Institute,* Martinus Nijhoff, Dordrecht, Netherlands, 1987.

54. P. Hariharan, "Interferometric Metrology: Current Trends and Future Prospects," *SPIE Proc.* **816** (1987).

55. R. J. Parker and D. G. Jones, "Holography in an Industrial Environment," *Opt. Eng.* **27**(1) (1988).

56. B. Ovryn, "Holographic Interferometry," *CRC Critical Reviews in Biomedical Engineering* **16**(4) (1989).

57. H. Rottenkolber and W. Juptner, "Holographic Interferometry in the Next Decade," *SPIE Proc.* **1162** (1990).

58. R. Jones and C. Wykes, *Holographic and Speckle Interferometry,* Cambridge University Press, Cambridge, 1983.

59. R. Jones and C. Wykes, *Holographic and Speckle Interferometry,* Cambridge University Press, Cambridge, 1983, p. 57.

60. J. N. Butters and J. A. Leendertz, "Holographic and Video Techniques Applied to Engineering Measurements," *J. Meas. Control* **4** (1971).

61. A. Macovski, D. Ramsey, and L. F. Schaefer, "Time Lapse Interferometry and Contouring Using Television Systems," *Appl. Opt.* **10**(12) (1971).

62. O. Schwomma, Osterreichisches, Patent No. 298830, 1972.

63. U. Kopf, in *Messtechnik* (in German), vol. 4, 1972, p. 105.

64. O. J. Lokberg, "Advances and Application of Electronic Speckle Pattern Interferometry (ESPI)," *SPIE Proc.* **215** (1980).

65. B. D. Bergquist, P. C. Montgomery, F. Mendoza-Santoyo, P. Henry, and J. Tyrer, "The Present Status of Electronic Speckle Pattern Interferometry (ESPI) With Respect to Automatic Inspection and Measurement," *SPIE Proc.* **654** (1986).

66. O. J. Lokberg, "The Present and Future Importance of ESPI," *SPIE Proc.* **746** (1987).

67. O. J. Lokberg, "Electronic Speckle Pattern Interferometry," *Proc. of the NATO Advanced Study Institute,* Martinus Nijhoff, Dordrecht, Netherlands, 1987, pp. 542–72.

68. O. J. Lokberg and G. A. Slettermoen, "Basic Electronic Speckle Pattern Interferometry," chap. 8 in R. R. Shannon and J. C. Wyant (eds.), *Applied Optics and Optical Engineering,* vol. X, Academic Press, Inc., 1987.

69. D. W. Robinson, "Holographic and Speckle Interferometry in the UK: A Review of Recent Developments," *SPIE Proc.* **814** (1988).

70. D. E. Parker, "Introductory Overview of Holography and Speckle," *SPIE Proc.* **1375** (1990).

71. O. J. Lokberg and S. Ellingsrud, "TV-Holography (ESPI) and Image Processing in Practical Use," *SPIE Proc.* **1332** (1990).

72. K. A. Stetson, W. R. Brohinsky, J. Wahid, and T. Bushman, "An Electro-Optic Holography System with Real-Time Arithmetic Processing," *J. Nondest. Eval.* **8**(2) (1989).

73. T. Bushman, "Development of a Holographic Computing System," *SPIE Proc.* **1162** (1989).

74. R. J. Pryputniewicz and K. A. Stetson, "Measurement of Vibration Patterns Using Electro-Optic Holography," *SPIE Proc.* **1162** (1989).

75. T. W. Stone and B. J. Thompson (eds.), "Selected Papers on Holographic and Diffractive Lenses and Mirrors, *SPIE Milestone Series, vol. MS 34,* 1991.

76. S. V. Pappu, "Holographic Optical Elements: State-of-the-Art Review Part 2," *Opt. Laser Technol.* **21**(6) (1989).

77. S. V. Pappu, "Holographic Optical Elements: State-of-the-Art Review Part 1," *Opt. Laser Technol.* **21**(5) (1989).

78. J. M. Lerner, J. Flamand, J. P. Laude, G. Passereau, and A. Thevenon, "Diffraction Gratings, Ruled and Holographic: A Review," *SPIE Proc.* **240** (1980).

79. E. G. Loewen, "Diffraction Gratings, Ruled and Holographic," chap. 2 in *Appl. Opt. and Opt. Eng.,* **IX** (1983).

80. "Interference (Holographic) Gratings," chap. 5, in C. Palmer and E. Loewen (eds.), *Diffraction Grating Handbook,* Milton Roy Company, 1991.

81. G. T. Sincerbox, "Holographic Scanners," chap. 1, in G. F. Marshall (ed.), *Laser Beam Scanning,* Marcel Dekker, Inc., New York, 1985.

82. W. Lee, "Holographic Optical Head for Compact Disk Applications," *Opt. Eng.* **28**(6) (1989).

83. K. M. Johnson, B. Cormack, A. Strasser, K. Dixon, and J. Carsten, "The Digital Holographic Sundial," *Topical Meeting on Holography Technical Digest* **86**(5), Opt. Soc. Am., Wash., D.C. (1986).

84. K. A. Arunkamar, J. D. Trolinger, S. Hall, and D. Cooper, "Holographic Inspection of Printed Circuit Board," *SPIE Proc.* **693** (1986).

85. Y. Lu, L. Jiang, L. Zou, X. Zhao, and J. Sun, "The Non-Destructive Testing of Printed Circuit Board by Phase Shifting Interferometry," *SPIE Proc.* **1332** (1990).

86. C. P. Wood and J. D. Trolinger, "The Application of Real-Time Holographic Interferometry in the Nondestructive Inspection of Electronic Parts and Assemblies," *SPIE Proc.* **1332** (1990).

87. D. Casasent, "Computer Generated Holograms in Pattern Recognition: A Review," *SPIE Proc.* **532** (1985).

88. L. S. Watkins, *Proc. IEEE* **57:**1634 (1969).

89. N. N. Axelrod, *Proc. IEEE* **60:**447 (1972).

90. R. A. Heinz, R. L. Odenweller, Jr., R. C. Oehrle, and L. S. Watkins, *Western Elect. Eng.* **17:**39 (1973).

91. R. L. Fusek, J. S. Harris, J. Murphy, and K. G. Harding, "Holographic Documentation Camera for Component Study Evaluation," in *High Power Lasers and Applications*: *Proceedings of the Meeting, SPIE Proc.,* Los Angeles, Calif., February 11–13, 1981.

92. L. Huff, "Holographic Documentation and Inspection," chap. 8, in J. Robillard and H. J. Caulfield (eds.), *Industrial Application of Holography,* Oxford University Press, 1990.

93. R. L. Fusek, L. H. Linn, K. Harding, and S. Gustafson, "Holographic Optical Processing for Submicrometer Defect Detection," *Opt. Eng.* **24**(5) (1985).

94. R. L. Fusek, J. S. Harris, and K. G. Harding, U.S. Patent 4,566,757, 28 January 1986.

95. L. H. Din, D. L. Cavan, R. B. Howe, and R. E. Graves, "A Holographic Photomask Defect Inspection System, "*SPIE Proc.* **538:**110–116 (1985).

96. D. L. Cavan, L. H. Lin, R. B. Howe, R. E. Graves, and R. L. Fusek, "Patterned Wafer Inspection Using Laser Holography and Spatial Frequency Filtering," *J. Vac. Sci. Technol.* **B6**(6) (1988).

97. E. Ochoa, J. W. Goodman, and L. Hesselink, "Real-Time Enhancement of Defects in a Periodic Mask Using Photorefractive $Bi_{12}SiO_{20}$," *Opt. Lett.* **10**(9) (1985).

98. J. Brook and R. Dandliker, "Submicrometer Holographic Photolithography," *Solid State Technology* (November 1989).

99. B. A. Omar, F. Clube, M. Hamidi, D. Struchen, and S. Gray, "Advances in Holographic Lithography," *Solid State Technology* (September 1991).

100. S. Gray and M. Hamidi, "Holographic Microlithography for Flat-Panel Displays," *SID 91 Digest* (1991).

101. K. Stetson, "Holography with Totally Internally Reflected Light," *Appl. Phys. Lett.* **11:**225 (1967).

102. B. Hill, "Holographic Memories and their Future," in N. H. Farhat (ed.), *Advances in Holography,* vol. 3, Marcel Dekker, New York, 1976, pp. 1–251.

103. S. V. Pappu, "Holographic Memories: A Critical Review," *Int. J. Optoelectronics* **5:**3 (1990).

104. G. A. Voskoboinik, I. S. Gibin, V. P. Koronkevich, E. S. Nezhevenko, P. E. Tverdokhleb, and Y. V. Ghugui, "Holographic Memory Device for Identifying Substances from Their Infrared Spectra," *Optika i Spektroskopiya* **30**(6) (1971), translated in *Optics and Spectroscopy* **30**(6) (1971).

105. I. S. Gibin, A. Gofman, S. K. Kibirev, E. F. Pen, and P. E. Tverdokhleb, "Holographic Memory Devices with Data Search Functions," *Avtometriya* **5:**37–51 (1977) translated in *Optoelectronics, Instrumentation and Data Processing* (1977).

106. E. F. Pen, P. E. Tverdokhleb, Y. N. Tishchenko, and A. V. Trubetskoi, "Acoustooptical Deflector for a Holographic Memory," *Optika i Spektroskopiya* **55**(1):148–55 (1983); translated in *Opt. Spectrosc.* **55**(1):86–90 (1983).

107. V. A. Dombrovskii, S. A. Dombrovskii, and E. F. Pen, "Investigation of Noise Stability of Holograms in a Holographic Memory," *Avtometriya* **4** (1985), translated in *Optoelectronics, Instrumentation and Data Processing* (1985).

108. A. A. Verbovetskii, A. P. Grammatin, V. N. Ivanov, V. G. Mityakov, A. A. Novikov, N. N. Rukavitsin, Y. S. Skvortsov, V. B. Fedorov, and V. V. Tsvetkov, "Holographic Memory for Archival Storage of Binary Information," *Optiko-Mekhanicheskaya Promyshlennost* **55:**5 (1988), translated in *Sov. J. Opt. Technol.* **55:**5 (1988).

109. V. A. Dombrovskii, S. A. Dombrovskii, and E. F. Pen, "Reliability of Data Readout in a Holographic Memory Channel with Constant Characteristics," *Avtometriya* **6** (1988), translated in *Optoelectronics, Instrumentation and Data Processing* **6** (1988).

110. Y. V. Vovk, L. V. Vydrin, N. N. V'yukhina, V. N. Zatolokin, P. E. Tverdokhleb, I. S. Shteinberg, and Y. A. Shchepetkin, "Fast Storage Device for Digital Data Based on a Pack of Optical Disks," *Avtometriya* **3:**82–94 (1989), translated in *Optoelectronics, Instrumentation and Data Processing* **3:**78–90 (1989).

111. Y. V. Vovk, L. V. Vydrin, P. E. Tverdokhleb, and Y. A. Shchepetkin, "Method for Multichannel Recording of Binary Data on Optical Disk," *Avtometriya* **2:**77–87 (1989), translated in *Optoelectronics, Instrumentation and Data Processing* **2:**79–89 (1989).

112. B. V. Vanyushev, N. N. V'yukhina, I. S. Gibin, A. P. Litvintseva, T. N. Mantush, B. N. Pankov, E. F. Pen, A. N. Potapov, I. B. Tatarnikova, and P. E. Tverdokhleb, "Architecture of Data System Based on Large Capacity Holographic Memory," *Avtometriya* **3:**74–82 (1989), translated in *Optoelectronics, Instrumentation and Data Processing* **3:**70–77 (1989).

113. A. A. Blok, R. S. Kucheruk, and E. F. Pen, "Diffraction Efficiency of Partially Superimposed Holograms," *Avtometriya* **3** (1989), translated in *Optoelectronics, Instrumentation and Data Processing* **3** (1989).

114. A. A. Blok, "Effect of Data Coding Methods in Holographic Memory on Characteristics of Reconstructed Images of Data Pages," *Avtometriya* **5** (1989), translated in *Optoelectronics, Instrumentation and Data Processing* **5** (1989).

115. V. A. Dombrovskii, S. A. Dombrovskii, and E. F. Pen, "Noise Immunity of Holographic Memory with Paraphrase Data Coding," *Avtometriya* **2** (1989), translated in *Optoelectronics, Instrumentation and Data Processing* **2** (1989).

116. P. E. Tverdokhleb and B. N. Pankov, "Parallel Associative VLSI Processor with Optical Input," *SPIE Proc.* **1230** (1990).

117. S. Redfield and L. Hesselink, "Data Storage in Photorefractives Revisited," *SPIE, Optical Computing 88* **963** (1988).

118. S. Redfield and L. Hesselink, "Enhanced Nondestructive Holographic Readout in Strontium Barium Niobate," *Opt. Lett.* **13**(10) (1988).

119. L. Hesselink and S. Redfield, "Photorefractive Holographic Recording in Strontium Barium Niobate Fibers," *Opt. Lett.* **13**(10) (1988).

P · A · R · T · 3

OPTICAL MEASUREMENTS

CHAPTER 24
RADIOMETRY AND PHOTOMETRY

Edward F. Zalewski
Hughes Danbury Optical Systems
Danbury, Connecticut

24.1 GLOSSARY

A	area
A_1, A_2, A_s, A_d	area of surface 1, surface 2, a source, a detector, respectively
A_r	area of an image on the retina of a human eye
A_p	area of the pupil of a human eye
$A_{\text{in}}, A_{\text{out}}, A_{\text{sph}}$	area of an input port, output port, and sphere surface, respectively
b	distance from optic axis
c	the speed of light in a vacuum
C_e	photon-to-electron conversion efficiency, i.e., quantum efficiency of a photodetector
D	diameter
dA	infinitesimal element of area
dA_1, dA_2, dA_s, dA_d	infinitesimal element of area of surface 1, surface 2, a source, a detector, respectively
dL_λ	infinitesimal change in radiance per wavelength interval
dT	infinitesimal change in temperature
$d\Phi_{12}$	infinitesimal amount of radiant power transferred from point 1 to point 2
$d\lambda$	infinitesimal wavelength interval
dv	infinitesimal frequency interval
$d\Omega$	infinitesimal change in solid angle
E	irradiance, the incident radiant power per the projected area of a surface
E_v	illuminance, the photometric equivalent of irradiance
E_r	average illuminance in an image on the retina of a human eye
E_T	retinal illuminance in units of trolands

$E_T(\lambda)$ — photopic retinal illuminance from a monochromatic source in trolands

$E_T'(\lambda)$ — scotoptic retinal illuminance from a monochromatic source in trolands

$E_{r\lambda}$ — retinal spectral irradiance in absolute units: $\mathrm{W\,nm^{-1}m^{-2}}$

f — focal length

$f\#$ — F-number

g — fraction of light lost through the input and output ports of an averaging sphere

h — Planck's constant

h_s, h_d — object (source) height, image (detector) height

I — radiant intensity, the emitted or reflected radiant power per solid angle

i — photoinduced current from a radiation detector

I_v — luminous intensity, the photometric equivalent of radiant intensity

k — Boltzmann's constant

K_m — luminous efficacy (i.e., lumen-to-watt conversion factor) for photopic vision

K_m' — luminous efficacy for scotopic vision

K_{ab} — nonlinearity correction factor for a photodetector

L — radiance, the radiant power per projected area and solid angle

L_{12} — radiance from point 1 into the direction of point 2

L_a, L_b — radiance in medium a, in medium b

L_e — radiance within the human eye

L_λ — radiance per wavelength interval

L_v — radiance per frequency interval

L_v — luminance, the photometric equivalent of radiance

$L_v(\lambda)$ — luminance of a monochromatic light source

M — exitance, the emitted or reflected radiant power per the projected area of a source

m — mean value

N — photon flux, the number of photons per second

n — index of refraction

n_a, n_b — index of refraction in medium a, in medium b

n_e — index of refraction of the ocular medium of the human eye

n_s, n_d — index of refraction in the object (i.e., source) region, in the image (i.e., detector) region

N_λ — photon flux per wavelength interval

N_v — photon flux per frequency interval

$N_{E\lambda}$ — photon flux irradiance on the retina of a human eye

Q — radiant energy

Q_λ	radiant energy per wavelength interval
Q_v	radiant energy per frequency interval
R	responsivity of a photodetector, i.e., electrical signal out per radiant signal in
r	radius
r_s, r_d, r_{sph}	radius of a source, detector, sphere, respectively
$R(\lambda)$	spectral (i.e., per wavelength interval) responsivity of a photodetector
s	distance
s_{12}	length of the light ray between points 1 and 2
s_{sd}	length of the light ray between points on the source and detector
s_{pr}	distance from the pupil to the retina in a human eye
T	absolute temperature
t	time
U	photon dose, the total number of photons
$V(\lambda)$	spectral luminous efficiency function (i.e., peak normalized human visual spectral responsivity) for photopic vision
$V'(\lambda)$	spectral luminous efficiency function for scotopic vision
w	width
x_i	the ith sample in a set of measurements
α	absorptance, fraction of light absorbed
β_a, β_b	angle of incidence or refraction
γ	absorption coefficient of a solute
δ	angle of rotation between crossed polarizers
ε	emittance of a blackbody simulator
E	étendue
η	total number of sample measurements
θ_s, θ_d	angle between the light ray and the normal to a point on the surface of a source, of a detector
θ_1, θ_2	angle between the light ray and the normal to a surface at point 1, at point 2
κ	concentration of a solute
λ	wavelength
ν	frequency
ρ	fraction of light scattered or reflected
σ	standard deviation
σ_m	standard deviation of the mean
τ	transmittance, radiant signal out per radiant signal into a material
$\tau(\lambda)$	spectral (i.e., per wavelength interval) transmittance
$\tau_e(\lambda)$	spectral transmittance of the ocular medium of the human eye

Φ radiant power or equivalently radiant flux

ϕ half angle subtended by a cone

Φ_{in}, Φ_{out} incoming radiant power, outgoing radiant power

Φ_r luminous flux at the retina of the human eye

Φ_λ radiant power per wavelength interval

Φ_v radiant power per frequency interval

Φ_v photopic luminous flux, radiant power detectable by photopic human vision

Φ_v' scotopic luminous flux, radiant power detectable by scotopic human vision

Ω solid angle, a portion of the area on the surface of a sphere per the square of the sphere radius

Ω_a, Ω_b solid angle in medium a, in medium b

24.2 INTRODUCTION

Radiometry is the measurement of the energy content of electromagnetic radiation fields and the determination of how this energy is transferred from a source, through a medium, and to a detector. The results of a radiometric measurement are usually obtained in units of power, i.e., in watts. However, the result may also be expressed as photon flux (photons per second) or in units of energy (joules) or dose (photons). The measurement of the effect of the medium on the transfer of radiation, i.e., the absorption, reflection, or scatter, is usually called *spectrophotometry* and will not be covered here. Rather, the assumption is made here that the radiant power is transferred through a lossless medium.

Traditional radiometry assumes that the propagation of the radiation field can be treated using the laws of geometrical optics. That is, the radiant energy is assumed to be transported along the direction of a ray and interference or diffraction effects can be ignored. In those situations where interference or diffraction effects are significant, the flow of energy will be in directions other than along those of the geometrical rays. In such cases, the effect of interference or diffraction can often be treated as a correction to the result obtained using geometrical optics. This assumption is equivalent to assuming that the energy flow is via an incoherent radiation field. This assumption is widely applicable since most radiation sources are to a large degree incoherent. For a completely rigorous treatment of radiant energy flow, the degree of coherence of the radiation must be considered via a formalism based on the theory of electromagnetism as derived from Maxwell's equations.[1,2] This complexity is not necessary for most of the problems encountered in radiometry.

In common practice, radiometry is divided according to regions of the spectrum in which the same experimental techniques can be used. Thus, vacuum ultraviolet radiometry, intermediate-infrared radiometry, far-infrared radiometry, and microwave radiometry are considered separate fields, and all are distinguished from radiometry in the visible and near-visible optical spectral region.

The reader should note that there is considerable confusion regarding the nomenclatures of the various radiometries. The terminology for radiometry that we have inherited is dictated not only by its historical origin,[3] but also by that of related fields of study. By the late 1700s, techniques were developed to measure light using the human eye as a null detector in comparisons of sources. At about the same time, radiant heating effects were studied with liquid-in-glass thermometers and actinic (i.e., chemical) effects of solar radiation were studied by the photoinduced decomposition of silver compounds into metallic silver. The discovery of infrared radiation in 1800 and ultraviolet radiation in 1801 stimulated a great deal of effort to study the properties of these radiations. However, the only practical detectors of ultraviolet radiation at that time were the actinic effects—for infrared radiation it was thermometers and for visible radiation it was human vision. Thus actinometry, radiometry, and photometry became synonymous with studies in the ultraviolet, infrared, and visible spectral regions. Seemingly independent fields of study evolved and even today there is confusion because the experimental methods and terminology developed for one field are often inappropriately applied to another. Vestiges of the confusion over what constitutes photometry and radiometry are to be found in many places. The problems encountered are not simply semantic, since the confusion can often lead to substantial measurement error.

As science progressed, radiometry was in the mainstream of physics for a short time at the end of the nineteenth century, contributing the absolute measurement base that led to Planck's radiation law and the discovery of the quantum nature of radiation. During this period, actinic effects, which were difficult to quantify, became part of the emerging field of photochemistry. In spite of the impossibility of performing an absolute physical measurement, it was photometry, however, that grew to dominate the terminology and technology of radiant energy measurement practice in this period. At the beginning of the nineteenth century, the reason photometry was dominant was that the most precise (not absolute) studies of radiation transfer relied on the human eye. By the end of the nineteenth century, the growth of industries such as electric lighting and photography became the economic stimulus for technological developments in radiation transfer metrology and supported the dominance of photometry. Precise photometric measurements using instrumentation in which the human eye was the detector continued into the last half of the twentieth century. The fact that among the seven internationally accepted base units of physical measurement there remains one unit related to human physiology—the candela—is an indication of the continuing economic importance of photometry.

Presently, the recommended practice is to limit the term photometry to the measurement of the ability of electromagnetic radiation to produce a visual sensation in a physically realizable manner, that is, via a defined simulation of human vision.[4–5] Radiometry, on the other hand, is used to describe the measurement of radiant energy independent of its effect on a particular detector. Actinometry is used to denote measurement of photon flux (photons per second) or dose (total number of photons) independent of the subsequent photophysical, photochemical, or photobiological process. Actinometry is a term that is not extensively used, but there are current examples where measurement of the "actinic effect of radiation" is an occasion to produce a new terminology for a specific photoprocess, such as for the Caucasian human skin reddening effect commonly known as sunburn. We do not attempt here to catalog the many different terminologies used in photometry and radiometry, instead the most generally useful definitions are introduced where appropriate.

This chapter begins with a discussion of the basic concepts of the geometry of radiation transfer and photon flux measurement. This is followed by several approximate methods for solving simple radiation transfer problems. Next is a discussion of radiometric calibrations and the methods whereby an absolute radiant power or photon flux measurement is obtained. The discussion of photometry that follows is restricted to measurements employing physical detectors rather than those involving a human observer. Because many esoteric terms are still in use to describe photometric measurements, the ones most likely to be encountered are listed and defined in the section on photometry.

It is not the intention that this chapter be a comprehensive listing or a review of the extensive literature on radiometry and photometry; only selected literature citations are made where appropriate. Rather, it is hoped that the reader will be sufficiently introduced to the conceptual basis of these fields to enable an understanding of other available material. There are many texts on general radiometry. Some of the recent books on radiometry are listed in the reference section.[6–9] In addition, the subject of radiometry or photometry is often presented as a subset of another field of study and can therefore be found in a variety of texts. Several of these texts are also listed in the reference section.[10–13] Finally, the reader will also find material related to radiometry and photometry in the chapters of this Handbook which treat colorimetry (Vol. I, Chap. 26), detectors (Vol. I, Chap. 15 and Chap. 19), and spectrophotometry (Vol. II, Chap. 25).

24.3   RADIOMETRIC DEFINITIONS AND BASIC CONCEPTS

Radiant Power and Energy

For a steadily emitting source, that is a radiation source with a continuous and stable output, radiometric measurement usually implies measurement of the power of the source. For a flashing or single-pulse source, radiometric measurement implies a measurement of the energy of the source.

Radiometric measurements are traditionally measurements of thermal power or energy. However, because of the quantum nature of most photophysical, photochemical, and photobiological effects, in many applications it is not the measurement of the thermal power in the radiation beam but measurement of the number of photons that would provide the most physically meaningful result. The fact that most radiometric measurements are in terms of watts and joules is due to the history of the field. The reader should examine the particular application to determine if a measurement in terms of photon dose or photon flux would not be more meaningful and provide insight for the interpretation of the experiment. (See section on "Actinometry" later in this chapter.)

Radiant Energy.   Radiant energy is the energy emitted, transferred, or received in the form of electromagnetic radiation.
Symbol: Q *Unit*: joule (J)

Radiant Power.   Radiant power or radiant flux is the power (energy per unit time t) emitted, transferred, or received in the form of electromagnetic radiation.
Symbol: Φ *Unit*: watt (W)

$$\Phi = \frac{dQ}{dt} \qquad (1)$$

Geometrical Concepts

The generally accepted terminology and basic definitions for describing the geometry of radiation transfer are presented below. More extensive discussions of each of these definitions and concepts can be found in the references.[4–13]

The concepts of irradiance, intensity, and radiance involve the density of the radiant power (or energy) over area, solid angle, and area times solid angle, respectively.

In situations where the density or distribution of the radiation on a surface is the required quantity, then it is the irradiance that must be measured. An example of where an irradiance measurement would be required is the exposure of a photosensitive surface such as the photoresists used in integrated circuit manufacture. The irradiance distribution over the surface determines the degree of exposure of the photoresist. A nonuniform irradiance distribution will result in overexposure and/or underexposure of regions across the piece and results in a defect in manufacture.

In an optical system where the amount of radiation transfer through the system is important, then it is the radiance that must be measured. The amount of radiation passing through the optical system is determined by the area of the source from which the radiation was emitted and the field of view of the optic, also known as the solid angle or collection angle. Radiance is often thought of as a property of a source, but the radiance at a detector is also a useful concept.

Both irradiance and radiance are defined for infinitesimal areas and solid angles. However, in practice, measurements are performed with finite area detectors and optics with finite fields of view. Therefore all measurements are in fact measurement of average irradiance and average radiance.

Irradiance and radiance must be defined over a projected area in order to account for the effect of area change with angle. This is easily seen from the observation that the amount of a viewed area diminishes as it is tilted with respect to the viewer. Specifically, the view of the area falls off as the cosine of the angle between the normal to the surface and the line of sight. This effect is sometimes called the *cosine law of emission* or the *cosine law of irradiation.*

Intensity is a term that is part of our common language and often a point of confusion in radiometry. Strictly speaking, intensity is definable only for a source that is a point. An average intensity is not a measurable quantity since the source must by definition be a point. All intensity measurements are an approximation, since a true point source is physically impossible to produce. It is an extrapolation of a series of measurements that is the approximation of the intensity. An accurate intensity measurement is one that is made at a very large distance and, consequently, with a very small signal at the detector and an unfavorable signal-to-noise ratio. Historically speaking, however, intensity is an important concept in photometry and, to a much lesser extent, it has some application in radiometry. Intensity is a property of a source, not a detector.

Irradiance. Irradiance is the ratio of the radiant power incident on an infinitesimal element of a surface to the projected area of that element, dA_d, whose normal is at an angle θ_d to the direction of the radiation.
Symbol: E Unit: watt/meter^2 (W/m^{-2})

$$E = \frac{d\Phi}{\cos \theta_d \, dA_d} \tag{2}$$

Exitance. The accepted convention makes a distinction between the irradiance, the surface density of the radiation incident on a radiation detector (denoted by the subscript d), and the exitance, the surface density of the radiation leaving the surface of a radiation source (denoted by the subscript s).

Exitance is the ratio of the radiant power leaving an infinitesimal element of a source to the projected area of that element dA_s whose normal is at an angle θ_s to the direction of the radiation.
Symbol: M Unit: watt/meter^2 (W/m^{-2})

$$M = \frac{d\Phi}{\cos \theta_s \, dA_s} \tag{3}$$

Intensity. Radiant intensity (often simply "intensity") is the ratio of the radiant power leaving a source to an element of solid angle $d\Omega$ propagated in the given direction.
Symbol: I Unit: watt/steradian (W/sr^{-1})

$$I = \frac{d\Phi}{d\Omega} \tag{4}$$

Note that in the field of physical optics, the word *intensity* refers to the magnitude of

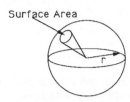

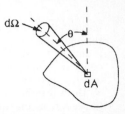

FIGURE 1 The solid angle at the center of the sphere is the surface area enclosed in the base of the cone divided by the square of the sphere radius.

FIGURE 2 The radiance at the infinitesimal area dA is the radiant flux divided by the solid angle times the projection of the area dA onto the direction of the flux.

the Poynting vector and thus more closely corresponds to irradiance in radiometric nomenclature.

Solid Angle. The solid angle is the ratio of a portion of the area on the surface of a sphere to the square of the radius r of the sphere. This is illustrated in Fig. 1.
Symbol: Ω *Unit*: steradian (sr)

$$d\Omega = \frac{dA}{r^2} \tag{5}$$

It follows from the definition that the solid angle subtended by a cone of half angle ϕ, the apex of which is at the center of the sphere, is given by

$$\Omega = 2\pi(1 - \cos\phi) = 4\pi \sin^2 \frac{\phi}{2} \tag{6}$$

Radiance. Radiance, shown in Fig. 2, is the ratio of the radiant power, at an angle θ_s to the normal of the surface element, to the infinitesimal elements of both projected area and solid angle. Radiance can be defined either at a point on the surface of either a source or a detector, or at any point on the path of a ray of radiation.
Symbol: L *Unit*: watt/steradian meter^2 (W sr^{-1} m^{-2})

$$L = \frac{d\Phi}{\cos\theta_s\, dA_s\, d\Omega} \tag{7}$$

Radiance plays a special role in radiometry because it is the propagation of the radiance that is conserved in a lossless optical system; see "Radiance Conservation Theorem, Homogeneous Medium." Radiance was often referred to as the brightness or the specific intensity, but this terminology is no longer recommended.

It should be emphasized that the above definitions are precisely accurate only for point sources and point detectors. In practice, a measurement cannot be obtained at a point or an infinitesimal area. Therefore, when the terms defined above are applied to actual measurements it is usually assumed that averages are being discussed.

Spectral Dependence of Radiometric Quantities

Polychromatic Radiation Definitions. For polychromatic radiation, the spectral distribution of radiant power (or radiant energy) is denoted as either radiant power (energy) per wavelength interval or radiant power (energy) per frequency interval.
Symbol: $\Phi_\lambda(Q_\lambda)$ *Unit*: watt/nanometer (W nm^{-1}) joule/nanometer (J nm^{-1});
or
Symbol: $\Phi_\nu(Q_\nu)$ *Unit*: watt/hertz (W Hz^{-1}) joule/hertz (J Hz^{-1})

It follows that $\Phi_\lambda \, d\lambda$ is the radiant power in the wavelength interval λ to $\lambda + d\lambda$, and $\varphi_v \, dv$ is the radiant power in the frequency interval v to $v + dv$. The total radiant power over the entire spectrum is therefore

$$\Phi = \int_0^\infty \Phi_\lambda \, d\lambda \tag{8a}$$

or

$$\Phi = \int_0^\infty \Phi_v \, dv \tag{8b}$$

Another way of stating the preceding is that monochromatic radiant power Φ_λ is the radiant power in an infinitesimal wavelength interval $d\lambda$.

If λ is the wavelength in the medium corresponding to the frequency v, and since $v = c/n\lambda$, where c is the speed of light in a vacuum and n is the index of refraction of the medium, then

$$dv = \frac{c}{n\lambda^2} d\lambda \tag{9}$$

and

$$\lambda \Phi_\lambda = v \Phi_v \tag{10}$$

Since the wavelength changes with the index of refraction of the medium, it is becoming more common to use the vacuum wavelength, $\lambda = c/v$. It is particularly important in high-accuracy applications to state explicitly whether or not the vacuum wavelength is being used.

Spectral versions of the other radiometric quantities, i.e., radiant energy, radiance, etc., are defined similarly.

Polychromatic Radiation Calculations. As an example of the application of the concept of the spectral dependence of a radiometric quantity, consider the calculation of the response of a radiometer consisting of a detector and a spectral filter. The spectral responsivity of a detector $R(\lambda)$ is the ratio of the output signal to the radiant input at each wavelength λ. The output is usually an electrical signal, such as a photocurrent i, and the input is a radiometric quantity, such as radiant power. The spectral transmittance of a filter $\tau(\lambda)$ is the ratio of the output radiant quantity to the input radiant quantity at each wavelength λ. For a spectral radiant power Φ_λ, the photocurrent i of the radiometer is

$$i = \int_0^\infty R(\lambda)\tau(\lambda)\Phi_\lambda \, d\lambda \tag{11}$$

In practice, either the responsivity of the detector or the transmittance of the filter are nonzero only within a limited spectral range. The integral need be evaluated only within the wavelength limits where the integrand is nonzero.

Photometry

The radiation transfer concepts, i.e., geometrical principles, of photometry are the same as those for radiometry. The exception is that the spectral responsivity of the detector, the human eye, is specifically defined. Photometric quantities are related to radiometric quantities via the spectral efficiency functions defined for the photopic and scotopic CIE

Standard Observer. The generally accepted values of the photopic and scotopic human eye response function are represented in the "Photometry" section in Table 2.

Luminous Flux.   The photometric equivalent of radiant power is luminous flux, and the unit that is equivalent to the watt is the lumen. Luminous flux is spectral radiant flux weighted by the appropriate eye response function. The definition of luminous flux for the photopic CIE Standard Observer is
Symbol: Φ_v *Unit*: lumen (lm)

$$\Phi_v = K_m \int \Phi_\lambda V(\lambda) \, d\lambda \tag{12}$$

where $V(\lambda)$ is the spectral luminous efficiency function and K_m is the luminous efficacy for photopic vision. The spectral luminous efficacy is defined near the maximum, $\lambda_m = 555$ nm, of the photopic efficiency function to be approximately 683 lm W^{-1}.

Definitions of the Density of Luminous Flux.

Illuminance.   Illuminance is the photometric equivalent of irradiance; that is, illuminance is the luminous flux per unit area.
Symbol: E_v *Unit*: lumen/meter^2 (lm m^{-2})

$$E_v = \frac{d\Phi_v}{\cos\theta_d \, dA_d} = \frac{d\left[K_m \int \Phi_\lambda V(\lambda) \, d\lambda\right]}{\cos\theta_d \, dA_d} \tag{13}$$

Luminous Intensity   Luminous intensity is the photometric equivalent of radiant intensity. Luminous intensity is the luminous flux per solid angle. For historical reasons, the unit of luminous intensity, the candela—not the lumen—is defined as the base unit for photometry. However, the units for luminous intensity can either be presented as candelas or lumens/steradian.
Symbol: I_v *Unit*: candela or lumen/steradian (cd or lm sr^{-1})

$$I_v = \frac{d\varphi_v}{d\Omega} = \frac{d\left[K_m \int \Phi_\lambda V(\lambda) \, d\lambda\right]}{d\Omega} \tag{14}$$

Luminance.   Luminance is the photometric equivalent of radiance. Luminance is the luminous flux per unit area per unit solid angle.
Symbol: L_v *Unit*: candela/meter^2 (cd m^{-2})

$$L_v = \frac{d\Phi_v}{\cos\theta_s \, dA_s \, d\Omega} = \frac{d\left[K_m \int \Phi_\lambda V(\lambda) \, d\lambda\right]}{\cos\theta_s \, dA_s \, d\Omega} \tag{15}$$

Actinometry

Radiant Flux to Photon Flux Conversion.   Actinometric measurement practice closely follows that of general radiometry except that the quantum nature of light rather than its thermal effect is emphasized. In actinometry, the amount of electromagnetic radiation being transferred is measured in units of photons per second (photon flux). The energy of a single photon is

$$Q = h\nu \tag{16}$$

where v is the frequency of the radiation and h is Planck's constant, 6.6262×10^{-34} J s. For monochromatic radiant power Φ_λ, measured as watts and wavelength λ, measured as nanometers, the number of photons per second N_λ in the monochromatic radiant beam is

$$N_\lambda = 5.0341 \times 10^{15} n\lambda\Phi_\lambda \tag{17}$$

Photon Dose and the Einstein. Dose is the total number of photons impinging on a sample. For a monochromatic beam of radiant power Φ_λ that irradiates a sample for a time t seconds, the dose U measured as Einsteins is

$$U = 8.3593 \times 10^{-9} n\lambda\Phi_\lambda t \tag{18}$$

The Einstein is a unit of energy used in photochemistry. An Einstein is the amount of energy in one mole (Avogadro's number, 6.0222×10^{23}) of photons.

Conversions Between Radiometry, Photometry, and Actinometry

Conversions between radiometric, photometric, and actinometric units is not simply one of determining the correct multiplicative constant to apply. As seen previously, the conversion between radiant power and photon flux requires that the spectral character of the radiation be known. It was also shown that, for radiometric to photometric conversions, the spectral distribution of the radiation must be known. Furthermore, there is an added complication for photometry where one must also specify the radiant power level in order to determine which CIE Standard Observer function is appropriate. Table 1, which summarizes the spectral radiation transfer terminology, may be helpful to guide the reader in determining the relationship between radiometric, photometric, and actinometric concepts. In Table 1, the power level is asusmed to be high enough to restrict the photometric measurements to the range of the photopic eye response function.

Basic Concepts of Radiant Power Transfer

Radiance Conservation Theorem, Homogeneous Medium. In a lossless, homogeneous isotropic medium, for a perfect optical system (i.e., having no aberrations) and ignoring interference and diffraction effects, the radiance is conserved along a ray through the optical system. In other words, the spectral radiance at the image always equals the spectral radiance at the source.

TABLE 1   Radiation Transfer Terminology, Spectral Relationships

	Radiometric	Photometric	Actinometric
Base quantity:	Radiant power (also radiant flux)	Luminous flux	Photon flux
Units:	Watts/nanometer	Lumens	Photons/second
Conversion:	—	$[W/nm] K_m V(\lambda)$	$[W/nm] \lambda (hc)^{-1}$
Surface density:	Irradiance	Illuminance	Photon flux irradiance
Solid angle density:	Radiant intensity	Luminous intensity	Photon flux intensity
Solid angle and surface density:	Radiance	Luminance	Photon flux radiance

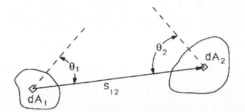

FIGURE 3 The radiant flux transferred between the infinitesimal areas dA_1 to dA_2.

It follows from Eq. (7), the definition of radiance, that for a surface A_1 with radiance L_{12} in the direction of a second surface A_2 with radiance L_{21} in the direction to a first surface, and joined by a light ray of length s_{12}, the net radiant power exchange between elemental areas on each surface is given by

$$\Delta\Phi = d\Phi_{12} - d\Phi_{21} = \frac{(L_{12} - L_{21}) \cos\theta_1 \cos\theta_2 \, dA_1 \, dA_2}{s_{12}^2} \tag{19}$$

where θ_1 and θ_2 are the angles between the ray s_{12} and the normals to the surfaces A_1 and A_2, respectively. The transfer of radiant power and the terminology used in this discussion is depicted in Fig. 3.

The total amount of radiation transferred between the two surfaces is given by the integral over both areas as follows:

$$\Phi = \int\int \frac{(L_{12} - L_{21}) \cos\theta_1 \cos\theta_2}{s_{12}^2} dA_1 \, dA_2 \tag{20}$$

This is the generalized radiant power transfer equation for net exchange between two sources. In the specialized case of a source and receiver, the radiant power emitted by a receiver is zero by definition. In this case, the term L_{21} in Eq. (20) is zero.

Refractive Index Changes. In the case of a boundary between two homogeneous isotropic media having indices of refraction n_a and n_b, the angles of incidence and refraction at the interface β_a and β_b are related by Snell's law. If the direction of the light ray is oblique to the boundary between n_a and n_b, the solid angle change at the boundary will be

$$d\Omega_a = \frac{n_b^2 \cos\beta_b}{n_a^2 \cos\beta_a} d\Omega_b \tag{21}$$

Therefore the radiance change across the boundary will be

$$\frac{L_a}{n_a^2} = \frac{L_b}{n_b^2} \tag{22}$$

This result is obtained directly by substituting the optical path for the distance in Eq. (19) and considering that the radiance transferred across the boundary between the two media is unchanged. Optical path is the distance within the medium times the index of refraction of the medium.

In the case of an optical system having two or more indices of refraction, the radiance conservation theorem is more precisely stated as: In a lossless, homogeneous isotropic medium, for a perfect optical system (i.e., having no aberrations) and ignoring interference and diffraction effects, at a boundary between two media having different indices of refraction the radiance divided by the square of the refractive index is conserved along a ray through the optical system.

Radiative Transfer Through Absorbing Media. For radiation transmitted through an absorbing and/or scattering medium, the radiance is not conserved. This is not only because of the loss due to the absorption and/or scattering but the medium could also emit radiation. The emitted light will be due to thermal emission (see the discussion on blackbody radiation later in this chapter). In some cases, the medium may also be fluorescent. Fluorescence is the absorption of radiant energy at one wavelength with subsequent emission at a different wavelength.

Historically, the study of radiative transfer through absorbing and/or scattering media dealt with the properties of stellar atmospheres. Presently, there is considerable interest in radiative transfer measurements of the earth and its atmosphere using instruments on board satellites or aircraft. An accurate measure of the amount of reflected sunlight (approximately 400 to 2500 nm) or the thermally emitted infrared (wavelengths >2500 nm) requires correction for the absorption, scattering, and, in the infrared, the emission of radiation by the atmosphere. This specialized topic will not be considered here. Detailed discussion is available in the references.[14–16]

24.4 RADIANT TRANSFER APPROXIMATIONS

The solution to the generalized radiant power transfer equation is typically quite complex. However, there are several useful approximations that in some instances can be employed to obtain an estimate of the solution of Eq. (20). We shall consider the simpler case of a source and a detector rather then the net radiant power exchange between two sources, since this is the situation commonly encountered in an optical system. In this case, Eq. (20) becomes

$$\Phi = \int \int \frac{L \cos \theta_s \cos \theta_d}{s_{sd}^2} dA_s \, dA_d \tag{23}$$

where the subscripts s and d denote the source and detector, respectively. Here it is assumed the detector behaves as if it were a simple aperture. That is, it responds equally to radiation at any point across its surface and from any direction. Such a detector is often referred to as a cosine corrected detector. Of course, deviations from ideal detection behavior within the spatial and angular range of the calculation reduces the accuracy of the calculation.

Point-to-point Approximation: Inverse Square Law

The simplest approximations are obtained by assuming radiant flux transfer between a point source emitting uniformly in all directions and a point detector. The inverse square law is an approximation that follows directly from the definitions of intensity, solid angle, and irradiance, Eqs. (2), (4), and (5), respectively. The irradiance (at an infinitesimal area whose normal is along the direction of the light ray) times the square of the distance from a point source equals the intensity of the source

$$I = \frac{\Phi}{A} s^2 = E s^2 \tag{24}$$

The relationship between the uniformly emitted radiance and the intensity of a point source is obtained similarly from Eqs. (4) and (7):

$$L = \frac{I}{A_s} \tag{25}$$

These point-to-point relationships are perhaps most important as a test of the accuracy of a radiation transfer calculation at the limit as the areas approach zero.

Lambertian Approximation: Uniformly Radiant Areas

Lambertian Sources.   A very useful concept for the approximation of radiant power transfer is that of a source having a radiance that is uniform across its surface and uniformly emits in all directions from its surface. Such a uniform source is commonly referred to as a lambertian source.

For the case of a lambertian source, Eq. (23) becomes

$$\Phi = L \int \int \frac{\cos \theta_s \cos \theta_d}{s_{sd}^2} dA_s \, dA_d \tag{26}$$

Configuration Factor.   The double integral in Eq. (26) has been given a number of different names: configuration factor, radiation interaction factor, and projected solid angle. There is no generally accepted terminology for this concept, although configuration factor appears most frequently. Analytical solutions to the double integral have been found for a variety of different shapes of source and receiver. Tabulations of these exact solutions to the integral in Eq. (26) are usually found in texts on thermal engineering,[17–18] under the heading of radiant heat transfer or configuration factor.

Radiation transfer between complex shapes can often be determined by using various combinations of configuration factors. This technique is often referred to as configuration factor algebra.[17] The surfaces are treated as pieces, each with a calculable configuration factor, and the separate configuration factors are combined to obtain the effective configuration factor for the complete surface.

Étendue.   The double integral in Eq. (26) is often used as a means to characterize the flux-transmitting capability of an optical system in a way that is taken to be independent of the radiant properties of the source. Here the double integral is written as being over area and solid angle:

$$\Phi = L \int \int \cos \theta_d \, dA_s \, d\Omega \tag{27}$$

In this case, the surface of the lambertian source is assumed perpendicular to the optic axis and to lie in the entrance window of the optical system. The solid angle is measured from a point on the source to the entrance pupil. The étendue E of an optical system of refractive index n is defined as

$$E = n^2 \int \int \cos \theta_d \, dA_s \, d\Omega \tag{28}$$

Equation (28) is sometimes referred to as the throughput of an optical system.

Total Flux into a Hemisphere.   The total amount of radiation emitted from a lambertian source of area dA_s into the hemisphere centered at dA_s (or received by a hemispherical, uniform detector centered at dA_s) is obtained from integrating Eq. (26) over the area A_d. Note that the ray s_{sd} is everywhere normal to the surface of the hemisphere; i.e., $\cos \theta_d = 1$.

$$\Phi = L\pi \int dA_s \tag{29}$$

Using Eq. (3), the definition of the exitance, the radiance at each point on the surface of the source is

$$L = \frac{M}{\pi} \tag{30}$$

Because of the relationship expressed in Eq. (30), Eq. (26) is often written in terms of the exitance.

$$\Phi = M\pi \int \int \frac{\cos \theta_s \cos \theta_d}{s_{sd}^2} dA_s \, dA_d \tag{31}$$

In this case, the factor π is considered to be part of the configuration factor. Note again that there is no generally accepted definition of the configuration factor.

Radiation Transfer Between a Circular Source and Detector. The particular case of radiation transfer between circular apertures, the centers of which are located along the same optical axis as shown in Fig. 4, is a configuration common to many optical systems and is therefore illustrated here. The radius of the source (or first aperture) is r_s, the detector (second aperture) radius is r_d, and the distance between the centers is s_{sd}. The exact solution of the integral in Eq. (26) yields

$$\Phi = \frac{2L(\pi r_s r_d)^2}{r_s^2 + r_d^2 + s_{sd}^2 + [(r_s^2 + r_d^2 + s_{sd}^2)^2 - 4r_s^2 r_d^2]^{1/2}} \tag{32}$$

This result can be approximated for the case where the sum of the squares of the distance and radii is large compared to the product of the radii, that is, $(r_s^2 + r_d^2 + s_{sd}^2) \gg 2r_s r_d$, so that Eq. (32) reduces to

$$\Phi \cong \frac{L(\pi r_s r_d)^2}{r_s^2 + r_d^2 + s_{sd}^2} \tag{33}$$

From this expression the irradiance at the detector can be obtained

$$E = \frac{\Phi}{A_d} \cong \frac{LA_s}{r_s^2 + r_d^2 + s_{sd}^2} \cong \frac{LA_s}{s_{sd}^2} \tag{34}$$

where A_s is the area of the lambertian disk and A_d is the detector area. The approximation at the extreme right is obtained by assuming that the radii are completely negligible with respect to the distance. This is the same result that would be obtained from a point-to-point approximation.

Off-axis Irradiance: Cosine-to-the-Fourth Approximation. Equation (34) describes the irradiance from a small lambertian disk to a detector on the ray axis and where both surfaces are perpendicular to the ray. If the detector is moved off-axis by a distance b as depicted in Fig. 5, the ray from A_s to A_d will then be at an angle with respect to the normal at both surfaces as follows

$$\theta_s = \theta_d = \theta = \tan^{-1}\left(\frac{b}{s_{sd}}\right) \tag{35}$$

The projected areas are then $(A_s \cos \theta)$ and $(A_d \cos \theta)$. In addition, the distance from the source to the detector increases by the factor $(1/\cos \theta)$. The radiant power at a distance b away from the axis therefore decreases by the fourth power of the cosine of the angle formed between the normal to the surface and the ray.

$$\Phi \cong \frac{LA_s A_d}{s_{sd}^2} \cos^4 \theta \tag{36}$$

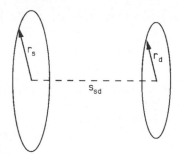

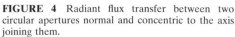

FIGURE 4 Radiant flux transfer between two circular apertures normal and concentric to the axis joining them.

FIGURE 5 illustration of the cosine-fourth effect on irradiance, displacement of the receiving surface by a distance b.

Since the radiance is conserved for propagation in a lossless optical system, Eq. (36) also approximates the radiant power from an off-axis region of a large lambertian source received at a small detector. The approximate total radiant power received at the detector would then be the sum of the radiant power contributed by each region of the source.

Spherical Lambertian Source.   In order to compute the radiant power at a point at a distance s_{sd} from the center of a spherical lambertian source of radius r_{sph}, it is not necessary to explicitly solve the integrals in the radiation transfer equation. The solution is readily obtained from the symmetry of the lambertian sphere. Using the relationship between the exitance and radiance of a lambertian source [Eq. (30)], the total radiation power emitted by the source is obtained from the product of the surface area of the source times the exitance.

$$\Phi = 4\pi^2 r_{sph}^2 L \tag{37}$$

The radiant power is isotropically emitted. Therefore, the irradiance at any point on an enclosing sphere of radius s_{sd} is the total radiant power divided by the area of the enclosing sphere.

$$E = \frac{\pi r_{sph}^2 L}{s_{sd}^2} \tag{38}$$

Note that the irradiance from a spherical lambertian source follows the inverse square law at all distances from the surface of the sphere. The intensity of a spherical lambertian source is

$$I = \pi r_{sph}^2 L \tag{39}$$

Radiant Flux Transfer Through a Lambertian Reflecting Sphere.   A lambertian reflector is a surface that uniformly scatters a fraction ρ of the radiation incident upon it.

$$L = \frac{\rho E}{\pi} \tag{40}$$

where E is the irradiance.

A spherical enclosure whose interior is coated with a material that approximates a lambertian reflector is a widely used tool in radiometry and photometry.[19] Such spheres are used either for averaging a nonuniform radiant power distribution (averaging sphere) or for measuring the total amount of radiant power emitted from a source (integrating sphere).

The sphere has the useful property whereby the solid angle subtended by any one section of the wall times the projected area is constant over all other points on the inside surface of the sphere. Therefore, if radiation falling on any point within the sphere is uniformly reflected, the reflected radiation will be uniformly distributed, i.e., produce uniform irradiance, throughout the interior. This result follows directly from the symmetry of the sphere.

Consider a sphere of radius r_{sph} and the radiant power transfer between two points on the inner surface. The normals to the two points are radii of the sphere and form an isosceles triangle when taken with the ray joining the points. Therefore, the angles between the ray and the normals to each point are equal. From Eq. (26)

$$\Phi = L \int \int \frac{\cos^2 \theta}{s_{sd}^2} \, dA_s \, dA_d \tag{41}$$

The length of the ray joining the points is $2r_{\text{sph}} \cos \theta$. The irradiance is therefore

$$E = \frac{\Phi}{A_d} = \frac{LA_s}{4r_{\text{sph}}^2} \tag{42}$$

which is independent of the angle θ. If Φ_{in} is the radiant power entering the sphere, the irradiance at any point on the sphere after a single reflection will be

$$E = \frac{\rho \Phi_{\text{in}}}{4\pi r_{\text{sph}}^2} \tag{43}$$

A fraction ρ of the flux will be reflected and again uniformly distributed over the sphere. After multiple reflections the irradiance at any point on the wall of the sphere is

$$E = \frac{(\rho + \rho^2 + \rho^3 + \cdots)\Phi_{\text{in}}}{4\pi r_{\text{sph}}^2} = \frac{\rho \Phi_{\text{in}}}{(1 - \rho)A_{\text{sph}}} \tag{44}$$

where A_{sph} is the surface area of the sphere. The flux Φ_{out} exiting the sphere through a port of area A_{out} is

$$\Phi_{\text{out}} = \frac{\rho \Phi_{\text{in}} A_{\text{out}}}{(1 - \rho)A_{\text{sph}}} \tag{45}$$

In Eq. (45) it is assumed that the loss of radiation at the entrance and exit ports is negligible and does not affect the symmetry of the radiation distribution.

The effect of the radiation lost through the entrance and exit ports is approximated as follows. After the first reflection, the fraction of radiation lost in each subsequent reflection is equal to the combined areas of the ports divided by the sphere area.

$$g = \frac{A_{\text{in}} + A_{\text{out}}}{A_{\text{sph}}} \tag{46}$$

Using this in Eq. (44) yields

$$\Phi_{\text{out}} = \frac{\rho \Phi_{\text{in}} A_{\text{out}}}{(1 - \rho g)A_{\text{sph}}} \tag{47}$$

Since the sphere is approximately a lambertian source, the radiance at the exit port is

$$L = \frac{\rho \Phi_{\text{in}}}{(1 - \rho g)\pi A_{\text{sph}}} \tag{48}$$

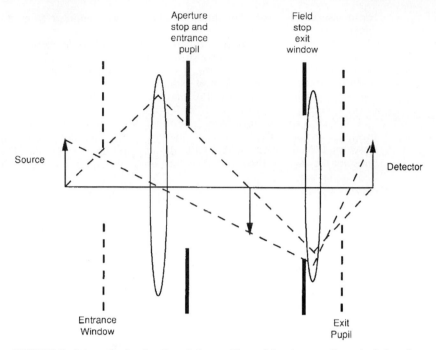

FIGURE 6   Schematic showing the relative positions of the stops, pupils, and windows in a simple optical system.

Radiometric Effect of Stops and Vignetting

Refer to Fig. 6 for an illustration of these definitions. The *aperture stop* of an optical system is an aperture near the entrance to the optical system that determines the size of the bundle of rays leaving the source that can enter the optical system.

The *field stop* is an aperture within the optical system that determines the maximum angle of the rays that pass through the aperture stop that can reach the detector. The position and area of the field stop determines the field of view of the optical system. The field stop limits the extent of the source that is represented in its image at the detector.

The image of the aperture stop in object space, i.e., in the region of the source, is the *entrance pupil*. The image of the aperture stop in image space, i.e., in the region of the detector, is the *exit pupil*. Light rays that pass through the center of the aperture stop also pass through the centers of the images of the aperture stop at the entrance and exit pupils. Since all of the light entering the optical system must pass through the aperture stop, all of the light reaching the detector appears to pass through the exit pupil.

The field stop defines the solid angle within the optical system, the system field of view. When viewed from the image, the field stop of an optical system takes on the radiance of the object being imaged. This is a useful radiometric concept since a complex optical system can often be approximated as an exit pupil having the same radiance as the object being imaged (modified by the system transmission losses). The direction in which the radiation in the image appears to be emitted is, of course, limited by the aperture at the field stop. A word of caution: if the object is small, its image will be limited by diffraction effects and its radiance will depart even further from the extended source (large area) approximations used here.

The *entrance window* is the image of the field stop at the source and the *exit window* is the image of the field stop at the detector. If the field stop coincides with the detector, i.e., the detector is in the image plane of the optical system, then the entrance window will correspond with the object plane on the source. If the field stop does not coincide with the image plane at the detector, then because of parallax, different portions of the source will be visible from different points within the exit pupil. This condition, known as *vignetting*, causes a decrease in the irradiance at the off-axis points on the detector or image plane.

Approximate Radiance at an Image

Aplanatic Optical Systems. Except for rays that lie on the optic axis, the radiance of an image must be based on a knowledge of the image quality since any aberrations introduced by the optical system divert some of the off-axis rays away from the image.

Consider a well-corrected optical system that is assumed to be aplanatic for the source and image points. That is, the optical system obeys Abbe's sine condition which is

$$n_s h_s \sin \theta_s = n_d h_d \sin \theta_d \tag{49}$$

where n_s and n_d are the refractive indices of the object (source) and image (detector) spaces, h_s and h_d are the object and image heights, and θ_s and θ_d are the angles between the off-axis rays and the optic axis in object and image space. From Eq. (27) the flux radiated by a small lambertian source of area A_s into the solid angle of the optical system is

$$\Phi \cong 2\pi L A_s \int_0^{\theta_s} \cos \theta \sin \theta \, d\theta = \pi L A_s \sin^2 \theta_s \tag{50}$$

The differential of the solid angle is obtained from Eq. (6). Since Φ is the radiant flux at the image and A_d is the area of the image, the irradiance at the image is

$$E = \frac{\Phi}{A_d} = \frac{\pi L A_s}{A_d} \sin^2 \theta_s \tag{51}$$

If h_s and h_d are the radii of circular elements A_s and A_d, then according to Abbe's sine condition

$$\frac{A_s}{A_d} = \frac{n_d^2 \sin^2 \theta_d}{n_s^2 \sin^2 \theta_s} \tag{52}$$

The irradiance at the image is

$$E = \frac{\pi L n_d^2}{n_s^2} \sin^2 \theta_d \tag{53}$$

Numerical Aperture and F-number. The quantity $n_d \sin \theta_d$ in Eq. (53) is called the *numerical aperture* of the imaging system. The irradiance of the image is proportional to the square of the numerical aperture. Geometrically speaking, the image irradiance increases with the angle of the cone of light converging on the image.

Another approximate measure of the image irradiance of an optical system is the F-number, $f\#$ (sometimes called the focal ratio) defined by the ratio of focal length (in the image space) f to the diameter D of the entrance pupil. For a source at a very large distance

$$f\# = \frac{f}{D} = \frac{1}{2 \tan \theta_d} \cong \frac{1}{2 \sin \theta_d} \tag{54}$$

The approximate image irradiance expressed in terms of the F-number is

$$E \cong \pi L \left(\frac{n_d}{2 n_s f\#} \right)^2 \tag{55}$$

24.5 ABSOLUTE MEASUREMENTS

An absolute measurement, often referred to as an absolute calibration, is a measurement that is based upon, i.e., derived from, one of the internationally recognized units of physical measurement. These units are known as the SI units (Système International d'Unités[20]). The absolute SI base units are the meter, second, kilogram, kelvin, ampere, candela, and mole. The definitions of the SI units, the methods for their realization, or their physical embodiment are a matter of international agreement under the terms of the 1875 Treaty of the Meter. A convenient method (but often not a sufficient condition) for achieving absolute accuracy is to obtain traceability to one of the SI units via a calibration transfer standard issued by one of the national standards laboratories. The United States standards laboratory is the National Institute of Standards and Technology (NIST, formerly the National Bureau of Standards).

A relative measurement is one that is not required to be traceable to one of the SI units. Relative measurements are usually obtained as the ratio of two measurements. An example of a relative measurement is the determination of the transmittance of an optical material wherein the ratio of the output radiant power to the input radiant power is measured; the measurement result is independent of SI units.

Absolute Accuracy and Traceability

Establishment of legal traceability to an SI unit requires that one obtain legally correct documentation, i.e., certification, of the device that serves as the calibration transfer standard and sometimes of the particular measurement process in which the device is to be used. Certification of legal traceability within each nation is obtained from the national standards laboratory of that nation. Often another nation's standards laboratory can be used to establish legal traceability, provided that there exists the legal framework for mutual recognition of the legality of each other's standards.

In order to establish accurate traceability to an SI unit, one needs to determine the total accumulated error arising from: (1) the realization of the base SI unit; (2) if applicable, the derivation of an associated measurement quantity; (3) if applicable, scaling to a higher or lower value; and $(4, 5 \cdots)$ transfer of the calibration from one device to another. The last entries must include the instability of the calibration transfer devices; the others may or may not involve a transfer device.

Legal traceability to SI units does not guarantee accurate traceability and vice versa. In order to obtain accurate traceability, it is not necessary to prove traceability to a national standards laboratory. Instead, the measurement must trace back to one of the SI units. However, it is usually convenient to establish accurate traceability via one of the national standards laboratories. The degree of convenience and accuracy will depend upon the accuracy of the measurement method and type of calibration transfer device available from the particular national standards laboratory.

As an example of an absolute radiometric measurement chain consider the following method for determining absolute spectral irradiance (based on the method used by NIST[21]). The steps that lead to the calibration of a spectral irradiance meter are: (1) determination of the absolute temperature of a simulated planckian radiator (i.e., a blackbody; see later discussion); (2) derivation of the spectral radiance output of the blackbody simulator (by evaluation of the deviations from the ideal radiator, i.e., deviations from Planck's law); (3) scaling up to the radiance level produced by an incandescent lamp; (4) transfer of spectral radiance from the blackbody simulator to the incandescent lamp; (5) derivation of spectral irradiance from radiance via the measurement of solid angle; (6) transfer of spectral irradiance to another type of incandescent lamp; (7) spectral irradiance transfer from primary standard lamps to secondary standard lamps; $(8, 9 \cdots x\text{-}1)$ other lamp-to-lamp transfers (to get to the user laboratory); and (x) calibration of the response of the spectral irradiance meter via the final calibration transfer lamp. The absolute accuracy of the spectral irradiance meter is established by evaluating and summing the errors accumulated in all the measurements listed. Measurements 1 through 7 are performed at NIST and NIST estimates the accuracy of their traceability to the SI temperature unit. Note that it is up to the user to determine the accuracy of the remaining x-7 measurements. The major uncertainty is typically the instability (noise and drift) of the incandescent lamps used at several steps. Lamp drift is often not quantifiable.

Other methods for the determination of absolute spectral irradiance exist,[22–24] and some are employed by other national standards laboratories. Each method has a unique set of errors associated with it. A comparison of the absolute measurements obtained via independent methods is the best means of verifying absolute accuracy. If the comparison shows that the two calibrations agree to within their combined uncertainty, then the hypothesis that the calibration and the true value are the same within the uncertainty estimation is strengthened. Comparison of independent methods of calibration provides the best available verification of absolute accuracy. However, a note of caution is in order because compensating mistakes can occur and, even in the case of good agreement between independent methods, there is no 100 percent guarantee of accurate traceability.

Relative measurements do not require traceability to one of the SI base units. Since a relative measurement is the ratio of two measurements, accuracy is assured by the linearity of the measuring instrument (or by a precise knowledge of its nonlinearity function) and by the elimination of differences (or a knowledge of the effect of such differences) in the two measurements being ratioed. For relative measurements the best available assurance of accuracy is also obtained by comparison to a measurement obtained by an independent method.

Types of Errors, Uncertainty Estimates, and Error Propagation

It is almost pointless to state a value for an absolute or relative measurement without an estimate of the uncertainty and the degree of confidence to be placed in the uncertainty estimate. Verification of the accuracy and the confidence limits is not only desirable but is often a legal requirement.

The accuracy and the uncertainty of a measurement are synonymous. The usual terminology is that a measurement is "accurate to within $\pm x$" or "uncertain to within $\pm x$", where x is either a fraction (percent) of the measured value or an interval within which the true value is known to within some degree of confidence. The degree of confidence in the uncertainty estimate is the confidence interval or σ-level.

Errors are classified as type A errors, also known as random errors, and type B, or systematic errors. Type A errors are the variations due to the effects of uncontrolled variables. The magnitude of these effects is usually small and successive measurements form a random sequence. Type B errors are not detectable as variations since they do not change for successive measurements with a given apparatus and measurement method. Type B errors arise because of differences between the ideal behavior embodied in fundamental laws of physics and real behavior embodied in an experimental simulation of the ideal. A type B error could also be a function of the quantity being measured; for example, in a blackbody radiance standard using the freezing point of a metal and its defined temperature instead of the true absolute temperature.

Type A errors are estimated using standard statistical methods. If the distribution of the measurements is known (e.g., either Gaussian, which is often called a normal distribution, or Poisson), then one uses the formalism appropriate to the distribution. Unless enough data is obtained to establish that the distribution is not Guassian, it is usual to assume a gaussian distribution. A brief discussion of Gaussian statistical concepts and terminology is given here to guide the reader in interpreting or determining the uncertainty in a radiometric or photometric calibration. One of the many texts on statistics,[25–26] or experimental methods,[27] should be consulted for a more thorough discussion.

The mean value m, the standard deviation σ, and the standard deviation of the mean σ_m, of a set of measurements x_i, are estimated for a small sample from a gaussian distribution of measurements as follows:

$$m = \sum_i \frac{x_i}{\eta} \qquad (56)$$

$$\sigma^2 = \frac{\sum (x_i - m)^2}{\eta - 1} \qquad (57)$$

$$\sigma_m = \frac{\sigma}{\sqrt{\eta}} \qquad (58)$$

where $i = 1$ to η, and η is the total number of measurements.

The standard deviation is an estimate of the spread of the individual measurements within a sample, and it approaches a constant value as η is increased.

The standard deviation of the mean is an estimate of the spread of the values of the mean that would be obtained from several different sets of sample measurements. The standard deviation of the mean decreases as the number of samples in a set increases, since the estimate of the mean approaches the true mean for an infinite data set. The standard deviation of the mean is used in the estimate of the confidence interval assigned to the reported value of the mean.

The degree of confidence to which a reported value of the mean is valid is known as the *confidence interval* (CI). If it is assumed that a very large set of measurements has been sampled, then the CI is often given in terms of the number of standard deviations of the mean (one-σ level, two-σ level, etc.) within which the type A error of a reported value is known.

The CI is the probability that the mean from a normal distribution will be within the estimated uncertainty. That is, for a z-percent confidence interval, z-percent of the measurements will fall inside and $(100 - z)$ percent will fall outside of the uncertainty estimate. For small measurement samples from a gaussian distribution, Student's t-distribution is used to estimate the CI. Tables of Student's t-distribution along with discussions concerning its use are presented in most statistics textbooks. For large sets of measurements, a one-σ level corresponds approximately to a 68-percent CI, a two-σ level to a 95-percent CI, and a three-σ level to a 99.7-percent CI.

The reader is cautioned about using the σ level designation to describe the CI for a small sample of measurements. As an example of the small versus large sample difference, consider two data sets, one consisting of three samples and the other ten. Using a Student's t-distribution to estimate the CI for the three sample set, the one-σ, two-σ, and three-σ levels correspond to CIs of 61 percent, 86 percent, and 94 percent, respectively. For the ten-sample set, the respective CIs are 66 percent, 93 percent, and 99 percent. It can be quite misleading to state only the σ level of the uncertainty estimate without an indication of the size of the measurement set from which it was drawn. In order to avoid misleading accuracy statements, it is recommended that, instead of simply reporting the σ level, either the estimated CI be reported or the standard deviation of the mean be reported along with the number of measurement samples obtained.

Type B error estimates are either educated guesses of the magnitude of the difference between the real and the ideal or they are the result of an auxiliary measurement. If an appropriate auxiliary experiment can be devised to measure a systematic or type B error, then it need no longer be considered an error. The result obtained from the auxiliary measurement can usually be used as a correction factor. If a correction factor is applied, then the uncertainty is reduced to the uncertainty associated with the auxiliary experiment.

Most of the effort in high-accuracy radiometry and photometry is devoted to reducing type B errors. The first rule for reducing type B errors is to insure that the experiment closely simulates the ideal. The second rule is that the differences between real and ideal should be investigated and that a correction be applied. Unlike type A errors for which an objective theory exists, the educated guess for a type B error is often subjective. For type B errors, neither a confidence interval nor a σ level is objectively quantifiable.

Error propagation, error accumulation, or a combined uncertainty analysis is the summation of all the type A and type B uncertainties that contribute to the final measurement in the chain. Because type A errors are truly random, they are uncorrelated and the accumulated type A error is obtained from the square-root of the sum of the squares (also known as root-sum-square, RSS) of the several type A error estimates. Type B uncertainties, however, may be either correlated or uncorrelated. If they are uncorrelated, the total uncertainty is the RSS of the several estimates. Type B uncertainties that are correlated must be arithmetically summed in a way that accounts for their correlation. Therefore, it is usually desirable to partition type B uncertainties so that they are uncorrelated.

Absolute Sources

Planckian or Blackbody Radiator. A blackbody, or planckian, radiator is a thermal radiation source with a predictable absolute radiance output. An ideal blackbody is a uniform, i.e., lambertian, source of radiant power having a predictable distribution over area, solid angle, and wavelength. It is used as a standard radiance source from which the other radiometric quantities, e.g., irradiance, intensity, etc., can be derived.

Blackbody simulators are in widespread use not only at national standards laboratories but also in many other industrial, academic, and government laboratories. Blackbody simulators are commercially available from a number of manufacturers and cover a wide range of temperatures and levels of accuracy. Because they are in such widespread use as absolute standard sources for a variety of radiometric applications, particularly in the infrared, they are discussed here in some detail. Furthermore, since many practical sources of radiation can be approximated as a thermal radiation source, a blackbody function is often used in developing the radiometric model of an optical system.

An ideal blackbody is a completely enclosed volume containing a radiation field which is in thermal equilibrium with the isothermal walls of the enclosure that is at a known absolute temperature. The radiation in equilibrium with the walls does not depend upon the shape or constitution of the walls provided that the cavity dimensions are much larger than the wavelengths involved in the spectrum of the radiation.

Since the radiometric properties of a blackbody source are completely determined by its temperature, the SI base unit traceability for blackbody-based radiometry is to the kelvin.

Since the radiation field and the walls are in equilibrium, the energy in the radiation field is determined by the temperature of the walls. The relationship between the absolute temperature T and the spectral radiance L_λ is given by Planck's law:

$$L_\lambda = \frac{2hc^2}{n^2\lambda^5}[e^{(hc/n\lambda kT)} - 1]^{-1} \tag{59}$$

Here h is Planck's constant, c is the speed of light in a vacuum, k is Boltzmann's constant, λ is the wavelength, and n is the index of refraction of the medium. Incorporating the values of the constants in this equation yields,

Spectral radiance units: $W\,m^{-2}\,sr^{-1}\,\mu m^{-1}$

$$L_\lambda = \frac{1.1910 \times 10^8}{n^2 \lambda^5} [e^{(1.4388 \times 10^4 / n\lambda T)} - 1]^{-1} \tag{60}$$

It follows that the peak of the spectrum of a blackbody is determined by its temperature (Wein displacement law).

$$n\lambda_{max} T = 2898 \; \mu m\,K \tag{61}$$

It is often useful to measure blackbody spectral radiance in units of photons per second N_λ. The form of Planck's law in this case is
Spectral radiance units: $photons\,s^{-1}\,m^{-2}\,sr^{-1}\,\mu m^{-1}$

$$N_\lambda = \frac{2c}{n\lambda^4} [e^{(hc/n\lambda kT)} - 1]^{-1} \tag{62}$$

The peak of this curve is not at the same wavelength as in the case of radiance measured in units of power. Wein's displacement law for blackbody radiance measured in photons per second is

$$n\lambda_{max} T = 3670 \; \mu m\,K \tag{63}$$

In other applications, the spectral distribution of the blackbody radiation may be required in units of photons per second per frequency interval (symbol: N_v). This form of Planck's law is
Spectral radiance units: $photons\,s^{-1}\,m^{-2}\,sr^{-1}\,Hz^{-1}$

$$N_v = \frac{2n^2 v^2}{c^2} [e^{(hv/kT)} - 1]^{-1} \tag{64}$$

and that of Wein's displacement law is

$$\frac{T}{v_{max}} = 1.701 \times 10^{-11} \; K\,Hz^{-1} \tag{65}$$

Planck's law integrated over all wavelengths (or frequencies) leads to the Stefan-Boltzmann law which describes the temperature dependence of the total radiance of a blackbody. For blackbody radiance measured as radiant power, the Stefan-Boltzmann law is
Radiance units: $W\,m^{-2}\,sr^{-1}$

$$L = 1.8047 \times 10^{-8} \, n^2 T^4 \tag{66}$$

Equation (66) is the usual form of the Stefan-Boltzmann law; however, it can also be derived for blackbody radiance measured as photon flux.
Radiance units: $photons\,s^{-1}\,m^{-2}\,sr^{-1}$

$$N = 4.8390 \times 10^{14} \, n^2 T^3 \tag{67}$$

The preceding expressions are valid provided that the cavity dimensions are much larger than the wavelengths involved in the spectrum of the radiation. The restriction imposed by the cavity dimension may lead to significant errors in very high accuracy radiometry or very long wavelength radiometry. For example, in a cube 1 mm on a side and at a wavelength of 1 μm, the approximate correction to Planck's equation is only 3×10^{-7}; however, if the measurement is made within a 1-nm bandwidth or less, the root mean square fluctuation of the signal is about 2×10^{-3} which may not be negligible.

Recent work describes how well the Planck and Stefan-Boltzmann equations describe the radiation in small cavities and at long wavelengths.[28-30]

Blackbody Simulators.   An ideal blackbody, being completely enclosed, does not radiate into its surrounds and therefore cannot serve as an absolute radiometric source. A blackbody simulator is a device that does emit radiation but only approximates the conditions under which Planck's law is valid. In general, a blackbody simulator is an enclosure at some fixed temperature with a hole in it through which some of the radiation is emitted. Some low-accuracy blackbody simulators are fabricated as a flat surface held at a fixed temperature.

A blackbody simulator can be used as an absolute source provided that the type B errors introduced by the deviations from the ideal Planck's-law conditions are evaluated and the appropriate corrections are applied. In a blackbody simulator there are three sources of type B error: inaccurate surface temperature, nonequilibrium between the radiant surface and the radiation field due to openings in the enclosure, and nonuniformity in the temperature of the radiant surface.

Calculation of the effect of a temperature error on the spectral radiance is obtained from the derivative of Planck's law with respect to temperature.

$$\frac{dL_\lambda}{L_\lambda} = \frac{hc}{n\lambda kT}[1 - e^{-(hc/n\lambda kT)}]^{-1}\frac{dT}{T} \tag{68}$$

Since the radiation field is in equilibrium with the surface of the cavity, it is the absolute temperature of the surface that must be measured. It is usually impractical to have the thermometer located on the emitting surface and it is the temperature within the wall that is measured. The difference between the temperature within the wall and the surface must therefore be measured, or calculated from a thermal model, and the correction applied.

The error due to nonequilibrium occurs because a practical radiation source cannot be a completely closed cavity. The correction factor for the effect on the radiance due to the escaped radiation is obtained from application of Kirchhoff's law. Simply stated, Kirchhoff's law states that the absorptive power of a material is equal to its emissive power. According to the principle of detailing balancing, for a body to be in equilibrium in a radiation field, the absorption of radiation by a given element of the surface for a particular wavelength, state of polarization, and in a particular direction and solid angle must equal the emission of that same radiation. If this were not true, the body would either emit more than it absorbs or vice versa, it would not be in equilibrium with the radiation, and it would either heat up or cool off.

Radiation impinging upon a body is either reflected, transmitted, or absorbed. The fraction of the incident radiation that is reflected ρ (reflectance), plus the fraction absorbed α (absorptance), plus the fraction transmitted τ (transmittance), is equal to one.

$$1 = \rho + \alpha + \tau \tag{69}$$

From Kirchhoff's law for a surface in radiative equilibrium, the fraction of absorbed radiation equals the fraction emitted ε (emittance or emissivity). Therefore, the sum of the reflectance, transmittance, and emittance must also be equal to one. If the body is opaque, the transmittance is zero and the emittance is just equal to one minus the reflectance.

$$\varepsilon = 1 - \rho \tag{70}$$

For a body not in an enclosed volume to be in equilibrium with a radiation field, it must absorb all the radiation impinging upon it, because any radiation lost through reflection will upset the equilibrium. An emittance less than one is the measure of the departure from a perfect absorber and, therefore, it is a measure of the radiance change due to the departure from closed-cavity equilibrium. In general, cavities with an emittance nearly equal to unity are those for which the size of the whole is very small in comparison to the size of the cavity.

Temperature nonuniformity modifies the radiant flux over the whole cavity in much the same way as the presence of a hole in that it is a departure from equilibrium. Radiation loss from the region of the cavity near the hole is typically larger than from other regions and this loss produces a temperature change near the hole and a nonuniformity along the cavity wall. In addition, the temperature nonuniformity is another source of uncertainty in the absolute temperature. In practice, the limiting factor in the accuracy of a high-emittance blackbody simulator is typically the nonuniformity of the temperature.

Accurate calculation of the emittance of a cavity radiator requires a detailed knowledge of the geometry of the cavity and the viewing system. This is a radiance transfer calculation and, in order to perform it accurately, one must know the angular emitting or reflecting properties of the cavity surface. The regions that contribute most to the accuracy of the calculation are those that radiate directly out the hole into the direction of the solid angle of the optical detection system.

There are many methods of calculating the emittance. The most popular are based upon the assumption of uniform emission that is independent of direction, i.e., lambertian emission. One can calculate the spectral emittance and temperature of each element along the cavity wall and sum the contribution from each element to the cavity radiance. Extensive discussion of the diffuse emittance and temperature nonuniformity calculation methods can be found elsewhere.[31-35]

Instead of calculating the emittance of a cavity directly, the problem may be transformed into one of calculating the absorptance for a ray incident from the direction in which the emittance is required.[36-38] The quantity to be calculated in this case is that fraction of the radiation entering the hole from a particular direction which is subsequently reflected out of the hole into a hemisphere.

Real surfaces are not perfectly diffuse reflectors and often have a higher reflectance in the specular direction. A perfect specularly reflecting surface is at the other extreme for calculating the emittance of a blackbody simulator. In some applications, a specular black surface might perform better than a diffusely reflecting one, particularly if the viewing geometry is highly directional and well known. The calculation of the emittance of a cavity made from a perfectly specular reflector is obtained in terms of the number of reflections undergone by an incident ray before it leaves the cavity.[39]

One can reduce the error due to temperature nonuniformity by reducing the emittance of those regions along the cavity wall that do not contribute radiation directly to that emitted from the cavity.[40] That is, by fabricating the "hidden" portions of the cavity wall from a specular, highly reflecting material and by proper orientation of these surfaces, the highly reflecting surfaces absorb almost none of the radiation but reflect it back to the highly absorbing surfaces. Since the highly reflective surfaces absorb and emit very little radiation, their temperature will have a minimal effect on the equilibrium within the cavity.

In high-accuracy applications, it is preferable to measure rather than calculate the emittance of the blackbody cavity. This can be done either by comparison of the radiance of the device under test to that of a higher-quality blackbody simulator (emittance closer to unity) or by a direct measurement of the reflectance of the cavity.[41] Accurate measurement of thermal nonuniformity by measurement of the variations in the radiance from different regions within the cavity is made difficult by the fact that radiance variations depend not only on the local temperature but also upon the emittance of the region.

Synchrotron Radiation. A synchrotron is an electronic radiation source that if well-characterized has a predictable absolute radiance output. A synchrotron source is a very nonuniform, i.e., highly directional and highly polarized, radiance standard in contrast to a blackbody which is uniform and unpolarized. However, like a blackbody, a synchrotron has a predictable spectral output and it is useful as a standard radiance source from which the other radiometric quantities, e.g., irradiance, intensity, etc., can be derived.

Classical electrodynamic theory predicts that an accelerated charged particle will emit radiation. A synchrotron is a type of electron accelerator where the electron beam is accelerated in a closed loop and synchrotron radiation is the radiation emitted by the electrons undergoing acceleration. The development of these and other charged particle accelerators led to closer experimental and theoretical scrutiny of the radiation emitted by an accelerated charged particle. These studies culminated in Schwinger's complete theoretical prediction, including relativistic effects, of the spectral and angular distribution of the radiation emitted by a beam in a particle accelerator.[42] The accuracy of Schwinger's predictions have been verified in numerous experimental studies.[43–45]

Schwinger's theoretical model of the absolute amount of radiation emitted by an accelerated charged particle is analogous to the Planck equation for blackbody sources in that both predict the behavior of an idealized radiation source. Particle accelerators, when compared to even the most elaborate blackbodies, are, however, far more expensive. Furthermore, in order to accurately predict the spectral radiance of the beam in a particle accelerator, much detailed information is required of the type not found for most accelerators. Accurately predictable radiometric synchrotron sources are consequently found only in a few laboratories throughout the world.

The magnitude of the radiant power output from a synchrotron source is proportional to the number of electrons in the beam and their velocity, i.e., the number of electrons per second or current and their energy. Therefore, synchrotron radiometry is traceable to the SI unit of electricity, the ampere.

Because absolute synchrotron sources are so rare, a detailed discussion of Schwinger's model of synchrotron radiation and the various sources of uncertainty will not be presented here. It is generally useful, however, to know some of the characteristics of synchrotron radiation. For example, the radiance from a synchrotron beam is highly polarized and very nonuniform: radiant power is almost entirely in the direction of the electron velocity vector and tangent to the electron beam. The peak of the synchrotron radiation spectrum varies from the vacuum ultraviolet to the soft x-ray region depending upon the energy in the beam. Higher-energy beams have a shorter wavelength peak: 1 GeV peaks near 10 nm, 6 GeV peaks near 0.1 nm. Radiant power decreases to longer wavelengths by very roughly two decades for every decade increase of wavelength, so that for the typical radiometric-quality synchrotron source, there is usually sufficient energy to perform accurate radiometric measurements in the visible for intercomparison to other radiometric standards.[45]

Absolute Detectors

Electrical Substitution Radiometers. An electrical substitution radiometer, often called an electrically calibrated detector, is a device for measuring absolute radiant power by comparison to electrical power.[8] As a radiant power standard, an electrical substitution radiometer can be used as the basis for the derivation of the other radiometric quantities (irradiance, radiance, or intensity) by determining the geometrical distribution (either area and/or solid angle) of the radiation.[22,24,46–49] Since an electrical substitution radiometer measures the spectrally total radiant power, it is used primarily for the measurement of monochromatic sources or those with a known relative spectral distribution.

An electrical substitution radiometer consists of a thermal detector (i.e., a thermometer) that has a radiation-absorbing surface and an electrical heater within the surface, or the heater is in good thermal contact with the surface. When the device is irradiated, the thermometer senses the temperature of the radiantly heated surface. The radiation source is then blocked and the power to the electrical heater adjusted to reproduce the temperature of the radiantly heated surface. The electrical power to the heater is measured and equated to the radiant power on the surface. The absolute base for this measurement is the electrical power measurement which is traceable to the SI ampere. In order for the measurement to be accurate, differences between the radiant and electrical heating modes must be evaluated and the appropriate corrections applied.

Electrical substitution radiometers predate the planckian radiator as an absolute radiometric standard.[50–51] They were the devices used to quantify the radiant power output of the experimental blackbody simulators studied at the end of the nineteenth century. Electrical substitution radiometers are in widespread use today and are commercially available in a variety of forms that can be classified either as to the type of thermometer, the type of radiant power absorber, or the temperature at which the device operates.

Early electrical substitution radiometers operated at ambient temperature and used either a thermocouple, a thermopile, or a bolometer as the detector. Thermopile- and bolometer-based radiometers are presently used in a variety of applications. They have been refined over the years to produce devices of either greater accuracy, sensitivity, and/or faster response time. Thermopile-based, ambient temperature electrical substitution radiometers used for radiant power (and laser power, see later discussion) measurements at about the 1-mW level at several national standards laboratories have estimated uncertainties reported to be within ±0.1 percent.[52–54] Electrical substitution radiometers have also been used for very high-accuracy absolute radiant power measurements of the total solar irradiance both at the surface of the earth,[52,54] and above its atmosphere.[53] The type of high-accuracy radiometer used at various national standards laboratories is a custom-built device and is not commercially available in general. On the other hand, electrical substitution radiometers for solar and laser power measurements at a variety of accuracy levels are available commercially.

An ambient temperature electrical substitution radiometer based on a pyroelectric as the thermal detector was developed in the 1970s.[55–56] A pyroelectric detector is a capacitor containing a dielectric with a temperature-sensitive spontaneous electrical polarization; a change in temperature results in a change in polarization. Small and rapid changes of polarization are readily detectable, making the pyroelectric a sensitive and fast thermal detector. It is most useful as a detector of a pulsed or chopped radiant power signal. During the period when the radiant power signal is blocked, electrical power can be introduced to a heater in the absorptive surface of the radiometer. As in the method for a thermopile- or bolometer-based electrical substitution radiometer, the electrical power is adjusted to equal the heating produced by the radiant power signal. Chopping can be done at a reasonable frequency, hence the electrical heating can be adjusted to achieve a balance in a comparatively short time. Because the radiant-to-electrical heating balance is more rapidly obtained, a pyroelectric radiometer is often more convenient to use than a thermopile radiometer. Pyroelectric electrical substitution radiometers are generally more sensitive but are usually less accurate than the room temperature thermopile or bolometer electrical substitution radiometers.

Electrical substitution radiometers are further distinguished by two types of radiant power absorber configurations: a flat surface coated with a highly absorbing material or a cavity-shaped, light-trapping detector. Cavity-shaped radiometers are usually more accurate over a greater spectral range than flat-surface radiometers. However, a flat-surface receiver can usually be fabricated with less thermal mass than a cavity-shaped receiver and therefore may have greater sensitivity and/or a faster response time.

Electrical substitution radiometers are further distinguished by the temperature at which the electrical-to-radiant-power comparison is performed. In the last two decades there have been significant advancements[57–58] made in instruments that perform the radiant-to-electrical comparison at a temperature near to that of liquid helium (4.2 K). Such devices are known as cryogenic electrical substitution radiometers or electrically calibrated cryogenic detectors and they are commercially available. Cryogenic electrical substitution radiometers are presently the most accurate absolute radiometric devices; the uncertainty of some measurements of radiant power has been estimated to be within ± 0.005 percent.[48]

Sources of Error in Electrical Substitution Radiometers. The relative significance of each of the possible sources of error and the derived correction factor depends upon the type of radiometer being used and the particular measurement application. It is possible to determine the total error occurring in equating radiant to electrical power and thence the accuracy of the traceability to the absolute electrical SI unit of measurement. Most manufacturers provide extensive characterization of their instruments. In such cases, the traceability to SI units is independent of radiometric standards such as blackbodies, hence electrical substitution radiometers are sometimes called absolute detectors. A commercially produced electrical substitution radiometer is capable of far greater accuracy (within ± 0.01 percent for the cryogenic instruments) than any of the typical radiometric transfer devices available from a national standards laboratory. Hence, establishing traceability through a radiometric standard from a national standards laboratory is almost pointless for a cryogenic electrical substitution radiometer.

The sources of error in an electrical substitution radiometer can be divided into three categories: errors in traceability to the absolute base unit, errors due to differences in the radiant-versus-electrical heating modes of operation, and errors arising in a particular application. The major error sources common to all electrical substitution radiometers as well as some of the less common are briefly described here. An extensive listing and description of all of these errors is given in Ref. 8, Chap. 1.

Electrical power measurement accuracy is first determined by the accuracy of the voltage and resistance standards (or voltmeter and resistance meter) used to measure voltage and current. Electrical power measurement accuracy within ± 0.01 percent is readily achievable and if needed it can be improved by an order of magnitude or better. Additional error is possible due to improper electrical measurement procedures such as those giving rise to ground-loops (improper connection to earth).

Differences between electrical-versus-radiant heating appear as differences in radiative, conductive, or convective losses. Most of these differences can be measured and a correction factor applied to optimize accuracy. The most obvious example is probably that of the radiative loss due to reflection from the receiver surface. Less obvious perhaps is the effect due to extraneous heating in the portion of the electrical conductors outside the region defined by the voltage sensing leads.

Differences between electrical heating and radiant heating may also arise due to spatial nonuniformity of the thermal sensor and/or differences in the heat conduction paths in the electrical-versus-radiant heating modes. These effects are specific to the materials and design of each radiometer. The electrical heater is typically buried within the device, whereas radiant heating occurs at the surface, so that the thermal conductivity paths to the sensor may be very different. Also, the distribution of the radiant power across the receiver is usually quite different compared to the distribution of the electrical heating. A detailed thermal analysis is required to create a design which minimizes these effects, but for optimum accuracy, the measurement of the magnitude of the nonuniformity effects is required to test the thermal model. Nonuniformity can be measured either by placing small auxiliary electrical heaters in various locations or by radiative heating of the receiver in several regions by moving a small spot of light across the device.

It should be noted that the thermal conduction path differences may also be dependent upon the environment in which the radiometer is to be operated. For example, atmospheric-pressure-dependent differences between the electrical-to-radiant power correction factor have been detected for many radiometers. These differences are, of course, greatest for a device for which the correction factors have been characterized in a normal atmosphere and which is then used in a vacuum.

Application-dependent errors arise from a variety of sources. Some examples are window transmission losses if a window is used, the accuracy of the aperture area and diffraction corrections are critical for measurements of irradiance; and, if a very intense source such as the sum is measured, heating of the instrument case and the body of the aperture could be an important correction factor. The last effect might also be very sensitive to atmospheric pressure changes.

Photoionization Devices. Another type of absolute detector is a photoionization detector which can be used for absolute photon flux, i.e., radiant power, measurements of high-energy photon beams. Since a photoionization detector is a radiant power standard like the electrical substitution radiometer, it can in principle be used as the basis for the derivation of the other radiometric quantities (irradiance, radiance, or intensity) by determining the geometrical distribution (either area and/or solid angle) of the radiation.

A photoionization detector is a low-pressure gas-filled chamber through which a beam of high-energy (vacuum ultraviolet) photons is passed between electrically charged plates, the electrodes. The photons absorbed by the gas, if of sufficient energy, ionize the gas and enable a current to pass between the electrodes. The ion current is proportional to the number of photons absorbed times the photoionization yield of the gas and is, therefore, proportional to the photon flux.

The photoionization yield is the number of electrons produced per photon absorbed. If the photon is of sufficiently high energy, the photoionization yield is 100 percent for an atomic gas. The permanent atomic gases are the rare gases: helium, neon, argon, krypton, and xenon. Their photoionization yields have been measured relative to each other and shown to be 100 percent over specific wavelength ranges.[59–60] If an ionization chamber is constructed properly and filled with the appropriate gas so that all of the radiation is absorbed, then the number of photons per second incident on the gas is simply equal to the ion current produced. If instead of measuring the ion current one were to measure each pulse produced by a photon absorption, then one would have a photon counter.

Carefully constructed ion current measurement devices have been used as absolute detectors from 25 to 102.2 nm and photon counters from 0.2 to 30 nm. Careful construction implies that all possible systematic error sources have either been eliminated or can be estimated, with an appropriate correction applied. Because of the difficulty in producing accurate and well-characterized devices, ion chambers and high-energy photon counters are not claimed to be high-accuracy radiometric devices. Furthermore, they are limited to applications in vacuum ultraviolet radiometry and are consequently of restricted interest.

Predictable Quantum Efficiency Devices. A useful and quite economical type of absolute detector is a predictable quantum efficiency (PQE) device using high-quality silicon photodiodes. Quantum efficiency is the photon flux-to-photocurrent conversion efficiency. Because there have been many technological advancements made in the production of solid-state electronics, it is now possible to obtain very high quality silicon photodiodes whose performance is extremely close to that of the theoretical model.[61–62] The technique for predicting the quantum efficiency of a silicon photodiode is also known as the self-calibration of a silicon photodiode.[63–64] It is a relatively new absolute radiometric technique, quite simple to implement and of very high accuracy.[54,65–66]

Conversion of a detector calibration from spectral responsivity $R(\lambda)$, in units of A/W, to quantum efficiency, i.e., photon-to-electron conversion efficiency, is as follows:

$$C_e = 1239.85 \frac{R(\lambda)}{\lambda} \tag{71}$$

where λ is the in-vacuum wavelength in nm and C_e is in units of electrons per photon.

As in the case of the other absolute detectors discussed previously, a PQE device is used for absolute photon flux, i.e., radiant power, measurements. It can also be used as the basis for the derivation of the other radiometric quantities such as irradiance, radiance, or intensity.[23] The extension to other radiometric measurements is by the determination of the geometrical distribution (area and/or solid angle) of the radiation. Also, like other absolute detectors, it measures spectrally total flux (within its spectral response range) and is therefore used primarily for the measurement of monochromatic sources or those with a known relative spectral distribution.

In a solid-state photodiode, the process for the conversion of a photon to an electronic charge is as follows. Photons not lost through reflection or by absorption in a coating at the front surface are absorbed in the semiconductor—if the photon is of high enough energy. To be absorbed, the photon energy must be greater than the band gap; the band gap for silicon is 1.11 eV (equivalent wavelength, 1.12 μm). In silicon, the absorption of a photon causes a promotion of a charge carried to the conduction band. Absorption of very high energy photons will create charge carriers with sufficient energy to promote a second, third, or possibly more charge carriers into the conduction band by collision processes. However, for silicon, the photon energy throughout the visible spectral range is insufficient for such impact ionization processes to occur. Therefore, in the visible to near-ir spectral region (about 400 to 950 nm), one absorbed photon produces one electron in the conduction band of silicon.

In a photodiode, impurity atoms diffused into a portion of the semiconductor material create an electric field. The internal electric field causes the newly created charge carriers to separate, eventually promoting the flow of an electron in an external measurement circuit. The efficiency with which the charge carriers are collected depends upon the region of the photodiode in which they are created. In the electric field region of a high-quality silicon photodiode, this collection efficiency has been demonstrated to approach 100 percent to within about 0.01 percent. Outside the field region, the collection efficiency can be determined by simple electrical bias measurements.

For the spectral regions in which the collection efficiency is 100 percent, the only loss in the photon-to-electron conversion process is due to reflection from the front surface of the detector. Several silicon photodiodes can be positioned to more effectively collect the radiation, acting as a light trap.[67–68] If the radiation reflected from the first photodiode is directed to a second photodiode, then onto a third photodiode, etc., almost all the radiation will be collected in a small number of reflections. The photocurrents from all of the photodiodes are then summed and the total current (electrons per second) will be nearly equal, within 0.1 percent or less, to the photon flux (photons per second).

The more common type of silicon photodiode is the pn-type (positive charge impurity diffused into negative charge impurity starting material). High-quality pn-type detectors have their high collection efficiency in the long wavelength visible to near-ir spectral region. On the other hand, np-type silicon photodiodes have high collection efficiency in the short wavelength spectral region. At this time, the silicon photodiodes with the highest quantum efficiency (closest to ideal behavior) in the blue spectral region are the np-type devices, while nearly ideal red region performance is obtained with pn-type devices. The predictable quantum efficiency technique for silicon photodiodes has been demonstrated[54,65–68] to be absolutely accurate to within ±0.1 percent from about 400 nm to 900 nm.

A disadvantage of the light-trap geometry is the limited collection angle (field of view) of the device. Light-trap silicon photodiode devices are now commercially available using large area devices and a compact light-trap configuration that maximizes the field of view.

An np-type silicon photodiode trap detector optimized for short-wavelength performance and a pn-type silicon photodiode trap detector optimized for long-wavelength performance can be used as an almost ideal radiometric standard. The pair covers the 400- to 900-nm spectral range, has direct absolute SI base unit traceability via convenient electrical standards, and they are sufficiently independent to be meaningfully cross-checked to verify absolute accuracy and long-term stability. These detectors are not only useful radiometric standards by themselves but can be used with various source standards to either verify the absolute accuracy or to correct for the instabilities in the source standards.

The concept of a PQE light-trapping device is extendable to other high-quality photodiodes. Very recently, InGaAs devices with nearly 100-percent collection efficiency in the 1000- to 1600-nm spectral range have been developed. A light-trapping device employing these new detectors is now commercially available.

Calibration Transfer Devices

The discussion to this point focused on absolute radiometric measurements using methods that in themselves can be made traceable to absolute SI units. It is often more convenient (and sometimes required by contractual agreements) to obtain a device that has been calibrated in radiometric units at one of the national standards laboratories. Specific information as to the type and availability of various calibration transfer devices and calibration services may be obtained by directly contacting any of the national standards laboratories in the world. The products and services offered by the various standards laboratories cover a range of applications and accuracies, and differ from country to country.

Radiometric calibration transfer devices are either sources or detectors. The calibration transfer sources are either incandescent, tungsten filament lamps, deuterium lamps, or argon arc discharge sources.[21,69–70] Generally, calibration transfer detectors are photodiodes of silicon, germanium, or indium gallium arsenide. The most prevalent calibration transfer sources are incandescent lamps and the typical calibration transfer detector is a silicon photodiode.[71]

The commonly available spectral radiance calibration transfer devices that span the 250- to 5000-nm region are typically tungsten strip filament lamps. Lamps calibrated in the 250- to 2500-nm region by a national standards laboratory are available. Lamps calibrated in the 2.5- to 5-μm region by comparison to a blackbody are commercially available. These devices are calibrated within specific geometrical constraints: the area on the filament, and the direction and solid angle of observation. The calibration is reported at discrete wavelengths, for a specified setting of the current through the filament and the ambient laboratory temperature. The optimum stability of spectral radiance is obtained with vacuum rather than gas-filled lamps, and with temperature controlled, i.e., water-cooled electrodes. Vacuum lamps cannot be operated at high filament temeratures and consequently do not have sufficient uv output. Gas-filled lamps cover a broader spectral and dynamic range and are the more commonly available calibration transfer device.

The commonly available spectral irradiance calibration transfer devices that span the 250- to 2500-nm region are tungsten coiled filament lamps. These are usually gas-filled lamps that have a halogen additive to prolong filament life and enable higher-temperature operation. Lamps calibrated in the 250- to 2500-nm region by a national standards laboratory are available. These devices are calibrated within the specific geometrical constraints of the distance and the direction with respect to a location on the lamp base or the filament. The calibration is reported at discrete wavelengths, for a specified setting of the current through the filament and the ambient laboratory temperature. Because the filament is operated at a higher temperature, the spectral irradiance lamps are usually less stable than the radiance lamps.

The drift of an incandescent lamp's radiance or irradiance output is not reliably predictable. It is for this reason that the calibration is most reliably maintained not by an individual lamp but by a group of lamps. The lamps are periodically intercompared and the average radiance (irradiance) of the group is considered to be the calibration value. The calculated differences between the group average and the individual lamps is used as a measure of the performance of the individual lamp. Lamps that have drifted too far from the mean are either recalibrated or replaced.

Spectral radiance and irradiance calibration transfer devices for the vacuum to near-uv (from about 160 to 400 nm) are typically available as deuterium lamps.

The commonly available calibrated transfer detectors for the 250- to 1100-nm spectral region are silicon photodiodes and for the 1000- to 1700-nm region, they are either germanium or indium gallium arsenide photodiodes. The calibration is reported at discrete wavelengths in absolute responsivity units (A/W) or irradiance response units (A cm^2/W). In the first case, the calibration of the detector is performed with its active area underfilled, while in the second case, it is overfilled. If the detector is fitted with a precision aperture and if its spatial response is acceptably uniform, then the area of the aperture can be used to calculate the calibration in either units. The conditions under which the device was calibrated should be reported. The critical parameters are the location and size of the region within the active area in which it was calibrated, the radiant power in the calibration beam (alternately the photocurrent), and the temperature at which the calibration was performed. The direction in which the device was calibrated is usually assumed to be normal to its surface and the irradiation geometry is usually that from a nearly collimated beam. Significant departures from normal incidence or near collimation should be noted.

Lasers

Power and Energy Measurement.   Lasers are highly coherent sources and the previous discussion of radiometry has been limited to the radiometry of incoherent sources. Nevertheless, the absolute power (or energy) in a laser beam can be determined to a very high degree of accuracy (within ±0.01 percent in some cases) using some of the detector standards discussed here. The most accurate laser power measurements are made with cryogenic and room temperature electrical substitution radiometers and with predictable quantum efficiency devices. In order to measure the laser power (energy) it is necessary to insure that all the radiation is impinging on the sensitive area of the detector and, if the absolute detector characterization was obtained at a different power (energy) level, that the detector is operating in a linear fashion. For pulsed lasers, the peak power may substantially exceed the dynamic range of the detector's linear performance. (A discussion of detector linearity is presented later in this chapter.) Furthermore, caution should be exercised to insure that the detector not be damaged by the high photon flux levels achieved with many lasers.

In addition to ensuring that the detector intercept all of the laser beam, it is necessary to determine that all coherence effects have been eliminated (or minimized and corrected).[72–73] The predominant effects of coherence are, first, interference effects at windows or beam splitters in the system optics and, second, diffraction effects at aperture edges. The use of wedged windows will minimize interference effects, and proper placement of apertures or the use of specially designed apertures[74] will minimize diffraction effects.

Lasers as a Radiometric Characterization Tool.   It should be noted that lasers, particularly the cw (continuous wave) variety, are particularly useful as characterization tools in a radiometric laboratory. Some of their applications are instrument response uniformity mapping, detector-to-detector spectral calibration transfer, polarization sensitivity, linearity verifications, and both diffuse and specular reflectance measurements.

Lasers are highly polarized and collimated sources of radiation. It is usually simple to construct an optical system as required for each measurement using mostly plane and spherical mirrors and to control scattered light with baffles and apertures. Lasers are high-power sources so that the signal-to-noise levels obtained are very good. If the power level is excessive it can usually be easily attenuated. Also, care must be taken to avoid local saturation of a detector at the peak of a laser's typical gaussian beam profile. They are highly monochromatic so that spectral purity, i.e., out-of-band radiation, is not usually a problem. However, in very high accuracy, within <0.1 percent, measurements, lasing from weaker lines may be significant and additional spectral blocking filters could be required.

Lasers are not particularly stable radiation sources. This problem is overcome by putting a beam splitter and stable detector into the optical system near the location of the measurement. The detector either serves to monitor the laser beam power and thereby supplies a correction factor to compensate for the instability, or its output is used to actively stabilize the laser.[23,49] In the latter case, an electronically controllable attenuator, such as an electro-optical, acousto-optical, or a liquid crystal system, is used to continuously adjust the power in the laser beam at the beam splitter. Feedback stabilization systems for cw lasers, both the electro-optical and liquid crystal type, are commercially available. For the highest-accuracy measurements, i.e., optimum signal-to-noise ratios, it is necessary both to actively stablize the laser source and also to monitor the beam power close to the measurement in order to correct for the residual system drifts.

Various Type B Error Sources

Offset Subtraction. One common error source, which is often simply an oversight, is the incorrect (or sometimes neglected) adjustment of an instrument reading for electronic and radiometric offsets. This is often called the dark signal or dark current correction since it is obtained by shutting off the radiation source and reading the resulting signal. The shuttered condition needs to be close to radiant zero, at least within less than the expected accuracy of the measurement.

A dark signal reading is usually easy to achieve in the visible and near-visible spectral regions. However, in the long-wavelength infrared a zero radiance source is one that is at a temperature of ideally 0 K. Often an acceptably cold shutter is not easily obtained so that the radiance, i.e., temperature, of the "zero" reference source must be known in order to determine the true instrument offset.

Scattered Radiation and Size of Source Effect. An error associated with the offset correction is that of scattered radiation from regions outside the intended optical path of the measurement system. Often, by judicial placement of the shutter, the principal light path can be blocked while the scattered light is not. In this case, the dark signal measurement includes the scattered light which is then subtracted from the measurement of the unshuttered signal. It is not possible to formulate a general scattered light elimination method so that each radiometric measurement system needs to be evaluated on an individual basis. The effects of scattered radiation can often be significantly reduced by using an optical chopper, properly placed, and lock-in amplifier system to read the output of the photodetector.

No optical element will produce a perfect image and there will be an error due to geometrically introduced stray light. Sharp edges between bright and dark regions will be blurred by aberrations, instrument fabrication errors, scattering due to roughness and contamination of the optical surfaces, and scattered light from baffles and stops within the instrument enclosure. Diffraction effects will also introduce stray light. Light originating from the source will be scattered out of the region of the image and light from the area surrounding the source will be scattered into the image. The error resulting from scattering at the objective lens or mirror is related to the size of the source since the scattering is proportional to the irradiance of the objective element. Thus, the error introduced by the lack of image quality is commonly referred to as the size-of-source effect.

The effect of the aberrations on the radiant power both into and out of an image can, in principle, be calculated. The diffraction-related error can also be calculated in some situations.[73-75] However, the effect due to scattering is very difficult to model accurately and usually will have to be measured. In addition, the amount of scattering can be expected to change in time due to contamination of the optics, baffles, and stops. It is often more practical to measure the size of source effect and determine a correction factor for the elimination of this systematic error.

There are two different methods for measuring the size-of-source effect. The first method measures the response of the instrument as the size of the source is increased from the area imaged to the total area of the source. In the second method, a dark target of the same size as the image is placed at the imaged region on the source, and the surrounding area is illuminated. The second method has the advantage in that the effect being measured is the error signal above zero, whereas in the first method a small change in a large signal is being sought. In either case, the total error signal is measured; it includes aberrations, diffraction, and scattering effects.

Polarization Effects.   These are often significant perturbations of radiant power transfer due to properties of the radiation field other than its geometry. One such possible error is that due to the polarization state of the radiation field. The signal from a photodetector that is polarization-sensitive will be dependent upon the relative orientation of the polarization state of the radiation with respect to the detector orientation. Examples of polarization-dependent systems are grating monochromators and radiation transfer through a scattering medium or at a reflecting surface. In principle, the polarization state of the radiation field may be included in the geometrical transfer equation as a discrete transformation that occurs at each boundary or as a continuous transformation occurring as a function of position in the medium. Often it is sufficient to perform a calibration at two orthogonal rotational positions of the instrument or its polarization-sensitive components. However, it is recommended that other measurements at rotations intermediate between the two orthogonal measurements be included to test if the maximum and minimum polarization sensitivities have been sampled. The average of the maximum and minimum polarization measurements is then the calibration factor of the instrument for a nonpolarized radiation source.

Detector Nonlinearity.

Nonlinearity Measurement by Superposition of Sources.   Another possibly significant error source is photodetector and/or the electronic signal processing system nonlinearity. If the calibration and subsequent measurements are performed at the same radiant power level, then nonlinearity errors are avoided. Often conditions require that the measurements be performed over a range of power levels. In general, a separate measurement is required either to verify the linearity of the photodetector (and/or the electronics) or to deduce the form of the nonlinearity function in order to apply the appropriate correction.[72,76–78]

The typical form of a nonlinearity appears as a saturation of the photoelectronic process at high irradiance levels. At low radiant flux levels what often appears to be a nonlinearity may be the result of failing to apply a dark signal or offset correction. There are, of course, other effects that will appear as a nonlinearity of the photodetector and/or electronics.[79]

Either the linearity of the detector and electronics can be directly verified by experiment or it can be determined by comparison to a photodetector/electronics system of verified linear performance. It is useful to note that several types of silicon photodiodes using a transimpedance or current amplifier have been demonstrated to be linear within ± 0.1 percent over up to eight decades for most of its principal spectral range.[78]

The fundamental experimental method for determining the dynamic range behavior of a photodetector is the superposition of sources method.[76–78] The principle of the method is as follows. If a photodetector/electronics system is linear, then the arithmetic sum of the individual signals obtained from different radiant power sources should equal the signal obtained when all the sources irradiate the photodetector at the same time. There are many variations of the multiple source linearity measurement method using combinations of apertures or beam splitters. A note of caution: Interference effects must be avoided when combining beams split from the same source or when combining highly coherent sources such as lasers.

The difference between the arithmetic sum and the measured signal from the combined sources is used as the nonlinearity correction factor. Consider the superposition of two sources having approximately equal radiant powers ϕ_a and ϕ_b, which when combined have a radiant power of $\phi_{(a+b)}$. The signals from the photodetector when irradiated by the individual and combined sources is i_a, i_b, and $i_{(a+b)}$. The following equation would be equal to unity for a linear detector:

$$K_{ab} = \frac{i_{(a+b)}}{i_a + i_b} \tag{72}$$

For a calibration performed at the radiant power level ϕ_a (or ϕ_b), the detector responsivity is R and

$$i_a = R\phi_a \tag{73}$$

For a measurements at the higher radiant power level $\phi_{(a+b)}$,

$$i_{(a+b)} = K_{ab} R \phi_{(a+b)} \tag{74}$$

Scaling up to much higher radiant power levels (or down to lower levels) requires repeated application of the superposition-of-sources method. For example, in order to scale up to the next higher radiant power level, the source outputs from the first level are increased to match the second level (e.g., by using larger apertures). The increased source outputs are then combined to reach a third level and a new correction factor calculated. The process is repeated to cover the entire dynamic range of a photodetector/electronics system in factor-of-two steps.

Note that when type B errors, such as the interference effects noted above, are eliminated, the accumulated uncertainty in the source superposition method is the accumulated imprecision of the individual measurements.

Various Nonlinearity Measurement Methods. Other techniques for determining the dynamic range behavior of a photodetector are derivable from predictable attenuation techniques.[72] One such method is based upon chopping the radiation signal using apertures of known area in a rotating disk. This is often referred to as Talbot's law: the average radiant power from a source viewed through the apertures of a rotating disk is given by the product of the radiant power of the source and the transmittance of the disk. The transmittance of the disk is the ratio of the open area to the blocked area of the disk. The accuracy of this technique depends upon the accuracy with which the areas are known and may also be limited by the time dependence of the photodetector and/or electronics.

Another predictable attenuation technique is based upon the transmittance obtained when rotating, i.e., crossing two polarizers.

$$\tau = \tau_0 \cos^4 \delta \tag{75}$$

Here δ is the angle of rotation between the linear polarization directions of the two polarizers and τ_0 is the transmittance at $\delta = 0°$. This technique, of course, assumes ideal polarizers that completely extinguish the transmitted beam at $\delta = 90°$, and its accuracy is limited by polarization efficiency of the polarizers.

A third predictable attenuation technique is the application of Beer's law which states that the transmittance of a solution is proportional to the concentration κ of the solute

$$\tau = e^{-\gamma\kappa} \tag{76}$$

Here γ is the absorption coefficient of the solute. The accuracy of this technique depends upon the solubility of the solute and the absence of chemical interference, i.e., concentration-dependent chemical reactions.

Time-dependent Error.   For measurements of pulsed or repeatedly chopped sources of radiation, the temporal response of the detector could introduce a time-dependent error.

A photodetector that has a response that is slow compared to the source's pulse width or the chopping frequency will not have reached its peak signal during the short time interval. Time-dependent error is avoided by determining if the detector's frequency response is suitable before undertaking the calibration and measurement of pulsed or chopped radiation sources.

Nonuniformity. The nonuniformity of the distribution of radiation over an image or within the area sampled in an irradiance or radiance measurement may lead to an error if the response of the instrument is nonuniform over this area. The calibration factor for a nonuniform instrument will be different for differing distributions of radiation. The size of the error will depend upon the relative magnitudes of the source and instrument nonuniformities and it is a very difficult error to correct. This type of error is usually minimized either by measuring only sources that are uniform or by insuring that the instrument response is uniform. It is usually easier to assure that the instrument response is uniform.

Nonideal Aperture. For very high-accuracy radiometric calibrations, the error due to the effect of the land on an aperture must be correctly taken into account. An ideal aperture is one that has an infinitesimally thin edge that intercepts the radiation beam. In practice an aperture will have a surface of finite thickness at its edge. This surface is referred to as the land; see Fig. 7. The effective radius of the aperture will be slightly reduced by vignetting caused by the land on its edge, assuming that the vignetting is small compared to the aperture radius and that all the radiation reflected by the land eventually falls on the detector (i.e., the land has a highly reflective surface). The effective radius of the aperture r' is

$$r' = r\left[1 - \frac{(1-\rho)w}{s}\right] \tag{77}$$

Here ρ is the reflectance, s is the distance between this aperture and the mirror or lens (or other aperture), and w is the width of the land.

Spectral Errors.

   Wavelength Error. In those cases where a spectrally selective element, such as a monochromator or a filter, is included in the radiometric instrument, spectral errors must be taken into account.[24] The first type of error is called the wavelength error and is due to misassignment of the wavelength of the spectrum of the filter or the monochromator in the instrument. That is, either the monochromator used to calibrate the filter transmission or the

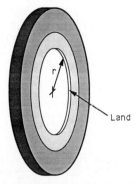

FIGURE 7 The nonideal aperture showing the location of the land.

monochromator within the radiometric instrument has an error in its wavelength setting. This error is eliminated by calibration of the monochromator wavelength setting using one or more atomic emission lines from either a discharge lamp (usually a mercury and/or a rare gas lamp) or one of the many spectra of the elements available as a hollow cathode lamp. The wavelengths of most atomic emission lines are known with an accuracy that exceeds the requirements of radiometric calibrations.

A special note regarding the wavelength error and the use of interference filters in a radiometric instrument. The typical angular sensitivity of an interference filter is 0.1 nm per angular degree of rotation. If the transmission of the interference filter is measured in a collimated beam and then used in a convergent beam there will be an error due to the angularly dependent shift in the spectral shape of the transmission. For an accurate radiometric instrument it is important to measure interference filter transmission in nearly the same geometry as it will be used. Furthermore, the temperature coefficient of the transmission of an interference filter is about 0.2 nm K^{-1}. Therefore, it is also important in subsequent measurements to assure that the filter remains at the same temperature as that used during the calibration.

It should also be noted that in order to accurately determine the spectral transmission of a monochromator it is necessary to completely fill the aperture of the dispersing element in the monochromator. This usually means that the field of view of the monochromator must be filled by the beam from the spectral calibration instrument.

Out-of-band Radiation Error. The second type of spectral error is called the out-of-band or spectrally stray radiation error. This error is due to the radiation transmitted at both longer and shorter wavelengths that are beyond the edges of the principle transmission band of the filter or monochromator. This radiation is not taken into account if the limits in the integral in Eq. (11) are restricted to the edges of the principle transmission band. Although the relative amount of radiation transmitted at any wavelength beyond the edges of the principle transmission band may appear to be negligible with respect to the amount of radiation within the band, it is the spectrally total radiation that "leaks through" that is the error signal. It is therefore necessary to determine the transmission of the filter or monochromator over the entire spectrum of either the detector's response and/or the light source's output, whichever is greater. If the out-of-band radiation effect is small, it is possible to determine the correction factor from nominal values or limited accuracy measurements of the out-of-band spectra of the source, detector, and filter (monochromator).

Temperature Dependence. The effect of temperature on the various elements in a radiometric instrument must not be overlooked. Unless the temperature of the instrument at the time of its calibration is maintained during subsequent applications, there may be substantial changes introduced in the instrument calibration factor that could well be above the uncertainty of its traceability to absolute SI units. The simple solution is to control the temperature of the system from the calibration to the subsequent measurements. A more practical solution is often to measure the relative change in the instrument calibration factor as a function of temperature and then apply this as a correction factor to account for the temperature difference in the subsequent measurements.

24.6 PHOTOMETRY

Definition and Scope

Photometry is the measurement of radiation in a way that characterizes its effectiveness in stimulating the normal human visual system.[4–5,80–82] Since visual sensation is a subjective experience, it is not directly quantifiable in absolute physical units. Attempts to quantify human visual sensation, therefore, were by comparison to various specified sources of light. The first sources used as standards were candles and later flames of prescribed construction. About the turn of the century, groups of incandescent lamps were selected as photometric standards and eventually a planckian radiator at a specific temperature was adopted by international agreement as the standard source. At present, the SI base unit for photometry, the candela, is no longer defined in terms of a given light source but is related to the radiant intensity by a multiplicative constant. Therefore, either an absolute source or detector can be used to establish an internationally recognized photometric calibration. Furthermore, there is no need for a human observer to effect a quantitative photometric measurement.

Photometry, as discussed here, is more precisely referred to as physical photometry to distinguish it from psychophysical photometry. Early photometric calibrations relied on human observers to compare an unknown light source to a standard. Presently, photometric calibrations are based on measurements using physical instruments. The instrument simulates human visual response either by having a detector with a spectral response that approximates that of the CIE standard observer or by using the CIE standard observer spectral response function in the data analysis.

Psychophysical photometry is the measurement of the effectiveness of light in individual observers and is more generally referred to as visual science. An individual's visual system may differ from that of the CIE standard observer defined for physical photometry, and these differences are sometimes important in experiments in visual science.

Photometry is restricted to the measurement of the magnitude of the visual sensation without regard to color, although it is well known that the perception of brightness is highly dependent on color in many circumstances. Measurement of the human response to color in terms of color matching is known as colorimetry. See Vol. I, Chap. 26.

Under reasonable light levels, the human eye can detect a difference of as little as 0.5 percent between two adjacent fields of illumination. For fields of illumination which are not adjacent, or are viewed at substantially different times, the eye can only detect differences of 10 to 20 percent. A discussion of the performance of the human visual system can be found in Vol. I, Chap. 25. Extensive treatments of photometry can be found in Walsh,[77] and Wyszecki and Stiles.[78]

Photopic, Scotopic, and Mesopic Vision

Electromagnetic radiation of sufficient power and in the wavelength range from about 360 to 830 nm, will stimulate the human visual system and elicit a response from an observer. The spectral range given here is the range over which measurements in physical photometry are defined. The range of reasonably perceptible radiation is usually given as about 400 to 700 nm. After light enters through the optical system of the eye—the cornea, iris or pupil, lens, and vitreous humor—the next stage of the visual response occurs in the retina. The retina contains two types of receptor cells: cones, which are the dominant sensors when the eye is adapted to higher radiance levels of irradiation (*photopic* vision), and rods, the dominant sensors at lower radiance levels (*scotopic* vision). Between the higher and lower levels of light adaptation is the region of *mesopic* vision, the range of radiance levels where both cones and rods contribute in varying degrees to the visual process.

Three types of cones having different spectral sensitivity functions exist in the normal human eye. The brain is able to distinguish colors by comparison of the signals from the three cone types. Of the three cone types, only the middle- and long-wavelength-sensitive cones contribute to the photopic sensation of the radiation entering the eye. The relative spectral sensitivity functions of the photopically and scotopically adapted human eye have been measured for a number of observers. From averages of these measurements, a set of values has been adopted by international agreement as the spectral efficiency for the CIE

TABLE 2 Photopic and Scotopic Spectral Luminous Efficiency Functions

Wavelength	Photopic	Scotopic	Wavelength	Photopic	Scotopic
375	0.00002	—	575	0.91540	0.1602
380	0.00004	0.00059	580	0.87000	0.1212
385	0.00006	0.00111	585	0.81630	0.0899
390	0.00012	0.00221	590	0.75700	0.0655
395	0.00022	0.00453	595	0.69490	0.0469
400	0.00040	0.00929	600	0.63100	0.03315
405	0.00064	0.01852	605	0.56680	0.02312
410	0.00121	0.03484	610	0.50300	0.01593
415	0.00218	0.0604	615	0.44120	0.01088
420	0.00400	0.0966	620	0.38100	0.00737
425	0.00730	0.1436	625	0.32100	0.00497
430	0.01160	0.1998	630	0.26500	0.00334
435	0.01684	0.2625	635	0.21700	0.00224
440	0.02300	0.3281	640	0.17500	0.00150
445	0.02980	0.3931	645	0.13820	0.00101
450	0.03800	0.455	650	0.10700	0.00068
455	0.04800	0.513	655	0.08160	0.00046
460	0.06000	0.567	660	0.06100	0.00031
465	0.07390	0.620	665	0.04458	0.00021
470	0.09098	0.676	670	0.03200	0.00015
475	0.11260	0.734	675	0.02320	0.00010
480	0.13902	0.793	680	0.01700	0.00007
485	0.16930	0.851	685	0.01192	0.00005
490	0.20802	0.904	690	0.00821	0.00004
495	0.25860	0.949	695	0.00572	0.00003
500	0.32300	0.982	700	0.00410	0.00002
505	0.40730	0.998	705	0.00293	0.00001
510	0.50300	0.997	710	0.00209	0.00001
515	0.60820	0.975	715	0.00148	0.00001
520	0.71000	0.935	720	0.00105	0.00000
525	0.79320	0.880	725	0.00074	0.00000
530	0.86200	0.811	730	0.00052	0.00000
535	0.91485	0.733	735	0.00036	0.00000
540	0.95400	0.650	740	0.00025	0.00000
545	0.98030	0.564	745	0.00017	0.00000
550	0.99495	0.481	750	0.00012	0.00000
555	1.00000	0.402	755	0.00008	0.00000
560	0.99500	0.3288	760	0.00006	0.00000
565	0.97860	0.2639	765	0.00004	0.00000
570	0.95200	0.2076	770	0.00003	0.00000

standard observer for photopic vision and another set for the CIE standard observer for scotopic vision (CIE, Commission Internationale de l'Eclairage). Because of the complexity of the spectral sensitivity of the eye at intermediate irradiation levels, there is no standard of spectral efficiency for mesopic vision. Values of the photopic and scotopic spectral efficiency functions are listed in Table 2.

Photometric quantities can be calculated or measured as either photopic or scotopic quantities. Adaptation to luminance levels of $\geq 3 \, \text{cd m}^{-2}$ (see further discussion) in the visual field usually leads to photopic vision, whereas adaptation to luminance levels of $\leq 3 \times 10^{-5} \, \text{cd m}^{-2}$ usually leads to scotopic vision. Photopic vision is normally assumed in photometric measurements and photometric calculations unless explicitly stated to be otherwise.

Basic Concepts and Terminology

As noted in the earlier section on "Photometry," the principles of photometry are the same as those for radiometry with the exception that the spectral responsivity of the detector is defined by general agreement to be specific approximations of the relative spectral response functions of the human eye. Photometric quantities are related to radiometric quantities via the spectral efficiency functions defined for the photopic and scotopic CIE standard observers.

Luminous Flux. If physical photometry were to have been invented after the beginning of the twentieth century, then the physical basis of measurement might well have been the relationship between visual sensations and the energy of the photons and their flux density. It would follow naturally because vision is a photobiological process that is more closely related to the quantum nature of the radiation rather than its thermal heating effects. However, because of the weight of historical precedent, the basis of physical photometry is defined as the relationship between visual sensation and radiant power and its wavelength. The photometric equivalent of radiant power is luminous flux, and the unit that is equivalent to the watt is the lumen.

Luminous flux, Φ_v, is the quantity derived from spectral radiant power by evaluating the radiation according to its action upon the CIE standard observer.

$$\Phi_v = K_m \int \Phi_\lambda V(\lambda)\,d\lambda \qquad (78)$$

where $V(\lambda)$ is the spectral efficiency function for photopic vision listed in Table 2, and K_m is the luminous efficacy for photopic vision. The spectral luminous efficacy is defined near the maximum, $\lambda_m = 555$ nm, of the photopic efficiency function to be

$$K_m = 683 \frac{V(\lambda_m)}{V(555.016 \text{ nm})} \cong 683 \text{ lm W}^{-1} \qquad (79)$$

The definitions are similar for scotopic vision

$$\Phi'_v = K'_m \int \Phi_\lambda V'(\lambda)\,d\lambda \qquad (80)$$

where $V'(\lambda)$ is the spectral luminous efficiency function for scotopic vision listed in Table 2, and K'_m is the luminous efficacy for scotopic vision. The scotopic luminous efficiency function maximum occurs at $\lambda_m = 507$ nm. The defining equation for K'_m is

$$K'_m = 683 \frac{V'(\lambda_m)}{V'(555.016 \text{ nm})} \cong 1700 \text{ lm W}^{-1} \qquad (81)$$

The spectral shifts indicated in Eqs. (79) and (81) are required in order to obtain the precise values for the photopic and scotopic luminous efficacies. The magnitudes of the shifts follow from the specification of an integral value of frequency instead of wavelength in the definition of the SI base unit for photometry, the candela.

Luminous Intensity, Illuminance, and Luminance. The candela, abbreviated cd, is defined by international agreement to be the luminous intensity in a given direction of a source that emits monochromatic radiation of frequency 540×10^{12} Hz (equal to 555.016 nm) and that has a radiant intensity of $1/683$ W sr^{-1} in that direction. The spectral luminous efficacy of radiation at 540×10^{12} Hz equals 683 lm W^{-1} for all states of visual adaptation.

Because of the long history of using a unit of intensity as the basis for photometry, the candela was chosen as the SI base unit instead of the lumen, notwithstanding the fact that intensity is, strictly speaking, measurable only for point sources.

The functional form of the definitions of illuminance, luminance, and luminous intensity were presented in Eqs. (13), (14), and (15). The concepts are briefly reviewed here for the sake of convenience.

Luminous intensity is the photometric equivalent of radiant intensity, that is, luminous intensity is the luminous flux per solid angle. The symbol for luminous intensity is I_v. The unit for luminous intensity is the candela.

Illuminance is the photometric equivalent of irradiance, that is, illuminance is the luminous flux per unit area. The symbol for illuminance is E_v. The typical units for illuminance are lumens/meter^2.

Luminance is the photometric equivalent of radiance. Luminance is the luminous flux per unit area per unit solid angle. The symbol for luminance is L_v. The units for luminance are typically candelas/meter^2. In many older treatises on photometry, the term brightness is often taken to be equivalent to luminance, however, this is no longer the accepted usage.

In present usage, luminance and brightness have different meanings. In visual science (psychophysical photometry), two spectral distributions that have the same luminance typically do not have the same brightness. Operationally, spectral distributions of equal luminance are established with a psychophysical technique called heterochromatic flicker photometry. The observer views two spectral distributions that are rapidly alternated in time at the same spatial location, and the radiance of one is adjusted relative to the other to minimize the appearance of flicker. Spectral distributions of equal brightness are established with heterochromatic brightness matching, in which the two spectral distributions are viewed side-by-side and the radiance of one is adjusted relative to the other so that the fields appear equally bright. Though repeatable matches can easily be set with each technique, flicker photometric matches and brightness matches differ for many pairs of spectral distributions.

Photometric radiation transfer calculations and measurements are performed using the same methods and approximations that apply to the radiometric calculations discussed earlier. The exception, of course, is that the spectral sensitivity of the detector is specified.

Retinal Illuminance

In vision research it is frequently required to determine the effectiveness of a uniform, extended field of light (i.e., a large lambertian source that overfills the field of view of the eye) by estimating the illuminance on the retina. If it is assumed that the cornea, lens, and vitreous humor are lossless, then the luminous flux Φ_v in the image on the retina can be approximated from the conservation of the source luminance L_v as follows [see Eq. (22)],

$$L_v = \frac{L_e}{n_e^2} = \frac{\Phi_r s_{pr}^2}{n_c^2 A_r A_p} = E_r \frac{s_{pr}^2}{n_c^2 A_p} \tag{82}$$

where L_e is the radiance within the eye, n_e is the index of refraction of the ocular medium (the index of refraction of air is 1), Φ_r is the luminous flux at the retina, A_r is the area of the image of the retina, A_p is the area of the pupil, s_{pr} is the distance from the pupil to the retina, and E_r is the average illuminance in the image. Therefore, the average illuminance on the retina is

$$E_r = L_v \frac{n_e^2 A_p}{s_{pr}^2} \tag{83}$$

The luminance can, of course, be in units of either photopic or scotopic $\mathrm{cd\,m^{-2}}$. The area of the pupil is measurable, but the distance between the pupil and retina is typically not available. Therefore, a unit of retinal illuminance that avoids the necessity of determining this distance has been defined in terms of just the source luminance and pupil area. This unit is the troland, abbreviated td, and is defined as the retinal illumination for a pupil area of $1\ \mathrm{mm}^2$ produced by a radiating surface having a luminance of $1\ \mathrm{cd\,m^{-2}}$.

$$E_T = L_v A_p \tag{84}$$

Although it may be construed as an equivalent unit, one troland is *not equal* to one microcandela. The source is not a point but is infinite in extent. The troland is useful for relating several vision experiments where sources of differing luminance levels and pupil areas have been used.

The troland is, furthermore, not a measure of the actual illuminance level on the retina since the distance, index of refraction, and transmittance of the ocular medium are not included. For a schematic eye, which is designed to include many of the optical properties of the typical human eye, the effective distance between the pupil and the retina including the effect of the index of refraction is $16.7\ \mathrm{mm}$[83] (see also Vol. I, Chap. 26). For the schematic eye with a $1\text{-}\mathrm{mm}^2$ pupil area, the effective solid angle at the retina is approximately $0.0036\ \mathrm{sr}$. The retinal illuminance equivalent to one troland is therefore $0.0036\ \mathrm{lm\,m^{-2}}$ times the ocular transmittance.

Recall from the section on "Radiometric Effects of Stops and Vignetting" the effect of the aperture stop on the light entering an optical system; that is, all of the light entering the optical system appears to pass through the exit pupil, and the image of the aperture stop on the retina is the exit pupil. If the source is uniform and very large so that it overfills the field of view of the eye, the illuminance on the retina is independent of the distance between the source and the eye. If there is no intervening optic between the eye and the source, then the pupil is the aperture stop. However, if one uses an optical system to image the source into the eye, then the aperture stop need not be the pupil. An external optical system enables both the use of a more uniform, smaller source and the precise control of the retinal illumination by adjustment of an external aperture. The first configuration is called the newtonian view and the second is the maxwellian view of a source[80] (see also Vol. I, Chap. 28).

Though the troland is a very useful and a commonly used photometric unit among vision researchers, it should be interpreted with some caution in situations where one wishes to draw quantitative inferences about the effect of light falling on the retina. The troland is not a precise predictor because, besides not including transmission losses, no angular information is conveyed. The photoreceptors exhibit directional sensitivity where light entering through the center of the pupil is more effective than light entering through the pupil margin (the Stiles-Crawford effect, see Vol. I, Chap. 24). Finally, specifying retinal illuminance in photometric units of trolands does not completely define the experimental conditions because the spectral distribution of the light on the retina is unspecified. Rather, the relative spectral responsivity of the eye (including the spectral dependence of the transmittance) is assumed by the inclusion of the $V(\lambda)$ or $V'(\lambda)$ functions. Experiments performed under mesopic conditions will be particularly prone to error.

If one measures the absolute spectral radiance of the light source, then Eq. (83) may be used in the radiometric form; that is, one substitutes $E_{r\lambda}$ and L_λ for E_r and L_v. The retinal spectral irradiance will then be in absolute units: $\text{W nm}^{-1}\text{m}^{-2}$.

Because the process of vision is a photobiological effect determined by the number and energy of the incident photons, the photon flux irradiance may be a more meaningful measure of the effect of the light on the retina. Using the radiometric form of Eq. (83) and the conversion to photon flux in Eq. (17), the photon flux irradiance $N_{E\lambda}$ on the retina is as follows.

$$N_{E\lambda} = 5.03 \times 10^{15} \lambda \tau_e(\lambda) L_\lambda \frac{n_e^2 A_p}{s_{pr}^2} = 1.80 \times 10^{13} \lambda \tau_e(\lambda) L_\lambda A_p \tag{85}$$

The ocular transmittance is included in this expression as $\tau_e(\lambda)$. The wavelength is in nm and, for radiance in units of $\text{W m}^{-2}\text{sr}^{-1}\text{nm}^{-1}$, the photon flux irradiance is in units of photons $\text{s}^{-1}\text{m}^{-2}\text{nm}^{-1}$.

If one uses a monochromatic light source, then a relationship between a monochromatic troland $E_T(\lambda)$ and the photon flux irradiance may be derived.

$$N_{E\lambda} = 1.80 \times 10^{13} \lambda A_p \tau_e(555) \frac{L_v(\lambda)}{K_m V(\lambda)} = 1.53 \times 10^{10} \lambda \frac{E_T(\lambda)}{V(\lambda)} \tag{86}$$

Only the transmittance at the peak of the $V(\lambda)$ curve, $\tau_e(555) = 0.58^{83}$ needs to be included since the spectral dependence of the ocular transmittance is already included in the $V(\lambda)$ function. The term $L_v(\lambda)$ is the luminance of a monochromatic light source.

An equivalent expression can be derived for the scotopic form of the monochromatic troland $E'_T(\lambda)$. Here, an ocular transmittance at 505 nm of 0.55^{83} has been used.

$$N_{E\lambda} = 5.82 \times 10^9 \lambda \frac{E'_T(\lambda)}{V'(\lambda)} \tag{87}$$

The reader is reminded that Eqs. (86) and (87) are valid only for a monochromatic source.

Absolute Photometric Calibrations

Photometric calibrations are in principle derived from the SI base unit for photometry, the candela. However, as one can see from the definitions of the candela and the other photometric quantities, photometric calibrations are in fact derived from absolute radiometric measurements using either a planckian radiator or an absolute detector. Typically, the relationship between illuminance and irradiance, Eq. (13), is used as the defining equation in deriving a photometric calibration.

The photometric calibration transfer devices available from national standards laboratories are usually incandescent lamps of various designs.[84] The photometric quantities commonly offered as calibrations are luminous intensity and total luminous flux.

The luminous intensity of a lamp, at a specified minimum distance and in a specified direction, is derived from a calibration of the spectral irradiance of the lamp, in the specified direction and at measured distance(s). The radiometric-to-photometric conversion [see Eq. (13)] is used to convert from spectral irradiance to illuminance. The inverse-square-law approximation, Eq. (24), is then used to derive luminous intensity.

Total luminous flux is a measure of all the flux emitted in every direction from a lamp. Total luminous flux is derived from illuminance (or luminous intensity) by measuring the flux emitted in all directions around the lamp. This procedure is known as goniophotometry. For an illuminance-based derivation, the total flux is the average of all the illuminance measurements times the surface area of the sphere described by the locus of the points at which the average illuminance was sampled. In the case of an intensity-based derivation, the total flux is the average of all the intensity measurements times 4π steradians. These are, in principle, the calculation methods for goniophotometry. In practice, the average illuminance (or intensity) is measured in a number of zones of fixed area (or solid angle) around the lamp. The product of the illuminance times the area of the zone (or the intensity times the solid angle of the zone) is the flux. The flux from each of the zones is then summed to obtain the total flux from the lamp.

A number of national standards laboratories provide luminance calibration transfer devices. These are typically in the form of a translucent glass plate that is placed at a specified distance and direction from a luminous intensity standard. One method of deriving the luminance calibration of the lamp/glass unit is to restrict the area of the glass plate with an aperture of known area. The intensity of the lamp/glass combination is then calibrated by comparison to an intensity standard lamp and the average luminance calculated by dividing the measured intensity by the area of the aperture.

Some national standards laboratories also offer calibrations of photometers,[85] also known as illuminance meters. A photometer is a photodetector that has been fitted with a filter to tailor its relative (peak normalized) spectral responsivity to match that of the CIE standard photopic observer. Calibration of a photometer is usually obtained by reference to a luminous intensity standard positioned at a measured distance from the detector aperture. The inverse-square-law approximation is invoked to obtain the value of the illuminance at the measurement distance.

Other Photometric Terminology

Foot-candles, Foot-lamberts, Nits, etc.   The following units of illuminance are often used in photometry, particularly in older texts:

lux (abbreviation: lx) = lumen per square meter

phot (abbreviation: ph) = lumen per square centimeter

meter candle = lumen per square meter

footcandle (abbreviation: fc) = lumen per square foot

One foot-candle = 0.0929 lux.

The following units of luminance are often used:

nit (abbreviation: nt) = candela per square meter

stilb (abbreviation: sb) = candela per square centimeter

It is sometimes the practice, particularly in illuminating engineering, to express the luminance of an actual surface in any given direction in relation to the luminance of a uniform , diffuse, i.e., lambertian, source that emits one lumen per unit area into a solid angle of π steradians [see Eq. (30)]. This concept is one of relative luminance and its units (given following) are not equatable to the units of luminance. Furthermore, in spite of what may appear to be a similarity, this concept is not, strictly speaking, the photometric equivalent of the exitance of a source because, in its definition, the integral of the flux over the entire hemisphere is referenced. In other words, the equivalence to emittance is true only for a perfectly uniform radiance source. For all other sources it is the luminance in a particular direction divided by π. The units for luminance normalized to a lambertian source are:

1 apostilb (abbreviation: asb) = $(1/\pi)$ candela per square meter

1 lambert (abbreviation: L) = $(1/\pi)$ candela per square centimeter

1 foot-lambert (abbreviation: fL) = $(1/\pi)$ candela per square foot

Sometimes the total flux of a source is referred to as its *candlepower* or *spherical candlepower*. This term refers to a point source that uniformly emits in all directions, that is, into a solid angle of 4π steradians. Such a source does not exist, of course, so that the terminology is more precisely stated as the *mean spherical candle power*, which is the mean value of the intensity of the source averaged over the total solid angle subtended by a sphere surrounding the source.

Distribution Temperature, Color Temperature. Distribution temperature is an approximate characterization of the spectral distribution of the visible radiation of a light source. Its use is restricted to sources having relative spectral outputs similar to that of a blackbody such as an incandescent lamp. The mathematical expression for evaluating distribution temperature is

$$\int_{\lambda_1}^{\lambda_2} \left[1 - \frac{\phi_x(\lambda)}{a\phi_b(\lambda, T)} \right]^2 d\lambda \Rightarrow \text{minimum} \tag{88}$$

where $\phi_x(\lambda)$ is the relative spectral radiant power distribution function of the test source, $\phi_b(\lambda, T)$ is the relative spectral radiant power distribution function of a blackbody at the temperature T, and a is an arbitrary constant. The limits of integration are the limits of visible radiation. Since distribution temperature is only an approximation, the exact values of the integration limits are arbitrary: typical limits are 400 and 750 nm. Values of a and T are adjusted simultaneously until the value of the integral is minimized. The temperature of the best-fit blackbody function is the distribution temperature.

Color temperature and correlated color temperature are defined in terms of the perceived color of a source and are obtained by determining the chromaticity of the radiation rather than its relative spectral distribution. Because they are not related to physical photometry they are not defined in this section of the Handbook. These quantities do not provide information about the spectral distribution of the source except when the source has an output that closely approximates a blackbody. Although widely, and mistakenly, used in general applications to characterize the relative spectral radiant power distribution of light sources, color temperature and correlated color temperature relate only to the three types of cone cell receptors for photopic human vision and the approximate manner in which the human brain processes these three signals. As examples of incompatibility, there are the obvious differences between the spectral sensitivity of the human eye and physical receptors of visible optical radiation, e.g., photographic film, TV cameras. In addition, ambiguities occur in the ability of humans to distinguish the perceived color from different spectral distributions (metameric pairs). These ambiguities and the spectral sensitivity functions of the eye are not replicated by physical measurement systems. Caution must be exercised when using color temperature or correlated color temperature to predict the performance of a physical measurement system.

24.7 REFERENCES

1. M. Born and E. Wolf, *Principles of Optics,* 6th ed., Pergamon, Oxford, 1980.

2. E. Wolf, "Coherence and Radiometry," *J. Opt. Soc. Am.* **68:**6 (1978).

3. J. Geist, *McGraw-Hill Encyclopedia of Science and Technology,* McGraw-Hill, New York, 1987, p. 156.

4. W. Blevin et al., "Principles Governing Photometry," *Metrologia* **19:**97 (1983).

5. "The Basis of Physical Photometry," 2d ed., *Commission International de L'Eclairage Publ. No. 18.2,* Central Bureau of the CIE, Vienna, 1983.

6. R. W. Boyd, *Radiometry and the Detection of Optical Radiation,* Wiley, New York, 1983.

7. F. Grum and R. J. Becherer, *Optical Radiation Measurements,* Academic Press, New York 1979.

8. F. Hengstberger, ed. *Absolute Radiometry,* Academic Press, New York, 1989.

9. C. L. Wyatt, *Radiometric Calibration: Theory and Methods,* Academic Press, New York, 1978.

10. E. L. Dereniak and D. G. Crowe, *Optical Radiation Detectors,* Wiley, New York, 1984, pp. 1–14, and Apps. A and C.

11. M. V. Klein and T. E. Furtak, *Optics,* 2d ed., Wiley, New York, 1986, pp. 203–222.

12. T. J. Quinn, *Temperature,* Academic Press, New York, 1983, pp. 284–363.

13. W. S. Smith, *Modern Optical Engineering,* 2d ed., McGraw-Hill, New York, 1990, pp. 135–136, 142–145, and 205–231.

14. M. E. Chahine, D. J. McCleese, P. W. Rosenkranz, and D. H. Staelin, in *Manual of Remote Sensing,* 2d ed., American Society of Photogrammetry, Falls Church, VA, 1983, pp. 172–179.

15. J. E. Hansen and L. D. Travis, "Light Scattering in Planetary Atmospheres," *Space Science Reviews* **16:**527 (1974).

16. Y. J. Kaufman, in Ghassem Asrar (ed.), *Theory and Applications of Optical Remote Sensing,* Wiley, New York, 1989, pp. 350–378.

17. H. Y. Wong, *Handbook of Essential Formulae and Data on Heat Transfer for Engineers,* Longman, London, 1977, pp. 89–128.

18. E. M. Sparrow and R. D. Cess, *Radiation Heat Transfer,* Brooks-Cole, Belmont, CA, 1966.

19. D. G. Goebel, "Generalized Integrating Sphere Theory," *Appl. Opt.* **6:**125 (1967).

20. "Le Système International d'Unités," 3d ed., Bureau International des Poids et Mesures, Sèvres, France, 1977.

21. J. H. Walker, R. D. Saunders, J. K. Jackson, and D. A. McSparron, *Spectral Irradiance Calibrations,* National Bureau of Standards Special Publication No. 250-20, U.S. Government Printing Office, Washington, D.C., 1987.

22. L. P. Boivin, "Calibration of Incandescent Lamps for Spectral Irradiance by Means of Absolute Radiometers," *Appl. Opt.* **19:**2771 (1980).

23. E. F. Zalewski and W. K. Gladden, "Absolute Spectral Irradiance Measurements Based on the Predicted Quantum Efficiency of a Silicon Photodiode," *Opt. Pura y Aplicada* **17:**133 (1984).

24. L. P. Boivin and A. A. Gaertner, "Realization of a Spectral Irradiance Scale in the Near Infrared at the National Research Council of Canada," *Appl. Opt.* **28:**6082 (1992).

25. A. M. Mood and F. A. Graybill, *Introduction to the Theory of Statistics,* McGraw-Hill, New York, 1963.

26. G. W. Snedecor and W. G. Cochran, *Statistical Methods,* Iowa State Univ. Press, Ames, Iowa, 1967.

27. E. B. Wilson, Jr., *An Introduction to Scientific Research,* McGraw-Hill, New York, 1952.

28. H. P. Baltes and E. R. Hilf, *Spectra of Finite Systems,* Bibliographisches Institut Manheim, Vienna, Zurich, 1976.

29. H. P. Baltes, "Deviations from the Stefan-Boltzmann Law at Low Temperatures," *Appl. Phys.* **1:**39 (1973).

30. H. P. Baltes, "Planck's Law for Finite Cavities and Related Problems," *Infrared Phys.* **16:**1 (1976).

31. E. M. Sparrow, L. U. Ulbers, and E. R. Eckert, "Thermal Radiation Characteristics of Cylindrical Enclosures," *J. Heat Transfer* **C84:**188 (1962).

32. B. A. Peavy, "A Note on the Numerical Evaluation of Thermal Radiation Characteristics of Diffuse Cylindrical and Conical Cavities," *J. Res. Natl. Bur. Stds.* **70C:**139 (1966).

33. R. E. Bedford and C. K. Ma, "Emissivities of Diffuse Cavities: Isothermal and Non-isothermal Cones and Cylinders," *J. Opt. Soc. Amer.* **64:**339 (1974).

34. R. E. Bedford and C. K. Ma, "Emissivities of Diffuse Cavities, II: Isothermal and Non-isothermal Cylindro-cones," *J. Opt. Soc. Amer.* **65:**565 (1975).

35. R. E. Bedord and C. K. Ma, "Emissivities of Diffuse Cavities, III: Isothermal and Non-isothermal Double Cones", *J. Opt. Soc. Amer.* **66:**724 (1976).

36. J. C. de Vos, "Evaluation of the Quality of a Blackbody," *Physica* **20:**669 (1954).

37. T. J. Quinn, "The Calculation of the Emissivity of Cylindrical Cavities Giving Near Blackbody Radiation," *Brit. J. Appl. Phys.* **18:**1105 (1967).

38. E. M. Sparrow and V. K. Johnson, "Absorption and Emission Characteristics of Diffuse Spherical Enclosures," *J. Heat Transfer* **C84:**188 (1962).

39. T. J. Quinn, "The Absorptivity of a Specularly Reflecting Cone for Oblique Angles of View," *Infrared Phys.* **21:**123 (1981).

40. T. J. Quinn and J. E. Martin, "Blackbody Source in the −50 to +200°C Range for the Calibration of Radiometers and Radiation Thermometers," *Appl. Opt.* **30:**4486 (1991).

41. E. F. Zalewski, J. Geist, and R. C. Willson, "Cavity Radiometer Reflectance," *Proc. of the SPIE* **196:**152 (1979).

42. J. Schwinger, "On the Classical Radiation of Accelerated Electrons," *Phs. Rev.* **75:**1912 (1949).

43. D. H. Tomboulian and P. L. Hartman, "Spectral and Angular Distribution of Ultraviolet Radiation from the 300-Mev Cornell Synchrotron," *Phys. Rev.* **102:**1423 (1956).

44. D. Lemke and D. Labs, "The Synchrotron Radiation of the 6-Gev DESY Machine as a Fundamental Radiometric Standard," *Appl. Opt.* **6:**1043 (1967).

45. N. P. Fox, P. J. Key, P. J. Riehle, and B. Wende, "Intercomparison Between Two Independent Primary Radiometric Standards in the Visible and near Infrared: a Cryogenic Radiometer and the Electron Storage Ring BESSY," *Appl. Opt.* **25:**2409 (1986).

46. T. J. Quinn and J. E. Martin, "A Radiometric Determination of the Stefan-Boltzmann Constant and Thermodynamic Temperatures Between −40°C and +100°C," *Phil. Trans. Roy. Soc. London* **316:**85 (1985).

47. V. E. Anderson and N. P. Fox, "A New Detector-based Spectral Emission Scale," *Metrologia* **28:**135 (1991).

48. N. P. Fox, J. E. Martin, and D. H. Nettleton, "Absolute Spectral Radiometric Determination of the Melting/freezing Points of Gold, Silver and Aluminum," *Metrologia* **28:**357 (1991).

49. L. Jauniskis, P. Foukal, and H. Kochling, "Absolute Calibration of an Ultraviolet Spectrometer Using a Stabilized Laser and a Cryogenic Radiometer," *Appl. Opt.* **31:**5838 (1992).

50. F. Kurlbaum, "Über eine Methode zur Bestimmung der Strahlung in Absolutem Maass un die Strahlung des schwarzen Köpers zwischen 0 und 100 Grad," *Wied. Ann.* **65:**746 (1898).

51. K. Ångstrom, "The Absolute Determination of the Radiation of Heat with the Electrical Compensation Pyrheliometer, with Examples of the Application of this Instrument," *Astrophys. J.* **9:**332 (1899).

52. W. R. Blevin and W. J. Brown, "Development of a Scale of Optical Radiation," *Austr. J. Phys.* **20:**567 (1967).

53. R. C. Willson, "Active Cavity Radiometer Type V," *Appl. Opt.* **19:**3256 (1980).

54. L. P. Boivin and F. T. McNeely, "Electrically Calibrated Absolute Radiometer Suitable for Measurement Automation," *Appl. Opt.* **25:**554 (1986).

55. J. Geist and W. R. Blevin, "Chopper-stabilized Radiometer Based on an Electrically Calibrated Pyroelectric Detector," *Appl. Opt.* **12:**2532 (1973).

56. R. J. Phelan and A. R. Cook, "Electrically Calibrated Pyroelectric Optical-radiation Detector," *Appl. Opt.* **12:**2494 (1973).

57. J. E. Martin, N. P. Fox, and P. J. Key, "A Cryogenic Radiometer for Absolute Radiometric Measurements," *Metrologia* **21:**147 (1985).

58. C. C. Hoyt and P. V. Foukal, "Cryogenic Radiometers and Their Application to Metrology," *Metrologia* **28:**163 (1991).

59. J. A. R. Samson, "Absolute Intensity Measurements in the Vacuum Ultraviolet," *J. Opt. Soc. Amer.* **54:**6 (1964).

60. F. M. Matsunaga, R. S. Jackson, and K. Watanabe, "Photoionization Yield and Absorption Coefficient of Xenon in the Region of 860–1022Å," *J. Quant. Spectrosc. Radiat. Transfer* **5:**329 (1965).

61. J. Geist, "Quantum Efficiency of the p–n Junction in Silicon as an Absolute Radiometric Standard," *Appl. Opt.* **18:**760 (1979).

62. J. Geist, W. K. Gladden, and E. F. Zalewski, "The Physics of Photon Flux Measurements with Silicon Photodiodes," *J. Opt. Soc. Amer.* **72:**1068 (1982).

63. E. F. Zalewski and J. Geist, "Silicon Photodiode Absolute Spectral Response Self-calibration," *Appl. Opt.* **19:**1214 (1980).

64. J. Geist, E. F. Zalewski, and A. R. Schaefer, "Spectral Response Self-calibration and Interpolation of Silicon Photodiodes," *Appl. Opt.* **19:**3795 (1980).

65. J. L. Gardner and W. J. Brown, "Silicon Radiometry Compared to the Australian Radiometric Scale," *Appl. Opt.* **26:**2341 (1987).

66. E. F. Zalewski and C. C. Hoyt, "Comparison Between Cryogenic Radiometry and the Predicted Quantum Efficiency of Silicon Photodiode Light Traps," *Metrologia* **28:**203 (1991).

67. E. F. Zalewski and C. R. Duda, "Silicon Photodiode Device with 100 Percent External Quantum Efficiency," *Appl. Opt.* **22:**2867 (1983).

68. N. P. Fox, "Trap Detectors and Their Properties," *Metrologia* **28:**197 (1991).

69. J. H. Walker, R. D. Saunders, and A. T. Hattenburg, *Spectral Radiance Calibrations,* National Bureau of Standards Special Publication No. 250-1, U.S. Government Printing Office, Washington, D.C., 1987.

70. J. Z. Klose, J. M. Bridges, and W. R. Ott, *Radiometric Standards in the Vacuum Ultraviolet,* National Bureau of Standards Special Publication No. 250-3, U.S. Government Printing Office, Washington, D.C., 1987.

71. E. F. Zalewski, *The NBS Photodetector Spectral Response Calibration Transfer Program,* National Bureau of Standards Special Publication No. 250-17, U.S. Government Printing Office, Washington, D.C., 1987.

72. W. Budde, *Physical Detectors of Optical Radiation*, Academic Press, New York, 1983.

73. L. P. Boivin, "Some Aspects of Radiometric Measurements Involving Gaussian Laser Beams," *Metrologia* **17:**19 (1981).

74. L. P. Boivin, "Reduction of Diffraction Errors in Radiometry by Means of Toothed Apertures," *Appl. Opt.* **17:**3323 (1978).

75. W. R. Blevin, "Diffraction Losses in Photometry and Radiometry," *Metrologia* **6:**31 (1970).

76. C. L. Sanders, "A Photocell Linearity Tester," *Appl. Opt.* **1:**207 (1962).

77. C. L. Sanders, "Accurate Measurements of and Corrections for Non-linearities in Radiometers," *J. Res. Natl. Bur. Stand.* **A76:**437 (1972).

78. W. Budde, "Multidecade Linearity Measurements on Silicon Photodiodes," *Appl. Opt.* **18:**1555 (1979).

79. A. R. Schaefer, E. F. Zalewski, and J. Geist, "Silicon Detector Non-linearity and Related Effects," *Appl. Opt.* **22:**1232 (1983).

80. J. W. T. Walsh, *Photometry,* Dover, New York, 1965.

81. G. Wyszecki and W. S. Stiles, *Colour Science*: *Concepts and Methods,* Wiley, New York, 1967.

82. "Light as a True Visual Quantity: Principles of Measurement," *Commission International de L'Eclairage Publ. No. 41*, Central Bureau of the CIE, Vienna, 1978.

83. E. N. Pugh, "Vision: Physics and Retinal Physiology," in R. C. Atkinson, R. J. Herrnstein, G. Lindsey, and R. D. Luce, eds., Steven's Handbook of Experimental Psychology, 2d ed., Wiley, New York, 1988, pp. 75–163.

84. R. L. Booker and D. A. McSparron, *Photometric Calibrations,* National Bureau of Standards Special Publication No. 250–15, U.S. Government Printing Office, Washington, D.C., 1987.

85. "Methods of Characterizing the Performance of Radiometers and Photometers," *Commission International de L'Eclairage Publ. No. 53*, Central Bureau of the CIE, Vienna, 1982.

CHAPTER 25
THE MEASUREMENT OF TRANSMISSION, ABSORPTION, EMISSION, AND REFLECTION

James M. Palmer
Optical Sciences Center
University of Arizona
Tucson, Arizona

25.1 GLOSSARY

A	area
a	absorption
bb	blackbody
c_2	second radiation constant
E	irradiance
f	bidirectional scattering distribution function
i	internal
L	radiance
P	electrical power
R	reflectance factor
r	reflection
T	temperature
t	transmission
α	absorptance
α'	absorption coefficient
ε	emittance (emissivity)
θ, ϕ	angles
ρ	reflectance
σ	Stefan Boltzmann constant
τ	transmittance
Φ	power (flux)
Ω	projected solid angle

25.2   *INTRODUCTION AND TERMINOLOGY*

When radiant flux is incident upon a surface or medium, three processes occur: transmission, absorption, and reflection. Figure 1 shows the ideal case, where the transmitted and reflected components are either specular or perfectly diffuse. Figure 2 shows the transmission and reflection for actual surfaces.

The symbols, units, and nomenclature employed in this chapter follow the established

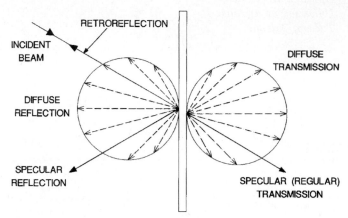

FIGURE 1   Idealized reflection and transmission.

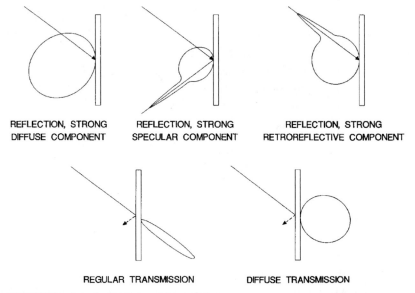

FIGURE 2   Actual reflection and transmission.

usage as defined in *ISO Standards Handbook 2* (1982), Cohen and Giacomo (1987), and Taylor (1991). Additional general terminology applicable to this chapter is from ASTM (1971), IES (1979), IES (1986), Drazil (1983), and CIE (1987). The prefix *spectral* is used to denote a characteristic at a particular wavelength and is indicated by the symbol (λ). The absence of the *spectral* prefix implies integration over all wavelengths with a source function included (omission of the source function is meaningless except where the characteristic is constant with wavelength.)

There has been a continuing dialog over terminology, particularly between the suffixes *-ance* and *-ivity* (Worthing, 1941; Richmond, 1962; Richmond et al., 1963; Wolfe, 1982; Richmond, 1982). The suggested usage reserves terms ending with *-ivity* (such as transmissivity, absorptivity, and reflectivity) for properties of a pure material and employs the suffix *-ance* for the characteristics of a specimen or sample. For example, one can distinguish between the reflec*tivity* of pure aluminum and the reflec*tance* of a particular sample of 6061-T6 aluminum with a natural oxide layer. This distinction can be extended to differentiate between emissivity (of a pure substance) and emittance (of a sample). This usage of *emittance* should not be confused with the older term *radiant emittance*, now properly called *radiant exitance*. In this chapter, the suffix *-ance* will be used exclusively inasmuch as the measurement of radiometric properties of materials is under discussion.

25.3 TRANSMITTANCE

Transmission is the term used to describe the process by which incident radiant flux leaves a surface or medium from a side other than the incident side, usually the opposite side. The spectral transmittance $\tau(\lambda)$ of a medium is the ratio of the transmitted spectral flux $\Phi_{\lambda t}$ to the incident spectral flux $\Phi_{\lambda i}$, or

$$\tau(\lambda) = \frac{\Phi_{\lambda t}}{\Phi_{\lambda i}} \tag{1}$$

The transmittance τ is the ratio of the transmitted flux Φ_t, to the incident flux Φ_i, or

$$\tau = \frac{\int_0^\infty \tau(\lambda)\Phi_{\lambda i}\, d\lambda}{\int_0^\infty \Phi_{\lambda i}\, d\lambda} \neq \int_\lambda \tau(\lambda)\, d\lambda \tag{2}$$

Note that the integrated transmittance is *not* the integral over wavelength of the spectral transmittance, but must be weighted by a source function Φ_λ as shown.

The transmittance may also be described in terms of radiance as follows:

$$\tau = \frac{\int_0^\infty \int_\Omega L_{\lambda i}\, d\Omega_t\, d\lambda}{\int_0^\infty \int_\Omega L_{\lambda i}\, d\Omega_i\, d\lambda} \tag{3}$$

where $L_{\lambda i}$ represents the spectral radiance $L_{\lambda i}(\lambda; \theta_i, \phi_i)$ incident from direction (θ_i, ϕ_i), $L_{\lambda t}$ represents the spectral radiance $L_{\lambda t}(\lambda; \theta_t, \phi_t)$ transmitted in direction (θ_t, ϕ_t), and $d\Omega$ is the elemental projected solid angle $\sin\theta \cos\theta\, d\theta\, d\phi$.

The bidirectional transmittance distribution function (BTDF, symbol f_t) relates the transmitted radiance to the radiant incidence as:

$$f_t(\lambda; \theta_i, \phi_i) \equiv \frac{dL_{\lambda t}}{dL_{\lambda i} d\Omega_i} = \frac{dL_{\lambda t}}{dE_{\lambda i}} \quad (\text{sr}^{-1}) \tag{4}$$

Geometrically, transmittance can be classified as specular, diffuse, or total, depending upon whether the specular (regular) direction, all directions other than the specular, or all directions are considered.

25.4 ABSORPTANCE

Absorption is the process by which incident radiant flux is converted to another form of energy, usually heat. Absorptance is the fraction of incident flux that is absorbed. The absorptance α of an element is defined by $\alpha = \Phi_a/\Phi_i$. Similarly, the spectral absorptance $\alpha(\lambda)$ is the ratio of spectral power absorbed $\Phi_{\lambda a}$ to the incident spectral power $\Phi_{\lambda i}$,

$$\alpha = \frac{\displaystyle\int_0^\infty \alpha(\lambda)\Phi_{\lambda i}\, d\lambda}{\displaystyle\int_0^\infty \Phi_{\lambda i}\, d\lambda} \neq \int_\lambda \alpha(\lambda)\, d\lambda \tag{5}$$

An absorption coefficient α' (cm^{-1} or km^{-1}) is often used in the expression $\tau_i = e^{-\alpha' t}$, where τ_i is internal transmittance and t is pathlength (cm or km).

25.5 REFLECTANCE

Reflection is the process where a fraction of the radiant flux incident on a surface is returned into the same hemisphere whose base is the surface and which contains the incident radiation. The reflection can be specular (in the mirror direction), diffuse (scattered into the entire hemisphere), or a combination of both. Table 1 (CIE, 1977) shows a wide range of materials that have different goniometric (directional) reflectance characteristics.

The most general definition for reflectance ρ is the ratio of the radiant flux reflected Φ_r to the incident radiant flux Φ_i, or

$$\rho = \frac{\Phi_r}{\Phi_i} \tag{6}$$

Spectral reflectance is similarly defined at a specified wavelength λ as

$$\rho(\lambda) = \frac{\Phi_{\lambda r}}{\Phi_{\lambda i}} \tag{7}$$

(Spectral) reflectance factor (symbol R) is the ratio of (spectral) flux reflected from a sample to the (spectral) flux which would be reflected by a perfect diffuse (lambertian) reflector.

No single descriptor of reflectance will suffice for the wide range of possible geometries.

TABLE 1 Goniometric Classification of Materials (CIE, 1977)

Material classification	Scatter	σ	γ	Structure	Example
Exclusively reflecting materials	None	0	$\cong 0$	None	Mirror
	Weak	≤ 0.4	$\leq 27°$	Micro	Matte aluminum
				Macro	Retroreflectors
$\tau = 0$	Strong	>0.4	$>27°$	None	Laquer & enamel coatings
				Micro	Paint films, $BaSO_4$, Halon
				Macro	Rough tapestries, road surfaces
Weakly transmitting, strongly reflecting materials	None	0	$\cong 0$	None	Sunglasses, color filters cold mirrors
	Weak	≤ 0.4	$\leq 27°$	Micro	Matte-surface color filters
$\tau \leq 0.35$				Macro	Glossy textiles
	Strong	>0.4	$>27°$	None	Highly turbid glass
				Micro	Paper
				Macro	Textiles
Strongly transmitting materials	None	0	$\cong 0$	None	Window glass
	Weak	≤ 0.4	$\leq 27°$	None	Plastic film
				Micro	Ground glass
$\tau > 0.35$				Macro	Ornamental glass prismatic glass
	Strong	>0.4	$>27°$	None	Opal glass
				Micro	Ground opal glass
				Macro	Translucent acrylic plastic with patterned surface

[1] Structure refers to the nature of the surface. In a microscattering structure, the scatterers cannot be resolved with the unaided eye. The macrostructure scatterers can be readily seen.

[2] σ is a diffusion factor, the ratio of the mean of radiance measured at 20° and 70° to the radiance measured at 5° from the normal, when the incoming radiation is normal. $\sigma = [L(20) + L(70)]/[2L(5)]$. It gives an indication of the spatial distribution of the radiance, and is unity for a perfect (lambertian) diffuser.

[3] γ is a half-value angle, the angle from the normal where the radiance has dropped to one-half the value at normal.

[4] It is suggested that the diffusion factor is appropriate for strongly diffusing materials and that the half-angle is better for weakly diffusing materials.

The fundamental geometric descriptor of reflectance is the bidirectional reflectance distribution function (BRDF, symbol f_r). It is defined as the differential element of reflected radiance dL_r in a specified direction per unit differential element of radiant incidence dE_i, also in a specified direction (Nicodemus et al., 1977), and carries unit of sr^{-1}:

$$f_r(\theta_i, \phi_i, \theta_r, \phi_r) = \frac{dL_r(\theta_i, \phi_i; \theta_r, \phi_r; E_i)}{dE_i(\theta_i, \phi_i)} \quad [sr^{-1}] \tag{8}$$

The polar angle θ is measured from the surface normal and the azimuth angle ϕ is measured from an arbitrary reference in the surface plane, most often the plane containing the incident beam. The subscripts i and r refer to the incident and reflected beams, respectively.

By integrating over varying solid angles, Nicodemus et al. (1977), based upon earlier work by Judd (1967), defined nine goniometric reflectances, and by extension, nine goniometric reflectance factors. These are shown in Tables 2 and 3 and Fig. 3. In these

TABLE 2   Nomenclature for Nine Types of Reflectance (Nicodemus et al., 1977)

1. Bidirectional reflectance

$$d\rho(\theta_i, \phi_i; \theta_r, \phi_r) = f_r(\theta_i, \phi_i; \theta_r, \phi_r) \cdot d\Omega_r$$

2. Directional-conical reflectance

$$\rho(\theta_i, \phi_i; \omega_r) = \int_{\omega_r} f_r(\theta_i, \phi_i; \theta_r, \phi_r) \cdot d\Omega_r$$

3. Directional-hemispherical reflectance

$$\rho(\theta_i, \phi_i; 2\pi) = \int_{2\pi} f_r(\theta_i, \phi_i; \theta_r, \phi_r) \cdot d\Omega_r$$

4. Conical-directional reflectance

$$d\rho(\omega_i; \theta_r, \phi_r) = (d\Omega_r/\Omega_i) \cdot \int_{\omega_i} f_r(\theta_i, \phi_i; \theta_r, \phi_r) \cdot d\Omega_i$$

5. Biconical reflectance

$$\rho(\omega_i; \omega_r) = (1/\Omega_i) \cdot \int_{\omega_i} \int_{\omega_r} f_r(\theta_i, \phi_i; \theta_r, \phi_r) \cdot d\Omega_r \cdot d\Omega_i$$

6. Conical-hemispherical reflectance

$$\rho(\omega_i; 2\pi) = (1/\Omega_i) \cdot \int_{\omega_i} \int_{2\pi} \cdot \int_{2\pi} f_r(\theta_i, \phi_i; \theta_r, \phi_r) \cdot d\Omega_r \cdot d\Omega_i$$

7. Hemispherical-directional reflectance

$$d\rho(2\pi; \theta_r, \phi_r) = (d\Omega_r/\pi) \cdot \int_{2\pi} f_r(\theta_i, \phi_i; \phi_r, \phi_r) \cdot d\Omega_i$$

8. Hemispherical-conical reflectance

$$\rho(2\pi; \omega_r) = (1/\pi) \cdot \int_{2\pi} \int_{\omega_r} f_r(\theta_i, \phi_i; \theta_r, \phi_r) \cdot d\Omega_r \cdot d\Omega_i$$

9. Bihemispherical reflectance

$$\rho(2\pi; 2\pi) = (1/\pi) \cdot \int_{2\pi} \int_{2\pi} f_r(\theta_i, \phi_i; \theta_r, \phi_r) \cdot d\Omega_r \cdot d\Omega_i$$

TABLE 3   Nomenclature for Nine Types of Reflectance Factor (Nicodemus et al., 1977)

1. Bidirectional reflectance factor

$$R(\theta_i, \phi_i; \theta_r, \phi_r) = \pi f_r(\theta_i, \phi_i; \theta_r, \phi_r)$$

2. Directional-conical reflectance factor

$$R(\theta_i, \phi_i; \omega_r) = (\pi/\Omega_r) \cdot \int_{\omega_r} f_r(\theta_i, \theta_j; \theta_r, \phi_r) \cdot d\Omega_r$$

3. Directional-hemispherical reflectance factor

$$R(\theta_i, \phi_i; 2\pi) = \int_{2\pi} f_r(\theta_i, \phi_i; \theta_r, \phi_r) \cdot d\Omega_r$$

4. Conical-directional reflectance factor

$$R(\omega_i; \theta_r, \phi_r) = (\pi/\Omega_i) \cdot \int_{\omega_i} f_r(\theta_i, \phi_i; \theta_r, \phi_r) \cdot d\Omega_i$$

5. Biconical reflectance factor

$$R(\omega_i; \omega_r) = [\pi/(\Omega_i \cdot \Omega_r)] \cdot \int_{\omega_i} \int_{\omega_r} f_r(\theta_i, \phi_i; \theta_r, \phi_r) \cdot d\Omega_r \cdot d\Omega_i$$

6. Conical-hemispherical reflectance factor

$$R(\omega_i; 2\pi) = (1/\Omega_i) \cdot \int_{\omega_i} \int_{2\pi} f_r(\theta_i, \phi_i; \theta_r, \phi_r) \cdot d\Omega_r \cdot d\Omega_i$$

7. Hemispherical-directional reflectance factor

$$R(2\pi; \theta_r, \phi_r) = \int_{2\pi} f_r(\theta_i, \phi_i; \theta_r, \phi_r) \cdot d\Omega_i$$

8. Hemispherical-conical reflectance factor

$$R(2\pi; \omega_r) = (1/\Omega_r) \cdot \int_{2\pi} \int_{\omega_r} f_r(\theta_i, \phi_i; \theta_r, \phi_r) \cdot d\Omega_r \cdot d\Omega_i$$

9. Bihemispherical reflectance factor

$$R(2\pi; 2\pi) = (1/\pi) \cdot \int_{2\pi} \int_{2\pi} f_r(\theta_i, \phi_i; \theta_r, \phi_r) \cdot d\Omega_r \cdot d\Omega_i$$

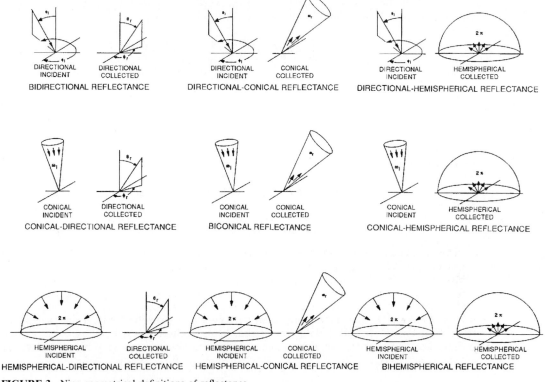

FIGURE 3   Nine geometrical definitions of reflectance.

tables, the term *directional* refers to a differential solid angle $d\omega$ in the direction specified by (θ, ϕ). *Conical* refers to a cone of finite extent centered in direction (θ, ϕ); the solid angle ω of the cone must also be specified.

   Details on these definitions and further discussion can be found in ASTM STP475 (1971), ASTM E808 (1981), Judd (1967), Nicodemus (1965), Nicodemus (1970), and Nicodemus et al. (1977).

25.6   *EMITTANCE*

Emittance (ε) is the ratio of the radiance of an object or surface to the radiance of a blackbody (planckian radiator) at the same temperature. It is therefore dimensionless and can assume values between 0 and 1 for thermal radiators at equilibrium. Spectral emittance $\varepsilon(\lambda)$ is the emittance at a given wavelength. If a radiator is neutral with respect to wavelength, with a constant spectral emittance less than unity, it is called a graybody.

$$\varepsilon = \frac{L}{L^{bb}} \qquad \varepsilon(\lambda) = \frac{L_\lambda}{L_\lambda^{bb}} \tag{9}$$

Directional emittance $\varepsilon(\theta, \phi)$ is defined by

$$\varepsilon(\theta, \phi) = \frac{L(\theta, \phi)}{L^{bb}} \tag{10}$$

Note that if the body is nongray, its emittance is dependent upon temperature inasmuch as the integral must be weighted by the source (Planck) function.

$$\varepsilon = \frac{\displaystyle\int_0^\infty \varepsilon(\lambda) L_\lambda^{bb}\, d\lambda}{\displaystyle\int_0^\infty L_\lambda^{bb}\, d\lambda} = \frac{1}{\pi} \cdot \frac{\displaystyle\int_0^\infty \varepsilon(\lambda) L_\lambda^{bb}\, d\lambda}{\sigma T^4} \tag{11}$$

25.7   KIRCHHOFF'S LAW

In a closed system at thermal equilibrium, conservation of energy necessitates that emitted and absorbed fluxes are equal. Since the radiation field in such a system is isotropic (the same in all directions), the directional spectral emittance and the directional spectral absorptance must be equal, i.e.,

$$\varepsilon(\lambda;\, \theta,\, \phi) = \alpha(\lambda;\, \theta,\, \phi) \tag{12}$$

This statement was first made by Kirchhoff (1860). Strictly, this equation holds for each orthogonal polarization component, and for it to be valid as written, the total radiation must have equal orthogonal polarization components. Kirchhoff's law is often simplified to the declaration $\alpha = \varepsilon$; however, this is not a universal truth; it may only be applied under a limited set of conditions. The geometrical and spectral averaging (integration) is governed by a specific set of rules as demonstrated by Siegel and Howell (1981). Table 4, adapted from Siegel and Howell (1981) and Grum and Becherer (1979), shows the various geometrical and spectral conditions under which the absorptance may be related to the emittance.

25.8   RELATIONSHIP BETWEEN TRANSMITTANCE, REFLECTANCE, AND ABSORPTANCE

Radiant flux incident upon a surface or medium undergoes transmission, reflection, and absorption. Application of conservation of energy leads to the statement that the sum of the transmission, reflection, and absorption of the incident flux is equal to unity, or

$$\alpha + \tau + \rho = 1 \tag{13}$$

In the absence of nonlinear effects (i.e., Raman effect, etc.),

$$\alpha(\lambda) + \tau(\lambda) + \rho(\lambda) = 1 \tag{14}$$

If the situation is such that one of the above Kirchhoff-type relations is applicable, then emittance ε may be substituted for absorptance α in the previous equations, or

$$\varepsilon = 1 - \tau - \rho \qquad \varepsilon(\lambda) = 1 - \tau(\lambda) - \rho(\lambda) \tag{15}$$

25.9   MEASUREMENT OF TRANSMITTANCE

A knowledge of the transmission of optical materials and elements, gaseous atmospheres, and various liquids is necessary throughout the realm of optics. Most of these measure-

TABLE 4 Summary of Absorptance-Emittance Relations (Siegel and Howell, 1981)

Quantity	Equality	Required conditions
Directional spectral	$\alpha(\lambda; \theta, \phi, T_a) = \varepsilon(\lambda; \theta, \phi, T_a)$	None other than thermal equilibrium
Directional total	$\alpha(\theta, \phi, T_a) = \varepsilon(\theta, \phi, T_a)$	(1) Spectral distribution of incident energy proportional to blackbody at T_a, or
		(2) $\alpha(\lambda; \theta, \phi, T_a) = \varepsilon(\lambda; \theta, \phi, T_a)$ independent of wavelength
Hemispherical spectral	$\alpha(\lambda, T_a) = \varepsilon(\lambda, T_a)$	(1) Incident radiation independent of angle, or
		(2) $\alpha(\lambda; \theta, \phi, T_a) = \varepsilon(\lambda; \theta, \phi, T_a)$ independent of angle
Hemispherical total	$\alpha(T_a) = \varepsilon(T_a)$	(1) Incident energy independent of angle *and* spectral distribution proportional to blackbody at T_a, or
		(2) Incident energy independent of angle *and* $\alpha(\lambda; \theta, \phi, T_a) = \varepsilon(\lambda; \theta, \phi, T_a)$ independent of wavelength, or
		(3) Incident energy at each angle has spectral distribution proportional to blackbody at T_a and $\alpha(\lambda; \theta, \phi, T_a) = \varepsilon(\lambda; \theta, \phi, T_a)$ independent of angle, or
		(4) $\alpha(\lambda; \theta, \phi, T_a) = \varepsilon(\lambda; \theta, \phi, T_a)$ independent of angle and wavelength

ments are made with commercial spectrophotometers. It is beyond the scope to discuss the design and operation of spectrophotometric equipment beyond sample-handling practices. For further discussion, see Grum and Becherer (1979), ASTM E275 (1989), and ASTM E409 (1990).

Conventional spectrophotometers are of the double-beam configuration, where the output is the ratio of the signal in the sample beam to the signal in the reference beam plotted as a function of wavelength. It is incumbent upon the experimenter to ensure that the only difference between the two beams is the unknown. Therefore, if liquid or gas cells are employed, one should be placed in each beam. For gas cells, an equal amount of carrier gas should be injected into each cell, with the unknown to be sample placed in only one cell, destined for the sample beam. For liquids, an equal amount of solute should be placed in each cell. A critical issue with liquid and solid samples is the beam geometry. Most spectrophotometers feature converging beams in the sample space. If the optical path (the product of index of refraction and actual distance) is not identical for each beam, a systematic difference is presented to either the entrance slit or the detector. In addition, some specimens (e.g., interference filters) are susceptible to errors when measured in a converging beam. Most instruments also have a single monochromator which is susceptible to stray radiation, the limiting factor when trying to make measurements of samples that are highly absorbing in one spectral region and transmitting in another. Some recent instruments feature linear detector arrays along with single monochromators to allow the acquisition of the entire spectrum in several milliseconds; these are particularly applicable to reaction rate studies.

Conventional double-beam instruments are limited by these factors to uncertainties on the order of 0.1 percent. For lower uncertainties, the performance deficiencies found in double-beam instruments can largely be overcome by the use of a single-beam architecture. The mode of operation is sample-in–sample-out. If the source is sufficiently stable with time, the desired spectral range can be scanned without the sample, then rescanned with the sample in place. Otherwise, the spectrometer can be set at a fixed wavelength and alternate readings with and without the sample in place must be made. Care should be taken to ensure that the beam geometry is not altered between sequential readings.

To achieve the ultimate in performance from conventional spectrophotometry, several design characteristics should be included. A double monochromator is essential to minimize stray light. The beam geometry in the sample compartment should be highly collimated to avoid focus shifts with optically thick samples. Some form of beam integration, such as an integrating sphere or other diffuser, should be employed to negate the effects of nonuniform detectors and beam shifts. An exemplary instrument is the high-accuracy spectrophotometer developed by the National Institute for Standards and Technology (NIST), described in Mielenz et al. (1972), Mielenz et al. (1973), Venable et al. (1976), Eckerle (1976), and Eckerle et al. (1983). A particularly useful review is Eckerle et al. (1987). Similar laboratory instruments have also been built elsewhere by Clarke (1972), Freeman (1986), and Zwinkles and Gignac (1992).

Numerous other instruments have been described in the literature; some have been designed for singular or limited purposes while others have a more universal appeal. Use of integrating spheres is common, both for the averaging effects and for the isolation of the specular and diffuse components, as shown in Fig. 4 (CIE, 1977; Roos, 1991). Several useful instruments are described by Karras (1921), Taylor (1922), Zerlaut and Anderson (1981), Clarke and Larkin (1985), and Kessel (1986).

Conventional instruments lack a wide dynamic range because there are simply not enough photons available in a narrow bandpass in a reasonable time. Solutions include Fourier transform spectrometers with a large multiplex advantage, the use of tunable lasers, and heterodyne spectrometry (Migdall et al., 1990).

Simple instruments can be purchased or constructed for specific purposes. For example, solar transmittance can be determined using either the natural sun (if available) or simulated solar radiation as the source. A limited degree of spectral isolation can be achieved with an abridged spectrophotometer using narrow bandpass interference or glass absorption filters.

Several publications have suggested methods for making accurate and repeatable measurements including Hughes (1963), Mielenz (1973), Venable and Hsia (1974), Burke and Mavrodineanu (1983), ASTM F768 (1987), ASTM E971 (1988), ASTM E903 (1988), and ASTM E179 (1990). Calibration and performance assessment of spectrophotometers includes photometric accuracy, linearity, stray light analysis, and wavelength calibration. Particular attention should be paid to luminescent samples that absorb radiant energy in

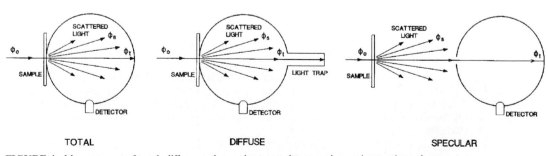

FIGURE 4 Measurement of total, diffuse, and specular transmittance using an integrating sphere.

one spectral region and re-emit it at longer wavelengths. Pertinent references include Hawes (1971), Bennett and Ashley (1972), ASTM E387 (1984), ASTM E275 (1989), and ASTM, E409 (1990).

Standards of spectral transmittance are available as Standard Reference Materials from NIST. These take the form of metal-on-glass, metal-on-quartz, and solid glass filters. Most are used for verifying the photometric scale or for checking the wavelength calibration of a recording spectrophotometer. Descriptions of their development and use are given in Mavrodineanu and Baldwin (1975), Mavrodineanu and Baldwin (1980), Eckerle et al. (1987), and Hsia (1987). The standardization laboratories of several countries occasionally conduct international intercomparisons of traveling standards. Recent intercomparisons have been reported in Eckerle et al. (1990) and Fillinger and Andor (1990).

25.10 *MEASUREMENT OF ABSORPTANCE*

In most cases, absorptance is not directly measured, but is inferred from transmission measurements, with appropriate corrections for reflection losses. These corrections can be calculated from the Fresnel equations if the surfaces are polished and the index of refraction is known. For materials where the absorption is extremely small, this method is unsatisfactory, as the uncertainties are dominated by the reflection contribution. In this case, direct measurements (such as laser calorimetry) must be made as discussed by Lipson et al. (1974) and Hordvik (1977).

25.11 *MEASUREMENT OF REFLECTANCE*

Instrumentation for the measurement of reflectance takes many forms. Only a few of the definitions for reflectance (Table 2) and reflectance factor (Table 3) have been adopted as standard configurations. The biconical configuration with small solid angles is most suited to a measurement of specular (regular, in the mirror direction) reflection. A simple reflectometer for the absolute measurement of specular reflectance was devised by Strong (1938, 1989) and is shown schematically in Fig. 5. Numerous detail improvements have

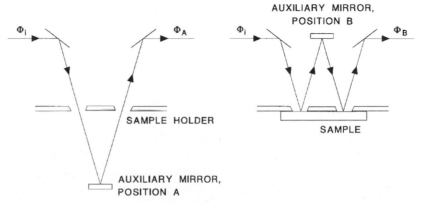

FIGURE 5 The Strong "VW" reflectometer.

been made on this fundamental design, including the use of averaging spheres. Designs range from simple (Castellini et al., 1990; Ram et al., 1991; Weeks, 1958; Weidner and Hsia, 1980; Zhuang and Yang, 1989) to complex (Bennett and Koehler, 1960). Some reflectometers have been built specifically to measure at normal incidence (Bittar and Hamlin, 1984; Boivin and Theriault, 1981; Shaw and Blevin, 1964). Measurement methods and data interpretation are also given in ASTM F768 (1987), ASTM D523 (1988), ASTM F1252 (1990), Hernicz and DeWitt (1973), and Snail et al. (1986).

The characterization of appearance of materials involves measurements of reflectance, both diffuse and specular. Numerous procedures and instruments have been devised for goniophotometry, the measurement of specular gloss with biconical geometry. Measurements are made at several angles from normal (20°, 30°, 45°, 60°, 75°, and 85°) depending upon the material under scrutiny. Further details can be found in ASTM C347 (1983), ASTM E167 (1987), ASTM D523 (1988), ASTM E1349 (1990), ASTM E179 (1990), ASTM E430 (1983), Erb (1980, 1987), and Hunter (1975).

The measurement of diffuse reflectance can be accomplished using any one of the nine definitions from Table 2 and integrating where necessary. One could, for example, choose to measure the bidirectional reflectance distribution function as a function of incident beam parameters and to integrate over the hemisphere, but this would be a tedious process, and the large amount of data generated would be useful only to those involved with detailed materials properties research. Most practical measurements of diffuse reflectance involve the use of an integrating sphere. Several papers have discussed the general theory of the integrating sphere (Jacques and Kuppenheim, 1955; Goebel, 1967, and references therein).

In the visible and near-IR spectral regions, the integrating sphere is the instrument of choice for both specular and diffuse specimens. Many papers have been written detailing instruments, methods, and procedures, some of which are shown in Fig. 6. The specular

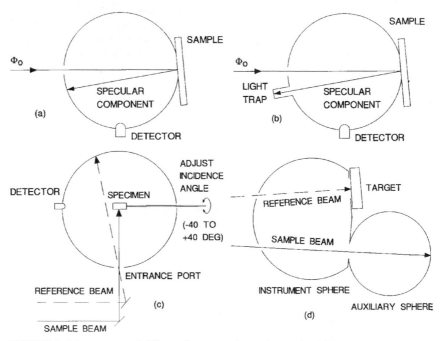

FIGURE 6 Measurement of diffuse reflectance using an integrating sphere.

component of the reflected flux can be included to determine the total reflectance (Fig. 6*a*) or excluded to measure just the diffuse component (Fig. 6*b*). The angle of incidence can be varied by placing the sample at the center of the sphere (Fig. 6*c*, Edwards et al., 1960). Others making contributions include Clarke and Compton (1986), Clarke and Larkin (1988), Dunkle, (1960), Egan and Hilgeman (1975), Goebel et al. (1966), Hisdal (1965a, 1965b), Karras (1921), McNicholas (1928), Richter and Erb (1987), Sheffer et al. (1987, 1991), Taylor (1935), and Venable et al. (1976). Some of these methods have been incorporated into standard methods and practices, such as ASTM C523 (1984), ASTM E429 (1987), ASTM E903 (1988), CIE (1979a), and IES (1974). Most integrating sphere measurements require reference to some form of an artifact standard, but the double-sphere method (Fig. 6*d*) produces absolute diffuse reflectance (Van den Akker et al., 1966; Venable et al., 1977; Zerlaut and Anderson, 1981; Goebel et al., 1966). Lindberg (1987) has demonstrated a method to scale relative measurements to absolute.

Alternative forms of hemispherical irradiation and/or collection have been described, several of which are shown in Fig. 7. Specular hemispherical (Fig. 7*a*), paraboloidal (Fig. 7*b*), and ellipsoidal (Fig. 7*c*) collectors have been used, particularly in those spectral regions where integrating sphere coatings are difficult to obtain (Barnes and Hsia, 1989; Blevin and Brown, 1965; Dunkle, 1960; Dunn et al., 1966b; Hanssen & Snail, 1987; Hartman and Logothetis, 1964; Neu et al., 1987; Wood et al., 1976). The Helmholtz reciprocity principle has been invoked to demonstrate the reversibility of the source and

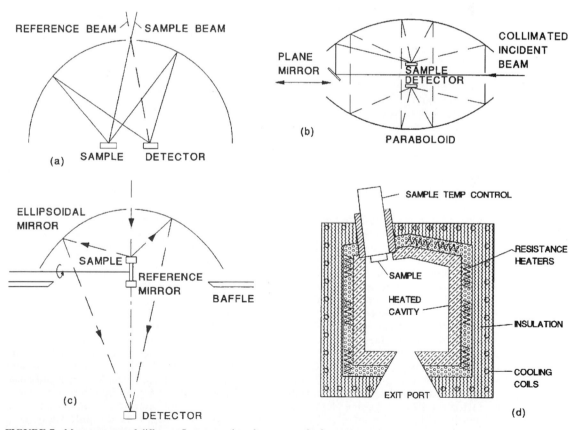

FIGURE 7   Measurement of diffuse reflectance using alternate methods.

the collector (Clarke and Parry, 1985). Hemispherical irradiation has also been employed by placing a cooled sample coplanar with the wall of a furnace and comparing the radiance of the sample with that of the wall of the furnace as shown in Fig. 7d (Agnew and McQuistan, 1953; Gier et al., 1954; Reid and McAlister, 1959; Dunkle, 1960).

The procedures and instrumentation for the measurement of reflectance factor are identical with those for diffuse reflectance using the 0°/45° or 45°/0° geometry with annular, circumferential, or uniplanar illumination or viewing. Reference is made to a white reflectance standard characterized for reflectance factor, which must be compared with a perfect diffuse reflector. Pertinent references are ASTM E1349 (1990), ASTM E97 (1987), ASTM E1348 (1990), Hsia and Weidner (1981), and Taylor (1920, 1922).

Orbiting sensors measure the radiance of the earth-atmosphere system with some known geometry (generally nadir) and in well-defined wavelength bands. The quantity of interest is reflectance, as it is related to factors such as crop assessment, mineralization, etc. Corrections must be made for the atmospheric absorption, emission, and scattering and for the BRDF (bidirectional reflectance distribution function) of the target. BRDF has been characterized in the field using the sun as the source, as described by Duggin (1980, 1982).

Laboratory measurements of BRDF are made using goniometers where the sample-source angle and sample-receiver angle are independently adjustable. Coherent sources (lasers) are employed for the characterization of smooth specimens where a large source power is necessary for adequate SNR for off-specular angles and where speckle is not a concern. Incoherent sources (xenon arcs, blackbody simulators, or tungsten-halogen) sources are often employed with spectral filters for more diffuse specimens. Similar measurements and techniques are employed to characterize bidirectional transmittance distribution function (BTDF) and bidirectional scattering distribution function (BSDF). For further information, see Asmail (1991), ASTM E1392 (1990), and Bartell et al. (1980).

The measurement of retroreflection poses the situation that the return beam coincides with the incident beam. The usual solution is to employ a beam splitter in the system, allowing the incident beam to pass and the return beam to be reflected. This immediately imposes both a significant loss in flux and the situation where the beam reaching the sample is partially polarized, unless a nonpolarizing beamsplitter is employed. In addition, it is imperative that the reflected component of the incident beam be well-trapped, as the radiometer is looking in the same direction. The special vocabulary for retroflection is given in CIE (1982) and ASTM E808 (1981). Test methods are given in ASTM E810 (1981), ASTM E809 (1991), and Venable and Johnson (1980). An instrument specifically designed to measure retroflection is detailed in Eckerle et al. (1980).

Most measurements of reflectance are relative and require artifact standards. Specular reflectance measurements are occasionally made using the absolute technique shown in Fig. 5, but are more commonly done using a simple reflectance attachment for a commercial spectrophotometer, requiring a calibrated reference. Freshly deposited metallic films have been used, with the assumption that an individual coating is the same as accepted data as shown in Table 5 for Al and Au (Hass and Waylonis 1961; Bennett et al., 1962; Bennett et al., 1963; Hass and Thun 1964; Bennett et al., 1965; Hass, 1982). Standards for specular reflectance are also available (Richmond and Hsia, 1972; Richmond et al., 1982; Verrill 1987; and Weidner and Hsia, 1982).

Diffuse reflectance standards exist in several forms. Ideally, a perfect ($\rho = 1$, lambertian) diffuser would be used, particularly for measuring reflectance factor (Erb, 1975, Erb and Budde, 1979). Certain materials approach the ideal over a limited angular and wavelength range. Historically, MgO was used in the visible spectrum (Middleton and Sanders, 1951; Agnew and McQuistan, 1953). It was replaced first by $BaSO_4$ (Grum and Luckey, 1968) and more recently by PTFE (Grum and Saltzman, 1976; Weidner and Hsia, 1981) and ASTM E259 (1992). Table 5 also shows a typical 6°/hemispherical reflectance factor for PTFE (Weidner and Hsia, 1987). This fine, white powder, when pressed to a density of $1\ g/cm^3$, is close to ideal over a wide spectral range. It is not quite lambertian, showing a falloff of BRDF at angles far removed from the specular direction (Fairchild

TABLE 5 Reflectance Standards

Wavelength (nm)	Aluminum	Gold	PTFE (6°/hemi)
250		0.295	0.973
300	0.921	0.346	0.984
350	0.921	0.330	0.990
400	0.919	0.360	0.993
450	0.918	0.358	0.993
500	0.916	0.453	0.994
550	0.916	0.800	0.994
600	0.912	0.906	0.994
650	0.906	0.947	0.994
700	0.898	0.963	0.994
750	0.886	0.970	0.994
800	0.868	0.973	0.994
850	0.868	0.973	0.994
900	0.891	0.974	0.994
950	0.924	0.974	0.994
1000	0.940	0.974	0.994
1100		0.975	0.994
1200	0.964	0.975	0.993
1300		0.975	0.992
1400		0.975	0.991
1500	0.974	0.975	0.992
1600		0.975	0.992
1700		0.976	0.990
1800		0.976	0.990
1900		0.976	0.985
2000	0.978	0.976	0.981
2100		0.976	0.968
2200		0.976	0.977
2300		0.976	0.972
2400		0.976	0.962
2500	0.979	0.977	0.960

and Daoust, 1988) and exhibiting a slight amount of retroreflection. It may also be slightly luminescent when excited by far-ultraviolet (Saunders, 1976).

In the infrared, two materials have proven useful. Flowers of sulfur (Kronstein, et al., 1963; Dunn, 1965; Tkachuk and Kuzina, 1978) is suitable over the spectral range 1–15 μm. Gold is highly reflective and very stable. To be useful as a diffuse reflectance standard, it must be placed on top of a lambertian surface. Several substrates for gold have been suggested, including sandpaper (Stuhlinger et al., 1981) and flame-sprayed aluminum.

PTFE is a satisfactory laboratory standard but is not well-suited for field use as it is not particularly rugged and is highly adsorbant and therefore subject to contamination. Several solutions have been proposed for working standards, including Eastman integrating sphere paint ($BaSO_4$), Vitriolite tile, and the Russian MS20 and MS14 opal glasses. These materials are, in general, more rugged, stable, and washable than PTFE. Further details can be found in CIE (1979b) and Clarke et al. (1983).

Discussion on the fabrication, calibration, and properties of various diffuse reflectance standards can be found in ASTM E259 (1987), Budde (1960), Budde (1976), Egan and Hilgeman (1976), Fairchild and Daoust (1988), Morren et al. (1972), TAPPI (1990), and Weidner (1983, 1986). International intercomparisons of laboratory standards of diffuse reflectance have been reported in Budde et al. (1982) and Weidner and Hsia (1987).

There are no standards available for retroflection. However, NIST offers a Measurements Assurance Program (MAP) to enable laboratories to make measurements consistent with other national standards.

25.12 MEASUREMENT OF EMITTANCE

Measurements of emittance can be done in several ways. The most direct method involves forming a material into the shape of a cavity in such a way that near-blackbody radiation is emitted. A measurement then compares the radiation from a location within the formed cavity to radiation from a flat, outside surface of the material, presumably at the same temperature (Sparrow et al., 1973). The cavity can take the form of a cylinder, cone, or sphere. Similarly, a small-diameter, deep hole can be drilled into a specimen and radiation from the surface compared to radiation from the hole. Care must be taken that the specimen is isothermal and that the reflected radiation is considered. The definitive measurements of several materials, such as tungsten (DeVos, 1954), were determined in this fashion. The significant advantage in this direct method is that it is relative, depending on neither absolute radiometry or thermometry, but only requiring that the radiometer or spectroradiometer be linear over the dynamic range of the measurement. This linearity is also determinable by relative measurements.

If a variable-temperature blackbody simulator and a suitable thermometer are available, the specimen can be heated to the desired temperature T_s and the blackbody simulator temperature T_{bb} can be adjusted such that its (spectral) radiance matches that of the specimen. Then the (spectral) emittance is calculable using the following equations for spectral emittance $\varepsilon(\lambda)$ and emittance ε (for a graybody only).

$$\varepsilon(\lambda) = \frac{e^{c_2/\lambda T_s} - 1}{e^{c_2/\lambda T_{bb}} - 1} \qquad \varepsilon = \frac{T_{bb}^4}{T_s^4} \tag{16}$$

If an absolutely calibrated radiometer and a satisfactory thermometer are available, a direct measurement can be made, as L_b is calculable if the temperature is known. Again, the reflected radiation must be considered.

Simple "inspection meter" techniques have been developed, and instrumentation is commercially available to determine the hemispherical emittance over a limited range of temperatures surrounding ambient. These instruments provide a single number, as they integrate both spatially and spectrally. A description of the technique can be found in ASTM E408 (1990).

Measurements of spectral emittance are most often made using spectral reflectance techniques, invoking Kirchhoff's law along with the assumption that the transmittance is zero. A review of early work is found in Dunn et al (1966a) and Millard and Streed (1969). The usual geometry of interest is the directional-hemispherical. This can be achieved by either hemispherical irradiation-directional collection, or, using Helmholtz reciprocity (Clarke and Perry, 1985), directional irradiation-hemispherical collection. Any standard reflectometry technique is satisfactory.

A direct method for the measurement of total (integrated over all wavelengths) hemispherical emittance is to use a calorimeter as shown in Fig. 8. A heated specimen is suspended in the center of a large, cold, evacuated chamber. The vacuum minimizes gaseous conduction and convection. If the sample suspension is properly designed, the predominant means of heat transfer is radiation. The chamber must be large to minimize the configuration factor between the chamber and the specimen. The chamber is cooled to

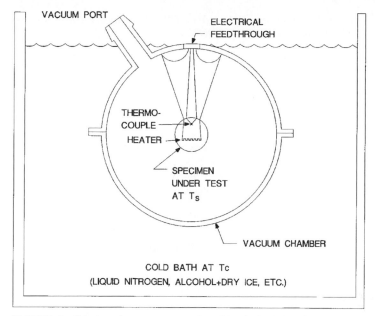

FIGURE 8 Calorimetric measurement of total hemispherical emittance.

T_c to reduce radiation from the chamber to the specimen. The equation used to determine emittance ε is

$$\varepsilon = \frac{P}{\sigma A(T_s^4 - T_c^4)} \tag{17}$$

where P is the power input to the specimen heater necessary to maintain an equilibrium specimen temperature T_s and A is the specimen area. The equation has been simplified with the aid of the following assumptions: (1) no thermal conduction from the specimen to the chamber, (2) no convective losses, (3) equilibrium has been achieved, and (4) the specimen area is much less than the chamber area. The power can be supplied electrically by means of a known heater or optically via a window in the chamber. In the latter case, a direct measurement of the ratio of solar absorptance α_s to thermal emittance ε_T can be directly obtained if the optical source simulates solar radiation. By varying the input power, the emittance can be determined as a function of temperature. There are numerous small corrections to account for geometry, lead conduction, etc. Details can be found in ASTM C835 (1988), ASTM E434 (1990), Edwards (1970), and Richmond and Harrison (1960).

Several attempts have been made to define and characterize artifact standards of spectral emittance for direct measurements (Richmond, 1962; Richmond et al., 1963). These were specimens of a thermally stable metal (i.e., Inconel) which were calibrated for emittance as a function of wavelength at several temperatures. No such standards are currently available. Interlaboratory comparisons have been made and reported (Willey, 1987).

Special problems include measurements at cryogenic temperatures (Weber, 1959) and effects of partially transparent materials (Gardon, 1956). Some additional references relating to emittance and its measurement are ASTM E307 (1990), ASTM E423 ((1990), Clarke and Larkin (1988), DeWitt (1986), DeWitt and Richmond (1970), Hornbeck

(1966), Millard and Streed (1969), Redgrove (1990), Sparrow et al. (1973), Stierwalt (1966), and Wittenberg (1968).

25.13 REFERENCES

Agnew, J. T. and R. B. McQuistan, "Experiments Concerning Infrared Diffuse-Reflectance Standards in the Range 0.8 to 20.0 Micrometers," *J. Opt. Soc. Am.* **43:**999 (1953).

Asmail, C., "Bidirectional Scattering Distribution Function (BSDF): A Systematized Bibliography," *J. Res. Natl. Inst. Stand. Technol.* **96:**215 (1991).

ASTM, "Estimating Stray Radiant Energy of Spectrophotometers," *ASTM E387,* ASTM, Philadelphia, (1984).

ASTM, "Measurement and Calculation of Reflecting Characteristics of Metallic Surfaces Using Integrating Sphere Instruments," *ASTM E429,* ASTM, Philadelphia (1987).

ASTM, "Measurement of Gloss of High-Gloss Surfaces by Goniophotometry," *ASTM E430,* ASTM, Philadelphia, (1983).

ASTM, "Nomenclature and Definitions Applicable to Radiometric and Photometric Characteristics of Matter," *ASTM Special Technical Publication 475,* ASTM, Philadelphia, (1971).

ASTM, "Practice for Describing and Measuring Performance of UV/VIS/near-IR Spectrophotometers," *ASTM E275,* ASTM, Philadelphia (1989).

ASTM, "Practice for Preparation of Reference White Reflectance Standards," *ASTM E259,* ASTM, Philadelphia (1987).

ASTM, "Preparation of Pressed Power White Reflectance Factor Transfer Standards for Hemispherical Geometry," *ASTM E259,* ASTM, Philadelphia (1992).

ASTM, "Procedure for Description and Performance in the Spectrophotometer," *ASTM E409,* ASTM, Philadelphia (1990).

ASTM, "Recommended Practice for Goniophotometry of Objects and Materials," *ASTM E167,* ASTM, Philadelphia (1987).

ASTM, "Specular Reflectance/Transmittance of Optically Flat Coated/Non-coated Specimens," *ASTM F768,* ASTM, Philadelphia (1987).

ASTM, "Standard Guide for Selection of Geometric Conditions for Measurement of Reflectance and Transmittance Properties of Materials," *ASTM E179,* ASTM, Philadelphia (1990).

ASTM, "Standard Practice for Angle Resolved Optical Scatter Measurements on Specular or Diffuse Surfaces," *ASTM E1392,* ASTM, Philadelphia (1990).

ASTM, "Standard Practice for Measuring Photometric Characteristics of Retroreflectors," *ASTM E809,* ASTM, Philadelphia (1991).

ASTM, "Standard Test Method for Reflectance Factor and Color by Spectrophotometry Using Bidirectional Geometry," *ASTM E1349,* ASTM, Philadelphia (1990).

ASTM, "Standard Test Method for Reflectance Factor and Color by Spectrophotometry Using Hemispherical Geometry," *ASTM E1348,* ASTM, Philadelphia (1990).

ASTM, "Standard Test Method for Solar Absorptance, Reflectance and Transmittance of Materials Using Spectrophotometers with Integrating Spheres," *ASTM E903,* ASTM, Philadelphia (1988).

ASTM, "Terminology for Retroreflection and Retroreflectors," *ASTM E808,* ASTM, Philadelphia (1981).

ASTM, "Test for Calorimetric Determination of Hemispherical Emittance and the Ratio of Solar Absorptance to Hemispherical Emittance Using Solar Simulation," *ASTM E434,* ASTM, Philadelphia (1990).

ASTM, "Test for Light Reflectance of Acoustical Materials by the Integrating Sphere Method," *ASTM C523,* ASTM, Philadelphia (1987).

ASTM, "Test for Normal Spectral Emittance at Elevated Temperatures," *ASTM E307,* ASTM, Philadelphia (1990).

ASTM, "Test for Normal Spectral Emittance at Elevated Temperatures of Non-Conducting Specimens," *ASTM E423,* ASTM, Philadelphia (1990).

ASTM, "Test for Photometric Transmittance/Reflectance of Materials to Solar Radiation," *ASTM E971,* ASTM, Philadelphia (1988).

ASTM, "Test for Reflectivity and Coefficient of Scatter of White Porcelain Enamels," *ASTM C347,* ASTM, Philadelphia (1983).

ASTM, "Test for Total Normal Emittances of Surfaces Using Inspection Meter Techniques," *ASTM E408,* ASTM, Philadelphia (1990).

ASTM, "Test Method for Coefficient of Retroreflection on Retroreflective Sheeting,"*ASTM E810,* ASTM, Philadelphia (1981).

ASTM, "Test Method for Measuring Optical Reflectivity of Transparent Materials," *ASTM F1252,* ASTM, Philadelphia (1990).

ASTM, "Test Method for Specular Gloss," *ASTM D523,* ASTM, Philadelphia (1988).

ASTM, "Test Method for (45-0) Directional Reflectance Factor of Opaque Specimens by Broad-Band Filter Reflectometry," *ASTM E97,* ASTM, Philadelphia (1987).

ASTM, "Total Hemispherical Emittance of Surfaces from 20 to 1400C," *ASTM C835,* ASTM, Philadelphia (1988).

Barnes, P. Y. and J. J. Hsia, "45°/0° Bidirectional Reflectance Distribution Function Standard Development," *Proc. SPIE* **1165:**165 (1989).

Bartell, F. O., E. L. Dereniak, and W. L. Wolfe, "Theory and Measurement of Bidirectional Reflectance Distribution Function (BRDF) and Bidirectional Transmittance Distribution Function (BTDF)," *Proc. SPIE* **257:**154 (1980).

Bennett, H. E., J. M. Bennett, and E. J. Ashley, "Infrared Reflectance of Evaporated Aluminum Films," *J. Opt. Soc. Am.* **52:**1245 (1962).

Bennett, H. E. and W. F. Koehler, "Precision Measurements of Absolute Specular Reflectance with Minimized Systematic Errors," *J. Opt. Soc. Am.* **50:**1 (1960).

Bennett, H. E., M. Silver, and E. J. Ashley, "Infrared Reflectance of Aluminum Evaporated in Ultra-high Vacuum," *J. Opt. Soc. Am.* **53:**1089 (1963).

Bennett, J. M. and E. J. Ashley, "Calibration of Instruments Measuring Reflectance and Transmittance," *Appl. Opt.* **11:**1749 (1972).

Bennett, J. M. and E. J. Ashley, "Infrared Reflectance and Emittance of Silver and Gold Evaporated in Ultra-high Vacuum," *Appl. Opt.* **4:**221 (1965).

Bittar, A. and J. D. Hamlin, "High-accuracy True Normal-incidence Absolute Reflectometer," *Appl. Opt.* **23:**4054 (1984).

Blevin, W. R. and W. J. Brown, "An Infrared Reflectometer with a Spheroidal Mirror," *J. Sci. Instrum.* **42:**1 (1965).

Boivin, G. and J. M. Theriault, "Reflectometer for Precise Measurement of Absolute Specular Reflectance at Normal Incidence," *Rev. Sci. Instrum.* **52:**1001 (1981).

Budde, W., "Calibration of Reflectance Standards," *J. Res. Natl. Bur. Stand* **A80:**585 (1976).

Budde, W., "Standards of Reflectance," *J. Opt. Soc. Am.* **50:**217 (1960).

Budde, W., W. Erb, and J. J. Hsia, "International Intercomparison of Absolute Reflectance Scales," *Color Res. Appl.* **7:**24 (1982).

Burke, R. W. and R. Mavrodineanu, *Standard Reference Material: Accuracy in Analytical Spectrophotometry,* NBS Special Publication SP260-81, U.S. National Bureau of Standards, Washington, D.C., 1983.

Castellini, C., G. Emiliani, E. Masetti, P. Poggi, and P. P. Polato, "Characterization and Calibration of a Variable-angle Absolute Reflectometer, *Appl. Opt.* **29:**538 (1990).

C.I.E., "A Review of Publications on Properties and Reflection Values of Material Reflection Standards", *C.I.E. Publication 46,* C.I.E., Paris (1979b).

C.I.E., "Absolute Methods for Reflection Measurements," *C.I.E. Publication 44,* C.I.E., Paris (1979a).

C.I.E., "International Lighting Vocabulary", *C.I.E. Publ. 17.4,* C.I.E., Paris (1987).

C.I.E., Radiometric and Photometric Characteristics of Materials and their Measurement", *C.I.E. Publication 38,* C.I.E., Paris (1977).

C.I.E., "Retroflection: Definition and Measurement," *C.I.E. Publ. 54,* CIE, Paris (1982).

Clarke, F. J. J., "High-Accuracy Spectrophotometry at the National Physical Laboratory," *J. Res. Natl. Bur. Stand.* **A76:**375 (1972).

Clarke, F. J. J. and J. A. Compton, "Correction Methods for Integrating Sphere Measurement of Hemispherical Reflectance," *Color. Res. Appl.* **11:**253 (1986).

Clarke, F. J. J., F. A. Garforth, and D. J. Parr, "Goniophotometric and Polarization Properties of White Reflection Standard Materials," *Light. Res. Technol.* **15:**133 (1983).

Clarke, F. J. J. and J. A. Larkin, "Improved Techniques for the NPL Hemispherical Reflectometer," *Proc. SPIE* **917:**7 (1988).

Clarke, F. J. J. and J. A. Larkin, "Measurement of Total Reflectance, Transmittance and Emissivity over the Thermal IR Spectrum," *Infrared Phys.* **25:**359 (1985).

Clarke, F. J. J. and D. J. Parry, "Helmholtz Reciprocity: Its Validity and Application to Reflectometry," *Light. Res. Technol.* **17:**1 (1985).

Cohen, E. R. and P. Giacomo, *Symbols, Units, Nomenclature and Physical Constants in Physics,* Document IUPAP-25, International Union of Pure and Applied Physics, 1987.

DeVos, J. C., "A New Determination of the Emissivity of Tungsten Ribbon," *Physica* **20:**690 (1954).

DeWitt, D. P., "Inferring Temperature from Optical Radiation Measurements," *Opt. Eng.* **25:**596 (1986).

DeWitt, D. P. and J. C. Richmond, "Theory and Measurement of the Thermal Radiative Properties of Metals," in *Techniques of Metals Research,* vol. VI, Wiley, NY, 1970.

Drazil, J. V., *Quantities and Units of Measurement*: *A Dictionary and Handbook,* Mansell, London, (1983).

Duggin, M. J., "The Field Measurement of Reflectance Factors," *Photogram. Eng. Rem. Sens.* **46:**643 (1980).

Duggin, M. J. and W. R. Philipson, "Field Measurement of Reflectance: Some Major Considerations," *Appl. Opt.* **21:**2833 (1982).

Dunkle, R. V., "Spectral Reflectance Measurements," in F. J. Clauss (ed.), *Surface Effects on Spacecraft Materials,* Wiley, NY, 1960, p. 117.

Dunn, S. T., "Application of Sulfur Coatings to Integrating Spheres," *Appl. Opt.* **4:**877 (1965).

Dunn, S. T., J. C. Richmond, and J. F. Parmer, "Survey of Infrared Measurement Techniques and Computational Methods in Radiant Heat Transfer," *J. Spacecraft Rockets* **3:**961 (1966a).

Dunn, S. T., J. C. Richmond, and J. C. Weibelt, "Ellipsoidal Mirror Reflectometer," *J. Res. Nat. Bur. Stand.* **70C:**75 (1966b).

Eckerle, K. E., E. Sutter, G. H. Freeman, G. Andor, and L. Fillinger, "International Intercomparison for Transmittance," *Metrologia* **27:**33 (1990).

Eckerle, K. L., *Modification of an NBS Reference Spectrophotometer,* NBS Technical Note TN913, U.S. National Bureau of Standards, Washington, D.C., 1976.

Eckerle, K. L., J. J. Hsia, K. D. Mielenz, and V. R. Weidner, *Regular Spectral Transmittance,* NBS Special Publication SP250-6, U.S. National Bureau of Standards, Washington, D.C., 1987.

Eckerle, K. L., J. J. Hsia, V. R. Weidner, and W. H. Venable, "NBS Reference Retroreflectometer," *Appl. Opt.* **19:**1253 (1980).

Eckerle, K. L., V. R. Weidner, J. J. Hsia, and Z. W. Chao, *Extension of a Reference Spectrophotometer into the Near Infrared,* NBS Technical Note TN1175, U.S. National Bureau of Standards, Washington, D.C. (1983).

Edwards, D. K., "Thermal Radiation Measurements," Chap. 9 in E. R. G. Eckert and R. J. Goldstein (eds), *Measurement Techniques in Heat Transfer,* AGARD 130, Technivision, Slough, England, 1970, p. 353.

Edwards, D. K., J. T. Gier, K. E. Nelson, and R. D. Roddick, "Integrating Sphere for Imperfectly Diffuse Samples," *J. Opt. Soc. Am.* **51:**1279 (1961).

Egan, W. G. and T. Hilgeman, "Integrating Spheres for Measurements Between 0.185 micrometers and 12 micrometers," *Appl. Opt.* **14:**1137 (1975).

Egan, W. G. and T. Hilgeman, "Retroflectance Measurements of Photometric Standards and Coatings," *Appl. Opt.* **15:**1845 (1976).

Erb, W., "Computer-controlled Gonioreflectometer for the Measurement of Spectral Reflection Characteristics," *Appl. Opt.* **19:**3789 (1980).

Erb, W., "High Accuracy Gonioreflectance Spectrometry," Chap. 2.2, in C. Burgess and K. D. Mielenz (eds), *Advances in Standards and Methodology in Spectrophotometry,* Elsevier, Amsterdam, 1987.

Erb, W., "Requirements for Reflection Standards and the Measurement of their Reflection Value," *Appl. Opt.* **14:**493 (1975).

Erb, W. and W. Budde, "Properties of Standard Materials for Reflection," *Color Res. Appl.* **4:**113 (1979).

Fairchild, M. D. and D. J. O. Daoust, "Goniospectrophotometric Analysis of Pressed PTFE Powder for Use as a Primary Transfer Standard," *Appl. Opt.* **27:**3392 (1988).

Fillinger, L. and G. Andor, "International Intercomparison of Transmittance Measurement," *C.I.E. Journal* **7:**21 (1988).

Freeman, G. H. C., "The New Automated Reference Spectrophotometer at NPL," in C. Burgess and K. D. Mielenz (eds.), *Advances in Standards and Methodology in Spectrophotometry,* Elsevier, Amsterdam, 1986, p. 69.

Gardon, R., "The Emissivity of Transparent Materials," *J. Am. Ceram. Soc.* **39:**278 (1956).

Gier, J. T., R. V. Dunkle, and J. T. Bevans, "Measurement of Absolute Spectral Reflectivity from 1.0 to 15 microns," *J. Opt. Soc. Am.* **44:**558 (1954).

Goebel, D. G., "Generalized Integrating Sphere Theory," *Appl. Opt.* **6:**125 (1967).

Goebel, D. G., B. P. Caldwell, and H. K. Hammond III, "Use of an Auxiliary Sphere with a Spectroreflectometer to Obtain Absolute Reflectance," *J. Opt. Soc. Am.* **56:**783 (1966).

Grum, F. and R. J. Becherer, "Radiometry," in *Optical Radiation Measurements* vol. 1, Academic, NY, p. 115 (1979).

Grum, F. and G. W. Luckey, "Optical Sphere Paint and a Working Standard of Reflectance," *Appl. Opt.* **7:**2289 (1968).

Grum, F. and M. Saltzman, "New White Standard of Reflectance," *C.I.E. Publication 36,* C.I.E., Paris (1976).

Hanssen, L. M. and K. A. Snail, "Infrared Diffuse Reflectometer for Spectral, Angular and Temperature Resolved Measurements," *Proc. SPIE* **807:**148 (1987).

Hartman, P. L. and E. Logothetis, "An Absolute Reflectometer for Use at Low Temperatures," *Appl. Opt.* **3:**255 (1964).

Hass, G., "Reflectance and Preparation of Front-Surface Mirrors for Use at Various Angles of Incidence from the Ultraviolet to the Far Infrared," *J. Opt. Soc. Am.* **72:**27 (1982).

Hass, G. and J. E. Waylonis, "Optical Constants and Reflectance and Transmittance of Evaporated Aluminum in the Visible and Ultraviolet," *J. Opt. Soc. Am.* **51:**719 (1961).

Hass, G. and R. E. Thun, *Physics of Thin Films,* vol. 2, Academic, NY, 1964, p. 337.

Hawes, R. C., "Technique for Measuring Photometric Accuracy," *Appl. Opt.* **10:**1246 (1971).

Hernicz, R. S. and D. P. DeWitt, "Evaluation of a High Accuracy Reflectometer for Specular Materials," *Appl. Opt.* **12:**2454 (1973).

Hisdal, B. J., "Reflectance of Nonperfect Surfaces in the Integrating Sphere," *J. Opt. Soc. Am.* **55:**1255 (1965b).

Hisdal, B. J., "Reflectance of Perfect Diffuse and Specular Samples in the Integrating Sphere," *J. Opt. Soc. Am.* **55:**1122 (1965a).

Hordvik, A., "Measurement Techniques for Small Absorption Coefficients: Recent Advances," *Appl. Opt.* **16:**2827 (1977).

Hornbeck, G. A., "Optical Methods of Temperature Measurement," *Appl. Opt.* **5:**179 (1966).

Hsia, J. J., "National Scales of Spectrometry in the U.S.," in C. Burgess and K. D. Mielenz (eds.), *Advances in Standards and Methodology in Spectrophotometry,* Elsevier, Amsterdam, 1987.

Hsia, J. J. and V. R. Weidner, "NBS 45-degree/Normal Reflectometer for Absolute Reflectance Factors," *Metrologia* **17:**97 (1981).

Hughes, H. K., "Beer's Law and the Optimum Transmittance in Absorption Measurements," *Appl. Opt.* **2:**937 (1963).

Hunter, R. S., *The Measurement of Appearance,* Wiley, NY, 1975.

IES Nomenclature Committee, *American National Standard Nomenclature and Definitions for Illuminating Engineering,* ANSI/IES RP-16-1986. Illuminating Engineering Society of North America, NY, 1986.

IES Nomenclature Committee, "Proposed American National Standard Nomenclature and Definitions for Illuminating Engineering (proposed revision of Z7.1-R-1973)," *J. Illum. Eng. Soc.* **8:**2 (1979).

IES Testing Procedures Committee, "IES Approved Method for Total and Diffuse Reflectometry," IES LM-44-1985, *J. Illum. Eng. Soc.* **19:**195 (1985).

ISO, *Units of Measurement,* ISO Standards Handbook 2, International Organization for Standardization, Geneva, 1982.

Jacquez, J. A. J. and H. F. Kuppenheim, "Theory of the Integrating Sphere," *J. Opt. Soc. Am.* **45:**460 (1955).

Judd, D. B., "Terms, Definitions and Symbols in Reflectometry," *J. Opt. Soc. Am.* **57:**445 (1967).

Karras, E., "The Use of the Ulbricht Sphere in Measuring Reflection and Transmission Factors," *J. Opt. Soc. Am.* **11:**96 (1921).

Kessell, J., "Transmittance Measurements in the Integrating Sphere," *Appl. Opt.* **25:**2752 (1986).

Kirchhoff, G., "On the Relation Between the Radiating and Absorbing Powers of Different Bodies for Light and Heat," *Phil. Mag.* **20:**1 (1860).

Kronstein, M., R. J. Kraushaar, and R. E. Deacle, "Sulfur as a Standard of Reflectance in the Infrared," *J. Opt. Soc. Am.* **53:**458 (1963).

Lindberg, J. D., "Absolute Diffuse Reflectance from Relative Reflectance Measurements," *Appl. Opt.* **26:**2900 (1987).

Lipson, H. G., L. H. Skolnik, and D. L. Stierwalt, "Small Absorption Coefficient Measurement by Calorimetric and Spectral Emittance Techniques," *Appl. Opt.* **13:**1741 (1974).

Mavrodineanu, R. and J. R. Baldwin, *Standard Reference Materials*: *Glass Filters as a Standard Reference Material for Spectrophotometry—Selection, Preparation, Certification, Use,* NBS Special Publication SP260-51, U.S. National Bureau of Standards, Washington, D.C., 1975.

Mavrodineanu, R. and J. R. Baldwin, *Standard Reference Materials*: *Metal-on-Quartz Filters as a Standard Reference Material for Spectrophotometry,* NBS Special Publication SP260-68, U.S. National Bureau of Standards, Washington, D.C., 1980.

McNicholas, H. J., "Absolute Methods in Reflectometry," *J. Res. Natl. Bur. Stand.* **1:**29 (1928).

Middleton, W. E. K. and C. L. Sanders, "The Absolute Spectral Diffuse Reflectance of Magnesium Oxide," *J. Opt. Soc. Am.* **41:**419 (1951).

Mielenz, K. D., "Physical Parameters in High-Accuracy Spectrophotometry," in R. Mavrodineanu, J. I. Schultz, and O. Menis (eds), *Accuracy in Spectrophotometry and Luminescence Measurements,"* NBS SP378, U.S. National Bureau of Standards, Washington, D.C., 1973.

Mielenz, K. D. and K. L. Eckerle, *Design, Construction, and Testing of a New High Accuracy Spectrophotometer,* NBS Tech Note 729, U.S. National Bureau of Standards, Washington, D.C., 1972.

Mielenz, K. D., K. L. Eckerle, R. P. Madden, and J. Reader, "New Reference Spectrophotometer," *Appl. Opt.* **12:**1630 (1973).

Migdall, A. L., B. Roop, Y. C. Zheng, J. E. Hardis, and G. J. Xia, "Use of Heterodyne Detection to Measure Optical Transmittance over a Wide Range," *Appl. Opt.* **29:**5136 (1990).

Millard, J. P. and E. R. Streed, "A Comparison of Infrared Emittance Measurements and Measurement Techniques," *Appl. Opt.* **8:**1485 (1969).

Morren, L., G. Vandermeersch, and P. Antoine, "A Study of the Reflection Factor of Usual Photometric Standards in the Near Infrared," *Light. Res. Technol.* **4:**243 (1972).

Neu, J. T., R. S. Dummer, and O. E. Myers, "Hemispherical Directional Ellipsoidal Infrared Spectroreflectometer," *Proc. SPIE* **807:**165 (1987).

Nicodemus, F. E., "Directional Reflectance and Emissivity of an Opaque Surface," *Appl. Opt.* **4:**767 (1965).

Nicodemus, F. E., "Reflectance Nomenclature and Directional Reflectance and Emissivity," *Appl. Opt.* **9:**1474 (1970).

Nicodemus, F. E., J. C. Richmond, and J. J. Hsia, *Geometrical Considerations and Nomenclature for Reflectance,* NBS Monograph 160, U.S. National Bureau of Standards, Washington, D.C., 1977.

Ram, R. S., O. Prakash, J. Singh, and S. P. Varma, "Simple Design for a Reflectometer," *Opt. Eng.* **30:**467 (1991).

Redgrove, J. S., "Measurement of the Spectral Emissivity of Solid Materials," *Measurement* (UK) **8:**90 (1990).

Reid, D. C. and E. D. McAlister, "Measurement of Spectral Emissivity from 3µ to 15µ," *J. Opt. Soc. Am.* **49:**78 (1959).

Richmond, J. C., "Physical Standards of Emittance and Reflectance," in H. H. Blau and H. Fischer (eds), *Radiative Transfer from Solid Material,* Macmillan, NY, 1962.

Richmond, J. C., "Rationale for Emittance and Reflectivity," *Appl. Opt.* **21:**1 (1982).

Richmond, J. C. and W. N. Harrison, "Equipment and Procedures for Evaluation of Total Hemispherical Emittance," *Am. Ceram. Soc. Bull.* **39:**668 (1960).

Richmond, J. C., W. N. Harrison, and F. J. Shorten, "An Approach to Thermal Emittance Standards, in J. C. Richmond (ed.), *Measurement of Thermal Radiation Properties of Solids,* NASA SP-31, NASA, Washington, DC, 1963.

Richmond, J. C. and J. J. Hsia, *Preparation and Calibration of Standards of Spectral Specular Reflectance,* NBS Special Publication SP260-38, U.S. National Bureau of Standards, Washington, D.C., 1972.

Richmond, J. C., J. J. Hsia, V. R. Weidner, and D. B. Wilmering, *Second-Surface Mirror Standards of Spectral Specular Reflectance,* NBS Special Publication SP260-79, U.S. National Bureau of Standards, Washington, D.C., 1982.

Richter, W. and W. Erb, "Accurate Diffuse Reflectance Measurements in the IR Spectral Range," *Appl. Opt.* **26:**4620 (1987).

Roos, A., "Interpretation of Integrating Sphere Signal Output for Nonideal Transmitting Samples," *Appl. Opt.* **30:**468 (1991).

Saunders, R. D. and W. R. Ott, "Spectral Irradiance Measurements: Effect of UV-produced Luminescence in Integrating Spheres," *Appl. Opt.* **15:**827 (1976).

Shaw, J. E. and W. R. Blevin, "Instrument for the Absolute Measurement of Direct Spectral Reflectances at Normal Incidence," *J. Opt. Soc. Am.* **54:**334 (1964).

Sheffer, D., U. P. Oppenheim, D. Clement, and A. D. Devir, "Absolute Reflectometer for the 0.8–2.5 micrometer region," *Appl. Opt.* **26:**583 (1987).

Sheffer, D., U. P. Oppenheim, and A. D. Devir, "Absolute Measurements of Diffuse Reflectances in the x°/d Configuration", *Appl. Opt.* **30:**3181 (1991).

Scheffer, D., U. P. Oppenheim, and A. D. Devir, "Absolute Reflectometer for the Mid-Infrared Region," *Appl. Opt.* **29:**129 (1990).

Siegel, R. and J. R. Howell, *Thermal Radiation Heat Transfer,* 2d ed., Hemisphere, NY, p. 63 (1981).

Snail, K. A., A. A. Morrish, and L. M. Hanssen, "Absolute Specular Reflectance Measurements in the Infrared," *Proc. SPIE* **692:**143 (1986).

Sparrow, E. M., P. D. Kruger, and R. P. Heinisch, "Cavity Methods for Determining the Emittance of Solids," *Appl. Opt.* **12:**2466 (1973).

Stierwalt, D. L., "Infrared Spectral Emittance Measurements on Optical Materials," *Appl. Opt.* **5:**1911 (1966).

Strong, J., *Procedures in Applied Optics,* Dekker, NY, 1989, p. 162.

Strong, J., *Procedures in Experimental Physics,* Prentice-Hall, NY, 1938, p. 376.

Stuhlinger, T. W., E. L. Dereniak, and F. O. Bartell, "Bidirectional distribution function of gold-plated sandpaper," *Appl. Opt.* **20:**2648 (1981).

TAPPI, "Calibration of Reflectance Standards for Hemispherical Geometry," TAPPI Standard TIS 0804-07 in *1990 TAPPI Test Methods,* TAPPI, Atlanta, 1990.

Taylor, A. H., "Errors in Reflectometry," *J. Opt. Soc. Am.* **25:**51 (1935).

Taylor, A. H., "The Measurement of Diffuse Reflection Factors and a New Absolute Reflectometer," *J. Opt. Soc. Am.* **4:**9 (1920).

Taylor, A. H., "A Simple Portable Instrument for the Absolute Measurement of Reflection and Transmission Factors," *Sci. Papers Bur. Standards* **17:**1 (1922).

Taylor, B. N., "The International System of Units (SI)," *NIST Special Publication 330,* National Institute of Standards and Technology, Washington, D.C., 1991.

Tkachuk, R. and F. D. Kuzina, "Sulfur as a Proposed Near Infrared Reflectance Standard," *Appl. Opt.* **17:**2817 (1978).

Van den Akker, J. A., L. R. Dearth, and W. M. Shilcox, "Evaluation of Absolute Reflectance for Standardization Purposes," *J. Opt. Soc. Am.* **56:**250 (1966).

Venable, W. H. and J. J. Hsia, *Describing Spectrophotometric Measurements,* NBS Technical Note TN594-9, U.S. National Bureau of Standards, Washington, D.C., 1974.

Venable, W. H., J. J. Hsia, and V. R. Weidner, *Development of an NBS Reference Spectrophotometer for Diffuse Transmittance and Reflectance,* NBS Tech Note TN594-11, U.S. National Bureau of Standards, Washington, D.C., 1976.

Venable, W. H., J. J. Hsia, and V. R. Weidner, "Establishing a Scale of Directional-Hemispherical Reflectance Factor I: The Van den Akker Method," *J. Res. Natl. Bur. Stand.* **82:**29 (1977).

Venable, W. H. and N. L. Johnson, "Unified Coordinate System for Retroflectance Measurements," *Appl. Opt.* **19:**1236 (1980).

Verrill, J. F., "Physical Standards in Absorption and Reflection Spectrometry," chap. 3.1 in C. Burgess and K. D. Mielenz (eds), *Advances in Standards and Methodology in Spectrophotometry,* Elsevier, Amsterdam, 1987.

Weber, D., "Spectral Emissivity of Solids in the Infrared at Low Temperatures," *J. Opt. Soc. Am.* **49:**815 (1959).

Weeks, R. F., "Simple Wide Range Specular Reflectometer," *J. Opt. Soc. Am.* **48:**775 (1958).

Weidner, V. R., "Gray Scale of Diffuse Reflectance for the 250–2500 nm Wavelength Range," *Appl. Opt.* **25:**1265 (1986).

Weidner, V. R., *Standard Reference Materials*: *White, White Opal Glass Diffuse Spectral Reflectance Standards for the Visible Spectrum,* NBS Special Publication SP260-82, U.S. National Bureau of Standards, Washington, D.C., 1983.

Weidner, V. R. and J. J. Hsia, "NBS Specular Reflectometer-Spectrophotometer," *Appl. Opt.* **19:**1268 (1980).

Weidner, V. R. and J. J. Hsia, "Reflection Properties of Pressed Polytetrafluorethylene Powder," *J. Opt. Soc. Am.* **71:**856 (1981).

Weidner, V. R. and J. J. Hsia, *Spectral Reflectance,* NBS Special Publication SP250-8, U.S. National Bureau of Standards, Washington, D.C., 1987.

Weidner, V. R. and J. J. Hsia, *Standard Reference Materials*: *Preparation and Calibration of First Surface Aluminum Mirror Specular Reflectance Standard,* NBS Special Publication SP260-75, U.S. National Bureau of Standards, Washington, D.C., 1982.

Willey, R. R., "Results of a Round-Robin Measurement of Spectral Emittance in the Mid-Infrared," *Proc. SPIE* **807:**140 (1987).

Wittenberg, A. M., "Determination of Total Emittance of a Nongray Surface," *J. Appl. Phys.* **39:**1936 (1968).

Wolfe, W. L., "Proclivity for Emissivity," *Appl. Opt.* **21:**1 (1982).

Wood, B. E., J. G. Pipes, A. M. Smith, and J. A. Roux, "Hemi-ellipsoidal Mirror Infrared Reflectometer," *Appl. Opt.* **15:**940 (1976).

Worthing, A. G., "Temperature Radiation Emissivities and Emittances," in *Temperature, Its Measurement and Control in Science and Industry,* Reinhold, NY, p. 1164 (1941).

Zerlaut, G. A. and T. E. Anderson, "Multiple-Integrating Sphere Spectrophotometer for Measuring Absolute Spectral Reflectance and Transmittance," *Appl. Opt.* **20:**3797 (1981).

Zhuang, D. K. and T. L. Yang, "Spectral Reflectance Measurements Using a Precision Multiple Reflectometer in the UV and VUV Range," *Appl. Opt.* **28:**5024 (1989).

Zwinkles, J. C. and D. S. Gignac, "Design and Testing of a New High-Accuracy Ultraviolet-Visible-Near-Infrared Spectrophotometer," *Appl. Opt.* **31:**1557 (1992).

25.14 FURTHER READING

ASTM, *ASTM Standards on Color and Appearance Measurement,* 3d ed., ASTM, Philadelphia, 1991.

Blau, Jr, H. H. and H. Fischer (eds.), *Radiative Transfer from Solid Materials,* Macmillan, NY, 1962.

Burgess, C. and K. D. Mielenz (eds.), *Advances in Standards and Methodology in Spectrophotometry,* Elsevier, Amsterdam, 1987.

Clauss, F. J. (ed.), *First Symposium, Surface Effects on Spacecraft Materials,* Wiley, NY, 1960.

Frei, R. W. and J. D. MacNeil, *Diffuse Reflectance Spectroscopy in Environmental Problem-Solving,* CRC Press, Cleveland, 1973.

Grum, F. and R. J. Becherer, "Radiometry," in *Optical Radiation Measurements* vol. 1, Academic, NY, 1979.

Hammond III, H. K., and H. L. Mason (eds.), "Selected NBS Papers on Radiometry and Photometry," NBS Special Publication SP300-7, *Precision Measurement and Calibration,* U.S. National Bureau of Standards, Washington, D.C., 1971.

Hunter, R. S., *The Measurement of Appearance,* Wiley, NY, 1975.

Kortum, G., *Reflectance Spectroscopy,* Springer-Verlag, NY, 1969.

Nimeroff, L. (ed.), "Selected NBS Papers on Colorimetry," NBS Special Publication SP300-9, *Precision Measurement and Calibration,* U.S. National Bureau of Standards, Washington, D.C., 1972.

Richmond, J. C. (ed.), "Measurement of Thermal Radiation Properties of Solids," *NASA Special Publication SP-31,* National Aeronautics and Space Administration, Washington, D.C., 1963.

Walsh, J. W. T., *Photometry,* 3d ed., Dover, NY, 1958.

Wendlandt, W. W. and H. G. Hecht, *Reflectance Spectroscopy,* Interscience, NY, 1966.

CHAPTER 26
SCATTEROMETERS

John C. Stover
TMA Technologies
Bozeman, Montana

26.1 GLOSSARY

BRDF	bidirectional reflectance distribution function
BTDF	bidirectional transmittance distribution function
BSDF	bidirectional scatter distribution function
f	focal length
FN	focal ratio
L	distance
P	power
R	length
r	radius
TIS	total integrated scatter
θ	angle
θ_N	vignetting angle
θ_{spec}	specular angle
λ	wavelength
σ	rms roughness
Ω	solid angle

26.2 INTRODUCTION

The measurement of optical scatter has received increased attention in the last decade. In addition to being a serious source of noise, scatter reduces throughput, limits resolution, and has been the unexpected source of practical difficulties in many optical systems. On the other hand, its measurement has proved to be an extremely sensitive method of providing metrology information for components used in many diverse applications. Measured scatter is a good indicator of surface quality and can be used to characterize surface roughness as well as locate and size discrete defects. It is also used to measure the quality of optical coatings and bulk optical materials. It is emerging as a valuable noncontact measurement technique outside the optics industry as well. Point sources of scatter imaged onto position-sensitive detectors are used to measure displacement. Doppler-shifted scatter is used to measure velocity, and polarization changes in scattered light can be used to reveal reflector material properties, such as the optical constants.

The instrumentation required for many of these measurements has to be fairly sophisticated. Scatter signals are generally small compared to the specular beam and can vary by several orders of magnitude in just a few degrees. Complete characterization may require measurement over a large fraction of the sphere surrounding the scatter source. For many applications, a huge array of measurement decisions (incident angle, wavelength, source and receiver polarization, scan angles, etc.) faces the experimenter. The instrument may faithfully record a signal, but is it from the sample alone? Or, does it also include light from the instrument, the wall behind the instrument, and the experimenter's shirt? These are not easy questions to answer at nanowatt levels in the visible and get even harder in the infrared and ultraviolet. It is easy to generate scatter data—lots of it. Obtaining accurate values of appropriate measurements and communicating them requires know-ledge of the instrumentation as well as insight into the problem being addressed.

In 1961, Bennett and Porteus[1] reported measurement of signals obtained by integrating scatter over the reflective hemisphere. They defined a parameter called *total integrated scatter* (TIS) and, using a scalar diffraction theory result drawn from the radar literature,[2] related it to reflector root mean square (rms) roughness. By the mid-1970s, several scatterometers had been built at various university, government, and industry labs that were capable of measuring scatter as a function of angle; however, instrument operation and data manipulation were not always well automated.[3-6] Scattered power per unit solid angle (sometimes normalized by the incident power) was usually measured. Analysis of scatter data to characterize sample surface roughness was the subject of many publications.[7-11] Measurement comparison between laboratories was hampered by instrument differences, sample contamination, and confusion over what parameters should be compared. A derivation of what is commonly called BRDF (for bidirectional reflectance distribution function) was published by Nicodemus and coworkers in 1970, but did not gain common acceptance as a way to quantify scatter measurements until after publication of their 1977 NBS monograph.[12] With the advent of small powerful computers in the 1980s, instrumentation became more automated. Increased awareness of scatter problems and the sensitivity of many end-item instruments increased government funding for better instrumentation.[13-14] As a result, instrumentation became available that could measure and analyze as many as 50 to 100 samples a day instead of just a handful. Scatterometers became commercially available and the number (and sophistication) of measurement facilities increased.[15-17] Two ASTM standards[18,19] were published (TIS in 1987 and BRDF in 1991). The 1990s will be characterized by less dramatic increases in instrumentation capabilities, but a large increase in applications. Further instrumentation improvements will include more out-of-plane capability, extended wavelength control, and polarization control at both source and receiver. These systems will find applications outside the optical industry in increasing numbers.

This review gives basic definitions, instrument configurations and components, scatter specifications, measurement techniques, and briefly discusses calibration and error analysis.

26.3 *DEFINITIONS AND SPECIFICATIONS*

One of the difficulties encountered in comparing measurements made on early instruments was getting participants to calculate the same quantities. There were problems of this nature as late as 1988 in a measurement round-robin run at 0.63 micrometers.[20] But, there are other reasons for reviewing these basic definitions before discussing instrumentation. The ability to write useful scatter specifications (i.e., the ability to make use of quantified scatter information) depends just as much on understanding the defined quantity as it does

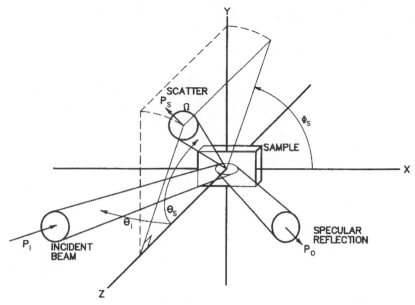

FIGURE 1   Geometry for the definition of BRDF.

on understanding the instrumentation and the specific scatter problem. In addition, definitions are often given in terms of mathematical abstractions that can only be approximated in the lab. This is the case for BRDF.

$$\text{BRDF} = \frac{\text{differential radiance}}{\text{differential irradiance}} \approx \frac{dP_s/d\Omega}{P_i \cos \theta_s} \approx \frac{P_s/\Omega}{P_i \cos \theta_s} \tag{1}$$

BRDF has been strictly defined as the ratio of the sample differential radiance to the differential irradiance under the assumptions of a collimated beam with uniform cross section incident on an isotropic surface reflector (no bulk scatter allowed). Under these conditions, the third quantity in Eq. (1) is found, where power P in watts instead of intensity I in watts/m^2 has been used. The geometry is shown in Fig. 1. The value θ_s is the polar angle in the scatter direction measured from reflector normal and Ω is the differential solid angle (in steradians) through which dP_s (watts) scatters when P_i (watts) is incident on the reflector. The cosine comes from the definition of radiance and may be viewed as a correction from the actual size of the scatter source to the apparent size (or projected area) as the viewer rotates away from surface normal.

The details of the derivation do not impact scatter instrumentation, but the initial assumptions and the form of the result do. When light power is measured, it is through a finite diameter aperture, and the resulting calculation is for an average BRDF over the aperture. This is expressed in the final term of Eq. (1) where P_s is the measured power through the finite solid angle Ω defined by the receiver aperture and the distance to the scatter source. Thus, when the receiver aperture is swept through the scatter field to obtain angle dependence, the measured quantity is actually the convolution of the aperture over the differential BRDF. This does not cause serious distortion unless the scatter field has abrupt intensity changes, as it does near specular or near diffraction peaks associated with periodic surface structure. But there are even more serious problems between the strict definition of BRDF (as derived by Nicodemus) and practical measurements. There are no such things as uniform cross-section beams and isotropic samples that scatter only from surface structure. So, the third term of Eq. (1) is not exactly the differential radiance/irradiance ratio for the situations we create in the lab with our instruments. However, it makes perfect sense to measure normalized scattered power density as a function of direction [as defined in the fourth term of Eq. (1)] even though it cannot be exactly expressed in convenient radiometric terms.

A slightly less cumbersome definition (in terms of writing scatter specifications) is realized if the cosine term is dropped. This is referred to as "the cosine-corrected BRDF," or sometimes, "the scatter function." Its use has caused some of the confusion surrounding measurement differences found in scatter round robins. In accordance with the original definition, accepted practice, and the ASTM Standard,[19] the BRDF contains the cosine, as given in Eq. (1), and the cosine-corrected BRDF does not. It also makes sense to extend the definition to volume scatter sources and even make measurements on the transmissive side of the sample. The term BTDF (for bidirectional transmission distribution function) is used for transmissive scatter, and BSDF (bidirectional scatter distribution function) is all-inclusive.

The BSDF has units of inverse steradians and, unlike reflectance and transmission, can take on very large values as well as very small values.[1,21] For near-normal incidence, a measurement made at the specular beam results in a BSDF value of approximately $1/\Omega$, which is generally a large number. Measured values at the specular direction on the order of $10^6\,\text{sr}^{-1}$ are common for a HeNe laser source. For low-scatter measurements, large apertures are generally used and values fall to the noise equivalent BSDF (or NEBSDF). This level depends on incident power and polar angle (position) as well as aperture size and detector noise, and typically vary from $10^{-4}\,\text{sr}^{-1}$ to $10^{-10}\,\text{sr}^{-1}$. Thus, the measured BSDF can easily vary by over a dozen orders of magnitude in a given angle scan. This large variation results in challenges in instrumentation design as well as data storage, analysis, and presentation, and is another reason for problems with comparison measurements.

Instrument signature is the measured background scatter signal caused by the instrument and not the sample. It is caused by a combination of scatter created within the instrument and by the NEBSDF. Any instrument scatter that reaches the receiver field of view (FOV) will contribute to it. Common causes are scatter from source optics and the system beam dump. It is typically measured without a sample in place; however, careful attention has to be paid to the receiver FOV to ascertain that this is representative of the sample measurement situation. It is calculated as though the signal came from the sample (i.e., the receiver/sample solid angle is used) so that it can be compared to the measured sample BSDF. Near specular, it is dominated by scatter (or diffraction) contributions from the source. At high scatter angles it can generally be limited to NEBSDF levels. Sample measurements are always a combination of desired signal and instrument signature. Reduction of signature, especially near specular, is a prime consideration in instrument design and use.

BSDF specifications always require inclusion of incident angle, source wavelength, and polarization as well as observation angles, scatter levels, and sample orientation. Depending on the sample and the measurement, they may also require aperture information to account for convolution effects. Specifications for scatter instrumentation should include instrument signature limits and the required NEBSDF. Specifications for the NEBSDF must include the polar angle, the solid angle, and the incident power to be meaningful.

TIS measurements are made by integrating the BSDF over a portion of the sphere surrounding the scatter source. This is usually done with instrumentation that gathers (integrates) the scattered light signal. The TIS can sometimes be calculated from BSDF data. If an isotropic sample is illuminated at near-normal incidence with circularly polarized light, data from a single measurement scan is enough to calculate a reasonably accurate TIS value for an entire hemisphere of scatter. The term "total integrated scatter" is a slight misnomer in that the integration is never actually "total", as some scatter is not measured. Integration is commonly performed from a few degrees from specular to polar angles approaching 90° (less than 2.5° to more than 70° in the ASTM Standard).[18] Measurements can be made of either transmissive or reflective scatter. TIS is calculated by ratioing the integrated scatter to the reflected (or transmitted) specular power as shown below. The conversion to rms roughness (σ) under the assumption of a smooth, clean, reflective surface, via Davies' scalar theory,[2] is also given. This latter calculation does not require gaussian surface statistics (as originally assumed by Davies) or even surface isotropy, but will work for other distributions, including gratings and machined optics.[8,11] There are other issues (polarization and the assumption of mostly near specular scatter) that cause some error in this conversion. Comparison of TIS-generated roughness to profile generated values is made difficult by a number of issues (bandwidth limits, one-dimensional profiling of a two-dimensional surface, etc.) that are beyond the scope of this section.

$$\text{TIS} = \frac{\text{integrated scatter}}{\text{reflected specular power}} = \left(\frac{4\pi\sigma}{\lambda}\right)^2 \qquad (2)$$

TIS is one of three ratios that may be formed from the incident power, the specular reflected (or transmitted) power, and the integrated scatter. The other two ratios are the diffuse reflectance (or transmittance) and the specular reflectance (or transmittance). Typically, all three ratios may be obtained from measurements taken in TIS or BSDF instruments. Calculation, or specification, of any of these quantities that involve integration of scatter, also requires that the integration limits be given, as well as the wavelength, angle of incidence, source polarization, and sample orientation.

26.4 INSTRUMENT CONFIGURATIONS AND COMPONENT DESCRIPTIONS

The scatterometer shown in Fig. 2 is representative of the most common instrument configuration in use. The source is fixed in position. The sample is rotated to the desired

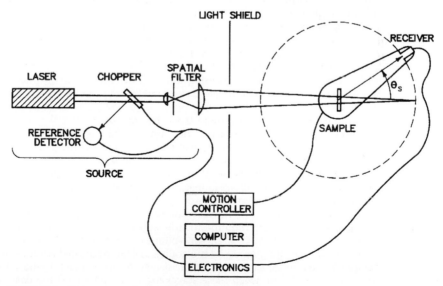

FIGURE 2 Components of a typical BSDF scatterometer.

incident angle, and the receiver is rotated about the sample in the plane of incidence. Although dozens of instruments have been built following this general design, other configurations are in use. For example, the source and receiver may be fixed and the sample rotated so that the scatter pattern moves past the receiver. This is easier mechanically than moving the receiver at the end of an arm, but complicates analysis because the incident angle and the observation angle change simultaneously. Another combination is to fix the source and sample together, at constant incident angle, and rotate this unit (about the point of illumination on the sample) so that the scatter pattern moves past a fixed receiver. This has the advantage that a long receiver/sample distance can be used without motorizing a long (heavy) receiver arm. It has the disadvantage that heavy (or multiple) sources are difficult to deal with. Other configurations, with everything fixed, have been designed that employ several receivers to merely sample the BSDF and display a curve fit of the resulting data. This is an economical solution if the BSDF is relatively uniform without isolated diffraction peaks.

Computer control of the measurement is essential to maximize versatility and minimize measurement time. The software required to control the measurement plus display and analyze the data can be expected to be a significant portion of total instrument development cost. The following paragraphs review typical design features (and issues) associated with the source, sample mount, and receiver components.

The source in Fig. 2 is formed by a laser beam that is chopped, spatially filtered, expanded, and finally brought to a focus on the receiver path. The beam is chopped to reduce both optical and electronic noise. This is accomplished through the use of lock-in detection in the electronics package which suppresses all signals except those at the chopping frequency. Low-noise, programmable gain electronics are essential to reducing NEBSDF. The reference detector is used to allow the computer to ratio out laser power fluctuations and, in some cases, to provide the necessary timing signal to the lock-in electronics. Polarizers, wave plates, and neutral density filters are also commonly placed prior to the spatial filter when required in the source optics. The spatial filter removes scatter from the laser beam and presents a point source which is imaged by the final focusing element to the detector zero position. Although a lens is shown in Fig. 2, the use of a mirror, which works over a larger range of wavelengths and generally scatters less light, is more common. For most systems the large FN of the final focusing element allows use of a spherical mirror with only minor aberration. Low-scatter spherical mirrors are easier to obtain than other conic sections. The incident beam is typically focused at the receiver to facilitate near specular measurement. Another option (a collimated beam at the receiver) is sometimes used and will be considered in the discussion on receivers. In either case, curved samples can be accommodated by adjusting the position of the spatial filter with respect to the final focusing optic. The spot size on the sample is obviously determined by elements of the system geometry and can be adjusted by changing the focal length of the first lens (often a microscope objective). The source region is completed by a shield that isolates stray laser light from the detector.

Lasers are convenient sources, but are not necessary. Broadband sources are often required to meet a particular application or to simulate the environment where a sample will be used. Monochromators and filters can be used to provide scatterometer sources of arbitrary wavelength.[22] Noise floor with these tunable incoherent sources increases dramatically as the spectral bandpass is narrowed, but they have the advantage that the scatter pattern does not contain laser speckle.

The sample mount can be very simple or very complex. In principal, six degrees of mechanical freedom are required to fully adjust the sample. Three translational degrees of freedom allow the sample area (or volume) of interest to be positioned at the detector rotation axis and illuminated by the source. Three rotational degrees of freedom allow the sample to be adjusted for angle of incidence, out-of-plane tilt, and rotation about sample normal. The order in which these stages are mounted affects the ease of use (and cost) of the sample holder. In practice, it often proves convenient to either eliminate, or occasionally duplicate, some of these degrees of freedom. Exact requirements for these stages differ depending on whether the sample is reflective or transmissive, as well as with size and shape. In addition, some of these axes may be motorized to allow the sample area to be raster-scanned to automate sample alignment or to measure reference samples. The order in which these stages are mounted affects the ease of sample alignment. As a general rule, the scatter pattern is insensitive to small changes in incident angle but very sensitive to small angular deviations from specular. Instrumentation should be configured to allow location of the specular reflection (or transmission) very accurately.

The receiver rotation stage should be motorized and under computer control so that the input aperture may be placed at any position on the observation circle (dotted line in Fig. 2). Data scans may be initiated at any location. Systems vary as to whether data points are taken with the receiver stopped or "on the fly." The measurement software is less complicated if the receiver is stopped. Unlike many TIS systems, the detector is always approximately normal to the incoming scatter signal. In addition to the indicated axis of rotation, some mechanical freedom is required to assure that the receiver is at the correct height and pointed (tilted) at the illuminated sample. Sensitivity, low noise, linearity, and dynamic range are the important issues in choosing a detector element and designing the receiver housing. In general, these requirements are better met with photovoltaic detectors than photoconductive detectors. Small area detectors reduce the NEBSDF.

Receiver designs vary, but changeable apertures, bandpass filters, polarizers, lenses, and field stops are often positioned in front of the detector element. Figure 3 shows two receiver configurations, one designed for use with a converging source and one with a collimated source. In Fig. 3*a*, the illuminated sample spot is imaged on a field stop in front of the detector. This configuration is commonly used with the source light converging on the receiver path. The field stop determines the receiver FOV. The aperture at the front of the receiver determines the solid angle over which scatter is gathered. Any light entering this aperture, that originates from within the FOV, will reach the detector and become part of the signal. This includes instrument signature contributions scattered through small

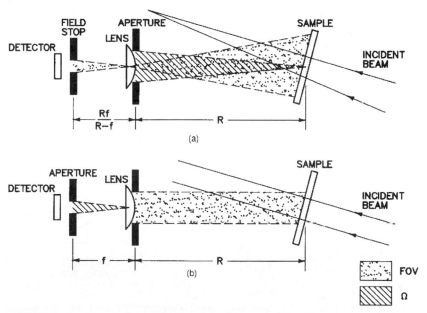

FIGURE 3   Receiver configurations: (*a*) converging source; (*b*) collimated source.

angles by the source optics. It will also include light scattered by the receiver lens so that it appears to come from the sample. The configuration in Fig. 3a can be used to obtain near specular measurements by bringing a small receiver aperture close to the focused specular beam. With this configuration, reducing the front aperture does not limit the FOV. The receiver in Fig. 3b is in better accordance with the strict definition of BRDF in that a collimated source can be used. An aperture is located one focal length behind a collecting lens (or mirror) in front of the detector. The intent is to measure bundles of nearly parallel rays scattered from the sample. The angular spread of rays allowed to pass to the detector defines the receiver solid angle, which is equal to the aperture size divided by the focal length of the lens. This ratio (not the front aperture/sample distance) determines the solid angle of this receiver configuration. The FOV is determined by the clear aperture of the lens, which must be kept larger than the illuminated spot on the sample. The Fig. 3b design is unsuitable for near specular measurement because the relatively broad collimated specular beam will scatter from the receiver lens for several degrees from specular. It is also limited in measuring large incident angle situations where the elongated spot may exceed the FOV. If the detector (and its stop) can be moved in relation to the lens, receivers can be adjusted from one configuration to the other. Away from the specular beam, in low instrument signature regions, there is no difference in the measured BSDF values between the two systems.

The two common methods of approaching TIS measurements are shown in Fig. 4. The first one, employed by Bennett and Porteus in their early instrument, uses a hemispherical mirror (or Coblentz Sphere) to gather scattered light from the sample and image it onto a detector. The specular beam enters and leaves the hemisphere through a small circular hole. The diameter of that hole defines the near specular limit of the instrument. The reflected beam (not the incident beam) should be centered in the hole because the BSDF will be symmetrical about it. Alignment of the hemispherical mirror is critical in this approach. The second approach involves the use of an integrating sphere. A section of the sphere is viewed by a recessed detector. If the detector FOV is limited to a section of the sphere that is not directly illuminated by scatter from the sample, then the signal will be proportional to total scatter from the sample. Again, the reflected beam should be centered on the exit hole. The Coblentz Sphere method presents more signal to the

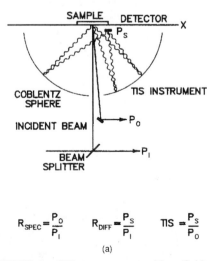

$$R_{SPEC} = \frac{P_o}{P_I} \qquad R_{DIFF} = \frac{P_s}{P_I} \qquad TIS = \frac{P_s}{P_o}$$

(a)

FIGURE 4a TIS measurement with a Coblentz Sphere.

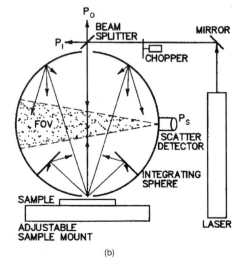

(b)

FIGURE 4b TIS measurement with a diffuse integrating sphere.

detector; however, some of this signal is incident on the detector at very high angles. Thus, this approach tends to discriminate against high-angle scatter (which is not a problem for many samples). The integrating sphere is easier to align, but has a lower signal-to-noise ratio (less signal on the detector) and is more difficult to build in the IR where uniform diffuse surfaces are harder to obtain. A common mistake with TIS measurements is to assume that for near-normal incidence, the orientation between source polarization and sample orientation is not an issue. TIS measurements made with a linearly polarized source on a grating at different orientations will quickly demonstrate this dependence.

TIS measurements can be made over very near specular ranges by utilizing a diffusely reflecting plate with a small hole in it. A converging beam is reflected off the sample and through the hole. Scatter is diffusely reflected from the plate to a receiver designed to uniformly view the plate. The reflected power is measured by moving the plate so the specular beam misses the hole. Measurements starting closer than 0.1° from specular can be made in this manner, and it is an excellent way to check incoming optics or freshly coated optics for low scatter.

26.5 INSTRUMENTATION ISSUES

Measurement of near specular scatter is often one of the hardest requirements to meet when designing an instrument and has been addressed in several publications.[23–25] The measured BSDF may be divided into two regions relative to the specular beam, as shown in Fig. 5. Outside the angle θ_N from specular, is a low-signature region where the source optics are not in the receiver FOV. Inside θ_N, at least some of the source optics scatter directly into the receiver and the signature increases rapidly until the receiver aperture reaches the edge of the specular beam. As the aperture moves closer to specular center, the measurement is dominated by the aperture convolution of the specular beam, and there is no opportunity to measure scatter. The value θ_N is easily calculated (via a small

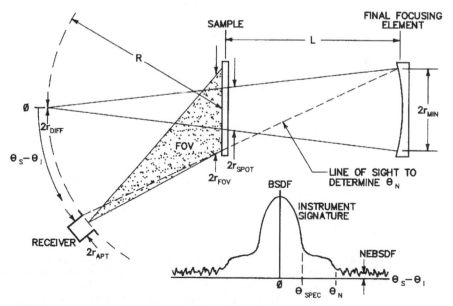

FIGURE 5 Near specular geometry and instrument signature.

angle approximation) using the instrument geometry and parameters identified in Fig. 5, where the receiver is shown at the θ_N position. The parameter F is the focal length of the sample.

$$\theta_N = (r_{MIR} + r_{FOV})/L + (r_{FOV} + r_{apt})/R - r_{spot}/F \tag{3}$$

It is easy to achieve values of θ_N below 10° and values as small as 1° can be realized with careful design. The offset angle from specular, θ_{spec}, at which the measurement is dominated by the specular beam, can be reduced to less than a tenth of a degree at visible wavelengths and is given by

$$\theta_{spec} = \frac{r_{diff} + r_{apt}}{R} \approx \frac{3\lambda}{D} + \frac{r_{apt}}{R} \tag{4}$$

Here, r_{diff} and r_{apt} are the radius of the focused spot and the receiver aperture, respectively (see Fig. 5 again). The value of r_{diff} can be estimated in terms of the diameter D of the focusing optic and its distance to the focused spot, $R + L$ (estimated as $2.5R$). The diffraction limit has been doubled in this estimate to allow for aberrations.

To take near specular measurements, both angles and the instrument signature need to be reduced. The natural reaction is to "increase R to increase angular resolution." Although a lot of money has been spent doing this, it is an unnecessarily expensive approach. Angular resolution is achieved by reducing r_{apt} and by taking small steps. The radius r_{apt} can be made almost arbitrarily small so the economical way to reduce the r_{apt}/R terms is by minimizing r_{apt}—not by increasing R. A little thought about r_{FOV} and r_{diff} reveals that they are both proportional to R, so nothing is gained in the near specular game by purchasing large-radius rotary stages.

The reason for building a large-radius scatterometer is to accommodate a large FOV. This is often driven by the need to take measurements at large incident angles, which creates a large spot on the sample. When viewing normal to the sample, the FOV requirements can be stringent. Because the maximum FOV is proportional to detector diameter (and limited at some point by minimum receiver lens FN), increasing R is the only open-ended design parameter available. It should be sized to accommodate the smallest detector likely to be used in the system. This will probably be in the mid-IR where, as of this writing, uniform high-detectivity photovoltaic detectors larger than 2 mm are difficult to obtain. On the other hand, a larger detector diameter means increased electronic noise and a larger NEBSDF.

Scatter sources of instrument signature can be reduced by these techniques.

1. Use the lowest-scatter focusing element in the source that you can afford and learn how to keep it clean. This will probably be a spherical mirror.

2. Keep the source area as "black" as possible. This especially includes the sample side of the spatial filter pinhole which is conjugate with the receiver aperture. Use a black pinhole.

3. Employ a specular beam dump that rides with your receiver and additional beam dumps to capture sample reflected and transmitted beams when the receiver has left the near specular area. Use your instrument to measure the effectiveness of your beam dumps.[26]

4. Near specular scatter caused by dust in the air can be significantly reduced through the use of a filtered air supply over the specular beam path.

Away from specular, reduction of NEBSDF is the major factor in measuring low-scatter samples and increasing instrument quality. Measurement of visible scatter from a clean semiconductor wafer will take many instruments right down to instrument signature levels. Measurement of cross-polarized scatter requires a low NEBSDF for even high-scatter optics. For a given receiver solid angle, incident power, and scatter direction, the NEBSDF is limited by the noise equivalent power of the receiver (and associated electronics), once

TABLE 1.

Detector (2 mm dia.)	NEP (W/Hz)	Wavelength (nanometers)	P_i (W)	NEBSDF (sr^{-1})
PMT	10^{-15}	633	0.005	10^{-10}
Si	10^{-13}	633	0.005	10^{-8}
Ge	3×10^{-13}	1320	0.001	10^{-7}
InSb	$10^{-12}*$	3390	0.002	5×10^{-7}
HgMgTe	$10^{-11}*$	10600	2.0	10^{-8}
Pyro	10^{-8}	10600	2.0	10^{-5}

* Detector at 77° Kelvin.

optical noise contributions are eliminated. The electronic contributions to NEBSDF are easily measured by simply covering the receiver aperture during a measurement. Because the resulting signal varies in a random manner, NEBSDF should be expressed as an rms value (roughly equal to one-third of the peak level). An absolute minimum measurable scatter signal (in watts) can be found from the product of three terms: the required signal-to-noise ratio, the system noise equivalent power (or NEP given in watts per square root hertz), and the square root of the noise bandwidth (BW_n). The system NEP is often larger than the detector NEP and cannot be reduced below it. The detector NEP is a function of wavelength and increases with detector diameter. Typical detector NEP values (2-mm diameter) and wavelength ranges are shown as follows for several common detectors in Table 1. Notice that NEP tends to increase with wavelength. The noise bandwidth varies as the reciprocal of the sum of the system electronics time constant and the measurement integration time. Values of 0.1 to 10 Hz are commonly achieved. In addition to system NEP, the NEBSDF may be increased by contributions from stray source light, room lights, and noise in the reference signal. Table 1 also shows achievable rms NEBSDF values that can be realized at unity cosine, a receiver solid angle of 0.003 sr, one-second integration, and the indicated incident powers. This column can be used as a rule of thumb in system design or to evaluate existing equipment. Simply adjust by the appropriate incident power, solid angle, etc., to make the comparison. Adjusted values substantially higher than these indicate there is room for system improvement (don't worry about differences as small as a factor of two). Further reduction of the instrument signature under these geometry and power conditions will require dramatically increased integration time (because of the square root dependence on noise bandwidth) and special attention to electronic dc offsets. Because the NEP tends to increase with wavelength, higher powers are needed in the mid-IR to reach the same NEBSDFs that can be realized in the visible. Because scatter from many sources tends to decrease at longer wavelengths, a knowledge of the instrument NEBSDF is especially critical in the mid-IR.

As a final configuration comment, the software package (both measurement and analysis) is crucial for an instrument that is going to be used for any length of time. Poor software will quickly cost work-years of effort due to errors, increased measurement and analysis time, and lost business. Expect to expend one to two work-years with experienced programmers writing a good package—it is worth it.

26.6 MEASUREMENT ISSUES

Sample measurement should be preceded (and sometimes followed) by a measurement of the instrument signature. This is generally accomplished by removing the sample and measuring the apparent BSDF from the sample as a transmissive scan. This is not an exact measure of instrument noise during sample measurement, but if the resulting BSDF is multiplied by sample reflectance (or transmission) before comparison to sample data, it can define some hard limits over which the sample data cannot be trusted. The signature should also be compared to the NEBSDF value obtained with the receiver aperture blocked. Obtaining the instrument signature also presents an opportunity to measure the incident power, which is required for calculation of the BSDF. The ability to see the data displayed as it is taken is an extremely helpful feature when it comes to reducing instrument signature and eliminating measurement setup errors.

Angle scans, which have dominated the preceding discussion, are an obvious way to take measurements. BSDF is also a function of position on the sample, source wavelength, and source polarization, and scans can also be taken at fixed angle (receiver position) as a function of these variables. Obviously, a huge amount of data is required to completely characterize scatter from a sample.

Raster scans are taken to measure sample uniformity or locate (map) discrete defects. A common method is to fix the receiver position and move the sample in its own $x = y$ plane recording the BSDF at each location. Faster approaches involve using multiple detectors (array cameras for example) with large area illumination, and scanning the source over the sample. Results can be presented using color maps or 3-D isometric plots. Results can be further analyzed via histograms and various image-processing techniques.

There are three obvious choices for making wavelength scans. Filters (variable or discrete) can be employed at the source or receiver. A monochromator can be used as a source.[22] Finally, there is some advantage to using a Fourier transforming infrared spectrometer (FTIR) as a source in the mid-IR.[22] Details of these techniques are beyond the scope of this discussion; however, a couple of generalities will be mentioned. Even though these measurements often involve relatively large bandwidths at a given wavelength (compared to a laser), the NEBSDF is often larger by a few orders because of the smaller incident power. Further, because the bandwidths change differently between the various source types given above, meaningful measurement comparisons between instruments are often difficult to make.

Polarization scans are often limited to SS, SP, PS, and PP (source/receiver) combinations. However, complete polarization-dependence of the sample requires the measurement of the sample Mueller matrix. This is found by creating a set of Stokes vectors at the source and measuring the resulting Stokes vector in the desired scatter direction.[10,11,27–31] This is an area of instrumentation development that is the subject of increasing attention.

Speckle effects in the BSDF from a laser source can be eliminated in several ways. If a large receiver solid angle is used (generally several hundred speckles in size) there is not a problem. The sample can be rotated about its normal so that speckle is time averaged out of the measurement. This is still a problem when measuring very near the specular beam because sample rotation unavoidably moves the beam slightly during the measurement. In this case, the sample can be measured several times at slightly different orientations and the results averaged to form one speckle-free BSDF.

Scatter measurement in the retrodirection (back into the incident beam) has been of increasing interest in recent years and represents an interesting measurement challenge. Measurement requires the insertion of a beam splitter in the source. This also scatters light and, because it is closer to the receiver than the sample, dramatically raises the NEBSDF. Diffuse samples can be measured this way, but not much else. A clever (high tech) Doppler-shift technique, employing a moving sample, has been reported[32] that allows separation of beam-splitter scatter from sample scatter and allows measurement of mirror scatter. A more economical (low tech) approach simply involves moving the source chopper to a location between the receiver and the sample. Beam-splitter scatter is now dc and goes unnoticed by the ac-sensitive receiver. Noise floor is now limited by scatter from the chopper which must be made from a low-scatter, specular, absorbing material. Noise floors as low as $3 \times 10^{-8}\,\mathrm{sr}^{-1}$ have been achieved.[33]

26.7 INCIDENT POWER MEASUREMENT, SYSTEM CALIBRATION, AND ERROR ANALYSIS

Regardless of the type of BSDF measurement, the degree of confidence in the results is determined by instrument calibration, as well as by attention to the measurement limitations previously discussed. Scatter measurements have often been received with considerable skepticism. In part, this has been due to misunderstanding of the definition of BSDF and confusion about various measurement subtleties, such as instrument signature or aperture convolution. However, quite often the measurements have been wrong and the skepticism is justified.

Instrument calibration is often confused with the measurement of P_i, which is why these topics are covered in the same section. To understand the source of this confusion, it is necessary to first consider the various quantities that need to be measured to calculate the BSDF. From Eq. (1), they are P_s, θ_s, Ω and P_i. The first two require measurement over a wide range of values. In particular, P_s, which may vary over many orders of magnitude, is a problem. In fact, linearity of the receiver to obtain a correct value of P_s, is a key calibration issue. Notice that an absolute measurement of P_s is not required, as long as the P_s/P_i ratio is correctly evaluated. P_i and Ω generally take on only one (or just a few) discrete values during a data scan. The value of Ω is determined by system geometry. The value of P_i is generally measured in one of two convenient ways.[11,19]

The first technique, sometimes referred to as *the absolute method* makes use of the scatter detector (and sometimes a neutral density filter) to directly measure the power incident upon the sample. This method relies on receiver linearity (as does the overall calibration of BSDF) and on filter accuracy when one is used. The second technique, sometimes referrred to as *the reference method* makes use of a known BSDF reference sample (usually a diffuse reflector and unfortunately often referred to as the "calibration sample") to obtain the value of P_i. Scatter from the reference sample is measured and the result used to infer the value of P_i via Eq. (1). The $P_i\Omega$ product may be evaluated this way. This method depends on knowing the absolute BSDF of the reference. Both techniques become more difficult in the mid-IR, where "known" neutral density filters and "known" reference samples are difficult to obtain. Reference sample uniformity in the mid-IR is often the critical issue and care must be exercised. Variations at 10.6 micrometers as large as 7 : 1 have been observed across the face of a diffuse gold reference "of known BRDF." The choice of measurement methods is usually determined by whether it is more convenient to measure the BSDF of a reference or the total power P_i. Both are equally valid methods of obtaining P_i. However, neither method constitutes a system calibration, because calibration issues such as an error analysis and a linearity check over a wide range of scatter values are not addressed over the full range of BSDF angles and powers when P_i is measured (or calculated). The use of a reference sample is an excellent system check regardless of how P_i is obtained.

System linearity is a key part of system calibration. In order to measure linearity, the receiver transfer characteristic, signal out as a function of light in, must be found. This may be done through the use of a known set of neutral density filters or through the use of a comparison technique[34] that makes use of two data scans—with and without a single filter. However, there are other calibration problems than just linearity. The following paragraph outlines an error analysis for BSDF systems.

Because the calculation of BSDF is very straightforward, the sources of error can be examined through a simple analysis[11,35] under the assumption that the four defining parameters are independent.

$$\frac{\Delta \text{BSDF}}{\text{BSDF}} = \left[\left(\frac{\Delta P_s}{P_s} \right)^2 + \left(\frac{\Delta P_i}{P_i} \right)^2 + \left(\frac{\Delta \Omega}{\Omega} \right)^2 + \left(\frac{\Delta \theta_s \sin \theta_s}{\cos^2 \theta_s} \right)^2 \right]^{1/2} \tag{5}$$

In similar fashion, each of these terms may be broken into the components that cause errors in it. When this is done, the total error may be found as a function of angle. Two high-error regions are identified. The first is the near specular region (inside one degree), where errors are dominated by the accuracy to which the receiver aperture can be located in the cross-section direction. Or, in other words, did the receiver scan exactly through the specular beam, or did it just miss it? The second relatively high error region is near the sample plane where $\cos \theta_s$ approaches zero. In this region, a small error in angular position results in a large error in calculated BSDF. These errors are often seen in BSDF data as an abrupt increase in calculated BSDF in the grazing scatter direction, the result of division by a very small cosine into the signal gathered by a finite receiver aperture (and/or a dc offset voltage in the detector electronics). This is another example where use of the cosine-corrected BSDF makes more sense.

Accuracy is system-dependent; however, at signal levels well above the NEBSDF, uncertainties less than ± 10 percent can be obtained away from the near specular and grazing directions. With expensive electronics and careful error analysis, these inaccuracies can be reduced to the ± 1 percent level.

Full calibration is not required on a daily basis. Sudden changes in instrument signature are an indication of possible calibration problems. Measurement of a reference sample that varies over several orders of magnitude is a good system check. It is prudent to take such a reference scan with data sets in case the validity of the data is questioned at a later time. A diffuse sample, with nearly constant BRDF, is a good reference choice for the measurement of P_i but a poor one for checking system calibration.

26.8 SUMMARY

The art of scatter measurement has evolved to an established form of metrology within the optics industry. Because scatter measurements tend to be a little more complicated than many other optical metrology procedures, a number of key issues must be addressed to obtain useful information. System specifications and measurements need to be given in terms of accepted, well-defined (and understood) quantities (BSDF, TIS, etc.). All parameters associated with a measurement specification need to be given (such as angle limits, receiver solid angles, noise floors, wavelength, etc.). Measurement of near specular scatter and/or low BSDF values are particularly difficult and require careful attention to instrument signature values; however, if the ASTM procedures are followed, the result will be repeatable, accurate data.

TIS and BSDF are widely accepted throughout the industry and their measurement is defined by ASTM standards. Scatter measurements are used routinely as a quality check on optical components. BSDF specifications are now often used (as they should be) in place of scratch/dig or rms roughness, when scatter is the issue. Conversion of surface scatter data to other useful formats, such as surface roughness statistics, is commonplace. The sophistication of the instrumentation (and analysis) applied to these problems is still increasing. Out-of-plane measurements and polarization-sensitive measurements are two areas that are experiencing rapid advances. Measurement of scatter outside the optics community is also increasing. Although the motivation for scatter measurement differs in industrial situations, the basic measurement and instrumentation issues encountered are essentially the ones described here.

26.9 REFERENCES

1. H. E. Bennett and J. O. Porteus, "Relation Between Surface Roughness and Specular Reflectance at Normal Incidence," *J. Opt. Soc. Am.* **51:**123 (1961).
2. H. Davies, "The Reflection of Electromagnetic Waves from a Rough Surface," *Proc. Inst. Elec. Engrs.* **101:**209 (1954).

3. J. C. Stover, "Roughness Measurement by Light Scattering," in A. J. Glass, A. H. Guenther (eds.), *Laser Induced Damage in Optical Materials,* U.S. Govt. Printing Office, Washington, D.C., 1974, p. 163.

4. J. E. Harvey, *Light Scattering Characteristics of Optical Surfaces,* Ph.D. dissertation, Univ. of Arizona, 1976.

5. E. L. Church, H. A. Jenkinson and J. M. Zavada, "Measurement of the Finish of Diamond-Turned Metal Surfaces By Differential Light Scattering," *Opt. Eng.* **16:**360 (1977).

6. J. C. Stover and C. H. Gillespie, "Design Review of Three Reflectance Scatterometers," *Proc. SPIE* **V362** (Scattering in Optical Materials):172 (1982).

7. J. C. Stover, "Roughness Characterization of Smooth Machined Surfaces by Light Scattering," *Appl. Opt.* **V14**(N8):1796 (1975).

8. E. L. Church and J. M. Zavada, "Residual Surface Roughness of Diamond-Turned Optics," *Appl. Opt.* **14:**1788 (1975).

9. E. L. Church, H. A. Jenkinson and J. M. Zavada, "Relationship Between Surface Scattering and Microtopographic Features," *Opt. Eng.* **18**(2):125 (1979).

10. E. L. Church, "Surface Scattering," in M. Bass (ed.), *Handbook of Optics,* vol. I, 2d ed., McGraw-Hill, New York, 1994.

11. J. C. Stover, *Optical Scattering Measurement and Analysis,* McGraw-Hill, NY, 1990.

12. F. E. Nicodemus, J. C. Richmond, J. J. Hsia, I. W. Ginsberg, and T. Limperis, *Geometric Considerations and Nomenclature for Reflectance,* NBS Monograph 160, U.S. Dept. of Commerce, 1977.

13. W. L. Wolfe and F. O. Bartell, "Description and Limitations of an Automated Scatterometer," *Proc. SPIE* **362:**30 (1982).

14. D. R. Cheever, F. M. Cady, K. A. Klicker, and J. C. Stover, "Design Review of a Unique Complete Angle-Scatter Instrument (CASI)," *Proc. SPIE* **818** (Current Developments in Optical Engineering II):13 (1987).

15. P. R. Spyak and W. L. Wolfe, "Cryogenic Scattering Measurements," *Proc. SPIE* **967:**15 (1989).

16. W. L. Wolfe, K. Magee, and D. W. Wolfe, "A Portable Scatterometer for Optical Shop Use," *Proc. SPIE* **V525:**160 (1985).

17. J. Rifkin, "Design Review of a Complete Angle Scatter Instrument," *Proc. SPIE* **V1036:**15 (1988).

18. ASTM Standard #F1048-87, "Standard Test Method for Measuring the Effective Surface Roughness of Optical Components by Total Integrated Scattering," August 1987.

19. ASTM Standard #E 1392-90, "Standard Practice for Angle Resolved Scatter Measurements on Specular or Diffuse Surfaces," November 1991.

20. T. A. Leonard and M. A. Pantoliano, "BRDF Round Robin," *Proc. SPIE* **967:**22 (1988).

21. T. F. Schiff, J. C. Stover, D. R. Cheever, and D. R. Bjork, "Maximum and Minimum Limitations Imposed on BSDF Measurements," *Proc. SPIE* **V967** (1988).

22. F. M. Cady, M. W. Knighton, D. R. Cheever, B. D. Swimley, M. E. Southwood, T. L. Hundtoft, and D. R. Bjork, "Design Review of a Broadband 3-D Scatterometer," *Proc. SPIE* **1753:**21 (1992).

23. K. A. Klicker, J. C. Stover, D. R. Cheever, and F. M. Cady, "Practical Reduction of Instrument Signature in Near Specular Light Scatter Measurements," *Proc. SPIE* **V818:**26 (1987).

24. S. J. Wein and W. L. Wolfe, "Gaussian Apodized Apertures and Small Angle Scatter Measurements," *Opt Eng.* **28**(3):273–280 (1989).

25. J. C. Stover and M. L. Bernt, "Very Near Specular Measurement via Incident Angle Scaling, *Proc. SPIE* **1753:**16 (1992).

26. F. M. Cady, D. R. Cheever, K. A. Klicker, and J. C. Stover, "Comparison of Scatter Data From Various Beam Dumps," *Proc. SPIE* **V818:**21 (1987).

27. W. S. Bickle and G. W. Videen, "Stokes Vectors, Mueller Matrices and Polarized Light:

Experimental Applications to Optical Surfaces and All Other Scatterers," *SPIE Proc.* **1530:**02 (1991).

28. T. F. Schiff, D. J. Wilson, B. D. Swimley, M. E. Southwood, D. R. Bjork, and J. C. Stover, "Design Review of a Unique Out-of-Plane Polarimetric Scatterometer," *Proc. SPIE* **1753:**33 (1992).

29. T. F. Schiff, D. J. Wilson, B. D. Swimley, M. E. Southwood, D. R. Bjork, and J. C. Stover, "Mueller Matrix Measurements with an Out-Of-Plane Polarimetric Scatterometer," *Proc. SPIE* **1746:**33 (1992).

30. T. F. Schiff, B. D. Swimley, and J. C. Stover, "Mueller Matrix Measurements of Scattered Light," *Proc. SPIE* **1753:**34 (1992).

31. J. M. Bennett, "Polarization," in M. Bass (ed.), *Handbook of Optics,* vol. I, 2d ed., McGraw-Hill, New York, 1994.

32. Z. H. Gu, R. S. Dummer, A. A. Maradudin, and A. R. McGurn, "Experimental Study of the Opposition Effect in the Scattering of Light From a Randomly Rough Metal Surface," *Appl. Opt.* **V28:**(N3):537 (1989).

33. T. F. Schiff, D. J. Wilson, B. D. Swimley, M. E. Southwood, D. R. Bjork, and J. C. Stover, "Retroreflections on a Low Tech Approach to the Measurement of Opposition Effects, *Proc. SPIE* **1753:**35 (1992).

34. F. M. Cady, D. R. Bjork, J. Rifkin, and J. C. Stover, "Linearity in BSDF Measurement," *Proc. SPIE* **1165:**44 (1989).

35. F. M. Cady, D. R. Bjork, J. Rifkin, and J. C. Stover, "BRDF Error Analysis," *Proc. SPIE* **1165:**13 (1989).

CHAPTER 27
ELLIPSOMETRY

Rasheed M. A. Azzam

Department of Electrical Engineering
College of Engineering
University of New Orleans
New Orleans, Louisiana

27.1 GLOSSARY

A	instrument matrix
D_ϕ	$\lambda/2S_1$
E	electrical field
$\mathbf{E}_0$	constant complex vector
$f(\ )$	function
I	interface scattering matrix
k	extinction coefficient
L	layer scattering matrix
N	complex refractive index $= n - jk$
n	real part of the refractive index
R	reflection coefficient
r	reflection coefficient
S_{ij}	scattering matrix elements
s, p	subscripts for polarization components
X	$\exp\left(-j2\pi d/D_\phi\right)$
Δ	ellipsometric angle
ϵ	dielectric function
$\langle\epsilon\rangle$	psuedo dielectric function
ρ	χ_i/χ_r
ϕ	angle of incidence
χ_i	E_{is}/E_{ip}
χ_r	E_{rs}/E_{rp}
ψ	ellipsometric angle

27.2 *INTRODUCTION*

Ellipsometry is a nonperturbing optical technique that uses the change in the state of polarization of light upon reflection for the in-situ and real-time characterization of surfaces, interfaces, and thin films. In this chapter we provide a brief account of this subject with an emphasis on modeling and instrumentation. For extensive coverage, including applications, the reader is referred to several monographs,[1–3] user's guides,[4–5] collected reprints,[6] conference proceedings,[7–12] and general and topical reviews.[13–32]

In ellipsometry, a collimated beam of monochromatic or quasi-monochromatic light, which is polarized in a known state, is incident on a sample surface under examination, and the state of polarization of the reflected light is analyzed. From the incident and reflected states of polarization, ratios of complex reflection coefficients of the surface for two incident orthogonal polarization states (commonly the linear polarizations parallel and perpendicular to the plane of incidence) are determined. These ratios are subsequently related to the structural and optical properties of the ambient-sample interface region by invoking an appropriate model and the electromagnetic theory of reflection. Finally, model parameters of interest are determined by solving the resulting inverse problem.

In ellipsometry, one of the two copropagating orthogonally polarized waves can be considered to act as a reference for the other. Inasmuch as the state of polarization of light is determined by the superposition of the orthogonal components of the electric field vector, an ellipsometer may be thought of as a common-path polarization interferometer. And because ellipsometry involves only relative amplitude and relative phase measurements, it is highly accurate. Furthermore, its sensitivity to minute changes in the interface region, such as the formation of a submonolayer of atoms or molecules, has qualified ellipsometry for many applications in surface science and thin-film technologies.

In a typical scheme, Fig. 1, the incident light is linearly polarized at a known but arbitrary azimuth and the reflected light is elliptically polarized. Measurement of the ellipse of polarization of the reflected light accounts for the name ellipsometry, which was first coined by Rothen.[33] (For a discussion of light polarization, the reader is referred to Vol. I, Chap. 5. For a historical background on ellipsometry, see Rothen[34] and Hall.[35])

For optically isotropic structures, ellipsometry is carried out only at oblique incidence. In this case, if the incident light is linearly polarized with the electric vector vibrating parallel p or perpendicular s to the plane of incidence, the reflected light is likewise p- and s-polarized, respectively. In other words, the p and s linear polarizations are the

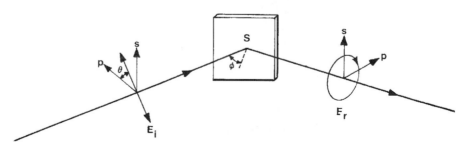

FIGURE 1 Incident linearly polarized light of arbitrary azimuth θ is reflected from the surface S as elliptically polarized. p and s identify the linear polarization directions parallel and perpendicular to the plane of incidence and form a right-handed system with the direction of propagation. ϕ is the angle of incidence.

eigenpolarizations of reflection.[36] The associated eigenvalues are the complex amplitude reflection coefficients R_p and R_s. For an arbitrary input state with phasor electric-field components E_{ip} and E_{is}, the corresponding field components of the reflected light are given by

$$E_{rp} = R_p E_{ip} \qquad E_{rs} = R_s E_{is} \qquad (1)$$

By taking the ratio of the respective sides of these two equations, one gets

$$\rho = \chi_i / \chi_r \qquad (2)$$

where

$$\rho = R_p / R_s \qquad (3)$$

$$\chi_i = E_{is} / E_{ip} \qquad \chi_r = E_{rs} / E_{rp} \qquad (4)$$

χ_i and χ_r of Eqs. (4) are complex numbers that succinctly describe the incident and reflected polarization states of light;[37] their ratio, according to Eqs. (2) and (3), determines the ratio of the complex reflection coefficients for the p and s polarizations. Therefore, ellipsometry involves pure polarization measurements (without account for absolute light intensity or absolute phase) to determine ρ. It has become customary in ellipsometry to express ρ in polar form in terms of two *ellipsometric angles* ψ and Δ ($0 \leq \psi \leq 90°$, $0 \leq \Delta < 360°$) as follows

$$\rho = \tan \psi \exp (j\Delta) \qquad (5)$$

$\tan \psi = |R_p|/|R_s|$ represents the relative amplitude attenuation and $\Delta = \arg (R_p) - \arg (R_s)$ is the differential phase shift of the p and s linearly polarized components upon reflection.

Regardless of the nature of the sample, ρ is a function,

$$\rho = f(\phi, \lambda) \qquad (6)$$

of the angle of incidence ϕ and the wavelength of light λ. Multiple-angle-of-incidence ellipsometry[38–43] (MAIE) involves measurement of ρ as a function of ϕ, and spectroscopic ellipsometry[3,22,27–31] (SE) refers to the measurement of ρ as a function of λ. In variable-angle spectroscopic ellipsometry[43] (VASE) the ellipsometric function ρ of the two real variables ϕ and λ is recorded.

27.3 CONVENTIONS

The widely accepted conventions in ellipsometry are those adopted at the 1968 Symposium on Recent Developments in Ellipsometry following discussions of a paper by Muller.[44] Briefly, the electric field of a monochromatic plane wave traveling in the direction of the z axis is taken as

$$\mathbf{E} = \mathbf{E}_0 \exp (-j2\pi Nz/\lambda) \exp (j\omega t) \qquad (7)$$

where $\mathbf{E}_0$ is a constant complex vector that represents the transverse electric field in the $z = 0$ plane, N is the complex refractive index of the optically isotropic medium of propagation, ω is the angular frequency, and t is the time. N is written in terms of its real and imaginary parts as

$$N = n - jk \qquad (8)$$

where $n > 0$ is the refractive index and $k \geq 0$ is the extinction coefficient. The positive

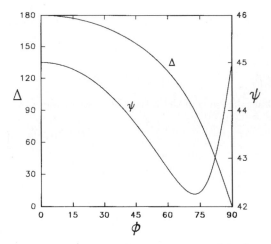

FIGURE 2 Ellipsometric parameters ψ and Δ of an air/Au interface as functions of the angle of incidence ϕ. The complex refractive index of Au is assumed to be $0.306 - j2.880$ at 564-nm wavelength.

directions of p and s before and after reflection form a right-handed coordinate system with the directions of propagation of the incident and reflected waves, Fig. 1. At normal incidence ($\phi = 0$), the p directions in the incident and reflected waves are antiparallel, whereas the s directions are parallel. Some of the consequences of these conventions are:

1. At normal incidence, $R_p = -R_s$, $\rho = -1$, and $\Delta = \pi$.

2. At grazing incidence, $R_p = R_s$, $\rho = 1$, and $\Delta = 0$.

3. For an abrupt interface between two homogeneous semi-infinite media, Δ is in the range $0 \leq \Delta \leq \pi$, and $0 \leq \psi \leq 45°$.

As an example, Fig. 2 shows ψ and Δ vs. ϕ for light reflection at the air/Au interface, assuming $N = 0.306 - j2.880$ for Au[45] at $\lambda = 564$ nm.

27.4 MODELING AND INVERSION

The following simplifying assumptions are usually made or implied in conventional ellipsometry: (1) the incident beam is approximated by a monochromatic plane wave; (2) the ambient or incidence medium is transparent and optically isotropic; (3) the sample surface is a plane boundary; (4) the sample (and ambient) optical properties are uniform laterally but may change in the direction of the normal to the ambient-sample interface; (5) the coherence length of the incident light is much greater than its penetration depth into the sample; and (6) the light-sample interaction is linear (elastic), hence frequency-conserving.

Determination of the ratio of complex reflection coefficients is rarely an end in itself. Usually, one is interested in more fundamental information about the sample than is conveyed by ρ. In particular, ellipsometry is used to characterize the optical and structural properties of the interfacial region. This requires that a stratified-medium model (SMM) for the sample under measurement be postulated that contains the sample physical parameters of interest. For example, for visible light, a polished Si surface in air may be modeled as an optically opaque (semi-infinite) Si substrate which is covered with a SiO_2 film, with the Si and SiO_2 phases assumed uniform, and the air/SiO_2 and SiO_2/Si interfaces considered as parallel planes. This is often referred to as the three-phase model. More complexity (and more layers) can be built into this basic SMM to represent such finer details as the interfacial roughness and phase mixing, a damage surface layer on Si caused by polishing, or the possible presence of an outermost contamination film. Effective medium theories[46–54] (EMTs) are used to calculate the dielectric functions of mixed phases based on their microstructure and component volume fractions; and the established theory of light reflection by startified structures[55–60] is employed to calculate the ellipsometric function for an assumed set of model parameters. Finally, values of the model parameters are sought that best match the measured and computed values of ρ. Extensive data (obtained, e.g., using VASE) is required to determine the parameters of more complicated samples. The latter task, called the inverse problem, usually employs linear regression analysis,[61–63] which yields information on parameter correlations and confidence limits. Therefore, the full practice of ellipsometry involves, in general, the execution and integration of three tasks: (1) polarization measurements that yield ratios of complex reflection coefficients, (2) sample modeling and the application of electromagnetic theory to calculate the ellipsometric function, and (3) solving the inverse problem to determine model parameters that best match the experimental and theoretically calculated values of the ellipsometric function.

Confidence in the model is established by showing that complete spectra can be described in terms of a few wavelength-independent parameters, or by checking the predictive power of the model in determining the optical properties of the sample under new experimental conditions.[27]

The Two-phase Model

For a single interface between two homogeneous and isotropic media, 0 and 1, the reflection coefficients are given by the Fresnel formulas[1]

$$r_{01p} = (\epsilon_1 S_0 - \epsilon_0 S_1)/(\epsilon_1 S_0 + \epsilon_0 S_1) \tag{9}$$

$$r_{01s} = (S_0 - S_1)/(S_0 + S_1) \tag{10}$$

in which

$$\epsilon_i = N_i^2, \qquad i = 0, 1 \tag{11}$$

is the dielectric function (or dielectric constant at a given wavelength) of the ith medium,

$$S_i = (\epsilon_i - \epsilon_0 \sin^2 \phi)^{1/2} \tag{12}$$

and ϕ is the angle of incidence in medium 0 (measured from the interface normal). The ratio of complex reflection coefficients which is measured by ellipsometry is

$$\rho = [\sin \phi \tan \phi - (\epsilon - \sin^2 \phi)^{1/2}]/[\sin \phi \tan \phi + (\epsilon - \sin^2 \phi)^{1/2}] \tag{13}$$

where $\epsilon = \epsilon_1/\epsilon_0$. Solving Eq. (13) for ϵ gives

$$\epsilon_1 = \epsilon_0\{\sin^2 \phi + \sin^2 \phi \tan^2 \phi[(1 - \rho)/(1 + \rho)]^2\} \tag{14}$$

For light incident from a medium (e.g., vacuum, air, or an inert ambient) of known ϵ_0, Eq. (14) determines, concisely and directly, the complex dielectric function ϵ_1 of the reflecting second medium in terms of the measured ρ and the angle of incidence ϕ. This accounts for an important application of ellipsometry as a means of determining the

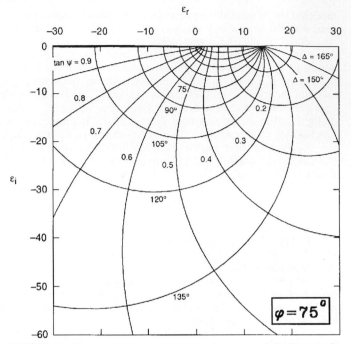

FIGURE 3 Contours of constant tan ψ and constant Δ in the complex plane of the relative dielectric function ϵ of a transparent medium/absorbing medium interface.

optical properties (or *optical constants*) of bulk absorbing materials and opaque films. This approach assumes the absence of a transition layer or a surface film at the two-media interface. If such a film exists, ultrathin as it may be, ϵ_1 as determined by Eq. (14) is called the pseudo dielectric function and is usually written as $\langle \epsilon_1 \rangle$. Figure 3 shows lines of constant ψ and lines of constant Δ in the complex ϵ plane at $\phi = 75°$.

The Three-phase Model

This often-used model, Fig. 4, consists of a single layer, medium 1, of parallel-plane boundaries which is surrounded by two similar or dissimilar semi-infinite media 0

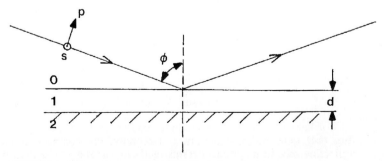

FIGURE 4 Three-phase, ambient-film-substrate system.

and 2. The complex amplitude reflection coefficients are given by the Airy-Drude formula[64,65]

$$R = (r_{01v} + r_{12v}X)/(1 + r_{01v}r_{12v}X), \qquad v = p, s \tag{15}$$

$$X = \exp\left[-j2\pi(d/D_\phi)\right] \tag{16}$$

r_{ijv} is the Fresnel reflection coefficient of the ij interface ($ij = 01$ and 12) for the v polarization, d is the layer thickness, and

$$D_\phi = (\lambda/2)(1/S_1) \tag{17}$$

where λ is the vacuum wavelength of light and S_1 is given by Eq. (12). The ellipsometric function of this system is

$$\rho = (A + BX + CX^2)/(D + EX + FX^2) \tag{18}$$

$$
\begin{aligned}
A &= r_{01p} & B &= r_{12p} + r_{01p}r_{01s}r_{12s} & C &= r_{12p}r_{01s}r_{12s} \\
D &= r_{01s} & E &= r_{12s} + r_{01p}r_{01s}r_{12p} & F &= r_{12s}r_{01p}r_{12p}
\end{aligned} \tag{19}
$$

For a transparent film, and with light incident at an angle ϕ such that $\epsilon_1 > \epsilon_0 \sin^2 \phi$ so that total reflection does not occur at the 01 interface, D_ϕ is real, and X, R_p, R_s, and ρ become periodic functions of the film thickness d with period D_ϕ. The locus of X is the unit circle in the complex plane and its multiple images through the conformal mapping of Eq. (18) at different values of ϕ give the constant-angle-of-incidence contours of ρ. Figure 5 shows a family of such contours[66] for light reflection in air by the SiO_2–Si system at 633-nm wavelength at angles from 30 to 85° in steps of 5°. Each and every value of ρ, corresponding to all points in the complex plane, can be realized by selecting the appropriate angle of incidence and the SiO_2 film thickness (within a period).

If the dielectric functions of the surrounding media are known, the dielectric function ϵ_1 and thickness d of the film are obtained readily by solving Eq. (18) for X,

$$X = \{-(B - \rho E) \pm [(B - \rho E)^2 - 4(C - \rho F)(A - \rho D)]^{1/2}\}/2(C - \rho F) \tag{20}$$

and requiring that[66,67]

$$|X| = 1 \tag{21}$$

Equation (21) is solved for ϵ_1 as its only unknown by numerical iteration. Subsequently, d is given by

$$d = [-\arg(X)/2\pi]D_\phi + mD_\phi \tag{22}$$

where m is an integer. The uncertainty of an integral multiple of the film thickness period is often resolved by performing measurements at more than one wavelength or angle of incidence and requiring that d be independent of λ or ϕ.

When the film is absorbing (semitransparent), or the optical properties of one of the surrounding media are unknown, more general inversion methods[68–72] are required which are directed toward minimizing an error function of the form

$$f = \sum_{i=1}^{N} [(\psi_{im} - \psi_{ic})^2 + (\Delta_{im} - \Delta_{ic})^2] \tag{23}$$

where ψ_{im}, ψ_{ic} and Δ_{im}, Δ_{ic} denote the ith measured and calculated values of the ellipsometric angles, and N is the total number of independent measurements.

Multilayer and Graded-index Films

For an outline of the matrix theory of multilayer systems, refer to Vol. I, Chap. 42. For our purposes, we consider a multilayer structure, Fig. 6, that consists of m plane-parallel

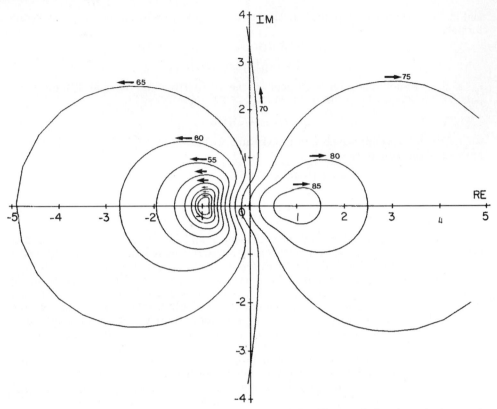

FIGURE 5   Family of constant-angle-of-incidence contours of the ellipsometric function ρ in the complex plane for light reflection in air by the SiO_2/Si film-substrate system at 633-nm wavelength. The contours are for angles of incidence from 30° to 85° in steps of 5°. The arrows indicate the direction of increasing film thickness.[66]

layers sandwiched between semi-infinite ambient and substrate media (0 and $m + 1$, respectively). The relationships between the field amplitudes of the incident (i), reflected (r), and transmitted (t) plane waves for the p or s polarizations are determined by the scattering matrix equation[73]

$$\begin{bmatrix} E_i \\ E_r \end{bmatrix} = \begin{bmatrix} S_{11} & S_{12} \\ S_{21} & S_{22} \end{bmatrix} \begin{bmatrix} E_t \\ 0 \end{bmatrix} \tag{24}$$

The complex amplitude reflection and transmission coefficients of the entire structure are given by

$$R = E_r/E_i = S_{21}/S_{11}$$
$$T = E_t/E_i = 1/S_{11} \tag{25}$$

The scattering matrix $\mathbf{S} = (S_{ij})$ is obtained as an ordered product of all the interface $\mathbf{I}$ and layer $\mathbf{L}$ matrices of the stratified structure,

$$\mathbf{S} = \mathbf{I}_{01}\mathbf{L}_1\mathbf{I}_{12}\mathbf{L}_2 \cdots \mathbf{I}_{(j-1)j}\mathbf{L}_j \cdots \mathbf{L}_m\mathbf{I}_{m(m+1)} \tag{26}$$

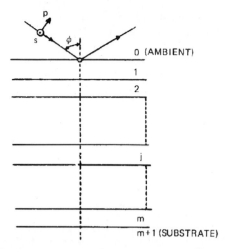

FIGURE 6 Light reflection by a multilayer structure.[1]

and the numbering starts from layer 1 (in contact with the ambient) to layer m (adjacent to the substrate) as shown in Fig. 6. The interface scattering matrix is of the form

$$\mathbf{I}_{ab} = (1/t_{ab})\begin{bmatrix} 1 & r_{ab} \\ r_{ab} & 1 \end{bmatrix} \tag{27}$$

where r_{ab} is the local Fresnel reflection coefficient of the $ab[j(j+1)]$ interface evaluated [using Eqs. (9) and (10) with the appropriate change of subscripts] at an incidence angle in medium a which is related to the external incidence angle in medium 0 by Snell's law. The associated interface transmission coefficients for the p and s polarizations are

$$t_{abp} = 2S_a/(S_a + S_b)$$
$$t_{abs} = 2(\epsilon_a \epsilon_b)^{1/2}S_a/(\epsilon_b S_a + \epsilon_a S_b) \tag{28}$$

where S_j is defined in Eq. (11). The scattering matrix of the jth layer is

$$\mathbf{L}_j = \begin{bmatrix} Y_j & 0 \\ 0 & 1/Y_j \end{bmatrix} \tag{29}$$

$$Y_j = X_j^{1/2} \tag{30}$$

and X_j is given by Eqs. (16) and (17) with the substitution $d = d_j$ for the thickness, and $\epsilon_1 = \epsilon_j$ for the dielectric function of the jth layer.

Except in Eqs. (28), a polarization subscript $v = p$ or s has been dropped for simplicity. In reflection and transmission ellipsometry, the ratios $\rho_r = R_p/R_s$ and $\rho_t = T_p/T_s$ are measured. Inversion for the dielectric functions and thicknesses of some or all of the layers requires extensive data, as may be obtained by VASE, and linear regression analysis to minimize the error function of Eq. (23).

Light reflection and transmission by a graded-index (GRIN) film is handled using the scattering matrix approach described here by dividing the inhomogeneous layer into an adequately large number of sublayers, each of which is approximately homogeneous. In fact, this is the most general approach for a problem of this kind because analytical closed-form solutions are only possible for a few simple refractive-index profiles.[74–76]

Dielectric Function of a Mixed Phase

For a microscopically inhomogeneous thin film that is a mixture of two materials, as may be produced by coevaporation or cosputtering, or a thin film of one material that may be porous with a significant void fraction (of air), the dielectric function is determined using EMTs.[46–54] When the scale of the inhomogeneity is small relative to the wavelength of light, and the domains (or grains) of different dielectric functions are of nearly spherical shape, the dielectric function of the mixed phase ϵ is given by

$$\frac{\epsilon - \epsilon_h}{\epsilon + 2\epsilon_h} = v_a \frac{\epsilon_a - \epsilon_h}{\epsilon_a + 2\epsilon_h} + v_b \frac{\epsilon_b - \epsilon_h}{\epsilon_b + 2\epsilon_h} \tag{31}$$

where ϵ_a and ϵ_b are the dielectric functions of the two component phases a and b with volume fractions v_a and v_b and ϵ_h is the host dielectric function. Different EMTs assign different values to ϵ_h. In the Maxwell Garnett EMT,[47,48] one of the phases, say b, is dominant $(v_b \gg v_a)$ and $\epsilon_h = \epsilon_b$. This reduces the second term on the right-hand side of Eq. (31) to zero. In the Bruggeman EMT,[49] v_a and v_b are comparable, and $\epsilon_h = \epsilon$, which reduces the left-hand side of Eq. (31) to zero.

27.5 TRANSMISSION ELLIPSOMETRY

Although ellipsometry is typically carried out on the reflected wave, it is possible to also monitor the state of polarization of the transmitted wave, when such a wave is available for measurement.[77–81] For example, by combining reflection and transmission ellipsometry, the thickness and complex dielectric function of an absorbing film between transparent media of the same refractive index (e.g., a solid substrate on one side and an index-matching liquid on the other) can be obtained analytically.[79,80] Polarized light transmission by a multilayer was discussed previously under "Multilayer and Graded-index Films." Transmission ellipsometry can be carried out at normal incidence on optically anisotropic samples to determine such properties as the natural or induced linear, circular, or elliptical birefringence and dichroism. However, this falls outside the scope of this chapter.

27.6 INSTRUMENTATION

Figure 7 is a schematic diagram of a generic ellipsometer. It consists of a source of collimated and monochromatic light L, polarizing optics PO on one side of the sample S,

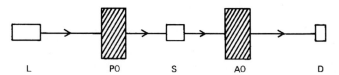

FIGURE 7 Generic ellipsometer with polarizing optics PO and analyzing optics AO. L and D are the light source and photodetector, respectively.

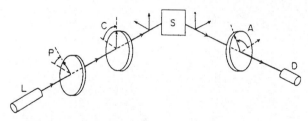

FIGURE 8 Polarizer-compensator-sample-analyzer (PCSA) ellipsometer. The azimuth angles P of the polarizer, C of the compensator (or quarter-wave retarder), and A of the analyzer are measured from the plane of incidence, positive in a counterclockwise sense when looking toward the source.[1]

and polarization analyzing optics AO and a (linear) photodetector D on the other side. An apt terminology[25] refers to the PO as a polarization state generator (PSG) and the AO plus D as a polarization state detector (PSD).

Figure 8 shows the commonly used polarizer-compensator-sample-analyzer (PCSA) ellipsometer arrangement. The PSG consists of a linear polarizer with transmission-axis azimuth P and a linear retarder, or compensator, with fast-axis azimuth C. The PSD consists of a single linear polarizer, that functions as an analyzer, with transmission-axis azimuth A followed by a photodetector D. All azimuths $P, C,$ and $A,$ are measured from the plane of incidence, positive in a counterclockwise sense when looking toward the source. The state of polarization of the light transmitted by the PSG and incident on S is given by

$$\chi_i = [\tan C + \rho_c \tan (P - C)]/[1 - \rho_c \tan C \tan (P - C)] \qquad (32)$$

where $\rho_c = T_{cs}/T_{cf}$ is the ratio of complex amplitude transmittances of the compensator for the incident linear polarizations along the slow s and fast f axes. Ideally, the compensator functions as a quarter-wave retarder (QWR) and $\rho_c = -j$. In this case, Eq. (32) describes an elliptical polarization state with major-axis azimuth C and ellipticity angle $-(P - C)$. (The tangent of the ellipticity angle equals the minor-axis-to-major-axis ratio and its sign gives the handedness of the polarization state, positive for right-handed states.) All possible states of total polarization χ_i can be generated by controlling P and C. Figure 9 shows a family of constant C, variable P contours (continuous lines) and constant $P - C$, variable C contours (dashed lines) as orthogonal families of circles in the complex plane of polarization. Figure 10 shows the corresponding contours of constant P and variable C. The points R and L on the imaginary axis at $(0, +1)$ and $(0, -1)$ represent the right- and left-handed circular polarization states, respectively.

Null Ellipsometry

The PCSA ellipsometer of Fig. 8 can be operated in two different modes. In the null mode, the output signal of the photodetector D is reduced to zero (a minimum) by adjusting the azimuth angles P of the polarizer and A of the analyzer with the compensator set at a fixed azimuth C. The choice $C = \pm 45°$ results in rapid convergence to the null. Two independent nulls are reached for each compensator setting. The two nulls obtained with $C = +45°$ are usually referred to as the nulls in zones 2 and 4; those for $C = -45°$ define zones 1 and 3. At null, the reflected polarization is linear and is crossed with the transmission axis of the analyzer; therefore, the reflected state of polarization is given by

$$\chi_r = -\cot A \qquad (33)$$

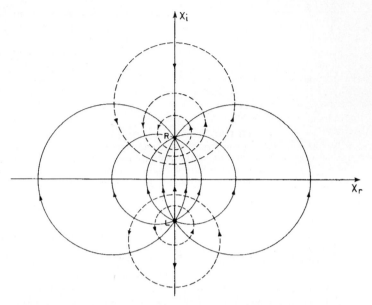

FIGURE 9 Constant C, variable P contours (continuous lines), and constant $P - C$, variable C contours (dashed lines) in the complex plane of polarization for light transmitted by a polarizer-compensator (PC) polarization state generator.[1]

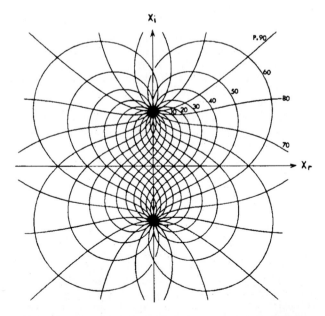

FIGURE 10 Constant P, variable C contours in the complex plane of polarization for light transmitted by a polarizer-compensator (PC) polarization state generator.[1]

where A is the analyzer azimuth at null. With the incident and reflected polarizations determined by Eqs. (32) and (33), the ratio of complex reflection coefficients of the sample for the p and s linear polarizations ρ is obtained by Eq. (2). Whereas a single null is sufficient to determine ρ in an ideal ellipsometer, results from multiple nulls (in two or four zones) are usually averaged to eliminate the effect of small component imperfections and azimuth-angle errors. Two-zone measurements are also used to determine ρ of the sample and ρ_c of the compensator simultaneously.[82–84] The effects of component imperfections have been considered extensively.[85]

The null ellipsometer can be automated by using stepping or servo motors[86–87] to rotate the polarizer and analyzer under closed-loop feedback control; the procedure is akin to that of nulling an ac bridge circuit. Alternatively, Faraday cells can be inserted after the polarizer and before the analyzer to produce magneto-optical rotations in lieu of the mechanical rotation of the elements.[88–90] This reduces the measurement time of a null ellipsometer from minutes to milliseconds. Large ($\pm 90°$) Faraday rotations would be required for limitless compensation. Small ac modulation is often added for the precise localization of the null.

Photometric Ellipsometry

The polarization state of the reflected light can also be detected photometrically by rotating the analyzer[91–95] of the PCSA ellipsometer and performing a Fourier analysis of the output signal I of the linear photodetector D. The detected signal waveform is simply given by

$$I = I_0(1 + \alpha \cos 2A + \beta \sin 2A) \tag{34}$$

and the reflected state of polarization is determined from the Fourier coefficients α and β by

$$\chi_r = [\beta \pm (1 - \alpha^2 - \beta^2)^{1/2}]/(1 + \alpha) \tag{35}$$

The sign ambiguity in Eq. (35) indicates that the rotating-analyzer ellipsometer (RAE) cannot determine the handedness of the reflected polarization state. In the RAE, the compensator is not essential and can be removed from the input PO (i.e., the PSA instead of the PCSA optical train is used). Without the compensator, the incident linear polarization is described by

$$\chi_i = \tan P \tag{36}$$

Again, the ratio of complex reflection coefficients of the sample ρ is determined by substituting Eqs. (35) and (36) in Eq. (2). The absence of the wavelength-dependent compensator makes the RAE particularly qualified for SE. The dual of the RAE is the rotating-polarizer ellipsometer which is suited for real-time SE using a spectrograph and a photodiode array that are placed after the fixed analyzer.[31]

A photometric ellipsometer with no moving parts, for fast measurements on the μs time scale, employs a photoelastic modulator[96–100] (PEM) in place of the compensator of Fig. 8. The PEM functions as an oscillating-phase linear retarder in which the relative phase retardation is modulated sinusoidally at a high frequency (typically 50–100 KHz) by establishing an elastic ultrasonic standing wave in a transparent solid. The output signal of the photodetector is represented by an infinite Fourier series with coefficients determined by Bessel functions of the first kind and argument equal to the retardation amplitude. However, only the dc, first, and second harmonics of the modulation frequency are usually detected (using lock-in amplifiers) and provide sufficient information to retrieve the ellipsometric parameters of the sample.

Numerous other ellipsometers have been introduced[25] that employ more elaborate

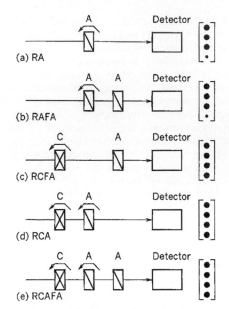

FIGURE 11 Family of rotating-element photo-polarimeters (REP) and the Stokes parameters that they can determine.[25]

PSDs. For example, Fig. 11 shows a family of rotating-element photopolarimeters[25] (REP) that includes the RAE. The column on the right represents the Stokes vector and the fat dots identify the Stokes parameters that are measured. (For a discussion of the Stokes parameters, see Vol. I, Chap. 5 of this Handbook.) The simplest complete REP, that can determine all four Stokes parameters of light, is the rotating-compensator fixed-analyzer (RCFA) photopolarimeter originally invented to measure skylight polarization.[101] The simplest handedness-blind REP for totally polarized light is not the RAE but the rotating-detector ellipsometer[102–103] (RODE), Fig. 12, in which the tilted and partially reflective front surface of a solid-state (e.g., Si) detector performs as a polarization analyzer.

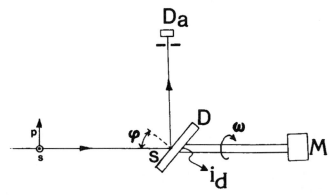

FIGURE 12 Rotating-detector ellipsometer (RODE).[102]

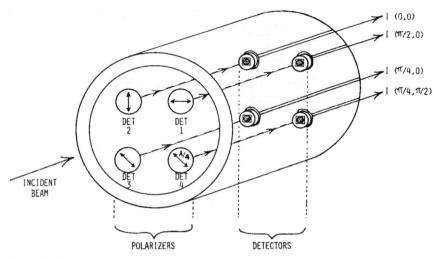

FIGURE 13   Division-of-wavefront photopolarimeter for the simultaneous measurement of all four Stokes parameters of light.[104]

Ellipsometry Using Four-detector Photopolarimeters

A new class of fast PSDs that measure the general state of partial or total polarization of a quasi-monochromatic light beam is based on the use of four photodetectors. Such PSDs employ the division of wavefront, the division of amplitude, or a hybrid of the two, and do not require any moving parts or modulators. Figure 13 shows a division-of-wavefront photopolarimeter (DOWP)[104] for performing ellipsometry with nanosecond laser pulses. The DOWP has been adopted recently in commercial automatic polarimeters for the fiber-optics market.[105,106]

Figure 14 shows a division-of-amplitude photopolarimeter[107,108] (DOAP) with a coated

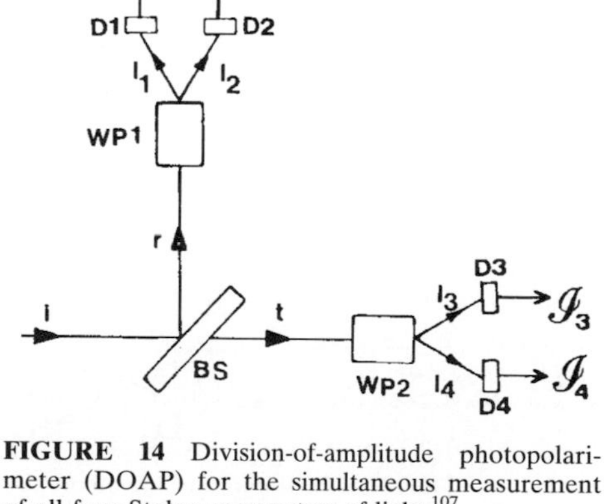

FIGURE 14   Division-of-amplitude photopolarimeter (DOAP) for the simultaneous measurement of all four Stokes parameters of light.[107]

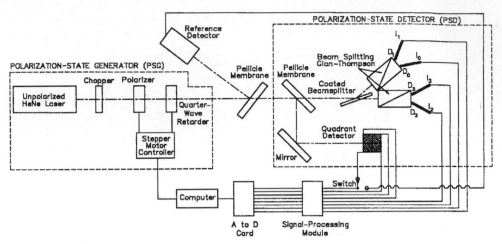

FIGURE 15 Recent implementation of DOAP.[109]

beam splitter BS and two Wollaston prisms WP1 and WP2, and Fig. 15 represents a recent implementation[109] of that technique. The multiple-beam-splitting and polarization-altering properties of grating diffraction are also well-suited for the DOAP.[110-111]

The simplest DOAP consists of a spatial arrangement of four solid-state photodetectors, Fig. 16, and no other optical elements. The first three detectors (D_0, D_1, and D_2) are partially specularly reflecting and the fourth (D_3) is antireflection-coated. The incident light beam is steered in such a way that the plane of incidence is rotated between successive oblique-incidence reflections, hence the light path is nonplanar. In this four-detector photopolarimeter[112-117] (FDP), and in other DOAPs, the four output signals of the four linear photodetectors define a current vector $\mathbf{I} = [I_0 \ I_1 \ I_2 \ I_3]^t$ which is linearly related,

$$\mathbf{I} = \mathbf{AS} \tag{37}$$

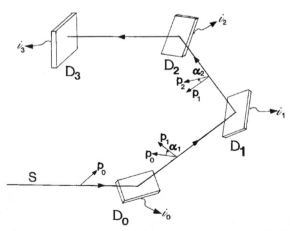

FIGURE 16 Four-detector photopolarimeter for the simultaneous measurement of all four Stokes parameters of light.[112]

to the Stokes vector $\mathbf{S} = [S_0 \ S_1 \ S_2 \ S_3]^t$ of the incident light, where t indicates the matrix transpose. The 4×4 instrument matrix $\mathbf{A}$ is determined by calibration[115] (using a PSG that consists of a linear polarizer and a quarter-wave retarder). Once $\mathbf{A}$ is determined, $\mathbf{S}$ is obtained from the output signal vector by

$$\mathbf{S} = \mathbf{A}^{-1}\mathbf{I} \tag{38}$$

where $\mathbf{A}^{-1}$ is the inverse of $\mathbf{A}$. When the light under measurement is totally polarized (i.e., $S_0^2 = S_1^2 + S_2^2 + S_3^2$), the associated complex polarization number is determined in terms of the Stokes parameters as[118]

$$\chi = (S_2 + jS_3)/(S_0 + S_1) = (S_0 - S_1)/(S_2 - jS_3) \tag{39}$$

Ellipsometry Based on Azimuth Measurements Alone

Measurements of the azimuths of the elliptic vibrations of the light reflected from an optically isotropic surface, for two known vibration directions of incident linearly polarized light, enable the ellipsometric parameters of the surface to be determined at any angle of incidence. If θ_i and θ_r represent the azimuths of the incident linear and reflected elliptical polarizations, respectively, then[119–121]

$$\tan 2\theta_r = (2 \tan \theta_i \tan \psi \cos \Delta)/(\tan^2 \psi - \tan^2 \theta_i) \tag{40}$$

A pair of measurements $(\theta_{i1}, \theta_{r1})$ and $(\theta_{i2}, \theta_{r2})$ determines ψ and Δ via Eq. (40). The azimuth of the reflected polarization is measured precisely by an ac-null method using an ac-excited Faraday cell followed by a linear analyzer.[119] The analyzer is rotationally adjusted to zero the fundamental-frequency component of the detected signal; this aligns the analyzer transmission axis with the minor or major axis of the reflected polarization ellipse.

Return-path Ellipsometry

In a return-path ellipsometer (RPE), Fig. 17, an optically isotropic mirror M is placed in, and perpendicular to, the reflected beam. This reverses the direction of the beam, so that it retraces its path toward the source with a second reflection at the test surface S and second passage through the polarizing/analyzing optics P/A. A beam splitter BS sends a sample of the returned beam to the photodetector D. The RPE can be operated in the null or photometric mode.

In the simplest RPE,[122,123] the P/A optics consists of a single linear polarizer whose azimuth and the angle of incidence are adjusted for a zero detected signal. At null, the angle of incidence is the principal angle, hence $\Delta = \pm 90°$, and the polarizer azimuth equals the principal azimuth, so that the incident linearly polarized light is reflected circularly polarized. Null can also be obtained at a general and fixed angle of incidence by adding a compensator to the P/A optics. Adjustment of the polarizer azimuth and the compensator azimuth or retardance produces the null.[124–125] In the photometric mode,[126] an element of the P/A is modulated periodically and the detected signal is Fourier-analyzed to extract ψ and Δ.

The RPEs have the following advantages: (1) the same optical elements are used as polarizing and analyzing optics; (2) only one optical port or window is used for light entry

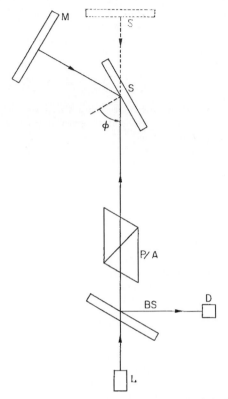

FIGURE 17 Return-path ellipsometer. The dashed lines indicate the configuration for perpendicular incidence ellipsometry on optically anisotropic samples.[126]

into and exit from the chamber in which the sample may be mounted; and (3) the sensitivity to surface changes is increased because of the double reflection at the sample surface.

Perpendicular-incidence Ellipsometry (PIE)

Normal-incidence reflection from an optically isotropic surface is accompanied by a trivial change of polarization due to the reversal of the direction of propagation of the beam (e.g., right-handed circularly polarized light is reflected as left-handed circularly polarized). Because this change of polarization is not specific to the surface, it cannot be used to determine the properties of the reflecting structure. This is why ellipsometry of isotropic surfaces has to be performed at oblique incidence. However, if the surface is optically anisotropic, PIE is possible and offers two significant advantages: (1) simpler single-axis instrumentation of the return-path type with common polarizing/analyzing optics, and (2) simpler inversion for the sample optical properties, because the equations that govern the reflection of light at normal incidence are much simpler than those at oblique incidence.[127–128]

Like RPE, PIE can be performed using null or photometric techniques.[126–132] For

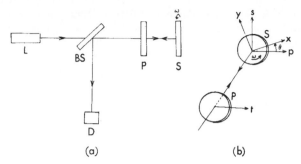

FIGURE 18 Normal-incidence rotating-sample ellipsometer (NIRSE).[128]

example, Fig. 18 shows a simple normal-incidence rotating-sample ellipsometer[128] (NIRSE) that is used to measure the ratio of the complex principal reflection coefficients of an optically anisotropic surface S with principal axes x and y. (The incident linear polarizations along these axes are the eigenpolarizations of reflection.) If we define

$$\eta = (R_{xx} - R_{yy})/(R_{xx} + R_{yy}) \tag{41}$$

then

$$\eta = \{a_2 \pm j[8a_4(1 - a_4) - a_2^2]^{1/2}\}/2(1 - a_4) \tag{42}$$

R_{xx} and R_{yy} are the complex-amplitude principal reflection coefficients of the surface, and a_2 and a_4 are the amplitudes of the second and fourth harmonic components of the detected signal normalized with respect to the dc component. From Eq. (41), we obtain

$$\rho = R_{yy}/R_{xx} = (1 - \eta)/(1 + \eta) \tag{43}$$

PIE can be used to determine the optical properties of bare and coated uniaxial and biaxial crystal surfaces.[127–130,133]

Interferometric Ellipsometry

Ellipsometry using interferometer arrangements with polarizing optical elements has been suggested and demonstrated.[134 136] Compensators are not required because the relative phase shift is obtained by the unbalance between the two interferometer arms; this offers a distinct advantage for SE. Direct display of the polarization ellipse is possible.[134–136]

27.7 JONES-MATRIX GENERALIZED ELLIPSOMETRY

For light reflection at an anisotropic surface, the p and s linear polarizations are not, in general, the eigenpolarizations of reflection. Consequently, the reflection of light is no longer described by Eqs. (1). Instead, the Jones (electric) vectors of the reflected and incident waves are related by

$$\begin{bmatrix} E_{rp} \\ E_{rs} \end{bmatrix} = \begin{bmatrix} R_{pp} & R_{ps} \\ R_{sp} & R_{ss} \end{bmatrix} \begin{bmatrix} E_{ip} \\ E_{is} \end{bmatrix} \tag{44}$$

or, more compactly,

$$\mathbf{E}_r = \mathbf{R}\mathbf{E}_i \tag{45}$$

where $\mathbf{R}$ is the nondiagonal reflection Jones matrix. The states of polarization of the incident and reflected waves, described by the complex variables χ_i and χ_r of Eqs. (4), are interrelated by the bilinear transformation[85,137]

$$\chi_r = (R_{ss}\chi_i + R_{sp})/(R_{ps}\chi_i + R_{pp}) \qquad (46)$$

In generalized ellipsometry (GE), the incident wave is polarized in at least three different states $(\chi_{i1}, \chi_{i2}, \chi_{i3})$ and the corresponding states of polarization of the reflected light $(\chi_{r1}, \chi_{r2}, \chi_{r3})$ are measured. Equation (46) then yields three equations that are solved for the normalized Jones matrix elements, or reflection coefficients ratios,[138]

$$R_{pp}/R_{ss} = (\chi_{i2} - \chi_{i1}H)/(-\chi_{r1} + \chi_{r2}H)$$

$$R_{ps}/R_{ss} = (H - 1)/(-\chi_{r1} + \chi_{r2}H)$$

$$R_{sp}/R_{ss} = (\chi_{i2}\chi_{r1} - \chi_{i1}\chi_{r2}H)/(-\chi_{r1} + \chi_{r2}H) \qquad (47)$$

$$H = (\chi_{r3} - \chi_{r1})(\chi_{i3} - \chi_{i2})/(\chi_{i3} - \chi_{i1})(\chi_{r3} - \chi_{r2})$$

Therefore, the nondiagonal Jones matrix of any optically anisotropic surface is determined, up to a complex constant multiplier, from the mapping of three incident polarizations into the corresponding three reflected polarizations. A PCSA null ellipsometer can be used. The incident polarization χ_i is given by Eq. (32) and the reflected polarization χ_r is given by Eq. (33). Alternatively, the Stokes parameters of the reflected light can be measured using the RCFA photopolarimeter, the DOAP, or the FDP, and χ_r is obtained from Eq. (39). More than three measurements can be taken to overdetermine the normalized Jones matrix elements and reduce the effect of component imperfections and measurement errors. GE can be performed based on azimuth measurements alone.[139] The main application of GE has been the determination of the optical properties of crystalline materials.[138–143]

27.8 MUELLER-MATRIX GENERALIZED ELLIPSOMETRY

The most general representation of the transformation of the state of polarization of light upon reflection or scattering by an object or sample is described by[1]

$$\mathbf{S}' = \mathbf{MS} \qquad (48)$$

where $\mathbf{S}$ and $\mathbf{S}'$ are the Stokes vectors of the incident and scattered radiation, respectively, and $\mathbf{M}$ is the real 4×4 Mueller matrix that succinctly characterizes the linear (or elastic) light-sample interaction. For light reflection at an optically isotropic and specular (smooth) surface, the Mueller matrix assumes the simple form[144]

$$\mathbf{M} = r \begin{bmatrix} 1 & a & 0 & 0 \\ a & 1 & 0 & 0 \\ 0 & 0 & b & c \\ 0 & 0 & -c & b \end{bmatrix} \qquad (49)$$

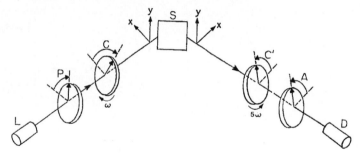

FIGURE 19 Dual rotating-retarder Mueller-matrix photopolarimeter.[145]

In Eq. (49), r is the surface power reflectance for incident unpolarized or circularly polarized light, and a, b, c are determined by the ellipsometric parameters ψ and Δ as:

$$a = -\cos 2\psi, \qquad b = \sin 2\psi \cos \Delta, \qquad \text{and} \qquad c = \sin 2\psi \sin \Delta \qquad (50)$$

and satisfy the identity $a^2 + b^2 + c^2 = 1$.

In general (i.e., for an optically anisotropic and rough surface), all 16 elements of **M** are nonzero and independent.

Several methods for Mueller matrix measurements have been developed.[25,145–149] An efficient and now popular scheme[145–147] uses the PCSC′A ellipsometer with symmetrical polarizing (PC) and analyzing (C′A) optics, Fig. 19. All 16 elements of the Mueller matrix are encoded onto a single periodic detected signal by rotating the quarter-wave retarders (or compensators) C and C' at angular speeds in the ratio 1:5. The output signal waveform is described by the Fourier series

$$I = a_0 + \sum_{n=1}^{12} (a_n \cos nC + b_n \sin nC) \qquad (51)$$

where C is the fast-axis azimuth of the slower of the two retarders, measured from the plane of incidence. Table 1 gives the relations between the signal Fourier amplitudes and

TABLE 1 Relations Between the Signal Fourier Amplitudes and the Elements of the Scaled Mueller Matrix M′

n	0	1	2	3	4	5	6
a_n	$m'_{11} + \frac{1}{2}m'_{12}$ $+ \frac{1}{2}m'_{21} + \frac{1}{4}m'_{22}$	0	$\frac{1}{2}m'_{12} + \frac{1}{4}m'_{22}$	$-\frac{1}{4}m'_{43}$	$-\frac{1}{2}m'_{44}$	0	$\frac{1}{2}m'_{44}$
b_n		$m'_{14} + \frac{1}{2}m'_{24}$	$\frac{1}{2}m'_{13} + \frac{1}{4}m'_{23}$	$-\frac{1}{4}m'_{42}$	0	$-m'_{41} - \frac{1}{2}m'_{42}$	0

n	7	8	9	10	11	12
a_n	$\frac{1}{4}m'_{43}$	$\frac{1}{8}m'_{22} + \frac{1}{8}m'_{33}$	$\frac{1}{4}m'_{34}$	$\frac{1}{2}m'_{21} + \frac{1}{4}m'_{22}$	$-\frac{1}{4}m'_{34}$	$\frac{1}{8}m'_{22} - \frac{1}{8}m'_{33}$
b_n	$-\frac{1}{4}m'_{42}$	$-\frac{1}{8}m'_{23} + \frac{1}{8}m'_{32}$	$-\frac{1}{4}m'_{24}$	$\frac{1}{2}m'_{31} + \frac{1}{4}m'_{32}$	$\frac{1}{4}m'_{24}$	$\frac{1}{8}m'_{23} + \frac{1}{8}m'_{32}$

The transmission axes of the polarizer and analyzer are assumed to be set at 0 azimuth, parallel to the scattering plane or the plane of incidence.

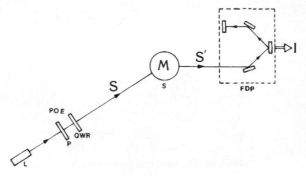

FIGURE 20 Scheme for Mueller-matrix measurement using the four-detector photopolarimeter.[152]

the elements of the Mueller matrix **M′** which differs from **M** only by a scale factor. Inasmuch as only the normalized Mueller matrix, with unity first element, is of interest, the unknown scale factor is immaterial. This dual-rotating-retarder Mueller-matrix photopolarimeter has been used to characterize rough surfaces[150] and the retinal nerve-fiber layer.[151]

Another attractive scheme for Mueller-matrix measurement is shown in Fig. 20. The FDP (or equivalently, any other DOAP) is used as the PSD. Fourier analysis of the output current vector of the FDP, $\mathbf{I}(C)$, as a function of the fast-axis azimuth C of the QWR of the input PO readily determines the Mueller matrix **M**, column by column.[152,153]

27.9 APPLICATIONS

The applications of ellipsometry are too numerous to try to cover in this chapter. The reader is referred to the books and review articles listed in the bibliography. Suffice it to mention the general areas of application. These include: (1) measurement of the optical properties of materials in the visible, IR, and near-UV spectral ranges. The materials may be in bulk or thin-film form and may be optically isotropic or anisotropic.[3,22,27–31] (2) Thin-film thickness measurements, especially in the semiconductor industry.[2,5,24] (3) Controlling the growth of optical multilayer coatings[154] and quantum wells.[155,156] (4) Characterization of physical and chemical adsorption processes at the vacuum/solid, gas/solid, gas/liquid, liquid/liquid, and liquid/solid interfaces.[26,157] (5) Study of the oxidation kinetics of semiconductor and metal surfaces in various gaseous or liquid ambients.[158] (6) Electrochemical investigations of the electrode/electrolyte interface.[18,19,32] (7) Diffusion and ion implantation in solids.[159–160] (8) Biological and biomedical applications.[16,20,151]

27.10 REFERENCES

1. R. M. A. Azzam and N. M. Bashara, *Ellipsometry and Polarized Light,* North-Holland, Amsterdam, 1987.
2. K. Riedling, *Ellipsometry for Industrial Applications,* Springer-Verlag, New York, 1988.
3. R. Röseler, *Infrared Spectroscopic Ellipsometry,* Akademie-Verlag, Berlin, 1990.
4. R. J. Archer, *Manual on Ellipsometry,* Gaertner, Chicago, 1968.

5. H. G. Tompkins, *A User's Guide to Ellipsometry,* Academic, Orlando, 1992.

6. R. M. A. Azzam (ed.), *Selected Papers on Ellipsometry,* vol. MS 27 of the Milestone Series, SPIE, Bellingham, Wash., 1991.

7. E. Passaglia, R. R. Stromberg, and J. Kruger (eds.), *Ellipsometry in the Measurement of Surfaces and Thin Films,* NBS Misc. Publ. 256, USGPO, Washington, D.C., 1964.

8. N. M. Bashara, A. B. Buckman, and A. C. Hall (eds.), *Recent Developments in Ellipsometry,* Surf. Sci. vol. 16, North-Holland, Amsterdam, 1969.

9. N. M. Bashara and R. M. A. Azzam (eds.), *Proceedings of the Third International Conference on Ellipsometry,* Surf. Sci. vol. 56, North-Holland, Amsterdam, 1976.

10. R. H. Muller, R. M. A. Azzam, and d. E. Aspnes (eds.), *Proceedings of the Fourth International Conference on Ellipsometry,* Surf. Sci. vol. 96, North-Holland, Amsterdam, 1980.

11. *Proceedings of the International Conference on Ellipsometry and Other Optical Methods for Surface and Thin Film Analysis,* J. de Physique, vol. 44, Colloq. C10, Les Editions de Physique, Paris, 1984.

12. A. C. Boccara, C. Pickering, and J. Rivory (eds.), *Proceedings of the First International Conference on Spectroscopic Ellipsometry,* Thin Solid Films, Vols. 233 and 234, Elsevier, Amsterdam, 1993.

13. A. B. Winterbottom, "Optical Studies of Metal Surfaces," in *The Royal Norwegian Sci. Soc. Rept.,* No. 1, F. Bruns, Trondheim, 1955, pp. 9–81.

14. F. L. McCrackin, E. Passaglia, R. Stromberg, and H. L. Steinberg, "Measurement of the Thickness and Refractive Index of Very Thin Films and the Optical Properties of Surfaces by Ellipsometry," *J. Res. Natl. Bur. Std.* **67A:**363–377 (1963).

15. K. H. Zaininger and A. G. Revesz, "Ellipsometry—A Valuable Tool in Surface Research," *RCA Review* **25:**85–115 (1964).

16. G. Poste and C. Moss, "The Study of Surface Reactions in Biological Systems by Ellipsometry," in S. G. Davison (ed.), *Progress in Surface Science,* vol. 2, pt. 3, Pergamon, New York, 1972, pp. 139–232.

17. R. H. Muller, "Principles of Ellipsometry," in R. H. Mueller (ed.), *Advances in Electrochemistry and Electrochemical Engineering,* vol. 9, Wiley, New York, 1973, pp. 167–226.

18. J. Kruger, "Application of Ellipsometry in Electrochemistry," in R. H. Muller (ed.), *Advances in Electrochemistry and Electrochemical Engineering,* vol. 9, Wiley, New York, 1973, pp. 227–280.

19. W.-K. Paik, "Ellipsometric Optics with Special Reference to Electrochemical Systems," in J. O'M. Bockris (ed.), *MTP International Review of Science, Physical Chemistry,* series 1, vol. 6, Butterworths, Univ. Park, Baltimore, 1973, pp. 239–285.

20. A. Rothen, "Ellipsometric Studies of Thin Films," in D. A. Cadenhead, J. F. Danielli, and M. D. Rosenberg (eds.), *Progress in Surface and Membrane Science,* vol. 8, Academic, New York, 1974, pp. 81–118.

21. R. H. Muller, "Present Status of Automatic Ellipsometers," *Surf. Sci.* **56:**19–36 (1976).

22. D. E. Aspnes, "Spectroscopic Ellipsometry of Solids," in B. O. Seraphin (ed.), *Optical Properties of Solids: New Developments,* North-Holland, Amsterdam, 1976, pp. 799–846.

23. W. E. J. Neal, "Application of Ellipsometry to Surface Films and Film Growth," *Surf. Technol.* **6:**81–110 (1977).

24. A. V. Rzhanov and K. K. Svitashev, "Ellipsometric Techniques to Study Surfaces and Thin Films," in L. Marton and C. Marton (eds.), *Advances in Electronics and Electron Physics,* vol. 49, Academic, New York, 1979, pp. 1–84.

25. P. S. Hauge, "Recent Developments in Instrumentation in Ellipsometry," *Surf. Sci.* **96:**108–140 (1980).

26. F. H. P. M. Habraken, O. L. J. Gijzeman, and G. A. Bootsma, "Ellipsometry of Clean Surfaces, Submonolayer and Monolayer Films," *Surf. Sci.* **96:**482–507 (1980).

27. D. E. Aspnes, "Characterization of Materials and Interfaces by Visible-Near UV Spectrophotometry and Ellipsometry," *J. Mat. Educ.* **7:**849–901 (1985).

28. P. J. McMarr, K. Vedam, and J. Narayan, "Spectroscopic Ellipsometry: A New Tool for

Nondestructive Depth Profiling and Characterization of Interfaces," *J. Appl. Phys.* **59:**694–701 (1986).

29. D. E. Aspnes, "Analysis of Semiconductor Materials and Structures by Spectroellipsometry," *SPIE Proc.* **946:**84–97 (1988).

30. R. Drevillon, "Spectroscopic Ellipsometry of Ultrathin Films: From UV to IR," *Thin Solid Films* **163:**157–166 (1988).

31. R. W. Collins and Y.-T. Kim, "Ellipsometry for Thin-Film and Surface Analysis," *Anal. Chem.* **62:**887A–900A (1990).

32. R. H. Muller, "Ellipsometry as an In Situ Probe for the Study of Electrode Processes," in R. Varma and J. R. Selman (eds.), *Techniques for Characterization of Electrode Processes,* Wiley, New York, 1991.

33. A. Rothen, *Rev. Sci. Instrum.* **16:**26–30 (1945).

34. A. Rothen, in Ref. 7, pp. 7–21.

35. A. C. Hall, *Surf. Sci.* **16:**1–13 (1969).

36. R. M. A. Azzam and N. M. Bashara, Ref. 1, sec. 2.6.1.

37. R. M. A. Azzam and N. M. Bashara, Ref. 1, sec. 1.7.

38. M. M. Ibrahim and N. M. Bashara, *J. Opt. Soc. Am.* **61:**1622–1629 (1971).

39. O. Hunderi, *Surface Sci.* **61:**515–520 (1976).

40. J. Humlíček, *J. Opt. Soc. Am.* A **2:**713–722 (1985).

41. Y. Gaillyová, E. Schmidt, and J. Humlíček, *J. Opt. Soc. Am.* A **2:**723–726 (1985).

42. W. H. Weedon, S. W. McKnight, and A. J. Devaney, *J. Opt. Soc. Am.* A **8:**1881–1891 (1991).

43. P. G. Snyder, M. C. Rost, G. H. Bu-Abbud, and J. A. Woollam, *J. Appl. Phys.* **60:**3293–3302 (1986).

44. R. H. Muller, *Surf. Sci.* **16:**14–33 (1969).

45. E. D. Palik, *Handbook of Optical Constants of Solids,* Academic, New York, 1985, p. 294.

46. L. Lorenz, *Ann. Phys. Chem.* (Leipzig) **11:**70–103 (1880).

47. J. C. Maxwell Garnett, *Philos. Trans. R. Soc. London* **203:**385–420 (1904).

48. J. C. Maxwell Garnett, *Philos. Trans. R. Soc. London* **205:**237–288 (1906).

49. D. A. G. Bruggeman, *Ann. Phys.* (Leipzig) **24:**636–679 (1935).

50. C. G. Granqvist and O. Hunderi, *Phys. Rev.* **B18:**1554–1561 (1978).

51. C. Grosse and J.-L. Greffe, *J. Chim. Phys.* **76:**305–327 (1979).

52. D. E. Aspnes, J. B. Theeten, and F. Hottier, *Phys. Rev.* **B20:**3292–3302 (1979).

53. D. E. Aspnes, *Am. J. Phys.* **50:**704–709 (1982).

54. D. E. Aspnes, *Physica* **117B/118B:**359–361 (1983).

55. F. Abelès, *Ann. de Physique* **5:**596–640 (1950).

56. P. Drude, *Theory of Optics,* Dover, New York, 1959.

57. J. R. Wait, *Electromagnetic Waves in Stratified Media,* Pergamon, New York, 1962.

58. O. S. Heavens, *Optical Properties of Thin Solid Films,* Dover, New York, 1965.

59. Z. Knittl, *Optics of Thin Films,* Wiley, New York, 1976.

60. J. Lekner, *Theory of Reflection,* Marinus Nijhoff, Dordrecht, 1987.

61. E. S. Keeping, *Introduction to Statistical Inference,* Van Nostrand, Princeton, 1962, chap. 12.

62. D. W. Marquardt, *SIAM J. Appl. Math.* **11:**431–441 (1963).

63. J. R. Rice, in *Numerical Solutions of Nonlinear Problems,* Computer Sci. Center, University of Maryland, College Park, 1970.

64. G. B. Airy, *Phil. Mag.* **2:**20 (1833).

65. P. Drude, *Annal. Phys. Chem.* **36:**865–897 (1889).

66. R. M. A. Azzam, A.-R. M. Zaghloul, and N. M. Bashara, *J. Opt. Soc. Am.* **65:**252–260 (1975).

67. A. R. Reinberg, *Appl. Opt.* **11:**1273–1274 (1972).

68. F. L. McCrackin and J. P. Colson, in Ref. 7, pp. 61–82.

69. M. Malin and K. Vedam, *Surf. Sci.* **56:**49–63 (1976).

70. D. I. Bilenko, B. A. Dvorkin, T. Y. Druzhinina, S. N. Krasnobaev, and V. P. Polyanskaya, *Opt. Spectrosc.* **55:**533–536 (1983).

71. G. H. Bu-Abbud, N. M. Bashara, and J. A. Woollam, *Thin Solid Films* **138:**27–41 (1986).

72. G. E. Jellison, Jr., *Appl. Opt.* **30:**3354–3360 (1991).

73. R. M. A. Azzam and N. M. Bashara, Ref. 1, sec. 4.6, and references cited therein.

74. F. Abelès, Ref. 7, pp. 41–58.

75. J. C. Charmet and P. G. de Gennes, *J. Opt. Soc. Am.* **73:**1777–1784 (1983).

76. J. Lekner, Ref. 60, chap. 2.

77. R. M. A. Azzam, M. Elshazly-Zaghloul, and N. M. Bashara, *Appl. Opt.* **14:**1652–1663 (1975).

78. I. Ohlídal and F. Lukeš, *Thin Solid Films* **85:**181–190 (1981).

79. R. M. A. Azzam, *J. Opt. Soc. Am.* **72:**1439–1440 (1982).

80. R. M. A. Azzam, Ref. 11, pp. 67–70.

81. I. Ohlídal and F. Lukeš, *Thin Solid Films* **115:**269–282 (1984).

82. F. L. McCrackin, *J. Opt. Soc. Am.* **60:**57–63 (1970).

83. J. A. Johnson and N. M. Bashara, *J. Opt. Soc. Am.* **60:**221–224 (1970).

84. R. M. A. Azzam and N. M. Bashara, *J. Opt. Soc. Am.* **62:**222–229 (1972).

85. R. M. A. Azzam and N. M. Bashara, Ref. 1, secs. 3.7 and 3.8 and references cited therein.

86. H. Takasaki, *Appl. Opt.* **5:**759–764 (1966).

87. J. L. Ord, *Appl. Opt.* **6:**1673–1679 (1967).

88. A. B. Winterbottom, Ref. 7, pp. 97–112.

89. H. J. Mathiu, D. E. McClure, and R. H. Muller, *Rev. Sci. Instrum.* **45:**798–802 (1974).

90. R. H. Muller and J. C. Farmer, *Rev. Sci. Instrum.* **55:**371–374 (1984).

91. W. Budde, *Appl. Opt.* **1:**201–205 (1962).

92. B. D. Cahan and R. F. Spanier, *Surf. Sci.* **16:**166–176 (1969).

93. R. Greef, *Rev. Sci. Instrum.* **41:**532–538 (1970).

94. P. S. Hauge and F. H. Dill, *IBM J. Res. Develop.* **17:**472–489 (1973).

95. D. E. Aspnes and A. A. Studna, *Appl. Opt.* **14:**220–228 (1975).

96. M. Billardon and J. Badoz, *C. R. Acad. Sci.* (Paris) **262:**1672 (1966).

97. J. C. Kemp, *J. Opt. Soc. Am.* **59:**950–954 (1969).

98. S. N. Jasperson and S. E. Schnatterly, *Rev. Sci. Instrum.* **40:**761–767 (1969).

99. J. Badoz, M. Billardon, J. C. Canit, and M. F. Russel, *J. Opt.* (Paris) **8:**373–384 (1977).

100. V. M. Bermudez and V. H. Ritz, *Appl. Opt.* **17:**542–552 (1978).

101. Z. Sekera, *J. Opt. Soc. Am.* **47:**484–490 (1957).

102. R. M. A. Azzam, *Opt. Lett.* **10:**427–429 (1985).

103. D. C. Nick and R. M. A. Azzam, *Rev. Sci. Instrum.* **60:**3625–3632 (1989).

104. E. Collett, *Surf. Sci.* **96:**156–167 (1980).

105. Lightwave Polarization Analyzer Systems, Hewlett Packard Co., Palo Alto, California 94303.

106. Polarscope, Electro Optic Developments Ltd, Basildon, Essex SS14 3BE, England.

107. R. M. A. Azzam, *Opt. Acta* **29:**685–689 (1982).

108. R. M. A. Azzam, *Opt. Acta* **32:**1407–1412 (1985).

109. S. Krishnan, *J. Opt. Soc. Am.* A **9:**1615–1622 (1992).

110. R. M. A. Azzam, *Appl. Opt.* **31:**3574–3576 (1992).

111. R. M. A. Azzam and K. A. Giardina, *J. Opt. Soc. Am.* A **10:**1190–1196 1993.

112. R. M. A. Azzam, *Opt. Lett.* **10:**309–311 (1985); U.S. Patent 4,681,450, July 21, 1987.

113. R. M. A. Azzam, E. Masetti, I. M. Elminyawi, and F. G. Grosz, *Rev. Sci. Instrum.* **59:**84–88 (1988).

114. R. M. A. Azzam, I. M. Elminyawi, and A. M. El-Saba, *J. Opt. Soc. Am.* A **5:**681–689 (1988).

115. R. M. A. Azzam and A. G. Lopez, *J. Opt. Soc. Am.* A **6:**1513–1521 (1989).

116. R. M. A. Azzam, *J. Opt. Soc. Am.* A **7:**87–91 (1990).

117. The Stokesmeter, Gaertner Scientific Co., Chicago, Illinois 60614.

118. P. S. Hauge, R. H. Muller, and C. G. Smith, *Surf. Sci.* **96:**81–107 (1980).

119. J. Monin and G.-A. Boutry, *Nouv. Rev. Opt.* **4:**159–169 (1973).

120. C. Som and C. Chowdhury, *J. Opt. Soc. Am.* **62:**10–15 (1972).

121. S. I. Idnurm, *Opt. Spectrosc.* **42:**210–212 (1977).

122. H. M. O'Bryan, *J. Opt. Soc. Am.* **26:**122–127 (1936).

123. M. Yamamoto, *Opt. Commun.* **10:**200–202 (1974).

124. T. Yamaguchi and H. Takahashi, *Appl. Opt.* **15:**677–680 (1976).

125. R. M. A. Azzam, *Opt. Acta* **24:**1039–1049 (1977).

126. R. M. A. Azzam, *J. Opt.* (Paris) **9:**131–134 (1978).

127. R. M. A. Azzam, *Opt. Commun.* **19:**122–124 (1976).

128. R. M. A. Azzam, *Opt. Commun.* **20:**405–408 (1977).

129. R. H. Young and E. I. P. Walker, *Phys. Rev.* B **15:**631–637 (1977).

130. D. W. Stevens, *Surf. Sci.* **96:**174–201 (1980).

131. R. M. A. Azzam, *J. Opt.* (Paris) **12:**317–321 (1981).

132. R. M. A. Azzam, *Opt. Eng.* **20:**58–61 (1981).

133. R. M. A. Azzam, *Appl. Opt.* **19:**3092–3095 (1980).

134. A. L. Dmitriev, *Opt. Spectrosc.* **32:**96–99 (1972).

135. H. F. Hazebroek and A. A. Holscher, *J. Phys. E: Sci. Instrum.* **6:**822–826 (1973).

136. R. Calvani, R. Caponi, and F. Cisternino, *J. Light. Technol.* **LT4:**877–883 (1986).

137. R. M. A. Azzam and N. M. Bashara, *J. Opt. Soc. Am.* **62:**336–340 (1972).

138. R. M. A. Azzam and N. M. Bashara, *J. Opt. Soc. Am.* **64:**128–133 (1974).

139. R. M. A. Azzam, *J. Opt. Soc. Am.* **68:**514–518 (1978).

140. D. J. De Smet, *J. Opt. Soc. Am.* **64:**631–638 (1974).

141. D. J. De Smet, *J. Opt. Soc. Am.* **65:**542–547 (1974).

142. P. S. Hauge, *Surf. Sci.* **56:**148–160 (1976).

143. M. Elshazly-Zaghloul, R. M. A. Azzam, and N. M. Bashara, *Surf. Sci.* **56:**281–292 (1976).

144. R. M. A. Azzam and N. M. Bashara, Ref. 1, p. 491.

145. R. M. A. Azzam, *Opt. Lett.* **2:**148–150 (1978).

146. P. S. Hauge, *J. Opt. Soc. Am.* **68:**1519–1528 (1978).

147. D. H. Goldstein, *Appl. Opt.* **31:**6676–6683 (1992).

148. A. M. Hunt and D. R. Huffman, *Appl. Opt.* **17:**2700–2710 (1978).

149. R. C. Thompson, J. R. Bottiger, and E. S. Fry, *Appl. Opt.* **19:**1323–1332 (1980).

150. D. A. Ramsey, *Thin Film Measurements on Rough Substrates using Mueller-Matrix Ellipsometry,* Ph.D. thesis, The University of Michigan, Ann Arbor, 1985.

151. A. W. Dreher, K. Reiter, and R. N. Weinreb, *Appl. Opt.* **31:**3730–3735 (1992).

152. R. M. A. Azzam, *Opt. Lett.* **11:**270–272 (1986).

153. R. M. A. Azzam, K. A. Giardina, and A. G. Lopez, *Opt. Eng.* **30:**1583–1589 (1991).

154. Ph. Houdy, *Rev. Phys. Appl.* **23:**1653–1659 (1988).

155. J. B. Theeten, F. Hottier, and J. Hallais, *J. Crystal Growth* **46:**245–252 (1979).

156. D. E. Aspnes, W. E. Quinn, M. C. Tamargo, M. A. A. Pudensi, S. A. Schwarz, M. J. S. Brasil, R. E. Nahory, and S. Gregory, *Appl. Phys. Lett.* **60:**1244–1247 (1992).

157. R. M. A. Azzam and N. M. Bashara, Ref. 1, sec. 6.3.

158. R. M. A. Azzam and N. M. Bashara, Ref. 1, sec. 6.4.

159. D. E. Aspnes and A. A. Studna, *Surf. Sci.* **96:**294–306 (1980).

160. M. Erman, P. Chambon, B. Prévot, and C Schwab, in Ref. 11, pp. 261–265.

CHAPTER 28
SPECTROSCOPIC MEASUREMENTS

Brian Henderson
*Department of Physics
and Applied Physics
University of Strathclyde
Glasgow, United Kingdom*

28.1 GLOSSARY

A_{ba} Einstein coefficient for spontaneous emission

a_o Bohr radius

B_{if} Einstein coefficient between initial state $|i\rangle$ and final state $|f\rangle$

e charge on the electron

ED   electric dipole term

E_{DC} Dirac Coulomb term

$\mathbf{E}_{hf}$ hyperfine energy

E_n eigenvalues of quantum state n

EQ   electric quadrupole term

$\mathbf{E}(t)$ electric field at time t

$\mathbf{E}(\omega)$ electric field at frequency ω

g_a degeneracy of ground level

g_b degeneracy of excited level

g_N gyromagnetic ratio of nucleus

h Planck's constant

H_{SO} spin-orbit interaction hamiltonian

I nuclear spin

$I(t)$ the emission intensity at time t

$\mathbf{j}$ total angular momentum vector given by $\mathbf{j} = 1 \pm \frac{1}{2}$

l_i orbital state

m mass of the electron

MD   magnetic dipole term

M_N mass of nucleus N

$n_\omega(T)$	equilibrium number of photons in a blackbody cavity radiator at angular frequency ω and temperature T
QED	quantum electrodynamics
$R_{nl}(r)$	radial wavefunction
R_∞	Rydberg constant for an infinitely heavy nucleus
s	spin quantum number with the value $\frac{1}{2}$
s_i	electronic spin
T	absolute temperature
W_{ab}	transition rate in absorption transition between states a and b
W_{ba}	transition rate in emission transition from state b to state a
Z	charge on the nucleus
$\alpha = e^2/4\pi\varepsilon_0\hbar c$	fine structure constant
$\Delta\omega$	natural linewidth of the transition
$\Delta\omega_D$	Doppler width of transition
ε_0	permittivity of free space
$\zeta(r)$	spin-orbit parameter
μ_B	Bohr magneton
$\rho(\omega)$	energy density at frequency ω
τ_R	radiative lifetime
ω	angular velocity
ω_k	mode k with angular frequency ω
$\langle f| \, v^1 \, |i\rangle$	matrix element of perturbation V

28.2 INTRODUCTORY COMMENTS

The conceptual basis of optical spectroscopy and its relationship to the electronic structure of matter as presented in the article entitled "Optical Spectroscopy and Spectroscopic Lineshapes" in Vol. I, Chap. 8 of this Handbook. The article entitled "Optical Spectrometers" in Vol. II, Chap. 20 of this Handbook discusses the operating principles of optical spectrometers. This article illustrates the underlying themes of the earlier ones using the optical spectra of atoms, molecules, and solids as examples.

28.3 OPTICAL ABSORPTION MEASUREMENTS OF ENERGY LEVELS

Atomic Energy Levels

The interest in spectroscopic measurements of the energy levels of atoms is associated with tests of quantum theory. Generally, the optical absorption and luminescence spectra of atoms reveal large numbers of sharp lines corresponding to transitions between the stationary states. The hydrogen atom has played a central role in atomic physics because of the accuracy with which relativistic and quantum electrodynamic shifts to the one-electron energies can be calculated and measured. Tests of quantum electrodynamics usually involve transitions between low-lying energy states (i.e., states with small principal quantum number). For the atomic states $|a\rangle$ and $|b\rangle$, the absorption and luminescence lines occur at exactly the same wavelength and both spectra have the same gaussian lineshape. The $1s \rightarrow 2p$ transitions on atomic hydrogen have played a particularly prominent role, especially since the development of sub-Doppler laser spectroscopy.[1] Such techniques resulted in values of $R_\infty = 10973731.43$ m^{-1}, 36.52 m^{-1} for the spin-orbit splitting in the $n = 2$ state and a Lamb shift of 3.53 m^{-1} in the $n = 1$ state. Accurate isotope shifts have been determined from hyperfine structure measurements on hydrogen, deuterium, and tritium.[2]

Helium is the simplest of the multielectron atoms, having the ground configuration ($1s^2$). The energy levels of helium are grouped into singlet and triplet systems. The observed spectra arise *within* these systems (i.e., singlet-to-singlet and triplet-to-triplet); normally transitions between singlet and triplet levels are not observed. The lowest-lying levels are 1^1S, 2^3S, 2^1S, 2^3P, and 2^1P in order of increasing energy. The $1^1S \rightarrow 2^1S$ splitting is of order 20.60 eV and transitions between these levels are not excited by optical photons. Transitions involving the 2^1S and 2^3S levels, respectively, and higher-lying spin singlet and spin triplet states occur at optical wavelengths. Experimental work on atomic helium has emphasized the lower-lying triplet levels, which have long excited-state lifetimes and large quantum electrodynamic (QED) shifts.

As with hydrogen, the spectra of He atoms are inhomogeneously broadened by the Doppler effect. Precision measurements have been made using two-photon laser spectroscopy (e.g., $2^3S \rightarrow n^3S$ ($n = 4$–6) and n^3D ($n = 3$–6), or laser saturation absorption spectroscopy ($2^3S \rightarrow 2^3P$ and $3^3P \rightarrow 3^3D$).[3-6] The $2^1S \rightarrow 3^1P$ and two photon $2^1S \rightarrow n^1D$ ($n = 3$–7) spectra have been measured using dye lasers.[7,8] The wide tune ranging of the Ti-sapphire laser and the capability for generating frequencies not easily accessible with dye lasers using frequency-generation techniques makes it an ideal laser to probe transitions starting on the 2S levels of He.[9] Two examples are the two-photon transition $2^3S \rightarrow 3^3S$ at 855 nm and the $2^3S \rightarrow 3^3P$ transition of 389 nm. The power of Doppler-free spectroscopy is shown to advantage in measurements of the $2^3S \rightarrow 2^3P$ transition.[10] Since both ^3S and ^3P are excited levels, the homogeneous width is determined by the sum of the reciprocal lifetimes of the two levels. Since both 2^3S and 2^3P levels are long-lived, the resulting homogeneous width is comparatively narrow. Figure 1a shows the Doppler-broadened profile of the $2^3S \rightarrow 2^3P$ transition of ^{4}He for which the FWHM is about 5.5 GHz. The inhomogeneously broadened line profile shown in Fig. 1a also shows three very weak "holes" corresponding to saturated absorption of the Ti-sapphire laser radiation used to carry out the experiment. These components correspond to the $2^3S_1 \rightarrow 3^3P_2$ and $2^3S_1 \rightarrow 3^3P_1$ transitions and their crossover resonance. The amplitude of the saturated signal is some 1–2 percent of the total absorption. The relativistic splittings including spin-orbit coupling of $^3P_0$–$^3P_1$ and $^3P_1$–$^3P_2$ are 8.1 GHz and 658.8 MHz, respectively. Frequency modulation of the laser (see Vol. 2, Chap. 20 of this Handbook) causes the "hole" to be detected as a first derivative of the absorption line, Fig. 1b. The observed FWHM of the Doppler-free signals was only 20 MHz. The uncertainty in the measured $2^3S \rightarrow 3^3P$ interval was two parts in 10^9 (i.e., 1.5 MHz), an improvement by a factor of sixty on earlier measurements. A comparison of the experimental results with recent calculations of the non-QED terms[11] gives a value for the one-electron Lamb shift of -346.5 (2.8) MHz, where the uncertainty in the quoted magnitude is in parentheses. The theoretical value is -346.3 (13.9) MHz. Finally, the frequencies of the $2^3S_1 \rightarrow 3^3P_1$, 3^3P_2 transitions were determined to be 25708.60959 (5) cm^{-1} and 25708.58763 (5) cm^{-1}, respectively.

The H$^-$ ion is another two-electron system, of some importance in astrophysics. It is the simplest quantum mechanical three-body species. Approximate quantum mechanical techniques give a wave function and energy eigenvalue which are exact, for all practical purposes. Experimentally, the optical absorption spectrum of H$^-$ is continuous, a property of importance in understanding the opacity of the sun. H$^-$ does not emit radiation in characteristic emission lines. Instead the system sheds its excess (absorbed) energy by

(a)

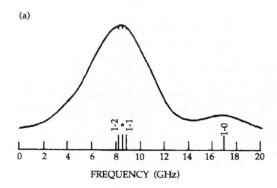

(b)

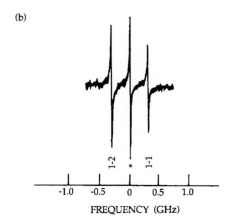

FIGURE 1 (*a*) Showing the inhomogeneously broadened line profile of the $2^3S \rightarrow 2^3P$ absorption in ^{4}He, including the weak "holes" due to saturated absorption and the position of the $2^3S \rightarrow 2^3P_2$, $^3P_1$ and $^3P_0$ components: (*b*) Doppler-free spectra showing the $2^3S \rightarrow 2^3P_2$, $^3P_1$ transitions and the associated crossover resonance. (*After Adams, Riis, and Ferguson*).[10]

ejecting one of the electrons. The radiant energy associated with the ejected electron consists of photons having a continuous energy distribution. Recent measurements with high-intensity pulsed lasers, counterpropagating through beams of 800-MeV H$^-$ ions, have produced spectacular "doubly excited states" of the He$^-$ ion. Such ions are traveling at 84 percent of the velocity of light and, in this situation, the visible laboratory photons are shifted into the vacuum ultraviolet region. At certain energies the H$^-$ ion is briefly excited into a system in which both electrons are excited prior to one of the electrons being ejected. Families of new resonances up to the energy level $N = 8$ have been observed.[12] These resonances are observed as windows in the continuous absorption spectrum, at energies given by a remarkably simple relation reminiscent of the Bohr equation for the Balmer series in hydrogen.[13] The resonance line with the lowest energy in each family corresponds to both electrons at comparable distances from the proton.

These experiments on H$^-$ are but one facet of the increasingly sophisticated measurements designed to probe the interaction between radiation and matter driven by experiments in laser technology. "Quantum jump" experiments involving single ions in an electromagnetic trap have become almost commonplace. Chaos has also become a rapidly growing subfield of atomic spectroscopy.[14] The particular conditions under which chaos may be observed in atomic physics include hydrogenic atoms in strong homogeneous magnetic fields such that the cyclotron radius of the electron approaches the dimensions of the atomic orbitals. A more easily realizable situation using magnetic field strengths of only a few tesla uses highly excited orbitals close to the ionization threshold. Iu et al.[15] have reported the absorption spectrum of transitions from the 3s state of Li to bound and continuum states near the ionization limit in a magnetic field of ca. six tesla. There is a remarkable coincidence between calculations involving thousands of energy levels and experiments involving high-resolution laser spectroscopy.

Atomic processes play an important role in the energy balance of plasmas, whether they be created in the laboratory or in star systems. The analysis of atomic emission lines gives much information on the physical conditions operating in a plasma. In laser-produced plasmas, the densities of charged ions may be in the range 10^{20}–10^{25} ions cm^{-3}, depending on the pulse duration of the laser. The spectra of many-electron ions are complex and may have the appearance of an unresolved transition array between states belonging to specific initial and final configurations. Theoretical techniques have been developed to determine the average ionization state of the plasma from the observed optical spectrum. In many cases, the spectra are derived from ionic charge states in the nickel-like configuration containing 28 bound electrons. In normal nickel, the outershell configuration is $(3d^8)(4s^2)$, the 4s levels having filled before 3d because the electron–electron potentials are stronger than electron-nuclear potentials. However, in highly ionized systems, the additional electron-nuclear potential is sufficient to shift the configuration from $(3d^8)(4s^2)$ to the closed shell configuration $(3d^{10})$. The resulting spectrum is then much simpler than for atomic nickel. The Ni-like configuration has particular relevance in experiments to make x-ray lasers. For example, an analog series of collisionally pumped lasers using Ni-like ions has been developed, including a Ta^{45+} laser operating at 4.48 nm and a W^{46+} laser operating at 4.32 nm.[16]

Molecular Spectroscopy

The basic principles of gas-phase molecular spectroscopy were also discussed in "Optical Spectroscopy and Spectroscopic Lineshapes", Vol. I, Chap. 8 of this Handbook. The spectra of even the simplest molecules are complicated by the effects of vibrations and of rotations about an axis. This complexity is illustrated elsewhere in this Handbook in Fig. 8 of Vol. I, Chap. 8, "Optical Spectroscopy and Spectroscopic Lineshapes," which depicts a photographically recorded spectrum of the $2\Pi \rightarrow 3\Sigma$ bands of the diatomic molecule NO, which was interpreted in terms of progressions and line sequences associated with the P-, Q-, and R-branches. The advent of Fourier transform spectroscopy has led to great improvements in resolution and greater efficiency in revealing all the fine details that characterize molecular spectra. Figure 2a is a Fourier-transform infrared spectrum of nitrous oxide, N_2O, which shows the band center at 2462 cm^{-1} flanked by the R-branch and a portion of the P-branch; the density of lines in the P- and R-branch is evident. On an expanded scale, in Fig. 2b, there is a considerable simplification of the rotational-vibrational structure at the high-energy portion of the P-branch. The weaker lines are the so-called "hot bands."

More precise determinations of the transition frequencies in molecular physics are measured using Lamb dip spectroscopy. The spectrum shown in Fig. 3a is a portion of the laser Stark spectrum of methyl fluoride measured using electric fields in the range 20–25 KV cm^{-1} with the 9-μm $P(18)$ line of the CO_2 laser, which is close to the ν_3 band origin of CH_3F.[17] The spectrum in Fig. 3a consists of a set of $\Delta M_J = \pm 1$ transitions, brought into resonance at different values of the static electric field. Results from the alternative

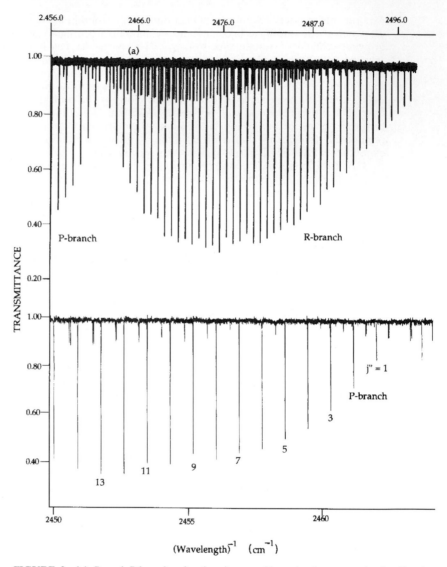

FIGURE 2 (*a*) *P*- and *R*-branches for the nitrous oxide molecule measured using Fourier-transform infrared spectroscopy. The gas pressure was 0.5 torr and system resolution 0.006 cm^{-1}, (*b*) on an expended scale, the high-energy portion of the *P*-branch up to $J'' = 15$.

high-resolution technique using a supersonic molecular beam and bolometric detector are shown in Fig. 3*b*: this spectrum was obtained using CH_3F in He mixture expanded through a 35-μm nozzle. The different $\mathbf{M_J}$ components of the $Q(1, 0)$, $Q(2, 1)$, and $Q(3, 3)$ components of the *Q*-branch are shown to have very different intensities relative to those in Fig. 3*a* on account of the lower measurement temperature.

There is considerable interest in the interaction of intense laser beams with molecules. For example, when a diatomic molecule such as N_2 or CO is excited by an intense (10^{15} W cm^{-2}), ultrashort (0.6 ps) laser pulse, it multiply ionizes and then fragments as a

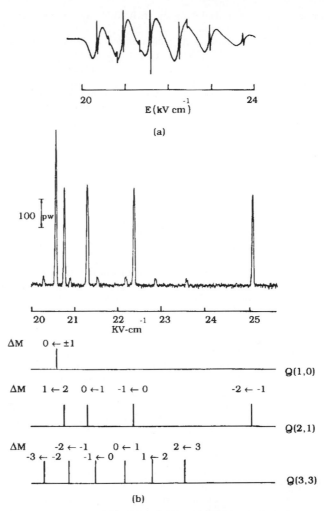

FIGURE 3 (*a*) Laser Stark absorption spectrum of methyl fluoride measured at 300 K using Lamb dip spectroscopy with a gas pressure of 5 m Torr; (*b*) the improved resolution obtained using molecular-beam techniques with low-temperature bolometric detection and CH_3F in He expanded through a 35-μm nozzle. (*After C. Douketic and T. E. Gough*).[17]

consequence of Coulomb repulsion. The charge and kinetic energy of the resultant ions can be determined by time-of-flight (TOF) mass spectrometry. In this technique, the daughter ions of the "Coulomb explosion" drift to a cathode tube at different times, depending on their weight. In traditional methods, the TOF spectrum is usually averaged over many laser pulses to improve the signal-to-noise ratio. Such simple averaging procedures remove the possibility of correlations between particular charged fragments. This problem has been overcome by the *covariance mapping* technique developed by Frasinski and Codling.[18] Experimentally, a linearly polarized laser pulse with **E**-vector pointing towards the detector is used to excite the molecules, which line up with their internuclear axis parallel to the **E**-field. Under Coulomb explosion, one fragment heads toward the detector and the other away from the detector. The application of a dc electric field directs the "backward" fragment ion to the detector, arriving at some short time after the forward fragment. This temporal separation of two fragments arriving at the detector permits the correlation between molecular fragments to be retained. In essence, the TOF spectrum, which plots molecular weight versus counts, is arranged both horizontally (forward ions) and vertically (backward ions) on a two-dimensional graph. A coordinate point on a preliminary graph consists of two ions along with their counts during a single pulse. Coordinates from 10^4 pulses or so are then assembled in a final map. Each feature on the final map relates to a specific fragmentation channel, i.e., the pair of fragments and their parent molecule. The strength of the method is that it gives the probability for the creation and fragmentation of the particular parent ion. Covariance mapping experiments on N_2 show that 610-nm and 305-nm pulses result in fragmentation processes that are predominantly charge-symmetric. In other words, the Coulomb explosion proceeds via the production of ions with the same charge.

Optical Spectroscopy of Solids

One of the more fascinating aspects of the spectroscopy of electronic centers in condensed matter is the variety of lineshapes displayed by the many different systems. Those already discussed in this Handbook include Nd^{3+} in YAG (Vol. I, Chap. 8, Fig. 6), O_2^- in KBr (Vol. I, Chap. 8, Fig. 11), Cr^{3+} in YAG (Vol. I, Chap. 8, Fig. 12), and F centers in KBr (Vol. I, Chap. 8, Fig. 13). The very sharp Nd^{3+} lines (in Vol. I, Chap. 8, Fig. 6) are zero-phonon lines, inhomogeneously broadened by strain. The abundance of sharp lines (in Vol. I, Chap. 8) is characteristic of the spectra of trivalent rare-earth ions in ionic crystals. Typical low-temperature linewidths for Nd^{3+} : YAG are $0.1–0.2 \ cm^{-1}$. There is particular interest in the spectroscopy of Nd^{3+} because of the efficient laser transitions from the $^4F_{3/2}$ level into the ^4I_J-manifold. The low-temperature luminescence transitions between $^4F_{3,2} \rightarrow {}^4I_{15/2}, {}^4I_{13/2}, {}^4I_{11/2}$ and $^4I_{9/2}$ levels are shown in Fig. 6 ("Optical Spectroscopy and Spectroscopic Lineshapes"): all are split by the effects of the crystalline electric field. Given the relative sharpness of these lines, it is evident that the Slater integrals $F^{(k)}$, spin-orbit coupling parameters ζ, and crystal field parameters, B_t^k, may be measured with considerable accuracy. The measured values of the $F^{(k)}$ and ζ vary little from one crystal to another.[19] However, the crystal field parameters, B_t^k, depend strongly on the rare-earth ion-ligand ion separation. Most of the $4f^n$ ions have transitions which are the basis of solid-state lasers. Others such as Eu^{3+} and Tb^{3+} are important red-emitting and green-emitting phosphor ions, respectively.[20]

Transition-metal ion spectra are quite different from those of the rare-earth ions. In both cases, the energy-level structure may be determined by solving the Hamiltonian

$$H = H_o + H' + H_{so} + H_c \tag{1}$$

in which H_o is a sum of one-electron Hamiltonians including the central field of each ion, H' is the interaction between electrons in the partially-filled $3d^n$ or $4f^n$ orbitals, H_{so} is the spin-orbit interaction, and H_c is the interaction of the outer shell electrons with the crystal field. For rare-earth ions H', $H_{so} \gg H_c$, and the observed spectra very much reflect the free-ion electronic structure with small crystal field perturbations. The spectroscopy of the transition-metal ions is determined by the relative magnitudes of $H' \simeq H_c \gg H_{so}$.[19,21] The simplest of the transition-metal ions is Ti^{3+}: in this $3d^1$ configuration a single 3d electron resides outside the closed shells. In this situation, $H' = 0$ and only the effect of H_c needs be considered (Fig. 4a). The Ti^{3+} ion tends to form octahedral complexes, where the 3d configuration is split into ^2E and $^2T_2$ states with energy separation $10Dq$. In cation sites with weak, trigonally symmetric distortions, as in Al_2O_3 and $Y_3Al_5O_{12}$, the lowest-lying

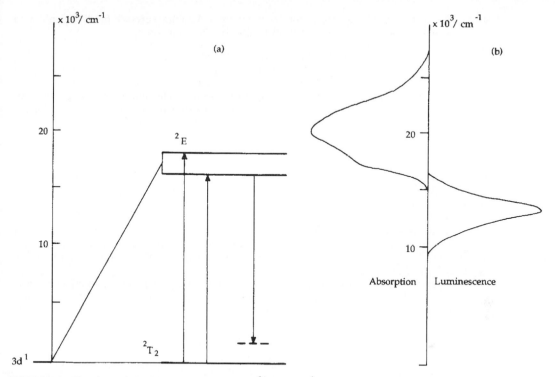

FIGURE 4   The absorption and emission spectra of Ti^{3+} ions (($3d^1$) configuration) in Al_2O_3 measured at 300 K.

state, $^2T_2$, splits into $^2A_1$ and ^2E states (using the C_{3v} group symmetry labels). In oxides, the octahedral splitting is of order $10Dq \approx 20,000\ cm^{-1}$ and the trigonal field splitting $v \simeq 700-1000\ cm^{-1}$. Further splittings of the levels occur because of spin-orbit coupling and the Jahn-Teller effect. The excited ^2E state splits into $2\bar{A}$ and $\bar{E}$ (from $^2A_1$), and $\bar{E}$ and $2\bar{A}$ (from ^2E). The excited state splitting by a *static* Jahn-Teller effect is large, $\sim 2000-2500\ cm^{-1}$, and may be measured from the optical absorption spectrum. In contrast, ground-state splittings are quite small: a *dynamic* Jahn-Teller effect has been shown to strongly quench the spin-orbit coupling ζ and trigonal field splitting v parameters.[22] In $Ti^{3+}:Al_2O_3$ the optical absorption transition, $^2T_2 \rightarrow {}^2E$, measured at 300 K, Fig. 4b, consists of two broad overlapping bands separated by the Jahn-Teller splitting, the composite band having a peak at ca. 20,000 cm^{-1}. Luminescence occurs only from the lower-lying excited state $2\bar{A}$, the emission band peak occurring at ca. 14,000 cm^{-1}. As Fig. 4 shows, both absorption and emission bands are broad because of strong electron-phonon coupling. At low temperatures the spectra are characterized by weak zero-phonon lines, one in absorption due to transitions from the $2\bar{A}$ ground state and three in emission corresponding to transitions in the $\bar{E}$, $\bar{E}$, and $2\bar{A}$ levels of the electronic ground state.[22,23] These transitions are strongly polarized.

For ions with $3d^n$ configuration it is usual to neglect H_c and H_{so} in Eq. (1), taking into account only the central ion terms and the Coulomb interaction between the 3d electrons. The resulting energies of the free-ion LS terms are expressed in terms of the Racah parameters A, B, and C. Because energy differences between states are measured in spectroscopy, only B and C are needed to categorize the free-ion levels. For pure d-functions, $C/B = 4.0$. The crystal field term H_c and H_{so} also are treated as perturbations. In many crystals, the transition-metal ions occupy octahedral or near-octahedral cation sites. The splittings of each free-ion level by an octahedral crystal field depend in a complex manner on B, C, and the crystal field strength Dq given by

$$Dq = \left(\frac{Ze^2}{24\pi\varepsilon_0} \right) \frac{\langle r^4 \rangle 3d}{a^5} \qquad (2)$$

The parameters D and q always occur as a product. The energy levels of the $3d^n$ transition-metal ions are usually represented on Tanabe-Sugano diagrams, which plot the energies $\mathbf{E}(\Gamma)$ of the electronic states as a function of the octahedral crystal field.[19,21] The crystal field levels are classified by irreducible representations Γ of the octahedral group, O_h. The Tanabe-Sugano diagram for the $3d^3$ configuration, shown in Fig. 5a, was constructed using a C/B ratio = 4.8: the vertical broken line drawn at $Dq/B = 2.8$ is appropriate for Cr^{3+} ions in ruby. If a particular value of C/B is assumed, only two variables, B and Dq, need to be considered: in the diagram $\mathbf{E}(\Gamma)/B$ is plotted as a function of Dq/B. The case of ruby, where the ^2E level is below $^4T_2$, is referred to as the *strong field* case. Other materials where this situation exists include $YAlO_3$, $Y_3Al_5O_{12}$ (YAG), and MgO. In many fluorides, Cr^{3+} ions occupy *weak field* sites, where $\mathbf{E}(^4T_2) < \mathbf{E}(^2E)$ and Dq/B is less than 2.2. When the value of Dq/B is close to 2.3, the *intermediate* crystal field, the $^4T_2$ and ^2E states almost degenerate. The value of Dq/B at the level crossing between $^4T_2$ and ^2E depends slightly on the value of C.

The Tanabe-Sugano diagram represents the static lattice. In practice, electron-phonon coupling must be taken into account: the relative strengths of coupling to the states involved in transitions and the consequences may be inferred from Fig. 5a. Essentially ionic vibrations modulate the crystal field experienced by the central ion at the vibrational frequency. Large differences in slope of the $\mathbf{E}$ versus Dq graphs indicate large differences

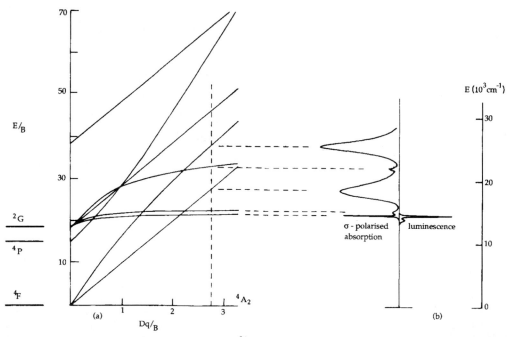

FIGURE 5 The Tanabe-Sugano diagram for Cr^{3+} ions with $C/B = 4.8$, appropriate for ruby for which $Dq/B = 2.8$. On the right of the figure are shown the optical absorption and photoluminescence spectrum of ruby measured at 300 K.

in coupling strengths and hence large homogeneous bandwidths. Hence, absorption and luminescence transitions from the $^4A_2$ ground state to the $^4T_2$ and $^4T_1$ states will be broadband due to the large differences in coupling of the electronic energy to the vibrational energy. For the $^4A_2 \rightarrow {}^2E, {}^2T_1$ transition, the homogeneous linewidth is hardly affected by lattice vibrations, and sharp line spectra are observed.

The Cr^{3+} ion occupies a central position in the folklore of transition-metal ion spectroscopy, having been studied by spectroscopists for over 150 years. An extensive survey of Cr^{3+} luminescence in many compounds was published as early as 1932.[24] The Cr^{3+} ions have the outer shell configuration, $3d^3$, and their absorption and luminescence spectra may be interpreted using Fig. 5. First, the effect of the octahedral crystal field is to remove the degeneracies of the free-ion states ^4F and ^2G. The ground term of the free ion, $^4F_{3/2}$, is split by the crystal field into a ground-state orbital singlet, $^4A_{2g}$, and two orbital triplets $^4T_{2g}$ and $^4T_{1g}$, in order of increasing energy. Using the energy of the $^4A_2$ ground state as the zero, for all values of Dq/B, the energies of the $^4T_{2g}$ and $^4T_{1g}$ states are seen to vary strongly as a function of the octahedral crystal field. In a similar vein, the ^2G free-ion state splits into $^2E, {}^2T_1, {}^2T_2$, and $^2A_1$ states, the two lowest of which, ^2E and $^2T_1$, vary very little with Dq. The energies $\mathbf{E}(^2T_2)$ and $\mathbf{E}(^2A_1)$ are also only weakly dependent on Dq/B. The free-ion term, ^4P, which transforms as the irreducible representation $^4T_1$ of the octahedral group is derived from (e^2t_2) configuration: this term is not split by the octahedral field although its energy is a rapidly increasing function of Dq/B. Low-symmetry distortions lead to strongly polarized absorption and emission spectra.[25]

The σ-polarized optical absorption and luminescence spectra of ruby are shown in Fig. 5b. The expected energy levels predicted from the Tanabe-Sugano diagram are seen to coincide with appropriate features in the absorption spectrum. The most intense features are the vibronically broadened $^4A_2 \rightarrow {}^4T_1, {}^4T_2$ transitions. These transitions are broad and characterized by large values of the Huang-Rhys factor ($S \approx 6$–7). These absorptions occur in the blue and yellow-green regions, thereby accounting for the deep red color of ruby. Many other Cr^{3+}-doped hosts have these bands in the blue and orange-red regions, by virtue of smaller values of Dq/B; the colors of such materials (e.g., MgO, $Gd_3Sc_2Ga_3O_{12}$, $LiSrAlF_6$, etc.) are different shades of green. The absorption transitions, $^4A_2 \rightarrow {}^2E, {}^2T_1, {}^2T_2$ levels are spin-forbidden and weakly coupled to the phonon spectrum ($S < 0.5$). The spectra from these transitions are dominated by sharp zero-phonon lines. However, the low-temperature photoluminescence spectrum of ruby is in marked contrast to the optical absorption spectrum since only the sharp zero-phonon line (R-line) due to the $^2E \rightarrow {}^4A_2$ transition being observed. Given the small energy separations between adjacent states of Cr^{3+}, the higher excited levels decay nonradiatively to the lowest level, ^2E, from which photoluminescence occurs across a band gap of ca. $15{,}000 \text{ cm}^{-1}$. Accurate values of the parameters Dq, B, and C may be determined from these absorption data. First, the peak energy of the $^4A_2 \rightarrow {}^4T_2$ absorption band is equal to $10Dq$. The energy shift between the $^4A_2 \rightarrow {}^4T_2, {}^4T_1$ bands is dependent on both Dq and B, and the energy separation between the two broad absorption bands is used to determine B. Finally, the position of the R-line varies with Dq, B, and C: in consequence, once Dq and B are known, the magnitude of C may be determined from the position of the $^4A_2 \rightarrow {}^2E$ zero-phonon line.

This discussion of the spectroscopy of the Cr^{3+} ion is easily extended to other multielectron configurations. The starting points are the Tanabe-Sugano diagrams collected in various texts.[19,21] Analogous series of elements occur in the fifth and sixth periods of the periodic table, respectively, where the $4d^n$ (palladium) and $5d^n$ (platinum) groups are being filled. Compared with electrons in the 3d shell, the 4d and 5d shell electrons are less tightly bound to the parent ion. In consequence, charge transfer transitions, in which an electron is transferred from the cation to the ligand ion (or vice versa), occur quite readily. The charge transfer transitions arise from the movement of electronic charge over a typical interatomic distance, thereby producing a large dipole moment and a concomitant large oscillator strength for the absorption process. For the Fe-group ions ($3d^n$ configuration), such charge transfer effects result in the absorption of ultraviolet photons.

For example the Fe^{3+} ion in MgO absorbs in a broad structureless band with peak at 220 nm and half-width of order 120 nm (i.e., 0.3 eV). The Cr^{2+} ion also absorbs by charge transfer process in this region. In contrast, the palladium and platinum groups have lower-lying charge transfer states. The resulting intense absorption bands in the visible spectrum may overlap spectra due to low-lying crystal field transitions. Rare-earth ions also give rise to intense charge transfer bands in the ultraviolet region.

Various metal cations have been used as broadband visible region phosphors. For example, transitions between the $4f^n$ and $4f^{(n-1)}$ 5d levels of *divalent* rare-earth ions give rise to intense broad transitions which overlap many of the sharp $4f^n$ transitions of the *trivalent* rare-earth ions. Of particular interest are Sm^{2+}, Dy^{2+}, Eu^{2+}, and Tm^{2+}. In Sm^{2+} $(4f^6)$ broadband absorption transitions from the ground state $^7F_1$ to the $4f^5$ 5d level may result in either broadband emission (from the vibronically relaxed $4f^5$ 5d level) or sharp line emission from $^5D_0$ $(4f^6)$ depending upon the host crystal. The $4f^5$ 5d level, being strongly coupled to the lattice, accounts for this variability. There is a similar material-by-material variation in the absorption and emission properties of the Eu^{2+} ($4f^7$ configuration), which has the $^8S_{7/2}$ ground level. The next highest levels are derived from the $4f^6$ 5d state, which is also strongly coupled to the lattice. This state is responsible for the varying emission colors of Eu^{2+} in different crystals, e.g., violet in $Sr_2P_2O_7$, blue in $BaAl_{12}O_{19}$, green in $SrAl_2O_4$, and yellow in Ba_2SiO_5.

The heavy metal ions Tl^+, In^+, Ga^+, Sn^{2+}, and Pb^{2+} may be used as visible-region phosphors. These ions all have two electrons in the ground configuration ns^2 and excited configurations $(ns)(np)$. The lowest-lying excited states, in the limit of Russell Saunders coupling, are then $^1S_0(ns^2)$, $^3P_{0,1,2}$, and $^1P_1$ from $(ns)/(np)$. The spectroscopy of Tl^+ has been much studied especially in the alkali halides. Obviously $^1S_0 \rightarrow {}^1P_1$ is the strongest absorption transition, occurring in the ultraviolet region. This is labeled as the C-band in Fig. 6. Next in order of observable intensity is the A-band, which is a spin-forbidden

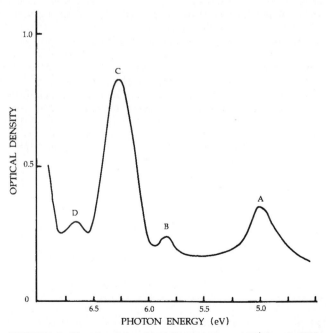

FIGURE 6   The ultraviolet absorption spectrum of Tl^+ ions in KCl measured at 77 K. (*After C. J. Delbecq, et al.*)[26]

absorption transition $^1S_0 \rightarrow {}^3P_1$, in which the relatively large oscillator strength is borrowed from the $^1P_1$ state by virtue of the strong spin-orbit interaction in these heavy metal ions. The B and D bands, respectively, are due to absorption transitions from $^1S_0$ to the $^3P_2$ and $^3P_0$ states induced by vibronic mixing.[26] A phenomenological theory[26,27] has been developed which quantitatively accounts for both absorption spectra and the triplet state emission spectra.[28,29]

The examples discussed so far have all concerned the spectra of ions localized in levels associated with the central fields of the nucleus and closed-shells of electrons. There are other situations which warrant serious attention. These include electron-excess centers in which the positive potential of an anion vacancy in an ionic crystal will trap one or more electrons. The simplest theory treats such a *color center* as a particle in a finite potential well.[19,27] The simplest such center is the F-center in the alkali halides, which consist of one electron trapped in an anion vacancy. As we have already seen (e.g., Fig. 13 in "Optical Spectroscopy and Spectroscopic Lineshapes", Vol. I, Chap. 8 of this Handbook), such centers give rise to broadbands in both absorption and emission, covering much of the visible and near-infrared regions for alkali halides. F-aggregate centers, consisting of multiple vacancies arranged in specific crystallographic relationships with respect to one another, have also been much studied. They may be positive, neutral, or negative in charge relative to the lattice depending upon the number of electrons trapped by the vacancy aggregate.[30]

Multiquantum wells (MQWs) and strained-layer superlattices (SLSs) in semiconductors are yet another type of *finite-well* potential. In such structures, alternate layers of two different semiconductors are grown on top of each other so that the bandgap varies in one dimension with the periodicity of the epitaxial layers. A modified Kronig-Penney model is often used to determine the energy eigenvalues of electrons and holes in the conduction and valence bands, respectively, of the narrower gap material. Allowed optical transitions between valence band and conduction band are then subject to the selection rule $\Delta n = 0$, where $n = 0$, 1, 2, etc. The example given in Fig. 7 is for SLSs in the II-VI family of semiconductors ZnS/ZnSe.[31] The samples were grown by metalo-organic vapor phase epitaxy[32] with a superlattice periodicity of ca. 6–8 nm while varying the thickness of the narrow gap material (ZnSe) between 0.8 nm and 7.6 nm. The splitting between the two

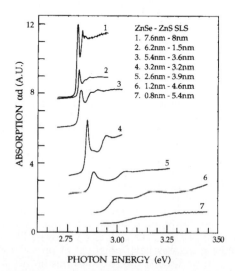

FIGURE 7 Optical absorption spectra of SLSs of ZnS/ZnSe measured at 14 K. (*After Fang et al.*)[31]

sharp features occurs because the valence band states are split into "light holes" (lh) and "heavy holes" (hh) by spin-orbit interaction. The absorption transitions then correspond to transitions from the $n = 1$ lh- and hh-levels in the valence band to the $n = 1$ electron states in the conduction band. Higher-energy absorption transitions are also observed. After absorption, electrons rapidly relax down to the $n = 1$ level from which emission takes place down to the $n = 1$, lh-level in the valence band, giving rise to a single emission line at low temperature.

28.4 THE HOMOGENEOUS LINESHAPE OF SPECTRA

Atomic Spectra

The homogeneous widths of atomic spectra are determined by the uncertainty principle, and hence by the radiative decaytime, τ_R (as discussed previously in this Handbook in Vol. I, Chap. 8), "Optical Spectroscopy and Spectroscopic Lineshapes"). Indeed, the so-called natural or homogeneous width of the transition, $\Delta\omega$, is given by the Einstein coefficient for spontaneous emission, $A_{ba} = (\tau_R)^{-1}$. The homogeneously broadened line has a lorentzian lineshape with FWHM given by $(\tau_R)^{-1}$. In gas-phase spectroscopy, atomic spectra are also broadened by the Doppler effect: random motion of atoms broadens the lines inhomogeneously leading to a guassian-shaped line with FWHM proportional to $(T/M)^{-1/2}$, where T is the absolute temperature and M the atomic mass. Saturated laser absorption or optical hole-burning (OHB) techniques are among the methods which recover the true homogeneous width of an optical transition. Experimental aspects of these types of measurement were discussed in this Handbook in Vol. II, Chap. 20, "Optical Spectroscopy", and examples of Doppler-free spectra (Figs. 1 and 3, Vol. I, Chap. 8, "Optical Spectroscopy and Spectroscopic Lineshapes") were discussed in terms of the fundamental tests of the quantum and relativistic structure of the energy levels of atomic hydrogen. Similar measurements were also discussed for the case of He (Fig. 1) and in molecular spectroscopy (Fig. 3). In such examples, the observed lineshape is very close to a true lorentzian, typical of a lifetime-broadened optical transition.

Zero-phonon Lines in Solids

Optical hole-burning (OHB) reduces the effects of inhomogeneous broadening in solid-state spectra. For rare-earth ions, the homogeneous width amounts to some 0.1–1.0 MHz, the inhomogeneous widths being determined mainly by strain in the crystal. Similarly improved resolution is afforded by fluorescence line narrowing (FLN) in the R-line of Cr^{3+} (Vol. I, Chap. 8, Fig. 12). However, although the half-width measured using OHB is the true homogeneous width, the observed FLN half-width, at least in resonant FLN, is a convolution of the laser width and twice the homogeneous width of the transition.[33] In solid-state spectroscopy, the underlying philosophy of OHB and FLN experiments may be somewhat different from that in atomic and molecular physics. In the latter cases, there is an intention to relate theory to experiment at a rather sophisticated level. In solids, such high-resolution techniques are used to probe a range of other dynamic processes than the natural decay rate. For example, hole-burning may be induced by photochemical processes as well as by intrinsic lifetime processes.[34,35] Such photochemical hole-burning processes have potential in optical information storage systems. OHB and FLN may also be used to study distortions in the neighborhood of defects. Figure 8 is an example of Stark spectroscopy and OHB on a zero-phonon line at 607 nm in irradiated NaF.[34] This line had been attributed to an aggregate of four F-centers in nearest-neighbor

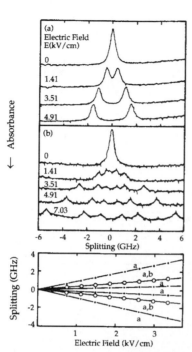

FIGURE 8 The effects of an applied electric field in a hole burned in the 607-nm zero-phonon line observed in irradiated NaF. (*After Macfarlane et al.*)[22]

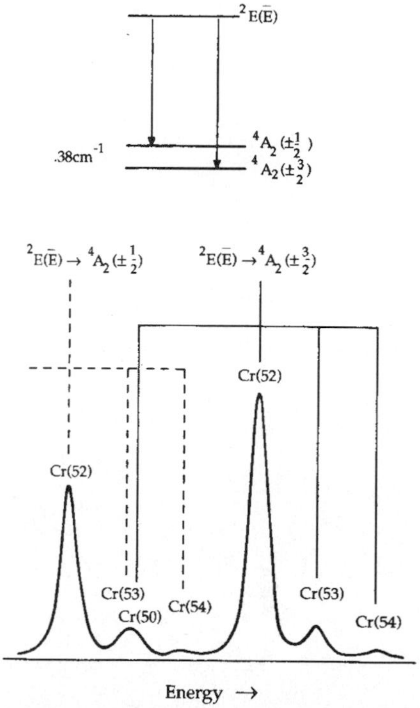

FIGURE 9 Showing the fine structure splitting in the $^4A_2$ ground state and the isotope shifts of Cr^{3+} in the R_1-line of ruby measured using FLN spectroscopy. (*After Jessop and Szabo.*)[36]

anion sites in the rocksalt-structured lattice on the basis of the polarized absorption/emission measurements. The homogeneous width in zero electric field was only 21 MHz in comparison with the inhomogeneous width of 3 GHz. Interpretation of these results is inconsistent with the four-defect model.

The FLN technique may also be used to measure the effects of phonon-induced relaxation processes and isotope shifts. Isotope and thermal shifts have been reported for $Cr^{3+} : Al_2O_3$[36] and $Nd^{3+} : LaCl_3$.[37] The example given in Fig. 9 shows both the splitting in the ground $^4A_2$ state of Cr^{3+} in ruby and the shift between lines due to the Cr(50), Cr(52), Cr(53), and Cr(54) isotopes. The measured differential isotope shift of $0.12 \, cm^{-1}$ is very close to the theoretical estimate.[19] Superhyperfine effects by the 100 percent abundant Al isotope with $I = 5/2$ also contribute to the homogeneous width of the FLN spectrum of Cr^{3+} in Al_2O_3 (Fig. 12 in Vol. I, Chap. 8, "Optical Spectroscopy and Spectroscopic Lineshapes").[36] Furthermore, in antiferromagnetic oxides such as $GdAlO_3$, $Gd_3Ga_5O_{12}$, and $Gd_3Sc_2Ga_3O_{12}$, spin-spin coupling between the Cr^{3+} ions $(S = 3/2)$ and nearest-neighbor Gd^{3+} ions $(S = 3/2)$ contributes as much to the zero-phonon R-linewidth as inhomogeneous broadening by strain.[38]

Configurational Relaxation in Solids

In the case of the broadband $^4T_2 \rightarrow \, ^4A_2$ transition of Cr^{3+} in YAG (Fig. 12 — Optical Spectroscopy and Spectroscopic Lineshapes) and MgO (Fig. 6), the application of OHB

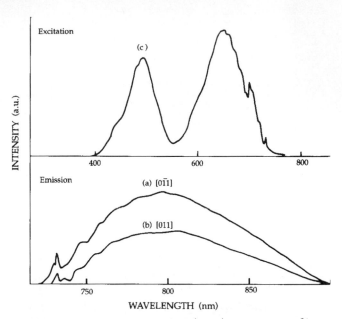

FIGURE 10   Polarized emission of the $^4T_2 \to {}^4A_2$ band from Cr^{3+} ions in orthorhombic sites in MgO. Shown also, (c) is the excitation spectrum appropriate to (a). (*After Yamaga et al.*)[48]

and FLN techniques produce no such narrowing because the vibronic sideband is the homogeneously broadened shape determined by the phonon lifetime rather than the radiative lifetime. It is noteworthy that the vibronic sideband emission of Cr^{3+} ions in orthorhombic sites in MgO, Fig. 10, shows very little structure. In this case, the Huang-Rhys factor $S \simeq 6$, i.e, the strong coupling case, where the multiphonon sidebands tend to lose their separate identities to give a smooth bandshape on the lower-energy side of the peak. By way of contrast, the emission sideband of the R-line transition of Cr^{3+} ions in octahedral sites in MgO is very similar in shape to the known density of one-phonon vibrational modes of MgO^{39} (Fig. 11), although there is a difference in the precise positions of the peaks, because the Cr^{3+} ion modifies the lattice vibrations in its neighborhood relative to those of the perfect crystal. Furthermore, there is little evidence in Fig. 11 of higher-order sidebands which justifies treating the MgO R-line process in the weak coupling limit. The absence of such sidebands suggests that $S < 1$, as the earlier discussion in "Optical Spectroscopy and Spectroscopic Lineshapes" (Vol. I, Chap. 8 of this Handbook) showed. That the relative intensities of the zero-phonon line and broadband, which should be ca. e^{-s}, is in the ratio 1 : 4 shows that the sideband is induced by odd parity phonons. In this case it is partially electric-dipole in character, whereas the zero-phonon line is magnetic-dipole in character.[19]

There has been much research on bandshapes of Cr^{3+}-doped spectra in many solids. This is also the situation for F-centers and related defects in the alkali halides. Here, conventional optical spectroscopy has sometimes been supplemented by laser Raman and sub-picosecond relaxation spectroscopies to give deep insights into the dynamics of the optical pumping cycle. The F-center in the alkali halides is a halide vacancy that traps an electron. The states of such a center are reminiscent of a particle in a finite potential well[19] and strong electron-phonon coupling. Huang-Rhys factors in the range $S = 15$–40 lead to

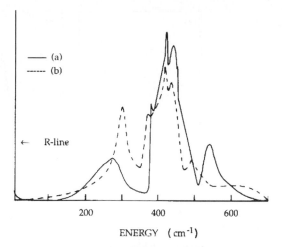

FIGURE 11 A comparison of (*a*) the vibrational side-band accompanying the $^2E \to {}^4A_2$ *R*-line of Cr^{3+} : MgO with (*b*) the density of phonon modes in MgO as measured by Peckham et al.,[39] using neutron scattering. (*After Henderson and Imbusch.*)[19]

broad, structureless absorption/luminescence bands with large Stokes shifts (see Vol. I, Chap. 8, Fig. 13). Raman-scattering measurements on *F*-centers in NaCl and KCl (Fig. 23 in Vol. II, Chap. 20, "Optical Spectrometers" in this Handbook), showed that the first-order scattering is predominantly due to defect-induced local modes.[40]

The F_A-center is a simple variant on the *F*-center in which one of the six nearest cation neighbors of the *F*-center is replaced by an alkali impurity.[41] In KCl the K^+ may be replaced by Na^+ or Li^+. For the case of the Na^+ substituent, the F_A(Na) center has tetragonal symmetry about a $\langle 100 \rangle$ crystal axis, whereas in the case of Li^+ an off-axis relaxation in the excited state leads to interesting polarized absorption/emission characteristics.[19,41] The most dramatic effect is the enormous Stokes shift between absorption and emission bands, of order $13,000 \text{ cm}^{-1}$, which has been used to advantage in color center lasers.[42,43] For F_A(Li) centers, configurational relaxation has been probed using picosecond relaxation and Raman-scattering measurements. Mollenauer et al.[44] used the experimental system shown in this Handbook, (Vol. II, Chap. 20, Fig. 10), to carry out measurements of the configurational relaxation time of F_A(Li) centers in KCl. During deexcitation many phonons are excited in the localized modes coupled to the electronic states which must be dissipated into the continuum of lattice modes. Measurement of the relaxation time constitutes a probe of possible phonon damping. A mode-locked dye laser producing pulses of 0.7-ps duration at 612 nm was used both to pump the center in the F_{A2}-absorption band and to provide the timing beam. Such pumping leads to optical gain in the luminescence band and prepares the centers in their relaxed state. The probe beam, collinear with the pump beam, is generated by a CW F_A(Li)-center laser operating at 2.62 μm. The probe beam and gated pulses from the dye laser are mixed in a nonlinear optical crystal (lithium iodate). A filter allows only the sum frequency at 496 nm to be detected. The photomultiplier tube then measures the rise in intensity of the probe beam which signals the appearance of gain where the F_A(Li)-centers have reached the relaxed excited state. The pump beam is chopped at low frequency to permit phase-sensitive detection. The temporal evolution of F_A(Li)-center gains (Figs. 12*a* and *b*) was measured

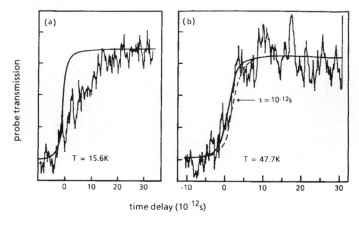

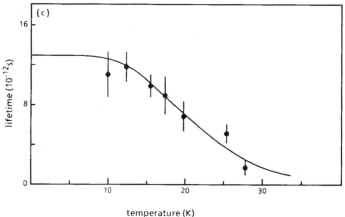

FIGURE 12   (*a*) Temporal evolution of gain in the F_A(Li) center emission in picosecond pulse-probe measurements, and (*b*) the temperature-dependence of the gain process. (*After Mollenauer et al.*)[44]

by varying the time delay between pump and gating pulses. In this figure, the solid line is the instantaneous response of the system, whereas in *b* the dashed line is the instantaneous response convolved with a 1.0-ps rise time.

Measurements of the temperature dependence of the relaxation times of F_A(Li)- in potassium chloride (Fig. 12*c*) show that the process is very fast, typically of order 10 ps at 4 K. Furthermore, configurational relaxation is a multiphonon process which involves mainly the creation of some 20 low-energy phonons of energy $E_P/hc \simeq 47$ cm^{-1}. That only ca. $(20 \times 47/8066)$ eV = 0.1 eV is deposited into the 47 cm^{-1} mode, whereas 1.6 eV of optical energy is lost to the overall relaxation process, indicates that other higher-energy modes of vibrations must be involved.[44] This problem is resolved by Raman-scattering experiments. For F_A(Li)-centers in potassium chloride, three sharp Raman-active local modes were observed with energies of 47 cm^{-1}, 216 cm^{-1}, and 266 cm^{-1}, for the ^{7}Li isotope.[45] These results and later polarized absorption/luminescence studies indicated that the Li$^+$ ion lies in an off-center position in a $\langle 110 \rangle$ crystal direction relative to the z axis of the F_A-center. Detailed polarized Raman spectroscopy resonant and nonresonant with the

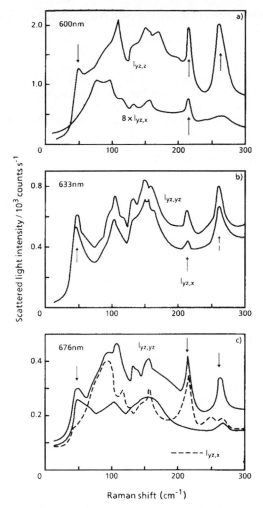

FIGURE 13 The Raman spectra of F_A(Li)-centers in potassium chloride measured at 10 K for different senses of polarization. In (a) the excitation wavelength $\lambda = 600$ nm is midway between the peaks of the F_{A1}-bands; (b) $\lambda = 632.8$ nm is resonant with the F_{A1}-band, and (c) $\lambda = 676.4$ nm is nonresonant. (*After Joosen et al.*)[44]

F_A-center absorption bands are shown in Fig. 13.[46] These spectra show that under resonant excitation in the F_{A1} absorption band, each of the three lines due to the sharp localized modes is present in the spectrum. The polarization dependence confirms that the 266 cm^{-1} mode is due to Li$^+$ ion motion in the mirror plane and parallel to the defect axis. The 216 cm^{-1} mode is the stronger under nonresonant excitation, reflecting the off-axis vibrations of the Li$^+$ ion vibrating in the mirror plane perpendicular to the z axis. On the other hand, the low-frequency mode is an amplified band mode of the center which hardly involves the motion of the Li$^+$ ion.

28.5 *ABSORPTION, PHOTOLUMINESCENCE, AND RADIATIVE DECAY MEASUREMENTS*

The philosophy of solid-state spectroscopy is subtly different from that of atomic and molecular spectroscopies. It is often required not only to determine the nature of the absorbing/emitting species but also the symmetry and structure of the local environment. Also involved is the interaction of the electronic center with other neighboring ions, which leads to lineshape effects as well as time-dependent phenomena. The consequence is that a combination of optical spectroscopic techniques may be used in concert. This general approach to optical spectroscopy of condensed matter phenomena is illustrated by reference to the case of Al_2O_3 and MgO doped with Cr^{3+}.

Absorption and Photoluminescence of Cr^{3+} in Al_2O_3 and MgO

The absorption and luminescence spectra may be interpreted using the Tanabe-Sugano diagram shown in Fig. 5, as discussed previously. Generally, the optical absorption spectrum of $Cr^{3+}:Al_2O_3$ (Fig. 14) is dominated by broadband transitions from the $^4A_2 \rightarrow {}^4T_2$ and $^4A_2 \rightarrow {}^4T_1$. The crystal used in this measurement contained some $10^{18}\ Cr^{3+}$ ions cm^{-3}. Since the absorption coefficient at the peak of the $^4A_2 \rightarrow {}^4T_2$ band is only $2\ cm^{-1}$, it is evident from Eq. (6) in Chap. 20, "Optical Spectrometers", in Vol. II of this Handbook, that the cross section at the band peak is $\sigma_o \cong 5 \times 10^{-19}\ cm^2$. The spin-forbidden absorption transitions $^4A_2 \rightarrow {}^2E$, $^2T_1$ are just distinguished as weak absorptions ($\sigma_o \sim 10^{21}\ cm^2$) on the long-wavelength side of the $^4A_2 \rightarrow {}^4T_2$ band. This analysis strictly applies to the case of octahedral symmetry. Since the cation site in ruby is distorted from

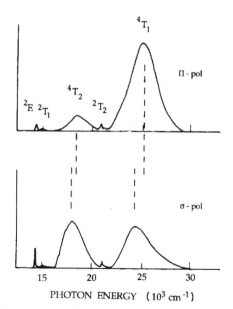

FIGURE 14 Showing the polarized optical absorption spectrum of a ruby $Cr^{3+} : Al_2O_3$ crystal containing $2 \times 10^{18}\ Cr^{3+}$ ions cm^{-3} measured at 77 K.

perfect octahedral symmetry, there are additional electrostatic energy terms associated with this reduced symmetry. One result of this distortion, as illustrated in Fig. 14, is that the absorption and emission spectra are no longer optically isotropic. By measuring the peak shifts of the $^4A_2 \rightarrow {}^4T_2$ and $^4A_2 \rightarrow {}^4T_1$ absorption transitions between π and σ senses of polarization, the trigonal field splittings of the $^4T_2$ and $^4T_1$ levels may be determined.[25]

The Cr^{3+} ion enters the MgO substitutionally for the Mg^{2+} ion. The charge imbalance requires that for every two impurity ions there must be one cation vacancy. At low-impurity concentrations, charge-compensating vacancies are mostly remote from the Cr^{3+} ions. However, some 10–20 percent of the vacancies occupy sites close to individual Cr^{3+} ions, thereby reducing the local symmetry from octahedral to tetragonal or orthorhombic.[19] The optical absorption spectrum of Cr^{3+} : MgO is also dominated by broadband $^4A_2 \rightarrow {}^4T_2$, $^4T_1$ transitions; in this case, there are overlapping contributions from Cr^{3+} ions in three different sites. There are substantial differences between the luminescence spectra of Cr^{3+} in the three different sites in MgO (Fig. 15), these overlapping

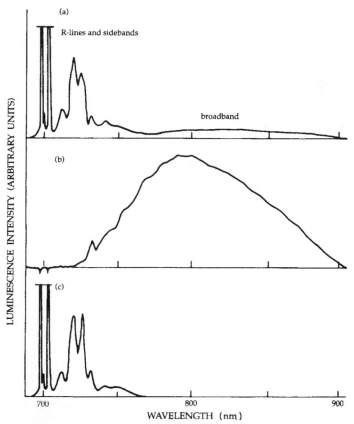

FIGURE 15 Showing photoluminescence spectra of Cr^{3+} : MgO using techniques of phase-sensitive detection. In (a) the most intense features are sharp R-lines near 698–705 nm due to Cr^{3+} ions at sites with octahedral and tetragonal symmetry; a weak broadband with peak at 740 nm is due to Cr^{3+} ions in sites of orthorhombic symmetry. By adjusting the phase-shift control on the lock-in amplifier (Fig. 4 in Vol. 2, Chap. 20 of this Handbook), the relative intensities of the three components may be adjusted as in parts (b) and (c).

spectra being determined by the ordering of the $^4T_2$ and ^2E excited states. For strong crystal fields, $Dq/B > 2.5$, ^2E lies lowest and nonradiative decay from $^4T_1$ and $^4T_2$ levels to ^2E results in very strong emission in the sharp R-lines, with rather weaker vibronic sidebands. This is the situation from Cr^{3+} ions in octahedral and tetragonal sites in MgO.[19] The $^2E \rightarrow {}^4A_2$ luminescence transition is both spin- and parity-forbidden (see this Handbook, Vol. I, Chap. 8) and this is signaled by relatively long radiative lifetimes— 11.3 ms for octahedral sites and 8.5 ms for tetragonal sites at 77 K. This behavior is in contrast to that of Cr^{3+} in orthorhombic sites, for which the $^4T_2$ level lies below the ^2E level. The stronger electron-phonon coupling for the $^4T_2 \rightarrow {}^4A_2$ transition at orthorhombic sites leads to a broadband luminescence with peak at 790 nm. Since this is a spin-allowed transition, the radiative lifetime is much shorter—only 35 μs.[47]

As noted previously the decay time of the luminescence signals from Cr^{3+} ions in octahedral and tetragonal symmetry are quite similar, and good separation of the associated R-lines using the phase-nulling technique are then difficult. However, as Fig. 15 shows, good separation of these signals from the $^4T_2 \rightarrow {}^4A_2$ broadband is very good. This follows from the applications of Eqs. (8)–(12) in this volume of the Handbook, Chap. 20. For Cr^{3+} ions in cubic sites, the long lifetime corresponds to a signal phase angle of 2° : the R-line intensity can be suppressed by adjusting the detector phase angle to $(90° + 2°)$. In contrast, the Cr^{3+} ions in orthorhombic sites give rise to a phase angle of 85°: this signal is reduced to zero when $\phi_D = (90° + 85°)$.

Excitation Spectroscopy

The precise positions of the $^4A_2 \rightarrow {}^4T_2$, $^4T_1$ absorption peaks corresponding to the sharp lines and broadbands in Fig. 15 may be determined by excitation spectroscopy (see Vol. II, Chap. 20). An example of the application of this technique is given in Fig. 10, which shows the emission band of the $^4T_2 \rightarrow {}^4A_2$ transition at centers with orthorhombic symmetry, Figs. 10a and b, and its excitation spectrum, Fig. 10c.[47] The latter was measured by setting the wavelength of the detection spectrometer at $\lambda = 790$ nm, i.e., the emission band peak, and scattering the excitation monochromator over the wavelength range 350–750 nm of the Xe lamp. Figure 10 gives an indication of the power of excitation spectroscopy in uncovering absorption bands not normally detectable under the much stronger absorptions from cubic and tetragonal centers. Another example is given in Fig. 16—in this case, of recombining excitons in the smaller gap material (GaAs) in GaAs/AlGaAs quantum wells. In this case, the exciton luminescence peak energy, $h\nu_x$, is given by

$$h\nu_x = E_G + E_{1e} + E_{1h} + E_b \qquad (3)$$

where E_G is the band gap of GaAs, E_{1e} and E_{1h} are the $n = 1$ state energies of electrons (e) and holes (h) in conduction and valence bands, respectively, and E_b is the electron-hole binding energy. Optical transitions involving electrons and holes in these structures are subject to the $\Delta n = 0$ selection rule. In consequence, there is a range of different absorption transitions at energies above the bandgap. Due to the rapid relaxation of energy in levels with $n > 1$, the recombination luminescence occurs between the $n = 1$ electron and hole levels only, in this case at 782 nm. The excitation spectrum in which this luminescence is detected and excitation wavelength varied at wavelengths shorter than 782 nm reveals the presence of absorption transitions above the bandgap. The first absorption transition shown is the $1lh \rightarrow 1e$ transition, which occurs at slightly longer wavelength than the $1hh \rightarrow 1e$ transition. The light hole (lh)-heavy hole (hh) splitting is caused by spin-orbit splitting and strain in these epilayer structures. Other, weaker transitions are also discernible at higher photon energies.

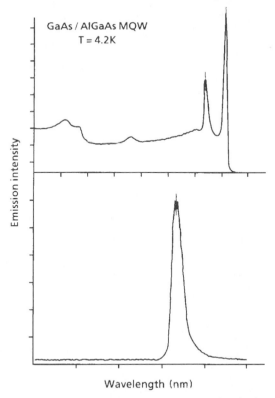

FIGURE 16 The luminescence spectrum and excitation spectrum of multiple quantum wells in GaAs/AlGaAs samples measured at 6 K. (*P. Dawson, 1986 private communication to the author.*)

Polarization Spectroscopy

The discussions on optical selection rules, in this Handbook in Vol. 1, Chap. 8 and Vol. II, Chap. 20, showed that when a well-defined axis is presented, the strength of optical transitions may depend strongly on polarization. In atomic physics the physical axis is provided by an applied magnetic field (Zeeman effect) or an applied electric field (Stark effect). Polarization effects in solid-state spectroscopy may be used to give information about the site symmetry of optically active centers. The optical properties of octahedral crystals are normally isotropic. In this situation, the local symmetry of the center must be lower than octahedral so that advantage may be taken of the polarization-sensitivity of the selection rules (see , e.g., Vol. II, Chap. 20 of this Handbook). Several possibilities exist in noncubic crystals. If the local symmetry of all centers in the crystal point in the same direction, then the crystal as a whole displays an axis of symmetry. Sapphire (Al_2O_3) is an example, in which the Al^{3+} ions occupy trigonally distorted octahedral sites. In consequence, the optical absorption and luminescence spectra of ions in this crystal are naturally polarized. The observed π- and σ-polarized absorption spectra of ruby shown in Fig. 14 are in general agreement with the calculated selection rules,[25] although there are undoubtedly vibronic processes contributing to these broadband intensities.[47] The other important ingredient in the spectroscopy of the Cr^{3+} ions in orthorhombic symmetry sites in MgO is that the absorption and luminescence spectra are strongly polarized. It is then

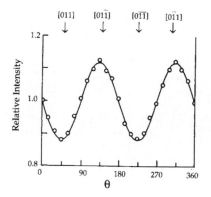

FIGURE 17 The polarization characteristics of the luminescence spectrum of Cr^{3+} ions in orthorhombic sites in Cr^{3+} : MgO. (*After Henry et al.*)[47]

quite instructive to indicate how the techniques of polarized absorption/luminescence help to determine the symmetry axes of the dipole transitions. The polarization of the $^4T_2 \rightarrow {}^4A_2$ emission transition in Fig. 10 is clear. In measurements employing the "straight-through" geometry, Henry et al.[47] report the orientation intensity patterns shown in Fig. 17 for the broadband spectrum. A formal calculation of the selection rules and the orientation dependence of the intensities shows that the intensity at angle θ is given by

$$I(\theta) = (A_\pi - A_\sigma)(E_\pi - E_\sigma) \sin^2 \left(\theta + \frac{\pi}{4} \right) + \text{constant} \qquad (4)$$

where A and E refer to the absorbed and emitted intensities for π- and σ-polarizations.[48] The results in Fig. 17 are consistent with the dipoles being aligned along $\langle 110 \rangle$ directions of the octahedral MgO lattice. This is in accord with the model of the structure of the Cr^{3+} ions in orthorhombic symmetry, which locates the vacancy in the nearest neighbor cation site relative to the Cr^{3+} ion along a $\langle 110 \rangle$ direction.

Zeeman spectroscopy

The Zeeman effect is the splitting of optical lines by a static magnetic field due to the removal of the spin degeneracy of levels involved in the optical transitions. In many situations the splittings are not much larger than the optical linewidth of zero-phonon lines and much less than the width of vibronically broadened bands. The technique of optically detected magnetic resonance (ODMR) is then used to measure the Zeeman splittings. As we have already shown, ODMR also has the combined ability to link inextricably, an excited-state ESR spectrum with an absorption band and a luminescence band. The spectrum (Fig. 16 in Vol. II, Chap. 20 of this Handbook) is an example of this unique power, which has been used in such diverse situations as color centers, transition-metal ions, rare-earth ions, phosphor- and laser-active ions (e.g., Ga^+, Tl°), as well as donor-acceptor and exciton recombination in semiconductors.[19] We conclude this review by illustrating the relationship of the selection rules and polarization properties of the triplet-singlet transitions.

The F-center in calcium oxide consists of two electrons trapped in the Coulomb field of a negative-ion vacancy. The ground state is a spin singlet, $^1A_{1g}$, from which electric dipole absorption transitions are allowed into a $^1T_{1u}$ state derived from the $(1s2p)$ configuration. Such $^1A_{1g} \rightarrow {}^1T_{1u}$ transitions are signified by a strong optical absorption band centered at a wavelength $\lambda \simeq 400$ nm (Vol. II, Chap. 20, Fig. 16). De-excitation of this $^1T_{1u}$ state does not proceed via $^1T_{1u} \rightarrow {}^1A_{1g}$ luminescence. Instead, there is efficient nonradiative decay from $^1T_{1u}$ into the triplet $^3T_{1u}$ state also derived from the $(1s2p)$ configuration (Henderson et al., 1969).[49] The spin-forbidden $^3T_{1u} \rightarrow {}^1A_{1g}$ transition gives rise to a striking orange fluorescence, which occurs with a radiative lifetime $\tau_R = 3.4$ ms at 4.2 K. The ODMR spectrum of the F-center and its absorption and emission spectral dependences were depicted in Fig. 16 in Chap. 20, "Optical Spectrometers" in Vol. II of this Handbook; other details are shown in Fig. 18. With the magnetic field at some general orientation in the (100) plane there are six lines. From the variation of the resonant fields with the orientation of the magnetic field in the crystal, Edel et al. (1972)[50] identified the spectrum with the $S = 1$ state of tetragonally distorted F-center. The measured orientation dependence gives $g_{\parallel} \simeq g_{\perp} = 1.999$ and $D = 60.5$ mT.

Figure 18 shows the selection rules for emission of circularly polarized light by $S = 1$ states in axial crystal fields. We denote the populations of the $M_s = 0, \pm 1$ levels as N_o and $N_{\pm 1}$. The low-field ESR line, corresponding to the $M_s = 0 \rightarrow M_s = +1$ transition, should be observed as an increase in σ_+-light because $N_o > N_{+1}$ and ESR transitions enhance the

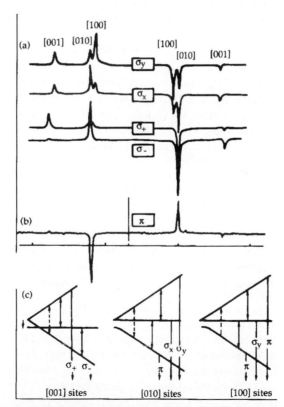

FIGURE 18 The polarization selection rules and the appropriately detected ODMR spectra of the $^3T_{1u} \rightarrow {}^1A_{1g}$ transition of F-centers in CaO. (*After Edel et al.*)[50]

$M_s = \pm 1$ level. However, the high-field line is observed as a change in intensity of σ_--light. If spin-lattice relaxation is efficient (i.e., $T_1 < \tau_R$), then the spin states are in thermal equilibrium, $N_o < N_{-1}$, and ESR transitions depopulate the $|M_s = -1\rangle$ level. Thus, the high-field ODMR line is seen as a decrease in the F-center in these crystals (viz., that for the lowest $^3T_1$ state, D is positive and the spin states are in thermal equilibrium). It is worth noting that since the $|M_s = 0\rangle \rightarrow M_s = \pm 1\rangle$ ESR transitions occur at different values of the magnetic field, ODMR may be detected simply as a change in the emission intensity at resonance; it is not necessary to measure specifically the sense of polarization of the emitted light. The experimental data clearly establish the tetragonal symmetry of the F-center in calcuim oxide: the tetragonal distortion occurs in the excited $^3T_{1u}$ state due to vibronic coupling to modes of E_g symmetry resulting in a static Jahn-Teller effect.[50]

28.6 REFERENCES

1. T. W. Hänsch, I. S. Shakin, and A. L. Shawlow, *Nature* (London), **225:**63 (1972).

2. D. N. Stacey, private communication to A. I. Ferguson.

3. E. Giacobino and F. Birabem, *J. Phys.* B**15:**L385 (1982).

4. L. Housek, S. A. Lee, and W. M. Fairbank Jr., *Phys. Rev. Lett.* **50:**328 (1983).

5. P. Zhao, J. R. Lawall, A. W. Kam, M. D. Lindsay, F. M. Pipkin, and W. Lichten, *Phys. Rev. Lett.* **63:**1593 (1989).

6. T. J. Sears, S. C. Foster, and A. R. W. McKellar, *J. Opt. Soc. Am.* B**3:**1037 (1986).

7. C. J. Sansonetti, J. D. Gillaspy, and C. L. Cromer, *Phys. Rev. Lett.* **65:**2539 (1990).

8. W. Lichten, D. Shinen, and Zhi-Xiang Zhou, *Phys. Rev.* A**43:**1663 (1991).

9. C. Adams and A. I. Ferguson, *Opt. Commun.* **75:**419 (1990) and **79:**219 (1990).

10. C. Adams, E. Riis, and A. I. Ferguson, *Phys. Rev.* A (1992) and A**45:**2667 (1992).

11. G. W. F. Drake and A. J. Makowski, *J. Opt. Soc. Amer.* B**5:**2207 (1988).

12. P. G. Harris, H. C. Bryant, A. H. Mohagheghi, R. A. Reeder, H. Sharifian, H. Tootoonchi, C. Y. Tang, J. B. Donahue, C. R. Quick, D. C. Rislove, and W. W. Smith, *Phys. Rev. Lett.* **65:**309 (1990).

13. H. R. Sadeghpour and C. H. Greene, *Phys. Rev. Lett.* **65:**313 (1990).

14. See, for example, H. Friedrich, *Physics World* **5:**32 (1992).

15. Iu et al., *Phys. Rev. Lett.* **66:**145 (1991).

16. B. MacGowan et al., *Phys. Rev. Lett.* **65:**420 (1991).

17. C. Douketic and T. E. Gough, *J. Mol. Spectrosc.* **101:**325 (1983).

18. See, for example, K. Codling et al., *J. Phys.* B. **24:**L593 (1991).

19. B. Henderson and G. F. Imbusch, *Optical Spectroscopy of Inorganic Solids,* Clarendon Press, Oxford, 1989.

20. G. Blasse, in B. Di Bartolo (ed.), *Energy Transfer Processes in Condensed Matter,* Plenum Press, New York, 1984).

21. Y. Tanabe and S. Sugano, *J. Phys. Soc. Jap.* **9:**753 (1954).

22. R. M. MacFarlane, J. Y. Wong, and M. D. Sturge, *Phys. Rev.* **166:**250 (1968).

23. B. F. Gachter and J. A. Köningstein, *J. Chem. Phys.* **66:**2003 (1974).

24. See O. Deutschbein, *Ann. Phys.* **20:**828 (1932).

25. D. S. McClure, *J. Chem. Phys.* **36:**2757 (1962).

26. After C. J. Delbecq, W. Hayes, M. C. M. O'Brien, and P. H. Yuster, *Proc. Roy. Soc.* A**271:**243 (1963).

27. W. B. Fowler, in Fowler (ed.), *Physics of Color Centers,* Academic Press, New York, 1968. See

also G. Boulon, in B. Di Bartolo (ed), *Spectroscopy of Solid State Laser-Type Materials,* Plenum Press, New York, 1988.

28. Le Si Dang, Y. Merle d'Aubigné, R. Romestain, and A. Fukuda, *Phys. Rev. Lett.* **38:**1539 (1977).

29. A. Ranfagni, D. Mugna, M. Bacci, G. Villiani, and M. P. Fontana, *Adv. in Phys.* **32:**823 (1983).

30. E. Sonder and W. A. Sibley, in J. H. Crawford and L. F. Slifkin (eds), *Point Defects in Solids,* Plenum Press, New York, vol. 1, 1972.

31. Y. Fang, P. J. Parbrook, B. Henderson, and K. P. O'Donnell, *Appl. Phys. Letts.* **59:**2142 (1991).

32. P. J. Parbrook, B. Cockayne, P. J. Wright, B. Henderson, and K. P. O'Donnell, *Semicond. Sci. Technol.* **6:**812 (1991).

33. T. Kushida and E. Takushi, *Phys. Rev.* B**12:**824 (1975).

34. R. M. Macfarlane, R. T. Harley, and R. M. Shelby, *Radn. Effects* **72:**1 (183).

35. W. Yen and P. M. Selzer, in W. Yen and P. M. Selzer (eds.), *Laser Spectroscopy of Solids,* Springer-Verlag, Berlin, 1981.

36. P. E. Jessop and A. Szabo, *Optics Comm.* **33:**301 (1980).

37. N. Pelletier-Allard and R. Pelletier, *J. Phys.* C **17:**2129 (1984).

38. See Y. Gao, M. Yamaga, B. Henderson, and K. P. O'Donnell, *J. Phys.* (Cond. Matter) (1992) in press (and references therein).

39. G. E. Peckham, *Proc. Phys. Soc.* (Lond.) **90:**657 (1967).

40. J. M. Worlock and S. P. S. Porto, *Phys. Rev. Lett.* **15:**697 (1965).

41. F. Luty, in W. B. Fowler (eds.), *The Physics of Color Centers,* Academic Press, New York, 1968.

42. L. F. Mollenauer and D. H. Olson, *J. App. Phys.* **24:**386 (1974).

43. F. Luty and W. Gellerman, in C. B. Collins (ed.), *Lasers '81,* STS Press, McClean, 1982.

44. L. F. Mollenauer, J. M. Wiesenfeld, and E. P. Ippen, *Radiation Effects* **72:**73 (1983); see also J. M. Wiesenfeld, L. F. Mollenauer, and E. P. Ippen, *Phys. Rev. Lett.* **47:**1668 (1981).

45. B. Fritz, J. Gerlach, and U. Gross, in R. F. Wallis (ed.), *Localised Excitations in Solids,* Plenum Press, New York, 1968, p. 496.

46. W. Joosen, M. Leblans, M. Vahimbeek, M. de Raedt, E. Goovaertz, and D. Schoemaker, *J. Cryst. Def. Amorph. Solids* **16:**341 (1988).

47. M. O. Henry, J. P. Larkin, and G. F. Imbusch, *Phys. Rev.* B**13:**1893 (1976).

48. M. Yamaga, B. Henderson, and K. P. O'Donnell, *J. Luminescence* **43:**139 (1989); see also *ibid.* **46:**397 (1990).

49. B. Henderson, S. E. Stokowski, and T. C. Ensign, *Phys. Rev.* **183:**826 (1969).

50. P. Edel, C. Hennies, Y. Merle d'Aubigné, R. Romestain, and Y. Twarowski, *Phys. Rev. Lett.* **28:**1268 (1972).

CHAPTER 29
OPTICAL METROLOGY

Daniel Malacara and Zacarias Malacara
Centro de Investigaciones en Optica, A.C.
León, Gto, Mexico

29.1 GLOSSARY

B	baseline length
f	focal length
f	signal frequency
f_b	back focal length
I	irradiance
N	average group refractive index
R	radius of curvature of an optical surface
r	radius of curvature of a spherometer ball
R	range
α	attenuation coefficient
λ	wavelength of light
τ	delay time

In "Optical Metrology" the purpose is to measure some physical parameters using optical methods. In this chapter we describe the most common procedures for the measurements of length and straightness, angles between plane optical surfaces, and curvature and focal length of lenses and mirrors. We also describe some optical procedures using Doppler shifts to measure velocities. The reader may obtain some more details in the book *Optical Shop Testing*, D. Malacara (ed.), 2d ed., John Wiley and Sons, New York, 1992, or in Chap. 21 on "Interferometers" in Vol. II of this Handbook.

29.2 INTRODUCTION AND DEFINITIONS

Metrology is an activity in which we quantify a physical variable to differentiate a system from another or to analyze the same system under different circumstances. Once a *fundamental unit* of measurement is defined, several *derived units* may be obtained. In

---

Note: Figures 10, 11, 12, 17, 18, 22, 23, 24, 25, 26, 27, and 29 are from *Optical Shop Testing*, 2d ed., edited by D. Malacara. (*Reprinted by permission John Wiley and Sons, Inc., New York, 1992.*)

TABLE 1 SI Fundamental Units

Quantity	Name	Symbol
Length	meter	m
Mass	kilogram	kg
Time	second	s
Electric current	ampere	A
Quantity of a substance	mole	mol
Temperature	kelvin	K
Luminous intensity	candela	cd

order to have a common measurement comparison, a *standard* has to be established, and it must be accessible and invariable.

The generally accepted measurement system of units is the International System or Système International (SI). Table 1 lists the fundamental units for the SI system.

There are two kinds of standards, a *primary standard* which is a measurement reference that has been built according to the defined standard, and a *secondary standard* which is a measurement scale built by comparison with a primary standard. Time is a fundamental standard and is defined as follows: "The second is the duration of 9,192,631,770 periods of the radiation corresponding to the transition between the two hyperfine levels of the ground state of the cesium 133 atom." Formerly, the meter was a fundamental standard defined as: 1,650,763.73 wavelengths in vacuum of the orange-red spectral line from the $2p_{10}$ and $5d_5$ levels of the krypton 86 atom. Shortly after the invention of the laser, it was proposed to use a laser line as a length standard (Bloom, 1971). In 1986, the speed of light was defined as 299,792,458 m/sec, thus the meter is now a derived unit defined as: "the distance traveled by light during 1/299,792,458 of a second." The advantages of this definition, compared with the former meter definition, lie in the fact that it uses a relativistic constant, not subjected to physical influences, and is accessible and invariant. To avoid a previous definition of a time standard, the meter could be defined as: "The length equal to 9,192,631,770/299,792,458 wavelengths in vacuum of the radiation corresponding to the transition between the two hyperfine levels of the ground state of the cesium 133 atom" (Goldman, 1980; Giacomo, 1980).

On the SI fundamental units, the candela is defined as: "The luminous intensity, in the perpendicular direction, of a surface of 1/600,000 square meter of a blackbody at the temperature of freezing platinum under a pressure of 101,325 newton per square meter." Details about primary and secondary standards are described by Larocca (1994) in Vol. I, Chap. 10 and by Zalewski (1994) in Vol. II, Chap. 24 of this Handbook.

The measurement process is subjected to errors. They may be systematic or stochastic. a *systematic error* occurs in a poorly calibrated instrument. An instrument low in systematic error is said to be accurate. *Accuracy* is a measure of the amount of systematic errors. The accuracy is improved by adequate tracing to a primary standard. *Stochastic errors* appear due to random noise and other time-dependent fluctuations that affect the instrument. Stochastic errors may be reduced by taking several measurements and averaging them. An instrument is said to be precise when the magnitude of stochastic errors is low. *Precision* is a term used to define the repeatability for the measurements of an instrument. In measurements, a method for data acquisition and analysis has to be developed. Techniques for experimentation planning and data reduction can be found in the reference (Baird, 1962; Gibbins, 1986).

29.3   *LENGTH AND STRAIGHTNESS MEASUREMENTS*

Length measurements may be performed by optical methods, since the definition of the meter is in terms of a light wavelength. Most of the length measurements are, in fact, comparisons to a secondary standard. Optical length measurements are made by comparisons to an external or internal scale a light time of flight, or by interferometric fringe counting.

Stadia and Range Finders

A *stadia* is an optical device used to determine distances. The principle of the measurement is a bar of known length W set at one end of the distance to be measured (Fig. 1). At the other end, a prism superimposes two images, one coming directly and the other after a reflection from a mirror, and are thus observed through a telescope. At this point, the image of one end of the bar is brought in coincidence with that of the other end by rotating the mirror an angle θ. The mirror rotator is calibrated in such a way that for a given bar length W, a range R can be read directly in a dial, according to the equation (Patrick, 1969):

$$R = \frac{W}{\theta} \tag{1}$$

where θ is small and expressed in radians.

Another stadia method uses a graduated bar and calibrated reticle in the telescope. For a known bar length W imaged on a telescope with focal length f, the bar on the focal plane will have a size i, and the range can be calculated approximately by (Dickson and Harkness, 1969):

$$R = \left(\frac{f}{i}\right)W \tag{2}$$

Most surveying instruments have stadia markings usually in a 1 : 100 ratio, to measure distances from the so-called anallatic point, typically the instrument's center. A theodolite

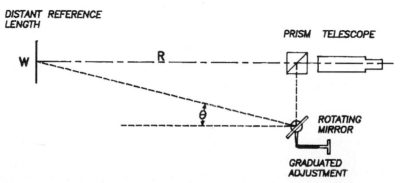

FIGURE 1   Stadia range meter. (*From Patrick, 1969.*)

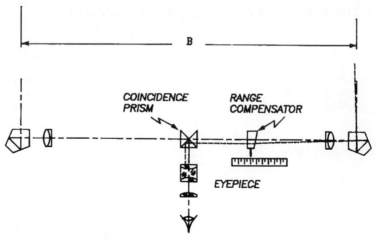

FIGURE 2 A range finder.

may be used for range measurements using the subtense bar method. In this method, a bar of known length is placed at the distance to be measured from the theodolite. By measuring the angle from one end to another end of the bar, the range can be easily measured.

A range finder is different from the stadia, in that the reference line is self-contained within the instrument. A general layout for a range finder is shown in Fig. 2. Two pentaprisms are separated a known baseline B; two telescopes form an image, each through a coincidence prism. The images from the same reference point are superimposed in a split field formed by the coincidence prism. In one branch, a range compensator is adjusted to permit an effective coincidence from the reference images.

Assuming a baseline B and a range R, for small angles θ (large distances compared with the baseline), the range is:

$$R = \frac{B}{\theta} \tag{3}$$

For an error $\Delta\theta$ in the angle measurement, the error ΔR in the range determination would be,

$$\Delta R = -B\theta^{-2}\Delta\theta \tag{4}$$

and by substituting in Eq. (3),

$$\Delta R = -\frac{R^2}{B}\Delta\theta \tag{5}$$

From this last equation, it can be seen that the range error increases with the square of the range. Also, it is inversely proportional to the baseline. The angle error $\Delta\theta$ is a function of the eye's angular acuity, about 10 arcsec. or 0.00005 radians (Smith, 1990).

Pentaprisms permit a precise 90° deflection, independent of the alignment, and are like mirror systems with an angle of 45° between them. The focal length for the two telescopes in a range finder must be closely matched to avoid any difference in magnification. There are several versions of the range compensator. In one design, a sliding prism makes a variable deviation angle (Fig. 3a); in another system (Fig. 3b), a sliding prism displaces the image, without deviating it. A deviation can be also made with a rotating glass block (Fig. 3c). The Risley prisms (Fig. 3d) are a pair of counterrotating prisms located on the entrance pupil for one of the arms. Since the light beam is collimated, there is no astigmatism contribution from the prisms (Smith, 1990). A unit magnification galilean telescope (Fig. 3e) is made with two weak lenses. A sliding lens is used to form a variable wedge to deviate the image path (Patrick, 1969).

Time-based and Optical Radar

Distance measurements can also be done by the time-of-flight method. An application of the laser, obvious at the time of its advent, is the measurement of range. Distance determination by precise timing is known as optical radar. Optical radar has been used to measure the distance to the moon. Since a small timing is involved, optical radar is applicable for distances from about 10 km. For distances from about a meter up to 50 km, modulated beams are used.

Laser radars measure the time of flight for a pulsed laser. Since accuracy depends on the temporal response of the electronic and detection system, optical radars are limited to distances larger than one kilometer. Whenever possible, a cat's eye retroreflector is set at the range distance, making possible the use of a low-power pulsed laser. High-power lasers can be used over small distances without the need of a reflector. Accuracies of the order of 10^{-6} can be obtained (Sona, 1972).

The beam modulation method requires a high-frequency signal, about 10–30 MHz (modulating wavelength between 30 and 10 m) to modulate a laser beam carrier. The amplitude modulation is usually applied, but polarization modulation may also be used. With beam modulation distance measurements, the phase for the returning beam is compared with that of the output beam. The following description is for an amplitude modulation distance meter, although the same applies for polarizing modulation. Assuming a sinusoidally modulated beam, the returning beam can be described by:

$$I_R(t) = \alpha I_o[1 + A \sin \omega(t - \tau)] \tag{6}$$

where α is the attenuation coefficient for the propagation media, I_o is the output intensity for the exit beam, ω is 2π times the modulated beam frequency, and τ is the delay time. By measuring the relative phase between the outgoing and the returning beam, the quantity ω is measured. In most electronic systems, the delay time τ is measured, so, the length in multiples of the modulating wavelength is;

$$L = \frac{c}{2N_g} \tau \tag{7}$$

where N_g is the average group refractive index for the modulating frequency (Sona, 1972).

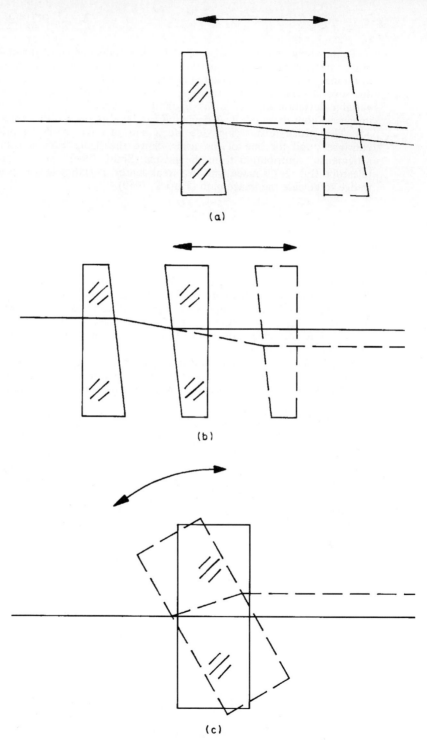

FIGURE 3 Range compensators for range finders (*a*) and (*b*): sliding prism; (*c*) rotating glass block; (*d*) counterrotating prisms; (*e*) sliding lens.

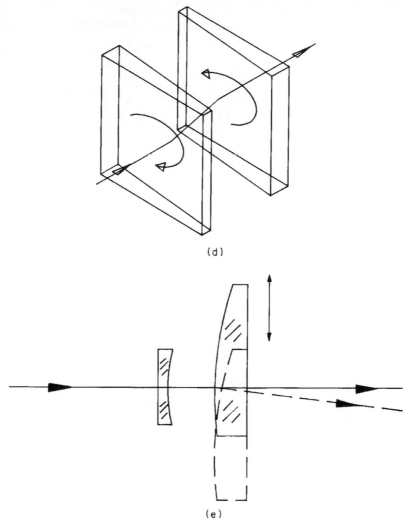

(d)

(e)

FIGURE 3 (*Continued*)

Since the measured length is a multiple of the modulating wavelength, one is limited in range to one-half of the modulating wavelength. To measure with acceptable precision, and at the same time measure over large distances, several modulating frequencies are used, usually at multiples of ten to one another.

The traveling time is measured by comparing the phase of the modulating signal for the exiting and the returning beams. This phase comparison is sometimes made using a null method. In this method, a delay is introduced in the exiting signal, until it has the same phase as the returning beam, as shown in Fig. 4. Since the introduced delay is known, the light traveling time is thus determined.

Another method uses the down conversion of the frequency of both signals. These signals, with frequency f, are mixed with an electrical signal with frequency f_o, in order to obtain a signal with lower frequency f_L, in the range of a few kHz. The phase difference

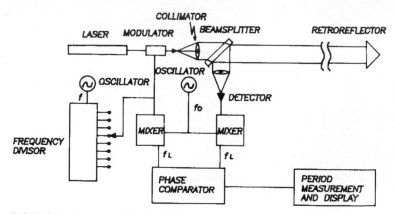

FIGURE 4 A wave modulation distance meter.

between the two low-frequency signals is the same as that between the two original signals. The lowering of the frequencies permits us to use conventional electronic methods to determine the phase difference between the two signals. A complete study on range finders has been done by Stitch (1972).

Interferometric Measurement of Small Distances

Interferometry may be used to measure small distances with a high degree of accuracy. This is usually done with a Michelson interferometer by comparing the thickness of the lens or glass plate with that of a calibrated reference glass plate. Both plates must have approximately the same thickness and should be made with the same material (Tsuruta and Ichihara, 1975). The two mirrors of a dispersion-compensated Michelson interferometer are replaced by the glass plate to be measured and by a reference plane-parallel plate of the same material as the lens. (The next step is to adjust the interferometer, to produce white-light Newton rings with the front surfaces of the lens and the plate.) Then, the plate is translated along the arm of the interferometer until the rear surface produces white-light rings. The displacement is the optical thickness Nt of the measured plate.

Interferometric Measurement of Medium Distances

Long and medium distances may also be measured by interferometric methods (Bruning, 1978; Steinmetz, Burgoon, and Herris, 1987; Massie and Caulfield, 1987). Basically, the method counts the fringes in an interferometer while increasing or decreasing the optical path difference. The low temporal coherence or monochromaticity of most light sources limits this procedure to short distances. Lasers, however, have a much longer coherence length, due to their higher monochromaticity. Using lasers, it has been possible to make interferometric distance measurements over several meters.

In these interferometers, three things should be considered during their design. The first is that the laser light illuminating the interferometer should not be reflected back to the laser because that would cause instabilities in the laser, resulting in large irradiance fluctuations. As the optical path difference is changed by moving one of the mirrors, an irradiance detector in the pattern will detect a sinusoidally varying signal, but the direction of this change cannot be determined. Therefore, the second thing to be considered is that there should be a way to determine if the fringe count is going up or down; that is, if the distance is increasing or decreasing. There are two basic approaches to satisfy this last requirement. One is by producing two sinusoidal signals in phase quadrature (phase difference of 90° between them). The direction of motion of the moving prism may be sensed by determining the phase of which signal leads or lags the phase of the other signal. This information is used to make the fringe counter increase or decrease. The alternative method uses a laser with two different frequencies. Finally, the third thing to consider in the interferometer design is that the number of fringes across its aperture should remain low and constant while moving the reflector in one of the two interferometer arms. This last condition is easily satisfied by using retroreflectors instead of mirrors. Then the two interfering wavefronts will always be almost flat and parallel. A typical retroflector is a cube corner prism of reasonable quality to keep the number of fringes over the aperture low.

One method of producing two signals in quadrature is to have only a small number of fringes over the interferogram aperture. One possible method is by deliberately using an imperfect retroreflector. Then, the two desired signals are two small slits parallel to the fringes and separated by one-fourth of the distance between the fringes (Sona, 1965). To avoid illuminating back the laser, the light from this laser should be linearly polarized. Then, a $\lambda/4$ phase plate is inserted in front of the beam, with its slow axis set at 45° to the interferometer plane to transform it into a circularly polarized beam.

Another method used to produce the two signals in quadrature is to take the signals from the two interference patterns that are produced in the interferometer. If the beam splitters are dielectric (no energy losses), the interference patterns will be complements of each other and, thus, the signals will be 90° apart. By introducing appropriate phase shifts in the beam splitter using metal coatings, the phase difference between the two patterns may be made 90°, as desired (Peck and Obetz, 1953). This method was used by Rowley (1966), as illustrated in Fig. 5. In order to separate the two patterns from the incident light beam, a large beam splitter and large retroreflectors are used. This configuration has the advantage that the laser beam is not reflected back. This is called a nonreacting interferometer configuration.

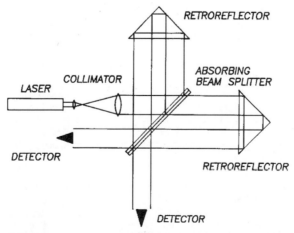

FIGURE 5   Two-interference-pattern, distance-measuring interferometer.

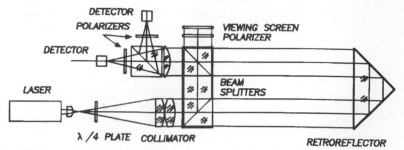

FIGURE 6   Minkowitz distance-measuring interferometer.

One more method, illustrated in Fig. 6, is the nonreacting interferometer designed at the Perkin Elmer Corporation by Minkowitz et al. (1967). A circularly polarized light beam, produced by a linearly polarized laser and a $\lambda/4$ phase plate, illuminates the interferometer. This beam is divided by a beam splitter into two beams going to both arms of the interferometer. Upon reflection by the retroreflector, one of the beams changes its state of polarization from right to left circularly polarized. The two beams with opposite circular polarization are recombined at the beam splitter, thus producing linearly polarized light. The angle of the plane of polarization is determined by the phase difference between the two beams. The plane of polarization rotates 360° if the optical path difference is changed by $\lambda/2$, the direction of rotation being given by the direction of the displacement. This linearly polarized beam is divided into two beams by a beam splitter. On each of the exiting beams, a linear polarizer is placed, one at an angle of +45° and the other at an angle of −45°. Then the two beams are in quadrature to each other.

In still another method, shown in Fig. 7, a beam of light, linearly polarized at 45° (or circularly polarized), is divided at a beam splitter, the p and s components. Then, both beams are converted to circular polarization with a $\lambda/4$ phase plate in front of each of them, with their axis at 45°. Upon reflection on the retroreflectors, the handedness of the polarization is reversed. Thus, the linearly polarized beams exiting from the phase plates

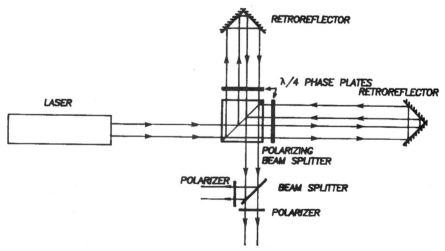

FIGURE 7   Bruning distance-measuring interferometer.

on the return to the beam splitter will have a plane of polarization orthogonal to that of the incoming beams. It is easily seen that no light returns to the laser. Here, the nonreacting configuration is not necessary but it may be used for additional protection. After recombination on the beam splitter, two orthogonal polarizations are present. Each plane of polarization contains information about the phase from one arm only so that no interference between the two beams has occurred. The two desired signals in quadrature are then generated by producing two identical beams with a nonpolarizing beam splitter with a polarizer on each exiting beam with their axes at $+45°$ and $-45°$ with respect to the vertical plane. The desired phase difference between the two beams is obtained by introducing a $\lambda/4$ phase plate after the beam splitter but before one of the polarizers with its slow axis vertical or horizontal. These two polarizers may be slightly rotated to make the two irradiances equal, at the same time preserving the $90°$ angle between their axes. If the prism is shifted a distance x, the fringe count is

$$\Delta_{\text{count}} = \pm \left[\frac{2x}{\lambda} \right] \tag{8}$$

where $[\,]$ denotes the integer part of the argument.

These interferometers are problematic in that any change in the irradiance may be easily interpreted as a fringe monitoring the light source. A more serious problem is the requirement that the static interference pattern be free of, or with very few, fringes. Fringes may appear because of multiple reflections or turbulence.

A completely different method uses two frequencies. It was developed by the Hewlett Packard Co. (Burgwald and Kruger, 1970; Dukes and Gordon, 1970) and is illustrated in Fig. 8. The light source is a frequency-stabilized He-Ne laser whose light beam is Zeeman split into two frequencies f_1 and f_2 by application of an axial magnetic field. The frequency difference is several megahertz and both beams have circular polarization, but with

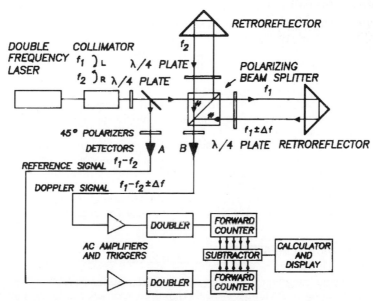

FIGURE 8 Hewlett-Packard double-frequency distance-measuring interferometer.

opposite sense. A $\lambda/4$ phase plate transforms the signals f_1 and f_2 into two orthogonal linearly polarized beams, one in the horizontal and the other in the vertical plane. A sample of this mixed signal is deviated by a beam splitter and detected at photodetector A, by using a polarizer at 45°. The amplitude modulation of this signal, with frequency $f_1 - f_2$, is detected and passed to a counter. Then, the two orthogonally polarized beams with frequencies f_1 and f_2 are separated at a polarizing beam splitter. Each is transformed into a circularly polarized beam by means of $\lambda/4$ phase plates. After reflection by the prisms, the handedness of these polarizations is changed. Then they go through the same phase plates where they are converted again to orthogonal linearly polarized beams. There is no light reflecting back to the laser. After recombination at the beam splitter, a polaroid at 45° will take the components of both beams in this plane. This signal is detected at the photodetector B. As with the other signal, the modulation with frequency $f_1 - f_2 + \Delta f$ is extracted from the carrier and sent to another counter. The shift Δf in the frequency of this signal comes from a Doppler shift due to the movement of one of the retroreflectors,

$$\Delta f = \frac{2}{\lambda_2} \frac{dx}{dt} \tag{9}$$

where dx/dt is the cube corner prism velocity and λ_1 is the wavelength corresponding to the frequency f_1. The difference between the results of the two counters is produced by the displacement of the retroreflector. If the prism moves a distance x, the number of pulses detected is given by

$$\Delta_{\text{count}} = \pm \left[\frac{2x}{\lambda_2} \right] \tag{10}$$

The advantage of this method compared with the first is that fringe counting is not subject to drift. These signals may be processed to obtain a better signal-to-noise ratio and higher resolution (Dukes and Gordon, 1970).

Straightness Measurements

Light propagation is assumed to be rectilinear in a homogeneous medium. This permits the use of a propagating light beam as a straightness reference. Besides the homogeneous medium, it is necessary to get a truly narrow pencil of light to improve accuracy. Laser light is an obvious application because of the high degree of spatial coherence. Beam divergence is usually less than 1 mrad for a He-Ne laser. One method uses a position-sensing detector to measure the centroid of the light spot despite its shape. In front of the laser, McLeod (1954) used an axicon as an aligning device. When a collimated light beam is incident on an axicon, it produces a bright spot on a circular field. An axicon can give as much as 0.01 arcsec.

Another method to measure the deviation from an ideal reference line uses an autocollimator. A light beam leaving an autocollimator is reflected by a mirror. The surface slope is measured at the mirror. By knowing the distance to the mirror, one can determine the surface's profile by integration, as in a curvature measurement (see "Optical Methods for Measuring Curvature," later in this chapter). A method can be designed for measuring flatness for tables on lathe beds (Young, 1967).

29.4 ANGLE MEASUREMENTS

Angle measurements, as well as distance measurements, require different levels of accuracy. For cutting glass, the required accuracy can be as high as several degrees, while for test plates, an error of less than a second of arc may be required. For each case, different measurement methods are developed.

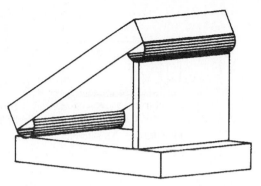

FIGURE 9   Sine plate.

Mechanical Methods

The easiest way to measure angles with medium accuracy is by means of mechanical nonoptical methods. These are:

Sine plate: Essentially it is a table with one end higher than the other by a fixed amount, as shown in Fig. 9. An accuracy close to 30 arcmin may be obtained.

Goniometer: This is a precision spectrometer. It has a fixed collimator and a moving telescope pointing to the center of a divided circle. Accuracies close to 20 arcsec may be obtained.

Bevel gauge: Another nonoptical method is by the use of a bevel gauge. This is two straight bars hinged at their edges by a pivot, as shown in Fig. 10. This device may be used to measure angle prisms whose angle accuracies are from 45 to about 20 arcsec (Deve, 1945). For example, if the measured prism has a 50-mm hypotenuse, a space of 5 μm at one end represents an angle of 0.0001 radians or 20 arcsec.

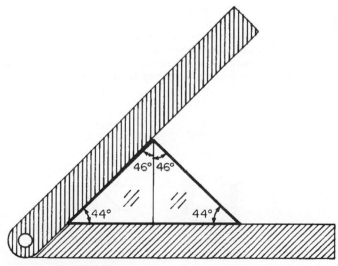

FIGURE 10   Bevel gauge.

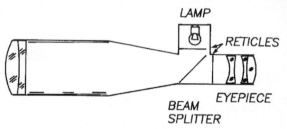

FIGURE 11   An autocollimator.

Autocollimators

As shown in Fig. 11, an autocollimator is essentially a telescope focused at infinity with an illuminated reticle located at the focal plane. A complete description of autocollimators is found in Hume (1965). A flat reflecting surface, perpendicular to the exiting light beam, forms an image of the reticle on the same plane as the original reticle. Then, both the reticle and its image are observed through the eyepiece. When the reflecting surface is not exactly perpendicular to the exiting light beam, the reticle image is laterally displaced in the focal plane with respect to the reticle. The magnitude of this of displacement d is:

$$d = 2\alpha f \tag{11}$$

where α is the tilt angle for the mirror in radians and f is the focal length of the telescope.

Autocollimator objective lenses are usually corrected doublets. Sometimes a negative lens is included to form a telephoto lens to increase the effective focal length while maintaining compactness. The collimating lens adjustment is critical for the final accuracy. Talbot interferometry can be used for a precise focus adjustment (Konthiyal and Sirohi, 1987).

The focal plane is observed through an eyepiece. Several types of illuminated reticles and eyepieces have been developed. Figure 12 (Noble, 1978) illustrates some illuminated eyepieces, in all of which the reticle is calibrated to measure the displacement. Gauss and Abbe illuminators show a dark reticle on a bright field. A bright field (Carnell and Welford, 1971) may be more appropriate for low reflectance surfaces. Rank (1946) modified a Gauss eyepiece to produce a dark field. In other systems, a drum micrometer displaces a reticle to position it at the image plane of the first reticle. To increase sensitivity some systems, called microoptic autocollimators, use a microscope to observe the image.

Direct-reading autocollimators have a field of view of about one degree. Precision in an autocollimator is limited by the method for measuring the centroid of the image. In a

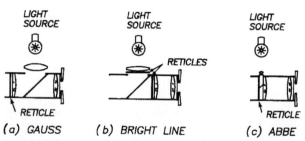

FIGURE 12   Illuminated eyepieces for autocollimators and traveling microscopes.

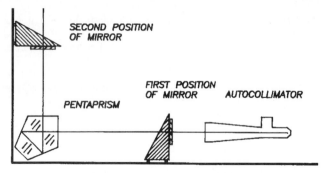

FIGURE 13   Perpendicularity measurement with an autocollimator.

diffraction-limited visual system, the diffraction image size sets the limit of precision. In a precision electronic measuring system, the accuracy of the centroid measurement is limited by the electronic detector, independent of the diffraction image itself, and can exceed the diffraction limit. In some photoelectric systems, the precision is improved by more than an order of magnitude.

Autocollimators are used for angle measurements in prisms and glass polygons. But they also have other applications; for example, to evaluate the parallelism between faces in optical flats, or to manufacture divided circles (Horne, 1974). By integrating measured slope values with an autocollimator, flatness deviations for a machine tool for an optical bed can also be evaluated (Young, 1967).

The reflecting surface in autocollimation measurements must be kept close to the objective in order to make the alignment easier and to be sure that all of the reflected beam enters the system. The reflecting surface must be of high quality. A curved surface is equivalent to introducing another lens in the system with a change in the effective focal length (Young, 1967).

Several accessories for autocollimators have been designed. For single-axis angle measurement, a pentaprism is used. An optical square permits angle measurements for surfaces at right angles. Perpendicularity is measured with a pentaprism and a mirror as shown in Fig. 13. A handy horizontal reference can be produced with an oil pool, but great care must be taken with the surface stability.

Theodolites

Theodolites are surveying instruments to measure vertical and horizontal angles. A telescope with reticle is the reference to direct the instrument axis. In some theodolites, the telescope has an inverting optical system to produce an erect image. The reticle is composed of a crosswire and has a couple of parallel horizontal lines, called stadia lines, used for range measurements (see "Stadia and Range Finders" earlier in this chapter). The telescope has two focus adjustments: one to sharply focus the reticle according to the personal setting for each observer, and the other to focus the objective on the reticle. This later focus adjustment is performed by moving the whole eyepiece.

The theodolite telescope has a tree-screw base altitude-azimuth mounting on a base made horizontal with a spirit level. Divided circles attached to both telescope axes measure vertical and horizontal angles. In older instruments, an accuracy of 20 arcmin was standard. Modern instruments are accurate to within 20 arcsec, the most expensive of which can reach an accuracy of one arcsec. To remove errors derived from eccentric scales as well as orthogonality errors, both axes are rotated 180° and the measurement repeated. Older instruments had a provision for reading at opposite points of the scale. Scales for theodolites can be graduated in sexagesimal degrees or may use a centesimal system that divides a circle into 400 grades.

Some of the accessories available for theodolites include:

1. An *illuminated reticle* that can be used as an autocollimator when directed to a remote retroreflector. The observer adjusts the angles until both the reflected and the instrument's reticle are superposed. This increases the pointing accuracy.

2. A *solar filter,* which can be attached to the eyepiece or objective side of the telescope. This is used mainly for geographic determination in surveying.

3. An *electronic range meter,* which is superposed to the theodolite to measure the distance. Additionally, some instruments have electronic position encoders that allow a computer to be used as an immediate data-gathering and reducing device.

4. A *transverse micrometer* for measuring angular separation in the image plane.

Accuracy in a theodolite depends on several factors in its construction. Several of these errors can be removed by a careful measuring routine. Some of the systematic or accuracy limiting errors are:

1. *Perpendicularity-deviation* between vertical and horizontal scales. This error can be nulled by plunging and rotating the telescope, then averaging.

2. *Concentricity deviation* of scales. When scales are not concentric, they are read at opposite ends to reduce this error. Further accuracy can be obtained by rotating the instrument 90, 180, and 270° and averaging measurements.

Level

Levels are surveying instruments for measuring the deviation from a horizontal reference line. A level is a telescope with an attached spirit level. The angle between the telescope axis and the one defined by the spirit level must be minimized. It can be adjusted by a series of measurements to a pair of surveying staffs (Fig. 14). Once the bubble is centered in the tube, two measurements are taken on each side of the staffs (Kingslake, 1983). The level differences between the two staffs must be equal if the telescope axis is parallel with the level horizontal axis.

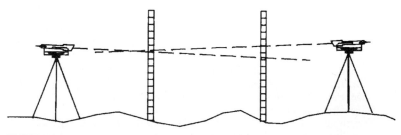

FIGURE 14   Level adjustment. (*After Kingslake, 1983.*)

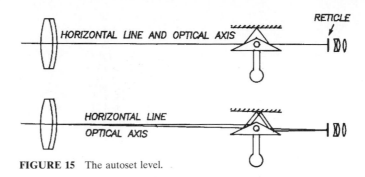

FIGURE 15   The autoset level.

The autoset level (Fig. 15) uses a suspended prism and a fixed mirror on the telescope tube. The moving prism maintains the line aimed at the horizon and passing through the center of the reticle, despite the tube orientation, as long it is within about 15 arcmin. Typical precision for this automatic level can go up to 1 arcsec (Young, 1967).

Interferometric Measurements

Interferometric methods find their main applications in measuring very small wedge angles in glass slabs (Met, 1966; Leppelmeier and Mullenhoff, 1970) and in parallelism evaluation (Wasilik et al., 1971) by means of the Fizeau or Haidinger interferometers (Malacara, 1992).

Interferometric measurements of large angles may also be performed. In one method, a collimated laser beam is reflected from the surfaces by a rotating glass slab. The resulting fringes can be considered as coming from a Murty lateral shear interferometer (Malacara and Harris, 1970). This device can be used as a secondary standard to produce angles from 0 to 360° with an accuracy within a second of arc. Further analysis of this method has been done by Tentori and Celaya (1986). In another system, a Michelson interferometer is used with an electronic counter to measure over a range of $\pm 5°$ with a resolution of 10° (Stijns, 1986, Shi and Stijns, 1988). An interferometric optical sine bar for angles in the milliseconds of arc was built by Chapman (1974).

Angle Measurements in Prisms

A problem frequently encountered in the manufacture of prisms is the precise measurement of angle. In most cases, prism angles are 90°, 45°, and 30°. These angles are easily measured by comparison with a standard but it is not always necessary.

An important aspect of measuring angles in a prism is to determine if the prism is free of pyramidal error. Consider a prism with angles A, B, and C (Fig. 16a). Let OA be perpendicular to plane ABC. If line AP is perpendicular to segment BC, then the angle AOP is a measurement of the pyramidal error. In a prism with pyramidal error, the angles between the faces, as measured in planes perpendicular to the edges between these faces, add up to over 180°. To simply detect pyramidal error in a prism, Johnson (1947) and Martin (1924) suggest that both the refracted and the reflected images from a straight line be examined (Fig. 16b). When pyramidal error is present, the line appears to be broken. A remote target could be graduated to measure directly in minutes of arc. A sensitivity of up to 3 arc min may be obtained.

During the milling process in the production of a prism, a glass blank is mounted in a jig collinear with a master prism (Fig. 17). An autocollimator aimed at the master prism

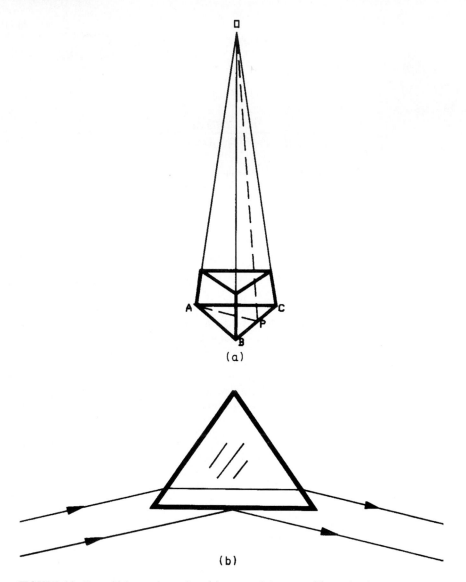

FIGURE 16   Pyramidal error in a prism: (*a*) nature of the error; (*b*) test for the error.

accurately sets the position for each prism face (Twyman, 1957; DeVany, 1971). With a carefully set diamond lap, pyramidal error is minimized. In a short run, angles can be checked with a bevel gauge. Visual tests for a prism in a bevel gauge can measure an error smaller than a minute of arc (Noble, 1978).

A 90° angle in a prism can be measured by internal reflection as shown in Fig. 18*a*. At the autocollimator image plane, two images are seen with an angular separation of $2N\alpha$, where α is the magnitude of the prism angle error, and its sign is unknown. Since the hypotenuse face has to be polished and the glass must be homogeneous, the measurement of the external angle with respect to a reference flat is preferred (Fig. 18*b*). In this case, the sign of the angle error is determined by a change in the angle by tilting the prism. If the

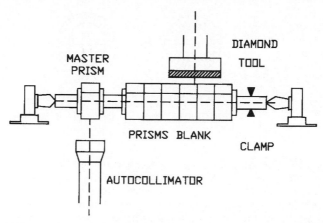

FIGURE 17 Milling prisms by replication.

external angle is decreased and the images separate further, then the external angle is less than 90°. Conversely, if the images separate by tilting in such way that the external angle increases, then the external angle is larger than 90°.

To determine the sign of the error, several other methods have been proposed. DeVany (1978) suggested that when looking at the double image from the autocollimator, the image should be defocused inward. If the images tend to separate, then the angle in the prism is greater than 90°. Conversely, an outward defocusing will move the images closer to each other for an angle greater than 90°. Another way to eliminate the sign of the error in the angle is by introducing, between the autocollimator and the prism, a glass plate with a small wedge whose orientation is known. The wedge should cover only one-half of the prism aperture. Ratajczyk and Bodner (1966) suggested a different method using polarized light.

Right-angle prisms can be measured using an autocollimator with acceptable precision (Taarev, 1985). With some practice, perfect cubes with angles more accurate than two arcsec can be obtained (DeVany, 1978). An extremely simple test for the 90° angle in

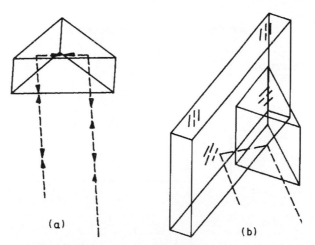

FIGURE 18 Right-angle measurement in prisms: (*a*) internal measurements, (*b*) external measurements.

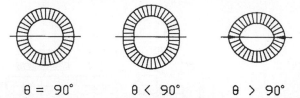

$\theta = 90°$ $\theta < 90°$ $\theta > 90°$

FIGURE 19 Retroreflected images of the observer's pupil on a 90° prism.

prisms (Johnson, 1947) is performed by looking to the retroreflected image of the observer's pupil without any instrument. The shape of the image of the pupil determines the error, as shown in Fig. 19. The sensitivity of this test is not very great and may be used only as a coarse qualitative test. As shown by Malacara and Flores (1990), a small improvement in the sensitivity of this test may be obtained if a screen with a small hole is placed in front of the eye, as in Fig. 20a. A cross centered on the small hole is painted on the front face of the screen. The observed images are as shown in the same Fig. 20b. As opposed to the collimator test, there is no uncertainty in the sign of the error in the tests just described, since the observed plane is located where the two prism surfaces intersect. An improvement described by Malacara and Flores (1990), combining these simple tests with an autocollimator, is obtained with the instrument in Fig. 21. In this system, the line defining the intersection between the two surfaces is out of focus and barely visible while the reticle is in perfect focus at the eyepiece.

29.5 CURVATURE AND FOCAL LENGTH MEASUREMENTS

The curvature of a spherical optical surface or the local curvature of an aspherical surface may be measured by means of mechanical or optical methods. Some methods measure the sagitta, some the surface slope, and some others directly locate the position of the center of curvature.

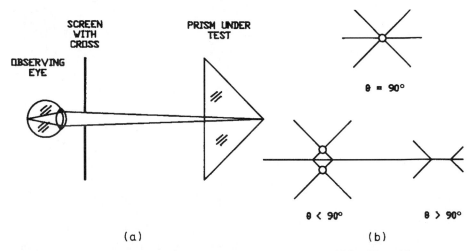

FIGURE 20 Testing a right-angle prism: (*a*) screen in front of the eye; (*b*) its observed images.

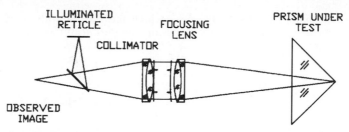

FIGURE 21   Modified autocollimator for testing the right angle in prisms without sign-uncertainty in measured error.

Mechanical Methods for Measuring Curvature

Templates.   The simplest and most common way to measure the radius of curvature is by comparing it with metal templates with known radii of curvature until a good fit is obtained. The template is held in contact with the optical surface with a bright light source behind the template and the optical surface. If the surface is polished, gaps between the template and the surface may be detected to an accuracy of one wavelength. If the opening is very narrow, the light passing through the gap becomes blue due to diffraction.

Test Plates.   This method uses a glass test plate with a curvature opposite to that of the glass surface to be measured. The accuracy is much higher than in the template method, but the surface must be polished.

Spherometers.   This is probably the most popular mechanical device used to measure radii of curvature. It consists of three equally spaced feet with a central moving plunger. The value of the radius of curvature is calculated after measuring the sagitta, as shown in Fig. 22. The spherometer must first be calibrated by placing it on top of a flat surface. Then it is placed on the surface to be measured. The difference in the position of the central plunger for these two measurements is the sagitta of the spherical surface being measured. Frequently, a steel ball is placed at the end of the legs as well as at the end of the plunger

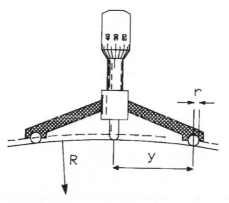

FIGURE 22   Three-leg spherometer.

to avoid the possibility of scratching the surface with sharp points. In this case, if the measured sagitta is z, the radius of curvature R of the surface is given by

$$R = \frac{z}{2} + \frac{y^2}{2z} \pm r \tag{12}$$

where r is the radius of curvature of the balls. The plus sign is used for concave surfaces and the minus sign for convex surfaces. The precision of this instrument in the measurement of the radius of curvature for a given uncertainty in the measured sagitta may be obtained by differentiating Eq. (12):

$$\frac{dR}{dz} = \frac{1}{2} - \frac{y^2}{2z^2} \tag{13}$$

obtaining:

$$\Delta R = \frac{\Delta z}{2} \left(1 - \frac{y^2}{z^2} \right) \tag{14}$$

This precision assumes that the spherometer is perfectly built and that its dimensional parameters y and r are well known. The uncertainty comes only from human or instrumental errors in the measurement of the sagitta. Noble (1978) has evaluated the precision for a spherometer with $y = 50$ mm and an uncertainty in the sagitta reading equal to 5 μm, and has reported the results in Table 2 where it can be seen that the precision is better than 2 percent. An extensive analysis of the precision and accuracy of several types of spherometers is given by Jurek (1977).

A ring may be used instead of the three legs in a mechanical spherometer. A concave surface contacts the external edge of the cup, and a convex surface is contacted by the internal edge of the ring. Thus, Eq. (5) may be used if a different value of y is used for concave and convex surfaces, and r is taken as zero. Frequently in spherometers of this type, the cups are interchangeable in order to have different diameters for different surface diameters and radii of curvature. The main advantage of the use of a ring instead of three legs is that an astigmatic deformation of the surface is easily detected, although it cannot be measured.

A spherometer that permits the evaluation of astigmatism is the *bar spherometer*, shown in Fig. 23. It can measure the curvature along any diameter. A commercial version of a small bar spherometer for the specific application in optometric work is the Geneva gauge, where the scale is directly calibrated in diopters assuming that the refractive index of the glass is 1.53.

Automatic spherometers use a differential transformer as a transducer to measure the

TABLE 2 Spherometer Precision

Radius of sphere R (mm)	Sagitta z (mm)	Fractional precision ΔR (mm)	Precision R/R
10,000	0.125 –	400 –	0.040
5,000	0.250 –	100 –	0.020
2,000	0.625 –	16 –	0.008
1,000	1.251 –	4 –	0.004
500	2.506 –	1 –	0.002

$y = 50$ mm; $\Delta z = 5$ μm (*from Noble, 1978*).

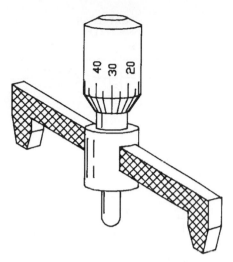

FIGURE 23   Bar spherometer.

plunger displacement. This transformer is coupled to an electronic circuit and produces a voltage that is linear with respect to the plunger displacement. This voltage is fed to a microprocessor which calculates the radius of curvature or power in any desired units and displays it.

Optical Methods for Measuring Curvature

Foucault Test.   Probably the oldest and easiest method to measure the radius of curvature of a concave surface is the knife-edge test. In this method, the center of curvature is first located by means of the knife edge. Then, the distance from the center of curvature to the optical surface is measured.

Autocollimator.   The radius of curvature may also be determined through measurements of the slopes of the optical surface with an autocollimator as described by Horne (1972). A pentaprism producing a 90° deflection of a light beam independent of small errors in its orientation is used in this technique, as illustrated in Fig. 24, where the pentaprism travels over the optical surface to be measured along one diameter. First the light on the reticle of the autocollimator is centered on the vertex of the surface being examined. Then the

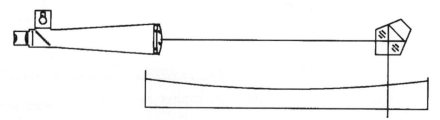

FIGURE 24   Autocollimator and pentaprism used to determine radius of curvature by measuring surface slopes.

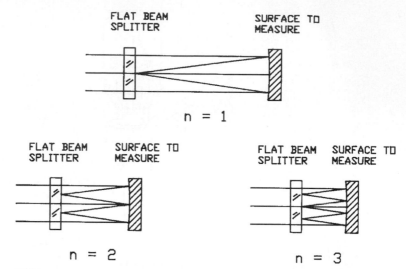

FIGURE 25 Confocal cavity arrangements used to measure radius of curvature.

pentaprism is moved towards the edge of the surface in order to measure any slope variations. From these slope measurements, the radius of curvature may be calculated. This method is useful only for large radii of curvature for either concave or convex surfaces.

Confocal Cavity Technique. Gerchman and Hunter (1979 and 1980) have described the so-called optical cavity technique that permits the interferometric measurement of very long radii of curvature with an accuracy of 0.1 percent. The cavity of a Fizeau interferometer is formed as illustrated in Fig. 25. This is a confocal cavity of nth order, where n is the number of times the path is folded. The radius of curvature is equal to approximately $2n$ times the cavity length Z.

Traveling Microscope. This instrument is used to measure the radius of curvature of small concave optical surfaces with short radius of curvature. As illustrated in Fig. 26, a

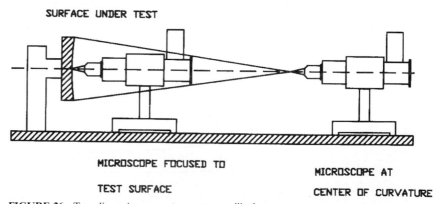

FIGURE 26 Traveling microscope to measure radii of curvature.

point light source is produced at the front focus of a microscope objective. This light source illuminates the concave optical surface to be measured near its center of curvature. Then this concave surface forms an image which is also close to its center of curvature. This image is observed with the same microscope used to illuminate the surface. During this procedure, the microscope is focused both at the center of curvature and at the surface to be measured. A sharp image of the light source is observed at both places. The radius of curvature is the distance between these two positions for the microscope.

This distance traveled by the microscope may be measured on a vernier scale, obtaining a precision of about +0.1 mm. If a bar micrometer is used, the precision may be increased by an order of magnitude. In this case, two small convex buttons are required, one fixed to the microscope carriage and the other to the stationary part of the bench. They must face each other when the microscope carriage is close to the optical bench fixed component.

Carnell and Welford (1971) describe a method that requires only one measurement. The microscope is focused only at the center of curvature. Then the radius of curvature is measured by inserting a bar micrometer with one end touching the vertex of the optical surface. The other end is adjusted until it is observed to be in focus on the microscope. Accuracies of a few microns are obtained with this method.

In order to focus the microscope properly, the image of an illuminated reticle must fall, after reflection, back on itself, as in the Gauss eyepiece shown in Fig. 12. The reticle and its image appear as dark lines in a bright field. The focusing accuracy may be increased with a dark field. Carnell and Welford obtained a dark field with two reticles, as in Fig. 12, one illuminated with bright lines and the other with dark lines.

A convex surface may also be measured with this method if a well-corrected lens with a conjugate longer than the radius of curvature of the surface under test is used. Another alternative for measuring convex surfaces is by inserting an optical device with prisms in front of the microscope, as described by Jurek (1977).

Some practical aspects of the traveling microscope are examined by Rank (1946), who obtained a dark field at focus with an Abbe eyepiece which introduces the illumination with a small prism. This method has been implemented using a laser light source by O'Shea and Tilstra (19881).

Additional optical methods to measure the radius of curvature of a spherical surface have been described. Evans (1971, 1972a, and 1972b) determines the radius by measuring the lateral displacements on a screen of a laser beam reflected on the optical surface when this optical surface is laterally displaced. Cornejo-Rodriguez and Cordero-Dávila (1980), Klingsporn (1979), and Diaz-Uribe et al. (1986) rotate the surface about its center of curvature on a nodal bench.

Focal Length Measurements

There are two focal lengths in an optical system: the *back focal length* and the *effective focal length*. The back focal length is the distance from the last surface of the system to the focus. The effective focal length is the distance from the principal plane to the focus. The back focal length is easily measured, following the same procedure used for measuring the radius of curvature, using a microscope and the lens bench. On the other hand, the effective focal length requires the previous location of the principal plane.

Nodal Slide Bench.   In an optical system in air, the principal points (intersection of the principal plane and the optical axis) coincide with the nodal points. Thus, to locate this point we may use the well-known property that small rotations of the lens about an axis perpendicular to the optical axis and passing through the nodal point do not produce any

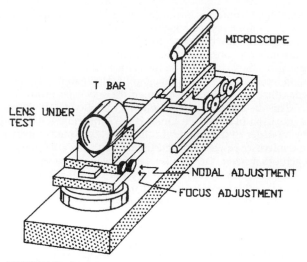

FIGURE 27 Nodal slide bench.

lateral shift of the image. The instrument used to perform this procedure, shown in Fig. 27, is called an optical nodal slide bench (Kingslake, 1932). This bench has a provision for slowly moving the lens under test longitudinally in order to find the nodal point.

The bench is illuminated with a collimated light source and the image produced by the lens under test is examined with a microscope. The lens is then displaced slightly about a vertical axis as it is being displaced longitudinally. This procedure is stopped until a point is found in which the image does not move laterally while rotating the lens. This axis of rotation is the nodal point. Then, the distance from the nodal point to the image is the effective focal length.

Focimeters. A focimeter is an instrument designed to measure the focal length of lenses in a simple manner. The optical scheme for the classical version of this instrument is shown in Fig. 28. A light source illuminates a reticle and a convergent lens, with focal length *f*, is

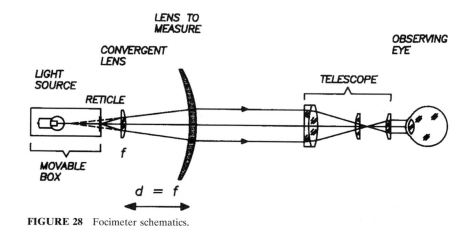

FIGURE 28 Focimeter schematics.

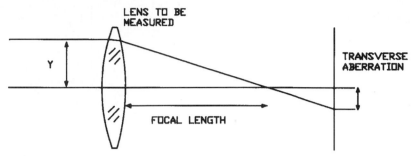

FIGURE 29 Focal length determination by transverse aberration measurements.

placed at a distance x from the reticle. The lens to be measured is placed at a distance d from the convergent lens. The magnitude of x is variable and is adjusted until the light beam going out from the lens under test becomes collimated. This collimation is verified by means of a small telescope in front of this lens focused at infinity. The values of d and the focal length f are set to be equal. Then, the back focal length f_B of the lens under test is given by

$$\frac{1}{f_B} = \frac{1}{d} - \frac{x}{d^2} \tag{15}$$

As can be seen, the power of the lens being measured is linear with respect to the distance x. There are many variations of this instrument. Some modern focimeters measure the lateral deviation of a light ray from the optical axis (transverse aberration), as in Fig. 29, when a defocus is introduced (Evans 1971, 1972a, and 1972b; Bouchard and Cogno, 1982). This method is mainly used in some modern automatic focimeters for optometric applications. To measure the transverse aberration, a position-sensing detector is frequently used.

Other Focal Length Measurements. A clever method used to automatically find the position of the focus has been described by Howland and Proll (1970). They used optical fibers to illuminate the lens in an autocollimating configuration, and the location of the image was also determined using optical fibers.

29.6 VELOCITY MEASUREMENTS

Optical measurement of particles can be done by differentiating their position with respect to time. Any instrument for position measurement can be used also for velocity measurement. But two other optical methods are available for velocimetry: particle image velocimetry (PIV) and laser Doppler velocimetry. The former method is useful for tangential displacements, while the latter is applicable for radial displacements.

In particle image velocimetry, the tangential component of velocity for small particles is found by using a double-exposure photograph for the moving particles. Later, the plate is analyzed using laser speckle interferometry (Dudderar and Simpkins, 1977; Coupland et al., 1987).

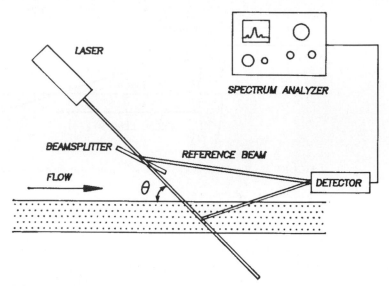

FIGURE 30 Reference beam laser Doppler velocimeter.

Laser Doppler Velocimetry

Laser Doppler velocimetry measures the frequency shift from the scattered or reflected light. This method is applicable to a reflecting surface or an ensemble of scattering particles. Typical frequency shifts are too small to be measured with spectrometric methods, so heterodyne interferometers and spectrum analysis are applied to the heterodyned signal. Figure 30 shows a basic setup for laser Doppler velocimetry with a reference beam. A laser beam is incident at an angle θ on the sample to be measured. A beam splitter takes a sample from the laser and mixes it with the scattered light. By heterodyning the reference beam with the reflected signal, the frequency shift is measured with a frequency-measuring device. The resulting frequency shift is:

$$\nu_d = \frac{2\nu}{\lambda} \sin\left(\frac{\theta}{2}\right) \tag{16}$$

A simpler setup can be done by the differential Doppler technique. On this method, a reference beam is no longer used, but two beams are incident on the sample with a different incidence angle. The reference method is common for solid objects, while for small particles, the differential Doppler method is used. Several differential laser Doppler configurations are discussed by Drain (1980).

In a third method, a rotating diffraction grating with radial slits is placed in front of the laser beam. This grating generates two beams, one (order of diffraction +1) with a shifted-up frequency and another (order of diffraction −1) with its frequency shifted down. These two beams permit sensing not only the magnitude but also the direction of the liquid flow. A stop after the first lens blocks out the strong nondiffracted beam. See Fig. 31.

The signal from the photodetector is processed in several ways. Among these we can mention:

1. *Spectrum analysis*: An electronic spectrum analyzer can be used for data reduction with equipment that is commercially available, although expensive. The maximum measurable frequency can go up to 1 GHz. Some systems, especially those for low

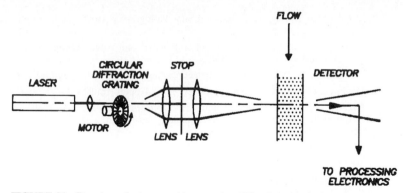

FIGURE 31 Doppler velocimeter with a rotating diffraction grating.

frequencies, are rather slow in response. Basically, it is a sweeping heterodyne radio receiver. The detected signal is fed to an oscilloscope. Some spectrum analyzers can be attached to a computer for digital processing.

2. *Phase-locked loop*: An electronic phase-locked loop, as those used in FM detection, can also be used. This method is useful in the presence of noise—its response is fast, but intermittent signals are not easily tracked. The main disadvantage as compared with the spectrum analyzer is that the phase-locked loop can only detect one signal at the time.

3. *Frequency meters*: An inexpensive means to measure a Doppler shift is to measure the frequency shift with a frequency meter. During a sampling interval, the incoming signal is compared with a precise reference signal. This system can be very fast and is appropriate for intermittent signals, but is sensitive to noise.

4. *Filter bank*: For a system to measure several frequencies and at the same time respond to signals with poor signal-to-noise ratios, a filter bank can be used. Filter banks can be cheap for low-resolution systems, but expensive for high-resolution ones. Their accuracy is low compared with other systems.

29.7 REFERENCES

Baird, D. C., *Experimentation,* Prentice-Hall, New Jersey, 1962.

Bloom, A. L., "Gas Lasers and Their Application to Precise Length Measurements," in E. Wolf (ed.), *Progress in Optics,* vol. IX, North Holland, Amsterdam, 1971.

Bouchaud, P. and J. A. Cogno, "Automatic Method for Measuring Simple Lens Power," *Appl. Opt.* **21:**3068 (1982).

Bruning, J., "Fringe Scanning," in D. Malacara (ed.), *Optical Shop Testing,* 1st ed, John Wiley and Sons, New York, 1978.

Burgwald, G. M. and W. P. Kruger, "An Instant-On Laser for Length Measurements," *Hewlett Packard Journal* **21:**2 (1970).

Carnell, K. H. and W. T. Welford, "A Method for Precision Spherometry of Concave Surfaces," *J. Phys.* **E4:**1060–1062 (1971).

Chapman, G. D., "Interferometric Angular Measurement," *Appl. Opt.* **13:**1646–1651 (1971).

Cornejo-Rodriguez, A. and A. Cordero-Dávila, "Measurement of Radii of Curvature of Convex and Concave Surfaces Using a Nodal Bench and a He–Ne Laser," *Appl. Opt.* **19:**1743–1745 (1980).

Coupland, J. M., C. J. D. Pickering, and N. A. Halliwell, "Particle Image Velocimetry: Theory of

Directional Ambiguity Removal Using Holographic Image Separation," *Appl. Opt.* **26:**1576–1578 (1987).

DeVany, A. S., "Reduplication of a Penta-Prism Angle Using Master Angle Prisms and Plano Interferometer," *Appl. Opt.* **10:**1371–1375 (1971).

DeVany, A. S., "Testing Glass Reflecting-Angles of Prisms," *Appl. Opt.* **17:**1661–1662 (1978).

Deve, C., *Optical Workshop Principles,* T. L. Tippell (transl.), Hilger and Watts, London, 1945.

Diaz-Uribe, R., J. Pedraza-Contreras, O. Cardona-Nuñez, A. Cordero-Dávila, and A. Cornejo Rodriguez, "Cylindrical Lenses: Testing and Radius of Curvature Measurement," *Appl. Opt.* **25:**1707–1709 (1986).

Dickson, M. S. and Harkness, D., "Surveying and Tracking Instruments," chap. 8 in R. Kingslake (ed.), *Applied Optics and Optical Engineering,* vol. V, Academic Press, New York, 1969.

Drain, L. E., "The Laser Doppler Technique," Wiley, New York, 1980.

Dudderar, T. D. and Simpkins, "Laser Speckle Photography in a Fluid Medium," *Nature* (London) **270:**45 (1977).

Dukes, J. N. and G. B. Gordon, "A Two-Hundred-Foot Yardstick with Graduations Every Microinch," *Hewlett Packard Journal* **21:**2 (1970).

Evans, J. D., "Equations for Determining the Focal Length of On-Axis Parabolic Mirrors by He-Ne Laser Reflection," *Appl. Opt.* **11:**712–714 (1972a).

Evans, J. D., "Error Analysis to: Method for Approximating the Radius of Curvature of Small Concave Spherical Mirrors Using a He-Ne Laser," *Appl. Opt.* **11:**945–946 (1972b).

Evans, J. D., "Method for Approximating the Radius of Curvature of Small Concave Spherical Mirrors Using a He-Ne Laser," *Appl. Opt.* **10:**995–996 (1971).

Gerchman, M. C. and G. C. Hunter, "Differential Technique for Accurately Measuring the Radius of Curvature of Long Radius Concave Optical Surfaces," *Opt. Eng.* **19:**843–848 (1980).

Gerchman, M. C. and G. C. Hunter, "Differential Technique for Accurately Measuring the Radius of Curvature of Long Radius Concave Optical Surfaces," *Proc. SPIE* **192:**75–84 (1979).

Giacomo, P., *Metrology and Fundamental Constants.* Proc. Int. School of Phys. "Enrico Fermi," course 68, North-Holland, Amsterdam, 1980.

Gibbins, J. C., *The Systematic Experiment,* Cambridge Univ. Press, Cambridge, 1986.

Goldman, D. T. "Proposed New Definition of the Meter," *J. Opt. Soc. Am.* **70:**1640–1641 (1980).

Horne, D. F., *Dividing, Ruling and Mask Making,* Adam Hilger, London, 1974, chap. VII.

Horne, D. F., *Optical Production Technology,* Adam Hilger, London, and Crane Russak, New York, 1972, chap. XI.

Howland, B. and A. F. Proll, "Apparatus for the Accurate Determination of Flange Focal Distance," *Appl. Opt.* **11:**1247–1251 (1970).

Hume, K. J., *Metrology with Autocollimators,* Hilger and Watts, London, 1965.

Johnson, B. K., *Optics and Optical Instruments,* Dover, New York, 1947, chaps. II and VIII.

Jurek, B., *Optical Surfaces,* Elsevier Scient. Pub. Co., New York, 1977.

Kingslake, R., "A New Bench for Testing Photographic Lenses," *J. Opt. Soc. Am.* **22:**207–222 (1932).

Kingslake, R., *Optical System Design,* Academic Press, New York, 1983, chap. 13.

Klingsporn, P. E., "Use of a Laser Interferometric Displacement-Measuring System for Noncontact Positioning of a Sphere on a Rotation Axis Through its Center and for Measuring the Spherical Contour," *Appl. Opt.* **18:**2881–2890 (1983).

Kothiyal, M. P. and R. S. Sirohi, "Improved Collimation Testing Using Talbot Interferometry," *Appl. Opt.* **26:**4056–4057 (1987).

Larocca, A. J., "Artificial Sources," Chap. 10 in *Handbook of Optics,* 2d ed., vol. I. McGraw-Hill, New York, 1994.

Leppelmier, G. W. and D. J. Mullenhoff, "A Technique to Measure the Wedge Angle of Optical Flats," *Appl. Opt.* **9:**509–510 (1970).

Malacara, D. (ed.), *Optical Shop Testing,* 2d ed., John Wiley and Sons, New York, 1992.

Malacara, D. and O. Harris, "Interferometric Measurement of Angles," *Appl. Opt.* **9:**1630–1633 (1970).

Malacara, D. and R. Flores, "A Simple Test for the 90 Degrees Angle in Prisms," *Proc. SPIE* **1332:**678 (1990).

Martin, L. C., *Optical Measuring Instruments,* Blackie and Sons Ltd., London, 1924.

Massie, N. A. and J. Caulfield, "Absolute Distance Interferometry," *Proc. SPIE* **816:**149–157 (1987).

McLeod, J. H., "The Axicon: A New Type of Optical Element," *J. Opt. Soc. Am.* **44:**592–597 (1954).

Met, V., "Determination of Small Wedge Angles Using a Gas Laser," *Appl. Opt.* **5:**1242–1244 (1966).

Minkowitz, S. and W. Reid Smith Vanir, *Proceed. 1st Congress on Laser Applications* (Paris) and *J. Quantum Electronics* **3:**237 (1967).

Noble, R. E., "Some Parameter Measurements," in D. Malacara (ed.), *Optical Shop Testing,* 1st ed., John Wiley and Sons, New York, 1978.

O'Shea, D. C. and S. A. Tilstra, "Non-Contact Measurements of Refractive Index and Surface Curvature," *Proc. SPIE* **966:**172–176 (1988).

Peck, E. R. and S. W. Obetz, *J. Opt. Soc. Am.* **43:**505 (1953).

Patrick, F. B., "Military Optical Instruments," Chap. 7, in R. Kingslake (ed.), *Applied Optics and Optical Engineering,* vol. V, Academic Press, New York, 1969.

Rank, D. H., "Measurement of the Radius of Curvature of Concave Spheres," *J. Opt. Soc. Am.* **36:**108–110 (1947).

Ratajczyk, F. and Z. Bodner, "An Autocollimation Measurement of the Right Angle Error with the Help of Polarized Light," *Appl. Opt.* **5:**755–758 (1966).

Rowley, W. R. C., *IEEE Trans on Instr. and Measur.* **15:**146 (1966).

Shi, P. and E. Stijns, "New Optical Method for Measuring Small Angle Rotations," *Appl. Opt.* **27:**4342–4344 (1988).

Smith, W. J., *Modern Optical Engineering,* 2d ed., McGraw-Hill, New York, 1990.

Sona, A., "Lasers in Metrology," in F. T. Arecchi and E. O. Schulz-Dubois (eds.), *Laser Handbook,* vol. 2, North Holland, Amsterdam, 1972.

Steinmetz, C., R. Burgoon, and J. Herris, "Accuracy Analysis and Improvements for the Hewlett-Packard Laser Interferometer System" *Proc. SPIE* **816:**79–94 (1987).

Stijns, E., "Measuring Small Rotation Rates with a Modified Michelson Interferometer," *Proc. SPIE* **661:**264–266 (1986).

Stitch, M. L., "Laser Rangefinding," in F. T. Arecchi and E. O. Schulz-Dubois (eds.), *Laser Handbook,* vol. 2, North Holland, Amsterdam, 1972.

Tareev, A. M., "Testing the Angles of High-Precision Prisms by Means of an Autocollimator and a Mirror Unit," *Sov. J. Opt. Technol.* **52:**50–52 (1985).

Tentori, D. and M. Celaya, "Continuous Angle Measurement with a Jamin Interferometer," *Appl. Opt.* **25:**215–220 (1986).

Tsuruta, T. and Y. Ichihara, "Accurate Measurement of Lens Thickness by Using White-Light Fringes," *Jap. J. Appl. Phys. Suppl.* **14-1:**369–372 (1975).

Twyman, F., *Prisms and Lens Making,* 2d. ed., Hilger and Watts, London, 1957.

Wasilik, J. H., T. V. Blomquist, and C. S. Willett, "Measurement of Parallelism of the Surfaces of a Transparent Sample Using Two-Beam Non-Localized Fringes Produced by a Laser," *Appl. Opt.* **10:**2107–2112 (1971).

Young, A. W., "Optical Workshop Instruments," in R. Kingslake (ed.), *Applied Optics and Optical Engineering,* vol. 4, chap. 7, Academic Press, New York, 1967.

Zalewsky, E., "Radiometry and Photometry," chap. 25 in *Handbook of Optics,* 2d ed., vol. II, McGraw-Hill, New York, 1992.

CHAPTER 30
OPTICAL TESTING

Daniel Malacara
Centro de Investigaciones en Optica, A.C.
León, Gto, Mexico

30.1 GLOSSARY

E	electric field strength
k	radian wave number
r	position
t	time
λ	wavelength
ϕ	phase
ω	radian frequency

30.2 INTRODUCTION

The requirements for high-quality optical surfaces are more demanding every day. Thus, they should be tested in an easier, faster, and more accurate manner. Optical surfaces usually have a flat or a spherical shape, but they also may be toroidal or generally aspheric. Frequently, an aspherical surface is a conic of revolution. An aspherical surface can only be made as good as it can be tested. Here, the field of optical testing will be reviewed. There are some references that the reader may consult for further details (Malacara, 1991).

30.3 CLASSICAL NONINTERFEROMETRIC TESTS

Some classical tests will never be obsolete, because they are cheap, simple, and provide almost instantly qualitative results about the shape of the optical surface or wavefront. These are the Foucault or knife-edge test, the Ronchi test, and the Hartmann test. They will be described next.

Foucault Test

The Foucault or knife-edge test was invented by Leon Foucault (1852) in France, to evaluate the quality of spherical surfaces. This test detects the presence of transverse aberrations by intercepting the reflected rays deviated from their ideal trajectory, as Fig. 1 shows. The observer is behind the knife, looking at the illuminated optical surface, with

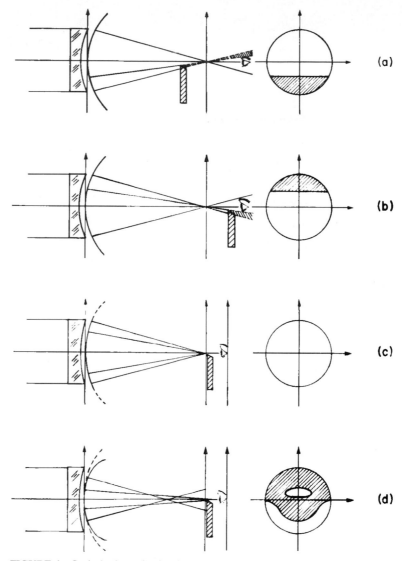

(a)

(b)

(c)

(d)

FIGURE 1 Optical schematics for the Foucault test of a spherical mirror, at several positions of the knife edge.

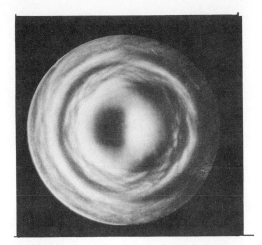

FIGURE 2 An optical surface being examined by the Foucault test. (*From Ojeda-Castañeda, 1978.*)

the reflected rays entering the eye. The regions corresponding to the intercepted rays will appear dark, as in Fig. 2.

This test is extremely sensitive. If the wavefront is nearly spherical, irregularities as small as a fraction of the wavelength of the light may be easily detected. This is the simplest and most powerful qualitative test for observing small irregularities and evaluating the general smoothness of the spherical surface under test. Any other surface or lens may be tested, as long as it produces an almost spherical wavefront, otherwise, an aberration compensator must be used, as will be described later. Very often a razor blade makes a good, straight, sharp edge that is large enough to cover the focal region.

Ronchi Test

Vasco Ronchi (1923) invented his famous test in Italy in 1923. A coarse ruling (50–100 lines per inch) is placed in the convergent light beam reflected from the surface under test, near its focus. The observer is behind the ruling, as Fig. 3 shows, with the light entering the eye. The dark bands in the ruling intercept light, forming shadows on the illuminated optical surface. These shadows will be straight and parallel only if the reflected wavefront is perfectly spherical. Otherwise, the fringes will be curves whose shape and separation depends on the wavefront deformations. The Ronchi test measures the transverse aberrations in the direction perpendicular to the slits on the grating. The wavefront deformations $W(x, y)$ are related to the transverse aberrations $TA_x(x, y)$ and $TA_y(x, y)$ by the following well-known relations:

$$TA_x(x, y) = -r\frac{\partial W(x, y)}{\partial x} \tag{1}$$

and

$$TA_y(x, y) = -r\frac{\partial W(x, y)}{\partial y} \tag{2}$$

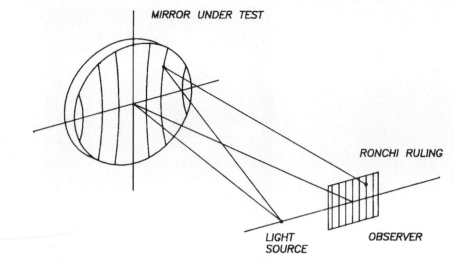

FIGURE 3   Testing a concave surface by means of the Ronchi test.

where r is the radius of curvature of the wavefront $W(x, y)$. Thus, if we assume a ruling with period d, the expression describing the mth fringe on the optical surface is given by

$$\frac{\partial W(x, y)}{\partial x} = -\frac{md}{r} \tag{3}$$

Each type of aberration wavefront has a characteristic Ronchi pattern, as shown in Fig. 4; thus, the aberrations in the optical system may be easily identified, and their magnitude estimated. We may interpret the Ronchi fringes not only as geometrical shadows, but also as interferometric fringes, identical with those produced by a lateral shear interferometer.

Hartmann Test

J. Hartmann (1900) invented his test in Germany. It is one of the most powerful methods to determine the figure of a concave spherical or aspherical mirror. Figure 5 shows the optical configuration used in this test, where a point light source illuminates the optical surface, with its Hartmann screen in front of it. The light beams reflected through each hole on the screen are intercepted on a photographic plate near the focus. Then, the position of the recorded spots is measured to find the value of the transverse aberration on each point. If the screen has a rectangular array of holes, the typical Hartmann plate image for a parabolic mirror looks like that in Fig. 6. The wavefront $W(x, y)$ may be obtained from integration of Eqs. (1) and (2) as follows:

$$W(x, y) = -\frac{1}{r} \int_0^x TA_x(x, y) \, dx \tag{4}$$

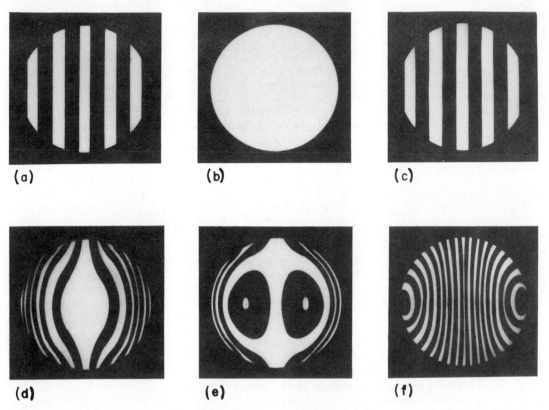

FIGURE 4 Typical Ronchi patterns for a spherical and a parabolic mirror for different positions of the Ronchi ruling.

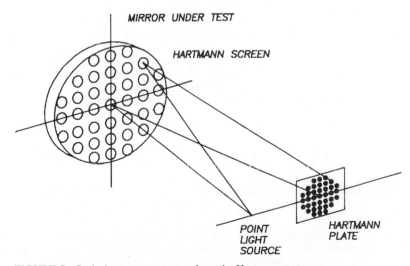

FIGURE 5 Optical arrangement to perform the Hartmann test.

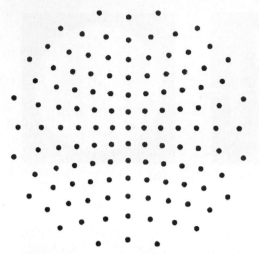

FIGURE 6 Array of spots in a Hartmann plate of a parabolic mirror.

and

$$W(x, y) = -\frac{1}{r} \int_0^y \mathrm{TA}_y(x, y) \, dy \tag{5}$$

After numerical integration of the values of the transverse aberrations, this test provides the concave surface shape with very high accuracy. If the surface is not spherical, the transverse aberrations to be integrated are the difference between the measured values and the ideal values for a perfect surface. Extended, localized errors, as well as asymmetric errors like astigmatism, are detected with this test. The two main problems of this test are that small, localized defects are not detected if they are not covered by the holes on the screen. Not only is this information lost, but the integration results will be false if the localized errors are large. The second important problem of the Hartmann test is that it is very time consuming, due to the time used in measuring all the data points on the Hartmann plate. These problems are avoided by complementing this test with the Foucault test, using an Offner compensator, in order to be sure about the smoothness of the surface (discussed under "Measuring Aspherical Wavefronts"). Various stratagems are available to speed the process. These include modulating the light at different frequencies at each of the holes. Variations also include measuring in front of, behind, or at the focus to get slope information. This technique can be considered an experimental ray trace.

30.4 INTERFEROMETRIC TESTS

Classical geometrical tests are very simple, but they do not provide the accuracy of the interferometric tests. Quite generally, an interferometric test produces an interferogram by producing the interference between two wavefronts. One of these two wavefronts is the

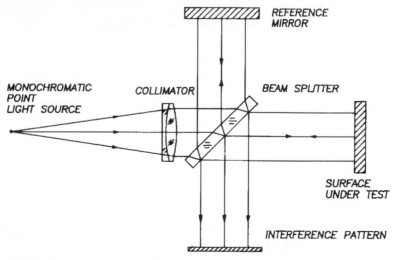

FIGURE 7 Twyman-Green interferometer.

wavefront under test. The other wavefront is either a perfectly spherical or flat wavefront, or a copy of the wavefront under test.

Interferometers with a Reference Wavefront

When the second wavefront is perfectly spherical or flat, this wavefront acts as a reference. The separation between the two wavefronts, or optical path difference $OPD(x, y)$, is a direct indication of the deformations $W(x, y)$ of the wavefront under test. Then, we may simply write $W(x, y) = OPD(x, y)$. There are many types of interferometers producing interferograms of these types, for example, the Twyman-Green, Newton, Fizeau, Point Diffraction, Burch interferometers, and many others that will not be described. Two of these interferometers are in Figs. 7 and 8. Figure 9 shows some typical interferograms made with these interferometers (Malacara, 1991).

Shearing Interferometers

When the second wavefront is not perfectly flat or spherical, but a copy of the wavefront under test, its relative dimensions or orientation must be changed (sheared) in some way with respect to the wavefront under test. Otherwise, no informaton about the wavefront deformations is obtained, because the fringes will always be straight and parallel independent of any aberrations. There are several kinds of shearing interferometers, depending on the kind of transformation applied to the reference wavefront.

The most popular of these instruments is the lateral shearing interferometer, with the reference wavefront laterally displaced with respect to the other, as Fig. 10 shows. The optical path difference $OPD(x, y)$ and the wavefront deformations $W(x, y)$ are related by

$$OPD(x, y) = W(x, y) - W(x - S, y) \qquad (6)$$

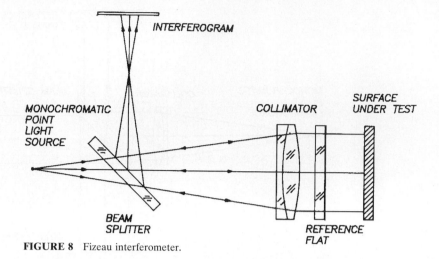

FIGURE 8   Fizeau interferometer.

where S is the lateral shear of one wavefront with respect to the other. If the shear is small with respect to the diameter of the wavefront, this expression may be approximated by

$$OPD(x, y) = -S\frac{\partial W(x, y)}{\partial x} = -\frac{S}{r}TA_x(x, y) \tag{7}$$

This relation suggests that the parameter being directly measured is the slope in the x direction of the wavefront (x component TA_x of the transverse aberration). An example of a lateral shear interferometer is the Murty interferometer, illustrated in Fig. 11.

There are also radial, rotational, and reversal shearing interferometers, where the interfering wavefronts are as Fig. 12 shows. A radial shear interferometer with a large shear approaches an interferometer with a perfect reference wavefront.

30.5   INCREASING THE SENSITIVITY OF INTERFEROMETERS

The sensitivity of interferometers is a small fraction of the wavelength being used (about $\lambda/20$). There are several methods to increase this sensitivity, but the most common methods will now be described.

Multiple-reflection Interferometers

A method to increase the sensitivity of interferometric methods is to use multiple reflections, as in the Fabry-Perot interferometer. The Newton as well as the Fizeau interferometers can be made multiple-reflection interferometers by coating the reference surface and the surface under test with a high-reflection film. Then, the fringes are

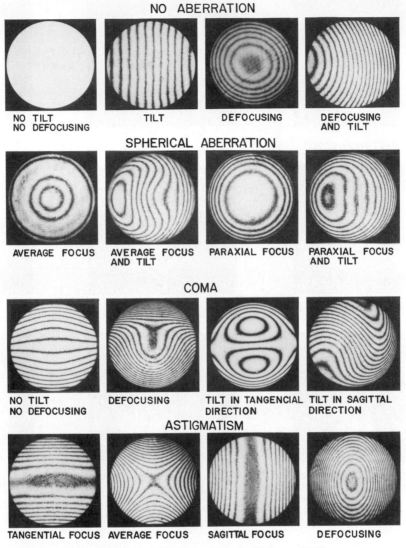

FIGURE 9 Twyman-Green interferograms. (*From D. Malacara, 1978.*)

greatly narrowed and their deviations from straightness are more accurately measured (Roychoudhuri, 1991).

Multiple-pass Interferometers

Another method to increase the sensitivity of interferometers is by double, or even multiple, pass. An additional advantage of double-pass interferometry is that the symmetrical and antisymmetrical parts of the wavefront aberration may be separated. This makes their identification easier, as Hariharan and Sen (1961) have proved. Several arrangements have been devised to use multiple pass (Hariharan, 1991).

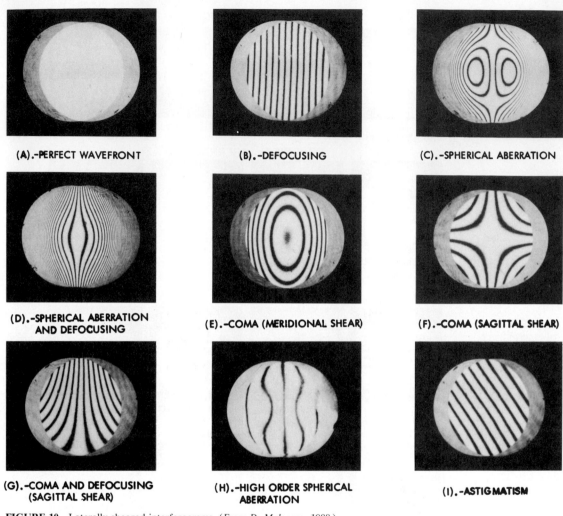

(A).-PERFECT WAVEFRONT

(B).-DEFOCUSING

(C).-SPHERICAL ABERRATION

(D).-SPHERICAL ABERRATION AND DEFOCUSING

(E).-COMA (MERIDIONAL SHEAR)

(F).-COMA (SAGITTAL SHEAR)

(G).-COMA AND DEFOCUSING (SAGITTAL SHEAR)

(H).-HIGH ORDER SPHERICAL ABERRATION

(I).-ASTIGMATISM

FIGURE 10 Laterally sheared interferograms. (*From D. Malacara, 1988.*)

Zernike Tests

The Zernike phase-contrast method is another way to improve the sensitivity of an interferometer to small aberrations. It was suggested by Zernike as a way to improve the knife-edge test (Zernike, 1934a). There are several versions of this test. The basic principle in all of them is the introduction of a phase difference equal to $\lambda/2$ between the wavefront under test and the reference wavefront. To understand why this phase difference is convenient, let us consider two interfering beams and irradiances $I_1(x, y)$ and $I_2(x, y)$ and a phase $\phi(x, y)$ between them. The final irradiance $I(x, y)$ in the interferogram is given by

$$I(x, y) = I_1(x, y) + I_2(x, y) + 2\sqrt{I_1(x, y)I_2(x, y)} \cos \phi(x, y) \tag{8}$$

Thus, the irradiance $I(x, y)$ of the combination would be a sinusoidal function of the phase, as illustrated in Fig. 13. If the phase difference is zero for a perfect wavefront,

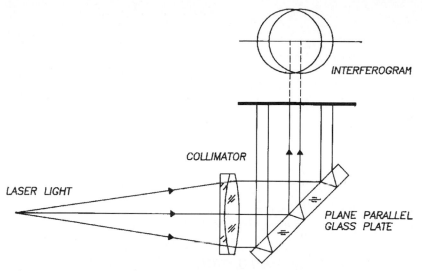

FIGURE 11 Murty's lateral shear interferometer.

deformations of the wavefront smaller than the wavelength of the light will not be easy to detect, because the slope of the function is zero for a phase near zero. The slope of this function is larger and linear for a phase value of 90°. Thus, the small wavefront deformations are more easily detected if the interferometer is adjusted, so that the wavefronts have a phase difference equal to 90° when the wavefront under test is perfect.

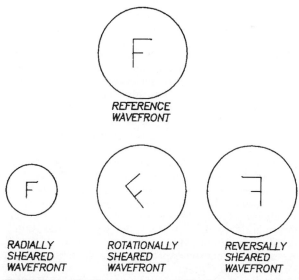

FIGURE 12 Wavefronts in radial, rotational, and reversal shear interferometers.

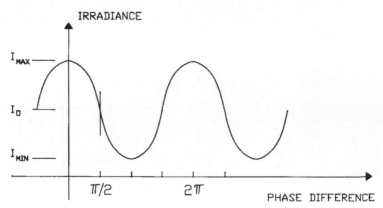

FIGURE 13 Irradiance in an interference pattern, as a function of the phase difference between the two interfering waves.

30.6 INTERFEROGRAM EVALUATION

An interferogram may be analyzed in several manners. One way begins by measuring several points on the interferogram, on top of the fringes. Then, the wavefront values between the fringes are interpolated. Another way uses a Fourier analysis of the interferogram. A third method interprets the fringe deformations as a phase modulation.

Fixed Interferogram Evaluation

Once the interferogram has been formed, a quantitative evaluation of it is a convenient method to find the wavefront deformations. The fixed interferogram evaluation by fringe measurements is done by measuring the position of several data points located on top of the fringes. These measurements are made in many ways, for example, with a measuring microscope, with a digitizing tablet, or with a video camera connected to a computer.

The fringe centers can be located either manually, using a digitizing tablet, or automatically, with the computer directly examining a single fringe image that has been captured using a digital frame grabber. After locating the fringe centers, fringe order numbers must be assigned to each point. The wavefront can then be characterized by direct analysis of the fringe centers. If desired, instead of global interpolation, a local interpolation procedure may be used.

To analyze the fringes by a computer, they must first be digitized by locating the fringe centers, and assigning fringe order numbers to them. The optical path difference (OPD) at the center of any fringe is a multiple m of the wavelength λ (OPD $= m\lambda$), where m is the fringe order. To obtain the wavefront deformation, only the relative values of the fringe order are important. So any value of the fringe order may be assigned to the first fringe being measured. However, for the second fringe, it may be increased or decreased by one. This choice affects only the sign of the OPD. An important disadvantage of the fixed interferogram analysis is that the sign of the OPD cannot be obtained from the interferogram alone. This information can be retrieved if the sign of any term in the wavefront deformation expression, like defocusing or tilt, is previously determined when taking the interferogram.

Fringes have been digitized using scanners (Rosenzweig and Alte, 1978), television cameras (Womack et al., 1979), photoelectric scanners, and digitizing tablets. Review articles by Reid (1986, 1988) give useful references for fringe digitization using television cameras.

Global and Local Interpolation of Interferograms

After the measurements are made, the wavefront is computed with the measured points. The data density depends on the density of fringes in the interferogram. Given a wavefront deformation, the ratio of the fringe deviations from straightness to the separation between the fringes remains a constant, independently of the number of fringes introduced by tilting of the reference wavefront. If the number of fringes is large due to a large tilt, the fringes look more straight than if the number of fringes is small. Thus, the fringe deviations may more accurately be measured if there are few fringes in the interferogram. Thus, information about many large zones is lost. A way to overcome this problem is to interpolate intermediate values by any of several existing methods. One method is to fit the wavefront data to a two-dimensional polynomial with a least-squares fitting, as described by Loomis (1978) and Malacara et al. (1990) or by using splines as described by Hayslett and Swantner (1980) and Becker et al. (1982). Unfortunately, this procedure has many problems if the wavefront is very irregular. The values obtained with the polynomial may be wrong, especially near the edge, or between fringes if the wavefront is too irregular.

The main disadvantage of global fits is that they smooth the measured surface more than desired. Depending on the degree of the polynomial, there will be only a few degrees of freedom to fit many data points. It is even possible that the fitted surface will pass through none of the measured points. If the surface contains irregular features that are not well described by the chosen polynomial, such as steps or small bumps, the polynomial fit will smooth these features. Then, they will not be visible in the fitted surface.

Global interpolation is done by least-squares fitting the measured data to a two-dimensional polynomial in polar coordinates. The procedure to make the least-squares fitting begins by defining the variance σ of the discrete wavefront fitting as follows:

$$\sigma = \frac{1}{N} \sum_{i=1}^{N} [W_i' - W(\rho_i, \theta_i)]^2 \tag{9}$$

where N is the number of data points, W_i' is the measured wavefront deviation for data point i, and $W(\rho_i, \theta_i)$ is the functional wavefront deviation after the polynomial fitting. The only requirement is that this variance or fit error is minimized. It is well known that the normal least-squares procedure leads to the inversion of an almost singular matrix. Then, the round-off errors will be so large that the results will be useless. To avoid this problem, the normal approach is to fit the measured points to a linear combination of polynomials that are orthogonal over the discrete set of data points. Thus, the wavefront is represented by

$$W(\rho_i, \theta_i) = \sum_{n=1}^{L} B_n V_n(\rho_i, \theta_i) \tag{10}$$

$V(\rho, \theta)$ are polynomials of degree r and not the monomials x^r. These polynomials satisfy the orthogonality condition

$$\sum_{i=1}^{N} V_n(\rho_i, \theta_i) V_m(\rho_i, \theta_i) = F_n \delta_{nm} \tag{11}$$

where $F_n = \sum V_n^2$

The advantage of using these orthogonal polynomials is that the matrix of the system becomes diagonal and there is no need to invert it.

The only problem that remains is to obtain the orthogonal polynomials by means of the Gram-Schmidt orthogonalization procedure. It is important to notice that the set of orthogonal polynomials is different for every set of data points. If only one data point is removed or added, the orthogonal polynomials are modified. If the number of data points tends to infinity and they are uniformly distributed over a circular pupil with unit radius, these polynomials V_r approach the Zernike polynomials (Zernike, 1934b).

Several properties of orthogonal polynomials make them ideal for representing wavefronts, but the most important of them is that we may add or subtract one or more polynomial terms without affecting the fit coefficients of the other terms. Thus, we can subtract one or more fitted terms—defocus, for example—without having to recalculate the least-squares fit. In an interferometric optical testing procedure the main objective is to determine the shape of the wavefront measured with respect to a best-fit sphere. Nearly always it will be necessary to add or subtract some terms.

The only problem with these orthogonal polynomials over the discrete set of data points is that they are different for every set of data points. A better choice for the wavefront representation is the set of Zernike polynomials, which are orthogonal on the circle with unit radius, as follows.

$$\int_0^1 \int_0^{2\pi} U_n(\rho, \theta) U_m(\rho, \theta) \rho \, d\rho \, d\theta = F_{nm} \delta_{nm} \tag{12}$$

These polynomials are not exactly orthogonal on the set of data points, but they are close to satisfying this condition. Therefore, it is common to transform the wavefront representation in terms of the polynomials V_n to another similar representation in terms of Zernike polynomials U_n, as

$$W(\rho, \theta) = \sum_{n=1}^{L} A_n U_n(\rho, \theta) \tag{13}$$

Fourier Analysis of Interferograms

A completely different way to analyze an interferogram without having to make any interpolation between the fringes is by a Fourier analysis of the interferogram. An interpolation procedure is not needed because the irradiance at a fine two-dimensional array of points is measured and not only at the top of the fringes. The irradiance should be measured directly on the interferogram with a two-dimensional detector or television camera, and not on a photographic picture. Womack (1983, 1984), Macy (1983), Takeda et al. (1982), and Roddier and Roddier (1987) have studied in detail the Fourier analysis of interferograms to obtain the wavefront deformations.

Consider an interferogram produced by the interference of the wavefront under test and a flat reference wavefront, with a large tilt between them. The tilt is about the y axis, increasing the distance between the wavefronts in the x direction. The picture of this interferogram may be thought of as a hologram reconstructing the wavefront. Thus, three wavefronts (images) are generated when this hologram is illuminated with a flat wavefront. In order to have complete separation between these images, the tilt between the wavefronts must be large enough, so that the angle between them is not zero at any point over the interferogram. This is equivalent to saying that the fringes must be open, and never cross any line parallel to the x axis more than once. One image is the wavefront under test and another is the conjugate of this wavefront.

If the tilt between the wavefront is θ, and the wavefront shape is $W(x, y)$, the irradiance, from Eq. 5, is given by

$$I(x, y) = I_1(x, y) + I_2(x, y) + 2\sqrt{I_1(x, y)I_2(x, y)} \cos\left(\phi_0 + kx \sin\theta + kW(x, y)\right) \qquad (14)$$

where $k = 2\pi/\lambda$. This expression may be rewritten as

$$I = [I_1 + I_2] + \sqrt{I_1 I_2}\, e^{i(kx\sin\theta + kW)} + \sqrt{I_1 I_2}\, e^{-i(kx\sin\theta + kW)} \qquad (15)$$

The first term represents the zero order, the second is the real image, and the third is the virtual image. We also may say that the Fourier transform of the interferogram is formed by a Dirac impulse $\delta(f)$ at the origin and two terms shifted from the origin, at frequencies $+f_o$ and $-f_o$. The quantity f is the spatial frequency, defined by the tilt between the reference wavefront and the wavefront under test $(f = \sin\theta/\lambda)$. These terms may be found by taking the Fourier transform of the interferogram. The term at $+f_o$ is due to the wavefront under test. This wavefront may be obtained by taking the Fourier transform of this term, mathematically isolated from the others. This method is performed in a computer by using the fast Fourier transform. The undesired terms are simply eliminated before taking the second fast Fourier transform in order to obtain the wavefront.

Direct Interferometry

This is another method to obtain the wavefront from an interferogram without the need of any interpolation. As in the Fourier method, the image of the interferogram is directly measured with a two-dimensional detector or television camera. The interferogram must have many fringes, produced with a large tilt between the wavefronts. The requirements for the magnitude of this tilt are the same as in the Fourier method.

Consider the irradiance in the interferogram along a line parallel to the x axis. This irradiance plotted versus the coordinate x is a perfectly sinusoidal function only if the wavefront is perfect, that is, if the fringes are straight, parallel, and equidistant. Otherwise, this function appears as a wave with a phase modulation. The phase-modulating function is the wavefront shape $W(x, y)$. If the tilt between the wavefronts is θ, the irradiance function is described by Eq. (14). If ϕ_o is a multiple of 2π, this expression may be rewritten as

$$I = I_1 + I_2 + 2\sqrt{I_1 I_2} \cos\left(kx \sin\theta + kW\right) \qquad (16)$$

Multiplying this phase-modulated function by a sinusoidal signal with the same frequency as the carrier $\sin(kx \sin\theta)$ a new signal S is obtained. Similarly, multiplying by a cosinusoidal signal $\cos(kx \sin\theta)$ a new signal C is obtained. If all terms in the signals S and C with frequencies equal to or greater than the carrier frequency are removed with a low pass filter, they become

$$S(x, y) = -\sqrt{I_1 I_2} \sin kW(x, y) \qquad (17)$$

$$C(x, y) = \sqrt{I_1 I_2} \cos kW(x, y) \qquad (18)$$

then, the wavefront $W(x, y)$ is given by

$$W(x, y) = -\frac{1}{k} \tan^{-1}\left[\frac{S(x, y)}{C(x, y)}\right] \qquad (19)$$

which is our desired result.

30.7 *PHASE-SHIFTING INTERFEROMETRY*

All the methods just described are based on the analysis of a single static interferogram. Static fringe analysis is generally less precise than phase-shifting interferometry, by more than one order of magnitude. However, fringe, analysis has the advantage that a single image of the fringes is needed. On the other hand, phase-shifting interferometry requires several images, acquired over a long time span during which the fringes must be stable. This is the main reason why phase-shifting interferometry has seldom been used for the testing of astronomical optics.

Phase-shifting interferometry (Bruning, 1974; Greivenkamp and Bruning, 1991) is possible, thanks to modern tools like array detectors and microprocessors. Figure 14 shows a Twyman-Green interferometer adapted to perform phase-shifting interferometry. Most conventional interferometers, like the Fizeau and the Twyman-Green, have been used to do phase shifting. A good review about these techniques may be found in the review article by Creath (1988).

In phase-shifting interferometers, the reference wavefront is moved along the direction of propagation, with respect to the wavefront under test, changing in this manner their phase difference. This phase shifting is made in steps or in a continuous manner. Of course this relative displacement of one wavefront with respect to the other may only be achieved through a momentary or continuous change in the frequency of one of the beams, for example, by Doppler shift, moving one of the mirrors in the interferometer. In other words, this change in the phase is accomplished when the frequency of one of the beams is modified in order to form beats.

By measuring the irradiance changes for different values of the phase shifts, it is possible to determine the initial difference in phase between the wavefront under test and

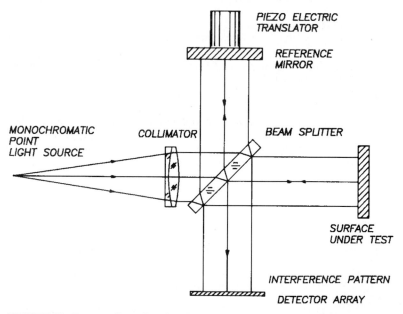

FIGURE 14 Twyman-Green interferogram adapted to do phase shifting.

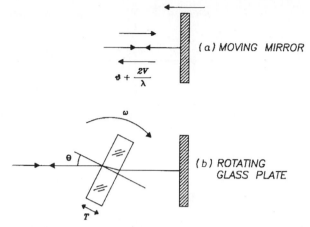

FIGURE 15 Obtaining the phase shift by means of a moving mirror or a rotating glass plate.

the reference wavefront, for that measured point over the wavefront. By obtaining this initial phase difference for many points over the wavefront, the complete wavefront shape is thus determined.

If we consider any fixed point in the interferogram, the initial phase difference between the two wavefronts has to be changed in order to make several measurements.

One method that can be used to shift this phase is by moving the mirror for the reference beam along the light trajectory, as in Fig. 15. This can be done in many ways, for example, with a piezoelectric crystal or with a coil in a magnetic field. If the mirror moves with a speed V, the frequency of the reflected light is shifted by an amount equal to $\Delta v = 2V/\lambda$.

Another method to shift the phase is by inserting a plane parallel glass plate in the light beam (see Fig. 15). Then the plate is rotated about an axis perpendicular to the optical axis.

The phase may also be shifted by means of the device shown in Fig. 16. The first

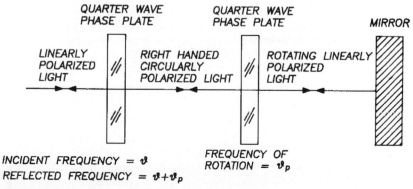

FIGURE 16 Obtaining the phase shift by means of phase plates and polarized light, with a double pass of the light beam.

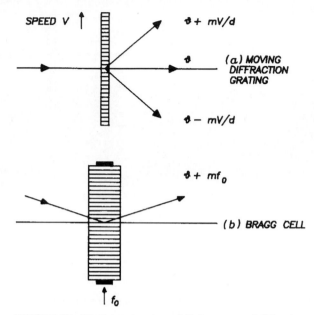

FIGURE 17   Obtaining the phase shift by means of diffraction: (*a*) with a diffraction grating; (*b*) with an acousto-optic Bragg cell.

quarter-wave retarding plate is stationary, with its slow axis at 45° with respect to the plane of polarization of the incident linearly polarized light. This plate also transforms the returning circularly polarized light back to linearly polarized. The second phase retarder is also a quarter-wave plate, it is rotating, and the light goes through it twice, therefore it is acting as a half-wave plate.

Still another manner to obtain the shift of the phase is by a diffraction grating moving perpendicularly to the light beam, as shown in Fig. 17*a*, or with an acousto-optic Bragg cell, as shown in Fig. 17*b*. The change in the frequency is equal to the frequency f of the ultrasonic wave times the order of diffraction m. Thus: $\Delta v = mf$.

The nonshifted relative phase of the two interfering wavefronts is found by measuring the irradiance with several predefined and known phase shifts. Let us assume that the irradiance of each of the two interfering light beams at the point x, y in the interference patterns are $I_1(x, y)$ and $I_2(x, y)$ and that their phase difference is $\phi(x, y)$. It was shown before, in Eq. (5), that the resultant irradiance $I(x, y)$ is a sinusoidal function describing the phase difference between the two waves. The basic problem is to determine the nonshifted phase difference between the two waves, with the highest possible precision. This may be done by any of several different procedures.

Phase Stepping

This method (Creath, 1988) consists of measuring the irradiance values for several known increments of the phase. There are several versions of this method, which will be described later. The measurement of the irradiance for any given phase takes some time, since there is a time response for the detector. Therefore, the phase has to be stationary during a

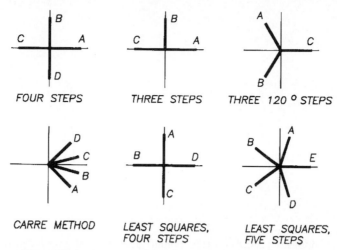

FIGURE 18 Six different ways to shift the phase using phase steps.

short time in order to take the measurement. Between two consecutive measurements, the phase is changed by an increment α_i. For those values of the phase, the irradiance becomes

$$I = I_1 + I_2 + 2\sqrt{I_1 I_2} \cos{(\phi + \alpha_i)} \tag{20}$$

There are six different algorithms, as shown in Fig. 18, with different numbers of measurements of the phase. As we see, the minimum number of steps needed to reconstruct this sinusoidal function is three. As an example with four steps,

$$I_A = I_1 + I_2 + 2\sqrt{I_1 I_2} \cos{\phi} \tag{21}$$

$$I_B = I_1 + I_2 - 2\sqrt{I_1 I_2} \sin{\phi} \tag{22}$$

$$I_C = I_1 + I_2 - 2\sqrt{I_1 I_2} \cos{\phi} \tag{23}$$

$$I_D = I_1 + I_2 + 2\sqrt{I_1 I_2} \sin{\phi} \tag{24}$$

From these relations the desired phase is

$$\phi(x, y) = \tan^{-1}\left\{\frac{I_D(x, y) - I_B(x, y)}{I_A(x, y) - I_C(x, y)}\right\} \tag{25}$$

Integrating Bucket

In the integrating phase-shifting method the detector continuously measures the irradiance during a fixed time interval, without stopping the phase. Since the phase changes continuously, the average value of the irradiance during the measuring time interval is measured. Thus, the integrating phase-stepping method may be mathematically considered a particular case of the phase-stepping method if the detector has an infinitely short time

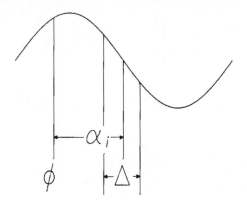

FIGURE 19 Averaged signal measurements with the integrating phase-shifting method.

response. Then, the measurement time interval is reduced to zero. If the measurement is taken as in Fig. 19, from $\alpha_i - \Delta/2$ to $\alpha_i + \Delta/2$ with center at α_i, then

$$I = \frac{1}{\Delta} \int_{\alpha - \Delta/2}^{\alpha + \Delta/2} [I_1 + I_2 + 2\sqrt{I_1 I_2} \cos{(\phi + \alpha_i)}]\, d\alpha \tag{26}$$

$$I = I_1 + I_2 + 2\sqrt{I_1 I_2} \sin{c(\Delta/2)} \cos{(\phi + \alpha_i)} \tag{27}$$

In general, in the phase-stepping as well as in the integrating phase-shifting methods, the irradiance is measured at several different values of the phase α_i, and then the phase is calculated.

Two Steps Plus One Method

As pointed out before, phase-shifting interferometry is not useful for testing systems with vibrations or turbulence because the three or four interferograms are taken at different times. An attempt to reduce this time is the so-called two steps plus one method, in which only two measurements separated by 90° are taken (Wizinowich, 1989). A third reading is taken any time later, of the sum of the irradiance of the beams, independently of their relative phase. This last reading may be taken using an integrating interval $\Delta = 2\pi$. Thus

$$I_A = I_1 + I_2 + 2\sqrt{I_1 I_2} \cos{\phi} \tag{28}$$

$$I_B = I_1 + I_2 + 2\sqrt{I_1 I_2} \sin{\phi} \tag{29}$$

$$I_C = I_1 + I_2 \tag{30}$$

Therefore:

$$\phi = \tan^{-1}\left\{\frac{I_B - I_C}{I_A - I_C}\right\} \tag{31}$$

Simultaneous Measurement

It has been said several times that the great disadvantage of phase-shifting interferometry is its great sensitivity to vibrations and atmospheric turbulence. To eliminate this problem, it has been proposed that the different interferograms corresponding to different phases be taken simultaneously (Bareket, 1985 and Koliopoulos, 1991). To obtain the phase-shifted interferogram, they have used polarization-based interferometers. The great disadvantage of these interferometers is their complexity. To measure the images these interferometers have to use several television cameras.

Heterodyne Interferometer

When the phase shift is made in a continuous manner rather than in steps, the frequency of the shifting beam is permanently modified, and a beating between the two interferometer beams is formed (Massie, 1987).

The phase of the modulated or beating wave may be determined in many ways. One way is by electronic analog techniques, for example, using leading-edge detectors. Another way is by detecting when the irradiance passes through zero, that is, through the axis of symmetry of the irradiance function.

Phase Lock

The phase-lock method (Johnson et al., 1977, 1979; Moore, 1979) can be explained with the help of Fig. 20. Assume that an additional phase difference is added to the initial phase $\phi(x, y)$. The additional phase being added has two components, one of them with a fixed value and the other with a sinusoidal time shape. Both components can have any predetermined desired value. Thus:

$$\phi = \phi(x, y) + \delta(x, y) + a \sin \omega t \tag{32}$$

then, the irradiance $i(x, y)$ would be given by

$$I = I_1 + I_2 + 2\sqrt{I_1 I_2} \cos \left[\phi + \delta + a \sin \omega t \right] \tag{33}$$

The amplitude of the phase oscillations $a \sin \omega t$ is much smaller than π. We may now adjust the fixed phase δ to a value such that $\phi + \delta = \pi/2 + n\pi$. Then the value of $\cos (\phi + \delta)$ is zero. The curve is antisymmetric at this point; hence, only odd harmonics remain on the irradiance signal. This is done in practice by slowly changing the value of the phase δ, while maintaining the oscillation $a \sin \omega t$, until the maximum amplitude of the first harmonic, or fundamental frequency, is obtained. At this point, then, we have $\delta + \phi = \pi/2 + n\pi$, and since the value of δ is known, the value of ϕ has been determined.

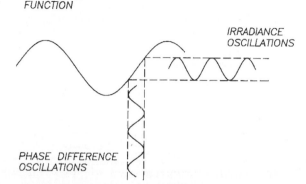

IRRADIANCE FUNCTION

IRRADIANCE OSCILLATIONS

PHASE DIFFERENCE OSCILLATIONS

FIGURE 20 Phase-lock method to find the phase with a small sinusoidal modulation of the phase.

30.8 *MEASURING ASPHERICAL WAVEFRONTS*

The most common type of interferometer, with the exception of lateral or rotational shearing interferometers, produces interference patterns in which the fringes are straight, equidistant, and parallel, when the wavefront under test is perfect and spherical with the same radius of curvature as the reference wavefront.

If the surface under test does not have a perfect shape, the fringes will not be straight and their separations will be variable. The deformations of the wavefront may be determined by a mathematical examination of the shape of the fringes. By introducing a small spherical curvature on the reference wavefront (focus shift) or by changing its angle with respect to the wavefront under test (tilt), the number of fringes in the interferogram may by changed. This is done to reduce the number of fringes as much as possible, since the greater the number of fringes, the smaller the sensitivity of the test. However, for aspherical surfaces this number of fringes cannot be smaller than a certain minimum. The larger the asphericity is, the greater is this minimum number of fringes. Since the fringe separations are not constant, in some places the fringes will be widely spaced, but in some others the fringes will be too close together.

The sensitivity of the test depends on the separation between the fringes, because an error of one wavelength in the wavefront distorts the fringe shape by an amount equal to the separation between the fringes. Thus, the sensitivity is directly proportional to the fringe separation. When the fringes are widely separated, the sampled points will be quite separated from each other, leaving many zones without any information. On the other hand, where the fringes are very close to each other, there is a high density of sampled data points, but the sensitivity is low.

Then, it is desirable that the spherical aberration of the wavefront under test is compensated in some way, so that the fringes appear straight, parallel, and equidistant, for a perfect wavefront. This is called a null test and may be accomplished by means of some special configurations. These special configurations may be used to conduct a null test of a conic surface. These are described in several books (Malacara, 1991). Almost all of these surfaces have rotational symmetry.

If no testing configuration can be found to get rid of the spherical aberration, additional optical components, called null compensators, have to be used. Many different types of compensators have been invented. The compensators may be refractive (lenses), reflective (mirrors), or diffractive (real or computer-generated holograms).

Refractive or Reflective Compensators

The simplest way to compensate the spherical aberration of a paraboloid or a hyperboloid tested at the center of curvature is a single convergent lens placed near the point of convergence of the rays, as Fig. 21 shows. This lens is called a Dall compensator.

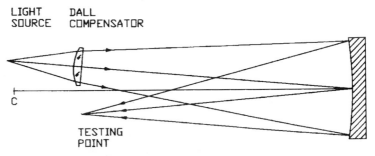

FIGURE 21 The Dall compensator.

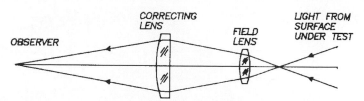

FIGURE 22 The Offner compensator. Only the reflected beam is shown.

Unfortunately, the correction due to a single lens is not complete, so a system of two lenses must be used to obtain a better compensation. This system is called an Offner compensator and is shown in Fig. 22. The field lens L is used to image the surface under test on the plane of the compensating lens L. Mirrors may also be used to design a null compensator.

As the sad experience of the Hubble space telescope proves, the construction parameters in a lens compensator have to be very carefully measured and adjusted, otherwise an imperfect correction is obtained either by undercorrection or overcorrection. The distance from the compensator to the surface under test is one of those parameters to be carefully measured. A way around this problem would be to assume that the compensator detects smoothness imperfections but not the exact degree of asphericity. This degree of asphericity may then be measured with some independent measurement like the Hartmann test.

Holographic Compensators

Diffractive holographic elements also may be used to compensate the spherical aberration of the system and to obtain a null test. The hologram may be real, produced by photographing an interferometric pattern. This pattern has to be formed by superimposing on the screen a wavefront like the one we have to test and a perfectly flat or spherical wavefront. The only problem with this procedure is that a perfect wavefront with the same shape as the wavefront to be tested has first to be produced. This is not always easy.

A better approach is to simulate the holographic interference pattern in a computer (Wyant, 1978) as in Fig. 23. Then this image is transferred to a small photographic plate, with the desired dimensions. There are many experimental arrangements to compensate the aspherical wavefront aberration with a hologram. One of these is illustrated in Fig. 24.

Infrared Interferometry

Another simple approach to reduce the number of fringes in the interferogram is to use a long infrared wavelength. Light from a CO_2 laser has been used with this purpose. It can also be used when the surface is still quite rough.

Two-wavelength Interferometry

In phase-shifting interferometry, each detector must have a phase difference smaller than π from the closest neighboring detector, in order to avoid 2π phase ambiguities and ensure phase continuity. In other words, there should be at least two detector elements for each fringe. If the slope of the wavefront is very large, the fringes will be too close together and the number of detector elements would be extremely large (Wyant et al., 1984).

A solution to this problem is to use two different wavelengths λ_1 and λ_2

FIGURE 23 Computer-generated hologram for testing an aspherical wavefront. (*From Wyant, 1978.*)

simultaneously. The group wavelength or equivalent wavelength λ_{eq} is longer than any of the two components and is given by:

$$\lambda_{eq} = \frac{\lambda_1 \lambda_2}{|\lambda_1 - \lambda_2|} \tag{34}$$

Under these conditions, the requirement in order to avoid phase uncertainties is that there should be at least two detectors for each fringe produced if the wavelength is λ_{eq}.

The great advantage of this method is that we may test wavefronts with large

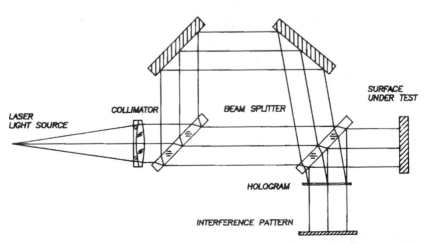

FIGURE 24 An optical arrangement for testing an aspherical wavefront with a computer-generated hologram.

asphericities, limited in asphericity by the group wavelength, and accuracy limited by the shortest wavelength of the two components.

Moiré Tests

An interferogram in which a large amount of tilt has been introduced is an ideal periodic structure to form moiré patterns. A moiré pattern represents the difference between two periodic structures. Thus, a moiré formed by two interferograms represents the difference between the two interferograms. There are several possibilities for the use in optical testing of this technique, as shown by K. Patorski (1988).

Let us assume that the two interferograms are taken from the same optical system producing an aspherical wavefront, but with two different wavelengths λ_1 and λ_2. The moiré obtained represents the interferogram that would be obtained with an equivalent wavelength λ_{eq} given by Eq. (31). If the tilt is of different magnitude in the two interferograms, the difference appears as a tilt in the moiré between them. Strong aspheric wavefronts may be tested with this method.

A second possibility is to produce the moiré between the ideal interferogram for an aspheric wavefront and the actual wavefront. Any differences between both would be easily detected.

Another possibility of application is for eliminating the wavefront imperfections in a low-quality interferometer. One interferogram is taken with the interferometer alone, without any optical piece under test. The second interferogram is taken with the optical component being tested. The moiré represents the wavefront deformations due to the piece being tested, without the interferometer imperfections.

Sub-Nyquist Interferometry

It was pointed out before that in phase-shifting interferometry each detector must have a phase difference smaller than π from the closest neighboring detector, in order to avoid 2π phase ambiguities and to ensure phase continuity. In other words, there should be at least two detector elements for each fringe. This condition is known as the Nyquist condition.

Since there is a minimum practical distance between detectors, the maximum asphericity in a surface to be tested by phase-shifting interferometry is only a few wavelengths. This condition may be relaxed (Greivenkamp, 1987) if the wavefront and its slope are assumed to be continuous on the whole aperture. Then, optical surfaces with larger asphericities may be tested.

30.9 REFERENCES

Bareket, N. "Three-Phase Phase Detector for Pulsed Wavefront Sensing," *Proc. SPIE* **551:**12 (1985).

Becker, F., G. E. A. Maier, and H. Wegner, "Automatic Evaluation of Interferograms," *Proc. SPIE* **359:**386 (1982).

Bruning, J. H., D. J. Herriott, J. E. Gallagher, D. P. Rosenfeld, A. D. White, and D. J. Brangaccio, "Digital Wavefront Measurement Interferometer," *Appl. Opt.* **13:**2693 (1974).

Creath, K., "Phase-Measurement Interferometry Techniques," in E. Wolf (ed.), *Progress in Optics* vol. XXVI, Elsevier Science Publishers, Amsterdam, 1988.

Foucault, L. M., "Description des Procedes Employes pour Reconnaitre la Configuration des Surfaces Optiques," *C.R. Acad. Sci. Paris* **47:**958 (1852); reprinted in Armand Colin, *Classiques de al Science,* vol. II.

Greivenkamp, J. E., "Sub-Nyquist Interferometry," *Appl. Opt.* **26:**5245 (1987).

Greivenkamp, J. and J. H. Bruning, "Phase Shifting Interferometers," in D. Malacara (ed.), *Optical Shop Testing,* 2d ed., John Wiley and Sons., New York, 1991.

Hariharan, P. and D. Sen, "The Separation of Symmetrical and Asymmetrical Wave-Front Aberrations in the Twyman Interferometer," *Proc. Phys. Soc.* **77:**328 (1961).

Hariharan, P., "Multiple-Pass Interferometers," in D. Malacara (ed.), *Optical Shop Testing,* 2d ed., John Wiley and Sons, New York, 1991.

Hartmann, J., "Bemerkungen über den Bann und die Justirung von Spektrographen," *Zt. Instrumentenkd.* **20:**47 (1990).

Hayslett, C. R. and W. Swantner,"Wave Front Derivation from Interferograms by Three Computer Programs," *Appl. Opt.* **19:**3401 (1980).

Johnson, G. W., D. C. Leiner, and D. T. Moore, "Phase Locked Interferometry," *Proc. SPIE* **126:**152 (1977).

Johnson, G. W., D. C. Leiner, and D. T. Moore, "Phase Locked Interferometry," *Opt. Eng.* **18:**46 (1979).

Koliopoulos, C. I.., "Simultaneous Phase Shift Interferometer," *Proc. SPIE* (in press) (1991).

Loomis, J. S., "Analysis of Interferograms from Waxicons," *Proc. SPIE* **171:**64 (1979).

Malacara, D., *Optical Shop Testing,* 2d ed., John Wiley and Sons, New York, 1991.

Malacara, D., "Interference," in *Methods of Experimental Physics,* Academic Press, New York, 1988.

Malacara, D., J. M. Carpio-Valadéz, and J. J. Sánchez-Mondragón, "Wavefront Fitting with Discrete Orthogonal Polynomials in a Unit Radius Circle," *Opt. Eng.* **29:**672 (1990).

Macy, W. W., Jr., "Two Dimensional Fringe Pattern Analysis," *Appl. Opt.* **22:**3898 (1983).

Massie, N. A., "Digital Heterodyne Interferometry," *Proc. SPIE* **816:**40 (1987).

Moore, D. T., "Phase-locked Moire Fringe Analysis for Automated Contouring of Diffuse Surfaces," *Appl. Opt.* **18:**91 (1979).

Ojeda-Castañeda, J., "Foucault, Wire and Phase Modulation Tests," in D. Malacara (ed.), *Optical Shop Testing,* 2d ed., John Wiley and Sons, New York, 1991.

Patorski, K., "Moiré Methods in Interferometry," *Opt. and Lasers in Eng.* **8:**147 (1988).

Reid, G. T., "Automatic Fringe Pattern Analysis: A Review," *Opt. and Lasers in Eng.* **7:**37 (1986).

Reid, G. T., "Image Processing Techniques for Fringe Pattern Analysis," *Proc. SPIE* **954:**468 (1988).

Roddier, C. and F. Roddier, "Interferogram Analysis Using Fourier Transform Techniques," *Appl. Opt.* **26:**1668 (1987).

Ronchi, V., "Le Franque di Combinazione Nello Studio Delle Superficie e Dei Sistemi Ottici," *Riv. Ottica mecc. Precis.* **2:**9 (1923).

Roychoudhuri, C., "Multiple-Beam Interferometers," in D. Malacara (ed.), *Optical Shop Testing,* 2d ed., John Wiley and Sons., New York, 1991.

Takeda, M., H. Ina, and S. Kobayashi, "Fourier Transform Method of Fringe-Pattern Analysis for Computer-Based Topography and Interferometry," *J. Opt. Soc. Am.* **72:**156 (1982).

Wizinowich, P. L., "Systems for Phase Shifting Interferometry in the Presence of Vibration," *Proc. SPIE* **1164:**25 (1980).

Womack, K. H., J. A. Jonas, C. L. Koliopoulos, K. L. Underwood, J. C. Wyant, J. S. Loomis, and C. R. Hayslett, "Microprocessor-Based Instrument for Analysis of Video Interferograms," *Proc. SPIE* **192:**134 (1979).

Womack, K. H., "Frequency Domain Description of Interferogram Analysis," *Opt. Eng.* **23:**396 (1984).

Wyant, J. C., "Holographic and Moire Techniques," in D. Malacara (ed.), *Optical Shop Testing,* John Wiley and Sons, New York, 1978.

Wyant, J. C., B. F. Oreb, and P. Hariharan, "Testing Aspherics Using Two-Wavelength Holography: Use of Digital Electronic Techniques," *Appl. Opt.* **23:**4020 (1984b).

Zernike, F., "Diffraction Theory of Knife Edge Test and its Improved Form, The Phase Contrast," *Mon. Not. R. Astron. Soc.* **94:**371 (1934a).

Zernike, F., "Begünstheorie des Schneidener-Fahrens und Seiner Verbasserten Form, der Phasenkontrastmethode," *Physica.* **1:**689 (1934b).

CHAPTER 31
USE OF COMPUTER-GENERATED HOLOGRAMS IN OPTICAL TESTING

Katherine Creath
Optical Sciences Center
University of Arizona
Tucson, Arizona

and

James C. Wyant
Optical Sciences Center
University of Arizona
Tucson, Arizona
and
WYKO Corporation
Tucson, Arizona

31.1 GLOSSARY

CGH	computer-generated hologram
M	linear, lateral magnification
N	diffracted order number
n	integers
P	number of distortion-free resolution points
r	radius
S	maximum wavefront slope (waves/radius)
$x, \Delta x$	distance
$\Delta \theta$	rotational angle error
$\Delta \phi$	wavefront phase error
θ	rotational angle
λ	wavelength
$\phi(\ )$	wavefront phase described by hologram

31.2   *INTRODUCTION*

Holography is extremely useful for the testing of optical components and systems. If a master optical component or optical system is available, a hologram can be made of the wavefront produced by the component or system and this stored wavefront can be used to perform null tests of similar optical systems. If a master optical system is not available for making a hologram, a synthetic or a computer-generated hologram (CGH) can be made to provide the reference wavefront.[1–10] When an aspheric optical element with a large departure from a sphere is tested, a CGH can be combined with null optics to perform a null test.

There are several ways of thinking about CGHs. For the testing of aspheric surfaces, it is easiest to think of a CGH as a binary representation of the ideal interferogram that would be produced by interfering the reference wavefront with the wavefront produced by a perfect sphere. In the making of the CGH the entire interferometer should be ray traced to determine the so-called perfect aspheric wavefront at the hologram plane. This ray trace is essential because the aspheric wavefront will change as it propagates, and the interferometer components may change the shape of the perfect aspheric wavefront.

31.3   *TYPES OF CGHs*

While there are several types of CGHs, the three most common types used in optical testing are (1) binary detour-phase,[11] (2) Lee hologram,[12,13] and (3) binary synthetic interferogram.[2,3] Generally, the holograms are binary amplitude rather than grayscale because they are easier to produce, and the only disadvantage is that additional orders are produced. These orders can be eliminated by spatial filtering, as explained later. Sometimes the holograms are bleached to give additional light in the order of interest.

Binary Detour-phase

The first hologram used for the testing of aspheric surfaces was the binary detour-phase hologram. In the binary detour-phase hologram the hologram is divided into equispaced resolution cells. A small rectangular aperture is placed in each cell. The area of each rectangle is made proportional to the amplitude of the wavefront at the position of the cell. Generally, in the case of testing aspheric optical elements, the irradiance across the wavefront is constant, so all rectangles in the hologram can be made of equal area. The cell is displaced from the center of the cell an amount proportional to the phase of the wavefront at the cell. Thus, for zero phase, the rectangle would be placed at the center of the cell; for −180 degrees, the rectangle would be at the left edge of the cell; for +180 degrees, the rectangle would be at the right edge of the cell, and so forth. The cell size would be selected such that, in going across any cell, the phase difference between the reference wave and the wavefront being produced would change by less than 180°. The accuracy of the wavefront produced by the hologram is limited by the accuracy with which the rectangles can be placed in the cells. If the cells are placed to the correct position to within 1/5 the cell width, the accuracy of the produced wavefront is at best 1/5 wave. A positional accuracy of 1/10 cell width would give 1/10 wave, and so forth.

Lee Hologram

The Lee hologram is also made up of a collection of cells. The phase of the aspheric wavefront is calculated at the center of each cell. Each cell contains four equally spaced rectangular apertures. For binary holograms, the length of the four rectangles is proportional to the positive real portion of the wavefront, positive imaginary portion, negative real portion, and negative imaginary portion of the wavefront, respectively. Any given cell would contain at most two rectangles. An advantage of the Lee hologram compared to the detour-phase hologram is that positional accuracy of the rectangles is not as critical; however, the size accuracy of the rectangles is critical, so it is no easier to make good-quality wavefronts with a Lee hologram than with a detour-phase hologram. The Lee approach is of particular interest if several gray levels, rather than strictly binary apertures, are available.

Binary Synthetic Hologram

The most common way to produce a CGH for optical testing is to make a binary representation of the actual hologram that would be produced if the reference wave were interfered with the aspheric wavefront under test. Figure 1 shows an example of a binary synthetic hologram. Since the amplitude of the aspheric wavefront is constant across the wavefront, best results are obtained if the lines making up the hologram have approximately one-half the spacing of the lines (i.e., fringe spacing) at the location of the lines. Thus, the line width will vary across the hologram. The major difference between the binary synthetic hologram and the real hologram that would be produced by interfering a reference wavefront and the aspheric wavefront is that additional diffraction orders are produced. These additional diffraction orders can be eliminated by spatial filtering, as shown subsequently.

31.4 PLOTTING CGHs

The largest problem in making CGHs is the plotting. The accuracy of the plot determines the accuracy of the wavefront. It is easier to see the plotting accuracy by comparing a binary synthetic hologram with an interferogram. In an interferogram, a wavefront error of $1/n$ waves causes a fringe to deviate from the ideal position by $1/n$ the fringe spacing. The same is true for CGHs. A plotting error of $1/n$ the fringe spacing will cause an error in the produced aspheric wavefront of $1/n$ wave.

To minimize wavefront error due to the plotter, the fringe spacing in the CGH should be as large as possible. The maximum acceptable fringe spacing is set by the requirement to separate the diffraction orders so spatial filtering can be used to select out only the first order. Figure 2a shows a drawing of the diffracted orders from a hologram in the Fourier plane, where the maximum slope of the aspheric wavefront is S. An actual photograph of the diffracted orders is shown in Fig. 2b. As shown in Fig. 2, to insure no overlapping of the first and second orders in the Fourier plane, the tilt angle of the reference beam needs to be greater than three times the maximum slope of the aberrated wave.[14] This means that, in general, the maximum slope difference between the reference and test beams is four times the maximum slope of the test beam. Thus, the error produced by plotter distortion is proportional to the slope of the aspheric wavefront being produced.

A convenient unit for slope of the aspheric wavefront is waves/radius. By units of waves/radius we mean that if a wavefront having a slope of S waves/radius is interfered with a nontilted plane wave, the minimum fringe spacing is such that if we had this fringe spacing across the hologram there would be S fringes across the radius of the hologram. If

FIGURE 1   Sample binary computer-generated hologram.

S is the maximum slope of the aspheric wavefront, measured in waves/radius, then the maximum slope difference between the reference and test waves is $4S$. If the plotter has P distortion-free resolution points across the hologram plot, then the maximum error due to plotter distortion in the first diffracted wave is $4S/P$ waves.[15,16] This maximum error is a worst-case situation and it occurs only if, in the region of the hologram where the fringe spacing is the smallest, the plotter distortion is the largest.

Many plotters have been used to plot holograms. Early work utilized pen plotters to make an enlarged version of the hologram which was then photographically reduced to the appropriate size.[2,3,17,18] The large format enabled a high-resolution CGH to be formed. However, problems due to plotter irregularities such as line thickness, pen quality, plotter distortion, and quantization caused errors in the reconstructed wavefront. Nonlinearities inherent in the photographic process and distortion in the reduction optics caused further degradation. With the advent of laser-beam recorders, resolution improved due to machine speed and an increased number of distortion-free recording points. The most recent

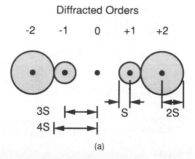

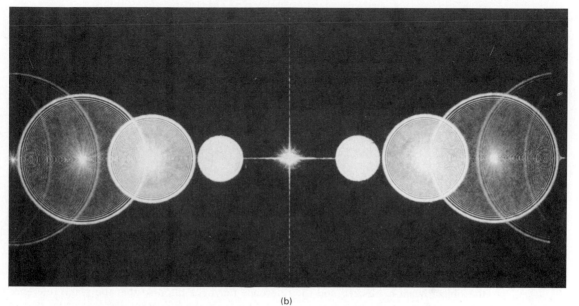

FIGURE 2 Diffracted orders in Fourier plane of CGH: (*a*) drawing; (*b*) photograph.

advances in the recording of CGHs have been made using the electron-beam (e-beam) recorders used for producing masks in the semiconductor industry.[19–21] These machines write onto photoresist deposited on an optical-quality glass plate and currently produce the highest-quality CGHs. Patterns with as many as 10^8 data points can be produced in a hologram of the desired size. Typical e-beam recorders will write a 1-mm area with a resolution of 0.25 μm. Large patterns are generated by stitching a number of 1-mm scans together. Errors in this technique are due to aberrations in the electron optics, beam drift, instabilities in the controlling electronics, and positioning of the stepper stage. Many of these errors are reproducible and can be compensated for in the software controlling the transfer.

As a conservative estimate of the amount of asphericity that can be measured using a CGH, consider that using an e-beam recorder with 0.25-micron resolution over a 10-mm diameter hologram (i.e., $P = 40{,}000$) would enable the measurement of an aspheric wavefront with a maximum wavefront slope of 1000 waves per radius (i.e., $4S = 4000$) to be tested to an accuracy of 1/10 wave.

Plotter distortion can be measured and calibrated out in the making of the

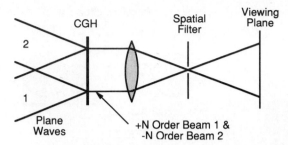

FIGURE 3 Test setup with $+/-N$ orders of hologram interfering to test the quality of a CGH plotter.

hologram.[18,22] The easiest way of determining plotter distortion is to draw straight lines and then treat this plot as a diffraction grating. If the computer-generated grating is illuminated with two plane waves as shown in Fig. 3, and the $-N$ order of beam 1 is interfaced with the $+N$ order of beam 2, the resulting interferogram gives us the plotter distortion. If the lines drawn by the plotter are spaced a distance Δx, a fringe error in the interferogram corresponds to a distortion error of $\Delta x/2N$ in the plot.

31.5 *INTERFEROMETERS USING COMPUTER-GENERATED HOLOGRAMS*

Many different experimental setups can be used for the holographic testing of optical elements. Because a hologram is simply an interferogram with a large tilt angle between the reference and object wavefronts, holographic tests can be performed either with standard interferometers or with setups having a larger angle between the object and reference beams. Figure 4 shows an interferometer that can be used for making a hologram of a concave mirror. The hologram is made in a plane conjugate to the test mirror. Once the hologram is made, it can be replaced in the same location, and reconstructed by illuminating with a plane wave and imaging onto a viewing screen. When the object beam is blocked and the reference mirror is tilted so that the plane reference wave interferes with the first-order diffraction from the hologram, the wavefront due to

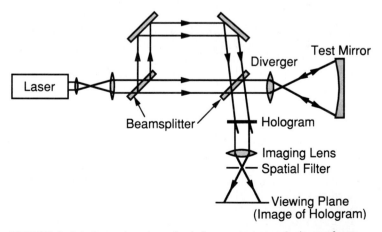

FIGURE 4 Interferometer setup using holograms to test aspheric wavefronts.

the test mirror will be reconstructed. Because several diffraction orders are produced by the hologram, it is usually necessary to select one of the diffraction orders using a spatial filter. The imaging lens and spatial filter are only necessary for the reconstruction of the hologram.

The same setup shown here for the making of a real hologram can be used if a CGH is used. The setup must be ray traced so the aberration in the hologram plane is known. While in theory there are many locations where the hologram can be placed, generally the hologram is made in a plane conjugate to the asphere under test. One reason is that in this plane the intensity across the image of the asphere is uniform, which simplifies the making of the hologram. A second reason is that if the hologram is made in a plane conjugate to the exit pupil of the mirror under test, the amount of tilt in the resulting interferogram can be selected by simply changing the tilt of the wavefront used in the hologram reconstruction process. If the hologram is not made in a plane conjugate to the exit pupil of the test surface, tilting the reference beam not only adds tilt to the interferogram, it also causes a displacement between the image of the exit pupil of the test surface and the exit pupil stored by the hologram. The longitudinal positional sensitivity for the hologram is reduced if the hologram is made in a region where the beams are collimated.

While there are many interferometric setups that can be used with CGHs, it is best if both the test and the reference beams pass through the hologram because a very serious error resulting from hologram substrate thickness variations is eliminated without the need of putting the hologram in a fluid gate or index matching the two surfaces to good optical flats.

The CGH interferometer can also be thought of in terms of moiré patterns.[23] Interference fringes resulting from the wavefront stored in the hologram and the wavefront coming from the optics under test can be regarded as the moiré pattern between the synthetic interference fringes recorded on the hologram plate, and the real-time interference fringes formed by the wavefront under test and a plane wavefront. The contrast in this moiré pattern is increased with spatial filtering by selecting only the wavefront produced by the mirror under test and the diffraction order from the hologram giving the stored wavefront produced by the master optics.

CGHs can be combined with partial null optics to test much more complicated aspherics than can be practically tested with either a CGH or null optics. This combination gives the real power of computer-generated holograms.

31.6 ACCURACY LIMITATIONS

The largest source of error is the error due to plotter distortion as discussed previously. The other large sources of error are improper positioning of the hologram in the interferometer, and incorrect hologram size.

Any translation of rotation of the hologram produces error.[3] If the hologram is made conjugate to the exit pupil of the master optical system, the exit pupil of the system under test must coincide with the hologram. If the test wavefront in the hologram plane is described by the function $\phi(x, y)$, a displacement of the hologram a distance Δx in the x direction produces an error

$$\Delta\phi(x, y) \approx \frac{\partial\phi(x, y)}{\partial x}\Delta x \tag{1}$$

where $\partial\phi/\partial x$ is the slope of the wavefront in the x direction. Similarly, for a wavefront described by $\phi(r, \theta)$, the rotational error $\Delta\theta$ is given by

$$\Delta\phi(r, \theta) \approx \frac{\partial\phi(r, \theta)}{\partial\theta}\Delta\theta \tag{2}$$

Another source of error is incorrect hologram size. If the aberrated test wavefront in the plane of the hologram is given by $\phi(r, \theta)$, a hologram of incorrect size will be given by $\phi(r/M, \theta)$, where M is a magnification factor. The error due to incorrect hologram size will be given by the difference $\phi(r/M, \theta) - \phi(r, \theta)$, and can be written in terms of a Taylor expansion as

$$\phi\left(\frac{r}{M}, \theta\right) - \phi(r, \theta) = \phi\left[r + \left(\frac{1}{M} - 1\right)r, \theta\right] - \phi(r, \theta)$$

$$= \left[\frac{\partial \phi(r, \theta)}{\partial r}\right]\left(\frac{1}{M} - 1\right)r + \cdots, \tag{3}$$

where terms higher than first order can be neglected if M is sufficiently close to 1, and a small region is examined. Note that this error is similar to a radial shear. When the CGH is plotted, alignment aids, which can help in obtaining the proper hologram size, must be drawn on the hologram plot.

31.7 EXPERIMENTAL RESULTS

Figure 5 shows the results of using the setup shown in Fig. 4 to measure a 10-cm-diameter F/2 parabola using a CGH generated with an e-beam recorder.[20] The fringes obtained in a Twyman-Green interferometer using a helium-neon source without the CGH present are shown in Fig. 5a. After the CGH is placed in the interferometer, a much less complicated interferogram is obtained as shown in Fig. 5b. The CGH corrects for about 80 fringes of spherical aberration, and makes the test much easier to perform.

To illustrate the potential of a combined CGH/null-lens test, results for a CGH/null-lens test of the primary mirror of an eccentric Cassegrain system with a departure of approximately 455 waves (at 514.5 nm) and a maximum slope of approximately 1500 waves per radius are shown.[17] The mirror was a 69-cm-diameter off-axis segment whose center lies 81 cm from the axis of symmetry of the parent aspheric surface. The null optics was a

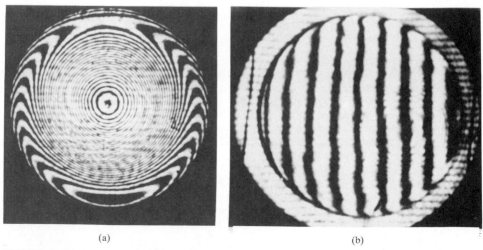

(a) (b)

FIGURE 5 Results obtained testing a 10-cm-diameter F/2 parabola: (*a*) without using CGH; (*b*) using CGH made using an e-beam recorder.

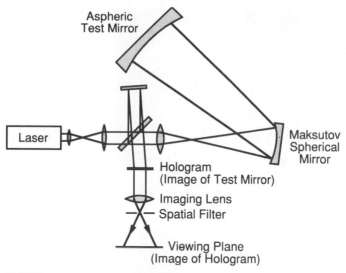

FIGURE 6 Setup to test the primary mirror of a Cassegrain telescope using a Maksutov sphere as a partial null and a CGH.

Maksutov sphere (as illustrated in Fig. 6), which reduces the departure and slope of the aspheric wavefront from 910 to 45 waves, and 300 to 70 waves per radius, respectively. A hologram was then used to remove the remaining asphericity.

Figure 7a shows interferograms of the mirror under test obtained using the CGH-Maksutov test. Figure 7b shows the results when the same test was performed using a rather expensive refractive null lens. When allowance is made for the fact that the interferogram obtained with the null lens has much more distortion than the CGH-Maksutov interferogram, and for the difference in sensitivity ($\lambda = 632.8$ nm for the null-lens test and 514.5 nm for the CGH-Maksutov test), the results for the two tests are seen to be very similar. The "hills" and "valleys" on the mirror surface appear the same for both tests, as expected. The peak-to-peak surface error measured using the null lens was 0.46 waves (632.8 nm), while for the CGH-Maksutov test it was 0.39 waves (514.5 nm). The rms surface error measured was 0.06 waves (632.8 nm) for the null lens, while the CGH-Maksutov test gave 0.07 waves (514.5 nm). These results certainly demonstrate that expensive null optics can be replaced by a combination of relatively inexpensive null optics and a CGH.

The difficult problem of testing aspheric surfaces, which are becoming increasingly popular in optical design, is made easier by the use of CGHs. The technology has reached the point that commercial interferometers using computer-generated holograms are now available.[24,25] The main problem with testing aspheric optical elements is reducing the aberration sufficiently to ensure that light gets back through the interferometer. Combinations of simple null optics with a CGH to perform a test enable the measurement of a wide variety of optical surfaces. The making and use of a CGH are analogous to using an interferometer setup that yields a large number of interference fringes, and measuring the interferogram at a large number of data points. Difficulties involved in recording and analyzing a high-density interferogram and making a CGH are very similar. In both cases, a large number of data points are necessary, and the interferometer must be ray traced so that the aberrations due to the interferometer are well known. The advantage of the CGH technique is that once the CGH is made, it can be used for testing a single piece of optics many times or for testing several identical optical components.

(a)

(b)

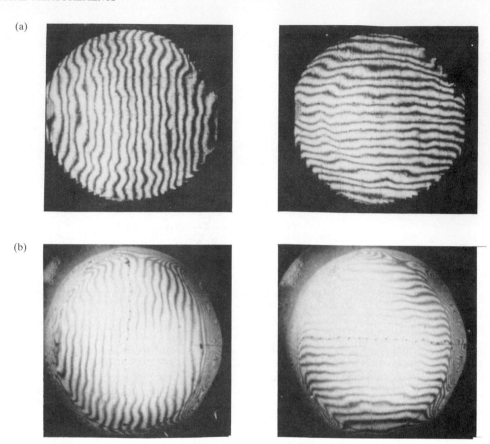

FIGURE 7   Results obtained using Fig. 6: (*a*) CGH-Maksutov test ($\lambda = 514.5\,\text{nm}$); (*b*) using null lens ($\lambda = 632.8\,\text{nm}$).

31.8   REFERENCES

1. J. Pastor, "Hologram Interferometry and Optical Technology," *Appl. Opt.* **8**(3):525–531 (1969).

2. A. J. MacGovern and J. C. Wyant, "Computer Generated Holograms for Testing Optical Elements," *Appl. Opt.* **10**(3):619–624 (1971).

3. J. C. Wyant and V. P. Bennett, "Using Computer Generated Holograms to Test Aspheric Wavefronts," *Appl. Opt.* **11**(12):2833–2839 (1972).

4. A. F. Fercher and M. Kriese, "Binare Synthetische Hologramme zur Prüfung Aspharischer Optischer Elemente," *Optik* **35**(2):168–179 (1972).

5. Y. Ichioka and A. W. Lohmann, "Interferometric Testing of Large Optical Components with Circular Computer Holograms," *Appl. Opt.* **11**(11):2597–2602 (1972).

6. J. Schwider and R. Burrow, "The Testing of Aspherics by Means of Rotational-Symmetric Synthetic Holograms," *Optica Applicata* **6**:83 (1976).

7. T. Yatagai and H. Saito, "Interferometric Testing with Computer-Generated Holograms: Aberration Balancing Method and Error Analysis," *Appl. Opt.* **17**(4):558–565 (1978).

8. W. H. Lee, "Computer-Generated Holograms: Techniques and Applications," in E. Wolf (ed.), *Progress in Optics,* vol. XVI, North Holland Publishing Co., Inc., Amsterdam, 1978, pp. 121–232.

9. J. Schwider, R. Burow, and J. Grzanna, "CGH—testing of Rotational Symmetric Aspheric in Compensated Interferometers," *Optica Applicata* **9:**39 (1979).

10. W. H. Lee, "Recent Developments in Computer-generated Holograms," *Proc. SPIE* **25:**52–58 (1980).

11. A. W. Lohmann and D. P. Paris, "Binary Fraunhofer Holograms, Generated by Computer," *Appl. Opt.* **6**(10):1739–1748 (1967).

12. Lee, W. H., "Sampled Fourier Transform Hologram Generated by Computer," *Appl. Opt.* **9**(3):639–641 (1970).

13. W. H. Lee, "Binary Synthetic Holograms," *Appl. Opt.* **13**(7):1677–1682 (1974).

14. J. R. Goodman, *Introduction to Fourier Optics,* McGraw-Hill, New York, 1968.

15. J. S. Loomis, "Applications of Computer-generated Holograms in Optical Testing," Ph.D. dissertation, Optical Sciences Center, University of Arizona, Tucson, AZ (University Microfilms, Ann Arbor, MI, 1980a).

16. J. S. Loomis, "Computer-generated Holography and Optical Testing," *Opt. Eng.* **19**(5):679–685 (1980b).

17. J. C. Wyant and P. K. O'Neill, "Computer Generated Hologram; Null Lens Test of Aspheric Wavefronts," *Appl. Opt.* **13**(12):2762–2765 (1974).

18. J. C. Wyant, P. K. O'Neill, and A. J. MacGovern, "Interferometric Method of Measuring Plotter Distortion," *Appl. Opt.* **13**(7):1549–1551 (1974).

19. P. M. Emmel and K. M. Leung, "A New Instrument for Routine Optical Testing of General Aspherics," *Proc. SPIE* **171:**93–99 (1979).

20. K. M. Leung, J. C. Lindquist, and L. T. Shepherd, "E-beam Computer Generated Holograms for Optical Testing," *Proc. SPIE* **215:**70–75 (1980).

21. K. M. Leung, S. M. Arnold, and J. C. Lindquist, "Using E-beam Written Computer-generated Holograms to Test Deep Aspheric Wavefronts," *Proc. SPIE* **306:**161–167 (1981).

22. A. F. Fercher, "Computer Generated Holograms for Testing Optical Elements: Error Analysis and Error Compensation," *Opt. Acta* **23**(5):347–365 (1976).

23. J. Pastor, G. E. Evans, and J. S. Harris, "Hologram-Interferometry: A Geometrical Approach," *Opt. Acta* **17**(2):81–96 (1970).

24. S. M. Arnold, "How to Test an Asphere with a Computer Generated Hologram," *Proc. SPIE* **1052:**191–197 (1989).

25. S. M. Arnold and A. K. Jain, "An Interferometer for Testing of General Aspherics Using Computer Generated Holograms," *Proc. SPIE* **1396:**473–480 (1990).

CHAPTER 32
TRANSFER FUNCTION TECHNIQUES

Glenn D. Boreman
*The Center for Research and
Education in Optics and Lasers (CREOL)
University of Central Florida
Orlando, Florida*

32.1 GLOSSARY

B	spot full width
CTF	contrast transfer function (square wave response)
$e(x)$	edge response
FN	focal ratio
$F(\xi, \eta)$	Fourier transform of $f(x, y)$
$f(x, y)$	object function
$G(\xi, \eta)$	Fourier transform of $g(x, y)$
$g(x, y)$	image function
$H(\xi, \eta)$	Fourier transform of $h(x, y)$
$h(x, y)$	impulse response
$\ell(x)$	line response
$S(\xi, \eta)$	power spectrum
W	detector dimension
$\delta(x)$	delta function
$\theta(\xi, \eta)$	phase transfer function
$**$	two-dimensional convolution

32.2 INTRODUCTION

Transfer functions are a powerful tool for analyzing optical and electro-optical systems. The interpretation of objects and images in the frequency domain makes available the whole range of linear-systems analysis techniques. This approach can facilitate insight, particularly in the treatment of complex optical problems. For example, when several optical subsystems are combined, the overall transfer function is the multiplication of the individual transfer functions. The corresponding analysis, without the use of transfer functions, requires convolution of the corresponding impulse responses.

32.3 DEFINITIONS

The image quality of an optical or electro-optical system can be characterized by either the system's impulse response or its Fourier transform, the transfer function. The impulse response $h(x, y)$ is the two-dimensional image formed in response to a delta-function object. Because of the limitations imposed by diffraction and aberrations, the image quality produced depends on the following: the wavelength distribution of the source; the F-number (FN) at which the system operates; the field angle at which the point source is located; and the choice of focus position.

A continuous object $f(x, y)$ can be decomposed, using the sifting property of delta functions, into a set of point sources, each with a strength proportional to the brightness of the object at that location. The final image $g(x, y)$ obtained is the superposition of the individually weighted impulse responses. This result is equivalent to the convolution of the object with the impulse response:

$$f(x, y)**h(x, y) = g(x, y) \tag{1}$$

where the double asterisk denotes a two-dimensional convolution.

The validity of Eq. (1) requires shift invariance and linearity. Shift invariance is necessary for the definition of a single impulse response and linearity is necessary for the superposition of impulse responses. These assumptions are often violated in practice, but the convenience of a transfer-function analysis dictates that we preserve this approach if possible. While most optical systems are linear, electro-optical systems that include a receiver (such as photographic film, detector arrays, and xerographic media) are often nonlinear. A different impulse response (and hence transfer function) is obtained for inputs of different strengths. In optical systems with aberrations that depend on field angle, separate impulse responses are defined for different regions of the image plane.

Although $h(x, y)$ is a complete specification of image quality (given a set of optical parameters), additional insight is gained by use of the transfer function. A transfer-function analysis considers the imaging of sinusoidal objects, rather than point objects. It is more convenient than an impulse-response analysis because the combined effect of two or more subsystems can be calculated by a point-by-point multiplication of the transfer functions, rather than by convolving the individual impulse responses. Using the convolution theorem of Fourier transforms, we can rewrite the convolution of Eq. (1) as a multiplication of the corresponding spectra:

$$F(\xi, \eta) \times H(\xi, \eta) = G(\xi, \eta) \tag{2}$$

where the uppercase variables denote the Fourier transforms of the corresponding lowercase variables: $F(\xi, \eta)$ is the object spectrum; $G(\xi, \eta)$ is the image spectrum; $H(\xi, \eta)$ is the spectrum of the impulse response. As a transfer function, $H(\xi, \eta)$ multiplies the object spectrum to yield the image spectrum. The variables ξ and η are spatial frequencies in the x and y directions. Spatial frequency is the reciprocal of the crest-to-crest distance of a sinusoidal waveform used as a basis function in the Fourier analysis of an object or image. In two dimensions, a sinusoid of arbitrary orientation has a spatial period along both the x and y axes. The reciprocals of these spatial periods are the spatial frequencies ξ and η. Typical units of spatial frequency are cycles/mm when describing an image, and cycles/milliradian when describing an object at a large distance. For an object located at infinity, these two representations are related through the focal length of the image-forming optical system:

$$\xi_{\text{angular}}[\text{cycles/mrad}] = 0.001 \times \xi \, [\text{cycles/mm}] \times f[\text{mm}] \tag{3}$$

The function $H(\xi, \eta)$ in Eq. (2) is usually normalized to have unit value at zero frequency. This yields a transfer function relative to the response at low frequency, and ignores frequency-independent attenuations, such as losses caused by Fresnel reflection or by obscurations. This normalization is appropriate for most optical systems, because[1] the transfer function of an incoherent optical system is proportional to the two-dimensional autocorrelation of the exit pupil, which is maximum at zero frequency. For more general imaging systems (for example, the human eye, photographic film, and electronic imaging systems), the transfer function is not necessarily maximum at the origin, and may be more useful in an unnormalized form.

With the above normalization, $H(\xi, \eta)$ is called the optical transfer function (OTF). In general, OTF is a complex function, having both a magnitude and a phase portion:

$$\text{OTF}(\xi, \eta) = H(\xi, \eta) = |H(\xi, \eta)| \exp\{-j\theta(\xi, \eta)\} \tag{4}$$

The magnitude of the OTF, $|H(\xi, \eta)|$, is referred to as the modulation transfer function (MTF), while the phase portion of the OTF, $\theta(\xi, \eta)$, is referred to as the phase transfer function (PTF).

MTF is the magnitude response of the imaging system to sinusoids of different spatial frequencies. This response is described in terms of the modulation depth, a measure of visibility or contrast:

$$M = \frac{A_{\max} - A_{\min}}{A_{\max} + A_{\min}} \tag{5}$$

where A refers to a value of the waveform (typically W/cm^2 vs position) that describes the object or image. These quantities are nonnegative, so the sinusoids always have a dc bias. Modulation depth is thus a number between 0 and 1. The effect of the finite-size impulse response is that the modulation depth in the image is less than that in the object. This attenuation is usually more severe at high frequencies. MTF is the ratio of image modulation to object modulation, as a function of spatial frequency:

$$\text{MTF}(\xi, \eta) = \frac{M_{\text{image}}(\xi, \eta)}{M_{\text{object}}(\xi, \eta)} \tag{6}$$

PTF describes the relative phases with which the various sinusoidal components recombine in the image. A linear phase such as $\text{PTF} = x_0\xi$ corresponds to a shift of the image by an amount x_0, each frequency component being shifted the amount required to reproduce the original waveform at the displaced location. For impulse responses that are symmetric about the ideal image point, the PTF exhibits phase reversals, with a value of either 0 or π radians as a function of spatial frequency. A general impulse response that is real but not even yields a PTF that is a nonlinear function of frequency, resulting in image degradation. Linearity of PTF is a sensitive test for aberrations (such as coma) which produce asymmetric impulse responses, and is often a design criterion.

32.4 MTF CALCULATIONS

OTF can be calculated from wave-optics considerations. For an incoherent optical system, the OTF is proportional to the two-dimensional autocorrelation of the exit pupil. This calculation can account for any phase factors across the pupil, such as those arising from aberrations or defocus. A change of variables is required for the identification of an autocorrelation (a function of position in the pupil) as a transfer function (a function of image-plane spatial frequency). The change of variables is

$$\xi = \frac{x}{\lambda d_i} \qquad (7)$$

where x is the autocorrelation shift distance in the pupil, λ is the wavelength, and d_i is the distance from the exit pupil to the image. A system with an exit pupil of full width D has an image-space cutoff frequency consistent with Eq. (7):

$$\xi_{\text{cutoff}} = \frac{1}{(\lambda \text{ FN})} \qquad (8)$$

where FN equals (focal length)/D for a system with the object at infinity, and d_i/D for a system operating at finite conjugates.

A diffraction-limited system has a purely real OTF. Diffraction-limited MTFs represent the best performance that a system can achieve, for a given FN and λ, and accurately describe systems with negligible aberrations, whose impulse-response size is dominated by diffraction effects. A diffraction-limited system with a square exit pupil of dimensions $D \times D$ has a linear MTF along ξ or η:

$$\text{MTF}\left(\frac{\xi}{\xi_{\text{cutoff}}}\right) = 1 - \frac{\xi}{\xi_{\text{cutoff}}} \qquad (9)$$

For a system with a circular exit pupil of diameter D, the MTF is circularly symmetric, with ξ profile[2]:

$$\text{MTF}\left(\frac{\xi}{\xi_{\text{cutoff}}}\right) = \frac{2}{\pi}\left\{\cos^{-1}\left(\frac{\xi}{\xi_{\text{cutoff}}}\right) - \frac{\xi}{\xi_{\text{cutoff}}}\left[1 - \left(\frac{\xi}{\xi_{\text{cutoff}}}\right)^2\right]^{1/2}\right\}, \text{ if } \xi \leq \xi_{\text{cutoff}}$$

$$= 0, \text{ if } \xi > \xi_{\text{cutoff}} \qquad (10)$$

Equation (10) is plotted in Fig. 1, along with MTF curves obtained for annular pupils, which arise in obscured systems such as Cassegrain telescopes. The plots are functions of the obsuration ratio, and the emphasis at high frequencies has been obtained by an overall decrease in flux reaching the image, proportional to the obscured area. If the curves in Fig. 1 were plotted without normalization to 1 at $\xi = 0$, they would all be contained under the envelope of the unobscured diffraction-limited curve.

A system exhibiting effects of both diffraction and aberrations has an MTF curve

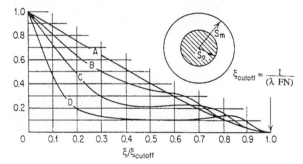

FIGURE 1 (*a*) Diffraction-limited MTF for system with circular pupil (no obscuration: $S_o/S_m = 0$). (*b*) through (*d*) are diffraction-limited MTF for a system with an annular pupil: (*b*) $S_o/S_m = 0.25$; (*c*) $S_o/S_m = 0.5$; (*d*) $S_o/S_m = 0.75$. (*Adapted from W. J. Smith,* Modern Optical Engineering, *McGraw-Hill, New York, 1966, p. 322.*)

bounded by the diffraction-limited MTF curve as the upper envelope. Aberrations broaden the impulse response, resulting in a narrower and lower MTF, with less integrated area. The area under the MTF curve relates to a figure of merit called the Strehl ratio S, a number between 0 and 1, defined as the irradiance at the center of the actual impulse response divided by that at the center of a diffraction-limited impulse response. Using the central-ordinate theorem for Fourier transforms, S can be written as the ratio of the area under the actual MTF curve to that under the diffraction-limited MTF curve:

$$S = \frac{h_{\text{actual}}(0,0)}{h_{\text{diffraction-limited}}(0,0)} = \frac{\iint \text{MTF}_{\text{actual}}(\xi, \eta)\, d\xi\, d\eta}{\iint \text{MTF}_{\text{diff ltd}}(\xi, \eta)\, d\xi\, d\eta} \tag{11}$$

The effect of defocus on the MTF is shown in Fig. 2. The MTF curves resulting from

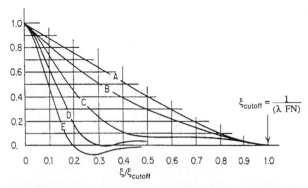

FIGURE 2 Diffraction MTF for a defocused system: (*a*) in focus, OPD = 0.0; (*b*) defocus = $\lambda/2N \sin^2 u$, OPD = $\lambda/4$; (*c*) defocus = $\lambda/N \sin^2 u$, OPD = $\lambda/2$; (*d*) defocus = $3\lambda/2N \sin^2 u$, OPD = $3\lambda/4$; (*e*) defocus = $2\lambda/N \sin^2 u$, OPD = λ. (*Adapted from W. J. Smith,* Modern Optical Engineering, *McGraw-Hill, New York, 1966, p. 320.*)

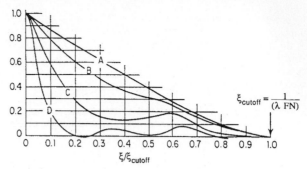

FIGURE 3 Diffraction MTF for system with third-order spherical aberration (image plane midway between marginal and paraxial foci): (*a*) $LA_m = 0.0$, $OPD = 0$; (*b*) $LA_m = 4\lambda/N \sin^2 u$, $OPD = \lambda/4$; (*c*) $LA_m = 8\lambda/N \sin^2 u$, $OPD = \lambda/2$; (*d*) $LA_m = 16\lambda/N \sin^2 u$, $OPD = \lambda$. (*Adapted from W. J. Smith*, Modern Optical Engineering, *McGraw-Hill, New York, 1966, p. 322.*)

small amounts of third-order spherical aberration are shown in Fig 3. MTF results for specific cases of other aberrations, including higher-order spherical, coma, astigmatism, and chromatic, are contained in Ref. 3.

A geometrical-aberration OTF can be calculated from ray-trace data, without regard for diffraction effects. Optical-design computer programs typically yield a diagram of ray-intersection density in the image plane, a geometrical-optics spot diagram. A geometrical-aberration OTF is calculated by Fourier transforming the spot-density distribution. The OTF thus obtained is accurate if the impulse-response size is dominated by aberration effects. A one-dimensional uniform blur spot of full width B has the following OTF in the ξ direction:

$$OTF(\xi) = \frac{\sin(\pi\xi B)}{\pi\xi B} \tag{12}$$

which has a zero at $\xi = 1/B$, and also exhibits the phase reversals mentioned above. When an MTF has been calculated from ray trace data, an approximation to the total system MTF may be made[4] by multiplying the diffraction-limited MTF of the proper FN and λ with the ray-trace data MTF. This is equivalent to a convolution of the spot profiles from diffraction and geometrical aberrations.

In electronic imaging systems, an electronics subsystem performs signal-handling and signal-processing functions. The performance characterization of electronic networks by transfer-function techniques is well established. The usual independent variable for these time-domain transfer functions is the temporal frequency f(Hz). To interpret the electronics transfer function in the same units as the image-plane spatial frequency (cycles/mm), the temporal frequencies are divided by the scan velocity (mm/s). For a scanning system, this is the velocity of the instantaneous field of view, referred to as image coordinates. For a staring system, an effective scan velocity is the horizontal dimension of the image plane divided by the video line time. With this change of variables from temporal frequencies to spatial frequencies, the electronics can be analyzed as simply an additional subsystem, with its own transfer function that will multiply the transfer functions of the other subsystems. It should be noted that an electronics transfer function is not bounded by a pupil autocorrelation the way an optical transfer function is. Thus, it need not be maximum at the origin, and can amplify certain frequencies and have sharp cutoffs at others. Thus, the usual normalization of MTF may not be appropriate for analysis of the electronics subsystems, or for the entire imaging system including the electronics.

An unavoidable impact of the electronics subsystem is the contribution of noise to the image. This limits the amount of electronic amplification that is useful in recovering modulation depth lost in other subsystems. A useful figure of merit, which has been validated to correlate with image visibility,[5] is the area between two curves: the MTF and the noise power spectrum. To facilitate comparison on the same graph, the noise power spectrum is expressed in modulation depth units, and is interpreted as a noise-equivalent modulation depth (the modulation needed for unit signal-to-noise ratio) as a function of spatial frequency.

The detector photosensitive area has finite size, rather than being a true point. It thus performs some spatial averaging[6] on any irradiance distribution that falls on it. Large detectors exhibit more attenuation of high spatial frequencies than do small detectors. For a detector of dimension W in the x direction, the MTF is

$$\text{MTF}(\xi) = \left| \frac{\sin(\pi \xi W)}{\pi \xi W} \right| \tag{13}$$

which has a zero at $\xi = 1/W$. This MTF component applies to any system with detectors, and will multiply the MTFs of other subsystems.

In electronic imaging systems, the image is typically sampled in both directions. The distance between samples will determine the image-plane spatial frequency at which aliasing artifacts will occur. Care must be taken in the calculation of MTF, because different impulse responses are possible depending on the location of the impulse response with respect to the sampling positions. This violates the assumption of shift-invariance needed for a transfer-function analysis.[7] One approach for defining a generalized MTF is to average over all possible positions of the impulse response with respect to the sampling lattice (Eq. 4 in Ref. 8). Research is still underway on the specification of MTF for sampled-image systems.

32.5 MTF MEASUREMENTS

In any situation where the measurement of MTF involves the detection of the image-plane flux, one component of the measurement-system MTF is caused by the finite aperture of the detector, which can be accounted for in the calibration of the instrument by dividing out the detector MTF seen in Eq. (13).

When OTF is measured with a point-source object, the image formed by the system under test is the impulse response. The two-dimensional impulse response can be Fourier transformed in two dimensions to yield $\text{OTF}(\xi, \eta)$. If an illuminated pinhole is used, it should be as small as possible. However, flux-detection considerations dictate a finite size for any source. The object is small enough not to affect the measurement if its angular subtense is much smaller than the angular subtense of the impulse response, when both are viewed from the aperture stop of the system. For sources of larger extent, a Fourier analysis can be made of the object, and an OTF can be calculated using Eq. (2), over the range of spatial frequencies provided by the source.

If higher flux levels are needed to maintain signal-to-noise ratio, a line response can be measured. The system under test is presented with an illuminated line source, which acts as a delta function in one direction and a constant in the other: $\delta(x)1(y)$. The system forms an image, the line response $\ell(x)$, which is a summation of vertically displaced impulse responses. In general $\ell(x) \neq h(x, 0)$.[2] The line response only yields information about one profile of $\text{OTF}(\xi, \eta)$. The one-dimensional Fourier transform of the line response produces the corresponding profile of the two-dimensional OTF: $\mathscr{F}\{\ell(x)\} = \text{OTF}(\xi, 0)$. To obtain other profiles of the OTF, the line source is reoriented. Line response data are also available from the response of the system to a point source, using a receiver that integrates the impulse response along one direction: a detector that is long in one dimension and is scanned perpendicularly, or a long slit that is scanned in front of a large-area detector.

Another measurement of OTF uses the edge response $e(x)$, which is the response of the system to an illuminated knife edge. Each line in the open part of the aperture produces a displaced line response, so $e(x)$ is a cumulative distribution, related to the line response as follows: $d/dx\{e(x)\} = \ell(x)$, which Fourier transforms to the ξ profile of the OTF. The derivative operation increases the effect of noise. Any digital filter used for data smoothing has its own impulse response, and hence its own OTF contribution. The edge response can also be measured by using a scanning knife edge in front of a detector in the image plane, with a point-source or a line-source object.

An MTF calculated from a measured profile is the product of a diffraction MTF and a geometrical-aberration MTF. When combining the separately-measured MTFs of several optical subsystems, care should be taken to ensure that the diffraction MTF (determined by the aperture stop of the combined system) contributes only once to the calculation. The geometrical-aberration MTFs for each subsystem will cascade if each subsystem operates independently on an irradiance basis, with no partial coherence effects.[9] The major exception to this condition occurs when two subsystems are designed to correct for each other's aberrations, and the MTF of the combined system is better than the individual MTFs would indicate.

MTF can also be obtained by the system's response to a sine-wave target, where the image modulation depth is measured as a function of spatial frequency. PTF can also be measured from the position of the waveform maxima as a function of frequency. Sine-wave targets are available as photographic prints or transparencies, which are suitable for testing visible-wavelength systems. Careful control in their manufacture is exercised[10] to avoid harmonic distortions, including a limitation to relatively small modulation depths. Sine-wave targets are difficult to fabricate for testing infrared systems, and require the use of half-tone techniques.

A more convenient target to manufacture is the three- or four-bar target of equal line and space width, with a binary transmission or reflection characteristic. These are widely used for testing both visible-wavelength and infrared systems. The square-wave response is called the contrast transfer function (CTF) and is not equivalent to the sine-wave response for which MTF is defined. CTF is a function of the fundamental spatial frequency ξ_f inverse of the bar-to-bar spacing) and is measured on the peak-to-valley variation of image (irradiance. For any particular fundamental frequency, the measured response to bar targets will be higher than that measured for sinewaves of the same frequency, because additional harmonic components contribute to the modulation. For a square-wave pattern of infinite extent, an analytical relationship exists[11] between CTF(ξ_f) and MTF(ξ). Each Fourier component of the square wave has a known transfer factor given by MTF(ξ), and the modulation depth as a function of ξ_f of the resultant waveform can be calculated by Eq. (5). This process yields the following series:

$$CTF(\xi_f) = \frac{4}{\pi}\{MTF(\xi_f) - \tfrac{1}{3}MTF(3\xi_f) + \tfrac{1}{5}MTF(5\xi_f) - \tfrac{1}{7}MTF(7\xi_f) + \tfrac{1}{9}MTF(9\xi_f) - \cdots\} \quad (14)$$

CTFs for the practical cases of three- and four-bar targets are slightly higher than the CTF curve for an infinite square wave. Figure 4 compares the MTF for a diffraction-limited circular-aperture system with CTFs obtained for infinite, three-, and four-bar targets. Because of the broad spectral features associated with bar patterns of limited extent, a finite level of modulation is present in the image, even when the fundamental frequency of the bar pattern equals the cutoff frequency of the system MTF.[12] The inverse process of expressing the MTF in terms of CTFs is more difficult analytically, since square waves are not an orthogonal basis set for the expansion of sinusoids. A term-by-term series subtraction[11] yields the following:

$$MTF(\xi_f) = \frac{\pi}{4}\{CTF(\xi_f) + \tfrac{1}{3}CTF(3\xi_f) - \tfrac{1}{5}CTF(5\xi_f) + \tfrac{1}{7}CTF(7\xi_f) + \tfrac{1}{11}CTF(11\xi_f) - \cdots\} \quad (15)$$

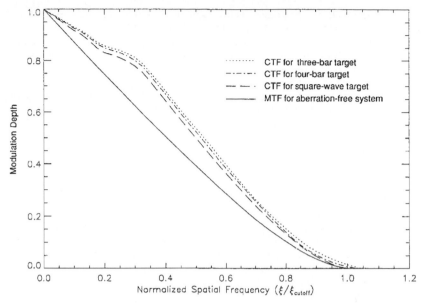

FIGURE 4 Comparison of MTF to CTFs obtained with infinite square wave, four-bar, and three-bar targets for a diffraction-limited system with circular pupil.

Narrowband electronic filtering can be used to isolate the fundamental spatial-frequency component for systems where the image data are available as a time-domain waveform. These systems do not require the correction of Eq. (15), because the filter converts bar-target data to sinewave data.

The MTF can also be measured by the response of the system to a random object. Laser speckle provides a convenient means to generate a random object distribution of known spatial-frequency content. The MTF relates the input and output spatial-frequency power spectra of the irradiance waveforms:

$$S_{\text{output}}(\xi, \eta) = |\text{MTF}(\xi, \eta)|^2 \times S_{\text{input}}(\xi, \eta) \tag{16}$$

This method is useful in the measurement of an average MTF for sampled-image systems,[13] since the speckle pattern has a random position with respect to the sampling sites.

A number of interferometric methods have been developed for measuring MTF.[14] An interferogram of the wavefront exiting the system is reduced to find the phase map. The distribution of amplitude and phase across the exit pupil contains the information necessary for calculation of OTF by pupil autocorrelation.

32.6 REFERENCES

1. J. W. Goodman, *Introduction to Fourier Optics,* McGraw-Hill, New York, 1968, pp. 116–120.
2. J. D. Gaskill, *Linear Systems, Fourier Transforms, and Optics,* Wiley, New York, 1978, pp. 305–307.
3. C. S. Williams and O. A. Becklund, *Introduction to the Optical Transfer Function,* Wiley-Interscience, New York, 1989, chap. 9 and app. A.
4. W. J. Smith, *Modern Optical Engineering,* McGraw-Hill, New York, 1966, p. 323.

5. H. L. Snyder, "Image Quality and Observer Performance," in L. M. Biberman (ed.), *Perception of Displayed Information,* Plenum, New York, 1973.

6. G. D. Boreman and A. E. Plogstedt, "Spatial Filtering by a Line-Scanned Nonrectangular Detector-Application to SPRITE Readout MTF," *Appl. Opt.* **28:**1165–1168, 1989.

7. W. Wittenstein, J. C. Fontanella, A. R. Newbery, and J. Baars, "The Definition of the OTF and the Measurement of Aliasing for Sampled-Imaging Systems," *Optica Acta* **29**(1):41–50 (1982).

8. S. K. Park, R. Schowengerdt, and M. Kaczynski, "Modulation-Transfer-Function Analysis for Sampled Image Systems," *Appl. Opt.* **23:**2572 (1984).

9. J. B. DeVelis and G. B. Parrent, "Transfer Function for Cascaded Optical Systems," *J. Opt. Soc. Am.* **57:**1486–1490 (1967).

10. R. L. Lamberts, "The Production and Use of Variable-Transmittance Sinusoidal Test Objects," *Appl. Opt.* **2:**273–276 (1963).

11. J. W. Coltman, "The Specification of Imaging Properties by Response to a Sine Wave Input," *J. Opt. Soc. Am.* **44:**468 (1954).

12. D. H. Kelly, "Spatial Frequency, Bandwidth, and Resolution", *Appl. Opt.* **4:**435 (1965).

13. G. D. Boreman and E. L. Dereniak, "Method for Measuring Modulation Transfer Function of Charge-Coupled Devices Using Laser Speckle," *Opt. Eng.* **25:**148 (1986).

14. D. Malacara, *Optical Shop Testing,* Wiley, New York, 1978, chap. 3.

OPTICAL AND PHYSICAL PROPERTIES OF MATERIALS

CHAPTER 33
PROPERTIES OF CRYSTALS AND GLASSES

William J. Tropf, Michael E. Thomas, and Terry J. Harris
Applied Physics Laboratory
Johns Hopkins University
Laurel, Maryland

33.1 GLOSSARY

A_i, B, C, D, E, G	constants
a, b, c	crystal axes
B	inverse dielectric constant
B	bulk modulus
C	heat capacity
c	speed of light
$\mathbf{c}$	elastic stiffness
$\mathbf{D}$	electric displacement
d	piezoelectric coefficient
$d_{ij}^{(2)}$	nonlinear optical coefficient
E	Young's modulus
E	energy
$\mathbf{E}$	electric field
$\mathbf{e}$	strain
G	shear modulus
g	degeneracy
Hi	Hilbert transform
h	heat flow
k	extinction coefficient
k_B	Boltzmann constant
ℓ	phonon mean free path
MW	molecular weight
m	integer
$N(\ )$	occupation density

n	refractive index
$\tilde{n}$	complex refractive index $= n + ik$
$\mathbf{P}$	electric polarization
$P_{x,y}$	relative partial dispersion
$\mathbf{p}$	elasto-optic tensor
p	elasto-optic compliance
p	pyroelectric constant
$\mathbf{q}$	piezo-optic tensor
r	electro-optic coefficient
r	amplitude reflection coefficient
r_{ij}	electro-optic coefficient
$S(\ )$	line strength
$\mathbf{s}$	elastic compliance
T	temperature
t	amplitude transmission coefficient
U	enthalpy
u	atomic mass unit
V	volume
v	velocity of sound
x	displacement
x	variable of integration
Z	formulas per unit
α	linear expansion coefficient
α	intensity absorption
α	thermal expansion
α_m	macroscopic polarizability
α, β, γ	crystal angles
β	power absorption coefficient
$\gamma(\ )$	line width
γ	Gruneisen parameter
ϵ	dielectric constant, permittivity
ϵ	emittance
θ_D	Debye temperature
κ	thermal conductivity
$\Lambda(\ )$	complex function
μ	permeability
ν	wave number $(\omega/2\pi c)$
ρ	density
ρ	intensity reflectivity
σ	stress
τ	intensity transmission
τ	power transmittance

χ	susceptibility
$\chi^{(2)}$	second-order susceptibility
Ω	solid angle
ω	radian frequency

Subscripts

ABS	absorptance
bb	blackbody
c	656.3 nm
d	587.6 nm
EXT	extinctance
F	486.1 nm
i	integers
0	vacuum, $T = 0$, or constant terms
P	constant pressure
p, s	polarization component
r	relative
SCA	scatterance
V	constant volume

33.2 INTRODUCTION

Nearly every nonmetallic crystalline and glassy material has a potential use in optics. If a nonmetal is sufficiently dense and homogeneous, it will have good optical properties. Generally, a combination of desirable optical properties, good thermal and mechanical properties, and cost and ease of manufacture dictate the number of readily available materials for any application. In practice, glasses dominate the available optical materials for several important reasons. Glasses are easily made of inexpensive materials, and glass manufacturing technology is mature and well-established. The resultant glass products can have very high optical quality and meet most optical needs.

Crystalline solids are used for a wide variety of specialized applications. Common glasses are composed of low-atomic-weight oxides and therefore will not transmit beyond about 2.5 μm. Some crystalline materials transmit at wavelengths longer (e.g., heavy-metal halides and chalcogenides) or shorter (e.g., fluorides) than common glasses. Crystalline materials may also be used for situations that require the material to have very low scatter, high thermal conductivity, or high hardness and strength, especially at high temperature. Other applications of crystalline optical materials make use of their directional properties, particularly those of noncubic (i.e., uni- or biaxial) crystals. Phasematching (e.g., in wave mixing) and polarization (e.g., in wave plates) are example applications.

This chapter gives the physical, mechanical, thermal, and optical properties of selected crystalline and glassy materials. Crystals are chosen based on availability of property data and usefulness of the material. Unfortunately, for many materials, property data are imprecise, incomplete, or not applicable to optical-quality material. Glasses are more accurately and uniformly characterized, but their optical property data are usually limited to wavelengths below 1.06 μm. Owing to the preponderance of glasses, only a representative small fraction of available glasses are included below. SI derived units, as commonly applied in material characterization, are used.

Property data are accompanied with brief explanations and useful functional relationships. We have extracted property data from past compilations[1-11] as well as recent literature. Unfortunately, property data are somewhat sparse. For example, index data may be available for only a portion of the transparent region or the temperature dependence of the index may not be known. Strength of many materials is poorly characterized. Thermal conductivity is frequently unavailable and other thermal properties are usually sketchy.

33.3   OPTICAL MATERIALS

Crystalline and amorphous (including glass) materials are differentiated by their structural (crystallographic) order. The distinguishing structural characteristic of amorphous substances is the absence of long-range order; the distinguishing characteristics of crystals are the presence of long-range order. This order, in the form of a periodic structure, can cause directional-dependent (anisotropic) properties that require a more complex description than needed for isotropic, amorphous materials. The periodic features of crystals are used to classify them into six crystal systems,* and further arrange them into 14 (Bravais) space lattices, 32 point groups, and 230 space groups based on the characteristic symmetries found in a crystal.

Glass is by far the most widely used optical material, accounting for more than 90 percent of all optical elements manufactured. Traditionally, glass has been the material of choice for optical systems designers, owing to its high transmittance in the visible-wavelength region, high degree of homogeneity, ease of molding, shaping, and machining, relatively low cost, and the wide variety of index and dispersion characteristics available.

Under the proper conditions, glass can be formed from many different inorganic mixtures. Hundreds of different optical glasses are available commercially. Primary glass-forming compounds include oxides, halides, and chalcogenides with the most common mixtures being the oxides of silicon, boron, and phosphorous used for glasses transmitting in the visible spectrum. By varying the chemical composition of glasses (glasses are not fixed stoichiometrically), the properties of the glass can be varied. Most notably for optical applications, glass compositions are altered to vary the refractive index, dispersion, and thermo-optic coefficient.

Early glass technologists found that adding BaO offered a high-refractive-index glass with lower than normal dispersion, B_2O_3 offered low index and very low dispersion, and by replacing oxides with fluorides, glasses could be obtained with very low index and very low dispersion. Later, others developed very high index glasses with relatively low dispersions by introducing rare-earth elements, especially lanthanum, to glass compositions. Other compounds are added to silica-based glass mixtures to help with chemical stabilization, typically the alkaline earth oxides and in particular Al_2O_3 to improve the resistance of glasses to attack by water.

To extend the transmission range of glasses into the ultraviolet, a number of fluoride and fluorophosphate glasses have been developed. Nonoxide glasses are used for infrared applications requiring transmission beyond the transmission limit of typical optical glasses (2.4 to 2.7 μm for an absorption coefficient of 1 cm^{-1}). These materials include chalcogenides such as As_2S_3 glass and heavy-metal fluorides such as ZrF_4-based glasses.

Crystalline materials include naturally occurring minerals and manufactured crystals. Both single crystals and polycrystalline forms are available for many materials. Polycrystalline optical materials are typically composed of many small (cf., 50 μm) individual

---

* Cubic (or isometric), hexagonal (including rhombohedral), tetragonal, orthorhombic, monoclinic, and triclinic are the crystallographic systems.

crystals with random orientations and *grain boundaries* between them. These grain boundaries are a form of material defect arising from the lattice mismatch between individual grains. Polycrystalline materials are made by diverse means such as pressing powders (usually with heat applied), sintering, or chemical vapor deposition. Single crystals are typically grown from dissolved or molten material using a variety of techniques. Usually, polycrystalline materials offer greater hardness and strength at the expense of increased scatter.

Uniformity of the refractive index throughout an optical element is a prime consideration in selecting materials for high-performance lenses, elements for coherent optics, laser harmonic generation, and acousto-optical devices. In general, highly pure, single crystals achieve the best uniformity, followed by glasses (especially those selected by the manufacturer for homogeneity), and lastly polycrystalline materials. Similarly, high-quality single crystals have very low scatter, typically one-tenth that of glasses.

Applications requiring optical elements with direction-dependent properties, such as polarizers and index-matching materials for harmonic generation, frequently use single crystals.

Purity of starting materials is a prime factor in determining the quality of the final product. High material quality and uniformity of processing is required to avoid impurity absorption, index nonuniformity, voids, cracks, and bubbles, and excess scatter. Practical manufacturing techniques limit the size of optics of a given material (glasses are typically limited by the moduli, i.e., deformation caused the weight of the piece). Some manufacturing methods, such as hot pressing, also produce significantly lower quality material when size becomes large (especially when the thinnest dimension is significantly increased). Cost of finished optical elements is a function of size, raw material cost, and the difficulty of machining, polishing, and coating the material. Any one of these factors can dominate cost.

General information on the manufacturing methods for glasses and crystalline materials is available in several sources.[9-11] Information on cutting and polishing of optical elements can be found in the literature.[2,12,13]

33.4 PROPERTIES OF MATERIALS

Symmetry Properties

The description of the properties of solids depends on structural symmetry. The structural symmetry crystalline materials dictate the number of appropriate directional-dependent terms that describe a property.[14] *Neumann's principle* states that physical properties of a crystal must possess at least the symmetry of the point group of the crystal. Amorphous (glassy) materials, having no long-range symmetry, are generally considered isotropic,* and require the least number of property terms. Macroscopic material properties are best described in tensor notation; the rank of the property tensors can range from zero (scalar property) up to large rank. Table 1 summarizes common material properties and the rank of the tensor that describes them.

The rank of the property tensor and the symmetry of the material (as determined by the point group for crystals) determines the number of terms needed to describe a property. Table 2 summarizes both the number of terms needed to describe a property and the number of unique terms as a function of property tensor rank and point group. Another important term in the definition of tensor characteristics is the *principal values* of

---

* Glass properties are dependent on cooling rate. Nonuniform cooling will result in density variation and residual stress which cause anisotropy in all properties. Such nonisotropic behavior is different from that of crystals in that it arises from thermal gradients rather than periodic structure order. Hence, the nature of anisotropy in glasses varies from sample to sample.

TABLE 1 Tensor Characteristics and Definitions of Properties

Tensor rank	Property	Symbol	Units	Relationship
0	Enthalpy (energy)	U	J/mole	—
	Temperature	T	K	—
	Heat capacity	C	J/(mole · K)	$C = \partial U / \partial T$
1	Displacement	x	m	—
	Heat flow	h	W/m^2	—
	Electric field	$\mathbf{E}$	V/m	—
	Electric polarization	$\mathbf{P}$	C/m^2	$\mathbf{P} = \epsilon_0 \chi \mathbf{E}$
	Electric displacement	$\mathbf{D}$	C/m^2	$\mathbf{D} = \epsilon_0 \mathbf{E} + \mathbf{P}$
	Pyroelectric constant	p	C/(m^2 · K)	$\Delta \mathbf{P} = p \, \Delta T$
2	Stress	σ	Pa	—
	Strain	e	—	—
	Thermal expansion	α	K^{-1}	$e = \alpha T$
	Thermal conductivity	κ	W/(m · K)	$h = -\kappa(\partial T / \partial x)$
	Dielectric constant (relative permittivity)	ϵ_r	—	$\mathbf{D} = \epsilon_0 \epsilon_r \mathbf{E}$
	Inverse dielectric tensor	$\mathbf{B}$	—	$\mathbf{B} = \epsilon_r^{-1}$
	Susceptibility	χ	—	$\epsilon_r = \chi + 1$
3	Piezoelectric coefficient (modulus)	d	m/V $\equiv$ C/N	$\mathbf{P} = d \cdot \sigma$
	(converse piezoelectric effect)	d	m/V $\equiv$ C/N	$e = d \cdot \mathbf{E}$
	Electro-optic coefficient (linear)	r	m/V	$\Delta \mathbf{B} = r \cdot \mathbf{E}$
	Second-order susceptibility	$\chi^{(2)}$	m/V	$\mathbf{P} = \epsilon_0 \chi^{(2)} \mathbf{E}_1 \mathbf{E}_2$
4	Elastic stiffness	$\mathbf{c}$	Pa	$\sigma = c \cdot e; c = 1/s$
	Elastic compliance	$\mathbf{s}$	Pa^{-1}	$e = s \cdot \sigma; s = 1/c$
	Elasto-optic tensor	$\mathbf{p}$	—	$\Delta \mathbf{B} = p \cdot e; p = q \cdot c$
	Piezo-optic tensor	$\mathbf{q}$	Pa^{-1}	$\Delta \mathbf{B} = q \cdot \sigma; q = p \cdot s$

a second-rank tensor property. The principal values of a property are those values referenced to (measured along) the crystal axes as defined in Table 2. For example, a second-rank tensor of a triclinic crystal has nine nonzero coefficients, and because of symmetry, six independent coefficients. These six coefficients can be separated into (1) three principal values of the quantity (e.g., thermal expansion coefficient, dielectric constant, refractive index, and stress), and (2) the three angles describing the orientation of the crystal axes (α, β, γ).

Properties of materials depend on several fundamental constants that are listed in Table 3.[15] These constants are defined as follows.

Optical Properties: Introduction

Refractive Index. Important optical properties, definitions, formulas, and basic concepts are derived from a classical description of propagation based on the macroscopic Maxwell's equations. The standard wave equation for the electric field $\mathbf{E}$ is obtained from the Faraday, Gauss, and Ampere Laws in Maxwell's form:*

$$\nabla^2 \mathbf{E} = (-i\omega\mu\sigma - \omega^2\mu\epsilon)\mathbf{E} \qquad (1)$$

---

* The definition of the dielectric constant and refractive index in this section is based on a harmonic field of the form $\exp(-i\omega t)$. Other definitions lead to different sign conventions (e.g., $\bar{n} = n - ik$) and care must be taken to insure consistency.

TABLE 2 Crystal Classes and Symmetries

Crystal system	Crystal axes	Space lattice		Point group		Space group nos.	Tensor coefficients*			
		Types	Symmetry	Schönflies	Internat'l		Rank 1	Rank 2	Rank 3	Rank 4
Triclinic (—)	$a \neq b \neq c$ $\alpha \neq \beta \neq \gamma$	P	$\bar{1}$	C_1	1	1	3 (3)	9 (6)	18 (18)	36 (21)
				C_i	$\bar{1}$	2	0	9 (6)	0	36 (21)
Monoclinic (2-fold axis)	$a \neq b \neq c$ $\alpha = \beta = 90°$ $\gamma \neq 90°$	P, I (or C)	$2/m$	C_2	2	3–5	1 (1)	5 (4)	8 (8)	20 (13)
				C_s	m	6–9	2 (2)	5 (4)	10 (10)	20 (13)
				C_{2h}	$2/m$	10–15	0	5 (4)	0	20 (13)
Orthorhombic ($3 \perp$ 2-fold axes)	$a \neq b \neq c$ $\alpha = \beta = \gamma = 90°$	P, I, C, F	mmm	D_2	222	16–24	0	3 (3)	3 (3)	12 (9)
				C_{2v}	$2mm$	25–46	1 (1)	3 (3)	5 (5)	12 (9)
				D_{2h}	mmm	47–74	0	3 (3)	0	12 (9)
Tetragonal (4-fold axis)	$a = b \neq c$ $\alpha = \beta = \gamma = 90°$	P, I	$4/mmm$	C_4	4	75–80	1 (1)	3 (2)	7 (4)	16 (7)
				S_4	$\bar{4}$	81–82	0	3 (2)	7 (4)	16 (7)
				C_{4h}	$4/m$	83–88	0	3 (2)	0	16 (7)
				D_4	422	89–98	0	3 (2)	2 (1)	12 (6)
				C_{4v}	$4mm$	99–110	1 (1)	3 (2)	5 (3)	12 (6)
				D_{2d}	$\bar{4}2m$	111–122	0	3 (2)	3 (2)	12 (6)
				D_{4h}	$4/mmm$	123–142	0	3 (2)	0	12 (6)
Hexagonal (3-fold axis)	$a = b \neq c$ $\alpha = \beta = 90°$ $\gamma = 120°$	P, R	$\bar{3}$	C_3	3	143–146	1 (1)	3 (2)	13 (6)	24 (7)
				C_{3i}	$\bar{3}$	147–148	0	3 (2)	0	24 (7)
			$\bar{3}m$	D_3	32	149–155	0	3 (2)	5 (2)	18 (6)
				C_{3v}	$3m$	156–161	1 (1)	3 (2)	8 (4)	18 (6)
				D_{3d}	$\bar{3}m$	162–167	0	3 (2)	0	18 (6)
(6-fold axis)		P	$6/m$	C_6	6	168–173	1 (1)	3 (2)	7 (4)	12 (5)
				C_{3h}	$\bar{6}$	174	0	3 (2)	6 (2)	12 (5)
				C_{6h}	$6/m$	175–176	0	3 (2)	0	12 (5)
			$6/mmm$	D_6	622	177–182	0	3 (2)	2 (1)	12 (5)
				C_{6v}	$6mm$	183–186	1 (1)	3 (2)	5 (3)	12 (5)
				D_{3h}	$\bar{6}2m$	187–190	0	3 (2)	3 (1)	12 (5)
				D_{6h}	$6/mmm$	191–194	0	3 (2)	0	12 (5)
Cubic (isometric) (4 3-fold axes)	$a = b = c$ $\alpha = \beta = \gamma = 90°$	P, I, F	$m3$	T	23	195–199	0	3 (1)	3 (1)	12 (3)
				T_h	$m3$	200–206	0	3 (1)	0	12 (3)
			$m3m$	O	432	207–214	0	3 (1)	0	12 (3)
				T_d	$\bar{4}3m$	215–220	0	3 (1)	3 (1)	12 (3)
				O_h	$m3m$	221–230	0	3 (1)	0	12 (3)
Isotropic	Amorphous	—	—	—	—	—	0	3 (1)	0	12 (2)

*Values are the number of nonzero coefficients in (equilibrium) property tensors and the values in parentheses are the numbers of independent coefficients in the tensor. Note that the elasto-optic and piezo-optic tensors have lower symmetry than the rank-4 tensors defined in this table and therefore have more independent coefficients than shown. Second-, third-, and fourth-rank tensors are given in the usual reduced index format (see text).

TABLE 3   Fundamental Physical Constants (1986 CODATA Values)

Constant	Symbol	Value	Unit
Atomic mass unit (amu)	u	$1.660\,540\,2 \cdot 10^{-27}$	kg
Avogadro constant	N_A	$6.022\,136\,7 \cdot 10^{23}$	mole^{-1}
Boltzmann constant	k_B	$1.380\,658 \cdot 10^{-23}$	J/K
Elementary charge	e	$1.602\,177\,33 \cdot 10^{-19}$	C
Permeability of vacuum	μ_0	$4\pi \cdot 10^{-7} =$ $12.566\,370\,614 \cdot 10^{-7}$	N/A^2 or H/m
Permittivity of vacuum	ϵ_0	$8.854\,187\,187 \cdot 10^{-12}$	F/m
Planck constant	h	$6.626\,075\,5 \cdot 10^{-34}$	$\text{J} \cdot \text{s}$
Speed of light	c	$299\,792\,458$	m/s

where σ, ϵ, and μ are the frequency-dependent conductivity, permeability, and permittivity, respectively. These quantities are scalars in an isotropic medium. To simplify notation, a generalized permittivity is sometimes defined as:

$$\epsilon_c(\omega) = \epsilon_r(\omega)\left[1 + i\,\frac{\sigma(\omega)}{\omega\varepsilon(\omega)}\right] \tag{2}$$

where $\epsilon_c(\omega)$ is a generalized relative permittivity of dielectric constant (i.e., with ε_0 removed from $\epsilon = \epsilon_0\epsilon_c$) that includes contributions from free charges [via the conductivity $\sigma(\omega)$] and bound charges [via the relative permittivity, $\epsilon_r(\omega)$]. Assuming a nonmagnetic material ($\mu_r = 1$), and using the preceding generalized dielectric constant, the plane-wave solution to the wave equation is

$$\mathbf{E}(z, \omega) = \mathbf{E}(0)\exp\left[i(\omega z/c)\sqrt{\epsilon_c(\omega)}\right] \tag{3}$$

In optics, it is frequently convenient to define a *complex refractive index* $\bar{n}$ the square root of the complex dielectric constant (henceforth, the symbol ϵ will be used for the relative complex dielectric constant):

$$\begin{aligned}
\bar{n} &= n + ik = \sqrt{\epsilon} = \sqrt{\epsilon' + i\epsilon''} \\
\epsilon' &= n^2 - k^2; \qquad \epsilon'' = 2nk \\
n^2 &= \tfrac{1}{2}[\epsilon' + \sqrt{\epsilon'^2 + \epsilon''^2}] \\
k^2 &= \tfrac{1}{2}[-\epsilon' + \sqrt{\epsilon'^2 + \epsilon''^2}]
\end{aligned} \tag{4}$$

where n is the (real) *index of refraction* and k is the *index of absorption* (or imaginary part of the complex refractive index). (The index of absorption is also called the absorption constant, index of extinction, or some other combination of these terms.) Using this definition of the complex index of refraction and the solution of the wave equation [Eq. (3)], the optical power density (proportional to $\tfrac{1}{2}|\mathbf{E}|^2$ from Poynting's vector) is

$$\text{Power density} = \tfrac{1}{2}\sqrt{\mu_0/\epsilon_0}\,|\mathbf{E}(z)|^2 = \tfrac{1}{2}\sqrt{\mu_0/\epsilon_0}\,|\mathbf{E}(0)|^2\exp\left(-2\omega kz/c\right) \tag{5}$$

where the exponential function represents the attenuation of the wave. The meanings of n and k are clear: n contributes to phase effects (time delay or variable velocity) and k contributes to attenuation by absorption. In practice, attenuation is conveniently described by a power absorption coefficient, β_{ABS}, which describes the internal transmittance over a distance z, i.e.,

$$\tau = \frac{|\mathbf{E}(z)|^2}{|\mathbf{E}(z=0)|^2} = e^{-2\omega kz/c} = e^{-\beta_{\text{ABS}}z} \tag{6a}$$

and β_{ABS} (with units of reciprocal length, usually cm^{-1}) is

$$\beta_{\text{ABS}} = 2\omega k/c = 4\pi\nu k \tag{6b}$$

Kramers–Krönig and Sum Rule Relationships. The principal of causality—that a material cannot respond until acted upon—when applied to optics, produces important symmetry properties and relationships that are very useful in modeling and analyzing optical properties. As a consequence of these symmetry properties, the real and imaginary parts of the dielectric constant (and of the complex index of refraction) are Hilbert transforms of each other. The Hilbert transform, Hi[$\Lambda(\omega)$], of the complex function $\Lambda(\omega)$ is defined as (the symbol P denotes the principal value of the integral)

$$\text{Hi}\,[\Lambda(\omega)] = \frac{i}{\pi}\,P\int_{-\infty}^{\infty}\frac{\Lambda(\omega')}{\omega - \omega'}\,d\omega' \tag{7}$$

and the relationships between the components of the dielectric constant or index of refraction are

$$\begin{aligned}
\epsilon' - 1 &= \text{Hi}\,[\epsilon''] \\
\epsilon'' &= \text{Hi}^{-1}\,[\epsilon' - 1] \\
n - 1 &= \text{Hi}\,[k] \\
k &= \text{Hi}^{-1}\,[n - 1]
\end{aligned} \tag{8}$$

These are the *Kramers–Krönig relationships* (abbreviated KK). Usually the Hilbert transforms for the refractive index are written in single-sided form:

$$n(\omega) - 1 = \frac{2}{\pi}\,P\int_{0}^{\infty}\frac{\omega' k(\omega')}{\omega'^2 - \omega^2}\,d\omega' \tag{9a}$$

with the inverse transform given by

$$k(\omega) = \frac{2\omega}{\pi}\,P\int_{0}^{\infty}\frac{n(\omega') - 1}{\omega^2 - \omega'^2}\,d\omega' \tag{9b}$$

These are fundamental relationships of any causal system.

A number of useful integral relationships, or *sum rules* result from Fourier transforms and the Kramers–Krönig relationship.[16] For example, the real part of the refractive index satisfies

$$\int_{0}^{\infty}[n(\omega) - 1]\,d\omega = 0 \tag{10a}$$

and

$$\begin{aligned}
n(\omega = 0) - 1 &= \frac{2}{\pi}\int_{0}^{\infty}\frac{k(\omega')}{\omega'}\,d\omega' \\
&= \frac{c}{\pi}\int_{0}^{\infty}\frac{\beta_{\text{ABS}}(\omega')}{\omega'^2}\,d\omega'
\end{aligned} \tag{10b}$$

The dielectric constant and the refractive index also have the following symmetry properties:

$$\begin{aligned}
\epsilon(\omega) &= \epsilon^*(-\omega) \\
\bar{n}(\omega) &= \bar{n}^*(-\omega)
\end{aligned} \tag{10c}$$

Practical models of the dielectric constant or refractive index must satisfy these fundamental symmetry properties and integral relationships.

Optical Properties: Origin and Models

Intrinsic optical properties of a material are determined by three basic physical processes: electronic transitions, lattice vibrations, and free-carrier effects.[5,6,17,18] However, the dominant physical process depends on the material and spectral region of interest. All materials have contributions to the complex index of refraction from electronic transitions. Insulators and semiconductors also require the characterization of the lattice vibrations (or phonons) to fully understand the optical properties. Transparency of semiconductors, particularly those with small band gaps, are additionally influenced by free-carrier effects. The strength of free-carrier influence on transmission and absorption depends on the free-carrier concentration; thus free-carrier effects dominate the optical properties of metals in the visible and infrared.

In the range of transparency of a bulk material, more subtle effects such as multiphonon processes (see later discussion), impurity and defect absorption, and scattering become the important loss mechanisms. Intrinsic atomic (Rayleigh) scattering is a very weak effect, but is important in long-path optical fibers and ultraviolet materials. Extrinsic scattering, caused by density (local composition) variations, defects, or grains in polycrystalline solids, is typically much larger than intrinsic scattering in the visible and infrared spectral regions. Impurity and defect (electronic or vibrational) absorption features can be of great concern depending on the spectral region, incident radiation intensity, or material temperature required by the application.

Figure 1 illustrates the frequency dependence of n and k for an insulating polar crystal.[19] The value of $n(\omega, T)$ is essentially the sum of the contributions of all electronic and lattice vibration resonances, and is dominated by those with fundamental oscillation frequencies above ω. Figure 1a indicates regions of validity for the popular Sellmeier model (see discussion under "Electronic Transitions"). Frequency dependence of the imaginary part of the index of refraction $k(\omega, T)$ requires consideration of not only the dominant physical processes but also higher-order processes, impurities, and defects as illustrated in Fig. 1b. The spectral regions of the fundamental resonances are opaque. The infrared edge of transparency is controlled by multiphonon transitions. Transparent regions for insulators are divided in two regions: microwave and visible/infrared.

Lattice Vibrations. Atomic motion, or lattice vibrations, accounts for many material properties, including heat capacity, thermal conductivity, elastic constants, and optical and dielectric properties. Lattice vibrations are quantized; the quantum of lattice vibration is called a *phonon*. In crystals, the number of lattice vibrations is equal to three times the number of atoms in the primitive unit cell (see further discussion); three of these are acoustic vibrations (translational modes in the form of sound waves), the remainder are termed optical vibrations (or modes of phonons). For most practical temperatures, only the acoustic phonons are thermally excited because optical phonons are typically of much higher frequency, hence acoustic modes play a dominant role in thermal and elastic properties. There are three types of optical modes: infrared-active, Raman-active, and optically inactive. Infrared-active modes, typically occurring in the region from 100 to 1000 cm^{-1}, are those that (elastically) absorb light (photon converted to phonon) through an interaction between the electric field and the light and the dipole moment of the crystal. Raman modes* (caused by phonons that modulate the polarizability of the crystal to induce a dipole moment) weakly absorb light through an inelastic mechanism (photon converted to phonon and scattered photon) and are best observed with intense (e.g.,

---

* Brillouin scattering is a term applied to inelastic scattering of photons by acoustic phonons.

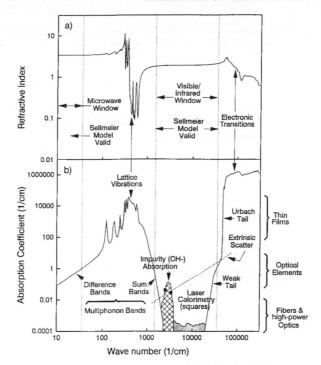

FIGURE 1 The wave number (frequency) dependence of the complex refractive index of Yttria:[19] (*a*) shows the real part of the refractive index, $n(\omega)$. The real part is high at low frequency and monotonically increases, becomes oscillatory in the lattice vibration (phonon) absorption bands, increases monotonically (normal dispersion) in the optical transparent region, and again becomes oscillatory in the electronic absorption region; (*b*) shows the imaginary part of the refractive index, in terms of the absorption coefficient, $\beta(v) = 4\pi v k(v)$. The absorption coefficient is small in the transparent regions and very high in the electronic and vibrational (phonon) absorption bands. The optical transparent region is bounded by the "Urbach tail" absorption at high frequency and by multiphonon absorption at low frequency (wave number). In between, loss is primarily due to impurities and scatter. (*Reprinted by permission of McGraw-Hill, Inc.*)

laser) light. Optically inactive modes have no permanent or induced dipole moment and therefore do not interact with light. Optical modes can be both infrared- and Raman-active and are experimentally observed by infrared or Raman spectroscopy, as well as by x-ray or neutron scattering.

Crystal symmetries reduce the number of unique lattice vibrations (i.e., introduce vibrational degeneracies). A group theory analysis will determine the number of optical modes of each type for an ideal material. Defects and impurities will increase the number of observed infrared-active and Raman modes in a real material. As structural disorder increases (nonstoichiometry, defects, variable composition), the optical modes broaden and additional modes appear. Optical modes in noncrystalline (amorphous) materials such as glasses are very broad compared to those of crystals.

Lattice vibration contributions to the static dielectric constant, $\epsilon(0)$, are determined from the longitudinal- (LO) and transverse-mode (TO) frequencies of the optical modes using the Lyddane-Sachs-Teller (LST) relationship[20] as extended by Cochran and Cowley[21] for materials with multiple optical modes:

$$\epsilon(0) = \epsilon(\infty) \prod_{j}^{j_{max}} \frac{\omega_j^2(\text{LO})}{\omega_j^2(\text{TO})} \tag{11}$$

where ω is frequency (usually in wave numbers) and $\epsilon(\infty)$ is the high-frequency (electronic transition) contribution to the dielectric constant (not the dielectric constant at infinite frequency). This relationship holds individually for each principal axis. The index j denotes infrared-active lattice vibrations with minimum value usually found from group theory (discussed later). This LST relationship has been extended to include the frequency dependence of the dielectric constant:

$$\epsilon(\omega) = \epsilon(\infty) \prod_{j} \frac{\omega_j^2(\text{LO}) - \omega^2}{\omega_j^2(\text{TO}) - \omega^2} \tag{12a}$$

A modified form of this fundamental equation is used by Gervais and Piriou[22] and others to model the dielectric constant in the infrared:

$$\epsilon(\omega) = \epsilon(\infty) \prod_{j}^{j_{max}} \frac{\omega_j^2(\text{LO}) - \omega^2 - i\omega\gamma_j(\text{LO})}{\omega_j^2(\text{TO}) - \omega^2 - i\omega\gamma_j(\text{TO})} \tag{12b}$$

where γ is the line width of the longitudinal and transverse modes as denoted by the symbol in parentheses. This form of the dielectric constant is known as the *semiquantum four-parameter model*.

Frequently, the infrared dielectric constant is modeled in a three-parameter classical oscillator form (or Maxwell–Helmholtz–Drude[23] dispersion formula), namely

$$\epsilon(\omega) = \epsilon(\infty) + \sum_{j}^{j_{max}} \frac{S_j \omega_j^2(\text{TO})}{\omega_j^2(\text{TO}) - \omega^2 - i\omega\gamma_j} \tag{13}$$

where $S_j(=\Delta\epsilon_j)$ is the strength and γ_j is the full width of the jth mode. This model assumes no coupling between modes and provides a good representation of the dielectric constant, especially if the modes are weak [small separation between $\omega(\text{TO})$ and $\omega(\text{LO})$] and uncoupled. The static dielectric constant $\epsilon(0)$ is merely the sum of the high-frequency dielectric constant $\epsilon(\infty)$ and strengths S_j of the individual, IR-active modes. This formulation has been widely used to model infrared dispersion. This model can also be used to represent the high-frequency dielectric constant using additional models (i.e., $\epsilon(\infty)$ is replaced by 1, the dielectric constant of free space, plus additional modes, see later discussion). Strengths and line widths for the classical dispersion model can be derived from the values of the four-parameter model.[24]

Both the classical and four-parameter dispersion models satisfy the Kramers–Krönig relationship [Eqs. (7) and (8)] and are therefore physically realizable. The frequencies $\omega(\text{TO})$ and $\omega(\text{LO})$ arise from the interaction of light with the material, correspond to solutions of the Maxwell wave equation, $\nabla \cdot \mathbf{D} = 0$ (no free charges), and are obtained from measurements. The transverse frequency corresponds to an electromagnetic wave with $\mathbf{E}$ perpendicular to the wave vector or $\mathbf{E} = 0$. At higher frequencies [$\omega > \omega(\text{TO})$], the external electric field counters the internal polarization field of the material until the real part of the dielectric constant is zero, hence $\mathbf{D} = 0$ at $\omega(\text{LO})$ [when $\gamma_j = 0$, i.e., case of Eq. (12a)]. The longitudinal frequency $\omega(\text{LO})$ is always greater than the transverse frequency

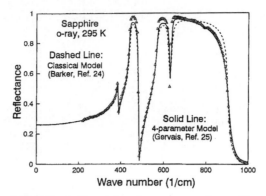

FIGURE 2 The infrared spectrum of sapphire (Al_2O_3) showing the fit to data of the three-parameter classic oscillator model of Barker (dashed line[25]) and the four-parameter model of Gervais (solid line[26]). The four-parameter model better fits experimental data (triangles) in the 650–900 cm^{-1} region.

ω(TO). The separation of ω(LO) and ω(TO) is a measure of the strength (S_j or $\Delta\epsilon_j$) of the optical mode. Raman modes are, by their nature, very weak and therefore ω(LO) $\approx$ ω(TO), hence Raman modes do not appreciably contribute to dielectric properties.

In Eq. (12a), the ω(TO) frequencies correspond to the poles and the ω(LO) frequencies correspond to the zeros of the dielectric constant. The dielectric constant continuously rises with frequency [except for a discontinuity at ω(TO)] and the dielectric constant is real and negative [i.e., highly absorbing, Eq. (3)] between the transverse and longitudinal frequencies. When damping is added to the dielectric constant model to represent the response of real materials [e.g., Eq. (12b) or (13)], the transverse and longitudinal frequencies become the maxima and minima of the dielectric constant. With damping, a negative dielectric constant is not a necessary condition for absorption, and the material will also absorb outside the region bounded by ω(TO) and ω(LO). Furthermore, damping allows the real part of the dielectric constant (and also the refractive index) to decrease with frequency near ω(TO). This dispersive condition is called *anomalous dispersion* and can only occur in absorptive regions.

Figure 2 shows a typical crystal infrared spectrum and the corresponding classical oscillator[25] and four parameter model[26] fits to the data.

As an example of lattice vibrations, consider the simple case of crystalline sodium chloride (NaCl) which has four molecules (eight atoms) per unit cell. Since the sodium chloride structure is face-centered cubic, a primitive cell has one molecule or two atoms. The number of unique vibrations is further reduced by symmetry: for sodium chloride, the lattice vibrations consist of one (triply degenerate) acoustic mode and one (triply degenerate) optical mode. Many metals have one atom per primitive unit cell and therefore have no optical modes. Group IV cubic materials (diamond, silicon, germanium) have two atoms per unit cell with one (triply degenerate) optical mode that is Raman-active only. Therefore, to first order, these nonpolar materials are transparent from their band gap to very low frequencies. In practice, multiphonon vibrations, defects, impurities, and free carriers can introduce significant absorption.

Electronic Transitions. Electronic transitions in a solid begin at the material's band gap. This point generally marks the end of a material's useful transparency. Above the band gap, the material is highly reflective. The large number of possible electronic transitions

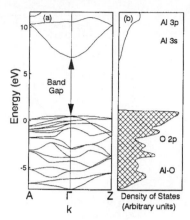

FIGURE 3   Electronic energy band diagram for sapphire at room temperature:[27] (*a*) shows the complex k-space energy levels of the electrons of sapphire. The arrow denotes the direct band gap transition; (*b*) shows the density of electronic states as a function of energy level. The many direct and indirect electronic transitions give rise to a broad electronic absorption with few features. (*Reprinted by permission of the American Ceramic Society.*)

produces broad-featured spectra. However, electronic structure is fundamental to understanding the nature of the bonds forming the solid and thus many of the material properties. In three dimensions, the band structure becomes more complicated, because it varies with direction just as lattice vibrations do. This point is illustrated in Fig. 3 for the case of the electronic k-space diagram for sapphire.[27] Also included in this figure is the corresponding electronic density of states that determines the strength of the absorption.

In polar insulators, no intraband electronic transitions are allowed, and the lowest frequency electronic absorption is frequently caused by creation of an *exciton,* a bound electron-hole pair. The photon energy required to create this bound pair is lower than the band-gap energy. Other lower-frequency transitions are caused by interband transitions between the anion valence band and the cation conduction band. For sapphire (see Fig. 3), the lowest energy transitions are from the upper valence band of oxygen ($2p^6$) to the conduction band of aluminum ($3s + 3p$). There are typically many of these transitions which appear as a strong, broad absorption feature. Higher energy absorption is caused by surface and bulk plasmons (quanta of collective electronic waves), and still higher energy absorption is attributable to promotion of inner electrons to the conduction band and ultimately liberation of electrons from the material (photoemission).

Excitons have many properties similar to that of a hydrogen atom. The absorption spectra of an exciton is similar to that of hydrogen and occurs near the band gap of the host material. The bond length between the electron-hole pair, hence the energy required to create the exciton, depends on the host medium. Long bond lengths are found in semiconductors (low-energy exciton) and short bond lengths are found in insulating materials.

Classical electronic polarization theory produces a model of the real part of the dielectric constant similar to the model used for the real dielectric constant of lattice vibrations [Eq. (13)]. General properties of the real dielectric constant (and real refractive index) can be deduced from this model. Bound electrons oscillate at a frequency proportional to the square root of the binding energy divided by the electronic mass.

Oscillator strength is proportional to the inverse of binding energy. This means that insulators with light atoms and strong bonding have large band gaps, hence good UV transmission (cf., LiF). Furthermore, high bonding energy (hence high-energy band gap) means low refractive index (e.g., fluorides).

When both electronic and lattice vibrational contributions to the dielectric constant are modeled as oscillations, the dielectric constant in the transparent region between electronic and vibrational absorption is (mostly) real and the real part takes the form

$$\epsilon'(\omega) - 1 = n^2(\omega) - 1 = \sum_j \frac{S_j \omega_j^2}{\omega_j^2 - \omega^2} \tag{14}$$

which is the widely-used *Sellmeier dispersion formula.* The sum includes both electronic (*UV*) and vibrational (IR or *ionic*) contributions. Most other dispersion formulas (such as the Schott glass power series) are recast or simplified forms of the Sellmeier model. The refractive index of most materials with good homogeneity can be modeled to a few parts in 10^5 over their entire transparent region with a Sellmeier fit of a few terms. The frequency (ω_j) term(s) in a Sellmeier fit are not necessarily TO modes, but are correlated to the strong TOs nearest the transparent region, with adjustments to the constants made to account for weaker modes, multiphonon effects, and impurities. The Sellmeier model works well because it is (1) based on a reasonable physical model and (2) adjusts constants to match data. The relationship between the Sellmeier equation and other dispersion formulas is discussed later.

The *Urbach tail* model is successfully applied to model the frequency and temperature dependence of the ultraviolet absorption edge in a number of materials, particularly those with a direct band gap, over several orders of magnitude in absorption. Urbach[28] observed an exponential absorption edge in silver halide materials (which have an indirect band gap). Further development added temperature dependence in the form:

$$\beta_{\mathrm{ABS}}(E, T) = \beta_0 \exp\left[-\sigma(T)\frac{(E_0 - E)}{k_B T}\right] \tag{15a}$$

where

$$\sigma(T) = \sigma_0(T)\frac{2k_B T}{E_p} \tanh\frac{E_p}{2k_B T} \tag{15b}$$

where E_0 is the band gap energy at $T = 0\,\mathrm{K}$ (typically the energy of an exciton), T temperature in Kelvins, and E_p a characteristic phonon energy. The interpretation of the Urbach tail is a broadening of the electronic band gap by phonon interactions, and several detailed theories have been proposed.[18,29]

Below the Urbach tail, absorption continues to decrease exponentially, albeit much slower than predicted by the Urbach formula. This region of slowly decreasing absorption is sometimes called the "weak tail," and has been observed in both semiconductor[30] and crystalline materials.[31,32] Typically, the weak tail begins at the point when the absorption coefficient falls to 0.1 cm^{-1}.

Free Carriers. Free carriers, such as electrons in metals, or electrons or holes in semiconductors, also affect the optical properties of materials. For insulators or wide-band-gap semiconductors (i.e., band gap greater than 0.5 eV) with a low number of free carriers at room temperature (low conductivity), the effect of free carriers on optical absorption is small [see Eq. (2)]. For nonmetals, the free-carrier concentration grows with temperature so that even an "insulator" has measurable conductivity (and free-carrier absorption) at very high temperature. Commonly used optical materials such as silicon and germanium

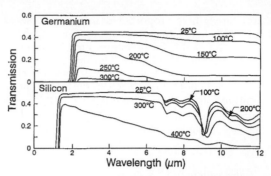

FIGURE 4 Decreased transmission of germanium and silicon with temperature is attributable to an increase in free-carrier concentration resulting in increased absorption.[33] The absorption is greater at longer wavelengths; see Eq. (16). (*Reprinted by permission of the Optical Society of America.*)

have a significant increase in free-carrier absorption at moderately high temperature as illustrated in Fig. 4.[33]

Free-carrier effects can be modeled as an additional contribution to the dielectric constant model. For example, the classical model [Eq. (13)] takes the form

$$\epsilon(\omega) = \epsilon(\infty) + \sum_{j}^{j_{max}} \frac{S_j \omega_j^2(\text{TO})}{\omega_j^2(\text{TO}) - \omega^2 - i\omega\gamma_j} - \frac{\omega_p^2}{\omega^2 + i\omega\gamma_c} \tag{16}$$

where ω_p is the plasma frequency, proportional to the square root of the free-carrier density, and γ_c is the damping frequency (i.e., determines the effective width of the free-carrier influence). Such a model is well known to accurately predict the far-infrared ($\geq 10\,\mu$m) refractive index of metals[34] and also has been used to model the free-carrier contribution to optical properties of semiconductors.

Multiphonon Absorption and Refraction. Absorption at the infrared edge of insulators is principally caused by anharmonic terms in the lattice potential leading to higher harmonics of the lattice resonances. This phenomenon is called multiphonon absorption because the frequencies are harmonics of the characteristic lattice phonons (vibrations). For absorption in the infrared, each successively higher multiple of the fundamental frequency is weaker (and broader) leading to decreasing absorption beyond the highest fundamental absorption frequency (maximum transverse optical frequency). At about three times $\omega(\text{TO})$ the absorption coefficient becomes small and a material with thickness of 1 to 10 mm is reasonably transparent. The infrared absorption coefficient of materials (especially highly ionic insulators) can be characterized by an exponential absorption coefficient[35] β_{ABS} of the form

$$\beta_{\text{ABS}} = \beta_0 \exp\left(-\gamma \frac{\omega}{\omega_o}\right) \tag{17a}$$

where β_0 is a constant (dimensions same as the absorption coefficient, typically cm^{-1}), γ is a dimensionless constant (typically found to be near 4), ω_o is frequency or wave number of the maximum transverse optical frequency (units are cm^{-1} for wave numbers; values are given in the property data tables), and ω is the frequency or wave number of interest (units are cm^{-1} for wave numbers). This formula works reasonably well for ionic materials at room temperature for the range of absorption coefficients from 0.001 to 10 cm^{-1}.

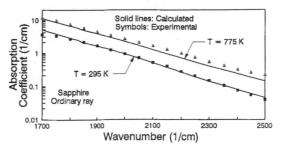

FIGURE 5 Temperature-dependent change of absorption in insulators is principally confined to the absorption edges, especially the infrared multiphonon absorption edge. This figure shows measured and predicted absorption coefficients at the infrared edge of transparency for the ordinary ray of crystalline sapphire. Increasing temperature activates higher multiphonon processes, resulting in a rapid increase in absorption. The multiphonon model of Thomas et al.,[38] accurately predicts the frequency- and temperature-dependence of infrared absorption in highly ionic materials such as oxides and halides.

In the classical (continuum) limit, the temperature dependence of multiphonon absorption has T^{n-1} dependence where n is the order of the multiphonon process,[36] i.e., $n \approx \omega / \omega_o$. At low temperature, there is no temperature dependence since only transitions from the ground state occur. Once the temperature is sufficiently high (e.g., approaching the Debye temperature), transitions that originate from excited states become important and the classical temperature dependence is observed. Bendow[37] has developed a simple model of the temperature dependence of multiphonon absorption based on a Bose–Einstein distribution of states:

$$\beta_{\text{ABS}}(\omega, T) = \beta_0 \frac{[N(\omega_o, T) + 1]^{\omega/\omega_o}}{N(\omega, T) + 1} \exp\left(-A\omega/\omega_o\right) \tag{17b}$$

where $N(\omega, T)$ is phonon occupation density from Bose–Einstein statistics.

Thomas et al.,[38] have successfully developed a semiempirical, quantum mechanical model of (sum band) multiphonon absorption based on the Morse interatomic potential and a Gaussian function for the phonon density of states. Use of the Morse potential leads to an exact solution to the Schrödinger equation and includes anharmonic effects to all orders. The model contains parameters derived from room-temperature measurements of absorption, is computationally efficient, and has been applied to many ionic substances. Figure 5 shows a typical result compared to experimental data.

Multiphonon absorption modeling also contributes to the real refractive index. Although multiphonon contributions to the real index are small compared to one-phonon contributions, they are important for two cases in the infrared: (1) when the refractive index must be known beyond two decimal places, or (2) at high temperature. In the first case, multiphonon contributions to the index are significant over a large spectral region. In the second case, the contribution of multiphonon modes to real refractive index grows rapidly at high temperature because of the T^{n-1} dependence of the nth mode strength.

Absorption in the Transparent Region. In the transparent region, away from the electronic and vibrational resonances, absorption is governed by impurities and defects. The level of absorption is highly dependent on the purity of the starting materials, conditions of manufacture, and subsequent machining and polishing. For example, OH impurities are common in oxides,* occurring at frequencies below the fundamental (nonbonded) OH vibration at $3735\ cm^{-1}$. OH can be removed by appropriate treatment.

Low-level absorption coefficient measurements are typically made by laser calorimetry or photoacoustic techniques. Data are available for a number of materials in the visible,[31,39] at $1.3\ \mu m$,[40] and at 2.7 and $3.8\ \mu m$.[41]

Optical Properties: Applications

Dielectric Tensor and Optical Indicatrix. Many important materials are nonisotropic (i.e., crystals Al_2O_3, SiO_2, and MgF_2) and their optical properties are described by tensor relationships (see earlier section, "Symmetry Properties"). The dielectric constant ϵ, a second-rank tensor, relates the electric field **E** to the electric displacement **D**:

$$\begin{vmatrix} \mathbf{D}_x \\ \mathbf{D}_y \\ \mathbf{D}_z \end{vmatrix} = \epsilon_0 \cdot \begin{vmatrix} \epsilon_{xx} & \epsilon_{xy} & \epsilon_{xz} \\ \epsilon_{yx} & \epsilon_{yy} & \epsilon_{yz} \\ \epsilon_{zx} & \epsilon_{zy} & \epsilon_{zz} \end{vmatrix} \cdot \begin{vmatrix} \mathbf{E}_x \\ \mathbf{E}_y \\ \mathbf{E}_z \end{vmatrix} \qquad (18a)$$

From the symmetry of properties, this is a symmetric tensor with $\epsilon_{ab} = \epsilon_{ba}$. Usually, the dielectric constant components are given as principal values, i.e., those values along the unit cell of the appropriate crystal class. In this case, the principal dielectric constants are

$$\epsilon_{x'x'} \equiv \epsilon_1$$

$$\epsilon_{y'y'} \equiv \epsilon_2 \qquad (18b)$$

$$\epsilon_{z'z'} \equiv \epsilon_3$$

where the primes on the subscripts denote principal values (i.e., along the crystallographic axes, possible in a nonorthogonal coordinate system) and the subscripts 1, 2, and 3 denote reduced notation for these values (see "Elastic Properties"). The relationship between dielectric constant and refractive index, Eq. (4), means there are similarly three principal values for the (complex) refractive index. Also, the components of the dielectric tensor are individually related to the corresponding components of the refractive index. (Subscripts a, b, and c or x, y, and z as well as others may be used for the principal values of the dielectric constant or refractive index.)

Three important cases arise:

1. Isotropic and cubic materials have only one dielectric constant, ϵ, (hence one refractive index, $\bar{n}$). Therefore $\epsilon_1 = \epsilon_2 = \epsilon_3 = \epsilon = \bar{n}^2$.

2. Hexagonal (including trigonal) and tetragonal crystals have two principal dielectric constants, ϵ_1 and ϵ_3 (hence two refractive indices, $\bar{n}_1$ and $\bar{n}_3$). Therefore $\epsilon_1 = \epsilon_2 \neq \epsilon_3$. Such materials are called *uniaxial*; the unique crystallographic axis is the c axis, which is also called the *optical axis*. One method of denoting the two unique principal axes is to state the orientation of the electric field relative to the optical axis. The dielectric constant for $\mathbf{E} \perp c$ situation is ϵ_1 or $\epsilon_\perp$. This circumstance is also called the *ordinary ray*, and the corresponding symbol for the refractive index is $\bar{n}_1$, $\bar{n}_\perp$, $\bar{n}_o$ (for ordinary ray), and $\bar{n}_\omega$ or ω

---

* The OH vibrational impurity absorption in oxides is known for Al_2O_3, ALON, $MgAl_2O_4$, MgO, SiO_2, Y_2O_3, and Yb_2O_3.

(primarily in the older literature). The dielectric constant for $\mathbf{E} \parallel c$ situation is ϵ_3 or $\epsilon_\parallel$. This condition is called the *extraordinary ray,* and the corresponding symbol for the refractive index is $\bar{n}_3$, $\bar{n}_\parallel$, $\bar{n}_e$ (for extraordinary ray), and $\bar{n}_\epsilon$ or ϵ (again, primarily in the older literature). Crystals are called *positive uniaxial* when $n_e - n_o > 0$, and *negative uniaxial* otherwise. Since the dispersions of the ordinary and extraordinary wave are different, a crystal can be positive uniaxial in one wavelength region and negative uniaxial in another ($AgGaS_2$ is an example).

3. Orthorhombic, monoclinic, and triclinic crystals have three principal dielectric constants, ϵ_1, ϵ_2, and ϵ_3 (hence, three refractive indices, $\bar{n}_1$, $\bar{n}_2$, and $\bar{n}_3$). Therefore $\epsilon_1 \neq \epsilon_2 \neq \epsilon_3 \neq \epsilon_1$. These crystals are called *biaxial.* Confusion sometimes arises from the correlation of the principal dielectric constants with the crystallographic orientation owing to several conventions in selecting the crystal axes. [The optical indicatrix (see following discussion) of a biaxial material has two circular sections that define optical axes. The orientation of these axes are then used to assign a positive- or negative-biaxial designation.]

The existence of more than one dielectric constant or refractive index means that, for radiation with arbitrary orientation with respect to the crystal axes, two plane-polarized waves, of different speed, propagate in the crystal. Hence, for light propagating at a random orientation to the principal axes, a uniaxial or biaxial crystal exhibits two effective refractive indices different from the individual principal values. The refractive index of the two waves is determined from the *optical indicatrix* or *index ellipsoid,* a triaxial ellipsoidal surface defined by

$$\frac{x_1^2}{n_1^2} + \frac{x_2^2}{n_2^2} + \frac{x_3^2}{n_3^2} = 1 \tag{19a}$$

where the x_1, x_2, and x_3 are the principal axes of the dielectric constant. The indicatrix is illustrated in Fig. 6: for a wave normal in an arbitrary direction (OP), the two waves have

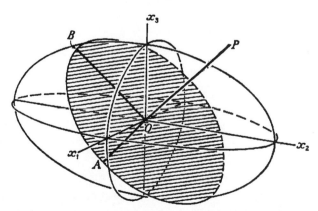

FIGURE 6 The *optical indicatrix* or *index ellipsoid* used to determine the effective refractive index for an arbitrary wave normal in a crystal. The axes of the ellipsoid correspond to the principal axes of the crystal, and the radii of the ellipsoid along the axes are the principal values of the refractive indices. For propagation along an arbitrary wave normal (OP), the effective refractive indices are the axes of the ellipse whose normal is parallel to the wave normal. In the illustrated case, the directions OA and OB define the effective refractive indices.[14] (*Reprinted by permission of Oxford University Press.*)

refraction indices equal to the axes of the ellipse perpendicular to the wave normal (OA and OB). The directions represented by OA and OB are the vibrational planes of the electric displacement vector **D** of the two waves. When the wave normal is parallel to an optic axis, the two waves propagate with principal refractive indices. For uniaxial material and the wave normal parallel to the x_3 (or z or c crystallographic or optical) axis, the vibrational ellipsoid is circular and the two waves have the same refractive index (n_o), and there is no double refraction. Equation (19a) can also be written

$$B_1 x_1^2 + B_2 x_2^2 + B_3 x_3^2 = 1 \qquad (19b)$$

where $B_i = 1/\bar{n}_i^2 = 1/\epsilon_i$ is called the *inverse dielectric tensor*. The inverse dielectric tensor is used in defining the electro-optic, piezo-optic, and elasto-optic effects.

Reflection: Fresnel Formulas. Fresnel formulas for reflection and transmission at an interface are presented here. Results for two cases are given. The first is for cubic materials, and the second is for uniaxial (tetragonal and hexagonal) that have the crystallographic c axis normal to the interface surface. Orientation of the electric field in the plane of incidence is the case of *vertical polarization* or ***p**-polarization*; the electric field perpendicular to the plane of incidence is the case of *horizontal polarization* or ***s**-polarization*.

For a cubic medium $\epsilon_{xx} = \epsilon_{yy} = \epsilon_{zz} = \bar{n}^2$, and the field (amplitude) reflection r_{12} and transmission t_{12} (Fresnel) coefficients are, for **p**-polarization propagating from medium 1 to medium 2,

$$r_{p12} = \frac{\dfrac{\bar{n}_2}{\cos \theta_t} - \dfrac{\bar{n}_1}{\cos \theta_i}}{\dfrac{\bar{n}_2}{\cos \theta_t} + \dfrac{\bar{n}_1}{\cos \theta_i}} = \frac{\bar{n}^2 \cos \theta_i - \sqrt{\bar{n}^2 - \sin^2 \theta_i}}{\bar{n}^2 \cos \theta_i + \sqrt{\bar{n}^2 - \sin^2 \theta_i}} = -r_{p21} \qquad (20a)$$

and

$$t_{p12} = \frac{2\bar{n}_1 \cos \theta_1}{\bar{n}_1 \cos \theta_i + \bar{n}_2 \cos \theta_t} = \frac{\bar{n}_1 \cos \theta_i}{\bar{n}_2 \cos \theta_t} t_{p21} \qquad (20b)$$

and the corresponding formulas for **s**-polarization are

$$r_{s12} = \frac{\bar{n}_1 \cos \theta_i - \bar{n}_2 \cos \theta_t}{\bar{n}_1 \cos \theta_i + \bar{n}_2 \cos \theta_t} = \frac{\cos \theta_i - \sqrt{\bar{n}^2 - \sin^2 \theta_i}}{\cos \theta_i + \sqrt{\bar{n}^2 - \sin^2 \theta_i}} = -r_{s21} \qquad (20c)$$

and

$$t_{s12} = \frac{2\bar{n}_1 \cos \theta_i}{(\bar{n}_1 \cos \theta_i + \bar{n}_2 \cos \theta_t)} = \frac{\bar{n}_1 \cos \theta_i}{\bar{n}_2 \cos \theta_t} t_{s21} \qquad (20d)$$

where $\bar{n} = \bar{n}_2/n_1$ for n_1 real. Snell's law is then given by

$$n_1 \sin \theta_1 = \bar{n}_2 \sin \theta_2 \qquad (20e)$$

where θ_1 is real and θ_2 is complex. The terms r_{21} and t_{21} are the field reflection and transmission coefficients for propagation from medium 2 to medium 1, respectively. The relationships between r_{12} and r_{21} and between t_{12} and t_{21} are governed by the *principle of reversibility.* The single-surface (Fresnel) *power* coefficients for reflection (R) and transmission (T) are directly obtained from the field coefficients:

$$R_{p,s} = |r_{p,s}|^2 \tag{20f}$$

and

$$T_{p,s} = \frac{\bar{n}_2 \cos \theta_t}{\bar{n}_1 \cos \theta_i} |t_{p,s}| \tag{20g}$$

Using the formulas of Eq. (20), and assuming medium 1 is vacuum $(n_1 = 1)$ and medium 2 is absorbing $(\bar{n}_2 = n + ik)$, the power coefficients become

$$R_s = \frac{(a - \cos \theta_i)^2 + b^2}{(a + \cos \theta_i)^2 + b^2} \tag{21a}$$

and

$$R_p = R_s \left[\frac{(a - \sin \theta_i \tan \theta_i)^2 + b^2}{(a + \sin \theta_i \tan \theta_i)^2 + b^2} \right] \tag{21b}$$

where the terms a and b are

$$a^2 = \tfrac{1}{2}([(n^2 - k^2 - \sin^2 \theta_1)^2 + 4n^2 k^2]^{1/2} + (n^2 - k^2 - \sin^2 \theta_i)) \tag{21c}$$

and

$$b^2 = \tfrac{1}{2}([(n^2 - k^2 - \sin^2 \theta_1)^2 + 4n^2 k^2]^{1/2} - (n^2 - k^2 - \sin^2 \theta_i)) \tag{21d}$$

The principle of reversibility and Snell's law require that

$$R_{12} = R_{21} \quad \text{and} \quad T_{12} = T_{21} \tag{22a}$$

Also, from the preceding definitions, it can be shown that

$$R_{s,p} + T_{s,p} = 1 \tag{22b}$$

For unpolarized light, the single-surface reflection coefficient becomes

$$R = \tfrac{1}{2}(R_s + R_p) \tag{22c}$$

For uniaxial media, $\epsilon_{xx} = \epsilon_{yy} = \bar{n}_1^2 = \bar{n}_2^2 = \bar{n}_o^2$ and $\epsilon_{zz} = \bar{n}_3^2 = \bar{n}_e^2$. For the special case with the surface normal parallel to the crystallographic c axis (optical axis), the Fresnel formula becomes

$$R_p = \left| \frac{n_1 n_3 \cos \theta_i - [n_3^2 - \sin^2 \theta_i]^{1/2}}{n_1 n_3 \cos \theta_i + [n_3^2 - \sin^2 \theta_i]^{1/2}} \right|^2 \tag{23a}$$

and

$$R_s = \left| \frac{\cos \theta_i - (n_1^2 - \sin^2 \theta_i)^{1/2}}{\cos \theta_i + (n_1^2 - \sin^2 \theta_i)^{1/2}} \right|^2 \tag{23b}$$

If $n_1 = n_3 = n$, the isotropic results are obtained.

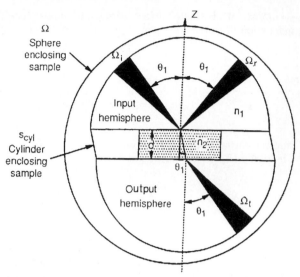

FIGURE 7 Geometry of incident, transmitted, and reflected beams for a plane transparent slab of thickness d. The power equals the reflected, refracted, and absorbed, assuming no scatter.

Total Power Law. Incident light on a material is reflected, transmitted, or absorbed. Scattering is a term used to describe diffuse reflectance (surface scatter) and diffuse transmittance (bulk scatter). Conservation of energy dictates that the fractional amount reflected ρ, absorbed α_{ABS}, transmitted τ, and scattered α_{SCA} total to unity, hence

$$1 = \rho(\Omega_i, \omega) + \alpha_{\mathrm{SCA}}(\Omega_i, \omega) + \alpha_{\mathrm{ABS}}(\Omega_i, \omega) + \tau(\Omega_i, \omega) \tag{24}$$

where these time-averaged quantities, illustrated in Fig. 7, are

$$\rho(\Omega_i, \omega) = \frac{\Phi_r(\Omega_i, \omega)}{\Phi_i(\omega)} = \text{total integrated reflectance}, \tag{25a}$$

$$\alpha_{\mathrm{SCA}}(\Omega_i, \omega) = \frac{\Phi_s(\Omega_i, \omega)}{\Phi_i(\omega)} = \text{total integrated scatterance}, \tag{25b}$$

$$\alpha_{\mathrm{ABS}}(\Omega_i, \omega) = \frac{\Phi_s(\Omega_i, \omega)}{\Phi_i(\omega)} = \text{total integrated absorptance}, \tag{25c}$$

and

$$\tau(\Omega_i, \omega) = \frac{\Phi_t(\Omega_i, \omega)}{\Phi_i(\omega)} = \text{total integrated transmittance} \tag{25d}$$

Notice that these quantities are functions of the angle of incidence and frequency only.

The sum of total integrated scatterance and total integrated absorptance can be defined as the total integrated extinctance α_{EXT},

$$\alpha_{\mathrm{EXT}}(\Omega_i, \omega) = \alpha_{\mathrm{ABS}}(\Omega_i, \omega) + \alpha_{\mathrm{SCA}}(\Omega_i, \omega) \tag{26}$$

and the total power law becomes

$$1 = \rho(\Omega_i, \omega) + \alpha_{\mathrm{EXT}}(\Omega_i, \omega) + \tau(\Omega_i, \omega) \tag{27}$$

Another useful quantity is emittance, which is defined as

$$\epsilon(\Omega_i, \omega) = \frac{\Phi_e(\Omega_i, \omega)}{\Phi_{bb}(\omega)} = \text{total integrated emittance} \tag{28a}$$

where Φ_{bb} is the blackbody function representing the spectral emission of a medium which totally absorbs all light at all frequencies. When $\Phi_i(\omega) = \Phi_{bb}(\omega)$, then the total integrated emittance equals the total integrated absorptance:

$$\varepsilon(\Omega_i, \omega) = \alpha_{ABS}(\Omega_i, \omega) \tag{28b}$$

Dispersion Formulas for Refractive Index. The dielectric constant and refractive index are functions of frequency, hence wavelength. The frequency or wavelength variation of refractive index is called *dispersion*. Disperson is an important property for optical design (i.e., correction of chromatic aberration) and in the transmission of information (i.e., pulse spreading). Other optical properties are derived from the change in refractive index with other properties such as temperature (*thermo-optic* coefficient), stress or strain (*piezo-optic* or *elasto-optic* coefficients), or applied field (*electro-optic* or *piezo-electric coefficients*). Since the dielectric constant is a second-order tensor with three principal values, the coefficients defined here are also tensor properties (see Table 1).

Precise refractive index measurements give values as functions of wavelength. Frequently, it is desirable to have a functional form for the dispersion of the refractive index (i.e., for calculations and value interpolation). There are many formulas used for representing the refractive index. One of the most widely used is the Sellmeier (or Drude or Maxwell-Helmholtz-Drude) dispersion model [Eq. (14)], which arises from treating the absorption like simple mechanical or electrical resonances. Sellmeier proposed the following dispersion formula in 1871 (although Maxwell had also considered the same derivation in 1869). The usual form of this equation for optical applications gives refractive index as a function of wavelength rather than wave number, frequency, or energy. In this form, the Sellmeier equation is:

$$n^2(\lambda) - 1 = \sum_{i=1} \frac{A_i \cdot \lambda^2}{\lambda^2 - \lambda_i^2} \tag{29a}$$

An often-used, slight modification of this formula puts the wavelength of the shortest wavelength resonance at zero ($\lambda_1 = 0$), i.e., the first term is a constant. This constant term represents contributions to refractive index from electronic transitions at energies far above the band gap. Sellmeier terms with small λ_i (representing electronic transitions) can be expanded as a power series,

$$\frac{A_i \cdot \lambda^2}{\lambda^2 - \lambda_i^2} = A_i \cdot \sum_{j=0}^{\infty} \cdot (\lambda_i^2/\lambda^2)^j$$

$$= A_i + \frac{A_i \cdot \lambda_i^2}{\lambda^2} + \frac{A_i \cdot \lambda_i^4}{\lambda^4} + \cdots \tag{29b}$$

and the terms with large λ_i (representing vibrational transitions) are expanded as:

$$\frac{A_i}{\lambda^2 - \lambda_i^2} = -A_i \cdot \sum_{j=1}^{\infty} (\lambda^2/\lambda_i^2)^j$$

$$= -A_i \cdot \frac{\lambda^2}{\lambda_i^2} - A_i \cdot \frac{\lambda^4}{\lambda_i^4} - \cdots \tag{29c}$$

The first term of this expansion is occasionally used to represent the long-wavelength contributions to the index of refraction (see Schott dispersion formula, following).

A generalized form of the short-wavelength approximation to the Sellmeier equation is the Cauchy formula, developed in 1836. This was the first successful attempt to represent dispersion by an equation:

$$n = A_0 + \sum_{i=1} \frac{A_i}{\lambda^{2i}} \quad \text{or} \quad n^2 = A_0' + \sum_{i=1} \frac{A_i'}{\lambda^{2i}} \tag{30}$$

Power series approximations to the Sellmeier equation are expressed in many forms. One common form is the Schott glass formula used for glasses:

$$n^2 = A_0 + A_1\lambda^2 + A_2\lambda^{-2} + A_3\lambda^{-4} + A_4\lambda^{-6} + A_5\lambda^{-8} \tag{31}$$

For typical high-quality glasses, this equation is accurate to $\pm 3 \cdot 10^{-6}$ in the visible (400 to 765 nm) and within $\pm 5 \cdot 10^{-6}$ from 365 to 1014 nm. A comparison of the Schott power series formula with a three-term Sellmeier formula showed equivalent accuracy of the range of the Schott fit, but that the Sellmeier model was accurate over a much wider wavelength range.[42] A number of other power series dispersion formulas (e.g., Ketteler–Neumann[43]) are occasionally used.

Frequently, Sellmeier terms are written in altered fashion such as this form used by Li:[44]

$$\frac{A_i}{(\lambda^2 - \lambda_i^2)} = \frac{A_i(\lambda^2/\lambda_i^2)}{(\lambda^2 - \lambda_i^2)} - \frac{A_i}{\lambda_i^2} \tag{32a}$$

which is the combination of two Sellmeier terms, one located at zero wavelength and the other at λ_i. The Zernike formula[45] also uses a term in this form. Another way to modify Sellmeier terms is to convert the wavelength of the resonances to wave number or energy [see Eq. (14)].

Another common formula for the index of refraction is the Hartmann or Cornu equation:

$$n = A + \frac{B}{\lambda - \lambda_0} \tag{32b}$$

This equation is more distantly related to the Sellmeier formulation. Note that a two-term Sellmeier formula (with $\lambda_1 = 0$) can be written as

$$(n^2 - n_0^2) \cdot (\lambda^2 - \lambda_0^2) = (n - n_0) \cdot (\lambda - \lambda_0) \cdot (n + n_0) \cdot (\lambda + \lambda_0) = \text{constant} \tag{32c}$$

and the Hartmann formula can be written as a hyperbola:

$$(n - n_0) \cdot (\lambda - \lambda_0) = \text{constant} \tag{32d}$$

Note that in a limited spectral region, the difference terms of the Sellmeier formula of Eq. (33c) vary much more rapidly than do the sum terms, hence the Hartmann and Sellmeier forms will have the same shape in this limited spectral range.

Other equations that combine Sellmeier and power series tersms (cf., Wemple formula) are often used. One such formulation is the Herzberger equation, first developed for glasses[46] and later applied to infrared crystalline materials:[47]

$$n = A + \frac{B}{(\lambda^2 - 0.028)} + \frac{C}{(\lambda^2 - 0.028)^2} + D\lambda^2 + E\lambda^4 \tag{32e}$$

where the choice of the constant $\lambda_o^2 = 0.028$ is arbitrary in that it is applied to all materials.

The Pikhtin-Yas'kov formula[48] is nearly the same as the Sellmeier form with the addition of another term representing a broadband electronic contribution to index:

$$n^2 - 1 = \frac{A}{\pi} \ln \frac{E_1^2 - (\hbar\omega)^2}{E_0^2 - (\hbar\omega)^2} + \sum_i \frac{G_i}{E_i^2 - (\hbar\omega)^2} \qquad (32f)$$

This formulation has been applied to some semiconductor materials. The unique term arises from assuming that the imaginary part of the dielectric constant is a constant between energies $\mathbf{E}_0$ and $\mathbf{E}_1$ and that infinitely narrow resonances occur at $\mathbf{E}_i$. The formula is then derived by applying the Kramers–Krönig relationship to this model.

Typically for glasses, the *Abbe number,* or constringence v_d, is also given. The Abbe number is a measure of dispersion in the visible and is defined as $v_d = (n_d - 1)/(n_F - n_C)$ where n_d, n_F, and n_C are refractive indices at 587.6 nm, 486.1 nm, and 656.3 nm. The quantity $(n_F - n_C)$ is known as the *principle dispersion.* A relative partial dispersion $P_{x,y}$ can be calculated at any wavelengths x and y from

$$P_{x,y} = \frac{(n_x - n_y)}{(n_F - n_C)} \qquad (33a)$$

For so-called "normal" glasses, the partial dispersions obey a linear relationship, namely

$$P_{x,y} = a_{x,y} + b_{x,y} v_d \qquad (33b)$$

where $a_{x,y}$ and $b_{x,y}$ are empirical constants characteristic of normal glasses. However, for correction of secondary spectrum in an optical system (that is, achromatization for more than two wavelengths), it is necessary to employ a glass that does not follow the glass line. Glass manufacturers usually list $\Delta P_{x,y}$ for a number of wavelength pairs as defined by:

$$P_{x,y} = a_{x,y} + b_{x,y} v_d + \Delta P_{x,y} \qquad (33c)$$

The deviation term $\Delta P_{x,y}$ is a measure of the dispersion characteristics differing from the normal glasses. Schott glasses F2 and K7 define the normal glass line.

The Sellmeier formula has the appropriate physical basis to accurately represent the refractive index throughout the transparent region in the simplest manner. The Sellmeier constants have physical meaning, particularly for simple substances. Most other dispersion formulas are closely related to (or are a disguised form of) the Sellmeier equation. Many of these other dispersion formulas are unable to cover a wide spectral region, and unlike the Sellmeier form, do not lend themselves to extrapolation outside the region of available measurements. For these reasons, we strongly urge that the Sellmeier model be universally used as the standard representation of the refractive index.

Modifications of the Sellmeier terms that include composition variation[49,50] and temperature dependence[51] have been applied to successfully model refractive index. The variation of the Sellmeier A_i and λ_i constants is usually modeled as linearly dependent on the mole fraction of the components and temperature.

In the transparent region, refractive index decreases with wavelength, and the magnitude of $dn/d\lambda$ is a minimum between the electronic and vibrational absorptions. The wavelength of minimum $dn/d\lambda$, called the zero-dispersion point, is given by

$$\frac{d^2 n}{d\lambda^2} = 0 \qquad (34)$$

which is the desired operating point for high-bandwidth, information-carrying optical fibers as well as the optimum wavelength for single-element refractive optical systems. For glassy silica fibers, the zero dispersion point is 1.272 μm. One approach to reducing both dispersion and loss is to use a material with a wide transparent region, i.e., widely separated electronic and vibrational absorptions, hence the interest in materials such as heavy-metal fluoride glasses for fiber applications.

Thermo-optic and Photoelastic Coefficients. Temperature is one of the main factors influencing the refractive index of solids. The thermo-optical coefficients $\partial n/\partial T$ (or $\partial \epsilon/\partial T$) can be estimated from a derivation of the Clausius-Mossotti relationship:[52]

$$\frac{1}{(\epsilon-1)(\epsilon+2)}\left(\frac{\partial \epsilon}{\partial T}\right) = -\alpha\left[1 - \frac{V}{\alpha_m}\left(\frac{\partial \alpha_m}{\partial V}\right)_T\right] + \frac{1}{3\alpha_m}\left(\frac{\partial \alpha_m}{\partial T}\right)_v \qquad (35)$$

where α_m is the macroscopic polarizability. The first two terms are the principal contributors in ionic materials: a positive thermal expansion coefficient α results in a negative thermo-optic coefficient and a positive change in polarizability with volume results in a positive thermo-optic coefficient. In ionic materials with a low melting point, thermal expansion is high and the thermo-optic coefficient is negative (typical of alkali halides); when thermal expansion is small (indicated by high melting point, hardness, and high elastic moduli), the thermo-optic coefficient is positive, dominated by the volume change in polarizability (typical of the high-temperature oxides).

Thermal expansion has no frequency dependence but polarizability does. At frequencies (wavelengths) near the edge of transparency, the polarizability (and $\partial \alpha_m/\partial V$) rises, and $\partial n/\partial T$ becomes more positive (or less negative). Figure 8 shows the variation of refractive index for sodium chloride as a function of frequency and temperature.

A formalism similar to the preceding for the thermo-optic coefficient can be used to estimate the photoelastic constants of a material. The simplest photoelastic constant is that produced by uniform pressure, i.e., dn/dP. More complex photoelastic constants are tensors whose components define the effect of individual strain (elasto-optic coefficients) or stress (piezo-optic coefficients) tensor terms. Bendow et al.,[53] calculate dn/dP and the

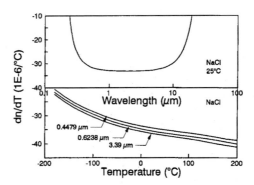

FIGURE 8 The thermo-optic coefficient (dn/dT) of sodium chloride (NaCl): (*a*) shows the wavelength dependence of room-temperature thermo-optic coefficient. The thermo-optic coefficient is nearly constant in the transparent region, but increases significantly at the edges of the transparent region; (*b*) illustrates the temperature dependence of the thermo-optic coefficient. The thermo-optic coefficient decreases (becomes more negative) with increasing temperature, primarily as the density decreases.

elasto-optic coefficients for a number of cubic crystals and compare the results to experiment.

Nonlinear Optical Coefficients. One of the most important higher-order optical coefficients is the nonlinear (or second-order) susceptibility. With the high electric fields generated by lasers, the nonlinear susceptibility gives rise to important processes such as second-harmonic generation, optical rectification, parametric mixing, and the linear electro-optic (Pockels) effect. The second-order susceptibility, $\chi^{(2)}$, is related to the polarization vector **P** by

$$\mathbf{P}_i(\omega) = \epsilon_0[\chi_{ij}\mathbf{E}_j + g\chi_{ijk}^{(2)}\mathbf{E}_j(\omega_1)\mathbf{E}_k(\omega_2)] \tag{36}$$

where g is a degeneracy factor arising from the nature of the electric fields applied. If the two frequencies are equal, the condition of optical rectification and second-harmonic generation (SHG) arises, and $g = \frac{1}{2}$. When the frequencies of $\mathbf{E}_j$ and $\mathbf{E}_k$ are different, parametric mixing occurs and $g = 1$. If $\mathbf{E}_k$ is a dc field, the situation is the same as the linear electro-optic (or Pockels) effect, and $g = 2$. The value of nonlinear susceptibility is a function of the frequencies of both the input fields and the output polarization ($\omega = \omega_1 \pm \omega_2$).

The nonlinear susceptibility is a third-order (3 by 3 by 3) tensor. A nonlinear optical coefficient $d^{(2)}$, frequently used to describe these nonlinear properties, is equal to one-half of the second-order nonlinear susceptibility, i.e., $d^{(2)} = \frac{1}{2}\chi^{(2)}$. Nonlinear optical coefficients are universally written in reduced (matrix) notation, d_{ij}, where the index $i = 1$, 2, or 3 and the index j runs from 1 to 6.[54] (Both the piezo-electric coefficient and the nonlinear optical coefficient are given the symbol d, and the resulting confusion is enhanced because both coefficients have the same units.) The relationship between the electro-optic coefficient r and the nonlinear optical coefficient $d^{(2)}$ is

$$r_{ij} = \frac{2gd_{ji}^{(2)}}{\epsilon^2} \tag{37}$$

Units of the second-order nonlinear optical coefficient are m/V (or pm/V, where pm = 10^{-12} m) in mks units.

Typical values of the nonlinear optical coefficients are listed in Table 4.[55] Additional nonlinear optical coefficients are given in reviews.[54,56,57]

Scatter. Scatter is both an intrinsic and extrinsic property. Rayleigh, Brillouin, Raman,

TABLE 4 Typical Nonlinear Optical Coefficients

Crystal	Nonlinear optical coefficient (pm/V)
β-BaB_2O_4	$d_{11} = 1.60$
KH_2PO_4	$d_{36} = 0.39$
LiB_3O_5	$d_{32} = 1.21$
$LiNbO_3$	$d_{31} = 5.07$
$LiIO_3$	$d_{31} = 3.90$
$KTiOPO_4$	$d_{31} = 5.85$
Urea	$d_{14} = 1.17$

and stoichiometric (index variation) contributions to scatter have been derived in simple form and used to estimate scatter loss in several fiber-optic materials.[58] Rayleigh scattering refers to elastic scatter from features small compared to the wavelength of light. In highly pure and defect-free optical crystals, Rayleigh scatter is caused by atomic scale in-homogeneities (much smaller than the wavelength) in analog to Rayleigh scatter from molecules in the atmosphere. In most materials, including glasses, Rayleigh scatter is augmented by extrinsic contributions arising from localized density variations (which also limit the uniformity of the refractive index). Attenuation in high-quality optical materials is frequently limited by Rayleigh scatter rather than absorption.

Brillouin and Raman scatter are forms of inelastic scattering from acoustic and optical phonons (vibrations). The frequency of the scattered light is shifted by the phonon frequency. Creation of phonons results in longer-wavelength (low-frequency) scattered light (Stokes case) and annihilation of phonons results in higher-frequency scattered light (anti-Stokes case). Rayleigh, Brillouin, and Raman scatter all have a λ^{-4} wavelength dependence. Polycrystalline and translucent materials have features such as grain boundaries and voids whose size is larger than the wavelength of light. This type of scatter is often called Mie scatter because the scattering features are larger than the wavelength of the light. Mie scatter typically has a measured λ^{-m} dependence where the parameter m typically lies between 1 and 2.[59,60] Rayleigh and Mie scatter may arise from either surface roughness or bulk nonuniformities.

Other Properties of Materials

Characterization of Crystals and Glasses. All materials are characterized by name(s) for identification, a chemical formula (crystalline materials) or approximate composition (glasses, amorphous substances), and a density (ρ, in kg/m^3). Crystalline materials are further identified by crystal class, space group, unit-cell lattice parameters, molecular weight (of a formula unit in atomic mass units, amu), and number of formula units per unit cell (Z). (See standard compilations of crystallographic data.[61,62])

Material Designation and Composition. Crystals are completely identified by both the chemical formulation and the space group. Chemical formulation alone is insufficient for identification because many substances have several structures (called polymorphs) with different properties. Properties in the data tables pertain only to the specific structure listed. Materials in the data tables having several stable polymorphs at room temperature include SiO_2 (eight polymorphs), C (diamond, graphite, and amorphous forms), and SiC and ZnS (both have cubic and hexagonal forms).

The space group also identifies the appropriate number of independent terms (see Tables 1 and 2) that describe a physical quantity. Noncubic crystals require two or more values to fully describe thermal expansion, thermal conductivity, refractive index, and other properties. Often, scalar quantities are given in the literature when a tensor characterization is needed. Such a characterization may be adequate for polycrystalline materials, but is unsatisfactory for single crystals that require knowledge of directional properties.

Optical glasses have been identified by traditional names derived from their composition and their dispersion relative to their index of refraction. Crown glasses have low dispersion (typically with Abbe number $v_d > 50$) and flint glasses have higher dispersion (typically, $v_d < 50$). More than a century ago, the crown and flint designations evolved to indicate primarily the difference between a standard soda-lime crown glass and that of flint glass which had a higher index and higher dispersion because of the addition of PbO to a silica base.

A more specific glass identifier is a six-digit number (defined in military standard

MIL-G-174) representing the first three digits of $(n_d - 1)$ and the first three digits of v_d. Each manufacturer also has its own designator, usually based on traditional names, that uniquely identifies each glass. For example, the glass with code 517624 has the following manufacturer's designations:

Manufacturer	Designation for glass 517624
Schott	BK-7
Corning	B-16-64
Pilkington	BSC-517642
Hoya	BSC-7
Ohara	BSL-7 (glass 516624)

Properties of glass are primarily determined from the compositions, but also depend on the manufacturing process, specifically the thermal history. In fact, refractive index specifications in a glass catalog should be interpreted as those obtained with a particular annealing schedule. Annealing removes stress (and minimizes stress-induced birefringence) and minimizes the effect of thermal history, producing high refractive-index uniformity. Special (*precision*) annealing designed to maximize refractive index homogeneity may, however, increase refractive index slightly above a nominal (catalog) value.

In general, both glass composition and thermal processing are proprietary (so the compositions given are only illustrative). However, manufacturers' data sheets on individual glasses can provide detailed and specific information on optical and mechanical properties. Also, the data sheets supply useful details on index homogeneity, climate resistance, stain resistance, and chemical (acid and alkaline) resistance for a particular glass type. For very detailed work or demanding applications, a glass manufacturer can supply a *melt data sheet* providing accurate optical properties for a specific glass lot.

Unit Cell Parameters, Molecular Weight, and Density. The structure and composition of crystals can be used to calculate density. This calculated (theoretical or x-ray) density should closely match that of optical-quality materials. Density ρ is mass divided by volume:

$$\rho = \frac{Z \cdot (\text{MW}) \cdot u}{a \cdot b \cdot c \cdot \sqrt{\sin^2 \alpha + \sin^2 \beta + \sin^2 \gamma - 2(1 - \cos \alpha \cdot \cos \beta \cdot \cos \gamma)}} \tag{38}$$

where Z is the number of formula units in a crystal unit cell, MW is the molecular weight of a formula unit in amu, u is weight of an amu (Table 3), a, b, and c are unit cell axes lengths, and α, β, and γ are unit cell axes angles.

Typically, pure amorphous materials have lower density than the corresponding crystalline materials. Density of glasses and other amorphous materials is derived from measurements.

Elastic Properties. Elastic properties of materials can be described with a hierarchy of terms. On the atomic scale, interatomic force constants or potential energies can be used to predict the vibrational modes, thermal expansion, and elastic properties of a material. On the macroscopic scale, elastic properties are described using elastic moduli (or constants) related to the directional properties of a material. The tensor relationships between stress (σ, a second-order tensor) and strain (**e**, a second-order tensor) are

$$\sigma_{ij} = \mathbf{c}_{ijkl} \cdot e_{kl}$$
$$e_{ij} = \mathbf{s}_{ijkl} \cdot \sigma_{kl} \tag{39a}$$

where the fourth-rank tensors $\mathbf{c}_{ijkl}$ and $\mathbf{s}_{ijkl}$ are named elastic stiffness **c** and elastic compliance **s**, respectively. This is the tensor form of Hooke's Law. Each index ($i, j, k,$ and l) has three values (i.e., $x, y,$ and z), hence the **c** and **s** tensors have 81 terms.

The stiffness and compliance tensors are usually written in a matrix notation made possible by the symmetry relationship of the stress and strain tensors. Symmetry reduces the number of independent terms in the stiffness and compliance tensors from 81 to 36. The usual notation for the reduced (matrix) notation form of the stiffness and compliance tensors is

$$\sigma_i = \mathbf{c}_{ij} \cdot e_j$$
$$e_i = \mathbf{s}_{ij} \cdot \sigma_j \tag{39b}$$

where the indices, an abbreviation of the ij or kl components, run from 1 to 6. Table 5 shows the conversion from tensor to matrix notation. Thus, the stiffness and compliance tensors are written as 6 by 6 matrices which can again be shown to be symmetric, given 21 independent terms. Virtually all data will be found in matrix notation. These tensors (matrices) that relate stress and strain are sometimes called *second-order* stiffness and compliance. Higher-order tensors are used to describe nonlinear elastic behavior (i.e., third-order stiffness determines the stress tensor from the square of the strain tensor).

Stiffness and compliance tensors are needed to completely describe the linear elastic properties of a crystal. Even a completely amorphous material has two independent constants that describe the relationship between stress and strain. Usually, the elastic properties of materials are expressed in terms of engineering (or technical) moduli: Young's modulus (E), shear modulus (or modulus of rigidity G), bulk modulus (B, compressibility^{-1}), and Poisson's ratio (v). For example, Young's modulus is defined as the ratio of the longitudinal tension to the longitudinal strain for tension, a quantity which is anisotropic (i.e., directionally dependent) for all crystal classes (but is isotropic for amorphous materials). Therefore the engineering moduli only accurately describe the elastic behavior of isotropic materials. The engineering moduli also approximately describe the elastic behavior of polycrystalline materials (assuming small, randomly distributed grains). Various methods are available to estimate the engineering moduli of crystals.

TABLE 5 Matrix Notation for Stress, Strain, Stiffness, and Compliance Tensors

Tensor-to-matrix index conversion		Tensor-to-matrix element conversion	
Tensor indices ij or kl	Matrix indices m or n	Notation	Condition
11	1	$\sigma_m = \sigma_{ij}$ $e_m = e_{ij}$	$m = 1, 2, 3$
22	2	$\sigma_m = \sigma_{ij}$ $e_m = 2e_{ij}$	$m = 4, 5, 6$
33	3	$\mathbf{c}_{mn} = \mathbf{c}_{ijkl}$	all m, n
23 or 32	4	$\mathbf{s}_{mn} = \mathbf{s}_{ijkl}$	$m, n = 1, 2, 3$
13 or 31	5	$\mathbf{s}_{mn} = 2\mathbf{s}_{ijkl}$	$m = 1, 2, 3$ and $n = 4, 5, 6$ $m = 4, 5, 6$ and $n = 1, 2, 3$
12 or 21	6	$\mathbf{s}_{mn} = 4\mathbf{s}_{ijkl}$	$m, n = 4, 5, 6$

Values of the engineering moduli for crystalline materials given in the data tables are estimated from elastic moduli using the Voigt and Reuss methods (noncubic materials) or the Haskin and Shtrickman method[63] (cubic materials) to give shear and bulk moduli. Young's modulus and Poisson's ratio are then calculated assuming isotropy using the following relationships:

$$E = \frac{9 \cdot G \cdot B}{G + 3 \cdot B} \qquad v = \frac{3 \cdot B - 2 \cdot G}{6 \cdot B + 2 \cdot G} \tag{39c}$$

Hardness and Strength. Hardness is an empirical and relative measure of a material's resistance to wear (mechanical abrasion). Despite the qualitative nature of the result, hardness testing is quantitative, repeatable, and easy to measure. The first measure of hardness was the Mohs scale which compares the hardness of materials to one of 10 minerals. Usually, the Knoop indent test is used to measure hardness of optical materials. The test determines the resistance of a surface to penetration by a diamond indenter with a fixed load (usually 200 to 250 grams). The Knoop hardness number (in kg/mm^2) is the indenter mass (proportional to load) divided by the area of the indent. Figure 9 compares the Mohs and Knoop scales.[9]

Materials with Knoop values less than $100\,kg/mm^2$ are very soft, difficult to polish, and susceptible to handling damage. Knoop hardness values greater than 750 are quite hard. Typical glasses have hardness values of 350 to $600\,kg/mm^2$. Hardness qualitatively correlates to Young's modulus and to strength. Hardness of crystals is dependent on the orientation of the crystal axes with respect to the tested surface. Coatings can significantly alter hardness.

Strength is a measure of a material's resistance to fracture (or onset of plastic deformation). Strength is highly dependent on material flaws and therefore on the method of manufacture, as well as the method of measurement. For optical materials, strength is most conveniently measured in flexure; tensile strengths are typically 50 to 90 percent of those measured in flexure. Because of high variability in strength values, quoted strength values should only be used as a guide for comparision of materials. Strength of crystals

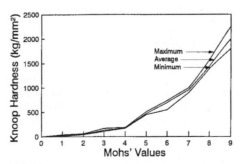

FIGURE 9 A comparison of Mohs and Knoop hardness scales.[9] The Mohs scale is qualitative, comparing the hardness of a material to one of 10 minerals: talc (Moh $\equiv$ 1), gypsum ($\equiv$ 2), calcite ($\equiv$ 3), fluorite ($\equiv$ 4), apatite ($\equiv$ 5), orthoclase ($\equiv$ 6), quartz ($\equiv$ 7), topaz ($\equiv$ 8), sapphire ($\equiv$ 9), and diamond (Moh $\equiv$ 10). The Knoop scale is determined by area of a mark caused by an indenter; the Knoop value is the indenter mass divided by the indented area. The mass of the indenter (load) is usually specified; 200 or 500 grams are typical. (*Reprinted by permission of Ashlee Publishing Company.*)

TABLE 6 Fracture Toughness of Some Materials

Material	Fracture toughness, $MPa \cdot m^{1/2}$	Material	Fracture toughness, $MPa \cdot m^{1/2}$
Al_2O_3	3	$MgAl_2O_4$	1.5
ALON	1.4	MgF_2	1.0
AlN	3	Si	0.95
C, diamond	2.0	fused SiO_2	0.8
CaF_2	0.5	Y_2O_3	0.7
$CaLa_2S_4$	0.68	ZnS	0.5
			0.8 (CVD)
GaP	0.9	ZnSe	0.33
Ge	0.66	$ZrO_2 : Y_2O_3$	2.0

also is dependent on the orientation of the crystal axes with respect to the applied stress. Applications requiring high strength to avoid failure should use large safety margins (typically a factor of four) over average strength whenever possible.

Fracture toughness is another measure of strength, specifically, a material's ability to resist crack propagation. Fracture toughness measures the applied stress required to enlarge a flaw (crack) of given size and has units of $MPa \cdot m^{1/2}$. Values for representative materials are given in Table 6.

Characteristic Temperatures. Characteristic temperatures of crystalline materials are those of melting (or vaporization or decomposition) and phase transitions. Of particular importance are the phase-transition temperatures. These temperatures mark the boundaries of a particular structure. A phase transition can mean a marked change in properties. One important phase-transition point is the Curie temperature of ferroelectric materials. Below the Curie temperature, the material is ferroelectric; above this temperature it is paraelectric. The Curie temperature phase transition is particularly significant because the change in structure is accompanied by drastic changes in some properties such as the static dielectric constant which approaches infinity as temperature nears the Curie temperature. This transition is associated with the lowest transverse optical frequency (the *soft mode*) approaching zero [hence the static dielectric constant approaches infinity from the Lyddane-Sachs-Teller relationship, Eq. (11)].

The term *glass* applies to a material that retains an amorphous state upon solidification. More accurately, glass is an undercooled, inorganic liquid with a very high viscosity at room temperature and is characterized by a gradual softening with temperature and a hysteresis between glass and crystalline properties. The gradual change in viscosity with temperature is characterized by several temperatures, especially the glass transition temperature and the softening-point temperature. The glass transition temperature defines a second-order phase transition analogous to melting. At this temperature, the temperature dependence of various properties changes (in particular, the linear thermal expansion coefficient) as the material transitions from a liquid to glassy state. Glasses can crystallize if held above the transition temperature for sufficient time. The annealing point is defined as the temperature resulting in a glass viscosity of 10^{13} poise and at which typical glasses can be annealed within an hour or so. In many glasses, the annealing point and the glass transition temperature are close. In an optical system, glass elements need to be kept 150 to 200°C below the glass transition temperature to avoid significant surface distortion. At the *softening temperature,* viscosity is $10^{7.6}$ poise and glass will rapidly deform under its own weight; glasses are typically molded at this temperature. Glasses do not have a true melting point; they become progressively softer (more viscous) with increased temperature. Other amorphous materials may not have a well-defined glass transition; instead they may have a conventional melting point. Glasses that crystallize at elevated temperature also have a well-defined mellting point.

Glass-ceramics have been developed which are materials with both glasslike and crystalline phases. In particular, low-thermal-expansion ceramics comprise a crystalline phase with a negative thermal expansion and a vitreous phase with a positive thermal expansion. Combined, the two phases result in very high dimensional stability. Typically, the ceramics are made like other glasses, but after stresses are removed from a blank, a special heat-treatment step forms the nuclei for the growth of the crystalline component of the ceramic. Although not strictly ceramics, similar attributes can be found in some two-phase glasses.

Heat Capacity and Debye Temperature.   Heat capacity, or specific heat, a scalar quantity, is the change in thermal energy with a change in temperature. Units are typically $J/(gm \cdot K)$. Debye developed a theory of heat capacity assuming that the energy was stored in acoustical photons. This theory, which assumes a particular density of states, results in a Debye molar heat capacity (units $= J/(mole \cdot K)$) of the form

$$C_V(T) = 9mN_A k_B \left(\frac{T}{\theta_D}\right)^3 \int_0^{\theta_D/T} \frac{x^4 e^x}{(e^x - 1)^2}\, dx \qquad (40a)$$

where C_V is the molar heat capacity in $J/(mole \cdot K)$ per unit volume, θ_D is the Debye temperature, m the number of atoms per formula unit, N_A is Avogadro's number, and k_B is Boltzmann's constant. At low temperatures ($T \to 0$ K), heat capacity closely follows the T^3 law of Debye theory

$$C_V(T) = \frac{12\pi^4}{5} mN_A k_B \left(\frac{T}{\theta_D}\right)^3 = 1943.76\, J/(mole \cdot K) \cdot m \left(\frac{T}{\theta_D}\right)^3 \qquad (T \ll \theta_D) \quad (40b)$$

and the high-temperature (classical) limit is

$$C_V(T) = 3mN_A k_B = 24.943\, J/(mole \cdot K) \cdot m \qquad (T \gg \theta_D) \qquad (40c)$$

If heat capacity data are fit piecewise to the Debye equation, a temperature-dependent θ_D can be found. Frequently, a Debye temperature is determined from room-temperature elastic constants, and is therefore different from the low-temperature value. The Debye temperature given in the tables is, when possible, derived from low-temperature heat capacity data.

The Debye equations can be used to estimate heat capacity C_V over the entire temperature range, typically to within 5 percent of the true value using a single Debye temperature value. Usually, however, C_P, the constant pressure heat capacity, rather than C_V, is desired. At low temperatures, thermal expansion is small, and $C_P \approx C_V$. At elevated temperature, the relationship between C_V and C_P is given by the thermodynamic relationship

$$C_P(T) = C_V(T) + 9\alpha^2 TVB \qquad (41)$$

where T is temperature (K), V is the molar volume ($m^3/mole$), α is the thermal expansion coefficient, and B is the bulk modulus ($Pa = Nt/m^2$).

Molar heat capacity can be converted to usual units by dividing by the molecular weight (see the physical property tables for crystals) in $g/mole$. Since molar heat capacity

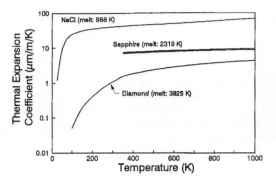

FIGURE 10 Thermal expansion of several materials. Expansion arises from the anharmonicity of the interatomic potential. At low temperature, expansion is very low and the expansion coefficient is low. As temperature increases, the expansion coefficient rises, first quickly, then less rapidly.

approaches the value of 24.943 J/(mole · K), the heat capacity per unit weight is inversely proportional to molecular weight at high temperature (i.e., above the Debye temperature).

Thermal Expansion. The linear thermal expansion coefficient α is the fractional change in length with a change in temperature as defined by

$$\alpha(T) = \frac{1}{L}\frac{dL}{dT} \qquad (42a)$$

and units are $1/K$. The units of length are arbitrary. Thermal expansion is a second-rank tensor; nonisometric crystals have a different thermal expansion coefficient for each principal direction. At low temperature, thermal expansion is low, and the coefficient of thermal expansion approaches zero as $T \rightarrow 0$. The expansion coefficient generally rises with increasing temperature; Fig. 10 shows temperature dependence of the expansion coefficient for several materials. Several compilations of data exist.[64,65]

The volume expansion coefficient α_V is the fractional change in volume with an increase in temperature. For a cubic or isotropic material with a single linear thermal expansion coefficient,

$$\alpha_V(T) = \frac{1}{V}\frac{dV}{dT} = 3\alpha \qquad (42b)$$

which can be used to estimate the temperature change of density.

The Grüneisen relationship relates the thermal expansion coefficient to molar heat capacity

$$\alpha(T) = \frac{\gamma C_V(T)}{3B_0 V_0} \qquad (42c)$$

where B_0 is the bulk modulus at $T = 0\,K$, V_0 is the volume at $T = 0\,K$, and γ is the Grüneisen parameter. This relationship shows that thermal expansion has the same temperature dependence as the heat capacity. Typical values of γ lie between 1 and 2.

Thermal Conductivity. Thermal conductivity κ determines the rate of heat flow through a material with a given thermal gradient. Conductivity is a second-rank tensor with up to

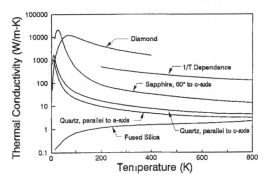

FIGURE 11 Thermal conductivity of several materials. The thermal conductivity of materials initially rises rapidly as the heat capacity increases. Peak thermal conductivity of crystals is high due to the long phonon mean free path of the periodic structure which falls with increasing temperature as the phonon free path length decreases ($\approx 1/T$). The phonon mean free path of amorphous materials is small and nearly independent of temperature, hence thermal conductivity rises monotonically and approaches the crystalline value at high temperature.

three principal values. This property is especially important in relieving thermal stress and optical distortions caused by rapid heating or cooling. Units are $W/(m \cdot K)$.

Kinetic theory gives the following expression for thermal conductivity κ

$$\kappa = \tfrac{1}{3} C_V v \ell \qquad (43)$$

where v is the phonon (sound) velocity and ℓ is the phonon mean free path. At very low temperature ($T < \theta_D/20$), temperature dependence of thermal conductivity is governed by C_V, which rises as T^3 [see Eq. (40b)]. At high temperature ($T > \theta_D/10$), the phonon mean free path is limited by several mechanisms. In crystals, scattering by other phonons usually governs ℓ. In the high-temperature limit, the phonon density rises proportional to T and thermal conductivity is inversely proportional to T. Figure 11 illustrates the temperature dependence of thermal conductivity of several crystalline materials.

Thermal conductivity in amorphous substances is quite different compared to crystals. The phonon mean free path in glasses is significantly less than in crystals, limited by structural disorder. The mean free path of amorphous materials is typically the size of the fundamental structural units (e.g., 10 Å) and has little temperature dependence; hence the temperature dependence of thermal conductivity is primarily governed by the temperature dependence of heat capacity. At room temperature, the thermal conductivity of oxide glasses is a factor of 10 below typical oxide crystals. Figure 11 compares the thermal conductivity of fused silica to crystalline silica (quartz).

Thermal conductivity of crystals is highly dependent on purity and order. Mixed crystals, second-phase inclusion, nonstoichiometry, voids, and defects can all lower the thermal conductivity of a material. Values given in the data table are for the highest-quality material. Thermal conductivity data are found in compilations[65] and reviews.[66]

Correlation of Properties. All material properties are correlated to a relatively few factors, e.g., constituent atoms, the bonding between the atoms, and the structural symmetry. The binding forces, or chemical bonds, play a major role in properties. Tightly bonded materials have high moduli, high hardness and strength, and high Debye temperature (hence high room-temperature thermal conductivity). Strong bonds also mean lower thermal expansion, lower refractive index, higher-frequency optical vibrational modes (hence less infrared transparency), and higher energy band gaps (hence more ultraviolet transparency). Increased mass of the constituent atoms lowers the frequency of both electronic and ionic resonances. Similarly, high structural symmetry can increase hardness and eliminate (to first order) ionic vibrations (cf., diamond, silicon, and germanium).

Combinations of Properties. A given material property is influenced by many factors. For example, the length of a specimen is affected by stress (producing strain), by electric fields (piezo-electric effect), and by temperature (thermal expansion). The total strain is then a combination of these linear effects and can be written as

$$\Delta e_{ij} = \left(\frac{\partial e_{ij}}{\partial \sigma_{kl}}\right)_{E,T} \Delta \sigma_{kl} + \left(\frac{\partial e_{ij}}{\partial E_k}\right)_{\sigma,T} \Delta E_k + \left(\frac{\partial e_{ij}}{\partial T}\right)_{\sigma,E} \Delta T$$

$$\Delta e_{ij} = s_{ijkl} \Delta \sigma_{kl} + d_{ijk} \Delta E_k + \alpha_{ij} \Delta T \tag{44}$$

Often, these effects are interrelated, and frequently dependent on measurement conditions. Some properties of some materials are very sensitive to measurement conditions (the subscripts in the preceding equation denote the variable held constant for each term). For example, if the measurement conditions for the elastic contribution were adiabatic, stress will cause temperature to fall, which, in turn, decreases strain (assuming positive thermal expansion coefficient). Thus the elastic contribution to strain is measured under isothermal (and constant **E** field) conditions so as not to include the temperature effects already included in the thermal expansion term.

The conditions of measurement are given a variety of names that may cause confusion. For example, the mechanical state of "clamped," constant volume, and constant strain all refer to the same measurement condition which is paired with the corresponding condition of "free," "unclamped," constant pressure, or constant stress. In many cases, the condition of measurement is not reported and probably unimportant (i.e., different conditions give essentially the same result). Another common measurement condition is constant **E** field ("electrically free") or constant **D** field ("electrically clamped").

Some materials, particularly ferroelectrics, have large property variation with temperature and pressure, hence measurement conditions may greatly alter the data. The piezoelectric effect contributes significantly to the clamped dielectric constant of ferroelectrics. The difference between the isothermal clamped and free dielectric constants is

$$\epsilon_0 \cdot (\epsilon_i^e - \epsilon_i^\sigma) = -d_{ij} d_{ik} c_{jk}^{\mathbf{E}} \tag{45}$$

where d_{ij} is the piezo-electric coefficient and $c_{jk}^{\mathbf{E}}$ is the (electrically free) elastic stiffness. If the material structure is centrosymmetric, all components of d_{ij} vanish, and the two dielectric constants are the same.

33.5 PROPERTIES TABLES

The following tables summarize the basic properties for representative crystals and glasses. In general, the presented materials are (1) of general interest, (2) well-characterized (within the limitations imposed by general paucity of data and conflicting property values), and (3) represent a wide range of representative types and properties. Few materials can be regarded as well-characterized. Crystalline materials are represented by alkali halides, oxides, chalcogenides, and a variety of crystals with nonlinear, ferroelectric, piezoelectric, and photorefractive materials.

The physical property tables define the composition, density, and structure (of crystalline materials). Table 7 gives data for over 90 crystalline materials. Table 8 gives similar data for 30 representative "optical" glasses intended for visible and near-infrared use (typically to 2.5 μm). Table 9 gives physical property data for 24 specialty glasses and substrate materials. The specialty glasses include fused silica and germania, calcium aluminate, fluoride, germanium-, and chalcogenide-based glasses, many of which are intended for use at longer wavelengths. The three substrate materials, Pyrex, Zerodur, and ULE, are included because of their widespread use for mirror blanks.

Mechanical properties for crystals are given in two forms, room-temperature elastic constants (or moduli) for crystals (Tables 10 through 15), and engineering moduli, flexure strength, and hardness for both crystals (Table 16) and glasses (Table 17). Engineering moduli for crystalline materials should only be applied by polycrystalline forms of these materials. Accurate representation of the elastic properties of single crystals requires the use of elastic constants in tensor form. Strength is highly dependent on manufacture method and many have significant sample-to-sample variability. These characteristics account for the lack of strength data. For these reasons, the provided strength data is intended only as a guide. Glasses and glass-ceramics flexure strengths typically range from 30 to 200 MPa, although glass fibers with strength exceeding 1000 MPa have been reported.

Thermal properties are given in Tables 18 and 19 for crystals and glasses, respectively. Characteristic temperatures (Debye, phase change, and melt for crystals; glass transitions, soften, and melt temperatures for glasses), heat capacity, thermal expansion, and thermal conductivity data are included. Directional thermal properties of crystals are given when available. Only room-temperature properties are reported except for thermal conductivity of crystals, which is also given for temperatures above and below ambient, if available.

Optical properties are summarized in Tables 20 and 21 for crystals and glasses, respectively. These tables give the wavelength boundaries of the optical transparent region (based on a 1-cm^{-1} absorption coefficient), $\bar{n}_\infty$, the electronic contribution to the refractive index, and values of dn/dT at various wavelengths. Tables 22 and 23 give dispersion formulas for crystals and glasses. Tables 24, 25, 26, 27, and 28 give tabular refractive index data for some crystals.

Vibrational characteristics of many optical materials are summarized in Tables 29 through 44 for a number of common optical crystal types. These tables give number and type of zone-center (i.e., the wave vector ≈ 0, where Γ is the usual symbol denoting the center of the Brillouin zone) optical modes predicted by group theory (and observed in practice) as well as the frequency (in wave number) of the infrared-active and Raman modes. Mulliken notation is used. Table 45 summarizes the available lattice vibration dispersion models for many crystals.

Table 46 summarizes Urbach tail parameters [Eq. (15)] for several crystals. Table 47 gives room-temperature ultraviolet and infrared absorption edge parameters for a number of glasses. These parameters are given in a form similar to Eqs. (15a) and (17a) for exponential absorption edges:

$$\beta_{ABS} = \beta_0(UV) \exp(\sigma_{UV} E) = \beta_0(UV) \exp\left[\frac{1.24\sigma_{UV}}{\lambda}\right]$$

$$\beta_{ABS} = \beta_0(IR) \exp(-\sigma_{IR} \nu)$$

(46)

where **E** is energy in eV, λ is wavelength in micrometers, and ν is wave number in cm^{-1}. Table 48 summarizes available room-temperature infrared absorption edge parameters for a number of crystalline and amorphous materials in the format of Eq. (17a).

TABLE 7 Composition, Structure, and Density of Crystals

Material	Crystal system & space group	Unit cell dimension (Å)	Molecular weight (amu)	Formulae/ unit cell	Density (g/cm^3)
Ag_3AsS_3 (proustite)	Hexagonal R3c (C_{3v}^6) #161	a = 10.80 c = 8.69	494.72	6	5.615
AgBr (bromyrite)	Cubic Fm3m (O_h^5) #225	5.7745	187.77	4	6.477
AgCl (cerargyrite)	Cubic Fm3m (O_h^5) #225	5.547	143.32	4	5.578
$AgGaS_2$	Tetragonal I$\bar{4}$2d (D_{2d}^{12}) #122	a = 5.757 c = 10.304	241.72	4	4.701
$AgGaSe_2$	Tetragonal I$\bar{4}$2d (D_{2d}^{12}) #122	a = 5.973 c = 10.88	335.51	4	5.741
β-AgI (iodyrite)	Hexagonal P6$_3$mc (C_{6v}^4) #186	a = 4.5924 c = 7.5104	234.77	2	5.684
AlAs	Cubic F$\bar{4}$3m (T_d^2) #216	5.6611	101.90	4	3.731
AlN	Hexagonal P6$_3$mc (C_{6v}^4) #186	a = 3.1127 c = 4.9816	40.988	2	3.257
Al_2O_3 (sapphire, alumina)	Hexagonal R$\bar{3}$c (D_{3d}^6) #167	a = 4.759 c = 12.989	101.96	6	3.987
$Al_{23}O_{27}N_5$, ALON	Cubic Fd3m (O_h^7) #227	7.948	1122.59	1	3.713
$Ba_3[B_3O_6]_2$, BBO	Hexagonal R3 (C_3^4) #146	a = 12.547 c = 12.736	668.84	6	3.838
BaF_2	Cubic Fm3m (O_h^5) #225	6.2001	175.32	4	4.886
$BaTiO_3$	Tetragonal P4$_2$/mnm (D_{4h}^{14}) #136	a = 39920 c = 4.0361	233.21	1	6.021
BeO (bromellite)	Hexagonal P6$_3$mc (C_{6v}^4) #186	a = 2.693 c = 4.395	25.012	2	3.009
$Bi_{12}GeO_{20}$, BGO	Cubic I23 (T^3) #197	10.143	2900.43	2	9.231
$Bi_{12}SiO_{20}$, BSO (sellenite)	Cubic I23 (T^3) #197	10.1043	2855.84	2	9.194
BN	Cubic F$\bar{4}$3m (T_d^2) #216	3.6157	24.818	4	3.489
BP	Cubic F$\bar{4}$3m (T_d^2) #216	4.538	41.785	4	2.970
C (diamond)	Cubic Fd3m (O_h^7) #227	3.56696	12.011	8	3.516
$CaCO_3$ (calcite)	Hexagonal R$\bar{3}$c (D_{3d}^6) #167	a = 4.9898 c = 17.060	100.09	6	2.711
CaF_2 (fluorite)	Cubic Fm3m (O_h^5) #225	5.46295	78.07	4	3.181
$CaLa_2S_4$	Cubic I$\bar{4}$3d (T_d^6) #220	8.685	446.15	4	4.524
$CaMoO_4$ (powellite)	Tetragonal I4$_1$/a (C_{4h}^6) #88	a = 5.23 c = 11.44	200.02	4	4.246
$CaWO_4$ (scheelite)	Tetragonal I4$_1$/a (C_{4h}^6) #88	a = 5.243 c = 11.376	287.93	4	6.116
$CdGeAs_2$	Tegragonal I$\bar{4}$2d (D_{2d}^{12}) #122	a = 5.9432 c = 11.216	334.86	4	5.614
CdS (greenockite)	Hexagonal P6$_3$mc (C_{6v}^4) #186	a = 4.1367 c = 6.7161	144.48	2	4.821
CdSe	Hexagonal P6$_3$mc (C_{6v}^4) #186	a = 4.2972 c = 7.0064	191.37	2	5.672

TABLE 7 Composition, Structure, and Density of Crystals (*Continued*)

Material	Crystal system & space group	Unit cell dimension (Å)	Molecular weight (amu)	Formulae/ unit cell	Density (g/cm^3)
CdTe	Cubic $F\bar{4}3m$ (T_d^2) #216	6.4830	240.01	4	5.851
CsBr	Cubic $Pm3m$ (O_h^5) #221	4.286	212.81	1	4.488
CsCl	Cubic $Pm3m$ (O_h^5) #221	4.121	168.36	1	3.995
CsI	Cubic $Pm3m$ (O_h^5) #221	4.566	259.81	1	4.532
CuCl (nantokite)	Cubic $F\bar{4}3m$ (T_d^2) #216	5.416	98.999	4	4.139
CuGaS$_2$	Tetragonal $I\bar{4}2d$ (D_{2d}^{12}) #122	a = 5.351 c = 10.480	197.40	4	4.369
GaAs	Cubic $F\bar{4}3m$ (T_d^2) #216	5.65325	144.64	4	5.317
GaN	Hexagonal $P6_3mc$ (C_{6v}^4) #186	a = 3.186 c = 5.178	83.73	2	6.109
GaP	Cubic $F\bar{4}3m$ (T_d^2) #216	5.4495	100.70	4	4.133
Ge	Cubic $Fd3m$ (O_h^7) #227	5.65741	72.61	8	5.327
InAs	Cubic $F\bar{4}3m$ (T_d^2) #216	6.0584	189.74	4	5.668
InP	Cubic $F\bar{4}3m$ (T_d^2) #216	5.8688	145.79	4	4.791
KBr	Cubic $Fm3m$ (O_h^5) #225	6.600	119.00	4	2.749
KCl	Cubic $Fm3m$ (O_h^5) #225	6.293	74.55	4	1.987
KF	Cubic $Fm3m$ (O_h^5) #225	5.347	58.10	4	2.524
KH$_2$PO$_4$, KDP	Tetragonal $I\bar{4}2d$ (D_{2d}^{12}) #122	a = 7.452 c = 6.959	136.09	4	2.339
KI	Cubic $Fm3m$ (O_h^5) #225	7.065	166.00	4	3.127
KNbO$_3$	Orthorhombic $Bmm2$ (C_{2v}^{14}) #38	a = 5.6896 b = 3.9692 c = 5.7256	180.00	2	4.621
KTaO$_3$	Cubic $Pm3m$ (O_h^5) #221	3.9885	268.04	1	7.015
KTiOPO$_4$, KTP	Orthorhombic $Pna2_1$ (C_{2v}^9) #33	a = 12.8172 b = 6.4029 c = 10.5885	197.95	8	3.025
LaF$_3$	Hexagonal $P\bar{3}c1$ (D_{3v}^4) #165	a = 7.183 c = 7.352	195.90	6	5.941
LiB$_3$O$_5$, LBO	Orthorhombic $Pna2_1$ (C_{2v}^9) #33	a = 8.4473 b = 7.3788 c = 5.1395	238.74	4	2.475
LiF	Cubic $Fm3m$ (O_h^5) #225	4.0173	25.939	4	2.657
α-LiIO$_3$	Hexagonal $P6_3$ (C_6^6) #173	a = 5.4815 c = 5.1709	181.84	2	4.488
LiNbO$_3$	Hexagonal $R3c$ (C_{3v}^6) #161	a = 5.1483 c = 13.8631	147.85	6	4.629

TABLE 7 Composition, Structure, and Density of Crystals (*Continued*)

Material	Crystal system & space group	Unit cell dimension (Å)	Molecular weight (amu)	Formulae/ unit cell	Density (g/cm^3)
LiYF$_4$, YLF	Tetragonal I4$_1$/a (C$_{4h}^6$) #88	a = 5.175 c = 10.74	171.84	4	3.968
MgAl$_2$O$_4$ (spinel)	Cubic Fd3m (O$_h^7$) #227	8.084	142.27	8	3.577
MgF$_2$ (sellaite)	Tetragonal P4$_2$/mnm (D$_{4h}^{14}$) #136	a = 4.623 c = 3.053	62.302	2	3.171
MgO (periclase)	Cubic Fm3m (O$_h^5$) #225	4.2117	40.304	4	3.583
NaBr	Cubic Fm3m (O$_h^5$) #225	5.9732	102.89	4	3.207
NaCl (halite, rock salt)	Cubic Fm3m (O$_h^5$) #225	5.63978	58.44	4	2.164
NaF (valliaumite)	Cubic Fm3m (O$_h^5$) #225	4.6342	41.99	4	2.802
NaI	Cubic Fm3m (O$_h^5$) #225	6.475	149.89	4	3.668
[NH$_4$]$_2$CO (urea, carbamide)	Tetragonal I$\bar{4}$2$_1$m (D$_{2d}^3$) #113	a = 5.661 c = 4.712	60.056	2	1.321
NH$_4$H$_2$PO$_4$, ADP	Tetragonal I$\bar{4}$2d(D$_{2d}^{12}$) #122	a = 7.4997 c = 7.5494	115.03	4	1.799
PbF$_2$	Cubic Fm3m (O$_h^5$) #225	5.951	245.20	4	7.728
PbMoO$_4$ (wulfenite)	Tetragonal I4$_1$/a (C$_{4h}^6$) #88	a = 5.4312 c = 12.1065	367.14	4	6.829
PbS (galena)	Cubic Fm3m (O$_h^5$) #225	5.935	239.26	4	7.602
PbSe (clausthalite)	Cubic Fm3m (O$_h^5$) #225	6.122	286.16	4	8.284
PbTe (altaite)	Cubic Fm3m (O$_h^5$) #225	6.443	334.80	4	8.314
PbTiO$_3$	Tetragonal P4$_2$/mnm (D$_{4h}^{14}$) #136	a = 3.8966 c = 4.1440	303.08	1	7.999
Se	Hexagonal P3$_1$21 (D$_3^4$) #152	a = 4.35448 c = 4.94962	78.96	3	4.840
Si	Cubic Fd3m (O$_h^7$) #227	5.43085	28.0855	8	2.329
β-SiC (3C)	Cubic F$\bar{4}$3m (T$_d^2$) #216	4.3596	40.097	4	3.214
α-SiC (2H)	Hexagonal P6$_3$mc (C$_{6v}^4$) #186	a = 3.0763 c = 5.0480	40.097	2	3.219
SiO$_2$ (α-quartz)	Hexagonal P3$_2$21 (D$_3^6$) #154	a = 4.9136 c = 5.4051	60.084	3	2.648
SrF$_2$	Cubic Fm3m (O$_h^5$) #225	5.7996	125.62	4	4.277
SrMoO$_4$	Tetragonal I4$_1$/a (C$_{4h}^6$) #88	a = 5.380 c = 11.97	247.56	4	7.746
SrTiO$_3$	Cubic Pm3m (O$_h^5$) #221	3.9049	183.50	1	5.117
Te	Hexagonal P3$_1$21 (D$_3^4$) #152	a = 4.44693 c = 5.91492	127.60	3	6.275
TeO$_2$ (paratellurite)	Tetragonal P4$_1$2$_1$2 (D$_4^4$) #92	a = 4.810 c = 7.613	159.60	4	6.019
TiO$_2$ (rutile)	Tetragonal P4$_2$/mnm (D$_{4h}^{14}$) #136	a = 4.5937 c = 2.9618	79.879	2	4.245

TABLE 7 Composition, Structure, and Density of Crystals (*Continued*)

Material	Crystal system & space group	Unit cell dimension (Å)	Molecular weight (amu)	Formulae/ unit cell	Density (g/cm^3)
Tl_3AsSe_3, TAS	Hexagonal R3m (C_{3v}^5) #160	a = 9.870 c = 7.094	924.95	3	7.70
TlBr	Cubic Pm3m (O_h^5) #221	3.9846	284.29	1	7.462
Tl[Br, I], KRS-5	Cubic Pm3m (O_h^5) #221	4.108	307.79	1	7.372
TlCl	Cubic Pm3m (O_h^5) #221	3.8452	239.84	1	7.005
Tl[0.7Cl, 0.3Br], KRS-6	Cubic Pm3m (O_h^5) #221		253.17	1	
$Y_3Al_5O_{12}$, YAG	Cubic Ia3d (O_h^{10}) #230	12.008	593.62	8	4.554
Y_2O_3 (yttria)	Cubic Ia3 (T_h^7) #206	10.603	225.81	16	5.033
$ZnGeP_2$	Tetragonal I$\bar{4}$2d (D_{2d}^{12}) #122	a = 5.466 c = 10.722	199.95	4	4.146
ZnO (zincite)	Hexagonal P6$_3$mc (C_{6v}^4) #186	a = 3.242 c = 5.176	81.39	2	5.737
β-ZnS (zincblende)	Cubic F$\bar{4}$3m (T_d^2) #216	5.4094	97.456	4	4.090
α-ZnS (wurtzite)	Hexagonal P6$_3$mc (C_{6v}^4) #186	a = 3.8218 c = 6.2587	97.456	2	4.088
ZnSe	Cubic F$\bar{4}$3m (T_d^2) #216	5.6685	144.34	4	5.264
ZnTe	Cubic F$\bar{4}$3m (T_d^2) #216	6.1034	192.99	4	5.638
$ZrO_2{:}0.12Y_2O_3$ (cubic zirconia)	Cubic Fm3m (O_h^5) #225	5.148	121.98	4	5.939

TABLE 8 Physical Properties of Optical Glasses

Glass type	Selected glass code	Density (g/cm^3)	Example composition (for the general type)
Deep crown	479587 TiK1	2.39	Alkali alumo-borosilicate glass
Fluor crown	487704 FK5	2.45	(Boro)phosphide glass w/high fluoride content
Titanium flint	511510 TiF1	2.47	Titanium alkali alumoborosilicate glass
Borosilicate	517642 BK7	2.51	70%SiO_2, 10%B_2O_3, 8%Na_2O, 8%K_2O, 3%BaO, 1%CaO
Phosphate crown	518651 PK2	2.51	70%P_2O_5, 12%K_2O, 10%Al_2O_3, 5%CaO, 3%B_2O_3
Crown	522595 K5	2.59	74%SiO_2, 11%K_2O, 9%Na_2O, 6%CaO
Crown flint	523515 KF9	2.71	67%SiO_2, 16%Na_2O, 12%PbO, 3%ZnO, 2%Al_2O_3
Light barium crown	526600 BaLK1	2.70	Borosilicate glass
Antimony flint	527511 KzF6	2.54	Antimony borosilicate glass
Zinc crown	533580 ZK1	2.71	71%SiO_2, 17%Na_2O, 12%ZnO
Extra light flint	548458 LLF1	2.94	63%SiO_2, 24%PbO, 8%K_2O, 5%Na_2O
ULTRAN 30	548743	4.02	
Dense phosphate crown	552635 PSK3	2.91	60%P_2O_5, 28%BaO, 5%Al_2O_3, 3%B_2O_3
Barium crown	573575 BaK1	3.19	60%SiO_2, 19%BaO, 10%K_2O, 5%ZnO, 3%Na_2O, 3%B_2O_3
Light barium flint	580537 BaLF4	3.17	51%SiO_2, 20%BaO, 14%ZnO, 5%K_2O, 4%PbO
Light flint	581409 LF5	3.22	53%SiO_2, 34%PbO, 8%K_2O, 5%Na_2O
Special long crown	586610 LgSK2	4.15	Alkali earth aluminum fluoroborate glass
Fluor flint*	593355 FF5	2.64	

TABLE 8 Physical Properties of Optical Glasses (*Continued*)

Glass type	Selected glass code	Density (g/cm^3)	Example composition (for the general type)
Dense barium crown	613586 SK4	3.57	$39\%SiO_2, 41\%BaO, 15\%B_2O_3, 5\%Al_2O_3$
Special short flint	613443 KzFSN4	3.20	Aluminum lead borate glass
Extra-dense barium crown	618551 SSK4	3.63	$35\%SiO_2, 42\%BaO, 10\%B_2O_3, 8\%ZnO, 5\%Al_2O_3$
Flint	620364 F2	3.61	$47\%SiO_2, 44\%PbO, 7\%K_2O, 2\%Na_2O$
Dense barium flint	650392 BaSF10	3.91	$43\%SiO_2, 33\%PbO, 11\%BaO, 7\%K_2O, 5\%ZnO, 1\%Na_2O$
Barium flint*	670472 BaF10	3.61	$46\%SiO_2, 22\%PbO, 16\%BaO, 8\%ZnO, 8\%K_2O$
Lanthanum crown	720504 LaK10	3.81	Silicoborate glass w. rare earth oxides
Tantalum crown*	741526 TaC2	4.19	$B_2O_3/La_2O_3/ThO_2/RO$
Niobium flint*	743492 NbF1	4.17	
Lanthanum flint	744447 LaF2	4.34	Borosilicate glass w. rare earth oxides
Dense flint	805254 SF6	5.18	$33\%SiO_2, 62\%PbO, 5\%K_2O$
Dense tantalum flint*	835430 TaFD5	4.92	$B_2O_3/La_2O_3/ThO_2/Ta_2O_5$

* Hoya glasses; others are Schott glasses.

TABLE 9 Physical Properties of Specialty Glasses and Substrate Materials

Glass type	Density (g/cm^3)	Typical composition
Fused silica (SiO_2) (e.g., Corning 7940)	2.202	$100\%SiO_2$
Fused germania (GeO_2)	3.604	$100\%GeO_2$
BS-39B (Barr & Stroud)	3.1	$50\%CaO, 34\%Al_2O_3, 9\%MgO$
CORTRAN 9753 (Corning)	2.798	$29\%SiO_2, 29\%CaO, 42\%Al_2O_3$
CORTRAN 9754 (Corning)	3.581	$33\%GeO_2, 20\%CaO, 37\%Al_2O_3, 5\%BaO, 5\%ZnO$
IRG 2 (Schott)	5.00	Germanium glass
IRG 9 (Schott)	3.63	Fluorophosphate glass
IRG 11 (Schott)	3.12	Calcium aluminate glass
IRG 100 (Schott)	4.67	Chalcogenide glass
HTF-1 (Ohara) [443930]	3.94	Fluoride glass
ZBL	4.78	$62\%ZrF_4, 33\%BaF_2, 5\%LaF_3,$
ZBLA	4.61	$58\%ZrF_4, 33\%BaF_2, 5\%LaF_3, 4\%AlF_3$
ZBLAN	4.52	$56\%ZrF_4, 14\%BaF_2, 6\%LaF_3, 4\%AlF_3, 20\%NaF$
ZBT	4.8	$60\%ZrF_4, 33\%BaF_2, 7\%ThF_4$
HBL	5.78	$62\%HfF_4, 33\%BaF_2, 5\%LaF_3$
HBLA	5.88	$58\%HfF_4, 33\%BaF_2, 5\%LaF_3, 4\%AlF_3$
HBT	6.2	$60\%HfF_4, 33\%BaF_2, 7\%ThF_4$
Arsenic trisulfide (As_2S_3)	3.198	$100\%As_2S_3$
Arsenic triselenide (As_2Se_3)	4.69	$100\%As_2Se_3$
AMTRI-1/TI-20	4.41	$55\%Se, 33\%Ge, 12\%As$
AMTIR-3/TI-1173	4.70	$60\%Se, 28\%Ge, 12\%Sb$
Pyrex (e.g., Corning 7740)	2.23	$81\%SiO_2, 13\%B_2O_3, 4\%Na_2O, 2\%Al_2O_3$ [two-phase glass]
Zerodur (Schott)	2.53	$56\%SiO_2, 25\%Al_2O_3, 8\%P_2O_5, 4\%Li_2O, 2\%TiO_2$ $2\%ZrO_2, ZnO/MgO/Na_2O/As_2O_3$ [glass ceramic]
ULE (Corning 7971)	2.205	$92.5\%SiO_2, 7.5\%TiO_2$ [glass ceramic]

TABLE 10 Room-temperature Elastic Constants of Cubic Crystals

Material	Stiffness (GPa)			Compliance (TPa^{-1})			Refs.
	c_{11}	c_{12}	c_{44}	s_{11}	s_{12}	s_{44}	
AgBr	56.3	32.8	7.25	31.1	−11.5	138	57, 67
AgCl	59.6	36.1	6.22	31.1	−11.7	161	57, 68
AlAs	116.3	57.6	54.1	12.8	−4.24	18.5	69
ALON	393	108	119	2.89	−0.62	8.40	70
BaF$_2$	90.7	41.0	25.3	15.2	−4.7	39.6	57
Bi$_{12}$GeO$_{20}$ (BGO)	125.0	32.4	24.9	8.96	−1.84	40.4	71
Bi$_{12}$SiO$_{20}$ (BSO)	129.8	29.7	24.7	8.42	−1.57	40.2	72
BN	783	146	418	1.36	−0.21	2.39	73
BP	315	100	160	3.75	−0.90	6.25	74
C (diamond)	1040	170	550	1.01	−0.14	1.83	57
CaF$_2$	165	46	33.9	6.94	−1.53	29.5	57
CaLa$_2$S$_4$	98	47	50	15	−5	20	75
CdTe	53.8	37.4	20.18	43.24	−17.73	49.55	76
CsBr	30.7	8.4	7.49	36.9	−7.9	134	57
CsCl	36.6	9.0	8.07	30.2	−6.0	124	57
CsI	24.5	6.6	6.31	46.1	−9.7	158	57
CuCl	45.4	36.3	13.6	76.1	−33.8	73.5	77
GaAs	118	53.5	59.4	11.75	−3.66	16.8	57
GaP	142	63	71.6	9.60	−2.93	14.0	57
Ge	129	48	67.1	9.73	−2.64	14.9	57
InAs	83.4	45.4	39.5	19.46	−6.86	25.30	78
InP	102	58	46.0	16.4	−5.9	21.7	57
KBr	34.5	5.5	5.10	30.3	−4.2	196	57
KCl	40.5	6.9	6.27	25.9	−3.8	159	57
KF	65.0	15.0	12.5	16.8	−3.2	79.8	57
KI	27.4	4.3	3.70	38.2	−5.2	270	57
KTaO$_3$	431	103	109	2.7	−0.63	9.2	57
LiF	112	46	63.5	11.6	−3.35	15.8	57
MgAl$_2$O$_4$	282.9	155.4	154.8	5.79	−2.05	6.49	79
MgO	297.8	95.1	155.8	3.97	−0.96	6.42	79
NaBr	40.0	10.6	9.96	28.1	−5.8	100	57
NaCl	49.1	12.8	12.8	22.9	−4.8	78.3	57
NaF	97.0	24.2	28.1	11.5	−2.3	35.6	57
NaI	30.2	9.0	7.36	38.3	−8.8	136	57
PbF$_2$	96.37	46.63	21.04	15.3	−4.9	47.6	80
PbS	126	16.2	17.1	8.16	−0.93	58.5	81
PbSe	117.8	13.9	15.53	8.71	−0.92	64.4	82
PbTe	105.3	7.0	13.22	9.58	−0.60	75.6	83
Si	165	64	79.2	7.74	−2.16	12.6	57
β-SiC	350	142	256	3.18	−0.85	3.91	84
SrF$_2$	124	45	31.7	9.89	−2.59	31.6	57
SrTiO$_3$	315.6	102.7	121.5	3.77	−0.93	8.23	85
TlBr	37.6	14.8	7.54	34.2	−9.6	133	57
Tl[0.5Br, 0.5I], (KRS-5)	34.1	13.6	5.79	38.0	−10.8	173	57
TlCl	40.3	15.5	7.69	31.6	−8.8	130	57
Tl[0.7Cl, 0.3Br], (KRS-6)	39.7	14.9	7.23	31.9	−8.8	139	57
Y$_3$Al$_5$O$_{12}$ (YAG)	328.1	106.4	113.7	3.62	−0.89	8.80	86
Y$_2$O$_3$	233	101	67	5.82	−1.76	14.93	70
ZnS	101	64.4	44.3	19.7	−7.6	22.6	57
ZnSe	85	50.2	40.7	21.1	−7.8	24.6	57
ZnTe	71.5	40.8	31.1	23.9	−8.5	32.5	57
ZrO$_2$:Y$_2$O$_3$	405.1	105.3	61.8	2.77	−0.57	16.18	87

TABLE 11 Room-temperature Elastic Constants of Tetragonal Crystals (Point Groups 4mm, $\bar{4}2m$, 422, and 4/mmm)

Material	c or s	Subscript of stiffness (GPa) or compliance (TPa^{-1})						Refs.
		11	12	13	33	44	66	
AgGaS$_2$	c	87.9	58.4	59.2	75.8	24.1	30.8	88
	s	26.5	−7.7	−14.5	35.9	41.5	32.5	
AgGaSe$_2$	c							
	s							
BaTiO$_3$	c^E	275	179	152	165	54.4	113	57
	s^E	8.05	−2.35	−5.24	15.7	18.4	8.84	
	c^E	211	107	114	160	56.2	127	89
	s^E	8.01	−1.57	−4.60	12.8	17.8	7.91	
CdGeAs$_2$	c	94.5	59.6	59.7	83.4	42.1	40.8	90
	s	21.6	−7.04	−10.4	26.9	23.8	24.5	
CuGaS$_2$	c							
	s							
KH$_2$PO$_4$	c	70.9	−5.5	14.3	56.7	12.7	6.24	57
(KDP)	s	15.1	2.1	−4.3	19.8	78.7	160	
MgF$_2$	c	137	87	61.5	199	56.4	95.5	57
	s	12.6	−7.2	−1.7	6.1	17.7	10.5	
[NH$_4$]$_2$CO	c	21.7	8.9	24	53.2	6.26	0.45	91
(Urea)	s	95	16	−50	64	160	2220	
NH$_4$H$_2$PO$_4$	c	67.3	5.0?	19.8	33.7	8.57	6.02	57
(ADP)	s	18.3	2.2	−12.0	43.7	117	166	
PbTiO$_3$	c							57
	s	7.2	−2.1		32.5	12.2	7.9	
TeO$_2$	c	56.12	51.55	23.03	105.71	26.68	66.14	92, 93
	s	114.5	−104.3	−2.3	10.5	37.5	15.1	
TiO$_2$	c	270	176	147	480	124	193	57
	s	6.71	−3.92	−0.85	2.60	8.04	5.20	
ZnGeP$_2$	c							
	s							

TABLE 12 Room-temperature Elastic Constants of Tetragonal Crystals (Point Groups 4, $\bar{4}$, and 4/m)

Material	c or s	Subscript of stiffness (GPa) or compliance (TPa^{-1})							Refs.
		11	12	13	16	33	44	66	
CaMoO$_4$	c	144	64.8	44.8	−14.2	126	36.9	46.1	94
	s	9.92	−4.3	−2.0	4.4	9.4	27.1	24.4	
CaWO$_4$	c	146	62.6	39.2	−19.1	127	33.5	38.7	95
	s	10.1	−5.1	−1.7	7.7	8.8	29.8	33.5	
PbMoO$_4$	c	109	68.0	53.0	−14.0	92.0	26.7	33.7	96
	s	21.0	−12.4	−4.9	13.3	16.6	37.5	40.6	
SrMoO$_4$	c	119	62.0	48.0	−12.0	104	34.9	42.0	97
	s	13.6	−6.3	−3.4	5.7	12.7	28.7	27.1	
YLiF$_4$	c	121	60.9	52.6	−7.7	156	40.9	17.7	98
	s	12.8	−6.0	−2.3	8.16	7.96	24.4	63.6	

TABLE 13 Room-temperature Elastic Constants of Hexagonal Crystals (Point Groups 6, $\bar{6}$, 6/m, 622, 6mm, $\bar{6}2m$, and 6/mmm)

Material	c or s	Subscript of stiffness (GPa) or compliance (TPa^{-1})					Refs.
		11	12	13	33	44	
β-AgI	c^E	29.3	21.3	19.6	35.4	3.73	99
	s^E	79	−46	−19	49	268	
AlN	c	345	125	120	395	118	100
	s	3.53	−1.01	−0.77	3.00	8.47	
BeO	c	470	168	119	494	153	57
	s	2.52	−0.80	−0.41	2.22	6.53	
CdS	c	87.0	54.6	47.5	94.1	14.9	57
	s	20.8	−10.1	−5.4	16.0	66.8	
CdSe	c	74.1	45.2	38.9	84.3	13.4	57
	s	23.2	−11.2	−5.5	16.9	74.7	
GaN	c	296	130	158	267	241	101
	s	5.10	−0.92	−2.48	6.68	4.15	
LiIO$_3$	c^E	81.24	31.84	9.25	52.9	17.83	102
	s^E	14.7	−5.6	−1.6	19.5	56.1	
α-SiC	c	502	95	56	565	169	57
	s	2.08	−0.37	−0.17	1.80	5.92	
ZnO	c	209	120	104	218	44.1	57
	s	7.82	−3.45	−2.10	6.64	22.4	
ZnS	c	122	58	42	138	28.7	57
	s	11.0	−4.5	−2.0	8.6	34.8	

TABLE 14 Room-temperature Elastic Constants of Hexagonal (Trigonal) Crystals (Point Groups 32, 3m, and $\bar{3}m$)

Material	c or s	Subscript of stiffness (GPa) or compliance (TPa^{-1})						Refs.
		11	12	13	14	33	44	
Ag$_3$AsS$_3$	c	57.0	31.8			36.4	9.0	103
	s							
Al$_2$O$_3$	c	495	160	115	−23	497	146	57
	s	2.38	−0.70	−0.38	0.49	2.19	7.03	
β-Ba$_3$B$_6$O$_{12}$	c	123.8	60.3	49.4	12.3	53.3	7.8	104
(BBO)	s	25.63	−14.85	−9.97	−63.97	37.21	331.3	
CaCO$_3$	c	144	53.9	51.1	−20.5	84	33.5	57
(Calcite)	s	11.4	−4.0	−4.5	9.5	17.4	41.4	
LaF$_3$	c	180	88	59	<0.5	222	34	105
	s	7.6	−3.3	−1.1		5.1	29.4	
LiNbO$_3$	c	202	55	71	8.3	242	60.1	57
	s	5.81	−1.15	−1.36	−0.96	4.94	16.9	
Se	c	19.8	6.6	20.2	|6.9|	83.6	18.3	106
	s	92.6	−32.5	−14.5	|47.2|	19.0	90.2	
α-SiO$_2$	c	86.6	6.7	12.6	−17.8	106.1	57.8	57
	s	12.8	−1.74	−1.32	4.48	9.75	20.0	
Te	c	32.57	8.45	25.7	|12.38|	71.7	30.94	106
	s	57.3	−13.1	−15.9	|28.2|	25.3	54.9	
Tl$_3$AsSe$_3$	c							
	s							

TABLE 15 Room-temperature Elastic Constants of Orthorhombic Crystals

Material	c or s	\multicolumn{9}{c}{Subscript of stiffness (GPa) or compliance (TPa^{-1})}	Refs.
		11	12	13	22	23	33	44	55	66	
KNbO$_3$	c^E	224	102	182	273	130	245	75	28.5	45	107
	s^E	11.3	−0.3	−8.2	4.9	−2.4	11.5	13.3	35.1	10.5	
KTiOPO$_4$	c	159			154		175				108
(KTP)	s										
LiB$_3$O$_5$	c										
(LBO)	s										

TABLE 16 Mechanical Properties of Crystals

Material	\multicolumn{3}{c}{Moduli (GPa)}	Poisson's ratio	Flexure strength (MPa)	Knoop hardness (kg/mm^2)
	Elastic	Shear	Bulk			
Ag$_3$AsS$_3$	[28]	[10]	[37]	[0.38]		
AgBr	24.7	8.8	40.5	0.39$_9$		7.0
AgCl	22.9	8.1	44.0	0.41	26	9.5
AgGaS$_2$	52	19	67	0.37		320
AgGaSe$_2$			[60]			230
β-AgI	12	4.4	24	0.4		
AlAs	108	42.4	77.2	0.27		490
AlN	294	117	202	0.26	225	1230
Al$_2$O$_3$	400	162	250	0.23	1200	2250
ALON	317	128	203	0.24	310	1850
Ba$_3$B$_6$O$_{12}$, BBO	30	11	60.6	0.41		
BaF$_2$	65.8	25.1	57.6	0.31	27	78
BaTiO$_3$	145	53	174	0.36		580
BeO	395	162	240	0.23	275	1250
Bi$_{12}$GeO$_{20}$, BGO	82	32	63.3	0.28		
Bi$_{12}$SiO$_{20}$, BSO	84	33	63.1	0.28		
BN	833	375	358	0.11		>4600
BP	324	136	172	0.19		4700
C, diamond	1100	500	460	0.10	2940	9000
CaCO$_3$, calcite	83	32	73.2	0.31		100
CaF$_2$	110	42.5	85.7	0.29	90	170
CaLa$_2$S$_4$	96	[38.4]	[64]	0.25	81	570
CaMoO$_4$	103	40	80	0.29		250
CaWO$_4$	96	37	78	0.29		300
CdGeAs$_2$	74	28	70	0.32		470
CdS	42	15	59	0.38	28	122
CdSe	42	15.3	53	0.37	21	65
CdTe	8.4	14.2	42.9	0.35	26	50
CsBr	22	8.8	15.8	0.27	8.4	18
CsCl	25	10.0	18.2	0.27		
CsI	18	7.3	12.6	0.26	5.6	
CuCl	24.8	8.9	39.3	0.30$_5$		
CuGaS$_2$			[94]			430
GaAs	116	46.6	75.0	0.24	55	710
GaN	294	118	195	0.25	70	750
GaP	140	56.5	89.3	0.24	100	875

TABLE 16 Mechanical Properties of Crystals (*Continued*)

Material	Moduli (GPa)			Poisson's ratio	Flexure strength (MPa)	Knoop hardness (kg/mm^2)
	Elastic	Shear	Bulk			
Ge	132	54.8	75.0	0.20_6	100	850
InAs	74	28	61	0.30		390
InP	89	34	72.7	0.30		520
KBr	18	7.2	15.2	0.30	11	6.5
KCl	22	8.5	18.4	0.29	10	8
KF	41	16	31.8	0.28		
KH$_2$PO$_4$, KDP	[38]	[15]	[28]	[0.26]		
KI	14	5.5	11.9	0.30		5
KNbO$_3$	[250]	[71]	[95]	[0.22]		500
KTaO$_3$	316	124	230	0.27		
KTiOPO$_4$, KTP						
LaF$_3$	120	46	100	0.32	33	450
Li$_2$B$_6$O$_{10}$, LBO						600
LiF	110	45	65.0	0.22_5	27	115
LiIO$_3$	55	22.4	33.5	0.23		
LiNbO$_3$	170	68	112	0.25		~5
LiYF$_4$, YLF	85	32	81	0.32	35	300
MgAlO$_4$	276	109	198	0.26_8	170	1650
MgF$_2$	137	53.9	99.1	0.26_9	100	500
MgO	310	131	163	0.18	130	675
NaBr	29	11.6	19.9	0.26		
NaCl	37	14.5	25.3	0.26	9.6	16.5
NaF	76	30.7	48.5	0.24		
NaI	22	8.4	16.1	0.28		
[NH$_4$]$_2$CO, urea	~9	~3	17	0.41		
NH$_4$H$_2$PO$_4$, ADP	29	11	27.9	0.32_5		
PbF$_2$	59.8	22.4	60.5	0.33_5		200
PbMoO$_4$	66	24	72	0.35		
PbS	70.2	27.5	52.8	0.28		
PbSe	64.8	25.4	48.5	0.28		
PbTe	56.9	22.6	39.8	0.26		
PbTiO$_3$						
Se	24	9	17	0.27		
Si	162	66.2	97.7	0.22_4	130	1150
β-SiC	447	191	224	0.17	250	2880
α-SiC	455	197	221	0.16		3500
SiO$_2$, α-quartz	95	44	38	0.08		740
SrF$_2$	89	34.6	71.3	0.29		150
SrMoO$_4$	87	33	73	0.30		
SrTiO$_3$	283	115	174	0.23		600
Te	35	14	24	0.25	11	18
TeO$_2$	45	17	46	0.33		
TiO$_2$	293	115	215	0.27		880
Tl$_3$AsSe$_3$, TAS						
TlBr	24	8.9	22.4	0.32		12
Tl[Br, I], KRS-5	19.6	7.3	20.4	0.34	26	40
TlCl	25	9.3	23.8	0.33		13
Tl[0.7Cl, 0.3Br], KRS-6	24	9.0	32.2	0.33	21	30
Y$_3$Al$_5$O$_{12}$, YAG	280	113	180	0.24		1350
Y$_2$O$_3$	173	67	145	0.30	150	700

TABLE 16   Mechanical Properties of Crystals (*Continued*)

Material	Moduli (GPa)			Poisson's ratio	Flexure strength (MPa)	Knoop hardness (kg/mm^2)
	Elastic	Shear	Bulk			
ZnGeP$_2$			[86]			980
ZnO	127	47	144	0.35		
β-ZnS	82.5	31.2	76.6	0.32	60	175
α-ZnS	87	33	74	0.30	69	
ZnSe	75.4	29.1	61.8	0.30	55	115
ZnTe	61.1	23.5	51.0	0.30	24	82
ZrO$_2$:12% Y$_2$O$_3$	233	88.6	205	0.31	(200)	1150

TABLE 17   Mechanical Properties of Optical and Specialty Glasses and Substrate Materials

Selected glass code or designation	Moduli (GPa)			Poisson's ratio	Flexure strength (MPa)	Knoop hardness (kg/mm^2)
	Elastic	Shear	Bulk			
479587 TiK1	40	16	27	0.254		330
487704 FK5	62	26	35	0.205		450
511510 TiF1	58	23	37	0.239		440
517642 BK7	81	34	46	0.208		520
518651 PK2	84	35	48	0.209		520
522595 K5	71	29	43	0.227		450
523515 KF9	67	28	37	0.202		440
526600 BaLK1	68	28	43	0.234		430
527511 KzF6	52	21	30	0.212		380
533580 ZK1	68	27	44	0.240		430
548458 LLF1	60	25	34	0.210		390
548743 Ultran-30	76	29	62	0.297		380
552635 PSK3	84	34	51	0.226		510
573575 BaK1	74	30	50	0.253		460
580537 BaLF4	76	31	49	0.244		460
581409 LF5	59	24	36	0.226		410
586610 LgSK2	76	29	60	0.290		340
593355 FF5	[65]	[26]	[41]	[0.238]		500
613586 SK4	82	32	59	0.268		500
613443 KzFSN4	60	24	45	0.276		380
618551 SSK4	79	31	56	0.265		460
620364 F2	58	24	35	0.225		370
650392 BaSF10	67	27	46	0.256		400
670472 BaF10	95	37	71	0.277	104	610
720504 LaK10	111	43	87	0.288		580
741526 TaC2	117	45	97	0.299		715
743492 NbF1	108	41	94	0.308		675
744447 LaF2	93	36	73	0.289		480
805254 SF6	56	22	37	0.248		310
835430 TaFD5	126	48	104	0.299		790
Fused silica	72.6	31	36	0.164	110	635
Fused germania	43.1	18	23	0.192		
BS-39B	104	40	83	0.29	90	760
CORTRAN 9753	98.6	39	75	0.28		600
CORTRAN 9754	84.1	33	67	0.290	44	560

TABLE 17 Mechanical Properties of Optical and Specialty Glasses and Substrate Materials (*Continued*)

Selected glass code or designation	Moduli (GPa)			Poisson's ratio	Flexure strength (MPa)	Knoop hardness (kg/mm^2)
	Elastic	Shear	Bulk			
IRG 2	95.9	37	73	0.282		481
IRG 9	77.0	30	61	0.288		346
IRG 11	107.5	42	83	0.284		610
IRG 100	21	8	15	0.261		150
HTF-1	64.2	25	49	0.28		320
ZBL	60	23	53	0.31		228
ZBLA	60.2	24	40	0.25	11	235
ZBLAN	60	23	53	0.31		225
ZBT	60	23	45	0.279	62	250
HBL	55	21	46	0.3		228
HBLA	56	22	47	0.3		240
HBT	55	21	46	0.3	62	250
Arsenic trisulfide	15.8	6	13	0.295	16.5	180
Arsenic triselenide	18.3	7	14	0.288		120
AMTIR-1/TI-20	21.9	9	16	0.266	18.6	170
AMTIR-3/TI-1173	21.7	9	15	0.265	17.2	150
Pyrex	62.8	26	35	0.200		
Zerodur	91	37	58	0.24		630
ULE	67.3	29	34	0.17	50	460

TABLE 18 Thermal Properties of Crystals

Material	CC*	Temperature (K)		Heat capacity (J/g · K)	Thermal expansion (10^{-6}/K)	Thermal conductivity (W/m · K)		
		Debye	Melt†			@ 250 K	@ 300 K	@ 500 K
Ag$_3$AsS$_3$	H		763 m					
AgBr	C	145	705 m	0.2790	33.8	1.11	0.93	0.57
AgCl	C	162	728 m	0.3544	32.4	1.25	1.19	
AgGaS$_2$	T	255	1269 m		28.5 ∥ a			
					−18.7 ∥ c			
AgGaSe$_2$	T	156	1129 m		35.5 ∥ a			
					−15.0 ∥ c			
β-AgI	H	116	423 p	0.242			0.4	
AlAs	C	416	2013 m	0.452	3.5		(80)	
AlN	H	950	3273 m	0.796	5.27 ∥ a	500	320	150
					4.15 ∥ c			
Al$_2$O$_3$	H	1030	2319 m	0.777	6.65 ∥ a	58	46	24.2
					7.15 ∥ c			
ALON	C		2323 m	0.830	5.66		12.6	7.0
Ba$_3$B$_6$O$_{12}$, BBO	H		900 p		0.5 ∥ a		0.08 ∥ a	
					33.3 ∥ c		0.80 ∥ c	
BaF$_2$	C	283	1553 m	0.4474	18.4	7.5	12?	
BaTiO$_3$	T	—	278 p	0.439	16.8 ∥ a	—	6	—
			406 p		−9.07 ∥ c			
BeO	H	1280	2373 p	1.028	5.64 ∥ a	420	350	200
			2725 m		7.47 ∥ c			

TABLE 18 Thermal Properties of Crystals (*Continued*)

Material	CC*	Temperature (K) Debye	Temperature (K) Melt†	Heat capacity (J/g · K)	Thermal expansion (10^{-6}/K)	Thermal conductivity (W/m · K) @ 250 K	@ 300 K	@ 500 K
$Bi_{12}GeO_{20}$ (BGO)	C				16.8			
$Bi_{12}SiO_{20}$ (BSO)	C							
BN	C	1900	1100 p	0.513	3.5		760	
BP	C	985	1400 d	0.71	3.65	460	360	
C, diamond	C	2240	1770 p	0.5169	1.25	2800	2200	1300
$CaCO_3$, calcite	H		323 p	0.8820	−3.7 ∥ a	5.1 ∥ a	4.5 ∥ a	(3.4) ∥ a
			3825 m		25.1 ∥ c	6.2 ∥ c	5.4 ∥ c	(4.2) ∥ c
CaF_2	C	510	1424 p	0.9113	18.9	13	9.7	5.5
$CaLa_2S_4$	C		2083 m	(0.36)	14.6		1.7	1.5
$CaMoO_4$	T		1730 m	0.573	7.6 ∥ a		4.0 ∥ a	
					11.8 ∥ c		3.8 ∥ c	
$CaWO_4$	T		1855 m	0.396	6.35 ∥ a		16	9.5
					12.38 ∥ c			
$CdGeAs_2$	T	253	900 p		8.4 ∥ a			
			943 m		0.25 ∥ c			
CdS	H	215	1560 m	0.3814	4.6 ∥ a		27	13
					2.5 ∥ c			
CdSe	H	181	1580 m	0.272	4.9 ∥ a		(9)	
					2.9 ∥ c			
CdTe	C	160	1320 m	0.210	5.0	8.2	6.3	
CsBr	C	145	908 m	0.2432	47.2		0.85	
CsCl	C	175	918 m	0.3116	45.0			
CsI	C	124	898 m	0.2032	48.6		1.05	
CuCl	C	179	700 m	0.490	14.6	1.0	0.8	0.5
$CuGaS_2$	T	356	1553 m	0.452	11.2 ∥ a			
					6.9 ∥ c			
GaAs	C	344	1511 m	0.345	5.0	(65)	54	27
GaN	H		1160 d		3.17 ∥ a			
					5.59 ∥ c		130 ∥ c	
GaP	C	460	1740 m	0.435	5.3	120	100	(45)
Ge	C	380	1211 m	0.3230	5.7	74.9	59.9	33.8
InAs	C	251	1216 m	0.2518	4.4	(50)	27.3	15
InP	C	302	1345 m	0.3117	4.5	90	68	32
KBr	C	174	1007 m	0.4400	38.5	5.5	4.8	2.4
KCl	C	235	1043 m	0.6936	36.5	8.5	6.7	3.8
KF	C	336	1131 m	0.8659	31.4		8.3	
KH_2PO_4, KDP	T	—	123 p	0.88	22.0 ∥ a	2.0	2.1	
			450 p		39.2 ∥ c			
			526 m					
KI	C	132	954 m	0.3192	40.3		2.1	
$KNbO_3$	O	—	223 p		(37)			
			476 p					
$KTaO_3$	C	311				0.2	0.17	
$KTiOPO_4$, KTP	O		1209 c	0.728	11 ∥ a		2 ∥ a	
			1423 d		9 ∥ b		3 ∥ b	
					0.6 ∥ c		3 ∥ c	
LaF_3	H	392	1700 m	0.508	15.8 ∥ a	5.4	5.1	
					11.0 ∥ c			
LiB_3O_5, LBO	O		1107 p					
LiF	C	735	1115 m	1.6200	34.4	19	14	7.5

TABLE 18 Thermal Properties of Crystals (*Continued*)

Material	CC*	Temperature (K) Debye	Temperature (K) Melt†	Heat capacity (J/g · K)	Thermal expansion (10^{-6}/K)	Thermal conductivity (W/m · K) @ 250 K	Thermal conductivity (W/m · K) @ 300 K	Thermal conductivity (W/m · K) @ 500 K
LiIO$_3$	H		520 p		28 ∥ a			
			693 m		48 ∥ c			
LiNbO$_3$	H	560	1470 c	0.63	14.8 ∥ a		5.6	
			1523 m		4.1 ∥ c			
LiYF$_4$, YLF	T		1092 m	0.79	13.3 ∥ a		6.3	
					8.3 ∥ c			
MgAlO$_4$	C	850	2408 m	0.8191	6.97	30	25	
MgF$_2$	T	535	1536 m	1.0236	9.4 ∥ a			
					13.6 ∥ c		30 ∥ a	
							21 ∥ c	
MgO	C	950	3073 m	0.9235	10.6	73	59	32
NaBr	C	225	1028 m	0.5046	41.8		5.6	
NaCl	C	321	1074 m	0.8699	41.1	8	6.5	4
NaF	C	492	1266 m	1.1239	33.5	22		168
NaI	C	164	934 m	0.3502	44.7			4.7
[NH$_4$]$_2$CO, urea	T		408 m	1.551				
NH$_4$H$_2$PO$_4$, ADP	T	—	148 p	1.26	27.2 ∥ a		1.26 ∥ a	
			463 m		10.7 ∥ c		0.71 ∥ c	
PbF$_2$	C	225	422 p	0.3029	29.0		(28)	
			1094 m					
PbMoO$_4$	T		1338 m	0.326	8.7 ∥ a			
					20.3 ∥ c			
PbS	C	227	1390 m	0.209	19.0		2.5	
PbSe	C	138	1338 m	0.175	19.4	2	1.7	1
PbTe	C	125	1190 m	0.151	19.8	2.5	2.3	1.8
PbTiO$_3$	T		763 p				4	2.8
Se	H	151	490 m	0.3212	69.0 ∥ a	1.5 ∥ a	1.3 ∥ a	—
					−0.3 ∥ c	5.1 ∥ c	4.5 ∥ c	—
Si	C	645	1680 m	0.7139	2.62	191	140	73.6
β-SiC (3C)	C	(1000)	3103 d	0.670	2.77		490	
α-SiC (6H)	H		3000 v	0.690			450 ∥ a	
SiO$_2$, α-quartz	H	271	845 p	0.7400	12.38 ∥ a	7.5 ∥ a	6.2 ∥ a	3.9 ∥ a
					6.88 ∥ c	12.7 ∥ c	10.4 ∥ c	6.0 ∥ c
SrF$_2$	C	378	1710 m	0.6200	18.1	11	8.3	
SrMoO$_4$	T		1763 m	0.619			4.0 ∥ a	
							4.2 ∥ c	
SrTiO$_3$	C	—	110 p	0.536	8.3	12.5	11.2	
			2358 m					
Te	H	152	621 p	0.202	27.5 ∥ a	2.5 ∥ a	2.1 ∥ a	1.5 ∥ a
			723 m		−1.6 ∥ c	4.9 ∥ c	3.9 ∥ c	2.5 ∥ c
TeO$_2$	T		1006 m	[0.41]	15.0 ∥ a			
					4.9 ∥ c			
TiO$_2$	T	760	2128 m	0.6910	6.86 ∥ a	8.3 ∥ a	7.4 ∥ a	(5.5) ∥ a
					8.97 ∥ c	11.8 ∥ c	10.4 ∥ c	(8.0) ∥ c
Tl$_3$AsSe$_3$, TAS	H		583 m	0.19	28 ∥ a		0.35	
					18 ∥ c			
TlBr	C	116	740 m	0.1778	51		0.53	
Tl[Br, I], KRS-5	C	(110)	687 m	(0.16)	58		0.32	
TlCl	C	126	703 m	0.2198	52.7		0.74	
Tl[Cl, Br], KRS-6	C	(120)	697 m	0.201	51		0.50	

TABLE 18 Thermal Properties of Crystals (*Continued*)

Material	CC*	Temperature (K)		Heat capacity (J/g · K)	Thermal expansion (10^{-6}/K)	Thermal conductivity (W/m · K)		
		Debye	Melt†			@ 250 K	@ 300 K	@ 500 K
$Y_3Al_5O_{12}$, YAG	C	754	2193 p	0.625	7.7		13.4	
Y_2O_3	C	465	2640 p	0.4567	6.56		13.5	
$ZnGeP_2$	T	428	1225 p		7.8 ‖ a		18	
			1300 m		5.0 ‖ c			
						16.7 (30)		9.2 7.6
ZnO	H	416	2248 m	0.495	6.5 ‖ a 3.7 ‖ c		30	15
β-ZnS	C	340	1293 p	0.4732	6.8		16.7	10
α-ZnS	H	351	2100 m	0.4723	6.54 ‖ a 4.59 ‖ c			
ZnSe	C	270	1790 m	0.339	7.1		13	8
ZnTe	C	225	1510 m	0.218	8.4		10	
ZrO_2:12%Y_2O_3	C	563	3110 m	0.46	10.2		1.8	1.9

* CC = crystal class; C = cubic; H = hexagonal; O = orthorhombic; T = tetragonal.
† *Temperature codes*: m = melt temperature; c = Curie temperature; d = decomposition temperature; p = phase change (to different structure) temperature; v = vaporization (sublimation) temperature.

TABLE 19 Thermal Properties of Optical and Specialty Glasses and Substrate Materials

Selected glass code	Temperature (K)			Heat capacity (J/g · K)	Thermal expansion (10^{-6}/K)	Thermal conductivity (W/m · K)
	Glass	Soften	Melt*			
479587 TiK1	613			0.842	10.3	0.773
487704 FK5	737	945		0.808	9.2	0.925
511510 TiF1	716	[892]		[0.81]	9.1	[0.953]
517642 BK7	836	989		0.858	7.1	1.114
518651 PK2	841	994		[0.80]	6.9	1.149
522595 K5	816	993		0.783	8.2	0.950
523515 KF9	718	934		[0.75]	6.8	[1.01]
526600 BaLK1	782	954		0.766	9.1	1.043
527511 KzF6	717			[0.82]	5.5	[0.946]
533580 ZK1	835	1005		[0.77]	7.5	[0.894]
548458 LLF1	721	901		[0.71]	8.1	[0.960]
548743 Ultran-30	786	873		0.58	11.9	0.667
552635 PSK3	875	1009		[0.72]	8.6	[1.004]
573575 BaK1	875	1019		0.687	7.6	0.795
580537 BaLF4	842	1004		0.670	6.4	0.827
581409 LF5	692	858		0.657	9.1	0.866
586610 LgSK2	788			[0.51]	12.1	0.866
593355 FF5	788	843		[0.80]	8.6	[0.937]
613586 SK4	916	1040		0.582	6.4	0.875
613443 KzFSN4	765	867		[0.64]	5.0	[0.769]
618551 SSK4	912	1064		[0.57]	6.1	[0.806]
620364 F2	705	866		0.557	8.2	0.780
650392 BaSF10	757	908		[0.54]	8.6	[0.714]
670472 BaF10	853	908		0.569	7.2	0.967
720504 LaK10	893	976		[0.53]	5.7	[0.814]
741526 TaC2	928	958		[0.48]	5.2	[0.861]

TABLE 19 Thermal Properties of Optical and Specialty Glasses and Substrate Materials (*Continued*)

Selected glass code	Temperature (K)			Heat capacity $(J/g \cdot K)$	Thermal expansion $(10^{-6}/K)$	Thermal conductivity $(W/m \cdot K)$
	Glass	Soften	Melt*			
743492 NbF1	863	898		[0.48]	5.3	[0.845]
744447 LaF2	917	1013		[0.47]	8.1	[0.695]
805254 SF6	696	811		0.389	8.1	0.673
835430 TaFD5	943	973		[0.41]	6.4	
Fused silica	1273		1983	0.746	0.51	1.38
Fused germania	800		1388		6.3	
BS-39B		[970]		0.865	8.0	1.23
CORTRAN 9753	1015	1254		0.795	6.0	2.3
CORTRAN 9754	1008	1147		0.54	6.2	0.81
IRG 2	975			0.495	8.8	0.91
IRG 9	696			0.695	16.1	0.88
IRG 11	1075			0.749	8.2	1.13
IRG 100	550	624			15.0	0.3
HTF-1	658				16.1	
ZBL	580		820	0.538	18.8	
ZBLA	588		820	0.534	18.7	
ZBLAN	543		745	0.520	17.5	0.4
ZBT	568	723		0.511	4.3	
HBL	605		832	0.413	18.3	
HBLA	580		835	0.414	17.3	
HBT	593		853	0.428	6.0	
Arsenic trisulfide	436	573		0.473	26.1	0.17
Arsenic triselenide		345		0.349	24.6	0.205
AMTIR-1/TI-20	635	678		0.293	12.0	0.25
MATIR-3/TI-1173	550	570		0.276	14.0	0.22
Zerodur				0.821	0.5 (20–300°C)	1.64
Pyrex	560	821		1.05	3.25	1.13
ULE	1000	1490		0.776	±0.03 (5–35°C)	1.31

* Or liquidus temperature.

TABLE 20 Summary Optical Properties of Crystals

Material	Transparency (μm) UV	IR	Refractive index n_∞	Thermo-optic coefficients (10^{-6}/K) λ (μm)	dn/dT	λ (μm)	dn/dT	λ (μm)	dn/dT	Refs.
Ag$_3$AsS$_3$	0.63(o) / 0.61(e)	12.5(o) / 13.5(e)	2.736(o) / 2.519(e)							109
AgBr	0.49	35	2.166	0.633		3.39	−61	10.6	−50	110
AgCl	0.42	23	2.002		−61	3.39		10.6, 10.0	−35	109
AgGaS$_2$	0.50(o) / 0.52(e)	12.0(o) / 12.5(e)	2.408(o) / 2.355(e)	0.6	258(o) / 255(e)	1.0	176(o) / 179(e)		153(o) / 155(e)	111
AgGaSe$_2$	0.75	17.5	2.617(o) / 2.586(e)							
β-AgI			[2.1(o)] / [2.1(e)]							
AlAs			2.857							
AlN			2.17(o) / 2.22(e)							
Al$_2$O$_3$	0.19(o)	5.0(o) / 5.2(e)	1.7555(o) / 1.7478(e)	0.458	11.7(o) / 12.8(e)	0.589	13.6(o) / 14.7(e)	0.633	12.6	112, 113, 114
ALON	0.23	4.8	1.771		11.7					115
BBO	0.205	3.0	1.540(o) / 1.655(e)	0.4047	−16.6(o) / −9.8(e)	0.5790	−16.4(o) / −9.4(e)	1.014	−16.8(o) / −8.8(e)	104
BaF$_2$	0.14	12.2	1.4663	0.633	−16.0	3.39	−15.9	10.6	−14.5	113
BaTiO$_3$			2.277(o) / 2.250(e)							
BeO	0.21	3.5	1.709(o) / 1.724(e)	0.458	8.2(o) / 13.4(e)	0.633	8.2(o) / 13.4(e)			116
BGO	0.50	3.1	2.37	0.51	−34.5	0.65	−34.9			117
BSO	0.52		[2.15]							
BN	[0.2]		2.12							
BP	[0.5]		2.78							
Diamond	0.24	2.7	2.380	0.546	10.1			30	9.6	118, 119
CaCO$_3$	0.24(o) / 0.21(e)	2.2(o) / 3.3(e)	2.942(o) / 2.850(e)	0.365	3.6(o) / 14.4(e)	0.458	3.2(o) / 13.1(e)	0.633	2.1(o) / 11.9(e)	4
CaF$_2$	0.135	9.4	1.4278	0.254	−7.5	0.663	−10.4	3.39	−8.1	120
CaLa$_2$S$_4$	0.65	14.3	2.6							
CaMoO$_4$			1.945(o) / 1.951(e)	0.588	−9.6(o) / −10.0(e)					121
CaWO$_4$	(0.2)	5.3	1.884(o) / 1.898(e)	0.546	−7.1(o) / −10.2(e)					122

Material										Ref
CdGeAs$_2$	2.5	15	3.522(o) 3.608(e)							123
CdS	0.52(o) 0.51(e)	14.8(o) 14.8(e)	1.7085(o) 1.7234(e)	10.6	58.6(o) 62.4(e)					
CdSe	0.75	20	2.448(o) 2.467(e)							
CdTe	0.85	29.9	2.6829	1.15	147	3.39	98.2	10.6	98.0	124
CsBr	0.230	43.5	1.669	0.254	-82.0	0.633	-84.7	30.0	-75.8	125
CsCl	0.41	18.5	1.620	0.365	-78.7	0.633	-77.4	20.0	-70.0	125
CsI	0.245	62	1.743	0.365	-87.5	0.633	-99.3	30.0	-88.0	126
CuCl	0.45	(15)	1.974							127
CuGaS$_2$			2.493(o) 2.487(e)	0.55	130(o) 173(e)	1.0	59(o) 60(e)	10.0	56(o) 57(e)	128
GaAs	0.90	17.3	3.32	1.15	250	3.39	200	10.6	200	129
GaN	1.15		2.35(o) 2.31(e)	1.15	61					
GaP	0.54	10.5	3.015	0.546	200	0.633	160	20.0	401	130
Ge	1.8	15	4.001	2.5	462	5.0	416	5.0	5.8	131
HfO$_2$:Y$_2$O$_3$	0.35	6.5	2.074	0.365	14.1	0.436	11.0	1.01	300	132
InAs	3.9	20	3.44	4.0	500	6.0	400	10.0	77	129
InP	0.93	20	3.09	5.0	83	10.6	82	20.0	-41.1	109
KBr	0.200	30.2	1.537	0.458	-39.3	1.15	-41.9	10.6	-34.8	113
KCl	0.18	23.3	1.475	0.458	-34.9	1.15	-36.2	10.6	-17.0	113
KF	0.14	15.8	1.357	0.254	-19.9	1.15	-23.4	10.6		113
KH$_2$PO$_4$	0.176	1.42	1.502(o) 1.460(e)	0.624	-39.6(o) -38.2(e)					133
KI	0.250	38.5	1.629	0.458	-41.5	1.15	-44.7	30	-30.8	113
KNbO$_3$	0.4	5.0	2.199(x) 2.233(y) 2.102(z)	0.436	67(x) -26(y) 125(z)	1.064	23(x) -34(y) 63(z)	3.00	21(x) -23(y) 55(z)	134
KTaO$_3$			2.14							
KTiOPO$_4$	0.35	4.5	1.733(x) 1.740(y) 1.822(z)		11(x) 13(y) 16(z)					135
LaF$_3$	0.15	10	1.593(o) 1.586(e)	0.532	-0.9	1.064				
LiB$_3$O$_5$	0.17	2.5	1.571(x) 1.594(y) 1.612(z)		-0.9 -13.5 -7.4		-1.9 -13.0 -8.3			136
LiF	0.120	6.60	1.388	0.458	-16.0	1.15	-16.9	3.39	-14.5	113
LiIO$_3$	0.38	5.5	1.846(o) 1.711(e)	0.4	-74.5(o) -63.5(e)	1.0	-84.9(o) -69.2(e)			54

TABLE 20 Summary Optical Properties of Crystals (*Continued*)

Material	Transparency (μm)		Refractive index n_∞	Thermo-optic coefficients (10^{-6}/K)						Refs.
	UV	IR		λ (μm)	dn/dT	λ (μm)	dn/dT	λ (μm)	dn/dT	
LiNbO$_3$	0.5	5.0	2.214(o) 2.140(e)	0.66	4.4(o) 37.9(e)	3.39	0.3(o) 28.9(e)			137
LiYF$_4$	0.18	6.7	1.447(o) 1.469(e)	0.436	−0.54(o) −2.44(e)	0.546	−0.67(o) −2.30(e)	0.578	−0.91(o) −2.86(e)	138
MgAl$_2$O$_4$	0.2	5.3	1.701	0.589	9.0					112
MgF$_2$	0.13(o) 0.13(e)	7.7(o) 7.7(e)	1.3734(o) 1.3851(e)	0.633	1.12(o) 0.58(e)	1.15	0.88(o) 0.32(e)	3.39	1.19(o) 0.6(e)	113
MgO	0.35	6.8	1.720	0.365	19.5	0.546	16.5	0.768	13.6	139, 140
NaBr	0.20	24	1.615	0.365	−30.4	1.15	−40.2	10.6	−37.9	125
NaCl	0.174	18.2	1.555	0.458	−34.2	0.633	−35.4	3.39	−36.3	113
NaF	0.135	11.2	1.320	0.458	−11.9	1.15	−13.2	3.39	−12.5	113
NaI	0.26	24	1.73	0.458	−38.4	1.15	−49.6	10.6	−48.6	125
[NH$_4$]$_2$CO	0.21	1.4	1.481(o) 1.594(e)							
NH$_4$H$_2$PO$_4$	0.185	1.45	1.516(o) 1.470(e)	0.624	−47.1(o) −4.3(e)					133
PbF$_2$	0.29	12.5	1.731							
PbMoO$_4$	0.5	5.4	2.132(o) 2.127(e)	0.588	−75(o) −41(e)					121
PbS	0.29		4.1	3.39	−2100	5.0	−1900	10.6	−1700	141
PbSe	4.6		4.7	3.39	−2300	5.0	−1400	10.6	−860	141
PbTe	4.0	20	5.67	3.39	−2100	5.0	−1500	10.6	−1200	141
PbTiO$_3$			2.52(o) 2.52(e)							
Se	(1)	(30)	2.65(o) 3.46(e)							
Si	1.1	6.5	3.415	2.5	166	5.0	159	10.6	157	131

Material										Ref.
β-SiC	0.5	4	2.563							
α-SiC	0.5	4	2.560(o) 2.596(e)							
α-SiO$_2$	0.155	4.0	1.5352(o) 1.5440(e)	0.254	−2.9(o) −4.0(e)	0.365	−5.4(o) −6.2(e)	0.546	−6.2(o) −7.0(e)	140
SrF$_2$	0.13	11.0	1.4316							
SrMoO$_4$			1.867(o) 1.869(e)	0.633	−16.0	1.15	−16.2	10.6	−14.5	113
SrTiO$_3$	0.5	5.1	2.283							
Te	3.5	32	4.778(o) 6.222(e)							
TeO$_2$	0.34	4.5	2.18(o) 2.32(e)	0.436	30(o) 25(e)	0.644	9(o) 8(e)			142
TiO$_2$	0.42	4.0	2.432(o) 2.683(e)	0.405	4(o) −9(e)					143
Tl$_3$AsSe$_3$	1.3	16	3.35(o) 3.16(e)	2–10	−45(o) 36(e)					144
TlBr	0.44	38	2.271							
KRS-5	0.58	42	2.38	0.633	−250	10.6	−233	30	−195	145
TlCl	0.38	25	2.136							
KRS-6	0.42	27	2.196							
Y$_3$Al$_5$O$_{12}$	0.21	5.2	1.815	0.458	11.9	0.633	9.4	1.06	9.1	146
Y$_2$O$_3$	0.29	7.1	1.892	0.663	8.3					115
ZnGeP$_2$	0.8	12.5	3.121(o) 3.158(e)	0.64	359(o) 376(e)	1.0	212(o) 230(e)	10	165(o) 170(e)	128, 147
ZnO	0.38(o) 0.37(e)		1.922(o) 1.936(e)							
β-ZnS	0.4	12.5	2.258	0.633	63.5	1.15	49.8	10.6	46.3	124
α-ZnS			2.271(o) 2.275(e)							
ZnSe	0.51	19.0	2.435	0.633	91.1	1.15	59.7	10.6	52.0	124
ZnTe	0.6	25	2.70							
ZrO$_2$:Y$_2$O$_3$	0.38	6.0	2.0892	0.365	16.0	0.458	10.0	0.633	7.9	148

TABLE 21 Summary Optical Properties of Optical and Specialty Glasses

Material	Transparency (μm) UV	Transparency (μm) IR	Refractive index n_d	Abbe number v_d	Thermo-optic coeff. $(10^{-6}/K)$* λ (μm)	dn/dT	λ (μm)	dn/dT	Refs.
TiK1	0.35		1.47869	58.70	0.4358	−1.8	1.060	−2.6	149
FK5	0.29	(2.5)	1.48749	70.41	0.4358	−1.1	1.060	−1.8	149
TiF1	0.35		1.51118	51.01	0.4358	−0.1†	1.014	−1.5†	149, 150
BK7	0.31	2.67	1.51680	64.17	0.4358	3.4	1.060	2.3	149
PK2	0.31	(2.5)	1.51821	65.05	0.4358	3.7	1.060	2.3	149
K5	0.31	2.7	1.52249	59.48	0.4358	2.4	1.060	1.1	149
KF9	0.33	(2.5)	1.52341	51.49	0.4358	5.1	1.060	3.3	149
BaLK1	0.32	(2.6)	1.52642	60.03	0.4358	1.1	1.060	0.1	149
KzF6	0.32		1.52682	51.13	0.4358	5.8†	1.014	4.2†	149, 150
ZK1	0.31		1.53315	57.98	0.4358	4.4†	1.014	2.8†	149, 150
LLF1	0.32		1.54814	45.75	0.4358	4.4	1.060	2.1	149
Ultran 30	0.23	3.95	1.54830	74.25	0.4358	−5.5	1.060	−6.2	149
PSK3	0.30	(2.5)	1.55232	63.46	0.4358	3.5	1.060	2.3	149
BaK1	0.31	2.7	1.57250	57.55	0.4358	3.3	1.060	1.9	149
BaLF4	0.335	(2.6)	1.57957	53.71	0.4358	6.3	1.060	4.3	149
LF5	0.32	(2.5)	1.58144	40.85	0.4358	4.4	1.060	1.6	149
LgSK2	0.35		1.58599	61.04	0.4358	−2.5	1.060	−4.0	149
FF5	0.36		1.59270	35.45	0.6328	0.7			151
SK4	0.325	2.68	1.61272	58.63	0.4358	3.5	1.060	2.1	149
KzFSN4	0.325	2.38	1.61340	44.30	0.4358	6.2	1.060	4.4	149
SSK4	0.325	2.9	1.61765	55.14	0.4358	4.0	1.060	2.2	149
F2	0.33	2.73	1.62004	36.37	0.4358	5.9	1.060	2.8	149
BaSF10	0.34	(2.6)	1.65016	39.15	0.4358	5.5	1.060	2.1	149
BaF10	0.35	2.7	1.67003	47.20	0.6328	4.7			151
LaK10	0.34	(2.5)	1.72000	50.41	0.4358	5.8	1.060	3.8	149
TaC2	0.325		1.74100	52.29	0.6328	6.8			151
NbF1	0.325		1.74330	59.23	0.6328	7.9			151
LaF2	0.355	(2.5)	1.74400	44.72	0.4358	2.2	1.060	0.2	149
SF6	0.37	2.74	1.80518	25.43	0.4358	16.2	1.060	6.9	149
TaFD5	0.34		1.83500	42.98	0.6328	4.6			151
SiO$_2$	0.16	3.8	1.45857	67.7	0.5893	10			152
GeO$_2$	0.30	4.9	$n_D = 1.60832$	41.2					153
BS-39B	0.38	4.9	$n_D = 1.6764$	44.5	0.5893	7.4			154
Corning 9753	0.38	4.3	$n_D = 1.60475$	54.98					155
Corning 9754	0.36	4.8	$n_D = 1.6601$	46.5					155
Schott IRG 2	0.44	5.1	1.8918	30.03					149
Schott IRG9	0.38	4.1	1.4861	81.02					149
Schott IRG 11	0.44	4.75	1.6809	44.21					149
Schott IRG 100	0.93	13	$n_1 = 2.7235$	—	2,5	103	10.6	56	149
Ohara HTF-1	0.21	6.9	1.44296	92.46					150
ZBL	0.25	7.0	$n_D = 1.523$						156, 157
ZBLA	0.29	7.0	$n_D = 1.521$	62					156, 157, 158
ZBLAN	0.25	6.9	$n_D = 1.480$	64	0.6328	−14.5			158, 159
ZBT	0.32	6.8	$n_D = 1.53$						160, 161
HBL	0.25	7.3	$n_D = 1.498$						158
HBLA	0.29	7.3	$n_D = 1.504$						158, 162
HBT	0.22	7.7	$n_D = 1.53$						160
As$_2$S$_3$	0.62	11.0	$n_1 = 2.47773$	—	0.6	85	1.0	17	163, 164
As$_2$Se$_3$	0.87	17.2	$n_{12} = 2.7728$	—	0.83	55	1.15	33	165, 166
AMTIR-1/TI-20	0.75	(14.5)	$n_1 = 2.6055$	—	1.0	101	10.0	72	166, 167
AMTIR-3/TI-1173	0.93	16.5	$n_3 = 2.6366$	—	3.0	98	12.0	93	166, 167

* Thermo-optic coefficient in air: $(dn/dT)_{rel}$.
† Data from comparable Ohara glass.

TABLE 22   Room-temperature Dispersion Formulas for Crystals

Material	Dispersion formula (wavelength, λ, in μm)	Range (μm)	Ref.
Ag_3AsS_3	$n_o^2 = 7.483 + \dfrac{0.474}{\lambda^2 - 0.09} - 0.0019\lambda^2$	0.63–4.6(o) 0.59–4.6(e)	168
	$n_e^2 = 6.346 + \dfrac{0.342}{\lambda^2 - 0.09} - 0.0011\lambda^2$		
AgBr	$\dfrac{n^2 - 1}{n^2 + 2} = 0.452505 + \dfrac{0.09939\lambda^2}{\lambda^2 - 0.070537} - 0.00150\lambda^2$	0.49–0.67	169
AgCl	$n^2 - 1 = \dfrac{2.062508\lambda^2}{\lambda^2 - (0.1039054)^2} + \dfrac{0.9461465\lambda^2}{\lambda^2 - (0.2438691)^2} + \dfrac{4.300785\lambda^2}{\lambda^2 - (70.85723)^2}$	0.54–21.0	110
$AgGaS_2$	$n_o^2 = 3.6280 + \dfrac{2.1686\lambda^2}{\lambda^2 - 0.1003} + \dfrac{2.1753\lambda^2}{\lambda^2 - 950}$	0.49–12	170
	$n_e^2 = 4.0172 + \dfrac{1.5274\lambda^2}{\lambda^2 - 0.1310} + \dfrac{2.1699\lambda^2}{\lambda^2 - 950}$		
$AgGaSe_2$	$n_o^2 = 4.6453 + \dfrac{2.2057\lambda^2}{\lambda^2 - 0.1879} + \dfrac{1.8377\lambda^2}{\lambda^2 - 1600}$	0.73–13.5	170
	$n_e^2 = 5.2912 + \dfrac{1.3970\lambda^2}{\lambda^2 - 0.2845} + \dfrac{1.9282\lambda^2}{\lambda^2 - 1600}$		
β-AgI	$n_o = 2.184;\quad n_e = 2.200 @ 0.659\ \mu$m	—	171
	$n_o = 2.104;\quad n_e = 2.115 @ 1.318\ \mu$m		
AlAs	$n^2 = 2.0792 + \dfrac{6.0840\lambda^2}{\lambda^2 - (0.2822)^2} + \dfrac{1.900\lambda^2}{\lambda^2 - (27.62)^2}$	0.56–2.2	172
AlN	$n_o^2 = 3.1399 + \dfrac{1.3786\lambda^2}{\lambda^2 - (0.1715)^2} + \dfrac{3.861\lambda^2}{\lambda^2 - (15.03)^2}$	0.22–5.0	173
	$n_e^2 = 3.0729 + \dfrac{1.6173\lambda^2}{\lambda^2 - (0.1746)^2} + \dfrac{4.139\lambda^2}{\lambda^2 - (15.03)^2}$		
Al_2O_3	$n_o^2 - 1 = \dfrac{1.4313493\lambda^2}{\lambda^2 - (0.0726631)^2} + \dfrac{0.65054713\lambda^2}{\lambda^2 - (0.1193242)^2} + \dfrac{5.3414021\lambda^2}{\lambda^2 - (18.028251)^2}$	0.2–5.5	174
	$n_e^2 - 1 = \dfrac{1.5039759\lambda^2}{\lambda^2 - (0.0740288)^2} + \dfrac{0.55069141\lambda^2}{\lambda^2 - (0.1216529)^2} + \dfrac{6.5927379\lambda^2}{\lambda^2 - (20.072248)^2}$		
ALON	$n^2 - 1 = \dfrac{2.1375\lambda^2}{\lambda^2 - 0.10256^2} + \dfrac{4.582\lambda^2}{\lambda^2 - 18.868^2}$	0.4–2.3	175
BBO	$n_o^2 = 2.7405 + \dfrac{0.0184}{\lambda^2 - 0.0179} - 0.0155\lambda^2$	0.22–1.06	104
	$n_e^2 = 2.3730 + \dfrac{0.0128}{\lambda^2 - 0.0156} - 0.0044\lambda^2$		
BaF_2	$n^2 - 1 = \dfrac{0.643356\lambda^2}{\lambda^2 - (0.057789)^2} + \dfrac{0.506762\lambda^2}{\lambda^2 - (0.10968)^2} + \dfrac{3.8261\lambda^2}{\lambda^2 - (46.3864)^2}$	0.27–10.3	176
$BaTiO_3$	$n_o^2 - 1 = \dfrac{4.187\lambda^2}{\lambda^2 - (0.223)^2}$	0,4–0.7	177
	$n_e^2 - 1 = \dfrac{4.064\lambda^2}{\lambda^2 - (0.211)^2}$		
BeO	$n_o^2 - 1 = \dfrac{1.92274\lambda^2}{\lambda^2 - (0.07908)^2} + \dfrac{1.24209\lambda^2}{\lambda^2 - (9.7131)^2}$	0.44–7.0	178
	$n_e^2 - 1 = \dfrac{1.96939\lambda^2}{\lambda^2 - (0.08590)^2} + \dfrac{1.67389\lambda^2}{\lambda^2 - (10.4797)^2}$		

TABLE 22 Room-temperature Dispersion Formulas for Crystals (*Continued*)

Material	Dispersion formula (wavelength, λ, in μm)	Range (μm)	Ref.
$Bi_{12}SiO_{20}$, BSO	$n^2 = 2.72777 + \dfrac{3.01705\lambda^2}{\lambda^2 - (0.2661)^2}$	0.48–0.7	72 179
$Bi_{12}GeO_{20}$, BGO	$n^2 - 1 = \dfrac{4.601\lambda^2}{\lambda^2 - (0.242)^2}$	0.4–0.7	117 180
BN	$n \approx 2.117$	0.589	181
BP	$n^2 - 1 = \dfrac{6.841\lambda^2}{\lambda^2 - (0.267)^2}$	0.45–0.63	74
C, diamond	$n^2 - 1 = \dfrac{4.3356\lambda^2}{\lambda^2 - (0.1060)^2} + \dfrac{0.3306\lambda^2}{\lambda^2 - (0.1750)^2}$	0.225–∞	182
$CaCO_3$, calcite	$n_o - 1 = \dfrac{0.8559\lambda^2}{\lambda^2 - (0.0588)^2} + \dfrac{0.8391\lambda^2}{\lambda^2 - (0.141)^2} + \dfrac{0.0009\lambda^2}{\lambda^2 - (0.197)^2} + \dfrac{0.6845\lambda^2}{\lambda^2 - (7.005)^2}$	0.2–2.2	4
	$n_e - 1 = \dfrac{1.0856\lambda^2}{\lambda^2 - (0.07897)^2} + \dfrac{0.0988\lambda^2}{\lambda^2 - (0.142)^2} + \dfrac{0.317\lambda^2}{\lambda^2 - (11.468)^2}$	0.2–3.3	4
CaF_2	$n^2 - 1 = \dfrac{0.5675888\lambda^2}{\lambda^2 - (0.050263605)^2} + \dfrac{0.4710914\lambda^2}{\lambda^2 - (0.1003909)^2} + \dfrac{3.8484723\lambda^2}{\lambda^2 - (34.649040)^2}$	0.23–9.7	120
$CaLa_2S_4$	$n \approx 2.6$	—	75
$CaMoO_4$	$n_o^2 - 1 = \dfrac{2.7840\lambda^2}{\lambda^2 - (0.1483)^2} + \dfrac{1.2425\lambda^2}{\lambda^2 - (11.576)^2}$	0.45–3.8	183
	$n_e^2 - 1 = \dfrac{2.8045\lambda^2}{\lambda^2 - (0.1542)^2} + \dfrac{1.0055\lambda^2}{\lambda^2 - (10.522)^2}$		
$CaWO_4$	$n_o^2 - 1 = \dfrac{2.5493\lambda^2}{\lambda^2 - (0.1347)^2} + \dfrac{0.9200\lambda^2}{\lambda^2 - (10.815)^2}$	0.45–4.0	183
	$n_e^2 - 1 = \dfrac{2.6041\lambda^2}{\lambda^2 - (0.1379)^2} + \dfrac{4.1237\lambda^2}{\lambda^2 - (21.371)^2}$		
$CdGeAs_2$	$n_o^2 = 10.1064 + \dfrac{2.2988\lambda^2}{\lambda^2 - 1.0872} + \dfrac{1.6247\lambda^2}{\lambda^2 - 1370}$	2.4–11.5	170
	$n_e^2 = 11.8018 + \dfrac{1.2152\lambda^2}{\lambda^2 - 2.6971} + \dfrac{1.6922\lambda^2}{\lambda^2 - 1370}$		
CdS	$n_o^2 - 1 = \dfrac{3.96582820\lambda^2}{\lambda^2 - (0.23622804)^2} + \dfrac{0.18113874\lambda^2}{\lambda^2 - (0.48285199)^2}$	0.51–1.4	42
	$n_e^2 - 1 = \dfrac{3.97478769\lambda^2}{\lambda^2 - (0.22426984)^2} + \dfrac{0.26680809\lambda^2}{\lambda^2 - (0.46693785)^2} + \dfrac{0.00074077\lambda^2}{\lambda^2 - (0.50915139)^2}$		
CdSe	$n_o^2 = 4.2243 + \dfrac{1.7680\lambda^2}{\lambda^2 - 0.2270} + \dfrac{3.1200\lambda^2}{\lambda^2 - 3380}$	1–22	170
	$n_e^2 = 4.2009 + \dfrac{1.8875\lambda^2}{\lambda^2 - 0.2171} + \dfrac{3.6461\lambda^2}{\lambda^2 - 3629}$		
CdTe	$n^2 - 1 = \dfrac{6.1977889\lambda^2}{\lambda^2 - (0.317069)^2} + \dfrac{3.2243821\lambda^2}{\lambda^2 - (72.0663)^2}$	6–22	184
CsBr	$n^2 - 1 = \dfrac{0.9533786\lambda^2}{\lambda^2 - (0.0905643)^2} + \dfrac{0.8303809\lambda^2}{\lambda^2 - (0.1671517)^2} + \dfrac{2.847172\lambda^2}{\lambda^2 - (119.0155)^2}$	0.36–39	185
CsCl	$n^2 = 1.33013 + \dfrac{0.98369\lambda^2}{\lambda^2 - (0.119)^2} + \dfrac{0.00009\lambda^2}{\lambda^2 - (0.137)^2} + \dfrac{0.00018\lambda^2}{\lambda^2 - (0.145)^2} + \dfrac{0.30914\lambda^2}{\lambda^2 - (0.162)^2} + \dfrac{4.320\lambda^2}{\lambda^2 - (100.50)^2}$	0.18–40	125
CsI	$n^2 - 1 = \dfrac{0.34617251\lambda^2}{\lambda^2 - (0.0229567)^2} + \dfrac{1.0080886\lambda^2}{\lambda^2 - (0.1466)^2} + \dfrac{0.28551800\lambda^2}{\lambda^2 - (0.1810)^2} + \dfrac{0.39743178\lambda^2}{\lambda^2 - (0.2120)^2} + \dfrac{3.3605359\lambda^2}{\lambda^2 - (161.0)^2}$	0.29–50	126

TABLE 22 Room-temperature Dispersion Formulas for Crystals (*Continued*)

Material	Dispersion formula (wavelength, λ, in μm)	Range (μm)	Ref.
CuCl	$n^2 = 3.580 + \dfrac{0.03162\lambda^2}{\lambda^2 - 0.1642} + \dfrac{0.09288}{\lambda^2}$	0.43–2.5	186
CuGaS$_2$	$n_o^2 = 3.9064 + \dfrac{2.3065\lambda^2}{\lambda^2 - 0.1149} + \dfrac{1.5479\lambda^2}{\lambda^2 - 738.43}$	0.55–11.5	127
			128
	$n_e^2 = 4.3165 + \dfrac{1.8692\lambda^2}{\lambda^2 - 0.1364} + \dfrac{1.7575\lambda^2}{\lambda^2 - 738.43}$		
GaAs	$n^2 = 3.5 + \dfrac{7.4969\lambda^2}{\lambda^2 - (0.4082)^2} + \dfrac{1.9347\lambda^2}{\lambda^2 - (37.17)^2}$	1.4–11	187
GaN	$n_o^2 = 3.60 + \dfrac{1.75\lambda^2}{\lambda^2 - (0.256)^2} + \dfrac{4.1\lambda^2}{\lambda^2 - (17.86)^2}$	<10	188
	$n_e^2 - 5.35 \qquad\qquad + \dfrac{5.08\lambda^2}{\lambda^2 - (18.76)^2}$		
GaP	$n^2 - 1 = \dfrac{1.390\lambda^2}{\lambda^2 - (0.172)^2} + \dfrac{4.131\lambda^2}{\lambda^2 - (0.234)^2} + \dfrac{2.570\lambda^2}{\lambda^2 - (0.345)^2} + \dfrac{2.056\lambda^2}{\lambda^2 - (27.52)^2}$	0.8–10	189
Ge	$n^2 = 9.28156 + \dfrac{6.72880\lambda^2}{\lambda^2 - 0.44105} + \dfrac{0.21307\lambda^2}{\lambda^2 - 3870.1}$	2–12	190
HfO$_2$: 9.8% Y$_2$O$_3$	$n^2 - 1 = \dfrac{1.9558\lambda^2}{\lambda^2 - (0.15494)^2} + \dfrac{1.345\lambda^2}{\lambda^2 - (0.0634)^2} + \dfrac{10.41\lambda^2}{\lambda^2 - (27.12)^2}$	0.365–5	132
InAs	$n^2 = 11.1 + \dfrac{0.71\lambda^2}{\lambda^2 - (2.551)^2} + \dfrac{2.75\lambda^2}{\lambda^2 - (45.66)^2}$	3.7–31.3	191
InP	$n^2 = 7.255 + \dfrac{2.316\lambda^2}{\lambda^2 - (0.6263)^2} + \dfrac{2.765\lambda^2}{\lambda^2 - (32.935)^2}$	0.95–10	192
KBr	$n^2 = 1.39408 + \dfrac{0.79221\lambda^2}{\lambda^2 - (0.146)^2} + \dfrac{0.01981\lambda^2}{\lambda^2 - (0.173)^2} + \dfrac{0.15587\lambda^2}{\lambda^2 - (0.187)^2}$	0.2–40	125
	$\qquad\qquad + \dfrac{0.17673\lambda^2}{\lambda^2 - (60.61)^2} + \dfrac{2.06217\lambda^2}{\lambda^2 - (87.72)^2}$		
KCl	$n^2 = 1.26486 + \dfrac{0.30523\lambda^2}{\lambda^2 - (0.100)^2} + \dfrac{0.41620\lambda^2}{\lambda^2 - (0.131)^2} + \dfrac{0.18870\lambda^2}{\lambda^2 - (0.162)^2} + \dfrac{2.6200\lambda^2}{\lambda^2 - (70.42)^2}$	0.18–35	125
KF	$n^2 = 1.55083 + \dfrac{0.29162\lambda^2}{\lambda^2 - (0.126)^2} + \dfrac{3.60001\lambda^2}{\lambda^2 - (51.55)^2}$	0.15–22.0	125
KH$_2$PO$_4$	$n_o^2 - 1 = \dfrac{1.256618\lambda^2}{\lambda^2 - 0.0084478168} + \dfrac{33.89909\lambda^2}{\lambda^2 - 1113.904}$	0.4–1.06	193
KDP	$n_e^2 - 1 = \dfrac{1.131091\lambda^2}{\lambda^2 - 0.008145980} + \dfrac{5.75675\lambda^2}{\lambda^2 - 811.7537}$		
KI	$n^2 = 1.47285 + \dfrac{0.16512\lambda^2}{\lambda^2 - (0.129)^2} + \dfrac{0.41222\lambda^2}{\lambda^2 - (0.175)^2} + \dfrac{0.44163\lambda^2}{\lambda^2 - (0.187)^2}$	0.25–50	125
	$\qquad\qquad + \dfrac{0.16076\lambda^2}{\lambda^2 - (0.219)^2} + \dfrac{0.33571\lambda^2}{\lambda^2 - (69.44)^2} + \dfrac{1.92474\lambda^2}{\lambda^2 - (98.04)^2}$		
KNbO$_3$	$n_x^2 - 1 = \dfrac{2.49710\lambda^2}{\lambda^2 - (0.12909)^2} + \dfrac{1.33660\lambda^2}{\lambda^2 - (0.25816)^2} - 0.025174\lambda^2$	0.40–3.4	134
	$n_y^2 - 1 = \dfrac{2.54337\lambda^2}{\lambda^2 - (0.13701)^2} + \dfrac{1.44122\lambda^2}{\lambda^2 - (0.27275)^2} - 0.028450\lambda^2$		
	$n_z^2 - 1 = \dfrac{2.37108\lambda^2}{\lambda^2 - (0.11972)^2} + \dfrac{1.04825\lambda^2}{\lambda^2 - (0.25523)^2} - 0.019433\lambda^2$		

TABLE 22 Room-temperature Dispersion Formulas for Crystals (*Continued*)

Material	Dispersion formula (wavelength, λ, in μm)	Range (μm)	Ref.
KTaO$_3$	$$n^2 - 1 = \frac{3.591\lambda^2}{\lambda^2 - (0.193)^2}$$	0.4–1.06	194
KTiOPO$_4$ KTP	$$n_x^2 = 2.16747 + \frac{0.83733\lambda^2}{\lambda^2 - 0.04611} - 0.01713\lambda^2$$	0.4–1.06	195
	$$n_y^2 = 2.19229 + \frac{0.83547\lambda^2}{\lambda^2 - 0.04970} - 0.01621\lambda^2$$		
	$$n_z^2 = 2.25411 + \frac{1.06543\lambda^2}{\lambda^2 - 0.05486} - 0.02140\lambda^2$$		
LaF$_3$	$$n_o^2 - 1 = \frac{1.5376\lambda^2}{\lambda^2 - (0.0881)^2}$$	0.35–0.70	196
	$$n_e^2 - 1 = \frac{1.5149\lambda^2}{\lambda^2 - (0.0878)^2}$$		
LiB$_3$O$_5$ LBO	$$n_x^2 = 2.45768 + \frac{0.0098877}{\lambda^2 - 0.026095} - 0.013847\lambda^2$$	0.29–1.06	197
	$$n_y^2 = 2.52500 + \frac{0.017123}{\lambda^2 - 0.0060517} - 0.0087838\lambda^2$$		
	$$n_z^2 = 2.58488 + \frac{0.012737}{\lambda^2 - 0.016293} - 0.016293\lambda^2$$		
LiF	$$n^2 - 1 = \frac{0.92549\lambda^2}{\lambda^2 - (0.07376)^2} + \frac{6.96747\lambda^2}{\lambda^2 - (32.79)^2}$$	0.1–10	125
LiIO$_3$	$$n_0^2 = 2.03132 + \frac{1.37623\lambda^2}{\lambda^2 - 0.0350823} + \frac{1.06745\lambda^2}{\lambda^2 - 169.0}$$	0.5–5	56
	$$n_e^2 = 1.83086 + \frac{1.08807\lambda^2}{\lambda^2 - 0.0313810} + \frac{0.554582\lambda^2}{\lambda^2 - 158.76}$$		
LiNbO$_3$	$$n_o^2 = 2.39198 + \frac{2.51118\lambda^2}{\lambda^2 - (0.217)^2} + \frac{7.1333\lambda^2}{\lambda^2 - (16.502)^2}$$	0.4–3.1	198
	$$n_e^2 = 2.32468 + \frac{2.25650\lambda^2}{\lambda^2 - (0.210)^2} + \frac{14.503\lambda^2}{\lambda^2 - (25.915)^2}$$		
LiYF$_4$	$$n_o^2 = 1.38757 + \frac{0.70757\lambda^2}{\lambda^2 - 0.00931} + \frac{0.18849\lambda^2}{\lambda^2 - 50.99741}$$	0.23–2.6	138
	$$n_e^2 = 1.31021 + \frac{0.84903\lambda^2}{\lambda^2 - 0.00876} + \frac{0.53607\lambda^2}{\lambda^2 - 134.9566}$$		
MgAl$_2$O$_4$	$$n^2 - 1 = \frac{1.8938\lambda^2}{\lambda^2 - (0.09942)^2} + \frac{3.0755\lambda^2}{\lambda^2 - (15.826)^2}$$	0.35–5.5	199
MgF$_2$	$$n_o^2 - 1 = \frac{0.48755108\lambda^2}{\lambda^2 - (0.04338408)^2} + \frac{0.39875031\lambda^2}{\lambda^2 - (0.09461442)^2} + \frac{2.3120353\lambda^2}{\lambda^2 - (23.793604)^2}$$	0.20–7.04	200
	$$n_e^2 - 1 = \frac{0.41344023\lambda^2}{\lambda^2 - (0.03684262)^2} + \frac{0.50497499\lambda^2}{\lambda^2 - (0.09076162)^2} + \frac{2.4904862\lambda^2}{\lambda^2 - (12.771995)^2}$$		
MgO	$$n^2 - 1 = \frac{1.111033\lambda^2}{\lambda^2 - (0.0712465)^2} + \frac{0.8460085\lambda^2}{\lambda^2 - (0.1375204)^2} + \frac{7.808527\lambda^2}{\lambda^2 - (26.89302)^2}$$	0.36–5.4	139
NaBr	$$n^2 = 1.06728 + \frac{1.10463\lambda^2}{\lambda^2 - (0.125)^2} + \frac{0.18816\lambda^2}{\lambda^2 - (0.145)^2} + \frac{0.00243\lambda^2}{\lambda^2 - (0.176)^2} + \frac{0.24454\lambda^2}{\lambda^2 - (0.188)^2} + \frac{3.7960\lambda^2}{\lambda^2 - (74.63)^2}$$	0.21–34	125

TABLE 22 Room-temperature Dispersion Formulas for Crystals (*Continued*)

Material	Dispersion formula (wavelength, λ, in μm)	Range (μm)	Ref.
NaCl	$n^2 = 1.00055 + \dfrac{0.19800\lambda^2}{\lambda^2 - (0.050)^2} + \dfrac{0.48398\lambda^2}{\lambda^2 - (0.100)^2} + \dfrac{0.38696\lambda^2}{\lambda^2 - (0.128)^2} + \dfrac{0.25998\lambda^2}{\lambda^2 - (0.158)^2}$ $+ \dfrac{0.08796\lambda^2}{\lambda^2 - (40.50)^2} + \dfrac{3.17064\lambda^2}{\lambda^2 - (60.98)^2} + \dfrac{0.30038\lambda^2}{\lambda^2 - (120.34)^2}$	0.2–30	125
NaF	$n^2 = 1.41572 + \dfrac{0.32785\lambda^2}{\lambda^2 - (0.117)^2} + \dfrac{3.18248\lambda^2}{\lambda^2 - (40.57)^2}$	0.15–17	125
NaI	$n^2 = 1.478 + \dfrac{1.532\lambda^2}{\lambda^2 - (0.170)^2} + \dfrac{4.27\lambda^2}{\lambda^2 - (86.21)^2}$	0.25–40	125
[NH$_4$]$_2$CO Urea	$n_o^2 = 2.1823 + \dfrac{0.0125}{\lambda^2 - 0.0300}$ $n_e^2 = 2.51527 + \dfrac{0.0240}{\lambda^2 - 0.0300} + \dfrac{0.020(\lambda - 1.52)}{(\lambda - 1.52)^2 + 0.8771}$	0.3–1.06	201
NH$_4$H$_2$PO$_4$ ADP	$n_o^2 - 1 = \dfrac{1.298990\lambda^2}{\lambda^2 - 0.0089232927} + \dfrac{43.17364\lambda^2}{\lambda^2 - 1188.531}$ $n_e^2 - 1 = \dfrac{1.162166\lambda^2}{\lambda^2 - 0.0085932421} + \dfrac{12.01997\lambda^2}{\lambda^2 - 831.8239}$	0.4–1.06	193
PbF$_2$	$n^2 - 1 = \dfrac{0.66959342\lambda^2}{\lambda^2 - (0.00034911)^2} + \dfrac{1.3086319\lambda^2}{\lambda^2 - (0.17144455)^2} + \dfrac{0.01670641\lambda^2}{\lambda^2 - (0.28125513)^2} + \dfrac{2007.8865\lambda^2}{\lambda^2 - (796.67469)^2}$	0.3–11.9	202
PbMoO$_4$	$n_o^2 - 1 = \dfrac{3.54642\lambda^2}{\lambda^2 - (0.18518)^2} + \dfrac{0.58270\lambda^2}{\lambda^2 - (0.33764)^2}$ $n_e^2 - 1 = \dfrac{3.52555\lambda^2}{\lambda^2 - (0.17950)^2} + \dfrac{0.20660\lambda^2}{\lambda^2 - (0.32537)^2}$	0.44–1.08	203
PbS	$n^2 - 1 = \dfrac{15.9\lambda^2}{\lambda^2 - (0.77)^2} + \dfrac{133.2\lambda^2}{\lambda^2 - (141)^2}$	3.5–10	141
PbSe	$n^2 - 1 = \dfrac{21.1\lambda^2}{\lambda^2 - (1.37)^2}$	5.0–10	141
PbTe	$n^2 - 1 = \dfrac{30.046\lambda^2}{\lambda^2 - (1.563)^2}$	4.0–12.5	204
PbTiO$_3$	$n_o^2 - 1 = \dfrac{5.363\lambda^2}{\lambda^2 - (0.224)^2}$ $n_e^2 - 1 = \dfrac{5.366\lambda^2}{\lambda^2 - (0.217)^2}$	0.45–1.15	205
Se	$n_o = 2.790; \quad n_e = 3.608 @ 1.06\ \mu$m $n_0 = 2.64; \quad n_e = 3.41 @ 10.6\ \mu$m	—	206
Si	$n^2 - 1 = \dfrac{10.6684293\lambda^2}{\lambda^2 - (0.301516485)^2} + \dfrac{0.003043475\lambda^2}{\lambda^2 - (1.13475115)^2} + \dfrac{1.54133408\lambda^2}{\lambda^2 - (1104.0)^2}$	1.36–11	42
β-SiC	$n^2 - 1 = \dfrac{5.5705\lambda^2}{\lambda^2 - (0.1635)^2}$	0.47–0.69	207
α-SiC	$n_o^2 - 1 = \dfrac{5.5515\lambda^2}{\lambda^2 - (0.16250)^2}$ $n_e^2 - 1 = \dfrac{5.7382\lambda^2}{\lambda^2 - (0.16897)^2}$	0.49–1.06	208

TABLE 22 Room-temperature Dispersion Formulas for Crystals (*Continued*)

Material	Dispersion formula (wavelength, λ, in μm)	Range (μm)	Ref.
α-SiO$_2$ quartz	$n_o^2 - 1 = \dfrac{0.663044\lambda^2}{\lambda^2 - (0.060)^2} + \dfrac{0.517852\lambda^2}{\lambda^2 - (0.106)^2} + \dfrac{0.175912\lambda^2}{\lambda^2 - (0.119)^2} + \dfrac{0.565380\lambda^2}{\lambda^2 - (8.844)^2} + \dfrac{1.675299\lambda^2}{\lambda^2 - (20.742)^2}$	0.18–0.71	140
	$n_e^2 - 1 = \dfrac{0.665721\lambda^2}{\lambda^2 - (0.060)^2} + \dfrac{0.503511\lambda^2}{\lambda^2 - (0.106)^2} + \dfrac{0.214792\lambda^2}{\lambda^2 - (0.119)^2} + \dfrac{0.539173\lambda^2}{\lambda^2 - (8.792)^2} + \dfrac{1.807613\lambda^2}{\lambda^2 - (197.70)^2}$		
SrF$_2$	$n^2 - 1 = \dfrac{0.67805894\lambda^2}{\lambda^2 - (0.05628989)^2} + \dfrac{0.37140533\lambda^2}{\lambda^2 - (0.10801027)^2} + \dfrac{3.8484723\lambda^2}{\lambda^2 - (34.649040)^2}$	0.21–11.5	113
SrMoO$_4$	$n_o^2 - 1 = \dfrac{2.4839\lambda^2}{\lambda^2 - (0.1451)^2} + \dfrac{0.1015\lambda^2}{\lambda^2 - (4.603)^2}$	0.45–2.4	183
	$n_e^2 - 1 = \dfrac{2.4923\lambda^2}{\lambda^2 - (0.1488)^2} + \dfrac{0.1050\lambda^2}{\lambda^2 - (4.544)^2}$		
SrTiO$_3$	$n^2 - 1 = \dfrac{3.042143\lambda^2}{\lambda^2 - (0.1475902)^2} + \dfrac{1.170065\lambda^2}{\lambda^2 - (0.2953086)^2} + \dfrac{30.83326\lambda^2}{\lambda^2 - (33.18606)^2}$	0.43–3.8	80
Te	$n_o^2 = 18.5346 + \dfrac{4.3289\lambda^2}{\lambda^2 - 3.9810} + \dfrac{3.7800\lambda^2}{\lambda^2 - 11813}$	4–14	170
	$n_e^2 = 29.5222 + \dfrac{9.3068\lambda^2}{\lambda^2 - 2.5766} + \dfrac{9.2350\lambda^2}{\lambda^2 - 13521}$		
TeO$_2$	$N_o^2 - 1 = \dfrac{2.584\lambda^2}{\lambda^2 - (0.1342)^2} + \dfrac{1.157\lambda^2}{\lambda^2 - (0.2638)^2}$	0.4–1.0	142
	$n_e^2 - 1 = \dfrac{2.823\lambda^2}{\lambda^2 - (0.1342)^2} + \dfrac{1.542\lambda^2}{\lambda^2 - (0.2631)^2}$		
TiO$_2$	$n_o^2 = 5.913 + \dfrac{0.2441}{\lambda^2 - 0.0803}$	0.43–1.5	143
	$n_e^2 = 7.197 + \dfrac{0.3322}{\lambda^2 - 0.0843}$		
Tl$_3$AsSe$_3$ TAS	$n_o^2 - 1 = \dfrac{10.210\lambda^2}{\lambda^2 - (0.444)^2} + \dfrac{0.522\lambda^2}{\lambda^2 - (25.0)^2}$	2–12	144
	$n_e^2 - 1 = \dfrac{8.993\lambda^2}{\lambda^2 - (0.444)^2} + \dfrac{0.308\lambda^2}{\lambda^2 - (25.0)^2}$		
TlBr	$\dfrac{n^2 - 1}{n^2 + 2} = 0.48484 + \dfrac{0.10279\lambda^2}{\lambda^2 - 0.090000} - 0.0047896\lambda^2$	0.54–0.65	169
Tl[Br, I] KRS-5	$n^2 - 1 = \dfrac{1.8293958\lambda^2}{\lambda^2 - (0.150)^2} + \dfrac{1.6675593\lambda^2}{\lambda^2 - (0.250)^2} + \dfrac{1.1210424\lambda^2}{\lambda^2 - (0.350)^2} + \dfrac{0.04513366\lambda^2}{\lambda^2 - (0.450)^2} + \dfrac{12.380234\lambda^2}{\lambda^2 - (164.59)^2}$	0.58–39.4	145
TlCl	$\dfrac{n^2 - 1}{n^2 + 2} = 0.47856 + \dfrac{0.07858\lambda^2}{\lambda^2 - 0.08277} - 0.00881\lambda^2$	0.43–0.66	169
Tl[Cl, Br] KRS-6	$n^2 - 1 = \dfrac{3.821\lambda^2}{\lambda^2 - (0.02234)^2} - 0.000877\lambda^2$	0.6–24	209
Y$_3$Al$_5$O$_{12}$ YAG	$n^2 - 1 = \dfrac{2.293\lambda^2}{\lambda^2 - (0.1095)^2} + \dfrac{3.705\lambda^2}{\lambda^2 - (17.825)^2}$	0.4–4.0	183
Y$_2$O$_3$	$n^2 - 1 = \dfrac{2.578\lambda^2}{\lambda^2 - (0.1387)^2} + \dfrac{3.935\lambda^2}{\lambda^2 - (22.936)^2}$	0.2–12	210
ZnGeP$_2$	$n_o^2 = 4.4733 + \dfrac{5.2658\lambda^2}{\lambda^2 - 0.1338} + \dfrac{1.4909\lambda^2}{\lambda^2 - 662.55}$	0.4–12	128
	$n_e^2 = 4.6332 + \dfrac{5.3422\lambda^2}{\lambda^2 - 0.1426} + \dfrac{1.4580\lambda^2}{\lambda^2 - 662.55}$		

TABLE 22 Room-temperature Dispersion Formulas for Crystals (*Continued*)

Material	Dispersion formula (wavelength, λ, in μm)	Range (μm)	Ref.
ZnO	$n_o^2 = 2.81418 + \dfrac{0.87968\lambda^2}{\lambda^2 - (0.3042)^2} - 0.00711\lambda^2$ $n_e^2 = 2.80333 + \dfrac{0.94470\lambda^2}{\lambda^2 - (0.3004)^2} - 0.00714\lambda^2$	0.45–4.0	183
β-ZnS	$n^2 - 1 = \dfrac{0.33904026\lambda^2}{\lambda^2 - (0.31423026)^2} + \dfrac{3.7606868\lambda^2}{\lambda^2 - (0.1759417)^2} + \dfrac{2.7312353\lambda^2}{\lambda^2 - (33.886560)^2}$	0.55–10.5	113
α-ZnS	$n_o^2 - 1 = 3.4175 + \dfrac{1.7396\lambda^2}{\lambda^2 - (0.2677)^2}$ $n_e^2 - 1 = 3.4264 + \dfrac{1.7491\lambda^2}{\lambda^2 - (0.2674)^2}$	0.36–1.4	211
ZnSe	$n^2 - 1 = \dfrac{4.2980149\lambda^2}{\lambda^2 - 0.1920630^2} + \dfrac{0.62776557\lambda^2}{\lambda^2 - 0.37878260^2} + \dfrac{2.8955633\lambda^2}{\lambda^2 - 46.994595^2}$	0.55–18	113
ZnTe	$n^2 = 9.92 + \dfrac{0.42530}{\lambda^2 - (0.37766)^2} + \dfrac{2.63580}{\lambda^2/(56.5)^2 - 1}$	0.55–30	212
ZrO$_2$:12% Y$_2$O$_3$	$n^2 - 1 = \dfrac{1.347091\lambda^2}{\lambda^2 - (0.062543)^2} + \dfrac{2.117788\lambda^2}{\lambda^2 - (0.166739)^2} + \dfrac{9.452943\lambda^2}{\lambda^2 - (24.320570)^2}$	0.36–5.1	148

TABLE 23 Room-temperature Dispersion Formula for Glasses

Material	Dispersion formula (wavelength, λ, in μm)	Range (μm)	Ref.
TiK1	$n^2 = 2.1573978 - 8.4004189 \cdot 10^{-3}\lambda^2 + 1.0457582 \cdot 10^{-2}\lambda^{-2}$ $\quad + 2.1822593 \cdot 10^{-4}\lambda^{-4} - 5.5063640 \cdot 10^{-6}\lambda^{-6} + 5.4469060 \cdot 10^{-7}\lambda^{-8}$	0.37–1.01	149
FK5	$n^2 = 2.1887621 - 9.5572007 \cdot 10^{-3}\lambda^2 + 8.9915232 \cdot 10^{-3}\lambda^{-2}$ $\quad + 1.4560516 \cdot 10^{-4}\lambda^{-4} - 5.2843067 \cdot 10^{-6}\lambda^{-6} + 3.4588010 \cdot 10^{-7}\lambda^{-8}$	0.37–1.01†	149
	$n^2 - 1 = \dfrac{1.036330719\lambda^2}{\lambda^2 - (0.0776227030)^2} + \dfrac{0.152107703\lambda^2}{\lambda^2 - (0.138959626)^2} + \dfrac{0.913166269\lambda^2}{\lambda^2 - (9.93162512)^2}$	0.37–2.33	42
TiF1	$n^2 = 2.2473124 - 8.9044058 \cdot 10^{-3}\lambda^2 + 1.2493525 \cdot 10^{-2}\lambda^{-2}$ $\quad + 4.2650638 \cdot 10^{-4}\lambda^{-4} - 2.1564809 \cdot 10^{-6}\lambda^{-6} + 2.6364065 \cdot 10^{-6}\lambda^{-8}$	0.37–1.01	149
BK7	$n^2 = 2.2718929 - 1.0108077 \cdot 10^{-2}\lambda^2 + 1.0592509 \cdot 10^{-2}\lambda^{-2}$ $\quad + 2.0816965 \cdot 10^{-4}\lambda^{-4} - 7.6472538 \cdot 10^{-6}\lambda^{-6} + 4.9240991 \cdot 10^{-7}\lambda^{-8}$	0.37–1.01†	149
PK2	$n^2 = 2.2770533 - 1.0532010 \cdot 10^{-2}\lambda^2 + 1.0188354 \cdot 10^{-2}\lambda^{-2}$ $\quad + 2.9001564 \cdot 10^{-4}\lambda^{-4} - 1.9602856 \cdot 10^{-5}\lambda^{-6} + 1.0967718 \cdot 10^{-6}\lambda^{-8}$	0.37–1.01	149
K5	$n^2 = 2.2850299 - 8.6010725 \cdot 10^{-3}\lambda^2 + 1.1806783 \cdot 10^{-2}\lambda^{-2}$ $\quad + 2.0765657 \cdot 10^{-4}\lambda^{-4} - 2.1314913 \cdot 10^{-6}\lambda^{-6} + 3.2131234 \cdot 10^{-7}\lambda^{-8}$	0.37–1.01†	149
KF9	$n^2 = 2.2824396 - 8.5960144 \cdot 10^{-3}\lambda^2 + 1.3442645 \cdot 10^{-2}\lambda^{-2}$ $\quad + 2.7803535 \cdot 10^{-4}\lambda^{-4} - 4.9998960 \cdot 10^{-7}\lambda^{-6} + 7.7105911 \cdot 10^{-7}\lambda^{-8}$	0.37–1.01†	149
BaLK1	$n^2 = 2.2966923 - 8.2975549 \cdot 10^{-3}\lambda^2 + 1.1907234 \cdot 10^{-2}\lambda^{-2}$ $\quad + 1.9908305 \cdot 10^{-4}\lambda^{-4} - 2.0306838 \cdot 10^{-6}\lambda^{-6} + 3.1429703 \cdot 10^{-7}\lambda^{-8}$	0.37–1.01	149
KzF6	$n^2 = 2.2934044 - 1.0346122 \cdot 10^{-2}\lambda^2 + 1.3319863 \cdot 10^{-2}\lambda^{-2}$ $\quad + 3.4833226 \cdot 10^{-4}\lambda^{-4} - 9.9354090 \cdot 10^{-6}\lambda^{-6} + 1.1227905 \cdot 10^{-6}\lambda^{-8}$	0.37–1.01	149
ZK1	$n^2 = 2.3157951 - 8.7493905 \cdot 10^{-3}\lambda^2 + 1.2329645 \cdot 10^{-2}\lambda^{-2}$ $\quad + 2.6311112 \cdot 10^{-4}\lambda^{-4} - 8.2854201 \cdot 10^{-6}\lambda^{-6} + 7.3735801 \cdot 10^{-7}\lambda^{-8}$	0.37–1.01	149

TABLE 23 Room-temperature Dispersion Formula for Glasses (*Continued*)

Material	Dispersion formula (wavelength, λ, in μm)	Range (μm)	Ref.
LLF1	$n^2 = 2.3505162 - 8.5306451 \cdot 10^{-3}\lambda^2 + 1.5750853 \cdot 10^{-2}\lambda^{-2}$ $+ 4.2811388 \cdot 10^{-4}\lambda^{-4} - 6.9875718 \cdot 10^{-6}\lambda^{-6} + 1.7175517 \cdot 10^{-6}\lambda^{-8}$	0.37–1.01†	149
Ultran 30	$n^2 = 2.3677942 - 6.0818906 \cdot 10^{-3}\lambda^2 + 1.0509568 \cdot 10^{-2}\lambda^{-2}$ $+ 1.3105575 \cdot 10^{-4}\lambda^{-4} - 4.9854380 \cdot 10^{-7}\lambda^{-6} + 1.0473652 \cdot 10^{-7}\lambda^{-8}$	0.37–1.01†	149
	$n^2 - 1 = \dfrac{1.36689\lambda^2}{\lambda^2 - (0.089286)^2} + \dfrac{0.34711\lambda^2}{\lambda^2 - (7.9103)^2}$	0.37–2.3	149*
PSK3	$n^2 = 2.3768193 - 1.0146514 \cdot 10^{-2}\lambda^2 + 1.2167148 \cdot 10^{-2}\lambda^{-2}$ $+ 1.1916606 \cdot 10^{-4}\lambda^{-4} + 6.4250627 \cdot 10^{-6}\lambda^{-6} - 1.7478706 \cdot 10^{-7}\lambda^{-8}$	0.37–1.01	149
BaK1	$n^2 = 2.4333007 - 8.4931353 \cdot 10^{-3}\lambda^2 + 1.3893512 \cdot 10^{-2}\lambda^{-2}$ $+ 2.6798268 \cdot 10^{-4}\lambda^{-4} - 6.1946101 \cdot 10^{-6}\lambda^{-6} + 6.2209005 \cdot 10^{-7}\lambda^{-8}$	0.37–1.01†	149
BaLF4	$n^2 = 2.4528366 - 9.2047678 \cdot 10^{-3}\lambda^2 + 1.4552794 \cdot 10^{-2}\lambda^{-2}$ $+ 4.3046688 \cdot 10^{-4}\lambda^{-4} - 2.0489836 \cdot 10^{-5}\lambda^{-6} + 1.5924415 \cdot 10^{-6}\lambda^{-8}$	0.37–1.01†	149
	$n^2 - 1 = \dfrac{1.25385390\lambda^2}{\lambda^2 - (0.0856548405)^2} + \dfrac{0.198113511\lambda^2}{\lambda^2 - (0.173243878)^2} + \dfrac{1.01615191\lambda^2}{\lambda^2 - (10.8069635)^2}$	0.37–2.33	42
LF5	$n^2 = 2.4441760 - 8.3059695 \cdot 10^{-3}\lambda^2 + 1.9000697 \cdot 10^{-2}\lambda^{-2}$ $+ 5.4129697 \cdot 10^{-4}\lambda^{-4} - 4.1973155 \cdot 10^{-6}\lambda^{-6} + 2.3742897 \cdot 10^{-6}\lambda^{-8}$	0.37–1.01†	149
LgSK2	$n^2 = 2.4750760 - 5.4304528 \cdot 10^{-3}\lambda^2 + 1.3893210 \cdot 10^{-2}\lambda^{-2}$ $+ 2.2990560 \cdot 10^{-4}\lambda^{-4} - 1.6868474 \cdot 10^{-6}\lambda^{-6} + 4.3959703 \cdot 10^{-7}\lambda^{-8}$	0.37–1.01	149
FF5	$n^2 = 2.4743324 - 1.0955338 \cdot 10^{-2}\lambda^2 + 1.9293801 \cdot 10^{-2}\lambda^{-2}$ $+ 1.4497732 \cdot 10^{-3}\lambda^{-4} - 1.1038744 \cdot 10^{-4}\lambda^{-6} + 1.1136008 \cdot 10^{-5}\lambda^{-8}$	0.37–1.01	151
SK4	$n^2 = 2.5585228 - 9.8824951 \cdot 10^{-3}\lambda^2 + 1.5151820 \cdot 10^{-2}\lambda^{-2}$ $+ 2.1134478 \cdot 10^{-4}\lambda^{-4} - 3.4130130 \cdot 10^{-6}\lambda^{-6} + 1.2673355 \cdot 10^{-7}\lambda^{-8}$	0.37–1.01†	149
KzFSN4	$n^2 = 2.5293446 - 1.3234586 \cdot 10^{-2}\lambda^2 + 1.8586165 \cdot 10^{-2}\lambda^{-2}$ $+ 5.4759655 \cdot 10^{-4}\lambda^{-4} - 1.1717987 \cdot 10^{-5}\lambda^{-6} + 2.0042905 \cdot 10^{-6}\lambda^{-8}$	0.37–1.01†	149
	$n^2 - 1 = \dfrac{1.38374965\lambda^2}{\lambda^2 - (0.0948292206)^2} + \dfrac{0.164626811\lambda^2}{\lambda^2 - (0.201806158)^2} + \dfrac{0.85913757\lambda^2}{\lambda^2 - (8.28807544)^2}$	0.37–2.33	42
SSK4	$n^2 = 2.5707849 - 9.2577764 \cdot 10^{-3}\lambda^2 + 1.6170751 \cdot 10^{-2}\lambda^{-2}$ $+ 2.7742702 \cdot 10^{-4}\lambda^{-4} + 1.2686469 \cdot 10^{-7}\lambda^{-6} + 4.5044790 \cdot 10^{-7}\lambda^{-8}$	0.37–1.01	149
F2	$n^2 = 2.5554063 - 8.8746150 \cdot 10^{-3}\lambda^2 + 2.2494787 \cdot 10^{-2}\lambda^{-2}$ $+ 8.6924972 \cdot 10^{-4}\lambda^{-4} - 2.4011704 \cdot 10^{-5}\lambda^{-6} + 4.5365169 \cdot 10^{-6}\lambda^{-8}$	0.37–1.01†	149
BaSF10	$n^2 = 2.6531250 - 8.1388553 \cdot 10^{-3}\lambda^2 + 2.2995643 \cdot 10^{-2}\lambda^{-2}$ $+ 7.3535957 \cdot 10^{-4}\lambda^{-4} - 1.3407390 \cdot 10^{-5}\lambda^{-6} + 3.6962325 \cdot 10^{-6}\lambda^{-8}$	0.37–1.01†	149
BaF10	$n^2 = 2.7324621 - 1.2490460 \cdot 10^{-2}\lambda^2 + 1.8562334 \cdot 10^{-2}\lambda^{-2}$ $+ 9.9990536 \cdot 10^{-4}\lambda^{-4} - 6.8388552 \cdot 10^{-5}\lambda^{-6} + 4.9257931 \cdot 10^{-6}\lambda^{-8}$	0.37–1.01	151
LaK10	$n^2 = 2.8984614 - 1.4857039 \cdot 10^{-2}\lambda^2 + 2.0985037 \cdot 10^{-2}\lambda^{-2}$ $+ 5.4506921 \cdot 10^{-4}\lambda^{-4} - 1.7297314 \cdot 10^{-5}\lambda^{-6} + 1.7993601 \cdot 10^{-6}\lambda^{-8}$	0.37–1.01	149
TaC2	$n^2 = 2.9717137 - 1.4952593 \cdot 10^{-2}\lambda^2 + 2.0162868 \cdot 10^{-2}\lambda^{-2}$ $+ 9.4072283 \cdot 10^{-4}\lambda^{-4} - 8.8614104 \cdot 10^{-5}\lambda^{-6} + 5.3191242 \cdot 10^{-6}\lambda^{-8}$	0.37–1.01	151
NbF1	$n^2 = 2.9753491 - 1.4613470 \cdot 10^{-2}\lambda^2 + 2.1096383 \cdot 10^{-2}\lambda^{-2}$ $+ 1.1980380 \cdot 10^{-3}\lambda^{-4} - 1.1887388 \cdot 10^{-4}\lambda^{-6} + 7.3444350 \cdot 10^{-6}\lambda^{-8}$	0.37–1.01	151
LaF2	$n^2 = 2.9673787 - 1.0978767 \cdot 10^{-2}\lambda^2 + 2.5088607 \cdot 10^{-2}\lambda^{-2}$ $+ 6.3171596 \cdot 10^{-4}\lambda^{-4} - 7.5645417 \cdot 10^{-6}\lambda^{-6} + 2.3202213 \cdot 10^{-6}\lambda^{-8}$	0.37–1.01†	149
SF6	$n^2 = 3.1195007 - 1.0902580 \cdot 10^{-2}\lambda^2 + 4.1330651 \cdot 10^{-2}\lambda^{-2}$ $+ 3.1800214 \cdot 10^{-3}\lambda^{-4} - 2.1953184 \cdot 10^{-4}\lambda^{-6} + 2.6671014 \cdot 10^{-5}\lambda^{-8}$	0.37–1.01†	149

TABLE 23 Room-temperature Dispersion Formula for Glasses (*Continued*)

Material	Dispersion formula (wavelength, λ, in μm)	Range (μm)	Ref.
TaFD5	$n^2 = 3.2729098 - 1.2888257 \cdot 10^{-2}\lambda^2 + 3.3451363 \cdot 10^{-2}\lambda^{-2}$ $+ 6.8221381 \cdot 10^{-5}\lambda^{-4} + 1.1215427 \cdot 10^{-4}\lambda^{-6} - 4.0485659 \cdot 10^{-6}\lambda^{-8}$	0.37–1.01	151
Fused silica	$n^2 - 1 = \dfrac{0.6961663\lambda^2}{\lambda^2 - (0.0684043)^2} + \dfrac{0.4079426\lambda^2}{\lambda^2 - (0.1162414)^2} + \dfrac{0.8974794\lambda^2}{\lambda^2 - (9.896161)^2}$	0.21–3.71	152
Fused germania	$n^2 - 1 = \dfrac{0.80686642\lambda^2}{\lambda^2 - (0.06897261)^2} + \dfrac{0.71815848\lambda^2}{\lambda^2 - (0.1539661)^2} + \dfrac{0.85416831\lambda^2}{\lambda^2 - (11.841931)^2}$	0.36–4.3	153
BS-39B	$n^2 - 1 = \dfrac{1.7441\lambda^2}{\lambda^2 - (0.1155)^2} + \dfrac{1.6465\lambda^2}{\lambda^2 - (14.981)^2}$	0.43–4.5	154*
CORTRAN 9753	$n = 1.61251 @ 0.4861 \ \mu m$ $n = 1.60475 @ 0.5893 \ \mu m \qquad n = 1.60151 @ 0.6563 \ \mu m$	—	155
CORTRAN 9754	$n^2 - 1 = \dfrac{1.66570\lambda^2}{\lambda^2 - (0.10832)^2} + \dfrac{0.04059\lambda^2}{\lambda^2 - (0.23813)^2} + \dfrac{1.31792\lambda^2}{\lambda^2 - (13.57622)^2}$	0.4–5.5	155
IRG 2	$n^2 - 1 = \dfrac{2.07670\lambda^2}{\lambda^2 - (0.11492)^2} + \dfrac{0.35738\lambda^2}{\lambda^2 - (0.23114)^2} + \dfrac{2.88166\lambda^2}{\lambda^2 - (17.48306)^2}$	0.365–4.6	149*
IRG 9	$n^2 - 1 = \dfrac{1.1852\lambda^2}{\lambda^2 - (0.084353)^2} + \dfrac{0.66877\lambda^2}{\lambda^2 - (11.568)^2}$	0.365–4.6	149*
IRG 11	$n^2 - 1 = \dfrac{1.7531\lambda^2}{\lambda^2 - (0.1185)^2} + \dfrac{0.4346\lambda^2}{\lambda^2 - (8.356)^2}$	0.48–3.3	149*
IRG 100	$n^2 = 4.5819 + \dfrac{2.2693\lambda^2}{\lambda^2 - (0.447)^2} - 0.000928\lambda^2$	1–14	149*
HTF-1	$n^2 = 2.0633034 - 3.2906345 \cdot 10^{-3}\lambda^2 + 7.1765160 \cdot 10^{-3}\lambda^{-2}$ $- 1.9110559 \cdot 10^{-4}\lambda^{-4} + 3.8123441 \cdot 10^{-5}\lambda^{-6} - 2.0668501 \cdot 10^{-6}\lambda^{-8}$	0.37–1.01	150
	$n^2 - 1 = \dfrac{1.06375\lambda^2}{\lambda^2 - (0.078958)^2} + \dfrac{0.80098\lambda^2}{\lambda^2 - (15.1579)^2}$	0.37–5	150*
ZBL	$n^2 = 2.03 + \dfrac{0.265\lambda^2}{\lambda^2 - (0.182)^2} + \dfrac{1.22\lambda^2}{\lambda^2 - (20.7)^2} + \dfrac{1.97\lambda^2}{\lambda^2 - (37.9)^2}$	1–5	213
ZBLA	$n^2 - 1 = \dfrac{1.291\lambda^2}{\lambda^2 - (0.0969)^2} + \dfrac{2.76\lambda^2}{\lambda^2 - (26.0)^2}$	0.64–4.8	158
ZBLAN	$n^2 - 1 = \dfrac{1.168\lambda^2}{\lambda^2 - (0.0954)^2} + \dfrac{2.77\lambda^2}{\lambda^2 - (25.0)^2}$	0.50–4.8	158
ZBT			
HBL	$n^2 = 1.96 + \dfrac{0.299\lambda^2}{\lambda^2 - (0.172)^2} + \dfrac{0.86\lambda^2}{\lambda^2 - (20.8)^2} + \dfrac{2.22\lambda^2}{\lambda^2 - (41.5)^2}$	1–5	213
HBLA			
HBT			
Arsenic trisulfide	$n^2 - 1 = \dfrac{1.8983678\lambda^2}{\lambda^2 - (0.150)^2} + \dfrac{1.9222979\lambda^2}{\lambda^2 - (0.250)^2} + \dfrac{0.8765134\lambda^2}{\lambda^2 - (0.350)^2} + \dfrac{0.1188704\lambda^2}{\lambda^2 - (0.450)^2} + \dfrac{0.9569903\lambda^2}{\lambda^2 - (27.3861)^2}$	0.56–12	163
Arsenic triselenide	$n^2 - 1 = \dfrac{6.6906\lambda^2}{\lambda^2 - (0.3468)^2}$	0.83–1.15	165
AMTIR-1/TI-20	$n^2 - 1 = \dfrac{5.298\lambda^2}{\lambda^2 - (0.29007)^2} + \dfrac{0.6039\lambda^2}{\lambda^2 - (32.022)^2}$	1–14	167*
AMTIR-3/TI-1173	$n^2 - 1 = \dfrac{5.8505\lambda^2}{\lambda^2 - (0.29192)^2} + \dfrac{1.4536\lambda^2}{\lambda^2 - (42.714)^2}$	3–14	167*
	$n^2 - 1 = \dfrac{5.8357\lambda^2}{\lambda^2 - (0.29952)^2} + \dfrac{1.064\lambda^2}{\lambda^2 - (38.353)^2}$	0.9–14	214

* Out dispersion equation from referenced data.
† Schott dispersion formula range; data available to 2.3 μm.

TABLE 24   Refractive Index of Calcite $(CaCO_3)^4$

λ, μm	n_o	n_e	λ, μm	n_o	n_e
0.198		1.57796	0.768	1.64974	1.48259
0.200	1.90284	1.57649	0.795	1.64886	1.48216
0.204	1.88242	1.57081	0.801	1.64869	1.48216
0.208	1.86733	1.56640	0.833	1.64772	1.48176
0.211	1.85692	1.56327	0.867	1.64676	1.48137
0.214	1.84558	1.55976	0.905	1.64578	1.48098
0.219	1.83075	1.55496	0.946	1.64480	1.48060
0.226	1.81309	1.54921	0.991	1.64380	1.48022
0.231	1.80233	1.54541	1.042	1.64276	1.47985
0.242	1.78111	1.53782	1.097	1.64167	1.47948
0.257	1.76038	1.53005	1.159	1.64051	1.47910
0.263	1.75343	1.52736	1.229	1.63926	1.47870
0.267	1.74864	1.52547	1.273	1.63849	
0.274	1.74139	1.52261	1.307	1.63789	1.47831
0.291	1.72774	1.51705	1.320	1.63767	
0.303	1.71959	1.51365	1.369	1.63681	
0.312	1.71425	1.51140	1.396	1.63637	1.47789
0.330	1.70515	1.50746	1.422	1.63590	
0.340	1.70078	1.50562	1.479	1.63490	
0.346	1.69833	1.50450	1.497	1.63457	1.47744
0.361	1.69316	1.50224	1.541	1.63381	
0.394	1.68374	1.49810	1.609	1.63261	
0.410	1.68014	1.49640	1.615		1.47695
0.434	1.67552	1.49430	1.682	1.63127	
0.441	1.67423	1.49373	1.749		1.47638
0.508	1.66527	1.48956	1.761	1.62974	
0.533	1.66277	1.48841	1.849	1.62800	
0.560	1.66046	1.48736	1.909		1.47573
0.589	1.65835	1.48640	1.946	1.62602	
0.643	1.65504	1.48490	2.053	1.62372	
0.656	1.65437	1.48459	2.100		1.47492
0.670	1.65367	1.48426	2.172	1.62099	
0.706	1.65207	1.48353	3.324		1.47392

TABLE 25   Refractive Index of Calcium Molybdate $(CaMoO_4)$ and Lead Molybdate $(PbMoO_4)^{203}$

Calcium molybdate (19.5°C)			Lead molybdate (19.5°C)		
λ, μm	n_o	n_e	λ, μm	n_o	n_e
0.40466	2.04452	2.06156	0.40466		2.44317
0.43584	2.02553	2.04029	0.43584	2.60487	2.39011
0.46782	2.01091	2.02409	0.46782	2.53415	2.35258
0.47999	2.00629	2.01898	0.47999	2.51398	2.34119
0.50858	1.99697	2.00873	0.50858	2.47589	2.31877
0.54607	1.98730	1.99810	0.54607	2.43929	2.29618
0.57696	1.98089	1.99109	0.57696	2.41651	2.28162
0.57906	1.98049	1.99066	0.57906	2.41515	2.28073
0.58756	1.97893	1.98897	0.58756	2.40977	2.27729
0.64385	1.97033	1.97961	0.64385	2.38119	2.25835
0.66781	1.96737	1.97639	0.66781	2.37170	2.25197
0.70652	1.96321	1.97189	0.70652	2.35878	2.24311
1.0830	1.94317	1.95030	1.0830	2.30177	2.20267

TABLE 26 Refractive Index of Lead Fluoride (PbF_2)[202,203]

λ, μm	n
0.30	1.93665
0.35	1.85422
0.40	1.81804
0.45	1.79644
0.50	1.78220
0.55	1.77221
0.60	1.76489
0.65	1.75934
0.70	1.75502
0.75	1.75158
0.80	1.74879
0.85	1.74648
0.90	1.74455
0.95	1.74291
1.00	1.74150
1.50	1.73371
2.00	1.72983
2.50	1.72672
3.00	1.72363
3.50	1.72030
4.00	1.71663
4.50	1.71255
5.00	1.70805
5.50	1.70310
6.00	1.69769
6.50	1.69181
7.00	1.68544
7.50	1.67859
8.00	1.67125
8.50	1.66340
9.00	1.65504
9.50	1.64615
10.00	1.63674
10.50	1.62679
11.00	1.61629
11.50	1.60523
12.00	1.59597

TABLE 27 Refractive Index of New Oxide Materials[215]

Wavelength (μm)	Material and supplier		
	ALON* (Raytheon)	Spinel* (Alpha Optical)	$Y_2O_3:La_3O_3$† (GTE Laboratories)
0.40466	1.81167	1.73574	1.96939
0.43583	1.80562	1.73054	1.95660
0.54607	1.79218	1.71896	1.92941
0.85211	1.77758	1.70728	1.90277
1.01398	1.77375	1.70300	1.89643
1.52958	1.76543	1.69468	1.88446
1.97009	1.75838	1.68763	1.87585
2.32542	1.75268	1.68194	1.86822

* Accuracy: $\pm 1 \cdot 10^{-5}$.
† Accuracy: $\pm 1 \cdot 10^{-4}$.

TABLE 28 Refractive Index of Strontium Titanate $(SrTiO_3)$[203]

λ, μm	n
0.40	2.66386
0.44	2.56007
0.48	2.49751
0.52	2.45553
0.56	2.40285
0.60	2.40285
0.64	2.38532
0.68	2.37135
0.72	2.35998
0.76	2.35055
0.80	2.34260
0.84	2.33583
0.88	2.32997
0.92	2.32486
0.96	2.32035
1.00	2.31633
1.40	2.29073
1.80	2.27498
2.20	2.26109
2.60	2.24676
3.00	2.23111
3.40	2.21370
3.80	2.19428
4.20	2.17265
4.60	2.14865
5.00	2.12212
5.40	2.09290

TABLE 29 Optical Modes of Crystals with Diamond Structure; Space Group: Fd3m (O_h^7) #227; $\Gamma = F_{2g}(R)$

Material	Raman mode location (cm^{-1})	Reference
	F_{2g}	
C (diamond)	1332.4	216
Si	519.5	217
Ge	300.6	218

TABLE 30 Optical Modes of Crystals with Cesium Chloride Structure; Space Group: Pm3m (O_h^5) #221; $\Gamma = F_{1u}(IR)$

Material	Mode location (cm^{-1})		Refs.
	ω_{TO}	ω_{LO}	
CsBr	75	113	219
CsCl	99	160	219, 220
CsI	62	88	219
TlBr	43	101	221
TlCl	63	158	221
TlI	52		222

TABLE 31 Optical Modes of Crystals with Sodium Chloride Structure; Space Group: Fm3m (O_h^1) #225; $\Gamma = F_{1u}(IR)$

| | Mode location (cm^{-1}) | | |
Material	ω_{TO}	ω_{LO}	Refs.
AgBr	79	138	221
AgCl	106	196	221
BaO	132	425	223
BaS	158	230	224
CaO	295	557	223, 225
CaS	256	376	224
CdO	270	380	226
KBr	113	165	221
KCl	142	214	221
KF	190	326	221
KI	101	139	221
LiF	306	659	221, 227
MgO	401	718	221, 227
MgS	237	430	224
NaBr	134	209	221
NaCl	164	264	221
NaF	244	418	221
NaI	117	176	221
NiO	401	580	221
SrO	227	487	223, 225
SrS	194	284	224
PbS	71	212	228
PbSe	39	116	229
PbTe	32	112	230

TABLE 32 Optical Modes of Crystals with Zincblende Structure; Space Group: F$\bar{4}$3m (T_d^2) #216; $\Gamma = F_2(R, IR)$

| | Mode location (cm^{-1}) | | |
Material	ω_{TO}	ω_{LO}	Refs.
AlAs	364	402	231
AlSb	319	340	232
BN	1055	1305	233, 234
BP	799	829	233, 234
CdTe	141	169	235
CuCl	172	210	236
GaAs	269	292	232
GaP	367	403	232, 237
GaSb	230	240	238
InAs	217	239	238, 239
InP	304	345	232
InSb	179	191	239, 240
β-SiC	793	970	218, 241
ZnS	282	352	235, 242
ZnSe	206	252	235
ZnTe	178	208	235, 243

TABLE 33 Optical Modes of Crystals with Fluorite Structure; Space Group: Fm3m (O_h^1) #225; $\Gamma = F_{1u}(IR) + F_{2g}(R)$

Material	Ref.	Mode locations (cm^{-1})		
		$F_{1u}(\omega_{TO})$	$F_{1u}(\omega_{LO})$	F_{2g}
BaCl$_2$	244			185
BaF$_2$	245, 246, 247	184	319	241
CaF$_2$	245, 246, 247	258	473	322
CdF$_2$	247, 248	202	384	317
EuF$_2$	244	194	347	287
β-PbF$_2$	244, 248	102	337	256
SrCl$_2$	244, 249, 250	147	243	182
SrF$_2$	245, 246, 247	217	366	286
ThO$_2$	251	281	568	
UO$_2$	251, 252	281	555	445
ZrO$_2$	253	354	680	605

TABLE 34 Optical Modes of Crystals with Corundum Structure; Space Group: R$\bar{3}$c (D_{3d}^6) #167; $\Gamma = 2A_{1g}(R) + 2A_{1u}(-) + 3A_{2g}(-) + 2A_{2u}(IR, E \parallel c) + 5E_g(R) + 4E_u(IR, E \perp c)$

Material	Mode locations (cm^{-1})												
	Infrared modes (LO in parentheses)						Raman modes						
	E_u	E_u	E_u	E_u	A_{1u}	A_{1u}	E_g	E_g	E_g	E_g	E_g	A_{1g}	A_{1g}
Al$_2$O$_3$	385	422	569	635	400	583	378	432	451	578	751	418	645
(Refs: 253, 254)	(388)	(480)	(625)	(900)	(512)	(871)							
Cr$_2$O$_3$	417	444	532	613	538	613	—	351	397	530	609	303	551
(Refs: 255, 256)	(420)	(446)	(602)	(766)	(602)	(759)							
Fe$_2$O$_3$	227	286	437	524	299	526	245	293	298	413	612	226	500
(Refs: 256, 257)	(230)	(368)	(494)	(662)	(414)	(662)							

TABLE 35 Optical Modes of Crystals with Wurtzite Structure. Space Group: $P6_3mc$ (C_{6v}^4) #186; $\Gamma = A_1(R, IR\ E \parallel c) + 2B_1(-) + E_1(R, IR\ E \perp c) + 2E_2(R)$

| | Mode locations (cm^{-1}) | | | | |
| | Infrared modes (LO) | | Raman modes | | |
Material	E_1	A_1	E_2	E_2	Refs.
β-AgI	106	106	17	112	258
	(124)	(124)			
AlN	672	659		655	234, 259
	(895)	(888)			
BeO	725	684	340	684	260, 261
	(1095)	(1085)			
CdS	235	228	44	252	261, 263
	(305)	(305)			
CdSe	172	166	34	?	264
	(210)	(211)			
GaN	560	533	145	568	188, 265
	(746)	(744)			
α-SiC	797	788	149	788	266
	970	964			
ZnO	407	381	101	437	261, 262
	(583)	(574)			
ZnS	274	274	55	280	261
	(352)	(352)			

TABLE 36 Optical Modes of Crystals with Trigonal Selenium Structure; Space Group: $P3_121$ (D_3^4) #152; $\Gamma = A_1(R) + A_2(IR, E \parallel c) + 2E(IR, E \perp c, R)$

| | Mode locations (cm^{-1}) | | | | |
| | Raman | Infrared, $E \perp c$ | | Infrared, $E \parallel c$ | |
Material	A_1	E	E	A_2	References
Se	237	144	225	102	267, 268
		(150)	(225)	(106)	
Te	120	92	144	90	269
		(106)	(145)	(96)	

TABLE 37 Optical Modes of Crystals with α-Quartz Structure; Space Group: P3$_2$21 (D$_3^6$) #154; $\Gamma = 4A_1(R) + 4A_2(IR, E \parallel c) + 8E(IR, E \perp c, R)$

Material	Infrared modes, $E \perp c$							
	E	E	E	E	E	E	E	E
SiO$_2$	128	265	394	451	697	796	1067	1164
(Refs: 270, 271, 272)	(128)	(270)	(403)	(511)	(699)	(809)	(1159)	(1230)
GeO$_2$	121	166	326	385	492	583	857	961
(Refs: 273)	(121)	(166)	(372)	(456)	(512)	(595)	(919)	(972)

Material	Infrared modes, $E \parallel c$				Raman modes			
	A$_2$	A$_2$	A$_2$	A$_2$	A$_1$	A$_1$	A$_1$	A$_1$
SiO$_2$	364	500	777	1080	207	356	464	1085
(Refs: 270, 271, 272)	(388)	(552)	(789)	(1239)				
GeO$_2$					212	261	440	880
(Ref: 273)								

TABLE 38 Optical Modes of Crystals with Rutile Structure; Space Group: P4$_2$/mnm (D$_{4h}^{14}$) #136; $\Gamma = A_{1g}(R) + A_{2g}(-) + A_{2u}(IR, E \parallel c) + B_{1g}(R) + B_{2g}(R) + 2B_{1u}(-) + E_g(R) + 3E_u(IR, E \perp c)$

Material	Mode locations (cm^{-1})							
	Infrared modes (LO in parentheses)				Raman modes			
	E$_u$	E$_u$	E$_u$	A$_{2u}$	B$_{1g}$	E$_g$	A$_{1g}$	B$_{2g}$
CoF$_2$	190	270	405	345	68	246	366	494
(Refs: 274, 275, 276	(234)	(276)	(529)	(506)				
FeF$_2$	173	244	405	307	73	257	340	496
(Refs: 275, 277, 278)	(231)	(248)	(530)	(487)				
GeO$_2$	300	370	635	455	97	680	702	870
(Refs: 273, 279, 280)	(345)	(470)	(815)	(755)				
MgF$_2$	247	410	450	399	92	295	410	515
(Refs: 277, 278, 281, 282, 283)	(303)	(415)	(617)	(625)				
SnO$_2$	243	284	605	465	123	475	634	776
(Refs: 284, 285, 286)	(273)	(368)	(757)	(703)				
TiO$_2$	189	382	508	172	143	447	612	826
(Refs: 22, 26, 277, 287)	(367)	(444)	(831)	(796)				
ZnF$_2$	173	244	380	294	70	253	350	522
(Refs: 277, 278, 281, 283)	(227)	(264)	(498)	(488)				

TABLE 39 Optical Modes of Crystals with Scheelite Structure; Space Group: $I4_1/a$ (C_{4h}^6) #88; $\Gamma = 3A_g(R) + 4A_u(IR, E \parallel c) + 5B_g(R) + 3B_u(-) + 5E_g(R) + 4E_u(IR, E \perp c)$

Material	Refs.	Infrared modes, $E \perp c$				Infrared modes, $E \parallel c$			
		E_u	E_u	E_u	E_u	A_u	A_u	A_u	A_u
CaMoO$_4$	288, 289, 290	146	197	322	790	193	247	420	772
		(161)	(258)	(359)	(910)	(202)	(317)	(450)	(898)
CaWO$_4$	288, 289, 290	142	200	313	786	177	237	420	776
		(153)	(248)	(364)	(906)	(181)	(323)	(450)	(896)
SrMoO$_4$	289	125	[181]	[327]	[830]	153	[282]	[404]	[830]
SrWO$_4$	289	[140]	[168]	[320]	[833]	[150]	[278]	[410]	[833]
PbMoO$_4$	289, 290	90	105	301	744	86	258	373	745
		(99)	(160)	(318)	(886)	(132)	(278)	(387)	(865)
PbWO$_4$	289, 290, 291	73	104	288	756	58	251	384	764
		(101)	(137)	(314)	(869)	(109)	(278)	(393)	(866)
LiYF$_4$	292, 293	143	292	326	424	195	252	396	490
		(173)	(303)	(367)	(566)	(224)	(283)		

Material	Refs.	Raman modes												
		A_g	A_g	A_g	B_g	B_g	B_g	B_g	B_g	E_g	E_g	E_g	E_g	E_g
CaMoO$_4$	294	205	333	878	110	219	339	393	844	145	189	263	401	797
CaWO$_4$	294, 295	218	336	912	84	210	336	401	838	117	195	275	409	797
SrMoO$_4$	294	181	327	887	94	157	327	367	842	111	137	231	381	797
SrWO$_4$	294, 295	187	334	925	75		334	370	839	101	131	238	378	797
PbMoO$_4$	296	164	314	868	64	75	317	348	764	61	100	190	356	744
PbWO$_4$	296	178	328	905	54	78	328	358	766	63	78	192	358	753
LiYF$_4$	292, 293	[150]	264	425	177	248	329	382	427	153	199	329	368	446

TABLE 40 Optical Modes of Crystals with Spinel Structure; Space Group: Fd3m (O_h^7) #227; $\Gamma = A_{1g}(R) + E_g(R) + F_{1g}(-) + 3F_{2g}(R) + 2A_{2u}(-) + 2E_u(-) + 4F_{1u}(IR) + 2F_{2u}(-)$

Material	Refs.	Mode locations (cm^{-1})								
		Infrared modes				Raman modes				
		F_{1u}	F_{1u}	F_{1u}	F_{1u}	F_{2g}	F_{2g}	F_{2g}	E_g	A_{1g}
MgAl$_2$O$_4$	199, 297, 298	305	428	485	670	311	492	611	410	722
		(311)		(497)	(800)					
CdIn$_2$S$_4$	299, 300	68	171	215	307	93	247	312	185	366
		(69)	(172)	(270)	(311)					
ZnCr$_2$O$_4$	301	186	372	506	624	186	515	610	457	692
		(194)	(377)	(522)	(711)					
ZnCr$_2$S$_4$	302, 303, 304	115	249	340	388	116	[290]	361	249	403
		(117)	(250)	(360)	(403)					

TABLE 41   Optical Modes of Crystals with Cubic Perovskite Structure; Space Group: Pm3m (O_h^1) #221; $\Gamma = 3F_{1u}(IR) + F_{2u}(-)$

Material	Infrared mode locations (cm^{-1})			References
	$F_{1u}*$	F_{1u}	F_{1u}	
KTaO$_3$	85	199	549	305
	(88)	(200)	(550)	
SrTiO$_3$	88	178	544	205, 306, 307
	(173)	(473)	(804)	
KMgF$_3$	168	299	458	308
	(197)	(362)	(551)	
KMnF$_3$	119	193	399	308
	(144)	(270)	(483)	

* "Soft" mode with strong temperature dependence.

TABLE 42   Optical Modes of Crystals with Tetragonal Perovskite Structure; Space Group: P4$_2$/mnm (D_{4h}^{14}) #136; $\Gamma = 3A_1(IR, E \parallel c, R) + B_1(R) + 4E(IR, E \perp c, R)$

Material	Mode locations (cm^{-1})								Refs.
	Infrared (E $\perp$ c)				Raman	Infrared (E $\parallel$ c)			
	E	E	E	E	B$_1$	A$_1$	A$_1$	A$_1$	
BaTiO$_3$	34	181	306	482	305	180	280	507	305, 309, 310
	(180)	(306)	(465)	(706)		(187)	(469)	(729)	
PbTiO$_3$	89	221	250	508	415	127	351	613	311
	(128)		(445)	(717)		(215)	(445)	(794)	

TABLE 43   Optical Modes of Crystals with the Chalcopyrite Structure; Space Group: I$\bar{4}$2d (D_{2d}^{12}) #122; $\Gamma = A_1(R) + 2A_2(-) + 3B_1(R) + 3B_2(IR, E \parallel c, R) + + 6E(IR, E \perp c, R)$

Material	Refs.	Mode locations (cm^{-1})						
		Raman modes				Infrared modes (E $\parallel$ c)		
		A$_1$	B$_1$	B$_1$	B$_1$	B$_2$	B$_2$	B$_2$
AgGaS$_2$	312, 313	295	118	179	334	195	215	366
						(199)	(239)	(400)
AgGaSe$_2$	312, 314	179	12.5				152	246
							(159)	(272)
CdGeAs$_2$	315, 316	[188]	[84]	[167]	[245]	[85]	203	270
						[(85)]	(210)	(278)
CuGaS$_2$	312	312	138	203	243	259	339	371
						(284)	(369)	(402)
ZnGeP$_2$	317, 318	328	120	247	389	[140]	361	411
							(341)	(401)
ZnSiP$_2$	319	344	131				352	511
							(362)	(535)

TABLE 43 Optical Modes of Crystals with the Chalcopyrite Structure; Space Group: I$\overline{4}$2d (D$_{2d}^{12}$) #122; $\Gamma = A_1(R) + 2A_2(-) + 3B_1(R) + 3B_2(IR, E \parallel c, R) + +6E(IR, E \perp c, R)$ (*Continued*)

		Infrared mode location (E $\perp$ c)					
Material	Refs.	E	E	E	E	E	E
AgGaS$_2$	312, 313	63	93	158	225	323	368
		(64)	(96)	(160)	(230)	(347)	(392)
AgGaSe$_2$	312, 314	78	133		160	208	247
		(78)	(135)	(112)	(163)	(213)	(274)
CdGeAs$_2$	315, 316		95	159	200	255	272
			(98)	(161)	(206)	(258)	(280)
CuGaS$_2$	312	75	95	147	260	335	365
		(76)	(98)	(167)	(278)	(352)	(387)
ZnGeP$_2$	317, 318	94	141	201	328	369	386
		(94)	(141)	(206)	(330)	(375)	(406)
ZnSiP$_2$	319	105	185	270	335	477	511
		(105)	(185)	(270)	(362)	(477)	(535)

TABLE 44 Optical Modes of Other Crystals

Material/space group	Irreducible optical representation and optical modes locations (cm^{-1})
Orpiment, As$_2$S$_3$ P2$_1$/b (C$_{2h}^5$) #14 (Ref. 320)	$\Gamma = 15A_g(R) + 15B_g(R) + 14A_u(IR, E \parallel b) + 13B_u(IR, E \parallel a, E \parallel c)$ $[7A_1(IR, E \parallel c, R) + 7A_2(R) + 7B_1(IR, E \parallel b, R) + 6B_2(IR, E \parallel a, R)$ for a noninteracting molecular layer structure] A$_g$ = 136, 154, 204, 311, 355, 382 B$_u$ = 140, 159, 198, 278, 311, 354, 383
Calcite, CaCO$_3$ R$\overline{3}$c (D$_{3d}^6$) #167 (Refs. 321, 322)	$\Gamma = A_{1g}(R) + 2A_{1u}(-) + 3A_{2g}(-) + 3A_{2u}(IR, E \parallel c) + 4E_g(R)$ $+ 5E_u(IR, E \perp c)$ A$_{1g}$ = 1088 E$_g$ = 156, 283, 714, 1432 A$_{2u}$ = 92(136), 303(387), 872(890) E$_u$ = 102(123), 223(239), 297(381), 712(715), 1407(1549)
BBO, Ba$_3$[B$_3$O$_6$]$_2$ R3 (C$_3^4$) #146 (Ref: 323)	$\Gamma = 41A(IR, E \parallel c, R) + 41E(IR, E \perp c, R)$
BSO, sellenite, Bi$_{12}$SiO$_{20}$ I23 (T^3) #197 (Ref: 324)	$\Gamma = 8A(R) + 8E(R) + 24F(IR)$ A = 92, 149, 171, 282, 331, 546, 785 E = 68, 88, 132, 252, —, 464, 626 F = 44, 51, 59, 89, 99, 106, —, 115, 136, 175, 195, 208, 237, 288, 314, 353, 367, 462, 496, 509, 531, 579, 609, 825
Iron Pyrite, FeS$_2$ Pa3 (T$_h^6$) #205 (Refs: 325, 326)	$\Gamma = A_g(R) + E_g(R) + 3F_g(R) + 2A_u(-) + 2E_u(-) + 5F_u(IR)$ A$_g$ = 379 E$_g$ = 343 F$_g$ = 435, 350, 377 F$_u$ = 293(294), 348(350), 401(411), 412(421), 422(349)

TABLE 44 Optical Modes of Other Crystals (*Continued*)

Material/space group	Irreducible optical representation and optical modes locations (cm^{-1})
KDP, KH_2PO_4 I$\bar{4}$2d (D_{2d}^{12}) #122 (Ref: 327)	$\Gamma = 4A_1(R) + 5A_2(-) + 6B_1(R) + 6B_2(IR, E \parallel c, R) + 12E(IR, E \perp c, R)$ $A_1 = 360, 514, 918, 2700$ $B_1 = 156, 479, 570, 1366, 1806, 2390$ $B_2 = 80, 174, \text{—}, 386, 510, 1350$ $E = 75, 95, 113, 190, 320, 490, 530, 568, 960, 1145, \text{—}, 1325$
KTP Pna2$_1$ (C_{2v}^9) #33 (Ref: 328)	$\Gamma = 47A_1(IR, E \parallel c, R) + 48A_2(R) + 47B_1(IR, E \parallel a, R) + 47B_2(IR, E \parallel b, R)$
Lanthanum fluoride, LaF_3 P$\bar{3}$c1 (D_{3d}^4) #165 (Refs: 329, 330, 331)	$\Gamma = 5A_{1g}(R) + 12E_g(R) + 6A_{2u}(IR, E \parallel c) + 11E_u(IR, E \perp c)$ $A_{1g} = 120, 231, 283, 305, 390$ $E_g = 79, 145, 145, 163, 203, 226, 281, 290, 301, 315, 325, 366$ $A_{2u} = 142(143), 168(176), 194(239), \text{—}, 275(296), 323(468)$ $E_u = 100(108), 128(130), 144(145), 168(183), 193(195), 208(222),$ $\qquad 245(268), 272(316), 354(364), 356(457)$
Lithium iodate, $LiIO_3$ P6$_3$ (C_6^6) #173 (Refs: 332, 333)	$\Gamma = 4A(IR, E \parallel c, R) + 5B(-) + 4E_1(IR, E \perp c, R) + 5E_2(R)$ $A = 148(148), 238(238), 358(468), 795(817)$ $E_1 = 180(180), 330(340), 370(460), 764(848)$ $E_2 = 98, 200, 332, 347, 765$
Lithium niobate, $LiNbO_3$ R3c (C_{3v}^6) #161 (Refs: 334, 335)	$\Gamma = 4A_1(IR, E \parallel c, R) + 5A_2(-) + 9E(IR, E \perp c, R)$ $A_1 = 255(275), 276(333), 334(436), 633(876)$ $E = 155(198), 238(243), 265(295), 325(371), 371(428), 431(454),$ $\qquad 582(668), 668(739), 743(880)$
Potassium niobate, $KNbO_3$ Bmm2 (C_{2v}^{14}) #38 (Ref: 336)	$\Gamma = 4A_1(IR, E \parallel z, R) + 4B_1(IR, E \parallel x, R) + 3B_2(IR, E \parallel y, R) + A_2(R)$ $A_1 = 190(193), 290(296), 299(417), 607(827)$ $B_1 = 187(190), 243(294), 267(413), 534(842)$ $B_2 = 56(189), 195(425), 511(838)$ $A_2 = 283$
Paratellurite, TeO_2 P4$_1$2$_1$2 (D_4^4) #92 (Refs: 337, 338)	$\Gamma = 4A_1(R) + 4A_2(IR, E \parallel c) + 5B_1(R) + 4B_2(R) + 8E(IR, E \perp c, R)$ $A_1 = 148, 393, 648$ $A_2 = 82(110), 259(263), 315(375), 575(775)$ $B_1 = 62, 175, 216, 233, 591$ $B_2 = 155, 287, 414, 784$ $E = 121(123), 174(197), 210(237), 297(327), 330(379), 379(415), 643(720),$ $\qquad 769(812)$
Yttria, Y_2O_3 Ia3 (T_h^7) #206 (Ref: 339)	$\Gamma = 4E_g(R) + 4A_g(R) + 14F_g(R) + 5E_u(-) + 5A_u(-) + 16F_u(IR)$ $E_g = 333, 830, 948$ $A_g = 1184$ $F_g = (131), 431, 469, 596$ $F_u = 120(121), 172(173), 182(183), 241(242), 303(315), 335(359), 371(412),$ $\qquad 415(456), 461(486), 490(535), 555(620)$
Yttrium aluminum garnet (YAG), $Y_3Al_2(AlO_4)_3$ Ia3d (O_h^{10}) #230 (Refs: 340, 341)	$\Gamma = 5A_{1u}(-) + 3A_{1g}(R) + 5A_{2u}(-) + 5A_{2g}(-) + 10E_u(-) + 8E_g(R)$ $\qquad + 14F_{1g}(-) + 17F_{1u}(IR) + 14F_{2g}(R) + 16F_{2u}(-)$ $A_{1g} = 373, 561, 783$ $E_g = 162, 310, 340, 403, 531, 537, 714, 758$ $F_{2g} = 144, 218, 243, 259, 296, 408, 436, 544, 690, 719, 857$ $F_{1u} = 122(123), 163(172), 177(180), 219(224), 290(296), 330(340), 373(378),$ $\qquad 387(388), 395(403), 428(438), 446(472), 472(511), 516(549), 569(585),$ $\qquad 692(712), 723(765), 782(841)$

TABLE 45 Summary of Available Lattice Vibration Model Parameters

Material	Classical	Four-parameter	Material	Classical	Four-parameter
	Dispersion model reference			Dispersion model reference	
AgBr		342	KI	351	
AgCl		342	$KNbO_3$	336	
$AgGaS_2$	312		$KTaO_3$	305	
$AgGaSe_2$	314		$KTiOPO_4$	328	
Al_2O_3	25	26	LaF_3		330
ALON	175		La_2O_3	352	
As_2S_3 (cryst)	320		LiF	227	
As_2S_3 (glass)	343		$LiIO_3$	332, 333	
As_2S_3 (cryst)	320		$LiNbO_3$	353	
As_2Se_3 (glass)	343		$YLiF_4$	292	
BaF_2	245		$MgAl_2O_4$	199	
$BaTiO_3$	305, 344	310	MgF_2	281, 283	278
BeO	260		MgO	227	
BN	233		NaF	354	
$CaCO_3$	322		PbF_2	248	
CaF_2	245		PbSe	355	
$CaMoO_4$	288		$PbWO_4$	291	
$CaWO_4$	288		Se	267, 345	
CdS	235		SiO_2	272, 356	
CdSe	264		SrF_2	245	
CdTe	235, 345		$SrTiO_3$	305	307
CsBr	219		Te	269	
CsCl	219, 220		TeO_2	338, 357	357
CsI	219		TiO_2	244	22
FeS_2	346, 347		TlBr		342
GaAs	348		TlCl		342
GaN	188, 265		$Y_3Al_5O_{12}$	358	341
GaP	237, 349		Y_2O_3	339	
GeO_2	280		ZnS	235	
$HfO_2:Y_2O_3$	132		ZnSe	235	
KBr	350		ZnTe	235	

TABLE 46 Urbach Tail Model Parameters

Material	α_0 (cm^{-1})	E_g (eV)	σ_s (−)	E_p (meV)	Temperature range (K)	Absorption range (cm^{-1})	Ref.
	Urbach model parameters						
AgBr	$[1.5 \cdot 10^4]$	[2.79]	1.0	—	>100		28
AgCl	$[1.6 \cdot 10^6]$	[3.44]	0.8	—	>100	$100{-}10^4$	28
$AgGaS_2$							
$E \perp c$	$[6.2 \cdot 10^6]$	2.92	1.09			$2{-}10^3$	359
$E \parallel c$	$[1.7 \cdot 10^4]$	2.69	1.14	—		$60{-}10^3$	359
Al_2O_3							
$E \perp c$	$6.531 \cdot 10^5$	9.1	0.559	—	RT	$1{-}100+$	360
ALON	$6.780 \cdot 10^5$	6.5	0.254	—	RT	$2{-}100+$	360
BaF_2	$4.17 \cdot 10^8$	10.162	0.58	40	78–573	$1{-}100$	361
$Bi_{12}SiO_{20}$	$1.0 \cdot 10^5$	3.54	0.47	25.0	41–293		362
CaF_2	$1.33 \cdot 10^{10}$	11.228	0.61	45	78–573	$1{-}100$	361

TABLE 46 Urbach Tail Model Parameters (*Continued*)

Material	Urbach model parameters				Temperature range (K)	Absorption range (cm^{-1})	Ref.
	α_0 (cm^{-1})	E_g (eV)	σ_s (−)	E_p (meV)			
CdS							
E ⊥ c	[2.7 · 10^9]	2.584	2.17	—	90–342	10^2–10^4	363
E ‖ c	[2.7 · 10^9]	2.608	2.17	—	90–300	10^2–10^4	363
CdSe							
E ⊥ c	[9 · 10^8]	[1.887]	2.2	28	110–340	10–10^3	364
E ‖ c	[5 · 10^8]	[1.902]	2.2	28	78–300	10–10^3	364
CdTe	3 · 10^{12}	1.65	4.39	—			365
CuCl	1 · 10^6	3.35	1.35	—	200–400	1–200	366
InP	8 · 10^7	1.43	1.35	—	6–298	10–10^4	367
KBr	6 · 10^9	6.840	0.774	10.5	70–536	3–10^5	368
KCl	1.26 · 10^{10}	7.834	0.745	13.5	10–573		368
KF		[10.0]		[17.1]			368
KI	6 · 10^9	5.890	0.830	4.5	65–573	1–10^6	368
LiF	3.72 · 10^{10}	13.09			400–600	1–10^3	361
LiNbO$_3$							
E ‖ c	[6 · 10^7]	[4.65]	0.75	60	10–667	2–10^4	369
MgAl$_2$O$_4$	4.931 · 10^5	8.0	0.267	—	RT	10–100+	360
NaBr	6 · 10^9	6.770	0.765	10.7			368
NaCl	1.2 · 10^{10}	8.025	0.741	9.5	10–573	3–500+	368
NaF	1.0 · 10^{10}	10.70	0.69	16.5	78–573		368
NaI	6 · 10^9	5.666	0.845	8.5			368
SrF$_2$	1.35 · 10^9	10.670	0.60	44	78–573		361
SrTiO$_3$	1.3 · 10^4	3.37	1.0	—	4–300	1–1000	370
TeO$_2$							371
E ‖ c		4.31	0.69	17	80–500	2–1000	
TlCl		3.44	1.1				372
	4 · 10^4	3.43	1.04	7			365
YAG	2.125 · 10^5	7.012	0.560	37.2	34–292	10–10^3	373
YAlO$_3$							
E ‖ a	3.30 · 10^6	8.056	0.553	33.5	10–300	10^2–10^5	374
E ‖ b	5.27 · 10^5	8.018	0.479	32.5			
E ‖ c	1.32 · 10^6	8.151	0.448	35.8			
Y$_2$O$_3$	8.222 · 10^6	6.080	0.688	18.6	10–297		375
ZnTe	1 · 10^{15}	2.556	2.8	50	77–300	10–300	376

TABLE 47 Ultraviolet and Infrared Room-temperature Absorption Edge Equation Constants

Selected glass code	Ultraviolet absorption edge			Infrared absorption edge		
	λ ($\beta = 1$ cm^{-1})	β_0 (cm^{-1})	σ_{UV} (eV^{-1})	λ ($\beta = 1$ cm^{-1})	β_0 (cm^{-1})	σ_{IR} (cm^{-1})
479587 TiK1	348	5.34 · 10^{-07}	4.059			
487704 FK5	287	1.78 · 10^{-13}	6.805			
511510 TiF1	352	4.63 · 10^{-14}	8.716			
517642 BK7	310	3.48 · 10^{-14}	7.753	2.69*	4.45 · 10^{62}	0.0388
518651 PK2	311	2.72 · 10^{-13}	7.248			
522595 K5	313	9.47 · 10^{-15}	8.143			
523515 KF9	325	8.97 · 10^{-15}	8.484			
526600 BaLK1	307	1.48 · 10^{-12}	6.749			

TABLE 47 Ultraviolet and Infrared Room-temperature Absorption Edge Equation Constants (*Continued*)

Selected glass code	Ultraviolet absorption edge			Infrared absorption edge		
	$\lambda\,(\beta = 1\,\text{cm}^{-1})$	$\beta_0\,(\text{cm}^{-1})$	$\sigma_{\text{UV}}\,(\text{eV}^{-1})$	$\lambda\,(\beta = 1\,\text{cm}^{-1})$	$\beta_0\,(\text{cm}^{-1})$	$\sigma_{\text{IR}}\,(\text{cm}^{-1})$
527511 KzF6	322	$3.21 \cdot 10^{-08}$	4.485			
533580 ZK1	313	$8.81 \cdot 10^{-15}$	8.165			
548458 LLF1	317	$2.49 \cdot 10^{-18}$	10.354			
548743 Ultran 30	228	$3.72 \cdot 10^{-05}$	1.877	3.98	$2.76 \cdot 10^{21}$	−0.0198
552635 PSK3	304	$1.48 \cdot 10^{-10}$	5.543			
573575 BaK1	305	$2.31 \cdot 10^{-09}$	4.890			
580537 BaLF4	335	$3.80 \cdot 10^{-15}$	8.973			
581409 LF5	321	$1.07 \cdot 10^{-17}$	10.125			
586610 LgSK2	346	$3.55 \cdot 10^{-10}$	6.073			
593355 FF5	359	$1.02 \cdot 10^{-19}$	12.678			
613586 SK4	323	$3.36 \cdot 10^{-11}$	6.274	2.70*	$2.75 \cdot 10^{48}$	0.0301
613443 KzFSN4	326	$3.86 \cdot 10^{-10}$	5.693	2.66*	$1.81 \cdot 10^{33}$	0.0204
618551 SSK4	325	$7.12 \cdot 10^{-13}$	7.327			
620364 F2	330	$4.15 \cdot 10^{-17}$	10.027	2.73	$1.20 \cdot 10^{30}$	0.0189
650392 BaSF10	341	$1.35 \cdot 10^{-13}$	8.146			
670472 BaF10	352	$2.90 \cdot 10^{-13}$	8.201			
720504 LaK10	341	$2.54 \cdot 10^{-10}$	6.083			
741526 TaC2	324	$7.63 \cdot 10^{-06}$	3.081			
743492 NbF1	326	$8.34 \cdot 10^{-08}$	4.281			
744447 LaF2	356	$1.83 \cdot 10^{-13}$	8.420			
805254 SF6	367	$4.42 \cdot 10^{-12}$	7.734	2.73	$7.73 \cdot 10^{28}$	0.0182
835430 TaFD5	340	$1.71 \cdot 10^{-09}$	5.542			

*Infrared edge description limited to absorption coefficients $\geq 2\,\text{cm}^{-1}$ and $\leq 10\,\text{cm}^{-1}$.

TABLE 48 Parameters for the Room-temperature Infrared Absorption Edge [Eq. (17a)]

Material	$A\,(\text{cm}^{-1})$	γ	$\nu_0\,(\text{cm}^{-1})$	Ref.
Al_2O_3, sapphire				
o-ray	55,222	5.03	900	35
	94,778	5.28	914	38
e-ray	33,523	4.73	871	377
ALON	41,255	4.96	969	38
BaF_2	49,641	4.5	344	35
BeO	36,550	5.43	1,090	†
$CaCO_3$, calcite				
o-ray	1,633	2.58	1,549	1
e-ray	456	1.88	890	1
CaF_2	105,680	5.1	482	35
$CaLa_2S_4$	3,910	3.58	314	†
CdTe	5,460	4.25	168	378*
CsI	12,800	3.88	85	379*
GaAs	2,985	3.69	292	380*
KBr	6,077	4.25	166	35, 381
KCl	8,696	4.19	213	35, 381
KI	8,180	3.86	139	382*
LiF	21,317	4.39	673	35, 381

TABLE 48 Parameters for the Room-temperature Infrared Absorption Edge [Eq. (17a)] (Continued)

Material	A (cm^{-1})	γ	ν_0 (cm^{-1})	Ref.
$MgAl_2O_4$, spinel	147,850	5.51	869	38,199*
MgF_2	11,213	4.29	617	35
MgO	41,420	5.29	725	383*
NaCl	24,273	4.79	268	35, 381
NaF	41,000	5.0	425	381
SiO_2, Quartz				
o-ray	107,000	4.81	1,215	1
e-ray	196,000	5.16	1,222	1
SrF_2	22,548	4.4	395	35
KRS-5	5,400	4.0	100	384
TlCl	6	1.58	158	384
Y_2O_3	184,456	5.36	620	38
β-ZnS	227,100	5.31	352	385*
ZnSe	179,100	5.33	250	386*
$ZrO_2:Y_2O_3$	226,390	4.73	658	387*
Fused SiO_2	54,540	5.10	1,263	†
ZBT	297,000	4.4	500	388
As_2S_3	4,900	2.93	350	389*
As_2Se_3	15,300	3.62	240	390*
AMTIR-3	10,320	3.13	235	391*

* Our fit to the referenced data.
† Estimated from our measurements.

33.6 REFERENCES

1. W. G. Driscoll (ed.), *Handbook of Optics,* McGraw-Hill, New York, 1978.

2. W. L. Wolfe and G. J. Zissis (eds.), *The Infrared Handbook,* Environmental Research Institute of Michigan, 1985.

3. M. J. Weber (ed.), *Handbook of Laser Science and Technology,* CRC Press, Boca Raton, 1986.

4. D. E. Gray (ed.), *American Institute of Physics Handbook,* 3d ed., McGraw-Hill, New York, 1972.

5. E. D. Palik (ed.), *Handbook of Optical Constants of Solids,* Academic Press, Orlando, 1985.

6. E. D. Palik (ed.), *Handbook of Optical Constants of Solids II,* Academic Press, Orlando, 1991.

7. B. O. Seraphin and H. E. Bennett, "Optical Constants," in R. K. Willardson and A. C. Beer (eds.), *Semiconductors and Semimetals, Vol. 3: Optical Properties of III-V Compounds,* Academic Press, New York, 1967.

8. S. Musikant (ed.), *Optical Materials: A Series of Advances,* Marcel Dekker, New York, 1990.

9. F. V. Tooley (ed.), *The Handbook of Glass Manufacture,* Ashlee Publishing Co., New York, 1985.

10. S. Musikant, *Optical Materials: An Introduction to Selection and Application,* Marcel Dekker, New York, 1985.

11. P. Klocek (ed.), *Handbook of Infrared Optical Materials,* Marcel Dekker, Inc., New York, 1991.

12. G. W. Fynn and W. J. A. Powell, *Cutting and Polishing Optical and Electronic Materials,* 2d ed., Adam Hilger, Bristol, 1988.

13. W. Zschommler, *Precision Optical Glassworking,* SPIE, Bellingham, WA, 1986.

14. J. F. Nye, *Physical Properties of Crystals,* Oxford University Press, Oxford, 1985.

15. E. R. Cohen and B. N. Taylor, "The 1986 CODATA Recommended Values for the Fundamental Physical Constants," *J. Phys. Chem. Ref. Data* **17:**1795–1801 (1988).

16. D. Y. Smith, "Dispersion Theory, Sum Rules, and Their Application to the Analysis of Optical Data," in E. D. Palik (ed.), *Handbook of Optical Constants of Solids,* Academic Press, New York, 1985.

17. T. Skettrup, "Urbach's Rule Derived From Thermal Fluctuations in the Band Gap Energy," *Phys. Rev. B* **18:**2622–2631 (1978).

18. F. K. Kneubühl, "Review of the Theory of the Dielectric Dispersion of Insulators," *Infrared Phys.* **29:**925–942 (1989).

19. W. J. Tropf, T. J. Harris, and M. E. Thomas, "Optical Materials: Visible and Infrared," in R. Waynant and W. Ediger (eds), *Electro-optics Handbook,* McGraw-Hill, New York, 1993.

20. R. H. Lyddane, R. G. Sachs, and E. Teller, "On the Polar Vibrations of Alkali Halides," *Phys. Rev.* **59:**673–676 (1941).

21. W. Cochran and R. A. Cowley, "Dielectric Constants and Lattice Vibrations," *J. Phys. Chem. Solids* **23:**447–450 (1962).

22. F. Gervais and B. Piriou, "Temperature Dependence of Transverse- and Longitudinal-optic Modes in TiO_2 (Rutile)," *Phys. Rev. B* **10:**1642–1654 (1974).

23. J. W. S. Rayleigh, "The Theory of Anomalous Dispersion," *Phil. Mag.* **48:**151–152 (1889).

24. L. Merten and G. Lamprecht, "Directional Dependence of Extraordinary Infrared Oscillator Parameters of Uniaxial Crystals," *Phys. Stat. Solidi (b)* **39:**573–580 (1970). [Also see J. L. Servoin, F. Gervais, A. M. Quittet, and Y. Luspin, "Infrared and Raman Response in Ferroelectric Perovskite Crystals: Apparent Inconsistencies," *Phys. Rev. B.* **21:**2038–2041 (1980).]

25. A. S. Barker, "Infrared Lattice Vibrations and Dielectric Dispersion in Corundum," *Phys. Rev.* **132:**1474–1481 (1963).

26. F. Gervais and B. Piriou, "Anharmonicity in Several-polar-mode Crystals: Adjusting Phonon Self-energy of LO and TO Modes in Al_2O_3 and TiO_2 to Fit Infrared Reflectivity," *J. Phys. C* **7:**2374–2386 (1974).

27. R. H. French, "Electronic Band Structure of Al_2O_3 with Comparision to ALON and AlN," *J. Am. Ceram. Soc.* **73:**477–489 (1990).

28. F. Urbach, "The Long-Wavelength Edge of Photographic Sensitivity and the Electronic Absorption of Solids," *Phys. Rev.* **92:**1324 (1953).

29. H. Sumi and A. Sumi, "The Urbach–Martienssen Rule Revisited," *J. Phys. Soc. Japan* **56:**2211–2220 (1987).

30. D. L. Wood and J. Tauc, "Weak Absorption Tails in Amorphous Semiconductors," *Phys. Rev. B* **5:**3144–3151 (1972).

31. H. Mori and T. Izawa, "A New Loss Mechanism in Ultralow Loss Optical Fiber Materials," *J. Appl. Phys.* **51:**2270–2271 (1980).

32. M. E. Innocenzi, R. T. Swimm, M. Bass, R. H. French, A. B. Villaverde, and M. R. Kokta, "Room-temperature Optical Absorption in Undoped α-Al_2O_3," *J. Appl. Phys.* **67:**7542–7546 (1990).

33. D. T. Gillespie, A. L. Olsen, and L. W. Nichols, "Transmittance of Optical Materials at High Temperature," *Appl. Opt.* **4:**1488–1493 (1965).

34. M. A. Ordal, R. J. Bell, R. W. Alexander, L. L. Long, and M. R. Querry, "Optical Properties of Fourteen Metals in the Infrared and Far Infrared: Al, Co, Cu, Au, Fe, Pb, Mo, Ni, Pd, Pt, Ag, Ti, V, and W," *Appl. Opt.* **24:**4493–4499 (1985).

35. T. F. Deutsch, "Absorption Coefficient of Infrared Laser Window Materials," *J. Phys. Chem. Solids* **34:**2091–2104 (1973). [Also see T. F. Deutsch, "Laser Window Materials—An Overview," *J. Electron. Mater.* **4:**663–719 (1975).]

36. D. L. Mills and A. A. Maradudin, "Theory of Infrared Absorption by Crystals in the High Frequency Wing of their Fundamental Lattice Absorption," *Phys. Rev. B* **8**:1617–1630 (1973). [Also see A. A. Maradudin and D. L. Mills, "Temperature Dependence of the Absorption Coefficient of Alkali Halides in the Multiphonon Regime," *Phys. Rev. Lett.* **31**:718–721 (1973).]

37. H. G. Lipson, B. Bendow, N. E. Massa, and S. S. Mitra, "Multiphonon Infrared Absorption in the Transparent Regime of Alkaline-earth Fluorides," *Phys. Rev. B* **13**:2614–2619 (1976).

38. M. E. Thomas, R. I. Joseph, and W. J. Tropf, "Infrared Transmission Properties of Sapphire, Spinel, Yttria, and ALON as a Function of Frequency and Temperature," *Appl. Opt.* **27**:239–245 (1988).

39. J. A. Harrington, B. L. Bobbs, M. Braunstein, R. K. Kim, R. Stearns, and R. Braunstein, "Ultraviolet-visible Absorption in Highly Transparent Solids by Laser Calorimetry and Wavelength Modulation Spectroscopy," *Appl. Opt.* **17**:1541–1546 (1978).

40. N. C. Frenelius, R. J. Harris, D. B. O'Quinn, M. E. Gangl, D. V. Dempsey, and W. L. Knecht, "Some Optical Properties of Materials Measured at 1.3 μm," *Opt. Eng.* **22**:411–418 (1983).

41. J. A. Harrington, D. A. Gregory, and W. F. Ott, "Infrared Absorption in Chemical Laser Window Materials," *Appl. Opt.* **15**:1953–1959 (1976).

42. B. Tatian, "Fitting Refractive-index Data with the Sellmeier Dispersion Formula," *Appl. Opt.* **23**:4477–4485 (1984).

43. P. G. Nutting, "Disperson Formulas Applicable to Glass," *J. Opt. Soc. Am.* **2–3**:61–65 (1919).

44. H. H. Li, "Refractive Index of ZnS, ZnSe, and ZnTe and Its Wavelength and Temperature Derivatives," *J. Phys. Chem. Ref. Data* **13**:103–150 (1984).

45. F. Zernike, "Refractive Indices of Ammonium Dihydrogen Phosphate and Potassium Dihydrogen Phosphate between 2000 Å and 1.5 μ," *J. Opt. Soc. Am.* **54**:1215–1220 (1964). [Errata: *J. Opt. Soc. Am.* **55**:210E (1965).]

46. M. Herzberger, "Colour Correction in Optical Systems and a New Dispersion Formula," *Opt. Acta* **6**:197–215 (1959).

47. M. Herzberger and C. D. Salzberg, "Refractive Indices of Infrared Optical Materials and Color Correction of Infrared Lenses," *J. Opt. Soc. Am.* **52**:420–427 (1962).

48. A. N. Pikhtin and A. D. Yas'kov, "Dispersion of the Refractive Index of Semiconductors with Diamond and Zinc-blende Structures," *Sov. Phys. Semicond.* **12**:622–626 (1978).

49. J. W. Fleming, "Dispersion in GeO_2-SiO_2 Glasses," *Appl. Opt.* **23**:4486–4493 (1984).

50. D. L. Wood, K. Nassau, and T. Y. Kometani, "Refractive Index of Y_2O_3 Stabilized Zirconia: Variation with Composition and Wavelength," *Appl. Opt.* **29**:2485–2488 (1990).

51. N. P. Barnes and M. S. Piltch, "Temperature-dependent Sellmeier Coefficients and Nonlinear Optics Average Power Limit for Germanium," *J. Opt. Soc. Am.* **69**:178–180 (1979).

52. A. J. Bosman and E. E. Havinga, "Temperature Dependence of Dielectric Constants of Cubic Ionic Compounds," *Phys. Rev.* **129**:1593–1600 (1963).

53. B. Bendow, P. D. Gianino, Y-F. Tsay, and S. S. Mitra, "Pressure and Stress Dependence of the Refractive Index of Transparent Crystals," *Appl. Opt.* **13**:2382–2396 (1974).

54. S. Singh, "Nonlinear Optical Properties," in *Handbook of Laser Science and Technology, Volume III Optical Materials: Part I,* CRC Press, Boca Raton, 1986, pp. 3–228.

55. D. F. Eaton, "Nonlinear Optical Materials," *Science* **253**:281–287 (1991).

56. M. M. Choy and R. L. Byer, "Accurate Second-order Susceptibility Measurements of Visible and Infrared Nonlinear Crystals," *Phys. Rev. B* **14**:1693–1706 (1976).

57. K.-H. Hellwege and A. M. Hellwege (eds.), *Landolt-Börnstein Numerical Data and Functional Relationships in Science and Technology, New Series, Group III: Crystal and Solid State Physics, Volume 11: Elastic, Piezoelectric, Pyroelectric, Piezooptic, Electrooptic Constants and Nonlinear Susceptibilities of Crystals,* Springer-Verlag, Berlin, 1979.

58. M. E. Lines, "Scattering Loss in Optic Fiber Materials. I. A New Parametrization," *J. Appl. Phys.* **55**:4052–4057 (1984); "II. Numerical Estimates," *J. Appl. Phys.* **55**:4058–4063 (1984).

59. J. A. Harrington and M. Sparks, "Inverse-square Wavelength Dependence of Attenuation in Infrared Polycrystalline Fibers," *Opt. Lett.* **8**:223–225 (1983).

60. D. D. Duncan and C. H. Lange, "Imaging Performance of Crystalline and Polycrystalline Oxides," *Proc. SPIE* **1326:**59–70 (1990).

61. J. D. H. Donnay and H. M. Ondik (eds.), *Crystal Data Determination Tables,* 3d ed., U. S. Department of Commerce, 1973.

62. R. W. G. Wyckoff, *Crystal Structures,* John Wiley & Sons, New York, 1963.

63. G. Simmons and H. Wang, *Single Crystal Elastic Constants and Calculated Aggregate Properties: A Handbook,* MIT Press, Cambridge, MA, 1971.

64. R. S. Krishnan, R. Srinivasan, and S. Devanarayanan, *Thermal Expansion of Crystals,* Pergamon Press, Oxford, 1979.

65. Y. S. Touloukian (ed.), *Thermophysical Properties of Matter,* IFI/Plenum, New York, 1970.

66. G. A. Slack, "The Thermal Conductivity of Nonmetallic Crystals," in H. Ehrenreich, F. Seitz, and D. Turnbull (eds.), *Solid State Physics, Vol. 34,* Academic Press, New York, 1979.

67. K. F. Loje and D. E. Schuele, "The Pressure and Temperature Derivative of the Elastic Constants of AgBr and AgCl," *J. Phys. Chem. Solids* **31:**2051–2067 (1970).

68. W. Hidshaw, J. T. Lewis, and C. V. Briscoe, "Elastic Constants of Silver Chloride from 4.2 to 300 K," *Phys. Rev.* **163:**876–881 (1967).

69. S. M. Kikkarin, A. V. Tsarev, V. V. Shashkin, and I. B. Yakovkin, "Elastic Properties of GaAlAs Solid Solutions," *Sov. Phys. Solid State* **30:**1689–1692 (1988). [Also see N. Chetty, A. Muñoz, and R. M. Martin, "First-principles Calculation of the Elastic Constants of AlAs," *Phys. Rev. B* **40:**11934–11936 (1989).]

70. Stiffness and compliance of ALON and yttria are estimated from the engineering moduli.

71. M. Gospodinov, P. Sveshtarov, N. Petkov, T. Milenov, V. Tassev, and A. Nikolov, "Growth of Large Crystals of Bismuth-Germanium Oxide and Their Physical Properties," *Bulgarian J. Phys.* **16:**520–522 (1989).

72. M. Gospodinov, S. Haussühl, P. Sveshtarov, V. Tassev, and N. Petkov, "Physical Properties of Cubic $Bi_{12}SiO_{20}$," *Bulgarian J. Phys.* **15:**140–143 (1988).

73. V. A. Pesin, "Elastic Constants of Dense Modification of Boron Nitride," *Sverkhtverd. Mater.* **6:**5–7 (1980) [in Russian].

74. W. Wettling and J. Windscheif, "Elastic Constants and Refractive Index of Boron Phosphide," *Solid State Commun.* **50:**33–34 (1984).

75. M. E. Hills, "Preparation, Properties, and Development of Calcium Lanthanum Sulfide as an 8- to 12-micrometer Transmitting Ceramic," Naval Weapons Center Report TP 7073, September 1989. [Engineering moduli used to estimate elastic constants.]

76. R. D. Greenough and S. B. Palmer, "The Elastic Constants and Thermal Expansion of Single-crystal CdTe," *J. Phys. D* **6:**587–592 (1973).

77. R. C. Hanson and K. Helliwell, "Anharmonicity in CuCl—Elastic, Dielectric, and Piezoelectric Constants," *Phys. Rev. B* **9:**2649–2654 (1984).

78. Yu. A. Burenkov, S. Yu. Davydov, and S. P. Nikanorov, "The Elastic Properties of Indium Arsenide," *Sov. Phys. Solid State* **17:**1446–1447 (1976).

79. A. Yoneda, "Pressure Derivatives of Elastic Constants of Single Crystal MgO and $MgAl_2O_4$," *J. Phys. Earth* **38:**19–55 (1990).

80. M. O. Manasrch and D. O. Pederson, "Elastic Constants of Cubic Lead Fluoride from 300 to 850 K," *Phys. Rev. B* **30:**3482–3485 (1984).

81. G. I. Peresada, E. G. Ponyatovskii, and Zh. D. Sokolovskaya, "Pressure Dependence of the Elastic Constants of PbS," *Phys. Stat. Sol. (a)* **35:**K177–K179 (1976).

82. N. J. Walker, G. A. Saunders, and N. Schäl, "Acoustic Mode Vibrational Anharmonicity of PbSe and other IV–VI Compounds," *J. Phys. Chem. Solids* **48:**91–96 (1987).

83. A. J. Miller, G. A. Saunders, and Y. K. Yoğurtçu, "Pressure Dependence of the Elastic Constants of PbTe, SnTe and $Ge_{0.08}Sn_{0.92}Te$," *J. Phys. C* **14:**1569–1584 (1981).

84. W. R. L. Lambrecht, B. Segall, M. Methfessel, and M. van Schilfgaarde, "Calculated Elastic Constants and Deformation Potentials of Cubic SiC," *Phys. Rev. B* **44**:3685–3694 (1991).

85. J. B. Wachtman, M. L. Wheat, and S. Marzullo, "A Method for Determining the Elastic Constants of a Cubic Crystal from Velocity Measurements in a Single Arbitrary Direction; Application to $SrTiO_3$," *J. Res. Nat. Bur. Stand.* **67A**:193–204 (1963).

86. Y. K. Yoğurtçu, A. J. Miller, and G. A. Saunders, "Elastic Behavior of YAG Under Pressure," *J. Phys. C* **13**:6585–6597 (1980).

87. H. M. Kandil, J. D. Greiner, and J. F. Smith, "Single Crystal Elastic Constants of Yttria-stabilized Zirconia in the Range 20° to 700°C," *J. Am. Ceram. Soc.* **67**:341–346 (1984).

88. M. H. Grimsditch and G. D. Holah, "Brillouin Scattering and Elastic Moduli of Silver Thiogallate ($AgGaS_2$)," *Phys. Rev. B* **12**:4377–4382 (1975).

89. Z. Li, S.-K. Chan, M. H. Grimsditch, and E. S. Zouboulis, "The Elastic and Electromechanical Properties of Tetragonal $BaTiO_3$ Single Crystals," *J. Appl. Phys.* **70**:7327–7332 (1991).

90. T. Hailing, G. A. Saunders, W. A. Lambson, and R. S. Feigelson, "Elastic Behavior of the Chalcopyrite $CdGeAs_2$," *J. Phys. C* **15**:1399–1418 (1982).

91. G. Fischer and J. Zarembowitch, "Elastic Properties of Single-crystal Urea," *C. R. Acad. Sc. Paris* **270B**:852–855 (1970).

92. Y. Ohmachi and N. Uchida, "Temperature Dependence of Elastic, Dielectric, and Piezoelectric Constants in TeO_2 Single Crystals," *J. Appl. Phys.* **41**:2307–2311 (1970).

93. I. M. Silvestrova, Y. V. Pisarevskii, P. A. Senyushenkov, A. I. Krupny, R. Voszka, I. F. Földvári, and J. Janszky, "Temperature Dependence of the Elastic Constants of Paratellurite," *Phys. Stat. Sol. (a)* **101**:437–444 (1987).

94. W. J. Alton and A. J. Barlow, "Acoustic-wave Propagation in Tetragonal Crystals and Measurements of the Elastic Constants of Calcium Molybdate," *J. Appl. Phys.* **38**:3817–3820 (1967).

95. J. M. Farley and G. A. Saunders, "Ultrasonic Study of the Elastic Behavior of Calcium Tungstate between 1.5 K and 300 K," *J. Phys. C* **5**:3021–3037 (1972). [Also see J. M. Farley and G. A. Saunders, "The Elastic Constants of $CaWO_4$," *Solid State Commun.* **9**:965–969 (1971).]

96. J. M. Farley, G. A. Saunders, and D. Y. Chung, "Elastic Properties of Scheelite Structure Molybdates and Tungstates," *J. Phys. C* **8**:780–786 (1975).

97. J. M. Farley, G. A. Saunders, and D. Y. Chung, "Elastic Properties of Strontium Molybdate," *J. Phys. C* **6**:2010–2019 (1973).

98. P. Blanchfield and G. A. Saunders, "The Elastic Constants and Acoustic Symmetry of $LiYF_4$," *J. Phys. C* **12**:4673–4689 (1979). [Also see P. Blanchfield, T. Hailing, A. J. Miller, G. A. Saunders, and B. Chapman, "Vibrational Anharmonicity of Oxide and Fluoride Scheelites," *J. Phys. C* **20**:3851–3859 (1983).]

99. T. A. Fjeldly and R. C. Hanson, "Elastic and Piezoelectric Constants of Silver-iodide: Study of a Material Covalent-ionic Phase Transition," *Phys. Rev. B* **10**:3569–3577 (1974).

100. K. Tsubouchi and N. Mikoshiba, "Zero-Temperature-Coefficient SAW Devices on AlN Epitaxial Films," *IEEE Trans. Sonics Ultrason.* **SU-32**:634–644 (1985).

101. V. A. Savastenko and A. U. Sheleg, "Study of the Elastic Properties of Gallium Nitride," *Phys. Stat. Sol. (a)* **48**:K135–K139 (1978).

102. S. Haussühl, "The Propagation of Elastic Waves in Hexagonal Lithium Iodate," *Acustica* **23**:165–169 (1970) [In German].

103. I. I. Zubrinov, V. I. Semenov, and D. V. Sheloput, "Elastic and Photoelastic Properties of Proustite," *Sov. Phys. Solid State* **15**:1921–1922 (1974).

104. D. Eimerl, L. Davis, and S. Velsko, "Optical, Mechanical, and Thermal Properties of Barium Borate," *J. Appl. Phys.* **62**:1968–1983 (1987).

105. R. Laiho, M. Lakkisto, and T. Levola, "Brillouin Scattering Investigation of the Elastic Properties of LaF_3, CeF_3, PrF_3, and NdF_3," *Phil. Mag. A* **47**:235–244 (1983).

106. D. Royer and E. Dieulesaint, "Elastic and Piezoelectric Constants of Trigonal Selenium and Tellurium Crystals," *J. Appl. Phys.* **50**:4042–4045 (1979).

107. A. G. Kalinichev, J. D. Bass, C. S. Zha, P. D. Han, and D. A. Payne, "Elastic Properties of Orthorhombic $KNbO_3$ Single Crystals by Brillouin Scattering," *J. Appl. Phys.* **74**:6603–6608 (1993).

108. V. V. Aleksandrov, T. S. Velichkina, V. I. Voronkona, L. V. Koltsova, I. A. Yakovlev, and V. K. Yanovskii, "Elastic Coefficients of $KTiOPO_4$, $RbTiOPO_4$, $TlTiOPO_4$ Crystals Determined from Mandelstamm-Brillouin Light Scattering Spectra," *Solid State Commun.* **69**:877–881 (1989).

109. Y. Tsay, B. Bendow, and S. S. Mitra, "Theory of the Temperature Derivative of the Refractive Index in Transparent Crystals," *Phys. Rev. B* **5**:2688–2696 (1972).

110. L. W. Tilton, E. K. Plyler, and R. E. Stephens, "Refractive Index of Silver Chloride for Visible and Infra-red Radiant Energy," *J. Opt. Soc. Am.* **40**:540–543 (1950).

111. G. C. Bhar, D. K. Ghosh, P. S. Ghosh, and D. Schmitt, "Temperature Effects in $AgGaS_2$ Nonlinear Devices," *Appl. Opt.* **22**:2492–2494 (1983).

112. K. Vedam, J. L. Kirk, and B. N. N. Achar, "Piezo- and Thermo-Optic Behavior of Spinel ($MgAl_2O_4$)," *J. Solid State Chem.* **12**:213–218 (1975).

113. A. Feldman, D. Horowitz, R. M. Walker, and M. J. Dodge, "Optical Materials Characterization Final Technical Report, February 1, 1978–September 30, 1978," NBS Technical Note 993, February 1979.

114. J. Tapping and M. L. Reilly, "Index of Refraction of Sapphire Between 24 and 1060°C for Wavelengths of 633 and 799 nm," *J. Opt. Soc. Am. A* **3**:610–616 (1986).

115. C. H. Lange and D. D. Duncan, "Temperature Coefficient of Refractive Index for Candidate Optical Windows," *SPIE Proc.* **1326**:71–78 (1990).

116. H. W. Newkirk, D. K. Smith, and J. S. Kahn, "Synthetic Bromellite. III. Some Optical Properties," *Am. Mineralogist* **51**:141–151 (1966).

117. K. Vedam and P. Hennessey, "Piezo- and Thermo-optical Properties of $Bi_{12}GeO_{20}$, II. Refractive Index," *J. Opt. Soc. Am.* **65**:442–445 (1975).

118. G. N. Ramachandran, "Thermo-optic Behavior of Solids, I. Diamond," *Proc. Ind. Acad. Sci.* **A25**:266–279 (1947).

119. J. Fontanella, R. L. Johnston, J. H. Colwell, and C. Andeen, "Temperature and Pressure Variation of the Refractive Index of Diamond," *Appl. Opt.* **16**:2949–2951 (1977).

120. I. H. Malitson, "A Redetermination of Some Optical Properties of Calcium Fluoride," *Appl. Opt.* **2**:1103–1107 (1963).

121. M. J. Dodge, "Refractive Index," in M. J. Weber (ed.), *Handbook of Laser Science and Technology, Volume IV, Optical Material: Part 2*, CRC Press, Boca Raton, 1986.

122. T. W. Houston, L. F. Johnson, P. Kisiuk, and D. J. Walsh, "Temperature Dependence of the Refractive Index of Optical Maser Crystals," *J. Opt. Soc. Am.* **53**:1286–1291 (1963).

123. R. Weil and D. Neshmit, "Temperature Coefficient of the Indices of Refraction and the Birefringence in Cadmium Sulfide," *J. Opt. Soc. Am.* **67**:190–195 (1977).

124. R. J. Harris, G. T. Johnson, G. A. Kepple, P. C, Krok, and H. Mukai, "Infrared Thermooptic Coefficient Measurement of Polycrystalline ZnSe, ZnS, CdTe, CaF_2 and BaF_2, Single Crystal KCl, and TI-20 Glass," *Appl. Opt.* **16**:436–438 (1977).

125. H. H. Li, "Refractive Index of Alkali Halides and Its Wavelength and Temperature Derivatives," *J. Phys. Chem. Ref. Data* **5**:329–528 (1976).

126. W. S. Rodney, "Optical Properties of Cesium Iodide," *J. Opt. Soc. Am.* **45**:987–992 (1955).

127. G. D. Boyd, H. Kasper, and J. H. McFee, "Linear and Nonlinear Optical Properties of $AgGaS_2$, $CuGaS_2$, and $CuInS_2$, and Theory of the Wedge Technique for the Measurement of Nonlinear Coefficients," *IEEE J. Quantum Electr.* **7**:563–573 (1971).

128. G. C. Bhar and G. Ghosh, "Temperature-dependent Sellmeier Coefficients and Coherence Lengths for some Chalcopyrite Crystals," *J. Opt. Soc. Am.* **69**:730–733 (1979). [Also see G. C. Bhar, "Refractive Index Dispersion of Chalcopyrite Crystals," *J. Phys. D* **13**:455–460 (1980).]

129. M. Bertolotti, V. Bogdanov, A. Ferrari, A. Jascow, N. Nazorova, A. Pikhtin, and L. Schirone, "Temperature Dependence of the Refractive Index in Semiconductors," *J. Opt. Soc. Am. B* **7:**918–922 (1990).

130. D. A. Yas'kov and A. N. Pikhtin, "Optical Properties of Gallium Phosphide Grown by Float Zone. I. Refractive Index and Reflection Coefficient," *Mat. Res. Bull.* **4:**781–788 (1969). [Also see D. A. Yas'kov and A. N. Pikhtin, "Dispersion of the Index of Refraction of Gallium Phosphide," *Sov. Phys. Sol. State* **9:**107–110 (1967).]

131. H. H. Li, "Refractive Index of Silicon and Germanium and Its Wavelength and Temperature Derivatives," *J. Phys. Chem. Ref. Data* **9:**561–658 (1980).

132. D. L. Wood, K. Nassau, T. Y. Kometani, and D. L. Nash, "Optical Properties of Cubic Hafnia Stabilized with Yttria," *Appl. Opt.* **29:**604–607 (1990).

133. C. S. Hoefer, "Thermal Variations of the Refractive Index in Optical Materials," *Proc. SPIE* **681:**135–142 (1986).

134. B. Zysset, I. Biaggio, and P. Günter, "Refractive Indices of Orthorhombic KNbO$_3$. I. Dispersion and Temperature Dependence," *J. Opt. Soc. Am. B* **9:**380–386 (1992). [Also see Y. Uematsu, "Nonlinear Optical Properties of KNbO$_3$ Single Crystal in the Orthorhombic Phase," *Jap. J. Appl. Phys.* **13:**1362–1368 (1974).]

135. J. D. Bierlein and H. Vanherzeele, "Potassium Titanyl Phosphate: Properties and New Applications," *J. Opt. Soc. Am. B* **6:**622–633 (1989).

136. S. P. Velsko, M. Webb, L. Davis, and C. Huang, "Phase-Matched Harmonic Generation in Lithium Triborate (LBO)," *IEEE J. Quantum Electron.* **27:**2182–2192 (1991).

137. D. S. Smith, H. D. Riccius, and R. P. Edwin, "Refractive Indices of Lithium Niobate," *Optics Commun.* **17:**332–335 (1976).

138. N. P. Barnes and D. J. Gettemy, "Temperature Variation of the Refractive Indices of Yttrium Lithium Fluoride," *J. Opt. Soc. Am.* **70:**1244–1247 (1980).

139. R. E. Stephens and I. H. Malitson, "Index of Refraction of Magnesium Oxide," *J. Nat. Bur. Stand.* **49:**249–252 (1952).

140. T. Radhakrishnan, "Further Studies on the Temperature Variation of the Refractive Index of Crystals," *Proc. Indian Acad. Sci.* **A33:**22–34 (1951).

141. J. N. Zemel, J. D. Jensen, and R. B. Schoolar, "Electrical and Optical Properties of Epitaxial Films of PbS, PbSe, PbTe, and SnTe," *Phys. Rev.* **140A:**330–342 (1965).

142. N. Uchida, "Optical Properties of Single Crystal Paratellurite (TeO$_2$)," *Phys. Rev. B* **4:**3736–3745 (1971).

143. J. R. DeVore, "Refractive Index of Rutile and Sphalerite," *J. Opt. Soc. Am.* **41:**416–419 (1951).

144. M. D. Ewbank, P. R. Newman, N. L. Mota, S. M. Lee, W. L. Wolfe, A. G. DeBell, and W. A. Harrison, "The Temperature Dependence of Optical and Mechanical Properties of Tl$_3$AsSe$_3$," *J. Appl. Phys.* **51:**3848–3852 (1980). [Also see J. D. Feichtner and G. W. Roland, "Optical Properties of a New Nonlinear Optical Material: Tl$_3$AsSe$_3$," *Appl. Opt.* **11:**993–998 (1972).]

145. W. S. Rodney and I. H. Malitson, "Refraction and Dispersion of Thallium Bromide Iodide," *J. Opt. Soc. Am.* **46:**956–961 (1956).

146. L. G. DeShazer, S. C. Rand, B. A. Wechsler, "Laser Crystals," in M. J. Weber (ed.), *Handbook of Laser Science and Technology, Volume V, Optical Material: Part 3,* CRC Press, Boca Raton, 1986.

147. G. D. Boyd, E. Beuhler, and F. G. Storz, "Linear and Nonlinear Optical Properties of ZnGeP$_2$ and CdSe," *Appl. Phys. Lett.* **18:**301–304 (1971).

148. D. L. Wood and K. Nassau, "Refractive Index of Cubic Zirconia Stabilized with Yttria," *Appl. Opt.* **21:**2978–2981 (1982).

149. Schott Glass Technologies, Duryea, Pa.

150. Ohara Corporation, Somerville, N.J.

151. Hoya Optics, Inc., Fremont, Calif.

152. W. S. Rodney and R. J. Spindler, "Index of Refraction of Fused-quartz Glass for Ultraviolet, Visible, and Infrared Wavelengths," *J. Res. Nat. Bur. Stand.* **53**:185–189 (1954). [Also see I. H. Malitson, "Interspecimen Comparison of the Refractive Index of Fused Silica," *J. Opt. Soc. Am.* **55**:1205–1209 (1965).]

153. J. W. Fleming, "Disperson in GeO_2–SiO_2 Glasses," *Appl. Opt.* **23**:4486–4493 (1984).

154. Barr & Stroud, Ltd., Glasgow, Scotland (UK).

155. Corning, Inc., Corning, N.Y. [Also see W. H. Dumbaugh, "Infrared Transmitting Germanate Glasses," *Proc. SPIE* **297**:80–85 (1981).]

156. C. T. Moynihan, M. G. Drexhage, B. Bendow, M. S. Boulos, K. P. Quinlan, K. H. Chung, and E. Gbogi, "Composition Dependence of Infrared Edge Absorption in ZrF_4 and HfF_4 Based Glasses," *Mat. Res. Bull.* **16**:25–30 (1981).

157. R. N. Brown, B. Bendow, M. G. Drexhage, and C. T. Moynihan, "Ultraviolet Absorption Edge Studies of Fluorozirconate and Fluorohafnate Glass," *Appl. Opt.* **21**:361–363 (1982).

158. R. N. Brown and J. J. Hutta, "Material Dispersion in High Optical Quality Heavy Metal Fluoride Glasses," *Appl. Opt.* **24**:4500–4503 (1985).

159. J. M. Jewell, C. Askins, and I. D. Aggarwal, "Interferometric Method for Concurrent Measurement of Thermo-optic and Thermal Expansion Coefficients," *Appl. Opt.* **30**:3656–3660 (1991).

160. M. G. Drexhage, C. T. Moynihan, and M. Saleh, "Infrared Transmitting Glasses Based on Hafnium Fluoride," *Mat. Res. Bull.* **15**:213–219 (1980).

161. B. Bendow, M. G. Drexhage, and H. G. Lipson, "Infrared Absorption in Highly Transparent Fluorozirconate Glass," *J. Appl. Phys.* **52**:1460–1461 (1981).

162. M. G. Drexhage, O. H. El-Bayoumi, C. T. Moynihan, A. J. Bruce, K.-H. Chung, D. L. Gavin, and T. J. Loretz, "Preparation and Properties of Heavy-Metal Fluoride Glasses Containing Ytterbium or Lutetium," *J. Am. Ceram. Soc.* **65**:C168–C171 (1982).

163. W. S. Rodney, I. H. Malitson, and T. A. King, "Refractive Index of Arsenic Trisulfide," *J. Opt. Soc. Am.* **48**:633–636 (1958).

164. A. R. Hilton and C. E. Jones, "The Thermal Change in the Nondispersive Infrared Refractive Index of Optical Materials," *Appl. Opt.* **6**:1513–1517 (1967).

165. Y. Ohmachi, "Refractive Index of Vitreous As_2Se_3," *J. Opt. Soc. Am.* **63**:630–631 (1973).

166. J. A. Savage, "Optical Properties of Chalcogenide Glasses," *J. Non-Cryst. Solids* **47**:101–116 (1982).

167. Amorphous Materials, Inc., Garland, Tex.

168. K. F. Hulme, O. Jones, P. H. Davies, and M. V. Hobden, "Synthetic Proustite (Ag_3AsS_3): A New Crystal for Optical Mixing," *Appl. Phys. Lett.* **10**:133–135 (1967).

169. H. Schröter, "On the Refractive Indices of Some Heavy-Metal Halides in the Visible and Calculation of Interpolation Formulas for Dispersion," *Z. Phys.* **67**:24–36 (1931) [in German].

170. G. C. Bhar, "Refractive Index Interpolation in Phase-matching," *Appl. Opt.* **15**:305–307 (1976).

171. B. F. Levine, W. A. Nordland, and J. W. Silver, "Nonlinear Optical Susceptibility of AgI," *IEEE J. Quantum Electron.* **QE-9**:468–470 (1973).

172. R. E. Fern and A. Onton, "Refractive Index of AlAs," *J. Appl. Phys.* **42**:3499–3500 (1971).

173. J. Pastrňák and L. Roskovcová, "Refractive Index Measurements on AlN Single Crystals," *Phys. Stat. Sol.* **14**:K5–K8 (1966). [Infrared term from D. A. Yas'kov and A. N. Pikhtin, "Refractive Index and Birefringence or Semiconductors with the Wurtzite Structure," *Sov. Phys. Semicond.* **15**:8–12 (1981).]

174. I. H. Malitson and M. J. Dodge, "Refractive Index and Birefringence of Synthetic Sapphire," *J. Opt. Soc. Am.* **62**:1405A (1972). [Also see M. J. Dodge, "Refractive Index," in *Handbook of Laser Science and Technology, Volume IV, Optical Materials: Part 2,* CRC Press, Boca Raton, 1986, p. 30.]

175. W. J. Tropf and M. E. Thomas, "Aluminum Oxynitride (ALON) Spinel," in E. D. Palik (ed.), *Handbook of Optical Constants of Solids II,* Academic Press, Orlando, 1991, pp. 775–785.

176. I. H. Malitson, "Refractive Properties of Barium Fluoride," *J. Opt. Soc. Am.* **54:**628–632 (1964).

177. S. H. Wemple, M. Didomenico, and I. Camlibel, "Dielectric and Optical Properties of Melt-Grown $BaTiO_3$," *J. Phys. Chem. Solids* **29:**1797–1803 (1968).

178. D. F. Edwards and R. H. White, "Beryllium Oxide," in Palik (ed.), *Handbook of Optical Constants of Solids II,* Academic Press, Orlando, 1991, pp. 805–814. [Our fit to the dispersion data.]

179. R. E. Aldrich, S. O. Hou, and M. L. Harvill, "Electrical and Optical Properties of $Bi_{12}SiO_{20}$," *J. Appl. Phys.* **42:**493–494 (1971).

180. E. Burattini, G. Cappuccio, M. Grandolfo, P. Vecchia, and Sh. M. Efendiev, "Near-infrared Refractive Index of Bismuth Germanium Oxide ($Bi_{12}GeO_{20}$)," *J. Opt. Soc. Am.* **73:**495–497 (1983).

181. P. J. Gielisse, S. S. Mitra, J. N. Plendl, R. D. Griffis, L. C. Mansur, R. Marshall, and E. A. Oascoe, "Lattice Infrared Spectra of Boron Nitride and Boron Monophosphide," *Phys. Rev.* **155:**1039–1046 (1967).

182. F. Peter, "In Refractive Indices and Absorption Coefficients of Diamond between 644 and 226 Micrometers," *Z. Phys.* **15:**358–368 (1923) [In German].

183. W. L. Bond, "Measurement of the Refractive Index of Several Crystals," *J. Appl. Phys.* **36:**1674–1677 (1965). [Our fit to the dispersion data.]

184. A. G. DeBell, E. L. Dereniak, J. Harvey, J. Nissley, J. Palmer, A. Selvarajan, and W. L. Wolfe, "Cryogenic Refractive Indices and Temperature Coefficients of Cadmium Telluride from 6 µm to 22 µm," *Appl. Opt.* **18:**3114–3115 (1979).

185. W. S. Rodney and R. J. Spindler, "Refractive Index of Cesium Bromide for Ultraviolet, Visible, and Infrared Wavelengths," *J. Res. Nat. Bur. Stand.* **51:**123–126 (1953).

186. A. Feldman and D. Horowitz, "Refractive Index of Cuprous Chloride," *J. Opt. Soc. Am.* **59:**1406–1408 (1969).

187. A. H. Kachare, W. G. Spitzer, and J. E. Fredrickson, "Refractive Index of Ion-implanted GaAs," *J. Appl. Phys.* **47:**4209–4212 (1976).

188. A. S. Barker and M. Ilegems, "Infrared Lattice Vibrations and Free-electron Dispersion in GaN," *Phys. Rev. B* **7:**743–750 (1973).

189. D. F. Parsons and P. D. Coleman, "Far Infrared Optical Constants of Gallium Phosphide," *Appl. Opt.* **10:**1683–1685 (1971).

190. N. P. Barnes and M. S. Piltch, "Temperature-dependent Sellmeier Coefficients and Nonlinear Optics Average Power Limit for Germanium," *J. Opt. Soc. Am.* **69:**178–180 (1979). [Also see H. W. Icenogle, B. C. Platt, and W. L. Wolfe, "Refractive Indexes and Temperature Coefficients of Germanium and Silicon," *Appl. Opt.* **15:**2348–2351 (1976).]

191. O. G. Lorimor and W. G. Spitzer, "Infrared Refractive Index and Absorption of InAs and CdTe," *J. Appl. Phys.* **36:**1841–1844 (1965).

192. G. D. Pettit and W. J. Turner, "Refractive Index of InP," *J. Appl. Phys.* **36:**2081 (1965). [Infrared term comes from A. N. Pikhtin and A. D. Yas'kov, "Disperson of the Refractive Index of Semiconductors with Diamond and Zinc-blende Structures," *Sov. Phys. Semicond.* **12:**622–626 (1978).]

193. K. W. Kirby and L. G. DeShazer, "Refractive Indices of 14 Nonlinear Crystals Isomorphic to KH_2PO_4," *J. Opt. Soc. Am. B* **4:**1072–1078 (1987).

194. Y. Fujii and T. Sakudo, "Dielectric and Optical Properties of $KTaO_3$," *J. Phys. Soc. Japan* **41:**888–893 (1976).

195. T. Y. Fan, C. E. Huang, B. Q. Hu, R. C. Eckardt, Y. X. Fan, R. L. Byer, and R. S. Feigelson, "Second Harmonic Generation and Accurate Index of Refraction Measurements in Flux-grown $KTiOPO_4$," *Appl. Opt.* **26:**2390–2394 (1987). [Also see H. Y. Shen et al., "Second Harmonic Generation and Sum Frequency Mixing of Dual Wavelength $Nd:YALO_3$ Laser in Flux Grown $KTiOPO_4$ Crystal," *IEEE J. Quantum Electron.* **28:**48–51 (1992).]

196. R. Laiho and M. Lakkisto, "Investigation of the Refractive Indices of LaF_3, CeF_3, PrF_3, and NdF_3," *Phil. Mag. B* **48:**203–207 (1983).

197. F. Hanson and D. Dick, "Blue Parametric Generation from Temperature-tuned LiB$_3$O$_5$," *Opt. Lett.* **16**:205–207 (1991). [Also see C. Chen, Y. Wu, A. Jiang, B. Wu, G. You, R. Li, and S. Lin, "New Nonlinear-optical Crystal: LiB$_3$O$_5$," *J. Opt. Soc. Am. B* **6**:616–621 (1989), K. Kato, "Tunable UV Generation to 0.2325 μm in LiB$_3$O$_5$," *IEEE J. Quantum Electron.* **QE-26**:1173–1175 (1990), and S. P. Velsko, M. Webb, L. Davis, and C. Huang, "Phase-Matched Harmonic Generation in Lithium Triborate (LBO)," *IEEE J. Quantum Electron.* **QE-27**:2182–2192 (1991).]

198. D. F. Nelson and R. M. Mikulyak, "Refractive Indices of Congruently Melting Lithium Niobate," *J. Appl. Phys.* **45**:3688–3689 (1974).

199. W. J. Tropf and M. E. Thomas, "Magnesium Aluminum Spinel (MgAlO$_4$), in E. D. Palik (ed.), *Handbook of Optical Constants of Solids II*, Academic Press, Orlando, 1991, pp. 881–895. [Improved fit with additional data.]

200. M. J. Dodge, "Refractive Properties of Magnesium Fluoride," *Appl. Opt.* **23**:1980–1985 (1984).

201. M. J. Rosker, K. Cheng, and C. L. Tang, "Practical Urea Optical Parametric Oscillator for Tunable Generation Throughout the Visible and Near-Infrared," *IEEE J. Quantum Electron.*, **QE-21**:1600–1606 (1985).

202. I. H. Malitson and M. J. Dodge, "Refraction and Dispersion of Lead Fluoride," *J. Opt. Soc. Am.* **59**:500A (1969). [Also see M. J. Dodge, "Refractive Index," in *Handbook of Laser Science and Technology, Volume IV, Optical Materials: Part 2*, CRC Press, Boca Raton, 1986, p. 31.]

203. I. H. Malitson, as quoted by W. L. Wolfe in W. G. Driscoll (ed.), *Handbook of Optics*, 1st ed., McGraw-Hill, New York, 1978. [Our fit to the dispersion data.]

204. F. Weiting and Y. Yixun, "Temperature Effects on the Refractive Index of Lead Telluride and Zinc Sulfide," *Infrared Phys.* **30**:371–373 (1990).

205. S. Singh, J. P. Remeika, and J. R. Potopowicz, "Nonlinear Optical Properties of Ferroelectric Lead Titanate," *Appl. Phys. Lett.* **20**:135–137 (1972).

206. L. Gample and F. M. Johnson, "Index of Refraction of Single-crystal Selenium," *J. Opt. Soc. Am.* **59**:72–73 (1969).

207. P. T. B. Schaffer, "Refractive Index, Dispersion, and Birefringence of Silicon Carbide Polytypes," *Appl. Opt.* **10**:1034–1036 (1971).

208. S. Singh, J. R. Potopowicz, L. G. Van Uitert, and S. H. Wemple, "Nonlinear Optical Properties of Hexagonal Silicon Carbide," *Appl. Phys. Lett* **19**:53–56 (1971).

209. G. Hettner and G. Leisegang, "Dispersion of the Mixed Crystals TlBr-TlI (KRS 5) and TlCl-TlBr (KRS-6) in the Infrared," *Optik* **3**:305–314 (1948). [In German] [Our fit to the dispersion data.]

210. Y. Nigara, "Measurement of the Optical Constants of Yttrium Oxide," *Jap. J. Appl. Phys.* **7**:404–408 (1968).

211. T. M. Bieniewski and S. J. Czyzak, "Refractive Indexes of Single Hexagonal ZnS and CdS Crystals," *J. Opt. Soc. Am.* **53**:496–497 (1963). [Our fit to the dispersion data.]

212. H. H. Li, "Refractive Index of ZnS, ZnSe, and ZnTe and Its Wavelength and Temperature Derivatives," *J. Phys. Chem. Ref. Data* **13**:103–150 (1984).

213. B. Bendow, R. N. Brown, M. G. Drexhage, T. J. Loretz, and R. L. Kirk, "Material Dispersion of Fluorozirconate-type Glasses," *Appl. Opt.* **20**:3688–3690 (1981).

214. P. Kloeck and L. Colombo, "Index of Refraction, Dispersion, Bandgap and Light Scattering in GeSe and GeSbSe Glasses," *J. Non-Cryst. Solids* **93**:1–16 (1987).

215. Measurements made by R. J. Scheller of Schott Glass Technologies, Inc, Duryea, Pa.

216. S. A. Solin and A. K. Ramdas, "Raman Spectrum of Diamond," *Phys. Rev. B* **1**:1687–1698 (1970). [Also see B. J. Parsons, "Spectroscopic Mode Grüneisen Parameters for Diamond," *Proc. Royal Soc. Lond. A* **352**:397–417 (1977).]

217. B. A. Weinstein and G. J. Piermarini, "Raman Scattering and Phonon Dispersion in Si and GaP at Very High Pressure," *Phys. Rev. B* **12**:1172–1186 (1975).

218. D. Olego and M. Cardona, "Pressure Dependence of Raman Phonons of Ge and 3C-SiC," *Phys. Rev. B* **25**:1151–1160 (1982).

219. P. Vergnat, J. Claudel, A. Hadni, P. Strimer, and F. Vermillard, "Far Infrared Optical Constants of Cesium Halides at Low Temperatures," *J. Phys.* **30:**723–735 (1969) [in French].

220. H. Shimizu, Y. Ohbayashi, K. Yamamoto, K. Abe, M. Midorikawa, and Y. Ishibashi, "Far-infrared Reflection Spectra of CsCl Single Crystals," *J. Phys. Soc. Japan* **39:**448–450 (1975).

221. S. S. Mitra, "Infrared and Raman Spectra Due to Lattice Vibrations," in *Optical Properties of Solids,* Plenum Press, New York, 1969.

222. G. O. Jones, D. H. Martin, P. A. Mawer, and C. H. Perry, "Spectroscopy at Extreme Infra-red Wavelengths II. The Lattice Resonances of Ionic Crystals," *Proc. Royal Soc. A* **261:**10–27 (1961).

223. M. Galtier, A. Montaner, and G. Vidal, "Optical Phonons of CaO, SrO, BaO at the Center of the Brillouin Zone at 300 and 17 K," *J. Phys. Chem. Solids* **33:**2295–2302 (1972) [in French].

224. R. Ramnarine and W. F. Sherman, "The Far-infrared Investigation of the Mode Frequencies of Some Alkaline Earth Sulfides," *Infrared Phys.* **26:**17–21 (1986).

225. J. L. Jacobson and E. R. Nixon, "Infrared Dielectric Response and Lattice Vibrations of Calcium and Strontium Oxides," *J. Phys. Chem. Solids* **29:**967–976 (1968).

226. Z. V. Popović, G. Stanišić, D. Stojanović, and R. Kostić, "Infrared and Raman Spectral of CdO," *Phys. Stat. Sol. (b)* **165:**K109–K112 (1991).

227. J. R. Jasperse, A. Kahan, J. N. Plendl, and S. S. Mitra, "Temperature Dependence of Infrared Dispersion in Ionic Crystals LiF and MgO," *Phys. Rev.* **146:**526–542 (1966). [Also see A. Kachare, G. Andermann, and L. R. Brantley, "Reliability of Classical Dispersion Analysis of LiF and MgO Reflectance Data," *J. Phys. Chem. Solids* **33:**467–475 (1972).]

228. R. Geick, "Measurement and Analysis of the Fundamental Lattice Vibration Spectrum of PbS," *Phys. Lett.* **10:**51–52 (1964).

229. H. Birkhard, R. Geick, P. Kästner, and K.-H. Unkelback, "Lattice Vibrations and Free Carrier Dispersion in PbSe," *Phys. Stat. Sol. (b)* **63:**89–96 (1974).

230. E. G. Bylander and M. Hass, "Dielectric Constant and Fundamental Lattice Frequency of Lead Telluride," *Solid State Commun.* **4:**51–52 (1966).

231. M. Ilegems and G. L. Pearson, "Infrared Reflection Spectra of $Ga_{1-x}Al_xAs$ Mixed Crystals," *Phys. Rev. B* **1:**1576–1582 (1970).

232. A. Mooradian and G. B. Wright, "First Order Raman Effect in III-V Compounds," *Solid State Commun.* **4:**431–434 (1966). [Also see W. J. Turner and W. E. Reese, "Infrared Lattice Bands in AlSb," *Phys. Rev.* **127:**126–131 (1962).]

233. P. J. Gielisse, S. S. Mitra, J. N. Plendl, R. D. Griffis, L. C. Mansur, R. Marshall, and E. A. Pascoe, "Lattice Infrared Spectra of Boron Nitride and Boron Monophosphide," *Phys. Rev.* **155:**1039–1046 (1967).

234. J. A. Sanjurjo, E. Lópex-Cruz, P. Vogl, and M. Cardona, "Dependence on Volume of the Phonon Frequencies and the IR Effective Charges of Several III-V Semiconductors," *Phys. Rev.* **B28:**4579–4584 (1983).

235. A. Manabe, A. Mitsuishi, and H. Yoshinga, "Infrared Lattice Reflection Spectra of II-VI Compounds," *Jap. J. Appl. Phys.* **6:**593–600 (1967).

236. G. R. Wilkenson, "Raman Spectra of Ionic, Covalent, and Metallic Crystals," in A. Anderson (ed.), *The Raman Effect, Vol. 2: Applications,* Marcel Dekker, New York, 1973.

237. A. S. Barker, "Dielectric Dispersion and Phonon Line Shape in Gallium Phosphide," *Phys. Rev.* **165:**917–922 (1968).

238. M. Hass and B. W. Henvis, "Infrared Lattice Reflection Spectra of III-V Compound Semiconductors," *J. Phys. Chem. Solids* **23:**1099–1104 (1962).

239. R. Carles, N. Saint-Cricq, J. B. Renucci, M. A. Renucci, and A. Zwick, "Second-order Raman Scattering in InAs," *Phys. Rev.* **B22:**4804–4815 (1980).

240. R. B. Sanderson, "Far Infrared Optical Properties of Indium Antimonide," *J. Phys. Chem. Solids* **26:**803–810 (1965).

241. L. Patrick and W. J. Choyke, "Lattice Absorption Bands in SiC," *Phys. Rev.* **123:**813–815 (1965).

242. W. G. Nilsen, "Raman Spectrum of Cubic ZnS," *Phys. Rev.* **182:**838–850 (1969).

243. J. C. Irwin and J. LaCombe, "Raman Scattering in ZnTe," *J. Appl. Phys.* **41:**1444–1450 (1970).

244. R. Shivastava, H. V. Lauer, L. L. Chase, and W. E. Bron, "Raman Frequencies of Fluorite Crystals," *Phys. Lett.* **36A:**333–334 (1971).

245. W. Kaiser, W. G. Spitzer, R. H. Kaiser, and L. E. Howarth, "Infrared Properties of CaF_2, SrF_2, and BaF_2," *Phys. Rev.* **127:**1950–1954 (1962).

246. I. Richman, "Longitudinal Optical Phonons in CaF_2, SrF_2, and BaF_2," *J. Chem. Phys.* **41:**2836–2837 (1966).

247. D. R. Bosomworth, "Far-infrared Optical Properties of CaF_2, SrF_2, BaF_2, and CdF_2," *Phys. Rev.* **157:**709–715 (1967).

248. J. D. Axe, J. W. Gaglianello, and J. E. Scardefield, "Infrared Dielectric Properties of Cadmium Fluoride and Lead Fluoride," *Phys. Rev.* **139:**A1211–1215 (1965).

249. R. Droste and R. Geick, "Investigation of the Infrared-Active Lattice Vibration in $SrCl_2$," *Phys. Stat. Sol.* (*b*) **62:**511–517 (1974).

250. A. Sadoc, F. Moussa, and G. Pepy, "The Lattice Dynamics of $SrCl_2$," *J. Phys. Chem. Solids* **37:**197–199 (1976).

251. J. D. Axe and G. D. Pettit, "Infrared Dielectric Dispersion and Lattice Dynamics of Uranium Dioxide and Thorium Dioxide," *Phys. Rev.* **151:**676–679 (1966).

252. P. G. Marlowe and J. P. Russell, "Raman Scattering in Uranium Dioxide," *Phil. Mag.* **14:**409–410 (1966).

253. S. Shin and M. Ishigame, "Defect-induced Hyper-Raman Spectra in Cubic Zirconia," *Phys. Rev. B* **34:**8875–8882 (1986).

254. S. P. S. Porto and R. S. Krishnan, "Raman Effect of Corundum," *J. Chem. Phys.* **47:**1009–1012 (1967).

255. D. R. Renneke and D. W. Lynch, "Infrared Lattice Vibrations and Dielectric Dispersion in Single-Crystal Cr_2O_3," *Phys. Rev.* **138:**A530–A533 (1965).

256. I. R. Beattle and T. R. Gibson, "The Single-crystal Raman Spectra of Nearly Opaque Materials. Iron(III) Oxide and Chromium(III) Oxide," *J. Chem. Soc.* (*A*), 980–986 (1970).

257. S. Onari, T. Arai, and K. Kudo, "Infrared Lattice Vibrations and Dielectric Dispersion in α-Fe_2O_3," *Phys. Rev. B* **16:**1717–1721 (1977).

258. G. L. Bottger and C. V. Damsgard, "Raman Scattering in Wurtzite-Type AgI Crystals," *J. Chem Phys.* **57:**1215–1218 (1972).

259. A. T. Collins, E. C. Lightowlers, and P. J. Dean, "Lattice Vibration Spectra of Aluminum Nitride," *Phys. Rev.* **158:**833–838 (1967).

260. E. Loh, "Optical Phonons in BeO Crystals," *Phys. Rev.* **166:**673–678 (1967).

261. C. A. Arguello, D. L. Rousseau, and S. P. S. Porto, "First-Order Raman Effect in Wurtzite-Type Crystals," *Phys. Rev.* **181:**1351–1363 (1969).

262. T. C. Damen, S. P. S. Porto, and B. Tell, "Raman Effect in Zinc Oxide," *Phys. Rev.* **142:**570–574 (1966).

263. B. Tell, T. C. Damen, and S. P. S. Porto, "Raman Effect in Cadmium Sulfide," *Phys. Rev.* **144:**771–774 (1966).

264. R. Geick, C. H. Perry, and S. S. Mitra, "Lattice Vibrational Properties of Hexagonal CdSe," *J. Appl. Phys.* **137:**1994–1997 (1966).

265. D. D. Manchon, A. S. Barker, P. J. Dean, and R. B. Zetterstrom, "Optical Studies of the Phonons and Electrons in Gallium Nitride," *Solid State Commun.* **8:**1227–1231 (1970).

266. D. W. Feldman, J. H. Parker, W. J. Choyee, and L. Patrick, "Raman Scattering in 6*H* SiC," *Phys. Rev.* **170:**698–704 (1968).

267. R. Geick, U. Schröder, and J. Stuke, "Lattice Vibrational Properties of Trigonal Selenium," *Phys. Stat. Sol.* **24:**99–108 (1967).

268. G. Locovsky, A. Mooradian, W. Taylor, G. B. Wright, and R. C. Keezer, "Identification of the Fundamental Vibrational Modes of Trigonal, α-monoclinic and Amorphous Selenium," *Solid State Commun.* **5:**113–117 (1967).

269. E. D. Palik, "Tellurium (Te)," in E. D. Palik (ed.), *Handbook of Optical Constants of Solids II*, Academic Press, Orlando, 1991, pp. 709–723.

270. B. D. Saksena, "Analysis of the Raman and Infra-red Spectra of α-Quartz," *Proc. Ind. Acad. Sci.* **12A:**93–139 (1940).

271. J. F. Scott and S. P. S. Porto, "Longitudinal and Transverse Optical Lattice Vibrations in Quartz," *Phys. Rev.* **161:**903–910 (1967).

272. S. M. Shapiro and J. D. Axe, "Raman Scattering from Polar Phonons," *Phys. Rev. B* **6:**2420–2427 (1972).

273. J. F. Scott, "Raman Spectra of GeO_2," *Phys. Rev. B* **8:**3488–3493 (1970).

274. A. S. Barker and J. A. Detzenberger, "Infrared Lattice Vibrations in CoF_2," *Solid State Commun.* **3:**131–132 (1965).

275. M. Balkanski, P. Moch, and G. Parisot, "Infrared Lattice-Vibration Spectra in NiF_2, CoF_2, and FeF_2," *J. Chem Phys.* **44:**940–944 (1966).

276. R. M. Macfarlane and S. Ushioda, "Light Scattering from Phonons in CoF_2," *Solid State Commun.* **8:**1081–1083 (1970).

277. S. P. S. Porto, P. A. Fleury, and T. C. Damen, "Raman Spectra of TiO_2, MgF_2, ZnF_2, FeF_2, and MnF_2," *Phys. Rev.* **154:**522–526 (1967).

278. J. Giordano and C. Benoit, "Infrared Spectra of Iron, Zinc, and Magnesium Fluorides: I. Analysis of Results," *J. Phys. C* **21:**2749–2770 (1988). C. Benoit and J. Giordano, "Dynamical Properties of Crystals of MgF_2, ZnF_2, and FeF_2: II. Lattice Dynamics and Infrared Spectral," *J. Phys. C* **21:**5209–5227 (1988).

279. A. Kahan, J. W. Goodrum, R. S. Singh, and S. S. Mitra, "Polarized Reflectivity Spectra of Tegragonal GeO_2," *J. Appl. Phys.* **42:**4444–4446 (1971).

280. D. M. Roessler and W. A. Albers, "Infrared Reflectance of Single Crystal Tetragonal GeO_2," *J. Phys. Chem. Solids* **33:**293–296 (1972).

281. A. S. Barker, "Transverse and Longitudinal Optic Mode Study in MgF_2 and ZnF_2," *Phys. Rev.* **136:**A1290–A1295 (1964).

282. R. S. Krishnan and J. P. Russell, "The First-order Raman Spectrum of Magnesium Fluoride," *Brit. J. Appl. Phys.* **17:**501–503 (1966).

283. J. Giordano, "Temperature Dependence of IR Spectra of Zinc and Magnesium Fluoride," *J. Phys. C* **20:**1547–1562 (1987).

284. R. Summit, "Infrared Absorption in Single-crystal Stannic Oxide: Optical Lattice-vibration Modes," *J. Appl. Phys.* **39:**3762–3767 (1967).

285. J. F. Scott, "Raman Spectrum of SnO_2," *J. Chem. Phys.* **53:**852–853 (1970).

286. R. S. Katiyar, P. Dawson, M. M. Hargreave, and G. R. Wilkerson, "Dynamics of the Rutile Structure III. Lattice Dynamics, Infrared and Raman Spectra of SnO_2," *J. Phys. C* **4:**2421–2431 (1971).

287. R. S. Katiyar and R. S. Krishnan, "The Vibrational Spectrum of Rutile," *Phys. Lett.* **25A:**525–526 (1967).

288. A. S. Barker, "Infrared Lattice Vibrations in Calcium Tungstate and Calcium Molybdate," *Phys. Rev.* **135:**A742–A747 (1964).

289. P. Tarte and M. Liegeois-Duyckaerts, "Vibrational Studies of Molybdates, Tungstates and Related Compounds—I. New Infrared Data and Assignments for the Scheelite-type Compounds $X^{II}MoO_4$ and $X^{II}WO_4$," *Spectrochim. Acta* **28A:**2029–2036.

290. V. M. Nagiev, Sh. M. Efendiev, and V. M. Burlakov, "Vibrational Spectra of Crystals with Scheelite Structure and Solid Solutions on Their Basis," *Phys. Stat. Sol. (b)* **125:**467–475 (1984).

291. J. M. Stencel, E. Silberman, and J. Springer, "Temperature-dependent Reflectivity, Dispersion Parameters, and Optical Constants for $PbWO_4$," *Phys. Rev. B* **12:**5435–5441 (1976).

292. S. A. Miller, H. E. Rast, and H. H. Caspers, "Lattice Vibrations of $LiYF_4$," *J. Chem. Phys.* **53:**4172–4175 (1970).

293. E. Schultheiss, A. Scharmann, and D. Schwabe, "Lattice Vibrations in $BiLiF_4$ and $YLiF_4$," *Phys. Stat. Sol.* (*b*) **138:**465–475 (1986).

294. S. P. S. Porto and J. F. Scott, "Raman Spectra of $CaWO_4$, $SrWO_4$, $CaMoO_4$, and $SrMoO_4$," *Phys. Rev.* **157:**716–719 (1967).

295. S. Desgreniers, S. Jandl, and C. Carlone, "Temperature Dependence of the Raman Active Phonons in $CaWO_4$, $SrWO_4$, and $BaWO_4$," *J. Phys. Chem. Solids* **45:**1105–1109 (1984).

296. R. K. Khanna, W. S. Brower, B. R. Guscott, and E. R. Lippincott, "Laser Induced Raman Spectra of Some Tungstates and Molybdates," *J. Res. Nat. Bur. Std.* **72A:**81–84 (1968).

297. M. P. O'Horo, A. L. Frisillo, and W. B. White, "Lattice Vibrations of $MgAl_2O_4$ Spinel," *J. Phys. Chem. Solids* **34:**23–28 (1973).

298. M. E. Strifler and S. I. Boldish, "Transverse and Longitudinal Optic Mode Frequencies of Spinel $MgAl_2O_4$," *J. Phys. C* **11:**L237–L241 (1978).

299. K. Yamamoto, T. Murakawa, Y. Ohbayashi, H. Shimizu, and K. Abe, "Lattice Vibrations in $CdIn_2S_4$," *J. Phys. Soc. Japan* **35:**1258 (1973).

300. H. Shimizu, Y. Ohbayashi, K. Yamamoto, and K. Abe, "Lattice Vibrations in Spinel-type $CdIn_2S_4$," *J. Phys. Soc. Japan* **38:**750–754 (1975).

301. H. D. Lutz, B. Müller, and H. J. Steiner, "Lattice Vibration Spectra. LIX. Single Crystal Infrared and Raman Studies of Spinel Type Oxides," *J. Solid State Chem.* **90:**54–60 (1991).

302. H. D. Lutz, G. Wäschenbach, G. Kliche, and H. Haeuseler, "Lattice Vibrational Spectra, XXXIII: Far-Infrared Reflection Spectra, TO and LO Phonon Frequencies, Optical and Dielectric Constants, and Effective Changes of the Spinel-Type Compounds MCr_2S_4 ($M = $ Mn, Fe, Co, Zn, Cd, Hg), MCr_2Se_4 ($M = $ Zn, Cd, Hg), and $MInr_2S_4$ ($M = $ Mn, Fe, Co, Zn, Cd, Hg)," *J. Solid State Chem.* **48:**196–208 (1983).

303. K. Wakamura, H. Iwatani, and K. Takarabe, "Vibrational Properties of One- and Two-Mode Behavior in Spinel Type Mixed Systems $Zn_{1-x}Cd_xCr_2S_4$," *J. Phys. Chem. Solids* **48:**857–861 (1987).

304. H. C. Gupta, G. Sood, A. Parashar, and B. B. Tripathi, "Long Wavelength Optical Lattice Vibrations in Mixed Chalcogenide Spinels $Zn_{1-x}Cd_xCr_2S_4$ and $CdCr_2(S_{1-x}Se_x)_4$," *J. Phys. Chem. Solids* **50:**925–929 (1989).

305. A. S. Barker and J. H. Hopfield, "Coupled-optical-phonon-mode Theory of the Infrared Dispersion in $BaTiO_3$, $SrTiO_3$, and $KTaO_3$," *Phys. Rev.* **135:**A1732–A1737 (1964).

306. A. S. Barker, "Temperature Dependence of the Transverse and Longitudinal Optic Mode Frequencies and Charges in $SrTiO_3$ and $BaTiO_3$," *Phys. Rev.* **145:**391–399 (1966).

307. J. L. Servoin, Y. Luspin, and F. Gervais, "Infrared Dispersion in $SrTiO_3$ at High Temperature," *Phys. Rev. B* **22:**5501–5506 (1980).

308. C. H. Perry and E. F. Young, "Infrared Studies of Some Perovskite Fluorides. I. Fundamental Lattice Vibrations," *J. Appl. Phys.* **38:**4616–4624 (1967).

309. A. Scalabrin, A. S. Chaves, D. S. Shim, and S. P. S. Porto, "Temperature Dependence of the A_1 and E Optical Phonons in $BaTiO_3$," *Phys. Stat. Sol.* (*b*) **79:**731–742 (1977).

310. J. L. Servoin, F. Gervais, A. M. Quittet, and Y. Luspin, "Infrared and Raman Responses in Ferroelectric Perovskite Crystals," *Phys. Rev. B* **21:**2038–2041 (1980).

311. G. Burns and B. A. Scott, "Lattice Modes in Ferroelectric Perovskites: PbTiO3," *Phys. Rev. B* **7:**3088–3101 (1973).

312. J. P. van der Ziel, A. E. Meixner, H. M. Kasper, and J. A. Ditzenberger, "Lattice Vibrations of $AgGaS_2$, $AgGaSe_2$, and $CuGaS_2$," *Phys. Rev. B* **9:**4286–4294 (1974).

313. W. H. Koschel and M. Bettini, "Zone-Centered Phonons in $A^IB^{III}S_2$ Chalcopyrites," *Phys. Stat. Sol.* (*b*) **72:**729–737 (1975).

314. L. Artus, J. Pascual, A. Goullet, and J. Camassel, "Polarized Infrared Spectra of $AgGaSe_2$," *Solid State Commun.* **69:**753–756 (1989).

315. G. D. Holah, A. Miller, W. D. Dunnett, and G. W. Isler, "Polarised Infrared Reflectivity of $CdGeAs_2$," *Solid State Commun.* **23:**75–78 (1977).

316. E. V. Antropova, A. V. Kopytov, and A. S. Poplavnoi, "Phonon Spectrum and IR Optical Properties of CdGeAs$_2$," *Opt. Spectrosc.* **64:**766–768 (1988).

317. A. Miller, G. D. Holah, and W. C. Clark, "Infrared Dielectric Dispersion of ZnGeP$_2$ and CdGeP$_2$," *J. Phys. Chem. Solids* **35:**685–693 (1974).

318. M. Bettini and A. Miller, "Optical Phonons and ZnGeP$_2$ and CdGeP$_2$," *Phys. Stat. Sol. (b)* **66:**579–586 (1974).

319. I. P. Kaminow, E. Buehler, and J. H. Wernick, "Vibrational Modes in ZnSiP$_2$," *Phys. Rev. B* **2:**960–966 (1970).

320. R. Zallen, M. L. Slade, and A. T. Ward, "Lattice Vibrations and Interlayer Interactions in Crystalline As$_2$S$_3$ and As$_2$Se$_3$," *Phys. Rev. B* **3:**4257–4273 (1971).

321. S. P. S. Porto, J. A. Giordmaine, and T. C. Damen, "Depolarization of Raman Scattering in Calcite," *Phys. Rev.* **147:**608–611 (1966).

322. K. H. Hellwege, W. Lesch, M. Plihal, and G. Schaack, "Two Phonon Absorption Spectra and Dispersion of Phonon Branches in Crystals of Calcite Structure," *Z. Physik* **232:**61–86 (1970). [Also see R. K. Vincent, "Emission Polarization Study on Quartz and Calcite," *Appl. Opt.* **11:**1942–1945 (1972).]

323. J. Q. Lu, G. X. Lan, B. Li, Y. Y. Yang, and H. F. Wang, "Raman Scattering Study of the Single Crystal β-BaB$_2$O$_4$ under High Pressure," *J. Phys. Chem. Solids* **49:**519–527 (1988).

324. W. Wojdowski, "Vibrational Modes in Bi$_{12}$GeO$_{20}$ and Bi$_{12}$SiO$_{20}$ Crystals," *Phys. Stat. Sol. (b)* **130:**121–130 (1985).

325. H. Vogt, T. Chattopadhyay, and H. J. Stolz, "Complete First-order Raman Spectra of the Pyrite Structure Compounds FeS$_2$, MnS$_2$, and SiP$_2$," *J. Phys. Chem. Solids* **44:**869–873 (1983).

326. H. D. Lutz, G. Schneider, and G. Kliche, "Far-infrared Reflection Spectra, TO- and LO-phonon Frequencies, Coupled and Decoupled Plasmon-phonon Modes, Dielectric Constants, and Effective Dynamical Charges of Manganese, Iron, and Platinum Group Pyrite Type Compounds," *J. Phys. Chem. Solids* **46:**437–443 (1985).

327. D. K. Agrawal and C. H. Perry, "The Temperature Dependent Raman Spectra of KDP, KD∗P, KDA, and ADP," in M. Balkanski (ed.), *Proceedings of the Second International Conference on Light Scattering in Solids,* Flammarion Sciences, Paris, 1971, pp. 429–435.

328. G. E. Kugel, F. Bréhat, B. Wyncke, M. D. Fontana, G. Marnier, C. Carabatos-Nedelec, and J. Mangin, "The Vibrational Spectrum of KTiOPO$_4$ Single Crystal Studied by Raman and Infrared Reflective Spectroscopy," *J. Phys. C* **21:**5565–5583 (1988).

329. R. P. Bauman and S. P. S. Porto, "Lattice Vibrations and Structure of Rare-earth Fluorides," *Phys. Rev.* **161:**842–847 (1967).

330. R. P. Lowndes, J. F. Parrish, and C. H. Perry, "Optical Phonons and Symmetry of Tysonite Lanthanide Fluorides," *Phys. Rev.* **182:**913–922 (1969).

331. E. Liarokapis, E. Anastassakis, and G. A. Kourouklis, "Raman Study of Phonon Anharmonicity in LaF$_3$," *Phys. Rev. B* **32:**8346–8355 (1985).

332. W. Otaguro, E. Weiner-Avnera, C. A. Arguello, and S. P. S. Porto, "Phonons, Polaritrons, and Oblique Phonons in LiIO$_3$ by Raman Scattering and Infrared Reflection," *Phys. Rev. B* **4:**4542–4551 (1971).

333. J. L. Duarte, J. A. Sanjurjo, and R. S. Katiyar, "Off-normal Infrared Reflectivity in Uniaxial Crystals: α-LiIO$_3$ and α-quartz," *Phys. Rev. B* **36:**3368–3372 (1987).

334. R. Claus, G. Borstel, E. Wiesendanger, and L. Steffan, "Directional Dispersion and Assignment of Optical Phonons in LiNbO$_3$," *Z. Naturforsch.* **27a:**1187–1192 (1972).

335. X. Yang, G. Lan, B. Li, and H. Wang, "Raman Spectra and Directional Dispersion in LiNbO$_3$ and LiTaO$_3$," *Phys. Stat. Sol. (b)* **141:**287–300 (1987).

336. D. G. Bozinis and J. P. Hurrell, "Optical Modes and Dielectric Properties of Ferroelectric Orthorhombic KNbO$_3$," *Phys. Rev. B* **13:**3109–3120 (1976).

337. A. S. Pine and G. Dresselhaus, "Raman Scattering in Paratellurite," *Phys. Rev. B* **5:**4087–4093 (1972).

338. D. M. Korn, A. S. Pine, G. Dresselhaus, and T. B. Reed, "Infrared Reflectivity of Paratellurite, TeO$_2$," *Phys. Rev. B* **8:**768–772 (1973).

339. W. J. Tropf and M. E. Thomas, "Yttrium Oxide (Y_2O_3)," in E. D. Palik (ed.), *Handbook of Optical Constants of Solids II,* Academic Press, Orlando, 1991, pp. 1081–1098.

340. M. Thirumavalavan, J. Kumar, F. D. Gnanam, and P. Ramasamy, "Vibrational Spectra of $Y_3Al_5O_{12}$ Crystals Grown from Ba- and Pb-based Flux Systems," *Infrared Phys.* **26:**101–103 (1986).

341. G. A. Gledhill, P. M. Nikolić, A. Hamilton, S. Stojilković, V. Blagojević. P. Mihajlovic, and S. Djurić, "FIR Optical Properties of Single Crystal $Y_3Al_5O_{12}$ (YAG)," *Phys. Stat. Sol. (b)* **163:**K123–K128 (1991).

342. R. P. Lowndes, "Anharmonicity in the Silver and Thallium Halides: Far-Infrared Dielectric Response," *Phys. Rev. B* **6:**1490–1498 (1972).

343. G. Lucovsky, "Optic Modes in Amorphous As_2S_3 and As_2Se_3," *Phys. Rev. B* **6:**1480–1489 (1972).

344. W. G. Spitzer, R. C. Miller, D. A. Kleinman, and L. E. Howarth, "Far Infrared Dielectric Dispersion in $BaTiO_3$, $SrTiO_3$, and TiO_2," *Phys. Rev.* **126:** 1710–1721 (1962).

345. E. J. Danielewicz and P. D. Coleman, "Far Infrared Optical Properties of Selenium and Cadmium Telluride," *Appl. Opt.* **13:**1164–1170 (1974).

346. J. L. Verble and R. F. Wallis, "Infrared Studies of the Lattice Vibrations in Iron Pyrite," *Phys. Rev.* **182:**783–789 (1969).

347. H. D. Lutz, G. Kliche, and H. Haeuseler, "Lattice Vibrational Spectra XXIV: Far-infrared Reflection Spectra, Optical and Dielectric Constants, and Effective Charges of Pyrite Type Compounds FeS_2, MnS_2, $MnSe_2$, and $MnTe_2$," *Z. Naturforsch.* **86a:**184–190 (1981).

348. C. J. Johnson, G. H. Sherman, and R. Weil, "Far Infrared Measurement of the Dielectric Properties of GaAs and CdTe at 300 K and 8 K," *Appl. Opt.* **8:**1667–1671 (1969).

349 D. A. Kleinman and W. G. Spitzer, "Infrared Lattice Absorption of GaP," *Phys. Rev.* **118:**110–117 (1960).

350. A. Hadni, J. Claudel, D. Chanal, P. Strimer, and P. Vergnat, "Optical Constants of Potassium Bromide in the Far Infrared," *Phys. Rev.* **163:**836–843 (1967).

351. A. Hadni, J. Claudel, G. Morlot, and P. Strimer, "Transmission and Reflection Spectra of Pure and Doped Potassium Iodide at Low Temperature," *Appl. Opt.* **7:**161–165 (1968) [in French].

352. J. Zarembowitch, J. Gouteron, and A. M. Lejus, "Raman Spectra of Lanthanide Sesquioxide Single Crystals with A-type Structure," *Phys. Stat. Sol. (b)* **94:**249–256 (1979).

353. A. S. Barker and R. Loudon, "Dielectric Properties and Optical Phonons in $LiNbO_3$," *Phys. Rev.* **158:**433–445 (1967).

354. I. F. Chang and S. S. Mitra, "Temperature Dependence of Long-Wavelength Optic Phonons of NaF Single Crystals," *Phys. Rev. B* **5:**4094–4100 (1972).

355. H. Burkhard, R. Geick, P. Kästner, and K.-H. Unkelbach, "Lattice Vibrations and Free Carrier Dispersion in PbSe," *Phys. Stat. Sol. (b)* **63:**89–96 (1974).

356. W. G. Spitzer and D. A. Kleinman, "Infrared Lattice Bands of Quartz," *Phys. Rev.* **121:**1324–1335 (1961).

357. B. Orel and V. Moissenko, "A Vibrational Study of Piezoelectric TeO_2 Crystals," *Phys. Stat. Sol. (b)* **165:**K37–K41 (1991).

358. A. M. Hofmeister and K. R. Campbell, "Infrared Spectroscopy of Yttrium Aluminum, Yttrium Gallium, and Yttrium Iron Garnets," *J. Appl. Phys.* **72:**638–646 (1992).

359. H. v. Campe, "Fundamental Absorption of $AgGaS_2$ Single Crystals and Thin Polycrystalline Films," *J. Phys. Chem. Solids* **44:**1019–1023 (1983).

360. M. E. Thomas, W. J. Tropf, and S. L. Gilbert, "Vacuum-Ultraviolet Characterization of Sapphire, ALON, and Spinel Near the Band Gap," to be published in *Opt. Eng.* **32:**1340–1343 (1993).

361. T. Tomiki and T. Miyata, "Optical Studies of Alkali Fluorides and Alkaline Earth Fluorides in VUV Region," *J. Phys. Soc. Japan* **27:**658–678 (1969).

362. T. Toyada, H. Nakanishi, S. Endo, and T. Irie, "The Fitting Parameters of Exponential Optical Absorption $Bi_{12}SiO_{20}$," *J. Phys. C* **19:**L259–L263 (1986).

363. D. Dutton, "Fundamental Absorption Edge in Cadmium Sulfide," *Phys. Rev.* **112:**785–792 (1958).

364. R. Nitecki and J. A. Gaj, "On the Urbach Rule in CdSe," *Phys. Stat. Sol (b)* **62:**K17–K19 (1974).

365. R. T. Williams and S. E. Schnatterly, "Magnetic Circular Dichroism of the Urbach Edge in KI, CdTe, and TlCl," in S. S. Mitra and B. Bendow (eds.), *Optical Properties of Transparent Solids,* Plenum, New York, 1975, pp. 145–160.

366. Y. Takubo and T. Koda, "The Urbach Tail of Cuprous Halides," *J. Phys. Soc. Japan* **39:**715–719 (1975).

367. W. J. Turner, W. E. Reese, and G. D. Pettit, "Exciton Absorption and Emission in InP," *Phys. Rev.* **136A:**1467–1470 (1964).

368. T. Tomiki, T. Mayata, and H. Tsukamoto, "The Urbach Rule for the Sodium- and Potassium-Halides," *Z. Naturforsch.* **29A:**145–157 (1974).

369. D. Redfield and W. J. Burke, "Optical Absorption Edge of $LiNbO_3$," *J. Appl. Phys.* **45:**4566–4571 (1974).

370. M. I. Cohen and R. F. Blunt, "Optical Properties of $SrTiO_3$ in the Region of the Absorption Edge," *Phys. Rev.* **168:**929–933 (1968).

371. T. Takizawa, "Optical Absorption and Reflection Spectra of Paratellurite, TeO_2," *J. Phys. Soc. Japan* **48:**505–510 (1980).

372. S. Tutihasi, "Ultraviolet Absorption in Thallous Halides," *J. Phys. Chem. Solids* **12:**344–348 (1960).

373. T. Tomiki, F. Fukudome, M. Kaminao, M. Fujisawa, Y. Tanahara, and T. Futemma, "Optical Spectra of $Y_3Al_5O_{12}$ (YAG) Single Crystals in the Vacuum Ultraviolet," *J. Phys. Soc. Japan* **58:**1801–1810 (1989).

374. T. Tomiki, M. Kaminao, Y. Tanahara, T. Futemma, M. Fujisawa, and F. Fukudome, "Anisotropic Optical Spectra of $YAlO_3$ (YAP) Single Crystals in the Vacuum Ultraviolet Region," *J. Phys. Soc. Japan* **60:**1799–1813 (1991).

375. T. Tomiki, J. Tamashiro, Y. Tanahara, A. Yamada, H. Fututani, T. Miyahara, H. Kato, S. Shin, and M. Ishigame, "Optical Spectra of Y_2O_3 Single Crystals in VUV," *J. Phys. Soc. Japan* **55:**4543–4549 (1986).

376. M. V. Kurik, "Urbach Rule," *Phys. Stat. Sol. (a)* **8:**9–45 (1971).

377. M. E. Thomas, "Infrared Properties of the Extraordinary Ray Multiphonon Processes in Sapphire," *Appl. Opt.* **28:**3277–3278 (1989).

378. E. D. Palik, "Cadmium Telluride (CdTe)," in E. D. Palik (ed.), *Handbook of Optical Constants of Solids,* Academic Press, Orlando, 1985, pp. 409–427.

379. J. E. Eldridge, "Cesium Iodide (CsI)," in E. D. Palik (ed.), *Handbook of Optical Constants of Solids II,* Academic Press, Orlando, 1991, pp. 853–874.

380. E. D. Palik, "Gallium Arsenide (GaAs)," in E. D. Palik (ed.), *Handbook of Optical Constants of Solids,* Academic Press, Orlando, 1985, pp. 429–443.

381. L. L. Boyer, J. A. Harrington, M. Hass, and H. B. Rosenstock, "Multiphonon Absorption in Ionic Crystals," *Phys. Rev. B* **11:**1665–1680 (1975).

382. J. I. Berg and E. E. Bell, "Far-infrared Optical Constants of KI," *Phys. Rev.* **B4:**3572–3580 (1971).

383. D. M. Roessler and D. R. Huffman, "Magnesium Oxide (MgO)," in E. D. Palik (ed.), *Handbook of Optical Constants of Solids II,* Academic Press, Orlando, 1991, pp. 919–955.

384. T. Hidaka, T. Morikawa, and J. Shimada, "Spectroscopic Small Loss Measurements on Infrared Transparent Materials," *Appl. Opt.* **19:**3763–3766 (1980).

385. E. D. Palik and A. Addamiano, "Zinc Sulfide (ZnS)," in E. D. Palik (ed.), *Handbook of Optical Constants of Solids,* Academic Press, Orlando, 1985, pp. 597–619.

386. P. A. Miles, "Temperature Dependence of Multiphonon Absorption in Zinc Selenide," *Appl. Opt.* **16:**2891–2986 (1977).

387. J. A. Cox, D. Greenlaw, G. Terry, K. McHenry, and L. Fielder, "Comparative Study of Advanced IR Transmissive Materials," *Proc. SPIE* **683:**49–62 (1988).

388. B. Bendow, R. N. Brown, M. G. Drexhage, T. J. Loretz, and R. L. Kirk, "Material Dispersion of Fluorozirconate-type Glasses," *Appl. Opt.* **20:**3688–3690 (1981).

389. D. J. Treacy, "Arsenic Sulfide (As_2S_3)," in E. D. Palik (ed.), *Handbook of Optical Constants of Solids,* Academic Press, Orlando, 1985, pp. 641–663.

390. D. J. Treacy, "Arsenic Selenide (As_2Se_3)," in E. D. Palik (ed.), *Handbook of Optical Constants of Solids,* Acadmic Press, Orlando, 1985, pp. 623–639.

391. A. R. Hilton, D. J. Hayes, and M. D. Rechtin, "Infrared Absorption of Some High-purity Chalcogenide Glasses," *J. Non-Cryst. Solids* **17:**319–338 (1975).

CHAPTER 34
POLYMERIC OPTICS

John D. Lytle
Advanced Optical Concepts
Santa Cruz, California

34.1 GLOSSARY

A_{H_2O} water absorption

K thermal conductivity

T_s maximum service temperature

α thermal expansion coefficient

ρ density

34.2 INTRODUCTION

A small number of carbon-based polymeric materials possesses some of those qualities which have made glass an attractive optical material. Most of these polymeric materials do exhibit certain physical deficiencies compared to glass. But, despite the fact that "plastic optics" has acquired an image as a low-end technology, it may nonetheless be a better choice, or even the best choice, in certain applications.

Selection Factors

Virtually all of the polymers having useful optical properties are much less dense than any of the optical glasses, making them worthy of consideration in applications where weight-saving is of paramount importance. Many of them exhibit impact resistance properties which exceed those of any silicate glass, rendering them well-suited to military applications (wherein high "g" loads may be encountered), or ideal for some consumer products in which safety may be a critical consideration.

Though the physical properties of the polymers may make them better matched to certain design requirements than glass, by far the most important advantage of polymeric optics is the considerable creative freedom they make available to the optical and mechanical design effort.[1] While the design constraints and guidelines governing glass optics design and fabrication are fairly well defined, the various replication processes which may be put to use in polymer optics fabrication make available unique opportunities for the creation of novel optical components and systems which would be unthinkable or unworkable in glass. Oftentimes, the differences in the engineering approach, or in the production processes themselves, may make possible very significant cost reductions in high-volume situations.[2]

34.3 FORMS

Thermoset Resins

Optical polymers fall into basically two categories—the *thermoset* resins and the *thermoplastic* resins. The thermoset resin group consists of chemistries in which the polymerization reaction takes place during the creation of the part, which may be produced by casting, or by transfer replication. The part which has been created at completion of the reaction may then be postprocessed, if desired, by machining. In general, the thermoset resins cannot be melted and reformed.

The most commonly encountered thermoset optical resin is that used to produce ophthalmic lenses for eyewear.[3] The monomer, which is stored in liquid form at reduced temperature, is introduced into a mold, where the polymerization reaction takes place, forming a part which assumes the shape of the cavity containing it. Alternatively, epoxy-based chemistries have been used with some success to form replicated reflecting surface shapes by a transfer process, and to produce aspheric figuring (at relatively modest expense) upon spherical refractive or reflective substrates.

Thermoplastic Resins

With the possible exception of eyewear, most polymeric optics are executed in thermoplastic materials which are supplied in already-polymerized form.[4] These materials are normally purchased in bulk as small pellets. These pellets are heated to a temperature beyond the softening point, so that they flow to become a single viscous mass. This mass is then formed to assume the shape desired in the final part.

Parts may be created by the injection molding process, in which the heated polymer is squirted into a mold at high pressure and allowed to cool in the shape of the desired component. Or the pellets may be directly heated between the two halves of a compression mold, and the mold closed to effect formation of the part. Hybrid molding technologies combining these two processes are recently experiencing increasing popularity in optical molding applications, and have produced optical surface figures of very high quality.

The capability of modern molding technology to produce optics having very good surface-figure quality has made possible the creation of polymeric optical components for a wide variety of applications. Among these are medical disposables, intraocular lenses, a host of consumer products, military optics, and a number of articles in which optical, mechanical, and electrical functions are combined in a single part.[5]

34.4 PHYSICAL PROPERTIES

Density

Optical glass types number in the hundreds (if all manufacturers worldwide are counted). The glass types available from the catalogs cover a wide range of optical, physical, thermal, and chemical properties. The density of these materials varies from about $2.3 \, \text{g/cm}^3$ to about $6.3 \, \text{g/cm}^3$. The heaviest optically viable polymer possesses a density of only about $1.4 \, \text{g/cm}^3$, whereas the lightest of these materials will readily float in water, having a density of $0.83 \, \text{g/cm}^3$.[6] All other things being equal, the total element count in an optical system may often be reduced (at modest cost penalty) by the inclusion of nonspherical surfaces. All things considered, then, polymeric optical systems may be made much less massive than their glass counterparts, especially if aspheric technology is applied to the polymer optical trains.

Hardness

Although cosmetic blemishes rarely impact final image quality (except in the cases of field lenses or reticles), optical surfaces are customarily expected to be relatively free of scratches, pits, and the like. Ordinary usage, especially cleaning procedures, are likely to result in some scratching with the passage of time. Most common optical glasses possess sufficient hardness that they are relatively immune to damage, if some modest amount of care is exercised.

The polymeric optical materials, on the other hand, are often so soft that a determined thumbnail will permanently indent them. The hardness of polymeric optics is difficult to quantify (in comparison to glass), since this parameter is not only material-dependent, but also dependent upon the processing. Suffice it to say that handling procedures which would result in little or no damage to a glass element may produce considerable evidence of abrasion in a polymeric surface, particularly in a thermoplastic. In fact, the compressibility of most thermoplastic polymers is such that the support for hard surface coatings is sufficiently low that protection provides immunity against only superficial abrasion. These deficiencies are of no particular consequence, however, if the questionable surfaces are internal, and thereby inaccessible.

Rigidity

A property closely related to hardness is the elastic modulus, or Young's modulus. This quantity, and the elongation factor at yield, are determinants of the impact resistance, a performance parameter in which the polymers outshine the glasses. These properties are, again, dependent upon the specified polymeric alloy, any additives which may be present, and processing history of the polymer, and cannot be dependably quoted.[7,8] The reader is referred to any of several comprehensive references listed herein for mechanical properties data. Those properties which create good impact resistance become liabilities if an optical part is subjected to some torsion or compressive stress. Since optical surface profiles must often be maintained to subwavelength accuracy, improper choice of the thickness/diameter ratio, or excessive compression by retaining rings, may produce unacceptable optical figure deformations.

Polymer chemistry is a complex subject probably best avoided in a discussion of polymer optics. Carbon-based polymers have been synthesized to include an extensive variety of chemical subgroups, however. Unfortunately, relatively few of these materials are actually in regular production, and only a handful of those possess useful optical properties for imaging purposes.

Service Temperature

Any decision involving a glass/plastic tradeoff should include some consideration of the anticipated thermal environment. While the optical glasses may exhibit upper service temperature limits of from 400 to 700°C, many of the glass types having the most interesting optical properties are quite fragile, and prone to failure if cooled too quickly. These failures are mostly attributable to cooling-induced shrinkage of the skin layer, which shatters because the insulating properties of the material prevent cooling (and shrinkage) of the bulk material at the same rate.

The polymeric materials, on the other hand, have much lower service temperature limits, in some cases no higher than about 60°C.[9] The limit may approach 250°C for some of the fluoropolymers. The thermal conductivity of many of these polymers may be as much as an order of magnitude lower than for the glasses and the thermal expansion coefficients

TABLE 1   Physical Properties

Material	ρ	α	T_s	K	A_{H_2O}
P-methylmethacrylate	1.18	6.0	85	4–6	0.3
P-styrene	1.05	6.4–6.7	80	2.4–3.3	0.03
NAS	1.13	5.6	85	4.5	0.15
Styrene acrylonitrile (SAN)	1.07	6.4	75	2.8	0.28
P-carbonate	1.25	6.7	120	4.7	0.2–0.3
P-methyl pentene	0.835	11.7	115	4.0	0.01
P-amide (Nylon)	1.185	8.2	80	5.1–5.8	1.5–3.0
P-arylate	1.21	6.3		7.1	0.26
P-sulfone	1.24	2.5	160	2.8	0.1–0.6
P-styrene co-butadiene	1.01	7.8–12			0.08
P-cyclohexyl methacrylate	1.11				
P-allyl diglycol carbonate	1.32		100	4.9	
Cellulose acetate butyrate	1.20			4.0–8.0	
P-ethersulfone	1.37	5.5	200	3.2–4.4	
P-chloro-trifluoroethelyne	2.2	4.7	200	6.2	0.003
P-vinylidene fluoride	1.78	7.4–13	150		0.05
P-etherimide	1.27	5.6	170		0.25

characterizing the polymers are often an order of magnitude larger than those associated with optical glasses. Consequently, subjecting any polymeric optical element to a significant thermal transient is likely to create more severe thermal gradients in the material, and result in significant thermally-induced optical figure errors.[10] Again, it is suggested that the interested reader consult the plastic handbooks and manufacturer's literature for a complete listing of this behavior, as additives and variation in molecular weight distribution may significantly affect all of these properties. Some of the most important physical properties of the more readily available optical polymers are tabulated in Table 1.

Conductivity (Thermal, Electrical)

Most materials which exhibit poor thermal conductivity are also poor electrical conductors. Since many unfilled polymers are very effective electrical insulators, they acquire static surface charge fairly easily, and dissipate it very slowly. Not surprisingly, these areas of surface charge quickly attract oppositely charged contaminants, most of which are harder than the plastic. Attempts to clear the accumulated particles from the surfaces by cleaning can, and usually do, result in superficial damage. Application of inorganic coatings to these surfaces may do double duty by providing a more conductive surface (less likely to attract contaminants), while improving the abrasion resistance.

Outgassing

In contrast to glass optical parts, which normally have very low vapor pressure when properly cleaned, most polymers contain lubricants, colorants, stabilizers, and so on, which may outgas throughout the life of the part. This behavior disqualifies most plastic optical elements from serving in space-borne instrumentation, since the gaseous products, once lost, surround the spacecraft, depositing upon solar panels and other critical surfaces. Some, but few, thermoset resins may be clean enough for space applications if their reaction stoichiometry is very carefully controlled in the creation of the part.

Water Absorption

Most polymers, particularly the thermoplastics, are hygroscopic. They absorb and retain water, which must, in most cases, be driven off by heating prior to processing. Following processing, the water will be reabsorbed if the surfaces are not treated to inhibit absorption. Whereas only a very small amount of water will normally attach to the surfaces of a glass optical element, the polymer materials used for optics may absorb from about 0.003 to about 2 percent water by weight. Needless to say, the trapped water may produce dimensional changes, as well as some minor alterations of the spectral transmission. Physical properties of some of the more familiar optical polymers are listed in Table 1. Density = ρ (g/cm^3); thermal expansion coefficient = α (cm/cm °C $\times 10^{-5}$); max. service temperature = T_s (°C); thermal conductivity = K (cal/sec cm °C $\times 10^4$); and water absorption (24 hr) = A_{H_2O} (%). Values are to be considered approximate, and may vary with supplier and processing variations.

Additives

Polymers are normally available in a variety of "melt flow" grades—each of which possesses viscosity properties best suited to use in parts having specific form factors. A number of additives are commonly present in these materials. Such additives may, or may not, be appropriate in an optical application. Additives for such things as flame retardancy, lubricants, lubrication, and mold release are best avoided if not included to address a specific requirement. Frequently, colorants are added for the purpose of neutralizing the naturally occurring coloration of the material. These additives create an artificial, but "clear," appearance. The colorants must, of course, absorb energy to accomplish this, resulting in a net reduction in total spectral transmission.

Radiation Resistance

Most of the optical polymers will be seen to exhibit some amount of fluorescence if irradiated by sufficiently intense high-energy radiation.[11] High-energy radiation of the ultraviolet and ionizing varieties will, in addition, produce varying amounts of polymer chain crosslinking, depending upon the specific polymer chemistry. Crosslinking typically results in discoloration of the material, and some amount of nonuniform energy absorption. Inhibitors may be added to the polymeric material to retard crosslinking, although, oddly enough, the polymers most susceptible to UV-induced discoloration are generally the least likely to be affected by ionizing radiation, and vice versa.

Documentation

Although polymeric materials suffer some shortcomings in comparison to glass (for optical applications), distinct advantages do exist. The major obstacle to the use of polymers, however, is the spotty and imprecise documentation of many of those properties required for good engineering and design. In general, the resin producers supply these materials in large quantity to markets wherein a knowledge of the optical properties is of little or no importance. With luck, the documentation of optical properties may consist of a statement that the material is "clear". In the rare case where refractive index is documented, the accuracy may be only two decimal places. In these circumstances, the optical designer or molder is left to investigate these properties independently—a complex task, since the processing itself may affect those properties to a substantial degree.

Unfortunately, optical applications may represent only a small fraction of a percent of the total market for a given resin formulation, and since these materials are sold at prices ranging from less than two dollars to a few dollars per pound, the market opportunity represented by optical applications seems minuscule to most polymer vendors.

34.5 OPTICAL PROPERTIES

Variations

It is only a fortuitous accident that some of the polymers exhibit useful optical behavior, since most all of these materials were originally developed for other end uses. The possible exceptions are the materials used for eyeglass applications (poly-diallylglycol), and the materials for optical information storage (specially formulated polycarbonate). Citation of optical properties for any polymeric material must be done with some caution and qualification, as different melt flow grades (having different molecular weight distribution) may exhibit slightly different refractive index properties. Additives to regulate lubricity, color, and so on can also produce subtle alterations in the spectral transmission properties.

Spectral Transmission

In general, the carbon-based optical polymers are visible-wavelength materials, absorbing fairly strongly in the ultraviolet and throughout the infrared.[12,13,14] This is not readily apparent from the absorption spectra published in numerous references, though. Such data are normally generated by spectroscopists for the purpose of identifying chemical structure, and are representative of very thin samples. One can easily develop the impression from this information that the polymers transmit well over a wide spectral range. Parenthetically, most of these polymers, while they have been characterized in the laboratory, are not commercially available. What is needed for optical design purposes is transmission data (for available polymers) taken from samples having sufficient thickness to be useful for imaging purposes.

Some specially formulated variants of poly-methylmethacrylate have useful transmission down to 300 nm.[15] Most optical polymers, however, begin to absorb in the blue portion of the visible spectrum, and have additional absorption regions at about 900 nm, 1150 nm, 1350 nm, finally becoming totally opaque at about 2100 nm. The chemical structure which results in these absorption regions is common to almost all carbon-based polymers, thus the internal transmittance characteristics of these materials are remarkably similar, with the possible exception of the blue and near-UV regions. A scant few polymers do exhibit some spotty narrowband transmission leakage in the far-infrared portion of the spectrum, but in thicknesses suitable only for use in filter applications.

Refractive Index

The chemistry of carbon-based polymers is markedly different from that of silicate glasses and inorganic crystals in common use as optical materials. Consequently, the refractive properties differ significantly. In general, the refractive indices are lower, extending to about 1.73 on the high end, and down to a lower limit of about 1.3. In practice, those materials which are readily available for purchase exhibit a more limited index range—from about 1.42 to 1.65. The Abbe values for these materials vary considerably, though, from about 100 to something less than 20. Refractive index data for a few of these polymers, compiled from a number of sources, is displayed in Table 2. In the chart,

TABLE 2 Refractive Index of Some Optical Polymers

Line ID	Wavl., nm	PMMA	P-styr	P-carb	SAN	PEI	PCHMA
	1014.0	1.4831	1.5726	1.5672	1.5519		
s	852.1	1.4850	1.5762	1.5710	1.5551		
r	706.5	1.4878	1.5820	1.5768	1.5601		
C	656.3	1.4892	1.5849	1.5799	1.5627		1.502
C′	643.9	1.4896	1.5858	1.5807	1.5634	1.651	
D	589.3	1.4917	1.5903	1.5853	1.5673		
d	587.6	1.4918	1.5905	1.5855	1.5674	1.660	1.505
e	546.1	1.4938	1.5950	1.5901	1.5713	1.668	
F	486.1	1.4978	1.6041	1.5994	1.5790		1.511
F′	480.0	1.4983	1.6052	1.6007	1.5800	1.687	
g	435.8	1.5026	1.6154	1.6115	1.5886		
h	404.7	1.5066	1.6253	1.6224	1.5971		
i	365.0	1.5136	1.6431	1.6432	1.6125		
Abbe number		57.4	30.9	29.9	34.8	18.3	56.1
$dn/dT \times 10^{-4}/°C$		−1.05	−1.4	−1.07	−1.1		

PMMA signifies polymethylmethacrylate; P = styr, polystyrene; p = care, polycarbonate; san, styrene acrylonitrile; PEI, polyetherimide; PCHMA, polycyclohexylmethacrylate. The thermo-optic coefficients at room temperature (change in refractive index with temperature) are also listed. Note that these materials, unlike most glasses, experience a reduction in refractive index with increasing temperature. Figure 1, a simplified rendition

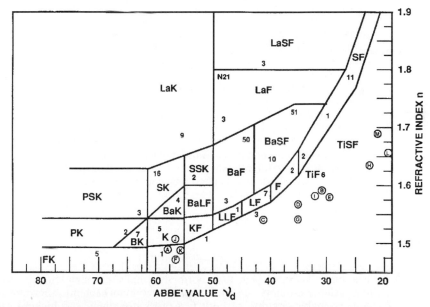

FIGURE 1 Optical glasses and polymers: (*a*) polymethylmethacrylate; (*b*) polystyrene; (*c*) NAS; (*d*) styrene acrylonitrile; (*e*) polycarbonate; (*f*) polymethyl pentene; (*g*) acrylonitrile-butadiene styrene (ABS); (*h*) polysulfone; (*i*) polystyrene co-maleic anhydride; (*j*) polycyclohexylmethacrylate (PCHMA); (*k*) polyallyl diglycol carbonte; (*l*) polyetherimide (PEI); (*m*) polyvinyl naphthalene.

of the familiar glass map (*n* vs. *v*), shows the locations of some of the more familiar polymers. Note that these materials all occupy the lower and right-hand regions of the map. In the Schott classification system, the polymers populate mostly the FK, TiK, and TiF regions of the map.[16]

Homogeneity

It must be kept constantly in mind that polymeric optics are molded and not mechanically shaped. The exact optical properties of a piece cannot, therefore, be quantified prior to manufacture of the element. In fact, the precise optical properties of the bulk material in an optical element are virtually certain to be a function of both the material itself, and of the process which produced the part. Some materials, notably styrene and butyrate resins, are crystalline to some degree, and therefore inherently birefringent. Birefringence may develop in amorphous materials, though, if the injection mold and process parameters are not optimized to prevent this occurrence. Likewise, the bulk scatter properties of a molded optical element are a function of the inherent properties of the material, but are also strongly related to the cleanliness of the processing and the heat history of the finished part.

34.6 OPTICAL DESIGN

Design Strategy

Virtually all optical design techniques which have evolved for use with glass materials work well with polymer optics. Ray-tracing formulary, optimization approaches, and fundamental optical construction principals are equally suitable for glass or plastic. The generalized approach to optical design with polymeric materials should be strongly medium-oriented, though. That is, every effort must be made to capitalize upon the design flexibility which the materials and manufacturing processes afford. Integration of form and function should be relentlessly pursued, since mechanical features may be molded integral with the optics to reduce the metal part count and assembly labor content in many systems.

Aberration Control

The basic optical design task normally entails the simultaneous satisfaction of several first-order constraints, the correction of the monochromatic aberrations, and the control of the chromatic variation of both first-order quantities and higher-order aberrations. It is well known that management of the Petzval sum, while maintaining control of the chromatic defects, may be the most difficult aspect of this effort.[17,18] It is also widely recognized that the choice of optical materials is key to success. While the available polymer choices cover a wide range of Abbe values, insuring that achromatization may be accomplished in an all-polymer system, the refractive index values for these materials are not well-positioned on the "glass" map to permit low Petzval sums to be easily achieved.

Material Selection

Simultaneous correction of the Petzval sum and the first-order chromatic aberration may, however, be nicely accomplished if the materials employed possess similar ratios of Abbe number to central refractive index. This implies that the *best* material combinations (involving polymers) should probably include an optical glass. Also implied is the fact that these hybrid material combinations may be inherently superior (in this respect) to all-glass combinations. Ideally, the chosen materials should be well-separated (in Abbe value) on the glass map, so that the component powers required for achromatization do not become unduly high. This condition is satisfied most completely with polymers which lie in the TiF sector of the glass map, coupled with glasses of the LaK, LaF, and LaSF families.

Most lens designers would prefer to utilize high-refractive-index materials almost exclusively in their work. Optical power must be generated in order to form images, and because the combination of optical surface curvature and refractive index creates this refractive power, these two variables may be traded in the lens design process. Since it is well known that curvature generates aberration more readily than does a refractive index discontinuity, one generally prefers to achieve a specified amount of refractive power through the use of low curvature and high refractive index. From this perspective, the polymers are at a distinct disadvantage, most of them being low-index materials.

Aspheric Surfaces

An offsetting consideration in the use of polymeric optical materials is the freedom to employ nonspherical surfaces. While these may be awkward (and very expensive) to produce in glass, the replication processes which create plastic optical parts do not differentiate between spherical and nonspherical surfaces.

As any lens designer can attest, the flexibility that aspheric surfaces make available is quite remarkable.[19,20] Spherical surfaces, while convenient to manufacture by grinding and polishing, may generate substantial amounts of high-order aberration if used in any optical geometry which departs significantly from the aplanatic condition. These high-order aberrations are often somewhat insensitive to substantial changes in the optical prescription. Thus, profound configurational alterations may be necessary to effect a reduction in these image defects.

On the other hand, the ability to utilize surface shapes which are more complex than simple spheres permits these high-order aberration components to be moderated at their point of origin, which may in turn reduce the amount of "transferred aberration" imparted to surfaces downstream in the optical train. In a multielement optical system, especially one employing cascaded aspheric surfaces, the required imagery performance may be achieved using fewer total elements. And due to the fact that the surface aberration contributions are diminished, the sensitivity to positioning errors may also be reduced, with the result that an aspheric optical system may actually be more forgiving to manufacture than its spherical counterpart.

In practice, the use of aspheric surfaces in polymer optical elements appears to more than compensate for the handicap imposed by low refractive index values. Using aspheric surfaces, it is possible to bend, if not break, many of the rules which limit design with spherical surfaces. Aspherics create extra leverage to deal with the monochromatic aberrations, and with the chromatic variation of these image defects. A designer experienced with aspherics, given a capable set of software tools, can frequently create optical constructions which deliver high performance, despite the fact that they appear odd to those accustomed to the more "classical" spherical surface configurations. Quite often, unfavorable design constraints such as an inconvenient aperture stop location, may be handled with less difficulty using aspherics.

Athermalization

The thermal behavior of the polymers, mentioned previously, may cast a shadow upon some applications where the temperature is expected to vary over a significant range, but the focal surface location must be fixed in space. In such cases, the variation of refractive index usually accounts for the largest share of the variation, with the dimensional changes playing a secondary role. In such situations, the thermally induced excursions of the focal surface may be compensated by modeling these functions and designing mechanical spacers of the proper material to stabilize the detector/image location.

Alternatively, the optical system may be designed to exhibit inherently athermal behavior over the operational temperature range.[21] Unfortunately, this is not strictly possible using only polymeric materials, as the thermo-optic coefficients display so little variation among themselves that the component powers would be absurdly high.

In combination with one or more glass elements, however, very nicely athermalized design solutions may be obtained with polymer elements.[22] Athermal designs may be generated by modeling the optical system in multiconfiguration mode in the lens design software, much as one would develop a zoom lens. The parameters to be "zoomed" in this case are the refractive indices at two or more temperatures within the operating range. The resulting designs frequently concentrate most of the refractive power in the glass elements, with the polymer elements functioning to achieve achromatism and control of the monochromatic aberrations. See also Vol. I, Chap. 39 of this Handbook.

Processing Considerations

In much the same manner that optical design with polymer materials is different from optical design with glass, the treatment of the fabrication and assembly issues are also quite different matters. The major issues requiring examination are those related to the materials themselves. While it is possible to characterize the glass for an optical system with complete certainty prior to performing any fabrication operations, with polymers, one's knowledge of the starting materials is only a rough indication of the properties of the finished optical parts.

When optical properties data are offered by the polymer supplier, it should be realized that these numbers apply *only* to measurement samples which have been predried to specification, have experienced a specified residence time in the extrusion barrel under specific temperature conditions, have been injected into the mold cavity at specific rates and pressures, and so on. Consequently, it is unlikely that the refractive properties of a polymer element will conform closely to catalog values (if such values are indeed supplied). Moreover, homogeneity, bubble content, scatter properties, and so on, are all process-dependent. So while the melt sheets may fix the optical properties of glass materials very precisely, the uncertainty associated with the polymers demands that refractive variations be allocated a significant portion of the fabrication and assembly error budget.

Manufacturing Error Budget

Other constructional parameters, conversely, may be implemented with great precision and repeatability in plastic. The molding process, executed by means of modern equipment, can be exceedingly stable. Vertex thickness, curvature, and wedge may often be maintained to a greater level of precision, with greater economy than is possible with glass fabrication technology. It is not unusual to see part-to-part variations in vertex thickness of less than 0.01 to 0.02 mm over a run of thousands of parts from a single cavity.

Multiple Cavities

The economic appeal of injection molding is the ability to create several parts in one molding cycle. In a multicavity scenario, the parts from different cavities may exhibit some small dimensional differences, depending upon the level of sophistication of the tool design and the quality of its construction. Cavity variations in axial thickness, fortunately, may be permanently minimized by implementing small tooling adjustments after the mold has been exercised. Consequently, part thickness variation rarely consumes a significant fraction of the constructional error budget.

Dimensional Variations

Surface radii, like axial thickness, may be replicated with great repeatability *if* the molding process is adjusted to a stable optimum. Radius errors, if they are present, are usually attributable to incorrect predictions of shrinkage, and may be biased out by correcting the radii of the mold inserts. Thus, the consistency of surface radii achievable with glass may often be equaled in plastic. Thus, radius errors, as well as axial thickness errors, frequently constitute a small portion of the polymer optics manufacturing error budget.

Element wedge, like axial thickness, may be minimized by careful attention to precision in the tool design and construction. It is quite possible to achieve edge-to-edge thickness variations of less than 0.01 mm in molded plastic lenses. With polymer lenses, the azimuthal location of the part gate may be used, if necessary, to define rotational orientation of the element in the optical train. Consequently, rotational alignment of plastic optical parts may be easily indexed.

Optical Figure Variations

Control of optical figure quality is obviously key to the successful execution of a good optical design. In glass, achievement of subfringe figure conformance is accomplished routinely, albeit at some cost penalty. In polymeric optics, the nonlinear shrinkage, surface tension, and other processing-related effects cause surface figure errors to scale with part size, sometimes at a rate proportional to some exponent of diameter. This limits the practical size range for polymeric optics, although capable optics molders may routinely produce elements in the 10-mm-diameter range to subfringe accuracy.[23]

On one hand, it can probably be stated that processing-induced variations in properties, and a dearth of dependable optical data, preclude any serious discussions of such things as apochromatic polymeric optics, or of large polymeric optics operating at the diffraction limit. On the other hand, the consistency with which some dimensional parameters may be reproduced in quantity, and the design freedom and flexibility afforded by molded aspherics, make possible the satisfaction of some design requirements which would be out of range for conventional glass optics.[24,25]

Specification

Given the fact that the guidelines and restrictions for design and implementation are very different for glass and polymeric optics, it is not surprising that the approach to specification of polymer optical parts and systems should be tailored to the materials and processes of polymer optics. Attempts to convert a glass optics concept to plastic are frequently unsuccessful if the translation overlooks the fundamental themes of the molding and tooling technologies involved. Much as optimum tube and solid-state electrical circuit topologies should be significantly different, so must the execution of a conceptual optical system, depending upon whether glass or polymer material is the medium.

It follows naturally that manufacturing drawings for polymer optics may contain annotations which seem unfamiliar to those versed in glass optics manufacture. Furthermore, some specifications which are universally present on all glass optics drawings may be conspicuously absent from a polymer optics print.

For example, thermal and cosmetic damage considerations preclude the use of the familiar test glasses in the certification of polymeric optics. Figure conformance, then, need only be specified in "irregularity" or asphericity terms, since the alternative method, use of a noncontacting interferometer, implies that the focus error (*fringe power* in test plate language) will be automatically removed in the adjustment of the test setup.

References to ground surfaces may be omitted from polymer optics drawings, since no such operation takes place. Discussions of "chips" inside the clear aperture, staining, and the like are also superfluous. Beauty defect specifications do apply, although such imperfections are almost always present in every sample from a specific cavity, probably implying the need to rework a master surface.

In general, the lexicon of optics, and that of the molding industry, do not overlap to a great extent. Molding terms like *flash* and *splay* are meaningless to most optical engineers. Those endeavoring to create a sophisticated polymeric optical system, anticipating a successful outcome, are advised to devote some time to the study of molding, and to discussions with the few experts in the arcane field of optics molding, before releasing a drawing package which may be unintelligible to or misunderstood by the vendor.

34.7   PROCESSING

Casting

As mentioned above, polymeric optics may be produced by any of several processes. These include fabrication, transfer replication, casting, compression molding, injection molding, and some combinations of the aforementioned.[26] The earliest polymeric optical parts were probably produced by fabrication or precipitation from solution. Large military tank prisms have been made by both processes. In the latter case, the polymer (typically PMMA) was dissolved, and the solvent then evaporated to produce a residue of polymer material in the shape of the mold—a very inefficient technique indeed.

Many of the polymers may be fabricated by cutting, grinding, and polishing, much as one would deal with glass materials. The thermoset resin tradenamed CR-39 (poly-diallylglycol) was formulated specifically to be processed using the same techniques and materials as those used to fabricate glass optics. And this material does indeed produce good results when processed in this manner. It is used extensively in the ophthalmic industry to produce spectacle lenses. The processing, in fact, usually involves casting the thermoset resin to create a lens blank which emerges from the mold with the optical surfaces polished to final form. More conventional fabrication techniques may then be utilized to edge the lens, or perhaps to add a bifocal portion.

Abrasive Forming

Unfortunately, the softness of most of the polymers, coupled with their poor thermal conductivity, complicates the achievement of a truly high quality polish using conventional methods. Even in the case of CR-39, which is relatively hard for a polymer material, some amount of "orange peel" in the polished surface seems unavoidable. Many thermoplastics, most of them softer than CR-39, may be conventionally ground and polished to give the appearance of an acceptable optical surface. Closer examination, however, reveals surface microstructure which probably does not fall within the standards normally associated with precision optics. Nonetheless, fabrication of optical elements from large slabs of plastic is often the only viable approach to the creation of large, lightweight refractive lenses, especially if cost is an issue.

In general, the harder, more brittle polymers produce better optical surfaces when ground and polished. PMMA and others seem to fare better than, say, polycarbonate, which is quite soft, exhibits considerable elongation at the mechanical yield point, but is in great demand due to its impact resistance.

Single-point Turning

An alternative approach to fabrication, one that is especially useful for the production of aspheric surfaces, is the computer numerical control (CNC) lathe turning of the bulk material using a carefully shaped and polished tool bit of single-crystal diamond or cubic boron nitride. See also Vol. I, Chap. 41, this Handbook. The lathe required to produce a good result is an exceedingly high precision tool, having vibration isolation, temperature control, hydrostatic or air bearings, and so on. On the best substrate materials (PMMA is again a good candidate), very good microroughness qualities may be achieved. With other materials, a somewhat gummy character (once more, polycarbonate comes to mind) may result in microscopic tearing of the surface, and the expected scatter of the incident radiation.

The diamond-turning process is often applied in conjunction with other techniques in order to speed progress and reduce cost. Parts which would be too large or too thick for economical stand-alone injection molding are frequently produced more efficiently by diamond-turning injection molded, stress-relieved preforms, which require minimal material removal and lathe time for finishing. Postpolishing, asymmetric edging, and other postoperations may be performed as necessary to create the finished part. Optics for illumination and TV projection applications are often produced by some combination of these techniques. Given the fact that the technology in most widespread use for the production of plastic optics involves some form of molding (a front-loaded process, where cost is concerned), diamond-turning is often the preferred production method for short production runs and prototype quantities.

Compression Molding

Most high-volume polymeric optics programs employ a manufacturing technology involving some form of molding to produce the optical surfaces, if not the entire finished part.[27] Of the two most widely used approaches, compression molding is best suited to the creation of large parts having a thin cross section. In general, any optical surface possessing relief structure having high spatial frequency is not amenable to injection molding, due to the difficulty of forcing the material through the cavity, and due to the fact that the relief structure in the mold disrupts the flow of the polymer. In addition, the relief structure in the master surfaces may be quite delicate, and prone to damage at the high pressures often present in the mold cavity.

The compression molding process is capable of producing results at considerably lower surface pressure than injection molding, and as long as the amount of material to be formed is small, this molding technology can replicate fine structure and sharp edge contours with amazing fidelity. Since the platens of a compression molding press are normally heated using steam or electrical heaters, most compression molded parts are designed to be executed in polymers having a relatively low temperature softening point, and materials like polyethersulfone are rarely utilized.

Injection Molding

Optical parts having somewhat smaller dimensions may be better suited to production by the injection molding process.[28] This is probably the preferred polymer manufacturing technology for optical elements having a diameter smaller than 0.1 m and a thickness not greater than 3 cm. Not only do the economics favor this approach in high production volume, but if properly applied, superior optical surfaces may be produced.[29]

It should be kept firmly in mind that the basic injection molded process (as it is known to most practitioners) requires a great deal of refinement and enhancement in order to produce credible optical parts.[30] Unfortunately, very few molders possess either the molding know-how, or the testing and measurement sophistication to do the job correctly. Given a supply of quality polymer material, the molding machine itself must be properly configured and qualified. Relatively new machinery is a must. The platens to which the mold halves are mounted must be very rigid and properly aligned. And this alignment must be maintainable on a shot-to-shot basis for long periods. The screw and barrel must be kept scrupulously clean, and must be carefully cleaned and purged when switching materials. The shot capacity, in ideal circumstances, should be more carefully matched to the part volume than for non-optical parts. The process control computer must be an inordinately flexible and accurate device, able to profile and servo a number of operational functions that might be of little importance if the molded part were not optical in nature.[31]

Since much of the heating of the injected polymer resin occurs as a result of physical shear and compression (due to a variable pitch screw), the selection of these machine characteristics is critical to success. In addition, the energy supplied to the machine barrel by external electric heaters must be controlled with more care than in standard industrial applications. A failure of a single heater, or a failure of one of the thermal measurement devices which close that servo loop, may result in many defective parts.

Vendor Selection

The injection mold itself requires special attention in both design and execution in order to produce state-of-the-art molded lenses. A number of closely held "trade tricks" normally characterize a mold designed to produce optical parts, and these subtle variations must be implemented with considerably greater accuracy than is normally necessary in ordinary molding. The mold and molding machine are often designed to operate more symbiotically than would be the case in producing non-optical parts. Control of the mold temperature and temperature gradients is extremely critical, as is the control bandwidth of those temperatures and the temperature of the molding room itself. The most important conclusion to be drawn from the preceding paragraphs is that *the molding vendor for polymer optical parts must be selected with great care*. A molding shop, no matter how sophisticated and experienced with medical parts, precision parts for electronics, and so on, will probably consume much time and many dollars before conceding defeat with optical parts.

Although success in molding optical elements is a strong function of equipment, process control, and engineering acumen, *attention to detail in the optical and mechanical design phases will consistently reduce the overall difficulty of manufacturing these items*. An awareness of the basic principles of injection molding procedures and materials is very helpful here, but it is necessary to be aware that, in the optical domain, we are dealing with micrometer-scale deformations in the optical surfaces. Thus, errors or oversights in design and/or molding technique which would totally escape notice in conventional parts can easily create scrap optics.

Geometry Considerations

The lens design effort, for best results, must be guided by an awareness of the basic physics of creating an injection molded part, and of the impact of part cross section, edge configuration, asymmetry, and so on. In general, any lens having refractive power will possess a varying thickness across its diameter. Unfortunately, meniscus-shaped elements may mold best due to the more uniform nature of the heat transfer from the bulk.[32] Positive-powered lens elements will naturally shrink toward their center of mass as they cool, and it may be difficult to fill the mold cavity efficiently if the edge cross section is only a small fraction of the center thickness.

Negative lenses, on the other hand, tend to fill in the outer zones more readily, since the thinner portion of the section (the center) tends to obstruct flow directly across the piece from the part gate. In extreme circumstances, it is possible that the outer zones of the lens element will be first to fill, trapping gases in the center, forming an obvious *sink* in molding terminology. Parts designed with molded-in bores may exhibit the '*weld-line*' phenomenon, which is a visible line in the part where the flow front of the molten plastic is divided by the mold cavity obstruction forming the bore. In the case of both negative and positive lens elements, it is good policy to avoid element forms wherein the center-to-edge thickness ratio exceeds three for positive elements, or is smaller than 0.3 for negative elements.

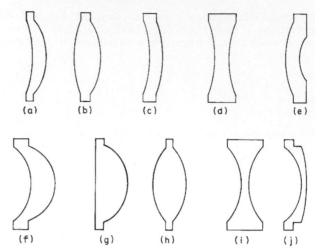

FIGURE 2 Some polymer lens element cross-sectional configurations.

Shrinkage

Surface-tension effects may play a significant role in the accuracy to which a precision optical surface may be molded.[33,34] Particularly in areas of the part where the ratio of surface area/volume is locally high (corners, edges), surface tension may create nonuniform shrinkage which propagates inward into the clear aperture, resulting in an edge rollback condition similar to that which is familiar to glass opticians. Surface tension and volumetric shrinkage may, however, actually aid in the production of accurate surfaces. Strongly curved surfaces are frequently easier to mold to interferometric tolerances than those having little or no curvature. These phenomena provide motivation to oversize optical elements, if possible, to a dimension considerably beyond the clear apertures. A buffer region, or an integrally molded flange provides the additional benefit of harmlessly absorbing optical inhomogeneities which typically form near the injection gate. Figure 2 depicts several optical element forms exhibiting favorable (*a–e*) and unfavorable (*f–j*) molding geometries. In some cases, a process combining injection and compression molding may be used to improve optical figure quality. Several variants of this hybrid process are in use worldwide, with some injection molding presses being specifically fitted at the factory to implement this procedure.[35]

Mechanical Assembly

In order to appreciate fully the design flexibility and cost-saving potential of polymer optics, it is necessary to modify one's approach to both optical and mechanical design. A fully optimized polymeric optical system not only makes use of aspheric technology and integrally molded features in the optical elements, but embodies an extension of this design philosophy into the lens housing concept and assembly strategy. These issues should ideally be considered in concert from the very beginning, so that design progress in one aspect does not preclude parallel innovation in other facets of the development process.

It is important to resist the urge to emulate glass-based optomechanical design approaches, since the polymer technology permits design features to be implemented which would be prohibitively expensive (or even impossible) in metal and glass. Spacers

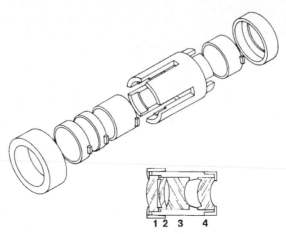

FIGURE 3 Collet-type lens housing. Joining by ultrasound eliminates the possibility of pinching lens elements.

required to separate elements may be molded as part of the elements (Fig. 2), reducing the metal part total, and simplifying assembly operations. See also Vol. I, Chap. 37, of this Handbook. Housings may be configurations which would be either improbable or unmanufacturable using machine tool technology. The collet-and-cap design shown in Fig. 3 is one such example. Joining might be accomplished by ultrasonic bonding. The clamshell concept shown in Fig. 4 may be designed so that the two halves of the housing are actually the same part, aligned by molded-in locating pins. Joining might be performed by a simple slip-on C ring.

Whereas lens assemblies in glass and metal are normally completed by seating threaded retaining rings, their plastic counterparts may be joined by snap-together pieces, ultrasonic bonding, ultraviolet-curing epoxies, expansion C rings, or even solvent bonding.[36] Solvent bonding is dangerous, however, since the errant vapors may actually attack the polymer optical surfaces.

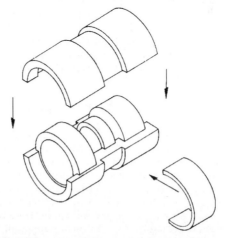

FIGURE 4 Clamshell lens housing. Possibility of lens jamming during assembly is minimized.

Following the basic polymeric optics philosophy, the lens element containment and assembly approach should probably not even consider the disassembly option in the event of a problem. In order to maximize assembly precision, and minimize unit cost, the design of the lens cell should evolve alongside that of the optical system, and this cell should be visualized as an extension of a fixture conceived to minimize the labor content of the assembly.

An in-depth treatment of optical mold design and tooling technology is obviously beyond the scope of this discussion. Many of the methods and procedures parallel those in use in the molding industry at large. However, a number of subtle and very important detail differences do exist, and these are not extensively documented in the literature. Issues having to do with metallurgy, heat treatment, chemical passivation, metal polishing, and so on, have little to do with the actual design and engineering of a polymeric optical system. In a modern tool design exercise, though, the flow behavior of the polymer material in the mold, and the thermal behavior of that mold, are carefully modeled in multinode fashion, so that part quality may be maximized, and cycle time minimized.[37] A nodding awareness of these methods, and the underlying physics, may be helpful to the person responsible for the engineering of the polymer optical system.

Testing and Qualification

In the process of implementing any optical system design, the matters of testing and certification become key issues. In molded optics, the master surfaces, whose shapes are ultimately transferred to the polymer optical parts, must be measured and documented. A convenient testing procedure for the optical elements replicated from these surfaces must likewise be contrived, in order to optimize the molding process and insure that the finished assembly will perform to specification. The performance of that assembly must itself be verified, and any disparities from specification diagnosed.

In general, mechanical dimensions of the polymer parts may be verified by common inspection tools and techniques used in the glass optics realm. The possibility of inflicting surface damage, however, dictates that noncontact interferometric techniques be used in lieu of test glasses for optical figure diagnosis. This is a straightforward matter in the case of spherical surfaces, but requires some extra effort in the case of aspherics. See also Vol. II, Chap. 30, of this Handbook.

Obviously, aspheric master surfaces must be scrupulously checked and documented, lest the molder struggle in vain to replicate a contour which is inherently incorrect. The verification of the aspheric masters and their molded counterparts may be accomplished in a variety of ways. Mechanical gauging, if properly implemented, works well, but provides reliable information through only one azimuthal section of the part. Measurement at a sufficient number of points to detect astigmatism is awkward, very time consuming, and expensive. And this is not exactly consistent with the spirit of polymer optics.

Null Optics

An optical *null corrector* permits the aspheric surface to be viewed in its entirety by the interferometer as if it were a simple spherical surface.[38,39] This is a rapid and convenient procedure. The null optics consist of very accurately manufactured (and precisely aligned) spherical glass elements designed to introduce aberration in an amount equal to, but of opposite sign from, that of the tested aspheric. Thus, interspersing this device permits aspherics to be viewed as if they were spherical. Since there exists no simple independent test of the null compensator, one must depend heavily upon the computed predictions of correction and upon the skill of the fabricator of the corrective optics. See also Vol. II, Chap. 31, of this Handbook.

The concept of greatest importance regarding the use of aspheric surfaces is that *successful production of the total system is cast into considerable doubt if a surface is present which is not amenable to convenient testing.* While some aspheric optics may be nulled fairly easily, those which appear in polymer optical systems are frequently strong, exhibiting significant high-order derivatives. If the base curves are strong, especially strongly convex, there may exist no practical geometry in which to create a nulling optical system. And if a favorable geometry does exist, several optical elements may be necessary to effect adequate correction. One can easily approach a practical limit in this situation, since the manufacturing and assembly tolerances of the cascaded spherical elements may themselves (in superposition) exceed the theoretical correction requirement. The bottom line is that one should not proceed with cell design, or any other hardware design and construction, until the aspheric testing issues have been completely resolved.

34.8 COATINGS

Reflective Coatings

Given the fact that optical polymers exhibit specular properties similar to those of glass, it is not surprising that optical coatings are often necessary in polymeric optical systems. The coatings deposited upon polymer substrates fall mostly into four general categories. These include coatings to improve reflectivity, to suppress specular reflection, to improve abrasion resistance, and to retard accumulation of electrostatic charge.

Reflective coatings may be applied by solution plating, or by vacuum-deposition. These are most often metallic coatings, usually aluminum if vacuum-deposited, and normally chromium if applied by plating. The abrasion resistance and general durability of such coatings is rather poor, and susceptibility to oxidation quite high, if no protective coating is applied over the metal film. In some applications, especially involving vacuum deposition, the overcoat may be a thin dielectric layer, deposited during the same process which applies the metal film. If the reflective coating has been applied by plating, the overcoat may be an organic material, perhaps lacquer, and may be deposited separately by spraying or dipping. Not surprisingly, the quality of a surface so treated will be poor by optical standards, and probably suitable only for toy or similar applications.

Antireflection Coatings

Antireflection coatings are frequently utilized on polymer substrates, and may consist of a single layer or a rudimentary multilayer stack yielding better reflection-suppression performance. Due to the stringent requirements for control of the layer thicknesses, such coating formulations may be successfully deposited only in high-vacuum conditions, and only if temperatures in the chamber remain well below the service temperature of the substrate material. Elevated temperatures, necessary for baking the coatings to achieve good adhesion and abrasion resistance, may drive off plasticizing agents, limiting the "hardness" of the chamber vacuum. Such temperatures can ultimately soften the optical elements, so that their optical figure qualities are compromised. Relatively recent developments in the area of ion beam-assisted deposition have made possible improvements in the durability of coatings on polymer materials without having to resort to significantly elevated chamber temperatures.[40] See also Vol. I, Chap. 42, of this Handbook.

Antiabrasion Coatings

In general, many polymeric optical systems which could benefit from application of coatings are left uncoated. This happens because the expense incurred in cleaning, loading, coating, unloading, and inspecting the optical elements may often exceed that of molding the part itself. Some optics, particularly those intended for ophthalmic applications, are constantly exposed to abuse by abrasion, and must be protected, cost notwithstanding. Antiabrasion coatings intended to provide immunity to scratching may be of inorganic materials (normally vacuum-deposited), or may be organic formulations.[41]

Inorganic antiabrasion coatings may be similar to those used for simple antireflection requirements, except that they may be deposited in thicknesses which amount to several quarter-wavelengths. The practical thickness is usually limited by internal stress buildup, and by differential thermal expansion between coating and substrate. In general, the inorganic coatings derive their effectiveness by virtue of their hardness, and provide protection only superficially, since sufficient pressure will collapse the underlying substrate, allowing the coating to fracture.

Organic coatings for abrasion resistance normally derive their effectiveness from reduction of the surface frictional coefficient, thereby minimizing the opportunity for a hard contaminant to gain the purchase required to initiate a scratch. These coatings are often applied by dipping, spraying, or spinning. Coatings thus deposited usually destroy the smoothness which is required if the piece is to be qualified as a precision optical element.

Antistatic Coatings

Coatings applied for the purpose of immunization against abrasion, or suppression of specular reflection, often provide a secondary benefit. They may improve the electrical conductivity of the host surface, thus promoting the dissipation of surface static charge, and the accumulation of oppositely charged contaminants. In circumstances where antireflection or antiabrasion coating costs cannot be justified, chemical treatments may be applied which increase conductivity. These materials typically leave a residue sufficiently thin that they are undetectable, even in interferometric testing.

34.9   REFERENCES

1. Richard M. Altman and John D. Lytle, "Optical Design Techniques for Polymer Optics," *S.P.I.E. Proc.* **237:**380–385 (1980).
2. Chuck Teyssier and Claude Tribastone, *Lasers & Optronics* **Dec:**50–53 (1990).
3. PPG Ind., Inc., Tech Bulletin-CR-39.
4. H. Dislich, *Angew. Chem. Int. Ed. Engl.* **18:**49–59 (1979).
5. H. D. Wolpert, *Photonics Spectra* **Feb:**68–71 (1983).
6. Plastics Desk Top Data Bank pp. 803–837 (1986).
7. *Modern Plastics Encyclopedia—Eng. Data Bank,* McGraw-Hill, New York, 1977, pp. 453–708.
8. *Plastics Technology Manufacturing Hdbk. and Buyer's Guide,* 1986, pp. 358–740.
9. *C.R.C. Hdbk. of Laser Science & Technology,* vol. IV, pp. 85–91.
10. *Encyclopedia of Polymer Science and Technology,* vol. 1, John Wiley & Sons, 1976.
11. *Space Materials Hdbk,* Lockheed Missiles and Space Corp., 1975.
12. John D. Lytle, Gary W. Wilkerson, and James G. Jaramillo, *Appl. Opt.* **18:**1842–1846 (1979).
13. D. C. Smith, Alpert, et al., *N.R.L. Report* 3924, 1951.
14. R. E. Kagarise and L. A. Weinberger, *N.R.L. Report* 4369, 1954.

15. Rohm & Haas Product Bulletin-PL 612d, 1979.

16. *Catalogue of Optical Glasses,* Schott Glass Technologies, 1989.

17. John D. Lytle, *S.P.I.E. Proc.* **1354:**388–394 (1990).

18. Jan Hoogland, *S.P.I.E. Proc.* **237:**216–221 (1980).

19. Atsuo Osawa et al., *S.P.I.E. Proc.* **1354:**337–343 (1990).

20. E. I. Betensky, *S.P.I.E. Proc.* **1354:**663–668 (1990).

21. Lee R. Estelle, *S.P.I.E. Proc.* **237:**392–401 (1980).

22. Kimball Straw, *S.P.I.E. Proc.* **237:**386–391 (1980).

23. M. Muranaka, M. Takagi, and T. Maruyama, *S.P.I.E. Proc.* **896:**123–131 (1988).

24. John D. Lytle, *S.P.I.E. Proc.* **181:**93–102 (1979).

25. Alice L. Palmer, "Practical Design Considerations for Polymer Optical Systems," *S.P.I.E. Proc.* **306,** (1981).

26. D. F. Horne, *Optical Production Technology,* 2d. ed., Adam Hilger, Ltd., 1983 pp. 167–170.

27. John R. Egger, *S.P.I.E. Proc.* **193:**63–69 (1979).

28. Roland Benjamin, *Plastics Design and Processing* **19**(3):39–49 (1979).

29. Richard F. Weeks, *Optical Workshop Notebook,* vol. I., O.S.A., 1974–1975, sect. XVII.

30. D. F. Horne, "Lens Mechanism Technology," Crane, Russack & Co., New York, 1975.

31. J. Sneller, *Modern Plastics Intl.* **11**(6):30–33 (1981).

32. John D. Lytle, "Workshop on Optical Fabrication and Testing," *Tech. Digest,* O.S.A., pp. 54–57 (1980).

33. Ernest C. Bernhardt and Giorgio Bertacchi, *Plastics Technology* **Jan:**81–85 (1986).

34. George R. Smoluk, *Plastics Engineering* **Jul:**107 (1966).

35. *Plastics News,* Aug. 1989, pp. 8–9.

36. *Plastics World,* July 1979, p. 34.

37. Matthew H. Naitove, *Plastics Technology* **Apr.** (1984).

38. D. Malacara (ed.), *Optical Shop Testing,* John Wiley & Sons, New York, 1978, chaps. 9, 14.

39. George W. Hopkins and Richard N. Shagam, *Appl. Opt.* **16**(10):2602 (1977).

40. James D. Rancourt, *Optical Thin Films-user's Hdbk.,* MacMillan, New York, 1987, pp. 197–199.

41. J. W. Prane, *Polymer News* **6**(4):178–181 (1980).

CHAPTER 35
PROPERTIES OF METALS

Roger A. Paquin
Advanced Materials Consultant
Tucson, Arizona
and
Optical Sciences Center
University of Arizona, Tucson

35.1 GLOSSARY

a	absorptance, absorptivity
a	plate radius (m)
B	support condition
C_{ij}	elastic stiffness constants (N/m^2)
C_p	specific heat (J/kg K)
CTE	coefficient of thermal expansion (K^{-1})
D	flexural rigidity (N m^2)
D	thermal diffusivity (m^2/s)
E	elastic modulus (Young's) (N/m^2)
$\mathbf{E}$	electromagnetic wave vector (J)
e	electron charge (C)
E_o	amplitude of electromagnetic wave at $x - 0$ (J)
G	load factor (N/kg)
G	shear modulus, modulus of rigidity (N/m^2)
g	acceleration due to gravity (m/s^2)
I	light intensity in medium (W/m^2)
i	$(-1)^{1/2}$
I_0	light intensity at interface (W/m^2)
I_0	section moment of inertia (m^4/m)
K	bulk modulus (N/m^2)
k	extinction coefficient
k	thermal conductivity (W/m K)

L	length (m)
M	materials parameter (kg/N m)
m	electron mass (kg)
$\mathbf{N}$	complex index of refraction
N	number of dipoles per unit volume (m^{-3})
n	index of refraction
P	plate size (m^4)
q	load (N/m^2)
r	reflectance, reflectivity
R_I	intensity reflection coefficient
S	structural efficiency (m^{-2})
T	temperature (K)
t	time (s)
t	transmittance, transmissivity
V_0	volume per unit area of surface (m)
x	distance (m)
α	coefficient of thermal expansion (K^{-1})
α	absorption coefficient (m^{-1})
β	deflection coefficient
β	dynamic deflection coefficient
Γ	damping constant
δ	skin depth (nm)
δ	deflection (m)
δ_{DYN}	dynamic deflection (m)
ϵ	emittance, emissivity (W/m^2)
$\boldsymbol{\varepsilon}$	complex dielectric constant
ε_0	permittivity of free space (F/m)
ε_1	real part of dielectric constant
ε_2	imaginary part of dielectric constant
$\ddot{\theta}$	angular acceleration (s^2)
λ	wavelength (m)
λ_0	wavelength in vacuum (m)
μ	magnetic susceptibility (H/m)
ν	frequency (s^{-1})
ν	Poisson's ratio

ρ mass density (kg/m^3)

σ conductivity (S/m)

ω radian frequency (s^{-1})

35.2 INTRODUCTION

Metals are commonly used in optical systems in three forms: (1) structures, (2) mirrors, and (3) optical thin films. In this article, properties are given for metal mirror substrate and structural materials used in modern optical systems. Many other materials have not been included due to their limited applicability. Metal film properties are discussed in the context of thick films (claddings) rather than optical thin films that are covered in Chap. 42, Optical and Physical Properties of Films and Coatings, in Vol. I. Since mirrors are structural elements, the structural properties are equally important as the optical properties to the designer of an optical system. Therefore, the properties addressed here include physical, mechanical, and thermal properties in addition to optical properties. Mechanical and thermal properties of silicon (Si) and silicon carbide (SiC) are included, but not their optical properties since they are given in the article entitled "Optical Properties of Semiconductors," Vol II, Chap. 36.

After brief discussions of optical properties, mirror design, and dimensional stability, curves and tables of properties are presented, as a function of temperature and wavelength, where available. For more complete discussions or listings, the reader should consult the references and/or one of the available databases.[1-3] A concise theoretical overview of the physical properties of materials is given by Lines.[4]

Nomenclature

The symbols and units used in this subsection are consistent with usage in other sections of this Handbook although there are some unavoidable duplications in the usage of symbols between categories of optical, physical, thermal, and mechanical properties. Definitions of symbols with the appropriate units are contained in the table at the beginning of this article.

Optical Properties

The definitions for optical properties given in this section are primarily in the geometric optics realm and do not go into the depth considered in many texts dealing with optical properties of solids.[5-8]

There is obviously a thickness continuum between thin films and bulk, but for this presentation, bulk is considered to be any thickness of material that has bulk properties. Typically, thin films have lower density, thermal conductivity, and refractive index than bulk; however, current deposition techniques are narrowing the differences. Optical properties of thin films are presented only when bulk properties have not been found in the literature.

The interaction between light and metals takes place between the optical electric field and the conduction band electrons of the metal.[9] Some of the light energy can be transferred to the lattice by collisions in the form of heat. The optical properties of metals are normally characterized by the two optical constants: index of refraction n and extinction coefficient k that make up the complex refractive index $\mathbf{N}$ where:

$$\mathbf{N} = n + ik \tag{1}$$

The refractive index is defined as the ratio of phase velocity of light in vacuum to the phase velocity of light in the medium. The extinction coefficient is related to the exponential decay of the wave as it passes through the medium. Note, however, that these "constants" vary with wavelength and temperature. The expression for an electromagnetic wave in an absorbing medium contains both of these parameters:

$$\mathbf{E} = E_o e^{-2\pi kx/\lambda_0} e^{-i(2\pi nx/\lambda_0 - \omega t)} \tag{2}$$

where E_o is the amplitude of the wave measured at the point $x = 0$ in the medium, E is the instantaneous value of the electric vector measured at a distance x from the first point and at some time t, ω is the angular frequency of the source, and λ_0 is the wavelength in vacuum.

The absorption coefficient α is related to the extinction coefficient by:

$$\alpha = 4k/\lambda_0 \tag{3}$$

and for the general case, the absorption coefficient also appears in the absorption equation:

$$I = I_0 e^{-\alpha x} \tag{4}$$

However, this equation implies that the intensities I and I_0 are measured within the absorbing medium. The complex dielectric constant $\boldsymbol{\varepsilon}$ for such a material is:

$$\boldsymbol{\varepsilon} = \varepsilon_1 + i\varepsilon_2 \tag{5}$$

where the dielectric constants are related to the optical constants by:

$$\varepsilon_1 = n^2 - k^2 \tag{6}$$

$$\varepsilon_2 = 2nk \tag{7}$$

Two additional materials properties that influence the light-material interaction are magnetic susceptibility μ and conductivity σ that are further discussed later.

The equations describing the reflection phenomena, including polarization effects for metals, will not be presented here but are explained in detail elsewhere.[5–8,10–11] After a brief description of Lorentz and Drude theories and their implications for metals, and particularly for absorption, the relationship among reflection, transmission, and absorption is discussed.[9]

The classical theory of absorption in dielectrics is due to H. A. Lorentz[12] and in metals to P. K. L. Drude.[13] Both models treat the optically active electrons in a material as classical oscillators. In the Lorentz model, the electron is considered to be bound to the nucleus by a harmonic restoring force. In this manner, Lorentz's picture is that of the

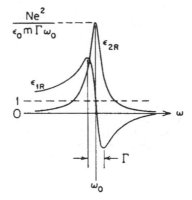

FIGURE 1 Frequency dependences of ε_{1R} and ε_{2R}.[9]

nonconductive dielectric. Drude considered the electrons to be free, and set the restoring force in the Lorentz model equal to zero. Both models include a damping term in the electron's equation of motion that in more modern terms is recognized as a result of electron-phonon collisions.

These models solve for the electron's motion in the presence of the electromagnetic field as a driving force. From this, it is possible to write an expression for the polarization induced in the medium and from that to derive the dielectric constant. The Lorentz model for dielectrics gives the relative real and imaginary parts of the dielectric constant ε_{1R} and ε_{2R} in terms of N, the number of dipoles per unit volume; e and m, the electron charge and mass; Γ, the damping constant; ω and ω_0, the radian frequencies of the field and the harmonically bound electron; and ε_0, the permittivity of free space. These functions are shown in Fig. 1. The range of frequencies where ε_1 increases with frequency is referred to as the *range of normal dispersion*, and the region near $\omega = \omega_0$ where it decreases with frequency is called the *range of anomalous dispersion*.

Since the ionic polarizability is much smaller than the electronic polarizability at optical frequencies, only the electronic terms are considered when evaluating optical absorption using the Lorentz model for dielectrics. The Drude model for metals assumes that the electrons are free to move. This means that it is identical to the Lorentz model except that ω_0 is set equal to zero. The real and imaginary parts of the dielectric constant are then given by

$$\varepsilon_{1R} = 1 - (Ne^2\varepsilon_0 m)\frac{1}{\omega^2 + \Gamma^2} \tag{8}$$

$$\varepsilon_{2R} = (Ne^2\varepsilon_0 m)\frac{\Gamma}{\omega(\omega^2 + \Gamma^2)} \tag{9}$$

The quantity Γ is related to the mean time between electron collisions with lattice vibrations, and by considering electronic motion in an electric field **E** having radian frequency ω, an expression for the average velocity can be obtained. An expression for the conductivity σ is then obtained and the parts of the dielectric constant can be restated. At electromagnetic field frequencies that are low, it can be shown that $\varepsilon_2 \gg \varepsilon_1$ and therefore it follows that:

$$\alpha = (\omega\mu\sigma/2)^{1/2} \tag{10}$$

In other words, the optical properties and the conductivity of a perfect metal are related through the fact that each is determined by the motion of free electrons. At high frequencies, transitions involving bound or valence band electrons are possible and there will be a noticeable deviation from this simple result of the Drude model. However, the experimental data reported for most metals are in good agreement with the Drude prediction at wavelengths as short as 1 μm.

From Eq. (10) it is clear that a field propagating in a metal will be attenuated by a factor of $1/e$ when it has traveled a distance:

$$\delta = (2/\omega\mu\sigma)^{1/2} \tag{11}$$

This quantity is called the *skin depth*, and at optical frequencies for most metals it is ~50 nm. After a light beam has propagated one skin depth into a metal, its intensity is reduced to 0.135 of its value at the surface.

Another aspect of the absorption of light energy by metals that should be noted is the fact that it increases with temperature. This is important because during laser irradiation the temperature of a metal will increase and so will the absorption. The coupling of energy into the metal is therefore dependent on the temperature dependence of the absorption. For most metals, all the light that gets into the metal is absorbed. If the Fresnel expression for the electric field reflectance is applied to the real and imaginary parts of the complex index for a metal-air interface, the field reflectivity can be obtained. When multiplied by its complex conjugate, the expression for the intensity reflection coefficient is obtained:

$$R_I = 1 - 2\mu\varepsilon_0\omega/\sigma \tag{12}$$

Since the conductivity σ decreases with increasing temperature, R_I decreases with increasing temperature, and at higher temperatures more of the incident energy is absorbed.

Since reflection methods are used in determining the optical constants, they are strongly dependent on the characteristics of the metallic surface. These characteristics vary considerably with chemical and mechanical treatment, and these treatments have not always been accurately defined. Not all measurements have been made on freshly polished surfaces but in many cases on freshly deposited thin films. The best available data are presented in the tables and figures, and the reader is advised to consult the appropriate references for specifics.

In this article, an ending of *-ance* denotes a property of a specific sample (i.e., including effects of surface finish), while the ending *-ivity* refers to an intrinsic material property. For most of the discussion, the endings are interchangeable.

Reflectance r is the ratio of radiant flux reflected from a surface to the total incident radiant flux. Since r is a function of the optical constants, it varies with wavelength and temperature. The relationship between reflectance and optical constants is:[5]

$$r = \frac{(n-1)^2 + k^2}{(n+1)^2 + k^2} \tag{13}$$

The reflectance of a good, freshly deposited mirror coating is almost always higher than that of a polished or electroplated surface of the same material. The reflectance is normally less than unity—some transmission and absorption, no matter how small, are always present. The relationship between these three properties is:

$$r + t + a = 1 \tag{14}$$

Transmittance t is the ratio of radiant flux transmitted through a surface to the total incident radiant flux and absorptance a is the ratio of the radiant flux lost by absorption to the total incident radiant flux. Since t and a are functions of the optical constants, they vary with wavelength and temperature. Transmittance is normally very small for metals except in special cases (e.g., beryllium at x-ray wavelengths). Absorptance is affected by surface condition as well as the intrinsic contribution of the material.

The thermal radiative properties are descriptive of a radiant energy-matter interaction that can be described by other properties such as the optical constants and/or complex dielectric constant, each of which is especially convenient for studying various aspects of the interaction. However, the thermal radiative properties are particularly useful since metallic materials are strongly influenced by surface effects, particularly oxide films, and therefore in many cases they are not readily calculated by simple means from the other properties.

For opaque materials, the transmission is near zero, so Eq. (14) becomes:

$$r + a = 1 \tag{15}$$

but since Kirchhoff's law states that absorptance equals emittance, ϵ, this becomes:

$$r + \epsilon = 1 \tag{16}$$

and the thermal radiative properties of an opaque body are fully described by either the reflectance or the emittance. Emittance is the ratio of radiated emitted power (in W/m^2) of a surface to the emissive power of a blackbody at the same temperature. Emittance can therefore be expressed as either *spectral* (emittance as a function of wavelength at constant temperature) or *total* (the integrated emittance over all wavelengths as a function of temperature).

Physical Properties

The physical properties of interest for metals in optical applications include density, electrical conductivity, and electrical resistivity (the reciprocal of conductivity), as well as crystal structure. Chemical composition of alloys is also included with physical properties.

For density, mass density is reported with units of kg/m^3. Electrical conductivity is related to electrical resistivity, but for some materials, one or the other is normally reported. Both properties vary with temperature.

Crystal structure is extremely important for stability since anisotropy of the elastic, electric, and magnetic properties and thermal expansion depend on the type of structure.[14] Single crystals of cubic metals have completely isotropic coefficient of thermal expansion (CTE), but are anisotropic in elastic properties—modulus and Poisson's ratio. Materials with hexagonal structures have anisotropic expansion and elastic properties. While polycrystalline metals with randomly oriented small grains do not exhibit these anisotropies they can easily have local areas that are inhomogeneous or can have overall oriented crystal structure induced by fabrication methods.

The combined influence of physical, thermal, and mechanical properties on optical system performance is described under "Properties Important in Mirror Design," later in this article.

Thermal Properties

Thermal properties of metals that are important in optical systems design include: coefficient of thermal expansion α, referred to in this section as CTE; thermal conductivity k; and specific heat C_p. All of these properties vary with temperature; usually they tend to decrease with decreasing temperature. Although not strictly a thermal property, the maximum usable temperature is also included as a guide for the optical designer.

Thermal expansion is a generic term for change in length for a specific temperature change, but there are precise terms that describe specific aspects of this material property. ASTM E338 Committee recommends the following nomenclature:[15]

Coefficient of linear thermal expansion (CTE or thermal expansivity):

$$\alpha \equiv \frac{1}{L}\frac{\Delta L}{\Delta T} \qquad (17)$$

Instantaneous coefficient of linear thermal expansion:

$$\alpha' \equiv \lim_{\Delta T \to 0} \left(\frac{1}{L}\frac{\Delta L}{\Delta T}\right) \qquad (18)$$

Mean coefficient of linear thermal expansion:

$$\bar{\alpha} \equiv \frac{1}{T_2 - T_1} \int_{T_1}^{T_2} \alpha' \, dT \qquad (19)$$

In general, lower thermal expansion is better for optical system performance, as it minimizes the effect of thermal gradients on component dimensional changes. CTE is the prime parameter in materials selection for cooled mirrors.

Thermal conductivity is the quantity of heat transmitted per unit of time through a unit of area per unit of temperature gradient with units of W/m K. Higher thermal conductivity is desirable to minimize temperature gradients when there is a heat source to the optical system.

Specific heat, also called heat capacity per unit mass, is the quantity of heat required to change the temperature of a unit mass of material one degree under conditions of constant pressure. A material with high specific heat requires more heat to cause a temperature change that might cause a distortion. High specific heat also means that more energy is required to force a temperature change (e.g., in cooling an infrared telescope assembly to cryogenic temperatures).

Maximum usable temperature is not a hard number. It is more loosely defined as the temperature at which there is a significant change in the material due to one or more of a number of things, such as significant softening or change in strength, melting, recrystallization, and crystallographic phase change.

Mechanical Properties

Mechanical properties are divided into elastic/plastic properties and strength, and fracture properties. The elastic properties of a metal can be described by a 6×6 matrix of constants called the elastic stiffness constants.[16–18] Because of symmetry considerations, there are a maximum of 21 independent constants that are further reduced for more symmetrical crystal types. For cubic materials there are three constants, C_{11}, C_{12}, and C_{44}, and for hexagonal five constants, C_{11}, C_{12}, C_{13}, C_{33}, and C_{44}. From these, the elastic properties of the material, Young's modulus E (the elastic modulus in tension), bulk modulus K, modulus of rigidity G (also called shear modulus), and Poisson's ratio ν can be calculated. The constants, and consequently the properties, vary as functions of temperature. The properties vary with crystallographic direction in single crystals,[14] but in randomly oriented polycrystalline materials the macroproperties are usually isotropic.

Young's modulus of elasticity E is the measure of stiffness or rigidity of a metal—the ratio of stress, in the completely elastic region, to the corresponding strain. Bulk modulus K is the measure of resistance to change in volume—the ratio of hydrostatic stress to the corresponding change in volume. Shear modulus, or modulus of rigidity, G is the ratio of shear stress to the corresponding shear strain under completely elastic conditions.

Poisson's ratio v is the ratio of the absolute value of the rate of transverse (lateral) strain to the corresponding axial strain resulting from uniformly distributed axial stress in the elastic deformation region.

For isotropic materials the properties are interrelated by the following equations:[18]

$$G = \frac{E}{2(1 + v)} \tag{20}$$

$$K = \frac{E}{3(1 - 2v)} \tag{21}$$

The mechanical strength and fracture properties are important for the structural aspect of the optical system. The components in the system must be able to support loads with no permanent deformation within limits set by the error budget and certainly with no fracture. For ductile materials such as copper, the yield and/or microyield strength may be the important parameters. On the other hand, for brittle or near-brittle metals such as beryllium, fracture toughness may be more important. For ceramic materials such as silicon carbide, fracture toughness and modulus of rupture are the important fracture criteria. A listing of definitions for each of these terms[19] follows:

creep strength: the stress that will cause a given time-dependent plastic strain in a creep test for a given time.

ductility: the ability of a material to deform plastically before fracture.

fatigue strength: the maximum stress that can be sustained for a specific number of cycles without failure.

fracture toughness: a generic term for measures of resistance to extension of a crack.

hardness: a measure of the resistance of a material to surface indentation.

microcreep strength: the stress that will cause 1 ppm of permanent strain in a given time; usually less than the microyield strength.

microstrain: a deformation of 10^{-6} m/m (1 ppm)

microyield strength: the stress that will cause 1 ppm of permanent strain in a short time; also called precision elastic limit (PEL).

ultimate strength: the maximum stress a material can withstand without fracture.

yield strength: the stress at which a material exhibits a specified deviation from elastic behavior (proportionality of stress and strain), usually 2×10^{-3} m/m (0.2 percent).

Properties Important in Mirror Design

There are many factors that enter into the design of a mirror or mirror system, but the most important requirement is optical performance. Dimensional stability, weight, durability, and cost are some of the factors to be traded off before an effective design can be established.[20–23] The loading conditions during fabrication, transportation, and use and the thermal environment play a substantial role in materials selection. To satisfy the cnd-use requirements, the optical, structural, and thermal performance must be predictable. Each of these factors has a set of parameters and associated material properties that can be used to design an optic to meet performance goals.

For optical performance, the shape or optical figure is the key performance factor followed by the optical properties of reflectance, absorptance, and complex refractive index. The optical properties of a mirror substrate material are only important when the mirror is to be used bare (i.e., with no optical coating).

To design for structural performance goals, deflections due to static (or inertial) and dynamic loads are usually calculated as a first estimate.[24] For this purpose, the well-known plate equations[25] are invoked. For the static case,

$$\delta = \beta q a^4 / D \tag{22}$$

where δ = deflection
β = deflection coefficient (depends on support condition)
q = normal loading (uniform load example)
a = plate radius (semidiameter)
D = flexural rigidity, defined as:

$$D = EI_0/(1 - v^2) \tag{23}$$

where, in turn: E = Young's modulus of elasticity
I_0 = moment of inertia of the section
v = Poisson's ratio.

But

$$q = \rho V_0 G \tag{24}$$

where ρ = material density
V_0 = volume of material per unit area of plate surface
G = load factor (g's)

After substitution and regrouping the terms:

$$\delta = \beta \frac{\rho(1 - v^2)}{E} \frac{V_0}{I_0} a^4 G \tag{25}$$

or:

$$\delta \times B = M \times S \times P \times G \tag{26}$$

where B = support condition
M = materials parameters
S = structural efficiency
P = plate size
G = load factor

This shows five terms, each representing a parameter to be optimized for mirror performance. B, P, and G will be determined from system requirements; S is related to the geometric design of the part; and M is the materials term showing that ρ, v, and E are the important material properties for optimizing structural performance.

For the dynamic case of deflection due to a local angular acceleration $\ddot{\theta}$ about a diameter (scanning applications), the equation becomes:

$$\delta_{\text{DYN}} = \beta_D \frac{\rho(1 - v^2)}{E} \frac{V_0}{I_0} a^5 \frac{\ddot{\theta}}{g} \tag{27}$$

The same structural optimization parameters prevail as in the static case. Note that in both cases maximizing the term E/ρ (specific stiffness) minimizes deflection.

The determination of thermal performance[26-27] is dependent on the thermal environment and thermal properties of the mirror material. For most applications, the most significant properties are the coefficient of thermal expansion CTE or α, and thermal conductivity k. Also important are the specific heat C_p, and thermal diffusivity D, a property related to dissipation of thermal gradients that is a combination of properties and

TABLE 1 Properties of Selected Mirror Materials

	ρ Density 10^3 kg/m^3	E Young's modulus GN/m^2	E/ρ Specific stiffness arb. units	CTE Thermal expansion 10^{-6}/K	k Thermal conductivity W/m K	C_p Specific heat J/kg K	D Thermal diffusivity 10^{-6} m^2/s	Distortion coefficient	
								CTE/k steady state μm/W	CTE/D transient s/m^2 K
Preferred	small	large	large	small	large	large	large	small	small
Fused silica	2.19	72	33	0.50	1.4	750	0.85	0.36	0.59
Beryllium: I-70	1.85	287	155	11.3	216	1925	57.2	0.05	0.20
Aluminum: 6061	2.70	68	25	22.5	167	896	69	0.13	0.33
Copper	8.94	117	13	16.5	391	385	115.5	0.53	0.14
304 stainless steel	8.00	193	24	14.7	16.2	500	4.0	0.91	3.68
Invar 36	8.05	141	18	1.0	10.4	515	2.6	0.10	0.38
Silicon	2.33	131	56	2.6	156	710	89.2	0.02	0.03
SiC: RB-30% Si	2.89	330	114	2.6	155	670	81.0	0.02	0.03
SiC: CVD	3.21	465	145	2.4	198	733	82.0	0.01	0.03

equal to $k/\rho C_p$. There are two important thermal figures of merit, the coefficients of thermal distortion α/k and α/D. The former expresses steady-state distortion per unit of input power, while the latter is related to transient distortions.

Typical room-temperature values for many of the important properties mentioned here are listed for a number of mirror materials in Table 1. It should be clear from the wide range of properties and figures of merit that no one material can satisfy all applications. A selection process is required and a tradeoff study has to be made for each individual application.[20]

Metal optical components can be designed and fabricated to meet system requirements. However, unless they remain within specifications throughout their intended lifetime, they have failed. The most often noted changes that occur to degrade performance are dimensional instability and/or environment-related optical property degradation. Dimensional instabilities can take many forms with many causes, and there are any number of ways to minimize them. Dimensional instabilities can only be discussed briefly here; for a more complete discussion, consult Refs. 28 to 31. The instabilities most often observed are:

- *temporal instability*: a change in dimensions with time in a uniform environment (e.g., a mirror stored in a laboratory environment with no applied loads changes figure over a period of time)

- *thermal/mechanical cycling* or *hysteresis instability*: a change in dimensions when the environment is changed and then restored, where the measurements are made under the same conditions before and after the exposure (e.g., a mirror with a measured figure is cycled between high and low temperatures and, when remeasured under the original conditions, the figure has changed)

- *thermal instability*: a change in dimensions when the environment is changed, but completely reversible when the original environment is restored (e.g., a mirror is measured at room temperature, again at low temperature where the figure is different, and finally at the original conditions with the original figure restored)

There are other types of instabilities, but they are less common, particularly in metals. The sources of the dimensional changes cited here can be attributed to one or more of the following:

- externally applied stress
- changes in internal stress
- microstructural changes
- inhomogeneity/anisotropy of properties

In general, temporal and cycling/hysteresis instabilities are primarily caused by changes in internal stress (i.e., stress relaxation). If the temperature is high enough, microstructural changes can take place as in annealing, recrystallization, or second-phase precipitation. Thermal instability is a result of inhomogeneity and/or anisotropy of thermal expansion within the component, is completely reversible, and cannot be eliminated by nondestructive methods.

To eliminate potential instabilities, care must be taken in the selection of materials and fabrication methods to avoid anisotropy and inhomogeneity. Further care is necessary to avoid any undue applied loads that could cause part deformation and subsequent residual stress. The fabrication methods should include stress-relief steps such as thermal annealing, chemical removal of damaged surfaces, and thermal or mechanical cycling. These steps become more critical for larger and more complex component geometries.

Instabilities can also be induced by attachments and amounts. Careful design to minimize induced stresses and selection of dissimilar materials with close thermal expansion matching is essential.[32]

35.3   SUMMARY DATA

The properties presented here are representative for the materials and are not a complete presentation. For more complete compilations, the references should be consulted.

Optical Properties

Thin films and their properties are discussed in Vol. I, Chap. 42, and therefore are not presented here except in the case where bulk (surface) optical properties are not available.

Index of Refraction and Extinction Coefficient. The data for the optical constants of metals are substantial, with the most complete listing available in the two volumes of *Optical Constants of Metals*,[33,34] from which most of the data presented here have been taken. Earlier compilations[35,36] are also available. While most of the data are for deposited films, the references discuss properties of polished polycrystalline surfaces where available. Table 2 lists room temperature values for n and k of Al,[37] Be,[38] Cu,[39] Cr,[40] Au,[39] Fe,[40] Mo,[39] Ni,[39] Pt,[39] Ag,[39] W,[39] and α-SiC.[41] Figures 2 to 14 graphically show these constants with the absorption edges shown in most cases.

Extensive reviews of the properties of aluminum[37] and beryllium[38] also discuss the effects of oxide layers on optical constants and reflectance. Oxide layers on aluminum typically reduce the optical constant values by 25 percent in the infrared, 10 to 15 percent in the visible, and very little in the ultraviolet.[37] As a result of the high values of n and k for aluminum in the visible and infrared, there are relatively large variations of optical constants with temperature, but they result in only small changes in reflectance.[37] The beryllium review[38] does not mention any variation of properties with temperature. The optical properties of beryllium and all hexagonal metals vary substantially with crystallographic direction. This variation with crystallography is shown for the dielectric constants

TABLE 2 n and k of Selected Metals at Room Temperature

Metal	eV	Wavelength Å	μm	n	k
Aluminum[37]	300.0	41.3		1.00	0.00
	180.0	68.9		0.99	0.01
	130.0	95.4		0.99	0.02
	110.0	113.0		0.99	0.03
	100.0	124.0		0.99	0.03
	95.0	131.0		1.00	0.04
	80.0	155.0		1.01	0.02
	75.0	165.0		1.01	0.02
	72.0	172.0		1.02	0.00
	50.0	248.0		0.97	0.01
	25.0	496.0		0.81	0.02
	17.0	729.0		0.47	0.04
	12.0	1,033.0	0.10	0.03	0.79
	6.00	2,066.0	0.21	0.13	2.39
	4.00	3,100.0	0.31	0.29	3.74
	3.10	4,000.0	0.40	0.49	4.86
	2.48	5,000.0	0.50	0.77	6.08
	2.07	6,000.0	0.60	1.02	7.26
	1.91	6,500.0	0.65	1.47	7.79
	1.77	7,000.0	0.70	1.83	8.31
	1.55	8,000.0	0.80	2.80	8.45
	1.10		1.13	1.20	11.2
	0.827		1.50	1.38	15.4
	0.620		2.00	2.15	20.7
	0.310		4.00	6.43	39.8
	0.177		7.00	14.0	66.2
	0.124		10.0	25.3	89.8
	0.062		20.0	60.7	147.0
	0.039		32.0	103.0	208.0
Beryllium[38]	300.0	41.3		1.00	0.00
	200.0	62.0		0.99	0.00
	150.0	82.7		0.99	0.01
	119.0	104.0		1.00	0.02
	100.0	124.0		0.99	0.00
	50.0	248.0		0.93	0.01
	25.0	496.0		0.71	0.10
	17.0	729.0		0.34	0.42
	12.0	1,033.0	0.10	0.30	1.07
	6.00	2,066.0	0.21	0.85	2.64
	4.00	3,100.0	0.31	2.47	3.08
		4,133.0	0.41	2.95	3.14
		5,166.0	0.52	3.03	3.18
		6,888.0	0.69	3.47	3.23
			1.03	3.26	3.96
			3.10	2.07	12.6
			6.20	3.66	26.7
			12.0	11.3	50.1
			21.0	19.9	77.1
			31.0	37.4	110.0
			62.0	86.1	157.0
Copper[39]	9,000.0	1.38		1.00	0.00
	4,000.0	3.10		1.00	0.00

TABLE 2 *n* and *k* of Selected Metals at Room Temperature (*Continued*)

Metal	eV	Wavelength Å	μm	*n*	*k*
Copper[39]	1,500.0	8.27		1.00	0.00
	1,000.0	12.4		1.00	0.00
	900.0	13.8		1.00	0.00
	500.0	24.8		1.00	0.00
	300.0	41.3		0.99	0.01
	200.0	62.0		0.98	0.02
	150.0	82.7		0.97	0.03
	120.0	103.0		0.97	0.05
	100.0	124.0		0.97	0.07
	50.0	248.0		0.95	0.13
	29.0	428.0		0.85	0.30
	26.0	477.0		0.92	0.40
	24.0	517.0		0.96	0.37
	23.0	539.0		0.94	0.37
	20.0	620.0		0.88	0.46
	15.0	827.0		1.01	0.71
	12.0	1,033.0	0.10	1.09	0.73
	6.50	1,907.0	0.19	0.96	1.37
	5.20	2,384.0	0.24	1.38	1.80
	4.80	2,583.0	0.26	1.53	1.71
	4.30	2,885.0	0.29	1.46	1.64
	2.60	4.768.0	0.48	1.15	2.5
	2.30	5,390.0	0.54	1.04	2.59
	2.10	5,904.0	0.59	0.47	2.81
	1.80	6,888.0	0.69	0.21	4.05
	1.50	8,265.0	0.83	0.26	5.26
	0.950		1.30	0.51	6.92
	0.620		2.00	0.85	10.6
	0.400		3.10	1.59	16.5
	0.200		6.20	5.23	33.0
	0.130		9.54	10.8	47.5
Chromium[40]	10,000.0	1.24		1.00	0.00
	6,015.0	2.06		1.00	0.00
	5,878.0	2.11		1.00	0.00
	3,008.0	4.12		1.00	0.00
	1.504.0	8.24		1.00	0.00
	992.0	12.5		1.00	0.00
	735.0	16.9		1.00	0.00
	702.0	17.7		1.00	0.00
	686.0	18.1		1.00	0.00
	403.0	30.8		1.00	0.00
	202.0	61.5		0.98	0.00
	100.0	124.0		0.94	0.03
	62.0	200.0		0.88	0.12
	52.0	238.0		0.92	0.18
	29.5	420.0		0.78	0.21
	24.3	510.0		0.67	0.39
	18.0	689.0		0.87	0.70
	14.3	867.0		1.06	0.82
	12.8	969.0		1.15	0.75
	11.4	1,088.0	0.109	1.08	0.69
	7.61	1,629.0	0.163	0.66	1.23

TABLE 2 *n* and *k* of Selected Metals at Room Temperature (*Continued*)

Metal	eV	Wavelength Å	μm	*n*	*k*
Chromium[40]	5.75	2,156.0	0.216	0.97	1.74
	4.80	2,583.0	0.258	0.86	2.13
	3.03	4.092.0	0.409	1.54	3.71
	2.42	5.123.0	0.512	2.75	4.46
	1.77	7,005.0	0.700	3.84	4.37
	1.26	9,843.0	0.984	4.50	4.28
	1.12		1.11	4.53	4.30
	0.66		1.88	3.96	5.95
	0.60		2.07	4.01	6.48
	0.34		3.65	2.89	12.0
	0.18		6.89	8.73	25.4
	0.09		13.8	11.8	33.9
	0.06		20.7	21.2	42.0
	0.04		31.0	14.9	65.2
Gold[39]	8,266.0	1.50		1.00	0.00
	2,480.0	5.00		1.00	0.00
	2,066.0	6.00		1.00	0.00
	1,012.0	12.25		1.00	0.00
	573.0	21.6		1.00	0.00
	220.0	56.4		0.99	0.01
	150.0	82.7		0.96	0.01
	86.0	144.0		0.89	0.06
	84.5	147.0		0.89	0.07
	84.0	148.0		0.89	0.06
	68.0	182.0		0.86	0.12
	60.0	207.0		0.86	0.16
	34.0	365.0		0.78	0.47
	30.0	413.0		0.89	0.60
	29.0	428.0		0.91	0.60
	27.0	459.0		0.90	0.64
	26.0	480.0		0.85	0.56
	21.8	570.0		1.02	0.85
	19.4	640.0		1.16	0.73
	17.7	700.0		1.08	0.68
	15.8	785.0		1.03	0.74
	12.4	1,000.0	0.10	1.20	0.84
	8.27	1,550.0	0.15	1.45	1.11
	7.29	1,700.0	0.17	1.52	1.07
	6.36	1,950.0	0.20	1.42	1.12
	4.10	3,024.0	0.30	1.81	1.92
	3.90	3,179.0	0.32	1.84	1.90
	3.60	3,444.0	0.34	1.77	1.85
	3.00	4.133.0	0.41	1.64	1.96
	2.60	4,769.0	0.48	1.24	1.80
	2.20	5,636.0	0.56	0.31	2.88
	1.80	6,888.0	0.69	0.16	3.80
	1.40	8,856.0	0.89	0.21	5.88
	1.20		1.03	0.27	7.07
	0.82		1.51	0.54	9.58
	0.40		3.10	1.73	19.2
	0.20		6.20	5.42	37.5
	0.125		9.92	12.2	54.7

TABLE 2 *n* and *k* of Selected Metals at Room Temperature (*Continued*)

Metal	eV	Wavelength Å	µm	*n*	*k*
Iron[36,40]	10,000.0	1.24		1.00	0.00
	7,071.0	1.75		1.00	0.00
	3,619.0	3.43		1.00	0.00
	1,575.0	7.87		1.00	0.00
	884.0	14.0		1.00	0.00
	825.0	15.0		1.00	0.00
	320.0	38.8		0.99	0.00
	211.0	58.7		0.98	0.01
	153.0	81.2		0.97	0.02
	94.0	132.0		0.94	0.05
	65.0	191.0		0.90	0.12
	56.6	219.0		0.98	0.19
	54.0	230.0		1.11	0.18
	51.6	240.0		0.97	0.05
	30.0	413.0		0.82	0.13
	22.2	559.0		0.71	0.35
	20.5	606.0		0.74	0.42
	18.0	689.0		0.78	0.51
	15.8	785.0		0.77	0.61
	11.5	1,078.0	0.11	0.93	0.84
	11.0	1,127.0	0.11	0.91	0.83
	10.3	1,200.0	0.12	0.87	0.91
	8.00	1,550.0	0.15	0.94	1.18
	5.00	2.480.0	0.25	1.14	1.87
	3.00	4.133.0	0.41	1.88	3.12
	2.30	5,390.0	0.54	2.65	3.34
	2.10	5,903.0	0.59	2.80	3.34
	1.50	8,265.0	0.83	3.05	3.77
	1.24		1.00	3.23	4.35
	0.496		2.50	4.13	8.59
	0.248		5.00	4.59	15.4
	0.124		10.0	5.81	30.4
	0.062		20.0	9.87	60.1
	0.037		33.3	22.5	100.0
	0.025		50.0	45.7	141.0
	0.015		80.0	75.2	158.0
	0.010		125.0	120.0	207.0
	0.006		200.0	183.0	260.0
	0.004		287.0	238.0	306.0
Molybdenum[39]	2,000.0	6.19		1.00	0.00
	1,041.0	11.6		1.00	0.00
	396.0	31.3		1.00	0.01
	303.0	40.9		1.00	0.01
	211.0	58.8		0.99	0.00
	100.0	124.0		0.93	0.01
	60.0	207.0		0.90	0.11
	37.5	331.0		0.81	0.29
	35.0	354.0		0.87	0.38
	33.8	367.0		0.91	0.33
	33.0	376.0		0.90	0.33
	31.4	394.0		0.92	0.31

TABLE 2 *n* and *k* of Selected Metals at Room Temperature (*Continued*)

Metal	eV	Wavelength Å	μm	*n*	*k*
Silver[39]	350.0	35.4		1.00	0.00
	170.0	72.9		0.97	0.00
	110.0	113.0		0.90	0.02
	95.0	131.0		0.86	0.06
	85.0	146.0		0.85	0.11
	64.0	194.0		0.89	0.21
	50.0	248.0		0.88	0.29
	44.0	282.0		0.90	0.33
	35.0	354.0		0.87	0.45
	31.0	400.0		0.93	0.53
	27.5	451.0		0.85	0.62
	22.5	551.0		1.03	0.62
	21.0	590.0		1.11	0.56
	20.0	620.0		1.10	0.55
	15.0	827.0		1.24	0.69
	13.0	954.0		1.32	0.60
	10.9	1,137.0	0.11	1.28	0.56
	10.0	1,240.0	0.12	1.24	0.57
	9.20	1,348.0	0.13	1.18	0.55
	7.60	1,631.0	0.16	0.94	0.83
	4.85	2,556.0	0.26	1.34	1.35
	4.15	2,988.0	0.30	1.52	0.99
	3.90	3,179.0	0.32	0.93	0.50
	3.10	4,000.0	0.40	0.17	1.95
	2.20	5,636.0	0.56	0.12	3.45
	1.80	6,888.0	0.69	0.14	4.44
	1.20		1.03	0.23	6.99
	0.62		2.00	0.65	12.2
	0.24		5.17	3.73	31.3
	0.125		9.92	13.1	53.7
Tungsten[39]	2,000.0	6.20		1.00	0.00
	1,016.0	12.2		1.00	0.00
	516.0	24.0		0.99	0.00
	244.0	50.8		0.99	0.02
	100.0	124.0		0.94	0.04
	43.0	288.0		0.74	0.27
	38.5	322.0		0.82	0.33
	35.0	354.0		0.85	0.31
	33.0	376.0		0.82	0.28
	32.0	388.0		0.79	0.30
	30.5	406.0		0.77	0.29
	23.8	521.0		0.48	0.60
	22.9	541.0		0.49	0.69
	22.1	561.0		0.49	0.76
	16.0	775.0		0.98	1.14
	15.5	800.0		0.96	1.12
	14.6	849.0		0.90	1.20
	11.8	1,051.0	0.11	1.18	1.48
	10.8	1,148.0	0.11	1.29	1.39
	10.3	1,204.0	0.12	1.22	1.33
	7.80	1,590.0	0.16	0.93	2.06
	5.60	2,214.0	0.22	2.43	3.70

TABLE 2   *n* and *k* of Selected Metals at Room Temperature (*Continued*)

Metal	eV	Wavelength Å	μm	*n*	*k*
Tungsten[39]	5.00	2,480.0	0.25	3.40	2.85
	4.30	2,883.0	0.29	3.07	2.31
	4.00	3,100.0	0.31	2.95	2.43
	3.45	3,594.0	0.36	3.32	2.70
	3.25	3,815.0	0.38	3.45	2.49
	3.10	4,000.0	0.40	3.39	2.41
	2.80	4.428.0	0.44	3.30	2.49
	1.85	6.702.0	0.67	3.76	2.95
	1.75	7,085.0	0.71	3.85	2.86
	1.60	7,749.0	0.77	3.67	2.68
	1.20		1.03	3.00	3.64
	0.96		1.29	3.15	4.41
	0.92		1.35	3.14	4.45
	0.85		1.46	2.80	4.33
	0.58		2.14	1.18	8.44
	0.40		3.10	1.94	13.2
	0.34		3.65	1.71	15.7
	0.18		6.89	4.72	31.5
	0.12		10.3	10.1	46.4
	0.07		17.7	26.5	73.8
	0.05		24.8	46.5	93.7
Silicon carbide[41]	30.0	413.0		0.74	0.11
	20.5	605.0		0.35	0.53
	13.1	946.0		0.68	1.41
	9.50	1,305.0	0.13	1.46	2.21
	9.00	1,378.0	0.14	1.60	2.15
	7.60	1,631.0	0.16	2.59	2.87
	6.40	1,937.0	0.19	4.05	1.42
	5.00	2,480.0	0.25	3.16	0.26
	3.90	3,179.0	0.32	2.92	0.01
	3.00	4,133.0	0.41	2.75	0.00
	2.50	4,959.0	0.50	2.68	0.00
	1.79	6,911.0	0.69	2.62	—
	1.50	8,266.0	0.83	2.60	—
	0.62		2.00	2.57	0.00
	0.31		4.00	2.52	0.00
	0.12		6.67	2.33	0.02
	0.11		9.80	1.29	0.01
	0.10		10.40	0.09	0.63
	0.10		10.81	0.06	1.57
	0.10		11.9	0.16	4.51
	0.09		12.6	8.74	18.4
	0.08		12.7	17.7	6.03
	0.05		13.1	7.35	0.27
			15.4	4.09	0.02
			25.0	3.34	—

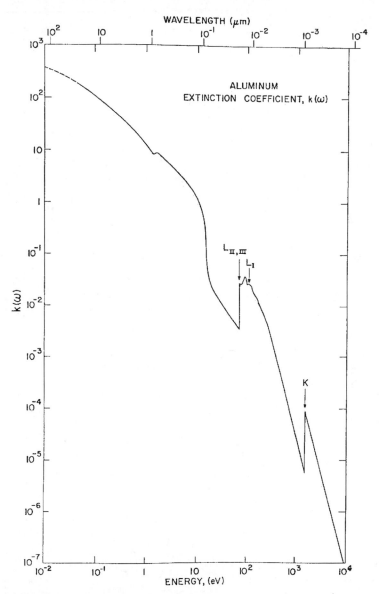

FIGURE 2 k for aluminum vs. photon energy.[37]

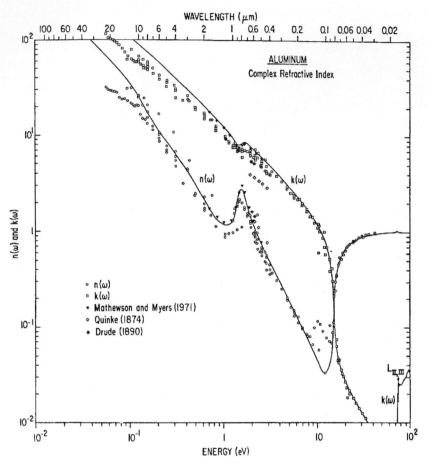

FIGURE 3 *n* and *k* for aluminum vs. photon energy.[37]

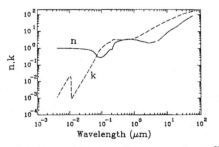

FIGURE 4 *n* and *k* for beryllium vs. wavelength.[38]

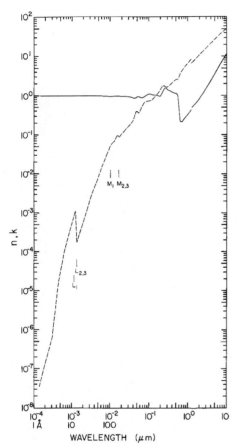

FIGURE 5 *n* and *k* for copper vs. wavelength.[39]

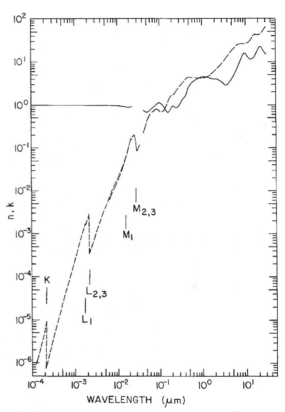

FIGURE 6 *n* and *k* for chromium vs. wavelength.[40]

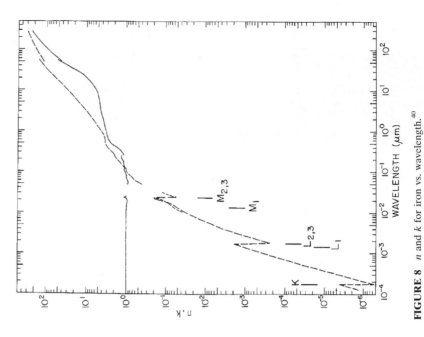

FIGURE 8 n and k for iron vs. wavelength.[40]

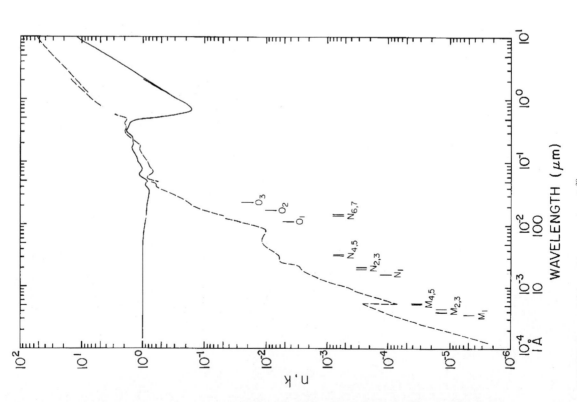

FIGURE 7 n and k for gold vs. wavelength.[39]

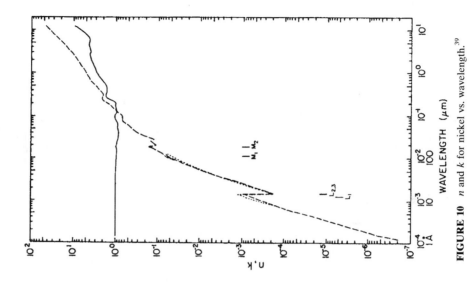

FIGURE 10 n and k for nickel vs. wavelength.[39]

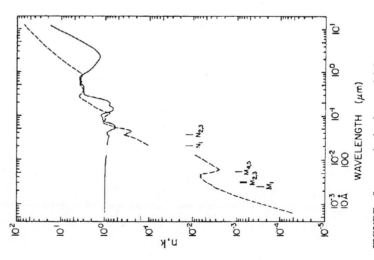

FIGURE 9 n and k for molybdenum vs. wavelength.[39]

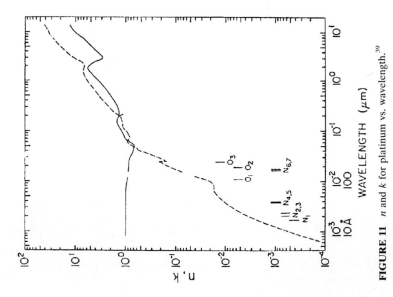

FIGURE 11 *n* and *k* for platinum vs. wavelength.[39]

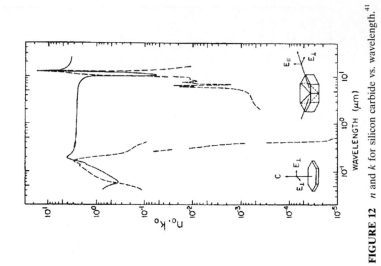

FIGURE 12 *n* and *k* for silicon carbide vs. wavelength.[41]

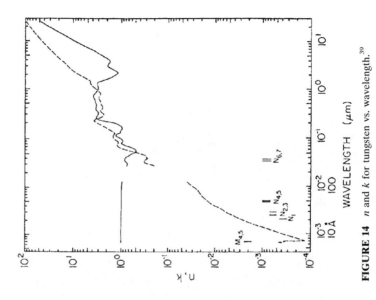

FIGURE 14 n and k for tungsten vs. wavelength.[39]

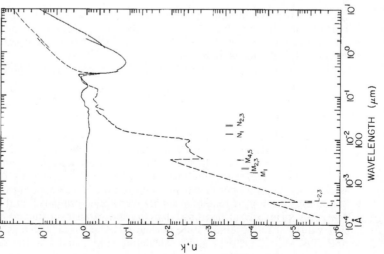

FIGURE 13 n and k for silver as a function of wavelength.[39]

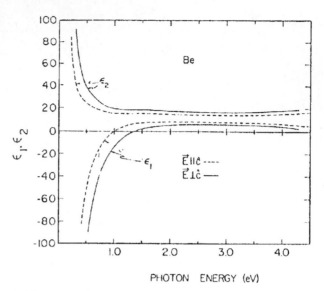

FIGURE 15 Dielectric function for beryllium vs. photon energy showing variation with crystallographic direction.[42]

of beryllium in Fig. 15.[42] The optical constants can be obtained from the dielectric constants using the following equations:[9]

$$n = \{[(\varepsilon_1^2 + \varepsilon_2^2)^{1/2} + \varepsilon_1]/2\}^{1/2} \tag{28}$$

$$k = \{[(\varepsilon_1^2 + \varepsilon_2^2)^{1/2} - \varepsilon_1]/2\}^{1/2} \tag{29}$$

This variation in optical properties results in related variations in reflectance and absorptance that may be the main contributors to a phenomenon called *anomalous scatter*, where the measured scatter from polished surfaces does not scale with wavelength when compared to the measured surface roughness.[43–47]

The optical constants reported for SiC are for single-crystal hexagonal material.

Reflectance and Absorptance

Reflectance data in the literature are extensive. Summaries have been published for most metals[35–36] primarily at normal incidence, both as deposited films and polished bulk material. Reflectance as a function of angle is presented for a number of metals in Refs. 48 and 49. Selected data are also included in Ref. 50. Temperature dependence of reflectance is discussed in a number of articles, but little measured data are available. Absorption data summaries are not as readily available, with one summary[35] and many articles for specific materials, primarily at laser wavelengths and often as a function of temperature. Table 3 lists values of room-temperature normal-incidence reflectance as a function of wavelength, and Figs. 16 to 26 show r and a calculated from n and k in the range of 0.015 to 10 μm.[35] Figure 27[35] shows reflectance for polarized radiation as a function of incidence angle for three combinations of n and k, illustrating the tendency toward total external reflectance for angles greater than about 80°.

Figures 28 to 34 show reflectance for polished surfaces and thin films of Al, Be, SiC, and Ni, including effects of oxide films on the surface.[51] One effect of absorption is to limit

TABLE 3 Reflectance of Selected Metals at Normal Incidence

Metal	eV	Wavelength Å	μm	R
Aluminum[36]	0.040		31.0	0.9923
	0.050		24.8	0.9915
	0.060		20.7	0.9906
	0.070		17.7	0.9899
	0.080		15.5	0.9895
	0.090		13.8	0.9892
	0.100		12.4	0.9889
	0.125		9.92	0.9884
	0.175		7.08	0.9879
	0.200		6.20	0.9873
	0.250		4.96	0.9858
	0.300		4.13	0.9844
	0.400		3.10	0.9826
	0.600		2.07	0.9806
	0.800		1.55	0.9778
	0.900		1.38	0.9749
	1.00		1.24	0.9697
	1.10		1.13	0.9630
	1.20		1.03	0.9521
	1.30	9537.0	0.95	0.9318
	1.40	8856.0	0.89	0.8852
	1.50	8265.0	0.83	0.8678
	1.60	7749.0	0.77	0.8794
	1.70	7293.0	0.73	0.8972
	1.80	6888.0	0.69	0.9069
	2.00	6199.0	0.62	0.9148
	2.40	5166.0	0.52	0.9228
	2.80	4428.0	0.44	0.9242
	3.20	3874.0	0.39	0.9243
	3.60	3444.0	0.34	0.9246
	4.00	3100.0	0.31	0.9248
	4.60	2695.0	0.27	0.9249
	5.00	2497.0	0.25	0.9244
	6.00	2066.0	0.21	0.9257
	8.00	1550.0	0.15	0.9269
	10.00	1240.0	0.12	0.9286
	11.00	1127.0	0.11	0.9298
	13.00	954.0		0.8960
	13.50	918.0		0.8789
	14.00	886.0		0.8486
	14.40	861.0		0.8102
	14.60	849.0		0.7802
	14.80	838.0		0.7202
	15.00	827.0		0.6119
	15.20	816.0		0.4903
	15.40	805.0		0.3881
	15.60	795.0		0.3182
	15.80	785.0		0.2694
	16.00	775.0		0.2326
	16.20	765.0		0.2031
	16.40	756.0		0.1789
	16.75	740.0		0.1460

TABLE 3 Reflectance of Selected Metals at Normal Incidence (*Continued*)

Metal	eV	Wavelength Å	μm	R
Aluminum[36]	17.00	729.0		0.1278
	17.50	708.0		0.1005
	18.00	689.0		0.0809
	19.00	653.0		0.0554
	20.0	620.0		0.0398
	21.0	590.0		0.0296
	22.0	564.0		0.0226
	23.0	539.0		0.0177
	24.0	517.0		0.0140
	25.0	496.0		0.0113
	26.0	477.0		0.0092
	27.0	459.0		0.0076
	28.0	443.0		0.0063
	30.0	413.0		0.0044
	35.0	354.0		0.0020
	40.0	310.0		0.0010
	45.0	276.0		0.0005
	50.0	248.0		0.0003
	55.0	225.0		0.0001
	60.0	206.0		0.0000
	70.0	177.0		0.0000
	72.5	171.0		0.0002
	75.0	165.0		0.0002
	80.0	155.0		0.0002
	85.0	146.0		0.0002
	95.0	131.0		0.0003
	100.0	124.0		0.0002
	120.0	103.0		0.0002
	130.0	95.4		0.0001
	150.0	82.7		0.0001
	170.0	72.9		0.0001
	180.0	68.9		0.0000
	200.0	62.0		0.0000
	300.0	41.3		0.0000
Beryllium[38]	0.020		61.99	0.989
	0.040		31.00	0.989
	0.060		20.66	0.988
	0.080		15.50	0.985
	0.100		12.40	0.983
	0.120		10.33	0.982
	0.160		7.75	0.981
	0.200		6.20	0.980
	0.240		5.17	0.978
	0.280		4.43	0.972
	0.320		3.87	0.966
	0.380		3.26	0.955
	0.440		2.82	0.940
	0.500		2.48	0.917
	0.560		2.21	0.887
	0.600		2.07	0.869
	0.660		1.88	0.841
	0.720		1.72	0.810

TABLE 3 Reflectance of Selected Metals at Normal Incidence (*Continued*)

Metal	eV	Wavelength Å	μm	R
Beryllium[38]	0.780		1.59	0.775
	0.860		1.44	0.736
	0.940		1.32	0.694
	1.00		1.24	0.667
	1.10		1.13	0.640
	1.20		1.03	0.615
	1.40	8856.0	0.89	0.575
	1.60	7749.0	0.77	0.555
	1.90	6525.0	0.65	0.540
	2.40	5166.0	0.52	0.538
	2.80	4428.0	0.44	0.537
	3.00	4133.0	0.41	0.537
	3.30	3757.0	0.38	0.536
	3.60	3444.0	0.34	0.536
	3.80	3263.0	0.33	0.538
	4.00	3100.0	0.31	0.541
	4.20	2952.0	0.30	0.547
	4.40	2818.0	0.28	0.558
	4.60	2695.0	0.27	0.575
Copper[36]	0.10		12.4	0.980
	0.50		2.48	0.979
	1.00		1.24	0.976
	1.50	8265.0	0.83	0.965
	1.70	7293.0	0.73	0.958
	1.80	6888.0	0.69	0.952
	1.90	6525.0	0.65	0.943
	2.00	6199.0	0.62	0.910
	2.10	5904.0	0.59	0.814
	2.20	5635.0	0.56	0.673
	2.30	5390.0	0.54	0.618
	2.40	5166.0	0.52	0.602
	2.60	4768.0	0.48	0.577
	2.80	4428.0	0.44	0.545
	3.00	4133.0	0.41	0.509
	3.20	3874.0	0.39	0.468
	3.40	3646.0	0.36	0.434
	3.60	3444.0	0.34	0.407
	3.80	3263.0	0.33	0.387
	4.00	3100.0	0.31	0.364
	4.20	2952.0	0.30	0.336
	4.40	2818.0	0.28	0.329
	4.60	2695.0	0.27	0.334
	4.80	2583.0	0.26	0.345
	5.00	2497.0	0.25	0.366
	5.20	2384.0	0.24	0.380
	5.40	2296.0	0.23	0.389
	5.60	2214.0	0.22	0.391
	5.80	2138.0	0.21	0.389
	6.00	2066.0	0.21	0.380
	6.50	1907.0	0.19	0.329
	7.00	1771.0	0.18	0.271
	7.50	1653.0	0.17	0.230

TABLE 3 Reflectance of Selected Metals at Normal Incidence (*Continued*)

Metal	eV	Wavelength Å	μm	R
Copper[36]	8.00	1550.0	0.15	0.206
	8.50	1459.0	0.15	0.189
	9.00	1378.0	0.14	0.171
	9.50	1305.0	0.13	0.154
	10.00	1240.0	0.12	0.139
	11.00	1127.0	0.11	0.118
	12.00	1033.0	0.10	0.111
	13.00	954.0		0.109
	14.00	886.0		0.111
	15.00	827.0		0.111
	16.00	775.0		0.106
	17.00	729.0		0.097
	18.00	689.0		0.084
	19.00	653.0		0.071
	20.00	620.0		0.059
	21.00	590.0		0.048
	22.00	564.0		0.040
	23.00	539.0		0.035
	24.00	517.0		0.035
	25.00	496.0		0.040
	26.00	477.0		0.044
	27.00	459.0		0.043
	28.00	443.0		0.039
	29.00	428.0		0.032
	30.00	413.0		0.025
	32.00	387.0		0.017
	34.00	365.0		0.014
	36.00	344.0		0.012
	38.00	326.0		0.010
	40.00	310.0		0.009
	45.00	276.0		0.006
	50.00	248.0		0.005
	55.00	225.0		0.004
	60.00	206.0		0.003
	70.00	177.0		0.002
	90.00	138.0		0.002
Chromium[36]	0.06		20.70	0.962
	0.10		12.40	0.955
	0.14		8.86	0.936
	0.18		6.89	0.953
	0.22		5.64	0.954
	0.26		4.77	0.951
	0.30		4.13	0.943
	0.42		2.95	0.862
	0.54		2.30	0.788
	0.66		1.88	0.736
	0.78		1.59	0.680
	0.90		1.38	0.650
	1.00		1.24	0.639
	1.12		1.11	0.631
	1.24	9998.0	1.00	0.629
	1.36	9116.0	0.91	0.631

TABLE 3 Reflectance of Selected Metals at Normal Incidence (*Continued*)

Metal	eV	Wavelength Å	μm	R
Chromium[36]	1.46	8492.0	0.85	0.632
	1.77	7005.0	0.70	0.639
	2.00	6199.0	0.62	0.644
	2.20	5635.0	0.56	0.656
	2.40	5166.0	0.52	0.677
	2.60	4768.0	0.48	0.698
	2.80	4428.0	0.44	0.703
	3.00	4133.0	0.41	0.695
	4.00	3100.0	0.31	0.651
	4.40	2818.0	0.28	0.620
	4.80	2583.0	0.26	0.572
	5.20	2384.0	0.24	0.503
	5.60	2214.0	0.22	0.443
	6.00	2066.0	0.21	0.444
	7.00	1771.0	0.18	0.425
	7.60	1631.0	0.16	0.378
	8.00	1550.0	0.15	0.315
	8.50	1459.0	0.15	0.235
	9.00	1378.0	0.14	0.170
	10.00	1240.0	0.12	0.120
	11.00	1127.0	0.11	0.103
	11.50	1078.0	0.11	0.100
	12.00	1033.0	0.10	0.101
	13.00	954.0		0.119
	14.00	886.0		0.135
	15.00	827.0		0.143
	16.00	775.0		0.139
	18.00	689.0		0.129
	19.00	653.0		0.131
	20.00	620.0		0.130
	22.00	563.0		0.112
	24.00	517.0		0.096
	26.00	477.0		0.063
	28.00	443.0		0.037
	30.00	413.0		0.030
Gold (electropolished)[36]	0.10		12.40	0.995
	0.20		6.20	0.995
	0.40		3.10	0.995
	0.60		2.07	0.994
	0.80		1.55	0.993
	1.00		1.24	0.992
	1.20		1.03	0.991
	1.40	8856.0	0.89	0.989
	1.60	7749.0	0.77	0.986
	1.80	6888.0	0.69	0.979
	2.00	6199.0	0.62	0.953
	2.10	5904.0	0.59	0.925
	2.20	5635.0	0.56	0.880
	2.30	5390.0	0.54	0.807
	2.40	5166.0	0.52	0.647
	2.50	4959.0	0.50	0.438
	2.60	4768.0	0.48	0.331

TABLE 3 Reflectance of Selected Metals at Normal Incidence (*Continued*)

Metal	eV	Wavelength Å	μm	R
Gold (electropolished)[36]	2.70	4592.0	0.46	0.356
	2.80	4428.0	0.44	0.368
	2.90	4275.0	0.43	0.368
	3.00	4133.0	0.41	0.369
	3.10	3999.0	0.40	0.371
	3.20	3874.0	0.39	0.368
	3.40	3646.0	0.36	0.356
	3.60	3444.0	0.34	0.346
	3.80	3263.0	0.33	0.360
	4.00	3100.0	0.31	0.369
	4.20	2952.0	0.30	0.367
	4.40	2818.0	0.28	0.370
	4.60	2695.0	0.27	0.364
	4.80	2583.0	0.26	0.344
	5.00	2497.0	0.25	0.319
	5.40	2296.0	0.23	0.275
	5.80	2138.0	0.21	0.236
	6.20	2000.0	0.20	0.203
	6.60	1878.0	0.19	0.177
	7.00	1771.0	0.18	0.162
	7.40	1675.0	0.17	0.164
	7.80	1589.0	0.16	0.171
	8.20	1512.0	0.15	0.155
	8.60	1442.0	0.14	0.144
	9.00	1378.0	0.14	0.133
	9.40	1319.0	0.13	0.122
	9.80	1265.0	0.13	0.124
	10.20	1215.0	0.12	0.127
	11.00	1127.0	0.11	0.116
	12.00	1033.0	0.10	0.109
	14.00	886.0		0.140
	16.00	775.0		0.123
	18.00	689.0		0.109
	20.00	620.0		0.133
	22.00	563.0		0.164
	24.00	517.0		0.125
	26.00	477.0		0.079
	28.00	443.0		0.063
	30.00	413.0		0.064
Iron[36]	0.10		12.40	0.978
	0.15		8.27	0.956
	0.20		6.20	0.958
	0.26		4.77	0.911
	0.30		4.13	0.892
	0.36		3.44	0.867
	0.40		3.10	0.858
	0.50		2.48	0.817
	0.60		2.07	0.783
	0.70		1.77	0.752
	0.80		1.55	0.725
	0.90		1.38	0.700
	1.00		1.24	0.678
	1.10		1.13	0.660

TABLE 3   Reflectance of Selected Metals at Normal Incidence (*Continued*)

Metal	eV	Wavelength Å	μm	R
Iron[36]	1.20		1.03	0.641
	1.30	9537.0	0.95	0.626
	1.40	8856.0	0.89	0.609
	1.50	8265.0	0.83	0.601
	1.60	7749.0	0.77	0.585
	1.70	7293.0	0.73	0.577
	1.80	6888.0	0.69	0.573
	1.90	6525.0	0.65	0.563
	2.00	6199.0	0.62	0.563
	2.20	5635.0	0.56	0.563
	2.40	5166.0	0.52	0.567
	2.60	4768.0	0.48	0.576
	2.80	4428.0	0.44	0.580
	3.00	4133.0	0.41	0.583
	3.20	3874.0	0.39	0.576
	3.40	3646.0	0.36	0.565
	3.60	3444.0	0.34	0.548
	4.00	3100.0	0.31	0.527
	4.33	2863.0	0.29	0.494
	4.67	2655.0	0.27	0.470
	5.00	2497.0	0.25	0.435
	5.50	2254.0	0.23	0.401
	6.00	2066.0	0.21	0.366
	6.50	1907.0	0.19	0.358
	7.00	1771.0	0.18	0.333
	7.50	1653.0	0.17	0.298
	8.00	1550.0	0.15	0.272
	8.50	1459.0	0.15	0.251
	9.00	1378.0	0.14	0.236
	9.50	1305.0	0.13	0.226
	10.00	1240.0	0.12	0.213
	11.00	1127.0	0.11	0.162
	11.17	1110.0	0.11	0.159
	11.33	1094.0	0.11	0.159
	11.50	1078.0	0.11	0.160
	12.00	1033.0	0.10	0.163
	12.50	992.0		0.165
	13.00	954.0		0.162
	13.50	918.0		0.159
	14.00	886.0		0.151
	15.00	827.0		0.135
	16.00	775.0		0.116
	17.00	729.0		0.102
	18.00	689.0		0.091
	20.00	620.0		0.083
	22.00	563.0		0.068
	24.00	517.0		0.045
	26.00	477.0		0.031
	28.00	443.0		0.021
	30.00	413.0		0.014
Molybdenum[36]	0.10		12.40	0.985
	0.20		6.20	0.985
	0.30		4.13	0.983

TABLE 3 Reflectance of Selected Metals at Normal Incidence (*Continued*)

Metal	eV	Wavelength Å	μm	R
Molybdenum[36]	0.50		2.70	0.971
	0.70		1.77	0.932
	0.90		1.38	0.859
	1.00		1.24	0.805
	1.10		1.13	0.743
	1.20		1.03	0.671
	1.30	9537.0	0.95	0.608
	1.40	8856.0	0.89	0.562
	1.50	8265.0	0.83	0.550
	1.60	7749.0	0.77	0.562
	1.70	7293.0	0.73	0.570
	1.80	6888.0	0.69	0.576
	2.00	6199.0	0.62	0.571
	2.20	5635.0	0.56	0.562
	2.40	5166.0	0.52	0.594
	2.60	4768.0	0.48	0.582
	2.80	4428.0	0.44	0.565
	3.00	4133.0	0.41	0.550
	3.20	3874.0	0.39	0.540
	3.40	3646.0	0.36	0.541
	3.60	3444.0	0.34	0.546
	3.80	3263.0	0.33	0.554
	4.00	3100.0	0.31	0.576
	4.20	2952.0	0.30	0.610
	4.40	2818.0	0.28	0.640
	4.60	2695.0	0.27	0.658
	4.80	2583.0	0.26	0.678
	5.00	2497.0	0.25	0.695
	5.20	2384.0	0.24	0.706
	5.40	2296.0	0.23	0.706
	5.60	2214.0	0.22	0.700
	6.00	2066.0	0.21	0.674
	6.40	1937.0	0.19	0.641
	6.80	1823.0	0.18	0.592
	7.20	1722.0	0.17	0.548
	7.40	1675.0	0.17	0.542
	7.60	1631.0	0.16	0.552
	7.80	1589.0	0.16	0.542
	8.00	1550.0	0.15	0.530
	8.40	1476.0	0.15	0.495
	8.80	1409.0	0.14	0.450
	9.20	1348.0	0.13	0.385
	9.60	1291.0	0.13	0.320
	10.00	1240.0	0.12	0.250
	10.40	1192.0	0.12	0.188
	10.60	1170.0	0.12	0.138
	11.20	1107.0	0.11	0.123
	11.60	1069.0	0.11	0.135
	12.00	1033.0	0.10	0.154
	12.80	969.0		0.178
	13.60	912.0		0.187
	14.40	861.0		0.182

TABLE 3 Reflectance of Selected Metals at Normal Incidence (*Continued*)

Metal	eV	Wavelength Å	μm	R
Molybdenum[36]	14.80	838.0		0.179
	15.00	827.0		0.179
	16.00	775.0		0.194
	17.00	729.0		0.233
	18.00	689.0		0.270
	19.00	653.0		0.284
	20.00	620.0		0.264
	22.00	563.0		0.207
	24.00	517.0		0.151
	26.00	477.0		0.071
	28.00	443.0		0.036
	30.00	413.0		0.023
	32.00	387.0		0.030
	34.00	365.0		0.034
	36.00	344.0		0.043
	38.00	326.0		0.033
	40.00	310.0		0.025
Nickel[36]	0.10		12.40	0.983
	0.15		8.27	0.978
	0.20		6.20	0.969
	0.30		4.13	0.934
	0.40		3.10	0.900
	0.60		2.07	0.835
	0.80		1.55	0.794
	1.00		1.24	0.753
	1.20		1.03	0.721
	1.40	8856.0	0.89	0.695
	1.60	7749.0	0.77	0.679
	1.80	6888.0	0.69	0.670
	2.00	6199.0	0.62	0.649
	2.40	5166.0	0.52	0.590
	2.80	4428.0	0.44	0.525
	3.20	3874.0	0.39	0.467
	3.60	3444.0	0.34	0.416
	3.80	3263.0	0.33	0.397
	4.00	3100.0	0.31	0.392
	4.20	2952.0	0.30	0.396
	4.60	2695.0	0.27	0.421
	5.00	2497.0	0.25	0.449
	5.20	2384.0	0.24	0.454
	5.40	2296.0	0.23	0.449
	5.80	2138.0	0.21	0.417
	6.20	2000.0	0.20	0.371
	6.60	1878.0	0.19	0.325
	7.00	1771.0	0.18	0.291
	8.00	1550.0	0.15	0.248
	9.00	1378.0	0.14	0.211
	10.00	1240.0	0.12	0.166
	11.00	1127.0	0.11	0.115
	12.00	1033.0	0.10	0.108
	13.00	954.0		0.105
	14.00	886.0		0.106

TABLE 3 Reflectance of Selected Metals at Normal Incidence (*Continued*)

Metal	eV	Wavelength Å	μm	R
Nickel[36]	15.00	827.0		0.107
	16.00	775.0		0.103
	18.00	689.0		0.092
	20.00	620.0		0.071
	22.00	564.0		0.055
	24.00	517.0		0.051
	27.00	459.0		0.042
	30.00	413.0		0.034
	35.00	354.0		0.022
	40.00	310.0		0.014
	50.00	248.0		0.004
	60.00	206.0		0.002
	65.00	191.0		0.002
	70.00	177.0		0.004
	90.00	138.0		0.002
Platinum[36]	0.10		12.40	0.976
	0.15		8.27	0.969
	0.20		6.20	0.962
	0.30		4.13	0.945
	0.40		3.10	0.922
	0.45		2.76	0.882
	0.50		2.50	0.813
	0.55		2.25	0.777
	0.60		2.07	0.753
	0.65		1.91	0.746
	0.70		1.77	0.751
	0.80		1.55	0.762
	0.90		1.38	0.765
	1.00		1.24	0.762
	1.20		1.03	0.746
	1.40	8856.0	0.89	0.725
	1.60	7749.0	0.77	0.706
	1.80	6888.0	0.69	0.686
	2.00	6199.0	0.62	0.664
	2.50	4959.0	0.50	0.616
	3.00	4133.0	0.41	0.565
	4.00	3100.0	0.31	0.472
	5.00	2497.0	0.25	0.372
	6.00	2066.0	0.21	0.276
	7.00	1771.0	0.18	0.230
	8.00	1550.0	0.15	0.216
	9.00	1378.0	0.14	0.200
	9.20	1348.0	0.13	0.198
	9.40	1319.0	0.13	0.200
	10.20	1215.0	0.12	0.211
	11.00	1127.0	0.11	0.199
	12.00	1033.0	0.10	0.173
	12.80	969.0		0.158
	13.60	912.0		0.155
	14.80	838.0		0.157
	15.20	816.0		0.155
	16.00	775.0		0.146

TABLE 3 Reflectance of Selected Metals at Normal Incidence (*Continued*)

Metal	eV	Wavelength Å	μm	R
Platinum[36]	17.50	708.0		0.135
	18.00	689.0		0.142
	20.00	620.0		0.197
	21.00	590.0		0.226
	22.00	564.0		0.240
	23.00	539.0		0.226
	24.00	517.0		0.201
	26.00	477.0		0.150
	28.00	443.0		0.125
	29.00	428.0		0.118
	30.00	413.0		0.124
Silver[36]	0.10		12.40	0.995
	0.20		6.20	0.995
	0.30		4.13	0.994
	0.40		3.10	0.993
	0.50		2.48	0.992
	1.00		1.24	0.987
	1.50	8265.0	0.83	0.960
	2.00	6199.0	0.62	0.944
	2.50	4959.0	0.50	0.914
	3.00	4133.0	0.41	0.864
	3.25	3815.0	0.38	0.816
	3.50	3542.0	0.35	0.756
	3.60	3444.0	0.34	0.671
	3.70	3351.0	0.34	0.475
	3.77	3289.0	0.33	0.154
	3.80	3263.0	0.33	0.053
	3.90	3179.0	0.32	0.040
	4.00	3100.0	0.31	0.103
	4.10	3024.0	0.30	0.153
	4.20	2952.0	0.30	0.194
	4.30	2883.0	0.29	0.208
	4.50	2755.0	0.28	0.238
	4.75	2610.0	0.26	0.252
	5.00	2497.0	0.25	0.257
	5.50	2254.0	0.23	0.257
	6.00	2066.0	0.21	0.246
	6.50	1907.0	0.19	0.225
	7.00	1771.0	0.18	0.196
	7.50	1653.0	0.17	0.157
	8.00	1550.0	0.15	0.114
	9.00	1378.0	0.14	0.074
	10.00	1240.0	0.12	0.082
	11.00	1127.0	0.11	0.088
	12.00	1033.0	0.10	0.100
	13.00	954.0		0.112
	14.00	886.0		0.141
	15.00	827.0		0.156
	16.00	775.0		0.151
	17.00	729.0		0.139
	18.00	689.0		0.124
	19.00	653.0		0.111

TABLE 3 Reflectance of Selected Metals at Normal Incidence (*Continued*)

Metal	eV	Wavelength Å	μm	R
Silver[36]	20.00	620.0		0.103
	21.00	590.0		0.112
	21.50	577.0		0.124
	22.00	564.0		0.141
	22.50	551.0		0.157
	23.00	539.0		0.163
	24.00	517.0		0.165
	25.00	496.0		0.154
	26.00	477.0		0.133
	28.00	443.0		0.090
	30.00	413.0		0.074
	34.00	365.0		0.067
	38.00	326.0		0.043
	42.00	295.0		0.036
	46.00	270.0		0.031
	50.00	248.0		0.027
	56.00	221.0		0.024
	62.00	200.0		0.016
	66.00	188.0		0.016
	70.00	177.0		0.021
	76.00	163.0		0.013
	80.00	155.0		0.012
	90.00	138.0		0.009
	100.00	124.0		0.005
Tungsten[36]	0.10		12.40	0.983
	0.20		6.20	0.981
	0.30		4.13	0.979
	0.38		3.26	0.963
	0.46		2.70	0.952
	0.54		2.30	0.948
	0.62		2.00	0.917
	0.70		1.77	0.856
	0.74		1.68	0.810
	0.78		1.59	0.759
	0.82		1.51	0.710
	0.86		1.44	0.661
	0.98		1.27	0.653
	1.10		1.13	0.627
	1.20		1.03	0.590
	1.30	9537.0	0.95	0.545
	1.40	8856.0	0.89	0.515
	1.50	8265.0	0.83	0.500
	1.60	7749.0	0.77	0.494
	1.70	7293.0	0.73	0.507
	1.80	6888.0	0.69	0.518
	1.90	6525.0	0.65	0.518
	2.10	5904.0	0.59	0.506
	2.50	4959.0	0.50	0.487
	3.00	4133.0	0.41	0.459
	3.50	3542.0	0.35	0.488
	4.00	3100.0	0.31	0.451
	4.20	2952.0	0.30	0.440

TABLE 3   Reflectance of Selected Metals at Normal Incidence (*Continued*)

Metal	eV	Wavelength Å	μm	R
Tungsten[36]	4.60	2695.0	0.27	0.455
	5.00	2497.0	0.25	0.505
	5.40	2296.0	0.23	0.586
	5.80	2138.0	0.21	0.637
	6.20	2000.0	0.20	0.646
	6.60	1878.0	0.19	0.631
	7.00	1771.0	0.18	0.607
	7.60	1631.0	0.16	0.556
	8.00	1550.0	0.15	0.505
	8.40	1476.0	0.15	0.449
	9.00	1378.0	0.14	0.388
	10.00	1240.0	0.12	0.287
	10.40	1192.0	0.12	0.270
	11.00	1127.0	0.11	0.290
	11.80	1051.0	0.11	0.318
	12.80	969.0		0.333
	13.60	912.0		0.325
	14.80	838.0		0.276
	15.60	795.0		0.246
	16.00	775.0		0.249
	16.80	738.0		0.273
	17.60	704.0		0.304
	18.80	659.0		0.340
	20.00	620.0		0.354
	21.20	585.0		0.331
	22.40	553.0		0.287
	23.60	525.0		0.252
	24.00	517.0		0.234
	24.80	500.0		0.191
	25.60	484.0		0.150
	26.80	463.0		0.105
	28.00	443.0		0.073
	30.00	413.0		0.047
	34.00	365.0		0.032
	36.00	344.0		0.036
	40.00	310.0		0.045

the penetration depth of incident radiation. Penetration depth is shown in Fig. 35 as a function of wavelength for Al, Be, and Ni.[51]

Absorption is a critical parameter for high-energy laser components, and is discussed in hundreds of papers as a function of surface morphology, angle of incidence, polarization state, and temperature. Only a few representative examples of this body of work can be cited here. When absorptance measurements were made of metal mirrors as a function of angle of incidence, polarization state, and wavelength,[52] it was found that measured values agreed with theory except at high angles of incidence where surface condition plays an undefined role. With the advent of diamond-turning as a mirror-finishing method, many papers have addressed absorptance characteristics of these unique surfaces as a function of surface morphology and angle of incidence, particularly on Ag and Cu mirrors.[53,54] It has been observed that mirrors have the lowest absorptance when the light is **s**-polarized and the grooves are oriented parallel to the plane of incidence.[54] The temperature dependence

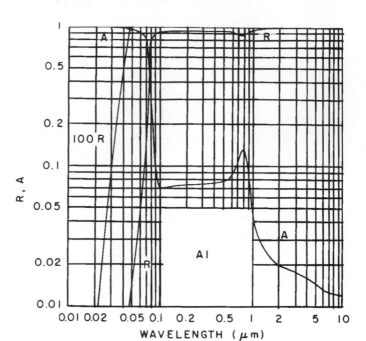

FIGURE 16 Reflectance and absorptance for aluminum vs. wavelength calculated for normal incidence.[35] (*With permission.*)

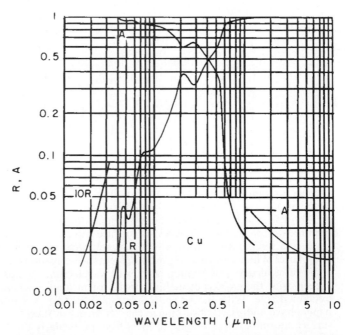

FIGURE 17 Reflectance and absorptance for copper vs. wavelength calculated for normal incidence.[35] (*With permission.*)

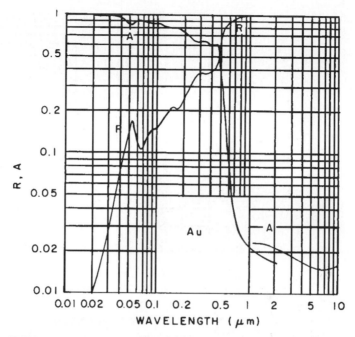

FIGURE 18 Reflectance and absorptance for gold vs. wavelength calculated for normal incidence.[35] (*With permission.*)

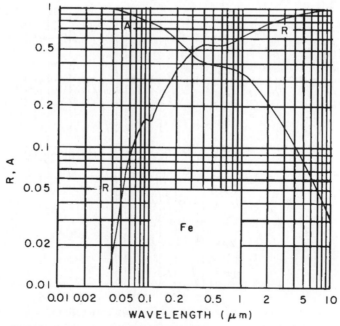

FIGURE 19 Reflectance and absorptance for iron vs. wavelength calculated for normal incidence.[35] (*With permission.*)

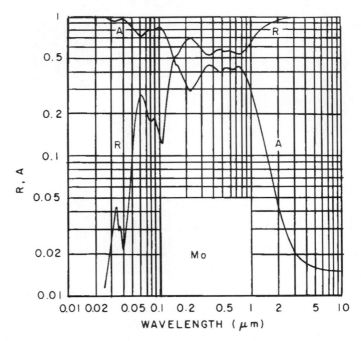

FIGURE 20 Reflectance and absorptance for molybdenum vs. wavelength calculated for normal incidence.[35] (*With permission.*)

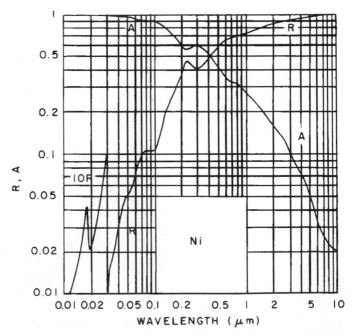

FIGURE 21 Reflectance and absorptance for nickel vs. wavelength calculated for normal incidence.[35] (*With permission.*)

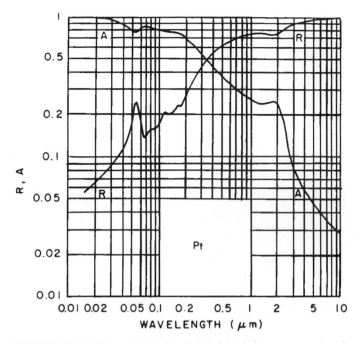

FIGURE 22 Reflectance and absorptance for platinum vs. wavelength calculated for normal incidence.[35] (*With permission.*)

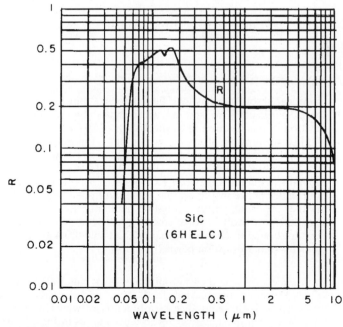

FIGURE 23 Reflectance for the basal plane of hexagonal silicon carbide vs. wavelength calculated for normal incidence.[35] (*With permission.*)

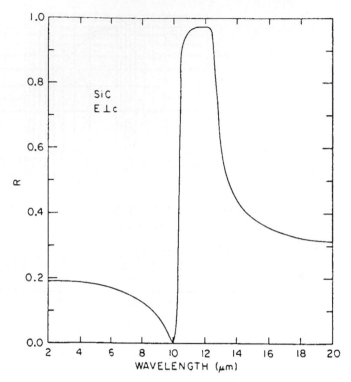

FIGURE 24 Infrared reflectance for the basal plane of hexagonal silicon carbide vs. wavelength calculated for normal incidence.[35] (*With permission.*)

of optical absorption has long been known,[55] but measurements and theory do not always agree, particularly at shorter wavelengths. Figure 36 shows absorptance of Mo as a function of temperature at a wavelength of 10.6 μm.[55]

Mass absorption of energetic photons[56] follows the same relationship as described in Eq. (4), but with the product mass attenuation coefficient μ and mass density ρ substituted for absorption coefficient α. Table 4 lists mass attenuation coefficients for selected elements at energies between 1 keV (soft x rays) and 1 GeV (hard gamma rays). Units for the coefficient are m²/kg, so that when multiplied by mass density in kg/m³, and depth x in m, the exponent in the equation is dimensionless. To a high approximation, mass attenuation is additive for elements present in a body, independent of the way in which they are bound in chemical compounds. Table 4 is highly abridged; the original[56] shows all elements and absorption edges.

Emittance. Where the transmittance of a material is essentially zero, the absorptance equals the emittance as described above and expressed in Eqs. (15) and (16). Spectral emittance ϵ_s is the emittance as a function of wavelength at constant temperature. These data have been presented as absorptance curves in Figs. 16 to 22, 25, and 26. For SiC, ϵ_s is given in Fig. 37.[57] Spectral emittance of unoxidized surfaces at a wavelength of 0.65 μm is given for selected materials in Table 5.[58]

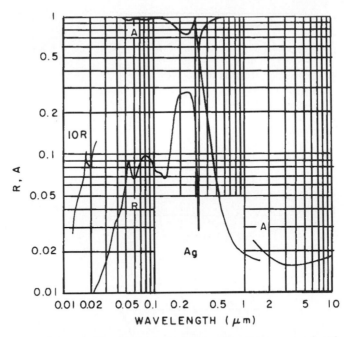

FIGURE 25 Reflectance and absorptance for silver vs. wavelength calculated for normal incidence.[35] (*With permission.*)

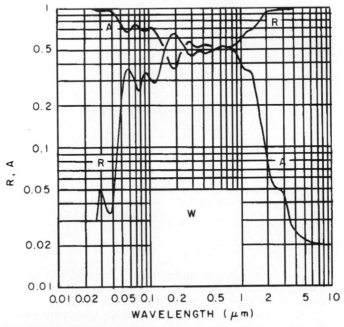

FIGURE 26 Reflectance and absorptance for tungsten vs. wavelength calculated for normal incidence.[35] (*With permission.*)

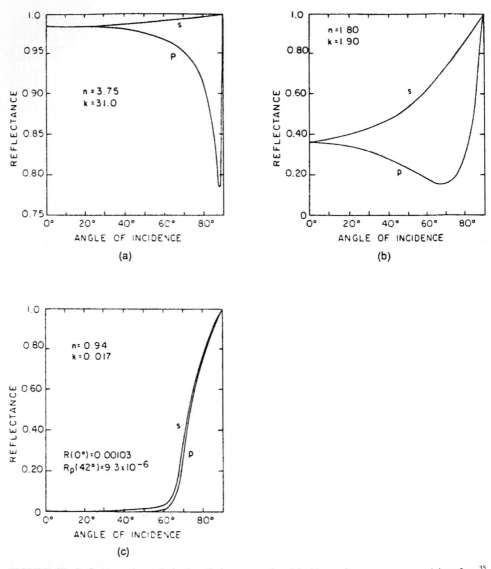

FIGURE 27 Reflectance for polarized radiation vs. angle of incidence for a vacuum-metal interface.[35] (*With permission.*) (*a*) $n = 3.75$, $k = 31.0$, approximate values for gold at $\lambda = 5\ \mu$m; (*b*) $n = 1.80$, $k = 1.90$, $\lambda = 0.3\ \mu$m; (*c*) $n = 0.94$, $k = 0.017$, approximate values for gold at $\lambda = 0.01\ \mu$m. Note the tendency toward total external reflectance for angle $\geq 80°$.

Total emittance ϵ_t is the emittance integrated over all wavelengths and usually given as a function of temperature. The total emittance of SiC is given in Fig. 38,[59] and for selected materials in Table 6.[60] Numerous papers by groups at the University of New Orleans (Ramanathan et al.[61–65]) and at Cornell University (Sievers et al.[66,67]) give high- and low-temperature data for the total hemispherical emittance of a number of metals including Ag, Al, Cu, Mo, W, and AISI 304 stainless steel.

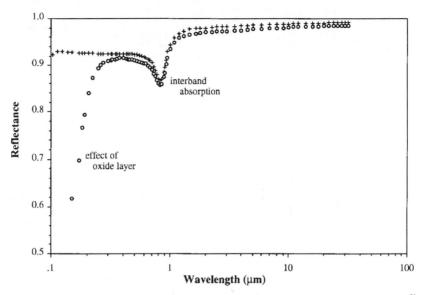

FIGURE 28 Effect of oxide layer on the reflectivity of aluminum vs. wavelength[51] calculated from n and k.[37]

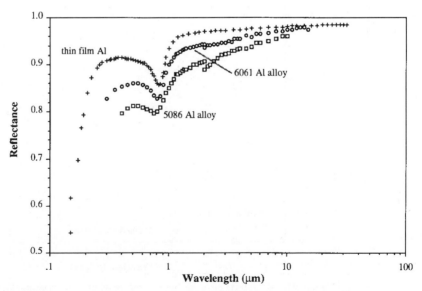

FIGURE 29 Reflectance of optical-grade aluminum alloys vs. wavelength.[51]

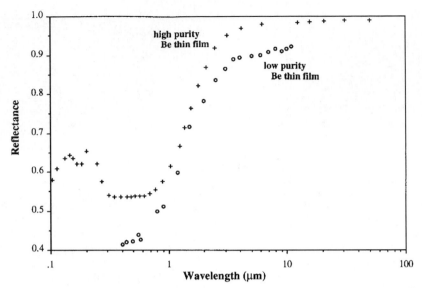

FIGURE 30 Effect of impurities on the reflectance of beryllium thin films vs. wavelength.[51]

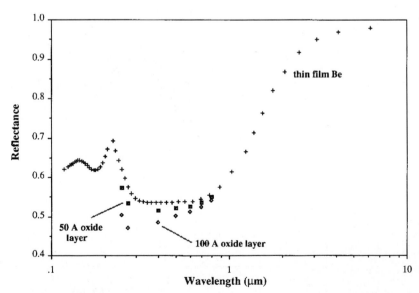

FIGURE 31 Effect of oxide layer thickness on the reflectance of beryllium vs. wavelength[51] calculated from n and k.[38]

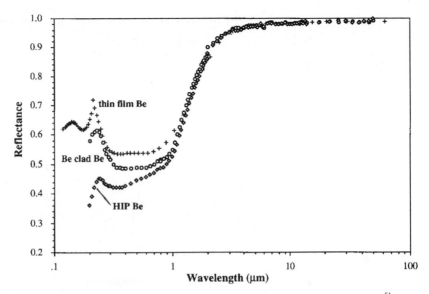

FIGURE 32 Reflectance of polished and evaporated beryllium vs. wavelength;[51] comparison of evaporated high-purity thin film,[38] polished high-purity thick film, and polished bulk beryllium (2 percent BeO).

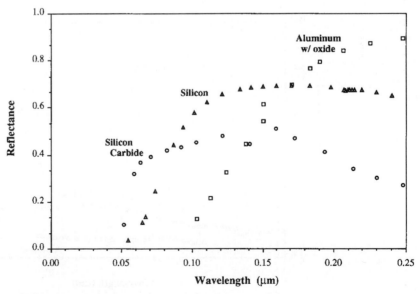

FIGURE 33 Ultraviolet reflectance of aluminum, silicon, and silicon carbide vs. wavelength.[51]

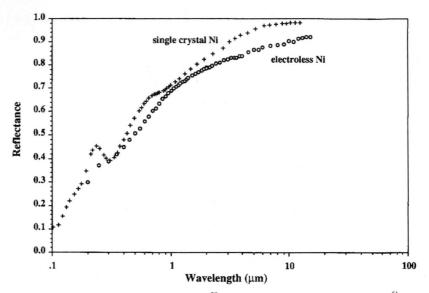

FIGURE 34 Reflectance of pure nickel[39] and electroless nickel (Ni-P alloy)[51] vs. wavelength calculated from *n* and *k*.

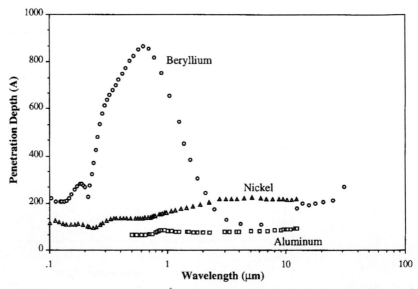

FIGURE 35 Penetration depth in Ångströms vs. wavelength for aluminum, beryllium, and nickel.[51]

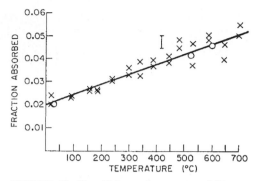

FIGURE 36 The 10.6-μm absorptance of Mo vs. temperature.[53] x is heating and o is cooling. The straight line is a least-squares fit to the data.

Physical Properties

The physical properties at room temperature of a number of metals are listed in Table 7. the crystal form does not appreciably affect the physical properties, but is a factor in the isotropy of thermal and mechanical properties. For most metals, resistivity is directly proportional to temperature and pure metals generally have increased resistivity with increasing amounts of alloying elements. This is shown graphically for copper in Fig. 39.[68] Resistivity for a number of pure, polycrystalline metals is listed as a function of temperature in Table 8.[69]

TABLE 4 Mass Attenuation Coefficients for Photons[56]

Mass attenuation coefficient, m^2/kg

		Photon energy, MeV						
	Atomic no.	0.001	0.01	0.1	1.0	10.0	100.0	1000.0
Be	4	6.04×10^1	6.47×10^{-2}	1.33×10^{-2}	5.65×10^{-3}	1.63×10^{-3}	9.94×10^{-4}	1.12×10^{-3}
C	6	2.21×10^{-2}	2.37×10^{-1}	1.51×10^{-2}	6.36×10^{-3}	1.96×10^{-3}	1.46×10^{-3}	1.70×10^{-3}
O	8	4.59×10^2	5.95×10^{-1}	1.55×10^{-2}	6.37×10^{-3}	2.09×10^{-3}	1.79×10^{-3}	2.13×10^{-3}
Mg	12	9.22×10^1	2.11	1.69×10^{-2}	6.30×10^{-3}	2.31×10^{-3}	2.42×10^{-3}	2.90×10^{-3}
Al	13	1.19×10^2	2.62	1.70×10^{-2}	6.15×10^{-3}	2.32×10^{-3}	2.52×10^{-3}	3.03×10^{-3}
Si	14	1.57×10^2	3.39	1.84×10^{-2}	6.36×10^{-3}	2.46×10^{-3}	2.76×10^{-3}	3.34×10^{-3}
P	15	1.91×10^2	4.04	1.87×10^{-2}	6.18×10^{-3}	2.45×10^{-3}	2.84×10^{-3}	3.45×10^{-3}
Ti	22	5.87×10^2	1.11×10^1	2.72×10^{-2}	5.89×10^{-3}	2.73×10^{-3}	3.71×10^{-3}	4.56×10^{-3}
Cr	24	7.40×10^2	1.39×10^1	3.17×10^{-2}	5.93×10^{-3}	2.86×10^{-3}	4.01×10^{-3}	4.93×10^{-3}
Fe	26	9.09×10^2	1.71×10^1	3.72×10^{-2}	5.99×10^{-3}	2.99×10^{-3}	4.33×10^{-3}	5.33×10^{-3}
Ni	28	9.86×10^2	2.09×10^1	4.44×10^{-2}	6.16×10^{-3}	3.18×10^{-3}	4.73×10^{-3}	5.81×10^{-3}
Cu	29	1.06×10^3	2.16×10^1	4.58×10^{-2}	5.90×10^{-3}	3.10×10^{-3}	4.66×10^{-3}	5.72×10^{-3}
Zn	30	1.55×10^2	2.33×10^1	4.97×10^{-2}	5.94×10^{-3}	3.18×10^{-3}	4.82×10^{-3}	5.91×10^{-3}
Ge	32	1.89×10^2	3.74	5.55×10^{-2}	5.73×10^{-3}	3.16×10^{-3}	4.89×10^{-3}	6.00×10^{-3}
Mo	42	4.94×10^2	8.58	1.10×10^{-1}	5.84×10^{-3}	3.65×10^{-3}	6.10×10^{-3}	7.51×10^{-3}
Ag	47	7.04×10^2	1.19×10^1	1.47×10^{-1}	5.92×10^{-3}	3.88×10^{-3}	6.67×10^{-3}	8.20×10^{-3}
W	74	3.68×10^2	9.69	4.44×10^{-1}	6.62×10^{-3}	4.75×10^{-3}	8.80×10^{-3}	1.08×10^{-2}
Pt	78	4.43×10^2	1.13×10^1	4.99×10^{-1}	6.86×10^{-3}	4.87×10^{-3}	9.08×10^{-3}	1.12×10^{-2}
Au	79	4.65×10^2	1.18×10^1	5.16×10^{-1}	6.95×10^{-3}	4.93×10^{-3}	9.19×10^{-3}	1.13×10^{-2}

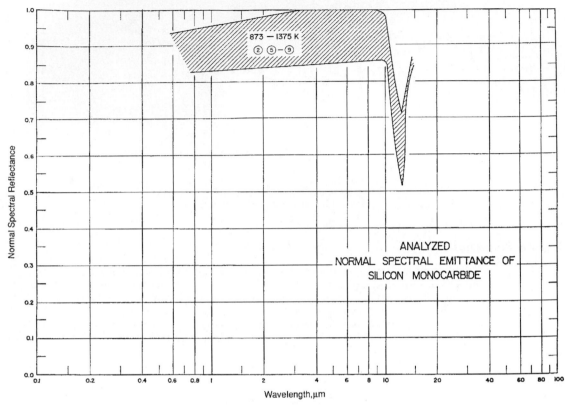

FIGURE 37   Analyzed normal spectral emittance of silicon carbide vs. wavelength.[57]

TABLE 5   Normal Spectral Emittance of Selected Metals ($\lambda = 0.65\ \mu m$)[58]

Metal	Emissivity
Beryllium	0.61
Chromium	0.34
Copper	0.10
Gold	0.14
Iron	0.35
Cast iron	0.37
Molybdenum	0.37
Nickel	0.36
80Ni-20Cr	0.35
Palladium	0.33
Platinum	0.30
Silver	0.07
Steel	0.35
Tantalum	0.49
Titanium	0.63
Tungsten	0.43

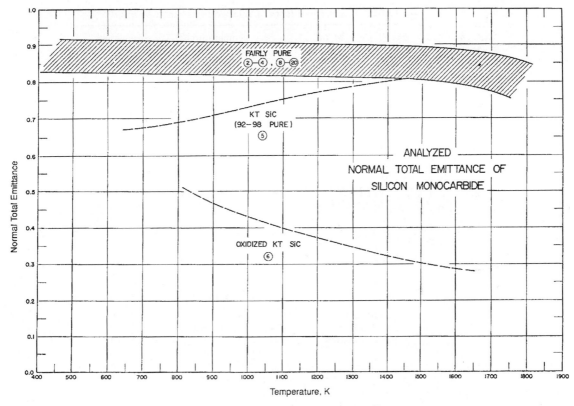

FIGURE 38 Analyzed normal total emittance of silicon carbide vs. temperature.[59]

TABLE 6 Total Emittance of Selected Materials[60]

Metal	Temperature (°C)	Emissivity
80 Ni-20Cr	100	0.87
	600	0.87
	1300	0.89
Aluminum		
Polished	50–500	0.04–0.06
Oxidized	200	0.11
	600	0.19
Chromium		
Polished	50	0.1
	500–1000	0.28–0.38
Copper		
Oxidized	50	0.6–0.7
	500	0.88
Polished	50–100	0.02
Unoxidized	100	0.02

TABLE 6 Total Emittance of Selected Materials (*Continued*)

Metal	Temperature (°C)	Emissivity
Glass	20–100	0.94–0.91
	250–1000	0.87–0.72
	1100–1500	0.7–0.67
Gold		
Carefully polished	200–600	0.02–0.03
Unoxidized	100	0.02
Iron, cast		
Oxidized	200	0.64
	600	0.78
Unoxidized	100	0.21
Molybdenum	600–1000	0.08–0.13
	1500–2200	0.19–0.26
Nickel		
Polished	200–400	0.07–0.09
Unoxidized	25	0.045
	100	0.06
	500	0.12
	1000	0.19
Platinum		
Polished	200–600	0.05–0.1
Unoxidized	25	0.017
	100	0.047
	500	0.096
	1000	0.152
Silver		
Polished	200–600	0.02–0.03
Unoxidized	100	0.02
	500	0.035
Steel		
304 SS	500	0.35
Unoxidized	100	0.08
Tantalum, unoxidized	1500	0.21
	2000	0.26
Tungsten, unoxidized	25	0.024
	100	0.032
	500	0.071
	1000	0.15

Thermal Properties

The thermal properties of materials were documented in 1970 through 1977 in the 13-volume series edited by Touloukian et al.[70] of the Thermophysical Properties Research Center at Purdue University. The properties database continues to be updated by the Center for Information and Numerical Data Analysis and Synthesis (CINDAS).[1]

Selected properties of coefficient of thermal expansion, CTE, thermal conductivity k, and specific heat C_p at room temperature, are listed in Table 9. Maximum usable temperatures are also listed in the table.

TABLE 7 Composition and Physical Properties of Metals

Metal	Mass density 10^3 kg/m^3	Electrical conductivity % IACS[a]	Electrical resistivity nohm·m[b]	Crystal form[c]	Chemical composition weight %, typical	Reference
Aluminum: 5086-O	2.66	31	56	fcc	4.0 Mg, 0.4 Mn, 0.15 Cr, bal. Al	84
Aluminum: 6061-T6	2.70	43	40	fcc	1.0 Mg, 0.6 Si, 0.3 Cu, 0.2 Cr, bal. Al	84
Beryllium: I-70-H	1.85	43	40	cph	99.0 Be min., 0.6 BeO, 0.08 Fe, 0.05 C, 0.03 Al, 0.02 Mg	85
Copper: OFC	8.94	101	17	fcc	99.95 Cu min.	84
Gold	19.3	73[d]	24	fcc	99.99 Au min.	84
Invar 36	8.1		820	bcc	36.0 Ni, 0.35 Mn, 0.2 Si, 0.02 C, bal. Fe	86
Molybdenum	10.22	34[e]	52[e]	bcc	99.9 Mo min., 0.015 C max.	84
Nickel: 200	8.9	18[d]	95	fcc	99.0 Ni min.	84
Nickel: electroless plate	7.75		900	fcc	10.5 P, bal. Ni	87
Silicon	2.33	[f]	[f]	dia. cubic	99.99 Si	84
Silicon carbide (SiC): CVD	3.21	[f]	[f]	cubic	99.99 SiC (beta)	88
SiC: reaction sintered	2.91	[f]	[f]	cph + dia. cubic	74.0 SiC (alpha), 26.0 Si	88
Silver	10.49	103[d]	15[e]	fcc	99.9 Ag min.	84
Stainless steel: 304	8.00	[d]	720	fcc	19.0 Cr, 9.0 Ni, 1.0 Mn, 0.5 Si, bal. Fe	89
Stainless steel: 416	7.80	[d]	570	distorted bcc	13.0 Cr, 0.6 Mn, 0.6 Mo, 0.5 Si, bal. Fe	89
Stainless steel: 430	7.80	[d]	600	bcc	17.0 Cr, 0.5 Mn, 0.5 Si, bal. Fe	89
Titanium: 6A14V	4.43	[d]	1710	bcc + cph	6.0 Al, 4.0 V, bal. Ti	90

[a] For equal volume at 293 K
[b] At 293 K
[c] fcc = face centered cubic; cph = close-packed hexagonal; bcc = body centered cubic
[d] Not available
[e] At 273 K
[f] Depends on impurity content

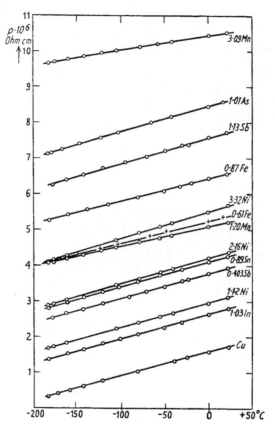

FIGURE 39 Electrical resistance of Cu and Cu alloys vs. temperature;[68] composition is in atomic percent.

The CTE of a material is a measure of length change at a specific temperature, useful for determining dimensional sensitivity to local temperature gradients. The total expansion (contraction) per unit length $\Delta L/L$ for a temperature change ΔT is the area under the CTE vs. T curve between the temperature extremes. Table 10 and Figs. 40 through 42 show recommended[71,72] CTE vs. T relationships for a number of materials. More recent expansion data have been published for many materials that are too numerous to list here, but those for beryllium[73] and beta silicon carbide[74] are included in Table 10.

Thermal conductivities of many pure polycrystalline materials have been published by the National Bureau of Standards[75,76] (now National Institute for Science and Technology) as part of the National Standard Reference Data System. Selected portions of these data, along with data from Touloukian et al.,[77,78] and specific data for beryllium[79] and beta silicon carbide,[80] are listed in Table 11 and shown in Figs. 43 through 46.

The specific heat of metals is very well documented.[73,81–83] Table 12 and Figs. 47 through 49 show the temperature dependence of this property. Table values are cited in J/kg K, numerically equal to W s/kg K.

Mechanical Properties

Mechanical properties are arbitrarily divided between the elastic properties of moduli, Poisson's ratio, and elastic stiffness, and the strength and fracture properties. All of these

TABLE 8 Electrical Resistivity (nohm m) of Pure, Polycrystalline Metals[69]

Temp. (K)	Aluminum	Beryllium	Chromium	Copper	Gold	Iron	Molybdenum	Nickel	Platinum	Silver	Tungsten
1	0.0010	0.332		0.020	0.220	0.225	0.0070	0.032	0.02	0.010	0.0002
10	0.0019	0.332		0.020	0.226	0.238	0.0089	0.057	0.154	0.012	0.0014
20	0.0076	0.336		0.028	0.350	0.287	0.0261	0.140	0.484	0.042	0.012
40	0.181	0.367		0.239	1.41	0.758	0.457	0.68	4.09	0.539	0.544
60	0.959	0.67		0.971	3.08	2.71	2.06	2.42	11.07	1.62	2.66
80	2.45	0.75		2.15	4.81	6.93	4.82	5.45	19.22	2.89	6.06
100	4.42	1.33	16.0	3.48	6.50	12.8	8.58	9.6	27.55	4.18	10.2
150	10.06	5.10	45.0	6.99	10.61	31.5	19.9	22.1	47.6	7.26	20.9
200	15.87	12.9	77.0	10.46	14.62	52.0	31.3	36.7	67.7	10.29	31.8
273	24.17	30.2	118.0	15.43	20.51	85.7	48.5	61.6	96.0	14.67	48.2
293	26.50	35.6	125.0	16.78	22.14	96.1	53.4	69.3	105.0	15.87	52.8
298	27.09	37.0	126.0	17.12	22.55	98.7	54.7	71.2	107.0	16.17	53.9
300	27.33	37.6	127.0	17.25	22.71	99.8	55.2	72.0	108.0	16.29	54.4
400	38.7	67.6	158.0	24.02	31.07	161.0	80.2	118.0	146.0	22.41	78.3
500	49.9	99.0	201.0	30.90	39.70	237.0	106.0	177.0	183.0	28.7	103.0
600	61.3	132.0	247.0	37.92	48.70	329.0	131.0	255.0	219.0	35.3	130.0
700	73.5	165.0	295.0	45.14	58.20	440.0	158.0	321.0	254.0	42.1	157.0
800	87.0	200.0	346.0	52.62	68.10	571.0	184.0	355.0	287.0	49.1	186.0
900	101.8	237.0	399.0	60.41	78.60		212.0	386.0	320.0	56.4	215.0

properties can be anisotropic as described by the elastic stiffness constants, but that level of detail is not included here. In general, cubic materials are isotropic in thermal properties and anisotropic in elastic properties. Materials of any of the other crystalline forms will be anisotropic in both thermal and elastic properties. For an in-depth treatment of this subject see, for example, Ref. 4.

TABLE 9 Thermal Properties of Metals at Room Temperature

Metal	Coeff. of thermal expansion ppm/K	Thermal conductivity W/m K	Specific Heat J/kg K	Maximum temperature K	References
Aluminum: 5086-O	22.6	127	900	475	84
Aluminum: 6061-T6	22.5	167	896	425	84
Beryllium: I-70-H	11.3	216	1925	800	85
Copper: OFC	16.5	391	385	400	84
Gold	14.2	300	130	400	84
Iron	11.8	81	450	900	84
Invar 36	1.0	10	515	475	86
Molybdenum	4.8	142	276	1100	84
Nickel: 200	13.4	70	456	650	84
Nickel: Electroless plate (11% P)	11.0	7	460	425	87
(8% P)	12.8			450	91
Silicon	2.6	156	710	725	84
Silicon Carbide (SiC): CVD	2.2	198	733	1200	88
	2.4	250	700		92
SiC: Reaction sintered	2.6	155	670	1100	92
Silver	19.0	428	235	400	84
Stainless steel: 304	14.7	16	500	700	89
Stainless steel: 416	9.5	25	460	500	89
Stainless steel: 430	10.4	26	460	870	89
Titanium: 6A14V	8.6	7	520	650	84

TABLE 10 Temperature Dependence of the Coefficient of Linear Thermal Expansion (ppm/K) of Selected Materials

Temp K	6061 Al	Be	Cu	Au	Fe	304 SS	416 SS	Mo	Ni	Ag	Si	alpha SiC	beta SiC
5		0.0003	0.005	0.03	0.01				0.02	0.015		0.01	
10		0.001										0.02	
20		0.005				9.8	4.3	0.3			0		
25		0.009	0.63	2.8	0.2			0.4	0.25	1.9	0	0.03	
50		0.096	3.87	7.7	1.3	10.5	4.9	1	1.5	8.2	−0.2	0.06	
75		0.47							4.3		−0.5	0.09	
100	12.2	1.32	10.3	11.8	5.6	11.4	6	2.8	6.6	14.2	−0.4	0.14	
125	18.7	2.55											
150	19.3	4.01				12.4	7				0.5	0.4	
175	20.3	5.54											
200	20.9	7.00	15.2	13.7	10.1	13.2	7.9	4.6	11.3	17.8	1.5	1.5	
225	21.5	8.32											
250	21.5	9.50				14.1	8.8				2.2	2.8	
293	22.5	11.3	16.5	14.2	11.8	14.7	9.5	4.8	13.4	18.9	2.6	3.3	3.26
300		11.5									3.4		3.29
350	23.8												3.46
400	25.0	13.6	17.6	14.8	13.4	16.3	10.9	4.9	14.5	19.7	3.2	4	3.62
450	26.3												3.77
500	27.5	15.1	18.3	15.4	14.4	17.5	12.1	5.1	15.3	20.6	3.5	4.2	3.92
600	30.1	16.6	18.9	15.9	15.1	18.6	12.9	5.3	15.9	21.5	3.7	4.5	4.19
700		17.8	19.5	16.4	15.7	19.5	13.5	5.5	16.4	22.6	3.9	4.7	4.42
800		19.1	20.3	17	16.2	20.2	13.8	5.7	16.8	23.7	4.1	4.9	4.62
900		20.0	21.3	17.7	16.4		13.9	6	17.1	24.8	4.3	5.1	4.79
1000		20.9	22.4	18.6	16.6	21.1	13.9	6.2	17.4	25.9	4.4	5.3	4.92
Reference	71	73	71	71	71	71	71	71	71	71	72	72	74

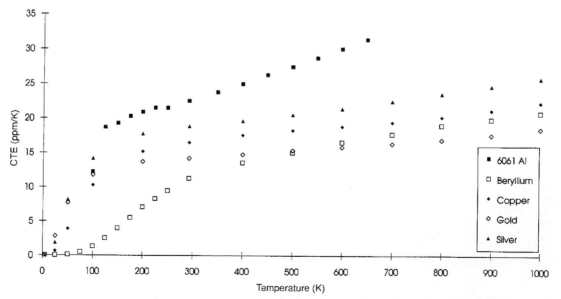

FIGURE 40 Coefficient of linear thermal expansion of 6061 aluminum alloy,[71] beryllium,[73] copper,[71] gold,[71] and silver[71] vs. temperature.

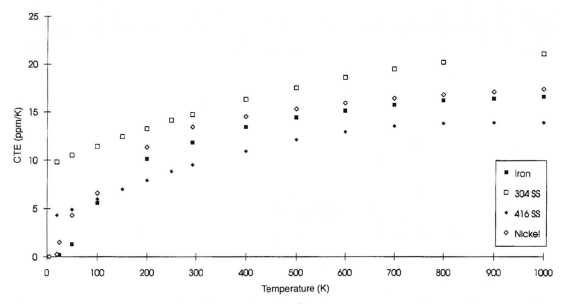

FIGURE 41   Coefficient of linear thermal expansion of iron,[71] stainless steel types 304[71] and 416,[71] and nickel[71] vs. temperature.

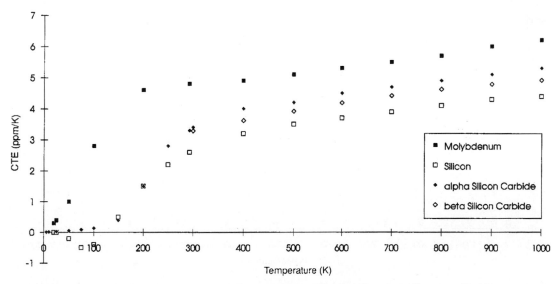

FIGURE 42   Coefficient of linear thermal expansion of molybdenum,[71] silicon,[72] and alpha[72] and beta[74] silicon carbide vs. temperature.

TABLE 11 Temperature Dependence of the Thermal Conductivity (W/m K) of Selected Materials

Temp K	Pure Al	6061 Al	5086 Al	Be	Cu	Au	Fe	304 SS	Mo	Ni	Ag	Si	alpha SiC	beta SiC
5	3,810				13,800	2,070	371		73	316	17,200	424		
10	6,610	87	8		19,600	2,820	705	1	145	600	16,800	2,110		
20	5,650	170	17	60	10,500	1,500	997		277	856	5,100	4,940	950	
50	1,000	278	40	140	1,220	420	936	6	300	336	700	2,680	2,872	
75	450			197			186		220	207	484	1,510	2,797	
100	300	213	64	268	463	345	132	10	179	158	450	884	2,048	
123														179
150		200	79	301	428	335	104	12	149	121	432	409		
167													970	
173														223
200	237	203	93	282	413	327	94	13	143	106	430	264		
250		209	103	232	404	320								
273	236		109		401	318	84	15	139	94	428	168		202
293		212												
298			115											193
300	237			200	398	315	80	15	138	90	427	148	420	
400	240			160	392	312	69	17	134	80	420	99		
500	237			139	388	309	61	18	130	72	413	76		
600	232			126	383	304	55	20	126	66	405	62		
700	226			115	377	298	49	21	122	65	397	51		
800	220			107	371	292	43	22	118	67	389	42		
900	213			98	364	285	38	24	115	70	382	36		
1000				89	357	278	33	25	112	72	374	31		
Reference	75, 76	77	77	79	77	77	77	77	77	77	77	77	78	80

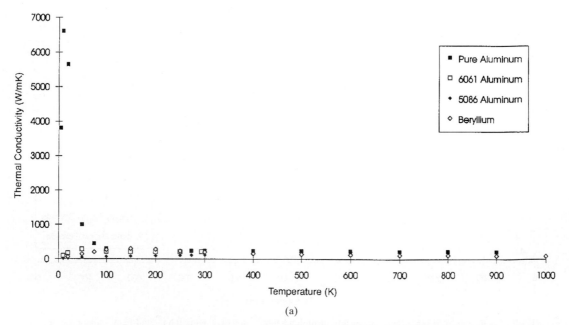

(a)

FIGURE 43 Thermal conductivity of three aluminum alloys[75-77] and beryllium[79] vs. temperature.

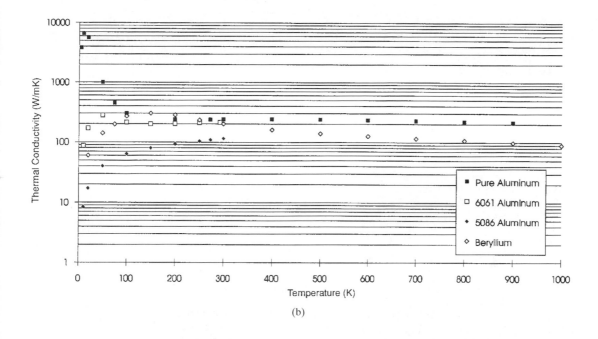

(b)

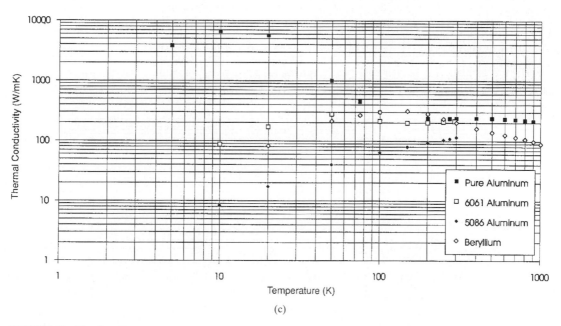

(c)

FIGURE 43 (*Continued*)

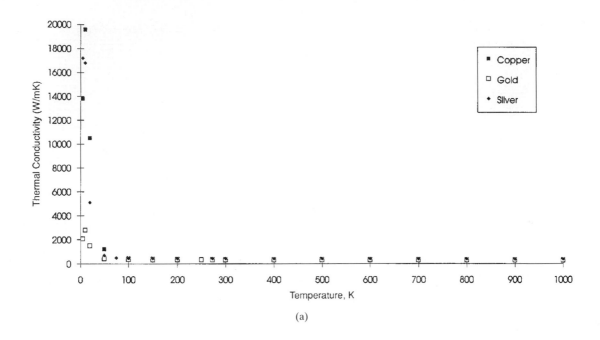

(a)

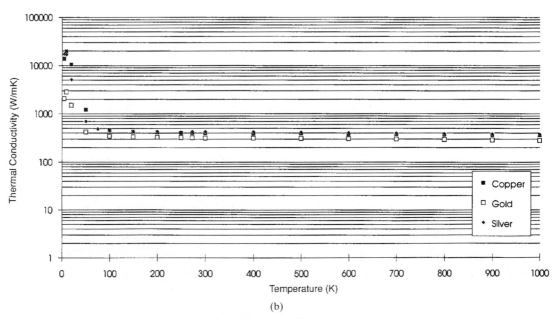

(b)

FIGURE 44   Thermal conductivity of copper,[77] gold,[77] and silver[77] vs. temperature.

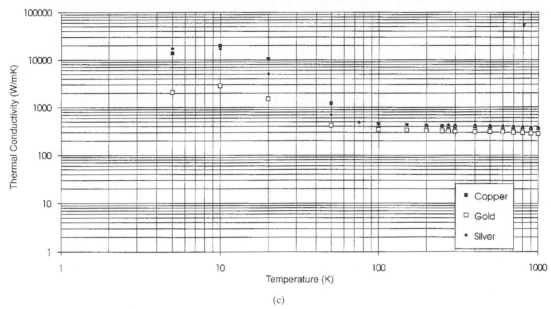

(c)

FIGURE 44   (*Continued*)

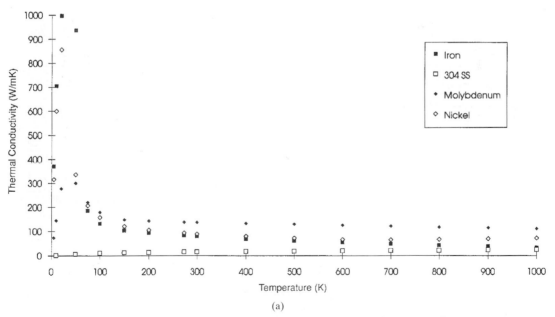

(a)

FIGURE 45   Thermal conductivity of iron,[77] type 304 stainless steel,[77] molybdenum,[77] and nickel[77] vs. temperature.

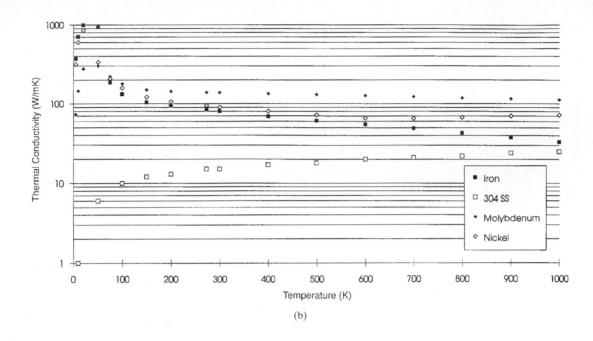

(b)

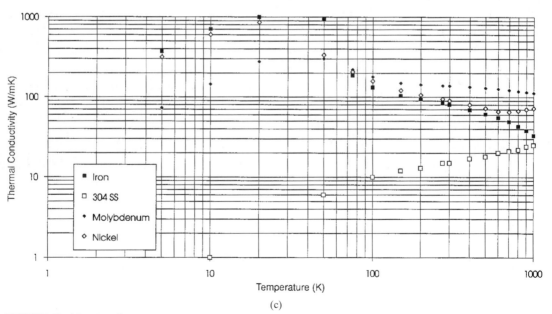

(c)

FIGURE 45   (*Continued*)

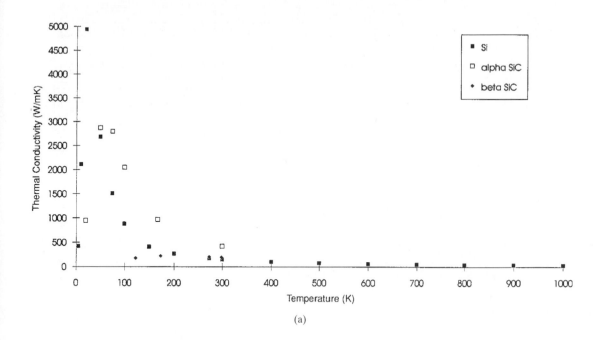

(a)

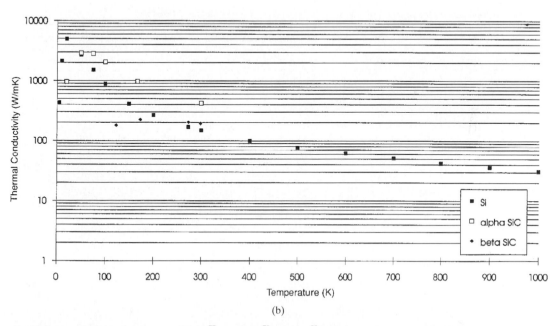

(b)

FIGURE 46 Thermal conductivity of silicon[77] and alpha[78] and beta[80] silicon carbide vs. temperature.

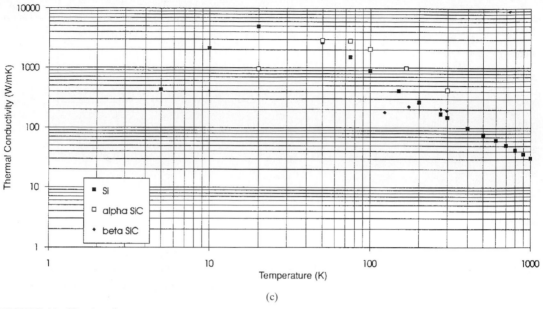

(c)

FIGURE 46 (*Continued*)

TABLE 12 Temperature Dependence of the Specific Heat (J/kg K) of Selected Materials

Temp K	Al	Be	Cu	Au	Fe	430 SS	Mo	Ni	Ag	Si	beta SiC
5	0.4	0.3	0.2		0.4				0.2		
10	1.4	0.4	0.9		1.2				1.7	0.3	
13		0.9					0.8	2.6	3.9		
20	8.9	1.7	7.5	16	4.6		2.2	5.4	15	3.4	
40	78	7.4	58	57	29		21	38	89	45	
60	214	29	137	84	84		61			115	
75		67			138		100	155	160	170	
100	481	177	256	109	213		141	232	187	259	
123											250
140			312					314			
150		636		119	324		197		214	426	
173											400
180			347					366			
200	791	1113		124	385		224	383	225	557	
220			368								
250	855	1536					241		232		
260			378								
273							255			663	700
293					445			441			
298			384	129					237		
300	899	1833								712	
323								453	236		
350	931										

TABLE 12 Temperature Dependence of the Specific Heat (J/kg K) of Selected Materials (*Continued*)

Temp K	Al	Be	Cu	Au	Fe	430 SS	Mo	Ni	Ag	Si	beta SiC
366			398								
373		2051					261	471		770	880
400	956			134	487						
473							266	514		825	1020
477			402								
500	995				526						
523								541	244		
573							271	573		848	1050
600	1034			142	568						
623		2574						626			
629								656			
630								669			
631								652			
673	1076	2658					277	530	251	864	1150
700			423		617	649					
733								526			
773								527	257	881	1200
800		2817		147	687	753					
873							288	542	262	898	
900		2918	452		786	862					
973							294	556		913	
1000		3022	467	151	1016	971			272		
Reference	81, 82	73, 83	83	83	83	83	83	83	83	83	80

Elastic Properties. The principal elastic stiffnesses C_{ij} of single crystals of some materials are given in Table 13. The three moduli and Poisson's ratio for polycrystalline materials are given in Table 14. These properties vary little with temperature, increasing temperature causing a gradual decrease in the moduli.

Strength and Fracture Properties The properties of tensile yield (at 0.2 percent offset), microyield strength, ductility (expressed as percent elongation in 50 mm), fracture toughness, flexural strength, and mechanical hardness are listed in Table 15. Most of these properties vary with temperature: strength and hardness decreasing, and fracture toughness and ductility increasing with temperature.

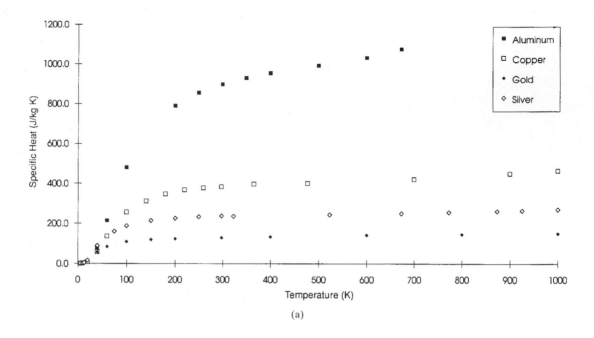

(a)

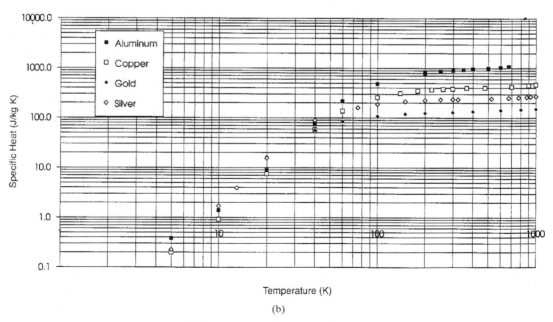

(b)

FIGURE 47   Specific heat of aluminum,[81,82] copper,[83] gold,[83] and silver[83] vs. temperature.

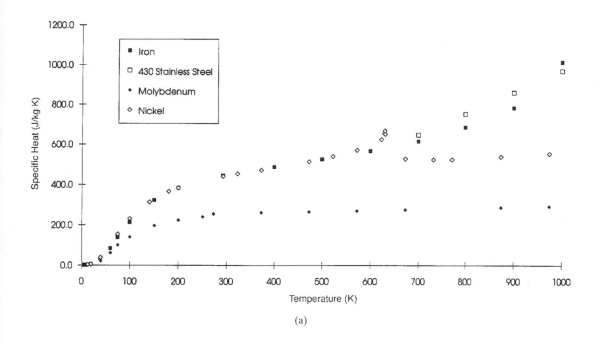

(a)

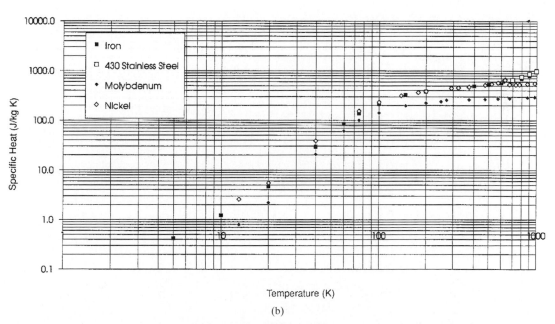

Temperature (K)

(b)

FIGURE 48   Specific heat of iron,[83] type 430 stainless steel,[83] molybdenum,[83] and nickel[83] vs. temperature.

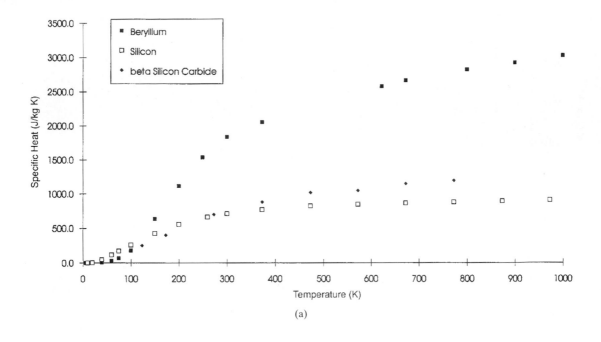

(a)

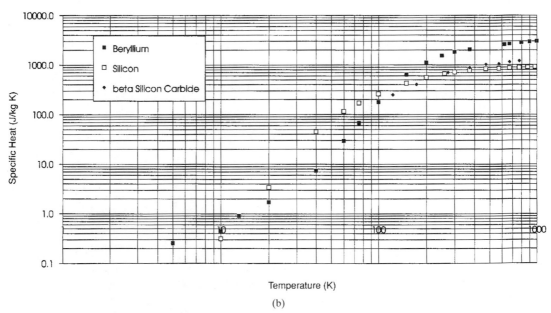

(b)

FIGURE 49 Specific heat of beryllium,[73,83] silicon,[83] and beta silicon carbide[80] vs. temperature.

TABLE 13 Elastic Stiffness Constants for Selected Single Crystal Metals

Cubic metals[93]	Elastic stiffness (GN/m^2)		
	C_{11}	C_{44}	C_{12}
Aluminum	108.0	28.3	62.0
Chromium	346.0	100.0	66.0
Copper	169.0	75.3	122.0
Germanium	129.0	67.1	48.0
Gold	190.0	42.3	161.0
Iron	230.0	117.0	135.0
Molybdenum	459.0	111.0	168.0
Nickel	247.0	122.0	153.0
Silicon	165.0	79.2	64.0
Silicon carbide[94]	352.0	233.0	140.0
Silver	123.0	45.3	92.0
Tantalum	262.0	82.6	156.0
Tungsten	517.0	157.0	203.0

Hexagonal metals	C_{11}	C_{33}	C_{44}	C_{12}	C_{13}
Beryllium[95]	288.8	354.2	154.9	21.1	4.7
Magnesium[93]	22.0	19.7	60.9	−7.8	−5.0
Silicon carbide[94]	500.0	521.0	168.0	98.0	

TABLE 14 Elastic Moduli and Poisson's Ratio for Selected Polycrystalline Materials

Material	Young's modulus GN/m^2	Shear modulus GN/m^2	Bulk modulus GN/m^2	Poisson's ratio	Reference
Aluminum: 5086-O	71.0	26.4		0.33	84
Aluminum: 6061-T6	68.9	25.9		0.33	84
Beryllium: I-701-H	315.4	148.4	115.0	0.043	96
Copper	129.8	48.3	137.8	0.343	97
Germanium	79.9	29.6		0.32	97
Gold	78.5	26.0	171.0	0.42	97
Invar 36	144.0	57.2	99.4	0.259	97
Iron	211.4	81.6	169.8	0.293	97
Molybdenum	324.8	125.6	261.2	0.293	97
Nickel	199.5	76.0	177.3	0.312	97
Platinum	170.0	60.9	276.0	0.39	97
Silicon	113.0	39.7		0.42	97
Silicon carbide: CVD	461.0			0.21	80
Silicon carbide: reaction sintered	413.0			0.24	88
Silver	82.7	30.3	103.6	0.367	97
Stainless steel: 304	193.0	77.0		0.27	97
Stainless steel: 416	215.0	83.9	166.0	0.283	97
Stainless steel: 430	200.0	80.0		0.27	97
Tantalum	185.7	62.2	196.3	0.342	97
Tungsten	411.0	160.6	311.0	0.28	97

TABLE 15 Strength and Fracture Properties for Selected Materials

Material	Yield strength MN/m^2	Microyield strength MN/m^2	Elongation (in 50 mm) %	Fracture toughness $MN\,m^{-3/2}$	Flexural strength MN/m^2	Hardness*	Reference
Aluminum: 5086-O	115.0	40.0	22.0	>25.0	—	55 HRB	84
Aluminum: 6061-T6	276.0	160.0	15.0	<25.0	—	95 HRB	84
Beryllium: I-70-H	276.0	30.0	4.0	12.0	—	80 HRB	84, 85
Copper	195.0	12.0	42.0	—	—	10 HRB	84
Germanium	—	—	—	1.0	110.0	800 HK	84
Gold	125.0	—	30.0	—	—	30 HK	84
Invar 36	276.0	37.0	35.0	—	—	70 HRB	86
Molybdenum	600.0	—	40.0	—	—	150 HK	84
Nickel	148.0	—	47.0	—	—	109 HRB	84
Platinum	150.0	—	35.0	—	—	40 HK	84
Silicon	—	—	—	1.0	207.0	1150 HK	84
Silicon carbide: CVD	—	—	—	3.0	595.0	2500 HK	80
Silicon carbide: reaction sintered	—	—	—	2.0	290.0	2326 HK	88
Silver	130.0	—	47.0	—	—	32 HK	84
Stainless steel: 304	241.0	—	60.0	—	—	80 HRB	89
Stainless steel: 416	950.0	—	12.0	—	—	41 HRC	89
Stainless steel: 430	380.0	—	25.0	—	—	86 HRB	89
Tantalum	220.0	—	30.0	—	—	120 HK	84
Tungsten	780.0	—	2.0	—	—	350 HK	84

* HK = Knoop (kg/mm^2); HRB = Rockwell B; HRC = Rockwell C.

35.4 REFERENCES

1. Center for Information and Numerical Data Analysis and Synthesis (CINDAS), Purdue Univ., 2595 Yeager Rd., W. Lafayette, IN 47906, (800) 428–7675.

2. Optical Properties of Solids and Liquids (OPTROP), Sandia National Laboratory, Div. 1824, P.O. Box 5800, Albuquerque, NM 87185, (505) 844–2109.

3. H. Wawrousek, J. H. Westbrook, and W. Grattideg (eds.), "Data Sources of Mechanical and Physical Properties of Engineering Materials," *Physik Daten/Physics Data,* No. 30–1, Fachinformationszentrum Karlsruhe, 1989.

4. M. E. Lines, "Physical Properties of Materials: Theoretical Overview," in Paul Klocek (ed.), *Handbook of Infrared Optical Materials,* Marcel Dekker, New York, 1991, pp. 1–69.

5. N. F. Mott and H. Jones, *The Theory of The Properties of Metals and Alloys,* Dover, New York, 1958, pp. 105–125.

6. A. V. Sokolov, *Optical Properties of Metals,* American Elsevier, New York, 1967.

7. F. Wooten, *Optical Properties of Solids,* Academic Press, New York, 1972.

8. M. Born and E. Wolf, *Principles of Optics,* 5th ed., Pergamon Press, London, 1975, pp. 611–627.

9. This discussion is adapted with permission from M. Bass, "Laser-Materials Interactions," in *Encyclopedia of Physical Science and Technology* **8**, Academic Press, New York, 1992, pp. 415–418.

10. F. Stern in F. Seitz and D. Turnbull (eds.), *Solid State Physics,* vol. 15, Academic Press, New York, 1963, pp. 300–324.

11. L. D. Landau and E. M. Lifshitz, *Electrodynamics of Continuous Media,* Addison-Wesley, Reading, Mass., 1965, pp. 257–284.

12. H. A. Lorentz, *The Theory of Electrons,* Dover, New York, 1952.

13. P. K. L. Drude, *Theory of Optics,* Dover, New York, 1959.

14. C. S. Barrett, *Structure of Metals,* 2d ed., McGraw-Hill, New York, 1952, pp. 521–537.

15. As reported by S. F. Jacobs in "Variable Invariables—Dimensional Instability with Time and Temperature," P. R. Yoder, Jr, ed., *Optomechanical Design,* Critical Reviews of Optical Science and Technology, **CR43,** SPIE Optical Engineering Press, Bellingham, Wash., 1992, p. 201.

16. C. Kittel, *Introduction to Solid State Physics,* 3d ed., Wiley, New York, 1966, pp. 111–129.

17. H. Reisman and P. S. Pawlik, *Elasticity,* Wiley, New York, 1980, pp. 128–135.

18. A. Kelly and N. H. Macmillan, *Strong Solids,* 3d ed., Clarendon Press, Oxford, 1986, pp. 382–393.

19. For more complete descriptions see, for example, *Metals Handbook,* 9th ed., **8**, Mechanical Testing, American Society for Metals, Metals Park, OH, 1985, pp. 1–15.

20. R. A. Paquin, "Selection of Materials and Processes for Metal Optics," in *Selected Papers on Optomechanical Design, Proc. SPIE,* Milestone Series, **770:**27–34 (1987).

21. D. Janeczko, "Metal Mirror Review," in R. Hartmann and W. J. Smith (eds.), *Infrared Optical Design,* Critical Reviews of Optical Science and Technology, **CR38**, SPIE Optical Engineering Press, Bellingham, Wash., 1991, pp. 258–280.

22. M. H. Krim, "Mechanical Design of Optical Systems for Space Operation," in P. R. Yoder, Jr. (ed.), *Optomechanical Design,* Critical Reviews of Optical Science and Technology, **CR43**, SPIE Optical Engineering Press, Bellingham, Wash., 1992, pp. 3–17.

23. P. R. Yoder, Jr. *Opto-Mechanical Systems Design,* 2d ed., Marcel Dekker, New York, 1993, pp. 1–41.

24. This analysis is the same as that used by many structural engineers such as the late G. E. Seibert of Perkin-Elmer and Hughes Danbury Optical Systems.

25. S. Timoshenko and S. Woinowsky-Kreiger, *Theory of Plates and Shells,* 2d ed., McGraw-Hill, New York, 1959, pp. 51–78.

26. P. K. Mehta,"Nonsymmetric Thermal Bowing of Curved Circular Plates," in A. E. Hatheway (ed.), *Structural Mechanics of Optical Systems II, Proc. SPIE,* **748** (1987).

27. E. Pearson, "Thermo-elastic Analysis of Large Optical Systems", in P. R. Yoder, Jr. (ed.), *Optomechanical Design,* Critical Reviews of Optical Science and Technology, **CR43**, SPIE Optical Engineering Press, Bellingham, Wash., 1992, pp. 123–130.

28. C. W. Marschall and R. E. Maringer, *Dimensional Instability, an Introduction,* Pergamon, New York, 1977.

29. R. A. Paquin (ed.), "Dimensional Stability," *Proc. SPIE* **1335** (1990).

30. R. A. Paquin and D. Vukobratovich (eds.), "Optomechanics and Dimensional Stability," *Proc. SPIE* **1533** (1991).

31. R. A. Paquin, "Dimensional Instability of Materials; How Critical Is It in the Design of Optical Instruments?," in P. R. Yoder (ed.), *Optomechanical Design,* Critical Reviews of Optical Science and Technology, **CR43**, SPIE Optical Engineering Press, Bellingham, Wash., 1992, pp. 160–180.

32. *op.cit.* Ref 23, pp. 271–320, 567–584.

33. E. D. Palik (ed.), *Handbook of Optical Constants of Solids,* Academic Press, Orlando, 1985.

34. E. D. Palik (ed.), *Handbook of Optical Constants of Solids II,* Academic Press, Orlando, 1991.

35. D. W. Lynch, "Mirror and Reflector Materials," in M. J. Weber (ed.), *CRC Handbook of Laser Science and Technology,* **IV**, Optical Materials, Part 2: Properties, CRC Press, Boca Raton, Florida, 1986.

36. J. H. Weaver, C. Krafka, D. W. Lynch, and E. E. Koch (eds.), "Optical Properties of Metals," Pts. 1 and 2, *Physik Daten/Physics Data,* Nos. 18-1 and 18-2, Fachinformationszentrum Karlsruhe, 1981.

37. D. Y. Smith, E. Shiles, and Mitio Inokuti, "The Optical Properties of Aluminum," in E. D. Palik (ed.), *Handbook of Optical Constants of Solids,* Academic Press, Orlando, 1985, pp. 369–406.

38. E. T. Arakawa, T. A. Callcott, and Y.-C. Chang, "Beryllium," in E. D. Palik (ed.), *Handbook of Optical Constants of Solids II,* Academic Press, Orlando, 1991, pp. 421–433.

39. D. W. Lynch and W. R. Hunter, in E. D. Palik (ed.), *Handbook of Optical Constants of Solids,* Academic Press, Orlando, 1985, pp. 275–367.

40. D. W. Lynch and W. R. Hunter, in E. D. Palik (ed.), *Handbook of Optical Constants of Solids II,* Academic Press, Orlando, 1991, pp. 341–419.

41. W. J. Choyke and E. D. Palik, "Silicon Carbide," in E. D. Palik (ed.), *Handbook of Optical Constants of Solids,* Academic Press, Orlando, 1985, pp. 587–595.

42. J. H. Weaver, D. W. Lynch, and R. Rossi, "Optical Properties of Single-Crystal Be from 0.12 to 4.5 eV," *Phys. Rev. B* **7:**3537–3541 (1973).

43. J. C. Stover, J. Rifkin, D. R. Cheever, K. H. Kirchner, and T. F. Schiff, "Comparison of Wavelength Scaling Data to Experiment," in R. P. Breault (ed.), *Stray Light and Contamination in Optical Systems, Proc. SPIE* **967:**44–49 (1988).

44. C. L. Vernold, "Application and Verification of Wavelength Scaling for Near Specular Scatter Predictions," in J. C. Stover (ed.), *Scatter from Optical Components, Proc. SPIE* **1165:**18–25 (1989).

45. J. E. Harvey, "Surface Scatter Phenomena: a Linear, Shift-invariant Process," in J. C. Stover (ed.), *Scatter from Optical Components, Proc. SPIE* **1165:**87–99 (1989).

46. J. C. Stover, M. L. Bernt, D. E. McGary, and J. Rifkin, "An Investigation of Anomalous Scatter from Beryllium Mirrors," in J. C. Stover (ed.), *Scatter from Optical Components, Proc. SPIE* **1165:**100–109 (1989).

47. See also papers in the "Scatter from Be Mirrors" session in J. C. Stover (ed.), *Optical Scatter: Applications, Measurement, and Theory, Proc. SPIE* **1530:**130–230 (1991).

48. Y. S. Touloukian and D. P. DeWitt, "Thermal Radiative Properties, Metallic Elements and Alloys," vol. 7 in Y. S. Touloukian and C. Y. Ho (eds.), *Thermophysical Properties of Matter,* IFI/Plenum, New York, 1970.

49. Y. S. Touloukian and D. P. DeWitt, "Thermal Radiative Properties, Nonmetallic Solids," vol. 8 in Y. S. Touloukian and C. Y. Ho (eds.), *Thermophysical Properties of Matter,* IFI/Plenum, New York, 1971.

50. J. S. Browder, S. J. Ballard, and P. Klocek in Paul Klocek (ed.), *Handbook of Infrared Optical Materials,* Marcel Dekker, New York, 1991, pp. 155–426.

51. C. M. Egert, "Optical Properties of Aluminum, Beryllium, Silicon Carbide (and more)" in *Proc. of Al, Be, and SiC Optics Technologies Seminar,* MODIL, Oak Ridge National Lab., 1993.

52. W. D. Kimura and D. H. Ford, Absorptance Measurement of Metal Mirrors at Glancing Incidence," *Appl. Optics* **25:**3740–3750 (1986).

53. M. Bass and L. Liou, "Calorimetric Studies of Light Absorption by Diamond Turned Ag and Cu Surfaces and Analyses Including Surface Roughness Contributions," *J. Appl. Phys.* **56:**184–189 (1984).

54. W. D. Kimura and T. T. Saito, "Glancing Incidence Measurements of Diamond Turned Copper Mirrors," *Appl. Optics* **26:**723–728 (1987).

55. M. Bass, D. Gallant, and S. D. Allen, "The Temperature Dependence of the Optical Absorption of Metals," in *Basic Optical Properties of Materials,* NBS SP574, U.S. Govt. Printing Office, Wash. D.C., 1980, pp. 48–50.

56. This discussion and Table 4 are based on the article: M. J. Berger and J. H. Hubbell, "Photon Attenuation Coefficients," in D. R. Lide (editor-in-chief), *CRC Handbook of Chemistry and Physics,* 74th ed., CRC Press, Boca Raton, Fla., 1993, pp. **10**-282–**10**-286.

57. Op. cit., Ref. 49, p. 798.

58. D. R. Lide (editor-in-chief), *CRC Handbook of Chemistry and Physics,* 74th ed., CRC Press, Boca Raton, Fla., 1993, p. **10**-299.

59. Op. cit., Ref. 49, p. 792.

60. Op. cit., Ref. 58, p. **10**-298.

61. K. O. Ramanathan and S. H. Yen, "High-temperature Emissivities of Copper, Aluminum, and Silver," *J. Opt. Soc. Am.* **67:**32–38 (1977).

62. E. A. Estalote and K. O. Ramanathan, "Low-temperature Emissivities of Copper and Aluminum," *J. Opt. Soc. Am.* **67:**39–44 (1977).

63. K. O. Ramanathan, S. H. Yen, and E. A. Estalote, Total Hemispherical Emissivities of Copper, Aluminum, and Silver," *Appl. Optics* **16:**2810–2817 (1977).

64. D. P. Verret and K. O. Ramanathan, "Total Hemispherical Emissivity of Tungsten," *J. Opt. Soc. Am.* **68:**1167–1172 (1978).

65. C. R. Roger, S. H. Yen, and K. O. Ramanathan, "Temperature Variation of Total Hemispherical Emissivity of Stainless Steel AISI 304," *J. Opt. Soc. Am.* **69:**1384–1390 (1979).

66. R. Smalley and A. J. Sievers, "The Total Hemispherical Emissivity of Copper," *J. Opt. Soc. Am.* **68:**1516–1518 (1978).

67. S. X. Cheng, P. Cebe, L. M. Hanssen, D. M. Riffe, and A. J. Sievers, "Hemispherical Emissivity of V, Nb, Ta, Mo, and W from 300 to 1000 K," *J. Opt. Soc. Am. B* **4:**351–356 (1987).

68. Op. cit. Ref. 5, p. 287.

69. Op. cit. Ref. 58, pp. **12**-32–**12**-33.

70. Y. S. Touloukian et al. (eds.), *Thermophysical Properties of Matter,* **1-13**, IFI/Plenum, New York, 1970–1977.

71. Y. S. Touloukian, R. K. Kirby, R. E. Taylor, and P. D. Desai, "Thermal Expansion, Metallic Elements and Alloys," vol. 12 in Y. S. Touloukian and C. Y. Ho (eds.), *Thermophysical Properties of Matter,* IFI/Plenum, New York, 1975, pp. 23 (Be), 77 (Cu), 125 (Au), 157 (Fe), 208 (Mo), 225 (Ni), 298 (Ag), 1028 (Al), 1138 (SS).

72. Y. S. Touloukian, R. K. Kirby, R. E. Taylor, and T. Y. R. Lee, "Thermal Expansion, Nonmetallic Solids," vol. 13 in Y. S. Touloukian and C. Y. Ho (eds.), *Thermophysical Properties of Matter* IFI/Plenum, New York, 1977, pp. 154 (Si), 873 (SiC).

73. C. A. Swenson, "HIP Beryllium: Thermal Expansivity from 4 to 300 K and Heat Capacity from 1 to 108 K," *J. Appl. Phys.* **70**(6):3046–3051 (Sep 1991).

74. Z. Li and C. Bradt, "Thermal Expansion of the Cubic (3C) Polytype of SiC," *J. Mater. Sci.* **21:**4366–4368 (1986).

75. C. Y. Ho, R. W. Powell, and P. E. Liley, *Thermal Conductivity of Selected Materials,* NSRDS-NBS-8, National Standard Reference Data System—National Bureau of Standards, Part 1 (1966).

76. C. Y. Ho, R. W. Powell, and P. E. Liley, *Thermal Conductivity of Selected Materials,* NSRDS-NBS-16, National Standard Reference Data System—National Bureau of Standards, part 2 (1968).

77. Y. S. Touloukian, R. W. Powell, C. Y. Ho, and P. G. Klemens, "Thermal Conductivity, Metallic Elements and Alloys," vol. 1 in Y. S. Touloukian and C. Y. Ho (eds.), *Thermophysical Properties of Matter,* IFI/Plenum, New York, 1970.

78. Y. S. Touloukian, R. W. Powell, C. Y. Ho, and P. G. Klemens, "Thermal Conductivity, Nonmetallic Solids," vol. 2 in Y. S. Touloukian and C. Y. Ho (eds.), *Thermophysical Properties of Matter,* IFI/Plenum, New York, 1970.

79. D. H. Killpatrick, private communication, Feb. 1993.

80. "CVD Silicon Carbide," Technical Bulletin #107, Morton International Advanced Materials, 1991.

81. Op. cit. Ref. 58, p. **12**-133.

82. E. A. Brandes and G. B. Brook (eds.), *Smithell's Metals Reference Book,* 7th ed., Butterworth Heinmann, Oxford, 1992, pp. **14**-3–**14**-5.

83. Y. S. Touloukian and E. H. Buyco, "Specific Heat, Metallic Elements and Alloys," vol. 4 in Y. S. Touloukian and C. Y. Ho (eds.), *Thermophysical Properties of Matter,* IFI/Plenum, New York, 1970.

84. *Metals Handbook,* **2**, 10th ed., Properties and Selection: Nonferrous Alloys and Special-Purpose Materials, ASM International, Metals Park, OH, 1990, pp. 93–94 & 102–103 (Al), 265 (Cu), 704–705 (Au), 1118–1129 (Fe), 1140–1143 (Mo), 441 (Ni), 1154–1156 (Si), and 699–700 & 1156–1158 (Ag).

85. I-70-H Optical Grade Beryllium Block, Preliminary Material Spec., Brush Wellman Inc., Nov. 1990.

86. *Carpenter Invar "36,"* Technical Data Sheet, Carpenter Technology Corp., Nov. 1980.

87. *Metals Handbook,* **5,** 9th ed., Surface Cleaning, Finishing, and Coating, American Society for Metals, Metals Park, OH, 1982, pp. 223–229.

88. *Engineered Materials Handbook,* **4,** Ceramics and Glasses, ASM International, Metals Park, OH, 1991, pp. 677, 806–808.

89. *Metals Handbook,* **1,** 10th ed., Properties and Selection: Irons, Steels, and High Performance Alloys, ASM International, Metals Park, OH, 1990, p. 871.

90. *Materials Engineering, Materials Selector 1993,* Dec. 1992, p. 104.

91. D. L. Hibbard, "Dimensional Stability of Electroless Nickel Coatings," in R. A. Paquin (ed.), *Dimensional Stability, Proc. SPIE* **1335:**180–185 (1990).

92. G. A. Graves, private communication, Feb. 1993.

93. Op. cit. Ref. 82, pp. **15**-5–**15**-7.

94. Z. Li and R. C. Bradt, "Thermal Expansion and Elastic Anisotropies of SiC as Related to Polytype Structure," in C. E. Selmer (ed.), *Proceedings of the Silicon Carbide 1987 Symposium* **2,** Amer. Ceram. Soc., Westerville, OH, 1989, pp. 313–339.

95. W. D. Rowland and J. S. White, "The Determination of the Elastic Constants of Beryllium in the Temperature Range 25 to 300°C," *J. Phys. F: Metal Phys.* **2:**231–236 (1972).

96. H. Ledbetter, private communication, Oct. 1987.

97. Op. cit., Ref. 82, pp. **15**-2–**15**-3.

CHAPTER 36
OPTICAL PROPERTIES OF SEMICONDUCTORS

Paul M. Amirtharaj and David G. Seiler
Materials Technology Group
Semiconductor Electronics Division
National Institute of Standards and Technology
Gaithersburg, Maryland

36.1 GLOSSARY

A	power absorption
$\mathbf{B}$	magnetic field
c	velocity of light
$\mathbf{D}$	displacement field
d	film thickness
E	applied electric field
E_c	energy, conduction band
E_{ex}	exciton binding energy
E_g	energy band gap
E_{H}	hydrogen atom ionization energy $= 13.6\,\text{eV}$
$\mathbf{E}$	electric field
$\mathbf{E}_n^{\pm}$	Landau level energy
E_v	energy, valence band
e_I	ionic charge
g^*	effective g-factor
$\mathbf{K}$	phonon wave vector
k	extinction coefficient
k_{B}	Boltzmann's constant
$\mathbf{k}$	electron/hole wave vector
$\text{L}_{\pm}$	coupled LO phonon–plasmon frequency
m_e^*	electron effective mass
m_h^*	hole effective mass

m_i	ionic mass
m_i'	reduced ionic mass
m_{imp}	impurity ion mass
m_l^*	longitudinal effective mass
m_o	electron rest mass
m_r	electron-hole reduced mass
m_t^*	transverse effective mass
N	volume density
n	refractive index (real part)
$\tilde{n} = (n + ik)$	complex index of refraction
$\mathbf{P}$	polarization field
$\mathbf{q}$	photon wave vector
R	power reflection
R_y	effective Rydberg
S	oscillator strength
T	power transmission
$\mathbf{T}$	temperature
V	Verdet coefficient
α	absorption coefficient
α_{AD}	absorption coefficient, allowed-direct transitions
α_{AI}	absorption coefficient, allowed-indirect transitions
α_{FD}	absorption coefficient, forbidden-direct transitions
α_{FI}	absorption coefficient, forbidden-indirect transitions
δ	skin depth or penetration depth
γ	phenomenological damping parameter
Δ	spin-orbit splitting energy
Γ	Brillouin zone center
ϵ	dielectric function
$\epsilon_{fc}(\omega)$	free-carrier dielectric function
$\epsilon_{imp}(\omega)$	impurity dielectric function
$\epsilon_{int}(\omega)$	intrinsic dielectric function
$\epsilon_{lat}(\omega)$	lattice dielectric function
$\epsilon(0)$	static dielectric constant
ϵ_0	free-space permittivity
ϵ_1	Real (ϵ)

ϵ_2 Im (ϵ)

ϵ_∞ high-frequency limit of dielectric function

η impurity ion charge

λ wavelength

λ_c cut-off wavelength

μ mobility

μ_B Bohr magneton

ν frequency

σ conductivity

τ scattering time

ϕ work function

χ susceptibility

$\chi^{(n)}$ induced nonlinear susceptibility

Ω phonon frequency

ω angular frequency

ω_c cyclotron resonance frequency

ω_{LO} longitudinal optical phonon frequency

ω_p free-carrier plasma frequency

ω_{pv} valence band plasma frequency

ω_{TO} transverse optical phonon frequency

36.2 INTRODUCTION

Rapid advances in semiconductor manufacturing and associated technologies have increased the need for optical characterization techniques for materials analysis and in-situ monitoring/control applications. Optical measurements have many unique and attractive features for studying and characterizing semiconductor properties: (1) They are contact-less, nondestructive, and compatible with any transparent ambient including high-vacuum environments; (2) they are capable of remote sensing, and hence are useful for in-situ analysis on growth and processing systems; (3) the high lateral resolution inherent in optical systems may be harnessed to obtain spatial maps of important properties of the semiconductor wafers or devices; (4) combined with the submonolayer sensitivity of a technique such as ellipsometry, optical measurements lead to unsurpassed analytical details; (5) the resolution in time obtainable using short laser pulses allows ultrafast phenomena to be investigated; (6) the use of multichannel detection and high-speed computers can be harnessed for extremely rapid data acquisition and reduction which is crucial for real-time monitoring applications such as in in-situ sensing; (7) they provide information that complements transport analyses of impurity or defect and electrical behavior; (8) they possess the ability to provide long-range, crystal-like properties and hence support and complement chemical and elemental analyses; and (9) finally, most optical techniques are "table-top" procedures that can be implemented by semiconductor device manufacturers at a reasonable cost. All optical measurements of semiconductors rely on a fundamental understanding of their optical properties. In this chapter, a broad overview of the optical properties of semiconductors is given, along with numerous specific examples.

The optical properties of a semiconductor can be defined as any property that involves the interaction between electromagnetic radiation or light and the semiconductor, including absorption, diffraction, polarization, reflection, refraction, and scattering effects. The electromagnetic spectrum is an important vehicle for giving an overview of the types of measurements and physical processes characteristic of various regions of interest involving the optical properties of semiconductors. The electromagnetic spectrum accessible for studies by optical radiation is depicted in Fig. 1a and b where both the photon wavelengths and photon energies, as well as the common designations for the spectral bands, are given.[1] Figure 1a shows the various techniques and spectroscopies and their spectral regions of applicability. Molecular, atomic, and electronic processes characteristic of various parts of the spectrum are shown in Fig. 1b. The high-energy x-ray, photoelectron, and ion desorption processes are important to show because they overlap the region of vacuum ultraviolet (VUV) spectroscopy. The ultraviolet (UV) region of the spectrum has often been divided into three rough regions: (1) the near-UV, between 2000 and 4500 Å; (2) the VUV, 2000 Å down to about 400 Å; and (3) the region below 400 Å covering the range of soft x-rays, 10 to 400 Å.[2] The spectrum thus covers a broad frequency range which is limited at the high-frequency end by the condition that $\lambda \gg a$, where λ is the wavelength of the light wave in the material and a is the interatomic distance. This limits the optical range to somewhere in the soft x-ray region. Technical difficulties become severe in the ultraviolet region (less the 100-nm wavelength, or greater than 12.3-eV photon energies), and synchrotron radiation produced by accelerators can be utilized effectively for ultraviolet and x-ray spectroscopy without the limitations of conventional laboratory sources. A lower limit of the optical frequency range might correspond to wavelengths of about 1 mm (photon energy of 1.23×10^{-3} eV). This effectively excludes the microwave and radio-frequency ranges from being discussed in a chapter on the optical properties of semiconductors.

From the macroscopic viewpoint, the interaction of matter with electromagnetic radiation is described by Maxwell's equations. The optical properties of matter are introduced into these equations as the constants characterizing the medium such as the dielectric constant, magnetic permeability, and electrical conductivity. (They are not real "constants" since they vary with frequency.) From our optical viewpoint, we choose to describe the solid by the dielectric constant or dielectric function $\epsilon(\omega)$. This dielectric constant is a function of the space and time variables and should be considered as a response function or linear integral operator. It can be related in a fundamental way to the crystal's refractive index n and extinction coefficient k by means of the Kramers-Krönig dispersion relations as discussed later. It is the values of the optical constants n and k that are usually directly measured in most optical experiments; they are real and positive numbers.

There are a number of methods for determining the optical constants n and k of a semiconductor as a function of wavelength. Five of the most common techniques are as follows.

1. Measure the reflectivity at normal incidence over a wide wavelength and use a Kramers-Krönig dispersion relation.

2. Measure the transmission of a thin slab of known thickness together with the absolute reflectivity at normal incidence or alternately observe the transmission over a wide spectral range and obtain n by a Kramers-Krönig analysis.

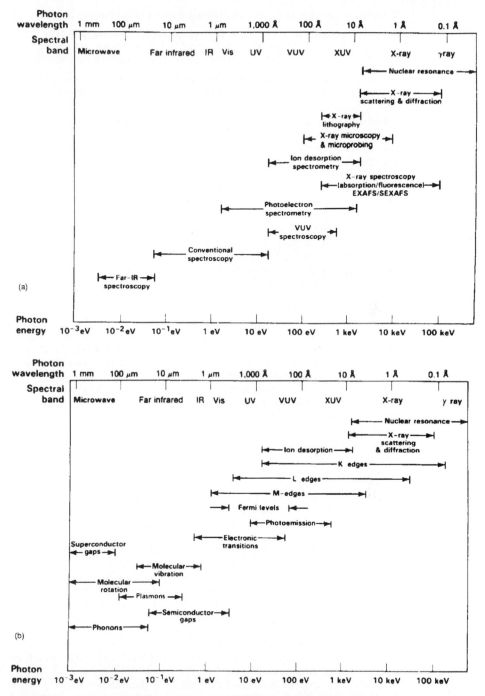

FIGURE 1 The electromagnetic spectrum comprising the optical and adjacent regions of interest: (*a*) characterization techniques using optical spectroscopy and synchrotron radiation; (*b*) molecular, atomic, and electronic processes characteristic in various parts of the electromagnetic spectrum plotted as a function of photon energy.[1]

3. Measure the reflection of unpolarized light at two or more angles of incidence.

4. Use a polarimetric method like ellipsometry which involves finding the ratio of reflectivities perpendicular and parallel to the plane of incidence at a nonnormal incidence together with the difference of phase shifts upon reflection.

5. Use detailed computer modeling and fitting of either reflection, transmission, or ellipsometric measurements over a large enough energy range.

These optical constants describe an electromagnetic wave in the medium of propagation; the refractive index n gives the phase shift of the wave, and the extinction coefficient or attenuation index k gives the attenuation of the wave. In practice, one often uses the absorption coefficient α instead of k because of the Beer's low formalism describing the absorption.

The field of optical spectroscopy is a very important area of science and technology since most of our knowledge about the structure of atoms, molecules, and solids is based upon spectroscopic investigations. For example, studies of the line spectra of atoms in the late 1800s and early 1900s revolutionized our understanding of the atomic structure by elucidating the nature of their electronic energy levels. Similarly for the case of semiconductors, optical spectroscopy has proven essential to acquiring a systematic and fundamental understanding of the nature of semiconductors. Since the early 1950s, detailed knowledge about the various eigenstates present in semiconductors has emerged including energy bands, excitonic levels, impurity and defect levels, densities of states, energy-level widths (lifetimes), symmetries, and changes in these conditions with temperature, pressure, magnetic field, electric field, etc. One of the purposes of this chapter is to review and summarize the major optical measurement techniques that have been used to investigate the optical properties of semiconductors related to these features. Specific attention is paid to the types of information which can be extracted from such measurements of the optical properties.

Most optical properties of semiconductors are integrally related to the particular nature of their electronic band structures. Their electronic band structures are in turn related to the type of crystallographic structure, the particular atoms, and their bonding. The full symmetry of the space groups is also essential in determining the structure of the energy bands. Group theory makes it possible to classify energy eigenstates, determine essential degeneracies, derive selection rules, and reduce the order of the secular determinants which must be diagonalized in order to compute approximate eigenvalues. Often, experimental measurements must be carried out to provide quantitative numbers for these eigenvalues. A full understanding of the optical properties of semiconductors is thus deeply rooted in the foundations of modern solid-state physics. In writing this chapter, the authors have assumed that the readers are familiar with some aspects of solid-state physics such as can be obtained from an undergraduate course.

Most semiconductors have a diamond, zinc-blende, wurtzite, or rock-salt crystal structure. Elements and binary compounds, which average four valence electrons per atom, preferentially form tetrahedral bonds. A tetrahedral lattice site in a compound AB is one in which each atom A is surrounded symmetrically by four nearest neighboring B atoms. The most important lattices with a tetrahedral arrangement are the diamond, zinc-blende, and wurtzite lattices. In the diamond structure, all atoms are identical, whereas the zinc-blende structure contains two different atoms. The wurtzite structure is in the hexagonal crystal class, whereas the diamond and zinc-blende structures are cubic. Other lattices exist which are distorted forms of these and others which have no relation to the tetrahedral structures.

Band structure calculations show that only the valence band states are important for predicting the following crystal ground-state properties: charge density, Compton profile, compressibility, cohesive energy, lattice parameters, x-ray emission spectra, and hole effective mass. In contrast, both the valence-band and conduction-band states are important for predicting the following properties: optical dielectric constant or refractive index, optical absorption spectrum, and electron effective mass. Further complexities arise because of the many-body nature of the particle interactions which necessitates understanding excitons, electron-hole droplets, polarons, polaritons, etc.

The optical properties of semiconductors cover a wide range of phenomena which are impossible to do justice to in just one short chapter in this Handbook. We have thus chosen to present an extensive, systematic overview of the field, with as many details given as possible. The definitions of the various optical properties, the choice of figures used, the tables presented, the references given all help to orient the reader to appreciate various principles and measurements that form the foundations of the optical properties of semiconductors.

The optical properties of semiconductors are often subdivided into those that are electronic and those that are lattice in nature. The electronic properties concern processes involving the electronic states of the semiconductor, while the lattice properties involve vibrations of the lattice (absorption and creation of phonons). Lattice properties are of considerable interest, but it is the electronic properties which receive the most attention in semiconductors because of the technological importance of their practical applications. Modern-day semiconductor optoelectronic technologies include lasers, light-emitting-diodes, photodetectors, optical amplifiers, modulators, switches, etc., all of which exploit specific aspects of the electronic optical properties.

Almost all of the transitions that contribute to the optical properties of semiconductors can be described as one-electron transitions. Most of these transitions conserve the crystal momentum and thus measure the vertical energy differences between the conduction and valence bands. In the one-electron approximation, each valence electron is considered as a single particle, moving in a potential which is the sum of the core potentials and a self-consistent Hartree potential of the other valence electrons.

The phenomena usually studied to obtain information on the optical properties of semiconductors are (1) absorption, (2) reflection, (3) photoconductivity, (4) emission, and (5) light scattering. Most of the early information on the optical properties of semiconductors was obtained from measurements of photoconductivity, but these measurements can be complicated by carrier trapping, making interpretation of the results sometimes difficult. Thus, most measurements are of the type (1), (2), (4), or (5). For example, the most direct way of obtaining information about the energy gaps between band extrema and about impurity levels is by measuring the optical absorption over a wide range of wavelengths. Information can also be obtained by (2) and (4).

The transient nature of the optical properties of semiconductors is important to establish because it gives insight to the various relaxation processes that occur after optical excitation. Because of the basic limitations of semiconductor devices on speed and operational capacity, ultrafast studies have become an extremely important research topic to pursue. The push to extend the technologies in the optoelectronic and telecommunication fields has also led to an explosion in the development and rise of ultrafast laser pulses to probe many of the optical properties of semiconductors: electrons, holes, optical phonons, acoustic phonons, plasmons, magnons, excitions, and the various coupled modes (polaritons, polarons, excitonic molecules, etc.). The time scale for many of these excitations is measured in femtoseconds (10^{-15}) or picoseconds (10^{-12}). Direct time measurements on ultrafast time scales provides basic information on the mechanisms, interactions, and dynamics related to the various optical properties. Some of the processes that have been investigated are the formation time of excitons, the cooling and thermalization rates of hot carriers, the lifetime of phonons, the screening of optical-phonon-carried interactions, the dynamics of ballistic transport, the mechanism of laser annealing, dephasing processes of electrons and excitons, optical Stark effect, etc. It is not possible in this short review chapter to cover these ultrafast optical properties of semiconductors. We refer the reader to the many fine review articles and books devoted to this field.[3,4]

The advent of the growth of artificially structured materials by methods such as molecular beam epitaxy (MBE) has made possible the development of a new class of materials and heterojunctions with unique electronic and optical properties. Most prominent among these are heterojunction quantum-wells and superlattices. The field of microstructural physics has thus been one of the most active areas of research in the past decade. The novel properties of structures fabricated from ultrathin layers of semiconductors of thicknesses <100 Å stem from microscopic quantum mechanical effects. The simplest case to visualize is that of a particle confined in a box which displays distinct quantum energy states, the equivalent of which are electrons and holes confined to a thin layer of a material such as GaAs sandwiched between two thick layers of AlAs. The new energy states produced by the confinement of the charges in the artificially produced potential well can be manipulated, by tailoring the size and shape of the well, to produce a wide variety of effects that are not present in conventional semiconductors. Microstructures formed from alternating thin layers of two semiconductors also lead to novel electronic and optical behavior, most notable of which is large anisotropic properties. The ability to "engineer" the behavior of these microstructures has led to an explosion of research and applications that is too large to be dealt with in this short review. The reader is referred to several review articles on their optical behavior.[5,6]

36.3 OPTICAL PROPERTIES

Background

The interaction of the semiconductor with electromagnetic radiation can be described, in the semiclassical regime, using response functions such as ϵ and χ which are defined in the following section. The task of the description is then reduced to that of building a suitable model of χ and ϵ that takes into account the knowledge of the physical characteristics of the semiconductor and the experimentally observed optical behavior. One example of a particularly simple and elegant, yet surprisingly accurate and successful, model of ϵ for most semiconductors is the linear-chain description of lattice vibrations.[7] This model treats the optical phonons, i.e., the vibrations that have an associated dipole moment, as damped simple harmonic motions. Even though the crystal is made up of $\sim 10^{23}$ atoms, such a description with only a few resonant frequencies and phenomenological terms, such as the damping and the ionic charge, accurately accounts for the optical behavior in the far-infrared region. The details of the model are discussed in the following section. Such simple models are very useful and illuminating, but they are applicable only in a limited number of cases, and hence such a description is incomplete.

A complete and accurate description will require a self-consistent quantum mechanical approach that accounts for the microscopic details of the interaction of the incident photon with the specimen and a summation over all possible interactions subject to relevant thermodynamical and statistical mechanical constraints. For example, the absorption of light near the fundamental gap can be described by the process of photon absorption resulting in the excitation of a valence-band electron to the conduction band. In order to obtain the total absorption at a given energy, a summation has to be performed over all the possible states that can participate, such as from multiple valence bands. Thermodynamic considerations such as the population of the initial and final states have to be taken into consideration in the calculation as well. Hence, a detailed knowledge of the specimen and the photon-specimen interaction can, in principle, lead to a satisfactory description.

Optical/Dielectric Response

Optical Constants and the Dielectric Function.   In the linear regime, the dielectric function ϵ and the susceptibility χ are defined by the following relations[8]:

$$\mathbf{D} = \epsilon_0 \mathbf{E} + \mathbf{P} \tag{1}$$

$$\mathbf{D} = \epsilon_0 (1 + \chi) \mathbf{E} \tag{2}$$

$$\mathbf{D} = \epsilon \mathbf{E} = (\epsilon_1 + i\epsilon_2) \mathbf{E} \tag{3}$$

where $\mathbf{E}$, $\mathbf{D}$, and $\mathbf{P}$ are the free-space electric field, the displacement field, and the polarization field inside the semiconductor; ϵ_0 is the permittivity of free space; and ϵ and χ are dimensionless quantities, each of which can completely describe the optical properties of semiconductors. The refractive index $\tilde{n}$ of the material is related to ϵ as shown below:

$$\tilde{n} = \sqrt{\epsilon} = n + ik \tag{4}$$

The real and imaginary parts of the refractive index, n and k, which are also referred to as the optical constants, embody the linear optical property of the material. The presence of k, the imaginary component, denotes absorption of optical energy by the semiconductor. Its relationship to the absorption coefficient α is discussed in the following section. In the spectral regions where absorptive processes are weak or absent, as in the case of the subband gap range, k is very small, whereas in regions of strong absorption, the magnitude of k is large. The optical constants for a large number of semiconductors may be found in Refs. 9 and 10. The variation in the real part n is usually much smaller. For example, in GaAs, at room temperature, in the visible and near-visible region extending from 1.4 to 6 eV, k varies from $<10^{-3}$ at 1.41 eV which is just below the gap, to a maximum of 4.1 at 4.94 eV.[11] In comparison, n remains nearly constant in the near-gap region extending from 3.61 at 1.4 eV to 3.8 at 1.9 eV, with the maximum and minimum values of 1.26 at 6 eV and 5.1 at 2.88 eV, respectively. The real and imaginary components are related by causal relationships that are also discussed in the following sections.

Reflection, Transmission, and Absorption Coefficients.   The reflection and transmission from a surface are given by:

$$\tilde{r} = \frac{(\tilde{n} - 1)}{(\tilde{n} + 1)} = |\tilde{r}| \cdot \exp(i\theta) \tag{5}$$

$$R = |\tilde{r}^2| \tag{6}$$

$$T = (1 - R) \tag{7}$$

where $\tilde{r}$ is the complex reflection coefficient and R and T are the power reflectance and transmission. For a thin slab, in free space, with thickness d and refractive index $\tilde{n}$, the appropriate expressions are[12]:

$$\tilde{r} = \frac{\tilde{r}_1 + \tilde{r}_2 \cdot \exp\left(i4\pi\tilde{n}d/\lambda\right)}{1 + \tilde{r}_1 \cdot \tilde{r}_2 \cdot \exp\left(i4\pi\tilde{n}d/\lambda\right)} \tag{8}$$

where $\tilde{r}_1$ and $\tilde{r}_2$ are the reflection coefficients at the first and second interfaces, respectively, and λ is the free-space wavelength.

For most cases of optical absorption, the energy absorbed is proportional to the thickness of the specimen. The variation of optical energy inside the absorptive medium is given by the following relationship:

$$I(x) = I(0) \cdot \exp\left(-\alpha \cdot x\right) \tag{9}$$

and α is related to the optical constants by:

$$\alpha = 4\pi k/\lambda \tag{10}$$

Here we note that α (measured in cm^{-1}) describes the attenuation of the radiation intensity rather than that of the electric field.

In spectral regions of intense absorption, all the energy that enters the medium is absorbed. The only part of the incident energy that remains is that which is reflected at the surface. In such a case, it is useful to define a characteristic "skin" thickness that is subject to an appreciable density of optical energy. A convenient form used widely is simply the inverse of α, i.e., $1/\alpha$. This skin depth is usually denoted by δ:

$$\delta = \frac{1}{\alpha} \tag{11}$$

The skin depths in semiconductors range from $>100\,\text{nm}$ near the band gap to $<5\,\text{nm}$ at the higher energies of $\sim 6\,\text{eV}$.

Kramers-Krönig Relationships.   A general relationship exists for linear systems between the real and imaginary parts of a response function as shown in the following:

$$\epsilon_1(\omega) = 1 + \frac{2}{\pi} P \int_0^\infty \frac{\omega' \epsilon_2(\omega')}{\omega'^2 - \omega^2}\, d\omega' \tag{12}$$

$$\epsilon_2(\omega) = -\frac{2\omega}{\pi} P \int_0^\infty \frac{\epsilon_1(\omega')\, d\omega'}{\omega'^2 - \omega^2} + \frac{\sigma_0}{\epsilon_0 \cdot \omega} \tag{13}$$

where σ_0 is the dc conductivity.

$$n(\omega) = 1 + \frac{2}{\pi} P \int_0^\infty \frac{\omega' k(\omega')}{\omega'^2 - \omega^2}\, d\omega' \tag{14}$$

$$k(\omega) = -\frac{2}{\pi} P \int_0^\infty \frac{n(\omega')}{\omega'^2 - \omega^2}\, d\omega' \tag{15}$$

where P denotes the principal part of the integral and σ_0 the conductivity. These are referred to as the Kramers-Krönig dispersion relationships.[13,14] An expression of practical utility is one in which the experimentally measured power reflection R at normal incidence is explicitly displayed as shown:

$$\theta(\omega) = -\frac{\omega}{\pi} P \int_0^\infty \frac{\ln (R(\omega')) \, d\omega}{\omega'^2 - \omega^2} \tag{16}$$

This is useful since it shows that if R is known for all frequencies, θ can be deduced, and hence a complete determination of both n and k can be accomplished. In practice, R can be measured only over a limited energy range, but approximate extrapolations can be made to establish reasonable values of n and k.

The measurement of the reflectivity over a large energy range spanning the infrared to the vacuum ultraviolet, 0.5- to 12-eV range, followed by a Kramers-Krönig analysis, used to be the main method of establishing n and k.[15] However, the advances in spectroscopic ellipsometry in the past 20 years have made this obsolete in all but the highest energy region. A discussion of the past methods follows for completeness.

The measured reflectivity range, in general, is not large enough to obtain accurate values of n and k. Hence, extrapolation procedures were used to guess the value of R beyond ~12 eV.[15] The most justifiable procedure, from a physical standpoint, assumed that the higher energy reflectivity was dominated by the valence-band plasma edge ω_{PV} and, hence, assumed the following forms for $\epsilon(\omega)$, $\tilde{n}(\omega)$, and $R(\omega)$:

$$\tilde{n} = \sqrt{\epsilon}(\omega) \approx -\frac{1}{2} \cdot \left(\frac{\omega_{PV}^2}{\omega^2} \right) \tag{17}$$

$$R(\omega) = \frac{(n(\omega) - 1)}{(n(\omega) + 1)} = \frac{1}{16} \cdot \left(\frac{\omega_{PV}^4}{\omega^4} \right) \tag{18}$$

Other less intuitive forms of extrapolations have also been used with an exponential falloff or a ω^{-p} fall where p is computer fit to get the most consistent results.

Sum Rules. Having realized the interrelationships between the real and imaginary parts of the response functions, one may extend them further using a knowledge of the physical properties of the semiconductor to arrive at specific equations, commonly referred to as sum rules.[13,14] These equations are useful in cross-checking calculations for internal consistency or reducing the computational effort. Some of the often-used relations are shown below:

$$\int_0^\infty \omega \epsilon_2(\omega) \, d\omega = \frac{\pi}{2} \omega_{PV}^2 \tag{19}$$

$$\int_0^\infty \omega \, \text{Im} \left[\frac{-1}{\epsilon(\omega)} \right] d\omega = \frac{\pi}{2} \omega_{PV}^2 \tag{20}$$

$$\int_0^\infty \omega k(\omega) \cdot d\omega = \frac{\pi}{4} \omega_{PV}^2 \tag{21}$$

$$\int_0^\infty [n(\omega) - 1] \, d\omega = 0 \tag{22}$$

where ω_{PV} is the valence-band plasma frequency.

The dc static dielectric constant, $\epsilon(0)$ may be expressed as:

$$\epsilon(0) = 1 + \frac{2}{\pi} \int_0^\infty \frac{\epsilon_2(\omega)}{\omega} \cdot d\omega \tag{23}$$

The reader is referred to Refs. 13 and 14 for more details.

Linear Optical Properties

Overview. The optical properties of semiconductors at low enough light levels are often referred to as linear properties in contrast to the nonlinear optical properties described later. There are many physical processes that control the amount of absorption or other optical properties of a semiconductor. In turn, these processes depend upon the wavelength of radiation, the specific properties of the individual semiconductor being studied, and other external parameters such as pressure, temperature, etc. Just as the electrical properties of a semiconductor can be controlled by purposely introducing impurity dopants (both p and n type) or affected by unwanted impurities or defects, so too are the optical properties affected by them. Thus, one can talk about *intrinsic* optical properties of semiconductors that depend upon their perfect crystalline nature and *extrinsic* properties that are introduced by impurities or defects. Many types of defects exist in real solids: point defects, macroscopic structural defects, etc. In this section we review and summarize intrinsic linear optical properties related to lattice effects, interband transitions, and free-carrier or intraband transitions. Impurity- and defect-related extrinsic optical properties are also covered in a separate section and in the discussion of lattice properties affected by them. Figure 2 schematically depicts various contributions to the absorption spectrum of a typical semiconductor as functions of wavelength (top axis) and photon energy (bottom axis). Data for a real semiconductor may show more structure than shown here. On the other hand, some of the structure shown may be reduced or not actually present in a particular semiconductor (e.g., impurity absorption, bound excitons,

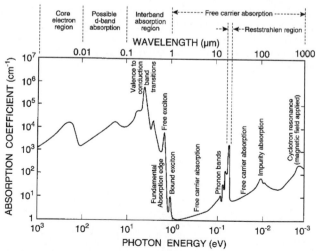

FIGURE 2 Absorption spectrum of a typical semiconductor showing a wide variety of optical processes.

TABLE 1 Classification by Wavelength of the Optical Responses for Common Semiconductors

Wavelength (nm)	Responses	Physical origin	Application	Measurement tech.
$\lambda > \lambda_{TO}$ Far-IR and micro-wave region	Microwave R and T Plasma R and T	Free-carrier plasma	Detectors Switches	R, T, and A* Microwave techniques Fourier Transform Spectrometry (FTS)
$\lambda_{LO} < \lambda < \lambda_{TO}$ Reststrahlen region	Reststrahlen R	Optical phonons in ionic crystals	Absorbers Filters	R, T, and A FTS & Dispersion spectrometry (DS)
$\lambda \sim \lambda_{LO}$, λ_{TO}, λ_P Far-IR region	Far-IR A	Optical phonons, impurities (vibration and electronic), free carriers, intervalence transitions	Absorbers Filters	R, T, and A FTS and DS
$\lambda_{LO} > \lambda > \lambda_G$ Mid-IR region	Mid-IR T and A	Multiphonon, multiphoton transitions, impurities (vibrational and electronic), intervalence transitions excitons, Urbach tail	Detectors Switches Absorbers Filters	R, T, and A Ellipsometry FTS and DS
$\lambda < \lambda_G$ IR, visible, and UV	R, T, and A	Electronic interband transitions	Reflectors Detectors	Reflection Ellipsometry
$\lambda \sim \lambda_W$ UV, far-UV	Photoemission	Fermi energy to vacuum-level electronic transitions	Photocathodes Detectors	High-vacuum, spectroscopy tech.
$\lambda_W > \lambda > \gtrsim a$	R, T, and A	Ionic-core transitions	Detectors	Soft x-ray and synchrotron-based analyses
	Diffraction	Photo—ionic-core interactions	X-ray optics and mono-chromators	X-ray techniques

* R, T, and A—Reflection, transmission, and absorption.
Note: P—Plasma; G—Energy gap; W—Work functions; (a lattice constant).
TO, LO: Transverse and longitudinal optical phonons.

d-band absorption). Table 1 shows the classification of the optical responses of the semiconductor to light in various wavelength regions showing the typical origin of the response and how the measurements are usually carried out. At the longest wavelengths shown in Fig. 2, cyclotron resonance may occur for a semiconductor in a magnetic field, giving rise to an absorption peak corresponding to a transition of a few meV energy between Landau levels. Shallow impurities may give rise to additional absorption at low temperatures and here a 10-meV ionization energy has been assumed. If the temperature was high enough so that $k_B\mathbf{T}$ was greater than the ionization energy, the absorption peak would be washed out. At wavelengths between 20 to 50 μm, a new set of absorption peaks arises due to the vibrational modes of the lattice. In ionic crystals, the absorption coefficient in the reststrahlen region may reach $10^5\,\text{cm}^{-1}$, whereas in homopolar semiconductors like Si and Ge, only multiphonon features with lower absorption coefficients are present (around 5 to 50 cm^{-1}).

Models of the Dielectric Function. The interaction of light with semiconductors can be completely described by the dielectric function, $\epsilon(\omega)$. The dielectric function $\epsilon(\omega)$ may be divided into independent parts to describe various physical mechanisms so long as the processes do not interact strongly with each other; this is an approximation, referred to as the adiabatic approximation which simplifies the task at hand considerably.[16] The major players that determine the optical behavior of an intrinsic semiconductor are the lattice, particularly in a nonelemental semiconductor; the free carriers, i.e., mobile electrons and holes; and the interband transitions between the energy states available to the electrons. These three mechanisms account for the intrinsic linear properties that lead to a dielectric function as shown:

$$\epsilon_{int}(\omega) = \epsilon_{lat}(\omega) + \epsilon_{fc}(\omega) + \epsilon_{inter}(\omega) \tag{24}$$

The addition of impurities and dopants that are critical to controlling the electronic properties leads to an additional contribution, and the total dielectric response may then be described as shown:

$$\epsilon(\omega) = \epsilon_{int}(\omega) + \epsilon_{imp}(\omega) \tag{25}$$

Lattice

Phonons. The dc static response of a semiconductor lattice devoid of free charges to an external electromagnetic field may be described by the single real quantity $\epsilon(0)$. As the frequency of the electromagnetic radiation increases and approaches the characteristic vibrational frequencies associated with the lattice, strong interactions can occur and modify the dielectric function substantially. The main mechanism of the interaction is the coupling between the electromagnetic field with the oscillating dipoles associated with vibrations of an ionic lattice.[7] The interactions may be described, quite successfully, by treating the solid to be a collection of damped harmonic oscillators with a characteristic vibrational frequency ω_{TO} and damping constant γ. The resultant dielectric function may be written in the widely used CGS units as:

$$\epsilon_{lat}(\omega) = \epsilon(\infty) + \frac{S\omega_{TO}^2}{(\omega_{TO}^2 - \omega^2 - i\omega\gamma)} \tag{26}$$

where S is called the oscillator strength and may be related to the phenomenological ionic charge e_i, reduced mass m_i', and volume density N, through the equation

$$S\omega_{TO}^2 = \frac{4\pi N e_1^2}{m_i'} \tag{27}$$

In the high frequency limit of $\epsilon(\omega)$, for $\omega \gg \omega_{TO}$,

$$\epsilon(\omega) \rightarrow \epsilon_\infty \tag{28}$$

The relationship may be easily extended to accommodate more than one characteristic vibrational frequency by the following relationship:

$$\epsilon_{lat}(\omega) = \epsilon_\infty + \sum_j \frac{S_j(\omega_{TO}^j)^2}{[(\omega_{TO}^j)^2 - \omega^2 - i\omega\gamma^j]} \tag{29}$$

It is worth noting some important physical implications and interrelations of the various parameters in Eq. (26).

For a lattice with no damping, it is obvious that $\epsilon(\omega)$ displays a pole at ω_{TO} and a zero at a well-defined frequency, usually referred to by ω_{LO}. A simple but elegant and useful relationship exists between these parameters as shown by

$$\frac{\epsilon(0)}{\epsilon_\infty} = \left(\frac{\omega_{LO}}{\omega_{TO}}\right)^2 \tag{30}$$

which is known as the Lydenne-Sachs-Tellers relationship.[17]

The physical significance of ω_{TO} and ω_{LO} is that these are the transverse and longitudinal optical phonon frequencies with zero wave vector, **K**, supported by the crystal lattice. The optical vibrations are similar to standing waves on a string. The wave pattern, combined with the ionic charge distribution, leads to oscillating dipoles that can interact with the incident radiation and, hence, the name optical phonons. $\epsilon(\omega)$ is negative for $\omega_{TO} \geq \omega \geq \omega_{LO}$ which implies no light propagation inside the crystal and, hence, total reflection of the incident light. The band of frequencies spanned by ω_{TO} and ω_{LO} is referred to as the reststrahlen band.

The reflectivity spectrum of AlSb[18] is shown in Fig. 3. It is representative of the behavior of most semiconductors. Note that the reflectivity is greater than 90 percent at ≈ 31 μm in the reststrahlen band spanned by the longitudinal and transverse optical phonons. The two asymptotic limits of the reflection tend to $(\sqrt{\epsilon(0)} - 1)/(\sqrt{\epsilon(0)} + 1)$ and $(\sqrt{\epsilon_\infty} - 1)/(\sqrt{\epsilon_\infty} + 1)$, respectively. The effects of the phonon damping are illustrated in Fig. 4.[19] For the ideal case with zero damping, the reflection in the reststrahlen band is 100 percent. Note that for the elemental semiconductors such as Si and Ge, the lack of a dipole moment associated with the optical vibrations of the lattice leads to the absence of any oscillator strength and hence no interaction with the radiation. The reflection spectrum will, therefore, show no change at or near the optical phonon frequencies. The optical phonon frequencies (ω_{LO} and ω_{TO}) and wavelengths and $\epsilon(0)$ and ϵ_∞ for the commonly known semiconductors are presented at the end of this chapter.

The optical phonons ω_{LO} and ω_{TO} are the frequencies of interest for describing the optical interactions with the lattice. In addition, the lattice is capable of supporting vibrational modes over a wide range of frequencies extending from 0 to $\omega_{LO}(\Gamma)$, the LO phonon frequency at the center of the Brillouin zone, Γ as discussed by Cowley in Ref. 20. The vibrational modes can be subdivided into two major categories. The first are optical phonons that possess an oscillating dipole moment and interact with light. The second group are the acoustic phonons, i.e., soundlike vibrations that do not possess a dipole

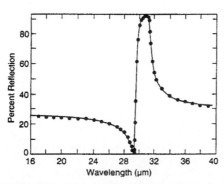

FIGURE 3   Reststrahlen reflection spectrum of AlSb.[18]

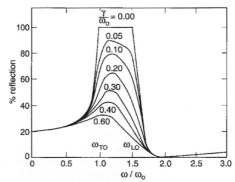

FIGURE 4   Reflection spectra of a damped oscillator for various values of the damping factor.[19]

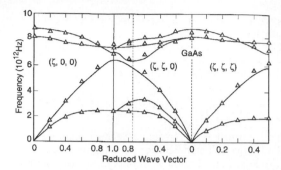

FIGURE 5 Phonon dispersion curves of GaAs. Experimental data are represented by triangles. Solid lines show the results of the calculation of Ref. 21.[21]

moment and hence are not of primary importance in determining the optical properties. For the optical band, only the $\mathbf{K} \approx 0$ modes are important since a strong interaction is precluded for other modes due to the large mismatch in the wave vectors associated with the light and vibrational disturbances.

The simple treatment so far, though very useful, is limited to the most obvious and strongest aspect of the interaction of light with the lattice. However, many weaker but important interaction mechanisms have not yet been accounted for. An attempt to produce a complete description starts with the consideration of the total number of atoms N that makes up the crystal. Each atom possesses three degrees of freedom, and hence one obtains $3N$ degrees of freedom for the crystal, a very large number of magnitude, $\sim 10^{24}$. The complexity of the description can be readily reduced when one realizes the severe restriction imposed by the internal symmetry of the crystal. The vibrational characteristics now break up into easily understandable normal modes with well-defined physical characteristics. The translational symmetry associated with the crystal makes it possible to assign a definite wave vector to each lattice mode. In addition, the phonons can be divided into two major classes: the optical vibrations, which we have already discussed, and the acoustic vibrations, which, as the name implies, are soundlike vibrations. The acoustic phonons for $|\mathbf{K}| \rightarrow 0$ are identical to the sound waves. Hence, specifying the type of phonon, energy, wave vector, and polarization uniquely describes each vibrational mode of the crystal. The conventional description of phonons is achieved by graphically displaying these properties in a frequency vs. $\mathbf{K}$ plot, as shown in Fig. 5 for GaAs.[21] The energies of the phonons are plotted as a function of the wave vector along the high-symmetry directions for each acoustic and optical branch. It can be shown that the number of acoustic branches is always three, two of transverse polarization and one longitudinal, leaving $3p$-3 optical branches, where p is the number of atoms in the primitive cell. The majority of the most important semiconductors fall into the three cubic classes of crystals—namely the diamond, zincblende, and the rock-salt structures that contain two atoms per primitive cell and hence possess three acoustic and three optical branches each.[22] However, more complicated structures with additional optical phonon branches can be found. The most important group among the second category is the wurtzite structure displayed by CdS and CdSe which contains four atoms in the primitive cell and hence two additional sets of optical phonon branches.[22] The specific symmetry associated with the vibrational characteristics of each mode is used to distinguish them as well as their energies at the high-symmetry points in the phonon dispersion curves.

Multiphonon Absorption. It has already been pointed out that the interaction of light with phonons is restricted to those with $\mathbf{K} \approx 0$. This is true only when single-phonon interactions are considered in ideal crystals. Higher-order processes, such as multiphonon absorption, can activate phonons with $\mathbf{K} \neq 0$.[22,23] Symmetry considerations and their implication on the multiphonon absorption are discussed by Birman in Ref. 24. In multiphonon processes, the total momentum of the interacting phonons will be 0, but many modes with $\mathbf{K} \neq 0$ can participate in the interaction.

The energy and momentum conservation conditions may be expressed as follows:

$$\hbar\omega = \sum_i \hbar\Omega_i \tag{31}$$

$$\hbar\mathbf{q} = \sum_i \hbar\mathbf{K}_i \approx 0 \tag{32}$$

where $\hbar\omega$ is the energy of the absorbed photon; $\hbar\Omega_i$ is the energy of the phonons; and $\hbar\mathbf{q}$ and $\hbar\mathbf{K}_i$ are the corresponding momenta. Any number of phonons can participate in the process. However, the strength of interaction between the incident photon and the higher-order processes falls off rapidly with increasing order, making only the lowest order, i.e., two- or three-phonon processes noteworthy in most semiconductors. The well-defined range that spans the phonon energies in most semiconductors, extending from 0 to the LO phonon energy at the center of the zone Γ, restricts the n-phonon process to a maximum energy of $n\hbar\omega_{LO}(\Gamma)$. Among the participating phonons, those with large values of $\mathbf{K}$ and those in the vicinity of the critical points are the most important owing to their larger populations. These factors are important in understanding the multiphonon absorption behavior, as we now discuss.[22,23]

Multiphonon processes may be subdivided into two major categories: (1) sum processes where multiple phonons are created, and (2) difference processes in which both phonon creation and annihilation occur with a net absorption in energy. The former process is more probable at higher-incident photon energies and the converse is true for the latter. The reduction of the equilibrium phonon population at low temperatures leads to a lower probability for the difference process, and hence it is highly temperature dependent.

The multiphonon interactions are governed by symmetry selection rules in addition to the energy and momentum conservation laws stated earlier. A list of the possible combination processes is presented in Table 2 for the diamond structure and Table 3 for the zincblende structure. Representative multiphonon absorption spectra are presented in Figs. 6 and 7 for Si[25] and GaAs.[26] The Si ω_{TO} and ω_{LO} frequency of ~522 cm^{-1} is indicated in the spectrum for reference. Note that the absence of a dipole moment implies that ω_{TO} and ω_{LO} are degenerate, and no first-order absorption is present. In contrast, for GaAs the very large absorption associated with the one-phonon absorption precludes the possibility

TABLE 2 Infrared Allowed Processes in the Diamond Structure[23]

Two-phonon processes
TO(X) + L(X)
TO(X) + TA(X)
L(X) + TA(X)
TO(L) + LO(L)
TO(L) + TA(L)
LO(L) + LA(L)
LA(L) + TA(L)
TO(W) + L(W)
TO(W) + TA(W)
L(W) + TA(W)

TABLE 3 Infrared Allowed Processes in the Zincblende Structure[23]

Two-phonon processes
2LO(Γ), LO(Γ) + TO(Γ), 2TO(Γ)
2TO(X), TO(X) + LO(X), TO(X) + LA(X), TO(X) + TA(X)
LO(X) + LA(X), LO(X) + TA(X)
LA(X) + TA(X)
2TA(X)
2TO(L), TO(L) + LO(L), TO(L) + LA(L), TO(L) + TA(L)
2LO(L), LO(L) + LA(L), LO(L) + TA(L)
2LA(L), LA(L) + TA(L)
2TA(L)
TO$_1$(W) + LO(W), TO$_1$(W) + LA(W)
TO$_2$(W) + LO(W), TO$_2$(W) + LA(W)
LO(W) + LA(W), LO(W) + TA$_1$(W), LO(W) + TA$_2$(W)
LA(W) + TA$_1$(W), LA(W) + TA$_2$(W)

of obtaining meaningful multiphonon absorption data in the reststrahlen band that spans the ~269- to 295-cm^{-1} (~34- to 37-meV) spectral region. The multiphonon spectra obtained from GaAs are displayed in Figs. 7a and b. The absorption associated with the multiphonon processes is much smaller than the single-phonon process. However, the rich structure displayed by the spectra is extremely useful in analyzing the lattice dynamics of the material. In addition, in applications such as windows for high-power lasers, even the small absorption levels can lead to paths of catastrophic failure. The multiphonon transmission spectrum of ZnS in the ω_{LO} to $2\omega_{LO}$ frequency range is presented in Fig. 8, and the frequencies and the assignments are presented in Table 4.[27]

Impurity-related Vibrational Optical Effects. The role of impurities is of primary importance in the control of the electrical characteristics of semiconductors and hence their technological applications. This section outlines the main vibrational features of impurities and the resulting modification of the optical properties of these semiconductors. Impurities can either lead to additional vibrational modes over and beyond that supported

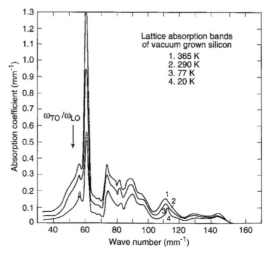

FIGURE 6 Multiphonon absorption of vacuum-grown Si.[25]

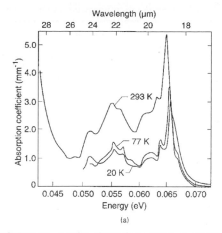

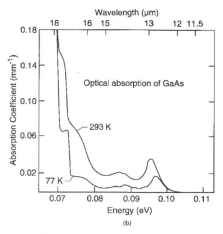

FIGURE 7a Lattice absorption coefficient of high-resistivity *n*-type GaAs vs. wavelength from 18 to 28 μm at 20, 77, and 293 K.[26]

FIGURE 7b Lattice absorption coefficient of high-resistivity *n*-type GaAs vs. wavelength from 10 to 18 μm at 77 and 293 K.[26]

by the unperturbed lattice or they can activate normally inactive vibrational modes.[28] The perturbation of the lattice by a substitutional impurity is a change in the mass of one of the constituents and a modification of the bonding forces in its vicinity. If the impurity is much lighter than the host atom it replaces, high-frequency vibrational modes above $\omega_{LO}(\Gamma)$, the maximum frequency supported by the unperturbed lattice, are introduced. These vibrational amplitudes are localized in the vicinity of the impurity and hence are known as local vibrational modes (LVM). For heavier impurities, the impurity-related vibrations can occur within the phonon band or in the gap between the acoustic and optical bands. These modes are referred to as resonant modes (RM) or gap modes (GM).[28]

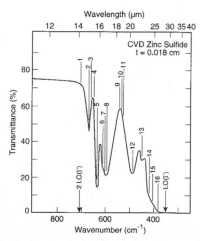

FIGURE 8 Optical transmittance of chemically vapor-deposited cubic ZnS in the two-phonon absorption regime.[27]

TABLE 4 Critical-point Analysis: Two-phonon Summation Spectrum in Cubic ZnS[27]

Feature*	Measured position (cm^{-1})	Phonon assignment	Calculated position (cm^{-1})	Comment†
1 (k)	704	2LO(Γ)	704	
2 (m)	668	2LO(L)	668	R
3 (s)	662	$2O_1$(W)	662	Q
4 (s)	650	LO(X) + TO(X)	650	
5 (m)	636	2TO(X)	636	R‡
6 (s)	612	$2O_2$(W)	612	Q, R
7 (m)	602	$2O_3$(W)	602	Q
8 (m)	596	2TO(L)	596	
9 (k)	544	LO(X) + LA(X)	544	
10 (s)	530	TO(X) + LA(X)	530	
11 (s)	526	LO(L) + LA(L)	526	
12 (m)	488	TO(L) + LA(L)	490	
13 (m)	450	O_1(W) + A_2(W)	450	R
14 (k)	420	LO(X) + TA(X)	420	R
15 (s)	406	TO(X) + TA(X)	406	also LO(L) + TA(L)
16 (s)	386	2LA(L)	384	R

* See Fig. 8 (k = kink, m = minimum, s = shoulder).
† R = Raman active and Q = quadrupole allowed.
‡ May also include the LO(L) + TO(L) summation at $632 \, cm^{-1}$.

The qualitative features of an impurity vibrational mode can be understood by considering a simple case of a substitutional impurity atom in a linear chain. The results of a numerical calculation of a 48-atom chain of GaP are presented in Figs. 9 and 10.[28] The highly localized character can be seen from Fig. 10. Note that the degree of localization reduces with increasing defect mass for a fixed bonding strength.

The absorption band produced by the LVM has been successfully used in impurity analyses. Figure 11 shows the IR absorption spectrum associated with interstitial oxygen in Si taken at NIST, and Fig. 12 displays the absorption spectrum from a carbon-related LVM in GaAs.[29,30] Note that the multiple peaks in the high-resolution spectrum are a consequence of the mass perturbations to the local environment resulting from the two naturally occurring isotopes of Ga. The fine structure in the spectrum helps identify the site occupied by C as belonging to the As sublattice. When accurate calibration curves are available, the concentration of the impurity can be determined from the integrated intensity of the LVM band. The LVM frequencies of a number of hosts and impurities are presented in Table 5.[28]

The absorption features produced by the LVM have been successfully employed in impurity and dopant analysis in semiconductors. The characteristic frequencies associated with specific impurities, as already discussed, can be used for chemical and structural identification, and the strength associated with each absorption may be exploited to obtain quantitative measurement accuracy. Two representative examples of LVM absorption are displayed in Figs. 11 and 12. The absorption of interstitial oxygen in Si is displayed in Fig. 11. The feature at $1107 \, cm^{-1}$ is due to the oxygen impurity, whereas the bands at 610 and $739 \, cm^{-1}$ arise from multiphonon absorption processes. For reference, the Si lattice degenerate ω_{TO}/ω_{LO} mode occurs at $522 \, cm^{-1}$ at room temperature. The spectrum was measured with a resolution of $1 \, cm^{-1}$.

In comparison, the LVM absorption from substitutional C is presented in Fig. 12. Note that the spectrum was recorded with a much higher resolution, namely, $0.06 \, cm^{-1}$,[29,30] and consequently displays a rich fine structure. The presence of the multiple peaks has been

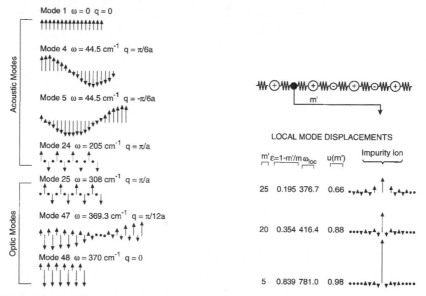

FIGURE 9 Linear-chain model calculations for GaP. A 48-atom chain is considered. Position along the chain is plotted horizontally and ion displacement vertically. Modes 24 and 25 occur on either side of the gap between the acoustic and optic bands.[28]

FIGURE 10 Eigenvectors for the highest-frequency (localized) mode for three isotopic substitutions on the $m = 31$ site. Note the extreme localization for a substituent of mass 5.[28]

interpreted to be a consequence of variations in the nearest-neighbor arrangement of the naturally occurring isotopes of the host lattice ions around the impurity. The results of calculation for the vibrational behavior of ^{12}C surrounded by the four possible configurations of combinations of ^{69}Ga and ^{71}Ga are displayed in the figure.[30] The agreement between theory and measurement is remarkable, which testifies to the power and

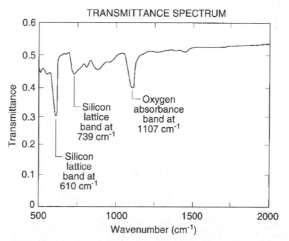

FIGURE 11 IR absorption due to interstitial oxygen in Si.

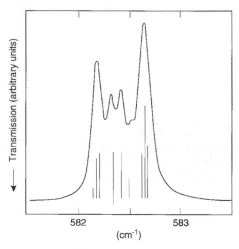

FIGURE 12 ^{12}C local modes in GaAs and the predicted fine structure. The height of each line is proportional to the strength of each mode.[30]

potential of the LVM analysis. Most noteworthy of the conclusion is the unambiguous assignment of the impurity site to be substitutional in the As sublattice. The use of LVM in the study of complex defects, particularly those that involve hydrogen, have led to a wealth of microscopic information, not easily attainable by any other means.[31]

The quantitative accuracy obtainable from LVM analysis may be illustrated by a simple

TABLE 5 Localized Modes in Semiconductors[28]

Host and impurity	Mode frequency (temp. K)	Defect symmetry, method of observation
Diamond		
N	1340(300)	T_d, A*
Silicon		
^{10}B	644(300), 646(80)	T_d, A
^{11}B	620(300), 622(80)	T_d, A
As	366(80) Reson.	T_d, A
P	441 Reson, 491(80) Reson.	T_d, A
^{14}C	570(300), 573(80)	T_d, A
^{13}C	586(300), 589(80)	T_d, A
^{12}C	605(300), 680(80)	T_d, A
O	Bands near 30, 500, 1100, 1200	
GaAs		
Al	362(80) ~ 371(4.2)	T_d, A, R
		T_d, T
P	355(80), 353(300) ~ 363(4.2)	T_d, A, R
		T_d, T
Si_{Ga}	384(80)	T_d, A
Si_{As}	399(80)	T_d, A

* A—absorption: T—transmission; R—reflection Reson.—resonant mode.

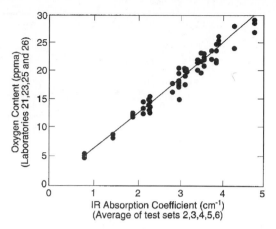

FIGURE 13 The absorption coefficient dependence on the concentration of interstitial oxygen in Si.[33]

harmonic model calculation. In such an approximation, the total integrated absorption over the entire band may be expressed as:[32]

$$\int \alpha \, d\omega = \frac{2\pi^2 N \eta^2}{\eta m_{\text{imp}} c} \tag{33}$$

where N is the volume density of the impurity and η and m_{imp} are the charge and mass of the impurity ion; c is the velocity of light. η is an empirically derived parameter that is specific to each center, i.e., a specific impurity at a specific lattice location. Once calibration curves are established, measurement of the intensity of absorption can be used to determine N. Figure 13 displays a calibration used to establish the density of interstitial oxygen.[33] Such analyses are routinely used in various segments of the electronic industry for materials characterization.

The effect of the impurities can alter the optical behavior in an indirect fashion as well. The presence of the impurity destroys translational symmetry in its vicinity and hence can lead to relaxation of the wave-vector conservation condition presented earlier in Eq. (33). Hence, the entire acoustic and optical band of phonons can be activated, leading to absorption bands that extend from zero frequency to the maximum $\omega_{\text{LO}}(\Gamma)$. The spectral distribution of the absorption will depend on the phonon density-of-states modulated by the effect of the induced dipole moment.[28] The latter is a consequence of the perturbation of the charge distribution by the impurity. The perturbation due to defects may be viewed in a qualitatively similar fashion. For instance, a vacancy may be described as an impurity with zero mass.

Interband

Absorption Near the Fundamental Edge. The fundamental absorption edge is one of the most striking features of the absorption spectrum of a semiconductor. Within a small fraction of an electron-volt at an energy about equal to the energy gap E_g of the material, the semiconductor changes from being practically transparent to completely opaque—the absorption coefficient changing by a factor of 10^4 or more. This increased absorption is caused by transitions of electrons from the valence band to the conduction band. This

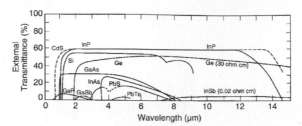

FIGURE 14   The transmission of CdS, InP, Si, Ge, GaAs, GaP, GaSb, InAs, InSb, PbTe, and PbS.[34]

characteristic optical property is clearly illustrated in Fig. 14, which shows the transmission versus wavelength for a number of major semiconductors.[34] At the lower wavelengths, the transmission approaches zero which defines a cut-off wavelength λ_c, for each material. For example, $\lambda_c \approx 7.1$ μm for InSb; $\lambda_c \approx 4.2$ μm, PbTe; $\lambda_c \approx 3.5$ μm, InAs; $\lambda_c \approx 1.8$ μm, GaSb and Ge; $\lambda_c \approx 1$ μm, Si; and $\lambda_c \approx 0.7$ μm for CdS. At much longer wavelengths than the edge at λ_c, lattice and free-carrier absorption become appreciable and the transmission drops. Studies of the fundamental absorption edge thus give values for the energy gap and information about the states just above the edge in the conduction band and below it in the valence band. Properties of these states are important to know since they are responsible for electrical conduction. Details of the band structure near the band extrema can be determined from the position and shape of the absorption edge and from its temperature, magnetic field, pressure, impurity concentration, and other parameters dependence. Finally, this fundamental gap region is important because usually it is only near the energy gap that phenomena such as excitons (both free and bound), electron-hole drops, donor-acceptor pairs, etc., are seen.

Interband transitions near the fundamental absorption edge are classified as (1) direct or vertical or (2) indirect or nonvertical. The momentum of light ($\hbar k = \hbar n\omega/c$) is negligible compared to the momentum of a **k**-vector state at the edge of the Brillouin zone. Thus, because of momentum conservation, electrons with a given wave vector in a band can only make transitions to states in a higher band having essentially the same wave vector. Such transitions are called vertical transitions. A nonvertical transition can take place, but only with the assistance of phonons or other entities which help preserve momentum.

Direct Transitions. The interband absorption coefficient depends upon the band structure and photon energy $\hbar\omega$. Use of quantum mechanics and, in particular, time-dependent perturbation theory, becomes necessary.[35] Direct transitions (with approximate conservation of the electron wave vector) can then be separated into "allowed" and "forbidden" transitions, depending on whether the dipole matrix element which determines the transition probability between the bands or the absorption coefficient is finite or vanishes in first approximation. For a nonzero momentum matrix element, a simple model gives for allowed direction transitions

$$\alpha_{AD} = C_{AD}(\hbar\omega - E_g)^{1/2} \tag{34}$$

and for forbidden direct transitions with a zero dipole matrix element

$$\alpha_{FD} = C_{FD}(\hbar\omega - E_g)^{3/2} \tag{35}$$

Both coefficients involve constants, valence- and conduction-band effective masses, and matrix elements, and only a slight dependence on photon energy. The absorption strengths of direct-gap semiconductors are related to their density of states and the momentum

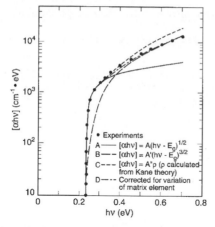

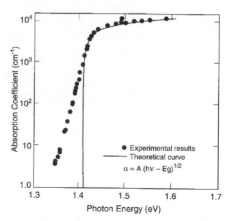

FIGURE 15 Theoretical fit to the experimental absorption edge of InSb at ~5 K.[36]

FIGURE 16 Absorption edge of GaAs at room temperature.[38]

matrix element that couples the bands of interest. Semiconductors such as AlAs, AlP, GaAs, InSb, CdS, ZnTe, and others have allowed direct transitions; many complex oxides such as Cu_2O, SiO_2, rutile, and others have forbidden direct absorption.

Figure 15 shows the spectral variation of the absorption coefficient for pure InSb at a temperature of 5 K compared to various theoretical predictions.[36] We note the extremely sharp absorption edge which is fit best by the $(\hbar\omega - E_g)^{1/2}$ dependence near the edge. However, a big deviation from the experimental data occurs at higher photon energies. Consideration of two more details allows a better fit: (1) use of a more complicated band model from Kane[37] which predicts a more rapidly increasing density of states than for the simple bands, and (2) taking into account a decrease in the optical matrix element at the higher photon energies because of the **k**-dependence of the wave functions. The calculated curves in Fig. 4 were arbitrarily shifted so that they look like a better fit than they are. The actual calculated absorption is a factor of about 15 too low at high energies. This discrepancy was attributed to the neglect of exciton effects which can greatly affect the absorption as discussed later.

Figure 16 shows the absorption behavior of GaAs at room temperature compared with calculations based on Kane's theory.[38] Below about 10^3 cm^{-1}, the absorption decreases much more slowly than predicted and absorption is even present for energies below E_g. In practice, there seems to exist an exponentially increasing absorption edge rule (called Urbach's rule) in most direct transition materials which is found to correlate reasonably well with transitions involving band tails. These band tails seem to be related to doping effects and phonon-assisted transitions.

Indirect Transitions. Semiconductors such as GaP, Ge, and Si have indirect gaps where the maximum valence-band energy and minimum conduction-band energy do not occur at the same **k** value. In this case, the electron cannot make a direct transition from the top of the valence band to the bottom of the conduction band because this would violate conservation of momentum. Such a transition can still take place but as a two-step process requiring the cooperation of another particle and which can then be described by second-order perturbation theory. The particle most frequently involved is an intervalley phonon of energy $\hbar\Omega_K$ which can be either generated or absorbed in the transition. (In some cases, elastic scattering processes due to impurity atoms or dislocations must be considered; they are less frequent than the phonon interactions.) The photon supplies the needed energy, while the phonon supplies the required momentum. The transition probability depends not only on the density of states and the electron-phonon matrix elements as in the direct case, but also on the electron-phonon interaction which is temperature dependent.

Calculations of the indirect-gap absorption coefficient give for the allowed indirect transitions

$$\alpha_{\mathrm{AI}} = C_{\mathrm{AI}}^{\mathrm{(abs)}}(\hbar\omega + \hbar\Omega_q - E_g)^2 + C_{\mathrm{AI}}^{\mathrm{(em)}}(\hbar\omega - \hbar\Omega_q - E_g)^2 \tag{36}$$

and for the forbidden indirect transitions

$$\alpha_{\mathrm{FI}} = C_{\mathrm{FI}}^{\mathrm{(abs)}}(\hbar\omega + \hbar\Omega_q - E_g)^3 + C_{\mathrm{FI}}^{\mathrm{(em)}}(\hbar\omega - \hbar\Omega_q - E_g)^3 \tag{37}$$

where the superscripts (abs) and (em) refer to phonon absorption and emission, respectively. These expressions are only nonzero when the quantities in parentheses are positive, i.e., when $\hbar\omega \pm \hbar\Omega_q > E_g$. We note that the phonon energies are usually small (≤ 0.05 eV) compared to the photon energy of about 1 eV and thus for the case of allowed indirect transitions with phonon absorption

$$\alpha_{\mathrm{AI}} \approx C_{\mathrm{AI}}^{\mathrm{(abs)}}(\hbar\omega - E_g)^2 \tag{38}$$

Thus the absorption increases as the second power of $(\hbar\omega - E_g)$, much faster than the half-power dependence of the direct transition as seen in Eq. (34).

Figure 17 shows the variation of the absorption coefficient of GaP with photon energy at room temperature near the indirect edge.[39] A reasonable fit to the experimental data of Spitzer et al.[40] is obtained indicating that, for GaP, allowed indirect transitions dominate. Further complications arise because there can be more than one type of phonon emitted or absorbed in the absorption process. Transverse acoustic (TA), longitudinal acoustic (LA), transverse optic (TO), and longitudinal optic (LO) phonons can be involved as shown in the absorption edge data of GaP, as seen in Fig. 18.[41] The phonon energies deduced from these types of experimental absorption edge studies agree with those found from neutron scattering.

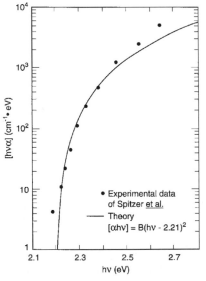

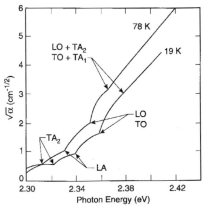

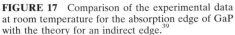

FIGURE 17 Comparison of the experimental data at room temperature for the absorption edge of GaP with the theory for an indirect edge.[39]

FIGURE 18 Absorption spectra at the edge of GaP, showing thresholds associated with the emission of each of several different phonons.[41]

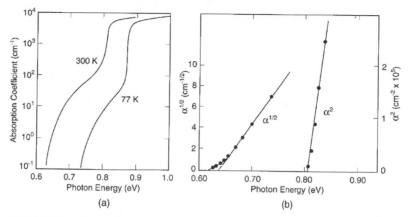

FIGURE 19 (*a*) α vs. $\hbar\omega$ for Ge; (*b*) the analysis of the 300 K experimental data.[42]

Both indirect and direct absorption edge data for Ge are shown in Fig. 19 while the analysis of the 300 K data is plotted in part *b*.[42] At the lowest energies, α rises due to the onset of indirect absorption as seen by the $\alpha^{1/2}$ dependence on photon energy. At higher energies, a sharper rise is found where direct transitions occur at the zone center and an α^2 dependence on energy is then seen. Note the large shifts of E_g with temperature for both the direct and indirect gaps. Also, the direct-gap absorption is much stronger than that of the indirect-gap absorption.

Excitons. Among the various optical properties of semiconductors, the subject of excitons has one of the dominant places because of their remarkable and diverse properties. Studies of exciton properties represent one of the most important aspects of scientific research among various solid-state properties. According to Cho,[43] there are a number of reasons for this: (1) excitonic phenomena are quite common to all the nonmetallic solids—semiconductors, ionic crystals, rare gas crystals, molecular crystals, etc.; (2) the optical spectra often consist of sharp structure, which allows a detailed theoretical analysis; (3) theories are not so simple as to be understood by a simple application of atomic theory or the Bloch band scheme, but still can be represented by a quasi-hydrogenlike level scheme; (4) sample quality and experimental techniques have continually been improved with subsequent experiments proving existing theories and giving rise to new ones; and (5) the exciton is an elementary excitation of nonmetallic solids, a quantum of electronic polarization. It has a two-particle (electron and hole) nature having many degrees of freedom, and along with the variety of energy-band structures, this leads to a lot of different properties from material to material or from experiment to experiment. Table 6 gives a definition of the major types of excitons in a glossary obtained from Hayes and Stoneham.[44] Many examples exist in the literature involving work on excitons in semiconductors to understand their nature and to determine their properties. Besides the references cited in this chapter, the authors refer the reader to the more detailed work presented in Refs. 45 and 46.

An electron, excited from the valence band to a higher energy state, can still be bound by the Coulomb attraction to the hole that the electron leaves in the valence band. This neutral bound-electron hole pair is called an exciton which can move throughout the crystal. Excitons are most easily observed at energies just below E_g using optical absorption or photoluminescence measurements. There are two models used for describing excitons in solids, named after Frenkel and Wannier. In a solid consisting of weakly interacting atoms, Frenkel considered excitons as described by excitations of a single atom or molecule.[47] An excited electron describes an orbit of atomic dimensions around an

TABLE 6 A Glossary of the Main Species of Excitons[44]

Exciton	In essence, an electron and hole moving with a correlated motion as an electron-hole pair.
Wannier exciton	Electron and hole both move in extended orbits; energy levels related to hydrogen-atom levels by scaling, using effective masses and dielectric constant; occurs in covalent solids such as silicon
Frenkel exciton	Electron and hole both move in compact orbits, usually essentially localized on adjacent ions; seen in ionic solids, such as KCl, in absorption
Self-trapped exciton	One or both carriers localized by the lattice distortion they cause; observed in ionic solids, such as KCl, in emission
Bound exciton	Only a useful idea when a defect merely prevents translational motion of an exciton and does not otherwise cause significant perturbation
Core exciton	Lowest-energy electronic excitation from a core state, leaving an unoccupied core orbital (e.g., the 1 s level of a heavy atom) and an electron in the conduction band whose motion is correlated with that of the core hole
Excitonic molecule	Complex involving two holes and two electrons
Multiple bound excitons	Complex of many holes and a similar number of electrons, apparently localized near impurities; some controversy exists, but up to six pairs of localized carriers have been suggested
Exciton gas	High concentration of electrons and holes in which each electron remains strongly associated with one of the holes (as insulating phase)
Electron-hole drops	High concentration of electrons and holes in which the motions are plasmalike (a metallic phase), not strongly correlated as in excitons

atom with a vacant valence state. The empty valence state acts as a mobile hole since the excitation can move from one atom to another. These tightly bound excitons are similar to an ordinary excited state of the atom, except that the excitation can propagate through the solid. The radius of the Frenkel exciton is on the order of the lattice constant. Frenkel excitons are useful to describe optical properties of solids like alkali halides and organic phosphors.

Wannier (or also called Mott-Wannier) excitons are also electrons and holes bound by Coulomb attraction.[48,49] In contrast to the Frenkel exciton, the electron and hole are separated by many lattice spacings producing a weakly bound exciton which is remarkably similar to a hydrogen-atom-like system. Since the electron and hole are, on the average, several unit cells apart, their Coulomb interaction is screened by the average macroscopic dielectric constant ϵ_∞, and electron and hole effective masses can be used. Their potential energy $-e^2/\epsilon_\infty r$ is just that of the hydrogen atom (except for ϵ_∞). The energy binding of the free exciton (relative to a free electron and free hole) is then given by the hydrogen-atom-like discrete energy levels plus a kinetic energy term due to the motion of the exciton:

$$E_{\text{ex}} = \frac{R_y}{n^2} - \frac{\hbar^2 k^2}{2(m_e^* + m_h^*)} \tag{39}$$

$$R_y = \left(\frac{m_r}{\epsilon_\infty^2}\right)\left(\frac{e^4}{2\hbar^2}\right) = \left(\frac{m_r}{m_H \epsilon_\infty^2}\right) E_H \tag{40}$$

where

$$m_r = \left(\frac{1}{m_e^*} + \frac{1}{m_h^*}\right)^{-1} \tag{41}$$

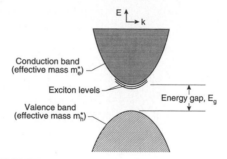

FIGURE 20 Exciton levels in relation to the conduction-band edge, for a simple band structure with both conduction- and valence-band edges at $k = 0$.

is the reduced mass of the exciton, m_H the reduced mass of the hydrogen atom, n is the principal quantum number $(1, 2, \ldots, \infty)$, E_H is the ionization energy of the hydrogen atom (13.6 eV), and R_y is the effective Rydberg energy. The lowest energy absorption transition of the semiconductor is thus $E_g - E_{ex}$. As an example, consider CdS where $m_e^* \approx 0.21\, m_o$, $m_h^* \approx 0.64\, m_o$, and $\epsilon_\infty \approx 8.9$; here $m_r \approx 0.158\, m_o$ and $R \approx 27$ meV. The Bohr radius for the $n = 1$ ground state is about 30 Å.

A series of excitonic energy levels thus exists just below the conduction band whose values increase parabolically with **k** and whose separation is controlled by n. Excitons are unstable with respect to radiative recombination whereby an electron recombines with a hole in the valence band, with the emission of a photon or phonons. These excitonic levels are shown in Figs. 20 and 21. For many semiconductors only a single peak is observed, as shown in Fig. 22 for GaAs.[50] However, even though only one line is observed, the exciton states make a sizable contribution to the magnitude of the absorption near and above the edge. At room temperatures, the exciton peak can be completely missing since the binding energy is readily supplied by phonons. In semiconductors with large enough carrier concentrations, no excitons exist because free carriers tend to shield the electron-hole interaction. Neutral impurities can also cause a broadening of the exciton lines, and, at large enough concentrations, cause their disappearance.

Extremely sharp exciton states can often be seen as shown in Fig. 23, which shows the absorption spectrum of a very thin, very pure epitaxial crystal of GaAs.[51] The $n = 1$, 2, and 3 excitons are clearly seen followed by excited states with $n > 3$ leading smoothly into the continuum. The dashed line, calculated neglecting the effects of excitons, illustrates how important exciton effects are in understanding the optical properties of semiconductors and confirms the qualitative picture of exciton absorption presented in Fig. 24.[50] Cu_2O exhibits a series of beautiful exciton absorption lines as shown in Fig. 25 where structure for $n = 2$

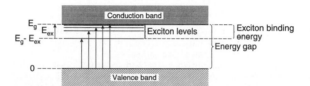

FIGURE 21 Energy levels of a free exciton created in a direct process. Optical transitions from the top of the valence band are shown by the arrows; the longest arrow corresponds to the energy gap.

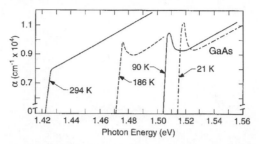

FIGURE 22 Observed exciton absorption spectra in GaAs at various temperatures between 21 K and 294 K. Note the decrease in the band edge with increasing temperatures.[50]

to $n = 9$ lines exists.[52] No $n = 1$ line is seen since Cu_2O has a direct but forbidden gap where the exciton emission is dipole-forbidden.

Excitons in direct-gap semiconductors such as GaAs are called direct excitons. For indirect-gap semiconductors like Si or GaP, the absorption edge is determined by the influence of indirect excitons as revealed by the shape of the absorption. Such indirect-exciton transitions have been observed in several materials including Ge, Si, diamond, GaP, and SiC.

Real semiconductor crystals contain impurities and defects which also can affect the optical properties related to excitonic features, in additon to their causing impurity/defect absorption. A bound exciton (or bound-exciton complex) is formed by binding a free

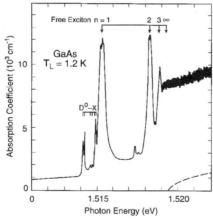

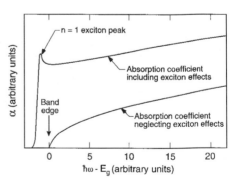

FIGURE 23 Absorption spectrum at 1.2 K of ultrapure GaAs near the band edge. The $n = 1, 2, 3$ free exciton peaks are shown; also the bandgap E_g, determined by extrapolation to $n = \infty$, and impurity lines (D_oX) from the excitons bound to $\sim 10^{15}\,cm^{-3}$ donors. (The rise at high energy is due to substrate absorption.) The dashed line shows the $(E - E_g)^{1/2}$ behavior expected in the absence of electron-hole interaction (the absolute magnitude is chosen to fit the absorption far from the band edge).[51]

FIGURE 24 Calculated optical absorption coefficient for direct transitions in the simple band model neglecting and including the $n = 1$ exciton peak below the band edge. The top trace represents the low-temperature data given in Fig. 22. (*Adapted from Ref. 50.*)

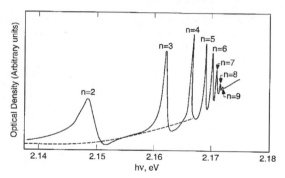

FIGURE 25 Absorption spectrum of the "yellow" exciton in Cu_2O at 1.8 K. Since transitions are allowed only to p states of this exciton, the $n = 1$ exciton is forbidden.[52]

exciton to a chemical impurity atom (ion), complex, or a host lattice defect. The binding energy of the exciton to the defect or impurity is generally weak compared to the free-exciton binding energy. These bound excitons are extrinsic properties of the semiconductor; the centers to which the free excitons are bound can be either neutral donors or acceptors or ionized donors or acceptors. They are observed as sharp-line (width ≈ 0.1 meV) optical transitions in both absorption and photoluminescence spectra. The absorption or emission energies of these bound-exciton transitions always appear below those of the corresponding free-exciton transitions. Bound excitons are very commonly observed because semiconductors contain significant quantities of impurities or defects which produce the required binding. These complexes are also of practical interest because they characterize the impurities often used to control the electrical properties of semiconductors as well as being able to promote radiative recombination near the band gap. Bound excitons exhibit a polarization dependence similar to the free-exciton states from which they originated.

At higher densities of free excitons and low temperatures, they can form an electron-hole droplet by condensing into a "liquid" phase. This condensed phase occurs for electron-hole concentrations of about 2×10^{17} cm^{-3} and can be thought of as an electron-hole plasma with a binding energy of several meV with respect to the free excitons. Table 7 shows some of the properties of these electron-hole drops in Si and Ge that must be considered in an understanding of their optical properties.[53]

TABLE 7 Some Properties of Electron-hole Drops in Si and Ge[53]

	Si	Ge
Droplet radius R	<1 μm	~4 μm
Binding energy or energy to remove one exciton at $T = 0$	8.2 ± 0.1 meV	1.8 ± 0.2 meV
Ground-state energy per e-h pair	22.8 ± 0.5 meV	6.0 ± 0.2 meV
Density of e-h pairs	$3.33 \pm 0.05 \times 10^{18}$ per cm^3	$2.38 \pm 0.05 \times 10^{17}$ per cm^3
Critical density of e-h pairs n_c, at $T = 0$	$1.2 \pm 0.5 \times 10^{18}$ per cm^3	$0.8 \pm 0.2 \times 10^7$ per cm^3
Critical temperature T_c	25 ± 5 K	6.5 ± 0.1 K
Principal luminescence band for e-h droplet	1.082 eV	0.709 eV
Principal luminescence band for single exciton	1.097 eV	0.713 eV

Polarizations. Interesting optical effects arise when one considers explicitly the influence of longitudinal and transverse optical phonons on a transverse electromagnetic wave propagating through the semiconductor. This influence can be taken into account via the dielectric function of the medium. Dispersion curves that arise do not conform either to the photon or to the phonon. The coupling between the photon and phonon becomes so strong that neither can continue to be regarded as an independent elementary excitation, but as a photon-phonon mixture! This mixture can be regarded as a single quantity which can be interpreted as a new elementary excitation, the *polariton.*[54] Similar couplings exist between an exciton and the photon. It is an important consideration for interpreting some optical processes involving Raman and luminescence measurements.

High-Energy Transitions Above the Fundamental Edge. The optical properties of most semiconductors have been thoroughly investigated throughout the visible and ultraviolet regions where transitions above the fundamental gap energy give rise to properties strongly dependent upon photon energy. This regime is dominated by optical absorption and reflection of a photon arising from both valence and core electron transitions from the ground state of the system into various bound, autoionizing, continuum, or other excited states. The sum of all excitations—both bound (nonionizing) and ionizing—gives the total absorption coefficient and the complex dielectric constant at each photon energy hv. Photoemission or photoelectron spectroscopy measurements in this high-energy regime provide an alternative to ultraviolet spectroscopy for providing detailed information on the semiconductor. (See the electromagnetic spectrum in Fig. 1 in the Introduction for a reminder that photoemission measurements overlap UV measurements.) Electrons may be ejected from the semiconductor by high-energy photons as shown in Fig. 26.[55] Their kinetic energies are measured and analyzed to obtain information about the initial electron states. Ionizing excitations involve electron excitations into unbound states above the vacuum level. These excitations result in photoemission as depicted for the two valence bands and the core level shown. For photoemission involving one-electron excitations (usually dominant), binding energies E_B of valence and core levels are given directly by the measured kinetic energies E (from $N(E)$ in the figure and energy conservation):

$$E_B = hv - E - \phi \tag{42}$$

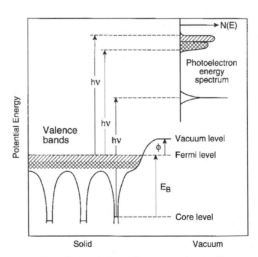

FIGURE 26 A schematic energy-level diagram showing photoemission from the valence bands and a core level in a solid.[55]

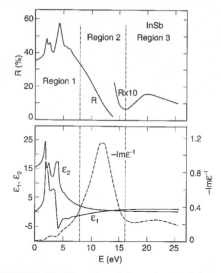

FIGURE 27 The spectral dependence of the reflectance R, the real and imaginary parts of the dielectric constant ϵ_1 and ϵ_2, and the energy-loss function $-Im(\epsilon^{-1})$ for InSb.[56,61]

Here ϕ is the work function (usually about 2 to 5 eV for most clean solids) which is known or easily measured. X-ray photoemission spectroscopy (XPS) is usually used to study core states and ultraviolet photoemission spectroscopy (UPS) to study valence-band states. Photoemission techniques are also used for the study of surface states. Synchrotron radiation provides an intense source of light over a large spectral region. By measuring the angular distribution of the emitted UPS electrons, a direct determination of the E versus k relation for the valence band of GaAs can be made.

There are several regions of importance that must be considered in describing the optical properties of this high-energy region. Figure 27 shows the regions for InSb that are representative of the results for other semiconductors.[56] InSb is a narrow gap material with a direct band gap of 0.17 eV at room temperature, so what is shown is at much greater energies than E_g. Sharp structure associated with transitions from the valence band to higher levels in the conduction band characterize the first region that extends to about 8 to 10 eV. That this behavior is characteristic of other III-V compounds can be seen in Fig. 28 which shows the reflectance for InAs, GaAs, and GaP as well as InSb.[57] To show how this type of optical spectra can be interpreted in terms of the materials energy band structure, consider Fig. 29 which shows the spectral features of ϵ_2 for Ge in a and the calculated energy bands for Ge in b.[58,59] Electronic transitions can take place between filled and empty bands subject to conservation of energy and wave vector. The initial and final electron wave vectors are essentially equal, and only vertical transitions between points separated in energy by $\hbar\omega = E_c(k) - E_v(k)$ are allowed.

The intensity of the absorption is proportional to the number of initial and final states and usually peaks when the conduction and valence bands are parallel in k-space. This condition is expressed by

$$\nabla_k[E_c(k) - E_v(k)] = 0 \qquad (43)$$

Places in k-space where this is true are called critical points or Van Hove singularities.

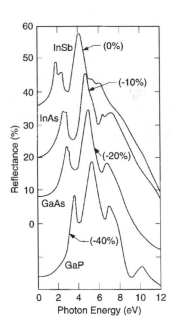

FIGURE 28 Reflectance of several III–V compounds at room temperature. For clarity, the spectra are offset along the y axis by the amounts shown in brackets.[57]

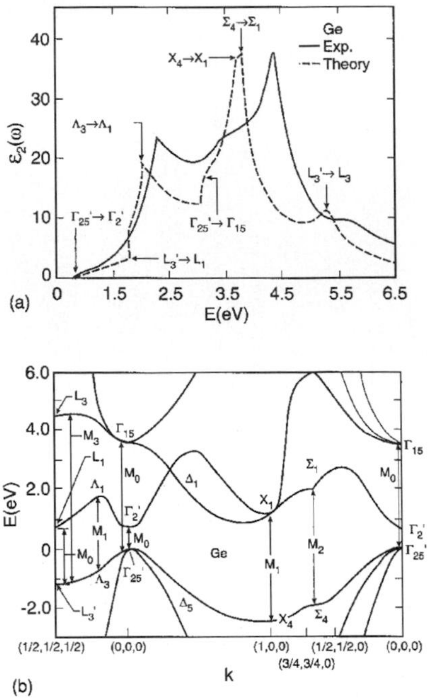

FIGURE 29 (*a*) Spectral features of ϵ_2 for Ge; (*b*) the calculated pseudopotential energy bands for Ge along some of the principal axes.[58,59]

The experimental peaks can be sharp as shown in Fig. 29*a* because interband transitions are not appreciably broadened by damping and thus the lineshapes are determined by the density of states. Much information is available from the data if a good theory is used. Figure 29*b* shows pseudopotential energy-band calculations that show the special points and special lines in the Brillouin zone that give rise to the data shown in *a*.

The second region in Fig. 27 extends to about 16 eV and shows a rapid decrease of reflectance due to the excitation of collective plasma oscillations of the valence electrons. The behavior in this second "metallic" region is typical of the behavior of certain metals in the ultraviolet. One can think of the valence electrons as being essentially unbound and able to perform collective oscillations. Sharp maxima in the function $-Im\,\epsilon^{-1}$, which describes the energy loss of fast electrons traversing the material, have been frequently associated with the existence of plasma oscillations.

In the third region, the onset of additional optical absorption is indicated by the rise in reflectance. This structure is identified with transitions between filled d bands below the valence band and empty conduction-band states. As shown in Fig. 30, the structure in region three is present in other III-V compounds, but is absent in Si which does not have a d-band transition in this region.[60] Other structure at higher photon energies is observed for Si as shown in Fig. 31 which shows the imaginary part of the dielectric constant of Si from 1 to 1000 eV.[61,62] The large peaks on the left are due to excitations of valence electrons, whereas the peaks on the right are caused by excitation of core electrons from L shell states. K shell electrons are excited at energies beyond the right edge of the graph.

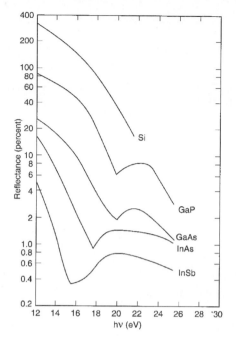

FIGURE 30 Reflectance of several semiconductors at intermediate energies. Starting from 12 eV the reflectance decreases, representing the exhaustion of bonding → antibonding oscillator strength at energies greater than $2E_g$. The rise in reflectivity in the 15–20 eV range in the Ga and In compounds is caused by excitation of electrons from Ga cores (3d states) or In cores (4d states). The ordinate should be multiplied by 2 for InSb, 1 for InAs, 1/2 for GaAs, 1/4 for GaP, and 1/10 for Si.[60]

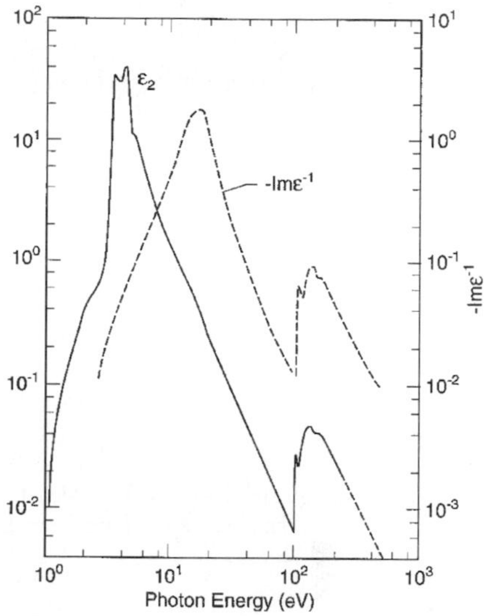

FIGURE 31 Imaginary part of the dielectric function for Si from 1 to 1000 eV.[2] [From reflectivity measurements of H. R. Philipp and H. Ehrenreich, Ref. 61, out to 20 eV and from transmission measurements of C. Gahwiller and F. C. Brown, Ref. 62, from 40 to 200 eV.]

Free Carriers

Plasmons. Semiconductors, in addition to a crystal lattice that may be ionic, may contain free charges as well.

The free-carrier contribution to the dielectric function is given by Maxwell's equation in CGS units

$$\epsilon_{\text{fc}}(\omega) = -i\frac{4\pi\sigma}{\omega} \tag{44}$$

The task of establishing the functional form of $\epsilon(\omega)$ hence reduces to one of determining the conductivity at the appropriate optical frequencies.

The response of a charge to an externally applied field may be described by classical methods, assuming a damping or resistive force to the charge that is proportional to the velocity of the charge. This simplification is known as the Drude approximation,[63] and it leads to the following relationship:

$$\sigma = \frac{Ne^2\tau}{m^*}\frac{1}{(1 - i\omega\tau)} \tag{45}$$

and is related to the dc conductivity by the relationship:

$$\sigma(0) = Ne^2\tau/m^* \tag{46}$$

where N is the free carrier density and $1/\tau$ is the constant of proportionality for the damping force and τ is a measure of the electron-electron collision time. Now,

$$\epsilon_{fc}(\omega) = -\frac{i\omega_P^2\epsilon_\infty}{\omega\left(\omega + \dfrac{i}{\tau}\right)} \tag{47}$$

ω_p is the plasma frequency that describes oscillations of the plasma, i.e., the delocalized charge cloud

$$\omega_P^2 = \frac{4\pi Ne^2}{m^*\epsilon_\infty} \tag{48}$$

against the fixed crystal lattice.

In an ideal plasma with no damping, the $\epsilon(\omega)$ reduces to:

$$\epsilon(\omega) = 1 - \frac{\omega_p^2}{\omega^2} \tag{49}$$

$\epsilon(\omega)$ is negative for $\omega < \omega_p$, which leads to total reflection and, hence, the term plasma reflectivity.

The behavior of plasma reflection and the relationship to the free-carrier density is illustrated in Fig. 32 using the far-infrared reflection spectrum from a series of PbTe

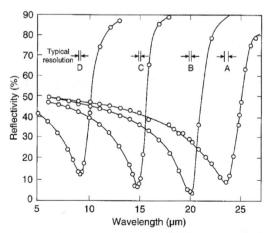

FIGURE 32 Reflectivity at 81 K and normal incidence of variously doped samples of p-type PbTe, showing the plasma resonance. Hole concentrations: A, $3.5 \times 10^{18}\,\text{cm}^{-3}$; B, $5.7 \times 10^{18}\,\text{cm}^{-3}$; C, 1.5×10^{19}; D, $4.8 \times 10^{19}\,\text{cm}^{-3}$.[64]

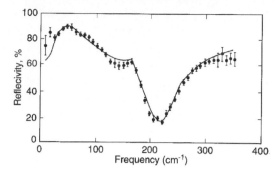

FIGURE 33 Far-infrared reflectivity of a 2-μm-thick epitaxial layer of PbSe on an NaCl substrate. Data points are shown with associated errors. The solid curve represents the best fit.[65]

samples with hole densities extending from 3.5×10^{18} cm^{-3} to 4.8×10^{19} cm^{-3}.[64] The plasma frequency increases with increasing carrier density as described by Eq. (48).

Coupled Plasmon-Phonon Behavior. Most semiconductor samples contain free carriers and phonons and the frequencies of both are comparable. Hence, a complete description of the far-infrared optical properties has to take both into account. This can be achieved readily using Eq. (47) to describe the free carriers. The combined $\epsilon(\omega)$ may then be expressed as:

$$\epsilon(\omega) = \epsilon_\infty + \frac{S\omega_{TO}^2}{\omega_{TO}^2 - \omega^2 - i\omega\gamma} - \frac{i\omega_p^2\epsilon_\infty}{\omega(\omega + i/\tau)} \tag{50}$$

A good example of the accurate description of the far-infrared behavior of a semiconductor is presented in Fig. 33.[65] The reflection spectrum from a 2-μm-thick PbSe film on an ionic substrate of NaCl is shown along with the results of a computer fit. The calculations included only an optical phonon contribution for the substrate and optical phonon and a plasmon contribution of the film. All the major features in the complicated spectrum can be well described using the simple oscillatory models described.

The coexistence of phonons and plasmons leads to a coupling between the two participants.[66] Of particular interest are the coupled plasmon-LO phonon modes denoted by L_+ and L_- that are exhibited as minimas in the reflection spectra. As explained earlier, the LO phonon frequencies occur at the zeros of the dielectric function $\epsilon(\omega)$. In the presence of plasmons, the zeros are shifted to the coupled mode frequencies L_+ and L_-. These frequencies can be determined directly for the case of no damping for both the phonon and the plasmon as shown below:

$$L_\pm = \tfrac{1}{2}\{(\omega_{LO}^2 + \omega_p^2) \pm [(\omega_{LO}^2 - \omega_p^2)^2 + 4\omega_{LO}^2\omega_p^2(1 - \epsilon_\infty/\epsilon(0)]^{1/2}\} \tag{51}$$

Note that the presence of the plasmon introduces an additional low frequency zero at L_-. The relationship of the L_+ and L_- frequencies with the carrier density is presented in Fig. 34.[67] The existence of the coupled modes were predicted by Varga[66] and later observed using Raman scattering[67,68] and far-infrared reflectivity.[69]

Impurity and Defect Absorption. The extended electronic states, excitons, lattice vibrations, and free carriers discussed thus far are all intrinsic to the pure and perfect

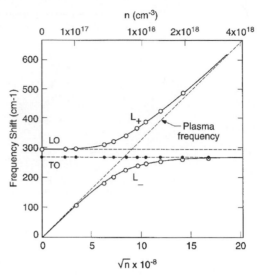

FIGURE 34 The solid curves labeled L_+ and L_- give the calculated frequencies of the coupled longitudinal plasmon-phonon modes and the measured frequencies are denoted by the open circles.[68]

crystal. In practice, real-life specimens contain imperfections and impurities. The characteristic optical properties associated with impurities and defects are the subject of discussion in this section. Two representative examples of the most widely observed effects, namely, shallow levels and deep levels in the forbidden gap, are considered in the following discussion.

Some of the effects due to impurities are considered in other parts of this chapter: impurity-related vibrational effects were considered under "Lattice"; excitons bound to impurity states were discussed under "Excitons"; and impurity-related effects in magneto-optical behavior are dealt with under "Magnetic-Optical Properties." In addition to these effects, optical absorption due to electronic transitions between impurity-related electronic levels may also be observed in semiconductors.

The presence of impurities in a semiconductor matrix leads to both a perturbation of the intrinsic electronic quantum states and the introduction of new states, particularly in the forbidden energy gap. The major classes of electronic levels are the shallow levels that form the acceptor and donor states and lie close to the valence and conduction band extremes, respectively, and those that occur deep in the forbidden gap. The former are well known and are critical in controlling the electrical behavior of the crystal, and the latter are less well known but are, nevertheless, important in determining the subband gap optical behavior.

Direct transitions from the shallow levels to the closest band extrema can be observed in the far-infrared transmission spectra of many semiconductors. An elegant example of this property is illustrated with the spectrum obtained from a high-purity Si wafer[70] as displayed in Fig. 35. Sharp, well-resolved absorption features from electronic transitions due to B, P, As, and Al are present, as are additional features, perhaps from unidentified impurities. Note that both acceptors and donor bands are observable due to the highly nonequilibrium state in which the specimen was maintained through the use of intense photoexcitation.

Electronic transitions from impurity- and defect-related levels deep in the forbidden gap can significantly alter the subband gap behavior. One of the best known examples of

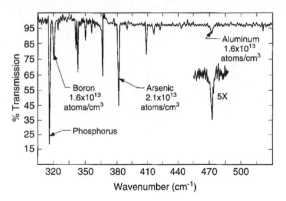

FIGURE 35 Total impurity spectrum of a 265-Ω-cm n-type Si sample obtained by the simultaneous illumination method. The input power to the illumination source was about 50 W.[70]

this is the native defect level known as EL2 in GaAs. The level occurs 0.75 eV below the conduction-band extremum, and when present, it can completely dominate the subband gap absorption. The absorption spectrum recorded from a GaAs sample containing EL2 is presented in Fig. 36.[71] The onset of the absorption at 0.75 eV is due to transitions to the conduction-band extremum at the direct gap, and the features at 1.2 and 1.4 eV are due to transitions to higher-lying extrema.[71] The figure also shows two spectra that exhibit the well-known photoquenching effect associated with EL2. When the specimen is subjected to intense white light radiation, the EL2 absorption is quenched, leaving only the band-to-band transitions with an onset at E_g which occurs at ~1.5 eV at 10 K.

Optical measurements of shallow impurities in semiconductors have been carried out by absorption (transmission) and photoconductivity techniques. The photoconductivity method is a particularly powerful tool for studying the properties of shallow impurity states, especially in samples which are too pure or too thin for precise absorption

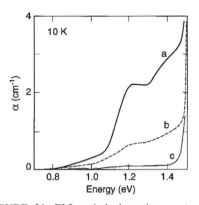

FIGURE 36 EL2 optical absorption spectra recorded at 10 K in the same undoped semi-insulating GaAs material. Curve a: after cooling in the dark; curves b and c: after white light illumination for 1 and 10 min, respectively.[71]

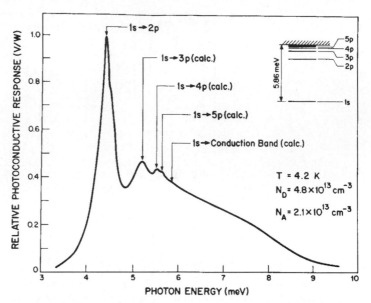

FIGURE 37 Far-infrared photoconductivity spectrum of a high-purity GaAs sample showing the measured transition energies and those calculated from the hydrogenic model using the (1s→2p) transition energy. The hydrogenic energy level diagram is shown in the inset.[72]

measurements. In most cases, this type of photoconductivity can only be observed in a specific temperature range, usually at liquid helium temperatures. Figure 37 shows the photoconductivity response of a high-purity GaAs sample with specific transition energies that correspond to hydrogenic-like transitions.[72] If the excited states of the impurity were really bound states, electrons in these states could not contribute to the conductivity of the sample, and the excited state absorption would not result in peaks in the photoconductivity spectrum. However, there have been several suggestions as to how the electrons that are excited from the ground state to higher bound excited states can contribute to the sample's conductivity. First, if the excited state is broadened significantly by interactions with neighboring ionized donor and acceptor states, it is essentially unbound or merged with the conduction band. Other mechanisms for impurity excited-state photoconductivity all involve the subsequent transfer of the electron into the conduction band after its excitation to the excited state by the absorption of a photon. Mechanisms that have been considered for this transfer include (1) impact ionization of the electrons in the excited state by energetic free electrons, (2) thermal ionization by the absorption of one or more phonons, (3) photoionization by the absorption of a second photon, and (4) field-induced tunneling from the excited state into the conduction band. All of these mechanisms are difficult to describe theoretically.

Magneto-optical Properties

Background. Phenomena occurring as a result of the interaction of electromagnetic radiation with solids situated in a magnetic field are called magneto-optical (MO) phenomena. Studies of MO phenomena began in 1845 when Michael Faraday observed that plane-polarized light propagating through a block of glass in a strong magnetic field has its plane of vibration of light rotated. By the 1920s, most MO effects were fairly well understood in terms of the classical dynamics of an electron in a magnetic field. However, when semiconductors were first investigated in the early 1950s, a quantum mechanical interpretation of the MO data in terms of the energy-band structure was found to be necessary. The two major limitations of the classical theory are that no effects depending on the density of states are predicted and no effects of electron spin are included.

Table 8 presents an overview of the typical types of magneto-optical phenomena observed in semiconductors and the information that can be determined from the experimental measurements. Four classes of MO phenomena can be distinguished: those arising from (1) interband effects, (2) excitonic effects, (3) intraband or free-carrier effects, and (4) impurity magnetoabsorption effects. Further clarification can then be made by determining whether the effect is absorptive or dispersive, resonant or nonresonant, and upon the relative orientation of the magnetic field to the direction of propagation of the electromagnetic radiation and its polarization components. Resonant experiments usually provide more detailed information about the band structure of a semiconductor and often are easier to interpret. There is thus a wide variety of effects as shown which can give different types of information about the crystal's energy-band structure, excitonic properties, and impurity levels. Before summarizing and discussing each of these magneto-optical effects, it is necessary to briefly describe the effects of a magnetic field on the enegy-band structure of a semiconductor.

Effect of a Magnetic Field on the Energy Bands. Magneto-optical experiments must be analyzed with specific energy-band models in order to extract the related band parameters and to emphasize the underlying physical concepts with a minimum of mathematical complexity. Most often, one deals with only the highest valence bands and the lowest conduction bands near the forbidden energy-gap region. If simple parabolic bands are assumed, a fairly complete analysis of the MO experiments is usually possible, including both the resonant frequencies and their line shapes. On the other hand, if more complicated energy bands (e.g., degenerate, nonparabolic) are needed to describe the solid, the detailed analysis of a particular experiment can be complicated.

The effect of a magnetic field on a free electron on mass m^* was determined in 1930 by Landau, who solved the Schroedinger equation. The free electrons experience a transverse Lorentz force which causes them to travel in orbits perpendicular to the magnetic field. The resulting energy eigenvalues corresponding to the transverse components of the wave vector are quantized in terms of harmonic oscillator states of frequency ω_c, while a plane-wave description characterizes the motion along the magnetic field. The allowed energy levels, referred to as Landau levels, are given by

$$E_n^\pm = (n + \tfrac{1}{2})\hbar\omega_c + \frac{\hbar^2 k_z^2}{2m^*} \pm \frac{1}{2} g^* \mu_B B \qquad (52)$$

where n is the Landau level number $(0, 1, 2, \ldots)$, ω_c (the cyclotron frequency) is equal to eB/m_c^*, and m_c^* is the cyclotron effective mass. The middle term represents the energy of an electron moving along the direction of the **B** field in the z direction; it is not quantized. The first term represents the quantized energy of motion in a plane perpendicular to the field. The last term represents the effect of the electron's spin; g^* is the effective spectroscopic g-factor or spin-splitting factor and

$$\mu_B = e\hbar/2m_0 = 5.77 \times 10^{-2}\,\text{meV/T} \qquad (53)$$

is the Bohr magneton. The g-factor in a semiconductor has values quite different from the usual value of two found for atomic systems (e.g., for narrow-energy gaps and a strong spin-orbit interaction, g^* can be large and negative).

TABLE 8 Magneto-optical Phenomena and Typical Information Obtainable

Magneto-optical effect	Some information or properties obtainable
Interband effects	
In transmission	
Band-to-band magnetoabsorption	Energy gaps, effective masses, g-factors, higher band parameters
Faraday rotation (resonant and nonresonant)	Energy gaps, effective masses
Faraday ellipticity (nonresonant)	Relaxation times
Voigt effect (resonant and nonresonant)	Effective masses
Cross-field magnetoabsorption	
In reflection	
Magnetoreflection	Studies of very deep levels or where absorption is high, similar information as in transmission magnetoabsorption
Kerr rotation	
Kerr ellipticity	
Magneto-excitonic effects	
	Diamagnetic and Zeeman shifts and splittings; energy gap; effective reduced-mass tensor; effective Rydberg; anisotropy parameter; dielectric tensor components; effective g-factors; effective masses; quality of materials, structures, alloys and interfaces
Intraband or free-carrier effects	
In transmission	
Cyclotron resonance (resonant)	Effective masses, relaxation times, nonparabolicity
Combined resonance	Same information as cyclotron resonance plus g-factors
Spin resonance	g-factors
Phonon-assisted cyclotron resonance harmonics	Same information as cyclotron resonance plus phonon information
Faraday rotation (resonant and nonresonant)	Carrier concentration, effective masses; use for impure materials, flexible, can be used at high temperature
Faraday ellipticity (nonresonant)	
Voigt effect (nonresonant)	
Interference fringe shift (nonresonant)	
Oscillatory variation of the Shubnikov-de Haas type	Carrier concentration
In reflection	
Magnetoplasma reflection	Effective masses, carrier concentration
Magnetoplasma rotation (Kerr effect)	
Magnetoplasma ellipticity	
Impurity magneto-absorption	
Zeeman and diamagnetic effect type behavior of impurities	Hydrogenic impurity information, binding energy, effective masses, effective Rydberg, central cell corrections, static dielectric constant; both impurity and Landau level information; impurity information as above plus Landau-level information related to effective masses and g-factors
Photoionization behavior (transitions from ground state of impurity to Landau levels)	

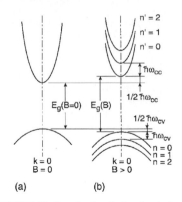

FIGURE 38 Landau levels for simple bands.

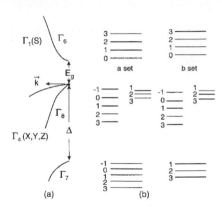

FIGURE 39 (*a*) Zincblende semiconductor energy bands (**H** = 0); and (*b*) in an applied magnetic field **H**.[73]

Figure 38 shows schematically the effect of a magnetic field on a simple parabolic direct gap extrema at $k = 0$. The quasi-continuous parabolic behavior of the nondegenerate conduction and valence bands at **B** = 0 (shown in *a*) is modified by the application of a magnetic field into Landau levels as shown in *b*. Each Landau level is designated by an integer $n = 0, 1, 2, 3, \ldots$ Finally, in Fig. 39, the Landau effects are shown schematically for zincblende energy bands when both spin and valence-band degeneracy are taken into account for both **B** = 0 (*a*) and **B** ≠ 0 (*b*).[73] With spin included, the valence band splits into the Γ_8 and Γ_7 bands with spin-orbit splitting energy Δ. The quantization of the bands into Landau levels is illustrated in the right-hand side of the figure. The Pidgeon and Brown[74] model has been successfully used to describe the magnetic field situation in many semiconductors, since it includes both the quantum effects resulting from the partial degeneracy of the p-like bands and the nonparabolic nature of the energy bands. The *a*-set levels are spin-up states, and the *b*-set, the spin-down states. These large changes in the **E** versus k relations of the bands when a magnetic field is applied also means large changes in the density of states which become periodic with a series of peaks at energies corresponding to the bottom of each Landau level. This oscillatory variation in the density of states is important for understanding the various oscillatory phenomena in a magnetic field.

Interband Magneto-optical Effects. Interband transitions in a magnetic field connect Landau-level states in the valence band to corresponding states in the conduction band. Thus, they yield direct information concerning energy gaps, effective masses, effective g-factors, and higher band parameters. The strongest allowed transitions are those that are proportional to the interband matrix element $p = -(i\hbar/m_o)\langle s|p_j|x_j\rangle$, where $j = x, y, z$.

This matrix element directly connects the p-like valence band (x, y, z) which is triply degenerate with the s-like conduction band through the momentum operator p_j. The transition energies can be calculated directly from a knowledge of the selection rules and use of an energy-band model. The selection rules are given by[75]

$$\sigma_L: \quad a(n) \to a(n-1),\ b(n) \to b(n-1)$$

$$\sigma_R: \quad a(n) \ > a(n+1),\ b(n) \to b(n+1)$$

$$\pi: \quad a(n) \to b(n+1),\ b(n) \to a(n-1)$$

where σ_L, σ_R, and π are left circular, right circular, and linear (**e** ∥ **B**) polarizations. As discussed earlier, a and b denote the spin-up and spin-down states, and n the Landau-level number. Often, sharp optical transitions between the Landau levels are observed, providing highly accurate information about the fundamental band parameters such as the energy gap E_g, effective masses of the electrons and holes, higher band parameters, etc.

InSb is a material in which interband effects have been studied very extensively; thus, it is a good representative example. Even though it is a narrow-gap semiconductor and thus has a small exciton binding energy, Weiler[76] has shown that excitonic corrections must be made to properly interpret the magnetic-field-dependent data. This shall be discussed in more detail in the next section. The band models discussed earlier predict that there is an increase in the energy of the absorption edge with magnetic field because of the zero point energy $1/2\hbar\omega_c$ of the lowest conduction band. At larger photon energies, transmission minima (absorption maxima) which are dependent upon magnetic field are observed. By plotting the photon energy positions of the transmission minima against magnetic field, converging, almost linear plots are obtained as shown in Fig. 40.[74] Extrapolation of the lines to zero field gives an accurate value for the energy gap. Use of a band model and specific transition assignments further allow the determination of other important band parameters, such as effective masses and g-factors.

Magnetoreflection. Besides the changes in absorption brought about by the magnetic field, there are changes in the refractive index. Since the reflectivity depends upon both the real and imaginary parts of the index, clearly it is affected by the field. Interband transitions are often observed in reflection because of the high absorption coefficients. Figure 41 shows the magnetoreflection behavior of InSb for σ_R and σ_L polarizations.[73] In addition, modulation spectroscopy techniques, in which a parameter such as stress, electric field, wavelength, magnetic field, etc. is periodically varied and the signal synchronously

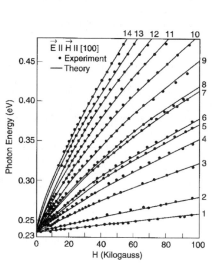

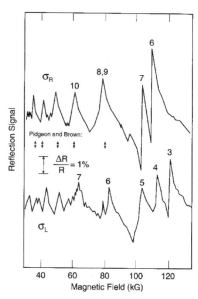

FIGURE 40 Energy values of transmission minima versus magnetic field for electron transitions between Landau levels of valence and conduction bands in InSb. Plot of the photon energy of the principal transmission minima as a function of magnetic field for **E** ∥**H**∥ [100]. The solid lines represent the best theoretical fit to the experimental data. The numeral next to each line identifies the quantum assignment.[74]

FIGURE 41 Magnetoreflection curves for InSb, compared to the magnetoabsorption results of Ref. 74 for InSb [001] with **E** ⊥ **H**. $\hbar\bar{\omega} = 387.1$ meV; $T = 24$ K.[73]

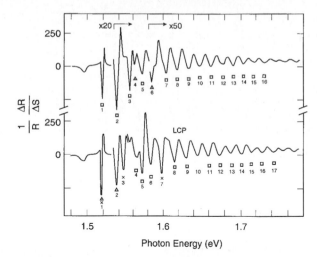

FIGURE 42 Stress-modulated magnetoreflectance spectra for the fundamental edge in epitaxially-grown high-purity ⟨211⟩ GaAs at $T \sim 30$ K, with $\Delta S = 5 \times 10^{-5}$, observed in the Faraday configuration with magnetic field **H** ∥ [11$\bar{2}$] and $H = 88.6$ kG. The number directly below each prominent transition refers to the identification of the transitions, Δ, LHa; x, LHb; □, HH(AV).[77]

detected, provide several orders of magnitude enhancement in the sensitivity for observing resonant transitions. This is especially important for the observation of higher energy transitions lying far away from the energy of the fundamental gap. Figures 42[77] and 43[78] show the stress-modulated magnetoreflectance spectra for the fundamental edge and the split-off valence to conduction-band edge in GaAs at $T \approx 30$ K and $B = 88.6$ kG. Again, quantitative information about the split-off can be obtained. Quantitative interpretation of the fundamental edge data must involve the effect of excitons on the transitions.

Faraday Rotation. A plane-polarized wave can be decomposed into two circularly polarized waves. The rotation of the plane of polarization of light as it propagates through the semiconductor in a direction parallel to an applied magnetic field is called the Faraday effect, or Faraday rotation. The amount of rotation is usually given by the empirical law $\Theta = V \mathbf{B} l$, where Θ is the angle of rotation, V is the Verdet coefficient, **B** is the magnetic field value, and l is the thickness. The Verdet coefficient is temperature-, wavelength-, and sometimes field-dependent. The Faraday effect can then be understood in terms of space anisotropy effects introduced by the magnetic field upon the right and left circularly polarized components. The refractive indices and propagation constants are different for each sense of polarization, and a rotation of the plane of polarization of the linearly polarized wave is observed. The sense of rotation depends on the direction of the magnetic field. Consequently, if the beam is reflected back and forth through the sample, the Faraday rotation is progressively increased. When measurements are thus made, care must be taken to avoid errors caused by multiple reflections. Faraday rotation may also be considered as birefringence of circularly polarized light.

If absorption is present, then the absorption coefficient will also be different for each sense of circular polarization, and the emerging beam will be elliptically polarized. Faraday ellipticity specifies the ratio of the axes of the ellipse.

Faraday rotation can be observed in the Faraday configuration with the light beam propagating longitudinally along the **B**-field direction. Light propagating transverse to the field direction is designated as the Voigt configuration. Two cases must be distinguished—

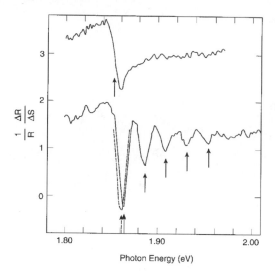

FIGURE 43 Stress-modulated magnetoreflectance spectra for the split-off valence-to-conduction band transitions in GaAs at $T \approx 30\,\text{K}$ with $\Delta S = 5 \times 10^{-5}$ in the Faraday configuration for the same conditions as for Fig. 42. The zero-field spectrum is displaced upward for clarity. For bottom curve, $H = 88.6\,\text{kG}$, RCP is solid line, and LCP is dashed line.[78]

the incident beam may be polarized so that its **E** field is either parallel or perpendicular to **B**. The Voigt effect is birefringence induced by the magnetic field and arises from the difference between the parallel and perpendicular indices of refraction in the transverse configuration. It is usually observed by inclining the incident plane-polarized radiation with the electric vector at 45° to the direction of **B**. The components resolved parallel and perpendicular to **B** then have different phase velocities, and recombine at the end of the sample to give emerging radiation which is elliptically polarized. Measurements of this ellipticity then determine the Voigt effect.

The interband Faraday effect is a large effect and is therefore useful for characterizing semiconductors. For frequencies smaller than the frequency corresponding to the energy gap, the interband Faraday effect arises from the dispersion associated with the interband magnetoabsorption. In this region, it has been used to determine energy gaps and their pressure and temperature dependence. Since the beam propagates through the crystal in a transparent region of the spectrum, it may be attractive to use in certain applications. Figure 44 shows the Verdet coefficient versus wavelength for GaAs at two temperatures, 298 and 77 K, and for **B** = 20 kG.[79] At long wavelengths it has a positive rotation, while near the gap the Verdet coefficient becomes negative. For frequencies equal to or larger than the frequency corresponding to the energy gap, the Faraday rotation is dominated by the nearest magneto-optical transition. Oscillatory behavior, like that seen in the magnetoabsorption, is also often observed.

Diluted magnetic semiconductors (DMS) are a class of materials that have been attracting considerable scientific attention. Any known semiconductor with a fraction of its constituent ions replaced by some species of magnetic ions (i.e., ions bearing a net magnetic moment) can be defined as a member of this group. The majority of DMS studied so far have involved Mn^{+2} ions embedded in various II–VI hosts. One important aspects of DMS controlling their optical properties is the interaction between the localized

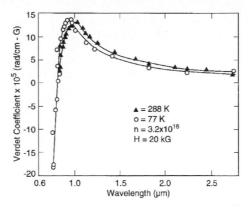

FIGURE 44   Interband Faraday rotation in GaAs.[79]

magnetic moments of Mn^{+2} and the conduction- and/or valence-band electrons (referred to as the sp-d interaction) which results in features unique to DMS. The best known (and quite spectacular) of these are the huge Faraday rotations of the visible and near-infrared light in wide-gap DMS. The origin of the large rotations is the sp-d exchange interaction which makes the band structure much more sensitive to the strength of external magnetic fields than in ordinary semiconductors. Figure 45 shows the Faraday effect in $Cd_{1-x}Mn_xSe(x = 0.25)$ at $T = 5$ K.[80] Each successive peak represents an additional Faraday rotation of 180°. The absolute Faraday rotation as a function of frequency is determined by monitoring the transmission at several fixed photon energies as the magnetic field is swept.

Excitonic Magneto-optical Effects.   As described earlier, a free exciton consists of an electron and hole bound together electrostatically. When the pair has an energy less than

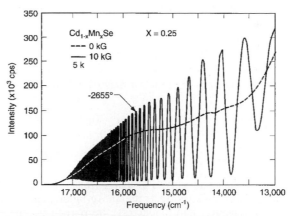

FIGURE 45   Faraday effect in $Cd_{1-x}Mn_xSe$ $(x = 0.25)$, $T = 5$ K. The dashed line represents the transmission of light as a function of its frequency for a 3-mm-thick sample located between two polaroids with their axes at 45° and at zero magnetic field. The light propagates along the optic axis, $\hat{c}$. The oscillations occur when a 10-kG magnetic field is applied along $\hat{c}$.[80]

that of the energy gap, they orbit around each other. If the orbital radius is large compared with the lattice constant, they can be approximately treated as two point charges having effective masses and being bound together by a Coulomb potential that gives rise to a hydrogen-atom-like behavior. In the presence of a magnetic field, excitons give rise to Zeeman and diamagnetic effects analogous to those in atomic spectra. Fine structure can occur due to motions other than the simple orbiting of an electron and hole—the carriers can have intrinsic motion, motion around an atom, spin motion, and motion of the complete exciton through the lattice. Some of these motions may even be coupled together.

Bound excitons (or bound-exciton complexes) or impurity-exciton complexes are extrinsic properties of materials. Bound excitons are observed as sharp-line optical transitions in both photoluminescence and absorption. The bound exciton is formed by binding a free exciton to a chemical impurity atom (or ion), a complex, or a host lattice defect. The binding energy of the exciton to the impurity or defect is generally smaller than the free-exciton binding energy. The resulting complex is molecularlike (hydrogen-molecule-like), and bound excitons have many spectral properties analogous to those of simple diatomic molecules.

The application of a magnetic field to samples where excitonic features are observed in the absorption spectra results in line splittings, energy shifts, and changes in linewidths. These arise from diamagnetic and Zeeman effects just as in atomic or molecular spectroscopy. The treatment of the problem of an exciton in a magnetic field in zincblende-type structures is difficult due to the complexity of the degenerate valence band. Often a practical solution is adopted that corrects the interband model calculations for exciton binding energies that are different for each Landau level. Elliott and Loudon showed that the absorption spectrum has a peak corresponding to the lowest $N = 0$ hydrogen-like bound state of the free exciton, which occurs below the free interband transition by the exciton binding energy E_B.[81] Weiler suggests that for transitions to the conduction-band Landau level n, the exciton binding energy E_B can be approximated by[73]

$$E_B(n) \approx 1.6R[\gamma_B/(2n+1)]^{1/3} \tag{54}$$

where R is the effective Rydberg,

$$R = R_0\mu/m_0\epsilon(0) \tag{55}$$

$R_o = 13.6$ eV, $\epsilon(0)$ is the static-dielectric constant, μ is the reduced effective mass for the transition, and γ_B is the reduced magnetic field

$$\gamma_B = m_o S/2\mu R \tag{56}$$

and

$$S = \hbar eB/m_o \tag{57}$$

Thus after calculating the interband transition energy for a particular conduction-band Landau level n, one subtracts the above binding energy to correct for exciton effects.

Nondegenerate semiconductors, in particular, materials belonging to the wurtzite crystal structure, have been extensively studied both with and without a magnetic field. Exciton states in CdS have been studied by high-resolution two-photon spectroscopy in a magnetic field using a fixed near-infrared beam and a tunable visible dye laser.[82,83] Figure 46 shows the photoconductive response versus total photon energy near the A-exciton region.[82] The two-photon transitions involve P states and both 2P and 3P states are clearly seen. As the field is increased, both Zeeman splittings and diamagnetic shifts occur. Figure

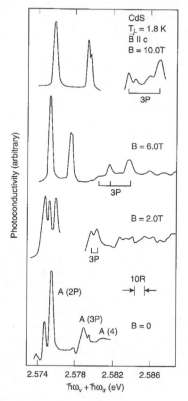

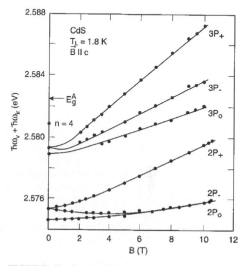

FIGURE 46 Photoconductivity vs. total photon energy $\hbar\omega_v + \hbar\omega_{ir}$ near the A-exciton region in CdS platelets for various magnetic fields. The magnetic field was parallel to the hexagonal c axis in a Voigt configuration with $\mathbf{E}$ perpendicular to c for the two photons at a lattice temperature of $T_L = 1.8$ K. The instrumental resolution $R = 0.1$ meV is narrower than the intrinsic linewidths.[82]

FIGURE 47 Peak positions, in total photon energy $\hbar\omega_v + \hbar\omega_{ir}$ for the $2P$ and $3P$ A excitons in CdS platelets as a function of applied $\mathbf{B}$ field. The solid points were determined experimentally and the solid curves are theoretically obtained from variational calculations of the diamagnetic shifts along with use of the experimental g factors.[82]

47 shows both experimental and theoretical transition energies versus $\mathbf{B}$ field.[82] Excellent agreement is obtained by using variational calculations that have been successfully used to describe impurity atoms in a magnetic field.

Intraband or Free-carrier Effects

Cyclotron Resonance. The simplest and most fundamental magneto-optical effect, cyclotron resonance absorption of free carriers, provides a direct determination of carrier effective masses. Classically, it is a simple phenomenon—charged particles move in circular orbits (in planes perpendicular to the direction of the magnetic field) whose radii increase as energy is absorbed from the applied electric fields at infrared or microwave frequencies. After a time τ, a collision takes place and the absorption process begins again. From the resonance relation $\omega_c = e\mathbf{B}/m^*$, extensive and explicit information about the effective masses and the shape of energy surfaces near the band extrema can be obtained. Excellent reviews of cyclotron resonance have been previously published by McCombe and Wagner[84,85] and Otsuka.[86]

A classical example of the explicit band structure information that can be obtained

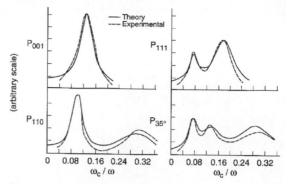

FIGURE 48 Microwave absorption in intrinsic n-type Ge at 4.2 K for four different directions of **B** in the $(\bar{1}10)$ plane as a function of the magnetic field. $P_{35°}$ represents the absorption with **B** at 35° to the [001] axis.[87]

from cyclotron resonance experiments is given for Ge. Figure 48 shows the microwave absorption for n-type germanium at 4.2 K for four different **B** field directions, each peak corresponding to a specific electron effective mass.[87] Figure 49 shows the orientation dependence of the effective masses obtained from the cyclotron resonance experiments which demonstrates that there is a set of crystallographically equivalent ellipsoids oriented along all $\langle 111 \rangle$ directions in the Brillouin zone.[88] Figure 50 shows the band structure of Ge that these measurements helped to establish[89]: a illustrates the conduction band minima along the $\langle 111 \rangle$ direction at the zone edge and b, the eight half-prolate ellipsoids of revolution or four full ellipsoids. The longitudinal and transverse masses are $m_l^* = 1.6 m_o$

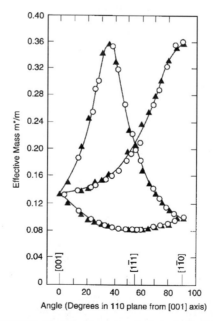

FIGURE 49 Effective mass of electrons in Ge at 4 K for magnetic field directions in a (110) plane.[88]

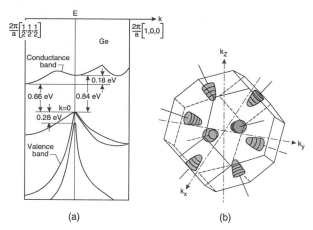

FIGURE 50 (*a*) Band structure of Ge plotted along the [100] and [111] directions; (*b*) ellipsoidal energy surface corresponding to primary valleys along the $\langle 111 \rangle$ directions.[89]

and $m_i^* = 0.082 m_o$, respectively. Both holes and electrons could be studied by using light to excite extra carriers. This illustrates the use of cyclotron resonance methods to obtain band structure information. Since that time, numerous experiments have been carried out to measure effective mass values for carriers in various materials.

The beginning of modern magneto-optics in which "optical" as opposed to "microwave" techniques were used, began in 1956 with the use of infrared frequencies at high magnetic fields. Far-infrared lasers are extremely important for modern-day measurements of cyclotron resonance as seen in Fig. 51 for *n*-type InSb.[90] At low temperatures, only the lowest CR transition C_1 (0^+ to 1^+) is seen. Raising the temperature populates the higher-lying Landau levels, and other CR transitions are seen at different fields because of the nonparabolicity of the conduction band which gives rise to an energy-dependent effective mass. At 13 K, a second transition C_2 (0^- to 1^-) is seen and at 92 K, C_3 (1^+ to 2^+). The low field feature denoted by *I* is called impurity cyclotron resonance because, although it is a neutral donor excitation, its appearance resembles that of the regular CR magnetoabsorption. It results from neutral donors exhibiting a Zeeman transition (1s→2p+), the transition energy of which is much larger than the ionization energy. The *I* signal gradually disappears as the temperature increases because the donors become ionized.

Cyclotron resonance also serves as a valuable tool for materials characterization through making use of the cyclotron resonance linewidth and intensity. Resonance linewidths can differ considerably from sample to sample. This difference in linewidth is attributed to differences in impurity content, with the higher-purity samples having the narrowest linewidths and largest intensities. Figure 52 shows the electron CR signals for both *n*-type and *p*-type GaAs crystals with low compensation.[86] The observed large difference in linewidths is primarily considered to reflect the difference in the electron-donor and electron-acceptor scattering rates.

An electron placed in the conduction band of a polar insulator or semiconductor surrounds itself with an induced lattice-polarization charge. The particle called a polaron consists of the electron with its surrounding lattice-polarization charge. The term *magnetopolaron* is also often referred to as a polaron in a magnetic field. Landau-level energies of these magnetopolarons are shifted relative to those predicted for band electrons. These energy shifts give rise to polaron effects that are most clearly evident in optical experiments such as cyclotron resonance. The recent review by Larsen provides an annotated guide to the literature on polaron effects in cyclotron resonance.[91]

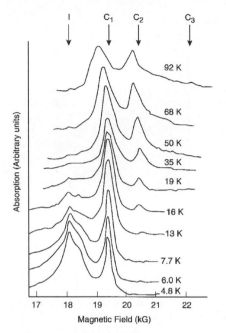

FIGURE 51 Thermal equilibrium resonance traces in n-type InSb at various temperatures. At 4.8 K, only the lowest cyclotron transition $C_1(0^+ \rightarrow 1^+)$ and impurity cyclotron resonance I (ICR) are visible. On raising the temperature, the signal I disappears on complete ionization of donors, while the second and third cyclotron transitions $C_2(0^- \rightarrow 1^-)$ and $C_3(1^+ \rightarrow 2^+)$ start to show up.[90]

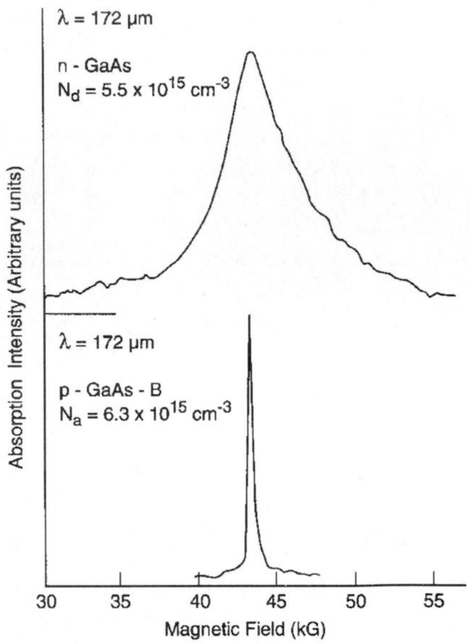

FIGURE 52 Difference in electron cyclotron resonance linewidth between n- and p-type GaAs crystals, having the same order of donor or acceptor concentrations.[86]

Free-carrier Faraday Rotation and Other Effects. Observation of free-carrier rotation was first reported in 1958. It is best understood as the differential dispersion of the cyclotron resonance absorption, and as such, it is an accurate method for determining carrier effective masses. It can be measured off resonance and detected under conditions which preclude the actual observations of cyclotron resonance absorption. Cyclotron resonance is a more explicit technique which enables carriers of different mass to be determined by measuring different resonant frequencies. The Faraday effect is an easier and more flexible technique, but is less explicit, since the dispersion involves the integral of all CR absorption. Also, it is unable to detect any anisotropy of the effective mass in a cubic crystal.

Other free-carrier effects are ellipticity associated with the Faraday rotation, the Voigt effect, and the magnetoplasma effect on the reflectivity minimum.

Impurity Magnetoabsorption. Impurity states forming shallow levels, i.e., those levels separated from the nearest band by an energy much less than the gap E_g, can be described by the effective mass approximation. This simple model for impurities is that of the hydrogen atom with an electron which has an effective mass m^* and the nuclear charge reduced to e/ϵ_∞, by the high-frequency dielectric constant of the crystal. This hydrogen-atom-like model leads to a series of energy levels leading up to a photoionization continuum commencing at an energy such that the electron (hole) is excited into the conduction (or valence) band.

The effect of a magnetic field on impurity levels and the related optical transitions is one of the most important tools for the study of the electronic states. The reason for this is that a magnetic field removes all degeneracies including the Kramers' degeneracy due to time reversal. It can produce new quantization rules, and it is a strong perturbation which is not screened like the Coulomb interactions.

For donor states with an isotropic conduction-band minimum, as in GaAs or InSb, the Hamiltonian acquires terms which are linear (Zeeman term) and quadratic (diamagnetic) in magnetic field. In this case, perturbation theory is appropriate to use until one considers transitions to very high quantum numbers or very high fields. Often complicated variational procedures must be used for accurate solutions. In the high field limit where $\hbar\omega_c/R \gg 1$ or for high energies, the impurity level quantum numbers do not provide a proper classification, and instead, the states should be related to the continuum Landau levels. The effect of a magnetic field on acceptor impurity states is more complicated because of band degeneracy and of spin-orbit coupling. Also for acceptors, one can be in the high field limit at relatively small values of the field for shallow levels and for excited states.

High-resolution photoconductivity measurements in high-purity GaAs indicate that the behavior of shallow donors in GaAs deviates from that predicted by the simple hydrogenic model. Figure 53 shows that under high resolution, the single ($1s \rightarrow 2p$) transition observed at lower resolution actually consists of several different unresolved transitions.[92] The photoconductive response is shown for two different high-purity GaAs samples. The magnetic field range shown covers about 1 kG at around 15 kG. The different transitions labeled A, B, C, and D each correspond to a different donor species, and variation of the amplitude of these transitions in different samples results from the different relative concentrations of the particular donor species. Thus, the use of magnetic fields in studying the absorption properties of shallow impurities has opened up a new analytical method of impurity characterization. Because of the central-cell corrections (chemical shifts), different impurities of the same type (donor or acceptors) can be distinguished and identified by back-doping experiments.

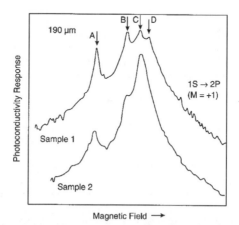

FIGURE 53 The photoconductivity due to the ($1s \rightarrow 2p, m = +1$) transition as a function of magnetic field for two different samples when excited by 190 μm laser radiation. For sample 1, $N_D = 2.0 \times 10^{14}$ cm^{-3}, $N_A = 4.0 \times 10^{13}$ cm^{-3}, $\mu_{77K} = 153{,}000$ cm^2/V-s. For sample 2, $N_D = 4.3 \times 10^{13}$ cm^{-3}, $N_A = 2.5 \times 10^{-13}$ cm^{-3}, and $\mu_{77K} = 180{,}000$ cm^2/V-s.[92]

Nonlinear Optical Properties of Semiconductors

Background. The study and characterization of nonlinear optical properties of semiconductors are increasingly important topics for research and development, with new effects being discovered and useful nonlinear optical devices being constructed. The underlying cause of these nonlinear optical effects lies in the interaction of electromagnetic radiation with matter. Each nonlinear optical process may be thought to consist of the intense light first inducing a nonlinear response in a medium and then the medium in reacting, modifying the optical fields in a nonlinear way. There is a wide variety of nonlinear optical phenomena in semiconductors, leading to many papers on the subject of nonlinear properties. It is thus not possible here to do justice to this field. We refer the reader to Tang's article on Nonlinear Optics in this Handbook (Vol. II, Chap. 38).[93] Here, we give a brief overview of the nonlinear optical processes starting from Maxwell's equations and describe and categorize some important second- and third-order nonlinear processes and properties.

Theoretical Overview of Nonlinear Optical Processes and Properties

Maxwell's Equations and Polarization Power Series Expansion. All electromagnetic phenomena are governed by Maxwell's equations for the electric and magnetic fields $\mathbf{E}(\mathbf{r}, t)$ and $\mathbf{B}(\mathbf{r}, t)$ or by the resulting wave equations

$$\left[\nabla \cdot \nabla + \frac{1}{c^2} \frac{\partial^2}{\partial t^2} \right] \mathbf{E}(\mathbf{r}, t) = \frac{-4\pi}{c^2} \frac{\partial^2}{\partial t^2} \mathbf{P}(\mathbf{r}, t) \tag{58}$$

$$\nabla \cdot \mathbf{E}(\mathbf{r}, t) = -4\pi \nabla \cdot \mathbf{P}(\mathbf{r}, t)$$

Here, $\mathbf{P}$ is the generalized electric polarization which includes not only the electric dipole part but all the multiple contributions. In general, $\mathbf{P}$ is a function of $\mathbf{E}$ which describes fully the response of the medium to the field. It is often known as the constitutive equation since all optical phenomena would be predictable and easily understood from it and its solution for the resulting set of Maxwell's equation with appropriate boundary conditions. Unfortunately, this equation is almost never possible to solve exactly, and physically reasonable approximations must be resorted to for progress to occur.

Most nonlinear optical properties can be described in terms of a power series expansion for the induced polarization. (This assumes that $\mathbf{E}$ is sufficiently weak.) Since lasers are most often used to observe nonlinear optical effects, one usually deals with the interaction of several monochromatic or quasi-monochromatic field components, and $\mathbf{E}$ and $\mathbf{P}$ can be expanded into their Fourier components as

$$\mathbf{E}(\mathbf{r}, t) = \sum_i \mathbf{E}(\mathbf{q}_i, \omega_i), \qquad \mathbf{P}(\mathbf{r}, t) = \sum_i \mathbf{P}(\mathbf{q}_i, \omega_i) \tag{59}$$

where

$$\mathbf{E}(\mathbf{q}_i \, \omega_i) = \mathbf{E}(\omega_i) \exp\left(i\mathbf{q}_i \cdot \mathbf{r} - i\omega_i t\right) + \text{c.c.} \tag{60}$$

The induced polarization is usually written as

$$\mathbf{P}(\mathbf{q}_i, \omega_i) = \chi^{(1)}(\mathbf{q}_i, \omega_i) \cdot \mathbf{E}(\mathbf{q}_i, \omega_i) + \sum_{j,k} \chi^{(2)}(\mathbf{q}_i = \mathbf{q}_j + \mathbf{q}_k, \omega_i = \omega_j + \omega_k) : \mathbf{E}(\mathbf{q}_j, \omega_j)\mathbf{E}(\mathbf{q}_k, \omega_k)$$

$$+ \sum_{j,k,l} \chi^{(3)}(\mathbf{q}_i = \mathbf{q}_j + \mathbf{q}_k + \mathbf{q}_l, \omega_i = \omega_j + \omega_k + \omega_l) : \mathbf{E}(\mathbf{q}_j, \omega_j)\mathbf{E}(\mathbf{q}_k, \omega_k)\mathbf{E}(\mathbf{q}_l, \omega_l) + \cdots \tag{61}$$

It is, however, sometimes more convenient to use $\mathbf{E}(\mathbf{r}, t)$ and $\mathbf{P}(\mathbf{r}, t)$ directly instead of their Fourier components, especially when dealing with transient nonlinear phenomena.

In the electric dipole approximation, $\chi^{(n)}(\mathbf{r}, t)$ is independent of $\mathbf{r}$, or $\chi^{(n)}(\mathbf{k}, \omega)$ is independent of $\mathbf{q}$, and the equations become simpler to write and to work with. These $\chi^{(n)}$ are the susceptibilities, with $\chi^{(1)} =$ linear electric dipole susceptibility and $\chi^{(2)}(\chi^{(3)}) =$ nonlinear second-order (third-order) susceptibility tensor. Both absorptive and refractive effects can be described in terms of these complex electric susceptibilities, which have real and imaginary parts for each tensor element. These linear and nonlinear susceptibilities characterize the optical properties of the medium and are related to the microscopic structure of the medium. Knowledge of $\chi^{(n)}$ allows, at least in principle, to predict the nth-order nonlinear optical effects from Maxwell's equations. Consequently, much effort (both experimentally and theoretically) has gone into determining the $\chi^{(n)}$.

The definitions of the nonlinear susceptibilities in the literature vary and have led to some confusion. Shen reviews these definitions and the reasons for the confusion.[94] In addition to some intrinsic symmetries, the susceptibilities must obey crystallographic symmetry requirements. The spatial symmetry of the nonlinear medium imposes restrictions upon the form of the various $\chi^{(n)}$ tensors. Butcher has determined the structure of the second- and third-order tensors for all crystals.[95] One important consequence is that for media with inversion symmetry, $\chi^{(2)} \equiv 0$, and thus $\chi^{(3)}$ represents the lowest-order nonlinearity in the electric-dipole approximation. Of the 12 nonzero elements, only 3 are independent. These susceptibility tensors must transform into themselves under the point group symmetry operations of the medium.

It is often convenient to discuss the various optical processes which might occur in terms of whether they are active or passive. Passive processes involve energy or frequency conservation, and the material medium acts basically as a catalyst. The susceptibilities are predominantly real for passive processes. Of course, as resonances are approached, susceptibilities become complex and may even become totally imaginary. These passive nonlinear optical phenomena are listed in Table 9.[96] Active nonlinear optical phenomena

TABLE 9 Passive Nonlinear Optical Phenomena[96]

Frequencies of incident fields	Frequencies of fields generated by the polarization of the medium	Susceptibility	Process (acronym)
ω_1	No polarization	$0(\epsilon = 1)$	Vacuum propagation (VP)
ω_1	ω_1	$\chi^{(1)}(\omega_1; \omega_1)$	Linear dispersion (LD)
ω_1, ω_2	$\omega_3[\omega_3 = \omega_1 + \omega_2]$	$\chi^{(2)}(\omega_3; \omega_1, \omega_2)$	Sum mixing (SM)
ω_1	$\omega_3[\omega_3 = 2\omega_1]$	$\chi^{(2)}(\omega_3; \omega_1, \omega_1)$	Second-harmonic generation (SHG)
$\omega_1, 0$	ω_1	$\chi^{(2)}(\omega_1; \omega_1, 0)$	Electro-optic linear Kerr effect (EOLKE)
ω_1	$\omega_2, \omega_3[\omega_1 = \omega_2 + \omega_3]$	$\chi^{(2)}(\omega_2; -\omega_3, \omega_1)$	Difference-frequency mixing (DFM)
ω_1	$\omega_2[\omega_1 = 2\omega_2]$	$\chi^{(2)}(\omega_2; -\omega_2, \omega_1)$	Degenerate difference-frequency (DDF)
ω_1	0	$\chi^{(2)}(0; -\omega_1, \omega_1)$	Inverse electro-optic effect (IEOE)
$\omega_1, \omega_2, \omega_3$	$\omega_4[\omega_4 = \omega_1 + \omega_2 + \omega_3]$	$\chi^{(3)}(\omega_4; \omega_1, \omega_1, \omega_1)$	Third-harmonic generation (THG)
ω_1, ω_2	$\omega_3, \omega_4[\omega_1 + \omega_2 = \omega_3 + \omega_4]$	$\chi^{(3)}(\omega_3; -\omega_4, \omega_1, \omega_2)$ $\chi^{(3)}(\omega_4; -\omega_3, \omega_1, \omega_2)$	Four-wave difference-frequency mixing
ω_1	$\omega_2, \omega_3, \omega_4[\omega_1 = \omega_2 + \omega_3 + \omega_4]$	$\chi^{(3)}(\omega_2; -\omega_3, -\omega_4, \omega_1)$	processes (FWDFMP)
ω_1	ω_1	$\chi^{(3)}(\omega_1; \omega_1, -\omega_1, \omega_1)$	Intensity-dependent refractive index (IDRI)
$\omega_1, 0$	ω_1	$\chi^{(3)}(\omega_1; 0, 0, \omega_1)$	Quadratic Kerr effect (QKE)

$\omega = 0$ indicates the presence of a uniform electric field.

TABLE 10 Active Nonlinear Optical Phenomena[96]

Susceptibility	Process
$\chi^{(1)}(\omega_1; \omega_1)$	Linear absorption ($\omega_1 \approx \omega_{10}$)
$\chi^{(3)}(\omega_2; \omega_1, -\omega_1, \omega_2)$	Raman scattering ($\omega_2 \approx \omega_1 \mp \omega_{10}$)
$\chi^{(3)}(\omega_1; \omega_1, -\omega_1, \omega_1)$	Two-photon absorption ($2\omega_1 \approx \omega_{10}$) or
	Saturable absorption ($\omega_1 \approx \omega_{10}$)
$\chi^{(5)}(\omega_2; \omega_1, \omega_1, -\omega_1, -\omega_1, \omega_2)$	Hyper-Raman scattering ($\omega_2 \approx 2\omega_1 \mp \omega_{10}$)

are listed in Table 10.[96] In general, energy is exchanged between the radiation and the material only for the active processes. We also note that second-order effects are always passive.

Second-order Nonlinear Optical Properties. Most existing nonlinear optical devices are based upon second-order nonlinear optical effects that are quite well understood. Here, we assume the presence of only three quasi-monochromatic fields

$$\mathbf{E} = \mathbf{E}(\omega_1) + \mathbf{E}(\omega_2) + \mathbf{E}(\omega_3) \tag{62}$$

and

$$\omega_1 = |\omega_2 \mp \omega_3| \tag{63}$$

Thus, Eq. (58) can be decomposed into three sets of equations for each $\mathbf{E}(\omega_i)$. They are then nonlinearity coupled with one another through the polarizations

$$\mathbf{P}(\omega_1) = \chi^{(1)}(\omega_1) \cdot \mathbf{E}(\omega_1) + \chi^{(2)}(\omega_1 = |\omega_2 \pm \omega_3|) : \mathbf{E}(\omega_2)\mathbf{E}(\omega_3) \tag{64}$$

The second-order nonlinear processes are then described by the solutions of the coupled-wave equations with the proper boundary conditions. $\chi^{(2)} = 0$ for materials with a center of inversion. The coefficient $\chi^{(2)}$ is a third-rank tensor. Some second-order processes include sum- and difference-frequency mixing, the electro-optic linear Kerr effect, the inverse electro-optic effect, parametric amplification and oscillation, and second-harmonic generation. The past emphasis has been on finding new nonlinear crystals with a large $\chi^{(2)}$. Semiconductor crystals have received much attention: III-V compounds like GaAs and InSb, II-VI compounds like ZnS and CdSe, I-III-VI compounds like $AgGaS_2$ and $CuInS_2$, and II-IV-V compounds like $CdSiAs_2$ and $ZnGeP_2$.

In most applications of second-order nonlinear optical effects, it is important to achieve phase-matching conditions

$$\Delta \mathbf{q} = \mathbf{q}_1 - \mathbf{q}_2 - \mathbf{q}_3 = 0 \tag{65}$$

where $\mathbf{q}_i$ is the wave vector of $\mathbf{E}(\omega_i)$. This ensures an efficient energy conversion between the pump field(s) and the signal field.

Third-order Nonlinear Optical Properties. In materials with inversion symmetry, third-order processes are the dominant nonlinearity. These processes are described by a fourth-rank nonlinear susceptibility tensor $\chi^{(3)}$ whose contribution to the polarization is given according to Eq. (61) by

$$\mathbf{P}(\omega_i) = \sum_{j,k,l} \omega^{(3)}(\omega_i = \omega_j + \omega_k + \omega_l) : \mathbf{E}(\omega_j)\mathbf{E}(\omega_k)\mathbf{E}(\omega_l) \tag{66}$$

In general, this nonlinearity will provide a coupling between four electromagnetic waves. Depending on whether the susceptibility tensor elements are real or imaginary and on whether some of the frequencies are identical or different, a large variety of physical phenomena can be understood and accounted for: third-harmonic generation, two-photon absorption, saturable absorption, intensity-dependent index of refraction, stimulated

Raman effect, anti-Stokes generation, stimulated Raleigh scattering, modulation of the index of refraction, and self-focusing of light. We concentrate on discussing only a few of the most pertinent cases of interest.

Third-harmonic Generation. Here, the output frequency $\omega_4 = \omega_1 + \omega_2 + \omega_3 = 3\omega_1$ since $\omega_1 = \omega_2 = \omega_3$. The polarization at the frequency ω_4 will generate radiation at the third-harmonic frequency. The quantum process responsible for the harmonic generation may be described as a scattering process in which three quanta at the fundamental frequency are annihilated and one quantum at the third-harmonic frequency is created. The system remains in the ground state, although three virtually excited states are involved in the scattering process. Since the phases are important, the process is actually an interference between many four-photon scattering processes.

Two-photon Absorption. Here, for example, $\omega_1 = -\omega_2 = \omega_3 = \omega_4$ and $\Delta \mathbf{k} = 0$, and the nonlinear polarization has components described by

$$\mathbf{P}(\omega_4) = \chi^{(3)}(\omega_4 = +\omega_1 - \omega_2 + \omega_3) : \mathbf{E}(\omega_1)\mathbf{E}^*(\omega_2)\mathbf{E}(\omega_3) \tag{67}$$

The nonlinear susceptibility $\chi^{(3)}$ is purely imaginary, but positive. One can define an absorption coefficient proportional to the intensity itself. In the important case of resonance, the sum of two frequencies of the exciting field is approximately equal to a transition frequency of the medium, $\omega_{ab} = 2\omega_1$, where $\hbar\omega_{ab}$ is the energy difference between two levels $|a\rangle$ and $|b\rangle$ with the same parity.

The TPA coefficient can be expressed in terms of a third-order nonlinear susceptibility tensor by solving the wave equation using the slowly varying amplitude approximation. The explicit expression for β is related to the imaginary part of the third-order electric dipole susceptibility tensor which depends upon the crystal class and laser electric field direction, for example, in crystals with $\bar{4}3$ m symmetry (e.g., $Hg_{1-x}Cd_xTe$, GaAs, InSb, etc.) and the electric field along the [001] direction,[97]

$$\beta = \frac{32\pi^2\omega}{n^2c^2}[3 \operatorname{Im} \chi^{(3)}_{1111}(-\omega, \omega, \omega, -\omega)] \tag{68}$$

where the convention used is that of Maker and Terhune (1965).[98]

Since $\chi^{(3)}$ is a second-rank tensor, there are, in general, nine terms contributing to β with magnitudes which vary with orientation. However, in most systems the symmetry is such that there are relations between certain of the terms. For example, in crystals with $\bar{4}3$ m symmetry, there are three possible values of χ for each β, and for crystals with m3m symmetry, there are four possible values. These can, in general, be measured by using both linearly and circularly polarized light. It is not always possible to sort out all the different TPA spectra simply by changing the sample orientation or the light polarization because of the competing process of absorption by second-harmonic-produced light.

Simulated Raman Scattering or Coherent Raman Spectroscopy. Levenson (1977) has written an excellent review of this area in *Physics Today*.[99]

Raman scattering provides a tool for the spectroscopic investigation of energy levels not accessible by the usual absorption or emission techniques. However, until the development of lasers, Raman scattering was a laborious and exotic technique. Now with coherent Raman techniques, the Raman modes are made to emit a beam of coherent radiation containing the details of the spectrum. The power in the beam can be many orders of magnitude larger than that in the spontaneously scattered radiation, and spatial filtering can be used to separate the output beam from the unwanted radiation.

Here, we can write

$$\mathbf{P}(\omega_4) = \chi^{(3)}(\omega_4 = \pm(\omega_1 - \omega_2) + \omega_3)\mathbf{E}(\omega_1)\mathbf{E}^*(\omega_2)\mathbf{E}(\omega_3) \tag{69}$$

The electric field with frequency components at ω_1 and ω_2 produces a force with Fourier components at $\pm(\omega_1 - \omega_2)$ which can then drive a Raman mode resonantly when

$|\omega_1 - \omega_2| \approx \omega_R$. The polarization in Eq. (66) acts as a source term in Maxwell's equation to produce the output beam at ω_4. Besides this resonant contribution to $\chi^{(3)}$, there also can be nonresonant contributions to $\chi^{(3)}$. The resonant Raman term interferes constructively and destructively with the nonresonant background which can become a nuisance.

There are a variety of coherent Raman spectroscopic techniques which can be used in observing Raman spectra by means of nonlinear optical mixing. Each technique has its own set of advantages and disadvantages, and it is important to match the technique properly to the investigation.

CARS (Coherent Anti-Stokes Raman Scattering, or Spectroscopy). This is the most widely used technique of coherent Raman spectroscopy. In this case, two incident laser frequencies ω_1 and ω_2 are employed with $\omega_3 = \omega_1$, giving an output frequency $\omega_4 = +(\omega_1 - \omega_2) + \omega_3 = 2\omega_1 - \omega_2$. With ω_1 corresponding to the laser frequency in a spontaneous-scattering experiment and ω_2 to the Stokes-scattered photon, the output occurs at the corresponding anti-Stokes frequency ($\omega_1 - \omega_2 > 0$). An analogous technique with $\omega_1 - \omega_2 < 0$ is called Coherent Stokes Raman Spectroscopy, or CSRS.

Two lasers are focused into a sample with an angle between the beams which best fulfills the wave-vector-matching conditions for the overall three-wave mixing process $\Delta \mathbf{q} = 0$. In solids, this depends upon $\omega_1 - \omega_2$ and the dispersion of the index of refraction. The beam emerging from the sample at ω_4 is selected by means of filters or a monochromator, and its intensity is then detected. The difference frequency $\omega_1 - \omega_2$ is varied by tuning one or both lasers which gives the Raman spectrum of the sample in the output plotted as a function of $\omega_1 - \omega_2$. This is a potentially powerful technique with resolution limited by the laser linewidth. However, in solids, to be able to see weak spectroscopically interesting modes by coherent techniques, it becomes necessary to suppress the nonresonant background. Without this suppression, coherent Raman techniques are actually less effective than spontaneous scattering in detecting Raman spectra of condensed phases.

RIKES (Raman-induced Kerr Effect Spectroscopy). The RIKES technique offers a simple method for suppressing the background.[100] Here, $\omega_3 = -\omega_1$ and the driven vibration produces an intensity- and frequency-dependent birefringence which alters the polarization condition of a wave probing the sample. The wave which results from the coherently driven vibration is at the same frequency as one of the inputs, $\omega_4 = -\omega_2$, but polarization selection rules are employed to ensure that the radiated field is in a different state of polarization than the input laser. The wave-vector-matching condition is automatically fulfilled, and if the ω_1 beam is circularly polarized, the nonresonant background is eliminated. Early RIKES experiments employed a linearly polarized probe beam at ω_2 and a circularly polarized pump at ω_1. The probe intensity was blocked by a crossed polarization analyzer placed after the sample except when $|\omega_1 - \omega_2| = \omega_R$, where a small transmission was detected.

In practice, some birefringence due to strains in the sample or the optics leads to a small background transmission of the probe frequency even when the pump beam is blocked. However, this background can be used to enhance the sensitivity of the technique since the radiated field resulting from the coherent Raman process interferes constructively or destructively with the background. The resulting small change in transmitted intensity can be detected by modulating the pump laser and using electronic signal-processing techniques. In this case, background-free Raman spectra are obtained with a sensitivity limited only by fluctuations in the probe-laser intensity.

This technique can be used to separately determine the real and imaginary parts of the complex nonlinear polarization as well as the real and imaginary parts of the total nonlinear susceptibility tensor. RIKES also has an advantage in studying low-frequency modes when $\omega_1 - \omega_2$ approaches zero. The CARS wave-vector-matching condition requires collinear propagation, making separation of the signal beam difficult. No such geometrical restriction applies to RIKES, where polarization selection and spatial filtering easily extract the signal.

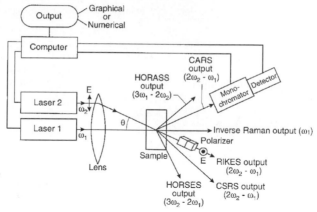

FIGURE 54 Typical layout for several coherent Raman experiments.[99]

Four-wave Mixing Techniques. Consider now the use of three lasers with $\omega_4 = \omega_1 - \omega_2 + \omega_3$ as before. Raman resonances can both occur when $|\omega_1 - \omega_2| = \omega_R$ and when $|\omega_3 - \omega_2| = \omega_R$. If $\omega_3 - \omega_2$ is set to a frequency at which the Raman contribution from one mode nearly cancels the nonresonant background, the sensitivity with which Raman modes are detected near $\omega_1 - \omega_2$ can be increased. Nonresonant background signals can also be completely eliminated with four-wave mixing using certain polarization configurations, like the "Asterisk." Thus, photons appear at the frequency ω_4 only when a Raman resonance condition exists.

A diagram of typical layouts for using coherent Raman techniques in nonlinear optical experiments is shown in Fig. 54. The directions of the output beams are given for CARS, Inverse Raman effect, RIKES, CSRS, HORASS (higher-order anti-Stokes scattering), and HORSES (higher-order Stokes effect scattering). In four-wave mixing, an additional beam would be added which is collinear with that from laser 1. The RIKES output has its electric-field vector polarized normal to the beam.

36.4 MEASUREMENT TECHNIQUES

Overview

The ability to measure the optical response of a semiconductor specimen precisely under well-controlled environmental conditions is of obvious importance in the determination of the optical properties of semiconductors. The significant resolution with which optical spectra can be measured makes it possible to perform precise determinations of the intrinsic properties such as the energy separation between electronic states, lattice vibration frequencies, as well as extrinsic properties due to impurities. The wealth of information that can be obtained and the direct relevance to device-related issues has led to much effort being expended in developing techniques and apparatus to perform a wide range of measurements. Over the past 30 years there has been an explosive growth in the scientific interest and technological applications of semiconductors. This chapter reviews the most widely used procedures and optical components employed in these investigations.

Most optical procedures are amenable for use on a small scale by a small group of investigators, in many cases by a single researcher. The costs and effort involved are

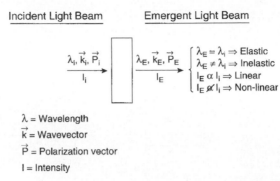

FIGURE 55 Schematic of interaction of light with semiconductors showing the linear, nonlinear, elastic, and inelastic processes.

reasonable when set in the proper perspective of scientific endeavors. Optical studies, except in rare instances, are contactless and noninvasive. These attractive features have led to widespread use of the techniques in both scientific analyses and, more recently, in manufacturing environments. Specific procedures and experimental apparatus and variations of them are too numerous to be dealt with in this brief chapter. The motivation here is, therefore, to set forth the essentials and dwell on the major aspects of each technique covered, as well as to provide some references which contain more details than could be given here.

The essence of spectroscopic investigations is to determine the interaction of a light beam, with a well-defined wavelength, intensity, polarization and direction, with a semiconductor specimen, in most cases from the point of view of the light beam. Figure 55 schematically displays this light-specimen interaction. Upon interacting, both the specimen and the light beam will change, and the experimental task is to precisely measure the change in the properties of the light beam. The changes may be classified into linear and the nonlinear regimes based on whether the response of the specimen is linear with respect to the incident power or not. The interactions may be elastic or inelastic; the light beam may also undergo a change in all aspects except its wavelength or photon energy in the former and the wavelength can also be modified in the latter. These terms arise from the elastic or inelastic interaction of the photon with the specimen where the incident photon energy is preserved or modified in the process. The incident light beam may be reflected, scattered, and transmitted by the specimen. In addition to these processes, a properly excited specimen may emit light as well. The last process, known as luminescence, may also be exploited to gain an insight into the physical behavior of the material.

The linear, elastic regime covers most of the procedures used to elucidate the equilibrium properties related to the optical constants, n and k, introduced previously. The techniques used are comprised of reflection and transmission spectroscopies where the energy reflection R and transmission T of the specimen are studied as function of the wavelength, polarization, and angle of incidence. The net energy absorption A may be determined from R and T as follows: $A = (1 - R - T)$.

As discussed earlier, a complete knowledge of R or T over a large range of energies is required to determine both n and k. Such a task is usually difficult. A more convenient and accurate procedure is to measure the change in polarization properties of an obliquely incident plane-polarized light beam after it interacts with the specimen. This procedure, called ellipsometry, uses the polarization change caused by the incident beam to extract n and k and, in many cases, information regarding overlayers.[101,102] The underlying principles and a more detailed discussion follow.

The nonlinear and inelastic spectroscopic procedures have been popular since the advent of lasers. The very large power densities achievable using lasers over the wide range of energies extending from the far-infrared (FIR) to the UV have driven the rapid developments in this field. At high excitation powers, the absorption of light energy can become superlinear due to the presence of higher-order interactions as discussed previously. Exploitation of these specific interactions as a means of gaining information regarding the specimen is the content of nonlinear spectroscopic techniques. On the other hand, the high degree of wavelength purity and coherence offered by lasers is exploited to perform inelastic spectroscopic analyses such as Raman[103] and Brillouin[103] spectroscopies. The crux of these techniques is to project a high-power highly coherent laser beam onto a specimen and observe the scattered part of the intensity. The scattered part will be dominated by light with the same wavelength as the incident laser beam; i.e., the elastic part, but a small part, usually $<10^{-8}$ of the original intensity, can be observed with well-defined frequency shifts. These additional frequency bands, similar to the sidebands that arise as a consequence of intensity modulation, can be analyzed to provide crucial information regarding the specimen. For instance, optical phonons can interact with the incoming laser photon and energy-shift it by an amount equal to a multiple of the phonon energy. Hence, an analysis of the frequency-shifted bands in the Raman spectrum can be used to establish phonon energies. The major task of Raman spectroscopy is to isolate the very weak frequency-shifted component in the scattered beam.[103]

Light emission is important for both spectroscopic analysis and device applications and, hence, has commanded a large amount of attention.[104,105] Any excited semiconductor will emit light as a means of relaxing to its equilibrium state. Under proper excitation, such as with above-band-gap radiation, the light emission can be made quite intense and can then be easily recorded and subjected to spectroscopic analyses. Within the specimen, the above-band-gap photoexcitation leads to a transition of the electron from a valence band to the conduction band, followed by a rapid process of thermalization whereby the excited electron and hole reach their respective band extrema and recombine from there radiatively, i.e., by emitting the potential energy in the form of a photon. The photoexcited free electrons and holes may also form an exciton, or interact with the impurity states in the forbidden band, or both. The final recombination can be mediated by a large number of such intermediate processes, and, hence, the luminescence spectra can display a very rich and complicated structure. The most important aspect of the luminescence spectrum is the fact that nearly all the interactions involve impurity and defect states in the forbidden band. Add to this the rapid thermalization and large self-absorption effects for emission of light with energies greater than the gap, and luminescence is almost entirely a subband-gap tool dominated by the impurities and defects present in the semiconductor.

The major categories of spectroscopic procedures are grouped by the energy range of photons used. The commonality of the instrumentation for a given wavelength range accounts for this categorization. The lowest energy region in the FIR spans the phonon energies and a variety of other possible excitations such as plasmons (free-carrier oscillations), magnons, and impurity-related electronic and vibrational excitations. The energy-band gap of semiconductors ranges from 0 for HgTe to $>5\,\text{eV}^{106}$ in diamond and can hence occur anywhere from the far-infrared to the vacuum ultraviolet. For the two most important electronic materials, namely Si and GaAs, the gaps occur in the near-infrared at 0.8 and 1.5 eV,[106] respectively. Hence, the mid- and near-infrared investigations have largely been confined to the study of impurity states within the forbidden gap. The electronic band transitions dominate the higher energies.

Instrumentation

The major components of a spectroscopic system are schematically displayed in Fig. 56. Light from the broadband source enters a monochromator fitted with a dispersion element

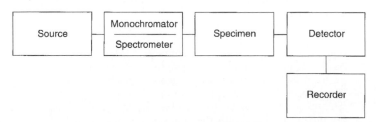

FIGURE 56   Schematic of a spectroscopic measurement apparatus.

that separates the various wavelength components and allows a chosen narrow spectral band of light to interact with the specimen. A detector converts the intensity information in the beam to an electrical or digital signal that can then be recorded by a computer or other recording device and then analyzed. Passive components such as lenses and mirrors, filters, polarizers, light pipes, and optical fibers, etc., that are needed to tailor the behavior of the system also form an integral part of the experimental apparatus. A short discussion of each of the major components follows.

Sources

Broadband.   The ideal broadband source should emit light with sufficient intensity in the wavelength band of interest, possess a stable output and exhibit a minimum amount of noise, and display a slowly varying spectral character, i.e., the source should not possess intense spectral features that will interfere with the measurement procedure.[107] All these characteristics can be satisfied by blackbodies, and if blackbodies with high enough temperatures can be fabricated, they would be ideal sources for any wavelength region. However, this is not possible since the operating temperatures required to obtain workable energy densities in the ultraviolet are extremely large. Hence, blackbody sources are usually restricted to wavelengths in the red-yellow region starting at ~500 nm or larger and ending at ~2000 μm in the FIR. The incandescent lamp with a hot filament is the best known BB source. Gas emission lamps such as high-pressure arc lamps and low-pressure discharge lamps are useful in the visible and ultraviolet region. Their main feature is the ability to produce a large intensity in the upper energy regions. However, the inherent atomic processes and associated line spectra that are present make using these somewhat complicated. Care should be exercised to avoid wavelength regions where intense spectroscopic features arise from the discharge medium. The unavoidable electronic activity in the discharge media can also be a source of noise.

Laser.   Laser sources are required in applications where large intensities are essential, as in the studies of nonlinear optical phenomena using Raman scattering or photoluminescence techniques. Lasers are currently available from the UV to the FIR and often both cw and pulsed operations are possible.[107] The argon and krypton ion lasers with emissions in the visible and near-infrared regions, the Nd:YAG with a 1.06-μm emission, and the CO_2 laser with emission in the 9.2- to 10.8-μm range have been the workhorses for a wide variety of semiconductor investigations.

Since efficient laser operation requires a set of excitable electronic or vibrational levels properly arranged to produce population inversion and sufficient amplification, intense laser emission is usually confined to specific wavelengths. However, the use of optically excited dye lasers and tunable solid-state lasers such as the Ti:sapphire lasers can be used to fill the wavelength regions in between most of the visible and near-infrared regions.

The semiconductor lasers that are currently available extend in wavelength from the red to the far-infrared. The III-V alloy-based double heterostructure lasers, fabricated from AL_xGa_{1-x} As and $In_{1-x}Ga_xAs$, are particularly efficient in the near-infrared region. The IV–VI alloy-based lasers, fabricated from $Pb_{1-x}Sn_xTe$, operate at considerably longer wavelengths of ~10 μm; a small range of emission wavelength tunability has been achieved based on the change of the band gap with the temperature. The intense interest in the development of blue-green laser emission is likely to lead to additional sources that will have substantial influence on future developments.

Spectrometers and Monochromators

Dispersion Spectrometers.   The monochromator or the spectrometer is the heart of the spectroscopic apparatus, and the dispersion element that analyzes the light is the main component of the monochromator.[108] The simplest and best known dispersion unit is the transparent prism. The physical mechanism that leads to the dispersion is the inherent dispersion in the optical refractive index n of the prism material over the wavelength range of interest. Prism-based monochromators work very well and are still employed. They possess a large throughput; i.e., they transmit a large fraction of the incident intensity but suffer from limited resolution since the degree of dispersion is restricted by the characteristics of the prism material.

The most widely used dispersion element is the diffraction grating which is a collection of finely spaced grooves or slits. Diffraction from the multiple slits leads to a dispersive action. The dispersion is given by the following relationship: $n\lambda = d \sin(\theta)$.

The fact that the degree of dispersion can be controlled by the slit spacing reduces the complexity of design and fabrication when compared to the prism. Gratings have a drawback in that they display multiple orders, and hence the throughput at any given order is likely to be very high. Blazing, i.e., control of the shape of the groove, can be used to increase energy in a given band in a particular order to reduce this shortcoming. Both transmission and reflection gratings can be fabricated as well as concave gratings that both disperse and focus the light beam. The major manufacturing flaw in grating fabrication that used mechanical devices was that groove spacing was not well controlled and hence the flawed grating led to the appearance of spectroscopic artifacts that were called "ghosts"! The advent of holographic grating fabrication procedures has eliminated these difficulties.

Fourier-transform Spectrometers.   An alternate method for performing spectroscopic measurements is the use of Michelson's interferometer.[109] Figure 57 displays the layout of the interferometer. The approach in this procedure is to divide the white-light beam from

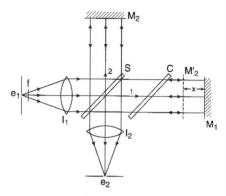

$$B(\omega) = \int_0^\infty [I(x) - I(\infty)] \cos(2\pi\omega x)\, dx$$

FIGURE 57 Schematic of the Fourier-transform interferometer.

the source into two wavefronts, introduce a path difference x between the two, recombine them, and record the interference-modulated intensity as a function of x. The recorded intensity variation $\mathbf{B}(x)$ is known as the interferogram, and a Fourier transformation of $I(x)$ will yield the spectral distribution of the white light. Experimentally, this is achieved as shown in Fig. 57. The incident beam from the source is collimated by lens L_1, split into two wavefronts by the beam splitter S. The two wave fronts are directed to a movable mirror, M_1 and a fixed mirror M_2. The reflected beams are combined at the detector where the intensity is recorded as a function of x, the path difference between M_1 and M_2. The spectrum $\mathbf{B}(\omega)$ is related to $I(x)$ by the following relationship:

$$B(\omega) = \int_0^\infty [I(x) - I(\infty)] \cos(2\pi\omega x) \, dx \tag{70}$$

This procedure of measuring the spectra is referred to as Fourier-transform spectroscopy and contains two major advantages: the throughput advantage and the multiplex advantage, both of which greatly add to the ultimate signal-to-noise ratios that can be achieved as compared to measurements performed with grating spectrometers under comparable conditions of illumination and recording times. The throughput advantage arises from the fact that improved resolution is not achieved at the expense of reducing slits and reduced throughput, and the multiplex advantage is a consequence of the fact that all wavelength channels are observed all the time as opposed to a one-channel-at-a-time measurement constriction in the dispersion-based instruments.

These significant advantages come with a price. The much larger signal intensities are likely to be seen by the detector and place stringent conditions on the detector performance in its dynamic range and linearity. Less-than-optimum performance may lead to significant distortions in the transformed spectrum that are not intuitively evident. The FT spectrometer was used first in the far-infrared, soon after the advent of high-speed computers that were capable of performing the Fourier transformations. However, the advances in the technology of designing and fabricating complicated optical elements and computer hardware have contributed greatly to the advancement of the field, and FT spectrometers are now available that cover a wide spectral region extending from FIR to the VUV.

Detectors. The photomultiplier (PMT)[110] is a widely used light detector that is a vacuum-tube-based device that uses a photoemitter followed by a large ($>10^5$) amplification stage so that very low signal levels can be detected. The wavelength band that the detector responds to is determined by the photocathode and window characteristics. Photomultipliers are particularly useful when low-light-level detection is needed as in Raman spectroscopy, but their use is largely confined to the visible and near-infrared as a consequence of the limitation in obtaining photoemitters for lower energies. Use of fluorescent phosphors can extend the upper working region of the PMTs to the VUV and beyond. Since they are vacuum-based devices, they are fragile and have to be handled with care.

Solid-state detectors[110] which are almost entirely fabricated from semiconductors have advanced to a state that, in many applications, they are preferable to PMTs. The simplest semiconductor detector is the photoconductor (PC) where absorption of an above-band-gap photon leads to an increase in conductivity. Semiconducting PC detectors are, therefore, sensitive to any radiation with an energy larger than the band gap. Since the band gap of semiconductors extends all the way from 0 in HgTe to >5 eV in diamond,[106] detectors that function over a very large energy range can be fabricated. Photovoltaic detectors, as the name implies, employ the photovoltaic effect in a p-n junction and can also detect above-band-gap radiation.

The explosive growth in semiconductor technology that has led to large-scale integration has benefited spectroscopic experimenters directly in the field of detectors. Imaging devices such as the CCD array have been incorporated in spectroscopy. The array detectors combined with a dispersion spectrometer can be used for observing multiple channels simultaneously. This has led to an advantage similar to the multiplex advantage in the Fourier spectrometer. In addition, the low noise levels present in these detectors, particularly when they are cooled, have had a large impact on high-sensitivity spectroscopic analysis. Commercially available CCDs are fabricated and, hence, have a lower energy limit of 0.8 eV, the band gap of Si. However, some linear and 2-D arrays, fabricated from InSb and HgCdTe, with longer wavelength response are beginning to become available.

Major Optical Techniques

Reflection and Transmission/Absorption. Measurements of the power reflection, transmission, and absorption, R, T, and A, respectively, are the simplest and most direct methods of spectroscopic analyses of semiconductor materials. The measurements are simple to perform so long as satisfactory spectroscopic apparatus is available in the wavelength region of interest. The major drawback is the less-than-satisfactory absolute accuracies with which the measurements can be performed.

The measurements are usually conducted at near-normal incidence for convenience; normal incidence measurements are difficult to perform and oblique incidence measurements are difficult to analyze. Once the major elements of the spectroscopic system—namely, the source, monochromator, and detector—have been chosen, the R measurements are obtained by directing the light beam on the specimen and measuring the incident and reflected intensities with the detector. The experimental R is determined by ratioing the incident and reflected power, I_0, and I_R, respectively.

$$R = \frac{I_R}{I_0} \tag{71}$$

The spectral behavior of the measurement apparatus will not affect the final result so long as they remain fixed during the observation of I_0 and I_R. A change in the source intensity or the response of the detector will be reflected in the measurement. Since the measurement conditions cannot be identical in the two measurements, this is an unavoidable source of error.

Several arrangements have been attempted to minimize such an error. The most direct approach is to hold the optical path fixed and instead of measuring I_0 and I_R, one measures I_{REF}, and I_R, where I_{REF} is the reflected intensity from a well-calibrated reference surface. A ratio of these two measurements leads to the following relationship:

$$\frac{I_R}{I_{REF}} = \frac{R}{R_{REF}} \tag{72}$$

A knowledge of R_{REF}, the known reflectivity of the reference surface, may then be used to extract R. This procedure works well and has been employed widely in the infrared region where reference surfaces from metal mirrors with ~100 percent reflectivity and stable blackbody sources are available. For the visible and higher-energy regions that use gas lamp sources that tend to be not as stable as blackbody sources, the reference method is not satisfactory. The procedure used to overcome the short-term variations in the lamp intensity was to perform the ratioing at each measurement wavelength. Such a "real-time" ratioing arrangement used a dynamical arrangement with a rotating optical element that directed the light beam alternately to the reference and the sample. Many such configurations were used, and a good example of a particularly ingenious arrangement is presented in Fig. 58. The essence of this rotating light pipe reflectometer[111] was the rotating light pipe that sampled the incident and reflected beam alternately at a frequency of ~70 Hz, a frequency large enough to remove any errors due to variations in the lamp intensity.

The use of reflectometers for the spectroscopic analysis of semiconductors in the visible and near-visible regions has been limited since the advent of the spectroscopic ellipsometer. However, they are still used at higher energies and in applications where their simplicity makes them attractive, for example, to monitor thin films in semiconductor device fabrication. The reflectivity spectra from several III-V semiconductors in the 0- to 12-eV energy range[57] were presented in Fig. 28. The energy range beyond 6 eV is not accessible by ellipsometers, but the data clearly display significant structure in the 6- to 12-eV VUV energy range.

Reflectometers are also currently used for the analysis of oxide overlayers on Si in the manufacture of integrated circuits and other semiconducting multilayer film measurements. The reflectivity spectrum obtained from three SiO_2 films on Si is presented in Fig. 59.[112]

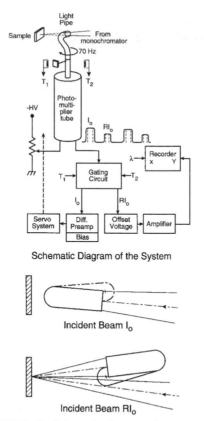

Schematic Diagram of the System

Incident Beam I_o

Incident Beam RI_o

FIGURE 58 Schematic diagram of the rotating light pipe reflectometer and expanded top view showing the geometry of the sample and bent light pipe.[111]

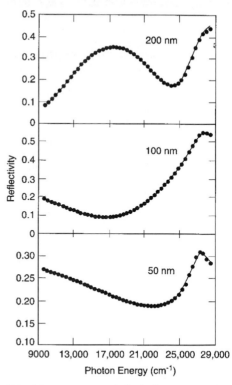

FIGURE 59 Measured (points) and calculated (line) optical reflectivity spectra of silicon dioxide films on silicon.[112]

The results of the computer fit are also displayed. The thicknesses determined from the reflection analyses agree to within ~0.5 percent of the values obtained from ellipsometric values. As illustrated in the figure, the change in the spectral shape from the 200-nm-thick film to 100-nm is much greater than that between the 100-nm and the 50-nm films which points to reducing sensitivity to film thickness determination as the films get thinner.

The techniques used to measure the transmission spectra are almost identical to those used for reflection except that nature provides a perfect reference, namely, the absence of the sample in the beam. The sample-in/sample-out reference intensity ratioing works satisfactorily. Since transmission measurements are mainly confined to energies below the forbidden gap and hence lower energies covered by blackbody sources, the difficulties faced by high-energy reflectivity analysis have not been as keenly felt.

The absorption of energy of a specimen may be determined from a knowledge of R and T. The most important absorption mechanism is that associated with the electronic transitions discussed previously. The absorption edge spectra for a number of semiconductors were reproduced in Fig. 14. The subband gap region has been studied extensively using absorption spectroscopy where the spectra are dominated by impurity-related effects. The absorption from the electronic transitions associated with a number of impurities in Si was presented earlier in Fig. 35. The impurity-related vibrational features and the lattice phonon bands can be observed as discussed under the section, "Lattice." The collective charge carrier oscillations and impurity to bandtype transitions may also be observed in this region.

Modulation Spectroscopy. A very useful variation of the reflection and transmission analysis is the use of modulation techniques that produce a periodic perturbation of the property of the specimen or the light beam and detect in-phase changes in R and T. The schematic diagram of a measurement apparatus is presented in Fig. 60.[113] The capacity of lock-in amplifiers for very large amplifications of greater than 10^5 with a concomitant reduction in broadband noise is the heart of the procedure. Modulations in R and T, namely, ΔR and ΔT, of 10^{-5} can be easily detected. This means the modulating perturbation can be quite small and accessible in routine experimental systems. In the simplest cases, the modulation response of a specimen to a property x may be expressed as follows:

$$\Delta R = \frac{dR}{dx} \cdot \Delta x \qquad (73)$$

where Δx is the intensity of modulation. The modulated property may be internal to the

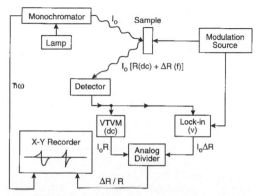

FIGURE 60 Block diagram of a typical modulation spectrometer.[114]

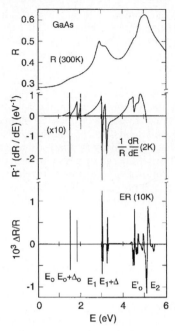

FIGURE 61 A comparison of three types of spectra from 0- to 6-eV for a typical semiconductor, GaAs. *top*: reflectance R; *middle*: energy-derivative reflectance; *bottom*: low-field electroreflectance.[114]

specimen, such as the temperature and pressure, or external, such as the wavelength or polarization of the probe. The most attractive feature, from a measurement point of view, is the fact that modulation spectra are derivative-like and hence suppress slowly varying background structure and emphasize features in the vicinity of the critical points in the electronic band structure. Figure 61 presents a comparison of the reflective spectrum of GaAs in the 0- to 6-eV region and a wavelength-modulated spectrum and an electric-field-modulated spectrum.[114] Note the intense oscillatory line shapes in both the WMR and ER spectra. For instance, the $E_0 + \Delta_0$, indistinguishable in R, is clearly observable in both WMR and ER. These techniques are crucial in providing accurate values of the interband transition energies, which results in a better understanding of the electronic band structure of a material.

A variety of modulation procedures, underlying mechanisms, measured specimen properties, and other salient features are presented in Table 11.[114] The techniques are broadly classified as internal or external modulation to signify if the perturbation is intrinsic to the measurement approach, as in the case of WMR, or is external, as in the case of ER where an externally applied electric-field modulation is needed at the specimen surface. Temperature, stress, and magnetic-field modulation are included for completion. Compositional modulation that compares two nearly identical alloy samples is less frequently used to investigate semiconductors. Spectroscopic ellipsometry, to be discussed later, may also be considered to be a polarization-modulation technique. Modulation techniques are useful in yielding crystal properties such as the electronic transition energies as well as information regarding the perturbation mechanism such as the electro-optic or magneto-optic effects.

A widely used and very useful form of a modulation technique for the study of

TABLE 11 Characteristics of Some Commonly Used Modulation Techniques[114]

Technique	Name	Type	Variable	Sample parameters affected	Lineshape type	Principal parameters measured	Principal advantages	Disadvantages
Wavelength modulation: energy derivative reflectance	Internal	Scalar	Wavelength λ; energy $\hbar\omega$	—	1st derivative	E_g, γ	Universal applicability; minimal sample preparation; fast, convenient	Measurement system can generate intrinsic structure, not easy to eliminate
Spectroscopic ellipsometry	Internal	Scalar	Wavelength λ; energy $\hbar\omega$	—	Absolute	Dielectric function	As energy derivative reflectance but on ϵ_1, ϵ_2 directly	Strongly influenced by surface preparation and thin films
Composition modulation	Internal	Scalar	Sample compared to control sample	—	Complicated	Doping, alloying effects	Obtains differences for parameters impossible to vary cyclically	Two samples/beam motion involved; alignment and surface preparation critical
Thermomodulation	External	Scalar	Temperature T	Threshold E_g; broadening parameters γ	1st derivative	E_g, γ, dE_g/dT, $d\gamma/dT$	Wide applicability; identifies Fermi-level transitions in metals	Slow response (1–40 Hz); broad spectra
Hydrostatic pressure	External	Scalar	Pressure P	Threshold E_g	1st derivative	Deformation potentials		Cannot be modulated; must be used in conjunction with another technique
Light modulation (photoreflectance) (photovoltage)	External	Scalar or tensor	Intensity I of secondary beam	Carrier concentration or surface electric field	Complicated	E_g	Convenient; minimal sample preparation required	Effect on material usually not well-defined
Uniaxial stress	External	Tensor	Stress X	Threshold E_g; matrix elements	1st derivative	E_g, γ symmetries; deformation potentials	Symmetry determination	Difficult to modulate; limited to high fracture/yield stress materials
Electric field	External	Tensor	Field E	Electron energy $E(k)$ oscillations	3rd derivative (low) Franz-Keldysh symmetries (high)	E_g, γ Effective masses in VUV; impurity concentrations	Very high resolution; only high-resolution technique usable	Requires certain resistivity ranges
Magnetic field	External	Tensor	Field **H**	Electron energy levels	Landau levels	E_g, effective masses	Extremely high resolution	Advantages realized only for lower conduction-band minima

semiconductors is the electric-field-modulated reflection spectroscopy, referred to as electroreflectance. The basis of the procedure is the electric-field-induced changes in the optical response.

The electric-field-induced perturbations can be treated in detail by considering the effects of the applied potential on the electronic band structure. For relatively weak fields, i.e., for field strengths not large enough to modify the band structure significantly, the major perturbation mechanism may be considered to be the acceleration of the electron to a successive set of momentum states. The perturbation to $\varepsilon(E, \mathbf{E})$, for such a simple case may be expressed as:[114]

$$\Delta\epsilon = \frac{(\hbar\Omega)^3}{3E^2}\frac{\partial^3}{\partial E^3}[E^2 \cdot \epsilon(E)] \tag{74}$$

where E is the energy and

$$(\hbar\Omega)^3 = (e^2\mathbf{E}^2\hbar^2)/(8\mu_{\parallel}\gamma^3) \tag{75}$$

e is the electronic charge, $\mu_{\parallel}$ the effective mass parallel to $\mathbf{E}$, and γ the energy broadening associated with the electronic level under consideration.

When the field is large, the electroreflectance spectra display oscillatory structure, known as Franz-Keldysh oscillations, in the energy vicinity of the electronic critical point transitions which may be expressed as[115]:

$$\frac{\Delta R}{R} \approx (\hbar\omega - E_g)^{-(d+1)/4} \cdot \exp\left[\frac{-\gamma(\hbar\omega - E_g)^{1/2}}{(\hbar\Omega)^{3/2}}\right] \times \cos\left[\frac{2}{3}\left(\frac{\hbar\omega - E_g}{\hbar\Omega}\right)^{3/2} - \frac{\pi}{4}(d-1)\right] \tag{76}$$

where d is an integer that is determined by the band structure near the critical point under consideration, and γ is the phenomenological broadening associated with the transition. In practical terms, the oscillations may be analyzed to determine $\hbar\Omega$ and, hence, the strength of the electric field that causes the perturbation.

A particularly attractive and widely used form of electroreflectance is the contactless form of electric-field-modulated reflectivity known as photoreflectance (PR).[116] In this procedure, one uses the electric-field modulation produced near the surface of the specimen by a chopped laser beam to modulate the reflectivity of a weak probe beam.

The sharp features in the PR spectrum associated with the critical points in the electronic density of states including the direct gap E_g may be employed to determine the transition's energies accurately. The information obtained using ER and PR was instrumental in leading to a detailed understanding of the band structure of semiconductors. The variation of the critical point energies with alloy composition has been used to understand the electronic behavior as well as characterize semiconductor alloys.[117] Similar studies have been used extensively in the study of microstructures, where PR is particularly useful since it allows the observation of the gap as well as several additional higher-energy transitions as shown in Fig. 62.[118] Distinct transition from both the well and the barrier region can be observed and, hence, a complete picture of the microstructure can be obtained. The electric-field-induced Franz-Keldysh oscillations in the ER and PR line shapes have been used to establish the electric field strengths,[115,119] and, more recently, a similar technique has been used to measure the electric field strength in the surface region of GaAs and the effects of passivation[119] as shown in Fig. 63. The electric field strength can be determined from the slope of the inset in Fig. 63. Since PR measurements can be performed at room temperature with minimal sample preparation, it is attractive for routine characterization. Commercial PR spectrometers are now available for use in semiconductor fabrication. Several excellent reviews of the applications of PR are available. A compilation of recent activity may be found in the Proceedings of the 1990

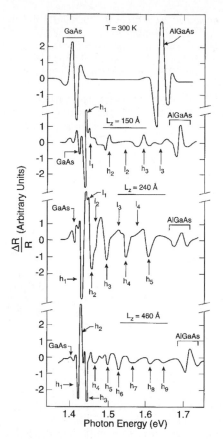

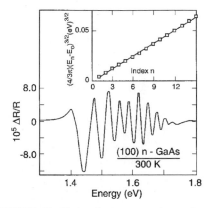

FIGURE 62 Room-temperature PR spectra for an undoped GaAs/Al$_x$Ga$_{1-x}$As heterojunction and three multiple quantum-well samples with $x \approx 0.2$, L$_z$ is the well thickness, and arrows labeled h$_1$, h$_2$, ... (l$_1$, l$_2$, ...) correspond to calculated values of interband transitions involving heavy (light) hole valence bands.[118]

FIGURE 63 Photoreflectance spectrum at room temperature with $I_{pump} = 3\,\mu W/cm^2$ and $I_{probe} = 2\,\mu W/cm^2$. The inset shows a plot of $(4/3\pi)(E_n - E_o)^{3/2}$ as a function of index n. The slope of the line in the inset yields the electrical field strength.[119]

International Conference on Modulation Spectroscopy[120] and the review by Glembocki and Shanabrook,[121] and the reader is referred to them for more details.

Ellipsometry. The reflection and transmission measurements discussed in the previous section consider the ratio of the incident to the reflected or transmitted optical power and hence ignore any information carried by the phase change suffered by the incident beam. A complete description of the reflection or transmission process will have to include the phase information as well. Ellipsometry attempts to obtain part of the phase information by measuring the phase difference, introduced upon reflection, between the normal components of obliquely incident plane-polarized light. Even though the absolute phase suffered by each component is not measured, the phase difference may be measured easily as an ellipticity in the polarization state of the reflected light and, hence, the name ellipsometry.

The intrinsic sensitivity of ellipsometry may be illustrated by considering a thin film with thickness d, refractive index n, and measurement wavelength λ. The phase change in the reflected component can be measured to a precision of 0.001° in ellipsometers which translates to a precision of $\sim 10^{-5}\lambda/2\pi n$ in thickness, i.e., $\sim 10^{-4}\lambda!$ or $\sim 10^{-4}d/\lambda$ in the refractive index. Such extraordinary precision may be used for the study of submonolayer films or as a real-time monitor for measuring very small changes in materials properties. The reader is referred to the chapter on ellipsometry[122] in this volume for more details.

Luminescence.   The process of luminescence, as described earlier, occurs in a suitably excited specimen. It is a mechanism through which the excited specimen relaxes to the equilibrium state.[104,105] Hence, unlike reflection and transmission spectroscopies, luminescence procedures concentrate on the relaxation of the specimen and often lead to complementary information. For instance, impurities in semiconductors (particularly at low concentrations) are impossible to detect through reflection and more difficult to detect by absorption than by luminescence. Luminescence spectroscopy is thus an important part of the analysis of the optical behavior of semiconductors.[104] Moreover, since one of the main applications of semiconductors is in the arena of light emitters including lasers, the study of luminescence provides direct access to device optoelectronic information.

Luminescence processes may be induced by excitations that produce free electron-hole (e-h) pairs that may recombine across the band gap or through defect- and impurity-related intermediate steps and emit a photon. The excitations employed most often are: (1) an incident intense above band-gap radiation such as from a laser source; (2) an incident electron beam; (3) electrical injection of electrons and/or holes through an appropriate contact; and (4) thermal excitation. These procedures are known as photoluminescence (PL), cathodoluminescence (CL),[123] electroluminescence (EL),[124] and thermoluminescence (TL), respectively. The most widely used technique for the analysis of semiconductor materials is PL. Cathodoluminescence measurements are usually conducted in a scanning electron microscope (SEM). The SEM electron beam can be focused to a spot size <1000 Å and can be scanned over the area of the sample. Hence, much work has been performed in CL imaging of wafers where one can obtain not only spectroscopic information but also spatial details. Electroluminescence is the most difficult to obtain because of the complexity of producing appropriate contacts. However, in terms of application, EL is the most important since a light emitter has to be able to produce light in an efficient manner under electrical excitation. Thermoluminescence is a technique used with insulators and wide-gap materials and not widely employed in the analysis of the commercially important materials such as GaAs and Si. The principles of luminescence and the optical information regarding the semiconductor that can be obtained are discussed in the next section using PL.

Photoluminescence.   Luminescence processes may be excited using an above-band-gap beam of light that leads to the creation of an electron-hole pair (e-h) that may recombine across the gap and emit a photon with energy equal to the gap E_g. This process is schematically displayed in Fig. 64.[105] Two possibilities are shown: namely, the direct-gap and the indirect-gap semiconductors. The recombination will be direct in the former and will have to involve an additional participant, mostly a phonon, to conserve momentum in the latter. This is the simplest possible recombination mechanism. Several additional routes exist for the relaxation that involves impurities. The photoexcited charges may recombine with ionized acceptors (A) and donors (D) with or without involving a phonon. The process may also be more complicated where, for example, an electron may be trapped by an ionized donor which may subsequently recombine with a hole at a neutral acceptor, leading to a donor-acceptor pair (DAP) transition. In addition, recombination may also occur through the annihilation of excitons considered further in the following section.

Excitons, as discussed earlier, are hydrogen-like two-particle electron-hole combinations that are not included in the one-electron energy band description of the solid. The strong Coulomb interaction between the electron and the hole leads to the excitonic coupling.

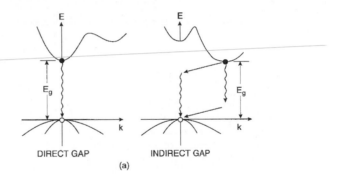

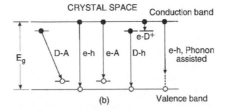

FIGURE 64 Schematic of photoexcitation and relaxation in semiconductors.[105]

As the excitation intensity increases, so does the population of the excitons. Higher-order interactions can occur between excitons and entities, such as biexcitons, and may be observed. Under intense excitation, the electrons and holes form a liquid state known as the electron-hole liquid. The presence of impurities and defects can also significantly alter the nature of the luminescence spectra, in particular the excitonic behavior. The electric field in the vicinity of the impurity can trap and localize excitons. Such an interaction leads to bound excitons; the binding energy of the bound exciton E_{BE} will, in addition to E_{ex} given in Eq. (39), contain a localization part δ as shown:

$$E_{BE} = E_{ex} + \delta \qquad (77)$$

The impurity potential that confines an exciton to a given center depends on both the impurity and its local environment. Hence, the impurity-bound excitonic features can be very rich and informative. A detailed knowledge of the PL excitonic spectrum may be used to identify both the chemical species and its environment, as is demonstrated later on. The observed luminescent photon energy at low temperatures may be written as:

$$E = E_g - E_{BE} \qquad (78)$$

Transitions from a free electron or hole to a neutral acceptor or donor, respectively, will occur at the following energies:

$$E = E_g - E_{ion} \qquad (79)$$

where E_{ion} is the relevant ionization energy. A complete description of the free-to-bound transition will have to take into account both the dispersion of the band in the vicinity of the minima and the population distribution. For the donor-to-acceptor transition, an additional electrostatic energy term $e^2/\epsilon(0)r_{sep}$, will have to be accounted for as shown:

$$E = E_g - (E_A + E_D) + \frac{e^2}{\epsilon(0)} r_{sep} \tag{80}$$

where r_{sep} is the distance that separates the two participating centers. The last term is needed to account for the electrostatic energy of the final ionized state of both centers. All of the transitions discussed may occur with phonon participation, and hence the emission energies should be reduced by the quantum of the energy of the phonon; multiple phonons may also be involved in more complex spectra.

The measurement of PL is, in most cases, routine and straightforward, which partly accounts for its popularity as a materials characterization technique. The laser provides the excitation and the dispersive spectrometer along with a sensitive detector, the detection. The sample should be cooled to ~5 K so that the temperature-induced broadening is kept to a minimum and the population of the processes with a small activation energy, such as the excitons, is sufficiently high to perform accurate measurements. Photoluminescence measurements may also be performed using Fourier transform (FT) spectrometers. The main advantage is the extraordinary resolution that can be achieved with the FT systems.

The extension of PL into a tool with quantitative accuracy, particularly in very high-purity Si has demonstrated the versatility of the analysis.[125] The paths of luminescence decay in high-purity Si are schematically presented in Fig. 65. The first step in the excitation process is the creation of electron-hole pairs, which subsequently can undergo a wide variety of processes before recombination. Figure 65 represents the possible intermediate states that eventually lead to radiative recombinations. A typical spectrum obtained from a sample containing 1.3×10^{13} cm^{-3} B, 1.8×10^{12} cm^{-3} P, and 3×10^{11} cm^{-3}, as is presented in Fig. 66.[125] The spectrum displays both the no-phonon (NP) component (shown as an inset) and TA, TO, and LO phonon-assisted features for the free exciton

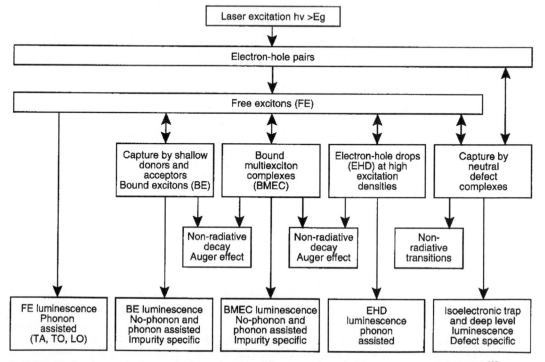

FIGURE 65   Luminescence decay processes stimulated by above-bandgap photoexcitation in high-purity Si.[125]

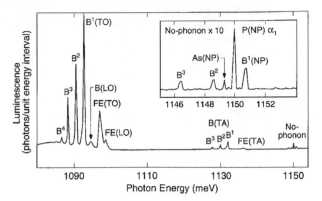

FIGURE 66 Photoluminescence spectrum from a Si sample doped with $1.3 \times 10^{13} \, \text{cm}^{-3}$ B, contaminated with $1.8 \times 10^{12} \, \text{cm}^{-3}$ P, and $3 \times 10^{11} \, \text{cm}^{-3}$ As.[125]

(FE) and the impurity-related bound exciton features. The measurement was performed at 4.2 K. Since B is an acceptor and P and As are donors in Si, electrical transport analyses are not sufficient to fully analyze the impurities. In contrast, all three impurities can be unambiguously identified using PL. In addition, when dependable calibration curves are available, the concentration of each species can also be established. A representative calibration curve is presented in Fig. 67 for the impurities B, Al, and P.[125] Note that quantitative measurements can be performed down to $10^{12} \, \text{cm}^{-3}$ for all three impurities. In the case of B, in ultrapure samples, measurements can be performed to levels as low as $\sim 10^{10} \, \text{cm}^{-3}$.

Next, we turn our attention to GaAs, which is a direct-band-gap material and hence exhibits efficient luminescence. Figure 68a displays a representative spectrum from a high-quality MOCVD-grown sample doped with C and Zn measured using an FT spectrometer.[126] The spectra shown are intense, sharp, and well-resolved near the band edge, with excitonic features appearing slightly below the band gap of $\sim 1.519 \, \text{eV}$ at 4.2 K. The near-edge features, labeled (A^0, X) and (D^0, X), are due to excitons bound to neutral donors and acceptors, and the deeper-lying features are the free-to-bound transition to the

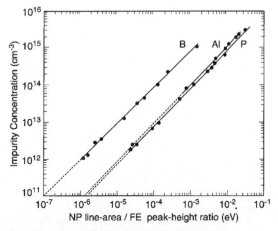

FIGURE 67 Calibration of the photoluminescence technique for measuring B, P, and Al concentrations in Si.[125]

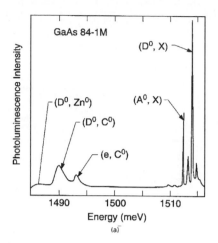

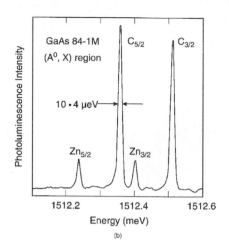

FIGURE 68a Representative PL spectrum from MOCVD-grown GaAs.[126]

FIGURE 68b High-resolution, near-band edge spectrum of the same sample as in previous figure.[126]

C acceptor, (e, C^0) and the donor-acceptor-pair transitions, (D^0, Zn^0), and (D^0, C^0), to the Zn and C centers, respectively. An expanded high-resolution version of the same spectrum, in the vicinity of (A^0, X) region, is displayed in Fig. 68b.[126] Note the impressive resolution obtainable in the FT system and the clear resolution of the excited states associated with the bound exciton.

The use of PL techniques may be extended deeper in the infrared region as well. Deep levels associated with impurities such as Fe, Cr, and Ag in GaAs are known to produce luminescence bands at ~ 0.35, ~ 0.6, and ~ 1.2 eV.[127] The narrow-gap semiconductors such as InAs, InSb, and HgCdTe luminesce in the 100- to 200-meV region farther into the infrared region. An example of the PL spectra observed from $Hg_{0.78}Cd_{0.22}Te$ narrow-gap alloy[128] is displayed in Fig. 69. The spectra, in comparison to those presented earlier, are rather featureless due to the fact that the effective masses in the narrow-gap semiconductors tend to be small, and hence the excitonic binding energies are small as well. Hence, sharp, well-resolved spectra are not normally observed. In addition, the Auger process, i.e., nonradiative decay processes, is much stronger, making the observation of luminescence much more difficult in narrow-gap semiconductors.

The luminescence processes described provide information on the relaxation mechanisms and hence are heavily weighted towards transitions that involve only the first excited state and the final ground state of a system. A complete study of a recombination process, for instance, the FE, requires information regarding the higher excited states. This can be achieved by a variation of the PL procedure known as photoluminescence excitation (PLE) spectroscopy. The crux of this technique is to concentrate on the PL response with respect to the excitation wavelength and thereby to determine the excitation resonances. Since the resonances occur when the excitation photon energy matches the excited energies, information regarding the excited states may be elucidated. The principle is illustrated with Fig. 70, where an attempt is made to determine the process that the photoexcited electron-hole pair undergoes before the formation of a free exciton in high-purity CdTe.[129] The spectrometer was set to the FE energy of 1.596 eV and the excitation photon energy was scanned from 1.6 to ~ 1.8 eV. The excitation wavelength may be scanned using a dye laser as shown in Fig. 70. Two PLE spectra obtained at 1.8 and 20.6 K are displayed in Fig. 70. The strong oscillatory behavior, with spacing of ~ 21 meV, demonstrates a cascade process through intermediate states separated by the LO phonon energy of 21 meV. This study illustrates the importance of the LO phonon in the

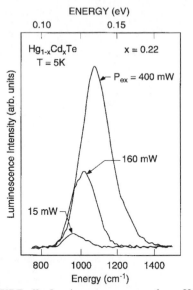

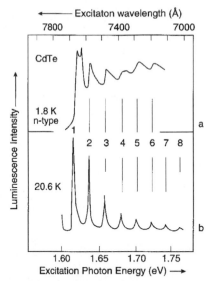

FIGURE 69 Luminescence spectra from $Hg_{0.78}$-$Cd_{0.22}Te$ measured at 6 K for different 1.06 μm excitation power.[128]

FIGURE 70 Excitation spectra of exciton luminescence of the emission line at 1.596 eV in the energy region higher than the bandgap of CdTe: (*a*) a pure *n*-type sample at 1.8 K; and (*b*) the sample at 20.6 K.[129]

electron-hole relaxation process and yields information regarding electronic states that lie above the conduction-band extremum, involving the hot electron behavior.

Inelastic Light Scattering (Raman and Brillouin). The elastic interaction of the incident photon with the specimen implies a process where the energy or wavelength of the photon is preserved. For instance, reflection and transmission spectroscopies involve the measurement of the fraction of the incident beam that is reflected or transmitted with no change in the incident wavelength. The vast majority of the incident photons undergo only elastic interactions, but a tiny fraction, of the order of $\sim 10^{-8}$, are subjected to inelastic scattering. Inelastic interactions, though very few, are extremely important as probes of the properties of the specimen. The advent of high-powered coherent laser sources and sensitive detectors has made the measurement of the scattered spectra straightforward, leading to an explosive growth in the last 20 years. Several excellent reviews may be found in the five volumes on light scattering, edited by Cardona and Guntherodt.[130] A good introduction to the procedure and theory may be found in Ref. 131.

The experimental procedure involves projecting a high-power laser beam onto a specimen, collecting the back-scattered light, and performing a spectrum analysis and detection. Multiple-dispersion stages are needed for improved resolution and filtering out of the unwanted elastically scattered beam. A typical set of spectra measured from GaAs[68] is presented in Fig. 71. The axes for the spectra are the detected intensity and the frequency shift suffered by the incident light. The intense peak at $0\,cm^{-1}$ is the elastically scattered peak. The spectrum at the room temperature of ~300 K displays both the frequency up-shifted and down-shifted components labeled as Stokes and anti-Stokes features corresponding to the TO phonon interaction; these correspond to microscopic processes where the incident photon loses or gains energy due to TO phonon emission or absorption, respectively. At lower temperatures, additional features labeled L_+ and L_- appear near the TO peak as well as a broad feature in the 0- to 100-cm^{-1} region. These features originate from charge carriers and were discussed in earlier sections.

The inelastic scattering techniques are usually divided into two categories, namely,

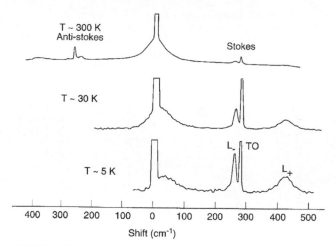

FIGURE 71 Typical Raman spectrum in GaAs, with $n = 1.4 \times 10^{18}\,\mathrm{cm}^{-3}$, at 300 K, 30 K, and 5 K showing the lineshape change with temperature.[68]

Brillouin and Raman scattering. The former covers low frequencies extending from 0 to $\sim 10\,\mathrm{cm}^{-1}$, while the latter spans the higher frequencies. The low-frequency Brillouin scattering measurements provide access to information on properties of acoustic phonons, spin waves, etc., and involve the use of specialized spectrometers needed to remove the very intense elastic peak.[132] The position of the elastic peak is taken to be the reference zero position, and the scattered spectrum is measured with respect to the zero position; i.e., the spectrum is recorded as a function of the frequency shift and not the absolute frequency. Raman scattering is, in general, easier to perform and more informative for the study of semiconductors and hence is emphasized in this section.

The techniques reviewed so far all involve first-order interactions that provide direct information regarding the electronic, vibrational, and impurity-related behavior of semiconductors. In contrast, inelastic light scattering such as Raman and Brillouin scattering involves the electron-phonon or other quasi-particle interactions and hence can provide additional information regarding these interactions. In addition, since inelastic scattering involves a higher-order interaction, processes that are inactive in first-order may be investigated. For instance, the optical phonons in Si do not possess a dipole moment and hence do not interact with infrared radiation but can be clearly observed in Raman scattering.

The response of the specimen may be expressed, through the susceptibility χ.[133] However, the presence of the inelastic interactions will give rise to additional contributions. Consider the example of lattice vibrations and their influence on χ in the visible-frequency range. Since the lattice vibrational frequencies correspond to far-infrared light frequencies, no direct contribution is expected. However, the lattice vibrations can influence χ even in the visible-frequency range in an indirect fashion by a small amount. The influence of this interaction may be expressed as follows:

$$\chi = \chi_0 + \sum_i \frac{\partial \chi}{\partial u_i} \cdot \mathbf{u}_i$$
$$+ \sum_{i,j} \frac{\partial^2 \chi}{\partial u_i\, \partial u_j} \mathbf{u}_i \mathbf{u}_j \qquad (81)$$
$$+ \cdots$$

where $\mathbf{u}_i$ and $\mathbf{u}_j$ are the displacements associated with the normal modes of lattice vibrations or phonons. The first term contains the elastic term, and the second and third terms denote the second- and third-order inelastic interactions with the phonons. Note that the form of χ shown here differs from that used previously in nonlinear optics because it is expressed explicitly in terms of the phonon coordinates. Assuming sinusoidal oscillations for the incident radiation $\mathbf{E}$ and the phonons $\mathbf{u}_j$, the polarization induced in the specimen may be expressed as follows:

$$\mathbf{P} = \chi_0 \mathbf{E}_0 \exp(i\omega_0 t) + \sum_i \chi' u_{i0} \exp\{i(\omega_0 \pm \omega_i)t\} + \sum_{ij} \chi'' u_{i0} u_{j0} \exp\{i(\omega_0 \pm \omega_i \pm \omega_j)\}t \quad (82)$$

The source of the inelastically scattered or frequency-shifted components is immediately apparent.

The scattered intensity, assuming only second-order interactions, is usually expressed in terms of a differential cross section, i.e., scattered energy per unit time in the solid angle $d\Omega$ as follows:

$$\frac{d\sigma}{d\Omega} = V^2 \left(\frac{\omega_s}{c}\right)^4 \hat{\mathbf{e}}_i \chi' \hat{\mathbf{e}}_s n_j \quad \text{Stokes}$$

$$(n_j + 1) \quad \text{Anti-Stokes} \quad (83)$$

where the $\hat{\mathbf{e}}_i$ and $\hat{\mathbf{e}}_s$ denote the polarization state of the scattered and incident light, ω_s the frequency of the scattered beam, and V the scattering volume.

χ' and χ'' are Raman tensors whose symmetry properties may be calculated from a knowledge of the structure of the crystals.[134] The notation used to express the combination of incident and scattered beam directions and polarizations is as follows: $a(b, c)d$ where a and d denote the incident and scattered directions, and b and c the respective polarizations. Using the cubic crystal axes x, y, and z, the selection rules may be expressed as follows.

Back scattering from (100) surface:

TO—disallowed for any combination of incident and scattered beam configurations

LO—allowed only for crossed polarizations, i.e., $z(x, y)\bar{z}$ and $z(y, x)\bar{z}$, where $\bar{z}$ is the opposite direction to z.

Similar selection rules may be derived for other orientations and crystal structures.

A microscopic description of the scattering event considers quantum mechanical details using perturbation theory. The incident photon with frequency ω_i and wave vector $\mathbf{k}_i$ produces a transition from the initial state i to a virtual state b where the incident photon is annihilated, followed by a transition to the final state f, accompanied by the emission of the "scattered" photon with frequency and wave vector ω_s and $\mathbf{q}_s$, respectively. The entire process is part of a complete quantum mechanical event and will be governed by momentum, energy, and symmetry conservation rules shown as:

$$\hbar\omega_i = \hbar\omega_s \pm \hbar\Omega$$
$$\hbar\mathbf{q}_i = \hbar\mathbf{q}_s \pm \hbar\mathbf{K} \quad (84)$$

where Ω and $\mathbf{K}$ denote the frequency and wave vector of the participating phonon. The exact forms of the scattering cross section will depend on the details of the interactions and are reviewed in Ref. 131.

One of the most powerful aspects of Raman scattering is the ability to perform resonance excitation. When the incident photon energy or the scattered photon energy matches an intrinsic excitation energy, a substantial increase in the scattered intensity is observed. This ability to resonate has been used for both an understanding of the details of the scattering process and applications in which the source of the scattering can be selected. For instance, the study of a particular impurity in a large matrix may be conducted by tuning the resonance to match that of the impurity. Examples of resonance studies are discussed in the following sections.

The power and versatility of Raman scattering to probe many important properties of semiconductors have been exploited for a wide variety of characterizations. Figure 72 schematically displays the broad areas where Raman scattering has been employed to

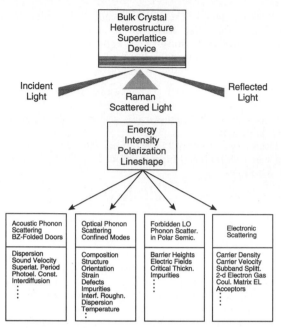

FIGURE 72 Schematics of inelastic light scattering and information which can be extracted from such measurements.[135]

date.[135] The study of materials devices and microstructural properties include chemical, structural, and electronic properties. The developments in microscopic measurements and analysis have opened up new applications as well. The measurement of the temperature in a spot ~1 μm in diameter is one such example that is discussed in the following sections.

The most dominant Raman-scattering mechanism in semiconductors is that due to phonons—in particular, the optical phonons at the center of the Brillouin zone—namely, TO (Γ) and LO (Γ). Even though a semiconductor crystal can support a variety of vibrations, the momentum conservation requirements restrict the interaction of the incident photon to only the TO (Γ) and LO (Γ). The position and shape of the spectra contain important information regarding the structural state of the material: the presence of strain will be reflected as shifts in the line position; degraded crystal quality due to multiple grains and concomitant distributed strain will lead to a broadening of the lineshape. The effects of crystal damage on the Raman spectrum as a result of ion implantation in GaAs is displayed in Fig. 73.[136] Note the large increase in the linewidth of the LO (Γ) feature. In addition, a series of new structures, not present in the undamaged sample, can also be observed. These have been interpreted to be the result of relaxing the momentum conservation laws and hence the activation of phonon-scattering processes not normally allowed in a good crystal. A simple interpretation of the broadening of the LO lineshape was provided using a phonon confinement model, where the LO phonon is described as being confined to small damage-free regions. The shift in the position of the LO peak and the full-width-at-half-maximum were related to the diameter of the undamaged region. The information obtained is useful in characterizing lattice damage, amorphous materials, and degrees of recovery during annealing.

As explained earlier, Raman processes can involve higher-order interactions that involve more than one phonon. Hence, multiphonon scattering can be performed and information complementary to that discussed in the section entitled "Lattice" on the infrared absorption properties can be obtained. The Raman spectra obtained from Si[137] are

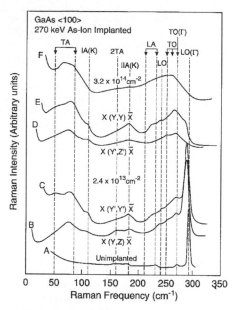

FIGURE 73 Raman spectra of $\langle 100 \rangle$ GaAs before implantation (A), implanted to a fluence of 2.4×10^{13} cm^{-2} for various polarization configurations (B, C, D, and E) and a fluence of 3.2×10^{14} cm^{-2} (F).[136]

displayed in Fig. 74. The spectra were recorded at room temperature where $X' = (100)$, $Y' = (0\bar{1}1)$, and $Z' = (0\bar{1}\bar{1})$. The irreducible representation of the phonons involved is also noted in the figure. Note the strong peak at ~ 522 cm^{-1} which is due to the degenerate TO(Γ)/LO(Γ) phonons that are not observable in infrared absorption measurements. Additional bands present in the 200- to 400-cm^{-1} and the 600- to 1000-cm^{-1} range are due to multiphonon scattering processes. The ability to employ polarization selection rules has been used effectively to isolate the symmetry character of the underlying vibrational features and can be used to eventually identify the source of the various features. Such studies are useful in understanding both the optical behavior of Si and in illuminating the lattice dynamical properties of the material.

In crystals that contain a substantial number of free carriers, the incident photon can scatter off collective charge oscillations, known as plasmons, discussed earlier. The Raman spectra measured from three GaAs samples[138] with electron densities ranging from 1.95×10^{18} cm^{-3} to 6.75×10^{18} cm^{-3} are presented in Fig. 75. The observed features in addition to the LO phonon peak at ~ 293 cm^{-1} are due to the coupled LO phonon-plasmon features, also discussed earlier. The variation of the L_+ mode frequencies with the carrier density may be calculated and compared to measurements, as was shown in Fig. 34. The shape of the L_+ and the L_- features can be used to deduce the mobility of the carriers as well. The uniqueness of the Raman results is that they can be employed to study the behavior of carriers near the surface. Since the incident laser light with a photon energy of ~ 2.5 eV penetrates only ~ 1000 Å into the sample, the Raman results represent only the behavior of this near-surface region.

The single-particle nature of the free carriers can also be probed by Raman scattering.[68] The mechanism responsible for the interaction is the scattering of the carriers from inside the Fermi-surface to momentum states that lie outside, the total change in momentum being equal to that imparted by the photons. The integrated effect in the case of the spherical Fermi-surface in GaAs is displayed in Fig. 76. The Fermi wave vector is denoted by $\mathbf{p}_F$ and that of the electron is $\mathbf{p}$. The wave-vector change as a result of the scattering is $\mathbf{q}$.

Two cases of small and large $\mathbf{q}$ and the resultant single-particle spectrum at 0 K are shown. The net effect in the first case will be a linear increase followed by a rapid fall and when $\mathbf{q}$ is large, the spectrum displays a bandlike behavior as shown. The single-particle spectrum measured from a sample of n-GaAs[139] is presented in Fig. 77. The measurements were performed using the 6471 Å line of the Kr^+ laser at 10 K. The ability to probe the Fermi sphere directly, using a spectroscopic technique, can lead to valuable insights into the electronic distributions that are complementary to that obtained from transport studies that usually provide only information regarding integrated effects of all the carriers.

The effects of resonance enhancement, mentioned earlier, is one of the most powerful features of Raman scattering. The effect can be illustrated with the variation of the scattering intensity of the optical phonons in CdS with the incident photon energy as discussed in Ref. 140. An order-of-magnitude enhancement was observed as the excitation energy approached the energy gap at ~2.6 eV for all the observed phonon modes. In addition, the TO modes both displayed a reduction before a large enhancement as E_g was approached. The reduction was interpreted as the result of a destructive interference between the resonant and nonresonant terms that contribute to the scattering cross section. Resonance Raman-scattering studies can therefore shed light on the microscopic details of the scattering process.

The application of Raman scattering to the localized vibration due to impurities is illustrated in Fig. 78 that present the data obtained from B-doped Si.[141] In most cases, the density of the impurities needs to be quite high to be observable in Raman scattering. However, rapid advances in the measurement procedures may improve the sensitivities. Direct measurement of the electronic transitions related to dopant ions can also be observed in Raman scattering. A good example is the electronic interbound state transitions from three donors in GaP (in Ref. 142). The normally symmetric, threefold degenerate 1S state of the donor, split into a $1S(A_1)$ singlet and a $1S(E)$ doublet due to the interaction with the conduction band valleys in the indirect-gap GaP was clearly observed as well-resolved peaks.[142] The Raman spectra are sensitive to the impurity electronic states and provide a tool to probe them as well.

36.5   ACKNOWLEDGMENTS

The authors express their deep appreciation for the help rendered by Ms. Tammy Clark and Ms. Jane Walters at NIST and Mrs. Nancy Seiler in preparing the manuscript.

36.6   SUMMARY AND CONCLUSIONS

An overview of the optical properties of semiconductors has been presented in this chapter. These properties form the foundation for understanding and utilizing the wide variety of optical devices manufactured today. A number of materials can be used together with electronic circuits to generate, detect, and manipulate light signals leading to the field of optoelectronics. Semiconductor materials are becoming increasingly important for use in optoelectronic devices: devices can be made very small, leading to a high degree of compactness and compatibility with other electronic and optical functions; they are robust and highly reliable; they are highly efficient as light-generating sources with internal efficiencies sometimes approaching 100 percent; they are capable of large power dissipation, of very high frequency performance, and can access an enormous range of wavelengths; performances can be turned over wavelength, frequency, and efficiency.[143] Table 12 lists some of the most important materials and their applications for optoelectronics.[143]

In conclusion, Table 13 presents some of the important parameters for the most common semiconductors that determine the optical behavior of each material. The forbidden-energy gap and higher-energy critical point energies are listed along

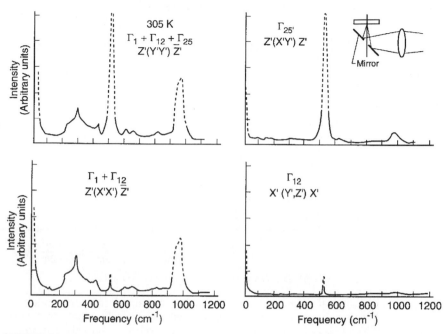

FIGURE 74 The Raman spectra of Si at room temperature showing the multiphonon contributions. The polarization configuration and the representations which were possible contributors to the spectra are shown for each of the four spectra.[137]

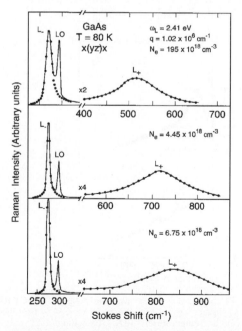

FIGURE 75 Raman spectra of three different n-GaAs samples obtained in backscattering geometry from (100) surfaces showing the coupled plasmon-LO phonon modes.[138]

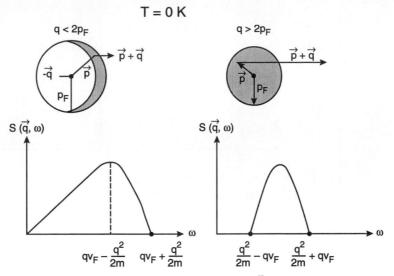

FIGURE 76   Single-particle excitation spectrum at 0 K.[68]

with optical phonon energies, dielectric constants, refractive index at energies below the energy gap, and free exciton binding energy. Closely related transport parameters such as the charge carrier effective masses and the mobilities are also included for completion. The reader is referred to Palik's compilation in Refs. 9 and 10 for refractive indices for several semiconductors over a wide range of energies. Additional information may also be obtained from many comprehensive collections of physical parameters such as those presented in the Landolt and Bornstein Tables.[144–147]

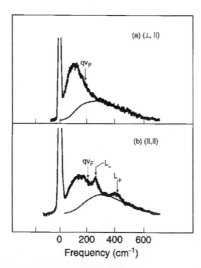

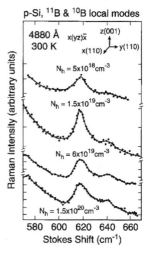

FIGURE 77   SIngle-particle spectra and coupled LO phonon-plasmon modes ($L_\pm$) for n-GaAs, $n = 1.3 \times 10^{18}$ cm^{-3}. Temperature: 10 K. Excitation: 6471 Å. The interband transition energy from the split-off valence band to the Fermi level is very close to the laser photon energy. Estimated luminescence background is shown by the dashed line.[139]

FIGURE 78   Scattering by local modes of boron (isotopes B^{10} and B^{11}) in p-type silicon.[141]

TABLE 12 Some Important Semiconductor Materials and Applications for Optoelectronics[143]

Material	Type	Substrate	Devices	Wavelength range (μm)	Applications
Si	IV	Si	Detectors, solar cells	0.5–1	Solar energy conversion, e.g., watches, calculators, heating, cooling, detectors
SiC	IV	SiC	Blue LEDs	0.4	Displays, optical disk memories, etc.
Ge	IV	Ge	Detectors	1–1.8	Spectroscopy
GaAs	III–V	GaAs	LEDs, lasers, detectors, solar cells, imagers, intensifiers, electro-optic modulators, optoisolators	0.85	Remote control TV, etc., video disk players, range-finding, solar energy conversion, optical fiber communication systems (local networks), image intensifiers
AlGaAs	III–V	GaAs	LEDs, lasers, solar cells, imagers	0.67–0.98	
GaInP	III–V	GaAs	Visible lasers, LEDs	0.5–0.7	Displays, control, compact disk players, laser printers/scanners, optical disk memories, laser medical equipment
GaAlInP	III–V	GaAs	Visible lasers, LEDs	0.5–0.7	
GaP	III–V	GaP	Visible LEDs	0.5–0.7	
GaAsP	III–V	GaP	Visible LEDs, optoisolators	0.5–0.7	
InP	III–V	InP	Solar cells	0.9	Space solar cells
InGaAs	III–V	InP	Detectors	1–1.67	Optical fiber communications (long-haul and local loop)
InGaAsP	III–V	InP	Lasers, LEDs	1–1.6	
InAlAs	III–V	InP	Lasers, detectors	1–2.5	
InAlGaAs	III–V	InP	Lasers, detectors	1–2.5	
GaSb/GaAlSb	III–V	GaSb	Detectors, lasers	2–3.5	
CdHgTe	II–VI	CdTe	IR detectors	3–5 and 8–12	Imaging Infrared imaging, night vision sights, missile seekers, and many other military applications
ZnSe	II–VI	ZnSe	Short LED, lasers	0.4–0.6	Commercial applications (R&D stage only)
ZnS	II–VI	ZnS	Short LED, lasers	0.4–0.6	
Pb compounds	IV–VI	Pb Compounds	IR lasers, detectors	3–30	Spectroscopy, pollution monitoring

TABLE 13 Materials Parameters[1]

Material	Type (I/D)	E_g 300 K (eV)	E_g 77 K (eV)	E_g ~0 K (eV)	dE_g/dT (10^{-4} eV/K)	E_0 (eV)	$E_0+\Delta_0$ (eV)	E_1 (eV)	$E_1+\Delta_1$ (eV)
Si	I	1.1242[a-1]	1.169[a-1]	1.170[a-1]	−2.8[a-2]	4.185(4.2 K)[a-3]	4.229(4.2 K)[a-3]	3.40[a-4]	
Ge	I	0.664[b-1]	0.734[b-1]	0.744[b-2]	−3.7[b-3]	0.888(10 K)[b-4]	1.184(10 K)[b-4]	2.05[b-5]	2.298[b-6]
α-Sn	D	0[c-1]		0[c-1]	0	−0.42(85 K)[c-2]	0.8(10 K)[c-2]	1.316[c-3]	1.798[c-3]
GaAs	I	1.424[d-1]	1.5115[d-2]	1.51914[d-3]	−3.9[d-4]	E_g	1.760[d-5]	2.915[d-6]	3.139[d-6]
AlAs	I	2.153[e-1]	2.223[e-1]	2.229[e-1]	−3.6[e-1]	3.02[e-2]	3.32[e-2]	~3.9[e-2]	~4.1[e-2]
InAs	D	0.354[f-1]	0.404[f-2]	0.418[f-3]	−3.5[f-4]	E_g	0.725[f-1]	2.5[f-5]	2.75[f-5]
InP	D	1.344[g-1]	1.4135[g-2]	1.4236[g-3]	−2.9[g-2]	E_g	1.45[g-4]	3.158[g-5]	3.28[g-6]
InSb	D	0.18[h-1]	0.23[h-2]	0.2368[h-2]	−2.7[h-2]	E_g	1.16[h-3]	1.88[h-4]	2.38[h-4]
GaP	I	2.272[i-1]	2.338[i-2]	2.350[i-1]	−3.7[i-2]	2.780[i-3]	2.860[i-3]	3.785(10 K)[i-3]	3.835(10 K)[i-3]
ZnS (cubic)	D	3.68[j-1]		3.78[j-1]	−4.7[j-2]	E_g	3.752(15 K)[j-3]	5.73[j-1]	
ZnSe	D	2.70[k-1]		2.8215[k-2]	−4.8[k-3]	E_g	3.12[k-1]	4.77[k-1]	4.97[k-1]
ZnTe	D	2.30[l-1]		2.3941[l-2]	−4.1[l-1]	E_g	3.18[l-3]	3.64[l-3]	4.22[l-3]
CdTe	D	1.505[m-1]	1.583[m-2]	1.6063[m-3]	−2.9[m-4]	E_g	2.1[m-5]	3.31[m-6]	3.87[m-5]
HgTe	D	0[n-1]	0[n-1]		0	−0.106[n-1]	1.08[n-2]	2.12[n-2]	2.78[n-2]
CdS (Hexagonal)	D	2.485[o-1]	2.573[o-1]	2.5825[o-2]	+4.1[o-1]	E_g			
PbS	D	0.42[p-1]	0.307[p-2]	0.286[p-3]	+5.2[p-3]	E_g		1.85[p-4]	
PbTe	D	0.311[q-1]	0.217[q-2]	0.188[q-1]	+4.5[q-3]	E_g		1.24[q-4]	
PbSe	D	0.278[r-1]	0.176[r-2]	0.1463[r-3]	+4.0[r-4]	E_g		1.59[r-5]	

Material	E_{ex} (FE) (eV)	$\epsilon(0)$	ϵ_∞	$n(\lambda)$ $\lambda > \lambda_c$	dn/dT (10^{-4}/K)	$\hbar\omega_{TO}$ (eV)	$\hbar\omega_{LO}$ (eV)	μ_{e2} (cm²/V·s)	μ_{h2} (cm²/V·s)	m_e^*/m_0	m_h^*/m_0
Si	0.014[a-5]	11.9[a-6]	11.9[a-6]	3.4179(10 μm)[a-7]	1.3[a-8]	0.0642[a-9]	0.0642[a-9]	1,500[a-10]	450[a-10]	0.98[3(a-10)], 0.19[4(a-10)]	0.16[5(a-10)], 0.49[6(a-10)]
Ge	0.00415[b-7]	16.2[b-8]	16.2[b-8]	4.00319(10 μm)[b-9]	4.0[b-10]	0.0373[b-11]	0.0373[b-11]	3,900[b-12]	1900[b-12]	1.64[3(b-12)], 0.082[4(b-13)]	0.045[5(b-12)], 0.28[6(b-12)]
α-Sn	0	24[c-4]	24[c-4]	—		0.0244[c-5]	0.0244[c-5]	1,400[c-5]	1200[c-6]	0.028[c-2]	0.195[c-2]
GaAs	0.0042[d-7]	13.18[d-8]	10.89[d-8]	3.298(5 μm)[d-9]	1.5[d-10]	0.0333[d-11]	0.0362[d-12]	8,500[d-12]	400[d-12]	0.067[d-12]	0.082[d-12]
AlAs	0.02[e-1]	10.06[e-3]	8.16[e-3]	2.87(2 μm)[e-4]	1.2[e-5]	0.04488[e-6]	0.05009[e-6]	300[e-7]	200[e-8]	1.1[3(e-2)], 0.19[4(e-2)]	0.153[5(e-9)], 0.409[6(e-9)]
InAs	0.0017[f-3]	15.15[f-6]	12.25[f-6]	3.42(10 μm)[f-7]		0.0269[f-8]	0.0296[f-8]	3,300[f-8]	460[f-8]	0.023[f-8]	0.4[f-8]
InP	0.0051[g-3]	12.61[g-7]	9.61[g-7]	3.08(5 μm)[g-8]	0.83[g-9]	0.0377[g-10]	0.0428[g-10]	4,600[g-11]	150[g-11]	0.077[g-11]	0.64[g-11]
InSb	0.00052[h-5]	16.8[h-6]	15.68[h-7]	3.953(10 μm)[h-8]	4.7[h-9]	0.0223[h-10]	0.0237[h-10]	80,000[h-11]	1250[h-11]	0.0145[h-11]	0.40[h-11]
GaP	~0.02[i-1]	11.11[i-4]	9.11[i-4]	2.90(10 μm)[i-5]		0.0455[i-6]	0.050[i-6]	110[i-7]	75[i-7]	0.82[i-7]	0.60[i-7]

ZnS (cubic)	0.036[i-4]	5.1[i-5]	2.95[j-6]	2.2014(10 μm)[j-7]	0.46[j-8]	0.03397[j-9]	0.04364[j-10]	165[j-11]	5[j-11]	0.28[j-12]	0.49[j-12]
ZnSe	0.018[k-4]	9.6[k-5]	6.3[k-5]	2.410(10 μm)[k-6]	0.52[k-7]	0.02542[k-8]	0.03099[k-8]	600[k-9]	100[k-10]	0.142[k-11]	0.57[k-12]
ZnTe	0.0132[l-2]	10.1[l-4]	7.28[l-5]	2.64(5 μm)[l-6]		0.02194[l-7]	0.02542[l-7]	330[l-8]		0.11[l-9]	0.6[l-10]
CdTe	0.0105[m-3]	10.2[m-7]	7.1[m-7]	2.684(5 μm)[m-8]		0.0174[m-7]	0.0208[m-9]	1,050[m-10]	100[m-10]	0.1[m-11]	0.4[m-12]
HgTe	0	21.0[n-3]	15.2[n-3]	—		0.01463[n-4]	0.0171[n-4]	35,000[m-5]		0.031[n-6]	0.42[n-7]
CdS (Hexagonal)	0.0274[o-3]	8.7($\epsilon_{11}(0)$)[o-4] / 9.25($\epsilon_{33}(0)$)[o-4]	5.53($\epsilon_{11\infty}$)[o-5] / 5.5($\epsilon_{33\infty}$)[o-5]	2.227(10 μm)$\perp$c[o-5] / 2.245(10 μm)$\parallel$c[o-5]	0.6[o-5] / 0.62[o-5]	See (o-6)	See (o-6)	340[o-7]	50[o-7]	0.21[o-7]	0.8[o-7]
PbS		169[p-5]	17.2[p-5]	3.68(16 μm)[p-6]		0.00810[p-7]	0.02539[p-7]	600[p-8]	700[p-8]	0.105[3][p-9] / 0.080[4][p-9]	0.105[3][p-9] / 0.075[4][p-9]
PbTe		414[q-5]	33[q-6]	5.66(10 μm)[q-7]		0.00399[q-5]	0.01414[q-8]	6,000[q-9]	4000[q-9]	0.24[3][q-10] / 0.024[4][q-10]	0.31[3][q-10] / 0.024[4][q-10]
PbSe		210[r-6]	22.9[r-7]	4.75(10 μm)[r-8]		0.00546[r-9]	0.0165[r-10]	1,000[r-11]	800[r-12]	0.07[3][r-13] / 0.04[4][r-13]	0.068[3][r-13] / 0.034[4][r-13]

[1] Most values quoted are at ~300 K, except where indicated.
[2] Highest values reported.
[3] Longitudinal effective mass.
[4] Transverse effective mass.
[5] Light-hole effective mass.
[6] Heavy-hole effective mass.
[7] Band extrema effective masses are obtained from low-temperature measurements. See quoted references for more details.

a-Si

(a-1) W. Bludau, A. Onton, and W. Heinke, *J. Appl. Phys.* **45**:1846 (1974).

(a-2) H. D. Barber, *Sol. St. Electronics* **10**:1039 (1967).

(a-3) D. E. Aspnes and A. A. Studna, *Solid State Communications* **11**:1375 (1972).

(a-4) A. Duanois and D. E. Aspnes, *Phys. Rev. B.* **18**:1824 (1978).

(a-5) K. L. Shaklee and R. E. Nahory, *Phys. Rev. Lett.* **24**:942 (1970).

(a-6) K. V. Rao and A. Smakula, *J. Appl. Phys.* **37**:2840 (1966).

(a-7) C. D. Salzberg and J. J. Villa, *J. Opt. Soc. Am.* **47**:244 (1957).

(a-8) M. Cardona, W. Paul, and H. Brooks, *J. Phys. Chem. Solids* **8**:204 (1959).

(a-9) G. Dolling, *Inelastic Scattering of Neutrons in Solids and Liquids*, vol. II, IAEA, Vienna, 1963, p. 37.

(a-10) S. M. Sze, *Physics of Semiconductor Devices*, 2d ed., John Wiley, New York, 1981, p. 849.

b-Ge

(b-1) G. G. Macfarlane, T. P. McLean, J. E. Quarinaton, and V. Roberts, *Phys. Rev.* **108**:1377 (1957).

(b-2) S. Zwerdling, B. Lax, L. M. Roth, and K. J. Button, *Phys. Rev.* **114**:80 (1959).

(b-3) T. P. McLean, in *Progress in Semiconductors*, vol. 5, A. F. Gibson (ed.), Heywood, London, 1960.

(b-4) D. E. Aspnes, *Phys. Rev. B* **12**:2297 (1975).

(b-5) A. K. Gosh, *Phys. Rev.* **165**:888 (1968).

(b-6) L. Vina and M. Cardona, *Physica* **117B** and **118B**:356 (1983).

(b-7) V. I. Sidorov and Ya. E. Pokrovski, *Sov. Phys. Semicond.* **6**:2015 (1973).

(b-8) F. A. D'Altroy and H. Y. Fan, *Phys. Rev.* **103**:1671 (1956).

(b-9) R. P. Edwin, M. T. Dudermel, and M. Lamare, *Appl. Opt.* **21**:878 (1982).

(b-10) See (a-8).

(b-11) G. Nilsson and G. Nelin, *Phys. Rev. B* **6**:3777 (1972).

(b-12) See (a-10).

TABLE 13 Materials Parameters *(Continued)*

c-αSn

(c-1) J. R. Chelikowsky and M. L. Cohen, *Phys. Rev. B* **14**:556 (1976).
(c-2) S. H. Groves, C. R. Pidgeon, A. W. Ewald, and R. J. Wagner, *J. Phys. Chem. Solids* **31**:2031 (1970).
(c-3) L. Vina, H. Hockst, and M. Cardona, *Phys. Rev. B* **31**:958 (1985).
(c-4) R. E. Lindquist and A. W. Ewald, *Phys. Rev.* **135**:A191 (1964).
(c-5) C. J. Buchenauer, M. Cardona, and F. H. Pollak, *Phys. Rev. B* **3**:1243 (1971).
(c-6) See (a-10).

d-GaAs

(d-1) D. D. Sell, H. C. Casey, and K. W. Wecht, *J. Appl. Phys.* **45**:2650 (1974).
(d-2) M. B. Panish and H. C. Casey, *J. Appl. Phys.* **40**:163 (1969).
(d-3) B. J. Skromme and G. E. Stillman, *Phys. Rev. B* **29**:1982 (1984).
(d-4) J. Camassel, D. Auvergne, and H. Mathieu, *J. Appl. Phys.* **46**:2683 (1975).
(d-5) C. Alibert, S. Gaillard, M. Erman, and P. M. Frijlink, *J. Phys. Paris* **44**:C10–229 (1983).
(d-6) P. Lautenschlager, M. Garriga, S. Logothetidis, and M. Cardona, *Phys. Rev. B* **35**:9174 (1987).
(d-7) D. D. Sell, *Phys. Rev. B* **6**:3750 (1972).
(d-8) G. A. Samara, *Phys. Rev. B* **27**:3494 (1983).
(d-9) K. G. Hambleton, C. Hilsum, and B. R. Holeman, *Proc. Phys. Soc.* **77**:1147 (1961).
(d-10) M. Cardona, *Proc. Int. Conf. Phys. Semicond.*, Prague, 1960, Publ. House of the Czech. Acad. of Sciences, Prague, 1960, p. 388.
(d-11) A. Mooradian and G. B. Wright, *Solid State Communications* **4**:431 (1966).
(d-12) See (a-10).

e-AlAs

(e-1) B. Monemar, *Phys. Rev. B* **8**:5711 (1983).
(e-2) S. Adachi, *J. Appl. Phys.* **58**:R1 (1985).
(e-3) R. E. Fern and A. Onton, *J. Appl. Phys.* **42**:3499 (1971).
(e-4) M. Hoch and K. S. Hinge, *J. Chem. Phys.* **35**:451 (1961).
(e-5) H. G. Grimmeiss and B. Monemar, *Phys. Stat. Sol. (a)* **5**:109 (1971).
(e-6) O. K. Kim and W. G. Spitzer, *J. Appl. Phys.* **50**:4362 (1979).
(e-7) A. G. Ettenberg, A. G. Sigai, A. Dreeben, and S. L. Gilbert, *J. Electrochem. Soc.* **119**:1355 (1971).
(e-8) J. D. Wiley, in *Semiconductors and Semimetals*, vol. 10, R. K. Willardson and A. C. Beer (eds.), Academic Press, New York, 1975, p. 91.
(e-9) M. Huang and W. Y. Ching, *J. Phys. Chem. Solids* **46**:977 (1985).

f-InAs

(f-1) F. Lukes, *Phys. Stat. Sol. (b)* **84**:K113 (1977).
(f-2) E. Adachi, *J. Phys. Soc. Jpn.* **2**:1178 (1968).
(f-3) A. V. Varfolomeev, R. P. Seisyan, and R. N. Yakimova, *Sov. Phys. Semicond.* **9**:560 (1975).
(f-4) F. Matossi and F. Stern, *Phys. Rev.* **111**:472 (1958).
(f-5) M. Cardona and G. Harbeke, *J. Appl. Phys.* **34**:813 (1963).
(f-6) M. Haas and B. W. Henvis, *J. Phys. Chem. Solids* **23**:1099 (1962).
(f-7) O. G. Lorimor and W. G. Spitzer, *J. Appl. Phys.* **36**:1841 (1965).
(f-8) R. Carles, N. Saint-Cricq, J. B. Renucci, M. A. Renucci, and A. Zwick, *Phys. Rev. B* **22**:4804 (1980).

g-InP

(g-1) M. Bugaski and W. Lewandowski, *J. Appl. Phys.* **57**:521 (1985).

(g-2) W. J. Turner, W. E. Reese, and G. D. Pettit, *Phys. Rev.* **136**:A1467 (1964).

(g-3) H. Mathieu, Y. Chen, J. Camassel, J. Allegre, and D. S. Robertson, *Phys. Rev. B* **32**:4042 (1985).

(g-4) K. L. Shaklee, M. Cardona, and F. H. Pollak, *Phys. Rev. Lett.* **16**:48 (1966).

(g-5) S. M. Kelso, D. E. Aspnes, M. A. Pollak, and R. E. Nahory, *Phys. Rev. B* **26**:6669 (1982).

(g-6) E. Matatagui, A. E. Thompson, and M. Cardona, *Phys. Rev.* **176**:950 (1968).

(g-7) See (f-6).

(g-8) F. Oswald, *Z. Naturforsch* **9a**:181 (1954).

(g-9) See (d-10).

(g-10) A. Mooradian and G. B. Wright, *Solid State Communications* **4**:431 (1966).

(g-11) See (a-10).

h-InSb

(h-1) F. Lukes and E. Schmidt, *Proc. Int. Conf. Phys. Semicond.*, Exeter, 1962, Inst. of Physics, London, 1962, p. 389.

(h-2) C. L. Littler and D. G. Seiler, *Appl. Phys. Lett.* **46**:986 (1985).

(h-3) S. Zwerdling, W. H. Kleiner, and J. P. Theriault, *MIT Lincoln Laboratory Report 8G-00M*, 1961.

(h-4) M. Cardona, K. L. Shaklee, and F. H. Pollak, *Phys. Rev.* **154**:696 (1967).

(h-5) A. Baldereschi and N. O. Lipari, *Phys. Rev. B* **3**:439 (1971).

(h-6) J. R. Dixon and J. K. Furdyna, *Solid State Communications* **35**:195 (1980).

(h-7) See (f-6).

(h-8) T. S. Moss, S. D. Smith, and T. D. F. Hawkins, *Proc. Phys. Soc. B* **70**:776 (1957).

(h-9) See (g-9).

(h-10) W. Keifer, W. Richter, and M. Cardona, *Phys. Rev. B* **12**:2346 (1975).

(h-11) See (a-10).

i-GaP

(i-1) R. G. Humphreys, U. Rossler, and M. Cardona, *Phys. Rev. B* **18**:5590 (1978).

(i-2) D. Auvergne, P. Merle, and H. Mathieu, *Phys. Rev. B* **12**:1371 (1975).

(i-3) S. E. Stokowski and D. D. Sell, *Phys. Rev. B* **5**:1636 (1972).

(i-4) G. A. Samara, *Phys. Rev. B* **27**:3494 (1983).

(i-5) H. Welker, *J. Electron* **1**:181 (1955).

(i-6) See (g-10).

(i-7) See (a-10).

j-ZnS

(j-1) D. Theis, *Phys. Stat. Sol. (b)* **79**:125 (1977).

(j-2) J. Camassel and D. Auvergne, *Phys. Rev. B* **12**:3258 (1975).

(j-3) B. Segall and D. T. F. Marple, *Physics and Chemistry of II-VI Compounds*, M. Aven and J. S. Prener (eds.), North-Holland, Amsterdam, 1967, p. 318.

(j-4) W. Walter and J. L. Birman, *Proc. Int. Conf on II-VI Semiconducting Compounds, 1967*, D. G. Thomas (ed.), W. A. Benjamin, New York, 1967, p. 89.

(j-5) G. Martinez, in *Handbook on Semiconductors*, vol. 2, T. S. Moss (ed.), North-Holland, Amsterdam, 1980, p. 210.

(j-6) M. Balkanski and Y. Petroff, *Proc. 7th Int. Conf. Physics of Semicond.* Paris, 1964, Dunod, Paris, 1964, p. 245.

(j-7) W. L. Wolfe, A. G. DeBell, and J. M. Palmer, *Proc. SPIE* **245**:164 (1980).

(j-8) R. J. Harris, G. T. Johnston, G. A. Kepple, P. C. Krock, and M. Mukai, *Appl. Opt.* **16**:436 (1977).

(j-9) M. Balkanski, M. Nusimovici, and R. Letoullec, *J. Phys. Paris* **25**:305 (1964).

(j-10) C. A. Klein and R. N. Donadio, *J. Appl. Phys.* **51**:797 (1980).

(j-11) See (a-10).

(j-12) J. C. Miklosz and R. G. Wheeler, *Phys. Rev.* **153**:913 (1967).

TABLE 13 Materials Parameters (*Continued*)

k-ZnSe

(k-1) See (j-1).
(k-2) P. J. Dean, D. C. Herbert, C. J. Werkhoven, B. J. Fitzpatrick, and R. M. Bhargava, *Phys. Rev. B* **23**:4888 (1981).
(k-3) L. Baillou, J. Daunay, P. Bugnet, Jac Daunay, C. Auzary, and P. Poindessault, *J. Phys. Chem. Solids* **41**:295 (1980).
(k-4) A. K. Ray and F. A. Kroger, *J. Appl. Phys.* **50**:4208 (1979).
(k-5) A. Hadni, J. Claudel, and P. Strimer, *Phys. Stat. Sol.* **26**:241 (1968).
(k-6) X. J. Jiang, T. Hisamura, Y. Nosua, and T. Goto, *J. Phys. Soc. Jpn* **52**:4008 (1983).
(k-7) See (j-8).
(k-8) M. Cardona, *J. Phys. Paris* **C8**:29 (1984).
(k-9) T. Yao, M. Ogura, S. Matsuoka, and T. Morishita, *J. Appl. Phys.* **43**:499 (1983).
(k-10) G. Jones and J. Woods, *J. Phys. D* **9**:799 (1976).
(k-11) T. Ohyama, E. Otsuka, T. Yoshida, M. Isshiki, and K. Igaki, *Jpn. J. Appl. Phys.* **23**:L382 (1984).
(k-12) M. Sondergeld, *Phys. Sta. Sol. (b)* **81**:253 (1977).

l-ZnTe

(l-1) See (j-1).
(l-2) M. Venghaus and P. J. Dean, *Phys. Rev. B* **21**:1596 (1980).
(l-3) See (h-5).
(l-4) D. Berlincourt, M. Jaffe, and L. R. Shiozawa, *Phys. Rev.* **129**:1009 (1983).
(l-5) D. T. F. Marple, *J. Appl. Phys.* **35**:539 (1964).
(l-6) T. L. Chu, S. S. Chu, F. Firszt, and C. Herrington, *J. Appl. Phys.* **59**:1259 (1986).
(l-7) See (k-8).
(l-8) A. G. Fisher, J. N. Carides, and J. Dresner, *Solid State Commun.* **2**:157 (1964).
(l-9) H. Venghaus, P. J. Dean, P. E. Simmonds, and J. C. Pfister, *Z. Phys. B* **30**:125 (1978).
(l-10) M. Aven and B. Segall, *Phys. Rev.* **130**:81 (1963).

m-CdTe

(m-1) P. M. Amirtharaj and D. Chandler-Horowitz, (to be published).
(m-2) P. M. Amirtharaj, R. C. Bowman, Jr., and R. L. Alt, *Proc. SPIE* **946**:57 (1988).
(m-3) N. Nawrocki and A. Twardowski, *Phys. Stat. Sol. (b)* **97**:K61 (1980).
(m-4) See (j-2).
(m-5) See (l-3).
(m-6) A. Moritani, K. Tamiguchi, C. Hamaguchi, and J. Nakai, *J. Phys. Soc. Jpn.* **34**:79 (1973).
(m-7) T. J. Parker, J. R. Birch, and C. L. Mok, *Solid State Communications* **36**:581 (1980).
(m-8) L. S. Ladd, *Infrared Phys.* **6**:145 (1966).
(m-9) J. R. Birch and D. K. Murray, *Infrared Phys.* **18**:283 (1978).
(m-10) See (a-10).
(m-11) K. K. Kanazawa and F. C. Brown, *Phys. Rev.* **135**:A1757 (1964).
(m-12) See (j-3).

n-HgTe

(n-1) W. Szuszkiewicz, *Phys. Stat. Sol. (b)* **81**:K119 (1977).
(n-2) See (m-6).
(n-3) J. Baars and F. Sorger, *Solid State Commun.* **10**:875 (1972).
(n-4) H. Kepa, T. Giebultowicz, B. Buras, B. Lebech, and K. Clausen, *Physica Scripta* **25**:807 (1982).
(n-5) T. C. Harmon, in *Physics and Chemistry of II-VI Compounds*, M. Aven and J. S. Prener (eds). North Holland Publishing, Amsterdam, 1967, p. 767.
(n-6) Y. Guldner, C. Rigaux, M. Grynberg, A. Mycielski, *Phys. Rev. B* **8**:3875 (1973).
(n-7) K. Shimizu, S. Narita, Y. Nisida, and V. I. Ivanov-Omskii, *Solid State Commun.* (eds.), **32**:327 (1979).

o-CdS

(o-1) V. V. Sobolev, V. I. Donetskina, and E. F. Zagainov, *Sov. Phys. Semicond.* **12**:646 (1978).

(o-2) D. G. Seiler, D. Heiman, and B. S. Wherrett, *Phys. Rev. B* **27**:2355 (1983).

(o-3) A series excitons: D. G. Seiler, D. Heiman, R. Fiegenblatt, R. Aggarwal, and B. Lax, *Phys. Rev. B* **25**:7666 (1982); B series excitons: see (o-2).

(o-4) A. S. Barker and C. J. Summers, *J. Appl. Phys.* **41**:3552 (1970).

(o-5) R. Weil and D. Neshmit, *J. Opt. Soc. Am.* **67**:190 (1977).

(o-6) Complex Phonon Structure with Nine Allowed Optical Modes. See B. Tel, T. C. Damen, and S. P. S. Porto, *Phys. Rev.* **144**:771 (1966).

(o-7) See (a-10).

p-PbS

(p-1) R. B. Schoolar and J. R. Dixon, *Phys. Rev. A* **137**:667 (1965).

(p-2) D. L. Mitchell, E. D. Palik, and J. N. Zemel, *Proc. 7th Int. Conf. Phys. Semicond.*, Paris, 1964, Dunod, Paris, 1964, p. 325.

(p-3) G. Nimtz and B. Schlicht, *Springer Tracts in Modern Physics*, vol. 98, Springer-Verlag, Berlin, 1983, p. 1.

(p-4) M. Cardona and D. L. Greenaway, *Phys. Rev. A* **133**:1685 (1964).

(p-5) R. Dalven, in *Solid State Physics*, vol. 28, H. Ehrenreich, F. Seitz, and D. Turnbull (eds.), Academic, NY, 1973, p. 179.

(p-6) R. B. Schoolar and J. N. Zemel, *J. Appl. Phys.* **35**:1848 (1964).

(p-7) M. M. Elcombe, *Proc. Soc. London,* **A300**:210 (1967).

(p-8) See (a-10).

(p-9) K. F. Cuff, M. R. Ellet, C. D. Kulgin, and L. R. Williams, in *Proc. 7th Int. Conf. Phys. Semicond.*, Paris, 1964, M. Hulin (ed.), Dunod, Paris, 1964, p. 677.

q-PbTe

(q-1) M. Preier, *Appl. Phys.* **20**:189 (1979).

(q-2) C. R. Hewes, M. S. Adler, and S. D. Senturia, *Phys. Rev. B* **7**:5195 (1973).

(q-3) See (p-3).

(q-4) See (p-4).

(q-5) W. E. Tennant, *Solid State Communications* **20**:613 (1976).

(q-6) J. R. Lowney and S. D. Senturia, *J. Appl. Phys.* **47**:1773 (1976).

(q-7) N. Piccioli, J. B. Beson, and M. Balkanski, *J. Phys. Chem. Solids* **35**:971 (1974).

(q-8) W. Cochran, R. A. Cowley, G. Dolling, and M. M. Elcombe, *Proc. R. Soc. London,* **A293**:433 (1966).

(q-9) See (a-10).

(q-10) See (p-9).

r-PbSe

(r-1) U. Schlichting, Dissertation Technische Universität Berlin, 1970.

(r-2) D. L. Mitchell, E. D. Palik, and J. N. Zemel, *Proc. 7th Int. Conf. Phys. Semicond., Paris, 1964,* M. Hulin (ed.), Dunod, Paris, 1964, p. 325.

(r-3) H. Pasher, G. Bauer, and R. Grisar, *Phys. Rev. B* **38**:3383 (1988).

(r-4) A. F. Gibson, *Proc. Phys. Soc.* (London) **B65**:378 (1952).

(r-5) See (p-4).

(r-6) See (p-5).

(r-7) J. N. Zemel, J D. Jensen, and R. B. Schoolar, *Phys. Rev. A* **140**:330 (1965).

(r-8) H. Burkhard, R. Geick, P. Kastner, and K. H. Unkelbach, *Phys. Stat. Sol. (b)* **63**:89 (1974).

(r-9) E. Burstein, R. Wheeler, and J. Zemel, *Proc. 7th Int. Conf. Phys. Semicond., Paris, 1964,* M. Hulin (ed.), Dunod, Paris, 1964, p. 1065.

(r-10) R. N. Hall and J. H. Racette, *J. Appl. Phys.* **32**:2078 (1961).

(r-11) J. N. Zemel, J. D. Jenson, and R. B. Schoolar, *Phys. Rev. A* **140**:330 (1965).

(r-12) U. Schlichting and K. H. Gobrecht, *J. Phys. Chem. Solids* **34**:753 (1973).

(r-13) See (p-9).

36.7 REFERENCES

1. D. Attwood, B. Hartline, and R. Johnson, *The Advanced Light Source: Scientific Opportunities,* Lawrence Berkeley Laboratory Publication 55, 1984.

2. F. C. Brown, in *Solid State Physics,* vol. 29, H. Ehrenreich, F. Seitz, and D. Turnbull (eds.), Academic Press, New York, 1974, p. 1.

3. A. V. Nurmikko, in *Semiconductors and Semimetals,* vol. 36, D. G. Seiler and C. L. Littler (eds.), Academic Press, New York, 1992, p. 85.

4. R. R. Alfano, (ed.), *Semiconductors Probed by Ultrafast Laser Spectroscopy,* vols. I and II, Academic Press, New York, 1984.

5. G. Bastard, C. Delalande, Y. Guldner, and P. Vosin, in *Advances in Electronics and Electron Physics,* vol. 72, P. W. Hawkes (ed.), Academic Press, New York, 1988, p. 1.

6. C. Weisbuch and B. Vinter, *Quantum Semiconductor Structures, Fundamentals and Applications,* Academic Press, New York, 1991, p. 57.

7. M. Born and K. Huang, *Dynamical Theory of Crystal Lattices,* chap. 2, Oxford University Press, London, 1954, p. 38.

8. W. K. H. Panofsky and M. Phillips, *Classical Electricity and Magnetism,* Addison-Wesley, New York, 1962, p. 29.

9. E. D. Palik (ed.), *Handbook of Optical Constants of Solids,* vol. 1., Academic Press, New York, 1985.

10. E. D. Palik (ed.), *Handbook of Optical Constants of Solids,* vol. 2., Academic Press, New York, 1991.

11. E. D. Palik (ed.), *Handbook of Optical Constants of Solids,* vol. 1., Academic Press, New York, 1985, p. 429.

12. M. Born and E. Wolf, *Principles of Optics,* Pergamon, London, 1970, p. 61.

13. J. S. Toll, *Phys. Rev.* **104:**1760 (1956).

14. D. Y. Smith, *J. Opt. Soc. Am.,* **66:**454 (1976).

15. D. L. Greenaway and G. Harbeke, *Optical Properties and Band Structure of Semiconductors,* Pergamon, London, 1968, p. 9.

16. J. M. Ziman, *Principles of the Theory of Solids,* Cambridge University Press, London, 1972, p. 200.

17. C. Kittel, *Introduction to Solid State Physics,* 4th ed., John Wiley, New York, 1971, p. 184.

18. W. J. Turner and W. E. Reese, *Phys. Rev.* **127:**126 (1962).

19. S. S. Mitra, in *Optical Properties of Solids,* S. Nudelman and S. S. Mitra (eds.), Plenum Press, New York, 1979, p. 333.

20. W. Cochran, *The Dynamics of Atoms in Crystals,* Crane, Rusak and Co., New York, 1973.

21. K. C. Rustagi and W. Weber, *Solid State Communications* **18:**673 (1976).

22. S. S. Mitra and N. E. Massa, in *Handbook on Semiconductors,* T. S. Moss and W. Paul (eds.), North Holland, Amsterdam, 1982, p. 81.

23. W. G. Spitzer, in *Semiconductors and Semimetals,* vol. 3., R. K. Willardson and A. C. Beer (eds.), Academic Press, New York, 1967, p. 17.

24. J. L. Birman, *Theory of Crystal Space Groups and Lattice Dynamics,* Springer-Verlag, Berlin, 1974, p. 271.

25. F. A. Johnson, *Proc. Phys. Soc.* (London) **73:**265 (1959).

26. W. Cochran, S. J. Fray, F. A. Johnson, J. E. Quarrington, and N. Williams, *J. Appl. Phys.* **32:**2102 (1961).

27. C. A. Klein and R. N. Donadio, *J. Appl. Phys.* **51:** 797 (1980).

28. A. S. Barker and A. J. Sievers, *Reviews of Modern Physics,* vol. 47, suppl. no. 2, S1 (1975).

29. W. M. Theis, K. K. Bajaj, C. W. Litton, and W. G. Spitzer, *Appl. Phys. Lett.* **41:**70 (1982).

30. R. S. Leigh and R. C. Newman, *J. Phys. C: Solid State Phys.* **15:** L1045 (1982).

31. M. Stavola and S. J. Pearton, in *Semiconductors and Semimetals,* vol. 34, J. I. Pankove and N. M. Johnson (eds.), Academic Press, New York, 1991, p. 139.

32. R. C. Newman, in *Growth and Characterization of Semiconductors,* R. A. Stradling and P. C. Klipstein (eds.), Adam Hilger, Bristol, 1990, p. 105.

33. A. Baghdadi, W. M. Bullis, M. C. Croarkin, Y. Li, R. I. Scace, R. W. Series, P. Stallhofer, and M. Watanabe, *J. Electrochem. Soc.* **136:**2015 (1989).

34. W. L. Wolfe, *The Infrared Handbook,* W. L. Wolfe and G. J. Zeiss (eds.), Environmental Research Institute, Ann Arbor, 1978, pp. 7–39.

35. A. Miller, *Handbook of Optics,* 2d ed., vol. I, chap. 9, McGraw-Hill, New York, 1994.

36. E. J. Johnson, in *Semiconductors and Semimetals,* vol. 3, R. K. Willardson and A. C. Beer (eds.), Academic Press, New York, 1967, p. 153.

37. E. O. Kane, *J. Phys. Chem. Solids* **1:**249 (1957).

38. T. S. Moss and T. D. F. Hawkins, *Infrared Phys.* **1:**111 (1961).

39. (op. cit.) E. J. Johnson, p. 191.

40. W. G. Spitzer, M. Gershenzon, C. J. Frosch, and D. F. Gibbs, *J. Phys. Chem. Solids* **11:**339 (1959).

41. M. Gershenzon, D. G. Thomas, and R. E. Dietz, *Proc. Int. Conf. Phys. Semicond. Exeter,* Inst. of Physics, London, 1962, p. 752.

42. W. C. Dash and R. Newman, *Phys. Rev.* **99:**1151 (1955). See also G. Burns, *Solid State Physics,* Academic Press, New York, 1985, p. 505.

43. K. Cho, *Excitons,* vol. 14 of *Topics in Current Physics,* K. Cho (ed.), Springer-Verlag, New York, 1979, p. 1.

44. W. Hayes and A. M. Stoneham, *Defects and Defect Processes in Nonmetallic Solids,* John Wiley and Sons, New York, 1985, p. 40.

45. D. C. Reynolds and T. C. Collins, *Excitons: Their Properties and Uses,* Academic Press, New York, 1981, pp. 1–291.

46. E. I. Rashba and M. D. Sturge (eds.), *Excitons,* North Holland, Amsterdam, 1982, pp. 1–865.

47. J. Frenkel, *Phys. Rev.* **37:**1276 (1931).

48. G. H. Wannier, *Phys. Rev.* **52:**191 (1937).

49. N. F. Mott, *Proc. Roy. Soc. A* **167:**384 (1938).

50. M. D. Sturge, *Phys. Rev.* **127:**768 (1962). For an excellent discussion of theoretical aspects see article by M. Sturge, "Advances in Semiconductor Spectroscopy," in B. DiBartolo (ed.), *Spectroscopy of Laser-Type Materials,* Plenum Press, New York, 1987, p. 267.

51. Adapted by permission from R. G. Ulbrich and C. Weisbuch, Contribution to the study of optical pumping in III-V Semiconductors, These de doctorat d'Etat, Univ. Paris **7,** 1977 (Unpublished).

52. K. T. Shindo, T. Gato, and T. Anzai, *J. Phys. Soc. Jpn.* **36:**753 (1974).

53. (op. cit.) W. Hayes and A. M. Stoneham, p. 51.

54. O. Madelung, *Introduction to Solid-State Theory,* Springer-Verlag, New York, 1978, p. 254.

55. G. Burns, *Solid State Physics,* Academic Press, New York, 1985, p. 969.

56. H. R. Philipp and H. Ehrenreich, in *Semiconductors and Semimetals,* vol. 3, R. K. Willardson and A. C. Beer (eds.), Academic Press, New York, 1967, p. 93. For a review of interband transitions in Semiconductors see M. L. Cohen and J. R. Chelikowski, *Electronic Structure and Optical Properties of Semiconductors,* Springer-Verlag, Berlin, 1988.

57. H. Ehrenreich, H. R. Philipp, and J. C. Phillips, *Phys. Rev. Lett.* **8:**59 (1962).

58. D. Brust, J. C. Phillips, and F. Bassani, *Phys. Rev. Lett.* **9:**94 (1962).

59. D. Brust, *Phys. Rev.* **134:**A1337 (1964).

60. H. R. Philipp and H. Ehrenreich, *Phys. Rev. Lett.* **8:**92 (1962).

61. H. R. Philipp and H. Ehrenreich, *Phys. Rev.* **129:**1550 (1963).

62. C. Gahwiller and F. C. Brown, *Phys. Rev. B* **2:**1918 (1970).

63. S. Perkowitz, in *Infrared and Millimeter Waves,* vol. 8, K. J. Button (ed.), Academic Press, New York, 1983, p. 71.

64. J. R. Dixon and H. R. Riedl, *Phys. Rev.* **138:**A873 (1965).

65. P. M. Amirtharaj, B. L. Bean, and S. Perkowitz, *J. Opt. Soc. Am.* **67:**939 (1977).

66. B. B. Varga, *Phys. Rev.* **137:**A1896 (1965).

67. A. Mooradian and G. B. Wright, *Phys. Rev. Lett.* **16:**999 (1966).

68. A. Mooradian, in *Advances in Solid State Physics,* vol. 9, O. Madelung (ed.), Pergamon Press, London, 1969, p. 74.

69. C. G. Olson and D. W. Lynch, *Phys. Rev.* **177:**1231 (1969).

70. S. C. Baber, *Thin Solid Films* **72:**201 (1980).

71. G. M. Martin, *Appl. Phys. Lett.* **39:**747 (1981).

72. C. M. Wolfe and G. E. Stillman, *Gallium Arsenide and Related Compounds,* Inst. Phys., London, 1971, p. 3.

73. M. H. Weiler, *Semiconductors and Semimetals,* vol. 16, R. K. Willardson and A. C. Beer (eds.), Academic Press, New York, 1981, p. 119.

74. C. R. Pidgeon and R. N. Brown, *Phys. Rev.* **146:**575 (1966).

75. M. H. Weiler, R. L. Aggarwal, and B. Lax, *Phys. Rev. B* **17:**3269 (1978).

76. M. H. Weiler, *J. Magn. Magn. Mater.* **11:**131 (1979).

77. M. Reine, R. L. Aggarwal, and B. Lax, *Phys. Rev B* **5:** 3033 (1972).

78. M. Reine, R. L. Aggarwal, B. Lax, and C. M. Wolfe, *Phys. Rev. B* **2:**458 (1970).

79. H. Piller, *Proc. 7th Int. Conf. Phys. Semicond.,* Dunod, Paris, 1964, p. 297.

80. E. Oh, D. U. Bartholomew, A. K. Ramdas, J. K. Furdyna, and U. Debska, *Phys. Rev. B* **38:**13183 (1988).

81. R. J. Eliott and R. Loudon, *J. Phys. Chem. Solids* **15:**196 (1960).

82. D. G. Seiler, D. Heiman, R. Feigenblatt, R. L. Aggarwal, and B. Lax, *Phys. Rev. B* **25:**7666 (1982).

83. D. G. Seiler, D. Heiman, and B. S. Wherrett, *Phys. Rev. B* **27:**2355 (1983).

84. B. D. McCombe and R. J. Wagner, *Adv. Electron. and Electron. Phys.* **37:**1 (1975).

85. B. D. McCombe and R. J. Wagner, *Adv. Electron. and Electron. Phys.* **38:**1 (1975).

86. H. Kobori, T. Ohyama, and E. Otsuka, *J. Phys. Soc. Jpn.* **59:**2164 (1990).

87. B. Lax, H. J. Zeiger, and R. N. Dexter, *Physica* **20:**818 (1954).

88. G. Dresselhaus, A. F. Kip, and C. Kittel, *Phys. Rev.* **98:**368 (1955).

89. M. A. Omar, *Elementary Solid State Physics,* Addison-Wesley, Reading, 1975, p. 285.

90. O. Matsuda and E. Otsuka, *J. Phys. Chem. Solids* **40:**809 (1979).

91. D. Larsen, in *Landau Level Spectroscopy,* vol. 27.1, chap. 3, G. Landwehr and E. I. Rashba (eds.), North Holland, Amsterdam, 1991, p. 109.

92. H. R. Fetterman, D. M. Larsen, G. E. Stillman, P. E. Tannenwald, and J. Waldman, *Phys. Rev. Lett.* **26:**975 (1971).

93. C. L. Tang, *Handbook of Optics,* 2d ed., vol. II, chap. 38, McGraw-Hill, New York, 1993.

94. Y. R. Shen, *The Principles of Nonlinear Optics,* John Wiley and Sons, New York, 1984, p. 38.

95. P. N. Butcher, *Nonlinear Optical Phenomena,* Ohio State Univ. Eng. Publications, Columbus, 1965, p. 1.

96. B. S. Wherrett, *Nonlinear Optics,* P. G. Harper and B. S. Wherrett (eds.), Academic Press, New York, 1977, p. 4.

97. J. H. Bechtal and W. L. Smith, *Phys. Rev. B* **8:**3515 (1976).

98. P. D. Maker and R. H. Terhune, *Phys. Rev.* **137:**A801 (1965).

99. M. D. Levenson, *Phys. Today* **30**(5):44 (1977).

100. D. Heiman, W. Hellwarth, M. D. Levenson, and G. Martin, *Phys. Rev. Lett.* **36:** 189 (1976).

101. R. M. A. Azzam and N. M. Bashara, *Ellipsometry and Polarized Light,* North-Holland, Amsterdam, 1987, p. 153. See also Azzam in *Handbook of Optics, 2d ed.,* vol. II, chap. 27, McGraw-Hill, New York, 1994.

102. D. E. Aspnes, *Proc. SPIE* **946:**84 (1988).

103. W. Hayes and R. Loudon, *Scattering of Light by Crystals,* John-Wiley, New York, 1978, p. 53.

104. P. J. Dean, *Prog. Crystal Growth and Characterization* **5:**89 (1982).

105. M. Voos, R. F. Lehney, and J. Shah, in *Handbook of Semiconductors,* vol. 2, T. S. Moss and M. Balkanski (eds.), North-Holland, Amsterdam, 1980, p. 329.

106. See Table 13.

107. See LaRocca in *Handbook of Optics, 2d ed.,* vol. I, chap. 10, McGraw-Hill, New York, 1994.

108. See Henderson in *Handbook of Optics, 2d ed.,* vol. II, chap. 20, McGraw-Hill, New York, 1994.

109. See Harinaran in *Handbook of Optics, 2d ed.,* vol. II, chap. 21, McGraw-Hill, New York, 1994.

110. See Norton in *Handbook of Optics, 2d ed.,* vol. I, chap. 15, McGraw-Hill, New York, 1994.

111. U. Gerhardt and G. Rubloff, *Appl. Opt.* **8:**305 (1969).

112. M. I. Bell and D. A. McKeown, *Rev. Sci. Instrum.* **61:**2542 (1990).

113. D. E. Aspnes and J. E. Fischer, in *Encyclopaedic Dictionary of Physics,* suppl. vol. 5, Thewlis (ed.), Pergamon, Oxford, 1975, p. 176.

114. D. E. Aspnes, in *Handbook on Semiconductors,* vol. 2, T. S. Moss and M. Balkanski (eds.), North-Holland, Amsterdam, 1980, p. 109.

115. D. E. Aspnes and A. A. Studna, *Phys. Rev. B* **7:**4605 (1973).

116. R. E. Nahory and J. L. Shay, *Phys. Rev. Lett.* **21:**1589 (1968).

117. F. H. Pollak, *Proc. SPIE* **276:** 142 (1981).

118. O. J. Glembocki, B. V. Shanabrook, N. Bottka, W. T. Beard, and J. Comas, *Appl. Phys. Lett* **46:**970 (1985).

119. X. Yin, H. M. Chen, F. Pollak, Y. Chan, P. A. Montano, P. D. Kichner, G. D. Pettit, and J. M. Woodall, *Appl. Phys. Lett.* **58:**260 (1991).

120. F. H. Pollak, M. Cardona, and D. E. Aspnes (eds.), *Proc. Int. Conf. on Modulation Spectroscopy,* Proc. SPIE, Bellingham, 1990, *1286.*

121. O. J. Glembocki and B. V. Shanabrook, in *Semiconductors and Semimetals,* vol. 36, D. G. Seiler and C. L. Littler (eds.), Academic Press, New York, 1992, p. 221.

122. R. M. A. Azzam, *"Ellipsometry,"* in M. Bass (ed.), *Handbook of Optics,* 2d ed., vol. II, chap. 27, McGraw-Hill, New York, 1994.

123. D. B. Holt and B. G. Yacobi, in *SEM Microcharacterization of Semiconductors,* D. B. Holt and D. C. Joy (eds.), Academic Press, New York, 1989, p. 373.

124. J. I. Pankove, in *Electroluminescence,* J. I. Pankove (ed.), Springer-Verlag, Berlin, 1977, p. 1.

125. E. C. Lightowlers, in *Growth and Characterization of Semiconductors,* R. A. Stradling and P. C. Klipstein (eds.), Adam Hilger, Bristol, 1990, p. 135.

126. M. L. W. Thewalt, M. K. Nissen, D. J. S. Beckett, and K. R. Lundgren, *Mat. Res. Soc. Symp.,* **163:**221 (1990).

127. *Properties of GaAs, 2d Ed., Inspec Data Reviews,* Series, no. 2, chap. 12, IEE, London, 1990, p. 229.

128. F. Fuchs, A. Lusson, P. Koidl, and R. Triboulet, *J. Crystal Growth* **101:**722 (1990).

129. P. Hiesinger, S. Suga, F. Willmann, and W. Dreybrodt, *Phys. Stat. Sol. (b)* **67:**641 (1975).

130. *Light Scattering in Solids,* M. Cardona (ed.), vol. 1., Springer-Verlag, Berlin, 1983; *Light Scattering in Solids,* vol. 2, M. Cardona and G. Guntherodt (eds.), Springer-Verlag, Berlin, 1982; *Light Scattering in Solids,* vol. 3, M. Cardona and G. Guntherodt (eds.), Springer-Verlag, Berlin, 1982; *Light Scattering in Solids,* vol. 4, M. Cardona and G. Guntherodt (eds.), Springer-Verlag, Berlin, 1984; *Light Scattering in Solids,* vol. 5, M. Cardona and G. Guntherodt (eds.), Springer-Verlag, Berlin, 1989.

131. (op. cit.) W. Hayes and R. Loudon, pp. 1–360.

132. A. S. Pine, in *Light Scattering of Solids,* vol. 1, M. Cardona (ed.), Springer-Verlag, Berlin, 1982, p. 253.

133. (op. cit.) W. Hayes and R. Loudon, p. 16.

134. (op. cit.) W. Hayes and R. Loudon, p. 44.

135. G. Abstreiter, *Applied Surface Science* **50:**73 (1991).

136. K. K. Tiong, P. M. Amirtharaj, F. H. Pollak, and D. E. Aspnes, *Appl. Phys. Lett.* **44:** 122 (1984).

137. P. A. Temple and C. E. Hathaway, *Phys. Rev. B* **7:**3685 (1973).

138. G. Abstreiter, R. Trommer, M. Cardona, and A. Pinczuk, *Solid State Communications* **30:**703 (1979).

139. A. Pinczuk, L. Brillson, E. Burstein, and E. Anastassakis, *Phys. Rev. Lett.* **27:**317 (1971).

140. J. M. Ralston, R. L. Wadsack, and R. K. Chang, *Phys. Rev. Lett.* **25:**814 (1970).

141. M. Chandrasekhar, H. R. Chandrasekhar, M. Grimsditch, and M. Cardona, *Phys. Rev. B* **22:**4285 (1980).

142. D. D. Manchon, Jr., and P. J. Dean, *Proc. 10th Int. Conf. on Physics of Semiconductors,* S. P. Keller, J. C. Hemsel, and F. Stern (eds.), USAEC, Cambridge, 1970, p. 760.

143. A. W. Nelson, in *Electronic Materials from Silicon to Organics,* L. S. Miller and J. B. Mullin (eds.), Plenum Press, New York, 1991, p. 67.

144. O. Madelung, M. Schulz, and H. Weiss (eds.), *Landolt-Bornstein Numerical Data and Functional Relationships in Science and Technology, Group III—Crystal and Solid State Physics,* vol. 17a, Springer-Verlag, Berlin, 1982.

145. O. Madelung, M. Schulz, and H. Weiss (eds.), *Landolt-Bornstein Numerical Data and Functional Relationships in Science and Technology, Group III—Crystal and Solid State Physics,* vol. 17b, Springer-Verlag, Berlin, 1982.

146. O. Madelung, M. Schulz, and H. Weiss (eds.), *Landolt-Bornstein Numerical Data and Functional Relationships in Science and Technology, Group III—Crystal and Solid State Physics,* vol. 17f, Springer-Verlag, Berlin, 1982.

147. O. Madelung and M. Schulz (eds.), *Landolt-Bornstein Numerical Data and Functional Relationships in Science and Technology, Group III—Crystal and Solid State Physics,* vol. 22a, Springer-Verlag, Berlin, 1987.

CHAPTER 37
BLACK SURFACES FOR OPTICAL SYSTEMS

Stephen M. Pompea
Pompea and Associates
Tucson, Arizona
and
Steward Observatory
University of Arizona
Tucson, Arizona

Robert P. Breault
Breault Research Organization, Inc.
Tucson, Arizona

37.1 INTRODUCTION

Optical instruments and telescopes rely on black baffle and vane surfaces to minimize the effect of stray light on overall system performance. For well-designed and well-baffled systems, the black surfaces chosen for the baffles and vanes can play a significant role in reducing the stray light on the detector (e.g., Wolfe, 1980; Breault, 1990; Bergener et al., 1984; Pompea et al., 1988, 1992a,b). Black surfaces are also used extensively in solar collector applications. Excellent reviews of spectrally selective surfaces for heating and cooling applications are found in Hahn and Seraphim (1978) and in Granqvist (1989). Black coatings are also used in radiometric detectors (Betts et al., 1985). Because the surface needed is often small, these surfaces may be even more specialized than the black coatings used for stray light reduction in optical instruments. This chapter will concentrate on the selection and characterization of black surfaces chosen for stray light suppression and suitable for application to relatively large areas of an optical system.

The optical system designer has a wide repertoire of baffle surfaces from which to choose. Summaries of optical properties of materials were given by Wolfe (1978), Pompea et al. (1988), and McCall et al. (1992a). Reviews of materials by McCall (1992b) and Smith and Howitt (1986) emphasized the ultraviolet/visible and infrared properties, respectively. A number of company databases of scattering data are available, including one using the same instrument (for bidirectional scatter distribution function or BSDF measurements at 0.5145 micrometers) for approximately 15,000 data runs (Schaub et al., 1990)! Large amounts of BSDF data are at Breault Research Organization and a BSDF database format has been proposed by Klicker et al. (1990). An organized effort to create specialized databases of optical properties of surfaces applicable to both ground- and space-based instruments has been undertaken and is proving to have great utility (McCall et al., 1993).

The choices of optical black surfaces are usually first narrowed by the nature of the application, the substrates available or possible, the wavelength or bandpasses of interest, the angles at which the surfaces must be nonreflective, and a host of system issues and environmental factors (Pompea and McCall, 1992a). As the system performance requirements have become more stringent, an array of surfaces has become available to meet these requirements. Many paints (e.g., Chemglaze Z-306, SolarChem) (Table 1); anodized surfaces (e.g., Martin Black, Infrablack, Tiodize) (Table 2); etched, electrodeposited, and plasma-sprayed metal surfaces (Table 3) are now available to meet quite specific optical and system performance and environmental requirements.

New classes of surfaces are also being developed to give selected performance at specific angles and wavelengths. Other materials are being developed for applications where "hardened" laser-resistant or radiation-resistant materials are needed. A third area where much development is currently taking place is in the area of materials that are able to withstand severe and unusual forms of environmental exposure for long periods of time.

The surface morphology of a "black" or diffusely scattering surface is only one determinant of its complex optical properties. However, for many surfaces, surface roughness plays a most significant role. This can be illustrated in Figs. 1 to 8, scanning electron micrographs of some important black or diffuse surfaces. The size and shapes of the surface features provide a valuable indication of how light will be absorbed and scattered or diffracted by such a surface. The creation and design of new surfaces will be

TABLE 1 Painted Surfaces

[Adapted from McCall (1992c).]

Surface name or designation	Manufacturer and/or distributor (contact person)	Historical notes	Surface type	Main literature references
Aeroglaze L300	Lord Corporation Chemical Products Division Industrial Coatings 2000 West Grandview Boulevard P.O. Box 10038 Erie, PA 16514–0038	Formerly called Chemglaze L300	Paint	Lord Corp., 1992
Chemglaze Z004	Lord Corporation Erie, Pa.		Paint	Lord Corp., 1992
Aeroglaze Z302	Lord Corporation Erie, Pa.	Formerly called Chemglaze Z302	Paint	Griner et al., 1979 Fernandez et al., 1988 Lord Corp., 1992
Aeroglaze Z306	Lord Corporation Erie, Pa.	Formerly called Chemglaze Z306	Paint	Brown et al., 1990 Lord Corp., 1992 Ames, 1990 Lompado et al., 1989 Muscari et al., 1981 Pompea et al., 1984a Schaub et al., 1990 Smith et al., 1982a

TABLE 1 Painted Surfaces (*Continued*)

Surface name or designation	Manufacturer and/or distributor (contact person)	Historical notes	Surface type	Main literature references
				Smith, 1982b
				Smith, 1984
				Smith and Howitt, 1986a
				Smith, 1990
				Viehmann et al., 1986
				CAL, 1986
				Nordberg et al., 1984
				Noll et al., 1980
				Smith, 1980
				Jelinsky et al., 1987
				Stierwalt, 1979
				Willey et al., 1983
Aeroglaze Z306 with microspheres	Lord Corporation Erie, Pa.	Formerly called Chemglaze Z306 with microspheres	Paint	Evans et al., 1984 Heslin et al., 1974 Evans, 1983 Ames, 1990 Lord Corp., 1992
Aeroglaze Z307	Lord Corporation Erie, Pa.	Formerly called Chemglaze Z307	Paint	Lord Corp., 1992
Aeroglaze Z313	Lord Corporation Erie, Pa.	Formerly called Chemglaze Z313	Paint	Viehmann et al., 1986 Lord Corp., 1992
Ames 24E Ames 24E2	NASA Ames Research Center Moffet Field, Calif. S. Smith Sterling Software 1121 San Antonio Rd. Palo Alto, CA 94303		Paint	Pompea et al., 1988 Smith, 1988 Nee et al., 1990 Smith, 1990
Cardinal 6450	Cardinal Industrial Finishes 1329 Potrero Ave., So. El Monte, CA 91733-3088	Formerly called "Cardinal 6550"	Paint	Cardinal, 1992 Viehmann et al., 1986
Cornell Black	Prof. J. Houck Department of Astronomy Cornell University Ithaca, NY 14853		Paint	Smith, 1980 Smith, 1982b Smith, 1984 Smith and Howitt, 1986a Houck, 1982
DeSoto Flat Black	DeSoto, Inc., Chemical Coatings Division		Paint	Smith, 1986a Viehmann et al., 1986
Electrically Conductive Black Optical Paint	Jet Propulsion Laboratory, Caltech, 4800 Oak Grove Dr. Pasadena, CA 91109	Has no trade name	Paint	Birnbaum et al., 1982 Metzler et al., 1988 Breault, 1988a

TABLE 1 Painted Surfaces (*Continued*)

Surface name or designation	Manufacturer and/or distributor (contact person)	Historical notes	Surface type	Main literature references
IIPRI Bone Black D-111 (IITRI D111)	IIT Research Institute 10 West 35th Street Chicago, IL 60616		Paint	Griver, 1979 IITRI, 1992 Smith et al., 1982a Smith et al., 1982b Smith, 1984 Grammar et al., 1980 Noll et al., 1980
LMSC Black	Lockheed Palo Alto Research Lab		Painted multilayer coating	Smith, 1982b Smith, 1984 Smith and Howitt, 1986a Grammar et al., 1980
MH21-1	IIT Research Institute Chicago, Ill.		Paint	IITRI, 1992
MH55	IIT Research Institute Chicago, Ill.		Paint	IITRI, 1992
MH2200	IIT Research Institute Chicago, Ill.	Formerly 3M's ECP 2200 paint, but sold to IIT	Paint	Fernandez et al., 1988 Brown et al., 1990 Driscoll, 1978 IITRI, 1992 Smith, 1983 Smith, 1984 Smith and Howitt, 1986a Smith, 1986b Viehmann et al., 1986 Wyman et al., 1975
Solarchem	Eastern Chem Lac Corporation 1080-T Eastern Ave. Malden, MA 02148		Paint	Pompea et al., 1988
463-3-8	Akzo Coatings, Inc. 434 W. Meats Avenue Orange, CA 92665	Formerly called "Cat-a-lac 463-3-8" diffuse black paint	Paint	Brown et al., 1990 Driscoll, 1978 Breault, 1983 Akzo, 1992 Muscari et al., 1981 Wyman et al., 1975 Willey et al., 1983 Grammar et al., 1980 Stierwalt, 1979
443-3-8	Akzo Coatings, Inc. Orange, Calif.	Formerly called "Cat-a-lac 443-3-8"	Paint	Freniere et al., 1986 Griner, 1979 Driscoll, 1978 Breault, 1983 Akzo, 1992 Wyman et al., 1975 Grammar et al., 1980 Stierwalt, 1979
443-3-17	Akzo Coatings, Inc. Orange, Calif.	Formerly called "Sikkens 443-3-17" glossy black	Paint	Fernandez et al., 1988 Akzo, 1992

TABLE 2 Anodized Surfaces
[*Adapted from McCall (1992c).*]

Surface name	Manufacturer	Notes	Surface type	Main literature references
Infrablack	Martin Marietta Astronautics Group P.O. Box 179, Denver CO 80201	For Al substrates only	Anodization process	Evans et al., 1984 Gull et al., 1986 Martin Marietta 1992 Bergener et al., 1984 Pompea et al., 1984 Pompea et al., 19884 Smith, 1988 Pompea, 1984b
Martin Black	Martin Marietta Astronautics Group Denver, CO	For Al substrates only	Anodization process	Evans et al., 1984 Griner, 1979 Geikas, 1983 Gull et al., 1986 Cady et al., 1987 Martin Marietta, 1992 Bartell et al., 1982 Bergener et al., 1984 Evans, 1983 Brooks et al., 1982 Breault, 1983 Breault, 1979 Young et al., 1980 Smith et al., 1982b Smith, 1984 Smith, 1988 Smith, 1990 Viehmann et al., 1986 Wyman et al., 1975 Grammar et al., 1980 Lompado et al., 1989 Noll et al., 1980 Pompea et al., 1984a Pompea et al., 1984b Pompea et al., 1988 Schaub et al., 1990 Shepard et al., 1988 Smith, 1980 Smith, 1981 Smith et al., 1982a Nordberg et al., 1989 Stierwalt, 1979 Noll et al., 1980
Martin Black, Enhanced	Martin Marietta Astronautics Group Denver, Colo.	For Al substrates only	Anodization process	Martin Marietta, 1992 Pompea et al., 1984a Pompea et al., 1988
Martin Black, Posttreated	Martin Marietta Astronautics Group Denver, Colo.	For Al substrates only	Anodization process	Martin Marietta, 1992

TABLE 3 Other Processes
[*Adapted from McCall (1992c).*]

Surface name	Manufacturer	Notes	Surface type	Main literature references
Black chrome-type surfaces	Martin Marietta Astronautics Group Denver, Colo.	For many kinds of metal substrates	Electrodeposition process	Martin Marietta, 1992 Pompea et al., 1988 Stierwalt, 1979
Black cobalt-type surfaces (cobalt black)	Martin Marietta Astronautics Group Denver, Colo. and many more companies	For many kinds of metal substrates; references to black copper, black, steel, etc., are covered by black cobalt	Electrodeposition processes, and can be followed by chemical or thermal oxidation	Martin Marietta, 1992 Pompea et al., 1988 Vitt, 1987 Hutchins et al., 1986
Black nickel (NBS black) (Ball black)	Ball Aerospace Electro-Optics Engineering Dept. P.O. Box 1062 Boulder CO 80306 and many others	For many kinds of metal substrate; Ball improved the patent developed by NBS	Deposition and etching process	Geikas, 1983 Smith and Howitt, 1986a Lompado et al., 1989 Smith, 1986a Nordberg et al., 1989 Johnson, 1979, 1980
Black Kapton film	DuPont Wilmington, DE 19898		Foil	Viehmann et al., 1986
Black Tedlar film	DuPont TEDLAR/Declar PPD, D-12082 1007 Market Street Wilmington, DE 19898		Foil	Viehmann et al., 1986
Boron black	Martin Marietta Astronautics Group Denver, Colo.	For many kinds of metals	Plasma spray deposition process	Pompea et al., 1988 Martin Marietta, 1992
Boron carbide	Martin Marietta Astronautics Group Denver, Colo.	For Ti substrates only	Proprietary process	Pompea et al., 1988
Silicon carbide	Martin Marietta Astronautics Group Denver, Colo. and more companies	For many kinds of metals	Chemical vapor deposition	Pompea et al., 1988
Textured surfaces	NASA Ames Research Center Moffet Field, Calif. (Sheldon Smith) Spire Corporation Patriots Park, Bedford, MA 01730 Optics MODIL, Oak Ridge National Lab Martin Marietta Energy Systems, Inc.	For many kinds of metal substrates	Either: —sputtered coated —ion beam-etched —sputter coated then etched	Cuomo et al., 1989 Banks, 1981 Mirtich et al., 1991 Lompado et al., 1989 Wollam et al., 1989 Von Benken et al., 1989 Murray et al., 1991 Nordberg et al., 1989 Spire, 1992 Egert et al., 1990 Mirtich et al., 1991 Blatchley et al., 1992

TABLE 3 Other Processes (*Continued*)

Surface name	Manufacturer	Notes	Surface type	Main literature references
	P.O. Box 2009, 9102-2, MS-8039 Oak Ridge, TN 37831-8039			
Black optical thin-film interference coatings	National Research Council of Canada Thin Films, Institute for Microstructural Sciences Montreal Road, Building M-36 Ottawa, Ontario, Canada, K1A OR6	For metal, dielectric, or other substrates	Vacuum deposition techniques; —Sputter deposition —ion vapor deposition —resistance-heated source —electron-beam gun deposition	Hass et al., 1956 Dobrowolski, 1981 Dobrowolski et al., 1989 Dobrowolski et al., 1992

touched upon later, in the section on design techniques for creating new surfaces for specific applications.

This chapter gives a summary of the materials currently (1994) available and describes their optical and material properties so that the optical designer can begin the material selection process. As such, it is rather an extreme condensation of the data available. However, even the data presented here cannot be considered very definitive. There are a number of reasons for inconsistencies and ambiguities in the data presented. First, many of these surfaces and the processes that create them are evolving, and improving continuously. Even though the actual surface may change, the name may not. Thus, optical measurements of the same named surface that are separated by several years may not be consistent, even if the measurement techniques are consistent.

A second cause of inconsistency among data sets comes from the remeasurement of "archival" samples. When new measurement techniques or improved instruments become available, or the needs of a program demand new measurements, archival samples are retrieved and remeasured. Sometimes these archival samples may not have been stored properly and may not be in pristine or original condition. Other times, these samples may never have been archival in quality. They may have been marketing samples made without specific quality control and distributed widely. These measurements still enter the body of literature with the reader usually unaware of the important circumstances. For robust samples, poor storage may be of little importance. For more exotic materials (e.g., specialized baffles for space applications) that must be handled carefully, it can be of great importance.

Wolfe (1982) compares the theory and experiments for bidirectional reflectance distribution function (BRDF) measurements of microrough surfaces. Measurements of optical quantities such as the BRDF, that are by definition measurement-device-independent, can also show large variations (Leonard and Pantoliano, 1988; Leonard et al., 1989). No attempt has been made in this chapter to reconcile measurement discrepancies; indeed, no attempt to even identify the areas where conflicting data occur is made. This would be a herculean task, but more importantly, it is probably unnecessary. The variations in the measurement process assure that the optical description of black surfaces is still an order-of-magnitude science, or, at best, a half-an-order-of-magnitude science. Optical practitioners should use their own safety factor or better yet, have their own measurements made (McCall, 1992a; Pompea and McCall, 1992a).

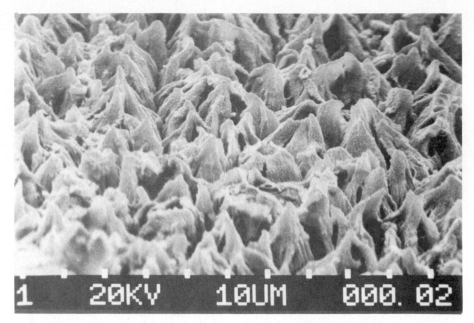

FIGURE 1   Scanning electron micrograph of Martin Optical Black, an anodized aluminum surface for ultraviolet, visible, and infrared use. *(Photo courtesy of Don Shepard, Martin Marietta Astronautics Group, Denver.)*

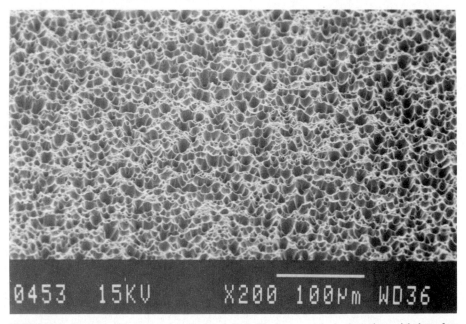

FIGURE 2   Scanning electron micrograph of Ball Black, an etched electroless nickel surface applicable to a variety of substrates. *(Photo courtesy of Arthur Olson, Ball Aerospace Systems Group, Boulder, Colorado.)*

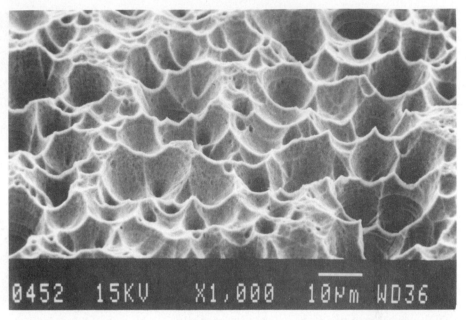

FIGURE 3 Scanning electron micrographs of Ball Black, an etched electroless nickel surface applicable to a variety of substrates. This surface is representative of a class of etched electroless nickel surfaces. *(Photo courtesy of Arthur Olson, Ball Aerospace Systems Group, Boulder, Colorado.)*

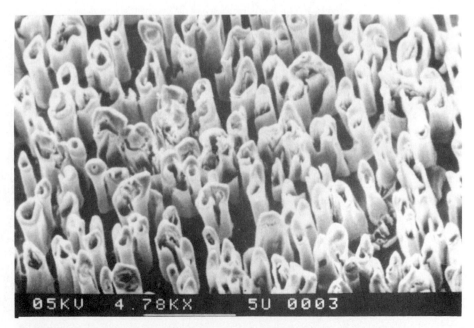

FIGURE 4 Scanning electron micrograph of a sputtered beryllium surface. *(Photo courtesy of Roland Seals, OPTICS MODIL, Oak Ridge National Laboratory.)*

FIGURE 5 Scanning electron micrographs of Ames Perfect Diffuse Reflector (PDR) at 24× magnification. *(Photo courtesy of Sheldon Smith, NASA Ames Research Center and Sterling Federal Systems, Palo Alto, Calif.)*

FIGURE 6 Scanning electron micrographs of Ames 24E at 24× magnification. *(Photo courtesy of Sheldon Smith, NASA Ames Research Center and Sterling Federal Systems, Palo Alto, Calif.)*

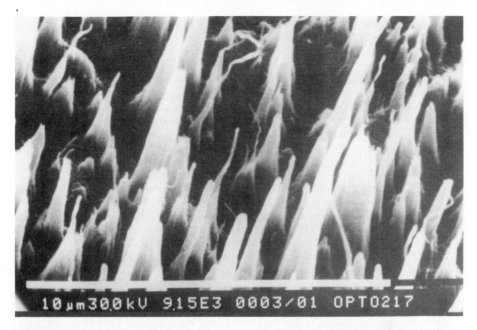

FIGURE 7 Scanning electron micrograph of a textured graphite surface created by bombarding a carbon surface with positive argon ions (Bowers et al., 1987; Culver et al. 1985.) *(Photo courtesy of Chuck Bowers, Hughes Aircraft, El Segundo, Calif.)*

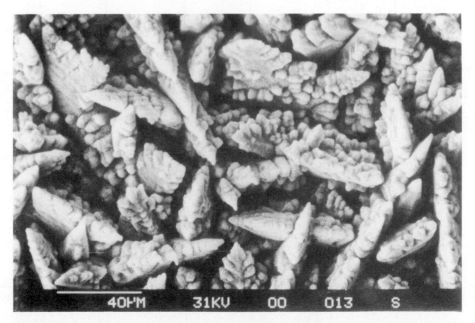

FIGURE 8 Scanning electron micrograph of Orlando Black surface, produced by electrodeposition of copper and subsequent oxidation in a proprietary process (Janeczko, 1992). *(Photo courtesy of D. Janeczko, Martin Marietta Electronic Systems, Orlando, Fla.)*

37.2 SELECTION PROCESS FOR BLACK BAFFLE SURFACES IN OPTICAL SYSTEMS

The selection of black surfaces is a systems issue; it must be addressed early and consider all aspects of system design and performance. It must also be examined from a total system performance perspective. Cost and schedule considerations have important effects on the decision and also need to be taken seriously. An unbuildable baffle structure with uncoatable surfaces is not a desirable state of affairs!

Similarly, proper financial and schedule support must be present. Black surfaces need to be taken as seriously as any other optical components, such as mirror surfaces or the thin films that act as filters. They deserve serious treatment in their design, fabrication, procurement, and testing. They are not last-minute design decisions or items to be created at the last minute. Since the baffle design and choice of surfaces often sets the final system performance (especially in infrared systems), it makes little sense to expend great energy and money to optimize other more traditional system components while ignoring stray light design and the surfaces so important in the design.

Figures 9 and 10 are a comparison of two processes that could lead to the selection of surfaces for an application. Figure 9 shows a process where design activities are done sequentially. In this process, stray light issues (including black surface issues) are left to the end of the program. In practice, with limited budgets, stray light and other fundamental system engineering issues are often not addressed at all in this process, or are addressed only at stages in the program where the system design is frozen, or cannot be changed without great expense. At this stage, when fundamental problems are found, they often

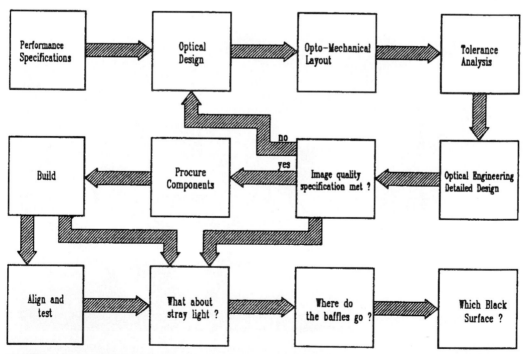

FIGURE 9 When selection of black surfaces is left to the end of a program, serious risks to the program are likely. (*From Pompea and McCall, 1992.*)

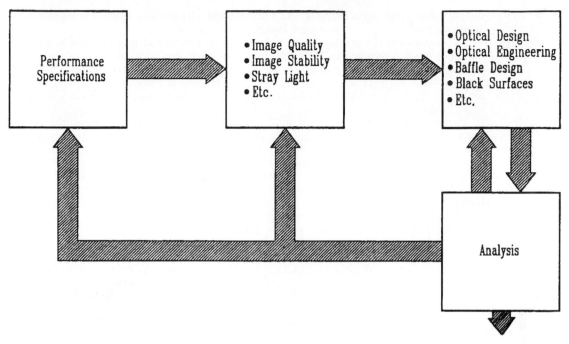

FIGURE 10   The most effective context for black surface selection is near the beginning of the program. This flowchart illustrates the early phase of a system-oriented design process. (*From Pompea and McCall, 1992a.*)

have severe performance, budget, and schedule penalties. That many stray light and black surface problems are rectified (with severe penalties to schedule and budget) at late stages in programs attests to the overall importance of black surfaces to the ultimate success of a system.

In an integrated systems engineering or concurrent engineering environment, illustrated in Fig. 10, the selection process begins early. There are many iterations of the design in the early stages, and close work among designers of different disciplines is essential. The level of detail increases with time, but no fundamental issues are decided without addressing their implication for the system and for each subsystem. This process is far superior to the one in the first flowchart, in which the surface selection process is treated as a trivial one and one waits until the end of the program to evaluate candidate surfaces. A review of general aspects of selection with emphasis on the system level issues inherent in the selection process is given by Pompea and McCall (1992a), McCall et al. (1992a), and Pompea (1994).

Once it has been decided to use black surfaces in an optical system, a whole host of system-level issues must be addressed to determine an appropriate coating or coatings. The high-level items include: (1) purpose and position of the surface in the instrument; (2) wavelength(s) and waveband(s) of interest; (3) general robustness of the surface; (4) nature of the installation process; (5) environment of the optical system; (6) availability; (7) cost; (8) substrate; and (9) other mission or system requirements.

For systems that operate over a wide wavelength range the selection process may be difficult because data may not be available over the entire wavelength range of interest. For example the next generation of infrared and submillimeter telescopes are being designed for wavelengths between about 10 and 800 micrometers. Baffle coatings must be adequate over this entire wavelength range.

The position of the coating in the system and its function at that location is a critical item. For positions where the incident radiation is nearly normal to the surface, the optical considerations become less critical, since the reflectance at near-normal incidence does not vary as significantly at this angle for the most often used surfaces. Also, the BRDF is usually at its lowest value. There is a wealth of data at these near-normal incident angles (e.g., Pompea et al., 1988).

The corollary is that if surfaces must be used close to grazing incidence, the optical properties become extremely important to understand (and quantify for system predictions) just as the amount of data becomes vanishingly small (Pompea et al., 1992b). Although there is sufficient optical data on specular reflections from specular surfaces, there is relatively little information on specular reflections from diffuse surfaces, let alone BRDFs at large angles of incidence. For these reasons, optical designers try to use lambertian-like surfaces at near to normal incidence light paths as possible.

The robustness of the surface in its environment is a complex subject. Some specific areas that will be addressed here are the atomic oxygen environment of low Earth orbit, outgassing in the vacuum of space, and particle generation by surfaces. The practical considerations of expense, schedule, ease of cleaning, exportability, and availability of specialized surfaces all play important roles. For a unique space instrument, there are different manufacturing considerations than for a large-volume production process. The installation process is often overlooked. It often plays a key role in the successful use of many surfaces.

Exposed spacecraft surfaces often must not contribute to spacecraft charging problems. Some nonconductive surfaces support the buildup of charges that can create large electrical potentials between spacecraft components, or the spacecraft and the plasma and subsequent damage may result. This is particularly important in the region near a detector, which is often very charge-sensitive. Surfaces such as Martin Black do not support charges in excess of 200 volts, since the coating can leak charge to the aluminum substrate. Textured metallic coatings, as a general rule, do not have surface charging problems.

Other considerations are the ability of a surface to withstand vibration, acceleration, elevated temperatures, thermal cycling, chemical attack (including atomic oxygen), solar and nuclear radiation, micrometeoroids, and moisture.

For the coating of baffle and (particularly) vane surfaces in ground-based telescopes, a further consideration may often be important. The baffle and vane surfaces in telescopes often have a large view angle or geometrical configuration factor (GCF) with the sky. This is particularly true for telescopes that operate without an enclosure or with a very large slit or opening to the night sky. In this case, the black baffle and vane surfaces must not radiatively couple strongly to the night sky. If there is a substantial view to the sky, and the emissivity of the surface is high, it will cool below the air temperature. This can create temperature induced "seeing" effects which degrade image quality. The requirement that a surface be strongly absorbent in the visible region, yet a poor emitter in the infrared is often a difficult one to meet. Table 4 gives the absorptivity and emissivity of some widely used surfaces.

In many laboratory or instrument applications, the use of one of a few standard black coatings or surfaces (such as the now unavailable 3M Black Velvet 101-C10) was entirely adequate to achieve the necessary performance of the optical system. In these applications, if the stray light performance was inadequate, it usually indicated that the stray light design, not the choice of black surface, was in error. The explanation was that the ignorance of some of the principles of stray light design could produce system performance five orders of magnitude worse than a poor choice of surfaces. The visibility of surfaces from the detector and the paths from those surfaces to the front of the instrument (not the optical properties or blackness of the surface) largely determine the total system performance. In stray light terminology, the GCF (geometrical configuration factor of

TABLE 4 Black Coatings

[*From Henniger (1984).*]

	$\bar{\alpha}_s$	$\bar{\varepsilon}_n$
Anodize Black	0.88	0.88
Carbon Black Paint NS-7	0.96	0.88
Cat-a-lac Black Paint	0.96	0.88
Chemglaze Black Paint Z306	0.96	0.91
Delrin Black Plastic	0.96	0.87
Ebanol C Black	0.97	0.73
Ebanol C Black-384 ESH* UV	0.97	0.75
GSFC Black Silicate MS-94	0.96	0.89
GSFC Black Paint 313-1	0.96	0.86
Hughson Black Paint H322	0.96	0.86
Hughson Black Paint L-300	0.95	0.84
Martin Black Paint N-150-1	0.94	0.94
Martin Black Velvet Paint	0.91	0.94
3M Black Velvet Paint	0.97	0.91
Paladin Black Lacquer	0.95	0.75
Parsons Black Paint	0.98	0.91
Polyethylene Black Plastic	0.93	0.92
Pyramil Black on Beryllium Copper	0.92	0.72
Tedlar Black Plastic	0.94	0.90
Velestat Black Plastic	0.96	0.85

* Note: Solar absorptivity (0.3 to 2.4 micrometers) and normal emittance (5–35 micrometers) of black coatings.

surfaces) is more important than the BRDF (bidirectional scatter characteristics of surfaces).

37.3 THE CREATION OF BLACK SURFACES FOR SPECIFIC APPLICATIONS

The optical properties in the visible (e.g., BRDF at 0.6328 micrometers) at near-normal incidence between painted black surfaces and more specialized high-performance black surfaces usually does not differ by more than a factor of ten. For infrared and ultraviolet wavelengths, the performance of more specialized surfaces is far superior to many paints by a much larger factor. The optical designer must take care not to overspecify the optical performance properties required for a baffle surface. The lowest reflectance surface is not necessary for all applications. On the other hand, the choice of low-performance coatings has important consequences.

Some systems have been designed with an excessive number of vanes on their baffles. In these cases, the baffle system is probably too heavy, costs more than necessary to manufacture, and the excessive number of vane edges may be introducing extra diffraction effects that degrade the performance. The advantage of the high-performance coatings often lies not only in their optical properties, but in the considerable experience of their manufacturers in working on baffle systems, the reliability and consistency in the application process, and the superior documentation of the optical performance characteristics. The environmental resistance of the high-performance coatings is generally much better and much better documented than paints that are used for a wide variety of applications.

Black Surfaces as Optical Design Elements

The systems approach to the choice of black surfaces in optical systems that has been stressed here has led to the rule of thumb in stray light design that the choice of a black surface's optical properties should not be the first design issue considered. If the performance of the preliminary design is inadequate, the designer should look at the design carefully, rather than look for a better black surface to boost the performance. A flaw in the general system design is probably more responsible for the performance than the "blackness" of surfaces.

One reason for this approach is that the choice of a different black surface may not improve the performance, if the surface is used in the system in certain ways (e.g., grazing incidence). An important equalizing factor among many black surfaces is that they become highly specular when used at angles significantly different from normal incidence. This optical performance characteristic makes it critical for the baffle designer to conceptualize the paths and ranges of angles that are required, and to avoid using a surface where its performance is requried at grazing incidence or close to it. The writing of scatter specifications is discussed by Stover (1990), and Breault (1990) discusses sources of BRDF data and facilities where BRDF measurements are routinely made.

Black surfaces can be contaminated and their absorbing or diffuse structure can be destroyed by improper handling. This can lead to a significant degradation in their optical properties. A margin of safety should always be included in the performance estimates.

Many specific high-performance applications today require that black surfaces must become critical design elements that are used with only small "safety" factors. First, all of the principles of stray light design are used. If the required performance is barely achievable without optimizing the choice of black surfaces, then considerable attention must be paid to the choice of surfaces. Sometimes, designers must often use several black surfaces in an instrument or telescope; each particular black surface is used where its performance is optimal. In the past few years, a variety of black surfaces have become available to supplement the old standbys such as 3M Black Velvet (the mainstay of laboratories ten years ago), and Chemglaze Z-306 (now called Aeroglaze Z306) and Martin Black (the standard bearers for space instruments in the 1980s). Since the applications for these blacks are so specific, we refer to them as *designer blacks*. This term refers not only to new classes of surfaces that have been developed, but also to the creativity of optical and material scientists in modifying existing surfaces to create special properties for specific applications.

This new array of designer optical blacks can still be subdivided into the traditional categories of specular blacks and diffuse (lambertian) blacks. Glossy or specular black surfaces have been used for vane cavities and their use is reviewed by Freniere (1980) and Freniere and Skelton (1986). The approach is to reflect unwanted radiation out of the system or to attenuate it while tightly controlling its reflected direction. The use of specular baffles (reflective black or shiny) requires considerable care to avoid sending specular beams further into the instrument at some particular narrow angles (Breault, 1983; Peterson and Johnston, 1992). The increasing sophistication of stray light analysis programs makes this analysis much more complete. The most widely used specular black surfaces are Chemglaze Z-302 and a paint formerly called Cat-a-lac Glossy Black, now available from Akzo as 463-3-8.

The lambertian black surface is often the most desirable type of surface to use in baffled telescopes and instruments. However, there is no *true* lambertian black, though several are extremely close for a near-normal angle of incidence. All diffuse blacks have larger BRDFs at large angles of incidence. Furthermore, these black surfaces can be further categorized by the wavelengths where they absorb best, the general categories being the ultraviolet, visible, near-infrared, middle-infrared, and far-infrared. They can also be distinguished by their degree of resistance to the atomic oxygen found in low Earth orbit, their resilience under laser illumination, and their performance at narrow laser wavelengths or specific angles of incidence. The method of manufacture and the performance of these specialized surfaces may be proprietary or classified. Even if the general qualities of these surfaces that perform in these areas are known, the full range of test results on them often are not readily available.

Design Techniques for Creating Black Surfaces

Four general tools are available to the black surface designer to tailor the optical properties. The first tool is to use absorbing compounds in the surface. Examples of this are the organic black dye, which is a good visible absorber, in the Martin Black anodized surface and the carbon black particles found in many paints. Similarly, for the infrared paints, the addition of silicon carbide particles provides good absorption near 12 micrometers. The addition of compounds may create quite different properties at visible and infrared wavelengths; a visual observation of blackness is no guarantee of its infrared properties. The converse is also true. Titanium dioxide is a good infrared absorber, but is very reflective in the visible region. "Black," therefore, means absorptive in the wavelength region being discussed.

The addition of large (relative to the wavelength of interest) cavities, craters, or fissures in the surface can aid in the absorption process by requiring the radiation to make multiple reflections within the material before it leaves the surface. The same effect is achieved by growing large angular projections from a flat substrate. Each reflection allows more of the radiation to be absorbed. For example, a surface with a reflectivity of 15 percent (a rather poor black) becomes a very good black if the light can scatter or reflect three times within the surface before it exits. After three internal reflections, the 15 percent reflectivity black has become a $(0.15)^3$ or 0.34-percent reflective surface. The NBS Black and Ball Black surfaces are good examples of this; they are shiny electroless nickel surfaces until etched. After etching, the surface is filled with a plethora of microscopic, cavities that absorb enough light to create a surface that is black to the eye.

The third phenomenon is the use of scattering from the surface structure, from particles in the surface coating, or from the substrate to diffuse the incoming beam over a hemispherical solid angle. Even without any absorption, this dilution factor can be very important in destroying specular paths, particularly in the far-infrared, where broad wavelength absorption is difficult. For an introduction to surface roughness and scattering phenomena, see Bennett and Mattsson (1989), Stover (1990), and Smith (1984, 1988). An in-depth, very readable treatment of scattering by small particles is given by Bohren and Huffman (1983).

The fourth phenomenon is based on optical interference of light in thin films. A black optical thin-film multilayer structure can significantly enhance, through optical interference, the amount of absorption of light in a multilayer structure over that of intrinsic absorption alone (Dobrowolski et al., 1992). The black layer system technology has been used successfully in a range of applications (Dobrowoski et al., 1992; Ludwig, 1984; Hass et al., 1956; Dobrowolski, 1981; Dobrowolski et al., 1989; Dobrowolski et al., 1991). Optical thin films provide tremendous design flexibility. The reflectance, absorptance, and transmittance can be tailored as a function of angle and wavelength. Black multilayers may have a higher cost per unit of surface area, are generally limited to small surfaces, and have a strong angular sensitivity. This last characteristic can be used to great advantage for specialized applications. The films (especially those deposited by ion-beam process) are durable and are suitable for space use.

There is no ideal black surface. For good absorption across a wide wavelength band (e.g., 0.5 to 20 micrometers), the surface might be excessively thicker than is needed for a narrowband application. Even a "perfect" absorbing surface has disadvantages associated with higher emissivity (e.g., the surface radiates to the detector). In practice, even specular surfaces can be acceptable as black baffle surfaces under the right circumstances, although they are much more difficult to use. By design they must direct the energy out of the system or in a direction that is not harmful to the system performance (Linlor, 1986; Greynolds, 1986; Peterson and Johnston, 1992). Their alignment is critical and there are caveats for their use (Breault, 1983). The "blackest" surface is not necessarily the best for a specific application.

37.4 ENVIRONMENTAL DEGRADATION OF BLACK SURFACES

Contamination control for terrestrial and space-borne sensors is a very rapidly developing and important area. A good introduction to the field with extensive references is given in Breault (1990) and in Glassford (1987, 1990, 1992). A good review of contamination assessment techniques is given by Heaney (1987), and Nahmet et al. (1989) discuss scattering from contaminated surfaces. The prevention, detection, and removal of contamination should be important considerations early in the design of the sensor.

Three areas of great interest for space-borne sensors are the effects of atomic oxygen, outgassing effects, and particle generation by surfaces. A large number of other effects can also influence the performance of surfaces, or of the systems where those surfaces are installed over short or long periods of time. These include adhesion of coatings, radiation effects on coatings, thermal cycling effects, vacuum ultraviolet effects, and electrostatic charging effects, to name a few. It would be helpful if each new coating developed could be tested by a suite of tests relevant to the area of concern, similar to what was done in cryomechanical tests of Ames 24E2 infrared black coating (Smith, 1990). All too often the tests are done on different formulations at different labs and with different degrees of care. This makes it extremely difficult for the optical designer to have the data necessary to make good decisions.

The surface contamination of a black surface as well as aging effects can affect its optical properties. However, the effect of contaminants from black surfaces on mirror scatter in the optical system is much more pronounced. Williams and Lockie (1979) use the BRDF as the sensitive parameter to judge the effects of dust, hydrocarbon oil, acrylic, and peelable coating residue on low-scatter mirror surfaces. Young (1976) describes the effect of particle scatter on mirrors. Shepard et al. (1988) have suggested that scattering properties of black baffle surfaces may be temperature-dependent. This may be a result of chemical changes in the surface or in changes to the surface as absorbed molecules are released at increased temperatures. Further work should be done at the temperatures of interest.

A large number of studies have been done to describe the behavior of black baffle surfaces to intense exposure to high-powered lasers and radiation, where damage to the surface is widespread. Accurate results can often be obtained using other vehicles to deliver energy into the surface. For example, rapid pulsed, low-energy electron beam irradiation of optical surfaces is described by Murray and Johnson (1990) as a cost effective way to test for radiation hardness. Black baffle surfaces are vulnerable even at low fluences because of their great absorptivity. The damage mechanisms are not well understood. Many of the damage mechanisms are related to the thermal shocks produced by rapid heating of a surface of thin layers on different materials. The destruction of surfaces not only reduces the ability of the surface to absorb stray radiation, but often also creates a contaminating cloud of particles that can greatly affect the optical system.

Atomic Oxygen Effects

Spacecraft surfaces on the early space-shuttle flights showed significant weight loss and aging effects. For example, on STS-1 the forward bulkhead camera blanket was milky yellow after the flight and the white paints on the Shuttle exhibited exposure-related degradation effects. On later flights, surfaces such as Kapton showed significant mass losses. These effects were attributed to the interaction of the surfaces with the atomic oxygen present in low Earth orbit (McCargo et al., 1983). The atomic oxygen attacks the binder materials in paints, thus removing mass, and causes glossy surfaces to become more lambertian (Whittaker, 1984). Carbon particles are released when the binder is removed. The weight loss is due both to chemical processes and to erosional (kinetic energy) processes.

A second problem is that many surfaces have a glow associated with them when exposed to the atomic oxygen. Both of these phenomena undermine the effectiveness of baffle surfaces on space instruments. The effect of exposure to atomic oxygen on anodized black surfaces (Martin Black and Enhanced Martin Black) has been measured and simulations of atomic oxygen exposure on Chemglaze (now called Aeroglaze) Z-306 paint has been done with a plasma etch chamber (Pompea et al., 1984b).

These tests showed that a carbon-black-based paint could be significantly degraded by atomic oxygen over time while anodized surfaces exhibited only small changes in their surface morphologies or in their visible and near-infrared reflectances. The graying of some smooth painted black surfaces is somewhat compensated for by an etching process that roughens these surfaces. If the exposure times are short, the etching action of the atomic oxygen may even improve the scattering properties of paints. However, removed material may show up on optical surfaces, degrading the optical performance, and it should be obvious that long exposure times would remove too much of the coating.

In general, however, the optical properties of surfaces degrade during longer exposure times and at higher atomic oxygen flux levels. Experiments in space on the Long Duration Exposure Facility (LDEF) have indicated how the properties of materials change after nearly six years of exposure to atomic oxygen. For Chemglaze (now called Aeroglaze) Z-306 paint, for example, this exposure has led to a loss of the binder material (Golden, 1991) and to loss of pigment. Some ion-textured surfaces were also flown as part of this experiment and showed high stability in the space environment (Mirtich et al., 1991).

Outgassing

For space applications, the amount of outgassed material and its composition are of great importance. One of the primary dangers is that the outgassed products may form a film with undesirable properties on optical components. This hazard of space exposure is particularly relevant for cooled systems. For cryogenic systems, the continued outgassing of materials as they break down after long exposure to the space environment becomes an additional source of contamination on optical surfaces. Outgassing is also of concern for severe environments on the ground, though the correlation between vacuum testing and a behavior of black surfaces under severe heat in air is not well understood. For the context of this chapter, only vacuum outgassing measures will be considered. In either case, the condensation of volatile materials on optical components can lead to catastrophic system failure.

The two primary measures or tests of outgassing are the *total mass loss* (TML) and the percent *collected volatile condensable materials* (CVCM). An American Society for Testing and Materials (ASTM) Standard Test Method was developed and is called E 595-77/84. Several NASA publications give values for TML and CVCM (Campbell and Scialdone, 1990, gives the most extensive listing; see also Predmore and Mielke, 1990). For example, measurements at Goddard Space Flight Center of Chemglaze (now Aeroglaze) Z-306 black urethane paint give a TML of 0.92 percent and a CVCM of 0.03 percent. Wood et al.

shows the near-infrared absorption due to the outgassing products of Aeroglaze Z-306 black paint. For the paint the total mass loss was 2.07 percent. Ames (1990) reports outgassing problems with Aeroglaze Z306 if exposed to temperatures greater than 40°C after a room temperature cure. However, many potential space material outgassing problems with this and other surfaces can be avoided completely with proper bakeout procedures. In 1993, a low volatile version of Z306 was being developed.

A third measure of outgassing relates to the water content of the material or surface. The *water vapor regained* (WVR) is a measure of the amount of water readsorbed/reabsorbed in 24 hours while the sample is exposed to 25°C and 50 percent relative humidity (for extensive listings, see Campbell and Scialdone, 1990). This determination is done after the vacuum tests for TML and CVCM.

Particle Generation

Materials for space use must be able to withstand launch vibration without generating an excessive number of particles. This is not simply a coating or surface problem—it is a dynamics and substrate problem as well. The details of the baffle design and assembly play an important role in the generation of particles. If a baffle surface is grossly distorted by the launch vibration, the adhesion of the black surface should not be blamed for the failure; the baffle design failed. A common source of particle generation is abrasion between coated baffle surfaces. For severe vibration environments and for fragile surfaces, special considerations are necessary (see for example, Young et al., 1980). Similarly, surfaces must be able to withstand thermal cycling without producing particles. As painted materials outgas and their composition is altered, particles in the paint can become dislodged.

Materials used in baffles can also get dusty or contaminated by exposure outgassing products from the spacecraft. For an accessible ground-based black surface, normal cleaning procedures appropriate to the surface may be used, if they do not damage the texture or optical properties of the surface. For a baffle system already in space, the cleaning procedures developed to clean mirrors in situ may apply. These techniques include laser cleaning, plasma or ion cleaning, and jet snow cleaning (see, for example, Peterson et al., 1992, and other papers in Glassford, 1992).

Baffle Material for Extreme Environments

One important form of degradation for Earth-based materials involves the exposure of the surface to the elements of temperature cycling, sunlight, humidity, and other factors of its operating environment. This combination is a severe test of many surfaces. Unfortunately, there is very little data published on the environmental degradation of black surfaces that are exposed to the outdoor environment, or to high temperatures, such as might be found in a closed automobile in the summer. Often, a black surface that is very ordinary in its optical properties is found to have been used in an extreme environment. The reason is that it was the first surface that did not peel or fade significantly.

Space-based surfaces must withstand severe launch vibrations, temperature extremes, collisions with space debris and micrometeoroids, and exposure to ultraviolet radiation. Military missions have also added the requirement that these materials be hardened against nuclear or laser threats. For example, these baffle surfaces and their associated optics must still perform after exposure to high-power lasers.

In the simplest sense, these black surfaces must be able to either selectively reject laser radiation (a difficult requirement) or be able to absorb significant amounts of energy and not degrade. Most of the coatings have been developed for application to substrates which can survive high temperatures, such as molybdenum, beryllium, tungsten, carbon/carbon, and titanium, because of their ability to withstand high flux densities of laser light. Generally, painted surfaces are completely inadequate. Even anodized surfaces on aluminum may not be adequate at these flux densities.

37.5 OPTICAL CHARACTERIZATION OF BLACK SURFACES

The baffle designer must be concerned about the optical properties of black surfaces over a large range of wavelengths and angles of incidence. Wolfe (1978) summarizes the relevant design data. Performance measurements, such as measurements of the specular reflectance and total hemispherical reflectance over a range of wavelengths are now supplemented by a more complete characterization of these optical surfaces: the bidirectional reflectance distribution function (BRDF) (Gunderson, 1977). The BRDF has units of inverse steradians and is discussed in detail elsewhere in this book. The BRDF is usually measured at several discrete wavelengths of the most common lasers. In particular, measurements at the HeNe laser wavelengths of 0.6328 micrometers and at the CO_2 wavelength of 10.6 micrometers are particularly valuable for visible and infrared applications. There is a notable shortage of BRDF data in the ultraviolet, from 1 to 4 micrometers, and for far-infrared wavelengths.

BRDF measurements at the same wavelength have the additional advantage that they are directly comparable. As Smith and Wolfe (1983) point out, the directional reflectivity of a surface is convolved with the instrument function (a weighted mean of the detector solid angle), a unique characteristic of each reflectometer. An approximate deconvolution of the instrument function can be easily performed by dividing the measured reflectance by the projected measurements are instrument independent as mentioned in ASTM Standard E12.09. However, simple specular reflectance measurements and spectra made on different instruments must first be divided by the projected detector solid angle before being compared. Several studies (Leonard and Pantoliano, 1988 and Leonard et al., 1989) have detailed the variation between BRDF measurements made in different laboratories. With the advent of the ASTM standard, greater congruence among BRDF measurements in the future is probable.

Because of the value of BRDF measurements in comparing the scattering characteristics of surfaces, a large set of BRDF data of black baffle materials that have traditionally been used in optical devices of all types, as well as the scattering functions of some new materials are presented in this chapter. These measurements may be compared to previous measurements made on other instruments (Bartell et al., 1982).

The specular and total hemispherical reflectance data thus become most valuable in describing some general characteristics of a surface over a wide wavelength range, while the BRDF data at a wavelength of particular interest allow the designer to characterize the angular scattering of the surface for angles of incidence of interest and for view angles from the detector. The total integrated scatter also is extremely useful for surface roughness determination (see Stover, 1990).

This chapter provides the optical designer a useful summary of data. Sufficient data are presented to allow a determination of which surfaces might be appropriate for a given task. The data generally can be divided into two forms, the first being specular and hemispherical reflectance data over a wavelength range. These data allow the designer to choose a surface with the right basic properties over the wavelength range of interest. BRDF measurements provides a second form of data. The data are usually presented for a few characteristic angles of incidence (5, 30, 45, and 60 degrees are the most common) at each wavelength. These BRDF measurements are particularly useful as input to optical modeling or stray light modeling programs.

In stray light analysis programs such as APART (Breault et al., 1980), ASAP (Greynolds, 1988), and GUERAP (Steadman and Likeness, 1977), as well as many other modeling programs, it is the mathematical value of the bidirectional reflectance distribution function that is used by the program models. In these programs, a library of BRDF functions for the most common surfaces is available for use by the analyst. For surfaces that do not explicitly have a BRDF associated with them in the code, the BRDF values can be input.

In most baffle designs, surfaces are used at angles not represented by the available BRDF measurements. Some surfaces are even used at angles close to grazing incidence where few measurements have been made. For these areas, the data available for the models are very limited, and often liberal extrapolations are needed. The assumption of a smooth transition between measurements at different angles or at very large angles of incidence, or from one wavelength to another is often assumed. The more unusual or nonrandom the surface, the more these assumptions or extrapolations should be questioned. Examples of how BRDF measurements are made and used in system designs and analysis are found in, for example, Breault (1986), Bergener et al. (1984), Evans and Breault (1984), Orazio et al. (1983), Bartell et al. (1984), and Silva et al. (1984).

Black surfaces are also used in radiators used for the thermal control of spacecraft. The hemispherical reflectivity can provide a general indication of the optical properties for a thermal analysis. The normal emittance at room temperature is also an important quantity for spacecraft design. Table 5 gives the hemispherical reflectance and normal emittance of some surfaces. Radiatively cooled surfaces are important for space infrared telescopes and, if designed properly, can provide a reliable means of reducing background radiation.

TABLE 5

[*Data courtesy of J. Heaney, Goddard Space Flight Center (Heaney, 1992).*]

Surface	Thick-ness	Hemispherical reflectance at 546 nm (incidence angle, degrees)					Normal emitt-ance (300°K)
		20	40	60	70	80	
Cat-a-Lac Black 463-3-8	0.002″	0.054	0.061	0.096	0.141	0.223	0.891
Black Velvet 401-C10	0.025″	0.035	0.036	0.047	0.058	0.085	0.911
Chemglaze Z306	0.005″	0.048	0.052	0.074	0.098	0.144	0.915
Chemglaze Z306 with silica powder		0.031	0.033	0.040	0.050	0.070	0.912
Chemglaze Z306 with 3M glass mic-roballoons left on 20-micron screen	0.006″	0.040	0.048	0.068	0.090	0.130	0.920
Chemglaze Z306 with 3M glass mic-roballoons left on 44-micron screen	0.010″	0.036	0.040	0.055	0.070	0.097	0.923
Chemglaze Z306 with 3M glass mic-roballoons left on 63-micron screen	0.011	0.037	0.040	0.053	0.064	0.084	0.923
Scotchlite Brand Reflective Sheeting #3285 Black		0.059	0.062	0.095	0.156	0.318	0.909
Scotchlite Brand Reflective Sheeting "C" Black W/A #234		0.146	0.130	0.129	0.135	0.157	0.846
NiS-dyed Anodized Al from Light Metal Coloring Corp.		0.056	0.061	0.097	0.151	0.333	0.912
Sandoz-Bk Organic Black anodized Al from Almag Co.		0.041	0.048	0.064	0.082	0.129	0.884
Black Chrome anodized Al from Goddard SFC		0.028	0.031	0.053	0.083	0.156	0.625
NiS-dyed anodized Al from Langley Research Center		0.050	0.053	0.083	0.138	0.253	0.915
NiS-dyed anodized Al, substrate blasted with glass shot, from Light Metals Coloring Corp.		0.045	0.049	0.073	0.100	0.146	0.920

Because of on-orbit degradation, the radiators must be sized for the appropriate performance at the end of the mission.

For these applications, the absorptivity and emissivity of surfaces as a function of wavelength and temperature must be well understood. Thus, room temperature measurements of normal emittance may not be very useful if the surface is to be used at cryogenic temperatures. Stierwalt (1963, 1979) has contributed greatly to the measurement of far-infrared emissivity of black materials. Clarke and Larkin (1985) describe measurements on a number of black surfaces suitable for radiometry detectors where a low reflectance is needed, independent of wavelength. For opaque surfaces, the spectral emittance at a given wavelength is equal to the spectral absorptance, which is equal to one minus the spectral reflectance.

These parameters are very sensitive to coating thickness and may change with exposure to ultraviolet radiation from the sun, with damage from solar electron and proton radiation, damage from space debris and micrometeoroids, and with outgassing. In practice, the solar absorptance is measured by obtaining the reflectance from 0.3 to 2.4 micrometers, the region containing 95 percent of the sun's energy. The normal emittance is determined using an infrared spectrophotometer over the wavelength range 5 to 35 micrometers, where close to 90 percent of the energy of a 300-K blackbody is emitted (see Table 4). While these measurements are useful for design of radiators that operate near room temperature, further measurements are required in order to predict a surface's thermal emittance at lower operating temperature.

37.6 SURFACES FOR ULTRAVIOLET AND FAR-INFRARED APPLICATIONS

Many surfaces are multipurpose and are useful for a variety of wavelengths. Before discussing the properties of these surfaces it may be instructive to look at several surfaces that are used for ultraviolet and far-infrared applications. In these wavelength regions, the performance of black surfaces is often critical to the performance of the optical instrument.

Black Surfaces for Ultraviolet Applications

For space telescopes or instruments operating in the ultraviolet, surfaces such as Martin Black have been used successfully. However, very few measurements of candidate baffle material have been made at wavelengths less than 0.3 micrometers (300 nanometers) and even fewer have been made at shorter ultraviolet wavelengths. Jelinsky and Jelinsky (1987) studied the performance characteristics of a variety of surfaces and surface treatments for use on baffle materials in the Extreme Ultraviolet Explorer. They determined some scattering characteristics of several materials at 30.4, 58.4, and 121.6 nanometers.

Heaney (1992) has made some pioneering bidirectional measurements at 15°, 45°, 75°, and 85° at 58.4 nanometers (584 Å) of candiate baffle materials for space instrument applications. Interestingly, two of his samples are "bare" aluminum with no applied coating. The natural oxide present on uncoated aluminum samples is strongly absorbing at this wavelength and, by itself, reduces the reflectances to below 5 percent at near-normal incidence. The baffle surface candidates are described here and their optical properties are presented in the figures.

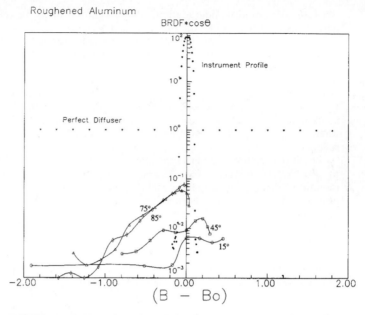

FIGURE 11   BRDF of roughened aluminum. Instrument profile and perfect diffuse reflector shown for reference. Wavelength is 58.4 nm. *(From Heaney, 1992.)*

Roughened Aluminum: Aluminum 6061 alloy, sheet finish, sandblasted to an un-specified degree of surface roughness. Uncoated. (Figure 11.)

Gold Iridite: Aluminum with a chromate surface coating produced in a room temperature chemical conversion process. (Figure 12.)

Black Nickel: Aluminum with a black nickel surface that was produced in an electroplating process with NiS and ZnS. (Figure 13.)

Copper Black: An aluminum substrate coated with approximately 80 to 100 nanometers of evaporated Cu to provide a conductive layer and then overcoated with an evaporated Cu black deposited at a chamber pressure in the 10^{-3} torr range. (Figure 14.)

Black Paint: An aluminum substrate painted with a carbon-loaded black silicone paint (NSB69-82). (Figure 15.)

Aluminum, grooved and blazed: An aluminum plate, 6061 alloy, with grooves cut at a blaze angle of about 20°, with a period of 1.1 mm, a depth of 0.4 mm, and otherwise uncoated. (Figures 16, 17, 18.)

Surfaces for Far-Infrared (>30 micrometers) Use

In the far-infrared wavebands, many surfaces such as paints which have good visible absorptance lose their absorbing and scattering characteristics. Although measurements have only been made at wavelengths longer than 15 micrometers on a few surfaces, the data suggest that most surfaces do not perform well as "blacks" for long wavelength operation (see Stierwalt, 1979; also Clarke and Larkin, 1985). Smith (1982) points out that many coatings become transparent at longer wavelength and the substrate roughness

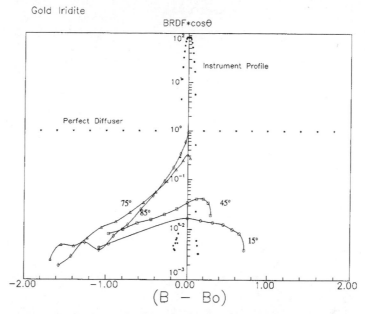

FIGURE 12 BRDF of gold iridite. Instrument profile and perfect diffuse reflector shown for reference. Wavelength is 58.4 nm. *(From Heaney, 1992.)*

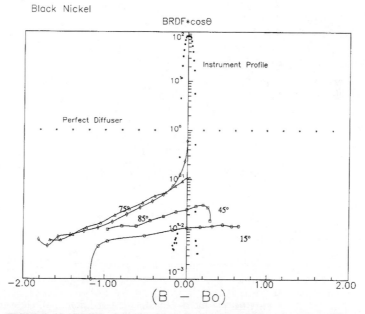

FIGURE 13 BRDF of black nickel. Instrument profile and perfect diffuse reflector shown for reference. Wavelength is 58.4 nm. *(From Heaney, 1992.)*

Evaporated Cu Black

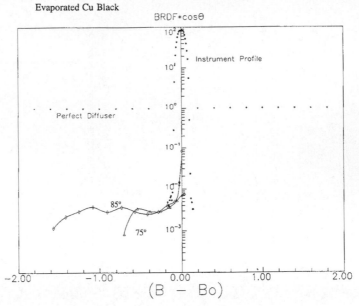

FIGURE 14 BRDF of evaporated Cu black. Instrument profile and perfect diffuse reflector shown for reference. Wavelength is 58.4 nm. *(From Heaney, 1992.)*

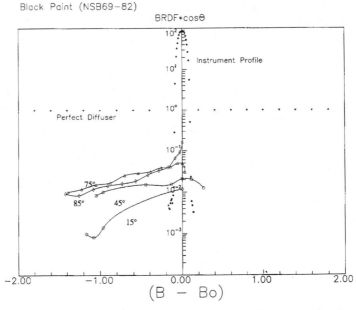

FIGURE 15 BRDF of NSB69-82 black paint. Instrument profile and perfect diffuse reflector shown for reference. Wavelength is 58.4 nm. *(From Heaney, 1992.)*

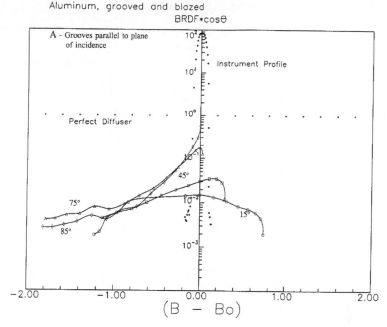

FIGURE 16 BRDF of aluminum, grooved and blazed, with grooves parallel to angle of incidence. Instrument profile and perfect diffuse reflector shown for reference. Wavelength is 58.4 nm. *(From Heaney, 1992.)*

enters strongly into the scattering characteristics. An example of this is Infrablack, a rough anodized surface that was previously described. The approach has been carried further in Ames 24E (Smith, 1991). A number of approaches and surfaces have been developed to create effective surfaces at these wavelengths.

A model has been constructed (Smith, 1984) to describe the effect of coating roughness and thickness on the specular reflectance. Smith and Howitt (1986) survey materials for infrared opaque coatings (2 to 700 micrometers). They highlight the value of silicon carbide and carbon black for far-infrared absorption. Smith (1980, 1981, 1982, 1983, 1984, 1988, 1991a, 1991b, 1992, 1993) provides much of the reference work in this area.

The optical properties of the several infrared coatings and surfaces are given in Figs. 19 to 24. For applications in laboratories and in calibration facilities, surfaces like velvet, commando cloth, and even neoprene may be used. The long-wavelength measurements on these surfaces are given in Figs. 25 to 32. The long-wavelength coatings may be grouped into a few basic categories discussed here. As this area is one of rapid development for space infrared telescopes and space radiators, there will likely be many new approaches in the next few years.

Multiple-layer Approach. For reducing the far-infrared reflectance of baffles, a technique of using antireflection surfaces composed of multiple layers with a different refractive index in each layer was discussed by Grammar et al. (1980). For such a coating to be effective it must match the optical constants of free space with the large constants of the baffle substrate.

Teflon Overcoat. A polytetrafluoroethylene (Teflon) spray-on lubricant was used as an antireflection overcoat for opaque baffle surfaces in the far-infrared and submillimeter by

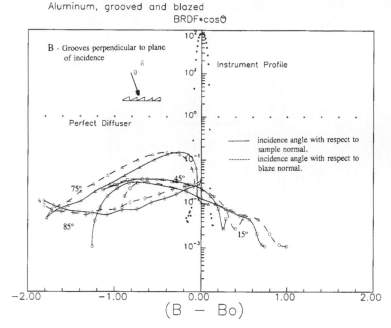

FIGURE 17 BRDF of aluminum, grooved and blazed, with grooves perpendicular to angle of incidence. Instrument profile and perfect diffuse reflector shown for reference. Wavelength is 58.4 nm. *(From Heaney, 1992.)*

Smith (1986). A thick Teflon overcoat created by spraying Teflon Wet Lubricant reduced the specular reflectance at millimeter and infrared wavelengths by a factor of two. The refractive index of the coating plays a more significant role than the thickness.

Cornell Black. Houck (1982) created a far-infrared black paint based on adding large particles of grit (silicon carbide #80 and #180 grit) to 3M Black Velvet Nextel 101-C10. This paint could be repeatedly cycled to cryogenic temperatures with no flaking or peeling. The far-infrared properties of Cornell Black are described by Smith (1982, 1984). Since this 3M paint is no longer available, Cornell Black is not now generally used. The ECP-2200 (now called MH2200) replacement for 3M Black Velvet was studied by Smith and Howitt (1986) in an effort to create a paint similar to Cornell Black. Houck (1982) gives complete instructions for making the Cornell Black paint.

Infrablack. Infrablack is a black anodized surface generally applied to 6061 Aluminum (Pompea et al., 1983b, 1984a,b). It is highly suitable for space radiators, infrared baffles, and for applications where a high emissivity is desired. The specular reflectance of the Infrablack is conservatively one order of magnitude less than that of Martin Black across the spectral region from 12 to 500 micrometers. The reduced reflectance of Infrablack is attributed to a large increase in the roughness of its substrate prior to anodization. The diffuse reflectance characteristics of the Infrablack are different from Martin Black at the longer wavelengths. At 12 and 50 micrometers, the BRDF of Infrablack is considerably less than that of Ames 24E, while at 100 and 200 micrometers they are comparable.

Ames 24E and Ames 24E2. Ames 24E is a new coating developed at NASA's Ames Research Center (Smith, 1988). The coating is a very rough, very thick, highly absorbing

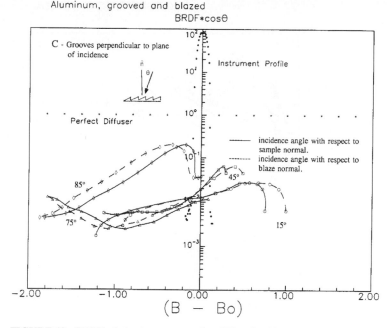

FIGURE 18 BRDF of aluminum, grooved and blazed, with grooves perpendicular to angle of incidence. Instrument profile and perfect diffuse reflector shown for reference. Wavelength is 58.4 nm. *(From Heaney, 1992.)*

coating composed largely of SiC grit and carbon black, and has very low far-infrared reflectance (Smith, 1988). BRDF measurements show that this surface is a model lambertian surface at a wavelength of 10.6 micrometers at near-normal incidence. Ames 24E is nearly lambertian at longer wavelengths, even at 200 microns.

Ames 24E2 is a formulation using ECP-2200, carbon black, silicone resin, and #80 silicon carbide grit. The coating has passed severe cryogenic cycling and vibration tests (Smith, 1990) and was developed for far-infrared space telescopes. The coating is considerably less reflective at 100 micrometers than Ames 24E and has excellent absorption out to about 500 microns. Smith (1992a) has also developed a highly lambertian, diffuse reflectance standard, which he calls PDR, for perfect diffuse reflector. It shows lambertian behavior from 10 to about 100 micrometers.

36.7 SURVEY OF SURFACES WITH OPTICAL DATA

A Note on Substrates and Types of Coatings

Many different substrates are employed for baffle and vane surfaces. In particular, for space or cryogenic use, the 6000 series of alloys of aluminum (particularly 6061) are popular. They represent the baseline by which other materials are judged. Titanium and graphite epoxy are also widely used, and beryllium and carbon fiber substrates are becoming more common. For high-temperature applications or for resistance to lasers, materials with high melting points such as nickel, molybdenum, and carbon/carbon are used.

PHOTOMETRIC SPECTRA

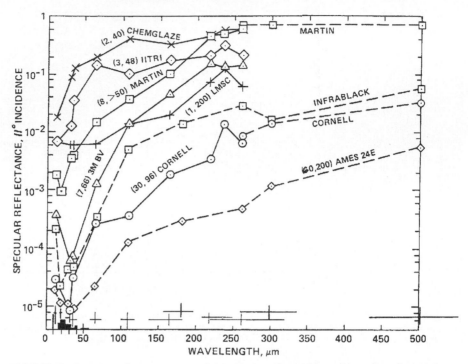

FIGURE 19   Specular reflectance spectra of eight black coatings. Measurements of coating roughness and thickness in micrometers are shown in parentheses before the name of each coating. Filter passbands are shown by horizontal error bars. The absorption bands of amorphous silicates are indicated by the solid histogram in lower left end of the graph. *(From Smith, 1982, 1984, 1988.)*

The distinction between substrate and surface or surface coating is blurry. When paints or other coatings are applied to a substrate, the distinction between the two is very clear. However, many surfaces today involve an extensive alteration of the substrate before the "coating" is applied. They may chemically alter the substrate to create the surface coating, which is an integral part of the substrate. In these cases, the use of the word substrate only serves as a description of the material that must be altered to create the black surface.

The classification of surfaces by substrate and/or type of coating or surface treatment is only one of the many ways available. However, it is a very useful form of classification, and will be used here.

37.8   PAINTS

Paints are easy to apply and their availability and usually low price make them ideal for laboratory experiments and instruments. However, their high outgassing rates and, in some cases, the large number of particles that can be liberated from them compromise their use in space-based applications. However, there are many space-qualified paints that have flown. Notable among them is Aeroglaze Z-306, and the derivatives of this basic paint system that have been developed at Goddard Space Flight Center.

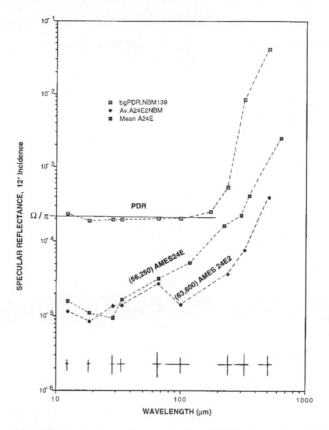

FIGURE 20   Specular reflectance spectra of PDR surface and two infrared black coatings. Dashed lines connect data points of a set, and the solid line indicates the theoretical value of reflectance for a perfect lambertian surface. The coating roughness and thickness are given in parentheses, and filter passbands are indicated by horizontal error bars. *(From Smith 1992a, 1992b.)*

There are many issues in paint application that affect the optical properties. The most obvious are the number of coats and the thickness; the substrate preparation also plays a critical role. Brown (1990) describes the effects of number of coats on optical performance for three diffuse paints. The importance of developing a procedure for painting that incorporates the paint manufacturers application recommendations cannot be overstated. This procedure must also include process control requirements, cleanliness requirements, application process and surface preparation requirements, safety requirements, quality assurance requirements, and storage shipping requirements, as McCall (1992b) describes in detail. A short description of a few of the more common paints is given here. Table 1 gives further reference data on these and other paints.

3M Paints and Their Current Derivatives

3M Nextel Black Velvet.   Several varieties of 3M Black Nextel paint from Minnesota Mining and Manufacturing Company have seen extensive use in ground-based instruments and were considered as standards for black coatings. These paints consisted of pigments with approximately 20 percent carbon black and 80 percent silicon dioxide. However,

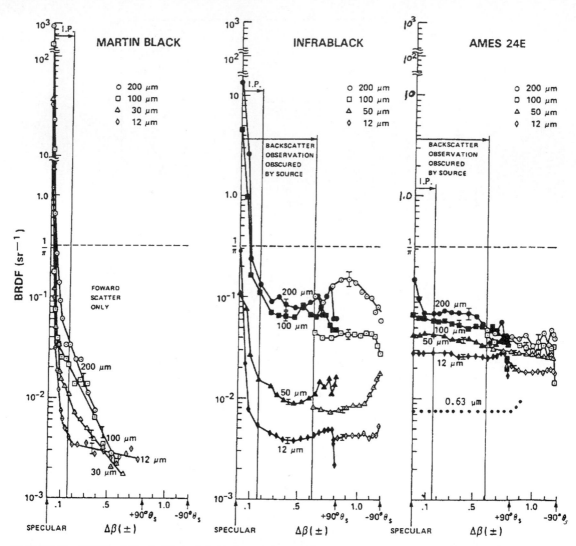

FIGURE 21 BRDFs of Martin Black, Infrablack, and Ames 24E. Note the different breaks in the ordinate scale of the Martin Black measurements used to compress its specular reflectance to fit the scale of the figure. Forward scatter data are solid symbols; backscatter data are open symbols. Forward and backscatter measurements at 0.6328 micrometers of Ames 24E are by D. Shepard of Martin Marietta (Denver) and are shown schematically by the dotted line in the right hand figure. *(From Smith, 1988.)* Smith (1993) recalibrated this data; the reletive comparison of these coatings remained essentially unchanged.

most of them are presently unavailable. Those that are available are manufactured under license by other companies. The optical properties of the original paint is given in Fig. 33.

The paint labeled 3M 101-C10 historically was one of the most widely used laboratory black paints; however, it is no longer made. Redspot Paint and Varnish (Evansville, Indiana) now makes a similar paint, but uses resin instead of glass microspheres, making it unsuitable for space applications according to Ames (1990). Two other paint varieties in the same line were 3M401-C10 and 3M3101 (see Figs. 34 and 35), apparently available from Red Spot Paint and Varnish.

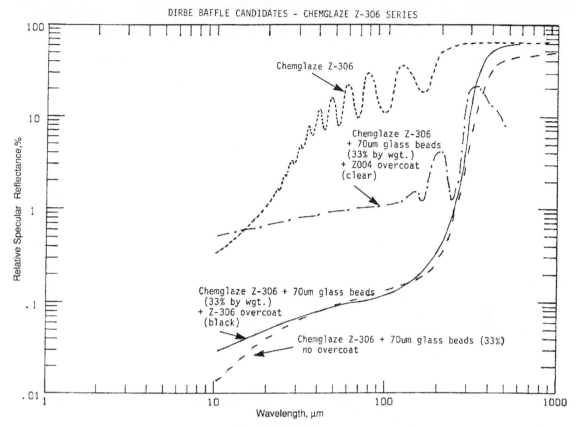

FIGURE 22 Specular reflectances of Diffuse Infrared Background Explorer candidate baffle surfaces based on Chemglaze (now Aeroglaze) Z-306, over the wavelength range 1–1000 micrometers. *(From Heaney, 1992.)*

The 3M Nextel Black Velvet was replaced by 3M SCS-2200 (3M Brand ECP-2200) (see Fig. 35). The optical differences between the original and its replacement are discussed by Willey et al. (1983). Illinois Institute of Technology now produces the ECP-2200 under the name MH2200 (see following).

Nextel 2010. A sprayable Nextel 2010 Velvet paint is available from the German company Mankiewicz Gebr. & Co., Hamburg, and has been used on several ground-based astronomical instruments, where cryogenic cycling durability is important. The 2010 surface has an outgassing rate of 1.3×10^{-4} grams/cm^2 at 84°F (29°C) in a 20-hour test at 10^{-5} to 10^{-6} torr. The solar alpha of this surface (from the company literature) is about 0.97 and the normal emittance is about 0.95.

The total emissivity of different black surfaces was investigated with regard to their being used as a total radiation standard in the temperature range −60°C to +180°C by Lohrengel (1987). The investigations showed that the matte varnish Nextel Velvet Coating 2010 black with an emissivity of 0.951 was best suited for these purposes. Additional changes to the surface (sprinkling of Cu globules, etching) could be made to increase the total emissivity by a further 0.007 or 0.016, respectively. The present availability of this paint from U.S. manufacturers is unknown.

Nextel Brand Suede Coating Series 3101-C10. C-10 is the black variety of this paint, available from Red Spot (see references). It is a low-gloss paint, with reasonable abrasion

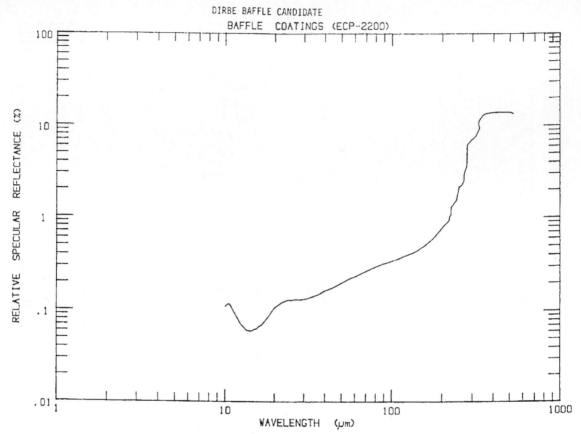

FIGURE 23 Specular reflectance of Diffuse Infrared Background Explorer candidate baffle surface based on ECP-2200 (now MH2200) black paint over the wavelength range 1–1000 micrometers. *(From Heaney, 1992.)*

resistance (see Fig. 36). It has a soft-cushioned feel and suedelike appearance, and is used on office equipment, furniture, and electronic equipment housings. The manufacturer recommends it for interior use only.

MH2200. MH2200 (formerly ECP-2200, which was derived from the original 3M Nextel Black Velvet) is a one-part, flat black, nonselective solar absorber coating designed for high-temperature service (Fig. 37). When applied to aluminum substrates, the coating integrity and its "solar absorption" of 0.96 are unaffected by temperatures to about 500°C. The coating is 43 percent solids in xylene and is intended for nonwear applications. ECP-2200 is available from IIT Research Institute. It is touted as the substitute for 3M Nextel Black, but does not have the low BRDF at visible wavelengths of its former relative. It has a normal emittance of 0.95 at 10.6 micrometers and an absorptance of 0.96 over the wavelength range of 0.35 to 2.15 micrometers. Its emittance is 0.86 on an aluminum substrate (normal emittance at 75°F from 2 to 25 micrometers).

Aeroglaze Z series. Aeroglaze (formerly Chemglaze) Z-306 diffuse black paints are part of the Z line of single-package, moisture-curing, ASTM Type II, oil-free polyurethanes (Lord, 1992). This paint has excellent chemical, solvent, and (if used with a primer) salt-spray resistance. This is a popular paint for aerospace use, especially after the

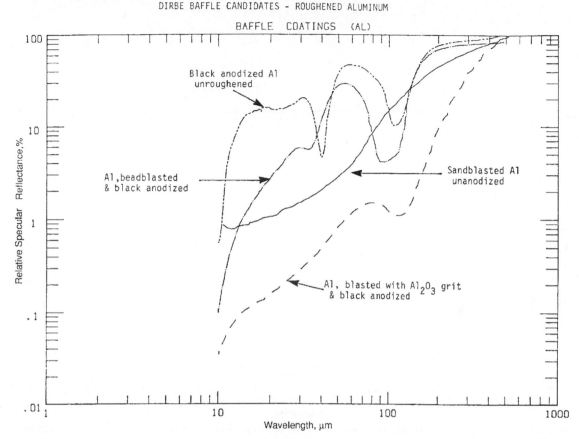

FIGURE 24 Specular reflectance of Diffuse Infrared Background Explorer candidate baffle surface based on roughened aluminum surfaces over the wavelength range 1–1000 micrometers. *(From Heaney, 1992.)*

discontinuation of 3M's 401-C10 Black Velvet Nextel (Ames, 1990). For space use, the Aeroglaze 9924 and Aeroglaze 9929 primers are recommended. The optical properties are given in Figs. 34 and 38 to 43.

A low volatile version of the paint is being developed. In a version developed at Goddard Space Flight Center (based on the Z-306) (Heslin et al., 1974) glass microballoons are added to reduce the specular component, though the particles may be a source of particulate contamination. Other versions were developed at Goddard based on the Z004 paint. Goddard Black is a generic term for several paints of composition and optical properties different from Z-306. This name is often incorrectly applied to the modified Z-306 described here.

Akzo Paints.   Three paints sold formerly under the label Cat-a-lac are now known as Akzo 463-3-8 (formerly Cat-a-lac 463-3-8 diffuse black), 443-3-8 (formerly Cat-a-lac 443-3-8) (see Figs. 34 and 43 to 46) and 443-3-17 (formerly called Sikkens 443-3-17). The Bostic name is also formerly associated with these same paints. The 463-3-8 is the most diffuse of the three blacks, and the other two are considered "glossy." It is an epoxy-based system with low outgassing and low reflectance.

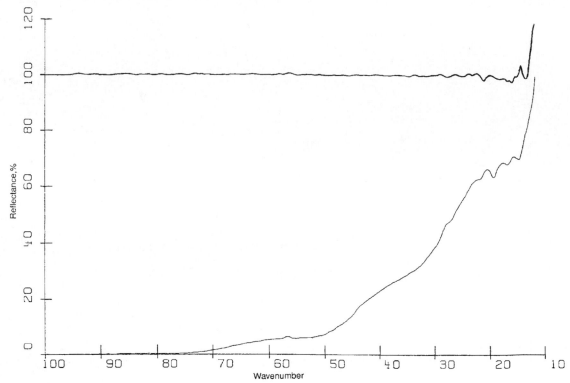

FIGURE 25   Reflectance of black felt contact paper from 100 to 10 wave numbers (100–1000 micrometers.) *(From Heaney, 1992.)*

Cardinal Black.   From Cardinal Industrial Finishes. Optical data are given in Fig. 47.

Cat-a-lac Black (Former Name).   This is an epoxy system with low outgassing and low reflectance. It is now manufactured by Sikkens Aerospace Finishes (see Akzo in the tables for address) and is described here.

DeSoto Black.   The DeSoto Black is a flat black paint most suitable for the range 0.2 to 2.5 micrometers. It has a reflectance of between 2 and 3 percent over this range. Its emissivity is 0.960 compared to a 300 K blackbody, while its absorptivity over the solar spectrum is 0.924. DeSoto Black can meet outgassing requirements for space applications when vacuum baked. The weight loss is typically below 0.5 percent and the volatile condensable material is below 0.5 percent, similar to 3M Black Velvet (Freibel, 1982). A BRDF curve is given in Fig. 35.

Floquil™: Black for Lens Edges.   Lewis et al. (1989) discuss experiments to find low-scatter blackening components for refractive elements. With index of refraction matching, gains of several magnitudes in lower BRDF can be achieved over other edge-blackening techniques. The paint Floquil™ (Polly S Corporation) has been used as

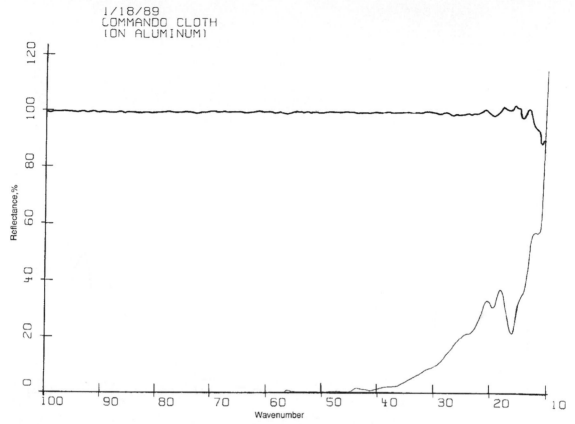

FIGURE 26   Reflectance of commando cloth on aluminum from 100 to 10 wave numbers (100–1000 micrometers). *(From Heaney, 1992.)*

an edge-blackening compound and is also discussed by Lewis. Smith and Howitt (1986) gives a specular reflectance curve for Floquil™ in the infrared from 20 to 200 micrometers.

Parson's Black.   Prior to 1970, Parson's Black was a standard black reference.

SolarChem.   SolarChem is a black paint with excellent visible absorbing properties and has been suggested as a replacement for the unavailable 3-M Nextel Black. The visible BRDF of SolarChem has a flat profile at near-normal angles of incidence (Fig. 48), while the 10.6-micrometer BRDF of SolarChem shows quite a different shape (Pompea et al., 1988). The surface is highly specular at 10.6 micrometers.

Anodized Processes

Anodized processes can produce fine black surfaces. The condition of the substrate plays an important role in the overall scattering properties of the surfaces. A common technique is to sandblast (vapor hone) the surface before anodization. Many varieties of grit may be used and each will produce a different kind of surface treatment. Sometimes small pieces of grit can become embedded in the surface and affect the anodizing process. The

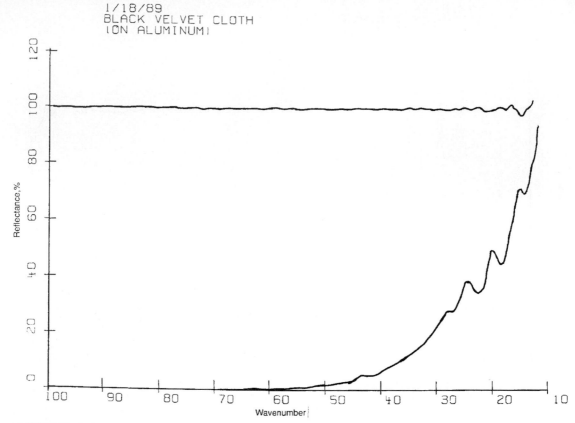

FIGURE 27 Reflectance of black velvet cloth on aluminum from 100 to 10 wave numbers (100–1000 micrometers.) *(From Heaney, 1992.)*

BRDF of sandblasted aluminum surfaces are given in Figs. 49 and 50. Table 2 gives reference data on anodized surfaces.

Martin Black. Martin Black is an anodized aluminum surface that is made microrough by a special anodization process developed by Martin Marietta Astronautics Group, Denver (Wade et al., 1978, Shepard, 1992). It is made black from the inclusion of an aniline dye which is sealed into the surface. It was developed for the Skylab program and has been used on a wide variety of space instruments operating from the vacuum ultraviolet to the far-infrared. The surface is rough, and scattering at several fundamentally different scale lengths occurs.

Special care in handling surfaces of this class is needed since the surface morphology consists of pyramids extending from the surface, and these features are easily crushed. The surface is still quite black, however, even if some surface damage occurs. The surface morphology works well for space applications as it does not support electrical potentials higher than 200 volts, is not affected by temperatures as high as 450°C, by hard vacuum, or by ultraviolet radiation. It also has very low outgassing rates and will not delaminate even under severe conditions (Wade and Wilson, 1975). It passes NASA SP-R-0022A specifications for outgassing.

The surface is also highly resistant to chemical attack by atomic oxygen prevalent in

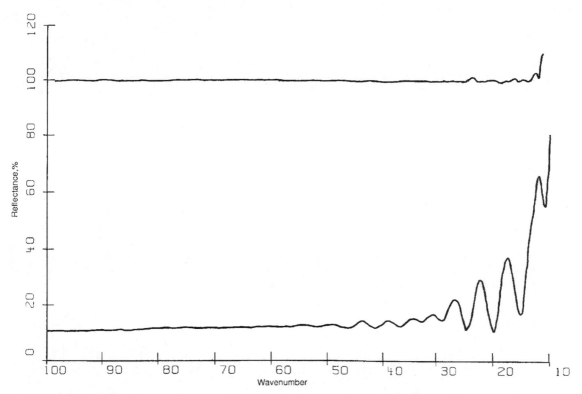

FIGURE 28 Reflectance of neoprene on aluminum from 100 to 10 wave numbers (100–1000 micrometers). *(From Heaney, 1992.)*

low Earth orbit (Pompea et al., 1984b). BRDF measurements of Martin Black show its lambertian character at 0.6328 and 10.6 micrometers. The reflectance spectrum shows significant reflectances at about 2.5 and 5 micrometers; at other wavelengths, the reflectance is low. The optical data are given in Figs. 36, 51, 52.

Enhanced Martin Black. Enhanced Martin Black is similar in its properties to Martin Black, but was created to provide an even more durable surface for long exposures to atomic oxygen in low Earth orbit. Experiments aboard the space shuttle confirm this superiority in the shuttle orbit environment (Pompea et al., 1984b). The process refinements also reduced the near- and middle-infrared reflectances from about 40 percent at 2.3 micrometers to about 25 percent and from about 15 percent at 5.5 micrometers to about 3.5 percent, while maintaining an absorption of about 99.6 percent in the visible region.

Post-treated Martin Black. Another variation of the Martin Black surface is designed for near-infrared applications. Posttreated Martin Black uses hydrogen fluoride to further reduce the near-infrared reflectance peak and eliminate the middle-infrared reflectance peak of Martin Black. The area under the 2.3-micrometer peak of Martin Black has been reduced by about two-thirds and the 5.8-micrometer peak of Martin Black has been eliminated.

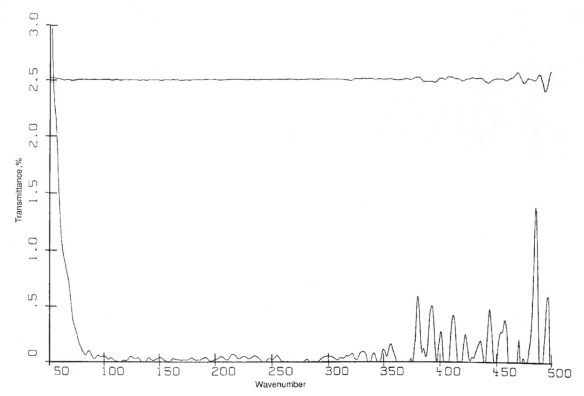

FIGURE 29 Transmittance of black felt contact paper from 50 to 500 wave numbers (200–20 micrometers). *(From Heaney, 1992.)*

Infrablack. Infrablack is another anodized surface from Martin Marietta which relies upon a very rough substrate to produce a diffusing and absorbing surface for wavelengths up to 750 micrometers (Pompea et al., 1983a; Pompea et al., 1983b). It was developed for the next generation of far-infrared NASA telescopes, but has seen applications in experiments to measure the Stefan-Boltzmann constant where a "blackbody" surface with emissivity near 1 was needed and in space radiator surfaces where high emissivity is very beneficial. Like other anodized surfaces in this class, the Infrablack surface is fragile and care in handling is needed to avoid crushing the surface structure. As with Martin Black, even a mistreated surface will be significantly blacker than most other black surfaces. Infrablack can be tuned for maximum absorption between 100 and 500 micrometers. The BRDF in the visible and infrared for Infrablack is similar to Martin Black and Enhanced Martin Black; all exhibit relatively lambertian behavior in the visible at near-normal angles of incidence (Figs. 53 and 54).

Tiodize V-E17 Surface. The Tiodize process is an electrolytic-conversion hard coating of titanium using an all-alkaline, room-temperature bath. It produces an antigall coating with good optical properties. The process can be used on all forms of titanium and its alloys and has been used in a wide variety of space vehicles and aircraft. The Ultra V-E17 coating is a black organic coating which changes the absorptivity and emissivity of

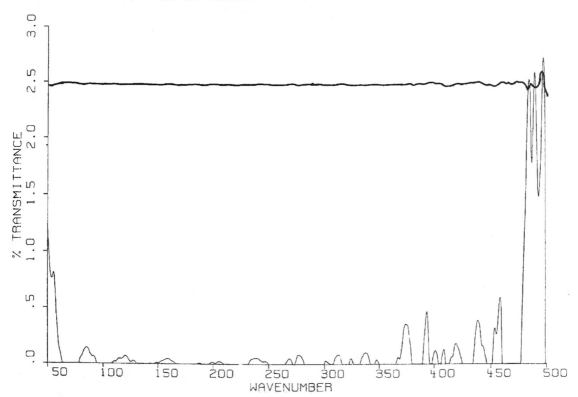

FIGURE 30   Transmittance of black velvet cloth from 50 to 500 wave numbers (200–20 micrometers). *(From Heaney, 1992.)*

titanium (0.62 and 0.89, respectively) to 0.89 and 0.91, respectively. The emittances were determined from a 25-point integration between 4.8 and 26.2 micrometers. The absorptance was determined by a 19-point integration between 0.32 and 2.1 micrometers. The surface has a total mass loss (TML) of 0.91 percent and no detectable volatile condensable material (VCM) in tests run at Ford Aerospace (see Friedrichs, 1992). Its interaction with atomic oxygen is unknown.

TRU-Color Diffuse Black.   An electrolytic coloring process from Reynolds Aluminum.

Hughes Airborne Optical Adjunct Coating.   A cleanable specular black for use at 1 to 10 micrometers.

Etching of Electroless Nickel

The etching of electroless nickel provides a black coating useful for many applications. Some reference data on several electroless nickel processes are presented in Table 3.

NBS Black.   This is a blackened electroless nickel surface developed for use as a solar

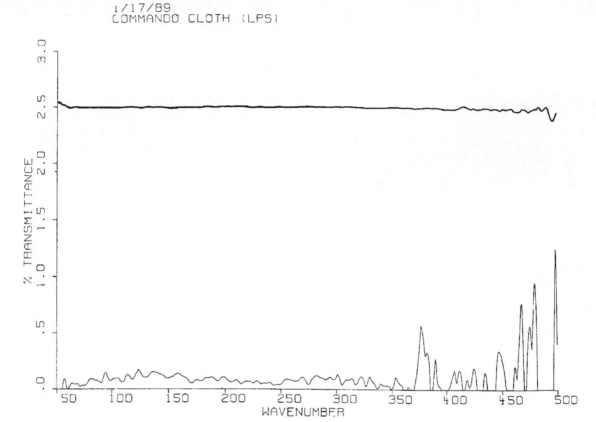

FIGURE 31 Transmittance of commando cloth (lps) from 50 to 500 wave numbers (200–20 micrometers). *(From Heaney, 1992.)*

collector at the National Bureau of Standards (now NIST) in Gaithersberg, MD (Johnson, 1979, 1980, 1982). It consists of an electroless nickel-phosphorous coating which can be plated onto a wide variety of substrates such as metals, glass, ceramics, and plastics. This coating is subsequently etched using nitric acid, creating conically shaped holes into the surface which act as light traps.

The specular reflectance is less than 1 percent over the range of 0.32 to 2.4 micrometers, making it an extremely black coating (Figs. 55 and 56). As this surface finish was developed for solar collectors, the surface has a high value for absorptivity (0.978), but a low value for room-temperature emissivity (about 0.5). The surface is moderately durable as the surface's relief consists of conical holes into the surface. The NBS Black process has been modified resulting in the next surface, which has greater infrared absorption.

Ball Black. Ball Black is an optical black surface produced with modifications to the techniques just described—the selective etching on an electroless nickel surface to produce a multiplicity of conical light traps in the surface (Geikas, 1983; Lompado et al., 1989). The surface appears intensely black to the eye. The surface, however, is still an unaltered nickel-phosphorous deposit with the same chemical properties as the unetched surface, which was shiny. Ball Black can be plated on aluminum, beryllium, copper, stainless steel,

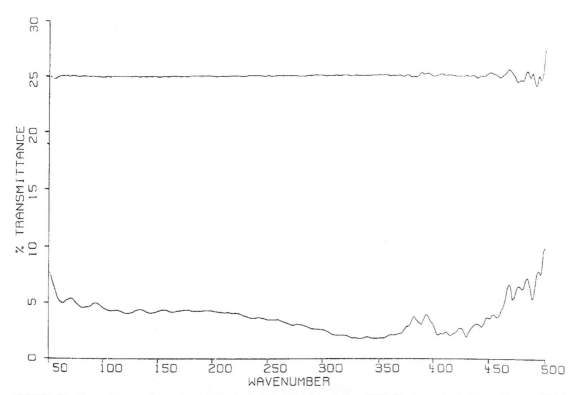

FIGURE 32 Transmittance of neoprene rubber from 50 to 500 wave numbers (200–20 micrometers). *(From Heaney, 1992.)*

invar, polycarbonate and ABS plastics, titanium, and some magnesium alloys. Since this process involves only deposition and etching of a metallic material, problems with outgassing or volatization of the surface in a vacuum are minimal.

The surface is able to withstand rapid changes in temperature, with immersion into boiling water after being at liquid nitrogen temperature. The surface is sensitive to handling, though less so than surfaces like Martin Black. Small blemishes in the surface can be restored by re-etching without replating. The solar absorptivity can be made to vary between 0.71 and 0.995 and its room-temperature emissivity can be tailored to values between 0.35 and 0.94 by changes in the etching parameters. The directional reflectance increases by less than a factor of two from 1.5 to 12 micrometers. It has a BRDF of about 6×10^{-2} at normal incidence at 0.6328 micrometers. See Figs. 57 and 58 for optical data.

Plasma-sprayed Surfaces

Boron Black. Martin Marietta Boron Black is a surface that is black in the visible (solar absorptivity between 0.89 and 0.97, and emissivity of at least 0.86) and black in the infrared (Shepard et al., 1991) (Fig. 59). Boron Black is a plasma-sprayed surface and can probably be applied to most metallic substrates. It has been applied to molybdenum,

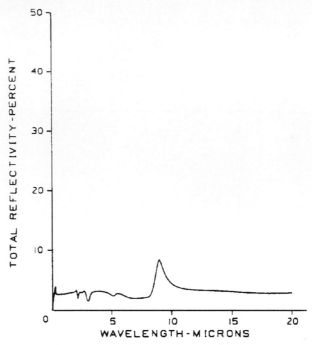

FIGURE 33 Total hemispherical reflectivity of 3M Nextel Black Velvet. *(From Willey et al., 1983.)*

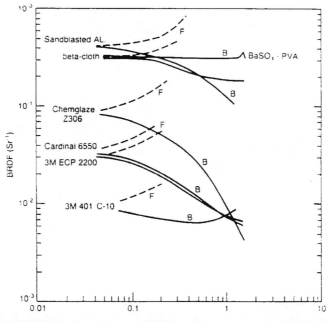

FIGURE 34 Comparison of BRDF profiles at an angle of 45° and 633-nm wavelength. *(From Viehmann and Predmore, 1986.)*

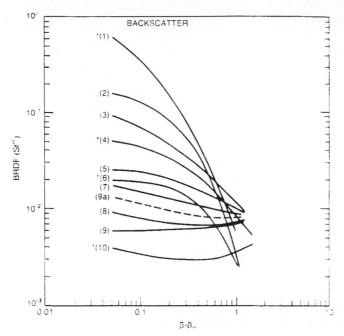

FIGURE 35 Comparison of BRDF profiles at an angle of 45° and 254-nm wavelength: (*a*) Reynolds Tru Color Diffuse; (*b*) Chemglaze Z306; (*c*) Sikkens 443; (*d*) Sanodal Fast Black GL; (*e*) Chemglaze Z306 plus microballoons; (*f*) Sanodal deep black MLW; (*g*) 3M 401-C10; (*h*) Chemglaze Z313; (*i*) DeSoto Flat; (*j*) DeSoto Flat after EUVSH; (*k*) Martin Black anodize. *(From Viehmann and Predmore, 1986).*

nickel, and titanium substrates. It offers a low-atomic-number surface with good optical properties and the ease of a plasma-spray deposition. The infrared BRDFs show a profile characteristic of a surface with lambertian profiles at near-normal angles of incidence.

Boron Carbide. This is a Martin Marietta proprietary process. The boron carbide surface is applied to a titanium substrate. A 10- to 15-micrometer layer of B_4C is applied in a proprietary plasma-spraying process. BRDF measurements at 30° angle of incidence at two wavelengths illustrate that this surface is more lambertian in the visible.

Beryllium Surfaces. Porous surfaces or those with steep-walled features which can trap radiation have been developed using plasma-sprayed beryllium and sputter-deposited beryllium by workers at Oak Ridge (Seals et al., 1990; Egert and Allred, 1990; Seals, 1991) and at Spire Corp. (e.g., Murray and Wollam, 1989) who are developing baffles that can operate in severe environments.

Plasma-sprayed beryllium samples appear visually rough with a matte gray finish. The specular reflectance is approximately constant at about 1.6 percent from 2 to 50 micrometers, while the BRDF is lambertian in character at near-normal angles of incidence with a value of about 5×10^{-2} sr^{-1}, equivalent to a total hemispherical reflectance of about 15 percent.

Beryllium surfaces with varying thicknesses and columnar grain sizes can also be

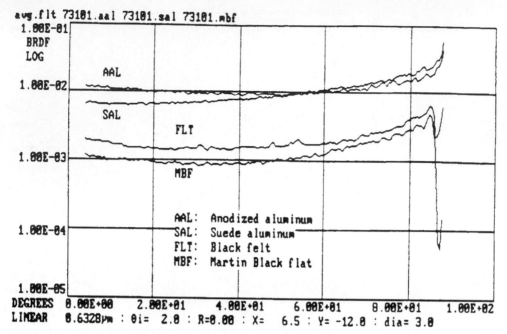

FIGURE 36   BRDF data for anodized aluminum, 3M Nextel Suede paint, black felt, and an aluminum flat with Martin Black wavelength is 633 nm, at near normal incidence. *(From Cady et al., 1987.)*

created by low-temperature magnetron sputtering. The surfaces can then be chemically etched to enhance their absorptive properties. These surfaces exhibit less than 2 percent specular reflectance in the near-infrared. Very thick surfaces (350 micrometers) exhibit specular reflectances of less than 0.5 percent up to 50 micrometers wavelength.

A sputtered coating of Be on Be can be made more absorptive through exposure to an oxygen plasma to form a layer of BeO on the surface of the coating (Lompado et al., 1989). Optical data are presented in Figs. 57, 58, 60 and 61.

Ion Beam-sputtered Surfaces

Ion beam-sputtering processes can roughen surfaces of a large variety of compositions. An excellent review is by Banks in Cuomo et al. (1989) (see also Banks, 1981). A variety of sputtering processes can create rough features of various geometries that decrease specular reflections and create a black-appearing surface from a light-appearing substrate. Since the roughness is integral to the surface, these surfaces often have desirable environmental resistance to atomic oxygen. However, these surfaces also have tremendous surface area for trapping of contaminants and thus a potential for large outgassing rates if preventative steps are not taken. These surfaces are expensive to make in large areas, though they are becoming less expensive as coating facilities are built to harness this technology. They do have tremendous potential as diffuse baffle surfaces and as high-emittance surfaces for high-temperature space radiators (Rutledge et al., 1987; Banks et al., 1988). They are also cleanable and very durable.

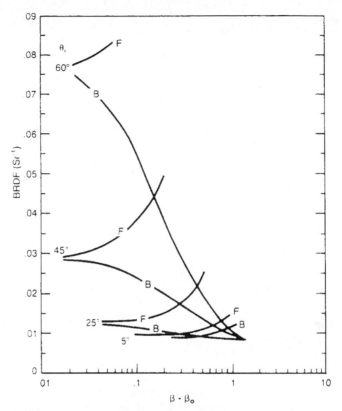

FIGURE 37 BRDF profiles of 3M ECP-2200 black paint 6550 at 633 nm. *(From Viehmann and Predmore, 1986.)*

Lompado et al. (1989) and Blatchley et al. (1992) describe several textured aluminum surfaces and give plots of their total hemispherical reflectance in the 0.4 to 1.0 micrometer range, as well as BRDF plots at 0.6328 and 10.6 micrometers (Fig. 60).

Electrodeposited Surfaces

Refer to Table 3 for further reference data on these surfaces.

Black Cobalt. Black cobalt is a patented surface developed at Martin Marietta (Shepard and Fenolia, 1990) which appears black in the visible, with a solar absorptivity of at least 0.96 and emissivity of at least 0.6. It is applied by an electrodeposition process to a substrate of nickel, molybdenum, or titanium; other substrates are also expected to work. It can be used at higher temperatures than anodized aluminum surfaces such as Martin Black. It is stable against loss of absorptivity at temperatures greater than 450°C for prolonged periods. The surface is not a lambertian surface at 10.6 micrometers (Fig. 62).

Black Chrome. A black chrome surface developed for aerospace use (Fenolia et al., 1990) is black in the visible and is created by an electroplating process on a conducting substrate. Many metallic substrates can be used and samples on molybdenum, nickel, and

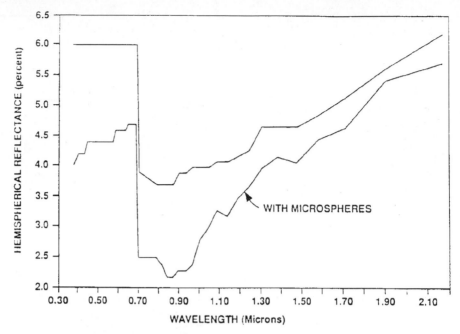

FIGURE 38 Diffuse visible and near-infrared hemispherical reflectance of Chemglaze (now called Aeroglaze) Z-306, and Z-306 with microspheres. *(From Ames, 1990.)*

titanium have been made. The preferred substrates are titanium and molybdenum. The coating process involves electrodeposition using chromium and chromium oxides. The visible BRDF is flat (Fig. 63), while the infrared profile shows the surface to be largely specular in its scattering profile. The solar absorptivity is at least 0.95 and the emissivity can be tailored in the range of 0.4 to 0.8. The surface has excellent adhesion to the substrate.

Orlando Black Optical Coating. A dendritic surface is produced by the electrodeposition of copper in a proprietary copper-plating formulation (Janeczko, 1992, and Engelhaupt). The dendritic structure is then oxidized to form smaller structures superimposed on the larger ones. The surface has excellent broadband absorption properties from 0.4 to 14 micrometers and good absorptance at non-normal angles of incidence (Fig. 64). It may be suitable for use at shorter wavelengths. The surface is usable in vacuum environments as well as a thermal reference in FLIR systems. A wide variety of materials can be coated including ABS plastic, nylon, Ultem, Noryl, fiberglass-reinforced epoxy, glass, stainless steel, copper, brass, nickel, aluminum, gold, and beryllium oxide. Dimensional allowances must be made for the thickness of the finish (approximately 0.001 in).

Other Specialty Surfaces

Refer to Table 3 for further data on these surfaces.

Electrically Conductive Black Paint. Birnbaum et al. (1982) developed an electrically conductive, flat black paint for space use in places such as Jupiter's radiation belts where spacecraft charging effects can be important. Its small resistivity prevents the buildup of

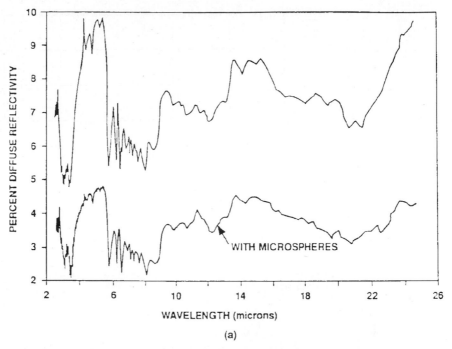

(a)

FIGURE 39a   Infrared diffuse reflectivity of Chemglaze Z-306 (now called Aeroglaze) and Z-306 with microspheres. *(From Ames, 1990.)*

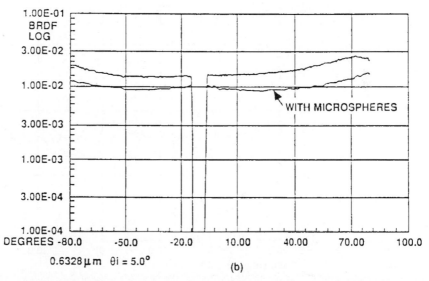

(b)

FIGURE 39b   BRDF for Chemglaze (now Aeroglaze) Z-306 at near-normal incidence (5°) at a wavelength of 0.6328 micrometers. *(From Ames, 1990.)*

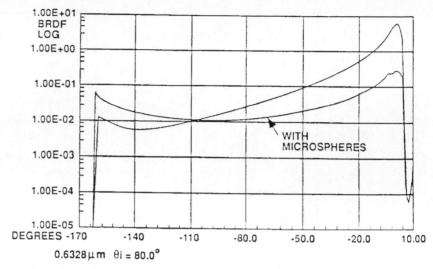

FIGURE 40 BRDF for Chemglaze (now Aeroglaze) Z-306 at near grazing incidence (80°) at a wavelength of 0.6328 micrometers. *(From Ames, 1990.)*

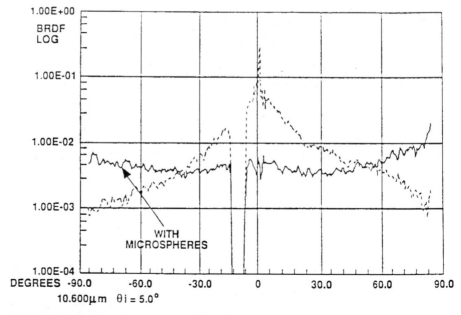

FIGURE 41 BRDF for Chemglaze (now Aeroglaze) Z-306 at near-normal incidence (5°) at a wavelength of 10.60 micrometers. *(From Ames, 1990.)*

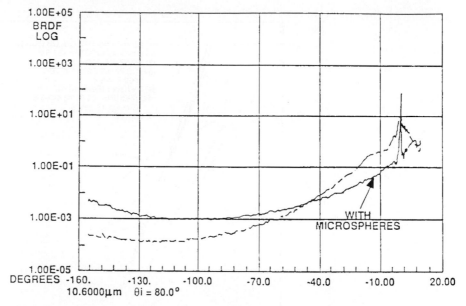

FIGURE 42 BRDF for Chemglaze (now Aeroglaze) Z-306 at near grazing incidence (80°) at a wavelength of 10.60 micrometers. *(From Ames, 1990.)*

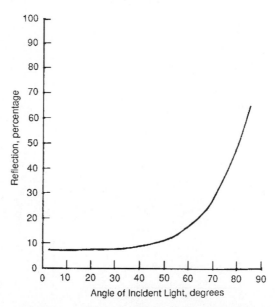

FIGURE 43 Specular reflection measured for Cat-A-Lac Glossy (now called Akzo 443-3-8) and for Chemglaze glossy (now called Aeroglaze Z-302). The wavelength is 0.6328 micrometers. The two lines overlap. *(From Griner, 1979.)*

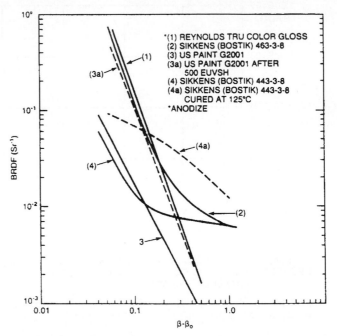

FIGURE 44 BRDFs of glossy surfaces at 0.254 micrometers and 45° angle of incidence. *(From Viehmann and Predmore, 1986.)*

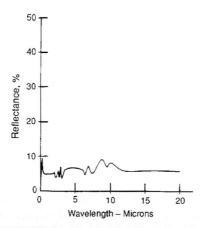

FIGURE 45 Total hemispherical reflectivity of Bostic 463-3-8 (now called Akzo 463-3-8) on bare aluminum. *(From Willey et al., 1983.)*

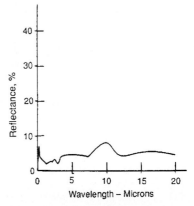

FIGURE 46 Total hemispherical reflectivity of Bostic 463-3-8 (now called Akzo 463-3-8) on ZnCr primer. *(From Willey et al., 1983.)*

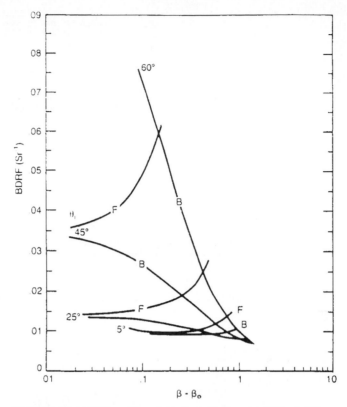

FIGURE 47 BRDF profiles of Cardinal black paint 6550 at 633 nm. *(From Viehmann and Predmore, 1986.)*

charge on the surface. It has a visual reflectance of less than 5 percent and other desirable optical properties.

High-resistivity Coatings. Strimer et al. (1981) describe a number of black surfaces with high electrical resistivity for visible and infrared applications.

Sputtered and CVD Surfaces. Carbon, quartz, and silicon surfaces have been modified by sputtering or by chemical vapor deposition (Robinson and Rossnagel, 1982; Culver et al., 1984; Bowers et al., 1987) to create surfaces that are black over the wavelength range of 1 to 15 micrometers. The textured surfaces are produced by sputtering with a low-energy (e.g., 500-eV) broadbeam ion source while adding impurities to the surface. The results from sputtering with this seeding process are structures in the form of cones, pyramids, and ridges. The exact nature of the surfaces created depends on the substrate temperature, ion and impurity fluxes, and the impurity species. The reflectivity of one modified silicon surface was below 1 percent throughout 1 to 25-micrometer wavelength range.

Silicon Carbide. A silicon carbide surface for high-temperature applications was developed at Martin Marietta for application to a carbon-carbon substrate. It is applied to a hot substrate by a chemical vapor deposition process. The visible BRDF is flat, while the

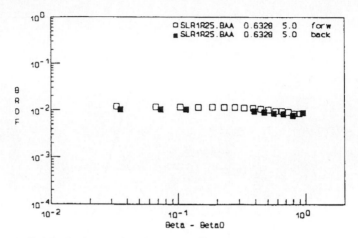

FIGURE 48 Visible (0.6328 micrometers) BRDF of SolarChem at 5° angle of incidence. *(From Pompea et al., 1988.)*

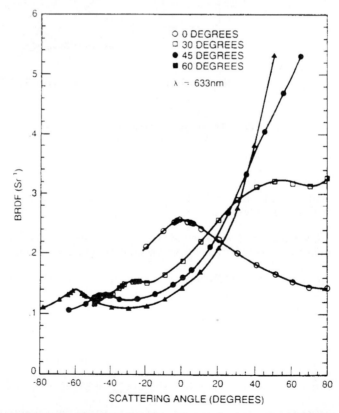

FIGURE 49 BRDF profiles of sandblasted aluminum. Open circle: 0°; open square: 30°; filled circle: 45°; filled square: 60°: wavelength of 633 nm. *(From Viehmann and Predmore, 1986).*

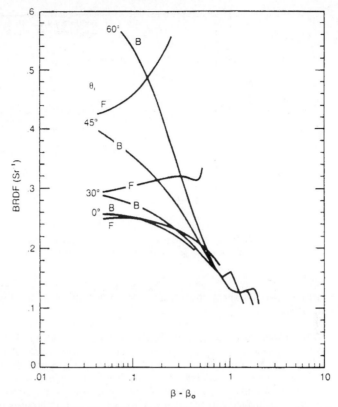

FIGURE 50 Forward (F) and back (B) scatter profiles of sandblasted aluminum. Wavelength of 633 nm. *(From Viehmann and Predmore, 1986.)*

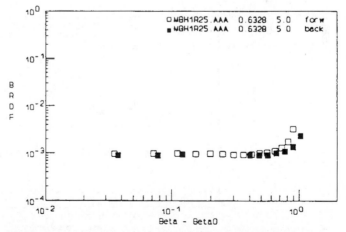

FIGURE 51 Visible (0.6328 micrometers) BRDF of Martin Black at 5° angle of incidence. *(From Pompea et al., 1988.)*

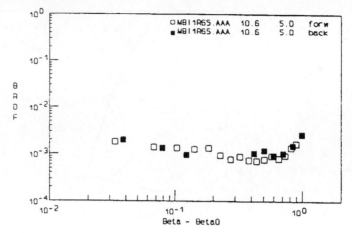

FIGURE 52   Infrared (10.6 micrometers) BRDF of Martin Black at 5° angle of incidence. *(From Pompea et al., 1988.)*

infrared profiles show the surface's specular nature at the longer wavelength (Pompea et al., 1988).

IBM Black (Tungsten Hexafluoride). This is a surface being produced at Martin Marietta, Orlando, using a process licensed from IBM. Any material that can tolerate 400°C can be used as a substrate for this process which involves vapor deposition of tungsten. The surface has dendritic structures which form light-absorbing traps. The surface looks like a collection of obelisks, does not outgas, and is fairly rugged to the touch. The surface can be given an anodize, which decreases its visible reflectance. Optical properties and an SEM photograph of the surface are given by Willey et al. (1983). The surface is very black in the 1- to 2-micrometer wavelength range.

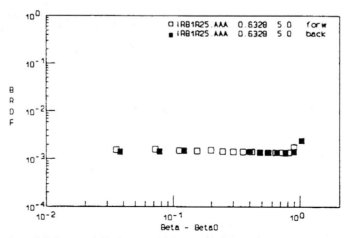

FIGURE 53   Visible (0.6328 micrometers) BRDF of Infrablack at 5° angle of incidence. *(From Pompea et al., 1988.)*

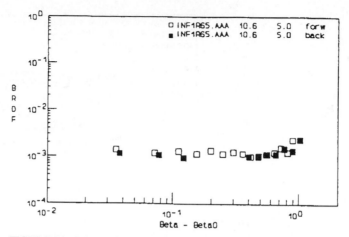

FIGURE 54   Infrared (10.6 micrometers) BRDF of Infrablack at 5° angle of incidence. *(From Pompea et al., 1988.)*

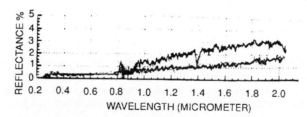

FIGURE 55   Total integrated scatter of NBS Black. The substrates are magnesium AZ31B-F (upper curve) and magnesium ZK60A-T5. Substrate is silica sandblasted. Alpha = 0.978. Epsilon = 0.697. *(From Geikas, 1983.)*

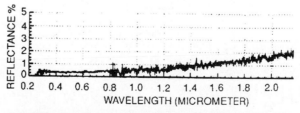

FIGURE 56   Diffuse scatter of NBS Black. The substrates are magnesium AZ31B-F and magnesium ZK60A-T5. The plots are overlaid and are virtually identical. Substrate is silica sandblasted. Alpha = 0.978. Epsilon = 0.697. *(From Geikas, 1983.)*

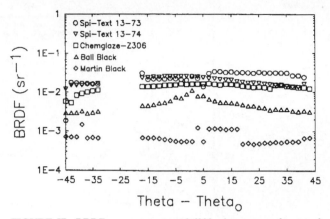

FIGURE 57 BRDF measurements at 0.6328 micrometers of textured metal surfaces from Spire Corporation and Chemglaze Z-306 paint, Ball Black electroless nickel coating, and Martin Black anodized coating. *(From Lompado et al., 1989.)*

ZO-MOD BLACK. ZO-MOD BLACK (ZYP Coatings, Oak Ridge, Tenn.) is a high-emissivity coating applied like paint to ceramic porous and fibrous structures. On heating, a hard abrasion- and chemical-resistant, calcia-stabilized zirconium oxide coating is formed. The coating has high emissivity. The coating has been used inside of ceramic furnaces.

Gold Black. Gold blacks are fragile surfaces made by evaporating gold in a low-pressure atmosphere of helium or nitrogen. The surface has good visible light absorption. The preparation and optical properties of gold blacks are given by Harris and McGinnies (1948), Blevin and Brown (1966), and Zaeschmar and Nedoluha (1972).

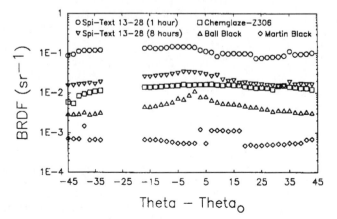

FIGURE 58 BRDF measurements at 0.6328 micrometers of two textured metal surfaces from Spire Corporation and Chemglaze Z-306 paint, Ball Black electroless nickel coating, and Martin Black anodized coating. *(From Lompado et al., 1989.)*

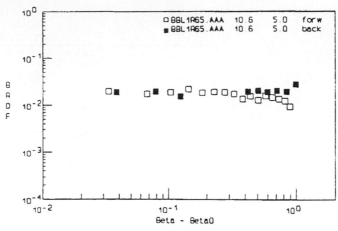

FIGURE 59 Infrared (10.6 micrometers) BRDF of Boron Black at 5° angle of incidence. *(From Pompea et al., 1988.)*

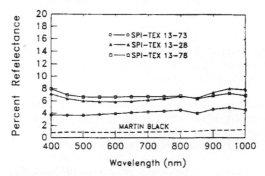

FIGURE 60 Total hemispherical reflectance in the visible bandpass for textured baffle materials from Spire Corp. The 13-73 surface is a textured aluminum surface. *(From Spire, 1992.)*

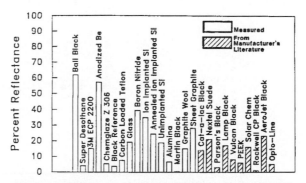

FIGURE 61 Total hemispherical reflectance data for baffle materials at 10.6 micrometers. The shaded measurements are from the manufacturer's literature and have not been independently verified. *(From E. Johnson, 1992.)*

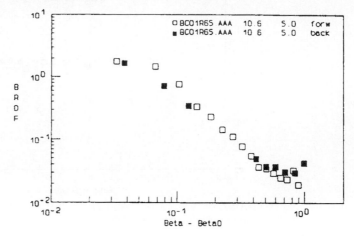

FIGURE 62 Infrared (10.6 micrometers) BRDF of Black Cobalt at 5° angle of incidence. *(From Pompea et al., 1988.)*

Flame-sprayed Aluminum. Flame-sprayed aluminum is a very durable, very rough surface which has use as an infrared diffuse reflectance standard. The parameters of the flame-spray process must be tightly controlled to create a similar surface each time.

Black Glass. Black glass is available from Schott Glass Technologies, Duryea, Pa., in a glass called UG1, which is an ionically colored glass. Star Instruments, Flagstaff, Ariz., produces a glossy black glass called LOX8. It is a low-expansion copper glass with a coefficient of expansion of about thirty times lower than Pyrex.

Black Kapton. The optical properties of Black Kapton are given in Fig. 65.

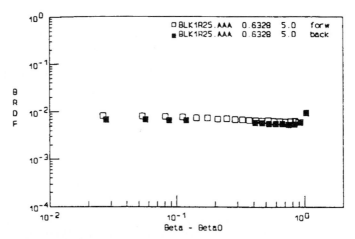

FIGURE 63 Visible (0.6328 micrometers) BRDF of Black Chrome at 5° angle of incidence. *(From Pompea et al., 1988.)*

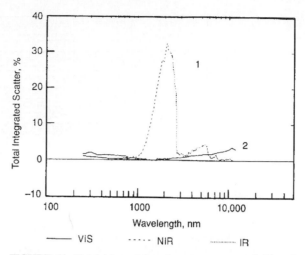

FIGURE 64 Total integrated scatter measurement of (1) and (2) Orlando Black commercially available anodized aluminum. *(From Janeczko, 1992.)*

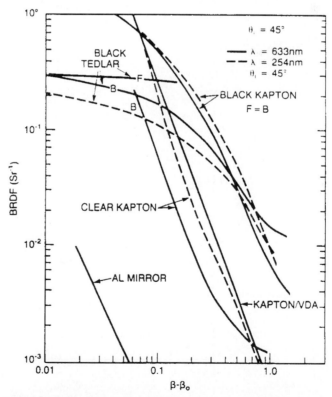

FIGURE 65 BRDFs of black Kapton and black Tedlar films. *(From Viehmann and Predmore, 1986.)*

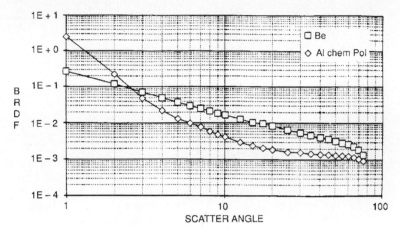

FIGURE 66 BRDF of specular baffle materials using Be and Al substrates that are polished and anodized. The measurements were made at 0.5145 micrometers. *(From Schaub et al., 1990.)*

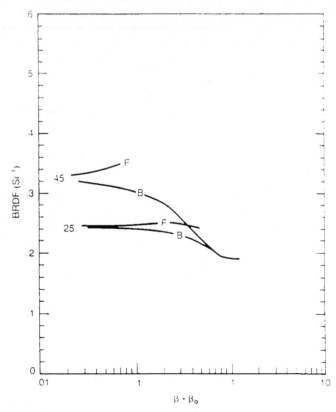

FIGURE 67 BRDF profiles of beta-cloth. Wavelength of 633 nm. *(From Viehmann and Predmore, 1986).*

Other Surfaces. Some other data is given for specular metallic anodized baffle surfaces (Fig. 66). This kind of basic treatment is adequate for some applications. The BRDF of Beta cloth is given in Fig. 67.

37.9 CONCLUSIONS

Black surfaces are used as optical elements in a variety of ways. There is a great choice of surfaces available, but the selection of the appropriate surface is often problematical. The consequences of choosing the wrong surface are often quite severe.

For the selection process to be most effective, it must be done as early as possible. While the optical data on surfaces is relatively extensive, the data are often inconsistent or are not available for the particular wavelength of interest. The creation of large databases of optical and material properties is a great aid to the optical designer and materials consultant. Of great concern for space materials is the lack of very long term environmental test data. The synergistic effect of a number of degrading influences (e.g., solar ultraviolet, charging effects, atomic oxygen) is unknown for many interesting and potentially effective space materials. The use of specialized surfaces that can be tailored for very specific applications is an emerging trend and is likely to continue.

When black surfaces are used effectively in optical systems, the performance of the system can be greatly enhanced. However, when more general considerations of stray light and other system-performance issues are ignored, even the "blackest" possible surfaces cannot restore the desired system performance. This cruel fact is a persuasive basis for an integration of black surface selection into the general stray light design studies, which today are nearly mandatory for the design of high-performance optical devices.

37.10 ACKNOWLEDGMENTS

Special thanks to J. Heaney (NASA Goddard Space Flight Center), S. H. C. P. McCall (Stellar Optics Laboratory, Inc.), and S. Smith (NASA ARC) for use of unpublished data. Thanks to R. Bonaminio (Lord Corp.), C. Bowers (Hughes Aircraft Co.), B. Banks (NASA Lewis Research Center), R. Harada (IIT Research Institute), D. Janeczko (Martin Marietta Corp.), A. Olson (Ball Aerospace), R. Seals (Oak Ridge National Laboratory), D. Shepard (Martin Marietta), and S. Smith (NASA Ames Research Center) for providing data and photographs. Thanks also to J. Martin, L. Bergquist, F. Bartko, R. Culver, and S. Russak for support in the early stages of this work, and to D. Shepard, S. McCall, S. Smith, and W. Wolfe for helpful suggestions on this manuscript.

37.11 REFERENCES

AKZO Coatings, Inc., product information and MSDSs on 443-3-17 and 443-3-8, Orange, Calif., 1992.

Anderson, S., S. M. Pompea, D. F. Shepard, and R. Castonguay, *Proc. SPIE* **967**:159 (1988).

Ames, A. J., "Z306 Black Paint Measurements," *Proc. SPIE: Stray Radiation in Optical Systems,* R. P. Breault (ed.) **1331**:299–304 (1990).

Banks, B. A., S. K. Rutledge, M. J. Mirtich, T. Behrend, D. Hotes, M. Kussmaul, J. Barry, C. Stidham, T. Stueber, and F. DiFillippo, "Arc-Textured Metal Surfaces for High Thermal Emittance Radiators," *NASA Technical Memorandum 100894*, 1988.

Banks, B. A., NASA Lewis Research Center, Cleveland, Oh., private communication.

Banks, B. A., *NASA Technical Memorandum 81721*, Lewis Research Center, 1981.

Banks, B. A., "Ion Beam Applications Research—A 1981 Summary of Lewis Research Center Programs," *NASA Technical Memorandum 81721,* 1981.

Bartell, F. O., E. L. Dereniak, and W. L. Wolfe, *Proc. SPIE* **257:**154 (1980).

Bartell, F. O., "BRDF Measurement Equipment: Intrinsic Design Considerations," *Proc. SPIE: Stray Radiation IV,* R. P. Breault (ed.) **511:**31–34 (1984).

Bartell, F. O., J. E. Hubbs, M. J. Nofziger, and W. L. Wolfe, *Appl. Opt.* **21**(17):3178 (1982).

Bennett, H. E., and E. L. Church, *Opt. Eng.* **18:**103 (1979).

Bennett, J. M., and L. Mattsson, *Introduction to Surface Roughness and Scattering,* Optical Society of America, 1989.

Bergener, D. W., S. M. Pompea, D. F. Shepard, and R. P. Breault, "Stray Light Performance of SIRTF: A Comparison," *Proc. SPIE: Stray Radiation IV,* R. P. Breault (ed.) **511:**65–72 (1984).

Bernt, M. L. and J. C. Stover, *Proc. SPIE* **1331:**261 (1990).

Betts, D. B., F. J. J. Clarke, L. J. Cox, and J. A. Larkin, "Infrared Reflection Properties of Five Types of Black Coating for Radiometric Detectors," *J. Physics E*: (*London*) *Scientific Instrum.* **18:**689–696 (August, 1985).

Birnbaum, M. B., E. C. Metzler, and E. L. Cleveland, "Electrically Conductive Black Optical Black Paint," *Proc. SPIE: Scattering in Optical Materials,* S. Musikant (ed.) **362:**60–70 (1982).

Blatchley, C. C., E. A. Johnson, Y. K. Pu, and C. Von Benken, "Rugged Dark Materials for Stray Light Suppression by Seeded Ion Beam Texturing," *Proc. SPIE: Stray Radiation II,* R. P. Breault (ed.) **1753** (1992).

Blevin, W. R. and W. J. Brown, "Black Coatings for Absolute Radiometers," *International Journal of Scientific Metrology* **2**(4):139–143 (October 1966).

Bohren, C. F., and D. F. Huffman, *Absorption and Scattering of Light by Small Particles,* John Wiley and Sons, New York, 1983.

Bonnot, A. M., H. Belkhir, D. Pailharey, and P. Mathiez, *Proc. SPIE* **562:**209 (1985).

Bonnot, A. M., H. Belkhir, and D. Pailharey, *Proc. SPIE* **653:**215 (1986).

Bowers, C. W., R. B. Culver, W. A. Solberg, and I. L. Spain, "Optical Absorption of Surfaces Modified by Carbon Filaments," *Appl. Opt.* **26:**4625 (1987).

Bowyer, S., Univ. of California at Berkeley, private communication, April 1992.

Breault, R. P., *Proc. SPIE* **181:**108 (1979).

Breault, R. P., *Proc. SPIE* **967:**2 (1988b).

Breault, R. P., "Stray Light Technology Overview of the 1980 Decade (And a Peek into the Future)," *Proc. SPIE: Stray Radiation in Optical Systems,* R. P. Breault (ed.) **1331:**2–11 (1990).

Breault, R. P., A. Greynold, and S. Lange, "APART/PADE Version 7: A Deterministic Computer Program Used to Calculate Scattered and Diffracted Energy," *Proc. SPIE: Radiation Scattering in Optical Systems,* G. Hunt (ed.) **257** (1980).

Breault, R. P., "Current Technology of Stray Light," *Proc. SPIE: Stray Radiation V,* R. P. Breault (ed.) **675:**4–13 (1986).

Breault, R. P., "Specular Vane Cavities," *Proc. SPIE: Generation, Measurement, and Control of Stray Radiation III,* R. P. Breault (ed.) **384:**90–97 (1983).

Brooks, L. D. and W. L. Wolfe, *Proc. SPIE* **257:**177 (1980).

Brooks, L. D., J. E. Hubbs, F. O. Bartell, and W. L. Wolfe, *Appl. Opt.* **21:**2465 (1982).

Brown, C. W., and D. R. Smith, "High-Resolution Spectral Reflection Measurements on Selected Optical-Black Baffle Coatings in the 5–20 Micrometer Region, *Proc. SPIE: Stray Radiation In Optical Systems,* R. P. Breault (ed.) **1331:**210–240 (1990).

Cady, F. M., D. R. Cheever, K. A. Klicker, and J. C. Stover, *Proc. SPIE* **818:**21 (1987).

CAL, "Space Systems Process Procedure for the Preparation and Application of Chemglaze Z306

Black Paint for Space Applications," *Document PQ-CAL-SR-10333* from Canadian Astronautics Ltd., 1986.

Campbell, W. A. and J. J. Scialdone, "Outgassing Data for Selecting Spacecraft Materials," *NASA Reference Publication 1134,* Revision 2, November 1990.

Cardinal, product information and MSDSs on Velvathane and 6450-01 paints, El Monte, Calif., 1992.

Clarke, F. J. J., and J. A. Larkin, "Measurements of Total Reflectance, Transmittance and Emissivity over the Thermal IR Spectrum," *Infrared Physics* **25**(1/2):359–367 (1985).

Culver, R. B., W. A. Solberg, R. S. Robinson, and I. L. Spain, "Optical Absorption of Microtextured Graphite Surfaces in the 1.1–2.4 Micrometer Wavelength Region," *Appl. Opt.* **24**:924 (1984).

Cuomo, J. J., S. M. Rossnagel, and H. R. Kaufman, *Handbook of Ion Beam Processing Technology: Principles, Deposition, Film Modification, and Synthesis,* Noyes Publications, Park Ridge, NJ, 1989. (See chap. 17, Banks, B. A., "Topography Texturing Effects," 1989.)

Davis, L., and J. G. Kepros, *Proc. SPIE* **675**:24 (1986).

Dobrowolski, J. A., F. C. Ho, and A. J. Waldorf, "Research on Thin Film Anticounterfeiting Coatings at the National Research Council of Canada," *Appl. Opt.* **28**:2702–2717 (1989).

Dobrowolski, J. A., "Versatile Computer Program for Absorbing Optical Thin Film Systems," *Appl. Opt.* **20** (1981).

Dobrowolski, J. A., E. H. Hara, B. T. Sullivan, and A. J. Waldorf, "A High Performance Optical Wavelength Multiplexer/Demultiplexer," *Appl. Opt.* **30** (1991).

Dobrowolski, J. A., B. T. Sullivan, and R. C. Bajcar, "An Optical Interference, Contrast Enhanced Electroluminescent Device," *Appl. Opt.* **31**:5988 (1992).

Driscoll, W. (ed.), and W. Vaughan (assoc. ed.), *Handbook of Optics,* McGraw-Hill, New York, 1978, pp. 7–112.

Duran, M., and C. Gricurt, *Proc. SPIE* **675**:134 (1986).

Egert, C. M., and D. D. Allred, "Diffuse Absorbing Beryllium Coatings Produced by Magnetron Sputtering," *Proc. SPIE: Stray Radiation in Optical Systems,* R. P. Breault (ed.) **1331**:170–178 (1990).

Engelhaupt, D. E., "Electrodeposited Anti-Reflective Surface Coatings and Method of Formulation," patent pending, Martin Marietta Corporation.

Evans, D. C., *Proc. SPIE* **384**:82 (1983).

Evans, D. C. and R. P. Breault, "APART/PADE Analytical Evaluation of the Diffuse Infrared Background Experiment for NASA's Cosmic Background Explorer," *Proc. SPIE: Stray Radiation IV,* R. P. Breult (ed.) **511**:54–64 (1984).

Fenolia, R. J., D. F. Shepard, and S. L. Van Loon, "Optically Black Pliable Foils," U.S. Patent 4,894,125, January 16, 1990, Martin Marietta Corporation.

Fernandez, R., R. G. Seasholtz, L. G. Oberle, and J. R. Kadambi, *Proc. SPIE* **967**:292 (1988).

Ferrel, V., Lord Corporation, Erie, PA, private communication, March 1992.

Freibel, V., *Colorado, M & P Information Bulletin No. 82.21,* Ball Aerospace, Boulder, 7/27/82.

Freniere, E. R. and D. L. Skelton, "Use of Specular Black Coatings in Well-Baffled Optical Systems," *Proc. SPIE: Stray Radiation V,* R. P. Breault (ed.) **675**:126–132 (1986).

Freniere, E. R., "First-Order Design of Optical Baffles," *Proc. SPIE: Radiation Scattering in Optical Systems,* G. Hunt (ed.) **257**:19–22 (1980).

Freniere, E. R., *Proc. SPIE* **675**:126 (1986a).

Friedrichs, Wade, product literature from Tiodize Co., Inc., 5858 Engineer Drive, Huntington Beach, CA 92649.

Funk, J., Ranbar Technology, Glenshaw, Pa., personal communication, Nov. 1990.

Garrison, J. D., J. C. Haiad, and A. J. Averett, *Proc. SPIE* **823**:225 (1987).

Geikas, G., "Scattering Characteristics of Etched Electroless Nickel Coatings," *Proc. SPIE: Generation, Measurement, and Control of Stray Radiation III,* R. P. Breault (ed.) **384**:10–18 (1983).

Geniere, Cardinal Industrial Finishes, El Monte, Calif., private communication, March 1992.

Glassford, A. P. (ed.), *Proc. SPIE: Optical System Contamination: Effects, Measurements, Control* **777** (1987).

Glassford, A. P. (ed.), *Proc. SPIE: Optical System Contamination: Effects, Measurements, Control II* **1329** (1990).

Glassford, A. P. (ed.), *Proc. SPIE: Optical System Contamination: Effects, Measurements, Control III* **1754** (1992).

Golden, J. L., "Results of an Examination of the A-276 White and Z-306 Black Thermal Control Paint Disks flown on LDEF," *NASA Conference Publication 10072,* First LDEF Post-Retrieval Symposium Abstracts, June 2–8, 1991.

Grammar, J. R., L. J. Balin, M. D. Blue, and S. Perkowitz, "Absorbing Coatings for the Far Infrared," *Proc. SPIE: Radiation Scattering in Optical Systems,* G. Hunt (ed.) **257:**192–195 (1980).

Granqvist, C. G., *Spectrally Selective Surfaces for Heating and Cooling Applications,* vol. TT 1, SPIE Optical Engineering Text, Bellingham, Wash., 1989.

Greynolds, A. W. and R. K. Melugin, "Analysis of an All-Specular Linlor Baffle Design," *Proc. SPIE: Stray Radiation IV,* R. P. Breault (ed.) **675:**240–248 (1986).

Greynolds, A. W., *Advanced Systems Analysis Package (ASAP) Users Manual,* Breault Research Organization, Inc., Tucson, Ariz., 1988.

Griner, D. B., *Proc. SPIE* **183:**98 (1979).

Gull, T. R., H. Hertzig, F. Osantowski, A. R. Toft, "Low Orbit Effects on Optical Coatings and Materials as Noted on Early Shuttle Flights," *RAL Workshop on Advanced Technology Reflectors for Space Instrumentation,* M. Grande (ed.) June 16–18, 1986.

Gunderson, John A., "Goniometric Reflection Scattering Measurements and Techniques at 10.6 Micrometers," thesis, Univ. of Arizona, 1977.

Hahn, R. E., and B. O. Seraphin, "Spectrally Selective Surfaces for Photothermal Solar Energy Conversion," in *Physics of Thin Films,* vol. 10, Academic Press, Orlando, Fla, 1978.

Harada, Y., IITRI, Chicago, Ill., private communication, March 1992.

Harris, L., R. T. McGinnies, and B. M. Siegel, "The Preparation and Optical Properties of Gold Black," *J. Opt. Soc. Am.* **38**(7) (1948).

Hass, G., H. H. Schroeder, and A. F. Turner, "Mirror Coatings for Low Visible and High Infrared Reflectance," *J. Opt. Soc. Am.* **46:**31–35 (1956).

Heaney, J. B., "A Comparative Review of Optical Surface Contamination Assessment Techniques", *Proc. SPIE: Optical System Contamination: Effects, Measurements, Control,* P. A. Glassford (ed.) **777:**179 (1987).

Heaney, J. B., Optics Branch, Code 717, Goddard Space Flight Center, Greenbelt, MD., private communication (unpublished data), 1992.

Henninger, J. H., "Solar Absorptance and Thermal Emittance of Some Common Spacecraft Thermal Control Coatings, *NASA Reference Publication 1121,* 1984, p. 7.

Heslin, T., J. Heaney, and M. Harper, "The Effects of Particle Size on the Optical Properties and Surface Roughness of a Glass-Balloon-Filled Black Paint," *NASA Technical Note TND-7643,* Goddard Space Flight Center, Greenbelt, MD., May 1974.

Houck, J. R., "New Black Paint for Cryogenic Infrared Applications," *Proc. SPIE,* S. Musikant (ed.) **362:**54–56 (1982).

Hunt, P. J., R. Noll, L. Andreozzi, and J. Hope, "Particle Contamination from Martin Optical Black," *Proc. SPIE: Radiation Scattering in Optical Systems,* G. Hunt (ed.) **257:**196 (1980).

Hutchins, M. G., P. J. Wright, and P. D. Grebenik, *Proc. SPIE* **653:**188 (1986).

IITRI, product information and MSDSs on paints MH-211, D111, and MH2200, from the Illinois Institute of Technology Research Institute, Chicago, Ill., 1992.

Janeczko, D. J., "Optics and Electro-Optics, Martin Marietta Corporation, Orlando, Fla., personal communication, 1992.

Jelinsky, P. and S. Jelinsky, "Low Reflectance EUV Materials: A Comparative Study, "*Appl. Opt.* **26**(4):613–615 (1987).

Johnson, E., Spire Corporation, Bedford, Mass., private communication, March 1992.

Johnson, C. E., "Unique Surface Morphology with Extremely High Light Absorption Capability," *Proc. Electro-less Nickel Conference,* No. 1, Cincinnati, Ohio, November 6–7, 1979.

Johnson, C. E., "Black Electro-less Nickel Surface Morphologies with Extremely High Light Absorption Capacity," *Metal Finishing,* July, 1980.

Johnson, C., U.S. Patent 4,361,630, 30, November 1982.

Kirchner, K., TMA Technologies, Inc., Bozeman, Mont., private communication, April 1992.

Klicker, K. A., D. M. Fuhrman, D. R. Bjork, "A BSDF Database," *Proc. SPIE: Stray Radiation In Optical Systems,* R. P. Breault (ed.) **1331**:270–279 (1990).

LaFolla, Surface Optics, San Diego, Calif., private communication, April 1992.

LDEF, "The Long Duration Exposure Facility (LDEF) Mission I Experiments," *NASA SP-473,* NASA Langley Research Center, Washington, D.C., November 1986.

Leger, L. J., J. T. Visentine, and J. F. Kuminecz, "Low Earth Orbit Oxygen Effects on Surfaces," presented at *AIAA 22nd Aerospace Sciences Meeting,* Reno, Nevada, January 9–12, 1984.

Leonard, T. A. and M. Pantoliano, "BRDF Round Robin," *Proc. SPIE: Stray Light and Contamination in Optical Systems,* R. P. Breault (ed.) **967**:226–235 (1988).

Leonard, T. A., M. Pantoliano, and J. Reilly, "Results of a CO_2 BRDF Round Robin," *Proc. SPIE: Scatter from Optical Components,* J. Stover (ed.) **1165**:444–449 (1989).

Lewis, I. T., A. R. Telkamp, and A. F. Ledebuhr, "Low Scatter Edge Blackening Compounds for Refractive Optical Elements," *Proc. SPIE: Scatter from Optical Components,* J. Stover (ed.) **1165**:227–236 (1989).

Linlor, W. I., "Reflective Baffle System with Multiple Bounces," *Proc. SPIE: Stray Radiation IV,* R. P. Breault (ed.) **675**:217–239 (1986).

Lohrengel, J., "Total Emissivity of Black Coatings," ("Gesamtemissionsgrad von Schwärzen"), *Wärme-Stoffübertrag* **21**(5):311–315 (27 July 1987).

Lompado, A., B. W. Murray, J. S. Wollam, and J. F. Meroth, "Characterization of Optical Baffle Materials," *Proc. SPIE: Scatter from Optical Components,* J. Stover (ed.) **1165**:212–226 (1989).

Lord Chemical Products Division, Industrial Coatings Office, 845 Olive Avenue, Novato, CA 94945. 1992.

Ludwig, R., "Antireflection Coating On a Surface With High Reflecting Power," U.S. Patent 4,425,022, Jan. 10, 1984.

Martin Marietta, product information from Martin Marietta, 1992.

McCall, S. H. C. P., S. M. Pompea, R. P. Breault, and N. L. Regens, "Reviews of Properties of Black Surfaces for Ground and Space-Based Optical Systems," *Proc. SPIE: Stray Radiation II,* R. P. Breault (ed.) **1753** (1992a).

McCall, S. H. C. P., "Optical Properties (UV and Visible) of Black Baffle Materials and Processes for Use in Space," *Proprietary Report from Stellar Optics Laboratories* (78 Normark Drive, Thornhill, Ontario, Canada, L3T 3R1) *to Space Astrophysics Laboratory,* ISTS, Ontario, Canada, June 17, 1992b.

McCall, S. H. C. P., *Black Materials Database of Stellar Optics Laboratories,* 78 Normark Drive, Thornhill, Ontario, Canada, L3T 3R1, unpublished, 1992c.

McCall, S. H. C. P., R. L. Sinclair, S. M. Pompea, and R. P. Breault, "Spectrally Selective Surfaces for Ground and Space-Based Instrumentation: Support for a Resource Base," *Proc. SPIE: Space Astronomical Telescopes and Instruments II,* P. Bely and J. B. Breckinridge (ed.) **1945** (1993).

McCargo, M., R. E. Dammann, J. C. Robinson, and R. J. Milligan, *Proceedings of the International Symposium on Environmental and Thermal Control Systems for Space Vehicles,* Toulouse, France, October, 1983, pp. 1–5.

McKowan, K., AKZO Coatings, Inc., Orange, Calif., private communication, March 1992.

Mende, S. B., P. M. Banks, and D. A. Klingelsmith III, *Geophys. Res. Lett.* **11**:527 (1984).

Metzler, E. C., "Application of Temperature Control Paints," *JPL Doc. No. FS501424 D,* 21 March 1988.

Metzler, E. C., JPL, Pasadena, Calif., private communication, March 1992.

Mirtich, M. J., S. K. Rutledge, N. Stevens, R. Olle, and J. Merrow, "Ion Beam Textured and Coated Surfaces Experiment (IBEX)," Presented at *LDEF First Postretrieval Symposium,* Orlando, Fla., June 3–8, 1991.

Murray, B. W., and J. S. Wollam, "Space Durable Beryllium Baffle Materials," *SPIE Symposium on Aerospace Sensing,* Orlando, Fla., March 1989.

Murray, B. W., and Johnson, E. A., "Pulsed Electron Beam Testing of Optical Surfaces," *Proc. SPIE: Conference on Optical Surfaces Resistant to Severe Environments,* paper no. 1330-01, San Diego, July 11, 1990.

Muscari, J. A., and T. O'Donnell, *Proc. SPIE* **287**:20 (1981).

Nahm, K., P. Spyak, and W. Wolfe, "Scattering from Contaminated Surfaces," *Proc. SPIE: Scatter from Optical Components,* J. Stove (ed.) **1165**:294–305 (1989).

Nee, S. F., and H. E. Bennett, *Proc. SPIE* **1331**:249 (1990).

Noll, R. J., R. Hrned, R. Breault, and R. Malugin, *Proc. SPIE* **257**:119 (1980).

Nordberg, M. M., C. Von Benken, and E. J. Johnson, *Proc. SPIE* **1050**:185 (1989).

Orazio, F. D., W. K. Stowell, and R. M. Silva, "Instrumentation of a Variable Angle Scatterometer (VAS)," *Proc. SPIE: Stray Radiation III,* R. P. Breault (ed.) **384**:123–131 (1983).

Peterson, Gary, and Steve Johnston, "Specular Baffles", *Proc. SPIE: Stray Radiation in Optical Systems II,* R. P. Breault (ed.) **1753** (1992).

Peterson, R. V., W. Krone-Schmidt, and W. V. Brandt, "Jet-Spray Cleaning of Optics," *Proc. SPIE: Optical System Contamination: Effects, Measurement, Control III,* A. P. Glassford (ed.) **1754** (1992).

Pompea, S. M., D. W. Bergener, and D. F. Shepard, "Reflectance Characteristics of an Infrared Absorbing Surface," *Proceedings 31st National Infrared Information Symposium,* 1983a, p. 487.

Pompea, S. M., D. W. Bergener, D. F. Shepard, S. L. Russak, and W. L. Wolfe, "Preliminary Performance Data on an Improved Optical Black for Infrared Use," *Proc. SPIE, New Optical Materials* **400**:128 (1983b).

Pompea, S. M., D. F. Bergener, D. W. Shepard, S. Russak, and W. L. Wolfe, "Reflectance Measurements on an Improved Optical Black for Stray Light Rejection from 0.3 to 500 Micrometers," *J. Opt. Eng.* **23**:149–152 (1984a).

Pompea, S. M., D. W. Bergener, D. F. Shepard, and K. S. Williams, "The Effects of Atomic Oxygen on Martin Black and Infrablack," *Proc. SPIE,* R. P. Breault (ed.) **511**:24–30 (1984b).

Pompea, S. M., D. F. Shepard, and S. Anderson, "BRDF Measurements at 6328 Angstroms and 10.6 Micrometers of Optical Black Surfaces for Space Telescopes," *Proc. SPIE: Stray Light and Contamination in Optical Systems,* R. P. Breault (ed.) **967**:236–248 (1988).

Pompea, S. M., and S. H. C. P. McCall, "Outline of Selection Process for Black Baffle Surfaces in Optical Systems," *Proc. SPIE: Stray Radiation in Optical Systems II,* R. P. Breault (ed.) **1753** (1992a).

Pompea, S. M., J. E. Mentzell, and W. A. Siegmund, "A Stray Light Analysis of the 2.5 Meter Telescope for the Sloan Digital Sky Survey," *Proc. SPIE: Stray Radiation in Optical Systems II,* R. P. Breault (ed.) **1753** (1992b).

Pompea, S. M., "Stray Radiation Issues on Adaptive Optics Systems," in *Adaptive Optics for Astronomy,* D. Alloin and J.-M. Mariotti (eds.), Kluwer Academic Publishers, Dordrecht, 1994.

Predmore, R. E. and E. W. Mielke, *Materials Selection Guide, Revision A,* Goddard Space Flight Center, August 1990.

Ranbar Technology, Inc., Glenshaw, Pa., product information and MSDSs on Ranbar G113 Paint, 1990.

Redspot Paint and Varnish Co., Inc., product information on Nextel Suede, P.O. Box 418, Evansville, In., 47703, 1992.

Robinson, R. S. and S. M. Rossnagel, "Ion-Beam Induced Topography and Surface Diffusion," *J. Vac. Sci. Technol.* **21**:790 (1982).

Rutledge, S. K., B. A. Banks, M. J. Mirtich, R. Lebed, J. Brady, D. Hotes, and M. Kussmaul, "High Temperature Radiator Materials for Applications in the Low Earth Orbital Environment", *NASA Technical Memorandum 100190,* 1987.

Schaub, C., M. Davis, G. Inouye, and P. Schaller, "Visible Scatter Measurements of Various Materials," *Proc. SPIE: Stray Radiation in Optical Systems,* R. P. Breault (ed.) **1331:**293–298 (1990).

Schiff, T. F., J. C. Stover, D. R. Cheever, and D. R. Bjork, *Proc. SPIE* **967:**50 (1988).

Seals, R. D., C. M. Egert, and D. D. Allred, "Advanced Infrared Optically Black Baffle Materials," *Proc. SPIE: Optical Surfaces Resistant to Severe Environments,* **1330:**164–177 (1990).

Seals, R. D., "Advanced Broadband Baffle Materials," *Proc. SPIE: Reflective and Refractive Materials for Earth and Space Applications,* **1485:**78–87 (1991).

Seraphin, B. O. (ed.), "Solar Energy Conversion—Solid State Physics Aspects," in *Topics in Applied Physics,* vol. 31, Springer Verlag, 1979.

Seraphin, B. O., Optical Sciences Center, Univ. of Arizona, private communication, April, 1992.

Shepard, D. F., S. M. Pompea, and S. Anderson, "The Effect of Elevated Temperatures on the Scattering Properties of an Optical Black Surface at 0.6328 and 10.6 Micrometers," *Proc. SPIE: Stray Radiation V.,* R. P. Breault (ed.) **967:**286–291 (1988).

Shepard, D. F., Martin Black and Infrablack Contact at Martin Marietta Astronautics Group, P.O. Box 179, Denver, Colorado 80201.

Shepard, D. F. and R. J. Fenolia, "Optical Black Cobalt Surface," U.S. Patent 4,904,353, February 27, 1990, Martin Marietta Corporation.

Shepard, D. F., R. J. Fenolia, D. C. Nagle, and M. E. Marousek, "High-Temperature, High-Emissivity, Optically Black Boron Surface," U.S. Patent 5,035,949, July 30, 1991, Martin Marietta Corporation.

Sikkens Aerospace Finishes Division of Akzo Coatings America, Inc., 434 W. Meats Avenue, Orange, CA 92665.

Silva, R. M., F. D. Orazio, and R. B. Sledge, "A New Instrument for Constant (Beta-Beta$_g$) Scatter Mapping of Contiguous Optical Surfaces of up to 25 Square Inches," *Proc. SPIE: Stray Radiation IV,* R. P. Breault (ed.) **511:**38–43 (1984).

Smith, S. M., "Far Infrared (FIR) Optical Black Bidirectional Reflectance Distribution Function (BRDF)," *Proc. SPIE: Radiation Scattering in Optical Systems* **362:**161–168 (1980).

Smith, S. M., "Bidirectional Reflectance Distribution Function (BRDF) Measurements of Sunshield and Baffle Materials for the Infrared Astronomy Satellite (IRAS) Telescope," *Proc. SPIE: Modern Utilization of Infrared Technology VII,* **304:**205–213 (1981).

Smith, S. M., "Far Infrared Reflectance Spectra of Optical Black Coatings," *Proc. SPIE: Scattering in Optical Materials,* S. Musikant (ed.) **362:**57–59 (1982).

Smith, S. M., "Analysis of 12–700 Micrometer Reflectance Spectra of Three Optical Black Samples," *Proc. SPIE: Generation, Measurement, and Control of Stray Radiation III,* R. P. Breault (ed.) **384:**32–36 (1983).

Smith, S. M., "Specular Reflectance of Optical-Black Coatings in the Far Infrared," *Appl. Opt.* **23**(14):2311–2326 (1984).

Smith, S. M., "A Simple Antireflection Overcoat for Opaque Coatings in the Submillimeter Region," *Proc. SPIE: Stray Radiation V,* R. P. Breault (ed.) **675:**55–60 (1986).

Smith, S. M., "The Reflectance of Ames 24E, Infrablack, and Martin Black," *Proc. SPIE: Stray Radiation V,* R. P. Breault (ed.) **967:**248–254 (1988).

Smith, S. M., "Cryo-Mechanical Tests of Ames 24E2 IR-Black Coating," *Proc. SPIE: Stray Radiation in Optical Systems,* R. P. Breault (ed.) **1331:**241–248 (1990).

Smith, S. M., "Formation of Ames 24E2 IR-Black Coatings," *NASA Tech. Memo 102864,* July (1991a).

Smith, S. M., "BRDFs of Ames 24E, Ames 24E2, and Ames 47A at Photometric Wavelengths of 220 and 350 micrometers," *Sterling Technical Note TN-91-8441-000-74,* October 1991b.

Smith, S. M., "An Almost 'Perfectly' Diffuse, 'Perfect' Reflector for Far-Infrared Reflectance Calibration," *Proc. SPIE: Stray Radiation in Optical Systems II,* Breault (ed.) **1753:**252–261 (1992a).

Smith, S. M., Sterling Software, NASA Ames Research Center, private communication, April 1992b.

Smith, S. M. and R. V. Howitt, "Survey of Materials for an Infrared-Opaque Coating," *Proc. SPIE: Infrared, Adaptive, and Synthetic Aperture Optical Systems* **643:**53–62 (1986).

Smith, S. M. and W. L. Wolfe, "Comparison of Measurements by Different Instruments of the Far-Infrared Reflectance of Rough, Optically Black Coatings," *Proc. SPIE: Scattering of Optical Materials,* S. Musikant (ed.) **362:**46–53 (1983).

Spire, product information from Spire Corporation (Appendix A–D of Spire Document FR 10106), Bedford, Mass., (1992).

Steadman, S. S., and B. K. Likeness, "GUERAP III Simulation of Stray Light Phenomena," *Proc. SPIE: Stray Light Problems in Optical Systems* **107** (1977).

Stierwalt, D. L., "Infrared Absorption of Optical Blacks," *Opt. Eng.* **18**(2):147–151 (1979).

Stierwalt, D. L., J. B. Bernstein, and D. D. Kirk, "Measurements of the Infrared Spectral Absorptance of Optical Materials," *Appl. Opt.* **2**(11):1169–1173 (1963).

Stover, J. C., C. Gillespie, F. M. Cady, D. R. Cheever, and K. A. Klicker, *Proc. SPIE* **818:**68 (1987b).

Stover, J., *Optical Scattering: Measurement and Analysis,* McGraw-Hill, New York, 1990.

Stover, J. C., and C. H. Gillespie, *Proc. SPIE* **362:**172 (1982).

Stover, J. C., C. H. Gillespie, F. M. Cady, D. R. Cheever, and K. A. Klicker, *Proc. SPIE* **818:**62 (1987a).

Strimer, P., X. Gerbaux, A. Hadni, T. Souel, "Black Coatings of Infrared and Visible, With High Electrical Resistivity," *Infrared Physics* **21:**7–39 (1981).

Sullivan, B. T., Institute for Microstructural Sciences, National Research Council of Canada, Ottawa, Ontario, Canada, private communication, March 1992.

Viehmann, W., and R. E. Predmore, *Proc. SPIE* **675:**67 (1986).

Visentine, J. T. (ed.), *Atomic Oxygen Effects Measurement for Shuttle Missions STS-8 and 41-G, Volume I–III,* NASA TM-100459, 1988.

Vitt, B., *Proc. SPIE* **823:**218 (1987).

Wade, J. F., and W. R. Wilson, *Proc. SPIE* **67:**59 (1973).

Wade, J. F., J. E. Peyton, B. R. Klitzky, and R. E. Groff, "Optically Black Coating and Process for Forming It," U.S. Patent 4,111,762, September 5, 1978.

Whittaker, A. F., "Atomic Oxygen Effects on Materials," *STS-8 Paint Data Summary,* Marshall Space Flight Center, January 1984.

Willey, R. R., R. W. George, J. G. Ohmart, J. W. Walvoord, "Total Reflectance Properties of Certain Black Coatings from 0.2 to 20 Micrometers," *Proc. SPIE: Generation, Measurement and Control of Stray Radiation III,* R. P. Breault (ed.) **384:**19–26 (1983).

Williams, V. L., and R. T. Lockie, "Optical Contamination Assessment by Bidirectional Reflectance-Distribution Function (BRDF) Measurement," *Opt. Eng.* **18**(2):152–156 (1979).

Wolfe, W. L., "Optical Materials," chap. 7 in W. L. Wolfe and G. J. Zissis (eds.), *The Infrared Handbook,* ERIM, Ann Arbor, Michigan, 1978.

Wolfe, W. L., "Scattered Thoughts on Baffling Problems," *Proc. SPIE: Radiation Scattering in Optical Systems* **257:**2 (1980).

Wolfe, W. L., and Y. Wang, "Comparison of Theory and Experiments for Bidirectional Reflectance Distribution Function (BRDF) of Microrough Surfaces," *Proc. SPIE: Scattering in Optical Materials,* S. Musikant (ed.) **362:**40 (1982).

Wollam, J. S., and B. W. Murray, *Proc. SPIE* **1118:**88 (1989).

Wood, B. E., W. T. Bertrand, E. L. Kiech, J. D. Holt, and P. M. Falco, *Surface Effects of Satellite Material Outgassing Products,* Final Report AEDC-TR-89-2, Arnold Engineering Development Center, 1989.

Wyman, C. L., D. B. Griner, G. H. Hunt, and G. B. Shelton, *Optical Engineering,* **14**(6):528 (1975).

Wyman, C. L., Marshall Space Flight Center, Alabama, private communication, March 1992.

Young, R. P., "Low Scatter Mirror Degradation by Particle Contamination," *Opt. Eng.* **15**(6):516–520 (1976).

Young, P. J., R. Noll, L. Andreozzi, and J. Hope, *Proc. SPIE* **257:**196 (1980).

Zaeschmar, G., and A. Nedoluha, "Theory of the Optical Properties of Gold Blacks," *J. Opt. Soc. Am.* **62**(3):348–352 (1972).

NONLINEAR AND PHOTOREFRACTIVE OPTICS

CHAPTER 38
NONLINEAR OPTICS

Chung L. Tang
School of Electrical Engineering
Cornell University
Ithaca, New York

38.1 GLOSSARY

c	velocity of light in free space
$\mathbf{D}$	displacement vector
d_{mn}	Kleinman's **d**-coefficient
$\mathbf{E}$	electric field in lightwave
$\tilde{\mathbf{E}}$	complex amplitude of electric field
e	electronic charge
f	oscillator strength
$\hbar$	Planck's constant
I	intensity of lightwave
$\mathbf{k}$	propagation vector
m	mass of electron
N	number of equivalent harmonic or anharmonic oscillators per volume
$n_{1,2}$	index of refraction at the fundamental and second-harmonic frequencies, respectively
$\mathbf{P}$	macroscopic polarization
$\mathbf{P}^{(n)}$	nth-order macroscopic polarization
$P_{0,2,+,-}$	power of lightwave at the fundamental, second-harmonic, sum-, and difference-frequencies, respectively
$\tilde{\mathbf{P}}$	complex amplitude of macroscopic polarization
$\mathbf{Q}$	amplitude of vibrational wave or optic phonons
$\mathbf{S}$	strain of acoustic wave or acoustic phonons

T_{mn}	relaxation time of the density matrix element ρ_{mn}
Γ_j	damping constant of jth optical transition mode
δ	Miller's coefficient
$\boldsymbol{\varepsilon}(\mathbf{E})$	field-dependent optical dielectric tensor
$\boldsymbol{\varepsilon}_n$	nth-order optic dielectric tensor
ε_0	optical dielectric constant of free space
$\boldsymbol{\eta}$	amplitude of plasma wave or plasmons
λ	wavelength
ρ_{mn}	density matrix element
$\boldsymbol{\chi}(\mathbf{E})$	field-dependent optic susceptibility tensor
$\boldsymbol{\chi}_1$ or $\boldsymbol{\chi}(1)$	linear optic susceptibility tensor
$\boldsymbol{\chi}_n$ or $\boldsymbol{\chi}(n)$	nth-order optic susceptibility tensor
ω_p	plasma frequency
$\langle a|\,\mathbf{p}\,|b\rangle$	dipole moment between states a and b

38.2 INTRODUCTION

For linear optical materials, the macroscopic polarization induced by light propagating in the medium is proportional to the electric field:

$$\mathbf{P} = \varepsilon_0 \boldsymbol{\chi}_1 \cdot \mathbf{E} \qquad (1)$$

where the linear optical susceptibility $\boldsymbol{\chi}_1$ and the corresponding linear dielectric constant $\boldsymbol{\varepsilon}_1 = \varepsilon_0(1 + \boldsymbol{\chi}_1)$ are field-independent constants of the medium.

With the advent of the laser, light intensities orders of magnitude brighter than what could be produced by any conventional sources are now possible. When the corresponding field strength reaches a level on the order of, say, 100 KV/m or more, materials that are normally "linear" at lower light-intensity levels may become "nonlinear" in the sense that the optical "constants" are no longer "constants" independent of the light intensity. As a consequence, when the field is not weak, the optical susceptibility χ and the corresponding dielectric constant ε of the medium can become functions of the electric field $\chi(\mathbf{E})$ and $\varepsilon(\mathbf{E})$, respectively. Such a field-dependence in the optical parameters of the material can lead to a wide range of nonlinear optical phenomena and can be made use of for a great variety of new applications.

Since the first experimental observation of optical second-harmonic generation by Franken[1] and the formulation of the basic principles of nonlinear optics by Bloembergen and coworkers[2] shortly afterward, the field of nonlinear optics has blossomed into a wide-ranging and rapidly developing branch of optics. There is now a vast literature on this subject including numerous review articles and books.[3,4,5,6] It is not possible to give a full review of such a rich subject in a short introductory chapter in this Handbook; only the basic principles underlying the lowest order, the second-order, nonlinear optical processes and some illustrative examples of related applications will be discussed here. The reader is referred to the original literature for a more complete account of the full scope of this field.

If the light intensity is not so weak that the field dependence can be neglected and yet not too strong, the optical susceptibility and the corresponding dielectric constant can be expanded in a Taylor series:

$$\chi(\mathbf{E}) = \chi_1 + \chi_2 \cdot \mathbf{E} + \chi_3 : \mathbf{EE} + \cdots \tag{2}$$

or

$$\varepsilon(\mathbf{E}) = \varepsilon_1 + \varepsilon_2 \cdot \mathbf{E} + \varepsilon_3 : \mathbf{EE} + \cdots \tag{3}$$

where

$$\varepsilon_1 = \varepsilon_0(1 + \chi_1) \tag{4}$$

$$\varepsilon_n = \varepsilon_0 \chi_n \quad \text{for} \quad n \ge 2 \tag{5}$$

and ε_0 is the dielectric constant of free space. When these field-dependent terms in the optical susceptibility are not negligible, the induced macroscopic polarization in the medium contains terms that are proportional nonlinearly to the field:

$$\mathbf{P} = \varepsilon_0 \chi_1 \cdot \mathbf{E} + \varepsilon_0 \chi_2 : \mathbf{EE} + \varepsilon_0 \chi_3 : \mathbf{EEE} + \cdots$$

$$= \mathbf{P}^{(1)} + \mathbf{P}^{(2)} + \mathbf{P}^{(3)} + \cdots \tag{6}$$

As the field intensity increases, these nonlinear polarization terms $\mathbf{P}^{(n>1)}$ become more and more important, and will lead to a large variety of nonlinear optical effects.

The more widely studied of these nonlinear optical effects are, of course, those associated with the lower-order terms in Eq. (6). The second-order nonlinear effects will be discussed in some detail in this chapter. Many of the higher-order nonlinear terms have been observed and are the bases of a variety of useful nonlinear optical devices. Examples of the third-order effects are: third-harmonic generation[7,8] associated with $|\chi^{(3)}(3\omega = \omega + \omega + \omega|^2$, two-photon absorption[9] associated with $\mathrm{Im}\, \chi^{(3)}(\omega_1 = \omega_1 + \omega_2 - \omega_2)$, self-focusing[10,11] and light-induced index-of-refraction[12] change associated with $\mathrm{Re}\, \chi^{(3)}(\omega = \omega + \omega - \omega)$, four-wave mixing[13] $|\chi^{(3)}(\omega_4 = \omega_1 + \omega_2 - \omega_3)|^2$, degenerate four-wave mixing or phase-conjugation[14,15] $|\chi^{(3)}(\omega = \omega + \omega - \omega)|^2$, optical Kerr effect[16] $\mathrm{Re}\, \chi^{(3)}(\omega = 0 + 0 + \omega)$, and many others.

There is also a large variety of dynamic nonlinear optical effects such as photon echo,[17] optical nutation[18] (or optical Rabi effect[19]), self-induced transparency,[20] picosecond[21] and femtosecond[22] quantum beats, and others.

In addition to the nonlinear optical processes involving only photons that are related to the nonlinear dependence on the $\mathbf{E}$-field as shown in Eq. (6), the medium can become nonlinear indirectly through other types of excitations as well. For example, the optical susceptibility can be a function of the molecular vibrational amplitude Q in the medium, or the stress associated with an acoustic wave S in the medium, or the amplitude η of any space-charge or plasma wave, or even a combination of these excitations as in a polariton, in the medium:

$$\mathbf{P} = \varepsilon_0 [\chi_1 + \chi_2 : \mathbf{E} + \chi_3 : \mathbf{EE} + \cdots]\mathbf{E}$$

$$+ \varepsilon_0 [\chi_q : \mathbf{Q} + \chi_a : \mathbf{S} + \chi_n : \boldsymbol{\eta} + \cdots]\mathbf{E} \tag{7}$$

giving rise to the interaction of optical and molecular vibrational waves, or optical and acoustic phonons, etc. Nonlinear optical processes involving interaction of laser light and molecular vibrations in gases or liquids or optical phonon in solids can lead to stimulated Raman[23,24,25] processes. Those involving laser light and acoustic waves or acoustic phonons lead to stimulated Brillouin[26,27,28] processes. Those involving laser light and mixed excitations of photons and phonons lead to stimulated polariton[29] processes. Again, there is a great variety of such general nonlinear optical processes in which excitations other than photons in the medium may play a role. It is not possible to include all such nonlinear optical processes in the discussions here. Extensive reviews of the subject can be found in the literature.[3–5]

38.3 BASIC CONCEPTS

Microscopic Origin of Optical Nonlinearity

Classical Harmonic Oscillator Model of Linear Optical Media. The linear optical properties, including dispersion and single-photon absorption, of optical materials can be understood phenomenologically on the basis of the classical harmonic oscillator model (or Drude model). In this simple model, the optical medium is represented by a collection of independent identical harmonic oscillators embedded in a host medium. The harmonic oscillator is characterized by four parameters: a spring constant k, a damping constant Γ, a mass m, and a charge $-e\sqrt{f}$ as shown schematically in Fig. 1. f is also known as the oscillator-strength and $-e$ is the charge of an electron. The resonance frequency ω_0 of the oscillator is then equal to $[k/m]^{1/2}$.

In the presence of, for example, a monochromatic wave:

$$\mathbf{E} = \tfrac{1}{2}[\tilde{\mathbf{E}}e^{-i\omega t} + \tilde{\mathbf{E}}^* e^{i\omega t}] \tag{8}$$

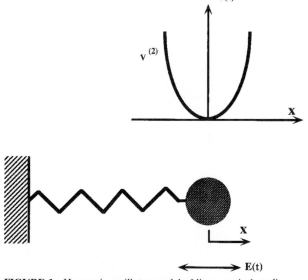

FIGURE 1 Harmonic oscillator model of linear optical media.

the response of the medium is determined by the equation of motion of the oscillator in the presence of the field:

$$\frac{\partial^2 X^{(1)}(t)}{\partial t^2} + \Gamma \frac{\partial X^{(1)}(t)}{\partial t} + \omega_0^2 X^{(1)}(t) = \frac{-e\sqrt{f}}{2m}[\tilde{\mathbf{E}}e^{-i\omega t} + \text{c.c.}] \cdot x \tag{9}$$

where $X^{(1)}(t)$ is the deviation of the harmonic oscillator from its equilibrium position in the absence of the field. The corresponding linear polarization in the steady state and linear complex susceptibility are from Eqs. (8) and (9):

$$\mathbf{P}^{(1)} = -NeX^{(1)}(t)x = \tfrac{1}{2}[\tilde{\mathbf{P}}^{(1)}e^{-i\omega t} + \tilde{\mathbf{P}}^{(1)*}e^{i\omega t}]$$

$$= \frac{Ne^2 f\tilde{\mathbf{E}}}{2mD(\omega)}e^{-i\omega t} + \text{c.c.} \tag{10}$$

and

$$\varepsilon_0 \chi^{(1)} = \frac{|\tilde{\mathbf{P}}|}{|\tilde{\mathbf{E}}|} = \frac{Ne^2 f}{mD(\omega)} \tag{11}$$

where N is the volume density of the oscillators and $D(\omega) = \omega_0^2 - \omega^2 - i\omega\Gamma$. The corresponding real and imaginary parts of the corresponding linear complex dielectric constant of the medium Re ε_1 and Im ε_1, respectively, describe then the dispersion and absorption properties of the linear optical medium. To represent a real medium, the results must be summed over all the effective oscillators (j):

$$\text{Re } \varepsilon_1 = \varepsilon_0 + \sum_j \frac{\omega_{pj}^2 f_j(\omega_{0j}^2 - \omega^2)}{(\omega_{0j}^2 - \omega^2)^2 + \omega^2\Gamma_j^2} \tag{12}$$

and

$$\text{Im } \varepsilon_1 = \sum_j \frac{\omega_{pj}^2 f_j \omega \Gamma_j}{(\omega_{0j}^2 - \omega^2)^2 + \omega^2\Gamma_j^2} \rightarrow \frac{\omega_{pj}^2 f_j}{2\omega} \frac{\Gamma_j/2}{(\omega - \omega_{0j})^2 - (\Gamma_j/2)^2} \qquad \text{for} \qquad \omega \approx \omega_{0j} \tag{13}$$

where $\omega_{pj}^2 = 4\pi N_j e^2/m$ is the plasma frequency for the jth specie of oscillators. Each specie of oscillators is characterized by four parameters: the plasma frequency ω_{pj}, the oscillator-strength f_j, the resonance frequency ω_{0j}, and the damping constant Γ_j. These results show the well-known anomalous dispersion and lorentzian absorption lineshape near the transition or resonance frequencies.

The difference between the results derived using the classic harmonic oscillator or the Drude model and those derived quantum mechanically from first principles is that, in the latter case, the oscillator strengths and the resonance frequencies can be obtained directly from the transition frequencies and induced dipole moments of the transitions between the relevant quantum states in the medium. For an understanding of the macroscopic linear optical properties of the medium, extended versions of Eqs. (12) and (13), including the tensor nature of the complex linear susceptibility, are quite adequate.

Anharmonic Oscillator Model of the Second-order Nonlinear Optical Susceptibility. An extension of the Drude model with the inclusion of suitable anharmonicities in the oscillator serves as a useful starting point in understanding the microscopic origin of the optical nonlinearity classically. Suppose the spring constant of the oscillator representing the optical medium is not quite linear in the sense that the potential energy of the oscillator

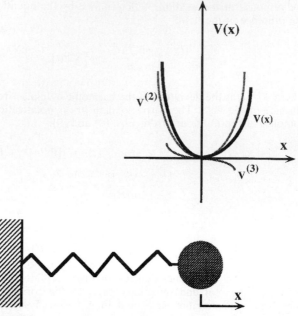

FIGURE 2 Anharmonic oscillator model of nonlinear optical media.

is not quite a quadratic function of the deviation from the equilibrium position, as shown schematically in Fig. 2. In this case, the response of the oscillator to a harmonic force is asymmetric. The deviation (solid line) from the equilibrium position is larger and smaller on alternate half-cycles than that in the case of the harmonic oscillator. This means that there must be a second-harmonic component (dark shaded curve) in the response of the oscillator as shown schematically in Fig. 3. It is clear, then, that the larger the

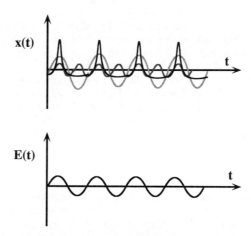

FIGURE 3 Response $[x(t)]$ of anharmonic oscillator to sinusoidal driving field $[\mathbf{E}(t)]$.

anharmonicity and the corresponding asymmetry in the oscillator potential, the larger the second-harmonic in the response. Extending this kind of consideration to a three-dimensional model, it implies that to have second-harmonic generation, the material must not have inversion symmetry and, therefore, must be crystalline. It is also clear that for the third and higher odd harmonics, the anharmonicity in the oscillator potential should be symmetric. Even harmonics will always require the absence of inversion symmetry. Beyond that, obviously, the larger the anharmonicities, the larger the nonlinear effects.

Consider first the second-harmonic case. The corresponding anharmonic oscillator equation is:

$$\frac{\partial^2 X(t)}{\partial t^2} + \Gamma \frac{\partial X(t)}{\partial t} + \omega_0^2 X(t) + v X(t)^2 = \frac{-e\sqrt{f}}{2m} [\tilde{E} e^{-i\omega t} + \text{c.c.}] \cdot x \tag{14}$$

Solving this equation by perturbation expansion in powers of the E-field:

$$X(t) = X^{(1)}(t) + X^{(2)}(t) + X^{(3)}(t) + \cdots \tag{15}$$

leads to the second-order nonlinear optical susceptibility

$$\varepsilon_0 \chi^{(2)} = \frac{|\tilde{P}|}{|\tilde{E}^2|} = \frac{Ne^3 f v}{2m^2 D^2(\omega) D(2\omega)} \tag{16}$$

Unlike in the linear case, a more exact expression of the nonlinear susceptibility derived quantum mechanically will, in general, have a more complicated form and will involve the excitation energies of, and dipole matrix elements between, all the states. Nevertheless, an expression like Eq. (16) obtained on the basis of the classical anharmonic oscillator model is very useful in discussing qualitatively the second-order nonlinear optical properties of materials. Expression (16) is particularly useful in understanding the dispersion properties of the second nonlinearity.

It is also the basis for understanding the so-called Miller's rule[30] which gives a very rough estimate of the order of magnitude of the nonlinear coefficient. We note that the strong frequency-dependence in the denominator of the $\chi^{(2)}$ involves factors that are of the same form as those that appeared in $\chi^{(1)}$. Suppose we divide out these factors and define a parameter which is called Miller's coefficient:

$$\delta = \chi^{(2)}(2\omega)/[\chi^{(1)}(\omega)]^2 \chi^{(1)}(2\omega)\varepsilon_0^2 = mv/2e^3 f^{1/2} N^2 \varepsilon_0^2 \tag{17}$$

from Eqs. (16) and (11). For many inorganic second-order nonlinear optical crystals, it was first suggested by R. C. Miller that δ was approximately a constant for all materials, and its value was found empirically to be on the order of 2–3×10^{-6} esu. If this were true, to find materials with large nonlinear coefficients, one should simply look for materials with large values of $\chi^{(1)}(\omega)$ and $\chi^{(1)}(2\omega)$. This empirical rule was known as Miller's rule. It played an important historical role in the search for new nonlinear optical crystals and in explaining the order of magnitude of nonlinear coefficients for many classes of nonlinear optical materials including such well-known materials as the ADP-isomorphs—for example, $KH_2PO_4(KDP)$, $NH_4PO_4(ADP)$, etc.—and the ABO_3 type of ferroelectrics—for example, $LiIO_3$, $LiNbO_3$, etc.—or III-V and II-VI compound semiconductors in the early days of nonlinear optics.

On a very crude basis, a value of δ can be estimated from Eq. (17) by assuming that the anharmonic potential term in Eq. (14) becomes comparable to the harmonic term when the deviation X is on the order of one lattice spacing in a typical solid, or on the order of an Angstrom. Thus, using standard numbers, Eq. (17) predicts that, in a typical solid, δ is on the order of 4×10^{-6} esu in the visible. It is now known that there are many classes of materials that do not fit this rule at all. For example, there are organic crystals with Miller's coefficients thousands of times larger than this value.

A more rigorous theory for the nonlinear optical susceptibility will clearly have to come from appropriate calculations based upon the principles of quantum mechanics.

Quantum Theory of Nonlinear Optical Susceptibility. Quantum mechanically, the nonlinearities in the optical susceptibility originate from the higher-order terms in the perturbation solutions of the appropriate Schrödinger's equation or the density-matrix equation.

According to the density-matrix formalism, the induced macroscopic polarization **P** of the medium is specified completely in terms of the density matrix:

$$\mathbf{P} = N \text{ Trace } [\mathbf{p}\rho] \tag{18}$$

where **p** is the dipole moment operator of the essentially noninteracting individual polarizable units, or "atoms" or molecules or unit cells in a solid, as the case may be, and N is the volume density of such units.

The density-matrix satisfies the quantum mechanical Boltzmann equation or the density-matrix equation:

$$\frac{\partial \rho_{mn}}{\partial t} + i\omega_{mn}\rho_{mn} + \frac{\rho_{mn} - \bar{\rho}_{mn}}{T_{mn}} = \frac{i}{\hbar} \sum_k [\rho_{mk}V_{kn} - V_{mk}\rho_{kn}] \tag{19}$$

where $\bar{\rho}_{mn}$ is the equilibrium density matrix in the absence of the perturbation V_{mn} and T_{mn} is the relaxation time of the density-matrix element ρ_{mn}. The nth-order perturbation solution of Eq. (19) in the steady state is:

$$\rho_{mn}^{(n)}(t) = \frac{i}{\hbar} \sum_k \int_{-\infty}^{t} [\rho_{mk}^{(n-1)}(t')V_{kn}(t') - V_{mk}(t')\rho_{kn}^{(n-1)}(t')] \exp\left[\left(i\omega_{mn} + \frac{1}{T_{mn}}\right)(t'-t)\right] dt' \tag{20}$$

The zeroth-order solution is clearly that in the absence of any perturbation or:

$$\rho_{mn}^{(0)} = \bar{\rho}_{mn} \tag{21}$$

In principle, once the zeroth-order solution is known, one can generate the solution to any order corresponding to all the nonlinear optical processes. While such solutions are formally complete and correct, they are generally not very useful, because it is difficult to know all the excitation energies and transition moments of all the states needed to calculate $\chi^{(n \geq 2)}$. For numerical evaluations of $\chi^{(n)}$, various simplifying approximations must be made.

To gain some qualitative insight into the microscopic origin of the nonlinearity, it can be shown on the basis of a simple two-level system that the second-order solution of Eq. (20) leads to the approximate result:

$$\chi^{(2)} \propto [(\omega_{ge} - \omega)(\omega_{ge} - 2\omega)]^{-1} |\langle g| \mathbf{p} |e\rangle|^2 [\langle e| \mathbf{p} |e\rangle - \langle g| \mathbf{p} |g\rangle] \tag{22}$$

It shows that for such a two-level system at least, there are three important factors: the resonance denominator, the transition-moment squared, and the change in the dipole moment of the molecule going from the ground state to the excited state. Thus, to get a large second-order optical nonlinearity, it is preferable to be near a transition with a large oscillator strength and there should be a large change in the dipole moment in going from the ground state to that particular excited state. It is known, for example, that substituted benzenes with a donor and an acceptor group have strong charge-transfer bands where the transfer of charges from the donor to the acceptor leads to a large change in the dipole moment in going from the ground state to the excited state. The transfer of the charges is mediated by the delocalized π electrons along the benzene ring. Thus, there was a great deal of interest in organic crystals of benzene derivatives. This led to the discovery of many organic nonlinear materials. In fact, it was the analogy between the benzene ring structure and the boroxal ring structure that led to the discovery of some of the best known recently discovered inorganic nonlinear crystals such as β-BaB_2O_4 (BBO)[31] and LiB_3O_5 (LBO).[32]

In general, however, there are few rules that can guide the search for new nonlinear optical crystals. It must be emphasized, however, that the usefulness of a material is not determined by its nonlinearity alone. Many other equally important criteria must be satisfied for the nonlinear material to be useful, for example, the transparency, the phase-matching property, the optical damage threshold, the mechanical strength, chemical stability, etc. Most important is that it must be possible to grow single crystals of this material of good optical quality for second-order nonlinear optical applications in bulk crystals. In fact, optical nonlinearity is often the easiest property to come by. It is these other equally important properties that are often harder to predict and control.

Form of the Second-order Nonlinear Optical Susceptibility Tensor

The simple anharmonic oscillator model shows that to have second-order optical nonlinearity, there must be asymmetry in the crystal potential in some direction. Thus, the crystal must not have inversion symmetry. This is just a special example of how the spatial symmetry of the crystal affects the form of the optical susceptibility. In this case, if the crystal contains inversion symmetry, all the elements of the susceptibility tensor must be zero. In a more general way, the form of the optical susceptibility tensor is dictated by the spatial symmetry of the crystal structure.[33]

For second-order nonlinear susceptibilities in the cartesian coordinate system:

$$P_i^{(2)} = \sum_{j,k} \varepsilon_0 \chi_{ijk}^{(2)} E_j E_k \tag{23}$$

$\chi_{ijk}^{(2)}$ in general has 27 independent coefficients before any symmetry conditions are taken into account. Taking into account the permutation symmetry condition, namely, the order E_j and E_k appearing in Eq. (23) is not important, or

$$\chi_{ijk}^{(2)} = \chi_{ikj}^{(2)} \tag{24}$$

the number of independent coefficients reduces down to 18. With 18 coefficients, it is sometimes more convenient to define a two-dimensional 3×6 tensor, commonly known as the Kleinman **d**-tensor:[34]

$$
\begin{pmatrix} P_x \\ P_y \\ P_z \end{pmatrix} = \varepsilon_0 \begin{pmatrix} d_{11} & d_{12} & d_{13} & d_{14} & d_{15} & d_{16} \\ d_{21} & d_{22} & d_{23} & d_{24} & d_{25} & d_{26} \\ d_{31} & d_{32} & d_{33} & d_{34} & d_{35} & d_{36} \end{pmatrix} \begin{pmatrix} E_x^2 \\ E_y^2 \\ E_z^2 \\ 2E_yE_z \\ 2E_xE_z \\ 2E_xE_y \end{pmatrix} \tag{25}
$$

rather than the three-dimensional tensor $\chi_{ijk}^{(2)} = \chi_{ikj}^{(2)}$. One obvious advantage of the $\mathbf{d}_{im}$ = tensor form is that the full tensor can be written in the two-dimensional matrix form, whereas it would be difficult to exhibit on paper any three-dimensional matrix.

An additional important point about the **d**-tensor is that it is defined in terms of the complex amplitudes of the **E**-field and the induced polarization with the 1/2 factor explicitly separated out in the front as shown in Eq. (8). In contrast, the definition of χ_{ijk} may be ambiguous in the literature because not all the authors define the complex amplitude with a 1/2 factor in the front. For linear processes, it makes no difference, because the 1/2 factors in the induced polarization and the **E**-field cancel out. In nonlinear processes, the 1/2 factors do not cancel and the numerical value of the complex susceptibility will depend on how the complex amplitudes of the **E**-field and polarization are defined.

For crystalline materials, the remaining 18 coefficients are, in general, not all independent of each other. Spatial symmetry requires, in addition, that they must satisfy the characteristic equation:

$$\chi_{ijk}^{(2)} = \sum_{\alpha\beta\gamma} \chi_{\alpha\beta\gamma}^{(2)} R_{\alpha i} R_{\beta j} R_{\gamma k} \tag{26}$$

where $R_{\alpha i}$, etc., represent the symmetry operations contained in the space group for the particular crystal structure and Eq. (26) must be satisfied for all the Rs in the group. For example, if a crystal has inversion symmetry, or $R_{\alpha i, \beta j, \gamma k} = (-1)\delta_{\alpha i, \beta j, \gamma k}$, Eq. (26) implies that $\chi_{ijk}^{(2)} = (-1)\,\chi_{ikj}^{(2)} = 0$ as expected. From the known symmetry elements of all 32 crystallographic point groups, the forms of the corresponding second-order nonlinear susceptibility tensors can be worked out and are tabulated. Equation (26) can in fact be generalized[33] to an arbitrarily high order n:

$$\chi_{ijk}^{(n)}\cdots = \sum_{\alpha\beta\gamma\cdots} \chi_{\alpha\beta\gamma}^{(n)}\cdots R_{\alpha i} R_{\beta j} R_{\gamma k}\cdots \tag{27}$$

for all the Rs in the group. Thus, the forms of any nonlinear optical susceptibility tensors can in principle be worked out once the symmetry group of the optical medium is known.

The **d**-tensors for the second-order nonlinear optical process for all thirty-two point groups derived from Eq. (26) are shown in, for example, Ref. 34. Similar tensors can in principle be derived from Eq. (27) for the nonlinear optical susceptibilities to any order for any point group.

Phase-matching Condition (or Conservation of Linear Photon Momentum) in Second-order Nonlinear Optical Processes

On a microscopic scale, the nonlinear optical effect is usually rather small even at relatively high light-intensity levels. In the case of the second-order effects, the ratio of the second-order term to the first-order term in Eq. (2), for example, is very roughly the ratio of the applied **E**-field strength to the "atomic **E**-field" in the material or:

$$\frac{\chi_2 E}{\chi_1} \approx \frac{E}{E_{\text{atomic}}} \tag{28}$$

which is on the order of 10^{-4} even at an intensity level of $1\ \text{MW/cm}^2$. The same ratio holds very roughly in each successively higher order. To see such a small effect, it is important that the waves generated through the nonlinear optical process add coherently on a macroscopic scale. That is, the new waves generated over different parts of the optical medium add coherently on a macroscopic scale. This requires that the phase velocities of the generated wave and the incident fundamental wave be "matched."[35]

Because of the inevitable material dispersion, in general the phases are not matched because the freely propagating second-harmonic wave will propagate at the phase velocity corresponding to the second-harmonic while the source polarization at the second-harmonic will propagate at the phase velocity of the fundamental. Phase matching requires that the propagation constant of the source polarization $2\mathbf{k}_1$ be equal to the propagation constant $\mathbf{k}_2$ of the second-harmonic or:

$$2\mathbf{k}_1 = \mathbf{k}_2 \tag{29}$$

Multiplying Eq. (29) by $\hbar$ implies that the linear momentum of the photons must be

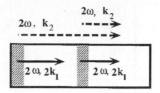

Phase-Matching Condition : $2\,\mathbf{k}_1 = \mathbf{k}_2$

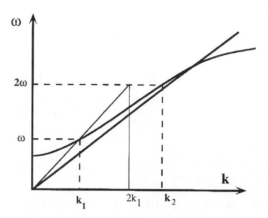

FIGURE 4   Phase-matching requirement and the effect of materials dispersion on momentum mismatch in second-harmonic process.

conserved. As shown in the schematic diagram in Fig. 4, in a normally dispersive region of an optical medium, $\mathbf{k}_2$ is always too long and must be reduced to achieve proper phase matching.

In bulk crystals, the most effective and commonly used method is to use birefringence to compensate for material dispersion, as shown schematically in Fig. 5. In this scheme, the **k**-vector of the extraordinary wave in the anisotropic crystal is used to shorten $\mathbf{k}_2$ or lengthen $2\mathbf{k}_1$ as needed. For example, in a negative uniaxial crystal, the fundamental wave is sent into the crystal as an ordinary wave and the second-harmonic wave is generated as an extraordinary wave in a so-called Type I phase-matching condition:

$$2\mathbf{k}_1^{(o)} = \mathbf{k}_2^{(e)} \tag{30}$$

or the fundamental wave is sent in both as an ordinary wave and an extraordinary wave while the second-harmonic is generated as an extraordinary wave in the so-called Type II phase-matching condition:

$$\mathbf{k}_1^{(o)} + \mathbf{k}_1^{(e)} = \mathbf{k}_2^{(e)} \tag{31}$$

In a positive uniaxial crystal, $\mathbf{k}_2^{(e)}$ in Eqs. (30) and (31) should be replaced by $\mathbf{k}_2^{(o)}$ and $\mathbf{k}_1^{(o)}$ in Eq. (30) should be replaced by $\mathbf{k}_1^{(e)}$. Crystals with isotropic linear optical properties clearly lack birefringence and cannot use this scheme for phase matching. Semiconductors of zinc-blende structure, such as the III-V and some of the II-VI compounds, have very large second-order optical nonlinearity but are nevertheless not very useful in the bulk crystal

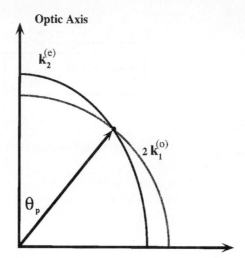

FIGURE 5 Phase matching using birefringence to compensate material dispersion in second-harmonic generation.

form for second-order nonlinear optical processes because they are cubic and lack birefringence and, hence, difficult to phase match. Phase matching can also be achieved by using waveguide dispersion to compensate for material dispersion. This scheme is often used in the case of III-V and II-VI compounds of zinc-blende structure. Other phase-matching schemes include the use of the dispersion of the spatial harmonics of artificial period structures to compensate for material dispersion.

These phase-matching conditions for the second-harmonic processes can clearly be generalized to other second-order nonlinear optical processes such as the sum- and difference-frequency processes in which two photons of different frequencies and momenta $\mathbf{k}_1$ and $\mathbf{k}_2$ either add or subtract to create a third photon of momentum $\mathbf{k}_3$. The corresponding phase-matching conditions are:

$$\mathbf{k}_1 \pm \mathbf{k}_2 = \mathbf{k}_3 \tag{32}$$

The practical phase-matching schemes for these processes are completely analogous to those for the second-harmonic process. For example, one can use the birefringence in a bulk optical crystal or the waveguide dispersion to compensate for the material dispersion in a sum- or difference-frequency process.

Conversion Efficiencies for the Second-harmonic and Sum- and Difference-frequency Processes

With phase matching, the waves generated through the nonlinear optical process can coherently accumulate spatially. The spatial variation of the complex amplitude of the generated wave follows from the wave equation:

$$\frac{\partial^2}{\partial z^2} E_i(z, t) - \frac{1}{c^2} \frac{\partial^2}{\partial t^2} E_i(z, t) = \frac{1}{c^2 \varepsilon_0} \frac{\partial^2}{\partial t^2} P_i(z, t) \tag{33}$$

where

$$E_i(z, t) = \tfrac{1}{2}[\tilde{E}_{0,i} e^{ik_0 z - i\omega_0 t} + \text{c.c.}] + \tfrac{1}{2}[\tilde{E}_{2,i}(z) e^{ik_2 z - i\omega_2 t} + \text{c.c.}] \tag{34}$$

and

$$P_i(z, t) = P_i^{(\omega_0)}(z, t) + [P_{\text{source},i}^{(2\omega_0)}(z, t) + P_i^{(2\omega_0)}(z, t)] \tag{35}$$

$$P_i^{(2\omega_0)}(z, t) = \tfrac{1}{2}\left[\sum_j \varepsilon_0 \chi_{ij}^{(1)}(2\omega_0) \tilde{E}_{2,j}(z) e^{ik_2 z - i\omega_2 t} + \text{c.c.} \right] \tag{36}$$

$$P_{\text{source},i}^{(2\omega_0)}(z, t) = \tfrac{1}{2}[\tilde{P}_{s,i}(z) e^{i2k_0 z - i\omega_2 t} + \text{c.c}] \tag{37}$$

$$\tilde{P}_{s,i}^{(2\omega_0)}(z, t) = \tfrac{1}{2} \sum_{jk} \varepsilon_0 \chi_{ijk}^{(2)}(2\omega_0) \tilde{E}_{0,j} \tilde{E}_{0,k} \tag{38}$$

The spatial variation in the complex amplitude of the fundamental wave $\tilde{\mathbf{E}}_{0,i}$ in Eq. (34) is assumed negligible and, in fact, we assume it to be that of the incident wave in the absence of any nonlinear conversion in the medium. It is, therefore, implied that the nonlinear conversion efficiency is not so large that the fundamental intensity is appreciably depleted. In other words, the small-signal approximation is implied. Solving Eq. (33) with the boundary conditions that there is no second-harmonic at the input and no reflection at the output end of the crystal, one finds the second-harmonic at the output end of the crystal $z = L$ to be:

$$I^{2\omega_0}(z = L) = \frac{2d^2 I_0^2}{c\varepsilon_0 n_2 (n_1 - n_2)^2} \sin^2 \left(\frac{L\omega_0}{c} \right)(n_1 - n_2) \tag{39}$$

where d is the appropriate Kleinman **d**-coefficient and the intensities refer to those inside the medium. When the phases of the fundamental and second-harmonic waves are not matched, or $n_1 \neq n_2$, it is clear from Eq. (39) that the second-harmonic intensity is an oscillating function of the crystal length. The maximum intensity is reached at a crystal length of:

$$L_{\max} = \frac{\lambda}{4 |n_2 - n_1|} \tag{40}$$

which is also known as the coherence length for the second-harmonic process. The maximum intensity that can be reached is:

$$I_{\max}^{2\omega_0}(z = L_{\max}) = \frac{2d^2}{c\varepsilon_0 n_2 (n_1 - n_2)^2} I_0^2 \tag{41}$$

regardless of the crystal length as long as it is greater than the coherence length. The coherence length for many nonlinear optical materials could be on the order of a few microns. Therefore, without phase matching, the second-harmonic intensity in such crystals corresponds to what is generated within a few microns of the output surface of the nonlinear crystal. A much more interesting or important case is clearly when there is phase matching or $n_1 = n_2$.

The second-harmonic intensity under the phase-matched condition is, from Eq. (39):

$$I_2 = \left(\frac{8\pi^2}{c\varepsilon_0 n_1^2 n_2} \right) \left(\frac{L}{\lambda_1} \right)^2 d_{\text{eff}}^2 I_0^2 \tag{42}$$

where d_{eff} is the effective **d**-coefficient which takes into account the projections of the

E-field and the second-harmonic polarization along the crystallographic axes and the form of the proper **d**-tensor for the particular crystal structure. The intensities in this equation refer to the intensities inside the nonlinear medium and the wavelength refers to the free-space wavelength. Equation (42) shows that the second-harmonic intensity under phase-matched conditions is proportional to the square of the length of the crystal measured in the wavelength, as expected for coherent processes. The second-harmonic intensity is also proportional to the effective **d**-coefficient squared and the fundamental intensity squared, as expected.

One might be tempted to think that, to increase the second-harmonic power conversion efficiency indefinitely, all one has to do is to focus the beam very tight since the left-hand side is inversely proportional to the beam cross section while the right-hand side is inversely proportional to the cross section squared. Because of diffraction, however, as the fundamental beam is focused tighter and tighter, the effective focal region becomes shorter and shorter. Optimum focusing is achieved when the Rayleigh range of the focal region becomes the limiting interaction length rather than the crystal length. A rough estimate assumes that a beam of square cross section doubles in width (w) due to diffraction in an "optimum focusing length," $L_{opt} \sim w^2/\lambda_2$, and that this optimum focusing length is equal to the crystal length L. Under such a nominally optimum focusing condition, the maximum second-harmonic power that can be generated in practice is, therefore, approximately:

$$P_2^{(opt)} = \left(\frac{2\pi^2}{c\varepsilon_0 n_1^2 n_2}\right)\left(\frac{L}{\lambda_2^3}\right)d_{eff}^2 P_0^2 \tag{43}$$

Note that this maximum power is linearly proportional to the crystal length. It must be emphasized, however, that this linear dependence is not an indication of incoherent optical process. It is because the beam spot size (area) under the optimum focusing condition is linearly proportional to the crystal length. Numerically, for example, approximately 3 W of second-harmonic power could be generated under optimum focusing in a 1-cm-long $LiIO_3$ crystal with 30 W of incident fundamental power at 1 μm.

Equation (43) can, in fact, be generalized to other three-photon processes such as the sum-frequency and difference-frequency processes:

$$P_+^{(opt)} = \left(\frac{2\pi^2}{c\varepsilon_0 n_1 n_2 n_+}\right)\left(\frac{L}{\lambda_+^3}\right)d_{eff}^2(\omega_+ = \omega_1 + \omega_2)P_1 P_2 \tag{44}$$

and

$$P_-^{(opt)} = \left(\frac{2\pi^2}{c\varepsilon_0 n_1 n_2 n_-}\right)\left(\frac{L}{\lambda_-^3}\right)d_{eff}^2(\omega_- = \omega_1 - \omega_2)P_1 P_2 \tag{45}$$

In using Eqs. (44) and (45), one must be especially careful in relating the numerical values of the d_{eff} coefficients for the sum- and difference-frequency processes to that measured in the second-harmonic process because the two low-frequency photons are degenerate in frequency in the latter process.

The Optical Parametric Process

A somewhat different, but rather important, second-order nonlinear optical process is the optical parametric process.[36,37] Optical parametric amplifiers and oscillators powerful solid-state sources of broadly tunable coherent radiation capable of covering the entire spectral range from the near-UV to the mid-IR and can operate down to the femtosecond time domain. The basic principles of optical parametric process were known even before the invention of the laser, dating back to the says of the masers. The practical development of the optical parametric oscillator had been impeded, however, due to the lack of suitable nonlinear optical materials. As a result of recent advances[38] in nonlinear optical materials research, these oscillators are now practical devices with broad potential applications in research and industry. The basic physics of the optical parametric process and recent developments in practical optical parametric oscillators are reviewed in this section as an example of wavelength-shifting nonlinear optical devices.

Studies of the optical parameters of materials clearly have always been a powerful tool to gain access to the atomic and molecular structures of optical materials and have played a key role in the formulation of the basic principles of quantum mechanics and, indeed, modern physics. Much of the information obtained through linear optics and linear optical spectroscopy came basically from just the first term in the expansion of the complex susceptibility, Eq. (2). The possibility of studying the higher-order terms in the complex susceptibility through nonlinear optical techniques greatly expands the power of such studies to gain access to the basic building blocks of materials on the atomic or molecular level. Of equal importance, however, are the numerous practical applications of nonlinear optics. Although there are now thousands of known laser transitions in all kinds of laser media, the practically useful ones are still relatively few compared to the needs. Thus, there is always a need to shift the laser wavelengths from where they are available to where they are needed. Nonlinear optical processes are the way to accomplish this. Until recently, the most commonly used wavelength-shifting processes were harmonic generation, sum-, and difference-frequency generation processes. In all these processes, the generated frequencies are always uniquely related to the frequencies of the incident waves. The parametric process is different. In this process, there is the possibility of generating a continuous range of frequencies from a single-frequency input.

For harmonic, sum-, and difference-frequency generation, the basic devices are nothing more than suitably chosen nonlinear optical crystals that are oriented and cut according to the basic principles already discussed in the previous sections and there is a vast literature on all aspects of such devices. The spontaneous optical parametric process can be viewed as the inverse of the sum-frequency process and the stimulated parametric process, or the parametric amplification process, can be viewed as a repeated difference-frequency process.

Spontaneous Parametric Process. The spontaneous parametric process, also known as the parametric luminescence or parametric fluorescence process, is described by a simple Feynman diagram as shown in Fig. 6. It describes the process in which an incident photon, called a pump photon, propagating in a nonlinear optical medium breaks down spontaneously into two photons of lower frequencies, called signal and idler photons using

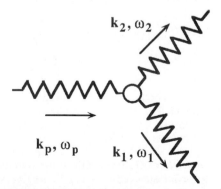

FIGURE 6 Spontaneous breakdown of a pump photon into a signal and an idler photon.

a terminology borrowed from earlier microwave parametric amplifier work, with the energy and momentum conserved:

$$\omega_p = \omega_s + \omega_i \tag{46}$$

$$\mathbf{k}_p = \mathbf{k}_s + \mathbf{k}_i \tag{47}$$

The important point about this second-order nonlinear optical process is that the frequency condition Eq. (46) does not predict a unique pair of signal and idler frequencies for each fixed pump frequency ω_p. Neglecting the dispersion in the optical material, there is a continuous range of frequencies that can satisfy this condition. Taking into account the dispersion in real optical materials, the frequency and momentum matching conditions Eqs. (46) and (47), in general, cannot be satisfied simultaneously. In analogy with the second-harmonic or the sum- or difference-frequency processes, one can use the birefringence in the material to compensate for the material dispersion for a set of photons propagating in the nonlinear crystal. By rotating the crystals, the birefringence in the direction of propagation can be tuned, thereby leading to tuning of the signal and idler frequencies. This tunability gives rise to the possibility of generating photons over a continuous range of frequencies from incident pump photons at one particular frequency, which means the possibility of constructing a continuously tunable amplifier or oscillator by making use of the parametric process.

A complete theory for the spontaneous parametric emission is beyond the scope of this introductory chapter because, as all spontaneous processes, it requires the quantization of the electromagnetic waves. Detailed descriptions of the process can be found in the literature.[4]

Stimulated Parametric Process, or the Parametric Amplification Process.   With only the pump photons present in the initial state, spontaneous emission occurs at the signal and idler frequencies under phase-matched conditions. With signal and pump photons present in the initial state, stimulated parametric emission occurs in the same way as in a laser medium, except here the pump photons are converted directly into the signal and the corresponding idler photons through the second-order nonlinear optical process and no exchange of energy with the medium is involved. The stimulated parametric process can also be viewed as a repeated difference-frequency process in which the signal and idler photons repeatedly mix with the pump photons in the medium, generating more and more signal and idler photons under the phase-matched condition.

The spatial dependencies of the signal and idler waves can be found from the appropriate coupled-wave equations under the condition when the pump depletion can be neglected. The corresponding complex amplitude of the signal wave at the output $\mathbf{E}_s(L)$ is proportional to that at the input $\mathbf{E}_s(0)$, as in any amplification process:[39]

$$\mathbf{E}_s(L) = \mathbf{E}_s(0) \cosh gL \tag{48}$$

where

$$g = \frac{d_{\text{eff}} |E_p| \sqrt{k_s k_i}}{2 n_s n_i} \tag{49}$$

is the spatial gain coefficient of the parametric amplification process. d_{eff} is the effective Kleinman **d**-coefficient for the parametric process. k_s and k_i are the phase-matched propagation constants of the signal and idler waves, respectively; n_s and n_i are the corresponding indices of refraction.

Optical Parametric Oscillator.   Given the parametric amplification process, a parametric oscillator can be constructed by simply adding a pair of Fabry-Perot mirrors, as in a laser, to provide the needed optical feedback of the stimulated emission. The optical parametric oscillator has the unique characteristic of being continuously tunable over a very broad

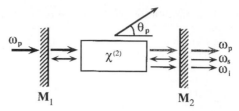

FIGURE 7 Schematic of singly resonant optical parametric oscillator.

spectral range. This is perhaps one of the most important applications of second-order nonlinear optics.

The basic configuration of an optical parametric oscillator (OP) is extremely simple. It is shown schematically in Fig. 7. Typically, it consists of a suitable nonlinear optical crystal in a Fabry-Perot cavity with dichroic cavity mirrors which transmit at the pump frequency and reflect at the signal frequency or at the signal and idler frequencies. In the former case, the OPO is a singly resonant OPO (SRO) and, in the latter case, it is a doubly resonant OPO (DRO). The threshold for the SRO is much higher than that for the DRO. The tradeoff is that the DRO tends to be highly unstable and, thus, not as useful.

Tuning of the oscillator can be achieved by simply rotating the crystal relative to the direction of propagation of the pump beam or the axis of the Fabry-Perot cavity. As an example of the spectral range that can be covered by the OPO, Fig. 8 shows the tuning

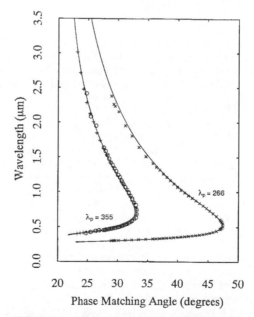

FIGURE 8 Tuning characteristics of BBO spontaneous parametric emission ($\times$ and $+$) and OPO (circles) pumped at the third-(355 nm) and fourth-(266 nm) harmonics of Nd-YAG laser output. Solid curves are calculated.

curve of a β-barium borate OPO pumped by the third-harmonic output at 355 μm and the fourth-harmonic at 266 μm of a Nd:YAG laser. Also shown are the corresponding spontaneous parametric emissions. The symbols correspond to the experimental data and the solid curves are calculated.[38] With a single set of mirrors to resonate the signal wave in the visible, the entire spectral range from about 400 nm to the IR absorption edge of the β-barium borate crystal can be covered. With $KTiO_2PO_4$ (KTP) or the more recently developed $KTiO_2AsO_4$ (KTA) crystals, the tuning range can be extended well into the mid-IR range to the 3- to 5-μm range. With $AgGaSe_2$, the potential tuning range could be extended to the 18- to 20-μm range.

The efficiency of the SRO that can be achieved in practice is relatively high, typically over 30 percent on a pulsed basis. Since the OPO is scalable, the output energy is only limited by the pump energy available and can be in the multijoule range.

A serious limitation at the early stage of development is the oscillator linewidth that can be achieved. Without rather complicated and special arrangements, the oscillator linewidth is typically a few Angstroms or more, which is not useful for high-resolution spectroscopic applications. The linewidth problem is, however, not a basic limitation inherent in the parametric process. It is primarily due to the finite pulse length of the pump sources, which limits the cavity length that can be used so that the number of passes by the signal through the nonlinear crystal is not too small. As more suitable pump sources are developed, various line-narrowing schemes[40] typically used in tunable lasers can be adapted for use in OPOs as well.

The OPO holds promise to become a truly continuously tunable powerful solid-state source of coherent radiation with broad applications as a research tool and in industry.

38.4 MATERIAL CONSIDERATIONS

The second-order nonlinearity is the lowest-order nonlinearity and the first to be observed as the intensity increases. As the discussion following Eq. (26) indicates, only materials without inversion symmetry can have second-order nonlinearity, which means that these must be crystalline materials. The lowest-order nonlinearity in a centrosymmetric system is the third-order nonlinearity.

To observe and to make use of the second-order nonlinear optical effects in a nonlinear crystal, an effective **d**-coefficient on the order of 10^{-13} m/V or larger is typically needed. In the case of the third-order nonlinearity, the effect becomes nonnegligible or useful in most applications when it is on the order of 10^{-21} MKS units or more.

Ever since the first observation of the nonlinear optical effect[1] shortly after the advent of the laser, there has been a constant search for new efficient nonlinear materials. To be useful, a large nonlinearity is, however, hardly enough. Minimum requirements in other properties must also be satisfied, such as transparency window, phase-matching condition, optical damage threshold, mechanical hardness, thermal and chemical stability, etc. Above all, it must be possible to grow large single crystals of good optical quality for second-order effects. The perfection of the growth technology for each crystal can, however, be a time-consuming process. All these difficulties tend to conspire to make good nonlinear optical materials difficult to come by.

The most commonly used second-order nonlinear optical crystals in the bulk form tend to be inorganic crystals such as the ADP-isomorphs $NH_4H_2PO_4$ (ADP), KH_2PO_4 (KDP), $NH_4H_2AsO_4$ (ADA), CsH_2AsO_4 (CDA), etc. and the corresponding deuterated version; the ABO_3 type of ferroelectrics such as $LiIO_3$, $LiNbO_3$, $KNbO_3$, etc.; and the borates such as β-BaB_2O_4, LiB_3O_5, etc. Although the III-V and II-VI compounds such GaAs, InSb, GaP, ZeTe, etc. generally have large **d**-coefficients, because their structures are cubic, there is no birefringence that can be used to compensate for material dispersion.

Therefore, they cannot be phase-matched in the bulk and are useful only in waveguide forms. Organic crystals hold promise because of the large variety of such materials and the potential to synthesize molecules according to some design principles. As a result, there have been extensive efforts at developing such materials for applications in nonlinear optics, but very few useful second-order organic crystals have been identified so far. Nevertheless, organic materials, especially for third-order processes, continue to attract a great deal of interest and remain a promising class of nonlinear materials.

To illustrate the important points in considering materials for nonlinear optical applications, a few examples of second-order nonlinear crystals with their key properties are tabulated in Tables 1 through 3. It must be emphasized, however, that because some

TABLE 1   Properties of Some Nonlinear Optical Crystals*

Crystal	LiB_3O_5	$\beta\text{-}BaB_2O_4{}^f$
Point group	$mm^{2\,a}$	3 m
Birefringence	$n_{x=a} = 1.5656^b$ $n_{y=c} = 1.5905$ $n_{z=b} = 1.6055$	$n_e = 1.54254$ $n_o = 1.65510$
Nonlinearity [pm/V]	$d_{32} = 1.16^b$	$d_{22} = 16$ $d_{31} = 0.08$
Transparency [μm]	0.16–2.6^c	0.19–2.5
Γ_{max} [GW/cm^2]	$\sim 25^b$	$\sim 5^g$
SHG cutoff [nm]	555^d	411
$\ell\,\Delta T$ [°C · cm]	3.9^e	55
$\ell\,\Delta\Theta$ [mrad · cm], CPM	31.3^e	0.52
$\ell^{1/2}\Delta\Theta$ [mrad (cm)$^{1/2}$]	71.9^e NCPM @ 148.0°C	Not available
$\ell\,\Delta\lambda$ [Å · cm]	Not available	21.1
Δv_g^{-1} @ 630 nm [fs/mm]	240^d	360
OPO tuning range [nm]	~ 415–2500^d $(\lambda_p = 355)$	~ 410–2500 $(\lambda_p = 355)$
Boule size	$20 \times 20 \times 15$ mm$^{3\,e}$	$\varnothing 84$ mm $\times$ 18 mm
Growth	TSSG^e @ ~ 810°C	TSSG from Na_2O @ ~ 900°C
Predominant growth defects	Flux^e inclusions	Flux and bubble inclusions
Chemical properties	Nonhygroscopic^e (m.p. ~ 834°C)	Slightly hygroscopic ($\beta \rightarrow \alpha \sim 925$°C)

* Data shown is at 1.064 μm unless otherwise indicated. Γ_{max}—surface damage threshold; $\ell\,\Delta T$—temperature-tuning bandwidth; $\ell\,\Delta\Theta$, CPM—critical phase-matching acceptance angle; $\ell^{1/2}\,\Delta\Theta$—noncritical phase-matching acceptance angle; $\ell\,\Delta\lambda$—SHG bandwidth; Δv_g^{-1}—group-velocity dispersion for SHG at 630 nm.

[a] Von H. Konig and A. Hoppe, *Z. Anorg. Allg. Chem.* **439:**71 (1978); M. Ihara, M. Yuge, and J. Krogh-Moe, *Yogyo-Kyokai-Shi* **88:**179 (1980); Z. Shuquing, H. Chaoen, and Z. Hongwu, *J. Cryst. Growth* **99:**805 (1990).

[b] C. Chen, Y. Wu, A. Jiang, B. Wu, G. You, R. Li, and S. Lin, *J. Opt. Soc. Am.* **B6:**616 (1989); S. Liu, Z. Sun, B. Wu, and C. Chen, *J. App. Phys.* **67:**634 (1989). On the basis of $\mathbf{d}_{32} = 2.69 \times \mathbf{d}_{36}$(KDP) and using the value $\mathbf{d}_{36}$(KDP) = 0.39 pm/V according to R. C. Ekart et al., *J. Quan. Elec.* **26:**922 (May 1990).

[c] 0.16–2.6 μm: C. Chen, Y. Wu, A. Jiang, B. Wu, G. You, R. Li, and S. Lin, *J. Opt. Soc. Am.* **B6:**616 (1989). 0.165–3.2 μm; S. Zhao, C. Huang, and H. Zhang, *J. Cryst. Growth.* **99:**805 (1990).

[d] Calculated by using Sellmeier equations reported in reference; B. Wu, N. Chen, C. Chen, D. Deng, and Z. Xu, *Opt. Lett.* **14:**1080 (1989).

[e] T. Ukachi and R. J. Lane, measurements carried out on Cornell LBO crystals grown by self-flux method.

[f] Reference sources given in: "Growth and Characterization of Nonlinear Optical Crystals Suitable for Frequency Conversion," by L. K. Cheng, W. R. Bosenberg, and C. L. Tang, review article in *Progress in Crystal Growth and Characterization* **20:**9–57 (Pergamon Press, 1990), unless indicated otherwise.

[g] Estimated surface damage threshold scaled from detailed bulk damage results reported by H. Nakatani et al., *Appl. Phys. Lett.* **53:**2587 (26 December, 1988).

TABLE 2 Properties of Several Visible Near-IR Nonlinear Optical Crystals*

Characteristics	$KNbO_3$†	$LiNbO_3$‡	$Ba_2NaNb_5O_{15}$
Point group	mm^2	$3m$	mm^2
Transparency [μm]	0·4–5.5	0.4–5.0	0.37–5.0
Birefringence	negative biaxial $n_{x=c} = 2.2574$ $n_{y=a} = 2.2200$ $n_{z=b} = 2.1196$	negative uniaxial $n^0 = 2.2325$ $n^e = 2.1560$	negative biaxial $n_{x=b} = 2.2580$ $n_{y=a} = 2.2567$ $n_{z=c} = 2.1700$
Second-order nonlinearity [pm/V]	$d_{32} = 12.9, d_{31} = -11.3$ $d_{24} = 11.9, d_{15} = -12.4$ $d_{33} = -19.6$	$d_{33} = -29.7$ $d_{31} = -4.8$ $\mathbf{d_{22}} = 2.3$	$d_{32} = -12.8, d_{31} = -12.8$ $d_{24} = 12.8, d_{15} = -12.8$ $d_{33} = -17.6$
$\partial(n^\omega - n^{2\omega})/\partial T$ [°C^{-1}]	1.6×10^{-4}	-5.9×10^{-5}	1.05×10^{-4}
T_{pm} [°C]	181, $\mathbf{d_{32}}$	$-8, d_{31}$	89, d_{32} 101, d_{31}
$\ell \Delta T$ [°C-cm]	0.3	0.8	0.5
λ_{SHG}(cutoff)[μm] @ 25°C	0.860	~1.08	1.01
Γ_{max} [MW/cm^2]	Not available	~120	40
Phase transition temperature (°C)	225 and 435	~1000	300
Growth technique	TSSG from K_2O @ ~1050°C	Czochralski @ ~1200°C	Czochralski @ ~1440°C
Predominant growth problems	Cracks, blue coloration, multidomains	Temp. induced compositional striations	Striations, microtwinning, multidomains
Postgrowth processing	Poling	Poling	Poling & detwinning
Crystal size	$20 \times 20 \times 20$ mm^3 (single domain)	⌀100 mm × 200 mm (as grown boule)	⌀20 mm × 50 mm (with striations)

* Unless otherwise specified, data are for $\lambda = 1.064$ μm. (Data taken from: a, e–i; a, b–c; and a, d, respectively.

† There is a disagreement on the sign of the nonlinear coefficients of $KNbO_3$ in the literature. Data used here are taken from Ref. e with the appropriate correction for the IRE convention.[a]

‡ Data are for congruent melting $LiNbO_3$.[b] Five-percent MgO doped crystals gives photorefractive damage threshold about 10–100 times higher.[k,l] The phase-matching properties for these crystals may differ due to the resulting changes in the lattice constants.[j]

[a] S. Singh in *CRC Handbook of Laser Science and Technology*, vol. 4, *Optical Materials*, part I, M. J. Weber (ed.), CRC Press, 1986, pp. 3–228.

[b] R. L. Byer, J. F. Young, and R. S. Feigelson, *J. Appl. Phys.* **41**:2320 (1970).

[c] R. L. Byer in *Quantum Electronics: A Treatise*, H. Rabin and C. L. Tang (eds), vol. 1, part A, Academic Press, 1975.

[d] S. Singh, D. A. Draegert, and J. E. Geusic, *Phys. Rev. B* **2**:2709 (1970).

[e] Y. Uematsu, *Jap. J. Appl. Phys.* **13**: 1362 (1974).

[f] P. Gunter, *Appl. Phys. Lett.* **34**:650 (1979).

[g] W. Xing, H. Looser, H. Wuest, and H. Arend, *J. Crystal Growth* **78**:431 (1986).

[h] D. Shen, *Mat. Res. Bull.* **21**: 1375 (1986).

[i] T. Fukuda and Y. Uematsu, *Jap. J. Appl. Phys.* **11**: 163 (1972).

[j] B. C. Grabmaier and F. Otto, *J. Crystal Growth* **79**: 682 (1986).

[k] D. A. Bryan, R. Gerson, and H. E. Tomaschke, *Appl. Phys. Lett.* **44**:847 (1984).

[l] G. Zhong, J. Jian, and Z. Wu, *11th International Quantum Electronics Conference*, IEEE Cat. No. 80 CH 1561-0, June 1980, p. 631.

of the materials are relatively new, some of the numbers listed are subject to confirmation and possibly revision. Discussions of other inorganic and organic nonlinear optical crystals can be found in the literature.[41]

As nonlinear crystals and devices become more commercialized, the issues of standardization of nomenclature and conventions and quantitative accuracy are becoming increasingly important. Some of these issues are being addressed[42] but much work remains to be done.

TABLE 3 Properties of Several UV, Visible, and Near-IR Crystals*

Crystal	KDP	KTP (II)†
Point group	42 m	mm^2
Birefringence	$n_e = 1.4599$ $n_o = 1.4938$	$n_{x=a} = 1.7367$ $n_{y=b} = 1.7395$ $n_{z=c} = 1.8305$
Nonlinearity [pm/V]	$d_{36} = 0.39$	$d_{32} = 5.0, d_{31} = 6.5$ $d_{24} = 7.6, \quad d_{15} = 6.1$ $d_{33} = 13.7$
Transparency [μm]	0.2–1.4	0.35–4.4
Γ_{max} [GW/cm^2]	~3.5	~15.0
SHG cutoff [nm]	487	~990
$\ell \Delta T$ [°C-cm]	7	22
$\ell \Delta \Theta$ [mrad-cm]	1.2	15.7
$\ell \Delta \lambda$ [Å-cm]	208‡	4.5
Δv_g^{-1} @ 630 nm [fs/mm]	185	Not applicable
OPO tuning range [nm] [nm]	~430–700 ($\lambda_p = 266$)	~610–4200 ($\lambda_p = 532$)
ΔT_F [°C]	12	Not available
Boule size	$40 \times 40 \times 100$ cm^3	~$20 \times 20 \times 20$ mm^3
Growth technique	Solution growth from H_2O	TSSG from $2KPO_3$-$K_4P_2O_7$ @ ~1000°C
Predominant growth defects	Organic impurities	Flux inclusions
Chemical properties	Hygroscopic (m.p. ~253°C)	Nonhygroscopic (m.p. ~1172°C)

* Unless otherwise stated, all data for 1064 nm. (Data taken from *c, e; a, b, f, m*; and *d, g–i,* respectively.)

† KTP Type I interaction gives $d_{eff} \sim d_{36}$ (KDP) or less for most processes.[m] The d_{ij} values[d] are for crystals grown by the hydrothermal technique.[j–l] Significantly lower damage thresholds were reported for hydrothermally grown crystals.[d]

‡ The anomalously large spectral bandwidth is a manifestation of the λ-noncritical phase matching.[n] This is equivalent to a very good group-velocity matching ($\Delta v_g^{-1} \sim 8$ fs/mm) for this interaction in KDP.

[a] D. Eimerl, *J. Quant. Elect.* **QE-23:**575 (1987).

[b] D. Eimerl, L. Davis, S. Velsko, E. K. Graham, and A. Zalkin, *J. Appl. Phys.* **62:**1968 (1987).

[c] D. Eimerl, *Ferroelectrics* **72:**95 (1987).

[d] Y. S. Liu, L. Drafall, D. Dentz, and R. Belt, *G. E. Technical Information Series Report,* 82CRD016, Feb. 1982.

[e] Y. Nishida, A. Yokotani, T. Sasaki, K. Yoshida, T. Yamanaka, and C. Yamanaka, *Appl. Phys. Lett.* **52:**420 (1988).

[f] A. Jiang, F. Cheng, Q. Lin, Z. Cheng, and Y. Zheng, *J. Crystal Growth* **79:**963 (1986).

[g] P. Bordui, in *Crystal Growth of KTiOPO$_4$ from High Temperature Solution,* Ph.D. thesis, Massachusetts Institute of Technology, 1987.

[h] *Information Sheet on KTiOPO$_4$,* Ferroxcube, Division of Amperex Electronic Corp., Saugerties, New York, 1987.

[i] P. Bordui, J. C. Jacco, G. M. Loiacono, R. A. Stolzenberger, and J. J. Zola, *J. Crystal Growth* **84:**403 (1987).

[j] F. C. Zumsteg, J. D. Bierlein, and T. E. Gier, *J. Appl. Phys.* **47:**4980 (1976).

[k] R. A. Laudis, R. J. Cava, and A. J. Caporaso, *J. Crystal Growth* **74:**275 (1986).

[l] S. Jia, P. Jiang, H. Niu, D. Li, and X. Fan, *J. Crystal Growth* **79:**970 (1986).

[m] L. K. Cheng, unpublished.

[n] J. Zyss and D. S. Chemla, in *Nonlinear Optical Properties of Organic Molecules and Crystals,* vol. 1, D. S. Chemla and J. Zyss (eds), Academic Press, 1987, pp. 146–159.

38.5   APPENDIX

The results in this article are given in the rationalized MKS systems. Unfortunately, many of the pioneering papers on nonlinear optics were written in the cgs gaussian system. In addition, different conventions and definitions of the nonlinear optical coefficients are used in the literature by different authors. These choices have led to a great deal of confusion. In this Appendix, we give a few key results to facilitate comparison of the results using different definitions and units.

First, in the MKS system, the displacement vector **D** is related to the **E**-field and the induced polarization **P** in the medium as follows:

$$\mathbf{D} = \varepsilon_0 \mathbf{E} + \mathbf{P}$$

$$= \varepsilon_0 \mathbf{E} + \mathbf{P}^{(1)} + \mathbf{P}^{(2)} + \cdots \tag{A-1}$$

The corresponding wave equation is given in Eq. (33). For the second-order polarization and the corresponding Kleinman **d**-coefficients, two definitions are in use. A more popular definition in the current literature is as follows:

$$\mathbf{P}^{(2)} = \varepsilon_0 \mathbf{d}_2 : \mathbf{EE} \tag{A-2}$$

In an earlier widely used reference,[34] Yariv defined his **d**-coefficient as follows:

$$\mathbf{P}^{(2)} = \mathbf{d}_2^{(\text{Yariv})} : \mathbf{EE} \tag{A-3}$$

The numerical values of $\mathbf{d}_2^{(\text{Yariv})}$ in this reference (e.g., Table 16.2)[34] are given in $(1/9) \times 10^{-22}$ MKS units. The numerical value of ε_0 in the MKS system is $10^7 \times (1/4\pi c^2)$ in MKS units. Thus, for example, a tabulated value of $\mathbf{d}_2^{(\text{Yariv})} = 0.5 \times (1/9) \times 10^{-22}$ MKS units in Ref. 34 converts to a numerical value of $\mathbf{d}_2 = 0.628$ pm/V in MKS units.

In the cgs gaussian system, the displacement vector **D** is related to the **E**-field and the induced polarization **P** in the medium as follows:

$$\mathbf{D} = \varepsilon_0 \mathbf{E} + 4\pi \mathbf{P}$$

$$= \varepsilon_0 \mathbf{E} + 4\pi \mathbf{P}^{(1)} + 4\pi \mathbf{P}^{(2)} + \cdots \tag{A-4}$$

The corresponding wave equation is:

$$\frac{\partial^2}{\partial z^2} E_i(z, t) - \frac{1}{c^2} \frac{\partial^2}{\partial t^2} E_i(z, t) = \frac{4\pi}{c^2} \frac{\partial^2}{\partial t^2} P_i(z, t) \tag{A-5}$$

The conventional definition of $\mathbf{d}_2$ is as follows:

$$\mathbf{P}^{(2)} = \mathbf{d}_2 : \mathbf{EE} \tag{A-6}$$

The numerical value of $\mathbf{d}_2$ in cgs gaussian units is, therefore, equal to $(3 \times 10^4/4\pi)$ times the numerical value of $\mathbf{d}_2$ in rationalized MKS units. Thus, continuing with the numerical example given in the preceding paragraph, $\mathbf{d}_2 = 0.628$ pm/V is equal to 1.5×10^{-9} cm/Stat-Volt or 1.5×10^{-9} esu.

As a final check, the expression Eq. (42) for the second-harmonic intensity in the MKS system becomes, in the cgs gaussian system:

$$I_2 = \left(\frac{512\pi^5}{cn_1^2 n_2}\right)\left(\frac{L}{\lambda_1}\right)^2 d_{\text{eff}}^2 I_0^2 \tag{A-7}$$

All the intensities refer to those inside the medium, and the wavelength is the free-space wavelength.

38.6 REFERENCES

1. P. A. Franken, A. E. Hill, C. W. Peters, and G. Weinreich, *Phys. Rev. Lett.* **7:**118 (1961).

2. J. A. Armstrong, N. Bloembergen, J. Ducuing, and P. S. Pershan, *Phys. Rev.* **127:**1918 (1962); N. Bloembergen and Y. R. Shen, *Phys. Rev.* **133:**A37 (1964).

3. N. Bloembergen, *Nonlinear Optics,* Benjamin, New York, 1965.

4. See, for example, H. Rabin and C. L. Tang (eds.), *Quantum Electronics: A Treatise,* vol. 1A and B *Nonlinear Optics,* Academic Press, New York, 1975, and the references therein.

5. See, for example, Y. R. Shen, *The Principles of Nonlinear Optics,* J. W. Wiley Interscience, New York, 1984.

6. M. D. Levenson and S. S. Kano, *Introduction to Nonlinear Laser Spectroscopy,* Academic Press, New York, 1988, and the references therein.

7. P. D. Maker and R. W. Terhune, *Phys. Rev. A* **137:**801 (1965).

8. See, for example, secs. 7.3 and 7.4 of Ref. 5.

9. H. Mahr, "Two-Photon Absorption Spectroscopy," in Ref. 4.

10. G. A. Askar'yan, *Sov. Phys. JETP* **15:**1088, 1161 (1962); M. Hercher, *J. Opt. Soc. Am.* **54:**563 (1964); R. Y. Chiao, E. Garmire, and C. H. Townes, *Phys. Rev. Lett.* **13:**479 (1964) [Erratum, **14:**1056 (1965)].

11. See, for example, Y. R. Shen, "Self-Focusing," chap. 17 in Ref. 5.

12. See, for example, R. W. Boyd, *Nonlinear Optics,* chap. 4, Academic Press, 1992.

13. See, for example, chap. 15 in Ref. 5.

14. Y. B. Zeldovich, V. I. Popoviecher, V. V. Ragul'skii, and F. S. Faizullov, *JETP Letters* **15:**109 (1972).

15. R. W. Hellwarth, *J. Opt. Soc. Am.* **68:**1050 (1978); A. Yariv, *IEEE J. Quant. Elect.* **QE-14:**650 (1978).

16. See, for example, A. Yariv and P. Yeh, *Optical Waves in Crystals,* Wiley, New York, 1984, p. 221.

17 N. A. Kurnit, I. D. Abella, and S. R. Hartmann, *Phys. Rev. Lett.* **13:**567 (1964); S. Hartmann, in R. Glauber (ed.), *Proc. of the Int. School of Phys. Enrico Fermi Course XLII,* Academic Press, New York, 1969, p. 532.

18. C. L. Tang and B. D. Silverman, "Physics of Quantum Electronics," P. Kelley, B. Lax, and P. E. Tannenwald (eds.), McGraw-Hill, 1966, p. 280. G. B. Hocker and C. L. Tang, *Phys. Rev. Lett.* **21:**591 (1969); *Phys. Rev.* **184:**356 (1969).

19. R. G. Brewer, *Phys. Today,* May 1977.

20. S. L. McCall and E. L. Hahn, *Phys. Rev. Lett.* **18:**908 (1967); *Phys. Rev.* **183:**457 (1969).

21. N. Bloembergen and A. H. Zewail, *J. Phys. Chem.* **88:**5459 (1984).

22. M. J. Rosker, F. W. Wise, and C. L. Tang, *Phys. Rev. Lett.* **57:**321 (1986); *J. Chem. Phys.* **86:**2827 (1987).

23. E. J. Woodbury and W. K. Ng, *Proc. IRE 50,* p. 2347 (1962); R. W. Hellwarth, *Phys. Rev.* **130:**1850 (1963).

24. E. Garmire, E. Pandarese, and C. H. Townes, *Phys. Rev. Lett.* **11:**160 (1963).

25. C. S. Wang, "The Stimulated Raman Process," chap. 7 in Ref. 4.

26. R. Y. Chiao, C. H. Townes, and B. P. Stoicheff, *Phys. Rev. Lett.* **12:**592 (1964); E. Garmire and C. H. Townes, *App. Phys. Lett.* **5:**84 (1964).

27. C. L. Tang, *J. App. Phys.* **37:**2945 (1966).

28. I. L. Fabellinskii, "Stimulated Mandelstam-Brillouin Process," chap. 5 in Ref. 4.

29. See, for example, sect. 10.7 in Ref. 5.

30. R. C. Miller, *App. Phys. Lett.* **5:**17 (1964).

31. C. Chen, B. Wu, A. Jiang, and G. You, *Sci. Sin. Ser. B* **28:**235 (1985).

32. C. Chen, Y. Wu, A. Jiang, B. Wu, G. You, R. Li, and S. Lin, *J. Opt. Soc. Am.* **B6:**616 (1989).

33. P. A. Franken and J. F. Ward, *Rev. of Mod. Phys.* **35:**23 (1963).

34. A. Yariv, *Quantum Electronics,* John Wiley, New York, 1975, pp. 410–411.

35. J. A. Giordmaine, *Phys. Rev. Lett.* **8:**19 (1962); P. D. Maker, R. W. Terhune, M. Nisenhoff, and C. M. Savage, *Phys. Rev. Lett.* **8:**21 (1962).

36. W. H. Louisell, *Coupled Mode and Parametric Electronics,* John Wiley, New York, 1960.

37. N. Kroll, *Phys. Rev.* **127:**1207 (1962).

38. See, for example, C. L. Tang, *Proc. IEEE* **80:**365 (March 1992).

39. Ref. 4, p. 428.

40. See, for example, L. F. Mollenauer and J. C. White (eds.), *Tunable Lasers,* Springer-Verlag, Berlin, 1987.

41. See, for example, S. K. Kurtz, J. Jerphagnon, and M. M. Choy, in *Landolt-Boerstein Numerical Data and Functional Relationships in Science and Technology,* New Series, K. H. Wellwege (ed.), Group III, vol. 11, Springer-Verlag, Berlin, 1979; *Nonlinear Optical Properties of Organic and Polymeric Materials,* D. Williams (ed.), Am. Chem. Soc., Wash., D.C., 1983; *Nonlinear Optical Properties of Organic Molecules,* D. Chemla and J. Zyss (eds.), Academic Press, 1987.

42. D. A. Roberts, *IEEE J. Quant. Elect.* **28:**2057 (1992).

CHAPTER 39
PHOTOREFRACTIVE MATERIALS AND DEVICES

Mark Cronin-Golomb
Electro-Optics Technology Center
Tufts, University
Medford, Massachusetts

Marvin Klein
Hughes Research Laboratories
Malibu, California

39.1 INTRODUCTION

The photorefractive effect is a real time holographic optical nonlinearity that is effective for low-power lasers over a wide range of wavelengths. It is relatively easy to use even in modestly equipped laboratories: all that is needed to get started is a sample of photorefractive material, almost any laser operating in the visible or near-infrared, and a few simple optics such as lenses and beam splitters. The diffraction efficiency of photorefractive holograms is roughly independent of the intensity of the writing beams and in many materials, the diffraction efficiency of these holograms can approach 100 percent, so that sophisticated detectors are not required. Its simplicity of use has been largely responsible for its widespread popularity during the last decade. As is often the case, however, attractive features such as these come only with associated disadvantages. For the photorefractive effect, the main disadvantage is one of speed. The nonlinearities come to steady state at a rate which is inversely proportional to intensity. The fastest high-diffraction efficiency materials have response times of the order of 1 ms at 1 W/cm². Even so, in certain applications, the characteristic slowness is not a disadvantage. In the first part of this chapter, we explain the basic mechanisms of the photorefractive effect. The second part deals with material selection considerations and the third part describes some typical applications. For the reader in need of an extensive overview of the photorefractive effect and its applications, we recommend a two-volume set edited by Gunter and Huignard.[1]

Grating Formation

The photorefractive effect is observed in materials which

1. exhibit a linear electro-optic effect
2. are photoconductive
3. have a low dark conductivity

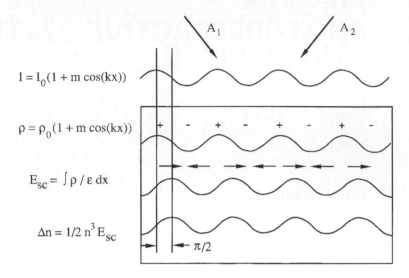

$$I = I_0(1 + m\cos(kx))$$

$$\rho = \rho_0(1 + m\cos(kx))$$

$$E_{sc} = \int \rho / \varepsilon \, dx$$

$$\Delta n = 1/2 \, n^3 E_{sc}$$

FIGURE 1 The photorefractive mechanism. Two laser beams intersect, forming an interference pattern $I(x)$. Charge is excited where the intensity is large and migrates to regions of low intensity. The electric field $\mathbf{E}_{sc}$ associated with the resultant space charge ρ_{sc} operates through the linear electro-optic coefficients to produce a refractive index grating Δn.

Two laser beams record a photorefractive hologram when their interference pattern is incident on a photorefractive crystal (Fig. 1). Charge carriers are preferentially excited in the bright fringes and are then free to drift and diffuse until they recombine with traps, most likely in the darker regions of the interference pattern. In this way, a space charge builds up inside the crystal in phase with the interference pattern. The electric field of this space charge acts through the linear electro-optic effect to form a volume holographic refractive index grating in real time. Real time means that the development occurs with a time constant of the order of the response time of the photorefractive crystal. The writing beams then diffract from the hologram into each other. Even though the process depends on the linear electro-optic effect, which is second-order, the whole process acts effectively as a third-order nonlinearity as far as the writing beams are concerned, and third-order coupled-wave equations can be written for their amplitudes. It is important to notice that the electric field is spatially shifted by 90° with respect to the interference pattern because of the Gauss's law integration that links the space charge to the electric field. The refractive index gradient is also shifted by ±90°. This shift is possible because the refractive index perturbation depends on the direction of the electric field, not just on its magnitude. The direction of the grating shift is determined by the sign of the electro-optic coefficient and crystal orientation. This dependence on crystal orientation is due to the lack of inversion symmetry associated with the linear electro-optic coefficient: if the crystal is inverted through its origin, the sign of the phase shift changes. In uniaxial crystals, inversion corresponds simply to reversing the direction of the optic axis. These symmetry effects are intimately related to the origin of photorefractive beam amplification.

The reason that use of the linear electro-optic effect is important is that a Bragg-matched volume hologram should have the same period as the optical interference pattern that wrote it. The refractive index change should be directly proportional to the space charge electric field. This is only possible if the material displays the linear electro-optic effect leading to a refractive index distribution $\Delta n \propto r_{eff} E_{sc}$, where r_{eff} is an effective electro-optic coefficient and E_{sc} is the space charge field. The lack of inversion symmetry needed by the linear electro-optic effect may be found in ferroelectric materials such as barium titanate, optically active materials such as bismuth silicon oxide and cubic compound semiconductors such as GaAs and InP.

When the charge transport is purely diffusive, the spatial phase shift is 90°. However, in certain circumstances the phase shift can depart from 90°. This occurs if electric fields are applied to the crystal or if the crystal exhibits the photovoltaic effect[2] so that drift mechanisms come into play, or, if the writing beams have different frequencies, so that the interference pattern moves in the crystal with the index grating lagging behind it.

The second requirement implies that the material should contain photoexcitable impurities. Direct band-to-band photoconductivity is usually not useful since it limits the optical interaction distances to rather small absorption depths.

The requirement for low dark conductivity ensures that the space charge can support itself against decay by leakage through background conduction.

The Standard Rate Equation Model

The simplest model, as formulated by Vinetskii and Kukhtarev,[3] involves optical excitation of charge carriers. For the purposes of this introduction, we will assume that the carriers are electrons which can be excited from a donor species such as Fe^{2+} and which can recombine into an acceptor such as Fe^{3+}. Let n be the local number density of mobile excited electrons and $N_D = N + N^+$ be the number density of the impurities or defects responsible for the photorefractive effect, where N^+ is the number density of acceptor dopants (e.g., Fe^{3+}) and N is the number density of donor dopants (e.g., Fe^{2+}). Let N_A be the number density of negative ions that compensate the excess positive charge of acceptor dopants when the charge is uniformly distributed in the dark. Neglecting the photovoltaic effect, we may write the following rate equations of generation and recombination, continuity, electric field (Poisson equation), and total drift and diffusion current:

$$\frac{\partial N^+}{\partial t} = (sI + \beta)N - \gamma n N^+$$

$$\frac{\partial N^+}{\partial t} = \frac{\partial n}{\partial t} - \frac{1}{e}\mathbf{\nabla} \cdot \mathbf{j} \qquad (1)$$

$$\mathbf{\nabla} \cdot \mathbf{E} = (N^+ - N_A - n)e/\varepsilon$$

$$\mathbf{j} = \mu e n \mathbf{E} + k_B T \mu \, \mathbf{\nabla} n$$

where s is proportional to the photoionization cross section σ ($s = \sigma/h\nu$), β is the dark generation rate, γ is a recombination coefficient, $\mathbf{j}$ is the electric current density. At steady state, the space charge is determined by a balance between charge diffusion away from bright fringes and electrostatic repulsion from charge concentrations. Extensions of these equations to include the effects of electron-hole competition,[4] multiple dopants,[5,6] photovoltaic effects,[7,8] and short pulse excitation[9] have been developed over the past few years. Nevertheless, the most important features of the photorefractive effect may be well-modeled by the simple equations shown here. Assuming that the number density of charge carriers is much less than the optically induced donor density perturbation ($N^+ - N_A$), a situation that almost always holds, the rate equations can be linearized to give the following solution for the fundamental spatial Fourier component of the space charge field E_{sc} induced by a sinusoidal optical fringe pattern of wave number k_g and fringe visibility m when a dc field E_0 is applied to the crystal:

$$E_{sc} = \frac{m/2}{1 + \beta/sI_0} \frac{E_q(iE_0 - E_d)}{E_0 + i(E_d + E_q)} \qquad (2)$$

where I_0 is the total average intensity of the interacting beams, E_q and E_d are characteristic fields of maximum space charge and diffusion, respectively $E_d = k_B T k_g / e$, $E_q = e N_A / \varepsilon k_g$. The response time τ is given by

$$\tau = \frac{N_A}{S N_D (I_0 + \beta/s)} \frac{E_0 + i(E_d + E_\mu)}{E_0 + i(E_d + E_q)} \tag{3}$$

where E_μ is the characteristic mobility field $E_\mu = \gamma N_A / \mu k_g$. These results show that the steady-state space charge field is approximately independent of total intensity and that the response time is inversely proportional to total intensity if the intensity exceeds the saturation intensity β/s.

In the absence of a dc applied field, the fundamental Fourier component of the space charge field is purely imaginary, indicating a 90° spatial phase shift between the interference pattern and the index grating. The effect of an applied dc field is to increase the magnitude of the space charge field and to move the spatial phase shift from 90°. The 90° phase shift is optimal for two-beam coupling amplifiers, as will be indicated, and can be restored by inducing a compensating phase difference through detuning the writing beams from each other to cause the grating to lag behind the interference pattern.[10] Alternatively, an ac applied field may be used to enhance the magnitude of the photorefractive grating and maintain the 90° phase shift.[11] For externally pumped four-wave mixing, the 90° phase shift is not optimum, so the applied field-induced deviation of the phase shift is advantageous.[13]

Wave Interactions

Two-beam Coupling.   Consider two laser beams writing a grating in a photorefractive medium as depicted in Fig. 2. The effect of the photorefractive nonlinearity on the writing beams can be described quite accurately by conventional coupled-wave theory. Coupled-wave equations for two-beam coupling can be found by taking the slowly varying envelope approximation and substituting the optical electric fields into the scalar wave equation.[14]

$$\nabla^2 E + k^2 E = 0 \tag{4}$$

where

$$k = \frac{\omega n}{c} = \frac{\omega (n_0 + \Delta n)}{c} \tag{5}$$

with Δn being the optically induced refractive index change.

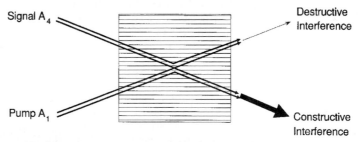

FIGURE 2   Two-beam coupling amplification. Beams 1 and 4 write a diffraction grating. Beam 1 diffracts from the grating to constructively interfere with and amplify beam 4. Beam 4 diffracts from the grating to destructively interfere with and attenuate beam 1.

The resultant coupled-wave equations read:

$$\frac{dA_1}{dz} = -\gamma \frac{A_1 A_4^*}{I_0} A_4$$

$$\frac{dA_4^*}{dz} = \gamma \frac{A_1 A_4^*}{I_0} A_1^*$$

(6)

where I_0 is sum of the intensities I_1 and I_4 of the interacting beams. The coupling constant γ is related to the space charge field by

$$\gamma = \frac{i\omega}{c \cos \theta} \frac{r_{\text{eff}} n^3 E_{\text{sc}}/m}{2}$$

(7)

It is proportional to the effective electro-optic coefficient r_{eff}[15] and is real when the index grating is 90° out of phase with the interference pattern (as in the case without an applied dc electric field, or with high-frequency ac applied fields). In that case, the coupled-wave equations show that beam 4 is amplified at the expense of beam 1 if γ is positive. The amplification effect can be explained in physical terms: the light diffracted by the grating from beam 1 interferes constructively with beam 4, so beam 4 is amplified. The light diffracted by the grating from beam 4 interferes destructively with beam 1, which consequently loses energy. At the same time, the phases of the interacting beams are preserved. If the spatial phase shift departs from 90° then the energy transfer effect becomes less, and phase coupling begins to appear.

The photorefractive beam coupling gain can be quite high in materials with large electro-optic coefficients. The intensity gain coefficient Γ which characterizes the transfer of energy between two beams ($\Gamma = \gamma + \gamma^*$; see Eq. 8) can exceed 60 cm^{-1} in BaTiO$_3$[16] and in SBN.[17] Such high gain makes possible the construction of devices such as photorefractive parametric oscillators and recursive image processors. In the high gain case, signal-to-noise issues become important. Generally, defects in photorefractive crystals scatter a small amount of light. In a process similar to amplified spontaneous emission, this scattered light can be very strongly amplified into a broad fan of light. This fanning effect[18,19] is a significant source of noise for photorefractive image amplifiers, and considerable effort has been devoted to lessening the effect by growing more uniform crystals and making device design modifications.[21,22] On the other hand, the fanning effect can be a useful source of seed beams for various oscillators[23,24] or as the basis for optical limiters.[25]

In the undepleted pump approximation ($I_1 \gg I_4$), theoretical analysis is extremely simple: the gain is exponential, with amplitude gain coefficient simply γ. But even in the pump depletion case, analysis is quite straightforward because, as in the case of second-harmonic generation, the nonlinear coupled-wave equations can be solved exactly.[14]

$$I_1(z) = \frac{I_0}{1 + (I_4(0)/I_1(0)) \exp{(\Gamma z)}}$$

$$I_4(z) = \frac{I_0}{1 + (I_1(0)/I_4(0)) \exp{(-\Gamma z)}}$$

(8)

$$\psi_1(z) = \psi_1(0) - \Gamma' z + \frac{\Gamma'}{\Gamma} \ln{(I_4(z)/I_4(0))}$$

$$\psi_4(z) = \psi_4(0) - \Gamma' z - \frac{\Gamma'}{\Gamma} \ln{(I_1(z)/I_1(0))}$$

where $\Gamma = \gamma + \gamma^*$ and $\Gamma' = (\gamma - \gamma^*)/2i$. The physical implications of these equations are clear: for $I_4(0) \exp{(\Gamma z)} \ll I_1(0)$, beam 4 is exponentially amplified; for $I_4(0) \exp{(\Gamma z)} \gg$

$I_1(0)$, beam 4 receives all of the intensity of both input beams, and beam 1 is completely depleted. There is phase transfer only if γ has an imaginary component. Linear absorption is accounted for simply by multiplying each of the intensity equations by $\exp(-\alpha z)$ where α is the intensity absorption coefficient.

The plane-wave transfer function of a photorefractive two-beam coupling amplifier may be used to determine thresholds (given in terms of $\gamma\ell$, where ℓ is the interaction length), oscillation intensities, and frequency pulling effects in unidirectional ring oscillators based on two-beam coupling.[26,27,28,29]

Four-wave Mixing. Four-wave mixing and optical phase conjugation may also be modeled by plane-wave coupled-wave theories for four interacting beams[13] (Fig. 3). In general, when all four beams are mutually coherent, they couple through four sets of gratings: (1) the transmission grating driven by the interference term $(A_1 A_4^* + A_2^* A_3)$, (2) the reflection grating driven by $(A_2 A_4^* + A_1^* A_3)$, (3) the counterpropagating pump grating $(A_1 A_2^*)$, and (4) the signal/phase conjugate grating $(A_3 A_4^*)$. The theories can be considerably simplified by modeling cases in which the transmission grating only or the reflection grating only is important. The transmission grating case may be experimentally realized by making beams 1 and 4 mutually coherent but incoherent with beam 2 (and hence beam 3, which is derived directly from beam 2). The four-wave mixing coupled-wave

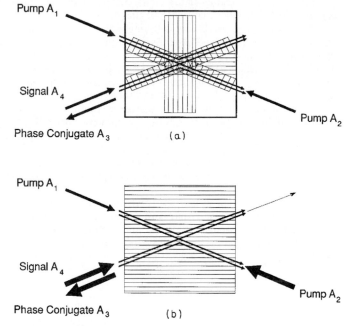

FIGURE 3 Four-wave mixing phase conjugation: (*a*) beams 1 and 2 are pump beams, beam 4 is the signal, and beam 3 is the phase conjugate. All four interaction gratings are shown: reflection, transmission, counterpropagating pump, and signal/phase conjugate. Beam 1 interferes with beam 4 and beam 2 interferes with beam 3 to write the transmission grating. Beam 2 interferes with beam 4 and beam 1 interferes with beam 3 to write the reflection grating. Beam 1 interferes with beam 2 to write a counterpropagating beam grating, as do beams 3 and 4. (*b*) If beams 1 and 4 are mutually incoherent, but incoherent with beam 2, only the transmission grating will be written. This interaction is the basis for many self-pumped phase conjugate mirrors.

equations can be solved analytically in several useful cases by taking advantage of conservation relations inherent in the four-wave mixing process,[30,31] and by using group theoretic arguments.[32] Such analytic solutions are of considerable assistance in the design and understanding of various four-wave mixing devices. But as in other types of four-wave mixing, the coupled-wave equations can be linearized by assuming the undepleted pump approximation. Considerable insight can be obtained from these linearized solutions.[13] For example, they predict phase conjugation with gain and self-oscillation. The minimum threshold for phase conjugation with gain in the transmission grating case is $\gamma \ell = 2 \ln (1 + \sqrt{2}) \approx 1.76$, whereas the minimum threshold for the usual $\chi^{(3)}$ nonlinearities is $\gamma \ell = i\pi/4 \approx 0.79$.

Self-oscillation (when the phase conjugate reflectivity tends to infinity) can only be achieved if the coupling constant is complex. In the $\chi^{(3)}$ case this is the normal state of affairs, but in the photorefractive case, as mentioned, it requires some additional effect such as provided by applied electric fields, photovoltaic effect, or frequency detuning. Photorefractive four-wave mixing thresholds are usually higher than the corresponding $\chi^{(3)}$ thresholds because photorefractive symmetry implies that either the signal or the phase conjugate tend to be deamplified in the interaction. In the $\chi^{(3)}$ case, both signals and phase conjugate can be amplified. This effect results in the fact that while the optimum pump intensity ratio is unity in the $\chi^{(3)}$ case, the optimum beam intensities are asymmetric in the photorefractive case. A consequence of the interplay between self-oscillation and coupling constant phase is that high-gain photorefractive phase conjugate mirrors tend to be unstable. Even if the crystal used is purely diffusive, running gratings can be induced which cause the coupling constant to become complex, giving rise to self-oscillation instabilities.[33,34]

Anisotropic Scattering. Because of the tensor nature of the electro-optic effect, it is possible to observe interactions with changing beam polarization. One of the most commonly seen examples is the anisotropic diffraction ring of ordinary polarization that appears on the opposite side of the amplified beam fan when a single incident beam of extraordinary polarization propagates approximately perpendicular to the crystal optic axis.[35,36,37] Since the refractive indices for ordinary and extraordinary waves differ from each other, phase matching for such an interaction can only be satisfied along specific directions, leading to the appearance of phase-matched rings. There are several other types of anisotropic scattering, such as broad fans of scattered light due to the circular photovoltaic effect,[38] and rings that appear when ordinary and extraordinary polarized beams intersect in a crystal.[39]

Oscillators with Photorefractive Gain and Self-pumped Phase Conjugate Mirrors

Photorefractive beam amplification makes possible several interesting four-wave mixing oscillators, including the unidirectional ring resonator and self-pumped phase conjugate mirrors (SPPCMs). The simplest of these is the linear self-pumped phase conjugate mirror.[40] Two-beam coupling photorefractive gains supports oscillation in a linear cavity (Fig. 4a). The counterpropagating oscillation beams pump the crystal as a self-pumped phase conjugate mirror for the incident beam. The phase conjugate reflectivity of such a device can theoretically approach 100 percent, with commonly available crystals. In practice, the reflectivity is limited by parasitic fanning loss. Other types of self-pumped phase conjugate mirrors include the following. The semilinear mirror, consisting of a linear mirror with one of its cavity mirrors removed[40] Fig. 4b. The ring mirror (transmission grating[41] and reflection grating[42] types). The transmission grating type is shown in Fig. 4c. The double phase conjugate mirror (Fig. 4d). This device is also sometimes known as a mutually pumped phase conjugator (MPPC). Referring to Fig. 4b, it can be seen that the double phase conjugate mirror is part of the semilinear mirror. Several variants involving combinations of the transmission grating ring mirror and double phase conjugate mirror:

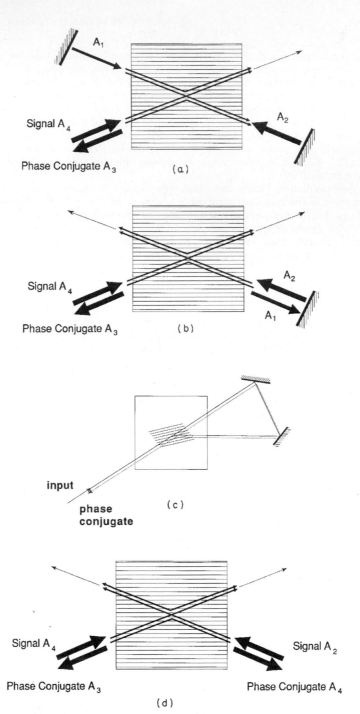

FIGURE 4 Self-pumped phase conjugate mirrors: those with external feedback: (*a*) linear; (*b*) semilinear; (*c*) ring; those self-contained in a single crystal with feedback (when needed) provided by total internal reflection: (*d*) double phase conjugate mirror; (*e*) cat; (*f*) frogs' legs; (*g*) bird-wing; (*h*) bridge; (*i*) mutually incoherent beam coupler.

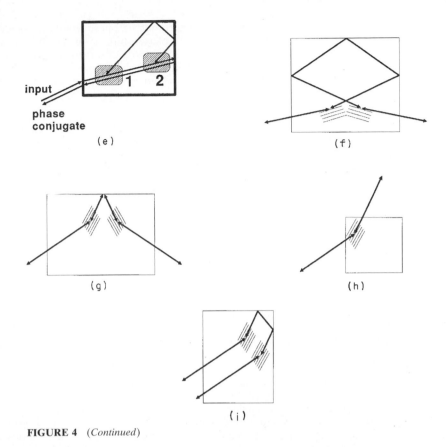

FIGURE 4 (*Continued*)

the cat mirror[23] (Fig. 4*e*) so named after its first subject, frogs legs[44] (Fig. 4*f*), bird-wing[45] (Fig. 4*g*), bridge (Fig. 4*h*), and mutually incoherent beam coupler[47] (Fig. 4*i*). The properties of these devices are sometimes influenced by the additional simultaneous presence of reflection gratings and gratings written between the various pairs of counterpropagating beams. The double phase conjugate mirror can be physically understood as being supported by a special sort of photorefractive self-oscillation in which beams 2 and 4 of Fig. 3 are taken as strong and depleted. The self-oscillation threshold for the appearance of beams 1 and 3 is $\gamma\ell = 2.0$ with purely diffusive coupling.[48] It can be shown that a transmission grating ring mirror with coupling constant $\gamma\ell$ is equivalent in the plane-wave theory to a double phase conjugate mirror with coupling constant $2\gamma\ell$.[49]

The matter of whether the various devices represent the results of absolute instabilities or convective instabilities has been the subject of some debate recently.[50,51]

Stimulated Photorefractive Scattering

It is natural to ask whether there is a photorefractive analogue to stimulated Brillouin scattering (SBS, a convective instability). In SBS, an intense laser beam stimulates a sound wave whose phase fronts are the same as those of the incident radiation. The Stokes wave reflected from the sound grating has the same phase fronts as the incident beam, and is traveling in the backward phase conjugate sense. The backward wave experiences gain because the sonic grating is 90° spatially out of phase with the incident beam. In the photorefractive case, the required 90° phase shift is provided automatically, so that stimulated photorefractive scattering (SPS) can be observed without any Stokes frequency shift[52,53] However, the fidelity of SPS tends to be worse than that of SBS because the intensity-gain discrimination mechanism is not as strong, as can be seen by examining the coupled-wave equations in each case.[54]

Time-dependent Effects

Time-dependent coupled-wave theory is important for studying the temporal response and temporal stability of photorefractive devices. Such a theory can be developed by including a differential equation for the temporal evolution of the grating. The spatiotemporal two-beam coupled-wave equations can be written, for example, as

$$\frac{\partial A_1}{\partial z} = -GA_4$$

$$\frac{\partial A_4^*}{\partial z} = GA_1^*$$

$$\frac{\partial G}{\partial t} + G/\tau = \frac{\gamma A_1 A_4^*}{\tau} \frac{1}{I_0}$$

(9)

where τ is an intensity dependent possibly complex response time determined from photorefractive charge transport models. Models like this can be used to show that the response time for beam amplification and deamplification is increased by a factor of $\gamma\ell$ over the basic photorefractive response time τ.[55] The potential for photorefractive bistability can be studied by examining the stability of the multiple solutions of the steady-state four-wave mixing equations. In general, there is no reversible plane-wave bistability except when other nonlinearities are included in photorefractive oscillator cavities.[56,57] (See the section on "Thresholding.") Temporal instabilities generate interesting effects such as deterministic chaos, found both experimentally[58] and theoretically[59,60] in high-gain photorefractive devices.

Influence of the Nonlinearity on Beam Spatial Profiles

In modeling the changes in the transverse cross section of beams as they interact, it is necessary to go beyond simple one-dimensional plane-wave theory. Such extensions are useful for analyzing fidelity of phase conjugation and image amplification, and for treating transverse mode structure in photorefractive oscillators. Several different methods have been used to approach the transverse profile problem.[61] In the quasi-plane-wave method, one assumes that each beam can be described by a single plane wave whose amplitude varies perpendicular to its direction of propagation. The resulting two-dimensional coupled partial differential equations give good results when propagative diffraction effects can be neglected.[62] Generalization of the one-dimensional coupled-wave equations to multi-coupled wave theory also works quite well,[63] as does the further generalization to coupling of continuous distributions of plane waves summed by integration.[20] This latter method has been used quite successfully to model beam fanning.[20] The split-step beam propagation method has also been used to include diffractive beam propagation effects as well as nonlinear models of the optically induced grating formation.[64]

Spatiotemporal instabilities have been studied in phase conjugate resonators.[65,66] These instabilities often involve optical vortices.[67]

39.2 MATERIALS

Introduction

Photorefractive materials have been used in a wide variety of applications, as will be discussed later. These materials have several features which make them particularly attractive.

- The characteristic phase shift between the writing intensity pattern and the induced space charge field leads to energy exchange between the two writing beams, amplified scattered light (beam fanning), and self-pumped oscillators and conjugators.
- Photorefractive materials can be highly efficient at power levels obtained using CW lasers. Image amplification with a gain of 4000[68] and degenerate four-wave mixing with a reflectivity of 2000 percent[69] have been demonstrated.
- In optimized bulk photorefractive materials, the required energy to write a grating can approach that of photographic emulsion ($50\,\mu J/cm^2$), with even lower values of write energy measured in photorefractive multiple quantum wells.
- The response time of most bulk photorefractive materials varies inversely with intensity. Gratings can be written with submillisecond response times at CW power levels and with nanosecond response times using nanosecond pulsed lasers. Most materials have a useful response with picosecond lasers.
- The high dark resistivity of oxide photorefractive materials allows the storage of holograms for time periods up to a year in the dark.

In spite of the great appeal of photorefractive materials, they have specific limitations which have restricted their use in practical devices. For example, oxide ferroelectric materials are very efficient, but are rather insensitive. Conversely, the bulk compound semiconductors are extremely sensitive, but suffer from low efficiency in the absence of an applied field. There are now active efforts in many laboratories to enhance the performance of these and other materials through a variety of approaches. In this section we will first review the figures of merit used to characterize photorefractive materials, and then discuss the properties of the different classes of materials.

Figures of Merit

The figures of merit for photorefractive materials can be conveniently divided into those which characterize the steady-state response, and those which characterize the early portion of the transient response[70–72] Most applications fall into one or the other of these regimes, although some may be useful in either regime. For example, aberration correction, optical limiting, and laser coupling are applications which generally require operation in the steady state. On the other hand, certain optical processing applications require a response only to a given level of index change or efficiency, and are thus better characterized by the initial recording slope of a photorefractive grating.

Steady-state Performance. The steady-state change in the refractive index is related to the space charge electric field by

$$\Delta n_{ss} = \tfrac{1}{2} n_b^3 r_{eff} E_{sc} \tag{10}$$

where n_b is the background refractive index, r_{eff} is the effective electro-optic coefficient (which accounts for the specific propagation direction and optical polarization in the sample), and E_{sc} is the space charge electric field. For large grating periods where diffusion limits the space charge field, the magnitude of the field is independent of material parameters and thus $\Delta n_{ss} \propto n_b^3 r_{eff}$. In this case (which is typical for many applications), the ferroelectric oxides are favored, because of their large electro-optic coefficients. For short grating periods or for very large applied fields, the space charge field is trap limited and $\Delta n_{ss} \propto n_b^3 r_{eff}/\varepsilon_r$, where ε_r is the relative dielectric constant.

The temporal behavior of the local space charge field in a given material depends on the details of the energy levels which contribute to the photorefractive effect. In many cases, the buildup or decay of the field is exponential. The fundamental parameter which characterizes this transient response is the write or erase energy W_{sat}. In many materials the response time τ at an average intensity I_0 is simply given by

$$\tau = W_{sat}/I_0 \tag{11}$$

This clearly points out the dependence of the response time on the intensity. As long as enough energy is provided, photorefractive gratings can be written with beams ranging in intensity from mW/cm^2 to MW/cm^2. The corresponding response time can be calculated simply from Eq. (11).

In the absence of an applied or photovoltaic field (assuming only a single charge carrier), the response time can be written as[70]

$$\tau = \tau_{di}(1 + 4\pi^2 L_d^2/\Lambda_g^2) \tag{12}$$

where τ_{di} is the dielectric relaxation time, L_d is the diffusion length, and Λ_g is the grating period. The diffusion length is given by

$$L_d = (\mu \tau_r k_B T/e)^{1/2} \tag{13}$$

where μ is the mobility, τ_r is the recombination time, k_B is Boltzmann's constant, T is the temperature, and e is the charge of an electron. The dielectric relaxation time is given by

$$\tau_{di} = \varepsilon_r \varepsilon_0/(\sigma_d + \sigma_p) \tag{14}$$

where σ_d is the dark conductivity and σ_p is the photoconductivity, given by

$$\sigma_p = \alpha e \mu \tau_r I_0/h\nu \tag{15}$$

where α is the absorption coefficient.

In as-grown oxide ferroelectric materials, the diffusion length is usually much less than the grating period. In this case, the response time is given by

$$\tau = \tau_{di} \tag{16}$$

In addition, the contribution from dark conductivity in Eq. (14) can be neglected for intensities greater than $\sim mW/cm^2$. In this regime, materials with large values of absorption coefficient and photoconductivity ($\mu\tau_r$) are favored.

In bulk semiconductors, the diffusion length is usually much larger than the grating period. In this case, the response time is given by

$$\tau \approx \tau_{di}(4\pi^2 L_d^2/\Lambda_g^2) \tag{17}$$

If we again neglect the dark conductivity in Eq. (14), then

$$\tau \approx \pi \varepsilon k_B T/e^2 \alpha I_0 \Lambda_g^2 \tag{18}$$

In this regime, materials with small values of dielectric constant and large values of absorption coefficient are favored. For a typical bulk semiconductor with $\varepsilon_r \approx 12$, $\alpha = 1$ cm^{-1}, and $\Lambda_g = 1$ μm, we find $W_{sat} \approx 100$ μJ/cm^2. This saturation energy is comparable to that required to expose high resolution photographic emulsion. In photorefractive multiple quantum wells (with $\alpha \approx 10^{13}$ cm^{-1}), the saturation energy can be much smaller. Note finally that an applied field leads to an increase in the write energy in the bulk semiconductors.[72]

Transient Performance. In the transient regime, we are typically concerned with the time or energy required to achieve a design value of index change or diffraction efficiency. This generally can be obtained from the initial *recording slope* of a photorefractive grating. One common figure of merit which characterizes the recording slope is the sensitivity,[70–72] defined as the index change per absorbed energy per unit volume:

$$S = \Delta n / \alpha I_0 \tau = \Delta n / \alpha W_{sat} \tag{19}$$

In the absence of an applied field and for large diffusion lengths, we find the limiting value of the sensitivity:

$$S \approx 1/4\pi (n_b^3 r_{eff}/h\nu\varepsilon)(me/\varepsilon_0)\Lambda_g \tag{20}$$

where m is the modulation index.

We will see that the only material dependence in the limiting sensitivity is through the figure of merit $n_b^3 r_{eff}/\varepsilon$. This quantity varies little from material to material. Using typical values of the materials parameters and assuming $\Lambda_g = 1$ μm, we find the limiting value $S = 400$ cm^3/kJ. Values in this range are routinely observed in the bulk semiconductors. In the as-grown ferroelectric oxides, in which the diffusion length is generally less than the grating period, the sensitivity values are typically two to three orders of magnitude smaller.

Comparison of Materials. In the following sections, we will briefly review the specific properties of photorefractive materials, organized by crystalline structure. To introduce this discussion, we have listed relevant materials parameters in Table 1. This table allows the direct comparison among BaTiO$_3$, Bi$_{12}$SiO$_{20}$ (BSO), and GaAs, which are representative of the three most common classes of photorefractive materials.

The distinguishing feature of BaTiO$_3$ is the magnitude of its electro-optic coefficients, leading to large values of steady-state index change. The sillenites are distinguished by

TABLE 1 Materials Parameters for BaTiO$_3$, BSO, and GaAs

(The parameters in bold type are particularly distinctive for that material.)

Material class	Ferroelectric oxide	Sillenite	Compound semiconductor
Material	BaTiO$_3$	BSO	GaAs
Wavelength range (μm)	0.4–1.1	0.45–0.65	0.9–1.3
Electro-optic coefficient e_{eff} (pm/V)	**100** (r_{33}) **1640** (r_{42})	4 (r_{41})	1.4 (r_{41})
Dielectric constant	135 (r_{33}) 3700 (r_{11})	56	13.2
$n_b^3 r_{eff}/\varepsilon$ (pm/V)	10 (r_{33}) 6 (r_{42})	1.4	3.3
Mobility μ (cm^2/V-s)	0.01	0.1	**6000**
Recombination time τ_r (s)	10^{-8}	**10^{-6}**	3 × 10^{-8}
Diffusion length L_d (μm)	0.01	0.5	20
Photoconductivity $\mu\tau_r$ (cm^2/V)	10^{-10}	10^{-7}	1.8 × 10^{-4}

their large value of recombination time, leading to a larger photoconductivity and diffusion length. The compound semiconductors are distinguished by their large values of mobility, leading to very large values of photoconductivity and diffusion length. Note also the different spectral regions covered by these three materials.

Ferroelectric Oxides

The photorefractive effect was first observed in ferroelectric oxides that were of interest for electro-optic modulators and second-harmonic generation.[73] Initially, the effect was regarded as "optical damage" that degraded device performance.[74] Soon, however, it became apparent that refractive index gratings could be written and stored in these materials.[75] Since that time, extensive research on material properties and device applications has been undertaken.

The photorefractive ferroelectric oxides can be divided into three structural classes: ilmenites ($LiNbO_3$, $LiTaO_3$), perovskites [$BaTiO_3$, $KNbO_3$, KTa_{1-x}, $Nb_xO_3(KTN)$], and tungsten bronzes [$Sr_{1-x}Ba_xNb_2O_6(SBN)$, $Ba_2NaNb_5O_{15}(BNN)$ and related compounds]. In spite of their varying crystal structure, these materials have several features in common. They are transparent from the band gap (~ 350 nm) to the intrinsic IR absorption edge near 4 μm. Their wavelength range of sensitivity is also much broader than that of other photorefractive materials. For example, useful photorefractive properties have been measured in $BaTiO_3$ from 442 nm[76] to 1.09 μm,[77] a range of a factor of $2\frac{1}{2}$. Ferroelectric oxides are hard, nonhygroscopic materials—properties which are advantageous for the preparation of high-quality surfaces. Their linear and nonlinear dielectric properties are inherently temperature-dependent, because of their ferroelectric nature. As these materials are cooled below their melting point, they undergo a structural phase transition to a ferroelectric phase. Additional transitions may occur in the ferroelectric phase on further lowering of the temperature. In general, samples in the ferroelectric phase contain regions of differing polarization orientation called domains, leading to a reduction in the net polar properties of the sample. To make use of the electro-optic and nonlinear optic properties of the ferroelectric oxides, these domains must be aligned to a single domain state. This process, called poling, can take place during the growth process, or more commonly, after polydomain samples have been cut from an as-grown boule.

Growth of large single crystals of ferroelectric oxides has been greatly stimulated by the intense interest in photorefractive and nonlinear optic applications. Currently, most materials of interest are commercially available. However, considerably more materials development is required before optimized samples for specific applications can be purchased.

Lithium Niobate and Lithium Tantalate. $LiNbO_3$ was the first material in which photorefractive "damage" was observed.[73] This material has been developed extensively for frequency conversion and integrated optics applications. It is available in large samples with high optical quality. For photorefractive applications, iron-doped samples are generally used. The commonly observed valence states are Fe^{2+} and Fe^{3+}. The relative populations of these valence states can be controlled by annealing in an atmosphere with a controlled oxygen partial pressure. In a reducing atmosphere (low oxygen partial pressure), Fe^{2+} is favored, while Fe^{3+} is favored in an oxidizing atmosphere. The relative Fe^{2+}/Fe^{3+} population ratio will determine the relative contributions of electrons and holes to the photoconductivity.[78] When Fe^{2+} is favored, the dominant photocarriers are electrons; when Fe^{3+} is favored, the dominant photocarriers are holes. In most oxides, electrons have higher mobilities, so that electron-dominated samples yield faster photorefractive response times. Even in heavily reduced $LiNbO_3$, the write energy is rarely lower than 10 J/cm^2, so this material has not found use for real-time applications. The properties of $LiTaO_3$ are essentially the same as those of $LiNbO_3$.

Currently, the most promising application of LiNbO$_3$ is for holographic storage. LiNbO$_3$ is notable for its very large value of dark resistivity, leading to very long storage times in the dark. In addition, the relatively large write or erase energy of LiNbO$_3$ makes this material relatively insusceptibe to erasure during readout of stored holograms.

Improved retention of stored holograms can be obtained by fixing techniques. The most common fixing approach makes use of complementary gratings produced in an ionic species which is not photoactive.[79,80] Typically, one or more holograms are written into the sample by conventional means. The sample is then heated to 150°C, where it is annealed for a few hours. At this temperature, a separate optically inactive ionic species is thermally activated and drifts in the presence of the photorefractive space charge field until it exactly compensates this field. The sample is then cooled to room temperature to "freeze" the compensating ion grating. Finally, uniform illumination washes out the photorefractive grating and "reveals" the permanent ion grating.

Another important feature of LiNbO$_3$ for storage applications is the large values of diffraction efficiency (approaching 100 percent for a single grating) which can be obtained. These large efficiencies arise primarily from the large values of space charge field, which, in turn, result from the very large value of photovoltaic field.

Barium Titanate. BaTiO$_3$ was one of the first ferroelectric materials to be discovered, and also one of the first to be recognized as photorefractive.[81] The particular advantage of BaTiO$_3$ for photorefractive applications is the very large value of the electro-optic coefficient r_{42} (see Table 1), which, in turn, leads to large values of grating efficiency, beam-coupling gain, and conjugate reflectivity. For example, four-wave mixing reflectivities as large as 20 have been observed,[68] as well as an image intensity gain of 4000.[69]

After the first observation of the photorefractive effect in BaTiO$_3$ in 1970, little further research was performed until 1980 when Feinberg et al.[82] and Krätzig et al.[83] pointed out the favorable features of this material for real-time applications. Since that time, BaTiO$_3$ has been widely used in a large number of experiments in the areas of optical processing, laser power combining, spatial light modulation, optical limiting, and neural networks.

The photorefractive properties of BaTiO$_3$ have been reviewed in Ref. 84. Crystals of this material are grown by top-seeded solution growth in a solution containing excess TiO$_2$.[85] Crystal growth occurs while cooling the melt from 1400°C to 1330°C. At the growth temperature, BaTiO$_3$ has the cubic perovskite structure, but on cooling through $T_c = 132$°C, the crystal undergoes a transition to the tetragonal ferroelectric phase. Several approaches to poling have been successfully demonstrated. In general, the simplest approach is to heat the sample to just below or just above the Curie temperature, apply an electric field, and cool the sample with the field present.

Considerable efforts have been expended to identify the photorefractive species in as-grown BaTiO$_3$. Early efforts suggested that transition metal impurities (most likely iron) were responsible.[86] In later experiments, samples grown from ultrapure starting materials were still observed to be photorefractive.[87] In this case, barium vacancies have been proposed as the dominant species.[88] Since that time, samples have been grown with a variety of transition metal dopants. All dopants produce useful photorefractive properties, but cobalt-doped samples[89] and rhodium-doped samples[90] appear particularly promising, because of their reproducible high gain in the visible and enhanced sensitivity in the infrared.[90]

One particular complication in developing a full understanding of the photorefractive properties of BaTiO$_3$ is the presence of shallow levels in the band gap, in addition to the deeper levels typically associated with transition-metal impurities or dopants. The shallow levels are manifested in several ways. Perhaps the most prominent of these is the observation that the response time (and photoconductivity) of as-grown samples does not scale inversely with intensity [see Eq. (11)], but rather has a dependence of the type $\tau \sim (I_o)^{-x}$ is observed, where $x = 0.6$–1.0.[81–84,91] Several models relating the sublinear behavior of the response time and photoconductivity to shallow levels have been reported.[92–95]

The characteristic sublinear variation of response time with intensity implies that the write or erase energy W_{sat} increases with intensity, which is a clear disadvantage for high-power, short-pulse operation. Nevertheless, useful gratings have been written in BaTiO$_3$ using nanosecond pulses[96,97] and picosecond pulses.[98] Another manifestation of the presence of shallow levels is intensity-dependent absorption.[99] The shallow levels have been attributed to oxygen vacancies or barium vacancies, but no unambiguous identification has been made to date.

The major limitation of BaTiO$_3$ for many applications is the relatively slow response time of this material at typical CW intensity levels. In as-grown crystals, typical values of response time are 0.1 to 1 s at 1 W/cm^2. These values are approximately three orders of magnitude longer than theoretical values determined from the band transport model (see "Steady-State Performance"), or from more fundamental arguments.[100,101]

Two different approaches have been studied to improve the response time of BaTiO$_3$. In the first, as-grown samples can be operated at an elevated temperature (but below the Curie temperature). In a typical experiment (see Fig. 5), an improvement in response time

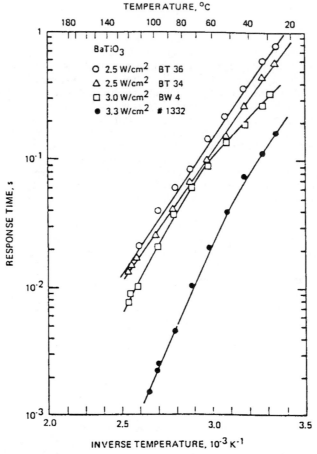

FIGURE 5 Measured response time as a function of temperature for four samples of BaTiO$_3$. The measurement wavelength was 515 nm and the grating period was 0.79 μm.

by two orders of magnitude was observed[102] when different samples were operated at 120°C. In some of these samples, the magnitude of the peak beam-coupling gain did not vary significantly with temperature. In these cases, the improvement in response time translates directly to an equivalent improvement in sensitivity.

While operation at an elevated temperature may not be practical for many experiments, the importance of the preceding experiment is that it demonstrates the *capability for improvement in response time,* based on continuing materials research.

Current materials research efforts are concentrated on studies of new dopants, as well as heat treatments in reducing atmospheres.[88,89,103,104] The purpose of the reducing treatments is to control the valence states of the dopants to produce beneficial changes of the trap density and the sign of the dominant photocarrier. While some success has been achieved,[89,104] a considerably better understanding of the energy levels in the $BaTiO_3$ band gap is required before substantial further progress can be made.

Potassium Niobate. $KNbO_3$ is another important photorefractive material with the perovskite structure. It undergoes the same sequence of phase transitions as $BaTiO_3$, but at higher transition temperatures. At room temperature it is orthorhombic, with large values of the electro-optic coefficients r_{42} and r_{51}.

$KNbO_3$ has been under active development for frequency conversion and photorefractive experiments since 1977 (Ref. 70). Unlike $BaTiO_3$, undoped samples of $KNbO_3$ have weak photorefractive properties. Iron doping has been widely used for photorefractive applications,[70] but other transition metals have also been studied.

Response times in as-grown $KNbO_3$ at 1 W/cm^2 are somewhat faster than those of $BaTiO_3$, but are still several orders of magnitude longer than the limiting value. The most common approach to improving the response time of $KNbO_3$ is electrochemical reduction. In one experiment at 488 nm,[101] a photorefractive response time of 100 μs at 1 W/cm^2 was measured in a reduced sample. This response time is very close to the limiting value, which indicates again the promise for faster performance in all the ferroelectric oxides.

Strontium Barium Niobate and Related Compounds. $Sr_{1-x}Ba_xNb_2O_6$ (SBN) is a member of the tungsten bronze family,[105] which includes materials such as $Ba_2NaNb_5O_{15}$ (BNN) and $Ba_{1-x}Sr_xK_{1-y}Na_yNb_5O_{15}$ (BSKNN). SBN is a mixed composition material with a phase transition temperature which varies from 60°C to 200°C as x varies from 0.75 to 0.25. Of particular interest is the composition SBN-60, which melts congruently,[106] and is thus easier to grow with high quality. SBN is notable for the very large values of the electro-optic tensor component r_{33}. In other materials such as BSKNN, the largest tensor component is r_{42}. In this sense it resembles $BaTiO_3$.

In general, the tungsten bronze system contains a large number of mixed composition materials, thus offering a rich variety of choices for photorefractive applications. In general, the crystalline structure is quite open, with only partial occupancy of all lattice sites. This offers greater possibilities for doping, but also leads to unusual properties at the phase transition, due to its diffuse nature.[105]

The photorefractive properties of SBN were first reported in 1969,[107] very soon after gratings were first recorded in $LiNbO_3$. Since that time, there has been considerable interest in determining the optimum dopant for this material. Until recently, the most common dopant has been cerium.[108–110] Cerium-doped samples can be grown with high optical quality and large values of photorefractive gain.[111,112] Another promising dopant is rhodium, which also yields high values of gain coefficient.[113]

As with $BaTiO_3$ and $KNbO_3$, as-grown samples of SBN and other tungsten bronzes are relatively slow at an intensity of 1 W/cm^2.[114] Doping and codoping has produced some improvement. In addition, the use of an applied dc electric field has led to improvement in the response time.[115]

The photorefractive effect has also been observed in fibers of SBN.[116] The fiber geometry has promise in holographic storage architectures.

TABLE 2 Material Properties of BSO, BGO, and BTO

Material	BSO	BGO	BTO
Wavelength range (μm)	0.5–0.65	0.5–0.65	0.6–0.75
Electro-optic coefficient r_{41} (pm/V)	4.5	3.4	5.7
$n_b^3 r_{41}$ (pm/V)	81	56	89
Dielectric constant	56	47	48
$n_b^3 r_{41}/\varepsilon$ (pm/V)	1.4	1.2	1.9
Optical activity at 633 nm (degrees/mm)	21	21	6

Cubic Oxides (Sillenites)

The cubic oxides are notable for their high photoconductivity, leading to early applications for spatial light modulation[117] and real-time holography using the photorefractive effect.[118] The commonly used sillenites are $Bi_{12}SiO_{20}$ (BSO), $Bi_{12}GeO_{20}$ (BGO), and $Bi_{12}TiO_{20}$(BTO). Some relevant properties of these materials are listed in Table 2.

The sillenites are cubic and noncentrosymmetric, with one nonzero electro-optic tensor component $\mathbf{r}_{41}$. The magnitude of $\mathbf{r}_{41}$ in the sillenites is small, ranging from approximately 4 pm/V to 6 pm/V in the visible. In addition, the sillenites are optically active, with a rotatory power (at 633 nm) of 21°/mm in BSO and BGO, and 6°/mm in BTO. These values increase sharply at shorter wavelengths. The optical activity of the sillenites tends to reduce the effective gain or diffraction efficiency of samples with normal thickness, but in certain experiments it also allows the use of an output analyzer to reduce noise.

The energy levels due to defects and impurities tend to be similar in each of the sillenites. In spite of many years of research, the identity of the photorefractive species is still not known. It is likely that intrinsic defects such as metal ion vacancies play an important role. With only one metal ion for each 12 bismuth ions and 20 oxygen ions, small deviations in metal ion stoichiometry can lead to large populations of intrinsic defects. In each of the sillenites, the effect of the energy levels in the band gap is to shift the fundamental absorption edge approximately 100 nm to the red.

BSO and BGO melt congruently and can be grown from stoichiometric melts by the Czochralski technique. On the other hand, BTO melts incongruently and is commonly grown by the top-seeded solution growth technique, using excess Bi_2O_3. BTO is particularly interesting for photorefractive applications (compared with BSO and BGO), because of its lower optical activity at 633 nm and its slightly larger electro-optic coefficient (5.7 pm/V).[119] It has been studied extensively at the Ioffe Institute in Russia, where both material properties and device applications have been examined.[120–122]

In the sillenites it is very common to apply large dc or ac electric fields to enhance the photorefractive space charge field, and thus provide useful values of gain or diffraction efficiency. A dc field will increase the amplitude of the space charge field, but the spatial phase will decrease from the value of 90° which optimizes the gain. In order to restore the ideal 90° phase shift, moving grating techniques are typically used.[123,124] By contrast, an ac field can enhance the amplitude of the space charge field, while maintaining the spatial phase at the optimum value of 90°.[12] In this case, the best performance is obtained when a square waveform is used, and when the period is long compared with the recombination time, and short compared with the grating formation time. Both dc and ac field techniques have produced large gain enhancements in sillenites and semiconductors, but only when the signal beam is very weak, i.e., when the pump/signal intensity ratio is large. As the amplitude of the signal beam increases, the gain decreases sharply, by an amount which cannot be explained by pump depletion. This effect is significant for applications such as self-pumped phase conjugation, in which the buildup of the signal wave will reduce the effective gain and limit the device performance.

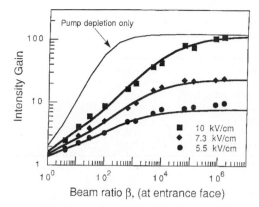

FIGURE 6 Measured two-wave mixing gain as a function of input pump-to-signal beam ratio in BTO. The measurement wavelength was 633 nm, the applied field was a 60-Hz ac square wave, and the grating period was 5.5 μm. The individual points are experimental data; the bold curves are fits using a large signal model. The thin curve is the standard pump depletion theory for the 10-kV/*cm* case.

A typical plot of intensity gain in BTO as a function of beam ratio for several values of ac square-wave voltage amplitude is given in Fig. 6.[121] Note that the highest gain is observed only for a beam ratio on the order of 10^5 (small signal limit). The simplest physical description of this nonlinearity is that the internal space charge field is clamped to the magnitude of the applied field; this condition only impacts performance for decreasing beam ratios (large signal limit). In a carefully established experiment using a very large beam ratio (10^5), gain coefficients approaching 35 cm^{-1} have been measured using an ac square wave field with an amplitude of 10 kV/cm (see Fig. 7).[125]

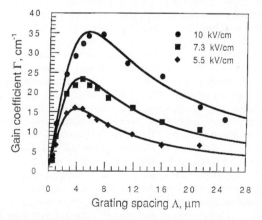

FIGURE 7 Measured gain coefficient as a function of grating spacing in BTO. The measurement wavelength was 633 nm, the applied field was a 60-Hz ac square wave, and the beam ratio was 10^5. The individual points are experimental data; the solid curves are fits using the basic band transport model.

TABLE 3 Relevant Materials Properties of Photorefractive Compound Semiconductors. (*Most of the values are taken from Ref.* 132.)

Material	GaAs	InP	GaP	CdTe	ZnTe
Wavelength range (μm)	0.92–1.3	0.96–1.3	0.63	1.06–1.5	0.63–1.3
EO coeff. r_{41} (pm/V)	1.2	1.45	1.1	6.8	4.5
$n_b^3 r_{41}$ (pm/V)	43	52	44	152	133
$n_b^3 r_{41}/\varepsilon$	3.3	4.1	3.7	16	13
Dielectric constant	13.2	12.6	12	9.4	10.1

Bulk Compound Semiconductors

The third class of commonly used photorefractive materials consists of the compound semiconductors (Si and Ge are cubic centrosymmetric materials, and thus have no linear electro-optic effect). Gratings have been written in CdS,[126] GaAs:Cr,[127] GaAs:EL2,[128] InP:Fe,[127] CdTe,[129] GaP,[130] and ZnTe.[131] These materials have several attractive features for photorefractive applications (see Table 3). First, many of these semiconductors are readily available in large sizes and high optical quality, for use as electronic device substrates. These substrates are generally required to be semi-insulating; the deep levels provided for this purpose are generally photoactive, with favorable photorefractive properties. Second, the semiconductors have peak sensitivity for wavelengths in the red and near-infrared. The range of wavelengths extends from 633 nm in GaP,[130] CdS,[133] and ZnTe[131] to 1.52 μm in CdTe:V.[129] Third, the mobilities of the semiconductors are several orders of magnitude larger than those in the oxides. There are several important consequences of these large mobilities. Most importantly, the resulting large diffusion lengths lead to fast response times [see Eqs. (12) and (18)]. The corresponding values of write/erase energies (10 to 100 μJ/cm^2) are very near the limiting values. These low values of write/erase energy have been observed not only at the infrared wavelengths used for experiments in InP and GaAs, but also at 633 nm in ZnTe.[131]

The large mobilities of the compound semiconductors also yield large values of dark conductivity (compared with the oxides), so the storage times in the dark are normally less than 1 sec. Thus, these materials are not suited for long-term storage, but may still be useful for short-term memory applications. Finally, the short diffusion times in the semiconductors yield useful photorefractive performance with picosecond pulses.[96]

The electro-optic coefficients for the compound semiconductors are quite small (see Table 3), leading to low values of beam-coupling gain and diffraction efficiency in the absence of an applied electric field. As in the sillenites, both dc and ac field techniques have been used to enhance the space charge field. Early experiments with applied fields produced enhancements in the gain or diffraction efficiency which were considerably below the calculated values.[134–136] The causes of these discrepancies are now fairly well understood. First, space charge screening can significantly reduce the magnitude of the applied field inside the sample. This effect is reduced by using an ac field, but even in this case the required frequencies to overcome all screening effects are quite high.[136] Second, the mobility-lifetime product is known to reduce at high values of electric field due to scattering of electrons into other conduction bands and cascade recombination. This effect is particularly prominent in GaAs.[137] Third, large signal effects act to reduce the gain when large fields are used.[138] As in the sillenites, the highest gains are only measured when weak signal beams (large pump/signal beam ratios) are used. Finally, when ac square-wave fields are used, the theoretical gain value is only obtained for sharp transitions in the waveform.[139,140]

Recently a new form of electric field enhancement has been demonstrated in iron-doped InP.[141,142] This material is unique among the semiconductors in that the operating temperature and incident intensity can be chosen so that the photoconductivity (dominated by holes) exactly equals the dark conductivity (dominated by electrons). In this case, an applied field will enhance the amplitude of the space charge field, while maintaining the ideal spatial phase of 90°. Gain coefficients as high as $11 \, cm^{-1}$ have been reported in InP:Fe at $1.06 \, \mu m$ using this technique.[142]

In early research on the photorefractive semiconductors, the wavelengths of operation were determined by available laser sources. Thus, all early experiments were performed at $1.06 \, \mu m$ (Nd:YAG), ~$1.3 \, \mu m$ (Nd:YAG or laser diode), and ~$1.5 \, \mu m$ (laser diode). Later, the He-Ne laser (633 nm) was used to study wider band-gap materials. As the Ti:sapphire laser became available, interest turned to investigating the wavelength variation of photorefractive properties, especially near the band edge. Near the band edge, a new nonlinear mechanism contributes to the refractive index change: the Franz-Keldysh effect.[143] In this case, the internal space charge field develops as before. This field slightly shifts the band edge, leading to characteristic electroabsorption and electrorefraction. These effects can be quite large at wavelengths near the band edge, where the background absorption is also high. However, the peak of the electrorefraction spectrum is shifted slightly to longer wavelengths, where the background absorption is smaller. This is generally the wavelength region where these effects are studied.

The electrorefractive photorefractive (ERPR) effect has different symmetry properties than the conventional electro-optic photorefractive effect. It is thus possible to arrange an experiment so that only the ERPR effect contributes, or both effects contribute to the gain. In addition, the ERPR effect is quadratic in applied electric field. Thus, energy transfer between two writing beams only occurs when a dc field is present. The direction of energy transfer is determined by the sign of the electric field; this allows switching energy between two output beams via switching of the sign of the applied field.

In the first report of the band-edge photorefractive effect,[143] a gain coefficient of $7.6 \, cm^{-1}$ was measured in GaAs:EL2 at 922 nm, for a field of 10 kV/cm. In this case, both nonlinear mechanisms contributed to the gain. When a moving grating was used to optimize the spatial phase of the grating, the gain coefficient increased to $16.3 \, cm^{-1}$.

In InP the temperature/intensity resonance can be used to optimize the spatial phase, thus eliminating the need for a moving grating. In the first experiment using band-edge resonance and temperature stabilization, gain coefficients approaching $20 \, cm^{-1}$ were measured in InP:Fe (see Fig. 8).[144] Later experiments on a thin sample using a beam ratio of 10^6 resulted in a measured gain coefficient of $31 \, cm^{-1}$.[145]

The photorefractive effect can also be used to measure basic materials properties of electro-optic semiconductors, without the need for electrical contacts.[146-148] Quantities which can be measured include the populations of filled and empty traps and the mobility-recombination time product. One particular feature of the photorefractive technique is the ability to map properties across a wafer.[148]

Multiple Quantum Wells

While the enhancement of the electro-optic effect near the band edge of bulk semiconductors is significant, much larger nonlinearities are obtained at wavelengths near prominent band-edge exciton features in multiple quantum wells (MQWs). In addition, the large absorption in these structures yields much faster response times than those in bulk semiconductors. Finally, the small device thickness of typical MQW structures (typically 1 to $2 \, \mu m$) provides improved performance of Fourier plane processors such as optical correlators.[149,150] One disadvantage of the small device thickness is that diffraction from gratings in these devices is in the Raman-Nath regime, yielding multiple diffraction orders.

In their early stages of development, MQWs were not optimized for photorefractive applications because of the absence of deep traps and the large background conductivity

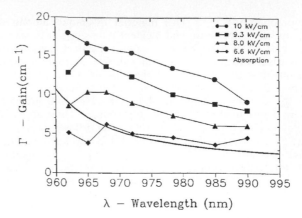

FIGURE 8 Measured gain coefficient as a function of wavelength in InP:Fe, for a grating period of 5 μm and four values of applied dc field. The beam rato was 1000, and the intensity was adjusted at each point to produce the maximum gain. The background absorption coefficient is also plotted.

within the plane of the structure. It was later recognized that defects resulting from ion implantation can provide the required traps and increase the resistivity of the structure.

The first photorefractive MQWs were GaAs/AlGaAs structures which were proton-implanted for high resistivity ($\rho = 10^9$/ohm-cm).[151,152] Two device geometries were considered (see Fig. 9), but only devices with applied fields parallel to the layers were studied. The principles of operation are initially the same for both device geometries. When two incident waves interfere in the sample, the spatially modulated intensity screens the applied field in direct proportion to the intensity, leading to a spatially modulated internal field. This spatially modulated field induces changes in both the refractive index and the absorption coefficient. The mechanism for these changes[152] is field ionization of excitons (Franz-Keldysh effect) in the parallel geometry and the quantum-confined Stark effect in the perpendicular geometry.

The magnitudes of the change in refractive index and absorption coefficient are strongly dependent on wavelength near the characteristic exciton peak. Both the index and the absorption grating contribute to the diffraction efficiency through the relationship

$$\eta = (2\pi \, \Delta n L / \lambda)^2 + (\Delta \alpha L / 2)^2 \tag{21}$$

where Δn and $\Delta \alpha$ are the amplitudes of the index and absorption gratings, respectively, and L is the device thickness. Although the device thickness of typical MQWs is much less than the thickness of a typical bulk sample (by 3 to 4 orders of magnitude), the values of Δn can be made larger, leading to practical values of diffraction efficiency (see following).

The first III-V MQWs using the parallel geometry[151,152] had rather small values of diffraction efficiency (10^{-5}). In later experiments, a diffraction efficiency of 3×10^{-4} and a gain coefficient of 1000 cm^{-1} were observed.[153] Still higher values of diffraction efficiency (on the order of 1.3 percent) were obtained using the perpendicular geometry in CdZnTe/ZnTe MQWs.[154] These II-VI MQWs have the added feature of allowing operation at wavelengths in the visible spectral region, in this case 596 nm.

In recent work on GaAs/AlGaAs MQWs in the perpendicular geometry, several device improvements were introduced.[155] First, Cr-doping was used to make the structure semi-insulating, thus eliminating the added implantation procedure and allowing separate control of each layer. Second, the barrier thickness and Al ratio were adjusted to give a reduced carrier escape time, leading to a larger diffraction efficiency. In these samples, a

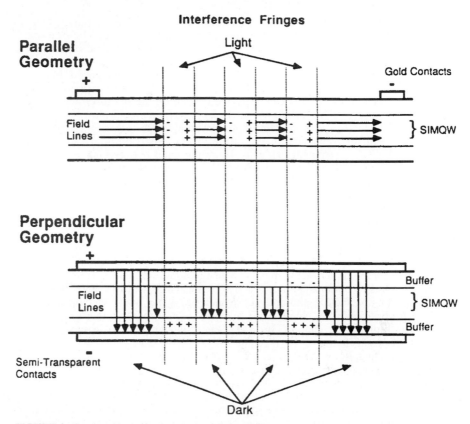

FIGURE 9 Device geometries for photorefractive MQWs.

diffraction efficiency of 3 percent was observed at 850 nm for an applied voltage of 20 V across a 2-μm-thick device. The response time was 2 μs at an intensity of 0.28 W/cm², corresponding to a very low write energy of 0.56 μJ/cm². The diffraction efficiency cited here was obtained at a grating period of 30 μm. For smaller values of grating period, the diffraction efficiency was smaller, due to charge smearing effects. The fast response time and small thickness of these structures make them ideal candidates for Fourier plane processors such as optical correlators. Competing bulk semiconductors or spatial light modulators have frame rates which are 2 to 3 orders of magnitude below the potential frame rate of ~10^6 sec⁻¹ which is available from photorefractive MQWs.

Future work on photorefractive MQWs would include efforts to grow thicker devices (so as to reduce the diffraction into higher grating orders) and to improve the diffraction efficiency at high spatial frequencies.

Organic Crystals and Polymer Films

In recent years, there has been increasing interest in the development of organic crystals for frequency conversion applications and polymer films for electro-optic waveguide devices. These materials are, in general, simpler to produce than their inorganic counterparts. In addition, the second-order nonlinear coefficients in these materials can be quite large, with values comparable for those of the well-known inorganic material

$LiNbO_3$. In most organic materials the nonlinearity is purely electronic in origin, and results from an extended system of π electrons produced by electron donor and acceptor groups. For a purely electronic nonlinearity, the dielectric constant ε is just the square of the refractive index, leading to much smaller values of ε than those in inorganic crystals. Thus, the electro-optic figure of merit $n_b^3 r/\varepsilon$ is enhanced in organic materials. This enhancement makes organic materials very appealing for photorefractive applications.

The only experiments reported to date on the photorefractive effect in organic crystals are those of Sutter et al.[156,157] on 2-cyclooctylamino-5-nitropyridine (COANP) doped with the electron acceptor 7,7,8,8,-tetracyanoquinodimethane (TCNQ). Pure COANP crystals (used for frequency doubling) are yellow, whereas the TCNQ-doped samples are green, due to a prominent extrinsic absorption band between 600 and 700 nm. In experiments at 676 nm with a grating period of 1.2 μm, both absorption and refractive index gratings were observed. Typical diffraction efficiencies were 0.1 percent, with a corresponding refractive index grating amplitude of $\sim 10^{-6}$. The recorded buildup times of the index gratings were on the order of 30 to 50 minutes at $3.2\ W/cm^2$. Clearly, the dopant species and concentration need to be optimized to provide greater photoconductivity, and thus faster response times.

The photorefractive effect in organic polymer films has also recently been reported.[158,159] From a materials science point of view, polymer films are much easier to prepare than organic single crystals. In addition, there is greater flexibility in modifying the films to optimize photorefractive performance. In this respect, there are four requirements for an efficient photorefractive material: (1) a linear electro-optic effect, (2) a source of photoionizable charges, (3) a means of transporting these charges, and (4) a means of trapping the charges. In an organic polymer, each of these functions can be separately optimized.

In spite of the large EO coefficients and the great flexibility in the materials engineering of organic polymers, there are also some practical problems which need to be addressed. First, polymers are most easily prepared as thin films. If propagation in the plane of the films is desired, then extremely high optical quality and low absorption is required. If propagation through the film is desired, then the grating efficiency is reduced. As the film thickness is made larger to enhance efficiency, the quality of the films is harder to maintain.

One important requirement of polymer films is that they must be poled to allow a linear electro-optic effect. If the poling voltage is applied normal to the film plane, then there is no electro-optic effect for light diffracted from gratings written with their normal in the plane of the film. In a practical sense, this means that the writing beams must enter the film at large angles to its normal. In addition, it becomes more difficult to provide large poling fields as the film thickness increases.

The first polymeric photorefractive material[158,159] was composed of the epoxy polymer bisphenol-A-diglycidylether 4-nitro-1,2-phenylenediamine (bisA-NPDA) made photo-conductive by doping with the hole transport agent diethylamino-benzaldehyde diphenyl-hydrazone (DEH). In this case, the polymer provides the nonlinearity leading to the electro-optic effect, as well as a mechanism for charge generation. The dopant provides a means for charge transport, while trapping is provided by intrinsic defects.

Films of this material with thicknesses between 200 and 500 μm were prepared. The material was not cross-linked, so a large field was required at room temperature to maintain the polarization of the sample. For an applied field of 120 kV/cm, the measured value of the electro-optic figure of merit $n_b^3 r$ was 1.4 pm/V. Using interference fringes with a spacing of 1.6 μm oriented 25° from the film plane, the measured grating efficiency at 647 nm was $\sim 2 \times 10^{-5}$. The grating buildup time was on the order of 100 seconds at an intensity of $25\ W/cm^2$. Analysis of the data showed that the photorefractive trap density had the relatively small value of $2 \times 10^{15}\ cm^{-3}$. In spite of the low value of trap density, relatively large values of space charge electric field were obtained, due to the low value of dielectric constant. In later experiments on related polymer films,[160] a grating buildup time of 1 sec was measured at an intensity of $1\ W/cm^2$.

It is clear from these efforts that considerable further improvements in materials properties will be required before organic polymers can compete with inorganic photorefractive materials in most applications. On the other hand, the materials tested to date are far from optimized, and the versatility of polymers may allow the engineering of optimized materials to meet future requirements.

39.3 DEVICES

Real-time Holography

The real-time phase holograms produced by the photorefractive effect can be used to perform any of the functions of regular holograms. In fact, many of the first applications of photorefractive nonlinear optics were replications of experiments and ideas first introduced in the 1960s in terms of the then new static holography. These include distortion correction by phase conjugation and one-way imaging through distortions, holographic interferometry for nondestructive testing, vibration mode visualization, and pattern recognition by matched filtering. A concise overview of these applications may be found in Goodman's classic text.[161] Their photorefractive realizations are well described in Huignard and Gunter.[1] Some of the advantages of photorefractive holography over conventional holography are:

- Photorefractive holograms are volume phase holograms. This results in high diffraction efficiencies.
- Photorefractive holograms are self-developing.
- Photorefractive holograms diffract light from the writing beams into each other during the writing process. This gives rise to a dynamic feedback process in which the grating and writing beams influence each other.
- Photorefractive holograms adapt to changing optical fields.

It is the last two features that most strongly differentiate the photorefractive effect from regular holography.

Pattern Recognition. As an example of the transfer of applications of conventional holography to photorefractive nonlinear optics we consider the case of pattern recognition by matched filtering.

A matched filter for pattern recognition is simply a Fourier transform hologram of the desired impulse response. It can be made in real time in a photorefractive crystal, a lens being used to produce the Fourier transform. The best-known system is in fact a nonlinear optical triple processor based on four-wave mixing[162] (Fig. 10). For a phase conjugate mirror, the coupled-wave equation for beam 3 which is the phase conjugate of beam 4 pumped by beams 1 and 2 is:

$$\frac{dA_3}{dz} = \gamma \frac{(A_1 A_4^* + A_2^* A_3)}{I_0} A_2$$

$$= \frac{\gamma}{I_0} I_2 A_3 + \frac{\gamma}{I_0} A_1 A_2 A_4^* \qquad (22)$$

where the A_j and I_j are the amplitudes and intensities of beams j respectively and I_0 is the total intensity of the interacting beams. The first term on the right-hand side simply corresponds to amplification of beam 3. The second term is the source for beam 3. Thus, the amplitude of beam 3 is proportional to the product $A_1 A_2 A_4^*/I_0$. As depicted in Fig. 10,

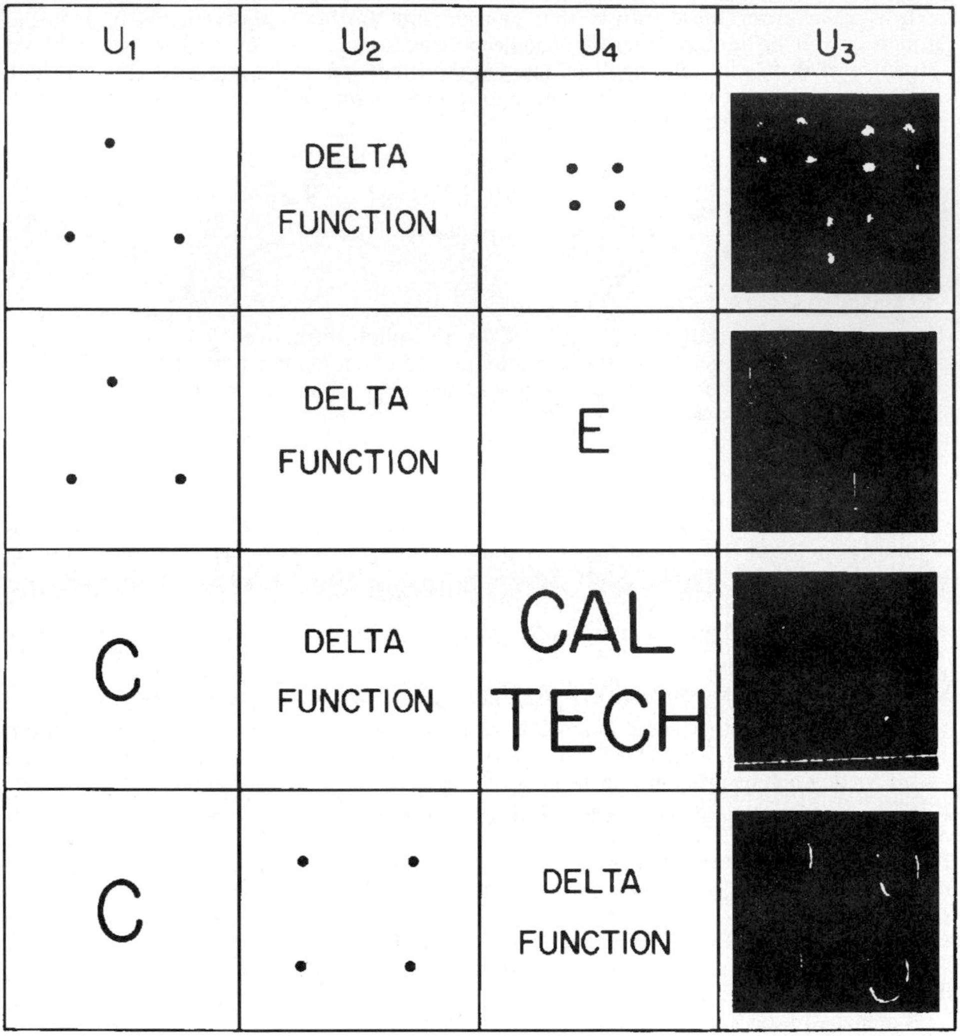

U_1	U_2	U_4	U_3
(image: dots)	DELTA FUNCTION	(image: four dots)	(image)
(image: dots)	DELTA FUNCTION	E	(image)
C	DELTA FUNCTION	CAL TECH	(image)
C	(image: four dots)	DELTA FUNCTION	(image)

(a)

FIGURE 10 (*a*) Results demonstrating real-time spatial convolution and correlation of two-dimensional images. The input fields are labeled $\mathbf{E}_1$, $\mathbf{E}_2$, and $\mathbf{E}_p$; the output is labeled $\mathbf{E}_c$. (*After White and Yariv.*[162])

A_1, A_2, and A_4 are the Fourier transforms of the spatially varying input fields a_1, a_2, and a_4. The output a_3 is proportional to the inverse Fourier transform of the product of the three Fourier transforms $A_1 A_2 A_4^* / I_0$. If I_0 is spatially constant the output is then none other than beam 1 convolved with beam 2 correlated with beam 4 ($a_3 \propto a_1 \otimes a_2 * a_4$) where $*$ represents the spatial correlation operation and $\otimes$ represents the convolution operation, all produced in real time. Modified filters such as phase-only filters can be produced by taking advantage of energy transfer during the filter writing process,[163–165] or by saturation induced by the presence of the possibly spatially varying intensity denominator I_0.[166–168]

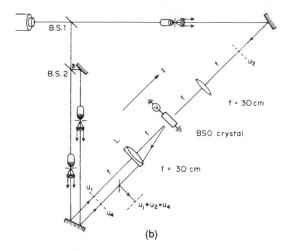

(b)

FIGURE 10 (b) Experimental apparatus for performing spatial convolution and correlation using four-wave mixing in photorefractive bismuth silicon oxide. Input and output planes are shown by dashed lines. (After White and Yariv.[162])

Sometimes, applications directly transferred from static holography inspire further developments made possible by exploitation of the physical mechanisms involved in the photorefractive effect. For example, acoustic signals can be temporally correlated with optical signals. An acoustic signal applied to a photorefractive crystal induces piezoelectric fields. If the crystal is illuminated by a temporally varying optical signal, then the photorefractive space charge generated by the photocurrent will be proportional to the product of the time-varying photoconductivity and the time-varying piezoelectric fields. Such correlators can be used to make photorefractive tapped delay lines with tap weights proportional to correlation values.[169–171] It is also possible to make acoustic filters which detect Bragg-matched retroreflection of acoustic waves from a photorefractive grating in a crystal such as lithium niobate that has low acoustic loss.[172]

Applications of Photorefractive Gain in Two-beam Coupling

Coherent Image Amplification. Coherent image amplification is especially important for coherent optical processors. Without it, the losses introduced by successive filtering operations would soon become intolerable. Practical considerations include maintenance of signal-to-noise ratio and amplification fidelity. The main contribution to noise introduced by photorefractive amplifiers is from the fanning effect. Although it can be reduced in a given crystal in a variety of different ways, such as by crystal rotation[21] and multiwavelength recording,[22] by far the best approach to this problem would be to undertake research to grow cleaner crystals. Amplification fidelity is determined by gain uniformity. In the spatial frequency domain, it is limited because photorefractive gain depends on the grating period. Images with a high spatial bandwidth write holograms with a wide range of spatial frequencies and grating periods. Optimal uniformity is obtained for reflection gratings, in which the image beam counterpropagates with respect to the pump. In that case, the change in grating period depends only to second order on the image spatial frequency. Any remaining first-order nonuniformity is due to the angular dependence of the effective electro-optic coefficient. In the space domain, gain uniformity is limited by pump depletion so that the most accurate results will be obtained if the pump beam is strong enough that it is not significantly depleted in the interaction.

Two-beam coupling amplification can also be used for beam cleanup:[173] a badly distorted beam, say from a laser diode, can be converted to a gaussian beam. A small sample of the beam is split off, spatially filtered and amplified in two-beam coupling by the remaining bulk of the distorted beam. The efficiency of the method can be quite high: fidelity limitations due to spatial variations in gain are usually quite small and can be removed by a second round of spatial filtering. A unidirectional ring resonator with an intracavity spatial filter can also be used for beam cleanup.[174]

Laser Power Combining.   An application related to two-beam coupling image amplification is coherent power combining, which would be especially useful for semiconductor lasers. The output of a single laser gain stripe is currently limited to values of the order of a few hundred milliwatts. Some applications require diffraction-limited beams containing many watts produced at high efficiency. Such a source can in principle be made by using two-beam coupling amplification of a diffraction-limited seed by the mutually injection locked outputs of many diode stripes. The injection locking can be achieved by evanescent coupling between laser gain stripes[175,176] or by retroreflecting a portion of the amplified beam with a partially transmitting mirror.[177] Another possibility involves forming a double phase conjugate mirror[178,179] or ring self-pumped phase conjugate mirror with a master laser providing one input, and the light from the gain elements loosely focused into the crystal providing the other inputs. Such a system will be self-aligning and will correct intracavity distortions by phase conjugation. Phase conjugate master oscillator/power amplifiers have also been used with some success.[180] Practical problems include the need to control the spectral effects of the associated multiple coupled cavities. Reference 181 gives an excellent exposition of these problems. Also, while self-pumped phase conjugate reflectivities and two-beam coupling efficiencies can theoretically approach 100 percent, in practice these efficiencies rarely exceed 80 percent. Among oxide ferroelectrics, barium titanate exhibits high gain at GaAs laser wavelengths. However, while some bulk semiconductors such as InP:Fe and CdTe:V are sensitive in the 1.3- to 1.5-μm wavelength range of interest for optical fiber communications, high gain requires the application of high electric fields.

Optical Interconnects.   Use of the double phase conjugate mirror for laser locking suggests another application. The beam-coupling crystal can be viewed as a device that provides optical interconnection of the laser gain elements to each other.[182] The basic idea exists in the realm of static holography in terms of computer interconnection by holographic optical elements[183,184] (HOEs). The use of photorefractive crystals should enable the construction of reprogrammable interconnects that would be self-aligning if phase conjugation were used:[185] the laser gain elements in the power combining case can be imagined as the input/output ports of an electronic chip.[186] Another way to go about the interconnection problem is to design in terms of an optical crossbar switch, or matrix vector multiplier.[187,188] The vector is an array of laser diode sources and the matrix describes connection patterns of the sources to a vector array of detectors (Fig. 11). The interconnection matrix is realized as a photorefractive hologram.

Applications of Photorefractive Loss in Two-beam Coupling

If the sign of the coupling constant is reversed (for example, by rotation of the uniaxial crystal by 180° so that the direction of the optic axis is reversed) the pump interferes destructively with the signal so that the output is reduced to a very low level. This resulting photorefractive loss can be used in a number of applications such as for optical limiters, optical bistability,[189] and novelty filters and achieved using a variety of different devices such as ring resonators and phenomena such as beam fanning.

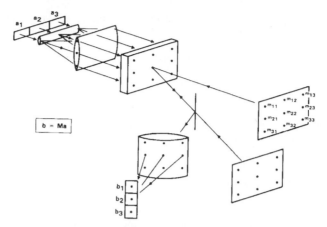

FIGURE 11 Schematic drawing of the basic principal of optical matrix vector multiplication through four-wave mixing in nonlinear media. Light from a linear source array a is fanned out by a cylindrical lens where it diffracts from an image plane hologram of the matrix $\mathbf{M}$ to produce a set of beams bearing the required products $\mathbf{M}_{ij}a_j$. A second cylindrical lens sums the diffracted beams to form $b = \mathbf{M}a$. (*After Yeh and Chiou.*[188])

Optical Limiters. The process of beam depletion in photorefractive materials forms the basis of their use as optical limiters. For this application, another important property of photorefractive materials is their ability to respond selectively to coherent optical inputs. Any portion of the input which is temporally incoherent is transmitted through a photorefractive material in a linear manner. Thus, photorefractive limiters will also selectively attenuate (or *excise*) a coherent beam while transmitting an incoherent beam; these devices have thus also been referred to as excisors.

The first studies of photorefractive limiters were published in 1985.[25] A number of device architectures involving resonators and self-pumped phase conjugate mirrors were discussed, but the primary emphasis was on beam fanning. Several features of limiters or excisors using beam fanning were pointed out: (1) the device design is very simple, with a single input beam and no separate external paths required; (2) the limiting mechanism is due to scattering (and not absorption), so that added heating is not present; (3) the device will operate at one or more wavelengths within the bandwidth of its photorefractive response, which (for ferroelectric oxides) extends over the entire visible band; and (4) the device will respond to sources with a relatively small coherence length, including mode-locked lasers producing picosecond pulses.

Experiments at 488 nm using $BaTiO_3$ in the beam fanning geometry[25] produced a steady-state device transmission of 2.5 percent, and a response time of 1.1 sec at 1 W/cm^2. Using the measured response time and intensity, we note that 1.1 J/cm^2 will pass through the device before it fully activates. In later measurements in the beam fanning geometry, attenuation values exceeding 30 dB and device activation energies as low as 1 to 10 mJ/cm^2 have been measured.

Two-beam coupling amplification of a second beam produced by beam splitters[190,191] or gratings in contact with the crystal[192,193] has also been studied as a mechanism for optical limiting. This mechanism is closely related to fanning, with the only difference that the second beam in a fanning device is produced internally by scattering.

Novelty Filters. Novelty filters are devices whose output consists of only the changing part of the input. The photorefractive effect can be used to realize the novelty filter operation in several different ways. The simplest way is to use two-beam coupling for image deamplification as was used in the fanning and two-beam coupling optical limiters. In that case, the pump interferes destructively with the signal so that the output is reduced to a very low level. Now if the signal suddenly changes, the output will be the difference between the new input signal and the reconstruction of the old signal by diffraction of the pump from the old grating. Thus, the output will show the changed parts of the scene until the grating adapts during the photorefractive response time to the new scene.[194,195] Such interferometers have been used, for example, to map turbulent flow,[196,197] to make photothermal measurements,[198] and to build acoustic spectrum analyzers.[199]

Phase Conjugate Interferometry

Another way to produce a novelty filter is to use a phase conjugate interferometer.[200,201] This is an interferometer in which some or all of the conventional mirrors are replaced by phase conjugate mirrors, thus achieving the benefit of self-alignment. The effects of phase objects inserted in the interferometer are canceled out by phase conjugation. One of the most common realizations is a phase conjugate Michelson interferometer (Figs. 12 and 13). If the phase conjugate mirrors have common pumping beams (this can be achieved by illuminating the same self-pumped phase conjugator with the beams in both arms), the phase of reflection for both beams will be the same and there will be a null at the output from the second port of the device. If a phase object inserted in one of the arms suddenly changes, the null will be disturbed, and the nonzero output will represent the changing parts of the phase object. The nonzero output persists until the gratings in the phase conjugate mirror adapt to the new input fields.[202] If two amplitude objects are inserted in the interferometer, one in each arm, the intensity of the output at the nulling port is the square of the difference between the squared moduli of the objects. This architecture thus gives rise to image subtraction.[203,204] A slightly modified version can be used to measure thin-film properties (refractive index, absorption coefficient, and thickness): the film under test on its substrate is used as the interferometer beam splitter.[205]

Associative Memories and Neural Networks

There are a number of adaptive processors that can be designed using photorefractive beam coupling. In addition to novelty filters, these include associative memories, neural network models, and other recursive image processors.

Photorefractive phase conjugate mirrors provide an elegant way to realize linear associative memories in which a fragment of an image can be used to recall the entire image from a bank of multiplexed holograms stored in a long-term storage photorefractive crystal such as lithium niobate.[206,207] Neural networks, on the other hand, are nets of interconnected signals with nonlinear feedback. They are currently under extensive investigation in the artificial intelligence community.[208] Typical applications are modeling of neural and cognitive systems and the construction of classification machines. As we have seen, there are a number of different ways to realize optical interconnections using the photorefractive effect, and nonlinear feedback can be introduced by using the nonlinear transfer properties of pumped photorefractive crystals. A pattern classifier can be built by recording a hologram for each image class (e.g. represented by a clear fingerprint) in a photorefractive crystal with a long time constant. Many holograms can be superposed if they are recorded with spatially orthogonal reference beams. As in the case of the linear associative memory, a smudged fingerprint introduced to the system will partially reconstruct each of the reference beams. The brightest reconstruction will be the reference

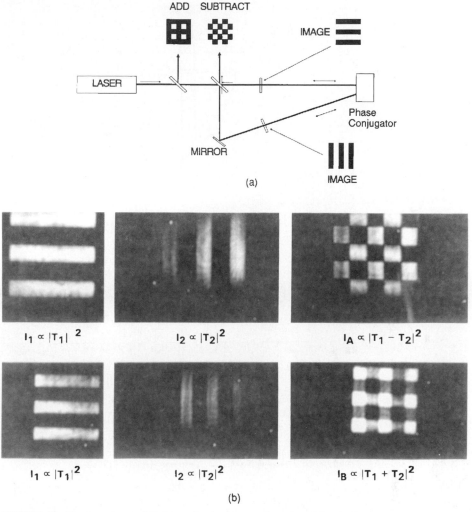

FIGURE 12 Phase conjugate interferometer for image subtraction: (*a*) amplitude images are placed in the interferometer arms. Their difference appears at the nulling output, the sum appears at the retroreflection output. (*b*) Real-time image subtraction and addition of images using above apparatus. (*After Chiou and Yeh.*[204])

associated with the fingerprint most like the smudged input. An oscillator with internal saturable absorption is built to provide competitive feedback of the reference reconstructions to themselves. The oscillator mode should be the mode associated with the proper fingerprint. That mode then reconstructs the clear fingerprint. The output of the device is the stored fingerprint which is most like the smudged input. We have described just one optical neural network model, but just as there are many theoretical neural network models, there are also many optical neural network models.[209–213] Each of them has its own practical difficulties, including those of reliability, suitability of available threshold functions, and stability.

In addition, the optical gain of photorefractive oscillators makes possible the design of

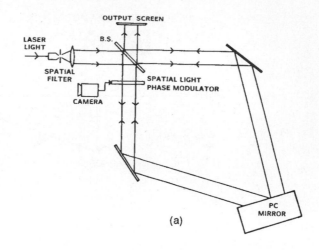

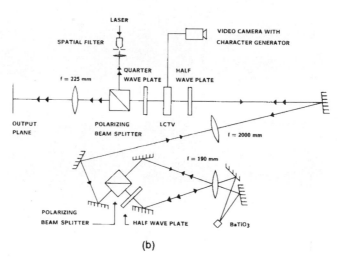

(b)

FIGURE 13 Phase conjugate interferometer as a novelty filter: any phase change in the object arm disturbs the null at the output until the phase conjugate mirror adjusts to the change. (*a*) Optical tracking novelty filter incorporating a spatial phase modulator. BS, beam splitter; PC, phase conjugate mirror. (*b*) Modification of preceding device to enable the use of a polarization modulating liquid crystal television (LCTV).

other recursive image processors, for example, to realize Gerschberg-Saxton-type algorithms in phase conjugate resonators.[214]

Thresholding

In optical data processing it is often necessary to determine if one or more elements of an optical pattern has an intensity above (or below) a set threshold value. For example, in optical associative memory applications, it is necessary to select the stronger modes among many in an optical resonator. In optical correlator applications, the output information

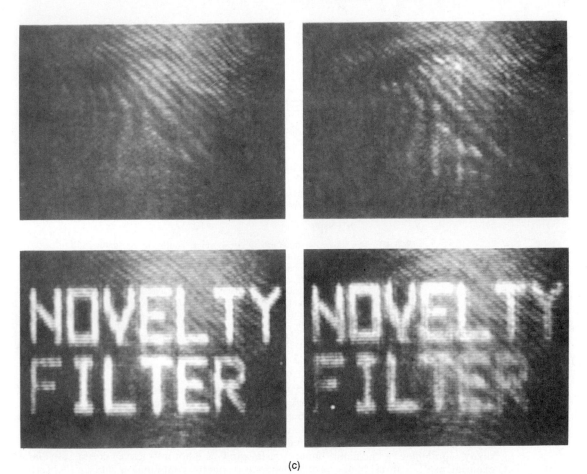

(c)

FIGURE 13 (c) Photograph of the output of the tracking novelty filter shown in (b). Input to the LCTV is taken from a character generator driving a video camera: (1) the character generator is off; the interferometer is essentially dark; (2) the character-generator display, showing the phrase NOVELTY FILTER is activated; (3) the filter adapts to the new scene and becomes nearly blank, as in (1). Some letters are visible; (4) the character-generator display is deactivated. The previous phrase appears at the output of the interferometer. Shortly thereafter it fades to (1). (*After Anderson, Lininger, and Feinberg.*[202])

plane may be thresholded to determine whether a correlation has been obtained, and to determine the location of the correlation peak(s). A closely related operation which is also useful for these applications is the Max or winner-take-all operation.

The ideal thresholding device should have the following properties: (1) the capability to process complex images with high resolution, (2) high sensitivity, (3) low crosstalk between pixels, (4) a sharp threshold, with constant output for intensities above the threshold value, (5) large signal-to-noise, and (6) the ability to control the threshold level by external means.

A large number of thresholding schemes using the photorefractive effect have been proposed or demonstrated, although only a few experimental demonstrations of thresholding of spatial patterns or images have been reported. Techniques used in early investigations include (1) uniform incoherent erasure of self-pumped phase conjugate mirrors[215] and of photorefractive end mirrors in phase conjugate resonators[216,217] and (2)

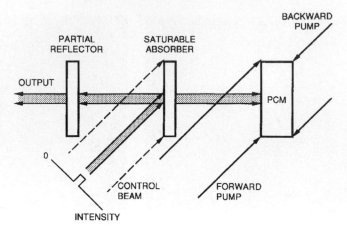

FIGURE 14 Schematic diagram of linear phase conjugate resonator containing an intracavity saturable absorber. Information is read into the resonator by means of a separate control beam incident on the saturable absorber. The control beam can be brought in through a beam splitter, or at an angle to the optic axis (as shown). The fluorescein-doped boric acid glass saturable absorber had a saturation intensity ($I_s = 20\,\mathrm{mW/cm^2}$) at its absorption peak (450 nm). The BaTiO$_3$ phase conjugate mirror was pumped at the neighboring Ar laser wavelength of 458 nm. The control beam was generated from the same laser, and cross-polarized to a void feedback into the resonator. (*After McCahon et al.*[220])

pump depletion in externally pumped phase conjugate mirrors and double phase conjugate mirrors.[55] A number of thresholding devices using ring resonators have also been demonstrated.[55,218] One way of increasing the sharpness of the threshold is to insert additional nonlinear media in the photorefractive resonator cavities.

Recently, Ingold et al.[219] have demonstrated winner-take-all behavior in a nonresonant cavity containing a nematic liquid crystal and a photorefractive crystal. In later experiments,[220] a thresholding phase conjugate resonator containing a saturable absorber consisting of a thin film of fluorescein-doped boric acid glass demonstrated with several performance improvements. In this architecture (shown schematically in Fig. 14) the input image was amplitude-encoded as a two-dimensional array of pixels on an incoherent control beam that was incident on the saturable absorber (a single pixel is shown in Fig. 14). If the intensity at a given pixel was above a threshold intensity, the saturable absorber bleached locally by an amount sufficient to switch on the phase conjugate resonator. The phase conjugate resonator continued to oscillate at these pixel locations even when the control beam was removed. The output was thus bistable and latching.

Photorefractive Holographic Storage

The neural networks described above rely on optical information storage in a material whose gratings are long-lived. One of the earliest potential applications for the photorefractive effect was for holographic information storage. One of the best materials for the purpose was and still is lithium niobate; it has storage times which can be as long as years. The principal concerns for holographic storage are information density, crosstalk minimization, fixing efficiency, signal-to-noise ratio, and development of practical readout/writing architectures. There has been a recent renewed interest in optical memory design, now that materials and computer technology have improved and schemes for rapid addressing have matured. The two main classes of addressing scheme use spatially orthogonal reference beams[221-223] and temporally orthogonal reference beams[224] (spatial multiplexing vs. frequency multiplexing). Spatial multiplexing has the advantage that relatively simple optical sources can be used. Frequency multiplexing has the advantage that spatial crosstalk is reduced compared to that associated with spatial multiplexing. However, it has the drawback of requiring frequency-tunable laser sources. An important consideration is the need to pack as much information as possible into each individual hologram. Here the frequency multiplexing approach is superior, because the information density can reach much higher values before crosstalk sets in.[64] Crosstalk can also be reduced by storing information in photorefractive fiber bundles instead of bulk, so that the information is more localized.[225,226]

Photorefractive Waveguides

There are three essentially different ways to prepare electro-optic (possibly photorefractive) waveguides. One is to produce local alterations in the chemistry of an electro-optic substrate, for example, by titanium in-diffusion into $LiNbO_3$ or ion implantation in B_0TiO_3.[227] Another way is to grow waveguides in layers by techniques such as RF sputtering, liquid phase epitaxy, laser ablation,[228] and metalorganic chemical vapor deposition (MOCVD).[229] A third way is to polish bulk material down to a thin wafer.[230]

The performance of electro-optic waveguide devices such as couplers and switches can be seriously compromised by the refractive index changes induced by the photorefractive effect in the host electro-optic materials such as lithium niobate. Therefore, one of the main motivations for understanding the photorefractive effect in waveguides is to develop ways to minimize its effects. MgO doping of lithium niobate is commonly used in attempts to reduce the photorefractive effect.[231]

Some researchers have taken advantage of waveguide photorefractivity. Optical confinement in waveguides enhances the effectiveness of optical nonlinearities. With conventional $\chi^{(3)}$ materials, waveguiding confinement increases the coupling constant-length product by maintaining high intensity over longer distances than would be possible in bulk interactions. In the photorefractive case, optical confinement reduces the response time.

An excellent review of earlier work on photorefractivity in waveguides may be found in Ref. 232. More recently Edson and co-workers have measured a response time improvement by a factor of 100 in an ion-implanted $BaTiO_3$ waveguide.[227] A bridge mutually pumped phase conjugator was also demonstrated.[233].

39.4 REFERENCES

1. P. Gunter and J.-P. Huignard (eds.), *Photorefractive Materials and their Applications I, Fundamental Phenomena,* Springer-Verlag, Berlin, 1988,: P. Gunter and J.-P. Huignard (eds.), *Photorefractive Materials and their Applications II. Survey of Applications.*
2. A. M. Glass, D. von der Linde, T. J. Negran, *Appl. Phys. Lett.* **25:**233 (1974).
3. V. L. Vinetskii and N. V. Khuktarev, *Sov. Phys. Solid State* **16:**2414 (1975).
4. F. P. Strohkendl, J. M. C. Jonathan, and R. W. Hellwarth, *Opt. Lett.* **11:**312 (1986).
5. N. V. Kukhtarev, G. E. Dovgalenko, and V. N. Starkov, *Appl. Phys. A* **33:**227 (1984).
6. G. C. Valley, *Appl. Opt.* **22:**3160 (1983).
7. N. V. Khuktarev, V. B. Markov, S. G. Odoulov, M. S. Soskin, and V. L. Vinetskii, *Ferroelectrics* **22:**949, 961 (1979).
8. B. Belinicher and B. Sturman, *Usp. Fiz. Nauk.* **130:**415 (1980).
9. G. C. Valley, *IEEE J. Quantum Electron.* **QE19:**1637 (1983).

10. J. P. Huignard and J. P. Herriau, *Appl. Opt.* **24:**4285 (1985).

11. S. I. Stepanov and M. P. Petrov, *Sov. Tech. Phys. Lett.* **10:**572 (1984).

12. M. Ziari, W. H. Steier, P. M. Ranon, M. B. Klein and S. Trivedi, *J. Opt. Soc. B* **9:**1461 (1992).

13. B. Fischer, M. Cronin-Golomb, J. O. White, and A. Yariv, *Opt. Lett.* **6:**519 (1981).

14. D. W. Vahey, *J. Appl. Phys.* **46:**3510 (1975).

15. J. Feinberg and K. R. MacDonald, "Phase Conjugate Mirrors and Resonators with Photorefractive Materials," in P. Gunter and J. P. Huignard (eds.), *Photorefractive Materials and their Applications II, Survey of Applications,* Springer Verlag, Berlin, 1989.

16. M. D. Ewbank, R. A. Vasquez, R. S. Cudney, G. D. Bacher, and J. Feinberg, Paper FS1, *Technical Digest, 1990 OSA Annual Meeting,* Boston, MA, Nov. 4–9 (1990).

17. R. A. Vasquez, R. R. Neurgaonkar, and M. D. Ewbank, *J. Opt. Soc. Am.,* **89:**1416 (1992).

18. V. V. Voronov, I. R. Dorosh, Yu. S. Kuz'minov, and N. V. Tkachenko, *Sov. J. Quantum Electron.* **10:**1346 (1980).

19. J. Feinberg, *J. Opt. Soc. Am.* **72:**46 (1982).

20. M. Segev, Y. Ophir, and B. Fischer, *Opt. Commun.* **77:**265 (1990).

21. H. Rajbenbach, A. Delboulbe, J. P. Huignard, *Opt. Lett.* **14:**1275 (1989).

22. W. S. Rabinovich, B. J. Feldman, and G. C. Gilbreath, *Opt. Lett.* **16:**1147 (1991).

23. J. Feinberg, *Opt. Lett.* **7:**486 (1982).

24. A. Zozulya, *Sov. J. Quantum Electron.* **22:**677 (1992).

25. M. Cronin-Golomb and A. Yariv, *J. Appl. Phys.* **57:**4906 (1985).

26. J. O. White, M. Cronin-Golomb, B. Fischer, and A. Yariv, *Appl. Phys. Lett.* **40:**450 (1982).

27. S. K. Kwong, A. Yariv, M. Cronin-Golomb, and I. Ury, *Appl. Phys. Lett.* **47:**460 (1985).

28. M. D. Ewbank and P. Yeh, *Opt. Lett.* **10:**496 (1985).

29. D. Anderson and R. Saxena, *J. Opt. Soc. Am. B* **4:**164 (1987).

30. M. Cronin-Golomb, B. Fischer, J. O. White, and A. Yariv, *IEEE J. Quantum Electron.* **QE20:**12 (1984).

31. A. A. Zozulya and V. T. Tikonchuk, *Sov. J. Quantum Electron.* **18:**981 (1988).

32. D. A. Fish, A. K. Powell, and T. J. Hall, *Opt. Commun.* **88:**281 (1992).

33. W. Krolikowski, K. D. Shaw, and M. Cronin-Golomb, *J. Opt. Soc. Am.* **B6:**1828 (1989).

34. W. Krolikowski, M. Belic, M. Cronin-Golomb, and A. Bledowski, *J. Opt. Soc. Am.* **B7:**1204 (1990).

35. R. A. Rupp and F. W. Drees, *Appl. Phys. B* **39:**223 (1986).

36. D. A. Temple and C. Warde, *J. Opt. Soc. Am. B* **3:**337 (1986).

37. M. D. Ewbank, P. Yeh, and J. Feinberg, *Opt. Commun.* **59:**423 (1986).

38. E. M. Avakyan, K. G. Belabaev, V. Kh. Sarkisov, and K. M. Tumanyan, *Sov. Phys. Solid State* **25:**1887 (1983).

39. L. Holtmann, E. Kratzig, and S. Odoulov, *Appl. Phys. B* **53:**1 (1991).

40. M. Cronin-Golomb, B. Fischer, J. O. White, and A. Yariv, *Appl. Phys. Lett.* **41:**689 (1982).

41. M. Cronin-Golomb, B. Fischer, J. O. White, and A. Yariv, *Appl. Phys. Lett.* **42:**919 (1983).

42. V. A. D'yakov, S. A. Korol'kov, A. Mamaev, V. V. Shkunov, and A. A. Zozulya, *Opt. Lett.* **16:**1614 (1991)

43. S. Weiss, S. Sternklar, and B. Fischer, *Opt. Lett.* **12:**114 (1987).

44. M. D. Ewbank, R. A. Vazquez, R. R. Neurgaonkar, and J. Feinberg, *J. Opt. Soc. B* **7:**2306 (1990).

45. M. D. Ewbank, *Opt. Lett.* **13:**47 (1988).

46. D. Wang, Z. Zhang, Y. Zhu, S. Zhang, and P. Ye, *Opt. Commun.* **73:**495 (1989).

47. A. M. C. Smout and R. W. Eason, *Opt. Lett.* **12:**498 (1987).

48. B. Fischer, S. Sternklar, and S. Weiss, *IEEE J. Quantum Electron* **QE25:**550 (1989).

49. M. Cronin-Golomb, *Opt. Lett.* **15:**897 (1990).

50. A. A. Zozulya, *Opt. Lett.* **16:**2042 (1991).

51. K. D. Shaw, *Opt. Commun.* **94:**458 (1992).

52. T. Y. Chang and R. W. Hellwarth, *Opt. Lett.* **10:**408 (1985).

53. R. A. Mullen, D. J. Vickers, L. West, and D. M. Pepper, *J. Opt. Soc. Am. B* **9:**1726 (1992).

54. G. C. Valley, *J. Opt. Soc. Am. B* **9:**1440 (1992).

55. M. Horowitz, D. Kliger, and B. Fischer, *J. Opt. Soc. Am. B* **8:**2204 (1991).

56. W. Krolikowski, K. D. Shaw, M. Cronin-Golomb, and A. Bledowski, *J. Opt. Soc. Am.* **B6:**1828 (1989).

57. W. Krolikowski and M. Cronin-Golomb, *Appl. Phys. B* **52:**150 (1991).

58. K. D. Shaw, *Opt. Commun.* **97:**148 (1993).

59. D. J. Gauthier, P. Narum, and R. W. Boyd, *Phys. Rev. Lett.* **58:**1640 (1987).

60. W. Krolikowski, M. Belic, M. Cronin-Golomb, and A. Bledowski, *J. Opt. Soc. Am.* **B7:**1204 (1990).

61. F. Vacchs and P. Yeh, *J. Opt. Soc. Am. B* **6:**1834 (1989).

62. W. Krolikowski and M. Cronin-Golomb, *Opt. Commun* **89:**88 (1992).

63. J. Hong, A. E. Chiou, and P. Yeh, *Appl. Opt.* **29:**3027 (1990).

64. M. Cronin-Golomb, *Opt. Commun.* **89:**276 (1992).

65. G. C. Valley and G. J. Dunning, *Opt. Lett.* **9:**513 (1984).

66. R. Blumrich, T. Kobialka, and T. Tschudi, *J. Opt. Soc. Am. B* **7:**2299 (1990).

67. S. R. Liu and G. Indebetouw, *J. Opt. Soc. Am. B* **9:**1507 (1992).

68. F. Laeri, T. Tschudi, and J. Libers, *Opt. Comm.* **47:**387 (1983).

69. J. Feinberg and R. W. Hellwarth, *Opt. Lett.* **5:**519 (1980).

70. P. Gunter, *Phys. Reports* **93:**200 (1983).

71. D. von der Linde and A. M. Glass, *Appl. Phys.* **8:**85 (1975).

72. G. C. Valley and M. B. Klein, *Opt. Engin.* **22:**704 (1983).

73. A. Ashkin, G. D. Boyd, J. M. Dziedzic, R. G. Smith, A. A. Ballman, J. J. Levinstein, and K. Nassau, *Appl. Phys. Lett.* **9:**72 (1966).

74. F. S. Chen, *J. Appl. Phys.* **38:**3418 (1967).

75. F. S. Chen, J. T. LaMacchia, and D. B. Fraser, *Appl. Phys. Lett.* **3:**213 (1968).

76. M. B. Klein and G. C. Valley, *J. Appl. Phys.* **57:**4901 (1985).

77. M. Cronin-Golomb, K. Y. Lau, and A. Yariv, *Appl. Phys. Lett.* **47:**567 (1985).

78. R. Orlowski and E. Kratzig, *Solid St. Comm.* **27:**1351 (1978).

79. J. J. Amodei and D. L. Staebler, *Appl. Phys. Lett.* **18:**540 (1971).

80. R. L. Townsend and J. T. LaMacchia, *J. Appl. Phys.* **41:**5188 (1970).

81. R. L. Townsend and J. T. LaMacchia, *J. Appl. Phys.* **41:**5188 (1970).

82. J. Feinberg, D. Heiman, A. R. Tanguay, Jr., and R. W. Hellwarth, *J. Appl. Phys.* **51:**1297 (1980).

83. E. Kratzig, F. Welz, R. Orlowski, V. Doorman, and M. Rosenkranz, *Solid St. Comm.* **34:**817 (1980).

84. M. B. Klein, "Photorefractive Properties of $BaTiO_3$" in P. Gunter and J.-P. Huignard (eds.), *Photorefractive Materials And Their Applications,* Springer-Verlag, Berlin, 1988.

85. V. Belruss, J. Kalnajs, A. Linz, and R. C. Folweiler, *Mater. Res. Bull.* **6:**899 (1971).

86. M. B. Klein and R. N. Schwartz, *J. Opt. Soc. Am.* **B3:**293 (1986).

87. P. G. Schunemann, T. M. Pollak, Y. Yang, Y. Y. Teng, and C. Wong, *J. Opt. Soc. Am.* **B5:**1702 (1988).

88. B. A. Wechsler and M. B. Klein, *J. Opt. Soc. Am.* B **5:**1713 (1988).

89. D. Rytz, R. R. Stephens, B. A. Wechsler, M. S. Keirstad, and T. M. Baer, *Opt. Lett.* **15:**1279 (1990).

90. G. W. Ross, P. Hribek, R. W. Eason, M. H. Garrett, and D. Rytz, *Opt. Comm.* **101:**60 (1993); B. A. Wechsler, M. B. Klein, C. C. Nelson, and R. N. Schwartz, *Opt. Lett.* 19, April 15 (1994).

91. S. Ducharme and J. Feinberg, *J. Appl. Phys.* **56:**839 (1984).

92. R. A. Rupp, A. Maillard, and J. Walter, *Appl. Phys.* **A49:**259 (1989).

93. L. Holtmann, *Phys. Status Solidi* **A113:**K89 (1989).

94. D. Mahgerefteh and J. Feinberg, *Phys. Rev. Lett.* **64:**2195 (1990).

95. G. A. Brost and R. A. Motes, *Opt. Lett.* **15:**1194 (1990).

96. L. K. Lam, T. Y. Chang, J. Feinberg, and R. W. Hellwarth, *Opt. Lett.* **6:**475 (1981).

97. N. Barry and M. J. Damzen, *J. Opt. Soc.* **B9:**1488 (1992).

98. A. L. Smirl, G. C. Valley, R. A. Mullen, K. Bohnert, C. D. Mire, and T. F. Boggess, *Opt. Lett.* **12:**501 (1987).

99. G. A. Brost, R. A. Motes, and J. R. Rotge, *J. Opt. Soc. Am.* **B5:**1879 (1988).

100. P. Yeh, *Appl. Opt.* **26:**602 (1987); A. M. Glass, M. B. Klein, and G. C. Valley, *Appl. Opt.* **26:**3189 (1987).

101. E. Voit, M. Z. Zha, P. Amrein, and P. Gunter, *Appl. Phys. Lett.* **51:**2079 (1987).

102. D. Rytz, M. B. Klein, R. A. Mullen, R. N. Schwartz, G. C. Valley, and B. A. Wechsler, *Appl. Phys. Lett.* **52:**1759 (1988).

103. S. Ducharme and J. Feinberg, *J. Opt. Soc. Am.* **3:**283 (1986).

104. M. H. Garrett, J. Y. Chang, H. P. Jenssen, and C. Warde, *Opt. Lett.* **17:**103 (1992).

105. M. E. Lines and A. M. Glass, *Principles And Applications of Ferroelectrics And Related Materials,* Clarendon Press, Oxford, 1977, pp. 280–292.

106. K. Megumi, N. Nagatshuma, Y. Kashiwada, and Y. Furuhata, *J. Matls. Sci.* **11:**1583 (1976).

107. J. B. Thaxter, *Appl. Phys. Lett.* **15:**210 (1969).

108. K. Megumi, H. Kozuka, M. Kobayashi, and Y. Furuhata, *Appl. Phys. Lett.* **30:**631 (1977).

109. V. V. Voronov, I. R. Dorosh, Y. S. Kuzminov, and N. V. Tkachenko, *Sov. J. Quantum Electron.* **10:**1346 (1980).

110. R. R. Neurgaonkar, W. K. Cory, J. R. Oliver, M. D. Ewbank, and W. F. Hall, *Opt. Engin.* **26:**392 (1987).

111. G. L. Wood and R. R. Neurgaonkar, *Opt. Lett.* **17:**94 (1992).

112. R A. Vazquez, F. R. Vachss, R. R. Neurgaonkar, and M. D. Ewbank, *J. Opt. Soc. Am.* **8:**1932 (1991).

113. R. A. Vasquez, R. R. Neurgaonkar, and M. D. Ewbank, *J. Opt. Soc. Am.* **B9:**1416 (1992).

114. M. D. Ewbank, R. R. Neurgaonkar, W. K. Cory, and J. Feinberg, *J. Appl. Phys.* **62:**374 (1987).

115. K. Sayano, A. Yariv, and R. R. Neurgaonkar, *Opt Lett.* **15:**9 (1990).

116. L. Hesselink and S. Redfield, *Opt. Lett.* **13:**877 (1988).

117. B. A. Horowitz and F. J. Corbitt, *Opt. Engin.* **17:**353 (1978).

118. J. P. Huignard and F. Micheron, *Appl. Phys. Lett.* **29:**591 (1976).

119. J. P. Wilde, L. Hesselink, S. W. McCahon, M. B. Klein, D. Rytz, and B. A. Wechsler, *J. Appl. Phys.* **67:**2245 (1990).

120. S. I. Stepanov and M. P. Petrov, *Opt. Comm.* **52:**292 (1985).

121. S. L. Sochava, S. I. Stepanov, and M. P. Petrov, *Sov. Tech. Phys. Lett.* **13:**274 (1987).

122. M. P. Petrov, S. L. Sochava, and M. P. Petrov, *Opt. Lett.* **14:**284 (1989).

123. S. I. Stepanov, V. V. Kulikov, and M. P. Petrov, *Opt. Comm.* **44:**19 (1982).

124. B. Imbert, H. Rajbenbach, S. Mallick, J. P. Herriau, and J.-P. Huignard, *Opt. Lett.* **13:**327 (1988).

125. J. E. Millerd, E. M. Garmire, M. B. Klein, B. A. Wechsler, F. P. Strohkendl, and G. A. Brost, *J. Opt. Soc. Am.* **B9:**1449 (1992).

126. R. Baltrameyunas, Yu. Vaitkus, D. Veletskas, and I. Kapturauskas, *Sov. Tech Phys. Lett.* **7:**155 (1981).

127. A. M. Glass, A. M. Johnson, D. H. Olson, W. Simpson, and A. A. Ballman, *Appl. Phys. Lett.* **44:**948 (1984).

128. M. B. Klein, *Opt. Lett.* **9:**350 (1984).

129. A. Partovi, J. Millerd, E. M. Garmire, M. Ziari, W. H. Steier, S. B. Trivedi, and M. B. Klein, *Appl. Phys. Lett.* **57:**846 (1990).

130. K. Kuroda, Y. Okazaki, T. Shimura, H. Okimura, M. Chihara, M. Itoh, and I. Ogura, *Opt. Lett.* **15:**1197 (1990).

131. M. Ziari, W. H. Steier, P. M. Ranon, S. Trivedi, and M. B. Klein, *Appl. Phys. Lett.* **60:**1052 (1992).

132. A. M. Glass and J. Strait, "The Photorefractive Effect in Semiconductors," in P. Gunter and J.-P. Huignard (eds.), *Photorefractive Materials and Their Applications I,* Springer-Verlag, Berlin, 1989, vol. 61, pp. 237–262.

133. P. Tayebati, J. Kumar, and S. Scott, *Appl. Phys. Lett.* **59:**3366 (1991).

134. J. Kumar, G. Albanese, and W. H. Steier, *J. Opt. Soc. Am.* **B4:**1079 (1987).

135. B. Imbert, H. Rajbenbach, S. Mallick, J. P. Herriau, and J.-P. Huignard, *Opt. Lett.* **13:**327 (1988).

136. M. B. Klein, S. W. McCahon, T. F. Boggess, and G. C. Valley, *J. Opt. Soc. Am.* **B5:**2467 (1988).

137. G. C. Valley, H. Rajbenbach, and H. J. von Bardeleben, *Appl. Phys. Lett.* **56:**364 (1990).

138. Ph. Refregier, L. Solymar, H. Rajbenbach, and J.-P. Huignard, *J. Appl. Phys.* **58:**45 (1985).

139. K. Walsh, A. K. Powell, C. Stace, and T. J. Hall, *J. Opt. Soc. Am.* **B7:**288 (1990).

140. M. Ziari, W. H. Steier, P. M. Ranon, M. B. Klein, and S. Trivedi, *J. Opt. Soc. Am.* **B9:**1461 (1992).

141. P. Gravey, G. Picoli, and J. Y. Labandibar, *Opt. Comm.* **70:**190 (1989).

142. G. Picoli, P. Gravey, C. Ozkul, and V. Vieux, *J. Appl. Phys.* **66:**3798 (1989).

143. A. Partovi, A. Kost, E. M. Garmire, G. C. Valley, and M. B. Klein, *Appl. Phys. Lett.* **56:**1089 (1990).

144. J. E. Millerd, S. D. Koehler, E. M. Garmire, A. Partovi, A. M. Glass, and M. B. Klein, *Appl. Phys. Lett.* **57:**2776 (1990).

145. J. E. Millerd, E. M. Garmire, and M. B. Klein, *Opt. Lett.* **17:**100 (1992).

146. G. C. Valley, S. W. McCahon, and M. B. Klein, *J. Appl. Phys.* **64:**6684 (1988).

147. A. Partovi, E. M. Garmire, G. C. Valley, and M. B. Klein, *Appl. Phys. Lett.* **55:**2701 (1989).

148. R. B. Bylsma, D. H. Olson, and A. M. Glass, *Appl. Phys. Lett.* **52:**1083 (1988).

149. D. M. Pepper, J. AuYeung, D. Fekete, and A. Yariv, *Opt. Lett.* **3:**7 (1978).

150. L. Pichon and J.-P. Huignard, *Opt. Comm.* **36:**277 (1981).

151. A. M. Glass, D. D. Nolte, D. H. Olson, G. E. Doran, D. S. Chemla, and W. H. Knox, *Opt Lett.* **15:**264 (1990).

152. D. D. Nolte, D. H. Olson, G. E. Doran, W. H. Knox, and A. M. Glass, *J. Opt. Soc. Am.* **B7:**2217 (1990).

153. Q. N. Wang, R. M. Brubaker, D. D. Nolte, and M. R. Melloch, *J. Opt. Soc. Am.* **B9:**1626 (1992).

154. A. Partovi, A. M. Glass, D. H. Olson, G. J. Zydzik, K. T. Short, R. D. Feldman, and R. F. Austin, *Opt. Lett.* **17:**655 (1992).

155. A. Partovi, A. M. Glass, D. H. Olson, G. J. Zydzik, H. M. O'Bryan, T. H. Chiu, and W. H. Knox, *Appl. Phys. Lett.* **62**:464 (1993).

156. K. Sutter, J. Hulliger, and P. Gunter, *Solid St. Comm.* **74**:867 (1990).

157. K. Sutter and P. Gunter, *J. Opt. Soc. Am.* **B7**:2274 (1990).

158. S. Ducharme, J. C. Scott, R. J. Twieg, and W. E. Moerner, *Phys. Rev. Lett.* **66**:1846 (1991).

159. W. E. Moerner, C. Walsh, J. C. Scott, S. Ducharme, D. M. Burland, G. C. Bjorklund, and R. J. Twieg, *Proc. SPIE* **1560**:278 (1991).

160. S. M. Silence, C. A. Walsh, J. C. Scott, T. J. Matray, R. J. Twieg, F. Hache, G. C. Bjorklund, and W. E. Moerner, *Opt. Lett.* **17**:1107 (1992).

161. J. W. Goodman, *Introduction to Fourier Optics,* McGraw-Hill, San Francisco, 1968.

162. J. O. White and A. Yariv, *Appl. Phys. Lett.* **37**:5 (1980).

163. J. Joseph, K. Singh, and P. K. C. Pillai, *Opt. Commun.* **85**:389 (1991).

164. T. Y. Chang, J. H. Hong, S. Campbell, and P. Yeh, *Opt. Lett.* **17**:1694 (1992).

165. J. Khoury, T. C. Fu, M. Cronin-Golomb, and C. Woods, "Nonlinear Optical Phase Rectifications," accepted for publication in *J. Opt. Soc. Am.* **B11** (1994).

166. J. P. Huignard and J. P. Herriau, *Appl. Opt.* **17**:2671 (1978).

167. J. Feinberg, *Opt. Lett.* **5**:330 (1980).

168. E. Ochoa, J. W. Goodman, and L. Hesselink, *Opt. Lett.* **10**:430 (1985).

169. J. J. Berg, J. N. Lee, M. W. Casseday, and B. J. Udelson, *Opt. Eng.* **19**:359 (1980).

170. H. Lee and D. Psaltis, *Opt. Lett.* **12**:459 (1987).

171. R. M. Montgomery and M. R. Lange, *Appl. Opt.* **30**:2844 (1991).

172. D. E. Oates, P. G. Gottschalk, and P. B. Wright, *Appl. Phys. Lett.* **46**:1125 (1985).

173. A. E. T. Chiou and P. Yeh, *Opt. Lett.* **10**:621 (1985).

174. S. K. Kwong and A. Yariv, *Appl. Phys. Lett.* **48**:564 (1986).

175. S. MacCormack and R. W. Eason, *Opt. Lett.* **15**:1212 (1990).

176. S. MacCormack and R. W. Eason, *J. Appl. Phys.* **67**:7160 (1990).

177. W. R. Christian, P. H. Beckwith, and I. McMichael, *Opt. Lett.* **14**:81 (1989).

178. M. Segev, S. Weiss, and B. Fischer, *Appl. Phys. Lett.* **50**:1397 (1987).

179. S. Weiss, M. Segev, and B. Fischer, *IEEE J. Quantum Electron.* **QE24**:706 (1988).

180. R. R. Stephens, R. C. Lind, and C. R. Guiliano, *Appl. Phys. Lett.* **50**:647 (1987).

181. P. D. Hillman and M. Marciniak, *J. Appl. Phys.* **66**:5731 (1989).

182. S. Weiss, M. Segev, S. Sternklar, and B. Fischer, *Appl. Opt.* **27**:3422 (1988).

183. J. W. Goodman, F. I. Leonberger, S. Y. Kung, and R. A. Athale, *Proc. IEEE* **72**:850 (1984).

184. J. W. Goodman, in *Optical Processing and Computing,* H. H. Arsenault, T. Szoplik, and B. Macukow (eds.), Academic Press, San Diego, 1989, chap. 1.

185. K. Wagner and D. Psaltis, *Appl. Opt.* **26**:5061 (1987).

186. M. Cronin-Golomb, *Appl. Phys. Lett.* **23**:2189 (1989).

187. A. E. T. Chiou and P. Yeh, *Appl. Opt.* **31**:5536 (1992).

188. P. Yeh and A. E. T. Chiou, *Opt. Lett.* **12**:138 (1987).

189. D. M. Lininger, P. J. Martin, and D. Z. Anderson, *Opt. Lett.* **14**:697 (1989).

190. S. W. McCahon and M. B. Klein, *Proc. SPIE* **1105**:119 (1989).

191. G. L. Wood, W. W. Clark III, G. J. Salamo, A. Mott, and E. J. Sharp, *J. Appl. Phys.* **71**:37 (1992).

192. M. B. Klein and G. J. Dunning, *Proc. SPIE* **1692**:73 (1992).

193. J. L. Schultz, G. J. Salamo, E. J. Sharp, G. L. Wood, R. J. Anderson, and R. R. Neurgaonkar, *Proc. SPIE* **1692**:78 (1992).

194. J. E. Ford, Y. Fainman, and S. H. Lee, *Opt. Lett.* **13**:856 (1988).

195. M. Cronin-Golomb, A. M. Biernacki, C. Lin, and H. Kong, *Opt. Lett.* **12:**1029 (1987).

196. G. F. Albrecht, H. F. Robey, and T. R. Moore, *Appl. Phys. Lett.* **57:**864 (1990).

197. H. F. Robey, *Phys. Rev. Lett.* **65:**1360 (1990).

198. S. D. Kalaskar and S. E. Bialkowski, *Anal. Chem.* **64:**1824 (1992).

199. G. Zhou, L. Bintz, and D. Z. Anderson, *Appl. Opt.* **31:**1740 (1992).

200. J. Feinberg, *Opt. Lett.* **8:**569 (1983).

201. M. D. Ewbank, P. Yeh, M. Khoshnevisan, and J. Feinberg, *Opt. Lett.* **10:**282 (1985).

202. D. Z. Anderson, D. M. Lininger, and J. Feinberg, *Opt. Lett.* **12:**123 (1987).

203. S. K. Kwong, G. A. Rakuljic, and A. Yariv, *Appl. Phys. Lett.* **48:**201 (1986).

204. A. E. Chiou and P. Yeh, *Opt. Lett.* **11:**306 (1986); J. Feinberg and K. R. MacDonald, in *Photorefractive Materials and their Applications II,* Springer Verlag, Berlin (1989).

205. E. Parshall and M. Cronin-Golomb, submitted to *Appl. Opt.,* **30:**5090 (1991).

206. G. J. Dunning, E. Marom, Y. Owechko, and B. H. Soffer, *Opt. Lett.* **12:**346 (1987).

207. Y. Owechko, *IEEE J. Quantum Electron.* **QE25:**619 (1989).

208. T. Kohonen, *Self-Organization and Associative Memory,* Springer-Verlag, New York, 1984.

209. D. Z. Anderson, *Opt. Lett.* **11:**56 (1986).

210. J. Hong, S. Campbell, and P. Yeh, *Appl. Opt.* **29:**3019 (1990).

211. D. Psaltis, D. Brady, and K. Wagner, *Appl. Opt.* **27:**1752 (1988).

212. A. V. Huynh, J. F. Walkup, and T. F. Krile, *Opt. Eng.* **31:**979 (1992).

213. E. G. Paek, P. F. Liao, and H. Gharavi, *Opt. Eng.* **31:**986 (1992).

214. K. P. Lo and G. Indebetouw, *Appl. Opt.* **31:**1745 (1992).

215. M. Cronin-Golomb and A. Yariv, *Proc. SPIE* **700:**301 (1986).

216. M. B. Klein, G. J. Dunning, G. C. Valley, R. C. Lind, and T. R. O'Meara, *Opt. Lett.* **11:**575 (1986).

217. S. W. McCahon, G. J. Dunning, K. W. Kirby, G. C. Valley, and M. B. Klein, *Opt. Lett.* **17:**517 (1992).

218. D. M. Lininger, P. J. Martin, and D. Z. Anderson, *Opt. Lett.* **14:**697 (1989).

219. M. Ingold, P. Gunter, and M. Schadt, *J. Opt. Soc. Am.* **B7:**2380 (1990).

220. S. W. McCahon, G. J. Dunning, K. W. Kirby, G. C. Valley, and M. B. Klein, *Opt. Lett.* **17:**517 (1992).

221. C. Denz, G. Pauilat, G. Roosen, and T. Tschudi, *Opt. Commun.* **85:**171 (1991).

222. F. Mok, M. C. Tackitt, and H. M. Stoll, *Opt. Lett.* **16:**605 (1991).

223. Y. Taketomi, J. E. Ford, H. Sasaki, J. Ma, Y. Fainman, and S. H. Lee, *Opt. Lett.* **16:**1774 (1991).

224. G. A. Rakuljic, V. Levya, and A. Yariv, *Opt. Lett.* **17:**1471 (1992).

225. L. Hesselink and S. Redfield, *Opt. Lett.* **13:**877 (1988).

226. F. Ito, K. Kitayama, and H. Oguri, *J. Opt. Soc. Am.* **B9:**1432 (1992).

227. K. E. Youden, S. W. James, R. W. Eason, P. J. Chandler, L. Zhang, and P. D. Townsend, *Opt. Lett.* **17:**1509 (1992).

228. K. E. Youden, R. W. Eason, M. C. Gower, and N. A. Vainos, *Appl. Phys. Lett.* **59:**1929 (1991).

229. Y. Nagao, H. Sakata, and Y. Mimura, *Appl. Opt.* **31:**3966 (1992).

230. B. Fischer and M. Segev, *Appl. Phys. Lett.* **54:**684 (1989).

231. J. L. Jackel, *Electron. Lett.* **21:**509 (1985).

232. V. E. Wood, P. J. Cressman, R. J. Holman, and C. M. Verber, in P. Gunter and J. P. Huignard (eds), *Photorefractive Materials and their Applications,* Springer-Verlag, Berlin, 1989.

233. S. W. James, K. E. Youden, P. M. Jeffrey, R. W. Eason, P. J. Chandler, L. Zhang, and P. D. Townsend, *Appl. Opt.* **32:**5299 (1993).

39.5 FURTHER READING

Books and Review Articles on Photorefractive Materials, Effects, and Devices

Gunter, P., "Holography, Coherent Light Amplification and Phase Conjugation in Photorefractive Materials," *Phys. Rep.* **93:**199 (1982).

Gunter, P., "Photorefractive Materials," in *CRC Handbook of Laser Science and Technology,* vol. IV, part 2, CRC Press Boca Raton, 1986.

Gunter, P. and J.-P. Huignard (eds.), *Photorefractive Materials and their Applications I and II,* Springer-Verlag, Berlin, 1988, 1989.

Klein, M. B., and G. C. Valley, "Optimal Properties of Photorefractive Materials for Optical Data Processing," *Opt. Engin.* **22:**704 (1983).

Odoulov, S., M. Soskin, and A. Khizniak, *Optical Oscillators with Degenerate Four-Wave Mixing (Dynamic Grating Lasers),* Harwood Academic Publishers, London, 1989.

Petrov, M. P., S. I. Stepanov, and A. V. Khomenko, *Photorefractive Crystals in Coherent Optical Systems,* Springer-Verlag, Berlin, 1991.

INDEX

Index note: The *f.* after a page number refers to a figure, the *n.* to a note, and the *t.* to a table.

ABOUT THE EDITORS

Michael Bass is Professor of Physics and Electrical and Computer Engineering at the University of Central Florida and is on the faculty of the Center for Research and Education in Optics and Lasers (CREOL). He received his B.S. in Physics from Carnegie-Mellon and his M.S. and Ph.D. in Physics from the University of Michigan.

Eric Van Stryland is a Professor of Physics and Electrical Engineering at the University of Central Florida Center for Research and Education in Optics and Lasers. He received his Ph.D. from the University of Arizona.

David R. Williams is a Professor of Psychology and Visual Science and the Director of the Center for Visual Science at the University of Rochester. He receive his B.S. in Psychology from Denison University, and his M.A. and Ph.D. in Psychology from the University of California, San Diego.

William L. Wolfe is a Professor at the Optical Sciences Center at the University of Arizona. He received his B.S. in Physics from Bucknell University, and his M.S. in Physics and M.S.E. in Electrical Engineering from the University of Michigan.